MW01197024

The Immunoassay Handbook

THIRD EDITION

The Immunoassay Handbook

THIRD EDITION

Edited by
DAVID WILD

2005

ELSEVIER

Amsterdam – Boston – Heidelberg – London – New York – Oxford – Paris
San Diego – San Francisco – Singapore – Sydney – Tokyo

ELSEVIER B.V.
Radarweg 29
P.O.Box 211, 1000 AE
Amsterdam, The Netherlands

ELSEVIER Inc.
525 B Street, Suite 1900
San Diego, CA 92101-4495
USA

ELSEVIER Ltd
The Boulevard, Langford Lane
Kidlington, Oxford OX5 1GB
UK

ELSEVIER Ltd
84 Theobalds Road
London WC1X 8RR
UK

Third edition 2005

Cover design by Dr. Rajeev Doshi

Library of Congress Cataloging in Publication Data
A catalog record is available from the Library of Congress.

British Library Cataloguing in Publication Data
A catalogue record is available from the British Library.

ISBN 10: 0 08 044526 8
ISBN 13: 9780080445267

Transferred to digital print 2007
Printed and bound by CPI Antony Rowe, Eastbourne

CONTENTS

PART 3 LABORATORY MANAGEMENT

CONTRIBUTORS

John Ardern
Alan S. Armstrong
Marie Arvola
Arthur L. Babson
Mary Beth Myers
Pierre Blockx
Doug Brandt
Kenneth F. Buechler
Bernard Capolaghi
Stephen J. Capper
Murielle Cazaubiel
Daniel W. Chan
Tim Chard
Nic D. Christofides
Penny Clark
Kay W. Colston
William A. Coty
Susan J. Danielson
Chris Davies
Derek Dawson
George J. Dawson
Alastair H. Dent
Bruce J. Dille
Theresa Donahoe
Roger Ekins
Gareth Evans
Steve Figard
Kent Ford
Gary Franklin
James K. Gimzewski
Elvio Gramignano
David A. Hilborn
Jeffrey K. Horton
Tom Houts
David Huckle
Kelvin T. Hughes
Alun D. Hutchings
Rhys John
Robert Karlsson
David F. Keren
Larry J. Kricka
Wlad Kusnezow
Tanguy Le Neel
Eryl Liddell
Rueyming Loor
P. Gordon Malan
Manuella Martin
Keith May
W.N. McLellan
Sami Medbak
Isa K. Mushahwar
James H. Nichols
Barry Nix
August C. Olivar
Michael Preece
Molly J. Price Jones
Jane Pringle
Frank A. Quinn
Jason Reed
Mats Rilvën
Robert F. Ritchie
Philip A. Routledge
John W. Safford Jr.
Colin H. Self
Dinesh Shah
Lori J. Sokoll
Carole A. Spencer
John C. Stevenson
Jim Stewart
Mavanur R. Suresh
Laura Taylor
Michael A. Teitelland
Stephen Thompson
Alain Truchaud
Edwin F. Ullman
Harry Waters
Michael J. Wheeler
Brian Widdop
David Wild
Colin Wilde
Larry A. Winger
Erwin Workman
Alan H.B. Wu
Lars Yman
Jean-Pierre Yvert

PREFACE

Immunoassay is a remarkable field, with a 50-year track record of continuous innovation. This book explains the science and technology, and many of the immense range of applications.

Immunoassays are extensively used, from over-the-counter pregnancy tests to proteomic research tools. Their sensitivity and specificity are remarkable. A minute quantity of an immunized animal's blood, mixed with a cocktail of chemicals, can be used to detect less than 1 part in 10^{12} of a hormone in a sample of blood. This sub-picomolar (10^{-12}) sensitivity is attained by some commercial hospital analyzers. Research immunoassays have detected concentrations as low as 10^{-21} mol. In theory this means that if a bucket of a new chemical was poured into the sea near my home in England, an immunoassay could eventually detect its presence in Australia.

The field has continued to advance at a breathtaking pace since the first edition of *The Immunoassay Handbook* was published in 1994. For example, pregnancy test technology is now used in multi-analyte, quantitative tests for cardiac conditions (Chapter 37), and discreet, miniature immunoassay analyzers accompany their owners, learn about their menstrual cycles, and advise when contraception is most needed (Chapter 40). Antibody libraries are made from human DNA using viruses as vectors and fine-tuned using DNA sequencers. A phage display library may contain more than 10^{12} different antibody clones in a single beaker and these can be used to develop immunoassays for chemicals too toxic to use for immunization (Chapter 8).

In the pages of this book you will discover a treasure chest of innovation, the result of individual brilliance and great teamwork, involving an unparalleled fusion of science and technology.

The first edition of *The Immunoassay Handbook* came about because there was a need for a comprehensive textbook on immunoassays and their applications. It was intended to demystify the science, reveal the technology behind commercial products, and summarize the more common clinical applications. The book was to be more complete than a collection of topical papers, and consistent in style and terminology from chapter to chapter. Readers responded well to the format and the first edition was reprinted twice. Since then, immunoassay has become a common experimental technique in diagnosis and research. It was inevitable in such a fast-moving field that a second edition would be necessary to bring the book up to date, but the decision was also taken to expand the scope of the book significantly.

The second edition was much larger than the first, with many new chapters and more contemporary material, and the book was awarded a prize in the British Medical Association's annual book competition in 2001. The award reflected the exceptionally talented team of contributing authors.

The third edition has been revised in the most fast-moving areas and has three new chapters from world leaders in their respective fields. At least half of the chapters have been revised.

Robert Ritchie has written a fascinating new account of the foundations of immunochemistry dating back to 1861, a sadly neglected subject (Chapter 5). This account of scientific endeavor in a complex field is a stark reminder of the challenges facing early researchers.

Looking forward, the third edition also includes major new contributions about advanced technology.

Robert Karlsson and colleagues from Biacore have provided an in-depth account of surface plasmon resonance (SPR) applications in kinetic, concentration and binding site analysis (Chapter 16). This revealing technique is now well established in many research laboratories.

The new chapter on SPR is complemented by a major revision of the immunobiosensor chapter by Professor Jim Gimzewski and his ground-breaking team at UCLA (Chapter 15).

Larry Kricka has written a comprehensive new chapter on lab-on-a-chip, micro- and nanoscale immunoassays (Chapter 17). It is packed with novel examples of miniaturized technology, but — as is a feature throughout the book — this chapter also *explains* the subject, especially the fundamental changes that apply when immunoassays are reduced in size. There is also a new section on microscale immunoassay development in the chapter on Separation Systems by Wlad Kusnezow, who is at the forefront in this field (Chapter 10).

Another first is a remarkable new technique for small analytes revealed by Professor Colin Self in Chapter 2. This method allows conventional antibodies from competitive immunoassays to be used in immunometric format assays, with dramatic improvements in performance.

Elsewhere in the book, new analytes have been added in Part 4, and new applications for existing analytes included. Product and marketing information has been updated in Part 2.

Part 1 (*Principles*) has been extended from 14 chapters to 17. It is a master class in immunoassay from leaders in the field. The opening four chapters explain the principles that apply to all immunoassays. The highly praised first chapter on competitive and immunometric assay theory includes an "Immunoassay for beginners" opening section, providing a basic explanation of immunoassay for newcomers. If you are new to immunoassay, start by reading this short section, then perhaps read the opening chapter in Part 2 (Chapter 18) before sampling the rest of the book. Chapter 2 introduces immunometric assays for small molecules, which until recently had proved impossible to achieve in practice, written by one of the great innovators and speakers in the field of immunoassay. This is an important scientific advance with great potential that has not yet been fully exploited. In Chapter 3, Professor Roger Ekins, the pioneer and innovative theorist who first introduced me to immunoassays, describes ambient analyte immunoassay. This format represents a whole new class of immunoassays, where neither sample nor reagent volumes have to be precisely measured. Although the compact disk assay presentation of ambient analyte immunoassay has deservedly received much attention, the revolutionary nature of the underlying principle is explained here, as there are other potential applications for this assay format that are also paradigm-breaking. The coverage of immunoassay formats concludes with a Chapter (4) on free analyte immunoassays, from a colleague who has achieved a series of breakthroughs in assay performance through a painstaking combination of experimentation and theoretical modeling.

Chapter 6 (*Concepts*) is really two chapters in one, covering a wide range of fundamental assay and clinical concepts. This chapter is essential reading for anyone responsible for applying immunoassays in clinical diagnostics or research. Then in Chapter 7, two experienced assay development project managers from the leading commercial supplier of clinical immunoassay equipment and reagents explain how commercial immunoassays are developed and optimized, using the latest experimental design software tools.

Antibodies, which are at the heart of every immunoassay, are described in Chapter 8. The scope of this chapter ranges from the basics to the latest developments in antibody engineering, which are exciting in the pharmaceutical world but have been somewhat neglected in diagnostics so far.

The chapter on separation systems (10) includes information about coating proteins onto plastic, and the principles behind assay separations, which are crucial to achieving high-performance immunoassays. Separation is much more complex in miniaturized formats, hence the new section mentioned earlier.

The coverage of signal generation and detection methods could have run to an entire volume, but Chapter 11 gives a whistle-stop tour of most of the methods, by the leading expert in this field. This edition includes new material on up-converting phosphor technology and new stable horseradish peroxidase chemiluminescent labels.

The chapter on homogeneous immunoassays (12), written by the pioneer who still dominates this field after 30 years, manages to compress an impressive amount of information into the pages allocated, in an accessible way. This is an excellent introduction to this complex subject.

The chapter on curve-fitting (13) includes the theory behind factory master calibrations and reduced calibration. Immunoassay manufacturers neglect an understanding of this important area at their peril.

Conjugation methods (chapter 14) are summarized by the author of *Bioconjugation*, previously a companion volume to this book.

As mentioned earlier there are now two chapters on immunobiosensors (15 and 16), reflecting the growth in this area.

Chapter 17 is the new chapter on lab-on-a-chip and other miniaturized immunoassay formats.

Part 2 encompasses the diversity of *Product Technology*. There is a short introduction to the wide range of products on the market (18) followed by an updated chapter on market trends (19), from the experienced author of a *Financial Times* publication on the diagnostics industry, including a well-argued glimpse into the future. There are two chapters on choosing automated systems (20, 21), written by well-respected teams of system reviewers in Europe and the US, followed by a chapter describing developments in over-the-counter pregnancy tests from the author of two systematic studies carried out in the UK in 1988 and 1997 (Chapter 22). The remainder of Part 2 comprises chapters describing 22 representative commercial immunoassay systems. Chapters describing some of the older systems have been left out of the third edition to make room for the new chapters in Part 1. Copies are available from me on request.

Part 3 (*Laboratory Management*) includes a comprehensive review of potential influences due to subject handling and sample collection (Chapter 41), and an introduction to Quality Assurance (Chapter 42). There is an authoritative chapter (43) on point-of-care testing by the leading authority on this subject, which has been revised for this edition. There is also a fully comprehensive troubleshooting guide (Chapter 44).

Part 4 (*Applications*) is a mine of information, with explanations for a wide range of clinical conditions, and information on the analytes relevant to clinical diagnosis. Each chapter is written by an expert (or experts) in that clinical field. The extensive chapter on immunoassays for drug-screening and life-science research by the GE Medical (previously Amersham Pharmacia Biotech) team (67) has been completely revised and updated, with new sections on Alzheimer's and Parkinson's Disease. Other chapters that have been significantly revised include the chapters on cardiac and cancer markers (55 and 56), diabetes mellitus (53), hematology (54), autoimmune diseases (58), growth and growth hormone deficiency (52), hirsutism (50) and veterinary applications (68), which now includes Foot and Mouth Disease and Bovine Spongiform Encephalopathy.

Just a brief note on nomenclature. In *The Immunoassay Handbook* most of the time assays are described either as *competitive* or *immunometric*. I prefer the term 'competitive' because it has a strong descriptive element that helps newcomers to remember how these assays work. However it is not universally accepted as a scientific description. I also have a personal bias towards 'immunometric' because '*non-competitive*' does not seem to do these assays justice, as they are typically far superior to their competitive counterparts. I also think the popular term for the immunometric format 'sandwich assay' is potentially misleading, although it is descriptive, as competitive immunoassays may also have three or more layers of molecules binding together. Having justified the use of these terms in the book, I must also point out that although *most* immunometric assays are also non-competitive, the term immunometric was introduced to describe any immunoassay that uses a labeled antibody, and there are some assays that are competitive and use labeled antibodies. In these cases the assays are described in the book as competitive rather than immunometric. I hope the experts will forgive these attempts to simplify and unify nomenclature in order to standardize the book across a wide range of fields that have sometimes developed independently.

This is about all we could possibly provide about immunoassays in one volume. The authors have packed an amazing amount of information into their chapters without making the subject indigestible. They have taken the time to tell the story of immunoassay in an accessible way, so that readers at all levels can be let into its secrets. My favorite story is told in the Foreword, in which Nobel Prize winner Rosalyn Yalow reflects on how immunochemistry was transformed by her work with Solomon Berson, but also sounds a warning bell to those who are tempted to sacrifice their scientific integrity in pursuit of commercial gain.

I hope you find this book a welcome source of information and insight, and discover for yourself something new and exciting in the field of immunoassay and its many applications. If you do, you will be continuing a long tradition of scientific fascination with this absorbing subject.

David Wild
david@davidwild.net

ACKNOWLEDGEMENTS

I would like to express my warmest thanks to the contributing authors, for taking time in their hectic lives to explain clearly a subject that is sometimes difficult to understand.

The striking science fiction images on the cover are by Rajeev Doshi, a rare individual who is both a scientist and an artist.

My family again accepted the inevitable interruptions to our leisure time and kept me in good spirits. I cannot thank them enough.

FOREWORD

While some of my girlfriends aspired to be movie stars, my great inspiration was Marie Curie. She set a wonderful example to all women scientists. Throughout my life I have been passionate about science, and felt the need for women to be treated equally to men, in science and in society as a whole. I still believe in the value of science as a powerful and positive influence for society. But over the years I have realized that great science is rare. Where it occurs, it should be cherished and nurtured.

Some readers may have assumed that I was a biologist or medical researcher but I was neither. I graduated in Physics in the College of Engineering in New York, not far from my home in the North Bronx, where I was raised as a child and still live. I was the first woman to take this course since World War I. There was a shortage of male students when I joined in September 1941; otherwise I doubt I would have been accepted as a mere woman! I received my PhD at the University of Illinois, Champaign-Urbana, and returned to work at the Veterans Administration Hospital near my home in 1947. Nearly 60 years later, I still work there for a few hours every week.

My collaborator, the late Solomon Berson, was a resident at the hospital in July 1950. He was a brilliant registrar. He was very good at mathematics and loved mathematical puzzles. The discovery of immunoassay was part mathematics, part physics and part biology. I learned about biology from Sol, and he learned about physics from me.

I believe that most scientific breakthroughs come out of painstaking research and experimentation; these are necessary precursors of those rare flashes of inspiration. But then, the most exciting new ideas can take a scientist into unexpected new territory. We were working on a medical problem involving diabetes, not trying to discover a new assay technique, but after we did, it became a tool that was used to investigate medical problems in a range of different fields much wider than we could ever have imagined.

Sol and I were investigating the suggestion of Dr I. Arthur Mirsky that maturity-onset diabetes might be due to excessive enzymic degradation of insulin, rather than an absolute deficiency of insulin secretion, which was the prevailing theory at the time. We administered ^{131}I-insulin to diabetic and non-diabetic subjects and observed a slower rate of disappearance of the labeled insulin from patients who had been treated with insulin, whether or not they were diabetic. We wondered whether antibodies had developed to the animal-derived insulin, changing the rate of disappearance. We used separation methods to prove that the protein that bound insulin in the treated patients was an antibody. This was something that immunologists could not accept at the time and our paper was rejected by *Science* and initially by the *Journal of Clinical Investigation*. The aim of science is to make new discoveries, so it is ironic that data that only confirms what is already thought is more likely to be published, and that new discoveries are often unwelcome to the scientific community. At my age I can say what I think and not worry about the consequences. Come to think of it, that is how I have always been and it is a necessary personality trait for a good scientist!

It is amazing that insulin had been given to patients since 1920, yet no one had realized that there was an immune reaction to it before. At the time, that was the discovery we thought was the most significant because of its medical application, and immunoassay was very much a spin-off.

We devised the principle of radioimmunoassay in 1956, but it took a couple of years to get it to work. We published our first paper describing the immunoassay technique in 1959, but research papers that we published in the preceding 2 years hint of our development work. The idea did not come in a flash of inspiration, but was the result of painstaking scientific method. We called the technique *immunoassay*, then *radioimmunoassay*, which is logical enough, but I have to admit that it has given people problems over the years as it is rather a tongue-twister. It quickly became RIA. I should have learned from Marie Curie that a simple name like X-ray would be easier to say and remember!

I am sometimes asked if I regret that we did not patent the principle of immunoassay. Not only do I not regret it, but I felt very strongly that we, as scientists, should not seek to exploit our scientific discoveries for our own financial gain. We realized that our new technique could have widespread applications, so we organized classes every year and invited scientists from all over the world to come and learn about our discovery, and how to develop their own assays. This may help to explain why the technique quickly spread to different analytes and many countries, and why so many companies eventually manufactured immunoassays.

I feel the same way about commercial links with manufacturers. I have declined lucrative offers to act as a paid consultant for immunoassay companies. How could I have retained my scientific integrity otherwise?

I had a wonderful time as a scientist. We had just a small laboratory, and nobody bothered us. I had to build upon my academic education, learning about medicine, so that I could make a useful contribution to scientific knowledge, then later on we had to set up educational courses to pass on what we had learned. This is how good science should be carried out and communicated. I believe that we were successful because we always worked at the laboratory bench, and were only interested in discovering the truth. I did not stop working at the bench until 1970, and I still carry out scientific research from my desk for a few hours most days. Over the years I carried out research into glycoprotein heterogeneity, and developed one of the first assays for Hepatitis B. Recently I have been studying the effects of radioactivity on the environment.

Of 10 women Nobel Prize winners in Science, I was the only one from an uneducated family. By coincidence one of the others, Gertrude Elian, also graduated from Hunter College in Manhattan.

I am immensely proud of my work with Sol. I hope that does not sound arrogant. But I was lucky enough to see our hard work amply rewarded by tangible results. I like to think that Marie Curie would have been proud of me had she known of my work. Perhaps she and Sol are having heated debates about a controversial new theory as I write this now. If so, there is indeed a heaven.

Rosalyn Yalow

Part 1

Principles

1 Introduction to Immunoassay Principles

Chris Davies

Immunoassays are sensitive analytical tests that harness the unique properties of antibodies. They proved to be one of the most productive technological contributions to medicine and fundamental life science research in the 20th century. Immunological techniques, and in particular immunoassay, have provided us with a sophisticated biochemical toolbox, which can be applied to investigate and manipulate minute concentrations of complex molecules. It seems appropriate that nature should provide us with the fundamental tools we use to study biological processes.

Immunoassays derive their unique characteristics from three important properties of antibodies:

- Their ability to bind to an extremely wide **range** of natural and man-made chemicals, biomolecules, cells and viruses. This is because antibodies are proteins, and the binding sites are derived from a huge number of potential combinations of amino acid sequences. Each of the 22 amino acids has its own unique properties for binding and orientation, and chains of amino acids can twist and fold to provide binding at multiple sites.
- Exceptional **specificity** for the substance to which each antibody binds. The remarkable specificity of antibodies enables minute concentrations of analyte to be assayed in the presence of many closely related substances, for example, the routine measurement of picomolar (10^{-12}) concentrations of hormones in blood samples.
- The **strength of the binding** between an antibody and its target. In an immunoassay test, this draws the analyte and antibody together, and creates a strong, noncovalent bond that survives the processing and signal generation stages. Immunoassays are accurate and precise, even at the low concentrations found in biological fluids.

Since the principles of immunoassay were first expounded by Rosalyn Yalow and Solomon Berson in 1959 there has been an exponential growth, not only in the range of applications to which it has successfully been applied, but also in the number of novel and ingenious assay designs. Some of these have resulted in more sensitive assays that have opened up new horizons of clinical research and diagnosis, while others have concentrated on simplifying the requirements for supporting technology, rendering it amenable to automation. Such technology has not been confined to medical diagnosis, but has also been applied to the pharmaceutical, veterinary, forensic, military and food sciences. In fundamental life science research, immunoassay remains an invaluable tool for elucidating basic biochemical phenomena, while substantial progress in simplifying the user protocols has enabled more prosaic applications such as pregnancy and ovulation tests for use in the home.

IMMUNOASSAY FOR BEGINNERS

Immunoassays use reagents to generate a signal from a sample. Imagine a magnet, securely tied to the end of a fishing line, resting in a stream. Several minute flecks of metal are attracted to the magnet and caught by the person fishing. In an immunoassay the magnet is replaced by an **antibody**, which is usually immobilized onto a plastic surface instead of a fishing line. Antibodies are very selective and only bind to their specific targets, even in the presence of a huge range of other materials in a sample. As the analytes are present in minute concentrations it is not enough simply to 'catch' them to know how much is there. Another reagent has to be used to generate a signal from the captured material. The level of signal indicates the concentration of the specific analyte under test.

The simplest type of immunoassay to understand is the **immunometric** design (Figure 1.1). An antibody immobilized onto a plastic surface (such as a polystyrene test tube) **captures** the test analyte from the sample, and a different antibody, specific for another part of the analyte molecule, is used as the basis of the signal

The Immunoassay Handbook. *Third Edition*
D. Wild (Ed.)

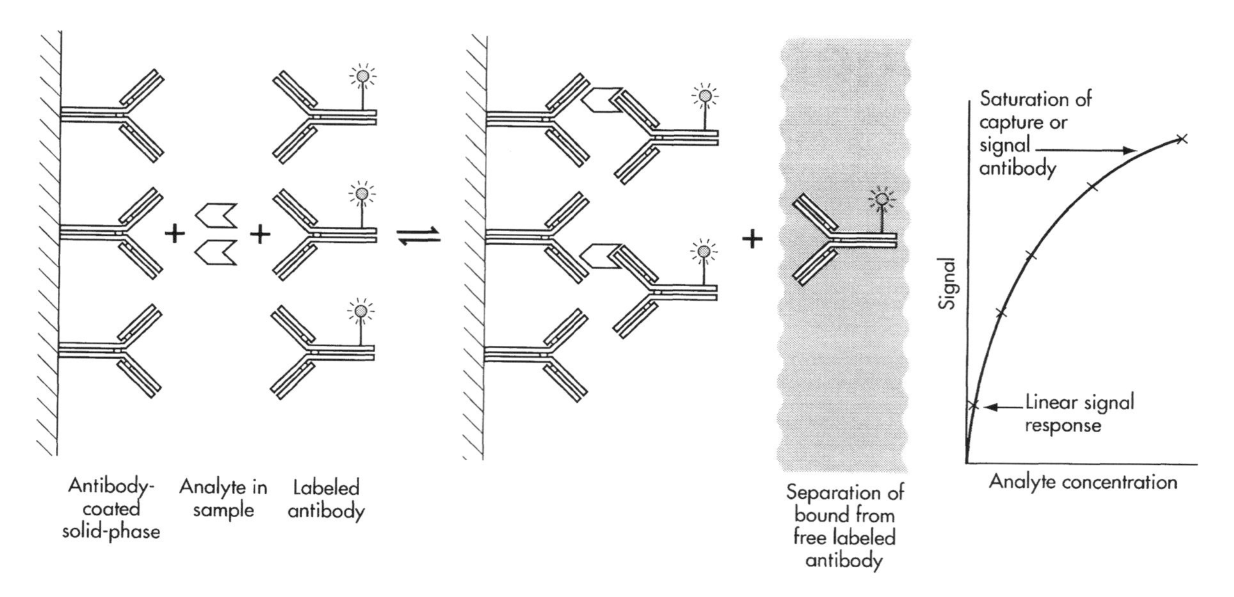

Fig. 1.1 Immunometric immunoassay.

generation system. This antibody is 'labeled', e.g. with a radioactive isotope. After an incubation to allow the antibodies to bind with the analyte, unbound labeled antibody is washed away. The final stage of the assay involves measurement of the signal level, in this example, measurement of the radioactivity. The signal level in this type of assay is clearly proportional to the analyte concentration in the sample.

The labeled component of an immunoassay is sometimes called the **tracer**. The analyte that the antibodies are specific for is known as the **antigen**. The efficient removal of unbound tracer by washing is a critical part of the assay, known as the **separation**. The material (normally plastic) that the antibody is irreversibly bound to is known as the **solid phase**. Because the antibodies form a sandwich around the analyte, immunometric assays are also known as **sandwich** assays.

In the next example the format is similar, but the immunoassay has been designed to detect *antibodies* in a blood sample using the appropriate antigen as the 'bait' (Figure 1.2). This application is useful for detecting previous exposure to a specific infectious disease. Proteins that occur on the surface of a virus can be immobilized onto plastic. They capture specific antibodies for that virus from the sample. As a tracer, labeled antibodies, raised in animals against the constant region of human antibody, can be used.

The next figure shows a similar assay with a different type of label used for the tracer (Figure 1.3). Instead of a radioactive isotope, an **enzyme** is chemically attached (**conjugated**) to the labeled antibody. Enzymes are proteins, like antibodies, that catalyze specific reactions. The starting material for an enzyme-catalyzed reaction is called a **substrate**. Enzyme labels, with the appropriate substrate, can be used to generate color, or create fluorescent or luminescent end-products, that can be readily measured by optical and electronic equipment. Each molecule of enzyme can convert many molecules of substrate, providing a very sensitive signal generation system. This format of assay is sometimes known as an **enzyme-linked immunosorbent assay (ELISA)**.

Immunometric assays work well when the target analyte is a large molecule with the available surface area to accommodate two molecules of antibody. However, many immunoassays are for small molecules such as drugs, and a different design is needed. This is illustrated in the next figure (Figure 1.4). Only one antibody is used, and it is present in a limited quantity. The other key reagent, the tracer, is the target analyte, labeled with a suitable signal generation material, such as a radioisotope or enzyme. The proportion of tracer that binds to the limited antibody sites is *indirectly* proportional to the concentration of analyte in the sample. This is known as a **competitive** immunoassay.

So far, each type of immunoassay described has depended on a separation of unbound tracer before the bound signal is measured. Without a separation (such as a thorough wash of the solid phase with buffer prior to signal generation) the level of signal would always be the same, regardless of the concentration of analyte. These assay formats are all examples of **heterogeneous** immunoassay. Some assays have been developed that do not require a separation, in which the signal is only generated by the tracer when it binds to the analyte. They are known as **homogeneous** immunoassays (Figure 1.5).

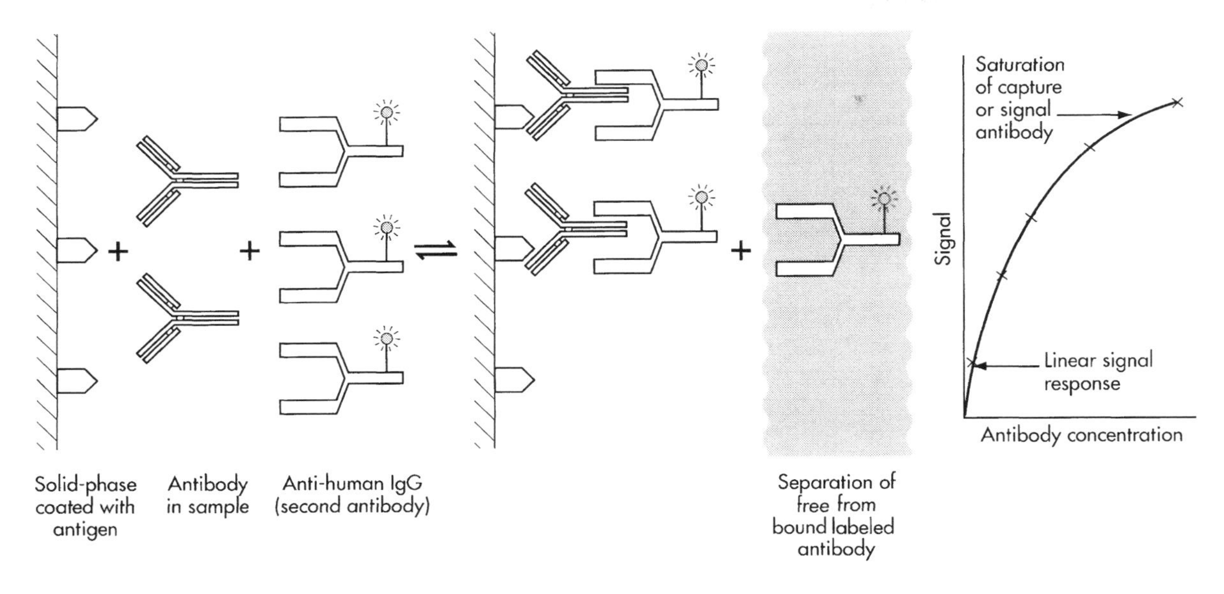

Fig. 1.2 Immunometric assay for antibody testing.

The diversity of immunoassay designs is due to the many possible permutations of sample handling, antibodies, solid phases, separation systems, signal generation systems and calibration methods, and the endless ingenuity of the many thousands of outstanding scientists and engineers that have worked with them. Every year, many new variations on these basic immunoassay formats are devised, and assays for hundreds of new analytes are developed.

KINETICS OF ANTIBODY–ANTIGEN INTERACTIONS

Immunoassays involve a binding reaction between an analyte and at least one antibody. This reaction takes from a few seconds to many hours to achieve equilibrium, depending on a range of different factors. The design of the assay influences the equilibration time in a number of ways. For example, if the antibody is immobilized on microparticles rather than the walls of a tube, the average distance that an analyte molecule has to move before coming into contact with a molecule of antibody is much reduced. The pH, ionic strength and temperature also affect the reaction time.

At the heart of every immunoassay is the binding reaction between the antibody and analyte. The nature of this reaction can vary considerably and is of profound significance to the development of an effective assay. Early immunoassays often had overnight incubations to allow the reaction to fully reach equilibrium. However, most current immunoassays involve comparatively short incubations, and many do not allow the reaction to go to completion. In these assay designs it is even more important to understand a few simple concepts and consequences of the kinetics of the antibody-analyte reaction.

The reaction between antibody and antigen may be simplistically described by the Law of Mass Action:

$$[\mathrm{Ag}] + [\mathrm{Ab}] \underset{k_d}{\overset{k_a}{\rightleftharpoons}} [\mathrm{Ag{-}Ab}] \qquad (1.1)$$

[Ag] = antigen concentration,
[Ab] = antibody concentration,
[Ag–Ab] = antigen–antibody complex,
k_a = association rate constant,
k_d = dissociation rate constant.

It is difficult at first to appreciate why the kinetics should be defined by two rate constants, rather than one. The importance of this distinction can be better understood by monitoring the binding reactions of different antibody-antigen combinations. Using biosensor technology it is possible to generate a signal in real time as analyte binds to an immobilized antibody on a sensor surface (*see* IMMUNOLOGICAL BIOSENSORS). In the experiment, a stream of buffer is run across the sensor surface, which has previously been coated with antibody. Then analyte is introduced into the device and it binds with the immobilized antibody. As more analyte binds, the signal response increases. After a few minutes the analyte solution is replaced by buffer, and the stability of the binding between the analyte and the antibody can

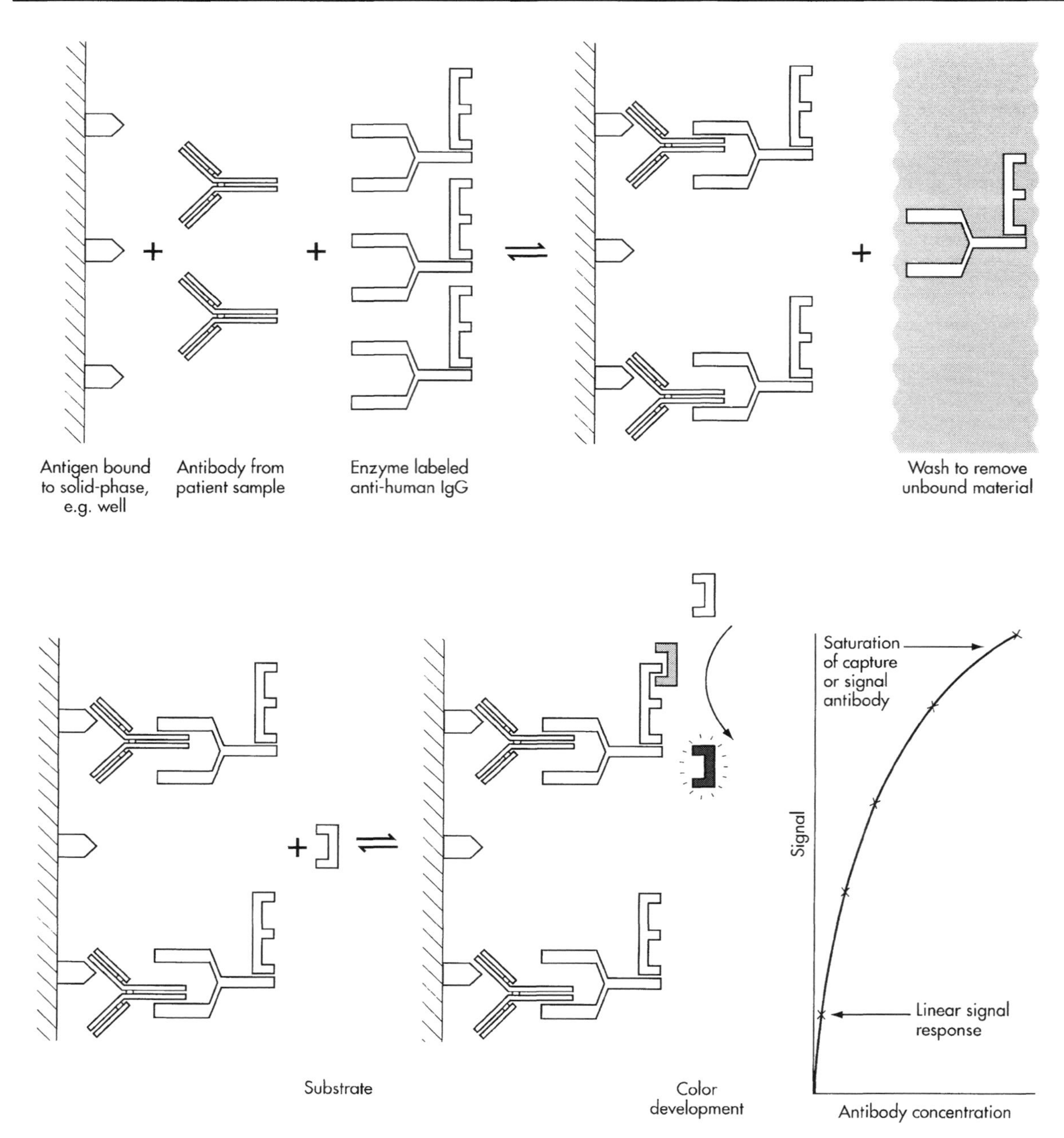

Fig. 1.3 Enzyme immunoassay for detection of antibodies (ELISA).

be observed. Figure 1.6 illustrates the signal responses for an analyte binding to three different monoclonal antibodies.

The three monoclonal antibodies vary in the rate with which they bind the analyte. This reflects the variation in the association constants. After the supply of analyte is cut off, very different responses follow. The antibody with the highest rate of association also suffers from significant dissociation. This antibody would be unsuitable for most immunoassay formats, because analyte would dissociate from the antibody during the separation of unbound label, e.g. using a buffer washing step.

This explains the importance of understanding k_a and k_d. However, it is also important that we recognize

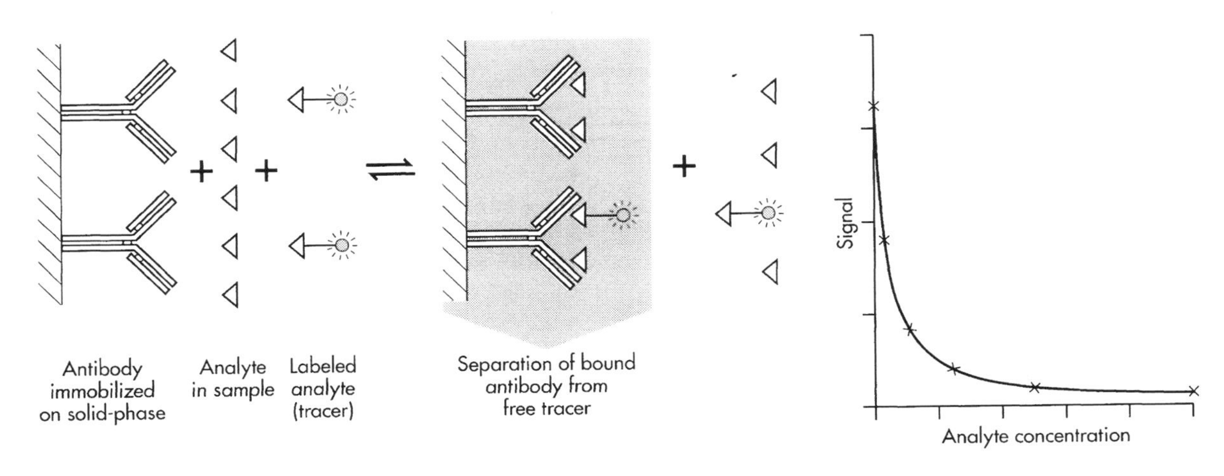

Fig. 1.4 Competitive immunoassay (solid-phase separation).

that the model in equation (1.1) represents an oversimplification of the usual situation. The assumptions of the model are

- The antigen is present in a homogeneous form consisting of only one chemical species.
- The antibody should also be homogeneous.
- The antigen possesses one epitope for binding.
- The antibody has a single binding site that recognizes one epitope of the antigen with one affinity.
- Binding should be uniform with no positive or negative allosteric effects (the binding of one antibody binding site should not influence the binding of the other site).
- The reaction must be at equilibrium.
- The separation of bound from free antigen must be complete.
- There should be no nonspecific binding, such as to the walls of the reaction vessel.

Although it is impossible for all of these assumptions to be completely met in practice, the Law of Mass Action does provide a useful framework on which to base a theoretical appreciation of the thermodynamic principles underlying immunoassay techniques.

The ratio of the two rate constants gives the **equilibrium constant** K_{eq}, which represents the ratio of bound to unbound analyte and antibody. This is also known as the **affinity constant**. This is a key measure of an antibody's ability to function well in an immunoassay.

From equation (1.1), at equilibrium:

$$K_{eq} = \frac{k_a}{k_d} = \frac{[Ag-Ab]}{[Ag][Ab]} \tag{1.2}$$

where

K_{eq} = equilibrium constant.

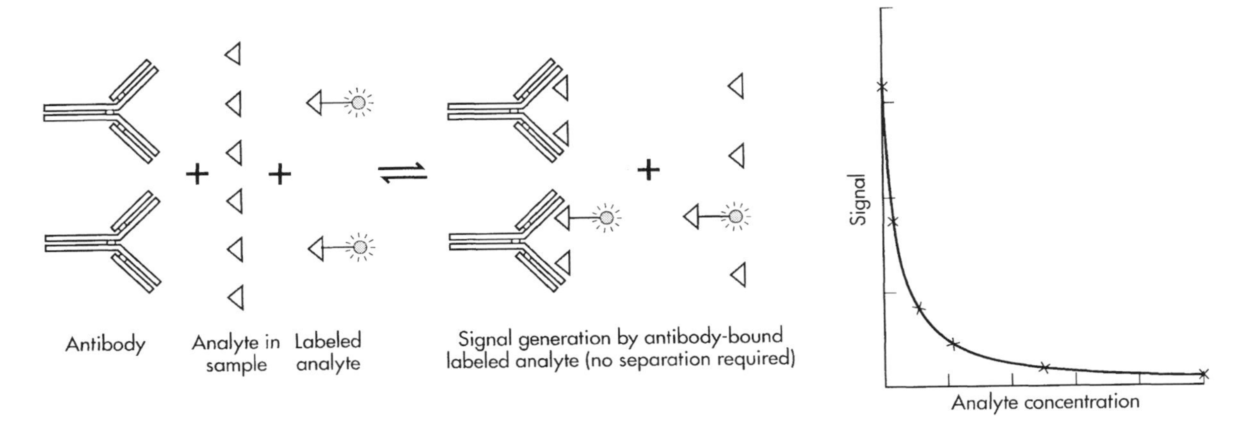

Fig. 1.5 Example of homogeneous immunoassay.

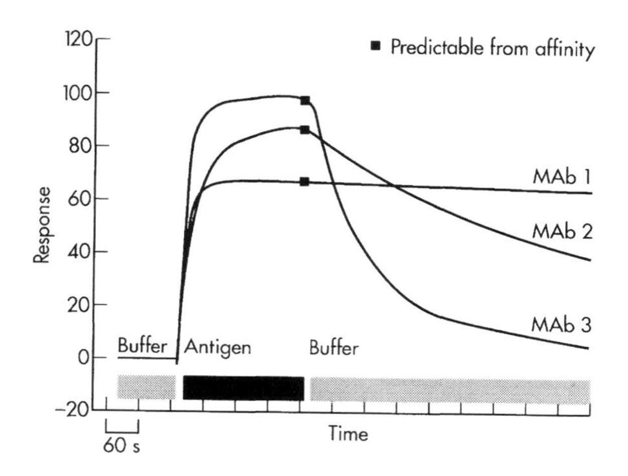

Fig. 1.6 Response curves illustrating the interaction of P24 antigen (125 nM) with three different monoclonal antibodies (MAbs).

Substituting

$$K_{eq} = \frac{[Ag\text{–}Ab]}{[Ab_t] - [Ag\text{–}Ab][Ag]} \qquad (1.3)$$

and rearranging

$$[Ab_t]K_{eq} - K_{eq}[Ag\text{–}Ab] = \frac{[Ag\text{–}Ab]}{[Ag]} \qquad (1.4)$$

$$[Ab_t]K_{eq} - K_{eq}[B] = \frac{[B]}{[F]} \qquad (1.5)$$

where

$[Ab_t]$ = the total concentration of antibody, $[Ab] + [Ab\text{–}Ag]$;
$[Ag_t]$ = the total concentration of antigen, $[Ag] + [Ab\text{–}Ag]$;
$[B]$ = concentration of bound antigen
$[F]$ = concentration of free antigen.

Equation (1.5) indicates a linear relationship between the ratio of [bound]/[free] antigen and the concentration of bound antigen. The graphic representation of this is known as a **Scatchard plot** (Scatchard, 1949) (see Figure 1.7).

Two useful parameters may be derived from the plot: the equilibrium constant, K_{eq}, from the slope of the line, and the total concentration of antibody binding sites, $[Ab_t]$, from the intersection on the x-axis, for as

$$\frac{[B]}{[F]} \rightarrow 0, \qquad [B] \rightarrow [Ab_t] \qquad (1.6)$$

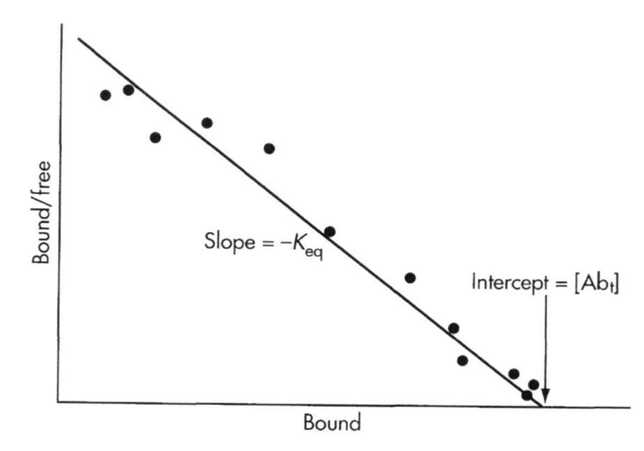

Fig. 1.7 Scatchard plot.

Figure 1.8 shows the effect of increasing the equilibrium constant while maintaining the antibody binding site concentration constant, and Figure 1.9 shows the converse, changing the amount of antibody but maintaining the equilibrium constant.

Not all the point estimates of [B]/[F] and [B] have equal weighting. Measurement errors have a disproportionate effect at both high and low [B]/[F] ratios. Care, therefore, needs to be exercised in constructing Scatchard plots to ensure that the concentrations of antigen are positioned to ensure maximum accuracy.

No separation of bound from free antigen can be 100% complete and there is always some residual free-antigen concentration associated with the bound fraction. This **nonspecific binding** needs to be determined with a considerable degree of accuracy as errors in its determination have a disproportionate effect on low [B]/[F] ratios, the region where the intercept with the bound fraction provides an estimate of the total antibody concentration.

Figure 1.10 shows a typical Scatchard plot using polyclonal antisera with a range of equilibrium constants.

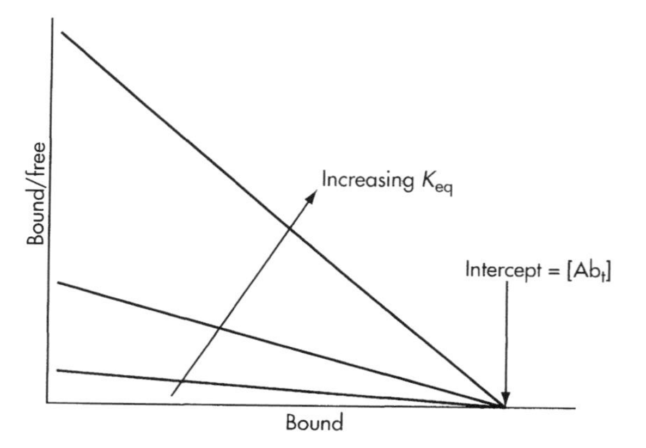

Fig. 1.8 Scatchard plot.

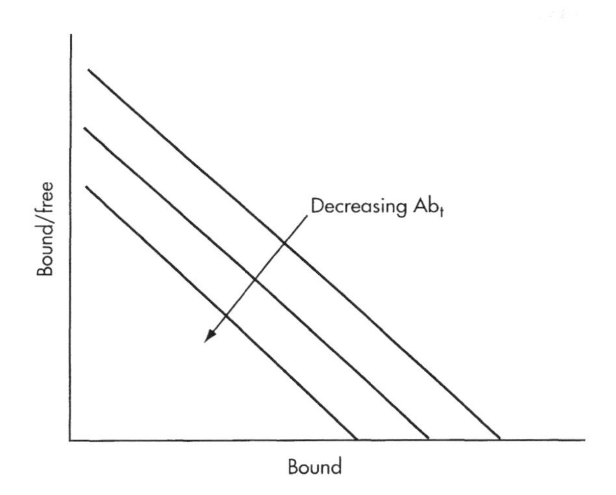

Fig. 1.9 Scatchard plot.

This gives rise to a curved line which in this example has been simplistically resolved into two populations of high- and low-affinity antibodies.

A reciprocal plot (also known as a Langmuir or Steward–Petty plot) may also be employed; this sometimes has the advantage of enabling the antibody concentration to be more accurately pinpointed.

Rearranging equation (1.3):

$$\frac{[Ab_t] - [Ag\text{–}Ab]}{[Ag\text{–}Ab]} = \frac{1}{K_{eq}[Ag]} \tag{1.7}$$

$$\frac{[Ab_t]}{[Ag\text{–}Ab]} = \frac{1}{K_{eq}[Ag]} + 1 \tag{1.8}$$

$$\frac{1}{[Ag\text{–}Ab]} = \frac{1}{K_{eq}[Ag]} \times \frac{1}{[Ab_t]} + \frac{1}{[Ab_t]} \tag{1.9}$$

substituting

$$\frac{1}{[B]} = \frac{1}{K_{eq}[F]} \times \frac{1}{[Ab_t]} + \frac{1}{[Ab_t]} \tag{1.10}$$

A typical reciprocal plot is shown in Figure 1.11, of 1/[B] versus 1/[F].

At infinite free antigen, and hence complete antibody saturation

$$\frac{1}{[F]} \rightarrow 0, \qquad \frac{1}{[B]} \rightarrow \frac{1}{[Ab_t]} \tag{1.11}$$

Ascertaining the most correct line on the Scatchard plot to estimate K_{eq} may be difficult, particularly with polyclonal antisera, where a mixture of antibodies of different affinities, are present. In such circumstances the average equilibrium constant may be the only useful measure of antibody affinity. There are two simple graphical solutions. The first approximates K_{eq} at half saturation of antibody binding sites.

From equation (1.2), when the antibody binding sites are half-saturated:

$$K_{eq} = \frac{1}{[Ag]} \tag{1.12}$$

The graphical solution is obtained by plotting [B] versus log[F] as shown in Figure 1.12 (logs are used purely because of the range of values found). A line is drawn at half the apparent maximal binding. The point on the x-axis where this corresponds to the intersection with the binding curve provides a reasonable average estimate of $1/K_{eq}$.

The second graphical solution is the **Sips plot** (Nisonoff and Pressman, 1958). An estimate of the total

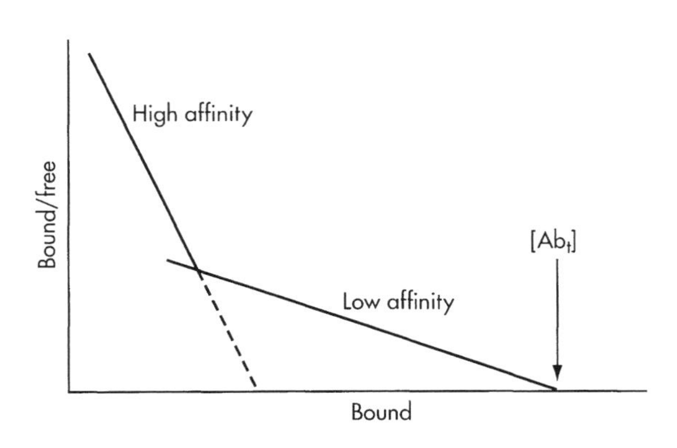

Fig. 1.10 Scatchard plot.

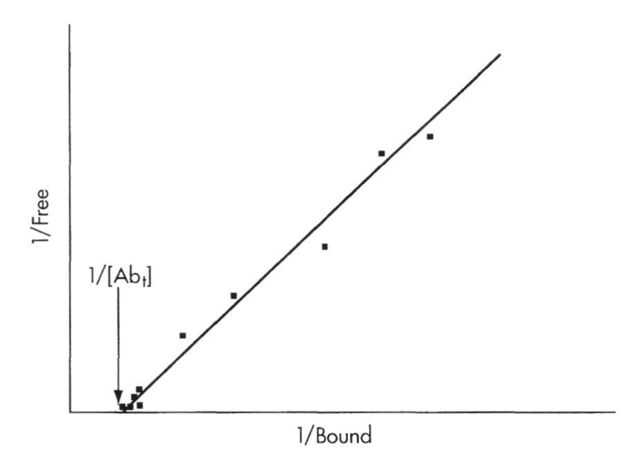

Fig. 1.11 Reciprocal plot.

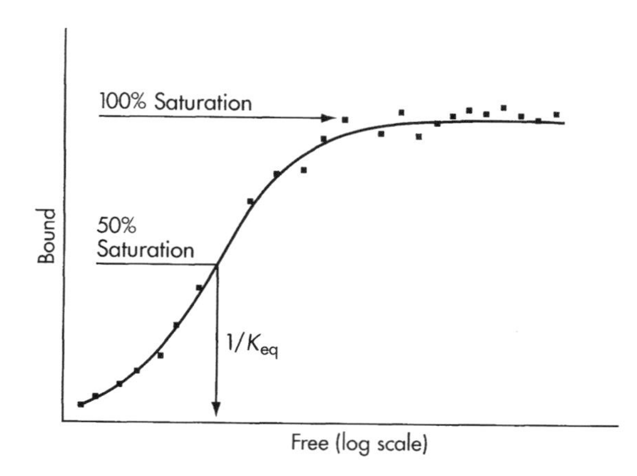

Fig. 1.12 Saturation plot.

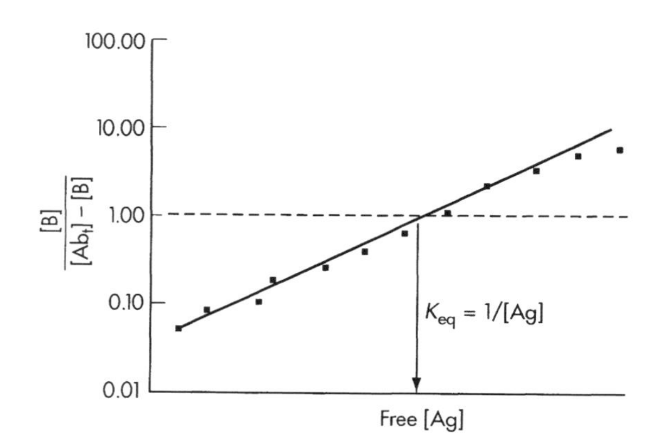

Fig. 1.13 Sips plot.

antibody binding sites is required from either the Scatchard or reciprocal plot.

Rearranging equation (1.2)

$$K_{eq}[Ag] = \frac{[Ag\text{–}Ab]}{[Ab_t] - [Ag\text{–}Ab]} \tag{1.13}$$

when

$$[Ag\text{–}Ab] = [Ab] = [Ab_t] - [Ag\text{–}Ab] \tag{1.14}$$

i.e. at half saturation

then

$$K_{eq}[Ag] = 1, \qquad K_{eq} = \frac{1}{[Ag]} \tag{1.15}$$

In practice a plot is constructed of

$$\log\left(\frac{[B]}{[Ab_t] - [B]}\right) \text{ versus } \log[F] \tag{1.16}$$

When

$$\frac{[B]}{[Ab_t] - [B]} = 1, \qquad \log\left(\frac{[B]}{[Ab_t] - [B]}\right) = 0 \tag{1.17}$$

and

$$K_{eq} = \frac{1}{[Ag]} \tag{1.18}$$

The graphical solution is shown in Figure 1.13. K_{eq} represents the estimate of the average affinity constant.

Commonly encountered antibody equilibrium constants range from 10^6 L/mol to, exceptionally, 10^{12} L/mol. Antibodies with equilibrium constants less than 10^8 L/mol are generally not useful in immunoassays. It should be stressed that concentrations of antibody refer to the total number of binding sites. For small haptens both IgG binding sites may bind to antigen: the **valence** is two. For some large proteins, steric restrictions may prevent binding of both antibody binding sites. Difficulties arise in the interpretation of Scatchard plots when the antigen is multivalent (with repeating epitopes). In this case the numbers of antibody molecules bound to each antigen follow a Poisson distribution and more complex modeling is required to estimate both K_{eq} and $[Ab_t]$.

The Law of Mass Action may also be used to predict the changes in [B]/[F] ratio with increasing concentration (or **dose**) of antigen [Ag], the so-called **dose-response curve**.

From equation (1.2), the ratio of bound to free antigen, [B]/[F] is

$$\frac{[B]}{[F]} = \frac{[Ag\text{–}Ab]}{[Ag]} \tag{1.19}$$

and the components [Ag] and [Ab] are

$$[Ag] = [Ag_t] - [Ag\text{–}Ab] \tag{1.20}$$

$$[Ab] = [Ab_t] - [Ag\text{–}Ab] \tag{1.21}$$

The ratio of bound to free antigen can be rearranged from equations (1.20) and (1.21) to

$$\frac{[B]}{[F]} = \frac{[Ag\text{–}Ab]}{[Ag_t] - [Ag\text{–}Ab]} \tag{1.22}$$

$$[Ag{-}Ab] = \frac{[Ag_t] \times \dfrac{[B]}{[F]}}{1 + \dfrac{[B]}{[F]}} \qquad (1.23)$$

Combining equations (1.2) and (1.19):

$$\frac{[B]}{[F]} = K_{eq}[Ab] \qquad (1.24)$$

Substituting equations (1.21) and (1.23) into equation (1.24):

$$\frac{[B]}{[F]} = K_{eq}\left([Ab_t] \frac{[Ag_t] \times \dfrac{[B]}{[F]}}{1 + \dfrac{[B]}{[F]}}\right) \qquad (1.25)$$

Multiplying by $1 + [B]/[F]$ to simplify

$$\left(\frac{[B]}{[F]}\right)\left(1 + \frac{[B]}{[F]}\right) - \left(1 + \frac{[B]}{[F]}\right)K_{eq}[Ab_t] + \frac{[B]}{[F]}K_{eq}[Ag_t] = 0 \qquad (1.26)$$

$$\left(\frac{[B]}{[F]}\right)^2 + \frac{[B]}{[F]} - K_{eq}[Ab_t] - \frac{[B]}{[F]}K_{eq}[Ab_t] + \frac{[B]}{[F]}K_{eq}[Ag_t] = 0 \qquad (1.27)$$

$$\left(\frac{[B]}{[F]}\right)^2 + \frac{[B]}{[F]}(K_{eq}[Ag_t] - K_{eq}[Ab_t] + 1) - K_{eq}[Ab_t] = 0 \qquad (1.28)$$

This reduces to the familiar quadratic, $ax^2 + bx + c = 0$. This can be solved:

for $ax^2 + bx + c = 0$:

$$x = \frac{-b \pm \sqrt{b^2 - 4ac}}{2a} \qquad (1.29)$$

It is, therefore, possible to calculate [B]/[F] for any given quantity of $[Ag_t]$, provided both K_a and $[Ab_t]$ are both accurately known, or conversely, and of more interest, $[Ag_t]$ can be solved given [B]/[F]. In other words the quantity of an unknown amount of antigen can be determined directly from the ratio of [B]/[F]. To ascertain the latter, all that is required is a method of separating the bound antigen from the free, and the ability to measure the relative proportion of each. The relationship is, however, more conveniently expressed as the percentage bound in relation to the total antigen present, because this more directly relates to what is measured in practice, as

$$\%\ \text{Bound} = \frac{\dfrac{[B]}{[F]}}{1 + \dfrac{[B]}{[F]}} \times 100\% \qquad (1.30)$$

A typical plot of per cent bound versus $[Ag_t]$ is shown in Figure 1.14.

Increasing the concentration of antibody shifts the percentage bound (%Bd) curve to the right, towards higher concentrations of antigen $[Ag_t]$ as shown in Figure 1.15.

Increasing the antibody equilibrium constant K_a acts to increase the slope of the response (see Figure 1.16).

IMMUNOASSAY DESIGN

A wide variety of assay designs and formats have been developed over the years. They can be considered under two main categories. The first includes what might be termed principal assay designs; those primarily concerned with exploiting the fundamental properties of immunoanalytical techniques. The second category comprises those that have built upon those fundamental principles in order to improve the precision of the assay, reduce assay incubation times, simplify the technology or render the assays amenable to automation. This section deals solely with the fundamental assay designs: a wide spectrum of assay designs built upon these principles are amply illustrated in the remaining chapters in PART 1 and in PART 2 – PRODUCTS.

COMPETITIVE (REAGENT LIMITED) ASSAYS

We have seen how, given the equilibrium constant and the antibody concentration, it is possible to derive the bound/free (B/F) ratio, and hence the percentage bound, for any given concentration of antigen. Conversely, if the percentage bound is measured then the antigen concentration of an unknown solution can be estimated. This principle formed the basis for the first type of immunoassay, described by Yalow and Berson (1959) for the assay of insulin in human serum.

All that is required is a method of separating the bound antigen from the free, and a means of determining the relative quantities of antigen in each. The former may simply be achieved by separation of the antibody component, and thus antibody–antigen complex, from the reaction mixture, leaving the free antigen in solution (see Figure 1.17).

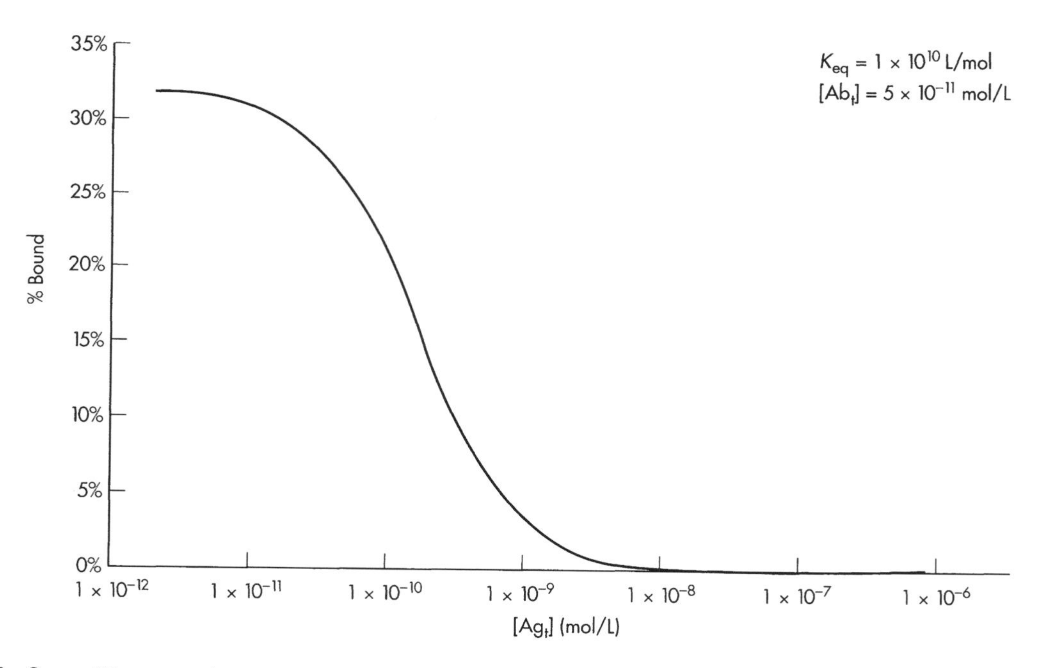

Fig. 1.14 Competitive assay dose–response curve.

In Yalow and Berson's assay the partition of insulin between the two fractions was determined by inclusion of a trace quantity of radiolabeled ^{125}I-insulin (the **tracer**) into the reaction, and subsequent monitoring of the distribution of activity between bound and free. In practice neither K_{eq} nor the antibody concentration are known with sufficient certainty to enable accurate predictions to be made of the concentration

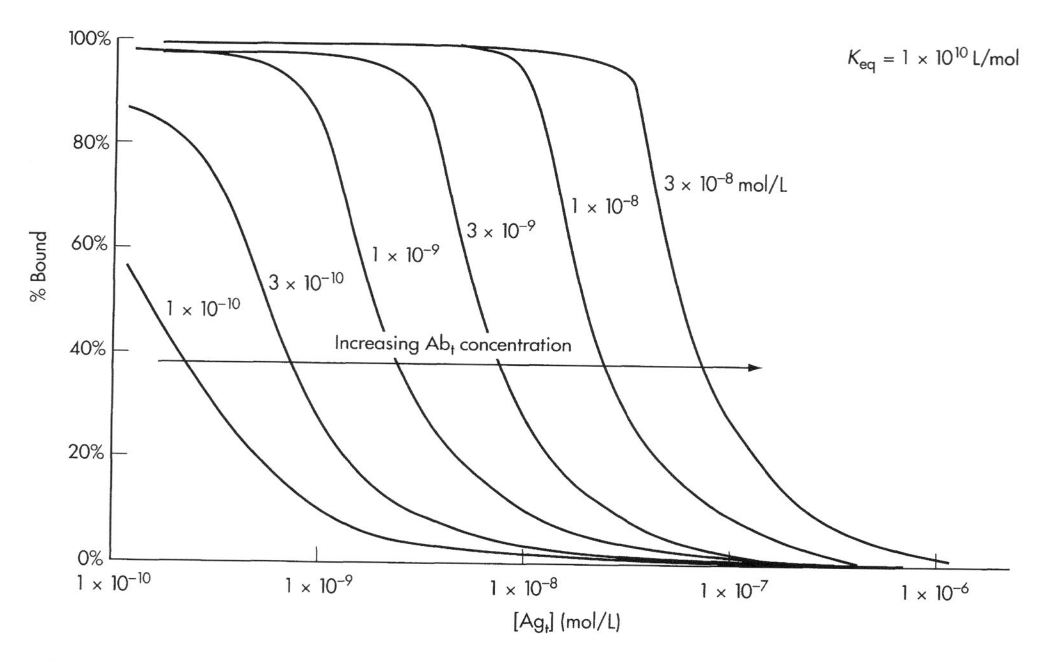

Fig. 1.15 Influence of antibody concentration.

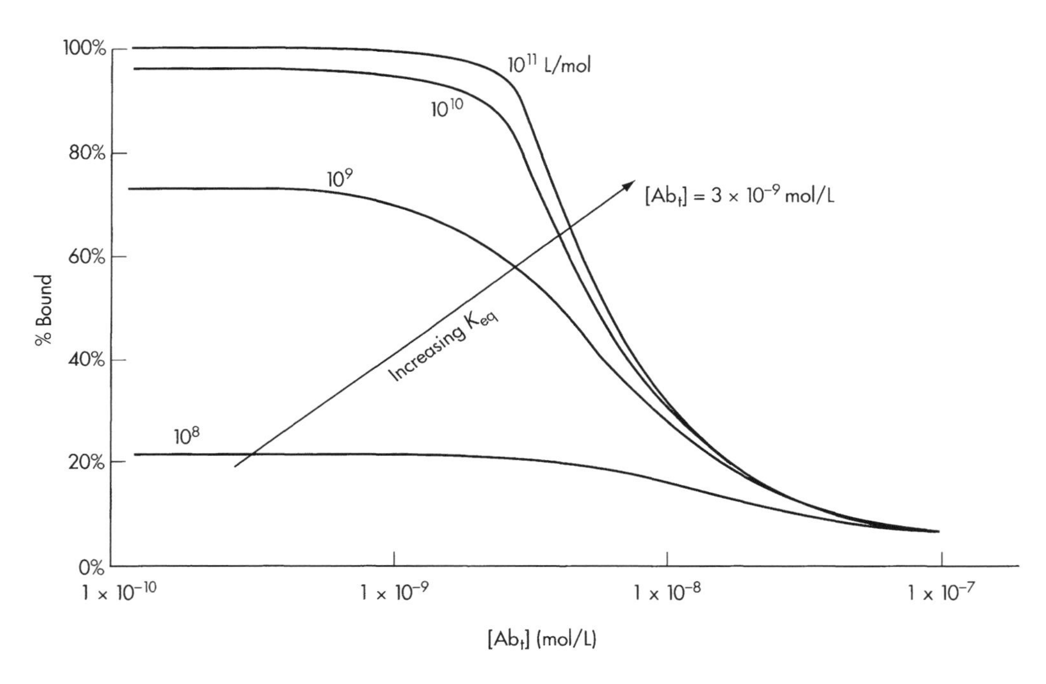

Fig. 1.16 Influence of K_{eq}.

of unknown antigen. Samples of known antigen concentrations are, therefore, included as standards (or **calibrators**) and a **calibration curve** drawn of percentage of activity in the bound fraction/total activity against the antigen concentration in the standards. The concentration of antigen in unknown samples may then be interpolated from the calibration curve (see Figure 1.18).

Such assays are commonly, but incorrectly, referred to as **competitive** assays. Competition can only be truly said to occur if the antibody binding sites are fully saturated. For competitive assays this is not the case and the function of the labeled antigen is simply to provide a means of assessing the relative partition of antigen between that which is bound and that which is free.

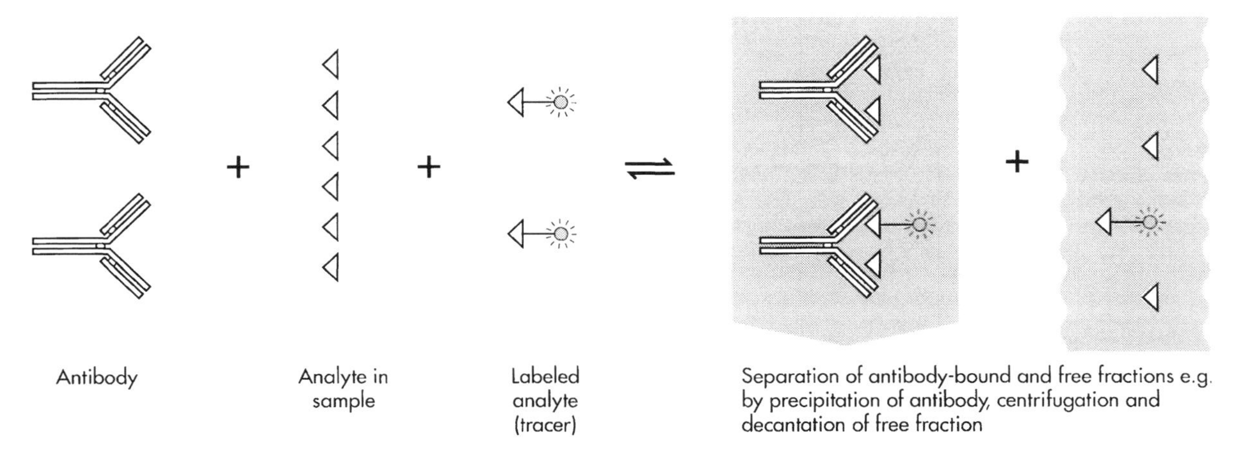

Fig. 1.17 Competitive (reagent-limited) immunoassay.

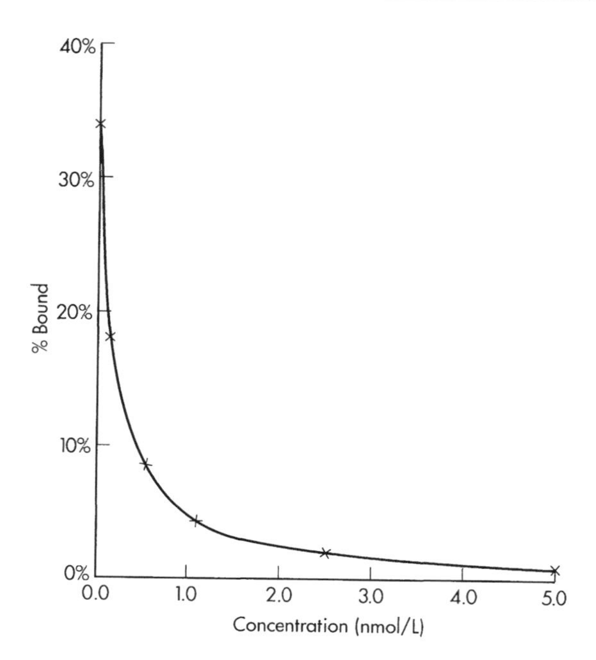

Fig. 1.18 Typical calibration (dose–response) curve.

Ekins has termed such assay designs as **reagent limited**, primarily and logically to distinguish them from **reagent excess** assays which are described later (Ekins, 1977). However, this terminology has not become established, and the term competitive assay has remained in common usage.

A variety of separation and detection systems have been employed in competitive assay designs. *See* SEPARATION SYSTEMS, SIGNAL GENERATION AND DETECTION SYSTEMS, AND HOMOGENEOUS IMMUNOASSAYS.

Assays where the antigen is labeled have some practical and theoretical disadvantages. First, the sensitivity of such assays is principally governed by the equilibrium constant of the antibody, and thus may not exploit to the full the potential that immunoanalytical techniques offer with alternative designs. Second, labeling the analyte may reduce or even abolish the recognition by antibody because a critical epitope has been affected or hidden. In some cases labeling may actually enhance recognition when it is directed to the same position that the hapten was originally conjugated to on the immunogen, a phenomenon known as **bridge recognition**. *See* ANTIBODIES.

SINGLE-SITE IMMUNOMETRIC ASSAYS

The first major advance in assay design was made in 1968, by Miles and Hales, using labeled antibody, rather than labeled antigen (Miles and Hales, 1968). Figure 1.19 shows the basic principle. Sample or standard antigen is incubated with labeled antibody. After the reaction, unbound labeled antibody is removed from solution by the addition of a large excess of solid-phase coupled antigen.

Empirically there are two major variants of this design although they do in fact represent extremes of the same continuum. The first uses antibody in concentrations that are limited and in this respect the assay is analogous to a competitive assay. A practical example of such an approach is in the assay of free thyroid hormones, where the labeled antibody is partitioned between free hormone in solution and hormone bound to the solid phase, both reactions occurring simultaneously. For the assay configuration, *see* FREE ANALYTE IMMUNOASSAYS.

The second variant uses labeled antibody in excess. On purely empirical grounds, the presence of a large excess of labeled antibody should drive the reaction between antigen and antibody to a far greater degree of completion (equation (1.1)) than is possible with competitive assays, and thus overcome some of the limitations of assay sensitivity dictated by the equilibrium constant in the latter type of design. In practice the sensitivity depends very much upon which fraction is measured: antibody bound to sample in solution, or extracted by the solid phase. Measurement of the former is greatly influenced by the presence of any immunologically inactive labeled tracer. Measurement of the latter, the amount of labeled antibody extracted by the solid phase, is seldom a practical proposition when the concentrations of labeled antibody are high, because the sensitivity in such cases depends upon the ability to detect with confidence a very small difference between two large measurements.

In practice one-site immunometric assays using large concentrations of labeled antibody seldom offer any advantages in terms of sensitivity over competitive designs, and hence have never achieved widespread popularity.

TWO-SITE IMMUNOMETRIC ASSAYS (REAGENT EXCESS)

Many proteins have multiple epitopes that are sufficiently well spatially separated that two antibodies may bind at the same time. This forms the principle of the two-site immunometric assay first described by Addison and Hales (1970). Binding of the two antibodies may occur sequentially or simultaneously. The principle is shown below. Sample is first incubated with the **capture** antibody, which reacts with the first epitope on the protein and binds it to the solid phase. The solid phase is then washed to remove unreacted components and further incubated with labeled **detecting** antibody, which binds to the antibody–antigen complex. Unreacted excess detecting antibody is then removed by washing

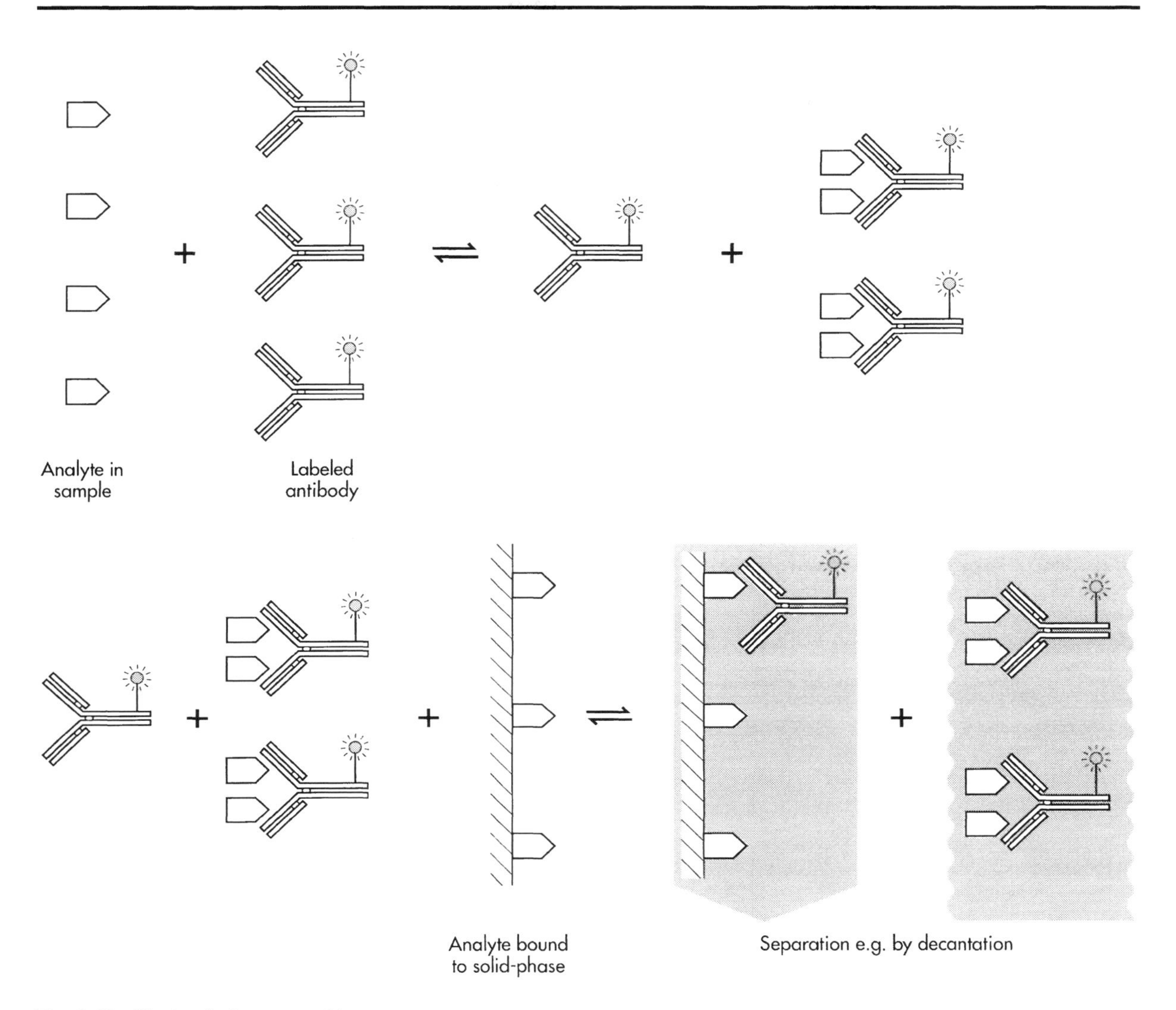

Fig. 1.19 Single-site immunometric assay.

and the signal level emanating from the solid phase determined (see Figure 1.20).

A typical dose–response (calibration) curve is shown below. If both capture and detecting antibody reactions proceeded to completion then the dose–response would approximate to a straight line. Many assays approach this, particularly at low concentrations, and some have sufficient linearity to enable single-point calibration of the standard curve (see Figure 1.21).

Such designs have the obvious advantage of increasing the specificity of the assay. For example, the glycoprotein hormones thyrotropin (TSH), human chorionic gonadotropin (hCG), follicle stimulating hormone (FSH) and luteinizing hormone (LH) share a common α-subunit, biological specificity being determined by the composition of relatively minor changes in the β-subunit. By targeting the capture antibody to a distinct TSH β-subunit epitope it is possible to capture only TSH, even in the presence of a vast excess of circulating hCG with the same α-subunit, as occurs in pregnancy. After removal of unreacted components, detection of the bound complex is achieved with an antibody specific for the common α-subunit. This type of design only really became practicable with the advent of **monoclonal antibodies** with single-epitope specificity. Many two-site immunometric assays have a single-stage incubation with binding of both capture and detecting antibody proceeding simultaneously.

Immunometric assays are also often described as **sandwich** assays. Immunometric assays that have antibody or antigen coated onto a solid phase, with

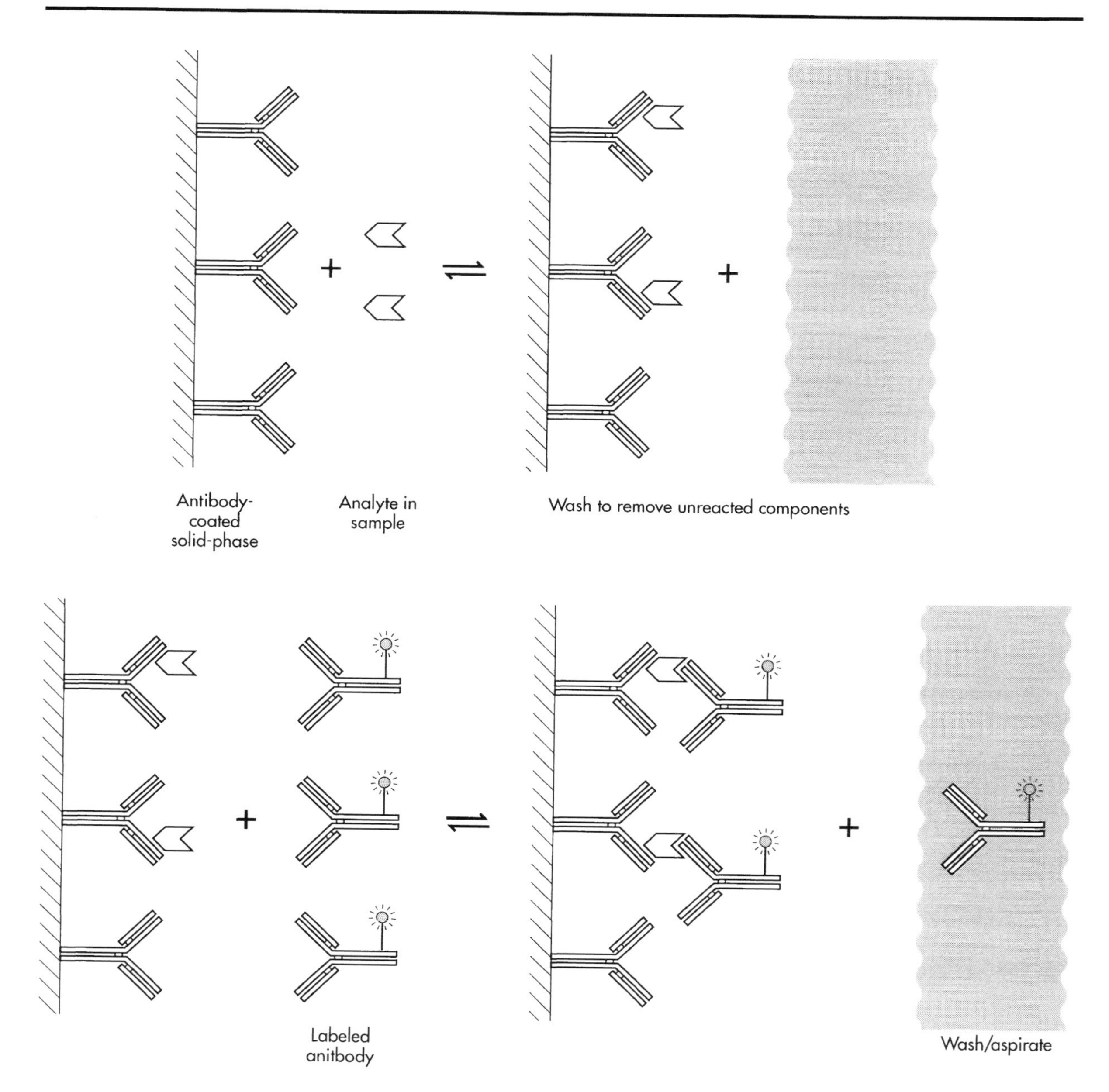

Fig. 1.20 Two-site immunometric assay (reagent excess).

an enzyme label, are also known as enzyme-linked immunosorbent assays (**ELISA**).

DETERMINANTS OF ASSAY SENSITIVITY

It seems obvious that an appreciation of the underlying principles governing assay design is an essential prerequisite for its application. It may seem surprising, therefore, that so little attention was given to these principles for many years following the initial description of the technique.

Roger Ekins has been one of the most vocal advocates for designs based on sound fundamental principles. Writing in 1979, and on many subsequent occasions, he expressed dismay that some assays appeared to be designed on the basis of certain empirical concepts that had been repeated in the literature to such an extent as to become accepted dogma.

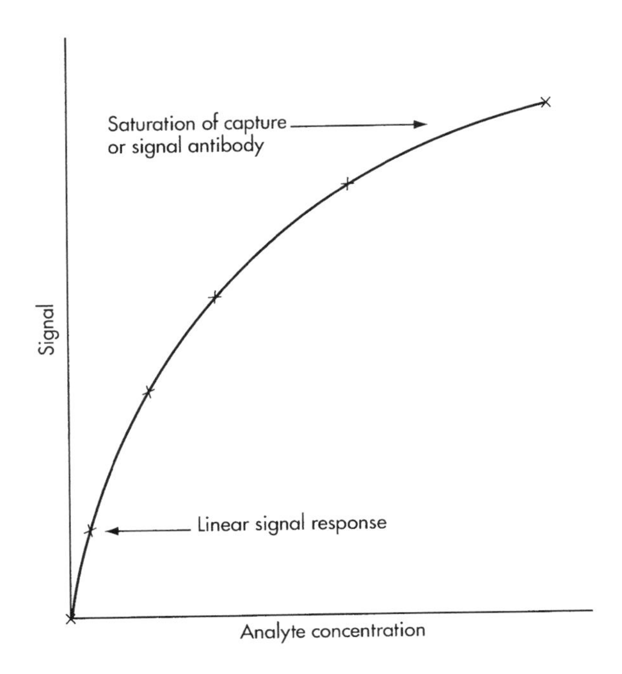

Fig. 1.21 Immunometric dose–response curve.

To some extent this is understandable given that initially many of the factors limiting performance were to a large extent practical ones, such as difficulties in the isolation, purification and labeling of antigens, and in obtaining antisera of sufficient specificity. There was also the widespread feeling that there was little incentive to resort to seemingly complex mathematical equations when a few simple experiments could achieve the same result.

Most of the theoretical treatments have focused on factors determining assay sensitivity. High sensitivity is an intrinsically desirable property of any analytical technique; indeed the high sensitivity that immunoassays offer is, and remains, the main reason for their continued use and further development. For some analytes, where the range of concentrations of interest is higher than the sensitivity available, it could be argued that higher sensitivity has little utility. There are, however, several other factors to consider:

- Higher-sensitivity assays may open up new opportunities in the diagnosis of disease that were previously unrecognized or unavailable. A good example is the development of so-called third generation thyrotropin (TSH) assays which for the first time open up the possibility of differentiating between the euthyroid and hyperthyroid state (*see* THYROID – THYROTROPIN).
- High sensitivity enables much smaller, or more easily obtainable, samples to be analyzed, for example, capillary blood samples from newborns or in saliva.
- Small sample size confers another important advantage. Few immunoassays are totally free from interference by what are often ill-defined factors in biological fluids. Different samples, each containing the same amount of analyte, may give different results. The lower the proportion of sample in the overall assay reaction, the less influence these **matrix effects** have on the accuracy of the assay. High-sensitivity assays tend, therefore, to be more robust.

Although the wide variety of assay designs dictates that optimal assay conditions and reagent concentrations differ, some observations are generally applicable.

The sensitivity of any analytical technique is defined as the minimal concentration that can be reliably estimated. It can be defined more formally as the minimal concentration of analyte that is statistically unlikely to form part of the range of signals seen in the absence of analyte, for example, the concentration corresponding to two standard deviations (s.d.) difference from the signal at zero concentration. The concept is familiar in practically all fields of measurement: in electronics it is commonly referred to as the **signal-to-noise ratio**. The latter analogy is particularly apt as it relates to music. The improvement in sound quality from a compact disc over cassette tape is governed not by the volume of music, but by the absence of background noise. Sensitivity, therefore, depends upon two factors: the increase in signal seen in the presence of analyte and the errors in measurement when no analyte is present (see Figure 1.22).

Yalow and Berson originally considered that the major determinant of assay sensitivity for competitive assays was the slope of the dose–response curve (Yalow and Berson, 1970; Berson and Yalow, 1973) and many workers concentrated on optimizing this parameter in isolation from the errors in measurement.

Ekins, however, focused attention on the errors in the measurement. He dismissed the influence of slope, arguing that if this is plotted in different frames of reference, for example, as bound or 1/bound against antigen concentration (Ekins, 1979; Ekins and Newman, 1970), the same assay could give different slopes and hence opposite conclusions regarding the optimal concentrations of reactants. To some extent this is true, as Yalow and Berson's arguments were based upon the slope of the percentage bound, whereas it is the rate of change of the measured signal that determines the difference between two measurements, not a derivative of the measurement in relation to the total amount added.

Any theoretical analysis must combine both thermodynamic and statistical elements, they cannot be used in isolation. It is worth noting that in the total absence of errors of any kind there would be no optimal design.

Assay optimization may, therefore, be regarded as simply an exercise in identifying the sources of error and determining, with any assay design, the appropriate concentrations of reagents that minimize their influence.

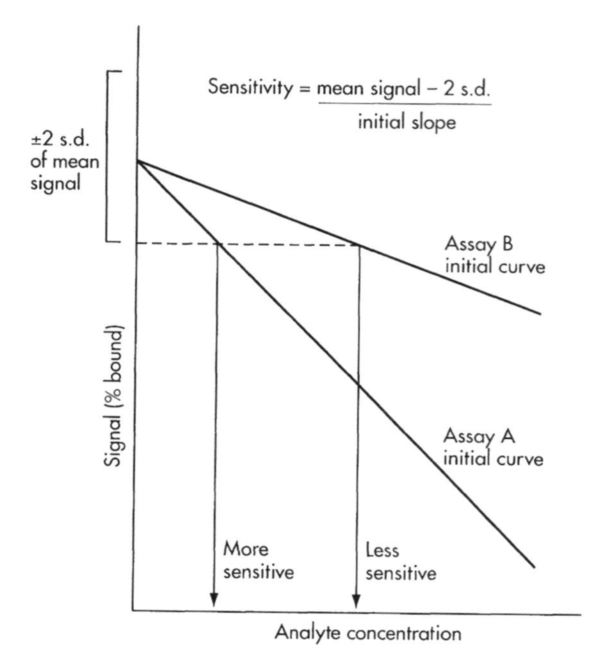

Fig. 1.22 Estimation of sensitivity.

Although only mathematical models can give quantitative estimates of the optimal concentrations of reagents, Ekins has demonstrated that the same qualitative conclusions may easily be drawn by examining the assay design in terms of whether the occupied or unoccupied antibody binding sites are being measured: the antibody occupancy principle (Ekins, 1991).

The Principle of Antibody Occupancy

All assays, regardless of which component is labeled, can essentially be described in one of two ways. Either they rely for measurement on estimates of the antibody binding sites occupied by antigen, or indirectly by measurement of unoccupied binding sites (see Figure 1.23).

In competitive assays the labeled antigen binds to antibody binding sites unoccupied by sample antigen. The addition of unlabeled (sample) antigen in such a system causes a reduction in the number of unoccupied binding sites.

In immunometric assays the labeled antibody measures binding sites occupied by antigen. Signal, therefore, rises with increasing antigen concentration from (one hopes) a small background signal in the absence of analyte. Generally speaking it is far better to measure a large signal against a small background than it is to measure the difference between two large signals. Purely on that basis alone immunometric assays should be capable of higher sensitivity than their corresponding competitive counterparts.

For competitive assay designs the rate of change of signal on addition of unlabeled sample antigen is greatest when the amount of labeled antigen is small by comparison.

In contrast, when the occupied binding sites are being estimated, as with immunometric designs, the change in binding is maximal when the concentration of labeled antibody is high.

In competitive designs, as a consequence of the small amount of labeled antigen, the concentration of antibody also needs to be low to minimize the errors in estimating the unoccupied sites. Because the fractional occupancy of the binding sites is primarily governed by the affinity of the antibody for antigen the equilibrium constant is the major limiting factor in determining sensitivity in competitive assay designs.

The conclusions from these qualitative analyses need to be judged in the context of the errors of estimation.

In competitive designs, as the concentration of labeled antigen tends to zero, so the errors in measurement become infinitely large. Clearly a limiting factor is the precision of measurement of low concentrations of labeled antigen. Nonspecific binding may also need to be considered. As this is related to the total concentration of labeled antigen, a low percentage binding of labeled antigen renders the nonspecific binding (and its error) large in relation to the specific binding. Consequently, variations in nonspecific binding form a major contribution to the overall imprecision.

In summary, the sensitivity of competitive assays is governed by three factors: the equilibrium constant, the precision of signal measurement and the level of nonspecific binding, where this forms a significant proportion of the specific binding. Each predominates in the absence of the others. If nonspecific binding and error in the measurement of labeled antigen are insignificant then optimal sensitivity can be achieved with a very low initial percentage binding, say, 1 or 2%, a figure very much at odds with accepted practice. In such cases the sensitivity of the assay is governed solely by residual errors (pipetting, for example) and the equilibrium constant. Where nonspecific binding is significant this necessitates higher antibody levels to increase the percentage of labeled antigen bound and hence signal-to-noise ratio. Sensitivity in such cases is a function of both K_{eq} and the level of nonspecific binding. Finally, where errors in signal measurement are significant then proportionately larger amounts of labeled antigen are required. Sensitivity in this case also depends on the specific activity of the detecting reagent.

The converse is true when occupied sites are measured directly as in immunometric designs. High concentrations of both capture and labeled antibody ensure that a high proportion of antigen is bound. Sensitivity should, therefore, improve as the antibody concentrations tend to infinity. Increasing the amount of labeled antibody,

Competitive assays measure unoccupied sites

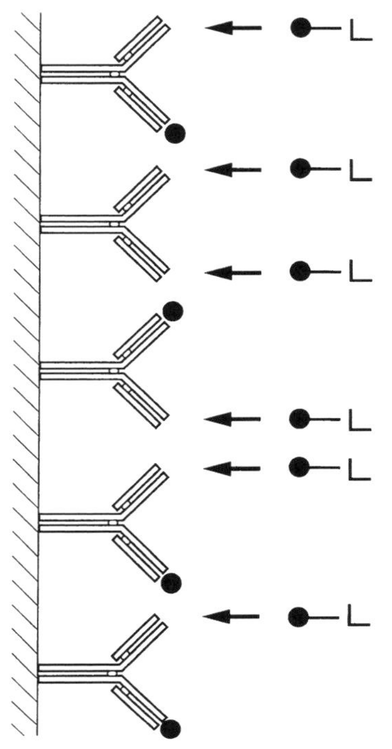

Immunometric assays measure occupied sites

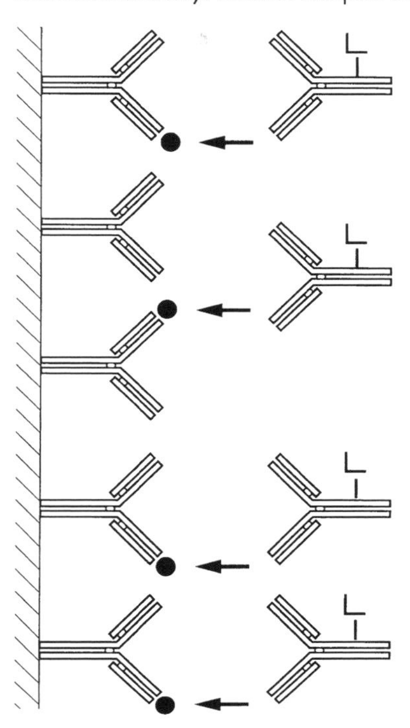

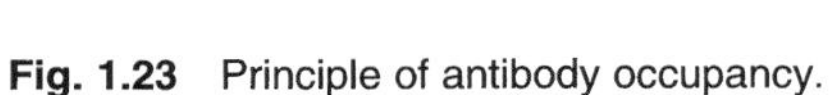

Fig. 1.23 Principle of antibody occupancy.

however, also increases the nonspecific binding. Because this is a function of the total amount of labeled antibody added, as more labeled antibody is added to the system, the nonspecific binding (and its error) also increases. Eventually a point is reached where the level of nonspecific binding increases faster than the rate of specific binding. In other words the signal-to-noise ratio has a maximum value. The point of maximum sensitivity depends upon the level of nonspecific binding and its associated error and the error in measurement of low levels of signal.

In summary, the equilibrium constant assumes less importance in determining assay sensitivity for immunometric assays. The major determining factors are the level and imprecision of the nonspecific binding and the error in measurement of signal. A high sensitivity assay, therefore, requires high concentrations of antibody in relation to the antigen, a low nonspecific binding and a high specific activity label. Where nonspecific binding is insignificant, error in signal measurement is the most important factor. When measurement errors and nonspecific binding are insignificant then the limiting factor is the equilibrium constant and other residual experimental errors. Highly sensitive immunometric assays, therefore, depend on achieving very low nonspecific binding in conjunction with a high specific activity detecting reagent.

Single-site immunometric assays can be considered as competitive type designs when the unoccupied binding sites are estimated using signal from the excess labeled antibody extracted by the solid phase. Alternatively, if the labeled antibody bound to analyte in solution is counted, then the occupied sites are estimated, and the assay is more akin to the immunometric design.

Theoretical Models and Quantitative Predictions

Any theoretical optimization model needs to accommodate several variables:

- the antibody equilibrium constant(s)
- the antibody concentration (both if two-site immunometric assay)
- the amount of labeled antigen added (if competitive)
- the sample and incubation volumes
- the activity of the detecting reagent (e.g. for radioactive assays, the specific activity)
- the level of nonspecific binding of detecting reagent

Optimization also requires models to be developed of the errors in measurement arising from

- the variation around the mean level of nonspecific binding
- the imprecision of pipetting
- additional variation as a result of manipulation errors (centrifugation, decantation, washing)
- the precision of measurement of the final signal.

Finally assumptions need to be made regarding the mechanism of reactions:

- The reactions are bimolecular and obey the Law of Mass Action.
- All of the reagents are chemically homogeneous, and react independently with no allosteric effects.
- All of the reactants are uniformly dispersed throughout the reaction mixture.
- The separation procedure does not disturb the equilibrium.

Competitive Assay Design

There are two different approaches in determining the theoretical assay sensitivity and optimal concentrations of reactants. The first assesses errors and dose–response (calibration curve) slope separately, the second relies on unified equations combining the errors with the slope.

The first method is conceptually the simplest, and gives values very close to those obtained by Ekins (Ekins and Newman, 1970) using unified response/error equations. Values for the reagent concentrations are substituted into equation (1.28) to yield estimates of the percentage bound. The percentage bound is then used to calculate the signal at zero dose of unlabeled antigen, and the standard deviation due to the sum of the contributory errors in its determination. The sensitivity in terms of percentage binding is then calculated and the figure substituted into the dose–response equation to obtain the corresponding concentration of antigen at the limit of sensitivity. Borth (1970) and Ezan *et al.* (1991) have proposed similar approaches.

Errors due to pipetting and other experimental manipulations are usually constant over a wide range of assay conditions and reactant concentrations. Their exact magnitude can only be determined by experimentation. However, they are likely to be of the order of 1–3% in terms of their coefficient of variation. In contrast, the imprecision in measurement of bound labeled antigen may increase dramatically at low concentrations of labeled antigen and/or low specific binding. For example, with radioactive or luminescent detection, the CV of 10,000 accumulated signal events would be 1% (the error is the square root of the total counts expressed as a percentage of the total counts), whereas the CV of 100 events would be 10% (10/100). Errors in nonspecific binding may also make a significant contribution to the overall error, particularly when it forms a significant proportion of the overall binding. Variations around the mean level of nonspecific binding may be quite large. For example, in cases where the antibody–antigen complex is separated by centrifugation, there may be considerable variation in the amount of free antigen left after decanting the tubes. The three sources of error need to be assessed separately and then combined in terms of their standard deviation.

There are several stages to the analysis.

1. Equation (1.28), restated below from earlier in this chapter, allows the bound/free ratio to be determined for any concentrations of reactant. Values for the equilibrium constant, and total antibody and labeled antigen concentration, may be substituted into the equation and the [B]/[F] ratio and the fraction bound (B_f) determined in the absence of added unlabeled antigen:

$$\left(\frac{[\mathrm{B}]}{[\mathrm{F}]}\right)^2 + \frac{[\mathrm{B}]}{[\mathrm{F}]} - (K_{eq}[\mathrm{Ag_t}] - K_{eq}[\mathrm{Ab_t}] + 1) - K_{eq}[\mathrm{Ab_t}] = 0 \qquad (1.28)$$

$$B_f = \frac{\frac{[\mathrm{B}]}{[\mathrm{F}]}}{1 + \frac{[\mathrm{B}]}{[\mathrm{F}]}} \qquad (1.31)$$

2. The total signal response (R_t) may be determined from the total concentration of antigen, the specific activity, the volume of the reaction and the time over which the signal is measured:

$$R_t = [\mathrm{Ag_t}]SVT \qquad (1.32)$$

where $\mathrm{Ag_t}$ is the total concentration of labeled antigen in mol/L, S is the specific activity of the labeled antigen, in terms of signal/mol/sec (for radioactive labels the specific activity in Bequerel/mol), V is the total volume of the reaction mixture in liters, T is the time the signal is counted in seconds.

3. The signal response due to specific antibody–antigen binding (R_o) is calculated:

$$R_o = R_t \times B_f \qquad (1.33)$$

4. The signal response due to nonspecific binding (R_{nsb}) is calculated as a function of the fractional binding of free antigen:

$$R_{nsb} = R_t(1 - B_f)\mathrm{NSB} \qquad (1.34)$$

where NSB is the fraction of free antigen bound nonspecifically.

5. The standard deviation of the specific binding signal in the absence of unlabeled antigen, due solely to experimental errors (s.d.$_{e}$), is calculated:

$$\text{s.d.}_{e} = \frac{CV_e}{100} \times R_o \quad (1.35)$$

where CV_e is the coefficient of variation due to experimental errors.

6. The standard deviation of the signal response due to nonspecific binding (s.d.$_{nsb}$) is calculated:

$$\text{s.d.}_{nsh} = \frac{CV_{nsb}}{100} \times R_{nsb} \quad (1.36)$$

where CV_{nsb} is the coefficient of variation around the mean level of nonspecific binding.

7. Radioactivity, chemiluminescence and fluorescence signals follow a Poisson distribution. The total observed signal response in the absence of unlabeled antigen is the sum of the specific binding, Ro and that due to nonspecific binding, R_{nsb}. The standard deviation of measurement (s.d.$_{m}$) is the square root of the total signal response:

$$\text{s.d.}_{m} = \sqrt{R_o + R_{nsb}} \quad (1.37)$$

8. The total combined standard deviation in terms of signal (s.d.$_{a}$) is, therefore,

$$\text{s.d.}_{a} = \sqrt{\text{s.d.}_{e}^{2} + \text{s.d.}_{nsb}^{2} + \text{s.d.}_{m}^{2}} \quad (1.38)$$

9. The signal corresponding to the 95% confidence level (by a one-sided test) for the distribution of signal in the absence of unlabeled antigen (R_{det}) is given by

$$R_{det} = R_o - 2\text{s.d.}_{a} \quad (1.39)$$

10. Hence the fractional binding (B_{det}), and [B]/[F] ratio at the limit of sensitivity ($[B]/[F]_{det}$) may be estimated:

$$B_{det} = \frac{R_{det}}{R_t} \quad (1.40)$$

$$\frac{[B]}{[F]}\text{det} = \frac{B_{det}}{1 - B_{det}} \quad (1.41)$$

11. Equation (1.28) can be rearranged to give the total antigen present at the limit of sensitivity in terms of the other reagent concentrations, the equilibrium constant and the [B]/[F] ratio:

$$[Ag_{det}] = [Ag_t]\left(1 + \frac{1}{\frac{[B]}{[F]}\text{det}}\right) - \frac{1}{K_{eq}}\left(1 + \frac{[B]}{[F]}\text{det}\right) \quad (1.42)$$

12. Substitution of $[B]/[F]_{det}$ permits calculation of the total antigen concentration present (labeled and unlabeled) at the limit of sensitivity, and hence the sensitivity can be derived as

$$\text{Sensitivity} = [Ag_{det}] - [Ag_t] \quad (1.43)$$

where $[Ag_{det}]$ is the total concentration of labeled and unlabeled antigen present at the detection limit, $[Ag_t]$ is the concentration of labeled antigen.

The sensitivity is that in the reaction mixture. If the sensitivity in the assay is 1 pmol/L but the sample is diluted 1/5 by the reagents, then it follows that the practical sensitivity is 5 pmol/L.

These series of simple equations are very useful in enabling 'what if' calculations to be performed by varying the reactant concentrations and equilibrium constant. Theoretical dose–response curves may also be produced by substitution of the relevant parameters into equation (1.28) and solving the quadratic to obtain [B]/[F] and hence percentage bound. Figure 1.24 below shows the percentage binding of antigen by dilutions of antibody with an average equilibrium constant of 1×10^{10} L/mol. The curves a, b, c, d and e represent decreasing antigen concentrations ranging from 10^{-9} to 10^{-11} (corresponding to $10/K_{eq}$ to $0.1/K_{eq}$).

The percentage binding of antigen increases with decreasing antigen concentration. As antigen concentrations decrease, the percentage binding tends towards

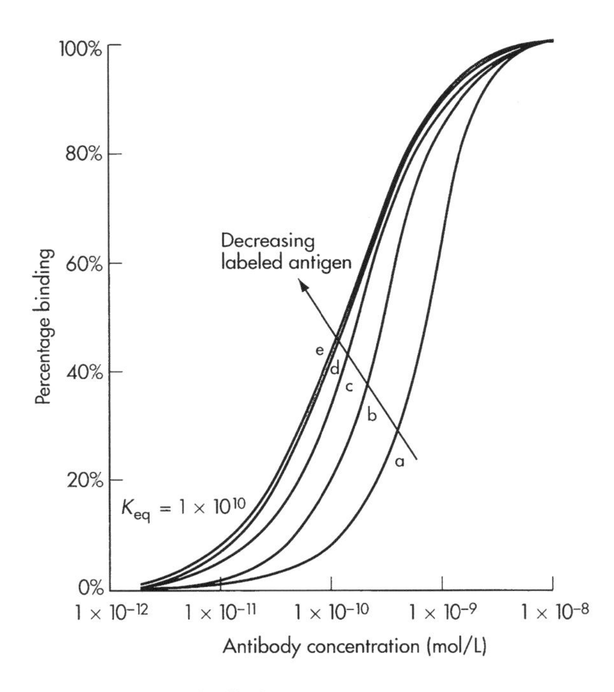

Fig. 1.24 Antibody dilution curves.

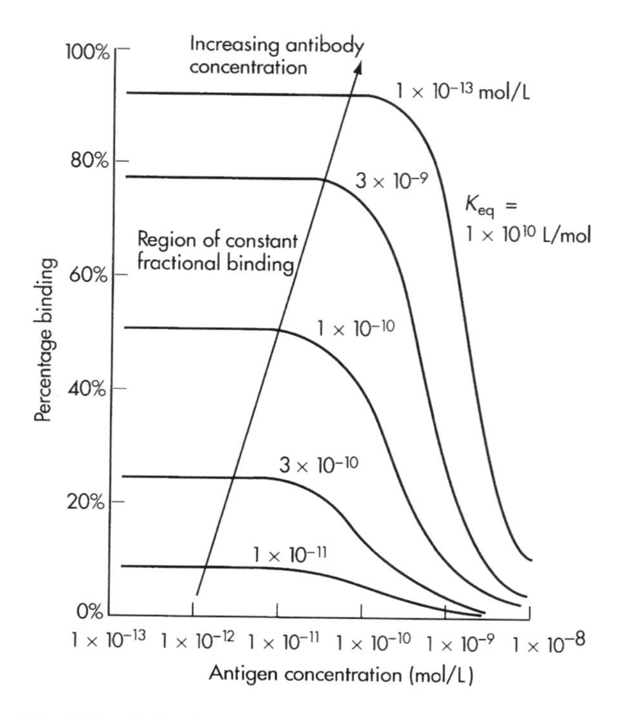

Fig. 1.25 Antibody dilution curves.

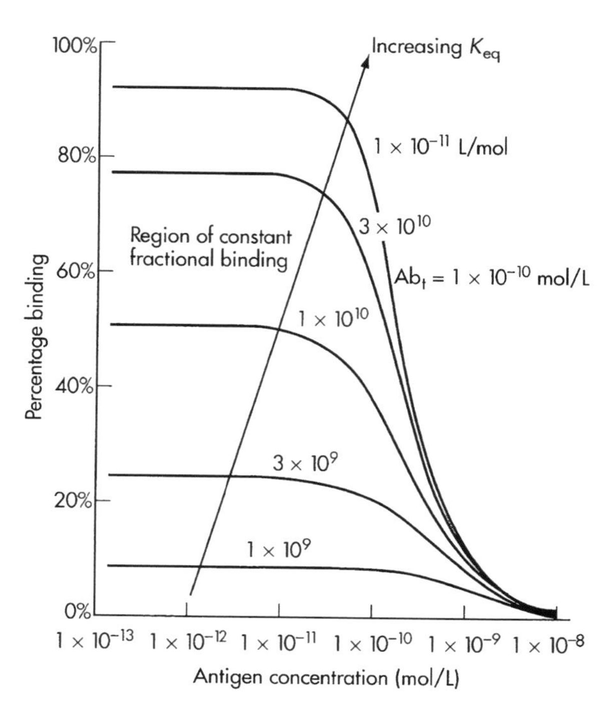

Fig. 1.26 Antigen dilution curves: effects of K_{eq}.

an upper limit for any given concentration of antibody. The implications can be seen more clearly in Figure 1.25 where the antigen concentration is varied for antibody concentrations a-e ranging from 10^{-9} to 10^{-11} ($10/K_{eq}$ to $0.1/K_{eq}$).

For any given concentration of antibody, increases in antigen concentration only alter the percentage binding above a critical region of antigen concentration. Below that point, in the region of constant proportional binding, changes in antigen concentration give only imperceptible changes in the percentage of total antibody bound. The position of this critical region decreases with decreasing concentrations of antibody. For competitive assays the use of low antigen concentrations, therefore, necessitates the use of correspondingly low antibody concentrations to maintain the assay in the region where further addition of antigen gives detectable displacement of the labeled antigen binding.

Figure 1.26 illustrates the influence of the equilibrium constant on percentage binding given an antibody concentration of 1×10^{-10} ($1/K_{eq}$). A high equilibrium constant increases the fractional occupancy of antibody binding sites and hence increases the rate of change of signal response with increasing addition of unlabeled antigen.

For a given equilibrium constant, the extent to which decreasing concentrations of labeled antigen and antibody increase assay sensitivity is bounded by the errors in measurement of low signal responses. As these decrease the optimal percentage binding decreases, eventually reaching a point at which errors in nonspecific binding impact upon the precision of measurement of the response. Any given combination of specific activity and nonspecific binding, therefore, has a unique combination of labeled antigen and antibody concentrations at which point sensitivity is maximal.

This can be best understood by using an example. Consider an assay of 1 mL final volume, with an antibody equilibrium constant of 1×10^{10} L/mol and where resulting signal is counted for 1 min. Let us assume first the situation where the antigen is labeled with an incorporation of 1 mol/mol of ^{125}I, giving a specific activity of 8.02×10^{16} signal events per second (Bequerels/mol: the counter efficiency is assumed to be 100%). The surface plot below shows the predicted sensitivity as a function of the antibody and labeled antigen concentrations (see Figure 1.27).

The maximum sensitivity, 4.1×10^{-12} mol/L, is bounded by inaccuracies of measurement on the one hand and mass action considerations on the other. Figures 1.28 and 1.29 show the effect of nonspecific binding at either 1 or 5% of the free labeled antigen, with a 5% coefficient of variation around the mean level of nonspecific binding.

Optimal sensitivity is achieved by decreasing the concentration of labeled antigen and increasing the concentration of antibody to raise the specific binding

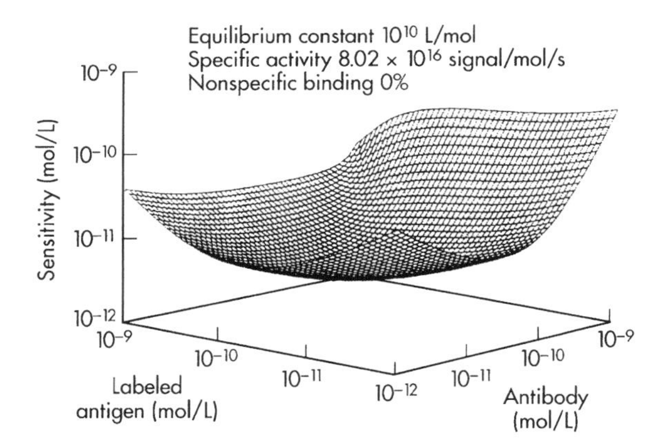

Fig. 1.27 Effect of labeled antigen and antibody concentrations on sensitivity.

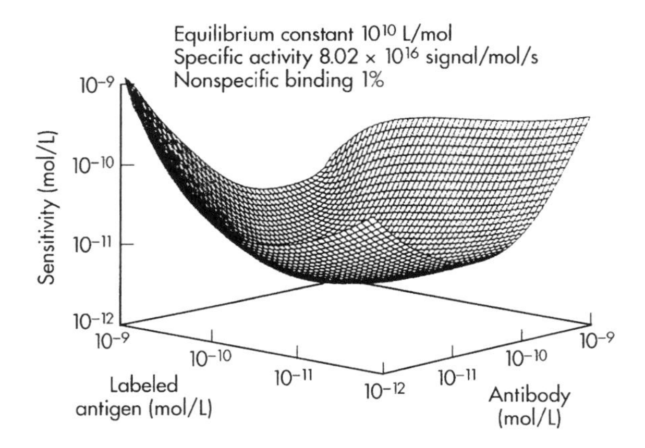

Fig. 1.28 Effect of 1% nonspecific binding on sensitivity.

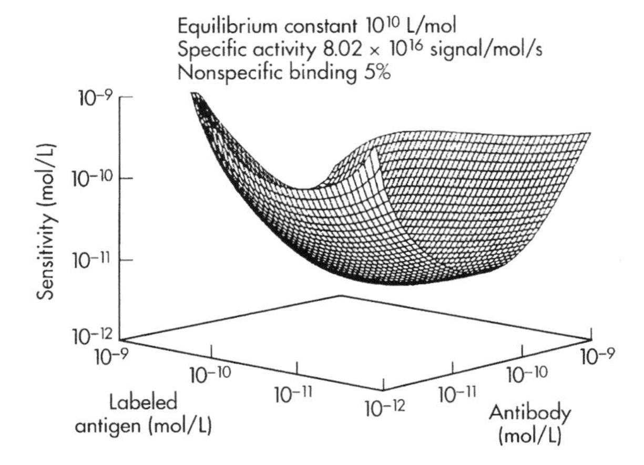

Fig. 1.29 Effect of 5% nonspecific binding on sensitivity.

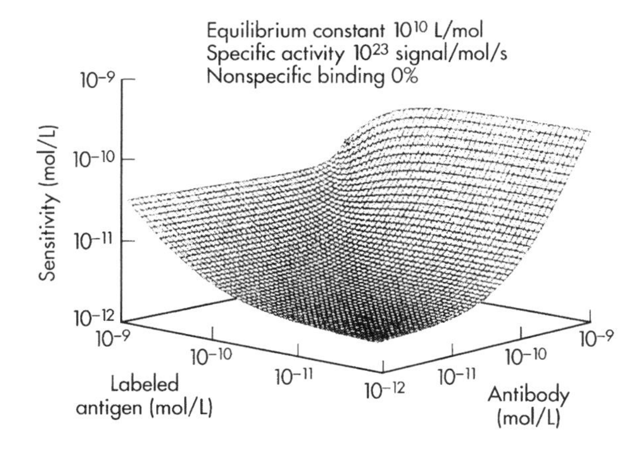

Fig. 1.30 Effect of high specific activity labeled antigen (0% nonspecific binding).

and thus partially compensate for the increase in nonspecific binding. Despite this adjustment in reagent concentrations the assay sensitivity is reduced to 4.3×10^{-12} mol/L at a nonspecific binding of 1% and to 5.8×10^{-12} mol/L at 5%.

Figures 1.30–1.32 similarly show the effects of nonspecific binding at 0, 1, and 5% but where the specific activity of the labeled antigen is much higher at 1×10^{23} signal events per second (approximately 1 out of every 6 molecules giving a detectable signal every second).

As before, reoptimization of the reagent concentrations only partially compensates for the increasing influence of nonspecific binding: the predicted sensitivities being 2.1×10^{-12}, 2.8×10^{-12} and 4.5×10^{-12} mol/L at nonspecific binding levels of zero, 1 and 5%, respectively.

Table 1.1 summarizes the optimal reagent concentrations, percentage binding and signal measurement errors for the above examples. Also shown are calcu-

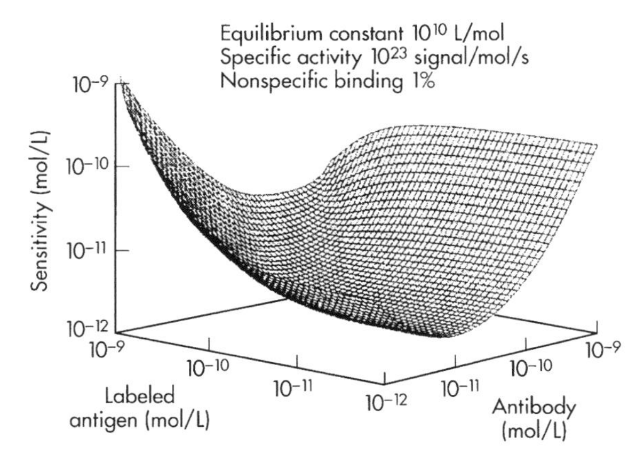

Fig. 1.31 Effect of high specific activity labeled antigen (1% nonspecific binding).

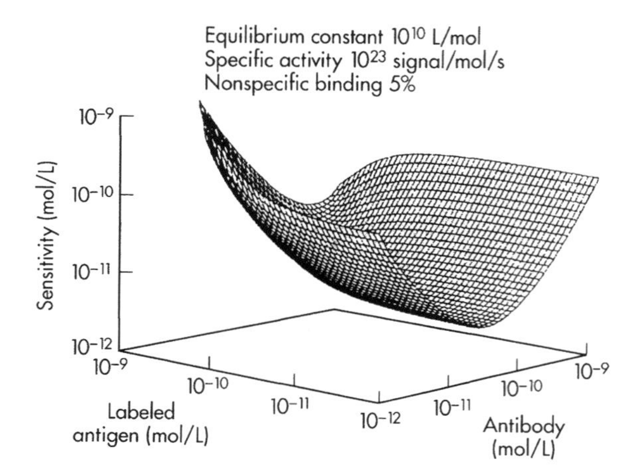

Fig. 1.32 Effect of high specific activity labeled antigen (5% nonspecific binding).

lations for an assay designed on a more 'traditional' basis, where the labeled antigen concentration has been chosen to give an acceptable signal response for an ^{125}I label (40,000 d.p.m.), and the antibody concentration adjusted to give 50% percentage binding (assay d). Nonspecific binding has been assumed to be 5%. There are many examples of such assay designs to be found in the literature and in commercially available kits.

The most striking observation is that in the absence of any nonspecific binding the optimal concentrations of labeled antigen and antibody are very much lower than those traditionally thought desirable. Indeed for a label of infinite specific activity the optimal concentrations of both antibody and labeled antigen tend to zero, at which point Jackson and Ekins (1983) have demonstrated theoretically that the sensitivity tends to

$$\text{Maximum sensitivity} = \frac{2 \times \frac{CV_e}{100}}{K_{eq}} \qquad (1.44)$$

where CV_e is the experimental error.

In the example above with no nonspecific binding the maximum sensitivity was 2.06×10^{-12} mol/L, close to the predicted 2.0×10^{-12} mol/L for labels of infinite specific activity.

Such low concentrations of both antigen and antibody are associated with percentage binding levels much lower than the 30–60% usually regarded as desirable, particularly where the level of nonspecific binding is negligible. For example, in the absence of nonspecific binding (assay a), the optimal percentage binding is predicted to be in the order of 11%.

The examples above may be used to illustrate two common misconceptions regarding assay design. The first is the undue focus placed on the slope of normalized expressions of the dose–response curve rather than that which is actually being measured: the signal response itself. The second is the failure to consider errors in the dose–response relationship. Figure 1.33 shows dose–response curves for assay c (optimized) and assay d (unoptimized) expressing the dose–response relationship in the familiar terms of percentage binding.

Based on traditional concepts of assay design, the unoptimized assay d might be considered to be more sensitive as the initial slope of the dose–response curve is slightly greater. Replotting the data in terms of what is actually measured reverses the interpretation (see Figure 1.34).

Both dose–response curves are shown in Figure 1.35, normalized in terms of the bound/bound at zero concentration (B/B_0), together with the errors in the estimation of the signal response at zero dose and the sensitivities: 5.8×10^{-12} mol/L for the optimized assay and 10.3×10^{-12} mol/L for the unoptimized.

For a competitive assay, where signal measurement error and the nonspecific binding are insignificant, the limiting factors governing sensitivity are the residual experimental errors and the equilibrium constant. Given that the maximum equilibrium constant for antibodies is about 10^{12} L/mol, and assuming that experimental errors can be restricted to a CV of 1%, this sets a practical limit to sensitivity of 2×10^{-14} ($2 \times 0.01/10^{12}$) for a label of infinite specific activity. Figure 1.36 illustrates the maximal sensitivities that could be achieved using ^3H (1.07×10^{15} Bq/mol), ^{125}I (8.02×10^{16} Bq/mol) or labels of nonfinite specific activity for a range of antibody equilibrium constants.

For antibodies with K_{eq} less than 10^{10} L/mol there is little advantage as far as sensitivity is concerned in changing to a nonisotopic label from ^{125}I. The specific activities of ^3H-labeled antigens (often used in steroid assays) are, however, far lower and in this case there would be an obvious and clear advantage in changing either to ^{125}I or a more active nonisotopic label. The calculations emphasize that any improvements to competitive assays need to focus primarily on nonspecific binding and experimental errors rather than the detecting reagent.

Caution needs to be exercised regarding the interpretation of Ekins' calculations regarding the ultimate sensitivities of competitive assays. These are often quoted in the literature without a full appreciation of the underlying assumptions used in their derivation (Ekins, 1991; Gosling, 1990). It should be remembered that the original calculations (Ekins *et al.*, 1968, 1970; Ekins and Newman, 1970) refer to one standard deviation of the error at zero dose, not the 95% confidence limit (2 × standard deviation). Second, it assumes that there is no nonspecific binding. Third, the sensitivity is that in the assay reaction, not in the sample (which is of more practical interest); and fourth it assumes that experimental errors are limited to a CV of 1%. The calculations are also based on other assumptions regarding homogeneity of labeled and unlabeled antigen, a single equilibrium constant, and that

Table 1.1. Optimal assay characteristics.

	^{125}I-label				High specific activity label		
	Assay a	**Assay b**	**Assay c**	**Assay d***	**Assay e**	**Assay f**	**Assay g**
Specific activity (signal/mol/s)	8.0×10^{16}	8.0×10^{16}	8.0×10^{16}	8.0×10^{16}	1×10^{23}	1×10^{23}	1×10^{23}
Nonspecific binding (%)	0	1	5	5	0	1	5
Labeled antigen concentration (mol/L)	3.1×10^{-11}	2.8×10^{-11}	2.1×10^{-11}	0.82×10^{-11}	26×10^{-14}	3.1×10^{-14}	2.0×10^{-14}
Antibody concentration (mol/L)	1.6×10^{-11}	2.0×10^{-11}	3.6×10^{-11}	10.0×10^{-11}	0.13×10^{-12}	11×10^{-12}	29×10^{-12}
Total signal (signal/min)	147,200	136,200	104,600	40,000	1.5×10^{9}	0.19×10^{9}	0.12×10^{9}
Signal at zero dose (B_0)	16,160	18,600	24,800	20,000	2.0×10^{6}	18×10^{6}	27×10^{6}
Percentage binding at zero dose (%)	11.0	13.7	23.8	50	0.13	9.5	22.3
Total error at zero dose (% CV)	1.3	1.3	1.5	1.3	1.0	1.1	1.3
Percentage bound at sensitivity (%)	10.7	13.3	23.1	48.7	0.13	9.3	21.7
Percentage B/B_0 at sensitivity (%)	97.5	97.4	97.1	97.5	98.0	97.8	97.3
Sensitivity (mol/L)	4.1×10^{-12}	4.3×10^{-12}	5.8×10^{-12}	10.3×10^{-12}	2.1×10^{-12}	2.8×10^{-12}	4.5×10^{-12}

Assumptions: $K_{eq} = 1 \times 10^{10}$ L/mol. Variation in nonspecific binding = 5% CV. Experimental errors = 1% CV. Counting time = 1 min. Volume of reaction = 1 mL.
*Assay d, assay unoptimized, reagent concentrations designed to give adequate count rate and 50% binding.

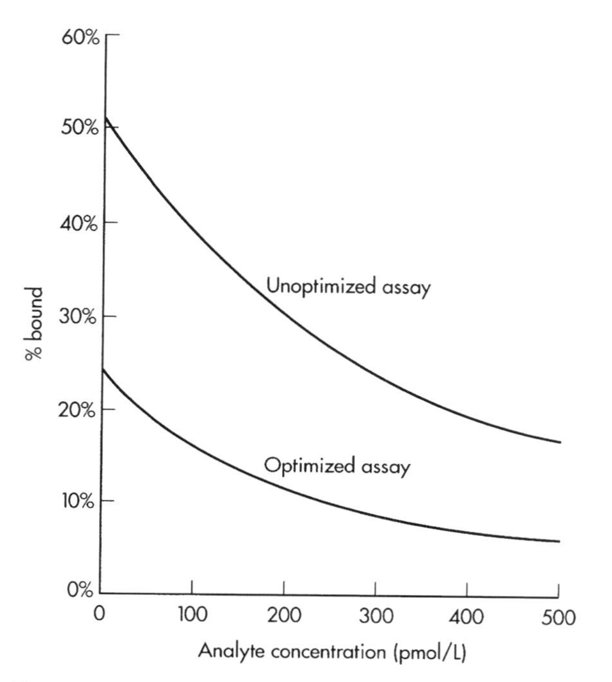

Fig. 1.33 Percentage bound.

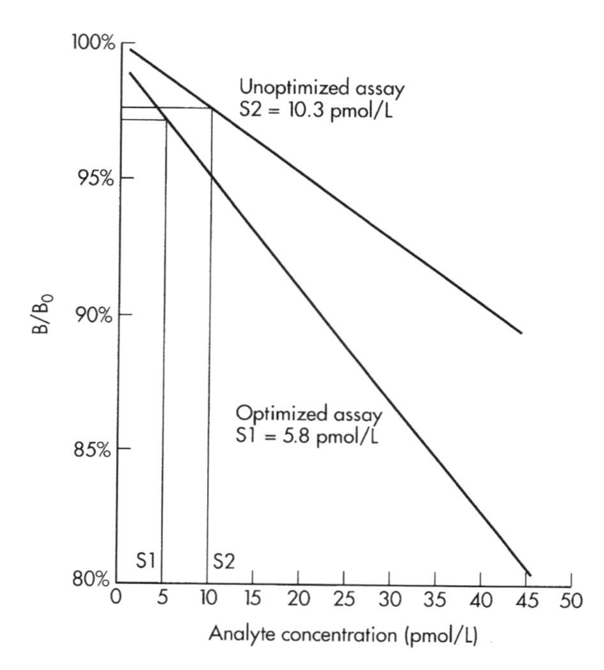

Fig. 1.35 Bound/bound at zero dose.

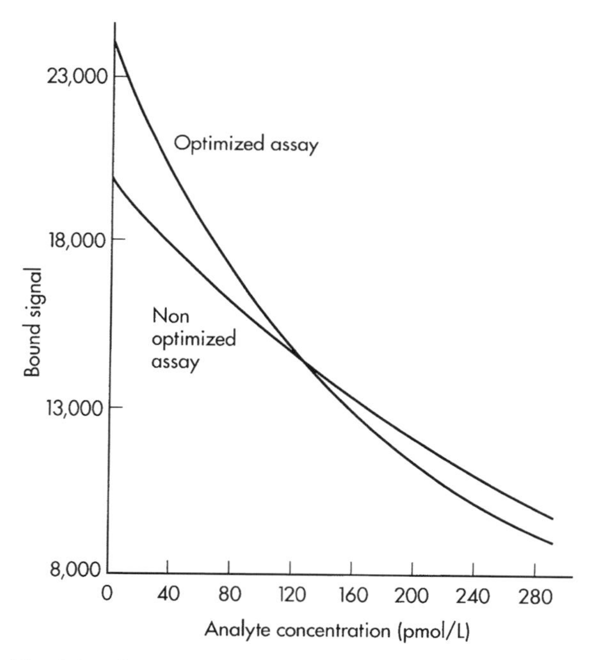

Fig. 1.34 Signal bound.

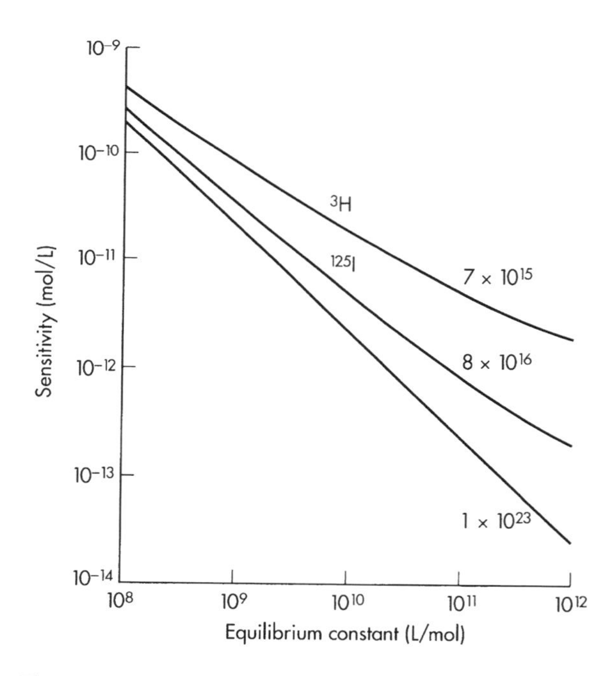

Fig. 1.36 Theoretical sensitivity.

the reaction proceeds to equilibrium. Few if any immunoassays comply with all of these assumptions in practice.

Immunometric Assay Design

As with the competitive assay design, there are two possible ways of producing theoretical models for the estimation and optimization of sensitivity in immunometric assays. Ekins and co-workers have derived unified models of dose–response relationships and errors for these assays (Jackson *et al.*, 1983), based on the same concepts of response/error relationships that were derived for competitive assays.

The unified model is complex, and simplifying assumptions are required to obtain algebraic solutions. An alternative approach is that described above for competitive assays, in that errors are considered separately and used to estimate the signal on the dose–response curve corresponding to the sensitivity. The signal is then substituted back into the dose–response equations to determine the corresponding dose. The simplest design to consider, used in both this and the models developed by Jackson *et al.*, is where the two reactions occur sequentially, with the following assumptions:

1. The first antibody is attached to a solid-phase support, incubated with antigen to equilibrium, and then washed to remove unreacted antigen.
2. The solid phase is then incubated with labeled antibody to equilibrium, washed to remove unbound labeled antibody and the bound signal measured.

With a perfect immunometric assay there would be no signal in the absence of antigen. In practice, however, there is always some signal, principally arising from

- The measuring instrument itself, assumed to be constant for a given counting time.
- The signal due to nonspecific adsorption of labeled antibody, assumed to vary as a simple proportion of the total concentration of labeled antibody.

The total signal arising, in the absence of antigen, R_o, may therefore, be expressed as

$$R_o = MT + \text{NSB}[\text{Ab2}_t]SVT \qquad (1.45)$$

where M is the background signal from the instrument, expressed as signal events per second, T is the signal measurement time in seconds, NSB is the fractional nonspecific binding of labeled antibody in the second incubation. For example, 0.01 represents a binding of 1% of the total labeled antibody, $[\text{Ab2}_t]$ is the total concentration of labeled antibody in the second assay incubation in mol/L, S is the specific activity of the labeled antibody, expressed in terms of signal events/mol/sec, V is the volume of the second assay incubation in liters.

Errors in the variables now need to be considered.

It is assumed that the signal conforms to a Poisson distribution, such that the standard deviation of the signal (s.d._m) is given by the square root of the accumulated signal.

Some variability also occurs in the actual fraction of nonspecific binding due to misclassification errors. It is assumed that these are normally distributed such that

$$\text{s.d.}_{nsh} = \frac{\text{CV}_{nsh}}{100} \times \text{NSB}[\text{Ab2}_t]SVT \qquad (1.46)$$

where s.d._{nsb} is the standard deviation of the misclassification errors, CV_{nsb} is the coefficient of variation of the NSB due to misclassification as a result of pipetting and manipulation errors.

The overall error, in terms of the standard deviation of R_o, can be estimated:

$$\text{s.d.}_a = \sqrt{\text{s.d.}_{nsb}^2 + R_o + \text{s.d.}_e^2} \qquad (1.47)$$

where s.d._a is the overall standard deviation of signal in the absence of antigen, s.d._e is the standard deviation of the experimental errors. The increase in signal level at the limit of sensitivity (R_{det}) is, therefore, defined as

$$R_{det} = 2\text{s.d}_a \qquad (1.48)$$

In other words the signal would have to increase by R_{det} above the sum of the instrument background and nonspecific binding to provide 95% confidence that antigen was present.

Assuming that the Law of Mass Action can be applied to solid-phase immunoassays, the concentration (the word is used loosely) of bound labeled antibody, $[\text{Ab2}_b]$, can be calculated from the signal response at the limit of sensitivity, the specific activity, the reaction volume and the period of signal measurement:

$$[\text{Ab2}_b] = \frac{R_{det}}{SVT} \qquad (1.49)$$

The free labeled antibody concentration, $[\text{Ab2}_f]$, is obtained by subtraction from the total concentration:

$$[\text{Ab2}_f] = [\text{Ab2}_t] - [\text{Ab2}_b] \qquad (1.50)$$

Equation (1.2) can be rearranged to give the total antigen concentration in terms of the equilibrium constant and the bound and free antibody:

$$[\text{Ag}_t] = \frac{[\text{B}]}{[\text{F}]K_{eq}} + [\text{B}] \qquad (1.51)$$

Substituting

$$[Ag2_t] = \frac{[Ab2_b]}{[Ab2_f]K2_{eq}} + [Ab2_b] \quad (1.52)$$

where $[Ag2_t]$ is the total concentration of antigen bound to the solid phase in the second incubation, $K2_{eq}$ is the equilibrium constant for the labeled antibody.

We can now turn to the first stage of the assay and calculate the concentration of antigen required in the first reaction to give the concentration of antigen in the second incubation at the sensitivity limit (calculated above). The concentration of antigen bound in the first reaction, $[Ag1_b]$, is the same as the concentration of occupied capture antibody binding sites in the first reaction, $[Ab1_b]$, and is equal to the total concentration of antigen present in the second reaction, $[Ag2_t]$:

$$[Ag1_b] = [Ab1_b] = [Ag2_t] \quad (1.53)$$

The free capture antibody concentration in the first incubation can be calculated by subtraction from the total concentration of capture antibody:

$$[Ab1_f] = [Ab1_t] - [Ab1_b] \quad (1.54)$$

The concentration of antigen that would need to be present in the first incubation to give the concentration previously calculated in the second can be found as before using equation (1.52). This corresponds to the sensitivity of the assay:

$$\text{Sensitivity} = [Ag1_t] = \frac{[Ab1_b]}{[Ab1_f]K1_{eq}} + [Ab1_b] \quad (1.55)$$

where $[Ag1_t]$ is the total concentration of antigen required in the first incubation to give the specific signal R_{det}, i.e. the sensitivity, $K1_{eq}$ is the equilibrium constant of the capture antibody in the first incubation.

Substituting typical values into the above equations enables 'what-if' models to be developed to examine the influence of reagent composition and errors in measurement.

Consider a two-stage immunometric assay, where the reaction volume is 0.2 mL, as typically would be the case for microtiter wells. It is assumed that both capture and labeled antibody have equilibrium constants of 1×10^{10} L/mol, the signal response is measured over a period of 1 min and the instrument background is one detectable signal event per second. Nonspecific binding is assumed to be a realistic figure for this type of assay of 0.1%, with a variation around this mean level due to experimental and misclassification errors of 5%. Figure 1.37 shows the calculated sensitivities as a function of both detecting and capture antibody concentrations, where the detecting antibody is labeled at 1 mol/mol with ^{125}I (specific activity 8.02×10^{16} Bq).

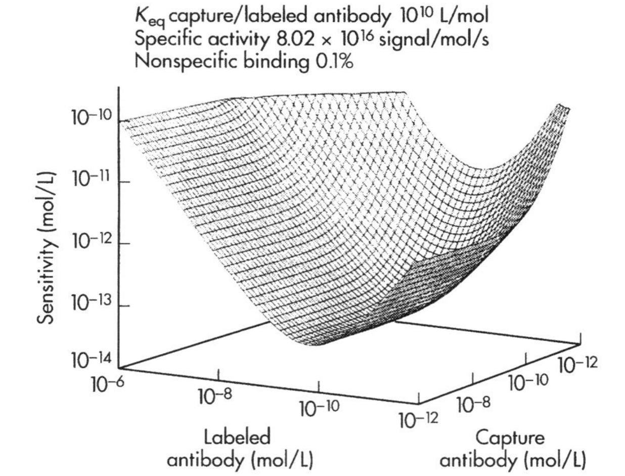

Fig. 1.37 Effect of antibody concentration on sensitivity.

There are two main features. First, there is an optimal concentration of labeled antibody giving maximal sensitivity. This region is bounded at high concentrations by the greater rate of increase in nonspecific binding than specific binding, in other words a reduction in signal-to-noise ratio. By contrast, for low concentrations of labeled antibody the sensitivity is constrained by errors arising from signal measurement and mass action considerations. Second, the sensitivity is governed by the concentration of labeled antibody rather than the capture antibody, which exerts little influence at concentrations greater than 1×10^{-9} mol/L. For the assay conditions outlined above the maximal sensitivity tended towards 5.5×10^{-14} mol/L with an optimal labeled antibody concentration of 1.24×10^{-10} mol/L, a 2000-fold excess over the concentration of antigen at the limit of sensitivity. For simplicity in subsequent examples the concentration of capture antibody is assumed to be 1×10^{-6} mol/L, a concentration sufficiently high enough to bind virtually all the antigen in the first incubation. Jackson (Jackson *et al.*, 1983; Jackson and Ekins, 1983) made similar assumptions in order to simplify the unified dose–response/error relationship model.

The limiting factors governing sensitivity are now examined in turn. Figure 1.38 shows the influence of the labeled antibody equilibrium constant on the concentration of labeled antibody required to attain maximal sensitivity.

For any given labeled antibody concentration, increasing the equilibrium constant increases the initial slope of the dose–response curve and hence the sensitivity. The optimal labeled antibody concentration increases from 1.1×10^{-11} to 1.3×10^{-9} as K_{eq} decreases from 1×10^{12} to 1×10^{8} L/mol. This is accompanied by a corresponding fall in sensitivity from 1.9×10^{-14} to 1.3×10^{-12} mol/L.

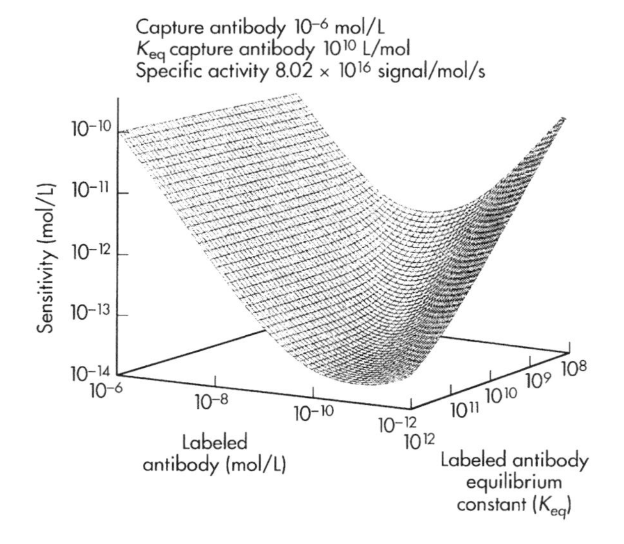

Fig. 1.38 Effect of labeled antibody equilibrium constant on sensitivity.

With competitive assays it was shown earlier that increasing the specific activity of the labeled reagent was of limited benefit in increasing the assay sensitivity, particularly where the equilibrium constant was below 1×10^{10} L/mol. The converse is true for immunometric assays where in Figure 1.39 sensitivity increases from approximately 2.2×10^{-13} to 1.0×10^{-14} as specific activity increases from 10^{16} to 10^{21}. Note how the optimum concentration of labeled antibody decreases as specific activity increases.

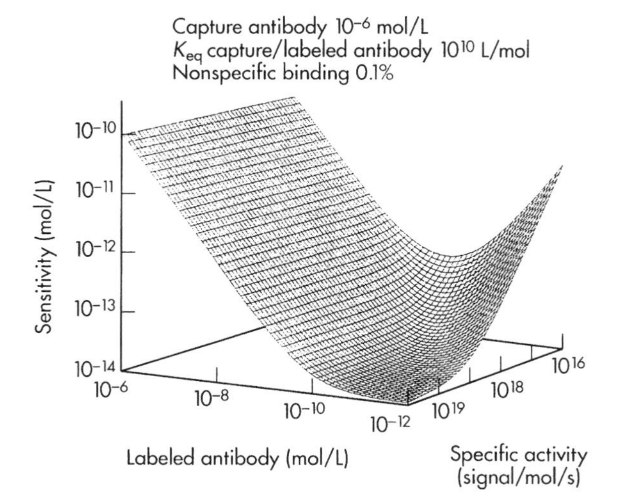

Fig. 1.39 Effect of specific activity of labeled antibody on sensitivity.

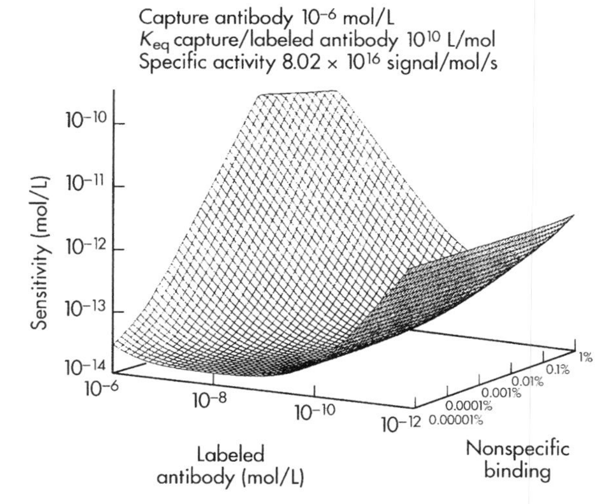

Fig. 1.40 Effect of nonspecific binding on sensitivity.

Figure 1.40 shows the effect of varying the level of nonspecific binding from 1 to 0.00001% using ^{125}I as label. Note how the optimal concentration of labeled antibody increases as nonspecific binding decreases.

Jackson and Ekins (1983) have demonstrated that the limiting sensitivity for immunometric assays, in the absence of signal measurement errors (infinite specific activity) can be described by the following relationship:

$$\frac{K_3 CV_{\mathrm{nsb}}}{K_2^*} \tag{1.56}$$

where K_3 is the fractional nonspecific binding (NSB in the terminology used above), CV_{nsb} is the relative error in the response at zero antigen concentration, K_2^* is the labeled antibody equilibrium constant.

For example, given an effective limit to antibody affinity of 1×10^{12} L/mol, nonspecific binding of 0.1% and an error in the signal response of 1%, this implies a maximum sensitivity of the order of 2×10^{-17} mol/L, three orders of magnitude greater than the corresponding maximal sensitivity for a competitive assay with comparable errors.

Figure 1.41a and b summarizes the various factors involved in determining the sensitivity that can be attained using immunometric designs, in a similar way to that presented for competitive assays. Figure 1.41a shows sensitivity as a function of the equilibrium constant for different levels of nonspecific binding for an assay using ^{125}I as label at maximum incorporation of 1 mol/mol. Figure 1.41b is with a labeled antibody of specific activity 1×10^{23} signal events/mol/sec.

Unlike competitive assays, the use of detection systems of higher activity than ^{125}I is of obvious benefit,

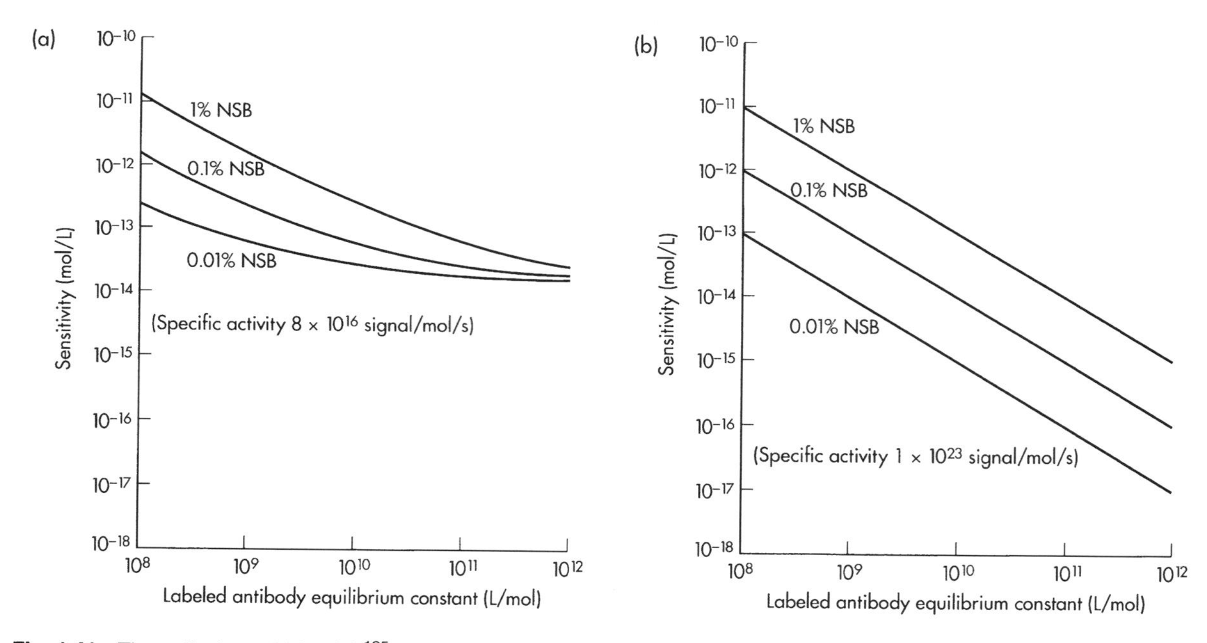

Fig. 1.41 Theoretical sensitivity: (a) ^{125}I; (b) very high specific activity labeled antibody.

particularly when combined with a high affinity antibody. Recognition of the benefits that high specific activity can bring to this type of assay has provided the main stimulus in the search for even more sensitive labeling and detection systems.

Applicability of Theoretical Models

Although theoretical models of the thermodynamics of antibody–antigen reactions have proved useful in determining the appropriate conditions and reagent concentrations to achieve maximum sensitivity, they do rely on the assumption that the reactions follow first-order mass-action laws, and make other assumptions regarding the sources of errors and their magnitude. Final optimization of any immunoanalytical technique must, therefore, be experimentally determined, although the reagent concentrations suggested as a result of theoretical calculations are an obvious starting point.

The major success of theoretical modeling, however, has been in identifying the critical determinants of sensitivity for each type of assay design, and hence directing attention to those components and processes where it is likely to prove of most benefit. A thorough understanding of the principles involved has also enabled the development of novel assay designs, such as the ambient analyte immunoassay, described by Ekins (1991). This exciting development may one day enable the measurement of many analytes simultaneously, by the use of small discrete antibody spots on a 'microanalytical compact disk' to sample the ambient concentration of analyte.

One other aspect of theoretical modeling deserves consideration, and that is as a means of training and education. It is hoped that the equations and models given in this section are sufficiently well detailed that they may be of practical use in demonstrating some of the fundamental principles of immunoassay design. Many of the equations have more elegant but complex derivation. They have been kept simple for two reasons, first so that the reader can follow the logic and algebra more easily, and second, because a number of simple formulae can be more easily and correctly placed in a spreadsheet than one large complex equation. Simple models can be constructed to enable the reader to perform 'what if' tests of the effects of varying component concentrations and antibody affinities, and the results can be easily displayed, for example, as dose–response curves. Such training with spreadsheet models has, in my experience, been of considerable value, both complementing and reinforcing theoretical aspects of immunoassay optimization. The spreadsheet models can of course also be of some practical use in assay design!

The models described above calculate errors in the signal response in the absence of unlabeled antigen and from this the precision at zero dose. By simply adding a further unlabeled antigen term to the equations above it is a relatively simple matter to predict errors at different points on the dose–response curve, and hence derive theoretical precision profiles for the assay across the whole range of antigen concentrations (Ekins, 1983; Jackson *et al.*, 1983).

DETECTION AND QUANTIFICATION OF ANTIBODIES

Most of the preceeding discussion has centered on the use of antibodies to measure antigens. Of course, immunoassays can also be designed to measure antibodies, for it is solely a question of semantics as to whether or not antibody is considered to bind to antigen or *vice versa*. Many of the thermodynamic and all of the statistical concepts governing assay design for antigens can equally and readily be applied to assays for antibody. There are, however, some important caveats relating both to assay design and the interpretation of the results.

- Unlike assays for antigens, the results of such assays estimate not antibody concentration but antibody activity: a combination of the concentration of functional binding sites and their affinities. From the clinical perspective, however, the biological activity may be of more importance than the mass concentration. Problems in interpretation also arise because different assay designs reflect differently the contributions of affinity and concentration. For example, using an immunometric design, sufficient antigen could be added to bind the majority of antibody. In this case the dose–response reflects more the concentration than the affinities of the antibodies. Conversely, the degree of binding in competitive assays is much more affected by the affinities of the antibodies. Assays of the same sample with different assay designs, therefore, only rarely give the same quantitative result even if each assay has been standardized against the same international reference preparation.
- Circulating antibodies are highly heterogeneous in terms of their epitope recognition and affinity. Assays using purified or recombinant antigens often give different quantitative, and occasionally different qualitative, results from those using whole lysates of bacteria or virus. It can be a moot point as to whether the increased specificity of such assays is unhelpful because they do not correlate well with established clinical profiles from assays based on whole lysates, or whether the clinical interpretation should change in order to take advantage of the micro-specificity that such techniques offer. For manufacturers of diagnostic tests it is sometimes easier to change the assay format to reflect established views than attempt to change (and possibly improve) clinical interpretation.
- A proportion of circulating antibodies may already be bound to antigen, especially in the acute phase of infection. The degree of prior saturation of antibody is, therefore, another factor that needs to be taken into account.
- Cross-reactivity may be a major issue if a particular isotype of antibody is being measured. In allergy testing in particular, IgG may interfere with binding by IgE.

Several assay designs are in routine use.

Liquid-Phase Assays

Liquid-phase assays represent the first type of assay described for the quantitative estimation of antibody activity, and are analogous to the liquid-phase competitive assay. Labeled antigen is incubated with dilutions of patient serum and the resulting immune complex or total antibody precipitated by physical or immunological techniques. Signal, invariably radioactive in these early assays, is measured in the precipitated complex. Unlike competitive assays, the dose–response curve increases with increasing antibody activity.

Liquid-phase assays are little used today and most immunoassays now rely upon some type of solid phase to effect the separation between bound and free.

Solid-Phase Immunoassays

Four main types of assay design are commonly used. In a competitive assay, sample antibody competes with labeled antibody for binding to antigen adsorbed onto the solid phase. The assay is directly analogous to the competitive assay of antigen and the dose–response curve falls similarly with increasing activity of sample antibody (see Figure 1.42).

One major disadvantage of this assay is the requirement to obtain a uniform amount of antigen bound to the solid phase. For some antigens and solid-phase combinations this may be difficult to achieve. However, one popular method of circumventing the problem is by linking them to the solid phase *via* an intermediary. Specific antibodies are often used, and, provided the coated linking antibodies do not obscure critical epitopes, have the added advantage that impure antigen can effectively be purified by such a technique. Other linkers involve the use of biotinylated antigen and streptavidin-coated plates. The latter method has the additional advantage that only one type of coated solid phase is required, and it is possible for the biotinylated antigen to be added at the start of the incubation rather than be precoated. A good example of such a design is in the allergy testing system developed by Diagnostic Products Corporation.

One type of immunometric assay design exploits the multivalent nature of antibodies. The **capture bridge** assay uses sample antibody to link antigen adsorbed to the solid phase to labeled antigen in solution (see Figure 1.43).

With simultaneous incubations, high-dose hook effects (*see* CONCEPTS – INTERFERENCE) may be problematic, with excessive amounts of antibody swamping the epitopes on both the solid phase and labeled antigen

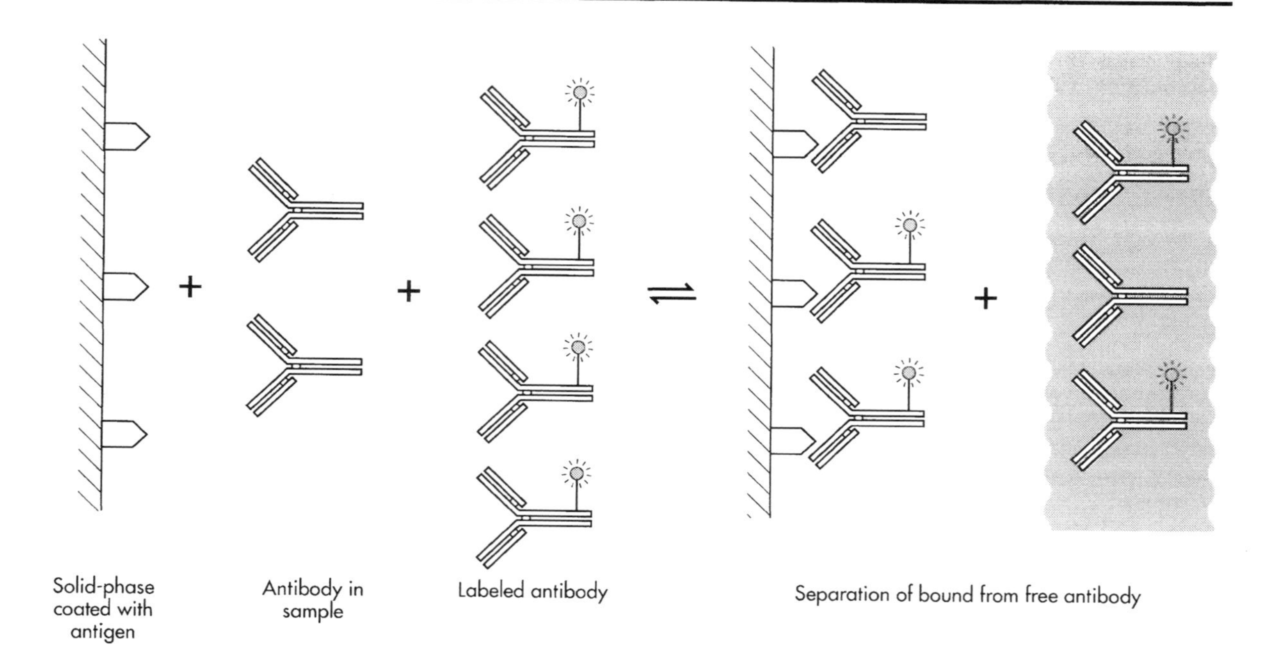

Fig. 1.42 Competitive assay for antibody testing.

and preventing coupling between them. When incubated sequentially very low antibody concentrations may bind bivalently to the solid-phase antigen and, therefore, be underestimated.

The sensitivity of both types of assay is determined, as with antigen assays, by the antibody affinities. Where these are low, which may often be the case, then little antibody is bound to the solid phase. If the supernatant

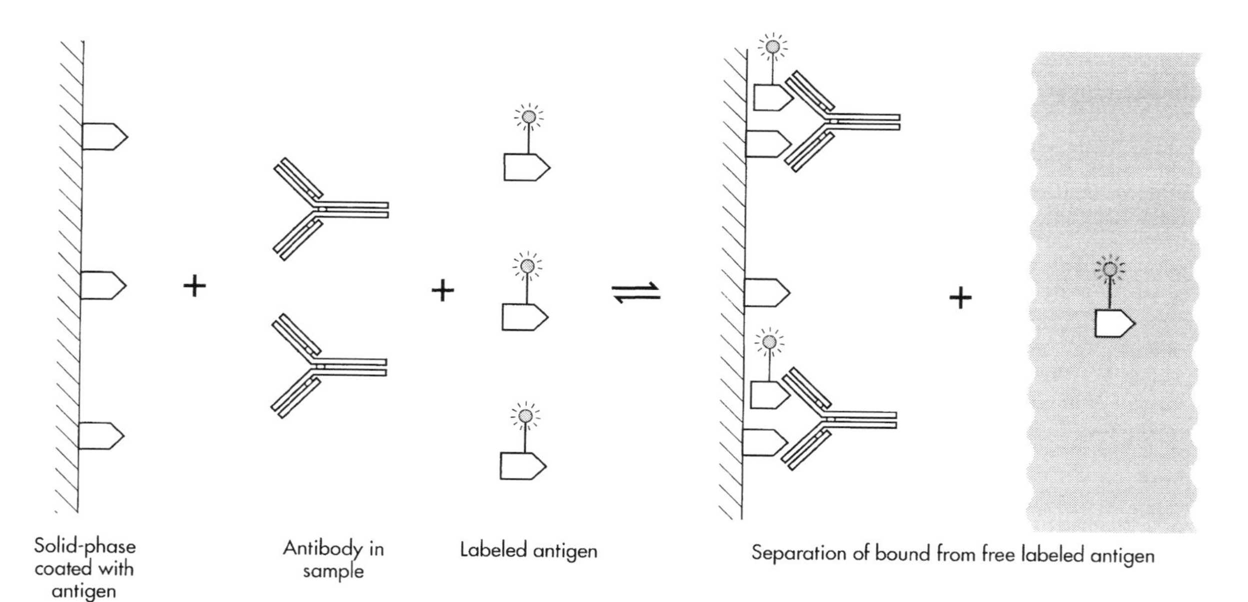

Fig. 1.43 Capture bridge assay.

from such assays is reassayed with fresh solid phase it is not unknown to obtain more or less the same signal on the second occasion, i.e. the uptake is too small to measure in relation to the total amount present in the sample. These techniques are, therefore, unlikely to have the sensitivity required for demanding applications, but do have some advantages in terms of simplicity and throughput. Neither of these designs are isotype specific.

The most popular type of immunometric assay involves binding the sample antibody to antigen immobilized onto a solid phase, washing to remove unreacted components, and then detecting the bound antibody with isotype-specific labeled second antibody in a second incubation. The design is directly analogous to the two-site immunometric assay ELISA of antigen, if one regards the solid-phase antigen as the capture reagent (see Figure 1.44).

Where the activity of one immunoglobulin isotype predominates this may swamp out the activity of other, possibly more relevant isotypes, and a good example of this is in allergy testing when IgE is the isotype of interest. Patients who are allergic often have IgG antibodies to the same antigen. The IgG may be present in much greater amounts and, owing to the process of affinity maturation, frequently has a higher affinity for the antigen than does IgE. Similar considerations apply when attempting to differentiate acute exposure to infection (by measuring IgM) from past infection (using IgG).

The fourth design satisfies the requirements of isotype specificity, even in the presence of large amounts of cross-reactive isotypes. In the **class-capture assay**, isotype-specific antibody immobilized onto a solid phase first captures the antibody type of interest. After washing, two methods of detection are commonly used. In the direct method, bound antibody is quantified using labeled antigen. In the indirect method unlabeled antigen is added and subsequently detected with specific labeled antibody. The latter method is more complex, but incorporates some degree of immunological purification, which may be particularly important when the antigen is not pure (see Figures 1.45 and 1.46).

By analogy with immunometric assays for antigen it might be considered that the sensitivities of the immunometric assays are governed primarily by the antibody affinities and the amount of nonspecific binding. In assays for antibodies, however, one of the main determinants of assay sensitivity is the surface density of functional epitopes on the solid phase.

In contrast to antigen immunoassay, where unified thermodynamic and statistical models have contributed much to assay design, relatively little attention has been paid to their application in the area of measurement of antibodies. Assay design in the latter area has often been characterized by a combination of empirical judgments and experimental optimization. To some extent this is not surprising, as there are other factors of a more practical nature that need to be considered:

- the clinical requirements, i.e. whether the assay is required principally for detection, quantification or for monitoring changes in concentration
- the availability, purity, stability and lot-to-lot variability of the antigen.

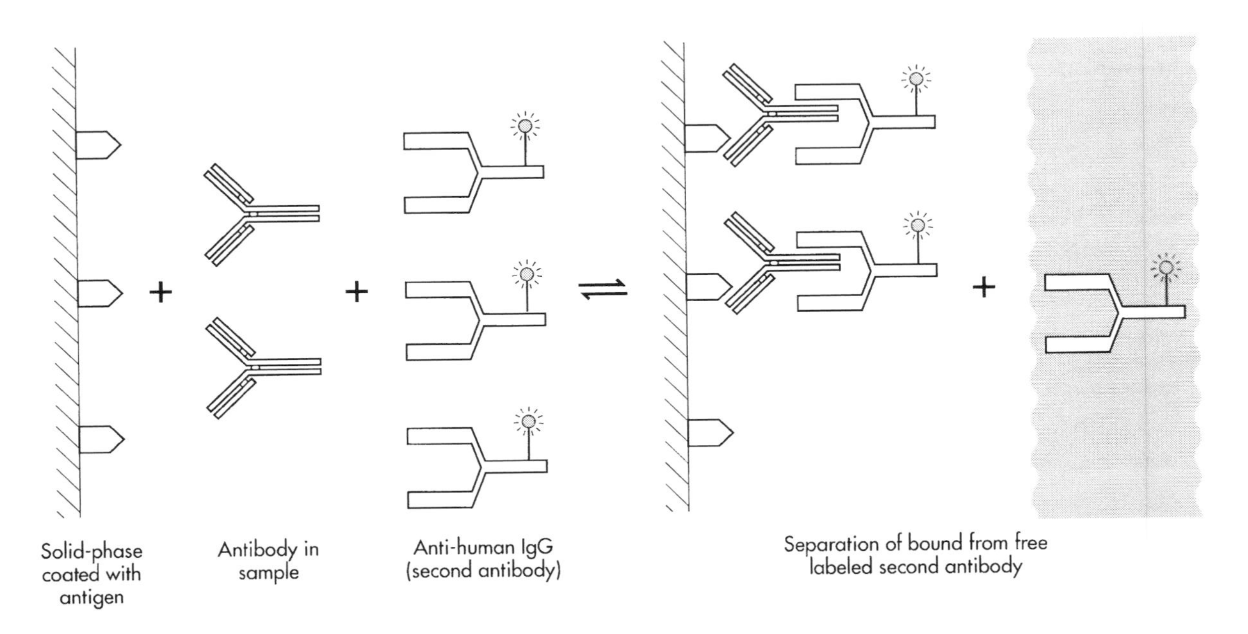

Fig. 1.44 Immunometric assay for antibody testing.

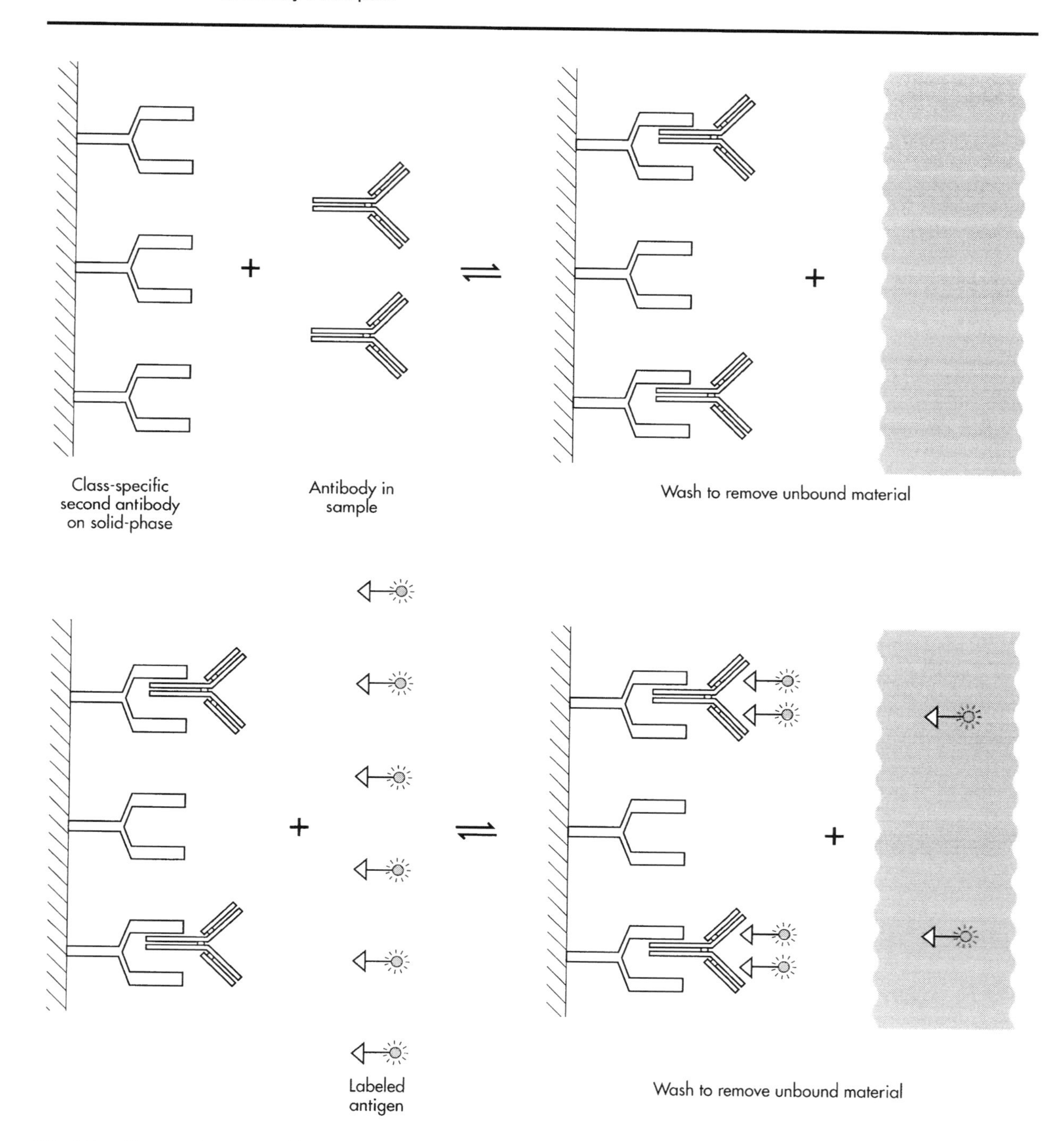

Fig. 1.45 Class capture assay (direct method).

- whether the antigen can be labeled directly without loss of critically important epitope expression.
- the ease with which the antigen can be coupled, either directly or indirectly, to the solid phase, and any losses of epitope expression when bound.

SPECIAL CONSIDERATIONS FOR SOLID-PHASE IMMUNOASSAYS

The ever-increasing popularity of solid-phase immunoassay has brought additional complexities, particularly in

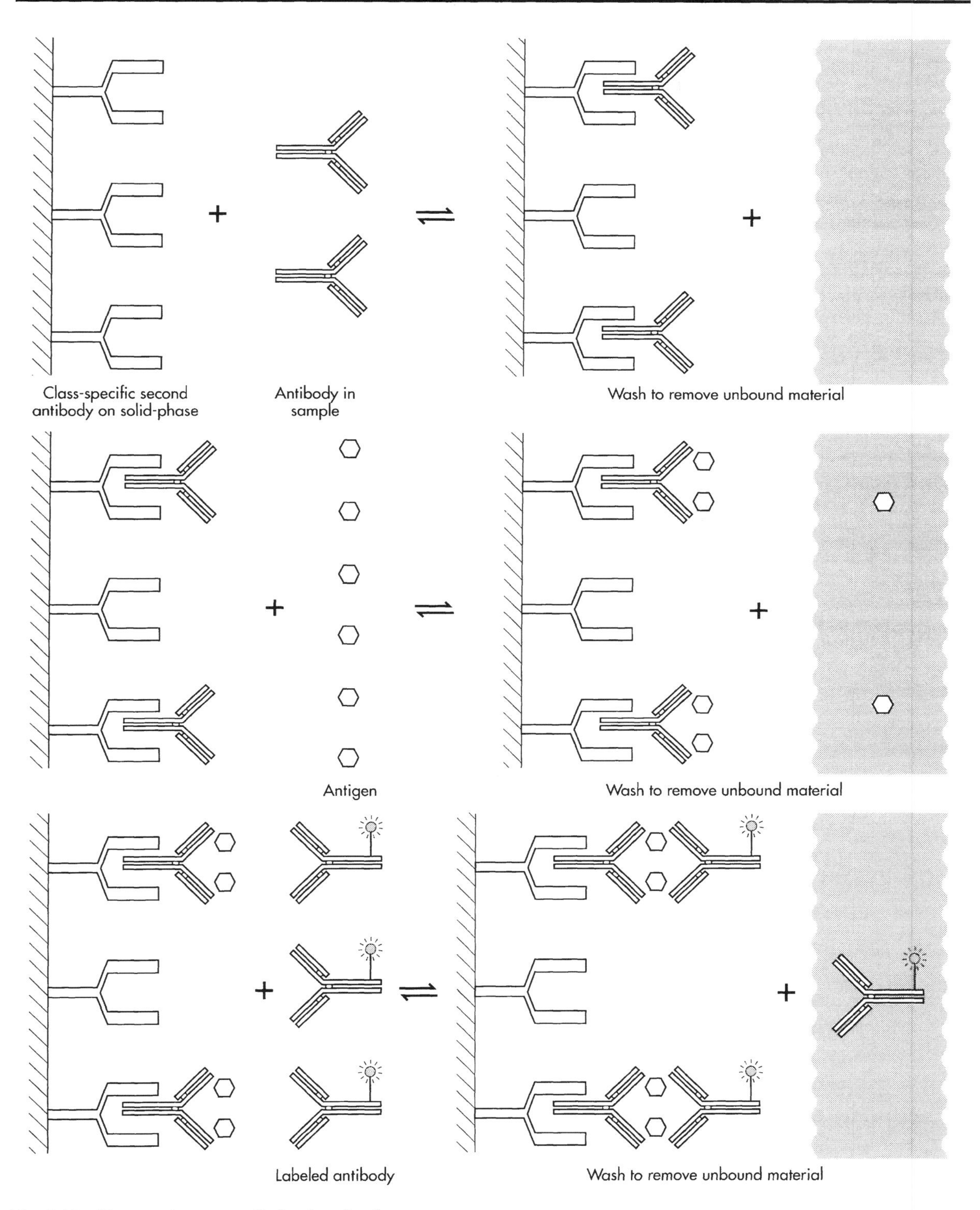

Fig. 1.46 Class capture assay (indirect method).

relation to an understanding of the thermodynamics of antibody–antigen interaction. Solid-phase assays differ in two important respects from their liquid-phase counterparts. The first relates to the reactivity of antibodies or antigens when adsorbed on solid surfaces. The second relates to the differences between reaction kinetics in solution and those occurring at the solid phase.

Surface Immobilization Effects

The solid phase should not be considered a passive component in the process of adsorption or coupling of antibody or antigen. Although chemical coupling of proteins to solid phases is possible, many solid-phase immunoassays rely upon noncovalent adsorption. The processes by which this occurs are poorly understood, yet what seems certain is that binding is essentially hydrophobic in nature. Given that most proteins tend to adopt a conformation in which hydrophilic groups tend to be on the outside of the protein and hydrophobic groups within it, it seems inevitable that hydrophobic binding to polymer surfaces causes conformational changes in the adsorbed protein. This phenomenon can be of benefit. Many antibodies show higher rates of uptake to polymer surfaces when they have been briefly exposed to a low pH glycine buffer prior to adsorption to render them more hydrophobic.

Antigens coated on to a solid phase may lose critical epitopes, either by conformational changes, because they are sterically hidden, or because the epitope is predominantly hydrophobic. New epitopes may also arise that were previously hidden. Microscopically, few polymer surfaces resemble their smooth pictorial representation and steric restrictions may have a major role in determining reaction kinetics.

Antibodies also undergo conformational changes, affecting not only the number of active binding sites, but also their affinity for antigen. It is by no means uncommon to find an antibody that works well in solution but not when bound to solid phase: monoclonal antibodies are affected more than polyclonal antibodies, most probably due to the diversity of antibodies in the latter.

Kinetic Constraints

Theoretical models of immunoassay design make various assumptions in order to retain the simplicity of the Law of Mass Action. In liquid-phase immunoassay most of those assumptions are not critical. In contrast, many of the assumptions are inappropriate when applied to solid-phase assays, for example, the concept of concentration. New constraints apply, and different assumptions need to be made.

Diffusion is a limiting factor in most solid-phase assays. The association rate constant is dependent upon the viscosity of the surrounding medium and a geometric factor that is minimal for reactants of equal size and uniform reactivity. It is not influenced by the strength of the bond between the antibody binding site and the antigen epitope. Where the reactants are of different size the association rate constant increases; for reactants with limited numbers of binding sites it decreases. In the absence of constant stirring, where diffusion is a function only of the size of the molecules and their viscosity, the association rate constant between antibody and hapten molecules should be in the order of 1×10^9/mol/sec (Stenberg and Nygren, 1988). Most association rate constants are substantially less than this and Stenberg and Nygren have proposed that an additional term is introduced into the equations, the **sticking coefficient**, which reflects that antibody and antigen can only bind when they are encountered in the correct alignment. It is likely that the sticking coefficient is radically altered when one of the components is bound, and hence significantly lowers the association rate constant.

A major factor in limiting the rate of forward reaction, and hence time to reach equilibrium, is the presence of the **boundary layer** at the solid phase. Here antigen is rapidly depleted by surface bound antibody, the replenishment of which is limited by diffusion. A quasi-stationary state is thus set up between bulk solution, boundary layer and solid-phase bound antigen (see Figure 1.47).

In the boundary layer the local concentration of antibody is very high. Dissociation kinetics may, therefore, be radically different from those found in free solution. It has been suggested that the binding reaction for solid-phase adsorbed antibody is essentially irreversible. This does not imply that the equilibrium constant *per se* is increased; it reflects instead the low probability that a dissociated antigen can 'escape' from the local high concentration of antibody.

One other interesting effect relates to the spatial orientation of antibody molecules when adsorbed to solid phase. It is widely assumed that the surface of the solid phase is homogeneous, and that bound antibodies are distributed evenly and randomly across the surface. In fact there is some evidence that antibodies may form **fractal clusters**, the solid phase consisting of islands of highly organized antibody, a process akin to the formation and growth of crystalline structures (see Figure 1.48).

Newly developed theories of heterogeneous fractal kinetics in the field of physical chemistry have had a major impact in the elucidation of complex reaction kinetics at solid surfaces, with such unusual discoveries as fractal orders for elementary bimolecular reactions, self-ordering and self-unmixing of reactants, and bimolecular rate coefficients which appear to be history dependent (Kopelman, 1988). It is likely that the same complex kinetics operate in solid-phase immunoassays. The tightly organized nature of such fractal clusters of antibody may lead to an apparent positive cooperative effect: indeed there is some evidence that the association rate constant in solid-phase immunoassays is variable (Werthén *et al.*, 1990). Extremely high concentrations of antibody in such clusters may effectively prevent dissociated antigen from leaving the local environment.

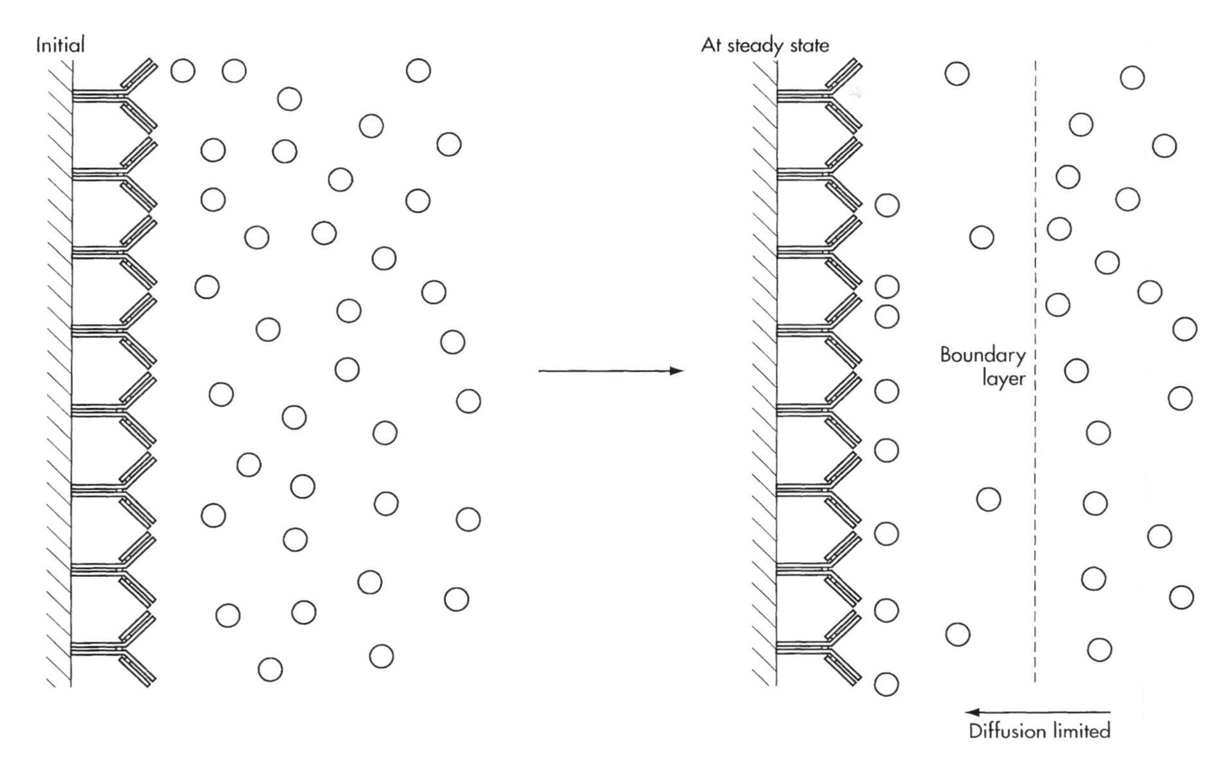

Fig. 1.47 Formation of boundary layer.

The observed dissociation rate constant from fractal clusters would, therefore, be much lower than when in free solution. This may in part explain the observation that solid-phase binding appears essentially irreversible.

COMPARISON OF EXPERIMENTAL AND THEORETICAL IMMUNOASSAY PERFORMANCE

A major problem that has plagued the literature concerning immunoassays is a lack of standardized terminology and practices when comparing the performance characteristics of various assay designs, and in particular estimates of their sensitivity. Differences in volumes, times of incubation, numbers of replicates, integrated time of signal detection, etc. can all lead to considerable differences in reported sensitivity for the same assay. Comparisons between different assays based on literature reports are, therefore, difficult.

It is important, however, to determine whether or not theoretical predictions can be translated quantitatively into practice. If theory and practice differ qualitatively then the underlying assumptions of the theory may be incorrect. Where theory and practice depart quantitatively then the theoretical models may be capable of numerical improvement.

For antibodies with realistic equilibrium constants of 10^9-10^{11} L/mol, high specific activity labeled antigen, and a 1% experimental error, the model (equation (1.44)) predicts sensitivities ranging from 0.2 to 20 pmol/L. Gosling (1990) has reviewed some of the most sensitive reported assays for steroids. All used either ^{125}I or peroxidase labels, the most sensitive being those with solid-phase antibody where errors in nonspecific binding could be minimized. The reported sensitivities in these assays ranged from 4 to 9 pmol/L, well within the range dictated by theory.

One possible reason for the close degree of agreement between theory and practice in competitive designs may lie in the use of solid-phase and nonisotopic detection systems, not just because of their intrinsic advantages of having low measurement errors and low nonspecific binding, but because of the constraints they impose on the more 'traditional' methods of assay optimization. Many competitive radioimmunoassays were developed to obtain the 30–50% zero-dose binding and a high signal, partly because of tradition and partly, in the case of diagnostic companies, because many of their customers still expected it. Estimates of percentage binding can rarely be obtained with solid-phase, nonisotopic assays:

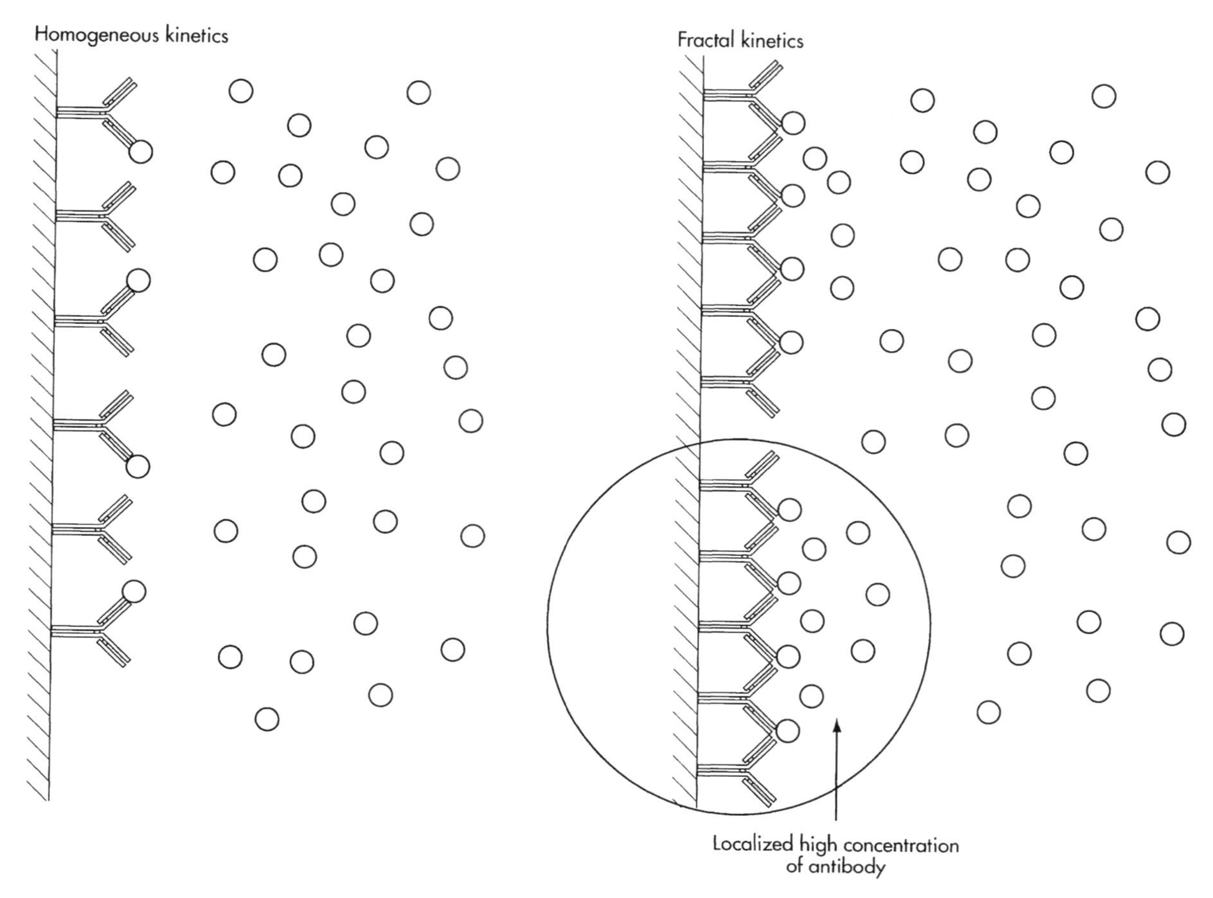

Fig. 1.48 Fractal kinetics.

they remain invisible to both assay development scientist and customer alike. Freed from these psychological constraints, assay optimization is free to focus on other aspects, such as minimizing errors, precisely the route that Ekins has advocated on theoretical grounds.

It is not possible to estimate the maximum sensitivity attainable with immunometric designs because this is determined mainly by experimental errors subject to a considerable degree of variation. Few published reports provide sufficient details of these errors to estimate how closely the practical sensitivities agree with those dictated by theory. However, making the assumption that nonspecific binding can be minimized to 0.1% and that the errors in nonspecific binding can be minimized to 1%, with an infinitely high specific activity label then the maximum sensitivity should be of the order of 0.20–20 fmol/L for commonly encountered antibody affinities ranging from 10^9 to 10^{11} L/mol (from equation (1.56)).

Precise measurement of very low concentrations of the hormone TSH is of great clinical importance. It is not surprising, therefore, that so much attention has been focused on increasing the sensitivity of this assay. Thonnart (Thonnart *et al.*, 1988) and McConway (McConway *et al.*, 1989) have reviewed the performance of 14 commercially available, highly sensitive assays for TSH: the reported sensitivities varied between 0.005 and 0.1 mIU/L, with the most sensitive being the nonisotopic assays. Taking the activity of human TSH as 5 IU/mg and a molecular mass of 28,000, these sensitivities translate in molar terms to 35–714 fmol/L. Using enzyme amplification and fluorescent detection Cook and Self (1993) have recently described an assay for proinsulin with a sensitivity of 17 fmol/L. Reported sensitivities are probably within an order of magnitude of those predicted on theoretical grounds, assuming nonspecific binding of 0.1%.

Given that detection systems of exquisite sensitivity are now available, understanding and optimizing the surface chemistry of such systems to give higher densities of active antibody and lower nonspecific binding may well pay dividends. In addition more sophisticated theoretical models, particularly those taking into account differences between solid- and liquid-phase kinetics, have a major role to play in enabling a better understanding of the fundamental properties of solid-phase interfacial kinetics and may help in the design of ever more sensitive assay configurations.

REFERENCES

Addison, G.M. and Hales, C.N. The immunoradiometric assay. In: *Radioimmunoassay Methods* (eds Kirkham, K.E. and Hunter, W.M.), 447–461 (Churchill-Livingstone, Edinburgh, 1970).

Berson, S.A. and Yalow, R.S. Radioimmunoassays. In: *Statistics in Endocrinology* (eds McArthur, J.W. and Colton, T.), 327–344 (MIT Press, Cambridge, MA, 1970).

Berson, S.A. and Yalow, R.S. Measurement of hormones – radioimmunoassay. In: *Methods in Investigative and Diagnostic Endocrinology*, vol. 2A (eds Berson, S.A. and Yalow, R.S.), 84–135 (Elsevier/North Holland, Amsterdam, 1973).

Borth, R. Discussion. In: *Karolinska Symposia on Research Methods in Reproductive Endocrinology: Second Symposium: Steroid Assay by Protein Binding* (ed Diczfalusy, E.), 30–36 (WHO/Karolinska Institut, Stockholm, 1970).

Cook, D.B. and Self, C.H. Determination of one thousandth of an attomole (1 Zeptomole) of alkaline phosphatase: application in an immunoassay of proinsulin. *Clin. Chem.* **39**, 965–971 (1993).

Ekins, R.P. The future development of immunoassay. In: *Radioimmunoassay and Related Procedures in Medicine* **1**, 241–268 (International Atomic Energy Agency Vienna, Vienna, 1977).

Ekins, R.P. Assay design and quality control. In: *Radioimmunoassay* (ed Bizollon, C.A.), 239–255 (Elsevier/North Holland, Amsterdam, 1979).

Ekins, R.P. The precision profile: its use in assay design, assessment and quality control. In: *Immunoassays for Clinical Chemistry*, 2nd edn (eds Hunter, W.M. and Corrie, J.E.T.), 76–105 (Churchill Livingstone, Edinburgh, 1983).

Ekins, R.P. Immunoassay design and optimisation. In: *Principles and Practice of Immunoassay* (eds Price, C.P. and Newman, D.J.), 96–153 (Macmillan, London, 1991).

Ekins, R.P. and Newman, B. Theoretical aspects of saturation analysis. In: *Karolinska Symposia on Research Methods in Reproductive Endocrinology: Second Symposium: Steroid Assay by Protein Binding* (eds Diczfalusy, E. and Dicfalusy, A.), 11–30 (WHO/Karoliska Institut, Stockholm, 1970).

Ekins, R.P., Newman, B. and O'Riordan, J.L.H. Theoretical aspects of 'saturation' and radioimmunoassay. In: *Radioisotopes in Medicine: In vitro Studies* (eds Hayes, R.L., Goswitz, F.A. and Murphy, B.E.P.), 59–100 (US Atomic Energy Commission, Oak Ridge, TN, 1968).

Ekins, R.P., Newman, B. and O'Riordan, J.L.H. Saturation assays. In: *Statistics in Endocrinology* (eds McArthur, J.W. and Colton, T.), 345–378 (MIT Press, Cambridge, MA, 1970).

Ezan, E., Tiberghien, C. and Dray, F. Practical method of optimising radioimmunoassay detection limits. *Clin. Chem.* **37**, 226–230 (1991).

Gosling, J.P. A decade of development in immunoassay methodology. *Clin. Chem.* **36**, 1408–1427 (1990).

Jackson, T.M. and Ekins, R.P. Theoretical limitations on immunoassay sensitivity: current practice and potential advantages of fluorescent Eu^{3+} chelates as nonradioactive tracers. *J. Immunol. Meth.* **87**, 13–20 (1983).

Jackson, T.M., Marshall, N.J. and Ekins, R.P. Optimisation of immunoradiometric assays. In: *Immunoassays for Clinical Chemistry*, 2nd edn (eds Hunter, W.M. and Corrie, J.E.T.), 557–575 (Churchill Livingstone, Edinburgh, 1983).

Kopelman, R. Fractal reaction kinetics. *Science* **241**, 1620–1626 (1988).

McConway, M.G., Chapman, R.S., Beastall, G.H., Brown, E., Tillman, J., Bonar, J.A., Hutchinson, A., Allinson, T., Finlayson, J., Weston, R., Beckett, G.J., Carter, G.D., Carlyle, E., Herbertson, R., Blundell, G., Edwards, W., Glen, A.C.A. and Reid, A. How sensitive are immunumetric assays for thyrotropin? *Clin. Chem.* **35**, 289–291 (1989).

Miles, L.E.M. and Hales, C.N. Labeled antibodies and immunological assay systems. *Nature* **219**, 186–189 (1968).

Nisonoff, A. and Pressman, D. Heterogeneity of antibody binding sites in their relative combining affinities for structurally related haptens. *J. Immunol.* **81**, 126 (1958).

Scatchard, G. The attractions of proteins for small proteins and molecules. *Ann. NY Acad. Sci.* **51**, 660–672 (1949).

Stenberg, M. and Nygren, H. Kinetics of antigen–antibody reactions at solid–liquid interfaces. *J. Immunol. Methods* **113**, 3–8 (1988).

Thonnart, B., Messian, O., Linhart, N.C. and Bok, B. Ten highly sensitive thyrotropin assays compared by receiver-operating characteristic curves analysis: results of a prospective multicenter study. *Clin. Chem.* **35**, 691–695 (1988).

Werthén, M., Stenberg, M. and Nygren, H. Theoretical analysis of the forward reaction of antibody binding to surface-immobilized antigen. *Progr. Colloid. Polym. Sci.* **82**, 349 (1990).

Yalow, R.S. and Berson, S.A. Assay of plasma insulin in human subjects by immunological methods. *Nature* **184**, 1648–1649 (1959).

FURTHER READING

Butler, J.E. *Immunochemistry of Solid-Phase Immunoassays* (CRC Press, Boca Raton, Florida, 1991).

Chappey, O., Debray, M., Niel, E. and Schermann, J.M. Association constants of monoclonal antibodies for hapten: heterogeneity of frequency distribution and possible relationship with hapten molecular weight. *J. Immunol. Meth.* **172**, 219–225 (1994).

Diamandis, E.P. and Christopoulos, T.K. (eds), *Immunoassay* (Academic Press, San Diego, 1996).

Harlow, E. and Lane, D. *Antibodies: A Laboratory Manual* (Cold Spring Harbor Laboratory Press, New York, 1988).

Hunter, W.M. and Corrie, J.E.T. (eds), *Immunoassays for Clinical Chemistry* 2nd edn, (Churchill Livingstone, Edinburgh, 1983).

Kricka, L.J. Principles of immunochemical techniques. In: *Tietz Textbook of Clinical Chemistry* 3rd edn, (eds Burtis, C.A. and Ashwood, E.R.), 205–225 (W.B. Saunders, Philadelphia, 1998).

Price, C.P. and Newman, D.J. (eds), *Principles and Practice of Immunoassay* 2nd edn, (Macmillan, London, 1997).

Roitt, I.M. and Delves, P.J. *Roitt's Essential Immunology* (Blackwell, Oxford, 2001).

Saunal, H., Karlsson, R. and Van Regenmortel, M.H.V. Antibody affinity measurements. In: *Immunochemistry, A Practical Approach*, vol. 2, (eds Johnstone, A.P. and Turner, M.W.), 1–30 (Oxford University Press, Oxford, 1997).

Sutton, B.J. Molecular basis of antibody–antigen reactions: structural aspects. In: *Methods of Immunological Analysis* (eds Masseyeff, R.F., Albert, W.A. and Stainess, N.A.), 66–79 (VCH, Weinheim, 1993).

Van Oss, C.J. Nature of specific ligand–receptor bonds, in particular the antibody–antigen bond. In: *Immunochemistry* (eds Van Oss, C.J. and Van Regenmortel, M.H.V.), 581–614 (Marcel Dekker, New York, 1994).

Van Oss, C.J. Hydrophobic, hydrophilic and other interactions in epitope–paratope binding. *Mol. Immunol.* **32**, 199–211 (1995).

2 Non-Competitive Immunoassays for Small Molecules – The Anti-Complex, Selective Antibody and Apposition Systems

Colin H. Self, Stephen Thompson and Larry A. Winger

INTRODUCTION

The history of immunoassay has been one of a trail of spectacular successes. High performance, fast and robust methods have been developed for a remarkable array of substances. It is clear, however, that outstanding performance has classically only been seen within the realm of heterogeneous assays for analytes of large enough molecular size such that they can be bound by both capture and detector antibodies simultaneously. These assays benefit from being 'reagent excess' systems and offer precise and fast assays equally at home on complex automated immunoassay machines or on dipsticks. Two-site immunometric assays for large molecules have found large international markets for qualitative home, clinic, or bedside diagnostic testing systems (May, 1994; Bonnar *et al.*, 1999).

In spite of the provision of high performance immunoassays for large analytes, simple physical constraints have precluded the use of such sandwich approaches for small molecule analytes, which are not large enough to bind two antibodies simultaneously. Instead, competitive-format assays have been widely used. In these approaches, it is the fraction of capture antibody that has not bound analyte that provides the signal to be measured. In effect these assays measure how much analyte is not present. This inverse approach brings a wealth of complexity to any assay systems thus based (Ekins, 1987; Self, 1993a,b).

The greatest problem of competitive-format systems is simply that the response at low analyte concentration is difficult, if not impossible, to distinguish from that at zero, as both conditions give rise to large signals. Also, by definition, competitive assays are not 'reagent excess' in nature. Limited concentrations of both capture antibody and competing partner must be employed, with the resulting limitations for assay speed. In addition, not only must reagent concentrations be limited, but they must also be precisely maintained across an assay to achieve acceptable results. These assay requirements demand high technical precision in the addition of reagents, besides posing clear constraints on shelf-life considerations. Even relatively modest changes of concentration of the competitive reagents can have a profound effect on the precision of the competitive assay. All of these factors taken together can lead to variable, relatively slow assays of limited sensitivity and range. Considering the very wide range of small molecule analytes of medical, forensic, environmental and quality assurance applications within the water, food and beverage industries, where fast, high performance assay systems are required, these are critical problems that required a solution.

It is for these reasons the anti-complex, Selective Antibody systems and subsequently, the Apposition System were developed. These systems provide direct measurement of the actual fraction of capture antibody bound by analyte and consequently share the operational advantages seen with large analyte sandwich methods.

THE ANTI-COMPLEX ASSAY

The **anti-complex immunoassay** system (Self, 1985) represents an extremely simple to use direct read-out system for small molecules. In this system, an antibody recognizes the changed characteristics of another antibody whose binding site is occupied by target analyte (Figure 2.1).

Whether these characteristics are a result of a new epitope composed of analyte and antibody, or an altered conformational state of the occupied antibody, is irrelevant to the utility of the assay. The point is that the analyte-bound fraction of antibody is determined and provides a signal commensurate with the amount of analyte occupying the binding sites on the capture antibody.

Application of the system to the determination of digoxin (Self, 1993a,b) has demonstrated the clear advantages of this anti-complex approach in terms of the critical parameters of sensitivity, specificity, precision and speed. For example, it has been shown that, if required, assays can be conducted with immuno-incubation times at least as short as 1 min. The assays exhibit high precision with a flat ('U-shaped') precision profile characteristic of two-site sandwich immunoassays rather than the more 'V-shaped' form typically found with competitive assays.

An example of the high performance of the system is given in Figure 2.2, which shows a standard curve for digoxin produced in our laboratory by an anti-complex antibody in conjunction with an anti-digoxin primary antibody. In this case, the primary antibody is labeled with surface-bound anti-complex antibody but as we have demonstrated, the assay can be used in the converse orientation with surface-bound primary antibody and labeled anti-complex antibody (Self, 1993a,b).

Other secondary antibody systems have been developed. Voss and colleagues (e.g. Carrero *et al.*, 1996) have developed antibodies binding anti-FITC antibody in the presence of its ligand. Ullman *et al.* (1993) have described a secondary antibody that binds an anti-cannabinoid antibody in the presence of THC and results in a system of increased specificity. Similarly, the work by Towbin *et al.* (1995) for the small peptide angiotensin II has resulted in an assay system of higher performance than obtainable by the conventional competitive assay format and a high-performance immune-complex assay for microcystin has been developed by Nagata *et al.* (1999).

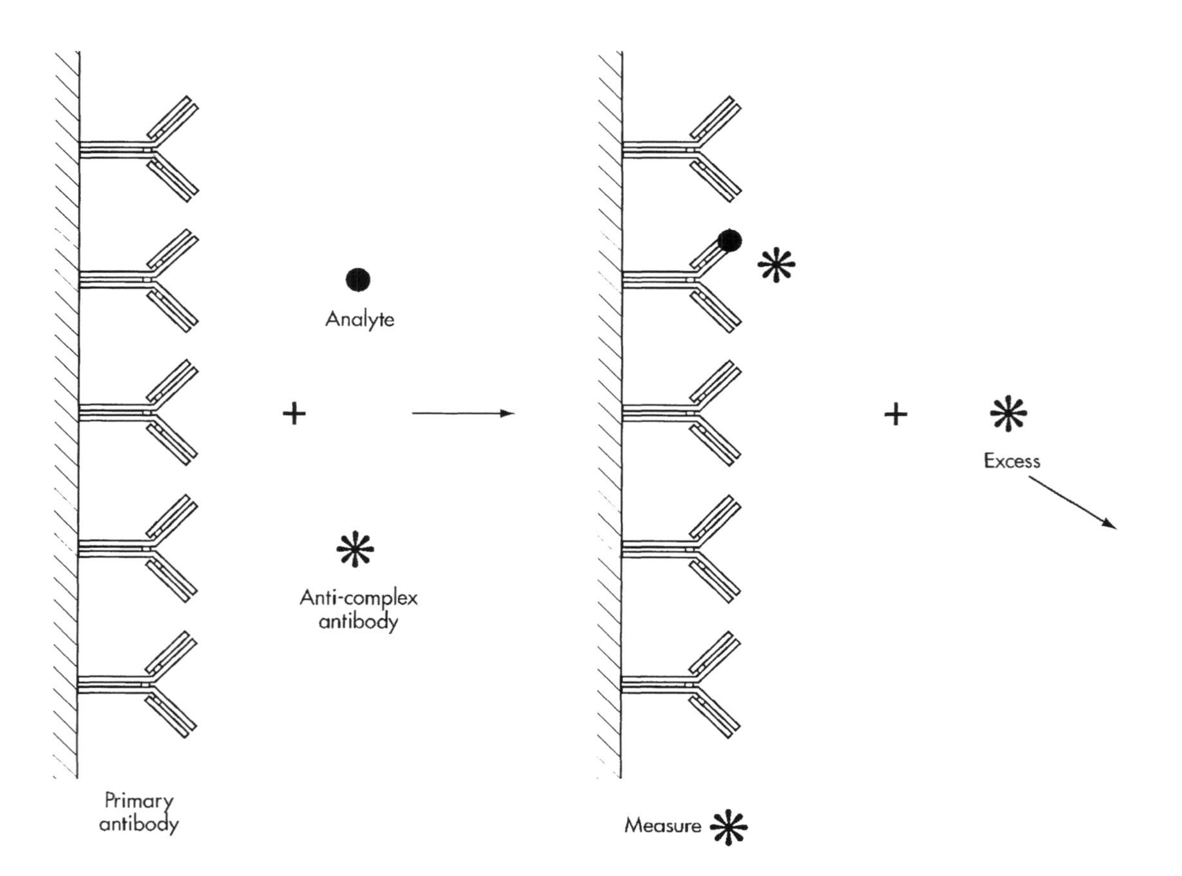

Fig. 2.1 Anti-immune complex assay for small molecules.

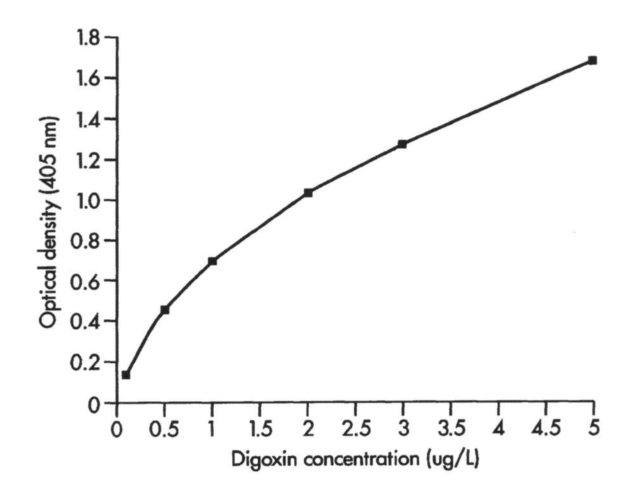

Fig. 2.2 Anti-complex assay.

ENHANCED SPECIFICITY

In addition to the other advantages of the anti-complex system our work has demonstrated the particular benefits of increased specificity resulting from the anti-complex antibody approach (Self *et al.*, 1994, 1996; Winger *et al.*, 1996). We have shown, for example, that the anti-complex assay for digoxin is less intrinsically susceptible to DLIF artifacts than competitive formats. Such increase in specificity is not surprising given the requirement of a hapten to participate as an active component of the assay (whether by contributing to a new determinant in the binding site or mediating a conformational change in the capture antibody) rather than passively competing with labeled competitor.

As noted above, this enhanced specificity is reflected also in the systems of Ullman *et al.* (1993), Towbin *et al.* (1995), and Carrero *et al.* (1996), which indicates that systems employing such secondary antibodies can elicit a better specificity profile from the primary antibody than is achievable by the use of the same primary antibodies used in the comparative competitive systems.

Specificity of anti-complex immunoassay can be greatly further increased over competitive format assay by the following simple maneuver. Two-site immunoassays benefit from a mechanical, technical feature that can be incorporated into their format. After analyte is exposed to the first analyte capture antibody on a solid phase, irrelevant and potentially cross-reactive analytes can be washed away, so that only the captured analyte is present to be recognized by labeled second antibody. In the same way, cross-reactive analytes (which might otherwise interfere with the assay if they remain when labeled anti-complex antibody is added, as in Figure 2.3a) can be washed away from the anti-complex digoxin assay, with no harm to the potential dynamic range of signal delivered by anti-complex antibody added secondary to such a wash (Figure 2.3b).

The dramatic enhancement of specificity by such a simple maneuver raises a fundamental question as to the limited specificity found for antibodies raised against a range of small molecules when these are determined by competitive immunoassay systems. It seems very likely indeed that, in many cases, the limited specificity is not a consequence of the particular antibody, or even a limitation of the immune system of the animal producing them, but rather of the competitive-format assay systems employed in testing them.

MULTIPLE BINDING ASSAY

As we have also demonstrated, specificity can be even further enhanced within an assay by sequential exposure of analyte to different analyte-capture antibodies. Thus in operation the **Multiple Binding Assay** (MBA) works, for example, by

- exposing the sample to a primary antibody;
- washing away unbound substances which may cross-react with subsequent antibodies;
- sequestration of analyte from the first antibody by a second antibody of different cross-reactivity profile to the first antibody;
- final determination of analyte (Self, 1993a,b). Particularly specific assays are shown to result when the determination step is that of an anti-complex system (Self, 1993a,b).

SELECTIVE ANTIBODY IMMUNOMETRIC ASSAY

The aim of determining the capture antibody-bound sites, with all of the advantages this brings, can also be achieved by a parallel technology that we have developed. This is the **Selective Antibody System** (Self, 1989). The principle of the selective antibody approach is quite clear. Essentially, while a small molecule hapten occupies a binding site on an array of analyte-capture primary antibodies, the remaining unoccupied binding sites can be hidden from recognition by a specific blocker that prevents the analyte-unbound sites from binding labeled selective antibody (Figure 2.4).

This masking of binding sites unoccupied by analyte in the test sample can be accomplished, for example, by the use of a protein that is covalently conjugated with hapten, such that when it is present its size prevents the binding of selective antibody. Alternatively, an anti-idiotypic antibody, which specifically recognizes the discrete binding site of the analyte, can be used as a masking reagent. In either situation, labeled selective antibody, which binds to

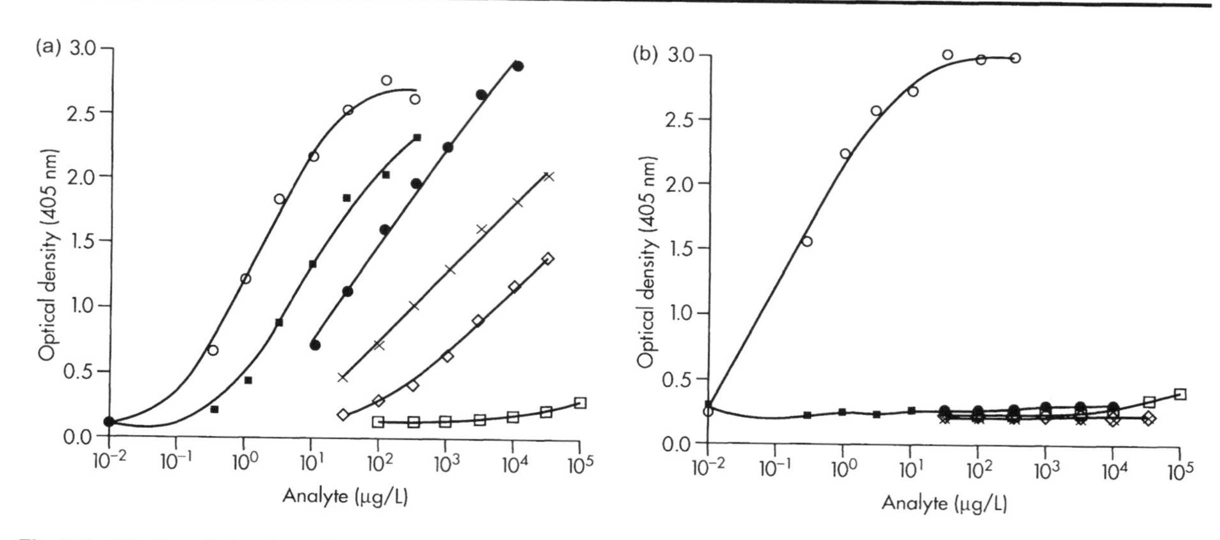

Fig. 2.3 Titration of digoxin or digoxin analogs in the anti-complex assay, either (a) simultaneously in the presence of labeled anti-complex antibody, or (b) prior to an intervening wash step before the addition of labeled anti-complex antibody. Digoxin (○), digoxigenin (■), digitoxin (●), digitoxigenin (×), acetylstrophanthidin (◇), ouabain (□).

the primary anti-analyte antibody even when analyte occupies its binding site, can then be used to probe for the presence of analyte.

We have produced a range of selective antibody systems that function not only with blocking anti-idiotypic antibodies, but also with hapten–protein conjugates working as useful blocking reagents. In addition, Barnard and Kohen and co-workers have used our approach, with anti-idiotypic blocking reagents in the system they subsequently termed the **idiometric assay**, for estradiol (Barnard and Kohen, 1990), and have followed up this demonstration with very interesting assays for

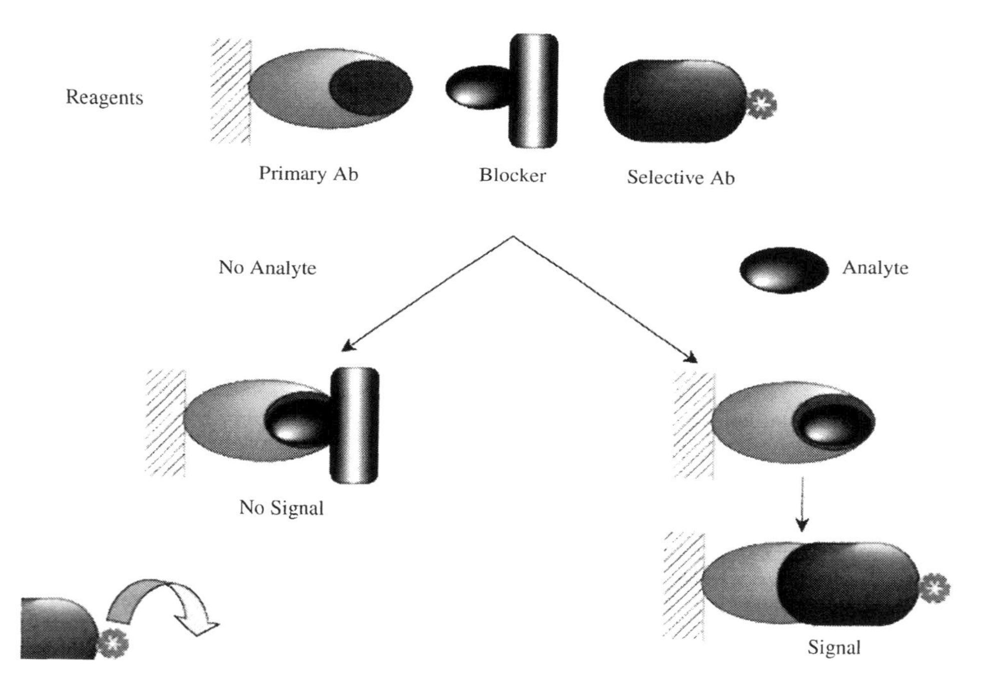

Fig. 2.4 The selective antibody immunometric assay for small molecules.

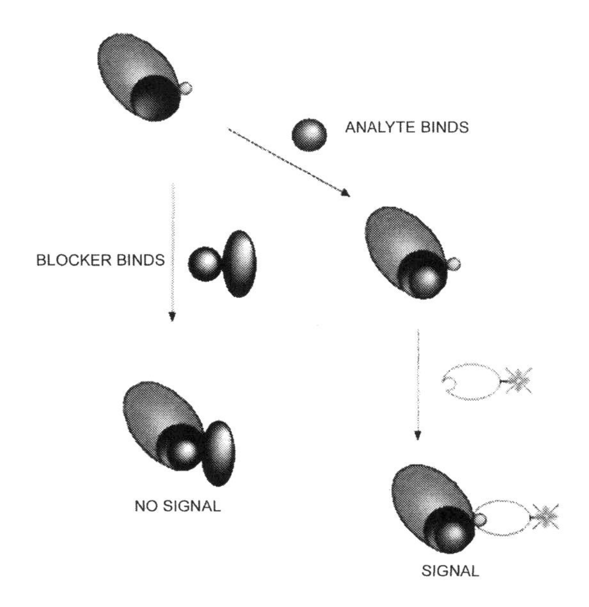

Fig. 2.5 The apposition immunometric assay for small molecules. This method does not require a specially developed antibody.

progesterone (Barnard *et al.*, 1995a) and estrone-3-glucuronide (Barnard *et al.*, 1995b). Similarly, Kobayashi has recently demonstrated selective antibody non-competitive anti-idiotypic assay systems for ursodeoxycholic acid 7-*N*-acetylglucosamine (Kobayashi *et al.*, 2003a) and 11-deoxycortisol (Kobayashi *et al.*, 2003b).

APPOSITION ASSAY

A further, alternative, system is our recently developed system, the **Apposition System** (Self *et al.*, 2003). In this a detectable moiety is synthetically placed close to the binding site of an anti-hapten receptor, such as a monoclonal antibody. This is positioned such that when hapten binds the receptor the detectable moiety can still be bound by a secondary antibody against it. In the absence of analyte the receptor binding site is bound by its specific blocker (such as a hapten–protein conjugate or an anti-idiotypic antibody) inhibiting the binding of the secondary antibody (Figure 2.5). The advantage of this system is that assays can be configured rapidly from an initial receptor preparation. Molecular biological means may be employed to specifically locate the detectable moiety, however, we have demonstrated that site-specific

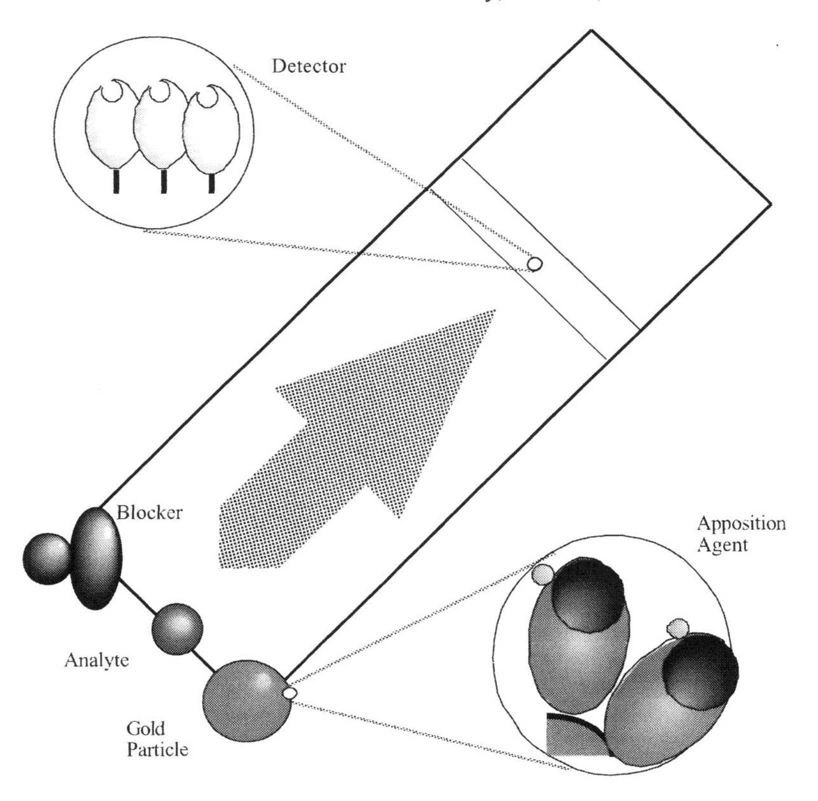

Fig. 2.6 Lateral flow apposition system.

affinity binding agents can also be employed to very good effect to provide suitably conjugated monoclonal antibodies rapidly. The detectable moiety itself can be chosen from substances for which antibodies already exist, for example, biotin. Such a detectable moiety can be used from assay to assay as a universal reagent or different substances can be used from assay to assay. The final apposition reagents can be used on a variety of systems from high throughput testing systems to lateral flow dip sticks (Figure 2.6) for small molecular weight analytes such as drugs of abuse. A main feature of the technology in such applications is that the presence of drug in a sample provides a visible red line clearly seen above a line-free background. This is in contrast to that seen with conventional competitive systems in which analyte negative samples produce a strong visible line that is somewhat reduced with analyte positive samples.

CONCLUSIONS

Both the anti-complex and the selective antibody systems offer obvious improvements in performance over that exhibited by competitive assay systems for a large range of small molecule analytes, including metabolites such as steroids, and extending to areas such as therapeutic drugs, drugs of abuse, toxins and pesticides. This has now been taken forward by the introduction of the apposition assay system with the particular advantage that it can be formulated rapidly using universal reagents if required. Primary antibodies or secondary anti-complex, selective or apposition antibodies may be labeled by any of the labels conveniently employed in immunoassay. These immunometric systems can also be readily adapted to the various formats utilized in two-site immunoassay systems. Importantly, our work has also shown that the strengths of these systems are particularly evident in dipstick formats where a positive signal developing from a clear zero background is very much easier to observe and interpret than the indirect signal obtained in competitive format systems, where a positive sample produces a reduction of a high zero signal such as a color. We believe that users as well as assay developers will appreciate and benefit from the clear advantages of these systems.

REFERENCES

Barnard, G. and Kohen, F. Idiometric assay: a noncompetitive immunoassay for small molecules typified by the measurement of estradiol in serum. *Clin. Chem.* **6**, 1945–1950 (1990).

Barnard, G., Osher, J., Lichter, S., Gayer, B., De Boever, J., Limor, R., Ayalon, D. and Kohen, F. The measurement of progesterone in serum by a non-competitive idiometric assay. *Steroids* **60**, 824–829 (1995a).

Barnard, G., Amir-Zaltsman, Y., Lichter, S., Gayer, B. and Kohen, F. The measurement of oestrone-3-glucuronide in urine by non-competitive idiometric assay. *J. Steroid Biochem. Mol. Biol.* **5**, 107–114 (1995b).

Bonnar, J., Flynn, A., Freundl, G., Kirkman, R., Royston, R. and Snowden, R. Personal hormone monitoring for contraception. *Br. J. Fam. Plann.* **24**, 128–134 (1999).

Carrero, J., Mallender, W.D. and Voss, E.W. Jr. Anti-metatype antibody stabilization of Fv 4-4-20 variable domain dynamics. *J. Biol. Chem.* **271**, 11247–11252 (1996).

Ekins, R.P. An overview of present and future ultrasensitive non-isotopic immunoassay development. *Clin. Biochem. Rev.* **8**, 12–23 (1987).

Kobayashi, N., Kubota, K., Oiwa, H., Goto, J., Niwa, T. and Kobayashi, K. Idiotype-anti-idiotype-based noncompetitive enzyme-linked immunosorbent assay of ursodeoxycholic acid 7-*N*-acetylglucosaminides in human urine with subfemtomole range sensitivity. *J. Immunol. Methods* **272**, 1–10 (2003a).

Kobayashi, N., Shibusawa, K., Kubota, K., Hsegawa, N., Sun, P., Niwa, T. and Goto, J. Monoclonal anti-idiotype antibodies recognizing the variable region of high-affinity antibody against 11-deoxycortisol. Production, characterization and application to a sensitive noncompetitive immunoassay. *J. Immunol. Methods* **274**, 63–75 (2003b).

May, K. Unipath Clearblue One Step™, Clearplan One Step™ and Clearview™. In: *The Immunoassay Handbook* (ed Wild, D.), (Stockton Press, New York, 1994).

Nagata, S., Tsutsumi, T., Yoshida, F. and Ueno, Y. A new type sandwich immunoassay for microcystin: production of monoclonal antibodies specific to the immune complex formed by micocystin and an anti-microcystin monoclonal antibody. *Nat. Toxins* **7**, 49–55 (1999).

Self, C.H. Antibodies manufacture and use. World Intellectual Patent Cooperation Treaty Publication No. 85/04422 (1985).

Self, C.H. Determination method, use and components. Patent Corporation Treaty Publication No. WO89/05453 (1989).

Self, C.H. The impact of new immunodiagnostic technologies. In: *Rapid Methods and Automation in Microbiology and Immunology* (eds Spencer, R.C. *et al.*), (Intercept, Andover, 1993a).

Self, C.H. Multiple binding patent. World Intellectual Patent Cooperation Treaty Publication No. WO 93/14404 (1993b).

Self, C.H., Dessi, J.L. and Winger, L.A. High-performance assays of small molecules: enhanced sensitivity, rapidity and convenience demonstrated with a noncompetitive immunometric anti-immune complex assay system for digoxin. *Clin. Chem.* **40**, 2035–2041 (1994).

Self, C.H., Dessi, J.L. and Winger, L.A. Ultra-specific immunoassays for small molecules: the roles of wash steps and multiple binding formats. *Clin. Chem.* **42**, 1527–1531 (1996).

Self, C.H., Winger, L.A., Dessi, J.L., Spoors, J.A. and Thompson, S. Synthetically changed receptors for high performance small molecule assay. Synthetic Receptors 2003, http://www.synthetic-receptors2003.com/oral.htm

Towbin, H., Motz, J., Oroszlan, O. and Zingel, O. Sandwich immunoassay for the hapten angiotensin II. A novel assay principle based on antibodies against immune complexes. *J. Immunol. Methods* **181**, 167–176 (1995).

Ullman, E.F., Milburn, G., Jelesko, J., Radika, K., Pirio, M., Kempe, T. *et al.* Anti-immune complex antibodies enhance affinity and specificity of primary antibodies. *Proc. Natl Acad. Sci. USA* **90**, 1184–1189 (1993).

Winger, L.A., Dessi, J.L. and Self, C.H. Enhanced specificity for small molecules in a convenient format which removes limitations imposed by competitive immunoassay. *J. Immunol. Methods* **199**, 185–191 (1996).

3 Ambient Analyte Assay

Roger Ekins

INTRODUCTION

Conventional immunoassays rely on the measurement of the amount of the target analyte in a specified volume of test fluid. Thus, if additional sample containing the analyte is added to the test incubation mixture (in immunoassays of either competitive or non-competitive assay design) the observed signal changes. The requirement that sample volume be carefully standardized is therefore an important feature of all conventionally designed immunoassay procedures. Constancy of sample volume can, of course, be readily achieved in large laboratory analyzers without the introduction of significant additional mechanical complexity; however, sample volume variations have long been known to constitute a common and significant source of analytical error in the case of manually performed tests.

It might therefore seem at first sight that use of a known sample volume is an inevitable requirement common to all immunoassays, and indeed to all other analytical procedures. Yet a pH meter gives the same pH reading regardless of the volume of fluid in which the electrode is immersed. In other words, the electrode 'senses' the ambient hydrogen ion concentration in the surrounding fluid. Clearly it would be advantageous, in certain circumstances, if immunoassays were available that functioned in a similar manner. In fact, the form of assay termed **ambient analyte ligand assay**[1] (Ekins, 1983) does exactly this.

This term embraces any type of 'binding' or 'ligand assay' (such as immuno- and DNA-assays) in which the analyte concentration in a test fluid is determined 'directly', i.e. not (as is generally the case) by measurements both of the amount of analyte in the sample and of sample volume. Ambient analyte assays rely on the use of extremely small amounts of antibody (or other specific analyte binder, such as an oligonucleotide) – amounts that are generally orders of magnitude lower than those previously regarded as obligatory in the immunoassay field. An extremely important practical illustration of this concept is provided by the miniaturized microarray methods first developed by my colleagues and myself (later in collaboration with Boehringer Mannheim GmbH, now Roche Diagnostics). These 'chip' technologies – now the subject of intensive development by many companies in the US and elsewhere – are widely predicted to dominate clinical chemistry in the next few years. However, similar concepts can also, in retrospect, be seen to govern some of the direct free hormone immunoassay methods already in routine diagnostic use (*see* FREE ANALYTE IMMUNOASSAY).

Ultimately, the ambient analyte assay principle is likewise likely to find expression in the form of miniaturized transducer-based 'immunosensor' technologies, such as that recently developed by Cornell *et al.* (1997). Moreover, the combination of this principle with sensor techniques opens the door to many new and exciting applications, such as the determination of (varying) analyte concentrations in body fluids *in vivo*, the on-line measurement of key constituents in industrial process control applications, and *in situ* environmental monitoring. In short, ambient analyte sensor systems of this kind are clearly advantageous in any situation where sample isolation and volume measurement are impossible, undesirable or inconvenient.

Immunoassays conforming to ambient analyte assay principles provide other important and surprising advantages. For example, the miniaturized antibody 'microspot' array systems referred to above, by minimizing diffusion constraints on analyte binding to antibody, are characterized by reaction kinetics that are faster than those observed using any other assay format. Likewise, contrary to widely accepted views, such systems yield higher assay sensitivities than all other immunoassay designs. But perhaps one of the most important long-term benefits may arise outside the immunoassay field *per se*, i.e in the

[1] It has been claimed that all binding assays measure the analyte concentration in test samples. But, as generally performed, assay results depend on the volume of sample (assuming the calibrants are unchanged), demonstrating their dependence on the amount of analyte in the sample.

The Immunoassay Handbook. *Third Edition*
D. Wild (Ed.)

field of DNA diagnostics, where the identification of genetic abnormalities may require the analysis of hundreds or thousands of individual polynucleotide sequences.

BASIC THEORETICAL PRINCIPLES

All immunoassays are implicitly based on measurement of antibody occupancy following reaction between a 'sensor' antibody and an antigen (the analyte) (Figure 3.1). 'Non-competitive' immunoassays rely on *direct* measurement of occupied 'sensor-antibody' binding sites (using, e.g. a second, labeled antibody directed against occupied sites). Conversely, so-called 'competitive' immunoassays depend on *indirect* determination of binding site occupancy by measurement of unoccupied sites (e.g. by the use of labeled analyte, analyte analog or anti-idiotypic antibody) (Figure 3.2).[2]

This general concept is expressed in the following equation, representing the fractional occupancy (F) by analyte of antibody binding sites at equilibrium (i.e. the fraction of sites that are occupied) as a function of analyte concentration ([An]):

$$F^2 - F(1/K[\mathrm{Ab}] + [\mathrm{An}]/[\mathrm{Ab}] + 1) + [\mathrm{An}]/[\mathrm{Ab}] = 0 \quad (3.1)$$

where [Ab] is the antibody concentration and K, the affinity constant governing the reaction.

Clearly F also depends on the values of K and [Ab]. Thus, in conventional immunoassay designs it is imperative that the total amount of antibody, the individual sample volume, the total incubation volume, and K, are identical in all the incubation tubes. Assuming this is the case, the only variable quantity in the system is the amount of analyte present in the test sample. Following measurement of the final analyte concentration in the incubation mixture (by reference to the known concentrations present in a set of standards), the amount of analyte in the sample is determined, and – since the original sample volume is known – the unknown analyte concentration in the sample may be derived.

In general, reduction of the amount of antibody used in the system results in increased antibody occupancy (assuming other factors remain constant). This conclusion is illustrated in Figure 3.3, in which F is plotted against antibody concentration (assuming the presence in incubation mixtures of the analyte concentrations shown).[3] However, as the amount of antibody (and hence the antibody concentration) employed approach 0, F tends to $[\mathrm{An}]/(1/K + [\mathrm{An}])$. Thus when the antibody concentration is low (less than approximately $0.01/K$–$0.05/K$) antibody fractional occupancy is dependent only on K and the analyte concentration in the medium to which the antibody is exposed, being independent both of the volume of the medium, and of the amount of antibody present.

In other words, when a small amount of antibody coupled to a solid support is exposed to analyte-containing fluid of such volume that the resulting antibody concentration is less than approximately $0.05/K$,[4] the fractional occupancy

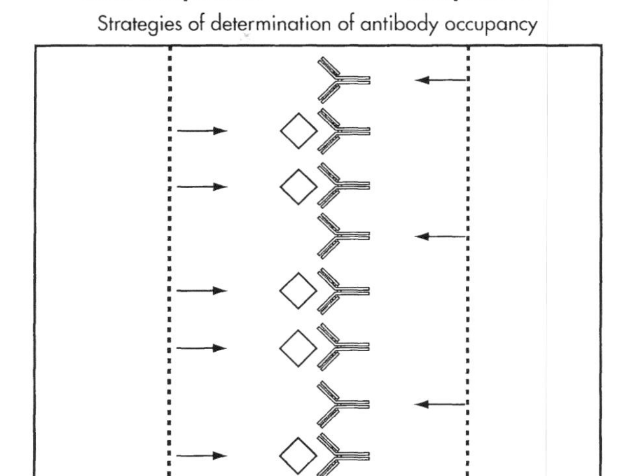

Fig. 3.1 The 'antibody binding site occupancy principle' of immunoassay. All immunoassays implicitly rely on the measurement of the fractional antibody binding site occupancy by analyte.

[2] These concepts implicitly assume sequential exposure of antibody first to the sample and subsequently to a labeled reagent (except in assays in which the sensor antibody is itself labeled (Figure 3.3)). However, in many circumstances the sensor antibody, test sample and labeled reagent are incubated together and the reactions occur simultaneously. Though the equations describing these situations therefore differ slightly, the concepts distinguishing competitive and non-competitive assays are unaffected.

[3] Note that antibody and analyte concentrations are expressed in units of $1/K$, thereby generalizing the figure.

[4] When antibody is coupled to a solid support, its concentration in the system is given by the number of effective antibody binding sites divided by the volume of fluid to which the antibody is exposed. In other words, the location of antibody within the medium is irrelevant under equilibrium conditions, only the kinetics of the binding reactions being affected when the antibody is restricted to a particular zone or compartment (such as – assuming the absence of secondary effects, such as antibody denaturation – by its location on a solid surface).

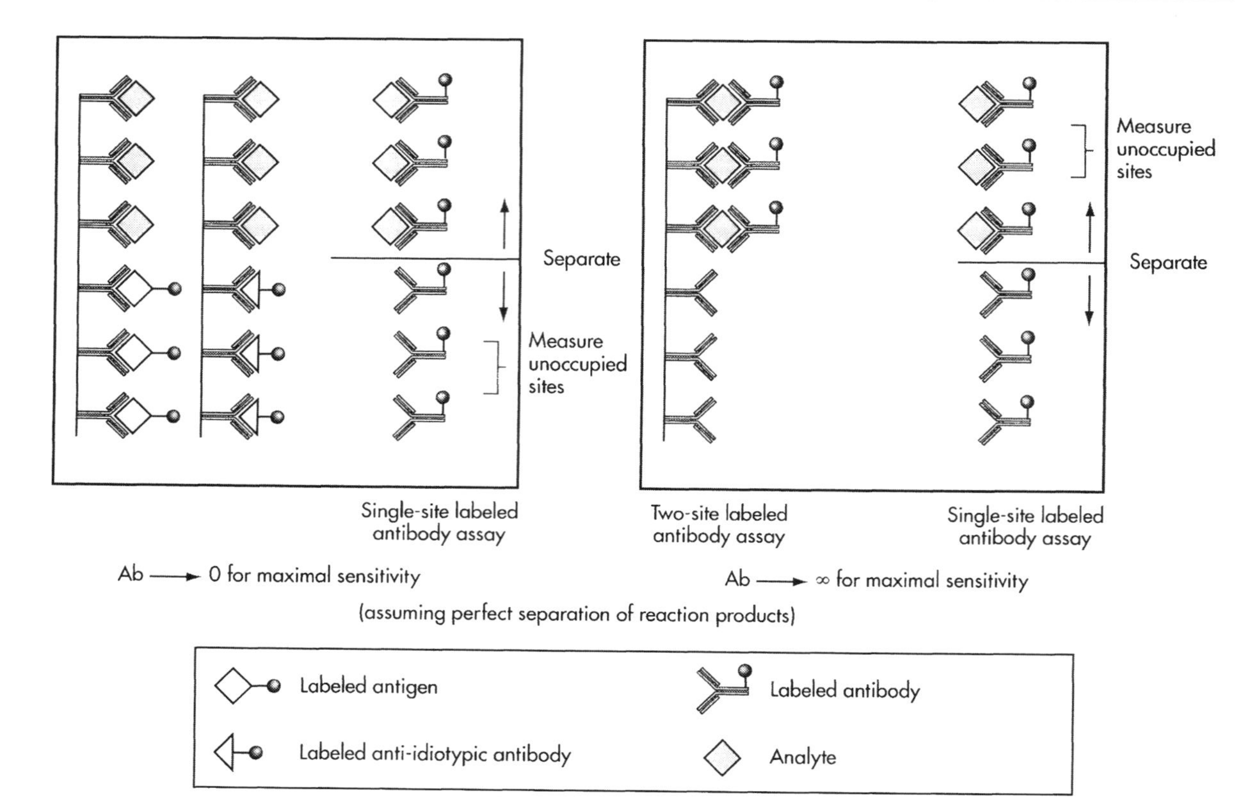

Fig. 3.2 Immunoassays differ in the way fractional occupancy of antibody binding sites is determined. So-called 'competitive' assays (left) rely on indirect measurement of occupancy, by observation of unoccupied sites. *Non-competitive* or *immunometric* assays (right) rely on direct measurement of binding site occupancy. Note that labeled antibody assays may be either of competitive or non-competitive design.

of antibody binding-sites solely reflects the pre-existing ambient analyte concentration *per se* (Figure 3.4), the total *amount* of analyte in the test sample being irrelevant. Analyte binding by antibody inevitably causes analyte depletion in the medium, but because the amount so bound is small, the reduction in the ambient analyte concentration caused by the introduction of antibody is insignificant. For example, if the sensor-antibody binding site concentration is less than $0.01/K$, the reduction in the analyte concentration is less than 1% (irrespective of the concentration's magnitude), (Figure 3.3).

A close similarity exists between these concepts and the operation of a simple mercury thermometer. A small thermometer – when placed in a liquid in a container – extracts heat from the liquid until it reaches thermal equilibrium, at which point the liquid's temperature is indicated. Provided the thermometer's thermal capacity is insignificant compared with that of the liquid, the indicated temperature is independent both of the thermometer's size and the volume of liquid in which it has been immersed. However, the introduction of a large thermometer of high thermal capacity into the liquid will alter the latter's temperature. The final recorded value will therefore depend both on the thermometer's size and the liquid volume (Figure 3.5).

In these circumstances the only practical way of determining the initial temperature of liquid samples is by ensuring that their volumes are standardized, that an identical thermometer (at the same initial temperature) is used, and that the system is calibrated using water standards of identical volume. By these means, a calibration curve may be drawn relating the true (initial) temperature of water samples to the final, observed, temperature.

Such a procedure is closely analogous to the steps adopted in conventional immunoassay protocols.

APPLICATIONS OF THE AMBIENT ANALYTE ASSAY PRINCIPLE

Two of the most important practical applications of the ambient analyte principle are 'microspot' assays (leading to the construction of multianalyte microarrays) and free

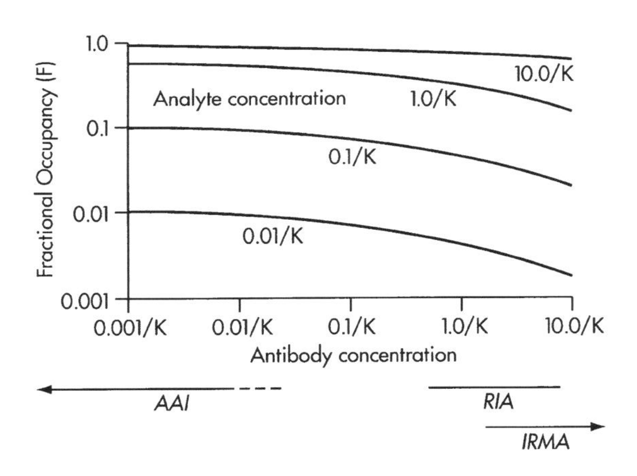

Fig. 3.3 Fractional antibody binding-site occupancy (F) plotted as a function of antibody binding-site concentration for different values of analyte (antigen) concentration. (All concentrations are expressed in multiples of 1/K, thereby generalizing the curves.) Note that for antibody concentrations $< 0.01/K$ (approx.), percentage binding of analyte is $<1\%$ for all analyte concentrations, and *F* is essentially unaffected by differences in antibody concentration, being governed solely by the analyte concentration (ambient analyte immunoassay (AAI)). Note that radioimmunoassays and other 'competitive' immunoassays conventionally rely on antibody concentrations approximating $0.5/K - 1/K$ or above (implying binding of labeled analyte alone (B_o) of at least 33%). Non-competitive (immunometric) assays are generally based on the use of considerably higher antibody concentrations.

hormone immunoassays. (Note that the latter were conceived of and developed before it was later recognized that they constituted a particular example of a more general principle.) These two areas of application each involve different additional concepts, and are therefore best examined separately.

MICROSPOT ASSAYS

As indicated above, ambient analyte assay conditions prevail if an amount of 'sensing' or 'capture' antibody is located on a solid support and exposed to an analyte containing medium of sufficient volume that the antibody concentration is less than approximately $0.05/K$ (or, preferably, less than about $0.01/K$). Moreover, the antibody can be confined within a 'microspot' such that the total number of antibody sites is ideally less than $\upsilon/K \times 10^{-5} \times N$ (where υ is the sample volume (mL) and N the Avogadro's number (6×10^{23})). Thus if $\upsilon = 1$ and $K = 10^{11}$ L/M, the number of binding sites causing negligible disturbance ($<1\%$) to the ambient analyte concentration is 6×10^7 (Figure 3.6). Following the microspot's exposure to an analyte-containing sample, occupancy of antibody within the spot may be determined by its exposure to a second (labeled) 'developing' antibody reactive with either occupied or unoccupied sites, these being equivalent to the non-competitive and competitive measurement strategies adopted in conventional immunoassays (Figure 3.7).

Since ambient analyte assays rely on measurement of the fractional occupancy of the sensor antibody (regardless of the exact amount of the latter present), a useful stratagem in this context is to label the sensor antibody itself with a second label, and to observe the ratio of signals emitted by the two labeled antibodies[5] (Figure 3.8). Fluorescent labels are particularly advantageous in that they possess very high specific activities, and permit an array of 'microspots' distributed on the surface of a 'chip' or sample holder (Figure 3.9) to be optically scanned using a confocal microscope or CCD camera (Figure 3.10). In principle, other labels, such as chemiluminescent labels, can also be used, though their lower specific activities would, in some circumstances, significantly limit assay sensitivities.

Intuitively, one might expect that the sensitivity of an immunoassay that uses a minute amount of antibody, confined to a spot so small as to be invisible, and binding only an insignificant fraction of the total analyte present in a sample, would be extremely low. However, microspot assays can yield higher sensitivities than other formats. This surprising finding requires explanation.

The reason may be readily understood without resort to mathematical theory by consideration of the effects of increasing the diameter of a microspot within which sensor antibody molecules are located at maximum surface density (Figure 3.11). If the amount of antibody in a small microspot is such that the antibody concentration in the system approximates $0.01/K$, then – as indicated above – less than 1% of analyte molecules in the solution to which the spot is exposed will, at equilibrium, be bound to the antibody. Enlarging the spot's size increases the amount of analyte bound (and hence the signal generated by labeled antibodies attached to the analyte, assuming use of a non-competitive 'sandwich assay' approach), but decreases antibody fractional occupancy. Thus the increase in signal emitted by analyte-bound labeled antibody molecules is less than that of the area of the spot. On the other hand, the background signal increases along with the size of the spot. The signal/background ratio therefore decreases with increase in spot area

[5] Measurement of the ratio of the two signals obviates (*inter alia*) problems arising from fluctuations in the intensity of the exciting light source, variations in coating density of the sensing antibody (or nucleotide) on the solid support, etc. Other advantages of the use of two labels are that it enables full quality control of manufactured microspot arrays, and – if necessary – exact positioning of the signal detector (e.g. confocal microscope) over each microspot in the array.

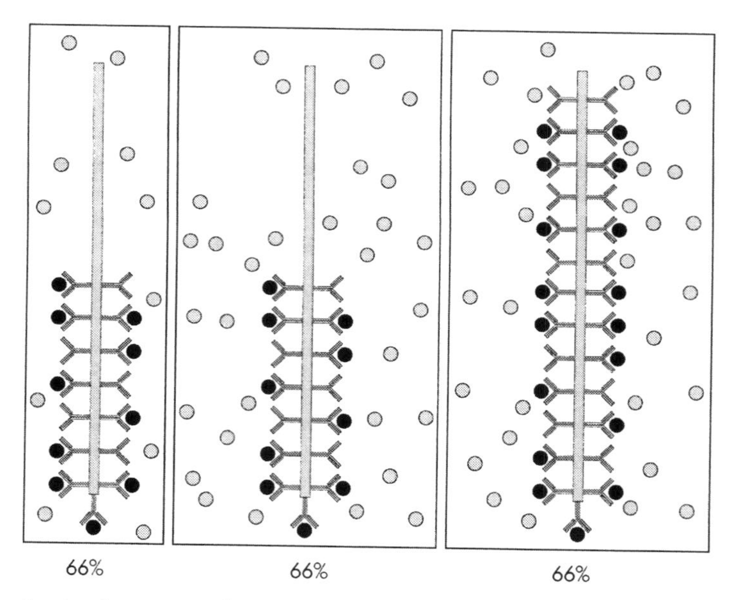

Fig. 3.4 If a probe bearing a small number of antibody molecules on its surface is placed in an analyte-containing medium of sufficient volume that ambient analyte conditions are fulfilled, then the fractional occupancy of antibody binding sites will be unaffected either by variations in the volume of the medium, or in the number of antibody molecules on the probe.

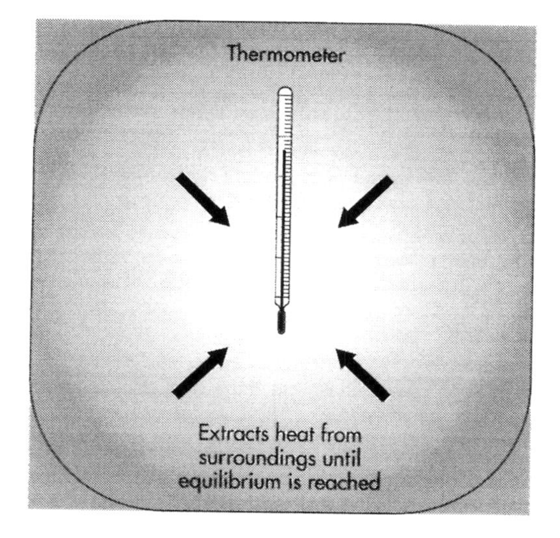

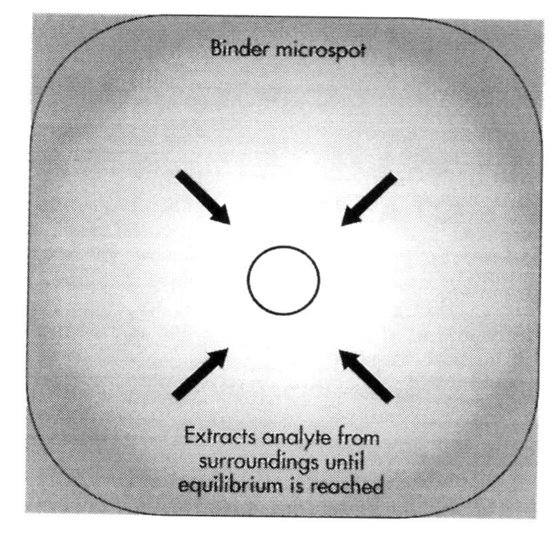

Fig. 3.5 The thermometer analogy. A (cold) thermometer absorbs heat until reaching thermal equilibrium with its surroundings. The temperature it records will not differ from the original ambient temperature if the amount of heat absorbed is an insignificant fraction of the total in the system.

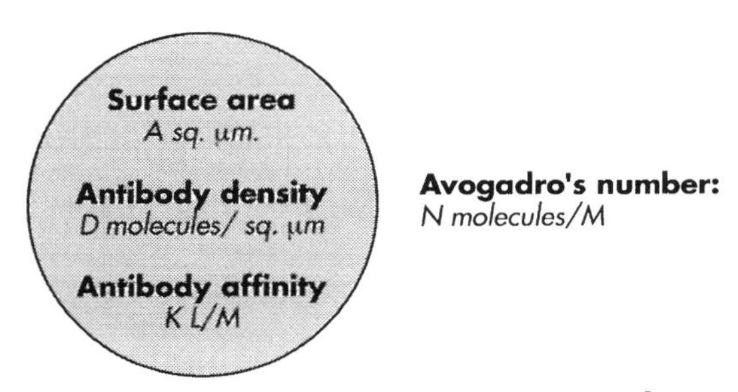

Minimum test sample volume (mL): $\frac{A \times D \times K \times 10^5}{N}$

Fig. 3.6 When exposed to an analyte-containing medium greater than shown, an antibody microspot assay conforms to ambient analyte conditions.

(Figure 3.12), this phenomenon being likely to reduce sensitivity.[6] Conversely reduction in spot size below that yielding an antibody concentration of 0.01/K will not significantly increase the signal/background ratio and hence not increase sensitivity.[7] Indeed progressive reduction in microspot area towards zero ultimately reduces the number of captured analyte molecules (and hence the signal emitted therefrom) to zero, in which event the system will clearly be totally insensitive.

Theoretical consideration of (non-competitive) microspot immunoassay sensitivity (Ekins, 1989) permits prediction of the sensitivities attainable using sensor antibodies of varying affinities as a function of the minimum detectable surface density of labeled antibody molecules (S^*_{min}) (Figure 3.13).[8] Such analysis indicates that the achievement of high microspot assay sensitivity requires a detector capable of accurately measuring low surface densities of labeled developing antibodies, and close packing of sensor–antibody molecules within the microspot area, maximizing the signal/background ratio. It also suggests that, assuming the use of very high specific activity non-isotopic labels, sensitivities yielded by microspot assays are unlikely to be inferior, and (depending on the measuring instrument used) may be considerably superior, to the sensitivities achievable in macroscopic assays of conventional design. For example, my colleagues and I – using antibodies labeled with fluorescent microspheres and a commercially available confocal microscope – achieved, in early studies, sensitivities in the order of 0.06 labeled antibody molecules/μm^2. Subsequently Boehringer Mannheim researchers, using improved scanning equipment incorporated into prototype analyzers, achieved detection limits of approximately 0.01 molecules/μm^2, suggesting that assay detection limits in the order of 10^{-17} mol/L (i.e. 10^3–10^4 analyte molecules/mL) are attainable using the ambient analyte microspot approach.

Predictions of high ambient analyte assay sensitivity can readily be verified in practice (Ekins and Chu, 1993), see Figure 3.14. However, it must be emphasized that the label used in microspot assays must be of such high specific activity that photon counting errors do not constitute the principal source of signal variation limiting sensitivity.

These conclusions assume the establishment of thermodynamic equilibrium in the system; however, the velocities of binding reactions are reduced with a reduction of concentrations of one or both of the reactants, and the time to reach equilibrium increases. Moreover, it is well known that diffusion constraints on binding reactions reduce reaction velocities when capture antibody molecules are linked to a solid support. Thus the suggestion that microspot assays are likely to be more rapid than conventional assays may likewise be counterintuitive. However, this prediction can again be readily understood by consideration of a series of antibody microspots of increasing diameter, each containing sensor antibody molecules at the same surface density. As indicated above, antibody fractional occupancy at equilibrium – and hence analyte surface density – is highest when the microspot area is small, the system therefore conforming to ambient analyte assay conditions. Moreover, it is intuitively evident, and may be confirmed theoretically (Crank, 1975) that the smaller the antibody microspot, the lower the diffusion constraints on analyte migration to it. Thus the surface density of analyte molecules within the microspot area increases more rapidly as the microspot area is decreased. Indeed, the thermometer analogy provides a good illustration of these concepts, it being evident that the smaller a thermometer, the faster it reaches thermal equilibrium with its surroundings.

Detailed theoretical analysis of the rate at which analyte molecules migrate towards, and bind to, an antibody microspot reveals that the (initial) antibody occupancy rate (OR) (per unit area of microspot) is given by (Ekins, 1995):

$$\text{OR} = 4r_{\text{m}}k_{\text{a}}D[\text{An}]d_{\text{Ab}}/(\pi r_{\text{m}}^2 k_{\text{a}} d_{\text{Ab}} + 4Dr_{\text{m}}) \quad \text{molecules/sec/cm}^2 \tag{3.2}$$

[6] Assuming the random variation in the background signal (noise) is approximately proportional to the background.
[7] This conclusion applies only when the reaction is allowed to proceed to equilibrium. It may be advantageous to reduce the antibody concentration below 0.01/K if the reaction is terminated before equilibrium is reached (see below).
[8] The analysis does not take into consideration the statistical problems that arise when analyte concentrations are so low that the probability of capturing a single analyte molecule within the microspot area becomes the major determinant of assay sensitivity.

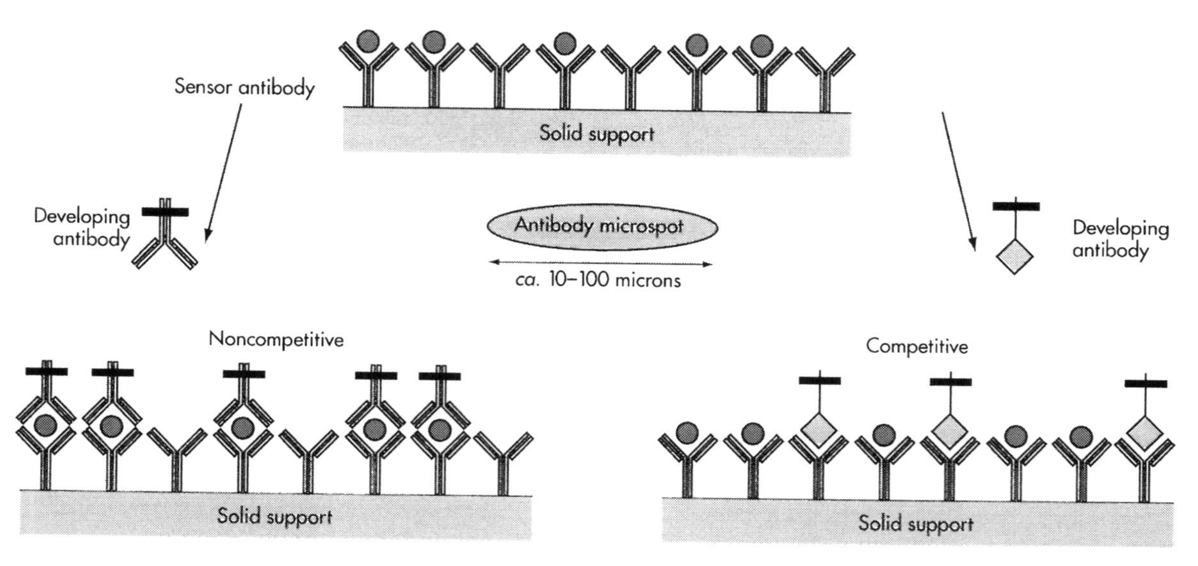

Fig. 3.7 Following exposure to the analyte-containing medium of the 'capture' (or 'sensor') antibody, a second labeled ('developing') antibody is used to determine sensor-antibody binding-site occupancy using a non-competitive approach (left) or competitive approach (right).

where k_a is the association rate constant (cm^3/sec/molecule); [An], the ambient analyte concentration (molecules/mL); D, the diffusion coefficient (cm^2/sec); r_m, the microspot radius; d_{Ab}, the antibody surface density (binding sites/cm^2).

Thus as r_m tends to zero, the term $\pi r_m^2 k_a d_{Ab}$ becomes small compared to $4Dr_m$, implying that OR approximates $k_a[An]d_{Ab}$. In other words, the velocity at which antigen molecules bind (per unit area) to antibodies attached to the solid support increases with reduction in r_m,

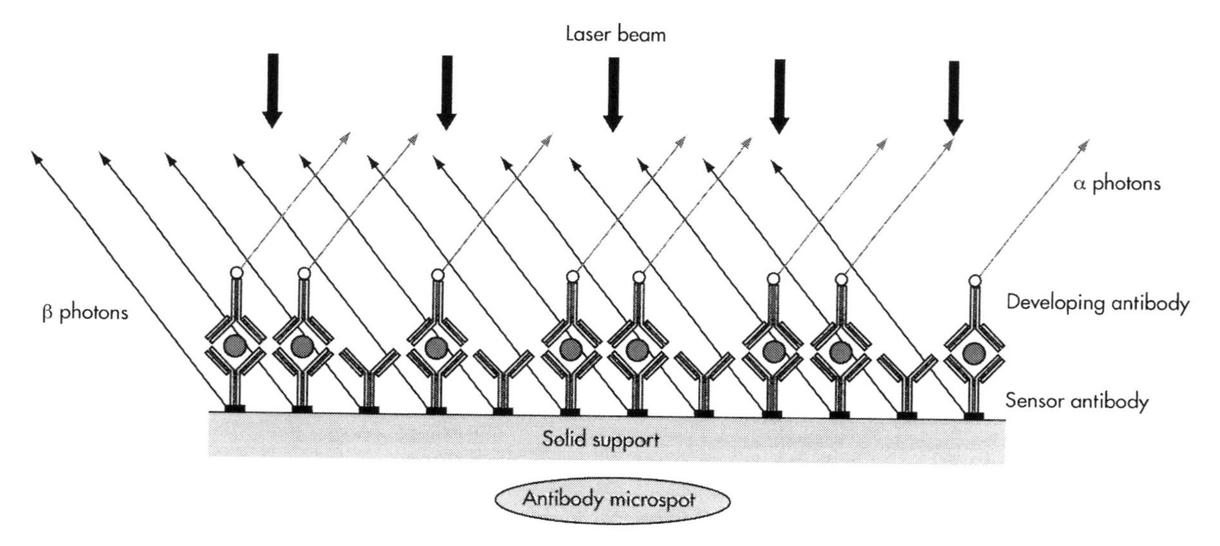

Fig. 3.8 Microspot immunoassay relying on fluorescent labeled antibodies. The ratio of α and β fluorescent photons emitted reflects the value of *F*. This ratio is solely dependent on the analyte concentration to which the probe has been exposed, being unaffected by the amount or distribution of antibody coated (as a monomolecular layer) on the probe surface.

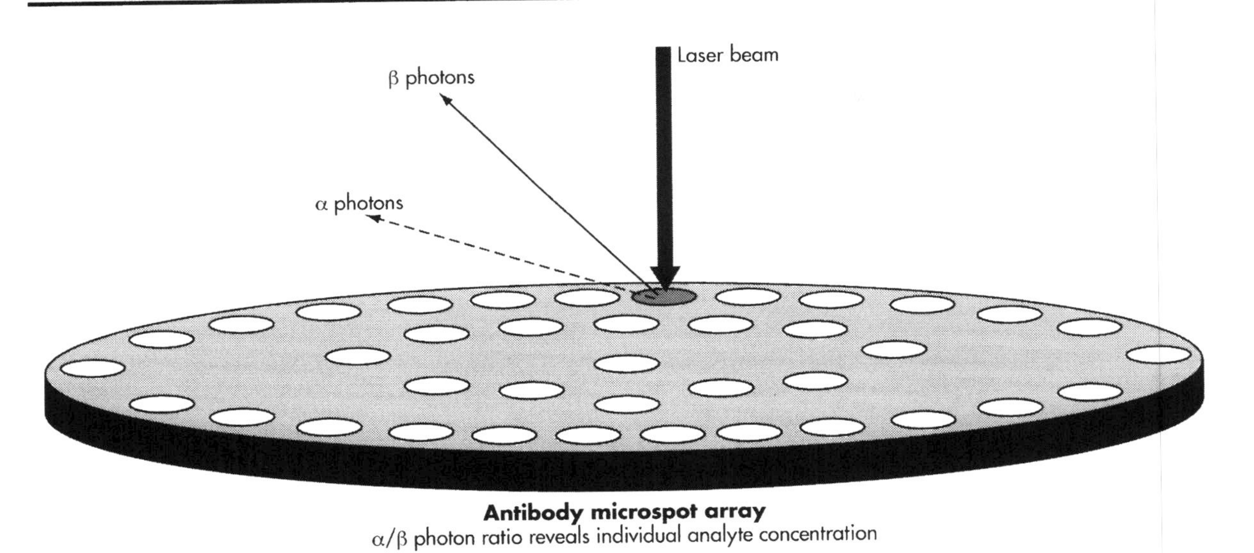

Fig. 3.9 Antibody microarray. Each microspot is interrogated to determine its fractional occupancy by the analyte against which it is directed.

ultimately approximating that seen in a homogeneous solution.

Detailed computer models illustrate the events following the introduction of a microspot into an analyte-containing medium (Figure 3.15), likewise demonstrating that the

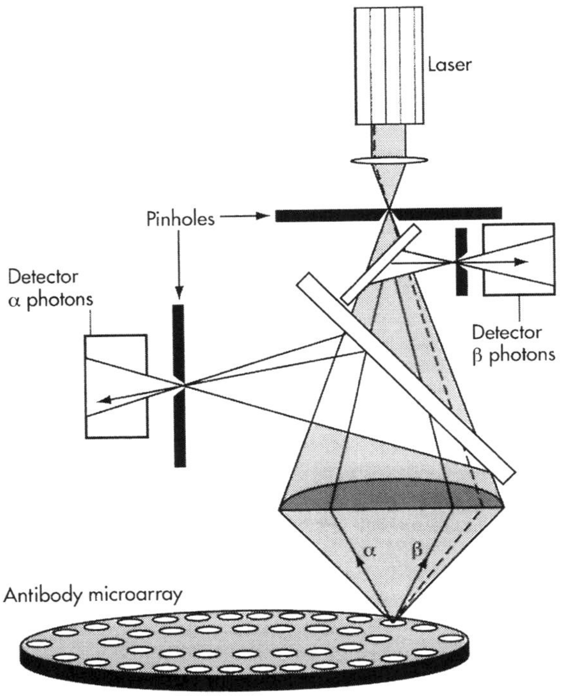

Fig. 3.10 A laser-based confocal microscope provides a sensitive method of determining fluorescent signals emitted from the spots.

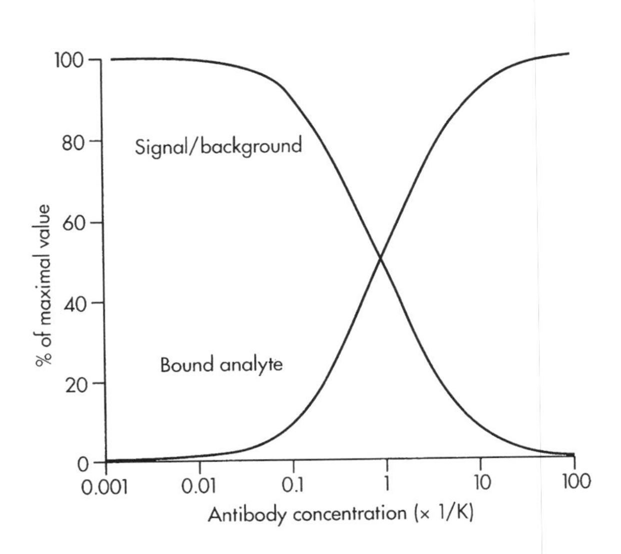

Fig. 3.11 Increase in 'capture' antibody-coated area results in increased analyte capture, but analyte surface density (and the corresponding signal generated by developing antibody) fall. Maximal analyte surface density (and a maximal signal/background ratio) is achieved when the antibody concentration falls below 0.01/K.

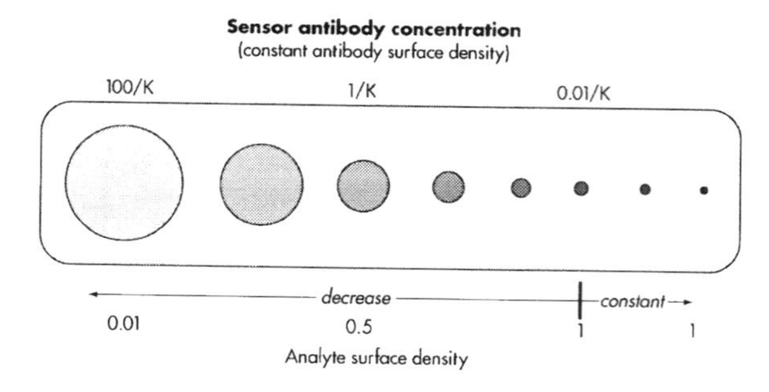

Fig. 3.12 As microspot size increases (implying an increase in sensor antibody concentration) the signal deriving from analyte molecules bound to the sensor antibody fades into the background.

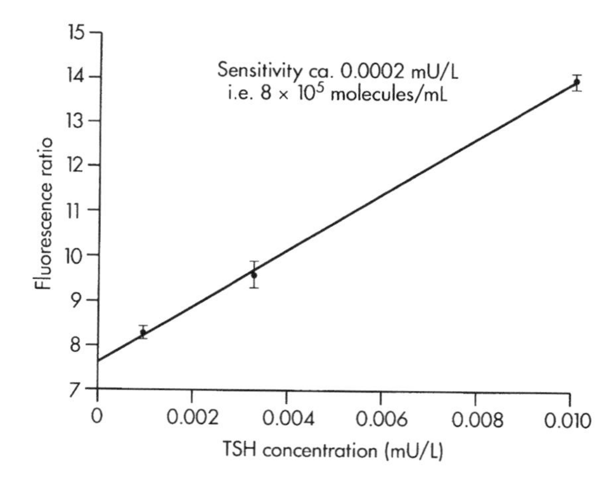

Fig. 3.14 Section of typical dose response curve falling below 0.01 μU/mL yielded in a two step TSH dual-labeled microspot 'ratiometric' assay using Texas Red-labeled solid-phase sensor antibody and a fluorescent microsphere-labeled developing antibody. Sensitivity (detection limit) derived from precision profile.

surface density of analyte molecules within the microspot area increases more rapidly (and equilibrium is reached sooner) the smaller the spot diameter (Figure 3.16).

In summary, microspot-based ambient analyte assays are potentially more sensitive, and can be performed in shorter times, than assays relying on conventional formats. Practical realization of these potentially important advantages nevertheless requires good instrumentation, and attention to key determinants of assay sensitivity

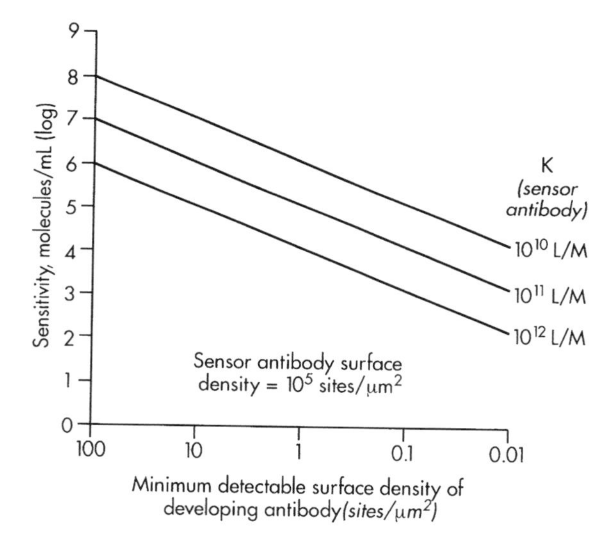

Fig. 3.13 Theoretically predicted non-competitive microspot immunoassay sensitivities plotted as a function of the minimum developing antibody surface density (S^*_{min}) detectable within the microspot area. Values of capture antibody surface density (S) of 10^5 binding sites/μm^2, and of developing antibody concentration of $1/K^*$ have been assumed. K and K^* are equilibrium constants of capture (sensor) and developing antibodies, respectively.

such as nonspecific binding of labeled reagents to solid supports, minimization of background signals from the supports themselves, etc. Likewise, industrial implementation of these ideas, and the development of reliable multianalyte microspot assays, presents considerably greater problems than that of conventional single analyte assay kits. Amongst these are the development of methods for manufacturing microarrays, for rapid and sensitive scanning of the arrays, and for fully automated analyzers incorporating the microfluidic and other sample processing systems required to ensure reliability of assay results.

Current methods of constructing microarrays have recently been reviewed by Schena *et al.* (1998). The industrial method evolved in the course of my own group's collaboration with Boehringer Mannheim relies on small disposable polystyrene carriers (or 'chips') onto which microspots are deposited using 'inkjet' technology. Arrays comprising 100–200 spots (each of diameter approximating 80 μm and spaced approximately 40 μm apart) are deposited in this manner on the flat bottom (*ca.* 3 mm diameter) of the carrier wells (Figure 3.17). Using prototype instrumentation, microspot arrays (each individually quality controlled) are produced at a speed of 10,000 arrays per hour.

A fundamentally different approach applicable to oligonucleotide (and polypeptide) microarray construction is that originally developed (to generate large numbers of candidate drugs) by Fodor *et al.* (1991) using combinatorial synthetic techniques. These permit construction of numerically large arrays, though (because the efficiency of synthesis is only in the order of 85–95%)

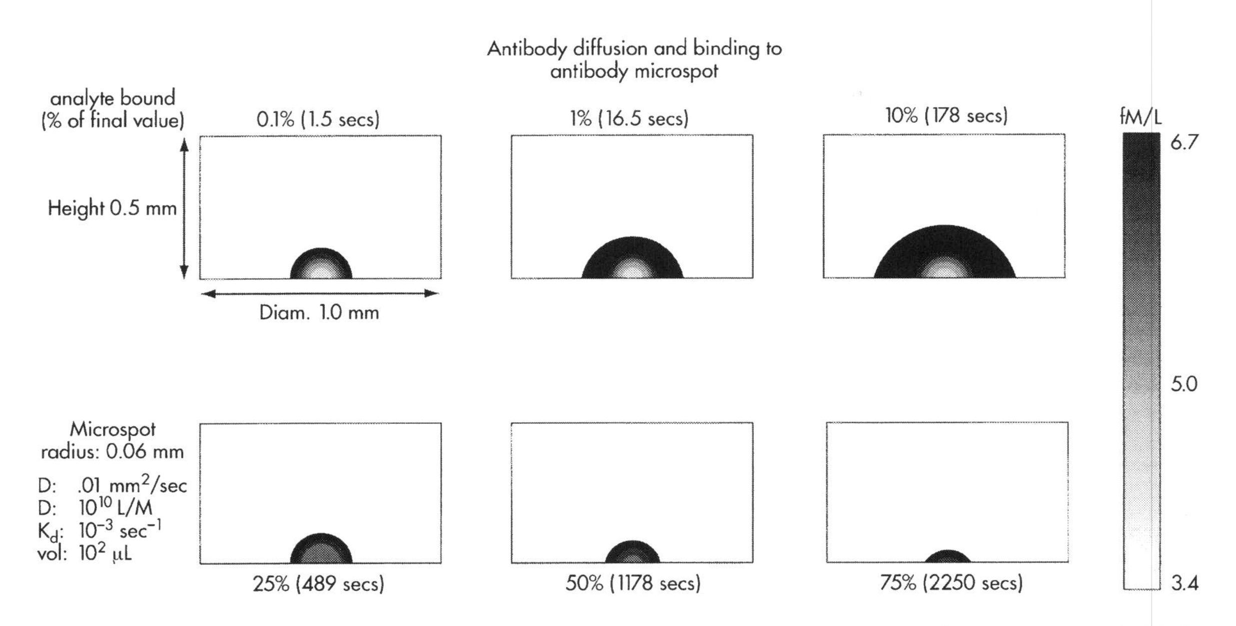

Fig. 3.15 Computer simulation of analyte binding to microspots (assuming typical analyte diffusion and antibody binding constants) shows that equilibrium is reached more rapidly, and that fractional occupancy of sensor antibody is at all times higher, with smaller spots.

oligonucleotides located within individual microspots are of lower purity than those produced using pre-synthesized material. The consequent need for 'redundant' spots implies that the potential information content of large arrays constructed in this manner is considerably less than might appear at first sight. Nevertheless, the realization that such arrays could potentially be used for diagnostic purposes led to the establishment of the US company Affymetrix in 1992, this having since assumed a prominent position amongst the many manufacturers

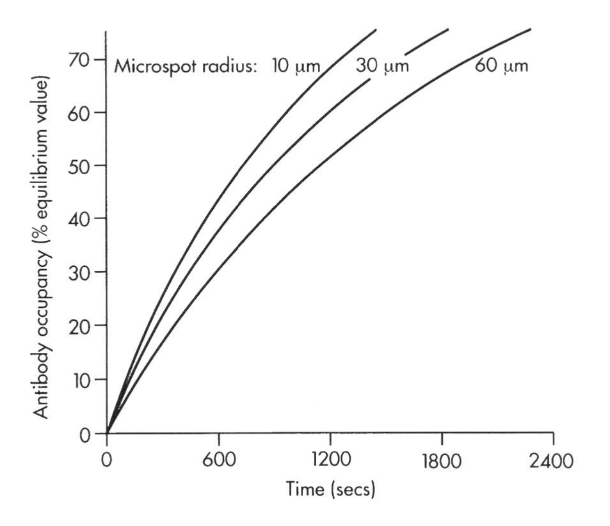

Fig. 3.16 Simulations of the kind portrayed in Figure 3.15 reveal that equilibrium is reached more rapidly, and that fractional occupancy of sensor antibody is at all times higher, with smaller spots.

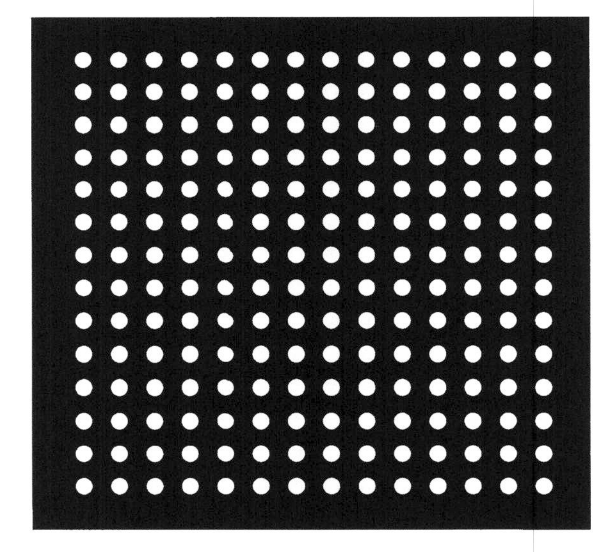

Fig. 3.17 Typical antibody (and oligonucleotide) microarray prepared by Boehringer Mannheim GmbH using ink-jet deposition technology.

now developing microarray-based technologies for DNA/RNA analysis. However, a number of other companies have developed equipment for the local production of oligonucleotide arrays (e.g. Bowtell, 1999) based on the use of inkjet dispensers, solid pens or 'quills'.

Although the principal use of oligonucleotide arrays is DNA analysis, they may also be used as standard array templates permitting individual researchers to construct antibody arrays of their own design, using antibodies to which complementary oligonucleotide sequences have been linked (Ekins, 1998a).

Meanwhile several manufacturers now market array-scanning equipment (Bowtell, 1999), their published sensitivities ranging from 0.5 molecules fluor/μm^2 (ScanArray, PerkinElmer Life and Analytical Sciences, Inc., Boston, MA) to <34.5 molecules fluor/μm^2 (BioChip Imager, previously manufactured by Packard Instrument Company, Meriden, CT).

The use of microarray technologies is nevertheless still in its infancy, and the number of published applications is as yet limited. Indeed, it has been said that the scientific literature contains more reviews about the technology than papers reporting its use (Bassett *et al.*, 1999). Nevertheless, in the course of my own group's collaborative studies with Boehringer Mannheim a variety of non-competitive and competitive 'immunoarray' systems have been developed (Finckh *et al.*, 1998), comprising multiple sandwich assays, labeled analyte back-titration assays for low molecular weight analytes, and capture-antigen assays for the determination of serum antibodies. Assays have primarily related to analytes within the fields of endocrinology, allergy and infectious disease, but similar techniques have also been employed for the screening of therapeutic drugs. Routine (15 min) TSH assays have been of high sensitivity (detection limits <0.01 μU/mL). Close correlations with the results obtained with the latest commercially available test kits have been demonstrated for a variety of allergens (e.g. birch, cat epithelia, house dust mite, α-amylase, bee venom and total IgE), assay precision and sensitivity being superior. For total IgE a detection limit of <0.01 IU/mL has been achieved. Microarray-based assays relating to a number of infectious diseases (e.g. HIV, HBsAg, anti-HBC, rubella) have likewise been developed and have been shown to be superior to the latest commercially available methods and instruments.

We have also carried out a limited number of studies exemplifying the technology's application to DNA analysis (Finckh *et al.*, 1998). For example, *Mycobacterium tuberculosis*, which is resistant to Rifampicin (an efficient first line drug) was specifically selected for study because of the technical challenges it poses (e.g. single-point mutations, formation of strong intra-strand secondary structures, extremely GC-rich segments) as well as of its clinical relevance. Rifampicin inhibits the RNA-polymerase by binding to its β-subunit (rpo-β); however, various single base transitions clustered in a 27 codon segment of the bacterium gene cause resistance. A study on 80 selected samples from two clinical centers specializing in tuberculosis diagnosis showed a high degree of concordance with a reference (culture) method.

In summary, miniaturized microarray-based assays constitute a ubiquitous technology, applicable to a wide range of analytes. Their need for small samples, their greater sensitivity, speed and reliability,[9] their reduced manufacturing costs, and the potential savings to clinical laboratories arising from the simultaneous determination of many different analytes in a single sample, are amongst important advantages that would in any event be likely to lead to their replacement of existing methodologies. But the most compelling factor currently driving microarray development is the perception of the potential diagnostic importance of the technology in the fields of genomics and – more recently – proteomics. Major pharmaceutical manufacturers are amongst the many that have realized the technology's implications, anticipating the future development of drugs tailored to individual patients according to their genetic make-up. There is, therefore, little doubt that the ligand assay field is presently on the brink of a revolution that is likely to totally transform diagnostic medicine, drug development, and other related areas within the next few years.

FREE (NON-PROTEIN BOUND) HORMONE IMMUNOASSAYS

The direct measurement by immunoassay of the 'free' (non-protein bound) concentrations of thyroid and steroid hormones has, in recent years, emerged as a standard diagnostic procedure in many clinical laboratories, it being widely accepted that the free hormone concentration measured under equilibrium conditions *in vitro* constitutes the determinant of the hormone's physiological activity. This concept, termed the 'free hormone hypothesis', derives primarily from observations that, in subjects in whom serum binding protein concentrations are 'abnormal', overall hormonal effects correlate closely with the free hormone concentration. Nevertheless, doubts regarding the hypothesis' validity remain. These have stemmed in part from the lack of explanation for the occurrence of specific binding proteins in mammalian blood, and the characteristic changes in their concentrations that accompany pregnancy in certain species (see, e.g. Seal and Doe, 1966). Such doubts have been reinforced by uncertainties regarding the underlying

[9] Note that microarray formats enable microspots to be included that enable detection in test samples of cross-reacting substances whose presence would be unnoticed in conventional assay formats.

physicochemical basis of the hypothesis (Ekins *et al.*, 1982; Ekins, 1985a), exemplified by conflicting views regarding the rate limitations on hormone efflux from the microcirculation held by thyroidologists (following Robbins and Rall (1979)) and by steroidologists (e.g. Tait and Burstein (1964)).[10]

Critics of the hypothesis have suggested that the bound hormone concentration determines hormone delivery to certain tissues, implying that serum binding proteins fulfill a specific tissue-targeting role. Changes in binding protein concentrations are thus postulated as redistributing the hormone supply between target organs in the body, albeit the suggested mechanisms underlying this putative phenomenon differ. For example, Keller *et al.* (1969) visualize that certain organs are permeable to bound hormone. In contrast, Pardridge and his co-workers (see, e.g. Pardridge, 1987) suggest that 'transient conformational changes about the ligand binding site within the microcirculation' cause changes in binding protein structure and hence in hormone binding affinities, resulting in enhanced hormone dissociation within certain tissues. Meanwhile – relying on an analysis of the kinetics of bound hormone dissociation, intracapillary hormone diffusion and capillary wall permeation – the present author has proposed that bound hormone concentrations influence the maternal hormone supply to the fetus in early pregnancy (Ekins, 1985b, 1990), the latter being postulated as of crucial importance to fetal brain development.

Notwithstanding continuing debate attaching to the physiological rôle (if any) of specific hormone binding proteins, the determination of serum free hormone (particularly free T_4) concentrations is of considerable diagnostic importance. Unfortunately, some of the immunoassay methods developed by kit manufacturers were based on fallacious physicochemical concepts, such methods yielding misleading results in certain clinical situations, and creating major controversy regarding the basic principles of free hormone measurement. However, only a brief summary of this topic (reviewed in greater detail elsewhere (Ekins, 1990, 1998b)) can be appropriately presented here.

All current free hormone immunoassay methods rely on the basic ambient analyte principle, i.e. that exposure of a small amount[11] of anti-hormone antibody to a test serum sample results in occupancy of antibody binding sites to an extent that reflects the ambient free hormone concentration in the sample (Figure 3.18). Occupancy of binding sites can be determined in three different ways, generally described as:

1. the 'labeled hormone, back-titration', approach ('two-step' free hormone immunoassay);
2. the 'labeled hormone analogue' approach ('single-step' free hormone immunoassay);
3. the 'labeled antibody' approach (likewise a 'single-step' method).

The first of these relies on determination of unoccupied antibody binding sites (the antibody being generally linked to a solid support) by their exposure to labeled hormone following removal of the test serum (thereby preventing reaction of the labeled hormone with serum binding proteins which, if permitted, would distort the measurement.)

The second obviates these sequential operations by the use of a labeled hormone analog that must, in principle, be totally unreactive with serum proteins (though retaining the ability to bind to antibody). However, the first commercial kits of this genre were based on a different (and erroneous) perception of these methods' underlying principle, this allowing a much higher degree of labeled analog binding to serum proteins (i.e. *ca.* 99%) than is permissible in valid methods.[12] Though – by the addition to kit reagents of albumin and other such artifices – free T_4 values yielded by these kits in normal and pregnant

[10] Robbins and Rall's view (1979) is that, as blood flows through target organ capillaries, the intracapillary free hormone concentration is maintained at its *in vitro* equilibrium value in the face of hormone loss into tissue by instantaneous hormone dissociation from binding proteins. In contrast, Tait and Burstein postulated (1964) that only hormone initially in the free state is available for tissue uptake, implying a decline in the intracapillary free hormone level as blood transits the target organ and free hormone molecules are lost into the extravascular compartment. This view is based on the supposition that release of hormone from bound hormone complexes is negligible during capillary transit, and implies, *inter alia*, that the rate of blood flow through the target organ constitutes a major determinant of its hormone supply.

[11] That is an amount that binds no more than 5% of the *total* hormone present in the sample.

[12] The developers of these kits considered that reduced analog binding to serum binding proteins was required solely to avoid displacement of endogenous hormone therefrom, thereby increasing the ambient free hormone concentration. They therefore postulated that – provided the affinity of the analog for endogenous binding proteins were sufficiently reduced as compared with that of the hormone itself (i.e. to less than 10%) that little or no displacement of hormone from binding proteins occurred – the analog could be described as not significantly bound (Midgeley and Wilkins, 1985). Amongst other implications of this entirely fallacious concept, major binding to serum albumin was permitted, on the grounds that such binding would not displace hormone because of this protein's high binding capacity. Ironically, though the analogs employed were almost entirely bound to the endogenous albumin present in test samples (causing significant errors when test sera contained unusual albumin levels, or abnormal albumins) such binding was essential for assays of this type to possess any superficial resemblance to a genuine free hormone assay. In other words, such success as early labeled analog assays enjoyed was based on an artifact of the system.

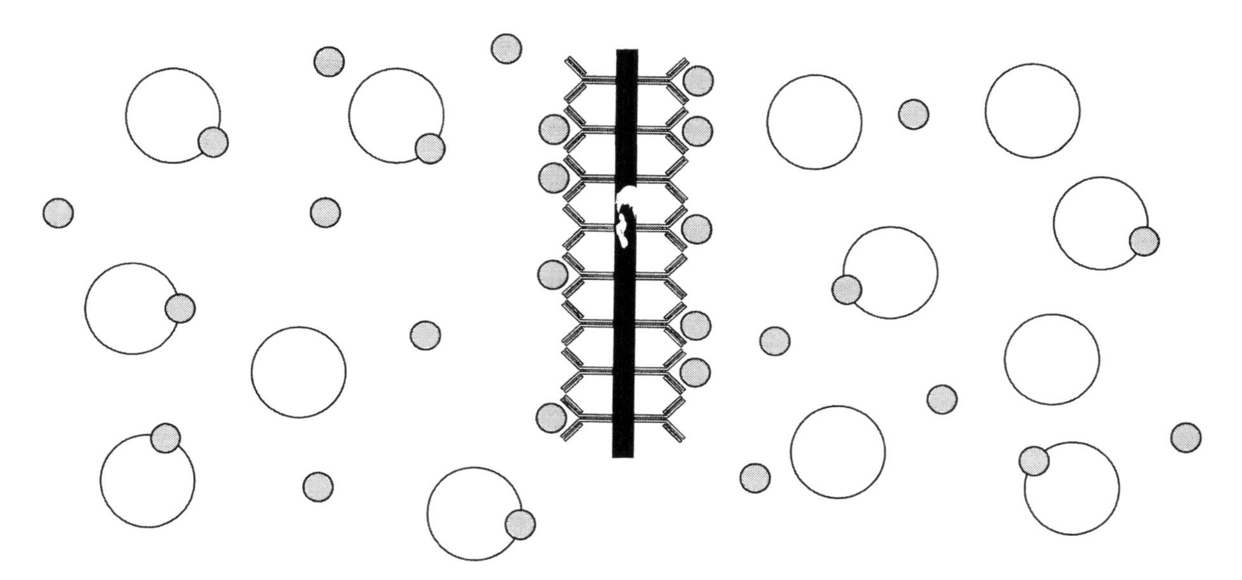

Fig. 3.18 Basic principle of free hormone immunoassay. A variety of different strategies may be used to determine occupancy of antibody binding sites.

subjects were 'engineered' to be closely comparable, incorrect and misleading results were frequently observed in other clinical situations. Labeled analog methods therefore fell into considerable disrepute, though if a genuinely unbound analog were to be used (i.e. one – in the case of T_4 – of an affinity vis-à-vis serum proteins some orders of magnitude less than analogs used in the original kits), assay results would be generally accurate.[13] Indeed certain manufacturers continue to market labeled analog kits, though the author has no recent information regarding their analytical validity or diagnostic reliability.

The third (labeled antibody) approach also relies on the use of a hormone analog, though kit manufacturers have clearly been somewhat reluctant to disclose this fact, presumably wishing to avoid the suspicion that attaches to analog-based methods. However, the analog used in labeled antibody techniques is coupled to a solid support, such attachment creating a 'macro-analog' and evidently contributing to a further major reduction of analog binding to serum proteins. For this and other reasons, labeled antibody-based kits appear to conform more closely to the principles governing valid analog-based free hormone immunoassays, and generally yield correct and clinically reliable results.

Though all three methods rely on the basic ambient analyte assay principle, only the two-step method is fully independent of incubation volume, because of the presence within single-step assay systems of another reagent (i.e. analog) whose reactions with antibody are concentration and volume dependent. However, if the *incubation* volume remains essentially unchanged, variations in the volume of sample added to the incubation mixture are largely irrelevant (provided that ambient analyte assay conditions are fulfilled), since differences in sample dilution have no affect on the ambient free hormone concentration in these circumstances.

Free hormone assays differ from those in which the analyte is totally unbound in so far as the reservoir of analyte maintaining near constancy of the ambient (free) analyte concentration in the face of antibody uptake comprises the rapidly dissociating pool of protein bound hormone present in the sample. Thus the total amount of hormone finally bound to antibody may greatly exceed the amount initially present in the free state. For example, in the case of free thyroxin measurements, up to *ca.* 5% of the total hormone in the sample may be bound to antibody at the termination of the assay, albeit only *ca.* 0.2% of the hormone in the incubation mixture is free (assuming a final 10-fold dilution of the serum sample by buffer and other assay reagents). The amount of hormone bound to antibody will nevertheless be proportional to the ambient free hormone concentration in the original sample.

[13] Note, however, that the presence of endogenous hormone antibodies in test samples can – as with most immunoassay methods – lead to incorrect results.

It should be perhaps be noted in this context that – according to Robbins and Rall (1979) – it is precisely this mechanism that operates during the delivery of thyroid hormones to target tissues and cells *in vivo*. In other words, such cells function as natural ambient analyte concentration detectors.

OTHER APPLICATIONS OF THE PRINCIPLE

As indicated in the introduction of this article, the ambient analyte assay principle is potentially applicable in many situations in which the measurement of sample volume is either impossible or inconvenient, such as the determination of analyte concentrations *in vivo*. A simple example of such an application was the subject of a study by my colleagues and myself some years ago, albeit it was abandoned before completion because of the competing demands on our time and resources consequent on the commencement of the collaboration with Boehringer Mannheim on microarray development.

As is well known, certain hormones, including steroid hormones, are found in saliva, the salivary concentration being claimed to reflect the free concentration present in serum. Salivary steroid assays have, therefore, attracted considerable attention in the past, particularly from participants in the WHO Human Reproduction Program, in the context of which steroid hormone assays on subjects reluctant to provide blood for religious and other reasons is frequently a complicating factor.

Nevertheless the collection of salivary samples also poses logistic problems, and is not without an attendant health risk. In principle, these could be obviated by the use of a small plastic probe bearing a small area of antibody at its tip, the probe being sucked by the subject for a specified time interval thereby permitting 'sensing' of the ambient salivary steroid concentration. It should be noted in this context that ambient analyte assay conditions are fulfilled if (assuming the presence of 10^5 molecules of antibody on the microspot surface; the antibody having an affinity constant of 10^{11} L/M) the fluid volume to which the antibody is exposed exceeds *ca.* 1.7 μL. Thus the presence of an extremely small amount of saliva suffices to permit measurements that are sample volume independent. However, though initial studies using this approach yielded encouraging results, the study was halted (for the reasons indicated earlier) before full validation and reliability tests could be completed.

This example nevertheless illustrates one potential use of what is, in effect, an ambient analyte sensor, albeit not one embodying a transduction system permitting continuous monitoring (by electronic or other means) of changing analyte concentrations.[14] Other such uses in medical practice and in other contexts can be readily envisaged.

SUMMARY AND CONCLUSION

Ambient analyte assay represents a concept that is not immediately apparent and often provokes initial disbelief. (Indeed, the author was once challenged to prove its validity experimentally by a well-known Nobel Laureate, a challenge which – given the concept's solid theoretical basis – it was not difficult to meet). In this article, some of its more important implications have been discussed, amongst which the emergence and widespread use of miniaturized multianalyte chip-based microarray methods (for DNA analysis, for the determination of the products of gene expression ('proteomics') and for conventional immunodiagnostic applications) are likely to have the most significant and enduring consequences. Indeed, for the various reasons indicated in this article, immunoanalyzers relying on conventionally formatted binding assays are ultimately likely to be replaced by much smaller instruments based on the use of microarrays, permitting the rapid determination of multiple or single analytes as required. It should be noted in this context that many hormones and other substances comprise heterogeneous mixtures, the only fully satisfactory solution to their assay being the determination of their principal components (Ekins, 1990). Methods adhering to this precept for the assay of human chorionic gonadotropin (hCG) are already under development by an expert group under the auspices of the International Federation of Clinical Chemistry (Sturgeon *et al.*, 1998), and pattern recognition methods of the kind made possible by microarray technology are ultimately likely to become the norm for the assay of analytes of this type.

Ambient analyte assay can, in short, be anticipated to revolutionize the entire medical diagnostics field within the next few years.

REFERENCES

Bassett, D.E., Eisen, B. and Boguski, M.S. Gene expression informatics – it's all in your mine. *Nature Genet.* **21**, 51–55 (1999).

Bowtell, D.D.L. Options available – from start to finish – for obtaining expression data by microarray. *Nature Genet.* **21**, 25–32 (1999).

[14] The slow kinetics of antibody–antigen reactions nevertheless preclude the monitoring of rapidly changing analyte concentrations notwithstanding the emergence of satisfactory and sensitive transduction systems permitting continuous measurement of antibody occupancy.

Cornell, B.A., Braach-Maksvitis, V.L.B., King, L.G. *et al.* A biosensor that uses ion-channel switches. *Nature* **387**, 580–583 (1997).

Crank, J. *The Mathematics of Diffusion*, 2nd edn (Oxford University Press, Oxford, 1975).

Ekins, R.P. Measurement of analyte concentration. British Patent No 8224600 (1983).

Ekins, R.P. The free hormone concept. In: *Thyroid Hormone Metabolism* (ed Hennemann, G.), 77–106 (Marcel Dekker, New York, 1985a).

Ekins, R.P. Hypothesis: the roles of serum thyroxine binding proteins and maternal thyroid hormones in foetal development. *Lancet* **i**, 1129–1132 (1985b).

Ekins, R.P. Development of microspot multi-analyte ratiometric immunoassay using dual fluorescent-labelled antibodies. *Anal. Chim. Acta* **227**, 73–96 (1989).

Ekins, R. Measurement of free hormones in blood. *Endocr. Rev.* **11**, 5–46 (1990).

Ekins, R. Immunoassay standardization. In: *Improvement of Comparability and Compatibility of Laboratory Assay Results in Life Sciences.* 3rd Bergmeyer Conference: Immunoassay Standardization (eds Kallner, A., Magid, E.D. and Albert, W.). *Scand. J. Clin. Lab. Invest.* **51** (Suppl 205), 33–46 (1991).

Ekins, R.P. Ultrasensitive ligand assays. In: *Principles of Nuclear Medicine* (eds Wagner, H.N., Szabo, Z. and Buchanan Murphy, J.W.), 247–266 (W.B. Saunders, Philadelphia, 1995).

Ekins, R.P. Ligand assays: from electrophoresis to miniaturized microarrays. *Clin. Chem.* **44**, 2015–2030 (1998).

Ekins, R. The science of free hormone measurement. *Proc. UK NEQAS Meeting* **3**, 35–59 (1998).

Ekins, R. and Chu, F. Multianalyte testing. *Clin. Chem.* **39**, 369–370 (1993).

Ekins, R.P., Edwards, P.R. and Newman, B. The role of binding proteins in hormone delivery. In: *Free Hormones in Blood* (eds Albertini, A. and Ekins, R.), 3–42 (Elsevier Biomedical Press, Amsterdam, 1982).

Finckh, P., Berger, H., Karl, J. *et al.* Microspot® – an ultrasensitive microarray-based ligand assay system. A practical application of ambient analyte assay theory. *Proc. UK NEQAS Meeting* **3**, 155–165 (1998).

Fodor, S.P.A., Read, J.L., Pirrung, M.C. *et al.* Light-directed, spatially addressable parallel chemical synthesis. *Science* **251**, 767–773 (1991).

Keller, N., Richardson, U.I. and Yates, F.E. Protein binding and the biological activity of corticosteroids: *in vivo* induction of hepatic and pancreatic alanine aminotransferases by corticosteroids in normal and estrogen treated rats. *Endocrinology* **84**, 49–92 (1969).

Midgeley, J.E.M. and Wilkins, T.A. A method for determining the free portions of substances in biological fluids. European Patent No 0026103 (1985).

Pardridge, W.M. Plasma protein-mediated transport of steroid and thyroid hormones. *Am. J. Physiol.* **252**, E157–E162 (1987).

Robbins, J. and Rall, J.E. Thyroid hormone transport in blood and extravascular fluids. In: *Hormones in Blood* (eds Gray, C.H. and James, V.H.T.), 575–688 (Academic Press, London, 1979).

Schena, M., Heller, R.A., Theriault, T.P. *et al.* Microarrays; biotechnology's discovery platform for functional genomics. *Trends Biotechnol.* **16**, 301–306 (1998).

Seal, V.S. and Doe, R.P. Corticosteroid-binding globulin: biochemistry, physiology and phylogeny. In: *Steroid Dynamics* (eds Pincus, G., Nadao, T. and Tait, J.F.), 63–90 (Academic Press, New York, 1966).

Sturgeon, C., Stenman, U.-H., Bidart, J.-M., Birken, S., Berger, P., Lequin, R.M., Norman, R.J. and Bristow, A. IFCC Working Group on standardisation of human chorionic gonadotropin measurements: progress report. *Proc. UK NEQAS Meeting* **3**, 134–139 (1998).

Tait, J.F. and Burstein, S. *In vivo* studies of steroid dynamics in man. In: *The Hormones*, vol. V, (eds Pincus, V., Thimann, K.V. and Astwood, E.B.), 441–557 (Academic Press, New York, 1964).

4 Free Analyte Immunoassay

Nic D. Christofides

The free hormone hypothesis, originally advanced by Robins and Rall (1957), states that, in the case of those hormones that exist in blood in free and protein-bound forms, the free hormone concentration constitutes the only determinant of physiological activity. This hypothesis is widely accepted for the thyroid hormones (thyroxine, T_4, and 3,3′,5-triiodothyronine, T_3) and some steroid hormones (e.g. testosterone). It may also apply to other analytes, such as vitamins and drugs, but evidence demonstrating a closer correlation of the free analyte concentration, rather than the total analyte concentration, to biological activity is scarce. This is primarily due to the very limited availability of accurate methods to measure free vitamins and drugs.

Development of methods to measure free analytes (e.g. free thyroxine (FT_4)) started in the 1970s, mainly through the pioneering work of Ekins and colleagues (e.g. Ekins and Ellis, 1975; Ekins *et al.*, 1980). Over the last two decades, a large number of methods have been developed and recently many have been automated (e.g. *see* THYROID or Demers, 1999). This makes their use in clinical practice very attractive. In Europe, a significantly higher proportion of laboratories measure FT_4, rather than Total T_4 (TT_4), whereas in the US, FT_4 measurement accounts for only 50% of the thyroid hormone testing at the time of writing. The main reason for the continued use of Total T_4 measurement in the US is the continuing controversy regarding the accuracy and validity of some commercial methods.

The primary aim of this chapter is to explain the basic mechanisms that dictate the free analyte concentration in the serum and how these mechanisms can be applied to develop valid free hormone assays. In addition, simple experimental designs will be described which can be used to challenge the validity of any free analyte assay.

BASIC PRINCIPLES GOVERNING THE FREE HORMONE CONCENTRATION

Following the *endogenous* release of certain hormones, such as thyroid and steroid hormones, into the blood circulation, they become bound to the **transport proteins**. Similarly, *exogenous* administration of some drugs and vitamins leads to the formation of binding protein–analyte complexes, with a small fraction of the drug or vitamin remaining unbound. The binding of the analyte to the serum proteins is reversible and, at equilibrium, the rate of dissociation of the analyte from the serum proteins is equal to its association rate. The proportion of analyte bound by the serum proteins is dictated by the relative affinity and concentration of each of the binding proteins. It is now widely accepted (at least in the case of thyroid and steroid hormones) that the free analyte fraction, rather than the fraction bound to the serum proteins, is the entity that binds to receptors and exerts biological activity. This does not mean that the hormone-binding protein complex has no function. Its role is to assure that a relatively constant supply of hormone to the organ tissues is maintained. This is achieved by dissociation of the hormone from the complex. Thus, when free hormone is removed from the pool as it is taken up by the tissues, sufficient free hormone becomes available by the dissociation of the complex to ensure a near constant supply of free hormone to the tissues. This property of the hormone–protein complex (i.e. dissociation of hormone from protein whilst maintaining a near constant free hormone concentration) has been utilized to develop methods that measure the free hormone concentration.

Throughout the remainder of this chapter, free thyroxine (FT_4) has been used as an example to illustrate the principles involved, both in the calculation of the free concentration and also in the development of methods for its measurement. This is because FT_4 has been the most extensively studied hormone. However, the principles discussed apply to all free analyte measurements.

CALCULATION OF FREE ANALYTE CONCENTRATION

In the case of T_4, following its release from the thyroid gland into the blood circulation, it becomes bound by three separate binding proteins. These are thyroxine-binding globulin (TBG), human serum albumin (HSA), and transthyretin (TTR). Typical concentrations of total thyroxine and the binding proteins, and the equilibrium constants (K_{eq}) of the binding proteins (to T_4) in a normal individual are shown below:

Concentration of $TT_4 = 1 \times 10^{-7}$ mol/L
Concentration of TBG $= 3.57 \times 10^{-7}$ mol/L, $K_{eq} = 2.2 \times 10^{10}$ L/M
Concentration of HSA $= 6.18 \times 10^{-4}$ mol/L, $K_{eq} = 1.3 \times 10^{6}$ L/M
Concentration of TTR $= 5.56 \times 10^{-6}$ mol/L, $K_{eq} = 3.9 \times 10^{7}$ L/M

The proportion of T_4 binding to the proteins is dictated by the **relative binding capacity** (affinity (K_{eq} or K) multiplied by concentration) of each protein. Thus, at equilibrium, the amount of T_4 bound by TBG will be equal to:

$$TT_4 \times \frac{K_{TBG}}{K_{TBG} + K_{HSA} + K_{TTR}} \quad (4.1)$$

The concentration of free T_4 is controlled by the equilibrium between the bound T_4 and the free binding sites of the proteins. The concentration of FT_4 can be calculated by equations derived from the Law of Mass Action. Thus, it can readily be shown that

$$FT_4 = PBT_4/K[P_{free}] \quad (4.2)$$

where PBT_4 denotes the concentration of protein-bound T_4, K the net affinity of the proteins (towards T_4), and $[P_{free}]$ the concentration of the unbound (free) binding sites of the proteins.

Using this equation, and knowing the total concentration and affinities of the binding proteins and of T_4, one can easily predict the *in vivo* serum FT_4 concentration. In the next section the construction of a simplified spreadsheet is described which can be used to calculate the *in vivo* FT_4 concentration in serum. The spreadsheet can also be expanded to include immunoassay reagents (such as antibody and other chemicals) and thus can be used to predict their influence on the FT4 concentration.

SPREADSHEET FOR CALCULATION OF FREE ANALYTE CONCENTRATION

This can be performed using a computer spreadsheet. Once again FT_4 is used as the example analyte. The program can be made available from the author. The program can be used to calculate both the *in vivo* and *in vitro* (e.g. by immunoassay) FT_4 concentrations.

Table 4.1 shows the calculations performed to derive the *in vitro* FT_4 concentration and Table 4.2 shows an example set of data. The *in vivo* FT_4 concentration can be calculated using the same program, by removing the reagent contribution (i.e. enter 'zero' for the reagent concentration and volume).

The calculation steps are also outlined below.

1. Enter the volumes of serum, antibody, and other reagents in column C (cells 4–6). The total reaction volume is calculated by adding the three individual volumes (in cell C7); the dilution factor incurred by the serum, in the reaction vessel, is calculated by dividing the total reaction volume by the serum volume (i.e. C7/C4). If one wants to estimate the *in vivo* FT_4 concentration, enter 0 μL for volume of antibody (C5) and for volume of other reagents (C6).
2. Enter the serum concentrations (in g/L) of the binding proteins (TBG, HSA, and TTR), TT_4, and the concentration of antibody used in the immunoassay reagents in column C (cells C13–C16); if any other binder is used in the immunoassay reagents (such as Bovine Serum Albumin (BSA)) enter the concentration in C20.
3. Enter the molecular weights of the binding proteins and of TT_4 in column D.
4. In column D calculate the molar concentrations of the binding proteins by dividing the inputs of column C by the molecular weights in column D.
5. The molar concentration of the binding proteins and TT_4 in the immunoassay 'tube' are calculated in column E.
6. Enter the affinity constants (K) of the individual binding proteins (including that of the antibody, if the program is to be used for simulations on immunoassay performance) in column G.
7. The binding capacity (i.e. $K[P_{total}]$) is calculated by multiplying column G by column F and placing in column H.
8. Calculate the sum of column H and place in cell H25.
9. The concentration of T_4 bound by each protein is calculated by multiplying column F by column H (divided by cell H25), and placing in column I.
10. The concentration of total protein-bound T_4 (PBT_4) is calculated by subtracting the fraction, cell F16/H25, from cell F16 (concentration of TT_4) and this is placed in cell H26.
11. The concentrations of free binding sites for each protein are calculated by subtracting column I from column F, placing in column J.
12. Calculate the product $K[P_{free}]$ by multiplying column G by column J and placing in column K.
13. Calculate the sum of column K and place in cell H27.
14. Calculate the FT_4 concentration (in M/L) by dividing H26 by H27 and placing in cell 28.
15. One can calculate the proportion of T_4 carried by each protein by dividing column H by the cell H25, and placing in column L.

Table 4.1. Spreadsheet for calculation of free analyte concentration.

	B	C	D	E	F	G	H	I	J	K	L	M
1	**B**	**C**	**D**	**E**	**F**	**G**	**H**	**I**	**J**	**K**	**L**	**M**
2												
3		μL										
4	*Sample volume*	*25*										
5	***Volume of antibody***	*1100*										
6	*Volume of other reagents*	*0*										
7	*Total volume*	= SUM(C4:C6)										
8	*Dilution*	= C7/C4										
9												
10												
11												
12		g/L (in serum)	*M.W*	*M/L (serum)*	*[P_{total}] M/L (tube)*	K_{eq}	$K[P_{total}]$	*Protein-bound T_4 (PBT_4)*	$[P_{free}]M/L$	$K[P_{free}]$	$K[P_{indiv}]/K[P_{total}]$	$[P_{free\ ab}]/[P_{total\ ab}]$
13	*TBG*	0.02	56,000	= C13/D13	= E13*(C4)*(1/C7)	2.20E + 10	= G13*F13	= F16*(H13/H25)	= IF(F13-I13 > 0,F13-I13,0)	= G13*J13	= H13/H25	
14	*HSA*	42	68,000	= C14/D14	= E14*(C4)*1/C7)	1.30E + 06	= G14*F14	= F16*(H14/H25)	= IF(F14-I14 > 0,F14-I14,0)	= G14*J14	= H14/H25	
15	*TTR*	0.3	54,000	= C15/D15	= E15*(C4)*(1/C7)	3.90E + 07	= G15*F15	= F16*(H15/H25)	= IF(F15-I15 > 0,F15-I15,0)	= G15*J15	= H15/H25	
16	*TT4*	8.89E − 05	889	= C16/D16	= E16)*(C4)*(1/C7)							
17												
18		g/L (in reagent)										
19	*Antibody*	0.000336	150,000	= C19/D19	= E19*(C5/1)*(1/C7)	1.00E + 10	= G19*F19	= F16*(H19/H25)	= IF(F19-I19 > 0,F19-I19,0)	= G19*J19	= H19/H25	= J19/F19
20	*BSA in assay reagents*	0	66,000	= C20/D20	= E20*(C5 + C6)*(1/C7)	1.30E + 06	= G20*F20	= F16*(H20/H25)	= IF(F20-I20 > 0,F20-I20,0)	= G20*J19	= H20/H25	
21												
22												
23												
24						*Calculations*						
25						Sum $K[P_{total}]$	= SUM(H13:H20)					
26						PBT_4	= F16-(F16/H25)					
27						Sum $K[P_{free}]$	= SUM(K13:K20)					
28						FT_4	= H26/H27					
29												
30												
31												

Table 4.2. Free analyte immunoassay.

1	B	C	D	E	F	G	H	I	J	K	L	M	N
2													
3		μL											
4	*Sample volume*	*25*											
5	*Volume of antibody*	*100*											
6	*Volume of other reagents*	*0*											
7	*Total volume*	*125*											
8	*Dilution*	*5*											
9													
10													
11													
12		g/L (in serum)	M.W	*M/L (serum)*	$[P_{total}]$ *M/L (tube)*	K_{eq}	$K[P_{total}]$	*Protein − bound* T_4 *(*PBT_4*)*	$[P_{free}]$ *M/L*	$K[P_{free}]$	$K[P_{indiv}]/K[P_{total}]$	$[P_{free\ ab}]/[P_{total\ ab}]$	
13	*TBG*	0.02	56,000	3.571E − 07	7.143E − 08	2.20E + 10	1.57E + 03	1.753E − 08	5.390E − 08	1.186E + 03	87.63%		
14	*HSA*	42	68,000	6.176E − 04	1.235E − 04	1.30E + 06	1.61E + 02	1.791E − 09	1.236E − 04	1.606E + 02	8.96%		
15	*TTR*	0.3	54,000	5.556E − 06	1.111E − 06	3.90E + 07	4.33E + 01	4.833E − 10	1.111E − 06	4.331E + 01	2.42%		
16	*TT4*	8.89E − 05	889	1.000E − 07	2.000E − 08								
17													
18		g/L (in reagent)											
19	*Antibody*	0.000336	150,000	2.240E − 09	1.792E − 09	1.00E + 10	1.79E + 01	2.00E − 10	1.592E − 09	1.592E + 01	1.00%	88.85%	
20	*BSA in assay reagents*	*0*	68,000	*0*	0.000E + 00	1.30E + 06	0.00E + 00	0.00E + 00	0.000E − 00	0.000E − 00	0.00%		
21													
22													
23													
24						*Calculations*							
25						Sum $K[P_{total}]$ =	1.79E + 03						
26						PBT_4	1.999E − 06						
28						Sum $K[P_{free}]$ =	1.406E + 03						
29						FT_4	1.422E − 11						
30													
31													

16. The fraction of the antibody which remains unbound can also be estimated by dividing J19 by F19 and placing the calculation in M19.

The spreadsheet program assumes that one substance (in this case T_4) is bound by the binding proteins, but could be expanded to include the binding of other substances, which may compete (e.g. T_3). The simulations presented in later sections have used the spreadsheet program outlined above, since in the case of FT_4, the contribution of T_3 does not greatly affect the FT_4 profiles obtained. However, the program will greatly underestimate the FT_4 concentration if large amounts of binding inhibitors (such as non-esterified fatty acids) are present in the serum. Another possible limitation of this and other software programs is that it assumes that full equilibrium between T_4 and any binding proteins has been reached, which may not be the case with some immunoassay methods. It is thus suggested that the results obtained are used in a qualitative fashion (i.e. to establish whether the hormone concentrations will increase or decrease) rather than quantitatively.

A similar program can be constructed for other free analytes (e.g. FT_3, cortisol, testosterone, etc.) following input of the relevant affinities and concentrations of the binding proteins (and concentration of the total analyte). It is, however, important to note that, in the case of FT_3, as the T_4 binding affinity is greater than that of T_3, the contribution of T_4 on the calculation of FT_3 concentration will be significant and will require the construction of a more complex program.

EFFECT OF SERUM PROTEINS ON FREE ANALYTE CONCENTRATION

As in the previous sections, free thyroxine is used as the example analyte.

Using the spreadsheet program, one can perform simulations to predict the concentration of FT_4 in any conceivable circumstance, e.g. when the concentration and affinity ($K[P_{total}]$) of individual proteins is altered or when the concentration of TT_4 is changed.

Figure 4.1 shows the effect, on FT_4, of altering the endogenous protein concentrations (whilst maintaining a constant TT_4 concentration).

In this example, the concentration of individual proteins (TBG, HSA, and TTR) was changed (from 1/4 to 4-fold of normal concentration) whilst the TT_4 concentration was kept at 100 nmol/L (euthyroid concentration). It is clear from the graph that an increase in the concentration of the proteins leads to a decrease in FT_4 concentration, whereas a reduction in the concentration of the proteins leads to an increase in FT_4. Also evident from this figure is that the FT_4 concentration is predominantly influenced (and controlled) by the TBG concentration (and affinity), rather than the other two

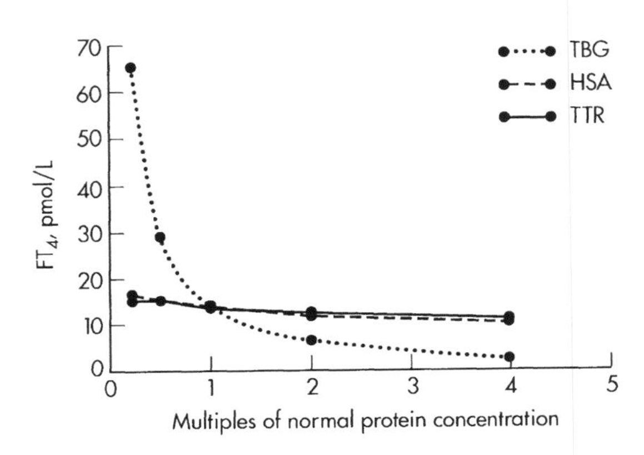

Fig. 4.1 FT_4 concentration following changes in the protein concentrations.

binding proteins. This is the reason why the T_4/TBG ratio has been used as an 'indirect' measure of FT_4.

However, it can be predicted that this ratio will not accurately describe the FT_4 concentration since the calculation uses the total TBG concentration rather than the concentration of the free binding sites. Figure 4.2 shows the predicted (using the spreadsheet program) FT_4 concentration and the FT_4 'index', calculated from the TT_4/TBG ratio (and calibrated in 'FT_4 units') in a euthyroid serum spiked by different concentrations of T_4.

The results show that the T_4/TBG ratio is linearly related to the TT_4 concentration whereas a curvilinear relationship is observed between FT_4 and TT_4. At high TT_4 concentrations (when the concentration of free sites on TBG is reduced), the T_4/TBG ratio becomes negatively biased (in comparison to FT_4) whereas at low TT_4 concentrations (when the concentration of free sites on TBG is increased) the T_4/TBG is positively biased. This is also illustrated (as a bias plot) in Figure 4.3. It can be predicted that biased T_4/TBG results will also be obtained

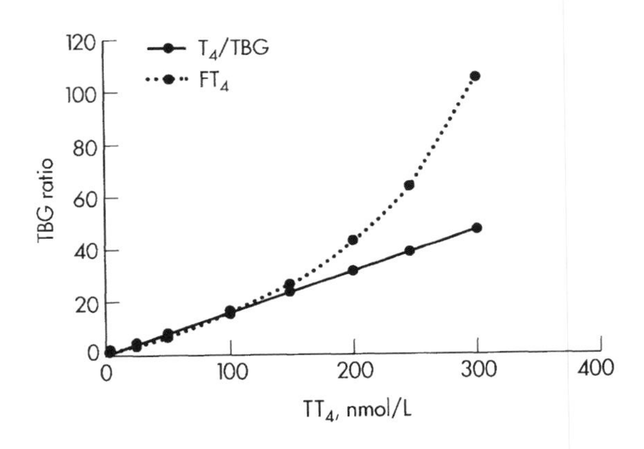

Fig. 4.2 Biasing effects of the T_4/TBG ratio.

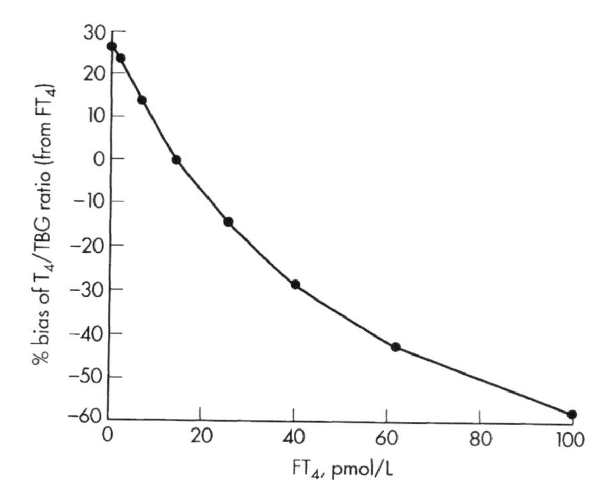

Fig. 4.3 Bias plot showing the expected difference (%) between the T_4/TBG ratio (calibrated in pmol/L) and FT_4, in a euthyroid serum spiked with increasing amounts of T_4.

in situations where the TBG affinity to T_4 is reduced, or in situations when the serum contains substances that can bind to TBG (and thus reduce the concentration of unoccupied binding sites). Of course, if the reduction in binding sites is not taken into consideration, the same will be true of estimates of FT_4 obtained by the spreadsheet program.

Nonetheless, the spreadsheet calculations can help in the understanding the mechanisms that may influence the FT_4 concentrations in certain clinical conditions. For example, the FT_4 concentration in severely ill patients is approximately 30% higher than the FT_4 concentration typically seen in ambulatory patients, despite the fact that the corresponding TT_4 concentration in these patients is approximately 50% lower than the ambulatory patients. A number of hypotheses have been put forward to explain the discordance between TT_4 and FT_4. These include:

- the presence of substances in the serum of ill patients that bind to albumin, reducing the concentration of unoccupied T_4 binding sites;
- decrease in albumin concentration;
- decrease in the binding affinity of albumin for T_4;
- decrease in the binding affinity of TBG for T_4 or reduction in the concentration of TBG.

Using the equations in the spreadsheet one can challenge these hypotheses by calculating the FT_4 concentration in different situations and establish the most likely cause of any TT_4/FT_4 discordance. It can be shown that, no matter how far one reduces the affinity and concentration of HSA and TTR, the presence of low TT_4 concentration (50% lower than normal) will not cause an elevation in the FT_4 concentration. However, the 30% increase in FT_4 (in the presence of a 50% reduction of TT_4) could result if either the concentration or affinity of TBG was reduced by 75–80% (or when the concentration and affinity are both reduced by 50%). Thus, one can show that the reason for the FT_4/TT_4 profile seen in ill patients is most likely due to a reduction in either the concentration or affinity of TBG. Both these possibilities have indeed been demonstrated to occur in some NTI patients (Csako *et al.*, 1989; Wilcox *et al.*, 1994).

IN VITRO MEASUREMENT OF FREE ANALYTE CONCENTRATION

There are a number of different methodologies that can be used to quantify free analyte concentrations in biological fluids. All the methods involve 'sampling' some of the free form in the serum sample and then quantitating the amount of free analyte sampled. The basic requirement, irrespective of methodology, is that the concentration of the free form sampled reflects the *in vivo* free analyte concentration. This section will examine whether this requirement has been met by the various methodologies.

DIRECT EQUILIBRIUM DIALYSIS

Direct equilibrium dialysis (ED) is a method that is considered by many investigators as the reference method for measuring free hormones. Figure 4.4 is an illustration of the mechanisms involved in this method.

The ED cell is made up of two compartments that are separated from each other by a semi-permeable membrane. This membrane allows small molecules to freely diffuse from one compartment to another, but prevents large molecules (such as proteins) from doing so. The serum sample is placed in one compartment and the buffer in the second compartment. During an incubation period T_4 (and other small molecules) diffuse through the membrane, from one compartment to the other. When equilibrium is reached, typically after 16–24 h, the concentrations of FT_4 and other small molecules in the two compartments are equal. However, as illustrated in Figure 4.4 the *number* of FT_4 molecules present in the buffer compartment is much larger than found in the serum compartment, although the FT_4 concentration (per mL) is similar. The number of FT_4 molecules is dependent on the ratio of the volumes in the two compartments, e.g. in Figure 4.4, 200 μL of serum is equilibrated against 2.4 mL buffer resulting in the presence of a 12-fold greater number of FT_4 molecules in the buffer compartment. This 'extra' FT_4 is derived from the binding proteins (i.e. T_4 which is normally bound to the serum proteins becomes dissociated and diffuses through to the buffer compartment).

A crucial requirement of an authentic FT_4 assay is that the amount of FT_4 sampled (or in the case of ED,

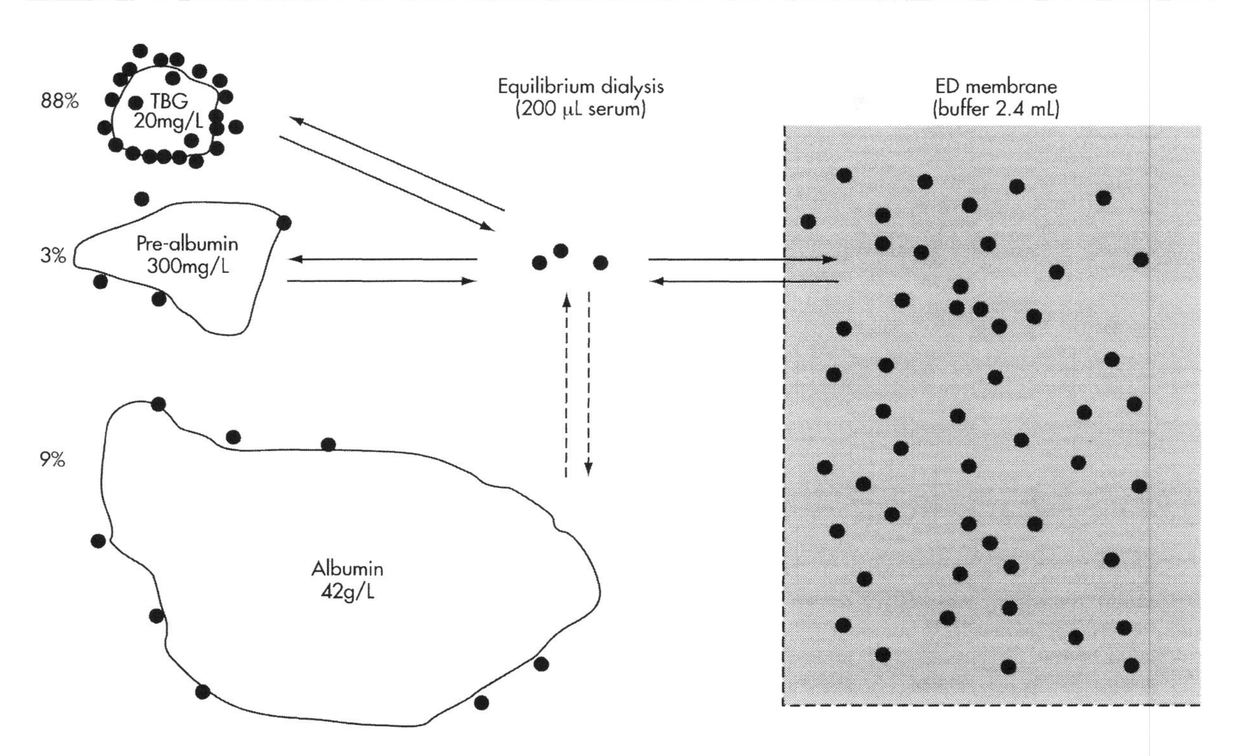

Fig. 4.4 Schematic representation of a direct dialysis FT_4 method. The serum (200 μL), containing normal concentrations of binding proteins, is placed in the serum compartment. Buffer (2.4 mL) is placed in a second compartment and is separated from the serum by a semi-permeable membrane. At equilibrium the concentration of FT_4 in the two compartments is equal. Measurement of the FT_4 concentration present in the buffer compartment should therefore reflect the FT_4 concentration present in the serum.

the amount of T_4 diffusing from the serum compartment to the buffer compartment) does not alter the *in vivo* FT_4 concentration, i.e. the concentration of FT_4 measured in the buffer compartment should be equal to the FT_4 concentration of the undialyzed serum. Whether the direct ED method used fulfills this requirement will depend on a number of factors. These include:

- the buffer composition and pH of the dialysis buffer (these will affect the affinity of the binding proteins);
- temperature used (affinity of proteins is temperature dependent);
- magnitude of non-specific binding (NSB) of T_4 (increased NSB will cause further dissociation of T_4 from its binding proteins);
- nature of the membrane (i.e. should only allow diffusion of small molecules);
- the volume of buffer relative to the volume of serum.

These factors are very important since they will dictate the concentration of T_4 dissociating from the buffer proteins and consequently the measured FT_4 concentration.

The Effect of a Reduced Protein-Bound T_4 Concentration on FT_4 Concentration

The FT_4 concentration expected following the removal of increasing amounts of T_4 can be calculated using the spreadsheet program (see Figure 4.5). As the amount of T_4 being removed from a euthyroid serum increases, the serum FT_4 becomes progressively decreased. It is, however, clear that one would need to remove a very large amount of T_4 from the serum before observing a large reduction in serum FT_4 concentration. For example, dissociation (and removal) of 1000 pmol/L of TT_4 will cause a reduction in FT_4 concentration of less than 2% (or <0.2 pmol/L).

Serum Dilution

In the example given above, I have considered diffusion of FT_4 from the serum compartment through the dialysis membrane to the buffer compartment. An identical effect, i.e. dissociation of T_4 from its binding proteins and near constancy of the FT_4 concentration, will also be seen if a serum is diluted in an inert buffer. Indeed, as far as the mechanisms involved, the presence or absence of

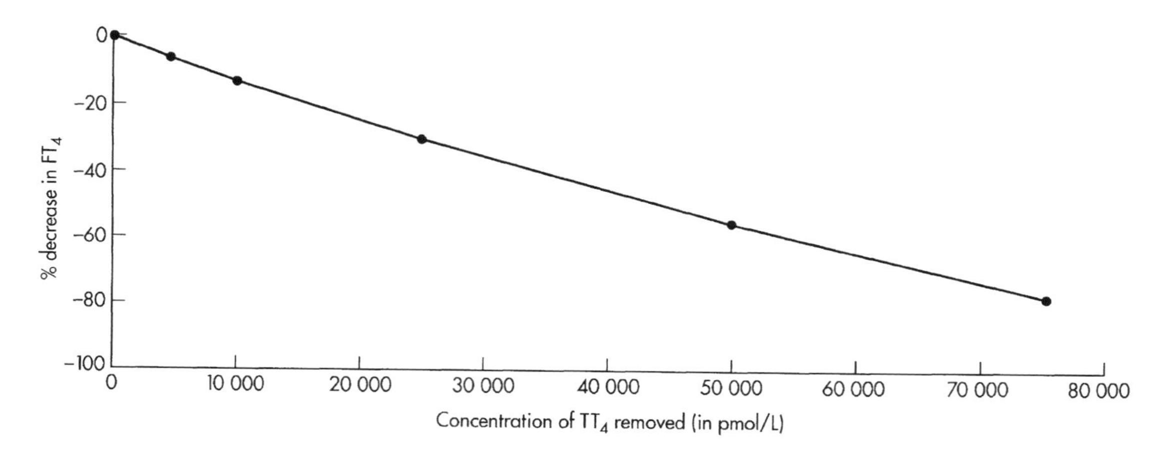

Fig. 4.5 The relationship between the amount of TT_4 removed from serum (in pmol/L) and reduction (%) in FT_4 concentration.

the dialysis membrane is irrelevant. Figure 4.6 shows the calculated (using the spreadsheet program) FT_4 concentrations of three serum samples diluted by an inert buffer (such as 10 mmol/L HEPES buffer, pH 7.4). One of the sera had a normal T_4 binding capacity (BC), another had a binding capacity which was 4-fold higher than normal and the final serum had a binding capacity which was 4-fold lower than normal. (The binding capacity is the affinity multiplied by the concentration of the binding proteins.) The results show that only when the dilution factor is increased more than 1000-fold does the FT_4 concentration decrease significantly (by more than 10%) in the normal and high binding capacity sera. However, dilution of the low binding capacity serum reduces the dilution window and causes a greater reduction of FT_4 concentration than those seen in the serum with normal binding capacity. These data suggest that in order to obtain unbiased FT_4 results in low binding capacity sera, the dilution of the assay used (equally applicable to all free hormone methodologies) should be kept to a minimum. Any assays that employ high serum dilutions will produce negatively biased results in such patients.

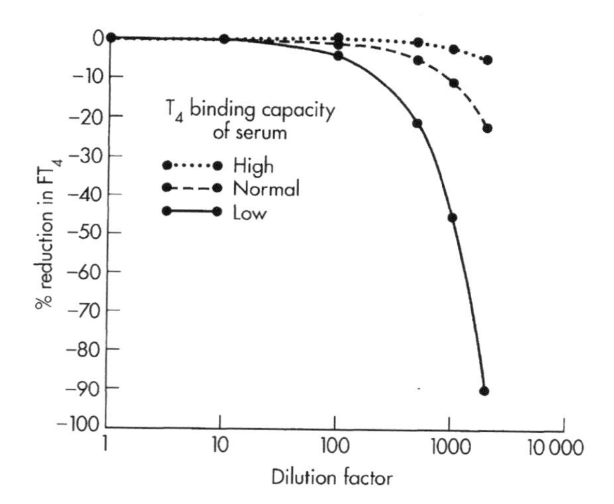

Fig. 4.6 Effect of serum dilution on FT_4 levels.

IMMUNOASSAYS FOR FREE ANALYTES

All immunoassays for FT_4 (and other free analytes), irrespective of assay architecture, have a number of common features. These are:

- a serum dilution step;
- addition of an antibody;
- quantification of free (unoccupied) binding sites of the antibody.

However, the assays do vary significantly, not only in architecture (i.e. the procedures used for quantification of the free binding sites of the antibody), but also on the level of disturbance of the T_4/protein equilibrium exerted by the assay reagents and protocols.

When an antibody is added to a diluted serum the following sequence of events occurs. As the antibody binds to the FT_4 more FT_4 becomes available by the dissociation of the protein–T_4 complex. The result is that T_4 becomes redistributed between the serum proteins and antibody. This redistribution is dictated not only by the concentration and affinity of the antibody used (relative to the serum binding capacity), but also by whether any other ingredients (e.g. BSA), included in the buffer formulation, can affect the binding of T_4 to the serum proteins. The reactions involved can best be described using two simple equations.

Equation (4.2) (also shown previously) describes the *in vivo* serum FT_4 concentration.

$$FT_4 = PBT_4/K[P_{free}] \qquad (4.2)$$

where PBT_4 denotes the concentration of protein-bound T_4, K the net affinity of the proteins (towards T_4), and $[P_{free}]$ the concentration of the unbound (free) binding sites of the proteins. Equation (4.3) describes the *in vitro* (i.e. in the immunoassay tube) serum FT_4 concentration.

$$FT_4 = (PB_{T_4} + IA_{T_4})/(K[P_{free}] + K[IA_{free}]) \quad (4.3)$$

where PB_{T_4} and IA_{T_4} denote the concentration of T_4 bound to the serum proteins and to the immunoassay reagents (including the antibody), respectively. $K[P_{free}] + K[IA_{free}]$ denote the binding capacities (i.e. affinity × concentration of free binding sites) of the serum proteins and immunoassay reagents, respectively.

The spreadsheet program can be used to calculate both the *in vitro* FT_4 (i.e. in the immunoassay tube) and the *in vivo* concentrations. It has been used to determine:

- the effects of adding antibodies (with different $K[P]$) on the FT_4 response to serum dilution;
- the biasing effects (on FT_4) of antibodies that have different binding capacities (K[Ab]) in different patient populations (i.e. sera having varying binding capacities ($K[P_{free}]$);
- the biasing effects (on FT_4) of exogenous binders (e.g. different amounts of BSA added to immunoassay reagents);
- the optimal affinity constant (K_{eq}) requirement for the antibody used in the immunoassay.

Effect of Antibody Addition on the Free Analyte Sample Dilution Profile

Figure 4.7 shows the serum dilution profiles expected when the affinity (K) and concentration of the antibody ($[P_{ab}]$) are varied from 0 (i.e. no antibody added) to situations where the $K[P_{ab}]$ is 0.2, 0.5, 1, and 15% of the total binding capacity in the immunoassay tube, i.e. the $K[P_{ab}]/(K[P_{ab}] + K[P_{total}])$ ratio was 0.002, 0.005, 0.1, and 0.15. In these situations, the antibody will sequestrate (or 'pull off') 0.2–15% of the serum total T_4. The concentrations of the binding proteins and of T_4 in the euthyroid serum sample used for these simulations were as described earlier in the chapter; the immunoassay protocol used 25 μL sample in a total reaction volume of 125 μL (with the only T_4 binder in the reagents being the antibody). The serum was used at dilution factors of 1 (i.e. no additional dilution above the one already used in the assay, which was a 5-fold dilution) to 160.

The results show that the FT_4 concentration is robust to serum dilution, as long as the combined effects of antibody concentration and affinity ($K[P_{ab}]$) are kept to a minimum, compared to the overall concentration and affinities of the native binding proteins, e.g. the $K[P_{ab}]/(K[P_{ab}] + K[P_{total}])$ ratio should be less than

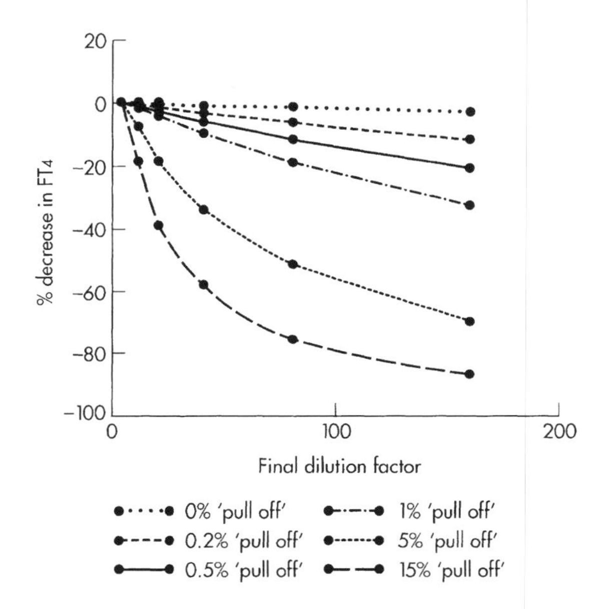

Fig. 4.7 Effect of antibody on free T_4 concentration.

0.5% in order to maintain robustness of FT_4 on serum dilution. At higher $K[P_{ab}]$ the FT_4 concentrations decrease in parallel to the dilution factor; the dilution-induced reduction in FT_4 becomes greater as the $K[P_{ab}]$ increases. The clinical significance of the serum dilution profile is discussed later in this chapter.

Effect of Antibody on the Free Analyte Concentration of Different Patient Populations

Figure 4.8 depicts the biasing effects of adding antibodies (of different $K[P_{ab}]$) in sera whose binding capacities span the range that would normally be seen in patients undergoing thyroid function testing. (Nelson and Wilcox (1996) suggested that a 30-fold range can be observed, with severely ill patients having binding capacities that are 8-fold lower and pregnancy sera (or patients with TBG excess) that are 4-fold higher than ambulatory subjects.) Other categories that have low serum binding capacities include patients with TBG deficiency, hyperthyroid patients, and newborns with respiratory distress syndrome.

The results show that the use of antibodies with high $K[P_{ab}]$ will cause a negative bias in sera with low binding capacities whereas a positive bias will be seen in patients whose sera have high binding capacities. Only assays that use antibodies with a low $K[P_{ab}]$, e.g. when the $K[P_{ab}]/(K[P_{ab}] + K[P_{total}])$ is less than 1%, will produce results that are close (i.e. $<10\%$ difference) to the true FT_4 concentration.

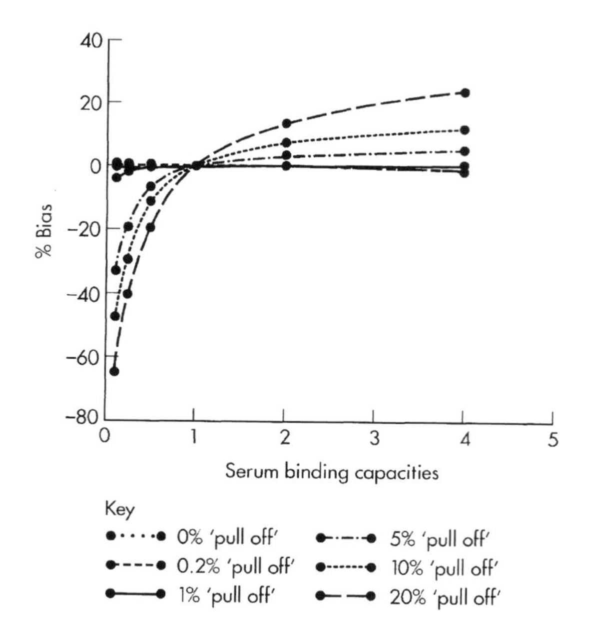

Fig. 4.8 Biasing effects of antibodies.

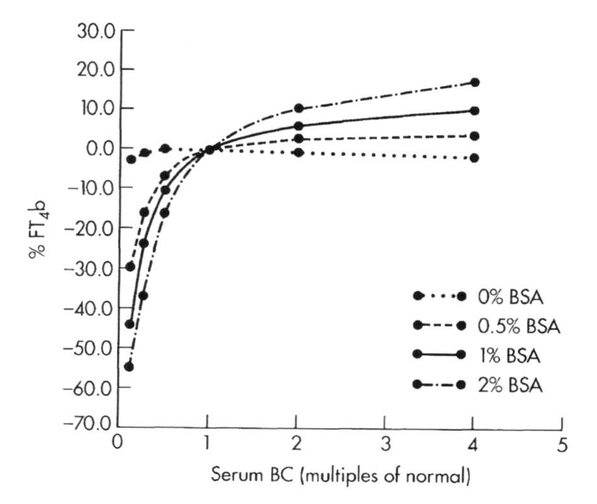

Fig. 4.9 Biasing effects of BSA.

Effect of BSA Present in Immunoassay Reagents on Free Analyte Concentration

Exogenous proteins, such as BSA, are commonly included in the reagents of most immunoassays, in order to reduce NSB of antibodies or of the analyte. In the case of the immunological measurement of FT_4 and FT_3, inclusion of BSA in the reagents was thought to make the assay robust to non-esterified fatty acids (NEFAs).[1] Figure 4.9 simulates the effects of including different concentrations of exogenous BSA in immunoassay reagents on the FT_4 concentration in sera having different T_4 binding capacities. The immunoassay protocol used 25 μL sample in a total reaction volume of 125 μL. The results in Figure 4.9 clearly show that inclusion of BSA in the immunoassay reagents of a free hormone assay will cause variable biases in results. Increasing the amount of BSA in the reagents causes increasingly negative biases in sera with low binding capacities whereas sera with high binding capacities become positively biased. As far as the notion that inclusion of BSA will protect against NEFA interference (since the NEFAs will be bound by the BSA) is concerned, BSA actually causes the FT_4 concentrations to be negatively biased, since the patients that are most likely to receive heparin are those with low serum T_4 binding capacities. So it is inadvisable for patients receiving heparin to have their FT_4 or FT_3 levels assayed, since the alterations occurring in the serum composition due to NEFA generation cause the *in vitro* FT_4 concentration to differ from the *in vivo* concentration.

Optimization of Antibody Affinity and Concentration

The spreadsheet program has so far been used to predict the free analyte concentrations when the analyte/protein equilibrium has been disturbed by exogenous addition of buffer (i.e. serum dilution), analyte binders (i.e. antibodies and BSA), or by alteration of the binding protein concentrations and affinities. The results of these simulations suggest that in order to develop a valid and accurate free analyte assay one should keep the disturbance of the analyte/protein equilibrium to a minimum. If the delicate equilibrium in the sample is upset, the advantages of free analyte measurement vs. total analyte measurement may be lost, as the assay results become like those of a total analyte.

The results presented in the above sections suggest that addition of the antibody to the serum causes a reduction in its FT_4 concentration. The magnitude of the reduction depends on the binding capacity of the antibody

[1] NEFAs are normally generated *in vitro* through the actions of lipoprotein lipases, and because they are able to bind to the thyroid-binding proteins, they cause a 'false' elevation in FT_3 and FT_4 concentrations. This is usually not a major problem, as the concentration of NEFAs in the serum is normally too low to have any significant effect. However, a significant *in vitro* generation of NEFAs, which can cause an elevation of FT_4, can occur when patients have been given heparin. Heparin is sometimes used to prevent clotting in infusion cannulae. *In vivo* administration of heparin stimulates the production of lipoprotein lipases and these act to release NEFAs.

($K[P_{ab}]$) and thus, in order to minimize bias the antibody binding capacity has to be kept at a low level. The simulations performed suggest that the optimum binding capacity of the antibody should be less than 1% of the binding capacity typically seen in a normal serum. Using the spreadsheet program, one can vary both the affinity and the concentration of antibody so that $K[P]$ is kept at 1% of the binding capacity of a normal serum. As explained previously, in these circumstances the antibody will sequestrate ('pull off') 1% of the serum TT_4. The percentage of free binding sites on the antibody (calculated in cell M19) can then be monitored, along with the FT_4 concentration, as the concentration of Total T_4 (cell C16 in the spreadsheet program) is altered. Figure 4.10 shows the results of these simulations. It is clear that the use of high-affinity antibodies (1×10^{11} L/mol) at a concentration (in the tube) of 1.79×10^{-10} mol/L will result in a FT_4 assay having excellent sensitivity but with a very reduced range, making it unsuitable for routine use.

As the affinity of the antibody is decreased (but keeping the $K[P]$ constant by adjusting the antibody concentration), the dose–response curve becomes shallower, but with an increased range. At affinities of less than 1×10^{10} L/M, the curve becomes too shallow making it unsuitable for routine use. It was thus surprising that two FT_4 assays, having curves with the required characteristics (in terms of curve shape and range), were claimed to have used anti-T_4 antibodies of affinities less than 1×10^{10} L/M (Christofides *et al.*, 1992; Christofides and Sheehan, 1995). This claim stimulated some debate in the literature, with Ekins (1992, 1998) suggesting that an error had been made in the measurement of the affinity constants of the antibodies and that only antibodies with affinity constants of more than 1×10^{11} L/M can produce the necessary curve shape. The data presented below show how antibodies with affinities of less than 1×10^{10} L/M can indeed be used in the measurement of FT_4. The first experiment used a T_4 antibody (K of 8×10^9 L/M, as measured by classical Scatchard plot analysis, using gravimetrically prepared T_4 standards, which were diluted in buffer). The assay protocol, based on a back titration format (*see* BACK-TITRATION (TWO-STEP) METHOD FOR FREE ANALYTE IMMUNOASSAY), was as follows. Twenty-five microliter aliquots of serum FT_4 standards (calibrated in ED) were pipetted into wells coated with a donkey anti-sheep antibody. One hundred microliters of sheep anti-T_4 antibody was added and the wells were incubated for 15 min at 37 °C. The wells were then washed and 100 μL of a solution containing T_3 conjugated to horseradish peroxidase (HRP) was added (T_3–HRP was used in order to reduce the possibility of dissociation of the antibody-bound T_4). This second incubation period was varied from 0.25 to 6 h. The wells were then washed and the HRP substrate added. The emitted luminescence was measured in a luminometer. The dose–response curves are shown in Figure 4.11. As predicted by Ekins (e.g. Ekins, 1998), the dose–response curve obtained with an antibody having an affinity of less than 1×10^{10} L/mol, when the assay was near equilibrium (after 6 h incubation), was too shallow to be a useful assay. However, as the incubation time was decreased the slope of the dose–response curve became progressively steeper with the curve produced after 0.25 h incubation period having the necessary slope and range.

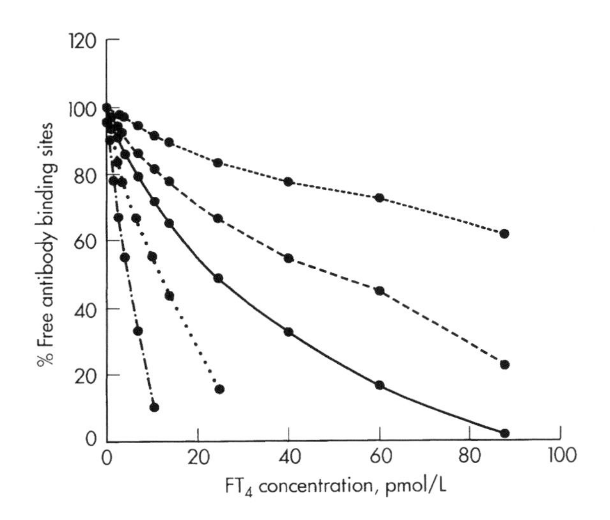

Fig. 4.10 Simulated FT_4 dose–response curves using antibodies of varying affinity constants •---• depicts an antibody with a K_{eq} of 1×10^{10} L/mol, • – – – – • depicts an antibody with a K_{eq} of 2×10^{10} L/mol, •—• depicts an antibody with a K_{eq} of 3×10^{10} L/mol, •···• depicts an antibody with a K_{eq} of 5×10^{10} L/mol and •-·-·-·• depicts an antibody with a K_{eq} of 1×10^{11} L/mol. The assumptions made for this simulation include that the antibody concentrations used are sufficient to sequestrate (i.e. pull off) 1% of the serum TT_4 and that the FT_4 assays proceeded to equilibrium.

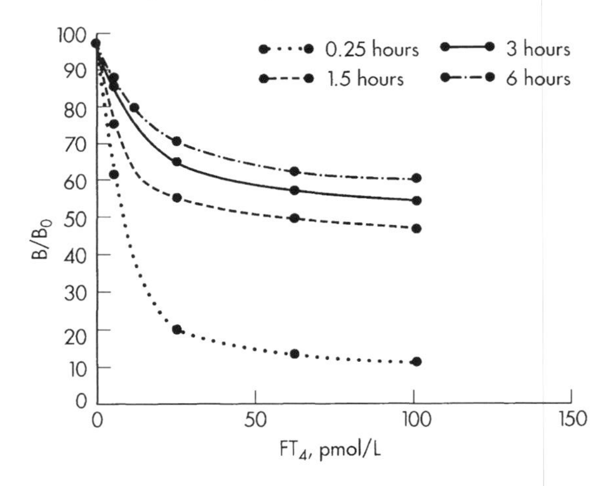

Fig. 4.11 Dose–response curves using the same antibody (over different incubation times).

In a second experiment, an anti-T_4 mouse monoclonal antibody with an affinity constant of 5×10^9 L/M was labeled with ^{125}I and used in a 'labeled' antibody method (*see* LABELED ANTIBODY METHODS). Fifty microliters of serum FT_4 calibrators were pipetted into polystyrene tubes, followed by 100 μL of the tracer antibody and 100 μL of a solution containing T_3, which was covalently linked to magnetizable cellulose separation suspension (SS). The concentration of the SS was varied from a 'neat' concentration to one that was 1000-fold lower. The tubes were incubated at 37 °C for 60 min, and then placed on a magnetic base for 20 min. The liquid supernatant was removed and the pellet counted in a gamma counter (NE1600). Figure 4.12 shows the dose–response curves (plotted as percentage of the total antibody binding to the SS vs. FT_4 concentration) of the assays using different SS concentrations. The data were also plotted (in Figure 4.13) as $\%B/B_0$ vs. FT_4 concentration. It is clear that the slopes obtained in the assays using high concentrations of T_3-cellulose are too shallow making the assays unsuitable for routine use. This outcome is in line with the prediction that FT_4 antibodies having affinities of $<1 \times 10^{10}$ L/M will produce curves that are too insensitive. However, reduction of the concentration of T_3-cellulose in the assay resulted in the generation of a dose–response curve that had the required (for a FT_4 assay) characteristics. The ED_{50} (i.e. the concentration of FT_4 required to reduce the amount of antibody bound by 50%) of the different assays presented in Figure 4.13 ranged from >100 pmol/L for the assay using 'neat' T_3-cellulose to 13 pmol/L for the assay using the T_3-cellulose at a concentration of 1 in 1000.

It is clear from the results of these two experiments that if the assays are taken to (near) equilibrium then the theoretical predictions (i.e. it is impossible to produce a workable FT_4 assay using antibodies with affinities of $<1 \times 10^{10}$ L/M) hold true. However, the use of non-equilibrium conditions and/or the optimization of the assay reactants, e.g. adjusting the concentration and affinity of the T_3-cellulose (for the separation of bound from free antibody) permits the development of FT_4 assays which have the necessary sensitivity and range.

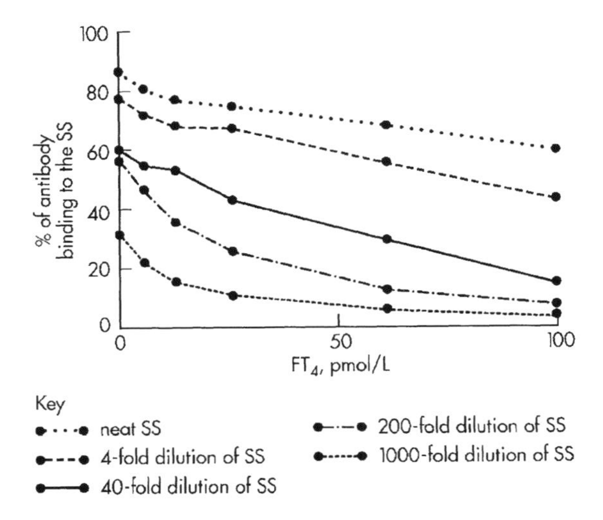

Fig. 4.12 Dose–response curves using different concentrations of separation suspension (SS).

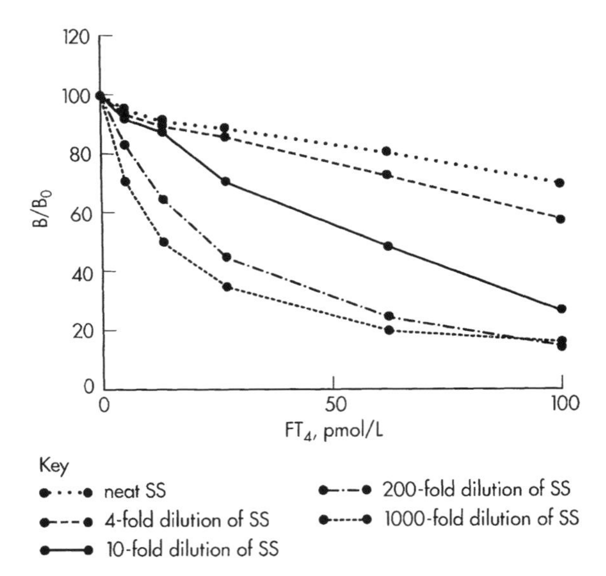

Fig. 4.13 Dose–response curves using different amounts of separation suspension (SS).

Back-Titration (Two-Step) Method for Free Analyte Immunoassay

In this method, the serum is allowed to react with an antibody that has been immobilized on a solid support. During this first incubation (which should be performed at 37 °C) the antibody binds to the analyte in the serum. After the first incubation is completed, the reaction mixture is removed by aspiration and the immobilized antibody washed. The unoccupied binding sites of the antibody can then be quantified by incubating it with labeled analyte. See Figure 4.14.

The use of a labeled analog of the analyte, having a lower affinity than the endogenous hormone towards the antibody, is preferable as this can reduce dissociation of the bound analyte from the antibody. The amount of tracer binding to the antibody can then be interpolated into concentration using the calibration curve, which is a plot of the amount of tracer bound by the antibody against free analyte concentration. The free analyte values assigned to the calibrators are normally derived by calibration in a direct equilibrium dialysis method.

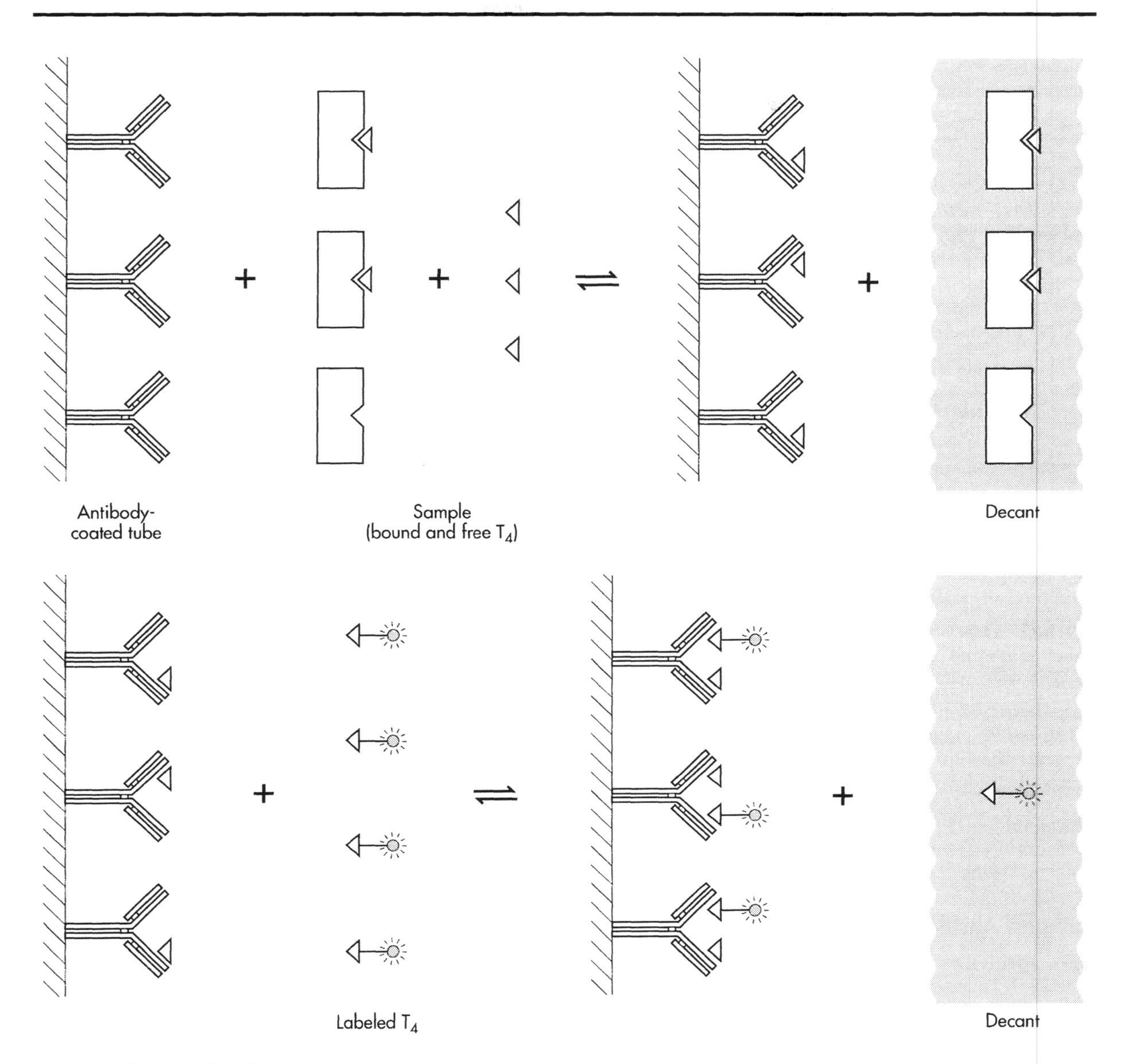

Fig. 4.14 Two-step free T_4.

Labeled Analog Tracer Method

In this method, the serum is incubated simultaneously with the antibody (this is usually immobilized on a solid surface) and a labeled derivative of the analyte (the labeled analog tracer). During a single incubation period (at 37 °C) the analog tracer competes with the free analyte for the limiting number of antibody binding sites. The amount of tracer binding to the antibody is inversely proportional to the concentration of the analyte. At the end of the incubation the antibody is separated from the rest of the reactants and the amount of bound tracer quantified (the measurement of the tracer depends on the nature of the label used, e.g. ^{125}I, enzyme, or fluorophore, etc.) and then converted into dose by interpolation from a calibration curve. The free analyte concentrations assigned to the calibrators are normally derived using direct equilibrium dialysis as the reference method. See Figure 4.15.

An important requirement for this methodology (and also the labeled antibody method) is that the analog used does not have an affinity towards any of the serum

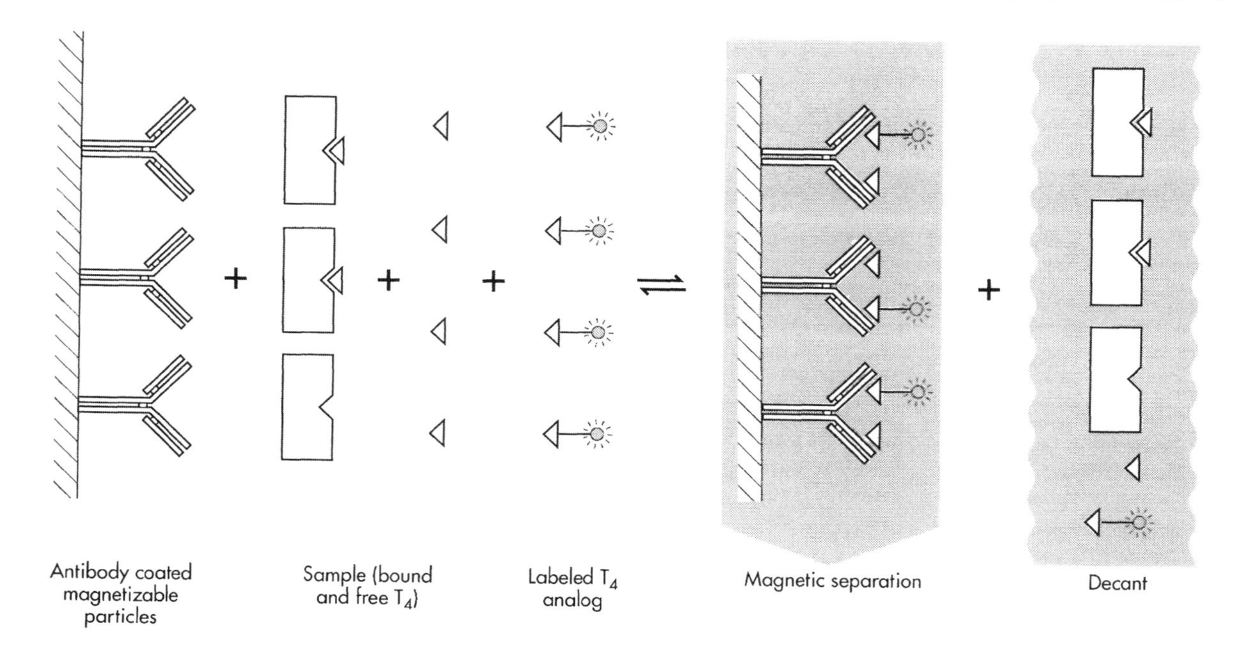

Fig. 4.15 Labeled analog free T_4 assay.

binding proteins. If the analog has affinity towards any of the serum binding proteins then the free analyte concentrations obtained with such an assay will be dependent on the protein concentration. Binding of the analog to proteins can be eliminated by conjugating it to large proteins (Georgiou and Christofidis, 1996; Ttsutsumi *et al.*, 1987). This conjugation causes a sufficient steric hindrance to eliminate binding to the serum proteins.

Labeled Antibody Methods

Once again, free thyroid hormones will be used as the example. Thyroid hormone (e.g. T_4 or a T_4–protein conjugate) is immobilized onto a solid surface (e.g. microtiter well surface). The serum sample and a labeled anti-T_4 antibody solution are added to the solid phase and the mixture incubated at 37 °C. During the incubation period the antibody partitions itself between the liquid phase (containing the endogenous FT_4) and the solid phase. The amount of labeled antibody binding to the solid phase (estimated after separating the liquid reactants from the solid phase) is thus inversely related to the amount of FT_4 in the serum, and can be quantified by interpolation from sera containing known concentrations of FT_4 (the values are commonly obtained from ED).

FT_3 assays can be developed using immobilized T_3 or a T_3-conjugate and a labeled anti-T_3 antibody tracer.

A variation of this method has been successfully employed (Christofides *et al.*, 1992, 1999a,b; Christofides and Sheehan, 1995) in developing commercial immunoassays for FT_4 (see Figure 4.16). This utilizes the weak cross-reactivity ($<1\%$) of the labeled anti-T_4 antibody to an immobilized T_3–protein conjugate. The weakly cross-reacting T_2–protein conjugate has been used in the development of a FT_3 assay.

The use of this 'heterologous assay' approach has a number of advantages. The first advantage, which is common with all heterologous assays, is that the dose–response curve becomes steeper. Other advantages include faster kinetics, a much higher signal and making the assay more robust to interference by endogenous anti-thyroid hormone antibodies. Note, however, that the presence of a high concentration of anti-T_3 autoantibodies with a very high affinity towards the immobilized antigen can still cause interference. Important requirements for these types of assays are lack of binding of thyroid-binding proteins to the immobilized antigen (this is generally met by linking the antigen to a large molecule) and, in common with all free thyroid hormone assays, minimizing the disturbance of the endogenous T_4/protein equilibrium.

TESTS OF VALIDITY (ACCURACY)

Using the Law of Mass Action model described earlier one can design a number of experiments to compare the performance of any free analyte assay with the ideal assay. Examples of experiments that can be performed are given in this section.

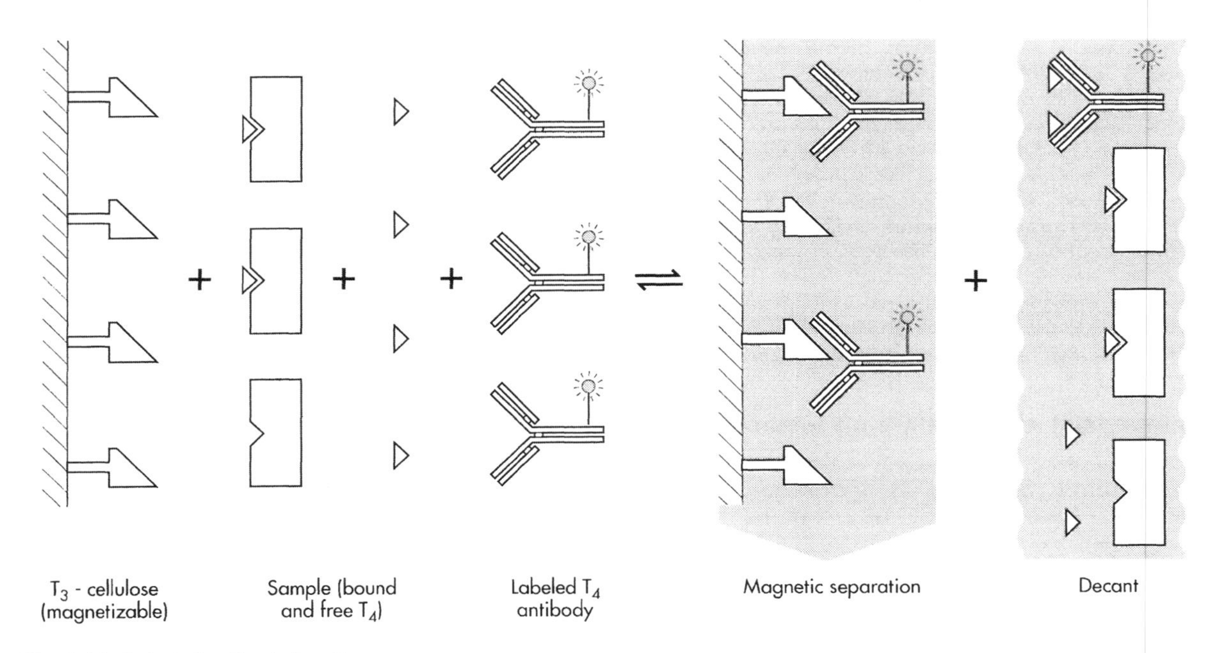

Fig. 4.16 Labeled antibody free T_4.

Spiking Serum Samples with Binding Proteins

The response expected is a gradual decrease in free analyte concentration, as more protein is added. In the case of FT_4, in the absence of any interference in the assay, one should expect that the percentage decrease in FT_4 would be greater following the addition of TBG than following the addition of equimolar concentrations of TTR or HSA. The problem with this test is that the magnitude of the decrease in free analyte concentration will depend not only on the concentration (and affinities) of the added proteins but also on the concentration and affinities of the endogenous proteins. Thus, one cannot readily compare the experimental results with those expected from theory, unless the concentration and affinities of both the exogenous and endogenous proteins are known. Nonetheless this test is useful in determining any gross problems with the assay, e.g. if the tracer used in the assay has an affinity towards any of the binding proteins then adding this protein in the serum will result in an apparent increase (or no change) in the FT_4 concentration rather than the expected decrease.

Spiking Serum Samples with Binding Blocking Agents

The response expected is a dose-dependent increase in free analyte concentration. Blockers that can be used for thyroid hormones include drugs such as furosemide, ketoprofen, phenylbutazone, mefenamic acid, diphenylhydantoin, probenecid, sulindac fenclofenac, and salicylic acid; other substances include anilino naphthalenesulfonic acid and non-esterified fatty acids (e.g. oleic acid). This test, like the protein spiking test (above), can be viewed as qualitative rather than quantitative, since the increase in free analyte concentration expected will depend on both the concentration (and affinity) of the spiked substance and the concentration and affinity of the endogenous proteins.

Dilution Test

A quantitative test that can readily be performed with any free analyte assay is the serum dilution test. Dilution of serum should produce near constant free analyte results if the assay is valid, but decreased free analyte values if the assay is invalid. In the case of thyroid hormones, it has been proposed (Christofides *et al.*, 1999a,b) that this test can be used to predict the performance of the assay in different patient categories. This is because serum dilution will, in effect, produce a panel of samples whose serum binding capacities reflect the spectrum of binding capacities seen in patients undergoing thyroid function tests. For example, it has been shown that there is a 20- to 30-fold span of binding capacities of patients having thyroid function tests; this span can be reproduced by diluting a third trimester pregnancy serum to 20-fold or 30-fold. Any decrease of FT_4 seen following dilution would indicate that this assay would underestimate the FT_4 concentrations in any patients that have low T_4 binding capacities, e.g. hospitalized patients. The buffer that is commonly used in the serum dilution experiment is 10 mmol/L HEPES (*N*-[2-hydroxyethyl]piperazine-*N*′-[2-ethane]sulfonic acid, which can be obtained from Sigma, catalogue No. H7523), pH 7.4. One possible problem with this approach is that dilution of sera to these extents (i.e. 20-fold to 30-fold) will excessively lower the protein content of the assay reagents and may introduce significant non-specific effects. It may thus be prudent to reduce the dilution window to no more than 4- to 8-fold dilution (to reflect severe hypo-proteinemia), to assure that there is sufficient protein present in the assay reagents to prevent these non-specific events from happening. The presence of a dilution-dependent reduction in FT_4 concentrations indicates that the assay will produce negatively biased results with sera having low T_4 binding capacities.

Comparison with a Reference Method

The composition of the patient panel used in such a comparison is of paramount importance, since an apparently excellent relationship can be obtained between an invalid free analyte assay and the reference method if the panel excludes patients with high or low analyte binding capacities. Thus, the panel chosen for this evaluation should include patient sera from severely ill, hospitalized patients, and pregnancy sera (preferably sera from the third trimester). More importantly, the reference method chosen should be one that has, itself, been proved to be a valid free hormone assay.

CONCLUDING REMARKS

To measure free hormones or drugs in patient samples, there must be minimal disturbance of the equilibrium between the analyte and its binding proteins. Assays should not be judged on their assay architecture but on the level of disturbance they exert on this equilibrium. The fact that an assay is based on accepted physico-chemical principles (e.g. equilibrium dialysis) does not necessarily make it a valid assay. Conversely, when a particular assay is shown to be invalid it does not necessarily mean that all assays based on this architecture are also invalid.

REFERENCES

Christofides, N.D. and Sheehan, C.P. Enhanced chemiluminescence labeled-antibody immunoassay (Amerlite MAB) for free thyroxine: design, development, and technical validation. *Clin. Chem.* **41**, 17–23 (1995).

Christofides, N.D., Sheehan, C.P. and Midgley, J.E.M. One-step, labeled antibody assay for measuring free thyroxine. I. Assay development and validation. *Clin. Chem.* **38**, 1118 (1992).

Christofides, N.D., Wilkinson, E., Stoddart, M., Ray, D.C. and Beckett, G.J. Assessment of serum thyroxine binding capacity-dependent biases in free thyroxine assays. *Clin. Chem.* **45**, 520–525 (1999a).

Christofides, N.D., Wilkinson, E., Stoddart, M., Ray, D.C. and Beckett, G.J. Serum T_4 binding capacity-dependent bias in the AXSYM FT_4 assay. *J. Immunoassay* **20**, 201–221 (1999b).

Csako, G., Zweig, M.H., Glickman, J., Ruddel, M. and Kestner, J. Direct and indirect techniques for free thyroxin compared in patients with nonthyroidal illness. II. Effect of prealbumin, albumin, and thyroxin-binding globulin. *Clin. Chem.* **35**, 1655–1662 (1989).

Demers, L.M. Thyroid function testing and automation. *J. Clin. Ligand Assay* **22**, 38–41 (1999).

Ekins, R. One-step, labeled antibody assay for measuring free thyroxin. I. Assay development and validation (letter). *Clin. Chem.* **38**, 2355–2357 (1992).

Ekins, R. The science of free hormone measurement. *Proc. UK NEQAS Meet.* **3**, 35–59 (1998).

Ekins, R.P. and Ellis, S. The radioimmunoassay of free thyroid hormones in serum. In: *Thyroid Research: Proceedings of the Seventh International Thyroid Conference, Boston* (eds Robbins, J. and Braverman, L.E.), 597–600 (Excerpta Medica, Amsterdam, 1975).

Ekins, R., Filetti, S., Kurtz, A.B. and Dwyer, K. A simple general method for the assay of free hormones (and drugs); its application to the measurement of serum free thyroxine levels and the bearing of assay results on the free thyroxine concept. *J. Endocrinol.* **85**, 29–30 (1980).

Georgiou, S. and Christofidis, I. Radioimmunoassay of free thyroxine (T_4) using ^{125}I-labeled T_4–IgG complex with very large molecular weight. *Clin. Chim. Acta* **244**, 209–220 (1996).

Nelson, J.C. and Wilcox, R.B. Analytical performance of free and total thyroxine assays. *Clin. Chem.* **42**, 146–154 (1996).

Robins, J. and Rall, J.E. The interaction of thyroid hormones and protein in biological fluids. *Recent Prog. Horm. Res.* **13**, 161–208 (1957).

Ttsutsumi, S., Ishibashi, K., Miyai, K., Nagase, S., Ito, M., Amino, N. and Endo, Y. A new radioimmunoassay of free thyroxine using ^{125}I-labelled thyroxine–protein complex uninfluenced by albumin and thyroxine-binding globulin. *Clin. Chim. Acta* **170**, 315–322 (1987).

Wilcox, R., Nelson, J.C. and Tomei, R.T. Heterogeneity in affinities of serum proteins for thyroxine among patients with non-thyroidal illness as indicated by the serum free thyroxine response to serum dilution. *Eur. J. Endocrinol.* **131**, 9–13 (1994).

5 The Foundations of Immunochemistry

Robert F. Ritchie

With the superb, readily available instrumentation we take for granted today, researchers at the beginning of the 21st century can focus their energies toward the end-point of their work. But we often, if not always, forget the underpinnings of our work. In an effort to refocus our view of the foundations of modern immunochemistry, I have attempted to identify and review those landmark publications that were increments towards new awareness and as a result, new investigative and clinically useful tools. In the collection of documents[1] establishing who did what, I have been fortunate in having powerful search tools that have only become available in the last two decades. Our predecessors did their searches with dogged perseverance and required a certain amount of luck to uncover relevant prior science. Today, our abilities are magnified enormously. Not all is easier, however. There is 'street knowledge' of what represents the founding publication for a given method or observation. The powerful tool of global searches and comprehensive archives that can produce copies of the actual document, which of course, must be read, presents serious problems. When the original publication is read carefully (because one has the document in its entirety) it can often be found that there is 'prior art'. With the help of my librarian and staff at various libraries and a personal dossier of articles, a collection spanning over a century has been assembled. There were some unsavory surprises, which, of course, I will prudently keep to myself. Plagiarism and outright academic and industrial espionage have occurred! But, more commonly, simple incomplete searches of the literature resulted in false-starts and misattribution of 'parenthood'. I make reference only to authors who have made significant contributions.

Identified are many authors who have made contributions to what we use routinely today. Inevitably, some papers will have been missed and for this I am truly sorry, but it has not been for the lack of effort. Those papers I have identified represent the basis for modern immunochemistry; those publications that detail technologically advanced assays are noted briefly, as these have been well covered in modern texts. Therefore, only those significant steps forward that lead up to modern methods are identified. The path is terminated there. It is with the archival methods, still very useful, though no longer recognized by many immunochemists, except by name, that I have worked. Known only to those involved, timing of publication, a feature that can lay claim to authorship of a method, work in progress by several groups can, and usually do, overlap. The first into print claims the 'prize' and not always fairly. There are many such instances in these pages, but timing can only be verified by the dates of publication.

Immunochemistry is inextricably meshed with protein chemistry; one cannot exist comfortably without the other. In this chapter I have elected to refer only to those papers that are clearly part of the foundation of immunochemistry. Also difficult to present in a serial manner is the parallel evolution of concepts, biological products, and instruments. Since these are interdependent phenomena, describing them separately is more confusing than informative, and never simple. However, in the pages to follow, I will attempt to develop, in chronologic blocks, the evolution with frequent reference to parallel topics.

INTRODUCTION

The concept that an immunochemical reaction could be examined both qualitatively and quantitatively became evident in the last years of the 19th century and in the first decade of the 20th century. But the potential that this apparently crude process could have the precision of traditional chemistries would wait until nearly the close of the last century, for full appreciation. During the last century, there have been great expectations, many of

[1] Many of the documents that were reviewed are old and much handled. To reduce artifact as a result, I have digitally repaired many of these images. However, the data as originally published remain unaltered.

which were quickly overrun by new developments. Many of the methods that have survived were not seen as great advances at the time. Others that showed great promise came to naught. If this narrative strikes a resonant chord with the current activity collectively referred to as 'Proteomics' (Anderson and Anderson, 2002), it is not without reason. Just possibly, our expectations might be closer to reality as a result of 'knowing history' rather than 'repeating it'.

THE IMMUNOLOGICAL REACTION (1895–1935)

The initial process evolved from the awareness that animals could be injected, or **immunized** – a new word then – with whole bacteria and survive. This stimulated several workers to search for factors that made this possible. Kraus (1897) made the observation that the serum of these immunized animals would agglutinate the introduced bacteria in 1895. Immune serum from these animals bound firmly and very specifically with the same bacteria to which the animal had been exposed (Bordet, 1900). This observation was the first qualitative immunochemical characterization. During the decade following the 1890s von Behring worked towards a solution of the diptheria epidemics that plagued Europe, especially among the military. Having worked during the era of Pasteur, Koch, Loeffler, Roux, and Yersin, the overwhelming importance of therapeutic antisepsis had a great effect on von Behring's work. He received the Nobel Prize in Physiology or Medicine in 1901, for his work in *Serum Therapy*, which he published in 1912 (von Behring, 1912) and appears as part of his Nobel Lecture (von Behring, 2001). During the two decades straddling the turn of the century several workers noted that the serum of the immunized animal could render the broth from the bacterial culture turbid. This was also specific to the species used, but the first suspicion that there were non-specific mechanisms at work also evolved. Should the culture broth from a related organism be used, turbidity developed, but to a much lesser degree, and often agglutination of the related organism did not occur. Bordet also noted that the antibody to rabbit red cells hemolyzed these cells, as did the serum from a naïve animal, unless the serum was heated to 55 °C. Perhaps this was first observation of the action of complement. During this time Paul Ehrlich, working at the new Frankfurt *Serum Institute*, was developing what was to become his 'side chain theory'. An illustration shown from Ehrlich's summary (Ehrlich, 1901) expanding his theoretical plan (Figure 5.1) was an attempt to organize what was known about the body's reaction to toxic (immunological) insult. While the nomenclature in the diagram is somewhat different from today's, the concept is easily recognized from the diagram. The cell and its surface structures, membrane proteins (antigens), are targeted by 'toxins' stimulating the cell to produce more receptors, if it survives. These receptors were believed to be shed into the serum to act as specific targets for free toxin and bind to it, i.e. an antitoxin. In Ehrlich's theory most of the terms appear that we associate with the immune reaction today. Paul Ehrlich received the Nobel Prize in Physiology or Medicine in 1908 shared with Ilya Ilyich Mechikov from the Institute Pasteur (Ehrlich, 2004).

Further insight was provided by Uhlenhuth (1903) who mixed reactants, antiserum, and culture broth, layered so as to create an interface at which precipitation, the **ring test**, developed. Subsequently flocculation was observed with the specific bacteria, or at least species closely related to that originally injected into the animal (Calmette and Massol, 1909). The use of a gelatin matrix as a medium for one of the reactants was introduced soon after (Nicolle *et al.*, 1920). The importance of time as a variable in the development of immunoprecipitates was appreciated when it was observed that with proper concentrations of both reactants, the development of precipitates was most rapid; the concept of 'equivalence' evolved (Ramon, 1922). Additionally, at high antibody concentration, reaction time diminished inversely with the concentration of antigen (Hooker and Boyd, 1935). Inhibition of reaction in situations of excess antibody and more so with excess antigen had been previously noted. The latter, however, was not to be fully appreciated for several decades as **antigen-excess**. Further understanding (Dean and Webb, 1926) came when time was measured against the concentration of the two reactants. The shortest time for a given amount of precipitate constituted the optimal proportion; an early assessment of immunological potency. Temperature, ionic strength, pH and agitation also had been observed to affect the degree of immune precipitation. Detailed work by Heidelberger and Kendall (1935a–c) and Goettsch and Kendall (1935), shown in Figure 5.2, amplified previous publications. These workers used a variety of antigen–antibody systems including a highly colored antigen (azo-albumin) that allowed careful observation of the antigen–antibody reaction under a variety of reactant-pair concentrations. They also gave credence to the theory that antigen–antibody reactions obeyed the 'laws of classical chemistry'. Also demonstrated was that as an animal is immunized, its antiserum changes in precipitating characteristics, believed to be due to an expansion in the number of antigenic sites on the antigen recognized by the animal's immune cells. Furthermore, they showed that there were several levels of what would now be termed **avidity**. It should be pointed out that at that time and until the early 1970s immunization was carried out by intramuscular injection of enormous quantities of antigen. The result was antisera of lower than expected affinity, and from what was believed to be pure antigen any number of ancillary antigen–antibody pairs developed (see below). As a result, this careful work supplied much of the theoretical information leading to today's quantitative analysis of serum proteins. These experiments used the laborious analysis of total nitrogen in a precipitate,

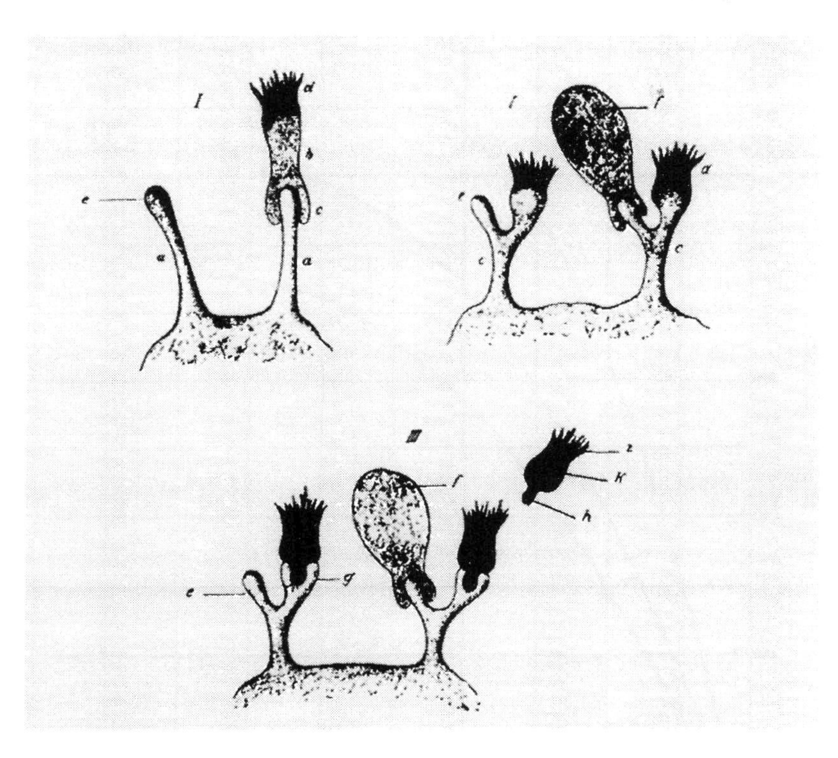

Fig. 5.1 Ehrlich's (1901) 'Side-Chain' theory. The terms Ehrlich used are different from today, however, the concept is clear. Cellular antigens, toxins, antitoxins (antibody) and complement are all represented.

the Kjeldahl reaction. Instrument analysis up to this point was used to indirectly measure the product of the reaction in milligrams of nitrogen.

Critical to what was to occur in the next few years, work in other disciplines was progressing. An early worker (McFarland, 1907) made use of a home-made comparator or 'nephelometer' (εφελή or 'the mist') as it was called, to measure the numbers of bacteria in a sample to measure the 'opsonic index' of vaccines. Perhaps this was the first effort to measure the quality of immune reagents.

The measurement of fat in blood was a laborious process in the early 1900s. A worker (Bloor, 1914, 1928) published a turbidimetric technique for measuring lipids in tissue and blood. After extracting tissue fats into an ether–alcohol mixture the resultant extract was evaporated to dryness and then dissolved in water. Measured in a Richards comparator the turbid solution gave, for then, acceptable results. The drawbacks of nephelometers of the time were recognized (nephelometric results were not proportional to the mass of the suspended precipitate) and a turbidimetric instrument (Dubosq colorimeter – a comparator) was substituted (Denis, 1921). These devices had all been used exclusively within the biochemical community. A new modality was introduced (Libby, 1938a,b): the numerical measurement of light scattered from an **immunoprecipitate**.

SUMMARY

During this early period in the evolution of immunochemistry, several major discoveries had been made:

- Animals react by producing serum factors (antibodies) when injected with either the whole noxious agent or extracts thereof (antigens).
- The animals producing these proteins can protect themselves against most of the effects of these agents.
- Serum from the injected animals causes flocculation of the agents in culture media.
- Serum produces turbidity when the filtered broth in which the agents have been grown is mixed with it.
- Related species of the immunizing agent also induce all the above reactions to a lesser degree.
- The precipitate produced *in vitro*, under controlled circumstances, can be quantified.
- The reaction of antibody and antigen produces a precipitate in a gel matrix that provides some information about the physical characteristics of the constituents.

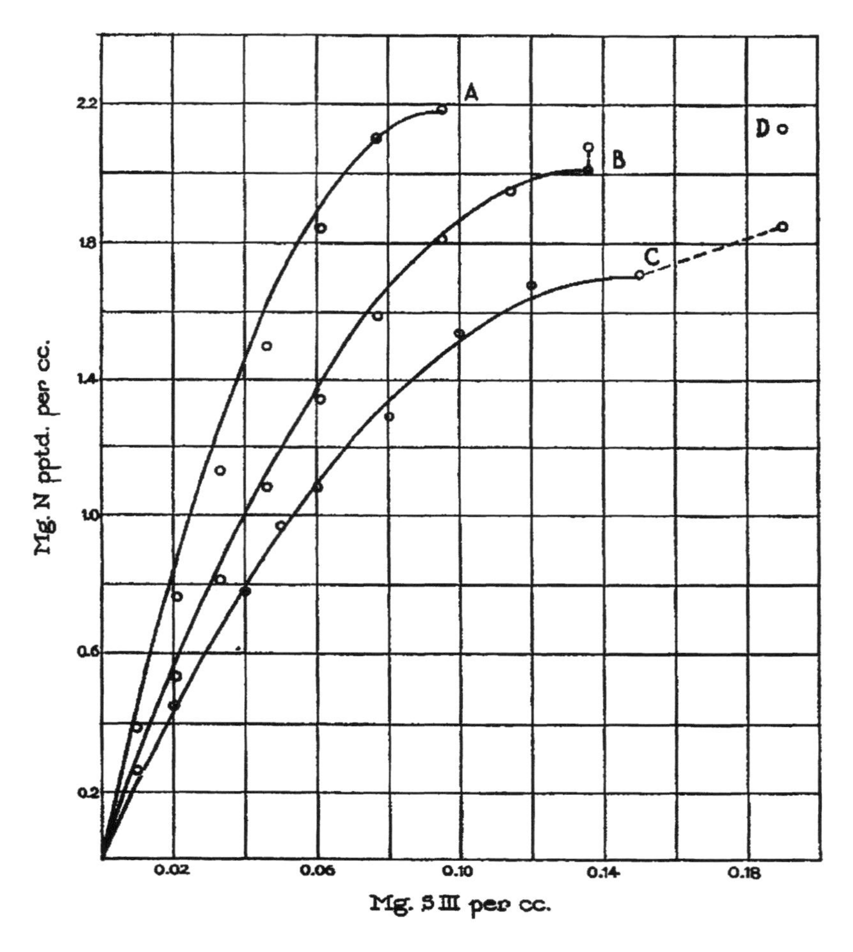

Fig. 5.2 The original Heidelberger curve as published in 1935. Milligrams of pneumococcal polysaccharide precipitated by homologous antibody on the ordinate and milligrams of nitrogen (N) on the abscissa: both per cubic centimeter.

- Simple instruments can be used to give numerical results for a reaction between antiserum and antigen.

QUALIFICATION BY DIFFUSION IN GEL (1861–1977)

The transition from reaction in free liquid to a semisolid matrix, gelatin, spawned new investigation that made use of considerable work that hearkened back 50 years. As early as the first decade of the 20th century, Bechhold (1905) and Bechnold and Ziegler (1906) described chemical reactions forming in a gelatin matrix between silver nitrate and NH_4O_2 including an image of multiple precipitin rings from the single chemical pair. Initially, it was assumed that the liquid phase of a gel presented no diffusion limitations. However, it was soon appreciated that even small molecules were inhibited during diffusion. This, therefore, presented complications to the quantitative measurements carried out at the time. Complex mathematical formulae (Friedman, 1930) incorporating all the characteristics of the molecules, medium, and physical constituents of the test were used to resolve these issues. Thus, the concept of a 'diffusion coefficient' was created (Stiles, 1920, 1923). Awareness that macromolecules such as proteins were affected by diffusion in gels had been previously described from a different point of view as far back as the mid-19th century (Graham, 1861). A curious phenomenon, which dogs workers in this field to this day, was described by Liesegang (1929) and bears his name. Liesegang rings are observed during precipitation in antigen–antibody systems forming in gels and also in certain geological formations. These bands are described as rhythmic precipitates formed from the reactants, which appear to become partially soluble only to form again at some distance further on. Certain geological formations present in the same manner from chemical interactions, as liquid solvents migrate through rock mobilizing solute. The observation of these multiple bands confused workers trying to dissect the protein constituents of animal serum. Were these bands evidence of related, but somehow different forms of a single species or were they several completely different species? Work over two decades later was to provide an explanation.

A visible ring of precipitate was observed (Petrie, 1932) when bacterial colonies on a culture medium reacted with incorporated antiserum (Figure 5.3). Petrie theorized that the precipitin rings or 'halos' were produced by soluble constituents (polysaccharides) of the bacterial colony (*Pneumococcus*). He also observed what were believed to be non-specific precipitates or 'Liesegang rings' under a microscope. The possibility that this ring phenomenon could be exploited to type certain bacteria proved to be useful (Sia and Chung, 1932). The precipitin reaction in agar gel format was combined with the tube method ('ring test') of Uhlenhuth by Oudin (1946, 1948, 1949). In this development, agar containing immune serum (at low concentration) fills the lower portion of a thin glass tube. Over this is layered an aliquot of soluble antigen. After incubation at constant temperature, precipitin bands develop that migrate from the gel–liquid interface into the gel column. The result is a complex system of antigen–antibody precipitates developing as they meet. Each pair acts independently of all others unless the antigens are related immunologically. After several hours of diffusion a series of bands develop. Many features can both complicate and enhance the results. These include:

- many overlapping bands
- the temperature at which the reaction occurs
- the mass of the antigen molecules (some types having several molecular sizes)
- the make-up of the solvent
- the characteristics of the gel.

For single antigen solutions semi-quantitation is possible. Beyond this, complexities make only crude quantification possible. Band migration occurred as the result of continuous immunoprecipitation and dissolution. The belief that this was the Liesegang phenomenon was dispelled when Oudin used a multiple antigen–antibody system, which produced several bands that were believed to be the result of several superimposed and specific precipitates migrating independently. This was not to dispel the true Liesegang phenomenon, which can develop with a pure system.

The Oudin method of assessing the concentrations of antigen and antibody became well recognized. Ouchterlony applied the one-dimensional, antigen from one well and antibody from the other (double diffusion technique) to double diffusion in two dimensions, when he allowed bacteria to grow in a gel matrix containing antiserum (Ouchterlony, 1948, 1949a,b). In this model, a sharp ring of precipitate developed around those colonies that produced toxic antigen (Figure 5.3). Thus, this simple technique became a quantitative means of assessing antiserum potency. In the ensuing years many publications appeared that explored permutations of this uncomplicated yet elegant means of using the exquisite sensitivity and specificity of the animal's immune system to explore the evolving topics of protein chemistry. Ouchterlony (1958) summed up the classification of serological reactions thus:

> ... simple, complex and multiple systems. A simple precipitation system is one in which an antigen or hapten reacts with the corresponding antibody. A complex system is one in which several closely related antigens of varying specificity react with a single antibody, originated by, and corresponding to, one of the antigens. Finally, a multiple system is one in which several different antigens react with corresponding antibodies.

This classification still stands although the physical dimension of the three types is lost, sometimes to our loss, in modern automated, one-dimensional, or 'Black Box' instruments.

An early experiment that was to become more detailed in work in thin gel films was described (Surgalla *et al.*, 1952). A four-part upper chamber in an *Oudin tube* format, each containing a different antigen, is sealed to the surface of a single gel containing antiserum. As the antigens diffuse downward, out of the compartments into the gel, they come into intimate contact as they diffuse laterally, in the presence of antibody. In this manner the relationships of one antigen to another can be assessed easily. The concept of identity/non-identity had begun.

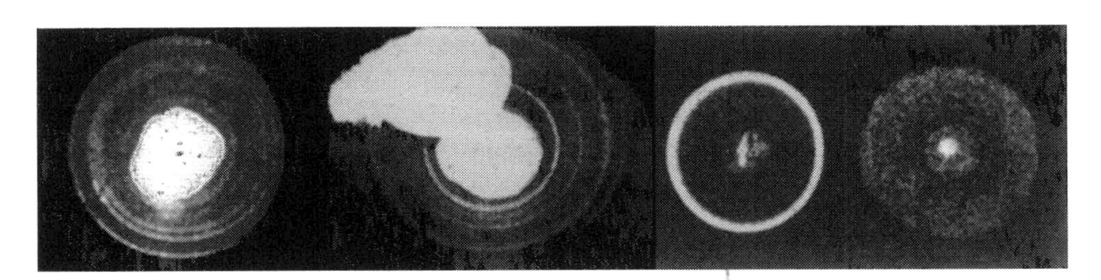

Fig. 5.3 The observation of Petrie (1932) that brought attention to the diffusion of antigen into an antibody-containing gel producing a 'halo' or ring.

What was soon to become a powerful new tool began as 'the simple diffusion in two dimensions' (Petrie, 1932). In this approach the familiar halo precipitates were visualized by indirect light, as we do today. It was believed that each ring corresponded to one antigen–antibody pair. The exception is the possible rhythmic precipitation-resolution representing Liesegang rings mentioned above, yet are produced by a single antigen–antibody pair. Perhaps the first use of the term '*radial diffusion*' appeared in this work. Important to these workers was the requirement that there be a balance of concentration between the reactants with the potential of no visible reaction at *either* extreme of antigen- or antibody-excess. This situation was to become a significant concern three decades later when automated instruments came into use.

Another permutation of the format in which reactants were introduced to each other came in the 'Basin Plate Technique' again introduced by Ouchterlony (1958). An infinite number of physical arrangements were possible and used to enhance the relationships between developing precipitin lines. In this way complex or multiple systems could be teased apart to fully describe the relationships in the reactant system (Figure 5.4). Elegant diffusion patterns were shown in this work enhanced by the physical arrangement and shapes of the wells. In this way diffusion time could be used to overcome overlapping precipitates with the result that the individual constituents could be visualized. Once again, new terms were introduced as a result: 'reaction type 1 – identity', 'reaction type 2 – non-identity', and 'reaction type 3 – partial identity' (Figure 5.5). Manipulating the positions of the wells allowed a plethora of sub-categories for each. By placing a series of wells around a central well, Wadsworth (1963) described a microtechnique to study antigen–antibody relationships and do so with much smaller amounts of reactants and also to reduce the time

(a)

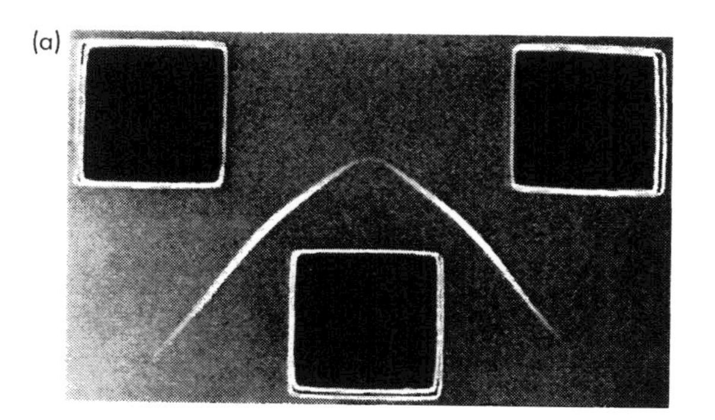

(b)

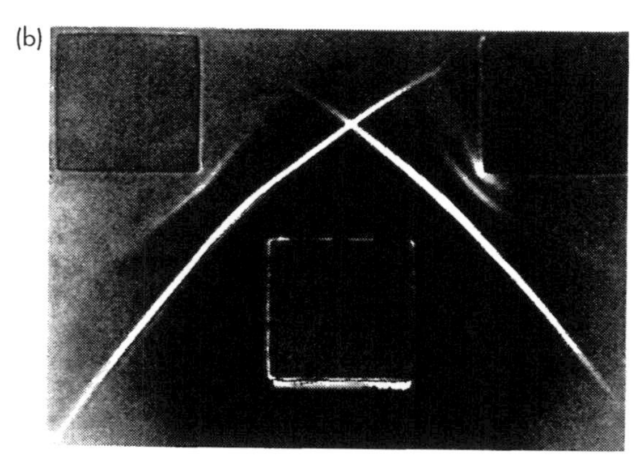

(c)

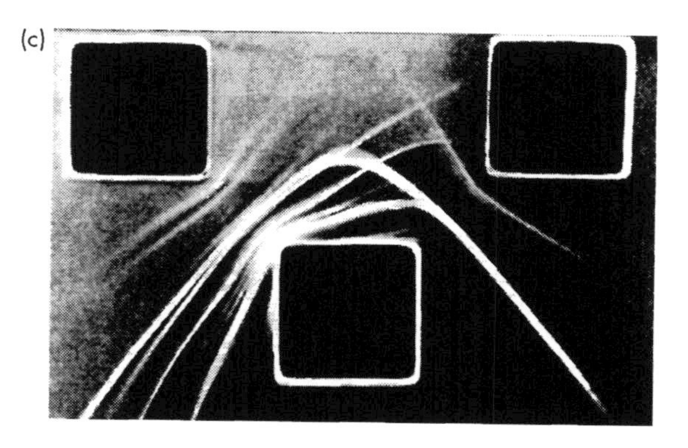

Fig. 5.4 (a) Actual image of a Type 1a precipitin pattern with precipitin arcs fusing, indicating immunological 'Identity' (Ouchterlony, 1958, Figure 23A). (b) Actual image of Type 2 precipitin pattern with individual arcs having no common precipitate, i.e. no fusion, indicating 'Non-identity' (Ouchterlony, 1958, Figure 28). (c) Pattern of immunodiffusion with upper left square well containing many different antigens, the center well containing a polyspecific antibody, and the upper right well containing two antigens, only one of which is present in the upper right-hand well (Ouchterlony, 1958, Figure 30F).

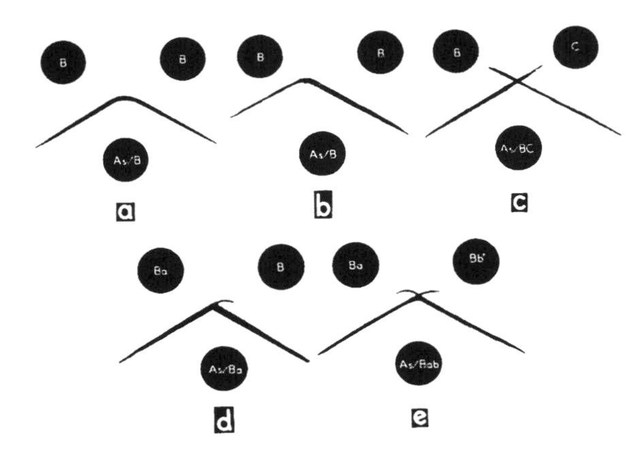

Fig. 5.5 Diagrammatic examples of double diffusion with: identity (a, b), non-identity (c), and partial identity (d, e) (Crowle, 1973, Figure 4.5).

to completion. Crowle published many papers describing how the physical arrangement of wells could be used better to tease apart the immunochemical characteristics of protein populations (Crowle and Atkins, 1974; Crowle, 1977; Crowle and Cline, 1977; Emmett and Crowle, 1982). He published informative texts on the topic of immunodiffusion in gel (Crowle, 1973, 1980). This uncomplicated procedure found use in the screening of rheumatic patients for specific anti-nuclear antibodies where different patterns of reactivity yielded valuable diagnostic information.

Another step improving interpretation of immunoprecipitates was the introduction of staining materials (basic fuchsin) and fixation (acid alcohol) by Björklund (1954), to enhance the visibility of immunoprecipitates. Previously, dark-field illumination for interpretation or photography was used. A year later, Sudan Black and Oil Red O were used to demonstrate certain precipitates, presumably lipoproteins (Burtin and Grabar, 1967). In 1956, a review of staining techniques was published (Uriel and Grabar, 1956; Uriel, 1964).

SUMMARY

During this time interval visual assessment of the reaction products of immunochemistry was the only means available. As a result, immunochemistry was still something of an art form.

- Reactants allowed to come into contact by diffusion in a transparent matrix gave information about the nature of the constituents by forming visible precipitates.
- Antigen and antibody reacting in this format produced infinitely variable patterns and the physical positioning of reactants could be adjusted to resolve questions.
- Analysis of these patterns could be used to understand what had previously only been seen as turbidity in the reaction vessel.

QUALITATIVE ANALYSIS – BY IMMUNOELECTROPHORESIS (1953–1978)

Immunoprecipitation in agar gel after first being separated by electrophoresis resulted in complex precipitates produced by reaction with polyspecific antiserum. The first publication of the technique was by Poulik (1952) where the combination of electrophoresis of protein was followed by diffusion of the appropriate antiserum. Traditionally the method has been ascribed to Grabar and Williams, who published their paper a year later in 1953 (Figure 5.6) without recognizing the prior publication. This paper also did not receive recognition in the excellent and encyclopedic review by Verbruggen (1975).

The new technique was called *immunoelectrophoresis* and was immediately used to study a wide variety of protein mixtures (Williams and Grabar, 1955) including allergens (Feinberg, 1957; Løwenstein, 1978), plant and microbiological proteins, and tissue extracts. The vast literature on the subject was reviewed, *in extensa*, by Verbruggen (1975). Immunoelectrophoresis became the standard for the clinical study of monoclonal immunoglobulins in human blood serum and remained so for over two decades in the microscope slide format proposed by Scheidegger (1955). Several workers attempted to extract some form of quantitative information for immunoelectrophoresis without success (Clarke and Freeman, 1967). Weime (1965) examined the use of agar and discussed the strong endosmotic flow of the material. This characteristic

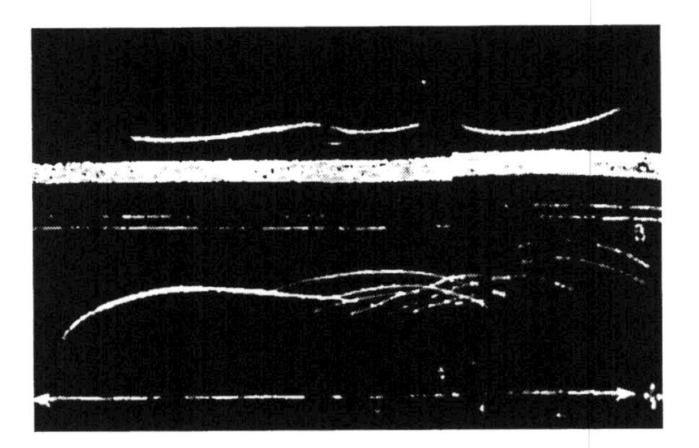

Fig. 5.6 The first published image of immunoelectrophoresis (Grabar and Williams, 1953). A polyspecific antibody has been placed in the central horizontal trough and two samples of human serum in each of the two wells above and below the center of the arcs. IgG is at the far left and albumin at the right.

interfered with the movement of proteins through the matrix despite the gel's large pore-size. Very large molecules could move as though in a free medium; however, interaction with the chemical matrix itself caused problems that would not be resolved for two decades.

The introduction of agarose, a fraction of traditional agar, followed the works of Hjerten (1961) and Brishammar *et al.* (1961). A miniaturized electrophoretic technique appeared 2 years later (Vanarkel *et al.*, 1963) and zone electrophoresis in agarose appeared about the same time (Hjerten, 1963). Further use of agarose appeared (Uriel and Avrameas, 1964) and soon after, its use for immunoelectrophoresis was published (Suellmann, 1964; Leise and Evans, 1965). Uriel (1966) proposed the use of an acrylamide–agarose mixture for electrophoresis laying the groundwork for sophisticated and high-resolution isoelectric techniques to follow later (Arnaud *et al.*, 1977). Cellulose acetate membranes were used (Marengo-Rowe and Leveson, 1972); however, the modification did not catch on.

Immunoelectrophoresis was used for close to two decades as the main method to identify and characterize monoclonal immunoglobulins for clinicians and to investigate new proteins in research. The technique, however, did require a considerable amount of experience to appreciate the nuances of the beautiful patterns. A modification of the procedure by which antiserum was layered over the electrophoretic pattern produced from a single well similar to immunoelectrophoresis began a new and more practical application (Wilson, 1964; Afonso, 1964, 1968). This technique was modified by applying antiserum over a conventional electrophoretic pattern using Laurell's technique in agarose (Jeppsson *et al.*, 1979; Alper and Johnson, 1969). After exposure to monospecific antisera, serum proteins that occurred in phenotypic patterns, and serum proteins that degraded, produced patterns that resembled a standard stained electrophoretic, except that only bands of the protein in question appeared, all others having been washed away. They coined the word **immunofixation**, which has remained in use. Our group (Ritchie and Smith, 1976a–c) described a modification of Alper's method which improved resolution and decreased the volume of antiserum required. Antiserum was applied with a strip of cellulose acetate saturated with whichever reagent was indicated. In this way several antisera could be applied to one sample electrophoresed in multiple lanes or one antiserum applied to multiple samples in adjacent lanes. The ability to interpret patterns became unambiguous since the immunofixation pattern had exactly the same proportions and configuration as the standard stained electrophoresis pattern except that the protein(s) of interest was the only band(s) visible. Some workers added the step of excising one lane for standard Amido Black staining. Subsequently immunofixation was adopted by several commercial suppliers and immunoelectrophoresis was relegated to the archives or used only for specialized research purposes by experienced workers. Immunoelectrofocussing (Catsimpoolas, 1969) was a further step towards improving resolution and characterization, and the use of antiserum overlays on polyacrylamide gels followed. Performed in plastic membranes, antigens were adsorbed in one dimension, in a slotted chamber, rotated by 90° and exposed to antibody in the same chamber, providing an efficient means of analyzing a number of antigens vs. any number of sera (Kazemi and Finkelstein, 1990; Ritchie *et al.*, 1992). The technique was called **checkerboard immunoassay**.

Interest in improving the visibility of immune precipitates in gel led to the introduction of several additives to the gels. Renn and Evans (1975) had been instrumental in developing several modified agaroses with tailored isoelectric points. During this work **brighteners** were explored, such as tannic, picric, and acetic acids. Other precipitants were tested including phosphotungstic and phosphomolybdic acids, sodium vanadate, sodium tungstate, and riboflavin. Phosphomolybdic acid was selected as being the most effective in differentiating immunoprecipitates from the background.

SUMMARY

Coincidentally with the development of more modern quantitative immunochemistry, qualitative techniques appeared aided by the application of an electric field.

- By simply applying a monospecific antiserum to an unfixed good quality serum protein electrophoresis, the immunochemical characteristics could be used to fix the reactants in place in an identical pattern. Immunoelectrophoresis was discovered.
- By using an electric field, antigen and antibody could be driven into contact thus reducing the time to test completion.
- Protein constituents of serum, tissues, and allergens could thus be separated and characterized by reaction with specific antisera.
- Staining agents enhanced the visualization of proteins.

QUANTIFICATION OF ANTIGENS BY IN-GEL IMMUNOCHEMISTRY (1963–PRESENT)

When precipitin analysis became available to clinical medicine, the focus was altered towards a more practical, and as a result, more widely used procedure, often by less skilled workers. Of paramount interest was the need for *numerical* rather than *qualitative* answers. Understanding the import of these words took several decades to be fully appreciated. To *quantitate* or to *quantify* incorporates the concept of *measurement* of a feature of an object (mass, length, hardness, color, electrophoretic mobility) or physical force (temperature, intensity, lumens, frequency, speed) familiar to the laboratorian. This is in contrast to non-numerical estimation (gender, phenotypic identity, race, taste, species). The former requires numerical representation while the latter speaks to other more imprecise descriptors that shade into judgment or artful interpretation, commonplace in clinical medicine. The requirement to derive a measurement for a parameter can only be expressed numerically, thus requiring several truths. A measurement must relate to a standard unit and device to derive that number precisely and accurately. Immunochemistry was for years denied access to measuring instruments, not because they were not available, but because they existed in separate and remote disciplines.

The observation that diffusion of some chemical reactions in gelatin produced rings of precipitate dates back to the work of Bechhold (1905, 1906). See comments above under QUALIFICATION BY DIFFUSION IN GEL. In 1932, Petrie also noted rings of precipitate when colonies of bacteria were grown on agar that contained antiserum to that species (Figure 5.3). However, these workers did not propose a quantitative relationship between the concentrations of the reactants and the diameter of the rings. Ouchterlony presented two figures in his treatise of 1958 showing 'halo reactions', noting that the time and concentration of antigen affected the size of the rings.

In a project to resolve some of the technical problems with the Oudin single-dimensional immunodiffusion method (e.g. the long time to completion and the non-linearity for large molecules), radial immunodiffusion (RID) was conceived. In 1963,[2] Heremans, Mancini, and co-workers (Mancini *et al.*, 1963, 1965) published work describing the **single-radial-immunodiffusion** method (Figure 5.7a,b). The method was immediately popular and made possible quantifying protein concentrations inexpensively, reasonably quickly (36–48 h) and even with antiserum that was of low potency.

No equipment was required other than illumination and a simple, yet accurate, measuring instrument. The method is still in use today in the developing world with commercially viable products.

When the question was asked of one of the original group that investigated ways of quantifying serum proteins in serum, 'How did it come about?', he wrote:

> Yes, I remember what led up to the Mancini method. At Joseph Heremans' lab, and with his advice, we were all trying to improve the 'Oudin' technique of specific immunological titration of antigens, which was using small glass tubes containing agar homogeneously mixed with specific antiserum against one protein. This was the 'linear immunodiffusion method'. If my memory is correct, it was initially the idea of Angelo Carbonara (now deceased) who, together with Giuliana Mancini (they were not yet married), were working with us at Heremans' lab in Louvain, in 1962; but G. Mancini devoted her Mémoire de Licence en Sciences Médicales to the subject, and obtained her scientific degree in 1964.
>
> Jean-Pierre Vaerman, co-worker, Brussels, Belgium, 2003

(a)

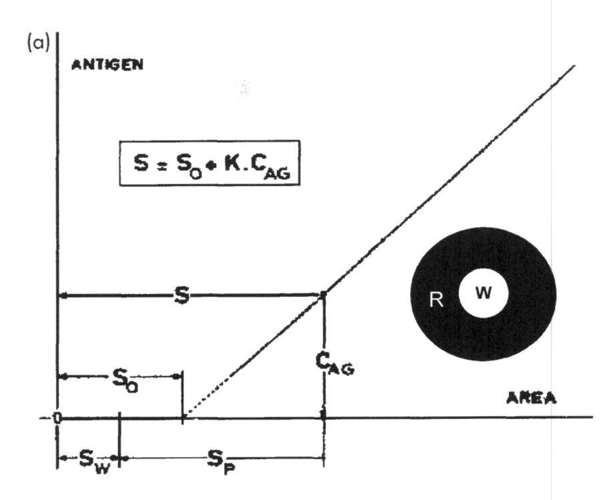

(b)

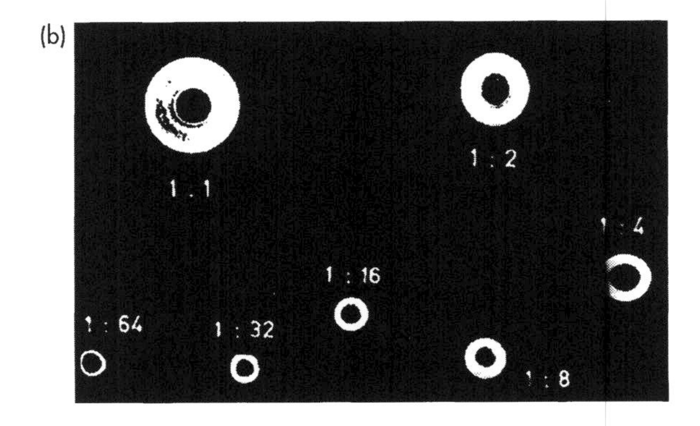

Fig. 5.7 (a) Relationship of antigen concentration (*Y*-axis) vs. diameter of precipitin ring (*X*-axis) (Mancini, 1963). (b) Image of actual unstained gel in Mancini *et al.* (1965). Dilutions of human serum albumin were added to each well as marked.

In 1965, a modification of Mancini's work was published (Fahey and McKelvey, 1965) that became the standard in the US. By shortening the time to completion of RID to less than 24 h the method was more attractive to laboratories pressed for timely results, although the precision of the shorter incubation time was less than the original method and unacceptable by today's standards. Commercial firms also adopted this modification. Fortunately, both formats can be used, short vs. long development time, to suit individual preferences.

Laurell (1965) presented a new and more sophisticated and complex method referred to as two-dimensional or crossed immunoelectrophoresis. A year later yet another simpler format for individually quantifying

[2] Through a publisher's error, the year was listed as 1964 when in fact it was the proceedings of the protides of the Biological Fluids for 1963.

many proteins was introduced, based on electroimmunodiffusion (EID). Eventually the technique would be appropriately called **rocket electrophoresis** (Laurell, 1966a,b) (Figure 5.8). This highly successful method had the advantage that antigen migrated several centimeters rather than a few millimeters as in other methods such as the recently introduced RID technique of Mancini. Like RID, EID only required a potent precipitating antiserum. This enabled any worker with good quality antiserum the ability to quantify any protein that was in reasonable concentration, about 10 mg/L, and which had an isoelectric point committing the protein to move in the electric field. Most serum proteins then identified, with the exception of the immunoglobulins, were accessible to quantitative analysis by almost any laboratory. The need for high-quality precipitating antisera was a great stimulus to manufacturers of commercial antiserum, some producing excellent material but others failing to do so, as will be discussed below. A close associate of C.-B. Laurell supplied this information for the same question of how this method came about:

'Joseph Heremans and Carl-Bertil Laurell agreed during the autumn of 1965 that there was a need for quantitative determinations of proteins as a complement to agarose electrophoresis. In early 1966 Mancini with Heremans reported about the success with the RID technique and CBL was stimulated and stressed to modify the crossed immunoelectrophoresis (for use with a) monovalent antiserum. The quantitation of alpha$_1$-antitrypsin, which up to then had usually been estimated from intensity of the electrophoretic band was successfully demonstrated only a short time before (Störiko and Schwick, 1963).

A series of diluted serum samples (3–4) was applied into punched holes in an antibody containing agarose gel. The same veronal buffer was used as before. The very first experiment showed precipitates like rockets and proportional to the antigen concentration. The first publication was in Analytical Biochemistry (Laurell, 1966b) and the findings were also presented in Brugges, May 1966 (Laurell, 1966a). A more detailed report of electroimmunoassay used for different proteins was published in a supplement in 1972 (Laurell, 1972).

Jan-Olof Jeppsson, Malmö, Sweden, 2003

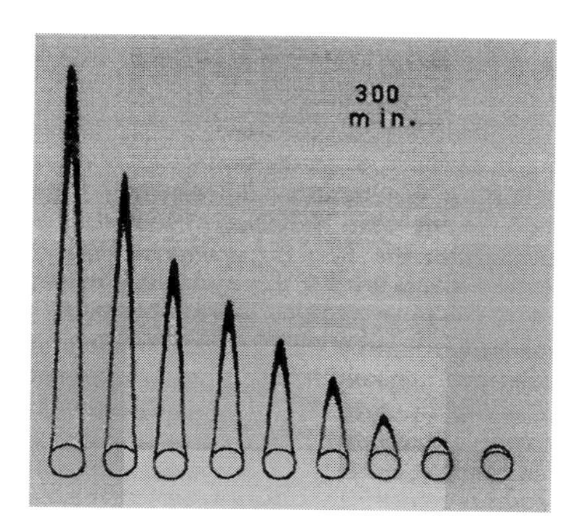

Fig. 5.8 The gel film contains a monospecific antiserum. Each well was filled with samples containing the homologous antigen. After an electric field was applied (anode above, cathode below) for a specific time, the height of the peaks are found to be proportional to the concentration of the antigen (Laurell, 1966a).

Various combinations of EID constituents were introduced to better characterize the proteins being investigated (Axelsen, 1973; Krøll, 1973; Weeke, 1973a,b).

Becker *et al.* (1968) used EID together with the photometric method (Schultze and Schwick, 1959) and RID (Mancini *et al.*, 1965) for determination of serum proteins, and prepared a table showing normal mean values, standard deviations, and normal ranges for the concentration of 25 proteins in human serum. Because EID required that the antibody move cathodally while the antigen moved anodally presented problems in using EID for assay of any protein in the gamma electrophoretic region which would not be expected to migrate at pH 8.6. While EID had been used by Laurell (1972) to quantify immunoglobulins by measuring the physical dimension of the bi-directional precipitates this was further resolved by modifying the antibody or the antigen, mostly IgG, by carbamylation before mixing with the agarose gel (Weeke, 1968; Bjerrum *et al.*, 1973). However, with the ongoing development of instrument-based immunoassays this modification did not receive a great deal of use.

SUMMARY

- Several scientific facts were combined to provide a means of easily and inexpensively measuring the concentrations of proteins in liquid solution. No complex instruments were required.
- The fact that reactants diffusing towards one another in transparent media produced visible halos, bands, and rings was known.
- It was appreciated that the position of these precipitates could be related to concentration.
- By correlating immunoprecipitin ring diameter to concentration, a simple, widely applicable method of quantifying proteins became available.
- The application of an electric field accelerated the reaction from days to hours meeting the practical needs of clinical applications.

QUANTIFICATION OF ANTIGENS BY IN-LIQUID IMMUNOCHEMISTRY (1935–PRESENT)

Measuring the amounts of a specific protein before the mid-1930s was an arduous task and carried out by investigators exclusively. Heidelberger and Kendall (1935a–c) published careful analyses of the antigen–antibody reaction using the amount of nitrogen as measured by the Kjeldahl reaction in, for example, the precipitated complexes of egg albumin and its homologous antibody. Few workers today have an appreciation for the effort expended in measuring dozens of such samples, but in 1935 there was no alternative. In 1938, Libby (1938a, b) recognized the need for more direct means of analyzing the immune reaction and published a method for determining the amount of precipitate between pneumococcal polysaccharides and the homologous antiserum using a new instrument, the Photonreflectometer. Measurement of the immunoprecipitate was given in galvanometer readings, without the intervention of secondary chemical analysis, for the first time. This instrument, basically a nephelometer, was an extension of that used to measure sample turbidity as noted above (Bloor, 1928).

The introduction of numerical quantities progressed slowly as listed above but in 1921 a Duboscq colorimeter, an early turbidimeter, was used for this purpose (Denis, 1921), to overcome the problems of non-proportional results in biological systems derived from nephelometers. The detailed work of Heidelberger and Kendall in 1935 (Figure 5.2) used laborious Kjehldal nitrogen analysis to give numerical values to their work. However, in 1938, the Photonreflectometer was used to measure the turbidity of $BaSO_4$ suspensions (Libby, 1938a, b) with answers given directly in galvanometric readings. In 1943, Boyden and DeFalco made the connection! They used the Photonreflectometer to quantitatively measure the turbidity resulting from the reaction of antiserum to lobster hemocyanin and turkey hemoglobin vs. the homologous antibody. In their work they avoided an issue that compromised many of the instruments measuring light scatter years later. They mixed their reactant in a single tube and followed the change in galvanometric reading, thus sidestepping the problems of variability induced by the differences among the reaction vessels. The effect of the concentration of reactants, nature of the antigen, pH, temperature, and reaction time was examined and it was felt that the antiserum was the greatest and most inconsistent variable. Different animals immunized in identical fashion gave reagents that reacted differently and unpredictably. Brief immunization produced more specific antisera than prolonged programs. They also noted that different antigens could show relative identity when tested against a single antiserum, which expanded on the extensive work of Landsteiner (1936). Using this principle they were able to show that species relationships could be demonstrated, sometimes even distant phyla could show a reaction of identity, albeit more weakly. They appear to be the first workers to observe the phenomenon of antigen-excess when they examined the effects of varying concentration ratios of reactants.

About the same time Pauling *et al.* (1944) applied similar experimental methods to the inhibition of precipitating systems by haptens and although instrument measurements were obtained after traditional chemical analysis, the concept of inhibition of immune precipitation was to bear commercial fruit only decades later. Further, very detailed, instrument-based investigation of the precipitin reaction by Bolton (1947) affirmed the theoretical value of quantitative methods when applied to immune reactions. He studied the reaction of pure bovine serum albumin and its homologous antibody and measured the amount of precipitate in solution with the Photonreflectometer (Libby, 1938a,b) and demonstrated that turbidity was directly proportional to the amount of precipitate in suspension. He further found that the proportions of each constituent affected the amount of precipitate. Gitlin (1949) reviewed the observations that the measured mass of purified proteins could vary as a result of the concentrations of three amino acids: phenylalanine, tyrosine, and tryptophan. Using ultraviolet spectroscopy it was shown that the dissolution of specific precipitates in sodium hydroxide or acetic acid produced results that compared well between species for purified proteins. He also applied ultraviolet absorption spectroscopy to a detailed investigation of the precipitin reaction presenting the now time-tested concept of measuring protein concentration by absorption at 280 nm (Figure 5.9).

Until 1948, most quantitative protein investigation had been at the academic level and published in the scientific literature. But in that year Kabat *et al.* (1948) published a study of albumin and gamma-globulin in patient serum and cerebrospinal fluid, in a clinical journal. There had been a previous paper approaching the topic (Goettsch and Kendall, 1935), but Kabat's paper was the first to approach a *pathological* state with immunochemical analysis confirming that antibody was in the gamma-globulin fraction. Schultze and Schwick (1959) published a paper on the quantitative analysis of nine human plasma proteins by measuring immunoprecipitates in the Heidelberger mode.

Clinical interest in quantifying plasma proteins began as early as 1948 (Kabat *et al.*, 1948; Kabat and Murray, 1950). The ability to quantitate antitoxins using early photometric devices was a forerunner of both quality assessment of antisera and of the specific antigens. Not fully recognized at that stage was the relative non-specificity of immune reagents. The non-specificity of antiserum could not be fully appreciated until the constituents of materials used to immunize animals were understood. Standing in the way was the method by which animals were immunized. See box for more information on changes in immunization practice, a crucial and often ignored issue.

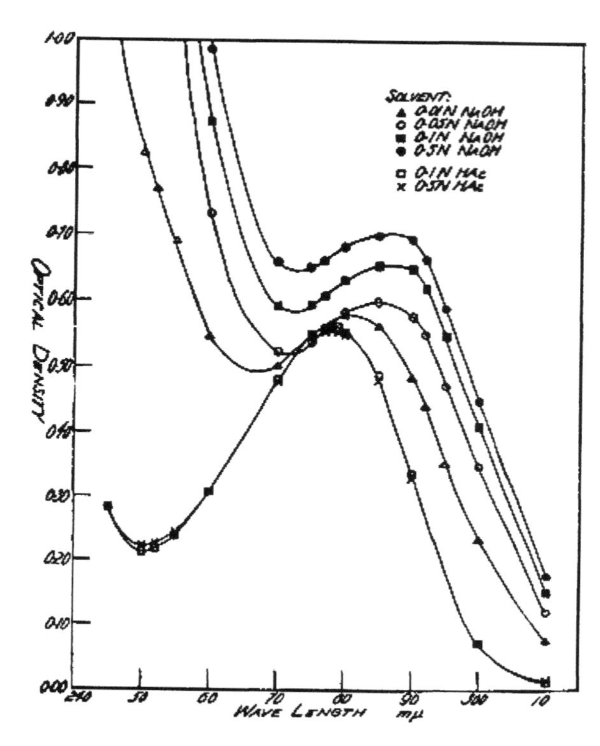

Fig. 5.9 The original Figure 2 from Gitlin (1949) illustrating the spectrophotometric spectra of crystalline human serum albumin hydrolyzed with NaOH or Hac. The ordinate is wavelength (240 nm at the left) and optical density (OD) on the abscissa. The peak value at OD 280 is the compromise wavelength commonly used today.

CHANGES IN IMMUNIZATION PRACTICE

Traditionally, immunization was effected by what we now know as truly enormous amounts of immunogen. Tens and even hundreds of milligrams of material were injected intramuscularly, intravenously, or into the footpads of rabbits. Not surprisingly, antiserum produced by this protocol contained massive quantities of irrelevant and unwanted antibodies, sometimes in concentrations that bore little relationship to the immunizing, multispecific (contaminated) antigens. In many ways the complex patterns developed by diffusion-in-gel methods (Figure 5.4c) led to the efforts to purify proteins further thus producing superior and less-expensive antisera. The development that forced major changes in antiserum production was the need for high potency, precipitating product in the mid-1960s.

At about the same time, protocols for purifying proteins with exchange resins and fractional precipitation with various salts and alcohol greatly improved the materials being used as immunogens. Hundred milligram amounts of protein were soon replaced by single milligram amounts and for certain materials even a few micrograms were sufficient in large animals. We have succeeded in making a usable rabbit antiserum with multiple intradermal injection of less than 1 μg of a rare antigen. Coupled to greater immunogen purity was the awareness that lower levels of contaminants made immunoabsorption less rigorous. Eventually solid-phase immunoadsorption became the rule making it possible to process antiserum in liter quantities with contaminant-bound columns. The subject of immunization schedules is a chapter in itself and need not concern us further here.

The use of commercial instruments to measure the immune reaction was extended to the immunoprecipitin reaction directly by Boyden and DeFalco (1943) using the Photonreflectometer. An extremely simple method using this inexpensive and commonly available turbidimeter was employed to quantify serum proteins in dilute solutions, with reasonable accuracy. They avoided the difficulty of sample concentration because of the increase in sensitivity. Using more potent antisera allowed us to measure protein concentrations as low as 0.5 μg/L in cerebrospinal fluid. While this end-point method (Ritchie, 1967) did not come into wide use, it was the forerunner of several commercial end-point systems (e.g. Behringwerke's BNA (Sieber and Gross, 1976), Hyland's Laser Nephelometer (Deaton *et al.*, 1976)). These and all end-point systems experienced the same serious problem; if the cuvette was rotated, clean or not, the results would vary unacceptably, because of variations in the cuvette's optical properties.

In 1961 an immunochemical method to determine β-lipoprotein using the immunocrit technique became available (Heiskell *et al.*, 1961). In this method serum was mixed with commercial anti-β-lipoprotein and aspirated into a small glass capillary tube to a defined height. The precipitate that resulted was centrifuged, measured, and compared against identically managed controls. Soon after, a new approach came under development by two groups, each without the knowledge of the other, and amazingly, without the knowledge of the personnel within the same company who supplied the equipment! Kahan and Sundblad (1966) adapted the immunoprecipitin reaction to the continuous flow Autoanalyzer equipment manufactured by Technicon Instruments Co. to quantify β-lipoproteins. Our group developed an almost identical system, for the analysis of several serum proteins in serum and in cerebrospinal fluid (Ritchie *et al.*, 1969, 1973; Ritchie and Graves, 1972; Ritchie and Stevens, 1972). The latter development prompted the company to introduce the Automated Immunoprecipitin (AIP) System to the market. This approach used the bubble-segmented continuous stream, introduced by Thiers and Ogelsby (1967) and Habig *et al.* (1969). In this mode a single aspirated sample is diluted to whatever ratio is desired and air bubbles are introduced producing an evenly spaced series of sub-aliquots of the original. Since flow dynamics and viscous drag quickly reduce a single unsegmented aliquot to a concentration distribution, introduction of bubbles halts further diffusion by viscous drag. Each subsegment reacts independently with identical doses of the provided antibody until just before analysis, when the bubbles are removed and immediately analyzed optically in a fixed cuvette (Figure 5.10). Those on the leading and trailing limbs of the set react appropriately (antibody-excess) while those in the center of the series may not react as expected if they are in antigen-excess (Figure 5.11). The shape of the recorded curves gives the operator a clue that the sample may have a high concentration and must be further diluted and reanalyzed. The AIP system was designed for rapid, high-volume testing (40 samples/h with total time per test of 20 min vs. 24–48 h for RID); it was automated, reasonably precise, and attracted worldwide attention. It did, however, require considerable mechanical expertise, but produced results that were significantly more precise than previous methods. Ebeling (1973) and Killingsworth *et al.* (1974) described optimization of buffers, ionic strength, enhancing additives, and other conditions for continuous flow immunoanalysis using the AIP system. In place of a conventional light source, a laser was introduced in a kinetic mode (Buffone *et al.*, 1974). The kinetic approach solved the problem of cuvette variability in previous static systems by analyzing the signal produced by a single sample's immunochemical reaction over time in a modified centrifugal analyzer. It appeared that assay performance in terms of precision was improved as their data demonstrated.

The addition of non-ionic polymers was a turning point for automated instrument-based immunoanalysis. Early attempts at light-scattering analyses were very successful for most analytes; however, certain proteins presented problems of an indolent reaction, most notably orosomucoid (α1-acid glycoprotein) and immunoglobulin M. These were both crucial to introducing viable immunochemistry to clinical laboratory programs. Hellsing and Laurent (1964) published work demonstrating that the addition of dextran to the immune reaction increased the amount of specific precipitate. In 1972, as AIP analysis was expanding to include pathologic samples and additional serum proteins, problems with the above-mentioned two analyses was of great concern. At a European Technicon Symposium in Brussels in 1972, the late Dr Kristofer Hellsing was introduced to the

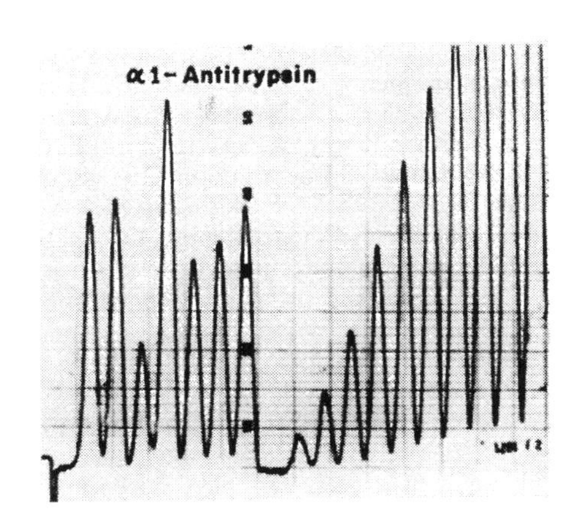

Fig. 5.10 AIP curve with unknowns to the left and a calibration curve to the right (Ritchie *et al.*, 1973).

dynamics of the AIP system. With his advice polyethylene glycol (PEG) at various percentages was added to the buffers (Hellsing, 1972). The results were dramatic and analysis of orosomucoid and IgM became as vigorous as for the other protein species. PEG M_m 6000 at 4–5% concentration was ideal (Hellsing, 1972, 1978). In theory PEG's effect could be explained by 'steric exclusion', i.e. each molecule of PEG is surrounded by a hydration-sphere which excludes other solute molecules to various degrees. Small species diffuse more deeply while large species, such as serum proteins,

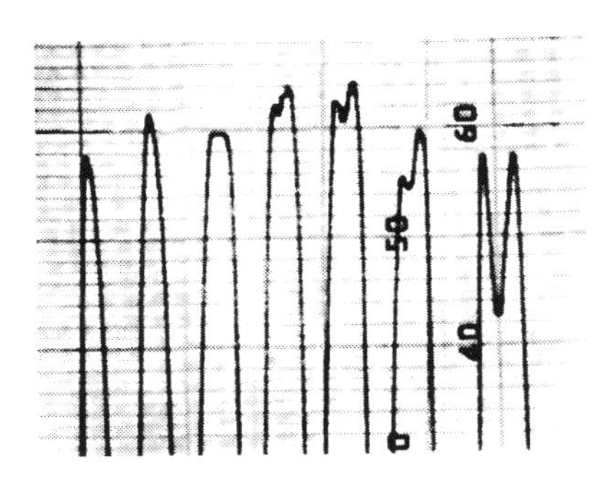

Fig. 5.11 The peak heights represent the antigen concentration of each sample except when the peak configuration becomes distorted. With further increases in concentration the peak height becomes flattened, then asymmetric, and finally it becomes forked. The explanation is that each peak represents many individual sub-samples. The early and late sub-samples contain concentrations below antigen-excess while those that become distorted are beyond the point of equivalence with the central samples in frank antigen-excess, hence diminished light-scattering ability (Ritchie *et al.*, 1973).

may be excluded completely, thus the total volume of solute exists as partitions, one freely available to salts and the other only available to proteins. As a result both antigen and antibody are forced into more intimate contact than in a PEG-free solution thus accelerating interaction.

The fact that the AIP was a large unit offered an opportunity for other manufacturers to design instruments expressly for low-volume users. By chance, the senior scientist of Beckman Instruments, Inc. and the author of the AIP system met at a symposium in 1974. As a result, after thorough research into the theory behind the method a device was designed and within 18 months the first operating rate-nephelometric system was delivered to our laboratory for a field trial. The SPA-1 (Serum Protein Analyzer-1) proved to be a remarkably precise and trouble-free instrument and was first stressed testing amniotic fluid alpha-fetoprotein. Precision was better than 5% CV, even in inexperienced hands (Haddow *et al.*, 1976). The architect of the instrument and the rate analyzer was the late James Sternberg (1977). The SPA-1 was followed by a family of ever more sophisticated devices from Beckman Instruments, Inc. and their competitors. Regrettably, the progress to more closed systems corresponded with a gradual decrease in technical expertise available in the majority of medical laboratories. At the same time requirements for greater precision and accuracy were coupled with the pressure to lower costs. For those laboratories with access to highly qualified staff, the development of cost-effective open systems has been rewarding (Hudson *et al.*, 1987; Ledue *et al.*, 2002).

All serum protein analyzers face a common and worrisome problem: antigen-excess. While it was a familiar phenomenon to the bench-workers of the past, in the routine delivery of serum protein values by automated instruments, there was no obvious evidence that such a problem existed. The AIP system presented a unique manner of demonstrating the situation. As each sample was divided into tiny aliquots by the bubble-segmentation, with the lowest concentration first, progressing to the highest concentration, and continuing to the lowest again, the possibility that the amount of serum protein antigen exceeded the antibody content was announced by the change in the shape of the recorded peak. In Figure 5.11, peak configuration becomes blunted and then bifid when antigen-excess supervenes. Keep in mind that the stream being analyzed has been held as multiple isolated segments until just before measurement. Those aliquots that are in antigen-excess have diminished amount of immunoprecipitate, hence less turbidity than expected. For systems that employ single cuvette analysis there is no detection device other than preparation of more than one dilution of each sample, as performed in an automated fashion by the AIP system. For rate analysis by turbidimetry or nephelometry, a more sophisticated approach is required (see Kusnetz and Mansberg (1978), for a complete description). It was recognized very early in the development of the Beckman analyzer that samples with high levels of an analyte would develop turbidity slowly, i.e. the rate was diminished yet usually reached high levels. However, in samples with extremely high Ag:Ab ratios, turbidity might not develop at all or be very limited. The concern was most acute with the immunoglobulins where the final result could be contrary to the actual fact, namely very high concentrations of protein recovered no value at all. The clinical concern was well placed, so an algorithm was devised to detect this. The automated rate nephelometric system included an internal dilution scheme triggered by the use of a 'gate' that instructed the instrument to insert a dilution different from the usual: if the recovered value was very low a more concentrated sample (less dilute) was submitted; for values higher, but still below a low cut-off, a more dilute sample was submitted. Eventually more sophisticated algorithms were developed that included the addition of an aliquot of antiserum and assessment of rate change.

Another interesting method is the kinetic turbidimetry developed by Metzmann (1985) and Tuengler *et al.* (1988). In this method a determination of protein concentrations is made through the simultaneous measurement on a highly automated analyzer of two (inter-related) reaction parameters: the maximum velocity (V_{max}) of the immunoprecipitation and the time required for the reaction to attain V_{max}.

Availability of such excellent devices, characterized as 'Black Box', 'Walk away', or similar marketing terms to advertise their sophistication, has allowed an increasing number of laboratories to offer high-quality serum protein analyses without the need for highly trained staff. As intimated above, the downside is that bench workers have been reduced to janitorial services as demanded by the machines. The positive aspect of course is that results are far superior than could be imagined two decades ago. A single laboratory with one technologist can produce accurate results on hundreds of tests a day and in many departments technical staff are never required to enter a result, it all having been managed by electronic data-transfer to central computing facilities. Inevitably, good comes with the bad and progress towards more automation, higher accuracy, lower per unit cost, and shorter turn-around time is taking its toll. The loss of technical expertise in the clinical laboratory is lamentable, but irreversible. The refuge of the expert technologist is now within industrial and research communities.

SUMMARY

While accurate measurement of protein concentration became available, results still required considerable time to complete. Electronics moved from crude, often home-made devices to commercially available instruments with consequent improvement in precision and standardization.

- It was known that immunochemistry in liquid yielded excellent meter-readable results; instruments needed to improve.
- Serious problems in the methods of preparing antisera became widely acknowledged and major changes in protocol were established largely by commercial endeavor.
- Several publications appeared demonstrating that instrument-based analyses of serum proteins were feasible.
- In clinical chemistry, early automation was extremely successful and made clear that immunochemistry in liquid could be designed for this application.
- The serious problem of erroneous results due to antigen-excess was appreciated and resolved operationally.
- End-point analyses were supplemented by kinetic methods. Conversion of laboratories to automated devices greatly improved performance.

QUANTIFICATION OF ANTIGENS BY PARTICLE-ENHANCED IMMUNOCHEMISTRY (1972–PRESENT)

Coincident with the development of automated immunochemical analysis, the potential for adding a layer of technical sophistication came early. During a seminar on the AIP system organized by Pierre Masson at the École de Santé Publique in University Catholique de Louvain in 1972, attended by among others, the director, Professor Joseph Heremans, I made a detailed description of how the mechanics of the device were viewed.

I observed with concern that Prof. Heremans appeared not to be listening, in fact I thought he was asleep. At the close of the presentation, Prof. Heremans came to the front of the room, fully alert, and in a rather disengaged fashion laid out his vision of the workings of the AIP system and then launched into a discussion of what became inhibition immunoassay, the inhibition of precipitation by the addition of non-precipitating antigen such as hapten.

Robert Ritchie, Scarborough, Maine, USA, 2003

Soon after, workers at the Unit of Experimental Medicine at Louvain, Belgium, were working with the theory and practice of nephelometric immunoassay and inhibition immunoassays for haptens (Riccomi *et al.*, 1972; Cambiaso *et al.*, 1974a,b). The practical application of this was grasped by Gauldie *et al.* (Gauldie *et al.*, 1975; Gauldie and Bienenstock, 1978) of McMaster University, Hamilton, Ontario, who published an abstract describing inhibition nephelometry of morphine and digoxin by this method. These workers recognized the difficulties of radioimmunoassay at the time, and applied the knowledge that antisera could be produced by immunization with antigen composed of an antigenic macromolecule conjugated to a low molecular weight non-precipitating compound. This haptenic antigen generates antibodies to the hapten, the base protein, and the complex. If the carrier protein is completely alien (e.g. horseshoe crab (*Limulus*) hemocyanin) to the species in which the hapten is to be measured (humans) it can be assumed that any reaction is due to the presence of the free hapten. By the addition of a concentration series of free hapten the light-scattering signal can be attenuated. In a similar fashion the addition of serum samples from the test species containing the same hapten attenuates the signal in proportion to the amount of hapten present. In 1978, this group described the analysis of dinitrophenol, progesterone, histamine, angiotensin, and other haptens.

The use of latex beads to enhance the sensitivity of immunoassays dates at least to the landmark paper by Singer and Plotz (1956) who observed that the serum of a co-worker had the ability to agglutinate latex particles. Further investigation by Lospalluto and Ziff (1959) provided the explanation for patients, sera agglutinating particles such as sensitized erythrocyte and latex particles. Further work (Hall *et al.*, 1958; Hall, 1961) set the stage for further enhancements of assays based upon the inhibition of particle agglutination. Building on this well-accepted diagnostic method and the awareness that inhibition by hapten was possible, the Louvain group (Cambiaso *et al.*, 1977, 1978; Masson and Holy, 1986; Masson, 1987) presented a unique approach of immunoanalysis. Using small polystyrene beads coated with antibody allowed measurement of the homologous antigen in serum samples. Called PArticle Counting ImmunoAssay (PACIA), they were able to use a particle-counter instead of a nephelometer. By the addition of diluted antibody to the surface of the particles, antigen in solution causes the particles to agglutinate, i.e. this reduces the absolute *number* of discrete particles counted. However, if the particles are coated with an antigen, i.e. non-specific IgG, rather than specific antibody, then incubated with a solution of the homologous antibody (rheumatoid factor – anti-IgG), agglutination again diminishes the absolute number of particles counted. In the former manner IgE, human placental lactogen, alpha-fetoprotein, T_3, and T_4 were effected. Later CRP was quantitated by the same technique in serum, cord serum, and cerebrospinal fluid (Collet-Cassart *et al.*, 1983). Assays for serum ferritin (Limet *et al.*, 1982), α_2-microglobulin in urine (Bernard *et al.*, 1981), theophylline (Liedtke, 1984), and pregnancy-specific α_1-glycoprotein (Fagnart *et al.*, 1985) were developed. Sensitivity was in the order of tens of micrograms/liter for serum proteins. A key issue with PACIA was the ability to decrease non-specific agglutination as a result of unwanted serum factors. The use of $F(ab')_2$ fragments for coating particles in the rheumatoid factor assay reduced non-specific agglutination as a result of patients having anti-rabbit or anti-goat antibodies. Various permutations of this system with haptens and inhibition of agglutination were successfully devised for over 100 applications. The instrument was commercially introduced, but because of non-scientific circumstances, this promising technology received only limited use.

Kapmeyer *et al.* (1988) described a particle-enhanced immunoassay that was further improved by developing 180 nm particles of novel composition. A polystyrene core surrounded a shell of chemically reactive material to which antibody could be easily bound in a more stable fashion. This method was used to develop an assay for C-reactive protein (CRP). Sensitivity was comparable to existing RID plates with a range from less than 1 to over 300 mg/L. Assay time was about 6 min. The assay for CRP with sufficient sensitivity to give precise answers in the low ranges (0.5–2.0 mg/L) spurred manufacturers to make available high-quality assays that were particle-enhanced. Ledue *et al.* (1989) compared a particle-enhanced assay to a fluorescence immunoassay and concluded that the automated particle-enhanced assay was superior in efficiency and precision to previously available methods. A CRP particle-enhanced assay for three serum proteins: CRP, serum amyloid A and mannose-binding protein was evaluated (Ledue *et al.*, 1998) and precision in a normal population was found to be equivalent to traditional chemical and colorimetric assays.

The subject was reviewed thoroughly by Galvin (1983).

SUMMARY

Available from the plastics industry, 'latex' beads of extraordinary constant size and a bewildering variety provided an investigative tool for immunochemists. Modification of the constitution of the beads, which measured only a few microns, also made the attachment of reactive elements routine. A large step forward was taken in the quest for increased sensitivity and precision.

- Polystyrene particles coated with either antibody or antigen provided a powerful tool for immunochemical analyses.
- Several instruments, including those used previously for traditional light-scattering analyses, were successful in lowering sensitivity by at least an order of magnitude to micrograms of antigen (or antibody) per liter.
- A wide array of low-level analytes of interest became easily and precisely analyzable.

LABELED IMMUNOASSAY (1959–PRESENT)

Progress towards immunoassays employing antibodies conjugated with ligands of a wide variety of materials such as enzymes, radioactive nuclides, and fluorescent tags is covered elsewhere in this book and is summarized *in extensa* in PRINCIPLES OF COMPETITIVE AND IMMUNOMETRIC ASSAYS by Davies, and will not be pursued to any degree here. As with all advances we stand on the shoulders of those before us. For these labeled immunoassays we have the pioneering work of the following scientists to thank for what we now take for granted:

- Avrameas and Uriel (1966): antibodies labeled with enzymes, Avrameas and Guilbert (1972): peroxidase-labeled antibodies, Avrameas (1976): method review
- Blanchard and Gardener (1982): solid-phase FIA of IgE
- Coons and Kaplan (1950): fluorescent-tagged antibody
- Coons (1956): fluorescent antibody and histochemistry
- Dandliker and de Saussure (1970): theory of polarization fluorescence assay, Dandliker *et al.* (1973): polarization fluorescence
- Masseyeff *et al.* (1973): rapid enzyme labeling of antibody, Masseyeff *et al.* (1981): immunoenzymatic analysis of immune complexes
- Nakane and Pierce (1967): localizing tissue antigens with enzyme-labeled antibody, Nakane (1971): peroxidase-labeled antibody
- Voller *et al.* (1974): Microplate ELISA method to detect malaria, Voller and Bidwell (1976): ELISA detection of viral infection, Voller *et al.* (1978): review
- Wang *et al.* (1980): immunoglobulins
- Yalow and Berson (1959, 1960): RIA assay of insulin.

In those days before grants-in-aid, sophisticated instruments and reagents, and the Internet with its powerful search engines, we can only be thankful that those gaps in knowledge were filled with intellectual brilliance, persistence and vision.

SUMMARY

The need for improved sensitivity to encompass many of the clinically important analytes prompted many workers to seek ways of modifying antibodies to provide higher detectability. Powerful contributions from radiochemistry, enzymology, and fluorochemistry provided these tools.

Instrument engineering has kept pace with changing needs and the scope of methods and materials seems to be endless.

COMPARATIVE STUDIES (1907–PRESENT)

Early work in immunochemistry, before the time when accuracy and precision were issues of paramount concern, proceeded in relative isolation. Each worker performed assays as best they could with what equipment and reagents were obtainable or could be made. The concept of quality assurance was not to develop fully in the developed world until the 1980s. As a result comparative studies for reference intervals, clinical cut-off points, and the like were not comparable. With the development of devices that could perform large numbers of assays under nearly identical conditions, this matter became addressable. Results from widely dispersed laboratories had to be harmonized for clinical utility. This was a desirable goal, but expensive. It was not until programs were introduced that required satisfactory performance in government mandated exercises for continuation of licensure to operate that the full value of automated testing was appreciated. Immunochemistry testing was early into the mandated evaluation.

By the 1970s, the availability of several systems of protein immunoassay raised the concern for reagent qualification. Implicit in several of the older papers was the comparison of one product with another and the semi-quantification of materials such as vaccines (McFarland, 1907; Libby, 1938a,b; Boyden and Defalco, 1943; Becker, 1969). As interest in automated immunochemistry was building, a group from the Centers for Disease Control (CDC) and Prevention published a comparison of commercial antisera (Phillips *et al.*, 1971). Using a panel of 20 different preparations of human serum IgA, the 13 major suppliers of anti-IgA were analyzed which showed at least a sevenfold difference in titer. Findings such as lack of specific antibody, presence of interclass specificity, other contaminant antibodies and low titer (one had none) by the CDC group resulted in a fairly rapid improvement in the quality of commercial antisera. A year later our concern for unrecognized contaminants in reagents used in the AIP system prompted a method to demonstrate these impurities (Ritchie and Stevens, 1972). Perhaps the most effective means of improving reagent antibody was the institution in the late 1970s by the College of American Pathologists of the Quality Assurance program for serum proteins. Manufacturers that supplied kits for these assays found themselves publicly compared with their competitors, sometimes with embarrassing results. To assist with this upgrading of products came a method of qualifying each batch of reagents in a system that performed in close parallel with others of the time (Hudson *et al.*, 1981, 1983). Some manufacturers still use this protocol while others have devised other, but equally demanding protocols. Hand in hand with improved antiserum requirements was the unavoidable realization that there were no easily available and authoritative reference materials. This last link in the harmonization of laboratories was to be the product of the European Union in the form of the Bureau de Communitaire de Référence (BCR), in Brussels. The European community, heavy users of serum protein assays, instituted a program to prepare and distribute a universally accepted and highly studied material, Certified Reference Preparation 470 (CRM 470), also known as Reference Preparation for Human Serum Proteins (RPPHS) in the US, distributed by the College of American Pathologists (Baudner *et al.*, 1992; Whicher *et al.*, 1994; Blirup-Jensen *et al.*, 1993, 2001; Blirup-Jensen, 2001a–c; Dati *et al.*, 1996). Crucial to the generation of these types of reference materials are immunoassays that can perform with a high degree of precision. Several highly automated instrument packages were used in this value transfer exercise (Johnson *et al.*, 1996) and to the surprise of all involved, immunoassays performed as well as, or better than, traditional chemistries. Precision with CVs of less than 5% were the norm.

SUMMARY

From the very earliest studies, workers were inevitably faced with differences in the quality and concentrations of their reagents. When improved instruments became available the effort to compare products, especially those on which assays depended, was a meaningful exercise.

- As a result of various reports demonstrating that product labels could be very misleading, product quality improved tremendously.
- Hand in hand with reagent verification and quality assurance came the need for a globally recognized reference material for human serum proteins. This was achieved.

ACKNOWLEDGEMENTS

I am indebted to our librarian, Linda Talamo, who has been tremendously helpful in finding and obtaining the necessary documents, many of which fall into the category of ancient archives. I am also indebted to colleagues, most of whom 'were there', for providing important insights that have faded from most of our memories, namely, Soren Blirup-Jensen, Cesar Cambiaso, Francesco Dati, Jan-Olof Jeppsson, A. Myron Johnson, Pierre Masson, Thomas

Ledue, Miroslav (David) Poulik, Jean-Pierre Vaerman, John Whicher. Lastly, I am grateful for the opportunity to assemble this narrative, which encompasses the better part of many of our lives, a privilege I attribute to David Wild.

REFERENCES

Afonso, E. Quantitative immunoelectrophoresis of serum proteins. *Clin. Chim. Acta* **10**, 114–122 (1964).

Afonso, E. Brief study of a myeloma protein. *Clin. Chim. Acta* **2**, 283–287 (1968).

Alper, C.A. and Johnson, A.M. Immunofixation electrophoresis: a technique for the study of protein polymorphism. *Vox Sang.* **17**, 445–452 (1969).

Anderson, N.L. and Anderson, N.G. The human plasma proteome: history, character, and diagnostic prospects. *Mol. Cell. Proteomics* **1.11**, 845–867 (2002).

Arnaud, P., Wilson, G.B., Koistinen, J. and Fudenberg, H.H. Immunofixation after electrofocusing: improved method for specific detection of serum proteins with determination of isoelectric points. I. Immunofixation print technique for detection of alpha-1-protease inhibitor. *J. Immunol. Methods* **16**, 221–231 (1977).

Avrameas, S. Immunoenzymic techniques for biomedical analysis. *Methods Enzymol.* **44**, 709–717 (1976).

Avrameas, S. and Guilbert, B. Enzyme-immunoassay for the measurement of antigens using peroxidase conjugates. *Biochimie* **54**, 837–842 (1972).

Avrameas, S. and Uriel, J. Méthode de marquage d'antigènes et d'anticorps avec les enzymes et son application en immunodiffusion. *C. R. Acad. Sci. Paris* **262**, 2543–3545 (1966).

Axelsen, N.H. Intermediate gel in crossed and in fused rocket immunoelectrophoresis. *Scand. J. Immunol. Suppl.* **1**, 71–77 (1973).

Baudner, S., Bienvenu, J., Blirup-Jensen, S., Carlström, A., Johnson, A.M., Milford-Ward, A., Ritchie, R.F., Svendsen, P.J. and Whicher, J. *Certification of a Matrix Reference Material for Immunochemical Measurement of 14 Human Serum Proteins*, CRM 470 (Bureau Communitaire de Reference, XII/268/93-EN, Brussels, Belgium, 1992).

Bechhold, H. Strukturbildung in gallerten. *Z. Phys. Chem.* **52**, 185–199 (1905).

Bechhold, H. and Ziegler, J. Die beeinflussbarkeit der diffusion in gallerten. *Z. Phys. Chem.* **56**, 105–121 (1906).

Becker, W. Determination of antisera titres using the single radial immunodiffusion method. *Immunochemistry* **6**, 539–546 (1969).

Becker, W., Rapp, W., Schwick, H.G. and Störiko, K. Methoden zur quantitativen bestimmung von plasmaproteinen durch immunpräzipitation. *Z. Klin. Chem. Klin. Biochem.* **6**, 113–122 (1968).

Bernard, A.M., Vyskocil, A. and Lauwerys, R.R. Determination of beta 2-microglobulin in human urine and serum by latex immunoassay. *Clin. Chem.* **27**, 832–837 (1981).

Bjerrum, O.J., Ingild, A., Løwenstein, H. and Weeke, B. Quantitation of human IgG by rocket immunoelectrophoresis at pH 5 by use of carbamylated antibodies. A routine laboratory method. *Clin. Chim. Acta* **46**, 337–343 (1973).

Björklund, B. Qualitative analysis of gel precipitates with the aid of chemical color reactions. *Proc. Soc. Exp. Biol.* **85**, 438–441 (1954).

Blanchard, G.C. and Gardner, R.E. A solid-phase fluoroimmunoassay for human IgE. *J. Immunol. Methods* **52**, 81–90 (1982).

Blirup-Jensen, S. Protein standardization I: protein purification procedure for the purification of human prealbumin, orosomucoid and transferrin as primary protein preparations. *Clin. Chem. Lab. Med.* **39**, 1076–1089 (2001a).

Blirup-Jensen, S. Protein standardization II: dry mass determination procedure for the determination of the dry mass of a pure protein preparation. *Clin. Chem. Lab. Med.* **39**, 1090–1097 (2001b).

Blirup-Jensen, S. Protein standardization III: method optimization basic principles for quantitative determination of human serum proteins on automated instruments based on turbidimetry or nephelometry. *Clin. Chem. Lab. Med.* **39**, 1098–1099 (2001c).

Blirup-Jensen, S., Carlström, A., Svendsen, P.J., Baudner, S., Bienvenu, J., Johnson, A.M., Ritchie, R.F., Ward, A.M. and Whicher, J.T. Ett nytt unternationellt referensmaterial för proteiner i humanserum. *Klin. Kem. Nord.* **5**, 9–12 (1993).

Blirup-Jensen, S., Johnson, A.M. and Larsen, M. Protein standardization IV: value transfer procedure for the assignment of serum protein values from a reference preparation to a target material. *Clin. Chem. Lab. Med.* **39**, 1110–1122 (2001).

Bloor, W.R. A method for the determination of fat in small amounts of blood. *J. Biol. Chem.* **17**, 377–384 (1914).

Bloor, W.R. The determination of small amounts of lipid in blood plasma. *J. Biol. Chem.* **77**, 53–73 (1928).

Bolton, E.T. Precipitin testing and its three-dimensional expression. *J. Immunol.* **57**, 391–394 (1947).

Bordet, J. Les sérum hémolytiques, leurs antitoxines et les thérories des sérums cytolotiques. *Ann. Inst. Pasteur* **14**, 257–296 (1900).

Boyden, A. and DeFalco, R.J. Report on the use of the Photonreflectometer in serological comparisons. *Physiol. Zool.* **16**, 229–241 (1943).

Brishammar, S., Hjerten, S. and von Hofsten, B. Immunological precipitates in agarose gels. *Biochim. Biophys. Acta* **53**, 518–521 (1961).

Buffone, G.J., Savory, J. and Cross, R.E. Use of a laser-equipped centrifugal analyzer for kinetic measurement of serum IgG. *Clin. Chem.* **20**, 1320–1323 (1974).

Burtin, P. and Grabar, P. In: *High Resolution Techniques in Electrophoresis – Theory, Methods and Applications* (ed Bier, M.), 110–156 (Academic Press, New York, 1967).

Calmette, A. and Massol, L. Les précipitines du sérum antivenimeux vis-à-vis du venin du cobra. *Ann. Inst. Pasteur* **23**, 155–165 (1909).

Cambiaso, C.L., Masson, P.L., Vaerman, J.P. and Heremans, J.F. Automated nephelometric immunoassay (ANIA) I. Importance of antibody affinity. *J. Immunol. Methods* **5**, 153–163 (1974a).

Cambiaso, C.L., Riccomi, H.A., Masson, P.L. and Heremans, J.F. Automated nephelometric immunoassay II. Its application to the determination of hapten. *J. Immunol. Methods* **5**, 293–302 (1974b).

Cambiaso, C.L., Leek, A.E., de Steenwinkel, F., Billen, J. and Masson, P.L. Particle counting immunoassay (PACIA). I. A general method for the determination of antibodies, antigens and haptens. *J. Immunol. Methods* **18**, 33–44 (1977).

Cambiaso, C.L., Riccomi, H., Sindic, C. and Masson, P.L. Particle counting immunoassay (PACIA). II. Automated determination of circulating immune complexes by inhibition of the agglutinating activity of rheumatoid sera. *J. Immunol. Methods* **23**, 29–50 (1978).

Catsimpoolas, N. Immunoelectrophoresis in agarose gels. *Clin. Chim. Acta* **23**, 237–238 (1969).

Clarke, H.G.M. and Freeman, T.A. A quantitative immunoelectrophoresis method (Laurell electrophoresis). In: *Protides Biol Fluids 14th Colloquium*, 503–509 (Pergamon Press, New York, 1967).

Collet-Cassart, D., Mareschal, J.C., Sindic, C.J., Tomasi, J.P. and Masson, P.L. Automated particle-counting immunoassay of C-reactive protein and its application to serum, cord serum and cerebrospinal fluid samples. *Clin. Chem.* **29**, 1127–1131 (1983).

Coons, A.H. Histochemistry with labeled antibody. *Int. Rev. Cytol.* **5**, 1–23 (1956).

Coons, A.H. and Kaplan, M.H. Localization of antigen in tissue cells II. Improvements in a method for detection of antigen by means of fluorescent antibody. *J. Exp. Med.* **91**, 1–13 (1950).

Crowle, A.J. Templates for antiserum application in immunoelectrophoresis and two-dimensional electroimmunodiffusion. *J. Immunol. Methods* **14**, 197–200 (1977).

Crowle, A.J. Precipitin and microprecipitin reactions in fluid medium and gels. In: *Manual of Clinical Laboratory Immunology*, 2nd edn (eds Rose, N.R. and Friedman, H.), (American Society for Microbiology, Washington, DC, 1980).

Crowle, A.J. and Cline, L.J. An improved stain for immunodiffusion tests. *J. Immunol. Methods* **17**, 379–381 (1977).

Crowle, A.J., Atkins, A.A. and Revis, G.J. Reversed electroimmunodiffusion for quantitating precipitins. *J. Immunol. Methods* **4**, 173–188 (1974).

Dandliker, W.B. and de Saussure, V.A. Fluorescence polarization in immunochemistry. *Immunochemistry* **7**, 799–828 (1970).

Dandliker, W.B., Kelly, R.J., Dandliker, J., Farquahar, J. and Levin, J. Fluorescence polarization immunoassay. Theory and experimental method. *Immunochemistry* **10**, 219–227 (1973).

Dati, F., Schumann, G., Thomas, L., Aguzzi, F., Baudner, S., Bienvenu, J., Blaabjerg, O., Blirup-Jensen, S., Carlström, A., Hyltoft-Petersen, P., Johnson, A.M., Milford-Ward, A., Ritchie, R.F., Svendsen, P.J. and Whicher, J. Consensus of a group of professional societies and diagnostic companies on guidelines for interim reference ranges for 14 proteins in serum based on the standardization against the IFCC/BCR/CAP reference material (CRM 470). *Eur. J. Clin. Chem. Clin. Biochem.* **34**, 517–520 (1996).

Dean, H.R. and Webb, R.A. The influence of optimal proportions of antigen and antibody in the serum precipitation reaction. *J. Pathol. Bacteriol.* **29**, 473–492 (1926).

Deaton, C.D., Maxwell, K.W., Smith, R.S. and Creveling, R.L. Use of laser nephelometry in the measurement of serum proteins. *Clin. Chem.* **22**, 1465–1471 (1976).

Denis, W. On the substitution of turbidimetry for nephelometry in certain biochemical analysis. *J. Biol. Chem.* **47**, 27–31 (1921).

Ebeling, H. Eine optimierte automatische methode zur quantitativen immunologischen antigen- und antikörperbestimmung. *Z. Klin. Chem. Klin. Biochem.* **11**, 209–214 (1973).

Ehrlich, P. Anemia. In: *Special Pathology and Therapy*, vol. 8 (ed Nothnagel, H.), 163–185 (1901).

Ehrlich, P., http://www.nobel.se/medicine/laureates/1908/index.html (2004).

Emmett, M. and Crowle, A.J. Crossed immunoelectrophoresis: qualitative and quantitative considerations. *J. Immunol. Methods* **50**, R65–R83 (1982).

Fagnart, O.C., Cambiaso, C.L., Sindic, C.J. and Masson, P.L. Particle-counting immunoassay of a fetuin-like antigen in serum and cerebrospinal fluid. *Clin. Chem.* **31**, 1820–1823 (1985).

Fahey, J.L. and McKelvey, E.M. Quantitative determination of serum immunoglobulins in antibody-agar plates. *J. Immunol.* **94**, 84–90 (1965).

Feinberg, J.G. Identification, discrimination and quantification in Ouchterlony gel plates. *Int. Arch. Allergy* **11**, 129–152 (1957).

Friedman, L. Diffusion of non-electrolytes into gelatin gels. *J. Am. Chem. Soc.* **52**, 1305–1310 (1930).

Galvin, J.P. Particle enhanced immunoassays – a review. In: *Diagnostic Immunology. Technology Assessment and Quality Assurance* (eds Rippy, J.H. and Nakamura, R.M.), 18–30 (College of American Pathologists, Skokie, IL, 1983).

Gauldie, J. and Bienenstock, J. Automated nephelometric analysis of haptens. In: *Automated Immunoanalysis* (ed Ritchie, R.F.), 321–333 (Marcel Dekker, New York, 1978).

Gauldie, J., Sherington, E. and Sircar, P.K. Automated nephelometric inhibition immunoassay of digoxin and morphine. *Fed. Proc. Soc.* **34**, 1047 (1975).

Gitlin, D. Use of ultraviolet spectroscopy in the quantitative precipitin reaction. *J. Immunol.* **62**, 437–451 (1949).

Goettsch, E. and Kendall, F.E. Analysis of albumin and globulin in biological fluids by the quantitative precipitin method. *J. Biol. Chem.* **109**, 221–231 (1935).

Grabar, P. and Williams, C.A. Méthode permettant l'étude conjugée des propriétés électrophorétique et immunochimiques d'un mélange de protéines; application au sérum sanguin. *Biochem. Biophys. Acta* **10**, 193–194 (1953).

Graham, T. Liquid diffusion applied to analysis. *Phil. Trans. R. Soc.* **151**, 183–224 (1861).

Habig, R.L., Schlein, B.W., Walters, L. and Thiers, R.E. A bubble-gating flow cell for continuous-flow analysis. *Clin. Chem.* **15**, 1045–1055 (1969).

Haddow, J.E., Macri, J.N., Munson, M.E., Baldwin, P. and Ritchie, R.F. Second trimester normal amniotic fluid protein levels as determined in the automated immunoprecipitin system. In: *Advances in Automated Analysis*, 270–273 (Technicon International Congress, New York, 1976).

Hall, A.P. Serologic tests in rheumatoid arthritis. *Med. Clin. North Am.* **45**, 1181–1196 (1961).

Hall, A.P., Mednis, A.D. and Bayles, T.B. The latex agglutination and inhibition reactions. Clinical experiences in the diagnosis of rheumatoid arthritis. *N. Engl. J. Med.* **258**, 731–735 (1958).

Heidelberger, M. and Kendall, F.E. A quantitative theory of the precipitin reaction II. A study of an azoprotein–antibody system. *J. Exp. Med.* **62**, 467–483 (1935a).

Heidelberger, M. and Kendall, F.E. The precipitin reaction between Type III pneumococcus polysaccharide and its homologous antibody. III. *J. Exp. Med.* **62**, 563–591 (1935b).

Heidelberger, M. and Kendall, F.E. A quantitative theory of the precipitin reaction. III. The reaction between crystalline egg albumin and its homologous antibody. *J. Exp. Med.* **62**, 697–720 (1935c).

Heiskell, C.L., Fisk, R.T., Florsheim, W.H., Tachi, A., Goodman, J.R. and Carpenter, C.M. A simple method for quantitation of serum beta-lipoproteins by means of the immunocrit. *Am. J. Clin. Pathol.* **35**, 222–226 (1961).

Hellsing, K. Influence of polymers on the antigen–antibody reaction in a continuous flow system. In: *Automated Immunoprecipitin Reactions. New Methods, Techniques and Evaluations* (ed Hamm, J.D.), 17–20 (Technicon Instruments Corp, Tarrytown, NY, 1972).

Hellsing, K. Enhancing effects of nonionic polymers on immunochemical reactions. In: *Automated Immunoanalysis* (ed Ritchie, R.F.), 67–112 (Marcel Dekker, New York, 1978).

Hellsing, K. and Laurent, T.C. The influence of dextran on the precipitin reaction. *Acta Chem. Scand.* **18**, 1303–1304 (1964).

Hjerten, S. Agarose as an anticonvection agent in zone electrophoresis. *Biochim. Biophys. Acta* **53**, 514–517 (1961).

Hjerten, S. Zone electrophoresis in columns of agarose suspensions. *J. Chromatogr.* **12**, 510–526 (1963).

Hooker, S.B. and Boyd, W.C. A formulation of the serological flocculation rate in the region of considerable antibody excess. *J. Gen. Physiol.* **19**, 373–378 (1935).

Hudson, G.A., Ritchie, R.F. and Haddow, J.E. Method for testing antiserum titer and avidity in nephelometric systems. *Clin. Chem.* **27**, 1838–1844 (1981).

Hudson, G.A., Haddow, J.E. and Ritchie, R.F. Disparity in antiserum avidity as measured by nephelometry, radial immunodiffusion and column-binding strength. *Clin. Chim. Acta* **130**, 239–244 (1983).

Hudson, G.A., Poulin, S.E. and Ritchie, R.F. Twelve-protein immunoassay profile on the Cobas FARA. *J. Clin. Lab. Anal.* **1**, 191–197 (1987).

Jeppsson, J.O., Laurell, C.-B. and Franzen, B. Agarose gel electrophoresis. *Clin. Chem.* **25**, 629–638 (1979).

Johnson, A.M., Sampson, E.J., Blirup-Jensen, S. and Svendsen, P.J. Recommendations for the selection and use of protocols for assignment of values to reference materials. *Eur. J. Clin. Chem. Clin. Biochem.* **34**, 279–285 (1996).

Kabat, E.A. and Murray, J.P. Comparison of human-globulins in their reactivity with rabbit anti-globulin by a quantitative precipitin method. *J. Biol. Chem.* **182**, 251–260 (1950).

Kabat, E.A., Glusman, M. and Knaub, V. Quantitative estimation of the albumin and gamma globulin in normal and pathologic cerebrospinal fluid by immunochemical methods. *Am. J. Med.* **4**, 653–662 (1948).

Kahan, J. and Sundblad, L. Automated immunochemical determination of β-lipoproteins. In: *Automation in Analytical Chemistry*, 361–364 (Technicon International Symposium, Paris, 1966).

Kapmeyer, W.H., Pauly, H.-E. and Tuengler, P. Automated nephelometric immunoassays with novel shell/core particles. *J. Clin. Lab. Anal.* **2**, 276–283 (1988).

Kazemi, M. and Finkelstein, R.A. Checkerboard immunoblotting (CBIB): an efficient, rapid and sensitive method of assaying multiple antigen/antibody cross-reactivities. *J. Immunol. Methods* **128**, 143–146 (1990).

Killingsworth, L.M., Buffone, G.J., Sonawane, M.B. and Lunsford, G.C. Optimizing nephelometric measurement of specific serum proteins: evaluation of three diluents. *Clin. Chem.* **20**, 1548–1552 (1974).

Kraus, R. Ueber specifische reaktionen in keimfreien filtraten aus cholera-, thyphus- und pestbouillonculturen, erzeugt durch homologes serum. *Wein. Klin. Wochenschr.* **10**, 736–738 (1897).

Krøll, J. Rocket-line immunoelectrophoresis. *Scand. J. Immunol.* **1** (Suppl 1), 83–87 (1973).

Kusnetz, J. and Mansberg, H.P. Optical considerations: nephelometry. In: *Automated Immunoanalysis* (ed Ritchie, R.F.), 1–43 (Marcel Dekker, New York, NY, 1978).

Landsteiner, K. *The Specificity of Serological Reactions* (Charles C. Thomas, Springfield, IL, 1936).

Laurell, C.-B. Antigen–antibody crossed electrophoresis. *Anal. Biochem.* **10**, 358–361 (1965).

Laurell, C.-B. Quantitative estimation of proteins by electrophoresis in agarose gel containing antibodies. *Anal. Biochem.* **15**, 45–52 (1966a).

Laurell, C.-B. Quantitative estimation of proteins by electrophoresis in antibody-containing agarose gel. In: *Protides of the Biological Fluids*, vol. 14, 499–502 (Elsevier, Amsterdam, 1966).

Laurell, C.-B. Electroimmunoassay. *Scand. J. Clin. Lab.* **29** (Suppl.), 21–37 (1972).

Ledue, T.B., Poulin, S.E., Leavitt, L.F. and Johnson, A.M. Evaluation of a particle-enhance immunoassay for quantifying C-reactive protein. *Clin. Chem.* **35**, 2001–2002 (1989).

Ledue, T.B., Weiner, D.L., Sipe, J.D., Poulin, S.E., Collins, M.F. and Rifai, N. Analytical evaluation of particle-enhanced immunonephelometric assays for C-reactive protein, serum amyloid A and mannose-binding protein in human serum. *Ann. Clin. Biochem.* **35**, 745–753 (1998).

Ledue, T.B., Collins, M.F. and Ritchie, R.F. Development of immunoturbidometric assays for 14 human serum proteins on the Hitachi 912™. *Clin. Chem. Lab. Anal.* **40**, 520–528 (2002).

Leise, E.M. and Evans, C.G. Immunoelectrophoretic analysis of serum proteins in agar and in agarose. *Proc. Soc. Exp. Biol. Med.* **120**, 310–313 (1965).

Libby, R.L. The Photonreflectometer – an instrument for the measurement of turbid systems. *J. Immunol.* **34**, 71–73 (1938a).

Libby, R.L. A new and rapid quantitative technic for the determination of the potency of types I and II antipneumococcal serum. *J. Immunol.* **34**, 269–279 (1938b).

Liedtke, R.J. Particle-enhanced turbidimetric inhibition immunoassay of theophylline adapted to the Cobas-Bio centrifugal analyzer. *Clin. Chem.* **30**, 1274–1275 (1984).

Liesegang, R. Spezielle methoden der diffusion in gallerten. In: *Handbuch der Biologischen Arbeitsmethoden* (ed Abderhalden, E.), 33 (IIIB Urban & Schwarzenberg, Berlin, 1929).

Limet, J.N., Collet-Cassart, D., Magnusson, C.G., Sauvage, P., Cambiaso, C.L. and Masson, P.L. Particle counting immunoassay (PACIA) of ferritin. *J. Clin. Chem. Clin. Biochem.* **20**, 141–146 (1982).

Lospalluto, J. and Ziff, M. Chromatographic studies of the rheumatoid factor. *J. Exp. Med.* **110**, 169–186 (1959).

Løwenstein, H. Quantitative immunoelectrophoretic methods as a tool for the analysis and isolation of allergens. *Prog. Allergy* **25**, 1–62 (1978).

Mancini, G., Vaerman, J.P., Carbonara, A.O. and Heremans, J.F. A single-radial-immunodiffusion method for the immunological quantitation of proteins. In: *Protides of the Biological Fluids, Proceedings of the 11th Colloquium, Bruges* (ed Peeters, H.), 370–373 (Elsevier, New York, 1964).

Mancini, G., Carbonara, A.O. and Heremans, J.F. Immunochemical quantitation of antigens by single radial immunodiffusion. *Immunochemistry* **2**, 235–254 (1965).

Marengo-Rowe, A.J. and Leveson, J.E. Immunoelectrophoresis of serum proteins on cellulose acetate. *Am. J. Clin. Pathol.* **58**, 300–302 (1972).

Masseyeff, R., Maiolini, R. and Bouron, Y. A method of enzyme immunoassay. *Biomedicine* **19**, 314–317 (1973).

Masseyeff, R., Maiolini, R., Roda, L. and Ferrua, B. Enzyme immunoassay for the demonstration and quantification of immune complexes. *Methods Enzymol.* **74**, 608–616 (1981).

Masson, P. Particle counting immunoassay – an overview. *J. Pharm. Biomed. Anal.* **5**, 113–117 (1987).

Masson, P.L. and Holy, H.W. Immunoassay by particle counting. In: *Manual of Clinical Laboratory Immunology*, 3rd edn, (eds Rose, N.R., Friedman, H. and Fahey, J.L.), (American Society for Microbiology, Washington, DC, 1986).

McFarland, J. The nephelometer: an instrument for estimating the number of bacteria in suspensions used for calculating the opsonic index and for vaccines. *JAMA* **69**, 1176–1178 (1907).

Metzmann, E. Protein quantitation on both branches of the Heidelberger curve by monitoring the kinetic of immunoprecipitation. *Behring Inst. Mitt.* **78**, 167–175 (1985).

Nakane, P.K. Peroxidase-labeled antibody method. *Adv. Biol. Skin* **11**, 283–286 (1971).

Nakane, P.K. and Pierce, G.B. Enzyme-labeled antibodies for the light and electron microscopic localization of tissue antigens. *J. Cell Biol.* **33**, 307–318 (1967).

Nicolle, M., Césari, E. and Debains, E. Etudes sur la précipitation mutuelles des anticorps et des antigènes. *Ann. Inst. Pasteur* **34**, 596–599 (1920).

Ouchterlony, Ő. In vitro method for testing the toxin-producing capacity of diptheria bacteria. *Acta Pathol. Microbiol. Scand.* **26**, 516–524 (1948).

Ouchterlony, Ő. Antigen–antibody reactions in gels II. Factors determining the site of the precipitate. *Ark. Kemi* **1**, 43–48 (1949a).

Ouchterlony, Ő. Antigen–antibody reactions in gels III. The time factor. *Ark. Kemi* **1**, 55–59 (1949b).

Ouchterlony, Ő. Diffusion-in-gel methods for immunological analysis. *Prog. Allergy* **5**, 1–78 (1958).

Oudin, J.C. Méthode d'analyse immunochimique par précipitation spécifique en milieux gélifié. *R. Acad. Sci.* **222**, 115–116 (1946).

Oudin, J.C. L'analyse immunochimique qualitative; méthode par diffusion des antigènes au sein de l'immunsérum précipitant gélosé. Première parties. *Ann. Inst. Pasteur* **75**, 30–52 (1948).

Oudin, J.C. La diffusion d'un antigène dans une colonne de gel contenant les anticorps précipitants homologues. Études quantitative des trois principales variables. *C. R. Acad. Sci.* **228**, 1890–1892 (1949).

Pauling, L., Pressman, D. and Grossberg, A.L. The serological properties of simple substances. VII. A quantitative theory of the inhibition by haptens of the precipitation of heterologous antisera with antigens, and comparison with experimental results for polyhaptenic simple substances and for azoproteins. *J. Am. Chem. Soc.* **66**, 784–792 (1944).

Petrie, G.F. A specific precipitin reaction associated with the growth on agar plates of meningococcus, pneumococcus and B. dysenteriæ (Shiga). *Br. J. Exp. Pathol.* **13**, 380–394 (1932).

Phillips, D.J., Shore, S.L., Maddison, S.E., Gordon, D.S. and Reimer, C.B. Comparative evaluation of commercial precipitating antisera against human IgA. *J. Lab. Clin. Med.* **77**, 639–644 (1971).

Poulik, M.D. Filter paper electrophoresis of purified diphtheria toxoid. *Can. J. Med. Sci.* **30**, 417–419 (1952).

Ramon, G. Flocculation dans une mélange neutre de toxine–antitoxine diphtérique. *C. R. Soc. Biol. (Paris)* **86**, 661–663 (1922).

Renn, D. and Evans, E. Use of heteropoly acids as immunological brighteners. *Anal. Biochem.* **64**, 620–623 (1975).

Riccomi, H., Masson, P.L., Vaerman, J.-P. and Heremans, J.F. An automated nephelometric inhibition immunoassay (NINIA) for haptens. In: *Automated Immunoprecipitin Reactions* (ed Hamm, J.H.), 9–11 (Technicon Instruments Corporation, Tarrytown, NY, 1972).

Ritchie, R.F. A simple, direct and sensitive technique for the measurement of specific proteins in cerebrospinal fluid. *J. Lab. Clin. Med.* **70**, 512–517 (1967).

Ritchie, R.F. and Graves, J. *Automated Analysis of Cerebrospinal Fluid Proteins* (Technicon International Congress, New York, 1972), 117–121.

Ritchie, R.F. and Smith, R. Immunofixation I. General principles and application to agarose gel electrophoresis. *Clin. Chem.* **22**, 497–499 (1976a).

Ritchie, R.F. and Smith, R. Immunofixation II. Application to typing of alpha$_1$-antitrypsin at acid pH. *Clin. Chem.* **22**, 1735–1737 (1976b).

Ritchie, R.F. and Smith, R. Immunofixation III. Application to the study of monoclonal proteins. *Clin. Chem.* **22**, 1982–1985 (1976c).

Ritchie, R.F. and Stevens, J. *Qualifications for Acceptable Antiserum Performance in the Automated Immunoprecipitin System: A Brief Review of Commercially Available Reagents*, 9–14 (Technicon International Congress, New York, 1972).

Ritchie, R.F., Alper, C.A. and Graves, J.A. Experience with a fully automated system for immunoassay of specific proteins. *Arthritis Rheum.* **12**, 693 (1969).

Ritchie, R.F., Alper, C.A., Graves, J., Pearson, N. and Larson, C. Automated quantitation of proteins in serum and other biological fluids. *Am. J. Clin. Pathol.* **59**, 151–159 (1973).

Ritchie, R.F., Collins, M. and Ledue, T.B. A flexible, efficient, checkerboard immunoblot system for the detection and semiquantitation of specific antinuclear antibodies. *J. Clin. Lab. Anal.* **6**, 391–398 (1992).

Scheidegger, J.J. Une micro-méthode de l'immunoélectrophorèse. *Int. Arch. Allergy* **7**, 103–110 (1955).

Schultze, H.E. and Schwick, G. Quantitative immunologische bestimmung von plasmaproteinen. *Clin. Chim. Acta* **4**, 15–25 (1959).

Sia, R.H. and Chung, S.F. Use of antipneumococcus-serum-agar for the identification of pneumococcal types. *Proc. Soc. Exp. Biol. Med.* **29**, 762–795 (1932).

Sieber, A. and Gross, J. Protein-bestimmung durch laser-nephelometrie. *Laborblätter* **26**, 117–123 (1976).

Singer, J.M. and Plotz, C.M. Latex fixation test. I. Application to serological diagnosis of rheumatoid arthritis. *Am. J. Med.* **21**, 888–892 (1956).

Sternberg, J.C. A rate nephelometer for measuring specific proteins by immunoprecipitin reaction. *Clin. Chem.* **23**, 1456–1464 (1977).

Stiles, W. The penetration of electrolytes into gels. I. The penetration of sodium chloride into gels of agar-agar containing silver nitrate. *Biochem. J.* **15**, 58–72 (1920).

Stiles, W. The penetration of electrolytes into gels. V. The diffusion of a mixture of chlorides into gels. *Biochem. J.* **17**, 530–534 (1923).

Störiko, K. and Schwick, G. Die quantitative immunologische bestimmung des α1-antitrypsins im menschlichen serum. In: *Protides of the Biological Fluids, Proceedings of the 11th Colloquium*, Bruges, (ed Peeters, H.), 411–414 (Elsevier, New York, 1963).

Suellmann, H. Immunoelectrophoresis with agarose gel as a carrier medium. *Clin. Chim. Acta* **10**, 569–571 (1964).

Surgalla, M.J., Bergdoll, M.S. and Dack, G.M. Use of antigen–antibody reactions in agar to follow the progress of fractionation of antigenic mixtures: application to purification of Staphylococcal enterotoxin. *J. Immunol.* **69**, 357–365 (1952).

Thiers, R.E. and Oglesby, K.M. Kinetic parameters of continuous flow analysis. *Clin. Chem.* **13**, 451–467 (1967).

Tuengler, P., Metzmann, E., Pauly, H.E. and Becker, W. New immunodiagnostic systems. *Behring Inst. Mitt.* **82**, 282–308 (1988).

Uhlenhuth, T. Demonstration der lactoserumreaction. *Dtsch. Med. Wochenschr.* **29**, 39 (1903).

Uriel, J. Les réaction de charactérization des constituants protéiques après électrophorèse ou immuno-électrophorèse en gélose. In: *Immunoélectrophorètique Analysis* (eds Grabar, P. and Burtin, P.), 33–57 (Elsevier, Amsterdam, 1964).

Uriel, J. Method of electrophoresis in acrylamide–agarose gels. *Bull. Soc. Chim. Biol. (Paris)* **48**, 969–982 (1966).

Uriel, J. and Avrameas, S. Demonstration of pancreatic hydrolase after agarose electrophoresis and immunoelectrophoresis. *Ann. Inst. Pasteur (Paris)* **106**, 396–407 (1964).

Uriel, J. and Grabar, P. Emploi de colorants dans l'analyse électrophorétique et immuno-électrophorétique en milieu gélifié. *Ann. Inst. Pasteur* **90**, 427–440 (1956).

Vanarkel, C., Ballieux, R.E. and Jordan, F.L. Micro-electrophoresis of mucopolysaccharides on agarose gel. *J. Chromatogr.* **11**, 421–423 (1963).

Verbruggen, R. Quantitative immunoelectrophoresis methods: a literature survey. *Clin. Chem.* **21**, 5–43 (1975).

Voller, A. and Bidwell, D.E. Enzyme-immunoassays for antibodies in measles, cytomegalovirus infections and after rubella vaccination. *Br. J. Exp. Pathol.* **57**, 243–247 (1976).

Voller, A., Voller, A., Bidwell, D., Huldt, G. and Engvall, E. A microplate method of enzyme-linked immunosorbent assay and its application to malaria. *Bull. WHO* **51**, 209–211 (1974).

Voller, A., Bartlett, A. and Bidwell, D.E. Enzyme immunoassays with special reference to ELISA techniques. *J. Clin. Pathol.* **31**, 507–520 (1978).

von Behring, E. *Einfuhrung in die Lehre von der Bekampfung von Infektionskrankheiten* (Hirschwald, Berlin, 1912).

von Behring, E., http://www.nobel.se/medicine/laureates/1901/index.html (Nobel Foundation, 2001).

Wadsworth, C. Comparative testing of a new photographic material for rapid registration of immunoprecipitates. Review. *Int. Arch. Allergy* **23**, 103–114 (1963).

Wang, R., Merrill, B. and Maggio, E.T. A simplified solid-phase immunoassay for measurement of serum immunoglobulins. *Clin. Chim. Acta* **28**, 169–177 (1980).

Weeke, B. Quantitation estimation of human immunoglobulins following carbamylation in antibody-containing agarose. *Scand. J. Clin. Lab. Invest.* **22**, 107–111 (1968).

Weeke, B. A manual of quantitative immunoelectrophoresis. Methods and applications. 1. General remarks on principle, equipment, reagents and procedures. *Scand. J. Immunol. Suppl.* **1**, 15–35 (1973a).

Weeke, B. Crossed immunoelectrophoresis. *Scand. J. Immunol. Suppl.* **1**, 47–56 (1973b).

Weime, R.J. *Agar Gel Electrophoresis* (Elsevier, Amsterdam, 1965).

Whicher, J.T., Ritchie, R.F., Johnson, A.M., Baudner, S., Bienvenu, J., Blirup-Jensen, S., Carlström, A., Dati, F., Milford-Ward, A. and Svendsen, P.J. A new international reference preparation for proteins in human serum (RPPHS). *Clin. Chem.* **41**, 1–5 (1994).

Williams, C.A. and Grabar, P. Immunoelectrophoretic studies on human serum proteins I. The antigens of human serum. *J. Immunol.* **74**, 158–168 (1955).

Wilson, A.T. Direct immunoelectrophoresis. *J. Immunol.* **92**, 431–434 (1964).

Yalow, R.S. and Berson, S.A. Assay of plasma insulin in human subjects by immunochemical methods. *Nature* **184**, 1648–1649 (1959).

Yalow, R.S. and Berson, S.A. Immunoassay of endogenous plasma insulin in man. *J. Clin. Invest.* **39**, 1157–1175 (1960).

FURTHER READING

Reviews

Johnson, A.M. Immunoprecipitation in gels. In: *Manual of Clinical Laboratory Immunology* 3rd edn, (eds Rose, N.R., Friedman, H. and Fahey, J.L.), (American Society for Microbiology, Washington, DC, 1987).

Johnson, A.M. Nephelometric immunoassay. *J. Pharm. Biomed. Anal.* **5**, 803–809 (1987).

Laurell, C.B. Electrophoresis, specific protein assays, or both in measurement of plasma proteins? *Clin. Chem.* **19**, 99–102 (1973).

Price, C.P., Spencer, K. and Whicher, J. Light-scattering immunoassay of specific proteins: a review. *Ann. Clin. Biochem.* **20**, 1–14 (1983).

Ritchie, R.F. New developments in light scattering immunoassay methods. In: *Clinical Laboratory Assays: New Technology and Future Directions* (eds Nakamura, R.M., Dito, W.R. and Tucker, E.S.), 1–15 (Masson Publishing, New York, 1983).

Books

Axelsen, N.H (ed) *Handbook of Immunoprecipitation-in-Gel Techniques*. *Scand. J. Immunol.* **17** (Suppl 10) (1983).

Axelson, N.H., Krøll, J. and Weeke, B (eds), *A Manual of Quantitative Immunoelectrophoresis: Methods and Applications*, *Scand. J. Immunol.* (Suppl 1) (1973).

Crowle, A.J. (ed), *Immunodiffusion* 2nd edn, (Academic Press, New York, 1973).

Marquardt, M. (ed), *Paul Ehrlich* (Henry Schuman, New York, 1951).

Wieme, R.J. (ed), *Agar Electrophoresis* (Elsevier, Amsterdam, 1965).

6 Concepts

Chris Davies

ASSAY CONCEPTS

ASSAY SENSITIVITY

Assay design determinants of sensitivity have been discussed previously (*see* PRINCIPLES – IMMUNOASSAY DESIGN). The concept of sensitivity is one of probability – at what concentration would one be uncertain that the sample was not actually zero? The classical statistical approach would be to assay low concentration samples and the zero calibrator repeatedly and perform a *t*-test of the null hypothesis that the sample result is not statistically different from zero. However, in this test the 'sensitivity' would decrease according to the number of replicates run, since the standard error is inversely proportional to the square root of the number of determinations. Clearly then, this test is unhelpful in estimating the inherent assay sensitivity.

The most common approach in practice is, therefore, to assay the zero standard repeatedly and define the limit of sensitivity as being the concentration corresponding to two or three standard deviations above the mean. This equates to the probability that a single replicate of sample does not form part of the distribution of zero analyte values. This is termed the **analytical sensitivity**.

Values at this limit are not necessarily reliable. The precision at this concentration may after all be considerably poorer than would be judged acceptable. There is one further, but not widely appreciated, problem with the estimation of sensitivity. Invariably the sensitivity lies below the concentration of the first analyte-containing calibrator, in some cases by an order of magnitude. One assumes, for the purpose of calculating analytical sensitivity, that the fitted calibration curve is a close reflection of the true dose–response curve below that calibrator. Such may not be the case, as shown in Figure 6.1. In such a case the calculated sensitivity (S_1) does not reflect the true sensitivity (S_2).

It is mainly for these reasons that the concept of **functional sensitivity** was introduced, which is of far more practical value. Functional sensitivity is defined as the lowest concentration in the assay for which the coefficient of variation is less than, say, 20%. (The figure of 20% is an arbitrary one and may be unsuitable for some applications.) The concept has, therefore, changed from being one of hypothesis testing to one of estimation. Functional sensitivity is derived from knowledge of the changes in precision over the assay range: the **precision profile**, the derivation of which is explained later. Functional sensitivity is often, but not invariably, at a higher concentration than analytical sensitivity.

The derivation of functional sensitivity is more complex than that of analytical sensitivity. Where the method of determination is not stated in the literature or manufacturer's package insert, it is more likely that the analytical sensitivity is being quoted than the functional sensitivity.

ACCURACY AND BIAS

Accuracy, when applied to the field of measurement, has a different meaning from the normal English usage, i.e. being of little error. In analytical measurements accuracy is defined as how close the *average* measured value is to the true value. **Bias** is a measure of the difference between the measured and the true value – in short

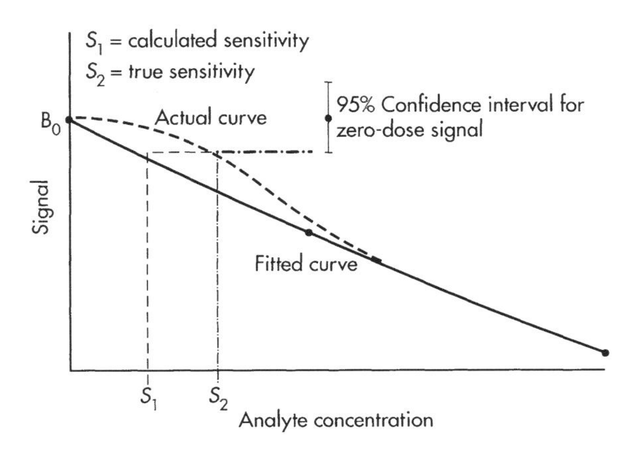

Fig. 6.1 Assay sensitivity.

the degree of inaccuracy. Bias may be **proportional**, where the assay reads a constant percentage higher or lower than the true value, or **constant**, where the assay reads a constant concentration higher or lower. Both types of bias can occur together, and therefore, the overall bias may vary over the assay range (see Figure 6.2a,b).

Although simple in concept, the interpretation of accuracy and bias can be difficult. For many analytes reference methods do not exist that are capable of measuring the true value. In such cases accuracy and bias may be relatively meaningless concepts and bias is usually assessed in such situations by comparison with the concensus results of other assay methodologies – a concept with obvious pitfalls. Recovery provides the most direct assessment of analytical accuracy where the analyte can be obtained in a pure homogeneous form, but the technique may be associated with other problems of interpretation (*see* RECOVERY).

PRECISION AND IMPRECISION

Precision describes the repeatability of an analytical technique. **Imprecision** is the opposite of precision. It is an estimate of the error in an analytical technique, expressed as the percentage coefficient of variation (%CV) or, less often, as the standard deviation (SD) at a particular analyte level. 'High precision' has a similar meaning to 'low imprecision'. The use of the term imprecision has been widely recommended rather than precision, on the grounds that it is imprecision that is actually measured (by %CV or SD). However, similar arguments have not yet been put forward for the use of (in)sensitivity, and the term 'precision' is widely accepted. In this book precision is used most often, to reflect common practice, but imprecision is also used selectively, where it improves the clarity of the text.

Assay imprecision is caused by the combined effects of several sources of variation, principally those determined by the antibody characteristics, separation, detection, and manipulation errors.

Effect of Antibody on Precision

Precision is much improved if true equilibrium conditions have been attained by the end of the assay. Higher antibody affinity and a faster rate of reaction make the achievement of equilibrium more likely at a given time and temperature.

Variation in the following parameters causes imprecision because they each affect the rate of the antibody: antigen binding reaction. Also, if incorrectly optimized for the assay, they can slow down the rate of reaction. They are

- antibody concentration
- temperature, and time:temperature profile
- the pH of the final reaction mixture
- the ionic strength of the final reaction mixture
- variation in effective antibody coating density if a solid-phase adsorbed antibody is used.

Incubation/Instrumentation Variations

Problems can arise where the incubation temperatures are poorly controlled or uneven. For example, there may be slight variations in temperature across the different positions in an analyzer incubator causing differences in binding at the end of the assay incubation. Variation in sample or reagent temperature prior to the assay may alter the time–temperature profile of an assay in an automated analyzer. The so-called **edge effects** may also present problems with microwell assays. This is due to differential heating at the edges of the plate and necessitates careful validation of incubators to ensure that heat distribution is evenly applied. Imprecision may also occur in solid-phase particle assays because of incomplete suspension of

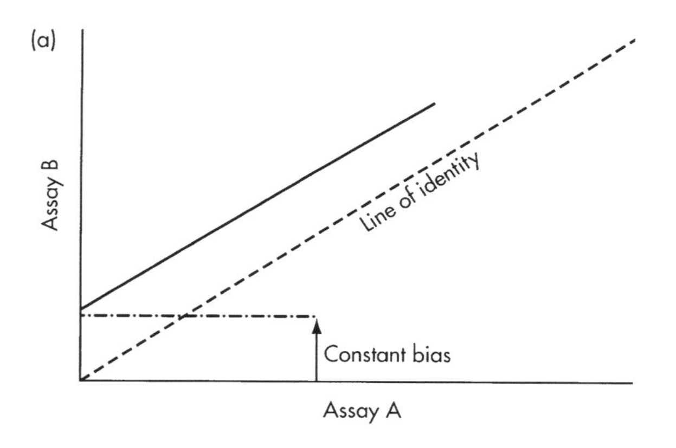

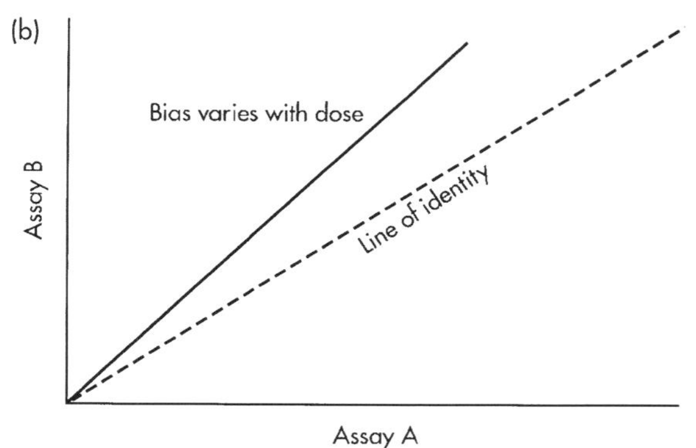

Fig. 6.2 (a) Constant bias. (b) Proportional bias.

the particles before use or differences in settling of the solid phase during incubation.

Effect of Separation on Precision

The free and bound fractions should be separated completely with the minimum of misclassification errors, and without perturbation of the specific binding reaction. An effective separation of unbound material from particles or wells requires a series of steps including washing and the processes involved (e.g. magnetic separation, resuspension, aspiration) may suffer from mechanical, fluidic or pressure fluctuations. Systems that handle solutions and particles can suffer from temporary blockages.

Effect of Detection Errors on Precision

Counting errors using radioactive tracers can be significant, particularly at low count rates. In radioactive detection the standard deviation is equal to the square root of the accumulated counts. The accumulation of 10,000 counts in each tube reduces the counting error to 1%.

Tritium and carbon-14 labels, which require scintillation counting, suffer from another source of imprecision, due to counting efficiency, although this may be partially reduced by the use of external or internal standardization. The use of glass tubes with varying wall thickness can add a further 1–2% imprecision in assays using iodine-125 as label, and should be avoided.

Multiwell gamma counters are commonly used for counting radioactive assays in order to increase throughput, and this may provide an additional source of imprecision if each counter well is not accurately calibrated and regularly checked for contamination. Similar problems can occur in microtiter plate readers which measure all wells through multiple fiber-optic transmission systems, and regular checks and maintenance are required to ensure that such readers are kept in good working order.

Optical systems can suffer from interferences that are sample-related, such as turbidity or the presence of fluorophores. Also lamps and detectors can deteriorate over a period of time. Reagents involved with signal generation may also lose activity over time, decreasing the signal-to-noise ratio.

Effect of Manipulation Errors on Precision

In addition to errors caused by poor pipetting, imprecision can be introduced by errors in the timing of additions. Metering probes can become blocked by salts. The most likely time for severe metering errors on an immunoassay analyzer is after a period of inactivity.

Precision Hierarchy

There are several hierarchical estimates of precision.

Within-run Precision

Within-run precision is defined as the precision of the same sample run on several occasions within the same assay. Twenty replicates of sample should be included and these should be spaced equidistantly throughout the assay rather than sequentially. The data may need to be assessed for outliers, preferably with a formal statistical technique such as that described by Healy (1979).

Between-run Precision

Between-run precision is an index of the ability of the assay to reproduce the same result on the same sample from run to run and from day to day. Where samples are routinely run as singleton determinations the between-run precision is due to a combination of the errors within and between assays. Where samples are run as replicates, precision is usually assessed using the mean results. Although this provides a meaningful estimate it is important to recognize that the mean result is inherently more precise. This is why, paradoxically, the between-run CV may be lower than the within-run CV if samples are replicated.

Between-lot Precision

Between-lot precision is an estimate of the variability of results using a variety of different lots of reagents.

Within-method, Within-method Group and All-Laboratory Precision

Within-method, within-method group and all-laboratory precision are estimates of precision derived by external quality assessment (proficiency testing) schemes for control samples run in many laboratories; only the within-method category is of particular relevance in assessing the precision of an individual assay method.

Combining Estimates of Precision

Precision estimates from discrete assays may be combined from several sources by using the **root mean square** (**RMS**) CV.

$$\text{RMS CV} = \sqrt{\frac{\text{CV}_1^2 N_1 + \text{CV}_2^2 N_2 + \cdots + \text{CV}_n^2 N_n}{N_1 + N_2 + \cdots + N_n}} \quad (6.1)$$

where CV is the coefficient of variation from individual assays 1 to n and N is the number of data points in each assay 1 to n.

Minimal Distinguishable Difference in Concentration

Knowledge of the precision of an immunoassay is important because it enables an assessment to be made of the probability that a given concentration differs from a specified value. It defines the **minimal distinguishable difference in concentration** (**MDDC**).

For example, consider an assay result for maternal serum alphafetoprotein (AFP) at 50 IU/mL, with a precision of 10% CV at this level. At what higher level would we be confident that the value was different from 50 IU/mL? At 55 IU/mL (one standard deviation higher than the mean) there is an 84% probability that the value differs from 50 IU/mL. At two standard deviations from the mean, 60 IU/mL, the probability is 97.7%. Other probabilities can easily be derived using a table of *Z*-scores (the *Z*-score is the number of standard deviations away from a given value) for the normal distribution. Note that a one-tailed statistical test is appropriate in this instance. For a 95% probability the *Z*-score is 1.645. For the example above, the MDDC at a 95% confidence level would be at 58.2 IU/mL. Sadler *et al.* (1992) have described the use of MDDCs in assessing the performance of assays for thyrotropin (TSH).

The MDDC provides a direct conceptual link between the technical and clinical performance of an assay, and is a useful method for relating analytical goals to clinical need.

The most basic concept in evaluating any assay is whether or not it gives the 'correct' result and it is unfortunate that there is no single scientific term which can be used to describe this 'correctness'. An ideal assay that gives perfectly 'correct' results would be one that gives the same result in samples containing the same level of analyte:

- from different patients;
- in different assay runs;
- in different laboratories;
- on different occasions;
- using different lots of reagents.

Perhaps it would have been better if the term accuracy had been used according to its common English usage, rather than specifically to describe bias. For an assay to give the correct result consistently it must be both precise and unbiased. An assay may be accurate yet imprecise, precise yet inaccurate, or both inaccurate and imprecise. In Figure 6.3, assay A is accurate but imprecise, assay B precise but inaccurate, and assay C imprecise and inaccurate. Paradoxically, individual results from assay B are more likely to be closer to the true value than assay A, despite the fact that assay A is formally defined as being the more accurate.

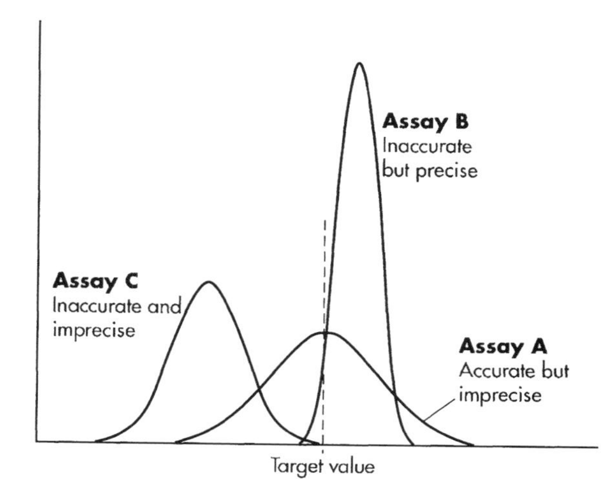

Fig. 6.3 Precision and bias.

Precision Profile

Assessment of precision in an analytical technique requires the measurement of variability at discrete concentrations using defined control solutions. There are several problems associated with such an approach. First, the precision is assessed at discrete levels of analyte; precision between these discrete intervals must be interpolated. Second, commercially available control solutions are prepared using processed blood products (usually delipidized and defibrinated plasma, with preservatives added). The matrix is, therefore, standardized and may show different matrix effects from routine samples. Even control sera made in-house from routine specimens suffer from the fact that pooling minimizes matrix differences between samples.

The ideal solution is, therefore, to estimate precision at many concentrations using routine samples. In this way the precision can be plotted across a wide range of concentrations to give the **precision profile** of the assay (Ekins, 1983).

The precision profile is calculated from the differences between duplicate measurements, grouped in sets of similar concentrations.

$$\mathrm{CV} = \frac{\left(\sqrt{\dfrac{d_1^2 + d_2^2 + \cdots + d_n^2}{N}}\right) \times 100}{\text{Mean concentration of samples in bin}} \qquad (6.2)$$

where d is the difference between duplicates for pairs 1 to n, and N the number of pairs of data points. The standard deviation divided by the mean of a group of values, and multiplied by 100, gives the %CV. Because the precision varies with concentration, only data within the same concentration range should be pooled.

A better method, and one that is very simple to calculate using a spreadsheet, is shown in Table 6.1 using, as an example, an assay for unconjugated estriol in maternal serum in early gestation.

The duplicate results are tabulated in columns a and b and the mean of the two results calculated in column c. In columns d and e the individual results are calculated as a percentage of the mean result. The mean sample results are sorted into ascending order and groups of >20 samples identified. The standard deviation for each group of 40 duplicate measurements in columns d and e is the CV at that mean concentration. The precision profile is then plotted against the mean result for the group sample mean, and the best-fit line drawn through the points, either subjectively or by fitting a suitable polynomial regression equation. Many immunoassay curve-fitting

Table 6.1. Calculation of precision profile.

Column A	Column B	Column C	Column D (%)	Column E (%)	Column F	Column G
0.57	0.66	0.62	93	107		
0.57	0.66	0.62	93	107		
0.64	0.64	0.64	100	100		
0.65	0.64	0.65	101	99		
0.58	0.73	0.66	89	111		
0.66	0.70	0.68	97	103		
0.70	0.70	0.70	100	100		
0.68	0.73	0.71	96	104		
0.74	0.71	0.73	102	98		
0.69	0.79	0.74	93	107		
0.74	0.75	0.75	99	101		
0.74	0.75	0.75	99	101		
0.70	0.81	0.76	93	107		
0.72	0.79	0.76	95	105		
0.80	0.74	0.77	104	96		
0.87	0.68	0.78	112	88		
0.85	0.77	0.81	105	95		
0.74	0.88	0.81	91	109		
0.84	0.87	0.86	98	102	Mean	CV
0.83	0.91	0.87	95	105	0.731	5.9%
0.91	0.90	0.91	101	99		
0.97	0.89	0.93	104	96		
0.96	0.96	0.96	100	100		
0.82	1.15	0.99	83	117		
1.11	1.07	1.09	102	98		
1.03	1.21	1.12	92	108		
1.27	1.10	1.19	107	93		
1.25	1.13	1.19	105	95		
1.17	1.25	1.21	97	103		
1.24	1.20	1.22	102	98		
1.22	1.29	1.26	97	103		
1.22	1.45	1.34	91	109		
1.36	1.51	1.44	95	105		
1.45	1.43	1.44	101	99		
1.43	1.48	1.46	98	102		
1.38	1.54	1.46	95	105		
1.48	1.48	1.48	100	100		
1.66	1.56	1.61	103	97		
1.72	1.62	1.67	103	97	Mean	CV
1.75	1.77	1.76	99	101	1.285	5.6%

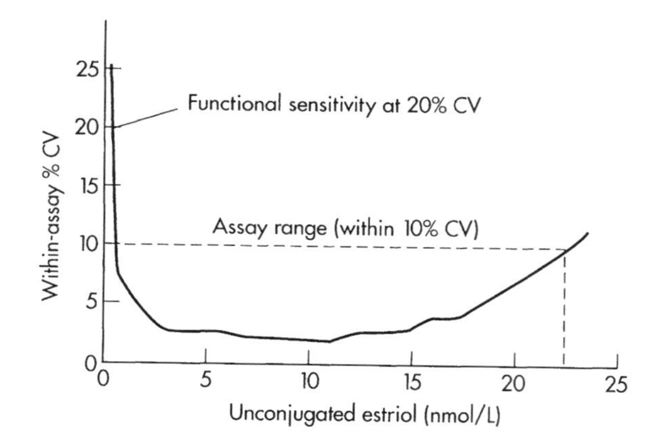

Fig. 6.4 Precision profile.

programs contain subroutines to calculate precision profiles from the assay data and some combine and compare these with previous data.

The precision profile contains useful information on the working range of the assay. A line can be drawn at a suitably acceptable CV, for example, 10%, and the intersection of the line with the fitted precision profile gives the concentration limits between which the assay precision is judged acceptable. If sufficient low-concentration samples are analyzed, information can be accumulated to permit estimation of the functional sensitivity.

A typical precision profile is shown in Figure 6.4.

One problem with conventional precision profile calculations is that the presence of just one outlying sample with a large spread in duplicates can grossly inflate the apparent CV for that particular set of binned data. Raggart (1989) has written an excellent review on the influence of such gross errors and how they can be dealt with.

A further variation of the technique described above, easily tabulated using a spreadsheet, is to calculate the mean and CV for the first 'bin' of, say, 20 samples, and then to copy the formulae down the entire data set, as shown in Table 6.2. This gives a moving average over the concentration range. One advantage of using the moving average method is that gross outliers are easily identified by the sudden increase in apparent precision when the outlying sample moves into the bin range and a sudden fall when it leaves it. Ideally, formal outlier tests should be employed on the binned data to check for their presence. One simple way is to use Healy's (1979) test to examine the ranked percentage differences from the mean result using the whole data set. This is usually successful in detecting gross outliers (see Figure 6.5).

It should be stressed that estimates of precision are themselves subject to sampling error, and an accurate assessment of the precision profile can only be achieved with sufficient numbers of samples. Sadler and Smith (1990) have reviewed some of the pitfalls associated with the calculation and interpretation of precision profiles and described a method for their derivation, together with the associated 95% confidence intervals.

In terms of clinical utility the precision should be optimal at the clinical decision cut-off points. The MDDC is then at its lowest at these values, and hence gives the least errors in clinical classification.

CROSS-REACTIVITY

Assay specificity describes the ability of an antibody to produce a measurable response only for the analyte of interest. **Cross-reactivity** is a measurement of antibody response to substances other than the analyte (see Figure 6.6). As it is the unique specificity of antibodies that renders them so useful in analytical techniques it is not surprising that the analysis of specificity and cross-reactivity should form an important part of the evaluation of any immunoanalytical technique. Many proteins have closely related structures with highly conserved epitopes, whereas steroid hormones have such closely related structures that evaluation of antibody cross-reactivity is inevitably the first stage in assay design. Similar problems occur in differentiating between drugs and their metabolites.

Cross-reactivity may be examined and expressed in several ways. In competitive assays the most common approach is to spike a pure sample of the cross-reacting substance into an analyte-free matrix to give a suitably wide range of concentrations. The cross-reactivity is often defined at the point where the reduction in signal corresponds to 50% of the signal achieved in the absence of analyte (B/B_0 of 50%), as a percentage of the analyte concentration giving the same fall in signal.

$$\%\text{ cross-reactivity} = \frac{\text{Concentration of analyte giving } 50\%\ B/B_0}{\text{Concentration of cross-reactant giving } 50\%\ B/B_0} = \frac{S1}{S2} \times 100\% \quad (6.3)$$

It is important to recognize, however, that the percentage cross-reactivity may vary across the assay range and this is particularly true when polyclonal antibodies are used. In such cases the observed cross-reactivity reflects the cross-reactivity of the particular antibody clone having most influence in the binding reaction at that analyte concentration. At low analyte concentrations this is predominantly due to high-affinity antibodies; at high concentrations the weaker-affinity clones have the most influence. One useful technique in the development of polyclonal steroid assays is to add deliberately a small quantity of the cross-reactive steroid in order to 'swamp' a minor but particularly cross-reactive antibody clone.

Table 6.2. Calculation of precision profile by moving averages.

Column A	Column B	Column C	Column D (%)	Column E (%)	Column F	Column G
0.57	0.66	0.62	93	107		
0.57	0.66	0.62	93	107		
0.64	0.64	0.64	100	100		
0.65	0.64	0.65	101	99		
0.58	0.73	0.66	89	111		
0.66	0.70	0.68	97	103		
0.70	0.70	0.70	100	100		
0.68	0.73	0.71	96	104		
0.74	0.71	0.73	102	98		
0.69	0.79	0.74	93	107		
0.74	0.75	0.75	99	101		
0.74	0.75	0.75	99	101		
0.70	0.81	0.76	93	107		
0.72	0.79	0.76	95	105		
0.80	0.74	0.77	104	96		
0.87	0.68	0.78	112	88		
0.85	0.77	0.81	105	95		
0.74	0.88	0.81	91	109		
0.84	0.87	0.86	98	102	Moving mean	Average CV (%)
0.83	0.91	0.87	95	105	0.73	5.9
0.91	0.90	0.91	101	99	0.75	5.6
0.97	0.89	0.93	104	96	0.76	5.5
0.96	0.96	0.96	100	100	0.78	5.5
0.82	1.15	0.99	83	117	0.79	6.7
1.11	1.07	1.09	102	98	0.82	6.2
1.03	1.21	1.12	92	108	0.84	6.4
1.27	1.10	1.19	107	93	0.86	6.6
1.25	1.13	1.19	105	95	0.89	6.6
1.17	1.25	1.21	97	103	0.91	6.7
1.24	1.20	1.22	102	98	0.93	6.5
1.22	1.29	1.26	97	103	0.96	6.5
1.22	1.45	1.34	91	109	0.99	6.8
1.36	1.51	1.44	95	105	1.02	6.7
1.45	1.43	1.44	101	99	1.06	6.6
1.43	1.48	1.46	98	102	1.09	6.6
1.38	1.54	1.46	95	105	1.13	6.1
1.48	1.48	1.48	100	100	1.16	6.0
1.66	1.56	1.61	103	97	1.20	5.7
1.72	1.62	1.67	103	97	1.24	5.7
1.75	1.77	1.76	99	101	1.28	5.6

Of more fundamental importance in competitive assay designs is that the measured cross-reactivity would be *expected* to vary at different positions on the dose–response curve. Ekins (1974) demonstrated some years ago that in competitive assays the **relative potency** (**RP**) can be expressed as

$$RP = \left(F\frac{K_{eq}}{K_{cr}}\right) + B \qquad (6.4)$$

where B and F represent the bound and free fractions, respectively, and K_{eq} and K_{cr} the equilibrium constants for the analyte and cross-reacting substances. If the equilibrium constant for the analyte is 10^{10} L/mol and that of the cross-reactant 10^{8} L/mol, then at 50% binding on the dose-response curve, the observed potency would be 50.5. At 10% binding the relative potency falls to 90.

In immunometric assay designs, assessment of cross-reactivity in the presence of endogenous analyte is essential. If the cross-reacting substance is added to

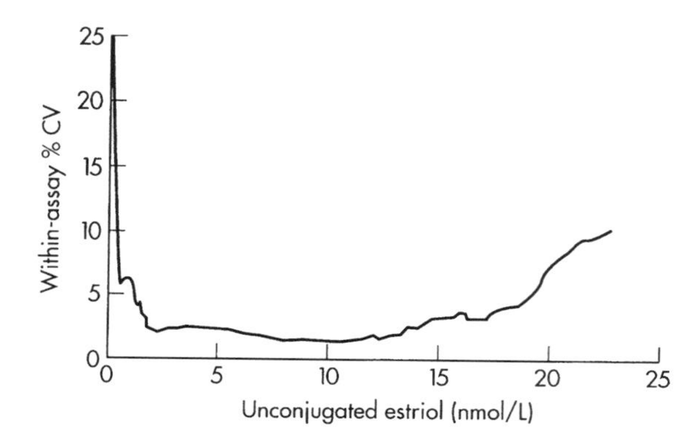

Fig. 6.5 Moving average precision profile.

analyte-free matrix, a signal increase is only observed if the substance binds to both antibodies. The absence of signal may imply that the substance does not cross-react. If cross-reaction is examined in the presence of endogenous analyte then a quite different picture can emerge. The cross-reactive substance may bind to the first capture antibody, the second detecting antibody or both, and give rise to marked differences in the measured cross-reactivity as shown in Figure 6.7.

If the cross-reactant binds to the capture antibody, but not the detecting antibody, then the measured concentration decreases, causing **negative interference**. This may also occur at high cross-reactant concentrations if the substance binds preferentially to the detecting antibody. Only if the cross-reactant binds to both antibodies does an increase in signal result, causing **positive interference**. True cross-reactivity can thus only be assessed in this type of assay by measuring the cross-reactivity of each antibody separately in a competitive assay design, although these individual measurements by themselves are of little practical use except as an assay design tool.

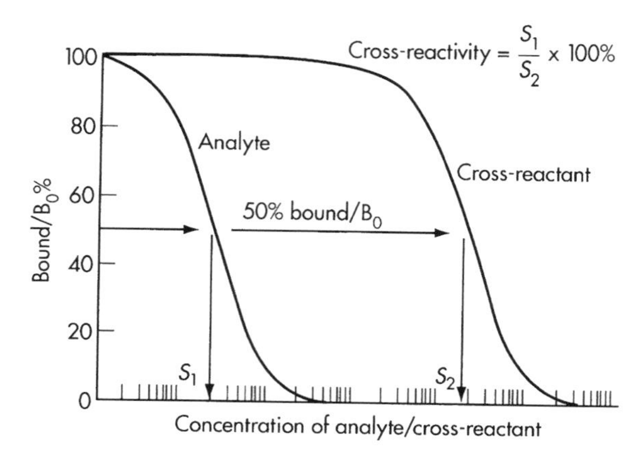

Fig. 6.6 Cross-reactivity.

The best method for examining cross-reactivity (and interference), regardless of assay design, is by spiking the substance into samples containing endogenous analyte. Routine samples provide a more realistic matrix than a chemically or physically stripped analyte-free matrix, and there may be differences in their behavior when the cross-reactive substance or analyte is bound to protein. The cross-reactivity should be calculated at some multiple of the upper reference limit for the cross-reacting substance, say twice, rather than at some arbitrary point on the dose–response curve, and expressed as the apparent percentage change in the endogenous analyte concentration. This provides a more clinically useful way of assessing the likely degree of interference that would be encountered in routine practice.

INTERFERENCE

A useful working definition of **interference** is any factor causing bias in an assay result, other than the presence of a true cross-reacting substance. Most of the manifestations of interference are classified as 'matrix' effects. This term is as useful in the field of immunoassay as 'idiopathic' is in the field of medicine. The tendency to categorize any unusual results as being matrix effects often provides little stimulus to investigate the problem further and possibly find a solution.

There are several mechanisms.

Alteration of Effective Analyte Concentration

Removal or Blocking of the Analyte

Removal or blocking of the analyte may be specific, such as the sequestration of steroids by sex hormone-binding globulin. In competitive assays this may be the labeled and unlabeled hormone, or the unlabeled hormone selectively. The effect is usually prevented by the inclusion of a non-antibody-reactive blocking agent to swamp the binding sites. Alternatively the blocking may be non-selective, such as the partitioning of hydrophobic drugs or steroids in lipid vesicles in lipemic serum samples. This is often corrected by the inclusion of selective detergents in the reagents. For cortisol, an ingenious approach is to incubate the assay at 50 °C and low pH, under which conditions binding to cortisol-binding globulin is minimal, and antibody binding remains relatively unaffected.

Displacement of Analyte from Physiological Binding Proteins

Displacement from binding proteins can be a major problem in assays of the free (non-protein-bound) fraction of small molecules such as thyroxine (T_4) or triiodothyronine (T_3). In these two examples, nearly all of the hormone is protein bound. Free hormone assays for these analytes may, therefore, be particularly sensitive to substances that displace them from their binding site,

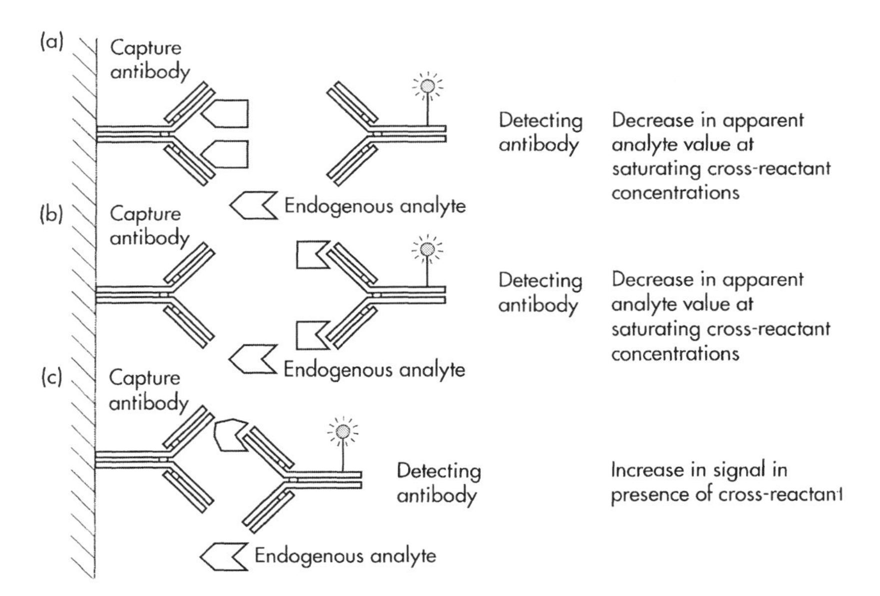

Fig. 6.7 Cross-reaction in immunometric assays. (a) Cross-reacting substance binds to capture antibody only. (b) Cross-reacting substance binds to detecting antibody only. (c). Cross-reacting substance binds to both capture and detecting antibodies.

for example non-esterified fatty acids, which may form by the action of lipases when samples are stored. Assay design in such cases is particularly challenging.

Alteration of Antigen Conformation

Some drugs and other haptens form complexes with divalent cations in solution. Many proteins also contain divalent cation-binding sites. The presence (in serum) or absence (in EDTA plasma) of magnesium or calcium may, therefore, result in changes in antigen conformation. *In vivo* the immunogen is complexed, and it is this conformation that provokes the immune response. The antibodies thus produced may have a higher affinity for the complexed form than the free. Although rare, this phenomenon should be borne in mind for certain analytes.

Interference with Antibody Binding

Physical Masking of the Antibody

Physical masking of the antibody is usually confined to solid-phase antibody systems, where the antibody is coated onto a predominantly hydrophobic polymer surface. Lipids and silicone oils used in some blood collection devices can sometimes cause problems in such systems by nonspecifically binding to the solid phase. Some loss of noncovalently bound solid-phase antibody may also result from displacement. The inclusion of specific surfactants and/or modification of the solid-phase surface to render it less hydrophobic can reduce or cure the problem. Surfactant concentrations do, however, require careful optimization: high concentrations may lead to direct loss of solid-phase antibody where this is noncovalently bound. Fibrin in plasma samples is another common cause of interference. The presence of high concentrations of protein may lead to protein overcoating of solid-phase antibody, particularly when this is the first component added to a solid-phase assay and hence at an undiluted protein concentration. Pre-blocking of active sites on the solid phase with a suitable inert substance is almost universally used to prevent nonspecific adsorption.

Alteration of the Antibody Binding Site Conformation

Alteration of the antibody binding site conformation may occur by sample-induced changes in the ionic strength or pH of the reaction medium, or by residual organic solvents where samples have been pre-extracted. The latter may be a particular problem with the assay of tissue, food, soil or forensic samples, since the precise composition of the sample is likely to vary. Most assays, however, have sufficient buffering and ionic strength capacity to minimize the effects of differences in sample pH and salt concentration.

Low-dose Hook

Low-dose hook effects are occasionally encountered in competitive assays, and in particular those radioactive assays where the antigen is labeled to a very high specific activity. At low concentrations of analyte, binding can

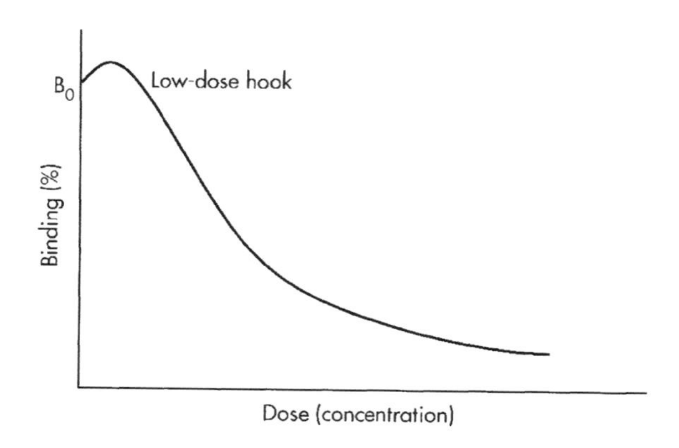

Fig. 6.8 Low-dose hook effect.

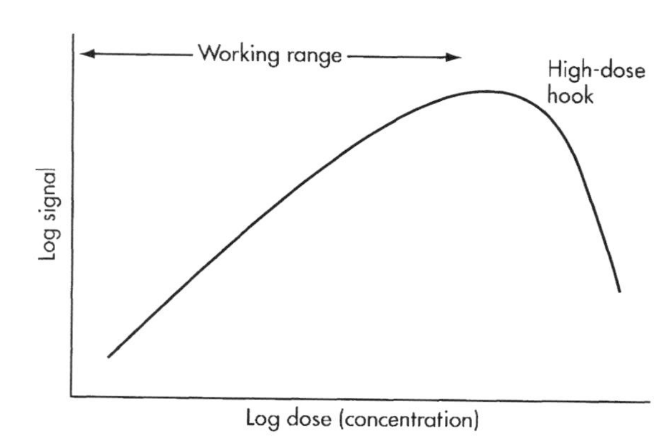

Fig. 6.10 High-dose hook effect.

exceed the binding found at zero concentration (see Figure 6.8). The phenomenon has been ascribed to positive cooperative binding of the antibody to the antigen.

High-dose Hook

The **high-dose hook** effect is confined to one-stage immunometric (sandwich) assays, giving a decrease in signal at very high concentrations of analyte (see Figure 6.9). It is caused by excessively high concentrations of analyte simultaneously saturating both capture and detecting antibodies. This prevents the formation of detectable capture antibody/analyte/detecting antibody complexes. It affects mainly solid-phase assays where the capture antibody concentrations may be limiting, and assays where the possible range of analyte concentrations is very wide, such as in assays for tumor markers. Careful assay design is necessary to ensure that the concentrations of both capture and detecting antibodies are sufficiently high to cope with levels of analytes over the entire pathological range. It is a common practice to re-assay samples in such assays at several dilutions as a check on the validity of the result (see Figure 6.10).

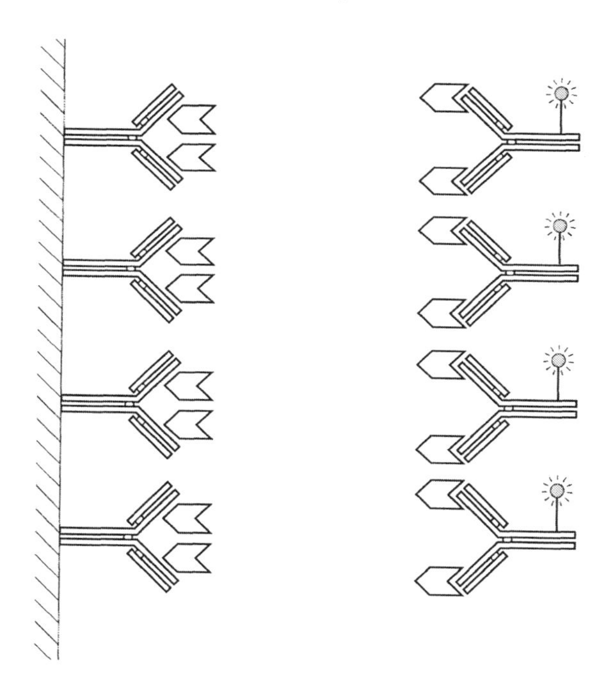

Fig. 6.9 High-dose hook effect.

Heterophilic Antibodies

Heterophilic antibodies are a well recognized source of interference in immunometric assays. Here antibodies to animal IgG in human samples may cross-react with reagent antibodies, particularly where these are from the same species, such as mouse monoclonal antibodies. Reaction with both capture and conjugate antibody leads to the formation of a stable, detectable antibody complex, mimicking that formed normally by the analyte. The prevalence of such heterophilic antibodies is probably higher than literature reports would tend to suggest, since such 'rogue' samples are only identified when the assay produces results that are widely at variance with the clinical picture. Small amounts of endogenous antibody leading to changes in concentration within the normal range are unlikely to arouse suspicion. The origin of such antibodies is the subject of some speculation. They are more common in people routinely handling animals, such as animal-house technicians. They may be due to the primary antibody having a similar epitope to the antigen originally responsible for the production of the endogenous antibody. Patients receiving monoclonal antibodies for therapy or imaging invariably produce an antibody response to mouse IgG after two or three exposures (see Figure 6.11).

Both IgG and IgM heterophilic antibodies have been described, the latter being particularly reactive as a result of their polyvalence and lower degree of steric restriction

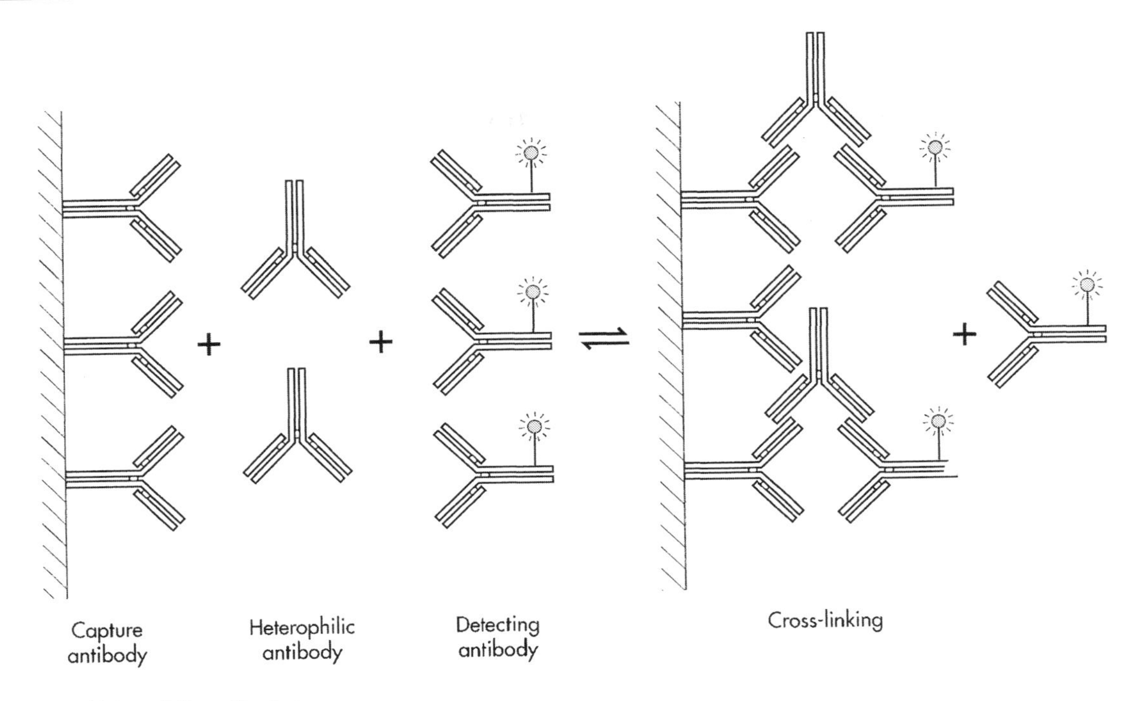

Fig. 6.11 Heterophilic antibodies.

when bound to a solid phase. Immunometric assays with simultaneous binding of capture and detecting antibody are the most prone to interference by heterophilic antibodies (Boscato and Stuart, 1986, 1988). The inclusion of serum or immunoglobulins from the same species as the antibody reagents in the assay is usually effective in preventing interference of this type. An alternative approach is the use of the Fab fragment of IgG rather than the whole IgG molecule, usually as the detecting antibody (Miller and Valdes, 1991).

Autoantibodies

Endogenous circulating autoantibodies have been described for a variety of analytes, notably T_3 and T_4 (Bendtzen *et al.*, 1990). In competitive assays this potential interference can lead to an apparent increase or decrease in measured concentration, depending on whether the autoantibody-analyte complex partitions into the free or bound fraction. In immunometric designs the net effect is almost invariably a decrease in measured concentration.

Complement and Rheumatoid Factor

Complement binds to the Fc portion of some subclasses of IgG, and can cause interference by blocking the antibody, leading to an effective reduction in concentration. Interference has been noted in both polyclonal and monoclonal antibody assays (Masson *et al.*, 1981; Käpyaho *et al.*, 1989).

Rheumatoid factor also reacts with the Fc portion of antibody–antigen complexes. In assays for antibody with solid-phase adsorbed antigen, this can lead to cross-linking of nonreactive IgG, and hence an increase in signal when the latter is detected with labeled anti-human IgG.

Interference in the Binding of the Capture Antibody to the Solid Phase

A popular technique is the use of biotinylated primary antibody with a streptavidin-coated solid phase (or alternatively biotin on the solid phase and streptavidin used as a linker). Interference here may occur in individuals with high circulating levels of endogenous biotin.

Interference in the Separation Stage

Interference in the separation procedure can occur when the antibody–antigen complex in solution is anchored to a solid phase *via* an intermediate linking reaction, for example, using a solid-phase adsorbed second antibody to anchor a primary rabbit IgG. Any endogenous heterophilic anti-rabbit antibodies in the sample compete with the bound second antibody. As with interference in immunometric assays the usual remedy is the inclusion of animal sera of the same or related species to swamp the effects, albeit at the expense of reducing the overall binding of the primary antibody.

Interference with Detection Systems

Endogenous Signal-generating Substances

The presence of endogenous signal-generating substances, such as europium in time-resolved fluorescence assays, or endogenous particulate enzymes such as membrane-bound peroxidases in horseradish peroxidase-based assays, can interfere in the signal detection stage, although they may be selectively depressed by the addition of suitable blocking agents. Interference by fluorescent drugs or metabolites may be a particular, though relatively uncommon, problem in homogeneous fluorescence assays.

Enzyme Inhibitors

Enzyme inhibitors may be chemical or immunological. Antibodies cross-reactive to horseradish peroxidase and alkaline phosphatase have been described. The presence of azide as preservative in some control sera may lead to suppression of enzyme activity in some assay designs using peroxidase as label.

Enzyme Catalysts or Cofactors

Enzyme-immunoassays may be affected by the presence of enzyme catalysts or cofactors as contaminants, for example, copper II ion contamination promotes luminol chemiluminescence in the presence of hydrogen peroxide.

RECOVERY

The test most often used to demonstrate the accuracy of an assay is the assessment of **recovery**. Accurate quantities of analyte are added to samples and the incremental increase in measured concentration determined. Recovery is correctly expressed as

$$\frac{\text{Measured increase in concentration}}{\text{Predicted increase in concentration}} \times 100\% \qquad (6.5)$$

Some workers have expressed recovery as the measured concentration divided by the expected concentration. This is incorrect: in this case the calculated recovery is a function of the measured concentration in the base sample as well as the amount added.

Recovery assesses more than just whether or not an assay is correctly calibrated. Calibration depends upon the preparation of an analyte-free base matrix to which known concentrations of analyte are added. This base matrix is often chemically or immunologically treated to remove endogenous analyte and this can result in a matrix that is sufficiently dissimilar from routine samples for matrix effects to occur in the assay. Recovery, therefore, assesses the calibration of the assay and the influence of differences between sample and calibrator matrices.

Recovery is normally performed using three or more base samples spiked with three concentrations of analyte. The samples and spiked samples should span the clinical range of the assay. The calculation of recovery is affected by errors in three independent measurements: the base matrix concentration, the spiked concentration and the preparation and addition of the recovery spikes. The smaller the recovery spike in relation to the concentration of the base sample, the greater the potential error in the calculated recovery. Several assays are, therefore, performed and the mean values across the assays calculated to minimize measurement errors.

The use of several assays and recovery spikes at different levels requires the use of relatively large aliquots of base sample. A common practice, therefore, is to pool patient samples to give a sufficient volume of material. This is not ideal. The calibration procedure itself probably uses pooled serum as a base. Use of a relatively homogeneous serum pool for recovery experiments may, therefore, fail to elicit any sample-to-sample matrix differences if these are present.

There are some pitfalls in using recovery as an objective criterion for assessing accuracy. First, the calculations are subject to considerable errors in measurement, and no recovery calculations can be assessed objectively without their accompanying 95% confidence intervals. Second, recovery only detects the presence of proportional bias. Constant bias is subtracted from both the base sample and recovery measurements equally. Constant bias can be assessed using samples with low or absent levels of analyte as a result of physiological or pathological processes (for example, for TSH using T_4-suppressed volunteers or patients with primary hyperthyroidism), or alternatively by comparison with a reference method, if one exists.

Assessment of recovery in assays for serum antibodies is of questionable value. Even where international standards are available, the increase in concentration observed is a function of both the concentration and avidity of the antibodies already present in the sample, as well as in the recovery spike. Accuracy in such assays is relative rather than absolute, and is best ascertained on the basis of clinical sensitivity and specificity rather than on technical criteria.

DILUTION

Dilution is a useful additional check on the accuracy of an assay. It provides information that is of more practical benefit than recovery as it answers the question: if a sample is diluted will it give the same result – and if not, which is the most accurate result? Dilution experiments are often referred to as **parallelism** studies because effectively one is judging whether or not dilutions of sample lie parallel to the calibration curve. Parallelism is assessed by assaying samples diluted in an appropriate analyte-free matrix. Several samples should be assessed, and the initial samples and dilutions chosen to fall evenly across the working range of the assay. Dilutions are often made serially, although this is not the best approach as any errors made with each dilution are compounded.

Results can be displayed by several methods, of which the simplest is to calculate the final concentrations by multiplying by the appropriate dilution factor and plotting these against dilution. Formal statistical tests can be used to test the significance of least-squares correlation of the measured value against dilution. However, one pitfall of such an approach is that the result is heavily biased by the lowest dilution if the dilution factor is plotted on the ordinal scale. For example, a 1/16 dilution has far more statistical weight than a 1/2 dilution but may be a less precise estimate of concentration in that region of the calibration curve.

A better method is to prepare fractional, rather than doubling, dilutions and assess dilution by performing least-squares regression on the corrected sample value versus the volume of original serum in the assay (e.g. 100, 90, 80, 70, 60, 50, 40, 30, 20, 10 μL, rather than 100, 50, 25, 12.5 μL), and examining the 95% confidence intervals for the slope of the line.

Dilution experiments are subject both to measurement and pipetting errors; the latter may cause systematic and accumulative errors in dilution. Gravimetric, independent dilutions are preferred to serial dilutions.

Assays of antibody only show parallelism if the affinity of the antibodies in the sample corresponds exactly with those in the assay calibrators, since such assays reflect both the concentration and affinity of the antibody. This is not a feasible proposition. Samples containing high-affinity antibody show over-recovery on dilution (apparently higher values). Samples with antibodies of low affinity show disproportionately lower levels on dilution. Such non-linearity can lead to difficulties in clinical interpretation where serial monitoring of antibody titers is performed and where the activity of the antibody is such as to require different dilutions on each occasion to bring the results into the working range of the assay.

Calibration of this type of assay is dictated by the availability of suitable standard reference materials. It is best if such reference antibodies contain antibodies of representative (average) affinity. This ensures that patient samples dilute correctly on *average*, that is to say approximately half of them show apparent under-recovery on dilution, and half over-recovery.

CORRELATION AND METHOD COMPARISON

Few published papers on new assays escape from the time-honored tradition of including a correlation study against an alternative method, usually quoting the Pearson product moment correlation coefficient, r, as an index of the degree of agreement. Strictly speaking the statistic is not particularly useful. Correlation examines whether or not two parameters are associated. The correlation coefficient assesses the strength of the relationship only in so far as to determine whether the parameters are statistically correlated. It is a very poor index of the degree of agreement between them. If r is significant, it simply means that there is a detectable relationship – the simplest way of ascertaining this is to read the label on the box, since it would be amazing if two methods for measuring the same substance were not associated!

Considerable weight is often placed upon the slope and intercept derived from the linear regression and on how close r is to a value of 1.000. There are several reasons why simple least-squares regression and correlation analysis can be highly misleading in method comparison studies.

- It assumes that the relationship is linear – it may not be.
- It assumes that the x variable is without error. This is not the situation when two immunoassays are compared. Deming's (1943) regression method is the more correct and is described later in this section.
- Correlation assumes that both sets of data are normally distributed. This is seldom the case, since most method comparison studies include disproportionately higher numbers of low and high samples in order to examine the whole assay range. The slope, intercept and regression coefficient are all heavily influenced by the presence of extreme values. A wide range gives better correlation but not necessarily better agreement. Even when the distribution of samples used is representative of the normal population, it is almost universal for the regression to be performed on the actual values even when the normal population results are known to be log-normally distributed (as are many biological substances).
- A high r value may provide reassurance that the methods agree, whereas in fact in the clinically important range the new method could be clinically invalid, and this is a particularly worrying aspect of the misuse of correlation in method comparison studies.
- The slope and intercept may be used to predict the new method result from the old. The converse is not true: the old method result may not be predicted from the new method result. Reversing the data and repeating least-squares correlation using the new method as the x data gives different estimates for the slope and intercept, although the regression coefficient remains the same. It is of note that Deming's regression method gives the same values for slope and intercept regardless of which assay method is used for the x data.

It is most improbable that two methods will agree exactly, giving identical results for all of the samples assayed, for the same reason that one method used twice on the same samples would not. The main aim of method comparison studies is, therefore, to judge how likely it is that a new method gives results that are so substantially at variance with the original method as to cause differences in clinical interpretation. The problem is not one of hypothesis testing but one of estimation.

What is required is an assessment of how closely two methods agree: this is a quite different question and requires a separate analytical technique. The preferred method is that of Bland and Altman (1986), modified by using the percentage rather than absolute differences between measurements, since in most immunoassay systems the standard deviation is concentration dependent. The calculation is described below, and may be simply performed using a spreadsheet, as illustrated in Table 6.3 for part of data taken from a comparison of two methods for the determination of maternal serum AFP using 220 second-trimester maternal sera.

- Assays should be performed using several lots of reagents and at least 100 routine samples. Some 'library' samples should also be included to assess results in the diagnostically abnormal range. Samples should be assayed at least in duplicate or higher (increasing the level of replicates increases the precision of the mean result and hence increases the confidence that differences between the methods are real and not just due to imprecision in one or both methods).
- Data for the reference (current) method, A, are tabulated in column A, and data for the new method, method B, in column B. The mean of the two results is placed in column C, and the difference between the new and reference method in column D. In column E the difference is expressed as a percentage of the mean result. Using this method a 10% difference between the results means that one result is approximately 1.1 times the other. Similarly a 66% difference means that one is twice the other and a 100% difference that one is three times the other.
- A scatter plot of the percentage differences versus the **mean** results (column E versus column C) can be produced. Plotting the percentage difference against either individual value is incorrect as the difference is dependent upon each value, a well known statistical artifact.
- A scatter plot of method B versus method A enables one to obtain a good 'feel' for the relationship and also provides a visual indication of any outliers.

Figure 6.12 shows the scatter plot of the data. Least-squares regression of this data gave the following relationship:

$$\text{method B} = 1.012 \times \text{method A} + 2.704, \quad r = 0.978$$

At first sight the agreement looks very good, with a slope close to unity, but with a small positive bias. The correlation coefficient, as judged by those cited in the other published method comparison studies, would be regarded as being more than acceptable.

The difference plot, however, reveals the true extent of the bias at low concentrations of AFP (see Figure 6.13). Both assays would be suitable for screening for neural tube defects, where a high result would warrant further investigation. But for Down's syndrome screening, where low values are more clinically relevant, the assays would show marked clinical disagreement. Knight *et al.* (1986) have reported the consequences of a low concentration

Table 6.3. Comparision of two methods.

Column A	Column B	Column C	Column D	Column E (%)
9.01	13.00	11.01	3.99	36.3
13.34	18.54	15.94	5.20	32.6
13.98	18.23	16.11	4.25	26.4
14.67	20.79	17.73	6.12	34.5
15.26	21.36	18.31	6.10	33.3
16.96	22.13	19.55	5.17	26.5
17.09	21.23	19.16	4.14	21.6
.	.	.	.	.
93.23	108.20	100.72	14.97	14.9
97.41	107.31	102.36	9.90	9.7
100.20	98.30	99.25	−1.90	−1.9
102.10	105.20	103.65	3.10	3.0
103.20	109.30	106.25	6.10	5.7
105.30	115.78	110.54	10.48	9.5
112.87	119.20	116.04	6.33	5.5

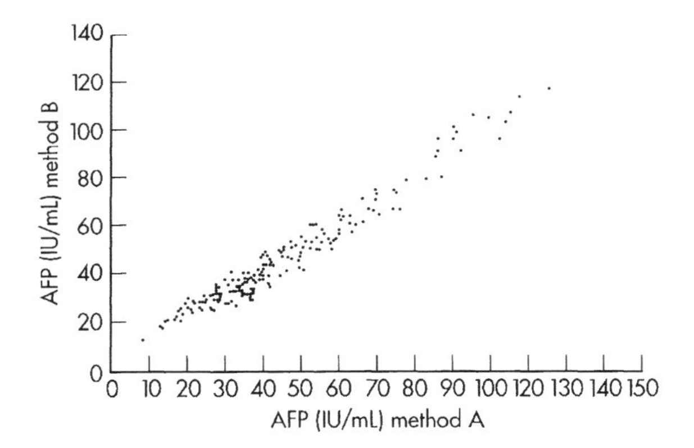

Fig. 6.12 Scalter plot.

bias when such an assay was used for Down's syndrome screening. In that case, the number of women deemed to be at increased risk for Down's syndrome increased by threefold. This example also underlines the necessity of checking the suitability of an existing assay when used for a different diagnostic purpose.

Least-squares regression is only appropriate in model I regression problems, that is where the independent variable, x, is measured without error, and the dependent variable, y, is subject to random error. Least-squares regression should only be used for method comparison studies in two specific circumstances. First where the x variable is a precise reference or 'gold standard' method, the results of which may be regarded as true. Second, in the so-called Berkson case (Sokal and Rohlf, 1969), where the x variables are target values, for example, where analyte has been added to a calibration matrix, even though such additions may be subject to error.

In practice both methods being examined are subject to error and there is often no reason to regard one or the other as being a reference method. Use of least-squares regression in such model II regression problems leads to different estimates of the slope depending on whether the x or y variable is regarded as independent. In reality the slope defining the true relationship between x and y lies somewhere in between.

The difference between Deming regression, sometimes termed functional relationship, and the more familiar least-squares method is that the latter seeks to minimize the squared differences in the y-direction, assuming that the x data are without error. Deming regression minimizes the sum of squares in both the y and x-directions. The calculations are shown below. First an estimate, λ, of the ratio of the standard deviations or CV is required for both assays to judge the relative imprecision of the methods.

$$\lambda = \frac{CV_x^2}{CV_y^2} \tag{6.6}$$

where CV_x and CV_y are the coefficients of variation for the two methods, x and y, respectively.

U is defined as follows:

$$U = \frac{\sum_{i=1}^{N}(y_i - \bar{y})^2 - (1/\lambda)\sum_{i=1}^{N}(x_i - \bar{x})^2}{2\sum(y_i - \bar{y})(x_i - \bar{x})} \tag{6.7}$$

where x_i and y_i are the individual values from 1 to N, and $\bar{x}$ and $\bar{y}$ are the respective means for the x and y data.

The slope of the regression line, Deming b_{yx} is given by

$$\text{Deming } b_{yx} = U + \sqrt{U^2 + \left(\frac{1}{\lambda}\right)} \tag{6.8}$$

The standard deviation, s_{yx}, of the residual error in regression in the y-direction can be calculated and estimates the degree of scatter of the points around the regression line.

$$\text{Deming } s_{yx} = \sqrt{\frac{\sum_{i=1}^{N}(y_i - \bar{y})^2 - b_{yx}(x_i - \bar{x})(y_i - \bar{y})}{N-2}} \tag{6.9}$$

The intercept, Deming a_{yx} is calculated conventionally as

$$\text{Deming } a_{yx} = y - b_{yx}x \tag{6.10}$$

The regression coefficient, r, is calculated in the same way, and gives the same value, as least-squares regression:

$$r = \frac{\sum_{i=1}^{N}(x_i - \bar{x})(y_i - \bar{y})}{\sqrt{\sum_{i=1}^{N}(x_i - \bar{x})^2\sum_{i=1}^{N}(y_i - \bar{y})^2}} \tag{6.11}$$

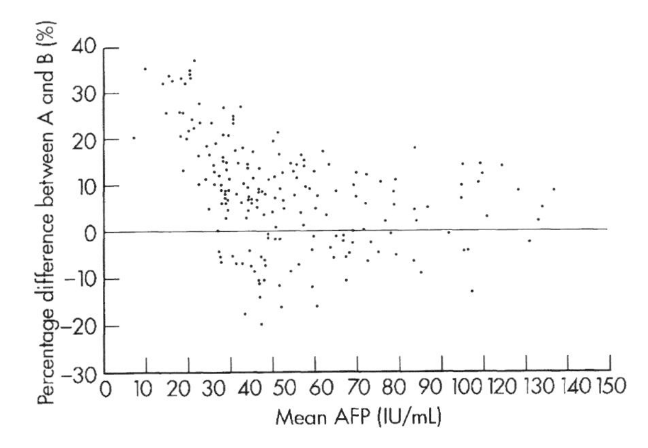

Fig. 6.13 Difference plot.

Linnet (1993) has written an excellent review of various regression procedures in method comparison studies and gives further modifications to allow for weighting of the data points.

It is unfortunate that many common statistical programs do not include Deming regression analysis as a matter of routine. The regression may, however, be easily programmed into a spreadsheet as shown in Table 6.4, using the data from the same AFP method comparison data as shown previously. It has been assumed, for simplicity, that the variance for both methods is equivalent, which generally is a reasonable assumption for most immunoassay methods.

1. Tabulate the results for method x in column A.
2. Tabulate the results for method y in column B.
3. Calculate the mean of the x data (mean x) and the mean of the y data (mean y).
4. In column C, calculate $x -$ mean x and in column D, $y -$ mean y.
5. Multiply column C by column D and place in column E.
6. Square column C and place in column F; square column D and place in column G.
7. Calculate the sums of columns E, F and G and the means of columns A and B.
8. Substitute the values into the above equations to calculate U and hence the slope and intercept.

Figure 6.14 shows the previous data overlayed with both the least-squares and Deming regression slopes. The Deming regression gives slightly different estimates of both slope and intercept from least-squares:

$$\text{Method B}(y) = 1.050 \times \text{method A}(x) + 1.087,$$

$$r = 0.978$$

A further, robust procedure, based on a structural relationship model, has been described by Passing and Bablock (1983, 1984). It requires no special assumptions with regard to the type of distribution of the sample-related expected values or the error terms. The calculations are, however, computationally complex, relying on the estimation of the median slope of the lines connecting every possible combination of pairs of data points.

Table 6.4. Deming regression.

	Column A A	Column B B	Column C A − mean A	Column D B − mean B	Column E C × D	Column F C^2	Column G D^2
	9.01	13.00	−33.78	−33.01	1114.93	1141	1089
	13.34	18.54	−29.45	−27.47	808.97	867	755
	13.98	18.23	−28.81	−27.78	800.21	830	771
	14.67	20.79	−28.12	−25.22	709.07	791	636
	15.26	21.36	−27.53	−24.64	678.44	758	607
	16.96	22.13	−25.83	−23.87	616.60	667	570
	17.09	21.23	−25.70	−24.78	636.73	661	614
	⋮	⋮	⋮	⋮	⋮	⋮	⋮
	93.23	108.20	50.44	62.19	3137.09	2544	3868
	97.41	107.31	54.62	61.30	3348.45	2983	3758
	100.20	98.30	57.41	52.29	3002.23	3296	2735
	102.10	105.20	59.31	59.19	3510.83	3518	3504
	103.20	109.30	60.41	63.29	3823.62	3649	4006
	105.30	115.78	62.51	69.78	4361.90	3907	4869
	112.87	119.20	70.08	73.19	5129.47	4911	5357
Mean	42.79	46.01			Sum 93,074	91,975	98,163
			By Deming regression		By linear regression		
U			0.033				
Slope			1.050		1.012		
Intercept			1.087		2.704		
Regression coefficient			0.997		0.997		

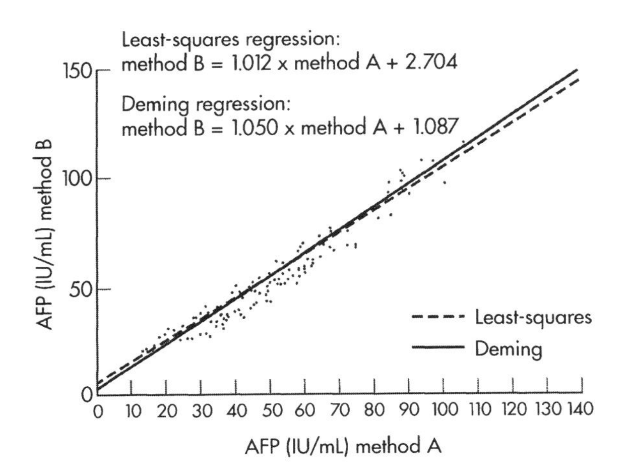

Fig. 6.14 Least-squares and Deming regression.

One further consideration relates to the relative precision of the assays. If the reference method is very imprecise then no new method will show good agreement with it, insofar as the reference method would not agree perfectly against itself if assayed on the same specimens twice (it can be very informative to do this experiment occasionally – the results can be surprising). If the level of imprecision is high, and for some assays this may be an inherent feature and one which is unlikely to be improved upon, then there is a strong case for repeatedly assaying the same samples and using the mean result to minimize the influence of imprecision.

Last, even where methods apparently agree well, this in itself would not justify the direct conversion from one assay to another without reassessing the reference interval. Correlation or direct method comparisons such as described above are not a substitute for that analysis, although good experimental design and judicious inclusion of the appropriate samples may enable both to be assessed in the same assays.

ASSAY DRIFT

As the name implies, **assay drift** refers to systematic rather than random changes in measured analyte concentrations, the magnitude of which depends upon the position within the assay. Drift is unlikely to be detected unless control sera or pooled patient samples are placed at regular intervals throughout the assay. The common practice of assessing the level of imprecision by running sequential replicates underestimates the true level of imprecision (and bias) when drift is present. Similar arguments apply when assessing sensitivity using sequential replicates of the zero calibrator. One consequential problem caused by drift may be in secondary calibration. Usually calibrators or control sera are assayed as unknowns following a series of reference standards. Once calibrated, their position in the assay changes when they are run routinely and this can lead to a systematic bias in sample results in the same direction as the original assay bias. The bias is thus compounded. There are several possible mechanisms.

- *Drift inherent in the assay design*. In many assays the immune reaction has not achieved the degree of completion necessary to be robust to variations in reagent addition between the beginning and the end of the assay. To some extent this may be mitigated by the corresponding delay in the addition of the other reagents. However, this may not always be the case, particularly for assays where separation is physical, and carried out on all samples at the same time (for example, batch washing, or centrifugation and decantation).
- *Incubation temperature variations*. Most immunoassay reagents are stored at 2–8 °C, yet few assays are performed at this temperature. If assays are set up using reagents straight from the fridge then those tubes pipetted first attain a higher temperature when the assay is subsequently incubated than the tubes assayed last. Given that the rate of reaction approximately doubles for each 10 °C rise in temperature it is not surprising that this may contribute to assay drift.
- *Settling of solid phase*. Few solid-phase suspensions remain in suspension on standing and the longer the time taken to add reagent, the more settling occurs, leading to changes in the concentration of solid-phase reagents. Gentle stirring of the solid phase during addition is recommended. Using a magnetic stirrer to mix magnetic particles is not recommended: an obvious point – but how many of us have done so – once?
- *Signal generation*. Many assays rely upon a final reagent to develop the signal. Although it may take a minute to add reagent to 100 microwells, many readers measure the signals from all of the wells in far less time than this. If the signal development time is 5 min this corresponds to a 20% longer period of signal development in the first well than the last. It is for this reason that colorimetric assays typically use a stopping reagent, although some laboratories economize by ignoring this step.

Drift is usually assessed using control sera equidistantly spaced throughout the assay run. Several assay runs should be performed and linear regression performed on the data with regard to assay position or time of addition. Controls at several levels should be run as drift is almost invariably concentration dependent. Taking the mean results from several assays increases the confidence that any differences seen are due to drift rather than imprecision. It is useful to run the assay at stressed conditions, for example, lower or higher temperatures than would normally be encountered, using slower reagent and sample addition times and shorter incubation periods. These stressed conditions usually elicit drift if present, providing both a positive control for the experiment and some reasonable boundaries in terms of reagent addition and incubation times, assay size, and temperatures outside which the assay does not perform satisfactorily. Both negative and positive drift can occur.

The tendency for assays to have ever shorter incubation times has exacerbated problems of drift and requires considerable ingenuity on the part of assay designers to minimize the problem.

Completely automated instruments seldom have problems with assay drift because of the high degree of control over reagent addition, incubation times and temperature. Problems are not uncommon, however, in manual assays converted to a semi-automated sampler protocol and drift should automatically be reassessed on such occasions. The timings of reagent addition are almost certainly different. For example, it is easy to add 100 μL of antibody reagent to 100 tubes manually using a repeating dispenser within 1 min whereas some automated instruments may take several minutes to complete the same task.

CLINICAL CONCEPTS

Although this section focuses on the relationship of immunoassays to clinical diagnosis, the concepts described are generally applicable to practically all fields of measurement.

When a clinical diagnostic test is carried out, each person affected views it from a different perspective. There is a hierarchy involving the patient, the physician, the laboratory and the kit manufacturer. Although perhaps occasionally forgotten, the patient is the most important, with each participant in the sequence of events effectively providing a service to the preceding 'customer'.

Patients seek medical assistance, not for a diagnosis, but for reassurance or treatment – will they die, will they get better, how long will the illness last and what will the quality of life be? Few people visit a physician hoping to have a disease! In other words the primary interests of the patient are reassurance, prognosis and treatment – not diagnosis.

For the physician, the primary aim is diagnosis, so that the appropriate action may be taken.

In busy laboratories, the primary requirements when selecting methods may be throughput, cost, correlation with other methods and performance in external quality control schemes.

For kit manufacturers, there is a focus on parameters that are easily measured and common to a range of products, such as control values, curve shape parameters and precision. These parameters tend to be selected for their ability to provide methods of quality control in an industrial setting, and may only rarely be considered from the patient's viewpoint.

In this complex situation, it is easy to see how judgments can be made that may be inappropriate to the clinical situation. For example, a particular TSH assay may be considered acceptable by a laboratory because the between-assay precision is less than 7% CV across the range. An alternative assay with precision of 30% CV above 60 mIU/L but <3% at 5 mIU/L could be considered unacceptable because of poor performance in an external quality assurance (proficiency testing) scheme, even though it has better performance in the region of clinical interest.

The best diagnostic assay will always be the one that gives the most reliable clinical diagnosis. If this premise is accepted, many of the technical requirements automatically follow. Some parameters usually considered important may become almost irrelevant by comparison. In particular, the great importance attached to numerical agreement with existing products is driven by manufacturers, laboratories, clinicians, regulatory authorities, and external quality assurance schemes and not by the ultimate needs of the patient. It matters not to the patient if his or her results are expressed in nmol/L or ng/mL, yet there may be several orders of magnitude of numerical difference between them, in the same way that two methods for measuring the same substance might disagree numerically.

Immunoassays are used in a number of ways, and are not limited to the identification of diseases. The principal uses can be categorized as follows.

Detection. The presence or absence of a particular substance, for example, in testing for infectious diseases or drugs of abuse.

Quantification. Accurately determining the level of a particular substance as an aid to differential diagnosis, for example, the assay of thyrotropin (TSH) in the diagnosis of thyroid disease.

Monitoring. Assessment of changes in concentration, for example, the use of specific tumor markers as an aid to the prognosis and therapy of cancer.

The use of tests for diagnosis is one of the most challenging applications, involving many important concepts that are easily misunderstood or underestimated. The remainder of this chapter concentrates on these concepts.

DIAGNOSIS

A diagnosis is a decision point. The decision is the intention to treat. It is the point at which sufficient evidence has been accumulated to state, beyond reasonable doubt, that the patient is or is not suffering from a particular disease. The English legal phrase 'beyond reasonable doubt' is apt as few diagnoses are absolutely certain. The scientific equivalent is 'high probability', and it is perhaps surprising that clinical diagnosis is rarely thought of routinely in such probabilistic terms.

REFERENCE INTERVAL

The most important concept is that of the **reference interval** (sometimes termed the **reference range** or **normal range**). This is the range of concentrations or values encompassing most, but not all, unaffected subjects. The term 'normal range' is misleading as it implies that all unaffected individuals fall within it, and the term 'reference interval' is the more proper, usually being defined as the range within which 95% of the population fall.

It is important to recognize the sources of variation that define the reference interval. Most biological substances are subject to some kind of homeostatic mechanism that ensures that their concentrations only vary within a particular range of values. The central point of this variation, the mean concentration, is referred to as the **homeostatic set point**, and is specific for the individual. The variation about this homeostatic set point is the within-subject biological variation, $\mathbf{CV_i}$ (intra-subject coefficient of variation). Variation between individual homeostatic set points, the between-subject biological variation, $\mathbf{CV_g}$ (group CV), comprises the second major component of variation. The third, usually minor, source of variation is in collection of the sample for analysis, the pre-analytical variation, $\mathbf{CV_p}$. The fourth source of variation is that inherent in the analysis: the between-batch analytical variation, $\mathbf{CV_a}$. The reference interval variation $\mathbf{CV_t}$, therefore, reflects the sum of these variations (Cotlove *et al.*, 1970; Harris, 1979):

$$CV_t = \sqrt{CV_a^2 + CV_p^2 + CV_i^2 + CV_g^2} \qquad (6.12)$$

There is a fifth source of error, applicable to some analytes, where the concentration varies physiologically with time, such as pregnancy hormones. In such cases the values are usually expressed in relation to the reference interval at that particular stage of gestation. Errors in the estimation of gestational age are relatively common, and often contribute to the overall variation to a considerably greater extent than the analytical error. Such variation may be considered semantically as part of the pre-analytical error.

The relationship between within- and between-subject variation is of critical importance, as it determines whether or not the whole concept of reference intervals is a valid one. Figure 6.15a illustrates a hypothetical reference interval defined from the observed Gaussian distribution of values. Figure 6.15b illustrates the components of that observed reference interval where the within-subject variation for the analyte is very small in relation to the between-subject variation. All subjects show different homeostatic set points, but the individual variation around them is small. Reference intervals for such an analyte are of little use – each subject needs their own reference interval! Figure 6.15c shows the more usual case, where the homeostatic set-points for individuals are relatively close together, and the within-subject variation forms a major component of the between-subject variation. Only in such cases is the reference interval valid.

It is generally accepted that reference intervals are useful provided that the **index of individuality** is greater than 1.4 (Harris, 1981). When this index is less than 0.6, reference intervals are of limited utility. The index is numerically expressed as

$$\sqrt{\frac{CV_a^2 + CV_i^2}{CV_g^2}} \qquad (6.13)$$

Inappropriate reference intervals are the commonest cause of impaired clinical performance, and it is, therefore, essential that considerable care is taken in their estimation. There are two major factors to consider: bias and statistical method.

BIAS

Bias is the major problem area, and there are several reasons why samples on which the reference interval has been based retrospectively may not reflect the population being tested prospectively.

- The analyte may show variation with age, sex or race. Reference intervals may need to be assigned separately for each group.
- The analyte may not be stable on storage or stored samples themselves may undergo changes in structure which may manifest themselves as 'matrix' effects in immunoassays.
- Analyte values may show marked fluctuations due to diurnal (e.g. cortisol), seasonal (e.g. vitamin D_3), cyclical (reproductive hormones), nutritional (insulin), gestational (pregnancy hormones) or episodic (growth hormone) effects. Some hormones are associated with stress (e.g. growth hormone) and levels found in volunteers who are frequently bled may be much lower than in anxious patients (*see* SUBJECT PREPARATION, SAMPLE COLLECTION, AND HANDLING).
- The reduction in physical activity in hospital in-patients may be reflected in the levels of some hormones, particularly those involved in bone metabolism, such as PTH.
- The samples used to determine the reference interval may have been subject to pre-selection bias. For example, requesting 300 euthyroid subjects from a thyroid clinic in order to set a reference interval for T_4 may result in samples being provided that have been classified as being euthyroid partly on the basis of their T_4 values. As an interesting aside this can cause considerable problems when comparing the diagnostic

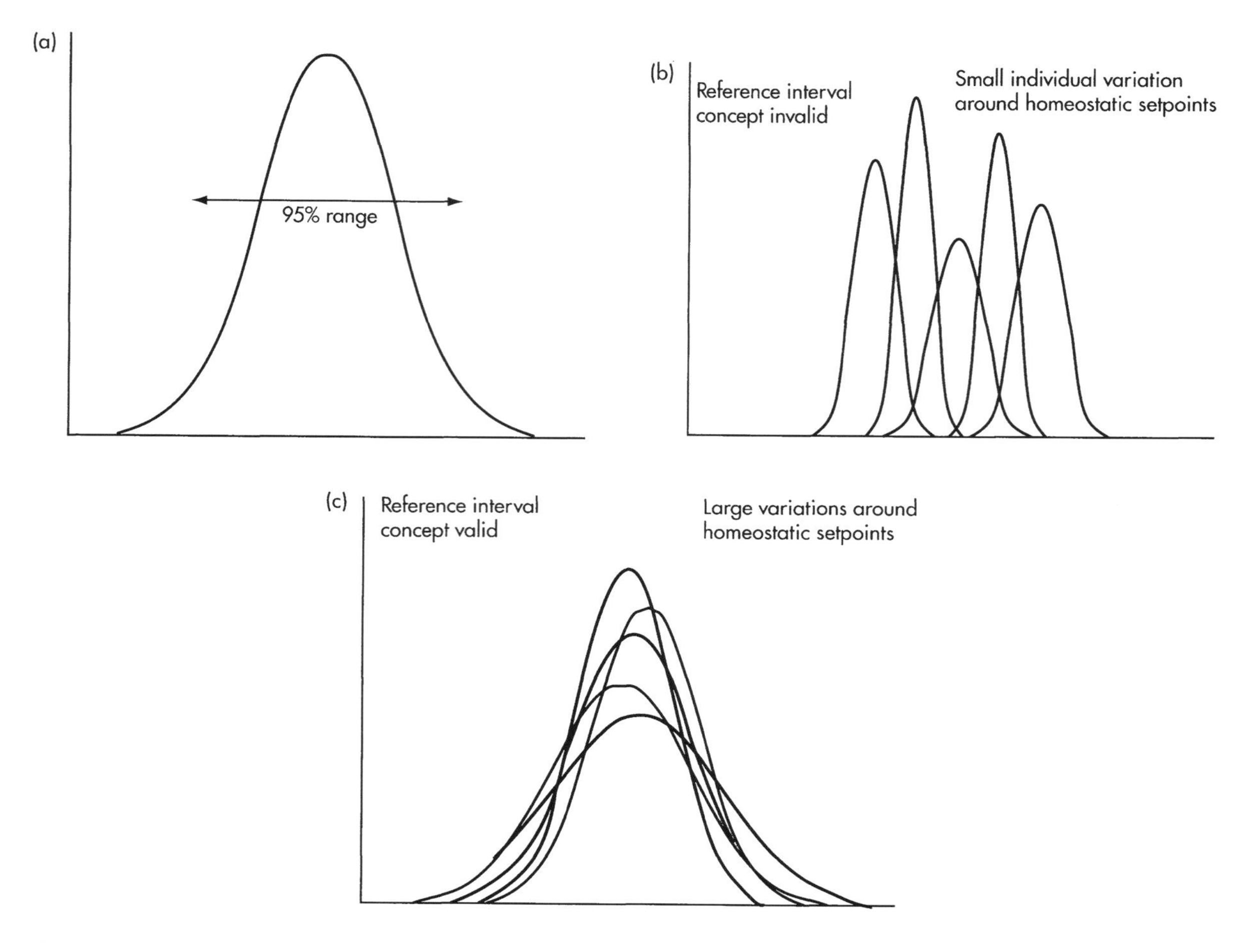

Fig. 6.15 (a) Reference interval. (b) Individual variation (i). (c) Individual variation (ii).

characteristics of two methods. If the samples used in the comparison have been diagnosed partly on the basis of their results with the existing method it would be impossible for the new method to show better sensitivity than the current one.

It can be argued that the most inappropriate samples on which to base reference intervals are those drawn from healthy, active and predominantly young volunteer subjects, such as blood donors. They are after all the least likely section of the population to see a physician (although for HIV and hepatitis blood donor screening they may be the most appropriate source). The preferred, and perhaps most obvious, source of subjects for a reference interval are those patients examined for the presence of disease and subsequently found to be unaffected. The assay in question should not be the only test used in making the diagnosis. Strictly speaking, it should not contribute in any way towards the final diagnosis. Great care also needs to be exercised as many markers of disease are correlated. A different marker used as part of the diagnostic process may act as surrogate marker for the analyte under examination.

Statistical Method and Statistical Power

There are two basic statistical methods for the determination of a reference interval. The first is by listing the data in ascending order of value, selecting cut-off values giving the desired percentiles, usually the 2.5 and 97.5 percentiles. This method has the advantage that it is numerically simple, does not rely on the data being normally distributed, and is relatively unaffected by the presence of outlying values. The great disadvantage is the high level of uncertainty associated with such estimates as they are based only on the few samples in the tails of the distribution.

The second approach is to calculate the reference intervals parametrically from the normal distribution, either on the data directly, or after suitable transformation to normality. The 95% confidence interval for a reasonably large sample is the mean ± 1.96 standard deviations. This method has the advantage that the uncertainty is much less, being based on estimates of the mean and standard deviation derived from the entire data set. It is, however, sensitive to outlying values, and to deviation from normality. The former may be checked by direct inspection of the frequency distribution of the data (this should always be done, no matter which method is used), or by a formal outlier test such as that of Healy (1979). Another robust method of determining the mean and standard deviation in the presence of outliers is to take the median as an estimator of the mean, and the standard deviation as the difference between the 90th and 10th centiles divided by 2.563.

Transformation to a normal distribution should not be confined to simple discrete methods such as taking logs. There are several continuous functional transforms to remove both skewness and kurtosis (Soldberg, 1987). Although there are several formal statistical tests of normality that can be employed (reviewed by Soldberg, 1987), an essential preliminary step is to plot the centiles of the data on normal probit paper, which has one axis scaled according to the normal standard deviate. If normal probability paper is unavailable, the standard deviates corresponding to the centiles may be obtained from standard statistical tables such as Geigy (Latner, 1982), or can easily be calculated using a good approximation. The following algorithm (Abramowitz and Stegun, 1972) gives a very good approximation (to four decimal places) for centiles up to 0.5 (for centiles greater than 0.5, subtract the centile from 1.0 and multiply Z by -1):

first calculate the quantity U:

$$U = \sqrt{-2\log_e(\text{centile})} \qquad (6.14)$$

then substitute into

$$Z = U - \frac{(a + bU + cU^2)}{(1 + dU + eU^2 + fU^3)} \qquad (6.15)$$

where:

$a = 2.515517$
$b = 0.802853$
$c = 0.010328$
$d = 1.432788$
$e = 0.189269$
$f = 0.001308$

Figure 6.16 shows a typical normal probability plot, in this case for TSH samples from 90 euthyroid individuals. The TSH data shows an approximately log–normal distribution, and hence the y-axis is logarithmically scaled.

Assessing a suitable transformation can be difficult and usually requires an iterative computer procedure in order to minimize the skewness and kurtosis. The procedure recommended by the International Federation of Clinical

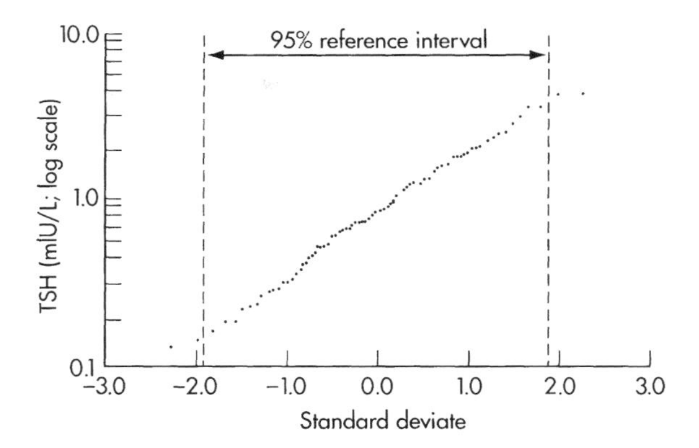

Fig. 6.16 Normal probability plot.

Chemistry is used in the program REFVAL (Soldberg, 1987). An example of program output is shown in Table 6.5, using the TSH data shown above as an example.

For the method of direct inspection of data a minimum of 240 samples should be used. For the parametric approach a minimum of 120 samples yield approximately the same level of precision for the reference interval 2.5 and 97.5 centiles.

The standard error of the reference interval cut-off points should be calculated to indicate the degree of uncertainty. The standard error at each limit (the 2.5 and 97.5 centiles) is approximately equal to the standard deviation × $\sqrt{(3/N)}$, where N is the number of samples (6.16).

CLINICAL SENSITIVITY AND SPECIFICITY

Consider an immunoassay with a single diagnostic application that measures a range of concentrations. For diagnostic purposes there are only two relevant categories of patients, those with and those without the disease, and two categories of results, positive or negative. A cut-off concentration needs to be set, above which samples are classified as positive. Suppose 100 patients with the disease and 100 without the disease are tested using this immunoassay. A 2 × 2 contingency table can be set up (Table 6.6).

The **clinical sensitivity** measures how well the test detects those patients with the disease (this term should not be confused with assay sensitivity, *see* ASSAY CONCEPTS). The sensitivity is the proportion of true positives that are correctly identified:

$$\text{Sensitivity} = a/(a + c); \qquad 95\% \text{ in the example}$$

The **clinical specificity** measures how well the test correctly identifies those patients who do not have the disease (this term should not be confused with assay

Table 6.5. Computerized normalization of data.

IFCC-RECOMMENDED METHODS ARE USED

HISTOGRAM

	UPPERLIMIT	N	NCUM	FREO	CUM
1	0.253	17	17	0.079	0.079
2	0.480	38	55	0.176	0.255
3	0.707	32	87	0.148	0.403
4	0.934	33	120	0.153	0.556
5	1.161	28	148	0.130	0.685
6	1.388	16	164	0.074	0.759
7	1.615	13	177	0.060	0.819
8	1.842	11	188	0.051	0.870
9	2.068	5	193	0.023	0.894
10	2.295	6	199	0.028	0.921
11	2.522	5	204	0.023	0.944
12	2.749	3	207	0.014	0.958
13	2.976	3	210	0.014	0.972
14	3.203	1	211	0.005	0.977
15	3.430	2	213	0.009	0.986
16	3.656	1	214	0.005	0.991
17	3.883	1	215	0.005	0.995
18	4.110	0	215	0.000	0.995
19	4.337	0	215	0.000	0.995
20	4.564	1	216	0.005	1.000

PERCENT: 1 2 3 4 5 6 7 8 9 10 1 2 3 4 15 6 7 8 9 20

DELETION OF OUTLIERS:
SUMMARY: 0 OUTLIERS DELETED

ANDERSON-DARLING TEST FOR GAUSSIAN DISTRIBUTION:
A-SQUARE STATISTIC: = 7.374
PROBABILITY = 0.00

CRAMER-VON MISES TEST FOR GAUSSIAN DISTRIBUTION:
W-SQUARE STATISTIC = 1.249
PROBABILITY = 0.00

KOLMOGOROV-SMIRNOV TEST FOR GAUSSIAN DISTRIBUTION:
DMAX STATISTIC = 0.149
PROBABILITY = 0.00

COEFFICIENTS-BASED TEST FOR GAUSSIAN DISTRIBUTION:
COEFFICIENT OF SKEWNESS (GS) = 1.537
PROBABILITY = 0.00
COEFFICIENT OF KURTOSIS (GK) = 2.883
PROBABILITY = 0.00

NON-PARAMETRIC DETERMINATION OF REFERENCE LIMITS:

FRACTION	REF. LIMIT	0.90 CONFIDENCE INTERVAL
0.025	0.146	0.125-0.196
0.975	3.119	2.716-3.850

PARAMETRIC DETERMINATION OF REFERENCE LIMITS:

FRACTION	REF. LIMIT	0.90 CONFIDENCE INTERVAL
0.025	0.167	0.137-0.200
0.975	2.958	2.607-3.449

TRANSFORMATION DETAILS:

MEAN OF ORIGINAL DATA	=	1.03807
ST.DEV. OF ORIGINAL DATA	=	0.76414
PARAMETER OF EXPO-TRANSFORM	=	−0.69938
MEAN OF EXPO-TRANSFORMED DATA	=	−0.29056
ST.DEV. OF EXPO-TRANSFORMED DATA	=	0.83254
SKEWNESS OF EXPO.TRANSF. DATA	=	0.00413
KURTOSIS OF EXPO-TRANSF. DATA	=	−0.85497
PARAMETER OF MODU-TRANSFORM	=	2.18932
MEAN OF MODU-TRANSFORMED DATA	=	0.00216
ST.DEV. OF MODU-TRANSFORMED DATA	=	1.91389
SKEWNESS OF MODU-TRANSF. DATA	=	−0.03505
KURTOSIS OF MODU-TRANSF. DATA	=	−0.00786
ANDERSON-DARLING A-SQUARE	=	0.683
PROBABILITY OF A-SQUARE	=	0.08

HISTOGRAM

	UPPERLIMIT	N	NCUM	FREQ	CUM
1	−5.183	1	1	0.005	0.005
2	−4.670	0	1	0.000	0.005
3	−4.158	3	4	0.014	0.019
4	−3.645	4	8	0.019	0.037
5	−3.133	5	13	0.023	0.060
6	−2.620	8	21	0.037	0.097
7	−2.108	6	27	0.028	0.125
8	−1.595	15	42	0.069	0.194
9	−1.083	16	58	0.074	0.269
10	−0.570	15	73	0.069	0.338
11	−0.058	30	103	0.139	0.477
12	0.454	39	142	0.181	0.657
13	0.966	15	157	0.069	0.727
14	1.479	12	169	0.056	0.782
15	1.991	15	184	0.069	0.852
16	2.504	8	192	0.037	0.889
17	3.016	8	200	0.037	0.926
18	3.529	7	207	0.032	0.958
19	4.042	5	212	0.023	0.961
20	4.554	4	216	0.019	1.000

PERCENT: 1 2 3 4 5 6 7 8 9 10 1 2 3 4 15 6 7 8 9 20

Table 6.6. A 2 × 2 contingency table.

	Disease status		
	Present	**Absent**	**Total**
Test positive	95	2	97
Test negative	5	98	103
Total	100	100	
Test positive	a	b	$a+b$
Test negative	c	d	$c+d$
Total	$a+c$	$b+d$	

specificity, *see* ASSAY CONCEPTS). The specificity is the proportion of true negatives that are correctly identified:

Specificity $= d/(b+d)$; 98% in the example

The **detection rate** is a term sometimes encountered which is synonymous with sensitivity.

The **false-positive rate** is the proportion of unaffected individuals falsely identified as having the disease:

False-positive rate = 1 − specificity

At first sight the above calculations appear to have defined the performance of the test, but the question has only been answered from one direction. In clinical practice the test result is all that is known (obviously the diagnosis is not), so the relevant measures are those that predict how good the test is at predicting the presence of disease.

POSITIVE AND NEGATIVE PREDICTIVE VALUES

Since the purpose of a diagnostic test in this situation is to provide diagnostic information, we need to establish the probability of the test giving the correct diagnosis *in the population being tested*. The sensitivity and specificity do not give this information. Going back to the previous table the **positive predictive value** (**PPV**) is the likelihood that a patient with a positive test result actually has the disease. The PPV is the proportion of patients with positive test results who are correctly diagnosed:

PPV $= a/(a+b)$; 97.9% in the example

The **negative predictive value** (**NPV**) is the likelihood that a patient with a negative test result does not have the disease. The NPV is the proportion of patients with negative test results who are correctly diagnosed:

NPV $= d/(c+d)$; 95.1% in the example

One other useful estimate of clinical utility is the **odds on being affected given a positive result** (**OAPR**). This is the odds ratio of affected:unaffected patients in the group with a positive test result.

OAPR = 1 : n, where $n = b/a$

In the example OAPR = 1:(2/95) = 1:0.02105.
Multiplying to obtain whole numbers = 95:2.

This means that in the population with a positive test result, 95 affected patients are correctly diagnosed as having the disease for every two unaffected patients incorrectly diagnosed as having the disease.

The PPV and NPV give a direct assessment of the usefulness of the test in the clinical trial reported, but only in that trial.

PREVALENCE AND CLINICAL UTILITY

The disadvantage of clinical sensitivity and specificity is that they do not assess the accuracy of a diagnostic test in a clinically useful way. They do, however, have the major advantage of not being affected by the proportion of patients with the abnormality, termed the **prevalence**. The predictive values (PPV and NPV), in contrast, are clinically useful but depend very strongly on the prevalence of the disorder. In the example quoted above the prevalence was very high, being 100 affected patients out of a total of 200 (50%). In routine practice it is likely that more unaffected than affected patients would be tested.

The prevalence of a disease is the proportion of people with the disease in a given population at a particular point in time. The prevalence varies with the population, for example, the prevalence of women with hypothyroidism is very different in the general population than it is in women referred to a thyroid clinic for clinically suspected thyroid disease.

The prevalence should not be confused with the **incidence** of a disease. The incidence is the number of new cases of a disease occurring within a given time period. For example, the incidence of insulin-dependent diabetes mellitus (IDDM) is the number of new cases reported per year per 1000 of the population, whereas the prevalence is the number of people with IDDM at any given point in time. For chronic diseases the incidence is almost always less than the prevalence.

The prevalence of a disease can markedly affect the clinical utility of a test in different settings. For example, let us consider a test where the sensitivity is found to be 90% and the specificity 98%, for a disease with a prevalence of 0.1 or 10%. Suppose 10,000 patients are tested.

The simplest way is to construct a 2 × 2 table as before, but this time arrange the relative proportions of affected and unaffected patients to reflect the prevalences (see Table 6.7).

Table 6.7. 2 × 2 tables for different levels of prevalence.

10,000 patients	Disease status (Prevalence 0.1%) Present	Absent	Disease status (Prevalence 10%) Present	Absent
Total tested	10	9990	1000	9000
Test positive	9	200	900	180
Test negative	1	9790	100	8820
Sensitivity (%)	90.00		90.00	
Specificity (%)	98.00		98.00	
False-positive rate (%)		2.00		2.00
Screen-positive rate (%)		2.09		10.80
Positive likelihood ratio		45		45
PPV (%)		4.31		83.33
NPV (%)		99.99		98.88
OAPR		1:22		5:1

If this test is used in a population with the lower prevalence of the disease, only about 4% of the patients who test positive actually have the disease. In contrast, 99.99% of patients with a negative test result do not have the disease (this is not surprising, since even in the absence of the test 99.9% are unaffected). If this is the only diagnostic information available, 22 unaffected individuals are misdiagnosed for every correct identification of a patient with the disease. In the higher prevalence population, only one unaffected individual is misdiagnosed for every five affected cases.

One term used almost exclusively in screening situations where the test is not diagnostic, but serves to identify a sub-population at increased risk of the disease is the **screen-positive rate** (**SPR**). Numerically it is the proportion of the entire population with a positive test result. By reference to Table 6.6:

$$\text{Screen-positive rate} = (a + b)/(a + b + c + d)$$

In populations with a low prevalence the screen-positive rate is numerically very close to the false-positive rate.

A further index of clinical utility that has been proposed is the **diagnostic efficiency**. This is the proportion of correct diagnoses. For example, using the coded 2 × 2 contingency table this would be

$$\text{Diagnostic efficiency} = (a + d)/(a + b + c + d)$$

This is not a particularly useful index as it gives the same 'value' to a false-positive as to a false-negative result.

LIKELIHOOD RATIO

For any test it is possible to calculate the probability of a positive result in patients with the disease, compared to the probability in patients not having the disease. The ratio of these probabilities is called the **positive likelihood ratio**, and is calculated as

$$\text{LR+} = \frac{\text{Probability (disease+, test+)}}{\text{Probability (disease−, test+)}} = \frac{\text{sensitivity}}{1 - \text{specificity}} \quad (6.17)$$

Similarly the **negative likelihood ratio** is the ratio of the probability of obtaining a negative test result in patients without the disease, compared to the probability in patients with the disease:

$$\text{LR−} = \frac{\text{Probability (disease−, test−)}}{\text{Probability (disease+, test−)}} = \frac{\text{specificity}}{1 - \text{sensitivity}} \quad (6.18)$$

The positive likelihood ratio measures the ability of the test to increase the certainty of a diagnosis. The prevalence of a disease is the probability of the disease being present in the group of patients being tested. The **odds** of having the disease are, therefore,

$$\text{Odds} = \frac{\text{Prevalence}}{(1 - \text{prevalence})} \quad (6.19)$$

For example, if the prevalence of the disease is 10%, i.e. 0.1, then the odds would be 0.11 or 9:1 against the disease being present. This is termed the **pre-test odds**. The odds of having the disease following the test are termed the **post-test odds**. By **Bayes' theorem** (Bayes, 1958):

Post-odds = pre-test odds × positive likelihood ratio

which demonstrates how the positive likelihood ratio measures the change in certainty about the diagnosis. For example, if the test result indicates a positive likelihood ratio of 2.0, this would modify the pre-test odds of 9:1–18:1. The calculation of likelihood ratios provides some measure of the degree of change in probability due to the test but, as with estimates of sensitivity and specificity, still gives proportional rather than numerical information, and takes no account of the prevalence of the disease. That is to say a high positive likelihood ratio might indicate a useful test but does not indicate whether or not the test is useful in predicting the presence or absence of disease in a population of low prevalence.

CONTINUOUS MEASUREMENT, ROC CURVES

So far the discussions have centered around the simplest case, where the test is either unambiguously positive or negative, and the disease either present or absent. This is rarely the case with assays where the result is a continuous measurement. The situation is, however, directly analogous to that of a positive and negative test result if a cut-off value is taken above which there is a high probability that the disease exists.

The choice of cut-off should not be based solely on statistical considerations, however. The cut-off must take into account the relative medical, ethical, psychological and financial costs associated with a false-positive and false-negative result. For example, quite different cut-offs are used when screening for open *spina bifida* using maternal serum AFP, where a positive result leads to the offer of amniocentesis and further tests, compared to the subsequent choice of cut-off used in assessing amniotic fluid AFP, where a positive result may lead to an offer of termination of pregnancy.

For a test in which a high concentration of analyte is associated with disease, raising the cut-off point increases the specificity of the test, while decreasing the sensitivity, and *vice versa*. Figure 6.17 shows this relationship for two overlapping Gaussian distributions of values for the affected and unaffected populations.

The two parameters are closely linked and are best represented by plotting the relationship between 1 – the specificity (false-positive rate) and the specificity (detection rate) at various cut-off levels. The curve obtained is known as a **receiver operating characteristic curve** (**ROC curve**), so called because it was first introduced during the second world war to optimize the discrimination of radar receivers to noise (false-positive results) and enemy aircraft (specificity).

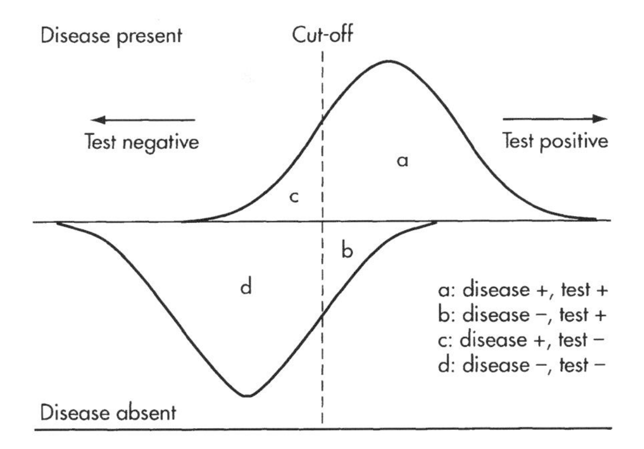

Fig. 6.17 Sensitivity and specificity.

Figure 6.18 shows theoretical ROC curves for a test with no discrimination (curve A at 45°), tests with increasing clinical utility (curves B, C and D), and a diagnostic test (curve E) with virtually 100% specificity and sensitivity. (It is impossible to prove that a test has precisely 100% specificity and 100% sensitivity in samples taken from a population – this can only be demonstrated by examining the entire population). If the 'cost' of a false-negative result is the same as a false-positive result, then the optimal cut-off point maximizes the sum of the specificity and the sensitivity, i.e. the point nearest the top left-hand corner.

For the majority of cases, where the consequences of false-negative and false-positive results are disproportionate, then the choice of cut-off is far more subjective, particularly when one takes into account the incidence of the disease in the population being tested. Curve B represents the type of performance expected from a typical screening test, the purpose of which would be to select a sub-population of increased risk of having the disease who would benefit from further testing with an appropriate

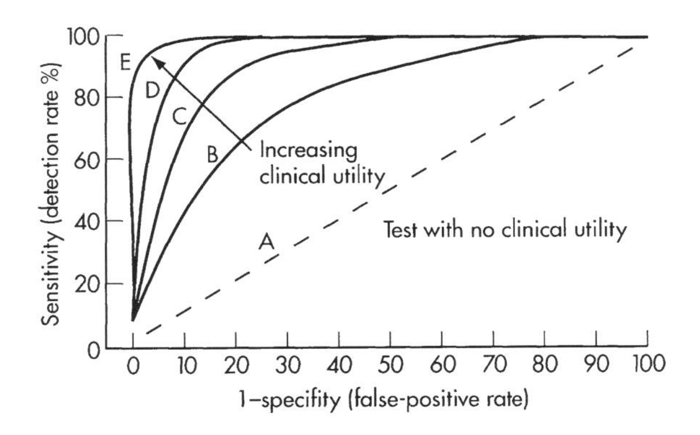

Fig. 6.18 ROC curves.

diagnostic procedure (use of which is restricted perhaps because it may be hazardous or expensive).

The ROC method is of limited applicability in assessing tests where the specificity is very high, but is extremely valuable in two other situations. First when assessing the utility of screening tests where the specificity may be relatively low in comparison to the sensitivity, and second in the comparison of two competing tests.

For small studies there is considerable noise in the relationship between the points. There are two solutions: either to draw the best fitting curve through them (non-parametric approach) or to use a parametric approach based on the mean and standard deviations of the observed Gaussian distributions in affected and unaffected subjects. The non-parametric method makes no assumptions regarding normality of the data, but random noise may lead to situations where two competing tests show a spurious apparent reversal of clinical utility at different cut-off points. The parametric approach requires fewer data to achieve the same degree of precision of the estimates, but requires an assumption of normality in both populations.

Zweig and Campbell (1993) and Henderson (1993) have recently reviewed the central role of ROC plots in the clinical evaluation and comparison of analytical methods.

CONFIDENCE INTERVAL ANALYSIS OF CLINICAL PERFORMANCE

Estimates of clinical performance are subject to sampling error, particularly if the number of patients is small, a likely situation when the disease or disorder is relatively uncommon. Claims are often made that one test gives better clinical performance than another, yet seldom in the clinical chemistry literature, and almost never in manufacturers' package inserts or promotional material, are such claims supported by confidence interval analysis. It is quite often possible for test A to be better than test B at one cut-off level, and for the situation to be apparently reversed at another.

Fortunately the confidence intervals are easy to calculate. Referring once again to the 2 × 2 contingency table shown previously, each of the values represents a simple proportion. The approximate standard error for any simple proportion, p, drawn from a sample of size N is

$$\mathrm{SE} = \sqrt{\frac{p(1-p)}{N}} \qquad (6.20)$$

The 95% confidence interval for the proportion in a population is given by

$$p - (1.96\mathrm{SE}) \text{ to } p(1.96\mathrm{SE}) \qquad (6.21)$$

The value 1.96 is the appropriate multiplier for the 95% confidence interval. For the 90% and the 99% confidence intervals the figures are 1.645 and 2.576, respectively.

Table 6.8. A 2 × 2 table.

	Disease status		
	Present	**Absent**	**Total**
Test positive	95	2	97
Test negative	5	98	103
Total	100	100	
Test positive	a	b	$a+b$
Test negative	c	d	$c+d$
Total		$a+c$	$b+d$
Sensitivity	$= a/(a+c)$	= 75/100	= 75.0%
Specificity	$= d/(c+d)$	= 90/100	= 90.0%
PPV	$= a/(a+d)$	= 75/85	= 88.2%
NPV	$= d/(c+d)$	= 90/115	= 78.3%

Table 6.9. Confidence intervals for 2 × 2 table.

	p	n	95% CI
Sensitivity	0.75	100 $(a+c)$	66.5 – 83.5%
Specificity	0.90	100 $(c+d)$	84.1 – 95.9%
PPV	0.88	85 $(a+b)$	81.4 – 95.1%
NPV	0.78	115 $(c+d)$	70.7 – 85.8%

Referring once again to the 2 × 2 table shown earlier. Each of the values in Table 6.8 represents a simple proportion such that the confidence intervals are as shown in Table 6.9.

The estimates are reasonably precise for samples greater than 30 in number and/or the proportions are less than 90%. A more detailed treatment of confidence intervals is given by Armitage and Berry (1987) and Gardner and Altman (1989). Simel *et al.* (1991) give general methods applicable to diagnostic tests with multiple outcomes.

PROBABILISTIC INTERPRETATION OF TEST RESULTS

Medicine is a science of uncertainty and an art of probability.

Sir William Osler (1849–1919)

Immunoanalytical techniques provide a wealth of information from a single measurement. Most of this information is lost in the diagnostic process by reduction to a dichotomous answer: is the value normal or abnormal, or if monitoring a patient's progress, has the value changed or stayed the same? The human brain is poor in evaluating numerical information – we reduce the information to manageable proportions by categorization of data (Hammond, 1990). In most cases this matters little. However, faced with a number of alternative diagnoses, the decision making process would be enhanced by an understanding not only of whether or not the levels of an analyte are raised, but precisely by how much, and in particular, how such information can be combined with other clinical and immunodiagnostic results in order to make the most optimal differential diagnosis (Dawson, 1993).

Patients with a test result above a defined cut-off value have an increased chance of having the disease, defined by the positive likelihood ratio.

However, the positive likelihood ratio refers to the group as a whole and not for any individual within that group, regardless of their individual analyte results. If the Gaussian distributions in the affected and unaffected populations are known then it is possible in the same way to calculate an individual likelihood ratio corresponding to the patient's actual result, and not just on the basis of whether or not it lies above some cut-off value. Such calculations are essential in cases where individuals within a group being tested have different pre-test odds of having the disease.

A good practical example is in the use of maternal serum AFP as a screening test for fetal Down's syndrome (Cuckle *et al.*, 1987). Here the pre-test probability of having an affected pregnancy ranges from 0.00065 (odds 1:1530) at age 20 to 0.0086 (odds 1:112) at age 40. Setting a single maternal serum AFP cut-off and offering the diagnostic test of amniocentesis and fetal chromosome analysis only on the basis of a low maternal serum AFP level, would mean that many younger women at low risk would be offered amniocentesis, whereas many older women with high pre-test odds of having an affected pregancy would be denied the opportunity. The only sensible way of using the test is by multiplying the pre-test odds by the individual likelihood ratio to derive the post-test odds (or risk) and to offer amniocentesis solely on the final odds or risk estimate. Figure 6.19 shows the basis for the calculation of individual positive likelihood ratios. Note that the point of neutrality (likelihood ratio equal to 1.0) is at the point of overlap of the two Gaussian distributions rather than at the mean of the unaffected distribution.

Two calculations are involved, first using the means and standard deviations for the affected distribution, and second using those for the unaffected distribution:

$$f_D(x) = \frac{1}{\sigma_D\sqrt{2\pi}} e^{-1/2((x-\bar{x}_D)/\sigma_D)^2} \qquad (6.22)$$

and

$$f_N(x) = \frac{1}{\sigma_D\sqrt{2\pi}} e^{-1/2((x-\bar{x}_N)/\sigma_N)^2} \qquad (6.23)$$

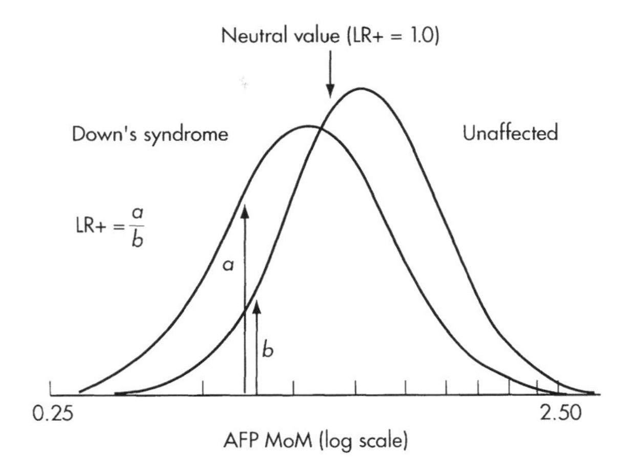

Fig. 6.19 Calculation of likelihood ratios.

where $\bar{x}_N$ and $\bar{x}_D$, σ_N and σ_D represent the means and standard deviations for unaffected and affected distributions, respectively, and x the analyte value.

The likelihood ratio is the ratio of the first calculation over the second:

$$\text{LR+} = \frac{f_D(x)}{f_N(x)} \qquad (6.24)$$

The risk of giving birth to a child with Down's syndrome is then calculated by multiplying the right-hand side of the odds ratio by the likelihood ratio. For example, for a 35-year-old woman with an AFP result of 0.5 multiple of the median (MoM) the right-hand side of the pre-test odds of 1:384 (one chance of being affected to 384 chances of being unaffected) would be divided by a likelihood ratio of 2.214 to give post-test odds of 1:173 (this is equivalent to multiplication of the left-hand side of the odds ratio by the likelihood ratio, but leads to simpler interpretation of low risks).

The calculation of likelihood ratios is not restricted to univariate distributions. Wald *et al.* (1989) described the use of likelihood ratios derived from trivariate distributions of maternal serum AFP, human chorionic gonadotropin and unconjugated estriol as markers, again for fetal Down's syndrome, and there is no theoretical limit to the number of test results that may be combined in such a way. Reynolds and Penney (1990) give a straightforward guide to the mathematical calculations.

ANALYTICAL GOALS

'A marksman without a target can never miss.'

This chapter has examined some of the major technical and clinical concepts associated with immunoanalytical techniques. The hierarchical nature of clinical diagnosis was briefly addressed with technical requirements arising from the needs of the patient for timely and appropriate action, and the needs of the physician for accurate and reliable information with which to assist in making a sound diagnosis. How may the needs of the patient and clinician define the required technical quality?

Quality standards for immunoassays have traditionally been driven by what has been available, rather than on what is required. Although there are many good reasons why this is so, in a sense it could be viewed as putting the cart before the horse. Quality control is centered around the numerous international, national and local QC schemes that have been established for many years and have proved invaluable in monitoring the technical performance standards of practically all clinically useful immunoassays. Nearly all, however, suffer from the disadvantage of defining unacceptable performance solely in relation to the performance of their peers. As eloquently stated by Gilbert (1977), "any QC scheme that declares a result as unacceptable must, at some time, face the question of defining acceptability."

Barnett (1989) has argued that at present:

> There is no evidence that patients or physicians are harmed by our present techniques as long as the techniques meet existing standards. Theoretical considerations aside, if we cannot do better, even in the best laboratories, it will not help to set goals requiring better performance.

Such a premise is difficult to prove one way or the other, but to the extent that the current assay quality standards already define current clinical performance, the argument could be viewed as somewhat circular.

Analytical goals based on clinical needs are a rational way forward. In the absence of target specifications there is no goal in sight for which to strive, nor an identifiable end-point beyond which further improvements in diagnostic technology will have little clinical impact. In addition, an understanding of the influence that different aspects of assay quality have on clinical performance enables priority to be focused on the most important.

How are clinical needs identified and quantified? Several proposals have been made.

Analytical Goals – State of the Art

This is defined as being the top 20% of current assays, on the grounds that if performance can be achieved by the top 20% of assays, then it should be attainable by all. This approach has some considerable appeal. There are, however, several serious disadvantages.

Advantages

- This is a simple approach.
- It leads to a continuous process of improvement, which fits in with current quality philosophies.
- In theory it should minimize the spectrum of performance between laboratories.

Disadvantages

- It provides a moving target.
- It does not define target in terms of need.
- It focuses on technical rather than clinical aspects (for example, between-method numerical differences may not be strictly relevant in the clinical setting, provided method-specific reference intervals are used).
- Quality control samples may not reflect true performance in the laboratory. Special attention may be placed on their analysis, particularly when continued accreditation may depend upon satisfactory performance in national quality control schemes (Rowan *et al.*, 1984).
- Quality control samples are often processed to add or remove analyte. Analytes added may not reflect the spectrum of biologically circulating variants of such analytes. In addition processing may interfere with the matrix of the sample so that it no longer reflects the performance of patient samples.
- Quality control samples measure performance at discrete intervals. These may not be at critical clinical decision points.
- It provides no end-point beyond which further improvement in precision has no clinical utility.

Analytical Goals – Clinicians' Opinions

At first sight it would appear logical that, since clinicians are the first customers of the laboratory, they should define the quality requirements. The most common approach has been to ask clinicians what difference in analyte concentrations would trigger a change in clinical action, usually by means of clinical vignettes (Thue *et al.*, 1991). Again there are several advantages and disadvantages.

Advantages

- The approach has immediate and logical appeal.
- It conforms to current quality philosophies.

Disadvantages

- Published studies indicate a wide range of responses to individual case studies. In addition, taking the median response as being the desired goal would only satisfy half of the clinicians.
- Clinicians' responses are colored by their experience of what the laboratory can already achieve.
- Clinical action depends not on the degree of change, but also on the concentration level from which the change is measured.
- Clinical responses in practice are likely to differ from 'paper exercises'.

- With the clinical vignette approach, clinicians' opinions may also be influenced by the background to the hypothetical patient and the format of the questions used to elicit an opinion.
- Clinical judgement is influenced by experience and 'gut-feeling'. Actions may well differ depending upon the circumstances of presentation and past history.
- Values are usually dichotomized into abnormal or normal, high or low, increasing or decreasing. The human brain is exceedingly poor at accurately assessing and weighting numerical information.

Analytical Goals Based on Reference Intervals

Tonks (1963) first addressed the issue of quality specifications derived from clinical need with the proposal that the total allowable limits of error, twice the coefficient of variation, should be set in percentage terms as

$$\mathrm{TAE}(\%) = \leq 0.25 \times \frac{\text{reference interval}}{\text{mean of reference interval}} \times 100 \tag{6.25}$$

However, the fraction 0.25 is empirical and the reference intervals depend to some extent upon the statistical method used to determine them. The main disadvantage, however, is that the analytical precision already contributes to the variance of the reference interval. Nonetheless the approach has advantages, in particular that the concept is simple and that reference intervals are readily available.

Analytical Goals Based on Biological Variation

The clinical performance of immunoanalytical techniques is defined by the separation between the mean values for healthy and diseased states and by the variation in values. The latter has three major sources: variation within an individual, variation between individuals and variation due to analytical error (see Figure 6.20).

Where CV_i is the within-subject coefficient of variation (that from day to day or week to week), CV_g is the between-individual, or group, biological variation (the variation in homeostatic set points between individuals), CV_a is the analytical precision, and CV_t the total imprecision.

In terms of defining quality specifications, attention should, therefore, focus on the relative contribution that analytical variation makes compared to the overall biological variation. The emphasis needs to be on relative contribution rather than the absolute. If one considers two analytes, one of which has a reference interval twice as wide as the other, then it is reasonable to argue that the precision requirements of the assays should reflect that difference in order to have the same degree of influence on clinical performance.

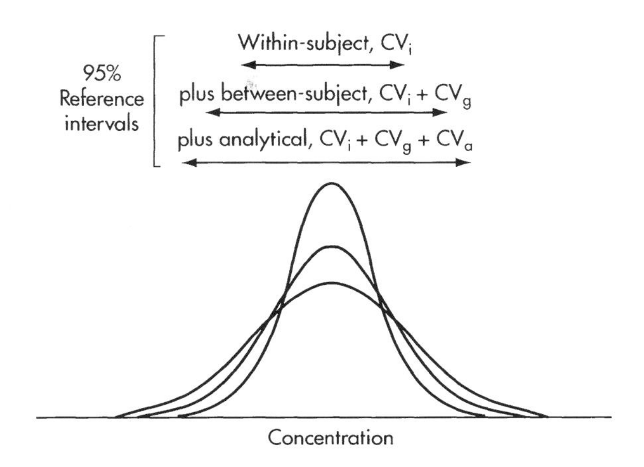

Fig. 6.20 Components of variation.

Analytical Goals for Precision

It has been recommended that in order to limit an increase in biological variation to approximately 10% (actually 11.8%), the maximum allowable analytical variability should be less than one half of the relevant biological variation (Harris, 1979):

$$CV_a \leq 0.5\sqrt{CV_i^2 + CV_g^2} \tag{6.26}$$

where CV_i, CV_g and CV_a are the within-subject, between-subject and analytical variation, respectively (Proceedings of the Subcommittee on Analytical Goals in Clinical Chemistry, 1979).

This concept is of most value in the diagnosis of disease in a population of patients of unknown status. The allowable imprecision takes into account both the within- and between-individual biological variation. Tests are, however, commonly used not only in the diagnosis of disease but also in monitoring. Here the biological variation is restricted to that within an individual subject, and hence the more appropriate analytical goal is that defined purely on the basis of the within-subject variability, CV_i:

$$CV_a \leq 0.5CV_i \tag{6.27}$$

As any particular assay is likely to be used for both purposes it has been recommended that this more stringent definition be adopted to ensure that assay performance meets the more demanding requirements of monitoring.

Analytical Goals for Accuracy

The analytical goal for accuracy is that there should be no bias. Although this must be the target, some definition of what constitutes an allowable tolerance is useful, if only as a means of quantifying the effects. It has been proposed (Fraser *et al.*, 1997) that the maximum allowable relative deviation should be less than one quarter of the overall biological variation (within- plus between-subject variation):

$$\text{Maximum allowable deviation} \leq 0.25\sqrt{CV_i^2 + CV_g^2} \qquad (6.28)$$

Analytical Goals for Drug Monitoring

Another common use of immunoanalytical techniques is in the field of therapeutic drug monitoring. Provided that absorption is rapid in relation to elimination, the elimination follows first-order kinetics, and the drug is distributed evenly throughout a single compartment, then steady-state conditions, where the concentration of drug varies around a set-point, is achieved after five half-lives. This fluctuation, expressed as four standard deviations, may be viewed as being analogous to the within-subject biological variation, and it has been proposed by Fraser (1987) that the desirable analytical goal for precision be expressed as

$$CV_a \leq 0.25\frac{(2^{T/t} - 1)}{(2^{T/t} + 1)} \times 100 \qquad (6.29)$$

where CV_a is the analytical CV, T is the dosing interval and t the elimination half life. Stewart and Fraser (1989) have derived analytical goals for a variety of common therapeutic drugs.

Analytical Goals for Interference

Few immunoanalytical techniques are free from some form or other of interference. Fuentes-Arderiu and Fraser (1991) have proposed that, since the total analytical error (TAE) should ideally be less than half the within-subject biological variation, the maximum allowable systematic error produced by an interfering substance, I, should be

$$I = CV_t - (1.96CV_a + SE) \qquad (6.30)$$

where CV_t is the within-subject biological variation, CV_a is the analytical imprecision, and SE is the systematic error.

These goals can equally be applied to cross-reactivity, matrix effects and carry-over of sample in automated immunoassay instruments and samplers.

Applicability

Much of the work on the derivation of analytical goals has been focused on clinical chemistry. In contrast, relatively little work has been performed in formulating quality standards for immunoanalytical techniques, although the concepts are both directly applicable and transferable. The main reason is that, for many commonly used immunoassays, data on within-subject biological variation has not been available until recently. Recent work by Browning (1989), Ricos and Arbos (1990), and Valero-Politi and Fuentes-Arderiu (1993) on 19 hormones has enabled some progress to be made in developing analytical goals for imprecision for the analytes listed in Table 6.10.

It is interesting to note the wide spectrum of allowable precision between these hormones. This reflects the differences in within-subject biological variation. For some analytes their current analytical performance comes very close to, or exceeds, these standards. Others, such as total T_4 and TBG, fall far short of meeting the proposed standards. It is important to recognize, however, that these are goals. The fact that the goal is unlikely to be met does not diminish its value as a target worth aiming for.

Table 6.10. Analytical goals for hormones.

Hormone	%CV A	%CV B	%CV C
11-Deoxycortisol	10.7		
17-OH-Progesterone	9.8		
Aldosterone	14.7		
Androstenedione	5.8		
Cortisol	7.6		
DHEA-Sulfate	0.6		
Estradiol	10.9		
Testosterone	4.2		5.4
C-peptide	4.7		
Insulin	7.6		
FSH	1.5 (women)		8.7 (men)
LH	6.2 (women)		12.0 (men)
Prolacin	3.5		
Total T_3	4.4	5.2	
Total T_4	1.8	2.5	
TBG	2.2		
TSH	9.7	8.1	
Free T_3		3.9	
Free T_4		4.7	

Analytical goals for imprecision derived from within-subject biological variation. A: data from Ricos and Arbos (1990); B: data from Browning (1989); C: data from Valero-Politi and Fuentes-Arderiu (1993).

SUMMARY

The preceding chapters have addressed some elements of the design and function of immunoassays, particularly in relation to meeting the ever more demanding technical requirements of sensitivity, precision and accuracy. This chapter has attempted to link those technical requirements to the clinical need.

Much work remains to be done, not only in accurately defining analytical goals for immunoassay, but also in developing the technology required to achieve them.

It is generally assumed that technological advances in immunodiagnostics are directed towards improving the standard of clinical diagnosis. It is important not to lose sight of the fact that developments in technology are also driven by the laboratory operational requirements for automation and high throughput. Such requirements may well conflict with fundamental principles in assay design leading to a trade-off between speed and clinical performance (Woodhead, 1991).

Some of the concepts related to clinical performance characteristics have been reviewed. Particular mention was made of the fact that much of the probabilistic information contained in the analyte values themselves is lost by reduction to a dichotomous outcome when formulating decisions. The combination of Bayesian probabilities and likelihood ratios may seem complex in a routine laboratory or clinical setting, but provides a far better mechanism for retaining information than commonly used rule systems. For diseases where multiple tests are utilized, likelihood ratios derived from multivariate or logistic regression models sometimes provide the only satisfactory way of combining the data in a statistically sound fashion.

The next revolution in immunoassay techniques may not lie with increasing automation, faster throughput, more sensitive assays or better precision. The next revolution may well be in how we interpret the results we already produce.

REFERENCES

Abramowitz, M. and Stegun, I.A. (eds), *Handbook of Mathematical Functions, 933* (Dover Publications, New York, 1972).

Armitage, P. and Berry, G. *Statistical Methods in Medical Research* 2nd edn, 117–120 (Blackwell, Oxford, 1987).

Barnett, R.N. Error limits and quality control. *Arch. Pathol. Lab. Med.* **113**, 829–830 (1989).

Bayes, T. An essay towards solving a problem in the doctrine of chances. *Biometrika* **45**, 293–315 (1958).

Bendtzen, K., Svenson, M., Jonsson, V. and Hippe, E. *Immunol. Today* **11**, 167–169 (1990).

Bland, J.M. and Altman, D.G. Statistical methods for assessing agreement between two methods of clinical measurement. *Lancet* **i**, 307–310 (1986).

Boscato, L.M. and Stuart, M.C. Incidence and specificity of interference in two-site immunoassays. *Clin. Chem.* **32**, 1491–1495 (1986).

Boscato, L.M. and Stuart, M.C. Heterophilic antibodies: a problem for all immunoassays. *Clin. Chem.* **34**, 27–33 (1988).

Browning, M.C.K. Analytical goals for quantities used to assess thyrometabolic status. *Ann. Clin. Biochem.* **26**, 1–12 (1989).

Cotlove, E., Harris, E.K. and Williams, G.Z. Biological and analytical components of variation in long-term studies of serum constituents in normal subjects. III. Physiological and medical complications. *Clin. Chem.* **16**, 1028–1032 (1970).

Cuckle, H.S., Wald, N.J. and Thompson, S.G. Estimating a woman's risk of having a pregnancy associated with Down's syndrome using her age and serum alpha-fetoprotein level. *Br. J. Obstet. Gynaecol.* **94**, 387–402 (1987).

Dawson, N.V. Physician judgement in clinical settings: methodological influences and cognitive performance. *Clin. Chem.* **39**, 1468–1480 (1993).

Deming, W.E. *Statistical Adjustment of Data*, 184 (Wiley, New York, 1943).

Ekins, R.P. Basic principles and theory. *Br. Med. Bull.* **30**, 3–11 (1974).

Ekins, R.P. The precision profile: its use in assay design, assessment and quality control. In: *Immunoassays for Clinical Chemistry*, 2nd edn (eds Hunter, W.M. and Corrie, J.E.T.), 76–105 (Churchill-Livingstone, Edinburgh, 1983).

Fraser, C.G. Desirable standards of performance for therapeutic drug monitoring. *Clin. Chem.* **33**, 387–390 (1987).

Fraser, C.G., Hyltoft Peterson, P.R., Libeer, J.C. and Ricos, C. Proposals for setting generally applicable quality goals solely based on biology. *Ann. Clin. Biochem.* **34**, 8–12 (1997).

Fuentes-Arderiu, X. and Fraser, C.G. Analytical goals for interference. *Ann. Clin. Biochem.* **28**, 393–395 (1991).

Gardner, M.J. and Altman, D.G. *Statistics with Confidence* 28–33 (British Medical Association, London, 1989).

Gilbert, R.K. CAP interlaboratory survey data and analytical goals. In: *College of American Pathologists Conference Report (1976): Analytical goals in Clinical Chemistry* (ed Elevitch, F.R.), 63–73 (College of American Pathologists, Skokie, IL, 1977).

Hammond, K.R. Intuitive and analytical cognition: information models. In: *Concise encyclopedia of information processing in systems and organisations* (ed Sage, A.), 306–312 (Pergamon Press, Oxford, 1990).

Harris, E.K. Statistical principles underlying analytical goal setting in clinical chemistry. *Am. J. Clin. Pathol.* **72**, 374–382 (1979).

Harris, E.K. Statistical aspects of reference values in clinical pathology. *Prog. Clin. Pathol.* **8**, 45–66 (1981).

Healy, M.J.R. Outliers in clinical chemistry. *Clin. Chem.* **25**, 675–677 (1979).

Henderson, A.R. Assessing test accuracy and its clinical consequences: a primer for receiver operating characteristics curve analysis. *Ann. Clin. Biochem.* **30**, 521–539 (1993).

Käpyaho, K., Tanner, P. and Weber, T. Effect of complement binding on a solid phase immunometric TSH assay. *Scand. J. Clin. Lab. Invest.* **49**, 211–215 (1989).

Knight, G.J., Palomaki, G.E. and Haddow, J.E. Maternal serum alphafetoprotein: a problem with a test kit. *N. Engl. J. Med.* **314**, 516 (1986).

Latner, C. (ed), *Geigy Scientific Tables* 8th edn (Geigy, Basle, Switzerland, 1982).

Linnet, K. Evaluation of regression procedures for methods comparison studies. *Clin. Chem.* **39**, 424–432 (1993).

Masson, P.L., Cambiaso, C.L., Cllet-Cassart, D., Magnusson, C.G.M., Richards, C.B. and Sindic, C.J.M. Particle counting immunoassay (PACIA). *Methods Enzymol.* **74**, 106–139 (1981).

Miller, I.J. and Valdes, R. Approaches to minimizing interference by cross-reacting molecules in immunoassays. *Clin. Chem.* **37**, 144–153 (1991).

Passing, H. and Bablock, W. A new biometrical procedure for testing the equality of measurements from two different analytical methods. *J. Clin. Chem. Clin. Biochem.* **21**, 709–720 (1983).

Passing, H. and Bablock, W. Comparison of several regression procedures for method comparison studies and determination of sample sizes. *J. Clin. Chem. Clin. Biochem.* **22**, 431–445 (1984).

Proceedings of the sub-committee on analytical goals in clinical chemistry. Analytical goals on clinical chemistry: their relationship to medical care. *Am. J. Clin. Pathol.* **79**, 624–630 (1979).

Raggart, P. Duplicates or singletons? – An analysis of the need for replication in immunoassay and a computer program to calculate the distribution of outliers, error rate and the precision profile from assay duplicates. *Ann. Clin. Biochem.* **26**, 26–37 (1989).

Reynolds, T.M. and Penney, M.J. The mathematical basis of multivariate risk screening: with special reference to screening for Down's syndrome associated pregnancy. *Ann. Clin. Biochem.* **27**, 452–458 (1990).

Ricos, C. and Arbos, M.A. Quality goals for hormone testing. *Ann. Clin. Biochem.* **27**, 353–358 (1990).

Rowan, R.M., Laker, M.F. and Alberti, K.G.M.M. The implications of assaying external quality control sera under 'special conditions'. *Ann. Clin. Biochem.* **21**, 64–68 (1984).

Sadler, W.A. and Smith, M.H. Use and abuse of imprecision profiles: some pitfalls illustrated by computing and plotting confidence intervals. *Clin. Chem.* **36**, 1346–1350 (1990).

Sadler, W.A., Murray, L.M. and Turner, J.G. Minimum distinguishable difference in concentration: a clinically orientated translation of assay precision summaries. *Clin. Chem.* **38**, 1773–1778 (1992).

Simel, D.L., Samsa, G.P. and Matchar, D.B. Likelihood ratios with confidence: sample size estimation for diagnostic test studies. *J. Clin. Epidemiol.* **44**, 763–770 (1991).

Sokal, R.R. and Rohlf, F.J. *Biometrics* 481–486 (W.H. Freeman, San Francisco, 1969).

Soldberg, H. The theory of reference values. *J. Clin. Chem. Clin. Biochem.* **25**, 545–656 (1987).

Stewart, M.J. and Fraser, C.G. Desirable performance standards for assays of drugs. *Ann. Clin. Biochem.* **26**, 220–226 (1989).

Thue, G., Sandberg, G. and Fugelli, P. Clinical assessment of haemoglobin values by general practitioners related to analytical and biological variation. *Scand. J. Clin. Lab. Invest.* **51**, 453–459 (1991).

Tonks, D.B. A study of the accuracy and precision of clinical chemistry determinations in 170 Canadian laboratories. *Clin. Chem.* **9**, 200–205 (1963).

Valero-Politi, J. and Fuentes-Arderiu, X. Within- and between-subject biological variations of follitropin, lutropin, testosterone and sex hormone-binding globulin in men. *Clin. Chem.* **39**, 1723–1725 (1993).

Wald, N.J., Cuckle, H.S., Densem, J.W., Nanchahal, K., Royston, P., Chard, T., Haddow, J.E., Knight, G.E., Palomaki, G.J. and Canick, J.A. Maternal serum screening for Down syndrome in early pregnancy. *Br. Med. J.* **297**, 883–887 (1989).

Woodhead, J.S. New Technology and improved analytical performance: do they always go together. *J. Clin. Immunoassay* **14**, 235–238 (1991).

Zweig, M.H. and Campbell, C.C. Receiver-operating characteristics (ROC) plots. A fundamental evaluation tool in clinical medicine. *Clin. Chem.* **39**, 561–577 (1993).

FURTHER READING

Altman, D.G. *Practical Statistics for Medical Research* (Chapman & Hall, London, 1991).

Fraser, C.G. Analytical goals are targets, not inflexible criteria of acceptability. *Am. J. Clin. Pathol.* **89**, 703–705 (1988).

Fraser, C.G. and Hyltoft Peterson, P.R. Proposed quality specifications for the imprecision and inaccuracy of analytical systems for clinical chemistry. *Eur. J. Clin. Chem. Clin. Biochem.* **30**, 311–317 (1992).

Fraser, C.G. and Hyltoft Peterson, P.R. Desirable standards for laboratory tests if they are to fulfil clinical needs. *Clin. Chem.* **39**, 1447–1455 (1993).

Fraser, C.G. and Hyltoft Peterson, P.R. Analytical performance characteristics should be judged against objective quality specifications. Editorial. *Clin. Chem.* **45**, 321–323 (1999).

Galen, R.S. and Gambino, S.R. *Beyond Normality: the Predictive Value and Efficiency of Medical Diagnoses* (Wiley, New York, 1975).

Gore, S. and Altman, D.G. *Statistics in Practice* (British Medical Association, London, 1982).

Haeckel, R. (ed), *Evaluation Methods in Laboratory Medicine* (VCH, Weinheim, Germany, 1993).

Klee, G.G. Tolerance limits for short-term analytical bias and analytical imprecision derived from clinical assay specificity. *Clin. Chem.* **39**, 1514–1518 (1993).

Linnet, K. Choosing quality-control systems to detect maximum clinically allowable errors. *Clin. Chem.* **35**, 834–837 (1989).

Magrid, E. (ed), Some concepts and principles of clinical test evaluation. Classification, analytical performance, monitoring and clinical interpretation. *Scand. J. Clin. Lab. Invest.* **52** (Suppl), 208 (1992).

1995 National Academy of Clinical Biochemistry Standards of Laboratory Practice Symposium on Thyroid Testing. *Clin. Chem.* **42** 119–192 (1996).

1995 National Academy of Clinical Biochemistry Standards of Laboratory Practice Symposium on Therapeutic Drug Monitoring. *Clin. Chem.* **44** 1072–1140 (1998).

Sebastian-Gambaro, M.A., Liron-Hernandez, F.J. and Fuentes-Arderiu, X. Intra- and inter-individual biological variability data bank. *Eur. J. Clin. Chem. Clin. Biochem.* **35**, 845–852 (1997).

Stockl, D., Baadenhuisjen, H., Fraser, C.G., Libeer, J.C., Petersen, P.H. and Ricos, C. Desirable routine analytical goals for quantities assayed in serum. Discussion paper from the members of the external quality assessment (EQA) Working Group A on analytical goals in laboratory medicine. *Eur. J. Clin. Chem. Clin. Biochem.* **33**, 157–169 (1995).

Thienpoint, L., Franzini, C., Kratochvila, J., Middle, J., Ricos, C., Siekmann, L. and Stockl, D. Analytical quality specifications for reference methods and operating specifications for networks of reference laboratories. *Eur. J. Clin. Chem. Clin. Biochem.* **33**, 949–957 (1995).

7 Immunoassay Development in the *In Vitro* Diagnostic Industry

Doug Brandt and Steve Figard

Developing an immunoassay entails both theoretical and empirical aspects. As discussed elsewhere in this book, a wealth of theoretical knowledge exists to guide the scientist in planning the assay format, selecting antibodies, determining optimal concentrations, and maximizing signal generation. For example, an understanding of antigen–antibody binding affinities will give the scientist a powerful tool for selecting the appropriate concentration of capture antibody in the assay, whether the format is competitive or immunometric (sandwich). In addition to theoretical considerations, there are an even greater number of empirical issues that confront the scientist. These include selecting solid phase and reporter molecules, reducing non-specific binding (NSB), maximizing reagent stability, developing an assay that is compliant with often diverse regulatory requirements across the globe, ensuring that design control practices are followed, such as those outlined by ISO 9000 and other regulatory guidance, and, most importantly, delivering an assay that the customer wants. Obviously, some of these will not apply to the academic scientist developing an immunoassay for in-house research only. In such cases, the assay developer is the customer, with fewer restrictions being placed on assay development and use.

Because of these diverse considerations, developing an immunoassay is a complex effort that requires careful planning, good scientific skills and insight, and coordination among the diverse groups involved in creating the assay. This complexity necessitates that a unified approach be utilized in order to gain understanding and lay the foundation for a solid immunoassay. Two issues are sometimes missing from this activity: ensuring first that development activities focus on what the customer needs, and second, that assay design and development result in a manufacturable product. The purpose of the chapter that follows is to discuss these issues and to present some guiding principles for early assay development primarily within the context of the *in vitro* diagnostics industry.

ASSAY DESIGN OVERVIEW

Essential to every immunoassay is the ability to detect an analyte in a physiologically based specimen. While there are a plethora of approaches to do this, the typical assay utilizes two important components: a solid phase and a conjugate. The solid phase serves to capture the analyte and thus allows complete separation from other components of the specimen; the conjugate serves to generate a signal, directly or indirectly, thereby detecting analyte. In addition to these two components, an assay may also use additional reagents to denature the specimen and allow analyte to be released, to passivate the solid phase to interfering substances present in the specimen, and/or to amplify the generation of signal.

Development of these reagents involves discovery, characterization, and manufacture. Discovery reveals that a particular reagent is important for assay performance. Characterization reveals the component(s), concentration, timing, or other process steps that are important in the production of this reagent. Manufacture selects the manufacturing ranges for final production. Within this process, the details of assay construction are up to the scientist and dependent on three key issues: the needs of the customer, the capability of the technology, and the requirements for consistent manufacture.

THE NEEDS OF THE CUSTOMER

A lack of sufficient customer input may well represent the single greatest deficiency in typical assay development efforts. This can result in the development of assays that do not meet the expectations, trade-off requirements, or market entry date needed by the customers. Expectations are not met because meaningful communication with the end-user customer does not occur in the early phases of product design. Thus, the characterization studies described above may not address the requirements that are essential to the customer. Trade-off requirements are

The Immunoassay Handbook. *Third Edition*
D. Wild (Ed.)

not met, not only because this communication did not occur, but also because clear performance priorities were not set. Thus, final manufacturing or operating ranges may not anticipate the trade-offs. For example, while exceptional functional sensitivity may be more desired or necessary than a large dynamic range for any given analyte, the reverse may be true for another. Finally, market entry dates are not met because of the technical complexity of modern immunoassay systems, poor planning, unforeseen interactions, and the multitude of changes in direction that sometimes occur during the product development cycle. More often than not, these changes are due to poorly defined customer input, or no input at all. Without clear input from real customers, opinion and speculation, rather than customer need, drive development. Stated another way, customer input should be a focus of early development of an assay. Those who develop the assay may move on to new challenges, but the needs of the customer, which should drive the design and development of immunoassays, usually change at a much slower pace.

THE CAPABILITY OF THE TECHNOLOGY

The capability of a given technology is an important aspect for the scientists who develop immunoassays. Unfortunately, this capability is often viewed too narrowly (i.e. on detection technology only) or is not driven by customer need. In the broadest sense, capability includes all aspects of a given technology that enable the technology to meet customer need. This includes reagents, instrument, and the interaction between reagents and instrument. More specifically, some other factors driving capability are detection technology, assay format (i.e. one-step versus two-step assay), antibody choice (analyte affinity and specificity), conjugation method, wash buffers, specimen diluent, pipetting accuracy and precision, mechanical movement, assay timing, and temperature control. Only as the scientist understands the impact of these components on assay performance can logical trade-offs be made during development. This is particularly important when considering instrument–reagent interactions.

Temperature control within an immunoassay analyzer is important to assay performance, but *how* important is it? No one would argue that pipetting accuracy is also critical, but *how* critical is it? These factors can only be known quantitatively as the scientist relates the variance of the instrument parameter, for example the temperature, to that of the assay, the control or panel value. As there are a number of instrument-controlled parameters that may impact a given assay's variance, the challenge to the assay development scientist is to identify which of these are more important and to determine just how important they are. We discuss methods that can yield such information in an efficient manner after we first briefly consider manufacturing issues.

THE REQUIREMENTS FOR CONSISTENT MANUFACTURING

Even if customer input is superb, and the capability of the technology extraordinary, all will be for nought if the manufacturing process is unreliable. Just as there are many factors that drive the capability of the technology, so there are also many that drive manufacturability. For both the solid phase and conjugate reagents, pH, protein concentration, ionic strength, and linker concentration are generally important factors, as well as the overcoat conditions used to passivate the solid phase and the coupling chemistry for conjugation. Reagent diluents, whether for solid phase or conjugate, are sometimes important for stability, reduction of NSB, and enhancement of signal. Important contributors to these effects include proteins, detergents, pH, ionic strength, and other agents.

Given the large number of variables typically encountered in the development of an immunoassay, a multivariate experimental approach is the method of choice to determine key drivers for assay performance, optimal concentrations and reaction conditions, and interactions between one or more key drivers. These factors form the 'knobs' that need to be 'turned' to influence the manufacturing output. Indeed, controlling these knobs is the key to consistent manufacture, whether of reagents or instrument. The rationale of the process of multivariate experimental design is given below.

EXPERIMENTAL DESIGN IN IMMUNOASSAY DEVELOPMENT

PREAMBLE

Experimental design in the context of immunoassay development requires the developer to go beyond the 'classical' change-one-and-only-one-thing-at-a-time approach. There are at least two reasons why it is important to do this. First, immunoassay development frequently occurs in an industrial setting, i.e. in the *in vitro* diagnostics industry, where product quality requires a thorough understanding of all aspects of a given product's composition and manufacturing processes. Second, immunoassays usually involve a multiplicity of interactions between reagent components, sample composition, and process parameters, making the entire affair far too complex and time consuming to dissect one step at a time, assuming it could be done at all by that methodology. Clearly there has to be a better way. The solution is frequently referred to as **Design of Experiments**, or **DOE**. DOE is a statistically based approach for planning and analyzing multivariate experiments with the goal of defining and/or optimizing a process. DOE designs allow the experimenter to simultaneously vary more than one input variable of a process and from the resulting data, either identify critical parameters for further study or construct a response surface model. This mathematical model defines the response relative to the input variables, allows the conditions for an optimum response to be determined, and permits the identification of those variables that interact with one another. All of these goals can be achieved much faster and with greater accuracy than with the traditional approach of changing one variable at a time. An additional advantage of DOE is the increased probability of finding the true optimum conditions to create the desired response, a goal that can be missed using traditional experimentation.

THE PROCESS IN OVERVIEW

In the optimization of a process or reagent formulation, DOE should be used in a two-step algorithm. First, a screening experiment should be done to identify the critical input parameters of the reagent formulation or process. Numerous input variables should be evaluated here. Second, a response surface modeling experiment of the critical parameters identified in the screening experiment should be executed. The model produced from these data may then be used to optimize the process or formulation. In some cases, preliminary non-DOE experiments may be carried out to define the region of experimentation (e.g. which components to evaluate in a diluent optimization). As always, model predictions should be verified prior to declaring the final process/formulation.

THE PROBLEMS ASSUAGED BY DOE

The investigator seeking to develop immunoassay reagents and methods will specifically face at least three difficulties which DOE methodology will help to overcome: first, experimental error, or noise; second, the possibility of confusing correlation with causation; last, as mentioned above, the complexity of the effects studied.

Experimental Error

Experimental error may be defined as variation in measured responses produced by perturbing influences, both known and unknown. Usually some small part of this variation can be directly attributed to error in measurement. Depending on its magnitude relative to the response being measured, important effects may be wholly or partially obscured by the experimental error. Worse still, the experimental error may cause the experimenter to reach erroneous conclusions.

Adequate experimental design and analysis using DOE methodology can greatly reduce the potential confusion caused by experimental error. In addition, statistical analysis generates measures of precision of the estimated quantities under study, permitting an objective evaluation of the presence of non-zero values for such quantities. The net effect is to increase greatly the probability that the investigator will come to true conclusions in the matter under study.

Confusion of Correlation with Causation

Correlation between two variables often occurs as a consequence of their each being associated with a third factor. Thus, correlation does not demonstrate that one variable causes the other. By using DOE principles of experimental design, and, in particular, randomization of data collection, data can be generated and analyzed in a way that will allow the valid deduction of causality to be achieved.

Complexity of Effects

When doing mathematical modeling of data as a function of input variables, the simplest relationship is a *linear* and *additive* one. Regrettably, *non-linear* and *interaction* effects abound in reality, and especially in immunoassay development. This leads to significantly more complex phenomena that can only be modeled by correspondingly more complex equations. DOE experimental designs allow the estimation of all four of these kinds of effects with the smallest possible transmission of error into the model.

THE PARTICULARS

Objectives

Generally speaking, the goal of any DOE endeavor will be to predict the performance of your process in a robust fashion with the least amount of experimental effort possible. Stated this way, the objectives encompass manufacturability (robustness), cost effectiveness (least amount of experimental effort), and understanding of the process (ability to predict performance). Because the answers arrived at are only as good as the questions asked, careful thought should be given to objectives, which should be clearly stated and outlined at the beginning. In the words of Henry David Thoreau, 'In the long run men hit only what they aim at.'

As has been mentioned previously, two specific objectives can be targeted by DOE methodology: first, screening, in which the experimenter sifts out of many control variables the critical few; second, response surface construction, in which an accurate map of the process is obtained. Screening designs contain more variables and assume a simpler surface for purposes of detection, whereas response surface designs evaluate fewer variables but permit more complex surface construction. In either case, the advent of computer software for design construction and analysis has greatly facilitated the use of this methodology, putting it into the hands of any experimenter with access to a computer.

Variables and Models

There are two types of variables: the dependent responses that are observed and the independent controls that affect the responses.

The dependent responses are, or should be the critical quality attributes of the processes that are measured to insure reproducible quality. They define the behavior characteristics of the product that are most important to the customer, or to some portion of the product performance that is critical to quality.

The model ties together the dependent responses and the independent control variables and defines the possible shape of the surface. Linear models (plane-linear in most cases) can only take the shape of a plane, whereas quadratic models add interaction terms to the equation to generate curvature. It is important to note that *the more complex the model, the more data will be required* to define the model, i.e. the greater the number of trials needed to evaluate. As the number of control variables increases, the human mind and eye rapidly lose the capacity to visualize or analyze the models, but fortunately, computers have no such limitations.

Resolving Power

Resolving power is one important factor that is usually overlooked despite the fact that most statistical software packages have the capability to evaluate it. The choice of resolving power is a trade-off between the desired resolution of the experiment and the number of trials needed to achieve that resolution. The analogy of a microscope is useful in visualizing this concept. The more trials run, the higher the magnification, and therefore, the higher the resolving power (resolution), or the ability to distinguish differences.

The resolution is the smallest signal or effect to be detected. The model describes the possible shape of the surface, with more terms required for more complex surfaces with twists and curves. The number of terms is chosen based on the level of predictive power you need for your process.

There are two ways to look at and calculate resolving power:

1. start with the size of the effect to be detected (resolution), and calculate the required number of trials (magnification);
2. start with the number of trials to be run (magnification) and calculate the size of the effect that can be detected (resolution).

Both require the selection of a model to determine the amount of data to be collected. In addition, an estimate of the replicate standard deviation (SD) must be known (a quantitation of the imprecision of the measurement process). Because this SD value is almost always an estimate at best, the results of the resolution calculation must not be viewed as ultimate and rigid determinations. The greatest value of this calculation is in eliminating the clearly impossible projects, those requiring hundreds of trials to obtain statistically significant results.

If the number of trials is too large, there are generally three options available to the experimenter:

1. choose a simpler model (works only if you are not already at the simplest model);
2. rethink the size of the least important difference;
3. reduce the replicate SD.

The latter two options focus on attempting to get bigger signal-to-noise ratios, either by increasing the signal (size of the least important difference), or by decreasing the noise (replicate SD).

Experimental Design

The experimental design is the specific collection of trials run and input conditions used to support a proposed model. The first critical step of choosing a design is choosing your variables for study. Four points can be made regarding the choice of variables:

1. choose them carefully;
2. set the ranges far enough apart to detect differences and avoid confusion;
3. experiment sequentially; and
4. make midcourse corrections only when absolutely necessary.

To choose a set of variables for study, use your expertise to prioritize those with the greatest likelihood of impacting the response(s). This is primarily for reasons of logistics and cost at this point. The mathematics can handle DOEs with large numbers of variables, but practical issues such as resources or time constraints may not provide the experimenter with the same freedom. It will probably always be possible to add lesser priority variables in the future.

Setting the ranges of your variables too close together can lead to wrong conclusions due to the experimental error in measurement. If the data at two points happen to fall in the opposing tails of statistical probability and the two points are too close, the relationship drawn between the two could actually be the opposite of the true relationship. This is avoided by having ranges between data collection points that are far enough apart.

Experimenting sequentially saves work and time. When the 'best' variables are chosen, it is necessary to assume that nature is simple. If this assumption turns out to be wrong, it is necessary to pick a more complex design. If the assumption turns out to be right, but there are still no answers, the variables are wrong and it is necessary to pick others. By approaching the process in a logical sequence, order is given to the search for understanding the process and success is made more likely.

There are two possible objectives for the experiments being designed; accordingly, there are two principle types of designs: screening and response surfaces.

Data Collection

There are three points to make in this section. First, get involved with the data collection. A career should not be risked on delivered data. It is necessary to be aware of everything that might be affecting response values. Anecdotes abound in which data were not collected properly because the person who did the design and analysis was not present for the collection of the data. One such story tells of an experimental design in a plant where the designer of the experiment arrived unannounced at the beginning of the shift that was supposed to collect the data and found the data collection sheet already filled out. Upon enquiry, he discovered that the 'old hands' of the plant already 'knew' what the results would be and so just filled out the data sheet based on their experience, not with the real data!

Which leads to the second point, communicate why and how things are done. Share the ownership of the design. When considering which variables to study, seek the input of others, especially those with experience of the process, or other like processes. Teamwork in scientific endeavor is a concept frequently missing, if not discouraged, in the typical academic laboratory. Yet it is indispensable for success in industry.

Third, randomize whenever possible to break correlations between the studied control variables and the unknown variables. Unknown effects will show up by inflating the residual SD (model is missing something).

Analysis

Analysis is just curve-fitting, finding the surface that comes closest to all the data using regression analysis. Two answers are needed quickly: does the model fit, and, what are the important variables? If a screening experiment is being done, the important variables will be carried forward to a response surface study. If a response surface experiment is being done, the important variables are the ones put on axes of a contour plot. Remember, the ultimate goal is to be able to predict the response for any given settings of the control variables.

The details of curve-fitting and analysis are usually handled by the software being used. Consequently, it is critical that the experimenter be well versed in the use of whatever software is chosen. Although many of the parameters and analysis outputs are standardized for DOE, most software has its own unique style of providing access to and presenting that output. Know where to find the appropriate results in your software package. Then be sure that the interpretation of the various outputs is understood. A complete discussion of this point is beyond the scope of this book, but abundant literature references are available, and most companies that sell this kind of software have training of some kind to aid the end-user (*see* THE SOFTWARE TOOLS).

Your experiment is *not* done until the model predictions have been verified with check point data. *Always* check the model's predictions. Verification points may be taken near the optimum, near suspicious behavior, at low cost settings, and inside versus outside the range of experiments.

The Software Tools

In the discussion above, the fact that software is the primary tool for design and analysis has been mentioned several times. Indeed, the very existence of this software is the only reason the methodology can be used effectively in any venue today. Two alternatives face the scientist in the market for DOE software: general statistics packages with DOE modules, or dedicated DOE software. Each has pros and cons, with the strengths of the one category being the weakness of the other, and *vice versa*. In the following discussion, some specific software packages are briefly mentioned and discussed, but this is neither an exhaustive list of such packages nor any kind of endorsement of the ones mentioned (although personal preferences are indicated).

General statistical packages have the primary advantage of greater cost effectiveness if you want or need more than just DOE capabilities. Many are becoming easier to use and more adept at DOE functions. Still, because the DOE functionality may be buried in a plethora of other tools, understanding how to use these modules effectively may have a steeper learning curve than that of a dedicated DOE package.

In this first category three packages can be mentioned: JMP® (SAS, www.jmp.com), Minitab® (Minitab, Inc., www.minitab.com), and Statgraphics Plus® (Manugistics, Inc., www.manugistics.com). Of these three, we are most familiar with and prefer to use JMP.

The second category, that of dedicated DOE software packages, shares the advantage of focus: because they are designed for only one primary task, their user interfaces are frequently more intuitive and the details of use more readily understandable. In addition, because of their focus, their power for their dedicated task may be greater than their general statistics counterparts. Their primary weakness, if it can be called that, is their lack of other statistical functions. But that is the nature of the beast.

Into this category two packages to be mentioned are ECHIP® (ECHIP, Inc., www.echip.com) and Design-Expert® and Design-Ease® (Stat-Ease, Inc., www.statease.com). Of these, we are most familiar with and prefer to use ECHIP. Comparing ECHIP and JMP for purposes of DOE, we would prefer to use ECHIP.

Regardless of which software package you choose, here are several characteristics that can be evaluated to narrow down the list of candidates:

- a user interface that you personally find easy-to-use and intuitive;
- well-written and thorough manuals with tutorials for hands-on instruction;
- a selection of designs that is wide enough to cover both screening and response surface experiments;
- data entry in a spreadsheet format that allows easy transfer from other applications such as spreadsheets (where the data is often tabulated during collection);
- the ability to randomize trials being run;
- design evaluations that flag problems with any given design (e.g. alias structures);
- data transformation capabilities;
- adequate reporting of diagnostics such as 'lack-of-fit' statistics and outlier identification to assess data and model quality;
- training workshops by the software designer to get hands-on experience and guidance in how to use the software to its greatest capacity.

Which of the above characteristics will be the most important depends on the individual situation of the one making the choice. Some may find ease-of-use to be most important, whereas others may choose based on the power of the package. Whichever software package is chosen, the primary concern is to master it thoroughly to avoid making costly mistakes as this powerful methodology is applied to your particular situation.

FOUR KEY ASSAY DESIGN PRINCIPLES

Given what has been discussed above, the following four principles are given to guide the development of immunoassays. These principles are not meant to provide detailed and specific guidance; rather, they are meant to outline key aids toward the planning and simplification required in the creation of a robust assay.

PLAN FIRST

Prior to starting the process of assay development, a list of critical customer requirements is essential. Furthermore, it is vital to understand the approach to be taken, the resources required, and the limitations of the approach. Part of the forethought involved here is to design for simplicity. This is a goal not only for the final product, but also for the product development process. Complex plans that include large numbers of 'performance panels' for evaluation of assay performance drastically limit the number of options and parameters that can be investigated. Conversely, such an approach may obscure relevant data, such that viable options are overlooked, or greatly increase the needed resources. On the other hand, a simplistic approach may miss or overlook key information. The foundation, among other things, involves identifying a very limited number of specimens or panels that can point the way for development. Evaluations that entail great numbers of specimens should be reserved for evaluation of two, or at the most three, options that initial investigations have revealed.

The principle of simplicity also applies to experimental design. If a number of negatives or positives are evaluated in an experiment, consider evaluating their responses via the mean and SDs of the negatives or positives. A few simple designed experiments may be easier to handle and evaluate than one complex experiment.

THINK FROM A THEORETICAL PERSPECTIVE

With all of the issues that need to be addressed in developing an assay, it is sometimes easy to forget what is important. For example, all assays measure analytes (antibodies or antigens) and it is their concentration, not signal generation, which is more important in assay performance. From a scientific perspective, the assay signal is a dependent variable; concentration is the independent variable. In other words, a theoretical understanding of what is going on in a given assay format is aided by a quantitative understanding. A 3% signal CV sounds good, but this can translate into a 20% concentration CV, depending on the dose–response relationship. For qualitative assays, thinking in this manner allows one to maximize the signal generated per concentration change at the cut-off, thereby minimizing imprecision. Stated another way, an increased uncertainty of measurement at the cut-off will directly translate into an increase in the number of sporadic false positives and false negatives among samples near the cut-off. It is important to understand how frequently this occurs for any given reagent formulation or assay protocol, but without quantitation of some kind, it is impossible to make such a determination.

Thinking from a theoretical perspective involves applying known scientific principles to problems where the mechanism of action is unknown. It involves visualizing the physical interactions that underlie the reasons for the behavior and performance of the immunoassay. This approach allows investigators to make predictions on the experimental outcome, which if proved true will allow them to mathematically model the experimental outcome. Not only does this allow the investigator to avoid unnecessary experimentation, it also provides a deeper understanding of how the immunoassay works. The key areas to apply such understanding include antigen–antibody binding and kinetics, suppression of sample-mediated interferences, design and manufacture of solid phase and conjugate reagents, limit of detection (LOD) and signal generation, and reagent–instrument interactions.

SIMPLIFY THE APPROACH

Clinical specificity and sensitivity are two important measures of immunoassay performance. The difficulty in determining these values is that hundreds, if not thousands, of samples must be evaluated before an accurate measure is obtained. Such studies are impractical during early assay development. A surrogate approach is outlined below.

Clinical Specificity

The *potential* for good clinical specificity can be evaluated by determining the mean and SD of a negative population (e.g. a 30-member panel, with one replicate of each) relative to that of an independent negative specimen (the number of replicates to match the number of members in the aforementioned panel). If an assay shows differing profiles among differing specimen types, or if the distribution of the 30-member panel is broader than that of a single specimen, then non-specific interactions are indicated. This is particularly true when the negative population displays tailing. Modifications of the assay protocol, conjugate, solid phase, diluents, etc. may be tested in order to minimize the difference. Identifying the loci of NSB may give insight as to which components to modify first. For example, for microparticle-based assays, if NSB decreases when particles are omitted, a conjugate–particle interaction is suspected; in contrast, if NSB is equivalent in the presence or absence of particles, then a conjugate–specimen interaction may be involved. Particle–specimen–conjugate interactions may also occur. Some approaches that have been used to resolve such problems are outlined below.

Particle–conjugate interactions may be eliminated by overcoating with protein, conjugate diluent changes, and/or elimination of unwanted cross-reactivity between the rare reagent components of conjugate and particles (through diluent changes or rare reagent purification and modification).

Particle–specimen–conjugate interactions may be eliminated by overcoating/co-coating with protein, addition of chelators, particle or conjugate diluent changes, and/or conjugate reformulation.

Conjugate–specimen interactions may be eliminated by the addition of protein or animal sera, and/or conjugate diluent modifications.

Clinical Sensitivity

Clinical sensitivity is related both to the inherent capability of the analyte to truly indicate disease or health and the ability of the assay to accurately measure analyte at the cut-off between these states. The analytical sensitivity, also known as the LOD, gives an idea of the *potential* clinical sensitivity of an assay (at least for analytes present in disease but absent in health). Some consider LOD to be a poor indicator of performance. Indeed, the functional sensitivity, defined as the concentration of analyte where the between run imprecision is 20%, is probably a better indicator of the ability of a given assay to quantify low concentration of analyte. However, LOD is useful in early assay development in that it is simple to determine and generally is proportional to the functional sensitivity. LOD is defined as the minimum concentration of analyte that can be detected with 95% confidence. In practice, LOD is determined according to the following formula:

$$\text{LOD} = (2\text{SD}/(B - A))[B]$$

where:

SD = the standard deviation of the zero or negative calibrator (OD, rate, or counts)
B = the output (OD, rate, or counts) for the B (low) calibrator
A = the output for the zero or negative calibrator
$[B]$ = the concentration of the B calibrator

This equation assumes a linear response between A and B and has several implications. First, at a given $B - A$, the lower the SD, the lower (better) the LOD (and potentially, the better the clinical sensitivity). Second, at a given SD, the higher the $B - A$, the lower the LOD. Third, the ratio of SD to $B - A$ defines an assay's analytical sensitivity. Last, the concentration of B does not have to be known to evaluate the LOD of potential assays (but a consistent specimen must be used as B).

ESTABLISH SPECIFICATIONS BEFORE STARTING

In other words, begin by planning first. Specifications should not be established based on what can be made; rather, they should be established by what the customer needs. Simplify specifications to metrics that can be measured: LODs can be related to B/A ratios; acceptable panel ranges to design specification imprecision targets or needed assay sensitivity. The number of specifications used in any component of the system should be the absolute minimum required.

While specifications should not be established based on the ability to manufacture, they should be evaluated against the manufacturing process. If the process is not robust relative to the specifications then one or more of the following are true.

(1) The process must be improved.
(2) The specifications are wrong.
(3) The test method is inappropriate.

If the process is clearly out of control, even though the test method is adequate, then (1) is likely. If the assay seems to be meeting the most critical design specifications even though a component is failing its component specification, then (2) is possible. In this case, rethink the specification based upon what is needed for the assay. If the test method is inappropriate, increased replicates or a new test method are required.

CONCLUDING COMMENTS

Development of immunoassays in the *in vitro* diagnostics industry has changed substantially in the last decade and this change has necessitated the use of a systematic approach to development, including multivariate analysis. While the approach must be clearly defined, it must also be flexible to allow rapid responses to new information to occur. Important in this endeavor is an aspect that historically has been overlooked in business, especially business in the US – people. Efforts that utilize the combined creative talents of the many talented people that work on a project are more successful than those that do not. True success, whether in the business, professional, or personal arenas, only occurs if everyone on a development project works together to get the job done and thinks beyond the conventional wisdom and past standards that no longer provide benefit to daily routines and jobs. Such thinking includes translating key customer needs into simple, measurable metrics, determining specifications based upon the need to achieve these key goals, thoroughly optimizing processes, evaluating manufacturing ranges by multivariate analysis, and eliminating non-value added testing and work.

FURTHER READING

Atkinson, A.C. and Donev, A.N. *Optimum Experimental Designs* (Oxford University Press, New York, 1992).

Box, G.E.P. and Draper, N. *Empirical Model-Building and Response Surfaces* (Wiley, New York, 1987).

Odeh, R.E. and Fox, M. *Sample Size Choice* (Marcel Decker, New York, 1975).

Ryan, T.P. *Statistical Methods for Quality Improvement* (Wiley, New York, 1989).

Schmidt, S.R. and Launsby, R.G. *Understanding Industrial Designed Experiments* 4th edn (Air Academy Press & Associates, Colorado Springs, CO, 1993).

Schmidt, S.R., Kiemele, M.J. and Berdine, R.J. *Knowledge Based Management* (Air Academy Press & Associates, Colorado Springs, CO, 1996).

Silvey, S.D. *Optimal Design* (Chapman & Hall, New York, 1980).

Snedecor, G.W. and Cochran, W.G. *Statistical Methods* 8th edn (Iowa State University Press, Ames, IN, 1989).

8 Antibodies

Eryl Liddell

INTRODUCTION

Antibodies are undoubtedly a crucial component of immunoassay performance. The selection of the correct specificity and binding affinity for the particular application is, clearly, important but it is also necessary to consider the choice between polyclonal, monoclonal or recombinant antibodies; purified or native sera; fragmented, bispecific or fusion proteins and whether to buy or make new reagents, taking into account the relative costs. The aim of this chapter is to provide an overview of the types of antibody reagent available, including how they are made and what one can expect from them in terms of degrees of precision, cross-reactivity, relative costs and production times, so that informed decisions can be made.

The design of the assay and its components also needs careful consideration, in order to make the best use of resources and to provide accurate information at the desired level. Sophisticated techniques that are expensive and take a long time to develop may not always be warranted when all that is required is a simple identification of a protein. Similarly, the production of monoclonal antibodies is pointless if they are to be used in insensitive precipitation assays. Indeed, certain monoclonal antibodies may not even work at all in this type of assay because of the difficulty in forming immune complexes. The interpretation of the assay result can also be misleading if the antigen samples have not been collected appropriately, if the antibody has not been properly characterized, or the assay is insufficiently sensitive or the statistical analysis is inaccurate. Other chapters in this volume address these issues and their interpretation.

ANTIBODY STRUCTURE

Antibodies are glycoproteins belonging to the immunoglobulin supergene family, which consists of many proteins involved in immune recognition and host defense systems.

There are five classes of immunoglobulin in most higher mammals, IgG, IgM, IgA, IgD, and IgE that differ on the basis of size, charge, amino acid composition, and carbohydrate content. In addition, there is some heterogeneity within classes, especially IgG, giving rise to four subclasses in humans (IgG1, IgG2, IgG3, IgG4) and also four in mice (IgG1, IgG2a, IgG2b, IgG3) although these subclasses are not synonymous between species.

The basic structure of an antibody consists of two identical 'heavy' polypeptide chains paired with two identical shorter 'light' chains, forming an approximate, flexible Y shape (Figure 8.1). The chains are subdivided into domains consisting of approximately 110 amino acids, which are linked by a variable number of disulfide bonds, giving a total molecular mass of approximately 150 kDa.

IgG antibodies, the most abundant antibody in serum, consist of one Y unit. The IgG subclasses differ mainly in the number of inter-heavy chain disulfide bonds.

The C_H and C_L domains are relatively constant in sequence with varying degrees of glycosylation and are responsible for the effector functions of the antibody, i.e. its binding to complement and cells involved in the immune response. The V_H and V_L domains, however, are more variable in sequence. Within the variable regions of each chain there are three areas of hypervariable sequence, known as complementarity determining regions (CDRs), each forming a loop structure supported by a framework of β sheets (Figure 8.2). The six CDRs on each arm of the antibody Y together form the antigen-binding site. Structural information about these sequences has been an important basis for the antibody engineering techniques discussed later in this chapter.

IgG is used almost exclusively in immunoassays. It is produced in the highest yield in response to immunization, binds with higher affinity to its epitope and is stable during isolation and purification processes, and has several functional sites that can be used for chemical coupling with minimal loss of antibody binding. It can be fractionated chemically into Fab and $F(ab')_2$ fragments, which can be an advantage in assays where the Fc region can interfere. Smaller antibody fragments such as single chain Fvs (scFv) are produced using molecular genetic

The Immunoassay Handbook. *Third Edition*
D. Wild (Ed.)

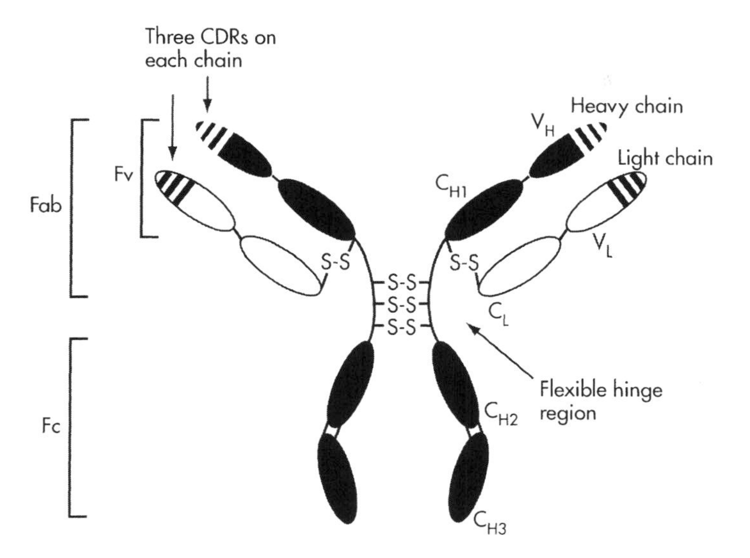

Fig. 8.1 Basic antibody structure (Human IgG1) illustrating a pair of identical heavy chains consisting of three constant domains (C_H1,C_H2,C_H3) and one variable domain (V_H) linked to a pair of identical light chains consisting of one constant domain (C_L) and one variable domain (V_L). The variable domains each have three hypervariable loops (complementarity determining regions or CDRs) which bind to antigen.

techniques and can be linked to other protein genes such as enzymes to make useful fusion proteins.

IgM antibodies are pentamers made up of five of the basic Y subunits joined by disulfide bonds and a J chain (*see* Figure 8.3). Each of the heavy chains is made up of five domains. The pentameric nature of IgM antibodies confers some benefit when amplification of the signal is required in assays such as immunohistochemistry but steric hindrance of the large molecule may pose problems. IgM tends to be of lower affinity than IgG since it is made predominantly during the early stages of immunization and it is less easy to purify and fractionate compared with IgG.

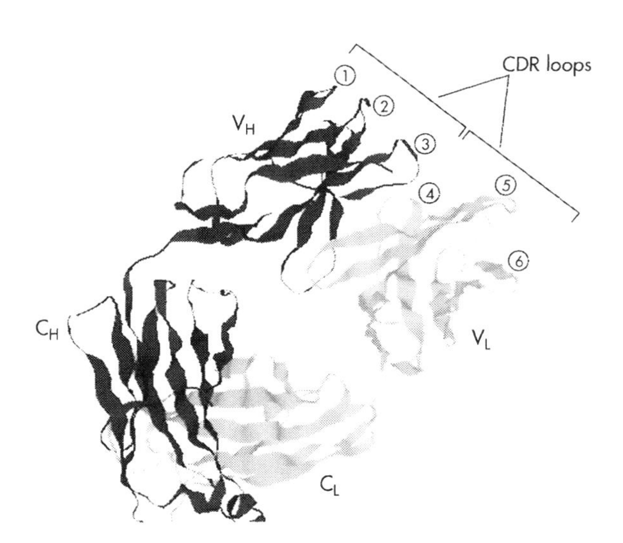

Fig. 8.2 Molecular model of antibody Fab region illustrating the CDR loops. The structure was obtained from the Brookhaven database and manipulated using Rasmol.

The other subclasses present *in vivo*, **IgA**, **IgD**, **IgE** are rarely used in immunoassay because of their low abundance or lower affinity. A summary of the essential differences between human immunoglobulin subclasses is given in Table 8.1.

THE ANTIBODY RESPONSE *IN VIVO*

Antibodies are produced by **B lymphocytes**, which develop in the fetal liver and subsequently in bone marrow. It is at these sites that immunoglobulin gene rearrangements occur (see below) and potentially harmful clones of cells producing antibodies to 'self' are eliminated. In the resting state, each B cell displays an antibody of one specificity on its surface. Foreign antigen therefore only becomes bound to a few of these resting B lymphocytes. However, binding to the antigen stimulates them to divide and to mature into plasma cells that secrete large amounts of the immunoglobulin that was originally displayed on their surface. During this proliferation there is a change in the main class of antibody produced from IgM to IgG, IgA or IgE, a complex process known as **class switching**, regulated by cellular signaling mechanisms from other cells

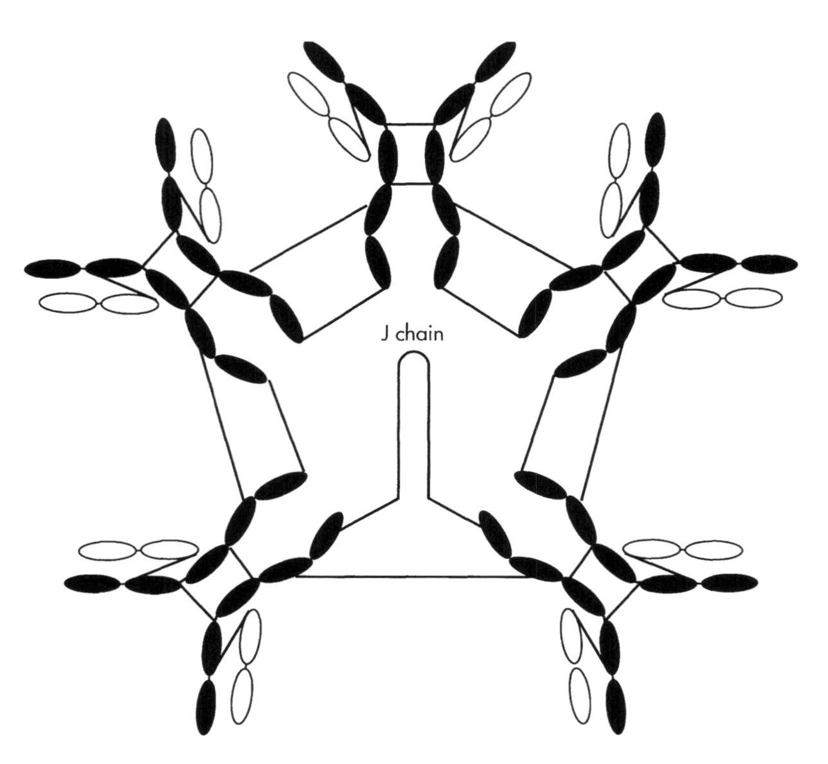

Fig. 8.3 Basic human IgM structure. IgM is essentially a pentamer of the basic IgG structure joined by disulfide bonds and a J chain. Each heavy chain consists of five domains. The carbohydrate side chains are not shown.

Table 8.1. Human immunoglobulin classes.

	IgG1	IgG2	IgG3	IgG4	IgM	IgA1	IgA2	sIgA	IgD	IgE
Normal adult serum concentration (mg/mL)	5–10	1.8–3.5	0.6–1.2	0.3–0.6	0.5–2.0	0.8–3.4	0.2–0.6	Trace	0.003–0.3	0.0001–0.0007
Molecular weight (kDa)	146	146	170	146	900	160	160	390	165	185
Molecular weight of heavy chain	51	51	60	51	67	56	53	53–56	60	70
No. of heavy chain domains	4	4	4	4	5	4	4	4	4	5
Intra-heavy chain disulfide bonds	2	4	11	2	1	2	2		2	2
% carbohydrate	3–4	3–4	3–4	3–4	12–16	10–15	10–15	10–15	14–17	13–16
Serum half-life (days)	12–21	12–21	7–8	11–21	10	5.8	5.8		1	2

Data selected from Kerr and Thorpe (1994).

involved in the immune response. B cells that secrete higher-affinity antibodies also begin to be preferentially selected, a process known as **affinity maturation**. Some of the proliferating class-switched B cells do not become secretory but continue to display their surface antibody. These **memory B cells** are responsible for the more rapid and greater immune response on subsequent antigen challenge. The consequence of these processes for immunoassay applications is that antisera will contain a heterogeneous mixture of antibody species, of varying specificity, affinity and isotype, the pattern of which will change with each bleed taken from the animal.

When raising antibodies, use can be made of the early elimination of self-reactive clones by exposing the newborn or fetal lymphocytes to an antigen in order to suppress subsequent production of antibodies to it, a process known as **tolerization**. For example, when raising antibodies to estradiol in a rabbit, the specificity can be increased by first tolerizing the newborn rabbit to potential cross-reactive substances. Antibody subsequently produced in the adult animal should not then produce clones that cross-react with the tolerizing cross-reactants.

ANTIBODY DIVERSITY

The unique diversity of the immunoglobulin repertoire is derived from the fact that each domain of the protein is coded for by different gene exons. The genes encoding the variable region of the heavy chain consist of V (variable), D (diversity), and J (joining) segments. In germ line DNA, one of approximately 100 V segments will combine with one of approximately 50 D segments and one of four J segments, to produce a functional **VDJ gene**. Similarly, light chain variable genes are made up from a choice of approximately 200 V segments and four κ J segments or two λ J segments to produce a functional **VJ gene**. The constant region genes are arranged downstream from the VDJ or VJ regions in such a way that any of the C genes can recombine with the VDJ or VJ genes, giving rise to antibodies of different isotypes. Additional diversity is conferred on the final antibody product by somatic mutation and combination of different light and heavy chains. Each B cell expresses a unique antibody from the original total of 10^{10} possible combinations. For further detail on antibody structure and the generation of diversity the reader is referred to standard immunology textbooks such as Roitt *et al.* (2001) and Janeway *et al.* (2005).

IMMUNIZATION

With the exception of the sophisticated methods of synthetic recombinant antibody production which will be discussed later in this chapter, virtually all antibodies are derived from active immunization of laboratory animals with an immunogen. In many countries, there is legislation governing the use of animals for research purposes. In the UK, for example, such work is strictly controlled by the Animals (Scientific Procedures) Act of 1986, under which the project, the scientist, and the premises have to be licensed individually and there are numerous regulations governing the breeding and housing of animals, record keeping and inspection of procedures, in order to protect the welfare of the animals. If in doubt over the regulations in a particular country, the technicians in charge of an Animal Facility will be familiar with local rules.

IMMUNOGEN

An **immunogen** is a molecule capable of eliciting an immune response when injected into an animal, not to be confused with an **antigen**, which is capable of binding to an antibody but not necessarily eliciting an immune response. An immunogen and antigen may have many different sites at which antibodies may bind. These are called **epitopes** or **antigenic determinants**. The main factors affecting the **immunogenicity** of the immunogen are the size and nature of the molecule, the length of exposure to the cells of the immune system, and its recognition by the host as foreign. The most potent immunogens are proteins and polysaccharides but lipids, nucleic acids, and synthetic polypeptides can also be immunogenic.

In general, proteins of molecular mass greater than 2000 will be immunogenic. However, molecules smaller than this which are not inherently immunogenic can be made so by being conjugated to carriers such as albumin or keyhole limpet hemocyanin (KLH). They are known as **haptens**. Synthetic peptides as small as 6–20 residues have produced antibody responses and they can be conveniently coupled to carriers such as polylysine during synthesis which obviates the need to purify and chemically conjugate them before injection. Table 8.2 shows some typical doses for different classes of immunogens.

In order to prolong the exposure of the immunogen *in vivo*, it is usually necessary to administer the immunogen with an **adjuvant** (large particulate or cellular material being the exception). There are three main classes of adjuvant:

1. antigen-depot adjuvants (Complete Freunds, Incomplete Freunds, and aluminum hydroxide) which act by slowly releasing antigen from the point of administration and also encourage granuloma formation rich in macrophages and other immunocompetent cells;
2. bacterial adjuvants *Bordetella pertussis* and muramyl dipeptide probably stimulate macrophages directly;
3. amphipathic and surface active agents such as saponin; liposomes can also be used particularly when toxic compounds need to be administered.

Table 8.2. Examples of typical immunogen doses.

Immunogen	Concentration per injection
Soluble or membrane proteins	10–100 μg (mice) 50–250 μg (rabbits) 250 μg–10 mg (sheep)
Conjugated peptides/haptens	100 μg (mice) 100–500 μg (rabbits)
Nucleic acids	200 μg
Eukaryotic cells	2–20 × 10^6
Bacterial cells	50 μg protein
Viruses	10^7 particles × 3 weekly
Fungal antigens	20–100 μg

These doses are a guide only. Previously published literature for immunogen/species should also be referred to where possible.

DNA immunization is a relatively new approach of particular interest to molecular biologists. DNA encoding a particular protein or peptide is cloned into an expression plasmid and injected into mice intramuscularly. This approach induces continuous secretion of antigen and high levels of circulating antibody (Davis *et al.*, 1993). It has the advantage of bypassing the need to express and purify relatively large amounts of protein *in vitro* prior to injection; a process which, in addition, may alter the structure of the protein. It therefore saves time, and only one injection of plasmid is necessary without conventional adjuvant, which is more favorable for animal welfare. The immune responses produced from what must be very small amounts of protein are surprisingly high, attributed to the recruitment of dendritic cells and a cytokine environment to augment the efficiency of antigen processing and epitope presentation (Donnelly *et al.*, 1997).

POLYCLONAL ANTISERA

Polyclonal antisera are a heterogeneous mixture of antibodies of varying binding affinities and isotype and also different specificities, recognizing epitopes both on the immunogen and any impurities injected with it. There will also be 'background' antibodies of unknown specificity that were present before the active immunization started. The antibody profile of each bleed of an individual animal will change especially with respect to affinity and isotype.

Antisera are relatively quick and inexpensive to make but their specificity is dependent on the purity of the immunogen and on the purification of the antiserum. This need not necessarily be a problem. Highly specific antisera have been developed to synthetic peptides, for instance, and when used in the right context the likelihood of obtaining false positive results is minimal. Antibodies are made in response to any immunogenic epitope in the injected material which includes any impurities in the preparation and any carrier protein or linker attaching them. It is, therefore, advisable to inject as pure a material as possible. In certain circumstances sufficient quantities of pure immunogen may not be available or perhaps the precise nature of the antigen is unknown (i.e. on a cell surface), in which case making monoclonal antibodies might yield better reagents.

For polyclonal antiserum production on a research scale, rabbits are the most common species used, which balances the low maintenance costs with useful volumes of serum (expected yields can be up to several hundred mL of serum per animal with repeated bleeds). It is advisable to use more than one individual animal since there can be significant differences between them even though they have been injected with the same material. This is especially noticeable when injecting synthetic peptides. For commercial scale production, larger domestic animals such as sheep, goats, horses, and donkeys are used, although theoretically any mammal or bird could be used. Antibodies can also be produced in eggs (IgY) following immunization of chickens (Jensenius *et al.*, 1981). In general, an immunogen is most immunogenic when injected into a species of further evolutionary distance from the donor.

The site of injection can also promote the slow release of immunogen or at least reduce its clearance from the body. Subcutaneous injection often at multiple sites near to lymph nodes is a preferred route that causes minimal discomfort to the animal. Other sites used less frequently are intradermal, intramuscular, intraperitoneal and footpad. Intravenous routes (without adjuvant) are only used as a boost of lymphocyte proliferation prior to monoclonal antibody production.

The timing of injections is based on what is known of the levels of antibody circulating in the serum. Following the first injection, predominantly IgM antibodies are produced, but if a second injection is given 3–4 weeks later, much greater levels of IgG antibodies of higher affinity circulate in the serum reaching a peak 10–14 days later with a gradual decline (see Figure 8.4). Therefore, a standard immunization protocol would be two injections, 4 weeks apart, followed by a test bleed after 10 days. If, on testing, the antiserum is not of sufficiently high titer or affinity repeated boosts at monthly intervals should improve this.

PURIFICATION OF ANTISERA

For many analytical techniques antisera can be successfully used as a straightforward dilution without the need for purification. Where the antibody is to be labeled with detecting reagent, some form of prior purification is essential to avoid the high nonspecific binding caused by other labeled contaminating proteins. There are four main techniques. The first uses simple physical separation by salting out, usually with ammonium or sodium sulfate, followed by dialysis or gel filtration to remove the salt. IgG can be obtained from antiserum in about 80–90% purity, which is sufficient for many purposes. IgG precipitates because the solubility decreases with increasing ionic strength. Some immunoglobulin will, however, always remain in solution and hence the yield is critically dependent upon the starting concentration of IgG. For neat antisera, where the IgG concentration is of the order of 25 mg/mL, recoveries of the order of 90% can easily be achieved. Care needs to be taken to remove all the ammonium sulfate as ammonium ions may reversibly compete with many amino-reactive coupling reagents used in antibody labeling procedures.

The second method is based on gradient ion-exchange techniques, which can give higher purity than salt fractionation, but the method is more time-consuming and may be limiting in scale. The yield is usually excellent, even with relatively low concentrations of antibody. The third method takes advantage of the specific and reversible binding of the Fc region to lectins such as Protein A and Protein G. Crude antibody is passed through a column containing lectin coupled to an inert support. After washing the column, IgG is eluted using low pH buffer to dissociate the complex. The technique is capable of producing antibody in high yield and purity, but is expensive and therefore of limited use when dealing with large-scale purification.

All of the above techniques nonspecifically isolate total IgG. **Immunoaffinity purification** depends on the immunological specificity of the antibodies and hence seeks to isolate only those IgG antibodies reactive with the antigen of interest. The antigen is covalently coupled to an inert support and the crude antibody passed through the column at a relatively low flow rate. After extensive washing of the solid phase, which removes everything that does not bind to the antigen, the specific antibody is eluted by disruption of immunological binding by either low pH or high concentrations of a chaotropic agent such as guanidinium chloride. A variant of the method, **immunoadsorption**, can also be used to remove unwanted antibody, for example, antibodies to rabbit IgG can first be purified using immunoaffinity chromatography with rabbit IgG, and then by passing down a column of coupled human IgG to remove any antibody strongly cross-reactive with human IgG.

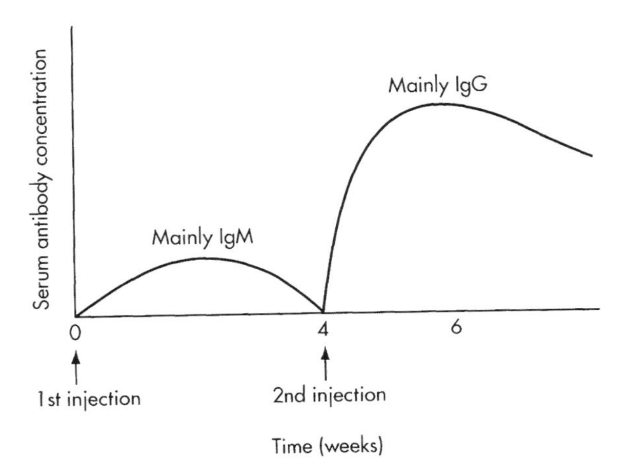

Fig. 8.4 Antibody production *in vivo*. Antibody produced following the first injection is predominantly IgM. After the second and subsequent injections, a greater concentration of higher affinity IgG antibodies is produced reflecting the maturing immune response. Antiserum should be collected at times appropriate for maximal antibody secretion.

MONOCLONAL ANTIBODIES

Polyclonal antisera, as described above, consist of the secreted products of thousands of different B lymphocytes, each of which produces a unique antibody recognizing a single epitope. If a single B lymphocyte could be isolated and propagated *in vitro*, a homogeneous antibody reagent could be obtained. In practice, this is not possible because lymphocytes cannot normally survive in these conditions.

However, in 1975, Köhler and Milstein were able to demonstrate the fusion of B lymphocytes with immortal **myeloma** cells to produce hybrid cells that inherited both the capacity to secrete specific antibodies and the ability to proliferate indefinitely in tissue culture. Through a series of selection procedures, described in this section, single antibody-secreting **hybridoma** cells can be isolated in individual tissue culture wells from which large colonies (monoclones) can develop through mitotic division, each secreting identical antibodies. This technique has given rise to reagents of superb specificity, able to distinguish very slight differences between molecules or cells or microorganisms. These **monoclonal antibodies** can be made available in limitless quantities since the hybridoma cells can be grown in tissue culture virtually indefinitely and at industrial scales. There is also the added advantage that the cells can be frozen for storage, and recovered when required without the need for recharacterization as would be necessary for a new batch of polyclonal antiserum.

This pioneering work, for which Köhler and Milstein were awarded the Nobel prize in 1986, has given rise to an explosion of new information in both fundamental biological research and in a wide spectrum of medical and diagnostic applications.

The vast majority of monoclonal antibodies are derived from mouse cells, although rat- and more rarely human-derived antibodies are also available. It is also possible to obtain antibodies derived from heterohybridomas that are products of parent cells from two of these species. The essential steps in the production of monoclonal antibodies are illustrated in Figure 8.5 and more practical detail on production can be found in Liddell and Cryer (1991) and Liddell (2003).

IMMUNE LYMPHOCYTES

The production of immune lymphocytes essentially follows the same principles as the production of polyclonal antisera as described previously. However, there are some important considerations. Any strain of mouse or rat can be chosen for immunization but if the intention is to propagate antibody *in vivo* by ascitic fluid production (*see* STORAGE AND PROPAGATION), the strain should be the same as that from which the parent myeloma was derived, i.e. for mice that strain is usually Balb/c, since the common myeloma cell lines are derived from this strain. Alternatively, ascitic fluid will have to be raised in F1 crosses of Balb/c and the immunized strain.

The purity of the immunogen is not quite so important as it is for polyclonal antiserum production since the selection of specific antibody secreting cell lines occurs after fusion and is dependent on the purity of the antigen used in the screening test. Indeed, it may be better to use an immunogen that has not been tampered with, if the requirement is to raise an antibody to a 'native' protein. Certain epitopes that monoclonal antibodies bind to are likely to be destroyed by alterations in the tertiary structure of the protein.

A conventional immunization protocol can be followed prior to test bleeding and the animals can be kept for up to a year before fusion. However, to enhance the success of fusion, animals should be boosted, preferably intravenously, four days before. This will ensure that the relevant lymphocytes are at their peak of proliferation rather than antibody secretion. The commonest source of lymphocytes for hybridoma production is the spleen but lymph nodes are also used. A mouse spleen contains approximately 10^8 cells, 50% of which are B cells. For most purposes, successful fusions can be achieved by using whole spleen suspensions without pre-selection of either B cells or those expressing specific antibodies.

MYELOMA CELL LINES

Myeloma or **plasmacytoma** cells are neoplastic antibody-producing cells. They can be generated experimentally in mice and rats (the Balb/c mouse strain being particularly susceptible), although it is very difficult to generate myeloma cells secreting antibody of predetermined specificity. A variety of lines adapted for survival in tissue culture and suitable for cell fusion (see Table 8.3) can be obtained from large international cell culture collections such as the American Type Culture Collection and the European Collection of Animal Cell Cultures, from which stocks can be grown and stored frozen for future use. There are two important characteristics of myeloma cell lines used in hybridoma production. First, they should have an enzyme deficiency to enable selection of fused from unfused cells and second they should ideally not secrete any immunoglobulin of their own.

Most myeloma cell lines are deficient in the enzyme **hypoxanthine guanine phosphoribosyl transferase** (**HGPRT**) although there are also others deficient in thymidine kinase (TK) and ouabain. HGPRT is essential for DNA and RNA synthesis. Lack of this enzyme prevents survival of myelomas in HAT medium, the culture medium in which post-fusion cell mixtures are grown, whereas fused cells are able to survive because they inherit the enzyme from the lymphocyte parent. The isolation of $HGPRT^-$ myeloma cells is relatively easy because the enzyme is coded for on the single active X chromosome present in each cell. This means that only a single mutation is required to result in the loss of HGPRT. Selection of $HGPRT^-$ cells is carried out in the presence of the toxic base analogs, 8-azaguanine or 6-thioguanine, which are incorporated into DNA via HGPRT. It is not usually necessary to select for enzyme deficiency in myeloma cells since there is a range of useful $HGPRT^-$ cell lines available commercially (see Table 8.3). The mechanism by which these enzyme deficiencies are exploited in hybridoma selection are described more fully in HYBRID SELECTION.

The myeloma line P3 that Köhler and Milstein first used secreted immunoglobulin, so when fused with lymphocytes the hybrid cells secreted a mixture of heavy and light chains derived from both parents. This obviously reduces the chances of obtaining a hybridoma secreting antigen-specific antibody derived solely from the immunized lymphocyte parent. Soon after, myeloma lines became available that synthesized but did not secrete κ light chains. Thus, after fusion, hybrids could secrete myeloma-derived κ chains in addition to immunoglobulin chains from the lymphocyte parent. These myeloma lines, of which NS-1 became very popular, then gave way to fully

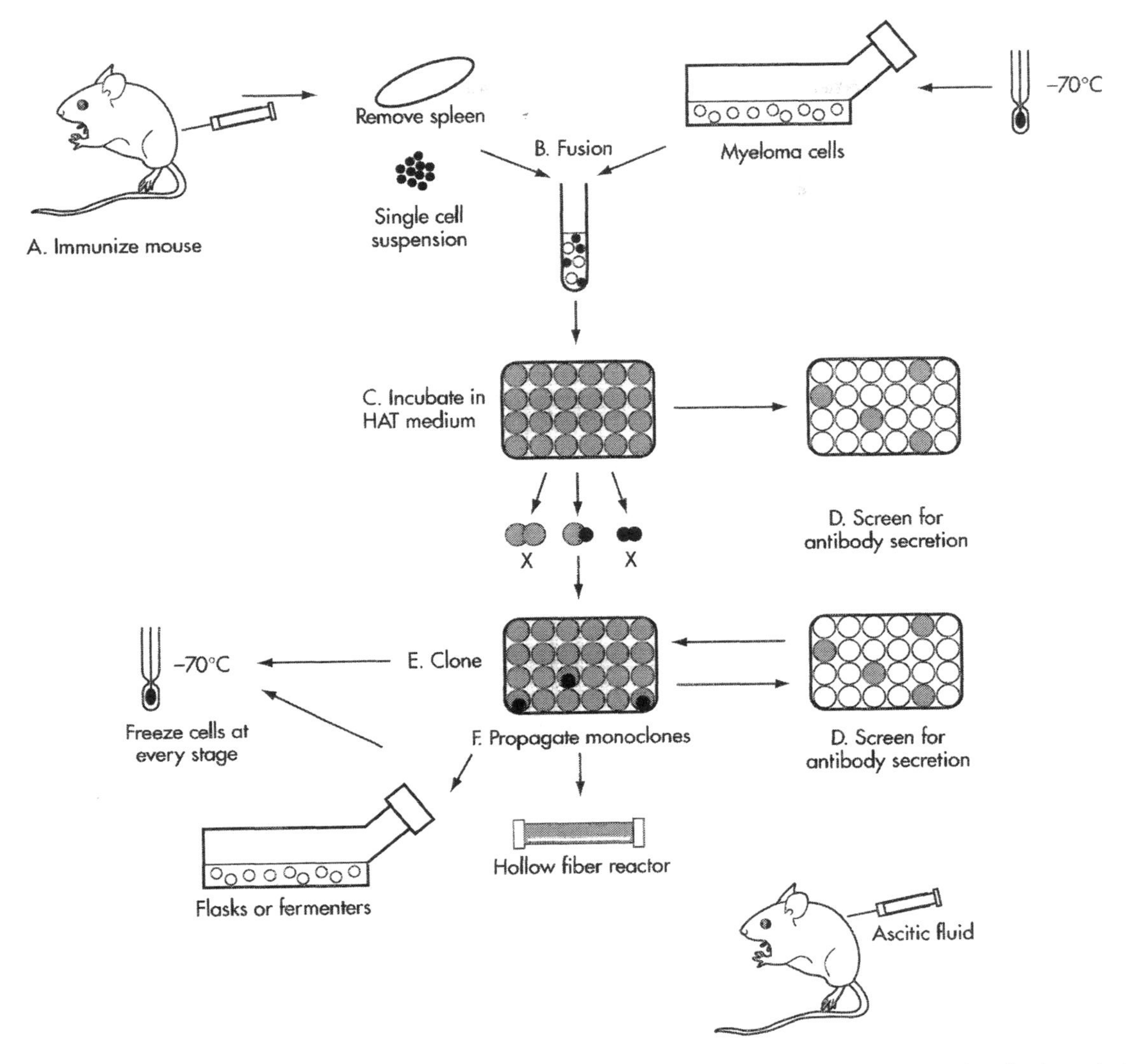

Fig. 8.5 Monoclonal antibody production. A flowchart showing the main sequence of events in monoclonal antibody production. NB: the production of ascitic fluid is now prohibited in some countries (see text).

'nonsecretor' cell lines, which did not synthesize any immunoglobulin chains (e.g. 653, SP-2) and these are the lines of choice when making new monoclonal antibodies.

Myeloma cells are very easy to grow in suspension tissue culture if certain characteristics are kept in mind. The growth medium to which they have been adapted will be recommended by the supplier (e.g. RPMI 1640 with 10% fetal calf serum), but they should not be grown continuously for long periods or at densities of more than 5×10^6 per mL in order to prevent overcrowding. When cells are put under stress, the likelihood of spontaneous mutations occurring increases. For instance, this could result in the line reverting back to be HGPRT positive, and indeed there are some human myeloma lines that are particularly prone to this problem. Reliable frozen stocks of cells should be kept and initiated at frequent intervals as an insurance against this and contamination problems or incubator breakdown. It also helps to culture the myeloma cells, periodically, in the presence of 8-azaguanine or 6-thioguanine, to eliminate any reverted HGPRT positive cells.

CELL FUSION

Although fusion is central to the whole process of hybridoma production, it requires no special equipment and takes approximately 30 min. Essentially, myeloma

Table 8.3. Selected myeloma cell lines for hybridoma production.

Species (and strain)	Cell line	Abbreviation	Ig chains secreted or synthesized	Original reference
Mouse (Balb/c)	P3-X63/Ag8	P3	IgG1,κ	Köhler and Milstein (1975)
	P3-NS1/1Ag4.1	NS-1	κ nonsecreted	Köhler and Milstein (1976)
	SP2/0Ag14	SP2	–	Shulman *et al.* (1978)
	P3-X63/Ag8.653	653	–	Kearney *et al.* (1979)
Rat (Lou)	Y3-Ag1.2.3	Y3	κ	Galfre (1979)
	YB2/3.0Ag20	Y0	–	Galfre *et al.* (1981)
				Kilmartin *et al.* (1982)
	IR983F	983	–	Bazin (1982)
Human	SK0-007		IgEλ	Olsson and Kaplan (1980)
	RPMI 8226		λ nonsecreted	Matsuoka *et al.* (1967)
	GM4672		IgG_2, κ	Croce *et al.* (1980)
	Karpas 707		λ nonsecreted	Karpas (1982)

The first two cell lines are no longer in common use but are included since many monoclonal antibodies commercially available may be derived from these lines.

cells (in the log growth phase) and a single cell suspension of immune lymphocytes from the spleen or lymph nodes are mixed together in a sterile centrifuge tube (in the proportion 10^7–10^8 myeloma: 10^8 splenocytes) and pelleted by gentle centrifugation. To the pellet a solution of polyethylene glycol (PEG, 40–50%; molecular weight 1500–4000) is added slowly over 1–2 min, then slowly diluted in culture medium, centrifuged and resuspended in fresh culture medium. The timing is crucial to avoid the potential toxic effects of PEG. The cell mixture is distributed between, typically, three 96-well microtiter plates and allowed to grow in a 37 °C/5% CO_2 incubator.

PEG fusion is not selective for cell types or antibody-secreting cells, so following fusion, a mixture of different cells will be present; unfused myeloma and spleen cells, hybrids of myeloma/myeloma, spleen/spleen and myeloma/spleen and also hybrids of more than two cells. The multiple hybrids will not survive and neither will the unfused spleen and spleen/spleen hybrids for more than a few weeks because of their inability to grow in tissue culture. The unfused myeloma and myeloma/myeloma hybrids must be disposed of so that the only remaining cells are the myeloma/spleen hybrids. This is achieved by growing the cells in a selection medium such as HAT.

HYBRID SELECTION

In order to understand the mechanism of **HAT selection**, one needs to appreciate that nucleic acid synthesis can follow one of two pathways; the *de novo* and the salvage pathways. All normal cells use the *de novo* pathway but if this is blocked they can bypass it and use the salvage pathway. The salvage pathway for purine synthesis uses the substrates hypoxanthine and guanine from which are formed inosinic-ribose phosphate and guanine acid-ribose phosphate respectively with the help of the enzyme HGPRT. Similarly, the salvage pathway for pyrimidine synthesis uses the substrate deoxythymidine to form thymidine monophosphate using the enzyme thymidine kinase. Obviously, if both pathways are blocked, cells cannot survive.

HAT medium consists of the normal culture medium with three additives: hypoxanthine, aminopterin, and thymidine. Aminopterin is an antibiotic that effectively blocks the *de novo* pathway, forcing all the cells to use the salvage pathway. Hypoxanthine and thymidine are the substrates converted by HGPRT and TK but if either of these enzymes is missing, as is the case with myeloma cells, purine or pyrimidine synthesis cannot occur and the cells will die. Hybrid cells, of enzyme-deficient myeloma and enzyme-positive lymphocytes, will be able to survive because they have inherited the enzyme from the lymphocyte parent. After five days of growth in HAT medium, most of the original cells will be dead, but the surviving cells amongst the debris, which are morphologically indistinguishable from myeloma cells, will be myeloma-spleen hybrids.

Natural **ouabain** sensitivity is a property of use particularly in the selection of heterohybridomas of human/mouse origin. Human cell lines normally die in the presence of ouabain at 10^{-7} M, whereas rodent lines are resistant up to 10^{-3} M. Unfused human cells can therefore be selected against.

ANTIBODY SCREENING TESTS

Within two weeks of fusion the hybrid cells will be growing rapidly (doubling times of 12–24 h) and colonies will have grown sufficiently to require transfer to larger containers. If this is not done, the cells will become overcrowded and start to die. Relatively few of these cells will be producing the desired antibodies. In order to identify which cells are relevant, it is necessary at this stage to test the culture supernatant for secreted antibody. Theoretically, most immunoassays can be adapted for use in monoclonal antibody detection. In practice, however, the most appropriate screening test is often a compromise based on the final proposed application of the antibody and the growth characteristics of the hybridoma cultures.

The screening test should be as similar as possible to the assay in which the antibody will be used, using antigen in the form it will be in, in the final application. The reason for this is that an epitope to which a monoclonal antibody will bind is not necessarily composed of a continuous sequence of amino acids; it is likely to be made up of sequences from neighboring and overlapping chains (**a discontinuous epitope**). If this conformation is altered, the epitope may be destroyed and the antibody will not bind. Thus, if a monoclonal antibody is selected on the basis of its binding to native protein, it may not react with denatured protein and vice versa. A monoclonal antibody that gives a positive signal in an enzyme linked immunosorbent assay (ELISA), therefore, may not recognize the same antigen that has been exposed to SDS electrophoresis.

It is preferable to use pure antigen in the screening test, but in practice the antigen may not be available in sufficient quantity or purity. In such cases, a secondary screening test for the major contaminant or alternative cell type may be of use to eliminate cross-reacting antibodies.

The rapid growth characteristics of the hybridoma cells require that the screening method should be quick to perform (24 h max.) and able to accommodate several hundred test samples at a time. Thus, assays that involve individual sample handling, centrifugation or observation of microscope slides or lengthy preparation of antigen or labeled probes, or parallel growth of antigen bearing cells are not ideal.

An ideal assay format for screening monoclonal antibodies that can fulfill most of the criteria suggested above is ELISA. Hundreds of samples can be analyzed within a few hours; labeled reagents are easily and cheaply available commercially; and relative concentrations can be determined visually, without the necessity for expensive equipment. This type of assay can be used to eliminate most of the irrelevant cell lines and when numbers of cultures have been reduced to more manageable levels, a secondary more specific screening test can be introduced if necessary.

CLONING

Once a cell culture well has been identified as containing specific antibody secreting cells, the cells need to be cloned in order to isolate a single colony from the mixture of other cells in the well. The most common method of achieving this is to **clone** by **limiting dilution**. This is done by counting the cell suspension and diluting it to levels that increase the likelihood of seeding single cells per well, which then grow into individual colonies. This can also be done more accurately using a fluorescence activated cell sorter (FACS) if available. Each cell within a colony is identical and so will secrete identical antibody. At the early stages these cultures must be supplemented with feeder cells or equivalent media supplements. **Feeder cells** are suspensions of thymocytes, peritoneal macrophages or splenocytes that secrete growth factors essential for clonal growth. Single clones in a well can be confirmed by careful observation under the microscope, which is relatively easy if the culture has remained undisturbed and not dispersed. Within two weeks of cloning, the colony will have grown sufficiently to produce enough antibody for testing. Those that are positive and also 'monoclonal' are transferred to larger culture wells and recloned. Recloning and antibody testing should be repeated until all the wells with growing cells are shown to secrete specific antibody when one can be reasonably sure that the cell line is monoclonal.

STORAGE AND PROPAGATION

At various stages during the production of hybridomas it is important to freeze samples of colonies to insure against infection or other events resulting in loss of the cell line.

Once a monoclonal cell line has been established, several vials of approximately 10^7 cells should be stored in several different − 70 °C or liquid nitrogen freezers. The freezing medium should contain 10% DMSO or equivalent to prevent crystallization of water within the cells.

Cells can be propagated by tissue culture in large flasks to produce culture supernatant containing antibody. If a static culture is allowed to overgrow until all the cells are dead, this will yield the highest concentration possible, which could reach 50 μg/mL and will be sufficient for preliminary characterization studies. Further concentration of this culture supernatant can be carried out but this will also concentrate other proteins derived from the serum supplement that might interfere with assays. Roller bottles and stirred flasks can be used to increase the concentration of antibody produced *in vitro* but can be very labor intensive and will require large volumes of culture medium containing expensive serum. There are totally defined serum-free media available but some hybridoma cell lines are unable to grow in total serum-free media. Most lines will need adaptation by gradual reduction of the serum percentage over several weeks.

Propagating antibody in **hollow fiber reactors** is becoming the preferred *in vitro* method. Hollow fiber reactors are essentially bundles of fine porous fibers, rather like kidney dialysis cartridges, through which culture medium is circulated. Cells are grown in a much smaller volume of medium in the extra capillary space. Nutrients and waste products can pass freely between the two spaces but larger molecules such as antibodies remain in the cell compartment until harvested. Large densities of cells can be maintained in this system for months producing potentially several hundred mg of antibody per day. The circulating culture medium does not contain serum, thus media costs are kept relatively low and the harvested antibody is already concentrated. The cartridges are the major expense so special care needs to be taken to avoid contamination.

Air-lift fermenters are also used for growing hybridoma cells. Cells are grown in suspension or on microcarriers in a vessel aerated from below in such a way that the gas bubbles gently mix the cells, with minimal damage from shear stresses that would be generated by conventional stirrers. They range in size from 5 L for research production but can be scaled up to commercial quantities of 1000–10,000 L capacity that can produce 40–500 mg of antibody per liter. The problem remains, however, of the expense of such large quantities of culture medium and the need to concentrate and purify the antibody.

The easiest way of obtaining very concentrated antibody (5–10 mg/mL) is to inject approximately 2×10^6 cells intraperitoneally into histocompatible mice that have been pretreated with Pristane (0.5 mL intraperitoneally, 10–60 days before cell injection). **Pristane** encourages the growth of plasmacytomas and also depresses the normal immunological function of the recipient animal. Within 2 weeks of cell injection, the abdomen should swell with **ascitic fluid** that is rich in antibody (5–15 mg/mL; approximately 4 mL per mouse). The amount of antibody and fluid development is very dependent on a particular cell line. Some cells only induce solid tumors with very little fluid and, of course, no tumor or fluid will develop in mice that are not histocompatible. If the mouse immune lymphocyte donor was not Balb/c, then an F1 hybrid of that strain and Balb/c must be used to propagate ascitic fluid. The resulting fluid will be 'contaminated' with other mouse proteins including general immunoglobulins but for many applications, at the typical working dilution of 1:100,000 or more, these become insignificant. In some countries, such as the UK and Germany, producing ascitic fluid in this way has been prohibited so there is no alternative but to propagate monoclonal antibodies *in vitro*.

HUMAN MONOCLONAL ANTIBODIES

The need to produce human monoclonal antibodies for diagnostic purposes is limited except perhaps for studies in autoimmunity. For therapeutic applications, however, antibodies of human origin are essential to avoid anti-mouse antibody reactions. In practice, however, it has proved very difficult to make them by this method, due in part to the instability of human myeloma cell lines, the difficulty in immunizing human volunteers against any but a small range of immunogens, and the necessity of using peripheral blood lymphocytes rather than spleen in most situations. Nevertheless, some human monoclonal antibodies have been raised by conventional hybridoma technology (see James and Bell (1987) for review; Shirahata *et al.*, 1998; Niedbala and Stott 1998). The hybridoma approach has now been largely superseded by humanization and recombinant DNA techniques to derive specific human antibody fragments from immunoglobulin gene libraries. However, more recently, the hybridoma approach has been revived by the creation of transgenic mice in which endogenous immunoglobulin genes have been replaced with the human equivalent (Mendez *et al.*, 1997). Both techniques will no doubt contribute to human therapeutics in the future.

PURIFICATION OF MONOCLONAL ANTIBODIES

The same purification methods apply to monoclonal antibodies as have been described for polyclonal. The need to purify monoclonal antibodies will depend on the source of antibody and the proposed use. Culture supernatant will contain bovine serum but if this poses a problem the hybridoma could be adapted to grow in reduced serum or serum-free media. Ascitic fluid will contain mouse serum proteins from the donor, but in practice since most ascitic fluid can be diluted 10^5 or even 10^6 fold and all the antibody is specific, the other mouse proteins will be too dilute to have an effect.

In many assays, specific antibodies are used directly, the signal being provided by a secondary anti-isotype antibody conjugated to an enzyme, fluorochrome or other label. A wide range of these secondary antibodies is available commercially, relatively cheaply so there is little need to prepare these oneself. If the antibody is to be conjugated directly with label, then it must be in concentrated form and of sufficient purity to minimize interference in the process by the contaminating proteins. Affinity purification of monoclonal antibodies is unnecessary and will reduce yields through damage caused during acid elution. Purification using Protein A or G should only be considered after isotyping the antibody since not all isotypes bind these reagents equally.

ANTIBODY FRAGMENTS

It is sometimes necessary to remove the Fc portion of an antibody in order to reduce nonspecific binding in an assay. For example, Fc receptors might be present in certain cells and tissues that could bind the Fc region of the antibody irrespective of the antibody specificity, thus giving false positive results. Similarly, complement, which may interfere in some assay designs, binds specifically to the Fc region. The smaller size of **Fab and F(ab′)$_2$** fragments may be beneficial in some stationary solid-phase immunoassays, where diffusion through boundary reaction layers is the rate-limiting step. However, set against these advantages are the greater difficulties encountered when antibody fragments are coated to solid-phase supports and the reduction in the number of accessible, noncritical, sites that can be used for labeling these fragments for use as detecting reagent.

IgG molecules can easily be fractionated without impairment of their specific binding to antigen, Fab or F(ab′)$_2$ fragments being the most useful. Fab fragments are made by proteolysis using the nonspecific thiol protease, papain. Papain is first activated by reducing the sulfydryl group in the active site with cysteine, 2-mercaptoethanol or dithiothreitol. Any heavy metals in the stock enzyme are removed by chelation with EDTA and the enzyme is mixed with antibody in a 1:100 weight ratio. The reaction is stopped by dialysis or irreversible alkylation of the thiol group with iodoacetamide. The resulting fragments, 2 Fab and an Fc (see Figure 8.6) are then separated by protein A-sepharose or ion-exchange chromatography. Fab fragments are, of course, monovalent and therefore the affinity of binding will be reduced in comparison to the original antibody. This might be critical if the original affinity was low.

Bivalent F(ab′)$_2$ fragments are more difficult to prepare using the enzyme pepsin, which causes cleavage at the C-terminal side of the inter-heavy chain disulfide bonds. There is considerable variation in the success of this process depending on the species and subclass of immunoglobulin, it being difficult to balance reasonable yields of F(ab′)$_2$ with complete degradation of the molecule. Mouse IgG2b is particularly susceptible so it is essential to isotype the antibody and tailor the conditions to produce the best results (Lamoyi, 1986).

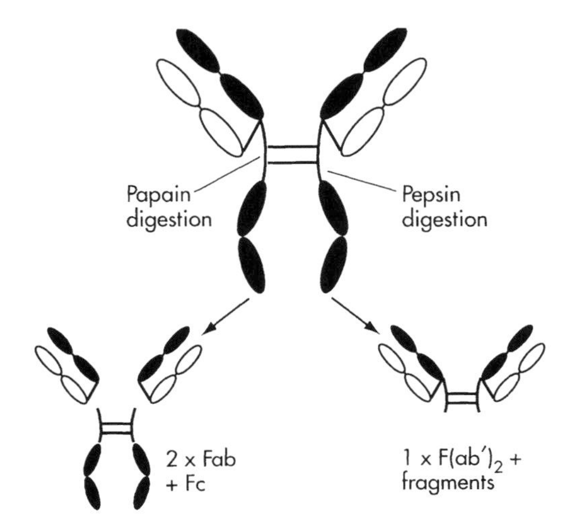

Fig. 8.6 Fab and F(ab′)$_2$ production by protease digestion.

When using Fab or F(ab′)$_2$ fragments in an assay with an indirect secondary antibody label, a specific anti-Fab or anti-F(ab′)$_2$ should be used, although some secondary antibodies (especially those of polyclonal origin) might still bind to the fragments.

BISPECIFIC ANTIBODIES

Bispecific antibodies are constructed from more than one source such that they retain both specificities of the original antibodies. In their simplest form, they lose their bivalency and therefore there will be a reduction in binding affinity to antigen. Consequently, their use in immunoassays has been fairly limited, there being more potential in therapeutic applications by redirection of cell targeting. Nevertheless, there may be some uses, for example, for one arm to bind peroxidase and the other the target antigen, which would help to reduce nonspecific binding of secondary reagents in immunohistochemistry and immunoblotting. They can be made in three ways: by chemical recombination; by cell fusion methods; and by genetic manipulation. A comprehensive review of bispecific antibodies was edited by Fanger in 1995.

The various antibody fragments generated by limited proteolysis of the inter chain disulfide bonds, are relatively stable and can be rejoined in different combinations without loss of their binding properties. Two or more Fabs can be joined together to form bi or trispecific F(ab′)$_2$s or bispecific trimeric molecules. Two homobifunctional reagents are commonly used to form the link: 5,5′-dithiobis(2-nitrobenzoic acid; DTNB), which regenerates the disulfide bonds and *o*-phenylenedimaleimide (*o*-PDM) which in turn creates a specific thioether bond between the two Fabs. Heterobifunctional linkers can also be used which introduce a reactive group into a Fab that will enable it to form a conjugate with another Fab. Succinimidyl 3-(2-pyridyldithio) proprionate (SPDP) and *N*-succinimidyl *S*-acetylthioacetate (SATA) react with primary amino groups to introduce free SH groups at these sites. Succinimidyl 4-(*N*-maleimidomethyl) cyclohexane-1-carboxylate (SMCC) reacts with primary amino groups to introduce a reactive maleimide group into the protein and 4-(4-*N*-maleimidophenyl) butyric acid hydrazide (MPBH) does so by reacting with carbohydrate groups on glycoproteins.

An alternative method of producing bispecific antibodies is to fuse two different hybridomas together to form a **quadroma** or **tetradoma** or to fuse a hybridoma with splenocytes secreting antibody of different specificity to form a **trioma**. Selection of the hybrid will depend on using different mechanisms for each partner. One of the hybridomas could be made HAT sensitive by growing it in

8-azaguanine to generate HGPRT deficiency or bromodeoxyuridine to generate tyrosine kinase deficiency. Other selection methods have been based on sensitivity to ouabain, actinomycin D, neomycin, methotrexate or emetine. Alternatively, antibiotic resistance can be introduced to one partner more reliably by transfection with a retrovirus-derived shuttle vector bearing genes for biochemical or antibiotic resistance such as G418 or methotrexate. There are also methods based on external labeling of the two fusion partners with the fluorescent labels FITC or TRITC followed by selection of cells with both labels by FACS.

Since H and L immunoglobulin chains are transcribed separately, antibodies secreted by such quadromas will contain random groupings of each, giving rise to 10 different combinations. Few of these will have the desired bispecific combination. There is evidence that this process is not completely random, there being preferential pairing of homologous chains, and fusion of hybridomas secreting the same class and subclass of antibody are likely to yield the highest levels of functional bispecific antibodies.

More reliable methods of making multivalent and multispecific antibodies are now based on recombinant genetic methods, which are discussed later in the chapter.

CHIMERIC AND HUMANIZED ANTIBODIES

Chimeric antibodies are molecules made up of domains from different species. For example, the Fc region or all the constant regions of a mouse monoclonal antibody may be replaced with those of a human or (any other species) antibody. The need to do this was prompted by the problems of using mouse monoclonals therapeutically. Human patients react to the foreign protein making chronic antibody treatment ineffective and causing complications due to immune complex formation known as the **human anti-mouse antibody (HAMA) response**. By replacing as much of the nonantigen binding part of the antibody as possible with human antibody (i.e. **humanization**), this response is reduced without affecting antigen binding. Nevertheless, even with all the constant regions replaced there can be enough of an anti-mouse response to cause problems.

The process can be taken one step further by replacing only the CDR regions and inserting them into a human framework, a technique known as **CDR grafting**. This is achieved by PCR techniques, site-directed mutagenesis or synthetic oligonucleotides. It takes advantage of the many well characterized mouse monoclonal antibodies already available that are otherwise unsuitable for therapy. Many of the therapeutic agents currently being trialed are constructs of this nature. It has been shown that transfer of the loops themselves may not be sufficient to retain binding affinity since the human framework residues can affect the orientation of the loop and may need alteration.

RECOMBINANT ANTIBODIES

The genetic approach to antibody production has gained momentum due to the failure of the hybridoma method to reliably produce human monoclonal antibodies for therapeutic purposes. Nevertheless, these techniques also have relevance in diagnostics in the production of improved reagents (*see* chapter 17, LAB-ON-A-CHIP SYSTEMS by Kricka and Wild). There are a number of advantages over the hybridoma methods, including greater speed of production, increased diversity of specificities obtainable, including those to poor immunogens and the facility to fine tune the affinities and specificities of antigen binding by manipulating the isolated genes. It is also now possible to bypass the immune system altogether and generate specific antibodies without the need to immunize animals.

Essentially, the process begins with the isolation of heavy and light chain gene fragments from immunized or naive lymphocytes. These are then cloned in random combinations into phage display vectors such that the antibody protein fragment representing the antigen binding site is expressed on the outer surface of the phage. This creates a **phage display library** of potentially 10^9 or more recombinant antibody fragments. The desired phage-attached antibody fragment (including the genes that encode it) can then be selected for on the basis of the antibody fragment binding to its antigen (a process known as **panning**). The phage antibody can be used as it is with anti-phage secondary antibodies, or the antibody fragment can be detached for use with alternative labels. The library can be repeatedly probed with any antigen to retrieve antibodies of many different specificities. Furthermore, the affinity and specificity of the antibodies can be further improved, if necessary, by various methods of genetic manipulation.

The phage display technique of expressing peptide fragments on the surface of phage was first demonstrated by Smith in 1985 and has made the selection of the desired clones from a library both easier and faster. This technological advance together with the polymerase chain reaction (PCR), and the demonstration that antigen-binding fragments (Fv and Fab) of antibody could be expressed in *E. coli* (Skerra and Plückthun, 1988; Better *et al.*, 1988) enabled cloned antibody genes to be quickly and easily amplified and analyzed. Several groups were among the first to isolate specific antibodies in this way (Huse *et al.*, 1989; McCafferty *et al.*, 1990; Barbas *et al.*, 1991; Hoogenboom *et al.*, 1991).

RECOMBINANT PHAGE ANTIBODY LIBRARY CONSTRUCTION

The general strategy of library construction is shown in Figure 8.7. More detail on the practical considerations can

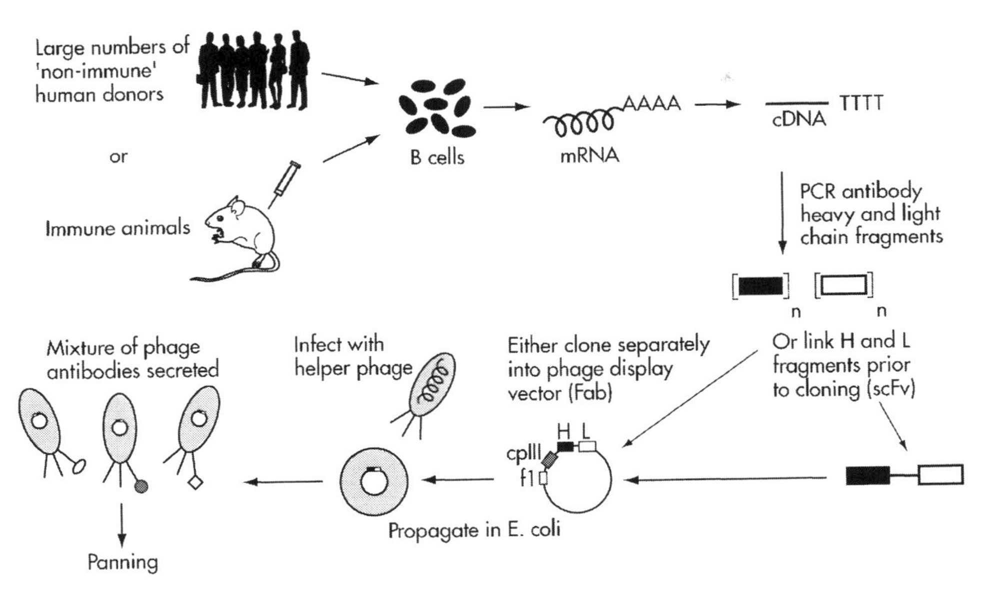

Fig. 8.7 Recombinant antibody gene library production by phage display.

be found in McCafferty *et al.* (eds), 1996; Kay *et al.* (eds), 1996; and Kontermann and Dübel, 2001.

Antibody Gene Amplification

First, RNA is isolated from immune lymphocytes such as the spleen, peripheral blood or bone marrow. From this template, cDNA is made using oligo-dT or antibody-specific primers, and amplified by PCR using pairs of specific primers complementary to either end of the desired antibody gene fragment. The heavy and light chain genes are amplified separately, and the size of the fragment will depend on whether whole Fab fragments or single chain variable fragments (scFv) are to be made (see Figure 8.8).

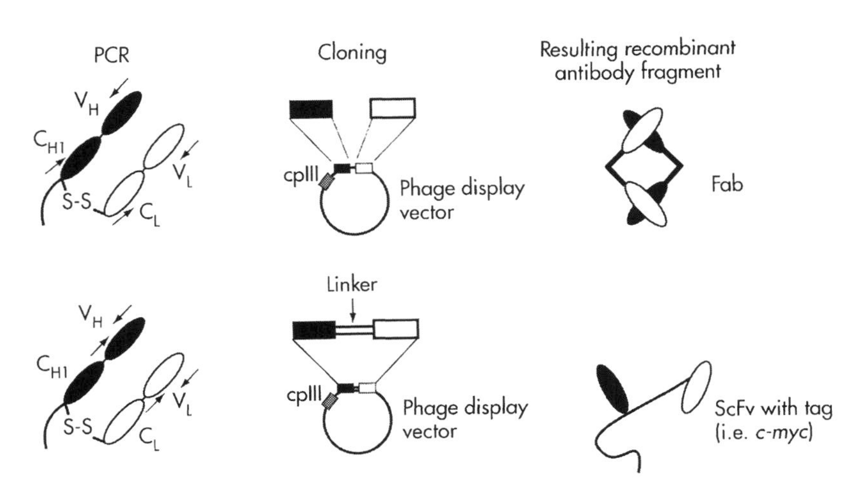

Fig. 8.8 Derivation of Fab and scFv antibody fragments. The antibody gene fragments amplified by PCR encode for the parts of the antibody protein shown by the arrows. To make a recombinant Fab, the heavy and light chain genes are cloned separately into the vector, necessitating two cloning steps. The Fab is formed by the natural formation of the disulfide bond. To make a scFv, the heavy and light chain genes are linked by a short piece of DNA before being cloned once. A tag such as *c-myc* or *his* has to be included to enable recognition by anti-tag antibodies.

In order to obtain **recombinant Fab fragments** the whole of the light chain is amplified and the V_H and C_H1 domains of the heavy chain (Barbas *et al.*, 1991). The 3′ PCR primer of the heavy chain hybridizes to a hinge region sequence which includes a Cys involved in the disulfide bond that naturally holds the two chains together. These primers (one for each isotype) are paired with several (typically 9) different primers for the 5′ end of the gene, therefore representing a large proportion of the different antibody variations in the original repertoire. The light chain is normally cloned into the vector first and amplified before insertion of the heavy chain into the same vector. The heavy chain is linked to the gIII gene in the vector so that, on expression, the gIII coat protein becomes anchored in the bacterial periplasm with the heavy chain attached. The light chain is secreted into the periplasm separately, but is able to associate with the heavy chain via the disulfide bond. No artificial linker is required. The Fab fragment structure, therefore, closely resembles the native form with its natural flexibility which is so important for binding to antigen.

The alternative antibody fragment, the **single chain Fv** (or **scFv**) is smaller (25 kDa) and represents only the variable regions (V_H and V_L) of both antibody chains (McCafferty *et al.*, 1990). More primers are needed for the 3′ end to account for the greater diversity in this region and there is no natural disulfide bond to hold the two chains together. Consequently, the two gene fragments are linked together by a small piece of DNA such as $(Gly_4Ser)_3$ before cloning into the vector. There is, therefore, only one cloning stage compared to that for Fab construction and a smaller fragment is produced which could facilitate intracellular penetration should it be required.

Cloning

The PCR amplified heavy and light chain gene fragments from each primer pair are analyzed by gel electrophoresis and two separate pools of heavy and light chains are isolated and purified. If Fab fragments are required the light and then the heavy chains are cloned separately into the phage display vector. If scFvs are to be made the heavy and light chain genes must first be linked by a short oligonucleotide linker but only one cloning step is required.

The original combinations of heavy and light chains forming the specific antibodies in the original repertoire are, of course, destroyed in the process of library construction and the separate chains recombined at random. Typically, 10^8 different recombinations and therefore potential specificities and binding affinities can be obtained. The size of the library, and the range of specificities, are directly dependent on the transformation efficiency of the bacteria with the vector, which is steadily improving as the technology improves. Following restriction enzyme digestion of the respective inserts, they are ligated into an appropriate expression vector linked to the gIII coat protein gene and transfected into *E. coli* by electroporation in order to maximize the transformation efficiency.

The expression vectors currently most in favor are phagemids, which combine the advantages of high transformation efficiency and control over the number of antibody fragments displayed. Phagemids are plasmids with an origin of replication for filamentous bacteriophage, thus they are incapable of packaging the DNA to produce free phage particles expressing the protein on the surface, without co-infection of the bacteria with helper phage. The helper phage has a defective origin of replication so the secretion of the phagemid package is favored. The experimental conditions can be adjusted so that only one of the gIII coat proteins carries the antibody fragment, thus the display is monovalent. Multivalent expression of antibody is not desirable for the phage selection process because several copies of poorer affinity binders will appear equivalent to high affinity binders and it will be difficult to select for the higher affinity binders.

SELECTION OF SPECIFIC PHAGE ANTIBODIES

Phage expressing the desired specificities can now be selected by affinity purification against specific antigens. The process is normally known as **panning** (see Figure 8.9), whereby antigen is attached to a solid phase, the mixture of phage display antibodies from the library is allowed to bind, and nonbinders are washed off. Binders are eluted with acid or excess antigen, and the phage is allowed to reinfect *E. coli*, which are allowed to grow and therefore amplify the selected phage. Secondary infection with helper phage releases more phage antibody particles, which are re-exposed to antigen and the process is repeated several times. Several rounds of panning are required because of some nonspecific adsorption of phage to the binding surface. With each round of panning the proportion of specific phage in the population can increase several thousand-fold. Ultimately, the bacterial colonies are plated out, and single colonies are selected and propagated for monoclonal production of phage antibodies. At this stage the antibody fragment is still attached to the phage and can be used in assays with a labeled antibody directed against the phage surface. Alternatively, free antibody fragments can be made by cutting out the gIII gene from the vector by various methods, depending on the particular vector used. Soluble Fab fragments can be detected by commonly available, inexpensive anti-Fab conjugate labels. Soluble scFv fragments cannot be detected with anti-Fab reagents so a marker such as a *c-myc* gene is included in the vector to enable the expressed scFv with a *c-myc* tag to be detected with anti-*c-myc* antibody labels.

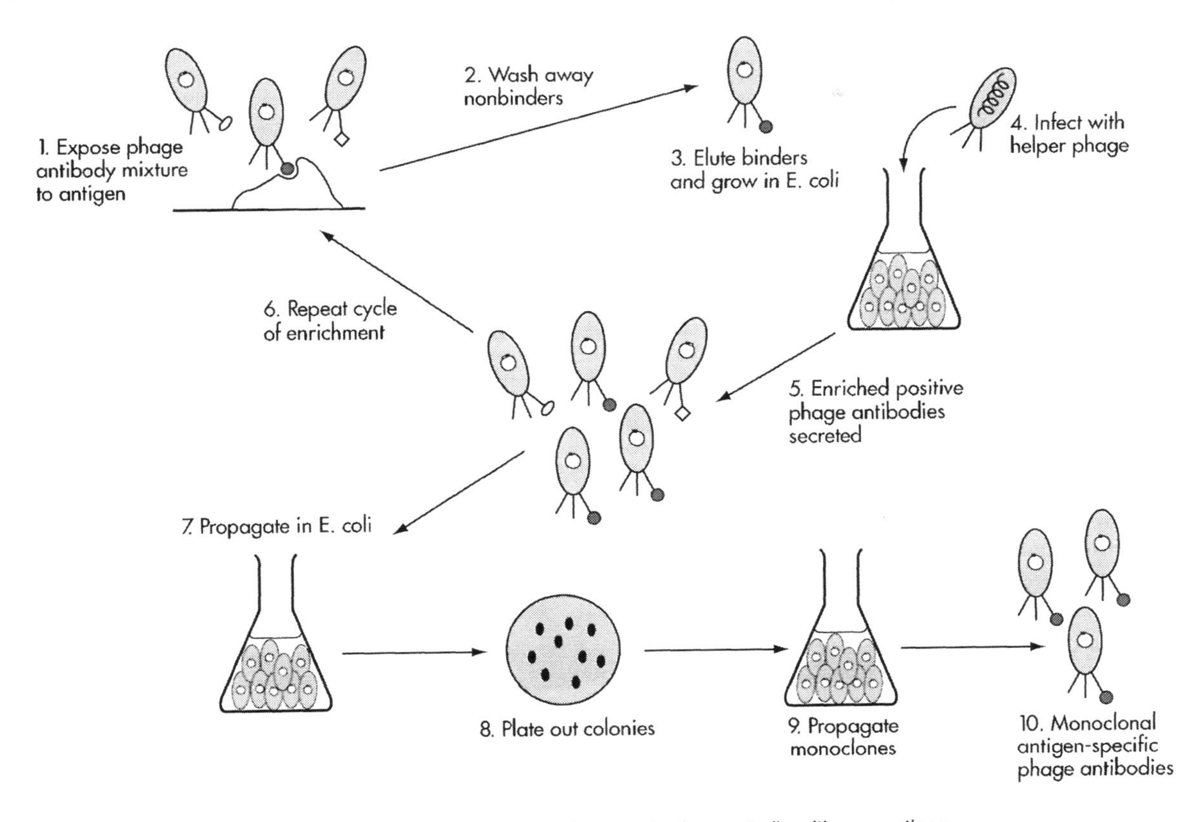

Fig. 8.9 Panning a phage antibody library. The library can be reprobed repeatedly with any antigen.

A novel method of selection of specific antibodies, selectively infective phage (SIP) technology, was reported in 1997 (Spada and Plückthun). The phage are rendered noninfective by removing the *N*-terminal domains (N1 and N2) of the gIII coat protein. These domains are responsible for docking and bacterial cell penetration. The antibody fragments are displayed on the remaining *C*-terminal domain attached to the phage. The *N*-terminal domains are attached to the antigen. When the desired pairing of antibody and antigen occurs, infectivity is restored to the gIII coat protein so only the desired phage antibodies are propagated.

ALTERING ANTIBODY CHARACTERISTICS

One important advantage of the combinatorial library approach over the hybridoma method is the ease and speed with which the initial antibody can be changed and improved upon without having to begin again. By genetic manipulation, antibody–antigen binding affinities can be improved, undesirable cross-reactivities can be eliminated, specificities can even be totally changed, bispecific antibodies can be constructed and structural analysis of the antibody combining site can assist in defining the antigen structure. In addition, once a library has been established, many other antibodies with different binding characteristics can be selected from it within days. The selection process can be biased towards eliminating undesirable cross-reactivities, broadening cross-reactivity in certain cases (for instance, to improve recognition of rapidly evolving viral antigens), and improving affinity.

The antibody–antigen binding affinities of recombinant antibodies obtained from immune sources tend to be in the range of 10^7–$10^9\,M^{-1}$, which are similar to those expected from an *in vivo* secondary immune response. These can be further improved upon *in vitro* by several different strategies, including **chain shuffling** (Barbas, 1993) and **mutagenesis** by various means (Myers *et al.*, 1985; Leung *et al.*, 1989; Schaaper, 1988). Chain shuffling involves retaining the original heavy chain and recombining it with the original light chain library, thus producing a family of specific antigen binders some of which may have improved binding characteristics (300 fold increases in affinity have been reported). Random or site directed mutagenesis of the variable region genes can be induced by standard chemical means, by PCR or by *in vivo* mutagenesis using mutator bacterial strains. Our knowledge of the most important targets to change or retain is accumulating as more antibody combining sites are being analyzed.

SYNTHETIC RECOMBINANT ANTIBODIES

One of the aims of producing human recombinant antibodies was to bypass immunization altogether, and this has been shown to be possible even to the extent that semi-synthetic antibodies have been made (Marks *et al.*, 1991; Barbas *et al.*, 1992). Antibody gene libraries derived from **naive** (i.e. nonimmunized) lymphocytes can be screened for virtually any antigen binding. The larger the library, the greater will be the chance of selecting a desired specificity. Antibodies have been obtained with binding affinities comparable to those found after primary immunization (10^6–10^7 M^{-1}). Antibodies derived from naive libraries are likely to have a different spectrum of specificities than those from immunized sources, where immunodominant epitopes are favored at the expense of the less immunogenic, including self-antigens that are very difficult to achieve by conventional methods. A similar approach has even led to the production of bacterial cell surface peptides that recognize small inorganic molecules such as Fe_3O_4 (Brown, 1992) that would never be possible using *in vivo* methods.

Semi-synthetic libraries have been constructed from pooled gene libraries, or single clones which have been subjected to recombination with randomized nucleotide sequences to replace the CDR3 loop of the heavy chain. Selection of alternative specificities is made by panning against different antigens. In this way, it has been possible to convert an anti-tetanus toxoid antibody to one specific to fluorescein (Barbas *et al.*, 1992). Several reports have demonstrated the wide variety of different antibody specificities that can be obtained from a single gene library (Lerner *et al.*, 1992; Williamson *et al.*, 1993; Nissim *et al.*, 1994).

The breadth of antibody diversity that can be obtained from a library is dependent on the size of the library, which has been restricted to a maximum of around 10^9 because of the limits of bacterial transformation efficiency. This limit was exceeded by applying the method of combinatorial infection (Waterhouse *et al.*, 1993). Antibody libraries of 6.5×10^{10} have been constructed (Griffiths *et al.*, 1994) by combining very diverse heavy chain plasmid libraries containing random CDR3 sequences, with a light chain phage library in the same bacteria. Lox P sites are contained within each of these vectors which can combine in the presence of Cre recombinase (introduced by co-infection with a phage containing the enzyme), thus the two vectors, and therefore the heavy and light chain genes, are randomly combined. Library sizes are potentially equivalent to the number of bacteria infected.

BIVALENT AND BISPECIFIC RECOMBINANT ANTIBODIES

The functional affinities of monovalent scFv and Fab fragments can be improved by constructing bi- and multivalent molecules and similar processes can also be employed to make bispecific molecules. An important aspect of therapeutic antibody development is not only improved affinity and flexibility but also the enhancement of tissue penetration, lengthening of serum half-life and reduction of immunogenicity. To this end, a variety of formats and protein designs have been attempted and continue to be the focus of much research (Fanger, 1995; Plückthun and Pack, 1997).

Certain scFvs can spontaneously dimerize or even multimerize. The percentage of monomer is directly proportional to the length of the linker connecting the two variable domains. Presumably with short linkers (5 amino acids), the V_H or V_L domain cannot reach the partner domain of the other scFv to form the dimer. This can only occur when the linker is long enough (20 amino acids) to span the distance, to form two V_H/V_L assemblies or **diabodies**. An alternative means of linking two scFvs or Fab fragments is to add extra *C*-terminal cysteine residues that, on oxidation, give rise to interdomain disulfide bridge formation. The linkage can also be effected by thioether formation using bis-maleimide derivatives.

A more directed approach is to engineer 'association domains' onto the *C*-terminus of each scFv. Association domains consist of a long flexible hinge sequence and a self-associating secondary structure and an optional cysteine tail, thus creating a bivalent or bispecific structure with similar flexibility to a normal Fab fragment. These are known as **miniantibodies.** Similarly, bivalent or bispecific Fab fragments can be constructed but are more difficult to produce because the V_H and V_L linkages are not fixed. ScFvs can be dimerized *in vivo* by incorporating a variety of self-associating secondary structures such as helix bundles or coiled-coils (leucine zippers). A 4-helix bundle containing a helix-turn-helix motif resulting in noncovalent dimerization has produced functional affinities and stability comparable to the whole parent antibody.

Multimeric constructs are possible if an appropriate association domain is used. A modified helix of the transcription factor, GCN4, streptavidin and a human IgG3 hinge/human p53 protein have all been used to produce tetrameric molecules with greatly enhanced functional affinities.

ANTIBODY CONJUGATES AND FUSION PROTEINS IN DIAGNOSTICS

Antibodies are used in immunoassays as whole molecules or fragments, as has been described in this chapter, and they are also used as conjugates with a wide variety of labels such as colorimetric enzymes, fluorescent and luminescent molecules, and radiolabels. These conjugates and the chemical methods of preparing them are described in Chapter 14. Use of the new recombinant antibody technology to form **antibody fusion proteins** genetically has the potential to improve production of conjugates in that the antigen-binding site should be unaffected and homogeneous reagents are produced directly. Several useful fusion proteins for immunoassay have been described (see Table 8.4) although none are yet widely available. It has sometimes been difficult to express fully functional fusion proteins, and the capacity to vary the stability of the linkage as with chemical conjugation is lost, but there is undoubtedly scope for development in this direction.

POLYCLONAL, MONOCLONAL OR RECOMBINANT?

In considering which type of antibody is appropriate for a particular application there are many factors to consider, e.g.

- is the antibody already available?
- cost
- ease of preparation
- equipment requirements
- time and costs of preparation, including downstream processing
- level of specificity and affinity required.

Polyclonal sera are the easiest, quickest, and cheapest to make but are limited in their degree of specificity, purity, and yield. These disadvantages can be overcome especially if pure antigen is available. For example, polyclonal sera to synthetic peptides can be of comparable specificity to monoclonals. There are assays in which a spectrum of antibodies to different epitopes may be an advantage although there are many applications where a polyclonal serum will never be able to produce the fine specificity of a monoclonal. See Table 8.5.

Unlike polyclonal sera, monoclonal antibodies can normally be used directly, without further purification.

Table 8.4. Recombinant antibody fusion proteins of use in diagnostics.

Fusion partner	Antibody fragment	Use	Reference
Alkaline phosphatase	ScFv, $F(ab')_2$	Produces colorimetric product with substrate	Wels *et al.* (1992); Ducancel *et al.* (1993)
Avidin	Fab, $F(ab')_2$, IgG	Binds biotin conjugated proteins	Shin *et al.* (1997)
Biotin-carboxyl carrier protein	Fab	Biotin attached during secretion	Weiss *et al.* (1994)
Protein A fragment	ScFv	Fc binding	Tai *et al.* (1990)
Metallothionein	$F(ab')_2$	Allows subsequent binding of ^{99M}Tc	Das *et al.* (1992)
Peptide	ScFv	Peptide chelates metal then binds to ^{99M}Tc	George *et al.* (1995)
Peptide	ScFv	Peptide can be enzymically labeled with ^{32}P	Neri *et al.* (1996)
Amino terminal of *E. coli* major lipoprotein	ScFv	Attaches lipid during expression	Laukkanen *et al.* (1993, 1995)
Streptavidin	ScFv	Binds biotin conjugated proteins. Also useful for producing tetramers	Dubel *et al.* (1995); Kipriyanov *et al.* (1995)

Table 8.5. Comparison of antibodies.

	Polyclonal antiserum	**Monoclonal antibody**	**Recombinant phage antibody**
Preparation			
Production time	Minimum six weeks	Minimum four months (including immunization)	Library prep.: 1–6 months. Panning from established library: a few weeks
Immunogen purity	Essential	Not necessary	Not necessary. 'Nonimmune' libraries available
Special facilities	Animal facility and licenses	Animal facility and licenses initially. Tissue culture	Molecular biology labs
Product characteristics			
Concentration	10–30 mg/mL (max 1 mg/mL of 'specific' often much less)	Culture supernatant (5–10 μg/mL static; 100–200 μg fermenter; 0.5–10 mg hollow fiber). Ascites 2–10 mg/mL	Up to 4 g/L in high cell-density fermenters
Quantity	~100 mL serum/rabbit	Limitless	Limitless
Purity	Dependent on immunogen. Affinity purification usually necessary	Epitope specific. Further purification not always necessary	Epitope specific. Detachment from phage required
Valency	IgG dimeric	IgG dimeric	Monomeric but can be multimerized genetically
Major contaminants	Ig to impurities; other serum proteins	Culture supernatant: bovine serum but can use serum-free media. Ascites: other mouse proteins	Bacterial proteins
Affinity	Heterogeneous	Homogeneous	Homogeneous
Specificity	Can be broad	Epitope specific though same epitope may be present in other proteins	Epitope specific though same epitope may be present in other proteins
Post-production versatility			
Batch variability	Each bleed slightly different	Antibody characteristics remain same, only concentration will vary	Antibody characteristics remain same, only concentration will vary
Affinity improvements	Can be improved slightly with boosts	Fixed affinity	Can be improved up to 100 fold genetically
Genetic manipulations	Not possible	Not possible although hybridoma genes could be	Possible
Patent restrictions	None	None, except for certain assay formats	Various on technology and libraries
Main advantages	Cheap, simple procedures	Superb specificity, unlimited quantities	Libraries can be reprobed for other specificities; potentially relatively fast production; human antibodies for therapy; 'new' specificities obtainable since immunization can be bypassed
Main disadvantages	Specificities and yield can be poor	Production labor intensive; human antibodies difficult	Initial antibodies can be of low affinity; some vector stabilities poor; technically demanding for beginners

When dealing with a monoclonal antibody, however, one should be aware that its unique specificity is for a particular epitope, not necessarily for the whole molecule. If that epitope is common to other molecules the antibody may also bind to apparently unrelated antigens. Having said that, however, it is not always necessary to know precisely what epitope the antibody recognizes, for instance, if it can distinguish between two different cell types or molecule.

A monoclonal antibody is raised using perhaps only one type of screening assay format (i.e. ELISA). That antibody may not recognize the 'same' antigen in a different type of assay because of possible conformational changes in the antigen. Some monoclonal antibodies (and indeed hybridomas) are sensitive to being frozen so their stability under different storage conditions may lead one to focus on a different hybridoma line of the same specificity.

Many monoclonal antibodies are now available commercially, in purified form or as native culture supernatant or ascitic fluid, and can be supplied with a variety of enzyme or fluorescent labels attached. Buying antibodies can be very expensive. If this is a serious problem, an option worth investigating is to obtain the hybridoma cell line from a cell culture collection or the originator (often free for research purposes) and make a batch of ascitic fluid oneself. Obviously, if after checking the commercial databases and scientific literature, the particular antibody specificity required is not available you must consider making your own monoclonal antibodies. In experienced hands, this can take a minimum of four months but it could take more than a year, especially if dealing with a poor immunogen. Also the work involved is fairly intensive once tissue culture has begun. Nevertheless, making ones own monoclonal antibodies can yield limitless quantities of useful, highly specific reagents that can also, if novel enough, be of commercial value. Alternatively, there are many companies and university-based units who offer custom production services.

The major advantages of the recombinant antibody approach to diagnostics are the speed of library production once the methods are established (theoretically, only a few weeks) and the diversity of specificities that can be obtained from that library (particularly, when the precise antigen target is unknown or of poor immunogenicity). Screening of the library for different specificities and affinities can be done in as little as a week. One then has all the advantages of having a range of different monoclonal antibodies plus the ability to manipulate the genes to eliminate any undesired cross-reactivities, improve affinities and create multivalent constructs. Propagation of the antibodies is also facilitated by the ability to grow in bacteria, which is both quicker and cheaper than *in vitro* methods of hybridoma propagation.

Making recombinant antibodies is becoming more accessible to nonspecialist laboratories and the equipment required is no more than that normally available in a general molecular biology laboratory. Although there have been problems with the stability of some phage display vectors, the technology is improving rapidly and will soon become routine. It is also sometimes possible to obtain ready-made naive libraries, which can be panned against the desired antigen, and some companies offer this as a service.

The characterization of any antibody reagent is, of course, important, and the procedures are similar whether the antibody is polyclonal, monoclonal or engineered. Thorough testing of the specificity and titer of an antibody in the context in which it will be used, with relevant controls, must be done before it can be applied confidently.

Knowing the isotype of a monoclonal antibody is important in order to determine the most appropriate purification method, the appropriate 'second' antibody in an immunoassay and whether it would have the appropriate effector function such as complement fixation. The affinity of the antibody–antigen bond is useful to know since those of highest affinity are required for good immunoassays whereas those of lower affinity are preferred for purification of antigen so that the bond can easily be broken without damaging the antigen. Measurement of binding on and off rates by surface plasmon resonance methodology (e.g. using BIAcore™, Biacore AB, Uppsala, Sweden) is the most effective way of providing this information. *See* SURFACE PLASMON RESONANCE IN KINETIC, CONCENTRATION AND BINDING SITE ANALYSES.

In summary, there are many different antibody products that have uses in a wide range of different applications, for which some will be more appropriate than others. The contribution of monoclonal antibodies to diagnostics has been phenomenal over the last 20 years. In therapeutics, progress has been comparatively slow but this has spurred rapid technological advances in recombinant antibody technology, which will no doubt produce novel and lucrative therapeutic reagents. In the coming years, this technology will also be applied with greater effect to improve the production and refinement of antibody reagents for immunoassay.

REFERENCES AND FURTHER READING

Barbas, C.F., III. Recent advances in phage display. *Current Opinion in Biotechnology* **4**, 526–530 (1993).

Barbas, C.F., Kang, A.S., Lerner, R.A. and Benkovic, S.J. Assembly of combinatorial antibody libraries on phage surfaces: the gene III site. *Proc. Natl. Acad. Sci.* **88**, 7978–7982 (1991).

Barbas, C.F. III, Bain, J.D., Hoekstra, D.M. and Lerner, R.A. Semisynthetic combinatorial antibody libraries: a chemical solution to the diversity problem. *Proc. Natl. Acad. Sci.* **89**, 4457–4461 (1992).

Bazin, H. Production of rat monoclonal antibodies with the Lou rat non-secreting IR983F myeloma cell line. In: *Protides of the Biological Fluids* (ed Peeters, H.), 615–618 (Pergamon Press, Oxford, 1982).

Bazin, H. *Rat Hybridomas and Rat Monoclonal Antibodies* (CRC Press, Boca Raton, FL, 1990).

Better, M., Chang, C.P., Robinson, R.R. and Horwitz, A.H. *Escherischia coli* secretion of an active chimeric antibody fragment. *Science* **240**, 1041–1043 (1988).

Brown, S. Engineered iron oxide-adhesion mutants of the *Escherichia coli* phage lambda receptor. *Proc. Natl. Acad. Sci.* **89**, 8651–8655 (1992).

Croce, C.M., Linnenbach, A., Hall, W., Steplewski, Z. and Koprowski, H. Production of human hybridomas secreting antibodies to measles virus. *Nature* **288**, 488–489 (1980).

Das, C., Kulkarni, P.V., Constantinescu, A., Antich, P., Blattner, F.R. and Tucker, P.W. Recombinant-antibody metallothionein: design and evaluation for radioimaging. *PNAS* **89**, 9749–9753 (1992).

Davies, H.L., Michel, M.-L. and Whalen, R.G. DNA-based immunization for Hepatitis B induces continuous secretion of antigen and high levels of circulating antibody. *Hum. Mol. Genet.* **2**, 1847–1851 (1993).

Donnelly, J.J., Ulmer, J.B., Shiver, J.W. and Liu, M.A. DNA vaccines. *Ann. Rev. Immunol.* **15**, 617–648 (1997).

Dubel, S., Breitling, F., Kontermann, R., Schmidt, T. and Skerra, A. Bifunctional and multimeric complexes of streptavidin fused to single chain antibodies (scFv). *J. Immunol. Methods* **178**, 201–209 (1995).

Ducancel, F., Gillet, D., Carrier, A., Lajeunesse, E., Menez, A. and Boulain, J.C. Recombinant colorimetric antibodies: construction and characterization of a bifunctional F(ab)2/ alkaline phosphatase conjugate produced in *Escherichia coli*. *Bio-technology* **11**, 601–605 (1993).

Fanger, M.W. (ed), *Bispecific Antibodies* (R.G. Landes, Austin, 1995).

Galfre, G., Milstein, C. and Wright, B. Rat × rat hybrid myelomas and a monoclonal anti-Fd portion of mouse IgG. *Nature* **277**, 131–133 (1979).

Galfre, G., Cuello, A.C. and Milstein, C. New tools for immunochemistry: internally labelled monoclonal antibodies. In: *Monoclonal Antibodies and Developments in Immunoassay* (eds Albertini, A. and Ekins, R.), 159–162 (Elsevier, Amsterdam, 1981).

George, A.J.T., Jamar, F., Tai, M.S., Heelan, B.T., Adams, G.P., McCartney, J.E., Houston, L.L., Weiner, L.M., Opperman, H., Peters, A.M. and Huston, J.S. Radiometal labeling of recombinant proteins by a genetically engineered minimal chelation site: techetium-99m coordination by single-chain Fv antibody fusion proteins through a C-terminal cysteinyl peptide. *PNAS* **92**, 8358–8362 (1995).

Griffiths, A.D., Williams, S.C., Hartley, O., Tomlinson, I.M., Waterhouse, P., Crosby, W.L., Kontermann, R.E., Jones, P.T., Low, N.M., Allison, T.J., Prospero, T.D., Hoogenboom, H.R., Nissim, A., Cox, J.P.L., Harrison, J.L., Zaccolo, M., Gherardi, E. and Winter, G. Isolation of high affinity human antibodies directly from large synthetic repertoires. *EMBO J.* **13**, 3245–3260 (1994).

Hoogenboom, H.E., Griffiths, A.D., Johnson, K.S., Chiswell, D.J., Hudson, P. and Winter, G. Multi-subunit proteins on the surface of filamentous phage: methodologies for displaying antibody (Fab) heavy and light chains. *Nucleic Acids Res.* **19**, 4133–4137 (1991).

Houston, L.L., Weiner, L.M., Oppermann, H., Peters, A.M. and Huston, J.S. Radiometal labelling of recombinant proteins by a genetically engineered minimal chelation site. *PNAS* **92**, 8358–8362 (1995).

Huse, W.D., Sastry, L., Iverson, S.A., Kang, A.S., AltingMees, M., Burton, D.R., Benkovic, S.J. and Lerner, R.A. Generation of a large combinatorial library of the immunoglobulin repertoire in phage lambda. *Science* **246**, 1275–1281 (1989).

James, K. and Bell, G.T. Human monoclonal antibody production. *J. Immunol. Methods* **100**, 5–40 (1987).

Janeway, C.A., Travers, P., Walport, M., Shlomchik, M., *Immunobiology*, 6th edn. (Garland Publishing, New York, 2005).

Jensenius, J.C., Andersen, I., Hau, J., Crone, M. and Koch, C. Eggs: conveniently packaged antibodies. Methods for purification of yolk IgG. *J. Immunol. Methods* **46**, 63–68 (1981).

Kay, B., Winter, J. and McCafferty, J. (eds), *Phage Display of Peptides and Proteins: A Laboratory Manual* (Academic Press, San Diego, 1996).

Karpas, A., Fischer, P. and Swirsky, D. Human myeloma cell line carrying a Philadelphia chromosome. *Science* **216**, 997–999 (1982).

Kearney, J.F., Radbruch, A., Leisegang, B. and Rajewsky, K. A new mouse myeloma cell line that has lost immunoglobulin expression but permits the construction of antibody-secreting hybrid cell lines. *J. Immunol.* **123**, 1548–1550 (1979).

Kilmartin, J.V., Wright, B. and Milstein, C. Rat monoclonal antibodies derived by using a new non-secreting rat cell line. *J. Cell Biol.* **93**, 576–582 (1982).

Kipriyanov, S.M. Generation of bispecific and tandem diabodies. *Methods Mol. Biol.* **178**, 317–331 (2002).

Kipriyanov, S.M., Breitling, F., Little, M. and Dubel, S. Single-chain antibody streptavidin fusions; Tetrameric bifunctional scFv-complexes with biotin binding activity and enhanced affinity to antigen. *Hum. Antibodies Hybridomas* **6**, 93–101 (1995).

Köhler, G. and Milstein, C. Continuous cultures of fused cells secreting antibody of predefined specificity. *Nature* **256**, 495–497 (1975).

Kohler, G. and Milstein, C. Derivation of specific antibody-producing tissue culture and tumour lines by cell fusion. *Eur. J. Immunol.* **6**, 511–579 (1976).

Kontermann, R. and Dübel, S. (eds), *Antibody Engineering* (Springer, Berlin, 2001).

Lamoyi, E. Preparation of F(ab′)2 fragments from mouse IgG of various subclasses. *Methods Enzymol.* **121**, 652–663 (1986).

Laukkanen, M.J., Teeri, T.T. and Keinanen, K. Lipid-tagged antibodies: bacterial expression and characterisation of a lipoprotein-single chain antibody fusion protein. *Protein Engng.* **6**, 449–454 (1993).

Laukkanen, M.J., Orellana, A. and Keinanen, K. Use of genetically engineered lipid-tagged antibody to generate functional europium chelate loaded liposomes. Application in fluoroimmunoassay. *J. Immunol. Meth.* **185**, 95–102 (1995).

Lerner, R.A., Kang, A.S., Bain, J.D., Burton, D.R. and Barbas, C.F. III Antibodies without immunisation. *Science* **258**, 1313–1314 (1992).

Leung, D.W., Chen, E. and Goeddel, D.V. A method for random mutagenesis of a defined DNA segment using a modified polymerase chain reaction. *J. Meth. Cell. Mol. Biol* **1**, 11–15 (1989).

Liddell, J.E. Production of monoclonal antibodies. In: *Nature Encyclopedia of Life Sciences* (Nature Publishing Group, London, 2003), http://www.els.net.

Liddell, J.E. and Cryer, A. *A Practical Guide to Monoclonal Antibodies* (Wiley, Chichester, 1991).

Marks, J.D., Hoogenboom, H.R., Bonnert, T.P., McCafferty, J., Griffiths, A.D. and Winter, G. By-passing immunization: Human antibodies from V-gene libraries displayed on phage. *J. Mol. Biol.* **222**, 581–597 (1991).

Matsuoka, Y. *et al.* Production of free light chains of immunoglobulin by a hematopoietic cell line derived from a patient with multiple myeloma. *Proc. Soc. Exp. Biol. Med* **125**, 1246–1250 (1967).

McCafferty, J., Griffiths, A.D., Winter, G. and Chiswell, D.J. Phage antibodies: filamentous phage displaying antibody variable domains. *Nature* **348**, 552–554 (1990).

McCafferty, J., Hoogenboom, H.R. and Chiswell, D.J. (eds), *Antibody Engineering: A Practical Approach* (IRL Press, Oxford, 1996).

Mendez, M.J., Green, L.L., Corvalan, J.R.F. *et al.* Functional transplant of megabase human immunoglobulin loci recapitulates human antibody response in mice. *Nat. Genet.* **15**, 147–156 (1997).

Myers, R.M., Lerman, L.S. and Maniatis, T. A general method for saturation mutagenesis of cloned DNA fragments. *Science* **229**, 242–247 (1985).

Neri, D., Petrul, H., Winter, G., Light, Y., Marais, R., Britton, K.E. and Creighton, A.M. Radioactive labelling of antibody fragments by phosphorylation using human casein kinase II and g32P-ATP. *Nat. Biotechnol.* **14**, 485–490 (1996).

Niedbala, W.G. and Stott, D.I. A comparison of three methods for production of human hybridomas secreting autoantibodies. *Hybridoma* **17**, 299–304 (1998).

Nissim, A., Hoogenboom, H.R., Tomlinson, I.M., Flynn, G., Midgley, C., Lane, D. and Winter, G. Antibody fragments from a single pot phage display library as immunochemical reagents. *EMBO J.* **13**, 692–698 (1994).

Olsson, L. and Kaplan, H.S. Human-human hybridomas producing monoclonal antibodies of predefined antigenic specificity. *PNAS* **77**, 5429–5431 (1980).

Plückthun, A. and Pack, P. New protein engineering approaches to multivalent and bispecific antibody fragments. *Immunotechnology* **3**, 83–105 (1997).

Roitt, I., Brostoff, J. and Male, D. (eds), *Immunology* 6th edn (Edinburgh, 2001).

Schaaper, R.M. Mechanisms of mutagenesis in the *Escherichia coli* mutator mutD5: role of DNA mismatch repair. *PNAS* **85**, 8126–8130 (1988).

Sharon, J., Kao, C.-Y.Y. and Sompuram, S.R. Oligonucleotide-directed mutagenesis of antibody combining sites. *Intern. Rev. Immunol.* **10**, 113–127 (1993).

Shin, S.U., Wu, D., Ramanthan, R., Pardridge, W.M. and Morrison, S.L. Functional and pharmacokinetic properties of antibody–avidin fusion proteins. *J. Immunol.* **158**, 4797–4804 (1997).

Shirahata, S., Katakura, Y. and Teruya, K. Cell hybridization, hybridomas, and human hybridomas. *Methods Cell Biol.* **57**, 111–145 (1998).

Shulman, M., Wilde, C.D. and Kohler, G. A better cell line for making hybridomas secreting specific antibodies. *Nature* **276**, 269–270 (1978).

Skerra, A. and Plückthun, A. Assembly of a functional immunoglobulin Fv fragment in *Escherichia coli*. *Science* **240**, 1038–1041 (1988).

Smith, G.P. Filamentous fusion phage: Novel expression vectors that display cloned antigens on the surface of the virion. *Science* **228**, 1315–1317 (1985).

Spada, S. and Plückthun, A. Selectively infective phage (SIP) technology: a novel method for in vivo selection of interacting protein–ligand pairs. *Nat. Med.* **3**, 694–696 (1997).

Tai, M.S., Mudgett-Hunter, M., Levinson, D., Wu, G.M., Haber, E., Oppermann, H. and Huston, J.S. A bifunctional fusion protein containing Fc binding fragment B of Staphylococcal protein A amino terminal to antidigoxin single-chain Fv. *Biochemistry* **29**, 8024–8030 (1990).

Tijssen, P. *Practice and Theory of Enzyme Immunoassays* (ed Tijssen, P.), (Elsevier, Amsterdam, 1985).

Waterhouse, P., Griffiths, A.D., Johnson, K.S. and Winter, G. Combinatorial infection and *in vivo* recombination: a strategy for making large phage antibody repertoires. *Nucleic Acids Res.* **21**, 2265–2266 (1993).

Weiss, E., Chatallier, J. and Orfanoudakis, G. *In vivo* biotinylated recombinant antibodies: construction, characterization and application of a bifunctional Fab-BCCP fusion protein produced in E. coli. *Protein Expression Purification* **5**, 509–517 (1994).

Wels, W., Harwerth, I.M., Zwickl, M., Hardman, N., Groner, B. and Hynes, N.E. Construction, bacterial expression and characterisation of a bifunctional single-chain antibody-phosphatase fusion protein targeted to the human erb-2 receptor. *Bio-technology* **10**, 1128–1132 (1992).

Williamson, R.A., Burioni, R., Sanna, P.P. and Partridge, L.J. Human monoclonal antibodies against a plethora of viral pathogens from single combinatorial libraries. *PNAS* **90**, 4141–4145 (1993).

Winter, G., Griffiths, A.D., Hawkins, R.E. and Hoogenboom, H.R. Making antibodies by phage display technology. *Ann. Rev. Immunol.* **12**, 433–455 (1994).

9 Standardization and Calibration

David Wild

As in all forms of measurement there is a need for immunoassay standardization. **Standardization** is the process of ensuring that all methods for determining the concentration of a particular analyte give the same results. **Calibration** is the process of assigning values to unknown samples using a **standard**. Immunoassays do not provide direct measurements (unlike a meter rule, for example). They estimate analyte concentration in unknown samples by comparing signal strength (originating from a labeled reagent) with that from similarly treated standard samples. However, for most of the analytes determined by immunoassay, there are no reference methods with which to calibrate standards. This presents a dilemma that is far from resolved for many immunoassay analytes.

The importance of standardization is often underestimated. It is usual practice to state the reference interval (*see* CONCEPTS) for a particular method alongside patient results, but once a reference interval becomes lodged in the memory it can be difficult to avoid it being used subconsciously to interpret results. This can have serious implications for patient diagnosis if a laboratory changes method and the reference interval is significantly different.

Lack of standardization hinders communication and scientific progress. For example, it reduces the value of comprehensive age and sex-related reference intervals in the scientific literature, which could otherwise be used to improve the diagnostic capability of different manufacturers' tests for the same analyte. There are also knock-on effects. For example, some publications stated that a luteinizing hormone:follicle stimulating hormone (LH:FSH) ratio of greater than 3 is diagnostic of polycystic ovarian syndrome. However, subsequent immunometric assays for LH gave much lower values than the methods that had been used when this ratio was established, making the use of a value of 3 as a diagnostic cut-off inappropriate.

Standardization requires:

- Value assignment in meaningful units.
- A standard that is identical to the analyte in the test samples.
- Absence of interference from the test sample matrix.
- A reference method.
- Demonstration of inter-method agreement.

As will be seen in this chapter, these criteria are rarely met in the immunoassay field and standardization is beset with problems. Diligent efforts by many different organizations have led to a form of consensus, but there have had to be some compromises, caused primarily by the limitations of immunoassay technology and the heterogeneity of patient samples. However, there is great scope for improved standardization and this is a fertile area for development. During a course of treatment, individual patients are more likely than ever before to have tests carried out for the same analyte by different methods (e.g. point-of-care and laboratory analyzer), and to have their results checked against reference intervals that were set elsewhere, and perhaps not using the same method. Therefore, there is now an even greater need for standardization, and this field is still one of the great challenges facing immunoassay scientists.

STANDARDIZATION

THE ROLE OF EXTERNAL QUALITY ASSESSMENT (PROFICIENCY TESTING) SCHEMES

Historically, the first step in achieving the standardization of different immunoassays for the same analyte has been the initiation of a regular external quality assessment (proficiency testing) scheme. Pooled samples are sent to a number of laboratories and the results analyzed by the scheme coordinator. Often the first results show a wide variation. However, it has been shown for many analytes that the strategy of feeding back biases from the all-laboratory trimmed mean (ALTM) to scheme participants

The Immunoassay Handbook. *Third Edition*
D. Wild (Ed.)

and manufacturers leads eventually to a reduction in the overall variation. Methods that have consistent bias from other methods are clearly highlighted, and laboratories that are not in consensus with their peer group are made aware of their relative position. Some scheme organizers send out spiked samples to estimate recovery and this can help to pinpoint fundamental problems with individual methods. As collaborative efforts to achieve standardization occur, the improvement can be tracked in the scheme and aberrant methods become more clearly identifiable. Studies have shown that the ALTM tends to be close to the 'true' concentration, but this cannot be taken for granted in the absence of a reference method, although recovery experiments may help to identify inappropriately standardized methods.

INTERNATIONAL STANDARDS

Ultimately, to achieve agreement between different methods, a single recognized standard is needed. To achieve widespread credibility, it is best prepared under the overall management of an international body. Some analytes are available in a highly purified form, e.g. cortisol, available from the NIST, Washington DC, USA, at 98.9 ± 0.2% purity. However, most **International Standards** for immunoassay analytes are prepared by the National Institute for Biological Standards and Control in the UK (www.nibsc.ac.uk).

Preparation of International Standards is a specialized activity and validation of each step is required, with support from several independent laboratories. One of the frequent problems of standard preparation is heterogeneity. Many immunoassay analytes are proteins with varying degrees of glycosylation (attachment of sugar residues). Some of this glycosylation needs to be retained if the standard is to be of use. Gentle methods of purification are used to try and preserve the structural integrity of the analyte. The starting material must also be from a reliable source and, if the analyte is a protein, of human origin. If the protein is extracted from organs, an organ is selected with a protein structure as near to that found in the blood as possible.

The purified protein is mixed with an inert carrier compound, otherwise the amount would be too small to be visible. Then the solution is divided into aliquots in glass ampoules, freeze-dried, and the ampoules sealed under nitrogen.

After demonstrating reproducibility across the ampoules (better than ±0.25%) and ensuring a low residual moisture content has been achieved, samples are sent to laboratories for independent estimates of concentration and to demonstrate that the standard analyte in the preparation behaves similarly to the naturally occurring analyte in human blood samples.

To be designated as an International Standard (IS), the preparation has to be authenticated by the Expert Committee on Biological Standardization (ECBS) of the World Health Organization (WHO). An IS is a preparation to which an **International Unit (IU)** has been assigned on the basis of an international collaborative study involving several assay systems in different laboratories. The term 'IRP' (International Reference Preparation) is no longer used for new materials. It used to be assigned to preparations that did not meet the demanding criteria for an IS but were useful for method to method standardization.

The designation of IS status is sometimes portrayed as the end of the road for standardization. It is, without doubt, a significant achievement to achieve worldwide agreement on a common standard, but IUs are still arbitrary units. To some extent, this is a vestige from enzyme determinations, which were expressed in terms of substrate conversion activity units, and from bioassays, which detected hormones in samples through their biological activity. Where protein concentrations are not stated in terms of IU, they are sometimes presented in terms of mass, e.g. ng/L. The limitations of arbitrary and mass units were illustrated by the unitage of the International Standard for hCG (IS 75/537), and the International Reference Preparations for the α subunit (IRP 75/569) and β subunit (IRP 75/551) of hCG. The IS value assignment was based on bioassay units, and the subunit IRPs were based on mass, so they were not easily comparable, except by using crude conversion factors. (Note that the latest versions of these standards are: 75/589 (intact hCG), 99/720 (α subunit), and 99/650 (β subunit), and there is a special 1st WHO Reference Reagent (99/688) for intact hCG for immunoassays.) At first sight, mass concentration units might seem to be ideal as units of measurement for immunoassay analytes but they also have limitations, because of the variation in the molecular weights of different analytes. For example, the presence of the same concentration of β subunit *molecules* of hCG in one sample as intact protein in another sample would result in a mass concentration that is approximately half as much, because of the lower molecular weight. Ultimately, immunoassay standardization should be in more meaningful terms, and it is to be hoped that one day we will see concentrations of proteins expressed in molar concentration. To achieve this, much work needs to be done on characterization of the proteins concerned. Progress has been made for insulin and adrenocorticotrophic hormone (ACTH) and their concentrations are often expressed as moles per liter.

The problems associated with assigning molar concentrations to existing International Standards are very significant. A designation of a molar concentration to no more than two significant figures by one laboratory could soon be followed by another announcing a different estimate, and this would not further the cause of standardization at all. The problems associated with assigning molar concentrations to heterogeneous analytes are particularly complex and it has even been argued that this goal is scientifically invalid. However, a pragmatic approach to these difficulties is emerging, as demonstrated by the current IS for FSH for immunoassay (92/510), which is a recombinant DNA preparation, giving it a defined structure, and the standard is accompanied by a statement of the approximate content in moles based on UV absorbance measurements and the theoretical extinction coefficient of the peptide chain.

The IS designation has helped to resolve potential differences between European and US attempts to standardize locally. Many organizations are active in the area of standardization, including the Bureau Communautaire de Référence (BCR) in Brussels, and in the US the Centers for Disease Control (CDC), College of American Pathologists (CAP), National Committee for Clinical Laboratory Standardization (NCCLS), and the National Institute of Health (NIH), and these organizations have subtle differences in approach. The International Federation of Clinical Chemists (IFCC) has been active in encouraging standardization worldwide.

Long-term stability of International Standards is tested using parallel storage at a number of different temperatures. Some indication can be given using an Arrhenius plot of activity vs. time for each temperature (accelerated degradation testing), and there is a rough rule-of-thumb that shelf-life doubles for each 10 °C reduction in temperature. A pragmatic approach that has been adopted by the WHO is to store samples of the standards at −20 and −80 °C. If no differences are observed in activity, it is a safe assumption that both stocks are secure and have not lost integrity. In this way a stable international standardization for insulin has been maintained for over 70 years, through several generations of standard, each calibrated from its predecessor.

DEFINITIVE AND REFERENCE METHODS

In the field of clinical chemistry certain well-established methods have been chosen as **definitive** or **reference methods** (NCCLS document NRSCL12-P), against which all methods can be calibrated. (NCCLS, 940 West Valley Road, Suite 1400, Wayne, Pennsylvania 19087-1898, USA, Tel.: 610-688-0100, fax: 610-688-0700, www.nccls.org). According to NCCLS terminology, a definitive method is the most desirable. It will have been thoroughly investigated and evaluated for sources of inaccuracy, including non-specificity, and uses accurate primary reference materials. A definitive method is considered to be accurate and precise, and suitable for calibration of all methods and standards in the US. A reference method may be a subsidiary method, calibrated from the definitive method, used locally to calibrate field methods and reference materials, or it can be a future candidate for definitive method status, still undergoing evaluation. Reference methods have been thoroughly investigated, and the methods have been very concisely documented to allow repeatability. For some analytes (classified as Class C), reference methods are considered the highest category, because experts consider that a definitive method is unlikely to become available.

An immunoassay for one analyte, digoxin, has been submitted as a candidate reference method. Yet, for many analytes, it is difficult to prove that one method is more accurate than another, as immunoassay is an indirect measurement technique, and depends on the use of standards with predetermined or arbitrary values assigned to them, and an ill-defined biological reagent: the antibody. Immunoassays do not measure defined physical or chemical parameters, such as moles or grams, directly. However, they have the advantage of a level of analytical sensitivity that is unmatched by technologies that would otherwise be preferred for standardization.

For a few immunoassay analytes, independent reference methods do exist. Isotope dilution-gas chromatography mass spectrometry (ID-GCMS) is suitable as a reference method for cortisol. It involves separation of cortisol by gas–liquid chromatography, specific detection by mass spectrometry, and determination of the recovery of a heavy isotope label added to the sample. A reference preparation of cortisol was prepared with a high purity and authenticated by international agreement. Several specialist laboratories around the world that had developed ID-GCMS methods validated them independently, using the reference preparation. As a final test, patient samples were submitted to each reference laboratory and their estimates of concentration were confirmed to agree within tight tolerances. Only then was the method assigned the status of 'reference method'. For an account of the use of ID-GCMS reference methods and other aspects of the standardization of cortisol assays (see Gosling *et al.*, 1993).

The availability of the reference method for cortisol enabled the UK National External Quality Assessment Scheme (UK NEQAS) to lead the standardization process across the industry, by providing to scheme participants data on their bias on a range of control samples over a period of many months. Serum pools containing cortisol were calibrated using ID-GCMS by a reference laboratory in Cardiff and sent to testing laboratories around the UK. Several immunoassay methods were found to be biased with respect to ID-GCMS, some by as much as 25%. One method was biased by 20%, even though the kit calibrators had been calibrated using ID-GCMS at the same reference laboratory in Cardiff. This large discrepancy was because the calibrators were made using charcoal-stripped human serum as a base matrix and, therefore, did not closely mimic patient samples. Although the calibrators were correctly assigned cortisol values, the immunoassay did not behave identically when producing signal from patient samples. (This is a classic example of a **matrix effect**.) Over a period of time, the various commercial immunoassay methods for cortisol were recalibrated by the manufacturers to align them with ID-GCMS. This exercise had the beneficial effect of reducing variation in the UK, and linking the immunoassay methods to a reference method, improving accuracy. Similar campaigns have been successful in Germany for steroids such as estradiol.

It is sometimes erroneously assumed that all ID-GCMS methods are reference methods. However, they can only be designated as reference methods when several highly controlled laboratories validate their methods internally, to a high standard, then agree with each other within close limits in an international study.

Unfortunately, ID-GCMS is only suitable for small homogeneous molecules such as steroids. There are few reference methods for proteins and, to make matters more difficult, many proteins, especially glycoproteins, are heterogeneous (i.e. they exist in different isoforms). In order to calibrate an immunoassay in the absence of a reference method, it is necessary to use pure analyte or a purified preparation that has been agreed to as an International Standard.

OTHER REFERENCE MATERIALS

International Standards are supplied as highly purified protein extracts, without contamination by serum or plasma. Calibration of immunoassays using these materials as primary standards does not guarantee that patient samples will give the same results in different assays. For this reason international reference materials provided in a human serum based matrix have been prepared. These include cortisol and progesterone certified reference materials (CRM) from BCR, the Serum Protein Standard (with values for 14 analytes), and the Apolipoprotein Standard.

HETEROGENEITY OF STANDARD MATERIAL

Many analytes, especially glycoproteins, are naturally heterogeneous. The sources of heterogeneity include glycosylation, sialylation, polymerization, aggregation, and decomposition. Different levels of prohormones (e.g. ACTH), subunits (e.g. α- and β-hCG), fragments (e.g. PTH), complexes (e.g. PSA), and subtypes (e.g. CEA) create additional diversity that may affect one immunoassay method more than another. Natural variation between individuals occurs, and certain forms can increase or decrease with different physiological states, sex or age.

It is important to use purified materials for calibration purposes but the purification process may further alter the structure or properties of the analyte. Because the IS represents only one sample of a heterogeneous analyte, methods standardized against the same IS may vary in the values they provide for a given series of patient samples. This is because each immunoassay method may have a unique set of antibodies that recognize specific epitopes on the analyte. Dual monoclonal antibody assays tend to be more selective than competitive polyclonal assays and may give lower values for glycoproteins.

Some analytes have different properties, depending on the tissue or body fluid from which they originate. For example, LH and FSH extracted from pituitaries are more like the forms of the hormones that circulate in serum than urinary preparations. The 3rd International Standard for hCG (IS 75/537) was widely used for assay standardization, but it was prepared from pregnancy urine and therefore contained nicks (cleavages in the amino acid sequence) that may result in decreased immunological and biological potency. This standard has been replaced by the 4th International Standard (75/589).

The Canadian Society of Clinical Chemists published a useful review of diversity and its effects on standardization of assays for growth hormone, prolactin, hCG, LH, FSH, and TSH (1992). They made a number of constructive recommendations including the following:

- Serum or plasma concentrations of structurally defined polypeptides (e.g. ACTH, PTH, insulin, gastrin, anti-diuretic hormone, glucagon, and somatostatin) should be reported in molar units.
- Two-site immunometric assays should be used for polypeptide hormone measurement.
- A consensus should be sought regarding the epitopes to which antibodies should be directed for each of the polypeptide hormones.
- Polypeptide hormone assay calibrators should be manufactured using recombinant DNA technology if possible.

Determination of the molar concentration of heterogeneous glycoproteins cannot be determined simply by dividing the weight by the molecular weight because of the variation in molecular weight and the difficulty in removing residual water. However, an alternative approach is to measure the amino acid composition. This is planned for the next generation International Standard for hCG.

METHOD-RELATED CAUSES OF STANDARDIZATION DIFFERENCES

Differences between methods may occur even when pure, homogeneous analyte is available for the preparation of standards. Although the antibodies used in immunoassays are remarkably specific and have high affinity for their target antigens, the lack of absolute specificity and adequate affinity in the presence of other substances are at the root cause of much method variation. This indicates that ultimately, it is the quality of the antibody that is most important in the development of a robust and unbiased assay.

Sample and Calibrator Matrices

The potential for variability in the molecular structure of the analyte between reference standard and patient sample has been explained above. This, coupled with antibody heterogeneity, can give rise to differences between immunoassays standardized using the same reference material. However, such differences can also be caused by the non-analyte constituents of the standard and samples (the **matrix**). The influence of matrix on the measurement of analyte concentrations is well known, and there is a thin line between the concepts of *matrix effect*, *sample interference*, and *cross-reactivity*. A **matrix effect** is a consistent bias in analyte determinations between two sources of matrix, such as between serum and plasma, or serum and charcoal-stripped serum. It is often used to describe a known source of bias with an unknown cause. The most important type of matrix effect is any that occurs between the matrix used to prepare the calibration curve, and the matrix of the test samples. **Sample interference** describes an action of a known substance that can be isolated and used to cause bias if added to a previously unaffected sample. Sample interference may be the root cause of a matrix effect. Many years ago I discovered that the consistent bias of an assay in an external quality assurance (proficiency testing) scheme was caused by the presence of excessive levels of non-esterified fatty acids in the control samples, due to sample decomposition. Until the cause was identified, this was best described as a matrix effect. But in reality, it was sample interference.

As another example, many immunoassays give different analyte concentrations in *paired* serum and plasma samples (i.e. each blood sample is split into two aliquots and separately processed into serum and plasma). This is typical of a matrix effect. Yet the actual percentage difference may be sample or patient-specific. This is a characteristic of sample interference. Complement in samples may bind to antibodies, especially IgG2 subclass monoclonal antibodies, interfering with the binding to analyte. Differences in results between serum and plasma may be due to inhibition of the complement activity by chelating agents in the anticoagulant in plasma collection tubes. Urine sample measurements tend to be significantly biased when serum or plasma standards are used for the calibration curve. This is a more extreme example of a matrix effect. Yet individual urine samples also vary enormously in terms of salt concentration.

The third type of effect is **cross-reactivity**. This is distinguishable from sample interference by the nature of the interfering molecule. If it binds to the analyte-binding site of the antibody because of structural similarity to the test analyte, giving rise to a false elevation of the test result, it is a cross-reactant. So it is conceivable that a *matrix effect* with an unknown cause may subsequently be characterized as a *sample interference*, when it is found to be sample-specific and transferable, and finally be traced to *cross-reactivity*, once the identity of the interfering substance has been identified.

From the above explanation, it follows that individual samples can vary within one type of sample matrix. Serum and plasma may be affected by the presence of rheumatoid factor, autoantibodies, human anti-mouse antibodies (HAMA) and cross-reacting drugs and metabolites. Fibrinogen in plasma may also interfere, and can displace proteins from solid phases. The chemicals used as anticoagulants for plasma collection (e.g. EDTA) may directly interfere in signal generation. Serum proteins, bilirubin, and NADH can cause background fluorescence and hemolyzed samples may contain peroxidases from ruptured red blood cell membranes. Homogeneous assays are most susceptible to interferences in signal generation, because of the absence of a washing step to remove unbound interfering molecules. Other potential sources of interference are complement, phospholipids, heparin, non-esterified (free) fatty acids, and chemicals used in – or leaching from – sample collection devices.

Samples and standards should always be carefully stored, to avoid further problems due to instability. Although sample stability may be experimentally verified with one immunoassay method, the presence of breakdown products may affect another.

Sample interferences cause discrepant results for *individual* samples between methods. But susceptibility to the non-analyte constituents of the sample – whether matrix effects, sample interferences, or cross-reactivity – can also result in consistent bias between two methods standardized against the same International Standard. This is due to differences in the nature of the non-analyte components between the standard matrix and the samples. For example, charcoal-stripped plasma, which is a convenient base matrix for preparing standards for steroids has had most small molecules completely removed and therefore differs from the population of patient samples for which the assay is intended. Potential cross-reactants are absent in the standard, but may be present in the patient samples. Less significant differences occur between serum samples and defibrinated, delipidized plasma, the usual source of a matrix for the manufacture of kit calibrators. Another example concerns the loss of dissolved carbon dioxide during freeze-drying of serum or plasma standards and controls, increasing the pH of the reconstituted matrix.

Every part of an assay, and every step of the calibration process between the IS and the test samples, via secondary standards and the kit calibrator set, may be influenced by matrix effects. The binding between antibody and analyte may be affected by the local environment, for example, pH, ionic strength, presence of protein, and level of hydrophobic material. Enzyme activity during signal generation may also be affected, although most enzyme label based assays involve a wash step that removes interfering substances. Each type of signal system has unique sample-based interferences, e.g. presence of natural fluorophores in samples can interfere in fluorescence measurement. The structure and integrity of the solid phase may also be affected.

The influence of the sample matrix on antigen–antibody binding can be minimized by using a low ratio of sample to assay reagents in the incubation – although this reduces the sensitivity of the assay – and increasing protein and ionic concentration, and buffering capacity, through the assay reagents. Animal sera or immunoglobulins derived from the same species as the antiserum, can eliminate some sample-specific effects, such as HAMA (human anti-mouse antibody) interference. The effects of potential interferences on signal generation can be much reduced by an efficient separation system, including an effective wash step, particularly if a short soak stage is included, to allow loosely bound material to diffuse away from the solid phase into the wash solution.

Homogeneous assays do not have a wash step and therefore require meticulous attention to minimize matrix effects.

To summarize, it is important to have a good understanding of a particular immunoassay system's inherent weaknesses, and ensure that assays are designed to compensate for them. Select the highest affinity and most specific antibodies wherever the application allows it and take particular care over the selection of the matrix used for standards and calibrators. Attention to these aspects of assay design can significantly reduce method-related bias, and sample interferences.

Buffers

Buffer standards are very different from patient samples and are therefore unreliable for achieving accuracy in method standardization. There are also buffer components in the assay reagents, but as these are added to standards and patient samples alike, they are unlikely to cause method bias directly. However, the nature of the buffer may have an influence on the conformation of proteins, and this in turn could affect antibody binding. So, careful choice of buffer, and optimization of pH, ionic strength and other active constituents can help to minimize method bias in standardization.

Buffers may also include chemicals that release analyte from binding sites on carrier proteins (known as **blocking agents**). It is important that the concentration is chosen to release the analyte completely in a wide variety of samples and the standard matrix. All buffer constituents should be of a high purity, and free of potentially interfering contaminants.

Antibodies

The antibodies are at the heart of an immunoassay. Selection of the highest affinity and most specific antibody available will reduce the influence of matrix effects considerably. However, this strategy will *increase* the susceptibility of the assay to *analyte* heterogeneity. Thus, selection of the most suitable antibodies for a particular immunoassay involves a series of compromises, depending on a detailed understanding of the analyte and the components of the immunoassay system (sample/standard matrix, separation system, and signal generation system). Where understanding is limited, there is no substitute for experimentation.

Successful antibody production is dependent on a well-characterized immunogen. It can be difficult to obtain a purified source of analyte that is exactly like the analyte in the sample. As mentioned previously, hCG is normally derived from pregnancy urine, contains nicks between amino acids, and lacks some of the carbohydrate side-chains. This favors generation of antibodies that have a higher affinity with analyte in a standard or control, than in a patient sample, causing standardization problems.

Two-site immunometric assays are more specific. Monoclonal antibodies provide specificity but this may be at the expense of affinity. Low affinity antibodies are particularly susceptible to matrix effects and sample interferences. Monoclonal antibodies are also more likely to change their binding properties significantly if directly immobilized onto a solid phase, much reducing affinity.

If antibodies can be selected to epitopes that are least likely to vary between different sources of analyte, the immunoassay will be less sensitive to analyte heterogeneity. A promising area for standardization involves choosing a recombinant source of protein for preparation of standards, then utilizing monoclonal antibodies specifically chosen to bind epitopes common to the recombinant and naturally occurring protein.

Labeled Analyte or Antibody

In the context of standardization, the aim of labeling should be to modify the binding region of the labeled molecule (analyte in a competitive assay, and antibody in an immunometric assay) as little as possible. This is a particular challenge with competitive assays for small molecules. For example, peroxidase-labeled triiodothyronine (T_3) is 62 times larger than T_3 alone.

Separation

Some separation systems are susceptible to matrix interferences. The use of solid phases with efficient wash processes helps to minimize the effect. Problems tend to occur at low analyte concentrations in immunometric assays. Inefficient washing can lead to spurious high signal levels due to sample constituents. Wash efficiency is influenced by the physical nature of the washing process, the formulation of the wash solution, and the temperature of the solution during the wash. Over-washing may remove bound material.

SPECIAL CONSIDERATIONS FOR ASSAY OF ANTIBODIES

Although International Standards exist for some antibodies, the heterogeneity inherent in naturally occurring antibody populations represents a major barrier to consistent standardization between methods. It is essential that each laboratory determines its own reference interval or cut-off based on a normal population. In the long term, standardization may be improved by the use of recombinant proteins as capture antigens, as is already happening in the field of infectious disease tests for blood donor screening. In this field there is also widespread use of accredited gray-zone samples and seroconversion sample sets. This helps to define the standardization of qualitative methods at the negative:positive interface.

CALIBRATION

ANALYTE

Calibration requires a source of pure or purified analyte of the highest grade available. Analytes that can be synthesized, such as steroids, drugs and small polypeptides, do not normally represent a problem, although they must be handled with extreme care, avoiding contamination and storing them exactly according to the manufacturer's instructions. Particular care must be taken with hygroscopic materials, and allowance may have to be made for water of crystallization for some chemicals. Pure analytes used for primary calibration should be stored in a desiccator and protected from light. Many immunoassay analytes of this type are dissolved in an organic solvent, such as ethanol. Such solutions should be diluted with an appropriate aqueous matrix, to reduce the final concentration of the organic solvent to less than 1%.

Most proteins of clinical importance exhibit some degree of heterogeneity in patient samples, although there are exceptions, such as alphafetoprotein (AFP). Several are glycoproteins, e.g. carcino-embryonic antigen (CEA), FSH, LH, and human chorionic gonadotropin (hCG). The heterogeneous nature of proteins has created a need for International Standards, which represent a single source of standard material. While they may not always be homogeneous, they at least provide a common, unitized source of analyte for all the methods available.

INTERNATIONAL STANDARDS

To calibrate a new method for a protein, it is usually necessary to obtain an ampoule of an IS or IRP.

Preparation of primary standard solutions from ampoules requires great care and should be entrusted only to experienced personnel. Ideally, two ampoules should be reconstituted by different individuals on different occasions, with a check that the results from standard solutions made from the two ampoules are in good agreement. The standard solutions should be stored in aliquots at −70 °C.

The availability of ampoules of International Standards is strictly limited. A catalogue of biological standards and reference materials is available from the NIBSC.

SECONDARY STANDARDS

In practice it is useful to prepare multiple sets of secondary standards, from which future lots of calibrators can be assigned values. They act as an intermediate between the IS primary standard, which has a very limited availability, and new lots of calibrators for routine use. A range of 6–10 working concentration solutions should be prepared in bulk, covering the full range of the assay, by diluting a reliable source of the analyte in an appropriate matrix. The solutions should be subdivided into aliquots to prevent repeated thawing and re-freezing. These should either be stored at −70 °C, or freeze-dried. The matrix should be the same as the matrix intended for the final calibrators (e.g. defibrinated, delipidized human plasma). The secondary standards are then calibrated from the IS (or pure analyte) by making up a standard curve consisting of about 10 dilutions of the IS solution and determining the concentrations of the secondary standards from the IS standard curve. Great attention should be paid to calibration curve-fitting error, by checking the actual vs. fitted concentrations of the IS dilutions. It may be necessary to hand-draw the calibration curves to minimize bias. Consistent 'wobbles' in the curves may indicate incorrect dilutions (immunoassay curves should only have a single point of inflection). Doubling dilutions should not be used, as cumulative errors may cause bias at lower concentrations. At least 20 assays should be run to obtain mean values for the secondary standards. Different reagent lots and instruments should be used if possible, in an ANOVA (Analysis of Variance) experimental design. Any outlier data should be rejected. Any significant variability between reagent lots or analyzers, detected in an ANOVA analysis, should be investigated.

CALIBRATORS

Immunoassays require the use of calibrators in order to assign values or concentrations to unknown samples. In a classical immunoassay, a set of about six calibrators is run prior to the unknown samples, a calibration curve of signal vs. concentration is plotted, and the concentrations of the unknown samples determined by interpolation (*see* CALIBRATION CURVE-FITTING). This is usually carried out using a computer. The calibrator sets are made in bulk, and the values may be assigned by reference to the secondary standards (in about 20 assays). Many immunoassay reagent kits contain **fixed-value** calibrators, which always have the same values assigned, regardless of the lot. This requires careful process control against the secondary standards during manufacture.

Calibrators should ideally be prepared by using a base matrix identical to that in the test samples. A preferred material for many clinical tests is defibrinated, delipidized plasma, from a pool of donated human blood. Each donated blood sample must be separately tested for HIV and HCV antibodies, and HBsAg.

One of the difficulties facing manufacturers is that a range of calibrators must be made, including one with an analyte concentration of zero. Many analytes are present in normal human serum, and preparation of a zero calibrator requires the use of a different matrix. Possible alternatives include animal serum or a buffer solution containing protein, but both of these may give rise to matrix effects. Horse serum is particularly variable and should be avoided. Possibly the best option is to use human serum stripped of analyte by affinity chromatography, using antibody immobilized onto a column. However, trace amounts of antibody may leach from the column. Some low molecular weight analytes may be removed from serum by stripping using charcoal or ion-exchange resin, but this removes other small molecules, changing the matrix considerably.

STANDARD AND CALIBRATOR MATRICES

The matrix of a calibrator needs to behave in a similar way to the sample matrix, keeping less soluble analytes in solution or bound to a carrier protein, and providing a background level of proteins that may play a role in the incubation or separation stages of the assay. There are two conflicting requirements for a calibrator matrix:

1. The matrix should be consistent from lot to lot.
2. The matrix should reflect any non-analyte constituents of patient samples that have a background effect in the assay.

The ideal way to achieve (1) would be to use buffer containing, for example, 1% bovine serum albumin, but this is often inadequate in meeting the second requirement. Animal serum is occasionally used, but can suffer from lot-lot variability. For clinical applications, human serum (or defibrinated, delipidized plasma) is the preferred base matrix. The effects of the natural variation between donors are best minimized by using pooled collections from a number of individuals.

REDUCED AND STORED CALIBRATION CURVES

Unless many samples are being run at a time, the need to run a set of six calibrators in duplicate, to derive a calibration curve, is wasteful of reagents. In small assay runs, or on random access analyzers, more reagents could be used up for the calibrators than for the test samples. For this reason, a variety of reduced calibration curve methods have been developed. Some immunometric assays with linear calibration curves require only one or two calibrators to be run. The linearity is achieved by immobilizing high levels of capture antibody on the solid phase; however, this approach only works for some assays. A major step forward: **stored calibration curves**, came with the introduction of stable, fully automated analyzers with stable reagents that only required calibration at periodic intervals. For example, six calibrators are run in duplicate and the calibration curve computed. The curve is stored in the analyzer's memory and used to determine the concentration of any samples tested using the same lot of reagents. Typically these stored calibration curves are stable for at least 14 days, but some systems have calibration stabilities of several months.

The next step was to extend the life of the full calibration with a reduced set of one or two calibrators, used to correct the main calibration curve, e.g. for long-term analyzer fluctuations. A good example is the Abbott IMx™ single point 'MODE 1' correction.

The ACS™:180 took this a stage further, with the introduction of encoded master calibration curves. They are derived from a series of assays carried out at the manufacturer's QC laboratory on each new lot, and encoded in a bar code that is read by a bar code reader on the analyzer. Periodic recalibration is accomplished by running two multianalyte calibrators (or *adjusters*). The computer uses the signal levels for the two calibrators to adjust the master curve, compensating for analyzer–analyzer differences and changes in the kit reagents over the shelf-life.

Encoded master calibration curves are now used on many random access analyzers. The master calibration data, and other information, are transferred from the manufacturer to the analyzer using conventional bar codes, two-dimensional bar codes (which can encode more data and are more robust), magnetic cards, smart cards, or floppy disks.

Normally the adjusting calibrator values are assigned by the manufacturer using several lots of reagents, then used in the clinical laboratory with later lots. However, small variations between reagent lots can cause inconsistencies in calibrator value assignment. The Vitros ECi™ system has lot-specific calibrator values, linking the value assignment of the calibrators with the determination of the master curve for each lot. These values are encoded with the master curve data on a magnetic card.

To achieve stable calibration curves the analyzer must maintain consistency and stability in incubation temperature (including warm-up rate), signal measurement and other variables that can affect signal levels. It must also withstand changes in ambient laboratory temperature and humidity within specified limits. Kits within a lot must give consistent results even if they have been shipped at different times and thus subjected to different conditions while in transit. To ensure consistency, some manufacturers have introduced cooled shipment for distribution. Any generic reagents (e.g. substrate or wash reagent) must also give constant signal levels across multiple lots and throughout their shelf-lives.

Many immunoassays for drugs of abuse and infectious diseases are run using a single calibrator or 'control'. This is run to determine a cut-off level between 'negative' and 'positive'. Normally, positive and negative controls are also run as unknowns to check the assay calibration.

Tests for use in doctors' offices, pharmacies, and in the home are self-calibrating, typically generating color in the presence of a clinically significant concentration of analyte.

RECOVERY AND DILUTION

Once a method has initially been calibrated, recovery efficiency and dilution characteristics should be checked. These are key indicators of calibration quality. These and other checks are described in chapter 6, CONCEPTS.

REFERENCES AND FURTHER READING

Apple, F.S. Clinical and analytical standardization issues confronting cardiac troponin I. *Clin. Chem.* **45**, 18–20 (1999).

Canadian Society of Clinical Chemists, Position paper: standardization of selected polypeptide hormone measurements. *Clin. Biochem.* **25**, 415–424 (1992).

Dikkeschei, L.D., de Ruyter-Buitenhuis, A.W., Nagel, G.T., Schade, J.H., Wolthers, B.G., Kraan, G.P.B. and van der Slik, W. GC–MS as a reference method in immunochemical steroid hormone analyses. *J. Clin. Immunoassay* **14**, 37–43 (1991).

Gosling, J.P., Middle, J., Siekmann, L. and Read, G. Standardization of hapten immunoprocedures: total cortisol. *Scand. J. Clin. Lab. Invest.* **53** (Suppl 216), 3–41 (1993).

Hilgers, J., von Mensdorff-Pouilly, S., Verstraaten, A. *et al.* Quantitation of polymorphic epithelial mucin: a challenge for biochemists and immunologists. *Scand. J. Clin. Lab. Invest.* **55** (Suppl 221), 81–86 (1995).

Jeffcoate, S.L. Analytical and clinical significance of peptide hormone heterogeneity with particular reference to growth hormone and luteinizing hormone in serum. *Clin. Endocrinol.* **38**, 113–121 (1993).

Kallner, A., Magid, E. and Albert, W (eds). Improvement of comparability and compatibility of laboratory assay results in life sciences, 3rd Bergmeyer Conference: Immunoassay Standardization. *Scand. J. Clin. Lab. Invest.* **51**, Suppl 205 (1991).

Kallner, A., Magid, E. and Ritchie, R (eds). Improvement of comparability and compatibility of laboratory assay results in life sciences, 4th Bergmeyer Conference: proposals for two immunomethod reference systems: cortisol and human chorionic gonadotropin. *Scand. J. Clin. Lab. Invest.* **53**, Suppl 216 (1993).

Marcovina, S.M., Albers, J.J., Henderson, L.O. *et al.* International Federation of Clinical Chemistry standardisation project for measurements of A-1 and B. III. Comparability of apolipoprotein A-1 values by use of international reference material. *Clin. Chem.* **39**, 773–781 (1993).

Mire-Sluis, A.R., Das, R.G. and Padilla, A. WHO cytokine standardization: facilitating the development of cytokines in research, diagnosis and as therapeutic agents. *J. Immunol. Methods* **216**, 103–116 (1998).

NCCLS. A candidate reference method for serum digoxin. I/LA9-T (NCCLS, Wayne, PA).

NCCLS. National Reference System for the Clinical Laboratory. Terminology and definitions for use in NCCLS documents. NRSCL8-A (NCCLS, Wayne, PA).

NCCLS. National Reference System for the Clinical Laboratory. Reference methods, materials and related information for the clinical laboratory. NRSCL12-P (NCCLS, Wayne, PA).

NCCLS. National Reference System for the Clinical Laboratory. Reference system: clinical laboratory. NRSCL13-P (NCCLS, Wayne, PA).

NIBSC. *Catalogue of Biological Standards and Reference Materials* (National Institute for Biological Standards and Control, Potters Bar, UK, 2004, www.nibsc.ac.uk).

Panteghini, M. Recent approaches in standardization of cardiac markers. *Clin. Chim. Acta* **311**, 19–25 (2001).

Seth, J. Standardization of protein hormone immunoassays. *Ann. Clin. Biochem.* **33**, 482–485 (1996).

Stamey, T.A. 2nd Stanford Conference on International Standardization of Prostate-Specific Antigen Immunoassays, 1994. *Urology* **45**, 173–184 (1995).

Stenman, U.-H., Bidart, J.-M., Birken, S., Mann, K., Nisula, B. and O'Connor, J. Standardization of protein immunoprocedures: choriogonadotropin. *Scand. J. Clin. Lab. Invest.* **53** (Suppl 216), 42–78 (1993).

Stenman, U.-H. Standardisation of immunoassays. In: *Principles and Practice of Immunoassay* (eds Price, C.P. and Newman, D.J.), 243–268 (Macmillan, London, 1997).

Storring, P.L. Assaying glycoprotein hormones – the influence of glycosylation on immunoreactivity. *Trends Biotechnol.* **10**, 427–432 (1992).

Sturgeon, C.M. and McAllister, E.J. Analysis of hCG: clinical applications and assay requirements. *Ann. Clin. Biochem.* **35**, 460–491 (1998).

Thienpont, L., Siekmann, L., Lawson, A., Colinet, E. and De Leenheer, A. Development, validation and certification by isotope dilution gas chromatography–mass spectrometry of lyophilized human serum reference materials for cortisol (CRM 192 and 193) and progesterone (CRM 347 and 348). *Clin. Chem.* **37**, 540–546 (1991).

Vihko, P. and Wagener, C. Structure and genetic engineering of antigens and antibodies: applications in immunoassays. *Ann. Biol. Clin.* **50**, 607–611 (1992).

Whicher, J.T., Ritchie, R.F., Johnson, A.M. *et al.* New international reference preparation for proteins in human serum (RPPHS). *Clin. Chem.* **40**, 934–938 (1994).

Wood, W.G. 'Matrix effects' in immunoassays. *Scand. J. Clin. Lab. Invest.* **51** (Suppl. 205), 105–112 (1991).

10 Separation Systems

David Wild and Wlad Kusnezow

Immunoassays that require separation of the antibody-bound and free fractions of the tracer are known as **heterogeneous**. This chapter provides an insight into the wide range of separation techniques that have been used. **Homogeneous** immunoassays do not require separations. They are described later in the book (*See* HOMOGENEOUS IMMUNOASSAYS). Some immunoassay tests incorporate separations other than of bound and free fractions. For example, plasma may be separated from blood cells in a whole-blood test device. There are also dual technology analytical systems in which a separation (such as electrophoresis) is followed by an immunoassay test. These specialized types of separation are not included in this chapter, which addresses the many possible methods for separating bound label from unbound label, a fundamental requirement of heterogeneous immunoassays.

In a **competitive** assay the separation is responsible for removing the unbound analyte and tracer so that only the antibody-bound fraction remains at the signal generation and detection stage. (Occasionally the bound fraction is removed instead, and the signal measured in the free fraction). In an **immunometric** assay the separation removes the unbound labeled antibody. In both assay designs the efficiency of the separation has direct effects on the quality of the assay (*see* PRINCIPLES). Separation efficiency is reduced by:

- failure to remove all of the unbound tracer, i.e. the labeled moiety associated with signal generation after the separation. Residual unbound tracer causes high background signal;
- partial removal of bound tracer;
- interference in the reaction between antibody and analyte;
- inconsistent separation due to variability in the sample matrix.

The quality of the separation in immunometric assays has a very direct effect on assay sensitivity. Failure to remove all of the unbound labeled antibody reduces assay sensitivity and causes poor precision, most evident at low concentrations. The separation in an immunometric assay is much improved by washing off unbound, residual labeled antibody from the solid phase with wash fluid. The best systems have a separation efficiency of >99.99%.

In competitive assays nonspecific binding is the most common problem, as it adds a level of background noise that can reduce precision, particularly at high concentrations. This can be due to binding of the tracer to the assay tube, or chemical impurities in the tracer. It can be minimized by using solid-phase or second antibody separation systems. Separation efficiency should be greater than 99%. Beyond 99.9%, other factors have a more direct effect on assay sensitivity and precision.

Separation efficiency in an immunometric assay is the percentage of total tracer removed in an assay of a sample with zero concentration of analyte. In a competitive assay, separation efficiency is the percentage of total tracer removed in an assay of a sample with an infinitely high concentration of analyte. In competitive assays, the separation efficiency can be estimated indirectly by measuring the percentage of total tracer removed in an assay in which the antibody is omitted.

LIQUID-PHASE SEPARATIONS

Liquid-phase (sometimes called **solution-phase**) separations occur in solution. The walls of the assay vessel are not actively involved in the reactions and there are no antibody-coated particles. They are sometimes used in competitive immunoassays. Although they have been largely superseded by solid-phase separation systems, second antibody separations are still used for some liquid-phase radioimmunoassays.

ELECTROPHORESIS

Electrophoresis may be used to separate the bound and free fractions, but it is too expensive and time-consuming for routine use.

The Immunoassay Handbook. *Third Edition*
D. Wild (Ed.)

GEL FILTRATION

Gel filtration chromatography involves the use of a cross-linked gel matrix in the form of small particles, such as Sephadex®. Within each particle there is a three-dimensional net through which only small molecules (such as unbound tracer) can pass. The method is sometimes referred to as size-exclusion chromatography. If the solution remaining at the end of the incubation is passed down a column packed with gel, the antibody-bound tracer passes through quickly, as it cannot penetrate the particles of gel.

This method is too expensive and time consuming for routine use.

DEXTRAN-COATED CHARCOAL

Powdered **charcoal** has the ability to adsorb small molecules of free tracer into small crevices on its surface. Antibody-bound tracer is too large to be adsorbed in this way. Dextran-coated charcoal is added to the assay following the incubation. It may be supplied in tablet or powder form, and has to be suspended in water, e.g. by vortex mixing. Following addition of the charcoal, the tubes are centrifuged. Charcoal assays are unusual in that the free fraction is pelleted and the bound fraction stays in solution. The supernatant is decanted into another tube for counting.

Time and temperature are critical during the separation and centrifugation, and great care needs to be taken to ensure that a consistent amount of charcoal is added to each tube. Charcoal can in some cases compete with the primary antibody for analyte, interfering with the equilibrium. Charcoal assays are only applicable to assays for small analytes where a low molecular weight label is used to make the tracer. The main remaining application of charcoal is in radioassays for vitamin B_{12} and folate.

PRECIPITATION BY SALTS, ORGANIC SOLVENTS AND POLYETHYLENE GLYCOL (PEG)

Immunoglobulins are not particularly soluble proteins at neutral pH. It is therefore not difficult to precipitate antibodies in solution by increasing the concentration of salt. The salt attracts water molecules away from the antibody, and because water is essential for the maintenance of its structure, the antibody denatures, falling out of solution. **Ammonium sulfate** is the most common salt used in immunoassays. At the end of the incubation, ammonium sulfate solution is added and the antibodies precipitate quickly. Centrifugation pellets the antibody precipitate, including the bound tracer. Free tracer in solution is removed by decantation of the supernatant fluid. Ammonium sulfate precipitation is simple, cheap, effective and virtually devoid of lot-to-lot variation. However, it results in a high level of nonspecific binding (5–20%) and is seldom used now.

Ethanol may be used in exactly the same way as ammonium sulfate, and has much the same advantages and disadvantages.

Precipitation by **polyethylene glycol** (**PEG**) has a similar principle. PEG is a polymer that soaks up water and provides a well-controlled antibody precipitation system. PEG 6000 (which has an average molecular mass of approximately 6000) is the most frequently used type.

DOUBLE (SECOND) ANTIBODY PRECIPITATION

Second antibody precipitation was introduced to increase the specificity of the separation of antibody, and hence the bound fraction, from the free fraction. At the end of the incubation, an anti-IgG antibody is added that is specific to the species in which the primary antibody was raised. For example if sheep anti-AFP is the primary antibody, donkey anti-sheep is a suitable second antibody. The divalent nature of antibodies is responsible for the formation of a lattice consisting of cross-linked antibodies and analyte. More than one molecule of antibody may be bound to each molecule of analyte, at different epitopes, each molecule of primary antibody may bind to two molecules of analyte, and each molecule of second antibody can bind to two molecules of primary antibody. The lattice of cross-linked molecules precipitates and separation is by centrifugation and decantation.

The main benefit of second antibody separations is the lower nonspecific binding (approximately 2%). The disadvantage is the need for a second incubation that can more than double the overall time it takes to perform the assay.

PEG-ASSISTED SECOND ANTIBODY PRECIPITATION

PEG-assisted second antibody precipitation combines the benefits of the second antibody and PEG methods: specificity and speed. The PEG, which is used at the lower concentration of 4%, hastens the precipitation of the cross-linked matrix. This is probably the best of all liquid-phase separation methods, taking performance and convenience into account.

ASPIRATION AND DECANTATION METHODS

Liquid-phase separations depend on separation of a precipitate from the supernatant. Aspiration may be used but it is difficult to remove all of the supernatant liquid without disturbing the pellet. Separation is best

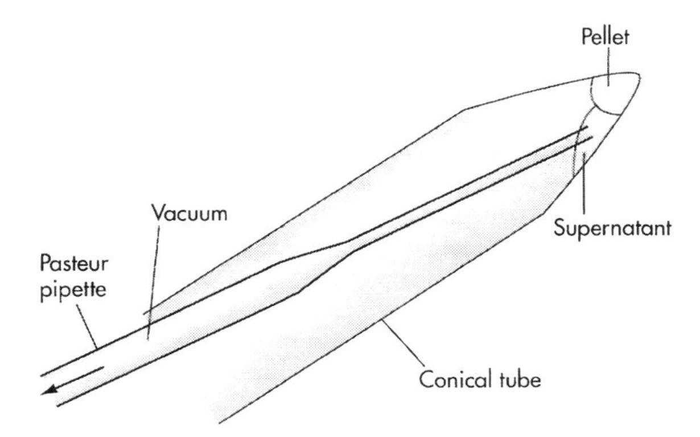

Fig. 10.1 Decantation with aspiration of excess supernatant.

accomplished using decantation in decanting racks, or aspiration, or a combination of both. One of the most effective methods is to decant the supernatants at a steep angle (but not vertical) in decanting racks, then press them onto a wad of tissues, rocking them gently to remove droplets adhering to the rims. While holding the rack inverted at an angle of 45°, use a fine drawn-out Pasteur pipette attached to a vacuum line to remove the last droplets by aspiration. This procedure works even better using conical assay tubes. The precision of most assays can be improved dramatically using this procedure (Figure 10.1).

It is important to carry out the decantation stage promptly and consistently. Tubes may be left to drain on a wad of tissues with a final manipulation to remove droplets at the tube rims. Tubes should not be drained into a sink, reinverted, then placed upside-down again on tissues elsewhere in the laboratory, or the pellets may be displaced.

Each type of separation system requires a different decanting method to get the best performance. Some need quite strong tapping on tissues to remove the droplets adhering to the precipitate, as the solution may be viscous.

SURFACE-COATED SOLID PHASES

Classical separation methods, such as ammonium sulfate and PEG precipitation, are nonspecific and sometimes affected by the level of protein in the sample. Double antibody separations are more specific but require large amounts of antibody. A simple way to improve the efficiency of the separation is to attach the primary or capture antibody to an insoluble support, such as the surface of the assay tube or well, or small plastic particles. This method was introduced by Catt and Tregear (1967) and the significant improvement in separation efficiency transformed assay performance. High sensitivity immunometric assays would not be possible without their pioneering work. Because the antibody is now no longer freely in solution, it becomes much easier to separate it from the supernatant after the incubation. Tubes, microwells, beads, and particles are examples of **solid phases**.

Solid-phase immunoassays have low nonspecific binding, particularly if a washing protocol is used. Nonspecific separation reagents, such as PEG, cause a dense precipitate to form, trapping unbound labeled material, whereas solid phases have an impenetrable and inert matrix, and binding only occurs at the surface. Pellets of solid-phase particles can be resuspended in wash fluid, which streams around the particles removing loosely bound material. The same effect can be achieved in coated tubes and wells by vigorous dispensing and aspiration with wash solutions containing a small amount of detergent. The principle is similar to that of a dishwasher. For a detailed discussion on solid-phase theory *see* PRINCIPLES—SPECIAL CONSIDERATIONS FOR SOLID-PHASE IMMUNOASSAYS.

GENERAL PRINCIPLES OF PROTEIN BINDING TO PLASTIC SURFACES

In most solid-phase immunoassays, protein is bound to a plastic surface. In a conventional immunoassay format, antibody is bound, but if the assay is for the detection of antibodies, the bound protein is normally the appropriate antigen. In either case the immobilized protein is known as the **capture protein**. Direct binding of the capture protein to plastic can result in conformational changes that reduce its affinity for the analyte. To avoid this problem, a spacer protein (such as streptavidin) may be bound to the plastic instead, and the capture protein modified to have a high affinity for the spacer protein (in this case by biotinylation).

Proteins will passively adsorb onto glass or plastic surfaces. The proportion of protein bound can range from 5–95%, so careful optimization of the coating process is important. Low coating efficiency is wasteful of antibody, but over-saturation can produce assays with inferior performance. Binding of proteins to plastic can be very strong, primarily through hydrophobic forces. The binding occurs because water molecules have a much stronger affinity for each other than for hydrophobic materials, such as the plastic, or nonpolar parts of the protein. It is therefore exclusion of hydrophobic sites from the solution that causes these parts of proteins to stick to the plastic.

Large proteins tend to bind more strongly, presumably because they have more hydrophobic regions, resulting in multiple sites of attachment. Perhaps also they have greater flexibility, and are able to refold to expose the normally hidden hydrophobic regions to the plastic. This results in an increase in avidity, even though the affinity of the binding at each site may be no stronger than for small proteins. Most plasma proteins bind strongly, with fibrinogen being a particularly high affinity binder. Antibodies have comparatively hydrophobic properties (as illustrated by the fact that they can be selectively precipitated out of aqueous solution by PEG or ammonium sulfate) and therefore bind well to plastic. Some proteins are highly glycosylated and this gives the surface a high degree of polarity. Pretreatment of proteins with exposure to a low pH (pH 2–3) can increase hydrophobicity, presumably by altering the conformation, exposing more hydrophobic sites.

The concentration of the protein in the coating solution has an effect on the speed of the coating process. Protein levels of 10–100 μg/mL are typically used, with the higher concentrations requiring shorter coating times. However high concentrations of protein may not bind with the same final conformation because of the effect of neighboring protein molecules on the protein–plastic interactions. At very high concentrations proteins may aggregate.

Optimum binding of most antibodies occurs at neutral or slightly alkaline pH, and for proteins in general a pH near the isoelectric point is most effective. This may be because it minimizes charges on the protein, and hence maximizes its hydrophobicity. For antibody coating a pH slightly above the isoelectric point is often chosen, as this offsets the slightly acidic environment near the surface of the plastic and also favors binding of the Fc region of the immunoglobulin molecules. The pH has the most effect on dilute coating solutions. The ionic strength has only a minor effect on the binding rate, although the presence of certain ions may be necessary to maintain protein structural stability. Temperature can have a significant effect on binding rate, and increased temperature is useful in speeding up particle coating reactions. A temperature of 37 °C is suitable. For wells and tubes, the reaction is slowed because of the rate of diffusion of protein towards the plastic (the force between plastic and protein only occurs in a region equivalent to the thickness of a few molecules of protein). Therefore agitation is probably the most effective way of increasing the rate of reaction, although high speeds of up to 12,000 rpm need to be used to overcome the resistance of the small volume of coating solution due to surface tension. This increases the rate of diffusion of protein in solution towards the solid phase.

The amount of protein that can be bound is affected by the size of the protein and the extent of the coverage of the plastic surface. As a rule of thumb, approximately 0.5–1 μg of IgG may be bound per cm^2 in an efficient coating process.

Time is a key variable in the coating process. At high protein concentrations, much of the binding can take place in just a few minutes, although additional protein may be deposited on the surface for many hours.

However binding of proteins to plastics can alter their shape and reduce their binding affinity. This is an important consideration when capture antibodies or polypeptide antigens (in ELISA tests for antibodies) are being bound to solid phases. These conformational changes are hardly surprising, as hydrophobic sites are more likely to be inside the protein, and charged sites at the outer surface, when in aqueous solution. On nearing the plastic surface, the hydrophobic sites are pulled towards it, at first gently, then with some force, as the distance reduces. Some proteins retain their activity when bound to plastic, but most suffer some loss of binding affinity because of the change in shape. A small minority benefit from improved affinity when bound to plastic. In addition to problems due to conformational changes, the active binding site may be in or near the hydrophobic area of the protein, and become unavailable due to steric hindrance by the plastic. Experiments with monoclonal antibodies have shown that most suffer from a loss of at least 90% of their functional binding capability, and many become completely denatured. Lower concentrations of proteins are more susceptible to denaturation, and albumin is sometimes added to low concentration protein solutions to reduce this effect (Figure 10.2).

The disruption of protein conformation is highly protein specific, as is the percentage of protein that becomes bound under a given set of conditions.

The nature of the binding reaction has been studied extensively but the full picture is still unclear. Under

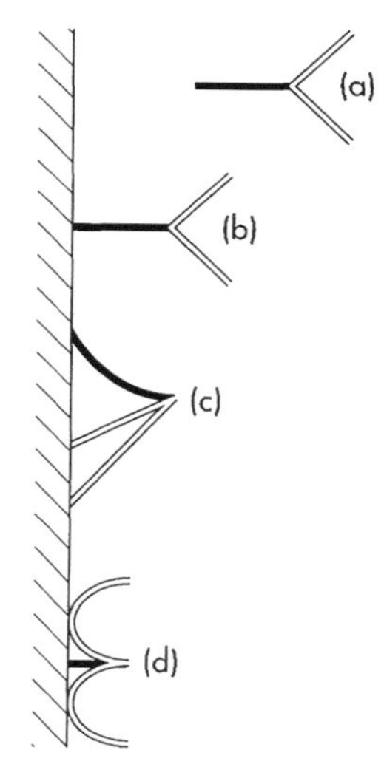

Fig. 10.2 Antibody binding to plastic: (a) Unbound antibody. (b) The idealized image of an antibody bound at the base of the Fc region, with the antigen-specific Fv regions available for binding. This never happens in practice. (c) If the Fv regions are hydrophobic, they may bind to the plastic and not be available for antigen binding because of steric hindrance. (d) The reality of successful antibody immobilization. Several hydrophobic regions of the antibody molecule bind tightly to the plastic, but the antigen-specific paratopes in the Fv regions of the antibodies remain intact.

certain conditions, the hydrophobic binding between protein and plastic follows the kinetics of a first order reaction, indicating that it is far from a haphazard or nonspecific reaction. However after a certain time, or above a defined concentration, the linear reaction kinetics no longer apply, and multiple reactions occur, e.g. involving clustering of protein into clumps. In one study, the linear reaction proceeded until about 20% of the plastic surface was covered, then second and third order effects occurred. The binding of protein can best be visualized as a pattern of individual molecules of protein as islands on the sea of plastic. As concentrations of protein are increased, some of the islands become clumps containing more than one molecule of protein. If the plastic has a carefully optimized pretreatment process, and a suitable protein is chosen, it is possible to coat the plastic with an almost uniform single layer of protein molecules, but this is an exceptional situation. Possibly the individual molecules of protein bind to the plastic in such a way that they repel other molecules in the near vicinity, but there is also evidence that nonuniformity of the plastic may play a role. Some studies have indicated that coating with high concentration protein solutions may lead to proteins adopting different conformations when binding to the plastic.

The strength of the binding between protein and plastic gradually increases, and it becomes difficult to remove bound protein. However, a very strongly binding protein like fibrinogen can displace weaker binding proteins, and there is also some evidence that bound protein can creep along the plastic surface.

As most monoclonal antibodies lose some or all of their binding affinity when adsorbed onto plastic, there is an advantage in coating with polyclonal antibodies, because at least some of the population of antibodies retain binding activity when adsorbed. However coating polyclonal antibodies to plastic is not as effective as coating a single protein, assuming that the protein retains its functionality, because the plastic can only adsorb about 1 $\mu g/cm^2$, and some of the binding site capacity is taken up by low affinity antibodies. In practice only about 1% of the polyclonal antibody is effective in an immunoassay. To some extent this limitation can be overcome by affinity purifying the polyclonal antibodies prior to coating. As a minimum, a partial fractionation with 40% ammonium sulfate will separate out immunoglobulins from other serum proteins.

Because of the above problems with coating polyclonal and monoclonal antibodies, most current systems use microparticles (to increase available surface area) or have a common capture protein coated onto the plastic (typically streptavidin or second antibody) to which monoclonal antibody (biotinylated in the case of streptavidin coated wells) can become bound during the assay incubation.

After antibody (or antigen) has been bound to the plastic surface, the remaining available binding sites are blocked off by washing the particles or wells with buffer containing BSA. This binds to any exposed plastic remaining, preventing nonspecific binding, e.g. of conjugates, later on during the assay.

Polystyrene is an ideal material for use as a solid phase. It is cheap, and can easily be molded into a wide variety of shapes, or prepared as latex particles of a reasonably uniform size distribution. Its surface characteristics are well suited for protein adsorption. Although protein binding occurs through hydrophobic interactions, hydrophilic sites are also necessary to retain functional activity. Studies have shown that optimum binding occurs when there are similar proportions of polar and nonpolar sites. Polystyrene contains some oxygen atoms near the surface, absorbed from the atmosphere during the molding process. This adds some polar sites to the otherwise hydrophobic polymeric structure. The amount of surface oxygen can be increased by exposing the plastic to γ-irradiation or plasma discharge, which oxidize the surface in the presence of air. For many proteins this process increases the surface binding capacity significantly. However over-treatment reduces the binding capacity.

Control of the plastic onto which antibody is coated is critical, especially where primary antibody is being coated directly for competitive assays (as variation in antibody concentration directly affects the signal level in a competitive assay). Consistency of antibody coating needs to be better than 2% for this application. Some major manufacturers own the plastic raw material formulation, the stocks of plastic mix and the production equipment for production of particles, beads or wells to maintain consistency. Immunometric assays are less critical because the capture antibody is present in excess. However they can be affected by the presence of contaminants in the plastic or mold release agents, which can cause release of bound material during the assay. Also, variation in antibody concentration may affect the kinetics of an immunometric assay, which can lead to variation in signal levels in nonequilibrium assays.

There is little difference between the protein-binding activities of polypropylene and polystyrene. However polystyrene is much more common, both for particles and wells. This is due to the availability of uniform polystyrene latex particles and because polystyrene can be injection molded very precisely into useful shapes such as microtitration plates or wells.

Traditional coating of tubes and wells was carried out with an overnight incubation at room temperature. Recently coating times have been reduced to a matter of minutes, for example by increasing the temperature or the concentration of binding protein. The drying stage used to be something of a black art, although it was well known from early on that drying has to be carefully controlled to remove surplus moisture without completely desiccating the coating protein. Drying used to be accomplished in drying rooms with controlled humidity over several days. However experimental work has shown that surplus water can be removed quickly by applying a current of temperature and humidity controlled air across the surface of the tubes or wells, immediately after an efficient aspiration step. We routinely coat and dry streptavidin coated wells in our Cardiff factory in less than 2 h. This involves 3 consecutive coatings, 8 washes and tunnel drying. The precision of these coated wells when tested in immunoassays is approximately 1.6%.

Particles and beads are coated in large vessels. After coating, magnetic particles can be separated from the coating solution by the application of large electromagnets and resuspended in wash fluid to remove loosely bound material. Beads and nonmagnetic particles may be separated using filters. Alternative methods for separating and cleaning particles are cross-flow filtration and mixed ion-exchange chromatography.

When developing or troubleshooting a coating method it is best to try to understand what happens to the solid phase (well, bead, particle etc.) from the raw material plastic formulation, through the solid phase manufacturing process, during each stage of the coating process, and finally through the immunoassay protocol. Each step has key process variables that have to be understood, and there are delicate relationships between coating and assay steps. For example, the strength of the wash/aspirate step to remove loosely bound material during the coating at the factory has to be compatible with the strength of the wash/aspirate stage of the assay protocol performed on an immunoassay analyzer.

An often overlooked source of problems is the water used to make coating and washing solutions. The water must be very highly purified as the presence of unwanted ions can interfere with the coating process, and organic impurities can be bound to the plastic. Impurities present in the ingredients of the coating and washing solutions can have the same effect.

COVALENT ATTACHMENT OF PROTEINS TO SOLID PHASES

In practice, passive adsorption of proteins onto plastic surfaces has been the most common technique used to manufacture solid-phase separation systems for commercial immunoassays. However there are situations where covalent attachment of proteins (antibodies, antigens or secondary capture proteins such as streptavidin) can provide a greater level of control in assay design. It should not be assumed that a higher concentration of protein can always be bound this way, as a well-optimized passive adsorption can produce a coating approaching a uniform monolayer on the plastic. However, covalent attachment of a protein to plastic may lessen the reduction in activity due to conformational change or steric hindrance around the antigen binding site, if an appropriate covalent binding site on the protein is selected. Covalent immobilization of proteins also provides an irreversible method of binding. Although desorption is not a problem for most passively adsorbed proteins, some proteins have only a weak binding to plastic, and covalent attachment prevents the bound material in an immunoassay becoming detached during the assay incubation or washing stages.

The surfaces of most common plastics are too inert to facilitate direct covalent attachment. Chemical treatment may be used to modify the plastic surface to create active groups. For example, treatment of polystyrene with concentrated nitric acid will create NO_2 groups that can be converted to amine groups using dithionite. Irradiation of polystyrene in the presence of ammonia gas also introduces amine groups. Conjugation of protein to the amine groups may be achieved using common enzyme–antibody conjugation methods. If the conjugation reagent (e.g. a homobifunctional molecule with two amine-reactive groups) is pre-reacted with the activated solid phase, it may then be used as a common, reactive solid phase, for subsequent covalent immobilization of a wide range of proteins for immunoassays. Carboxyl, hydrazide, hydroxyl, and thiol-reactive groups can also be created by chemical treatment. Surface carboxyl groups can be conjugated with proteins by preactivating the solid phase with an *N*-substituted carbodiimide, which creates highly reactive and short-lived *O*-acylisourea derivatives that react with amines in the protein to form amide bonds.

Surface coatings with specific active groups may be applied to previously made plastic wells or particles. Alternatively, special plastics may be created using appropriate monomers or additives in the raw materials for the polymerization process. For example styrene may be copolymerized with acrylic acid yielding particles with carboxylate groups covering their surface. Many activated solid phases (particles and microtitration plates) are available commercially from suppliers such as Amersham Pharmacia Biotech, BioRad, Corning, Costar, Isco, Nunc, Pierce, Toso Haas and Waters.

GLASS AND PLASTIC PARTICLES

Particulate solid phases include glass, polystyrene latex, Sepharose®, Sephadex®, Sephacryl® and nylon. Of these, latex is the most common. Polystyrene latex particles used in immunoassays have a wide range of particle sizes, from 0.015–20 μm in diameter (larger particles or beads are available up to 25 mm). They are manufactured by polymerizing styrene under carefully controlled conditions, resulting in a uniform particle size distribution. The distribution of particle size is typically 1–3%. To form a stable colloidal solution it is important that the particles do not attract each other forming clumps. The natural charge on the surface of latex particles helps by repelling other particles.

The large number of small, spherical particles provides a high surface area for capture antibody binding. Latex particles have a similar density to water, so they stay in suspension during the incubation if the particle size is less than 0.5 μm. Because many small particles are in suspension, molecular diffusion distances are short, and the kinetics of the assay are faster, minimizing incubation times.

Separation is usually by centrifugation and decantation, but in microparticle-capture enzyme-immunoassay (MEIA), the latex particles are captured by a filter consisting of a glass–fiber matrix. Particles may also be captured by membrane filtration. An advantage of the MEIA and membrane-based systems is that the particles may be easily washed, reducing the background signal (*see* IMX).

Incubation times tend to be longer when antibody is bound to a well or a tube, because the antibody is not free to diffuse into the solution and diffusion distances are therefore longer. However small particles are more difficult to separate out. Most particle-based assays have longer separation stages than well-based assays. So there is a trade-off and the total assay time tends to be about the same. Well-based assays have faster kinetics if the assay volume is kept small. This also allows the reagents to warm up more quickly in an incubator, increasing reaction rates further.

MAGNETIZABLE PARTICLES

A natural progression from latex particles led to the inclusion of iron oxide in cellulose or latex particles to form superparamagnetic microparticles. A surface coating on the particles prevents the iron oxide from coming into contact with the assay reagents. These are commonly used as a solid phase with the advantage of a separation that does not require centrifugation. Separation is achieved by the application of a magnetic field, drawing the particles to the side or base of the assay vessel or tubing. The supernatant solution is removed by aspiration or decantation.

Chromium dioxide has also been employed in particulate solid phases as it has less residual magnetism than iron oxide, and hence is easier to resuspend during the important washing process.

Not all magnetic separation systems are based on suspensions of microparticles. The solid phase in the AIA®-PACK range of assays manufactured by TOSOH consists of twelve 1.5 mm polymer beads coated with ferrite. An external magnetic field keeps the beads in motion during the incubation.

TUBES, WELLS AND MICROTITRATION PLATES

Antibody binds passively to polystyrene tubes, wells or microtitration plates. The optimum conditions are similar to those described for particles. A wide range of competitive and immunometric immunoassays have been developed that rely on antibody- or antigen-coated tubes or microtitration plates. Microtitration plates are more convenient than coated tubes, because individual positions are identified by the row and column label, avoiding the need for hand-labeling, although the fixed 96-well format can result in waste. This has been improved upon using strips of wells that fit into a plate holder. Some types of strips can be broken so that the exact number of wells required can be included in the assay, and the spare spaces in a row blocked off with blank wells. Recently microtitration plates with many more, smaller wells have been introduced for automated batch assays. These are particularly economical for high volume users, such as pharmaceutical researchers. Vitros ECi® uses tapered, round-bottomed wells that are stacked inside the reagent pack.

For competitive assays, it is essential that all of the wells (or tubes) in a lot are coated with an identical amount of antibody. This is achieved by careful control of the manufacturing conditions, including the processes used to manufacture the polystyrene wells. Unoccupied binding sites are blocked using protein, such as bovine serum albumin or casein. The coated wells or tubes are protected by packaging them in air-tight packs containing desiccant. In theory the exact amount of antibody coated should be less critical for immunometric or second antibody assays because the antibody is in excess in the immunoreaction. However most coated tube or well immunoassays are nonequilibrium assays and the concentration of coated antibody has an effect on the speed of the binding reaction, and hence on the final signal level.

The main disadvantage of coated wells and tubes is the small surface area for antibody binding compared to coated particles. The antibody–antigen reaction may only take place at the interface between the solid phase and liquid, slowing down the rate of the assay. Wells only have about one-tenth of the capacity for the combined volume of sample and reagent, compared to tubes, although the smaller size reduces diffusion distances and enables faster warming of the contents to an optimal incubation temperature. Irradiation and chemical pretreatments of polystyrene plates have been used to improve binding capacity. The assay kinetics can also be improved by agitation of tubes or plates during the incubation.

Incubation at 37 °C is usually required for microtitration plate assays. This must be carried out in an air incubator, as a water bath could splash water into the tiny wells. In a conventional air incubator, the wells at the corners warm up more quickly than the wells in the center, because of the insulating effect of surrounding wells. This shows up as a variation in binding. To avoid this, special plate incubators have been developed that provide consistent warming of all of the wells. In analyzers, consistency of incubation temperature is an engineering challenge. For 15 min incubations, temperature may need to be controlled to within ± 0.2 °C, not only from one position to the next, but also from one calibration to the next, which can be several weeks later.

In spite of their theoretical limitations, and the difficulties associated with their manufacture, immunoassays based on coated tubes and microtitration plates have been very successful. The success of coated tube radioimmunoassays was due to the avoidance of one reagent addition stage (the antibody) and the reduced nonspecific binding. The widespread use of microtitration plates is because of their added convenience and compact size, and the availability of a wide range of processing and signal measurement equipment.

BEADS

Until the advent of fully automated analyzers, bead assays were commonly used for immunometric assays. The beads are much larger than microparticles, so that just one bead is used for each test. The surface area offered by the bead is similar to that of a coated tube. Coated beads are easier to manufacture reproducibly than tubes as they can be coated in a bulk suspension rather than individually.

COMMON SOLID PHASE

There are distinct commercial advantages in using 'universal' solid phases, because development of an industrial coating process requires a great deal of investment of time and money. It is therefore better not to have to do this for each individual antiserum. The coating of a comparatively cheap and easily obtainable substance, such as streptavidin, also avoids the wastage of up to 90% of the antiserum, as can occur with passive antibody coating. Finally, the antiserum affinity is unaffected by the coating process. The use of an intermediary molecule to form a bridge to the plastic surface allows the use of many high affinity monoclonal antibodies that lose activity if immobilized directly onto plastic.

Streptavidin is commonly used in conjunction with biotinylated primary antibodies (*see* SIGNAL GENERATION AND DETECTION SYSTEMS—STREPTAVIDIN/AVIDIN–BIOTIN). Streptavidin may be directly adsorbed onto plastic or via a biotinylated carrier protein (streptavidin has four binding sites for biotin, allowing biotin to be used to anchor the streptavidin to a solid phase, and to bind the capture antibody). Direct coating of streptavidin reduces its biotin binding activity. The biotinylated capture antibody for the assay-specific analyte is provided in the liquid reagents and it becomes bound to the solid phase during the assay incubation. Many monoclonal antibodies that lose activity or specificity when directly adsorbed onto plastic, because of conformational changes, retain much of their functionality using this approach. Biotinylation of antibodies can, however, also affect binding affinity through steric hindrance, depending on the positions of the biotinylation sites.

As an alternative to streptavidin, a species-specific **polyclonal anti-IgG** antibody may be coated onto plastic. This provides a highly specific immobilizer for the capture antibody while preserving the capture antibody conformation. Steric hindrance may be less likely as the anti-IgG binds to the constant region of the capture antibody.

It is possible to use other antibody binding agents coated onto plastic, such as protein A or protein G, which are of bacterial origin and bind to the Fc region of immunoglobulins, but it may be necessary to chemically cross-link them with the capture antibody, as the affinity is less strong than streptavidin–biotin or anti-IgG–IgG. Otherwise some of the capture antibody will desorb from the solid phase during the assay incubation. Cross-linking may also be used to 'fix' proteins that weakly bind to certain types of plastic or membranes to prevent desorption.

Experimentally, chemical linkers have been used to provide a bridge between protein and plastic, with a hydrophobic region to bind the plastic and amine-reactive groups to capture protein.

MAIAclone® assays were liquid-phase immunometric assays in which the capture antibody is conjugated to **fluorescein isothiocyanate** (**FITC**). Following the first incubation, which occurs in solution, the generic solid phase (magnetizable particles that have an attached antibody to FITC) is added and, following a second incubation, the particles are sedimented in a magnetic field. This design uniquely combines the fast kinetics of a liquid-phase assay with the low background signal of a solid-phase assay.

MEMBRANE FILTRATION

Membrane filtration may be used as a separation method. In the Hybritech ICON® test, capture antibody (or antigen) was immobilized either onto a nylon membrane or onto polymer particles dried onto the surface of the membrane. The sample passes through the membrane onto an absorbent pad. Bound material is held on the membrane and unbound material passes through. Addition of washing solution ensures that the membrane is purged of any unbound material.

IMMUNOCHROMATOGRAPHY

In many point-of-care and home-use tests the sample is transported by **chromatography**. In the Unipath Clearblue One Step®, Clearplan One Step® and Clearview® tests, the sample moves along an absorbent matrix such as a porous membrane. The capture antibody is fixed in a band and immobilizes analyte and bound labeled antibody (attached to colored latex particles), as the sample moves past. The unbound material is carried away from the capture antibody band by the chromatographic effect.

The Baxter Stratus® system used glass fiber tabs, coated with immobilized antibody. The unbound fraction is washed away by radial diffusion, as successive reagents are added to the center.

The AccuLevel® tests are unique in that they utilize capillary flow to produce a vertical colored stripe, the height of which is proportional to the concentration of analyte. The solid phase is a strip of chromatography paper 4 mm wide and 90 mm long. It is impregnated with the immobilized capture antibody. Whole blood is mixed with the conjugate (horseradish peroxidase-labeled analyte) and glucose oxidase before the paper strip is immersed in the mixture, which migrates upwards. Because the concentration of labeled analyte in the mixture is always the same, the migration distance (before it has all bound to the immobilized antibody) depends on the concentration of the analyte in the sample. The strip has a serrated edge to avoid a natural tendency for liquid at the edge of the strip to migrate at a faster rate.

WESTERN BLOT

The western blot method is an important test for specific antibodies. It allows subtypes of infectious diseases to be distinguished, depending on the exact viral antigens that have generated antibodies in the patient.

Viral lysate is applied to a gel and the various proteins are separated by electrophoresis. The gel is left in contact with a nitrocellulose membrane that absorbs the proteins from the gel. Sample is incubated with the membrane and any bands with bound antibody are visualized by enzyme-labeled second antibody and substrate.

WASHING

The process of **washing** improves the quality of the separation of bound and unbound material in heterogeneous immunoassays. Washing has the most effect on precision at low signal levels, where the signal:noise ratio is highest. For this reason, assay performance is most improved at high concentrations in competitive assays and at low concentrations in immunometric assays. The impact of washing is also more dramatic with immunometric assays

COMPETITIVE ASSAYS

In a competitive immunoassay the lowest level of binding occurs at the highest concentration and is normally at least 2% of the total level of signal that can be generated (although there are exceptions, which we will ignore for now). The **total** is the signal if no separation is carried out and all of the labeled material is left in the test unit, whether it is bound or unbound. If the lowest level of bound signal is 2% of the total, then 98% of the signal must be removed from the bound material. Let us assume our separation method removes 99% of the *unbound* material. The remaining 1% would increase the lowest level of signal from the bound fraction (2%) to 3%, a gross error of 50% in the bound signal. If 99.9% of the signal can be removed, the error is now 0.1%, which would change the bound fraction from 2% to 2.1%, an error in the bound signal of 5%. Washing is simply a process of improving the separation. If a wash step improves the removal of unbound material from 99 to 99.9% it can be seen from this example that it would significantly improve the assay precision at high concentrations. However there comes a point where the error in the signal from other sources become more significant, so a separation efficiency beyond 99.9% may not bring about much additional improvement. In a competitive assay at high concentrations the flattening of the curve results in only small changes in signal level as analyte concentration increases, and this results in a source of error that is independent of the quality of the separation.

IMMUNOMETRIC ASSAYS

In immunometric assays (and some competitive assays with very low % binding) efficient removal of unbound tracer is even more critical. Whereas in a typical competitive immunoassay, other factors limit the ultimate level of precision and sensitivity, in an immunometric assay with a high specific activity tracer and a sensitive signal detection system, the efficiency of the separation has a very direct effect on sensitivity. From the example of the competitive immunoassay above, we can see that efficiency of separation is most important when the signal level is a small proportion of the total signal. At low signal levels, presence of a minute trace of the unbound label can cause a large percentage bias in the bound signal. In an immunometric assay the lowest level of signal occurs at zero analyte concentrations. In an ideal assay the signal generated would be zero in the absence of analyte.

As an example, let us consider a hypothetical immunoassay for TSH with a sensitivity of 0.005 μU/mL. The range of the assay is 0–50 μU/mL. In order to protect against hook effects at high analyte concentrations, the assay has enough capture and labeled antibody to be able to bind 500 μU/mL. In order to measure the signal at 0.005 μU/mL the separation system must remove at least 99.999% of the labeled antibody. It is not enough for the separation to remove exactly 99.999% of the unbound material, because the remaining 0.001% would double the bound signal. So a separation system that can achieve an efficiency of 99.9999% is required, to leave no more than 1 part of unbound material in 1 million. Clearly, a simple one-off removal of the supernatant by decantation, aspiration or filtration is unlikely to achieve this. Washing is the process that enables the achievement of these incredible separation efficiencies, in some of the best immunoassay analyzers available.

REMOVAL OF INTERFERING SUBSTANCES

Washing not only removes unbound label. It also removes substances that could interfere with the signal generation stage. These may include fluorophores and enzyme inhibitors from patient samples, or constituents of the assay reagents that can affect the signal generation step. Enzyme and fluorimetric assays are most susceptible to interferences of this type.

THE MECHANICS OF WASHING

To understand the process of washing let us start by taking a simple example: a tube coated with antibody in which an immunometric assay has been carried out. At the end of the assay incubation the tube contains liquid from the sample and the assay reagents. If we decant this liquid, droplets remain in the tube containing unbound material. If we add water to the tube and decant again, the droplets disperse, and although new droplets form after the second decantation, they contain a much lower concentration of the unbound material. This illustrates the first function of washing: **dilution**. If we also leave the wash water in the tube before we decant, say for 30 sec, we find that the level of unbound material remaining is decreased significantly. This process is called **soaking**. Soaking allows time for the diffusion of loosely bound material into the wash fluid. If we add a trace of detergent, the removal of sticky unbound material improves, through **solubilization**. We should of course include a buffer at a neutral pH in the wash fluid formulation so as to provide to the bound material a level of **stabilization**. We could also warm the wash fluid (**thermal enhancement**) and provide some **agitation** to improve the wash efficiency. We could improve the wash efficiency further by repeating the addition and decantation of wash fluid (**multiple washing**).

Let us take a closer look at the process of wash fluid removal. Decantation is a gentle process that leaves large droplets attached to the tube wall, but the inverted tube can be tapped onto tissues to achieve a good separation. This is the traditional manual method, but is difficult to automate, so many analyzers rely on **aspiration** to provide efficient removal of surplus droplets. The **vacuum depression** of the aspiration has to be carefully optimized to avoid stripping off or damaging the bound material. The **shape** of the tube or well also has an affect. A tapered tube bottom used in conjunction with a narrow aspirator nozzle positioned centrally works well. Microtitration wells, with their square bottoms, present a more challenging shape, so wide bore aspirator nozzles are often used to suck solution away from the edge of the well base. Round-bottomed wells and tubes are a good compromise for aspiration systems. The parameters that influence the way the wash fluid is added to the tube are also important (e.g. **nozzle diameter**, **flow rate**). A spray nozzle may remove more of the loosely bound material than a gentle dripping of wash fluid into the center of the tube.

Particle based solid phases have an advantage and a disadvantage compared to tubes and wells. The advantage is that when particles are **sedimented** in wash fluid, e.g. by application of a magnetic field to paramagnetic particles, the wash fluid flows over the surface of every particle, providing a very efficient washing process. The main disadvantage of particles is that **resuspension** into fresh wash fluid is required after each sedimentation, to achieve the high separation efficiencies demanded by high performance immunometric assays. However MEIA assays used an absorbent blotter placed under the filter that trapped the particles. A supply of warmed wash fluid can be applied to the particles on the filter without resuspension. There are other membrane based assays that trap particles on the membrane while wash fluid is flushed through the filter.

Beads are more like wells to wash than particles. Special wash/aspirate heads have been designed to create a flow of wash fluid over the bead.

In immunochromatography, the bound phase is held in position on a solid phase, e.g. a membrane, while the unbound material is carried through a matrix, e.g. by **capillary action**. Provision of additional wash buffer improves the separation considerably.

MICROARRAY IMMUNOASSAY SEPARATION

There is much interest in miniaturized immunoassay tests, including microarrays. (*see* LAB-ON-A-CHIP, MICRO- AND NANOSCALE IMMUNOASSAY SYSTEMS). Microarray immunoassays, as distinct from DNA microarrays, may provide information of great consequence about modulation and regulation of cellular activity at the proteome level, which is only slightly correlated with the level of mRNAs (Kusnezow and Hoheisel, 2002). The applicability of antibody and antigen microarrays for detection of serum proteins has been extensively demonstrated. It would be however also be very useful to be able to apply the antibody microarray techniques for protein expression profiling in tumor biopsy samples (Huang *et al.*, 2001).

The concept of the microspot multi-analyte immunoassay was proposed in the late 1980s by Roger Ekins (Ekins, 1998). In contrast to classical immunoassays, the microspot assay of Ekins represents a system where analyte is in excess compared to the number of binding sites and the formation of receptor–ligand complexes is greatly facilitated, since the analyte is not significantly depleted from the fluid to be analyzed. Consequently, the microspot assays under so-called **ambient analyte condition** may be many orders of magnitude more sensitive than conventional ELISAs or radioimmunoassays, for example. It is however clear that this gain in sensitivity is only attainable under thermodynamic equilibrium in microarray systems and is increased with a higher affinity of applied antibody (Kusnezow *et al.*, 2004). Based on these two assumptions, Ekins' theory suggests a detection limit of approximately 10^{-17} M for a microarray with spots of 100–1000 μm^2.

In contrast to competitive assay, Ekins proposed the separation principles for such immunoassay as **non-competitive.** They are based on direct measurement of **fractional occupancy of binding sites (FOB)**. Generally, this strategy should be more sensitive and reproducible. Clearly, it is easier to measure low signal arising from low analyte concentration directly than to obtain this value indirectly by subtraction from much higher signals with correspondingly larger errors on an absolute scale. An often applied strategy for microarrays is the typical **two-site immunometric assay**, which is also known as a **sandwich** assay. This is however very costly, since two antibodies are required for every antigen. Therefore, **direct analyte detection** by labeling with dyes or biotin-containing reagents is the next most widespread strategy. However, the competitive approach is not completely excluded from the microarray area and is also applied for detection of small analytes.

In contrast to the theory, performance of antibody microarrays in protein profiling has only been moderately productive thus far. Kezevic *et al.* analyzed protein expression in squamous cell carcinoma of the oral cavity using 368 antibodies in a microarray format. The data obtained could not be presented in quantified form (e.g. ratios) and only eleven proteins were found to have a different expression level. Another example is a microarray system with 146 antibodies created by Sreekumar *et al.* to detect protein level alternations in a colon carcinoma line that had been treated with ionizing radiation. The data obtained with only 20 of the 146 antibodies could be presented in this publication.

A number of technical challenges need to be overcome in this technology, to fully validate the concept for commercial application. Most of the issues stem from well-known biophysical protein properties, which may negatively influence the microarray detection process. Also, diversity and complexity of proteins represents a further hindrance for successful microarray immunoassay development. To understand the problematic nature of microarray detection, we will try to circumscribe below some biophysical basics underlying separation by means of this technique.

BASIC PRINCIPLES OF RECEPTOR LIGAND INTERACTION ON MICROARRAY

Reaction Kinetics Limitations

Due to the very small binding area on the surface, analyte concentration determines the time required to achieve the thermodynamic equilibrium (Kusnezow *et al.*, 2004). Seen only from the point of view of ideal kinetics, the time required to achieve the thermodynamic equilibrium in ambient analyte assays may vary from a few seconds or few minutes at high analyte concentrations to many hours at low analyte concentrations (Kusnezow *et al.*, in preparation). It is clear that the optimal incubation time depends on the affinity of particular antibodies. Therefore considering only the impact of ideal kinetics, a complex microarray should be incubated 10–20 h to achieve the maximal sensitivity.

Mass Transport Limitations

The situation described above is additionally complicated by strong diffusion constraints. In a conventional solid-phase immunoassay using microparticles, a large surface area is in contact with the sample and reagents, minimizing diffusion distances between analyte and antibodies captured on the solid phase. In nonparticle solid-phase assays for single analytes, encouraging diffusion is a challenge as the analyte has to be brought into close proximity with a continuous surface at the perimeter of the solution, such as a microtiter well. However in microarrays, only a very small part of this continuous surface is coated with each antibody, and diffusion is a major issue.

To achieve the thermodynamic equilibrium in a typical microarray experiment under nonmixing conditions is possible only at very high nanomolar analyte concentrations. At lower concentrations, it is impossible to reach thermodynamic equilibrium in a typical antibody microarray experiment even after many days of incubation (Kusnezow *et al.*, in preparation). Therefore most if not all antibody microarray systems, incubated usually for 1–2 h, are strongly nonequilibrium assays reaching only a few percent of the maximum possible signal intensities.

It has been widely recognized that mixing is very important for high association rates on DNA microarrays, leading to manifestly higher signal intensities, especially for low abundance transcripts. The importance of mixing is even greater for protein microarrays due to the generally lower affinity of antibody–antigen interaction in comparison to DNA–DNA duplexes and consequently much lower reaction rates on an antibody spot. Additionally, proteins vary strongly in their size and shape, resulting in significant differences in diffusion velocities for particular analytes. To demonstrate this let us consider two very different protein molecules: insulin, which is 5.8 kDa and KLH, about 6000 kDa. Their diffusion coefficients are roughly related in the same fluid as $D_1/D_2 = \sqrt[3]{(M_2/M_1)}$, where D_1, D_2 are diffusion coefficients and M_2, M_1 the molecular weights of two different proteins. Consequently, the diffusion velocity of insulin will be ten times higher than KLH and the reaction velocity on an anti-KLH spot will be at least 10 times slower even if the affinities of both the anti-KLH and anti-insulin antibodies are similar. Additionally, mobility of proteins in interface areas is complicated by their amphiphatic properties leading to a continuous adsorption/desorption of a protein on a surface.

Therefore, nonlaminar, very effective mixing is indispensable for protein microarrays. It not only enables the reaction rates to be increased but also improves reproducibility in this technology. Under strongly mixing condition, sensitivities in the low femtomolar range are attainable (Kusnezow *et al.*, in preparation). Recently, companies such as Advalytix (Munich, Germany) and BioMicro® Systems (Salt Lake City, USA) have demonstrated agitation systems specially designed for microarray systems.

Binding to Solid Phase

At first sight antibody or antigen microarrays appear deceptively similar to DNA microarrays, because they involve polypeptide chains instead of polynucleotides. However polypeptides are much more diverse than polynucleotides, and predicting and controlling their behavior for immobilization is very difficult. Proteins are also less robust than DNA and easily change shape, lose their binding properties and denature if exposed to dehydrating or other destabilizing environments. In general, binding affinities of antibodies are reduced and become heterogeneous by immobilization (Vijayendron and Leckband, 2001). This may result again in reduction of reaction rates on the spot. To reduce denaturation of proteins on the microarray surface, the addition of protective substances such as glycerol, disaccharides (trehalose or sucrose), or low molecular weight PEG to the spotting buffer is required.

Nonspecific Binding

It is a fundamental requirement of a microarray immunoassay that the binding protein be immobilized in a specific location and not elsewhere on the capture surface. However, large amphiphatic protein molecules exhibit abundant surface activities. The high degree of the protein adsorption is caused by electrostatic, Van der Waals and Lewis acid–base forces, and hydrophobic interactions as well as conformational changes and restricted lateral diffusion in the vicinity of a surface. Nonspecific protein adsorption on a microarray occurs mostly in the spaces between the spots, which are not occupied by large antibodies. Nonspecific binding increases with greater sample complexity or total number of secondary antibodies applied. Since the adsorption process has a competitive character, a low level of nonspecific binding is extremely difficult to achieve if a complex protein sample with thousands of molecules is incubated over a long time, even under optimal blocking conditions.

Even if this is achieved, cross-reactivity can be an issue, particularly when the microarray target analytes are related in structure. Interfering substances that may be present in the sample increase the likelihood of interference.

SOLID-PHASE MEDIUM

Filters and Microtiter Plates

Conventional nitrocellulose filters and microtiter plates have been used for microarray immunoassays. The limitations of all filter arrays are the relatively low resolution, the considerable background signal leading to limitation of further miniaturization, and difficulties in automating the analysis process. Due to the relatively large reaction volume required, it is also impracticable to use filter arrays in applications, such as protein expression profiling of tumor biopsies, where only limited sample quantities are available.

The maximum capacity of arrays in wells is up to a few hundred features per well. The achievable sensitivity and dynamic range of detection of these arrays are considerable better than conventional ELISA assays. They have the advantage that sample and reagent handling, incubation and mixing equipment are widely available.

Chip Format

The advantage of the chip format is nonlimited miniaturization of assays leading to increased multiplexing of detection and a decrease in sample volume. The chip format includes a number of systems, according to the different modes of applications and the manufacturing and detection possibilities of protein microarrays. The most widespread strategy is based on robotic instrumentation adapted from the production of cDNA microarrays. Currently, this format involves a large variety of protein microarray applications including antigen and antibody microarrays, microarray ELISAs or microarray Western analysis on a chip. Use of glass slide supports enables a variety of chemical surface derivatizations, which are described below.

PROTEIN ATTACHMENT ON MICROARRAYS

Microarray manufacture requires a generic binding method, yet despite the common underlying IgG structure involved in most immunoassays, there is an enormous diversity of structure in the variable part of antibody molecules, inherently necessary for their natural function. Since the variable regions of antibodies are constructed from the same amino acids as the common parts of these molecules, it is a challenge to find a generic method of immobilization that does not interfere with the wide range of binding sites present in a microarray. If *antigens* are immobilized in microarrays to test for the presence of antibodies, the problem of diversity simply adds to the complexity of the problem.

Another complication is the enormous variation in the size, structure, and charge of analyte molecules in the sample. For this reason generic immobilization methods need to include a neutralization step after binding, to provide an inert surface during the immunoassay procedure.

Good spot quality is also essential to the subsequent signal generation and detection stage of the immunoassay. Hydrophobic surfaces tend to produce small but inhomogeneous spots, whereas most hydrophilic surfaces yield homogeneous spots, which are often of irregular shape (Kusnezow and Hoheisel, 2003). Consistency of antibody coating needs to be very high to avoid statistical variations coming from a relatively small number of signal events generated on a small spot. Use of optimal concentration of antibody in the spotting solution ensures the maximal possible rates on the spots as well as better reproducibility and homogeneity of the signal intensity. The optimal concentration usually varies from several hundred μg up to many mg/mL.

Covalent binding of proteins to a surface is the preferred method for microarrays. It is advantageous to use a method that results in alignment of the bound proteins, and to bind the antibody to the surface using a region of the antibody remote from the analyte-binding site. It is also an advantage to use a spacer between the antibody and immobilization surface. However, chemical modification of antibodies adds another stage to the production process, multiplied by the number of antibodies in the assay. Methods include immobilization via amino, thiol and carbohydrate groups and the use of Protein A and biotin-streptavidin linkers. For more details see Kusnezow and Hoheisel, 2003.

Suitable surface chemistry is one of the first keys to the protein microarray technology. Many single analyte immunoassays use passive techniques to allow antibodies to bind nonspecifically to a plastic surface, followed by a secondary protein wash to inactivate binding sites on the plastic. The development of microarrays has forced assay developers to re-examine specific binding methods. However, applying DNA microarray surface chemistry directly to proteins was not very feasible because of many problems due to fundamental biophysical and biochemical differences between these two classes of biomolecules. Therefore, it is clear that there is a need for new sophisticated immobilization chemistries for proteins in consideration of the special features of protein microarrays listed above. An optimal protein microarray surface should meet the following criteria:

- high density and nondenaturing protein attachments;
- low protein adsorption;
- attachment chemistry that minimally disturbs different kinds of receptor–ligand interactions;
- compatibility with manufacturing devices;
- spot uniformity.

There is unlikely to be a single immobilization strategy that can be applied to all microarray systems. Inevitably, compromises have to be made and the implications fully understood and communicated to users. The interest in microarray-based immunoassays has stimulated the development of more elegant and controllable immobilization methods.

Conventional Coatings

Typical DNA microarray surfaces like poly-L-lysine and aminosilane coatings have been applied for a variety of protein microarray assays. However, since the binding on these surfaces occurs by adsorption, this strategy results in a lower protein binding capacity and a lower signal intensity compared with covalent binding. Glass may also be coated with epoxy, mercapto- and aminosilanes and optionally activated further using a bifunctional crosslinker (i.e. thiol- and aminoreactive crosslinkers). The main advantages of these attachment strategies are the simplicity and the low cost of the derivatization (Kusnezow *et al.*, 2003). One problem caused by most of the surfaces is protein denaturation due to the relatively high surface hydrophobicity. Hydrophobic surfaces also exhibit a higher degree of nonspecific binding when compared to hydrophilic support media. Additionally, there may be a larger steric influence on binding events due to the close proximity of the surface and the sensor molecules.

Polyethylene Glycol Coated Surfaces

PEG-derivatives provide a two-dimensional matrix, isolating the antibodies from the support matrix, increasing the potential surface area for binding sites and strongly decreasing nonspecific binding. A usual strategy is to bind linear PEG-derivatives containing amino-, thiol- or other functional groups to a reactive surface. However, the grafting efficiency of PEGs is often relatively low in such cases. An alternative strategy to obtain high-density PEG modified surfaces is graft-copolymerization of PEG-side chains with a polymer backbone. In particular poly-L-lysine grafted PEG copolymer (PLL-g-PEG) has excellent properties exhibiting more than hundred times lower nonspecific protein binding compared with untreated surfaces. Extremely low nonspecific binding of nonbiotinylated proteins to PLL-g-PEG-biotin coated with streptavidin even enables purification of biotinylated proteins from bacterial lysates directly onto a microarray.

Self-Assembled Monolayers

Self-assembled monolayers (SAMs) are formed by absorption of alkanethiols, typically onto a gold surface. Alkanethiols that terminate in short PEG groups provide improved nonspecific absorption properties. Functional and nonfunctional polymers are usually mixed to optimize the density of reactive groups. SAM provides a controlled way to achieve a well-characterized topography, suitable for use with atomic force microscopy, where binding may be observed as an increase in the height of the spot. It is also useful for planar waveguide techniques, where excitation of fluorescent labels only occurs within a short distance of the actual surface.

Three-Dimensional Solid-Phase Surfaces

In order to measure a wide dynamic range of analyte concentrations it is necessary to increase the number of binding sites, while minimizing nonspecific binding. The principles are exactly the same as for a conventional single-analyte immunoassay. Three-dimensional, dendrimeric matrices have been considered for DNA microarrays.

Gel pockets attached to glass slides have been tried, and achieved the goals of wide dynamic range, low nonspecific binding and limited protein denaturation. However their use impeded diffusion and long incubation times were necessary to achieve equilibrium (Arenkow *et al.*, 2000).

A simpler method involves the use of HydroGel polyacrylamide slides, from Perkin–Elmer Life Sciences, or nitrocellulose-coated FAST slides from Schleicher and Schuell. They have enormous binding capacity, and are useful for spotting protein lysates for probing with specific antibodies. However, molecules trapped in the three-dimensional matrix cannot be easily washed away and produce enormous background on this surface. Consequently, it is impractical to use such coatings for long incubation times. Application of these surfaces for immobilized antibody assays with unpurified samples of biological materials is therefore limited.

Binding of Recombinant Antibodies on Microarray

Recombinant antibodies (*see* ANTIBODIES) offer a unique way to solve one of the fundamental problems of microarray-based immunoassays: how to immobilize a wide range of antibody (or antigen) molecules using a generic method, while minimizing nonspecific binding from complex mixtures of proteins in the test sample. Additionally, because of the small molecular weight of recombinant proteins, dense attachment to support surface is facilitated.

The isolation and enhancement of recombinant antibodies, using techniques such as phage display, result in a strand of DNA coding for an antibody fragment with the desired binding specificity. Using routine molecular biology techniques it is possible to add to the DNA sequence an additional section coding for a unique polypeptide linker that can act as a specific tag for immobilization. This results in a **fusion protein**. The affinity tags are widely applied in the purification of recombinant proteins and are applicable to promote correct oriented binding of recombinant sensor molecules onto microarray surfaces. The specific orientation may also lead to the improvement of the stability of attached proteins and increase the sensitivity of the assay. The tag may be designed to bind uniquely to a common generic binder such as streptavidin for immobilization of the *in vivo* biotinylated antibody fragment or nickel coated surface His-tag fusions. Tags can be used to immobilize fusion proteins covalently on activated surfaces. Cutinase is a 22 kDa serine esterase that forms a site specific covalent adduct with phosphonate ligands. Cutinase fusion protein can be immobilized covalently on the monolayer of a phosphanate ligand.

REFERENCES AND FURTHER READING

Andrade, J.D. and Hlady, V. Protein adsorption and materials biocompatibility: a tutorial review and suggested hypotheses. *Adv. Polym. Sci.* **79**, 1–63 (1986).

Arenkov, P., Kukhtin, A., Gemmell, A., Voloshchuk, S., Chupeeva, V. and Mirzabekov, A. Protein microchips: use for immunoassay and enzymatic reactions. *Anal. Biochem.* **278**, 123–131 (2000).

Avseenko, N.V., Morozova, T.Y., Ataullakhanov, F.I. and Morozov, V.N. Immobilization of proteins in immunochemical microarrays fabricated by electrospray deposition. *Anal. Chem.* **73**, 6047–6052 (2001).

Avseenko, N.V., Morozova, T.Y., Ataullakhanov, F.I. and Morozov, V.N. Immunoassay with multicomponent protein microarrays fabricated by electrospray deposition. *Anal. Chem.* **74**, 927–933 (2002).

Ball, V., Huetz, P., Elaissari, A., Cazenave, J.P., Voegel, J.C. and Schaaf, P. Kinetics of exchange processes in the absorption of proteins on solid surfaces. *Proc. Natl Acad. Sci. USA* **91**, 7330–7334 (1994).

Benters, R., Niemeyer, C.M. and Wohrie, D. Dendrimer-activated solid supports for nucleic acid and protein microarrays. *ChemBioChem* **2**, 686–694 (2001).

Bernard, A., Fitzli, D., Sonderegger, P., Delamarche, E., Michel, B., Bosshard, H.R. and Biebuyck, H. Affinity capture of proteins from solution and their dissociation by contact printing. *Nat. Biotechnol.* **19**, 866–869 (2001).

Butler, J.E. (ed) *Immunochemistry of Solid-phase Immunoassay* (CRC Press, Boca Raton, 1991).

Butler, J.E. The behavior of antigens and antibodies immobilized on a solid-phase. In: *Structure of Antigens*, vol. 1, (ed van Regenmortel, M.H.V.), 209–259 (CRC Press, Boca Raton, 1992).

Catt, K.J. and Tregear, G.W. Solid-phase radioimmunoassay in antibody-coated tubes. *Science* **158**, 1570–1572 (1967).

Dent, A.H., and Aslam, M. Other categories of protein coupling. In *Bioconjugation: Protein Coupling Techniques for the Biomedical Sciences* (Macmillan, London, 1997).

Ekins, R.P. Ligand assays: from electrophoresis to miniaturized microarrays. *Clin. Chem.* **44**, 2015–2030 (1998).

Harris, T.M., Massimi, A. and Childs, G. Injecting new ideas into microarray printing. *Nat. Biotechnol.* **18**, 384–385 (2000).

Huang, J.X., Mehrens, D., Wiese, R., Lee, S., Tam, S.W., Daniel, S., Gilmore, J., Shi, M. and Lashkari, D. High-throughput genomic and proteomic analysis using microarray technology. *Clin. Chem.* **47**, 1912–1916 (2001).

Kusnezow, W. and Hoheisel, J.D. Antibody microarrays: promises and problems. *Biotechniques* **33** (Suppl), 14–23 (2002).

Kusnezow, W. and Hoheisel, J.D. Solid supports for microarray immunoassays. *J. Mol. Recognit.* **16**, 165–176 (2003).

Kusnezow, W., Jacob, A., Walijew, A., Diehl, F. and Hoheisel, J.D. Antibody microarrays: an evaluation of production parameters. *Proteomics* **3**, 254–264 (2003).

Kusnezow, W., Pulli, T., Witt, O. and Hoheisel, J.D. Solid support for protein microarrays and related devices. In: *Protein Microarrays* (ed Schena, M.), 247–283 (Jones and Bartlett Publishers, Sudbury, 2004).

Kusnezow W., Syagailo, Y.V., Rüffer, S., Klenin, K., Sebald, W., Hoheisell, J.D., Gauer, C. and Goychuk, I. Kinetics of antigen binding to antibody microarrays: limitation by analyte concentration and mass transport to the surface. (in preparation).

Masseyeff, R.F., Delaage, M., Barbet, J. *et al.* Separation (distribution) methods. In: *Methods of Immunological Analysis*, vol. 1, (eds Masseyeff, R.F., Albert, W.H. and Staines, N.A.), 475–533 (VCH, Basel, 1993).

Moody, V.D., Van Arsdell, S.W., Murphy, K.P., Orencole, S.F. and Burns, C. Array-based ELISAs for high-throughput analysis of human cytokines. *Biotechniques* **31**, 186–190 (2001).

Morozov, V.N. and Morozova, T. Electrospray deposition as a method for mass fabrication of mono- and multicomponent microarrays of biological and biologically active substances. *Anal. Chem.* **71**, 3110–3117 (1999).

Newman, D.J. and Price, C.P. Separation techniques. In: *Principles and Practice of Immunoassay*, 2nd edn (eds Price, C.P. and Newman, D.J.), 153–172 (Macmillan, London, 1997).

Norde, W. Adsorption of proteins from solution at the solid–liquid interface. *Adv. Colloid Interface Sci.* **25**, 267–340 (1986).

Okamoto, T., Suzuki, T. and Yamamoto, N. Microarray fabrication with covalent attachment of DNA using bubble jet technology. *Nat. Biotechnol.* **18**, 438–441 (2000).

Pandey, A. and Mann, M. Proteomics to study genes and genomes. *Nature* **405**, 837–846 (2000).

Robinson, W.H., DiGennaro, C., Hueber, W., Haab, B.B. *et al.* Autoantigen microarray for multiplex characterization of autoantibody responses. *Nat. Med.* **8**, 295–301 (2002).

Schetters, H. Avidin and streptavidin in clinical diagnostics. *Biomol. Eng.* **16**, 73–78 (1999).

Stillman, B.A. and Tonkinson, J.L. FAST slides: a novel surface for microarrays. *Biotechniques* **29**, 630–635 (2000).

Tijssen, P. *Practice and Theory of Enzyme Immunoassays* (Elsevier, Amsterdam, 1985).

Vijayendran, R.A. and Leckband, D.E. A quantitative assessment of heterogeneity for surface-immobilized proteins. *Anal. Chem.* **73**, 471–480 (2001).

Zhu, H., Bilgin, M., Bangham, R. *et al.* Global analysis of protein activities using proteome chips. *Science* **293**, 2101–2105 (2001).

11 Signal Generation and Detection Systems (Excluding Homogeneous Assays)

Larry J. Kricka and David Wild

A wide range of labels and label detection systems are used in immunological assays. Radiolabels, such as iodine 125, were among the first to be used in immunoassay. However, concern over exposure to radioactivity, the problems associated with regulation and disposal of radioactive waste, and the inherent instability of radiolabeled reagents, have stimulated the development of non-radioactive labels (Table 11.1). There has also been extensive development of strategies to achieve ultrasensitivity, either by increasing signal or reducing background in the immunoassay (Kricka, 1994). The prime requirement of the signal generation and detection system is to allow detection of the label above background noise and minimize non-specific binding of the labeled reagent. This is particularly important in immunometric assays where very high specific activity labels have been used to design immunoassays with zeptomole sensitivities (10^{-21}), equivalent to just 1000 molecules.

RADIOACTIVE LABELS

Radioactive isotopes are unstable variants of atoms that spontaneously transform to a more stable state, emitting energy in the form of particles or electromagnetic pulses. The rate of decay is specific to each isotope and described by the **half-life**, the time elapsed before half of the radioisotope has decayed. Radioactive labels can be thought of as powerful beacons. Use of a radioactively labeled antigen as a tracer enables the bound or free fractions to be estimated in a competitive assay, whereas a radioactively labeled antibody enables the amount of bound analyte to be determined in an immunometric assay.

The most commonly used radioisotope is iodine 125 (^{125}I). It has a half-life of 60 days. Tritium (^{3}H) is also used in some applications. It has a half-life of 12.26 years. Cobalt 57 (^{57}Co, used for vitamin B_{12} assays) and carbon 14 (^{14}C) are also used occasionally.

The level of activity of a sample of radioactive isotope gradually decreases as it decays. Also some of the energy emitted may cause damage to the labeled compound (**radiolysis**).

RADIOACTIVITY MEASUREMENT

The level of radioactivity is measured using a scintillation counter. Ionizing radiation is detected by a **scintillant**, which is a material that emits flashes of light in the presence of ionizing radiation. The β particles emitted by tritium are weak and easily absorbed by the vessel walls, requiring the scintillant to be mixed in solution with each sample to be counted. As the scintillation fluid is organic, and the sample is aqueous, detergent has to be included to allow mixing to occur. γ-Rays, from ^{125}I or ^{57}Co, are not absorbed by the sample container, and the scintillant, a crystal of sodium iodide coated with thallium, is located within the counting well. For both β- and γ-counting, a photomultiplier tube converts the light flashes from the scintillant into electric pulses, which trigger electronic counting circuits.

Most laboratories that perform radioimmunoassays (RIAs) use multiwell γ-counters that can count a number of tubes at once.

Radioactivity is measured by counting the number of **disintegrations per minute** (**d.p.m.**). Not all of the disintegrations are detected by the counter. The proportion detected as **counts per minute** (**c.p.m.**) is the **counting efficiency**. For ^{125}I the counting efficiency is typically about 80%. Because the counts per minute are not continuous variables (unlike absorbance in enzyme-immunoassay (EIA)), low counts are subject

The Immunoassay Handbook. *Third Edition*
D. Wild (Ed.)

Table 11.1. Examples of labels and fusion conjugates for immunoassay.

Chemiluminophore	Acridinium ester, acridinium sulphonamide, isoluminol
Coenzyme	ATP, FAD, NAD
Electrochemiluminophore	Ruthenium tris(bipyridyl)
Enzyme	Acetate kinase, alkaline phosphatase, β-lactamase, glucose oxidase, firefly luciferase, β-D-galactosidase, horseradish peroxidase, glucose 6-phosphate dehydrogenase, laccase, Renilla luciferase, xanthine oxidase
Fluorophore	Europium trisbipyridine cryptate (and other lanthanide cryptates), fluorescein, β-phycoerythrin, rhodamine, umbelliferone derivatives, Texas Red, semiconductor nanocrystal quantum dots
Free radical	Nitoxide
Fusion conjugates	Alkaline phosphatase – anti-phytochrome single chain antibody; alkaline phosphatase – basic fibroblast growth factor receptor; apoaequorin – IgG heavy chain; bacterial alkaline phosphatase – IgG Fc binding protein; bacterial alkaline phosphatase – synthetic octapeptide; bacterial alkaline phosphatase – anti-HIV 1 gp 41 single chain antibody; bacterial alkaline phosphatase – human proinsulin; β-galactosidase – interferon α_2; β-galactosidase – B19 specific oligopeptide; core streptavidin – single chain antibody (scFv); firefly luciferase – protein A; human placental alkaline phosphatase – 4-1 BB ligand; marine bacterial luciferase (β-subunit) – protein A; metapyrocatechase – protein A; protein A – antiphytochrome single chain antibody; *Pyrophorus plagiophthalamus* luciferase – protein A
Gene	Firefly luciferase
Metal and metalloid	Gold, silver, selenium
Metal complex	Cyclopentadienylmanganese(I) tricarbonyl, gold cluster
Microparticle	Latex, erythrocytes, liposomes
Nucleic acid	pUC19 DNA
Phosphor	Europium-activated yttrium oxisulfide, Up-converting Phosphor Technology (UPT)
Photoprotein	Aequorin
Radioisotope	^{125}I
Substrate	Galactosyl umbelliferone
Virus	Bacteriophage T4

to considerable error. Sufficient counting time needs to be allowed to accumulate at least 10,000 counts to achieve a counting error of 1%. (The standard deviation is approximately equal to the square root of the total number of counts).

Radioactive counters, and the sample carriers placed in them, should be regularly checked for backgrounds, which should be 50 c.p.m. or less. This is particularly important with multiwell counters where each well needs to give a consistent measurement. High backgrounds may be caused by contamination. Decay of ^{125}I is a two photon event, and the background in RIA can be reduced by setting the detection system to register only two photon events as signal. In this way background counts fall to 0.5 counts per week, and ^{125}I can be detected down to 0.1 amol. This is the principle behind **multi-photon detection** (**MPD**), in which twin detectors are used with pulse shape and height detection (BioTraces, Inc., Herndon, Virginia, USA).

PREPARATION OF RADIOACTIVE TRACERS

In some cases it is possible to replace one or more atoms in a potential analyte with radioisotopes to make a tracer. For example, the iodine in thyroxine (T_4) may be replaced by ^{125}I, and tritium can be used to replace hydrogen in most analytes. However, this is very much the exception and the most commonly used isotope, ^{125}I, is normally tagged onto the analyte to make a tracer. Iodine is a large atom, comparable in size to a benzene ring, so the position of the iodine label relative to the antibody binding site on the antigen is very important, if steric hindrance is to be avoided.

There are two main methods for labeling antigens with ^{125}I. Most proteins may be labeled by the direct substitution of ^{125}I onto the aromatic ring of tyrosine, or to a lesser extent, histidine. Where this is not practical, such as for polypeptides without a suitable tyrosine residue, or haptens such as drugs or steroids, tracers are synthesized by iodinating a carrier that includes a phenol or imidazole group, and coupling the carrier to the antigen. The carrier chosen has a reactive group that provides a covalent link to the antigen when the two are mixed together.

In theory, higher specificity tracers lead to more sensitive assays, but if proteins are labeled with too much ^{125}I they become damaged by radiolysis. Typically proteins are labeled with approximately one atom of ^{125}I per molecule.

Direct labeling methods involve mixing sodium iodide (^{125}I) with protein and an oxidizing agent, which converts the iodide into free iodine, or positively charged iodine radicals. This reactive form of iodine substitutes hydrogen atoms on the phenolic ring of any tyrosine residues on the protein. The most common oxidizing agent is chloramine T, which yields hypochlorous acid in aqueous solution (Greenwood *et al.*, 1963). The reaction is terminated after 10–15 sec by adding a reducing agent, sodium metabisulfite.

A simpler procedure involves the use of a solid-phase catalyst, iodogen (Fraker and Speck 1978; Salacinski *et al.*, 1979).

Lactoperoxidase, in the presence of a low concentration of hydrogen peroxide, provides a gentler method that is useful for sensitive antigens that are damaged by chloramine T, iodogen, or sodium metabisulfite.

There are a number of conjugation methods for haptens. They may also be used for proteins where direct iodination is not possible or interferes with the antibody binding sites. Bolton and Hunter (1973) described a method for labeling amino groups in any lysine residues or at the *N* terminus of a polypeptide. Using chloramine T, 3-(*p*-hydroxyphenyl) propionic acid *N*-hydroxysuccinimide ester is iodinated and then added to the polypeptide. The reaction is terminated by the addition of glycine, which inactivates any unconjugated ester. Other compounds that are used for conjugation include tyrosine methyl ester, tyramine, and histamine.

The selection of conjugates requires care, skill, and experience. The antigen labeled by conjugation is quite different structurally to the natural analyte, particularly in the region where the analyte is conjugated to the label. The conjugated tracer may be more than twice the size of the analyte. If the binding site recognized by the antibody is well away from the site of the conjugation, the antibody may not be able to distinguish between the analyte and the tracer, but this is often not the case. The main body of the conjugate (the iodinated ring) may be distanced from the antigen by means of a molecular **bridge**.

In hapten assays, the relationship between the conjugate and the immunogen is very important. If the same bridge is used, in the same position, to conjugate the carrier protein to the analyte to make an immunogen, as is used to make a conjugated tracer, antibodies to the bridge may be produced, so that the tracer binds much more strongly than the natural analyte, making the immunoassay insensitive. Normally, different bridges are used for the immunogen and tracer to avoid this effect, with the bridge for the tracer often being bulkier.

Following iodination, the reaction mixture is quickly diluted (to minimize chemical and radiolytic damage) and purified, to remove free iodine, and oxidizing and reducing agents. The most common method for proteins is gel-filtration chromatography. By measuring the counts per minute in each fraction eluted from the column, and determining the total activity in the protein and iodide peaks, it is possible to calculate the percentage incorporation of iodide into the protein. The **specific activity** of the tracer can be calculated from the amount of protein iodinated and the total activity of iodide used, corrected for the incorporation. It is the radioactivity per unit mass or mole of analyte, expressed for example as mCi/μg.

For a detailed explanation of iodination methods see Chard, 1990.

ENZYME LABELS

Enzymes are now used more widely than any other type of label. An enzyme is measurable at very low concentrations by utilizing its catalytic properties to generate colored, fluorescent, or luminescent compounds from a neutral substrate. A single molecule of enzyme may cause the conversion of 10^7 molecules of substrate per minute, increasing the strength of the signal, and hence the sensitivity of the assay, over a million-fold compared to a label that produces just one signal event. The activity of the enzymes most often used in immunoassays is more stable than ^{125}I, so that reagent shelf-lives are longer, and recalibration may only be necessary every few days or weeks (Gosling, 1990; Wisdom, 1976).

The main disadvantage of EIAs is that they have a susceptibility to interferences and changes in assay conditions during the signal generation stage, whereas RIA is more robust. In an RIA, the incubation between the antibody and the antigen needs careful attention and optimization during assay development, but signal generation is normally a simple and trouble-free process. In EIA the signal generation step must be controlled, optimized and kept free from interferences, in much the same way as the antibody–antigen incubation. Each of these two incubations has individual requirements, and the need for care, skill, and experience during assay development is even greater than for RIA. The enzyme-substrate incubation is sensitive to time, temperature, and pH, and inhibitory substances should be absent. Some enzymes require cofactors, such as Mg^{2+}. Substrate is normally provided in excess, except where high concentrations of substrate inhibit enzyme activity, as is the case with horseradish peroxidase and hydrogen peroxide.

The two enzymes most commonly used in immunoassays are alkaline phosphatase and horseradish peroxidase. **Horseradish peroxidase** (HRP) is a 44 kDa glycosylated hemoprotein. It consists of a protoporphyrin IX hemin prosthetic group and 308 amino acids, including four disulfide bridges, and two calcium ions. It contains approximately 20% carbohydrate by weight. It has six lysine residues available for conjugation without any loss of enzyme activity. HRP is an oxidoreductase that can be used with a wide variety of hydrogen donors to reduce hydrogen peroxide. This property has been utilized to generate colored, fluorescent, or luminescent derivatives depending on the substrate used. It has a high catalytic rate, resulting in amplification of the signal. The pH range of HRP activity is 4.0–8.0.

Alkaline phosphatase is a 140 kDa dimeric glycoprotein containing many free amino groups that can be used for conjugation. This enzyme catalyzes the hydrolysis of phosphate esters of primary alcohols, phenols, and amines. The form used in immunoassays is purified from calf intestine and contains zinc. It is inhibited by orthophosphate, zinc chelators, borate, carbonate, and urea. Activity is optimal at pH 9.5–10.5.

Many other enzymes have been tried over the years in immunoassays (e.g. acetate kinase, firefly luciferase, xanthine oxidase), with β-D-galactosidase, glucose oxidase, and glucose 6-phosphate dehydrogenase the only ones to find application in commercial products.

Enzymes are covalently linked to the antigen or antibody (depending on the format of the assay) to form a conjugate, using techniques similar to those described for radioisotopes. Horseradish peroxidase and alkaline phosphatase may be conjugated to antigens or antibodies using glutaraldehyde. There are two basic methods: the one-step method is more prone to the formation of polymers (conjugates containing more than one molecule of either enzyme or protein) than the two-step method. Horseradish peroxidase is a glycoprotein and may also be conjugated via the carbohydrate residues using the periodate method (Tijssen, 1985). Horseradish peroxidase in the conjugate may be quantified by measuring its absorption at 403 nm. There are also other conjugation methods, including molecular biological techniques in which the gene for the label and the gene for an antigen or binding protein are fused and expressed to form a fusion conjugate (e.g. human proinsulin–alkaline phosphatase, protein A-firefly luciferase). This is a useful technique for sensitive proteins that may be damaged during conjugation reactions. A list of fusion conjugates used for research immunoassays is included in Table 11.1.

COLORIMETRY

Enzyme labels may give rise to several different signal sources, depending on the substrate. The simplest signal to measure is color. Figure 11.1 illustrates how a conventional competitive immunoassay may be designed using an enzyme. In this example, antibody is coated onto microtitration wells, the solid phase. Analyte labeled with enzyme competes with analyte in the sample for a limited number of antibody binding sites. After the incubation, unbound material is aspirated or decanted, leaving the bound enzyme attached to the well.

The signal generation stage involves the addition of substrate in buffer to the wells. During a second incubation the enzyme gradually converts a proportion of the substrate to its colored end-product. The enzyme is a catalyst, which brings about the change in the substrate without changing chemically and can therefore keep on converting substrate until the reaction is terminated. At the end of the incubation a stopping reagent is added which stops enzyme activity, for example by changing the pH. The strength of the color depends on the amount of enzyme present, which in a competitive assay is inversely proportional to the concentration of analyte in the original sample.

To determine the strength of the signal, and hence estimate the analyte concentration using a calibration curve, a spectrophotometer is required.

Enzyme labels work equally well in immunometric (sandwich) formats, and with numerous different separation systems. Special immunoassay plate readers have been developed that quickly determine the color through the transparent base of each well in a microtitration plate, and compute the calibration curve, patient sample values and various QC parameters.

Colorimetric assays are limited by the working range of the spectrophotometer. Typically this is from 0.10 to 2.00 absorbance units. Although this is adequate for competitive assays, immunometric assays are capable of a wider working range than this. The range can be extended by reading high levels of absorbance at a wavelength offset from the center of the main peak. Alternatively, the rate of color development can be monitored over a period of time in a kinetic reader. This method works well with peroxidase and TMB as the substrate, as the rate of generation of the color is comparatively fast.

Common peroxidase substrates are ABTS® (2,2′-azino-bis(ethylbenzothiazoline-6-sulfonate)), OPD (*o*-phenylenediamine) and TMB (3,3′,5,5′-tetramethylbenzidine). Each of these substrates also requires the presence of hydrogen peroxide. ABTS is a good all-purpose reagent with a wide working range. TMB is often preferred, as it gives the highest absorbance values, low backgrounds and, unlike OPD, is not mutagenic. The commonest alkaline phosphatase substrate is *p*-nitrophenyl phosphate. Indoxyl phosphate has been used in point-of-care tests to produce an insoluble indigo dye.

Horseradish peroxidase-based assays are capable of greater sensitivity than alkaline phosphatase with colorimetric substrates, as the signal intensity (under identical conditions) is an order of magnitude greater.

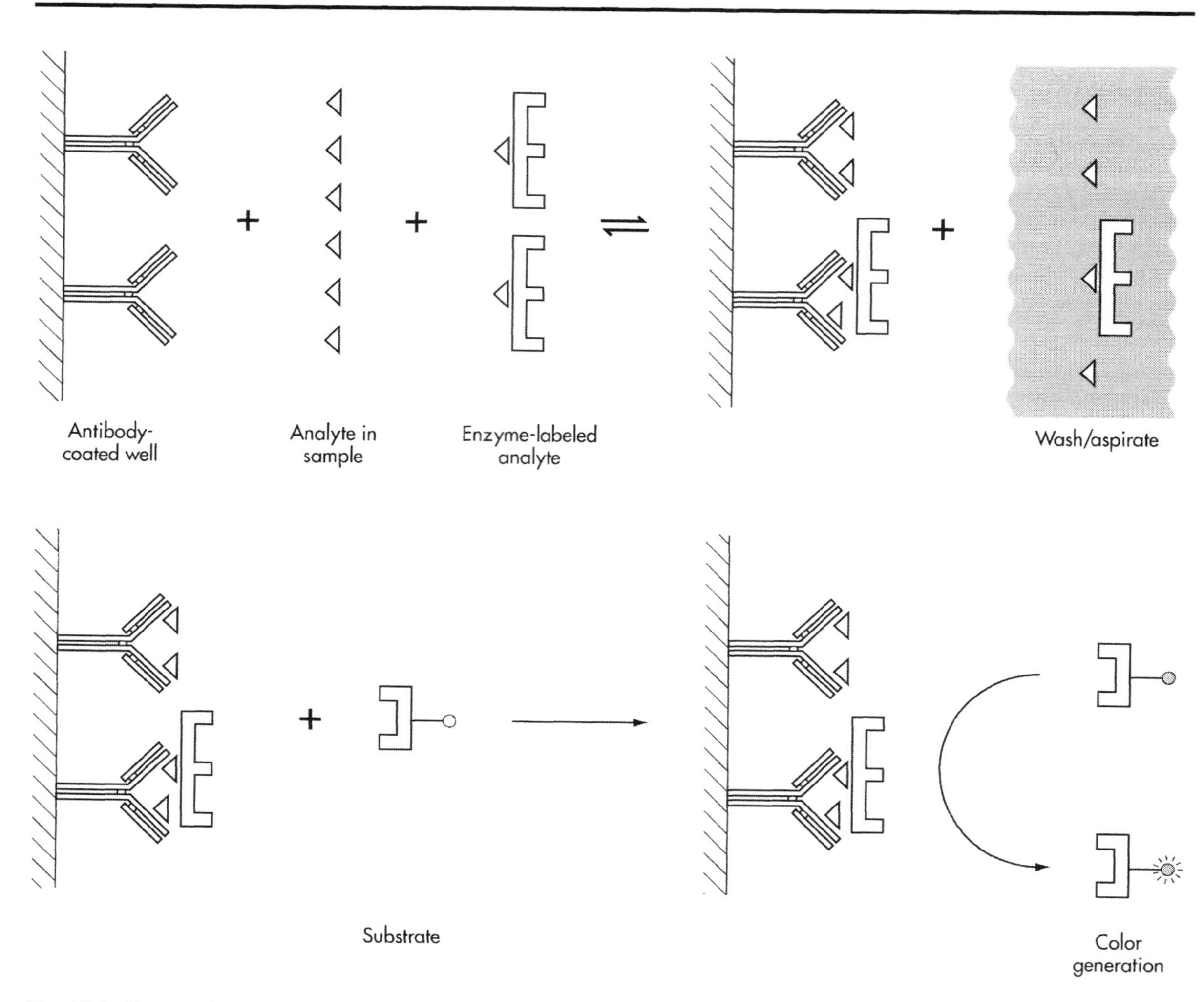

Fig. 11.1 Enzyme-immunoassay with colorimetric end-point (competitive).

FLUOROMETRY

In theory, colorimetric assays are capable of sensitivity that is as good as for RIA, but this is seldom achieved in practice. However even radioactive labels have limited sensitivity, because each radioactive atom only gives off its energy on one occasion, and only half of the atoms emit this energy during the half-life, which may be many weeks. In the case of ^{125}I, only one atom disintegrates per second in every 7.5×10^6 radioisotopic atoms. Colorimetric assays are primarily limited by the lower and upper limits of spectrophotometric measurements; colorimetry (absorption spectrometry) involves measuring the difference between two large light signals. However, fluorometric EIAs are capable of sensitivities several orders of magnitude better than colorimetric assays using the same enzymes (in the absence of extraneous interferences). This is partly because fluorescent compounds may be repeatedly excited by incident radiation to produce a signal in a short space of time. There is a greater relative improvement in sensitivity with alkaline phosphatase than with horseradish peroxidase, and fluorescence EIAs that utilize this enzyme have a level of sensitivity approaching those based on horseradish peroxidase.

There are two types of fluoroimmunoassays: those that employ a fluorophore label directly (in the same way that ^{125}I is used), and EIAs that use substrates that are converted to fluorescent end-products. This section describes the principles of fluorescence for both assays but the assay design of direct fluorescence immunoassays is described later in the chapter. *see* DIRECT FLUORESCENCE and TIME-RESOLVED FLUORESCENCE.

The principle of fluorescence is shown in Figure 11.2. Certain molecules, called **fluorophores**, absorb light at one wavelength and emit light at a longer wavelength.

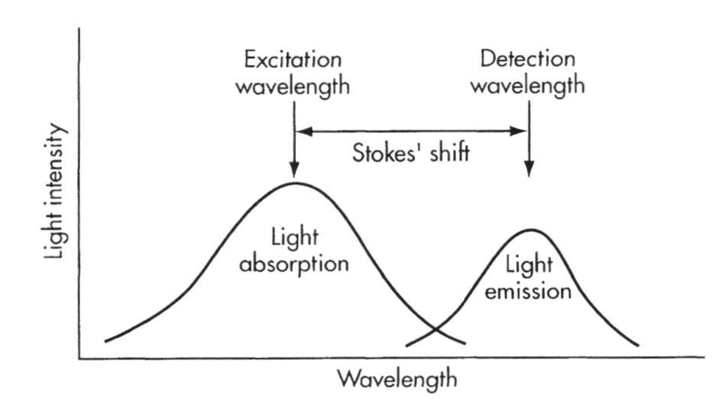

Fig. 11.2 The principle of fluorescence measurement.

If the appropriate wavelength of light is applied, the fluorophore is excited, resonating with the same frequency as the incident light. As the molecule returns to the ground state, it emits a photon of light at a lower energy, i.e. a longer wavelength and slower frequency. The difference between the excitation and emission wavelengths is known as the **Stokes' shift** and the proportion of energy re-emitted is referred to as the **quantum efficiency**. If the Stokes' shift is large, it is easier to measure the emitted light without interference from the incident light. Also, if the quantum efficiency is high, i.e. close to a value of 1, the signal is stronger and the assay more sensitive.

The most common substrate used in fluorometric EIAs is 4-methylumbelliferyl phosphate (4-MUP) with alkaline phosphatase as the enzyme label. The enzyme dephosphorylates 4-MUP to form 4-methylumbelliferone (4-MU), the fluorophore. The incident light is at 365 nm and the emitted light is at 448 nm. It is important that the incident light is filtered to avoid excitation of 4-MUP at wavelengths below 350 nm.

In a conventional fluorometer, the detector is a photomultiplier tube, set at right angles to the incident light source. However, front-surface fluorometers are now commonly used, in which a dichroic beamsplitter is positioned at an angle of 45°, directly above (or below) the solution containing the fluorophore. The incident light, produced by a lamp above, and to one side of the assay solid phase, passes through a filter and strikes the dichroic beamsplitter. This is reflective at 365 nm and transmissive at 448 nm. The incident light is reflected down onto the solid phase through a focusing lens. A small proportion of the incident light passes through the dichroic beamsplitter and strikes a photodiode, which measures the light level and controls the current supplied to the lamp, compensating for variation due to the age of the light source bulb. Light emitted from fluorophore bound to the solid phase passes up, through the dichroic filter and another bandpass filter, to reduce unwanted fluorescent signals from 4-MUP or from fluorophores in the original sample (Figure 11.3).

Fluorescence is directly affected by temperature, polarity, pH, and the dissolved oxygen content. If washing is not effective before the signal generation stage, interferences may arise from light scattering, background fluorescence, and quenching from the sample. Light scattering arises from particulate matter and causes an increase in background. Background fluorescence is naturally present in serum and plasma samples, caused by proteins at shorter wavelengths, and NADH and bilirubin at longer wavelengths. The peak fluorescence emission of plasma is around 350 nm (excitation at around 280 nm), with the weaker (but sometimes more critical) bilirubin peak at 520 nm. Quenching can occur when molecules in the sample absorb excitation light from the fluorometer or emitted light from the fluorophore. At extreme light intensities, photodecomposition can occur in the sample.

CHEMILUMINESCENCE

Well designed fluorometric EIAs are generally more sensitive than colorimetric EIAs, but their sensitivity is limited by interference from background fluorescence and quenching effects. Chemiluminescent immunoassays depend on the use of luminescent compounds (e.g. acridinium esters, acridinium sulphonamides, isoluminol), which emit light during the course of a chemical reaction.

Unlike fluorescence measurement, there is no incident light, and the only signal emanates from the luminescent molecules. If the luminometer can be made light-tight, the minimum background achievable is theoretically zero. Unlike radioactive disintegrations, all the photons from the luminescent compounds can be triggered in a very short time, much increasing the potential specific activity. Luminescent immunoassays have achieved levels of sensitivity several orders of magnitude better than radioisotopic and fluorometric immunoassays.

There are several types of luminescent immunoassays. This section describes conventional EIAs with a luminescent endpoint (Figure 11.4). Also *see* ENHANCED CHEMILUMINESCENCE and DIRECT CHEMILUMINESCENCE. There are substrates that give rise to luminescent end points for all of the commonly used enzyme labels, most notably horseradish peroxidase and alkaline phosphatase.

Adamantyl 1,2-dioxetane arylphosphate (AMPPD) provides a sensitive and versatile substrate for alkaline phosphatase labels. Cleavage of the phosphate group produces an unstable anion that decomposes with the emission of light (Figure 11.5). The detection limit for alkaline phosphatase is 1 zeptomole (10^{-21} moles, 602 molecules). Light emission is a protracted glow (>1 h)

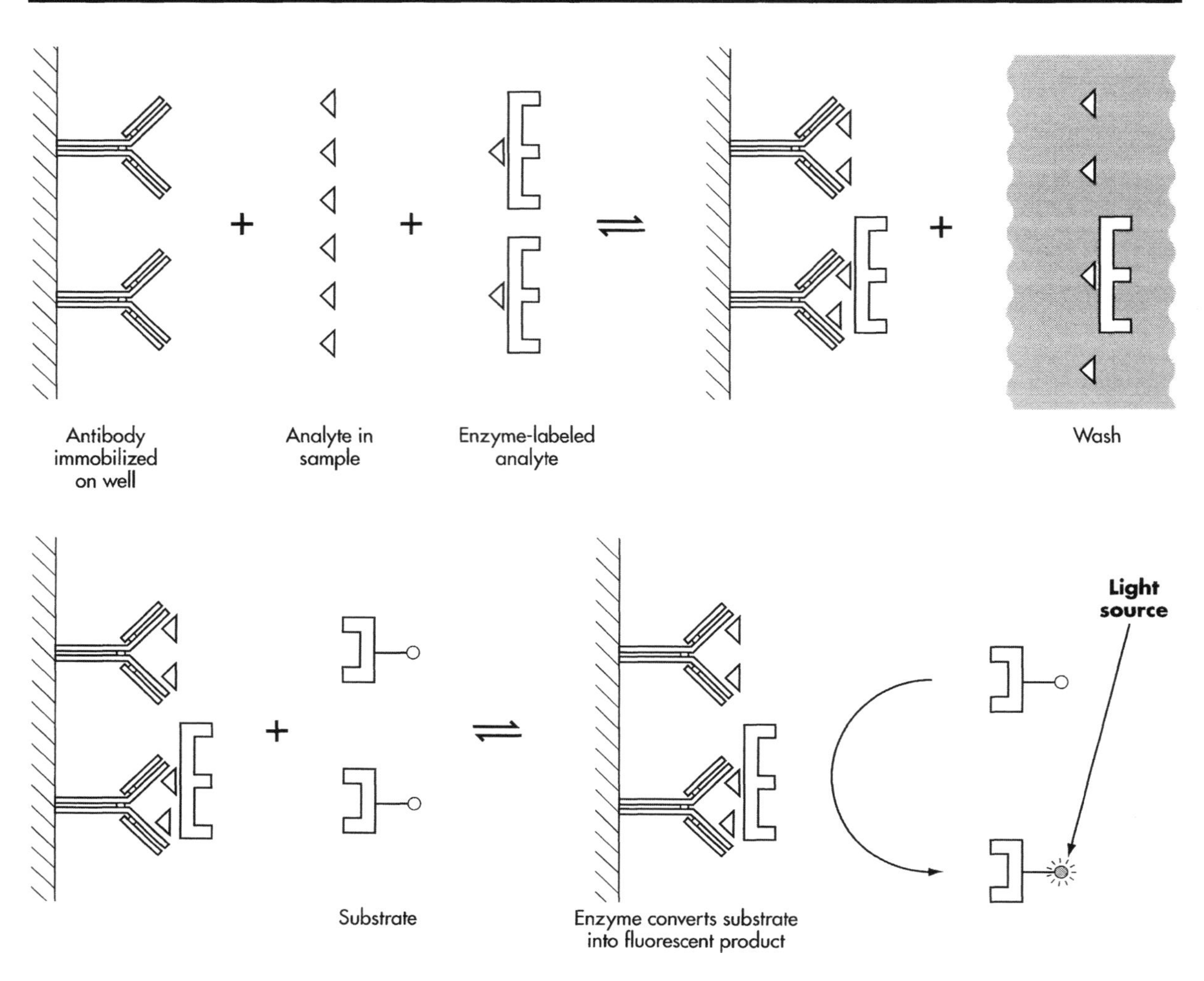

Fig. 11.3 Enzyme-immunoassay with fluorometric end-point (competitive).

and this simplifies initiation of the chemiluminescent reaction and measurement of the light emission.

Chemiluminescent detection reagents are also available for horseradish peroxidase. Lumigen® PS-1 is based on the oxidation of an acridan substrate. Lumigen PS-atto™ is intended for solution-based assays and has extended stability of the working solution, which can be used for at least a year. Lumigen TMA-6, which is optimized for membrane-based assays, is used by Amersham Biosciences for Western Blot kits. The working solution is stable for several months. Lumigen chemiluminescent reagents are used in several commercial immunodiagnostic systems.

Families of dioxetane substrates are available for assay of a range of enzyme labels (e.g. β-D-galactosidase) (Bronstein and Kricka, 1992).

ENHANCED CHEMILUMINESCENCE

Most enzyme-mediated luminescent reactions have low quantum yields, producing a weak light emission that rapidly decays. However, the addition of another chemical can enhance the output of light by several orders of magnitude. Horseradish peroxidase, in the presence of hydrogen peroxide, causes oxidation of luminol with the emission of light. The output of light is increased more than 1000-fold in the presence of certain substituted phenols and naphthols (e.g. 4-iodophenol). Background light emission from luminol and peroxide is also reduced, improving the signal-to-noise ratio (Thorpe and Kricka, 1986). Enhanced luminescence is a glow rather than a flash of light which persists for hours, although signal reading

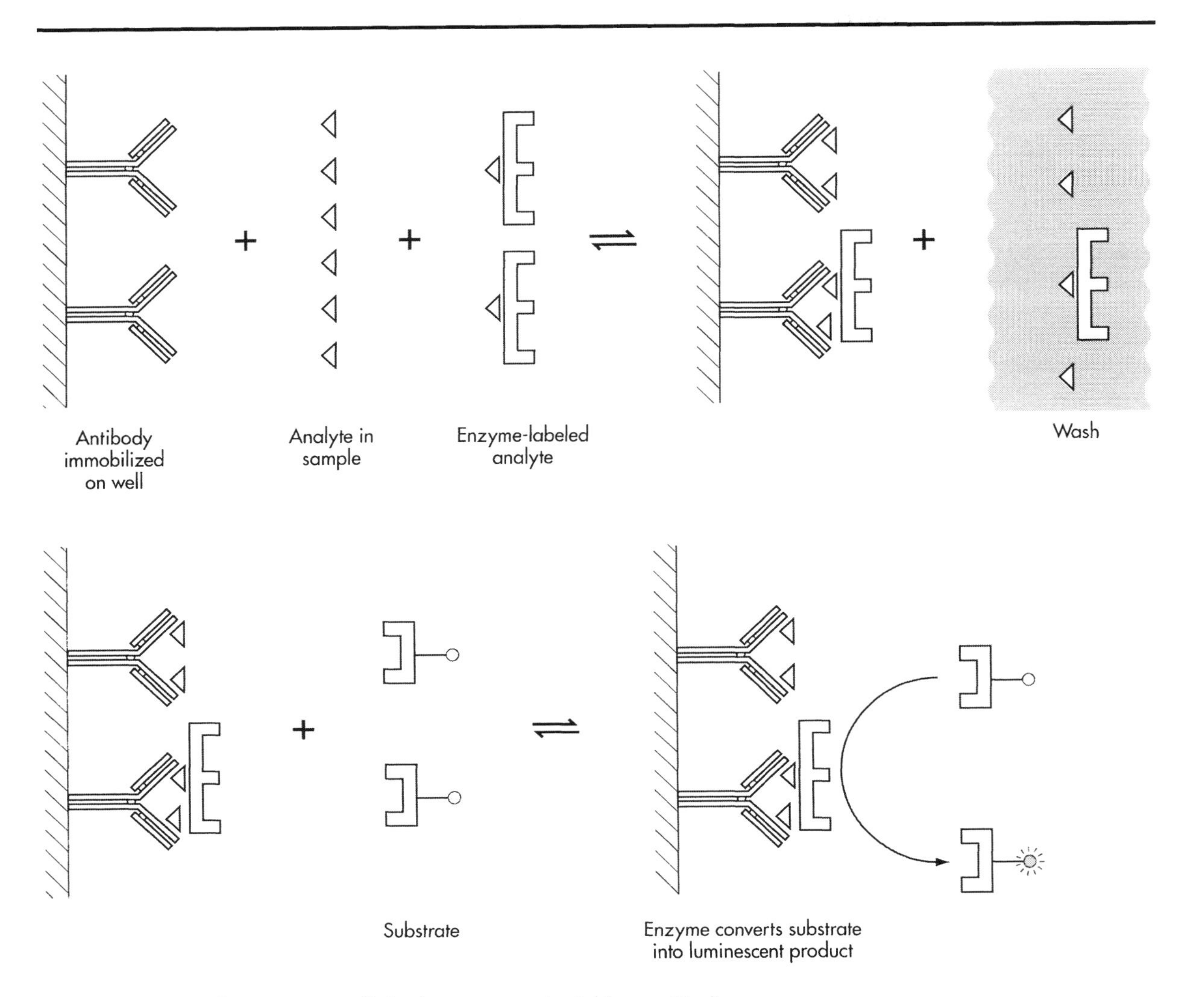

Fig. 11.4 Enzyme-immunoassay with luminescence end-point (competitive).

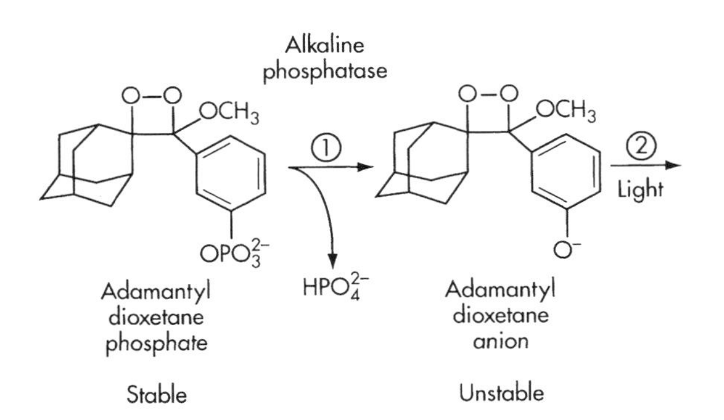

Fig. 11.5 Action of alkaline phosphatase in the IMMULITE system.

normally takes place within 2–20 min (Figure 11.6). This type of signal generation system is used in Vitros® ECi immunodiagnostic products (Ortho-Clinical Diagnostics).

New enhancers and substrates for horseradish peroxidase have been discovered that increase the sensitivity of this signal generation system.

DIRECT FLUORESCENCE

The principles of fluorescence measurement are explained in ENZYME LABELS – FLUOROMETRY above. Fluorophores are not only applicable to EIAs, but may also be used as direct labels, substituting for radioisotopes in competitive or immunometric assay designs

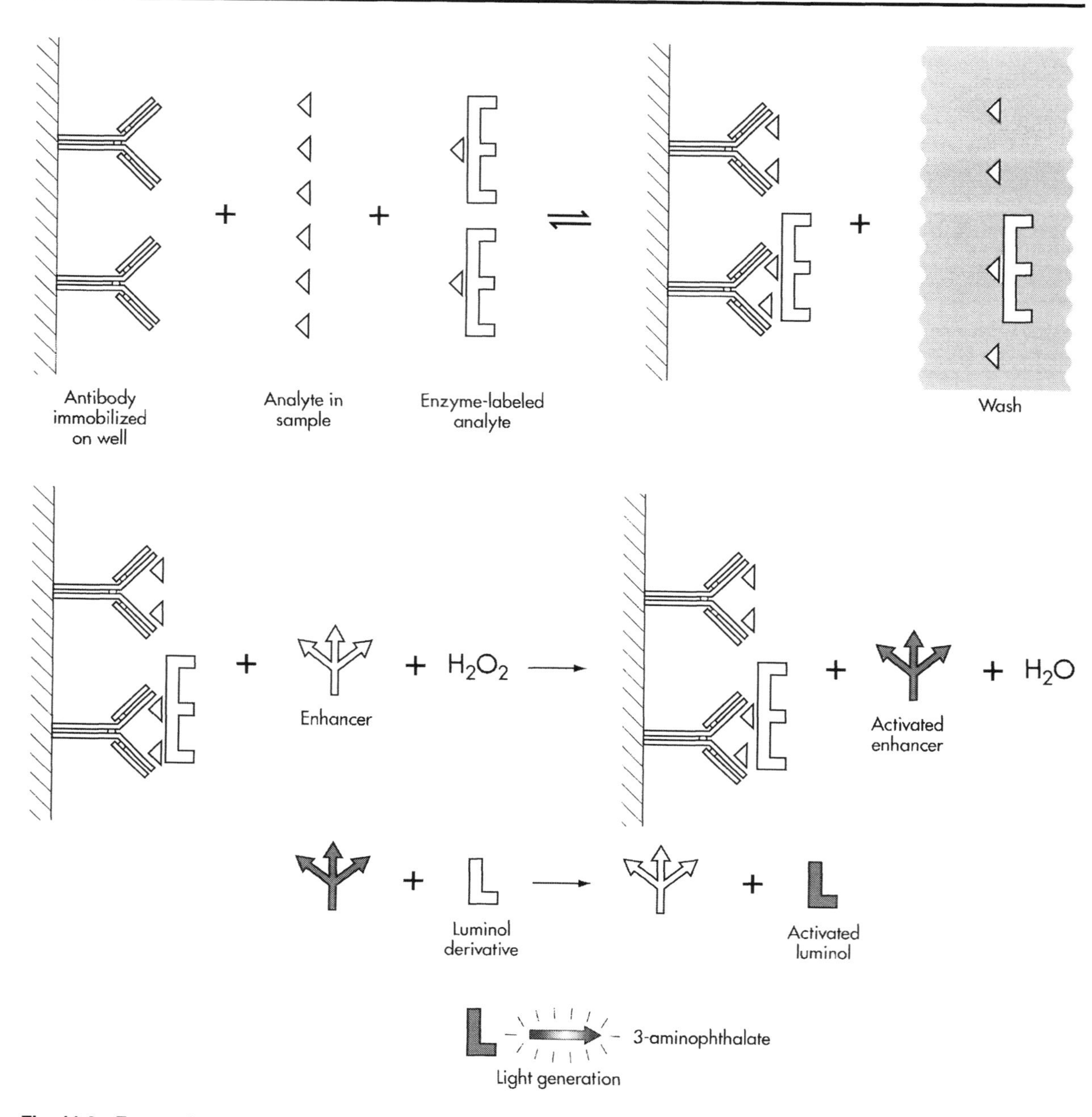

Fig. 11.6 Enzyme-immunoassay with enhanced luminescence end-point (competitive).

(Figure 11.7). Fluorescein or rhodamine derivatives may be used, for example.

The use of fluorescein-labeled antibodies in **immunofluorescence** has long been established to visualize antigens in microscopic sections of tissue.

Nanocrystalline, tunable **quantum dots** are available from Evident Technologies (EviTags™). These nanoparticles are 30–50 nm in diameter. There are a range of colors, with narrow emissions and non-interfering excitation properties. Each color is associated with a specific size of particle, with less than 5% variation in diameter. One ultraviolet light source may be used to excite all the colors, as the particles are excited by any wavelength shorter than the emission wavelength, which

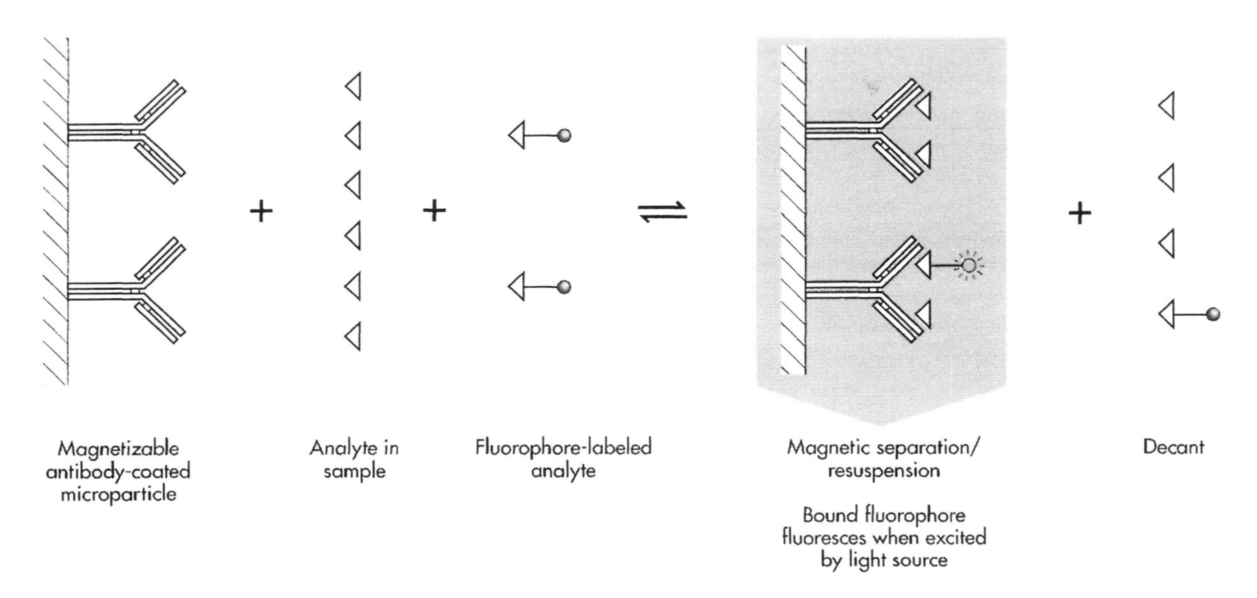

Fig. 11.7 Direct fluorometric detection system.

lies within the visible spectrum. They are supplied with amine or carboxyl terminating groups, suitable for conjugation.

LONG WAVELENGTH FLUORESCENCE

The term **long wavelength fluorescence** is sometimes applied to emission wavelengths of 600–1000 nm, encompassing the long wavelength end of the visible spectrum, and the very near infrared region. At these wavelengths, many fewer natural organic compounds are fluorescent, and background fluorescence from sample constituents is therefore much reduced. Also, Rayleigh scattering (scattered photons at the same wavelength as the incident light due to the solvent, i.e. water), and Raman scattering (a lesser effect, causing fluorescence from water) are much reduced. Long wavelength fluorimeters can be constructed from solid state light sources, such as diode lasers, and charge coupled devices (CCDs) which are available cheaply and miniaturized, from the electronics industry. There are a number of suitable fluorophore labels, such as the oxazine group of dyes.

TIME-RESOLVED FLUORESCENCE

Assay detection limits for conventional fluorescent tracers are limited by background interference, such as fluorescence from some of the components in blood samples (e.g. serum proteins, NADH and bilirubin). This interference can be avoided if there is a time gap between excitation and measurement of the emitted light from the fluorophore, particularly if the fluorophore has a comparatively long decay time. This is the principle behind time-resolved fluorescence. Some lanthanides, such as europium, form highly fluorescent chelates with certain organic ligands that absorb the excitation light and transfer its energy. The efficiency of the energy transfer between the ligand and the lanthanide ion can be as high as 100%. These chelates have exceptionally large Stokes' shifts (>200 nm), long decay times (>500 ns) and high quantum yields (30–100%). The long decay times make these fluorophores ideal for time-resolved fluorescence and the large Stokes' shift minimizes interference from fluorescent molecules in the sample, and from light scattering due to proteins or colloids. These properties reduce background noise, whilst the high quantum yield increases signal strength, so these assays are particularly sensitive. Figure 11.8 shows the principle as used in the DELFIA® (Wallac) system.

The most commonly used labeling reagent is N^1-(*p*-isothiocyanatobenzyl)-diethylene triamine tetra-acetic acid-Eu^{3+}. It binds to amine groups in target proteins at alkaline pH. Following the main assay incubation and separation the europium ions are dissociated from the labeling chelate into an acidic solution and enhancement reagent is added. The europium ions form a new fluorescent complex with the ligand in the enhancement reagent, a fluorinated β-diketone. The chelate would have limited solubility and suffer from quenching in aqueous solution, because water molecules quench the fluorescence due to oscillation of the OH bonds. To overcome this problem, micelles are formed

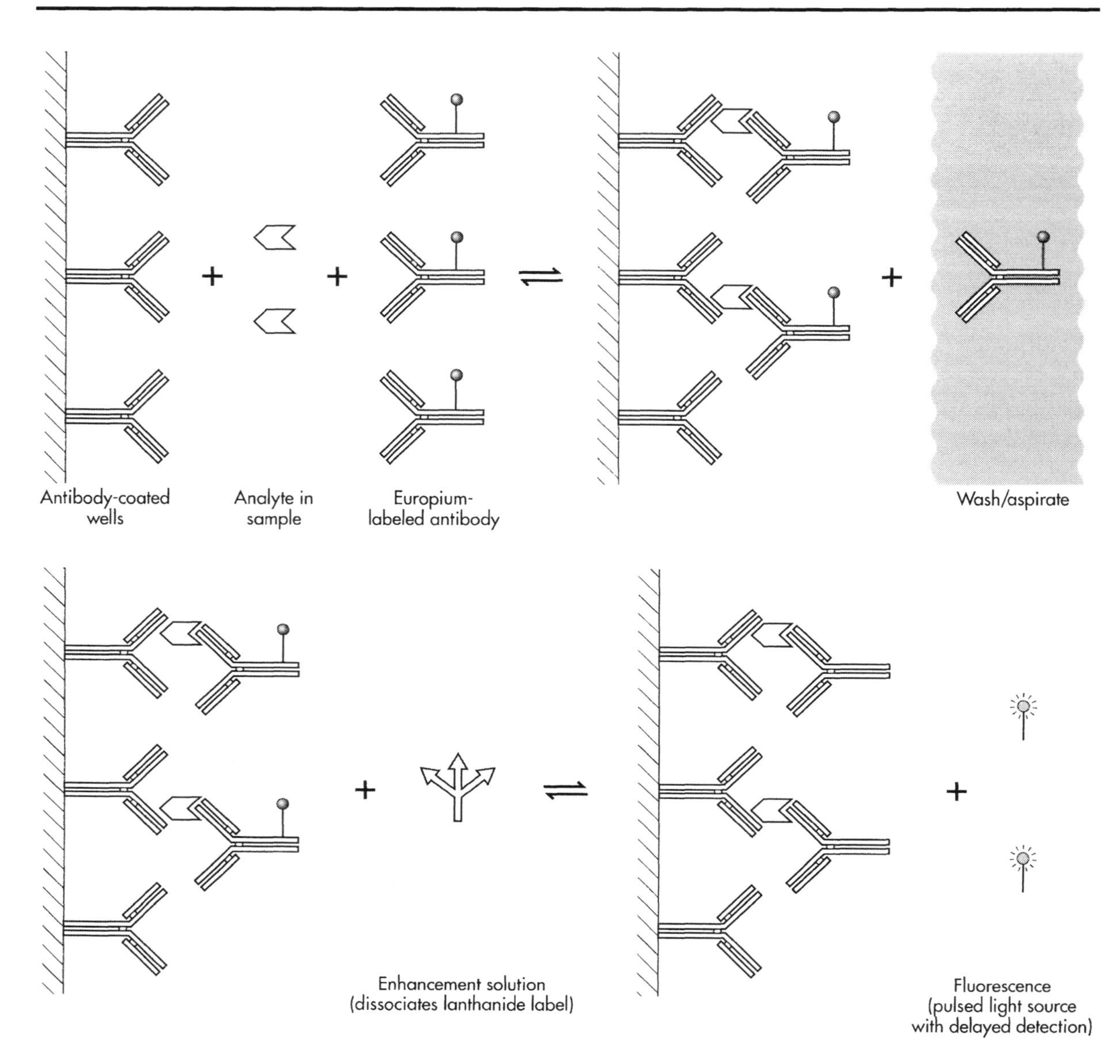

Fig. 11.8 Time-resolved fluorometric detection system (immunometric).

by incorporation of a detergent, trioctylphosphine oxide, and 2-naphthoyl-trifluoroacetone in the enhancement solution, protecting the chelate within a hydrophobic shell.

Recently a new lanthanide chelate system has been developed in Finland that is inert and stable, and intrinsically fluorescent in aqueous solution, avoiding the need for the special enhancement solution (Lövgren *et al.*, 1996). In this case the lanthanide chelate is used directly as a label and signal is measured following the wash step. The chelate is 4-[2-(4-isothiocyanatophenyl)ethynyl]-2,6,-bis([*N*,*N*-bis(carboxymethyl)amino]-methyl)pyridine. Another intrinsically fluorescent europium chelate that can be used to label proteins, with an exceptionally high signal-to-noise ratio, has been developed in Japan (Yuan and Matsumoto, 1998).

In the DELFIA system, the fluorometer supplies 1000 pulses of light per second, each lasting for less than 1 μs. Fluorescence is measured between 400 and 800 μs after each pulse.

Assays have been developed that use enzyme-generated fluorophores for time-resolved fluorescence signal detection. A range of salicyl phosphates have been synthesized as substrates for alkaline phosphatase that produce suitable fluorophores when chelated with lanthanides in the presence of EDTA. This combines the amplification capability of an enzyme with this highly sensitive signal generation system.

DIRECT CHEMILUMINESCENCE

The advantages of chemiluminescence as a signal generation system have been explained for EIAs (ENZYME LABELS – CHEMILUMINESCENCE). Acridinium esters and sulphonamides are chemiluminescent labels that can be attached directly to antigens or antibodies, in much the same way that ^{125}I is used in RIAs and immunoradiometric assays (Pringle, 1993). At the signal generation stage of the assay the ester linkage is cleaved under alkaline conditions to release the unstable compound *N*-methylacridone, which decomposes with the emission of a flash of light. The maximum intensity occurs 0.4 sec after initiation and the decay half-time is 0.9 sec, so it is essential that the addition of the alkaline reagent, and the luminometric reading, are carried out in one specially designed automated device. This is the principle behind the ACS™:180 and ACS:Centaur™ systems (Chiron Diagnostics) (Figure 11.9).

Electrochemiluminescent substances provide another option for a label in luminescent immunoassays, used in the ELECSYS™ system (Roche Diagnostics). Cations such as ruthenium tris(bipyridyl) undergo an electrochemical reaction at the surface of an electrode to form excited state species that decay to the electronic ground state with the emission of light (electrochemiluminescence). The detection step is usually performed in the presence of tripropylamine in a flow-through electrochemical cell (*see* ELECSYS).

BIOLUMINESCENCE

Aequorin is also a luminescent label, obtained from the jellyfish (*Aequorea victoria*). It is a 22 kDa protein complex that contains bound oxygen, the luciferin coelenterazine and three calcium binding sites. Recombinant aequorin and assays based on aequorin (AquaLite) have been developed in the past. Upon completion of the assay, following separation, the addition of calcium ions triggers a flash of blue light, with a peak at 469 nm. During the incubation, the label is prevented from premature activation by calcium ions using chelators, such as EDTA and EGTA. This is another very sensitive label.

PHOSPHORESCENCE

Phosphors produce an intense and long-lived light following excitation by incident radiation. Proteins may be adsorbed onto phosphor particles and, at the end of the immunoassay, ultraviolet light is used to activate the phosphorescent emission of light, which occurs at a wavelength in the visible spectrum, due to the loss of energy. The process of visible spectrum light production from an ultra-violet source is known as **down-conversion**.

OraSure Technologies, Inc. has reversed this process using **Up-Converting Phosphor Technology** (**UPT**™), based on lanthanide-containing, submicron-sized, ceramic, phosphor particles. Infrared light is used to excite phosphor particles that produce light in the visible color spectrum, a process known as **up-conversion**. Because up-conversion does not occur naturally, this method avoids interference from naturally fluorescent biomolecules in samples. Phosphors that emit light at different wavelengths are available for multi-analyte assays. So far, nine different colored phosphors have been developed.

MICROPARTICLE LABELS

Another method of labeling involves the use of latex, colloidal gold or selenium microparticles.

Latex microparticles are widely used as a solid phase in immunoassays (*see* SEPARATION SYSTEMS), and they are also useful as labels in quantitative turbidimetric and nephelometric assays (*see* HOMOGENEOUS IMMUNOASSAYS). Their use as labels in heterogeneous immunoassays is based on the use of very small **dyed latex particles**. For example, in the Clearblue One Step® pregnancy test, blue particles are coated with antibody to hCG and dried on a nitrocellulose membrane. Urine is applied at one end of the test strip and flows along by capillary action. The particles are picked up and carried along by the urine until they reach an immobilized band of anti-hCG antibody. In the presence of hCG in the urine a sandwich forms between the immobilized and particle-labeled antibodies, giving rise to a blue stripe that can easily be seen by the user without the need for any instrumentation.

Colloidal gold particles have been used to label antibodies for visualization by electron microscopy, as gold is extremely electron dense. Colloidal gold carries a negative charge that allows binding to positively charged proteins. **Silver enhancement** creates a dense black deposit around each gold particle, enabling visualization of antibodies by light microscopy. To apply these techniques to immunometric assays, antibodies are labeled with gold particles. At the end of the assay silver enhancer solution is added. The gold catalyzes the reduction of silver ions (silver lactate) to metallic silver in the presence of hydroquinone, a reducing agent. The silver forms a dense black deposit around each particle of gold. The density of the deposit may be determined colorimetrically in a conventional microtitration plate reader. The use of particles with a chemical enhancement avoids many of the potential interferences that can affect EIAs during the signal development stage, and simplifies

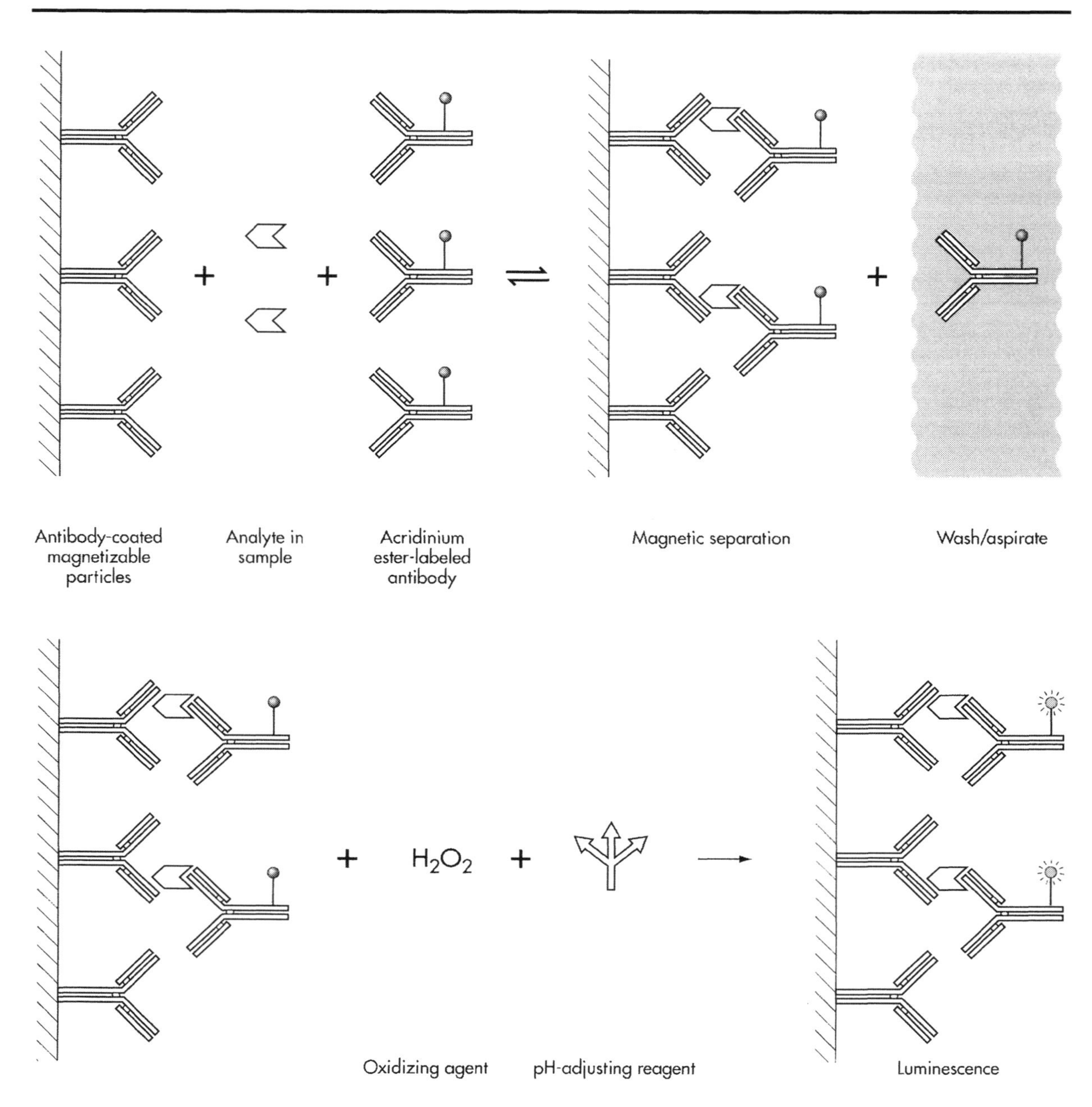

Fig. 11.9 Direct chemiluminescent detection system.

labeling of the antibodies. The resulting black coloration is also permanent.

Gold clusters contain a core of gold atoms held in a specific molecular structure by either a tris(aryl) phosphine ligand or by a halide ion. They are covalently linked to specific functional groups on the target molecule to be used as label. They do not carry a charge, and give clearer labeling of cellular structure in immunofluorescence, as they suffer less from non-specific binding.

Colloidal selenium is also used as an alternative to colored latex particles in simple qualitative immunoassays, e.g. for hCG, that form the basis of point-of-care and home pregnancy tests. The particles are red in color, simplifying the read-out of the assay result.

One extremely sensitive assay format reported for AFP (detection limit 1.6 amol) used antibodies labeled with carboxylated acrylate microparticles less than 1 μm in diameter. The surface of the well was photographed at the completion of the assay through a microscope and the number of particles in each photograph counted.

STREPTAVIDIN/AVIDIN–BIOTIN

Biotin is a small, water-soluble vitamin with a molecular mass of 244. It has an exceptionally high affinity for the egg-white protein, avidin (affinity constant 10^{15} L/mol) in free solution. One of the biological functions of biotin is to prevent absorption of avidin from raw eggs in the gut, which has harmful properties if taken into the circulation. The strength and speed of the binding between avidin and biotin may be used to link molecules together to create signal generation systems. In general a bacterial source of the protein, streptavidin, is used in preference to avidin. Unlike avidin, streptavidin has a neutral isoelectric point and does not contain carbohydrates. These properties make streptavidin more inert in assay systems, resulting in lower non-specific binding and hence greater sensitivity. Streptavidin has a molecular mass of 60 kDa.

Each molecule of streptavidin has four binding sites for biotin. Biotin is normally used as the label, in place of a radioisotope or enzyme. Biotinylation of protein is a gentle reaction that usually does not reduce the biological activity. At the end of the assay, a conjugate of streptavidin linked to a signal-generating substance is added. Examples of suitable conjugates are streptavidin–alkaline phosphatase, streptavidin–horseradish peroxidase, streptavidin-^{125}I, streptavidin–fluorescein and streptavidin–rhodamine. The conjugate and biotin quickly bind together, labeling the bound fraction with the signal source (Figure 11.10). In an immunometric assay, each molecule of the labeled antibody may have many of the small biotin molecules attached to it. This provides a means of attaching extra molecules of enzyme to amplify the final level of signal. Extra amplification may be achieved by using streptavidin conjugates that are labeled with two or three molecules of enzyme. One advantage of using a streptavidin-enzyme conjugate is that it can act as a generic signal generation reagent that can be applied to a range of analyte tests, simply by biotinylation of the relevant analyte-specific antibodies (immunometric format) or analyte (competitive format).

Streptavidin has been incorporated into macromolecular complexes containing large numbers of chelation sites for europium ions, providing several thousand fold signal amplification of each biotin label (**streptavidin-based macromolecular complexes** (**SBMC**)).

Perhaps the greatest benefit of streptavidin–biotin systems has been for the speed and simplicity of assay development. Use of generic signal generation reagents labeled with streptavidin avoids the need to develop individual conjugation methods for each assay. Streptavidin and biotin have also been used by some companies to develop generic capture systems. The solid phase (e.g. microtitration plate) is coated with streptavidin, and the capture antibody is biotinylated. This minimizes the need for new coating methods and facilitates the use of antibodies with high affinities for analyte but poor coating properties. (*see* SEPARATION SYSTEMS). For a review of the use of biotin–streptavidin systems in immunoassays, see Diamandis and Christopoulos, 1991.

PROTEIN A

Protein A is derived from the cell wall of *Staphylococcus aureus*. It has a high affinity for the Fc part of IgG. When conjugated with a suitable enzyme such as peroxidase or fluorescein, it may be used as a universal label for immunometric assays, western blots and immunocytochemistry. It may also be used as a generic capture method for solid-phase separation systems.

AMPLIFICATION STRATEGIES

Several amplification strategies have been developed in order to increase the signal from a label in an immunoassay.

In a conventional EIA, the enzyme label converts a substrate into a product that can be readily detected, either through its color, or some other property such as fluorescence or luminescence. At very low enzyme concentrations, hardly any product is generated, resulting in a weak signal that is difficult to measure because of background noise. It is possible to increase the amount of detectable product generated by using a multiple label, e.g. polylysine to which multiple molecules of 4-methylumbelliferyl groups have been attached via the amine groups.

Another amplification strategy that has been applied commercially uses several enzymes and substrates in a coupled cycle, as shown in Figure 11.11. The first stage of the signal generation process is conventional: the enzyme label catalyzes the conversion of substrate 1 into product. However, it is not this product that is detected. Instead, two recycling enzymes in the signal reagent convert the product to another form and back, over and over again. During each cycle, products from substrates 2 and 3 (present in excess) are generated. One of these final products is colored, fluorescent or luminescent.

By recycling the product of substrate 1 over and over, many molecules of colored product are generated for each molecule of substrate 1 converted by the enzyme used as label. This gives a stronger signal and hence a

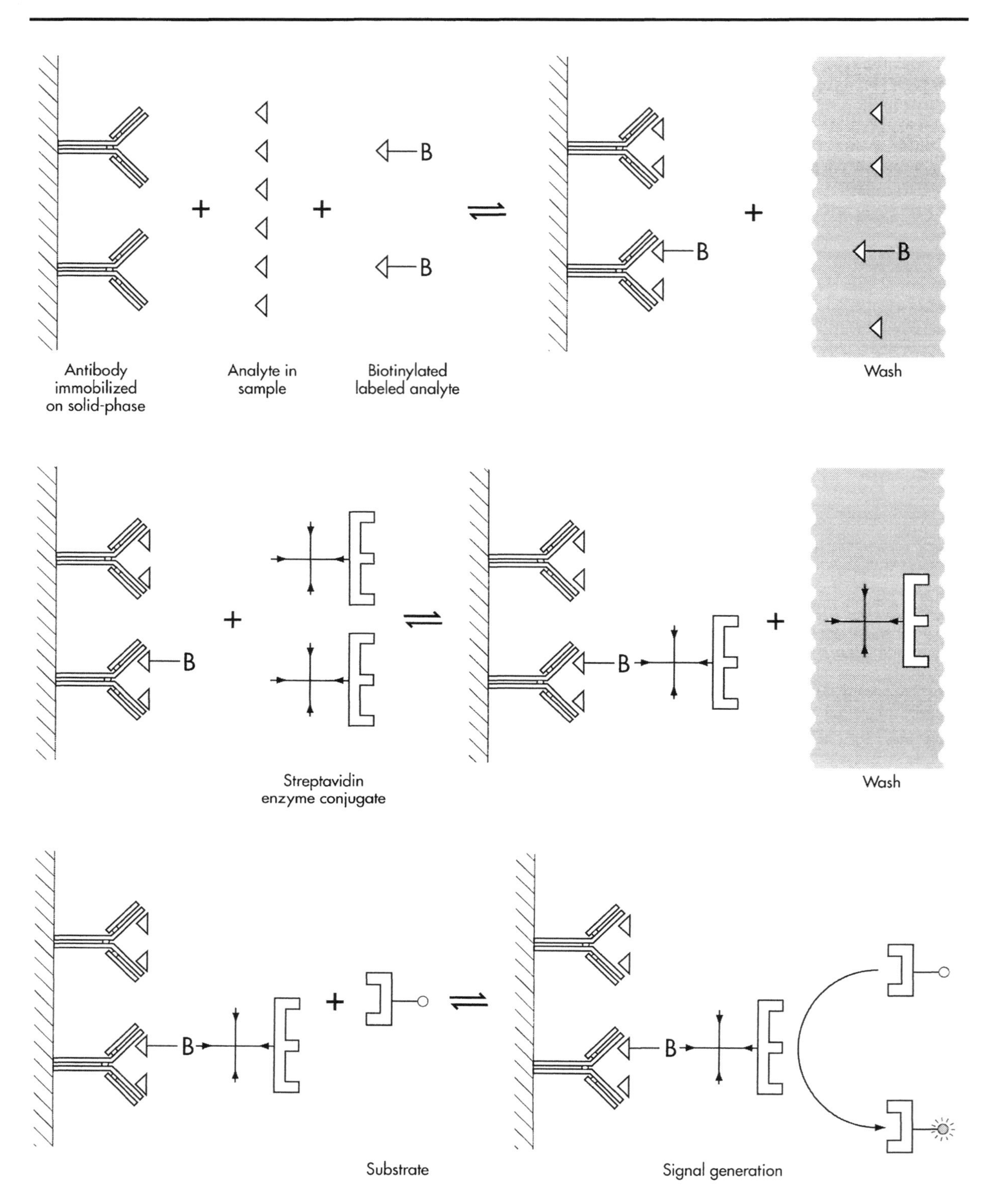

Fig. 11.10 Streptavidin-biotin detection system.

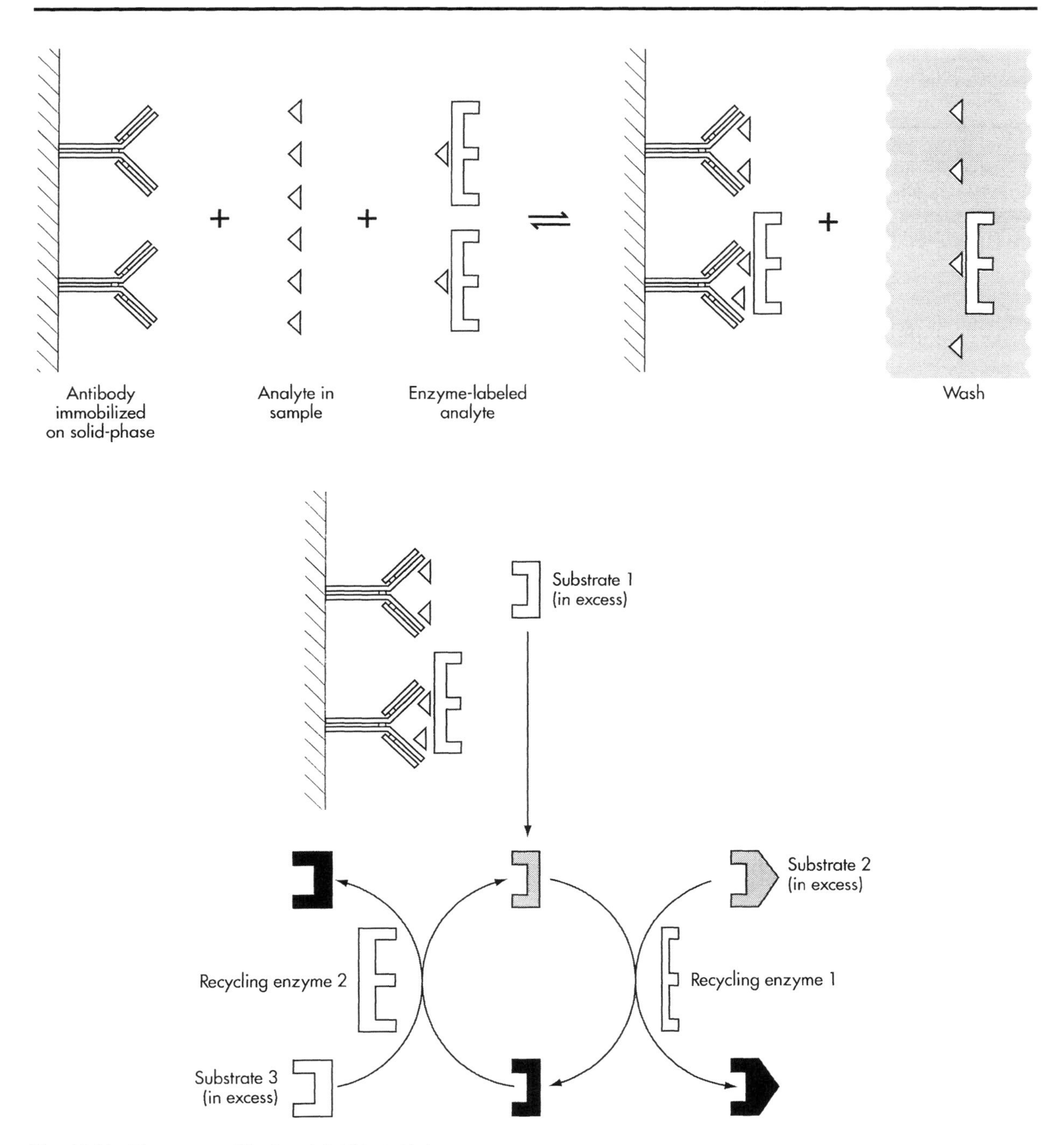

Fig. 11.11 Enzyme amplification detection system.

more sensitive assay. This remarkable system has been used successfully commercially. It is a genuine biomolecular amplifier, resulting in as much as 1000-fold amplification of the signal (Self, 1985).

Alkaline phosphatase may be used as the label, and NADPH as substrate 1. It is converted to NADH in the initial reaction and then to NAD and back in the cycle. Diaphorase and alcohol dehydrogenase are recycling enzymes 1 and 2. Formazan is the colored product of the diaphorase-catalyzed reaction.

An extremely sensitive assay system has been reported with acetate kinase as the label. The enzyme produces

ATP from acetyl phosphate and ADP. The ATP is detected using firefly luciferase and firefly luciferin in the presence of magnesium ions. The detection limit for acetate kinase was 9 zeptomoles (9×10^{-21} mol), less than 6000 molecules of the enzyme.

Liposome labels have also been made, containing many thousands of molecules capable of signal generation, such as an enzyme.

The **catalyzed reporter deposition (CARD)** assay is another example of an amplification scheme designed to increase assay sensitivity. It has been used in immunocytochemistry and immunoassays to increase sensitivity. In an immunometric (sandwich) assay format, after capture of the analyte onto an antibody-coated solid phase, and reaction of the captured analyte with a peroxidase-antibody conjugate, the peroxidase is reacted with a biotin–tyramine substrate. This leads to the formation of biotin–tyramine radicals, and these highly reactive species react with amino acids in the protein close to the label. This results in the deposition of biotin groups and these are then reacted with a streptavidin–peroxidase conjugate in a secondary incubation step, and bound peroxidase label quantitated using a substrate. The eventual number of peroxidase molecules is up to 30-fold more than were initially bound.

Other experimental amplification strategies include the use of a sequence of DNA as a label, which is then amplified using the polymerase chain reaction (**Immuno-PCR**). The DNA is replicated in 25 cycles, resulting in a million-fold increase in the number of copies of DNA (Sano *et al.*, 1992).

Alternatively a sequence of DNA coding for an enzyme can be used as a label (**expression immunoassay**). This acts as a factory that produces many enzyme molecules hence providing a significant amplification factor in an immunoassay. For example, the 'labeled' antibody in the assay is biotinylated, and after the immunoreaction, a DNA fragment encoding firefly luciferase is added. The DNA has previously been biotinylated and complexed with streptavidin. After a final wash to remove unbound DNA, a cocktail of reagents is added to allow cell-free transcription and translation, resulting in the production of luciferase. This in turn generates light in the presence of luciferin, oxygen and ATP. There is effectively a double stage amplification, as every DNA fragment can result in the synthesis of multiple molecules of enzyme, which can each turn over many molecules of substrate. As few as 3000 molecules of DNA label can be detected.

One of the most sensitive immunoassays reported to date relied on purification of the antibody-label complex to remove the background noise. The incubation between capture antibody, enzyme-labeled antibody and analyte was carried out in solution, then the complex was immobilized onto a polystyrene bead. This was achieved by using anti-dinitrophenyl (DNP) antibody coated beads and dual labeling the capture antibody with DNP and biotin. After washing, the immune complex was displaced from the bead using dinitrophenyl-L-lysine. The complex was finally captured using a streptavidin coated polystyrene bead, and the enzyme-labeled antibody quantified using substrate. The assay detected 1 zeptomole of ferritin (602 molecules).

MULTIPLE ANALYTES AND MINIATURIZATION

The availability of extremely sensitive labels and signal detection techniques, and the reduction of non-specific binding and background signal, has produced assays that in many cases offer performance characteristics that are well in excess of the basic requirements. However there are two areas where there is scope for exploiting recent advances to improve the practical benefits of immunoassays. These are the ability to measure several analytes in the same test, and miniaturization of the test. Simultaneous analyte assays have been available for many years but have never really caught on, for a variety of reasons. However, with the identification of new analytes, particularly from the human genome project, and more powerful algorithms to interpret the results, multiple test formats are likely to become more important.

Advances in coating techniques, and the introduction of the concept of ambient analyte immunoassay (*see* AMBIENT ANALYTE IMMUNOASSAY), offer the possibility of localizing discrete microscopic spots of immobilized antibodies for different analytes. Novel signal detection equipment is required to measure the small spots of signal involved. Fluorescence microscopy involves flooding a sample exposed to a fluorescent label with light at the excitation wavelength. This is commonly used in immunocytochemistry. The field of view is limited by the microscope optics. This provides sufficient contrast for localization of fluorescent label. However, for accurate quantification of fluorescent label, as is required by immunoassays, **confocal microscopy** is much superior. In a scanning confocal microscope, the incident light comes from a laser, which is focused on a single spot on the solid phase to which the immunocomplex is bound. Emitted light is directed, via a mirror, to a detector. The lens system co-focuses the detection system to measure light from the same spot. Two detectors may be used for α and β photons, emitted from the same microspot, but multianalyte capability derives from the ability of the microscope to scan the surface of the solid phase, measuring the fluorescence in different spots. An assay for thyroid stimulating hormone reached a detection limit of 0.0002 mU/L (Ekins and Chu, 1993).

MICROARRAYS

For a comprehensive review of signal generation systems used in miniaturized immunoassays *see* LAB-ON-A-CHIP, MICRO- AND NANOSCALE IMMUNOASSAY SYSTEMS.

Fluorophores are useful for multi-analyte assays. This is illustrated by a 2-color assay using fluorescein and Cy3 for avidin and sheep anti-digoxigenin IgG (Joos *et al.*, 2000). Three-color assays have also been developed, for example using BODIPY-FL, Cy3 and Cy5 dyes, and $Alexa_{488}$, Cy3, and Cy5 dyes (Sreekumar *et al.*, 2001). Time resolved fluorescence detection has also been adapted for use in microarray immunoassay (Scorilas *et al.*, 2000; Luo and Diamandis, 2000). Biotinylated antibodies can be detected using a streptavidin:biotinylated-polyvinylamine 4,7-bis(chlorosulfophenyl)-1, 10-phenanthroline-2,9-dicarboxylic acid (BCPDA)–europium chelate complex (50–100 BCPDA–europium complexes per polyvinylamine molecule) at levels of approximately 0.25 pg/spot (Scorilas *et al.*, 2000).

Increased sensitivity can be achieved in microarray immunoassay by using the Rolling Circle Amplification technique to amplify an oligonucleotide primer covalently linked to a detection antibody. When combined with fluorescent detection of the amplified oligonucleotide, it has proved possible to detect signals from individual antigen:antibody complexes on a microarray and achieve highly sensitive assays (Schweitzer *et al.*, 2000).

Another detection technique applied in microarray immunoassay is qualitative **matrix-assisted laser desorption/ionization time-of-flight mass spectrometric analysis (MALDI-TOF)** of the analytes bound on an array surface (Nedelkov and Nelson, 2001). Atomic force microscopic detection of the height increase when an antibody binds to an arrayed antigen is also used in microarrays (Silzel *et al.*, 1998).

As well as the multicolor detection format, it is also possible to test multiple samples simultaneously against an array of antigens by patterning lines of antigens onto a surface and then delivering solutions to be analyzed by the channels of a microfluidic network across the lines of antigens. This type of assay has been named **mosaic-format immunoassay** because specific binding of target antibodies with immobilized antigens on the surface produces a mosaic of binding events that can be readily visualized in a single assay (Bernard *et al.*, 2001).

REFERENCES AND FURTHER READING

Akhavan-Tafti, H., DeSilva, R., Eickholt, R., Handley, R., Mazelis, M. and Sandison, M. Characterization of new fluorescent peroxidase substrates. *Talenta* **60**, 345–354 (2003).

Avseenko, N.V., Morozova, T.Y., Ataullakhanov, F.I. and Morozov, V.N. Immobilization of proteins in immunochemical microarrays fabricated by electrospray deposition. *Anal. Chem.* **73**, 6047–6052 (2001).

Bernard, A., Michel, B. and Delamarche, E. Micromosaic immunoassays. *Anal. Chem.* **73**, 8–12 (2001).

Beverloo, H.B., van Schadewijk, A., Zijlmans, H.J.M.A.A. and Tanke, H.J. Immunochemical detection of proteins and nucleic acids on filters using small luminescent inorganic crystals as markers. *Anal. Biochem.* **203**, 326–334 (1992).

Bobrow, M.N., Harris, T.D., Shaughnessy, K.J. and Litt, G.J. Catalyzed reporter deposition, a novel method of signal amplification. *J. Immunol. Methods* **125**, 279–285 (1989).

Bobrow, M.N., Shaughnessy, K.J. and Litt, G.J. Catalyzed reporter deposition, a novel method of signal amplification. II. Application to membrane immunoassays. *J. Immunol. Methods* **137**, 103–112 (1991).

Bolton, A.E. and Hunter, W.M. The labelling of proteins to high specific radioactivities by conjugation to a ^{125}I-containing acylating agent. *Biochem. J.* **133**, 529–539 (1973).

Briggs, J. Sensor-based system for rapid and sensitive measurement of contaminating DNA and other analytes in biopharmaceutical development and manufacturing. *J. Parenteral Sci. Technol.* **45**, 7–12 (1991).

Briggs, J. and Fanfili, P.R. Quantitation of DNA and protein impurities in biopharmaceuticals. *Anal. Chem.* **63**, 850–859 (1991).

Bronstein, I. and Kricka, L.J. Chemiluminescence: properties of 1,2-dioxetanes. In: *Nonradioactive Labeling and Detection of Biomolecules* 168–175 (Springer, Berlin, 1992).

Case, J.F., Herring, P.J., Robison, B.H., Haddock, S.H., Kricka, L.J. and Stanley, P.E. (eds), *Bioluminescence and Chemiluminescence* (Wiley, Chichester, 2000).

Chard, T. *An Introduction to Radioimmunoassay and Related Techniques*, 4th edn (Elsevier, Amsterdam, 1990).

Christopoulos, T.K. and Chiu, N.H. Expression immunoassay. Antigen quantitation using antibodies labeled with enzyme-coding DNA fragments. *Anal. Chem.* **67**, 4290–4294 (1995).

Diamandis, E.P. and Christopoulos, T.K. The biotin-(strept)-avidin system: principles and applications in biotechnology. *Clin. Chem.* **37**, 625–636 (1991).

Diamandis, E.P. and Christopoulos, T.K. (eds), *Immunoassay* (Academic Press, San Diego, 1997).

Diamandis, E.P., Kitching, R. and Christopoulos, T.K. Enzyme amplified time-resolved fluoroimmunoassays. *Clin. Chem.* **37**, 1038 (1991).

Ekins, R.P. Ligand assays: from electrophoresis to miniaturized microarrays. *Clin. Chem.* **44**, 2015–2030 (1998).

Ekins, R.P. and Chu, F.W. Multianalyte microspot immunoassay microanalytical 'compact disk' of the future. *Clin. Chem.* **37**, 1955–1967 (1991).

Ekins, R. and Chu, F.W. Multianalyte testing. *Clin. Chem.* **39**, 369–370 (1993).

Ekins, R.P. and Chu, F. Microspot®, array-based, multianalyte binding assays: the ultimate microanalytical technology? In: *Principles and Practice of Immunoassay* 2nd edn (eds Price, C.P. and Newman, D.J.), 625–646 (Macmillan, London, 1997).

Ekins, R. and Chu, F.W. Microarrays: their origins and applications. *Trends Biotechnol.* **17**, 217–218 (1999).

Ekins, R., Chu, F. and Biggart, E. Multispot, multianalyte, immunoassay. *Ann. Biol. Clin.* **48**, 655–666 (1990).

Evangelista, R.A., Wong, H.E., Templeton, E.F.G., Granger, T., Allore, B. and Pollak, A. Alkyl- and aryl-substituted salicyl phosphates as detection reagents in enzyme-amplified fluorescence DNA hybridization assays on a solid support. *Anal. Biochem.* **203**, 218–226 (1992).

Exley, D. and Ekeke, G.I. Fluoroimmunoassay for 5-dihydrotestosterone. *J. Steroid Biochem.* **14**, 1297–1302 (1981).

Fraker, P.J. and Speck, J.C. Protein and cell membrane iodinations with a sparingly soluble chloramide, 1,3,4,6-tetrachloro-3α,6α-diphenylglycoluril. *Biochem. Biophys. Res. Commun.* **80**, 849–857 (1978).

Gosling, J.P. A decade of development of immunoassay methodology. *Clin. Chem.* **36**, 1408–1427 (1990).

Greenwood, F.C., Hunter, W.M. and Glover, J.S. The preparation of ^{131}I-labelled human growth hormone of high specific radioactivity. *Biochem. J.* **89**, 114–123 (1963).

Guifeng, J., Attiya, S., Ocvirk, G., Lee, W.E. and Harrison, D.J. Red diode laser induced fluorescence detection with a confocal microscope on a microchip for capillary electrophoresis. *Biosens. Bioelectron.* **14**, 10–11 (2000).

Hadd, A.G., Raymond, D.E., Halliwell, J.W., Jacobson, S.C. and Ramsey, J.M. *Anal. Chem.* **69**, 3407–3412 (1997).

Handley, R.S., Akhavan-Tafti, H. and Schaap, A.P. Chemiluminescent detection of DNA in low- and medium-density arrays. *Clin. Chem.* **44**, 2065–2066 (1998).

Hashida, S. and Ishikawa, E. Detection of one milliattomole of ferritin and ultrasensitive enzyme immunoassay. *J. Biochem.* **108**, 960–964 (1990).

Hastings, J.W., Kricka, L. and Stanley, P. (eds), *Bioluminescence and Chemiluminescence Reporting with Photons* (Wiley, Chichester, 1997).

Hayat, M.A. (ed), *Immunogold–Silver Staining: Principles, Methods and Applications* (CRC Press, Boca Raton, 1995).

Hemmilä, I. *Applications of Fluorescence in Immunoassays* (Wiley, New York, 1991).

Hiller, R., Laffer, S., Harwanegg, C. *et al.* Microarrayed allergen molecules: diagnostic gatekeepers for allergy treatment. *FASEB J.* **16**, 414–416 (2002).

Huang, R.P., Huang, R., Fan, Y. and Lin, Y. Simultaneous detection of multiple cytokines from conditioned media and patient's sera by an antibody-based protein array system. *Anal. Biochem.* **294**, 55–62 (2001).

Johannsson, A., Ellis, D.H., Bates, D.L., Plumb, A.M. and Stanley, C.J. Enzyme amplification for immunoassays. Detection limit of one hundredth of an attomole. *J. Immunol. Methods* **87**, 7–11 (1986).

Joos, T.O., Schrenk, M., Hopfl, P. *et al.* A microarray enzyme-linked immunosorbent assay for autoimmune diagnostics. *Electrophoresis* **21**, 2641–2650 (2000).

Kerstens, H.M.J., Poddighe, P.J. and Hanselaar, A.G.J.M. A novel in situ hybridisation signal amplification method based on the deposition of biotinylated tyramine. *J. Histochem. Cytochem.* **43**, 347–352 (1995).

Kricka, L.J. Chemiluminescent and bioluminescent techniques. *Clin. Chem.* **37**, 1472–1481 (1991).

Kricka, L.J. Ultrasensitive immunoassay techniques. *Clin. Biochem.* **26**, 325–331 (1993).

Kricka, L.J. Selected strategies for improving sensitivity and reliability of immunoassays. *Clin. Chem.* **40**, 347–357 (1994).

Kricka, L.J. Strategies for immunoassay. *Pure Appl. Chem.* **68**, 1825–1830 (1996).

Kricka, L.J. Microchips, microarrays, biochips and nanochips: personal laboratories for the 21st century. *Clin. Chim. Acta.* **307**, 219–223 (2001).

Kricka, L.J. and Wilding, P. Microfabricated immunoassay devices. In: *Principles and Practice of Immunoassay* 2nd edn (eds Price, C.P. and Newman, D.J.), 605–624 (Stockton, New York, 1997).

Law, S.J., Miller, T., Piran, U., Klukas, C., Chang, S. and Unger, J. Novel poly-substituted aryl acridinium esters and their use in immunoassay. *J. Biolumin. Chemilumin.* **4**, 88–98 (1989).

Lövgren, T., Meriö, L., Mitrunen, K., Mäkinen, M.-L., Mäkelä, M., Blomberg, K., Palenius, T. and Pettersson, K. One-step all-in-one dry reagent immunoassays with fluorescent europium chelate label and time-resolved fluorometry. *Clin. Chem.* **42**, 1196–1201 (1996).

Luo, L.Y. and Diamandis, E.P. Preliminary examination of time-resolved fluorometry for protein array applications. *Luminescence* **15**, 409–413 (2000).

Mangru, S.D. and Harrison, D.J. Chemiluminescence detection in integrated post-separation reactors for microchip-based capillary electrophoresis and affinity electrophoresis. *Electrophoresis* **19**, 2301–2307 (1998).

Nedelkov, D. and Nelson, R.W. Analysis of human urine protein biomarkers via biomolecular interaction analysis mass spectrometry. *Am. J. Kidney Dis.* **38**, 481–487 (2001).

Nguyen, N.-T. and Nwereley, S.T. *Fundamentals and Applications of Microfluidics* (Artech House, Norwood, Massachusetts, 2002).

Niedbala, R.S., Feindt, H., Kardos, K., Vail, T., Burton, J., Bielska, B., Li, S., Milunic, D., Bo, P. and Vallejo, R. Detection of analytes by immunoassay using up-converting phosphor technology. *Anal. Biochem.* **293**, 22–30 (2001).

Okano, K., Takahashi, S., Yasuda, K., Tokinaga, D., Imai, K. and Koga, M. Using microparticle labeling and counting for attomole-level detection in heterogeneous immunoassay. *Anal. Biochem.* **202**, 120–125 (1992).

Papanastasiou-Diamandi, A., Christopoulos, T.K. and Diamandis, E.P. Ultrasensitive thyrotropin assay based on enzymatically amplified time-resolved fluorescence with a terbium chelate. *Clin. Chem.* **38**, 545–548 (1992).

Price, C.P. and Newman, D.J. (eds), *Principles and Practice of Immunoassay* 2nd edn (Macmillan, London, 1997).

Pringle, M.J. Analytical applications of chemiluminescence. *Adv. Clin. Chem.* **30**, 89–183 (1993).

Ryan, O., Smyth, M.R. and Fagain, C.O. Horseradish peroxidase: the analyst's friend. *Essays Biochem.* **28**, 129–146 (1994).

Salacinski, P., Hope, J., McLean, C., Clement-Jones, V., Sykes, J., Price, J. and Lowry, P.J. A new simple method which allows theoretical incorporation of radio-iodine into proteins and peptides without damage. *J. Endocrinol.* **81**, 131P (1979).

Sano, T., Smith, C.L. and Cantor, C.R. Immuno-PCR: very sensitive antigen detection by means of specific antibody-DNA conjugates. *Science* **258**, 120–122 (1992).

Sato, K., Tokeshi, M., Kimura, H. and Kitamori, T. Integration of immunoassay system into a microchip. *Jpn. J. Electrophoresis* **44**, 73–77 (2000).

Sato, K., Tokeshi, M., Odake, T., Kimura, H., Ooi, T., Nakao, M. and Kitamori, T. Integration of an immunosorbent assay system: analysis of secretory human immunoglobulin A on polystyrene beads in a microchip. *Anal. Chem.* **72**, 1144–1147 (2000).

Sato, K., Tokeshi, M., Kimura, H. and Kitamori, T. Determination of carcinoembryonic antigen in human sera by integrated bead immunoassay in a microchip for cancer diagnosis. *Anal. Chem.* **73**, 1213–1218 (2001).

Schweitzer, B., Wiltshire, S., Lambert, J. *et al.* Inaugural article: immunoassays with rolling circle DNA amplification: a versatile platform for ultrasensitive antigen detection. *Proc. Natl. Acad. Sci. USA* **97**, 10113–10119 (2000).

Scorilas, A., Bjartell, A., Lilja, H., Moller, C. and Diamandis, E.P. Streptavidin–polyvinylamine conjugates labeled with a europium chelate: applications in immunoassay, immunohistochemistry, and microarrays. *Clin. Chem.* **46**, 1450–1455 (2000).

Self, C.H. Enzyme amplification: a general method applied to provide an immunoassisted assay for placental alkaline phosphatase. *J. Immunol. Methods* **76**, 389–393 (1985).

Silzel, J.W., Cercek, B., Dodson, C., Tsay, T. and Obremski, R.J. Mass-sensing, multianalyte microarray immunoassay with imaging detection. *Clin. Chem.* **44**, 2036–2043 (1998).

Smith, D.F., Stults, N.L., Rivera, H., Gehle, W.D., Cummings, R.D. and Cormier, M.J. Applications of recombinant bioluminescent proteins in diagnostic assays. In: *Bioluminescence and Chemiluminescence: Current Status* (eds Stanley, P.E. and Kricka, L.J.), 529–532 (Wiley, Chichester, 1991).

Sreekumar, A., Nyati, M.K., Varambally, S. *et al.* Profiling of cancer cells using protein microarrays: discovery of novel radiation-regulated proteins. *Cancer Res.* **61**, 7585–7593 (2001).

Thorpe, G.H.G. and Kricka, L.J. Enhanced chemiluminescent reactions catalyzed by horseradish peroxidase. *Methods Enzymol.* **133**, 331–353 (1986).

Thorpe, G.H.G., Kricka, L.J., Moseley, S.B. and Whitehead, T.P. Phenols as enhancers of the chemiluminescent reactions catalyzed by the horseradish peroxidase reaction: application in luminescence monitored enzyme immunoassays. *Clin. Chem.* **31**, 1335–1341 (1985).

Tijssen, P. *Practice and Theory of Enzyme Immunoassays* (Elsevier, Amsterdam, 1985).

Wang, J., Ibanez, A., Chatrathi, M.P. and Escarpa, A. Electrochemical enzyme immunoassays on microchip platforms. *Anal. Chem.* **73**, 5323–5327 (2001).

Wilson, R., Akhaven-Tafti, H., DeSilva, R. and Schaap, A. Comparison between acridan ester, luminol and ruthenium chelate electrochemiluminescence. *Electroanalysis* **13**, 1083–1092 (2001).

Wisdom, G.B. Enzyme-immunoassay. *Clin. Chem.* **22**, 1243–1255 (1976).

Yakovleva, J., Davidsson, R., Lobanova, A., Bengtsson, M., Eremin, S., Laurell, T. and Emneus, J. Microfluidic enzyme immunoassay using silicon microchip with immobilized antibodies and chemiluminescence detection. *Anal. Chem.* **74**, 2994–3004 (2002).

Yuan, J. and Matsumoto, K. A new tetradentate β-diketonate-europium chelate that can be covalently bound to proteins for time-resolved fluoroimmunoassay. *Anal. Chem.* **70**, 596–601 (1998).

12 Homogeneous Immunoassays

Edwin F. Ullman

INTRODUCTION

Homogeneous immunoassays require only mixing of a sample and immunochemical reagents followed by detection. Immunochemical binding produces a physically detectable signal that obviates the need to separate bound from free label. Because the rate of the binding reaction is not limited by slow diffusion to a surface, incubation times are fast, usually only a few seconds to a few minutes; and the non-separation assay protocols minimize the requirements for automation. Homogeneous methods are also, at least theoretically, more sensitive than heterogeneous immunoassays. This is because the separation and washing steps of heterogeneous methods are inherently more error prone and tend to reverse weak binding reactions, with an attendant decrease in sensitivity. However, because sample constituents are not removed by a wash step, variations in signal caused by non-specific effects of the sample matrix can prevent realization of this potential advantage. Only in the past few years have methods been developed that partially circumvent the sample interference problem and offer sensitivities equal to or better than the best heterogeneous methods.

Homogeneous methods have been developed for large and small analytes using both competitive and non-competitive protocols. Immunochemical binding can be followed either kinetically or after attainment of binding equilibrium. The unique and common characteristic of each of these methods is that they provide a mechanism for modifying the signal produced by a label as a function of an immunochemical binding event. This contrasts with heterogeneous methods which depend on identifying the location of an inert label after a separation step. The majority of labels are therefore environmentally sensitive sensors that are responsive to local variations in pH, solute concentration, electric field, steric constraints, radiation intensity, solvation, etc.

The first reported immunoassays were homogeneous but did not employ a label. They are attributed to Kraus, who in 1897 coined the term precipitin for the precipitate formed upon mixing an antigen and an antibody (Kraus, 1897). Likewise, the earliest immunoassays that employed a label were homogeneous (Meyer, 1922). Sheep erythrocytes serving as a label were coated with human immunoglobulin, and anti-immunoglobulin antibodies appearing in rheumatoid arthritis patients were shown to cause readily visible clumping of the cells. This method became known as hemagglutination. Subsequently, with the advent of spectrophotometry, it was shown that the visibly scattered light in Kraus' precipitin reaction could be measured by turbidometry to quantitate precipitin formation (Boyden *et al.*, 1947). However, the method was unsuitable for routine use. Not only was it very sensitive to other components in the sample but also the signal was biphasic, increasing and then decreasing upon adding antigen to a fixed amount of antibody (frequently referred to as the prozone phenomenon or hook effect). This occurs because only near equimolar concentrations can antibody bridge between antigens and produce a polymeric complex. At low antigen concentrations the excess antibody coats every antigen molecule and at high antigen concentrations the excess antigen occupies every antibody binding site.

PARTICLE AGGLUTINATION

ERYTHROCYTES AND LATEX

An important improvement in the evolution of homogeneous immunoassays was the replacement of Meyer's erythrocytes with more readily controllable latex particles (Singer and Plotz, 1956). Typically, antibodies are bound to the surface of the latex particles or erythrocytes, which form aggregates when a polyvalent antigen is present. Alternatively, the particles can be coated with an antigen to permit detection of antibodies. All the same assay architectures as in heterogeneous assays can be used. A non-competitive 'sandwich' assay can be set up with two antibody-coated particles that bind a multivalent antigen. The more antigen that is present, the more agglutination occurs. Although not as troublesome as the precipitin reaction, a sufficient excess of antigen will produce a biphasic response. Alternatively, added antigen can

The Immunoassay Handbook. *Third Edition*
D. Wild (Ed.)

inhibit antibody-induced aggregation of antigen-coated particles. This 'agglutination inhibition' method avoids the biphasic response but is generally less sensitive. In each case it is only necessary to combine the sample and reagents and optically measure the agglutination. One particularly innovative variant of this method was introduced by Agen Corp., which uses the patient's own erythrocytes as the particles in an assay for HIV antibodies. An HIV antigen conjugated to antibodies to an erythrocyte surface antigen is added to a whole blood sample. The antigen immediately becomes bound to the erythrocytes, which subsequently agglutinate if HIV antibodies are present (see Figure 12.1, Wilson *et al.*, 1991).

Visual observation of agglutination is moderately sensitive but of course is not quantitative. Most instrumental measurements rely on **turbidometry** or **nephelometry**. Turbidometry measures the intensity of a beam of light transmitted through the sample and nephelometry measures the light that is scattered at an angle away from the beam. Nephelometry is more sensitive but is more subject to interference from particulate matter in the sample. Simple subtraction of sample background is not possible because light scattering is not linearly dependent on the concentration of the particles.

Various strategies have been developed to avoid this problem. By monitoring the rate of change in a nephelometric signal it is possible to reduce the problem sufficiently that standard urine assays (Abuscreen®, Roche Diagnostics) for drugs of abuse and common serum proteins (ICS-II™, Beckman) can be quantitatively measured. Another approach is to detect individual particles in a flowing stream. These '**particle counting immunoassays**' (**PACIA**) estimate the change in the number of unaggregated particles during an immunochemical reaction (Masson *et al.*, 1981). This method is found to be most reliable for lower concentration analytes.

High sensitivity can be achieved by measuring the angular anisotropy of laser light scattered from a suspension of particles (**laser nephelometry**). The light scattering is a sensitive function of the particle size, particularly when the wavelength of light is in the order of the size of the particles. The method provides greater sensitivity than simple nephelometry, but interference from particulate material in the sample and reagents is also increased (Von Schulthess *et al.*, 1976).

GOLD SOLS

A related method for detecting agglutination uses particles that have greatly enhanced light scattering relative to latex or erythrocytes. Colloidal metals have high indices of refraction and produce particularly strong light scattering. The scattering intensity is strongly dependent on particle size. For gold sols relatively sharp apparent absorption and scattered light maxima are seen that increase from about 520 to 620 nm as the size of the particles increases from about 50 to 120 nm. Gold particle concentrations in the low femtomolar range can be detected. The theoretical basis for this phenomenon has recently been described (Yguerabide and Yguerabide, 1998).

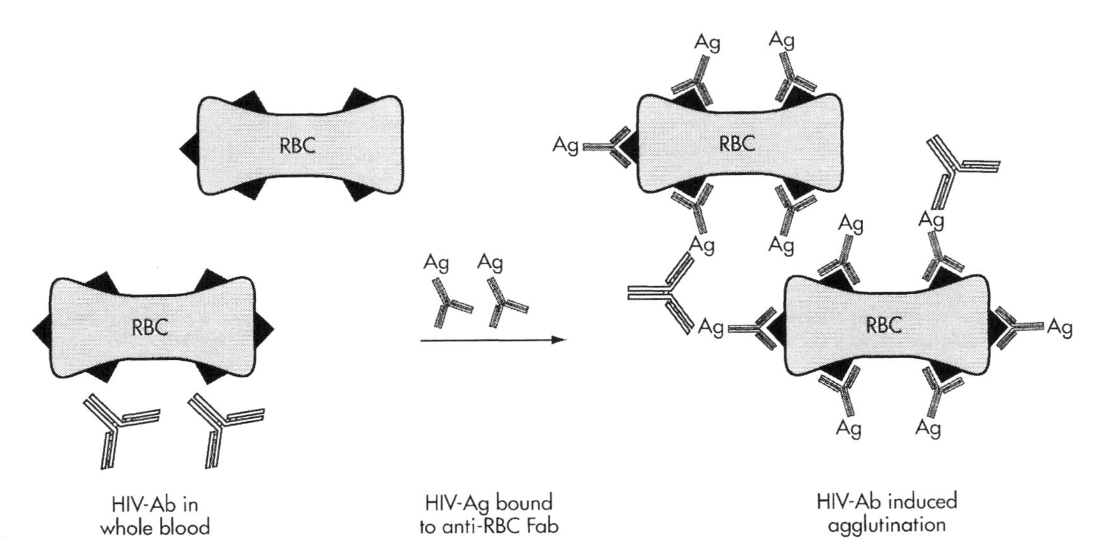

Fig. 12.1 Erythrocytes in anti-HIV positive blood are agglutinated upon addition of HIV antigen conjugated to antibodies to the cell surface antigen, glycophorin.

Agglutination of gold sols changes the effective particle size, which is reflected in a change in color of the suspension. A number of '**sol particle immunoassays**' (**SPIA**) based on this phenomenon have been demonstrated. For example, a sandwich SPIA capable of detecting 5.4 pM human placental lactogen (HPL) was carried out by mixing the sample with gold particles conjugated to anti-HPL antibodies and measuring the change in color with a colorimeter. Other assays, including competitive immunoassays for hormones, have also been demonstrated (Leuvering *et al.*, 1980). The relatively modest sensitivity of this method is probably related to difficulty in detecting small fractions of agglutinated particles, interference from sample components such as hemoglobin, and non-specific binding interactions.

Another useful property of gold particles is their ability to absorb light without photobleaching. This property was used in an assay for alpha-fetoprotein based on **photothermal beam deflection** (**PBD**). Antibodies were bound to both 20 nm gold particles and to large 50 μm glass particles. In the presence of a multivalent antigen, many gold particles aggregate around each glass particle. Irradiation of the suspension with an intense light beam produced highly localized heating near each aggregate. The resulting change in the index of refraction of the solution near the aggregates caused an angular deflection of the beam. This assay provided an order of magnitude increase in sensitivity relative to agglutination of latex particles although rapid settling of the large glass particles detracts from the convenience of the method (Sakashita *et al.*, 1995).

LYSIS IMMUNOASSAYS

Cell lysis initiated by binding of immunoglobulins to cellular antigens is a fundamental part of the body's defense mechanism. The process depends on the action of complement, a complex mixture of proteins that undergoes a cascade of events triggered by binding to antibodies on the cell surface and culminating in lysis of the cell. This process is the basis for a much explored but little utilized homogeneous immunoassay, **hemolysis immunoassay**. Antigens covalently conjugated to erythrocytes are allowed to bind to an antibody in the presence of a sample to be analyzed. The amount of antibody available for binding to the cells is affected by competitive binding to free antigen present in the sample. Following binding, serum that contains complement is added. After further incubation to complete lysis the remaining intact cells are removed and the concentration of hemoglobin released into the solution is measured by its light absorption (see Figure 12.2). Because many molecules of hemoglobin are released as a result of relatively few binding events the assay sensitivity approaches that of radioimmunoassay (Arquilla and Stavitsky, 1956).

Despite this early success, hemolysis immunoassay is now of little more than historical interest. Erythrocytes are difficult to store; the protocol is complex; complement is unstable; and precautions must be taken to deactivate complement that may be present in the sample. Numerous attempts to overcome these problems and improve assay sensitivity have met with some degree of success. A simpler protocol was devised by measuring the peroxidase activity of the released hemoglobin rather than its optical absorption. Hemoglobin catalyzes the formation of fluorescent or chemiluminescent products. Since luminescent substrates can be used that do not enter the cells, removal of the cells is avoided. However, reagent stability and interference from sample components remain problematic (Tatsu and Yoshikawa, 1990; Tatsu *et al.*, 1992).

A major advance in lysis immunoassays came when it was shown that **liposomes**, which are more stable than erythrocytes, can also be lysed by complement when antibodies are bound to a surface antigen (Kataoka *et al.*, 1971; Kinsky, 1972). The substance that is released need no longer be limited to hemoglobin. Nearly any compound that is in solution during formation of the liposomes can be encapsulated. Proteins and other large molecules can be retained nearly indefinitely and sufficiently hydrophilic smaller molecules leak out only slowly. Many encapsulated labels have been investigated. Encapsulated fluorescent compounds at sufficiently high concentration are usually quenched and become fluorescent only when released (Yasuda *et al.*, 1988), and dyes showing concentration dependence perform similarly (Frost *et al.*, 1994). Solutions of entrapped stable nitroxide radicals show a change in their EPR spectra upon dilution (Chan *et al.*, 1978). Encapsulated chelators can bind to rare earth ions upon their release to form fluorescent complexes (Ius *et al.*, 1996). Encapsulated enzyme substrates when released can be converted to a detectable product by an enzyme present in the bulk solution (Thompson and Gaber, 1985), and coenzymes can similarly be released to react with the corresponding apo-enzyme (Haga *et al.*, 1990).

Much attention has been given to the development of commercial liposome lysis immunoassays based on the use of **encapsulated enzymes** (Canova-Davis *et al.*, 1986; Yu *et al.*, 1987). Not only are enzymes resistant to leakage, but they also offer the opportunity for enormous amplification because many enzyme molecules can be released from each liposome and each molecule can then generate many molecules of a detectable product. Unfortunately, practical applications of this potentially powerful method have been difficult to develop. Methods for stabilizing standardized solutions of complement remain elusive and the presence of complement activity

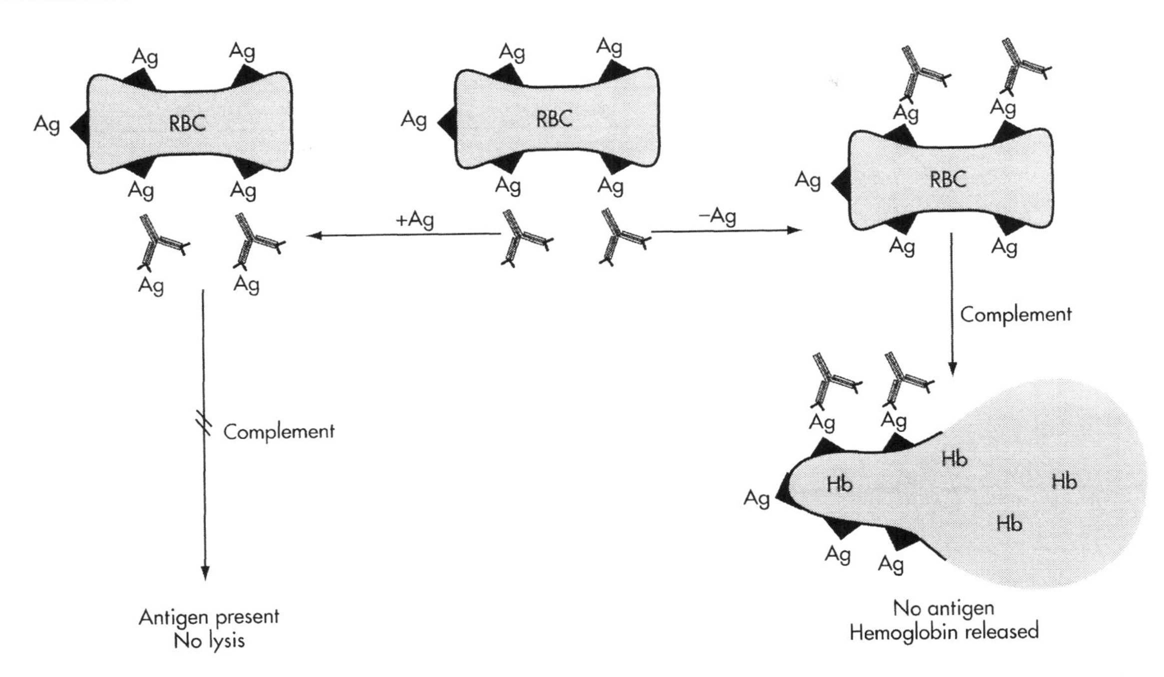

Fig. 12.2 Complement mediated lysis immunoassay. Erythrocytes labeled with an antigen become bound by antibodies and lysed only in the absence of competing antigen.

and inhibitors in serum samples has not been successfully overcome.

SPIN IMMUNOASSAYS

The use of molecular reporter groups in place of particles in immunoassays was first described by Coons who used antibodies labeled with fluorescein to visualize binding to tissue slices (Coons *et al.*, 1942). Although fluorescent labels were eventually found to be useful for homogeneous immunoassays, stable nitroxide radicals became the first reporter groups commonly used in homogeneous immunoassays. Nitroxides have one unpaired electron, which is detectable by **electron paramagnetic resonance (EPR) spectroscopy**. Electrons have a spin of 1/2 which permits two $(2S + 1)$ orientations in a magnetic field. Transitions between these states are stimulated at a precise microwave frequency depending upon the field strength. The closely associated nitrogen nucleus has $S = 1$ which imparts three possible local fields and thus there are three separate resonance frequencies. This hyperfine coupling is anisotropic with respect to the orientation of the nitroxide group in the magnetic field. The many possible orientations in a solution of a nitroxide produce a broad spectrum when the molecules are relatively immobile. However, sharp three line spectra are produced if the molecules are small enough to cause averaging of the hyperfine coupling due to rapid tumbling.

Binding of a small nitroxide-labeled drug such as morphine to anti-morphine antibodies sharply reduces its rate of tumbling leading to line broadening (Figure 12.3). In the presence of free morphine, binding of the nitroxide labeled drug is inhibited and the spectral lines become narrow. This '**spin immunoassay**' was the first widely used immunoassay, homogeneous or heterogeneous, to be employed commercially (Leute *et al.*, 1972). Sold under the trade name FRAT®, it was adopted by the US Army for screening its personnel for abused drugs during the Vietnam War. The method required simply drawing a urine sample into a capillary, expelling the contents into a mixture of the labeled drug and the antibody, plugging one end of the capillary and dropping it into an EPR cavity. The whole process could be done manually in about 30 sec with little training. Following the war, the method fell into disuse because of its poor sensitivity ($\sim 10^{-7}$ M) and the costly instrumentation.

FLUORESCENT LABELS

FLUORESCENCE POLARIZATION IMMUNOASSAY

An alternative method of monitoring changes in tumbling rates is **fluorescence polarization**. Light is differentially

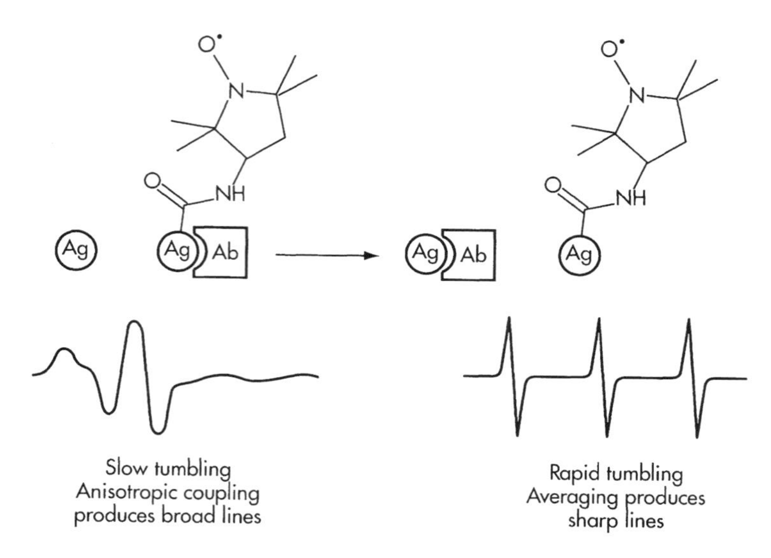

Fig. 12.3 Spin immunoassay. The line broadened EPR spectrum of a slowly tumbling spin-labeled drug bound to an antibody is replaced by the sharp line spectrum of the free label when drug from a sample competes for antibody binding sites.

absorbed by molecules as a function of their orientation relative to the direction and polarization of the exciting light. The light subsequently emitted as fluorescence by each of the resulting electronically excited molecules will usually be polarized. Rotation during the lifetime of the excited states randomizes the orientation of the excited molecules leading to a net reduction in fluorescence polarization. The more rapid the tumbling the less polarization is observed (see Figure 12.4). This phenomenon was first applied in a homogeneous immunoassay by Dandliker and Feigen (1961) and Dandliker *et al.* (1973) but the method was initially little more than a laboratory curiosity because of the primitive state of development of commercial spectrofluorometers and the requirement for two separate measurements differing by a 90° rotation of a polarizing lens.

Fluorescence polarization immunoassay (FPIA), like spin immunoassay, has been applied almost exclusively to small molecule analytes. The sample and antibody are combined and the antigen in the sample competes with fluorescer-labeled antigen for binding to the antibody. Increasing concentrations of the antigen produce decreased polarization. Abbott Laboratories uses FPIA primarily for therapeutic drug monitoring and drug abuse

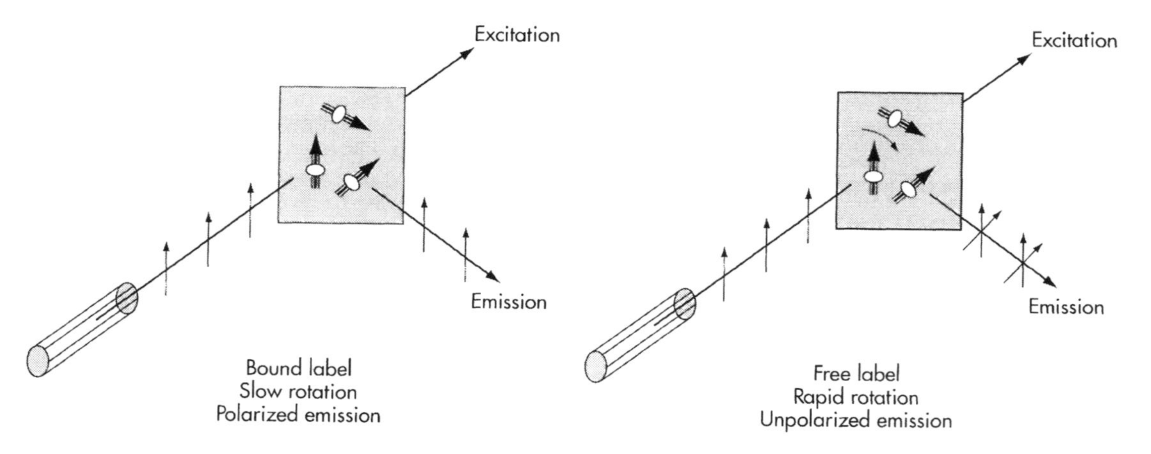

Fig. 12.4 Fluorescence polarization immunoassay. Excitation of fluorescent labels leads to selective absorption of light by appropriately oriented molecules. Polarized emission of these oriented molecules occurs when their rate of rotation is low relative to the rate of fluorescence emission.

testing on their TDx immunochemistry system. The success of the method after years of disuse stemmed in larger measure from the development of improved solid state methods for analyzing polarized light.

High molecular weight analytes are difficult to analyze by FPIA for two reasons. First, binding of a large antigen to an antibody that may be of similar mass produces a smaller relative change in the tumbling rate and thus a smaller change in polarization. Secondly, most fluorescent labels have excited state lifetimes in the range of 10^{-9}–10^{-7} sec, which is too short to permit rotational reorientation of larger bio-polymers. Recently, a rhenium based fluorophore has been identified that shows polarized fluorescence with a lifetime of 3 msec. This extends the theoretically accessible molecular weight range for special applications (Guo *et al.*, 1998). However, interference from quenching impurities becomes increasingly problematic with long lived excited states because quenching affects the measured polarization.

Interference from adventitious fluorophores and non-specific binding of the label to proteins in biological samples have historically restricted the detection limit of FPIA to concentrations of about 100 pM. By using a highly hydrophilic long wavelength dye and time-delayed measurements that permit discrimination between the emission from the background and the label, detection of concentrations down to 1 fM has been claimed (Devlin *et al.*, 1993).

FLUORESCENCE RESONANCE ENERGY TRANSFER

Fluorescence resonance energy transfer (FRET) refers to the transfer of energy from an electronically excited donor molecule to a nearby acceptor molecule which has an energetically accessible excited state. The excited states of one or both of the donor and acceptor can decay with fluorescence emission. When both are fluorescent, energy transfer is observed by a reduction in the emission intensity of the donor with a concomitant increase in the intensity of the longer wavelength emission from the acceptor. The energy transfer mechanism of most relevance in immunoassays depends on through-space coupling of the electronic transition dipoles (Förster, 1948). The rate of energy transfer is inversely dependent on the sixth power of the distance between the donor and acceptor and is directly related to the spectral overlap of the donor emission and acceptor absorption spectra. Distances at which the donor fluorescence is reduced by 50%, R_0, can be as high as 80 Å depending on the dyes that are used.

Fluorescence energy transfer immunoassays (**FETI**) using this phenomenon provide an attractive alternative to FPIA (Ullman *et al.*, 1976; Ullman and Khanna, 1981). Only a simple fluorometer is needed and the method is applicable to both small and large molecules. Competitive immunoassays are constructed by conjugating a fluorescent donor to an antigen and multiple acceptor molecules to an antibody. This assures that there will be at least one acceptor at an efficient energy transfer distance within the immune complex. In sandwich immunoassays, multiple donors and acceptors are used to label the two antibodies (see Figure 12.5). Usually the assays are carried out by following the rate of change in fluorescence during the initial phase of immunochemical binding. This reduces interference from fluorescence by the sample, which normally does not change over the course of the assay. In competitive immunoassays, the rate of decrease in donor fluorescence is reduced by antigen competing for antibody binding sites. In sandwich immunoassays increasing concentrations of antigen accelerate quenching of the donor emission due to accelerated binding of the two antibodies to the antigen. As in the precipitin and latex agglutination methods, at antigen excess, antibody binding to separate antigen molecules leads to a prozone phenomenon in which the rate of quenching again decreases.

The first commercial use of FETI was in Syva's Advance® immunochemistry system. In this system, the reduction in fluorescence of the donor was monitored rather than the theoretically more sensitive appearance of fluorescence of the acceptor. This was necessary because it is difficult to find a fluorescent acceptor that is not directly excited by the light used to excite the donor. Nevertheless, by using phycoerythrin as the donor, a highly fluorescent protein with an extinction coefficient near $2 \times 10^6\ M^{-1}\ cm^{-1}$, sufficient sensitivity could be obtained for a homogeneous serum digoxin assay that could quantitate 500 pM of the drug in serum (25 pM in the assay).

Attempts to achieve higher sensitivity by using a chemiluminescent donor in place of a fluorescent donor have not been successful. Several haptens and protein antigens labeled with isoluminol were shown to sensitize fluorescein emission when bound to fluorescein-labeled antibodies (Patel and Campbell, 1983). The assays were equivalent to radioimmunoassay in sensitivity but the high susceptibility of the chemiluminescent reaction to differences in sample composition led to unacceptable performance, at least for clinical applications.

An important improvement in the method came with the development of fluorescent rare earth chelates that permit the use of time-resolved fluorescence (*see* SIGNAL GENERATION AND DETECTION SYSTEMS). These labels, which have fluorescent lifetimes in the order of milliseconds, are used as energy donors. The acceptors are conventional dyes that absorb strongly at the emission wavelengths of the chelate. Irradiation with a brief pulse of light is followed by rapid decay of the short-lived fluorescence of most sample impurities and any acceptor molecules that are directly excited by the light. The delayed fluorescence that is subsequently measured is associated only with emission from those acceptor

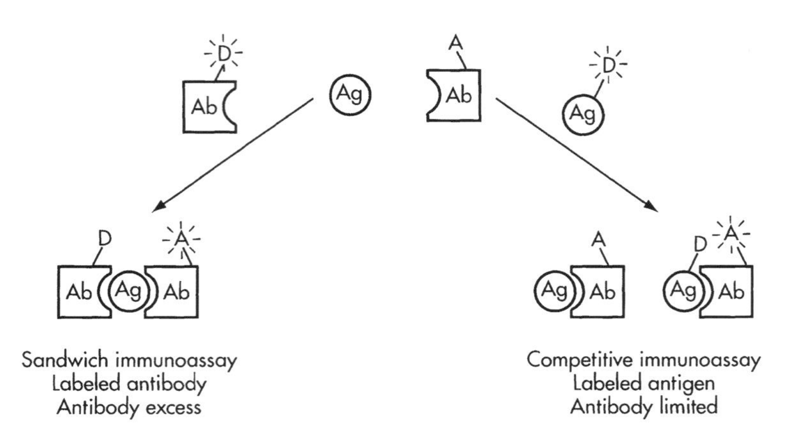

Fig. 12.5 Fluorescence energy transfer immunoassay (FETI). Fluorescence resonance energy transfer (FRET) produces a reduction in donor fluorescence and an increase in acceptor fluorescence upon binding of the two labels within an immune complex. Competitive and sandwich immunoassay protocols are illustrated.

molecules that are excited by resonance energy transfer together with emission from the donor.

Reduction in background emission and the ability to measure the appearance of induced acceptor emission provide greater sensitivity than when using steady-state fluorescence measurements. However, differentiation of immunochemically modulated emission from the acceptor from analyte-independent emission from the donor requires measurement at more than one wavelength and use of a deconvolution algorithm. This negatively impacts precision and increases instrument complexity. The best reported sensitivity in a sandwich immunoassay for human chorionic gonadotropin in serum is about 10 pM (Blomberg *et al.*, 1999) which is at least an order of magnitude more sensitive than can be obtained using steady-state fluorescence.

The method is used by Cis Bio under the trade name TRACE™ (**time-resolved amplified cryptate emission**) which employs a europium chelate donor together with a derivative of allophycocyanin, an energy-accepting fluorophore obtained from red algae (Mathis, 1993). Homogeneous binding assays for drug discovery are carried out with Cis Bio's Kryptor™ time-resolved fluorescence instrument.

FLUORESCENCE PROTECTION ASSAYS

An alternative method for modulating fluorescence in an immunoassay is to arrange the system so that those fluorescent labels that are not bound in an immune complex are quenched. This is a direct way to obviate the need to separate free from bound label and can be used to convert any heterogeneous fluorescence immunoassay into a homogeneous format. A competitive assay, for example, can then be run by first incubating the analyte, antibody, and a fluorescent labeled antigen. The antibody can be in solution or bound to a surface. A quenching reagent that reacts selectively only with unbound label is then added and the residual fluorescence associated with the bound label is measured.

The most general fluorescence protection assay strategy is to use an anti-fluorescer antibody as the quenching reagent. Some fluorophores such as fluorescein are quenched simply on binding to an antibody. When binding provides insufficient quenching a non-fluorescent energy acceptor can be attached to the antibody. Provided the fluorescent label is conjugated to a low molecular weight hapten by a short chain, simultaneous binding of antibodies to the hapten and the label is sterically hindered. Fluorescer-labeled haptens are therefore quenched unless they are protected by being bound to an anti-hapten antibody (see Figure 12.6). Fluorescer-labeled proteins are less efficiently quenched unless some way is provided to enhance the steric interactions. This can be readily accomplished, for example, by simply complexing the anti-fluorescer antibody with anti-immunoglobulin antibodies (Zuk *et al.*, 1979).

Fluorescence protection can similarly be applied to assays that use fluorescer-labeled antibodies provided the binding components are rendered sufficiently bulky. For example, binding of fluorescer-labeled antibodies to a protein antigen on agarose particles can protect the fluorescer from subsequent binding to anti-fluorescer antibodies. The more free antigen that is present the less labeled antibody becomes bound to the agarose. The unbound antibody is then selectively quenched by addition of carbon particles coated with anti-fluorescer antibody, and the fluorescence associated with antibodies bound to the agarose can then be directly measured.

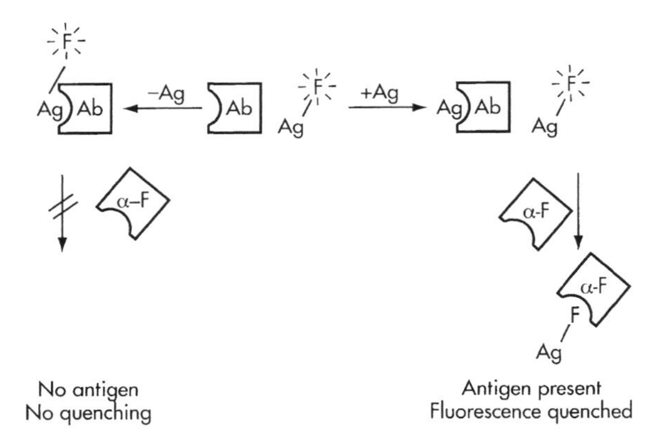

Fig. 12.6 Fluorescence protection immunoassay. Binding of antibody to the antigen in an antigen–fluorophore conjugate prevents anti-fluorescer antibody from binding and quenching the fluorescer.

The carbon particles provide steric constraints that inhibit binding to fluorescent labels on the agarose and also serve to enhance the quenching of fluorophores on unbound antibody by masking of the light beam (see Figure 12.7, Ullman, 1981a,b).

Particularly for competitive immunoassays with small analytes, fluorescence protection offers an advantage over FETI because only one label is required. It is therefore possible to multiplex several assays by using different fluorescent labels. Additionally, up to 98% of the fluorescent signal can be modulated as compared to 30–80% modulation by FETI. The sensitivity of both methods is limited by the fluorescent background from the sample except where time-resolved fluorescence is employed. As in FETI, this problem can be minimized by measuring the quenching rate rather than the end point after addition of the final reagent. This effectively subtracts out the static sample background, although with lipemic serum samples changes in light scattering due to dissolution of lipids following dilution can be a problem.

Surprisingly, fluorescence protection immunoassays have received little attention despite their advantages over FETI. The main application has been in a non-immunochemical assay for the binding of ligands to cellular receptors. A fluorescein-labeled ligand and an unlabeled test compound are allowed to compete for binding to receptors on a cell surface. Addition of anti-fluorescein antibodies quenches only the unbound conjugate so that higher fluorescence intensity relates to less effective competition by the test compound (Sklar *et al.*, 1982).

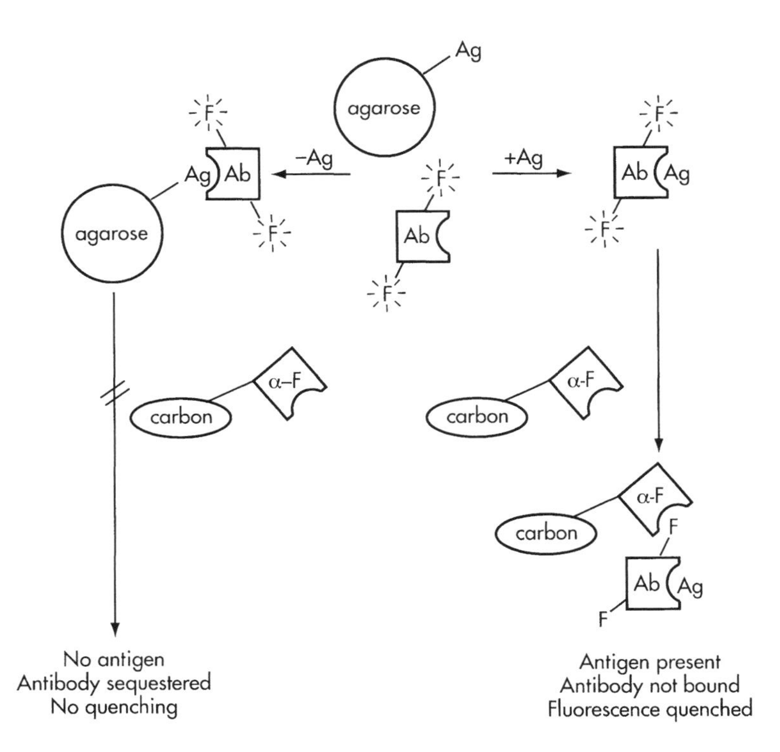

Fig. 12.7 Fluorescence protection immunoassay (as in Figure 12.6) employing carbon particles to enhance quenching.

FLUORESCENCE FLUCTUATION

A number of different approaches have been developed for homogenous assays based on detecting immunochemical binding by monitoring temporal variations in fluorescence intensity. Binding of a drug–fluorescer conjugate to latex particles coated with an antibody to a drug will of course cause the particles to become fluorescent. Provided the number and affinity of the antibodies on a particle is sufficiently high, the concentration of the conjugate within the volume defined by the particle will be higher than in the bulk solution. Binding can be detected by interrogating many small volumes of a suspension of these particles. Using a fluorescence fluctuation–correlation method, increased fluctuations are observed upon binding due to the presence of fluorescent particles in some volumes and not in others (Elings *et al.*, 1983).

The method has been applied to a homogeneous blood typing system. Cells in whole blood are stained with a membrane-soluble fluorescent dye, and aggregation of the fluorescent erythrocytes by blood type-specific antibodies is detected by fluorescence fluctuation analysis. For this purpose a probe is immersed in the solution that has two optical fibers, one that delivers a narrow beam of light to the solution and the other that collects fluorescent emission from a short segment of the light path and delivers it to a PMT. By using an interrogation volume of about 1 nL, fluctuations produced by single cells can be distinguished from dimers without waiting for larger aggregates to form (see Figure 12.8, Ghazarossian *et al.*, 1988).

A related approach is currently being used for homogeneous detection of binding of ligands to cells or latex particles for use in high throughput screening. In PE Biosystem's FMAT® system cells or particles having a receptor or antibody on their surface are mixed with a fluorescer-labeled ligand and the test compounds in microtiter wells. The bottom surface of well is scanned with a laser to permit activation of only a small volume at a time. The magnitude of the fluorescence fluctuations permits estimation of the degree of binding of the label, which is inversely related to the ability of the test compound to bind.

A recent proposal carries this concept to the ultimate level of detection of single molecule binding (Winkler *et al.*, 1999). The association of two different fluorescent molecules can be detected in the presence of the unbound molecules by monitoring the coincidence of emission from both fluorescers. The method, called **confocal fluorescence coincidence analysis**, requires interrogation of solution volumes that are so small that they will normally be occupied by a single molecule. Using a 1 fL detection volume, 10 nM of intact DNA doubly labeled with Cy-5 and Rhodamine green can be distinguished from cut strands in which the fluorescent molecules are not associated. Binding assays have not yet been described but are expected to have limited sensitivity because at low concentrations occupancy of the small volumes by the bound labels will diminish and background from chance occupancy of unbound labels will not.

ENZYME IMMUNOASSAYS

ENZYME-MULTIPLIED IMMUNOASSAY TECHNIQUE (EMIT®)

Homogeneous enzyme immunoassays were first described in 1972, and assays for drugs of abuse in urine were introduced commercially the following year under the trade name EMIT® (Rubenstein *et al.*, 1972).

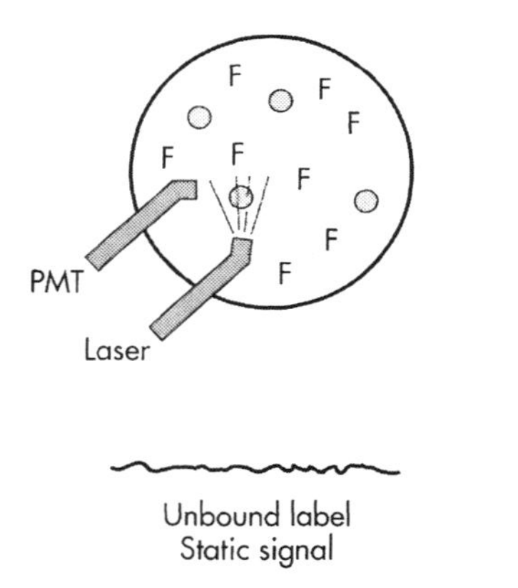

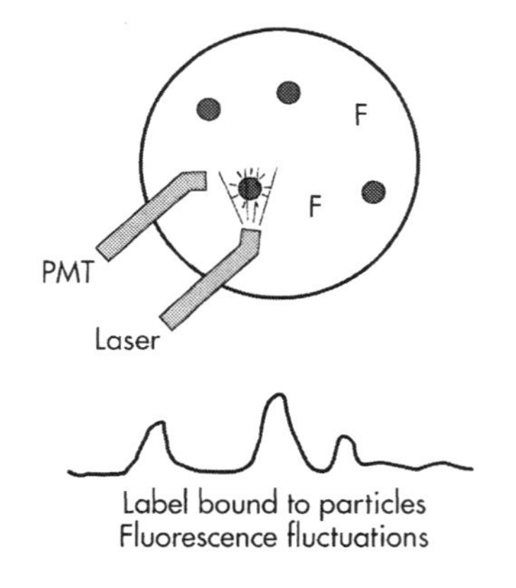

Fig. 12.8 Fluorescence fluctuation assay. Bent optical fibers immersed in a suspension of particles and capable of interrogating nanoliter volumes are used in fluorescence fluctuation analysis to distinguish between free fluorescent label from label bound to the particles.

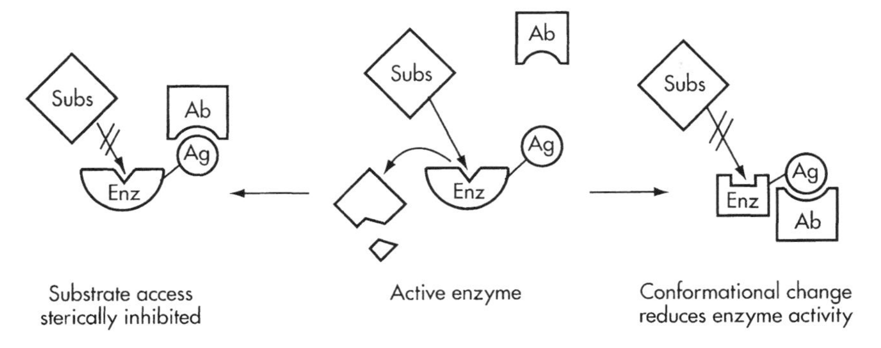

Fig. 12.9 EMIT. Antibodies can inhibit the activity of antigen-labeled enzymes either by sterically hindering approach of the enzyme substrate or by inducing a conformational change in the enzyme.

The assays used a simple mix and read protocol and had the advantage of using a general-purpose standard laboratory spectrophotometer. The method was based on the ability of anti-drug antibodies to inhibit the enzyme activity of a lysozyme–drug conjugate. The enzyme catalyzes bacterial cell wall hydrolysis, which is monitored by the clearing of a turbid bacterial suspension. In the presence of free drug, less antibody is available to bind to the conjugate, which results in an increase in the clearing rate. Lysozyme was chosen because reaction with its large substrate seemed likely to be susceptible to steric inhibition by the antibody (see Figure 12.9, left arrow). However, the light scattering assay was not very sensitive and the assays could not be used with serum because of serum-induced agglutination of the bacteria.

An ideal enzyme should not be present in the test sample, should be detectable at low concentrations by a general-purpose spectrometer, and the sample matrix should not affect its activity. Additionally, it is necessary for the enzyme to retain activity and stability following conjugation and, most importantly, the activity of the resulting conjugate must be modulated by antibody. Glucose-6-phosphate dehydrogenase (G6PDH) from the bacterium, *Leuconostoc mesenteroides*, was found to meet most of these requirements. This 109 kDa enzyme has two identical subunits and converts NAD to NADH, which is measured by its absorbance at 340 nm. Endogenous G6PDH in serum requires NADP and thus does not interfere with the measurement. Hapten conjugates of G6PDH are inhibited by anti-hapten antibodies when the hapten is bonded to lysine amino groups. Although the antibodies are unlikely to effect steric exclusion of the small substrates, molecular contact between the bound antibody and the enzyme produces an inhibitory effect by effecting a change in the enzyme conformation (see Figure 12.9, right arrow). Up to 80% inhibition by antibodies is occasionally observed, but the percent inhibitability depends strongly on the hapten structure and identity of the antibody. Increasing the number of haptens per enzyme molecule increases inhibitability but with concomitant loss in enzyme activity. Targeting of the hapten just to those lysines that effect inhibition is difficult because different constellations of sites are active with different haptens. As a result, extensive screening to identify suitable means of linking the conjugates and identifying effective antibodies is often needed.

Despite these limitations, serum EMIT G6PDH assays for a large number of therapeutic drugs and urine assays for abused drugs have been successfully developed. Several alternative enzymes utilizing macromolecular substrates have been described including amylase, phospholipase C, mitochondrial malate dehydrogenase, and dextranase, but the enzymes either are too sensitive to serum components or are difficult to measure at low concentrations.

Current EMIT assays use genetically modified G6PDH to achieve greater modulation with fewer haptens. The conjugates are more stable and provide greater assay response. Haptens are attached to a single cysteine per subunit (there are no natural cysteines) that is incorporated at a site that maximizes antibody-induced modulation. Suitable sites were identifying by mapping an enzyme epitope that is recognized by an inhibitory anti-G6PDH monoclonal antibody, which induces a conformational change of the enzyme. The amino acids within the epitope were then systematically replaced one at a time with cysteine (Ullman, 1999). The activity of conjugates prepared by attachment of drugs to the cysteine sulfhydryl group is not impaired. EMIT assays using recombinant G6PDH are currently used for assays such as digoxin that must provide accurate quantitation down to 500 pM.

The primary limitation of immunoassays based on simple modulation of the activity of enzyme conjugates is the difficulty in designing assays for larger molecules. The problem stems from the need to cause an antibody to bind close enough to the enzyme to have a strong effect on the enzyme activity. One method that has met with some success depends on the ability of antibodies to form aggregates with protein antigens. It has been found that horseradish peroxidase (HRP) is inhibited by high concentrations of hydrogen peroxide but is protected when it is polymerized as part of a large aggregate (Hoshino *et al.*, 1987). Homogeneous enzyme immunoassays for protein C, α-fetoprotein, and polymorphonuclear leukocyte elastase have been reported using this principle. In the assay, an antibody–HRP conjugate binds to the antigen to form an aggregate (precipitin) that has enhanced enzyme activity when high peroxide concentrations are used. Sufficient antibody must be used to avoid the prozone phenomenon and the sensitivity of the assays is limited by the susceptibility of the system to serum interference.

Another approach to protein assays is based on the previously discussed potential for large enzyme substrates to be sterically excluded from antibody–enzyme conjugate complexes. This was nicely demonstrated with β-galactosidase–antigen conjugates, which normally are not modulated by antibodies, even when the antigen is small. However, antibody-induced modulation of the enzyme conjugates of both small and large antigens was observed when the substrate, *o*-nitrophenyl-β-galactoside, was tethered to a large dextran polymer (Gibbons *et al.*, 1980). EMIT assays for human IgG and C-reactive protein (CRP) were demonstrated, and the latter used commercially in clinical applications. By using a dextran linked fluorogenic substrate a competitive ferritin assay was set up with a detection limit of about 4 pM in serum (Armenta *et al.*, 1985). Unfortunately, occasional low levels of antibodies to β-galactosidase in serum samples prevent application of this method for very high sensitivity serum assays.

CHARGE INDUCED ENZYME ACTIVATION

An alternative approach to immunochemical modulation of enzyme activity is to engineer a change in the microenvironment of an enzyme when it becomes bound in an immune complex. An assay for CRP in which the immune complex is highly charged is illustrative. Binding of CRP to a β-galactosidase conjugate of anti-CRP antibody does not significantly affect enzyme activity. Subsequent addition of excess unconjugated anti-CRP antibody increases congestion within the immune complex and reduces the enzyme activity when a macromolecular substrate is used. However, if the unconjugated antibody is succinylated and the macromolecular substrate carries an opposite positive charge, the enzyme activity is increased due to coulombic attraction of the substrate to the immune complex (see Figure 12.10). Enzyme activation immunoassays for both protein antigens and antibodies have been demonstrated based on this principle (Gibbons *et al.*, 1981). The prozone phenomenon is avoided simply by adding a large excess of the succinylated antibody.

Despite the apparent attractiveness of this approach the method has several pitfalls. Antigen concentrations as low as 60 fM in the assay mixtures can be detected but the use of greater than 0.1% serum affects quantitation. Hence the practical detection limit is closer to 60 pM. Additionally, while the assays do not show a prozone effect, data handling is difficult because the response curves of high sensitivity assays do not increase monotonically.

ENZYME CHANNELING

Enzyme channeling provides a method of detecting the proximity of two enzymes in an immune complex. The first enzyme catalyzes the formation of a substrate that is converted by the second enzyme into a detectable product. When both enzymes are independently dispersed in the same solution the rate of product formation is slow at first but accelerates as the concentration of the intermediate substrate builds up. This kinetic behavior changes when both enzymes are closely associated at a surface (Mosbach and Mattiasson, 1970). The local concentration of the intermediate in the vicinity of molecules of the first enzyme is determined by the rate of formation of the intermediate and its rate of diffusion away from the enzyme. A local

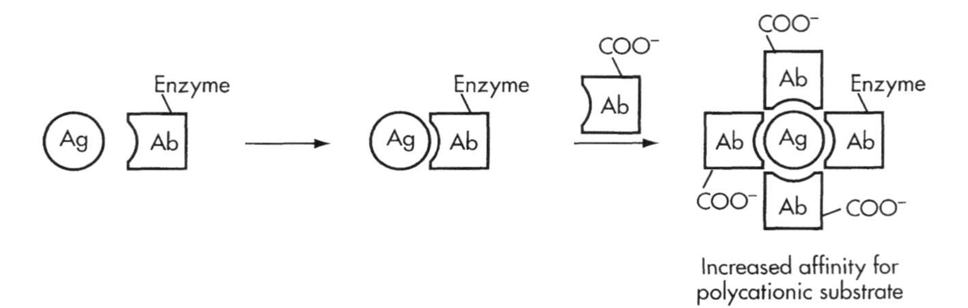

Fig. 12.10 Enzyme activation immunoassay. Binding of enzyme-labeled antibody and succinylated antibody to a multi-epitopic antigen produces increased catalytic activity because of increased charge attraction (lower K_M) between the resulting complex and a polycationic substrate.

steady state concentration is rapidly reached that is higher than the concentration in the bulk solution. Localization of several molecules of the first enzyme at a surface increases the rate of product formation and reduces the rate of product diffusion, and thus increases its local concentration. When the second enzyme becomes bound to this surface it experiences a relatively constant elevated concentration of its substrate leading to a rapid linear rate of formation of the final product.

Homogeneous enzyme channeling immunoassays take advantage of this phenomenon (Litman *et al.*, 1980). Various surfaces have been employed including agarose particles, latex beads, and the polystyrene surface of a microtiter well. One enzyme serves to label an antibody or antigen and an excess of the other enzyme is bound to the surface. Usually, the first enzyme is attached to the surface because more linear kinetics are obtained although channeling also occurs when the roles of the enzymes are reversed. A variety of enzyme pairs have been used including alkaline phosphatase/β-galactosidase, hexokinase/G6PDH, and glucose oxidase/HRP. When a natural substrate is not available, as in the case of alkaline phosphatase/β-galactosidase, synthetic constructs can be prepared that permit the sequential reaction to occur.

A competitive assay for HIgG can be carried out with agarose particles labeled with HIgG and glucose oxidase (GO) (see Figure 12.11). Upon reaction with glucose, these particles become surrounded by a halo of hydrogen peroxide. As the peroxide diffuses into the bulk solution it is diluted and the concentration is further reduced by catalase that is present in the reaction mixture. When HRP-labeled anti-HIgG antibodies bind to the particles in the presence of ABTS, an HRP substrate, there is a nearly constant rate of color formation that depends inversely on the concentration of the HIgG.

The most sensitive applications of enzyme channeling avoid the use of a pre-formed surface in favor of *in situ* formation of a colloidal precipitate. An assay for polyribose phosphate (PRP), a component of the cell wall of *Haemophilis influenzae*, was demonstrated using a reagent containing anti-PRP antibody labeled with GO (AB-GO), anti-PRP antibody labeled with HRP (Ab-HRP), and free GO (Ullman *et al.*, 1984). Combination of this reagent with a clinical sample to which anti-GO antibody had been added produced an Ab-GO:PRP:Ab-HRP sandwich complex that was incorporated into a colloidal GO:anti-GO immune complex (precipitin) (see Figure 12.12). Addition of glucose, ABTS, and catalase

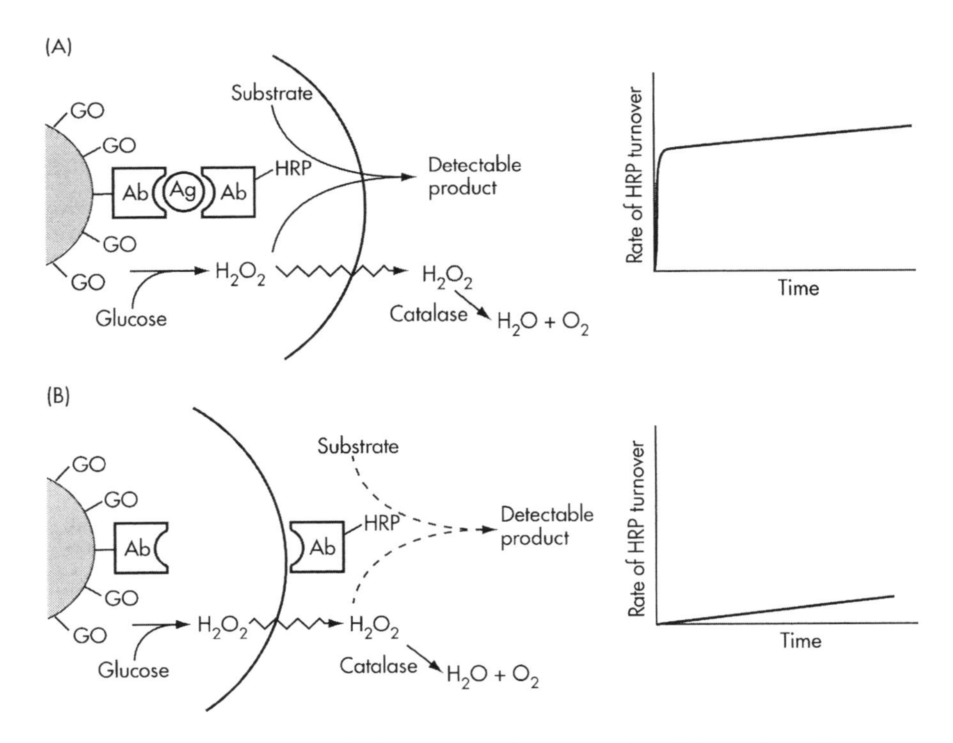

Fig. 12.11 Enzyme channeling immunoassay. Hydrogen peroxide concentrates at the surface of agarose particles that are labeled with glucose oxidase (GO): (A) horseradish peroxidase (HRP) labeled antibody bound to an antigen at the agarose surface catalyzes the oxidation of a leuco-dye; (B) low concentrations of hydrogen peroxide that escape into bulk solution permit only inefficient dye formation.

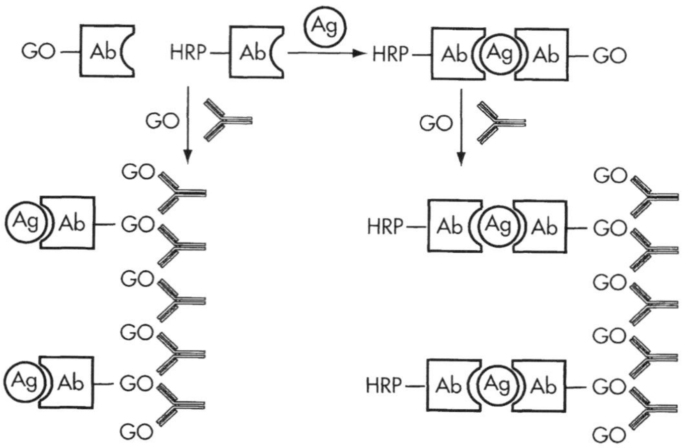

Fig. 12.12 Enzyme channeling immunoassay employing glucose oxidase and antibodies to glucose oxidase to create immune complex-containing particles *in situ*.

initiated the enzyme channeling reaction. The assay response was nearly linear with a detection limit of about 10 fM PRP in the assay mixture, sufficient for a cerebral spinal fluid assay for bacterial meningitis. Unfortunately, there has been little study to determine if similarly sensitive homogeneous enzyme channeling immunoassays can be carried out using serum samples.

ENZYME EFFECTOR IMMUNOASSAYS

Instead of using an intact enzyme, homogeneous immunoassays can be constructed with catalytically inert labels that can combine with an enzyme or enzyme fragment and effect a change in enzyme activity. Among the labels that have been studied include enzyme substrates, coenzymes and inhibitors, and enzyme fragments.

SUBSTRATE LINKED FLUORESCENCE IMMUNOASSAY (SLFIA)

Substrate labels have been used for monitoring therapeutic drug levels where there is a relatively high concentration of the drug in serum (Li *et al.*, 1981). Conjugation of β-galactosidylumbelliferone to theophylline does not affect its ability to act as a fluorogenic substrate for β-galactosidase. When anti-theophylline antibodies are present the reaction is inhibited because the substrate, when bound to the antibodies, is less sterically accessible to the enzyme. If theophylline is present it competes for antibody binding sites and there is an increase in the rate of appearance of the fluorescent umbelliferone product.

A major problem with this method is the necessarily low concentration of the substrate that must be used for the analyte to enter into effective competitive binding. At a low concentration the enzyme turn over is slow. Significant increases in rate can be achieved at very high enzyme concentrations but then the cost becomes prohibitive. Moreover, the method does not have the advantage of signal amplification that is provided by other enzyme immunoassay methods. That is, the number of molecules that provide a signal can never be greater than the number of molecules of the analyte.

One approach to overcome these limitations is to use a substrate that gives a very easily detectable signal. ATP, for example, can be used as the substrate. Reaction of the ATP–antigen conjugate with firefly luciferase produces a chemiluminescent signal that is reduced when antibody is present. The method has been applied to competitive assays for HIgG and 2,4-dinitrophenylalanine (Carrico *et al.*, 1976). While this overcomes some of the sensitivity limitations, sample to sample variations due to the susceptibility of chemiluminescent reactions to matrix effects more than offset the theoretical gain in sensitivity.

ENZYME COFACTOR IMMUNOASSAY

Cofactors or other prosthetic groups are more attractive labels for stimulation of enzyme activity. Unlike substrate labels, these groups provide the opportunity for signal

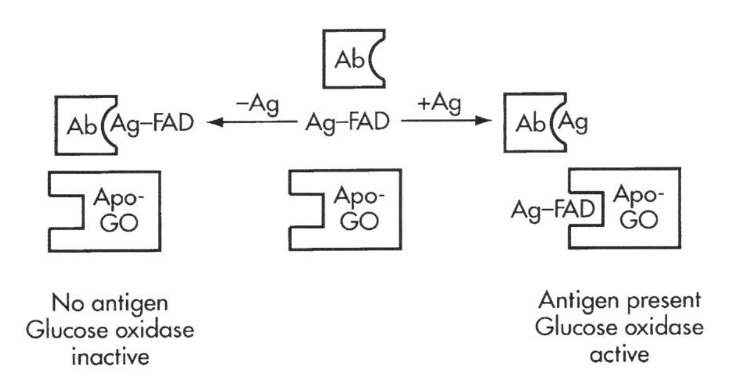

Fig. 12.13 Enzyme cofactor immunoassay. Free FAD-labeled antigen assembles with inactive apo-glucose oxidase (GO) to form the active holoenzyme. FAD-labeled antigen bound to antibody cannot react.

amplification. The most successful example is the use of FAD–analyte conjugates that can complex with inactive apo-glucose oxidase to form active glucose oxidase. When these conjugates are bound to an anti-analyte antibody, complexation is sterically inhibited and the enzyme is not activated. Competition by analyte from a sample increases the available conjugate concentration and thus increases the enzyme activity (see Figure 12.13). The method has been commercialized in the Miles Optimate system for protein analytes such as thyroid binding globulin. The reagents have also been dried onto strips for single step measurements of drugs in serum. Called ARIS™ (apo-enzyme reactivation immunoassay system), strips are impregnated with the dry conjugate, antibody, apo-enzyme, glucose and reagents for detecting hydrogen peroxide that is a product of the enzyme reaction. When contacted with a sample the intensity of color that is formed is used to determine the concentration of drugs such as phenytoin and theophylline. This is an ideal format for this type of assay because it permits the otherwise unstable apo-enzyme to be stored in a dry form.

ENZYME INHIBITOR IMMUNOASSAY

Enzyme inhibitors can be used as labels in much the same manner as coenzymes (see Figure 12.11). S-substituted ethoxymethylphosphonothioates are irreversible inhibitors of acetylcholinesterase. Conjugation of haptens such as thyroxine and theophylline to the sulfur does not affect the inhibition. However, the rate of reaction with the enzyme is reduced upon binding of the conjugate to an antibody to the hapten. Free hapten competes for antibody binding sites and increases the concentration of available inhibitor with a corresponding acceleration of inhibition and decrease in enzyme activity. The reaction is followed by the rate of color produced by reaction of 5,5′-dithiobis-(2-nitrobenzoic acid) with thiocholine formed upon enzymatic hydrolysis of acetylthiocholine (Blecka *et al.*, 1983). Methotrexate, a reversible enzyme inhibitor of dihydrofolate reductase has been used for a theophylline assay, in a similar manner (Place *et al.*, 1983).

The cholinesterase inhibition assay was sold briefly by Abbott using their Bichromatic Analyzers. The scope and limitations of the method have not been discussed. However, it is clear that assays using reversible inhibitors will be limited by the binding affinity of the inhibitor to the enzyme. Thus, only irreversible inhibitors have the potential for achieving high sensitivity assays. Further investigation of this method has undoubtedly been impeded by the difficulty of identifying an inhibitor/-enzyme pair that satisfies all of the requirements for high sensitivity serum assays.

ENZYME COMPLEMENTATION IMMUNOASSAY

Some enzymes can be disassembled into enzymatically inert fragments that become active only when they associate into the original holoenzyme. Complementation of the enzymatically inactive S-protein from ribonuclease A with its 20 amino acid *N*-terminal S-peptide leads to recovery of enzyme activity. Conjugation of thyroxine to the S-peptide can be achieved without blocking complementation with the larger S-protein fragment. However, the activity of the reassembled enzyme is reduced. Binding of anti-thyroxine antibodies to the reassembled enzyme produces a recovery in enzyme activity, which is the basis for a homogeneous immunoassay (Gonnelli *et al.*, 1981). Conversely, when anti-thyroxine antibodies are added to the S-peptide conjugate complementation is blocked and the enzyme is not formed. This process can similarly be used in a homogeneous assay for thyroxine (Farina and Gohlke, 1983). The former is simply an EMIT type assay in which the antibodies provide enhancement of activity instead of inhibition. The latter is based on a related but different phenomenon that is conceptually similar to an enzyme cofactor immunoassay. In both cases, competition by free thyroxine for the antibody leads to a change in enzyme activity.

Although ribonuclease has not proved to be attractive, possibly because there are no highly sensitive methods for the detection of ribonuclease activity, complementation immunoassays using an alternative enzyme, β-galactosidase, have been commercialized under the trade name CEDIA® (cloned enzyme donor immunoassay). The native enzyme has four identical sub-units, but an eight-subunit enzymatically active species can assemble spontaneously when two different genetically engineered inactive fragments are combined. These fragments include a smaller 'donor' peptide and a larger 'acceptor' protein. There are several donor–acceptor pairs that have this property. The members of each pair together contain the entire amino acid sequence of the native enzyme. In CEDIA, donor peptides are usually derived from the *N*-terminus and the acceptor has an amino acid deletion including some portion of the donor. The donor is modified to permit conjugation to an antigen at a specific site and genetically engineered to provide good recovery of enzyme activity (Engel and Khanna, 1992; Henderson *et al.*, 1986).

In CEDIA assays the analyte and the donor conjugate first compete for antibody binding sites. The substrate and excess acceptor are then added. Since both the acceptor and donor are monomeric and the donor is partially bound to antibody, the active enzyme is assembled relatively slowly. The higher the analyte concentration the less antibody is available to bind the conjugate and the faster the rate of assembly and final rate of product formation (see Figure 12.14).

CEDIA has been applied to a large variety of small molecules as well as to a few proteins. Like EMIT assays, a considerable amount of selection is required to identify antibodies that not only have the desired specificity and affinity but are also able to inhibit complementation. Donor conjugates of macromolecular antigens are difficult to design because bulky substituents can interfere with complementation, but the ability to genetically engineer the donor provides more flexibility than in EMIT. These assays offer somewhat higher sensitivity than EMIT because of the high turnover and good detectability of β-galactosidase. Their primary disadvantage arises from the need to reconstitute the relatively unstable enzyme fragments, which must be stored dry and the additional time required for the complementation reaction. Like EMIT

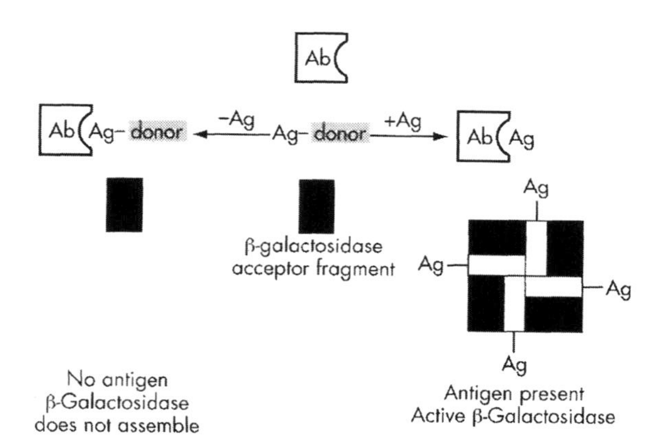

Fig. 12.14 Enzyme complementation immunoassay. Free peptide-labeled antigen (Ag-donor) assembles with a larger inactive acceptor fragment of β-galactosidase to form the active holoenzyme. Ag-donor bound to antibody cannot react.

there is no convenient way to construct practical non-competitive sandwich immunoassays.

ISOTOPIC LABELS

SCINTILLATION PROXIMITY ASSAY

One advantage of radioactive labels is that labeled ligands can be constructed that are chemically identical to the free ligand. This is of particular value in assays where chemical labeling of a ligand may interfere with binding to a natural receptor. However, this advantage must be weighed against the biohazard, stability, and disposal problems associated with radioactive isotopes, and the inconvenience of the separation and washing steps of standard radio binding assays.

Scintillation proximity assays (SPA) are radioisotopic assays that avoid separation and washing. SPA requires a radiolabel that emits alpha or weak beta particles that only traverse short distances in a condensed medium. A receptor or antibody is bound to the surface of a polymer such as a latex bead in which a fluorophore is dissolved. If the radiolabel becomes bound to the surface, radiation from the label passes through the polymer and produces pulses of light that are detected by scintillation counting. Most of the unbound labels will be more distant from the surface. The solvent will usually capture their radiation without light emission and will only infrequently encounter the polymer surface (see Figure 12.15). Instead of using fluorescent latex beads, fluorescent wells can also be used where the antibody or receptor is bound to the well surface. In either format it is only necessary to incubate the labeled antigen and sample together with the antibody coated surface and then measure the light emission (Hart and Greenwald, 1979).

Reagents for SPA have been developed by Amersham Biosciences and are used extensively, particularly for high throughput drug screening. In this application, natural receptors are normally used instead of antibodies, and competitive binding of drug candidates with a radiolabeled ligand is measured. However, the method is also applicable to sandwich assays in which an antibody or receptor is on the surface and a second antibody carries the radiolabel.

An important consideration in SPA is the choice of the label. The most light per decomposition event is produced by alpha particles, which have very high energy and short ionization tracks that have a high density of electrons. Additionally, the short pathlengths provide optimal discrimination between bound and unbound label. However, the high energies associated with alpha emitters render them particularly hazardous and not desirable for routine use as labels. Weak beta particles produce ionization tracks with lower electron density but they also have short pathlengths. Thus, weak β-decay isotopes such as ^{3}H, ^{33}P, ^{35}S, and ^{125}I, each of which emits

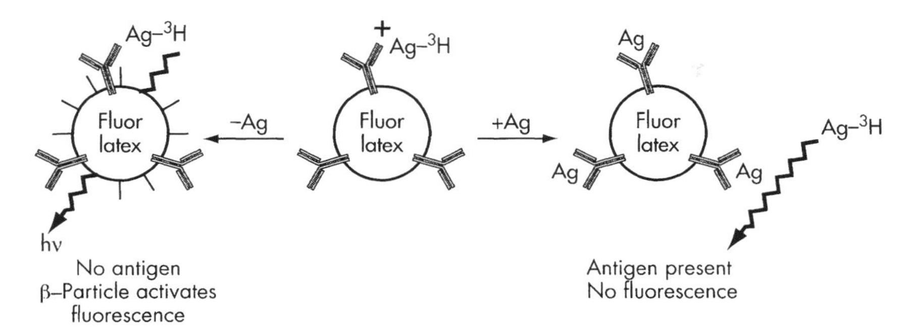

Fig. 12.15 Scintillation proximity assay. β particles emitted by tritium labeled antigen are intercepted by a scintillator loaded particle more efficiently when labeled antigen is bound to the particle. Free antigen inhibits binding.

β-particles with energies below 300 keV, are the most practical (Udenfriend *et al.*, 1987).

ELECTROACTIVE LABELS

ELECTROCHEMICAL DETECTION

Ferrocene can be readily oxidized at a voltage that does not affect most serum components (340 mV vs. SCE). Because electron transfer to an electrode requires a very close approach, the current falls off as the ferrocene is made more bulky. Homogeneous electrochemical immunoassays have been designed using this principle. Ferrocene is conjugated to a hapten such as thyroxine. Electrolytic oxidation of the conjugate is measured in the presence of an electrochemically inert agent designed to rapidly reduce the ferrocenium ions back to the neutral label. Glucose and glucose oxidase are used for this purpose. When an antibody to the thyroxine is present, the current is reduced because of decreased accessibility of the conjugate to the electrode (see Figure 12.16). Free thyroxine competes for the antibody and causes an increased current (Robinson *et al.*, 1986).

ELECTROCHEMILUMINESCENCE

Measurement of electrochemiluminescence (ECL) permits homogeneous immunoassays with much greater sensitivity. Pyrene, for example, can be oxidized at an electrode and the radical cation can then be rapidly reduced by cyclic voltammetry. At the appropriate voltage, the latter process is sufficiently exothermic that the pyrene ends up in an excited state which subsequently emits a photon. As in ampimetric detection, the process is hindered when the label is associated with a bulky complex. Thus, binding of anti-albumin antibodies reduces the emission produced by pyrene labeled albumin, and competitive binding by free albumin causes recovery of the signal (Ikariyama *et al.*, 1985).

Much improved sensitivity can be obtained by using a strongly fluorescent ruthenium II chelate in place of the pyrene (Blackburn *et al.*, 1991). As with pyrene, oxidation of the ruthenium takes place at the electrode. Reduction of the thus formed ruthenium III is accomplished without reversing the polarity of the electrodes by reaction with an amine free radical. This species forms during the oxidation of the ruthenium by electrochemical oxidation of tripropylamine which then loses a proton. The chemical process is sufficiently

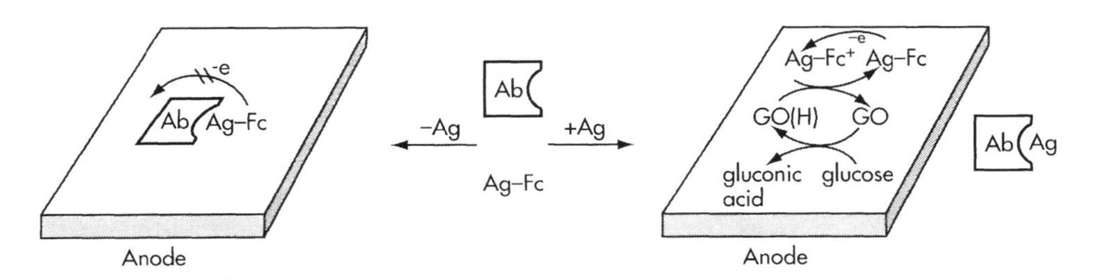

Fig. 12.16 Homogeneous electrochemical immunoassay. Ferrocene (Fc) labeled antigen produces a current by shuttling electrons between an anode and reduced glucose oxidase [GO(H)]. The ferrocene labeled antigen is inactive when bound to antibody.

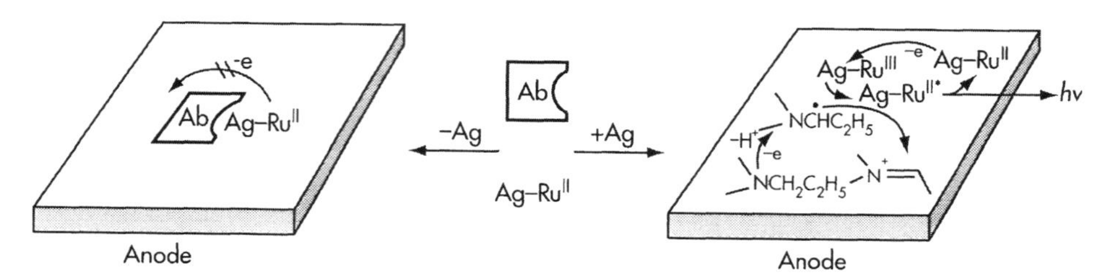

Fig. 12.17 Homogeneous electrochemiluminescence immunoassay. Ru(II) labeled antigen and tripropylamine are simultaneously oxidized at an anode. Reaction between the oxidation products, aminopropyl radical and Ru(III), populates an excited state of Ru(II) which subsequently emits light. The Ru(II) labeled antigen is less active when bound to antibody.

exothermic to leave the ruthenium in an electronically excited state which subsequently emits a photon (see Figure 12.17).

Immunoassays are carried out by causing the ruthenium label to bind to antibodies on suspendable beads. The beads further reduce the accessibility of the ruthenium to the electrode and enhance the assay response. Methods have also been worked out that cause the signal to increase when the label is bound to the beads. This is achieved by using magnetic beads that are concentrated on the electrode and washed prior to reading. This heterogeneous protocol provides higher sensitivity and extends the method to very low concentration analytes such as serum TSH. ECL assays for many clinically important analytes have been designed and are used in Roche's Elecsys® immunoassay system.

OXYGEN CHANNELING IMMUNOASSAYS

LUMINESCENT OXYGEN CHANNELING IMMUNOASSAY

In order to achieve ultra-high sensitivity in a homogenous immunoassay a label must be identified that not only is detectable at very low concentrations but also is able to produce a signal that can be modulated. With the exception of isotopic labels, which provide only intermediate sensitivity, all modulatable labels are affected to some degree by the sample matrix. High quantum yield fluorescent compounds with high extinction coefficients offer the best absolute sensitivity. Single copies of some fluorescent molecules can be detected because of the potential for multiple excitation and emission events. However, picoliter volumes must be interrogated, which requires pmol/L concentrations to assure occupancy of the interrogation volume. It is not possible to detect lower concentrations by simply increasing the interrogation volume because this increases the background fluorescence. Chemiluminescence detection is inherently much less sensitive for single molecule detection because a maximum of one photon can be produced by each molecule. However, the chemiluminescent background is immeasurable in most types of samples and chemiluminescent labels are therefore the best choice for measuring very low concentrations as distinct from very low absolute amounts. Unfortunately, the high susceptibility of chemiluminescent reactions to medium effects has thwarted most attempts to design practical homogeneous chemiluminescent immunoassays.

Luminescence oxygen channeling immunoassay (LOCI) is the only homogeneous immunoassay method developed to date that can take advantage of the exquisite sensitivity of chemiluminescence (Ullman *et al.*, 1996). This is accomplished by inducing the chemiluminescent reaction to take place within the interior of latex beads where it is completely isolated from the sample matrix. The method uses two types of ligand or receptor coated latex beads that are sufficiently small (~250 nm) that they do not settle out from a water suspension. One of the beads is a chemiluminescer bead in which is dissolved an olefin that can react rapidly with singlet oxygen ($^1\Delta_gO_2$). The reaction yields a dioxetane that spontaneously decomposes in about 1 sec with efficient emission of light. The other bead is a sensitizer bead in which is dissolved a photosensitizer such as a phthalocyanine that is capable of exciting oxygen to its singlet state upon exposure to light. Both the sensitizer and olefin are physically separated from contact with the sample and are therefore fully insulated from chemically associated non-specific medium effects.

The LOCI assay principle has similarities to enzyme channeling immunoassays (see Figure 12.18). Irradiation of a suspension of sensitizer beads results in energy transfer from excited sensitizer molecules to oxygen that is dissolved in the beads. The singlet oxygen that is formed diffuses into the aqueous solution. Because singlet oxygen has only a 4 μsec lifetime in water it exists only in the immediate vicinity of each sensitizer bead and is virtually undetectable beyond about 300 nm from the bead surface. When a chemiluminescer bead is bound to a sensitizer bead it is exposed to the singlet oxygen, which diffuses into the bead and initiates the chemiluminescent reaction. Unbound beads are unaffected. Because of rapid

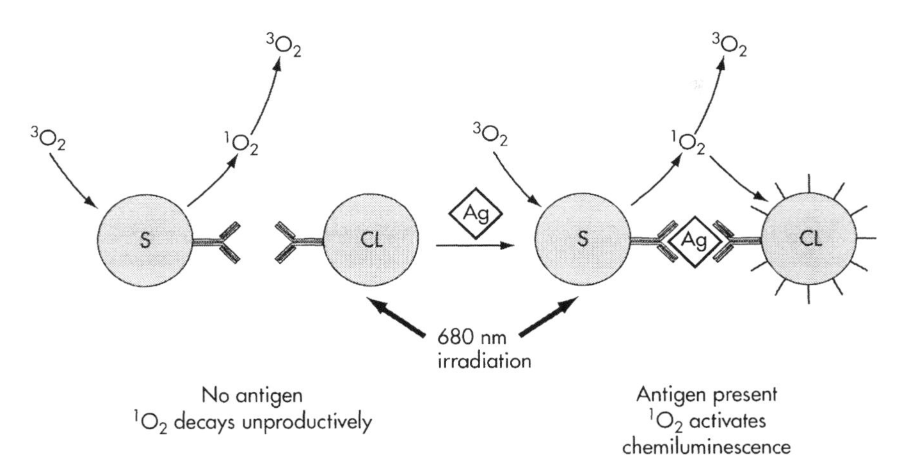

Fig. 12.18 Luminescence oxygen channeling immunoassay (LOCI). Singlet oxygen produced upon photoexcitation of a sensitizer particle (S) is intercepted by a bound chemiluminescent particle (CL) and induces a delayed chemiluminescent emission. In the absence of an antigen, binding does not occur and the singlet oxygen decays prior to encountering a chemiluminescent particle.

diffusion of oxygen in polystyrene, the latex does not impose a barrier for the transit of singlet oxygen between the beads or its reaction with the olefin.

Reagents bound to the bead surfaces are selected for the desired assay format. Competitive assays can have a ligand on one bead and an antibody on the other. Sandwich assays can have a different antibody on each bead. The assays are run by first incubating the beads and the sample to cause analyte dependent binding of sensitizer beads to chemiluminescer beads. The suspension is then irradiated using a 680 nm laser for 0.1–1 sec, and the ensuing chemiluminescent emission at 550–650 nm is integrated for a similar time period. Only a luminometer equipped with a laser and a mechanical shutter is required. Because the signal intensity is seldom limiting, the measurements are made well before binding equilibria are established. The signal is related to the number of bead pairs formed in a specific time period, which is dependent on the analyte concentration. Complications associated with bead aggregation are avoided, and rapid assay protocols can be used.

Assay sensitivity is at least equal and frequently superior to the most sensitive heterogeneous immunoassays. In a TSH sandwich immunoassay using different anti-TSH antibodies on each of the beads, concentrations as low as 0.001 mIU/L (4.1 fM or 120,000 molecules in 50 μL of serum) are detected with a total incubation period of 14 min. The absolute number of molecules detected can be reduced to under 25,000 by using a 20 μL total assay volume. Competitive assays for pmol/L concentrations of drugs such as digoxin can be run in under a minute.

Protocols for sandwich immunoassays have been developed that minimize the prozone phenomenon. The sample is first incubated with sensitizer beads coated with one antibody and an excess of biotin conjugate of another antibody. This leads to attachment of a number of biotins to each sensitizer bead that increases with increasing antigen concentration. Chemiluminescer beads coated with sufficient streptavidin to bind all of the biotinylated antibody are then added and the mixture incubated and measured. The measured bead/bead association rate increases with the number of biotins per bead but approaches a plateau when only about 10% of the maximum binding capacity has been reached. Thus, with a 10-fold antigen excess over biotinylated antibody there are 10-fold less biotins per sensitizer bead than antigen but only a slight reduction in the binding rate. Using these conditions, assay ranges in excess of five orders of magnitude in analyte concentration are obtained. Assays for drugs, proteins, antibodies, cell surface antigens, receptors, and nucleic acids have all been demonstrated using related protocols.

Interference from light scattering by the sample, a major limitation of conventional latex agglutination assays, is avoided in LOCI, nor does the sample affect the formation of singlet oxygen or its ensuing chemiluminescent reaction. Although singlet oxygen can in principle be intercepted by sample components during transit between beads its lifetime is so short that it can only be captured by compounds that are both very reactive and in high concentration. Proteins are the primary quenchers of singlet oxygen in serum (Wagner *et al.*, 1993). Accordingly, assays run with greater than 50% serum or plasma show some reduction in signal associated with the high protein concentrations, but the effect does not vary greatly between samples. High ascorbate concentrations in urine following megadose vitamin C therapy represent a more serious problem that necessitates relatively high dilution ($<1\%$ urine) or inclusion of an oxidant.

The effect of sample is thus largely confined to non-specific binding of sample components that can affect the rate of bead:bead binding. Non-specific binding to the beads increases linearly with time and can be minimized by measuring the signal during the initial rapid stage of the specific binding reaction.

LOCI assays are expected to be commercialized for diagnostic use by Dade Behring and assays for high throughput screening are being developed by Packard Instruments.

CONCLUSION

Homogeneous immunoassays offered significant advantages over heterogeneous immunoassays. Methods employing simpler protocols such as EMIT, CEDIA, FETI, latex agglutination, SPA, and FPIA have been available for many years, and LOCI now provides even greater sensitivity than heterogeneous assays. These advantages are reflected in the wide use of commercial products based on these methods, both for diagnostic and drug screening applications. Nevertheless, development of homogeneous methods for routine research applications has historically had to overcome a number of potential obstacles. These include limited sensitivity, lack of availability of suitable reagents, difficulty of assay development, and availability of instrumentation. Most non-commercial homogeneous methods are confined to fluorescence polarization. These assays are not highly sensitive and are unsuitable for large molecules, but they are relatively simple to set up and reagents and instruments are available. Until generic reagents and instrument systems are offered commercially, other homogeneous immunoassay methods are likely to remain used primarily in commercial kits for the assay of specific analytes.

REFERENCES AND FURTHER READING

Armenta, R., Tarnowski, T., Gibbons, I. and Ullman, E.F. Improved sensitivity in homogeneous enzyme immunoassays using a fluorogenic macromolecular substrate: an assay for serum ferritin. *Anal. Biochem.* **146**, 211–219 (1985).

Arquilla, E.R. and Stavitsky, A.B. *J. Clin. Invest.* **35**, 458 (1956).

Blackburn, G.F., Shah, H.P., Kenten, H., Leland, J., Kamin, R.A., Link, J., Peterman, J., Powell, M.J., Shah, A. *et al.* Electrochemiluminescence detection for development of immunoassays and DNA probe assays for clinical diagnostics. *Clin. Chem.* **37**, 1534–1539 (1991).

Blecka, L.J., Shaffer, M. and Dworschack, R. Inhibitor enzyme immunoassays for the quantification of various haptens: A review. In: *Immunoenzymatic Techniques* (eds Avrameas, S., Druet, P., Masseyeff, R. and Feldmann, G.), 207–214 (Elsevier, New York, 1983).

Blomberg, K., Hurskainen, P. and Hemmila, I. Terbium and rhodamine as labels in a homogeneous time-resolved fluorometric energy transfer assay of the β-subunit of human chorionic gonadotropin in serum. *Clin. Chem.* **45**, 855–861 (1999).

Boyden, A., Bolton, E. and Gemeroy, D. Precipitin testing with special reference to the measurement of turbidity. *J. Immunol.* **57**, 211 (1947).

Canova-Davis, E., Redemann, C.T., Vollmer, Y.P. and Kung, V.T. Use of a reversed-phase evaporation vesicle formulation for a homogeneous liposome immunoassay. *Clin. Chem.* **32**, 1687–1691 (1986).

Carrico, R.J., Yeung, K., Schroeder, H.R., Boguslaski, R.C., Buckler, R.T. and Christner, J.E. Specific protein-binding reactions monitored with ligand-ATP conjugates and firefly luciferase. *Anal. Biochem.* **76**, 95–110 (1976).

Chan, S.W., Tan, C.T. and Hsia, J.C. Clinical applications of ESR spin-labeling technique. Part 2. Spin membrane immunoassay: simplicity and specificity. *J. Immunol. Methods* **21**(1–2), 185–195 (1978).

Coons, A.H., Creech, H.J., Jones, R.N. and Berliner, E. The demonstration of pneumococcal antigen in tissues by the use of fluorescent antibody. *J. Immunol.* **45**, 159 (1942).

Dandliker, W.B. and Feigen, G. Quantification of the antigen–antibody reaction by the polarization of fluorescence. *Biochem. Biophys. Res. Commun.* **5**, 299 (1961).

Dandliker, W.B., Kelly, R.J., Dandliker, J., Farquhar, J. and Levin, J. Fluorescence polarization immunoassay. Theory and experimental method. *J. Immunochem.* **10**, 219–227 (1973).

Devlin, R., Studholme, R.M., Dandliker, W.B., Fahy, E., Blumeyer, K. and Ghosh, S.S. Homogeneous detection of nucleic acids by transient-state polarized fluorescence. *Clin. Chem.* **39**, 1939–1943, erratum 2343 (1993).

Elings, V.B., Nicoli, D.F. and Briggs, J. Fluorescence fluctuation immunoassay. *Methods Enzymol.* **92**, 458–472 (1983).

Engel, W.D. and Khanna, P.L. CEDIA *in vitro* diagnostics with a novel homogeneous immunoassay technique. Current status and future prospects. *J. Immunol. Methods* **150**, 99–102 (1992).

Farina, P. and Gohlke, J.R. Method for Carrying out Non-isotopic Immunoassays, Labeled Analytes and Kits for Use in such Assays. US Patent 4,378,428 (1983).

Förster, T. Zwischen molekulare energiewanderung und fluoreszenz. *Ann. Phys.* **2**, 55–75 (1948).

Frost, S.J., Firth, G.B. and Chakraborty, J. A novel colorimetric homogeneous liposomal immunoassay using sulforhodamine B. *J. Liposome Res.* **4**, 1159–1182 (1994).

Ghazarossian, V., Laney, M., Vorpahl, J., Pease, J., Skold, C., Watts, R., Jeong, H., Dafforn, A., Cook, R. and Ullman, E.F. A non-flow cytometric system for detecting antibodies and cellular antigens in blood. *Clin. Chem.* **34**, 1720–1725 (1988).

Gibbons, I., Skold, C., Rowley, G.L. and Ullman, E.F. Homogeneous enzyme immunoassay for proteins employing β-galactosidase. *Anal. Biochem.* **102**, 167–170 (1980).

Gibbons, I., Hanlon, T.M., Skold, C.N., Russell, M.E. and Ullman, E.F. Enzyme-enhancement immunoassay: a homogeneous assay for polyvalent ligands and antibodies. *Clin. Chem.* **27**, 1602–1608 (1981).

Gonnelli, M., Gabellieri, E., Montagnoli, G. and Felicioli, R. Complementing S-peptide as modulator in enzyme immunoassay. *Biochem. Biophys. Res. Commun.* **102**, 917–923 (1981).

Guo, X.-Q., Castellano, F.N., Li, L. and Lakowicz, J.R. Use of a long-lifetime Re(I) complex in fluorescence polarization immunoassays of high molecular weight analytes. *Anal. Chem.* **70**, 632–637 (1998).

Haga, M., Hoshino, S., Okada, H., Hazemoto, N., Kato, Y. and Suzuki, Y. An improved chemiluminescence-based liposome immunoassay involving apoenzyme. *Chem. Pharm. Bull.* **38**, 252–254 (1990).

Hart, H.E. and Greenwald, E.B. Scintillation proximity assay (SPA): a new method of immunoassay. Direct and inhibition mode detection with human albumin and rabbit antihuman albumin. *Mol. Immunol.* **16**, 265–267 (1979).

Hemmilä, I. Lanthanides as probes for time-resolved fluorometric immunoassays. *Scand. J. Clin. Lab. Invest.* **48**, 389–400 (1988).

Hemmilä, I., Malminen, O., Mikola, H. and Lövgren, T. Homogeneous time-resolved fluoroimmunoassay of thyroxin in serum. *Clin. Chem.* **34**, 2320–2322 (1988).

Henderson, D.R., Friedman, S.B., Harris, J.D., Manning, W.B. and Zoccoli, M.A. CEDIA®, A new homogeneous immunoassay system. *Clin. Chem.* **32**, 1637–1641 (1986).

Hoshino, N., Nakajima, R. and Yamazaki, I. The effect of polymerization of horseradish peroxidase on the peroxidase activity in the presence of excess hydrogen peroxide: a background for a homogeneous enzyme immunoassay. *J. Biochem. (Tokyo)* **102**, 785–791 (1987).

Ikariyama, Y., Kunoh, H. and Aizawa, M. Electrochemical luminescence-based homogeneous immunoassay. *Biochem. Biophys. Res. Commun.* **128**, 987–992 (1985).

Ius, A., Bacigalupo, M.A. and Meroni, G. A homogeneous time-resolved fluoroimmunoassay for haptens utilizing liposomes. *Anal. Biochem.* **238**, 208–211 (1996).

Kataoka, T., Inoue, K., Galanos, C. and Kinsky, S.C. Detection and specificity of lipid A antibodies using liposomes sensitized with lipid A and bacterial lipopolysaccharides. *Eur. J. Biochem.* **24**, 123–127 (1971).

Kinsky, S.C. Antibody-complement interaction with lipid model membranes. *Biochim. Biophys. Acta* **265**, 1–23 (1972).

Kraus, R. *Wien Klin. Wochenschr.* **10**, 736 (1897) as quoted by Kabat, E.A. and Mayer, M. *Experimental Immunochemistry* (Charles C. Thomas, Springfield, IL, 1961).

Leute, R., Ullman, E.F. and Goldstein, A. Spin immunoassay of opium narcotics in urine and saliva. *J. Am. Med. Assoc.* **221**, 1231–1234 (1972).

Leuvering, J.H.W., Thal, P.J.H.M., Van der Waart, M. and Schuurs, A.H.W.M. Sol particle immunoassay (SPIA). *J. Immunoassay* **1**, 77–91 (1980).

Li, T.M., Benovic, J.L., Buckler, R.T. and Burd, J.F. Homogeneous substrate-labeled fluorescent immunoassay for theophylline in serum. *Clin. Chem.* **27**, 22–26 (1981).

Litman, D.J., Hanlon, T.M. and Ullman, E.F. Enzyme channeling immunoassay: a new homogeneous enzyme immunoassay technique. *Anal. Biochem.* **106**, 223–229 (1980).

Liu, Y.P., de Keczer, S., Alexander, S., Pirio, M., Davalian, D., Kurn, N. and Ullman, E.F. Rapid luminescent oxygen channeling immunoassay (LOCI™) for homocysteine. *Clin. Chem.* **46**, 1506–1507 (2000).

Masson, P.L., Cambiaso, C.L., Collet-Cassart, D., Magnusson, C.G.M., Richards, C.B. and Sindic, C.J.M. *Methods Enzymol.* **74**, 106–139 (1981).

Mathis, G. Rare earth cryptates and homogeneous fluoroimmunoassays with human sera. *Clin. Chem.* **39**, 1953–1959 (1993).

Meyer, K. Über hämagglutininvermehrung und hämagglutinationfordernde wirkung bei menschlichen seren. *Z. Immunitaetsforsch. Exp. Ther.* **34**, 229 (1922).

Mikola, H., Takalo, H. and Hemmilä, I. Syntheses and properties of luminescent lanthanide chelate labels and labeled haptenic antigens for homogeneous immunoassays. *Bioconjug. Chem.* **6**, 235–241 (1995).

Mosbach, K. and Mattiasson, B. Matrix-bound enzymes. II. Matrix-bound two-enzyme-system. *Acta Chem. Scand.* **24**, 2093–2100 (1970).

Patel, A. and Campbell, A.K. Homogeneous immunoassay based on chemiluminescence energy transfer. *Clin. Chem.* **29**, 1604–1608 (1983).

Place, M.A., Carrico, R.J., Yeager, F.M., Albarella, J.P. and Boguslaski, R.C. A colorimetric immunoassay based on an enzyme inhibitor method. *J. Immunol. Methods* **61**, 209–216 (1983).

Robinson, G.A., Martinazzo, G. and Forrest, G.C. A homogeneous bioelectrochemical immunoassay for thyroxine. *J. Immunoassay* **7**, 1–15 (1986).

Rubenstein, K.E., Schneider, R.S. and Ullman, E.F. Homogeneous enzyme-immunoassay. A new immunochemical technique. *Biochem. Biophys. Res. Commun.* **47**, 846–851 (1972).

Sakashita, H., Tomita, A., Umeda, Y., Narukawa, H., Kishioka, H., Kitamori, T. and Sawada, T. Homogeneous immunoassay using photothermal beam deflection spectroscopy. *Anal. Chem.* **67**, 1278–1282 (1995).

Singer, J.M. and Plotz, R.M. The latex fixation test. I. Application to the serologic diagnosis of rheumatoid arthritis. *Am. J. Med.* **21**, 888–892 (1956).

Sklar, L.A., Jesaitis, A.J., Painter, R.G. and Cochrane, C.G. Ligand/receptor internalization: a spectroscopic analysis and a comparison of ligand binding, cellular response, and internalization by human neutrophils. *J. Cell. Biochem.* **20**, 193–202 (1982).

Tatsu, Y. and Yoshikawa, S. Homogeneous chemiluminescent immunoassay based on complement-mediated hemolysis of red blood cells. *Anal. Chem.* **62**, 2103–2106 (1990).

Tatsu, Y., Yamamura, S., Yamamoto, H. and Yoshikawa, S. Fluorimetry of hemolysis of red blood cells by catalytic reaction of leaked Hb: application to homogeneous fluorescence immunoassay. *Anal. Chim. Acta* **271**, 165–170 (1992).

Thompson, R.B. and Gaber, B.P. Improved fluorescence assay of liposome lysis. *Anal. Lett.* **18** (B15), 1847–1863 (1985).

Udenfriend, S., Gerber, L. and Nelson, N. Scintillation proximity assay: a sensitive and continuous isotopic method for monitoring ligand/receptor and antigen/antibody interactions. *Anal. Biochem.* **161**, 494–500 (1987).

Ullman, E.F. Recent advances in fluorescence immunoassay techniques. In: *Ligand Assay* (eds Langan, J. and Clapp), 113–136 (J. Masson, New York, NY, 1981).

Ullman, E.F. Homogeneous immunoassays. EMIT and beyond. *J. Clin. Ligand Assay* **22**, 221–227 (1999).

Ullman, E.F. and Khanna, P.L. Fluorescence excitation transfer immunoassay. *Methods Enzymol.* **74**, 28–60 (1981).

Ullman, E.F., Schwarzberg, M. and Rubenstein, K. Fluorescent excitation transfer immunoassay. *J. Biol. Chem.* **251**, 4172–4178 (1976).

Ullman, E.F., Bellet, N.F., Brinkley, J.M. and Zuk, R.F. Homogeneous fluorescence immunoassays. In: *Immunoassays: Clinical Laboratory Techniques for the 1980s* (eds Nakamura, R.M., Dito, W.R. and Tucker, E.S.), 13–43 (Alan R. Liss, New York, 1980).

Ullman, E.F., Gibbons, I., Weng, L., DiNello, R., Stiso, S.N. and Litman, D. Homogeneous immunoassays and immunometric assays. In: *Diagnostic Immunology: Technology Assessment and Quality Assurance* (eds Rippey, J.H. and Nakamura, R.M.), 31–46 (College of American Pathologists, Skokie, IL, 1984).

Ullman, E.F., Kirakossian, H., Switchenko, A.C., Ishkanian, J., Ericson, M., Wartchow, C.A., Pirio, M., Pease, J., Irvin, B.R. *et al.* Luminescent oxygen channeling assay (LOCI™): sensitive, broadly applicable homogeneous immunoassay method. *Clin. Chem.* **42**, 1518–1526 (1996).

Von Schulthess, G.K., Cohen, R.J. and Benedek, G.B. Laser light scattering spectroscopic immunoassay in the agglutination-inhibition mode for human chorionic gonadotropin (hCG) and human luteinizing hormone. *Immunochemistry* **13**, 963–966 (1976).

Wagner, J.R., Motchnik, P.A., Stocker, R., Sies, H. and Ames, B.N. The oxidation of blood plasma and low density lipoprotein components by chemically generated singlet oxygen. *J. Biol. Chem.* **268**, 18502–18506 (1993).

Wilson, K.M., Gerometta, M., Rylatt, D.B., Bundesen, P.G., McPhee, D.A., Hillyard, C.J. and Kemp, B.E. Rapid whole blood assay for HIV-1 seropositivity using an Fab-peptide conjugate. *J. Immunol. Methods* **138**, 111–119 (1991).

Winkler, T., Kettling, U., Koltermann, A. and Eigen, M. Confocal fluorescence coincidence analysis: an approach to ultra high-throughput screening. *Proc. Natl Acad. Sci. USA* **96**, 1375–1378 (1999).

Yasuda, T., Ishimori, Y. and Umeda, M. Immunoassay using fluorescent dye-trapped liposomes. Liposome immune lysis assay (LILA). In: *Nonisotopic Immunoassay* (ed Ngo, T.T.), 389–399 (Plenum, New York, NY, 1988).

Yguerabide, J. and Yguerabide, E.E. Light-scattering submicroscopic particles as highly fluorescent analogs and their use as tracer labels in clinical and biological applications, I and II. *Anal. Biochem.* **262**, 137–156, 157–176 (1998).

Yu, B.S., Choi, Y.K. and Chung, H.H. Development of immunoassay methods by use of liposomes. *Biotechnol. Appl. Biochem.* **9**, 209–216 (1987).

Zuk, R., Rowley, G.L. and Ullman, E.F. Fluorescence protection immunoassay: a new homogeneous assay technique. *Clin. Chem.* **25**, 1554–1560 (1979).

13 Calibration Curve-fitting

Barry Nix and David Wild

INTRODUCTION

Quantitative immunoassay relies on a calibration curve to determine the analyte concentration in samples from the strength of signal produced. The calibration curve is a plot of calibrator concentration against signal level. It can be drawn by hand with a pencil, flexicurve, and sheet of graph paper but the process is normally computerized to save time. The computer calculates a mathematical function for the calibration curve that allows it to calculate the concentration of analyte in each sample from the signal level without having to draw a curve, although a curve is normally produced by the curve-fit program to provide a visual graphic for the user. The mathematical function has to match the calibration curve as closely as possible across the concentration range, hence the term **curve-fitting**.

The curve-fitting routine used has the potential to introduce significant bias or imprecision into assay results. There are many different methods available, of varying complexity, and most work reasonably well with the majority of assays and individual data sets. But 100% success cannot be guaranteed, and curve-fit quality should always be checked in a number of assay runs, whenever a new system, test or curve-fitting algorithm is introduced.

DOSE–RESPONSE METAMETER

The two fundamental variables involved in curve-fitting are analyte concentration (known as **dose**) and signal level (or **response**). There are many different types of signal, such as radioactive counts in radioimmunoassay, and light intensity in luminescence immunoassays. It is sometimes useful to work with functions of these variables, for example, the logarithm of analyte concentration is sometimes plotted against the **percent bound**, which is the signal level expressed as a percentage of the total signal originally added to the tube. The **dose–response metameters** are the *functions* of dose and response that are being used.

RESPONSE–ERROR RELATIONSHIP

Precision varies according to the concentration of analyte. This can have important implications for the clinical use of the results, and the relationship between concentration and precision is therefore monitored by many quality control programs. The **response–error relationship** is the relationship between the standard deviation, coefficient of variation (CV) or variance of the signal, and the mean signal level.

HOMO- AND HETEROSCEDASTICITY

When designing a curve-fitting method, it is necessary to know if the error in the signal level is constant over the concentration range of the calibrators. If the error is constant over this range, the response variable is **homoscedastic**, whereas if it varies it is **heteroscedastic**. It is possible to transform a heteroscedastic variable into a homoscedastic variable. For example, a variable with a constant CV is heteroscedastic because the variance increases with concentration. A logarithmic transform changes it into a variable with roughly constant variance.

LINEARIZING TRANSFORMATIONS

As mentioned previously, it may be more convenient to work with functions of the concentration and signal variables. Some functions linearize the plot of signal against concentration. These are called **linearizing transformations**. These enable the use of simple methods, such as linear regression, to fit the calibration curve, but the transformations alter the error structure of the variables being transformed and, as a consequence, **weighted regression** should ideally be used. This is the term given to a regression that takes into account variation in precision across the curve. In weighted regression, data points that occur in areas of better precision are given a greater weighting in their influence on the final fitted curve function. As will be seen later, this can be counter-productive in some situations.

The Immunoassay Handbook. *Third Edition*
D. Wild (Ed.)

NORMALIZING TRANSFORMATIONS

Most curve-fitting programs require the calibrator replicates to be distributed in Gaussian form (i.e. a statistically normal distribution). This means that the measurement error is equally likely to make the signal higher than the mean as it is to make it lower, and 95% of values should fall within ±2 standard deviations of the mean. If the data have a non-Gaussian distribution, then an appropriate transformation may be sought. A useful general family of transformations is expressed in the following equation, where the transformed variable Y is defined in terms of the original variable X by

$$Y = ((X + C)^{\lambda} - 1)/\lambda \qquad (13.1)$$

where λ and C are parameters to be determined, possibly by a maximum likelihood criterion. This family of transformations is fairly general as it covers all power transforms and includes the popular logarithmic transform, this being achieved by letting λ tend to zero.

OUTLIERS

Outliers are atypical values that are distant from the main body of the data. There are many tests for detecting outliers, but as calibration curves are often established from duplicate responses they are normally detected as poor duplicates and great care should be exercised when deciding which of two values is an outlier. Outliers in calibration curves should always be a cause of concern, especially if samples are tested as singletons, as it may be difficult to detect outliers in singleton values. Ideally, all outliers should be logged and investigated.

EMPIRICAL AND THEORETICAL CURVE-FITTING METHODS

Box and Hunter (1962) categorized the various curve-fitting techniques in two ways:

1. empirical or interpolatory methods;
2. theoretical or regression methods.

The use of empirical methods assumes that the function connecting the points is continuous and smooth, but does not assume any overall functional form. A curve fitted in this way assumes that the calibrator signal levels are correct, even though their absolute accuracy is not possible in practice. These weaknesses in the interpolatory approach are highlighted when using linear interpolation techniques or simple splines because these forms of empirical curve-fitting always pass through the mean value of the signal or response metameter for each calibrator.

The advantage of theoretical methods is that they fit a given functional form or model to the data, so that errors in the calibrator points are partially corrected for, making the calibration curve more robust. They are usually based on the underlying physical–chemical mechanisms of immunoassays. Their only disadvantage is that each is based on a simplified model of immunoassay, which does not take all the possible factors into account. However, in practice this tends not to affect the quality of the curve-fitting adversely in most situations, and they are generally considered to be superior in performance. Use of a good theoretical method is essential if calibrators are run as singletons.

LEAST-SQUARES FITTING PROCEDURE

The statistical technique most commonly used to estimate the parameters in any regression method is the least-squares procedure (see Figure 13.1).

In the least-squares procedure, the vertical distance of each point from the curve is calculated and squared. The numbers are squared so that positive and negative values do not cancel each other out when they are added together. All of the squared distances are added to determine the sum of squares. The least-squares procedure selects the curve that gives the smallest sum of squares. This is fine in theory, but unexpected problems can arise. An exaggerated example is shown in Figure 13.2. Both curves pass through all of the points, and the sum of squares is zero, but clearly the straighter line (dashed) is a better fit of the calibrators. It is always sensible to inspect a good quality printout or plot

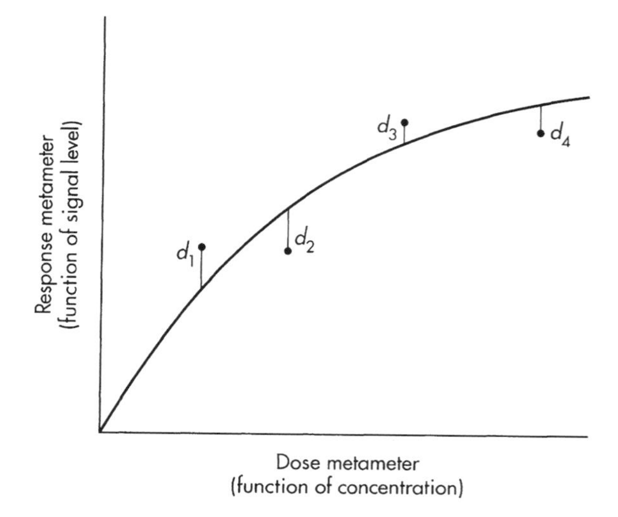

Fig. 13.1 Least-squares method. The least-squares procedure selects that curve which minimizes the residual sum of squares ($= d_1^2 + d_2^2 + d_3^2 + d_4^2$ in the above graph).

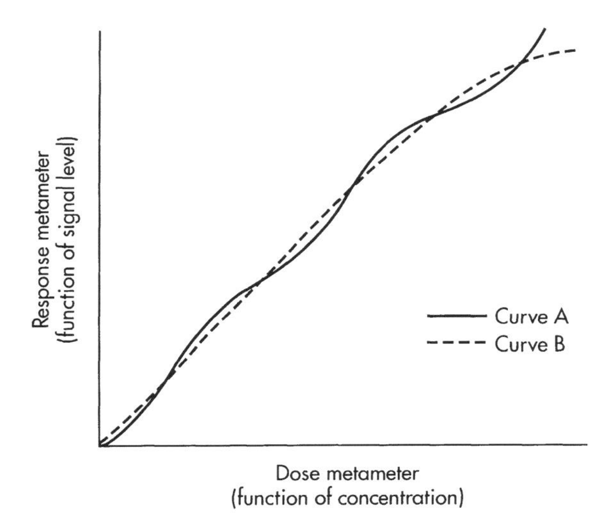

Fig. 13.2 Limitation of least-squares method. Two calibration curves with zero residual sum of squares. The least-squares procedure concludes they are equally good but clearly curve B is the preferred curve.

of the calibration curve to check for problems that may go undetected by the computer.

WEIGHTED LEAST-SQUARES PROCEDURE

Because precision varies according to concentration, it is better if a curve-fitting program weights the calibrator points so that the more precise ones influence the curve-fitting most. In most weighted regression procedures, the weight applied to each point is the reciprocal of the variance of the response metameter used. This means that the smaller the variance is, the larger the weight is, thereby forcing the curve-fitting routine to concentrate on those points on the calibration curve having greatest precision. Assuming there are no correlations between the signal levels from different calibrators, the weighted least-squares procedure involves minimizing the weighted sum of squares (WSS) defined by

$$\mathrm{WSS} = \frac{d_1^2}{\sigma_1^2} + \frac{d_2^2}{\sigma_2^2} + \frac{d_3^2}{\sigma_3^2} + \frac{d_4^2}{\sigma_4^2} = \sum_{i=1}^{4} \frac{d_i^2}{\sigma_i^2} \qquad (13.2)$$

where it is assumed that there are only four calibrator values, ${\sigma_i}^2$ is the variance of the response metameter for the ith calibrator and d_i is the vertical distance of the point from the curve. The values of σ_i^2 are often found from the response–error relationship described previously.

If the calibration curve is fitted to the means of replicate signal levels rather than to the individual points, a modified weighting function is required if the orders of replication for the calibrators differ. This is because the confidence in the mean is also dependent on the number of replicates.

PRECISION PROFILES

Precision profiles can be constructed by hand or by computer. Strictly speaking the term precision profile applies to a plot of within-assay error against concentration. However, the term is also sometimes applied to plots of between-assay error against concentration. To construct a within-assay precision profile, the calibration curve should be fitted by an appropriate statistical procedure. For each concentration X, the calibration curve gives the corresponding signal response Y.

A tangent to this curve, PT, is drawn at point P. From the point P, in the direction PX, a length PS is marked where S is chosen such that PS $= \sigma_y$, the standard deviation of the response at Y, this being obtained from the response–error relationship. A horizontal line is now constructed from S to meet PT at R. The distance SR then gives the standard deviation in the dose at the selected concentration, X (see Figure 13.3). The CV (standard deviation divided by concentration and multiplied by 100) is calculated and plotted against the corresponding concentration to give the precision profile (see Figure 13.4).

Two points are worth noting. First, it is customary to quote the precision profile for a singleton, i.e. the response error should be that for individual results and not for the means of replicates. The precision profile can be quoted for mean signal levels, but if this is done it

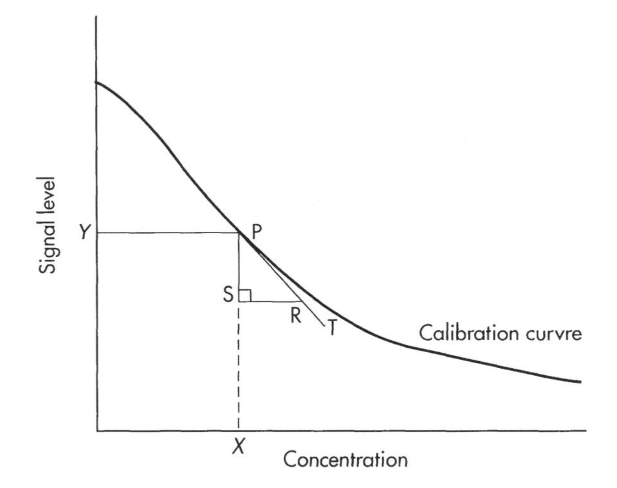

Fig. 13.3 Determination of error for the calculation of the precision profile. Precision at X = 100 × SR ÷ X%. PT is the tangent at P and PS is the standard deviation of the signal at a mean level of Y.

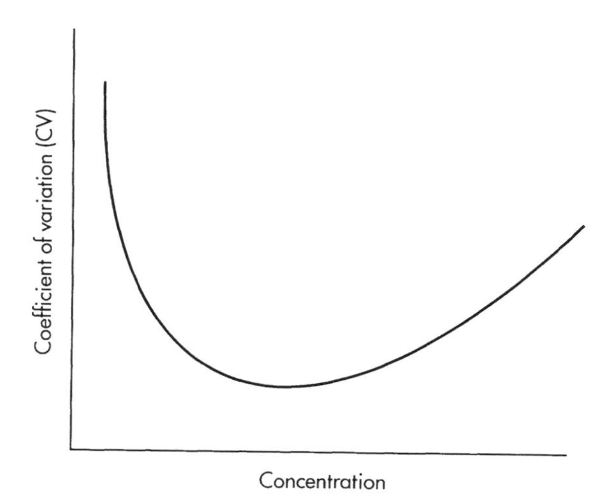

Fig. 13.4 A typical precision profile showing a decline, a plateau, and an increase with increasing analyte concentration.

should be clearly stated to avoid confusion. Second, the construction explained above assumes that the horizontal axis is the fundamental unit of concentration and not a dose metameter such as log concentration. Changing the scale of measurement clearly affects the shape of the curve and so could give the wrong estimate of the CV.

EXAMPLES OF CALIBRATION CURVE-FITTING METHODS

There is sometimes a lack of consensus for naming the different types of curve-fitting methods. If ambiguity does arise, check the mathematical basis of the method to try to identify it. The procedures described are the most commonly used.

HAND-PLOTS

A flexicurve and pencil may be used to draw carefully a curve through the data points. This can be done on linear–linear graph paper or specially prepared graph paper such as logit–log. The technique can be applied to any pair of dose and response metameters and is most easily done when the relationship is reasonably linear.

Strengths

This method is reasonably precise and unlikely to be biased. It is often used as a reference method for checking computerized curve-fitting techniques.

Limitations

The quality of the results depends on the skill and care of the user. It is difficult to smooth by eye if the error structure is heteroscedastic. It is time-consuming and impractical for routine use.

LINEAR INTERPOLATION

Any pair of dose and response metameters can be used. Linear interpolation involves drawing straight lines between the mean signal or response metameter for neighboring calibrators (see Figures 13.5 and 13.6).

If the selected dose and response metameters are such that the whole plot appears to be linear, then it may be better to fit all the points by linear regression rather than by linear interpolation.

Strengths

Linear interpolation is very simple to use and normally provides reasonably unbiased fitting around the calibrator concentrations.

Limitations

Considerable bias can occur between the calibrator concentrations when the curvature is high, causing consistent bias in controls. Point-to-point plots are significantly influenced by outliers and poor replicates, as there is no smoothing. Assay-to-assay reproducibility can therefore be poor.

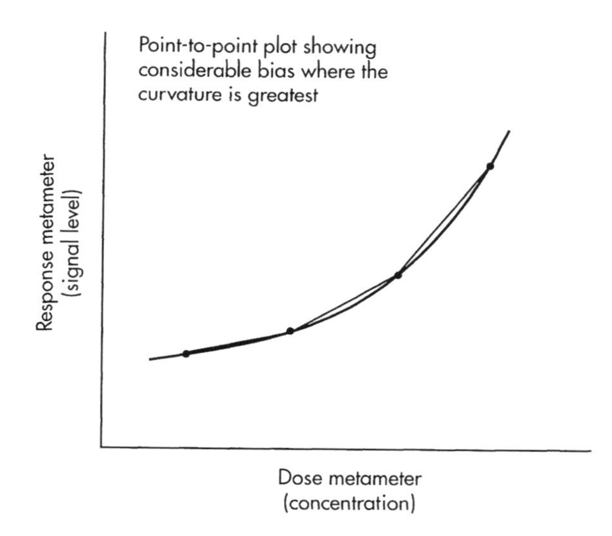

Fig. 13.5 Linear interpolation 1.

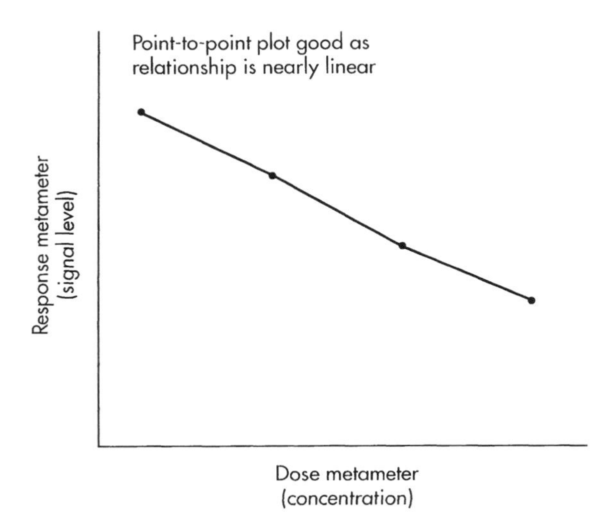

Fig. 13.6 Linear interpolation 2.

SPLINE FITS

The technique of spline-fitting is interpolatory, i.e. there is no physical–chemical structure to the model, and it can be used for any combination of dose and response metameters. In its simplest form, the cubic spline, which is often used, fits different cubic polynomials ($y = a + bx + cx^2 + dx^3$) through the mean values of the response metameter for neighboring calibrator values, imposing connectivity conditions making the curve value, slope, and curvature match at each point (see Figure 13.7).

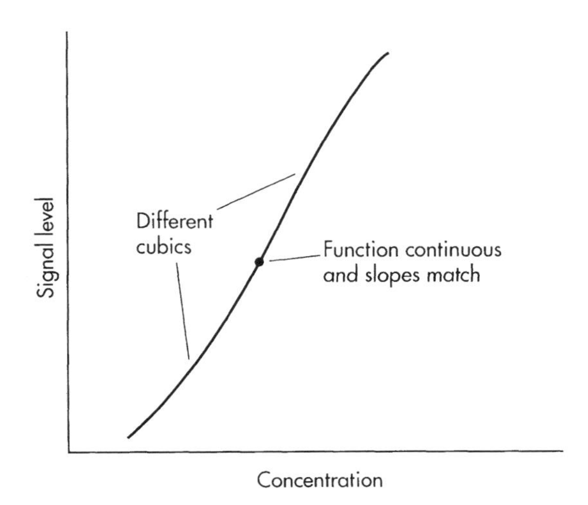

Fig. 13.7 How the simple spline function method imposes connectivity constraints at the knots (points where the component curves meet).

More sophisticated spline fits have a smoothing parameter which relaxes the condition that the curve has to pass through the mean responses of the calibrators, so that a straighter line can be produced. The most sophisticated spline programs allow for the number of curve turning points (points of inflection) to be restricted, typically to no more than one.

Strengths

Spline methods are applicable to a wide range of assays as they are the mathematical equivalent of a flexicurve. Smoothing gives some protection against outliers and poor duplicates.

Limitations

Less sophisticated spline programs fit a curve through the mean signal of each calibrator even if there are poor duplicates or outliers. In regions where the calibration curve flattens out (at asymptotes) the spline may oscillate badly, severely restricting the range of interpolation for which it can be used reliably. It is therefore particularly important that each spline fit is viewed on a high-resolution graphics screen or printout before the results are accepted.

Extrapolation beyond the calibration curve should be avoided, as spline fits do not have any built-in biochemical restrictions on the curve shape.

POLYNOMIAL REGRESSION

Any pair of dose and response metameters can be fitted by polynomial regression, although a near-linear plot often produces better fits. The method itself fits a polynomial of the form $y = a_0 + a_1x + \cdots + a_kx^k$ to the calibrator values. The order of the polynomial is k, which is normally ≤ 3 to avoid oscillations. The method of fitting is usually weighted least-squares, which takes account of the different levels of precision across the concentration range. It is important to truncate the curve at high concentrations as it may turn up in a competitive assay or down in an immunometric assay.

Strengths

Polynomial regression is a very flexible model with a wide application, as it works well with a range of dose and response metameters.

Limitations

High values of k should be avoided as higher order polynomials often cause oscillations in the calibration curve. Extrapolation beyond the calibration curve should also be avoided, as polynomial regression does not have any built-in biochemical restrictions on the curve shape.

LOGIT–LOG AND FOUR-PARAMETER LOG–LOGISTIC METHODS

Many texts describe logit–log and four-parameter log–logistic (4PL) curve-fitting methods separately, giving the impression that they are entirely different. However, the highly successful 4PL method is simply an extension of logit–log.

Logit–log is a mathematical conversion that comes close to linearizing the calibration curves of competitive and immunometric immunoassays because it mathematically models the main elements of the physicochemical situation. The logit function transforms an S-shaped curve with a single point of inflection, the characteristic shape shared by both competitive and immunometric assay curves, into a straight line. In competitive assays, the logarithm of the concentration is plotted against the logit of the signal, corrected for the binding at zero concentration (B_0) and the nonspecific binding (NSB), which is the signal generated when no antiserum is present (see Figure 13.8).

$$\text{logit}\, Y = \ln\left(\frac{Y}{1-Y}\right), \qquad (13.3)$$

$$\text{where } Y = \frac{\text{Bound} - \text{NSB (\% bound)}}{B_0 - \text{NSB (\% bound)}}$$

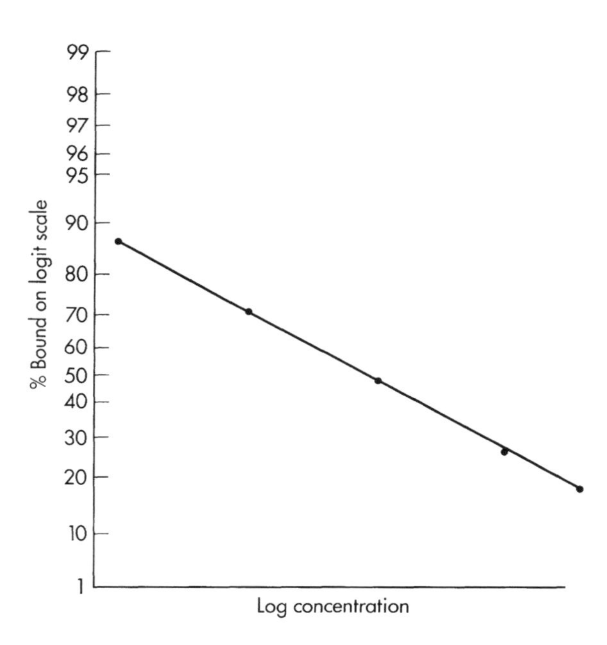

Fig. 13.8 Logit–log method.

The B_0 is obtained from the zero concentration calibrator, and the NSB is dealt with by one of the following methods:

- Determine the NSB by including in the assay extra zero calibrator tubes into which no antibody is added. This is only valid with liquid-phase antibody assays, which are now uncommon.
- Enter the NSB value provided for the assay by the manufacturer.
- Enter a value for the NSB that has been shown by experience to give good curve-fits.
- Ignore NSB (i.e. the computer assumes NSB is 0). In most automated laboratory assays, NSB is very low and this is not an unreasonable assumption.

In immunometric assays the mathematical model is the same except that the NSB is replaced by the maximum replicate signal of the highest concentration calibrator.

The converted data can be fitted by linear or polynomial regression, although polynomial regression is likely to give better results because the converted data often veer from a straight line at one or both ends of the calibration curve. There are a number of possible reasons for this, such as error in the experimentally derived values for B_0 and NSB.

When linear regression is used to fit a straight line to a plot of the log of the concentration against the logit of the signal, two parameters have to be derived to fit the linearized calibration curve: the slope of the line and the intercept.

In **four-parameter log–logistic (4PL)** curve-fitting, four parameters are derived by the computer program. Two parameters are equivalent to the intercept and slope derived for logit–log. In 4PL fits they relate to the point of inflection on the curve, and the slope at that point in a typical plot where signal is linear on the y-axis (if logit y is plotted instead, the slope is the slope of the straight line). The remaining two parameters define the upper and lower asymptotes of the curve (Figures 13.9 and 13.10). In a competitive assay these represent the B_0 and NSB true values (determined experimentally for logit–log). In a competitive assay the NSB is effectively the signal for infinite concentration so, in order to apply 4PL to immunometric assays, the NSB is replaced by a parameter that also represents the estimated signal for infinite concentration. The only difference is that this has the highest possible signal. Also the B_0 is used as a constant for immunometric assays, except that it has the lowest signal instead of the highest. The 4PL model is the same as the logit–log model, but it treats the B_0 and 'NSB' (signal for infinite concentration) as parameters that can be changed in value until the best fit is obtained. The equation for 4PL is

$$y = a + b\left(\frac{\exp(c - d\ln x)}{1 + \exp(c - d\ln x)}\right) \qquad (13.4)$$

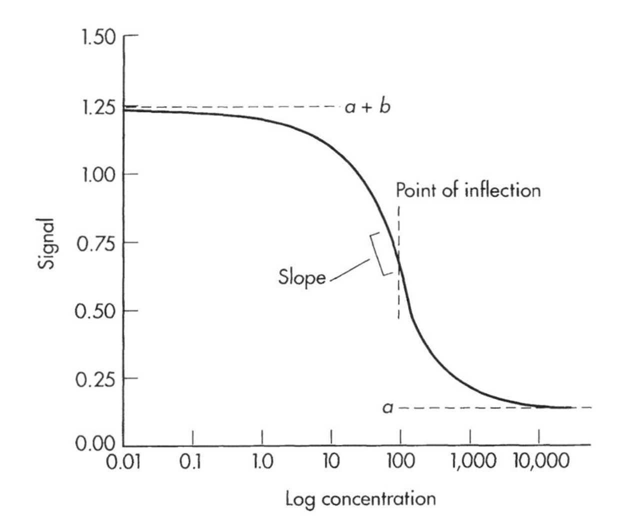

Fig. 13.9 Four-parameter log–logistic calibration curve fit (competitive assay). The point of inflection relates to parameter c. The slope relates to parameter d, but this parameter is also influenced by a and b.

where y is the signal and x is the analyte concentration. The constant a represents the NSB in a competitive assay. As mentioned above, this constant should theoretically represent the estimated signal for infinite concentration. However, in practice, to allow the best fit within the limitations of the symmetrical 4PL fitting model, this constant represents a signal level somewhere between

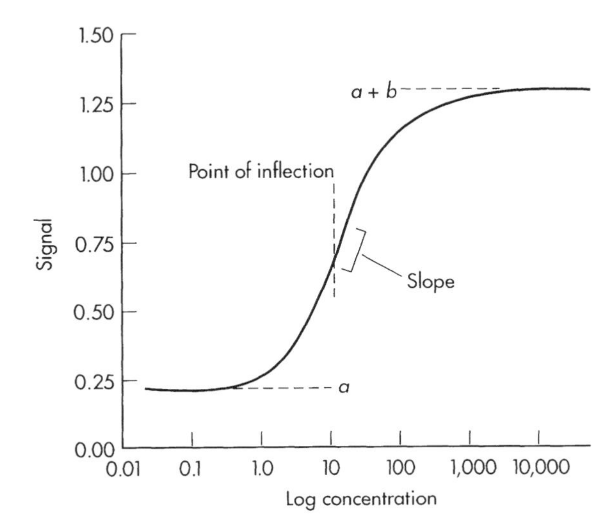

Fig. 13.10 Four-parameter log–logistic calibration curve fit (immunometric assay). The point of inflection relates to parameter c. The slope relates to parameter d, but this parameter is also influenced by a and b.

the upper concentration limit of the calibration curve and the signal at infinite concentration, assuming (in an immunometric assay) that no high dose hook effect occurs. B_0 is the upper asymptote in a competitive assay ($a + b$, *see* Figure 13.9). The two constants c and d are those defined in the logit–log model as the intercept and slope, respectively.

We have used Healy's (1972) representation of the 4PL; however, an alternative representation has been used by Rodbard and Hutt (1974).

Gerlach *et al.* (1993) show the effect on the curve shape of adjusting each of the four parameters in turn. This helps in understanding how the model can adapt to the curve within certain mathematical constraints.

The quality and flexibility of 4PL curve-fitting programs vary according to the skill, experience, and statistical knowledge of the programmer. Some commercial software ignores the zero calibrator, restricting the lower limit of the range of interpolation to the concentration range above the second nonzero calibrator. This is an unnecessary limitation. However, extrapolation beyond the positions of the highest and lowest concentration calibrators should be avoided. The 4PL method is one of the most reliable and flexible curve-fitting techniques for immunoassays so far developed, and it is the basis for the curve-fitting algorithms used in many automated systems.

Strengths

Logit–log is widely available and easy to fit by simple linear or polynomial regression. If weighted regression techniques are used, the model provides some protection against poor replicates or outliers.

Good-quality 4PL programs perform at least as well as logit–log, and do not require an experimentally determined NSB value or a calibrator with a true analyte concentration of zero.

Limitations

In logit–log the quality of fit depends, to a large extent, on the values used for B_0 and the NSB. If the logit–log transformation fails to linearize the plot, fitting a straight line can introduce considerable bias across the range of the calibration curve. Polynomial regression is much less prone to this weakness.

Depending on the weighting function used, the 4PL method can have a slight bias in regions of the calibration curve that have the greatest curvature. The fitting of a 4PL is an iterative procedure done on a computer. If the initial values assumed for the four parameters in the model are too far from the correct solution, some computer algorithms can fail to converge and do not provide a fit. This risk can be eliminated in well-designed software.

Although the 4PL method estimates the binding at zero and maximum dose in immunometric assays to give the best fit, there are additional sources of error that arise from differences between an 'ideal' immunoassay and reality. As an example, when horseradish peroxidase is used as an enzyme label, it does not respond to the presence of substrate consistently. At very low concentrations of analyte in an immunometric assay the amount of enzyme retained at the signal generation stage is correspondingly low. However, the concentration of substrate reagent added to every test is always the same. The ratio of enzyme to substrate concentration is therefore very low in this situation, and it is known that the enzyme is less efficient in the presence of very high excesses of substrate, which cause inhibition. This leads to a flattening of the dose–response curve near zero. There are other effects that cause departure from the theory on which 4PL models are based, such as the high-dose hook effect, inadequate washing, desorption (of the capture antibody, analyte, or labeled antibody) from the solid phase, and NSB of the detection system to the solid phase. Each of these causes a departure from the theoretical ideal immunometric assay curve, which should be near-linear from zero to a high dose, then gradually flatten out.

To some extent, the 4PL model can cope with departures from the ideal, but it cannot cope well with asymmetry in the curvature of the dose–response curve. All of the effects mentioned above affect either high- or low-dose regions of the curve, causing asymmetry.

Asymmetry problems can be dealt with by system manufacturers, by adding 'tweaks' tailored for a particular chemistry, or by using five-parameter log–logistic curve-fit software. However, if these are not available, it is sometimes possible to improve the fit by adjusting the weighting of the 4PL method. Although weighting is in principle an advantage, it may increase the weight applied to data in the middle of the curve, where these interferences are least apparent. This leads to even worse fits at the extremes of the curve, where departures from the theoretical ideal assay manifest themselves. Removal or adjustment of the weighting forces the 4PL to take more account of the extremes of the curve, where the key information exists that defines the departures from an ideal assay.

FIVE-PARAMETER LOG–LOGISTIC

The five-parameter log–logistic method is simply an extension of 4PL, including an additional parameter for asymmetry.

$$y = a + b\left(\frac{\exp(c - d\ln x)}{1 + \exp(c - d\ln x)}\right)^m \quad (13.5)$$

If the new parameter, m, has a value of 1, the curve is symmetrical around the ED_{50}. Alternative values allow asymmetry to get a better fit to the data, allowing for some of the effects that occur in real assays, as mentioned previously.

Strengths

The five-parameter log–logistic model is probably the most flexible for immunoassay data, retaining the constraints of theoretical models, while allowing for the departures from symmetry that often occur in practice.

Weaknesses

Use of the five-parameter model requires at least five calibrator points.

FOUR-PARAMETER LAW OF MASS ACTION

The four-parameter law of mass action method was developed specifically for competitive immunoassays. It is based on the molecular chemistry of competitive-binding assays. For example, the four-parameter model at one time available from Amersham International plc (now part of GEC Medical) took the form:

$$Y = \frac{2P(1 - B)}{K + P + A + X + [(K - P + A + X) + 4KP]^{1/2}} + B \quad (13.6)$$

where Y is the antibody-bound signal/total signal, X the analyte concentration, K the equilibrium dissociation constant, P the antibody concentration, A the labeled antigen concentration, and B the fraction of total signal due to NSB.

The four parameters to be estimated are K, P, A, and B. The fitting is again achieved by an iterative computer algorithm.

Strengths

The mathematical form of the model constrains the fit to look like a typical competitive immunoassay calibration curve and consequently has some protection against poor replication and outliers. The fitted values of the parameters can be used to reveal some of the underlying chemistry, although they cannot be directly equated because the fitting algorithm tends to change these values if a better fit results.

Limitations

If the mechanism of the assay is much different from that assumed by the model, the resulting fit can be poor and the parameters cannot be used to provide estimates of the molecular entities that they are meant to represent. This consideration has led to the development of a number of models, thereby creating further complications for the user. Some models suffer from bias effects in certain assays.

MONITORING THE QUALITY OF CURVE-FITTING

There are certain simple checks that can be made to monitor the quality of the fit produced by a curve-fitting program. Commercial packages normally provide a 'goodness-of-fit' number each time a calibration curve is fitted. This is clearly one parameter that can be monitored. It is, however, unreasonable to expect one number to adequately indicate all aspects of the quality of curve-fitting. A simple procedure, which gives more detail, is to record the back-calculated values of the calibrator signal levels. This information shows up concentration-dependent bias and the imprecision in terms of concentration, these being directly related to the particular assay that the curve-fitting method is being used for.

STORED CALIBRATION CURVES, FACTORY MASTER CURVES, AND ADJUSTERS

Traditionally, immunoassay tests were run in batches, with a full set of calibrators followed by samples and controls. But for samples to be analyzed on a continuous basis on random-access analyzers, calibration curves must be stable over a longer period of time. Most immunoassay systems on the market have calibration curves that are stable for at least 2 weeks, and across kits within one lot, although controls should still be run regularly. This reflects the level of stability and consistency now achievable with reagents and equipment. Recently, this stability has also led to a reduction in the number of calibrators required for user calibrations, through the provision of a master calibration curve from the manufacturer. A reduced number of calibrators are run by the user to adjust the master curve to take account of bias due to the user's analyzer. Hence, the *calibrators*, in this context, are really *adjusters*. It is beyond the scope of this text to comment on all the possible methods for using stored and reduced calibration curves, however, there are a few general principles.

There are two different stages to the process: establishing the initial calibration curve – sometimes referred to as the 'master curve' – at the manufacturer's laboratory, and the user adjustment in the local analyzer. This corrects for bias due to the analyzer or changes in the kit reagents since they were manufactured.

MASTER CALIBRATION CURVE

A master calibration curve is established by the manufacturer for each kit lot created. Its details are usually encoded in some way on the kits, by a conventional barcode, a two-dimensional barcode, a magnetic card or a smart card. The data may also be made available from a web site on the Internet, or communicated to analyzers via e-mail. For each kit lot around 6–10 calibrator concentrations are used with high order replication. The responses are obtained from at least two different analyzers. Generally, at least 20 'replicates' for each calibrator are used to establish the master calibration curve. The calibration curve-fitting method chosen is usually based on one of those that have already been described and so may have the strengths or weaknesses already discussed. Due to the closed nature of commercial systems, the curve-fitting process can be specially tailored to the particular assay or family of assays.

There are two major advantages of master curves, other than their obvious convenience for the user. First, the number of replicates is much greater than in conventional user calibrations. This is economically viable because the master calibration is only run once for all of the users of each kit lot. More concentration levels may be used, and many replicates run at each level, without excessive cost. The second advantage is that one set of master calibrator sets may be used to provide calibration curves to all the analyzers, possibly for many years, removing a potential source of bias (due to calibration error) when conventional *selling* calibrators are calibrated from the master set.

ADJUSTERS

No matter how many analyzers are used by the manufacturer to produce a master curve, any bias due to the user's analyzer will produce a bias if the master calibration curve is used. For this reason, some local calibration activity is necessary for each analyzer.

The user calibration consists of running two or three adjuster calibrators with known concentrations, and an algorithm to move the master calibration curve based on the signal levels for each adjuster. For example, if the master calibration curve gives a signal level of 1000 signal units at 100 concentration units, and the adjuster, which has a concentration of 100 units, gives a signal level of 950 signal units, the algorithm may lead to a shift in the master calibration curve of 5% at this point. Using multiple adjusters, the entire master calibration curve is moved to allow for bias in the user analyzer. However, this highlights the weakness of the adjustment stage. In this example, was the difference of 5% between the signal levels due to genuine bias of the user analyzer from the analyzers used for the master calibration, or was it just due to normal assay variation? The error in this type of calibration system thus derives largely from the assay imprecision when the adjuster calibrators are run at the user laboratory.

For the immunoassay system designer, the questions are: how many adjusters should be used, where should they be placed, and how should the user's analyzer process the information they produce? Clearly, the best results would be achieved by running at least four adjusters in duplicate, but this would negate the potential economic advantages of the master curve. The compromise solution is that two or three adjusters are usually run, sometimes in singleton and sometimes in duplicate.

Little has appeared in the literature about the theory of curve-fitting associated with factory calibration with user adjustment, but we can offer some guidelines in respect of the problems involved. One key principle applies throughout: *the total number of adjuster replicates must at least equal the number of parameters in the model that may change from the master curve determination.*

Linear Master Curves

It is helpful to distinguish between different types of linearity. There is the direct linear form where there are only the two parameters: slope (m) and intercept (c) that need to be determined. These parameters could well appear as the natural parameters after suitable transformations of response and dose. The other situation is where there is a pseudolinear curve shape, where the response and dose metameters are linear in the direct sense but there are other unknown parameters needed to specify the transformation. An example is the 4PL, where if it is assumed that the NSB and B_0 values are known, the logit–log transformation produces a linear plot. These two situations need to be differentiated.

Direct Linear Form

There are two cases to consider: fixed and nonfixed slope. If the master curve has a fixed slope, then there is only one parameter to determine, namely the intercept. The minimum number of adjusters is therefore 1. For a homoscedastic assay, if the slope of the line is m, the error in the response metameter for the adjuster σ_A, and the error in the response metameter for the unknowns σ_U, then the error in the interpolated dose is

$$\frac{1}{m}\sqrt{\sigma_U^2 + \sigma_A^2} \qquad (13.7)$$

σ_U and σ_A can be reduced by replication. So if the maximum error in prediction is specified, this will not only put constraints on m and σ_U, but also on σ_A, a result that could mean extra adjusters being needed in the form of replicates.

If the master curve has a nonfixed slope, then the minimum number of adjusters needed is 2 as there are the two unknown parameters: intercept (c) and slope (m). Standard statistical theory indicates that the placement of the adjusters should be such that they span the linear range, thus avoiding the increased loss of precision due to extrapolation.

Pseudolinear Form

As mentioned earlier, the use of only two adjusters could be challenged on theoretical grounds, for most immunoassays, since the linear relationship might be a consequence of a transformation from a mathematical form that had more than two parameters. For example, the logit–log transformation might well linearize a plot, but there are fundamentally four parameters to describe the response, the slope and the intercept in the logit–log domain, and the NSB and B_0. If the NSB and B_0 are 'known', then the logit–log plot is a truly linear one with the slope and intercept defining its properties. If NSB and B_0 are 'unknown', then extra information must be introduced into the process to infer their values. One familiar approach is now described. Suppose conventional calibration curves are established in a number of assay runs at the factory, then sometimes a plot of percent bound against dose produces a *profile* that is very stable across assays, even though the signal levels vary considerably between assays. This constant feature can be exploited to reduce the number of calibrators/adjusters needed. Suppose two adjusters have the fixed percentage bound B_1 and B_2, respectively, and for a particular assay run they have responses R_1 and R_2, respectively. Constant percent bound means we can write:

$$B_1 = \frac{(R_1 - \text{NSB})}{(B_0 - \text{NSB})}, \qquad B_2 = \frac{(R_2 - \text{NSB})}{(B_0 - \text{NSB})} \qquad (13.8)$$

Here B_1, B_2, R_1, and R_2 are known and so NSB and B_0 can be determined. This is all that is required, together with the master curve to run the assay. The statistical problem that has to be resolved by the manufacturer is to determine the optimum adjuster concentrations, taking account of the impact on the interpolated dose. Also, with any calibration curve with asymptotes (NSB and B_0), what procedure is adopted by the software when responses fall outside of the range of the calibration?

Integration of the Model, the Master Curve, and the Adjustment Process

Most manufacturers currently provide adjusters so that the curve shape can be adjusted empirically, by pulling it up and/or down, without taking account of the underlying model. This is validated by measuring precision on many assays, over the full reagent shelf lives. But this method is very prone to single point error and outliers. A better approach may be to record the key model parameters for the master curve, and then use the adjuster data to change the particular parameters in the model that have been shown, experimentally, to change from analyzer to analyzer, and across the reagent shelf life. The number of adjusters would then be determined by the number of parameters that can change.

Master Curve Model

We can start the process of integration with the master curve model. The increased number of calibrators and replicates run in the factory laboratory allows for more complex modeling. For example, a 4PL fit could be applied with additional constants included for variables unique to the chemistry and system, such as a change in enzyme efficiency in the presence of high substrate levels, as described previously. With as many as 200 replicates for the master calibration, there would be a more than adequate supply of data. So the master curve model could have many constants, reflecting the true variables in the assay. This would allow very precise, but constrained, modeling of the master calibration curve.

Fitting Process

The provision of many extra data points has another benefit. It is possible to systematically determine the optimum values of the parameters in the curve-fit model in a logical fashion. Conventional curve-fitting may involve the determination and fixing of parameters in a stepwise fashion to obtain best fit (least sum of squares) leaving one last parameter to be optimized. This may have to take on an extreme value to make the model fit, due to errors in the determination of the values of the other parameters. However, using the larger amount of data available in a factory calibration, it may be possible to determine each parameter independently, before finally making small adjustments to the values to obtain the best fit. Previous theoretical knowledge about the parameters that can be affected by reagent lot-to-lot variation can be used to restrain parameter changes from going outside previously established ranges.

Parameters Affected by Analyzer-to-Analyzer Variation

Analyzer variation needs to be investigated to identify which parameters in the curve-fit model are affected. For example, there could be a variation in absolute signal levels, variation in low signal sensitivity, or differences in high signal saturation characteristics. These should be incorporated in the model using as few parameters as possible. They may be assay-specific. For example, analyzer variation in the background noise level at low concentrations may only be a significant factor in a sensitive TSH assay.

Assay-Specific Parameters

During development of new kits, the model could be explored further. Is the assay competitive or immunometric? How linear is the dose–response? Are the zero reference calibrators, used for master curve generation, truly zero? Is the assay likely to involve very low levels of enzyme at the signal generation stage? The aim of this work would be to derive the basic model for the assay, perhaps chosen from a family of options for the system. Any parameters in the system model not relevant to the assay could be set to a fixed number or removed from the model.

Stability

During transport and stability studies, the model would be used to determine which parameters change in value with time. It may be that more than one parameter changes with time, but that two parameters vary according to a fixed ratio. Knowledge of this can be used to determine the number of adjusters and their concentrations.

Number of Adjusters

The number of adjusters can be determined from knowledge about analyzer–analyzer and stability effects. The minimum number depends on the number of parameters (or linked pairs of parameters) that can change.

Position of Adjuster Concentrations

Knowledge of the changes that can occur can be applied to the choice of adjuster concentration. For example, if background signal at zero concentration does not vary, the lowest adjuster does not have to be at zero.

Replication of Adjusters

This is a matter of trade-off between convenience and avoidance of adjustment bias. However, the replication is strongly influenced by the number of adjusters, the precision of the system in the user's laboratory, and the desired precision for the assay. As a rule of thumb, for assays with less than four replicates overall (e.g. four adjusters in singleton or two adjusters in duplicate) adjustment bias is a significant source of overall assay imprecision.

Method Used to Adjust Master Curve Using Adjuster Signal Levels

Using the accumulated information about the assay, obtained during development, it should be possible to use the adjuster data to modify the relevant parameters in the model, while retaining the parameter values that are not expected to change. In this way, the model is less likely to be forced into bias simply due to the error in the adjuster signal determination. However, the problem still remains that adjuster signal determination error can unduly influence the parameter(s) allowed to change and distort the curve. It is important that the system has some error checks to warn of signal changes that are outside of expected limits.

MODELING CALIBRATION CURVE CHANGES OVER SHELF LIFE

For reagents that are very stable, periodic recalibration within the shelf life may not be necessary. However, a concept that does not seem to have attracted much attention in the literature is that of modeling the time dependence of parameter values. As an example, suppose an appropriate calibration curve-fit model for a particular assay is the 4PL. Also suppose that a time series plot of the four-parameter values reveals a profile that can be accurately modeled over a period of time. If this turned out to be the case, then no adjusters would be required, since all that would be necessary is the date of use of the reagents. A periodic calibration of the analyzer may be required, but this may not need to be assay-specific. We are aware of this technique being used on one system, but it places great demands on the manufacturer to produce materials that have consistent and predictable changes during the shelf life.

USE OF ELECTRONIC DATA TRANSMISSION

As explained earlier, use of a master curve determined using a large number of replicates from a set of secondary reference standards at the manufacturer's laboratory has several advantages that are offset by the additional error due to use of very few adjuster concentrations and replicates in the user laboratory. In the future, we may see increasing use of master curve updating via the Internet, using modems in immunoassay analyzers. User analyzer calibration would comprise of a periodic determination of analyzer bias from the master analyzers at the manufacturer, which may not be assay-specific.

SUMMARY

The key issues about using stored calibration curves are stability and analyzer-to-analyzer variation. If, in a conventional assay, the underlying dose–response relationship requires k parameters to determine its form, then there should be at least k adjusters used, unless it has been shown experimentally that fewer parameters can change over time or between analyzers. It would be interesting to see if manufacturers could explain the fundamental theory of their stored calibration systems, as well as providing data to support stability and precision claims. There is no doubt that these 'black box' proprietary calibration systems should be approached with caution, with great attention paid to quality assurance and control schemes, to check that calibration integrity has been maintained.

SUITABLE CALIBRATION CURVE-FIT SOFTWARE

A well-established assay processing suite is MultiCalc™, from Perkin–Elmer Life Sciences (http://las.perkinelmer.com). This includes a number of curve-fit algorithms, quality control and interface capability for multiple workstations.

StatLIA®, by Brendan Scientific (www.brendan.com) offers a range of curve-fit algorithms for immunoassays, including a weighted five-parameter log–logistic program. It includes backfitted data for standards and good-quality graphs of the fitted curves. In addition, it accumulates information from up to 30 curves to define a reference curve for each assay. Each new curve is compared with the reference curve and any significant differences flagged. The program attempts to identify the cause of any changes in curve shape from a list including pipetting, separation, tracer, antibody, buffer, incubation, and standards. The package can be integrated with multiple workstations.

REFERENCES AND FURTHER READING

Baud, M., Mercier, M. and Chatelain, F. Transforming signals into quantitative values and mathematical treatment of data. *Scand. J. Clin. Lab. Invest.* **51** (Suppl 205), 120–130 (1991).

Box, F.E.P. and Hunter, W.G. A useful method for model-building. *Technometrics* **4**, 301–318 (1962).

Daniels, P.B. The fitting, acceptance, and processing of standard curve data in automated immunoassay systems, as exemplified by the Serono SR1 analyzer. *Clin. Chem.* **40**, 513–517 (1994).

Draper, N.R. and Smith, H. *Applied Regression Analysis* (Wiley, New York, 1966).

Dudley, R.A., Edwards, P., Ekins, R.P. *et al.* Guidelines for immunoassay data processing. *Clin. Chem.* **31**, 1264–1271 (1985).

Feldman, H. and Rodbard, D. Mathematical theory of radioimmunoassay. In: *Competitive Protein Binding Assays* (eds Daughaday, W.H. and Odell, W.D.), 158–203 (Lippincott, Philadelphia, PA, 1971).

Findlay, J.W., Smith, W.C., Lee, J.W., Nordblom, G.D., Das, I., DeSilva, B.S., Khan, M.N. and Bowsher, R.R. Validation of immunoassays for bioanalysis: a pharmaceutical industry perspective. *J. Pharm. Biomed. Anal.* **21**, 1249–1273 (2000).

Finney, D.J. *Statistical Method in Biological Assay* (Charles Griffin, London, 1978).

Gerlach, R.W., White, R.J., Deming, S.N., Palasota, J.A. and van Emon, J.M. An evaluation of five commercial immunoassay data analysis software systems. *Anal. Biochem.* **212**, 185–193 (1993).

Lynch, M.J. Extended standard curve stability on the CCD Magic Lite immunoassay system using a two-point adjustment. *J. Biolumin. Chemilumin.* **4**, 615–619 (1989).

Haven, M.C., Orsulak, P.J., Arnold, L.L. and Crowley, G. Data-reduction methods for immunoradiometric assays of thyrotropin compared. *Clin. Chem.* **33**, 1207–1210 (1987).

Healy, M.J.R. Statistical analysis of radioimmunoassay data. *Biochem. J.* **130**, 207–210 (1972).

Maciel, R.J. Standard curve-fitting in immunodiagnostics: a primer. *J. Clin. Immunoassay* **8**, 98–106 (1985).

Malan, P.G., Cox, M.G., Long, E., Wm, R. and Ekins, R.P. A multi-binding site model-based curve-fitting program for the computation of RIA data. In: *Radioimmunoassay and Related Procedures in Medicine* vol. I, 425–455 (IAEA, Vienna, 1973).

Plikaytis, B.D., Turner, S.H., Gheesling, L.L. and Carlone, G.M. Comparisons of standard curve-fitting methods to quantitate *Neisseria meningitidis* Group A polysaccharide antibody levels by enzyme-linked immunosorbent assay. *J. Clin. Microbiol.* **29**, 1439–1446 (1991).

Nisbet, J.A., Owen, J.A. and Ward, G.E. A comparison of five curve-fitting procedures in radioimmunoassay. *Ann. Clin. Biochem.* **23**, 694–698 (1986).

Nix, B. and Wild, D.G. Data processing. In: *Immunoassays – A Practical Approach* (ed Goshing, J.), (Oxford University Press, Oxford, 2000).

Peterman, J.H. Immunochemical considerations in the analysis of data from non-competitive solid-phase immunoassays. In: *Immunochemistry of Solid-Phase Immunoassay* (ed Butler, J.E.), (CRC Press, Boca Raton, FL, 1991).

Raggatt, P.R. Data manipulation. In: *Principles and Practice of Immunoassay*, 2nd edn (eds Price, C.P. and Newman, D.J.), 269–297 (Macmillan, London, 1997).

Rodbard, D. Statistical quality control and routine data processing for radioimmunoassay and immunometric assays. *Clin. Chem.* **20**, 1255–1270 (1974).

Rodbard, D. and Feldman, Y. Kinetics of two-site immunoradiometric ('sandwich') assays – I Mathematical models for simulation, optimization and curve-fitting. *Immunochemistry* **15**, 71–76 (1978).

Rodbard, D. and Hutt, D.M. Statistical analysis of radioimmunoassays and immunoradiometric (labeled antibody) assays: a generalized, weighted, iterative, least-squares method for logistic curve-fitting. In: *Radioimmunoassay and Related Procedures in Medicine* vol. I, 165–192 (IAEA, Vienna, 1974).

Rodbard, D., Munson, P.J. and De Lean, A. Improved curve-fitting, parallelism testing, characterization of sensitivity, validation and optimization for radioligand assays. In: *Radioimmunoassay and Related Procedures in Medicine, Proceedings of the Symposium, West Berlin, 1977* (IAEA, Vienna, 1978).

Rogers, R.P.C. Data analysis and quality control of assays: a practical primer. In: *Practical Immunoassay, the State of the Art* (ed Butt, W.R.), 253–308 (Marcel Dekker, New York, 1984).

Wilkins, T.A., Chadney, D.C., Bryant, J. *et al.* Non-linear least-squares curve fitting of a simple theoretical model using a mini-computer. *Ann. Clin. Biochem.* **15**, 123–135 (1978a).

Wilkins, T.A., Chadney, D.C., Bryant, J. *et al.* Non-linear least-squares curve fitting of a simple statistical model to radioimmunoassay dose–response data using a mini-computer. In: *Radioimmunoassay and Related Procedures in Medicine*, 399–423 (IAEA, Vienna, 1978b).

14 Conjugation Methods

Alastair H. Dent

As discussed in earlier chapters, immunoassay designs rely on the great specificity of antibodies, whose binding reactions can be related quantitatively to the concentration of an analyte of interest. Unfortunately, antibody molecules do not possess any intrinsic properties which allow them to be measured directly at low concentrations, and they must therefore be used in combination with a sensitive signal-generating technique such as radioactivity, fluorescence or luminescence (*see* SIGNAL GENERATION AND DETECTION).

This requirement is typically achieved through the use of **conjugates**, in which two or more different molecules of interest are coupled to yield a single reagent, and which therefore combine some of the properties of the individual constituents. Thus, many immunoassay conjugates consist of an antibody derivatized with a suitable **label**, which facilitates the generation of the signal. In some formats it is the antigen rather than the antibody which must be labeled, and these form a second category of immunoassay conjugates.

Thirdly, immunoassays also frequently require the immobilization of one of the binding molecules. For proteins this operation is sometimes carried out directly via chemisorption or physisorption to the surface of interest (see Butler, 1992, and Chapter 8 of Aslam and Dent, 1998), but it is often preferable to employ indirect techniques based, for example, on the biotin–avidin or biotin–streptavidin interaction. Frequently, these too require conjugate reagents to be prepared – a biotinylated antibody, for example.

CATEGORIES OF CONJUGATES EMPLOYED IN IMMUNOASSAY

The components that make up the conjugates for these three applications vary with the exact application, but in broad terms they can be categorized very simply (Figure 14.1). Sometimes (Figure 14.1a) both components of the conjugate are protein molecules, for example, where an antibody or antigen is labeled with an enzyme. Because proteins contain only a limited range of functional groups to act as coupling sites, there is a great deal in common between the methods used in the preparation of these **protein–protein conjugates**. Similarly, proteins are sensitive to many of the conditions typically employed in chemical synthesis; methods requiring nonaqueous solvent systems or extremes of temperature, pressure or pH are of little use in this field, so only a very small subset of the techniques of organic synthesis are applicable here.

Sometimes an antibody may instead be labeled with a small molecule such as biotin or a fluorophore, while in some immunoassay designs a low molecular weight antigen may be labeled with an enzyme. These are examples of **protein–small molecule conjugates** (Figure 14.1b). A great deal more variety may be encountered in the functional chemistry of these small molecules, often referred to as **haptens** – an immunological term implying that they are too small to elicit an immunogenic response in their own right. This property itself gives rise to another application of conjugation methods in the development of immunoassays – the need for protein–hapten conjugates as **immunogens** in the production of hapten-specific antibodies. Whether the product is a labeled conjugate or an immunogen, however, the small molecule must still be coupled to a protein target. The available reaction routes, therefore, are still constrained by most of the factors that impact protein–protein conjugation.

In contrast, where two small molecules are to be combined (for example, in the preparation of a biotinylated steroid), there are really very few constraints on the chemistry which can be employed. A wide range of reactive functionalities can be introduced if so desired, and the much greater stability of these small molecules to temperature, pH, etc. allows a broader choice of reaction conditions. Therefore, the preparation of these **small molecule–small molecule conjugates** (Figure 14.1c) really constitutes a branch of organic synthesis, and will not be discussed further here.

Finally, there are many other components which may be encountered in specific immunoassay applications. These include large, soluble carbohydrates such as dextrans, synthetic polymers with useful properties such as the polyethylene glycols (PEGs), the nucleic acids DNA

The Immunoassay Handbook. *Third Edition*
D. Wild (Ed.)

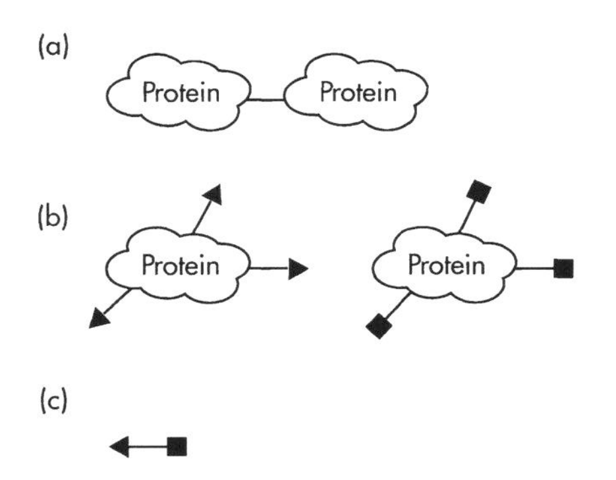

Fig. 14.1 Major categories of conjugates employed in immunoassays: (a) protein–protein; (b) protein–small molecule; (c) small molecule–small molecule.

and RNA, and macroscopic structures such as blood cells and their synthetic analogs, liposomes. These will not be discussed further here; neither will the coupling of antibodies or other soluble species to solid-phase components such as latex particles, although these reagents too are sometimes referred to as conjugates. *See* SEPARATION SYSTEMS for more information or see Aslam and Dent (1998), Chapters 7 and 8, for a comprehensive review of these miscellaneous applications of protein coupling methods.

The vast majority of immunoassay conjugations are protein–protein or protein–small molecule reactions, and the remainder of this chapter will concentrate on these two categories.

PROTEIN–PROTEIN COUPLING

FUNCTIONAL CHEMISTRY OF PROTEINS

In nature, proteins carry out an extraordinary range of functions, many of exquisite subtlety and specificity. It is a testament to the power of the evolutionary process that these properties have been achieved using a very small number of simple starting materials: the 20 common amino acids, some simple carbohydrates and metal ions, and the occasional organic moiety. As a consequence, it is relatively simple to list the useful reactive groups that are regularly encountered in protein molecules, and therefore provide the chemical basis for conjugation.

Amines

Each peptide chain within a protein molecule will provide a terminal amine group, although *in vivo* derivatization reactions can sometimes render this unreactive. More importantly though, lysine is one of the commoner amino acids, and each lysine side-chain provides an amine group. The presence of these in large numbers makes them the commonest target for conjugation reactions. They are reactive towards a wide range of functionalities, including activated carboxyls (e.g. succinimide esters, acid chlorides, and acid anhydrides), aldehydes, and isothiocyanates. These reactions require at least a fraction of the amine group to be present in the unprotonated (NH_2) rather than the protonated (NH_3^+) form, and are therefore typically carried out at a pH of 7 or above. Although the pK_a value for these groups (the pH value above which the unprotonated form predominates) may be as high as 9.5, there are enough unprotonated amines present two or even three pH units lower to allow reaction to proceed. Some useful reactions of protein amines are shown in Figure 14.2.

Thiols

The amino acid cysteine provides a side-chain thiol moiety, which has very useful reactivity. However, many of these are incorporated into intramolecular disulfide (cystine) links, and are therefore unavailable for derivatization. Where free thiols are available, they provide reactivity towards maleimide, disulfide, and haloacetyl groups; again high pH promotes the more reactive thiolate (S^-) form (pK_a typically 8.5 in proteins, though variable), and these conjugations are also generally carried out at pH 7 or above. Some useful reactions of protein thiols are shown in Figure 14.3.

Carbohydrates

Many proteins contain carbohydrate moieties, typically linked to the peptide chain via asparagine, serine or

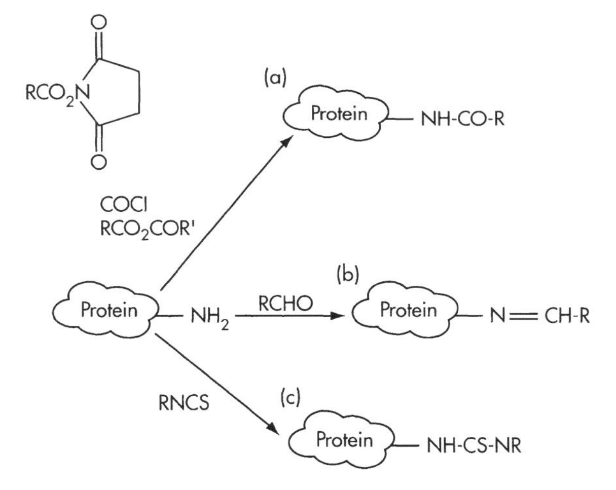

Fig. 14.2 Some useful reactions of protein amines: (a) with an activated carboxyl derivative to form an amide; (b) with an aldehyde to form an imine, and (c) with an isothiocyanate to form a thiourea.

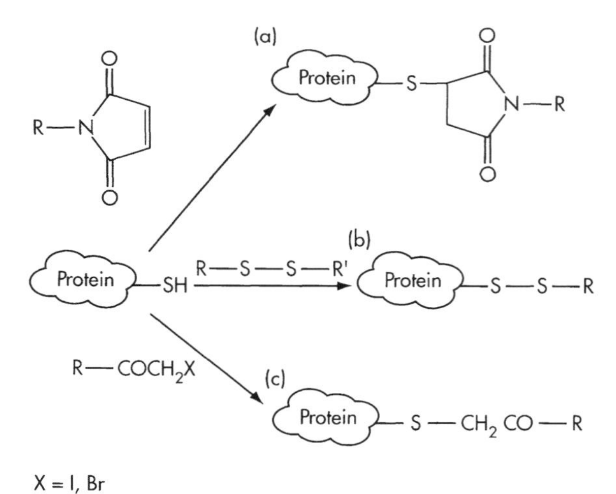

Fig. 14.3 Some useful reactions of protein thiols: (a) with a maleimide to form a succinimidyl thioether; (b) with a disulfide to form a second disulfide, and (c) with an acetyl halide to form a thioether.

threonine residues. The hydroxyl groups of these glycoproteins are not particularly useful *per se*, but where two such groups are present on adjacent carbon atoms, a useful cleavage reaction employing sodium periodate generates two amine-reactive aldehyde functions.

Others

Perhaps surprisingly, there are very few other groups in proteins that are exploited in common conjugation reactions. There are generally many **carboxylic acid** functions present from aspartic and glutamic acid groups and from the C-terminals of the peptide chains; however, these are unreactive without activation, which tends to be inefficient in the aqueous protein environment. Their use is therefore more or less limited to rare enzyme-catalyzed approaches. Where the acid group is amidated, as in asparagine and glutamine residues, the resulting **carboxamide** is even less reactive. **Tyrosine** and **histidine** contain ring structures that are susceptible to attack by electrophilic reagents, and this property can form the basis of conjugation via diazonium compounds; these routes were classically popular but are now mainly of historical interest. The major remaining use of these residues in labeling is in radioiodination, where many of the commonest techniques employ electrophilic iodine reagents. The hydroxyl functions of **serine** and **threonine** can occasionally possess anomalously high reactivity, and there are some very specific reactions of **arginine**, **methionine**, and **tryptophan**. In practical terms, however, the chemistry of proteins with respect to conjugation can be characterized adequately in terms of amines, thiols, and carbohydrates.

Table 14.1 summarizes the amino acid and carbohydrate content of a number of proteins commonly encountered in immunoassay work, namely labeling enzymes and antibodies.

Table 14.1. Amino acid and carbohydrate content of labeling enzymes and antibodies. The figures quoted for antibodies are necessarily broad estimates. There is a fairly high degree of homology between the immunoglobulin structures of different species. Therefore, the human-derived figures quoted here can give some indication of levels in other species.

Protein	Source	Molecular weight (kDa)	Number of lysines	Number of free thiols	Number of free carboxylic acids	Carbohydrate content (%)
Peroxidase (HRP)	Horseradish	44	6 (typically only around 2 accessible)	–	28	20
Alkaline phosphatase	*E. coli*	94	56	–	98	–
Alkaline phosphatase	Bovine	125	42	–	106	10
β-Galactosidase	*E. coli*	465	80	64 (typically only around 16 accessible)	508	–
IgG	Human	Around 150	Around 90	Variable	Around 120	Around 2
IgM	Human	Around 970	Around 350	Variable	Around 620	Around 12

CATEGORIES OF PROTEIN–PROTEIN COUPLING REACTION

There are a variety of general approaches to protein–protein coupling, and it is useful to consider these briefly before describing actual chemistries. The two protein molecules of interest can either be linked without the permanent introduction of any extra atoms (**direct coupling**), or via a bifunctional coupling agent. The two functional groups of the latter can be chosen to have the same reactivity (**homobifunctional coupling**), or to react with separate targets (**heterobifunctional coupling**).

These different approaches can be illustrated in the conjugation of two typical proteins (Figure 14.4). An amine group from one protein can react with a carboxyl group from the other (Figure 14.4a); the latter requires activation to the succinimide ester shown, but the final product contains only atoms present in the component proteins. In a homobifunctional approach (Figure 14.4b), a reagent combining two such succinimide esters can be used to couple the proteins via intrinsic amine functions. A heterobifunctional reagent may combine a succinimide ester with a functional group that reacts with thiols, such as a maleimide; this can be employed to create an amine–thiol link between native groups of the two proteins (Figure 14.4c). A fourth approach is to use two complementary heterobifunctional reagents, for example, one which introduces a thiol, and one which introduces a thiol-reactive maleimide (Figure 14.4d).

For protein–protein conjugates, direct coupling has many disadvantages. First, the approach is limited by the few classes of functional groups available in proteins, which possess mutual reactivity. Second, there are generally many of each functional group present on each molecule – therefore the reaction is unlikely to stop at the tidy 1:1 conjugate shown in Figure 14.4a, and may instead continue to form large cross-linked complexes. Finally, there is very little space between the protein molecules in the product, and steric hindrance may therefore interfere with their behavior.

Homobifunctional coupling gets round two of these problems: virtually all proteins contain enough amine groups to allow a bis-succinimide ester or similar amine-reactive homobifunctional linker to be employed, and the presence of the linker group minimizes steric hindrance issues. Cross-linking remains a problem, however, so that large and poorly defined conjugates will often be obtained.

Heterobifunctional reagents offer the most elegant approaches for protein–protein coupling. Free thiols are generally present only in proteins in small numbers, and the approach shown in Figure 14.4c can provide well-defined conjugates in such cases. In the ideal case, the degree of derivatization of Protein 1 is controlled such that there is only one maleimide group per molecule, and there is only one free thiol in Protein 2. Under these circumstances the only possible product of the reaction is a 1:1 conjugate.

In the majority of cases there is not a convenient single free thiol in one of the proteins of interest. However, this situation can be approximated to by the approach shown in Figure 14.4d, where thiol and maleimide functions are both introduced artificially via the protein's amine groups. Again if the incorporation of each group is controlled to one per molecule, a 1:1 conjugate is the only possible product. In practice, incorporation cannot be controlled quite so closely and there are bound to be some multiple-labeled molecules – and a higher mean incorporation may in any case be desirable from the point of view of immunoassay performance. However, these approaches can reproducibly yield well-defined conjugates: as an illustration, the size exclusion chromatograms obtained from a direct and a dual heterobifunctional approach to the preparation of an IgG-horseradish peroxidase conjugate are compared in Figure 14.5.

Because of this greater degree of control, heterobifunctional approaches have achieved great popularity in recent years – especially those based on mild chemistry such as the use of succinimide esters.

COMMON PROTEIN–PROTEIN COUPLING METHODS

Some of the commoner conjugation methods used in the preparation of protein–protein conjugates will now be outlined briefly, including examples from each of the categories above.

Periodate Method

The only direct coupling method which continues to enjoy widespread application is the periodate method (Nakane and Kawaoi, 1974), applicable if at least one of the proteins to be conjugated is a glycoprotein (Figure 14.6). As mentioned earlier, sodium periodate can be used to oxidize carbohydrate residues (Figure 14.6a), generating aldehyde groups at positions where there is a vicinal diol (hydroxyl residues on two adjacent carbon atoms). These aldehyde groups (Figure 14.6b) can be coupled to the amine groups of a second protein, generating an **imine** or **Schiff's base** (Figure 14.6c). These groups are considered somewhat unstable, and it is common practice to employ a reducing agent to generate a **secondary amine** link instead (Figure 14.6d). Sodium borohydride remains the most commonly employed reducing agent, but sodium cyanoborohydride exhibits greater selectivity.

Periodate is a powerful oxidizing agent and can cause some damage to protein structure. Therefore, low concentrations and short reaction times are typically used. Nevertheless, multiple aldehyde groups are bound to be generated from the glycoprotein, and as there are normally multiple amines on the protein to be conjugated, extensive cross-linking is often observed (see Figure 14.6).

In the immunoassay field, there have been two particularly significant applications of the periodate reaction. The first is in the preparation of conjugates of horseradish peroxidase (HRP), one of the commonest labeling enzymes, which is extensively glycosylated. Under mild oxidation conditions there is good retention of enzyme activity, and this has provided a convenient means for the preparation of HRP–IgG conjugates, etc.

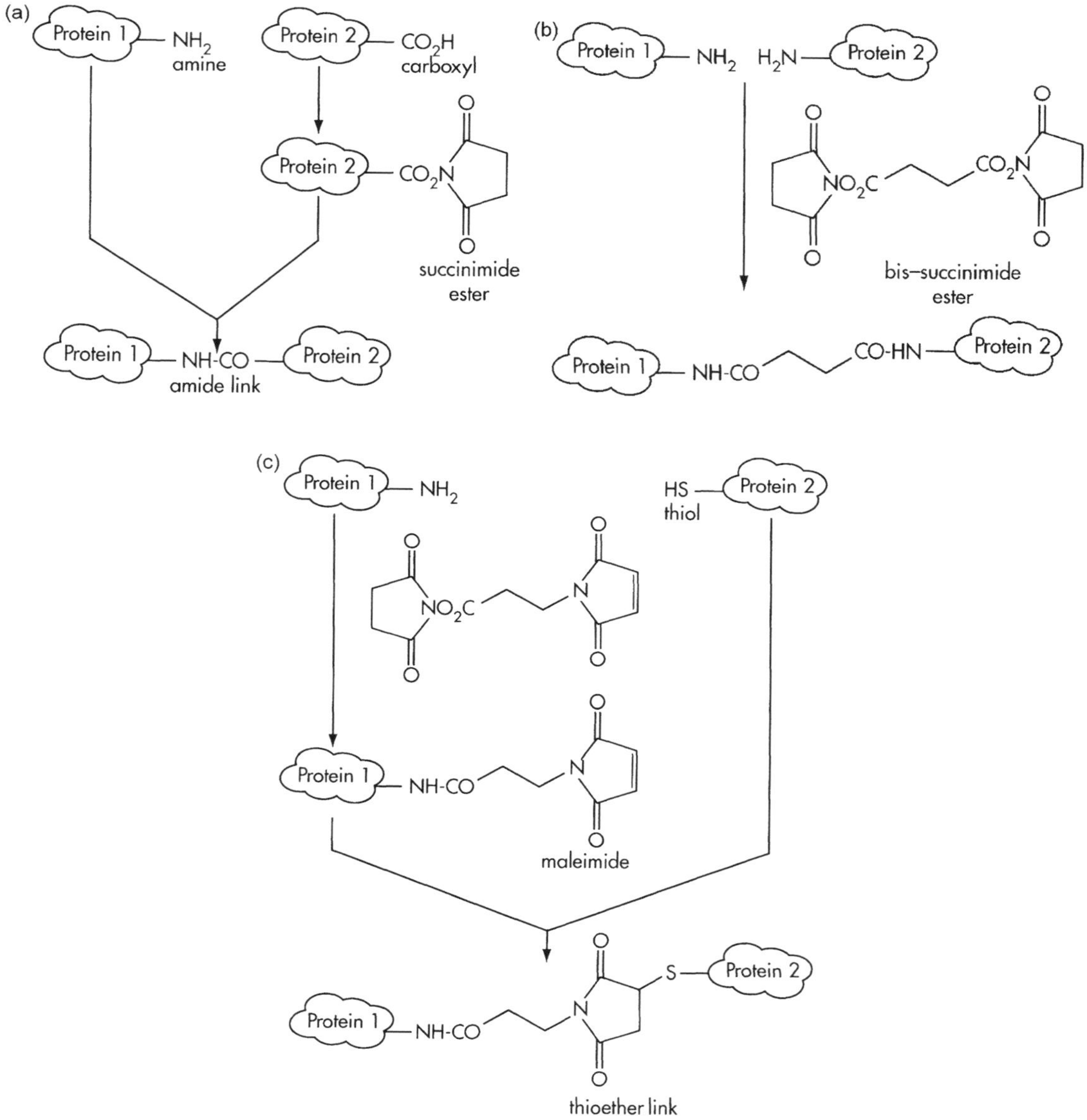

Fig. 14.4 Major categories of conjugation method: (a) direct coupling between the amine group of one protein and the carboxyl group of another, activated as a succinimide ester; (b) homobifunctional coupling of the amine groups of two proteins using a bis-succinimide ester; (c) use of an amine-reactive heterobifunctional reagent to introduce a maleimide group into one protein, allowing coupling to a thiol group in another; (d) use of two amine-reactive heterobifunctional reagents to introduce a maleimide group into one protein, and a thiol group into another, followed by coupling of these groups.

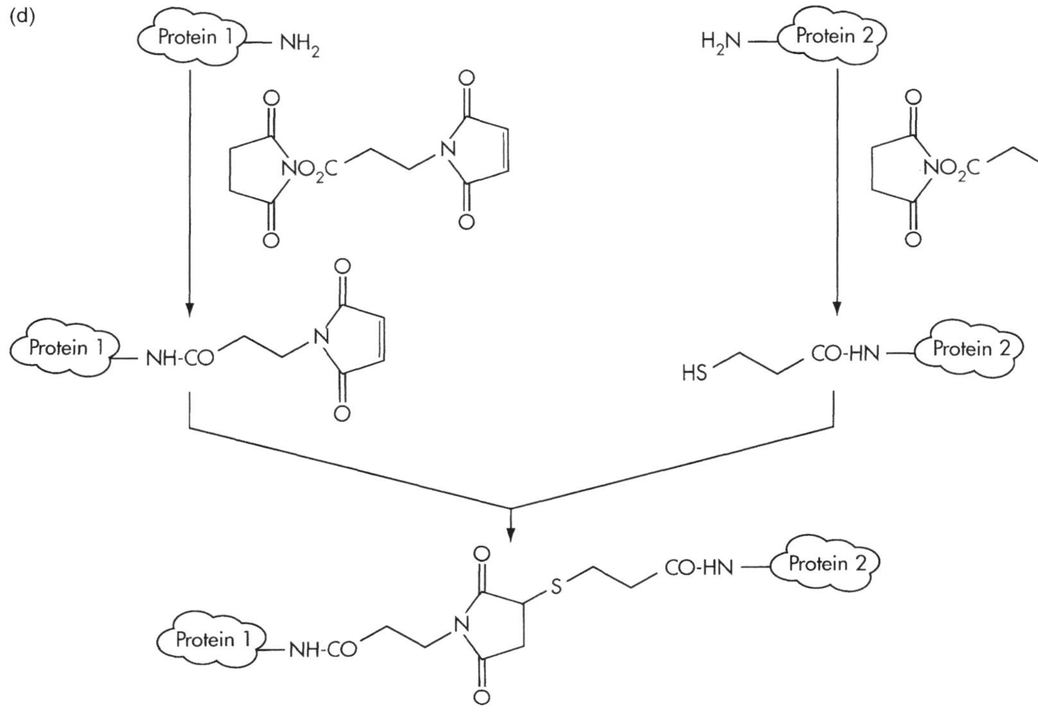

(d)

Fig. 14.4 Continued.

The second notable use is based on the fact that IgG molecules are glycosylated predominantly on the Fc region, distant from the binding site. Mild oxidation of these residues therefore allows conjugation to the amines of a labeling protein such as HRP or alkaline phosphatase, with minimal interference to the antigen-binding activity of the antibody.

Glutaraldehyde Method

Glutaraldehyde is in theory a simple bis-aldehyde, which can act as a homobifunctional reagent for coupling two proteins via their amine groups (Figure 14.7). However, the real situation is substantially more complex. Glutaraldehyde solutions as supplied commercially, generally contain a mixture of hydrates and cyclic acetals as well as the dialdehyde itself. Also, glutaraldehyde can self-condense to form α,β-unsaturated aldehydes; these may be present in small amounts in glutaraldehyde solutions, but more importantly can be formed in substantial quantities under the alkaline conditions often employed for coupling. These can still form protein–protein links, but multimeric examples may promote cross-linking of several protein molecules. Finally, glutaraldehyde which has partially reacted with protein amines can cyclize to form pyridinium compounds, which may or may not result in a protein–protein link. In summary, glutaraldehyde coupling features variable composition of the starting material, and a bewildering variety of possible reaction routes, which may or may not result in protein–protein coupling. Glutaraldehyde is also a toxic and unpleasant compound.

The main advantages of this approach, which have helped it to maintain some popularity in spite of the drawbacks just listed, are that it is rapid, simple, and almost always gives a product of some sort. Glutaraldehyde provided a significant step forward in this respect compared to the coupling agents used in the 1960s (bis-isothiocyanates, cyanuric chloride, bis-diazonium compounds). This approach was pioneered for histochemical purposes (Avrameas, 1969; Avrameas and Ternynck, 1971), but was soon widely reported for the preparation of immunoassay conjugates too. Glutaraldehyde was arguably the most widely used reagent for protein coupling in the 1970s.

Given the complex chemistry outlined above it is not surprising that glutaraldehyde coupling of proteins generates highly heterogeneous products, frequently containing large proportions of aggregated material. The use of two-step methods moderates this effect to some extent, one protein being activated with glutaraldehyde before the addition of the second. However, one-step approaches are also common, in which the glutaraldehyde

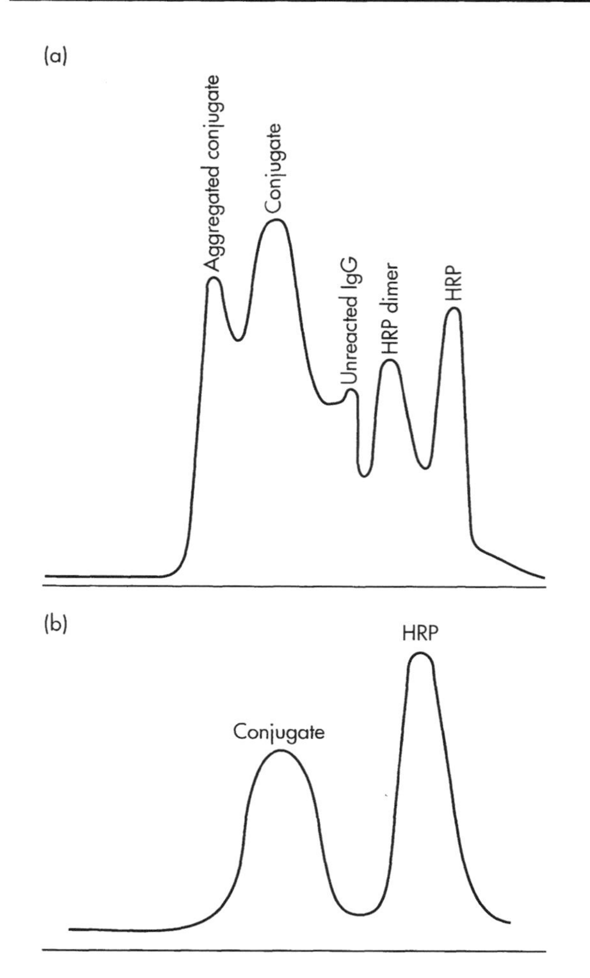

Fig. 14.5 Size exclusion chromatograms of the purification of IgG–HRP conjugates manufactured by (a) a direct periodate approach and (b) a dual heterobifunctional thiol–maleimide approach. Reproduced with permission from Aslam, M. and Dent, A.H. (eds.), *Bioconjugation: Protein Coupling Techniques for the Biomedical Sciences*, p. 90 (Macmillan, London, 1998).

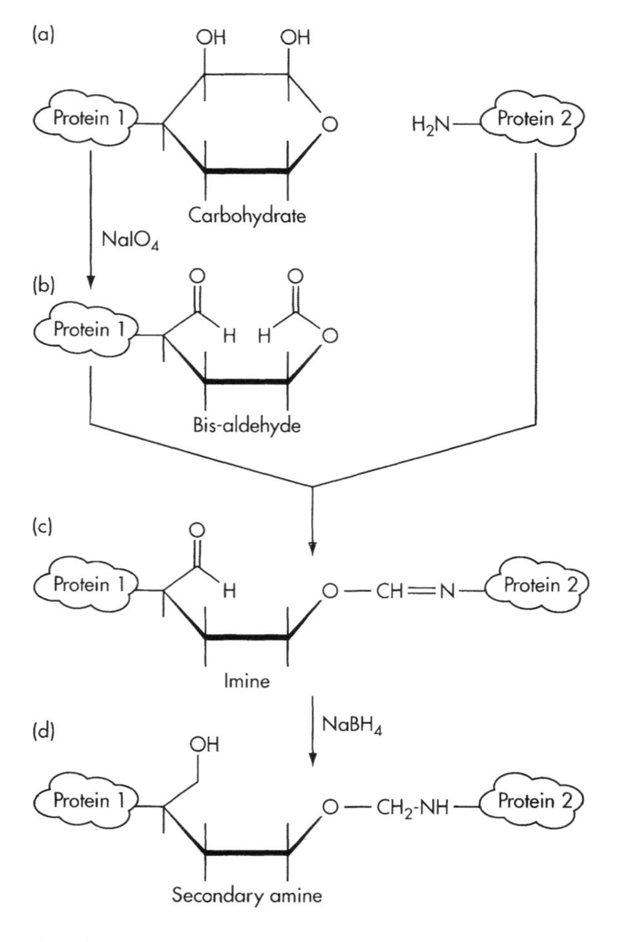

Fig. 14.6 Periodate coupling: the carbohydrate group of one protein (a), shown in a simplified form, contains a vicinal diol group; this is oxidized by periodate to yield a bis-aldehyde (b), which couples with the amine group of a second protein to form an imine link (c). This is frequently stabilized by reduction to a secondary amine (d).

at best provides a crude 'protein glue'. Reaction conditions are undemanding: so long as the pH is high enough to provide a reasonable quantity of deprotonated amine (roughly 7 or above) coupling will take place.

With the current availability of more controllable coupling reagents, and especially bearing in mind the toxicity issues, the use of glutaraldehyde should no longer be considered a first-line option for immunoassay conjugates.

Bis-maleimide Method

Where the proteins to be coupled both possess free thiol groups, homobifunctional coupling via these moieties is a feasible option. Obviously, a coupling agent bearing two thiol-reactive groups is called for, and *N,N'-o*-phenylenedimaleimide (PDM) has seen the most widespread use. The coupling reaction is shown in Figure 14.8.

In the immunoassay field a particularly elegant application of this approach has been to couple the labeling enzyme β-galactosidase to antibody fragments such as Fab or Fab′. Both of these proteins possess free thiols, and PDM coupling yields largely monomeric conjugates with excellent incorporation and little loss of enzyme activity (Ishikawa *et al.*, 1983). Thiol–maleimide coupling takes place readily at neutral to mildly basic pH.

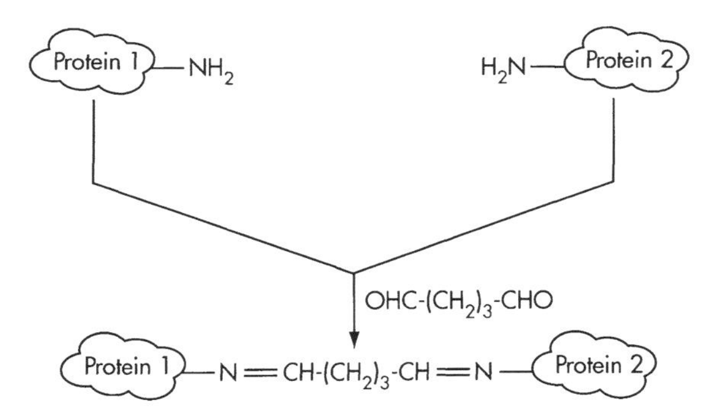

Fig. 14.7 Glutaraldehyde coupling: in this simplistic representation the amine groups of two proteins are cross-linked by this bis-aldehyde, but the real situation is considerably more complex.

The more widespread use of this approach has been prevented by the rarity of cases where the requisite free thiols are present on each of the proteins to be coupled.

Heterobifunctional Thiol–Maleimide Methods, and Related Approaches

As discussed earlier, heterobifunctional approaches can offer greater control over the product composition than the direct and homobifunctional procedures described so far. The commonest examples exploit the same thiol–maleimide chemistry just described, as shown in Figure 14.4d. Two coupling agents are required, one combining an amine-reactive group with a thiol, and one combining an amine-reactive group with a maleimide (Ishikawa *et al.*, 1983). In practice, the thiol is often provided in a protected form such as an *S*-acetyl compound, because free thiols tend to dimerize under the influence of atmospheric oxidation – an unwelcome side-reaction.

Numerous coupling agents of the types described are commercially available, some of which are shown in Tables 14.2 and 14.3. All the common amine-reactive maleimides are based on succinimide esters, sometimes derivatized with a sulfonic acid group, which confers water solubility. It is also important whether the maleimide group is attached to an aromatic or an aliphatic function, the latter being generally more stable. The length of the linker group is another parameter of interest, the very long chains of compounds such as SMTCC being claimed to give good performance in immunoassay conjugates by their inventors.

The thiolation reagents shown in Table 14.2 are interesting in terms of their effect on the charge of the protein – like other succinimide esters SATA reduces the charge at typical pH by one through the loss of an amine group. With SAMSA an extra

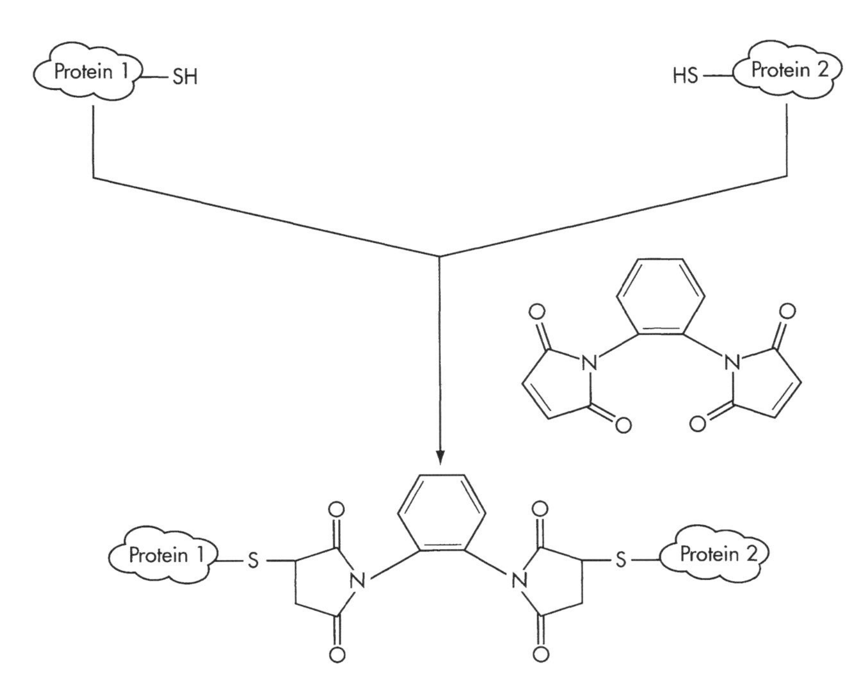

Fig. 14.8 Bis-maleimide reagents can be used for homobifunctional coupling of protein thiols.

Table 14.2. Amine-reactive heterobifunctional reagents for introduction of thiol groups into proteins.

Reagent
S-Acetylmercaptosuccinic anhydride (SAMSA) 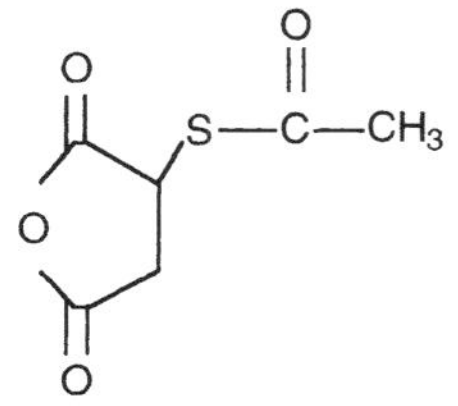
2-Iminothiolane hydrochloride (2-IT) 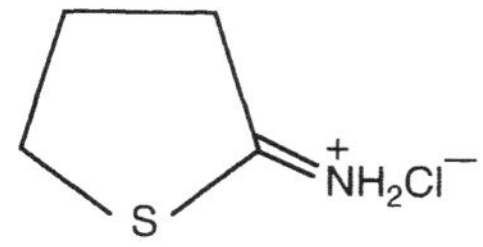
N-Succinimidyl-*S*-acetyl thioacetate (SATA) 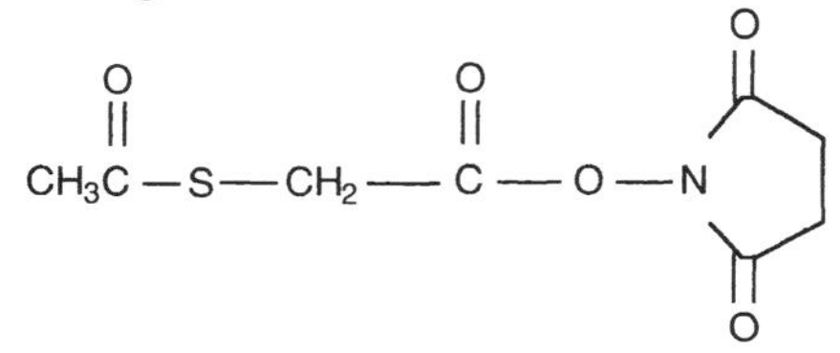

carboxylate function is introduced, reducing the net charge by two. With 2-IT the original charge of the protein is retained, as the amine group is replaced by a similarly charged amidine function. The *S*-acetylthiol yielded from SATA or SAMSA can be deprotected before use with hydroxylamine, while 2-IT yields the free thiol directly.

Subtle changes of conjugate performance can be obtained by using various combinations of the thiol and maleimide reagents listed above. The overall approach is highly satisfactory for the preparation of immunoassay conjugates, giving well-defined conjugates of controlled size, and generally high retention of protein functionality (e.g. enzyme activity). The thioether link formed in this method is very stable under typical conditions.

There are alternatives to maleimide as the thiol-reactive function in this approach. In particular, it is worth mentioning the reagent ***N*-Succinimidyl-3-(2-pyridyldithio)propionate, SPDP**, (Figure 14.9) which incorporates a disulfide moiety. This can be exploited in the same way as a maleimide group, yielding instead a conjugate with a disulfide link (see Figure 14.3). This may be unstable in the presence of thiols or other reducing agents – in fact, it is often used where the intention is to facilitate later cleavage of the protein–protein bond in this way. This is not generally the case for immunoassay applications, and the maleimide reagents described above predominate in this field.

GENETIC ENGINEERING APPROACHES TO PROTEIN CONJUGATION

There is a very different approach to the formation of protein–protein conjugates, which has become available over the last two decades – although it is still to have a significant impact on commercial immunoassays. This is the use of recombinant DNA techniques to construct a **fusion protein**, in which the DNA sequences coding for the two components of the conjugate are spliced together and the resulting sequence expressed in a suitable host organism. Well-defined 1:1 conjugates can be produced by this means, which can then be continually expressed in large quantities if required.

This approach has been used to express enzyme conjugates for immunoassay such as insulin coupled to alkaline phosphatase, and a malaria antigen coupled to β-galactosidase. Fusion proteins containing antibodies have also been produced as the relevant gene sequences have become available. Although still not widespread in commercial immunoassays, the use of recombinant antibody fragments such as **single chain variable fragments (ScFvs)** is an increasingly popular technology which can be tailored to conjugation applications (see, for example, Albrecht *et al.*, 2004).

See Bülow and Lindbladh (Chapter 9 in Aslam and Dent, 1998) for further discussion of this approach to conjugate preparation.

Table 14.3. Amine-reactive heterobifunctional reagents for introduction of maleimide groups into proteins.

Reagent
m-Maleimidobenzoyl succinimide ester (MBS) 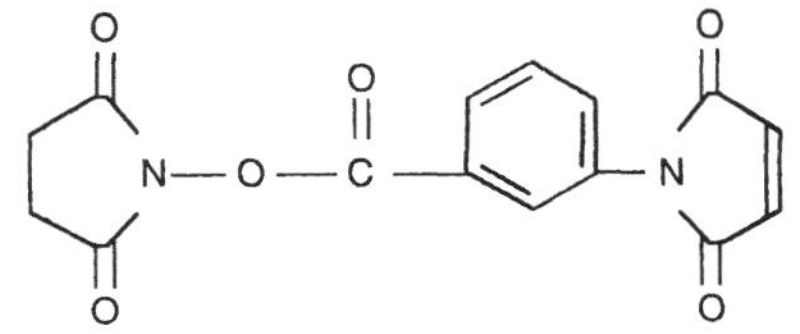
Succinimidyl 4-(*p*-maleimidophenyl) butyrate (SMPB) 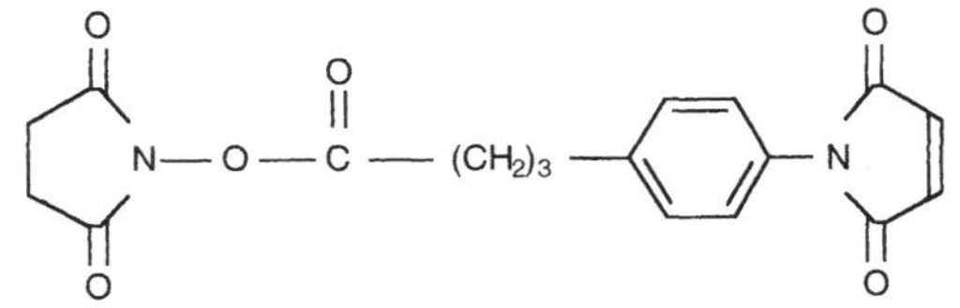
Succinimidyl 4-(*N*-maleimidomethyl) cyclohexane-1-carboxylate (SMCC) 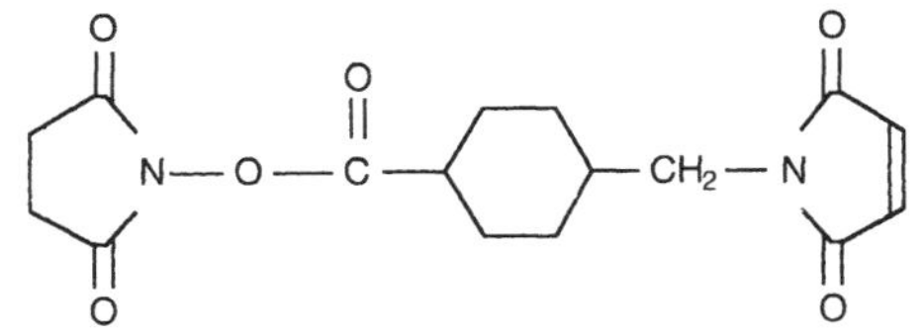
Succinimidyl 4-maleimidobutyrate (GMBS) 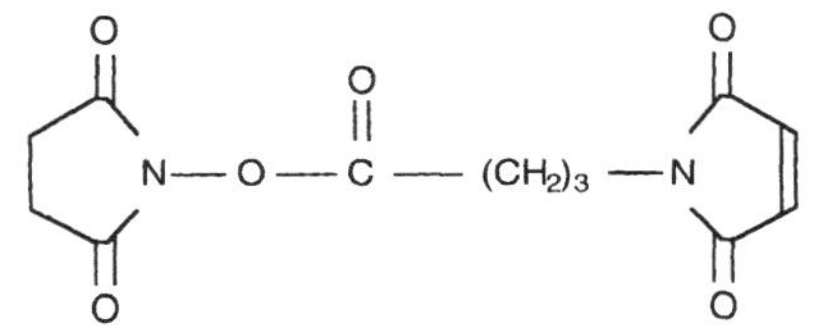
6-Maleimidohexanoyl-*N*-succinimide (MHS) 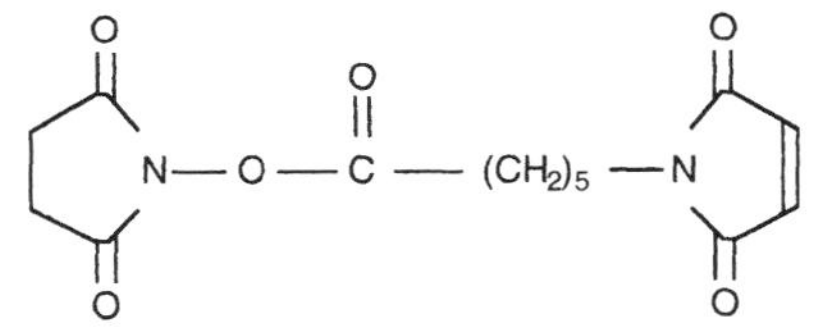
Succininimidyl 4-[(*N*-maleimidomethyl) tricaproamido]cyclohexane-1-carboxylate (SMTCC)] 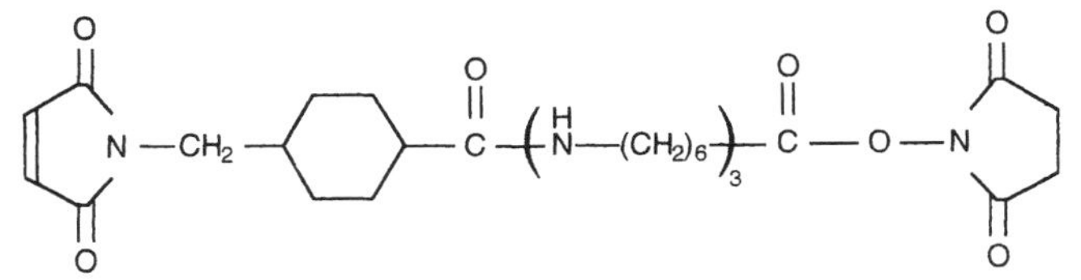

NB: most of these reagents are also available as the water-soluble sulfosuccinimidyl analogs (*cf.* Figure 14.16b, sulfo-NHS-biotin), e.g. sulfo-SMCC.

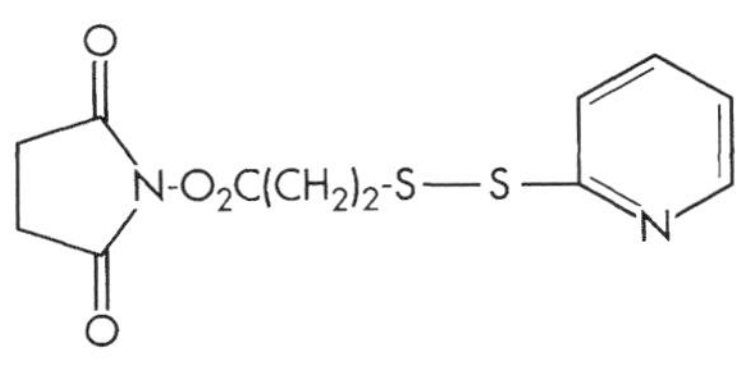

Fig. 14.9 *N*-Succinimidyl-3-(2-pyridyldithio)propionate, SPDP.

PROTEIN–SMALL MOLECULE COUPLING

INTRODUCTION

Where a small molecule is to be conjugated to a protein for immunoassay purposes, the small molecule concerned is almost always the analyte of interest or a close analog. This is true both for the production of conjugates for use in the assay itself, and for the preparation of immunogens used in the generation of antibodies.

The reactive groups available in protein molecules have already been summarized above. In some cases the small molecule of interest may already possess a group such as an amine, thiol, or carboxylic acid, which can form the basis of a coupling reaction, in which case no synthetic chemistry is necessary. Often however, there are no such groups present, and it can present a significant challenge to synthesize a derivative of the compound that possesses an appropriately reactive 'handle' for attachment to the protein, while retaining the ability to be recognized by the appropriate antibodies.

The choice of derivatization approach needs to be made with care. It is important to identify those parts of the molecule likely to be most important in antibody-binding terms, so that they can be avoided as potential derivatization sites. This choice is often driven primarily by cross-reactivity arguments. An antibody might be required, for example, to provide discrimination between a steroid analyte and a closely related analog differing only in the nature of single substituent on the A-ring. An immunogen in which the steroid is coupled to the carrier protein via this region of the molecule is unlikely to yield antibody of the necessary specificity. Similarly, once an appropriately specific antibody has been found, a conjugate coupled to a labeling protein through this site is likely to show poor binding. In both cases therefore, a derivatization site remote from the A-ring would probably give the best results.

(a)

$R—NH_2 \rightarrow R—NH\text{-}CO(CH_2)_2\text{-}CO_2H$

Succinate

$R—OH \rightarrow R—O\text{-}CO(CH_2)_2\text{-}CO_2H$

(b)

$R—C(H){=}O \xrightarrow{HCl\text{-}H_2NOCH_2CO_2H} R—C(H){=}N—O—CH_2CO_2H$

Aldehyde — Oxime

$R—C(R){=}O \xrightarrow{HCl\text{-}H_2NOCH_2CO_2H} R—C(R){=}N—O—CH_2CO_2H$

Ketone

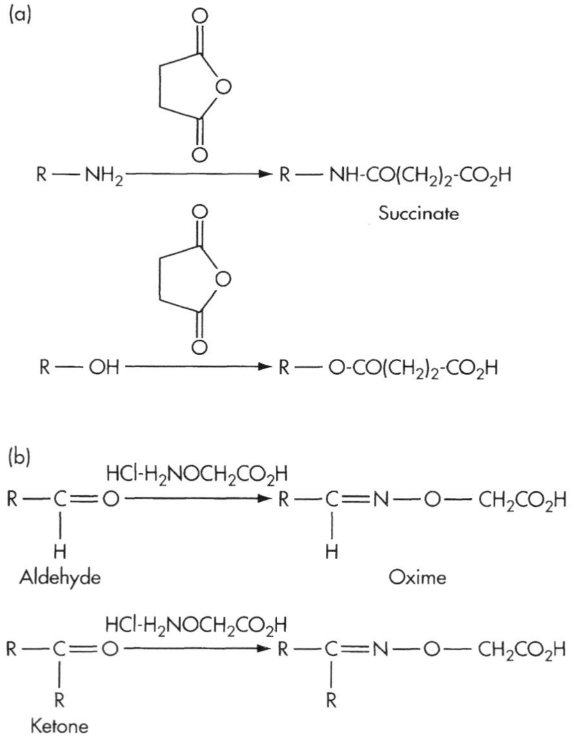

Fig. 14.10 Approaches for the introduction of carboxyl groups into small molecules through succinic anhydride activation of amine or hydroxyl functions (a), or derivatization of an aldehyde to form a carboxyl-bearing oxime (b).

A wide variety of 'handles' can provide appropriate reactivity for coupling to protein molecules, and some common examples reactive towards amines and thiols are shown in Figures 14.2 and 14.3. These offer attractive synthetic targets when the aim is to derivatize a small molecule for coupling to a protein. Routes which yield a carboxylic acid derivative are particularly popular, however, largely due to the well-established and controlled methods that can then be used for coupling to the protein's amine groups. A step commonly encountered in such a synthetic route, therefore, is one that exploits an existing functionality to introduce a carboxylic acid. Two common examples are shown in Figure 14.10 – the use of succinic anhydride to incorporate an acid at the site of an amine or (less efficiently) a hydroxyl group, and the use of carboxymethoxylamine hydrochloride to 'convert' an aldehyde or ketone to a carboxyl-containing **oxime** function.

Note that an amine group in the small molecule can itself be used as a handle. Homobifunctional coupling methods similar to those used for protein–protein applications are sometimes employed to attach such a molecule to a protein, using glutaraldehyde or a better-defined alternative such as a bis-succinimide ester. These approaches are often rather inefficient in terms of incorporation, however, and conversion to a carboxylic acid as just described is often preferred as a consequence. Additionally, the resulting succinate spacer group can yield better steric properties and hence improved assay performance.

COMMON PROTEIN–SMALL MOLECULE COUPLING METHODS

Because carboxyl-based approaches are so common, some of the commonest examples of this category are described in detail below. So too is the Mannich reaction, a useful standby for difficult cases where it is not straightforward to prepare a derivative bearing one of the standard handles.

Carbodiimide Methods

It is possible to activate carboxylic acids towards amines in many ways, but all of these involve

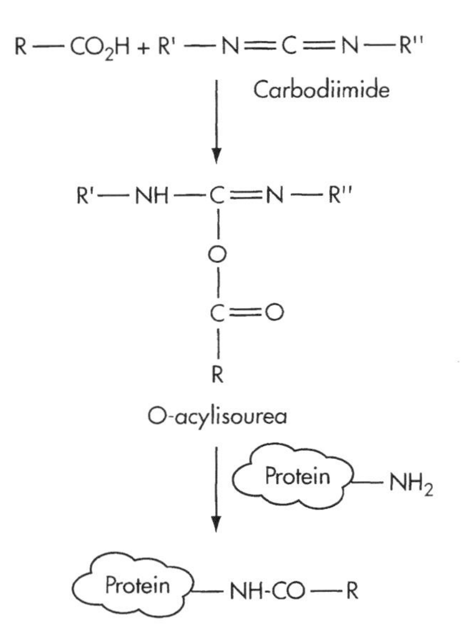

Fig. 14.11 Carbodiimide activation of a carboxylic acid yields an *O*-acylisourea, which is reactive towards amines.

intermediates that are unstable to some extent in an aqueous environment. For reaction with proteins it is therefore necessary to choose examples that strike a balance between reactivity towards amines and resistance towards hydrolysis. One such category of intermediate is the **O-acylisourea**, which is derived from the reaction of a carboxylic acid with a **carbodiimide** (Figure 14.11).

These compounds fall into two classes – those that are water-soluble and those that are not. The commonest example of the latter category is **dicyclohexylcarbodiimide (DCC)**. In a typical reaction scheme there are two steps: first the small molecule is derivatized with DCC in a water-miscible organic solvent, then the product is added to an aqueous protein solution for the coupling stage.

With a water-soluble carbodiimide, however, it is possible to carry out both reactions simultaneously – the isourea can be formed in the protein solution, so only one step is required and the need for organic solvents is also avoided. The commonest examples of water-soluble carbodiimides are **1-ethyl-3-(3-dimethylaminopropyl)carbodiimide hydrochloride (EDAC)** and **1-cyclohexyl-3-(2-(4-methylmorpholin-4-yl)ethyl)-carbodiimide *p*-toluenesulfonate (CMC)**.

Carbodiimide/*N*-hydroxysuccinimide Methods

A popular variation on the straightforward carbodiimide methods of the previous section is to convert the isourea *in situ* into a succinimide ester, which goes on to react with the protein amines (Figure 14.12); this can give improvements in yield. Thus, in a two-step approach it is common to activate a carboxyl-containing compound in an organic solvent with a mixture of DCC and ***N*-hydroxysuccinimide (NHS)** before addition to the aqueous protein solution. Similarly, the water-soluble ***N*-hydroxysulfosuccinimide (NHSS)**, can be used in one-step procedures with EDAC or CMC.

In terms of efficiency of incorporation, a two-step DCC/NHS protocol offers one of the best means of conjugating carboxylic compounds to proteins: yields are very high, and it is generally possible to expose the protein to the small quantity of organic solvent required without causing it much harm. Fully aqueous approaches avoid this potential issue, but there is invariably a loss of yield due to hydrolysis effects.

Mixed Anhydride Method

Acid anhydrides, formed by the loss of water between two carboxyl groups, can survive long enough in aqueous solutions to react with protein amines. The conditions required to create a symmetric anhydride by dehydration of the acid of interest are very harsh, so it is more common

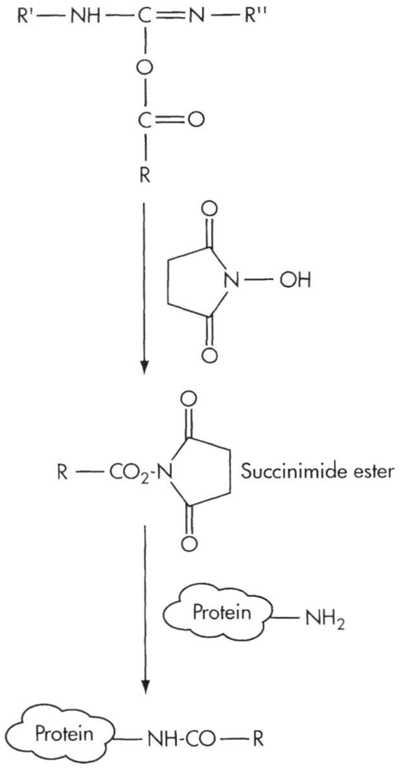

Fig. 14.12 Reaction of an *O*-acylisourea with *N*-hydroxysuccinimide yields an amine-reactive succinimide ester.

to form a **mixed anhydride** intermediate by reaction with a **chloroformate**. The commonest example of this class of reagents is **isobutyl chloroformate (IBCF)**. The conversion of the carboxylic compound to its mixed anhydride must be carried out in the absence of moisture, so a two-step approach is required (Figure 14.13), the product from the anhydrous activation reaction being added to the aqueous protein solution for coupling.

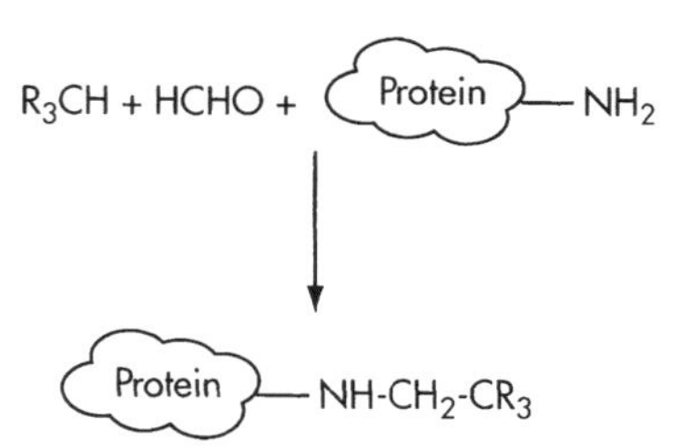

Fig. 14.14 The Mannich reaction allows a wide variety of compounds bearing a suitably 'active' hydrogen atom to be coupled to protein amines.

Mannich Reaction

In the Mannich reaction an amine (e.g. from a protein) is coupled in the presence of an aldehyde (usually formaldehyde, despite its toxic properties) to a suitable carbon atom in a wide range of organic compounds (Figure 14.14). The carbon atom must possess an 'active' hydrogen atom, loosely speaking one with a degree of acidic character. Examples of compounds that can take part in the Mannich reaction are shown in Figure 14.15. Reaction times can be slow and yields poor, but this approach can be useful where commoner approaches have failed.

Ready-Made Reagents for Biotinylation

The introduction of a biotin moiety into a protein is a common requirement of immunoassay systems based on the biotin–avidin or biotin–streptavidin interaction. This is generally achieved by rather simple means, as this is such a widespread need that one-step biotinylation reagents are readily available from commercial sources. The biotin molecule presents few challenges as a target for coupling to proteins: it possesses a side-chain ending in a carboxyl function, which only requires suitable activation for coupling to protein amines. The succinimide ester of biotin, **biotin-NHS**, is commercially available for this purpose. So too are longer chain analogs that incorporate extra spacer regions; some common biotinylation reagents are shown in Figure 14.16. Using these reagents, biotinylation is a simple one-step reaction at pH 7 or above. Yields are usually very high, and there is generally good retention of protein functionality.

Many other biotinylation reagents are commercially available, aimed at other functional groups: an example is ***N*-(6-biotinamidohexyl)-3′-(2′-pyridyldithio)propionamide, Biotin-HPDP** (Figure 14.17), which can be coupled to protein thiols. In practice, these are seldom required.

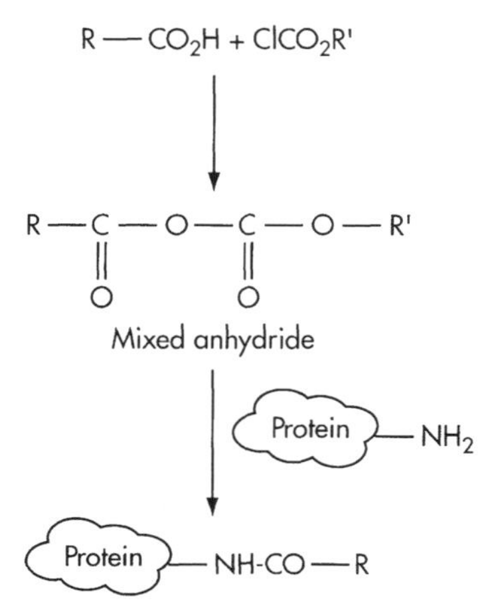

Fig. 14.13 Reaction of a carboxylic acid with a chloroformate yields an amine-reactive mixed anhydride.

Other Ready-Made Reagents for Protein Derivatization

Wherever there is sufficient interest in derivatization of proteins with a particular species, commercial suppliers market suitable reagents for this purpose. Thus, all the commonest fluorophores and luminophores can be purchased as amine-reactive derivatives, typically succinimide esters or isothiocyanates. Some examples are shown in Figure 14.18. Needless to say, these generally provide the simplest route for preparation of the relevant protein conjugate – a fluorescein-labeled IgG, for example.

PURIFICATION OF CONJUGATES

Conjugates are sometimes used in their crude form in immunoassays, but it is generally possible to achieve greater control by carrying out a purification step. Often this is essential to obtain the desired assay performance. By far the commonest approach to conjugate purification is chromatography.

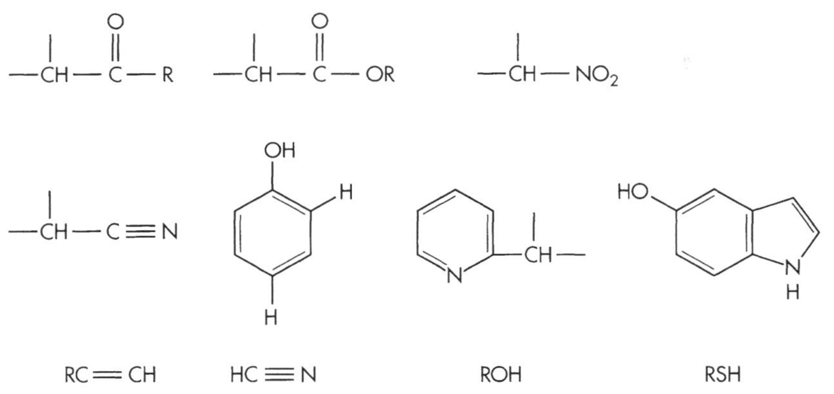

Fig. 14.15 Examples of moieties that can provide an active hydrogen for the Mannich reaction. Reproduced with permission from Aslam, M. and Dent, A.H. (eds.), *Bioconjugation: Protein Coupling Techniques for the Biomedical Sciences*, p. 449 (Macmillan, London, 1998).

CHROMATOGRAPHIC APPROACHES TO CONJUGATE PURIFICATION

All branches of chromatography are based on the different interactions of the components of a mixture in one phase with a second phase of some kind. In practical terms in this application, this nearly always means the interaction of solutes in an aqueous solution with a solid-phase column.

In **size-exclusion chromatography (SEC)** or **gel filtration**, the solid phase contains pores that are more accessible to smaller molecules than to large ones. The result is that large molecules pass through the column more quickly. This provides an excellent means for separating protein–small molecule conjugates from unreacted hapten, for example, or for separating a thiolated protein from excess thiolating reagent in a heterobifunctional protein–protein conjugation. In these 'de-salting' examples there is a huge difference in the molecular weights of the protein components (50–500 kDa) and the impurities (50–500 Da). However, high-performance SEC can also resolve separate protein components, allowing unreacted enzyme (44 kDa) to be removed from an IgG–HRP conjugate (200–300 kDa), for example.

In **ion exchange chromatography (IEC)**, the solid phase bears charged groups which interact electrostatically with the solutes, providing an alternative to the size-based separations of gel filtration. So long as the ionic strength of the solvent system is kept reasonably low, species of different charge can be eluted sequentially from a solid phase bearing positive (**anion exchange**) or negative (**cation exchange**) groups. Sometimes the pH of the eluent system is altered during the purification to effect elution of the more strongly bound components, but it is generally simpler to modulate ionic strength instead. As this is increased, the interactions of charged groups with the solid phase are outweighed by those with the solvent, and elution is again the result.

Hydrophobic interaction chromatography (HIC) and **reverse phase chromatography (RPC)** rely on the interactions of the hydrophobic regions of proteins with a hydrophobic solid phase. In HIC modulation of eluent, ionic strength is again typically employed to effect elution of bound solutes, but in this case it is a *low* ionic strength that is required; high salt content strengthens hydrophobic interactions. In RPC, which was adapted to protein applications from organic synthesis, the interactions tend to be stronger, and it is often necessary to employ organic solvents to disrupt the interactions of the bound solutes with the surface of the solid phase.

Affinity chromatography uses solid phases bearing reagents that have a specific ability to bind one component or class of components, thus separating them selectively from a mixture. This can provide highly specific purification regimes, often unachievable by the more generic techniques described above. There are countless examples of this approach in use, but some of the commoner applications in the immunoassay field use solid phases bearing **Protein A** or **Protein G**, bacterial proteins with a high affinity for most classes of immunoglobulin from most species. **Lectins** – naturally occurring proteins which bind carbohydrates – are also widely employed for the purification of glycoprotein conjugates. **Concanavalin A** is the commonest example of this category. Various means are used to elute bound proteins from affinity matrices: low pH is probably the commonest, but organic cosolvents or the structure-disrupting agents known as **chaotropes** are also employed.

An interesting variation on conjugate chromatography is to carry out the coupling reaction itself on proteins reversibly immobilized to a chromatographic support;

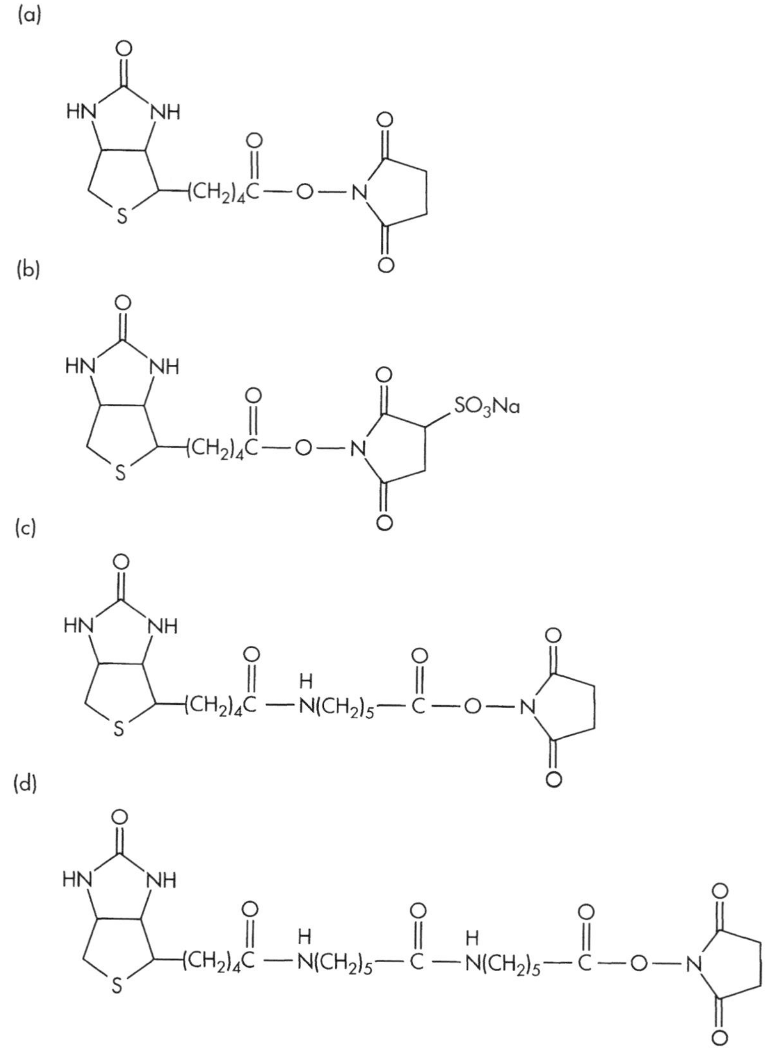

Fig. 14.16 Some common biotinylation reagents: (a) succinimidyl biotin (biotin-NHS); (b) sulfosuccinimidyl biotin, sodium salt (sulfo-NHS-biotin); (c) succinimidyl 6-(biotinamido)hexanoate (NHS-LC-Biotin II™ (Pierce), Biotin-X-NHS™ (Calbiochem)), (d) succinimidyl 6-[6-biotinamido(hexanamido)]hexanoate (Biotin-XX-NHS™ (Calbiochem)).

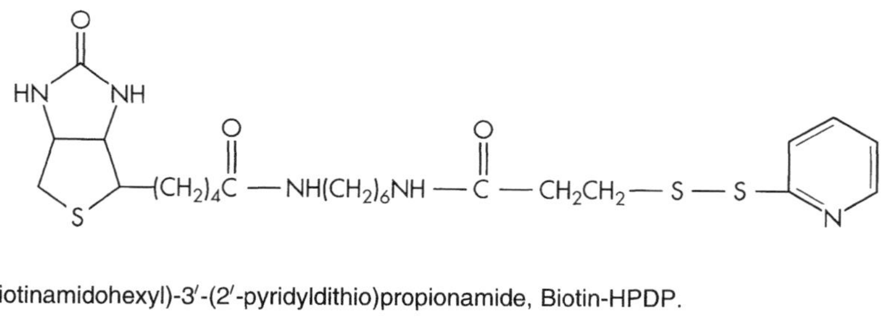

Fig. 14.17 *N*-(6-biotinamidohexyl)-3′-(2′-pyridyldithio)propionamide, Biotin-HPDP.

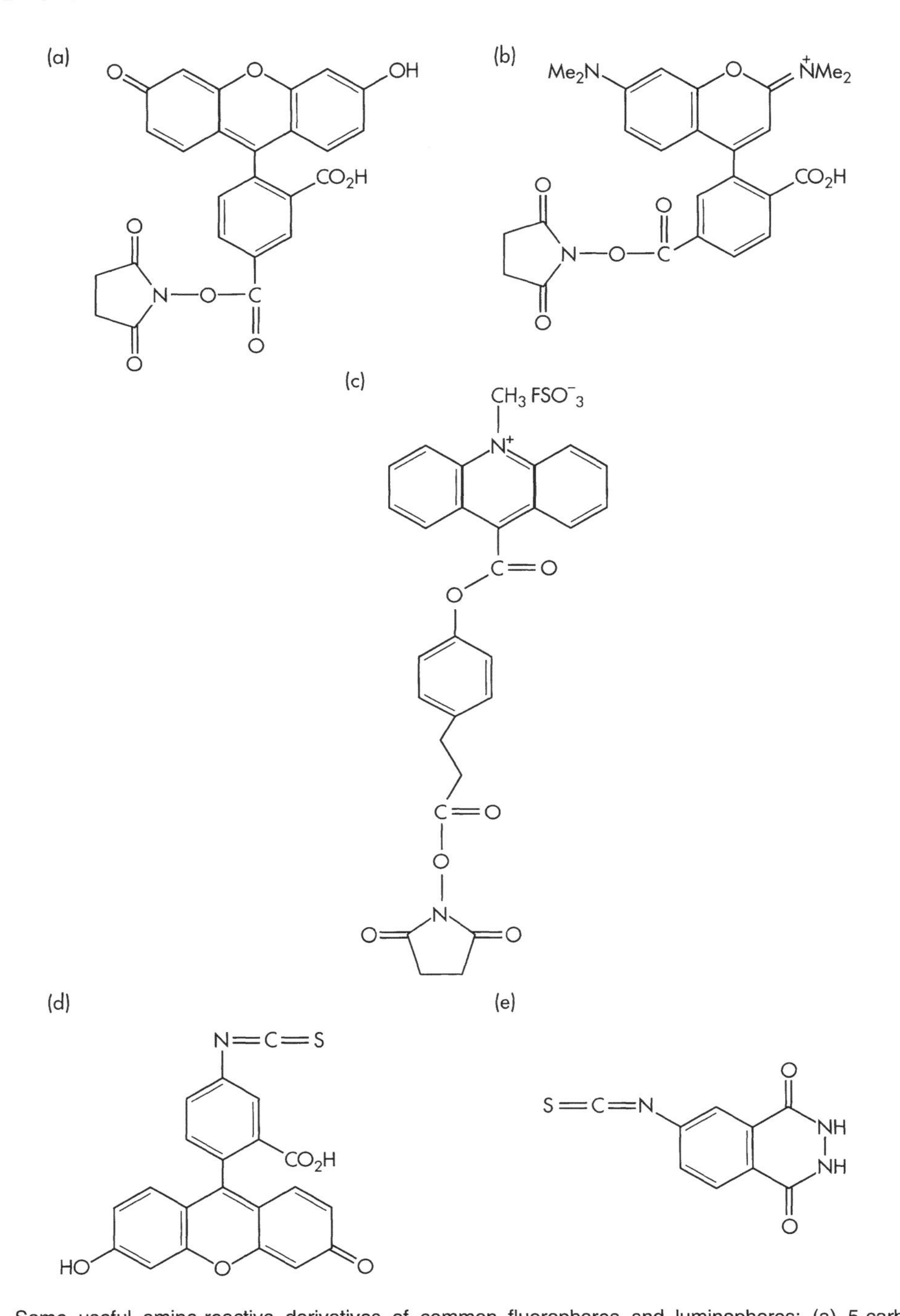

Fig. 14.18 Some useful amine-reactive derivatives of common fluorophores and luminophores: (a) 5-carboxyfluorescein succinimidyl ester; (b) 6-carboxytetramethylrhodamine succinimidyl ester; (c) 4-(2-succinimdyloxycarbonylethyl)phenyl-10-methylacridinium-9-carboxylate fluorosulfonate; (d) fluorescein-5-isothiocyanate (FITC Isomer I); (e) 4-isoluminol isothiocyanate.

good immunoassay performance has been claimed from the use of this approach (Russell *et al.*, 2002).

OTHER APPROACHES TO CONJUGATE PURIFICATION

There are a number of other common approaches for separating components that have a wide difference in size – alternatives to the de-salting applications of SEC. **Dialysis** is a time-consuming but well-established technique, where a semi-permeable membrane is employed to retain high molecular weight components while allowing smaller molecules to diffuse out. Good separations can be obtained by repeated buffer exchange. **Ultrafiltration** techniques use the same principle but employ pressure to drive the protein solution through the membrane – obviously a much quicker technique. Membranes with different molecular weight cut-offs are available, allowing some influence over which components are retained; it should be borne in mind, however, that these cut-offs are often rather approximate.

Electrophoresis techniques, where the solutes are driven through a suitable medium under the influence of an electric field, have excellent power to resolve protein components on the basis of their charge. The medium is typically a gelatinous matrix coated onto a flat support of some kind. However, the heat generated in the process puts serious constraints on the dimensions of the apparatus, and while powerful as an analytical tool, this approach is only rarely encountered in preparative applications.

CHARACTERIZATION OF CONJUGATES

An area which has received surprisingly little attention over the years is the study of the actual molecular composition of conjugates. Taking a typical reaction where a number of a protein's numerous amine groups are derivatized by a suitably reactive reagent – the succinimide ester of a small molecule, for example – it is commonplace to calculate a mean incorporation ratio such as '1.7 hapten groups per protein molecule'. This is generally established by techniques such as UV-visible spectrophotometry, which relies on the components of the conjugate bearing different absorbance characteristics at two or more wavelengths. Other techniques may be available in specific circumstances, for example, the use of radiometry to quantify a radiolabeled component, or the quantitation of biotin incorporation using a colorimetric assay based on the dye 4-hydroxyazobenzene-2′-carboxylic acid (HABA). However, the result is nearly always a *mean* incorporation.

Simple statistical calculations using Poisson or binomial distributions can be used to show the expected spread around these means: some examples are shown in Figure 14.19. These highlight the danger of taking mean incorporations too literally in understanding the behavior of protein conjugates.

To confirm these distributions of conjugate populations of differing individual stoichiometries, it is necessary to use techniques that can elucidate the composition of conjugate product mixtures at the molecular level. Three technologies have dominated such attempts. The first of these is **isoelectric focusing**, a branch of electrophoresis where fractions of the product are separated into tight bands based on their charge. As many small molecule conjugation procedures result in the replacement of positively charged amine groups, it is possible to track the extent of derivatization on this basis, and this approach has been used, for example, by Barbarakis and Bachas (1991) and Pham

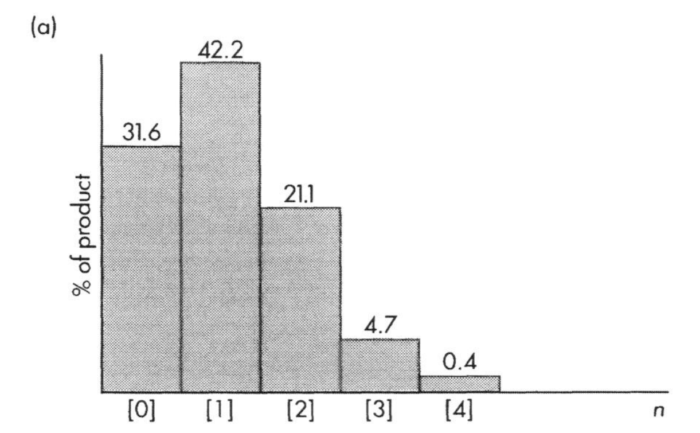

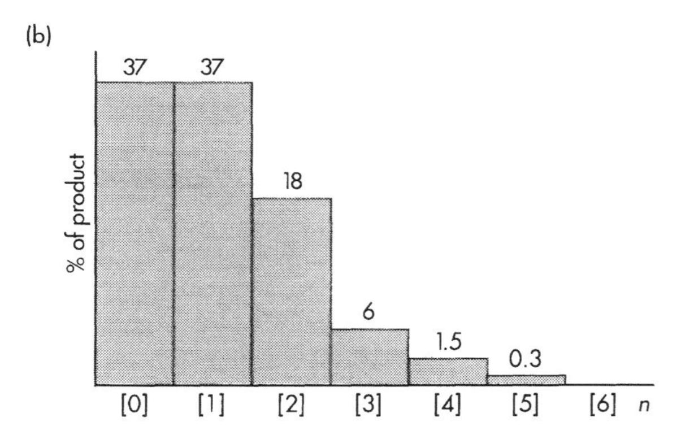

Fig. 14.19 (a) Binomial distribution of thiol incorporation (*n*) in a protein with four derivatization sites, where the mean incorporation is 1; (b) Poisson distribution of thiol incorporation (*n*) in a protein with a large, undefined number of derivatization sites, where the mean incorporation is 1. Reproduced with permission from Aslam, M. and Dent, A.H. (eds.), *Bioconjugation: Protein Coupling Techniques for the Biomedical Sciences*, pp. 92–93 (Macmillan, London, 1998).

et al. (1995). A second branch of electrophoresis, **SDS–PAGE**, allows resolution on the basis of size, and can therefore be used in a parallel fashion for the characterization of protein–protein conjugates; see, for example, Åkerblom *et al.* (1993). **Capillary electrophoresis** is now challenging the traditional gel-based formats for these methods, but to date has seen its widest application in the high-throughput screening arena.

Thirdly, the ability of **mass spectrometry** to give accurate mass determinations on large molecules is now well established, and techniques such as **electrospray** and **matrix-assisted laser desorption ionization (MALDI)** mass spectrometry are increasingly employed for protein conjugate characterization (see for example Singh *et al.*, 2004), or even simply as a tool to track the progress of the conjugation reaction (Safavy *et al.*, 2003). Indeed, the increasing ability of mass spectrometry to resolve protein populations from a biological milieu suggests that this approach may soon begin to compete with immunoassay as a clinical analysis tool in some applications (Petricoin and Liotta, 2003).

Given the central role of conjugates in immunoassay, it is certain that their composition can have a significant effect on assay behavior. More work is needed to increase the understanding of conjugate stoichiometry and its effect on immunoassay performance – and ideally to extend the capability of purification methods such that the individual populations can be resolved adequately at a preparative scale.

CONCLUSION

In conclusion, it can be observed that the chemistries employed in the preparation of immunoassay conjugates have reached a state of some maturity over the last two decades. Heterobifunctional methods now predominate in the production of protein–protein conjugates – those based on thiol–maleimide chemistry being particularly widespread – while the popularity of carbodiimide and mixed anhydride methods continues unabated for small molecule coupling.

Advances in chromatography continue to be driven primarily by commercial suppliers, and their regular introduction of better instrumentation and more efficient matrices is facilitating more effective purification of conjugates. Software for the electronic manipulation of chromatographic data has become much more powerful in recent years, providing valuable tools for monitoring separation efficiencies.

The last decade has seen big advances in the availability of techniques for the characterization of immunoassay conjugates – as observed earlier, this remains a rather neglected field. Mass spectrometry is likely to have a major part to play in increasing understanding in this area, and again commercial suppliers are playing a key role through the development of ever-better instrumentation.

Finally, the production of a conjugate for use in a commercial immunoassay is a manufacturing process, and as such should receive appropriate attention to ensure its robustness and reproducibility. See Chapter 13 of Aslam and Dent (1998) for a discussion of relevant approaches.

REFERENCES AND FURTHER READING

Åkerblom, E. *et al.* Preparation and characterization of conjugates of monoclonal antibodies and staphylococcal enterotoxin A using a new hydrophilic cross-linker. *Bioconjugate Chem.* **4**, 455–466 (1993).

Albrecht, H. *et al.* Production of soluble ScFVs with C-terminal free thiol for site-specific conjugation of stable dimeric ScFvs on demand. *Bioconjugate Chem.* **15**, 16–26 (2004).

*Aslam, M. and Dent, A.H.(eds), *Bioconjugation: Protein Coupling Methods for the Biomedical Sciences* (Macmillan, London, 1998).

Avrameas, S. Coupling of enzymes to proteins with glutaraldehyde: use of the conjugates for the detection of antigens and antibodies. *Immunochemistry* **6**, 43–52 (1969).

Avrameas, S. and Ternynck, T. Peroxidase labelled antibody and Fab fragments with enhanced intracellular penetration. *Immunochemistry* **8**, 1175–1179 (1971).

Barbarakis, M.S. and Bachas, L.G. Isoelectric focusing electrophoresis of protein–ligand conjugates: effect of the degree of substitution. *Clin. Chem.* **37**, 87–90 (1991).

Butler, J.E. The behaviour of antigens and antibodies immobilized on a solid phase. In: *Structure of Antigens*, (ed. Van Regenmortel, M. H. V.) vol. I, 209–259 (CRC Press, Boca Raton, FL, 1992).

Franks, F. (ed), *Protein Biotechnology. Isolation, Characterization and Stabilization* (Humana Press, Totowa, NJ, 1993).

*Hermanson, G.T. *Bioconjugate Techniques* (Academic Press, New York, 1995).

Ishikawa, E. *et al.* Enzyme-labeling of antibodies and their fragments for enzyme immunoassay and immunohistochemical staining. *J. Immunoassay* **4**, 209–325 (1983).

Kenney, A. and Fowell, S. *Practical Protein Chromatography* (Humana Press, Totowa, NJ, 1992).

Lundblad, R.L., Noyes, C.M. *Chemical Reagents for Protein Modification*, vols. I and II (CRC Press, Boca Raton, FL, 1984).

Nakane, P.K. and Kawaoi, A. Peroxidase-labeled antibody: a new method of conjugation. *J. Histochem. Cytochem.* **22**, 1084–1091 (1974).

Petricoin, E.F. and Liotta, L.A. Mass spectrometry-based diagnostics. The upcoming revolution in disease detection. *Clin. Chem.* **49**, 533–534 (2003).

Pham, D.T. *et al.* Electrophoretic method for the quantitative determination of a benzyl-DTPA ligand in DTPA monoclonal antibody conjugates. *Bioconjugate Chem.* **6**, 313–315 (1995).

Russell, J.C. *et al.* Solid phase assembly of defined protein conjugates. *Bioconjugate Chem.* **13**, 958–965 (2002).

Safavy, A. *et al.* Synthesis and biological evaluation of Paclitaxel-C225 conjugate as a model for targeted drug delivery. *Bioconjugate Chem.* **14**, 302–310 (2003).

Singh, K.V. *et al.* Synthesis and characterization of hapten–protein conjugates for antibody production against small molecules. *Bioconjugate Chem.* **15**, 168–173 (2004).

*Wong, S.S. *Chemistry of Protein Conjugation and Cross-Linking* (CRC Press, Boca Raton, FL, 1991).

NOTES

*Major publications providing detailed information on conjugation methods.

15 Immunological Biosensors

James K. Gimzewski, Jason Reed, Michael A. Teitell and P. Gordon Malan

A principal impetus for developing biosensor systems has been the need to produce a simple, very rapid, sensitive, and easy-to-use analytical system that does not need trained specialists to produce results. An aim for many applications is to develop a small, portable unit into which the test sample can be applied directly, without pre-processing. The result should be obtained within seconds, and the answer should not be subject to interference or modification by the test-sample matrix, the user, or environmental factors. Sensors that use immunological detection methods are among the most advanced type of biosensor technologies due to the ubiquity of traditional immunoassay techniques and because high-detection specificity is readily obtained with receptor–ligand interaction chemistry.

Traditional 'rapid' tests for blood glucose, fertility hormones, and drugs of abuse are well proven and cost effective. Immunosensors have become more important in recent years as a result of progress in **point-of-care testing (POC)**, most of which are 'rapid' immunoassay technologies, and due to an increased focus on monitoring environmental biohazards.

POC technology made notable strides with the introduction of emergency room rapid assays for cardiac stress proteins (troponins, B-type natriuretic peptide – BNP) and in-clinic tests for infectious agents. Biosite Inc. developed a commercially successful emergency room POC assay for BNP, a biomarker for heart failure, as one recent example for acute cardiac care. Manufacturers are now marketing quick, low complexity assays of this type for a battery of cardiac markers. Several *in vitro* diagnostic companies offer a rapid optical immunoassay test designed to diagnose influenza A and B infection in the doctor's office to aid in prescription of neuraminidase inhibitors. (Mahutte, 1998; Key *et al.*, 1999; Tucker *et al.*, 2001; Azzazy and Christenson, 2002; Rodriguez *et al.*, 2002).

Detection of chemical and biological warfare agents is a primary driver of sensor development and many prototype detectors in this area utilize receptor–ligand or enzyme–reporter formats. One significant example is the work of Ligler and colleagues, who have produced an automated, portable, multi-analyte optical array biosensor for real-time biohazard detection. This sensor can detect a variety of chemical toxins and infectious agents in a complex background such as human serum. Applications for immunosensors related to biohazard detection also include pollution monitoring, food safety and industrial process control. (Iqbal *et al.*, 2000; Billman *et al.*, 2002; Sapsford *et al.*, 2004).

Continuous, *in vivo* sensing for diagnostic monitoring and drug delivery is probably the most technically demanding application for immunosensors and as such this field is in its infancy. A notable commercially-produced *in vivo* sensor is the Medtronic MiniMed Continuous Glucose Monitoring System. This prototype system consists of a subcutaneous sensor and an external monitor; in clinical studies it improved patient's glycemic control resulting in lower hemoglobin A_{1C} values. A similar but non-invasive sensor is the Glucowatch Biographer marketed by Cygnus. This device uses reverse iontophoresis to extract sample through the skin and standard amperometric enzyme detection to measure glucose concentration. Both devices have received FDA approval but limitations in their stability, accuracy and longevity prevent widespread use. (Garg *et al.*, 2004; Kubiak *et al.*, 2004; Steil *et al.*, 2004; Abel and von Woedtke, 2002).

It is impossible to describe succinctly the rapidly developing field of immunosensors, so in this chapter we will focus on the core technical categories and provide a glimpse of especially promising nascent techniques and commercially significant efforts. Many excellent reviews exist for the reader who wishes a more in-depth discussion of a specific technology. Surface plasmon resonance (SPR) techniques, a very important sub-class of immunosensors, will not be addressed as they are treated at length in a separate chapter of this book (*see* CHAPTER 16, SURFACE PLASMON RESONANCE IN KINETIC, CONCENTRATION AND BINDING SITE ANALYSES) (Imoarai *et al.*, 2001; Fermann and Suyama, 2002; Hamilton *et al.*, 2002; Kratz *et al.*, 2002; Mastrovitch *et al.*, 2002; Phillips *et al.*, 2002;

The Immunoassay Handbook. *Third Edition*
D. Wild (Ed.)

Achyuthan *et al.*, 2003; Kabir, 2003; Murray *et al.*, 2003; Price, 2003; Armor and Britton, 2004; Donovan *et al.*, 2004; Gutierres and Welty, 2004).

OVERVIEW

A biosensor uses a biological system to measure a substance and differentiate this from other substances in a test sample. It is a measurement device that is comprised of three components: a **biological receptor** of appropriate specificity for the analyte (or test material to be measured); a **transducer** to convert the recognition event into a suitable physical signal; and a **detection system**, including analysis and processing, that is usually electrical. The physical signal can, e.g. be acoustic, electromagnetic or mechanical. To bring together these three components for development of a biosensor, therefore, requires an integrated, multidisciplinary team of biologists, chemists, physicists, engineers, and computer experts. This blend of skilled personnel is not found in every establishment, so biosensor development has resulted primarily from inter-institutional collaborative projects, or within industry.

Most of today's analyses of biological samples take place in laboratories that use relatively expensive equipment and skilled personnel. Tests that involve many manipulations are increasingly being automated. Where the result is not obtained for many minutes or even several hours, attempts are made to decrease the time of analysis, often in automated systems. Microbiological determinations using conventional culture techniques can take several days or sometimes weeks, so modern alternative tests are being examined as rapid screens, particularly in the food industry in response to regulatory pressure, to satisfy consumer concerns about safety. The pharmaceutical industry is perhaps the biggest user of automated analytical screening techniques in its quest to develop better drugs, faster.

It might have been expected that biosensors would be ideally suited to address some areas requiring rapid analysis. They offer the potential to use relatively inexpensive equipment to provide results almost instantaneously outside of the laboratory, where unskilled personnel may handle a sample. Over the past 10 years, however, the biosensors that have been commercialized have succeeded only in addressing niche markets; some of the reasons will be considered below.

There are many technologies that could potentially be used for biosensing. The range of these devices, with their many overlapping principles, adaptations and modifications, makes classification and discussion of all types quite difficult. Therefore, only those devices using **immunological or enzymatic components**, and which are likely to have some commercial potential, will be mentioned here.

Generally, in clinical chemistry analyses, a lower threshold of detection at around micromolar concentrations is satisfactory for most analytes. The demands for hormone measurements in the endocrinology clinic pushed the threshold for detection to around nanomolar concentrations and then below this to the picomolar range. These measurements were performed almost exclusively with biological binding assays, often immunoassays. For some more recent applications, lower detection limits are needed, for example with DNA fragments or pesticide residues.

The ability to detect a small amount of material in a test sample is determined by the signal-to-noise ratio. Jackson and Ekins (1983) have shown that the detection limit of an immunoassay system is directly proportional to the relative error in the signal, and inversely proportional to the equilibrium constant of the binding reagent. So, a decrease in the measurement error and a high affinity constant (a low-valued equilibrium constant) would contribute to a low detection-limit for the analysis (*see* CHAPTER 1 – PRINCIPLES). Most immunological methods of analysis are subject to more-or-less severe interference from other components in the complex sample matrix. Washing or other additional manipulations are often used to minimize interference effects.

The design features of a biosensor are little different from those of any modern laboratory instrumentation system. The concept of an ideal portable biosensor would probably feature most of the following characteristics:

- Instruments should be small, self-contained, cheap, and robust, capable of interfacing with existing central laboratory systems.
- The user interface should be simple, for use by unskilled operators.
- No volumetric measurement of the specimen, e.g. by pipette, should be necessary.
- The test specimen alone should be added, with no further reagent addition being required.
- Results should be unaffected by the test-specimen matrix, e.g. water, whole blood, serum, urine, or plasma.
- The time between presentation of the specimen and final result should be rapid, and ideally less than 5 min.
- Built-in standardization and controls are required.
- An easily understood record of the results should be available.
- The detection-limit should be appropriate to the analyte, and in the sub-picomolar range for the most sensitive systems.
- A wide analytical range is required if the same biosensor is to be adapted to many different analytes: for a generic analytical system, a capability for immunochemistry, clinical chemistry, enzymology, DNA probe measurements and a variety of other applications is desirable.
- The potential for simultaneous measurement of multiple analytes should be considered.

- There should be a good correlation of results with already-established test methods.
- Biosensor consumables must be cheap to manufacture in bulk and readily available (Vo-Dinh and Cullum, 2000; Luppa *et al.*, 2001; Raiteri *et al.*, 2001; Abel and von Woedtke, 2002; Cruz *et al.*, 2002; Kusnezow and Hoheisel, 2002; O'Sullivan, 2002; Peppas and Huang, 2002; Porwal *et al.*, 2002; Raman *et al.*, 2002; Schoning and Poghossian, 2002; Albers *et al.*, 2003; Dickert *et al.*, 2003; D'Orazio, 2003; Frederix *et al.*, 2003; Ganter and Zollinger, 2003; Hierlemann and Baltes, 2003; Hierlemann *et al.*, 2003; Ikai *et al.*, 2003; Robinson *et al.*, 2003; Selvaganapathy *et al.*, 2003; Yuqing *et al.*, 2003; Monk and Walt, 2004; Sadik *et al.*, 2004; Sapsford *et al.*, 2004; Turner and Magan, 2004).

ELECTROCHEMICAL SENSORS

AMPEROMETRIC SENSORS

Enzymes provide an attractive method of signal amplification. The continual turnover of a substrate generates a cascade signal that is large and can therefore be measured quite easily. Amperometric biosensors generally use reduction–oxidation (redox) enzyme systems. In the simplest case, a redox enzyme is immobilized by some convenient procedure at an electrode surface. The electrode is held at a fixed potential, adjusted so that electrons arising from an oxidized substrate are transferred to the electrode (or *vice versa* for a reduction reaction), and this regenerates the active form of a cofactor for another redox cycle by the enzyme. The specificity of the reaction is determined by the enzyme. Because the rate of enzymic reaction at a fixed temperature and pH is directly proportional to the substrate concentration, the current produced at the electrode is proportional to the rate of modification of the substrate by the enzyme (Figure 15.1).

The rate of an enzymic reaction is dependent on the temperature, pH, ions, cofactors, and competitive or non-competitive inhibitors (or activators) present in the test sample. Any redox compound present, such as oxygen, ascorbate, thiols or certain drugs can obviously interfere with the reaction. To circumvent some of these interference effects, chemical electron-acceptors are used as mediators. Thus, for example, the ferrocene–ferricinium redox system has been used in the mediation of electron transfer from glucose oxidase to graphite electrodes. The pen-sized glucose biosensor produced originally by MediSense Inc. is based on this system, and physiological concentrations of glucose in the millimolar concentration range may be measured in whole blood (Hill and Sanghera, 1990).

Immunoassay methods are normally required for measurements of analyte concentrations in the sub-millimolar range. Various methods of detection of antibody–antigen interactions using enzyme-labeled reagents have been tried, coupling the enzymic redox reaction to an amperometric detection system (Foulds *et al.*, 1990). However, the efficiency of coupling of the biological electron-generating steps to the electrode is not fully characterized. Additionally, the effects of interfering substances are greater when the substrate concentration falls below millimolar. These and other technical difficulties, such as attaching the reagents to the membrane or electrode, have hindered the development of systems based on this principle for immunosensing (Porter, 2000; Warsinke *et al.*, 2000; Dijksma *et al.*, 2001; Liu *et al.*, 2001a; Liu, *et al.*, 2001b; Pemberton *et al.*, 2001; Porter *et al.*, 2001; Lefeber *et al.*, 2002; Li, *et al.*, 2002b; Mittelmann *et al.*, 2002; Sarkar *et al.*, 2002; Albers *et al.*, 2003; Darain *et al.*, 2003; Fahnrich *et al.*, 2003; Lei *et al.*, 2003; Zeravik *et al.*, 2003; Zhou *et al.*, 2003; Dai *et al.*, 2004; Lei, *et al.*, 2004; Zacco *et al.*, 2004).

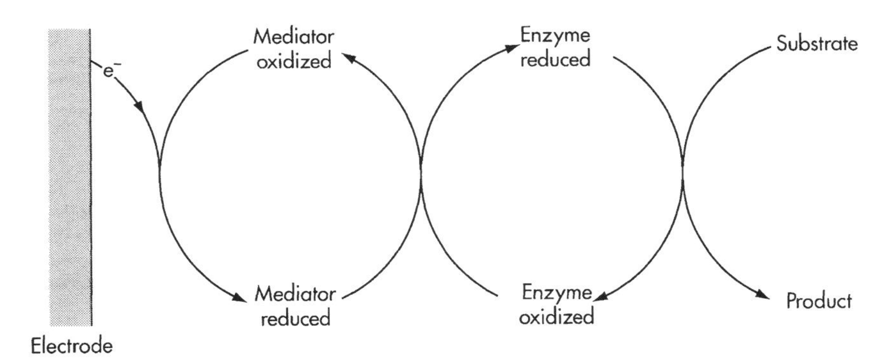

Fig. 15.1 An amperometric biosensor arrangement. A mediator is used to transfer electrons from an electrode to an enzyme-catalyzed redox reaction.

POTENTIOMETRIC SENSORS

Potentiometric devices rely on the measurement of changes in potential that arise from reaction of an analyte with a specific receptor. An extensive range of configurations has been described in which receptor molecules have been immobilized on ion-selective electrodes (Figure 15.2).

Advances in technology now allow silicon semiconductors to be coupled to a biological receptor, offering the potential of cheap, miniature, mass-produced biosensors. Several reports on the use of ion-sensitive field-effect transistors (ISFETs) as biosensors have appeared and Japanese workers are particularly active in this area. However, commercialization of these devices has been restricted because of technical difficulties associated with the reproducibility of depositing enzymatic material, its stability, and the relatively high cost of the devices compared with other systems.

When applied to immunoassay-based detection systems, this technology has also encountered problems similar to those of amperometric biosensors. Model systems can be shown to work in buffer solutions, but the interference effects that occur from materials present in the actual test specimens have restricted their more widespread application. (Holt *et al.*, 2002; Perez-Luna *et al.*, 2002; Schoning and Poghossian, 2002; Selvanayagam *et al.*, 2002; Zayats *et al.*, 2002; Besselink *et al.*, 2003; Hierlemann and Baltes, 2003; Hirano *et al.*, 2003; Yuqing *et al.*, 2003; Sadik *et al.*, 2004).

NANOMECHANICAL SENSORS

PIEZOELECTRIC MASS SENSORS

Mass detection sensors are among the most widely used microanalytic technologies. These methods rely, in general, on measuring the changes in vibrational resonant frequency of piezoelectric quartz oscillators that result from changes in mass on the oscillator's surface (see Figure 15.3).

Common configurations are the **quartz crystal microbalance (QCM)** and the **surface acoustic wave (SAW) device**. The QCM device consists of a quartz crystal disk driven by electrodes on either face. The mass of analytes that bind to the sensor is measured as a change in the crystal's resonant frequency. This type of sensor is also known as a **thickness-shear mode (TSM)** device. The mass determination in a TSM sensor is given in terms of the Sauerbrey relation:

$$\Delta M = (A\sqrt{(\rho\mu)}/2F^2)\Delta F$$

In the SAW sensor an acoustic wave is created by applying an alternating voltage to a metallized, interdigitated electrode plated onto one end of the thin piezoelectric planar substrate of the device. The acoustic wave acts on a symmetric transducer at the opposite end of the substrate, where the energy is converted back into an electrical signal. The SAW device can be altered for

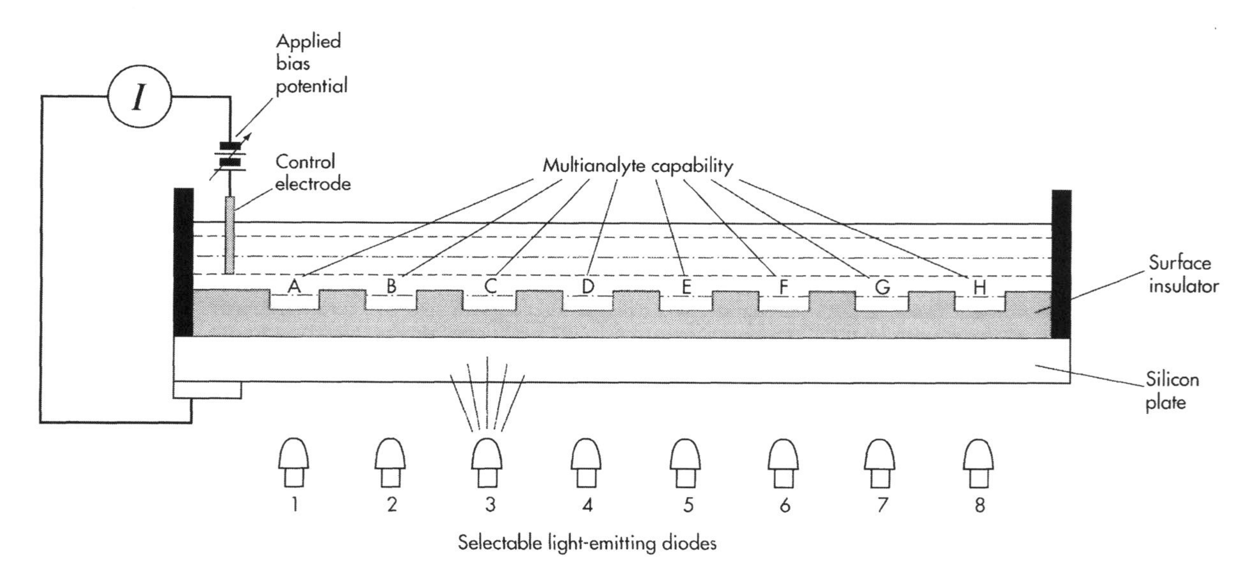

Fig. 15.2 Diagrammatic form of a light-addressable potentiometric sensor (LAPS). The underlying silicon plate has a surface insulator layer (shaded) of oxynitride. Different detection systems are located in the channels that become photo-responsive only when selectively illuminated by the light-emitting diodes. The alternating photocurrent (I) in the external circuit depends on the applied bias potential. (Redrawn from Hafeman *et al.* (1988).)

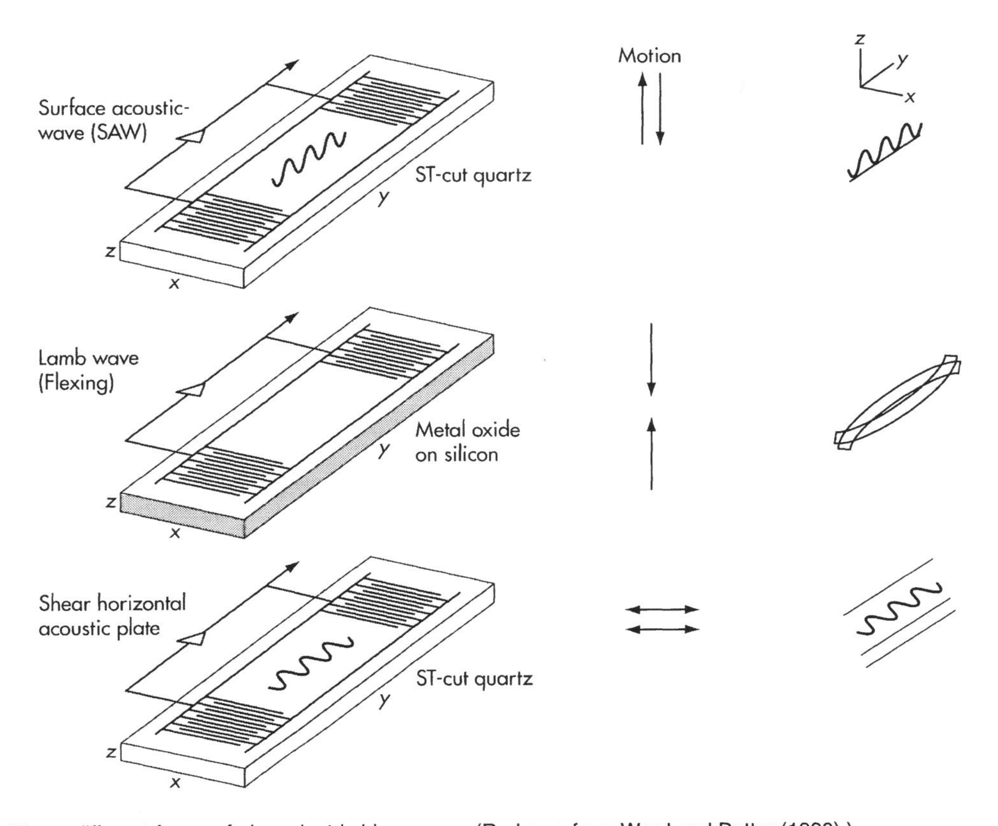

Fig. 15.3 Three different forms of piezoelectric biosensors. (Redrawn from Ward and Buttry (1990).)

particular applications by varying the interdigital spacing of the transducers, the distance between the transducers, and the thickness of the substrate.

SAW devices respond to changes in the surface wave amplitude or, more commonly, in the velocity of the acoustic wave when it interacts with molecules bound to the surface of the substrate. The thin layer of bound material alters the elasticity, density, viscosity, and conductivity of the SAW substrate. In some early work, it is apparent that the exquisite temperature sensitivity of these devices was not taken into account. There are many modes of acoustic wave that can propagate on the SAW devices, which can be used variously for sensing applications of gases, liquids or deposited solids.

Alternative acoustic modes induce particle motion on *both* surfaces of the substrate; these devices are sometimes referred to as **acoustic waveguide (AWG)** devices. The modes can be grouped into families and, by suitable selection of the type of mode, high sensitivity, and minimization of non-specific interferences can be achieved. Over the past decade, it has become clear that, for applications such as immunosensors, SAW devices should operate in shear horizontal (SH) or surface transverse wave (STW) modes (Collings and Caruso, 1997).

The so-called **Love wave device**, sometimes referred to as a surface-skimming bulk wave, is obtained when a layer of the appropriate thickness and acoustic properties is deposited over a conventional SH device. The energy of the bulk wave is concentrated in the guiding layer because the shear acoustic velocity in this layer is lower than in the substrate, leading to the alternative name of the surface-guided SH wave. The sensitivity to mass loading is increased by focusing the energy in this layer, which is dependent on the layer thickness and its acoustic properties.

The latter devices have the advantage of avoiding radiation losses in liquids, yet have much better sensitivity to mass loading by concentrating the energy near the surface. The SH wave may be guided along device surfaces by gratings as well as over-layers, as either SH waves or as STW. The improved sensitivity compared with simple SH devices, without much greater complexity in fabrication, holds much promise for their use in liquid-phase biosensing. This is because the viscous loading contribution of the solvent, in its attenuation of the shear wave in other modes, shifts the resonance of the transducer with consequent alteration in sensitivity.

Using conventional antibody–antigen interactions on thin substrates, detection limits of nanomolar down to picomolar concentrations have been claimed with 'ideal' test samples, though this is perhaps unlikely to be achieved with typical clinical specimens. Manufacture of the devices to produce cheap, disposable units with uniform characteristics may be relatively easy, although provision of multi-analyte analysis on a single unit would provide a challenge (Cavic *et al.*, 1999; O'Sullivan and Guilbault, 1999; O'Sullivan *et al.*, 1999; Kaiser *et al.*, 2000; Shen *et al.*, 2000; Uttenthaler *et al.*, 2001; Zhou and Cao, 2001; Chou *et al.*, 2002; Eun *et al.*, 2002; He and Zhang, 2002; Li *et al.*, 2002a; Liss *et al.*, 2002; Sota *et al.*, 2002; Wong *et al.*, 2002; Aizawa *et al.*, 2003; Kim *et al.*, 2003a; Kim *et al.*, 2003b; Kim and Park, 2003; Marx, 2003; Ruan *et al.*, 2003; Stubbs *et al.*, 2003; Tamarin *et al.*, 2003; Killard and Smyth, 2004; Laricchia-Robbio and Revoltella, 2004; Schaible *et al.*, 2004).

MICROCANTILEVER SENSORS

The microcantilever is an emerging and particularly versatile class of sensor which is unique in its combination of simplicity, sensitivity, and potential for miniaturization. Microcantilevers can sense and quantitate biological proteins, nucleic acids, and a variety of organic and inorganic chemical species. The principal mechanism of action for sensing is a nanoscopic deflection caused by receptor–ligand-induced stress on one face of the microcantilever. The deflection signal can be recorded with an optical lever, an interferometer or a piezoresistive element. When operated in 'active mode', cantilever sensors are induced to oscillate at their resonant frequency and function very much like the mass-detection sensors described above.

Gimzewski and colleagues produced the first cantilever array immunosensor, which could distinguish species-specific binding of protein A to rabbit IgG. The same sensor design could also detect single base-pair mismatches in DNA oligonucleotide hybridization experiments. Another iteration of this device, a 'nanomechanical nose', used an array of non-specific polymer probes to distinguish hydrogen, primary alcohols, natural flavors, and water vapor in air (Baller *et al.*, 2000; Fritz *et al.*, 2000; Lang *et al.*, 2002; Arntz *et al.*, 2003; Yue *et al.*, 2004).

Majumdar and co-workers produced a prototype POC cantilever immunosensor for prostate specific antigen (PSA). This device detected physiologic levels of free PSA in a high background of albumin and plasminogen serum proteins (Wu *et al.*, 2001; Majumdar, 2002; Yue *et al.*, 2004).

Thundat and colleagues have used single cantilever nanosensors to detect heavy metal ions, neurotoxins, and glucose. The latter application utilized cantilever-bound glucose oxidase enzyme as a detector–reporter element (Cherian *et al.*, 2002; Stevenson *et al.*, 2002; Cherian *et al.*, 2003; Yang *et al.*, 2003; Pei *et al.*, 2004).

Baltes and co-workers developed techniques for manufacturing cantilever sensors using a CMOS process, which provides the potential advantage of mass production for commercial applications. Their fully integrated device contains a Wheatstone bridge to sense cantilever bending, eliminating the need for external optics (Franks *et al.*, 2002; Hierlemann and Baltes, 2003; Hierlemann *et al.*, 2003).

Similarly, Boisen *et al.* have produced an active cantilever array system with completely integrated on-chip actuation and deflection sensors, highlighting the miniaturization potential that gives cantilever sensors an advantage over many competing technologies (Boisen *et al.*, 2000; Grogan *et al.*, 2002; Davis *et al.*, 2003).

Veeco offers the Scentis, an eight cantilever array, bench-top laboratory instrument. Startup companies Protiveris, Cantion, and Concentris are all attempting to perfect and commercialize cantilever technology (Luckham and Smith, 1998; Moulin *et al.*, 2000; Hansma, 2001; Ilic *et al.*, 2001; Allison *et al.*, 2002; Marie *et al.*, 2002; Sepaniak *et al.*, 2002; Arntz *et al.*, 2003; Dutta *et al.*, 2003; Kooser *et al.*, 2003; Liu *et al.*, 2003a; Rasmussen *et al.*, 2003; Ilic *et al.*, 2004; Zhang and Ji, 2004).

MICROMAGNETIC SENSORS

This sensor technology is less mature than others but has shown promise in proof-of-concept studies. Magnetically based nanomechanical sensors make use of the weak interaction of magnetic fields with the sample and are related to magnetic bead-based assays that are common in molecular biology. Experiments that measure receptor–ligand binding forces make use of magnetic particles to precisely control applied unbinding forces. The relatively inert magnetic particles can be translocated and rotated in a sample to measure force in a variety of directions and when coated with the appropriate sensor element can even probe receptor–ligand interactions inside single cells (Kausch and Bruce, 1994; Schalkhammer, 1998; Ptak *et al.*, 2001; Richardson *et al.*, 2001; Tanaka and Matsunaga, 2001; Graham *et al.*, 2003; Kim *et al.*, 2003a,b; Liu *et al.*, 2003b; Puckett *et al.*, 2003; Ruan *et al.*, 2003; Hong *et al.*, 2004; Weizmann *et al.*, 2004).

OPTICAL SENSORS

As with the other types of biosensor outlined above, there are now a large number of developments that make use of the physical properties of light in various ways to detect small changes in analyte concentration. In an immunosensor configuration, optical methods often rely on fiber-optic input and collection elements, with the receptor–ligand pair bound to the fiber surface or in the nearby fluid. Detection and quantitation typically make use of fluorescent tags or bio-/chemiluminecent reporters.

FLUORESCENT EVANESCENT WAVE SENSORS

This technique relies on the properties of light when it is reflected at a surface between two transparent media of different refractive indices. A light beam is totally internally reflected within an **optical waveguide** (e.g. an optical fiber) and, at the points of reflection, part of the light enters the external, lower refractive index medium before it returns to the waveguide. This results in an **evanescent field** which penetrates only a fraction of a wavelength of the light into the surrounding medium, and which decays exponentially with distance from the surface (Figure 15.4). Because there are multiple reflections within an optical fiber, it is essentially covered by this evanescent field, whose external influence is limited to within a few 100 nanometers from the fiber surface. Thus, if the optical fiber is dipped into a solution containing a fluorophore with an appropriate excitation wavelength, only those molecules in the evanescent field near the surface will fluoresce (Sutherland *et al.*, 1984).

When appropriate receptors or antibodies are attached to the fiber surface, then either competitive or immunometric assays can be performed using fluorophore-labeled reagents. An immunoglobulin G (IgG) antibody has dimensions of 5×15 nm, i.e. an average of about 10 nm. Although the influence of the evanescent field will extend a few 100 nanometers into the medium, the concentration of unbound fluorophore will be low, so a majority of the fluorescence will arise from specifically bound fluorophore at the fiber surface. Detection limits at nanomolar concentrations have been achieved with measurement times of only a few seconds.

A major deficiency with most fluorescent measurement techniques is that the fluorophore and exciting light are both influenced by the test specimen. When the specimen is serum, plasma or whole blood, light absorption and fluorescence quenching can significantly decrease the measured signal in a specimen-dependent manner. Alternatively, interference can occur from endogenous fluorophores present in the specimen; this is normally minimized by introducing a washing step. Although fluorescent techniques have the potential for great sensitivity, and hence even lower detection-limits, the interference effects have generally limited the attainment of very low detection-limits in real samples. Despite this limitation, commercial developments have been investigated by several companies (Squillante, 1998; Spiker and Kang, 1999; Cui *et al.*, 2000; DeLisa *et al.*, 2000; Marks *et al.*, 2000; Moreno-Bondi *et al.*, 2000; Blair and Chen, 2001; Anderson and Nerurkar, 2002; Balcer *et al.*, 2002; Neumann *et al.*, 2002; Vo-Dinh, 2002; Krioukov *et al.*, 2003; Liebermann and Knoll, 2003; Liu *et al.*, 2003b; Tedeschi *et al.*, 2003; Ekgasit *et al.*, 2004; Garden *et al.*, 2004; Monk and Walt, 2004).

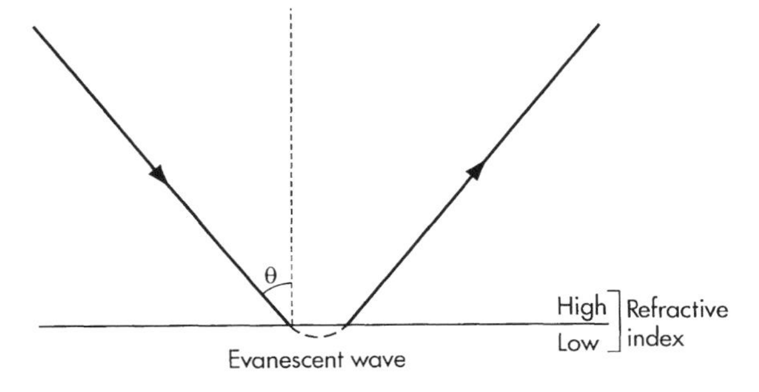

Fig. 15.4 Diagrammatic representation of the evanescent wave. The evanescent field decays exponentially with distance away from the high refractive-index surface.

INTEGRATED OPTICAL SENSORS

As mentioned above, Ligler and co-workers have developed and field tested an integrated optical immunosensor which is capable of detecting a variety of toxic substances and pathogens in real-world situations. This prototype is fully automated and portable, which is essential for continuous monitoring applications. Efforts in this area now focus on miniaturizing the optics and detector elements so that the entire device can be handheld. Further, most of the detection chemistry is antibody-based and new efforts are needed to expand the range of probe chemistries to accommodate a wider array of analytes (Rowe-Taitt *et al.*, 2000a; Rowe-Taitt *et al.*, 2000b; Rowe-Taitt *et al.*, 2000c; Sapsford *et al.*, 2001; Delehanty and Ligler, 2002; Holt *et al.*, 2002; Sapsford *et al.*, 2002; Taitt *et al.*, 2002; Sapsford *et al.*, 2003; Sapsford *et al.*, 2004).

One integrated optical sensor that has made it to the clinic is the **optode continuous intravascular blood gas monitor**. An example of this type of device is the FDA-approved Paratrend 7 from Diametrics Medical, which is a disposable, sterile, single-use fiber-optic sensor for continuous measurement of pH, pCO_2, pO_2 and temperature. This device is used in critical care situations to provide real-time oxygenation, ventilation, and metabolic data. Sensors of this type use a variety of biochemical reporter methods including chemically sensitive dyes and compound selective membranes; they are also compatible with affinity probe chemistries and enzyme-based reporter systems (Mahutte, 1998; Ganter and Zollinger, 2003).

QUANTUM DOTS

Quantum dots are luminescent semiconductor nanocrystals that behave similarly to fluorescent reporter molecules. They are unique in that the emitted light is confined to a very narrow frequency band making them ideal for use in multiplex assays. Also, quantum dots are much more physically robust than fluorescent molecules, which suffer from photobleaching and other undesirable photochemistry. Goldman and colleagues have conjugated antibodies to quantum dots for use in multiplex immunosensing. They detected and quantitated biotoxins including cholera toxin, ricin, shiga-like toxin 1, and staphylococcal enterotoxin B in a microtiter format (Cunin *et al.*, 2002; Medintz *et al.*, 2003; West and Halas, 2003; Goldman *et al.*, 2004).

CONCLUSIONS AND FUTURE DIRECTIONS

The benefits and deficiencies of the various types of biosensor systems discussed above are summarized in Table 15.1. The amperometric and potentiometric sensors have been intensively investigated over the past 35 years since Clark and Lyons (1962) described the first glucose sensor. It is notable that the most advanced commercial glucose sensors still use electrochemical detection methods. In studies to date electrochemical methods seem to be less versatile than other techniques, but they are competitive in applications where the proper chemistry exists, while ISFET efforts have great miniaturization potential.

Nanomechanical sensors such as microcantilevers, or one of the several optical devices described are the most likely candidates as general-purpose bioanalytical sensors of the future. They are easy to produce and have potential for use over a wide range of analyte concentrations. They also approach the low detection limits required for immunosensing, while the reproducibility of results is similar to that achieved by conventional immunoassays. These devices rely on relatively well-understood physical principles, and the associated instrumentation is quite simple to assemble. The analyte-detection elements in the optical or the piezoelectric sensors are relatively cheap and easy to construct.

The fluorescent methods used with the optical sensors, although intrinsically more sensitive than the other optical methods, suffer from potentially greater interference effects arising from the test-specimen matrix. These devices have not yet met all the criteria listed in this chapter for an 'ideal' biosensor, particularly with regard to miniaturization; but they have met most of these. Future developments are likely to extend their flexibility and deployment within the next 20 years.

A major development in the biosensor field is directed at designing a reaction cell in which both the kinetics of analytical reaction and flow characteristics at the sensor surface are optimized for ease of manufacture and use. The key to reproducible manufacture and consequent ease of use of these biosensor systems is the uniformity of physical and chemical properties of the biophysical interface. Instrumentation developments center around miniaturization, improved signal processing, and addressing the possibility of analyzing multiple analytes simultaneously on the same reaction-cell sensor surface. The chemistry of reagent deposition, assay format and other aspects related to the use of the reagents are, in general, not very different from those encountered with immunoassay systems. Attention has to be given to using

Table 15.1. Biosensors: a comparison of technologies.

Sensor technology	Main advantages	Main disadvantages
Amperometric	Proven technology Clinical chemistry analytes (micromolar)	Susceptible to interference Limited low detection limit Immunosensing difficult
Potentiometric	Potentially cheap microchip manufacturing technology	High set-up manufacture costs Limited low-detection limit Limited reproducibility
Light-addressable potentiometric	Proven technology Potentiometric stability Multi-analyte capability	Slow near lower detection limit Multiple steps, including washing
Fluorescence evanescent-wave	Rapid (<5 min)	Lower detection limit limited by sample interferences Non-homogeneous format
Surface acoustic wave	Theoretical lower detection limit is picomolar	High cost of manufacture of identical units Temperature sensitivity Non-specific effects

assay formats in which the signal change is large enough to measure. The assay system should be a homogeneous one, with no separation or measurement step required, so the design of the assay format becomes in some respects easier than for a conventional immunoassay.

Much of this discussion has focused on methods to detect a receptor–ligand binding event that is compatible with real-time, continuous monitoring. These technologies have become relatively sophisticated and each has strengths and weaknesses that vary by application. In the future we expect more research emphasis on design of the biochemical receptor elements themselves. Most current sensors use canonical antibody–antigen chemistry or something very similar. There is great need for probes which are more stable, more versatile in terms of recognition while retaining high specificity and which can be regenerated *in situ*. Some examples of new probe chemistries include molecularly imprinted polymers (MIPs), 'smart-polymers' and RNA aptamers (de Wildt *et al.*, 2000; Iqbal *et al.*, 2000; Kaiser *et al.*, 2000; Mishra and Schwartz, 2002; O'Sullivan, 2002).

We also expect more advanced probe design to be accompanied by signal processing informatics that compensate for the unavoidable non-specific background found in real-world biosensing situations (Baller *et al.*, 2000; Reder *et al.*, 2003; Turner and Magan, 2004).

Array-based approaches are already very popular in nucleic acid detection and much effort is being expended to produce reliable protein–protein recognition arrays. Arrays would allow 'super multiplex' analysis of samples and are needed to capture the complex patterns of disease states; this aspect of molecular diagnostics is severely limited by today's single- or few-analyte rapid immunoassay methods (Blank *et al.*, 2003; Niemeyer *et al.*, 1994; Rowe-Taitt *et al.*, 2000b; Delehanty and Ligler, 2002; Kusnezow and Hoheisel, 2002; Abedinov *et al.*, 2003; Albrecht, *et al.*, 2003; Arntz *et al.*, 2003; Pavlickova *et al.*, 2003; Peluso *et al.*, 2003).

Integrated immunosensors require some form of sample handling and for the miniaturization necessary, this will involve microfluidics. Microfluidics is a broad and promising field with many iterations of tiny fluid circuits, pumps, separation devices, etc. One successful commercial example is the Caliper Technologies LabChip, which allows very high density formats for traditional ligand–receptor and enzyme-relayed assays. The primary market for this device is drug discovery but in time this technology and its derivatives will move to the clinical laboratory and replace large immunoassay machines and other traditional instruments (Wells, 1998; Dutton, 1999; Nachamkin *et al.*, 2001; Christodoulides *et al.*, 2002; Schmut *et al.*, 2002; Wu *et al.*, 2003).

Cost per test/datapoint is very important when comparing immunosensors to traditional immunoassay technology. Standard ELISA assays are very sensitive and inexpensive, which creates a high hurdle for sensor technologies in standard applications. Thus, we expect immunosensors to emerge first in applications where real-time continuous sensing is a chief concern and cost/sensitivity vs. traditional techniques is less important. We have described a number of efforts to use semiconductor-like manufacturing technologies to produce immunosensors cheaply and in quantity. This, along with further miniaturization, will be necessary to make immunosensors competitive with current ELISA-type assays in many diagnostic applications (Hierlemann *et al.*, 2003; Lee, 2003).

Continuous, *in vivo* biosensing would revolutionize the diagnosis of disease and the controlled delivery of therapeutics. The blood glucose sensing technology already mentioned in this chapter and the arterial blood gas optode mark early but notable efforts. In reality, researchers have just begun to address the many substantive hurdles which exist in this area: these include complete sensor/readout integration, biocompatibility and long-term, *in vivo* sensor stability. Finally, the widespread use of biosensors is limited in many cases not by technology development but by questions about the utility of various biologic and environmental markers (Gerritsen *et al.*, 1999; Trettin *et al.*, 1999; Frost and Meyerhoff, 2002; Musham and Swanson, 2002; Nicolette and Miller, 2003; Yancy, 2003; Coradini and Daidone, 2004; Mark and Felker, 2004; Roongsritong *et al.*, 2004).

REFERENCES AND FURTHER READING

Abedinov, N., Popov, C. *et al.* Chemical recognition based on micromachined silicon cantilever array. *J. Vac. Sci. Technol. B* **21**, 2931–2936 (2003).

Abel, P.U. and von Woedtke, T. Biosensors for *in vivo* glucose measurement: can we cross the experimental stage. *Biosens. Bioelectron.* **17**, 1059–1070 (2002).

Achyuthan, K.E., Pence, L.M. *et al.* ZstatFlu (R)-II test: a chemiluminescent neuraminidase assay for influenza viral diagnostics. *Luminescence* **18**, 131–139 (2003).

Aizawa, M. Immunosensors. In: *Biosensor Principles and Applications* (eds Blum, L.J. and Coulet, P.R.), 249–266 (Marcel Dekker Inc., New York, 1991).

Aizawa, H., Kurosawa, S. *et al.* Conventional detection method of fibrinogen and fibrin degradation products using latex piezoelectric immunoassay. *Biosens. Bioelectron.* **18**, 765–771 (2003).

Albers, J., Grunwald, T. *et al.* Electrical biochip technology – a tool for microarrays and continuous monitoring. *Anal. Bioanal. Chem.* **377**, 521–527 (2003).

Albrecht, C., Blank, K. *et al.* DNA: a programmable force sensor. *Science* **301**, 367–370 (2003).

Allison, D.P., Hinterdorfer, P. *et al.* Biomolecular force measurements and the atomic force microscope. *Curr. Opin. Biotechnol.* **13**, 47–51 (2002).

Anderson, G.P. and Nerurkar, N.L. Improved fluoroimmunoassays using the dye Alexa Fluor 647 with the RAPTOR, a fiber optic biosensor. *J. Immunol. Methods* **271**, 17–24 (2002).

Armor, B.L. and Britton, M.L. Diabetes mellitus non-glucose monitoring: point-of-care testing. *Ann. Pharmacother.* **38**, 1039–1047 (2004).

Arntz, Y., Seelig, J.D. *et al.* Label-free protein assay based on a nanomechanical cantilever array. *Nanotechnology* **14**, 86–90 (2003).

Attridge, J.W., Daniels, P.B., Deacon, J.K., Robinson, G.A. and Davidson, G.P. Sensitivity enhancement of optical immunosensors by the use of a surface plasmon resonance fluoroimmunoassay. *Biosens. Bioelect.* **6**, 201–214 (1991).

Azzazy, H. and Christenson, R. Cardiac markers of acute coronary syndromes: is there a case for point-of-care testing? *Clin. Biochem.* **35**, 13–27 (2002).

Balcer, H.I., Kwon, H.J. *et al.* Assay procedure optimization of a rapid, reusable protein C immunosensor for physiological samples. *Ann. Biomed. Eng.* **30**, 141–147 (2002).

Baller, M.K., Lang, H.P. *et al.* A cantilever array-based artificial nose. *Ultramicroscopy* **82**, 1–9 (2000).

Besselink, G.A., Schasfoort, R.B. *et al.* Modification of ISFETs with a monolayer of latex beads for specific detection of proteins. *Biosens. Bioelectron.* **18**, 1109–1114 (2003).

Billman, G., Hughes, A. *et al.* Clinical performance of an in-line, *ex vivo* point-of-care monitor: a multicenter study. *Clin. Chem.* **48**, 2030–2043 (2002).

Blair, S. and Chen, Y. Resonant-enhanced evanescent-wave fluorescence biosensing with cylindrical optical cavities. *Appl. Opt.* **40**, 570–582 (2001).

Blank, K., Mai, T. *et al.* A force-based protein biochip. *Proc. Natl Acad. Sci. USA* **100**, 11356–11360 (2003).

Boisen, A., Thaysen, J. *et al.* Environmental sensors based on micromachined cantilevers with integrated read-out. *Ultramicroscopy* **82**, 11–16 (2000).

Cavic, B.A., Hayward, G.L. *et al.* Acoustic waves and the study of biochemical macromolecules and cells at the sensor–liquid interface. *Analyst* **124**, 1405–1420 (1999).

Cherian, S., Mehta, A. *et al.* Investigating the mechanical effects of adsorption of Ca^{2+} ions on a silicon nitride microcantilever surface. *Langmuir* **18**, 6935–6939 (2002).

Cherian, S., Gupta, R.K. *et al.* Detection of heavy metal ions using protein-functionalized microcantilever sensors. *Biosens. Bioelectron.* **19**, 411–416 (2003).

Chou, S.F., Hsu, W.L. *et al.* Determination of alpha-fetoprotein in human serum by a quartz crystal microbalance-based immunosensor. *Clin. Chem.* **48**, 913–918 (2002).

Christodoulides, N., Tran, M. *et al.* A microchip-based multianalyte assay system for the assessment of cardiac risk. *Anal. Chem.* **74**, 3030–3036 (2002).

Clark, L.C. and Lyons, C. Electrode systems for continuous monitoring in cardiovascular surgery. *Ann. NY Acad. Sci.* **102**, 29–45 (1962).

Collins, A.F. and Caruso, F. Biosensors: recent advances. *Rep. Prog. Phys.* **60**, 1397–1445 (1997).

Coradini, D. and Daidone, M.G. Biomolecular prognostic factors in breast cancer. *Curr. Opin. Obstet. Gynaecol.* **16**, 49–55 (2004).

Cruz, H.J., Rosa, C.C. *et al.* Immunosensors for diagnostic applications. *Parasitol. Res.* **88** (13 Suppl 1), S4–S7 (2002).

Cui, X.Q., Pei, R.J. *et al.* Detection of anti-human serum albumin antibody using surface plasmon resonance biosensor. *Chin. J. Anal. Chem.* **28**, 950–955 (2000).

Cunin, F., Schmedake, T.A. *et al.* Biomolecular screening with encoded porous-silicon photonic crystals. *Nat. Mater.* **1**, 39–41 (2002).

Dai, Z., Yan, F. *et al.* Novel amperometric immunosensor for rapid separation-free immunoassay of carcinoembryonic antigen. *J. Immunol. Methods* **287**, 13–20 (2004).

Darain, F., Park, S.U. *et al.* Disposable amperometric immunosensor system for rabbit IgG using a conducting polymer modified screen-printed electrode. *Biosens. Bioelectron.* **18**, 773–780 (2003).

Davis, Z.J., Abadal, G. *et al.* Monolithic integration of mass sensing nano-cantilevers with CMOS circuitry. *Sens. Actuators A (Phys.)* **105**, 311–319 (2003).

Delehanty, J.B. and Ligler, F.S. A microarray immunoassay for simultaneous detection of proteins and bacteria. *Anal. Chem.* **74**, 5681–5687 (2002).

DeLisa, M.P., Zhang, Z. *et al.* Evanescent wave long-period fiber bragg grating as an immobilized antibody biosensor. *Anal. Chem.* **72**, 2895–2900 (2000).

de Wildt, R.M., Mundy, C.R. *et al.* Antibody arrays for high-throughput screening of antibody–antigen interactions. *Nat. Biotechnol.* **18**, 989–994 (2000).

Dickert, F.L., Lieberzeit, P. *et al.* Sensor strategies for microorganism detection – from physical principles to imprinting procedures. *Anal. Bioanal. Chem.* **377**, 540–549 (2003).

Dijksma, M., Kamp, B. *et al.* Development of an electrochemical immunosensor for direct detection of interferon-gamma at the attomolar level. *Anal. Chem.* **73**, 901–907 (2001).

Donovan, B.J., Rublein, J.C. *et al.* HIV infection: point-of-care testing. *Ann. Pharmacother.* **38**, 670–676 (2004).

D'Orazio, P. Biosensors in clinical chemistry. *Clin. Chim. Acta.* **334**, 41–69 (2003).

Dutta, P., Tipple, C.A. *et al.* Enantioselective sensors based on antibody-mediated nanomechanics. *Anal. Chem.* **75**, 2342–2348 (2003).

Dutton, G. Hewlett Packard and Caliper launch microfluidics lab-on-chip – HP 2100 bioanalyzer debuts for biotech and pharmaceutical industries. *Genet. Eng. News* **19**, 1 (1999).

Ekgasit, S., Thammacharoen, C. *et al.* Evanescent field in surface plasmon resonance and surface plasmon field-enhanced fluorescence spectroscopies. *Anal. Chem.* **76**, 2210–2219 (2004).

Eun, A.J., Huang, L. *et al.* Detection of two orchid viruses using quartz crystal microbalance (QCM) immunosensors. *J. Virol. Methods* **99**, 71–79 (2002).

Fahnrich, K.A., Pravda, M. *et al.* Disposable amperometric immunosensor for the detection of polycyclic aromatic hydrocarbons (PAHs) using screen-printed electrodes. *Biosens. Bioelectron.* **18**, 73–82 (2003).

Fermann, G.J. and Suyama, J. Point of care testing in the Emergency Department. *J. Emerg. Med.* **22**, 393–404 (2002).

Foulds, N.C., Frew, J.E. and Green, M.J. Immunoelectrodes. In: *Biosensors: A Practical Approach* (ed Cass, A.E.G.), 97–124 (Oxford University Press, Oxford, 1990).

Franks, W., Lange, D. *et al.* Nanochemical surface analyzer in CMOS technology. *Ultramicroscopy* **91**, 21–27 (2002).

Frederix, P., Akiyama, T. *et al.* Atomic force bio-analytics. *Curr. Opin. Chem. Biol.* **7**, 641–647 (2003).

Fritz, J., Baller, M.K. *et al.* Translating biomolecular recognition into nanomechanics. *Science* **288**, 316–318 (2000).

Frost, M. and Meyerhoff, M.E. Implantable chemical sensors for real-time clinical monitoring: progress and challenges. *Curr. Opin. I Chem. Biol.* **6**, 633–641 (2002).

Ganter, M. and Zollinger, A. Continuous intravascular blood gas monitoring: development, current techniques, and clinical use of a commercial device. *Br. J. Anaesth.* **91**, 397–407 (2003).

Garden, S.R., Doellgast, G.J. *et al.* A fluorescent coagulation assay for thrombin using a fibre optic evanescent wave sensor. *Biosens. Bioelectron.* **19**, 737–740 (2004).

Garg, S.K., Hoff, H.K. *et al.* The role of continuous glucose sensors in diabetes care. *Endocrinol. Metab. Clin. N. Am.* **33**, 163 (2004).

Gerritsen, M., Jansen, J. *et al.* Performance of subcutaneously implanted glucose sensors for continuous monitoring. *Neth. J. Med.* **54**, 167–179 (1999).

Goldman, E.R., Clapp, A.R. *et al.* Multiplexed toxin analysis using four colors of quantum dot fluororeagents. *Anal. Chem.* **76**, 684–688 (2004).

Graham, D.L., Ferreira, H.A. *et al.* High sensitivity detection of molecular recognition using magnetically labelled biomolecules and magnetoresistive sensors. *Biosens. Bioelectron.* **18**, 483–488 (2003).

Grogan, C., Raiteri, R. *et al.* Characterisation of an antibody coated microcantilever as a potential immuno-based biosensor. *Biosens. Bioelectron.* **17**, 201–207 (2002).

Gutierres, S.L. and Welty, T.E. Point-of-care testing: an introduction. *Ann. Pharmacother.* **38**, 119–125 (2004).

Hafeman, D.G., Parce, J.W. and McConnell, H.M. Light-addressable potentiometric sensor for biochemical systems. *Science* **240**, 1182–1185 (1988).

Hamilton, M.S., Abel, D.M. *et al.* Clinical evaluation of the ZstatFlu-II test: a chemiluminescent rapid diagnostic test for influenza virus. *J. Clin. Microbiol.* **40**, 2331–2334 (2002).

Hansma, H.G. Surface biology of DNA by atomic force microscopy. *Ann. Rev. Phys. Chem.* **52**, 71–92 (2001).

He, F. and Zhang, L. Rapid diagnosis of M. tuberculosis using a piezoelectric immunosensor. *Anal. Sci.* **18**, 397–401 (2002).

Hierlemann, A. and Baltes, H. CMOS-based chemical microsensors. *Analyst* **128**, 15–28 (2003).

Hierlemann, A., Brand, O. *et al.* Microfabrication techniques for chemical/biosensors. *Proc. IEEE* **91**, 839–863 (2003).

Hill, H.A.O. and Sanghera, G.S. Mediated amperometric enzyme electrodes. In: *Biosensors: A Practical Approach* (ed Cass, A.E.G.), 19–46 (Oxford University Press, Oxford, 1990).

Hirano, A., Wakabayashi, M. *et al.* A single-channel sensor based on gramicidin controlled by molecular recognition at bilayer lipid membranes containing receptor. *Biosens. Bioelectron.* **18**, 973–983 (2003).

Holt, D.B., Gauger, P.R. *et al.* Fabrication of a capillary immunosensor in polymethyl methacrylate. *Biosens. Bioelectron.* **17**, 95–103 (2002).

Hong, X., Guo, W. *et al.* Preparation of nano-scaled magnetic biological probes of Fe_3O_4/dextran/antibody and chromatographic assay. *Chem. J. Chin. Univ (Chin.)* **25**, 445–447 (2004).

Ikai, A., Afrin, R. *et al.* Nano-mechanical methods in biochemistry using atomic force microscopy. *Curr. Protein Pept. Sci.* **4**, 181–193 (2003).

Ilic, B., Czaplewski, D. *et al.* Single cell detection with micromechanical oscillators. *J. Vac. Sci. Technol. B* **19**, 2825–2828 (2001).

Ilic, B., Craighead, H.G. *et al.* Attogram detection using nanoelectromechanical oscillators. *J. Appl. Phys.* **95**, 3694–3703 (2004).

Imoarai, T., Saitoh, N. *et al.* The development of POCTEM Flu A/B – a point of care test for the detection of influenza A and influenza B viruses. *Clin. Chem.* **47**, A-184 (2001).

Iqbal, S.S., Mayo, M.W. *et al.* A review of molecular recognition technologies for detection of biological threat agents. *Biosens. Bioelectron.* **15**, 549–578 (2000).

Jackson, T.M. and Ekins, R.P. Theoretical limits on immunoassay sensitivity: current practice and potential advantages of fluorescent Eu3 + chelates on nonradioactive tracers. *J. Immunol. Meth.* **87**, 13–20 (1983).

Kabir, S. Review article: clinic-based testing for *Helicobacter pylori* infection by enzyme immunoassay of faeces, urine and saliva. *Aliment. Pharmacol. Ther.* **17**, 1345–1354 (2003).

Kaiser, T., Gudat, P. *et al.* Biotinylated steroid derivatives as ligands for biospecific interaction analysis with monoclonal antibodies using immunosensor devices. *Anal. Biochem.* **282**, 173–185 (2000).

Kausch, A.P. and Bruce, B.D. Isolation and immobilization of various plastid subtypes by magnetic immunoabsorption. *Plant J.* **6**, 767–779 (1994).

Key, G., Schreiber, A. *et al.* Multicenter evaluation of an amperometric immunosensor for plasma fatty acid-binding protein: an early marker for acute myocardial infarction. *Clin. Biochem.* **32**, 229–231 (1999).

Killard, A.J. and Smyth, M.R. Biosensors. In: *Biomolecular Films*, vol. 111, Surfactant Series (ed Rusling, J.F.), 451–498 (Marcel Dekker, New York, 2003).

Kim, N. and Park, I.S. Application of a flow-type antibody sensor to the detection of *Escherichia coli* in various foods. *Biosens. Bioelectron.* **18**, 1101–1107 (2003).

Kim, G.H., Rand, A.G. *et al.* Impedance characterization of a piezoelectric immunosensor part II: *Salmonella typhimurium* detection using magnetic enhancement. *Biosens. Bioelectron.* **18**, 91–99 (2003a).

Kim, G.H., Rand, A.G. *et al.* Impedance characterization of a piezoelectric immunosensor. Part I: antibody coating and buffer solution. *Biosens. Bioelectron.* **18**, 83–89 (2003b).

Kooser, A., Manygoats, K. *et al.* Investigation of the antigen antibody reaction between anti-bovine serum albumin (a-BSA) and bovine serum albumin (BSA) using piezoresistive microcantilever based sensors. *Biosens. Bioelectron.* **19**, 503–508 (2003).

Kratz, A., Januzzi, J.L. *et al.* Positive predictive value of a point-of-care testing strategy on first-draw specimens for the emergency department-based detection of acute coronary syndromes. *Arch. Pathol. Lab. Med.* **126**, 1487–1493 (2002).

Krioukov, E., Greve, J. *et al.* Performance of integrated optical microcavities for refractive index and fluorescence sensing. *Sens. Actuators B (Chem.)* **90**, 58–67 (2003).

Kubiak, T., Hermanns, N. *et al.* Assessment of hypoglycaemia awareness using continuous glucose monitoring. *Diabetic Med.* **21**, 487–490 (2004).

Kusnezow, W. and Hoheisel, J.D. Antibody microarrays: promises and problems. *Biotechniques* **33** (Suppl.), 14–23 (2002).

Lang, H.P., Hegner, M. *et al.* Nanomechanics from atomic resolution to molecular recognition based on atomic force microscopy technology. *Nanotechnology* **13**, R29–R36 (2002).

Laricchia-Robbio, L. and Revoltella, R.P. Comparison between the surface plasmon resonance (SPR) and the quartz crystal microbalance (QCM) method in a structural analysis of human endothelin-1. *Biosens. Bioelectron.* **19**, 1753–1758 (2004).

Lee, L.J. BioMEMS and micro-/nano-processing of polymers – an overview. *J. Chin. Inst. Chem. Eng.* **34**, 25–46 (2003).

Lefeber, D.J., Gallego, R.G. *et al.* Isolation of oligosaccharides from a partial-acid hydrolysate of pneumococcal type 3 polysaccharide for use in conjugate vaccines. *Carbohydr. Res.* **337**, 819–825 (2002).

Lei, C.X., Gong, F.C. *et al.* Amperometric immunosensor for *Schistosoma japonicum* antigen using antibodies loaded on a nano-Au monolayer modified chitosan-entrapped carbon paste electrode. *Sens. Actuators B (Chem.)* **96**, 582–588 (2003).

Lei, C.X., Yang, Y. *et al.* Amperometric immunosensor for probing complement III (C-3) based on immobilizing C-3 antibody to a nano-Au monolayer supported by sol–gel-derived carbon ceramic electrode. *Anal. Chim. Acta* **513**, 379–384 (2004).

Li, J., Wu, Z.Y. *et al.* A novel piezoelectric biosensor for the detection of phytohormone beta-indole acetic acid. *Anal. Sci.* **18**, 403–407 (2002a).

Li, Z.Z., Gong, F.C. *et al.* Bacteria-modified amperometric immunosensor for a *Brucella melitensis* antibody assay. *Anal. Sci.* **18**, 625–630 (2002b).

Liebermann, T. and Knoll, W. Parallel multispot detection of target hybridization to surface-bound probe oligonucleotides of different base mismatch by surface-plasmon field-enhanced fluorescence microscopy. *Langmuir* **19**, 1567–1572 (2003).

Liss, M., Petersen, B. *et al.* An apatmer-based quartz crystal protein biosensor. *Anal. Chem.* **74**, 4488–4495 (2002).

Liu, G.D., Wu, Z.Y. *et al.* Renewable amperometric immunosensor for Schistosoma japonium antibody assay. *Anal. Chem.* **73**, 3219–3226 (2001a).

Liu, G.D., Zhong, T.S. *et al.* Renewable amperometric immunosensor for complement 3 assay based on the sol-gel technique. *Fresenius J. Anal. Chem.* **370**, 1029–1034 (2001b).

Liu, W., Montana, V. *et al.* Botulinum toxin type B micromechanosensor. *Proc. Natl Acad. Sci. USA* **100**, 13621–13625 (2003a).

Liu, Y., Ye, J. *et al.* Rapid detection of *Escherichia coli* O157:H7 inoculated in ground beef, chicken carcass, and lettuce samples with an immunomagnetic chemiluminescence fiber-optic biosensor. *J. Food Prot.* **66**, 512–517 (2003b).

Luckham, P.F. and Smith, K. Direct measurement of recognition forces between proteins and membrane receptors. *Faraday Discuss.* **111**, 307–320 (1998).

Luppa, P.B., Sokoll, L.J. *et al.* Immunosensors – principles and applications to clinical chemistry. *Clin. Chim. Acta* **314**, 1–26 (2001).

Mahutte, C. On-Line arterial blood gas analysis with optodes: current status. *Clin. Biochem.* **31**, 119–130 (1998).

Majumdar, A. Bioassays based on molecular nanomechanics. *Dis. Markers* **18**, 167–174 (2002).

Marie, R., Jensenius, H. *et al.* Adsorbtion kinetics and mechanical properties of thiol-modified DNA-oligos on gold investigated by microcantilever sensors. *Ultramicroscopy* **91**, 29–36 (2002).

Mark, D.B. and Felker, G.M. B-type natriuretic peptide – a biomarker for all seasons? *N. Engl. J. Med.* **350**, 718–720 (2004).

Marks, R.S., Margalit, A. *et al.* Development of a chemiluminescent optical fiber immunosensor to detect *Streptococcus pneumoniae* antipolysaccharide antibodies. *Appl. Biochem. Biotechnol.* **89**, 117–126 (2000).

Marx, K.A. Quartz crystal microbalance: a useful tool for studying thin polymer films and complex biomolecular systems at the solution–surface interface. *Biomacromolecules* **4**, 1099–1120 (2003).

Mastrovitch, T.A., Bithoney, W.G. *et al.* Point-of-care testing for drugs of abuse in an urban emergency department. *Ann. Clin. Lab. Sci.* **32**, 383–386 (2002).

Medintz, I.L., Mattoussi, H. *et al.* Prototype quantum dot FRET-based nanoscale biosensor. *Abs. Pap. Am. Chem. Soc.* **226**, U479–U480 (2003).

Mishra, B. and Schwartz, J. *Designer Molecules for Biosensor Applications* (Banbury Center, Cold Spring Harbor Laboratory, Long Island, NY, 2002).

Mittelmann, A.S., Ron, E.Z. *et al.* Amperometric quantification of total coliforms and specific detection of *Escherichia coli*. *Anal. Chem.* **74**, 903–907 (2002).

Monk, D.J. and Walt, D.R. Optical fiber-based biosensors. *Anal. Bioanal. Chem.* **379**, 931–945 (2004).

Moreno-Bondi, M.C., Mobley, J. *et al.* Antibody-based biosensor for breast cancer with ultrasonic regeneration. *J. Biomed. Opt.* **5**, 350–354 (2000).

Moulin, A.M., O'Shea, S.J. *et al.* Microcantilever-based biosensors. *Ultramicroscopy* **82**, 23–31 (2000).

Murray, C.K., Bell, D. *et al.* Rapid diagnostic testing for malaria. *Trop. Med. Int. Health* **8**, 876–883 (2003).

Musham, C., Trettin, L. *et al.* Before the storm: informing and involving stakeholder groups in workplace biomarker monitoring. *J. Public Health Policy* **20**, 319–334 (1999).

Nachamkin, I., Panaro, N.J. *et al.* Agilent 2100 bioanalyzer for restriction fragment length polymorphism analysis of the *Campylobacter jejuni* flagellin gene. *J. Clin. Microbiol.* **39**, 754–757 (2001).

Neumann, T., Johansson, M.L. *et al.* Surface-plasmon fluorescence spectroscopy. *Adv. Funct. Mater.* **12**, 575–586 (2002).

Nicolette, C.A. and Miller, G.A. The identification of clinically relevant markers and therapeutic targets. *Drug Discov. Today* **8**, 31–38 (2003).

Niemeyer, C.M., Sano, T. *et al.* Oligonucleotide-directed self-assembly of proteins – semisynthetic DNA streptavidin hybrid molecules as connectors for the generation of macroscopic arrays and the construction of supramolecular bioconjugates. *Nucleic Acids Res.* **22**, 5530–5539 (1994).

O'Sullivan, C.K. Aptasensors – the future of biosensing? *Anal. Bioanal. Chem.* **372**, 44–48 (2002).

O'Sullivan, C.K. and Guilbault, G.G. Commercial quartz crystal microbalances – theory and applications. *Biosens. Bioelectron.* **14**, 663–670 (1999).

O'Sullivan, C.K., Vaughan, R. *et al.* Piezoelectric immunosensors – theory and applications. *Anal. Lett.* **32**, 2353–2377 (1999).

Owicki, J.C., Bousse, L.J., Hafeman, D.G., Kirk, G.L., Olson, J.D., Wada, H.G. and Parce, J.W. The light-addressable potentiometric sensor: principles and biological applications. *Annu. Rev. Biomol. Struct.* **23**, 87–113 (1994).

Pathak, S.S. and Savelkoul, H.F.J. Biosensors in immunology: the story so far. *Immunol. Today* **18**, 464–467 (1997).

Pavlickova, P., Knappik, A. *et al.* Microarray of recombinant antibodies using a streptavidin sensor surface self-assembled onto a gold layer. *Biotechniques* **34**, 124–130 (2003).

Pei, J.H., Tian, F. *et al.* Glucose biosensor based on the microcantilever. *Anal. Chem.* **76**, 292–297 (2004).

Peluso, P., Wilson, D.S. *et al.* Optimizing antibody immobilization strategies for the construction of protein microarrays. *Anal. Biochem.* **312**, 113–124 (2003).

Pemberton, R.M., Hart, J.P. *et al.* An electrochemical immunosensor for milk progesterone using a continuous flow system. *Biosens. Bioelectron.* **16**, 715–723 (2001).

Peppas, N.A. and Huang, Y. Polymers and gels as molecular recognition agents. *Pharm. Res.* **19**, 578–587 (2002).

Perez-Luna, V.H., Yang, S. *et al.* Fluorescence biosensing strategy based on energy transfer between fluorescently labeled receptors and a metallic surface. *Biosens. Bioelectron.* **17**, 71–78 (2002).

Phillips, J.E., Ambrose, T.M. *et al.* Ligand assay technology in drugs of abuse screening point of care devices. *J. Clin. Ligand Assay* **25**, 342–347 (2002).

Porter, R.A. Investigation of electroplated conducting polymers as antibody receptors in immunosensors. *J. Immunoassay* **21**, 51–64 (2000).

Porter, R., van der Logt, P. *et al.* An electro-active system of immunoassay (EASI assay) utilising self assembled monolayer modified electrodes. *Biosens. Bioelectron.* **16**, 875–885 (2001).

Porwal, A., Narsude, M. *et al.* Microcantilever based biosensors. *IETE Tech. Rev.* **19**, 257–267 (2002).

Price, C.P. Point-of-care testing in diabetes mellitus. *Clin. Chem. Lab. Med.* **41**, 1213–1219 (2003).

Ptak, A., Takeda, S. *et al.* Modified atomic force microscope applied to the measurement of elastic modulus for a single peptide molecule. *J. Appl. Phys.* **90**, 3095–3099 (2001).

Puckett, L.G., Barrett, G. *et al.* Monitoring blood coagulation with magnetoelastic sensors. *Biosens. Bioelectron.* **18**, 675–681 (2003).

Purvis, D.R., Pollard-Knight, D. and Lowe, C.R. Direct immunosensors. In: *Principles and Practice of Immunoassay*, 2nd edn, (eds Price, C.P. and Newman, D.J.), 513–543 (Macmillan, London, 1997).

Raiteri, R., Grattarola, M. *et al.* Micromechanical cantilever-based biosensors. *Sens. Actuators B (Chem.)* **79**, 115–126 (2001).

Raman, S.C., Raje, M. *et al.* Immunosensors for pesticide analysis: antibody production and sensor development. *Crit. Rev. Biotechnol.* **22**, 15–32 (2002).

Rasmussen, P.A., Thaysen, J. *et al.* Optimised cantilever biosensor with piezoresistive read-out. *Ultramicroscopy* **97**, 371–376 (2003).

Reder, S., Dieterle, F. *et al.* Multi-analyte assay for triazines using cross-reactive antibodies and neural networks. *Biosens. Bioelectron.* **19**, 447–455 (2003).

Richardson, J., Hawkins, P. *et al.* The use of coated paramagnetic particles as a physical label in a magneto-immunoassay. *Biosens. Bioelectron.* **16**, 989–993 (2001).

Robinson, W.H., Steinman, L. *et al.* Protein arrays for autoantibody profiling and fine-specificity mapping. *Proteomics* **3**, 2077–2084 (2003).

Rodriguez, W.J., Schwartz, R.H. *et al.* Evaluation of diagnostic tests for influenza in a pediatric practice. *Pediatr. Infect. Dis. J.* **21**, 193–196 (2002).

Roongsritong, C., Warraich, I. *et al.* Common causes of troponin elevations in the absence of acute myocardial infarction – incidence and clinical significance. *Chest* **125**, 1877–1884 (2004).

Rowe-Taitt, C.A., Cras, J.J. *et al.* A ganglioside-based assay for cholera toxin using an array biosensor. *Anal. Biochem.* **281**, 123–133 (2000a).

Rowe-Taitt, C.A., Golden, J.P. *et al.* Array biosensor for detection of biohazards. *Biosens. Bioelectron.* **14**, 785–794 (2000b).

Rowe-Taitt, C.A., Hazzard, J.W. *et al.* Simultaneous detection of six biohazardous agents using a planar waveguide array biosensor. *Biosens. Bioelectron.* **15**, 579–589 (2000c).

Ruan, C.M., Zeng, K.F. *et al.* Magnetoelastic immunosensors: Amplified mass immunosorbent assay for detection of *Escherichia coli* O157: H7. *Anal. Chem.* **75**, 6494–6498 (2003).

Sadik, O.A., Wanekaya, A.K. *et al.* Advances in analytical technologies for environmental protection and public safety. *J. Environ. Monit.* **6**, 513–522 (2004).

Sapsford, K.E., Liron, Z. *et al.* Kinetics of antigen binding to arrays of antibodies in different sized spots. *Anal. Chem.* **73**, 5518–5524 (2001).

Sapsford, K.E., Charles, P.T. *et al.* Demonstration of four immunoassay formats using the array biosensor. *Anal. Chem.* **74**, 1061–1068 (2002).

Sapsford, K.E., Delehanty, J.B. *et al.* Array biosensor for homeland defense. *Abs. Pap. Am. Chem. Soc.* **225**, U-107 (2003).

Sapsford, K.E., Shubin, Y.S. *et al.* Fluorescence-based array biosensors for detection of biohazards. *J. Appl. Microbiol.* **96**, 47–58 (2004).

Sarkar, P., Pal, P.S. *et al.* Amperometric biosensors for detection of the prostate cancer marker (PSA). *Int. J. Pharm.* **238**, 1–9 (2002).

Schaible, U., Liss, M. *et al.* Affinity measurements of biological molecules by a quartz crystal microbalance (QCM) biosensor. *Methods Mol. Med.* **94**, 321–330 (2004).

Schalkhammer, T. Metal nano clusters as transducers for bioaffinity interactions. *Monatshefte Fur Chem.* **129**, 1067–1092 (1998).

Schmut, O., Horwath-Winter, J. *et al.* The effect of sample treatment on separation profiles of tear fluid proteins: qualitative and semi-quantitative protein determination by an automated analysis system. *Graefes Arch. Clin. Exp. Ophthalmol.* **240**, 900–905 (2002).

Schoning, M.J. and Poghossian, A. Recent advances in biologically sensitive field-effect transistors (BioFETs). *Analyst* **127**, 1137–1151 (2002).

Selvaganapathy, P.R., Carlen, E.T. *et al.* Recent progress in microfluidic devices for nucleic acid and antibody assays. *Proc. IEEE* **91**, 954–975 (2003).

Selvanayagam, Z.E., Neuzil, P. *et al.* An ISFET-based immunosensor for the detection of beta-Bungarotoxin. *Biosens. Bioelectron.* **17**, 821–826 (2002).

Sepaniak, M., Datskos, P. *et al.* Microcantilever transducers: a new approach to sensor technology. *Anal. Chem.* **74**, 568A–575A (2002).

Shen, J., Shu, N. *et al.* Development of a surface acoustic wave LB membrane immunity sensor. *J. Tongji Med. Univ.* **20**, 20–22 (2000).

Sota, H., Yoshimine, H. *et al.* A versatile planar QCM-based sensor design for nonlabeling biomolecule detection. *Anal. Chem.* **74**, 3592–3598 (2002).

Spiker, J.O. and Kang, K.A. Preliminary study of real-time fiber optic based protein C biosensor. *Biotechnol. Bioeng.* **66**, 158–163 (1999).

Squillante, E. Applications of fiber-optic evanescent wave spectroscopy. *Drug Dev. Ind. Pharm.* **24**, 1163–1175 (1998).

Steil, G.M., Panteleon, A.E. *et al.* Closed-loop insulin delivery – the path to physiological glucose control. *Adv. Drug Delivery Rev.* **56**, 125–144 (2004).

Stevenson, K.A., Mehta, A. *et al.* Nanomechanical effect of enzymatic manipulation of DNA on microcantilever surfaces. *Langmuir* **18**, 8732–8736 (2002).

Stubbs, D.D., Lee, S.H. *et al.* Investigation of cocaine plumes using surface acoustic wave immunoassay sensors. *Anal. Chem.* **75**, 6231–6235 (2003).

Sutherland, R.M., Dähne, C., Place, J.F. and Ringrose, A.R. Immunoassays at a quartz–liquid interface: theory, instrumentation and preliminary application to the fluorescent immunoassay of human immunolglobulin G. *J. Immunol. Methods* **74**, 253–265 (1984).

Swanson, B.N. Delivery of high-quality biomarker assays. *Dis. Markers* **18**, 47–56 (2002).

Taitt, C.R., Anderson, G.P. *et al.* Nine-analyte detection using an array-based biosensor. *Anal. Chem.* **74**, 6114–6120 (2002).

Tamarin, O., Comeau, S. *et al.* Real time device for biosensing: design of a bacteriophage model using love acoustic waves. *Biosens. Bioelectron.* **18**, 755–763 (2003).

Tanaka, T. and Matsunaga, T. Detection of HbA(1c) by boronate affinity immunoassay using bacterial magnetic particles. *Biosens. Bioelectron.* **16**, 1089–1094 (2001).

Tedeschi, L., Domenici, C. *et al.* Antibody immobilisation on fibre optic TIRF sensors. *Biosens. Bioelectron.* **19**, 85–93 (2003).

Tucker, S.P., Cox, C. *et al.* A flu optical immunoassay (Thermo BioStar's FLU OIA): a diagnostic tool for improved influenza management. *Philos. Trans. R. Soc. Lond. Ser. B (Biol. Sci.)* **356**, 1915–1924 (2001).

Turner, A.P. and Magan, N. Electronic noses and disease diagnostics. *Nat. Rev. Microbiol.* **2**, 161–166 (2004).

Uttenthaler, E., Schraml, M. *et al.* Ultrasensitive quartz crystal microbalance sensors for detection of M13-Phages in liquids. *Biosens. Bioelectron.* **16**, 735–743 (2001).

Vo-Dinh, T. Nanobiosensors: probing the sanctuary of individual living cells. *J. Cell. Biochem.* **39** (Suppl.), 154–161 (2002).

Vo-Dinh, T. and Cullum, B. Biosensors and biochips: advances in biological and medical diagnostics. *Fresenius J. Anal. Chem.* **366**, 540–551 (2000).

Ward, M.D. and Buttry, D.A. *In situ* interfacial mass detection with piezoelectric transducers. *Science* **249**, 1000–1007 (1990).

Warsinke, A., Benkert, A. *et al.* Electrochemical immunoassays. *Fresenius J. Anal. Chem.* **366**, 622–634 (2000).

Weizmann, Y., Patolsky, F. *et al.* Magneto-mechanical detection of nucleic acids and telomerase activity in cancer cells. *J. Am. Chem. Soc.* **126**, 1073–1080 (2004).

Wells, W.A. The next chip-based revolution – Caliper Technologies Corp. *Chem. Biol.* **5**, R115–R116 (1998).

West, J.L. and Halas, N.J. Engineered nanomaterials for biophotonics applications: improving sensing, imaging, and therapeutics. *Ann. Rev. Biomed. Eng.* **5**, 285–292 (2003).

Wong, Y.Y., Ng, S.P. *et al.* Immunosensor for the differentiation and detection of *Salmonella* species based on a quartz crystal microbalance. *Biosens. Bioelectron.* **17**, 676–684 (2002).

Wu, G.H., Datar, R.H. *et al.* Bioassay of prostate-specific antigen (PSA) using microcantilevers. *Nat. Biotechnol.* **19**, 856–860 (2001).

Wu, G., Irvine, J. *et al.* Assay development and high-throughput screening of caspases in microfluidic format. *Comb. Chem. High Throughput Screening* **6**, 303–312 (2003).

Yancy, C.W. Practical considerations for BNP use. *Heart Failure Rev.* **8**, 369–373 (2003).

Yang, Y., Ji, H.-F. *et al.* Nerve agents detection using a Cu^{2+}/l-cysteine bilayer-coated microcantilever. *J. Am. Chem. Soc.* **125**, 1124–1125 (2003).

Yue, M., Lin, H. *et al.* A 2-D microcantilever array for multiplexed biomolecular analysis. *J. Microelectromech. Syst.* **13**, 290–299 (2004).

Yuqing, M., Jianguo, G. *et al.* Ion sensitive field effect transducer-based biosensors. *Biotechnol. Adv.* **21**, 527–534 (2003).

Zacco, E., Pividori, M.I. *et al.* Renewable Protein A modified graphite-epoxy composite for electrochemical immunosensing. *J. Immunol. Methods* **286**, 35–46 (2004).

Zayats, M., Raitman, O.A. *et al.* Probing antigen–antibody binding processes by impedance measurements on ion-sensitive field-effect transistor devices and complementary surface plasmon resonance analyses: development of cholera toxin sensors. *Anal. Chem.* **74**, 4763–4773 (2002).

Zeravik, J., Ruzgas, T. *et al.* A highly sensitive flow-through amperometric immunosensor based on the peroxidase chip and enzyme-channeling principle. *Biosens. Bioelectron.* **18**, 1321–1327 (2003).

Zhang, J. and Ji, H.F. An anti *E. coli* O157: H7 antibody-immobilized microcantilever for the detection of *Escherichia coli* (*E. coli*). *Anal. Sci.* **20**, 585–587 (2004).

Zhou, X.C. and Cao, L. High sensitivity microgravimetric biosensor for qualitative and quantitative diagnostic detection of polychlorinated dibenzo-*p*-dioxins. *Analyst* **126**, 71–78 (2001).

Zhou, Y.M., Hu, S.Q. *et al.* An amperometric immunosensor based on an electrochemically pretreated carbon-paraffin electrode for complement III (C3) assay. *Biosens. Bioelectron.* **18**, 473–481 (2003).

16 Surface Plasmon Resonance in Kinetic, Concentration and Binding Site Analyses

Robert Karlsson, Marie Arvola and Gary Franklin

INTRODUCTION

The successful sequencing of the human genome has provided enormous advancements for basic life science research, while at the same time creating major new challenges. Probably, the biggest of these challenges will be to relate the large body of sequence data that has now been obtained to its corresponding biological functions. Since the genome functions primarily at the protein level, the focus of interest is now shifting towards protein studies and the technologies that will facilitate them.

OBJECTIVES OF PROTEIN STUDIES IN BASIC RESEARCH

The fundamental issues for protein research are expression and function. Expression studies establish the spatial and temporal patterns of the localization of a protein and while this can often provide some clues to function, this latter aspect is much more complex and requires information concerning interactions with other biomolecules. Expression can be studied using specific binding reagents, such as antibodies for example, to detect the protein of interest in a cell lysate or a tissue preparation. The more complex issue of protein interactions is typically addressed using yeast or mammalian two hybrid techniques (Coates and Hall, 2003; Ito *et al.*, 2002; Zhong *et al.*, 2003), or by a combination of affinity capture and mass spectrometry of proteins or peptide fragments (Nedelkov and Nelson, 2003; Turecek, 2002). Additional information may be obtained from computational methods that predict protein function from structure (Bader and Hogue, 2003; Turk and Cantley, 2003). Results from these methods can be translated into protein interaction maps and used to generate protein interaction databases (Hermjakob *et al.*, 2004; Peri *et al.*, 2004), which are emerging as new tools to access this type of data (Boulton *et al.*, 2001).

For each interaction, the concentrations of interacting biomolecules and the kinetics of the interaction determine the rate of complex formation and the extent of the reaction. This basic information can be complemented by comparing kinetic and affinity data obtained for wild type and mutant proteins, allowing the importance of a specific domain or amino acid residue for the interaction to be assessed.

PROTEIN STUDIES AND THERAPEUTICS

The proteins identified in protein interaction maps are likely to provide a rich source of future drug targets and therapeutic agents. It is also likely that the analysis of interactions between target proteins and small molecule inhibitors will provide further insights into protein function, as well as providing the basis for novel therapeutic agents. Detailed protein characterization will also be important during the development of new protein therapeutics, as well as in subsequent manufacturing quality control procedures, where binding activity and an evaluation of the structural integrity are important for batch-to-batch consistency.

TECHNOLOGY DEMANDS

The requirements of protein characterization studies, whether in basic life science or applied therapeutic

The Immunoassay Handbook. *Third Edition*
D. Wild (Ed.)

research, clearly create a demand for simple, direct methods to describe protein–protein interactions. These should enable the determination of the affinity and kinetic properties of an interaction and be suitable for topographical studies. In other research fields such as food and environmental analysis in particular, the monitoring and determination of potentially hazardous agents are of key interest. Here there is a need for rapid and robust analytical methods. The same requirements are valid for areas such as drug development and quality control applications, which are under the control of regulatory authorities. In these areas, the demands on, for example, reproducibility and accuracy of the assays are very high.

In this chapter, we will describe the use of Biacore's surface plasmon resonance (SPR) technology and examine how it can be used in kinetic analysis, binding site studies, and concentration analysis. As will now be discussed in detail, the main advantages of SPR analysis are that it provides direct, label-free, real-time analysis of the interactions of a very wide range of biomolecules. In contrast to many other technologies that provide only end point assays, SPR therefore gives a continuous readout of binding events over the entire course of the interaction.

SPR BIOSENSORS

TECHNOLOGY PRINCIPLES

SPR biosensors directly monitor the interaction between an analyte molecule introduced in solution and a second molecule or molecular complex present on a sensor surface. Detection of binding events is instantaneous and requires no labeling or auxiliary reagents. The sensor surface consists of a thin gold film deposited on a glass support. Normally, carboxymethylated dextran polymers are attached to the gold film, providing a convenient starting point for the immobilization of biomolecules (Johnsson *et al.*, 1995).

The gold side of the sensor surface is placed in contact with a flow system. This makes it possible to address samples to discrete spots or flow cells on the sensor surface. The glass side of the sensor surface is positioned on a prism and is thereby connec.e. › the optical unit of the biosensor. With this configuration, it is possible to exploit the phenomenon of SPR (Figure 16.1). Under conditions of total internal reflection, energy and momentum from light at a certain angle of incidence are transformed from photons into surface plasmons in the metal film of the sensor chip. At this angle, which depends

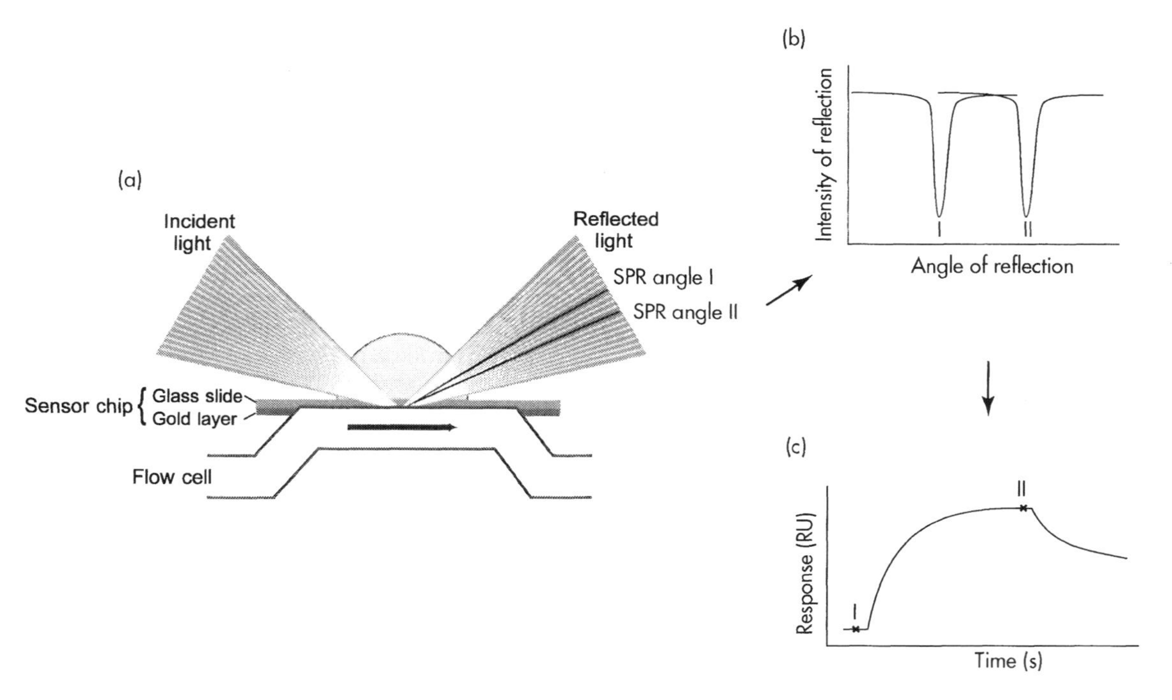

Fig. 16.1 (a) Changes in mass concentration at the sensor surface alter the refractive index and thereby alter the angle at which an SPR signal is generated. (b) This is observed as a shift in the angle at which the minimum of intensity of reflection is detected. (c) This shift is plotted as response units (RU) against time in a sensorgram.

on the refractive index close to the gold/dextran layers of the sensor surface, the intensity of the reflected light is therefore reduced. When an analyte binds to the sensor surface the resulting change in mass concentration causes a change in refractive index at the sensor surface, and the conditions for SPR will be fulfilled at a slightly different angle of incidence. This is monitored by the detector as a shift in the position at which the light intensity minimum is observed.

Continuous monitoring of the SPR angle as a function of time reveals the kinetic profile of interaction events at the surface and these are displayed in a sensorgram (Figure 16.1c). The SPR response on the y-axis of the sensorgram is expressed in resonance units (RU). One RU corresponds to a mass increase of approximately 1 pg mm^{-2} of protein.

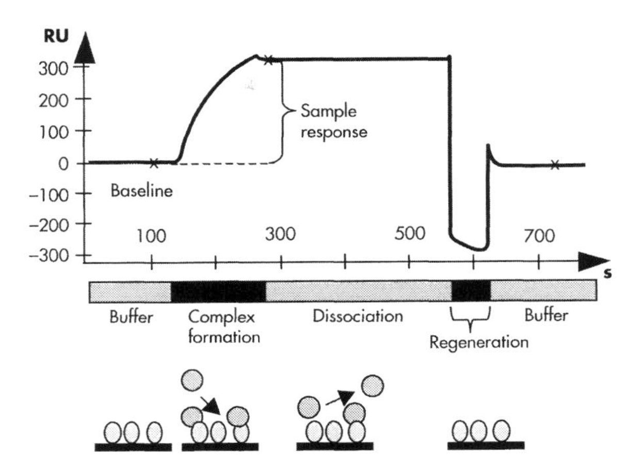

Fig. 16.2 In the beginning of this sensorgram, buffer flows over the surface with the immobilized interactant generating a baseline. As the analyte is injected, there is an increase in the response due to the change in mass generated by the complex formation on the surface. After sample injection, buffer flows over the surface again, allowing the dissociation of the complex to be monitored. Injection of a regeneration solution removes any remaining bound analyte and the response returns to the baseline level.

SENSORGRAMS AND REPORT POINTS

From a technical perspective, all SPR experiments can be viewed as a series of sample injections, where the duration of each injection and the time between injections are varied. Typical injection times are from 30 sec to 10 min and the flow rate can be varied during an experiment, with typical sample volumes ranging from 1 to 100 μL.

SPR biosensors are computer controlled and the control software guides the user through the various steps in an assay. Once the assay is programmed, automation allows analysis of more than a hundred samples without user intervention. Results are saved in a file and the data is analyzed using dedicated evaluation software. The interpretation of a sensorgram is relatively straightforward, as shown in the example presented in Figure 16.2. Here, sample is injected for 2 min and binds to immobilized biomolecules, giving a response close to 300 RU. The complex formed is stable, as can be seen by the slow release of analyte when the surface is washed with buffer. A regeneration solution is injected 5 min after the end of the sample injection, which removes analyte and returns the baseline level to a position close to the original position. Note that response levels can be negative. In this case, the regeneration solution has a lower bulk refractive index than the assay buffer and a negative signal is recorded.

In most cases, a control surface is employed to correct for the changes in bulk refractive index between different buffers, enabling a reference-subtracted response that depends specifically on molecular interactions at the surface to be viewed directly. Report points (SPR responses taken at specified time points during the analysis cycle) are set along the sensorgram. The result file gives access to both the full sensorgram and to a table listing the response values associated with each report point. Report points are useful for following trends from one sensorgram to another and for ranking purposes.

IMMOBILIZATION

Biomolecules can be immobilized on the sensor surface either by covalent coupling or via capture. These two principles are illustrated in Figure 16.3. Covalent coupling through amine groups, carboxyl groups, thiol groups, or sugar alcohols gives stable surfaces, and the level of immobilization can be controlled over a very wide range

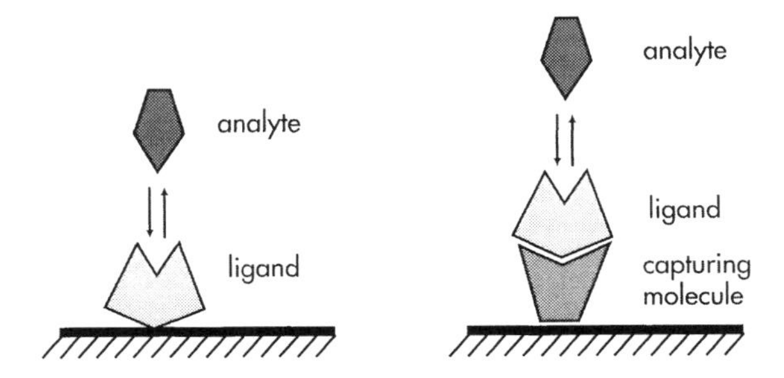

Fig. 16.3 The two major principles for immobilizing one interactant (the ligand in Biacore terms) on the sensor surface. In the covalent coupling approach (left panel), different chemical groups such as amine, thiol, or aldehyde groups on the ligand can be used for the covalent coupling. In the capture approach (right panel), a capturing molecule (for example an anti-Fc antibody) is first covalently coupled to the sensor surface, after which the ligand is injected and captured.

(less than 1 RU to over 10,000 RU). Covalent coupling involves activation of surface carboxyl groups, coupling of the biomolecule, and deactivation of unreacted groups on the surface.

This approach minimizes the consumption of the molecule for immobilization, but normally requires a purified preparation of the biomolecule. For protein immobilization, amine coupling is usually straightforward and is often the first method of choice. If this proves to be difficult, a capture approach can be tried if the biomolecule is appropriately tagged.

Capture is a two-step process in which an affinity capture molecule, such as an anti-histidine antibody or an anti-Fc antibody is first covalently immobilized to the sensor surface. In the second step, the biomolecule of interest is injected and captured ready for the analyte injection. With this approach, it is possible to use a single sensor surface for interaction studies with many different surface-bound biomolecules. Another advantage is that it is possible to capture biomolecules directly from culture media or cell lysates without prior purification. This approach can also be useful in cases where removal of analyte from a covalently immobilized ligand proves difficult, since in a capture assay, both analyte and ligand are removed from the sensor surface between each cycle. Capture is particularly useful for the analysis of tagged proteins and for antibodies. The potential drawbacks of capture techniques are reduced stability and lower binding capacity.

Combinations of capture and covalent coupling techniques are also possible, either by performing capture prior to surface deactivation or by reactivating the surface with the capturing molecule in place. Using these approaches, histidine-tagged molecules can be covalently coupled after capture on nitrilotriacetic acid (NTA) surfaces or on anti-histidine surfaces.

SURFACE ACTIVITY AND IMMOBILIZATION LEVELS

The binding activity of the immobilized biomolecule is conveniently determined by injecting increasing concentrations of the analyte, as illustrated in Figure 16.4.

A 10-fold dilution series of the analyte molecule is injected, starting with the lowest concentration. In this example, the surface becomes saturated at an analyte concentration of just over 80 μg mL^{-1} and the binding capacity is approximately 50 RU. This response can then be compared to the amount of immobilized biomolecule and the fraction of active biomolecules can be estimated from the following equation:

$$\text{Activity} = \frac{\text{Binding capacity}}{\text{Mw}_{\text{analyte}}} \times \frac{\text{Mw}_{\text{Immobilized biomolecule}}}{\text{Immobilization level}} \times \frac{1}{\text{Number of binding sites}}.$$

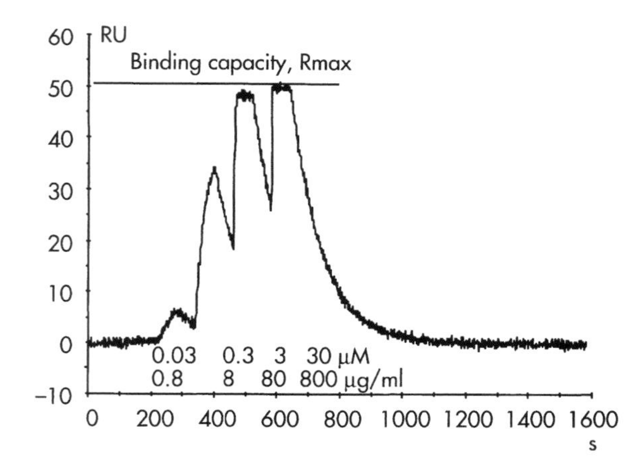

Fig. 16.4 The maximum binding response (R_{max}) is reached as the analyte binding capacity of the surface is saturated and can be determined by injecting increasing concentrations of analyte.

The number of binding sites refers to those on the immobilized biomolecule. Typical activity numbers range from 0.2 to 0.8 (i.e. 20–80%), but both lower and higher activities can be obtained. The surface activity can be influenced by the method of immobilization. If less than 20% of the immobilized biomolecules are active, alternative immobilization chemistry should be considered.

BINDING SITE ANALYSIS – EPITOPE MAPPING

Epitope mapping is commonly performed with the purpose of structural analysis or identification and characterization of antibody binding properties, for example, for finding pairs of antibodies that can be used in sandwich assays. Common techniques for epitope mapping experiments include ELISA and RIA. One advantage in using SPR biosensors for this type of experiments lies in the real-time measurements, which allow each step in the mapping experiment to be followed. In the case of structural analysis, direct mapping with antibodies as described here can be complemented with peptide inhibition mapping so that regions of the antigen can be linked to specific antibodies. An SPR variant of peptide inhibition mapping has been described by Fägerstam *et al.* (1990).

PAIR-WISE EPITOPE MAPPING

The design of a typical pair-wise mapping using mouse monoclonal antibodies (MAbs) is illustrated in Figure 16.5.

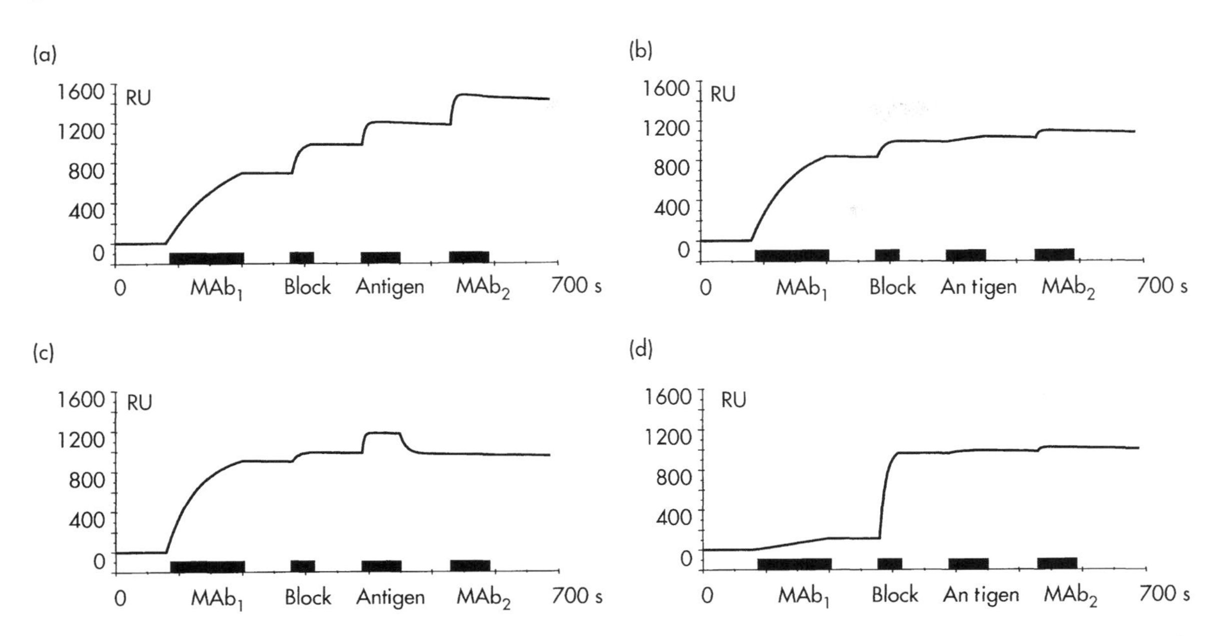

Fig. 16.5 Schematic illustrations of epitope mapping experiments. (a) MAb_2 generates a distinct response, indicating a binding epitope separate from the epitope of MAb_1. (b) A slow binding of antigen to MAb_1 results in a small binding of MAb_2, which is nevertheless relevant in comparison with antigen levels. The epitopes of these two MAbs, therefore, are likely to be separate from each other. (c) Because of the very rapid dissociation of antigen from MAb_1, no conclusions regarding epitope specificity can be drawn from these data. (d) Although the binding response of MAb_2 is low, it is relevant for the level of antigen on the surface and it is likely that it recognizes a second epitope on the antigen.

In this type of assay it is convenient to first immobilize an anti-mouse Fc antibody onto the sensor surface in order to capture the monoclonals of interest. Next, a first MAb is injected for 1–2 min and captured by the anti-mouse Fc, after which a blocking antibody with no relevance for the antigen is injected to block remaining antibody-binding sites on the surface. Antigen is then injected followed by injection of a second MAb. This injection cycle is terminated by injection of a regeneration agent (typically a low pH solution such as 10 mM glycine at pH 1.8) that removes captured antibodies and antigen (not shown). In this way, a single surface can often be reused over 100 times.

In contrast to other techniques where a response is obtained after the binding of the second antibody, SPR biosensors provide a detailed picture of all binding events. In Figure 16.5a, the first antibody is captured to a relatively high level and antigen binding proceeds rapidly. Bound antigen dissociates slowly and the second MAb can bind to another epitope of the antigen. In Figure 16.5b, the second antibody gives a very small response. By examining the sensorgram, this is clearly due to a very slow binding of antigen to the first MAb. In comparison to the level of antigen binding, the response from the second MAb is relevant and it can be concluded that the second antibody binds to another epitope on the antigen.

In Figure 16.5c, the second antibody does not give any response. This can be explained by a very rapid dissociation of antigen from the first MAb. This means that when the second antibody is injected, antigen is no longer present on the surface and therefore, no conclusions can be made regarding epitope specificity. In Figure 16.5d, the low response from the second MAb is due to the low captured level of the first MAb to the sensor surface. This indicates a low concentration of this antibody in the culture media. Again, the response of the second MAb is relevant for the level of antigen on the surface and it is likely that it recognizes a second epitope on the antigen.

Clearly, the complete picture of all binding events makes it possible to improve the interpretation of the mapping experiment. It also makes it possible to troubleshoot the assay and to adapt experimental conditions if necessary. For example, the results in Figure 16.5 b could be clarified by injecting a higher concentration of antigen. Earlier injection of the second MAb in Figure 16.5c could make this data interpretable. The data in Figure 16.5d could be improved by increasing the injection time or concentration of the first MAb. Control experiments include injection of antigen directly on the anti-mouse Fc antibody and over a surface with captured blocking antibody as well as injection of buffer instead of antigen in the mapping experiment.

As demonstrated by these examples, the mapping experiment provides data relevant to epitope specificity. At the same time it can give an approximate idea of the concentration of MAb in culture media and, more importantly, it provides data for ranking of antibody–antigen kinetics. Epitope mapping experiments using SPR biosensors often involve 10–50 antibodies, which are either purified or injected directly from cell culture media. Antigen is typically used at concentrations from 10 to 100 $\mu g\,mL^{-1}$. Recent examples include mapping of epitopes on *Botulinum* Neurotoxin Type A (Pless *et al.*, 2001), apolipoprotein B-100 (Robbio *et al.*, 2001), β-lactoglobulin (Clement *et al.*, 2002), and creatine kinase (Schlattner *et al.*, 2002). The last paper describes a variant of the mapping experiment using immobilized anti-FLAG antibody for capture of FLAG-tagged single chain Fv antibodies.

KINETIC ANALYSIS

For a bimolecular interaction, the affinity constant reflects the equilibrium of the interaction and can be expressed in the following terms:

$$K_D = \frac{A \times B}{AB}.$$

Here, K_D is the equilibrium dissociation constant, A and B are the concentrations of the interactants and AB is the concentration of the complex. The kinetics, on the other hand, provide information on the rates of the interaction and may be expressed in the following terms:

$$\frac{dAB}{dt} = k_a \times A \times B - k_d \times AB.$$

In this expression, $t =$ time and k_a and k_d represent the association and dissociation rate constants. At equilibrium, the following conditions apply:

$$\frac{dAB}{dt} = 0, \text{ and consequently } \frac{k_d}{k_a} = K_D.$$

Affinity and rate constants are operationally defined and depend on experimental conditions such as temperature, pH, and buffer composition. It is therefore important to state under what conditions affinity and rate constants have been determined. From the relationship between rate constants and the affinity constant, it is also clear that kinetic analysis will have a greater resolving power than affinity analysis. The affinity constant is the ratio between the two rate constants and it can therefore be obtained from the combination of an infinite number of rate constants.

When the kinetic experiment is repeated at different temperatures, reaction enthalpy and entropy can be determined. The affinity constant provides information on changes in free energy between the free and bound reaction states, while kinetic analysis provides additional information related to the energy changes between the free state, the transition state, and the bound state (Figure 16.6).

While affinity and rate constants, as well as changes in enthalpy and entropy are used to describe a single interaction, it is frequently the comparison of interaction properties for related interactions that is of interest. This is the case in antibody selection, where binding properties for a set of antibodies reacting with the same antigen are compared, in mutagenesis studies where wild type and mutants are compared, and in drug discovery where the interaction properties of drug candidates are investigated. Furthermore, two or more molecules that exhibit the same interaction properties at one set of interaction conditions may differ from each other when these conditions are changed.

When large sets of interaction pairs are compared, it is useful to display data in a plot of k_a versus k_d values (Markgren *et al.*, 2002). This is illustrated in Figure 16.7a, where a set of molecules with unknown interaction properties is first classified into three groups corresponding to low, intermediate, and high affinity. By kinetic analysis, a higher resolution is obtained and

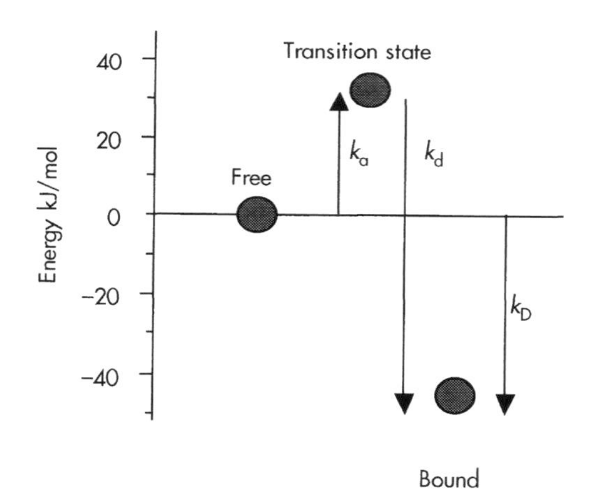

Fig. 16.6 The affinity constant (K_D) provides information on changes in free energy between the free and bound reaction states. Kinetic rate constants (k_a and k_d) give information related to energy changes between the free state, the transition state, and the bound state.

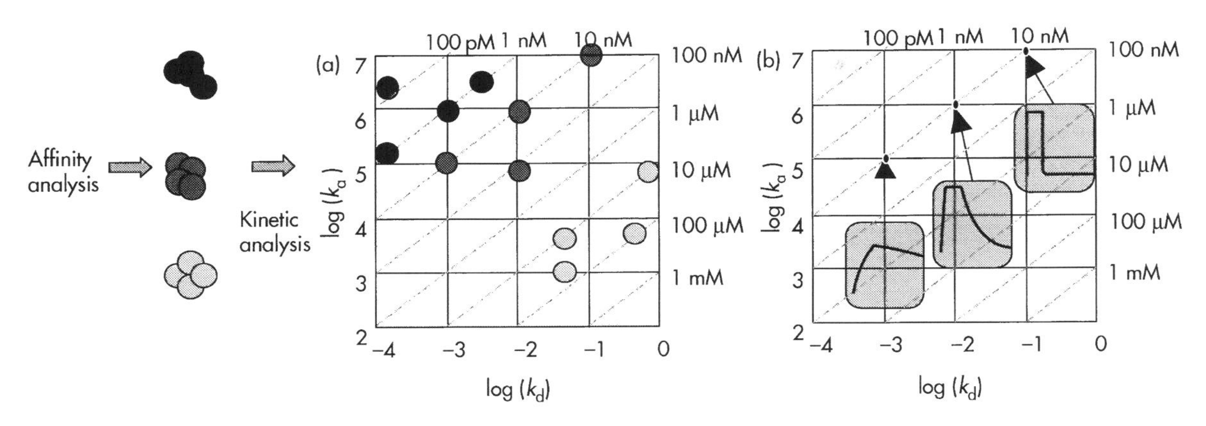

Fig. 16.7 Resolution of binding affinities using a k_a versus k_d plot. The y-axis indicates increasingly rapid association rates and the x-axis indicates increasingly rapid dissociation rates, producing isometric affinity diagonals (indicated by red dashed lines). On the plots, affinity increases from bottom-right to top-left. (a) Affinity analysis alone may discriminate several groups of analytes comprising of multiple molecules with very similar, or identical K_D values (shown schematically in the color-coded groups to the left of the plot). The k_a versus k_d plot, however, reveals very distinct kinetic characteristics for analytes falling on to the same affinity diagonal. (b) Insets to show the variation in kinetic profiles that may be seen among analytes with identical affinities.

the classification of these interactions becomes more detailed and informative. Figure 16.7b is a further clarification of the k_a versus k_d plot and illustrates the extent to which interactions with identical affinity actually can differ in binding profile when kinetic data is included in the comparison.

SPR IN KINETIC ANALYSIS

The fact that SPR measures mass changes at the sensor surface in real time makes it ideal for kinetic analysis. This type of universal detection makes it possible to apply almost the same experimental protocol to the study of interactions involving proteins, nucleic acids, and small molecules. Since the analyte is injected over the sensor surface under conditions of laminar flow, there will always be a balance between the rate at which the analyte is transported to the sensor surface by diffusion processes (mass transport) and the interaction-dependent binding rate at the sensor surface. The entire process can become transport limited and the observed binding rate may not reflect interaction properties when the transport rate is lower than the interaction-dependent binding rate. The binding rate at the sensor surface can be reduced by using low immobilization levels. Typically, the immobilization levels should be adjusted so that the binding capacity (R_{max}) falls between 10 and 100 RU. It is also possible to minimize mass transport effects by increasing the flow rate during the analyte injection, since the diffusion rate is related to the cube root of the liquid flow rate.

If the interaction-dependent binding rate is still higher than the transport rate despite optimization of assay conditions, the concentration of the analyte at the sensor surface will be depleted. Kinetics can still often be determined, however, as long as the concentration of the injected analyte is known. This is possible by introducing a mathematical formula by which the actual analyte concentration at the sensor surface is calculated. This leads to a set of rate equations that can be used in data analysis. For a reaction between analyte A and immobilized molecule B, therefore

$$dA/dt = k_t(\text{conc} - A) - (k_a \times A \times B - k_d \times AB) \quad |A(0) = 0(M)$$

$$dB/dt = -(k_a \times A \times B - k_d \times AB) \quad |B(0) = R_{max}(\text{RU})$$

$$dAB/dt = (k_a \times A \times B - k_d \times AB) \quad |AB(0) = 0(\text{RU}).$$

In these equations, 'conc' is the concentration of the injected analyte and k_t is a transport coefficient. To the right of each rate equation the starting values at time zero for each parameter are listed with the corresponding unit in parenthesis. The analyte concentration at the sensor surface and the concentration of the complex are both zero, while the concentration of the immobilized ligand is expressed in terms of the saturation response.

In a kinetic experiment, the analyte is typically injected at 5–6 concentrations. The injection times can be kept short (i.e. 1–2 min), whereas the dissociation phase, during which the complex dissociates under conditions of buffer flow, should be longer and is normally in the range of 15–20 min. With the experimental design and data analysis model described here, it should be possible to determine rate constants

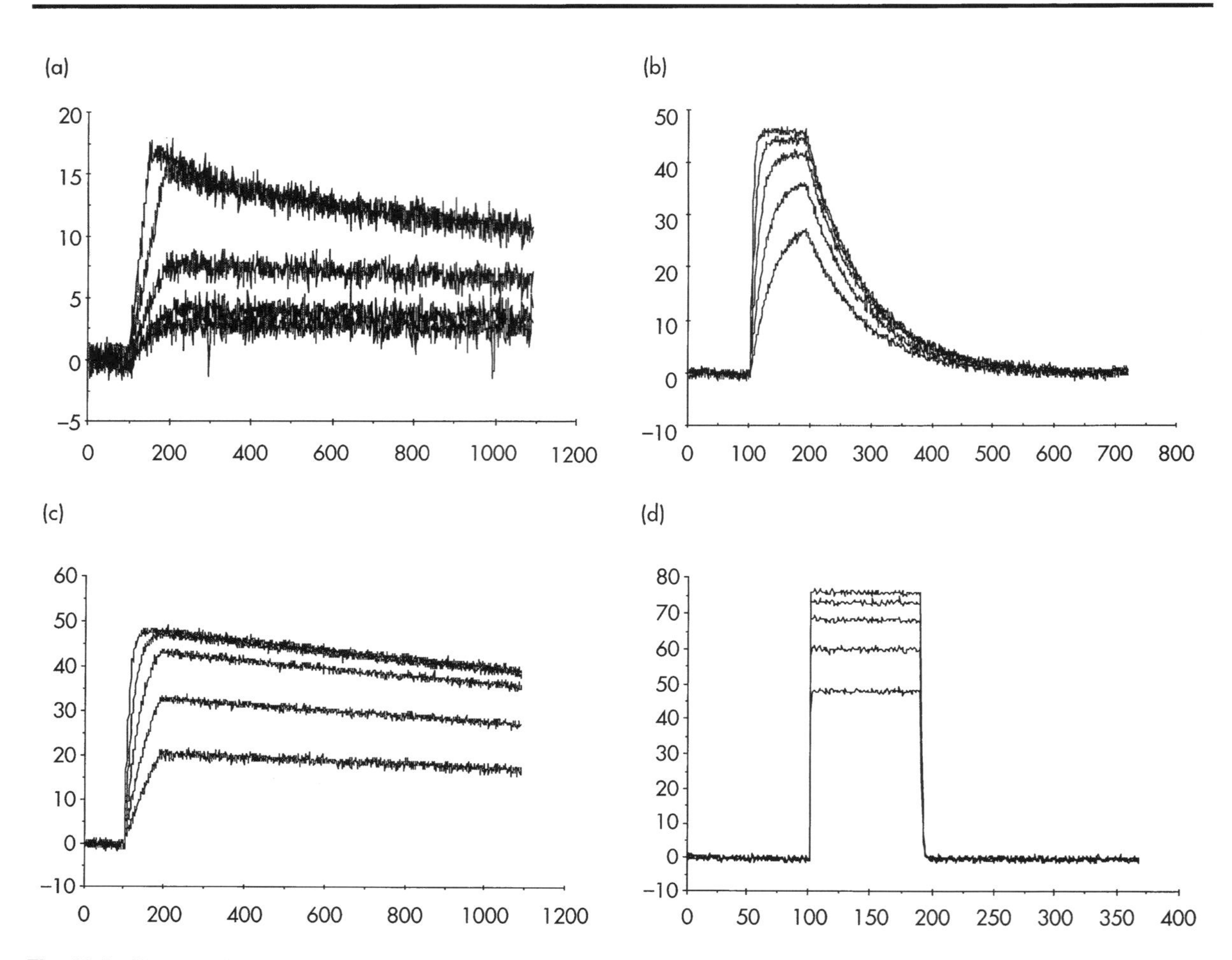

Fig. 16.8 Presentation of four sets of kinetic analysis data. (a) The lack of curvature of the sensorgram during the injection makes it difficult to obtain kinetic rate constants for the interaction. From the data in panels (b) and (c), rate constants for each interaction can be readily determined. (d) The data set mainly reflects steady-state levels and no kinetic information can be revealed.

in the following ranges:

$$k_a = 10^3 - 10^8 \text{ M}^{-1} \text{ sec}^{-1}$$

$$k_d = 10^{-4} - 10^{-1} \text{ sec}^{-1}.$$

These are basically the same ranges as illustrated in the k_a versus k_d plots in Figure 16.7. The ranges quoted are approximate, however, and can be extended up to one order of magnitude in either direction. For the association rate constant, the upper level depends on the balance between transport and binding rates and the lower level on the possibility of injecting very high concentrations of the analyte. Lower dissociation rate constants can be determined if the dissociation phase is prolonged. Dissociation events that occur faster than the time it takes to switch from analyte to buffer flow (approximately 1 sec at a high flow rate) are not possible to analyze.

In Figure 16.8, four sets of kinetic analysis data are presented. Sets b and c reflect interactions for which rate constants can be readily obtained. Note that the top curves approach almost the same response level. This indicates that sufficient information regarding the saturation level has been obtained. It is a clear characteristic of these sets that several analyte injections generate binding profiles with distinct curvature during the association phase. In both sets, dissociation profiles can readily be observed and these data therefore hold information on R_{max}, k_a, and k_d, which are the parameters calculated using the evaluation software.

In set a, the sensorgrams are almost straight during injection, with little or no curvature even when binding approaches saturation level. This indicates an unfavorable balance between transport rates and binding rates. Here it is difficult to obtain reliable values for the rate constants, but it is still possible to determine the affinity from these binding curves (Karlsson, 1999). In set d, dissociation is very rapid. The sensorgrams mainly reflect steady-state levels and no kinetic data can be obtained. In this case, the affinity constant can be derived from a plot of steady-state values versus analyte concentration.

Kinetic analysis is one of the most important applications for SPR technology, and even using a relatively limited database search, over 1300 publications citing SPR kinetic analysis can be identified. SPR has been used to shed light on the properties of cell adhesion molecules (van der Merwe and Barclay, 1994), to describe T-cell receptor interactions with MHC (Wu *et al.*, 2002), for screening and selection of antibodies (Canziani *et al.*, 2004), for identification of drug resistance (Shuman *et al.*, 2003), and in numerous other cases. These publications include a number of examples where the ability to perform detailed kinetic rate constant determinations by SPR analysis was indispensable in defining biological mechanisms (Katsamba *et al.*, 2001; Wu *et al.*, 2002) and where SPR-derived constants could predict *in vivo* behavior (Stahelin *et al.*, 2003).

CONCENTRATION ANALYSIS

Concentration measurements, particularly in areas under the control of regulatory authorities, put high demands on the reproducibility and accuracy of the techniques used. Validation experiments have demonstrated that Biacore concentration assays exhibit a high level of reproducibility with %CV values that are often below the corresponding assays in ELISA formats (Shelver and Smith, 2003). There are two principal assay formats for concentration analysis using SPR biosensors: direct binding assays and inhibition assays (see Figures 16.9 and 16.12, respectively).

DIRECT BINDING ASSAYS

In the direct binding assay (Figure 16.9), the sample is injected over the sensor surface and the analyte interacts with the immobilized biomolecule, which is often an antibody. The sensitivity in a direct binding assay depends on the interaction properties (k_a, k_d) of the reagents and on the injection time of the sample. With a typical antibody (with k_a close to $10^6\ M^{-1}\ sec^{-1}$ and k_d between 10^{-4} and $10^{-3}\ sec^{-1}$) immobilized and with an injection time of 10 min, it is normally possible to detect analyte present in buffer at 100 pM concentration.

The shape of the calibration curve in a direct binding assay, and hence the operating range, can be affected and optimized by adjusting several different parameters. In contrast to fixed volume assays, where the concentration of analyte is reduced as the reaction proceeds, the reaction here takes place under flow conditions and the analyte concentration is constant throughout the injection. This means that equilibrium responses will be higher than in fixed volume assays. Equilibrium is seldom reached during injection, however, and so the kinetic properties of the immobilized molecule for the analyte affect the position of the calibration curve. For the same dissociation rate constant, k_d $10^{-4}\ sec^{-1}$ (solid lines in Figure 16.10), a higher association rate constant moves the calibration curve so that lower concentrations can be determined. Note that the distance between curves representing a 10-fold change in affinity becomes smaller as

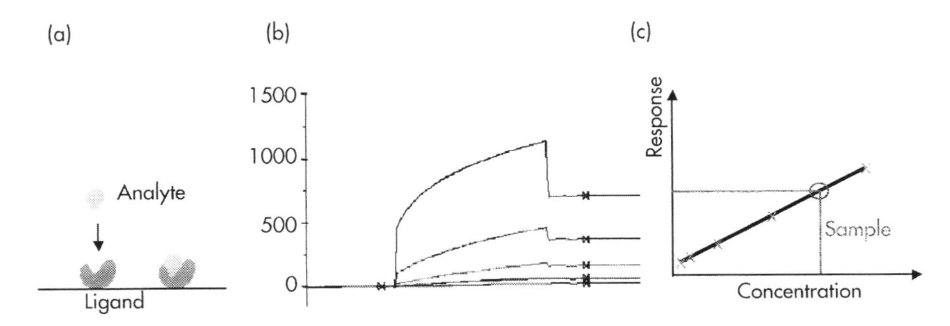

Fig. 16.9 Direct binding assay for concentration analysis. (a) Analyte concentration is measured via direct binding to a ligand immobilized on the sensor surface. (b) Overlay sensorgrams from the analysis of a series of samples the shows response levels that increase with higher analyte concentrations. (c) A calibration curve prepared from a series of known analyte concentrations is used to calculate analyte concentrations from the samples. In this assay format, the calibration curve has a positive slope relative to increasing concentration (compare with Figure 16.12).

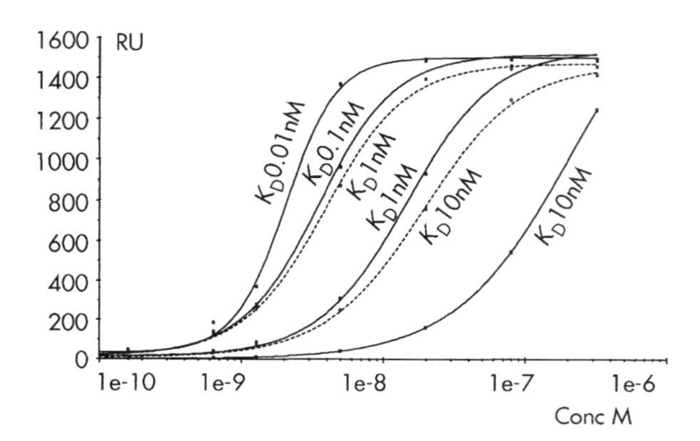

Fig. 16.10 Dose response curves obtained for immobilized ligands with different kinetic properties. Solid lines represent k_d values of 10^{-4} sec^{-1} and dashed lines represent k_d values of 10^{-3} sec^{-1} with affinities as indicated.

the affinity is increased, and that response curves start to overlap at lower analyte concentrations. This reflects transport effects. The dashed lines represent curves obtained where the dissociation rate constant is 10^{-3} sec^{-1}. The two calibration curves with 10 nM affinity are at clearly different positions and for the same affinity, a higher association rate constant shifts the calibration curve so that lower concentrations can be measured.

Another way of enabling measurements at lower concentrations is to increase the contact time (Figure 16.11). A higher immobilization level results in higher maximum analyte responses, which extends the range of the assay and gives greater precision in the measurements, although the effect on the shape of the calibration curve is limited. The direct binding assay is

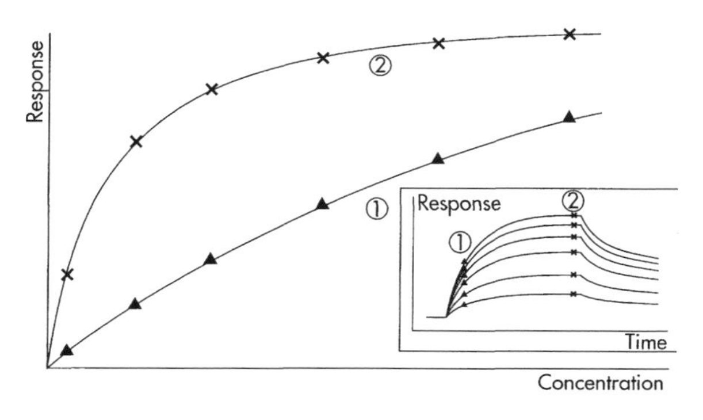

Fig. 16.11 Longer contact times allow measurements at lower analyte concentrations. The report points used for plotting calibration curve 1 are taken after a shorter contact time than the report points used for calibration curve 2, as is shown in the original sensorgrams in the inset.

often used for analysis of samples in complex matrices such as cell lysates (Liljeblad *et al.*, 2002) or serum (Mason *et al.*, 2003; Welt *et al.*, 2003), where SPR-based methods are emerging as new tools for the study of the antibody response against therapeutic proteins.

INHIBITION ASSAYS

The inhibition assay format (Figure 16.12) is often used for analysis of low molecular weight analytes. In this assay format, the small molecule (or an analogue or derivative thereof) is immobilized onto the sensor surface. A fixed amount of a detecting molecule (usually an antibody against the analyte) is mixed with and allowed to interact with the sample prior to injection into the instrument. The pre-incubated sample with antibody is injected and remaining free antibodies interact with the immobilized substance and generate a binding response. This means that the measured response is inversely related to the concentration of analyte in the samples. In an inhibition assay, the sensitivity is determined by the affinity of the interaction and to obtain a high sensitivity, it is an advantage to use a low concentration of the antibody. In practice the sensitivity of an inhibition assay is often close to 1 nM. An additional advantage of the inhibition format is that surfaces with small molecules immobilized are usually very stable and hundreds of samples can often be analyzed on the sensor surface.

As mentioned above, the range of an inhibition assay is determined by the affinity of the detecting molecule for analyte in solution together with the concentration of detecting molecule. A high affinity allows measurement at lower analyte concentrations, but also narrows the operating range (Figure 16.13). Changes in the amount of the detecting molecule affect the height of calibration curve, so that higher concentrations compress the curve at high response levels, while lower concentrations may allow measurement at lower analyte concentrations (Figure 16.13).

Changes in contact time affect the calibration curve in a similar way as for a direct binding assay, i.e. a longer contact time increases the response range. This, in combination with a lower concentration of detecting molecule, enables the measurement of lower analyte concentrations.

Inhibition assays are frequently used for the analysis of pathogens, toxins, veterinary drugs, and nutritional additives in food samples and typical sample matrixes include milk, jam, juice, honey, wheat, meat, liver, urine, and bile. In this context, SPR technology has been described as more flexible and reliable than ELISA (Johansson and Hellenas, 2001), and as more rapid and less tedious than HPLC (Samsonova *et al.*, 2002). Automation (Shelver and Smith, 2003) and the long-term stability of sensor surfaces (Gillis *et al.*, 2002) are also seen as advantages.

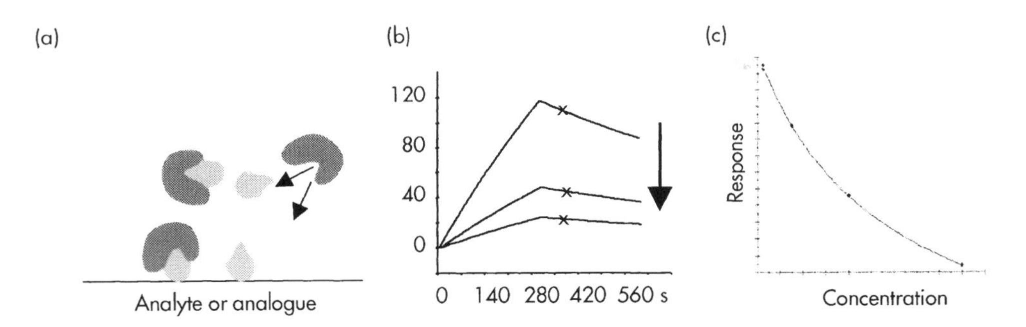

Fig. 16.12 The inhibition assay for concentration analysis. (a) Analyte (or analyte analogue) is immobilized on the sensor surface and the detecting molecule is injected at a fixed concentration together with samples. Increasing analyte concentration in the sample will reduce the amount of free detecting molecule available to bind to the surface. (b) Overlay sensorgrams from the analysis of a series of samples show response levels that are inversely proportional to the analyte concentrations. (c) A calibration curve prepared from a series of known analyte concentrations is used to calculate analyte concentrations from the samples. In this assay format, the calibration curve has a negative slope relative to increasing concentration (compare with Figure 16.9).

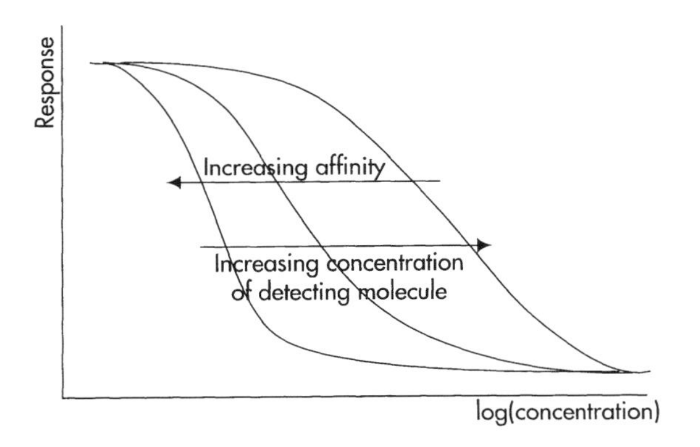

Fig. 16.13 Increasing the affinity of the detecting molecule moves the operating range to lower analyte concentrations and narrows the range. Increasing the concentration of detecting molecule moves the operating range to higher concentrations.

SUMMARY

In both basic life science and applied research, the ability to obtain high-quality characterization of biomolecular interactions is of the utmost importance. The examples described here demonstrate that SPR technology is an extremely valuable tool for the analysis of biomolecular interactions over a very wide range of research applications. The intrinsic advantages of the SPR detection principle, together with flexible surface chemistry options and a highly advanced microfluidics system have enabled this technology to be used for the analysis of interactions among a great diversity of biological interactants, from low molecular weight drug compounds, nucleic acids and proteins, through to whole cells. These features, together with the possibility to generate multiple data types including binding site analysis, concentration measurements, and affinity/kinetics, have resulted in the extensive use of this technology in fields as diverse as basic life science and biotechnology industry research, drug discovery, and in quality control applications. Biological research currently stands on the threshold of the much-anticipated explosion of the proteomics field, and the demands for the type of data provided by SPR biosensors can be expected to increase dramatically as a result of these developments.

REFERENCES

Bader, G.D. and Hogue, C.W. PreBIND and Textomy – mining the biomedical literature for protein–protein interactions using a support vector machine. *BMC Bioinformatics* **4**, 11 (2003).

Boulton, S.J., Vincent, S. and Vidal, M. Use of protein-interaction maps to formulate biological questions. *Curr. Opin. Chem. Biol.* **5**, 57–62 (2001).

Canziani, G.A., Klakamp, S. and Myszka, D.G. Kinetic screening of antibodies from crude hybridoma samples using Biacore. *Anal. Biochem.* **325**, 301–307 (2004).

Clement, G., Boquet, D., Frobert, Y., Bernard, H., Negroni, L., Chatel, J.M., Adel-Patient, K., Creminon, C., Wal, J.M. and Grassi, J. Epitopic characterization of native bovine beta-lactoglobulin. *J. Immunol. Methods* **266**, 67–78 (2002).

Coates, P.J. and Hall, P.A. The yeast two-hybrid system for identifying protein–protein interactions. *J. Pathol.* **199**, 4–7 (2003).

Fägerstam, L.G., Frostell, Å., Karlsson, R., Kullman, M., Larsson, A., Malmqvist, M. and Butt, H. Detection of antigen–antibody interactions by surface plasmon resonance. Application to epitope mapping. *J. Mol. Recognit.* **3**, 208–214 (1990).

Gillis, E., Gosling, J., Sreenan, J. and Kane, M. Development and validation of a biosensor-based immunoassay for progesterone in bovine milk. *J. Immunol. Methods* **267**, 131 (2002).

Hermjakob, H., Montecchi-Palazzi, L., Lewington, C., Mudali, S., Kerrien, S., Orchard, S., Vingron, M., Roechert, B., Roepstorff, P., Valencia, A., Margalit, H., Armstrong, J., Bairoch, A., Cesareni, G., Sherman, D. and Apweiler, R. IntAct: an open source molecular interaction database. *Nucleic Acids Res.* **32**, D452–D455 (2004), Database issue.

Ito, T., Ota, K., Kubota, H., Yamaguchi, Y., Chiba, T., Sakuraba, K. and Yoshida, M. Roles for the two-hybrid system in exploration of the yeast protein interactome. *Mol. Cell. Proteomics* **1**, 561–566 (2002).

Johansson, M.A. and Hellenas, K.E. Sensor chip preparation and assay construction for immunobiosensor determination of beta-agonists and hormones. *Analyst* **126**, 1721–1727 (2001).

Johnsson, B., Löfås, S., Lindquist, G., Edström, Å., Müller Hillgren, R.-M. and Hansson, A. Comparison of methods for immobilization to carboxymethyl dextran sensor surfaces by analysis of the specific activity of monoclonal antibodies. *J. Mol. Recognit.* **8**, 125–131 (1995).

Karlsson, R. Affinity analysis of non-steady-state data obtained under mass transport limited conditions using BIAcore technology. *J. Mol. Recognit.* **12**, 285–292 (1999).

Katsamba, P.S., Myszka, D.G. and Laird-Offringa, I.A. Two functionally distinct steps mediate high affinity binding of U1A protein to U1 hairpin II RNA. *J. Biol. Chem.* **276**, 21476–21481 (2001).

Liljeblad, M., Lundblad, A. and Pahlsson, P. Analysis of glycoproteins in cell culture supernatants using a lectin immunosensor technique. *Biosens. Bioelectron.* **17**, 883–891 (2002).

Markgren, P.O., Schaal, W., Hamalainen, M., Karlen, A., Hallberg, A., Samuelsson, B. and Danielson, U.H. Relationships between structure and interaction kinetics for HIV-1 protease inhibitors. *J. Med. Chem.* **45**, 5430–5439 (2002).

Mason, S., La, S., Mytych, D., Swanson, S.J. and Ferbas, J. Validation of the BIACORE 3000 platform for detection of antibodies against erythropoietic agents in human serum samples. *Curr. Med. Res. Opin.* **19**, 651–659 (2003).

Nedelkov, D. and Nelson, R.W. Surface plasmon resonance mass spectrometry: recent progress and outlooks. *Trends Biotechnol.* **21**, 301–305 (2003).

Peri, S., Navarro, J.D., Kristiansen, T.Z., Amanchy, R., Surendranath, V., Muthusamy, B., Gandhi, T.K., Chandrika, K.N., Deshpande, N., Suresh, S., Rashmi, B.P., Shanker, K., Padma, N., Niranjan, V., Harsha, H.C., Talreja, N., Vrushabendra, B.M., Ramya, M.A., Yatish, A.J., Joy, M., Shivashankar, H.N., Kavitha, M.P., Menezes, M., Choudhury, D.R., Ghosh, N., Saravana, R., Chandran, S., Mohan, S., Jonnalagadda, C.K., Prasad, C.K., Kumar-Sinha, C., Deshpande, K.S. and Pandey, A. Human protein reference database as a discovery resource for proteomics. *Nucleic Acids Res.* **32**, D497–D501 (2004), Database issue.

Pless, D.D., Torres, E.R., Reinke, E.K. and Bavari, S. High-affinity, protective antibodies to the binding domain of botulinum neurotoxin type A. *Infect. Immun.* **69**, 570–574 (2001).

Robbio, L.L., Uboldi, P., Marcovina, S., Revoltella, R.P. and Catapano, A.L. Epitope mapping analysis of apolipoprotein B-100 using a surface plasmon resonance-based biosensor. *Biosens. Bioelectron.* **16**, 963–969 (2001).

Samsonova, J.V., Baxter, G.A., Crooks, S.R., Small, A.E. and Elliott, C.T. Determination of ivermectin in bovine liver by optical immunobiosensor. *Biosens. Bioelectron.* **17**, 523–529 (2002).

Schlattner, U., Reinhart, C., Hornemann, T., Tokarska-Schlattner, M. and Wallimann, T. Isoenzyme-directed selection and characterization of anti-creatine kinase single chain Fv antibodies from a human phage display library. *Biochim. Biophys. Acta* **1579**, 124–132 (2002).

Shelver, W.L. and Smith, D.J. Determination of ractopamine in cattle and sheep urine samples using an optical biosensor analysis: comparative study with HPLC and ELISA. *J. Agric. Food Chem.* **51**, 3715–3721 (2003).

Shuman, C.F., Markgren, P.O., Hamalainen, M. and Danielson, U.H. Elucidation of HIV-1 protease resistance by characterization of interaction kinetics between inhibitors and enzyme variants. *Antiviral Res.* **58**, 235–242 (2003).

Stahelin, R.V., Rafter, J.D., Das, S. and Cho, W. The molecular basis of differential subcellular localization of C2 domains of protein kinase C-alpha and group IVa cytosolic phospholipase A2. *J. Biol. Chem.* **278**, 12452–12460 (2003).

Turecek, F. Mass spectrometry in coupling with affinity capture-release and isotope-coded affinity tags for quantitative protein analysis. *J. Mass Spectrom.* **37**, 1–14 (2002).

Turk, B.E. and Cantley, L.C. Peptide libraries: at the crossroads of proteomics and bioinformatics. *Curr. Opin. Chem. Biol.* **7**, 84–90 (2003).

van der Merwe, P.A. and Barclay, A.N. Transient intercellular adhesion: the importance of weak protein–protein interactions. *Trends Biochem. Sci.* **19**, 354–358 (1994).

Welt, S., Ritter, G., Williams, C. Jr., Cohen, L.S., John, M., Jungbluth, A., Richards, E.A., Old, L.J. and Kemeny, N.E. Phase I study of anticolon cancer humanized antibody A33. *Clin. Cancer Res.* **9**, 1338–1346 (2003).

Wu, L.C., Tuot, D.S., Lyons, D.S., Garcia, K.C. and Davis, M.M. Two-step binding mechanism for T-cell receptor recognition of peptide MHC. *Nature* **418**, 552–556 (2002).

Zhong, J., Zhang, H., Stanyon, C.A., Tromp, G. and Finley, R.L. Jr. A strategy for constructing large protein interaction maps using the yeast two-hybrid system: regulated expression arrays and two-phase mating. *Genome Res.* **13**, 2691–2699 (2003).

17 Lab-on-a-Chip, Micro-, and Nanoscale Immunoassay Systems

Larry J. Kricka and David Wild

In the Products section of this book, many different commercially available immunoassay systems are described. The instrumentation ranges in size from floor-standing laboratory analyzers to compact point-of-care tests. The first point-of-care tests (for human chorionic gonadotropin) were based on agglutination on a slide (Santomauro and Sciarra, 1967) and since then there has been a continuous expansion in the range available (Kasahara and Ashihara, 1997; Bayne, 1997; Graff, 1997; Price *et al.*, 1997; Price and Hicks, 1999; Kricka and Wilding, 1997). Although point-of-care tests are self-contained and comparatively small, the reagents and reaction vessels can easily be seen without the aid of magnification. Microscale immunoassays are still mostly at the development stage, but immunoassays involving microchip technology and some with elements of nanotechnology have been demonstrated at the laboratory scale, and several have significant commercial backing.

There are five primary objectives of miniaturization:

- Financial – reduction in the costs of biological materials consumed and the manufacturing processes.
- Environmental – reduction in biohazardous solid and liquid waste and packaging.
- Simplification – integration of all of the immunoassay steps (including sample preparation, analysis, data handling, and result presentation), which is essential for point-of-care, single use assays.
- Mobility – ease of use in field situations, e.g. detection of biological weapon materials (Iqbal *et al.*, 2000; Culpepper and Pratt, 2001).
- Scope – capability for multiple simultaneous testing for many different analytes, for example in proteomic studies.

Taking the concept of assay miniaturization further, the analyzer instrumentation may also be re-engineered to become integral with the test unit. Microchip-based analyzers have a number of potential advantages and benefits when compared with conventional macro-scale analyzers (Table 17.1). Total integration of all of the steps in the overall analytical process (i.e. sample preparation, analysis, data handling, and result presentation) is the ultimate objective (Kricka, 2002; Cheng and Kricka, 2001; Kricka, 2001).

Microchips are miniaturized assemblies, typically involving dimensions between 100 nm and 1 mm. Usually these chips are two-dimensional in appearance, and manufactured as a series of layers. **Microfluidic microchips** (also called **fluidic microchips**) are chips that contain chambers interconnected by narrow channels, along which sample and reaction fluids are transferred. Different assay stages are performed at different locations in the chip. Internal volumes depend on the cross-section and geometry of the particular structures but are usually in the nanoliter to microliter range. Microchips are fabricated from silicon, glass, or different types of plastic (e.g. polydimethylsiloxene (PDMS) or polymethylmethacrylate) (Anderson *et al.*, 2000; Bauer and Dietrich, 1999; Becker and Gartner, 2000) using various techniques including etching, hot embossing, wire imprinting, reactive ion etching, or laser ablation (Ning and Fitzpatrick, 2001). Microchips are typically 1–2 cm^2 in area and a few millimeters in thickness. The first microchip analyzer was a gas-chromatography system (Terry and Hawker, 1983; Terry *et al.*, 1979). Examples of fluidic microchip assays are shown in Table 17.2.

Bioelectronic chips have an interface between biomolecules (antibodies, antigens, or signal generating molecules) and non-biological materials, resulting in a transfer or modulation of signal from the biomolecule to the device, which can be amplified electronically. They contain built-in electrical components in combination with fluidic elements. The electrical components

The Immunoassay Handbook. *Third Edition*
D. Wild (Ed.)

Table 17.1. Selected advantages and disadvantages of miniaturized analyzers.

Advantages
High-volume low cost manufacture
Rapid low cost design cycles
Low sample volume
Rapid analysis
Simultaneous multi-analyte assays
Replicate analysis
Integration of analytical steps (lab-on-a-chip)
Small size facilitates extra-laboratory applications
Encapsulation for safe disposal
Disposable
Disadvantages
Non-representative sampling
Sensitivity limitations
Human interface with microchips

(e.g. electrodes) are located within, e.g. a microchamber, and used to manipulate a fluid or the components of a fluid contained within the chamber. The chip is usually fabricated on an electronic board that provides the connections to the electrical components within the chip, and plugs into a controller-monitor.

A **microarray** is an ordered collection of reagents immobilized on the surface of a small piece of silicon, glass, or plastic (a chip) (Schena, 1999; Schena, 2000). The array is formed by spotting, stamping, or depositing the reagents; or by *in situ* synthesis of the reagents on the surface of the chip. Microarrayed reagents have included complementary deoxyribonucleic acid (cDNA), oligonucleotides, antibodies, antigens, oligopeptides, and tissue sections. The size, density, and number of locations (**microspots**) on an array varies widely. Spot sizes are often less than 100 μm and arrays with hundreds of spots per square centimeter are in routine use.

Currently available **microfabrication** technology enables the construction of highly complex micron-sized interconnecting structures with diverse functions (e.g. filters, valves, and pumps) (Angell *et al.*, 1983; Bard, 1994) and the integration of electronic sensors, on a microchip of less than 1 cm^2, to form a **micro total analytical system** (**μTAS**) or **lab-on-a-chip** (Manz *et al.*, 1990; Manz *et al.*, 1991).

The small size of microchip devices facilitates rapid response times and production of small compact hand-held analyzers that are suitable for performing tests in a non-laboratory environment (e.g. point-of-care testing).

Table 17.2. Fluidic microchip immunoassays.

Analyte	Reference
Alpha-fetoprotein	Song *et al.* (1994)
Anti-DNP antibody	Yang *et al.* (2001)
Anti-estradiol antibody	Cheng *et al.* (2001)
Anti-HIV antibody	Karlsson *et al.* (1991)
Anti-p53 antibody	Askari *et al.* (2001)
Anti-secretory IgA antibody	Sato *et al.* (2000a,b)
Anti-theophylline antibody	Karlsson *et al.* (1991)
Atrazine	Yakovleva *et al.* (2002)
Blood group antigens	Wilding *et al.* (1997)
Carcinoembryonic antigen	Sato *et al.* (2001)
Cortisol	Koutny *et al.* (1996)
Human IgA	Sato *et al.* (2000a,b)
Human IgG	Linder *et al.* (2001)
Mouse IgG	Wang *et al.* (2001), Mangru and Harrison (1998) and Chiem and Harrison (1997)
Ovalbumin	Cheng *et al.* (2001) and Guifeng *et al.* (2000)
Phenytoin	Hatch *et al.* (2001)
Rabbit IgG	Dodge *et al.* (2001)
Secretory human immunoglobulin A	Sato *et al.* (2000a,b)
Theophylline	Chiem and Harrison (1997) and Von Heeren *et al.* (1996)
Thyroxine	Schmalzing *et al.* (1997)

A microchip-based analyzer can be encapsulated to provide extended operation over a wider range of environmental conditions of humidity and temperature.

Nanotechnology describes systems that operate at the atomic, molecular, or macromolecular levels with dimensions less than 100 nm. Although immunoassays involve molecular interactions, to qualify as nanotechnology the system must have an element of structure, fabrication, or control at this scale. It includes larger structures that are assembled from components at the nanometer scale and **nanoparticles**, which have a size of 100 nm or less.

Current immunoassays in commercial use are typically based on sample and reagent volumes of 10–200 μL. The size of a 10 μL droplet is approximately 2.15 mm^3, so it cannot truly be described as microscale. To be described as microscale, sample volumes of less than 1 nL are required. The dimension of a 1 nL droplet is 100 μm^3.

In electronics, miniaturization has been a way of life in industry for decades, and this has brought numerous benefits of cost reduction and increased product capability. Most of the conventional components of mechanical devices and fabrication processes have been replicated, experimentally, at microscale. The first products are already in widespread use. An example is the microcomb drive motor used in automobiles as an air-bag sensor.

It is perhaps inevitable that immunoassay scientists and engineers, who have already proved to be exceptionally innovative, find themselves drawn to nanotechnology. Since immunoassays are effectively molecular reactions, immunoassays appear to lend themselves to miniaturization, with potential advantages of cost reduction and greater capacity. Yet immunoassays are also cutting edge examples of science and technology at their current scale. Is it realistic to think that the delicate balance required to control immunoassay kinetics and measure faint signals could be miniaturized more than 1000-fold?

To understand the challenges and potential advantages of micro- and nanoscale immunoassays some basic theoretical principles need to understood. At the micro- and nanoscale, there are many differences from conventional immunoassays.

FUNDAMENTAL DIFFERENCES DUE TO MINIATURIZATION

There are a number of fundamental differences between the current macroscale immunoassays and micro- or nanoscale assays. For further details see Nguyen and Nwereley (2002) and Trimmer and Stroud (2002).

VISCOSITY AND SURFACE TENSION

The surface tension of water is strong, due to the hydrogen bonding between adjacent water molecules. Surface tension results from the intermolecular attractive forces at the surface of a liquid. Even at the centimeter scale surface tension can affect the distribution of sample and reagents in an assay vessel, such as a microtiter® well. In aqueous biological fluids such as blood, urine, and sputum, additional intermolecular attractive forces create higher surface tension and increase viscosity.

As size decreases, so does volume. Volume is proportional to weight and inertia. As the size of an object decreases, its volume decreases by a power of three, and so does inertia. Because surface tension decreases in direct proportion to the object's length, any reduction in surface tension is relatively small. As size goes down, surface tension decreases only by a power of one, while inertia drops by a power of three.

At very small scales, surface tension becomes the dominant force while inertia is insignificant. As a result the surface tension effects of the materials that come into contact with sample and reagents become very important.

Wettability is the preference of a solid to touch a liquid (or gas, such as air), known as the wetting phase, rather than another liquid or gas. The wetting phase will tend to spread on the solid surface and a porous solid will tend to imbibe the wetting phase. In the case of immunoassay, it is important that the channels along which fluids pass have a high wettability for aqueous solutions. Wettability is determined by measuring the **contact angle** of a drop of the aqueous solution on the solid, which may be plastic, glass, or metal.

Surface tension can be corrected for by using hydrophilic surfaces that attract the droplet. There are a number of methods used to increase the wettability of plastics used in assay devices. They include corona discharge and plasma treatment. Chemical surface treatments include cleaning, priming, coating, and etching. Increasing temperature decreases surface tension but this has limited potential with proteins that denature at higher temperatures. Surfactants may also be added, although they may interfere with binding reactions.

CAPILLARY FLOW

Capillary flow is fundamental to lateral flow immunoassays. It is due to surface tension. Water molecules are attracted by hydrophilic surfaces on the inside of a capillary, causing the leading edge of a narrow tube of water to move along, even against the force of gravity. Capillary action occurs when the adhesion to the walls of the capillary tube is stronger than the cohesive forces between the water molecules. The same attraction between water molecules that is responsible for surface tension causes the leading water molecules to pull the neighboring water molecules along behind them. Capillary action is restrained by the weight of the column of water. In a miniature system it is easier to achieve capillary flow because inertia and gravity have very little effect.

ELECTROOSMOSIS

Samples and reagents contain ions that will interact with a charged surface. Voltage may be applied to move them along a capillary. Electroosmosis is most effective for nanoliter samples, but at the microliter scale the use of pneumatics or hydrodynamic forces is preferable as a means of moving fluid around the structures in a microdevice.

THE EFFECTS OF REDUCED VOLUME ON LOW CONCENTRATION SAMPLES

The concentration of analyte in a sample is constant, regardless of the sample size. As the sample size is reduced, the number of molecules in the sample decreases. Immunoassay analytes are typically in the micro-, nano-, or picomolar range.

From Table 17.3 it can be seen that this places a fundamental restriction on scale for immunoassays. If the sample size is 1 fL, the scale of the sample is 1 μm^3, which is 10 times larger than a true nanotechnology component. Yet if the sample has a concentration of 1 nM, there is a reasonable chance that not one molecule of analyte will be in the sample.

The impact of low volume on the assay of low concentration samples is different for qualitative and quantitative determinations. If a theoretically perfect immunoassay could be developed with a detection efficiency of 100% (i.e. it detects every molecule that is in the sample) and a signal:noise ratio of greater than 1000 then a *qualitative* assay could be designed based on a sample volume of less than 200 fL in the example above. However, to reduce the error of a *quantitative* immunoassay to 1% the number of analyte molecules has to be 10,000 (because the standard deviation is approximately equal to the square root of the number of molecules), and the sample size would have to be 17 pL.

Table 17.3. Effect of volume on number of molecules in a sample containing 1 nmol analyte per liter.

Volume	Dimensions	Number of molecules
1 L	$100 \times 100 \times 100$ mm^3	6×10^{14}
1 mL	$10 \times 10 \times 10$ mm^3	6×10^{11}
1 μL	$1 \times 1 \times 1$ mm^3	6×10^{8}
1 nL	$100 \times 100 \times 100$ μm^3	600,000
1 pL	$10 \times 10 \times 10$ μm^3	600
1 fL	$1 \times 1 \times 1$ μm^3	0 or 1

One way to alleviate the problem of sample volume in a nanoscale device is to draw the required volume of sample across the device, so that the number of molecules brought into contact with the capture antibody or antigen is increased. Flow-through systems with antibody or antigen immobilized onto a surface allow analyte to be absorbed and concentrated at one point. However, it is important to remember that in both competitive and immunometric immunoassays the volume of sample needs to be known exactly to estimate the analyte concentration. The exception to this rule is ambient analyte immunoassay (*see* AMBIENT ANALYTE ASSAY).

Another complication with very low sample volumes containing low concentrations of analyte concerns the effect of removal of analyte molecules from the solution as the assay incubation progresses. Immunoassays rely on antibodies sampling the analyte. As the incubation progresses it is preferable that the analytes are not depleted in the solution by antibody binding. As explained by Ekins earlier in this book the effect would be like trying to measure the temperature of a small hot sample of water with a very cold thermometer, which cools the water on contact. Also as analyte concentration in the solution decreases, the amount of analyte dissociating from the antibody molecules increases. These complications can be compensated for to some extent by standardizing the protocol and volumes but an alternative way to avoid this situation is to use a minute amount of antibody, with a binding site concentration less than $0.01/K$. This reduces the analyte concentration by less than 1%, regardless of the analyte concentration.

EFFECT OF REDUCED VOLUMES ON KINETICS

In many respects, miniaturization presents additional challenges that need to be overcome. However, the use of a reduced scale has a significant advantage because it reduces the distances that molecules need to travel.

In traditional liquid-phase assays all the molecules move freely in solution. Unless the sample is excessively diluted by the assay reagents the volume does not directly affect the chance of antigen and antibody coming into contact. But as assay methodology has progressed, solid-phase assay formats are used almost exclusively for heterogeneous immunoassays, because of their advantages for the separation step. In solid-phase assays one of the key reactants is immobilized and this reduces the speed with which the assay progresses as only the reactant in the liquid-phase can freely move about. The use of very small volumes in microscale assays therefore increases the reaction rate and decreases the time required for equilibrium to be reached. Microscale assays can therefore be completed in a shorter time. This does not necessarily apply if the sample is moved across a detector for a period of time. In this type of assay design, it is important to minimize the diameter of the reaction vessel containing the immobilized antibody over which the sample is flowing.

Assay kinetics may be improved by using electrodes to attract appropriately charged biomolecules to the solid phase (*see* BIOELECTRONIC CHIPS AND IMMUNOASSAY).

IMMUNOASSAY DESIGN AT MICRO- AND NANOSCALE

ASSAY FORMAT

There are two commonly used immunoassay formats: competitive and immunometric, also referred to as competitive and non-competitive (*see* INTRODUCTION TO IMMUNOASSAY PRINCIPLES – IMMUNOASSAY FOR BEGINNERS). A fundamental limitation of competitive immunoassays is that the signal levels at zero and very low concentrations of analyte are relatively high, causing a low signal:noise ratio and impaired sensitivity. Since micro- and nanoscale immunoassays produce very low levels of signal it is advantageous to use the immunometric format to maximize the signal:noise ratio. For small molecules it is possible to develop immunometric format immunoassays using the anti-complex and apposition methods (*see* ANTI-COMPLEX AND SELECTIVE ANTIBODY IMMUNOMETRIC ASSAYS FOR SMALL MOLECULES and APPOSITION IMMUNOASSAY).

The ambient analyte assay format offers two additional advantages. First, it is unique in that the exact volume of sample does not need to be known, as the antibodies sample the analyte providing an estimate of concentration around the antibody spot. Second, this assay involves small microspots of antibody, typically less than 100 μm in diameter and spaced less than 50 μm apart, which lends itself to manufacture at the microscale. (*see* AMBIENT ANALYTE ASSAY).

ANTIBODIES

The requirement for antibodies with appropriate specificity is the same as with normal scale immunoassays. However, there is an increased need in microscale immunoassays for antibodies with high affinity (i.e. a high value of K_{eq}, $>10^{10}$). Monoclonal antibodies offer the highest potential concentration of active antibodies (advantageous for a micro- or nanoscale device) but most monoclonal antibodies lack the high affinities achievable with polyclonal antibodies. The ideal antibody would be monoclonal *and* have very high affinity. Such antibodies are possible using phage display to generate high affinity antibodies (typically $K_{eq} = 10^7 - 10^9$) from a large phage display library, followed by selective modification of the amino acid sequences of the complementarity determining regions using a sequencer to generate new phage DNA sequences. In this way monoclonal antibodies can be generated with affinities of $\geq 10^{11}$.

STANDARDIZATION AND CALIBRATION

The challenges of calibration in quantitative microscale immunoassays are much the same as at the normal scale. Microfabrication techniques are capable of producing identical assay channels and wells in a microchip, so that if calibrators and samples are run together the sample concentrations may be estimated by direct reference to a calibration curve. In microscale assays the kinetics are faster and this allows assays to progress to equilibrium in a shorter time. Assays that reach equilibrium are more forgiving of small differences in assay conditions between samples and calibrators. Also at very small scale it is easier to achieve equilibrium temperature quickly across a range of samples.

Immunometric assays with near-linear dose-response curves may require no more than two calibrators, and in theory an assay with very low non-specific binding and linear dose-response can be calibrated with just one calibrator.

Ultimately the ideal is to manufacture all test units identically so that a factory-generated calibration curve may be used.

Matrix effects such as those between buffer-based calibrators and blood samples may be exaggerated in microchannels, because of differences in surface tension. These effects can be reduced by diluting samples, but this lessens the number of molecules in the sample, which may impair assay sensitivity.

SEPARATION

Most homogeneous immunoassay formats tend to have less sensitivity than heterogeneous assays. So the use of antibody (or antigen) immobilized onto plastic is preferred at the microscale. However, homogeneous immunoassay formats have a distinct advantage in not requiring a washing stage to remove unbound signal generating material.

The effective removal of unbound signal generating material is crucial to assay performance. The ultra-low concentration of bound label can only be measured precisely if all unbound label is washed away. A washing stage is easily incorporated into microchannel structures and it is possible to provide a comparatively large excess of wash buffer from a small reservoir. However, it is not practical to provide vigorous washing of the type advocated to achieve >99.99% separation efficiency (*see* SEPARATION SYSTEMS).

For heterogeneous immunoassays, reagents can be immobilized on beads within microfluidic structures. Examples are:

- polystyrene (Sato *et al.*, 2000a,b),
- fluorescent latex (Wilding *et al.*, 1997),
- magnetic beads (Jin-Woo *et al.*, 2001).

Alternatively, internal surfaces within a microfluidic structure can be activated for immobilization of antigens or antibodies by coating the surfaces with reagents, for example:

- dinitrophenyl (DNP)-conjugated phospholipid bilayers coated on the surface of 50 μm wide PDMS microchannels (Yang *et al.*, 2001),
- protein A immobilized on glass microchannels (50 × 20 μm cross-section), (Dodge *et al.*, 2001),
- goat anti-mouse IgG-biotin:neutravidin complex immobilized on a glass-capped 20 μm deep PDMS microchannel (Linder *et al.*, 2001),
- direct coupling of antibodies to a porous silicon microchannel surface (235 μm deep × 25 μm wide) via chemical immobilization (Yakovleva *et al.*, 2002).

A further option is to position a microfluidic unit in contact with a dextran-coated sensor surface for surface plasmon resonance detection (Karlsson *et al.*, 1991).

Bound and free separation may be achieved using on-chip **capillary electrophoresis** (CE). The most extensively investigated type of fluidic microchip immunoassay format utilizes CE or CE variants (Von Heeren *et al.*, 1996; Schmalzing *et al.*, 2000) for separation and quantitation of bound and free fractions. In combination with laser-induced fluorescence (LIF) detection, CE is a fast and sensitive technique, and ideally suited for analysis of nanoliter–picoliter volumes of reaction mixture. It is also easily integrated with fluidic structures (e.g. mixing and reaction chambers), and electroosmotic pumping required to move fluid within an immunoassay chip. This is illustrated by a recent immunoassay chip that integrates six independent immunoassay manifolds on a single 4 × 4 in. thermally bonded glass chip (Cheng *et al.*, 2001). Each manifold comprises a multicomponent flow channel structure (50 μm wide × 10 μm deep). A sample and a reagent reservoir lead via channels to a 39 mm long reactor that in turn leads to a double-T injector for loading a sample from the reactor into the 9.3 cm long separation channel. The separation channels converge to a detection zone that is scanned to detect separated components by LIF. A model anti-estriol assay (limit of detection 4.3 nM) was completed in 30 sec (10 sec for loading and reaction plus 20 sec for CE analysis). Microchip-based CE has also been used to determine antibody affinity (Chiem and Harrison, 1998). Currently, the commercial equipment for microchip-based CE is bench-top in size (see Lab-on-a-Chip at http://www.chem.agilent.com), but further advances in miniaturization should effect further size-reduction to the hand-held size required for point-of-care applications.

SIGNAL GENERATION

In immunoassays the signal is potentially weak because of the very low concentration of analyte. In a microscale assay format the signal will be weaker still so it is essential that a very high specific activity label is used. A high signal:noise ratio must be achieved, so sources of noise must be eliminated. **Laser-induced fluorescence (LIF)** has proved to have excellent properties for microscale immunoassays, as the laser can be focused onto a small area, concentrating the light energy, providing a high specific activity signal. A laser-based confocal microscope enables the illumination of a microscopic area and the separation and quantification of emitted light at a different wavelength (*see* AMBIENT ANALYTE ASSAY).

Various assay strategies have been implemented on microchips including qualitative immunoassays based on visual inspection of agglutination with the aid of a microscope. On-chip label or immune complex detection options include:

- simple visual inspection,
- chemiluminescence detection of a horseradish peroxidase label (HRP), (Mangru and Harrison, 1998),
- fluorescence detection of a fluorophore label such as Cy-5 (Guifeng *et al.*, 2000),
- electrochemical detection of an alkaline phosphatase label (Wang *et al.*, 2001),
- potentiometric detection of urease labels (Briggs and Fanfili, 1991),
- surface plasmon detection of immune complexes (Karlsson *et al.*, 1991).

Simple detection via observation of agglutination is particularly effective within small microchannels on a chip. For example, a qualitative ABO typing agglutination immunoassay has been accomplished in microchambers (5 mm × 5 mm × 40 μm deep) in a glass-capped silicon microchip. A blood sample was injected into the chamber containing anti-A or anti-B antiserum, and agglutination of the red blood cells detected visually using a microscope (Wilding *et al.*, 1997). Similarly, IgG has been detected qualitatively in a 150 μm wide × 40 μm deep microchannel filled with anti-immunoglobulin G (IgG) coated fluorescent beads (4.55 μm diameter). The agglutination of the beads coated with anti-IgG was assessed using a fluorescence microscope (Wilding *et al.*, 1997). Agglutination immunoassays for alpha-fetoprotein have also been performed in microchannels (four 0.4 μL channels; 10 mm long, 80 μm deep, 500 μm wide) etched in a Pyrex glass microchip (Song *et al.*, 1994). An alternating electric field was used to accelerate the rate of the agglutination of the fluorescent latex beads (1.66 μm diameter) and agglutination assessed using a fluorescent microscope. This rapid (approximately 1 min) assay detected concentrations of alpha-fetoprotein as low as 10 pg/mL.

HOMOGENEOUS IMMUNOASSAY

Diffusion Immunoassay

Micronics Inc, Redmond, WA, USA has developed a competitive homogeneous immunoassay in a T-sensor fabricated on a glass-Mylar-glass hybrid chip (2.5 × 2.5 cm) that comprises two fluid inlets that lead to two channels that merge into a single 100 μm deep × 1200 μm wide channel, that connects to a fluid outlet (Hatch *et al.*, 2001; Weigl and Yager, 1998; Kamholz *et al.*, 1999). Under laminar flow conditions, the two fluid streams (the antibody reagent and the sample spiked with fluorescein-labeled antigen) flow side-by-side and the only mixing is by diffusion. Diffusion of antigen into the parallel stream is governed by the fraction that binds to antibody in the parallel flowing stream. Competition between antigen and labeled antigen for binding sites in the antibody stream provides the basis for antigen quantitation, and measurements are made after 18.5 sec of diffusive mixing. An advantage of this assay is that diluted whole blood can be assayed without the need to remove red cells. An assay for phenytoin in whole blood (10–400-fold dilution) required less than 1 min and detected 0.43 nM phenytoin.

ON-CHIP DETECTION METHODS

Chemiluminescence is a popular choice as a detection technology in routine immunoassay and, not surprisingly, it has also been adapted to microchip-based immunoassays (Yakovleva *et al.*, 2002; Mangru and Harrison, 1998). This is illustrated by an immunoassay for mouse IgG (Mangru and Harrison, 1998). The horseradish peroxidase conjugate was detected post-separation using the chemiluminescent HRP-catalyzed reaction of luminol with peroxide, in a CE microchip. All of the analytical components (injector, separator, and post-separation reactor) were integrated on a planar glass chip (10 or 40 μm etch depth, 1 or 8 nL injected sample plugs, respectively). An aluminum mirror fabricated onto the backside of the detection zone provided a reflective surface to enhance collection efficiency of the emitted light. The calibration curve for mouse IgG was linear in the range 0–60 μg/mL using an HRP-goat anti-mouse immunoglobulin G $F(ab')_2$ fragment conjugate.

There has been extensive development of microchip fluoroimmunoassays (Cheng *et al.*, 2001; Chiem and Harrison, 1997; Guifeng *et al.*, 2000; Mere *et al.*, 1999). CE is particularly effective for rapid separation (<30 sec) and quantitation of bound and free fractions in competitive immunoassays using liquid reagents (antibody and labeled antigen). These assays are performed on a microchip that includes a mixing channel that leads to an orthogonal CE channel arrangement (channel depth typically 20 μm). The rate of flow and length of the mixing channel determine the incubation time of the assay, and the reaction mixture is then sampled and the sample delivered into the separation channel for analysis. Laser-induced fluorescence reveals and quantifies the separated bound and free species in the channel. Various fluorophores have been used as labels in this type of immunoassay including Cy-5 (Guifeng *et al.*, 2000), and fluorescein (Chiem and Harrison, 1997).

Higher analytical throughput can be achieved using a six-channel microfluidic immunoassay, in which six independent mixing, reaction, and separation manifolds are integrated on one microfluidic wafer. Each manifold is operated simultaneously and data acquired using a scanned fluorescence detection system (Cheng *et al.*, 2001).

On-chip electrochemical detection of an alkaline phosphatase label is accomplished following separation of free antibody and antibody–antigen complex in a post-column reaction of the enzyme label with a 4-aminophenyl phosphate substrate, and downstream amperometric detection of the 4-aminophenol product (Wang *et al.*, 2001). A detection limit of 1.7 amol is achieved in a model assay for mouse IgG conducted in 50 μm wide and 20 μm deep channels in a glass microchip.

Surface plasmon resonance allows direct detection of immune complexes on a surface. This technique has been used in conjunction with a microfluidic unit in contact with a sensor surface to measure the kinetics of monoclonal antibody–antigen reactions in real time. The antibody or antigen was immobilized in a dextran matrix that was attached to the sensor surface and binding events monitored by surface plasmon resonance (Karlsson *et al.*, 1991).

A thermal lens microscope can detect colloidal gold labels and this detection option has been effectively applied in an immunometric immunoassay for carcinoembryonic antigen (CEA) (Sato *et al.*, 2001) and human secretory immunoglobulin A (Sato *et al.*, 2000a,b). The assays were conducted within microchannels in a transparent glass microchip using an antibody–colloidal gold conjugate and polystyrene beads coated with capture antibody. The beads were constrained within the 100 μm deep × 250 μm wide glass channel by a dam that had a 10 μm gap for fluidic connection between the two sides of the channel. Overall analysis time was 35–45 min, and this was attributed to the advantageous molecular behavior within the confines of the microchannel. Detection limit for CEA was 0.03 ng/mL in an assay that employed a 10 min incubation of sample-conjugate solution in the chip followed by a 10 sec wash step.

An early form of micromachined immunoassay device that can be considered as a microchip analyzer, utilized micromachined silicon pits incorporating a pH sensitive light addressable potentiometric sensor (LAPS) (Briggs and Fanfili, 1991; Owicki *et al.*, 1994). The sensor monitored the pH change that accompanied enzymatic action of bound urease labeled conjugate in immunocomplexes captured on a membrane. Assays for mouse IgG (linear range 25–5000 pg) and potential contaminants of recombinant protein products (e.g. *Escherichia coli* protein, protein A), hormones (e.g. hCG), and infectious agents (e.g. *Yersinia pestis*) were among the assays developed using this LAPS detector (Briggs and Fanfili, 1991; Briggs, 1991).

BIOELECTRONIC CHIPS AND IMMUNOASSAY

Electric fields generated by microelectrodes within microfluidic compartments of bioelectronic chips are effective for manipulating bioparticles and macromolecules. A direct current (dc) field provides electrophoretic control of charged species, and a non-uniform alternating current (ac) field provides dielectrophoretic control of polarized species (charged or neutral). These forces are exploited in immunoassay for sample preparation and for controlling antigen:antibody reactions (Table 17.4) (Huang *et al.*, 2001a–c; Ewalt *et al.*, 2001). In particular capture antibodies and conjugates can be specifically targeted to an electrode, a process known as **electronic addressing**.

The sample preparation capability of a biochip is an important advantage of this type of device. For example, *Bacillus globigii* spores can be separated and enriched from a stream of sample (40 μL) flowing over a 5 × 5 array of electrodes (80 μm diameter, 200 μm center to center spacing; dc current) in a biochip (internal volume 7.5 μL), and detected with fluorescent conjugate (Huang *et al.*, 2001a–c). The 6 min dc-driven capture of the spores from a 40 μL sample is followed by a dc-driven electrophoresis of a Texas Red-labeled anti *B. globigii* antibody conjugate to the positively biased electrodes. A biochip assay can also be formatted as an immunometric assay, as exemplified by an assay for *E. coli* bacteria using the chip described above. First, biotinylated anti-*E. coli* capture antibodies are addressed to the electrodes (dc-driven) where they become immobilized in the streptavidin-containing agarose permeation layer above the electrode. In the next ac-driven step, *E. coli* are collected by the electrodes from the applied sample during a 5 min period. Finally, fluorescein-labeled anti *E. coli* antibody is addressed to the positively biased electrodes (dc-driven) where it binds to the immunospecifically captured *E. coli*. Fluorescence microscopy reveals the presence and extent of binding of *E. coli* to the individual electrodes. Advantages of this technique include low nonspecific binding and faster reaction times compared to assays that rely on passive diffusion of reagents.

Table 17.4. Bioelectronic chip immunoassays.

Analyte	Reference
Bacillus globigii	Huang *et al.* (2001a–c)
Escherichia coli	Huang *et al.* (2001a–c)
Fluorescein-labeled staphylococcal enterotoxin B	Ewalt *et al.* (2001)
Fluorescein-labeled cholera toxin B	Ewalt *et al.* (2001)

MICROARRAYS AND IMMUNOASSAY

The protein microarray chip (protein chip) has proved useful for detecting and identifying proteins, profiling proteins in complex mixtures (e.g. cell lysates) (Huang *et al.*, 2001a–c), protein affinity studies, and for investigating protein–protein, protein–DNA, and protein–RNA interactions (Peluso *et al.*, 2003; Sydor *et al.*, 2003; Stoll *et al.*, 2002; Ge, 2000; Weng *et al.*, 2002; Haab, 2001; Zhu and Snyder, 2001; Albala, 2001; Cahill, 2001; MacBeath, 2001; Scorilas *et al.*, 2000; Walter *et al.*, 2000). It is also an effective solution for the vast number of immunoassays necessary for the high throughput testing required for proteomic studies. This format makes manageable thousands of simultaneous tests on a small compact device (Ekins and Chu, 1991, 1993, 1999; Ekins *et al.*, 1990; Ekins, 1998). Some examples of protein microarray-based assays are collected in Table 17.5 (Scorilas *et al.*, 2000; Ekins and Chu, 1993; Schweitzer *et al.*, 2000; Joos *et al.*, 2000; Huang *et al.*, 2001a–c; Avseenko *et al.*, 2001; Hiller *et al.*, 2002; Silzel *et al.*, 1998; Jones *et al.*, 1998).

Array Manufacture

Several protein microarray fabrication procedures have been developed including:

- spotting onto glass (Schweitzer *et al.*, 2000), gold-coated silicon (Silzel *et al.*, 1998), or plastic surfaces, e.g. polystyrene (Jones *et al.*, 1998),
- electrospray deposition onto aluminized plastic (Avseenko *et al.*, 2001),
- immobilization within arrays of small (e.g. 100 × 100 × 20 μm) polyacrylamide gel pads on a glass surface (Arenkov *et al.*, 2000).

Other ways of forming protein arrays include adsorption onto a self-assembled monolayer of *n*-octyldecyltrimethoxysilane on a silicon dioxide surface previously patterned by UV photolithography into an array of micrometer-sized features (Mooney *et al.*, 1996), and self-assembly via hybridization of RNA–protein fusions to

Table 17.5. Microarray immunoassays.

Analyte	Reference
Anti-digoxigenin antibody	Schweitzer *et al.* (2000)
Autoantigens	Joos *et al.* (2000)
Avidin	Schweitzer *et al.* (2000)
Biotinylated mouse IgG	Scorilas *et al.* (2000)
Cytokines	Huang *et al.* (2001a–c)
Human IgE	Schweitzer *et al.* (2000), Avseenko *et al.* (2001) and Hiller *et al.* (2002)
Goat anti-rabbit IgG	Silzel *et al.* (1998)
Mouse IgG	Avseenko *et al.* (2001)
IgG1	Jones *et al.* (1998)
IgG2	Jones *et al.* (1998)
IgG3	Jones *et al.* (1998)
IgG4	Jones *et al.* (1998)
PSA	Schweitzer et al. (2000)
Thyroid stimulating hormone (TSH)	Ekins and Chu (1993)

an array of surface bound DNA capture probes (Weng *et al.*, 2002).

Chemical immobilization methods used to form microarrays include attachment via UV-induced photolinking (Jones *et al.*, 1998); reaction with a crosslinker, e.g. *N*-[gamma-maleimidobutyrloxy]succinimide ester (Schweitzer *et al.*, 2000) or 1,4-phenylene-diisocyanate (Schweitzer *et al.*, 2000); and printing onto *N*-hydroxysuccinimide activated BSA-coated slides (MacBeath and Schreiber, 2000), or aldehyde activated glass slides (Sreekumar *et al.*, 2001). Arrays can also be formed by microstamping using a silicone elastomer stamp made by photolithography (individual elements in the stamp are 350×350 μm and 50 μm high). The protein-coated stamp is applied to an aminopropyltrimethoxysilane treated slide activated with bis-sulfosuccinimidyl suberate (Lin *et al.*, 2001), and the protein deposited by the stamp binds covalently via reaction with the protein coupling agent on the slide.

Zyomyx has developed a microarray based on a silicon wafer (Ruiz-Taylor *et al.*, 2001). Photolithography is used to create pillars 150 μm high and 50 μm across. The biochip has six 40 μL sample channels and each channel has 200 pillars (Figures 17.1 and 17.2). The surface is coated with an inorganic chemical to create a negative charge. This attracts poly(L-lysine)-grafted-poly(ethylene glycol)(PLL-g-PEG), which forms a monolayer. The PLL-g-PEG is biotinylated, and the biotin moieties are orientated into the sample stream following coating. This common capture system is used with streptavidin-labeled antibodies or antigens. Zyomyx has released a human cytokine biochip based on this technology, capable of the detection of 30 cytokines.

Assay Format and Detection

The detection methods employed with protein microarrays are analogous to those used in conventional immunoassays. In microarray-based enzyme immunoassays, enzyme labeled conjugates are detected using colorimetric (Avseenko *et al.*, 2001), fluorometric (Hiller *et al.*, 2002), and chemiluminescent signal detection systems (Huang *et al.*, 2001a–c).

Fluoroimmunoassay formats are popular because of the ease with which simultaneous assays can be implemented, thus allowing multiplexed or comparative assays on the same microarray. This is illustrated by a two-color assay using fluorescein and Cy3 for avidin, and sheep anti-digoxigenin IgG (Joos *et al.*, 2000). Three-color assays have also been developed using BODIPY-FL, Cy3, and Cy5 dyes for studying protein–protein interactions, and $Alexa_{488}$, Cy3, and Cy5 dyes for detecting binding of a digoxigenin derivative with anti-digoxigenin, FKBP12 (FK-506 binding-protein) with AP1497 (a synthetic pipecolyl alpha-ketoamide), and avidin with biotin, respectively (Sreekumar *et al.*, 2001). Time resolved fluorescence detection has also been adapted for use in microarray immunoassay (Scorilas *et al.*, 2000; Luo and Diamandis, 2000). Biotinylated antibodies immobilized by spotting (5 nL) as an array on a membrane coated slide can be detected using a streptavidin:biotinylated-polyvinylamine 4,7-bis(chlorosulfophenyl)-1,10-phenanthroline-2,9-dicarboxylic acid (BCPDA)–europium chelate complex (50–100 BCPDA– europium complexes per polyvinylamine molecule) at levels of approximately 0.25 pg/spot (Scorilas *et al.*, 2000).

Increased sensitivity can be achieved in microarray immunoassay by using the Rolling Circle Amplification

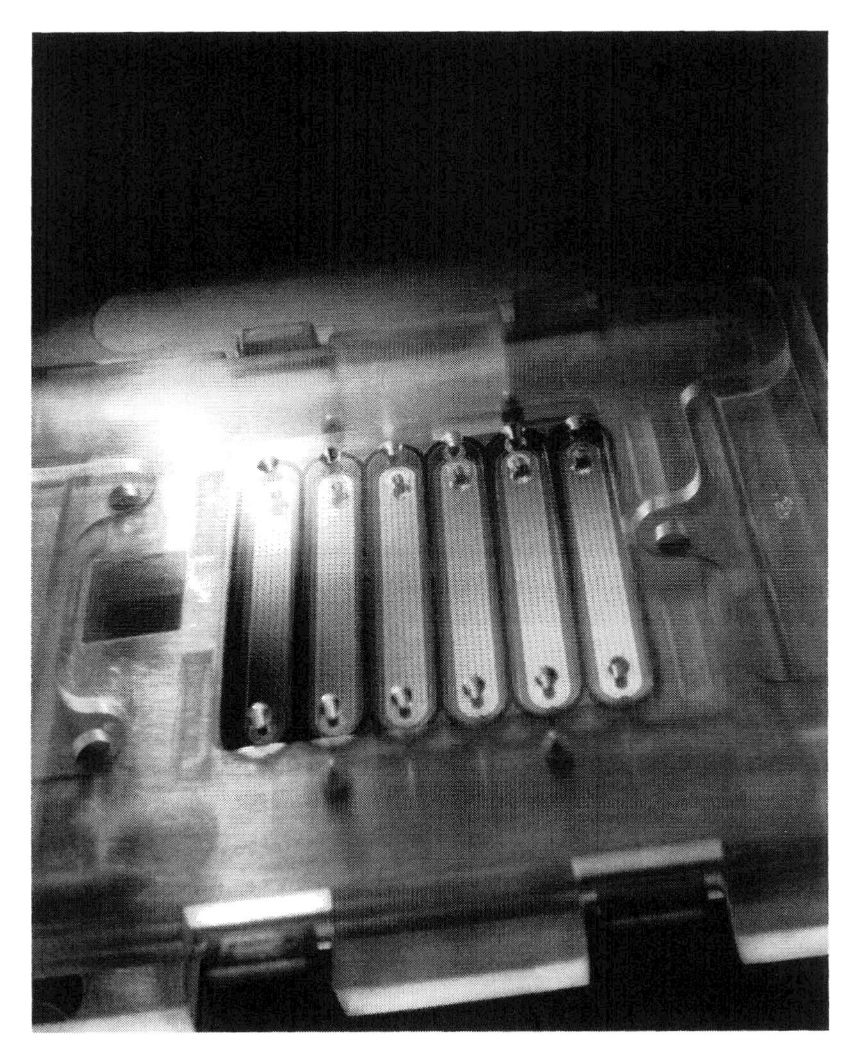

Fig. 17.1 The Zyomyx^{TM} Expression Profiling Biochip. The Biochip enables multianalyte expression profiling of 30 biologically relevant proteins. Each Biochip contains six channels, allowing for the simultaneous analysis of six different samples. A single channel contains an array of 200 discrete silicon pillars, each pillar coated with antibody to a protein. Replicate pillars are randomized across the silicon array. Filling each channel with only 16 μL of sample allows quantification of all 30 analytes. Human cytokine Biochips and murine cytokine Biochips are currently available.

technique to amplify an oligonucleotide primer covalently linked to a detection antibody. When combined with fluorescent detection of the amplified oligonucleotide, it has proved possible to detect signals from individual antigen:antibody complexes on a microarray and achieve highly sensitive assays (e.g. detection of 0.1 pg/mL prostate specific antigen (PSA), and 1 pg/mL IgE) (Schweitzer *et al.*, 2000). Another means of increasing sensitivity in a microarray immunoassay is to adopt the microspot strategy developed by Ekins (Ekins and Chu, 1991, 1993, 1999; Ekins *et al.*, 1990; Ekins, 1998) in which minute amounts of a capture antibody are spotted onto the array and captured antigen detected using a fluorophore-labeled detection antibody and confocal fluorescence microscopy. An assay for thyroid stimulating hormone reached a detection limit of 0.0002 mU/L (Ekins and Chu, 1993).

Another detection technique applied in microarray immunoassay is qualitative **matrix-assisted laser desorption/ionization time-of-flight mass spectrometric analysis** (**MALDI-TOF**) of the analytes bound on an array surface (Nedelkov and Nelson, 2001).

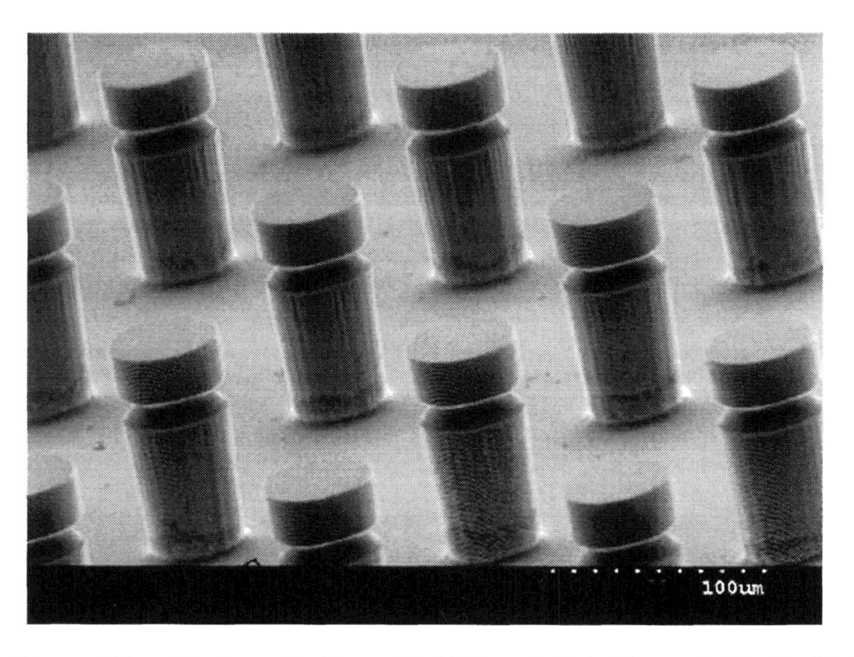

Fig. 17.2 Details of Zyomyx Silicon Pillar Array. The Zyomyx Biochip is used with the Zyomyx Profiling Biochip System™ which is based on double antibody immunoassay principles. Each 50 μm-diameter pillar is coated with a single specificity binding molecule, typically an antibody or antibody fragment, as a capture reagent. The detection reagent is labeled with a fluorescent probe and laser-induced fluorescence is measured.

Atomic force microscopic detection of the height increase when an antibody binds to an arrayed antigen is also used in microarrays, see below (Silzel *et al.*, 1998).

As well as the multicolor detection format, it is also possible to test multiple samples simultaneously against an array of antigens by patterning lines of antigens onto a surface and then delivering solutions to be analyzed by the channels of a microfluidic network across the lines of antigens. This type of assay has been named **mosaic-format immunoassay** because specific binding of target antibodies with immobilized antigens on the surface produces a mosaic of binding events that can be readily visualized in a single assay (Bernard *et al.*, 2001).

ATOMIC FORCE MICROSCOPY

Atomic Force Microscopy (**AFM**) is a nanoscale analytical technique, capable of resolving topographical features with a resolution of less than 1 nm. It may also be used to measure molecular forces of less than 1 nN.

In its simplest form, an antibody or antibody fragment is coupled to a tip on the end of a cantilever. The end of the tip needs to be at a molecular scale and the first AFM was made by gluing a tiny shard of diamond onto one end of a minute strip of gold foil. The tip-cantilever assembly is now usually microfabricated from silicon or silicon nitride. The tip is moved over the surface of a sample and, as the tip comes close to the complementary antigen, the binding forces between antibody and antigen pull the cantilever towards the sample. Alternatively the sample may be moved under the tip. The movement of the cantilever is detected using a diode laser beam deflected off the reflective upper surface of the cantilever onto a dual element photodiode. The photodetector measures the difference in voltage from the two detectors to indirectly estimate the strength of the binding between tip and sample. Feedback from the photodetector is used to maintain the tip at either a constant force or height from the sample.

AFM involves moving the tip (or the sample) in parallel lines so that a topographical image may be constructed. If an antibody is used, the image becomes a map of the locations of the antigen binding sites in the sample. Binding strength may also be mapped. The AFM equipment must be mounted on a special vibration isolation platform.

Initially AFM was applied in **contact mode**, where the tip scans the sample in close contact with the surface, but this can damage the sample, due to friction and adhesion. In **non-contact mode** the tip hovers 50–150 Å above the sample surface. Attractive Van der Waals forces between the tip and the sample are detected and the position of the tip adjusted during the scanning process. The forces are very weak so the tip is oscillated and the forces detected through changes in amplitude, phase, or frequency of the cantilever. However, sensitivity is limited due to the distance of the tip from the surface of the sample. A key advance in AFM was the introduction of **tapping mode**. In this mode the tip is made to touch and withdraw from the sample surface repeatedly, avoiding the sample damage that can occur if the tip is dragged, but enabling the high resolution associated with contact mode. The oscillation is over a comparatively wide amplitude (20 nm) at a frequency of 50–500 kHz, controlled by a piezoelectric crystal. As the tip begins to touch the sample surface the amplitude of the cantilever oscillation is reduced. The amplitude is increased over a depression. The sensitive electronics detect the changes and move the tip up and down to maintain constant amplitude over the bumps and dips, building a topographical model of the surface.

Biological sample preparation is a challenge for AFM because it is important to maintain structural integrity. The tissue or proteins must be immobilized (e.g. on salinized or carbon coated mica), and washed and bathed in appropriate salts and buffer.

AFM is often used in conjunction with optical microscopy, using bright-field or fluorescence labels to identify structures of interest prior to scanning.

MANUFACTURING AND BUSINESS CONSIDERATIONS

The small size of a microscale immunoassay necessitates appropriate packaging in order that it can be manipulated by a human operator. The user interfaces, primarily sample application, and result output, have to be on a human scale.

As the dimensions of immunoassays reduce there are several implications for manufacturers. The most obvious is that the cost of the materials used in the final product reduces. This is a clear benefit for immunoassays, which have complex cocktails of reagents, including some that are very expensive. The capacity of a factory can also be increased if the volume of each unit produced is reduced. However, manufacture of microscale assays requires greater investment in manufacturing and test equipment. Most commercial immunoassays start their life in the laboratory, where it is advantageous to be able to carry out the preparation of reagents manually. However, microscale technology requires automated equipment to be used throughout development. Part of the function of the development process is to understand and optimize process variables so that the eventual production process is as efficient as possible. This can result in several generations of expensive automated equipment. Also each immunoassay has unique characteristics, requiring the automated equipment to be designed with a high degree of flexibility. The cost of development of the current generation of hospital immunoassay systems was in the region of $100–200 million, and the cost of future microsystems is likely to be considerably greater. The benefits of reduced material costs are modest compared to the increased cost of development and capital investment.

However, microfabrication provides increased design flexibility and ease of manufacture compared to the manufacturing techniques used to produce macroscale devices. For microchips fabricated on silicon wafers, high device density allows large numbers of different microchip designs to be simultaneously fabricated on the same wafer. Consequently, many more design iterations can be tested than would be normally possible for a macroscale device fabricated using conventional methods. Once the design has been finalized, low cost manufacture is possible by adapting the same high volume techniques already employed in the electronics industry to produce microelectronic chips.

Also on the positive side of the cost equation, the current generation of hospital analyzers, with their large size and complexity, is very expensive to manufacture and maintain, and miniaturization offers the prospect of much smaller, simpler analyzers, with most of the complexity built into the disposable assay device. Miniaturization also offers more opportunity for commonality of point-of-care and laboratory test methods, and for a fusion between the technology used for tests involving antibody-binding, other proteins, DNA, and RNA.

It is quite possible that a revolutionary microscale immunoassay system could replace the current generation of ageing hospital analyzers and the plethora of point-of-care test technologies, but the investment required to develop the technology and manufacturing capability is probably in excess of $500 million. Alternatively new small companies with strong academic links may develop the underlying technology and commercialize it via a niche market where the advantages of miniaturization would fulfill an unmet clinical need. This may not be in diagnostics, but could be in pharmaceutical research, where large numbers of tests are carried out routinely and the speed and efficiency of the tests has a direct effect on the time to launch of a product with a limited life under patent protection.

CONCLUSIONS

The main thrust of the current range of microminiaturized immunoasssays is towards proteomic studies in which the massively parallel analysis capabilities of microarray chips matches the scope of testing required for the analysis of large numbers of complex protein mixtures. Some microarray-based immunoassay products are in development, most notably the Evidence analyzer (see http://www.randox.com), and this is directed towards routine testing in a clinical laboratory. Zyomyx has also developed a biochip-based microarray immunoassay system. This company has a cytokine test biochip available for research, capable of detection of 30 cytokines.

Less effort has been directed at the development of specific point-of-care chip-based immunoassay tests. The unique capabilities of microchips for integration of a complex analytical process onto a small device may offer the best option for increasing the scope of immunoassay at the point-of-care beyond the current assays such as hCG. Microchip-based immunoassay devices would facilitate an additional degree of sophistication not easily achieved with the current flow-through membrane-based immunoassay devices. Already, many of the functions and features required for either an immunometric or a competitive immunoassay have been formatted into disposable microchip devices, including sample preparation.

Microchips may well be the next step in the general trend towards qualitative and quantitative unitized, disposable immunoassay devices, but further development in this direction will depend on the continued growth of microchip technology, a favorable regulatory climate, economic viability, and demand by the health-care community and the general public.

REFERENCES

Albala, J.S. Array-based proteomics: the latest chip challenge. *Expert Rev. Mol. Diagn.* **1**, 145–152 (2001).

Angell, J.B., Terry, S.C. and Barth, P.W. Silicon micromechanical devices. *Sci. Am.* **248**, 44–55 (1983).

Anderson, J.R., Chiu, D.T., Jackman, R.J. *et al.* Fabrication of topologically complex three-dimensional microfluidic systems in PDMS by rapid prototyping. *Anal. Chem.* **72**, 3158–3164 (2000).

Arenkov, P., Kukhtin, A., Gemmell, A., Voloshchuk, S., Chupeeva, V. and Mirzabekov, A. Protein microchips: use for immunoassay and enzymatic reactions. *Anal. Biochem.* **278**, 123–131 (2000).

Askari, M., Alarie, J.P., Moreno-Bondi, M. and Vo-Dinh, T. Application of an antibody biochip for p53 detection and cancer diagnosis. *Biotechnol. Progr.* **17**, 543–552 (2001).

Avseenko, N.V., Morozova, T.Y., Ataullakhanov, F.I. and Morozov, V.N. Immobilization of proteins in immunochemical microarrays fabricated by electrospray deposition. *Anal. Chem.* **73**, 6047–6052 (2001).

A. Bard (ed), Integrated Chemical Systems, (Wiley, New York, 1994)

Bauer, C. and Dietrich, S. Design rules for polyimide solvent bonding. *Sens. Mater.* **11**, 269–278 (1999).

Bayne, C.G. Pocket-sized medicine. New POC technologies. *Nurs. Manag.* **28**, 30–32 (1997).

Becker, H. and Gartner, C. Polymer microfabrication methods for microfluidic analytical applications. *Electrophoresis* **21**, 12–26 (2000).

Bernard, A., Michel, B. and Delamarche, E. Micromosaic immunoassays. *Anal. Chem.* **73**, 8–12 (2001).

Briggs, J. Sensor-based system for rapid and sensitive measurement of contaminating DNA and other analytes in biopharmaceutical development and manufacturing. *J. Parenter. Sci. Technol.* **45**, 7–12 (1991).

Briggs, J. and Fanfili, P.R. Quantitation of DNA and protein impurities in biopharmaceuticals. *Anal. Chem.* **63**, 850–859 (1991).

Cahill, D.J. Protein and antibody arrays and their medical applications. *J. Immunol. Methods* **250**, 81–91 (2001).

J. Cheng and L.J. Kricka (eds), *Biochip Technology*, (Gordon Breach Publishers, Philadelphia, 2001)

Cheng, S.B., Skinner, C.D., Taylor, J. *et al.* Development of a multichannel microfluidic analysis system employing affinity capillary electrophoresis for immunoassay. *Anal. Chem.* **73**, 1472–1479 (2001).

Chiem, N.H. and Harrison, D.J. Microchip-based capillary electrophoresis for immunoassays: analysis of monoclonal antibodies and theophylline. *Anal. Chem.* **69**, 373–378 (1997).

Chiem, N.H. and Harrison, D.J. Microchip systems for immunoassay: an integrated immunoreactor with electrophoretic separation for serum theophylline determination. *Clin. Chem.* **44**, 591–598 (1998a).

Chiem, N.H. and Harrison, D.J. Monoclonal antibody binding affinity determined by microchip-based capillary electrophoresis. *Electrophoresis* **19**, 3040–3044 (1998b).

Culpepper, R.C. and Pratt, W.D. Advances in medical biological defense technology. *Clin. Lab. Med.* **21**, 679–689 (2001).

Dodge, A., Fluri, K., Verpoorte, E. and de Rooij, N.F. Electrokinetically driven microfluidic chips with surface-modified chambers for heterogeneous immunoassays. *Anal. Chem.* **73**, 3400–3409 (2001).

Ekins, R.P. Ligand assays: from electrophoresis to miniaturized microarrays. *Clin. Chem.* **44**, 2015–2030 (1998).

Ekins, R.P. and Chu, F.W. Multianalyte microspot immunoassay microanalytical 'compact disk' of the future. *Clin. Chem.* **37**, 1955–1967 (1991).

Ekins, R.P. and Chu, F.W. Multianalyte testing. *Clin. Chem.* **39**, 369–370 (1993).

Ekins, R.P. and Chu, F.W. Microarrays: their origins and applications. *Trends Biotechnol.* **17**, 217–218 (1999).

Ekins, R.P., Chu, F. and Biggart, E. Multispot, multianalyte, immunoassay. *Ann. Biol. Clin.* **48**, 655–666 (1990).

Ewalt, K.L., Haigis, R.W., Rooney, R., Ackley, D. and Krihak, M. Detection of biological toxins on an active electronic microchip. *Anal. Biochem.* **289**, 162–172 (2001).

Ge, H. UPA, a universal protein array system for quantitative detection of protein–protein, protein–DNA, protein–RNA, and protein–ligand interactions. *Nucleic Acids Res.* **28**, e3 (2000).

Graff, L.G. Bedside cardiac testing: emergency medicine leaps into the next millennium. *Acad. Emerg. Med.* **4**, 1015–1017 (1997).

Guifeng, J., Attiya, S., Ocvirk, G., Lee, W.E. and Harrison, D.J. Red diode laser induced fluorescence detection with a confocal microscope on a microchip for capillary electrophoresis. *Biosens. Bioelectron.* **14**, 10–11 (2000).

Haab, B.B. Advances in protein microarray technology for protein expression and interaction profiling. *Curr. Opin. Drug Discov. Dev.* **4**, 116–123 (2001).

Hatch, A., Kamholz, A.E., Hawkins, K.R. *et al.* A rapid diffusion immunoassay in a T-sensor. *Nat. Biotechnol.* **19**, 461–465 (2001).

Hiller, R., Laffer, S., Harwanegg, C. *et al.* Microarrayed allergen molecules: diagnostic gatekeepers for allergy treatment. *FASEB J.* **16**, 414–416 (2002).

Huang, Y., Ewalt, K.L., Tirado, M., Haigis, R., Forster, A., Ackley, D., Heller, M.J., O'Connell, J.P. and Krihak, M. Electronic manipulation of bioparticles and macromolecules on microfabricated electrodes. *Anal. Chem.* **73**, 1549–1559 (2001a).

Huang, R.P., Huang, R., Fan, Y. and Lin, Y. Simultaneous detection of multiple cytokines from conditioned media and patient's sera by an antibody-based protein array system. *Anal. Biochem.* **294**, 55–62 (2001b).

Huang, J.X., Mehrens, D., Wiese, R., Lee, S., Tam, S.W., Daniel, S., Gilmore, J., Shi, M. and Lashkari, D. High-throughput genomic and proteomic analysis using microarray technology. *Clin. Chem.* **47**, 1912–1916 (2001c).

Iqbal, S.S., Mayo, M.W., Bruno, J.G., Bronk, B.V., Batt, C.A. and Chambers, J.P. A review of molecular recognition technologies for detection of biological threat agents. *Biosens. Bioelectron.* **15**, 549–578 (2000).

Jin-Woo, C., Oh, K.W., Thomas, J.H. *et al.* An integrated microfluidic biochemical detection system with magnetic bead-based sampling and analysis capabilities. *Tech. Dig. MEMS*, 447–450 (2001).

Jones, V.W., Kenseth, J.R., Porter, M.D. *et al.* Microminiaturized immunoassays using atomic force microscopy and compositionally patterned antigen arrays. *Anal. Chem.* **70**, 1233–1241 (1998).

Joos, T.O., Schrenk, M., Hopfl, P. *et al.* A microarray enzyme-linked immunosorbent assay for autoimmune diagnostics. *Electrophoresis* **21**, 2641–2650 (2000).

Kamholz, A.E., Weigl, B.H., Finlayson, B.A. and Yager, B. Quantitative analysis of molecular interaction in a microfluidic channel: the T-Sensor. *Anal. Chem.* **71**, 5340–5347 (1999).

Karlsson, R., Michaelsson, A. and Mattsson, L. Kinetic analysis of monoclonal antibody–antigen interactions with a new biosensor based analytical system. *J. Immunol. Methods* **145**, 229–240 (1991).

Kasahara, Y. and Ashihara, Y. Simple devices and their possible application in clinical laboratory devices downsizing. *Clin. Chim. Acta* **267**, 87–102 (1997).

Koutny, L.B., Schmalzing, D., Taylor, T.A. and Fuchs, M. Microchip electrophoretic immunoassay for serum cortisol. *Anal. Chem.* **68**, 18–22 (1996).

Kricka, L.J. Microchips, microarrays, biochips, and nanochips: personal laboratories for the 21st century. *Clin. Chim. Acta* **307**, 219–223 (2001).

Kricka, L.J. *Hitchhikers Guide to Analytical Microchips* (AACC Press, Washington DC, 2002).

Kricka, L.J. and Wilding, P. Microfabricated immmunoassay devices. In: *Principles and Practice of Immunoassay* 2nd edn, (eds Price, C.P. and Newman, D.J.), 605–624 (Stockton, New York, 1997).

Lin, S.C., Tseng, F.G., Huang, H.M. *et al.* Microsized 2D protein arrays immobilized by micro-stamps and microwells for disease diagnosis and drug screening. *Fresenius' J. Anal. Chem.* **371**, 202–208 (2001).

Linder, V., Verpoorte, E., Thormann, W., de Rooij, N.F. and Sigrist, H. Surface biopassivation of replicated poly (dimethylsiloxane) microfluidic channels and application to heterogeneous immunoreaction with on-chip fluorescence detection. *Anal. Chem.* **73**, 4181–4189 (2001).

Luo, L.Y. and Diamandis, E.P. Preliminary examination of time-resolved fluorometry for protein array applications. *Luminescence* **15**, 409–413 (2000).

MacBeath, G. Proteomics comes to the surface. *Nat. Biotechnol.* **19**, 828–829 (2001).

MacBeath, G. and Schreiber, S.L. Printing proteins as microarrays for high-throughput function determination. *Science* **289**, 1760–1763 (2000).

Mangru, S.D. and Harrison, D.J. Chemiluminescence detection in integrated post-separation reactors for microchip-based capillary electrophoresis and affinity electrophoresis. *Electrophoresis* **19**, 2301–2307 (1998).

Manz, A., Graber, N. and Widmer, H.M. Miniaturized total analysis systems: a novel concept for chemical sensors. *Sens. Actuators* **B1**, 244–248 (1990).

Manz, A., Harrison, D.J., Verpoorte, E.M.J., Fettinger, J.C., Lüdi, H. and Widmer, H.M. Miniaturization of chemical analysis systems – a look into next century's technology or just a fashionable craze. *Chimia* **45**, 103–105 (1991).

Mere, L., Bennett, T., Coassin, P. *et al.* Miniaturized FRET assays and microfluidics: key components for ultra-high-throughput screening. *Drug Discov. Today* **4**, 363–369 (1999).

Mooney, J.F., Hunt, A.J., McIntosh, J.R. *et al.* Patterning of functional antibodies and other proteins by photolithography of silane monolayers. *Proc. Natl Acad. Sci. USA* **93**, 12287–12291 (1996).

Nedelkov, D. and Nelson, R.W. Analysis of human urine protein biomarkers via biomolecular interaction analysis mass spectrometry. *Am. J. Kidney Dis.* **38**, 481–487 (2001).

Nguyen, N.-T. and Nwereley, S.T. *Fundamentals and Applications of Microfluidics* (Artech House, Norwood, Massachusetts, 2002).

Ning, Y. and Fitzpatrick, G. Microfabrication processes for silicon glass chips. In: *Biochip Technology* (eds Cheng, J. and Kricka, L.J.), 17–38 (Harwood Academic Publishers, Philadelphia, 2001).

Owicki, J.C., Bousse, L.J., Hafeman, D.G. *et al.* The light-addressable poteniometric sensor. *Annu. Rev. Biophys. Biomol. Struct.* **23**, 87–113 (1994).

Peluso, P., Wilson, D.S., Do, D., Tran, H., Ventatasubbaiah, M., Quincy, D., Heidecker, B., Poindexter, K., Tolani, N., Phelan, M., Witte, K., Jung, L.S., Wagner, P. and Nock, S. Optimizing antibody immobilization strategies for the construction of protein microarrays. *Anal. Biochem.* **312**, 113–124 (2003).

Price, C.P. and Hicks, J.M. *Point-of-Care Testing* (AACC Press, Washington DC, 1999).

Price, C.P., Thorpe, G.H.G., Hall, J. and Bunce, R.A. Disposable integrated immunoassay devices. In: *Principles and Practice of Immunoassay* 2nd edn, (eds Price, C.P. and Newman, D.J.), 579–603 (Stockton, New York, 1997).

Ruiz-Taylor, L.A., Martin, T.L., Zaugg, F.G., Witte, K., Indermuhle, P., Nock, S. and Wagner, P. Monolayers of derivatized poly(L-lysine)-grafted poly(ethylene glycol) on metal oxides class of biomolecular interfaces. *Proc. Natl Acad. Sci. USA* **98**, 852–857 (2001).

Santomauro, A.G. and Sciarra, J.J. Comparative evaluation of a hemagglutination inhibition test and a latex agglutination inhibition test for HCG. *Obstet. Gynecol.* **29**, 520–525 (1967).

Sato, K., Tokeshi, M., Kimura, H. and Kitamori, T. Integration of immunoassay system into a microchip. *Jpn. J. Electrophor.* **44**, 73–77 (2000a).

Sato, K., Tokeshi, M., Odake, T., Kimura, H., Ooi, T., Nakao, M. and Kitamori, T. Integration of an immunosorbent assay system: analysis of secretory human immunoglobulin A on polystyrene beads in a microchip. *Anal. Chem.* **72**, 1144–1147 (2000b).

Sato, K., Tokeshi, M., Kimura, H. and Kitamori, T. Determination of carcinoembryonic antigen in human sera by integrated bead immunoassay in a microchip for cancer diagnosis. *Anal. Chem.* **73**, 1213–1218 (2001).

Schena, M. *DNA Microarrays. A Practical Approach* (Oxford University Press, Oxford, 1999).

Schena, M. *Microarray Biochip Technology* (Eaton Publishing, Natick, MA, 2000).

Schmalzing, D., Koutny, L.B., Taylor, T.A., Nashabeh, W. and Fuchs, M. Immunoassay for thyroxine (TL) in serum using capillary electrophoresis and micromachined devices. *J. Chromatogr. B Biomed. Sci. Appl.* **697**, 175–180 (1997).

Schmalzing, D., Buonocore, S. and Piggee, C. Capillary electrophoresis-based immunoassays. *Electrophoresis* **21**, 3919–3930 (2000).

Schweitzer, B., Wiltshire, S., Lambert, J. *et al.* Inaugural article: immunoassays with rolling circle DNA amplification: a versatile platform for ultrasensitive antigen detection. *Proc. Natl Acad. Sci. USA* **97**, 10113–10119 (2000).

Scorilas, A., Bjartell, A., Lilja, H., Moller, C. and Diamandis, E.P. Streptavidin-polyvinylamine conjugates labeled with a europium chelate: applications in immunoassay, immunohistochemistry, and microarrays. *Clin. Chem.* **46**, 1450–1455 (2000).

Silzel, J.W., Cercek, B., Dodson, C., Tsay, T. and Obremski, R.J. Mass-sensing, multianalyte microarray immunoassay with imaging detection. *Clin. Chem.* **44**, 2036–2043 (1998).

Song, M.I., Iwata, K., Yamada, M. *et al.* Multisample analysis using an array of microreactors for an alternating-current field-enhanced latex immunoassay. *Anal. Chem.* **66**, 778–781 (1994).

Sreekumar, A., Nyati, M.K., Varambally, S. *et al.* Profiling of cancer cells using protein microarrays: discovery of novel radiation-regulated proteins. *Cancer Res.* **61**, 7585–7593 (2001).

Stoll, D., Templin, M.F., Schrenk, M., Traub, P.C., Vohringer, C.F. and Joos, T.O. Protein microarray technology. *Front. Biosci.* **7**, c13–c32 (2002).

Sydor, J.R., Scalf, M., Sideris, S., Mao, G.D., Pandey, Y., Tan, M., Mariano, M., Moran, M.F., Nock, S. and Wagner, P. Chip-based analysis of protein–protein interactions by fluorescence detection and on-chip immunoprecipitation combined with microLC-MS/MS analysis. *Anal. Chem.* **75**, 6163–6170 (2003).

Terry, S.C. and Hawker, D.A. Automated high speed natural gas analysis using a new microcomputer controlled, high resolution GC analyzer. *Adv. Instrum.* **38**, 387–398 (1983).

Terry, S.C., Jerman, J.H. and Angell, J.B. A gas chromatographic air analyzer fabricated on a silicon wafer. *IEEE Trans. Electron Devices* **26**, 1880–1886 (1979).

Trimmer, W. and Stroud, R.H. Scaling of micromechanical devices. In: *The MEMS Handbook* (ed Gad-el-Hak, M.), 2.1–2.9 (CRC Press, Boca Raton, 2002).

Von Heeren, F., Verpoorte, E., Manz, A. and Thormann, W. Micellar electrokinetic chromatography separations and analyses of biological samples on a cyclic planar microstructure. *Anal. Chem.* **68**, 2044–2053 (1996).

Walter, G., Bussow, K., Cahill, D., Lueking, A. and Lehrach, H. Protein arrays for gene expression and molecular interaction screening. *Curr. Opin. Microbiol.* **3**, 298–302 (2000).

Wang, J., Ibanez, A., Chatrathi, M.P. and Escarpa, A. Electrochemical enzyme immunoassays on microchip platforms. *Anal. Chem.* **73**, 5323-5327 (2001).

Weigl, B.H. and Yager, P. Microfluidic diffusion-based separation and detection. *Science* **283**, 346–347 (1998).

Wilding, P., Kricka, L.J., and Zemel, J.N. Methods and Apparatus for the Detection of an Analyte Utilizing Mesoscale Flow Systems. United States Patent 5,637, 469, (1997)

Weng, S., Gu, K., Hammond, P.W., Lohse, P., Rise, C., Wagner, R.W., Wright, M.C. and Kuimelis, R.G. Generating addressable protein microarrays with PROfusion covalent mRNA-protein fusion technology. *Proteomics* **2**, 48–57 (2002).

Yakovleva, J., Davidsson, R., Lobanova, A., Bengtsson, M., Eremin, S., Laurell, T. and Emneus, J. Microfluidic enzyme immunoassay using silicon microchip with immobilized antibodies and chemiluminescence detection. *Anal. Chem.* **74**, 2994–3004 (2002).

Yang, T., Jung, S., Mao, H. and Cremer, P.S. Fabrication of phospholipid bilayer-coated microchannels for on-chip immunoassays. *Anal. Chem.* **73**, 165–169 (2001).

Zhu, H. and Snyder, M. Protein arrays and microarrays. *Curr. Opin. Chem. Biol.* **5**, 40–45 (2001).

FURTHER READING

Bilitewski, U., Genrich, M., Kadow, S. and Mersal, G. Biochemical analysis with microfluidic systems. *Anal. Bioanal. Chem.* **377**, 556–569 (2003).

Cahill, D.J. and Nordhoff, E. Protein arrays and their role in proteomics. *Adv. Biochem. Eng. Biotechnol.* **83**, 177–187 (2003).

Conrads, T.P., Zhou, M., Petricoin, E.F. III, Liotta, L. and Veenstra, T.D. Cancer diagnosis using proteomic patterns. *Expert Rev. Mol. Diagn.* **3**, 411–420 (2003).

Figeys, D. Adapting arrays and lab-on-a-chip technology for proteomics. *Proteomics* **2**, 373–382 (2003).

Hanash, S. Disease proteomics. *Nature* **422**, 226–232 (2003).

Huikko, K., Kostiainen, R. and Kotiaho, T. Introduction to micro-analytical systems: bioanalytical and pharmaceutical applications. *Eur. J. Pharm. Sci.* **20**, 149–171 (2003).

Ligler, F.S., Taitt, C.R., Shriver-Lake, L.C., Sapsford, K.E., Shubin, Y. and Golden, J.P. Array biosensor for detection of toxins. *Anal. Bioanal. Chem.* **377**, 469–477 (2003).

Liotta, L.A., Espina, V., Mehta, A.I., Calvert, V., Rosenblatt, K., Geho, D., Munson, P.J., Young, L., Wulfkuhle, J. and Petricoin, E.F. III Protein microarrays: meeting analytical challenges for clinical applications. *Cancer Cells* **3**, 317–325 (2003).

Mouradian, S. Lab-on-a-chip: applications in proteomics. *Curr. Opin. Chem. Biol.* **6**, 51–56 (2002).

Paegel, E.M., Blazej, R.G. and Mathies, R.A. Microfluidic devices for DNA sequencing: sample preparation and electrophoretic analysis. *Curr. Opin. Biotechnol.* **14**, 42–50 (2003).

Phizicky, E., Bastiaens, P.I., Zhu, H., Snyder, M. and Fields, S. Protein analysis on a proteomic scale. *Nature* **422**, 208–215 (2003).

Simon, R., Mirlacher, M. and Sauter, G. Tissue microarrays in cancer diagnosis. *Expert Rev. Mol. Diagn.* **3**, 421–430 (2003).

Valle, R.P. and Jendoubi, M. Antibody-based technologies for target discovery. *Curr. Opin. Drug Discov. Dev.* **6**, 197–203 (2003).

Weigl, B.H., Bardell, R.L. and Cabrera, C.R. Lab-on-a-chip for drug development. *Adv. Drug Delivery Rev.* **55**, 349–377 (2003).

Weigl, B.H. and Hedine, K. Lab-on-a-chip based separation and detection technology for clinical diagnostics. *Am. Clin. Lab.* **21**, 8–13 (2002).

Zhu, H., Bilgin, M. and Snyder, M. Proteomics. *Ann. Rev. Biochem.* **72**, 783–812 (2003).

Zieziulewicz, T.J., Unfricht, D.W., Hadjout, N., Lynes, M.A. and Lawrence, D.A. Shrinking the biologic world – nanobiotechnologies for toxicology. *Toxicol. Sci.* **74**, 235–244 (2003).

Part 2

Product Technology

18 Introduction to Product Technology in Clinical Diagnostic Testing

David Wild

Considering the specialized nature of the immunoassay field, there is a surprisingly wide range of technology available commercially, and many more assay variations have been reported in the scientific literature. Part 1 of this book covers the principles that lie behind this diverse array of immunoassays and Part 2 provides descriptions of a representative selection of commercial products. The immunoassay market is described in the next chapter MARKET TRENDS.

The diversity of commercial immunoassay products is due to

- the pace of discovery in the fundamental science of immunoassays. Examples of impressive scientific breakthroughs have been monoclonal antibodies, immunometric assay, homogeneous immunoassay, lateral flow immunoassay, anti-complex assay, ambient analyte immunoassay, and phage display libraries, but many more impressive breakthroughs could be added to this list, for example, in the field of signal generation and detection methods;
- advances in engineering, e.g. microelectronics, optical systems, sensors and actuators, and computer systems;
- demand for continuous improvements in automation;
- the availability of research funding for original immunoassay formats;
- the number of immunoassay product manufacturers competing in the market;
- major investment by well-funded multinational corporations;
- specialized technology for individual applications;
- differentiation of the market into multiple user groups, with unique product requirements.

It is difficult to find an area of biotechnology that has stimulated more innovation, and benefited more from scientists and engineers working closely together, than immunoassay.

IMMUNOASSAY PRODUCT TECHNOLOGIES

USE OF FLUORESCENCE-LABELED ANTIBODIES

Antibodies labeled with fluorescent molecules, such as fluorescein, are used in **fluorescent microscopy**, to show the location and quantity of specific proteins in tissue sections, or the presence of cell-surface marker proteins (*see* MICROTRAK®). **Fluorescent Activated Cell Sorters** (**FACS**) (not described in Part 2) are used to screen cells, using antibodies for cell-surface antigens labeled with latex particles incorporating fluorescent dyes. The cells are streamed through a read cell and the proportion of cells labeled with dye measured using laser fluorimetry.

AGGLUTINATION ASSAYS

Slide agglutination tests are qualitative tests used to detect the presence of antibodies in serology laboratories and blood banks. Treated red blood cells or colored latex beads, coated with antigen, clump in the presence of antibody to the antigen. The degree of clumping may be measured using absorbance at 600 nm and **latex agglutination** has been applied to quantitative assays. The Immuno 1™ analyzer uses latex agglutination as one of a number of assay formats.

The Immunoassay Handbook. *Third Edition*
D. Wild (Ed.)

Precipitin Assays

Precipitin assays, such as **radial immunodiffusion** and **immunoelectrophoresis**, are still used for certain applications, but these tend to be low-volume assays, in specialist centers. In these types of assays, the presence of antibody (or antigen) in the sample causes the formation of a precipitate in agar containing antigen (or antibody). Radial immunodiffusion involves adding samples to circular holes cut in agar plates and the formation of a circular ring. In immunoelectrophoresis, proteins are first separated by electrophoresis before incubation with antibodies in a parallel trough, with the formation of precipitin arcs (*see* AUTOIMMUNE DISEASES in Part 4 for examples).

Nephelometry and Turbidimetry

Nephelometric assays are used to determine proteins at relatively high concentrations on clinical chemistry analyzers and some immunoassay analyzers. They rely on the fact that antigen–antibody complexes are insoluble and cause solutions to scatter light (**nephelometry**) or become more opaque (**turbidimetry**).

RADIOIMMUNOASSAY AND IMMUNORADIOMETRIC ASSAY

Competitive **radioimmunoassays** (**RIA**) were once the gold standard for immunoassay, but two-site **immunoradiometric assays** (**IRMA**) brought about major improvements for large molecule analytes, including greater sensitivity, increased precision at low concentrations, and a wider dynamic range. RIA is occasionally used for some small molecule analytes, such as steroids, but non-radioactive methods have superseded them. IRMA was largely replaced by non-radioactive immunometric methods with higher specific activity signal generation systems. RIAs with a **liquid phase** (requiring the addition of a precipitating agent at the end of the incubation) are seldom used. Most are **solid-phase** assays, with antibodies most commonly bound to polystyrene assay tubes or magnetic particles.

HETEROGENEOUS ENZYME IMMUNOASSAY

RIA involves the handling of radioactive materials, restricting its use to licensed laboratories and causing waste handling problems. The counters are also very expensive. These disadvantages were overcome by replacing the radioactive label with an enzyme, although **enzyme immunoassays** (EIA) require an extra incubation of the bound fraction with a substrate, to generate a colorimetric signal. Enzyme immunoassays were initially most successful in an immunometric format, in the fields of infectious diseases and oncology. Antibody (or antigen) is bound to beads or microwells, which are amenable to automated washing. The Abbott Quantum™ reader, which was less expensive than a gamma counter, was widely used to read the colorimetric signal generated. These early enzyme immunoassays provided a bridge to later fully automated systems with simpler protocols. The first enzyme immunoassays were not as sensitive as RIA, which continued to be the test of choice in endocrinology, with primarily competitive assays, for some time. The Ortho-Clinical Diagnostics range of infectious disease assays and the SUMMIT™ processing system are enzyme immunoassays that are widely used in blood banks for screening donor plasma.

Initially, enzyme immunoassays were more susceptible to interferences than RIA, due to effects on the enzyme-mediated signal generation stage. However, various techniques were introduced, such as the addition of correction factors, to minimize interferences of this type.

HOMOGENEOUS ENZYME IMMUNOASSAY

One of the great breakthroughs in immunoassay technology was the introduction of **homogeneous immunoassays**, which did not require a physical separation of the bound and unbound fractions, much simplifying the assay protocol, and enabling automation to be introduced at a reasonable cost. The Emit® chemistry was very successfully applied to therapeutic drug monitoring at the end of the 1970s. Emit reagents are still widely used to run immunoassays on clinical chemistry analyzers.

AUTOMATED HOMOGENEOUS BATCH ANALYZER

Abbott's TDx® analyzer was not the first attempt at immunoassay automation, but it became the most successful, until it was eclipsed by the IMx®. Introduced in 1981, it eventually took over from Emit as the market leader in therapeutic drug monitoring. It is very simple to operate, giving rise to the term **black box analyzer**. The TDx is a batch analyzer, in which a series of samples can be tested for one analyte in a run. The reagent pack is barcoded, so that the analyzer can automatically carry out the appropriate protocol. Calibrators only have to be run periodically. The TDx signal generation system is **fluorescence polarization**, which does not require a separation, simplifying the design of the analyzer. The TDx is used to carry out immunoassays for small molecules, such as drugs and thyroid hormones (*see* IMX for an explanation of the technology).

AUTOMATED HETEROGENEOUS BATCH ANALYZER

Two years after the launch of the TDx, Baxter responded with the Stratus®, another automated batch analyzer, but with a heterogeneous format. Separation was by **radial partition immunoassay**. The bound antibody was immobilized in the center of a filter paper. Unbound material was washed away from the center by successive additions of reagent. This design enabled heterogeneous assays to be more easily automated. The assays were enzyme immunoassays in which the enzyme label converts a substrate into a fluorescent product. The Stratus (which is no longer available) included a minimum 14-day calibration curve stability, and introduced the concept of STAT capability, enabling urgent samples to be run immediately. It had the flexibility to perform immunoassays for large and small molecular weight analytes on the same analyzer.

NON-RADIOACTIVE SYSTEMS WITH INCREASED SENSITIVITY

The TDx and Stratus were primarily used for testing small molecules such as drugs, and were not considered to be as sensitive or widely applicable as RIA or IRMA methods, although sensitive assays for a number of proteins were developed for the Stratus. Their great advantages were automation and continual availability. In 1985, two new technologies, **time-resolved fluorescence** (Wallac DELFIA®) and **enhanced luminescence** (Ortho-Clinical Diagnostics Amerlite®) were incorporated in semi-automated systems with greater sensitivity than radioactive methods, and very wide dynamic ranges. The menus included analytes that were unavailable on the TDx or Stratus.

Amerlite used enzyme-mediated light generation with an enhancer to provide a strong signal. DELFIA is not an enzyme immunoassay system, but uses lanthanide chelates as labels, with a pulsed fluorometric system to minimize interference from background fluorescence. Another time-resolved fluorescence system was CyberFluor™, which used a Europium chelate label, bound to a thyroglobulin–streptavidin conjugate (EuroFluor S-streptavidin). The conjugate bound to any bound biotinylated antibody on the solid phase (microwell) at the end of the assay.

Another example was Ciba Corning's MagicLite™ system, which was based on **chemiluminescence**.

For an explanation of the luminescence chemistry used in Amerlite, *see* VITROS® ECi, and for MagicLite *see* ACS®:180 SE.

These systems introduced significant improvements in immunoassay signal generation and exploited the potential of immunometric assay formats more effectively than radioactive methods. Similar chemistry systems lie behind the fully automated analyzers that are in common use today.

SEMI-AUTOMATED SYSTEMS

Semi-automated, modular systems provide automated reagent addition, signal reading, and result processing, primarily for blood-bank screening and allergen-sensitivity testing. An example is the Ortho-Clinical Diagnostics SUMMIT™ system.

AUTOMATED DUAL TECHNOLOGY BATCH ANALYZER

In 1988, Abbott launched the IMx® system, which provides full automation in a batch assay system for a wide range of analytes. The IMx is capable of carrying out homogeneous fluorescence polarization assays, using TDx technology, but also includes a range of heterogeneous enzyme immunoassays, using microparticle capture to provide an automatable separation and signal generation system. The fluorescent signal generation system delivers sensitivity comparable to or better than RIA. The IMx soon became the most widely used immunoassay system ever (*see* IMX®).

AUTOMATED, MULTIANALYTE BATCH ANALYZERS

In 1990, Boehringer Mannheim introduced the ES 300™, which is an example of an automated, multianalyte batch analyzer. Based on established Enzymun-Test® reagents, the ES 300 could run up to 12 different tests on a load of up to 150 samples. Syva's Vista™ system offered up to 15 tests during a run. Other examples of multianalyte systems were the IMx® Select™ and the Amerlite® Processing Center, which ran panels of up to three analytes, and the TDxFLx™, which could run up to eight different drug tests. Multianalyte systems require advanced sample and test-handling software. Worklists that itemize the tests required for each sample are prepared by the user or obtained via a link to the main hospital or laboratory computer. In general laboratory work, these types of systems have now been largely superseded by the more flexible random-access analyzers. However, panel test systems are important in blood bank and allergen-screening applications (*see* UNICAP® 100 and PRISM™).

RANDOM-ACCESS ANALYZERS WITH BULK REAGENT PACKS

See CHOOSING AN AUTOMATED IMMUNOASSAY SYSTEM and AUTOMATED SYSTEM FEATURES in Part 2.

In 1987, TOSOH Corporation launched the AIA®-1200, a full random-access immunoassay analyzer, offering the continuous loading capabilities found on many clinical chemistry analyzers. The term **random**

access is used to describe analyzers on which samples may be loaded at any time, and usually the analyzer can perform several tests on a particular sample. The AIA-1200 is capable of testing up to 100 samples for as many as 21 analytes. Unlike the previous analyzers described, the reagents are stored on-board for weeks at a time, so that a wide range of tests is always available. This instrument has sophisticated reagent inventory monitoring systems in addition to patient sample worklist and result-handling software. The ACS:180, introduced by Ciba Corning in 1991, is also a random-access system, using luminescence as the signal source. It has **integrated reagent packs**, loaded onto the analyzer, containing liquids and suspensions to carry out a particular test. These analyzers were the first of many, and random-access machines have continued to evolve, becoming the main clinical laboratory systems in use today.

Both analyzers can be used in batch or random-access modes and have a STAT capability for urgent samples. The AIA-1200 has stored calibration curves, obtained by running the calibrator set every few weeks, or when the lot number changes. The ACS:180 has a **factory-determined calibration curve**, which is provided with the kit on a barcode. The user runs two calibrators periodically, which are used by the analyzer computer to make adjustments to the calibration curve.

Both these systems have fixed incubation times. Tests have to be designed to work on the analyzer protocols, to avoid the complicated timing problems that can occur if multiple protocols are used.

The ACS:180 was the first immunoassay system to allow the use of barcoded primary sample tubes. The AIA-1200 DX offered a similar capability. This feature is now present on most analyzers. Hospital staff may label the blood-sample tubes with barcodes when the blood is first collected. Test requests are entered into a terminal of the hospital's main computer. The blood is clotted in the tube, which is sent to the laboratory. The tube is loaded onto the analyzer, which identifies the sample from the barcode, and interrogates the hospital computer to determine the tests to be run. The analyzer sampling probe samples the serum from the top of the tube, well above the blood clot.

The term **primary tube** refers to the original tube into which whole blood is collected. A **worklist** is a list of test requests for a group of samples to be run on one analyzer.

The AxSYM® made IMx assays available in a random-access format. This instrument is able to run multiple protocols.

Subsequent analyzers have increased the level of performance and convenience further. For example, some analyzers allow loading of samples and reagents while the analyzer is running. The ability to handle a wide range of **sample tubes and cups** is also important, ideally within a standard sample rack.

There are sometimes trade-offs between performance and usability. For example, high-throughput analyzers may have limited numbers of protocols, as the arrangement of the analyzer subsystems involves a sequence of consecutive processing stages, each fully utilized for the time between tests. Shorter or longer incubation times, or a mixture of one- and two-step protocols, cause faster tests to catch up the ones in front, increasing the demand for the final assay processing modules at times of peak demand. Some systems have timing cycles, which divide up the time between tests (e.g. 30 sec) and give each analyzer process a part of the time available (e.g. pipetting of a particular reagent from 2–4 sec after the start of the sequence). This is important for rotor-based systems, as one rotor, full of test cuvettes, has to be able to move each test to the correct processing stage, at the appropriate time. Vitros ECi has a duological ring, which allows some test wells to be shuttled out of the main rotating incubator ring for longer duration processes, such as washing. It is able to run a wide range of protocols on a sequence of samples. The ability to run multiple protocols is important as a random-access immunoassay analyzer represents a major investment for any laboratory, and **breadth of menu** is essential, particularly with consolidation of many specialized tests into core laboratories (*see* MARKET TRENDS). Yet **test throughput** is also important, for larger laboratories, and to handle peaks in demand.

There have also been efforts to reduce disposable waste, for example, Vitros ECi and TOSOH systems do not use separate bulk-loaded reaction cuvettes, but instead utilize a well containing (or coated with) reagent.

Immunoassay random-access analyzers are masterpieces of engineering and science, involving advanced biochemistry, specialized injection molding, precision engineering, microfluidics, optoelectronics, and microprocessor and PC-based computer control systems. Working on the development of such a system is both exciting and challenging.

Many random-access immunoassay analyzers are described in this book (*see* ACS:180®SE, AXSYM®, ELECSYS®, VITROS® ECi, IMMULITE® 2000®, ADVIA CENTAUR®, and ARCHITECT i2000®).

UNITIZED RANDOM-ACCESS SYSTEMS

An alternative form of random-access system is sometimes referred to as unit dose, because the reagents for each individual sample test are provided in separate packs. Most unit dose systems can be continuously loaded with samples and reagents, one test at a time. Samples are either dispensed into the reagent test pack before loading, or placed onto a conveyor system in parallel with the appropriate reagent pack. The analyzers tend to be cheaper, and the reagents more expensive than random-access analyzers with bulk reagent packs. Examples are IMMULITE®, Dade Behring OPUS® and OPUS® PLUS, TOSOH AIA®-600 (described in the first edition of *The Immunoassay Handbook*) and BioMerieux VIDAS™ and mini-VIDAS™ (not described in this book).

CLINICAL CHEMISTRY ANALYZERS

Some immunoassays have been specially designed for clinical chemistry analyzers, which tend to have a high level of automation and fast test throughput, but lack the ability to carry out long incubations or separations. Examples are Syva's EMIT® range and Microgenics' CEDIA® assays which can be run on a range of systems. Ortho-Clinical Diagnostics market the VITROS® Immuno-Rate reagents for the VITROS range of clinical chemistry analyzers. The Dade Behring Dimension® RxL has a heterogeneous immunoassay module. Roche Diagnostics manufacture COBAS-FP™ reagents for the COBAS FARA™ II clinical chemistry analyzer based on fluorescence polarization.

See CEDIA®, and VITROS® IMMUNO-RATE.

NEAR-PATIENT TESTS

The use of immunoassay tests outside of the laboratory is described in more detail in the chapter on POINT-OF-CARE TESTING in Part 3. The Abbott VISION™ was an early example, designed for use in doctors' offices. Whole blood samples could be used (the instrument includes a centrifuge which spins down the blood cells leaving plasma for the test). Other systems are the Dupont Analyst™, Miles Seralyzer™, and Miles Clinimate-TDA™.

Many tests have been developed for physicians' offices that do not require any processing equipment. The simplest provide a qualitative result by means of a colored spot, and do not require any equipment. The TANDEM® ICON® QSR® device provided a quantitative result when used in conjunction with a small, low-cost reader. All are unitized and have simple protocols. Some method of internal control is also provided to confirm the validity of the result. In Part 2, the representative examples are Clearview™ and TRIAGE™ Cardiac System. Also the Stratus® CS STAT Fluorimetric analyzer, which is designed for Critical Care laboratories, is described.

HOME-USE TESTS

Several pregnancy (urinary hCG) tests are available for home use. They need to have very simple protocols, and a colorimetric end-point. Some have a wick at one end of the unitized device, which is simply placed in the stream of urine. One of the tests is made by Unipath (*see* UNIPATH CLEARBLUE PREGNANCY TEST™). Some of the tests are not as simple as this to use and users of pregnancy tests occasionally misinterpret the results (*see* OVER-THE-COUNTER PREGNANCY TEST KITS in Part 2).

Urinary LH tests, for the prediction of ovulation, are available (*see* UNIPATH CLEARBLUE OVULATION TEST™).

Carter-Wallace is a leading manufacturer of pregnancy and ovulation tests in the USA.

One of the most sophisticated immunoassay tests ever developed is Unipath's PERSONA™, which measures LH and estriol-3-glucuronide in urine, and applies algorithms to estimate the time of ovulation from multiple data points from both analytes. This alone makes it very advanced compared to the typical single analyte/single time application of immunoassay. Unlike many lateral flow tests, the measurement is quantitative, rather than qualitative. It uses a miniaturized, low-power spectrophotometer in a pocket-size unit. To initiate the device, the user presses a button at the time of a menstrual period. In the non-US version of the device, the user is prompted to run a test on appropriate days by showing a yellow light. A green light indicates that the user is unlikely to conceive on those days, and a red light indicates that contraception is advised. Over a period of up to six cycles, the device 'learns' about the owner's menstrual cycle and becomes more accurate, reducing the number of days on which a test is required. All the data from the six cycles is available for investigation purposes via an RS-232 interface hidden under a panel on the back of the unit. This masterpiece of design is sold in pharmacies for just $60.

There are a few home-use tests for the presence of cancer markers in urine, which are useful for the prompt detection of recurrence after treatment. However, home-use tests for HIV have not been particularly successful (so-called 'bad news' tests).

OTHER APPLICATIONS

Immunoassays are widely used in veterinary applications, scientific research, and pharmaceutical research and development (*see* IMMUNOASSAY APPLICATIONS IN VETERINARY DIAGNOSTICS and ASSAYS FOR DRUG-SCREENING APPLICATIONS AND RESEARCH for reviews of the technologies available). They are also used in food and agriculture, for example, to detect the presence of toxins, or to identify meat species. Immunoassays have detected pollutants in the environment and traces of explosives and biological agents in criminal investigations. Many non-clinical fields have the potential to further benefit from immunoassay science and technology in the future.

For a review of immunoassay technology in miniaturized formats, *see* LAB-ON-A-CHIP, MICRO- AND NANOSCALE IMMUNOASSAY SYSTEMS.

FUTURE DEVELOPMENTS

For a list of predicted trends for immunoassays in the future, *see* MARKET TRENDS, the following chapter.

FURTHER READING

Blick, K.E. Current trends in automation of immunoassays. *J. Clin. Ligand Assay*, **22**, 6–12 (1999).

Chan, D.W. *Immunoassay Automation: An Updated Guide to Systems* (Academic Press, San Diego, 1996).

Felder, R.A. Immunoassay automation. *J. Clin. Ligand Assay*, **22**, 13–24 (1999).

Kost, G.J. (ed.) *Handbook of Clinical Automation, Robotics and Optimization* (Wiley, New York, 1996).

Kricka, L.J., Nozarki, O., Wilding, P. Micromechanics and nanotechnology: implications and applications in the clinical laboratory. *JIFCC*, **6**, 54–59 (1994).

Price, C.P. The evolution of immunoassay as seen through the journal clinical chemistry. *Clin. Chem.*, **44**, 2071–2074 (1998).

Truchaud, A., Le Néel, T., Malvaux, S. *et al.* Laboratory engineering in immunoassays. *J. Clin. Ligand Assay*, **22**, 25–31 (1999).

Wheeler, M.J. Automated immunoassay analysers. *Ann. Clin. Biochem.*, **38**, 217–229 (2001).

19 Market Trends

David Huckle and David Wild

Immunoassays have had a major impact on diagnostic testing and scientific research. For many years they were unsurpassed in specificity and sensitivity leading to the introduction of a wide range of highly sensitive assays for diagnosis and patient monitoring. In the 1980s, automation made the technology accessible to non-specialized laboratories, and subsequently ingenious developments in reagent and hardware integration led to immunoassays on the pharmacy shelf. Most immunoassay tests are now simple to use and immunoassay has become the leading technology in use in the field of *in vitro* diagnostics (IVD). However, history teaches that past successes cannot guarantee anything about the future. The field of healthcare shows early signs of dramatic changes ahead, fueled by new technology and disenchantment with the current service providers. This chapter assesses why immunoassays have become so successful, and analyzes the potential of future immunoassay technologies to fulfill unmet needs in the market. Threats to the immunodiagnostics market from alternative technologies are also evaluated.

IMMUNOASSAY MARKET STATUS

Much of the diagnostics business is showing the characteristic signs of a mature market, although there are some exceptions. This has led to consolidation in the industry.

The quantitative picture of the market confirms that immunoassay is now the largest single technology in the IVD market (Table 19.1). The 2004 value of the IVD market was $27.6 billion, with a projected growth to $38 billion in 2008. The corresponding figures for immunoassay products are $12.4 billion for 2003 and $18.5 billion for 2008.

Commercial consolidation and globalization of the market has continued over the last few years. There are now a small number of major international healthcare corporations and several smaller international companies, which tend to be focused on specific areas of diagnostics.

The recent consolidation of medium-sized diagnostics suppliers into multi-billion dollar enterprises is partly caused by the maturity of IVD technology, as most companies have essentially comparable products, and to some extent this is true of the immunodiagnostics sector. The absence of true product differentiation gives an advantage to global companies supplying high throughput central laboratories because of economies of scale. With reduced market growth overall, companies can only achieve significant growth through acquisitions or mergers. The exception in the immunodiagnostics market is Abbott Diagnostics, which has grown organically, with few acquisitions in the immunoassay field, based on a combination of sound business, selling and technical strategies. They have the leading share of the business but with several other companies snapping at their heels.

A further group of primary manufacturers consists of essentially local suppliers. These feed the home base markets, with some exports supplied through local distributors, who make up the final group active in this industry. These distributors typically operate in single countries, build up a range of products from worldwide sources, and are the market entry point into Europe for many small companies based in North America, Australia, and Japan.

Table 19.1. Global diagnostics market figures for 2004 and projections for 2008 ($ billion).

Category	Region	2004	2008
Total diagnostics	Europe*	22.3	29.0
	USA	30.8	42.0
	World	76.0	105
In vitro diagnostics	Europe*	9.9	13.0
	USA	10.5	14.0
	World	27.6	38.0
Immunoassay	Europe*	4.0	5.5
	USA	4.7	6.5
	World	12.4	18.5

*Source: Adams Business Associates.
12 nation EU + Switzerland and Norway. Extended EU market will be relatively small in the short term, with low prices.

Innovative companies, particularly those using recombinant protein and DNA probe technology, have created niche markets and a tenable position in the market. A particularly successful example of this was Chiron, although the diagnostic applications have now been consolidated within the Roche Diagnostics business.

There are other factors at work in the immunodiagnostic market. The cost of product development has increased, partly because products have become more complex, and partly due to tighter regulatory requirements. There is also more centralized purchase control within the healthcare provider organizations, with the involvement of finance and management personnel as well as scientific and clinical staff, and this can increase companies marketing costs and reduce profit margins. These are all features of a mature market. The response of some companies, or of those with new potential to exploit, has been to avoid direct confrontation with the leading companies' by shifting the customer base. The most important of these initiatives is the drive toward Point-of-Care or near-patient testing.

The issues that face the diagnostics industry are typical of a mature business and these now apply to immunoassays as well as to clinical chemistry. The biggest issue for the industry is likely to be the impact of innovative new companies entering the market. Conventional wisdom is that market entry is not commercially viable in a mature market. However, the history of other industries shows that companies can and do enter such markets, but with new technological answers to problems unsolved by current suppliers. This is where expansion and change in the market is most likely to originate. To some extent, the diagnostics industry has been in this position since the early 1990s. Although major change may seem to be unlikely, market dynamics suggest that the industry is poised for a major, technology-led shift.

ESTABLISHED TRENDS

MARKET DRIVERS

Cost Containment

The economic issues of the last decade have brought the allocation of funds for healthcare throughout the world into critical focus. There is insufficient money to maintain the level of demand in any of the major markets.

Providers of healthcare resources, particularly fund managers, have major economic problems to resolve. Their response has been to try to reduce overall healthcare costs, directly through budget constraints, or indirectly through changes to reimbursement. The resulting trend is for a focus on the more clinically useful diagnostic tests rather than uncontrolled provision of a wide range of general tests. In the UK, rationing has applied in the National Health Service for some time. Budgets for all reagents are controlled, sometimes preventing useful new tests from being introduced widely. In the US, the concept of **managed care** has led to indirect rationing of healthcare resources through rigorous cost control and cost–benefit analysis. In many countries, reimbursement, approved test lists, or **disease-related groups** (**DRGs**) have been key control measures. In some cases, new products have been prevented from reaching the market effectively, even if approved by the regulatory authority, because government organizations have kept them off the reimbursement list.

The value of laboratory diagnostic tests in reducing long hospital stays and avoiding unnecessary drug prescriptions has not been widely recognized except by clinical chemists. However, the potential economic benefits of near-patient testing, through a reduction in the overall time taken between patient presentation and initiation of treatment, have been widely debated. Applications have grown in the USA and are progressing slowly in Europe led by the German market.

Safety

Healthcare managers and regulatory agencies have become heavily involved in the scrutiny of product quality and the safety of diagnostic procedures. The FDA has increased its requirements in the Quality System Regulation (QSR), which is enforced through site inspections, and now covers product design and development. Many companies have also been certified to ISO 9000 standard, the International Standard for quality systems, and the broad risk management standard, ISO 14971. The long awaited IVD Directive in Europe introduces another comprehensive set of requirements that companies planning to do business in Europe must comply with. New restrictions are also beginning to be applied to the internal production of reagents by hospitals. Pressures from both professionals and the public are behind the tightening of standards and regulations. Professional healthcare providers are particularly concerned about litigation, and members of the public have greater concerns about the reliability of healthcare processes. These pressures have increased the responsibilities being put on manufacturers, but favor large organizations that can invest in the necessary quality system infrastructure.

Controls that apply to testing laboratories, such as the CLIA 1988 regulations in the US, have prevented many tests being taken up outside of hospital laboratories. This temporarily stalled the growing Point-of-Care test market. However, with clear guidelines provided by CLIA for waived test classification, manufacturers have designed fully integrated products that meet these requirements. This regulatory pressure coincided with a customer preference for fully integrated systems. The overall effect is that responsibility for controlling the product during use has been partially transferred to the manufacturer from the user.

Companies now have to follow formal development systems that comply with regulations such as the QSR. Those R&D organizations, including universities and clinical laboratories, seeking to commercialize products through partnerships with major suppliers, will improve their chance of success if they work to the same standard.

Core Laboratories

The increased level of automation of immunoassays and other types of laboratory tests has lessened the need for separate specialized laboratories, e.g. for microbiology, immunology, and clinical chemistry. Studies have shown that in a typical medium size hospital, around 80% of the laboratory tests are automated. In many hospitals, traditional specialist laboratories have been consolidated into **core laboratories** that provide a wide range of services. Core laboratories are very efficient, as they can switch a small number of staff between different test technologies, operate in a comparatively small space and gain economies by combining common operations such as sample handling, data management, result transmission, and staff training. Core laboratories buy considerable quantities of reagents and are in a strong position to negotiate lower prices, especially if they are part of a large **Health Management Organization** (HMO). Core laboratories can be replicated and decentralized, with satellite operations in the suburbs of a large city performing common automated tests, backed up by a central laboratory offering a wider range of tests.

Reference Laboratories

Reference laboratories provide the widest range of services, often offering rare tests and bench tests that require special reagents and highly trained staff. Clinical biochemists are available to interpret complex and unusual data from individual patients. This aspect of reference laboratory operation seems to be secure for the future.

There has also been a move toward super-laboratories, with automated sample handling equipment that sends samples from a central collection point directly to individual analyzers via intelligent switching points. This trend started in Japan where there was already a high level of centralized testing.

Super-laboratories have the advantages of extensive automation, with its low labor costs, and great bargaining power with reagent and equipment suppliers. On the downside, they require high levels of capital investment and are sometimes located in high cost areas such as city centers. In a reference laboratory, many of the specialized tests offered may lack automation, limiting the benefit of the fast sample handling systems. Sample collection and transport logistics may extend the time before a result is received in distant patient locations but computerization can reduce associated administrative costs.

Innovative new tests from small companies often appear first in reference laboratories. DNA-based methods are an important new area in this segment. These tests parallel use of more routine markers tested for by immunoassay methods.

Point-of-Care

The fastest growing market sector is Point-of-Care. Point-of-Care tests are those that do not require laboratory facilities. A shift to rapid test results, which permit immediate action to be taken as part of the consultation process, has been slowly taking place over the last decade. Products allowing **non-invasive testing** are widely available, such as urinary tests for drugs of abuse. Further developments are taking place that will use direct but non-invasive patient contact.

There are issues raised by this response to demand from doctors and patients. For example, healthcare managers are faced with reallocations of budgets between cost centers, and both the immunodiagnostics companies and their laboratory-based customers recognize that much of their existing business could be moved out of laboratories.

An important stepping stone in this process is seen to be clinic testing using equipment with a computer link to the central laboratory. This allows the laboratory to control local instrument performance, result quality, and test interpretation.

The latest Point-of-Care tests can provide quantitative results. Some have a level of accuracy and precision comparable with laboratory tests. Sophistication is also increasing, for example immunoassays in Point-of-Care format for glycosylated hemoglobin (HbA_{1c}) are now in use as an alternative or complementary method to established glucose testing in diabetes management. Similar tests for cardiovascular disease are increasingly being used.

Home Tests

Pregnancy tests are widely used, and available through a broad range of distribution channels. Ovulation tests are also available, but their introduction to the consumer market has required a major investment in educational material, publicity and advertising, and telephone support. The main difference from other immunoassay markets is the very large number of potential end-customers. However, pregnancy tests are distributed in much the same way as the vast array of other consumer healthcare products.

ADVANCES IN TECHNOLOGY

Since the advent of the first commercial radioimmunoassays there have been many major advances in immunoassay technology. Some have still not been exploited in mainstream commercial products. The impact of micro- and nanotechnology is on the horizon and is likely to change many approaches to healthcare (*see* LAB-ON-A-CHIP, MICRO- AND NANOSCALE IMMUNOASSAY SYSTEMS). The very small scale of the technology will be able to reach the molecular level: it is projected that internal diagnostic devices could be developed as continuous monitors of health.

Homogeneous Assays

Classical (heterogeneous) immunoassays require a separation of bound and free material. However, a variety of ingenious methods have been developed to allow measurement of free or bound material without the use of a separation step. This simplifies the design of automated equipment and shortens the overall time required to carry out an assay. In the market, the most significant effect in recent years has been that homogeneous (non-separation) assays may be run on conventional clinical chemistry analyzers. Homogeneous assays have been most successful for small molecules with concentrations above the ng/mL range, and for comparatively high concentration proteins. Efforts have continued to try and develop a homogeneous assay format for a wider range of analytes with improved sensitivity. Experimental homogeneous assays have demonstrated remarkable sensitivity and a wide working range with large molecules, such as thyrotropin (TSH). However, the availability of automation for the separation step in heterogeneous assays has partially reduced the need. In the long term, a key application of a universally applicable homogeneous assay format may be in enabling fast, simple Point-of-Care tests (*see* HOMOGENEOUS IMMUNOASSAYS).

Immunometric Assays

Immunometric (sandwich) assays (*see* PRINCIPLES) transformed assay performance in the areas of sensitivity, specificity, and range. Until recently these benefits have only been available for large molecules, but recently anti-complex immunoassays have been described that offer the improved performance of immunometric assay formats for small molecule assays (*see* NON-COMPETITIVE IMMUNOASSAYS FOR SMALL MOLECULES).

Non-radioactive Labels

Non-radioactive immunoassays had a major effect on the market, allowing tests to be performed outside of the specialized nuclear medicine laboratory. As labels were developed with much greater specific activity than radioisotopes, assay sensitivity was increased, especially for immunometric assays (*see* SIGNAL GENERATION AND DETECTION SYSTEMS).

Automation

Automation of immunoassay tests has broadened the market enormously. The Abbott TDx™ was a breakthrough, offering 'black box' automation with minimal skill required from the operator. Automation was welcomed by laboratory managers on economic grounds because higher workloads could be handled with fewer, low-paid, less qualified staff. Random-access analyzers followed, reducing turnaround times in laboratories. Unipath's lateral flow technology, used in pregnancy tests, made reliable immunoassay technology available to the general public. This is a form of automation achieved, not with mechanics and electronics, but through the ingenious use of materials and physical forces such as capillary flow, requiring little intervention from the user. These test devices perform nearly all the steps without any intervention from the user. Even the mainstream random-access analyzers, designed for laboratory use, have become highly automated with the minimum of intervention required.

As less technical staff are given responsibility for immunoassay tests, manufacturers are taking a greater responsibility for the functions of the tests and the way they are used. Security can only be guaranteed if the user cannot influence the operation of the product, and this can only be achieved by fully integrated devices.

Relationship Between *In Vitro* and Other Types of Diagnostic Tests

Immunoassay and other *in vitro* tests, may look very different to *in vivo* tests and some of the electromedical equipment used in patient monitoring, but all serve a common purpose: the provision of information about a patient for diagnosis or monitoring. Convergences between diagnostic test technology are illustrated in Figure 19.1.

In the few situations where non-invasive, continuous monitoring types of tests have become available in areas that were once the province of immunoassays, the demise of the immunoassay test has been swift. An example is fetal heart-rate monitoring, which displaced immunoassay tests for human placental lactogen (hPL) and estriol.

This reflects the limitations of measuring an analytical quantity compared to a functional measurement and there is a trend towards functional tests throughout the diagnostics sector. Figure 19.2 illustrates how different diagnostic areas develop based on patient needs, and the availability of suitable technology to meet them.

One technology that has quickly grown from showing great promise to acceptance is DNA-based testing. This has already had an impact in some areas currently served by immunoassays, particularly infectious disease and cancer tests. DNA probe and amplification tests represent major breakthroughs, comparable in their significance to immunoassays, but the eventual displacement of immunoassay by these tests may have been overestimated.

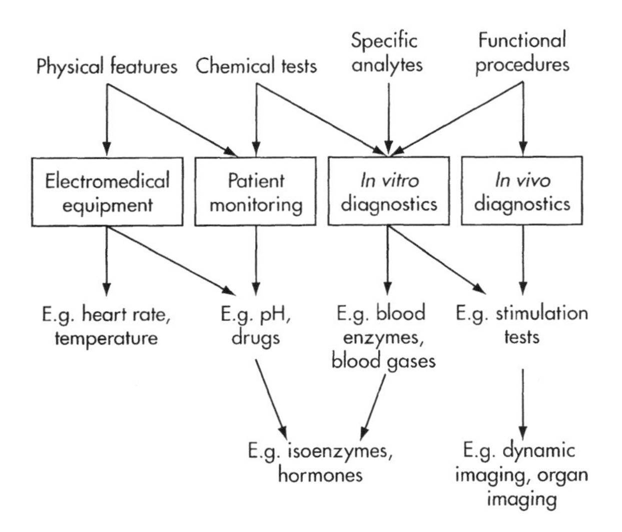

Fig. 19.1 Overall diagnostics market and analytical test measurements. This figure shows the central role of *in vitro* diagnostics in the overall diagnostics market, due to the importance of achieving specificity (versus non-specific monitoring) with minimal cost and invasiveness (versus *in vivo* diagnostics).

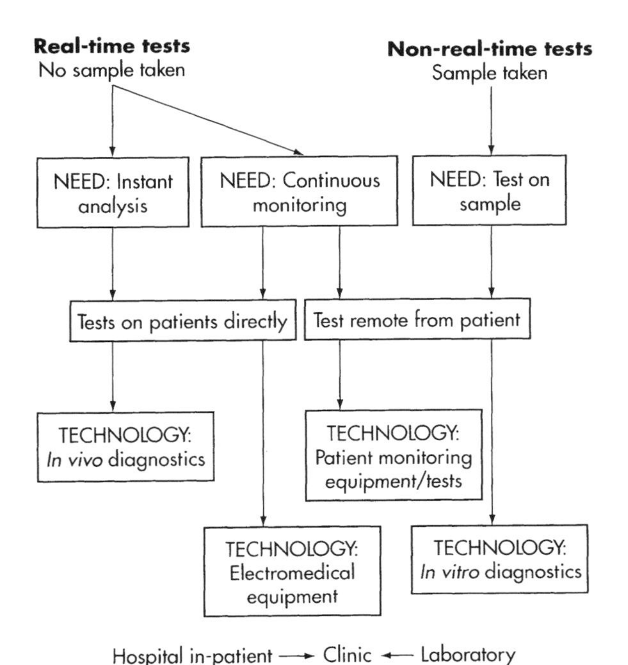

Fig. 19.2 Overall diagnostics market and patient needs. This figure shows that tests that can be carried out directly on the patient in the clinic most closely meet patient needs, subject to appropriate technology becoming available.

These products, however, have an important role and will open new areas of clinical investigation in a similar way to immunoassays 20–30 years ago. The main business opportunity will be in allowing totally new parameters to be measured with a resulting change in therapeutic options, e.g. viral load.

Point-of-Care Tests

Manufacturers have gradually simplified laboratory immunoassay tests through automation. The complexity and capital cost of random-access analyzers have required them to be located in a laboratory environment. However, disposable devices have been cleverly designed that require only very simple analyzers or even no analyzer at all. These tests do not require the support of a centralized laboratory. Their availability has coincided with an increased demand for immediate test results. Generally *Point-of-Care* or *near-patient* tests are defined as those that do not have to be sent to a laboratory. They are most likely to be performed by a nurse or paramedic.

There are some practical advantages of Point-of-Care tests, such as a reduced likelihood of mix-ups with samples or results, or sometimes, less travel for the patient. However, the primary difference from laboratory testing is the reduced overall time between test request and result availability. The patient is likely to receive the most appropriate treatment quickly and the doctor only has to deal with the test situation once. The increased test reagent cost may be more than offset by the decrease in administrative and treatment costs incurred sustaining the patient while awaiting the test result. The **cycle time** is the overall time taken by a connected series of events (such as those involved in making a single diagnosis). It is now well established in manufacturing industry that reduced cycle times can lead to lower costs, but in healthcare the economics of Point-of-Care testing have been argued both ways. As Point-of-Care tests fall in price, the argument will be resolved in favor of prompt and local testing in many common situations. There are also some acute clinical situations where speedier tests can reduce the risk of adverse effects on the patient. However, Point-of-Care tests must be just as clinically reliable as a laboratory test to be an acceptable substitute.

There is a trend away from laboratory tests that merely confirm diagnosis or reinforce actions already decided on. To be effective, Point-of-Care tests must give clear and unambiguous results that are simple to assess and have a clear course of action associated with their results. To be viable away from the laboratory environment they must involve minimal patient sampling or test processing.

Home Testing

Pregnancy tests, based on human chorionic gonadotropin immunoassay, are commonly used in the home. They may be purchased from pharmacies and, in some countries, in supermarkets. The technology required to deliver a foolproof test to an untrained user is similar to that used in Point-of-Care tests by medical professionals, based on lateral flow immunochromatography, with a visual end-point and a control indication. Home tests are also available for ovulation monitoring and cardiac assessment. The technology exists to provide home tests for a wider variety of analytes, subject to market demand.

Probabilistic Testing

Screening for Down's syndrome requires the use of multiple tests, and the results are fed into an algorithm along with other data, such as the subject's age. The end result provides a measure of the risk, or probability, of a condition occurring. This is a sensible approach to screening, wherever a reliable confirmatory test exists (via amniocentesis in this case). The same type of approach has been tried for cancer risk, using multiple, numerical sources of information. Down's syndrome is an area where DNA diagnostics have been developed for disease risk assessment and are gaining acceptance, often as a deciding step before more detailed patient work up that includes immunoassay tests.

NEW ANALYTES

Over the years there has been a steady flow of new analytes, such as *Helicobacter pylori* for gastritis, homocysteine for cardiovascular disease, and bone metabolism markers for osteoporosis. There are many innovative tests available from small, specialized suppliers, but few become mainstream diagnostic tests. Many new markers have been reported that have a correlation with a disease state but lack the necessary clinical specificity and sensitivity for diagnostic use.

THE FUTURE OF THE IMMUNODIAGNOSTICS BUSINESS

APPLICATION OF MARKETING THEORY TO IMMUNODIAGNOSTICS

The life cycles of technologies follow a pattern in the market that is known as an S-curve (Figure 19.3). S-curve analysis describes the whole life of any technology from its invention, to its first commercialization, and through to its demise, when an alternative technology provides improved performance or efficiency. In the early stages, much is spent on development before any income is received from sales. This is known as the **Functional** phase of development and matches the **Embryonic** phase of the market. As the first products come onto the market there is a slow uptake by a few pioneer customers and the cost of development is still high. This is the beginning of the **Feature** phase of development and represents the **Early Growth** phase of the market. There follows the most profitable period where sales increase rapidly as products improve in quality and features. Eventually the technology starts to lose momentum, leading to the **Late Growth** stage of the market. Now the cost of development of new features becomes higher as the technology is stretched to its limits. Much of the effort expended in development is aimed at reducing costs and this stage is therefore known as the **Process** phase of development. Eventually the product saturates its market in the **Mature** phase, where it is most vulnerable to replacement by newer technologies.

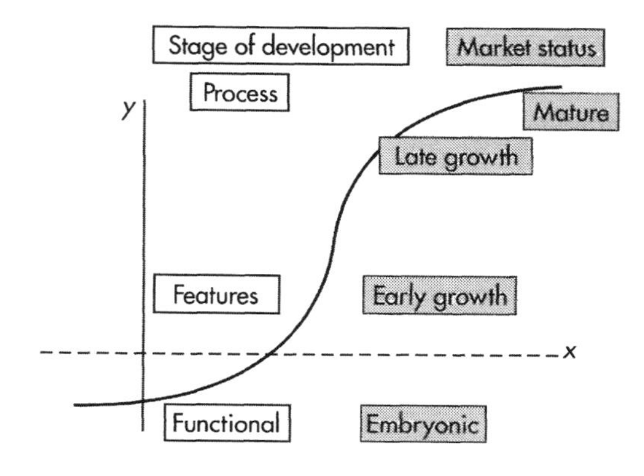

Fig. 19.3 S-curve model of technology development. Source: Adams Business Associates.

Using this principle the development status of key diagnostic technologies is compared: clinical chemistry, immunoassay, and molecular diagnostics (Figure 19.4). In practice, a scientific innovation, such as immunoassay, often proves most successful in the market when linked to a complementary technology. This was seen with both clinical chemistry and immunoassay products, where automation technology gave these markets a significant boost, allowing the techniques to be used in many more locations. Laboratory clinical chemistry products have reached the mature phase although there has been growth in near-patient applications of tests for certain analytes (e.g. home glucose testing). Laboratory immunoassays are in the late growth phase, possibly entering the mature phase. From this analysis it can be seen that further advances in laboratory applications of immunoassay technology will probably be expensive to develop, and limited in their scope. The majority of the most fundamental improvements in immunoassay design (homogeneous formats, immunometric assays, monoclonal antibodies, non-isotopic labels, and automation) date from the last decades of the 20th century. The new

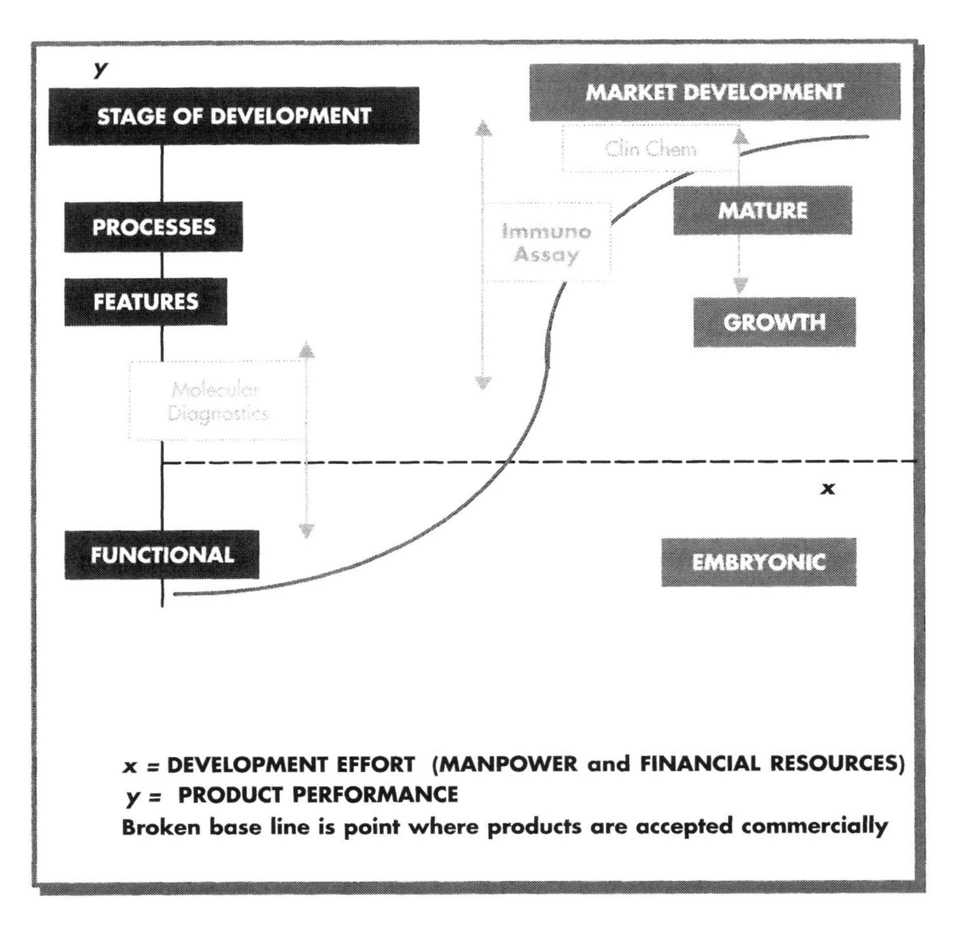

Fig. 19.4 S-curve model for diagnostic product groups, 2002.

systems of the early 21st century have added more features to this core of integrated technologies, but at considerable expense because of the complexity of the engineering required.

As the S-curve model predicts, new improvements are becoming very costly (at least $300 million for a completely new immunoassay system) with progressively smaller benefits to the user. With only a limited number of companies able to afford these high development costs, many have withdrawn from the market while others have focused on seeking the next technology. This focus was on biosensors for a period and is currently aimed at DNA probes, but there are areas where immunoassays could re-emerge in new formats, outside the laboratory. The challenge for immunoassay technology is that the near-patient situation has different requirements, and the factors that make immunoassays such an attractive option for the laboratory can actually be disadvantages for non-skilled personnel needing a result in a hurry.

Consolidation of the many companies in the immunodiagnostics business has been evident over the last few years. This is natural for any business and serves to confirm that the development of the market has reached a mature or late growth phase. Consolidation is not just limited to the industry. Laboratories specializing in different disciplines are also being consolidated into core laboratories. However, there is also an underlying trend bringing effective diagnostics closer to the patient. Point-of-Care testing is showing the highest growth rate at present but there are restraining forces at work, as it is not in the interests of the suppliers or their primary customers, the laboratories, to give up the current efficient manufacturing, selling and distribution systems, and there are cost issues to be resolved.

The increased internationalization of the market means that all the leading suppliers now have direct subsidiaries or regional headquarters in the major global markets. A highly active supply network of distributors remains, but they recognize that they are acting as temporary market developers for the larger companies. The trend is for major companies to move to direct sales in order to control the market in each country and to increase the profits at a time of pressure on costs.

The pressures from clinicians, administrators, and patients for more immediate results will slowly drive for change in the market. As the issues of ease of use and reliability of the technology continue to be resolved there will be a major impact on the distribution systems – the end customer will change from the laboratorian to the clinician.

Few of the major diagnostics companies have a suitable distribution channel to meet such a shift. Laboratories will also resist it, as they risk losing much of their existing business. As just one example of the size of the distribution problem, in the UK market there are some 450 laboratories but more than 60,000 doctors involved in primary healthcare diagnostics. Implementation of alternative strategies for marketing, selling, and distribution is critical to allow new, effective technologies to advance in the diagnostics market.

The action taken commercially to protect established businesses will be effective in the short term but the market forces for change will eventually overtake the situation. Change may well come from a company with no vested interest in the present diagnostics market as it will not be driven by past position and efforts to maintain profit levels in the laboratory sector.

FUTURE MARKET REQUIREMENTS

Cost Reduction

The pressures on healthcare budgets will continue to exert a major effect on the market. Larger companies with high market shares can benefit from economies of scale, as they can recoup the development costs of new products more quickly. Technology developments that reduce manufacturing costs will also enable prices for immunoassay reagents to fall in real terms.

There is an increased awareness that medical and clinical chemistry staff are being assessed for their effectiveness in the total healthcare cycle. Interest in 'outcomes' analysis is extending to all areas, and reduced bed occupancy is part of this critical assessment for those treating hospitalized patients.

The driving force is the effective use of limited budgets. There is gradual acceptance that rigid cost center budgets are not particularly useful. A moderate increase in expense in one area can greatly reduce the expense in another. Hopefully, the potential for diagnostic testing to be a positive investment will be more widely recognized in the context of the overall healthcare system. However, this depends on the diagnostic utility of the tests. Many are carried out to *confirm* diagnosis and do not provide much new clinical information. Also there are many hidden costs in a testing process that takes several days to accomplish (e.g. consultation, outpatient appointment, test, transmission of result to doctor, and follow-up appointment). For diagnostic tests to facilitate shorter hospital stays and more timely drug treatments, they must provide immediate results when and where they are needed. Finally, the increased costs associated with Point-of-Care tests must be more than offset by reductions in administrative and laboratory costs.

Even though greater investment in immunodiagnostic tests has the potential to reduce overall healthcare costs, budgetary constraints mean that it is unlikely that the value of the immunodiagnostics market will grow at much more than 5–8% per year.

Instantaneous Results

There is an underlying demand for testing to be moved closer to the patient. This has always been the ideal situation, but the complexity of most diagnostic tests and the difficulty of interpreting the results correctly have kept them in the province of influential and high-profile central laboratories and specialist consultants in hospitals. But tests are becoming much simpler to use, and the addition of computerized result *interpretation* for some tests requires only minor improvements in technology, so near-patient testing is becoming a more practical proposition.

As diagnostic tests progress from the laboratory toward the patient, immunoassays will face a greater challenge from direct physical methods, which give immediate results. For the laboratory, immunoassays have many advantages, but this is of little consequence when the doctor and patient need a result as quickly as possible. Immunoassays cannot easily compete with a sensor providing instant information. Electromedical equipment can monitor patient's status continuously, whereas current immunoassays cannot. The deciding factor will be the costs incurred.

For immunoassays to compete with new technologies, faster tests will be needed, with result generation in less than 5 min, and eventually in less than 1 min. Homogeneous assays are very fast, but not yet universally applicable. Capillary-based heterogeneous assays, which have very fast reaction kinetics, could potentially provide results in less than a minute.

The use of immunoassays for continuous monitoring will require a major shift in assay technology. Flow immunoassays allow a continuous series of tests to be carried out in rapid succession, but they are not instant, and require expensive equipment and a continuous supply of reagents. Atomic force microscopy has been used to measure attraction between antibodies and their targets reversibly, and this suggests that a continuous monitoring immunoassay device could be designed. This will require immunoassay technology way beyond that employed in current laboratory test kits and it will be a challenge to achieve the right balance between specificity, affinity and reversibility of binding.

Sample Type

Provision of immediate results will mean a shift away from processed blood (serum or plasma) as the sample medium of choice. Urine, whole blood, and capillary blood are the most likely candidates, although saliva has also been used. Determination of clinically meaningful results using whole or capillary blood is not devoid of technical difficulties. However, there are materials available that act as cell filters and such tests have been successfully developed for some analytes.

The use of whole blood samples eliminates the equipment and time required for the isolation of serum or plasma. Other alternatives, such as urine and saliva, offer less invasive ways of sampling.

More recently, experimental diagnostic devices for non-invasive testing have started to appear, especially targeted at glucose testing. Clearly, diabetic patients would prefer not to have a daily pinprick if it could be avoided. Use of devices that sample plasma through the peripheral vessels of the skin is a step forward, as are those that have attempted to use wave technology such as near-infra red spectroscopy. It is likely that the specificity and sensitivity of these tests will be enhanced using powerful data processing algorithms to reduce noise.

These non-invasive technologies remain in the *Functional Development* stage of technology development (see Figure 19.3). Their use in niche markets could well open the path to a wider acceptance once the key functional issues have been addressed.

User Convenience

Laboratory analyzers have become highly automated. Typically, technicians only have to load reagent packs and samples. However, most analyzers require a temperature-controlled, clean environment with skilled staff available for maintenance activities and to deal with the occasional problem. The main requirements for future laboratory analyzers will be wider menus of analytes, and improved reliability, simplicity of operation, sensitivity, specificity, range, and speed of result. Cheaper reagents will be sought by customers and the market will become very cost competitive. If a major manufacturer offered an open system with licensed third party suppliers it could probably make considerable gains in the market, but at the expense of own reagent sales, which normally provide most of the profit. Calibration could probably be simplified further, by provision of assay calibration data over a modem link from the manufacturer to the analyzer. Quality control, consumable ordering, and troubleshooting could all be improved using modem links and intranets. Some analyzers already have modems.

Point-of-Care tests have different requirements. In the absence of skilled laboratory technicians, they must be absolutely reliable. This can only be achieved if the device is totally self-contained and robust to application of the sample. Internal quality control to monitor every test is essential (whereas in an analyzer it is sufficient to run quality control samples no more than once a day). Point-of-Care tests tend to be used by nurses who have a more patient-oriented approach than laboratory staff. Charting of data from each patient is of greater interest to nurses, whereas QC charts are not a priority. Downloading of data to the hospital computer is important for Point-of-Care test equipment but the laboratory information system interface may not be like the conventional RS-232 interface used for laboratory analyzers. More likely an infrared or other wireless link will be needed, with patient tracking software in the hand-held test unit and base station. These types of products are already available.

Laboratory analyzer tests now have reagent packs for multiple tests, but Point-of-Care test reagents need to be unitized. Each test is likely to have a test tab that contains the necessary reagents and requires no preparation other than a simple loading step.

Provision of Clinical, Rather than Analytical Information

Historically, immunoassay systems have been designed primarily for the laboratory. They provide precise measurements of analyte concentration to the clinical chemist and specialist physician. Only a few existing immunoassays, such as home pregnancy tests for hCG, provide clinical information as the end-point (instead of analyte concentration). As testing moves nearer the patient, immunoassay products must be modified to provide the clinician with definitive clinical information. In some cases the test will have to measure the concentrations of more than just one analyte, and analyze the data to reach a clinically meaningful conclusion. The Unipath Persona™ system uses two analytes and an algorithm to provide the user with information about risk of conception. It can utilize data from up to six menstrual cycles to provide more reliable information. In practice this complexity has created some confusion for users resulting in an inappropriate use.

In order to interpret analytical data, future devices will need increased computing power. This is well within the capability of current microprocessor technology.

A new demand now emerging that will increase the importance of diagnostics is for **disease prevention**. Early diagnosis or even determination of susceptibility to disease is being seen as having major potential to reduce total health care costs.

Multiple-Analyte Testing

Pre-selected multiple-analyte panel tests are of little benefit to the laboratory as random-access analyzers already offer user-defined panel selection as an option, whilst retaining flexibility for different patient situations. However, carefully optimized multi-analyte tests have clear advantages in Point-of-Care applications, where diagnosis is the end-point, rather than the provision of an analytical result. An integrated test for more than one analyte is likely to provide more reliable clinical results for a comparatively small extra cost, as much of the manufacturing cost of a Point-of-Care test is due to the device and the packaging, rather than the immunoassay reagents.

Improved Clinical Specificity and Sensitivity

Achieving greater clinical specificity and sensitivity requires two objectives to be met:

1. Selection of an analyte or functional marker that reflects the clinical condition.
2. An assay method with sufficient specificity and sensitivity to measure the relevant concentrations of that analyte.

Immunoassays now dominate the *in vitro* market because they excel in the *second* category at moderate cost and with high convenience. They are unsurpassed in their ability to measure precisely a wide range of analytes in serum or plasma, without any need for a pre-test separation. However, there are areas where other, more complex methodologies can provide superior clinical sensitivity and specificity. One example is drugs of abuse testing. Gas chromatography is more accurate and specific than immunoassay because of the physical separation of the analytes. DNA probe tests may also be more specific because of precise base pair matching. The polymerase chain reaction (PCR) technique provides massive amplification, so these tests are also more sensitive. DNA tests may replace immunoassays for some infectious diseases, being capable of detecting extremely low concentrations of viral antigens, whereas immunoassays are mainly applied to determine the presence of antibodies, which are only detectable weeks after infection. Specific examples where this progress has been achieved are viral load tests for HIV, HCV, and HPV. Use of complementary immunoassay tests continues to assess seroconversion and progress of the infection. DNA probe tests are also useful for detecting specific gene sequences associated with a higher risk of a number of other clinical conditions, for example, cancer.

Where there is much scope for improvement in immunoassay tests is in the *first* category: the selection of a useful analyte. Many current markers cannot be completely relied upon to indicate the presence or absence of a particular clinical condition. The field of cancer screening provides many such examples. For immunoassays to continue to be the test of choice in a wide variety of clinical applications, better markers would be a distinct advantage. Otherwise, functional or more specific assay tests, based on alternative technologies, may replace them.

CHANGES IN THE CUSTOMER BASE

The Healthcare Budget Holder

The healthcare administrator has an overall responsibility to control expenditure. Diagnostics companies must provide convincing data that new tests will actually reduce overall healthcare costs. In the absence of product differentiation, reagents will be selected primarily on the basis of cost, although more astute administrators are seeking the lowest overall cost per delivered result, rather than simply buying the cheapest reagents.

The Laboratory Manager

Laboratory managers are increasing efficiency by consolidating automated tests in core laboratories. They need reliable instrumentation and reagent supplies. Automated immunoassay analyzers require wide menus of analytes and must be suitable for high and low throughputs of individual analytes, without excessive waste. Instrumentation must be easy to interface to computer systems handling samples and results. There is scope for standardization of QC, sample IDs, barcodes, sample tubes, and result data between different makes of analyzers and laboratory interface systems. Automated laboratories require interfaces between the analyzers and sample track systems.

The Clinician

Point-of-Care tests are primarily supplied to clinicians, although some are supplied direct to patients through pharmacies. The immunodiagnostics industry is primarily structured to supply to large laboratories. The growth of the Point-of-Care test market will have an enormous impact on the industry. The requirements for the products, support literature, promotion and distribution are quite different from those of central laboratories.

The Patient's Role as End-Customer

A major new trend is for patients to want a greater say in the use of diagnostics for their clinical assessment, due to their increased awareness of the benefits of early diagnostic procedures and a more open discussion of medical conditions. New breakthroughs in testing are publicized in newspapers and magazines, on television, and via the Internet. Patients are now demanding to be considered as consumers. For example, the consumer does not want to travel to a city center hospital from the suburbs, and may change to a different healthcare provider with a local laboratory. Consumers also prefer non-invasive tests. They are not, of course, concerned with the underlying technology behind a test.

The latest trend in healthcare is the provision of one-stop medical centers, e.g. in shopping malls, which provide a range of services on demand. Payment is for services received. This is a radical departure from the 'insurance' systems that take money primarily from the well, and distribute services to the sick, such as HMOs and the UK National Health Service. These locations could be a growth area for Point-of-Care tests. The Internet is also a source of a range of services to the consumer.

Patient concerns are many. The increasing proportion of elderly patients, and strong demands from younger patients who expect good service, will place extra demands on systems that are already under pressure. The healthy working population has shown an unwillingness to be generous in supporting the whole social structure from their earnings and taxes.

POTENTIAL IMPACT OF NEW TECHNOLOGIES

DNA Probes

DNA analysis can, in certain situations, provide clinical information that is difficult or impossible to obtain by other means. In these areas, new markets will be created and immunodiagnostic testing will be unaffected. However, in some areas, DNA probes provide more direct information than immunoassays. Diseases that are directly caused by genetic variation amongst the general population are in future likely to be tested for by DNA-based tests. However, where the incidence of disease is only partially related to genetic variation, or where only some patients are affected, immunoassays may continue to have an important role in monitoring the presence or status of disease.

Infectious diseases that are currently detected by the presence of antibodies may well be superseded or complemented by DNA or RNA detection tests, which have the potential to detect infection earlier. Immunoassays will still be useful to distinguish between different immunological states, e.g. after vaccination.

DNA probe tests are expensive, because of their complexity, and this will restrict their use if an immunoassay can be used to achieve the same clinical objective. The specialized nature of DNA diagnostics might provide central laboratories with a replacement technology if simpler tests move to Point-of-Care use. The application of DNA-array technology permits determination of hundreds of DNA defects at a low cost per test. The use of arrays also gives rapid screening for this large number of genetic defects with all results obtained at the same time. Costs of DNA procedures will fall as feature developments occur, as would be expected from the S-curve analysis of the technology.

Biosensors

Eventually, immunobiosensors may become firmly established as diagnostic tests. The concept of a small analyzer with disposable test strips fits well with the needs of the growing Point-of-Care market. Biosensors depend on a biomolecular detection system and for many analytes this is likely to be antibody based. In fact, biosensor based immunoassays could give the immunodiagnostics market a boost. This is because biosensors react instantly to binding between antibody and analyte. The inability of current Point-of-Care immunoassay tests to provide a near-instant result is a key limitation. Existing immunobiosensor systems are used primarily in research laboratories, because of the complex instrumentation and operator skills required, and also because of the sample pre-processing steps, which add considerably to the total assay time. However, this technology has long-term potential to deliver cheap and fast Point-of-Care tests if production costs and engineering issues can be overcome, possibly involving micro- and nanotechnology.

Flow Cytometry

An alternative technology for provision of multi-analyte data is the FlowMetrix® flow cytometry system from Luminex. This can analyze up to 64 different analytes from a single sample using a flow cytometer and digital signal processing of a range of color dyes. Polystyrene microspheres are dyed with various amounts of orange and red-emitting fluorophores to produce 64 different sets that each has unique orange-red emission profiles. Each set is coated with a different antibody. The signal analysis is set up so that interfering substances are effectively eliminated and whole blood can be used.

This approach provides cost savings through multi-analyte measurement. Instrument costs are being addressed to reduce this to some 20% of the current cost of flow cytometers. It has been suggested that dedicated profiles of relevant tests could be assembled. This has been achieved for assessment of myocardial infarction by combining tests for myoglobin, creatine kinase-MB, troponin-I, and digoxin, and with scope to add other related substances to develop a full cardiac function profile.

Flow cytometry is a laboratory based technology and falls short of providing non-invasive diagnostic tests, near-patient testing, and giving rapid test results to the clinician, despite being reasonably fast in operation (results are available within 30 min). However, the multi-analyte approach might still be exploited by immunoassay methods.

Capillary Electrophoresis

Capillary electrophoresis can be used to assess tissue well being. A minute sample of tissue is taken and inserted into a glass capillary tube containing a buffer. Electrophoresis, which takes about 40 min, is used to obtain the pattern of the metabolites in the tissue. This profile is compared with a reference or against a historical profile from the same patient. Changes can indicate early stages of disease, from profiles containing up to 30 different substances. This technique has been proposed for early detection of diabetes and for monitoring effectiveness of treatment.

Non-invasive Tests

Immunoassays have some potential disadvantages compared to non-invasive tests that can provide equivalent clinical information. Non-invasive tests not only avoid the need for blood samples, but some also have the capability to provide functional information continuously and in real time. The advent of highly sensitive ultrasonic scanners linked to powerful three-dimensional imaging software is likely to revolutionize the detection of diseased tissues. In some cases this will reduce or even eliminate the need for laboratory tests. However, determining the specific cause of disease may still require immunoassay tests, and early detection may not be possible with imaging equipment.

Several companies are working on, or placing in the market, glucose monitoring devices that are worn against the skin and draw material through the skin or use sensors embedded just below the surface layer of dead skin cells in the epidermis. Although such tests may be more convenient, they are not truly non-invasive.

As an alternative approach, wave technology and innovative signal detection systems could measure unique substances non-invasively through the skin. An example is the combination of laser diodes for direction of near infrared light via a fiber optic device held against the skin. The reflected light is analyzed for the substance of interest. A model system has been used for glucose detection. Whether this type of technology could measure a range of lower concentration analytes remains to be seen.

The skin is highly variable, and there are many obstacles to overcome before non-invasive tests for immunoassay analytes, which occur at very low concentrations, become feasible. Detection of analytes in the eye is one alternative that is being investigated. These technologies remain in the functional development stage.

In Vivo Imaging

A halfway house between *in vitro* tests and non-invasive tests does exist. This involves the patient taking some kind of imaging material that can be detected by an external signal detection instrument, such as a gamma camera for the detection of radioisotopically labeled material. Perhaps antibodies will find a role as highly specific binders in such tests, linked to signal generating molecules that can be detected externally. Experimental work has established that engineered antibody conjugates can be 'activated' to generate a signal in response to a binding event within the body. These are effectively *in vivo* immunoassays. However, such tests could be expensive and impractical to apply to large numbers of patients.

NEW ANALYTES

Many of the clinical fields in which immunoassays have been dominant, such as thyroid and obstetrics/gynecology, are now mature. This has meant that simple tests are used in carefully optimized test panels to minimize expenditure. Newer areas for growth have not been as lucrative as once expected, although there has been much emphasis on infectious and autoimmune diseases, osteoporosis, and markers for cardiac disease and cancer.

The human genome project is generating a large library of new proteins with hitherto unknown functions. Many of these have the potential to be clinically useful. This should give a boost to immunoassay test menus. However, some new test parameters are also likely to be addressed by alternative methodologies, such as gene amplification.

New analytes will not be accepted in the managed care environment unless there is proven clinical utility. This puts the onus on manufacturers to undertake comprehensive clinical trials and publish them in the literature.

The Role of the Clinical Laboratory

The level of automation of the latest generation of immunoassay analyzers has minimized the need to use a traditional laboratory. One of the functions of the laboratory has always been to provide an environment where complex chemical and biochemical tests can be developed and carried out, but most analyzers are effectively self-contained laboratories requiring nothing more than a power outlet. However, there are persuasive economic arguments for non-urgent testing in a hospital to be centralized. A core laboratory can be made highly efficient by investing in computerization, and making radical improvements in sample and result processing, and ancillary activities such as stock control and QC. The survival of individual core laboratories will therefore depend on economic considerations rather than technical capability.

The laboratory professional has lost some status in the area of assay technology expertise, now that the analyzers largely take care of the science and technology. In the future another challenge will have to be faced: clinical professionals (clinical chemists and clinicians) will find that test methodology often provides clinical (rather than analytical) results requiring little further interpretation. This will impact upon another province of clinical chemists: result interpretation.

Although change is inevitable, the laboratories, and the professionals who manage and operate them, will have an important and influential role to play in clinical chemistry in the future. As more testing is carried out close to the patient, many clinical chemists are seeking a role in organizing quality assurance and proficiency testing schemes in all the test locations, and this is welcomed by most healthcare administrators. Laboratorians can play a key role in evaluating new Point-of-Care tests, training personnel in their use, and overseeing test and data management strategy in the decentralized health centers. It has been shown that the enormous growth in home glucose testing has not led to a corresponding decrease in laboratory glucose testing. This is an important indication that Point-of-Care testing will not lead to laboratories running out of work. Instead, they will have to provide highly reliable confirmatory tests, a much wider menu of tests than can be managed in Point-of-Care locations, and guidance on the interpretation of test data that do not fit the simple rules of computerized expert systems.

Laboratory immunoassay services will have to provide a wide range of tests, and menu may therefore be the key determining factor for the ultimate winner(s) among the laboratory immunoassay manufacturers, now that automation is reaching its limits. An open immunoassay system, that allows laboratories to experiment with new markers as well as run made up reagent packs for a wide range of established analytes, would fit well in this laboratory of the future. While immunoassay testing may expand out of the laboratory, other, new test methodology may be limited to laboratories because of its complexity.

CONCLUSION

With immunoassay technology approaching the limits of its potential, as illustrated by the S-curve, there are likely to be new technologies to challenge immunoassays for the leading role in diagnostics. The commercialization process may be slow, but the new technology may already have been invented. It could be at the stage of functional development or embryonic market development. Probably, it will be a fusion of technologies that are already well known, but have not yet been applied to mainstream diagnostics. Whatever the technology, it will most likely be applied first in niche markets. The change in technology means that new companies will enter the industry and some of the major companies will either exit or be acquired by other companies more determined to establish and maintain their market position.

Immunoassay technology could be further exploited to provide very fast Point-of-Care tests but samples still have to be taken from the patient. Antibody-based tests can also be used in flow cytometry, biosensors, and *in vivo* tests. DNA and other diagnostic test formats will gain market acceptance where they offer improved clinical utility. However, the ultimate displacement of immunoassays from their leading position is likely to be due to non-invasive testing technology that provides functional or visual information in real time. This would probably only occur in limited clinical applications at first, but could widen in scope as technology develops. Even so, immunoassays will continue to have a key role in providing complementary information.

The apparent similarity between products from different manufacturers, reduced market growth, increased pressures to reduce the costs of diagnostic procedures, and the lack of differentiation of products all lead to the same conclusion. The market is mature and this will result in major technical and commercial changes from the market seemingly so strongly positioned in the 1990s.

The drivers for change are already present and the market is showing early signs of a response. Further evolution and development of Point-of-Care tests, and falling prices, will lead to a confrontation between the existing laboratory based market and the new near-patient segments. This is going to be just as disruptive for manufacturers as for laboratories. However, manufacturers can survive the change by offering wide menus on laboratory analyzers and more common tests in cheap, easy-to-use Point-of-Care formats. Laboratories should take a lead in managing changes in the healthcare services, introducing professional quality and data management systems. The changes will open up new markets and diagnostic testing will increase overall, in terms of the number of tests performed and market value.

Only non-invasive monitoring equipment can seriously upset the diagnostic test market, as it may not require any consumables or laboratory involvement. However, there are many obstacles to overcome and the impact is likely to be restricted to specific areas in the next few years. For those companies that wish to enter the market with new technology, non-invasive technology is probably the best target area. In assessing the likely areas where these niche markets will develop it may not be advisable to look at the largest market opportunities for tests. Breakthrough developments are more likely to be successful in areas where there are existing unmet needs and where new technology can solve a significant clinical problem. Subsequent exploitation of the technology will create new market dynamics.

SUMMARY OF LIKELY TRENDS IN IMMUNODIAGNOSTICS AND RELATED PRODUCTS

Here is a list of trends we predict based on current information.

- Continued growth of the immunodiagnostics test market.
- Significant growth of Point-of-Care test markets.
- Reduced rate of growth in laboratory test markets.
- Growth (but limited) in home testing markets.
- Reduction in cost per test within each market segment.
- Faster tests in each market segment.
- More robust tests with simpler user protocols.
- Increased use of whole blood and urine samples.

- Reduction in use of tests with unproved clinical utility.
- Increased proportion of fully automated laboratory tests.
- Increased assay sensitivity and specificity.
- Improved clinical sensitivity and specificity.
- New analytes with improved clinical utility.
- Fewer companies supplying the laboratory immunodiagnostics market.
- New entrants with new technology or more appropriate distribution chains for Point-of-Care market.
- New entrants that specialize in providing products to physicians in one clinical sector.
- Laboratory consolidation into core laboratories.
- Wider use of immunometric format assays for small molecules.
- Wider application of homogeneous assays.
- Recombinant proteins and monoclonal antibodies will be used routinely.
- Further use of antibody engineering techniques (phage display, recombinant antibody conjugates, etc.).
- Simpler assay calibration protocols.
- Improved assay standardization between manufacturers.
- More detailed reference interval information for patient sub-populations.
- Computer algorithms for diagnosis.
- Use of probability in decision algorithms to determine risk based on multiple sources of data.
- Greater use of patient-oriented trend analysis in diagnosis and monitoring.
- Use of multiple analyte testing in one test unit.
- Use of assay and non-assay data to compute risk.
- Standardization (e.g. sample ID, barcodes, QC, results, patient information).
- Increased level of regulation of manufacturers and users with common global standards.
- Greater influence of patients in choice of test.
- Smaller test disposables (reduced waste).
- Networks (intranets) between users and manufacturers.
- Test services available over the Internet.
- Information about tests available over the Internet.
- Growth in DNA-based, Molecular Diagnostic or Nucleic Acid Tests (NAT).
- Growth in biosensor-based tests.
- Immunoassays in some clinical areas will be replaced by non-invasive techniques.
- Progress towards pharmacogenomic testing, with drugs prescribed on a patient's probable response to that drug.
- Increased demand for tests that support disease prevention programs – wellness versus illness care.

20 Choosing an Automated Immunoassay System

Lori J. Sokoll and Daniel W. Chan

Choosing an automated immunoassay system, like choosing any new instrumentation, is not an easy task. There are currently over 25 immunoassay systems in the market, the majority of which are described in later chapters in this book. Although immunoassay analyzers have come a long way, with many systems now incorporating features that were previously found exclusively in high-throughput chemistry systems, there is no one perfect system. Furthermore, all laboratories, whether hospital, commercial, physician's office, satellite, etc., are different, each with unique requirements. The choice of an immunoassay analyzer therefore depends on the goals to be achieved with analyzer acquisition. This chapter will discuss automation goals, sources of system information, considerations and criteria, and briefly, system evaluation.

DEFINING AUTOMATION GOALS

The first and most important step in choosing an automated immunoassay system, or any automated system, is to define the goals and objectives to be accomplished with the system. It is then possible to survey systems in or soon to be in the marketplace to determine which systems can meet those goals. The advantages of automation are many. One of the most beneficial aspects of automation is the ability to consolidate testing and workstations, including manual testing, batch, selective, or random-access analyzers. Automation can therefore reduce labor requirements and hence testing costs. With shorter incubation and assay times and the ability to increase testing frequency, such as with random-access capabilities, turnaround times can be improved. Quality of testing can be achieved with immunoassay automation, with improved assay performance resulting from improved precision, sensitivity, and wide dynamic ranges. Errors due to sample handling and processing can be reduced by using primary tubes, avoiding the need for sample splitting. The use of barcoding and bidirectional interfaces can eliminate manual data entry errors. Automation also allows for increased capacity for growth and productivity. In summary, the overall goal of automation, in this case immunoassay automation, should be to improve overall testing efficiency with the previous issues discussed all contributing to that goal.

SOURCES OF INFORMATION

There are numerous sources of information and resources that can aid in gathering information about the attributes of the immunoassay analyzers currently available in the marketplace as well as those in development and soon to be released. One source is printed material. This book is of course an excellent resource with in-depth discussions of individual systems. We have also published several books and articles, listed at the end of this chapter, describing specific analyzers and their features. Considering the number of systems available, summarized information can be very useful. Several publications, including CAP (College of American Pathologists) Today and ADVANCE publish annual surveys of current instrumentation in tabulated format allowing for ease of comparison among systems. Similarly, the Laboratorian Desk Reference, a publication from the Clinical Ligand Assay Society, provides additional summarized information. Additional sources are journal articles as well as brochures and other promotional materials provided by individual manufacturers. The most up to date information may need to be obtained directly from individual companies. Manufacturers should also be able to provide performance data on all the analytes of interest including precision, linearity, correlations with other instruments, reference ranges, etc. Information to be obtained from the manufacturer also includes space, power, water, waste, LIS interface, and other specifications and requirements.

The Immunoassay Handbook. *Third Edition*
D. Wild (Ed.)

Another invaluable source of information is other users. Contacting colleagues at laboratories with similar test volumes and test mixes to learn which analyzers they are using and their experiences can be insightful. Manufacturers will also provide names of customers using their systems, although most likely their most satisfied users, as well as arrange visits to see their instruments in use out in the field. Current instrument users, including laboratory directors in addition to the end-users, can provide their experiences with instrument performance and operation. Users may also share certain documentation from their laboratory such as quality control records. Knowledge about systems from outside sources can also be obtained by employing industry consultants.

First-hand knowledge is unquestionably useful in selecting an instrument. Most manufacturers are now willing to bring a system into an institution for a demonstration as well as leave the system for an extended period to allow a partial or full evaluation. If the opportunity arises, participation in alpha and beta site testing as well as other research and development projects can also provide in-depth exposure to systems in development or to systems already available. It must be kept in mind, however, that early hardware and software versions may bear little resemblance to final production models. System evaluations allow the technologists running the analyzers to become closely involved in the selection procedure. Laboratory supervisors and administrators, as well as LIS administrators, should also play a role in the selection and evaluation process.

Finally, exhibitions are an obvious source of information at national and international meetings of organizations such as the American Association of Clinical Chemistry (AACC) and Clinical Laboratory Management Association (CLMA) in the US, the Association of Clinical Biochemists (ACB) in the UK, and the International Federation of Clinical Chemistry (IFCC) and at Medica in Germany. Time and the number of systems being exhibited may limit the number of instruments that can be viewed and investigated but these forums do allow the opportunity to have a first look at new systems with launch dates in the near future.

CONSIDERATIONS AND CRITERIA

Once automation goals have been determined for an individual laboratory, the next step is to identify systems that meet those goals. The previous section described sources of information on immunoassay analyzers while this section will discuss considerations and criteria for choosing a system. Laboratory environment, test menu, technical, clinical, financial, and operational issues will be covered.

LABORATORY ENVIRONMENT CONSIDERATIONS

The type of laboratory and the goals and plans for the organization of the laboratory are important considerations. The size and test volume of the lab in addition to the type of lab will dictate priorities. In a large hospital laboratory random-access analyzers with short times to first result, stat capabilities, and breadth of menu may be critical factors to consider, while commercial reference laboratories may consider throughput the most significant consideration. Laboratories with smaller volumes may be limited to smaller analyzers that may not have high throughput or walk-away capabilities.

The goal of the laboratory with respect to consolidation is another defining feature. It is now possible to combine testing from previously segregated disciplines by focusing on methodology as opposed to clinical pathology specialty. The standard immunoassay or special chemistry laboratory can now be combined with the drug assay area as well as testing from microbiology and diagnostic immunology. It is now possible to perform homogeneous and heterogeneous assays on the same instrument allowing chemistry and drug testing to be combined with standard immunoassays such as thyroid and cardiac markers. Immunoassays, encompassing all testing or high-volume stat tests, may also join other types of highly automated testing in a core laboratory. The ability to consolidate will therefore depend on the type of automation as well as specific tests available on the system.

Space is an important consideration. Where are the analyzers to be placed in the laboratory and what is the function of the laboratory? Are floor models appropriate or is a smaller bench top unit a better fit? In a satellite or emergency room setting space is at a premium. Satellite laboratories associated with the main laboratory or laboratories that are part of a network may choose to use the same instrumentation throughout the system. Therefore, choice of an instrument line with several models, such as a smaller bench top version, may be beneficial. It may also be beneficial to choose the same vendor for chemistry or hematology testing. Using one vendor will allow for better pricing and decreased training due to possible common user interfaces. Vendor choice may be limited by buying groups associated with the institution.

Variability in future testing volumes is also a consideration for the laboratory. Will the system acquired now be adequate in the near future? One possible solution is a modular system that can be expanded to adapt to changes in workflow. In these systems multiple analytical components can be linked together with a common specimen transport system and specimen and data management unit allowing for increased or decreased testing depending upon demand. Modular systems can also include modules for other types of testing such as high-volume chemistry testing as well as pre-analytical processing units.

Modular systems are one type of automation to take into consideration. Other types of automation currently present in the laboratory or planned for the future should also be kept in mind if the immunoassay system is to be incorporated. Automation may be a total, partial, or modular laboratory system, either closed or open to instruments from a number of specific manufacturers. A number of immunoassay analyzers are designed for use with specific systems while others are designed to be flexible in nature allowing the instrument to sample directly from any track system. Individual systems that allow stat specimens to be added at an additional entry point, and those that can operate independently from the automation system, are advantageous compared to those with inclusive track systems that are unusable when transport systems are inoperable.

TEST MENU CONSIDERATIONS

Test menu may in many cases be the most influential factor in choosing an immunoassay system. Again, the goal of the system will dictate required menu. If the goal is consolidation, assays in a large number of categories are available although no analyzer has a complete menu and the depth in each category may also be an issue. Categories currently include thyroid, fertility, cardiac, anemia, tumor markers, therapeutic drugs, drugs of abuse, adrenal/pituitary, reproductive, allergy, infectious disease, transplant, bone metabolism, cytokines, and other special proteins. Assays available worldwide typically outnumber those available in the United States due to regulatory requirements by the FDA for assays such as tumor markers and infectious disease tests. If a large number of platforms are consolidated resulting in an expanded menu, the number of tests on board and the quantity of reagents on board for each test should be examined carefully.

Menu is also of utmost importance if the function of the analyzer is to provide specific types of testing such as specialty endocrinology testing, hepatitis testing, tumor marker testing, etc. Specific testing can also be related to location such as those analyzers located in emergency departments or specialized laboratories such as those supporting IVF programs. There are also certain tests only available on one platform thereby limiting analyzer selection.

TECHNICAL AND CLINICAL CONSIDERATIONS

Technical issues such as precision, accuracy, sensitivity, and dynamic range should be considered when choosing an immunoassay analyzer. Automation has allowed for greatly improved precision of steps such as pipetting, washing, separating, and measuring such that specimens can be analyzed in singlicate as opposed to manual assays requiring duplicate analyses. Intra- and inter-assay precision, particularly at important medical decision points, and other technical parameters should be evaluated carefully when considering a system. Examining results from proficiency surveys can be useful to determine intra- and inter-method precision as well as compare absolute values to determine whether laboratory reference ranges may be affected and need adjustment. A contributing factor to inter-assay precision, which should be minimized, is lot-to-lot variability in reagents. Sensitivity and linear range can be affected by the type of label and detection method as well as by assay design, with improvements observed with chemiluminescent signals compared to older colorimetric and fluorometric detection methods. Accuracy of results has also improved with automation due to built-in quality checks such as sensors for malfunctions. There should be checks associated with pipetting of specimens and reagents and with timing of incubation and reading steps. Appropriate reagent addition as well as positive sample identification, including controls and calibrators, can be assured with barcoding. Specimen carry-over is a concern with many immunoassays that have results spanning a wide range. Therefore, some systems have taken the approach of using disposable tips for specimen pipetting. Some other accuracy-associated system features include clot and short sample detection, and autodilution and autorepeat capabilities.

In order to be useful clinically, assay results need to be accurate and precise, and be provided in a timely manner. Turnaround time is dependent upon how often the test is performed and how long it takes to complete the test. Random access allows tests to be performed upon receipt at the analyzer with testing time dependent upon incubation time, method of separation for heterogeneous immunoassays, and type of detection. Instrument throughput depends upon the time to first result, the time between results, number of assays on board the instrument as well as incubation time. Incubation times are fixed in some systems while variable in others. Specific test mix will greatly influence the throughput for systems with varying incubation times and therefore the optimal throughput may not be attainable. Other system features such as wide dynamic ranges (reducing dilutions required), use of primary and secondary tubes of different sizes, reflex testing capabilities, and barcoding can also increase testing efficiency. A final clinical consideration is specimen volume. Required assay volumes and dead volumes should be as small as possible.

FINANCIAL CONSIDERATIONS

There are a number of financial aspects to assess when considering the choice of an immunoassay analyzer. It must be determined if implementation of a specific analyzer will be cost effective. The total cost of testing should be considered. The following costs should be compared among systems: instrument, reagents, calibrators, controls, disposables, labor, service contract, maintenance and overhead, renovations required, and LIS-associated tasks such as interfacing. Features affecting cost effectiveness include long calibration stability (30–60 days), decreased maintenance requirements and training from instrument consolidation, decreased specimen processing, such as sample splitting or aliquotting, etc. Although difficult to quantify, cost savings resulting from the impact on patient care, such as reduced length of stay, and improvements in communication and reporting of results, should be included. Instrument acquisition may be accomplished through direct instrument purchase, instrument leasing, reagent rental contract, or on a cost per reportable result basis.

OPERATIONAL CONSIDERATIONS

Operational issues associated with immunoassay analyzers encompass a number of aspects including instrument, human, and manufacturer issues. The instrument itself should have features that allow for continued reliable operation. This includes a system that is easy to operate and requires minimal maintenance and service. Maintenance that is performed automatically by the system, including automatic startup and shutdown as well as automatic calibration, and the ability to add specimens and reagents, and disposal of waste without interruption are desirable features. Efficiency of operation is gained with automatic inventory of reagents, multiple stored calibrations, long-term calibration and reagent stability, and liquid reagents or lyophilized reagents reconstituted on board. Walk-away ability is realized with adequate on-board test availability and reagent capacity, on-board refrigeration or room temperature stable reagents, and a large reserve of disposables such as tips and reaction vessels.

Other operational issues related to the instrument include compatibility of the system with the laboratory LIS system and ability to easily interface the instrument. Barcoding in combination with a bidirectional interface operating in host query mode is a key feature for large laboratories.

Another aspect of the reliable operation of the analyzer is the inclusion of mechanisms to deal with system malfunctions. The system should alert the operator with audible prompts when there is a system failure, and provide error messages and troubleshooting assistance. The trend is for more operator involvement in system repair with help manuals and video clips either on-line or on CD-ROM or DVD. Modem connections to the manufacturer as well as video cameras for real time analysis can aid in troubleshooting. Error and system logs as well as instrument settings can be sent to the manufacturer via modem while messages and software upgrades can be automatically downloaded to the system. Help from the manufacturer should also be provided through a telephone hotline accessible 24 h a day as well as through on-site service. The response time for on-site technical service is a fundamental question to be asked when selecting a system. The availability of local technical service personnel may be a critical deciding factor for laboratories located in rural or remote areas. Even in locations with a local service representative, territories covered may be very large precluding a prompt response. Acceptable downtime, if any, must be determined. A back-up instrument may be required.

Human issues should also be considered when choosing an automated instrument. The system should be designed with the safety of the operator in mind with respect to mechanical aspects such as moving parts and exposure to infectious materials from both sampling and waste. The system should be easy and intuitive to operate with ease of training. Software should be user-friendly with a graphical user interface and on-line help and system manuals.

A final operational issue relates to instrument manufacturers and instrument placements. A consideration for systems on the market is the number of systems placed and track record of those systems. For systems in the market for several years, inquiries should be made as to the new systems being developed and time until launch. If a new system is imminent it may make sense to consider the new system or include trade-ins or upgrades in contracts if acquiring current systems. Another factor to consider is the number of immunoassay systems offered by a particular vendor. Recent mergers and acquisitions have resulted in companies with several product lines. It is clear that in the long term all these systems cannot be supported.

SUMMARY

This chapter has discussed the criteria for selecting an automated immunoassay system. It must be emphasized that choosing an immunoassay system is dependent upon the needs and requirements of the individual laboratory. Therefore it is necessary to first define the goals of the system and how those goals fit into the overall laboratory plans and then to determine which instrument features will help to achieve those goals. It is then possible to choose a system(s) with the required features. Once a system has been short-listed a thorough evaluation should be carried out to ensure that the system meets the specifications claimed by the manufacturer as well as the expectations of the laboratory.

System evaluation should include technical, clinical, operational, and economic elements. Briefly, a technical evaluation would assess within- and between-run precision, sensitivity, linearity, accuracy (including evaluation of recovery), dilutions, interferences, carry-over, calibration stability, lot-to-lot variability, and method comparisons. A clinical evaluation would assess the diagnostic accuracy of the methods with an emphasis on reference values and disease management issues. An operational evaluation would assess system operation including data management and interface capabilities, technical service issues, and an assessment of throughput capabilities from simulation studies with the expected test mix. Finally an economic evaluation should be performed to determine the system cost as well as the total cost of testing including pre- and post-analytical steps to determine the effects on labor requirements and productivity.

FURTHER READING

Auxter, S. What you should know before buying a lab instrument. *Clin. Lab. News* **24**, 16–47 (1998).

Blick, K.E. Specifications for the selection of automated immunoassay (IA) systems. *J. Clin. Ligand Assay* **19**, 220–228 (1996).

Blick, K.E. Current trends in automation of immunoassays. *J. Clin. Ligand Assay* **22**, 6–12 (1999).

Blick, K.E. Automated immunoassay selection. *Advance/Laboratory* **10**, 41–46 (2001).

Blick, K.E. Refinements in AIA technology. *Advance/Laboratory* **12**, 84–90 (2003).

Bock, J.L. The new era of automated immunoassay. *Am. J. Clin. Pathol.* **113**, 628–646 (2000).

Chan, D.W. (ed), *Immunoassay Automation: A Practical Guide* (Academic Press, San Diego, CA, 1992).

Chan, D.W. (ed), *Immunoassay Automation: An Updated Guide to Systems* (Academic Press, San Diego, CA, 1996).

Cook, T.M. Defining "best fit" features. *Advance/Laboratory* **7**, 24–28 (1998).

Felder, R.A. Immunoassay automation. *J. Clin. Ligand Assay* **22**, 13–24 (1999).

Li, D.J., Sokoll, L.J. and Chan, D.W. Automated chemiluminescent analyzers. *J. Clin. Ligand Assay* **21**, 377–385 (1999).

Pearlman, E.S., Swiss, S., Stauffer, J. and Bilello, L. Selection factors in the choice of immunoassay technology: implementing technology in the new era. *Clin. Lab. Man. Rev.* **12**, 27–30 (1998).

Phillips, J.E. Emerging trends in immunodiagnostics. *Advance/Laboratory* **12**, 46–50 (2003).

Sokoll, L.J. and Chan, D.W. Clinical instrumentation (immunoassay analyzers). *Anal. Chem.* **71**, 356–362 (1999).

Wheeler, M.J. Automated immunoassay analysers. *Ann. Clin. Biochem.* **38**, 217–229 (2001).

21 Automated System Features

Alain Truchaud, Tanguy Le Neel, Murielle Cazaubiel, Bernard Capolaghi and Jean-Pierre Yvert

INTRODUCTION

The field of immunoassay has shown enormous growth in the last two decades. This has caused laboratory workloads to increase at a time of diminishing resources, hence the importance of automation. The main requirements of laboratories performing clinical tests may be summarized as:

- user-friendly analyzer interface
- high test throughput
- large panel of analytes
- low detection limits
- sampling from primary tubes
- good precision
- absence of carryover
- infrequent calibration
- simultaneous multiselective access (**random access**)
- biosafety
- integrated **Total Quality Management** (**TQM**)
- continuous operation
- compatibility with **Total Laboratory Automation** (**TLA**) (Boyd *et al.*, 1996).

These requirements are adequately met by state-of-the-art chemistry analyzers. However, additional features are needed for immunoassays. They are described in the following section.

SPECIFIC REQUIREMENTS OF IMMUNOASSAYS

Immunoassays are used to measure or detect analytes with low concentrations, but the analytical range is very wide, thus inducing risks of **carryover** throughout the process. Immunoassay systems must be designed to minimize the risk of sample–sample interference.

Immunoassay analytes are often fragile molecules and require protection against temperature variations; most are more stable when the sample is stored or transported in a cool or frozen condition. **Analyte integrity** must be maintained by immunoassay systems.

Biosafety is a priority for any sample, but particularly for samples from patients with infectious diseases.

Immunoassay reagents, controls, and consumables are expensive. Radioisotopic labels are gradually being replaced by fluorescent and chemiluminescent labels that are more stable, resulting in longer reagent shelf lives. However, the **on-board reagent shelf-life** after opening and the **pack size** are also important to avoid wastage of unused reagent for the less common analytes.

Many automated immunoassay systems are available with a level of automation comparable to that of chemistry analytical systems. However, immunoassay automated systems often perform better in specific fields, such as tumor markers, infectious diseases, or endocrinology. The **clinical performance** of the tests relevant to the laboratory's main clinical customers is still an important consideration for immunoassays.

Some results of immunoassays are required urgently, e.g. '**stat**' tests for organ transplantation; outpatients requiring test results for adjustment of therapy. Most of the results are not necessarily needed urgently, but must be available within known timescales. An analyte that is not normally required urgently may be needed as a stat in specific clinical situations. The ability of a system to deliver results when needed has to be considered.

Laboratories performing immunoassays are merging to contain escalating costs, amplifying the need for efficient sample handling, laboratory automation, and computerized information processing. In busy laboratories, the quality of the results depends on the reliability of the system for sample handling, from blood collection through to the immunoassay test system

The Immunoassay Handbook. *Third Edition*
D. Wild (Ed.)

(Truchaud *et al.*, 1997). **Compatibility** with other laboratory systems and **breadth of menu** are therefore important considerations.

An increasing proportion of samples require long-term storage to comply with regulations, to be available in the case of malpractice lawsuits or other legal situations, and to provide effective long-term monitoring of diseases. These sample libraries can also be valuable for the validation of new analytical methods. It is essential that the immunoassay system does not adulterate the samples if they are sampled directly or subject them to long periods at ambient temperature.

In many cases, immunoassay results are crucial for diagnosis and monitoring patient condition. The interpretation of the results requires extensive knowledge and experience. Immunoassay systems should provide clinical biochemists with sufficient information to allow them to investigate unusual situations.

CONCEPTS IN IMMUNOASSAY AUTOMATION: WORKSTATIONS VS. INTEGRATED SYSTEMS

Most immunoassay systems are based on heterogeneous assay formats, i.e. they involve a separation of bound from unbound material. This increases the complexity of immunoassay analyzers, which are required to process multiple steps without user intervention (Figure 21.1). In the case of homogeneous immunoassays, there is no washing step.

Two automation concepts for immunoanalysis systems have evolved: a series of **workstations**, each responsible for a single step, such as washing, and **integrated systems**, where an automated analyzer is developed around a methodology. Workstations, or 'fully open robots' use microtiter® plates, and operators transfer the

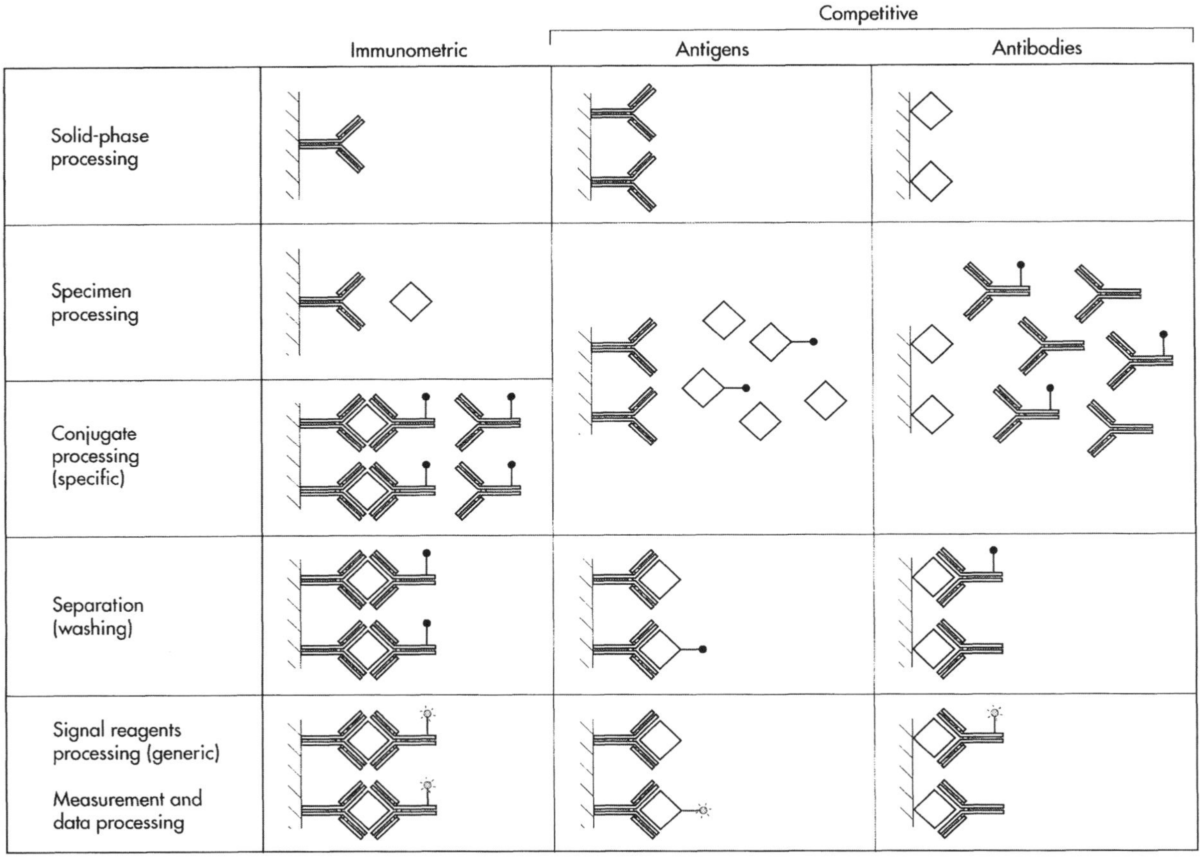

Fig. 21.1 Heterogeneous immunoassays.

plates manually, or using conveyors, through different modules for sample and reagent pipetting, incubation, washing, signal generation, and signal measurement. They are primarily used for high-throughput applications for a single analyte or a fixed panel of analytes, such as for blood bank screening or pharmaceutical research.

As the demands on integrated systems are perhaps the most challenging, we will focus on how they may best fulfill the requirements listed above.

REAGENT FEATURES

INFREQUENT CALIBRATION

To minimize the use of reagents for calibration, calibration curves must be stable. This creates a range of challenges for conjugates, the solid phase in the case of heterogeneous immunoassays, diluents, calibration solutions, signal reagents, and buffers.

The stability of the reagents depends on the nature of the raw materials, especially antibodies, antigens, and enzymes. This immunochemical stability may be improved by careful selection of the storage form, e.g. liquid, solid, suspension, lyophilized powder, or particles. Reagent packaging may be unitary or in multi-test packages.

The packaging plays a very active role in most immunoassay systems. For example many packs carry data for transmission to the analytical system such as the test identification, expiration date, and master calibration curve. Also many packs are responsible for carriage and storage of reagents inside the analyzer, once loaded. Such reagent packs must allow access, e.g. by automatic opening and closing or the use of elastic septums, and sometimes packs dispense the reaction cuvettes. To achieve longer term storage without recalibration, protection against evaporation is required. Cross-contamination of reagents must also be avoided.

TRACEABILITY

To fulfill TQM, users need automated tracking of the reagents to ensure complete traceability. Reagents introduced in the system must be automatically identified and the quantities tracked, and managed to ensure quality. The preservation of the quality of the reagents on board relies on the thermal and humidity control of the reagent inside the reagent tray or carousel. If the reagents are stabilized in a lyophilized form, reagents must be regenerated just before use. A reagent blank is not as easy to perform as in chemistry, and a dye or an internal control must be used to check this step.

MULTISELECTIVE TESTING AND CONTINUOUS OPERATION

Multiselective continuous testing is based on the number of reagents stored on board, and on the ability to load samples, reagents, and consumables while the system is performing assays without a pause.

SAMPLE FEATURES

SAMPLE INTEGRITY

Before loading the samples, especially for homogeneous assays with no separation steps, the optical quality of the sample must be checked (for hemolysis, turbidity, and hyperbilirubinemia). Also, the medical and biological history of the patient should be reviewed to anticipate the likelihood of a hook effect. This, and any risk of carryover, may be avoided by diluting the sample. These steps require a skilled operator and are usually neglected in robotic loading.

Samples need to be protected against evaporation and heat, e.g. by cooling the sample track or carousel to around 15 °C. Contamination of samples may be avoided by sampling from closed tubes.

In sample metering, there are two basic options: metering probes require a sophisticated washing system, generating extra liquid waste, but allowing sampling in primary tubes through capped tubes. Alternatively, the use of disposable tips avoids the risk of carryover or contamination of the original sample, but requires more self-checking and increases the amount of solid waste.

With the advent of robotic conveyors that have uncapping stations, the choice of an immunoassay system using disposable tips is the only way to really secure the sample from carryover if the user wants to perform immunoassays with a low detection limit or screen for infectious diseases.

The metering system must also be able to detect if there is a sample in the tube and control the entire pipetting and delivery process, detecting clots, bubbles, etc. by a pressure sensor.

BIOSAFETY

Concerning biosafety, the sample loading area must be screened to prevent any risk of mechanical injuries (Truchaud *et al.*, 1991). Biohazardous waste should be minimal, contained, and clearly identified.

DILUTION AND REFLEX TESTING

To decrease sample handling, if the result is out of the assay working range, dilutions should be performed automatically inside the system. Reflex testing may also include alternative analytes, according to retest algorithms. However, for immunoassay systems that have continuous sample loading and unloading, this feature is practical only if the system has the capability to transfer a sample or a rack from the unloading area back into the analytical area.

STAT TESTING

Stat immunoassays are now a requirement in many laboratories not only to support surgery, but also for outpatients needing a result that could result in a change in therapy. They can be handled in two ways: they can be manually loaded or presented on the front of the analyzer using a specific stat operating procedure, or, in TLA systems, after automatic or manual sample processing, the stat samples can be placed on the automation track just before the immunoassay system.

The loading of stat samples should not disturb the ongoing run of assays.

INCUBATION AND WASHING

To achieve a high test throughput within a compact footprint for the analyzer, it is necessary to minimize the duration of the incubation by optimizing the kinetics of the immunochemistry, using an elevated temperature and applying mixing or agitation.

The first generation of immunoassay systems often used so-called 'marching' incubation systems with a fixed incubation time. This allowed the automation to be kept simple and provided standardization of the analytical process, but the menu was limited. Subsequent systems have included flexible incubation times, with the option of one- or two-step protocols, to enable optimization of each immunoassay. This requires complex software for operations control.

In heterogeneous immunoassays, automation includes a separation step involving a solid phase (Truchaud *et al.*, 1991).

MEASUREMENT

Fluorescent and luminescent labels are now widely used in immunoassay systems; they provide not only a good signal for low concentrations but also a very wide linear range (Kricka, 1994).

In order to obtain reproducible results and low detection limits, weak signals must be measured very precisely, requiring accurate orientation of the reaction mixture emitting the signal against the detector. This requirement is less critical when an internal standard is used (Mathis, 1993).

TOTAL QUALITY MANAGEMENT/TRACEABILITY

The system needs to provide full traceability of calibration, quality controls, and sample results. Since some systems allow use of the same memorized calibration curve for several weeks, and, in random-access mode, quality controls are inevitably run less often than patient samples, it is crucial that each patient result can be associated with the corresponding calibration curve and quality control in order to assess the quality of the result.

MAINTENANCE

The classical definition of maintenance is to maintain or restore an instrument to a specified state where it is able to deliver a given performance (Truchaud *et al.*, 1998).

This definition implies that several requirements are fully understood. We recommend the following:

- Before purchasing the immunoassay system, the laboratory supervisors have defined the required level of performance from the medical needs.
- At the installation of a new system in the laboratory, key aspects of performances are checked and various indicators, including quality controls, are established to provide a benchmark against which performance can be assessed throughout the life of the instrument. Examples of indicators other than quality controls are precision tests, carryover tests, control of the optoelectronics, and control of the temperature of incubation.
- Periodically, but also after any instrument breakdown, the user must be able to check the performance by using the indicators.
- For any patient result, the user must be able to prove that the instrument has been properly maintained. The WINDOWS™-based maintenance tracking systems used in many immunoassay analyzers greatly assist traceability.

ERGONOMICS

Many improvements have been made to improve the ergonomics of immunoassay automated systems, for example:

- clear user interface, e.g. based on WINDOWS NT™, with attractive, colored icons to identify types of function and lead the user to the required information or control button; Quality Control diagrams and calibration curves that are easy to interpret, user manuals integrated in the software, etc.;
- safe loading and unloading, with continuous access to the system;

- quiet operation;
- improved user safety by limiting access to the robotic area while the instrument is performing assays.

INTEGRATED AUTOMATION

There are several different approaches to the integration of immunoassay systems into the macro-automation of the laboratory:

- Integrated modular systems: different analytical systems that have been combined into one analyzer, e.g. for clinical chemistry and immunoassays. Multiple tests are carried out using the same sample tube to avoid sample aliquoting. These integrated systems are very fast but are closed 'macro-analytical' systems.
- Linked modular systems, where a conveyor of tubes is linked to analytical systems from the same or different manufacturer by a conveyor. This approach is slower, consumes more space, but is more open. In this approach, electromechanical and computer interfaces are necessary to link with the conveyor system and the process control module for workload scheduling and possible retesting.
- Fully automated laboratory, in which a pre-analytical phase (sample sorting, centrifugation, and aliquoting) is included as a series of workstations at the beginning of a conveyor connected to the analytical systems. Alternatively, these steps can be performed in a sample processing 'island' of automation separated from analytical islands for the various tests.

CONCLUSION

Devising a suitable strategy for immunoassay automation requires a multi-disciplinary approach:

- Review the immunoassay methods and systems available; derive a bibliography of independent evaluations.
- Model the workload in one or several immunoassay work cells based on the existing equipment, monitor the level of performance and identify the limitations.
- Discuss the situation with the different teams affected and derive a consensus.
- Make a shortlist of immunoassay systems and carry out an objective, comparative evaluation based on the criteria described in this chapter.
- Derive different potential configurations including various instruments, computers (for technical and medical validation), and options for integration with existing or future TLA systems.
- Negotiate with manufacturers, and explore bundled options that include immunoassay systems, interfaces with TLA, maintenance, reagents, and consumables. The proposals should guarantee a level of performance corresponding to the laboratory's requirements and, if more than one manufacturer can meet those requirements, the system that provides the cheapest overall cost per test should be selected.

REFERENCES AND FURTHER READING

Boyd, J.C., Felder, R.A. and Savory, J. Robotics and the changing face of the clinical laboratory. *Clin. Chem.* **42**, 1901–1910 (1996).

Kricka, L. Selected strategies for improving sensitivity and reliability of immunoassays. *Clin. Chem.* **40**, 347–357 (1994).

Mathis, G. Rare earth cryptates and homogeneous fluoroimmunoassays with human sera. *Clin. Chem.* **39**, 1953–1959 (1993).

Truchaud, A., Barclay, J., Yvert, J.P. *et al.* Automated separation for heterogeneous immunoassays. *J. Autom. Chem.* **13**, 49–51 (1991).

Truchaud, A., Schnipelsky, P., Pardue, H.L. *et al.* Increasing the biosafety of analytical systems in the clinical laboratory. *Clin. Chim. Acta* **226**, 5–13 (1994).

Truchaud, A., Le Neel, T., Brochard, H. *et al.* New tools for laboratory design and management. *Clin. Chem.* **43**, 1709–1715 (1997).

Truchaud, A., Gouget, B., Gourmelin, Y. *et al.* Maintenance of biomedical equipment. *RBM* **11**, 111–114 (1998).

Wheeler, M.J. Automated immunoassay analysers. *Ann. Clin. Biochem.* **38**, 217–229 (2001).

22 Over-the-Counter Pregnancy Test Kits

Michael J. Wheeler

INTRODUCTION

Although pregnancy testing, especially home testing, might seem a relatively new development, it was in existence hundreds of years ago. An Egyptian papyrus described how a woman should pass her urine over the great Nile plant to determine if she was pregnant. If the plant was scorched the next morning, she was not pregnant but, if the plant continued to thrive, she was pregnant. Other Egyptian and Greek texts describe how women may determine the sex of the unborn child by methods considerably less complex and invasive than procedures practiced today! It was not until the 20th century that biological tests were developed, based on the presence of a 'pregnancy substance' in blood and urine. These tests used mice, rabbits, and toads, when control of the assay had a different connotation! The first immunological test was developed by Wide and Gemzell (1960). This was a hemagglutination-inhibition test that had a sensitivity of about 200 IU/L. The test took just over 90 min to complete. Early tests were quite complex, involved a number of steps, and were developed for laboratory use. These methods were gradually refined and developed so that they could be used for self-testing in the home. And so **over-the-counter (OTC) pregnancy tests** were developed.

Women from a wide age range and from all walks of life use OTC kits and these may be obtained from pharmacies, drug stores, and supermarkets. In the UK, it is known that pharmacists who offer a local pregnancy testing service also use these kits. However, since the pharmacist's choice of kit may be dictated more by an imminent expiry date than a knowledge of its reliability, the kit used by a pharmacist may vary from one month to another. In view of the above, it is critical that kits have similar analytical sensitivities, have simple instructions and are easy to use.

Twenty years ago this was far from true. Kits were fairly complex and some took a long time to complete. One of the early home methods used a column of concanavalin A Sepharose on to which the woman poured 3 mL urine (Shoham *et al.*, 1987). A blue dye reagent containing anti-hCG antibody was added and washed through the column with 2 mL buffer. A positive result was indicated by retention of the blue dye in the column. The test took only 10 min and women who evaluated the test found it easy to use and interpret. This laboratory carried out an evaluation of nine OTC kits in 1988 for the UK Department of Health. The number of steps in each kit varied from 2 to 6 with incubation times ranging from 5 to 90 min (Table 22.1). The incidence of false positives ranged from 2 to 8% and false negatives from 0 to 2%. One urine with a hCG concentration of 3000 IU/L gave a false negative result in four kits. The sensitivity of these kits ranged from about 100 to 200 IU/L. Non-pregnancy urines containing high concentrations of protein or hemoglobin gave positive results with some kits. Therefore, 17 years ago the sensitivity of the kits was not sufficient to detect pregnancy at the time of the first missed period, in many urines.

At about the same time, others also found that OTC kits were not very reliable. Latman and Bruot (1989) evaluated nine OTC pregnancy test kits and found a wide variation in accuracy, sensitivity, and ease of operation. Kits were found to have a high rate of false negatives (24.3%) which rose to 40% in those women who sought prenatal care in public clinics (Lee and Hart, 1990). After finding that lay persons, using OTC kits, could have up to a 12.5% inaccuracy in their results, Hicks and Iosofsohn (1989) queried "whether home pregnancy test kits should be on the market".

Nevertheless, the usage of these kits is high and a great deal of faith is put in them by both women and doctors. In a study of one UK health district, it was found that about seven pregnancy tests were performed for each live birth and about 44 pregnancy tests were performed for every 100 women of childbearing age each year. This was equivalent to 4.8 million pregnancy tests per annum in England and Wales (Voss, 1992). A study in the US of 438

The Immunoassay Handbook. *Third Edition*
D. Wild (Ed.)

Table 22.1. Details of OTC kits evaluated in 1988.

Kit	Number of steps	Incubation time (min)	Endpoint
Boots	2	60	Hemagglutination
ClearBlue	5	10 + 10 + 10	Blue color
Confirm	2	90	Hemagglutination
Discover Color	4	15 + 15	Blue color
Discover 2	2	60	Hemagglutination
Evatest Rapid	2	30	Hemagglutination
Evatest Blue	6	2 + 1 + 2	Blue color
Predictor Color	2	30	Loss of color
Predictor Test	4	30	Pink color

women entering a family planning clinic in the fall of 1987, found that nearly 40% of the respondents had used an OTC test at least once (Coons, 1989).

EVALUATION OF OTC KITS

The above evaluations took place about 17 years ago and no detailed evaluation has been carried out over recent years. Although some brand names have not changed over the last 17 years (cf Tables 22.1 and 22.2), the devices have. Because of this and the large commercial market for these products, we evaluated all the OTC kits available in the UK in 1997 (Foxon *et al.*, 1997). The sensitivity and specificity of the kits were examined in the laboratory, but they were also tested by 100 lay people. The kits evaluated are listed in Table 22.2 and shown in Figure 22.1.

The kits could be divided into two types of testing device: (a) a dipstick which may be placed either into the urine stream or into urine collected into a container and (b) a flat 'cassette' device which contains a wick sandwiched between two plastic layers. Windows in the top of the device allow urine to be added to the wick through one window and for results to be read in another window. A fuller description of examples of these devices is given later in this section of the book. An early kit of Early Bird™ had a sponge disc through which a wick was inserted. The sponge disc was floated on top of urine collected in a container. Although the wick was rather flimsy and the device fiddly, some women liked it for its novelty value. This kit has now been replaced with a stick test.

In the scientific part of the evaluation 200 urines were randomly collected from our routine pregnancy testing service. The concentration of hCG in each urine was

Table 22.2. Details of OTC kits evaluated in 1998.

Kit	Type	Incubation time (min)	Current format
Biocare One Step Midstream Pregnancy Test	Stick	5	Stick
Boots Home Pregnancy Test	Slide*	5	Stick
Cameo One-Step Home Pregnancy Test	Slide	5	NA
ClearBlue One-Step Pregnancy Test	Stick	3	Stick
Discover 2 Pregnancy Test	Stick	2	Stick
Early bird One-Step Pregnancy Test	Stick		Stick
First Response 1 Step Early Pregnancy Test	Stick	2	Stick
First Signal S Early Pregnancy Test	Stick	3	Stick
Lloyds Home Pregnancy Test	Slide	2–4	NA
Predictor Home Pregnancy Test	Stick	4	Stick
Unichem Home Pregnancy Test	Slide	3–5	Slide*

NA, No longer available.
*Now modified from the product evaluated.

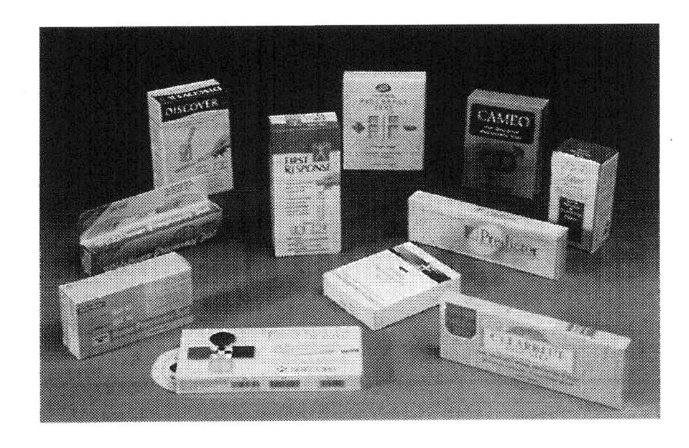

Fig. 22.1 Over-the-counter pregnancy test kits examined in an evaluation in 1997.

measured using the MAIAclone™ kit (Bio-stat Ltd, Stockport, Cheshire, UK) before testing with each OTC kit. All the kits had a clinical sensitivity and specificity of 100% in this part of the study. Urines with high concentrations of protein and hemoglobin, and urine from men and postmenopausal women, were tested to see if any false negatives were obtained. No false negatives were observed. The analytical sensitivity again was assessed by testing urines which previously had the hCG concentration measured by the MAIAclone kit. Sensitivity was also assessed by testing pregnancy urines, which had been serially diluted with non-pregnancy urine, and the hCG concentration measured. Finally, it was assessed by adding known amounts of the hCG reference preparation IS 75/537 to non-pregnancy urine (Table 22.3). The Discover 2 kit was particularly sensitive but no false negatives were obtained with this kit. All the kits had an analytical sensitivity of at least 50 IU/L and would be expected to detect a pregnancy at the time of the first missed period. We found a sensitivity that was better than that quoted by the manufacturer for Boots, ClearBlue™, Discover 2™, First Response™, and Predictor™, but we found First Signal S™ and Unichem™ to be less sensitive. The mean analytical sensitivities calculated from the results for the diluted urines generally were similar to the results for the spiked urines. The least detectable concentrations determined from serially diluted urines varied up to sevenfold between different urines. For instance for Predictor, this kit was able to detect 12 IU/L in one urine but only 87 IU/L in another urine although all the urines were diluted down to about 4 IU/L with the same non-pregnancy urine. Similar findings were found by Alfthan *et al.* (1993) in their evaluation of laboratory-based kits.

USER PREFERENCES

This laboratory also carried out a study to see which kits women preferred based on convenience, usability, clarity of instructions, and packaging. Five groups of 20 women were chosen to take part in the study:

- unskilled women employed as porters and domestic staff;
- secretarial staff and clerks;
- women in managerial positions, e.g. banks, publishing houses;
- university undergraduates; and
- high-school women aged 15–19 years.

Each woman was given two tests of each kit and asked to carry out the test on themselves following the instructions in the package. The second test was a spare in case they encountered problems with the first test. At the end of each test the women were also given a questionnaire to fill in. This asked about the clarity

Table 22.3 Sensitivity of each kit determined from a spiked urine and five diluted urines and compared with the manufacturer's quoted sensitivity.

Kit	Diluted urines	Spiked urine	Manufacturer
Biocare	18 ± 9 (9–30)	10	NS
Boots			
Cameo	48 ± 5 (44–53)	50	NS
ClearBlue			
Discover 2	10 ± 8 (5–24)	5	50
Early Bird			
First Response	21 ± 14 (9–44)	25	50
First Signal S	30 ± 8 (23–44)	50	25
Lloyds	46 ± 12 (24–53)	50	NS
Predictor	35 ± 30 (12–87)	25	50
Unichem	39 ± 12 (24–53)	50	25

NS, not specified.

of instructions and ease of use, and encouraged the women to share any particular comments or criticisms. When each woman had completed all the tests, they were visited by one of the investigators. At this visit, three urines were tested in front of them and they were asked to say whether the test result for each urine specimen was positive or negative. The three urines had been prepared so that one was negative, one positive, and the third adjusted to a concentration at the limit of sensitivity of the kit to give a 'borderline' result. The sensitivity of each kit had been predetermined in the laboratory. The three urines were tested in random order. At the end of the tests the women were asked which type of test they preferred, and which kit they would buy from a store if they had only the packaging from which to choose.

The user survey produced clear preferences and dislikes. The main criticisms are given in Table 22.4 with an overall preference for stick tests. These were found to be more convenient to use, and since they all had caps, it was suggested they could be placed in a handbag after use for either later disposal or for showing to the GP. However, some manufacturers do not recommend keeping the tests to show to a GP. This may be because some negative tests become positive with time, whereas positive tests might fade. Women preferred to place the dipstick into the urine stream rather than collect urine into a container. This was probably because no container was found to be ideal; the containers were usually found to be too small and were criticized for being messy and unhygienic. The Biocare kit did not contain a pot or tray to collect urine although it was suggested the test could be used either by placing directly into the urine stream or into a container of collected urine. The Cameo kit, which required urine to be collected before the test could be used, did not contain a collection vessel either. The Discover 2 kit only contained one collection container, although the kit had two tests. The users gave anecdotal reports of tests being carried out in toilets at work or college and, in such situations, a dipstick would be more convenient, due to the lack of a convenient flat surface for the use of a slide test, and the problem of disposal afterwards.

When the evaluators carried out tests in the users' homes, light conditions varied widely and it was found that the lighting conditions influenced the ease with which tests could be read. In poor light conditions faint positives were more likely to be scored negative, but more likely to be scored positive under bright lighting conditions. The positive result line in some tests was not very intense, so those tests could be difficult to read in poorly lit rooms.

The criticisms listed in Table 22.4 were common across the groups of women. The instructions in many kits were criticized for either being too lengthy or unnecessarily scientific. Women were not interested to know that antibodies in the tests had been produced in goats! Only First Signal S gave the instructions in more than one language but, as these were presented on a folded sheet, the women found it difficult to find the English version.

Most tests worked well and urines with concentrations above 50 IU/L gave a clear result line. The color that developed with the ClearBlue test was sometimes faint at the suggested reading time of 3 min, but became clearer after 4–5 min. The most difficult test to read was the Cameo test. During the test an intense blue color appeared at one end of the result window and moved across it. Often this color had not disappeared after 5 min when the test should be read. This led to many women either wrongly interpreting the test or repeating the test. Nearly every woman criticized this test because of this finding.

When the women were asked to choose a kit based on the packaging, the results were very interesting. Choices fell into two broad categories. The younger women chose the most attractive looking kit, with the Cameo kit as first choice in most cases. The packaging on this kit was a multicolored, pastel purple, and blue. The women also

Table 22.4. User comments about the kits they tested.

Kit	Criticism
Biocare	Instructions, lengthy, only one test, no urine container
Boots*	Collection tray, packaging
Cameo	Instructions, no urine container, test unclear, dropper
ClearBlue*	None
Discover 2	Urine container, only one container for two tests
First response*	Instructions, test result unclear
First signal S	Instructions
Lloyds	Test messy but generally positive comments. This was the only kit that had a urine collection container that users felt was of adequate size
Predictor	Test procedure
Unichem*	The size of the urine container seemed more ideal although some women found it too small and messy

*Modified since survey.

Table 22.5. The ideal OTC kit derived from the feedback from 100 women.

- Attractive eye-catching packaging
- Any outer cellophane covering should have an obvious, easy-open strip around it
- The instructions should be brief, clear, and unequivocal
- The test should be a dipstick device which can be placed in the urine stream
- The test device should have a cap that can be placed over the device when the wick has been wetted
- A collection tray, measuring about 10 cm × 5 cm and 2 cm deep, should be provided for those women who prefer to collect urine first and then place the device into the urine
- The result window and control windows should be separate and clearly labeled
- The colored band that appears for a positive result should show as a dark band at concentrations of hCG above 50 IU/L
- Two tests should be provided
- The test should have a sensitivity of 25 IU/L
- The antibodies should have minimal cross-reaction with β-core hCG, nicked hCG, or nicked β-hCG

thought that the device, which had a multicolored paper top, was attractive. The older women did not choose the Cameo kit at all but chose a kit either because of the reputation of the company (Boots) or because they were familiar with the name of the kit through advertising (ClearBlue and Predictor). Choice of the kit was not influenced by the way the test performed because all the kits had been tested and those kits preferred during testing were not necessarily first choice based on packaging. Certainly no woman liked the Cameo kit in use. Women were also critical of the main packaging of some of the kits, especially those that were covered in cellophane. Some were provided with an easily seen strip to pull, others either had nothing, so that the women needed sharp nails or scissors to pierce the cellophane, or the strip was not obvious.

Table 22.5 gives a description of the **ideal OTC kit** determined from the comments of the women in our study.

FACTORS AFFECTING THE MEASUREMENT OF hCG IN URINE

Our study showed that OTC kits currently available in the UK have high sensitivity and would detect pregnancy in the majority of women at the time of the first missed period. Although most kits showed a sensitivity close to the sensitivity stated by the manufacturer, there was variation, especially when sensitivity was assessed using diluted urines, but also with the spiked urine. The Discover 2 kit was found to be considerably more sensitive than other kits, although the manufacturers quoted a sensitivity of 50 IU/L. This kit was able to detect a concentration of hCG as low as 10 IU/L and might be suspected of detecting hCG in non-pregnant women. Nevertheless, we found no false positives in our study. Using diluted pregnancy urine, we found that the lowest concentration that could be detected varied by as much as fourfold. This was similar to the findings of Alfthan *et al.* (1993) for laboratory-based kits. For example, the First Response kit could detect hCG as low as 9 IU/L in one urine but only 44 IU/L in another urine. Stenman *et al.* (1991) and Alfthan *et al.* (1993) suggest that the pH and salt concentrations may affect the measurement of hCG although there appears to be no published data to support these statements. However, the preparation of the dilutions could affect the salt concentration. It is usual to ask women to collect the first urine passed in the morning; this is more concentrated than collections during the day. Our dilutions were made with a random non-pregnancy urine, collected during the day, which would have been less concentrated. If salt concentration does have an effect, then the time of day a urine is collected or, in our dilution experiments, the salt concentration of the urine used for dilution, could affect the sensitivity of the kits. An investigation of the variation in sensitivity observed with different urines could not be assessed with a number of urine specimens with added IRP as there was only enough reference preparation for one set of spiked samples prepared from one urine specimen. Manufacturers do not give details of how they establish the sensitivity of their kits, but it is likely that their experiments would be similar to ours.

Variation, both between kits and between urine samples, could be a result of differing cross-reactions with the different forms of hCG found in urine. β-core fragment is the principle degradation product in urine. By 6 weeks gestation, its concentration is similar to the total hCG concentration. Smaller amounts of α- and β-subunits as well as nicked hCG also occur in urine. Although the concentrations are lower than β-core hCG, concentrations of each of these products may reach at least 40% of the total hCG concentration (Cole *et al.*, 1993, 1997). These workers also noted that there was considerable variation in the amount of these products in different urines. Therefore, cross-reaction of the antibodies with any of the products could lead to variable sensitivities. Cole (1997) reported that the quantitative hCG kits have been designed to measure nicked hCG and non-nicked hCG, or free β-subunit. Details of the specificity of OTC kits is lacking but Alfthan *et al.* (1993)

found that of 10 laboratory-based qualitative hCG kits, five detected β-subunit and one detected β-core fragment. A similar study is required for OTC kits.

SUMMARY

When we first started this study, four methods used a flat plastic slide device, six were dipsticks, and one used a sponge disc that held a detection strip and floated on the collected urine. Several methods have now changed so that of the original 11 kits evaluated only nine remain available with the Cameo and Lloyds kits withdrawn. All but one of the remaining kits employ a dipstick device which reflects the preference we found for our users. From our conversations with our women volunteers, testing could take place at home, work, or school. The stick tests, therefore, were found to be more convenient since they did not need a flat surface on which to develop the test, and the wet end could be capped. The device could then be placed in a bag. Because location varied so did the lighting conditions and we found that this affected how easily faint positives could be read. In poor light, faint positives were more frequently reported as negative.

Packaging was only an important factor with the young women at school or college. In this study, the kit that was the first choice by this group of women (Cameo) proved to be the least liked in operation by all women. The older women, who chose kits according to advertising, chose kits that performed well in our evaluation, i.e. Predictor and ClearBlue. We also found the new Boots and First Response stick tests easy to use; both are easy to read and have a high-sensitivity. OTC pregnancy kits have now almost achieved the optimum specifications based on the preferences of our users. There is no need to improve sensitivity below 25 IU/L since this should be low enough to detect conception at the time of the first missed period but high enough not to result in false positives. Nevertheless, these very sensitive kits do have the risk that they will detect more pregnancies that end with an early spontaneous abortion. OTC pregnancy kits are now more sensitive than some tests used by GPs and hospital laboratories. Some of these latter tests have a sensitivity of 1000 IU/L and so it may be more than 10 days after the first missed period before one of these kits shows a positive result in a urine from a woman testing positive with an OTC kit. It is not known how widely this problem is appreciated. It would be wise for any GP who fails to confirm a positive OTC result reported by the patient, to either repeat his own test in about 2 weeks time, or to send a blood sample to the hospital laboratory. As most OTC kits have two tests, the patient could be asked to repeat the test with the second test from her kit. In these days of litigation, and given that many of the pregnancy test kits do not claim 100% sensitivity and specificity, these seem wise and inexpensive precautions to take.

REFERENCES AND FURTHER READING

Alfthan, H., Björses, U.-M., Tiitinen, A. and Stenman, U.-H. Specificity and detection limit of ten pregnancy tests. *Scand. J. Clin. Lab. Investig.* **53** (Suppl 210), 105–113 (1993).

Chard, T. Pregnancy tests: a review. *Hum. Reprod.* **7**, 701–710 (1992).

Cole, L. Immunoassay of human chorionic gonadotrophin, its free subunits, and its metabolites. *Clin. Chem.* **43**, 2233–2243 (1997).

Cole, L.A., Seifer, D.B., Kardana, A. and Braunstein, G.D. Selecting human chorionic gonadotrophin immunoassays: consideration of cross-reacting molecules in first trimester pregnancy serum and urine. *Am. J. Obstet. Gynecol.* **168**, 1580–1586 (1993).

Coons, S.J. A look at the purchase and use of home pregnancy-test kits. *Am. Pharm.* **NS29**, 46–48 (1989).

Foxon, L., Angel, P., D'Souza, A. and Wheeler, M.J. Evaluation of eleven over-the-counter pregnancy test kits. MDA Evaluation Report MDA/97/18 (1997).

Hicks, J.M. and Iosofsohn, M. Reliability of home pregnancy test kits in the hands of lay persons. *N. Engl. J. Med.* **320**, 320–321 (1989).

Latman, N.S. and Bruot, B.C. Evaluation of home pregnancy test kits. *Biomed. Instrum. Technol.* **23**, 144–149 (1989).

Lee, C. and Hart, L.L. Accuracy of home pregnancy tests. *Ann. Pharmacother.* **24**, 712–713 (1990).

Shoham, Z., Katz, E., Blinkstein, I., Katz, Z., Zosmer, A., Dgani, R. and Lancet, M. New immunological method for rapid detection of human chorionic gonadotrophin in urine. *Clin. Chem.* **33**, 800–802 (1987).

Stenman, U.-H., Alfthan, H. and Turpeinen, U. Method dependence of interpretation of immunoassay results. *Scand. J. Clin. Lab. Investig.* **51** (Suppl 205), 86–94 (1991).

Sturgeon, C.M. and McAllister, E.J. Analysis of hCG-assay requirements for clinical applications. *Ann. Clin. Biochem.* **35**, 460–491 (1998).

Wheeler, M.J. Home and laboratory pregnancy-testing kits. *Prof. Nurse* **14**, 571–576 (1999).

Wide, L. and Gemzell, C.A. An immunological pregnancy test. *Acta Endocrinol.* **35**, 261–267 (1960).

Voss, S. The usage of pregnancy tests in one health district. *Br. J. Obstet. Gynaecol.* **99**, 1000–1003 (1992).

23 Fluorescence Microscopy: MicroTrak®

Tom Houts

The Trinity Biotech MicroTrak® product line (as distinct from the MicroTrak EIA line) is a group of reagents to aid in the diagnosis of infection using fluorescence microscopy. The best-known product in the line, the *Chlamydia* direct specimen test, was the first nonculture test for the diagnosis of chlamydial infection. Fluorescein-labeled monoclonal antibodies are used to stain selectively the organism or its products, either directly in a patient specimen, or after culture amplification. The presence of characteristic fluorescent features provides a positive diagnosis. MicroTrak methods are available for the detection of *Chlamydia trachomatis*, herpes simplex virus (types I and II), cytomegalovirus and *Neisseria gonorrhoeae*.

MicroTrak reagents are manufactured by Trinity Biotech plc.

TYPICAL ASSAY PROTOCOL

- Inoculate cell monolayers with patient specimens, if culture amplification required. Incubate as appropriate (12–72 h).
- For direct specimen tests, transfer cellular material from patient specimen to glass slides. For culture-amplified tests, remove culture medium from cell monolayer.
- Fix cellular material to glass slide as appropriate (methanol, acetone, etc.). Allow to air dry.
- Add fluorescent-labeled antibody reagent and incubate for 15 min.
- Rinse slide to remove unbound reagent.
- Mount coverslip and examine slide using fluorescence microscope equipped for fluorescein.

PRODUCT FEATURES

- The reagents contain, in addition to fluorescein-labeled monoclonal antibodies, counterstaining agents that aid in the identification of mammalian cell types. The presence of appropriate cell types can be used to judge the adequacy of the specimens.
- Some reagents are designed for use as direct specimen tests, some as culture confirmation tests, and others to detect viral infection in culture prior to cytopathic effects (Pre-CPE).
- All the reagents are nonradioactive and have long shelf lives.
- The fluorescence microscope used for the test may serve other important purposes in the diagnostic laboratory.

ASSAY PRINCIPLE

The binding of fluorescein-labeled antibody to the fixed antigens causes them to appear apple green under the microscope. The size, shape, color, and distribution of the fluorescent areas are interpreted by the reader as a pattern. Training is provided to ensure that customers can distinguish positive from negative staining patterns with confidence.

Target antigens have been selected as being:

- specific to the pathogen of interest;
- consistently expressed in all strains of the pathogen;
- localized to provide easy-to-read staining patterns.

CONTROL MATERIALS

Both positive and negative control slides are available for each method. These are run periodically to confirm the performance of the reagents, and as a reference on the characteristic fluorescent features to be seen on positive results.

The Immunoassay Handbook. *Third Edition*
D. Wild (Ed.)

ANTIBODIES

Monoclonal antibodies with well-defined molecular targets are used.

SEPARATION

The target antigens in the specimens are fixed to glass slides during the specimen preparation procedure. Unbound fluorescent antibody is removed by rinsing the glass slide.

READING AND INTERPRETATION OF RESULTS

The technician views the prepared slide using a fluorescence microscope fitted with filters for fluorescein. The entire area (6–8 mm diameter wells) is scanned for the presence of fluorescent features characteristic of infection. The technician evaluates each specimen as positive or negative based upon its presence. In some instances, such as *Chlamydia* testing, interpretation errors are reduced by requiring a minimum number (e.g. 10) of the characteristic fluorescent features (*Chlamydia* elementary bodies) per slide.

FURTHER READING

Holmes, K.K., Mardh, P.E., Sparling, P.F., Weisner, P.J. (eds), *Sexually Transmitted Diseases* 2nd edn, (McGraw--Hill, New York, 1990), Chapters 73, 74, 77, 79.

24 Automated Batch Analyzers: IMx®

Kent Ford

The Abbott IMx® is a batch analyzer that incorporates three technologies, encompassing a wide range of analytes. It is a widely used immunoassay system with over 30,000 analyzers sold worldwide since its introduction in 1988. The assays for low molecular mass analytes are in a homogeneous format using **fluorescence polarization immunoassay** (**FPIA**), as in the TDx. For larger analytes, and increased sensitivity, Abbott developed a new technology, **microparticle capture enzyme-immunoassay** (**MEIA**). Recently, a third technology has been added called **ion capture immunoassay** (**ICIA**) for select assays (see Figures 24.1 and 24.2).

Reagents are available for a wide range of assays, including thyroid, anemia, cardiac, reproductive endocrinology, oncology, infectious disease and allergy analytes.

TYPICAL ASSAY PROTOCOL

An assay run is started by loading the appropriate number of reaction cells into a carousel. (Different carousels are required for FPIA and MEIA assays.) Approximately 150 μL of sample, control, or calibrator is added to the sample well of each of the reaction cells (see Figure 24.3). The carousel is then placed in the analyzer and the appropriate reagent pack loaded into the reagent heater block. The user then closes the door and presses the 'RUN' button.

The instrument first homes the stepper motors, checks the type of carousel, and reads the barcode on the reagent pack to identify the assay to be run. The reagents and carousel are warmed to the appropriate temperatures. An aliquot of each sample is pipetted from the sample well into the reaction well by the analyzer's dispensing probe (there is also an optional predilution stage).

Fig. 24.1 The IMx® analyzer.

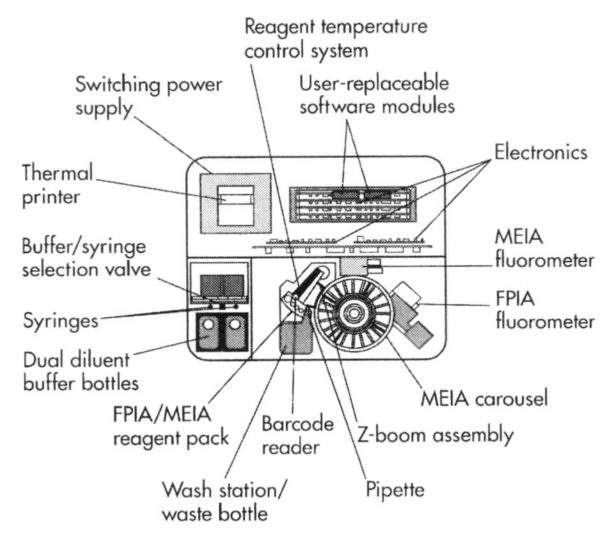

Fig. 24.2 Major components of the IMx analyzer.

The Immunoassay Handbook. *Third Edition*
D. Wild (Ed.)

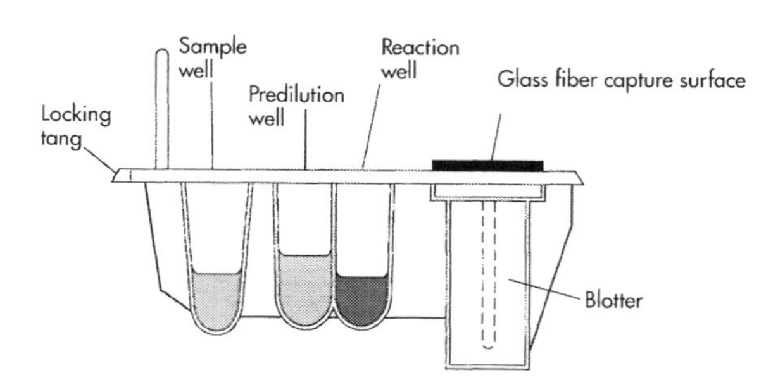

Fig. 24.3 MEIA reaction cell.

In a MEIA immunometric assay, the suspended antibody (or antigen)-coated microparticles are pipetted into the reaction well, mixed with the sample, and the well is incubated at 34 °C. After the incubation an aliquot of the well contents is transferred to the glass-fiber capture surface. The coated microparticles are irreversibly bound to the glass-fibers while unbound material passes through. Conjugate (alkaline phosphatase-labeled antibody or antigen) is added and incubated on the matrix. (This stage is omitted in competitive assays.) After incubation, the matrix is washed with prewarmed diluent. The substrate (4-methylumbelliferyl phosphate) is then added to the matrix, generating a fluorescent signal in the presence of alkaline phosphatase (Figure 24.4).

Ion capture immunoassay is similar to MEIA, except microparticles are replaced with a separation method based on polyelectrolyte interaction. In this approach, the glass-fiber matrix on the IMx disposable reaction cell is precoated with a high molecular mass quaternary ammonium compound. This imparts a net positive charge to the matrix, which enables electrostatic capture of anionic analyte–carrier complexes. Anionic complexes are formed between a selected polyanionic affinity reagent (e.g. polyacrylic acid and dihydroxyboronate) and the capture molecule, usually an antigen-specific antibody. The anionic–antibody complex is incubated with sample and transferred to the reaction cell as in MEIA. The anionic–antibody–analyte complex binds to the positively charged precoated reaction cell and unbound material passes through. Conjugate and substrate are the same material and are processed the same as described for MEIA.

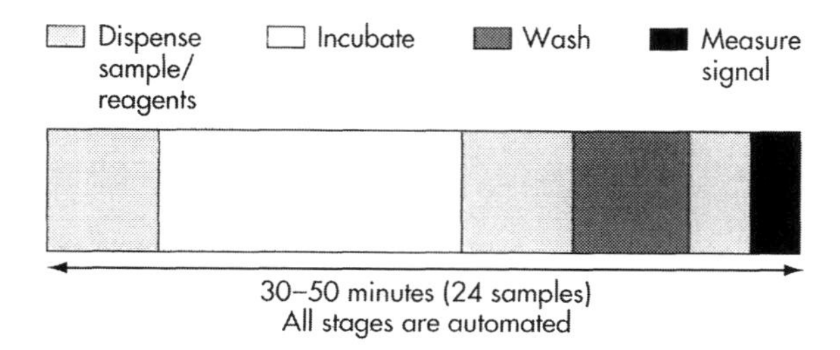

Fig. 24.4 IMx MEIA test sequence.

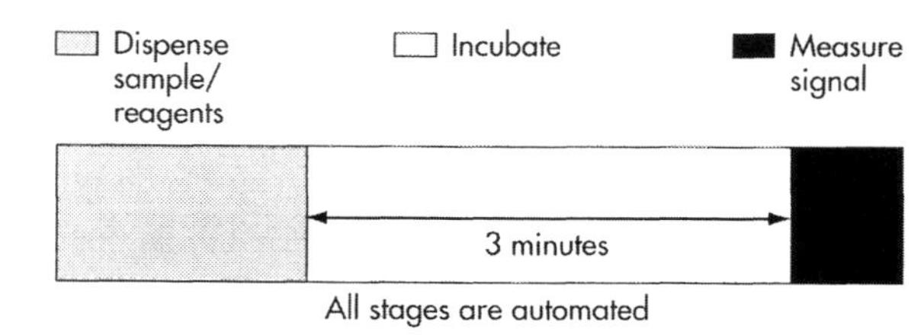

Fig. 24.5 IMx FPIA test sequence.

The FPIA assays are performed as on the TDx. The assays are homogeneous, i.e. do not require a separation. Sample, fluorescein-labeled analyte, and antibody are incubated in the reaction cell for 3 min prior to signal detection (Figure 24.5).

Two 1-L bottles of buffer are left on board the analyzer (one for MEIA and one for FPIA).

PRODUCT FEATURES

- The IMx analyzer is fully automated.
- The instrumentation is compact and fits on a bench top.
- The instrument can be upgraded to IMx Select, which performs up to three select assays in one run.
- The MEIA and ICIA carousel has a capacity for 24 samples. The FPIA carousel has a capacity for 20 samples.
- The typical turnaround time is approximately 30–50 min for 24 samples, and can be less than 20 min for five samples.
- By providing three technologies, the IMx has a broad menu capability, both for high and low molecular mass analytes.
- The IMx is capable of on-board sample dilution.
- The reagents are liquid and nonradioactive.
- The calibration curves for the assays are stable for a minimum of 14 days.
- The two-step format of the immunometric assays avoids high dose hook effects. In addition, assays have been optimized to eliminate these effects.

ASSAY PRINCIPLE

The IMx system uses three different technologies: MEIA, ICIA, and FPIA. Most MEIA assays are immunometric, with antibodies coated onto microparticles (although for some tests for antibodies, antigen is coated onto the microparticles). After binding occurs, and an enzyme-labeled antibody (conjugate) is added, a conjugate/analyte/antibody sandwich is formed, which is then bound to the glass-fiber matrix (see Figure 24.6).

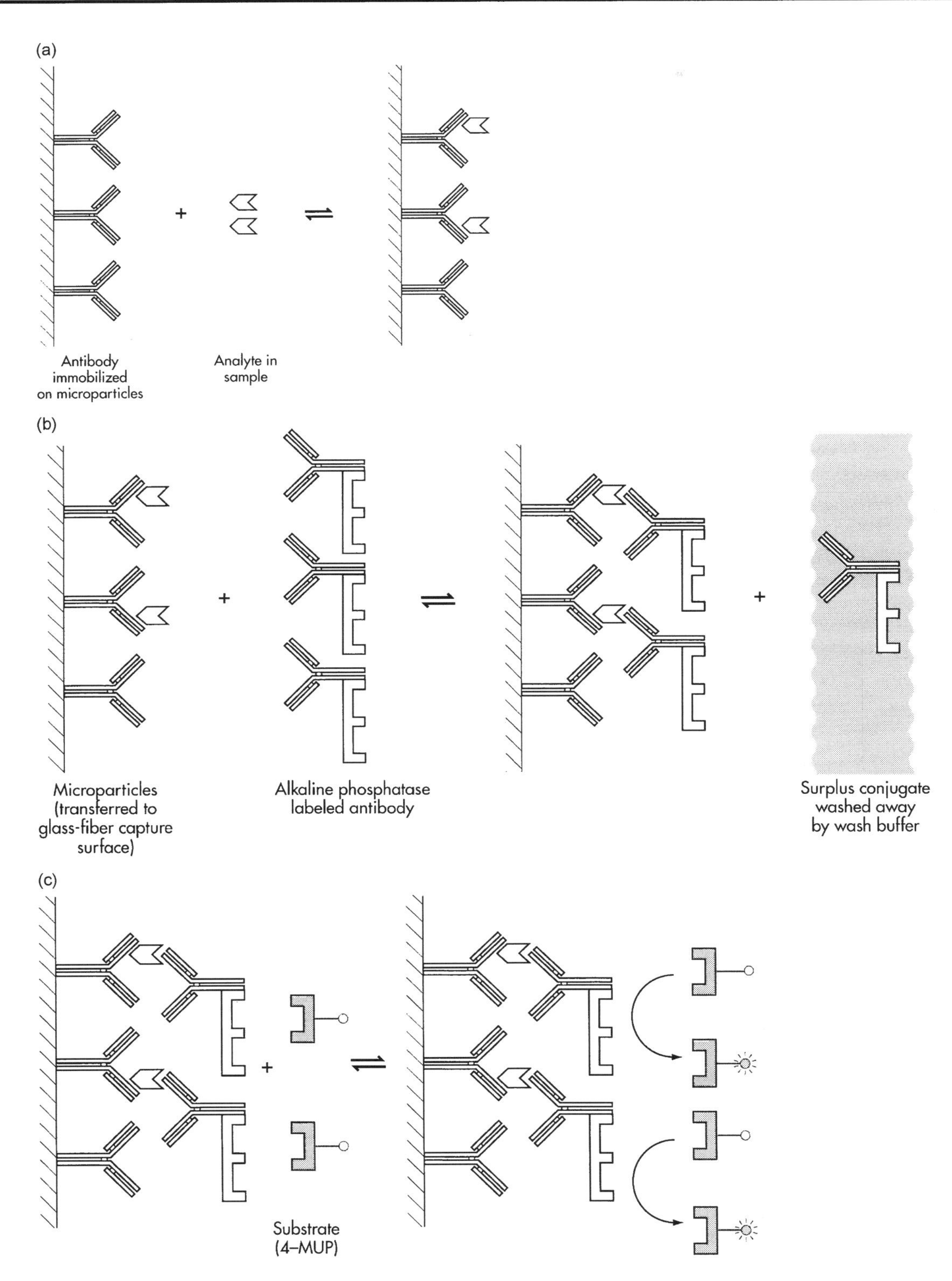

Fig. 24.6 IMx MEIA immunometric assay principle.

In the competitive format, antibody-coated microparticles are incubated with sample and enzyme-labeled analyte (conjugate). Removal of unbound conjugate and analyte is accomplished with the glass-fiber matrix (see Figure 24.7).

The ICIA assays are also immunometric, with antibody or capture molecule bound to a polyanionic affinity reagent. After binding of analyte occurs, the complex is transferred to the reaction cell, which is precoated with a positively charged quaternary ammonium reagent. As with the MEIA, conjugate is added to form a conjugate/analyte/antibody sandwich (see Figure 24.8).

FPIA assays are competitive: sample, fluorescein-labeled analyte, and antibody are incubated in the reaction cell for 3 min prior to signal detection. No separation is required (see Figure 24.9).

CALIBRATION

There are three calibration modes on the IMx: CAL, a full calibration of six calibrators run in duplicate; mode 1, a one-point adjustment of the stored calibration that is required for some MEIA assays; and mode 2, where the stored calibration curve is used.

ANTIBODIES

Both polyclonal and monoclonal antibodies are used, depending on the assay design.

SEPARATION

The MEIA and ICIA assays are heterogeneous, requiring a separation of the bound and unbound fractions during the procedure. With MEIA, the bound fraction is immobilized on uniform latex microparticles, which bind to a glass-fiber matrix. With ICIA, the bound fraction is immobilized by a polyanionic affinity reagent on the glass-fiber matrix. Unbound material is washed through the glass-fiber

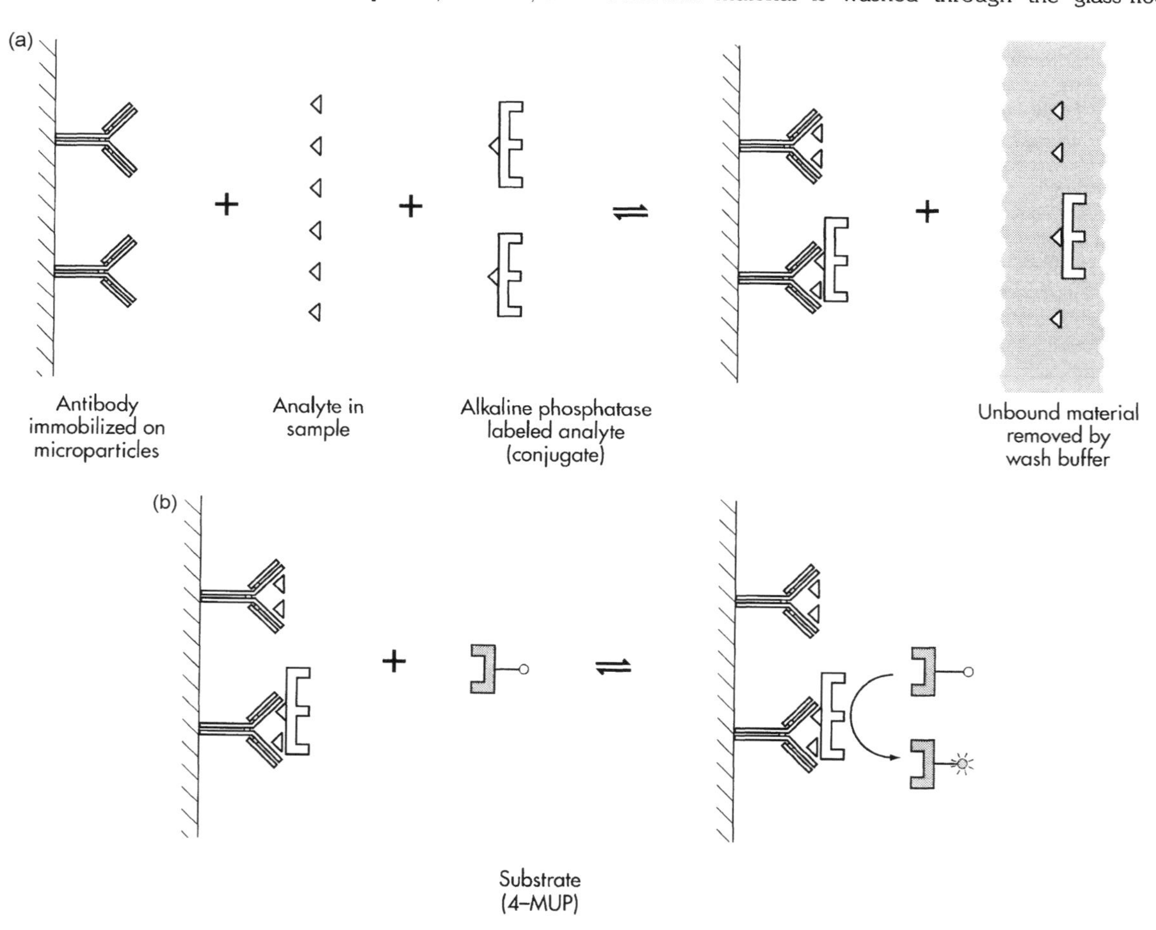

Fig. 24.7 IMx MEIA assay principle (competitive).

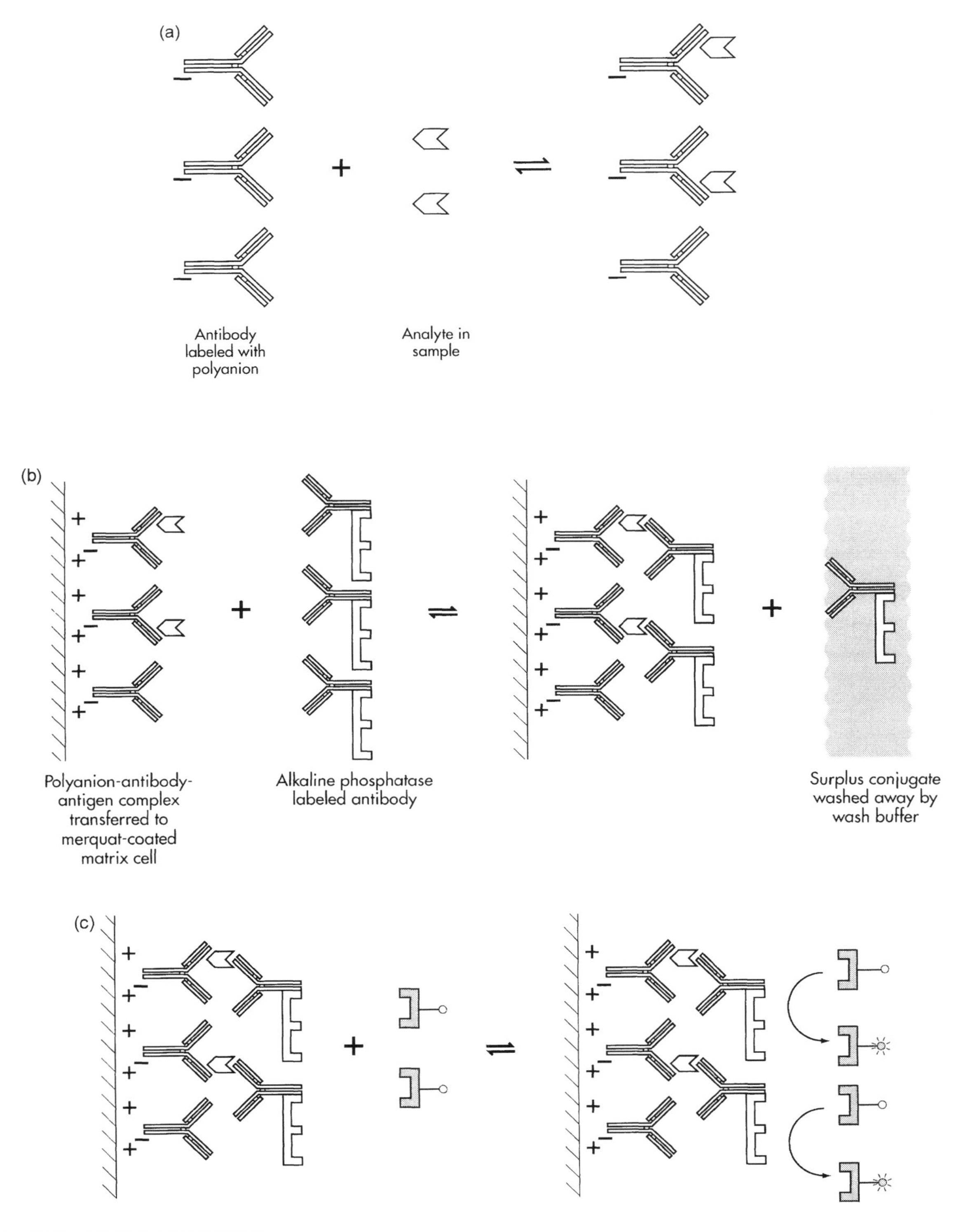

Fig. 24.8 IMx ICIA assay principle.

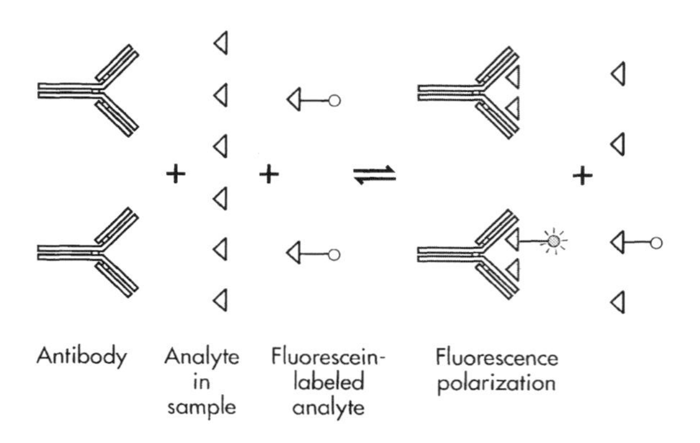

Fig. 24.9 IMx FPIA assay principle.

matrix into an absorbent, opaque blotter below, using prewarmed buffer.

The FPIA assays are homogeneous and do not require a separation.

SIGNAL GENERATION AND DETECTION

The MEIA and ICIA assays use alkaline phosphatase as the enzyme label, conjugated to an antibody or an antigen. The substrate for the enzyme-mediated reaction is 4-methylumbelliferyl phosphate (4-MUP), which is added to the glass-fiber capture surface immediately after the washing stage. Alkaline phosphatase present in the bound phase on the glass-fibers converts the 4-MUP to 4-methylumbelliferone (4-MU), which is fluorescent.

The analyzer measures the rate of fluorescence development. The front-surface fluorometer generates light at 365 nm which excites the molecules of 4-MU, which fluoresce at 448 nm. A photodiode detector and photomultiplier tube detect and amplify the emitted light, which is first filtered to remove stray light emissions. The analyzer takes eight consecutive fluorescence measurements of 500 msec each, with an interval between each measurement. A microprocessor uses linear regression analysis to convert the fluorescent measurement data to rates that are proportional to analyte concentration.

FPIA assays are homogeneous, i.e. the antibody-bound and free fractions of fluorescein-labeled tracer can be distinguished in solution without the need of a separation. Fluorescence polarization is a refined example of fluorescence. The fluorometer generates vertically (and horizontally) polarized light at 485 nm, the excitation frequency for fluorescein, and detects emitted light at 525–550 nm, through a vertical polarizing filter. The addition of polarization to the process moderates the strength of the emitted light, depending on whether the tracer is bound to the antibody or free in solution. The bound fraction generates a strong measurable signal whereas the free tracer produces only a weak signal. As the free tracer molecules rotate at very high speed, the emitted light is always in a different plane from the incident light, hence the amount of light passing through the polarized filter to the detector is minimal. However, the tracer bound to the much larger antibody molecules is restrained from rotating at such a high speed and the emitted light is in almost the same plane as the incident light. The speed of rotation of an antibody molecule is about one hundredth that of fluorescein (see Figure 24.10).

An additional refinement is that the analyzer measures the intensity of vertically polarized light after excitation with alternating pulses of vertically and horizontally polarized light. The signal detected after excitation with horizontally polarized light (I_{hv}) provides a measurement of the amount of emitted light which has been changed from the original plane of polarization. The intensity of the light which remains vertically polarized is designated as I_{vv}. To quantify the degree of polarization, P, the following equation is used:

$$P = (I_{vv} - I_{hv})/(I_{vv} + I_{hv})$$

DATA PROCESSING

The fluorescence or fluorescence polarization data are processed using point-to-point, linear regression or four-parameter logistic algorithms for calibration curve-fitting.

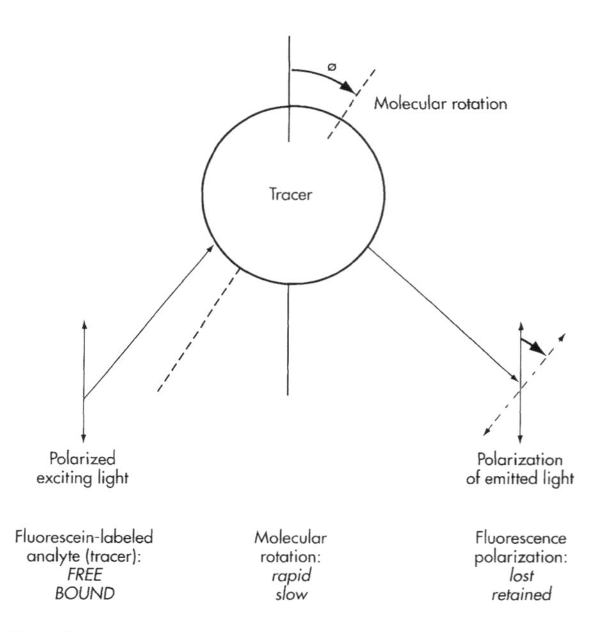

Fig. 24.10 Fluorescence polarization principle. For low molecular mass analytes, fluorescence polarization can distinguish free from protein-bound fluorescein-labeled analyte, without a separation step.

Results are presented on a 40-column printout at the end of each run.

INTERFACING TO LABORATORY INFORMATION SYSTEMS

The analyzers have an RS232-C serial interface for data output to a remote computer.

REFERENCES AND FURTHER READING

Chou, P.P. IMx system. In: *Immunoassay Automation: a Pratical Guide.* (ed Chan D.W.), 203–219 (Academic Press, San Diego, 1992).

Fiore, M., Mitchell, J., Doan, T., Nelson, R., Winter, G., Grandone, C., Zeng, K., Haraden, R., Smith, J., Harris, K., Leszczynski, J., Berry, D., Safford, S., Barnes, G., Scholnick, A. and Ludington, K. The Abbott IMx™ automated benchtop immunochemistry analyzer system. *Clin. Chem.*, **34**, 1726–1732 (1988).

Keller, C.H., Fitzgerald, K.L. and Barnes, A. The Abbott IMx® and IMx SELECT™ systems. *J. Clin. Immunoassay* **14**, 115–119 (1991).

Pennington, C., Cotter, S., Fabian, K., Hiltibran, R., Jenik, B. and Johnson, J. An automated digoxin assay on the IMx® system using soluble reagents and polyelectrolyte interaction as separation means (Abstract). *Clin. Chem.* **37**, 1046 (1991).

25 Bulk Reagent Random-Access Analyzers: UniCAP® 100

Gareth Evans and Mats Rilvën

UniCAP® is an integrated laboratory system, specially designed for allergy testing and related clinical fields. The reagents are based on the well-established ImmunoCAP format, and the instruments provide full automation and user-friendly software. UniCAP delivers fast and accurate test results that provide physicians with valuable clinical information. Over 450 different allergens are available for allergy testing, in addition to tests for monitoring asthma treatment, celiac disease and autoimmune disorders. UniCAP has been developed by Pharmacia & Upjohn Diagnostics AB in Sweden.

UniCAP 100 (Figure 25.1) provides smaller to medium sized laboratories with a fully automated allergy testing system incorporating sample and reagent handling, washing, reading, calculation, and integrated processing. The time from sample pipetting to first result is less than 2.5 h. Each run can yield up to 46 test results.

All processing steps have been combined into an efficient, simple to use, and extremely compact *in vitro* test system.

UniCAP 100 was introduced in 1995 and approximately 3000 instruments were in use worldwide at the beginning of 2000.

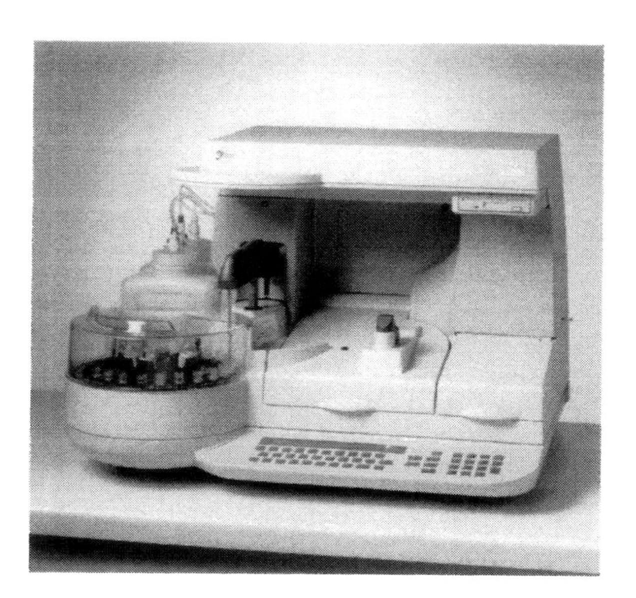

Fig. 25.1 UniCAP 100.

TYPICAL ASSAY PROTOCOL

UniCAP 100 is a batch analyzer designed to handle all the steps from request to result (Figure 25.2). The instrument is controlled from the panel at the front of the instrument. Reagents and samples are loaded into the Sample Carousel to the left of the instrument. Positions are marked to simplify the loading of the reagents.

In the automated assay protocol on the UniCAP 100, an ImmunoCAP carrier is loaded, then sample, ImmunoCAP and reagent are dispensed. There is an incubation, followed by washing, rinsing, and aspiration. The signal generation stage consists of substrate addition, stop reagent addition, and fluorescent signal measurement. ImmunoCAP carriers are loaded via the top, central

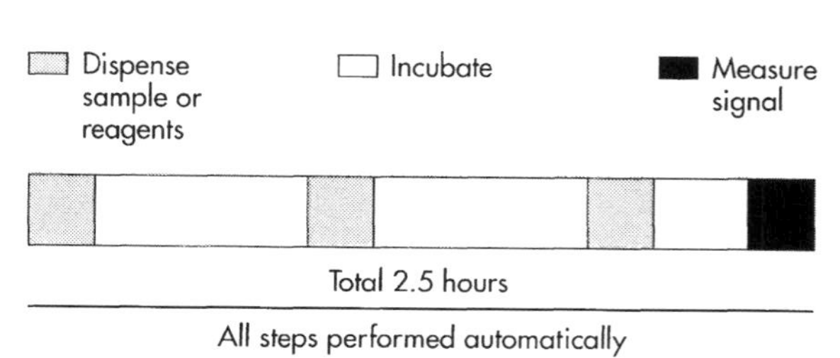

Fig. 25.2 Automated assay protocol.

The Immunoassay Handbook. *Third Edition*
D. Wild (Ed.)

section of the instrument. Required reagents, samples or ImmunoCAP are prompted on the display in the Load and Start mode. Each step is confirmed via the instrument keyboard. Wash, waste, and rinse bottles are placed just behind the carousel to the left of the instrument. The instrument is also equipped with a printer and a diskette reader.

The system is designed to use stored calibration curves. A new calibration curve is required approximately once a month and with every new lot number of conjugate or reference ImmunoCAP. Between calibrations, curve controls are used to check performance against the stored curve.

Requests, reagents, samples, and ImmunoCAP are loaded manually. The process is then started and takes 2.5 h. A laboratory report is automatically printed when the process is ended.

PRODUCT FEATURES

- A minimal set-up time;
- compact size;
- monthly calibration;
- up to 46 test results per run;
- four different analytes per run;
- results available in 2.5 h;
- logs and checks quality control samples against expected values;
- calibrated against WHO standards;
- bidirectional interfacing to central laboratory computer via PC;
- barcode reader optional.

ASSAY PRINCIPLE

See Figure 25.3.

CALIBRATION

Automation of the assay protocol steps, with good temperature control, has enabled the use of stored calibration curves. The high reproducibility of the reagents allows established calibration curves to be stored and used in future assay runs, within a lot. The calibration curve is checked by curve controls.

The stored calibration curve is valid for the specific calibration code for a maximum time of 28 days, as long as the curve controls are within limits.

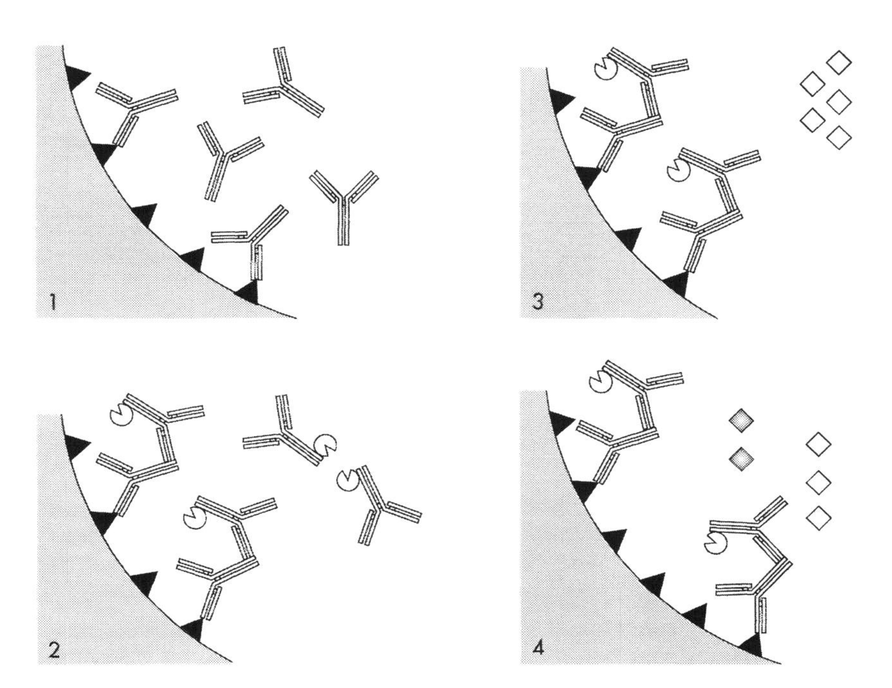

Fig. 25.3 Assay principle. 1. The antigen/antibody of interest, covalently bound to the ImmunoCAP, reacts with the analyte in the patient sample. 2. After washing away unbound analyte, enzyme-labeled antibodies are added to form a complex. 3. After incubation, unbound enzyme label is washed away and the bound complex is then incubated with a developing agent. 4. After stopping the reaction, the fluorescence of the eluate is measured. The higher the fluorescence the more analyte is present in the sample.

A FluoroC procedure checks the fluorometer and the pipette, and minimizes instrument variations. The response of the actual measurement is compared to the response of the FluoroC target value. The FluoroC procedure is automatically prompted by the software. A FluoroC run is also automatically prompted the first time the instrument is run, when there is no active calibration curve for the chosen method and when curve controls or calibration curve are not acceptable.

ANTIBODIES

Mixtures of monoclonal antibodies are used.

SEPARATION

The system is built around a new type of solid phase consisting of a flexible hydrophilic carrier polymer encased in a capsule, ImmunoCAP.

The carrier consists of a cyanogen bromide (CNBr)-activated cellulose derivative that can bind at least three times more protein than a paper disc and about 150 times more than can be adsorbed to the inner surface of a traditional coated tube or microtiter well (Figure 25.4). This solid phase is an excellent carrier of antigens (allergens) and provides favorable reaction conditions, including short diffusion distances. Equilibrium can be reached in less than 20 min. In fact, most of the antibodies of test samples or standards are bound to the immobilized antigens within 2 min of the initial incubation step (Figure 25.5).

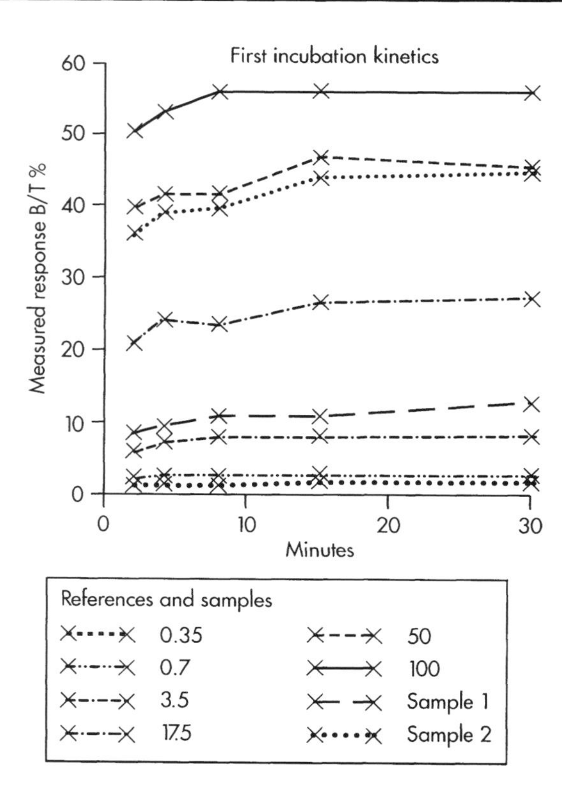

Fig. 25.5 Assay kinetics.

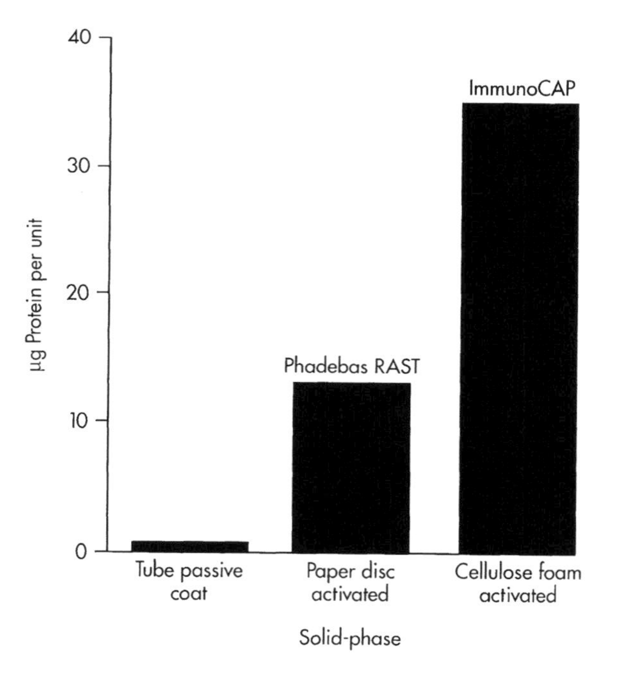

Fig. 25.4 ImmunoCAP binding capacity.

SIGNAL GENERATION AND DETECTION

Following the second wash of ImmunoCAP where the unbound enzyme-labeled conjugate is removed, development solution (4-methylumbelliferone-β-D-galactoside) is added. The bound enzyme reacts with the substrate forming a fluorescent product (4-methylumbelliferone). This product is eluted using a stop solution (sodium carbonate) and the fluorescent signal is measured.

DATA PROCESSING

UniCAP uses a four- or five-parameter log–logistic Rodbard curve model depending on the analyte. The Rodbard model constrains the shape of the curve, consistent with the assay chemistry, disregarding single replicate or calibration points that are totally out of limits in the actual calibration curve assay run.

Levey–Jennings plots of quality control data may be displayed or printed. An overlay plot of calibration curves is also available.

INTERFACING TO LABORATORY INFORMATION SYSTEMS

Bidirectional interfacing to a central laboratory computer is possible via a PC for downloading of worklists and uploading of completed results.

FURTHER READING

Costongs, G.M.P.J. and Bas, B.M. The first fully automated allergy analyzer UniCAP. Comparison with IMMULITE for allergy panel testing. *Eur. J. Clin. Chem. Clin. Biochem.* **35**, 885–888 (1997).

Paganelli, R., Ansotegui, I.J., Sastre, J., Lange, C.-E., Roovers, M.H.W.M., de Groot, H., Lindholm, N.B. and Ewan, P.W. Specific IgE antibodies in the diagnosis of atopic disease. Clinical evaluation of a new *in vitro* test system, UniCAP®, in six European allergy clinics. *Allergy* **53**, 763–768 (1998).

26 Automated Panel Analyzers PRISM™

Dinesh Shah and Jim Stewart

PRISM™ is a fully automated immunoassay analyzer utilizing chemiluminescent immunoassay technology (ChLIA) for performing high-volume screening (Figure 26.1). Solid-phase microparticles and chemiluminescent conjugates are used to simultaneously determine the presence of multiple specific infectious disease markers (antigens and antibodies). PRISM is manufactured internationally by the Diagnostics Division of Abbott Laboratories.

PRISM is uniquely designed for use in blood and plasma screening laboratories due to its high throughput, automation, and built-in process controls. The instrument has six independent temperature-controlled immunoassay processing channels, which are centrally computer controlled. Reagents have been developed for detection of hepatitis B surface antigen (HBsAg), antibody to hepatitis B core antigen (HBcore), antibodies to hepatitis B surface antigen (HBsAb), antibodies to hepatitis C virus

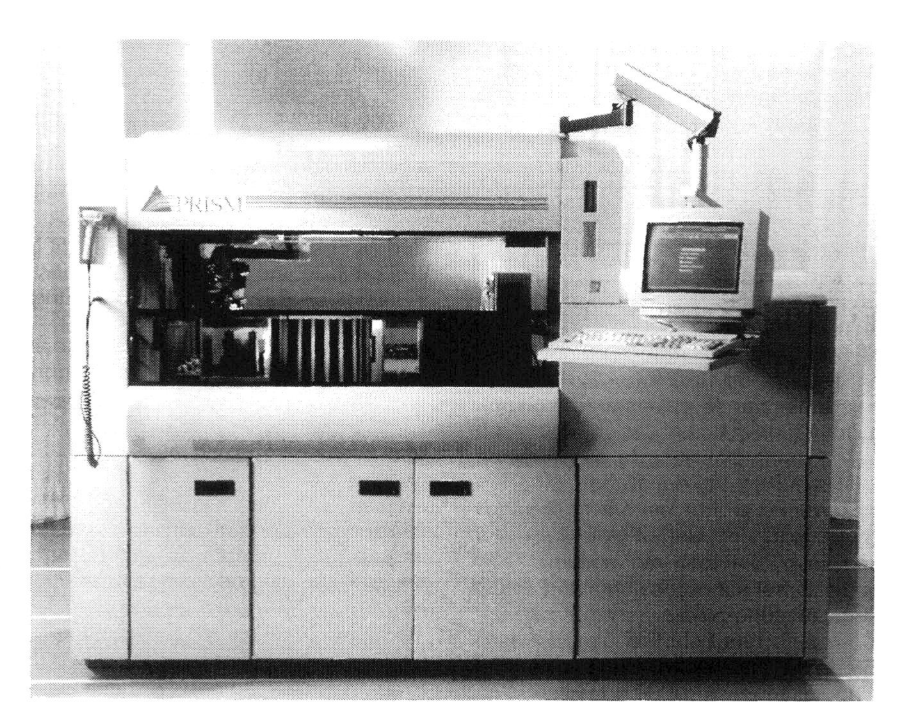

Fig. 26.1 PRISM analyzer.

The Immunoassay Handbook. *Third Edition*
D. Wild (Ed.)

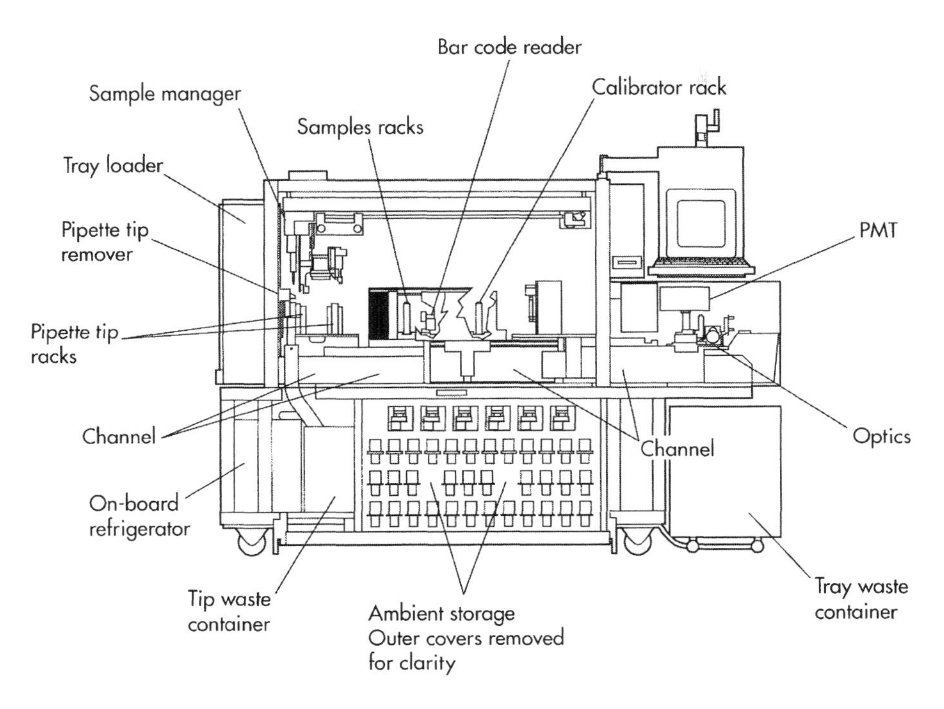

Fig. 26.2 Major components of the PRISM analyzer.

(HCV), antibodies to human T-lymphotropic viruses types I and II (HTLV-I/HTLV-II), and antibodies to human immunodeficiency viruses types I and II (HIV-1/HIV-2). Multiple subsystems within the instrument (see Figure 26.2) coordinate, control, and monitor all critical test steps including sample identification and pipetting, reagent dispensing, tray transport and incubation, result determination and reporting, and retest management.

The instrument design includes numerous electronic monitors, sensors and built-in redundant checks to ensure integrity of the testing process. Operators are alerted to problem conditions, but are restricted from tampering with assay results. Complete batch documentation is available through a variety of computer-generated reports.

TYPICAL ASSAY PROTOCOL

PRISM ChLIA assays use either immunometric or competitive/blocking protocols depending on the assay. Assays are carried out in 16-well reaction trays consisting of both an incubation well and a reaction well (see Figure 26.3). Microparticles coated with recombinant antigen, viral lysate, or monoclonal antibodies are incubated with the sample (plasma, serum, calibrator or control) in the incubation well of the reaction tray. During incubation, the analyte present in the sample binds to the microparticles. After this first incubation is complete, the reaction mixture is transferred to the glass-fiber membrane of the reaction well using a noncontact transfer wash. The microparticles are captured on the glass-fiber matrix while the remaining mixture flows through to an absorbent blotter. For two-step assays, an acridinylated antibody conjugate is added to the microparticles captured on the matrix and incubated. For three-step assays, biotinylated antigen (probe) is added prior to the addition of an acridinium-labeled anti-biotin conjugate and any unbound probe is washed off using a probe wash. The chemiluminescent signal is generated by the addition of an alkaline hydrogen peroxide solution. The resultant photons are counted and reported as assay counts. The intensity of chemiluminescent signal is proportional to the amount of analyte in the sample. A representative test sequence for a three-step PRISM assay channel is depicted in Figure 26.4.

PRISM requires minimal hands-on time. The operator primes the system, replenishes reaction trays and pipette tips, loads sample racks as needed, and purges the system on completion of the run. Up to 1380 samples can be processed in 8.5 h, requiring only 2.5 h of technician time.

PRODUCT FEATURES

- PRISM is self-contained and fully automated. All assays are performed entirely on board the analyzer.
- Laboratory efficiency is greatly enhanced by complete automation coupled with simultaneous throughput of all required assays on a single instrument.

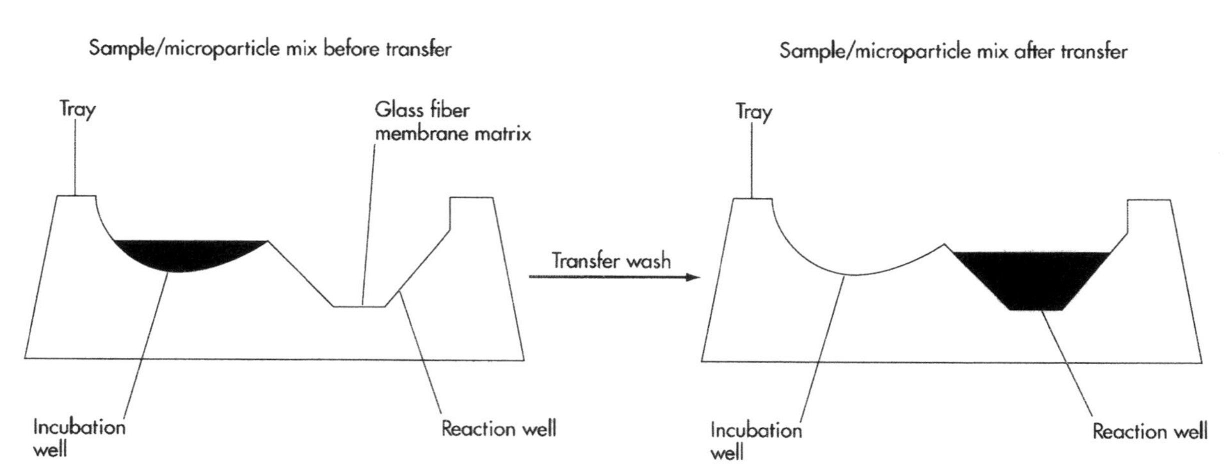

Fig. 26.3 Schematic of PRISM reaction tray.

- Sample throughput is 160 samples per hour or 960 tests per hour across the six PRISM assay channels. Throughput is not dependent on the test mix. Sample test results for all assays are available within approximately 54 min of pipetting.
- The sixth channel serves as a backup for any PRISM assay and can be used for menu expansion.
- PRISM can hold up to 280 samples at a time with continuous access and stat-processing capability. Stat sample requests can be made at any time and, once

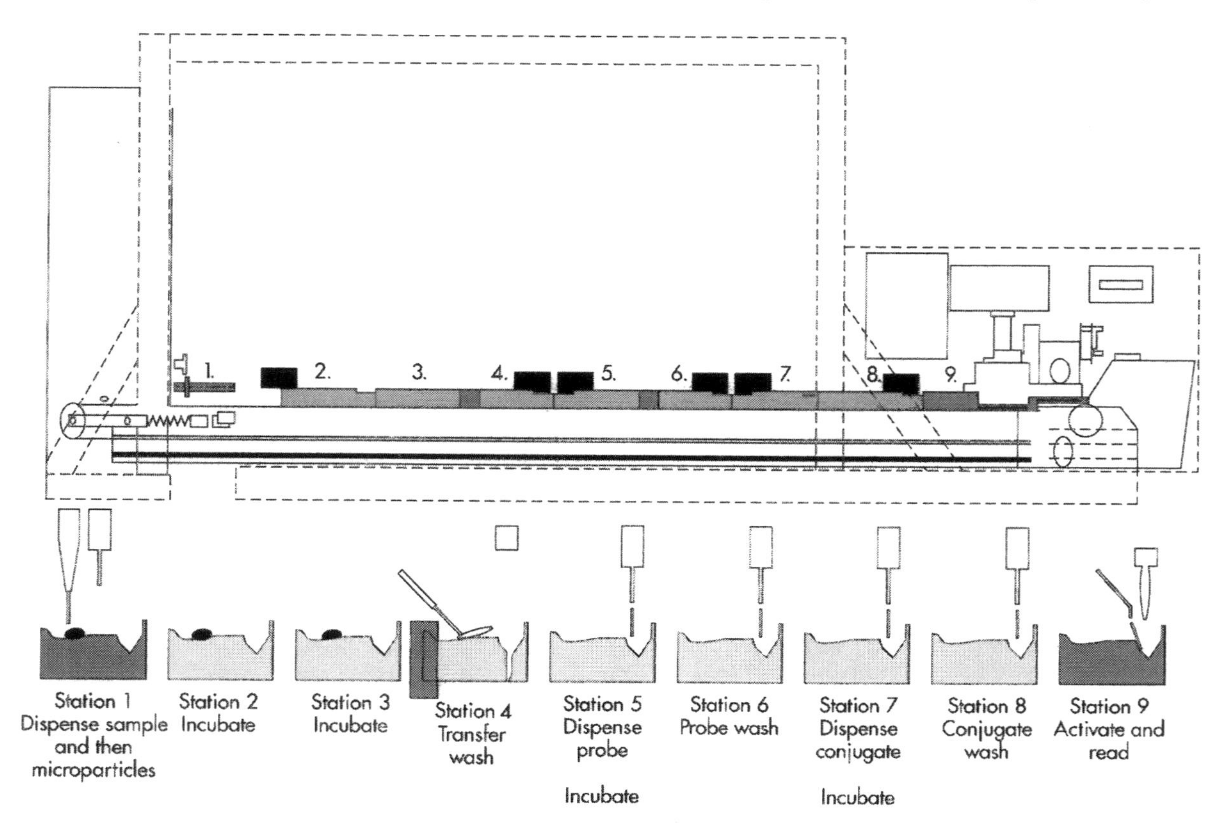

Fig. 26.4 Representative test sequence for a three-step PRISM assay channel.

loaded, stat samples will be processed in the next pipetting sequence.

- The same sample tube is used to pipette across all required PRISM channels. Typical sample volume is 100 μL per assay. Disposable pipette tips eliminate the risk of carryover between samples.
- Sample pipetting is electronically verified on both aspiration and dispense using pressure sensors. Sample barcodes are read automatically upon loading to ensure positive identification throughout the testing process.
- Refrigerated and ambient assay reagents are stored on board the instrument, up to approximately 5000 tests per assay. Barcode entry is required on reagent loading to ensure proper installation and to verify that matched and nonexpired reagents are used. PRISM monitors reagent usage and alerts the operator when kits need to be replenished. All reagents are nonradioactive with typical shelf-lives of 12 months.
- Critical reagent dispenses are monitored using patented optical verification sensors. All reagents are collected within the reaction tray; no liquid waste is generated.
- Trays are transported automatically through the temperature-controlled channels in timed increments, thus maintaining precise incubation time and temperature. No physical manipulation of the trays by the operator is required during processing. Trays are automatically disposed into waste bags within the instrument.
- PRISM employs an end-of-run release control, which is a low-level, multiconstituent positive sample to check system integrity. This control must pass specification in order for assay test results to be released from the instrument. Release controls can be placed throughout the batch to allow for efficient release of test results as needed.
- The instrument detects error conditions and alerts the operator via the computer monitor and/or audio alarms. Error codes are recorded on the batch report and in a comprehensive activity log. Errors detected that could affect the potential outcome of a specimen prevent the release of results for that sample.
- ChLIA technology allows for very sensitive and specific detection of viral markers. For example, HBsAg detectability on PRISM is ~0.1 ng/mL, which is very sensitive compared with enzymeimmunoassay (EIA) technology that is being currently used in whole blood and plasma screening. Acridinium-based chemiluminescence eliminates many of the temperature and pH effects associated with enzyme-based conjugate detection systems.
- Reliability of PRISM in releasing valid test results is equal to or better than current technology.
- The process control and documentation features built into PRISM can greatly enhance compliance with current good manufacturing practice (cGMP) guidelines for blood/plasma testing establishments.

ASSAY PRINCIPLE

The PRISM ChLIA assays are either immunometric or competitive/blocking, with either recombinant antigen or viral lysate coated onto microparticles (although for some tests for antigens, antibody is coated onto microparticles) and use either three-step or two-step formats (see Figures 26.5 and 26.6). Table 26.1 shows the assay configuration for the current test menu on PRISM.

The PRISM instrument can be configured to run new tests as they are developed either as a replacement to a current assay on channels 1–5 or on the back-up channel (channel 6). Future tests planned for PRISM include assays to detect antibodies to *Trypanosoma cruzi* (Chagas), combination detection of HIV-1 antigen and HIV-1/HIV-2 antibodies, and combination detection of HCV antigen and antibodies.

CALIBRATION

Three replicates each of assay specific, single analyte, negative and positive calibrators (antibody to HIV-1, antibody to HTLV-I, antibody to HBc, antibody to HCV, and reactive for HBsAg) are run at the beginning of each batch to validate reagent integrity and to establish the assay cutoff. In addition, supplemental controls are also run at the start of each batch for the three-step assays. These supplemental controls, which are not used in the assay cutoff determination, verify assay integrity for additional markers such as HTLV-II, HIV-2, and HIV-1 Group O.

ANTIBODIES

Both recombinant and lysate antigens as well as polyclonal and monoclonal antibodies are used, depending on the assay design.

SEPARATION

The ChLIA assays are heterogeneous, requiring a separation of the bound and unbound fractions during the procedure. The bound fraction is immobilized on uniform latex microparticles, which bind to a glass-fiber matrix. Unbound material is washed through the glass-fiber matrix into an absorbent black blotter using buffer (Wolf-Rogers *et al.*, 1990; Shah and Richerson, 1998; Shah *et al.*, 2003).

SIGNAL GENERATION AND DETECTION

The ChLIA assays use acridinium derivatives as the chemiluminescence label, conjugated to an antibody or an antigen. The acridinium is triggered by alkaline

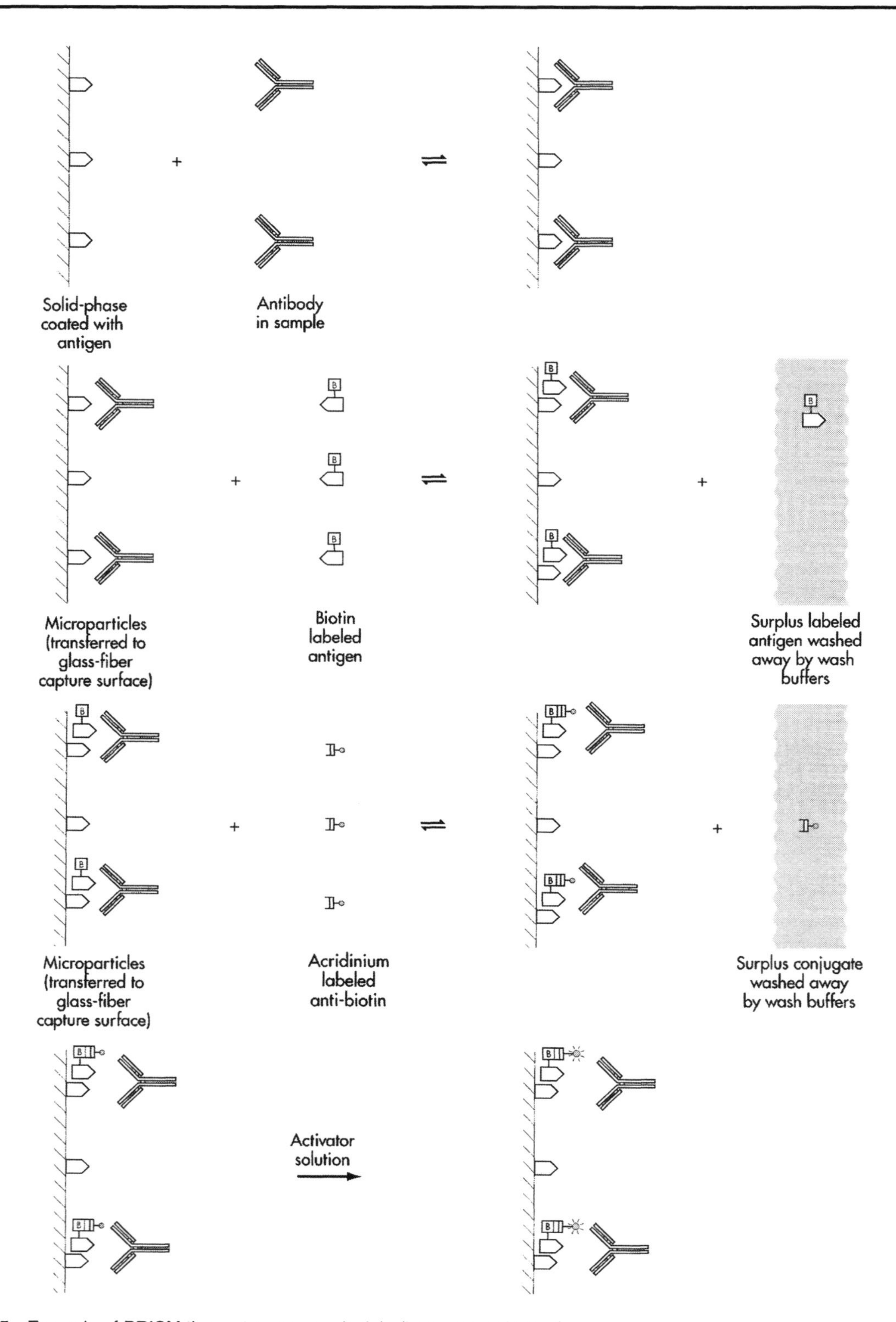

Fig. 26.5 Example of PRISM three-step assay principle (immunometric type).

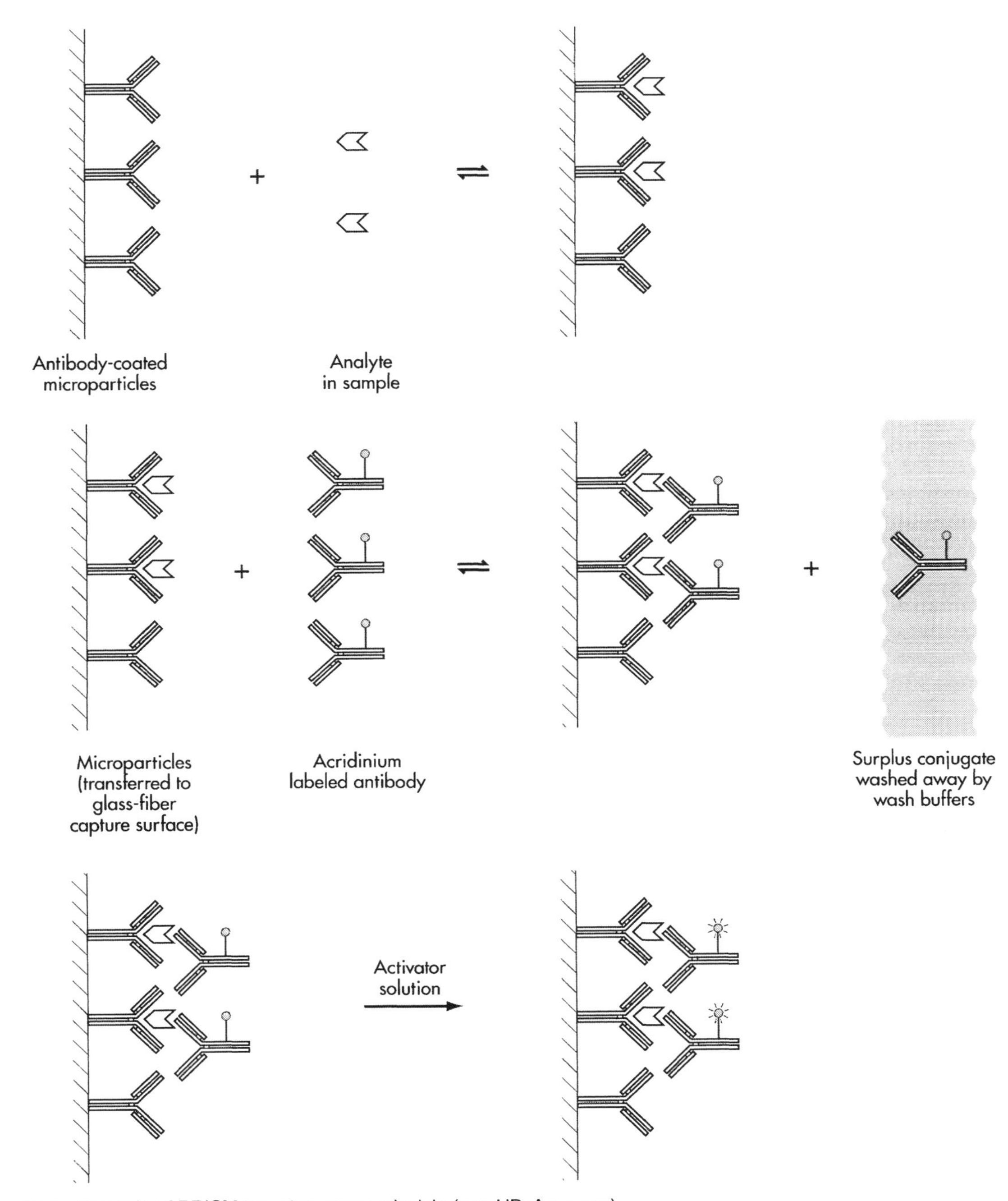

Fig. 26.6 Example of PRISM two-step assay principle (e.g. HBsAg assay).

peroxide activator solution, which is added to the glass-fiber capture surface after the conjugate washing stage. Acridinium present in the bound phase on the glass fiber generates photons, which are counted by a photomultiplier tube detector over a period of 2 sec (Shah *et al.*, 1994).

DATA PROCESSING

Photon counts, corrected for background (dark counts), are compared to a qualitative cutoff value generated from the calibrator values. Samples with calculated sample to

Table 26.1. PRISM assay configuration.

Channel configuration					
1	2	3	4	5	6
HTLV	HCV	HIV	HBsAg	HBc	Backup
DS	IDS	DS	DS	C/B	
Three-step	Two-step	Three-step	Two-step	Two-step	Two/three-step

DS, direct sandwich; IDS, indirect sandwich; C/B, competitive/blocking.

cutoff (S/CO) ratios of 1.0 or greater are reactive; samples with S/CO values less than 1.0 are considered negative (the inverse applies to these S/CO rules for competitive formats like PRISM HBcore). PRISM maintains a database of up to 14,000 samples. Using this database, PRISM automatically recognizes initially reactive samples and schedules them for duplicate retesting on the appropriate assay channels. This also prevents inadvertent retesting of negative samples. A computer server is available for managing larger sample databases across single or multiple PRISM instruments in a laboratory.

Printed results are available in a variety of reports providing detailed information on all test results, samples requiring retest, identified error codes, and complete documentation of assay reagents used including lot numbers and expiration dates.

INTERFACING TO LABORATORY INFORMATION SYSTEMS

PRISM data output can be transferred to a remote computer or data management system via an RS232-C serial interface.

PRISM can also be used in conjunction with the ABBOTT PRISM Retest Server (APRS), which is a computer system that assists laboratories by managing the sample test and retest functions (test management) for multiple ABBOTT PRISM Systems (Figure 26.7). The APRS allows the user to view sample results while the ABBOTT PRISM instrument is processing samples. The server also provides a central location to obtain ABBOTT PRISM instrument reports. The APRS can be used to receive ABBOTT PRISM result information and transfer

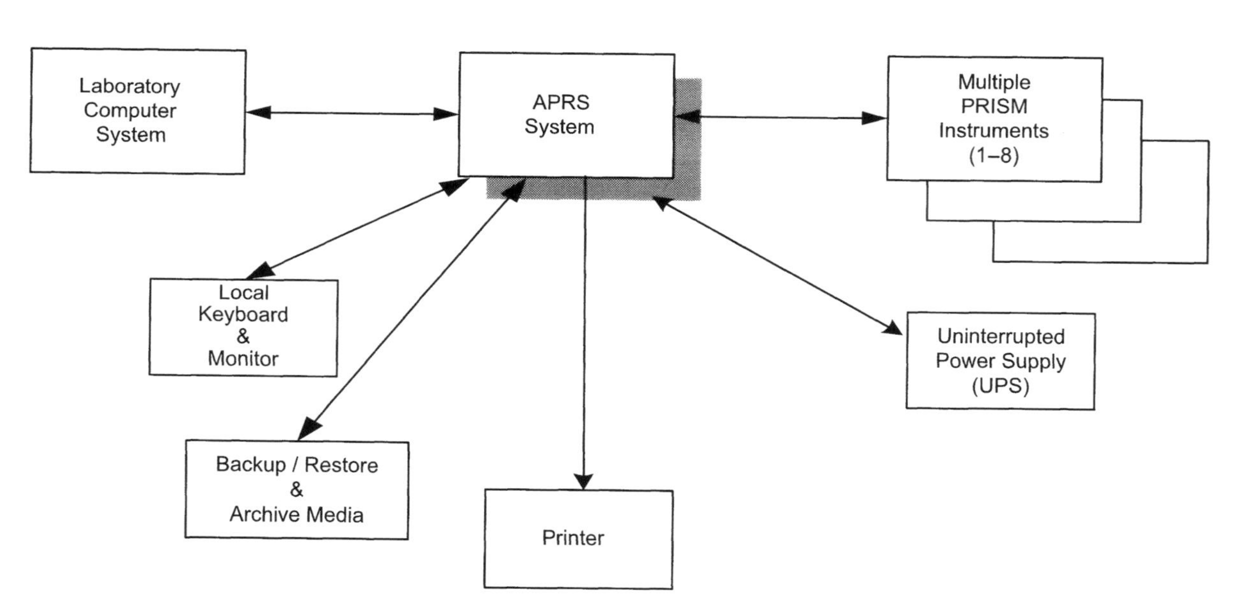

Fig. 26.7 APRS system.

this information to blood bank, hospital, and reference laboratory computers.

The APRS is capable of communicating with multiple ABBOTT PRISM instruments and a laboratory host computer. This system provides a centralized location for determining test replicates, reviewing sample results, and printing reports. The server also collects, tracks, and collates test results for samples tested on the interfaced PRISM instruments. Through communication with the laboratory host computer, the APRS can also obtain test order information and send test result information from the interfaced instruments.

REFERENCES AND FURTHER READING

Ekins, R.P. An overview of present and future ultrasensitive nonisotopic immunoassay development. *Clin. Biochem. Rev.* **8**, 12–23 (1987).

Khalil, O.S., Hanna, C.F., Huff, D., Zurek, T.F., Murphey, B., Pepe, C. and Genger, K. Reaction tray and noncontact transfer method for heterogeneous chemiluminescence immunoassays. *Clin. Chem.* **37**, 1612–1617 (1991a).

Khalil, O.S., Zurek, T.F., Tryba, J., Hanna, C.F., Hollar, R. and Pepe, C. Abbott Prism: a multichannel heterogeneous chemiluminescent immunoassay analyzer. *Clin. Chem.* **37**, 1540–1547 (1991b).

Shah, D.O. and Richerson, R.B. Chemiluminescent immunoassay for antibody detection. U.S. Patent 5,705,330 (1998).

Shah, D.O., Chandra, T., Chang, A., Klosterman, K., Richerson, R. and Keller, C. Acridinium-labeling to latex microparticles and application in chemiluminescence-based instrumentation. *Clin. Chem.* **40**, 1824–1825 (1994).

Shah, D.O., Smith, A.H., Blazejak, C. and Chandra, T. In-process testing of acridinium for conjugation with protein coated microparticles and application in chemiluminescence based analyzer, *Bioluminescence and Chemiluminescence. Proceedings of 10th International Symposium 1998* (Wiley, West Sussex, 1999).

Shah, D.O., Chang, C.D., Jiang, L.X., Cheng, K.Y., Stewart, J.L. and Dawson, G.J. Combined HCV core antigen and antibody assay on a fully automated chemiluminescence analyzer. *Transfusion* **43**, 1067–1074 (2003).

Wolf-Rogers, J., Wears, J.A., Rick, K., Robertson, E.F., Guidinger, P., Khalil, O.S. and Madson, G.A. A chemiluminescent, microparticle-membrane capture immunoassay for the detection of antibody to hepatitis B core antigen. *J. Immunol. Methods* **133**, 191–198 (1990).

27 Unitized Reagent Random-Access Analyzers: IMMULITE® and IMMULITE 1000

Arthur L. Babson

The IMMULITE® system is manufactured and distributed by Diagnostic Products Corporation (DPC). IMMULITE, a bench-top instrument, is illustrated in Figure 27.1a. It is a continuous, random-access, automated immunoassay analyzer designed around a proprietary assay tube called a test unit that allows for thorough and efficient washing of an integral antibody- or antigen-coated bead by rapidly spinning the tube on its vertical axis. All assays use an alkaline phosphatase enzyme label that is quantified with a sensitive chemiluminescent substrate. The assay menu includes 105 analytes and there are more than 5400 instruments in 70 countries around the world.

In 2001, IMMULITE was discontinued with the introduction of IMMULITE 1000 (Figure 27.1b), basically the same instrument but with new casework incorporating the computer and touch-screen monitor, as well as bulk fluids with automatic fluid-level indicators resulting in a consolidated footprint.

TYPICAL ASSAY PROTOCOL

The operator loads samples in barcoded sample cups into the loading chain of the IMMULITE instrument followed by up to five barcoded, assay-specific test units in any order. All subsequent steps are performed automatically by the IMMULITE immunoassay system. The IMMULITE dispenses the sample, selects and dispenses the reagent, incubates, separates bound from free, dispenses substrate and reads signal all automatically as follows:

- Sample and alkaline phosphatase-labeled reagent are added to a test unit.
- The reaction is incubated for 30 min at 37 °C and shaken every 10 sec.
- The bead is washed.
- Chemiluminescent substrate is added.
- The reaction is incubated for 10 min at 37 °C.
- The luminescence is read.

The time to first result is 42 min and the maximum throughput is 120 samples per hour. Throughput varies somewhat with assay mix and is reduced by sequential assays that require two 30-min incubations.

PRODUCT FEATURES

The layout of IMMULITE is illustrated schematically in Figure 27.2. Access to a chain transporting samples and test units to the various stations of the instrument is continuously available to the operator. Stat samples can be placed at the head of the queue without interrupting instrument operation. The pipettor and reagent carousel can be paused to allow for the addition of reagents at any time without interrupting those tests in progress. Up to 12 different reagents can be on board at any time.

Rapid and thorough washing of the coated bead captured in the test unit is achieved by spinning of the tube on its longitudinal axis. Excess sample, reagent, and wash solution are captured in a coaxial waste sump, which is integrated into the test unit (Figure 27.3).

Following the washing of the bead, the chemiluminescent substrate is added and after an additional 10-min incubation the signal is read by a photon counting PMT.

IMMULITE 1000 uses Microsoft Windows 2000 operating software for fast system processing, and the touch-screen monitor provides easier software access

The Immunoassay Handbook. *Third Edition*
D. Wild (Ed.)

(a)

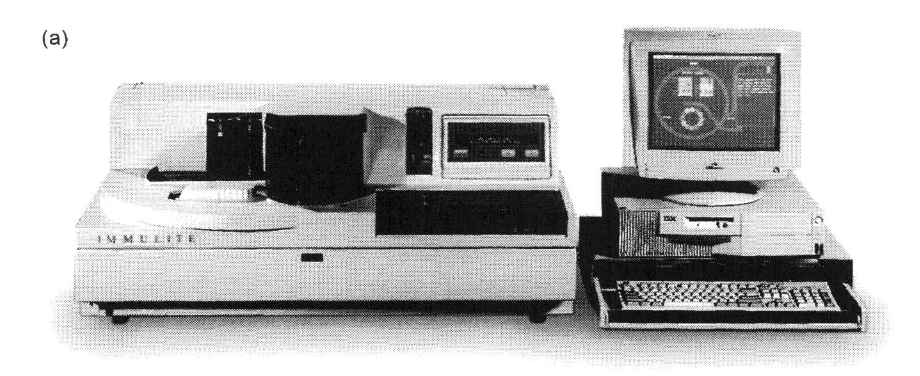

(b)

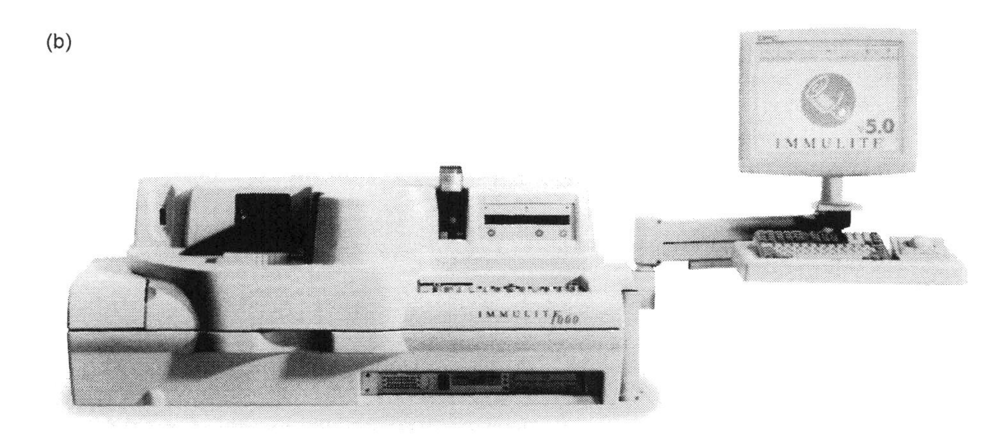

Fig. 27.1 (a) The IMMULITE system. (b) IMMULITE 1000.

and fewer keystrokes. On-board sample dilution is provided for selected assays, and positive sample identification using secondary sample tubes. Remote diagnostics connects the system via modem to DPC for faster technical support. Ten languages are supported: English, French, Italian, German, Spanish, Portuguese, Polish, Turkish, Russian, and Chinese.

ASSAY PRINCIPLE

Both competitive assays for haptens and sandwich assays for large molecules are run with either antibody- or antigen-coated beads. Alkaline phosphatase is the label for all assays. The patented axial centrifugal separation of bound and free label provides low background signal.

CALIBRATION

Each analyte has a stored master calibration curve in the external system computer which is described by a mathematical equation relating photon counts to concentration. Monthly calibration with two calibrators adjusts the relationship for the specific instrument and lot of reagents in use.

ANTIBODIES

The IMMULITE system uses both polyclonal and monoclonal antibodies.

SEPARATION

Separation is a key feature of the IMMULITE system (Figure 27.3). After incubation the test units are automatically transferred to the spin-wash station. The tube is spun rapidly on its longitudinal axis and the slight draft angle of the tube allows for instantaneous and complete transfer of sample and excess reagent from the tube to a coaxial sump chamber around the top of the test unit. Several 200 μL volumes of wash water are added to the tube to wash the bead, and the test unit is spun again after

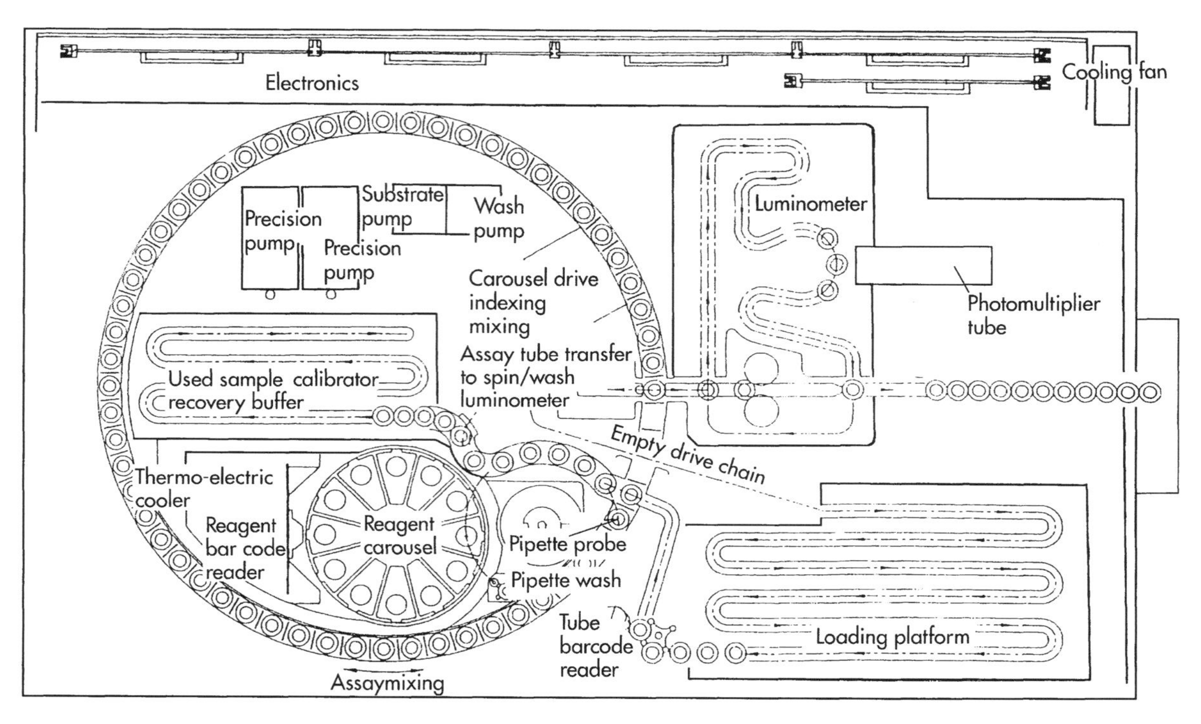

Fig. 27.2 The IMMULITE system, schematic.

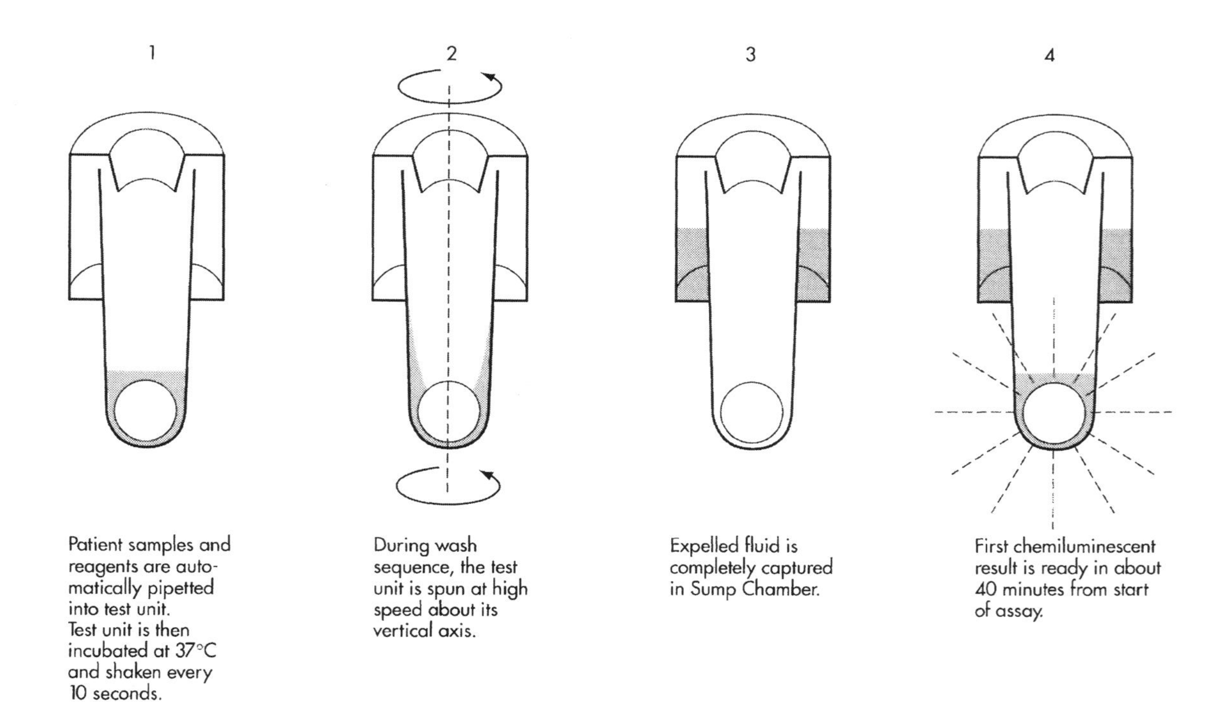

Fig. 27.3 The IMMULITE assay protocol.

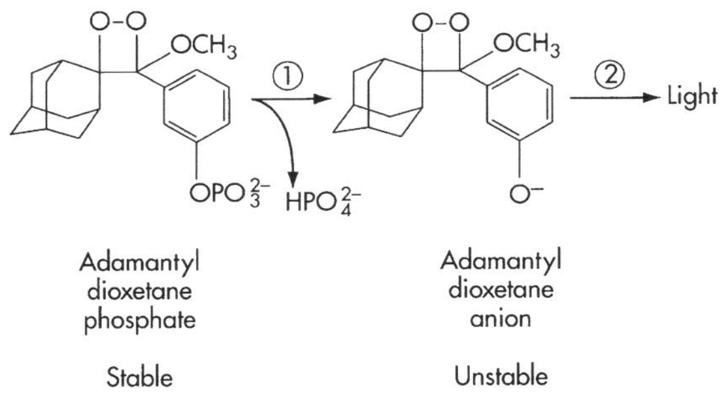

Adamantyl dioxetane phosphate

Stable

Adamantyl dioxetane anion

Unstable

Fig. 27.4 Action of alkaline phosphatase in the IMMULITE system.

each wash, leaving the tube containing the bead totally free of unbound alkaline phosphatase label.

SIGNAL GENERATION AND DETECTION

After the addition of substrate, the test units enter a section of the instrument that excludes ambient light and allows for a 10-min incubation at 37 °C for development of the luminescent signal.

The action of bound alkaline phosphatase on the adamantyl dioxetane phosphate creates the unstable adamantyl dioxetane anion. Breakdown of this unstable anion creates a prolonged 'glow' of light (see Figure 27.4).

When each tube reaches the read station in front of the photomultiplier tube, photon counts are measured for 1 sec through a 2A neutral density filter which attenuates the signal by a factor of 100. If counts are below a certain level, the attenuator is automatically removed. This automatic attenuation increases the dynamic range of the photomultiplier tube 100-fold. Counts are measured for 12 consecutive 1-sec intervals.

DATA PROCESSING

After discarding highest and lowest counts for each tube, the average counts per second are converted to analyte concentration by means of the stored calibration curves. The IMMULITE instrument has sufficient storage capacity to save data should any interruption of the communication with the external computer occur.

INTERFACING TO LABORATORY INFORMATION SYSTEM

The IMMULITE software allows for uploading all results to a laboratory information system.

FURTHER READING

Babson, A.L. The cirrus IMMULITE automated immunoassay system. *J. Clin. Immunoassay* **14**, 83–88 (1991).

Babson, A.L. An automated random-access immunoassay system. *Am. Clin. Lab.* Feb., 12 (1992).

Babson, A.L., Olson, D.R., Palmieri, T. *et al.* The IMMULITE assay tube: a new approach to heterogeneous ligand assay. *Clin. Chem.* **37**, 1521–1522 (1991).

Witherspoon, L.R., Babson, A.L. and Olson, D.R. IMMULITE chemiluminescent immunoassay system. In: *Immunoassay Automation: An Updated Guide to Systems* (ed Chan, D.W.), 102–130 (Academic Press, San Diego, CA, 1996).

28 Bulk Reagent Random-Access Analyzers: ACS:180® SE

Elvio Gramignano

The ACS:180 SE is a completely automated immunoassay system, using chemiluminescent tracers and paramagnetic particle solid-phase reagents, manufactured in the USA by Bayer HealthCare, Diagnostics Division (see Figure 28.1).

The system consists of a bench-top analyzer and several sets of assay reagents intended for measurement of over 49 analytes, including thyroid, fertility and adrenal hormones; anemia, oncology, allergy, cardiac damage and bone metabolism markers; and therapeutic drug levels. In 2004 the ACS:180 BNP assay was launched.

Introduced in October 1997, the ACS:180 SE is the enhanced generation of the ACS:180 Systems, which was originally introduced in 1991. There are more than 1500 such systems installed and operating in clinical laboratories worldwide.

The ACS:180 SE differs from the original ACS:180 and ACS:180 PLUS in that its user interface resides on a separate Pentium®-based PC, running Windows NT® software, and uses a larger color monitor. This new graphical user interface operates with 60% fewer screens compared to the original user interface. The system also has redesigned cabinetry, providing increased test cuvette and solid waste storage capacity for extended unattended operation.

TYPICAL ASSAY PROTOCOL

Assays are performed on the ACS:180 SE using either direct competitive binding or immunometric protocols depending on the assay. All of the steps are performed automatically, with the operator required only to load the samples and reagents, and to generate or download a worklist from an interfaced LIS. Total incubation time is the same for all assays, 7.5 min, and each assay is performed in a discrete disposable plastic cuvette, which

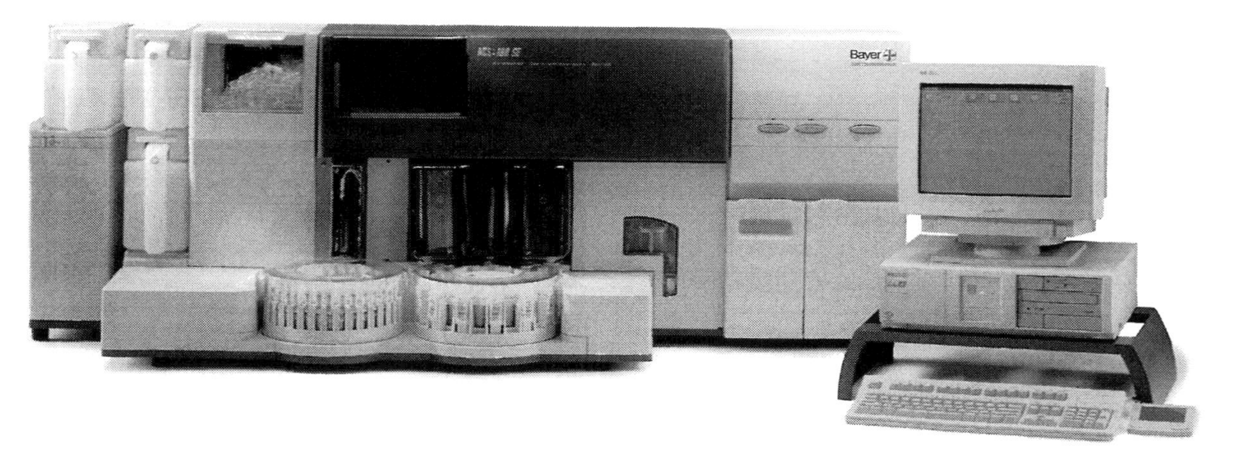

Fig. 28.1 ACS:180 SE.

The Immunoassay Handbook. *Third Edition*
D. Wild (Ed.)

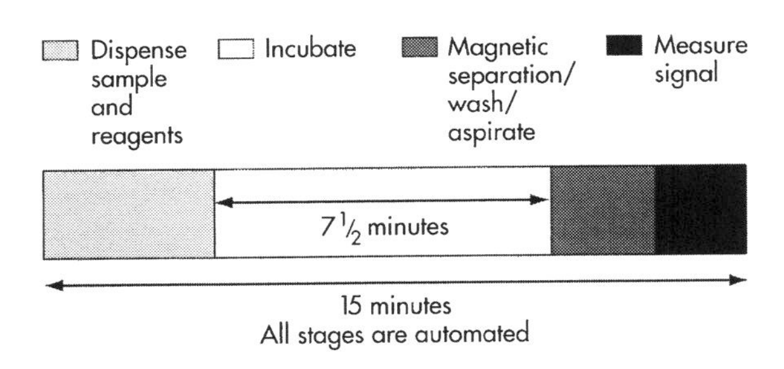

Fig. 28.2 ACS:180 SE test sequence.

travels linearly in 20-sec timed steps through a temperature-controlled reaction track. During this journey, which takes a total of 15 min, sample is dispensed, reagents added at up to four different time intervals for flexibility, separation of free from bound tracer is achieved by magnetic separation of the micron-sized paramagnetic particle solid-phase reagent, and bound tracer in the cuvette is measured by a luminometer component after the tracer is oxidized, yielding a light flash which is concentration dependent. The first result is reported in 15 min, each subsequent result in 20-sec intervals (see Figure 28.2).

PRODUCT FEATURES

- The ACS:180 and ACS:180 SE are self-contained and totally automated. All assays are performed entirely on board the analyzer (Dudley, 1991).
- Assays may be carried out in either batch or random-access mode. Random-access operation allows simple worklist generation, prompt reporting of multiple test results on any sample, and improved laboratory productivity (Weinzierl, 1993; Blick, 1997).
- Sample throughput is up to 180 results per hour, after start-up and processing of first sample, which takes 15 min.
- The ACS:180 SE has a sample tray capacity for 60 samples. After these samples have been accessed, subsequent sample trays may be added, following a brief instrument pause, with minimal delay.
- Stat samples may be added at any time without disrupting normal operation. These samples will be processed immediately, followed by automatic return to the normal worklist.
- Up to 13 different assay reagent sets may be loaded at one time, and any or all of these tests may be performed on any sample in true random-access operation.
- The sample probe is equipped with a patented pressure-sensing clog detection capability, ensuring full aspiration of sample. If a probe clog is detected, as from a clot or residual fibrin, the system notifies the operator. The system may be programmed to either wait for the operator to perform maintenance or to attempt washing the probe clear of the obstruction. If automatic clearing is selected, and the clog is successfully eliminated, the system will continue operation on subsequent samples without interruption, flagging the offending sample.
- The use of chemiluminescence for signal generation has been demonstrated to be at least as sensitive as radioimmunoassay and more sensitive than most non-isotopic alternatives (Ekins, 1987; Weeks, 1983; Woodhead, 1985). The use of acridinium ester tracers also eliminates the need for substrates or signal enhancing reagents because the oxidation process is direct and simple.
- All reagents are non-radioactive with shelf-lives of 3–12 months. Waste may be disposed of through the normal channels.
- Primary sample collection tubes may be loaded directly onto the sample tray, eliminating the need for pouring off serum into sample cups, secondary labeling, etc. Sample cups or secondary tubes may be used, however, if preferred. Barcoded labels may be used for positive sample identification by the system, and several barcode formats are recognized, including Code 39, Codabar, Code 128, and Interleaved 2 of 5.
- Bidirectional interfacing to a central laboratory computer is easily accomplished, allowing downloading of worklists in either batch or host-query mode, and uploading of test results. Data output and communication are organized in ASTM format to facilitate interfacing with a variety of LIS computers.
- The ACS:180 SE is capable of interfacing with robotic sample handling devices and also communicating with central remote control stations, allowing it to be compatible with a variety of totally automated laboratory systems.

ASSAY PRINCIPLE

Assay protocols, which are stored in memory, utilize either competitive binding or immunometric (sandwich) principles, depending on the specific assay. Quantification of results is achieved by oxidation of the acridinium ester tracer reagent, followed by detection and quantification of the resultant intense and short-lived light emission. Assays are performed in discrete, inexpensive, disposable cuvettes, which may be loaded and replenished randomly at any time without stopping or pausing the system (see Figure 28.3).

CALIBRATION

ACS:180 assays are calibrated using a stored, lot-specific master curve of up to 10 concentration points, established by the Quality Assurance Laboratory at Bayer HealthCare Diagnostics. These master curves are

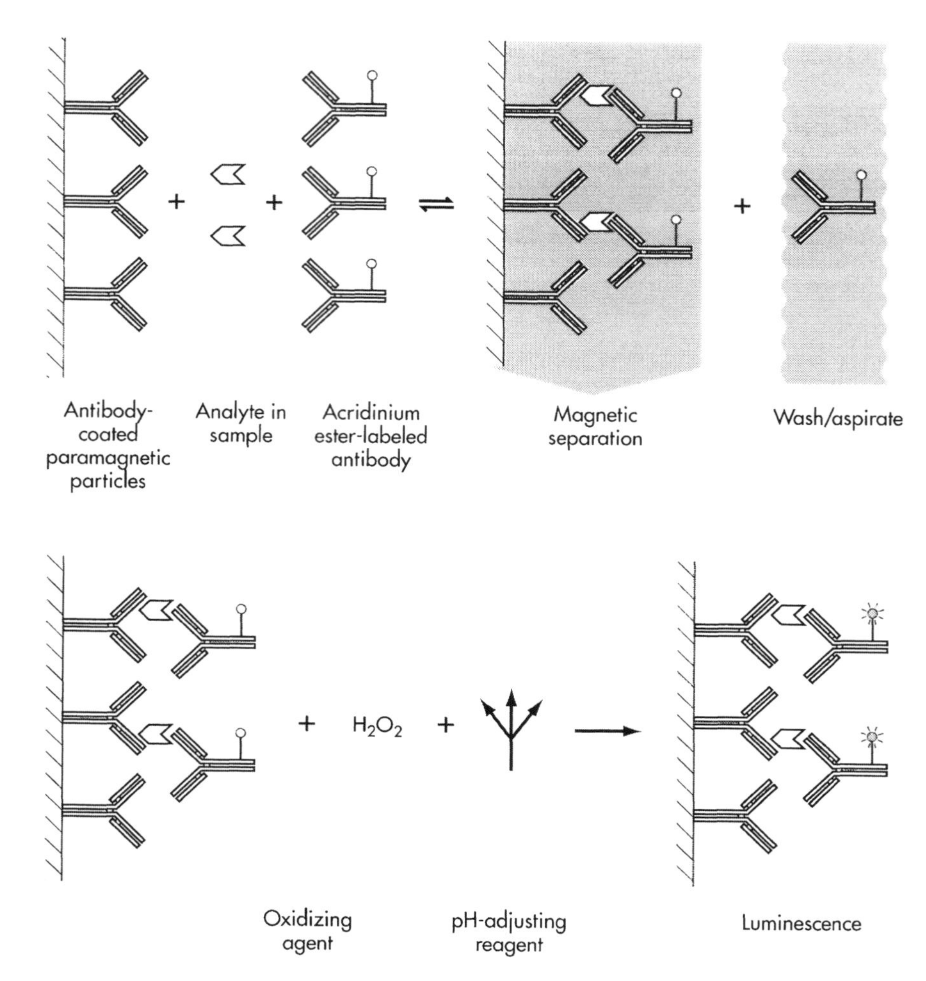

Fig. 28.3 Example of ACS:180 SE assay principle (immunometric type).

supplied with each lot of reagents on a barcoded card for rapid and accurate entry into the memory of the ACS:180 SE analyzer. The master curve is then used by the software to quantify each assay performed with that lot of reagents. Periodic recalibration is accomplished by inclusion of a set of two calibrators, many of which contain multi-analytes. Recalibration intervals vary from 7 to 28 days, depending on the assay, and are tracked and controlled by on-board software. Multiple master curves for each assay may be stored in system memory, simplifying moving from lot to lot.

ANTIBODIES

A mixture of polyclonal and mouse monoclonal antibodies is used in ACS:180 assays. In immunometric assays this multispecies technology minimizes the risk of aberrant results due to heterophilic binding by human anti-mouse antibodies (HAMA), which can occur in dual mouse monoclonal assays. Where dual mouse monoclonal antibodies are utilized, blocking reagents are added to minimize the risk of HAMA effects.

SEPARATION

The solid-phase reagents used in ACS:180 assays are suspensions of micron-sized, specially coated ferric oxide particles. These particles are small enough to provide an extremely high surface area for rapid binding kinetics, and stay in suspension for prolonged periods, but are rapidly and cleanly separated through their paramagnetic attraction when placed in a magnetic field. The ACS:180

SE reagent tray periodically agitates these reagents while on-board, and reagents typically have a minimum of 40 h stability on board at ambient temperature. Individual assay reagent bottles provide sufficient volume for 50 tests.

After dispensing the sample, tracer and solid-phase reagents into the test cuvette and completion of the incubation period, the particles are pulled to the flat wall of the cuvette by permanent magnets mounted on the back of the reaction track. The liquid contents of the cuvette are aspirated, water is injected to resuspend and remove non-specific bound tracer, and reseparation accomplished, automatically.

Water is provided from an on-board reservoir (2.6 L capacity) and liquid waste is collected into another on-board reservoir (2.8 L capacity), both sufficient for over 1 h of typical operation. An optional extended operation module is available, consisting of exterior water and waste reservoirs of 9 L capacity each, connected to the on-board reservoirs through a three-way valve and pumps. This module refills the on-board water reservoir and empties the on-board waste reservoir automatically, extending the unattended time for fluid management to over 4 h. The module also holds a reservoir for cleaning solution, providing the capability for programming automatic daily cleaning and flushing of tubing and probes.

SIGNAL GENERATION AND DETECTION

Following separation and washing of the solid-phase particles, they are resuspended in an oxidizing reagent at an acid pH, stored on-board (flash reagent 1). While held in front of a photomultiplier tube detector, a pH adjusting reagent, also stored on-board (flash reagent 2) is injected, causing the dose-related light emission which is quantified over a period of 5 sec.

DATA PROCESSING

The flash emission, corrected by background subtraction, is directly related to the stored master curve for quantification. Results are calculated and reported as mass concentration and flagged if out of range, based upon operator-controlled reporting formats. The software allows automatic conversion to SI units, if desired. Results may be reported individually as completed, or collated into patient reports containing all test results performed on a given sample.

If desired, the ACS:180 SE will automatically repeat a test on a sample, with or without dilution, if the initial result is outside an operator-programmed range.

Up to 10 different controls for each analyte may be tracked through a rolling 60-day history, and results displayed in tabular or Levey–Jennings format. Monthly summaries are also accessible for a period of 12 months. Up to 50,000 orders and/or test results are stored in on-board memory, and the most recent 10,000 test results per assay are displayed in an assay-specific histogram format, tracking assay performance and validation of stored expected ranges.

USER INTERFACE

The User Interface of the ACS:180 SE consists of a separate, connected Pentium-based computer, 15-inch color monitor, keyboard, and choice of standard mouse or Glidepoint® pointing device. The software is in Windows® NT format, allowing several screens to be accessed simultaneously. The use of intuitive icons and full on-line help throughout the software makes training and routine operation rapid and simple to master.

A built-in, high-speed CD-ROM reader allows rapid and easy loading of software and rapid access to the on-line operator's manual, including extensive graphics. Step-by-step instructions for activities such as maintenance, system diagnostics, and troubleshooting are immediately available from anywhere in the software.

INTERFACING TO LABORATORY INFORMATION SYSTEMS

Bidirectional interfacing to any central laboratory computer is achievable for downloading of worklists and uploading of test results, using barcoded samples, including host-query operation. Interfacing software is written in standardized ASTM format for consistency with a wide variety of LIS systems. Additionally, simple software selections may be made to allow robotic continuous sample loading/unloading, and also for remote operation from a central control station, allowing integration of the ACS:180 SE into totally automated laboratory systems.

REFERENCES AND FURTHER READING

Blick, K.E. Cost effective workstation consolidation using the Chiron ACS:180™ and Valunalysis process. *J. Clin. Lig. Assay* **20**, 265–268 (1997).

Dudley, R.F. The Ciba Corning ACS:180 automated immunoassay system. *J. Clin. Immunoassay* **14**, 77–82 (1991).

Ekins, R.P. An overview of present and future ultrasensitive non-isotopic immunoassay development. *Clin. Biochem. Rev.* **8**, 12–23 (1987).

Weeks, I., Behesti, I., McCapra, F. *et al.* Acridinium esters as high specific-activity labels in immunoassay. *Clin. Chem.* **29**, 1474–1479 (1983).

Weinzierl, C. Impact of random-access automation on the immunoassay laboratory. *Lab. Med.* **24**, 717–723 (1993).

Woodhead, J.S. Immunoassays based upon non-radioactive labels. Presented at Medica, Dusseldorf, Germany (1985).

29 Bulk Reagent Random-Access Analyzers: AxSYM®

Theresa Donahoe

The Abbott AxSYM® is a fully automated floor-standing analyzer that provides random- and continuous-access testing for more than 70 assays (Figure 29.1). Over 11,000 AxSYMs are in use worldwide and over 700 million tests have been run by customers since the system was introduced to the market in 1993.

AxSYM utilizes four assay technologies: microparticle enzyme immunoassay (MEIA), fluorescence polarization immunoassay (FPIA), radiative energy attenuation (REA®), and ion-capture immunoassay (ICIA) – which originated from the IMx® and TDx® batch analyzers. Tests on AxSYM may be processed in one- or two-step sandwich or competitive formats, with programmable pipetting sequences, incubation periods, washes and optical read formats. Menu capabilities for both systems include tests for hepatitis, retrovirus, congenital diseases, thyroid and metabolic function, reproductive endocrinology, tumor markers, cardiovascular markers, therapeutic drugs, and drugs of abuse.

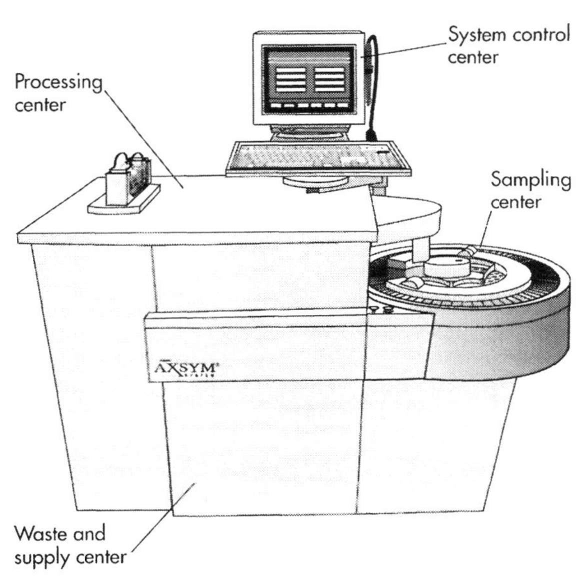

Fig. 29.1 AxSYM instrument.

The AxSYM has four main areas: a sampling center (Figure 29.2), a processing center (Figure 29.3), a supply center, and a system control center. The sampling center contains carousels for loading samples (including calibrators and controls), reagents, disposable reaction vessels and matrix cells that are used to process the tests, and an automated pipettor for transferring sample and reagents. Tests are completed in the temperature-controlled processing center. The supply center provides storage for four bulk solutions as well as liquid and consumable waste. The system control center consists of a color monitor with integrated touch screen, keyboard, printer, disk drives, interface ports, and barcode reader. The system control center is used for order entry and reviewing results, quality control functions, and system configuration, maintenance and diagnostics.

TYPICAL ASSAY PROTOCOL

All assays are run by ordering a test, loading the sample and the appropriate reagent pack onto the sampling center, verifying sufficient inventory for reaction vessels, matrix cells and bulk solutions, verifying that waste containers are not full and pushing the 'RUN' button. Tests may be ordered directly through the AxSYM system control center or downloaded from a host computer. If entered directly, the operator may control the position of the sample within the sampling carousel, or configure the system to automatically assign the sample location. Except for stat samples, which are given priority in both

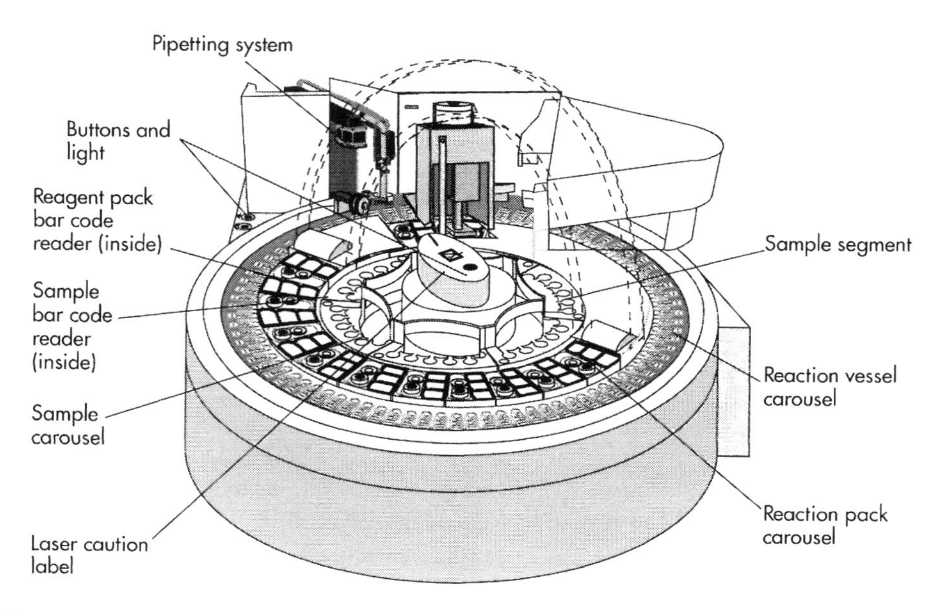

Fig. 29.2 AxSYM sampling center.

sampling and processing centers, samples are processed in the order in which they are presented to the pipettor.

When the sample tube arrives at the pipetting station, the system scans the sample barcode to positively identify the sample. The system checks the order list to determine the tests to be run and verifies reagent and consumable inventories. The system then schedules the activities needed to process the test as defined by the flexible assay protocol. The sampling center aligns the next reaction vessel (see Figure 29.4) and the correct

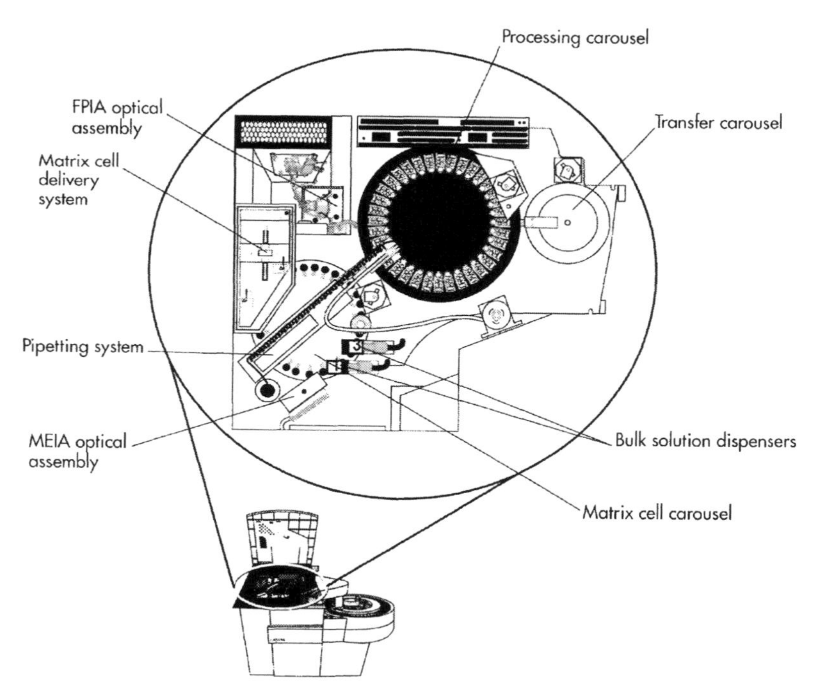

Fig. 29.3 AxSYM processing center.

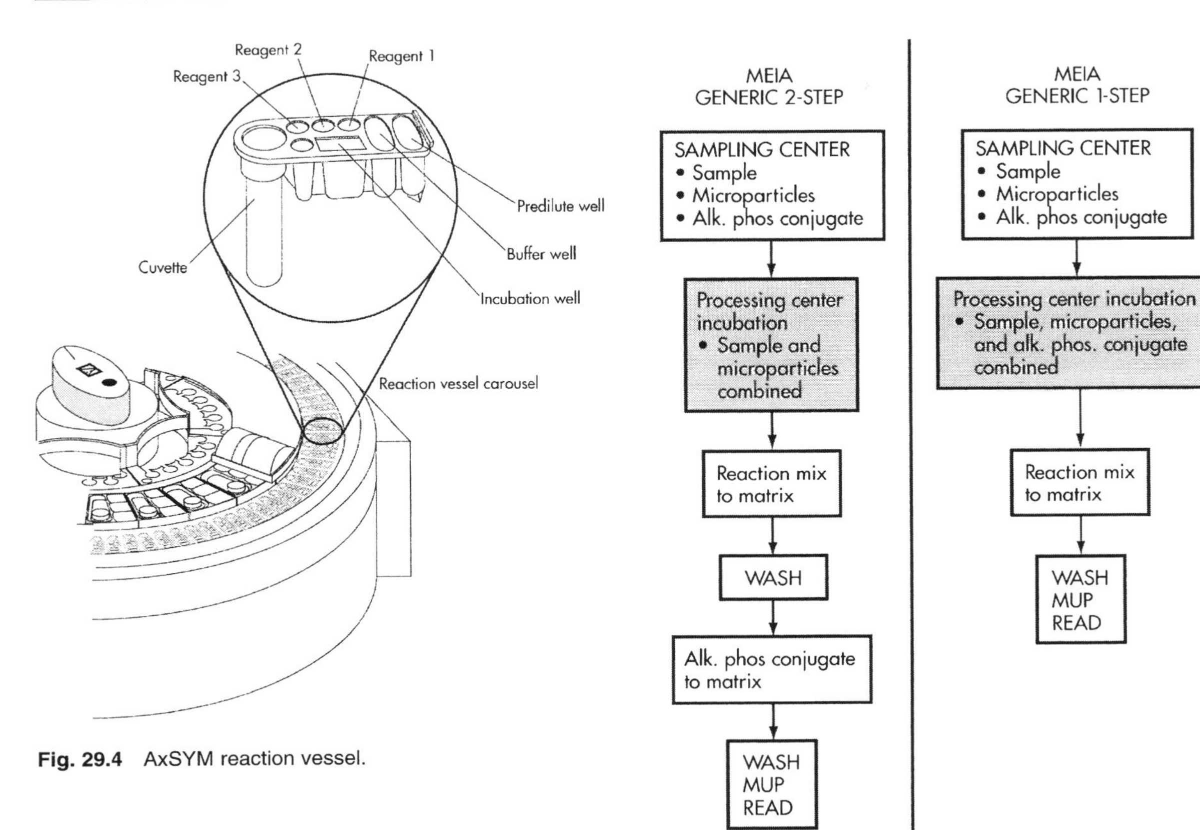

Fig. 29.4 AxSYM reaction vessel.

Fig. 29.6 Typical reaction sequences for MEIA assays.

Fig. 29.5 AxSYM matrix cell.

reagent pack with the pipetting station. A reagent pack actuator opens reagent bottles 1–3. The pipettor then transfers sample and reagents for the requested test into a reaction vessel, performing any mixing or dilution steps required by the assay protocol.

The reaction vessel is then automatically transferred to the processing center, where subsequent incubations, mixing, and wash steps occur. For FPIA and REA assays, the reaction is read through a cuvette by the fluorescence polarization reader. MEIA and ion capture assays require matrix cells to complete the assay sequence (Figure 29.5). The system automatically loads a matrix cell for each MEIA and ICIA test as dictated by the scheduler. MEIA and ICIA wash steps, substrate reactions, and optical reads occur on the matrix cells. Typical reaction sequences are shown in Figures 29.6–29.9.

PRODUCT FEATURES

- AxSYM has broad menu capabilities, including more than 70 assays spanning diverse clinical applications. Additional assays are planned or under development.

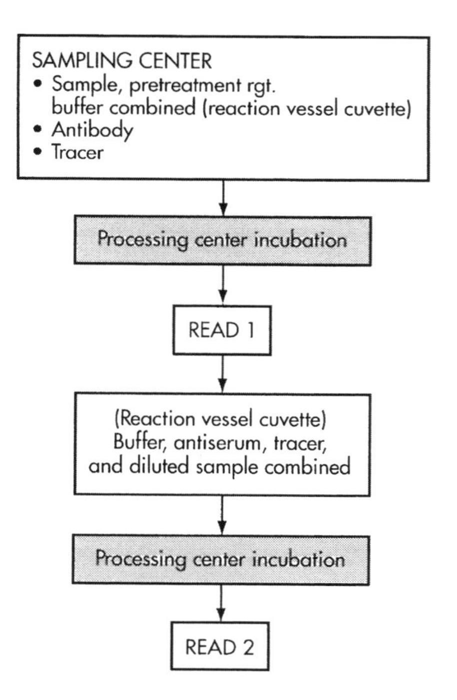

Fig. 29.7 Typical reaction sequence for FPIA assays.

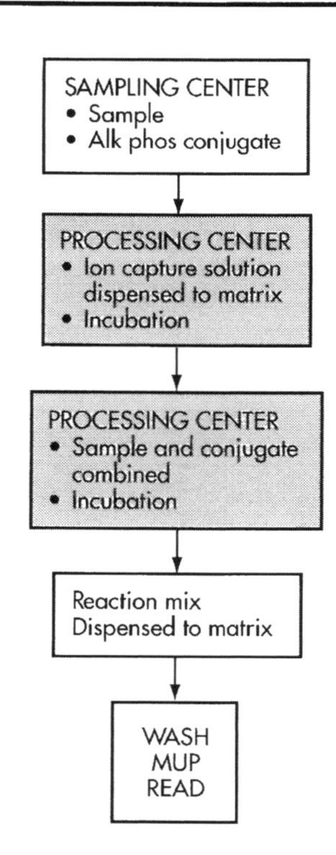

Fig. 29.8 Typical reaction sequence for ICIA assays.

- AxSYM improves laboratory work flow. It is fully automated and offers immediate, random, and continuous access. Samples and reagents may be added or removed at any time by simply pausing the sampling center movement and may be loaded in any available location on the respective carousel. Stat tests are prioritized upon order and stat results are available within 15 min. The system may also be configured for autodilution and reflex testing.
- AxSYM increases laboratory efficiency and productivity. AxSYM has on-board capacity for 20 barcoded reagent packs (100 tests each), and for up to 60 primary or aliquot sample tubes or for up to 90 sample cups. Average throughput for AxSYM is 80 (up to 120) tests per hour. Walkaway time is approximately 1 h. The system is easy to maintain and requires less than 15 min daily maintenance. AxSYM's user interface is logical and easy to use.
- AxSYM helps maintain laboratory quality. AxSYM has positive sample ID and supports Levey–Jennings QC and calibration review.
- AxSYM controls sample carryover by a proprietary, patented 'Smart Wash' system.

ASSAY PRINCIPLE

AxSYM incorporates four technologies for processing immunoassays. MEIA is a heterogeneous technology used for high molecular weight analytes as in IMx. FPIA, as in IMx and TDx, and REA, as in TDx, are homogeneous technologies used for low molecular weight analytes. ICIA is a heterogeneous technology that may be used for either low- or high-molecular-weight analytes and offers the potential for improved precision for some assays. (*see* IMx for details of MEIA, FPIA, and ICIA assay principles).

REA assays involve color development reactions. Their reaction systems use analyte to convert a chromogen (unreacted dye) to a chromophore (colored dye). A stable fluorescent substance (fluorophore) is also included in the reaction mixture. The light-absorbing properties of the chromophore produced cause a decrease of measured fluorescent light intensity from the fluorophore. The FPIA optics is used to measure this decrease in fluorescent intensity. Figure 29.10 illustrates the assay principle.

CALIBRATION

Quantitative assays employ a six-point calibration. However, many AxSYM Reagent Packs have a card with a barcode label containing a Master Curve generated in the factory using six calibrators for a particular reagent lot. An operator may elect to run two calibrator adjusters which are used to adjust the bar-coded Master Curve. This procedure is called a Master Cal. An operator may also

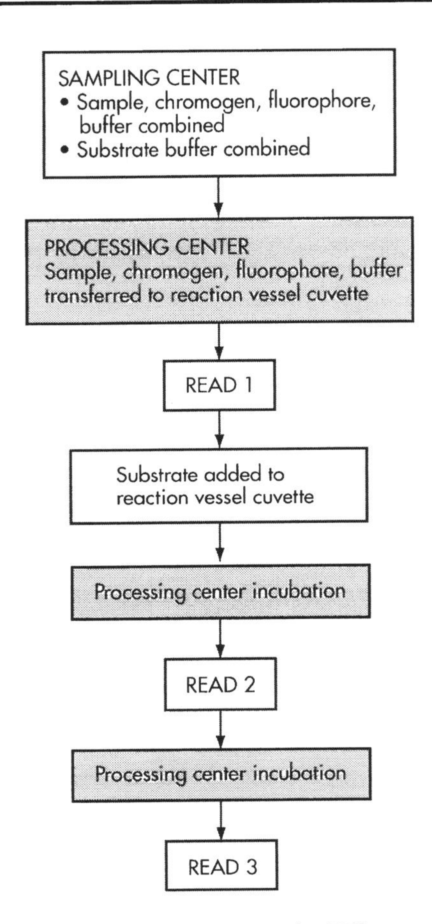

Fig. 29.9 Typical reaction sequence for REA assays.

choose to run a Standard Cal, in which six calibrators are run in duplicate to calibrate the assay. A Standard Cal is typically performed when the control values fall outside the specified range or in accordance with the policy of the individual laboratory.

Qualitative assays, also known as index or screening assays, use a single calibrator run in duplicate to establish a cut-off value. This is called an Index Cal.

There is no expiration time for validity of a calibration. However, assay controls must be run every 8 h as a quality assurance check in accordance with package insert instructions. Individual laboratories may elect to recalibrate if assay controls are out of specification, a new lot of bulk solution is used, a dispense system component is replaced, or specific system maintenance is performed. The AxSYM can store one active calibration curve for up to four different reagent lots for a given assay. If a fifth curve is added, it replaces the oldest active curve. Reagent packs have between 112 and 336 cumulative hours on-board stability on the AxSYM (between 14 and 42 eight-hour shifts).

ANTIBODIES

Both polyclonal and monoclonal antibodies are used, depending on the assay design.

SEPARATION

MEIA and ICIA assays are heterogeneous, requiring a separation of the bound and unbound fractions during the procedure. In MEIAs, the bound fraction is immobilized on uniform latex microparticles, which bind to a glass-fiber matrix. Unbound material is washed through the glass-fiber matrix into an absorbent, opaque blotter below, using prewarmed buffer. In ICIA assays, the negatively charged polyanion–analyte complex binds to the positively charged quaternary ammonium compound that coats the glass-fiber matrix. Unbound material is then washed through the glass-fiber matrix as in MEIAs.

FPIA and REA assays are homogeneous and do not require a separation step.

SIGNAL GENERATION AND DETECTION

Signal generation for MEIA, FPIA, and ICIA assays occurs as in IMx. (*see* IMx for details). The design of the electronic signal conversion in AxSYM, however, gives a dynamic range exceeding that of IMx.

REA assays use the FPIA optical reader to measure the decrease in fluorescent intensity at 448 nm (Figure 29.10). This signal follows Beer's Law and is calculated as percent transmission.

DATA PROCESSING

Analyte concentration for quantitative assays is determined from the rate of fluorescence, fluorescence

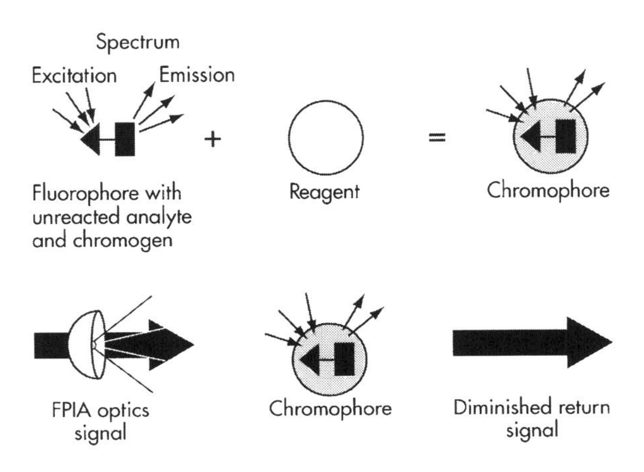

Fig. 29.10 AxSYM REA assay principle.

polarization, or percent transmission data. Calibration curve-fitting is based on point-to-point, linear regression, third-order polynomial regression or four-parameter logistic algorithms. Qualitative assays are evaluated using a signal-to-noise or sample-to-cutoff ratio formula defined for the individual assay.

Numerous assay error checks are performed before an assay result is generated. Assay calibrations include checks for excessive deviation between replicates of a calibrator, for calibrator rates that are out of specification, and for deviations in curve shape. Assay controls are checked against specifications. Laboratories can also configure error checking of controls based on their QC ranges. Error checking is also performed for patient samples. For MEIAs and ICIA tests, read precision and rate correlation are checked. The MUP (4-methylumbelliferyl phosphate) intercept value is also checked to prevent erroneous results due to the presence of red blood cells (low MUP intercepts) or degradation of MUP (high MUP intercepts). For FPIA assays, net intensity and polarization are checked. For REA assays, baseline intensity and deviation of the end point are also checked to establish acceptable background and sample integrity.

System error checks are also performed before assay results are generated. These include barcode read checks, robotics and sensor checks, and liquid level sense checks. All are designed to ensure the integrity of the assay result.

AxSYM provides storage and Levey–Jennings graphs for up to 31 days of assay controls.

INTERFACING TO LABORATORY INFORMATION SYSTEMS

AxSYM includes a bi-directional RS232 serial communication port operating at 1200, 2400, 4800, or 9600 baud rate in accordance with the ASTM standard.

FURTHER READING

Black, M., Schubert, D., Dawson, M., Clemens, J., Elmore, J., Fries, P., Hawkins, M., Osikowicz, G. and Ford, K. Minimization of sample carryover on Abbott AxSYM system. *Clin. Chem.* **40**, 1116 (1994).

Sanchez, B., Sobrino, J., Raymoure, W. and Ford, K. Evaluation of work flow with the Abbott AxSYM system, a random and continuous access immunoassay analyzer. *Clin. Chem.* **40**, 1047 (1994).

Shaffer, M. and Stroupe, S.D. A general method for routine clinical chemistry in the Abbott TDx® analyzer. *Clin. Chem.* **29**, 1251 (1983).

Smith, J., Osikowicz, G. *et al.* Abbott AxSYM random and continuous access immunoassay system for improved workflow in the clinical laboratory. *Clin. Chem.* **39**, 2063–2069 (1993).

30 Bulk Reagent Random-Access Analyzers: Elecsys® Immunoassay Systems

Mary Beth Myers

The Elecsys family of automated immunoassay systems was designed to meet the needs for cost efficiency and standardize testing across small-, medium-, and high-volume laboratories in integrated healthcare networks, hospitals, and independent laboratories. More than 40 analytes are currently available, including cardiac markers (including NT-proBNP), tumor markers, bone markers, infectious disease, thyroid function tests, anemia and fertility tests.

The Elecsys family includes the Elecsys 1010 System (Figure 30.1), the Elecsys 2010 System (Figure 30.2), and the E 170 Elecsys Immunoassay Module. The Elecsys 1010 System is designed for small- and medium-volume laboratories. It is also suitable for dedicated applications such as STAT testing. The Elecsys 2010 System is engineered for continuous, random-access operation in medium- to large-volume laboratories. For expanding workloads, the Elecsys 2010 System can be equipped for sample rack handling with the capacity for 100 samples. Based on the same five-position sample rack used on Roche/Hitachi chemistry systems, the Elecsys 2010 rack system facilitates sample flow by allowing sample racks to be transferred directly from the Elecsys 2010 System to Roche/Hitachi and Integra systems.

For high-volume laboratories, the E 170 Elecsys Immunoassay Module for MODULAR *SYSTEMS*™ offers a throughput of up to 510 tests per hour and 100 reagent packs onboard. The E 170 can be configured as a stand-alone system with up to three modules or as part of MODULAR *ANALYTICS* for integrated chemistry and immunodiagnostics testing (Figure 30.3). MODULAR Pre-Analytics *SYSTEMS* provides total lab integration, encompassing pre-analytics, chemistry and immunodiagnostics. This concept allows serum samples to be processed, from centrifugation and decapping to final results, without manual handling.

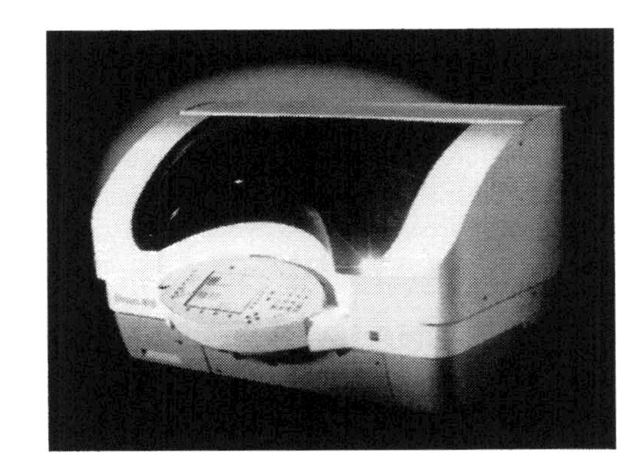

Fig. 30.1 Elecsys 1010.

Fig. 30.2 Elecsys 2010 rack system.

The Immunoassay Handbook. *Third Edition*
D. Wild (Ed.)

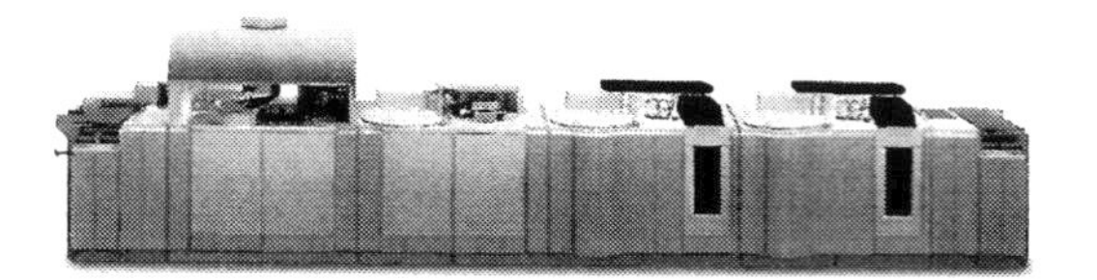

Fig. 30.3 Integrated chemistry and immunochemistry MODULAR *ANALYTICS*.

Every Elecsys system is designed for simplicity of operation, training, and maintenance. All systems use the same liquid, ready-to-use reagents, with the same packaging, making it easy to standardize operation procedures and compare patient results across different laboratories in an integrated health network.

The Elecsys systems use electrochemiluminescence (ECL) detection technology, which was developed by IGEN International, Inc. In 2004, Roche acquired IGEN International. Elecsys instrumentation was developed jointly by Roche Diagnostics and Hitachi. Hitachi manufactures the Elecsys 2010 and E 170 systems.

Elecsys reagents are developed and manufactured by Roche. The Elecsys family was launched globally in 1996. Approximately 4000 systems are currently in use worldwide.

TYPICAL ASSAY PROTOCOL

The Elecsys technology is very flexible and can be applied to sandwich assays for high-molecular-weight analytes, competitive assays for haptens, bridge assays for antibody measurement, and DNA/RNA probes.

PRODUCT FEATURES

- Representative incubation times are 9 min for critical assays (e.g. CARDIAC T® troponin T assay, CK-MB, myoglobin, and hCG) and 18 min for routine tests.
- The Elecsys 1010 System features six assays onboard and a throughput of approximately 50 tests per hour. There are 42 positions for primary tubes and 24 positions for secondary cups. STAT samples can be introduced into normal routine operation at any time and are assigned the highest priority for processing. Operator interface is via a graphical monitor with custom soft keys (Figure 30.4).
- The Elecsys 2010 System features up to 15 assays onboard and a throughput of 88 tests per hour. In addition to a 30-position sample disk, a 100-position rack is available. Primary tubes and secondary cups can be used without special adapters. STATs are processed with priority without interrupting the routine workflow. The operator interface consists of a color touch screen and customized keyboard (Figure 30.5).

Fig. 30.4 Elecsys 1010 System.

- E 170 Elecsys Module has a maximum sample throughput of 170 tests per hour, and up to three modules can be connected in one system for a total throughput of 510 tests per hour. As many as 25 different assays can be installed on each module. With three modules, the maximum number of reagent packs onboard is 75. STAT samples can be introduced at any time. Reruns are handled automatically. A single graphical interface controls up to six chemistry and/or E 170 modules plus ISE.
- Liquid level detection, clot detection, and automated sample dilution are standard for all Elecsys systems.
- Liquid, ready-to-use reagents are available in packages of 100- or 200-test packs. Only two points are needed to confirm calibration since the master curve is encoded in the two-dimensional barcode on each reagent pack. Shelf life of reagent kits is 18 months

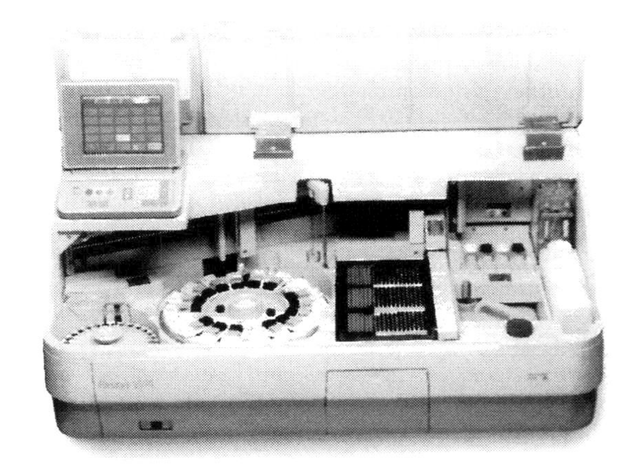

Fig. 30.5 Elecsys 2010 Disk System.

after manufacturing. Onboard stability for the majority of assays is 8 weeks.

- Elecsys systems provide programming-by-loading (PBL), a technology that initiates test and calibration selection when barcoded reagents and calibrators are placed on the system.
- All Elecsys systems support standard primary tube barcode symbologies.
- Disposable tips eliminate carryover.
- ECL technology provides a broad dynamic range, with a signal response that is linear over more than six orders of magnitude. Since the ECL reaction is initiated electrically, the imprecision inherent in enzyme detection systems is eliminated. The amplifying property of the label molecule permits the detection of extremely low concentrations of analyte.

ASSAY PRINCIPLE

The Elecsys assay combines conventional antigen–antibody reactions on the surface of a streptavidin-coated paramagnetic microparticle with electrochemical reaction on the surface of an electrode, which generates luminescence (Figure 30.6).

- A biotin-labeled antibody and a ruthenium-labeled antibody are incubated with the sample analyte. An antigen or nucleic acid probe may be substituted for either antibody to accommodate other assay types.
- The immune complex is captured by the streptavidin-coated microparticles, which have a high biotin binding capacity.

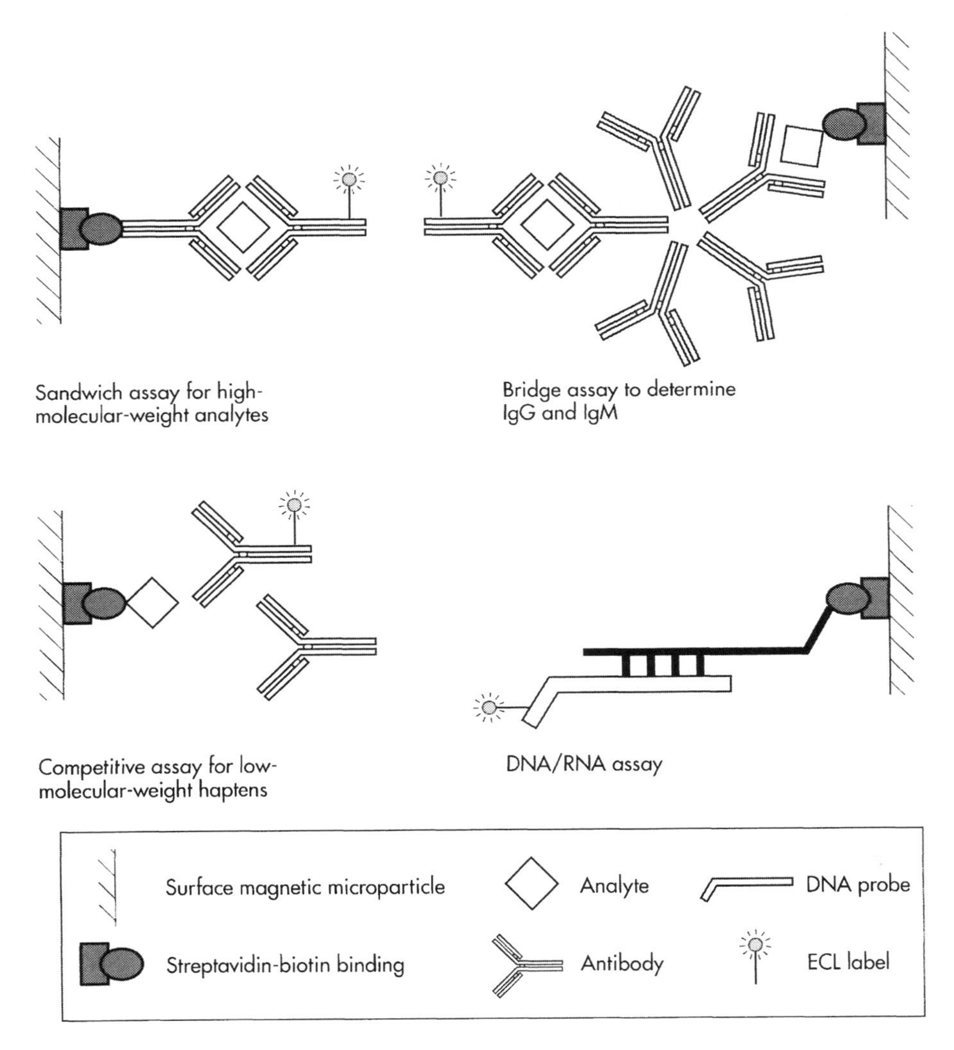

Fig. 30.6 Elecsys assay formats.

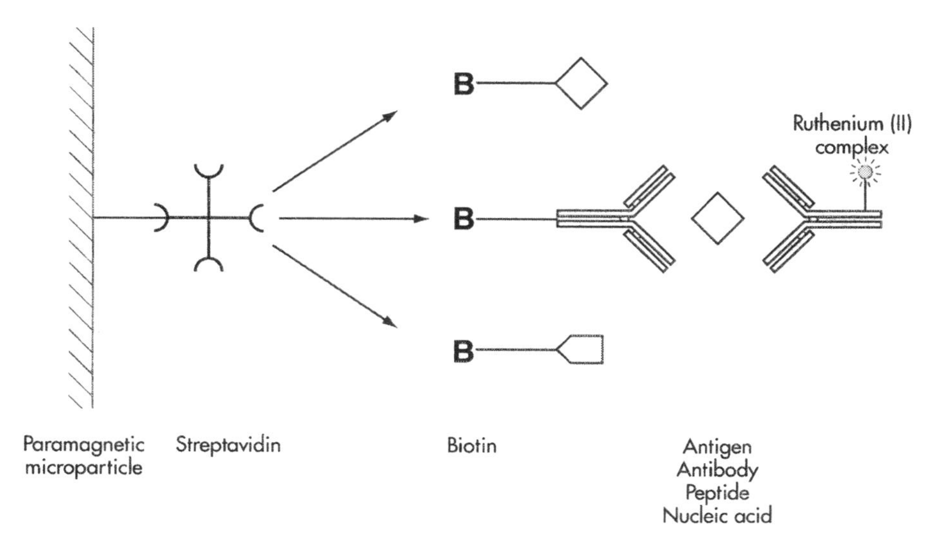

Fig. 30.7 Elecsys assay principle.

- Inside the ECL measuring cell, the microparticles with their bound immune complexes are uniformly deposited on the electrode. Unbound components are washed away.
- The sample analyte is quantitated by applying a voltage to the electrode and measuring the ECL signal.
- Once the ECL reaction is completed, the magnetic microparticles are released from the surface and washed away. The surface is thoroughly cleaned, and the cell is ready for another measurement (see Figure 30.7).

CALIBRATION

Only two points are needed to confirm calibration since the master curve is encoded in the two-dimensional barcode on the reagent pack.

A lot-specific master calibration curve ($n = 5$ or 6) is generated at the manufacturing site using lot-specific test kit reagents and master calibrators. The data characterizing this curve are stored in the reagent barcode. Calibrator values are assigned based on the master calibration curve and are encoded in the calibrator barcode.

At the customer site, results from the two-point calibration are combined mathematically with the encoded master curve data, based on which analyte concentration is calculated (Figure 30.8).

Calibration frequency varies but is typically once per lot, every month (same lot) or every 7 days (same kit on the system).

ANTIBODIES

Antibodies are carefully selected to optimize assay performance. About 75% of the Elecsys assays use

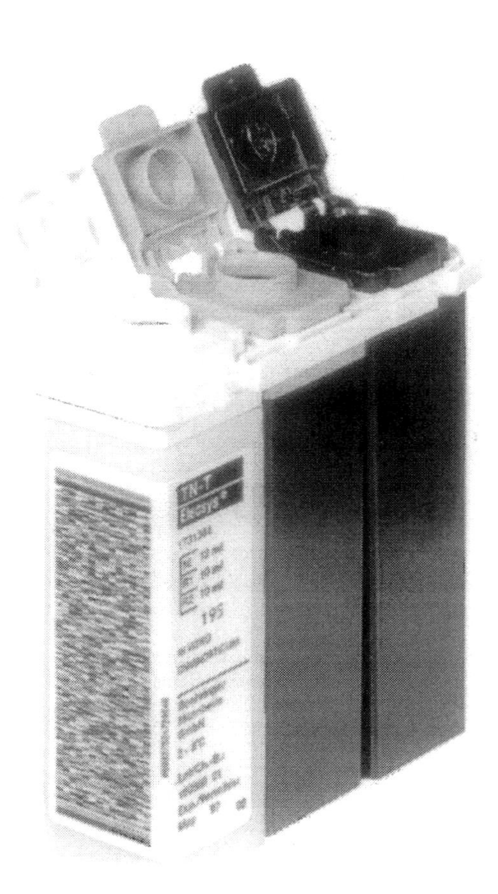

Fig. 30.8 Elecsys reagent pack, showing two-dimensional barcode.

monoclonal antibodies. Examples are troponin T, thyroid-stimulating hormone (TSH), and prostate-specific antigen (PSA). The remaining 25% use polyclonal antibodies selected for specificity to the analyte. Examples are triiodothyronine (T_3) and thyroxine (T_4). Monoclonal antibodies offer many advantages. Increased sensitivity and specificity along with consistency in manufacturing are primarily the reasons for the migration from polyclonal antibody use. Monoclonal antibodies, however, are more susceptible to heterophilic antibody interference than polyclonals. Elecsys assays that utilize monoclonal antibodies have additives to either minimize the effects of heterophilic antibodies or adsorb out the heterophilic reactive antibodies. In addition, the emergence of human/mouse chimeric antibodies has been shown to be more effective in neutralizing heterophilic interference than the addition of mouse serum or purified mouse IgG. The chimeric antibody, by its design, does not afford a binding site for the heterophilic antibody thereby eliminating the interference. The Elecsys CEA assay utilizes chimeric antibodies.

SEPARATION

The streptavidin-coated microparticles bound with the antigen–antibody complexes are magnetically captured onto the surface of the working electrode by the introduction of a magnetic field. A system buffer (ProCell) is used to wash the particles on the working electrode and to flush out the excess reagent and sample materials from the measuring cell.

SIGNAL GENERATION AND DETECTION

Two electrochemically active substances, the ruthenium label and tripropylamine (TPA), are involved in the reactions leading to the emission of light. The reactions occur at the electrode surface inside the measuring cell.

As an electrical potential is applied, the ruthenium label is oxidized at the electrode surface; simultaneously, TPA is oxidized to a radical cation that spontaneously loses a proton. The resulting TPA radical reacts with oxidized ruthenium, resulting in the excited state of the ruthenium label, which decays with the emission of a photon (620 nm). As the ruthenium label returns to ground state, it is regenerated and available to perform multiple light-generating cycles (Figure 30.9).

The subsequent light emission is then detected by the photomultiplier tube and converted to an electrical signal based on which analytes are quantitated.

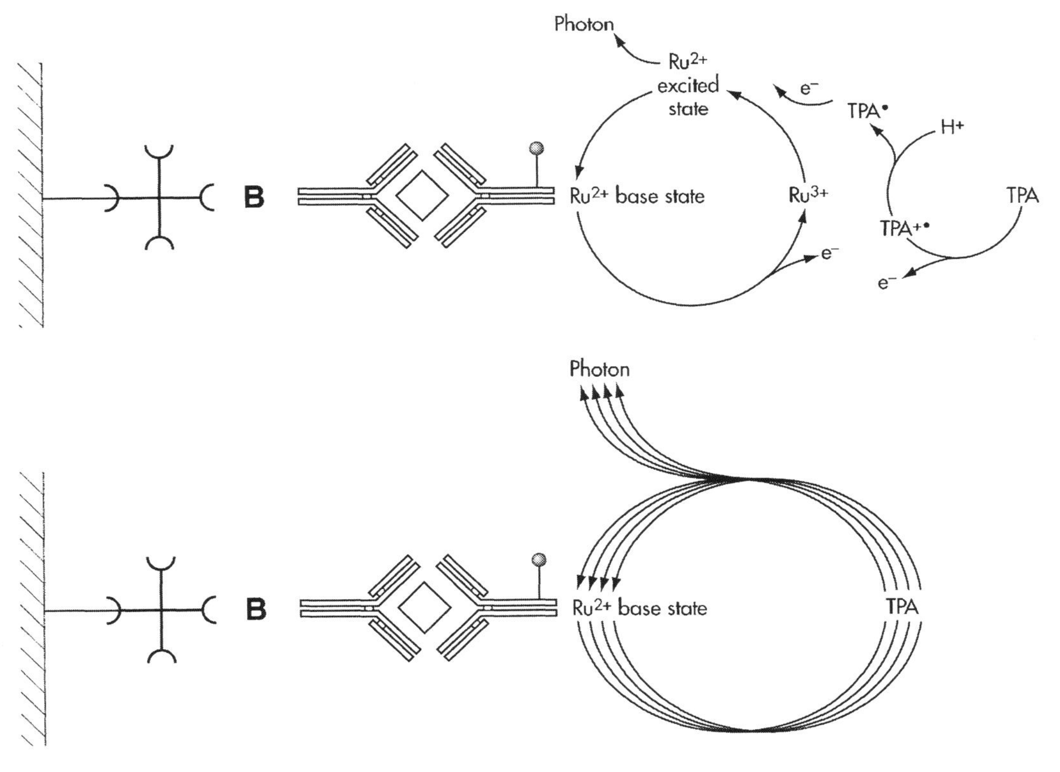

Fig. 30.9 Electrochemiluminescence (ECL) signal generation and detection.

DATA PROCESSING

Data transmission to and from the analyzer, results evaluation, documentation, and quality control are performed automatically by the system software, which also manages the data interface with the laboratory information system (LIS).

INTERFACING WITH LABORATORY INFORMATION SYSTEMS

Elecsys systems feature a bidirectional interface with host-query capability for connecting to LISs.

Elecsys and MODULAR and CARDIAC T are trademarks of a member of the Roche Group.

FURTHER READING

Alpert, N.L. Elecsys™ 2010 immunoassay system. *Clin. Instrum. Syst.* **15** (1998).

Blackburn, G.F., Shah, H.P., Kenten, J.H. *et al.* Electrochemiluminescence detection for development of immunoassays and DNA probe assays for clinical diagnostics. *Clin. Chem.* **37**, 1534–1539 (1991).

Leuther, M. Detecting ultrasensitive immunoassays. *Advance/Laboratory*, May (1997).

Kuroki, M., Matsumoto, Y., Arakawa, F. *et al.* Reducing interference from heterophilic antibodies in a two-site immunoassay for carcinoembryonic antigen (CEA) by using a human/mouse chimeric antibody to CEA as the tracer. *J. Immunol. Methods* **180**, 81–91 (1995).

31 Bulk Reagent Random-Access Analyzers: Vitros® ECi

David Wild

The Vitros® ECi Immunodiagnostic System is fully automated, providing random-access testing for a wide range of analytes with continuous access to the samples and reagents (Figure 31.1). It is manufactured by Ortho-Clinical Diagnostics, a Johnson & Johnson company (formerly the Clinical Diagnostics Division of Eastman Kodak Company). The immunoassay technology, with a signal generation system based on enhanced chemiluminescence, was first used on Amerlite™ (developed by Amersham International, now GE Healthcare). The system's engineering concepts originate from Ektachem™ (now VITROS 250, VITROS 950, and VITROS 5,1 FS Chemistry Systems), first developed by Eastman Kodak Company. However, the immunochemistry, reagent consumables, and system engineering were the result of a 6-year integrated development program, started by Eastman Kodak and completed by Ortho-Clinical Diagnostics, with the aim of providing exceptional assay performance with a high level of automation and a user-friendly interface. A key feature of the system is its ability to accommodate one and two-step assays, with multiple incubation times, enabling each assay to be separately optimized with respect to performance and speed.

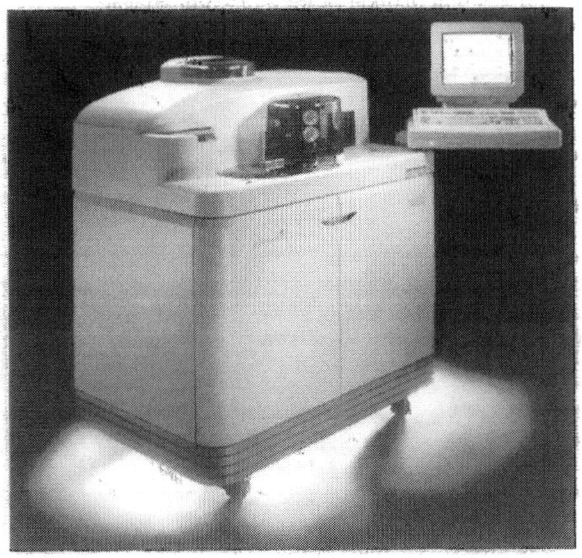

Fig. 31.1 Vitros ECi analyzer.

The Vitros ECi System (Figure 31.2) has four main areas:

Sample processing. The sampling center accommodates up to 60 samples in universal sample trays that can accept primary and secondary tube types as well as microcontainers. Sample trays may be loaded or unloaded during operation.

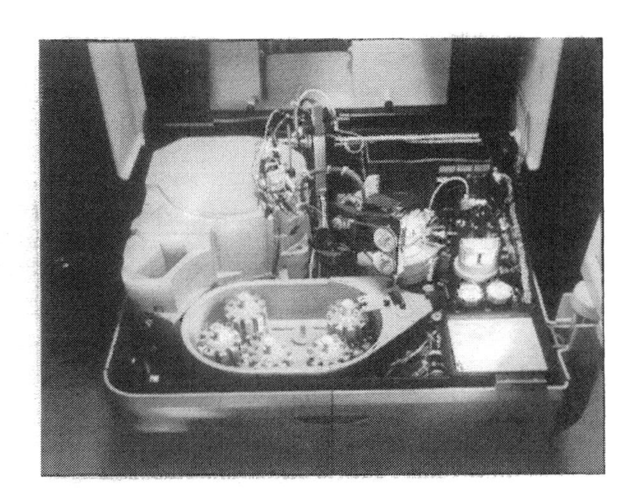

Fig. 31.2 Vitros ECi analyzer layout.

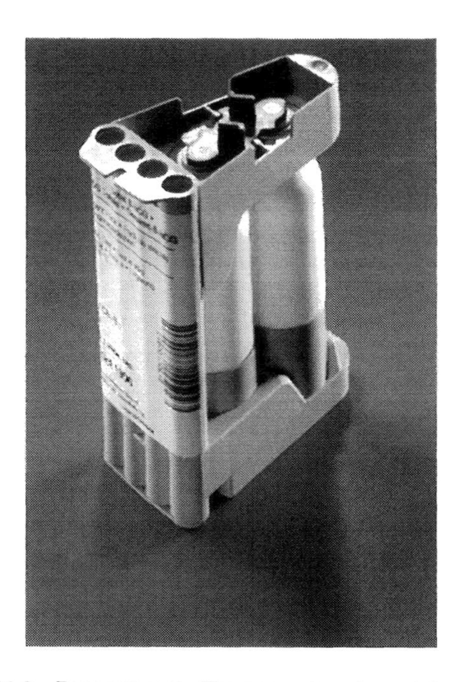

Fig. 31.3 Reagent pack. The reagent pack contains 50 or 100 coated wells and one or two reagents. The reagents are protected from evaporation loss by a closure that swivels to allow access to the reagents. Until first use, the reagents are also protected by a film seal that is punctured by the reagent metering probe. The coated wells are stored at the back of the pack. On-board stability is at least 2 months (although recalibration is typically required every 28 days).

Reagent management. Assay reagents and wells are supplied in special integrated reagent packs (Figure 31.3). Reagents for up to 2000 tests may be stored in the temperature and humidity controlled chamber, with access available at any time.

Assay processing. Dispensing, incubation, washing, and signal generation are carried out in a specially designed incubator that allows multiple protocols to be run simultaneously.

User interface. The system is operated by a touch-screen, graphical, color-coded icon-driven menu, and a keyboard with magnetic card reader. Magnetic cards are used to load in calibration or protocol data.

TYPICAL ASSAY PROTOCOL

Tests may be ordered directly or downloaded from a host computer. Assays are processed in coated microwells (Figure 31.4). A well is first dispensed into the circular incubator from the reagent pack. Sample is dispensed into the well, which rotates back to the reagent area for reagent addition. Incubation is at 37 °C. At the end of the incubation, the well is transferred from the outer to the inner ring. The contents are aspirated and the well washed. Signal reagent is added and the contents of the well are incubated for 4.5 min. Finally the luminescence is measured in a luminometer prior to the well being discarded to the waste container.

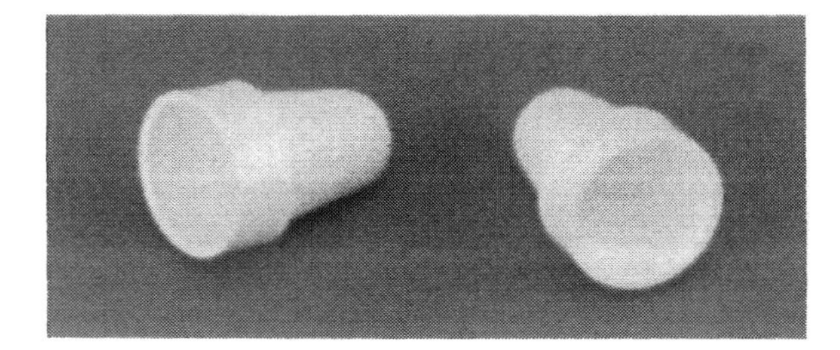

Fig. 31.4 Two of the coated wells. The tapered shape of the coated wells allows them to be nested at the back of the reagent pack. The shape also improves initial mixing and washing efficiency. The use of coated wells avoids the need for separate disposable reaction cuvettes, minimizing wastage and disposal.

PRODUCT FEATURES

- Broad menu capability. A dual ring incubator design allows assays of mixed incubation times to be run together (Figures 31.5 and 31.6).
- Fully automated, continuous random access. Samples and reagents may be loaded or removed at any time. Stat tests are prioritized upon order with results available within 15 min.
- Autodilution and reflex testing, i.e. Automatic re-assay at a different dilution or for a different analyte.
- Universal sample trays that can handle primary and secondary sample tubes and cups. There is no need for adaptors. Up to 60 samples can be loaded on-board.
- Positive sample identification. Barcode reader recognizes all major symbologies.
- Level sensing with clot and bubble detection including preserving the sample upon detection of a clot.
- Disposable tips avoid the risk of cross-contamination and carryover.
- On-board capacity for 20 barcoded integrated reagent packs (100 tests each). On- and off-board reagent inventory control. No additional components are required (stacked microwells are inside pack), minimizing waste. Throughput is up to 90 tests per hour. Walk-away time is approximately 1 h. Reagent packs may be loaded and unloaded safely while the system is running tests.
- On-board reagent stability (temperature and humidity-controlled) for up to 60 days.
- Calibration stability is 28 days. Ability to store multiple calibration curves per analyte. Magnetic card used to load master calibration data.
- New assay protocols can be loaded using a magnetic card.

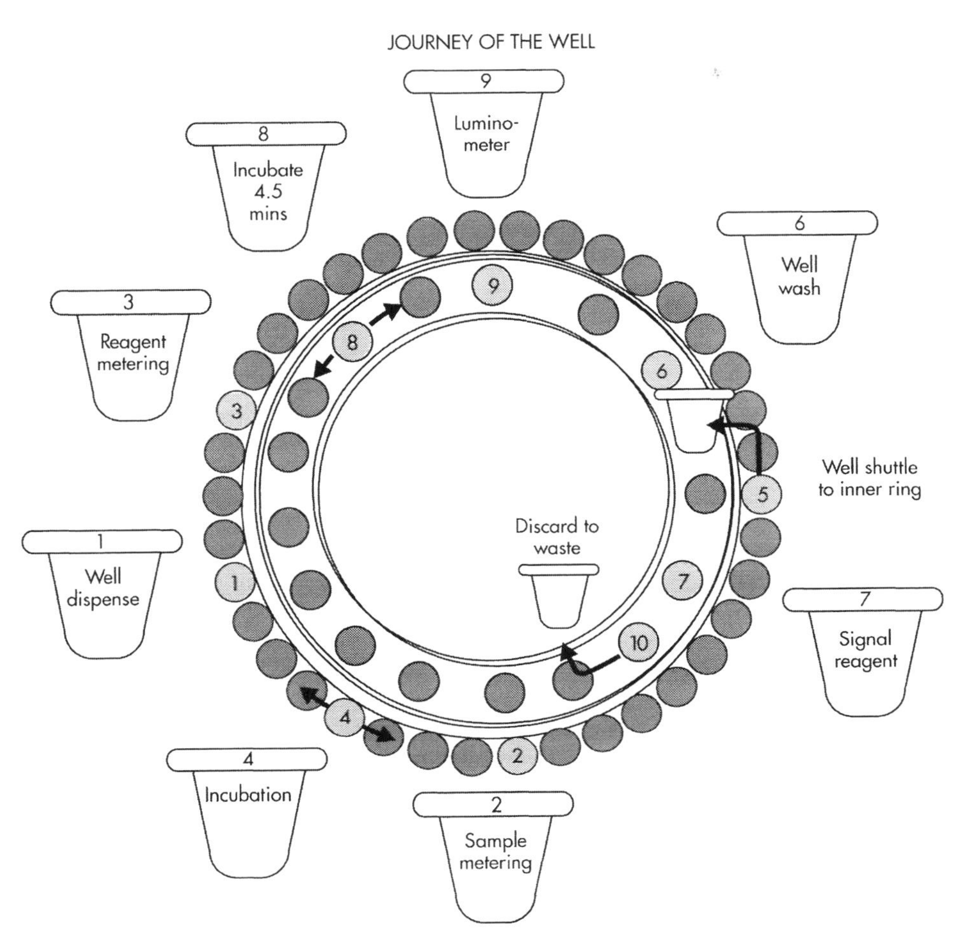

Fig. 31.5 Journey of the well. Coated wells are dispensed into the incubator ring via a shuttle, to keep the cooled reagent pack area separate from the warm incubator (1). Sample is metered into the well (2) followed by the reagents (3). The well is incubated for a fixed time, according to the test protocol (4). The well is then shuttled into the inner processing ring (5) for the longer processes of washing (6), signal reagent addition (7), incubation for 4.5 min (8) and signal reading (9). Finally the well is discarded into the waste collection box (10).

- Routine maintenance is minimal, guided by the software with an automated procedure via a special maintenance pack.
- Universal wash reagent for reagent dispensing probe and well wash.
- User-friendly user interface is intuitive, logical, and easy to use.
- Temperature control of every step, including wash cycle.

ASSAY PRINCIPLE

Competitive and immunometric formats are used for quantitative measurement of both antigens and antibodies. Most of the assays are designed for quantitative use except for those in the infectious diseases range

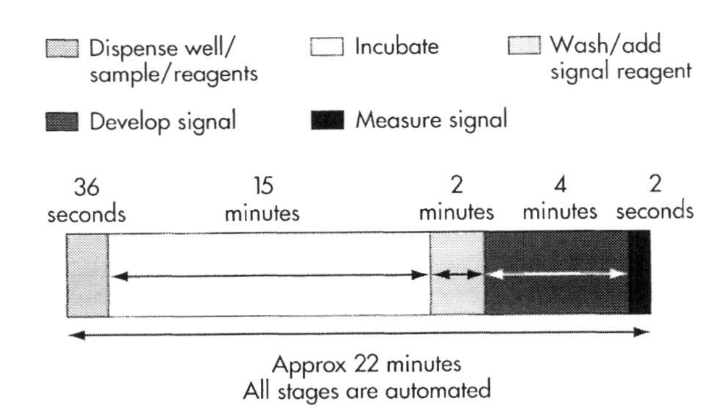

Fig. 31.6 Typical assay sequence.

where most are for qualitative testing (see Figures 31.7 and 31.8).

For qualitative tests other formats such as anti-globulin and antibody-class capture are also used.

CALIBRATION

A calibration pack is supplied with each kit lot. This includes a magnetic card, which is encoded with a master curve for the lot (typically based on at least seven reference standard concentrations) and either 2 or 3 calibrators that are run to adjust the master curve for that system. The calibrators are lot-specific, i.e. they are separately calibrated at the factory for each new kit lot coincident with the generation of the master curve data. This ensures that the master curve and adjusting calibrators are perfectly matched for each kit lot.

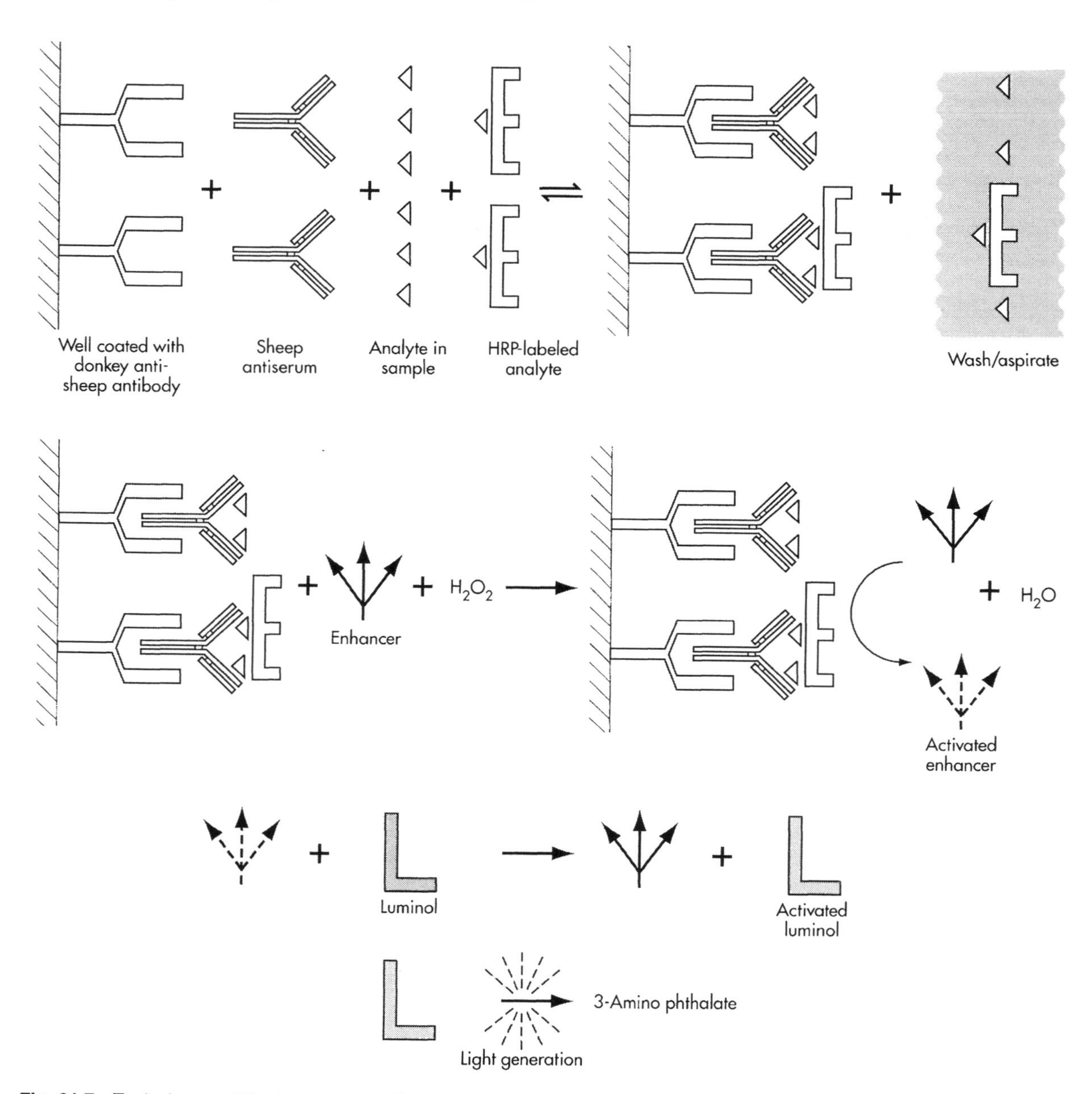

Fig. 31.7 Typical competitive immunoassay format.

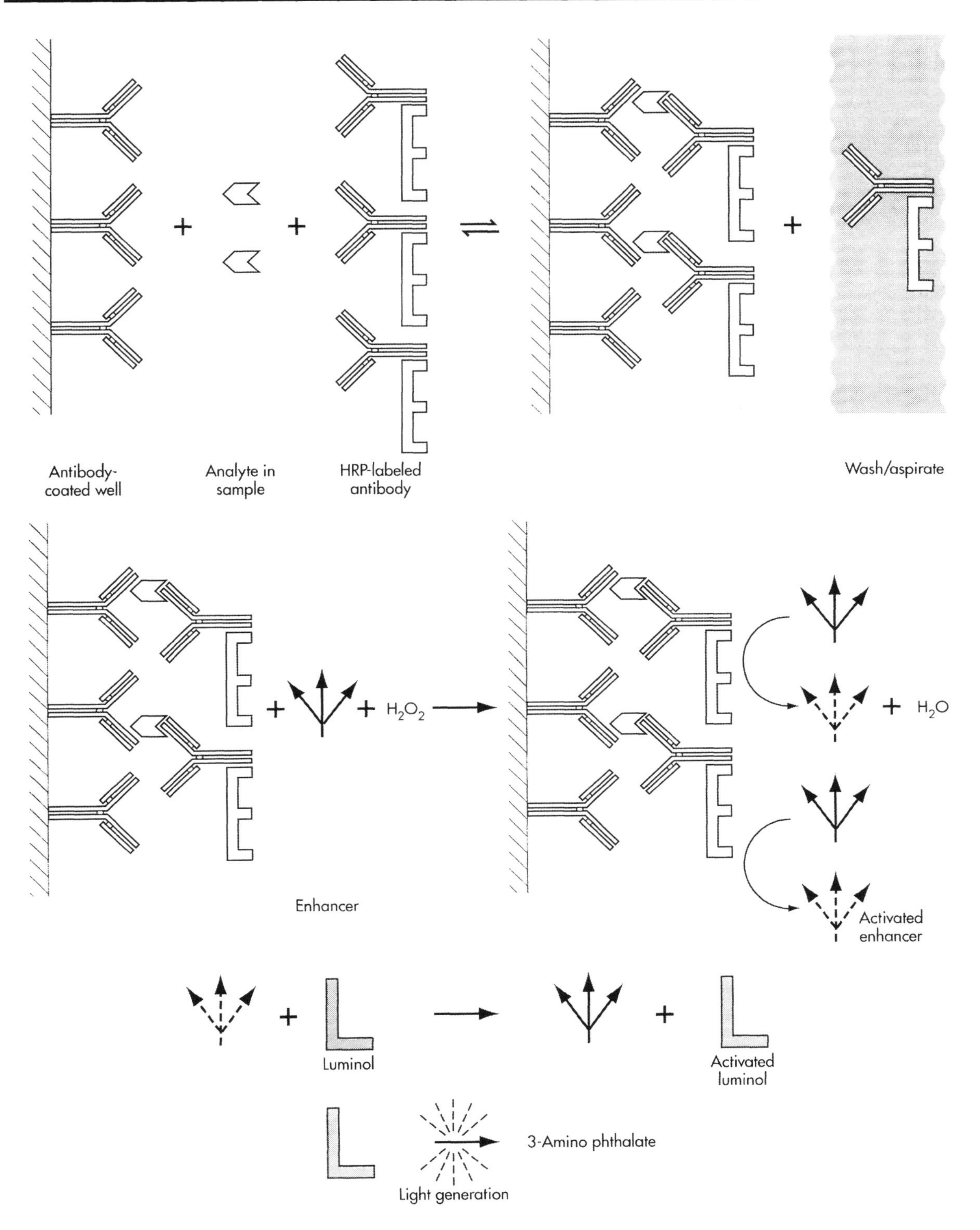

Fig. 31.8 Typical immunometric format.

Semi-quantitative assays use a single calibrator to establish a cut-off value.

Calibrations are stable for up to 28 days.

ANTIBODIES

Both polyclonal and monoclonal antibodies are used, depending on the assay design. In many assays the antibodies are not immobilized onto the plastic wells, but are biotinylated and provided in solution. Streptavidin coated onto the wells captures the biotinylated antibodies during incubation. This format prevents the deformation of antibody, with consequent loss of functionality, that can occur with protein immobilization.

SEPARATION

The solid phase is a polystyrene-based microwell, tapered to allow 100 wells to be stacked in the back of a reagent pack. The wells, which have exacting specifications, are manufactured in a purpose-designed facility. For the majority of the assays the wells are coated with streptavidin. In some cases the coating is donkey anti-sheep second antibody and various other protein coatings are used for specific tests.

The separation is performed by aspirating the well contents, then dispensing and aspirating temperature-controlled wash reagent four times.

SIGNAL GENERATION AND DETECTION

The enzyme label is horseradish peroxidase in all the assays. The bound fraction (conjugated antibody or antigen) is quantified by measuring the activity of horseradish peroxidase in the well after washing and aspiration. A signal reagent is added to the wells that contains a peracid salt, luminol, and an enhancer. It is supplied in a pack that contains two solutions. The peracid salt generates peroxide ions in solution, which oxidize the luminol, catalyzed by the horseradish peroxidase. In a series of reactions the oxidized form of luminol breaks down, with the generation of light. The enhancer increases the level of light produced several 100-fold and prolongs its emission. The light is read in the analyzer 4.5 min after the addition of the signal reagent.

DATA PROCESSING

For quantitative assays, calibration curves are fitted using a modified 4- or 5-parameter log–logistic program. The signal levels from the calibrators are used to adjust the master curve, which is transferred to the system on a magnetic card, supplied by the manufacturer. The system reads the luminescent signal for each sample and determines the analyte concentration from the stored calibration curve. Calibration curves and Levey–Jennings quality control charts may be viewed or printed.

INTERFACING TO LABORATORY INFORMATION SYSTEMS

The Vitros ECi System includes a bi-directional RS232 serial communication port operating for KERMIT and the ASTM standard.

FURTHER READING

Demers, L.M. Thyroid function testing and automation. *J. Clin. Ligand Assay* **22**, 38–41 (1999).

Thorpe, G.H.G., Kricka, L.J., Moseley, S.B. and Whitehead, T.P. Phenols as enhancers of the chemiluminescent horseradish peroxidase–luminol–hydrogen peroxide reaction: application in luminescence-monitored enzyme immunoassays. *Clin. Chem.* **31**, 1335–1341 (1985).

Thorpe, G.H.G. and Kricka, L.J. Enhanced chemiluminescent assays for horseradish peroxidase: characteristics and applications. In: *Bioluminescence and Chemiluminescence, New perspectives Proceedings of the Fourth International Bioluminescence and Chemiluminescence Symposium, Freiburg, September 1986* (eds Schölmerich, J., Andreesen, R., Kapp, A., Ernst, M. and Woods, W.G.) 199–208 (Wiley, Chichester 1986).

Whitehead, T.P., Thorpe, G.H.G., Carter, T.J.N., Groucutt, C. and Kricka, L.J. Enhanced luminescence procedure for sensitive determination of peroxidase-labelled conjugates in immunoassay. *Nature* **305**, 158–159 (1983).

32 Bulk Reagent Random-Access Analyzers: IMMULITE® 2000 and IMMULITE 2500

Arthur L. Babson

The IMMULITE 2000 is manufactured and distributed by Diagnostic Products Corporation (DPC) (see Figure 32.1). Like the DPC IMMULITE, it uses 0.25-in. antibody or antigen-coated polystyrene beads as the solid phase, and the separation of bound and free alkaline phosphatase label and washing of the bead are accomplished by spinning the reaction tube at high speed about its longitudinal axis. IMMULITE 2000 uses a generic reaction tube into which the appropriate beads are introduced by the instrument and the waste is collected by the instrument rather than the coaxial waste sump, which is integrated into the assay tube in IMMULITE. Since IMMULITE 2000 uses the same chemistry and sensitive chemiluminescent substrate as IMMULITE, transferring assay protocols to the new system is readily accomplished. Over 3200 IMMULITE 2000s have been placed in the US and around the world.

In 2004 DPC introduced the IMMULITE 2500, which is identical to the IMMULITE 2000 but with a random-access tube processor, allowing tests with any incubation time to be run simultaneously (Logic Driven Incubation™) and software to maximize throughput and minimize turnaround time.

TYPICAL ASSAY PROTOCOL

The instrument automatically delivers the appropriate bead to an assay tube, which is then advanced to the pipetting position. Separate liquid displacement pipettors deliver sample and liquid reagent to the assay tube to initiate the immune reaction. For IMMULITE 2000, the tube is incubated for 30 min with constant agitation at 37 °C. For IMMULITE 2500, the tube is incubated for whatever period is appropriate for the assay. Separation of bound and free label is accomplished by spinning the tube at 8000 rpm about its vertical axis. The bead is washed with four 400 μL volumes of water, spinning out the fluids after each wash. Chemiluminescent substrate is added and after a further 5-min incubation at 37 °C the light output is measured with a photon-counting photomultiplier tube. Time to first result is 35 min (as little as 10 min with IMMULITE 2500) and the throughput is 200 tests per hour maximum. Throughput varies somewhat with assay mix and is reduced by sequential assays, which require two incubations and washes. IMMULITE 2500 has two spin-wash stations to minimize throughput degradation from sequential assays.

PRODUCT FEATURES

IMMULITE 2000 and IMMULITE 2500 operate in a totally random-access mode with operator access to samples and reagents at any time. Its operation can best be understood by reference to the diagrammatic layout shown in Figure 32.2. It can accommodate up to 90 primary or secondary sample tubes of any size between 12–16 and 75–100 mm. The liquid reagent carousel is cooled to 4–8 °C and has a capacity of twenty-four 200-test reagent wedges. Each wedge has three separate compartments for assays that require more than one reagent. A patented cover that is actively opened by the instrument only when accessed by the reagent pipettor seals the compartments. A single laser scanning barcode reader interrogates both the samples and the reagents. The bead carousel has a capacity for 24 packs containing 200 beads each. The packs are hermetically sealed by a patented mechanism that is opened only when a bead is dispensed. The bead packs contain desiccant and, in addition, the bead carousel is actively dehumidified. A dedicated CCD barcode reader interrogates the bead

The Immunoassay Handbook. *Third Edition*
D. Wild (Ed.)

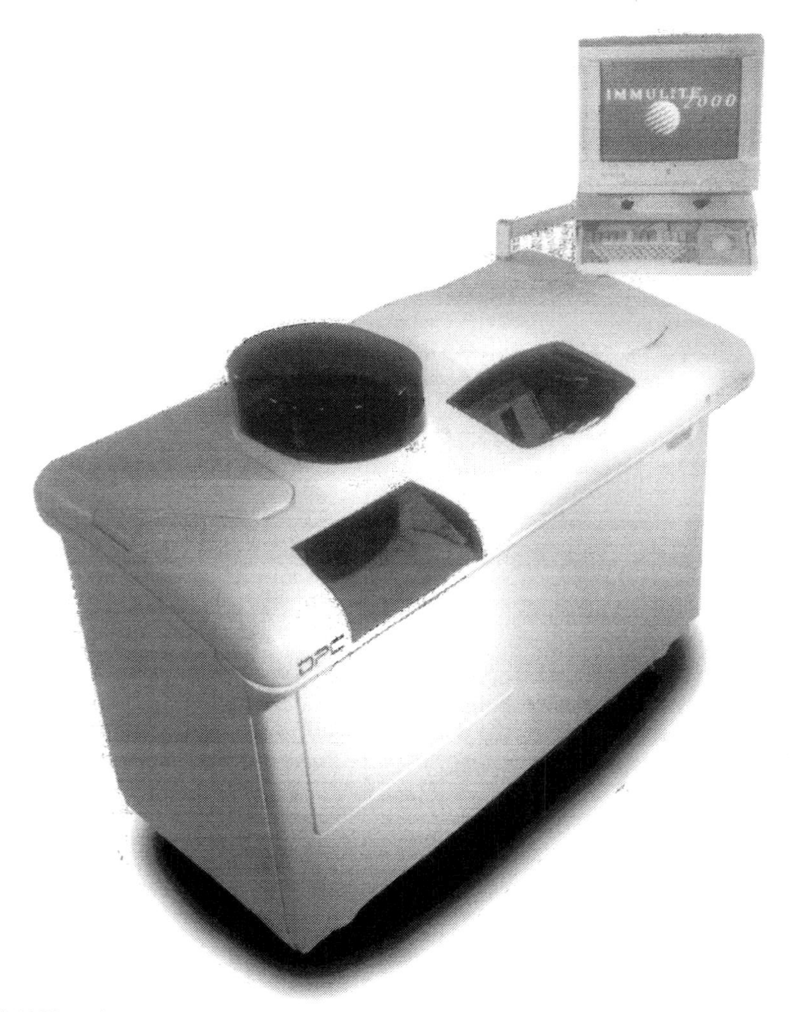

Fig. 32.1 IMMULITE 2000 analyzer.

barcode labels, which are serialized for reagent inventory by the software. The reagent and bead containers are shown in Figure 32.3. Reaction tubes are bulk loaded into a hopper, which has a capacity of 1000 tubes. The status of all reagents, bulk supplies, and waste is continuously monitored and displayed. A patented, reusable sample dilution well automatically prepares properly diluted sample when required. Samples giving out-of-range test results are automatically diluted and re-assayed.

The IMMULITE 2000 is specifically designed to support powerful, easy-to-use software running under Microsoft's Windows NT operating system, while the IMMULITE 2500 runs under Windows XP. They are driven by Intel Pentium microprocessors with 128 and 256 MB of RAM, respectively. A 40-GB hard drive provides online storage for several years of patient results. A second identical drive maintains a redundant copy of the data, providing the laboratory with reliable, on-board data storage backup. In addition, the unit includes an integrated floppy disk drive and a CD-ROM drive to facilitate data export and loading of software updates.

The operator interface is a 17-in., high-resolution color monitor with touch screen in addition to a keyboard and track ball. Software screens have large, customized icons, and navigating through the screens is very intuitive. The user can go directly from anywhere to anywhere without having to back out of several windows. Drop-down windows are never more than three deep. Many software features such as reflexive testing and quality control rules are user configurable. Eight languages are supported: English, German, Italian, French, Polish, Swedish, Spanish, and Portuguese.

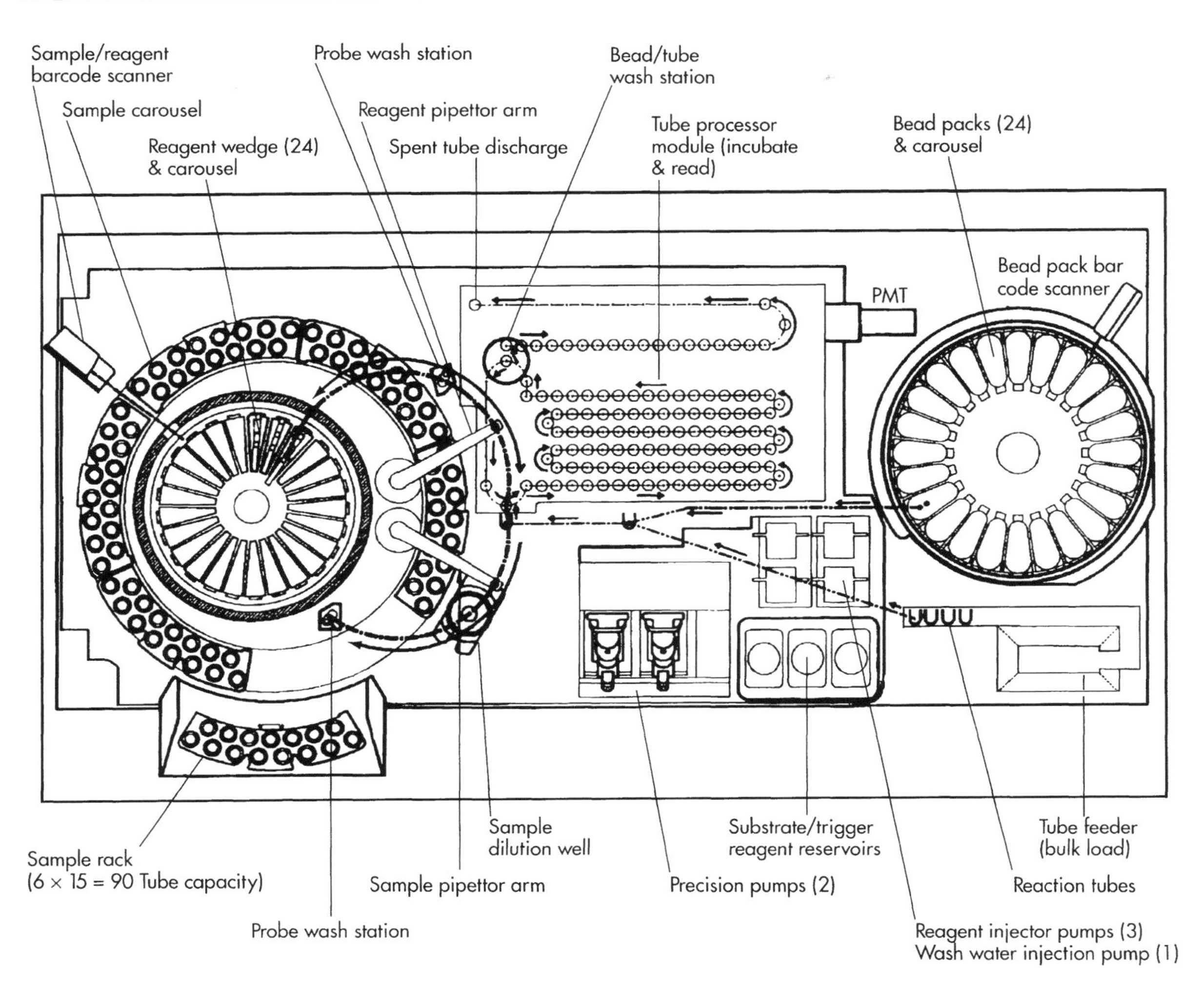

Fig. 32.2 IMMULITE 2000, schematic.

In addition to the customary help function, a unique function called 'guidance' allows the operator to immediately obtain an explanation or additional information for any item on the currently displayed screen.

The IMMULITE 2000 software includes a tutorial, which provides in-depth information on any aspect of the instrument. The information is provided in a variety of formats including text, photographs, video clips with sound, and animated sequences. Figure 32.4 shows the initial screen of the tutorial, and Figure 32.5 shows the initial screen of the instrument reference. The latter is an actual photograph of the instrument with the top cover open showing the major modules with callouts. Touching any callout button will immediately bring up another screen with additional callouts specific to that module. The arrow buttons can incrementally change the view, as if the viewer is walking around the instrument, with the callouts changing as different features are brought into view. The same multimedia approach is taken with the IMMULITE 2500 using a training CD that can be periodically updated to include new features.

ASSAY PRINCIPLE

Both competitive assays for haptens and immunometric assays for large molecules are run with antigen or antibody-coated beads. The patented axial centrifugal washing technique provides low background signal and the sensitive chemiluminescent substrate for alkaline

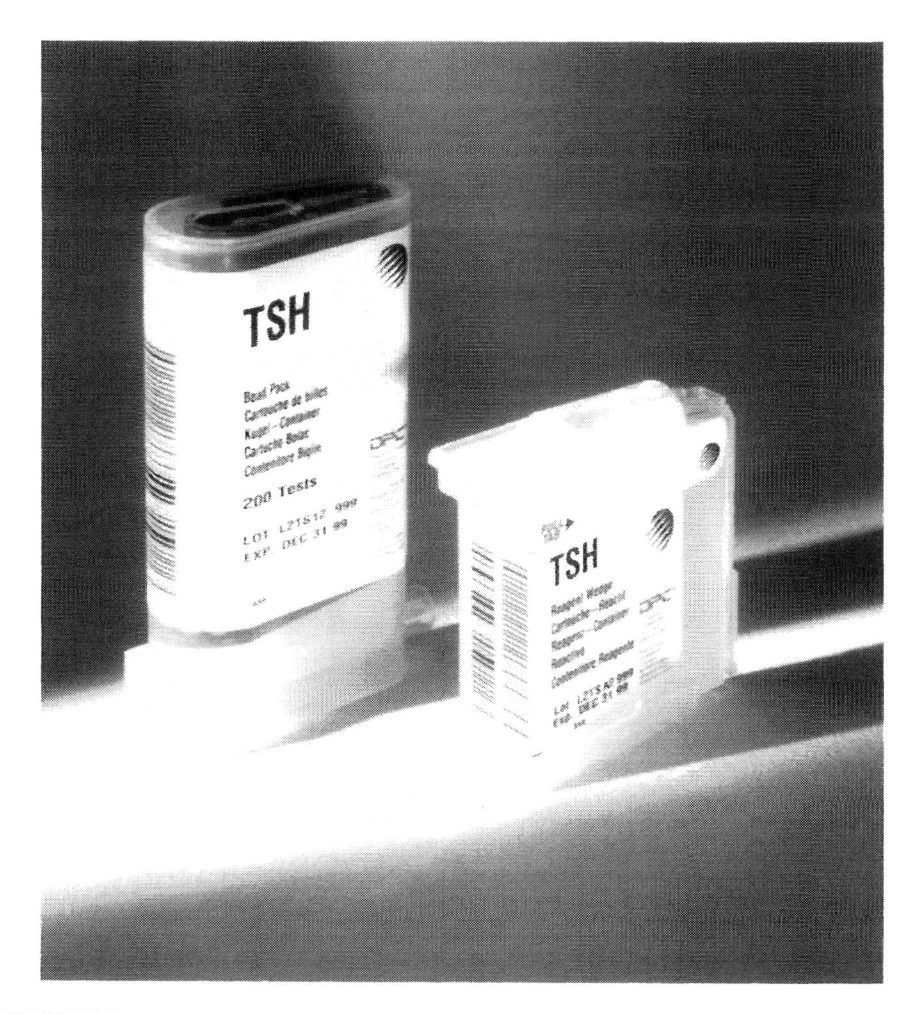

Fig. 32.3 IMMULITE 2000 reagent and bead containers.

phosphatase described in the IMMULITE chapter provides high signal levels.

CALIBRATION

IMMULITEs 2000 and 2500 use stored master curves generated by DPC and the parameters describing this curve as well as all the information the instrument requires to run the assay are encoded on a two-dimensional barcode label supplied with the kit. Before using a new lot this barcode is scanned with the hand-held scanner on the instrument. A 200-test kit consists of a reagent wedge, a bead pack, one or two adjustors, and several barcode labels defining the adjustors. Each new lot of reagents must be adjusted before use. This initial adjustment corrects for any variations in response between the user's instrument and the manufacturer's instrument that generated the master curve. Adjustment is performed by transferring aliquots of adjustors to barcoded tubes on the sample carousel. The instrument recognizes these adjustors and performs the adjustment automatically. Master curves for up to three different lots of any test can be stored in the instrument. The operator is informed by the system if and when any readjustment is required.

ANTIBODIES

Both polyclonal and monoclonal antibodies are used by the system.

SEPARATION

After incubation the tubes are shuttled to the patented spin-wash station where the tube is lifted to engage a chuck driven by a high-speed motor. Tube fluid contents are instantaneously spun out into a chamber connected to

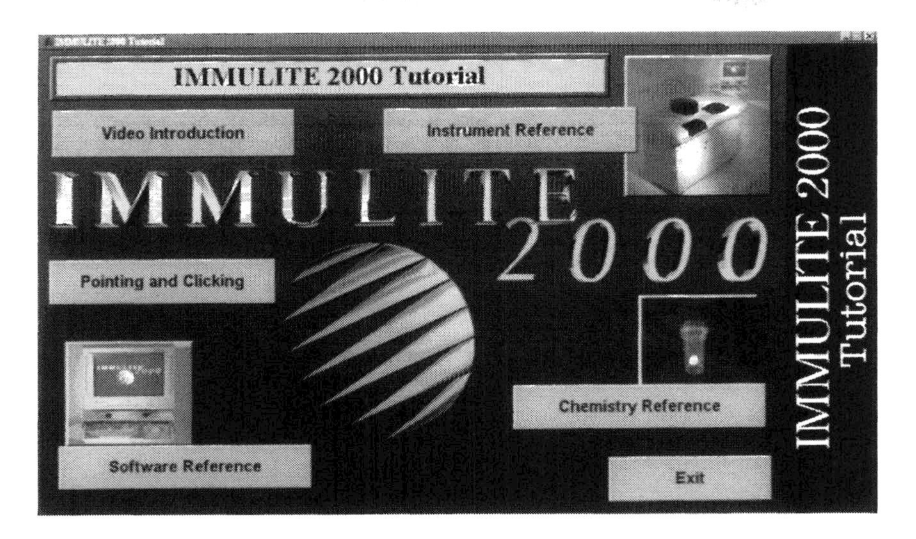

Fig. 32.4 IMMULITE 2000 tutorial – initial screen.

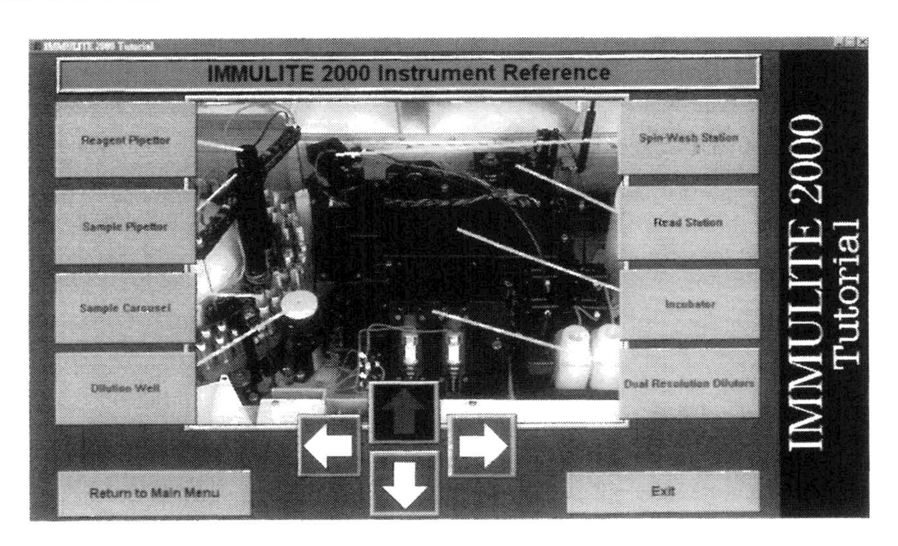

Fig. 32.5 IMMULITE 2000 tutorial – initial screen of instrument reference section.

the system liquid waste. Four 400 μL water washes with intervening spinning assure the bead has no unbound label.

SIGNAL GENERATION AND DETECTION

This is identical to the process described in the IMMULITE chapter.

DATA PROCESSING

This is identical to the process described in the IMMULITE chapter.

INTERFACING TO LABORATORY INFORMATION SYSTEM

The IMMULITEs 2000 and 2500 can be connected to the laboratory information system (LIS) through bidirectional batch download as well as bidirectional host query. In addition to an RS232 serial port to communicate with the LIS, it also has extra ports for connection to laboratory automation systems, an external modem for remote diagnostics, and other peripherals. An ethernet interface board is included, providing the capability to directly interface the instrument to the hospital LAN.

33 Bulk Reagent Random-Access Analyzers: ADVIA Centaur®

Laura Taylor and Elvio Gramignano

The ADVIA Centaur is a fully automated random-access immunoassay system, using chemiluminescent tracers and paramagnetic particle solid-phase reagents. The system uses the common reagents with other immunoassay systems from Bayer, along with an enhanced menu including allergy and infectious disease assays.

The system is a stand-alone floor model that includes a refrigeration unit holding up to 30 assay packs on-board. Introduced in March 1998, the ADVIA Centaur is the first of the next generation of immunoassay systems.

The ADVIA Centaur has a user-friendly interface that uses a dual-processor Unix platform. To allow for high system productivity all system supplies can be changed during operation. The ADVIA Centaur may be easily integrated into a fully automated laboratory with a sample transport system (see Figure 33.1).

TYPICAL ASSAY PROTOCOL

Assays are performed on the ADVIA Centaur using multiple architectures to obtain the most reliable results. All the steps are performed automatically, with the operator required only to load samples and reagents. Each assay is performed in a discrete disposable plastic cuvette, which travels in an incubation ring that indexes every 15 sec. The bidirectional incubation ring has the flexibility to support a wide variety of assay protocols. Incubation times vary per assay with a minimum of up to 8 min. During processing, sample is dispensed, reagents are added and separation of free from bound tracer is achieved by magnetic separation of the micron-sized paramagnetic particle solid-phase reagent and bound tracer in the cuvette. This reaction is measured by a luminometer component after the tracer is oxidized, yielding a light flash that is concentration dependent. The first result is reported in 19 min for most of the assays, each subsequent result in 15 sec intervals (see Figure 33.2).

PRODUCT FEATURES

- The ADVIA Centaur can perform up to 240 tests per hour with 840 tests of walk away time. The system holds 180 samples when fully loaded.
- Reagents, samples, and supplies may be loaded on the system at any time, without pausing.
- The ADVIA Centaur requires no daily startup. Due to the on-board refrigeration, 30 assay packs may be stored on-board for up to 6 weeks.
- The system has autodilution, repeat, and cascade reflexing capabilities. The samples will be held in the in-process queue if repeat tests or dilution is required, the operator does not have to find the samples and reload them.
- The ADVIA Centaur was designed with a dedicated STAT port, where STATs may be added to the system at any time and will be analyzed immediately without interruption to routine testing.
- The ADVIA Centaur uses disposable sample tips for every test, eliminating all carryover and the need for wash reagents. The system has clot detection and management capabilities.
- Primary sample tubes may be used on the system, eliminating the need for pouring off serum into sample cups. Barcoded samples may be utilized for positive sample identification by the system, and several barcode formats are recognized, including Code 39, Codabar, Code 128, and Interleaved 2 of 5.
- The daily maintenance consists of an automated cleaning procedure. Weekly and monthly maintenance requires minimal hands-on time.
- Bidirectional interfacing to a laboratory information system is easily accomplished, allowing downloading of information, operation in host query mode, and uploading of test results.
- The ADVIA Centaur is capable of interfacing with laboratory automation systems without additional robotics. The sample probe has the flexibility to sample directly from an automated track to

The Immunoassay Handbook. *Third Edition*
D. Wild (Ed.)

Fig. 33.1 The Centaur system.

the ADVIA Centaur, making the system efficient for managing workflow requirements.

ASSAY PRINCIPLE

Assay protocols, which are stored in memory, utilize either competitive binding, immunometric (sandwich) principles, antigen bridging, or μ-capture depending on the specific assay. The system can perform one-step, extended one-step, two-step, and extended two-step incubation protocols. Quantification of results is achieved by oxidation of the acridinium ester tracer reagent, followed by detection and quantification of the resultant intense quick flash emission. Assays are performed in discrete disposable cuvettes, which may be loaded and replenished randomly at any time. The cuvettes are the same as used on the other Bayer Immunoassay systems.

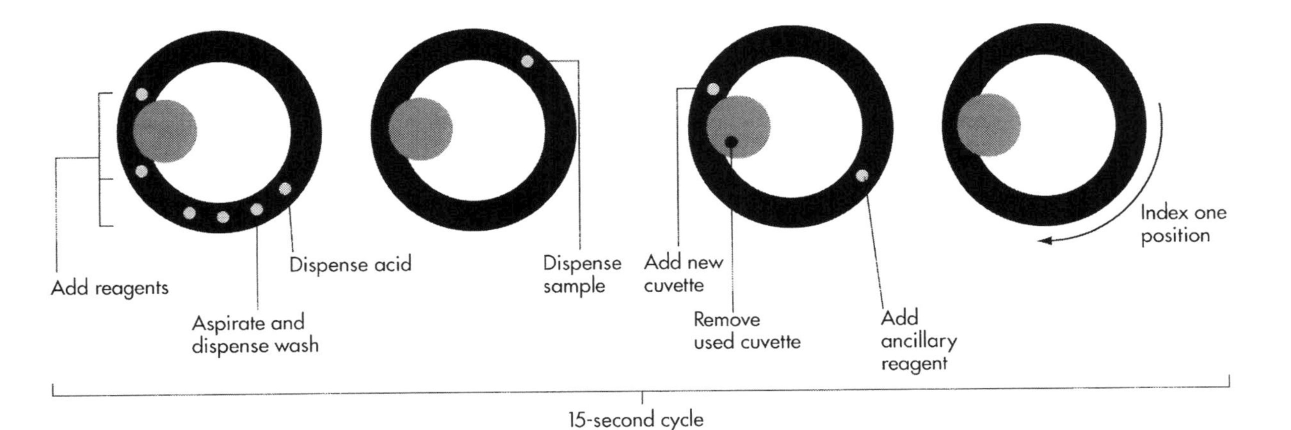

Fig. 33.2 The incubation ring.

CALIBRATION

ADVIA Centaur assays are calibrated using a stored, lot-specific or fixed master curve of up to 10 concentration points, depending on the assay. The curves are established by the Quality Control Laboratory at Bayer HealthCare. These master curves are supplied on a bar-coded card for rapid and accurate entry into the memory of the ADVIA Centaur user interface. The master curve is then used by the software to quantify each assay performed with that lot of reagents. Calibration Evaluation Ranges (CERs) which include slope, ratio, low and high calibrator deviation, and precision ensure a valid calibration. Periodic recalibration is accomplished by the inclusion of a set of two calibrators. Recalibration intervals vary depending on the assay, and are tracked and controlled by on-board software. Multiple master curves may be stored in system memory for each assay, simplifying lot-to-lot switching. Raw data (relative light units, RLUs) for controls and patients are retained by the system when a valid calibration is not present for the specified assay and can be quantified as soon as the calibration is valid.

ANTIBODIES

Reagents for the Centaur are identical to those used on the other Bayer Immunoassay systems. A mixture of polyclonal and mouse monoclonal antibodies are used in many ADVIA Centaur assays. In immunometric assays, this multispecies technology minimizes the risk of aberrant results due to heterophilic binding by human anti-mouse antibodies (HAMA), which can occur in dual mouse monoclonal assays. Where dual mouse monoclonal antibodies are utilized, blocking reagents are added to minimize the risk of HAMA effects.

SEPARATION

The solid-phase reagents used in the ADVIA Centaur assays are suspensions of micron-sized, specially coated ferric oxide particles. These particles are small enough to provide extremely high surface area for rapid binding kinetics, and stay in suspension for prolonged periods, but are rapidly and cleanly separated through their paramagnetic attraction when placed in a magnetic field. The ADVIA Centaur refrigerated reagent compartment gently rocks the reagent packs at periodic intervals. Reagents may be stored on-board up to 42 days, depending on the assay. The reagents are packaged in a Ready-Pack® that contains an S-shaped flow design for optimal resuspension of the particles. The Parafilm®-like coating covers the pack openings that are punctured during reagent probe aspiration. Individual assay packs provide sufficient volume for 50–200 tests (see Figure 33.3).

After dispensing the sample, tracer, and solid-phase reagents into the test cuvette and completion of the incubation period, the particles are magnetically pulled to the flat wall of the cuvette by magnets mounted on the back of a ring within the incubation chamber. This separate ring allows the magnets to move within the ring to any location they are required. The liquid contents of the cuvette are aspirated, water is injected to resuspend and remove the nonspecific bound tracer, and reseparation accomplished automatically.

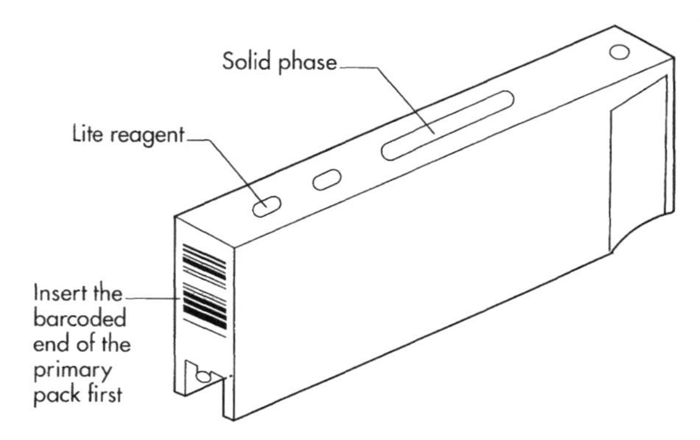

Fig. 33.3 The Ready-Pack, containing the test reagents.

Water is provided from an on-board reservoir (1.5 L capacity) fed by a bulk water bottle (9.5 L capacity) that may be filled at anytime during a run. The liquid waste area also contains a reservoir and a bulk bottle so that the waste may be emptied while the system is running. Both areas have capacity for 4 h of walk away time. The system may also be directly plumbed, so no operator intervention in these areas is required.

SIGNAL GENERATION AND DETECTION

Following separation and washing of the solid-phase particles, they are resuspended in an oxidizing reagent at an acid pH, stored on-board (acid reagent). While held in front of a photomultiplier tube detector, a pH adjusting reagent, also stored on-board (base reagent) is injected, causing the dose-related light emission, which is quantified over a period of 4 sec. The acid and base reagents have a bulk bottle and reservoir stored on-board, so that the reagents may be replenished at any time during a run.

DATA PROCESSING

The flash emission, corrected by background subtraction, is directly related to the stored master curve for quantification. Results are calculated and reported as mass concentration and flagged if out of range, based upon operator-controlled reporting formats. The software allows automatic conversion to SI Units, if desired.

Results may be reported individually as completed, or collated into patient reports containing all tests performed on a given sample.

If desired, the ADVIA Centaur will automatically hold a sample in the in-process queue for repeat, if the initial result is outside an operator-programmed range.

An unlimited number of controls can be defined for each analyte and may be tracked through a 90-day history, and results are displayed in a tabular or Levey–Jennings format. QC Rules can be also set to evaluate system performance. Monthly summaries are also accessible for a period of 12 months. Results from 15,000 tests are stored in a current or historical worklist view.

USER INTERFACE

The user interface of the ADVIA Centaur is a Sun computer system, 17 in. color monitor and keyboard, with a Glidepoint® pointing device. The software utilizes a Unix platform, allowing several screens to be accessed simultaneously. The use of intuitive icons and full online help throughout the software makes training and routine operation rapid and simple to master.

A built-in high-speed CD-ROM reader allows easy loading of software and test definitions. Step-by-step instructions and graphics for activities such as maintenance, system diagnostics, and troubleshooting are available through the help and online procedures in the event log.

A remote access feature has been built into the ADVIA Centaur to allow Bayer Technical Assistance groups access via modem to information in the user interface for troubleshooting.

INTERFACING TO LABORATORY INFORMATION SYSTEMS

Bidirectional interfacing to any central laboratory computer is achievable for downloading of testing requests, host query options, and uploading of test results, using barcoded samples. Interfacing software is written in standardized ASTM format for consistency with a wide variety of LIS systems. Additionally, simple software selections may be made to allow direct sampling from the back of the system off an automated track. Operators can still load routine and STAT samples manually at the front of the system without interfering with the track. This flexibility allows the ADVIA Centaur to easily integrate into totally automated laboratory systems.

FURTHER READING

Day, R., Horschke, W., Hallahan, M., Jackson, T. and Williams, J. System design solutions for high volume immunoassay analysis. *J. Clin. Ligand Assay* **22**, 184 (1999).

Geyer, S., Moore, S. and Michetti, D. A study of the potential impact of the ACS:Centaur in a hospital core laboratory. *Clin. Chem.* **44**, A57 (1998).

Girgensohn, S., Liedtke, R., Balzer, G., Castor, S. and Hauser, M. Performance of the ACS:Centaur high-capacity, random-access immunoassay system. *Clin. Lab.* **43**, 975–983 (1997).

Weinzierl, C. Productivity evaluation of random-access immunoassay platforms. *Am. Clin. Lab.* **16**, 4–5 (1997).

34 Architect® *i*2000® and *i*2000_{SR}® Analyzers

Frank A. Quinn

The ARCHITECT *i*-series is a family of immunoassay analyzers that combines flexible assay protocols with a patented chemiluminescent detection technology (CHEMIFLEX®). The first instrument in this family, the *i*2000 analyzer (Figure 34.1), was introduced in 1999. The *i*2000 analyzer is a modular, fully automated, random-access instrument that can process up to 200 immunoassays per hour. All assays use paramagnetic microparticles and chemiluminescent detection based on Abbott's patented acridinium derivative. The newest member of the ARCHITECT *i*-series, the *i*2000_{SR} (Figure 34.2), provides additional functionality by allowing STAT and automatic retest capabilities. STAT assays are processed ahead of routine immunoassay tests by means of a Retest Sample Handler (RSH) utilizing multidimensional sample handling (MDS) technology. All STAT tests are completed within 18 min, and retests are automatically scheduled without manual operator intervention.

The ARCHITECT family of systems also includes the *c*8000® clinical chemistry analyzer. The *c*8000 has a maximum throughput of 800 tests per hour (colorimetric assays) and 1200 tests per hour (ISE and colorimetric assays). STAT clinical chemistry profiles are completed and available in less than 15 min. Due to the unique design of the RSH, an *i*2000_{SR} analyzer and a *c*8000 analyzer may be fully integrated to form the *ci*8200™ analyzer. This allows consolidation of immunoassay and clinical chemistry testing into a single platform, run by one operator through a single computer interface. Because the *ci*8200 analyzer utilizes MDS, STAT sample processing is always prioritized, even during peak workload hours.

Future members of the ARCHITECT instrument family will include the *i*1000_{SR}™ analyzer for immunoassay testing, as well as the *c*4000™ and *c*16000™ clinical chemistry systems. The *i*1000_{SR} analyzer will have a maximum throughput of 100 tests per hour. The *c*4000 and *c*16000 systems will have throughputs of 400 and 16,000 (colorimetric) and 600 and 2000 (ISE and colorimetric) tests per hour, respectively.

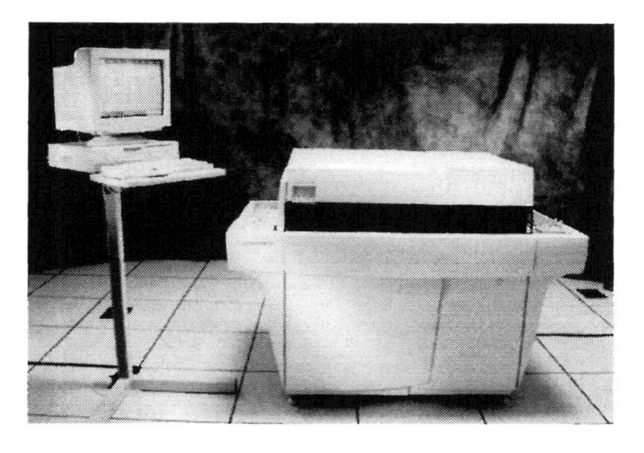

Fig. 34.1 ARCHITECT *i*2000 analyzer and system control center (SCC).

Fig. 34.2 The ARCHITECT *i*2000_{SR} analyzer and Retest Sample Handler (RSH).

The Immunoassay Handbook. *Third Edition*
D. Wild (Ed.)

For the ARCHITECT *i*2000, *i*2000$_{SR}$, and *ci*8200 analyzers, reagents are available for a wide range of assays including thyroid, fertility, oncology, cardiac, metabolic, and infectious disease analytes. These reagents can be run on all current and future ARCHITECT immunoassay analyzers. Although this chapter focuses on the *i*2000 and *i*2000$_{SR}$ analyzers, most of the features and details described below are common to all members of the ARCHITECT *i*-series of immunoassay analyzers.

TYPICAL ASSAY PROTOCOLS

The ARCHITECT *i*2000 and *i*2000$_{SR}$ analyzers are capable of simultaneously running two-step, one-step, and automated pretreatment protocols. In addition to these protocols, the *i*2000$_{SR}$ can also run STAT and automated re-run protocols. The availability of multiple protocols provides the flexibility required to address analyte-specific assay development needs. The process path for both analyzers is circular (see Figure 34.3), and maintained at a constant temperature of 37 °C. Reaction vessels progress through the process path by indexing to a new position every 18 sec. The majority of routine assays use a two-step protocol. In two-step routine assays, specimen and microparticles are added to a reaction vessel at positions 1 and 2 of the process path, respectively. Reaction components are then thoroughly mixed by a pop-up mixer (or, in-track vortexer, 'ITV') at position 3. Incubation continues until the reaction vessel reaches wash zone 1. In wash zone 1, magnets attract the microparticles to the inner wall of the reaction vessel, and unbound material is washed away in a series of three washes. After wash zone 1, conjugate or tracer is added to the reaction vessel, and reaction components are mixed

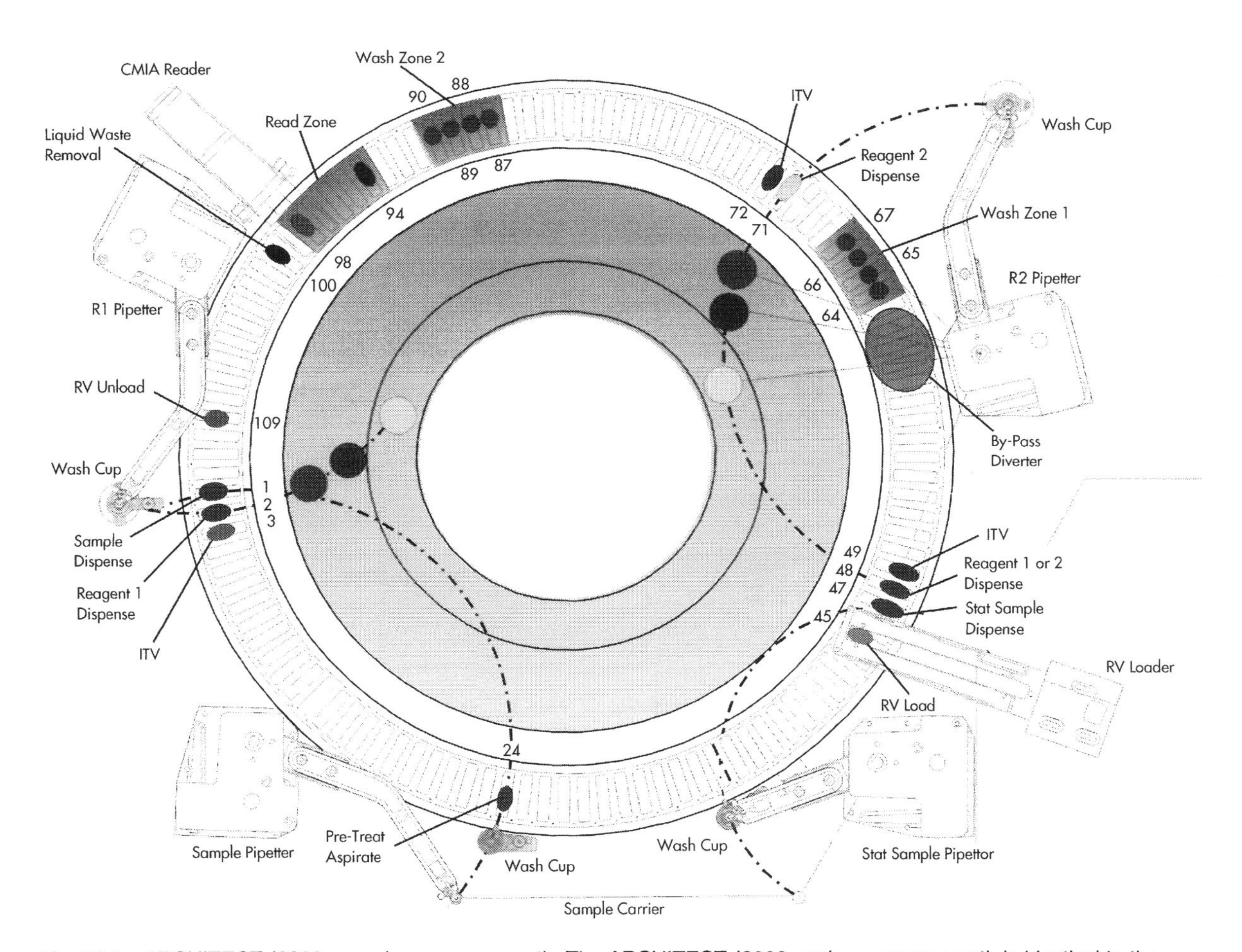

Fig. 34.3 ARCHITECT *i*2000$_{SR}$ analyzer process path. The ARCHITECT *i*2000 analyzer process path is identical to the *i*2000$_{SR}$ analyzer process path except it does not have the STAT sample pipettor and additional in-track vortexer (ITV, or pop-up mixer) at position 49.

using a pop-up mixer. Incubation continues until the reaction vessel reaches wash zone 2, where the microparticles undergo a second series of three washes to remove unbound material. After the reaction vessel leaves the second wash zone, a pretrigger reagent is added. This reagent releases the conjugate or tracer into solution, and prepares the label for the light generating reaction. Microparticles are then magnetically attracted to the inner wall of the reaction vessel, separating them from the label, and the trigger reagent is added. The trigger reagent initiates the final phase of the light generating reaction, and the resulting chemiluminescence is measured using a photomultiplier tube. The measured relative light units (RLUs) are processed by the data management software to yield the assay result. Finally, the reaction mixture is aspirated, and the empty reaction vessel is transferred to a waste container.

In routine one-step assays, specimen is added to the reaction vessel at position 1. Particles and conjugate or tracer are added at position 2. Reaction components are then thoroughly mixed by the pop-up mixer at position 3. As the reaction vessel approaches wash zone 1, it is diverted to the outer lane of the process path. This allows the reaction vessel to by-pass wash zone 1 and continues incubation without interruption. Incubation continues until the reaction vessel reaches wash zone 2. In wash zone 2, microparticles are magnetically attracted to the inner wall of the reaction vessel, and unbound material is washed away in a series of three washes. From this point, reaction vessel processing is as described for the two-step protocol.

A variety of fully automated pretreatment protocols are also available. In the pretreatment format, specimen and pretreatment reagent are combined in a reaction vessel at positions 1 and 2 of the process path, respectively. After an incubation of approximately 7 min, an aliquot of the pretreated specimen is removed by the sample pipetter and reintroduced into a new reaction vessel at position 1 of the process path. From this point, the assay runs in either a two-step or one-step protocol as described above.

On the $i2000_{SR}$ analyzer, STAT protocols are made possible by the addition of a second sample pipetter and pop-up mixer to the system hardware. These hardware modifications allow for shorter assay incubation times by providing for STAT sample introduction and mixing at a point farther along in the system sample processing path (refer to Figure 34.3). Reagents for STAT assays are optimized to allow for the shorter incubation time without compromising assay performance.

PRODUCT FEATURES

- The *i*2000 and $i2000_{SR}$ analyzers are fully automated, random-access instruments. Maximum throughput is 200 tests per hour.
- The *i*2000 and $i2000_{SR}$ analyzers can perform automated on-board sample dilution. The $i2000_{SR}$ can perform automatic repeat testing without manual operator intervention.
- For routine two-step and routine one-step assays, the time to first result is approximately 29 min (36 min for single pretreatment assays). STAT results ($i2000_{SR}$ analyzer) are produced in approximately 18 min. For all protocols, additional results are produced every 18 sec.
- Modular instrument design allows multiple *i*2000 analyzers to be combined to form *i*4000, *i*6000, or *i*8000 analyzers with maximum throughputs of 400, 600, and 800 tests per hour, respectively. In either format (individual module or multiple integrated modules) the instrument is run by one operator using a single system control center (SCC).
- The $i2000_{SR}$ may be integrated with a *c*8000 chemistry analyzer to form a *ci*8200 analyzer. The integrated system can be run by a single operator using the SCC.
- Using an available CD-ROM, interactive computer-based training (CBT) can be run on either the instrument SCC or a personal computer.
- The ARCHITECT *i*2000 and $i2000_{SR}$ analyzers have on-board refrigerated reagent storage with a capacity of 25 reagent kits per module.
- The sample handler can accommodate a wide variety of primary tubes as well as many different types of aliquot tubes and sample cups. The barcode reader recognizes a wide range of industry symbologies.
- The stainless steel sample probe uses pressure differential technology to detect short samples due to clots or bubbles. Electronic liquid level sensing is integrated with robotic pipetters, and automatically identifies foam or bubbles.
- The *i*2000 analyzer has direct track sampling capability and is lab automation compatible. Using the RSH, the $i2000_{SR}$ analyzer is capable of MDS. This ensures STAT tests are processed ahead of routine tests.
- For the *i*2000 and $i2000_{SR}$ analyzers, an automated reconstitution module (ARM) is also available. The ARM allows automated, continuous delivery of buffer concentrate to the instrument.

ASSAY PRINCIPLE

All assays use chemiluminescent magnetic microparticle immunoassay (CMIA) technology. The majority of CMIAs on the *i*2000 and $i2000_{SR}$ analyzers are two-step, immunometric assays with antibody attached to paramagnetic microparticles, and acridinium-derivative-labeled antibody/small molecule as conjugate or tracer (Figure 34.4). A few assays have antigens or other proteins coated on the microparticles. Using two-step formats wherever appropriate reduces assay nonspecific binding, eliminates exposure of the conjugate/tracer

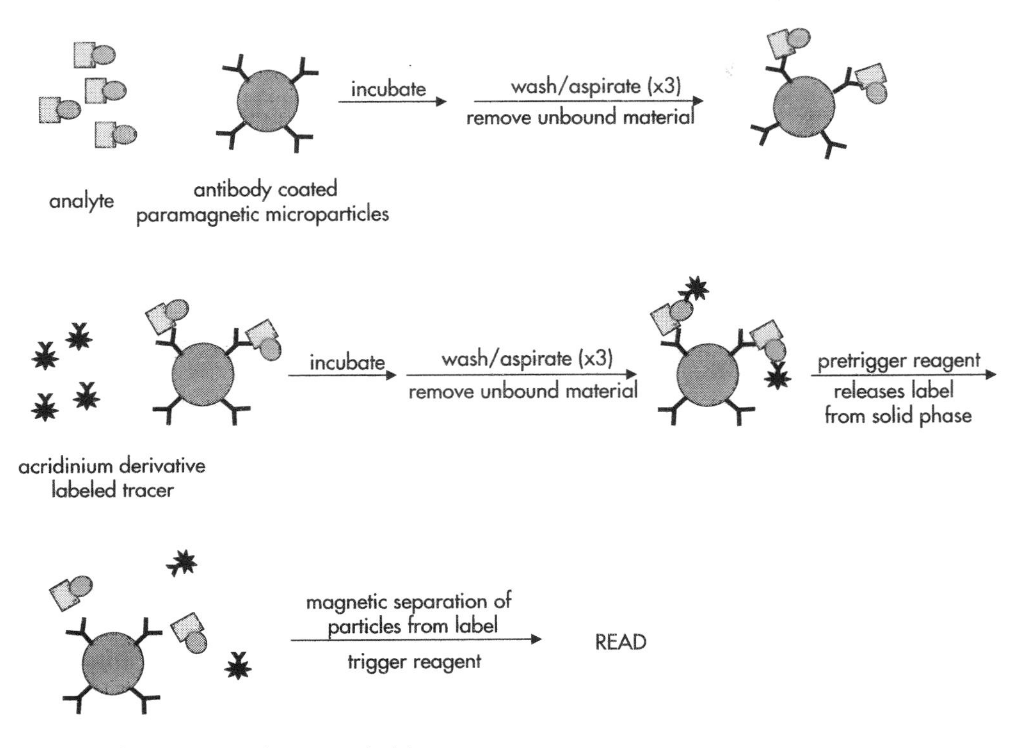

Fig. 34.4 ARCHITECT immunometric assay principle.

to potential interferents in the specimen (e.g. human anti-mouse antibodies, 'HAMA'), and minimizes the potential for high-dose hook effects. Low molecular weight analytes are predominantly measured using a two-step 'competitive' protocol. In this format, analyte is first extracted from specimen on to the paramagnetic microparticles. In step 2, tracer binds to unoccupied binding sites on the particles. A small number of assays use a one-step competitive protocol. This format essentially involves a competition between analyte and tracer for binding sites on the paramagnetic microparticles (Figure 34.5). Pretreatment assays utilize the same capture and signal generating principles as one- or two-step assays, except that the specimen is incubated with pretreatment reagent prior to the capture and detection phases of the assay. STAT assay protocols (*i*2000$_{SR}$ analyzer) operate under the same assay principles described above, but have a shorter first incubation time.

CALIBRATION

Assay calibration is dependent on the particular assay. Some assays utilize a six-point calibration, while others utilize a two-point adjustment of a master calibration curve. For the latter method, the master calibration curve is established by Abbott using six calibrators, and is stored within the two-dimensional barcode associated with each reagent lot. The statistical methodology for establishing six-point and master calibration curves, as well as the process for calibration curve adjustment, depends on the assay. The two-dimensional barcode associated with reagent and calibrator lots also contains expiration date and reagent inventory information. Where available, assays are standardized versus accepted international reference preparations (e.g. World Health Organization International Reference Preparations) or other highly purified commercially available material (e.g. USP grade steroid hormones).

ANTIBODIES

The majority of assays use monoclonal antibodies. However, there are a few assays that use polyclonal antibodies. Reagent storage buffers have been optimized with blocking agents to minimize potential interferences caused by heterophilic antibodies and HAMA.

SEPARATION

All assays use paramagnetic microparticles. Magnetic separation of the solid phase from unbound materials occurs in the wash zones. An optimized saline/surfactant buffer is used to perform the washing. Within each wash

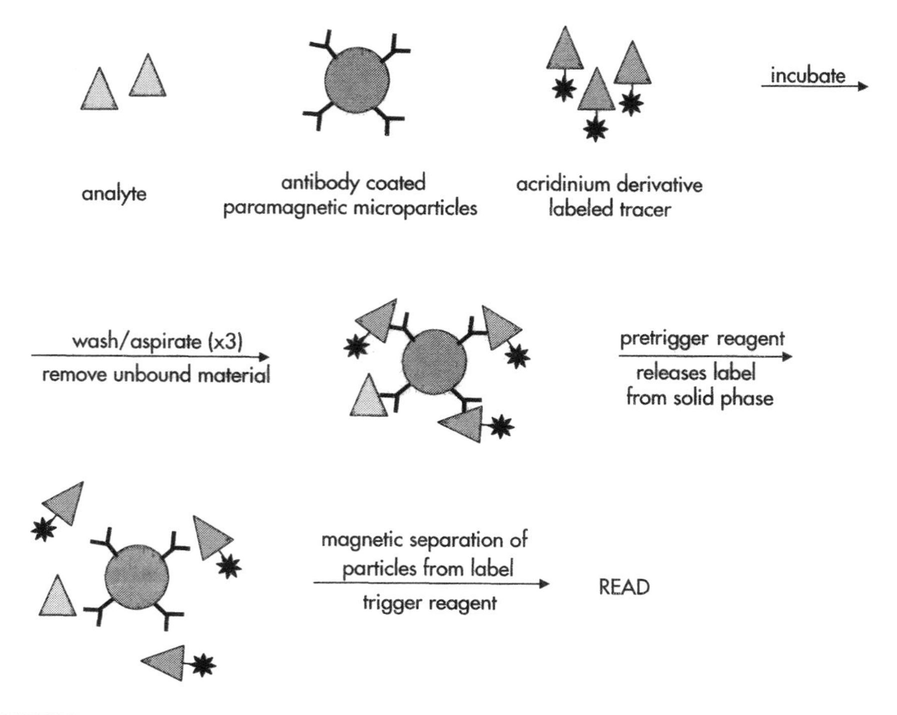

Fig. 34.5 ARCHITECT one-step competitive assay principle.

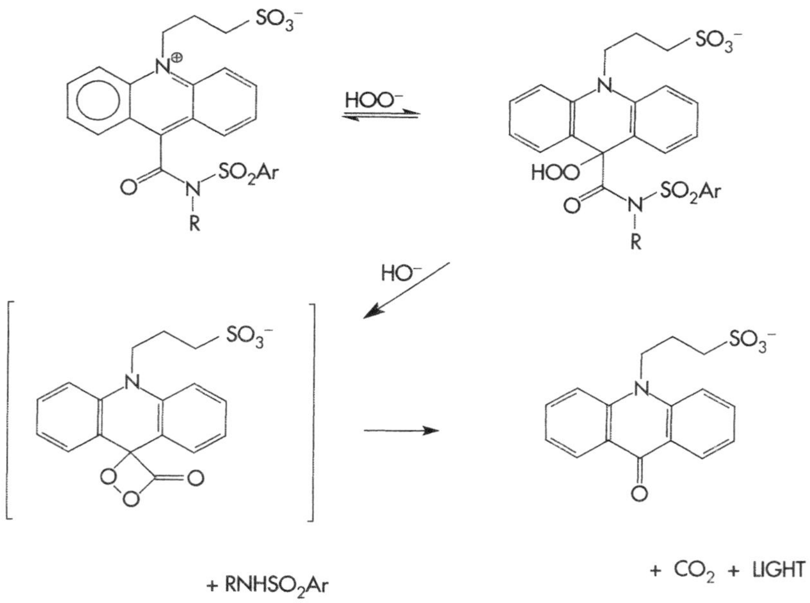

HOO^-

HO^-

$+ RNHSO_2Ar$

$+ CO_2 +$ LIGHT

Fig. 34.6 ARCHITECT immunoassay light generating reaction schematic.

zone, the washing event is composed of three distinct dispense/aspirate cycles of this system wash buffer.

SIGNAL GENERATION AND DETECTION

Signal generation is based on a patented acridinium derivative with improved chemical properties. This class of compounds (sulfopropyl acridinium carboxamides) has better aqueous solubility and stability than traditional N^{10}-methylacridinium-9-carboxylic acid phenyl esters. The light generating reaction is initiated by the addition of a pretrigger reagent containing acid and hydrogen peroxide. This pretrigger reagent causes the label to be released from the solid phase and into solution. The microparticles are then magnetically attracted to the inner wall of the reaction vessel, separating them from the label, which remains in solution. In the final phase of the reaction, a trigger reagent containing base is added. The resultant chemiluminescence is measured by a photomultiplier tube and translated by the data management software into the assay result. The light generating reaction is shown in Figure 34.6.

DATA PROCESSING

The SCC has a Windows NT-based data management software. This software is touch screen driven, with an average response time of less than 3 sec. The SCC has a data storage and retrieval capacity of up to 50,000 patient results and 25,000 quality control points.

INTERFACING TO LABORATORY INFORMATION SYSTEMS

The ARCHITECT *i*2000 and *i*2000$_{SR}$ analyzers have a standard ASTM RS232 port on the SCC for interfacing to other laboratory information systems.

FURTHER READING

Adamczyk, M., Chen, Y.Y., Mattingly, P.G., Pan, Y. and Rege, S. Neopentyl 3-trifloxypropanesulfonate. A reactive sulfopropylation reagent for the preparation of chemiluminescent labels. *J. Org. Chem.* **63**, 5636–5639 (1998).

Adamczyk, M., Fishpaugh, J.R., Gebler, J.C., Mattingly, P.G. and Shreder, K. Letter: detection of reaction intermediates by flow injection electrospray ionization mass spectrometry: reaction of chemiluminescent *N*-sulfonylacridinium-9-carboxamides with hydrogen peroxide. *Eur. Mass Spectrom.* **4**, 121–125 (1998).

Kimler, K. and Daly, D. A stepwise approach to 'flexible automation'. *J. Assoc. Lab. Autom.* **3**, 60–63 (1998).

35 CEDIA®, a Homogeneous Enzyme Immunoassay System☆

William A. Coty and Rueyming Loor

INTRODUCTION

Microgenics' CEDIA® (Cloned Enzyme Donor ImmunoAssay) product line consists of reagent kits, calibrators, and/or controls for performing homogeneous immunoassays on photometric clinical chemistry analyzers. The kits utilize Microgenics' patented technology (Henderson *et al.*, 1986; Henderson, 1997), CEDIA®, to allow highly sensitive detection of enzyme signal generated from complementation and antigen–antibody binding reaction without a separation step. CEDIA tests primarily measure low-molecular mass analytes such as hormones, therapeutic drugs (TDM), drugs of abuse, vitamins, and immunosuppressive drugs. However, the technology can also be used for low-concentration proteins such as ferritin (Shindelman *et al.*, 1992; Coty *et al.*, 1993).

TYPICAL ASSAY PROTOCOL

CEDIA products consist of two or three reagents; therefore, the assays are performed using two-reagent or three-reagent assay protocols. All reactions are carried out in the reaction cuvettes at 37 °C on clinical chemistry analyzers. A typical two-reagent assay protocol is carried out in the following steps.

1. Pipette and mix sample and reagent 1 (enzyme acceptor and antibody); incubate for 2–5 min.
2. Add reagent 2 (enzyme donor/ligand conjugate and substrate) and mix; incubate for 4–8 min.
3. Measure the rate of substrate hydrolysis photometrically for up to 3 min.

The three-reagent protocols are similar to the two-reagent protocol, except that reagent 1 contains only antibody. Enzyme acceptor is added as reagent 3, followed by a brief incubation prior to the photometric measurement.

PRODUCT FEATURES

Each CEDIA kit contains lyophilized reagents and their respective reconstitution buffers. The lyophilized reagents provide much longer shelf life of the kits for up to 36 months stored at 2–8 °C. Single or multi-constituent calibrators and controls are offered as separate components. Applications are available for a wide variety of clinical chemistry analyzers. All tests are fully automated and compatible with the random-access operation of many analyzers, and thus allow integration with special chemistry testing in routine profiles. This eliminates labor costs due to sample 'splitting', and errors due to sample identification and combining data from separate workstations.

The CEDIA assay (*see* ASSAY PRINCIPLE) provides the highest sensitivity of any homogeneous photometric immunoassay method. This allows the measurement of Vitamin B_{12} with a sensitivity of approximately 1×10^{-15} mol/test, and other low-concentration analytes such as digoxin, without pre-treatment or separation steps. There is no background β-galactosidase activity in human serum, and the CEDIA method is insensitive to endogenous interferences from hemolysis, icterus, or lipemia; these properties contribute to assay accuracy for sample quantitation. The CEDIA technology is also versatile, allowing measurement of a wide variety of analytes including low- and high-molecular weight analytes. The assay technology is also flexible toward sample matrices, i.e. the therapeutic drug monitoring assays use serum or plasma samples, drug of abuse assays measure urine or saliva samples, and the immunosuppressive drug assays quantify lysed or

☆ CEDIA is a registered trademark of Roche Diagnostic Systems.

The Immunoassay Handbook. *Third Edition*
D. Wild (Ed.)

extracted samples. For example, the CEDIA Cyclosporine Assay (Loor *et al.*, 2004) allows direct measurement of cyclosporine in lysed whole blood, thus eliminating organic solvent extraction required by other commercial cyclosporine assays. The new immunosuppressive drug assays, including sirolimus (rapamycin) (Luo *et al.*, 2004) and tacrolimus (FK506) (Charter *et al.*, 2004) quantify analytes in extracted samples because extraction is necessary to break the extremely high binding affinity of analyte toward FKBP (FK506 binding protein).

One of the key features of CEDIA tests is a linear relationship between kinetic enzyme rate and analyte concentration (*see* CALIBRATION). This allows two-point calibration, which maximizes the number of reportable results per kit, and eliminates the need for specialized curve-fitting software. CEDIA tests are non-isotopic, eliminating the costs and safety considerations associated with radioactive labels.

ASSAY PRINCIPLE

In the CEDIA technology, the polypeptide chain of the single subunit of *Escherichia coli* β-galactosidase has been split into two inactive fragments using genetic engineering methodology (Henderson *et al.*, 1986). The large fragment, termed enzyme acceptor (EA), contains a deletion near the amino terminus of approximately 5% of the β-galactosidase single subunit. The small fragment, termed enzyme donor (ED), contains the sequence missing from EA. When mixed together, these fragments spontaneously re-associate to form intact enzyme subunits, which then assemble to form tetramers and express β-galactosidase activity (Figure 35.1A); this process is termed complementation (Zabin, 1982).

For a precise conjugation position on the ED molecule, the sequence of ED has been modified to contain a cysteine or lysine residue as a specific site for covalent attachment of a hapten, which does not affect the complementation of EA with ED to form active enzyme (Figure 35.1B). However, the conjugation site was also chosen so that binding of antibody to the hapten blocks complementation of ED with EA (Figure 35.1C). Analyte present in a sample (calibrator, control or clinical sample) will compete for binding to the limited number of antibody sites, making ED–ligand conjugate available for formation of active enzyme (Figure 35.1D). Thus the amount of enzyme formed, as measured spectrophotometrically by the kinetic rate of chromogenic substrate hydrolysis, is directly proportional to the analyte concentration in the sample.

For the assay of protein analytes such as ferritin (Shindelman *et al.*, 1992), an alternative principle is used. ED is attached to anti-ferritin antibody so that complementation can still take place, although at a slightly reduced enzyme activity. Binding of the antibody to analyte present in a calibrator, control or unknown specimen will inhibit complementation, due to the formation of a large complex of ferritin and antibody. As a result, there is an inverse relationship between absorbance rate and ferritin concentration.

Different from the typical CEDIA assay design, a cyclic chimeric ED peptide was incorporated in the reagent kit for the detection of protease activity or for the screening of protease inhibitors (Coty *et al.*, 1999). The chimeric ED peptide was synthesized to contain ED peptide and a short sequence of a protease cleavage site such as HIV protease or caspase enzyme. Incubation of the linear chimeric ED peptide with EA resulted in the formation of active enzyme. When the chimeric ED peptide was cyclized by cross-linking agent, the cyclic chimeric ED peptide failed to complement with EA and exhibited no enzyme activity. Cleavage of cyclic chimeric ED peptide by corresponding protease yields a linear chimeric ED peptide and restored complementation activity, which can be measured in the same manner as typical CEDIA assays. This method can be utilized for the measurement of protease activity or for the screening of drug libraries for protease inhibitors.

Most CEDIA assays are designed for single analyte detection, but the technology also allows for simultaneous measurement of more than one analyte. The CEDIA Amphetamine/Methamphetamine assay measures both analytes by utilizing two antibodies and two ED–analyte conjugates in its reagents. By incorporating three antibodies and three ED–analyte conjugates in the kit reagents, a multiplex assay (Amphetamine/Ecstasy assay) is able to detect amphetamine, methamphetamine, and ecstasy drugs (Loor *et al.*, 2002). The CEDIA capability could be even further extended for detection of more than three analytes in one reagent kit.

CALIBRATION

CEDIA TDM and immunosuppressant immunoassays for low-molecular mass analytes are designed to give a linear calibration curve. Each kit is provided with two calibrators and tests are calibrated with two points in duplicate. Controls and unknown samples are quantified from the calibration curve by linear interpolation or regression analysis. Out-of-range samples can be diluted with low calibrator and retested; CEDIA tests do not give a high-dose ‘hook’ effect.

The CEDIA Drug of Abuse assays and Ferritin assay use a non-linear calibration curve to maximize detection sensitivity. The Drug of Abuse assays use a direct read out calibration curve that is different from the ferritin assay as described above (*see* ASSAY PRINCIPLE). In this case, five calibrators are provided, and analyzer

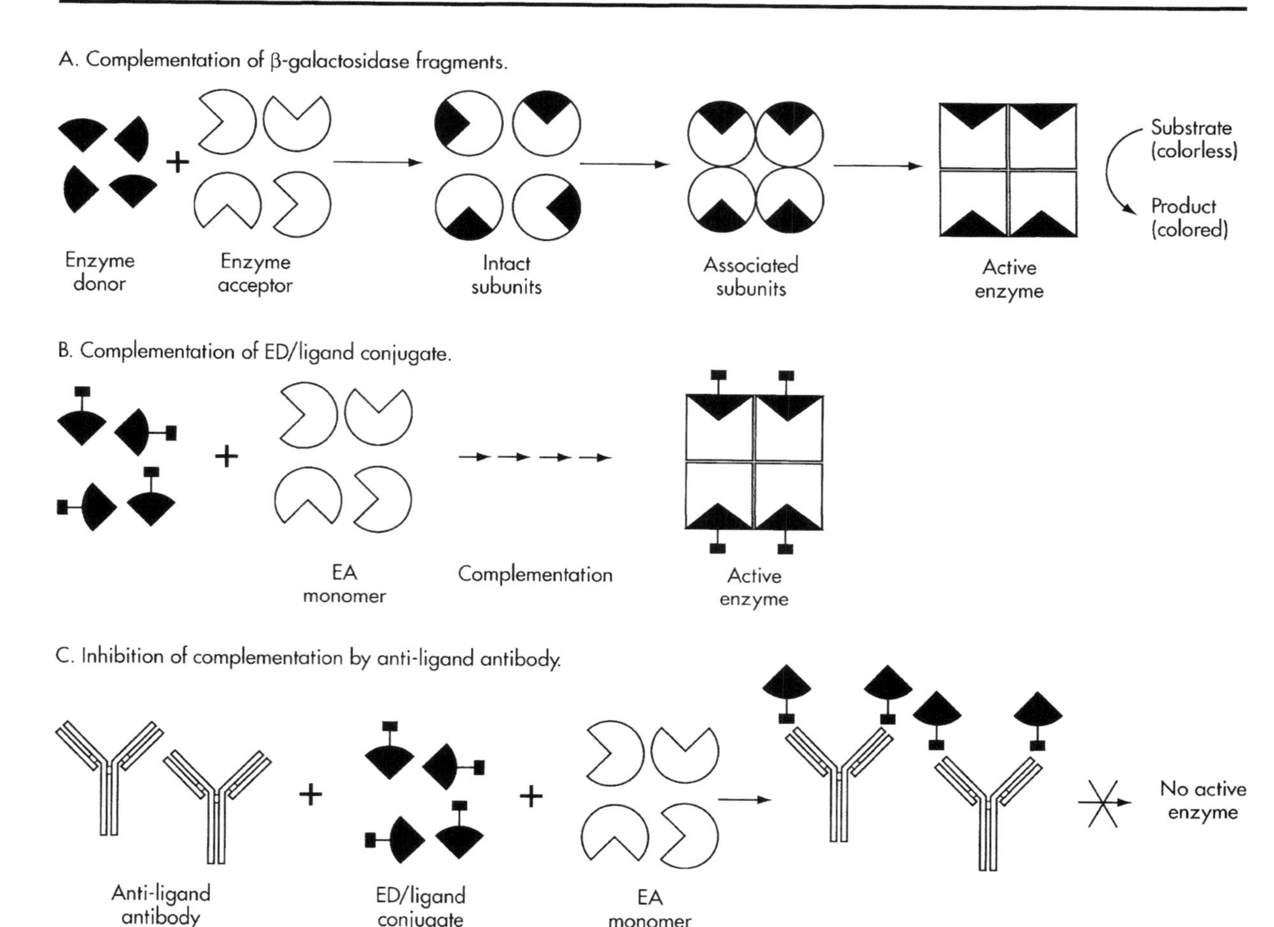

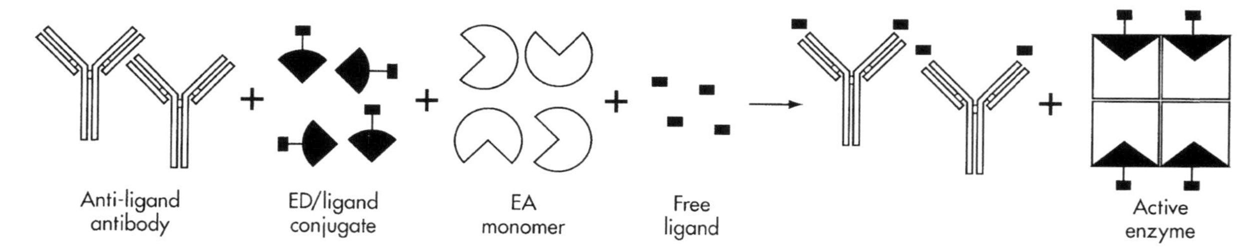

Fig. 35.1 CEDIA assay principle.

software is used to perform a four or five-parameter logistic curve-fit.

Calibration curves are stable for 3–7 days, depending on the test and the instrument used. In some cases, low calibrator updates are needed to provide maximum calibration curve stability.

ANTIBODIES

Monoclonal and polyclonal antibodies are used, depending on the nature of the test. Some assays use natural binding proteins instead of antibodies, such as intrinsic factor for the Vitamin B_{12} assay and folate-binding protein for the Folate assay. Intrinsic factor has a very high affinity toward Vitamin B_{12} and allows the assay to achieve an extremely low detection sensitivity, that antibody-based tests cannot reach, at approximately 1×10^{-15} mol/test Vitamin B_{12}.

SEPARATION

All CEDIA immunoassays are homogeneous; no separation step is required. An alkaline pre-treatment step is used in the Vitamin B_{12} and Folate assays to inactivate endogenous binding proteins; this step is fully automated using the Hitachi 911 and 912 analyzers. Due to high-binding affinity of immunosuppressant drugs toward FKBP, the Sirolimus assay (Luo *et al.*, 2004) and Tacrolimus assay (Charter *et al.*, 2004) require an extraction procedure including a centrifugation step. The extract is then placed in the sample cup and analyzed automatically on the analyzer.

SIGNAL GENERATION AND DETECTION

The label used in all CEDIA immunoassays is the ED peptide. Unbound ED–analyte conjugate reacts with EA to form active β-galactosidase; the kinetic properties of the assembled enzyme are identical to those of the native enzyme.

Chromogenic substrates such as *o*-nitrophenyl-β-D-galactopyranoside (ONPG) and chlorophenol red-β-D-galactopyranoside (CPRG) are used to measure enzyme activity. The hydrolysis of substrate is measured spectrophotometrically at 400–420 nm (ONPG) or 550–600 nm (CPRG). Some analyzers also correct for background rates measured at a secondary wavelength where neither the substrate nor the product have significant extinction coefficients.

Rate measurements are used to avoid interference due to sample background absorbance. Typical clinical chemistry analyzers make photometric measurements every 10–20 sec. Three or more readings are selected at the linear reaction phase and used to calculate a rate using linear regression analysis.

Besides the chromogenic substrates, fluorescence and chemiluminescence substrates can also be used to provide better sensitivity in the CEDIA assays (Loor *et al.*, 2000; Fung *et al.*, 2000) for high-throughput screening in drug discovery. The assays, including steroid receptor, cAMP, and tyrosine kinase, were miniaturized to 384- and 1536-well microtiter plates with simultaneous detection of all wells on a CCD-camera-based plate reader. Recently, a bioluminogenic substrate, D-luciferin-*O*-β-galactopyranoside, was utilized in the CEDIA assay and was shown to be more sensitive than chromogenic and chemiluminescence substrates (Yang *et al.*, 2005). The CEDIA assay using bioluminogenic substrate was further demonstrated to be suitable for drug and metabolite monitoring in point-of-care settings.

DATA PROCESSING AND INTERFACING TO LABORATORY INFORMATION SYSTEMS

For a given CEDIA assay, testing of the supplied calibrators produces a rate value. A linear calibration curve is created from the calibrator rates and their assigned concentrations. By linear interpolation or linear least squares regression, rates measured for controls and unknown samples are compared to the calibration curve to determine their respective analyte concentrations. These calculations are usually performed using standard software routines provided with the clinical chemistry analyzer, although they can also be performed using a host computer.

Most clinical chemistry analyzers provide for connection to a separate computer system via an RS232-C or similar interface. As with any clinical chemistry test, the CEDIA data can be transferred and analyzed in a report format for customers.

APPLICATION PROTOCOL FOR CLINICAL CHEMISTRY ANALYZERS

CEDIA tests are designed for use on automated clinical chemistry analyzers. Application of homogeneous immunoassays to clinical chemistry analyzers is more complex than for normal clinical chemistry tests. Therefore, only the instrument application protocols obtained from Microgenics, Roche Diagnostics or other licensed distributors of CEDIA reagents are warranted. Other than the specified protocols can be defined and approved by the user. Dilution or adulteration of the reagents is not recommended, as it will result in reduced performance. To obtain optimum performance of CEDIA reagents, the user must follow the analyzer care and maintenance instructions provided by the instrument manufacturer.

CONCLUSION

Since the first report by Henderson *et al.* (1986), the CEDIA assay system has been successfully used in the development of immunoassays for the measurement of hormones, therapeutic drugs, drugs of abuse, immunosuppressive drugs, and ferritin. Recent CEDIA immunoassays are summarized in Table 35.1.

Up to now, Microgenics has put forth tremendous effort toward CEDIA technology advancement and product

Table 35.1. Recent CEDIA immunoassays.

Testings	CEDIA assays
TDM (therapeutic drug monitoring)	Carbamazepine, Digoxin, Digitoxin, Gentamicin, Haloperidol/Bromperidol, Phenobarbital, Phenytoin, Procainamide, *N*-Acetyl-procainamide, Theophylline, Tobramycin, Valproic acid, Vancomycin
Immunosuppressive drugs	Cyclosporine, Sirolimus (Rapamycin), Tacrolimus (FK506) (under development), Mycophenolic acid (under development)
Hormones	T_4, T uptake
Anemia	Folate, Vitamin B_{12}
Drug of abuse	6-Acetyl-morphine, Amphetamine/Methamphetamine, Amphetamine/Methamphetamine/Ecstasy, Buprenorphine, Barbiturate, Benzodiazepine, Cocaine metabolite, Cannabinoids (THC), LSD, Methadone, Methadone metabolite, Opiate, PCP, Propoxphene, sample check
Calibrators and controls	Available for above assays

development as evidenced by the product list above. The CEDIA technology is continually utilized for the development of new assays such as immunosuppressive drugs, TDM drugs, and drugs of abuse. In the future, it is feasible to extend the CEDIA technology for the development of assays for high-molecular weight proteins or infectious disease agents.

REFERENCES AND FURTHER READING

Charter, L., Zhitnik, V., Caruso, A., Nguyen, T., Bodepudi, V. and Loor, R. Quantitative immunoassay for measuring tacrolimus (FK506). *Clin. Chem.* **50** (6), A132 (2004).

Coty, W.A., Loor, R., Bellet, N., Khanna, P.L., Kasper, P.L. and Baier, M. CEDIA® – homogeneous immunoassays for the 1990s and beyond. *Wien. Klin. Wochenschr.* **104** (Suppl 191), 5–11 (1992).

Coty, W.A., Shindelman, J., Sing, H., Loor, R. and Khanna, P.L. A homogeneous enzyme immunoassay for ferritin. *Am. Clin. Lab.* **12**, 10–11 (1993).

Coty, W.A., Loor, R., Powell, M. and Khanna, P.L. CEDIA® – homogeneous immunoassays: current status and future prospects. *J. Clin. Immunoassay* **17**, 144–150 (1994).

Coty, W.A., Shindelman, J., Rouhani, R. and Powell, M.J. Enzyme-fragment assay; a promising technology for HTS and drug screening. *Genet. Eng. News* **19**(7) (1999).

Fung, P., Panah, S., Araujo, J., Giertych, S., Lei, B., Rouhani, R., Loor, R. and Coty, W.A. Homogeneous, chemiluminescence and fluorescence assays for the high throughput screening. *The Society for Biomolecular Screening; 6th Annual Conference*, September, 198 (2000)

Henderson, D.R. Methods for Protein Binding Enzyme Complementation. United States Patent #5,604,091 (1997)

Henderson, D.R., Friedman, S.B., Harris, J.D., Manning, W.B. and Zoccoli, M.A. CEDIA™, a new homogeneous immunoassay system. *Clin. Chem.* **32**, 1637–1641 (1986).

Loor, R., Vainshtein, S., Deshpande, S., Fung, P., Panah, S., Zhao, S., Araujo, J., Lei, B., Rouhani, R., Coty, W.A., Donofrio, D. and Humphries, G. Homogeneous, chemiluminescence-based assay for receptor binding, cellular cyclic AMP and tyrosine kinase using the Clip CCD-camera-based Microplate reader. *The Society for Biomolecular Screening; 6th Annual Conference*, September, 225 (2000)

Loor, R., Lingenfelter, C., Wason, P.P., Tang, K. and Davoudzadeh, D. Multiplex assay of amphetamine, methamphetamine, and ecstasy drug using CEDIA technology. *J. Anal. Toxicol.* **26**, 267–273 (2002).

Loor, R., Pope, L., Boyd, R., Wood, K. and Bodepudi, V. Monitoring cyclosporine of pre-dose and post-dose samples using nonextraction homogeneous immunoassay. *Ther. Drug Monit.* **26**, 58–67 (2004).

Luo, W., Ye, L., Wang, J., Pai, V., Tsai, A., Bodepudi, V. and Loor, R. Development of CEDIA sirolimus assay for the automated clinical chemistry analyzer. *Clin. Chem.* **50** (6), A132 (2004).

Shindelman, J., Singh, H., Hertle, V., Davoudzadeh, F., Lingenfelter, D., Loor, R. and Khanna, P. Homogeneous enzyme immunoassay for measurement of human ferritin using recombinant enzyme fragments. *Clin. Chem.* **38**, 1078–1079 (1992).

Yang, X., Janatova, J. and Andrade, J. Homogeneous enzyme immunoassay modified for application to luminescence-based biosensors. *Anal. Biochem.* **336**, 102–107 (2005).

Zabin, I. β-Galactosidase α-complementation. A model of protein–protein interaction. *Mol. Cell. Biochem.* **49**, 87–96 (1982).

36 Clinical Chemistry Analyzers: VITROS™® Immuno-Rate and MicroTip™ Assays

Susan J. Danielson and David A. Hilborn

The VITROS 250, VITROS 950, and VITROS 5,1 FS Chemistry Systems are completely automated, random-access clinical chemistry systems that have the capability to perform immunoassays using Immuno-Rate technology developed by Ortho-Clinical Diagnostics (OCD), a Johnson & Johnson Company. The Immuno-Rate and clinical chemistry assays use dry slide technology (MicroSlides™), which was introduced in 1978 by the Eastman Kodak Company as an extension of its photographic technology. These coated, multilayer thin-film elements greatly simplify the number of operations performed by the customer since all the reagents required for an assay are contained within a postage stamp-sized MicroSlide. In addition, the VITROS 5,1 FS Chemistry System has extended immunoassay and special chemistry capability using new MicroTip technology.

The VITROS analyzers are floor-model instruments (Figures 36.1 and 36.2) with high throughput and sample capacity. The VITROS 250 and 950 analyzers can perform any of 45 + assays currently available. The VITROS 5,1 FS can accommodate up to 125 assays on board at one time. They use direct (primary) tube sampling and positive sample identification (barcode scanning), and can be incorporated into totally automated laboratory instrumentation systems. Results can be sent to central laboratory information systems. For the VITROS 250 and VITROS 950, reagents are provided in dry format in chemistry-specific thin-film MicroSlides. Only sample addition is required to initiate reactions for conventional clinical chemistry analytes. Immunoassays (Immuno-Rate assays) use an additional, automated wash step for bound-free separations prior to analyte detection, and potentiometric assays have a reference fluid added. In addition to the dry MicroSlides, the VITROS 5,1 FS also has ready-to-use liquid reagent packs supplied by OCD for its extended MicroTip assay menu.

Immuno-Rate immunoassays use either a competitive or immunometric mechanism. Competitive assays include those for digoxin, phenytoin, phenobarbital, and carbamazepine, while C-reactive protein (CRP) is measured using an immunometric assay. MicroTip immunoassays are liquid homogeneous assays which use either enzyme multiplied immunoassay technique (EMIT™) or turbidimetric technology for therapeutic drugs, serum proteins, and other assays (e.g. high sensitivity CRP, %A1c, and rheumatoid factor). These assays can be randomly ordered among the normal clinical assays, thus giving a laboratory the capability of running chemistries using either potentiometric, colorimetric, rate, Immuno-Rate, or MicroTip technology (on the VITROS 5,1 FS Chemistry System) in any order – at any time.

TYPICAL ASSAY PROTOCOLS

For Immuno-Rate assays, both competitive and immunometric assays are performed automatically starting with the application of 11 μL of a serum or plasma sample to a dry chemistry slide into which reagents specific for each assay have been incorporated. The slides are then incubated for 5 min at 37 °C to allow for equilibration of binding reactions. After incubation, 12 μL of a wash solution is applied to the slide. This wash solution serves a dual purpose. It separates bound reagents from free reagents using the flow of the fluid to move mobile (non-bound) reagents out of the detection area of the slide and also initiates the enzymatic reaction required for the detection reaction because it contains the enzyme substrate. Following the wash step, there is a second 2.5 min incubation at 37 °C during which reflectance densitometry measurements are made at regular intervals.

The Immunoassay Handbook. *Third Edition*
D. Wild (Ed.)

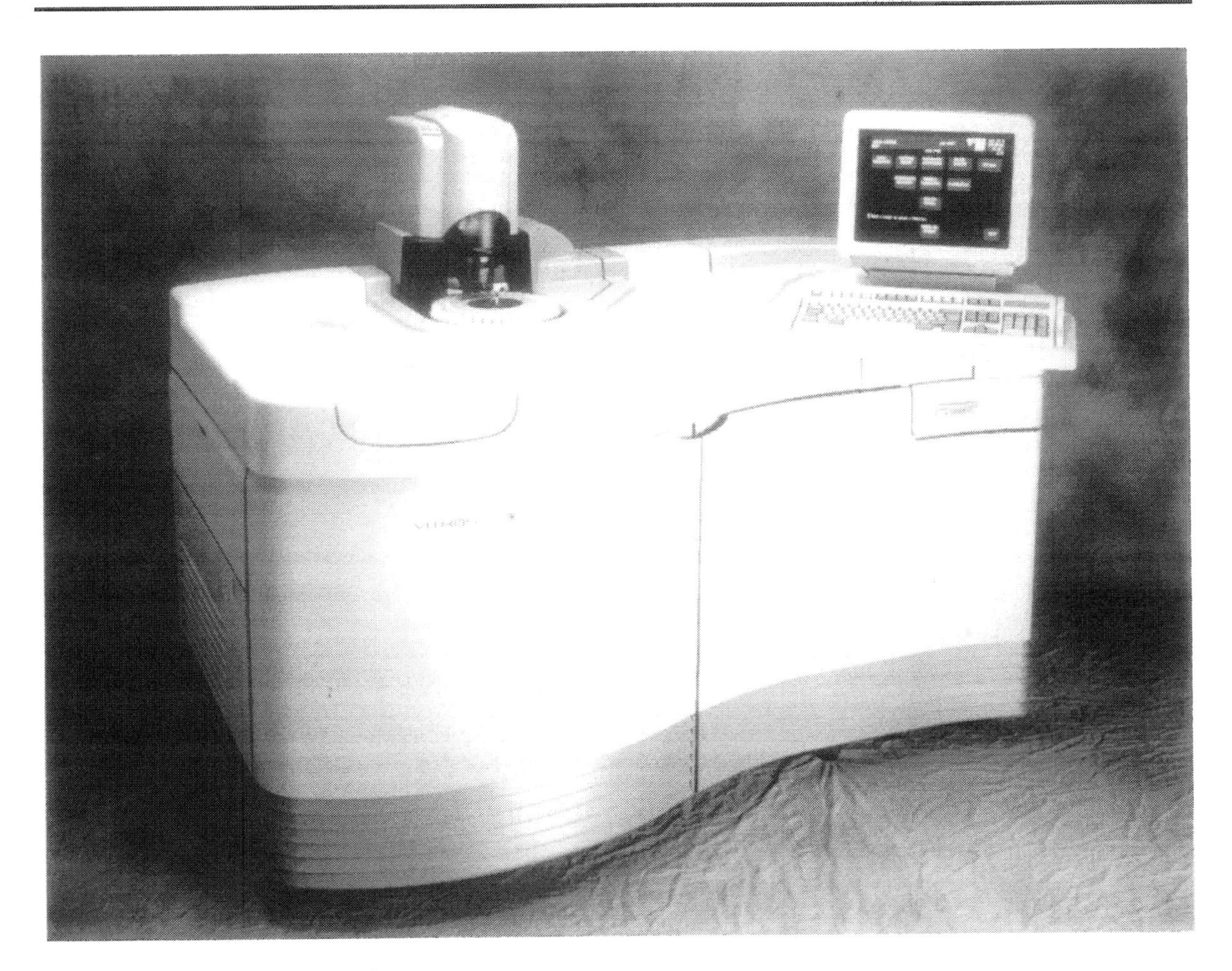

Fig. 36.1 VITROS 950 Chemistry System.

The rates of these reactions can be correlated to the concentration of analyte in the original serum sample. From the application of the first sample to its prediction requires approximately 9 min.

MicroTip assays are performed automatically with the application of 2–20 μL of a serum or plasma sample to a disposable cuvette into which reagents specific for each test have been added. The cuvette is then incubated at 37 °C to allow for thermal equilibration. After this time, a second reagent is added to the cuvette. Following the addition of the second reagent, there is a second incubation at 37 °C during which spectrophotometric measurements are made at regular intervals. The rates or endpoints of these reactions can be correlated to the concentration of analyte in the original serum sample. From the application of the first sample to its prediction typically requires about 11 min.

PRODUCT FEATURES

- The VITROS instruments are self-contained and fully automated. They are capable of performing both immunoassays and standard clinical chemistry tests in batch or random-access fashion.
- Less than 16 μL of sample is required for each of the immunoassays.
- For Immuno-Rate assays, MicroSlide reagents (other than a standard wash solution) are stored on the instrument for up to 1 week.
- For MicroTip immunoassays, reagents are stored in ready-to-use liquid reagent packs on the instrument for up to 4 weeks.
- There is no risk of carry-over from sample to sample or reagent to reagent. Disposable tips are used for individual metering steps for both Immuno-Rate and MicroTip assays.

Fig. 36.2 VITROS 5,1 FS Chemistry System.

- The integrity of patient results is maintained and sample repeats and redraws are minimized by the presence of Intellicheck™ Technology which includes liquid level sensing, sample clot, bubble detection, and sample and reagent aspirate and dispense verification features on the analyzers.
- In addition, the VITROS 5,1 FS Chemistry System utilizes new MicroSensor technology to measure sample quality indices for hemolysis, turbidity, and icterus without using additional samples, reagents, disposables, or slowing system throughput.
- The VITROS 5,1 FS Chemistry System also supports remote, real-time system diagnostics using e-Connectivity, which can reduce time spent on maintenance. In addition, the systems do not require the use of fixed probes or reagent mixing assemblies.
- Bidirectional interfacing with hospital laboratory information processing systems, supporting Broadcast Download (and Host Query on the VITROS 5,1 FS) is possible for increased laboratory productivity.
- The VITROS 250 system and the VITROS 5,1 FS system contain on-board dilution capability for improved labor optimization and error reduction.
- The VITROS 950, VITROS 250, and VITROS 5,1 FS systems do not require any plumbing to a water source or drain to execute testing and manage processing waste.

ASSAY PRINCIPLE

Immuno-Rate assays are based on heterogeneous competitive and immunometric methods adapted to thin-film formats (Figure 36.3). In general, the layers are formed as follows. Buffering agents are placed in a cross-linked gelatin layer, which comprises the lowest layer of the thin-film format. When sample is applied, the buffer is rehydrated and helps to control the pH at which the immunoassay binding reactions and subsequent enzyme reactions occur. Over the gelatin layer is an isotropically porous polymeric bead spreading layer whose function is to accept the serum or plasma sample and assure that it spreads to an area so that the surface density of analyte is constant. The non-porous beads that comprise this layer are about 30 μm in diameter, leaving ample interstitial capillary space so the sample is accepted rapidly (a few seconds). This layer may also contain buffer, stabilizers for

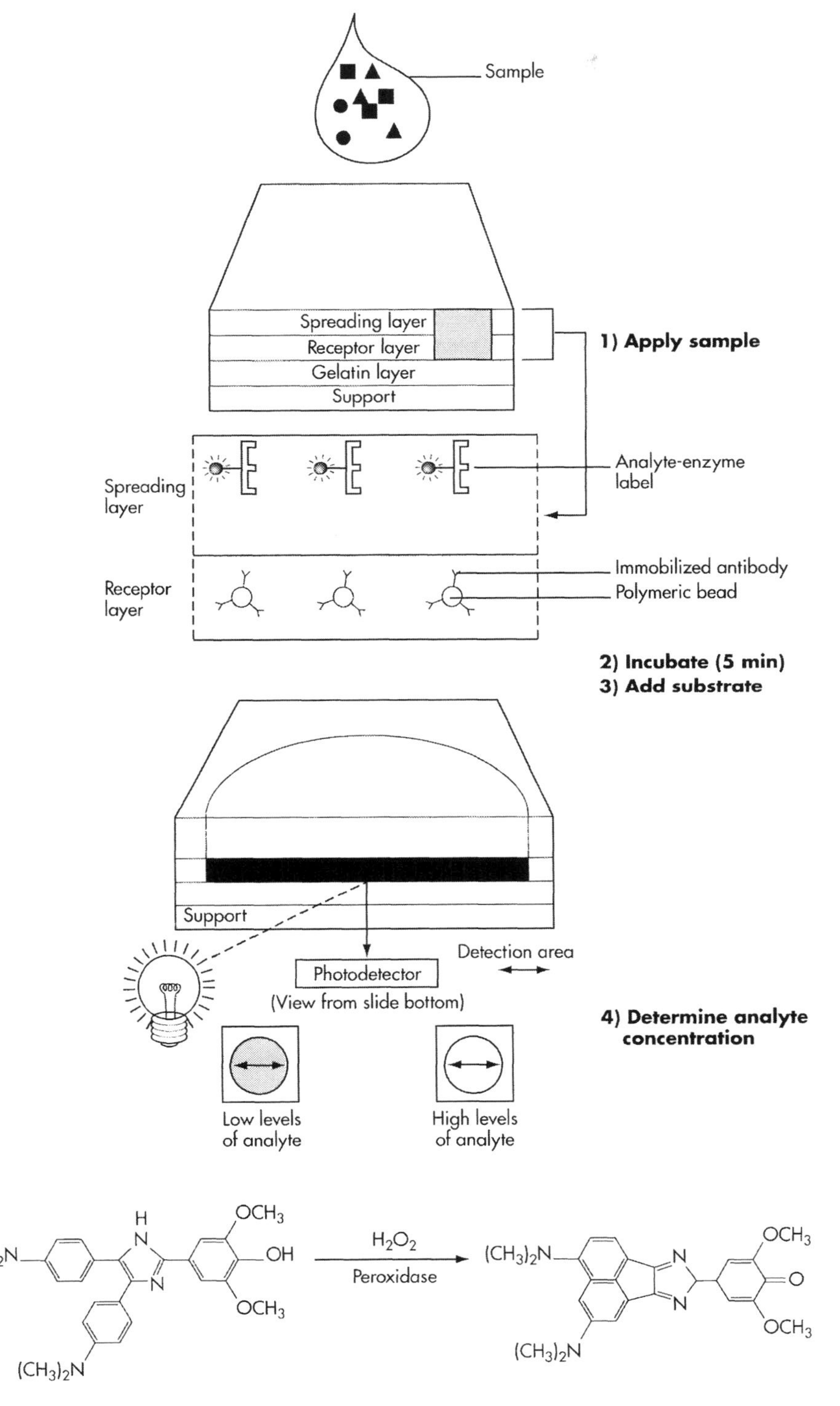

Fig. 36.3 Principle of competitive binding VITROS Immuno-Rate assays.

enzymes and antibodies, dyes that can be developed by the enzyme label, and other reactants as needed. Antibodies which have been covalently immobilized on polymeric beads (about 1 μm in diameter) may also be placed in this layer or alternatively in a thin receptor layer which lies between the gelatin layer and the spreading layer. For competitive assays, the enzyme-labeled drug (horseradish peroxidase) is coated in a thin layer at the top of the bead spread layer in order to prevent binding to the antibody before the serum sample is applied. For immunometric assays, where binding cannot occur in the absence of the analyte, the enzyme label is incorporated into a receptor layer.

Binding reactions are initiated upon sample addition. Because of the small spaces between the beads of the spread layer, diffusional distances are very small (a few microns) so that equilibrium is reached rapidly (within a few minutes) with no agitation necessary. After 5 min incubation at 37 °C, a wash solution is applied. The wash fluid flows by capillary action from the point of application through the read region of the slide to the periphery of the slide. Wash is efficient, a result of bulk fluid movement being the primary mode of molecular movement within the porous spreading layer and the small capillary structure within the spreading layer which directs fluid flow away from the point of application. Unbound molecules are carried away in the direction of fluid flow from their immobilized counterparts. There is little or no difference in movement between high and low molecular weight materials. Thus, a 12 μL volume is sufficient for the bound-free separation required by the heterogeneous assay format. The wash removes unbound material and initiates the enzyme reaction used for detection. The bound horseradish peroxidase produces a blue color. Rate measurements are made using reflectance densitometry at 670 nm during a 2.5 min incubation (at 37 °C) following the wash solution application. Movement of an inert magenta dye in the slide is automatically monitored to ensure that the wash step has been completed.

MicroTip immunoassays are based on homogeneous competitive and turbidimetric methods using traditional liquid reagents. In general, two liquid reagents in an integrated reagent pack are available for each of the MicroTip immunoassays. The reagent packs are stored refrigerated on the analyzer. The analyzer automatically uncaps and recaps the reagents when required. The assays themselves are carried out in disposable spectrophotometric cuvettes, which are incubated at 37 °C. A typical assay protocol uses a disposable pipette tip to add about 100 μL of the first liquid reagent to a cuvette. The sample is added (typically 2–16 μL) using a specially designed disposable VITROS 5,1 MicroTip, which provides the necessary metering precision. This tip also is used to mix the reagents and sample. After a typical 5 min incubation time to allow for thermal equilibration, the second reagent is added and mixed using a disposable tip. Incubation continues at 37 °C while spectrophotometric or turbidimetric readings are taken at prescribed times. Therapeutic drugs are measured using EMIT™ technology while serum protein and other assays (e.g. high-sensitivity CRP, %A1c, and rheumatoid factor) are measured turbidimetrically. Samples are automatically diluted if required by the specific assay protocol. Because MicroTip assays are processed using disposable tips, cuvettes, and reagents, the maintenance, cost, and carry-over associated with the use of reusable cuvettes, fixed probes, water, wash reagents, plumbing, drains, and mixing assemblies are eliminated. MicroTip technology also eliminates sample and reagent carryover that may occur with sample probes, mixing devices, and reusable cuvettes.

CALIBRATION

Immuno-Rate assays use a stored, lot-specific calibration function that has been factory determined and delivered to the instrument using a calibration disk supplied by OCD. Wet calibrations are performed on the analyzers using a three-level multianalyte calibration set available from OCD. A new lot of MicroSlides is calibrated when it is first used on an instrument. Subsequent re-calibrations are done every 3–6 months. Assays can be purchased in 18 or 50/60 slide cartridges. Unopened cartridges are stored refrigerated or frozen for up to 18 months.

MicroTip immunoassays use multiple level calibrators to generate a calibration curve on the analyzer. Calibrations are good for at least 4 weeks or until there is a change in the lot of assay reagents. Calibrator kits have a shelf life of greater than 1 year. The liquid reagent packs contain fluids for 50 assays. The typical shelf life of an unopened refrigerated reagent pack is greater than 1 year. Opened reagent packs can be stored on analyzer for typically 4 weeks.

ANTIBODIES

All VITROS Immuno-Rate assays use mouse monoclonal antibodies that have been covalently immobilized to small (1 μm), uniform-sized copolymeric latex beads. The CRP assay contains both a monoclonal anti-CRP antibody conjugated to horseradish peroxidase and a derivative of phosphorylcholine covalently bound to the 1 μm latex beads in place of a second monoclonal antibody.

MicroTip immunoassays use either monoclonal or polyclonal antibodies. Turbidimetric (agglutination) assays use uncomplexed antibodies or antibodies coupled to latex particles. EMIT assays use uncomplexed antibodies that have been selected for their ability to inhibit specific drug–enzyme conjugates in the absence of free drug.

SEPARATION

In Immuno-Rate assays, separation of bound enzyme label from free enzyme label is accomplished by application of a wash solution as described above. No separation is required in the homogeneous MicroTip assays.

SIGNAL GENERATION AND DETECTION

The Immuno-Rate assays use horseradish peroxidase as the enzyme label. Its reaction is initiated by the addition of the wash fluid, which contains hydrogen peroxide and an electron transfer agent, 4-hydroxyacetanilide. The reaction results in a blue color due to the oxidation of a leuco dye that is incorporated in one of the thin-film layers. Detection of the rate of blue color formation is by periodic reflectance densitometry reads at 670 nm. Only the color development at the center of the slide (washed region) is read.

The MicroTip immunoassays for therapeutic drugs (EMIT technology) use drug conjugates of glucose-6-phosphate dehydrogenase (G6P-DH). The reaction rate is monitored at 340 nm to detect the formation of NADH (NAD is a cofactor of G6P-DH). The turbidimetric MicroTip assays measure the agglutination reactions at a variety of wavelengths.

DATA PROCESSING

Once a rate or endpoint has been calculated by internal algorithms, it is transformed into a predicted concentration using the on-board calibrations described above. If the predicted concentration is outside the pre-determined calibration range either at the low or high end, the operator is alerted through an error message (flag). If the sample is beyond the high end of the concentration range, the sample can be diluted using an established protocol or automatically by the VITROS 250 and VITROS 5,1 Chemistry System and re-processed by the analyzer. Patient reports are flagged if there is not enough sample volume for the assay, and, for Immuno-Rate assays, if the sample is not applied to the thin-film or if the wash fluid is not applied.

Quality control fluids for immunoassays are also available from OCD. They can be run daily with the results tabulated and/or plotted in a variety of customer-specified formats.

INTERFACING TO LABORATORY INFORMATION SYSTEMS

An RS232-C serial interface with bidirectional capability, along with the necessary software, is available to link the VITROS chemistry systems with central laboratory information management systems. The instruments can be interfaced with lab automation consisting of a variety of transport systems that automate the pre- and post-analytical processes including centrifugation, aliquoting, sample integrity checking, stopper removal and recapping, sample sorting, and sample storage. The VITROS 950 AT, VITROS 250 AT, and VITROS 5,1 FS Chemistry System with an automation accessory can connect to any automation vendor's track. The design of these systems is centered around a concept in which the metering track is extended so that the disposable sample tip can sample from a tube while it is on the automation system transport track, thus eliminating the need for robotics to move tubes to and from the sample supply and for extra aliquot tubes. The VITROS 250 AT, VITROS 950 AT, and 5,1 FS Chemistry Systems retain the capability to process urgent STAT samples in a manual mode.

FURTHER READING

Curme, H.G., Columbus, R.L., Dappen, G.M. *et al.* Multilayer film elements for clinical analysis: general concepts. *Clin. Chem.* **24**, 1335–1342 (1978), .

Danielson, S.J. Thin-film immunoassays. In: *Immunoassays* (eds Diamandis, E.P. and Christopoulos, T.K.), 505–535 (Academic Press, San Diego, CA, 1996).

Danielson, S.J. and Hilborn, D.A. Single-layer and multilayer thin-film immunoassays. In: *Principles and Practice of Immunoassays* (eds Price, C.P. and Newman, D.J.), 545–577 (Stockton Press, New York, 1997).

Kwong, T., Meiklejohn, B., Bodman, V. *et al.* Performance of a new digoxin thin-film immunoassay in a hospital setting. *Clin. Chem.* **41**, S196(A) (1995).

Wu, A., Harmoinen, A., Chamber, D. *et al.* Comparison of a thin-film immunoassay for C-reactive protein with a turbidometric method. *Clin. Chem.* **40**, 1018(A) (1994).

37 Near-Patient Tests: The TRIAGE® System

Kenneth F. Buechler

The Biosite Incorporated TRIAGE® System is a quantitative, point-of-care immunoassay system for the detection of multiple analytes in whole blood. The products include TRIAGE BNP, Triage Cardiac, TRIAGE TOX, TRIAGE Profiler CP™ and TRIAGE SOB™. These products are intended to be used at the point of care as aids to the physician in the rapid diagnosis of heart failure, acute coronary syndromes (ACS), drug overdose, and delineating the cause of chest pain, and shortness of breath, respectively. The system is comprised of a portable instrument, the TRIAGE Meter, and a disposable assay device. The technology was developed and patented by Biosite Inc. The system is available in the United States, Canada, Europe, Central and South America, and Asia.

USER PROTOCOL

The assay is performed using a sample comprising whole blood or plasma collected in a sodium or lithium heparin tube. The test procedure involves addition of several drops of sample to the assay device, insertion of the assay device in the TRIAGE Meter and reading the results from the display screen. The quantitative results are generated within about 15 min of addition of the sample to the assay device.

PRODUCT FEATURES

- The TRIAGE System is easy to use. The product was designed for use in the laboratory and at the point of care.
- The assay system measures simultaneously and quantitatively, multiple markers from whole blood or plasma within about 15 min. Simultaneously measuring multiple markers improves the sensitivity and specificity for aiding in the identification of complex conditions and diseases because they have multiple etiologies and generally change as a function of time. The rapid turn around time allows for serial testing of samples at 20-min intervals.
- The system incorporates an end-point detection algorithm that allows the assay device to be run independent of the instrument. This enables an efficient throughput of testing which allows for up to 30 blood samples to be tested per hour per TRIAGE Meter.
- The system can provide a panel response when multiple markers are measured. The panel response can be considered a value that represents the measurements of the multiple markers on the panel and has an associated ROC curve, analogous to the measurement of a single analyte. One can select different cutoff values of the panel response to accompany the desired sensitivity and specificity of the diagnostic.
- Two products are designed to be aids in the differential diagnosis of the causes of chest pain and shortness of breath.
- The troponin I assay measures the oxidized and reduced forms of free troponin I, troponin I/C, troponin I/T and troponin I/C/T. The assay was designed so that each troponin I form is measured to approximately the same extent. This assures that the measured concentration of troponin I is independent of the form of troponin I in the sample.
- The assay system has multiple internal controls, external controls, and a QC lockout feature.

ASSAY PRINCIPLE

The immunoassay for the three cardiac markers is performed in a one-step assay device. The TRIAGE Meter detects the completion of the assay and measures the analyte concentrations. Several important aspects of the TRIAGE Cardiac System will be discussed: the fluorescence dye system, the assay device, and the TRIAGE Meter.

The Immunoassay Handbook. *Third Edition*
D. Wild (Ed.)

Fluorescence Energy Transfer Dye System

Fluorescence energy transfer in particles is used to achieve an excitation wavelength of 670 nm and an emission wavelength of 760 nm. At these wavelengths, scattering interferences, and intrinsic absorbance and fluorescence from the plasma are minimized. The fluorescent dyes are incorporated at optimal concentrations into latex particles to make fluorescent latex particles (fluorescence energy transfer latex, FETL) that exhibit fluorescence energy transfer. The fluorescent dyes are derivatives of silicon phthalocyanines (see Figure 37.1). The donor dye, silicon (IV) phthalocyanine bis (7-oct-1-enyldimethylsilyloxide), is excited at 670 nm and transfers its triplet state energy to the acceptor dye, [2^1,2^6,12^1,12^6-tetraphenyldinaphtho [b,l]-7,17-dibenzo [g,q]-5,10,15,20-tetraazoporphyrinato] silicon bis (7-oct-1-enyldimethylsilyloxide). The acceptor dye is a hybrid

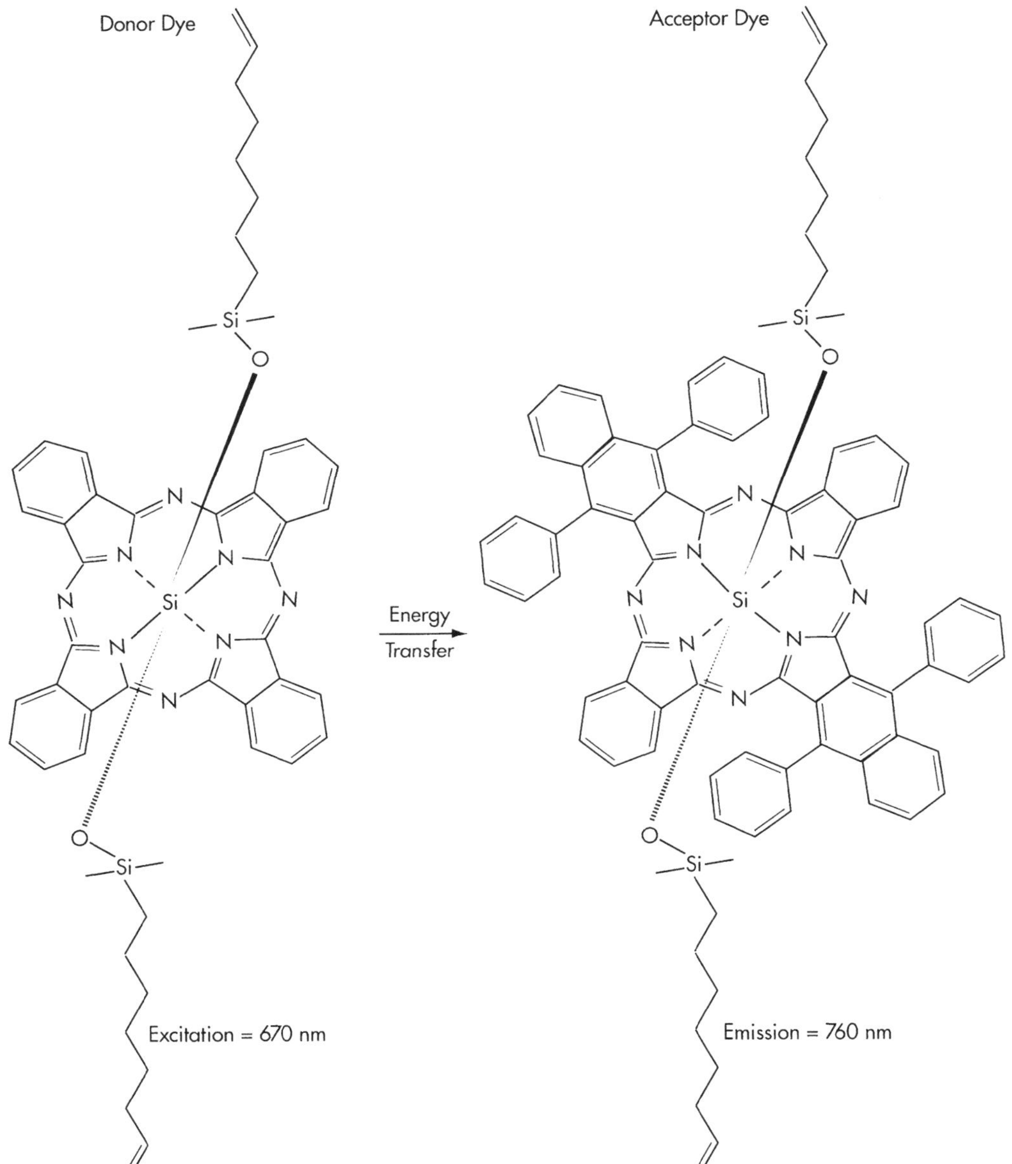

Fig. 37.1 Fluorescent energy dye system.

phthalocyanine derivative, having subunit characteristics of both phthalocyanine and naphthalocyanine dyes. The silicon-substituted dyes are derivatized with silyloxy aliphatic axial ligands, such that the ligands are transverse to the planar structure of the aromatic ring system. The axial ligands minimize stacking of the dyes in the particles through steric effects and thereby quenching of the dyes does not contribute to a decreased fluorescence yield. The Stokes shift of 90 nm allows the use of simple optics in the meter. The phthalocyanine dyes are photostable and exhibit high quantum and energy transfer efficiencies.

Assay Device

The assay device is comprised of a base, a blood filter, a lid, and a label (Figure 37.2). The base and lid are ultrasonically joined to create a series of precise capillaries through which the sample flows. A defined surface architecture of different textures in the base and the lid of the assay device control the movement of sample through the assay device. The use of capillarity to control fluid flow, based on the molded architecture of the device, is one of the unique features of the device.

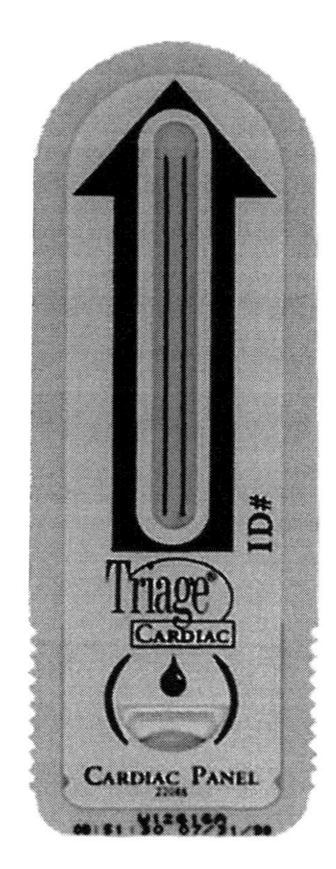

Fig. 37.2 Assay device.

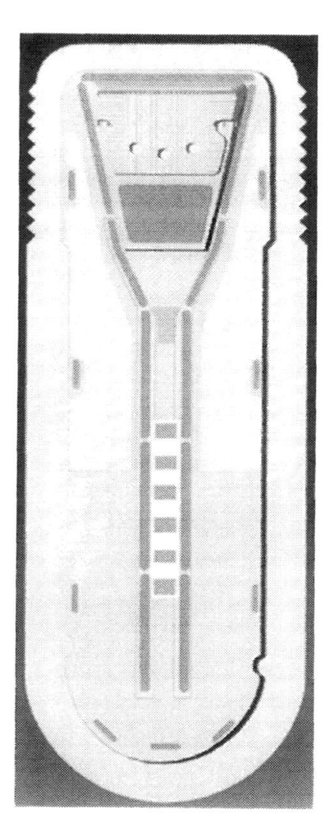

Fig. 37.3 Top view of device.

Figure 37.3 shows a skeleton diagram of the top view of the assay device. The blood sample is added at the sample addition zone. It enters a filter where red blood cells are separated from plasma by lateral flow in the filter. The plasma exits the filter and enters into a high capillarity region termed the sample reaction barrier (SRB). The SRB helps form the volume of the reaction chamber. In the walls of the SRB are grooves that promote the capillary flow of the sample into the reaction chamber and in doing so, the fluid moves from higher capillarity of the SRB to the lower capillarity of the reaction chamber. The sample enters and fills the reaction chamber and the fluid flow is delayed at the area of the time gate. The flow delays so that the cardiac marker analytes in the sample can react with the fluorescent antibody conjugates in the reaction chamber to form a reaction mixture. The time gate substantially delays fluid flow through the device because the surface of the time gate is hydrophobic. Proteins from the sample bind to the hydrophobic surface and make it hydrophilic. The rate of binding of the proteins and the relative hydrophobicity of the surface dictate the incubation time of the reaction mixture in the reaction chamber. This time is generally about 2 min, but can be varied by changing the relative hydrophobicity of the surface. The reaction mixture then flows from the time gate into the diagnostic lane. The change of the surface of the time gate to a hydrophilic surface allows an unimpeded flow of sample down the diagnostic lane. On the diagnostic lane are antibodies immobilized in discrete zones (detection zone antibodies). The zones include high- and low-positive controls, a negative control, and antibodies for the analytes to be measured. The analyte, already bound to the fluorescent antibody conjugate, binds to the immobilized antibodies in their respective zones as the reaction mixture flows through the diagnostic lane. The excess plasma that was added to the device acts as a wash to remove unbound fluorescent conjugates from the diagnostic lane. The wash solution fills the perimeter of the device in the capillary of the used reagent reservoir. The fluorescence of the analyte zones in the diagnostic lane is directly proportional to the analyte concentrations.

Figure 37.4 shows electron micrographs of the surface architecture. The surface texture and the relatively high capillarity of the SRB assure that sample will predominantly fill the capillary uniformly prior to the sample entering the reaction chamber (RC). The texture in the reaction chamber promotes uniform drying of the assay reagents during manufacture and aids in the filling of the reaction chamber with sample. The grooves at the time gate (TG) direct the fluid flow front to be essentially perpendicular to the direction of fluid flow. This minimizes the formation of air bubbles in the time gate capillary and promotes consistent time gate times. The point at the end of the time gate (TGP) assures that sample enters the diagnostic lane at a uniform location because the advancing sample front will interact first with the capillary at this point. The texture of the diagnostic lane (DL) and the used reagent reservoir is designed to promote uniform fluid flow through the long capillary where the fluorescent latex is captured at discrete zones and during the removal of the unbound fluorescent latex from the diagnostic lane.

Triage Meter

The TRIAGE Meter is a portable, battery powered fluorometer (see Figure 37.5). The excitation wavelength of 670 nm of the donor dye coincides with the output of the laser diode in the instrument. The detector is a silicon photodiode that measures the 760 nm emission light of the acceptor dye with a quantum efficiency of about 85%. A motor in the meter drives the assay device under the optic block for the measurement of fluorescence at the discrete assay zones. The fluorescence measurement is converted into an electrical signal which is then transformed into an analyte concentration using the calibration curve stored in the meter memory. The calibration curves and the meter software are downloaded into the instrument memory using a small code chip or EE-

(a)

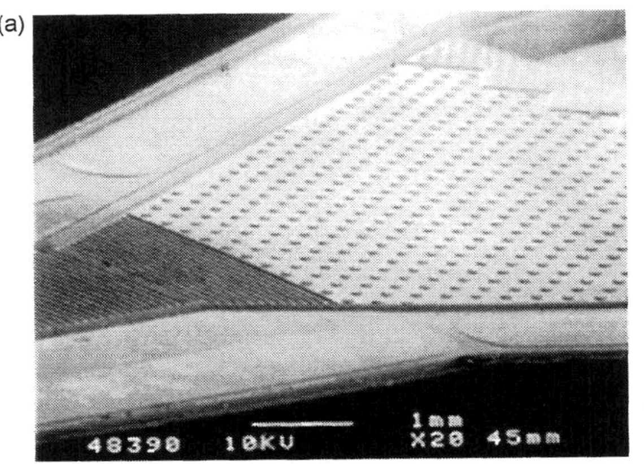

(b)

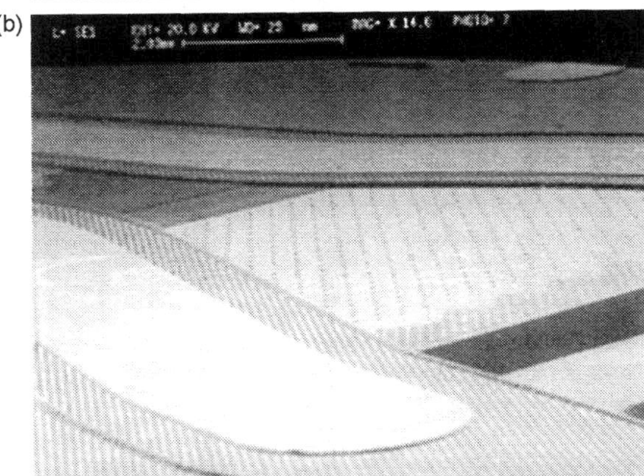

(c)

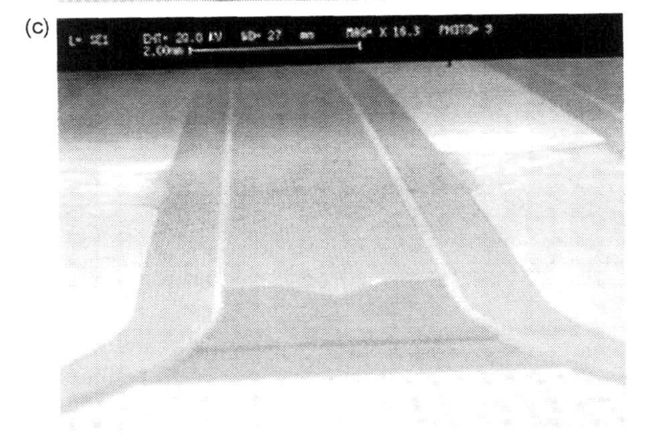

Fig. 37.4 Surface architecture.

prom. The assay results are displayed on a LCD or can be printed.

The meter is simple to use and the user interface is intuitive. The LCD screen prompts the user through the menu using the numeric keypad. There are software lockouts that allow only intended personnel to change

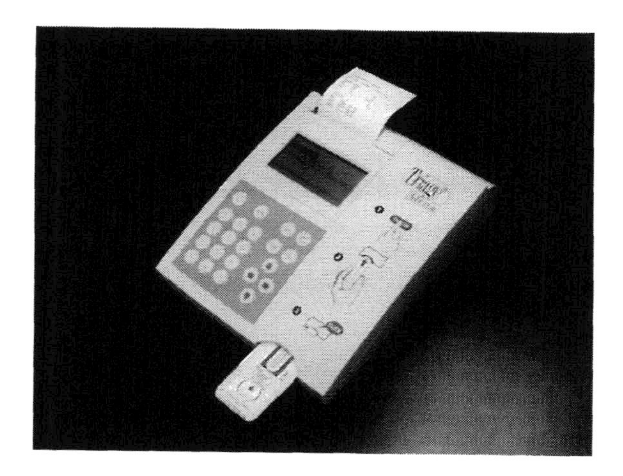

Fig. 37.5 TRIAGE Meter.

parameters such as frequency of calibration verification checks, cutoff concentrations, and deleting patient results.

MULTIANALYTE MEASUREMENTS

The TRIAGE TOX product, for example, has 10 different zones of immobilized antibody to measure overdosed drugs and assay controls. The diagnostic lane of the device is 2.5 cm in length and we have demonstrated the deposition of up to 100 discrete zones of antibody. See Figure 37.6 for an illustration of the intensity of a 100-array immunoassay as a function of the distance along the diagnostic lane. The measurement of the 100-array immunoassay device was performed by the TRIAGE Meter. This demonstrates that at least 100 different analytes can be measured by the TRIAGE System.

CALIBRATION

The TRIAGE Meter is calibrated against gold standard FETL. The meter has an internal standard FETL that is measured each time the instrument is turned on and each time an assay device is measured.

The assays are calibrated at Biosite using known concentrations of each analyte to be measured for each device lot. Each kit of assay devices comes with a code chip, which has specific calibration data for that lot. The calibration curves are loaded into the meter memory using the code chip. The meter memory can store calibration data for up to six different lots of product.

ANTIBODIES

The antibodies in the TRIAGE Cardiac Panel are comprised of monoclonal and polyclonal antibodies that perform sandwich assays in the assay device. The troponin I antibodies recognize with approximately equal affinity the various forms of troponin I (Buechler and McPherson, 1998). This is important because various forms of troponin I are released into the blood from damaged heart cells or change into different forms in the blood (Buechler and McPherson, 1998; Katrukha *et al.*, 1997). The majority of commercial assays measure the troponin I form to different extents (Wu *et al.*, 1998), which prevents the user from comparing the troponin I results of a patient obtained by different assays.

QUALITY CONTROL

Automatic checks of meter functionality, memory capacity, battery life, and software functionality are performed each time the TRIAGE Meter is turned on. An internal fluorescent standard, comprised of FETL in the instrument, verifies the proper function of the optics and the electronics. A QC simulator device is provided with the meter that confirms laser stability, alignment, and calibration. A supervisor code chip is supplied that allows access to special functions of the software, such as setting meter parameters and deleting results. A barcode reader in the TRIAGE Meter reads a barcode on the bottom of the assay device to protect against the use of expired product and to ensure that the TRIAGE Meter software has the appropriate calibration data for the specific lot of assay devices.

DATA PROCESSING

Data processing of the assay results occurs automatically after the user inserts the assay device into the TRIAGE Meter. Data for up to 600 patients are stored in the meter memory and the results can be downloaded via an RS232 port on the back of the meter.

THE PANEL RESPONSE

The diagnosis and prognosis of complex diseases will require the measurement of multiple protein markers because of the complex etiologies of the disease. This need can be seen today with the measurement of troponin I for aiding in the diagnosis of myocardial infarction. Because troponin I and its complexes rise 4–6 h after symptoms, an earlier marker is needed to improve sensitivity and specificity for detection of early myocardial infarction. The TRIAGE Cardiac and PROFILER CP products measure myoglobin and CKMB in addition to troponin I and its complexes because myoglobin and CKMB are earlier markers, although they are not specific to the heart. What is needed in the field of ACS diagnosis is a marker that can be measured earlier to rule in non-ST segment elevation myocardial infarction for patients who present within 0–4 h after symptom onset.

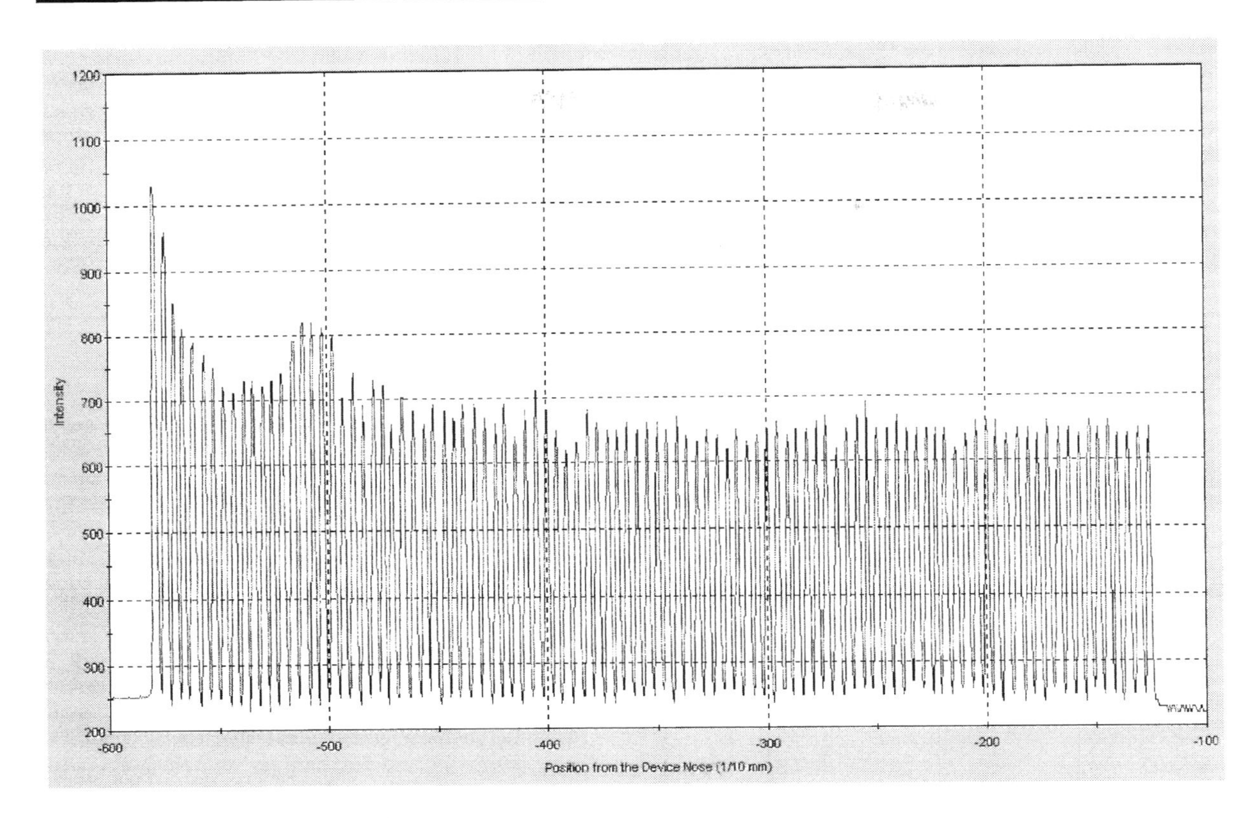

Fig. 37.6 TRIAGE device with 100 immunoassay array.

Stroke diagnostic biomarkers have been a major area of research at Biosite. During the course of our research, which included mining over 1000 patient samples and measuring over 50 biomarkers, we have found that one or two markers alone are not sufficient to optimize the identification of stoke. Through our research, we found six biomarkers that when measured together optimize the ROC curve area for stroke diagnosis. Our intention is to express the measurement of the six markers as a single value, which we term the 'panel response'.

The method of defining the panel response is a systematic iteration to find the optimal markers and panels of markers to distinguish or compare non-disease from disease, and it also optimizes the way in which the marker values can be used in a diagnostic product. A first step to simplify the problem of defining a marker or a panel of markers is to define an 'objective function'. An objective function is a scalar function and will represent the effectiveness of the test for diagnosis of non-disease from disease. For example, rather than requiring n elevated markers to define a positive state and then quantifying the effectiveness of this algorithm, one can generate a ROC curve from the number of elevated markers and use the area under the ROC curve to define the effectiveness of the test. By using the ROC curve area as the effectiveness of the test, the optimization problem has been simplified because the search space has been reduced since there is no need to calculate the effectiveness associated with each of the m values for n elevated markers. In this example, the number of elevated markers can be thought of as a concentration for the ROC curve, but the selection of the cutoff concentration is not required to determine if a test will be effective. Another step to simplify the problem of defining a marker or a panel of markers is to define a systematic method to find the best way to use the markers' values. Without this it is very difficult to find the best markers because one needs to distinguish the markers and how to use them. A systematic process to find the best way to use the markers is to combine all the values into one result, the 'panel response'. Functional forms of the panel response can be selected. Once this is done search routines can be employed to find the panel response function to maximize or minimize the objective function for a set of markers.

The panel response is calculated from an algorithm that was derived from the measurement of the panel markers in the non-diseased and diseased patient populations. The panel response behaves identically to a single analyte measurement, wherein the panel response has a ROC curve and associated sensitivities and specificities at specific cutoff panel responses. Specifically, the 'panel response' refers to a scalar function or its value, which is a function of the marker values of the panel. The panel response is a function of the marker values (M_{1-n}), written as PR$f(M_{1n})$. More specifically, the panel response is a summation over indicator values (I) of each marker. The indicator value is generally a function of the marker value. This can be represented as $PR \sum_{Markers} I_i(M_i)W_i$, where I_i is a function of the marker value M_i, W_i is a weighting coefficient that scales the indicator function. Furthermore, the panel response is scaled such that all values are between 0 and 1, but other increments can apply. For additional details relating to the panel response, see Anderberg *et al.* (2004).

SYMPTOM-BASED DIAGNOSTIC PANELS

Emergency room physicians make diagnostic decisions from the symptoms that the patient presents with. Different diseases and conditions can have the same symptom, thereby making the diagnosis more challenging for the physician. We have therefore launched two products that aid the physician in the differential diagnosis of conditions that present to the emergency department with a similar symptom.

The TRIAGE PROFILER CP product aids in the differential diagnosis of chest pain, which can be a result of heart failure, heart attack, or cardiac ischemia. The product measures BNP, CKMB, troponin I and its complexes, and myoglobin.

The TRIAGE PROFILER SOB product aids in the differential diagnosis of shortness of breath, which can be a result of heart failure, pulmonary embolism, or a silent heart attack, where the patient experiences no chest pain. The product measures D dimer, BNP, CKMB, troponin I and its complexes, and myoglobin.

REFERENCES

Anderberg, J., Buechler, K.F., McPherson, P.H., Kirchick, H. and Dahlen, J. Method and System for Disease Detection using Marker Combinations. International Patent Publication No. WO04/58055, July 15, 2004.

Buechler, K.F. and McPherson, P.H. Methods for the assay of troponin I and T and selection of antibodies for use in immunoassays. United States Patent 5,795,725, issued August 18, 1998.

Katrukha, A.G., Bereznikova, A.V., Esakova, T.V., Pettersson, K., Lövgren, T., Severina, M.E., Pulkki, K., Vuopio-Pulkki, L.-M. and Gusev, N.B. Troponin I is released in the bloodstream of patients with acute myocardial infarction not in free form but as complex. *Clin. Chem.* **43**, 1379–1385 (1997).

Product insert: Cardiac Panel for rapid quantification of CK-MB, myoglobin and troponin I. The TRIAGE Cardiac Panel. Biosite Incorporated, San Diego, CA.

Wu, A.H.B., Feng, Y.-J., Moore, R., Apple, F.S., McPherson, P.H., Buechler, K.F. and Bodor, G. Characterization of cardiac troponin subunit release into serum after acute myocardial infarction and comparison of assays for troponin T and I. *Clin. Chem.* **44**, 1198–1208 (1998).

38 Near-Patient Tests: Stratus® CS Acute Care™ Diagnostic System

W.N. McLellan

The Dade Behring Inc. Stratus® CS Acute Care™ Diagnostic System was introduced in 1998. It is a STAT, random-access analyzer (Figure 38.1) for the quantitative determination of analytes in whole blood or plasma. The initial menu focused on cardiac biomarkers. These are cardiac troponin I (cTnI), CK-MB mass and myoglobin. This menu has been expanded to include total βhCG and D-dimer. Assays under development are for other analytes that are significant to the triage of emergency patients such as NT pro B-type natriuretic peptide (NT pro-BNP). The Stratus® CS provides complete flexibility by allowing the selection of the most appropriate marker(s) specific to the individual patient. The analyzer is suitable for near-patient use or in the laboratory.

TYPICAL ASSAY PROTOCOL

There are six areas involved in the operation of the Stratus® CS analyzer. These are the Sample Manager, Pipette Manager, Pak Manager, Tip Manager, display/keypad, and waste container area. There are three types of reagent packs. These are the TestPak, DilPak, and CalPak.

Samples consist of whole blood or plasma. Processing whole blood involves its collection in a stoppered tube containing lithium or sodium heparin. An exception is that the D-dimer assay requires lithium heparin or sodium citrate whole blood or plasma. A barcoded tube can be passed by the barcode reader on the front of the analyzer. The patient ID can also be entered manually using the keypad. Whichever way the patient ID is entered, it is stored in the instrument, and displayed on the screen. The operator, after ensuring that the contents are well mixed, inverts the tube into a cannula without piercing the stopper. This cannula/tube assembly is placed in the Sample Manager and the operator closes the door.

The operator loads a rotor in the rotor table and TestPaks in the Pak Manager. A supply of tips is available in the Tip Manager.

Small red lights on the keypad are turned off when the necessary components have been loaded on the instrument. The system does not respond to the start key until all required components and data are present.

When the operator presses the start key, the Pak Manager moves the first TestPak under the barcode reader, which identifies the pack type and related information. The Pak Carousel moves the pack to the next position. All packs are inventoried to ensure that they can be used (i.e. TestPak lot is calibrated, DilPak has a corresponding TestPak, no expired Paks, etc.). While in the carousel, the TestPak is warmed to the testing temperature.

Next, the rotor moves under the sample manager. The cannula pierces the tube stopper and sample flows into the rotor. The rotor is mounted on the centrifuge, which spins it, preparing the sample for testing. A pipettor with a clean tip aspirates sample from the rotor and transfers it to the TestPak sample well. The pipettor then adds sample and reagents from TestPak wells to the glass fiber paper. To prevent fluid carryover, the tips are disposed of in the waste container after use.

The carousel moves the TestPak over the fluorometer, where the fluorescent signal is measured. The results are displayed on the screen as they are completed and printed on a report slip when all packs for the sample have been processed. The first result is reported within 15 min. This includes the time for the centrifugation of the whole blood. Subsequent test results are reported in 4-min intervals.

When processing is complete, the sample door opens and the operator removes the tube/cannula assembly. The TestPaks are automatically sent to the waste container. If the rotor has been used for four tests, it is automatically sent to the waste container. Otherwise,

The Immunoassay Handbook. *Third Edition*
D. Wild (Ed.)

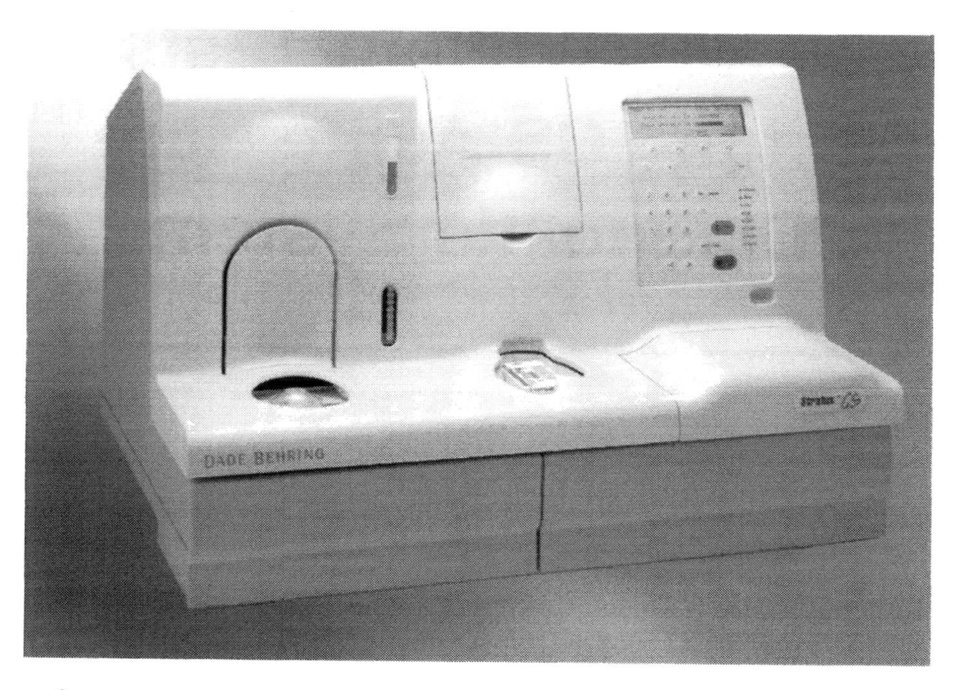

Fig. 38.1 Stratus® CS analyzer.

it can be held for up to 30 min in the rotor table, depending on the system configuration.

When a test requires a dilution, the rotor is not discarded automatically if fewer than four tests have been processed. The dilution can be processed on the centrifuged sample using a new TestPak and the method-specific DilPak. The DilPak provides the proper type and amount of diluent for the test.

When the waste container is full, as indicated by a warning light on the keypad, the operator removes it and disposes of it according to accepted local procedures for handling biohazardous waste.

A sample cup of plasma can be processed from a special position in the rotor table. Plasma cannot be tested through the tube/cannula assembly.

PRODUCT FEATURES

- The product has direct tube sampling, cap piercing, and barcode reading capability.
- A centrifuge is a built-in module of the analyzer that reduces the turnaround time with whole blood and ensures uniform centrifugation.
- The Stratus® CS analyzer fully complies with the Bloodborne Pathogens Preamble from the United States Occupational Safety and Health Administration (OSHA) (1998).
- The first result is available within 15 min with subsequent results at 4-min intervals thereafter.
- The operator has a choice of testing modes in that whole blood or plasma can be used with random access for one to four results.
- The unit uses reagent packs that are color-coded for easy recognition.
- There is a calibration curve stability of 60 days (cardiac assays and D-dimer) and 90 days (total βhCG).
- The analyzer can store up to three calibration curves per assay.
- There is an automated dilution protocol.
- The quality control requirement is a daily system check. Liquid quality control frequency is determined by the institution. There is a quality control lockout capability based on range and time that can be customized by authorized personnel.
- The product features dual monoclonal assay using radial partition immunoassay (RPIA) and dendrimer technology.
- The Acute Care™ troponin I assay is a high-sensitivity, second-generation assay with an analytical sensitivity of <0.03 ng/mL and imprecision of $<10\%$ CV at the 99th percentile of a healthy reference population as recommended by the joint committee of the European Society of Cardiology and the American College of Cardiology. The cTnI assay has an FDA-cleared risk stratification claim with a cutoff of 0.1 ng/mL.
- The results can be transmitted to a remote computer system (LIS/HIS) using a dedicated link as described

below. This capability, when used in conjunction with a POCT data management system, enables compliance with the NCCLS Standard POCT 1-A.
- Reprints of results for up to 20 samples can be obtained.
- The analyzer's dimensions are height (18 in./46 cm), width (27 in./69 cm) and depth (22 in./56 cm).
- Operator ID validation when configured prevents operation by untrained or unauthorized personnel.
- Sample/patient barcode reading capability ensures positive identification of the sample/patient.
- Reagent barcoding ensures proper identification of the assay, verification of expiration date, and verification of calibration.

ASSAY PRINCIPLE

The Stratus® CS analyzer procedure is a two-site sandwich assay based upon solid-phase RPIA (Giegel *et al.*, 1982) technology. In this procedure, dendrimer (Singh *et al.*, 1994) linked monoclonal antibody is added to the center portion of a square piece of glass fiber paper in the reagent TestPak. This antibody recognizes a distinct antigenic site on the analyte of interest. Sample is then added onto the paper where it reacts with the immobilized antibody. After a short incubation time, a conjugate consisting of enzyme-labeled monoclonal antibody directed against a second distinct antigenic site on the analyte of interest is pipetted onto the reaction zone of the paper. During this second incubation period, enzyme-labeled antibody reacts with the bound analyte of interest, forming an antibody–antigen–labeled antibody sandwich. The unbound labeled antibody is later eluted from the field of view of the Stratus® CS analyzer by applying a substrate wash solution to the center of the reaction zone.

CALIBRATION

Calibration update uses one CalPak and three TestPaks corresponding to the method being calibrated. Information needed by the instrument for the update is encoded on the barcoded labels of the CalPak and TestPaks.

When TestPaks are manufactured, calibration coefficients (C terms) are assigned to each lot using a six-level master calibration curve. The analyzer reads these coefficients from the TestPak barcode and stores them in memory.

In calibration mode, the instrument automatically processes the three TestPaks, using the contents of the CalPak. The mean of the three replicates is used to mathematically transform one of the calibration coefficients into a new, updated term. The instrument subsequently uses the updated term, along with the other coefficients from the TestPak barcode, to calculate test results when that lot of TestPaks is used on the Stratus® CS analyzer.

After calibration update, when a patient sample or quality control material is processed, the microprocessor within the instrument compares the test result to its internal calibration data for that TestPak lot, as based upon the stored coefficients. The test result is then displayed and printed.

ANTIBODIES

Monoclonal antibodies are used in these assays as described above.

SEPARATION

The Stratus® CS analyzer assays require a separation of the bound and unbound fractions during the procedure. The captured antibodies are immobilized on the glass fibers of the filter paper to form a reaction zone. The unbound materials are partitioned from the reaction zone, out of view of the detector, by a series of successive additions of reagent solutions, culminating in the addition of substrate wash solution.

SIGNAL GENERATION AND DETECTION

The Stratus® CS analyzer assays use alkaline phosphatase derived from mammalian intestine as the enzyme label conjugated to an antibody. The conjugate (consisting of enzyme-labeled monoclonal antibody directed against a second distinct antigenic site on the analyte of interest) is pipetted onto the reaction zone of the paper. During this second incubation period, enzyme-labeled antibody reacts with the bound analyte of interest, forming an antibody–antigen–labeled antibody sandwich. Next, the substrate wash solution is added to the tab. This is a dual function. First, it serves as an additional wash step (the first being the conjugate). Second, it is a substrate for the enzyme-mediated reaction because it contains 4-methylumbelliferyl phosphate (4-MUP). The addition of the substrate wash solution initiates the enzymatic reaction with the conversion of 4-MUP to 4-methylumbelliferone. The TestPak is transported to the fluorometer read station. This is a kinetic rather than an endpoint determination. The fluorescence of the 4-methylumbelliferone, a fluorogenic product, is detected by front surface fluorometry. Only the center of the tab, where the bound reaction has occurred, is read. On the periphery, any fluorescence due to the unbound fraction is not read by the detector because of its design. The results of the measurement are reported immediately on a thermal paper printer, shown on the display, and may be sent to a remote computer by way of the RS232-C serial port located on the analyzer. Thus, data can be sent because of the analyzer's connectivity capability to an LIS/HIS through a dedicated link. The options for the dedicated link are given in the 'Interfacing to LIS' section of this chapter.

DATA PROCESSING

The data reduction program has been designed to measure the rate of change of fluorescence for each reaction zone. The program has tests for various out-of-tolerance conditions such as excessive noise, substrate depletion, and excessive initial fluorescence. The raw data are expressed as millivolts per minute (MVM). The calibration routine is described above.

The quality control requirement is the processing of daily system check. This is a daily verification that interrogates four key areas of the system: optical detection, temperature, mechanical alignment, and fluid delivery. The frequency of running liquid quality controls is determined by each institution. Liquid controls are recommended after each calibration, with each new shipment of previously calibrated TestPak lot, as indicated by troubleshooting or part replacement procedures, or whenever the institution wishes to verify the performance of the system. The laboratory should run quality control materials in a manner consistent with local, state, or federal regulations.

Where appropriate, Daily System Check using Electronic QC provides liquid QC flexibility reducing the time and costs associated with processing daily liquid QC. The Daily System Check (Electronic QC) has a programmable time lockout.

INTERFACING TO LABORATORY INFORMATION SYSTEMS

Each analyzer has an RS232-C serial interface for data output to a remote computer. Data interface connectivity options are available between the Stratus® CS and a laboratory LIS or HIS, making it possible to monitor, manage, and accurately send data necessary for medical records and billing processes through a dedicated link. Many commercial providers of LIS/HIS are familiar with the Stratus® CS analyzer interface capabilities and formats.

FURTHER READING

Giegel, J.L., Brotherton, M.M., Cronin, P. *et al.* Radial partition immunoassay. *Clin. Chem.* **28**, 1894–1898 (1982).

Heeschen, C., Goldmann, B.U., Langenbrink, L., Matschuck, G. and Hamm, C.W. Evaluation of a rapid whole blood ELISA for quantification of troponin I in patients with acute chest pain. *Clin. Chem.* **45**, 1789–1796 (1999).

NCCLS. Point-of-Care Connectivity. Approved Standard. NCCLS Document POCT 1-A (ISBN1-56238-450-3), (2001).

Newby, L.K., Storrow, A.B., Gibler, W.B., Garvey, J.L., Tucker, J.F., Kaplan, A.L., Schreiber, D.H., Tuttle, R.H., McNulty, S.E. and Ohman, M. Bedside multimarker testing for risk stratification in chest pain units. *Circulation* **103**, 1832–1837 (2001).

Singh, P., Moll, F., Lin, S.H., Ferzli, C., Yu, K.S., Koski, R.K., Saul, R.G. and Cronin, P. Starburst™ dendrimers: enhanced performance and flexibility for immunoassays. *Clin. Chem.* **40**, 1845–1849 (1994).

Wu, A.H.B., Apple, F.S., Gibler, W.B., Jesse, R.L., Warshaw, M.M. and Valdes, R. Jr. National Academy of Clinical Biochemistry standards of laboratory practice: recommendations for the use of cardiac markers in coronary artery diseases. *Clin. Chem.* **45**, 1104–1121 (1999).

39 Over-the-counter Tests: Clearblue Pregnancy Test™, Clearblue Ovulation Test™ and Clearview™☆

Keith May

Clearblue Pregnancy Test (human chorionic gonadotropin, hCG), Clearblue Ovulation Test (luteinizing hormone, LH) and Clearview (hCG, *Chlamydia*, Strep A, IM, *Clostridium difficile* and Listeria) are unitized, rapid tests manufactured by Unipath in the United Kingdom and sold in over 80 countries. The tests are all based on immunochromatography using colored latex. The technology was developed and patented by Unipath, and the first product, Clearblue One Step, was the first truly one-step immunoassay to be used in the home (see Figure 39.1).

TYPICAL ASSAY PROTOCOL

To carry out the home tests for pregnancy and ovulation the device is simply held directly in the urine stream for 5 sec, then removed and the cap replaced. After 1 min the results can be read for Clearblue Pregnancy Test, and Clearblue Ovulation Test can be read after 3 min.

With tests in the Clearview format, a few drops of the sample or extract are added to the device and the result read after a few minutes. In the case of Clearview HCG, 120 μL of urine are added to the sample window using a one-shot pipette, and the results are read after 5 min. All of the tests contain an integral control which indicates when the test has been performed correctly.

PRODUCT FEATURES

- Clearblue Pregnancy Test and Clearblue Ovulation Test are provided in a pen-sized device for convenience when sampling from the urine stream.
- The one-step protocol is very simple. The device carries out the remaining steps automatically.
- The devices each have a built-in control.
- Clearblue Pregnancy Test has some additional features:

  (a) The wick changes from white to pink when enough urine has been applied.
  (b) The control line also acts as a timer to inform the user when the test is ready to read.
  (c) The result is displayed as a ‘ + ’ for ‘pregnant’ or ‘ − ’ for ‘not pregnant’.

ASSAY PRINCIPLES

The assay mechanisms of the two over-the-counter (OTC) tests (Clearblue Pregnancy Test and Clearblue Ovulation Test) and the Clearview devices are similar. The main difference is the means of sample addition. The OTC tests contain an integral urine collection device, a sampling wick, which protrudes from the plastic casing of the device and makes contact with a strip of porous membrane material internally. In the Clearview tests, sample is added onto the sample pad at one end of the device using a dropper.

☆Clearblue and Clearview are Trademarks.

The Immunoassay Handbook. *Third Edition*
D. Wild (Ed.)

Fig. 39.1 Clearblue Pregnancy Test.

The absorbent sample wick is held in contact with a porous membrane material. The membrane has three zones of antibody (see Figure 39.2). The first is mobilized by the sample, while the other two are immobilized. When the sample wick is saturated with urine, this passes along the membrane material until it reaches the mobile zone of the reagent. In the pregnancy test devices, the mobile reagent consists of colored latex particles sensitized

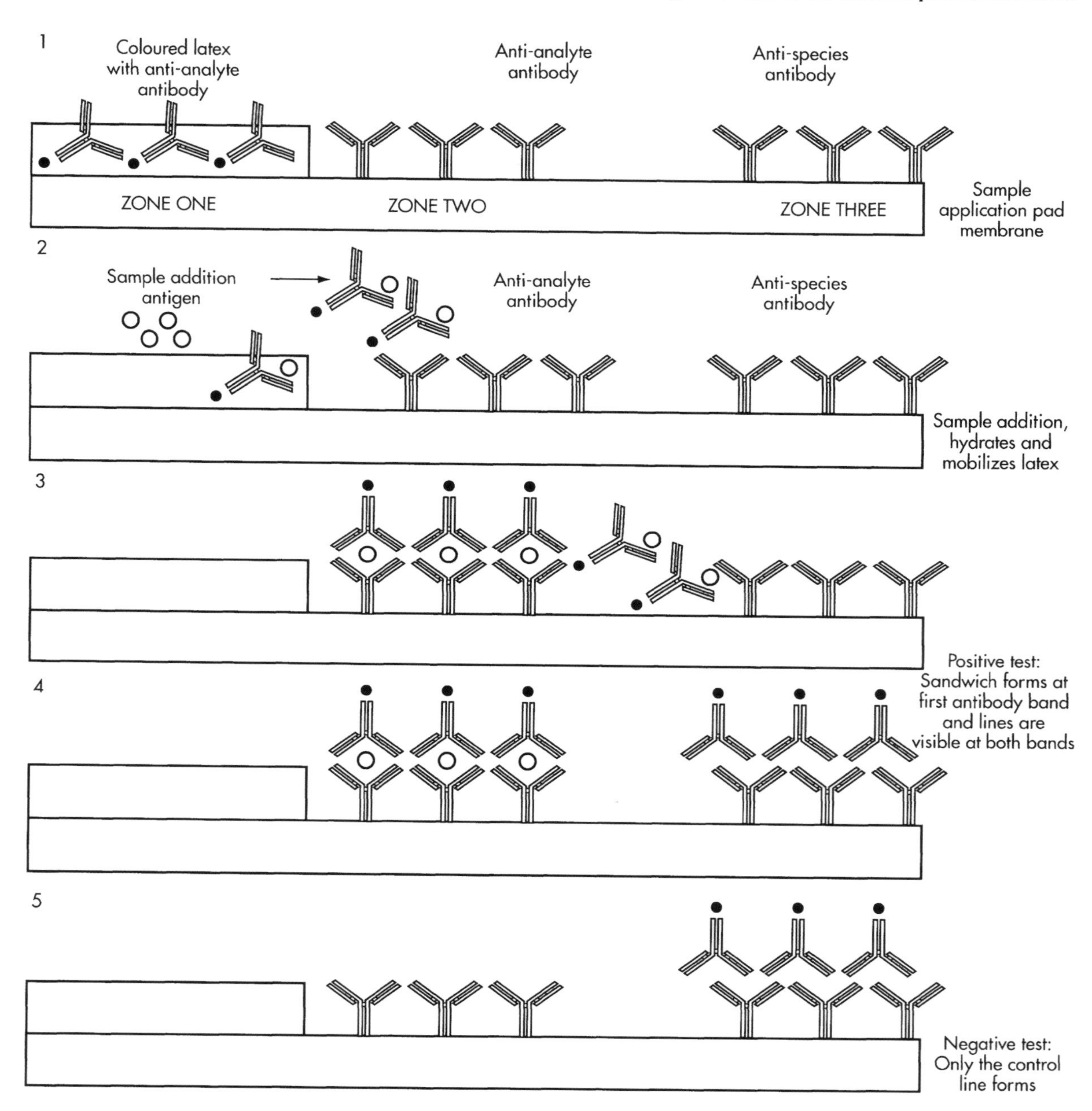

Fig. 39.2 Mechanism of Unipath's rapid assay.

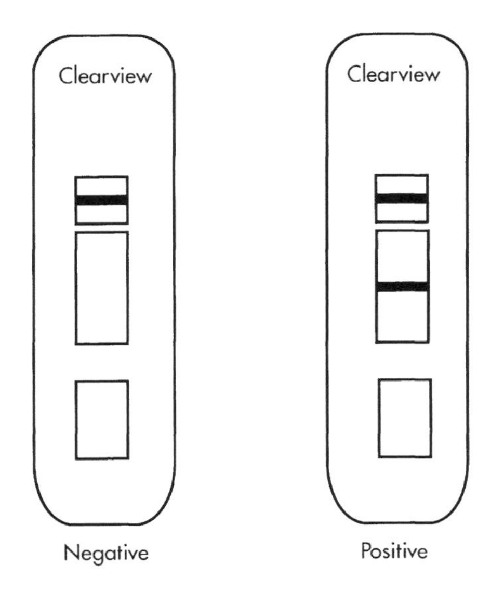

Fig. 39.3 Schematic diagram showing typical results obtained using Clearview.

with monoclonal antibody to the α subunit of hCG and colored latex particles sensitized with rabbit IgG. The urine picks up the colored latex and carries it to the second zone located directly below the large window. The second zone is composed of monoclonal antibody to the β subunit of the hCG immobilized in the membrane. If the urine contains hCG, the hCG reacts with the anti-α hCG on the latex and allows this to be trapped by the anti-β hCG zone, causing the formation of a blue line in the large window.

The unbound latex and urine continue to move along the strip by capillary action and come into contact with the third zone (below the small window). The third zone typically (but not for all the tests) consists of a goat anti-rabbit immunoglobulin antibody that is immobilized in the membrane. Latex sensitized with rabbit IgG is trapped, leading to the appearance of a blue line in the small window every time, whether hCG is present or not (see Figure 39.3).

The casing also contains a desiccant to remove moisture and the devices are sealed into foil pouches to ensure a long shelf-life at room temperature.

CALIBRATION

The quality control of Unipath tests is done by reference to the appropriate international standard (IS) where available, e.g. fourth IS for hCG.

ANTIBODIES

Monoclonal and polyclonal antibodies are used.

SEPARATION

Chromatography of the sample transports the label to the reaction zone and also washes unbound label away from the immunologically bound material.

SIGNAL GENERATION AND DETECTION

The rapid assay technology uses labels that can be directly visualized. The principal uses of the technology are for qualitative testing and the devices are designed to be read by eye.

LIMITATIONS

The way the assay system is currently configured, assays are qualitative with a visual end point and are therefore limited to the sensitivity of the human eye. The tests were also designed such that single or only a few samples are examined at one time.

In other respects the assay system is capable of handling most sample materials (urine, serum, swab samples, etc.) and can be applied to both macromolecular and hapten assays.

LATEST DEVELOPMENTS

Unipath have recently launched Clearblue Digital Pregnancy Test and Clearblue Digital Ovulation Test. These tests consist of test sticks, which are smaller but otherwise similar in design to those described above, and a holder. The test stick is inserted into the holder when the user is ready to test, this activates the holder. The absorbent sampler (of the test stick) is held in the urine stream for 5–7 sec. The holder displays digitally when the sample is detected, and displays the result in words or symbols within 3 min. The holder utilizes similar techniques to Persona to read the colored lines on the test stick (*see* PERSONA). This new technology makes it easier for the user to interpret the result of their test.

40 Over-the-Counter Tests: Persona

Keith May

Persona is a new method of contraception that works by monitoring changes in reproductive hormones to prospectively identify the fertile period. The information can then be used to modify behavior (abstain) to reduce the risk of pregnancy. The product was introduced in the UK in September 1996 and has since been introduced in Germany, Italy, France, Netherlands, and Scandinavia.

TYPICAL ASSAY PROCEDURE

On the first day of menstruation the user presses the 'M' button on the monitor which then registers this as day 1 of the cycle. For the first 5 days of the cycle, the monitor will display an 'M' symbol. Each day the monitor will indicate the fertility status by the use of a red or green light. On the days the monitor requires the user to do a test (16 tests in the first cycle and 8 in subsequent cycles) it will indicate this by a yellow light. The user then inserts a test stick into the urine stream for 3 sec, places the cap over the exposed end and then inserts it into the monitor. The monitor compares the levels of signal produced with those recorded from previous days, looking for significant changes in estrone-3-glucuronide (E-3-G) and luteinizing hormone (LH). When the monitor detects a significant change in E-3-G it changes from green to red. Detection of LH allows the system to indicate by the use of an 'O' symbol the expected time of ovulation and also uses this information to indicate the end of fertility by switching the light to green. As the monitor stores the information, it is able to target the tests to maximize the chances of locating the hormone rise and becomes personalized to the user's cycle lengths and patterns.

PRODUCT FEATURES

Persona consists of a small, hand-held, intelligent monitor and disposable test sticks (Figure 40.1). The test stick converts the level of the key reproductive hormones into colored signals which can be read by the monitor. The monitor compares the readings of the test sticks with those recorded on previous days to make decisions on the start and end of the fertile period.

ASSAY PRINCIPLE

Unipath Ltd has pioneered advances in rapid testing that have allowed the production of one-step pregnancy and ovulation tests that are simple and reliable. The immunochromatographic system used in these tests has now been extended to allow the rapid simultaneous detection of urinary hormones to predict the fertile period. The system consists of a series of disposable test sticks and a hand-held monitor. The test stick converts the concentrations of E-3-G and LH into a colored signal which can be read by the monitor. The monitor contains a sophisticated algorithm which uses the changes in E-3-G and LH levels together with the information obtained from previous cycles to determine the fertile period. The principles of the test stick operation are shown in Figures 40.2–40.5.

The device itself contains two key materials – a wicking material onto which the mobile latex reagent is deposited and a membrane onto which the other key reagents are immobilized. For simplicity, the assays are described here separately although they operate simultaneously with the test device. For the E-3-G assay, blue latex particles are coated with antibody specific for E-3-G and these are deposited onto the wick material in such a way that they can be mobilized by the sample (Figure 40.2). An E-3-G protein conjugate is immobilized in the test zone on the membrane that is furthest from the wick. When the test stick is introduced into the urine stream, urine rises up the wick by capillary action. The urine resuspends the latex and transports it to the test zone containing immobilized E-3-G. Any E-3-G in the sample is free to react with

The Immunoassay Handbook. *Third Edition*
D. Wild (Ed.)

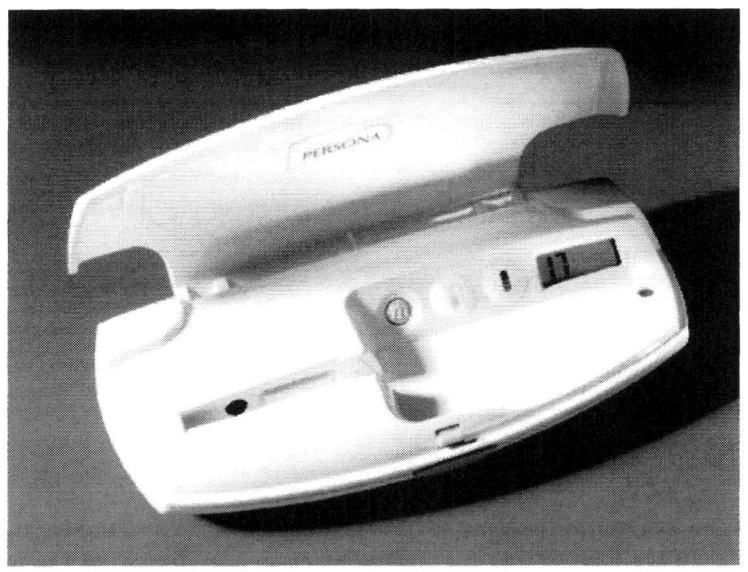

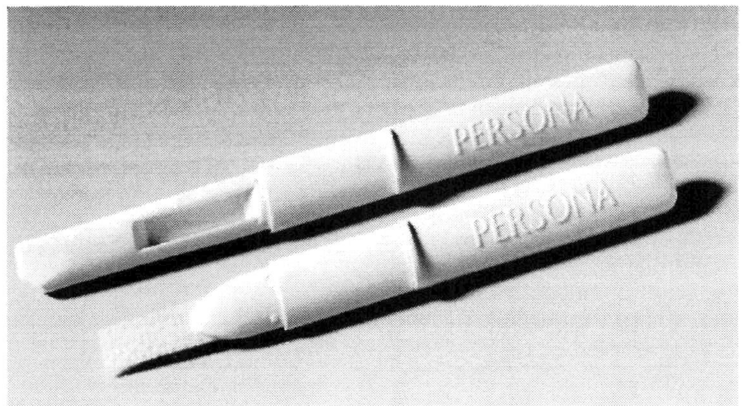

Fig. 40.1 Persona hand-held monitor and test strips.

Test stick set-up

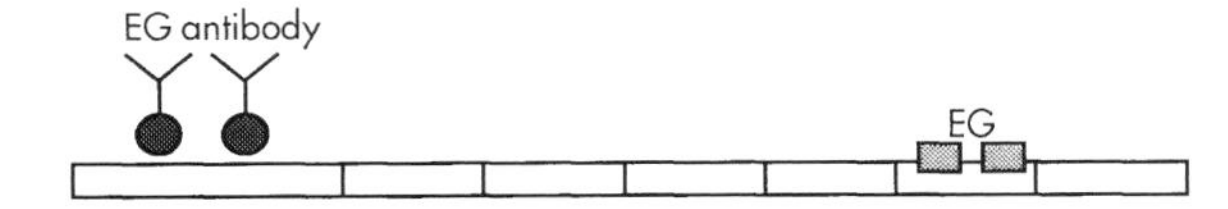

Fig. 40.2 Unipath rapid assay technology for the detection of estrone-3-glucuronide (E-3-G). E-3-G antibodies bound to blue latex particles are at the base of the strip and E-3-G is attached to the strip further along it (test zone furthest from the wick).

Test stick set-up

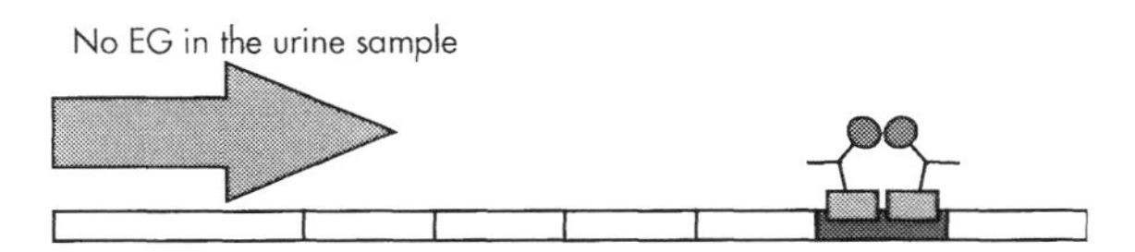

Fig. 40.3 Unipath rapid assay technology for the detection of E-3-G. When there is no E-3-G in the urine sample, the E-3-G antibodies bind to the E-3-G bound further along the strip (test zone furthest from the wick).

Test stick set-up

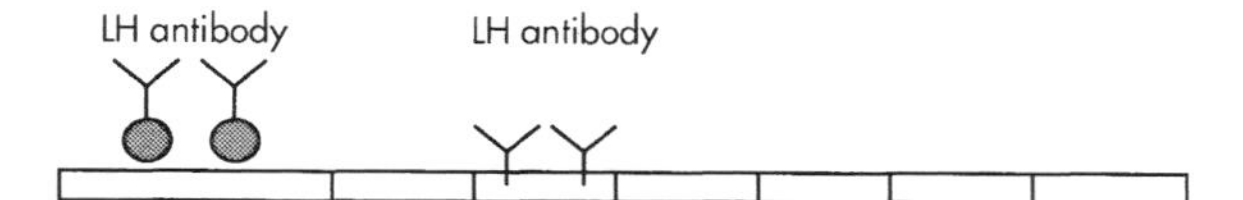

Fig. 40.4 Unipath rapid assay technology for the detection of LH. LH antibodies bound to blue latex particles are at the base of the strip and a band of LH antibodies is at a point further along the strip (test zone nearest to the wick).

antibody on the surface of the latex and residual sites on the latex are free to react with E-3-G which is immobilized on the membrane. The assay is a competitive system with decreasing color shown for increasing levels of E-3-G (Figure 40.3). The LH assay is shown in Figures 40.4 and 40.5. In this assay, blue latex is sensitized with antibody specific for the β-subunit of LH. When the urine sample makes contact with the latex, any LH in the sample is free to react with the anti-β LH antibody on the surface of the latex. With increasing amounts of LH in the sample, increasing amounts of latex are captured in the test zone that is nearest to the wick, and hence the intensity of the colored band rises.

The rapid assay technology uses colored labels that can be interrogated by the use of transmission measurements. The test stick design allows optical access to both sides of the test stick. Light at 635 nm is directed to one side of the membrane and the light that is transmitted through the strip is detected by a series of detectors. Increasing quantities of colored label lead to an increasing attenuation of the light emitted by the light source. The monitor compares the color levels on the test stick with those of previous tests from within that cycle and when it sees a significant decrease in the color level of the E-3-G test area (a rise in E-3-G concentration) it changes the fertility status from 'safe' to 'unsafe' (i.e. green light to red light). By subsequently detecting the rise in LH the system is able to indicate the end of the fertile period by switching the light back to green to indicate a 'safe' fertility status. In addition, the rise in LH is also used to indicate, by the use of an 'O' symbol, the days during which ovulation will occur. As the monitor stores information on the previous six cycles, it is able to target the tests to maximize the chances of locating the rises in both hormones. The system, therefore, adapts to the individual and becomes personalized to their hormone patterns and cycle lengths.

Test stick set-up

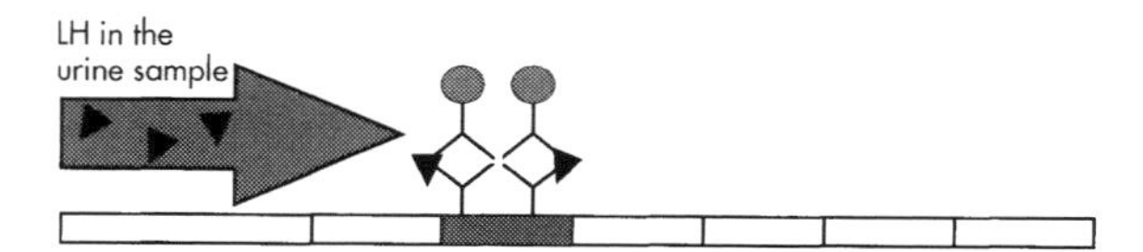

Fig. 40.5 Unipath rapid assay technology for the detection of LH. LH in the urine sample attaches to the latex-bound LH antibodies and moves by capillary action to the LH antibody bound to the strip (test zone nearest to the wick), to form a 'sandwich' of antibody producing a blue line.

CALIBRATION

The quality control of Persona test sticks is done by reference to the IS 71/264 for LH and by gravimetric processes for E-3-G. The Persona monitor is self-calibrating.

ANTIBODIES

All antibodies used in the reagents (anti-α LH, anti-β LH, and anti-E-3-G) are monoclonal.

DATA PROCESSING

The Persona monitor is able to store 6 months of test data. The monitor stores information on hormone changes, cycle length, the fertility status displayed on each day as well as information on its own internal functionality checks. The information on hormone levels is used by a sophisticated rule set (algorithm) coded within the microprocessor of the monitor to identify the fertile period. Data can be recovered from the monitor to provide a full record of its use over the last 6 months.

LATEST DEVELOPMENTS

Using the same technology Unipath has developed a system for use in conception that was launched in the USA in 1999, and has since been launched in Canada, Germany, Japan, Singapore, and Malaysia. This product is ideally suited for home and physician directed testing.

Part 3

Laboratory Management

41 Subject Preparation, Sample Collection and Handling

Colin Wilde

Although great emphasis is always placed on the accuracy and precision of immunoassays, some of the largest potential sources of error concern sample collection, handling methods, and the way the subject is prepared before the sample is taken. Unfortunately, this critical area is often neglected in laboratory investigations, and overlooked in quality control procedures and troubleshooting investigations.

This chapter describes collection methods for various types of samples, together with some of the types of problems to which they are susceptible. Examples are given to enable the reader to be aware of the types of difficulties that may arise, together with an understanding of how to avoid them. A comprehensive listing of published references can be found in Young (1997).

SUBJECT STATE AND PREPARATION

Certain aspects of the subject that may affect the interpretation of a test result are not under the control of the clinician or laboratory staff. These include age, sex, ethnic origin, pregnancy, and stage of menstrual cycle, and they must be well documented at the sample collection stage so that the test results may be interpreted accordingly. There are other variables, however, which often require active intervention and control if test results are to be meaningful.

STRESS

Stress, whether mental or physical, can alter bodily functions, including the secretion of various hormones into the body fluids that are to be tested. The subject should not therefore be anxious or tense when samples are collected, but should be made to feel relaxed and at ease in a comfortable environment. Fear and stress are potent stimuli of growth hormone, prolactin, cortisol, aldosterone, and plasma renin activity. Many other intermediary metabolites, and also carrier proteins such as transferrin, may be affected by longer term stress and, after a major stressful event, such as myocardial infarction, the results of assays of such analytes should be interpreted accordingly.

Postsurgical stress can be particularly severe, and immunoassay tests should be avoided if possible until the patient has stabilized. For example, transferrin can fall after about 3 h and ferritin starts to rise shortly afterwards. Thyroid hormone levels are also often depressed after surgery.

EXERCISE

Exercise, like mental stress, stimulates the production and secretion of a number of hormones including growth hormone, prolactin, cortisol, and plasma renin. The extent of the increase depends on the amount of exercise done and on the physical fitness of the individual. When resting-state levels of affected hormones are required it is important to ensure that the subject has not recently exercised, including running up the stairs to the collection center immediately before sample collection. Such stress, however, may be used in stimulation tests to assess a patient's reserve capacity for hormone production, during which controlled exercise is undertaken immediately before sample collection as, for instance, in the assessment of growth hormone deficiency.

FOOD AND DRINK

The dietary state of the subject may also be relevant to the investigation required. Many of the commonly measured constituents of plasma vary in concentration according to the time elapsed since the last meal, but these changes have been quantified for only a few analytes normally

The Immunoassay Handbook. *Third Edition*
D. Wild (Ed.)

measured by immunoassay. Serum insulin, gastrin, and calcitonin are examples of hormones with levels that significantly alter following food intake, and fasting samples are required unless a stimulation test is being done. The levels of circulating therapeutic drugs are also influenced by the timing of meals, because of slowing of the drug absorption from the gut.

Caffeine from coffee, tea, and soft drinks has a strong effect on some analytes, and can increase plasma cortisol levels by up to 50% after 3 h.

Malnutrition results in a lowering of levels of IGF_1, albumin, caeruloplasmin, transferrin, and prolactin, indeed IGF_1 and albumin are often used for monitoring nutritional status.

Lipids

There are well-documented changes in lipids after a fatty meal, when the blood serum or plasma may have a milky appearance due to the presence of triglycerides and chylomicrons. Such hyperlipidemic serum may interfere with antibody binding in immunoassay procedures and, although not always practicable, it is desirable to collect specimens from subjects after a period of fasting, usually overnight, for any type of investigation. Alternatively, ultracentrifugation or enzymatic cleavage are sometimes used to remove or break down lipids.

Alcohol

Alcohol ingestion induces changes to body fluid composition that vary depending on whether the subject is an abuser or casual drinker, and on the timing of collection after alcohol intake. Examples are ferritin and the liver enzymes alkaline phosphatase, aspartate aminotransferase, and γ-glutamyl transferase, although the latter are seldom measured in immunoassays.

Smoking

Smoking can influence the levels of a number of analytes commonly measured by immunoassay. Both cortisol and growth hormone rise as an acute response to smoking, and it is advisable not to take blood within 30 min of smoking. Long-term effects are seen as increases in IgE, androstenedione, insulin, C-peptide, placental alkaline phosphatase, and carcinoembryonic antigen, which can give false positives for the latter two tumor markers. Significantly lower levels of IgG and prolactin can be expected in smokers.

The metabolism of some drugs, e.g. theophylline and tricyclic antidepressants, is stimulated by smoking, resulting in reduced half-life and increased body clearance.

In pregnant smokers the levels of hCG and estradiol are significantly lower than those in nonsmokers.

POSTURE

When an individual assumes an upright position after a period of recumbency there is movement of ultrafiltrate from the intravascular to the extravascular compartment of the extracellular fluid. This produces a 10–20% hemoconcentration with a concomitant concentration of large molecules and those substances that are bound to them. Plasma concentrations of proteins, peptides, enzymes, and protein-bound substances, such as cortisol, thyroxine and drugs, are affected. When patients rise from a sitting position, similar changes occur but to a lesser degree.

For ambulatory patients, blood should be collected after the subject has been seated for 15 min, but results from seated subjects are not directly comparable to those from hospital in-patients whose blood is often collected after prolonged recumbency.

The renin–aldosterone–angiotensin system is strongly influenced by posture and it is normally recommended that blood samples are taken after overnight recumbency, without sitting or standing before collection. Even a short period of sitting will produce significant increases in aldosterone. Measurement before and after 4 h of ambulation may be used to differentiate between adrenal hyperplasia and adrenal adenoma.

MEDICAL PROCEDURES

Some medical procedures have short-term effects on levels of circulating analytes and it is important to allow sufficient time to elapse after such procedures before sample collection.

Surgery or intramuscular injection results in increased creatine kinase activity, and rectal examination or prostatic manipulation may cause an increase in circulating prostate-specific antigen. Care must be taken not to collect blood proximate to a site of intravenous injection as the concentrations of many components are likely to be misleadingly low.

Samples should never be taken from the arm in which a drip has been inserted because of the dilution effect on blood constituents.

Transfusion results in analyte concentrations that are a composite between recipient and donor. Also folate and ferritin concentrations may be increased. It is therefore preferable to avoid taking samples for testing for a few days after the transfusion if possible.

DRUGS

Drugs may cause difficulties in the interpretation of immunoassay results through their effect on *in vivo* physiological and biochemical mechanisms, or by *in vitro* effects in the analytical process. There are now many thousands of references in the literature testifying to these problems and over 80% of these relate to the *in vivo* type.

Drug Interactions

Oral contraceptives have a profound effect through their estrogenic activity, leading to increased levels of many binding proteins, including those for thyroxine, cortisol, and the sex hormones (sex hormone-binding globulin (SHBG)). Other drugs, such as barbiturate and phenytoin, cause hepatic enzyme induction with increased levels of liver enzymes. Amiodarone is one of the better-known drugs that interfere with thyroid hormone levels, raising the level of thyrotropin (TSH) and thyroxine. A novel type of interaction is the one caused by high concentrations of β-lactam antibiotics (penicillins and cephalosporins) which inactivate aminoglycosides both *in vivo* and *in vitro*. Samples for aminoglycoside assay that contain β-lactams should therefore either be assayed immediately or stored frozen.

The list of drug interactions is so large that it may only be accommodated on constantly updated computer data banks. Although nobody can be expected to know more than a small fraction of the documented drug–test interactions, it is important to record all drug therapy at the time of the request or sample collection so that anomalous results may be checked by reference to drug–test interaction data banks.

It is, of course, impractical to cease drug therapy before sample collection for all tests but for some it is necessary if the results are to be interpreted without ambiguity. For instance, because the levels of aldosterone and renin vary inversely with the activity of the sympathetic nervous systems, diuretic and hypertensive drugs must not be taken for 3 weeks before analysis if meaningful results are to be obtained.

PREGNANCY

The effects of pregnancy on a number of analytes are well documented elsewhere in this book. In addition to the well-known rises in human chorionic gonadotropin (hCG), estriol, human placental lactogen (hPL), etc. (*see* PREGNANCY), the serum concentrations of the binding proteins for cortisol and thyroid hormones also rise causing the total concentrations of these hormones to increase.

AGE

Changes occur in the levels of many components with aging; some of these are gradual, others occur quite rapidly at certain times of life such as the neonatal period, puberty and the menopause. Different reference ranges must be used for different age groups.

Thus, TSH and free thyroxine (FT_4) levels peak at birth, but settle to normal levels within 6 days. The concentration of 17-hydroxyprogesterone is very high at term, due to enzymic inhibition by placental steroids, but levels decline to normal within 48 h. Vitamin D levels are low at birth but rise in the first two days, this corresponds to an increase in parathyroid hormone (PTH) that often occurs in response to early hypocalcaemia.

At puberty there are well-known changes in all the sex hormone levels but this is also the time of peak levels of insulin-like growth factor-1 (IGF_1), and of growth hormone and prolactin in females.

The menopause in women results in increases in FSH and LH followed by a post-menopausal lowering of estradiol. Bone markers generally change with aging with steeper changes at the menopause in women; this is particularly true of osteocalcin levels. Thus, there are increases in PTH and urinary pyridinoline and decreases in 25-hydroxy vitamin D and calcitonin.

In older men there are decreases in testosterone levels and an increase in prostate specific antigen that is due to physiological benign prostatic hyperplasia.

RACE

There are ethnic differences for some analytes, the most notable being the lower levels of 25-hydroxy vitamin D and osteocalcin levels in colored races, and higher levels of prostatic specific antigen (PSA) in African and West Indian men. Asian–Pacific populations do not show the age-related increases in PSA described in the above section on AGE.

TIMING

BIOLOGICAL RHYTHMS

Changes that follow well-defined rhythms occur in a number of biological systems. The most common rhythms relating to biochemical substances are the menstrual cycle, with a periodicity of about 28 days in humans, and circadian rhythms, with a periodicity of about 24 h. It is important therefore to understand the rhythmic patterns of any biological substances measured by immunoassay, and to time the collection of samples carefully, making best use of this knowledge in the interpretation of the test results. It is also important to take samples at the same time in the rhythm if intra- or inter-individual results are to be compared.

Menstrual Cycle

Female reproductive hormones, e.g. luteinizing hormone (LH), follicle stimulating hormone (FSH), estradiol, and progesterone, follow a monthly cyclical pattern and it is important to know the timing in relation to the menstrual cycle and to have a series of two or more samples at known times in the cycle before any interpretation of the result is possible. Levels of SHBG are significantly higher during the luteal phase as are those of interleukin-1α, a finding which may be due to the rise in body temperature during this phase.

Circadian Rhythms

Of the circadian rhythms it is probably the hormonal ones that have been most studied. Those of adrenocorticotropic hormone (ACTH) and corticosteroids are perhaps the most striking with the peak levels in body fluids occurring between 06:00 and 08:00 with a nadir at midnight in humans. In the rat, a nocturnal animal, this rhythm is almost exactly 12 h out of phase with that of man. Prolactin, corticosteroid, and aldosterone levels follow a similar pattern of building up during sleep but a growth hormone (GH) peak occurs earlier, during the first few hours of sleep, and unlike other hormones the GH peak is abolished if the subject remains awake. A rhythm that is related to subject activity rather than photoperiodic effect occurs with plasma testosterone. This again normally peaks at 08:00 but with a much steadier fall during the day to a trough at 20:00. Interleukin-1α peaks early in the sleep cycle, being related to the onset of slow wave sleep, but perhaps the most striking circadian rhythm occurs with melatonin, which is almost entirely produced during the night with very low blood levels during the day.

Rhythms may vary at different stages of life and this is, of course, most pronounced with the female menstrual cycle, but a nocturnal rise in LH is characteristic of puberty in males. A loss of rhythm is often diagnostic of a disease process (e.g. cortisol in Cushing's disease), but sometimes the rhythm may become more marked, as with 17-hydroxyprogesterone in congenital adrenal hyperplasia in neonates.

Pulsatile Secretion

Within the well-defined rhythms there is sometimes a pronounced episodic or pulsatile secretion. This occurs with ACTH and the corticosteroids where strong pulses occur about 12 times every 24 h. Androstenedione, which shows a prominent circadian rhythm, also has marked episodic secretions with 25% swings in circulating levels. LH is another hormone with pulsed secretions but the peaks are much less pronounced than with corticosteroids.

DYNAMIC TESTS

Stimulation and suppression tests are often used to investigate the capacity of organs to respond to positive or negative stimuli. These usually involve the collection of timed samples following the stimulus. For reliable interpretation it is important that the prescribed timings are strictly adhered to and that each sample is clearly labeled with the time of collection. The dose of the stimulating substance should always be recorded.

PATHOLOGICAL CHANGES

Timing is sometimes important in the confirmation of a pathological event. A good example is that of the creatine kinase MB isoenzyme where, for reliable diagnostic information to be obtained, a blood sample should be collected within a predetermined time interval (usually between 6 and 30 h) following the suspected myocardial infarction.

NONTHYROIDAL ILLNESS

Severe illness or injury can induce changes in thyroid hormone levels. *See* THYROID.

THERAPEUTIC DRUG MONITORING

The timing of sample collection is especially important when measuring the circulating levels of therapeutic drugs. The appropriate timings during the dose interval vary for different drugs depending on their absorption and distribution characteristics. At the start of the therapy or following a change in dosage regimen there is an initial absorption phase with the average drug concentration continuing to rise after repeated doses until the steady-state condition is reached. This may not be until at least five half-lives and several dose intervals have elapsed.

As a general rule, samples should be taken when the concentration of drug is at the lowest point (or trough level) and this is usually immediately before the next dose. This trough level normally relates to the steady-state concentration. For digoxin, for instance, it is preferable to wait at least 8 h after administration because this drug has slow accumulation and excretion rates. Procainamide is a drug that is absorbed slowly, causing the highest blood levels at the end of the dose interval. Ideally, in such cases a number of samples should be taken at intervals or, at the very least, one at the end and one 2 h into the dose interval.

With some drugs where toxic levels are close to therapeutic levels, e.g. the aminoglycosides, it is recommended that both peak and trough concentrations within the dose interval are measured. This practice helps to prevent toxicity and ensure therapeutic efficiency. For similar reasons, theophylline is normally measured at the time of peak levels.

It should also be remembered that food intake may delay the attainment of peak concentrations by slowing absorption.

BLOOD COLLECTION BY VENEPUNCTURE

PRECAUTIONS RELATING TO THE PATIENT

Errors in interpretation of immunoassay results may occur if the sample is not collected correctly and under the best possible conditions. Taking a blood sample can be very stressful for the patient and stress causes major changes in some blood constituents (*see* SUBJECT PREPARATION). Considerable attention must therefore be paid to the environment, to the facilities available, and to reassuring the patient during the procedure.

Patient Identification

Correct patient identification is obviously essential but it is not unusual for errors to be made. The venepuncturist must always ensure that the blood specimen is being drawn from the patient designated on the request form and that it is placed in a correctly and unambiguously labeled container. The use of a single barcoded primary tube for sample collection and testing is becoming established now that automated immunoassay analyzers are being fitted with barcode readers. The test request can be logged onto the main hospital computer, which is connected to the analyzer via a two-way link. This avoids the risk of errors when transferring samples or test-request information.

Position

The patient must be positioned comfortably, preferably in a special phlebotomy chair with adjustable armrests or on a bed or examination couch. Patients should not be startled, such as by sudden wakening, because this may affect the levels of some analytes, as would sudden changes in posture, especially for investigations involving the renin–angiotensin–aldosterone axis (*see* SUBJECT PREPARATION – POSTURE).

Puncture Site

There are a number of veins in the arm that may be used but the larger median cubital and cephalic veins are used most frequently. The puncture site should be carefully chosen avoiding any area of hematoma and any area of extensive scarring. A specimen taken from the side on which a mastectomy has recently been done may not be truly representative because of lymphostasis; specimens should never be taken from an arm being used for intravenous therapy because hemodilution is likely. Lowering the arm over the side of the armrest or bed will cause the veins to distend. Stroking in an upward direction usually makes the veins more visible. The veins become more prominent and easier to enter when the patient forms a fist but vigorous hand exercise should not be allowed as this may change some blood constituent levels. The puncture site should be cleaned with an isopropanol swab and wiped dry to minimize contamination of the specimen.

Tourniquet

A tourniquet is often applied to enhance venous filling, increase vein distention, and aid vein location. Its use is contraindicated when analytes affected by hemoconcentration are being measured, because the occlusion of the upper arm results in ultrafiltration of blood in the forearm to produce spurious increases in the concentration of large molecular mass substances and anything bound to them. Such analytes include proteins and peptides, bound components, cellular elements, and enzymes. If a tourniquet has to be used in such instances for preliminary vein selection it should be released and reapplied after an interval of at least 2 min. The tourniquet must never be left on for longer than 1 min immediately before venepuncture and it should be removed as soon as the blood begins to flow, otherwise hemoconcentration will occur and local stasis is likely.

PHLEBOTOMY TECHNIQUES

For detailed descriptions of phlebotomy techniques see ECCLS Standard (1987) and NCCLS Standard (1984).

BLOOD COLLECTION

Blood-collection Systems

Blood collection may be achieved by the conventional needle and syringe method followed by transfer of the blood from the syringe to a suitable container, or alternatively by using an evacuated tube system. Evacuated or vacuum tubes are manufactured to withdraw a predetermined volume of blood in a closed system. The needle used is double ended, one end pierces the stopper or diaphragm that seals the tube, the other end is inserted into the vein. With these systems there is no transfer from syringe to tube and the blood comes into contact with anticoagulant or preservative immediately it is drawn. The likelihood of hemolysis and microclot formation is reduced and there is less deterioration of metabolically labile substances.

If the investigation requires multiple sampling, as in stimulation or suppression tests, it is often helpful to insert a cannula which can stay in position for the duration of the investigation and allow blood to be withdrawn at intervals with minimum stress to the subject. However, great care must be exercised to avoid sample contamination with heparin, which may be used to flush the cannula.

Common Types of Blood Container

The main types of blood container available are listed in Table 41.1. The codes and colors are those recommended by the National Committee for Clinical Laboratory Standards (NCCLS), the European Committee for Clinical Laboratory Standards (ECCLS) and the International Standards Organization (ISO), although many countries still use different codes and colors.

Withdrawal of Blood

The needle must be inserted carefully into the chosen vein bevel-side up. The desired amount of blood is drawn either by withdrawing the syringe plunger or by allowing the blood to flow into the evacuation tube until the vacuum is exhausted. If additional blood is required a further syringe or evacuated tube should be attached to the needle, which should remain in the vein. If the syringe method is used the blood should be transferred to an appropriate container after separation from the needle, and never through the needle as this causes hemolysis.

Filling-tubes Containing Additives

Whichever method is used, tubes containing anticoagulant must be filled to the mark, otherwise the concentration of anticoagulant will be too high and this may affect the assay system, particularly the antibody-binding characteristics. If several specimens are to be drawn at the same time the plain nonadditive tube should always be filled first, with additive-containing tubes being filled later, the recommended order of fill being plain tube, citrate, heparin, ethylene diamine tetra acetic acid (EDTA), and finally oxalate fluoride. Care must still be taken to avoid cross-contamination between different additive tubes as this may result in factitious test results.

Table 41.1. Blood sample containers: anticoagulant codes and colors.

Anticoagulant	Code	Color
Potassium	KE	
EDTA		Lavender
Sodium	NE	
Potassium oxalate	KX	
Trisodium citrate	9NC*	Blue
	4NC*	Black
Fluoride oxalate	FX	Grey
Ammonium and potassium oxalate	AKX	
Lithium heparin	LH	Green
Sodium heparin	NH	Green
ACD	ACD	Yellow
None	Z	Red

*Figures denote ratio between blood and anticoagulant.

PREPARATION OF SERUM

Blood serum is often preferred for immunoassay. For this the blood must be collected into a plain tube, and the serum separated from the blood cells by centrifugation following clot formation and retraction, which may normally take up to 1 h. This time can be reduced using axial centrifugation where the blood collection tube is rotated about its own longitudinal axis, separating the blood into three concentric cylinders, cells against the tube walls, serum inside the cells and air in the center. A separator may be introduced whilst the tube is spinning to permanently isolate the serum from the cells. The DuPont Axial Separation Technology (AST) system utilizes this technique.

Clotting Activators

Clotting activators are used to speed up the process to 10–15 min and these are present in many commercial blood-collecting devices. Minute silica or glass particles are commonly used. These may be introduced in the form of small beads or attached to the tube wall with a water-soluble silicone coating or to a 'carrier' such as a paper disc or polypropylene cup. Thromboplastin has been substituted for glass particles when very rapid clotting is required.

These materials accelerate the clotting process and help produce a clean, well-defined clot. They also diminish latent fibrin formation in the separated serum. If fibrin is allowed to form it may interfere with the accuracy of pipetting devices or with the efficiency of solid-phase binding in the assay system. Serum produced using clotting activators is less likely to be hemolyzed although the clotting process itself may cause the release of some erythrocyte contents, including erythrocyte enzymes, which are usually present in higher concentrations in serum than in plasma. There is unlikely to be interference with immunoassay from this source unless the enzymes themselves are being assayed using immunological techniques.

Serum samples are not always suitable. For instance, samples for antithrombin III assay should be taken into the anticoagulant EDTA, otherwise the analyte will be lost in the clotting process. The serum levels of thrombospondin are 100 times higher than plasma levels due to the release of this putative breast tumor marker from platelets during the clotting process.

PREPARATION OF PLASMA

Anticoagulants

When plasma is required blood must be collected into an anticoagulant. Anticoagulants work by interfering with the clotting process and the principal ones in general use are heparin and EDTA salts. Immediate gentle mixing of the anticoagulant with the blood after phlebotomy is essential if clotting is to be prevented but too vigorous a treatment will cause hemolysis. Separation of the plasma from the cells by centrifugation may be effected immediately. If the samples are centrifuged for a long period, or at too high a temperature, hemolysis will occur and cell contents leak into the plasma. A speed equivalent to 1000–1500g for 5–10 min, preferably at 20–25 °C in a temperature-controlled centrifuge, is recommended. Some analytes require cooled centrifugation at 2–8 °C, e.g. ACTH. Axial centrifugation (*see* PREPARATION OF SERUM) may also be used for preparation of plasma.

Interference by Anticoagulants

If plasma is to be used for immunoassay, care must be taken to select an appropriate anticoagulant. Chelation of essential metal ions by EDTA may inhibit the enzyme activity at the signal generation stage in an enzyme immunoassay, especially if alkaline phosphatase is used. Anticoagulants may also interfere with some antibody–antigen reactions. The use of heparin decreases the rate of reaction of some antibodies, particularly at the precipitation stage in second-antibody systems. Solid-phase systems and careful selection of antibodies have virtually eliminated this problem but some assays still exhibit interference by anticoagulants, including heparin, and care must be taken to identify these and use appropriate specimens. Heparin should not be used for the investigation of cryoproteins because the anticoagulant precipitates cryofibrinogen.

WHOLE BLOOD

If whole blood is required for a laboratory assay, an anticoagulant is used to prevent coagulation. EDTA is often preferred for this purpose because of its efficiency in preserving the integrity of the sample for blood cell investigations. In the laboratory, immunoassays are rarely carried out on whole blood but if they are, then the effect of anticoagulation on cellular components, as well as matrix effects on the immunological reaction, must be considered. In contrast, recently introduced point-of-care devices utilize fresh whole capillary blood applied directly to the device following skin puncture.

INTERFERENCE BY TUBE AND STOPPER COMPONENTS

Soluble substances may be leached out of tubes and their closures and in certain cases the substances may profoundly affect the results of assay procedures.

INTERFERENCE

There is much literature concerning such problems, particularly with the rubber stoppers used in some commercial blood-collection devices. Plastics such as tris(2-butoxy-ethyl) cause displacement of some drugs and other analytes from protein-binding sites with consequent redistribution between erythrocytes and plasma. They may also strip antibody from coated solid-phase systems such as coated tubes, interfere with the binding of antigen and antibody and inhibit enzyme activity at the signal generation stage of any assay. There has been considerable concern regarding interference with drug–protein interactions, particularly with the lipophilic drugs bound to acidic α_1-glycoprotein.

PRECAUTIONS

The manufacturers of blood-collection tubes recognize these problems and attempts have been made to reformulate the closure material using specially selected low-extractable rubbers to minimize interference. It is, however, good practice always to fill the tubes to their designated volume so that any material leached out is not concentrated in a low volume. Ensure also that the length of time the blood is in contact with the stopper is minimal. Inversion of the tube to mix the contents should be done gently not more than five times and roller mixing of samples should be avoided. The tubes and samples should always be stored upright to minimize contact with the stopper.

Storage in the tubes should be for no longer than 24 h and should be at low temperature (2–8 °C) because leaching is greater at higher temperatures. The tubes should always be stored upright. Such precautions will reduce but not eliminate the problem, particularly if a poor batch of stoppers is being used.

As it is almost impossible to predict interference with particular methodologies and because variation between batches is not uncommon, it is a wise precaution to test the batch of tubes before using them for a specific purpose, recognizing that a new analyte may invalidate the assumption that the system is appropriate for all assay purposes. It is also helpful to ask manufacturers for any data that they may hold and to consult references on the specific type of tube and stopper.

It is also necessary to consider possible adsorption of the analyte on to the tube. This is particularly pertinent with ACTH which is readily adsorbed on to glass and for which plastic tubes must be used.

THE USE OF SERUM SEPARATORS

Serum separators are now frequently used to assist the rapid and efficient separation of cell-free serum from clotted whole blood, although they may also be used for the separation of plasma from the cells in anticoagulated blood.

CHARACTERISTICS OF SEPARATORS

Separators are usually silicone gels or polyester formulations, although glass beads and plastic and fiber devices have been used. They all essentially have a specific gravity that is intermediate between that of serum and that of the cells or the clot. During centrifugation, while the cell–clot coagulum settles to the bottom of the tube, the gel viscosity decreases and the gel migrates to the surface of the clot. Following centrifugation, the original viscosity is restored and the gel forms a nonpermeable barrier at the serum–clot interface. The separators are generally best used at 20–25 °C because chilling may impair the flow characteristics and too high a temperature may cause breakdown of the gels as well as vitiation of the serum components.

Provided the gel barrier is visually checked for integrity it is generally safe to store the serum on the gel for up to 2–3 days without significant contamination from cellular components. The tubes should never be recentrifuged as this causes mixing of separated serum with serum that has remained in contact with the cells.

INTERFERENCE

Pieces of gel or droplets of oil may be seen on or within the separated serum with some batches of gel-containing tubes. There is evidence that this can interfere with liquid-sensing sampling devices and that it may coat tubes and cuvettes, possibly leading to physical interference with binding in solid-phase systems. Some laboratories routinely filter gel-separated serum before assay to prevent such contamination. Although undoubtedly some batches of gel have posed particular problems, it is important to follow the manufacturer's instructions and not use the gels at excessive temperatures or inappropriate centrifuge speeds, or to subject the tubes to rough handling or unusual orientations.

Despite the usefulness of separators in helping to provide a good clear serum specimen they may interfere with the measurement of some analytes. Progesterone shows a time-related absorption or adsorption phenomenon with decreases of up to 50% in measurable hormone when stored over a gel for 6 days. Other hormones investigated at the same time showed no significant change.

Lignocaine, pentobarbital, phenytoin, and carbamazepine are among the drugs that have shown similar reductions in levels on some gels even after only a few hours' storage. If there is any doubt about the storage characteristics of a particular analyte in gel-containing tubes, the serum should be decanted into a plain tube as soon as possible after separation.

OTHER ADDITIVES

Some of the low molecular mass polypeptide hormones such as ACTH, glucagon, gastrin, and other gastrointestinal hormones are rapidly destroyed by enzymes present in blood and may require protection by the addition of suitable antiproteolytic agents such as Trasylol (aprotinin) to the tubes into which the blood is collected. Even when antiproteolytic agents are used it is still necessary to centrifuge the sample at low temperature within 10 min of collection and to store the serum or plasma immediately at −20 °C, only thawing it out immediately before assay. If any transport is necessary to a distant laboratory, this must also be at −20 °C. FUTHAN-EDTA should be used as a sample stabilizer for some complement components such as the labile C2 and C5-9, which additionally should be stored at −70 °C. It must also be added to samples for complement breakdown products, e.g. C3d, otherwise spuriously high levels will be recorded due to *in vitro* destruction of C3 on storage.

HEMOLYSIS

Noticeable hemolysis will interfere with antibody–antigen reactions and with some signal generation stages. Heavily contaminated samples should always be discarded and not used for immunoassay. Even small degrees of hemolysis may be unacceptable, principally due to the release of proteolytic enzymes which destroy small peptides. Such peptides include insulin, glucagon, calcitonin, PTH, ACTH, and gastrin. Samples with any sign of hemolysis are not acceptable for such assays. Anomalous results may be obtained in other systems because of leakage of cellular components. Such an assay is serum folate where red cell values are approximately 30 times greater than those in serum.

COLLECTION OF BLOOD BY SKIN PUNCTURE

Small but adequate amounts of capillary blood may be obtained by skin puncture. Such collection procedures are especially important in babies but are also often useful for collection from adults where multiple sampling is required, when venesection may constitute a hazard to the patient or where no suitable superficial veins are available, for example in gross obesity or in cases of severe burns. It must be remembered, however, that capillary blood is not the same as venous blood – indeed, it more closely resembles arterial blood – and the differences in composition must be considered when interpreting test results. There is also a greater risk of hemolysis when capillary blood is taken.

SKIN PUNCTURE SITES

Usual sites for collecting blood by skin puncture are:

- the most lateral or most medial plantar heel surfaces (recommended in infants);
- the medial plantar surface of the big toe;
- the distal digit of a finger, preferably the third or fourth;
- the earlobe.

It is essential to obtain a good blood flow, otherwise constituents may be diluted with tissue fluid. A similar dilution will occur if the skin puncture site is edematous or if excess pressure is exerted at the site in an attempt to increase the flow rate.

Adequate flow rate is best obtained by covering the site with a warm, moist towel at a temperature no higher than 42 °C for at least 3 min. The site should be cleaned with an isopropyl alcohol swab and then thoroughly dried with a sterile gauze pad before being punctured. Residual alcohol causes rapid hemolysis. The skin is normally punctured with a special sterile lancet or puncture device that penetrates to a depth of 2.4 mm. Drops of blood should well out spontaneously without excessive rubbing or pressing. The first drop of blood must be discarded, as this is likely to contain excess tissue fluid.

COLLECTION INTO CAPILLARY TUBES

The blood may be conveniently collected into a capillary tube by allowing the tip of the tube to touch the drop forming over the puncture site. Blood will flow into the tube by capillary action. Capillary tubes with no additives or containing heparin or EDTA salts may be used depending on whether serum, plasma or whole blood is required. The capillary tube or tubes are then plugged with sealing clay and, if anticoagulant is used, must be inverted gently at least 10 times to prevent coagulation and hemolysis. Mixing may also be done by placing a small magnetic mixing bar in the barrel of the tube, and manipulating it from the outside. To prevent identification error, the tubes should be individually labeled or all tubes for each patient or timed sample placed in a single labeled test tube.

URINE COLLECTION

CONTAINERS

Collecting containers should be clean and made of inert disposable plastic for single use only. They should have a lid that can be tightened securely to prevent leakage of the contents and should be sterile if the sample is collected for microbiological studies.

PRESERVATIVE

If the sample is to be stabilized because of delayed analysis or an unstable constituent, chemical preservatives or stabilizers should be added to the container or the urine. Sodium merthiolate, 150 g per 24 h in the container, or boric acid, 27 g per 24 h in the container, are preferred preservatives for immunoassay. Their main function is to prevent bacterial destruction of the analyte. Whenever material is added it should be noted on the container as it may be hazardous or interfere with the assay system.

TYPES OF COLLECTION

There are several types of urine specimen:

- A random specimen may be collected at any unspecified time during a 24-h period and is generally used where only a qualitative result is required.
- Midstream specimen, normally required for microbiological investigations where precautions have to be taken to provide a clean specimen by careful washing of the genitalia beforehand, voiding the first portion of urine into the toilet and catching the mid portion in an appropriate sterile container.
- The early morning, overnight specimen is normally collected immediately after the patient rises from an overnight sleep. It is often used when more concentrated urine is necessary because of a low concentration of analyte.
- Timed specimens are collected between specified times and include those associated with dynamic tests where the urine is collected into a series of containers at specified times; 2 or 3 h specimens that are collected at specific times during the day or within a certain period following a meal or ingestion of pharmaceuticals; and the more usual 12 or 24-h urine collection which provides an average excretion value over a long period.

INSTRUCTIONS

The patient should always be given written instructions and these should also be explained orally. This is particularly important when a timed specimen is required.

The following are instructions for a 24-h sample:

- Always use the collection bottle provided.
- At the beginning of the 24-h collection the bladder should be voided and this specimen must be discarded. The exact time should be noted on the bottle label.
- All the urine passed during the next 24 h must be collected in the bottle, which should be kept in a cool place, preferably in a refrigerator.
- On the following day at exactly the time noted on the label on the previous day, the bladder should again be voided and the urine added to the bottle. This is the end of the 24-h collection.

- If any urine is accidentally discarded, a new 24-h collection should be started in an empty bottle.
- The urine must not be contaminated by bowel movements. If this happens a new 24-h collection should be started.

SAMPLE HANDLING

When the sample reaches the laboratory it must be thoroughly mixed before a specimen is taken for analysis. Urine is often cloudy, particularly if stored for any length of time, and should be centrifuged before use.

It may be necessary to take the pH of the sample and adjust it if appropriate. β_2-Microglobulin, for instance, is destroyed at above pH 6.0, and catecholamines and related substances must be collected at a pH less than 2.0.

It is often useful, particularly if there is doubt about the completeness of the collection, to measure the creatinine. Although creatinine excretion is related to body mass it is relatively constant in individuals from day to day and any gross changes throw doubt on the integrity of the sample.

SALIVA

Saliva may be collected repeatedly at regular intervals in reasonable quantities. Samples may be taken without stress while the patient continues normal activity, possibly at home or at work, without the need for specially trained personnel. This makes it an attractive alternative to blood for use in investigations involving immunoassay, especially when these are in neonates or infants.

Saliva has been most useful for analytes where it can be demonstrated that the level of analyte in saliva reflects that in the blood plasma. So far there is only a limited number of such substances as saliva is a complex fluid that is produced from different glands contributing different proportions of the total oral fluid depending on the type and intensity of stimulation. In resting conditions the rough proportions are: submandibular gland, 70%; parotid gland, 25%; and sublingual gland, 5%.

The transfer of plasma components to saliva may either be by intracellular or extracellular means. Passive diffusion is the main intracellular mechanism, and ultrafiltration, involving active transport, the main extracellular mechanism. Some factors that influence the saliva:plasma ratio are the lipophilic nature of the material, its degree of ionization, the pH of the saliva, the flow rate, and the extent of protein-binding of the substance measured.

Despite the potential variability of saliva composition due to the above factors, a number of analytes, principally hormones and drugs, have a low variability of saliva: plasma ratio because they are freely diffusible and have concentrations relatively independent of small changes in flow rate. Such analytes include free (rather than protein-bound) cortisol, progesterone, estradiol, 17-hydroxyprogesterone and testosterone, and among the drugs, paracetamol, digoxin, theophylline, ethanol, caffeine, diazepam, and amylobarbital. Secretory IgA has also been measured successfully in saliva as an aid to evaluating resistance against infectious diseases of the upper respiratory tract.

PROCEDURES FOR SAMPLING SALIVA

There are a variety of ways in which saliva may be collected. If saliva from a particular gland, e.g. the parotid, is required, a polythene catheter may be introduced into the opening of the gland (called Stenson's duct in the parotid). Such catheters are, however, easily dislodged and difficult to use. A special two-chambered cup (the Lashley cup) has been designed (initially by Lashley, but now used with little modifications) with an inner chamber that is placed over the opening of the Stenson's duct. When suction is applied to an outer concentric chamber, saliva flows into the inner chamber and may be channeled through an outlet tube to a collecting device.

For most purposes mixed saliva is used. One method of collection is by suction from the floor of the mouth using a syringe or other suction device. This is particularly useful in infants. The most common technique, however, is to have the patient spit directly into a container. Sufficient sample may normally be obtained without stimulation, but the flow rate may be increased by sucking or chewing inert material such as wax or polypropylene.

Mixed oral fluid contains much mucin which may make pipetting difficult by virtue of its viscosity and which may interfere with immunoassays by precipitating out of solution. It also contains bacteria, leucocytes, and oral squames. Most laboratories freeze the saliva to denature the mucin and to eliminate the froth associated with saliva samples, then after thawing, centrifuge the sample to leave a clear supernatant fluid suitable for analysis by immunoassay.

Parotid saliva is cell- and debris-free and is less viscous than oral fluid, but under resting conditions the flow rate is low and stimulation of the gland is necessary to obtain adequate quantities.

Saliva may also be collected by absorption onto a cotton-wool swab of the type used by dentists, which the patient chews for 30–90 sec. This forms the basis of commercially produced collection packs such as Salivette® (W. Sarstedt Corp., Germany) which may also incorporate citric acid to increase the flow rate. The saliva is retrieved by centrifugation of the cotton-wool swab in a special centrifuge vessel. Saliva collected in this way is not viscous and does not require a freeze–thaw treatment. Some authors have reported interference with several hormone and drug assays either because the material adsorbs the analyte giving a low recovery or because substances present in the material interfere directly with the dose–response curves of the assays. Other absorbent materials may be used for saliva collection but adsorption and assay interference should be carefully investigated before routine use for immunoassay purposes.

SPECIAL PRECAUTIONS

Saliva should be stored at low temperature before analysis as salivary bacteria may utilize certain analytes such as progesterone and estrogen.

As the concentration of analytes in saliva is often far less than in blood and may be as low as 1%, it is essential to prevent contamination of saliva by blood or plasma. It is therefore recommended that subjects should not brush their teeth or practice any other method of oral hygiene for at least 3 h before sampling.

Gingival fluid, which has a composition similar to blood plasma, may also contaminate saliva; indeed, it is probably always present in very small quantities. However, if the patient has obvious gingivitis, then the amount of contamination by plasma constituents may be significant and the result of immunoassay misleading.

If drug levels are being measured, the drug should not have recently been administered orally, or remnants of drug may contaminate the saliva. If this is possible the mouth should be rinsed before sampling and saliva should not be collected for a further 10 min to ensure there is no dilution by the rinse solution.

CEREBROSPINAL FLUID

FORMATION OF CEREBROSPINAL FLUID

Cerebrospinal fluid (CSF) is mainly produced by passive filtration from the plasma across the blood–CSF barrier formed by the blood vessels (choroid plexi) of the ventricles in contact with CSF. Proteins are filtered with some selectivity according to size so that the low molecular mass proteins pass more readily than those of high molecular mass, but all plasma proteins commonly measured have been found in CSF, demonstrating the lack of a molecular size exclusion limit by the blood–CSF barrier.

About 500 mL CSF is produced per day and the CSF occupies a volume of about 135 mL in the normal adult. Approximately two-thirds of this is produced by the choroid plexi lying in the lateral ventricles. The remainder is produced at extrachoroid sites. Outflow of CSF is by reabsorption into the superior sagittal and other adjacent sinuses following an upward flow over the cerebral cortex. Circulation downwards into the spinal cord is much slower as the spinal canal constitutes a 'dead-end'. Lumbar CSF has a protein concentration up to three times greater than the fluid found within the ventricles, which also has a higher proportion of the low molecular mass proteins than does lumbar fluid. This is due to the progressive equilibration of CSF with plasma through the capillary walls during passage down the spinal canal. Immobilized patients may exhibit increased protein levels in lumbar CSF as the reduced flow rate also allows increased equilibration with plasma.

COLLECTION OF CSF

CSF is normally collected by lumbar puncture which must only be done by an experienced medical practitioner. It is contraindicated in patients with increased intracranial pressure due to space-occupying lesions and in those with untreated blood clotting defects. It is normally collected with the patient in a lateral recumbent position under local procaine anesthesia. A sitting position may be preferable in some patients, especially neonates. The puncture should be done under sterile conditions with the needle inserted into the lumbar subarachnoid space, usually at the midline of the L3-4 interspace. When the needle is positioned correctly the CSF should slowly flow out.

Blood-stained fluid indicates contamination with plasma constituents and invalidates the results of immunoassays of protein fractions. This may be due to a traumatic tap such as transfixation of a vessel or because of a recent subarachnoid hemorrhage. A subarachnoid hemorrhage is characterized by xanthochromic CSF due to the accumulation of bilirubin.

AMNIOTIC FLUID

COLLECTION OF AMNIOTIC FLUID

Amniotic fluid bathes the fetus during its development in the amniotic sac within the womb. From early pregnancy, fetal skin cells are shed into the fluid and may be retrieved. At later stages the fluid consists mainly of fetal urine and lung fluid.

AMNIOCENTESIS

Amniocentesis, the procedure by which amniotic fluid is withdrawn from the amniotic sac using a needle through the mother's abdominal wall and uterus, must be done by an experienced medical practitioner. It is normally done under local anesthetic with ultrasound scanning control and not normally before 16 weeks' gestation, as the volume of amniotic fluid is small in early pregnancy but increases rapidly from about 16 weeks onwards.

SPECIAL PRECAUTIONS

There is a risk of spontaneous abortion after amniocentesis of about 1 in 200. This should be explained to the patient.

Amniotic fluid should be placed in a plain tube without additives and should be centrifuged before immunoassay. Care should be taken to avoid contamination with blood, particularly fetal blood, if this is likely to contain the analyte of interest. This is a particularly pertinent consideration in alphafetoprotein (AFP) assays, where blood-stained amniotic fluid samples may give factitious results. Conversely, if serum AFP is to be assayed in conjunction with amniocentesis, the blood sample must be taken before amniocentesis, as fetomaternal hemorrhage induced at amniocentesis may produce spuriously high maternal serum levels. It is possible, and not unknown, for urine to be obtained by percutaneous puncture in the mistaken belief that it is amniotic fluid. This should be borne in mind when interpreting apparently anomalous results, e.g. very low AFP levels.

It is useful to note the gestational age, preferably assessed by ultrasonography, as this is helpful in the interpretation of the results of most assays for amniotic fluid constituents, which generally change as pregnancy progresses.

SWEAT

Sweat samples are used for monitoring intake of drugs of abuse, particularly cocaine. Insensible perspiration may be collected over a period of days using a sweat patch. The PharmChek™ patch (Sudormed, Santa Ana, CA) is a nonocclusive dressing consisting of a cellulose blotter collection pad, covered by a thin polyurethane and acrylate adhesive membrane. The water component of the sweat passes through the polyurethane and solids, salts, and drugs excreted in the sweat are trapped on the collection pad, which will collect a minimum of 300 μL/day in a 22 °C environment. Exercise or higher temperatures will increase the amount collected. The drugs are eluted at the end of the collection period by placing the cellulose collection pad in a 5-mL capped tube together with 2.5 mL of a pH 5.0 acetate buffer with methanol.

Drugs deposited on the skin prior to wearing the patch may cause false positives.

SEMEN

Immunological tests on semen, such as the assessment of anti-sperm antibody, depend on a reasonable sperm count and on the integrity and viability of the sperm in the sample. It is, therefore, advisable for the subject to abstain from intercourse for 3 days before production of the specimen, which may be obtained by hand masturbation or *coitus interruptus* (withdrawal). A normal sheath or condom should not be used as these may contain spermicidal agents. However, there are special reusable condoms specifically designed for the collection of semen. The specimen should be delivered into a clean bottle provided by the laboratory.

It is essential that the sample is tested within 2 h of ejaculation and it is helpful to allow 30 min to elapse because the initial gel formation requires this time for dissolution, although the semen sample may be liquefied by warming in a sealed plastic container for 20 min at 37 °C.

The sample should be maintained at body or room temperature before assay because cooling damages the sperm and reduces motility. Frozen sperm must not be used as freezing alters the surface of the sperm and may interfere with antibody binding.

HAIR

Sectional hair analysis can be utilized for toxicological and illicit drug investigations as it can provide an assessment of the intake of certain analytes over a period of many months. It is increasingly being used to monitor drug abuse by sensitive immunoassay. There are a number of problems that have to be addressed. The first is surface contamination of the hair from the surrounding environment of the subject. The second is the release of the drug that was deposited in the hair structure during the keratinization process. A third is to ensure that drug has not been lost during cosmetic hair treatment as it is known that there can be an 80% loss during bleaching and that repeated shampooing can decrease the levels by up to 60%. The carefully selected hair sections undergo alkaline digestion before neutralization and buffering for analysis.

MILK

Milk, whether from humans or animals, may be collected either by expressing the milk manually or by use of a breast pump into plain, sterile, glass tubes. The milk should be thoroughly mixed before aliquots are taken for storage or for immediate analysis. If analysis is to be delayed, milk should be stored at −20 °C to suppress bacterial action and, after thawing, thorough mixing is again necessary before analysis. As an alternative to freezing, potassium dichromate may be used to preserve milk which may then be stored at 4 °C while awaiting analysis. This is a convenient way of storing bovine milk prior to assay for progesterone.

If the analytes are not lipid soluble, e.g. the immunoglobulins to food proteins, removal of the lipid layer after centrifugation minimizes lipid interference with the immunological reaction. If the analyte is lipid soluble, e.g. morphine, then extraction of the substance from the milk is usually necessary to eliminate possible matrix effects due to the high lipid and protein concentrations when the milk is assayed undiluted.

STORAGE AND TRANSPORTATION

Most analytes are more stable when the sample is maintained in a cool or frozen condition. For some, especially the small peptide hormones, storage at −20 °C and transportation in the frozen state is usually necessary for reliable results. Such hormones include insulin and insulin C-peptide, gastrin, glucagon, ACTH, and vitamin D. Some complement components need to be maintained at a temperature of −70 °C if spurious results are not to be obtained.

The need for low-temperature preservation of the sample depends not only on the analyte but on the method being used for assay, and in particular the antibodies and the epitopes to which they are directed. Thus, the sample for a particular assay may require different handling depending on the method to be used in the investigation.

Whilst it is strongly recommended that samples should be separated from red cells as soon as possible following venesection, and stored at low temperature, it has been reported (Diver *et al.*, 1994) that there was little change in the assayed levels of commonly measured hormones after storage at ambient temperature in whole blood for up to 1 week and that these changes were unlikely to be of clinical significance.

The specimen may be transported in the frozen state by packing in solid carbon dioxide (dry ice) in an insulated container or vacuum flask or, if such a low temperature is not necessary, in an insulated box containing a bottle of 20% sodium chloride which has been frozen solid over a number of days in a deep freeze capable of reaching −21.6 °C (the freezing point of the sodium chloride solution). The frozen sample, enclosed in a sealed polythene bag and placed next to a bottle of frozen sodium chloride solution, and cushioned with foam plastic in an insulated box, will stay at a temperature of below −8 °C for a period of 40 h, approximately twice as long as with the same volume of solid carbon dioxide.

For safety reasons, the transport of samples outside the laboratory is subject to legislation in many countries and local rulings should always be applied. Typically, they should ensure that every sample is enclosed in a primary container, which is securely sealed. This must be wrapped in sufficient absorbent material to absorb all possible leakage in the event of damage. The container and absorbent material must be sealed in a leakproof bag. This package should further be placed in a polypropylene, metal or strong cardboard container, which is securely closed and properly labeled for transit.

CONCLUSIONS

In this chapter, I have attempted to describe the procedures for the collection of samples and to draw attention to the many pre-analytical factors that may influence the results of immunoassays.

I have not tried to specifically define those factors that are clinically relevant as these will vary according to circumstances and the purpose for which the assay has been requested. But for reliable results, practitioners must know of these factors and set standards of control that minimize their effect on the test results and on the interpretation of the test outcome.

FURTHER READING

Diver, M.J., Hughes, J.G., Hutton, J.L., West, C.R. and Hipkin, L.J. The long term stability in whole blood of 14 commonly requested hormone analytes. *Ann. Clin. Biochem.* **31**, 561–565 (1994).

ECCLS Standard for Specimen Collection, Part 1, Blood Containers (European Committee for Clinical Laboratory Standards document 1 (1), 1984)

ECCLS Standard for Specimen Collection, Part 2, Blood Specimen by Venepuncture (European Committee for Clinical Laboratory Standards document 4 (1), 1987)

NCCLS Procedures for the Collection of Diagnostic Blood Specimens by Venepuncture, 2nd edn (National Committee for Clinical Laboratory Standards document H3-A2, 1984

NCCLS Physicians Office Laboratory Guidelines. Proposed Guidelines. Blood and Urine Collection, Preparation and Transport. 83–105 (National Committee for Clinical Laboratory Standards document POL 1-P, 1987)

Tietz, N.W (ed), *Clinical Guide to Laboratory Tests* (Saunders, Philadelphia, 1990).

Vining, R.F. and McGinley, R.A. Hormones in saliva. *CRC Crit. Rev. Clin. Lab. Sci.* **23**, 95–114 (1986).

Young, D.S. *Effects of Preanalytical Variables on Clinical Laboratory Tests* (American Association for Clinical Chemistry Press, Washington, DC, 1997).

42 Laboratory Quality Assurance

Pierre Blockx and Manuella Martin

The objective of a laboratory quality assurance program is to ensure that reliable results are provided for the laboratory's customers, whether they are clinicians, veterinarians, or researchers. The provision of good quality diagnostic information is essential for the treatment of the sick, reducing unnecessary suffering and helping to keep healthcare costs under control. The quality assurance program should include staff selection and training, equipment maintenance and testing, water purification, sample handling, and assay quality control.

Quality assurance is the term used to describe a wide range of activities that aim to prevent quality problems and optimize **precision** and **accuracy**. **Quality control (QC)** describes the tests that are applied to individual assays to check the validity of the results.

Monitoring the precision of assays is an important part of assay quality control. For each immunoassay, it is useful to know the precision, the optimum working range, and how the coefficient of variation (CV) varies with the sample concentration. Imprecision is due to random errors in the assay and these errors cause individual assay results for a particular sample to scatter around the mean value. If the precision is good, there is only a small amount of variation. If the precision is poor, the user has little confidence in the value of individual results, although a better estimate of the analyte concentration can be obtained by carrying out more determinations, and taking the mean of the values obtained (Figure 42.1).

Accuracy (or lack of bias) is a measure of the proximity of the mean of a series of values to the target and depends on the absence of systematic errors.

STAFF SELECTION AND TRAINING

The selection procedure for laboratory staff needs to identify people who will be capable of performing assays precisely. Although many procedures are automated, we still like to ensure that our staff are able to carry out practical steps competently. This provides an indication of manual dexterity and the ability to work with laboratory equipment. Previous experience is obviously advantageous but staff who have not worked in laboratories before may have had a job or hobby which involves precise handling of materials and tools, and this can be a useful indicator of future laboratory skills, following training. If time permits it may be possible for job applicants to perform practice assays to see if they have an aptitude for the work. Staff with previous computer experience have an advantage as much of the laboratory processes involve working with computer interfaces.

A quality assurance program needs to include the training, evaluation, and improvement of laboratory personnel. A training program should be set up to take

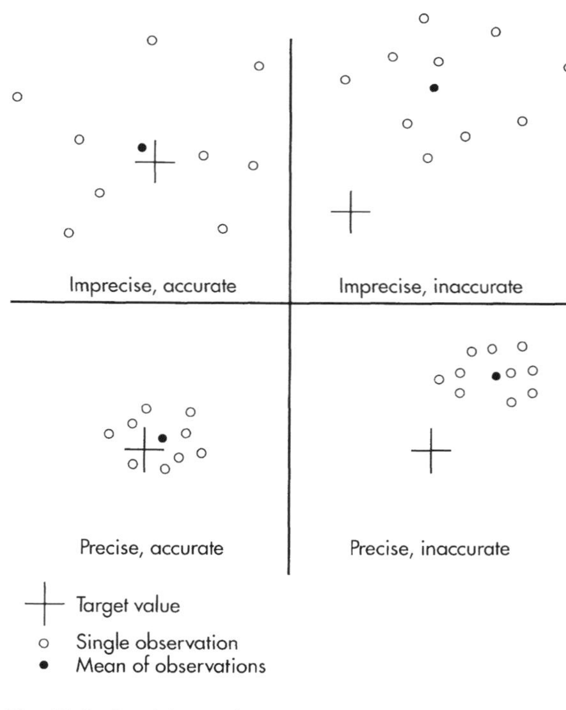

Fig. 42.1 Precision and accuracy.

The Immunoassay Handbook. *Third Edition*
D. Wild (Ed.)

technicians through the different techniques required, including an evaluation of competence at the end of each stage. In our laboratory the technician is not allowed to progress to the next stage without being successful in the previous one. Start with the simple procedures, such as the use of pipettes, then progress to semi-automated and automated equipment, and procedures such as sample dilution. The more background information included in the training program the better, as the technicians are then more likely to notice if anything is going wrong.

In the following sections we describe the tests that we apply when evaluating technicians undergoing training. Further on in the chapter, training for staff carrying out point-of-care (POC) testing will be discussed.

FIXED VOLUME PIPETTES

Volumes from 1000 down to 25 μL, both of water and of serum, are pipetted with 10 replicates of each. Evaluation of precision and accuracy is by weighing. The volume of 25 μL is also tested with a radioactive tracer solution, using 20 replicates to establish the precision. For each volume all replicates are dispensed with the same tip to eliminate tip variations. Precision should be better than 1% CV and accuracy between 99 and 101%. Laboratories not licensed to handle radioactivity could use colorimetric solutions for this and similar tests described below. (see Table 42.1).

REPEATING PIPETTES

Tests are performed with different volumes, e.g. 1000, 500, and 100 μL, and with solutions of different viscosities, e.g. a radioactive or colorimetric tracer solution and a polyethylene glycol (PEG) solution. For volumes smaller than 500 μL a disposable tip is used on the syringe to increase the dispensing precision (e.g. an Eppendorf® tip upon an Eppendorf Combitip® syringe if a Multipette® is used). Precision should be better than 0.5% CV (see Table 42.1).

SEMI-AUTOMATED INSTRUMENTS

Skill evaluation with semi-automated dispensers and diluters is performed using 10 replicates of a radioactive or colorimetric tracer solution, with a variety of volumes. To test dilution the tracer is diluted 1/10 with water. Precision should be better than 0.5% CV.

MANUAL DILUTIONS

Manual dilutions are tested by diluting a radioactive or colorimetric tracer solution with water in two different ways: serial dilutions and dilution series.

For serial dilutions, consecutive 1 in 2 dilutions are made out of a mother solution up to a 1 in 32 dilution: 1/1, 1/2, 1/4, 1/8, 1/16, 1/32. From each dilution, duplicates of 100 μL are pipetted. Precision on the means of the duplicates, after multiplication of the counts with the corresponding dilution factors, should be better than 2% CV.

For the dilution series the first three are made out of a mother solution: 1/2, 1/3, 1/5. These are further diluted to get 1/4, 1/8, 1/6, 1/9, and 1/10 dilutions. Precision on the means of the duplicates, after multiplication of the counts with the corresponding dilution factors, should be better than 2% CV.

It should be noted that, when using radioactive counts to evaluate the precision of a manipulation, errors due to insufficient counting statistics have to be eliminated. Collected counts should be high enough to yield a negligible statistical error, e.g. 100,000 counts represent

Table 42.1. Precision and accuracy tests for pipettes.

Volume (μL)	Liquid	Method	Replication	Precision (% CV)	Accuracy
Fixed volumes					
1000	Water	Weight	10	<1	99–101%
500					
200					
100	Water/	Weight/	10	<1	99–101%
50	serum/	weight/	10	<1	
25	tracer/	activity/	20	<1	
Repetitive pipettes					
1000	PEG	Weight	10	<0.5	
500	Tracer	Activity	20	<0.5	
100	Tracer	Activity	20	<0.5	

a counting error of 0.3% (%counting error = √counts/counts).

EQUIPMENT MAINTENANCE AND TESTING

It is important that the performance characteristics of laboratory equipment such as pipettes, dispensers, diluters, automated devices, and detection systems are checked. The precision and accuracy of each should be carefully tested, at least on the following occasions:

- when a new instrument or accessory is introduced, even a new type of pipette tip;
- when a repaired instrument is reintroduced;
- when an instrument is suspected of poor performance.

The methods that are used for testing equipment are the same as those described above for staff selection and training.

Three additional factors should be taken into consideration for automated pipetting stations: carry-over, dilution of sample by the system solution, and the influence of the difference in viscosity between calibrators and serum samples.

Carry-over of an automated pipetting station is evaluated by determining samples with a very low concentration of a parameter with a broad dynamic range (e.g. human chorionic gonadotropin (hCG), alphafetoprotein (AFP), hepatitis B surface antigen (HB_sAg)) directly after pathological samples. When the low samples give results higher than manually determined, carry-over is suspected. It should be noted that carry-over cannot be checked by pipetting water after a radioactive solution. Indeed, to demonstrate carry-over, dilutions should be detectable up to about 1/40,000. This is because of the broad positive ranges of some parameters (e.g. in some assays hCG has a cut-off value for nonpregnancy at 5 mIU/mL and a peak value of 200,000 mIU/mL). If carry-over is present, it should yield a signal significantly higher than the background signal of the detector. It is important to evaluate the carry-over with liquids similar to the ones examined routinely. Therefore, in most laboratories it is necessary to use serum samples to test for carry-over, because of the proteins present. These proteins influence the viscosity of the liquid and consequently the adhesion of the analyte onto the system tubing.

The **dilution** of a sample by the system solution is tested by pipetting 10 replicates of human serum, spiked with a tracer solution. Dilution with the system solution is proved when the counts in the tube sequence diminish successively.

The influence of the **difference in viscosity** is checked by comparing the serum concentration of a particular analyte obtained with manual and automated procedures. A difference in viscosity can lead to different results and consequently to inaccuracy.

Besides these three specific evaluations, automated pipetting stations should be checked daily for precision and accuracy. Five replicates of a mixture of tracer and serum pipetted by the equipment should have a precision better than 1% CV, and the accuracy should be between 99 and 101%, when compared to five replicates pipetted manually.

Detection systems such as gamma counters, colorimetric assay readers, fluorimeters, and luminometers have to be evaluated regularly. For example, for radioactive counters, background counting, detector normalization, and tray contamination should be checked once a week.

The accuracy and precision of **fully automated analyzers** can usually be tested by metering out a reference tracer (signal generation) solution dedicated to the instrument. Generally, the same solution can be used to check its detection system. Operator maintenance procedures should be performed according to a regular schedule either daily, weekly, monthly, or as required.

Depending on the system, daily procedures should include:

- emptying solid waste container and liquid waste bottle;
- removing outdated or empty specific and universal reagent packs;
- verifying inventory and if necessary loading reagents;
- inspecting and cleaning sample trays;
- verifying that QC samples have been processed.

Depending on the system, weekly procedures should include:

- cleaning metering probes and performing a probe rinse procedure;
- performing a system cleaning procedure;
- inspecting the incubator, if possible, for contamination or corrosion;
- performing accuracy and precision tests.

Depending on the system, monthly (or as required) procedures should include:

- backing up QC calibrations and configuration files;
- replacing filters;
- checking and, if necessary, replacing photometer/fluorimeter lamp;
- replacing syringes and needles;
- cleaning the incubator if possible.

There are a number of fully automated systems available to carry out immunoassays. These have mechanical, electrical, and liquid-handling components. Liquid-handling systems are prone to blockages and contamination, and mechanical components eventually wear out. Electrical connections can occasionally become damaged. For these reasons it is important that the manufacturer's instructions for maintenance and performance checking are carried out carefully.

The temperatures of fridges and freezers should also be checked at least once a week. Fridges should maintain a temperature between 2 and 8 °C and freezers should be below − 10 °C.

LABORATORY WATER

The modern laboratory requires readily available supplies of purified water. Untreated mains water is not suitable for most laboratory applications and should be purified to remove potential contaminants in order to meet the standards required for the particular application.

Many different contaminants can be present in a mains water supply: inorganics, organics, gases, suspended solids, colloids, microorganisms, and pyrogens. Depending upon the application, some or all of these will have to be removed.

The water needed in immunoassays is used to prepare, reconstitute, or dilute reagents, calibrators, and controls. Therefore, it should be as free from impurities as possible. The presence of interfering substances in the calibrators, for example, can cause inaccuracy in sample determinations.

The labware necessary for the preparation of calibrators and reagents must also be rinsed with purified water. Untreated tap water can typically contain dissolved inorganic salts in excess of 200 parts per million (ppm) and leaves residual deposits on glass tubes, pipettes, and other labware.

WATER-PURIFICATION METHODS

The two major purification techniques used are **deionization** and **reverse osmosis**.

Deionization is a rapid (and reversible) chemical reaction using cation- and anion-exchange resins to remove all dissolved ionizable substances from water supplies. Also the removal of organic or colloidal matter can be achieved by using specific ion-exchange resins. A filter is used to prevent particulates from the resin contaminating the purified water. An alternative method for removing organics is filtration through an activated carbon filter.

Reverse osmosis uses a semipermeable membrane, which separates the water from most of the impurities. This technique is based on the reverse principle of osmosis. A pressure, higher than the osmotic pressure, is applied to the higher-concentration solution (e.g. tap water). The water molecules pass through the semipermeable membrane from the higher-concentration to the lower-concentration solution. The rate of retention for ions dissolved in the water is approximately 95%, together with 97% of dissolved solids and up to 99% of bacteria, viruses, high molecular mass organics, and pyrogens.

Where necessary, both techniques (reverse osmosis and deionization) can be combined, reverse osmosis being used as pretreatment to deionization.

WATER PURITY MEASUREMENT

There are several methods for measuring water purity, each designed to measure the levels of particular types of contaminants.

Inorganics

All dissolved inorganic ions carry a charge. The net effect of these charges is measured using a resistivity or conductivity meter: the higher the resistivity (or the lower the conductivity), the better the water quality. The theoretical maximum purity of water with respect to inorganic ions is a resistivity of 18 MW/cm at 25 °C.

Organics

Numerous organic substances are present in water, e.g. humic and fulvic acid, or synthetics. Their levels can be measured by several techniques such as: total organic carbon (TOC), oxygen dissolved, biochemical oxygen demand, and chemical oxygen demand (Williams, 1984, 619–623). Typical TOC values for high purity water are in the range of 20–100 parts per billion (ppb).

Suspended Solids

These are measured by turbidimetry, i.e. by passing light through the water and measuring the amount of light scattered or absorbed.

Dissolved Gases

Carbon dioxide is a weak anion and will have a very slight effect on the conductivity value. However, when monitoring dissolved oxygen levels, special electrodes must be used.

Microorganisms

Microorganisms are measured by growing cultures of organisms in special nutrient media. The normal units of measurement are colony forming units (CFU) per mL.

Colloids

Colloids are substances in the twilight area between true solution and suspension. When present in water they cause haze or turbidity and are measured either by a fouling index test, in which the blockage rate of a standard filter is measured, or by turbidimetry.

Pyrogens

Pyrogens are biological substances that induce a rise in body temperature when injected into mammals. The pyrogens found in water are endotoxins, material from the cell walls of Gram-negative bacteria. Pyrogens are detected using the rabbit test, and endotoxins can be determined specifically using the *Limulus amoebocyte* lysate (LAL) test.

RECOMMENDED WATER QUALITY

The ideal water quality for making up reagents and calibrators is:

Resistivity: 10–18 MΩ/cm at 25 °C
Organics (TOC): 20–100 ppb
pH: 7

However, most immunoassays do not demand such high water quality. Most can be performed with water having a resistivity not lower than 5 MΩ/cm at 25 °C, with an organic content of less than 2 ppm. Depending upon the applications, it is therefore usually sufficient to have a deionization technique, controlled by a conductivity meter, to obtain the water purity necessary for assays and for rinsing labware.

SAMPLE CONTROL

Clinical information is only reliable if patient samples are collected and stored under optimal conditions. Three steps are very important in this process: sample collection, centrifugation, and storage.

SAMPLE COLLECTION

The time of sample collection can often affect the interpretation of the results. Moreover, for some parameters a fasting sample is required. Any advice given by the assay manufacturer concerning the choice of anticoagulant for blood collection should be strictly followed. It is well known that some analytes need to be determined in plasma, but not every anticoagulant is suitable, e.g. heparin interferes in some analyses. When heparin is used therapeutically, laboratory technicians should be on the lookout for small fibrin clots in the serum. In patients treated with heparin, the transformation of fibrinogen into fibrin takes place more slowly, so the formation of fibrin goes on even after separation of the blood cells. These tiny, sometimes almost invisible, clots may go unnoticed, especially in pipetting stations, and lead to inaccurate and imprecise results. This problem can be avoided by centrifuging serum samples before testing.

CENTRIFUGATION

Centrifugation temperature has to be considered, because some analytes, especially peptide hormones such as glucagon and adrenocorticotropic hormone (ACTH), break down rapidly. The recommended centrifugation temperature is 4 °C, controlled thermostatically.

Ideally, samples should be centrifuged as soon as they arrive at the laboratory. If this is not possible they should be stored at 4 °C for a maximum of a few hours only.

STORAGE

Never freeze blood samples before separation of the blood cells. When frozen whole blood is thawed, hemolysis occurs and the cell contents are released. Hemolysis may either lead to falsely low results due to dilution by the cell contents or to elevated results if the tested analyte is present at a higher concentration in the blood cells. The cell contents or membranes can cause interference in some immunoassays.

After centrifugation it is preferable to dispense the plasma or serum in aliquots of at least 0.5 mL. Multiple fractions are useful as they avoid repeated freezing and thawing cycles if the sample has to be diluted or re-analyzed. Multiple freezing and thawing may lead to erroneous results, due to molecular degradation or protein aggregation.

When the test is to be performed on the same day, the serum or plasma should be sealed and stored in the refrigerator. Polystyrene tubes can be used for storage at −20 °C for up to 1 month, but for longer periods, polypropylene tubes with a small capacity are preferred, because their very low permeability reduces the evaporation during storage to a minimum. Sometimes glass tubes are recommended, because certain molecules adsorb onto plastic, e.g. glucagon. After being stored frozen, the thawed samples should be mixed gently but thoroughly, because a concentration gradient of serum constituents may have been generated during storage.

REAGENT AND CONTROL PREPARATION

The control of assay quality begins with accurate preparation of the reagents and control sera. It is important to follow the instructions in the package insert carefully. In particular, the reconstitution of freeze-dried preparations is critical. Such vials should be opened carefully, allowing air to slowly enter the vial, which is at a negative pressure, to avoid material being forced out by the incoming rush of air.

The reconstitution should be carried out using a calibrated pipette, with the correct solvent at the appropriate temperature. The mixing of the solution has to be carried out cautiously (e.g. on a Coulter mixer or on a rotator), to avoid foaming. Extended contact between the solution and the vial stopper is best avoided as occasionally substances on rubber stoppers leak into the solution and can subsequently interfere with the assay. This may occur with commercial controls where the manufacturer of the control cannot test the effects of the stopper on all the assays with which the control may be used.

When subdivision of samples is required, e.g. for use as controls, it is best to use polypropylene tubes, because of their low permeability.

Before use on automated analyzers, liquid reagents should not be shaken as foaming may induce metering errors due to interference with level-sensing detectors. Some systems require reagents to be brought to room temperature before use whereas analyzers with cooled reagent storage areas may give the best results if reagents are transferred directly from the refrigerator. The status of fresh packs on loading is important if they are to be used immediately, particularly for calibrations.

When storing reagents and control preparations, it is important to adhere to the manufacturer's instructions. The following are general guidelines.

- Liquid and freeze-dried reagents should be stored at 2–8 °C.
- Reagents transported in dry ice should be stored in the deep-freeze at −20 °C.
- After reconstitution, freeze-dried reagents should be kept at 2–8 °C for short periods, or as aliquots in the freezer for longer periods.
- Thawed reagents should not be refrozen.

ASSAY QUALITY CONTROL

Quality control samples should be included in every assay or at regular intervals in random-access systems. They provide a retrospective check of each assay run using the following parameters:

- **within-assay precision**, through the evaluation of replicates;
- **between-assay precision**, from variation in the values of control samples obtained on different occasions;
- **bias**, by comparing average values of control samples with their target values;
- **trends**, **cycles**, and **patterns**, by careful monitoring of the data from each assay using control charts. This approach may also be applied to other parameters such as the percentage binding at zero concentration ($\%B_0$).

These checks involve comparing the run under evaluation with previous runs. Consistency of controls, calibration curve-shape, and within-assay precision indicates that the clinical results are also consistent.

WITHIN-ASSAY PRECISION

The within-assay precision of an assay is evaluated by calculating the coefficients of variation (CV) of the concentrations of replicates of control and patient samples.

These CVs are best interpreted by plotting them against concentration. This type of plot is known as a **precision profile**. When drawing a precision profile our criterion for acceptance is a CV of 5% for at least 90% of all the results between the lowest concentration calibrator (excepting zero calibrator) and the highest calibrator. The remaining 10% of the results may have a slightly higher CV but should be located at the ends of the calibration range, not in the midrange (see Figure 42.2).

This is the ideal target for analyses performed in a routine clinical laboratory. However, there are still some analytes for which it is difficult to obtain a within-assay precision of less than 5%. Some analytes are inherently difficult to measure reproducibly, stretching the available technology to the limit. Some determination methods are insufficiently sensitive to measure low concentrations of certain analytes in biological fluids precisely.

There are many other causes of poor precision. For example, some analytes are poorly immunogenic and rarely produce antibodies with a high enough affinity to provide precise assays. The size and shape of the analyte may also have an effect on the binding with antibody, causing steric hindrance. Alternatively, the assay may offer the user a short incubation time, or some other convenience feature, at the expense of precision.

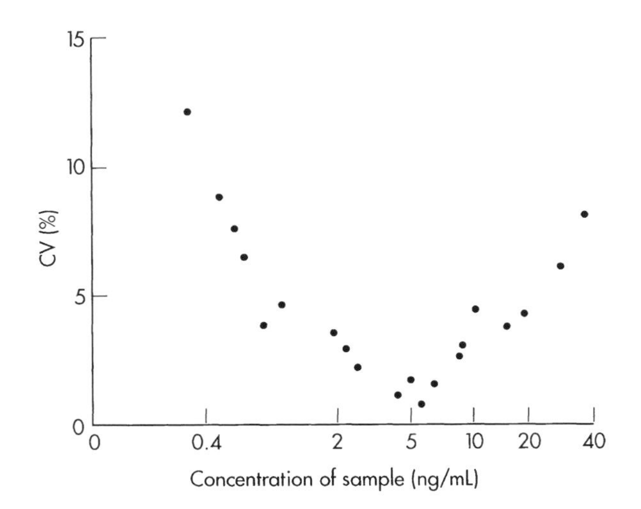

Fig. 42.2 Precision profile.

BETWEEN-ASSAY PRECISION

Control Preparations

Controls from two different sources are used in the evaluation of between-assay precision: internal controls are included in every run and external controls are analyzed periodically as part of an external quality control scheme (EQCS) (*see* EXTERNAL QUALITY CONTROL SCHEMES).

Three types of controls may be used for internal quality control: commercial control preparations, pools of patient samples, and controls supplied with the assay kit.

Commercial Control Preparations

Commercial control preparations have several advantages: they are independent of the determination method, freeze-dried, stable, and available in large quantities so that the same lot may be used for a long time. Moreover, a number of different analytes can be determined in each control and the concentrations are at different levels. However, although the controls are usually based on defibrinated human plasma, which is similar to serum, the plasma is stripped of most of its contents before the analytes are added back in to give the required concentrations. The manufacturers adjust protein and salt concentrations back to normal levels but commercial controls are not the same as unprocessed serum. Although these controls are suitable for routine QC they should not be used for comparisons of different assays for the same analyte. The presence of high concentrations of many different analytes in the same control preparation can cause a difference in control values between two methods that use different antisera, due to cross-reactivity. This can also result in a change in control value if a manufacturer changes the antibodies used in a kit.

Pools

Control preparations may also be 'home-made', using pools of patient samples. Pools have the advantages of being cheap, being more or less the same in composition as routine patient samples, and being available for unusual analytes for which no commercial preparations exist. However, they also have some important disadvantages which should not be overlooked: they may be positive for hepatitis B or HIV, they may not be very stable (unless freeze-dried), they are only available in limited quantities and they have unpredictable concentrations. Pools should also be excluded from dilution and correlation studies, because some analytes break down to fragments after the repeated cycles of freezing and thawing that are necessary for their preparation.

Controls from Reagent Manufacturer

The third type of internal control is supplied by manufacturers. These are frequently prepared in the same way as the calibrators and, therefore, may not behave in the same way as samples. Use of these controls to monitor reagent lot-to-lot variation over long periods is often not possible because the lot numbers of the manufacturers' controls change too often.

INTERNAL QC PROGRAMS

Internal QC programs have two main elements:

1. a method for recording the key parameters, such as control values, from each assay. This should include a graphical display of the values;
2. a system of rules for deciding whether to accept or reject results, and when to investigate changes in performance.

Originally, charts were plotted by hand in many laboratories; nowadays there are many computerized methods available which speed up this process and flag warnings of changes in performance automatically.

Control Sheets

The first evaluation system described may seem superficial, but it is easy to use without any computer help and gives enough practical information for a basic QC program (see Table 42.2).

Control preparations at three different concentrations are tested in each run to monitor the between-assay and between-lot variation. The control values obtained are recorded on a control sheet and for every 20 runs the mean, standard deviation, and %CV are calculated.

The control sheet provides information about between-assay and between-lot variation for the method used. It is an advantage to run as many assays as possible from each kit lot, as between-assay and between-lot variation can then be distinguished quite clearly. There are a number of reasons for between-lot variation. The commonest are changes in batches of critical raw materials and insufficient control of the production processes for the manufacture of the reagents. Some manufacturers do not pay enough attention to quality control. Many have introduced statistical process control (SPC), which has been used so successfully by many industries, to reduce end-product variation.

Shewhart/Levey-Jennings Control Chart

The daily control values collected on the control sheet are difficult to interpret for drift, cycles or other patterns that may indicate quality problems. These are much easier to recognize and interpret with a **Shewhart** or **Levey-Jennings** control chart (see Figure 42.3).

Control charts are constructed by plotting the control results from each assay on a graph with the mean value and the standard deviations marked on the y-axis. The run identification, which is generally the date or assay

Table 42.2. Control sheet.

QUALITY CONTROL

CONTROL PREPARATION: UVW
PARAMETER: Human Growth Hormone
EXPIRY DATE: October 2000
KIT: XYZ
MASTERLOT: 2785V
UNIT: mU/L

		Low	Medium	High	Lot no.	Expiry date
N, SD, %CV		**20, 0.12, 3.7**	**20, 0.54, 3.7**	**20, 2.2, 5.1**		
Mean $\bar{x}$		**3.26**	**14.6**	**42.5**		
$\bar{x} \pm 2SD$		**3.02–3.50**	**13.5–15.7**	**38.2–46.8**		
Date	**Techn.**					
10 April 99	CC	3.33	14.7	45.4	32051	9 May 99
17 April 99	KC	3.40	14.2	45.3	32051	9 May 99
24 April 99	CC	3.40	14.6	46.7	32051	9 May 99
30 April 99	MK	3.30	15.3	48.3	32051	9 May 99
8 May 99	AL	3.43	14.8	45.1	32051	9 May 99
14 May 99	SvA	3.48	14.6	45.5	32463	30 May 99
22 May 99	SvA	3.33	14.9	46.2	32463	30 May 99
29 May 99	AL	3.46	14.3	44.1	32463	30 May 99
5 June 99	RS	3.11	14.9	44.2	33064	25 Jul 99
12 June 99	GV	3.47	15.4	46.7	33064	25 Jul 99
19 June 99	MV	3.44	15.0	41.7	33064	25 Jul 99
26 June 99	LT	3.49	15.4	42.7	33064	25 Jul 99
3 July 99	LT	3.33	14.2	38.9	33064	25 Jul 99
10 July 99	LH	3.42	14.6	42.4	33064	25 Jul 99
17 July 99	GV	3.36	15.5	45.0	33560	06 Sep 99

number, is recorded under the *x*-axis. Sometimes the kit lot number is also written underneath. The warning limits used are typically ± 1, 2, and 3 standard deviations. The 2 standard deviations limit is approximately equal to the 95% confidence interval, which means that about 95% of the control values should lie within this limit if the assay is under control. About 99.7% of the values should lie within the 3 standard deviations limit and, for this reason, if a control lies outside the 3 standard deviations limit, the assay is normally rejected.

Control charts require an estimate of the mean and standard deviation for each control in use. The minimum number of assays required to estimate the mean is 10 (20 assays give a better estimate), so that the mean value provided by the control manufacturer is normally used for the first 10–20 assays after a new lot of controls is introduced. The minimum number of assays required to estimate the standard deviation is 20. However, once the assay is established, the same standard deviations may be applied across lots of controls if the values of the controls are very similar. Alternatively, the standard deviations for the new lot of controls may be recalculated for the new mean values so that they give the same %CVs. Control manufacturers normally provide target ranges for each method and these may be used in the absence of a reliable estimate of standard deviation.

Control charts are an excellent method for detecting problems such as bias or poor precision. Decision rules may be set to suit the laboratory. For example, exceeding the 2 standard deviations limit may require an experienced person to check the assay results carefully, and if two controls in the same assay both exceed the two standard deviations limit the assay may be rejected.

Westgard Analysis

There are a series of well-established and statistically sound decision rules that can be applied to control charts. These are collectively known as the **multi-rule Shewhart chart**, or **Westgard analysis**. Westgard analysis focuses

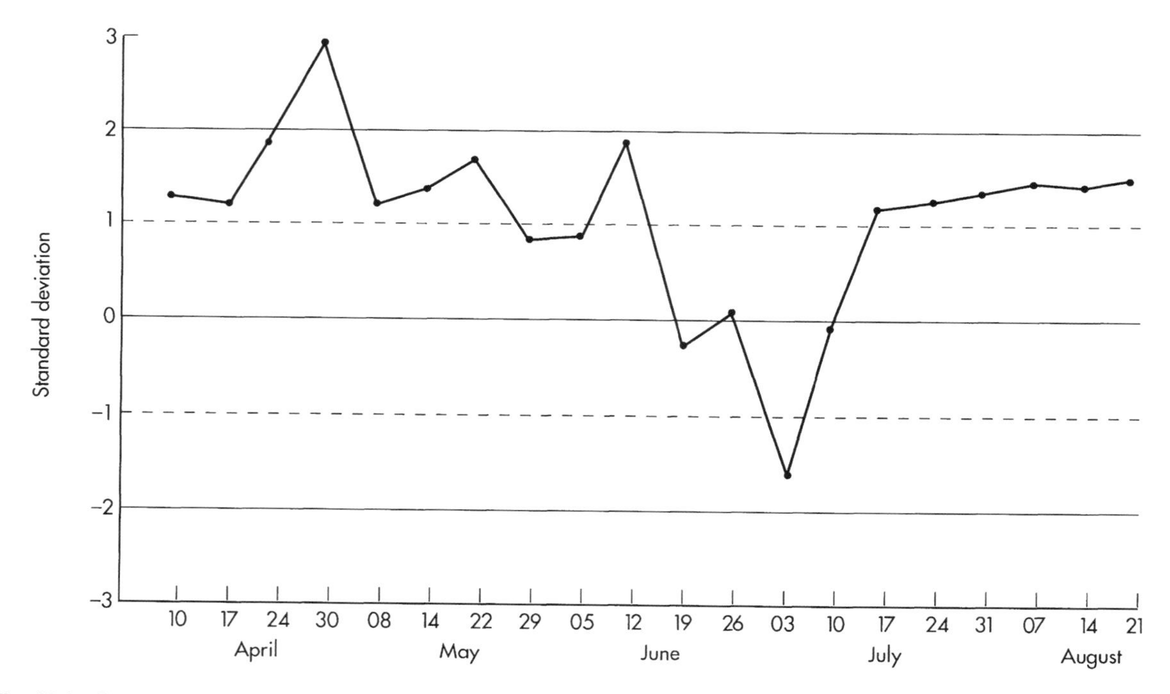

Fig. 42.3 Shewhart/Levey-Jennings control chart.

on the decision whether to accept or reject data from an assay run, in some cases taking data from previous runs into account. In this QC procedure several different control rules are used, hence the term 'multi-rule'.

Westgard's rules were carefully chosen to provide a low probability that normal assays would be rejected and a high probability that errors would be detected. This is difficult because of the background statistical variation. At least two decision rules need to be selected, one that detects random analytical error and the other that detects systematic analytical error. Westgard described six different decision criteria that can be applied to two or more control preparations with different concentrations, assayed in each run. A decision to accept the analytical run without further review requires that there be no violations of any of the rules.

For brevity and convenience, symbols are used to represent the different control rules. The symbol has the form A_L, where A is an abbreviation for a statistic or is the number of control observations and L is the control limit. Other abbreviations used are $\bar{x}$ for mean and s for standard deviation (see Figure 42.4).

$\mathbf{1}_{2s}$ represents the control rule where one control observation exceeds the control limits of $\bar{x} \pm 2s$. This is also the warning rule for a normal Shewhart/Levey-Jennings chart and it is interpreted as a requirement for additional inspection of the assay data and control charts, using the rules below, to judge whether the analytical run should be accepted or rejected.

$\mathbf{1}_{3s}$ symbolizes the control rule where one control observation exceeds the control limits of $\bar{x} \pm 3s$. In this case the run is rejected. This is also the usual action or rejection limit in a classical Shewhart control procedure.
$\mathbf{2}_{2s}$ is the control rule where the values of two controls in the same run, or two consecutive runs of the same control,

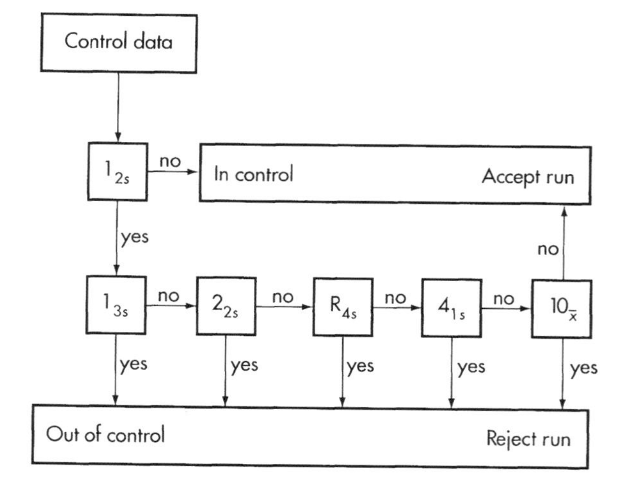

Fig. 42.4 Application of decision criteria in the multi-rule Shewhart/Westgard procedure (redrawn with permission of J.O. Westgard).

exceed the same 2s limit (either $\bar{x}+2s$ or $-2s$). Either situation results in the run being rejected.
R_{4s} is the control rule according to which the run is rejected if one control is outside its +2s limit and another is outside its −2s limit.
4_{1s} is the control rule where the run is rejected when four consecutive control observations exceed the same limit (either $\bar{x}+1s$ or $-1s$). These observations can occur within one control in four consecutive runs, or across two controls in the present run and the one before.
$10_{\bar{x}}$ is the control rule where the run is rejected when 10 consecutive control observations fall on the same side of the mean. These consecutive observations can occur within one control or across controls.

A practical set of decision rules for accepting or rejecting results based on control values is an essential part of a laboratory QC program. For example, the **1_{2s}** rule can be used to prompt a more detailed inspection of the data using the other control rules. If any of the other rules are violated, the run is rejected; otherwise the run is accepted. The particular rule violated gives an indication of the type of analytical error affecting the results. Random error is most often detected by the 1_{3s} and R_{4s} rules. Systematic error is usually detected by the 2_{2s}, 4_{1s} or $10_{\bar{x}}$ rules.

Cusum Chart

The Shewhart/Levey-Jennings type of control chart is not very sensitive to small changes in mean values. Consistent bias from the target mean shows up more quickly and clearly using cusum charts (see Table 42.3).

The cusum, or cumulative sum chart, uses a similar format to that of the Shewhart/Levey-Jennings chart. But instead of plotting the individual values for a control, the difference of each new value from the target mean is added to the cumulative sum of all the previous differences.

In a cusum chart a horizontal line signifies zero bias. Any systematic bias results in a slope away from the horizontal, upwards for a positive bias, and downwards for a negative bias. The steepness of the slope is proportional to the magnitude of the bias. Most cusum programs use a threshold value for the deviation, initiating the cusum calculation only when the deviation has exceeded the threshold value, which is typically 0.5 or 1 standard deviation. If the difference from the target value changes its sign, the cusum line is interrupted. A new plot is started when another observation exceeds the threshold value. The user can choose to set warning limits, e.g. at 2 or 3 standard deviations. When the cusum line crosses these warning limits, the assay is said to be out of control.

A useful property of cusum charts is that they can help pinpoint when a change in mean value occurred, because any change in mean results in a point of inflexion in the plotted line as the slope changes. This is helpful when troubleshooting the cause of bias (see Figure 42.5).

QC SOFTWARE ON AUTOMATED ANALYZERS

The QC software provided on automated analyzers is used to monitor the accuracy and the precision of the system and to collect QC data. Most of the latest analyzers can define multiple control samples, sometimes a virtually unlimited number. Control results are saved in a QC file in reverse chronological order by control sample and/or by analyte. QC files can save up to 1000 results for each analyte defined for each control sample. Some systems save up to 200 analytes across all defined control samples.

The QC results may be reviewed from various QC report options:

- QC report by analyte
- QC report by control sample
- QC daily report
- QC report by time interval.

Baseline and Record (updated) statistics are included in the reports, together with the control ID, and eventually the number of results (total/omitted) and codes or flags associated with the results. QC data can be presented in Levey-Jennings graphs. These graphs plot all the result data points or data for a selected time interval against the predefined mean and standard deviation. Intervals are user defined time periods that make it possible to group and save control data. For example, it may be useful to define a new interval when starting a new control lot.

NEW GENERATION QC SOFTWARE PROGRAMS

The primary purpose of a quality control system is to identify erroneous results before they are reported. Erroneous results are defined as results that are significantly different from the true result and can potentially shift a patient from one diagnostic sub-population to another. The latest QC software programs offer the operator more than the statistical calculation of control sample results.

Examples of these 'new generation' software programs are:

- QC Validator®.
- StatLIA® QCA.

QC Validator®

QC Validator offers the automated selection of statistical control rules and numbers of control measurements (*N*) that will assure the quality required by clinical decision interval (D_{int}) criteria or analytical total error (TE_a) criteria. It is available from Westgard QC, Madison, Wisconsin, USA.

Table 42.3. Calculations for cusum chart.

Date	Control concentration (mU/L)	Bias (value − $\bar{x}$)	Bias − 0.5s	Cusum	
10 April	45.6	0.6	"	0.0	
17 April	45.2	0.2	"	0.0	
24 April	46.8	1.8	0.9	0.9	
30 April	48.4	3.4	2.5	3.4	
8 May	45.6	0.6	"	3.4	
14 May	45.4	0.4	"	3.4	
22 May	46.6	1.6	0.7	4.1	
29 May	45.0	0	"	4.1	
5 June	44.8	−0.2	"	0.0	Reset
12 June	46.6	1.6	0.7	0.7	Reset
19 June	45.4	0.4	"	0.7	
26 June	45.0	0	"	0.7	
3 July	43.4	−1.6	−0.7	−0.7	Reset
10 July	44.8	−0.2	"	−0.7	
17 July	46.8	1.8	0.9	0.9	Reset
24 July	45.8	0.8	"	0.9	
31 July	45.5	0.5	"	0.9	
7 August	46.2	1.2	0.3	1.2	
14 August	45.7	0.7	"	1.2	
21 August	45.9	0.9	0.0	1.2	

Control mean ($\bar{x}$) = 45.0
Standard deviation (s) = 1.8
%CV = 4.0
Threshold (0.5s) = 0.9

To calculate the values to be plotted on a cusum chart, subtract the target mean for the control from each individual result, then subtract the threshold value. Then add each number to the 'cumulative sum' of the previous numbers. The cusum is reset to zero if the difference from the target value changes its sign.

The computer program allows the user to enter information about the characteristics of a method's performance (e.g. imprecision, inaccuracy, expected frequency of errors, within-subject biological variation) and the quality required (clinical important change or D_{int}, Te_a). The user then initiates automatic selection on the basis of the number of control materials to be analyzed (i.e. 1, 2, or 3), and the program constructs a chart of operating specifications (OPSpecs chart) that displays the selected rules and *N*. OPSpecs charts show the relation between precision and accuracy, as a function of the chosen combination of QC rules, ensuring detection of critical-sized errors that would otherwise cause method performance to be inferior, to defined analytical or clinical quality requirements.

StatLIA® CQA Software Program

StatLIA® CQA (available from Brendan Scientific, www.brendan.com) far surpasses conventional quality control methods in its ability to do what is really needed:

- assess the error in the testing process;
- monitor the quality of results over time;
- assess appropriate operator performance.

This methodology, called component quality analysis (CQA), improves the ability to measure the accuracy, precision, and reliability of a patient result and test methodology when compared to conventional methods.

The underlying principles of this methodology are simple and apply to all testing technologies in the clinical laboratory. They are as follows:

- Determine all of the components and steps in a testing methodology which introduce error (variation).
- Identify parameters for that testing technology which can measure these errors directly or indirectly.
- Compare each parameter in the assay statistically with the same parameter in a group of stable reference assays.

For example, in immunoassays there are nine components: instrument signal, tracer, antibody, buffer,

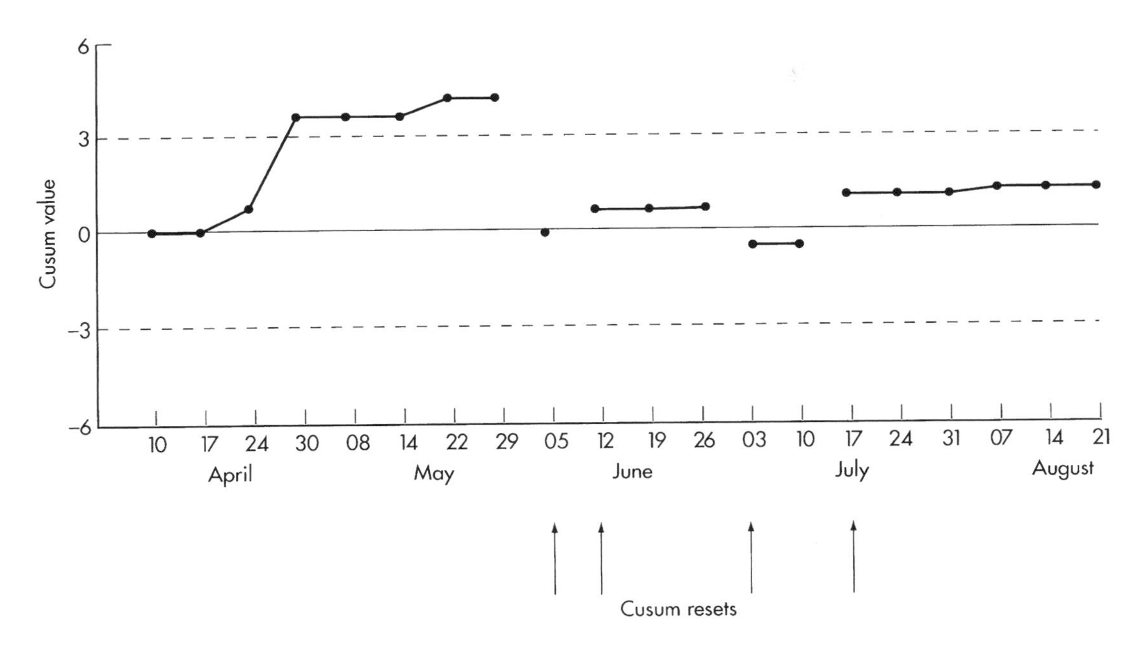

Fig. 42.5 Cusum chart.

incubation conditions (time, temperature, and pipetting), separation of bound and free tracer, standards, controls, and patients' samples. With each assay run, more than 50 control parameters can be compared statistically to a group of reference assays automatically, to measure the error generated by these nine components for that assay. The reference assays are a set of assays, typically 30, which the laboratory has validated as having acceptable performance. Each parameter of the assay being examined is compared statistically to the same parameter obtained from the reference assays. These tests indicate the statistical probabilities that there are no significant differences between the parameters of that assay and the reference assays. These statistical probabilities are combined into a probability index for the variation, between that assay and the reference assays. This offers an unbiased, stable method to measure the error in each individual assay.

In addition to quality control evaluations, which examine the performance of a single assay, quality assurance evaluations are also performed in CQA. A quality assurance evaluation compares a set of recent, or current, assays to the reference assays to detect changes in the testing process itself. The quality assurance evaluation is performed every 30 assays to detect significant differences between parameters of these two assay groups. Flagged parameter probabilities identify changes in reagent integrity or incubation conditions in the test methodology itself.

A CQA quality assurance evaluation can also be used to compare assays performed by a particular operator to the reference assays. This evaluation provides a statistical assessment of that operator's performance, with respect to the error, or variation, acceptable to the laboratory.

QC PARAMETERS OTHER THAN CONTROLS

Although controls are the most useful monitor of assay quality, there are many other parameters which, if charted in the same way as controls, can monitor assay performance and warn of assay problems. Examples are given below:

- percentage binding at zero concentration;
- percentage binding at the concentration of the highest calibrator;
- nonspecific binding (NSB);
- curve-fit parameters, e.g. goodness-of-fit, number of iterations, variance ratio, slope, and intercept;
- estimated dose (ED) at 20, 50, and 80% binding. The ED_{50} is the concentration that corresponds to 50% of the maximum binding, the ED_{20} is the concentration that corresponds to 20% of the maximum binding, etc.

Each of these variables can give an indication of the quality and behavior of the reagents, and the influence of the reaction conditions used. For example, a high NSB could indicate that the tracer has become degraded, or a low percent binding could be caused by an incubation at too low a temperature. This type of data should be recorded along with the control values from each assay. Ideally, computer-generated control charts of the data should also be available to help in troubleshooting problems such as poor precision or the deviation of control sera from the target means (see Table 42.4).

QC IN INFECTIOUS DISEASE TESTS

Infectious disease tests are mostly immunometric assays based on the enzyme-linked immunosorbent assay (ELISA) technique. They are performed either in batches (e.g. using microtitration plates) or on random-access or panel analyzers. These tests can involve several reagents and incubation steps in the assay protocol. So many variables have the opportunity to influence the test result. These variables include:

- reagent preparation and reagent stability;
- pipetting reproducibility and dilution accuracy;
- reagent additions;
- incubator temperature stability;
- washing station performance;
- photometer performance.

This process has to be controlled by:

- commercial controls and/or
- serum pools and/or
- commercial controls.

The composition of most kit controls is not comparable with human serum (*see* ASSAY QUALITY CONTROL–BETWEEN-ASSAY PRECISION–CONTROL PREPARATIONS). Their matrix is often the same as for the calibrators of the tested parameter. Therefore, kit controls cannot be used as an independent control to evaluate the quality of the test performed.

On the other hand, pools made of patient samples with known serum levels are reagent-independent and can be used to evaluate between-lot variations. The main disadvantage of these pools is their unpredictable stability over a longer period, unless they are lyophilized. Commercial controls are the most recommended as far as these controls are available for the desired parameters. They are stable for a long period and usually are available with different analyte levels so that the negative as well as the positive clinical ranges can be controlled covering several reagent lot numbers. The only disadvantage of commercial controls is the limited choice in serum level of the controls. The levels available for infectious disease tests are mostly tuned to the sensitivity and assay range of the US FDA approved methods, which are not always the same as those used in other countries.

External Quality Control Schemes

External quality control schemes (EQCS) provide a means for independent assessment of laboratory performance. In the UK they are known as **external quality assessment schemes (EQAS)**. (This is a more descriptive term as these schemes do not "control" quality in the same way as internal controls). In the USA such schemes are often termed **proficiency testing schemes**. At regular intervals the organizer sends the same controls to all the participating laboratories for testing. No target values are provided and the samples are tested in the same way as patient samples. The results of the tests are sent back to the organizer who collates and analyzes them. Finally, the compiled analysis of all the results is sent back to each laboratory, indicating how their results compare with other laboratories, using the same and different methods. Some of the schemes are conducted in 6-month cycles, with regular evaluations within each cycle, e.g. every 2 weeks. Many EQCS organizers act as a focal point for the exchange of information between laboratories and kit reagent manufacturers. They have undoubtedly had a beneficial effect on product and laboratory performance since their introduction, reducing the amount of variation in results between different laboratories performing the same tests. In some countries, continued unsatisfactory performance in an EQCS can result in the laboratory's license being revoked.

EQCS reports have the additional benefit of providing participating laboratories with data about the relative performance of different products on the market, which can help in choosing the best methods available.

Ideally, the control samples should be derived from human serum and be freeze-dried. Animal serum or plasma is not recommended because it has a different protein and lipid content compared to its human equivalent. It is important that the analyte used to make the higher concentration controls has a similar structure to that in the sample.

As an example, the reports issued during a 6-month cycle of the scheme could be as follows:

- 2-weekly individual laboratory reports in histogram form, together with a summary of the data and some interpretation;
- a 6-monthly survey of performance consisting of:
  - an individual laboratory bias/precision histogram together with a data summary;
  - an analysis of method performance with respect to precision;
  - an analysis of method performance with respect to bias;
  - summary tables that provide a 'ranking' system and a comparison with national, and even international performance.

Each laboratory should carefully analyze EQCS reports for signs of possible problems. A monthly report of the EQCS information for all the analytes under test is useful for the laboratory manager to

Table 42.4. Trend evaluation.

			Parameter: Free T_3 (pmol/L)						Kit: XYZ					
Batch number			**94**	”	”	”	**96**	”	”	”	”	”	”	”
Expiry date			**4 Sep**	”	”	”	**15 Oct**	”	”	”	”	”	”	”
Date test			**31 July**	**1 Aug**	**3 Aug**	**6 Aug**	**7 Aug**	**8 Aug**	**10 Aug**	**13 Aug**	**14 Aug**	**17 Aug**	**20 Aug**	**21 Aug**
Technician			**CC**	**CC**	**CC**	**AL**	**AL**	**AL**	**KC**	**RS**	**RS**	**RS**	**GV**	**GV**
Binding B/B_0%	1	2.55	77.9	77.8	78.4	77.7	75.3	74.8	75.3	75.5	74.7	75.3	74.7	74.3
Conc Std	2	6.24	60.1	61.3	61.0	60.5	57.1	57.3	57.3	57.7	56.8	56.7	57.5	56.7
	3	12.4	45.4	45.5	46.6	45.3	42.1	42.0	42.5	42.6	42.1	42.0	42.8	41.9
	4	25.6	30.4	30.6	31.2	30.1	27.3	28.1	28.0	27.8	27.5	27.5	28.1	27.7
	5	60.5	17.6	17.7	17.4	17.7	15.5	16.2	15.6	15.4	15.5	15.0	16.1	15.8
	6	138	9.7	9.8	9.7	10.3	8.7	9.1	8.5	8.3	8.2	8.0	9.0	8.6
	7													
	8													
	9													
	10													
NSB/(B0 − NSB)%			**1.1**	**0.94**	**1.1**	**1.6**	**2.9**	**2.1**	**0.87**	**0.96**	**1.1**	**1.1**	**1.2**	**1.6**
NSB/T%			**0.58**	**0.49**	**0.56**	**0.77**	**1.3**	**0.99**	**0.41**	**0.44**	**0.48**	**0.48**	**0.54**	**0.74**
B0/T%			**53**	**53**	**54**	**50**	**48**	**47**	**47**	**46**	**46**	**45**	**46**	**46**
Iterations			2	1	3	1	1	1	1	1	1	2	1	1
Variance ratio			2.1	0	3.3	0	0	0	0	0	0	3.6	0	0
ED 20%			49.8	50.2	50.5	49.5	41.0	44.0	42.0	41.6	41.7	40.4	43.5	42.6
ED 50%			10.0	10.2	10.5	10.1	8.7	8.7	8.7	8.9	8.6	8.5	8.9	8.5
ED 80%			2.2	2.2	2.3	2.2	1.9	1.9	1.9	1.9	1.8	1.9	1.8	1.7

Intercept y-axis.

monitor the performance of the laboratory. Performance in the EQCS should be widely publicized in the laboratory to give staff feedback about their laboratory's performance, whether good, bad, or average. If bias from other laboratories occurs, the internal QC data should also be reviewed for clues as to the cause. Continued poor performance in the EQCS indicates problems with the reagents, equipment, or staff concerned (*see* IMMUNOASSAY TROUBLESHOOTING GUIDE).

The maintenance of good quality assays depends upon all staff members being made aware of possible problems, anticipating them, and avoiding them. If a laboratory has quality problems, it can use the data in the EQCS reports to help solve them. One should differentiate between problems appearing consistently with all samples (systematic bias), with particular samples (matrix problems), with samples of particular analyte concentrations (dose-dependent bias) or with a set of data from a particular report (reconstitution or detection problems). Careful analysis should enable laboratory staff to suggest further courses of action, such as changing to an alternative method, determining if the matrix problem is also seen with clinical samples, or retraining laboratory technicians. In summary, a well-organized EQCS can provide a unified, reliable source of information, advice and education on all aspects of the assays, and an independent check of the effectiveness of the laboratory's internal QC program.

POINT-OF-CARE TESTS

New trends in health care have resulted in the need for laboratory testing at the point where patient care is rendered, commonly called POC or near-patient testing. Rapid advances in automated technology have produced instruments that are easy to operate and rapidly produce results using urine or whole blood specimens. Tests are now performed in operating theaters, patient rooms/ wards, out-patient units, and even patients' homes. The quality of the results, however, should be guaranteed. This is not self-evident, because advances in technology allow nonlaboratory personnel to perform the testing. It should be kept in mind, that Clinical Chemists and Medical Technologists are the only group of health care professionals who are trained in laboratory quality control processes and can guarantee quality testing regardless of where it is performed. Therefore, it should remain the clinical laboratory's own responsibility to develop appropriate training programs for nonlaboratory personnel in the use of ancillary testing equipment, test procedures, review of test results and performance, and monitoring of quality control, corrective action, proficiency testing, and equipment management. The assurance of personnel competence must be an integral part of these programs. The clinical laboratory, in collaboration with appropriate medical and nursing staff, must take the responsibility for defining appropriate parameters and establishing guidelines for repeat testing or rejection of results.

The training of nonlaboratory testing personnel should ensure:

- sufficient knowledge of the method to use it intelligently and safely;
- understanding the importance of checking instrument performance, running control tests, and recording results;
- accepting the responsibility for keeping any instrument in proper working order;
- recognition of instrument malfunction;
- basic knowledge of the significance of abnormal results.

There must be adequate and appropriate documentation at the point of testing. The minimum includes:

1. a list of personnel trained to perform the test;
2. a standard operating procedure that includes precise methodology;
3. a logbook for recording all data with regard to calibration, maintenance, quality control, reagent lot numbers and results.

Maintenance procedures, as established by the clinical laboratory, have to be carefully followed and documented. The protocols must be operated with national/ local quality assurance schemes as recommended by the local clinical laboratory and must comply with national/ local regulations related to *in vitro* diagnostic medical devices.

It remains the medical technologist's responsibility to maintain the quality of laboratory testing, and take a leading role in managing the quality control and quality assurance programs.

REFERENCES AND FURTHER READING

Dunn, J.R. A statistical-based quality control/quality assurance system for immunoassay analysis. *Am. Clin. Lab.* **12**, 7 (1993).

Maisonneuve, S.A. *European Pharmacopeia, Bacterial Endotoxins* (Sainte-Ruffine, France, 1987).

U.S. Pharmacopeia. *Bacterial Endotoxins Test* (Mack Publishing, Easton, PA, 1985).

Westgard, J.O. Internal quality control: planning and implementation strategies. *Ann. Clin. Biochem.* **40**, 593–611 (2003).

Westgard, J.O. and Stein, B. Automated selection of statistical quality control procedures to assure meeting clinical or analytical quality requirements. *Clin. Chem.* **43**, 400–403 (1997).

Westgard, J.O., Barry, P.L. and Hunt, M.R. A multi-rule Shewhart chart for quality control in clinical chemistry. *Clin. Chem.* **27**, 493–501 (1981).

Williams, S. *Official Methods of Analysis* (Association of Official Analytical Chemists, Arlington, VA, 1984).

Wright, J.D. Internal quality control in virology: validation of immunoassays using a personal computer spreadsheet. *Br. J. Biomed. Sci.* **57**, 302–306 (2000).

43 Point-of-Care Testing

James H. Nichols

The decision to conduct laboratory testing in a formal, main laboratory or to decentralize the testing closer to a patient care unit is a balance between clinical need and patient outcome. With recent economic pressures to decrease the cost of healthcare, the choice of where to conduct laboratory testing will ultimately be determined by the cost-effectiveness of the delivery model and the benefit to the patient. As hospitals and clinics consolidate to form more cost-effective partnerships, some main laboratories are being closed, resulting in the transportation of specimens to regional reference laboratories with delays in the reporting of test results. The clinical need for a rapid turnaround of critical care tests has increased the utility of point-of-care testing (POCT) at remote sites in these health systems.

POCT promises rapid results and earlier therapeutic intervention. However, the convenience of obtaining a fast result does not necessarily lead to improved patient care, particularly when the technical performance of POCT is not equivalent to that of a main laboratory result. Concerns over the quality of the test, incorrect test performance, and inadequate operator training can cause clinicians to doubt the validity of POCT and lead to duplicate and confirmatory testing that actually leads to higher healthcare costs and increased risk to the patient. In addition, there is a spectrum of delivery options for POCT from bedside testing, portable carts, home care, and even formal satellite laboratories. The optimal delivery model for any given institution is influenced by a variety of factors. Test menu, test volume, patient population, cost, number of operators, management structure of the institution, regulatory environment, educational level of the operators, difficulty level of performing the test, and management of test results are just some of the issues which must be considered before choosing to offer POCT.

DELIVERY OPTIONS

TERMINOLOGY

POCT refers to analyses conducted outside of a main laboratory. Other common terms for POCT are ancillary, bedside, decentralized, near-patient, patient-focused, peripheral, portable, and satellite testing. Some of the terms may be more general in meaning, as ancillary or peripheral testing can describe any testing outside of a main laboratory. Other terms are more specific, like bedside testing that describes testing conducted solely at the patient's bed. The multitude of words used to describe POCT adds confusion. In general, discussions of POCT should limit the number of terms and always define the exact meaning whenever changing POCT terminology.

SITES

POCT is not necessarily used in the same manner in every location, as each testing site involves different types of patients and clinical applications. The location determines the types of patients analyzed: the patient population. POCT in the ambulance/helicopter transport, emergency room, operating room, and intensive care units (ICUs) involves acutely ill patients. These patients present with extremes of hematocrit, blood gases, metabolic byproducts, and multiple drug therapies that place added technical demands on POCT performance. Devices calibrated to the average population may demonstrate significant bias when used on certain hospitalized patients. Chronically ill patients, on the other hand, present with more average profiles. Chronic patients are found in the general medicine, obstetrics/gynecology, psychiatric, dialysis, dental, outpatient clinics, and home nursing settings. Other areas for POCT involve the judiciary system, insurance industry, and even patient self-management.

Although POCT brings laboratory testing closer to the patient, there are many different options for conducting POCT. Testing can be performed at the patient's bedside on capillary whole blood collected by fingerstick. Another option is to collect a blood specimen from an indwelling line and carry the specimen to a utility closet on the ward where POCT can be conducted. The difference between bedside testing and testing in a spare utility closet down the hall may require anticoagulation of the specimen due to delays in testing, which may be minimal but sufficient to start specimen clotting. Differences in sample type and collection can further affect the technical performance of

The Immunoassay Handbook. *Third Edition*
D. Wild (Ed.)

a device. Thus, devices manufactured for home use on capillary blood may not perform equally well when used in a hospital setting on venous or arterial blood. Other delivery options for POCT include portable carts where POCT devices and reagents can be moved from room to room in an institution using a small staff of dedicated POCT operators. POCT has also been performed in patients' bathrooms where urine dipstick analysis and occult blood testing is common.

Given the variety of delivery options, immunoassay testing, whether conducted as POCT or in a main laboratory, must be equivalent throughout an institution or health system in order to manage clinical therapies. Patients may enter a hospital through the emergency room, have an operation and end up in an ICU, an intermediary care unit, a general medicine unit, and finally be treated as an outpatient through clinic visits and home healthcare. POCT results from each of these locations, performed by different operators on different devices or test kits, can be interspersed with main laboratory results. To achieve continuity of care, a test must be site independent and generate equivalent results, otherwise separate reference ranges need to be established for effective clinical interpretation. Maintaining testing equivalency is the goal of POCT quality assurance and management efforts.

OPERATORS

The quality of POCT is directly related to the ability of the operator to perform the test. POCT devices are marketed to be easy to use so that POCT can be conducted by a variety of individuals with minimal training. Quality results, however, require training on the specific POCT method with ongoing checks of operator competency. Operators can have varied levels of education from support technologists with only a high school education, to nurses with 2 or 4 years of college, and medical students and practicing physicians with postgraduate education. Research has demonstrated that for simple one-step devices containing internal controls and not requiring sophisticated volumetric pipetting, operators can perform equally well independent of educational level provided that they have successfully completed a standardized training program on that device. When devices are more complicated, involving multiple steps and technical expertise, training becomes even more important and performance is also dependent on educational level. Thus, training with periodic competency checks is fundamental to obtaining quality POCT results.

MENU

Many laboratory tests are currently available in a point-of-care format. Critical care whole blood analytes include blood gases, electrolytes, coagulation, glucose, and hemoglobin. Organ-specific tests can be conducted for cardiac enzymes (creatine kinase, myoglobin, and troponin), renal function (creatinine and urea nitrogen), pancreatic enzymes (amylase and lipase), and bone turnover (*N*-telopeptides). Occult blood is available for gastric and fecal specimens. Neonatal hypoxia can be assessed through scalp pH. Urine tests include pregnancy, specific gravity, urine dipsticks, and drugs of abuse. Rapid microbiologic screens are available for *Streptococcus*, *H. pylori* and influenza. Human immunodeficiency virus (HIV-1), infectious mononucleosis, and Lyme disease antibody testing can be performed on fingerstick capillary blood. Microscopic tests can also be conducted point-of-care by clinicians for wet mounts, pinworm exams, fern tests, semen, postcoital exams of cervical mucous, and urine sediment. The varied menu of tests available in a point-of-care format indicates the potential utility of POCT for patient care.

ECONOMICS

The cost of POCT is dependent on how the test is delivered and managed at each site. As each site is unique in the delivery of POCT, overall costs will be related to the number of testing devices, number of trained operators, test volume, and percentage of testing dedicated to quality control tests. Each device or test kit requires regular quality control to document minimal standards of performance. As the number of devices, opened test kits, or bottles of test reagents increase, additional quality control will be required. More labor and supplies will be dedicated to quality control testing, increasing cost. Limiting the number of devices is, thus, more cost-effective. Unless testing volumes at a particular site are high enough to necessitate the use of multiple devices, POCT management should seek to minimize the number of devices and open test kits at each site.

The number of operators at a site also affects the cost of testing. The time involved initially training operators and maintaining skills competency increases with the number of operators on a nursing unit. As the number of staff increases, more labor is required to manage staff competencies. Staff also need to perform quality control testing either as part of routine quality assurance or to document skills competency. More quality control leads to higher costs both in reagents and labor. Thus, decreasing the number of operators is a further means of minimizing POCT costs.

Test volume is another factor affecting POCT cost. POCT differs from testing conducted in a formal laboratory setting, as there is no dedicated space. For a large, automated analyzer in a main laboratory, there is a considerable initial investment in the instrument as well as ongoing facility overhead, for space, light, water, and electricity. Yet, the variable, unit reagent cost of each test is minimal. POCT is exactly the opposite, the capital and overhead costs for the analyzer are lower (or negligible with test reagents that are read visually) but the unit, variable costs are generally higher. POCT thus has higher variable costs but lower fixed instrument and facility costs when compared to testing in a main laboratory (Table 43.1).

The overall effect of these factors depends on the test volume. The cost per test for main laboratory analysis will drop faster than POCT as test volume increases, since high fixed costs can be distributed amongst more tests. For POCT, there are minimal fixed costs for instrumentation and facilities, but the labor involved in training and supervising operators is greater than in the main laboratory where there is a smaller group of staff. In comparing POCT to the main laboratory, fixed instrument costs and facilities in the main laboratory balance the cost of fixed labor for POCT involved in initial training and ongoing program supervision (Table 43.1). The greatest difference is in variable reagent costs. As testing volume increases for POCT, unit costs increase proportionally. Depending on the volume, cost for POCT can approach equivalence with the cost of testing in a main laboratory, particularly at those sites with high volumes (Table 43.1).

Quality control testing, however, can significantly affect test cost. Since POCT quality control is performed at

Table 43.1. Cost analysis of a hypothetical POCT ($).

	Main laboratory		**POCT**	
	10,000 pts/yr	**100,000 pts/yr**	**10,000 pts/yr**	**100,000 pts/yr**
Fixed costs				
Instrumentation	10,000	10,000	2000	2000
Facilities	2100	2100	–	–
Training	1200	1200	2800	2800
Supervision	4160	4160	10,400	10,400
Proficiency	500	500	500	500
Fixed total	17,960	17,960	15,700	15,700
Variable costs				
Reagent 10% QC	5500	55,000	11,000	110,000
Reagent 20% QC	6000	60,000	12,000	120,000
Labor 10% QC	3667	36,670	7370	73,700
Labor 20% QC	4000	40,000	8040	80,400
QC 10% calibration verification	1200	1200	1500	1500
QC 20% calibration verification	2000	2000	1500	1500
Variable total				
10% QC	10,367	92,870	19,870	185,200
20% QC	12,000	102,000	21,540	201,900
Total costs				
10% QC	28,327	110,830	35,570	200,900
20% QC	29,960	119,960	37,240	217,600
Cost/test				
10% QC	2.83/patient	1.11/patient	3.56/patient	2.01/patient
20% QC	3.00/patient	1.20/patient	3.72/patient	2.18/patient

Main laboratory costs based on: two laboratory instruments of $100,000/ea to perform four tests amortized over 5 years (2 × $100,000/4 × 5 = $10,000/yr); facility costs of $2100/yr for water, lighting, space rental; training of five technologists for 10 h at $20.00 (including benefits) and 10 h to write the procedure at $20/h (50 × $20 + 10 × $20 = $1200); supervision costs of 0.1 full-time equivalent (FTE) at $20/h or $41,600/yr (0.1 × $41,600 = $4160); proficiency survey $500; reagent costs $0.50/test (10,000 × $0.50 + 1000 × $0.50 for QC = $5500); labor of 1 min/test ($20/h/60 min = $0.33/test + 10 or 20% for QC); quality control and calibration verification based on utilization per year with a discount for higher volumes. POCT costs estimated as: four devices on two patient care units at $500 (4 × $500 = $2000); training of 100 operators for 1 h at $20/h nursing (including benefits) and writing standardized training program 40 h at $20 (100 × $20 + 40 × $20 = $2800); supervision of 0.25 laboratory FTE (0.25 × $41,600 = $10,400); proficiency survey $500; reagent cost $1.00/test (10,000 × $1.00 + 1000 × $1.00 for QC = $11,000); labor of 2 min/test at $20/h nursing ($20/60 min × 2 = $0.67/test + 10 or 20% for QC); quality control and calibration verification based on fixed price per year for up to 10 devices. Note how the price of POCT/patient is actually lower than the laboratory cost per patient when testing volume is high at the POCT site ($2.01/patient at 100,000 tests/yr) and low in the laboratory ($2.83/patient at 10,000 tests/yr). Cost analysis for POCT is therefore dependent on the organization and individual delivery options at each site.

higher unit cost, minimizing the amount of quality control significantly reduces the cost of POCT. Table 43.1 compares the price per test for main laboratory and POCT at 10 and 20% quality control to patient testing ratios. Increasing the percentage of quality control by only 10% leads to a change of $0.09–0.17/test, adding 4–8% to the total test cost. As more devices and operators are added to a POCT program, there is more quality control. The cost of testing increases, since the ratio of quality control to patient tests increases. Ultimately, in order to reduce the amount of quality control performed, a balance needs to be achieved at each site between the minimal number of staff required to meet clinical demands, and maintenance of a high, cost-effective volume of patient testing for each operator. Thus, keeping quality control testing to a minimum decreases POCT cost.

POCT costs are a summation of site-specific factors. Costs can vary, since each site has a different number of devices, operators, test volume, and frequency of quality control. Cost estimates for one site are, therefore, not entirely applicable to other sites. Published cost savings from one institution should be viewed cautiously when extrapolating to other institutions. Even with comparable methodology and clinical application, differences in POCT delivery and administrative policies can affect the amount and type of documentation and supervision required, and this can influence cost.

CLINICAL OUTCOME

POCT, as with any laboratory test, should always be used to answer specific clinical questions. Laboratory testing, in general, should reflect the clinical symptoms and assist in moving patients through clinical and diagnostic pathways. Universal, random POCT testing is as dangerous a practice as blindly ordering all available main laboratory tests on every patient in the hope of finding an abnormality. Laboratory testing is neither cost-effective nor helpful unless it is used judiciously. POCT is no exception. POCT should complement main laboratory testing by offering the clinician a comparable diagnostic test for specific patient care dilemmas where a faster result can improve patient outcome.

Despite the potential of POCT to enhance patient care, POCT is often used more for convenience than for defining a specific clinical situation. Overuse and use of POCT for convenience only duplicates testing efforts in the main laboratory. Important clinical reasons to justify implementation of POCT are:

- *Turnaround time*. Faster results and potential for rapid treatment.
- *Vascular entry*. Decreased risk for fingerstick compared to phlebotomy, especially with oncology patients and others with coagulation disorders.
- *Blood loss*. Less specimen is required for POCT which is beneficial for neonates.
- *Education*. Required as part of housestaff training in a teaching institution.
- *Practice trends*. The increased acuity of inpatient illnesses that require immediate clinical action on results.
- *Efficiency*. Clinicians treat other patients while waiting for test results. Time must then be spent refamiliarizing the case history when the result becomes available. POCT provides rapid results with immediate treatment and eliminates time wasted in rereading case histories.

The common theme sustaining POCT is its ability to resolve a clinical need. POCT blood gases and glucose, for instance, have been implemented in sites that have critically ill patients that cannot wait for remote testing. Immediate results and clinical treatment can move patients through an emergency room or ICU faster and result in shorter hospitalizations. Coagulation testing, for instance, has been used to adjust clotting times after cardiac bypass surgery. By providing faster coagulation results, POCT allows immediate titration of protamine doses, saving time spent in expensive postoperative recovery areas waiting for remote laboratory results and conserving the use of blood products. The use of pregnancy tests in the same day surgery area reduces patient wait times for those women who show up for outpatient surgery without a recent pregnancy test. By offering onsite pregnancy testing, POCT eliminates patient waiting, rescheduling surgery and the loss of expensive surgeon and operating room time. The use of hemoglobin testing in postsurgical recovery rooms has further allowed quicker assessment of postsurgical bleeding and helps to move patients through this area faster, opening the limited postoperative space for the monitoring of more critically ill patients and saving healthcare costs. Implementation of bedside blood gases in the ICU has decreased the frequency of testing and decreased the amount of blood drawn for each test, resulting in significantly less blood loss over an entire ICU stay. Therefore, POCT can offer beneficial patient outcomes when used to meet specific clinical needs. Outcomes can be measured in real units like time and money or in subjective units like patient satisfaction. In this manner, the cost of performing the test at the point-of-care can be balanced with the clinical effect, both justifying the need for POCT and documenting improvements in clinical service.

Although POCT may benefit some patient populations, these outcomes do not imply that the same test can be performed universally on all patients with the same effect. Publications touting the tremendous cost savings and significant advances in patient care provided by POCT need to be viewed critically for the administrative structure of the testing practices at that site, particularly with respect to the patient population, volume of testing, number of trained operators, quality control practices, and management oversight. Given the spectrum of possible situations, cost-effective POCT at one site on a specific patient population may not be cost-effective at another site on a different patient population. Each site needs to be viewed independently.

POCT devices also do not demonstrate the same accuracy and precision as core laboratory instrumentation. The US Food and Drug Administration (FDA), for instance, would like glucose meters to have a minimum performance of ±20% agreement to a laboratory comparative method more than 95% of the time for levels >100 and ±20 mg/dL agreement for levels <100 mg/dL. With variability of ±20 mg/dL, glucose meters would not be able to statistically distinguish 1.1 mmol/L (20 mg/dL) from 3.3 mmol/L (60 mg/dL) for a sample at a level of 2.2 mmol/L (40 mg/dL), a common cutoff for determining neonatal hypoglycemia. In comparison, core laboratory methods give precision of the order of <3% throughout the analytical range of physiologic glucose levels. While the FDA has recommended one set of performance goals for glucose meters, other organizations have different recommendations. There is currently no clear consensus among various international professional organizations. The American Diabetes Association (ADA) recommends glucose meter agreement within 15% of a laboratory comparative method, with the future goal of <5% deviation from comparative values. The National Academy of Clinical Biochemistry and the ADA have found that no published glucose comparison study has achieved the goals proposed by the ADA, and there are no published data to support a role for portable meters in the diagnosis of diabetes or for population screening. The imprecision of the meters, coupled with the substantial differences between meters, precludes their use in the diagnosis of diabetes and limits their usefulness in screening for diabetes. Differences between various POCT and core laboratory methods are thus an important consideration when physicians are ordering a test. The clinical situation should dictate the type of method that is appropriate for that point in a patient's care based on the method's limitations.

Once the decision to implement POCT has been made, clinicians must be informed of the limitations and advantage of POCT and educated on how to order the appropriate test. Criteria determined from outcomes assessment should assist clinicians in choosing whether to order a test as POCT, send a sample to a local stat laboratory or transport the specimen to a more distant main laboratory based on the method's overall performance in the specific patient care setting. The required turnaround time of results, clinical need, and when the result will be acknowledged and therapeutic action instituted by the clinician should be part of the decision of where to perform the test. Staff rushing to perform a POCT that sits in a patient's chart for hours before acknowledgement by the physician is as poor a use of clinical resources as if a patient suffers a bad clinical outcome due to the unavailability of a critical test result from delays at a distant laboratory. Optimal decision trees for the type of tests and where to send those tests for different patient treatments form the basis for current research in critical pathways. These pathways seek to optimize patient outcome and minimize healthcare costs by indicating which practices lead to the most beneficial outcome. Initial pathways for an institution tend to be generalized and only indicate types of tests ordered for particular disease groups. Later refinements of the initial pathways focus on specific outcomes from tests, including where and how testing is performed and used.

QUALITY ASSURANCE

COMPONENTS OF GOOD LABORATORY PRACTICE

Quality assurance of POCT, like other laboratory testing, requires control over the entire testing process, from physician order, patient preparation, and specimen collection, to transportation, analysis of the specimen, result reporting, clinical interpretation, and therapeutic intervention. Each step must be analyzed for potential sources of error, and control monitors instituted to determine when the testing process is successful and intervene when it fails. POCT shares similar quality issues with main laboratory testing, and POCT only eliminates the specimen transportation and processing steps. Problems with inappropriate testing, improper collection of specimens from unprepared patients, transposition of numbers in the test result, and incorrect clinical interpretation of the result are all areas for mishaps. Technical problems with the analysis must also be examined for instrument interferences and operator errors.

Historically, quality assurance of laboratory testing has focused on instrument errors. Systematic and random errors occur most frequently from the laboratory instrument and reagent interaction on the instrument. Performance of laboratory instrumentation is relatively independent of operator effects. Operators are only required to maintain reagent supplies, load samples, push a button to start the analyzer, and troubleshoot when problems happen. Quality assurance of laboratory testing is easily manageable given the limited number of instruments and technologists in a main laboratory. Each operator has narrow focus and considerable experience from repeated operation of the same instrumentation.

The performance of POCT devices, on the other hand, is more dependent on the operator. Visual results can change with differences in specimen loading, test timing, washing steps, and lighting conditions for manual tests like pregnancy, drug, and urine dipsticks. Even variations in color discrimination between operators can affect POCT results. Newer generations of devices have sought to make devices independent of operator effects by automating the test timing and detection steps. Yet, quality assurance of POCT is more difficult to manage than core laboratory testing, as there can be hundreds of POCT devices and operators distributed throughout a health system. The primary focus of these operators is patient care. Nurses, physicians, and other clinical staff are not attuned to laboratory practice, and POCT is only a minor part of their daily responsibilities. Some clinicians may only infrequently perform POCT. Quality assurance programs for POCT must therefore ensure consistent test performance by controlling both operators and devices.

Experience from institutions with POCT programs has linked several components of these programs to high initial quality of testing and consistent improvements in quality over time. These programs:

- involve laboratory personnel in the initial training of operators;
- include standardized training tools, like videotapes, as part of the training;
- repeat training and review performance at scheduled intervals;
- regularly compare POCT results to the main laboratory;
- use computerized data capture of POCT devices when available to store quality control and patient data.

Operator training with ongoing checks of competency are important aspects of obtaining good POCT results. Training specific to the particular POCT device guarantees correct initial performance, but periodic comparison of POCT results to the main laboratory with operator retraining guarantees high levels of continued performance. Automated capture of quality control and patient data assists the management of POCT quality by reducing the amount of labor required to review operator competencies and laboratory comparisons.

REGULATIONS

Regulation of POCT seeks to ensure that it is conducted under the direction of a structured quality assurance program. Currently, the United States of America has federally mandated POCT guidelines with monetary penalties for institutions that fail to meet the guidelines. Other countries are adopting quality regulations regarding POCT, and peer guidelines have appeared with recommendations for quality assurance practices for unit-use POCT devices.

In the US, POCT is subject to multiple regulations with varying levels of stringency. Home testing by patients on their own devices requires that the FDA approve the device for 'home use' before marketing. Hospital and healthcare use of POCT must meet both federal regulations and the guidelines for the state in which the test is conducted. State regulations for POCT vary, and some are more or less stringent than the federal regulations. Institutions are advised to consult their own local states for specific guidelines.

All healthcare use of POCT, however, is subject to the Department of Health and Human Services, Clinical Laboratory Improvement Amendments of 1988 (CLIA'88), which supersedes the previous 1967 regulations. CLIA divides laboratory testing into three levels of complexity: waived, moderate, and high complexity, depending on the difficulty of the testing process. These regulations are enforced through the threat of losing federal Medicare reimbursement for laboratory testing which can result in a significant loss of income to the institution as well as additional legal penalties. New updates to CLIA focus on controlling the entire testing process, from preanalytic sample collection, device operation, and analyst competency to result reporting and interpretation. The new regulations track the path that a specimen follows through the laboratory. Other changes include the required frequency of quality control for certain tests, establishing one set of quality control standards for all laboratories performing nonwaived complexity testing (i.e. moderate and high complexity tests).

Three accreditation agencies inspect and certify healthcare laboratories for CLIA compliance: the Joint Commission on the Accreditation of Healthcare Organizations (JCAHO) and the College of American Pathologists (CAP), which accredit hospitals, and the Commission on Office Laboratory Accreditation (COLA), which accredits private physicians' office laboratories (POLs). Since the CAP guidelines are stricter than the JCAHO and some state regulations, inspection and accreditation by CAP does not necessarily require reinspection by these other agencies provided that CAP has 'deemed status' with that agency. Other agencies and some states have similar equivalencies and 'deemed status' to the federal guidelines depending on their current standards. Laboratories may, therefore, have multiple certifications but only require a single inspection by the strictest of the accreditation agencies.

POCT under CLIA generally falls into waived and moderate complexity categories. Waived POCT includes simple test kits approved by the FDA for home use: glucose, urine dipsticks, hemoglobin, and pregnancy for instance. All other POCT not on the federal waived category lists are considered moderate complexity. The JCAHO has a separate set of guidelines for waived complexity tests. Waived POCT, under the JCAHO, must comply with six guidelines:

- The organization must define how the POCT is to be used, i.e. for diagnosis or patient management.
- A list of all personnel involved in POCT (director, supervisors, and operators) must be available.
- There must be documentation of initial training specific to the POCT device for all operators and demonstration of ongoing competency.
- Written policies covering all aspects of the testing process – patient preparation, sample collection,

analysis, reporting results, and interpretation – must be readily available to each operator.
- Quality control must be performed according to manufacturer specifications and at the recommended frequency.
- POCT results must be recorded so that there is an audit trail available between patient specimen, quality control performance records, operator competency, and training documentation.

The JCAHO guidelines are general recommendations, and the specifics of meeting the guidelines are left to each organization. For instance, the following can all demonstrate operator competency:

- routine performance of quality control;
- analysis of a blind specimen of known value (proficiency sample);
- visual inspection of the operator.

An institution only needs to document operator competency and that the institution follows its own policy in order to satisfy the JCAHO guidelines.

Moderate complexity POCT under the JCAHO, COLA, and CAP is considered comparable to main laboratory testing and is judged by the same standards. POCT must be conducted under the direction of the main laboratory. Moderate complexity testing requires a technical director, a supervisor, and a clinical consultant, and there are specific recommendations regarding the educational level of operators. Patient results must be verified on a regular basis (daily) by the laboratory supervisor. In addition, there are recommendations for method and reagent validation, storage, maintenance, laboratory environment, result documentation, and operator continuing education. At a minimum, two levels of quality control must be performed daily and calibration must be verified semiannually or with each shipment of reagent lots. POCT should further be part of an external proficiency testing program to compare POCT performance to other peer institutions. Oversight of POCT should involve a multidisciplinary team with nursing, clinician, and laboratory membership. Overall, the testing process should be part of a continuous quality improvement scheme that can document baseline performance, target areas for improvement, and indicate the result of systematic program changes to POCT quality.

Although the CAP follows moderate complexity CLIA standards, the CAP does not distinguish between different delivery options for laboratory testing. POCT and main laboratory testing, whether waived or moderately complex, are all treated the same. Waived complexity POCT under the CAP must therefore conform to all of the moderate complexity standards regarding laboratory direction, staff educational levels, daily result verification, external proficiency testing and the high levels of reagent, instrument, and operator documentation. In addition, the CAP requires POCT operators to be screened for color blindness.

In summary, the US regulations mandate laboratory involvement in the POCT process to ensure that testing is conducted under a defined quality assurance program. Although some of the regulations may seem trite by requiring excessive paperwork and oversight, the final consequence of these regulations has been to emphasize the assessment of clinical need for POCT within a healthcare organization. POCT cannot be viewed as mere convenience. By forcing institutions to realize that quality POCT has labor and supervision costs with regulatory impact, unnecessary redundant testing can be eliminated. Only through complementary use of POCT and testing in the main laboratory, with assessment of clinical outcomes, can overall healthcare be optimized.

DATA MANAGEMENT

Quality assurance of POCT generates volumes of data. Instrumentation records include initial validation, maintenance, troubleshooting, quality control, and routine testing. Operator records contain training and competency. Finally, reagent records document validation of shipments, use and storage. Managing these data becomes more difficult as the number of devices, operators, and reagent lots increase. Methods that automate the management of POCT data are an advantage by improving documentation compliance and saving labor in data review.

POCT data management is a threefold process involving capture of data by the device at the time of testing, transfer of data to a central information system, and statistical reduction to obtain quantitative parameters for review. Only computerized POCT devices are capable of automating data management at this time. In terms of testing volume, this includes about half of the current POCT: glucose, electrolytes, blood gas, hemoglobin, and coagulation testing. The other half is manual POCT: urine dipsticks, pregnancy, occult blood and drug testing. Management of manual testing continues to pose a challenge.

When available, computerized data management can assist the laboratory in supervising the POCT program. POCT devices with data management prompt the operator for required information. Operator and patient identification can be entered manually or by barcode reader. Test results are stored with the date, time, reagent lot, reagent expiration date, quality control lots, control expiration dates, control ranges, and operator and patient identification to provide a complete audit trail for quality assurance. Newer POCT devices also have lockouts, which only allow trained, certified operators to perform testing, remind the operators when to perform quality control, and prevent reporting of patient results if controls fail. These features ensure that the device only releases results within institutionally defined parameters. As POCT results have the potential to be used immediately in patient care, the POCT device lockouts are invaluable in checking the acceptability of results, since laboratory review would add to the result turnaround time and may not be practical with each result.

POCT devices, however, are limited in the amount of data that can be stored and they require periodic downloading. Data can be transferred via the serial port of the device, directly to a computer with larger memory capacity. Laptop computers are portable and provide a convenient means of collecting data from remote POCT devices. Yet, laptops require a person to manually connect the computer to each device, physically requiring someone to carry the laptop to the device or to periodically bring the device to the laptop. Telephone connections via modem are also available for downloading POCT data from remote sites, but the Internet provides a ready means of collecting data where Internet connections are available. Recent developments are exploring the ability to use radio, infrared, and other wireless technologies to transfer data from remote sites to a central computer. Wireless provides a real-time means of obtaining data. The amount of labor for a technologist to manually download POCT devices can be significant, particularly for POCT programs with hundreds of devices. Automating the transfer of data thus provides an opportunity for considerable labor savings and improved review of data.

Unfortunately, connecting POCT devices is costly and inefficient. Each manufacturer has developed their interfaces independently with proprietary hardware and unique communication protocols. For users, the custom nature of each POCT device translates into separate data manager computer systems for each device with the added expense of separate wiring and separate interfaces. This added expense discourages the implementation and conversion to new POCT. Efforts to standardize the communication protocols have led to a collaborative consortium of POCT users, organizations, and manufacturers, and the development of a POCT1-A standard for connectivity, now maintained by the National Committee for Clinical Laboratory Standards (NCCLS). This standard promotes seamless information exchange between POCT devices, electronic medical records, and laboratory information systems, and meets the key user requirements for bidirectional communication, standard device connections, use of existing hospital infrastructure, interoperability with commercial software, and data security.

Once POCT data is collected in a central database, the data can be statistically reduced and reviewed. Quality control can be tracked for trends, and problems identified, for targeting future program improvements. Operator performance can be monitored and compared. Patient data can be verified, trend reports printed for clinical management, and billing statements generated for reimbursement. Unfortunately, this type of analysis is only available for data stored on a computer. For manual POCT, comparable statistical review is only available if results are manually entered into a computer. Given the volume of manual testing, this is a laborious task that can generate other errors involved in numeric transposition and key entry mistakes. Compliance with manual data entry is also impractical on a busy patient care unit. Only a check on test use could indicate if all POCT manual testing is accurately being captured.

PRACTICAL MANAGEMENT

TECHNICAL VALIDATION

Ensuring that POCT gives accurate results that are comparable to a main laboratory, in all the situations likely to be encountered in an institution, requires knowledge of the technical limitations of the test. Before being used on patients, POCT should be validated to determine that results match manufacturers' specifications. Testing the devices under conditions in which they are planned to be used is important to accurately predict POCT performance once implemented. Hospitals, physician's office, and home nursing are different practices from the patient self-testing for which many POCT devices were intended. Hospitalized and acutely ill patients can have different physiologic parameters when compared to home patients (e.g. hematocrit). Differences in patient population can lead to POCT biases. These effects should be considered when selecting a possible POCT method.

Determination of acceptable performance, however, depends on the clinical application. More variability may be acceptable in some situations and unacceptable in others. A consensus of clinical tolerance needs to be established before selecting and validating a POCT method. Published standards can assist in setting tolerance limits for POCT, but ultimately institutions must determine the criteria for acceptable performance within their own clinical practice.

Method selection should consider the variables within the institution that can affect POCT. This involves determining how individual POCT is used at every testing site. Important variables to consider are:

- *Vibration*. Is the device sensitive to vibrations that could occur in mobile transport vehicles like ambulances?
- *Humidity*. Is the environment controlled or can seasonal variations occur?
- *Temperature*. Will the device reagents be exposed to ambient temperature extremes of summer heat or winter cold during storage, transportation, or analysis?
- *Lighting*. Will the POCT be exposed to direct sunlight during storage or analysis? Does indoor fluorescent lighting affect the development color of the test?
- *Hematocrit*. What are the acceptable tolerance limits of the POCT? What are the average and range limits

of hematocrit in the intended patient population? Do these match the device limits?

- *Altitude*. Is there an expected bias at different altitudes due to either barometric pressure or oxygen effects that could affect testing on an airplane or in a medical rescue helicopter?
- *Storage*. Is refrigeration required for any components of the POCT system? How will the temperature of storage refrigerators be monitored and documented?
- *Samples*. Can POCT be used on all types of samples like venous and arterial blood if the device was intended for home use with capillary fingersticks?
- *Specimen additives*. Do anticoagulants or glycolysis inhibitors affect the test? Is fresh urine required or should urine preservatives be used?

The possible combinations of variables are considerable and each practice needs to assess their unique situation. For instance, validation experiments, which use stored specimens with anticoagulants, may yield different results from fresh capillary specimens. Additionally, many POCT have narrow effective temperature ranges, particularly those containing protein enzymes and antibodies. Special attention must be given to home nursing practices where POCT may be stored at ambient temperature in cars and exposed to extremes of summer heat and winter cold. Validation of POCT should seek to mimic the actual conditions to which the POCT is subjected during routine use. In many situations, this may require concurrent use of POCT and correlation testing in the main laboratory for a limited period of time from the same patients.

Ultimately, decisions about POCT methodology should seek to meet the prime objective of standardizing results for a given analyte across an institution. As the number of different methods and manufacturers for any given test increases, there will be more problems encountered with test biases. POCT quality assurance programs are thus more easily managed when the numbers of different methods for a laboratory test are kept to a minimum. Minimizing the number of methods also simplifies the need for multiple policies and training programs.

QUALITY CONTROL

All laboratory testing requires a means of documenting that the method is stable and has not changed over time. This is often achieved through the analysis of liquid controls of known value that can be analyzed like patients. The control results can then be monitored and tracked over time to determine if changes have occurred that would affect patient results. This use of quality control has been perfected to monitor the systematic performance of laboratory instruments with large batches of reagent. However, standard quality control practices developed in the main laboratory do a poor job of assessing random, single test errors.

POCT tests are often manufactured in small batches or as single test cartridges. Routine quality control practice must therefore be modified to measure random failure rates of each test. Manufacturers have begun to include internal controls in some POCT to provide a means of monitoring single test failures. For manual tests, these internal controls can indicate problems with kit storage, sample application, and with analytical procedure by including a second antibody–antigen reaction for immunoassay-based kits. Internal controls can also signal appropriate timing for reading results. Activity of a separate area on guaiac occult blood tests can determine the potency of peroxide developing solutions. For POCT devices, spectrophotometric and electronic controls can detect the integrity of the device, especially if dropped or damaged, as well as manufacturing consistency of biosensor test cartridges. Electronic controls are beneficial, since there is no use of reagents or added cost associated with performance of the control. Yet, electronic controls do not assess the entire performance aspects of the test reagents and operator technique like liquid controls can. Recent regulatory changes in the JCAHO and CAP guidelines have allowed electronic controls to substitute for liquid controls provided that the institution can document that electronic controls detect the same frequency of errors as liquid controls on the POCT device. The latest update to CLIA takes a more global view of POCT quality assurance. This update allows the use of alternative control practices provided that labs also control the other components of the testing process, including areas that may not be covered by an electronic control.

ADMINISTRATION

POCT requires a defined administrative structure within an organization. The laboratory plays a key role in this structure, with a director and supervisory personnel assigned to review data generated by POCT and to oversee the testing practices on the patient care units. Clinical staff also need to be involved in the administrative structure with supervisors on the nursing unit who can address POCT problems as they arise. Educators need to be appointed to guarantee standardized training of staff and to handle the training of new staff as turnover occurs.

Oversight of the POCT program should be directed by a multidisciplinary committee that has the power to advise on policy issues and make recommendations regarding management issues as they arise. Membership of this committee should include representation from all parties involved in the POCT process. Laboratory and clinical staff are mandatory. Administrators should be included to handle organizational budgeting and to ensure the support of the organization trustees. Purchasing representation is also recommended to address inventory issues. Inventory control is particularly problematic given the variety of ways that POCT can enter an organization: pharmacy, physicians carrying in kits from other practices, patient devices, and corporate sales representatives leaving samples, to name a few of the ways. Overall, the multidisciplinary committee should provide guidance to the laboratory in selecting POCT methodologies and standardizing testing throughout the organization. The multidisciplinary committee thus provides the administrative power to allow testing to occur at a given site or to discontinue testing when problems occur and policies are not being followed.

SUMMARY

POCT is an alternate means of delivering laboratory testing closer to the patient that promises faster turnaround of results and improved clinical outcomes. Whether POCT actually delivers beneficial outcomes is determined by how the test is used at an individual site. Sites performing few tests with many POCT devices and operators may incur higher costs than other sites. Inefficient testing practices that result in redundant testing and values that sit in a medical record without clinical action do not necessarily lead to improved outcome. The effectiveness of POCT is thus intricately dependent on the clinical practices at a given site. Moreover, quality POCT requires deliberate efforts and a defined administrative structure to implement and maintain high standards of testing. Each testing site must be viewed as a unique situation where all practice variables come together to determine the most appropriate test delivery options for that particular patient population. POCT that is effective at one site may be very ineffective at another. Only through realization that POCT is an extension of the main laboratory with comparable quality assurance problems, and similar training and supervision requirements, will POCT find its optimal role within an institution.

REFERENCES AND FURTHER READING

American Diabetes Association, Self-monitoring of blood glucose. *Diabetes Care* **17**, 81–86 (1994).

Cohen, F.E., Sater, B. and Feingold, K.R. Potential danger of extending SMGB techniques to hospital wards. *Diabetes Care* **9**, 320–322 (1986).

College of American Pathologists Commission on Laboratory Accreditation. *Point-of-Care Testing Checklist* (College of American Pathologists, Northfield, IL, 2002).

Department of Health and Human Services Centers for Medicare & Medicaid Services Centers for Disease Control and Prevention 42 CFR Part 493; Medicare, Medicaid, and CLIA Programs; Laboratory Requirements Relating to Quality Systems and Certain Personnel Qualifications. *Fed. Regist.* **68**, 3640–3714 (2003), January 24.

Department of Health and Human Services Health Care Financing Administration Public Health Service 42 CFR; Final Rule; Medicare, Medicaid and CLIA Programs; Regulations Implementing Clinical Laboratories Improvement Amendments of 1988 (CLIA). *Fed. Regist.* **57**, 7001–7288 (1992), February 28.

Ehrmeyer, S.S. and Laessig, R.H. Regulatory requirements (CLIA'88, JCAHO, CAP) for decentralized testing. *Am. J. Clin. Pathol.* **104** (Suppl 1), S40–S49 (1996).

Handorf, C.R. Assuring quality in laboratory testing at the point of care. *Clin. Chim. Acta* **260**, 207–216 (1997).

International Organization for Standardization. *ISO/TC 212/DIS 15197 Determination of Performance Criteria for In Vitro Blood Glucose Monitoring Systems for Management of Human Diabetes Mellitus* (ISO, Geneva, 2000).

International Organization for Standardization. *ISO/TC 212/WG1/WD 22870 Clinical Laboratory Testing – Quality Management of Point-of-Care Testing (POCT)* (ISO, Geneva, 2002).

Jacobs, E. and Laudin, A.G. The satellite laboratory and point-of-care testing: integration of information. *Am. J. Clin. Pathol.* **104** (Suppl 1), S33–S39 (1995).

Jacobs, E., Vadasdi, E., Roman, S. and Colman, N. The influence of hematocrit, uremia and hemodialysis on whole blood glucose analysis. *Lab. Med.* **24**, 295–300 (1993).

Joint Commission on Accreditation of Healthcare Organizations. *CAMPCLS Comprehensive Accreditation Manual for Pathology and Clinical Laboratory Services* (Joint Commission on Accreditation of Healthcare Organizations, Oakbrook Terrace, IL, 2001).

Joint Commission on Accreditation of Healthcare Organizations. *CAMH Automated Comprehensive Accreditation Manual for Hospitals* (Joint Commission on Accreditation of Healthcare Organizations, Oakbrook Terrace, IL, 2003).

Jones, B.A. and Howanitz, P.J. Bedside glucose monitoring quality control practices: a College of American Pathologists, Q-Probes study of program quality control documentation, program characteristics, and accuracy performance in 544 institutions. *Arch. Pathol. Lab. Med.* **120**, 339–345 (1996).

Kost, G.J. Guidelines for point-of-care testing: improving patient outcomes. *Am. J. Clin. Pathol.* **104** (Suppl 1), S111–S127 (1995).

Kost, G.J. *Principles & Practice of Point-of-Care Testing* (Lippincott, Williams & Wilkins, Philadelphia, PA, 2002).

Kost, G.J., Vu, H.T., Lee, J.H., Bourgeois, P., Kiechle, F.L., Martin, C., Miller, S.S., Okorodudu, A.O., Podczasy, J.J., Webster, R. and Whitlow, K.J. Multicenter study of oxygen-insensitive handheld glucose point-of-care testing in critical care/hospital/ambulatory patients in the United States and Canada. *Crit. Care Med.* **26**, 581–590 (1998).

Lewandrowski, K., Cheek, R., Nathan, D.M., Godine, J.E., Hurxthal, K., Eschenbach, K. and Laposata, M. Implementation of capillary blood glucose monitoring in a teaching hospital and determination of program requirements to maintain quality testing. *Am. J. Med.* **93**, 419–426 (1992).

Macik, B.G. Designing a point-of-care program for coagulation testing. *Arch. Pathol. Lab. Med.* **119**, 929–938 (1995).

Marks, V. Essential considerations in the provision of near-patient testing facilities. *Ann. Clin. Biochem.* **25**, 220–225 (1988).

Nanji, A.A., Poon, R. and Hinberg, I. Quality of laboratory test results obtained by non-technical personnel in a decentralized setting. *Am. J. Clin. Pathol.* **89**, 797–801 (1988).

NCCLS. *Document EP18-P. Quality Management for Unit Use Testing; Proposed Guideline* (National Committee for Clinical Laboratory Standards, Wayne, PA, 1997).

NCCLS. *Document AST4-A. Blood Glucose Testing in Settings Without Laboratory Support; Approved Guideline* (National Committee for Clinical Laboratory Standards, Wayne, PA, 1999).

NCCLS. *Document POCT1-A. Point-of-Care Connectivity; Approved Standard* (National Committee for Clinical Laboratory Standards, Wayne, PA, 2001).

NCCLS. *Document C30-A2. Point-of-Care Blood Glucose Testing in Acute and Chronic Care Facilities; Approved Guideline – Second Edition* (National Committee for Clinical Laboratory Standards, Wayne, PA, 2002).

Nichols, J.H. Cost analysis of point-of-care testing. In: *Advances in Pathology and Laboratory Medicine*, vol. 9, (ed Weinstein, R.S.), 121–133 (Mosby-Year Book, St. Louis, MO, 1996).

Nichols, J.H. (ed), *Point-of-Care Testing: Performance Improvement and Evidence-Based Outcomes* (Marcel Dekker, New York, NY, 2003).

Nichols, J.H., Howard, C., Loman, K., Miller, C., Nyberg, D. and Chan, D.W. Laboratory and bedside evaluation of portable glucose meters. *Am. J. Clin. Pathol.* **103**, 244–251 (1995).

Parvin, C.A., Lo, S.F., Deuser, S.M., Weaver, L.G., Lewis, L.M. and Scott, M.G. Impact of point-of-care testing on patients' length of stay in a large emergency department. *Clin. Chem.* **42**, 711–717 (1996).

Sacks, D.B., Bruns, D.E., Goldstein, D.E., Maclaren, N.K., McDonald, J.M. and Parrott, M. Guidelines and recommendations for laboratory analysis in the diagnosis and management of diabetes mellitus. *Clin. Chem.* **48**, 436–472 (2002).

Seamonds, B. Medical, economic, and regulatory factors affecting point-of-care testing. *Clin. Chim. Acta* **249**, 1–19 (1996).

Wu, A.H.B. Reducing the inappropriate utilization of clinical laboratory tests. *Conn. Med.* **61**, 15–21 (1997).

44 Immunoassay Troubleshooting Guide

David Wild

INSTRUCTIONS

1. Identify the type of fault and find the most appropriate section of the troubleshooting guide. The CASE STUDY sections provide typical examples.
2. Review the POSSIBLE CAUSES and list all those that may apply.
3. Follow the diagnostic hints in SUGGESTED COURSE OF ACTION. Where possible, obtain the information listed in CHECKLIST OF USEFUL INFORMATION. Ask the manufacturer if other users have reported similar problems.
4. Gradually eliminate the possible causes. It may be necessary to carry out simple experiments or equipment tests to pinpoint the exact cause. Substitution of alternative equipment is the best way to check the operation of manual assays, although this may need to be over an extended period, especially if the problem is intermittent.
5. With automated immunoassay analyzers, start by drawing or copying a flow diagram with boxes for each separate instrument function. Concentrate on the paths of the samples and reagents, and the subsequent assay processing steps. Identify any functions that could, in theory, cause the specific fault. To test individual functions, use the test protocols supplied with the instrument or try to devise experiments that stress test the relevant assay processes on the instrument. It is surprising how many potential causes of problems can be eliminated, simply by carrying out experiments using standard clinical laboratory equipment. Characterize the data from the periods before and after the problem started, and from any experiments, carefully.
6. Note down signal levels for the zero calibrator and the routine controls from a number of occasions before and after the problem began. Any changes noted may help pinpoint exactly when the problem started and indicate possible causes. Even if there are no changes in signal, this information can be used to eliminate causes.
7. To check for calibration curve-fit bias, use the printout from a calibration curve. Compare the pre-assigned values of the calibrators (the *actual* values) with their concentrations determined by recalculating them from their signal levels using the equation for the fitted curve (the *fitted* values). This information is often included in the assay or calibration printouts or can be determined by re-entering the signal levels of the calibrators manually as unknowns.
8. Troubleshooters are taught to look for coincident differences between the two situations: when a problem occurs and when it does not. For example, if a problem started in June, what else changed in June? Although there may not be a connection, these coincident changes warrant further examination. Ask questions beginning with the words *who*, *what*, *why* and *when* to open up the investigation.
9. Look for similar problems with other immunoassays run on the same equipment. These indicate that the equipment is involved rather than the reagents. The way competitive and immunometric assays are affected by equipment faults can help pinpoint the exact cause of the problem.
10. Most determinations are run on single samples. Running duplicates, at least of controls, can be very useful for problem-solving.

TROUBLESHOOTING GUIDE

CONTROL BIAS – CONSISTENT CHANGE IN VALUES FROM ONE PERIOD OF TIME TO ANOTHER (OR FROM ONE REAGENT LOT TO ANOTHER)

Possible Causes

Reagent lot at fault – the change in control values should clearly coincide with a change in reagent lot number.

Error in stored calibration curve or manufacturer-generated calibration curve – the change in control value should coincide with the introduction of the new calibration curve. Rectify by running a new calibration curve. If the manufacturer-generated calibration curve is at fault, other users of the same reagent lot will also be affected.

Change in curve-fit program used – differences between *fitted* and *actual* calibrator values in the calibration data should reflect the control changes.

Incorrect reconstitution of freeze-dried control – if this is the cause the duration of the bias will be coincident with the use of aliquots from a specific bottle of control.

Bottle-to-bottle variation in controls – as above, changes will coincide with the bottle of control used.

Inadvertent change in control used – the technician may have changed from one batch or type of control to another without realizing. If so, the time of the change may not coincide with a change in the immunoassay reagent lot.

Change in protocol – check to see if the protocol has been changed.

Change in laboratory temperature if room temperature incubation used – other effects should be apparent, such as changes in the calibration curve shape or changes in the signal levels of all the controls. Seasonal extremes of temperature should be considered, as should the effects of cold or hot weather first thing in the morning if heating or air-conditioning is switched off at night.

Faulty pipette or automatic sampler/dispenser – changes in the controls may be mirrored by a change in signal levels, curve shape, or within-assay precision of duplicates. Test by substituting different dispensing equipment if possible or check the operation using the manufacturer's instructions. If competitive assays are affected more than immunometric assays, suspect the reagent dispenser. If all assays are equally affected, check the sample dispenser. The pipette used to reconstitute the calibrators or controls may be inaccurate. If this is the case all the controls will be biased by the same percentage from their target values.

Contamination of reusable reagent or distilled water container by dirt, disinfectant, or detergent – check the container appearance and cleaning procedure. Effective removal of cleaning agents may be more critical than the cleaning process itself.

Faulty mixing of sample and reagents – changes in the controls will be accompanied by poorer within-assay and between-assay precision.

Faulty incubator or water-bath – changes in the controls may be mirrored by a change in the curve shape. Test the temperature maintained by the equipment if possible.

Faulty centrifuge – insufficient centrifugation will also cause a change in curve shape, a worsening of within-assay precision and slipping pellets on draining (if a decantation assay).

Faulty magnetic separation – unlikely to cause a consistent change in the controls, but if so, it will be matched by poor within-assay precision.

Faulty washer – most likely to affect low concentration controls in immunometric assays and high concentration controls in competitive assays. Poor washing can show up differences between the calibrator and control matrices that can result in bias.

Change in technician – changes in the controls may be accompanied by other differences, e.g. in between-assay precision.

Suggested Course of Action

1. Analyze the control data and try to assess approximately when the change occurred. It is not normally possible to define this down to the nearest assay. In the example (Figure 44.1), the change could have started in the last assay using lot 40C or with the first assay using 42B. If there is any doubt about the statistical significance of the change, compare control values using an unpaired *t*-test.
2. If there have been similar reports before, analyze them to see if the problem is the same.
3. Identify any differences between your protocol and that in the package insert (manual assays).
4. Run any tests recommended by the system manufacturer.
5. Find out if the manufacturer has seen a similar bias in their control data or if any other customers have reported bias.
6. Check the assay printouts or calibration records for curve-fit bias before and after the change (by comparing *fitted* and *actual* calibrator values on the calibration record).

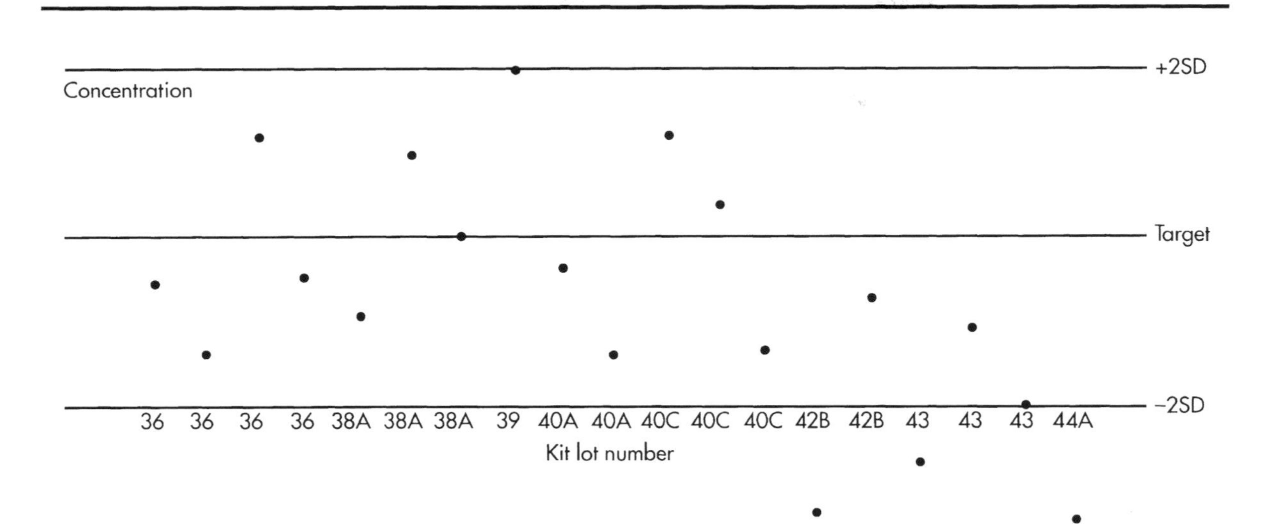

Fig. 44.1 Example of control bias (*see* CASE STUDY).

7. Check the signal levels of all the controls before and after the bias occurred. Check the signal levels of other immunoassays run on the same equipment over the same period. (This will help to differentiate between reagent and equipment problems.)
8. Identify any coincident changes in reagent lots. If the change is coincident with the introduction of a new reagent lot number, ask the manufacturer for further information about the nature of the lot change.

Checklist of Useful Information

It may be useful to have the following information available:

- reagent lot numbers immediately before and after the suspected change;
- control values and the corresponding signal levels should be plotted on control charts using data from at least 10 occasions before and as many as possible after the change occurred. Test whether the change is statistically significant using unpaired *t*-tests. Note reagent lot numbers on the graph;
- if the controls are made in-house, e.g. by pooling patient samples, find out how they were prepared;
- details of any other coincident changes in assay performance noticed, e.g. change in binding at zero dose (%B_0), curve shape, ED_{50}, or non-specific binding (NSB). Compare calibration curve shapes before and after the change in control values;
- assay printouts (or relevant calibration data) from at least one unbiased assay before the problem occurred and from at least one clearly biased assay after the change occurred;
- details of any differences between the protocol used and the protocol in the package insert;
- details of all the equipment used and a list of recent service operations;
- information on any coincident change in the protocol, equipment, technician, or curve-fit program.

Case Study

The 'hypothyroid' commercial control used by a laboratory for T_3 was within the target-expected range from January to May, then failures started to occur on the low concentration side of the target range. On looking at the control charts, the laboratory manager suspected that the mean value of the control in the assay had decreased around kit lot 40C (Figure 44.1). The manufacturer of the reagents did not observe control bias but several other laboratories using the same test reported bias. Those laboratories that experienced bias were all using spline curve-fit programs. (Cause – calibration error of one low calibrator, near to the control in concentration, caused bias with the laboratories' spline curve-fit programs but not with logit-log (used by the manufacturer) or other 'smoothed' curve-fits.)

GRADUAL CHANGE IN CONTROL VALUES (NOT CONSISTENT CHANGE AT ONE TIME OR WITH CHANGE IN REAGENT LOT)

Possible Causes

Gradual change in reagent lots – it is unlikely that consecutive reagent lots would suffer from a continual trend over a long period of time. This would be characterized by steps in control charts, the steps coinciding with changes in the reagent lot.

Reagents unstable or incorrectly stored – gradual change in the control values will be apparent from the first use of a new reagent lot until the last use, followed by a return to the initial values with the new reagent lot. If reagents are stored on board an analyzer, check that the storage system is working correctly and that the manufacturer's instructions are being followed. Instability of on-board reagents will result in drift in control values while individual reagent packs are used, with unbiased control values when a new pack is installed.

Controls unstable – this is the most likely cause if the change is gradual, over several months, and is independent of the reagent lot. Although the affected controls may change in value, no parallel changes in the reference interval or patient sample median should be apparent. Many analytes are unstable, even at −20 °C, over several months.

Automated analyzer requires recalibration – recalibrate analyzer.

Refrigerator or freezer not at correct temperature – it is not unknown for certain areas in refrigerators to be at +15 °C, and freezers which are being regularly opened during the day may only reach −20 °C at night. This can cause controls, calibrators, or assay reagents to deteriorate more quickly than expected.

Faulty equipment – gradual changes over a short period of time, such as a few days, may be due to faulty equipment. Other effects may also be apparent, such as changes in the control signal levels, calibration curve shape, or within-assay precision. Analyzer signal reading equipment and incubator temperature control are possible sources of drift. Check records or on-board database for evidence of drift in signal-related parameters.

Change in laboratory temperature – in extreme cases, cold or hot weather can affect assays with room temperature incubations if the laboratory is insufficiently heated or cooled. Automated analyzers may also be affected by extreme changes in laboratory temperature between calibrations. If either of these occurs, the gradual change in controls will coincide with the change in ambient temperature. Very low signal levels in enzyme immunoassays may be affected by air pollution.

Suggested Course of Action

1. Analyze the control data and the signal levels of the controls, and try to assess the type of change (e.g. gradual, in steps, or cyclic). Identify when the change started and any points of inflection, where the rate of change decreased or increased. If there have been similar reports before, analyze them to see if the problem is the same. If patterns are noted, review any parameters that may have changed in a corresponding way, e.g. laboratory temperature.
2. Find out if the manufacturer has seen corresponding changes in their control data, or has received any similar reports from other assay users.
3. Check equipment and consider where faults could arise. Check any analyzer parameters that could be correlated with drift in signal reading equipment. If other customers have not reported a similar effect, request a check of the equipment by a service engineer.
4. Also check refrigerator and freezer temperatures.
5. Replace controls if they are suspected as the cause.
6. If the assay has a room temperature incubation, review periods of extreme temperature against the changes in control values.

Checklist of Useful Information

It may be useful to have the following information available:

- control charts, with reagent lot numbers and dates, for controls (and for binding at zero dose, ED_{50}, NSB, etc. if available);
- signal levels of controls and calibrators;
- identity, description, and lot number of the controls concerned;
- storage details for the controls (i.e. temperature, liquid or freeze-dried, whether subaliquoted or repeatedly frozen and thawed, age of controls);
- storage details of reagents between assays;
- details of all the equipment used.

Case Study

Over a period of 12 months a laboratory noticed unconjugated estriol controls gradually decreasing in value, by about 1% per month. (Cause – the liquid controls, made from pools of patient samples, were stored at −20 °C. Estriol in serum is unstable for long periods of time at −20 °C.)

COMMERCIAL CONTROLS CONSISTENTLY BIASED FROM TARGET MEAN (NOT CHANGE IN VALUES ACROSS SAME LOT OF CONTROLS)

Possible Causes

Error by control manufacturer – check with the control manufacturer.

Error by immunoassay manufacturer when calibrating the controls for the control manufacturer – check with the assay manufacturer.

Imprecision or bias in calibration by assay manufacturer – check with the manufacturer.

Change in values obtained with product since calibration – check with the manufacturer.

Incorrect ranges being used – check the ranges in the package insert for the controls.

Use of ranges for different kit for same analyte – check the exact product name in the package insert against the name in the control package insert.

Use of different units to those for which the range is stated – check the units in the kit and control package inserts.

Use of a different protocol to that used in the calibration – if you do not use the primary protocol in the package insert, carry out an assay of the controls using this protocol to see if it corrects the bias.

Analyzer-to-analyzer variation – ask the manufacturer to run the same controls. If the control values are normal when the manufacturer runs them but biased in your laboratory, have the analyzer checked.

Error due to curve-fit program used – if this is the cause, differences between *fitted* and *actual* calibrator values should reflect the control bias from the target mean.

Defective pipette (especially pipette used to reconstitute calibrators or controls) or automatic sampler/dispenser – check the accuracy of dispensing or substitute with another pipette or dispenser.

Defective water-bath or incubator – check the temperature of the incubator or water-bath, or substitute with alternative equipment. If a stand-alone piece of equipment is suspected of causing bias try incubating in a water-bath.

Defective washer – have the washer checked or substitute with another washer.

Suggested Course of Action

1. Check that the controls have been consistently biased since they were first used.
2. Check the control and kit package inserts and check that the product name and units are the same.
3. Consult the control and kit manufacturers for advice. Ask if any other customers have reported a similar bias. If not, request a check of the equipment by a service engineer.
4. Reassay the controls using the primary protocol in the package insert and change the equipment used to see if the bias is corrected.
5. Check assay printouts (or calibration records) for curve-fit bias by comparing *fitted* and *actual* calibrator values.

Checklist of Useful Information

It may be useful to have the following information available:

- control lot number(s);
- package inserts for controls and kit;
- control values since control lot was first introduced;
- assay printouts (or relevant calibration records) from typical assays showing the bias;
- details of differences between the protocol used and the protocol in the package insert;
- details of all the equipment used;
- details of the storage conditions of the controls, including whether samples of controls are aliquoted, or are repeatedly frozen and thawed.

Case Study

A customer reported that she was consistently obtaining control values for a particular immunoassay in the range 70–90 ng/mL. The control had a calibrated value in the package insert of 100 ng/mL with a range of 80–120 ng/mL. (Cause – the customer had been using a new version of the assay, launched after the controls had been calibrated, that gave slightly different values.)

BIAS IN PROFICIENCY TESTING (EXTERNAL QUALITY ASSESSMENT) SCHEME FROM OTHER USERS OF THE SAME ASSAY

Possible Causes

Reagent lot at fault – unlikely unless you were the only laboratory using the lot concerned. Check the values of internal controls run at the same time as the scheme samples.

Reagents unstable or incorrectly stored; degradation of reagents stored on board analyzer – check for evidence of drift in internal control values during the storage period.

Error in stored calibration curve or manufacturer-generated calibration curve – internal control values at similar concentrations are likely to be biased if this is the cause. Check that calibrations are being carried out frequently enough. If the manufacturer-generated calibration curve is at fault, other users of the same reagent lot are also likely to be affected.

Bias due to curve-fit program used – this should be evident by curve-fit bias of the same magnitude at similar concentrations in the assay printout. Check for differences between *fitted* and *actual* calibrator values in recent calibrations.

Incorrect reconstitution of bottle of control – this is unlikely to occur on more than one occasion. All the analytes measured in the control will be affected by the same percentage bias.

Variation between different bottles of the same control – this should cause a wide variation in the scheme participant results.

Use of a different kit to the majority of other users classified under the same group – check the exact description of the product listed in the scheme.

Use of incorrect reporting units – check the units in the scheme against those used in the assay.

Reagent lot used after expiry – check the dates.

Use of incorrect kit calibrator values – check the values on the assay printouts or calibration records. Internal controls will also be biased.

Difference between your assay protocol and that used by other scheme participants – any major difference between your protocol and that in the package insert should be tested for effects on the control values.

Extreme laboratory temperature – if your laboratory is insufficiently heated or cooled, extreme weather may cause control bias if the assay has a room temperature incubation. Automated analyzers may also be affected by extreme laboratory temperatures and require recalibration more frequently.

Faulty pipette or sampler/dispenser; dirty pipette tips; contaminated sample or reagent probe; poor connection between probe and tubing; leaking gasket; carryover of analyte from high-concentration sample to low-concentration sample; carry-over of reagent by reagent-dispensing probe; splashing of reagents; tubing constricted – test by substituting different dispensing equipment or cleaning the system and rerunning the scheme samples.

Faulty mixing of sample and reagents – control bias will be mirrored by poorer within-assay and between-assay precision.

Faulty incubator or water-bath – test the temperature maintained by the equipment if possible.

Insufficient time allowed for incubator or water-bath to warm up – check procedure.

Faulty centrifuge – insufficient centrifugation may cause a change in curve shape, a worsening of within-assay precision, and slipping pellets on draining.

Faulty magnetic separation – this may be characterized by an occasional bad replicate if duplicates are run.

Faulty washer – this is most likely to affect low-concentration controls in immunometric assays and high-concentration controls in competitive assays. Poor washing can show up differences between calibrator and control matrices causing bias.

If the bias in the proficiency testing scheme varies considerably between assays, *see* POOR BETWEEN-ASSAY PRECISION.

Suggested Course of Action

1. Retest the proficiency testing scheme samples if they are still available.
2. Analyze the information from the scheme and try to assess approximately when the change occurred. Look for the extent of the bias with each proficiency testing scheme sample and assess whether there is a concentration dependency or other common factor.
3. Check whether the bias from other users of the same kit is consistent in different sample distributions.
4. Identify differences between your protocol and that in the package insert. If the protocol is suspected as the cause of the bias, retest the scheme samples using the primary protocol in the package insert.
5. Check control charts for changes coincident with any changes in bias in the proficiency testing scheme.
6. Identify coincident change in reagent lots. If a very new or nearly expired lot was used it may not have been used by other participants.
7. Check the assay printouts (or calibration records) for curve-fit bias (by comparing *fitted* and *actual* calibrator values).

Checklist of Useful Information

It may be useful to have the following information available:

- proficiency testing (external quality assessment) scheme printouts before and after the bias occurred;
- reagent lot numbers for the assays in which scheme samples were run;
- control data for all the assays concerned;
- assay printout(s) for all the assays concerned or calibration data for the calibration against which the sample was assayed;
- details of differences between the protocol used and the protocol in the package insert;
- details of all the equipment used;
- information on any coincident change in the protocol, equipment, technician, or curve-fit program.

Case Study

A participant reported that his laboratory's results had been negatively biased in the UK External Quality Assessment Scheme between October and February. (Cause – the customer's laboratory was poorly heated – sometimes the temperature was as low as 9 °C, and the assay, which had a room temperature incubation, was temperature-sensitive, giving rise to bias below about 16 °C.)

BIAS IN PROFICIENCY TESTING (EXTERNAL QUALITY ASSESSMENT) SCHEME OF ONE METHOD FROM ALL-LABORATORY MEAN OR REFERENCE METHOD

Possible Causes

Error in calibration of calibrator lot – the start of bias would have coincided with the introduction of the new calibrator lot and this should be apparent from laboratory control data.

Reagent lot at fault – the bias would have coincided with the introduction of a new reagent lot.

Incorrect standardization of assay during development – if this is the cause, the recovery of the assay will also be poor (*see* POOR RECOVERY).

Manufacturer's reference standards are unstable – this will be characterized by a stepwise increase in the bias from other methods with each new lot of kit calibrators, or manufacturer-generated master calibration curve.

Presence of cross-reacting substances in samples – immunoassays may give higher values than methods such as gas chromatography–mass spectrometry (GC–MS), which is used as a reference method, because of the presence of substances that cross-react in the assay but are not detected by GC–MS.

Specificity of immunoassay different to other products – there may be an indication of this in package insert cross-reactivity data. The manufacturer should be able to provide more information. In assays for proteins, international standards sometimes contain proteins with slightly different structures to those of the proteins that occur in patient samples. If this is the explanation, various products will give different results for patient or control samples, but all will give good recovery with the international standard.

Proficiency testing/external quality assessment (PT/EQA) scheme samples unstable – unlikely if samples are freeze-dried, but liquid samples, if used, may deteriorate in transit. If this occurs, there will be a wide spread of values among participants. Because of the varying specificity of different products, some may be more affected by sample instability than others. Check the package insert or ask the manufacturer for information on sample stability.

Change in collection method, treatment, or origin of PT/EQA scheme samples – may also affect certain other methods. Check with the scheme organizer.

Use of different units or international standards – check the PT/EQA scheme printouts for the correct units.

Apparent bias due to another manufacturer's kit, which is itself incorrectly standardized – if the majority of PT/EQA scheme participants use one method, the all-laboratory mean will be biased towards that method. To check which method is correct, data from recovery experiments are needed. Note that two different products may both be correctly calibrated against the same international standard but still give different results. *See* SPECIFICITY OF IMMUNOASSAY DIFFERENT TO OTHER PRODUCTS above.

Suggested Course of Action

1. Analyze the data in several PT/EQA scheme printouts and assess whether the bias is consistent or has started recently.
2. If the bias is recent look for coincident changes in control values on the internal QC charts. If such a change is evident, identify the reagent lots involved.
3. If the PT/EQA scheme has sent the same sample out more than once, check the values obtained for any samples distributed before and after the bias occurred.
4. Ask the manufacturer for advice.
5. In some countries, the organizers of the scheme provide advice about the performance of individual methods.

Checklist of Useful Information

It may be useful to have the following information available:

- PT/EQA scheme printouts covering the period before and after the bias commenced;
- QC charts for the reagent lots concerned.

Case Study

The users of a particular assay for luteinizing hormone (LH) complained that the method was slowly becoming positively biased from the all-laboratory mean in a national proficiency testing/external quality assessment scheme. (Cause – a new immunometric assay for LH had been launched that gradually became the most commonly used method in that country. This very specific method, which used dual monoclonal antibodies, gave lower values than any other methods in the scheme, and the all-laboratory mean changed downwards as the method became more established.)

POOR WITHIN-ASSAY PRECISION

Possible Causes

Reagent lot at fault – the change in precision should coincide with a change in the reagent lot number. This will only be noticeable if duplicates are run. Check whether the signal levels, control values, and curve shape are similar to those in the previous assays and in the package insert.

Poor reagent design or optimization – if this is the cause other users will have experienced poor precision and the manufacturer is likely to have received similar comments already.

Variation of coated beads, wells, or tubes – this is difficult to differentiate from other possible causes. It can only be confirmed by running replicates from the affected lot alongside a good lot at the same time. Equipment should be thoroughly checked first. Ask the manufacturer how much variability is allowed by their specifications.

Premature deterioration of reagent during shelf-life or exposure of reagent to extremes of temperature during transport – if reagents are deteriorating during their shelf-life the imprecision will increase as the reagents near expiry. If imprecision is due to extreme temperature during shipment, other reagents in the same shipment may also be affected. If it is likely that extreme temperatures were encountered, ask the manufacturer whether the reagent is sensitive to freezing or excessive warming.

Flat spots or assay insensitivity at extremes of calibration curve – if the imprecision is worse at either low or high concentrations, check whether the slope of the calibration curve is sufficient to give good discrimination and ask the manufacturer whether the precision is typical.

Incomplete mixing of sample – if controls are in good agreement between duplicates but some patient samples are not, try vortex mixing or centrifuging the samples before dispensing.

Dirty, shiny, or siliconized polystyrene assay tubes or cuvettes, or use of polypropylene tubes with microparticle or liquid-phase assays – test by using an alternative source of tubes.

Reagents not allowed to reach room temperature before use – ensure that reagents are allowed to reach room temperature before use. Large volumes (e.g. >100 mL) may take at least an hour to reach ambient temperature.

Incomplete mixing of reagent (if freeze-dried or suspension) – check the appearance of the reagents for uniform mixing. Follow the instructions in the package insert.

Pipette, dispenser, or probe not primed – check the procedure.

Defective pipettes or automatic sampler/dispenser; dirty pipette tips; contaminated sample or reagent probe; poor connection between pipette and pipette tips; poor connection between probe and tubing; leaking gasket; carryover of analyte from high-concentration sample to low-concentration sample; carryover of reagent by reagent-dispensing probe; splashing of reagents; tubing constricted – test by substituting different dispensing equipment if possible.

Use of reduced volumes or other difference from the recommended protocol – major differences between your protocol and that in the package insert should be tested for their effects on precision.

Insufficient or overaggressive mixing of sample and reagents – look for evidence of faulty operation, e.g. splashing.

Use of air incubator instead of water-bath (except purpose-designed incubators) – large air incubators are unsuitable for many immunoassays. Test by carrying out an assay in a water-bath.

Insufficient centrifugation (time or *g* force) or excessive centrifuge braking – check the *g* force from the speed and radius of the centrifuge rotor using a nomogram. Try using the centrifuge with the brake off.

Faulty magnetic separator or magnet not in close enough proximity to paramagnetic particles during separation – check the separation module of analyzer or method of use of manual separation device.

Faulty washer – characterized by poor precision at low concentrations in immunometric assays and high concentrations in competitive assays. Clean the washer thoroughly, using wire to clear the dispensing nozzles and flushing through with copious amounts of distilled water. Test the washer by dispensing intensely colored dye into the tubes or wells and monitoring the removal of the dye during washing.

Contamination of reusable reagent or distilled water container by dirt, disinfectant, or detergent – check appearance and cleaning procedures.

Interference due to sample collection device – if this is the cause, poor precision will be isolated to individual samples.

Poor decanting technique (manual separations) or insufficient draining time – if this is the cause, imprecision may vary between technicians.

Overaggressive tapping of microparticle-based assays after decanting (manual separations) – check the appearance of the tubes for slipping precipitates.

Coated tubes or tubes containing polyethylene glycol (PEG) or magnetizable cellulose separation systems not tapped hard enough – check the tubes for residual droplets of moisture.

Tubes or microtiter plates are being decanted, placed the right way up, and then reinverted to drain (manual separations) – check the technique used by the technician.

Contamination of outside of tube, well, or cuvette by tracer or conjugate (especially immunometric assays) – this is characterized by individual replicates with higher than expected signal levels which are normalized if the tube or well is wiped with a tissue and recounted.

Contamination by dust or fiber particles (especially enzyme assays at stage of signal generation) – check environment. If tissues are used for tapping (manual separations), check that they are not shedding fibers that interfere with the assay. Some latex gloves shed material that can interfere with enzyme-based signal generation systems.

Faulty or contaminated signal reader – check by reading the signal in an empty tube (or run a zero concentration TSH (thyrotropin) calibrator and check how close the signal is to zero).

Insufficient counting or signal reading time (manual counters or signal readers) – check whether sufficient signal is being generated.

Use of expired or poor-quality reagents (due to incorrect storage conditions, faulty refrigerator, or contamination) – check that the storage temperature is correct.

Poor technician technique – if imprecision is technician-dependent check the technique being used.

Suggested Course of Action

Within-assay imprecision is most likely to be noticed in manual assays where duplicates are run. For this reason the suggested course of action below is targeted at manual assay users.

1. Carefully check the assay data from several assays for patterns. If precision has deteriorated, when did the problem start? If duplicates for calibrators and controls are good and poor replicates only occur with patient samples, the problem may be due to sample preparation. Check the origin of the samples involved.
2. Check the position of the poor duplicates. For example, poor duplicates in the corners of microtiter plates may indicate uneven warming in an incubator, and poor duplicates at the beginning of an assay may indicate insufficient priming of pipettes or probes.
3. If poor precision only occurs in low-concentration samples in immunometric assays, the cause may be poor washing, insufficient draining, contamination with tracer or signal reagent, poor curve shape, or carryover. If the first replicate of a low-concentration sample is elevated following a high-concentration sample, the cause is likely to be carryover.
4. Check the signal levels and curve shape to see if they are normal. If they are not, check the expiry date, appearance, and odor of the reagents. If there have been similar reports before, analyze them to see if the problem is the same.
5. Check whether the precision varies between technicians, indicating that technique is involved.
6. Try using a completely different set of equipment, if possible, to test for equipment problems. If alternatives are not available, clean and check the equipment according to the manufacturer's instructions.
7. Identify any differences between your protocol and that in the package insert. Test whether the protocol is involved by repeating an assay with poor precision using the primary protocol in the package insert.
8. Consult the manufacturer for advice. Ask if any other customers have reported similar imprecision.

Checklist of Useful Information

It may be useful to have the following information available:

- kit and reagent lot numbers;
- when the problem started;
- details of any other coincident changes in immunoassay performance, e.g. change in signal levels, binding at zero dose, calibration curve shape, ED_{50}, or NSB;
- assay printout(s);
- details of any differences between the protocol used and the protocol in the package insert;
- details of all equipment used;
- information on any coincident change in the protocol, equipment, or technician.

Case Study

Staff in a laboratory noticed a gradually increasing incidence of poor within-assay precision with an immunometric assay based on coated wells. (Cause – a sealing gasket in the washer was worn and needed replacement.)

POOR BETWEEN-ASSAY PRECISION OR INDIVIDUAL CONTROL FAILURE

Possible Causes

Poor assay design or optimization – if this is the cause other users will have experienced poor precision and the manufacturer is likely to have received similar comments from other users.

Poor instrumentation design or optimization (dedicated systems) – multiple analytes may be affected. Other users will have experienced poor precision.

Poor controls – test by running a group of patient samples in several assays or using alternative controls, to assess precision independently.

Lot-to-lot variation – this can be detected on QC charts if more than one assay is carried out on each lot. Cusum charts show up lot-to-lot variation particularly clearly (*see* LABORATORY QUALITY ASSURANCE). The precision within a lot will be significantly better than across lots.

Error in stored calibration curve or manufacturer's calibration curve – the between-assay precision of control values should be acceptable between calibrations, but show significant changes coincident with new calibrations.

Change in values across shelf-life – note the reagent lot numbers and assay dates on the QC charts and look for trends with the age of the reagents, and clear differences when an old reagent lot is replaced by a new one.

Premature deterioration of reagent during shelf-life or exposure of reagent to extremes of temperature during transport – if reagents are deteriorating during their shelf-life the imprecision will increase as the reagents near expiry. If imprecision is due to extreme temperature during shipment, other reagents in the same shipment may also be affected. If it is likely that extreme temperatures were encountered, ask the manufacturer whether the reagent is sensitive to freezing or heating.

Variation of coated beads, wells, or tubes – this is difficult to differentiate from other possible causes. It can only be confirmed by running replicates from the affected lot alongside a good lot in the same assay. Equipment should be thoroughly checked first. Ask the manufacturer how much variability is allowed by their specifications.

Flat spots or assay insensitivity at extremes of calibration curve – if imprecision is worse at low or high concentrations, check whether the slope of the calibration curve is sufficient to give good discrimination and ask the manufacturer whether the precision is typical.

Use of reduced volumes or other difference from the recommended protocol – major differences between your protocol and that in the package insert should be tested for their effects on precision.

Variation in incubation temperature – check equipment and record the temperature daily if possible. If more than one unit of equipment is being used, check that each has the same incubation temperature.

Incubator or analyzer not allowed sufficient time to warm up – check manufacturer's instructions.

Dirty, shiny, or siliconized polystyrene assay tubes or cuvettes, or use of polypropylene tubes with microparticle or liquid-phase assays – test by using an alternative source of polystyrene assay tubes or cuvettes.

Reagents not allowed to reach room temperature before use – ensure that reagents are allowed to reach room temperature before use. Large volumes (e.g. >100 mL) may take at least an hour to reach ambient temperature.

Reagent heating plate ineffective; poor contact between reagent and heating plate; insufficient time allowed for reagent to be heated – some analyzers heat the reagents on board. Check the temperature of the reagent.

Incomplete mixing of reagent before use (if freeze-dried or suspension) – check appearance of reagents for uniform mixing. Check that magnetic particle suspensions are adequately mixed, particularly when kept on board automated analyzers.

Pipette, dispenser, or probe not primed – check procedure.

Defective pipettes or automatic sampler/dispenser; dirty pipette tips; contaminated sample or reagent probe; poor connection between pipette and pipette tips; poor connection between probe and tubing; leaking gasket; carryover of analyte from high-concentration sample to low-concentration sample; carryover of reagent by reagent-dispensing probe; splashing of reagents; tubing constricted – test by substituting different dispensing equipment if possible. Multiple assays may be affected.

Dilution of control by water in tubing – check the equipment.

Insufficient or overaggressive mixing of sample and reagents – look for evidence of faulty operation, e.g. splashing.

Use of air incubator instead of water-bath (except purpose-designed incubators) – large air incubators are unsuitable for many immunoassays. Test by carrying out assay incubations in a water-bath or purpose-designed microtitration plate incubator.

Change in laboratory temperature – in extreme cases, cold or hot weather can affect assays with room temperature incubations if the laboratory is insufficiently heated or cooled. Automated analyzers may also be affected by extreme laboratory temperatures and may require recalibration more often. If this occurs, changes in controls will coincide with changes in the ambient temperature.

Insufficient centrifugation (time or *g* force) or excessive centrifuge braking – check the *g* force from the speed and radius of the centrifuge rotor using a nomogram. Try using the centrifuge with the brake off.

Faulty magnetic separator or magnet not in close enough proximity to paramagnetic particles during

separation – check the separation module of analyzer or method of use of manual separation device.

Faulty washer – characterized by poor precision at low concentrations in immunometric assays and high concentrations in competitive assays. Clean the washer thoroughly, using wire to clear the dispensing nozzles and flushing through with copious amounts of distilled water. Test the washer by dispensing intensely colored dye into the tubes or wells and monitoring the removal of the dye during washing.

Contamination of reusable reagent or distilled water container by dirt, disinfectant, or detergent – check appearance and cleaning procedures.

Interference due to sample collection device – if this is the cause, poor precision will be isolated to individual samples.

Poor decanting technique or insufficient draining time (manual separations) – if this is the cause, imprecision may vary between technicians.

Overaggressive tapping of microparticle-based assays after decanting (manual separations) – check the appearance of the tubes for slipping precipitates.

Coated tubes or tubes containing PEG or magnetizable cellulose separation systems not tapped hard enough – check the tubes for residual droplets of moisture.

Tubes or microtiter plates are being decanted, placed the right way up, and then reinverted (manual separations) – check the technique used by the technician.

Contamination of outside of tube, well, or cuvette by tracer or conjugate (especially immunometric assays) – this is characterized by individual replicates with higher than expected signal levels which are normalized if the tube or well is wiped with a tissue and recounted.

Contamination by dust or fiber particles (especially enzyme assays at stage of signal generation) – check the environment. If tissues are used for tapping inverted tubes or microtitration plates, check that they are not shedding any fibers that interfere with the assay.

Faulty or contaminated signal reader – check by reading the signal in an empty tube (or run a zero concentration TSH (thyrotropin) calibrator and check how close the signal is to zero).

Insufficient counting or signal reading time (manual counters or signal readers) – check whether sufficient signal is being generated.

Time-critical stages, such as the addition of reaction stopping reagent, not being timed accurately (manual assays) – check times being used if a time-critical stage exists in the protocol (see the package insert).

Use of expired or poor-quality reagents (due to incorrect storage conditions, faulty refrigerator, or contamination) – check that the storage temperature is correct.

Reagents deteriorating while in the analyzer – make a note on the control charts each time a new pack is loaded onto the analyzer. A gradual change in the control values will be apparent from the first use of the pack until the last, followed by a return to the initial values with a new pack. Check that the manufacturer's instructions are being followed.

Poor technician technique (manual assays) – if imprecision appears to be technician-dependent check the technique being used.

Insufficient reagent in reagent container – check the reagent containers.

Defective data in reagent supply information – this could be due either to detector error or to the reagent container not being full when loaded onto the analyzer. Check the reagent containers.

Sample or reagent in incorrect position – in some multianalyte analyzers the test carried out on the sample (or control) is determined by the position relative to the position of the reagents. Check the sample and reagent positions.

Use of poor curve-fit method, e.g. linear interpolation without transformation of the data, or use of curve-fit method that is not suitable for the reagents – linear interpolation is unsuitable for most immunoassays unless the data are first linearized by using mathematical functions. To check the quality of curve-fit programs other than those that use linear interpolation, compare the *actual* and *fitted* values for the calibrators on the assay printouts.

Contamination of reagents during first use – replace the reagents.

Consistent use of reagents near to, or after, expiry – compare the assay dates with the reagent lot expiry dates.

Variation between different technicians – compare the control means and coefficients of variation (CV) for different technicians.

Suggested Course of Action

1. Calculate the between-assay CVs for all the controls and identify whether all, or some, of the controls are affected. If only some controls are affected, does the poor precision only occur at low or high

concentrations? If precision has deteriorated, approximately when did the problem start? Look for significant differences between new and old reagents. Check the within-assay precision by running replicates of the calibrators and the controls. Poor within-assay precision can be the sole cause of apparent poor between-assay CVs (*see* POOR WITHIN-ASSAY PRECISION).
2. Check the quality of the curve-fitting program (by comparing *fitted* and *actual* calibrator values).
3. If poor precision only occurs in low-concentration samples in immunometric assays the cause may be poor washing, insufficient draining, contamination with tracer or signal reagent, poor curve shape, or carryover. If the first replicate of a low-concentration sample is elevated following a high-concentration sample, the cause is likely to be carryover.
4. Check the signal level and the calibration curve shape to see if they are normal. If they are not, check the expiry date, appearance, and odor of the reagents. If there have been similar reports before, analyze them to see if the problem is the same.
5. Check whether the precision varies between technicians, indicating that technique is involved. For automated analyzers, check data for patterns, e.g. more extreme bias from norm when analyzer first used, e.g. after a weekend.
6. Try using a completely different set of equipment, if possible, to test for equipment problems. If alternatives are not available, clean and check the equipment according to the manufacturer's instructions.
7. Identify any differences between your protocol and that in the package insert. Test whether the protocol is involved by repeating assays with poor precision using the primary protocol in the package insert.
8. Consult the manufacturer for advice. Ask if any other customers have reported similar imprecision.

Checklist of Useful Information

It may be useful to have the following information available:

- reagent lot numbers;
- when the problem started;
- details of any other coincident changes in reagent performance, e.g. change in binding at zero dose, calibration curve shape, ED_{50}, or NSB;
- QC charts;
- assay printout(s);
- details of differences between the protocol used and the protocol in the package insert;
- details of all the equipment used;
- information on any coincident change in the protocol, equipment, or technician.

Case Study

A user of an adrenocorticotropic hormone (ACTH) immunoassay found that she was not achieving the between-assay CVs quoted in the package insert. (Cause – the laboratory received reagents on a standing order. Because fewer samples had been received than expected, there was an excess of reagents that were used near to, or after, their expiry date).

ASSAY DRIFT

There are two types of assay drift. In manual assays consisting of calibrators, samples, and controls, drift is characterized by changes in signal level at a given concentration across the assay. In automated random-access systems, assay drift describes changes in signal at a fixed concentration across any length of time between two consecutive calibrations.

Possible Causes

Poor immunoassay design or optimization – if this is the cause, other users may have reported drift or poor precision to the manufacturer.

Reagent lot at fault – this can be checked by running controls at the beginning and end of a batch run (manual assays), or on several occasions between calibrations (automated systems) with the affected and an unaffected reagent lot. For investigations into assay drift in manual assays, ensure that the assay length and addition times are consistent.

Variation between coated tubes, wells, or beads – if this is the cause, poor within-assay precision will be apparent if the controls are run in duplicate on multiple occasions.

Excessive number of tubes run in one assay (manual assays) – record the times taken to add the samples and reagents and check with the manufacturer.

Excessive time taken to add samples (manual assays) – record the time taken to add the samples and check with the manufacturer.

Excessive time taken to add reagents (manual assays) – record the time taken to add the reagents and check with the manufacturer.

Reaction stopping reagent added too quickly or slowly (manual assays) – check the timing in relation to the time taken to add the substrate.

Technician interrupted during pipetting stage (manual assays) – check with the technician.

Changes from the recommended protocol – major differences between your protocol and that in the package insert should be tested for their effects on drift.

Reagents not allowed to reach room temperature before use – ensure that the reagents are allowed to reach room temperature before use. Large volumes (e.g. >100 mL) may take at least an hour to reach ambient temperature.

Incomplete mixing of reconstituted reagents before use – check the mixing time, method, and effectiveness.

Unstable reagents – compare results obtained using fresh and stored reagents.

Low incubation temperature – check the temperature.

Incubator or analyzer not allowed sufficient time to warm up – check manufacturer's instructions.

Fluctuations in incubator temperature between calibrations – some systems have temperature sensor information available from the user interface. Otherwise, checking incubator temperature consistency is a service operation.

Incomplete mixing of separation reagent – check the mixing time, method, and effectiveness.

Variation in draining time after decantation within one assay – check the exact method used by the technician.

Defective pipettes or automatic sampler/dispenser; dirty pipette tips; contaminated sample or reagent probe; poor connection between pipette and pipette tips; poor connection between probe and tubing; leaking gasket; carryover of analyte from high-concentration sample to low-concentration sample; carryover of reagent by reagent-dispensing probe; splashing of reagents; tubing constricted – test by substituting different dispensing equipment. Multiple assays may be affected.

Use of air incubator instead of water-bath (except purpose-designed incubators) – large air incubators are unsuitable for many immunoassays. Test by carrying out an assay in a water-bath.

Faulty incubator – test by substituting a different incubator or using a water-bath.

Difference in water-bath temperature if more than one water-bath used – check the temperature.

Inconsistent separation, e.g. use of two different centrifuges for one assay – check with the technician.

Faulty washer – clean washer thoroughly, using wire to clear the dispensing nozzles and flushing through with copious amounts of distilled water. Test the washer by dispensing intensely colored dye into the tubes or wells and monitoring the removal of the dye during washing.

Faulty or contaminated signal reader – check by reading the signal in an empty tube (or run a zero concentration TSH (thyrotropin) calibrator in a TSH test and check how close the signal is to zero).

Suggested Course of Action

1. Analyze the data from more than one assay for consistent bias and check the within-assay precision if duplicates have been run.
2. Check whether the magnitude of the drift varies between technicians, indicating that technique is involved.
3. Plot signal level for controls against time using both a single-opened reagent pack and fresh reagent packs between calibrations (if drift occurs on random-access equipment). This experiment will show how stable reagents are in opened packs (using fresh packs as an experimental control). However, if signal also changes using fresh packs, it indicates either that the equipment is unstable or that the reagents are unstable within an unopened pack.
4. Try using a completely different set of equipment, if possible, to test for equipment problems. If alternatives are not available, clean and check the equipment according to the manufacturer's instructions.
5. Identify differences between your protocol and that in the package insert, including the times allowed for addition of each reagent. Test whether the protocol is involved by repeating assays in which drift is detected using the primary protocol in the package insert.
6. Consult the manufacturer for advice.

Checklist of Useful Information

It may be useful to have the following information available:

- reagent lot numbers;
- when the problem started;
- details of any other coincident changes in reagent performance, e.g. change in binding at zero dose, curve shape, ED_{50}, or NSB;
- QC charts;
- assay printout(s) or relevant calibration data;
- details of any differences between the protocol used and the protocol in the package insert;
- times taken to add the samples and reagents;
- details of all the equipment used;
- information on any coincident change in the protocol, equipment, or technician.

Case Study

A laboratory supervisor noticed that controls placed at the end of an assay of 200 tubes gave approximately 10% lower values than the same controls at the beginning of the assay. (Cause – the technician took longer to add the separation system than the maximum time stated in the package insert.)

LOW SIGNAL LEVEL

Possible Causes

Low concentration or deterioration of label; problem with tracer formulation, e.g. wrong pH, inactive cofactor – if the curve shape is normal, but the signal level is low this should be suspected. Some tracers can be checked independently (e.g. count rate of radioactive label). Try substitution of tracer from another lot. Return suspect enzyme-labeled conjugates to the manufacturer for checking.

Inactive enzyme or substrate or presence of enzyme inhibitor in enzyme assays – this may be characterized by a normal curve shape with a low signal level.

Defective signal generation reagent – characterized by a low signal and possibly by a non-linear calibration curve in immunometric assays. Similar effects may be observed in assays for different analytes when the same signal reagent is used.

Antigen concentration of tracer (in competitive assay) too high – this should be suspected if dilution of the tracer results in a more normal calibration curve.

Low specific activity in tracer – this is normally checked by the manufacturer.

Labeled antibody concentration (in immunometric assay) too low – this causes a non-linear and flattened calibration curve at higher concentrations.

Antiserum concentration too low – in an immunometric assay this causes a non-linear and flattened calibration curve at higher concentrations. In a competitive assay it causes low binding.

Defective separation system – likely to be accompanied by poor within-assay precision.

Premature deterioration of reagents within shelf-life or exposure of reagent to extreme temperature during transport – if reagents are deteriorating during their shelf-life the precision and signal level will deteriorate as the reagents near expiry. If the low signal is due to extreme temperature during shipment, other reagents in the same shipment may also be affected. If it is likely that extreme temperatures were encountered, ask the manufacturer whether the reagent is sensitive to freezing or heating. Check by replacing with new reagent from a different shipment.

Cross-contamination of one reagent by another – certain combinations of reagents can have an unexpectedly large effect on the signal, even if only a trace of one of the reagents is present. For example, a very small quantity of donkey anti-sheep second antibody reagent could reduce the binding of a sheep antiserum reagent. Check that the dispensing equipment is cleaned thoroughly and replace the reagents.

Unstable pellets in precipitation assays – these may be visible. Within-assay precision will be poor.

Incorrect storage of reagents or faulty refrigerator – check that the storage temperature is correct.

Reagents unstable while in analyzer – check the manufacturer's instructions. Record the signal level of a control at regular intervals during the life of the reagent pack to see whether there is a trend. Make sure that the laboratory temperature stays within the range allowed by the manufacturer.

Reduced reagent volumes used – check by using the volumes specified in the package insert.

Reagent at end of shelf-life – check using fresh reagents.

Contamination of reusable reagent or distilled water container by dirt, disinfectant, or detergent – check appearance and cleaning procedures.

Calibrators too concentrated in a competitive assay or too dilute in an immunometric assay due to, e.g. incorrect reconstitution volume or calibrator deterioration – check by replacing the calibrators. Controls will be biased.

Incorrect reconstitution of reagents or omission of dilution step, if required – check the protocol.

Accidental omission of tracer or addition of wrong tracer – check the protocol.

Accidental omission of antiserum or addition of wrong antiserum – check the protocol.

Accidental omission of separation system – check the protocol.

Changes from the recommended protocol (e.g. change in incubation time or temperature) or accidental incubation at too low or high a temperature (e.g. incubator not switched on) – major differences between your protocol and that in the package insert should be tested for their effects on precision.

Incubator or analyzer not allowed sufficient time to warm up – check manufacturer's instructions.

Insufficient centrifugation – precision will be poor. Check the protocol and the centrifuge.

Faulty magnet or magnet not in close enough proximity to paramagnetic particles during separation – check the separation module of analyzer or method of use of manual separation device.

Insufficient counting time (radioactive labels) – check counts per minute. Use the counting time appropriate to the assay.

Low counter efficiency or defective signal reader – check using a test signal source or an alternative assay.

Counter window setting incorrect for isotope used (radioactive tracers) – check the counter user manual and the settings.

Spectrophotometer set at incorrect wavelength – check the spectrophotometer settings.

Dirt on optics – check and clean if necessary.

Signal not measured within time limits specified in protocol (non-radioactive assays) – check the protocol.

Suggested Course of Action

1. Confirm that low signal levels only occur with one assay (otherwise the analyzer or assay equipment is at fault). Analyze the data from more than one occasion for the consistency of the low signal level and check the within-assay precision if duplicates have been run. Compare signal levels and curve shape with data from previous assays, prior to the occurrence of the problem. Low signal levels with a normal curve shape indicate a problem with the label or signal generation reagents. A change in curve shape indicates that another reagent may be the cause or that the binding moiety of the label or conjugate has changed. (See sections on changes in binding below).
2. Check whether there is any relationship between low counts or signal and other factors, such as the technician carrying out the assay or the age of the reagents.
3. Identify differences between your protocol and that in the package insert (manual assays). Test whether the protocol is involved by repeating assays with low counts or signal level using the primary protocol in the package insert.
4. Consult the manufacturer for advice, asking for the expected signal levels for the age of the reagents used and whether other users have reported similar findings.

Checklist of Useful Information

It may be useful to have the following information available:

- reagent lot numbers and the dates the assays were performed;
- when the problem started;
- details of any other coincident changes in performance, e.g. within-assay precision;
- QC charts;
- assay printout(s) or relevant calibration data;
- details of differences between the protocol used and the protocol in the package insert;
- details of all the equipment used;
- information on any coincident change in the protocol, equipment, or technician.

Case Study

Staff in a laboratory noticed that signal levels were consistently low in a specific lot of reagents for an enzyme-based immunoassay. However, the shape of the calibration curve seemed normal, and controls were in range, although precision was worse than normal. (Cause – the enzyme in the conjugate reagent had been partially inactivated, possibly by extreme temperature conditions during shipment.)

LOW BINDING IN A COMPETITIVE IMMUNOASSAY

The **% binding** is the percentage of total tracer available that is bound at a particular concentration, on completion of the assay protocol. This parameter is only apparent if the maximum signal level from all the tracers (the **total signal**) can be measured. In a radioimmunoassay (RIA) this can be measured by counting the radioactivity due to the tracer in the absence of sample or other reagents. For most assay systems this is difficult to measure or estimate.

In order to investigate any unexpected change in signal level where total signal cannot be measured, the following simple procedure is recommended. If signal level reduces by a constant percentage across the concentration range, *see* LOW SIGNAL LEVEL above. In other cases of signal levels increasing or decreasing, use the appropriate section from the following: LOW BINDING IN A COMPETITIVE IMMUNOASSAY, LOW BINDING IN AN IMMUNOMETRIC ASSAY, INCREASE IN BINDING IN A COMPETITIVE IMMUNOASSAY.

Possible Causes

Poor reagent design or optimization – compare the calibration curve with the typical curve shown in the package insert.

Premature deterioration of reagents within shelf-life or exposure of reagent to extreme temperature during transport – if reagents are deteriorating during their shelf-life the precision and signal level will deteriorate as the reagents near expiry. If low binding is due to extreme temperature during shipment, other reagents in the same shipment may also be affected. If it is likely that extreme temperatures were encountered, ask the manufacturer whether the reagent is sensitive to freezing or heating. Check by replacing with new reagent from a different shipment.

Concentration of antigen in tracer too high or other fault in tracer reagent – if the antigen concentration is too high, dilution may result in a more normal curve shape. Other faults in the tracer can be confirmed by replacing the tracer with fresh material.

Specific activity of tracer too low – this is normally checked by the manufacturer.

Inactive enzyme or presence of enzyme inhibitor in enzyme assays – this may be characterized by a normal curve shape with a low signal level.

Concentration of antibody too low or faulty antiserum reagent – if the antibody concentration is too low, using twice the quantity may result in a more normal curve shape. Other faults in the antiserum reagent may be confirmed by replacing with fresh material.

Faulty microparticles, beads, coated wells, or coated tubes – likely also to result in poor precision. Test by replacing with fresh material.

Concentration of separation reagent too low (if used) – if the separation reagent concentration is too low, using twice the quantity may result in a more normal curve shape. (This technique may be useful in diagnosing the cause of the problem, but it should not be used to correct the fault in clinical practice.) Other faults in the separation reagent may be confirmed by replacing with fresh material.

Faulty separation reagent (if used) – likely to be accompanied by poor precision. Test by replacing with fresh material.

Cross-contamination of one reagent by another – certain combinations of reagents can have an unexpectedly large effect on the signal, even if only a trace of one of the reagents is present. For example, a very small quantity of donkey anti-sheep second antibody reagent could reduce the binding of a sheep antiserum reagent. Check that the dispensing equipment is cleaned thoroughly and replace the reagents.

Unstable pellets in precipitation assays – these may be visible. Within-assay precision will be poor.

Defective signal generation reagent – similar effects may be observed in assays for different analytes when the same signal generation reagent is used. Test by replacing with fresh material.

Use of reagent near to or at expiry – check using fresh reagents.

Use of reagent after expiry date – check using fresh reagents.

Incorrect storage of reagents or faulty refrigerator – check that the storage temperature is correct.

Reagents unstable while in analyzer – check the manufacturer's instructions. Record the signal level of a control at regular intervals during the life of the reagent pack to see whether there is a trend. Make sure that the laboratory temperature stays within the range allowed by the manufacturer.

Contamination of reusable reagent or distilled water container by dirt, disinfectant, or detergent – check appearance and cleaning procedures.

Tracer reconstitution volume too low – if the bottle of reconstituted tracer is still available, check the reconstitution volume in the package insert, the number of tests carried out, and the remaining tracer volume to test whether the correct amount is left. (Allow for the use of tracer for pre-washing pipette tips or probes.)

Use of undiluted tracer if tracer reagent requires dilution – if the tracer used in the assay is still available, check the instructions in the package insert, the number of tests carried out, the remaining tracer concentrate volume, and the remaining diluted tracer volume to test whether the correct amounts are left. (Allow for the use of tracer for pre-washing pipette tips or probes.)

Volume of tracer dispensed too high – check the protocol and the dispensing equipment.

Antiserum reagent reconstitution volume too high – if the bottle of reconstituted antiserum is still available, check the reconstitution volume in the package insert, the number of tests carried out, and the remaining antiserum reagent volume to test whether the correct amount is left. (Allow for the use of antiserum for pre-washing pipette tips or probes.)

Inadequate mixing of antiserum reagent suspension before dispensing – check the protocol and the mixing equipment.

Volume of antiserum reagent dispensed too low or antiserum not dispensed – check the protocol and the dispensing equipment. Also check the amount of antiserum reagent left over after the assay.

Calibrator reconstitution volume too low – check the reconstitution volume in the package insert and the amount left after the assay is completed. Check the measuring equipment. Controls will be biased low.

Sample volume too high – check the protocol and the dispensing equipment.

Reduced incubation time – check the protocol used.

Incubation temperature too low or high – check the temperature setting used. Check that the incubation equipment was allowed sufficient time to warm up.

Inadequate mixing of separation reagent suspension before dispensing – check the protocol and the mixing equipment. Check that the separation reagent was mixed immediately before use, if this is required.

Volume of separation reagent dispensed too low or reagent not dispensed – check the protocol and the volume of separation reagent left over.

Insufficient centrifugation time – precision will be poor. Check the protocol.

Insufficient centrifugation *g* force – precision will be poor. Check the protocol.

Faulty magnetic separation – check the separation module of the analyzer or the method of use of a manual separation device.

Loss of precipitate on decanting because of over-vigorous tapping (manual assays) – precision will also be poor and slipping pellets may be visible on the tubes. Check the technique.

Defective pipettes or automatic sampler/dispenser – test by substituting different dispensing equipment.

Faulty incubator or water-bath (temperature too low or high) – test the temperature maintained by the equipment if possible.

Faulty centrifuge (*g* force too low or excessive braking) – check the centrifuge.

Faulty counter or signal reader – check using a test signal source or an alternative assay.

Suggested Course of Action

1. Analyze the data from more than one assay to evaluate the consistency of the low binding and check the within-assay precision if duplicates have been run. Try and assess approximately when the change occurred, or if there was a gradual change.
2. Check whether there is any relationship between low binding and other factors, such as the lot or age of the reagents, or the technician carrying out the assay.
3. Identify differences between your protocol and that in the package insert. Test whether the protocol is involved by repeating an assay with low binding using the primary protocol in the package insert.
4. Consult the manufacturer for advice, asking for the expected signal or binding for the age of the reagents used and whether other laboratories have reported similar findings.

Checklist of Useful Information

It may be useful to have the following information available:

- reagent lot numbers and the dates the assays were performed;
- when the problem started;
- details of any other coincident changes in performance, e.g. within-assay precision;
- QC charts;
- assay printout(s) or relevant calibration data;
- details of differences between the protocol used and the protocol in the package insert;
- details of all the equipment used;
- information on any coincident change in the protocol, equipment, or technician.

Case Study

A customer using a radioimmunoassay for parathyroid hormone (PTH) reported low percent bound at zero concentration ($\%B_0$) on several occasions. (Cause – on each occasion the reagents were at or near expiry. This assay was suffering from low binding at expiry due to radiolysis of the ^{125}I-PTH tracer.)

LOW BINDING IN AN IMMUNOMETRIC ASSAY

The **% binding** is the percentage of total tracer available that is bound at a particular concentration, on completion of the assay protocol. This parameter is only apparent if the signal level that arises from all the tracers (the **total signal**) can be measured. In a radioimmunoassay this can be measured by counting the radioactivity due to the tracer in the absence of sample or other reagents. For most assay systems this is difficult to measure or estimate.

In order to investigate any unexpected change in signal level where the total signal cannot be measured, the following simple procedure is recommended. If the signal level reduces by a constant percentage across the concentration range, *see* LOW SIGNAL LEVEL above. In all other cases, use the appropriate section from the following: LOW BINDING IN A COMPETITIVE IMMUNOASSAY, LOW BINDING IN AN IMMUNOMETRIC ASSAY, INCREASE IN BINDING IN A COMPETITIVE IMMUNOASSAY.

Possible Causes

Poor product design or optimization – compare the calibration curve with the typical curve shown in the package insert.

Premature deterioration of reagents within the shelf-life or exposure to extreme temperature during transport – if reagents are deteriorating during their shelf-life the precision and signal level will deteriorate as the reagents near expiry. If low binding is due to extreme temperature during shipment, other reagents in the same shipment may also be affected. If it is likely that extreme temperatures were encountered, ask the manufacturer whether the reagent is sensitive to freezing or heating. Check by replacing with new reagent from a different shipment.

Concentration of capture antibody too low or absence of capture antibody – a low concentration causes a non-linear calibration curve, flattening off at higher concentrations. Absence of capture antibody results in a lack of binding or a constant very low level of binding for all samples.

Concentration of labeled antibody too low – a low concentration causes a non-linear calibration curve, flattening off at higher concentrations. Absence of labeled antibody results in a lack of binding.

Inactive enzyme or substrate, or presence of enzyme inhibitor in enzyme assays (including colorimetric, fluorimetric, and luminometric enzyme-based assays) – this may be characterized by a normal curve shape with a low signal level.

Faulty microparticles, beads, coated wells, or coated tubes – likely also to result in poor precision. Test by replacing with fresh material.

Cross-contamination of one reagent by another – certain combinations of reagents can have an unexpectedly large effect on the signal, even if only a trace of one of the reagents is present. For example, a very small quantity of donkey anti-sheep second antibody reagent could reduce the binding of a sheep antiserum reagent. Check that the dispensing equipment is cleaned thoroughly and replace the reagents.

Unstable pellets in precipitation assays – these may be visible. Within-assay precision will be poor.

Defective signal generation reagent – similar effects may be observed in assays for different analytes when the same signal generation reagent is used. Test by replacing with fresh material.

Use of reagents near to or at expiry – check using fresh reagents.

Use of reagent after expiry date – check using fresh reagents.

Incorrect storage of reagents or faulty refrigerator – check that the storage temperature is correct.

Reagents unstable while in analyzer – check the manufacturer's instructions. Record the signal level of a control at regular intervals during the life of the reagent pack to see whether there is a trend. Make sure that the laboratory temperature stays within the range allowed by the manufacturer.

Contamination of reusable reagent or distilled water container by dirt, disinfectant, or detergent – check appearance and cleaning procedures.

Volume of capture antibody (if dispensed) too low or capture antibody not dispensed – check the protocol and the dispensing equipment. Also check the amount of antiserum reagent left over after the assay.

Volume of labeled antibody too low or labeled antibody not dispensed – check the protocol and the dispensing equipment. Also check the amount of labeled antiserum reagent left over after the assay.

Inadequate mixing of reagent suspension before dispensing – check the protocol and the mixing equipment. Check that the reagent was mixed immediately before use, if this is required.

Calibrator reconstitution volume too high – check the reconstitution volume in the package insert and the amount left after the assay is completed. Check the measuring equipment. Controls will be biased high.

Sample volume too low – check the protocol and the dispensing equipment.

Reduced incubation time (at any stage of multi-incubation assays, including signal generation in colorimetric, fluorimetric, or luminometric assays) – check the protocol used.

Incubation temperature too low or too high – check the temperature setting used. Check that the incubation equipment was allowed sufficient time to warm up.

Volume of separation reagent dispensed too low or reagent not dispensed – check the protocol and the volume of separation reagent remaining.

Insufficient centrifugation time – precision will be poor. Check the protocol.

Insufficient centrifugation *g* force – precision will be poor. Check the protocol.

Faulty magnetic separation – check separation module of analyzer or method of use of manual separation device.

Loss of precipitate on decanting because of overvigorous tapping (manual separations) – the precision will also be poor and slipping pellets may be visible on the tubes. Check the technique.

Defective pipettes or automatic sampler/ dispenser – test by substituting different dispensing equipment.

Faulty incubator or water-bath (temperature too high or too low) – test the temperature maintained by the equipment if possible.

Faulty centrifuge (*g* force too low or excessive braking) – check the centrifuge.

Faulty counter or signal reader – check using a test signal source or an alternative assay.

Suggested Course of Action

1. Analyze the data from more than one occasion to evaluate the consistency of the low binding and check within-assay precision if duplicates have been run. Try and assess approximately when the change occurred, or if there was a gradual change.
2. Check whether there is any relationship between low binding and other factors, such as the lot or age of the reagents, or the technician carrying out the assay.
3. Identify any differences between your protocol and that in the package insert. Test whether the protocol is involved by repeating an assay with low binding using the primary protocol in the package insert.

4. Consult the manufacturer for advice, asking for the expected signal or binding for the age of the reagents used and whether other users have reported similar findings.

Checklist of Useful Information

It may be useful to have the following information available:

- reagent lot numbers and the dates the assays were performed;
- when the problem started;
- details of any other coincident changes in reagent performance, e.g. within-assay precision;
- QC charts;
- assay printout(s) or relevant calibration data;
- details of differences between the protocol used and the protocol in the package insert;
- details of all the equipment used;
- information on any coincident change in the protocol, equipment, or technician.

Case Study

A laboratory using an immunometric assay kit for luteinizing hormone experienced low binding on a number of occasions. (Cause – the analyzer incubator had a fault and was running at too low a temperature.)

INCREASE IN BINDING IN A COMPETITIVE IMMUNOASSAY

Possible Causes

Premature deterioration of reagents within shelf-life or exposure of reagent to extreme temperature during transport – deterioration of the tracer can sometimes result in an increase rather than a decrease in binding, in a competitive immunoassay. If reagents are deteriorating during their shelf-life the precision may worsen as the reagents near expiry. If elevated binding is due to extreme temperature during shipment, other reagents in the same shipment may also be affected. If it is likely that extreme temperatures were encountered, ask the manufacturer whether the reagent is sensitive to freezing or heating. Check by replacing with new reagent from a different shipment.

Concentration of antigen in tracer lower than usual or tracer not present in tracer reagent – a low concentration causes elevated binding and, if the concentration is much reduced, a hooked calibration curve with a flat dose–response at low analyte concentrations.

Specific activity of tracer higher than usual – check with the manufacturer. If the specific activity of a radioactive tracer is much higher than normal it may be less stable. The calibration curve may be hooked, with a flat dose–response at low analyte concentrations.

Concentration of antibody higher than usual – a high concentration causes elevated binding and, if the concentration is much increased, a hooked calibration curve with a flat dose–response at low analyte concentrations.

Faulty microparticles, beads, coated wells, or coated tubes – likely also to result in poor precision. Test by replacing with fresh material.

Faulty separation system – likely to be accompanied by poor within-assay precision and an unusual calibration curve shape. Test by replacing with fresh material.

Incorrect storage of reagents or faulty refrigerator – check that the storage temperature is correct.

Use of newer reagent than usual – check the records.

Tracer reconstitution volume too high – check the reconstitution volume in the package insert and the amount left after the assay is completed. Check the measuring equipment.

Volume of tracer dispensed too low – check the protocol and the volume of tracer remaining.

Antiserum reagent reconstitution volume too low – check the reconstitution volume in the package insert and the amount left after the assay is completed. Check the measuring equipment.

Volume of antiserum reagent dispensed too high – check the protocol and the volume of antiserum remaining.

Calibrator reconstitution volume too high – check the control values for bias. Check the reconstitution volume in the package insert and the amount left after the assay is completed. Test the measuring equipment.

Sample volume too low – check the protocol and the measuring equipment.

Increased incubation time – check the protocol against the package insert.

Increased incubation temperature – check the temperature setting used.

Defective pipettes or automatic sampler/dispenser – test by substituting different dispensing equipment.

Faulty incubator or water-bath (temperature too high) – test the temperature maintained by the equipment if possible.

Suggested Course of Action

1. Analyze the data from more than one assay to check the consistency of the high binding and check the within-assay precision if duplicates have been run.

2. Check whether there is any relationship between the high binding and other factors, such as the lot or age of the reagents, or the technician carrying out the assay.
3. Identify any differences between your protocol and that in the package insert. Test whether the protocol is involved by repeating an assay with high binding using the primary protocol in the package insert.
4. Check the control values. If they are severely biased, the calibrators may have been incorrectly reconstituted.
5. Consult the manufacturer for advice, asking for the expected signal or binding for the lot and age of the reagents used and whether other laboratories have reported similar findings.

Checklist of Useful Information

It may be useful to have the following information available:

- reagent lot numbers and the dates the assays were performed;
- when the problem started;
- details of any other coincident changes in performance, e.g. within-assay precision;
- QC charts;
- assay printout(s) or relevant calibration data;
- details of differences between the protocol used and the protocol in the package insert;
- details of all the equipment used;
- information on any coincident change in the protocol, equipment, or technician.

Case Study

A technician reported a sudden increase in the binding in a T_3 immunoassay. (Cause – the change in binding was caused by the technician reconstituting the calibrators with 1 mL of water instead of 0.5 mL, the correct reconstitution volume.)

REDUCTION IN ED_{50} (ESTIMATED DOSE AT 50% OF BINDING AT ZERO CONCENTRATION) – COMPETITIVE IMMUNOASSAYS ONLY

Possible Causes

Premature deterioration of reagents within the shelf-life or exposure of the reagent to extreme temperature during transport – if reagents are deteriorating during their shelf-life the precision and signal level will deteriorate as the reagents near expiry. If low ED_{50} is due to extreme temperature during shipment, other reagents in the same shipment may also be affected. If it is likely that extreme temperatures were encountered, ask the manufacturer whether the reagent is sensitive to freezing or heating. Check by replacing with new reagent from a different shipment.

Concentration of antigen in tracer changed or other fault in tracer reagent – if the antigen concentration is too high, dilution may result in a more normal curve shape. Other faults in the tracer can be confirmed by replacing the tracer with fresh material.

Inactive enzyme or presence of enzyme inhibitor in enzyme assays – this may be characterized by a normal curve shape with a low signal level.

Concentration of antibody too low or faulty antiserum reagent – if the antibody concentration is too low, using twice the quantity may result in a more normal curve shape. Other faults in the antiserum reagent can be confirmed by replacing with fresh material.

Change in antiserum – check with the manufacturer.

Faulty microparticles, beads, coated wells, or coated tubes – likely also to result in poor precision. Test by replacing with fresh material.

Concentration of separation reagent too low (if used) – if the separation reagent concentration is too low, using twice the quantity may result in a more normal curve shape. (This technique may be useful in diagnosing the cause of the problem, but it should not be used to correct the fault in clinical practice.) Other faults in the separation reagent can be confirmed by replacing with fresh material.

Faulty separation reagent – likely to be accompanied by poor precision. Test by replacing with fresh material.

Unstable pellets in precipitation assays – these may be visible. Within-assay precision will be poor.

Defective signal generation reagent – similar effects may be observed in assays for different analytes when the same signal generation reagent is used. Test by replacing with fresh material.

Use of reagent near to or at expiry – check using fresh reagents.

Use of reagent after expiry date – check using fresh reagents.

Incorrect storage of reagents or faulty refrigerator – check that the storage temperature is correct.

Tracer reconstitution volume incorrect – if the bottle of reconstituted tracer is still available, check the reconstitution volume in the package insert, the number of tests carried out, and the remaining tracer volume to test whether the correct amount is left. (Allow for the use of tracer for pre-washing pipette tips or probes.)

Use of undiluted tracer if tracer reagent requires dilution – if the tracer used in the assay is still available, check the instructions in the package insert, the number of tests carried out, the remaining tracer concentrate volume, and the remaining diluted tracer volume to test whether the correct amounts are left. (Allow for the use of the tracer for pre-washing pipette tips or probes.)

Volume of tracer dispensed incorrect – check the protocol and the dispensing equipment.

Antiserum reagent reconstitution volume too high – if the bottle of reconstituted antiserum is still available, check the reconstitution volume in the package insert, the number of tests carried out, and the remaining antiserum reagent volume to test whether the correct amount is left. (Allow for the use of antiserum for pre-washing pipette tips or probes.)

Inadequate mixing of antiserum reagent suspension before dispensing – check the protocol and the mixing equipment.

Volume of antiserum reagent dispensed too low or antiserum not dispensed – check the protocol and the dispensing equipment. Also check the amount of antiserum reagent left over after the assay.

Calibrator reconstitution volume too low – check the reconstitution volume in the package insert and the amount left after the assay is completed. Check the measuring equipment. Controls will be biased low.

Sample volume too high – check the protocol and the dispensing equipment.

Increased incubation time – check the protocol used.

Incubation temperature too low or high – check the temperature setting used. Check that the incubation equipment was allowed sufficient time to warm up.

Inadequate mixing of separation reagent suspension before dispensing – check the protocol and the mixing equipment. Check that the separation reagent was mixed immediately before use, if this is required.

Volume of separation reagent dispensed too low or reagent not dispensed – check the protocol and the volume of separation reagent remaining.

Insufficient centrifugation time – precision will be poor. Check the protocol.

Insufficient centrifugation *g* force – precision will be poor. Check the protocol.

Defective pipettes or automatic sampler/dispenser – test by substituting different dispensing equipment.

Faulty incubator or water-bath (temperature too low or high) – test the temperature maintained by the equipment if possible.

Faulty centrifuge (*g* force too low or excessive braking) – check the centrifuge.

Faulty counter or signal reader – check using a test signal source or an alternative assay.

Suggested Course of Action

1. Analyze the data from more than one assay to evaluate the consistency of the low ED_{50} and check the within-assay precision if duplicates have been run. Try and assess approximately when the change occurred or if there was a gradual change.
2. Check whether there is any relationship between low ED_{50} and other factors, such as the lot or age of the reagents or the technician carrying out the assay.
3. Identify differences between your protocol and that in the package insert. Test whether the protocol is involved by repeating an assay with low ED_{50} using the primary protocol in the package insert.
4. Consult the manufacturer for advice, asking for the expected binding for the age of the reagents used and whether other users have reported similar findings.

Checklist of Useful Information

It may be useful to have the following information available:

- reagent lot numbers and the dates the assays were performed;
- when the problem started;
- details of any other coincident changes in performance, e.g. within-assay precision;
- QC charts;
- assay printout(s) or relevant calibration data;
- details of differences between the protocol used and the protocol in the package insert;
- details of all the equipment used;
- information on any coincident change in the protocol, equipment, or technician.

Case Study

A laboratory using an immunoassay with a magnetic separation noticed a reduction in ED_{50}. (Cause – the change in ED_{50} corresponded to the introduction of a new process control method for antibody-coated magnetic particles which resulted in a different concentration of antibody in the assay.)

INCREASE IN ED_{50} (ESTIMATED DOSE AT 50% OF BINDING AT ZERO CONCENTRATION) COMPETITIVE IMMUNOASSAYS ONLY

Possible Causes

Premature deterioration of reagents within shelf-life or exposure of reagent to extreme temperature

during transport – deterioration of the tracer can sometimes result in an increase rather than a decrease in ED_{50} in a competitive immunoassay. If reagents are deteriorating during their shelf-life the precision may worsen as the reagents near expiry. If elevated binding is due to extreme temperatures during shipment, other reagents in the same shipment may also be affected. If it is likely that extreme temperatures were encountered, ask the manufacturer whether the reagent is sensitive to freezing or heating. Check by replacing with new reagent from a different shipment.

Concentration of antigen in tracer different from usual or tracer not present in tracer reagent – a low concentration causes elevated binding and, if the concentration is much reduced, a hooked calibration curve with a flat dose–response at low analyte concentrations. Other faults in the tracer can be confirmed by replacing the tracer with fresh material.

Specific activity of tracer changed – check with the manufacturer. If the specific activity of a radioactive tracer is much higher than normal then the tracer may be less stable. The calibration curve may be hooked, with a flat dose–response at low analyte concentrations.

Concentration of antibody higher than usual – a high concentration causes elevated binding and, if the concentration is much increased, a hooked calibration curve with a flat dose–response at low analyte concentrations.

Change in antiserum – check with the manufacturer.

Faulty microparticles, beads, coated wells, or coated tubes – likely also to result in poor precision. Test by replacing with fresh material.

Incorrect storage of reagents or faulty refrigerator – check that the storage temperature is correct.

Use of newer or older reagent than usual – check the records.

Tracer reconstitution volume incorrect – check the reconstitution volume in the package insert and the amount left after the assay is completed. Check the measuring equipment.

Volume of tracer dispensed incorrect – check the protocol and the volume of tracer reagent remaining.

Antiserum reagent reconstitution volume too low – check the reconstitution volume in the package insert and the amount left after the assay is completed. Check the measuring equipment.

Volume of antiserum reagent dispensed too high – check the protocol and the volume of antiserum remaining.

Calibrator reconstitution volume too high – check the control values for bias. Check the reconstitution volume in the package insert and the amount left after the assay is completed. Test the measuring equipment.

Sample volume too low – check the protocol and the dispensing equipment.

Change in incubation time – check the protocol against the package insert.

Incubation temperature too low or high – check temperature setting used.

Incubator or analyzer not allowed sufficient time to warm up – check manufacturer's instructions.

Defective pipettes or automatic sampler/dispenser – test by substituting different dispensing equipment.

Faulty incubator or water-bath (temperature too low or high) – test the temperature maintained by the equipment if possible.

Insufficient centrifugation time – the precision will be poor. Check the protocol.

Insufficient centrifugation *g* force – the precision will be poor. Check the protocol.

Faulty counter or signal reader – check using a test signal source or an alternative assay.

Suggested Course of Action

1. Analyze the data from more than one assay and check the consistency of the high ED_{50}. Check the within-assay precision if duplicates have been run.
2. Check whether there is any relationship between the high ED_{50} and other factors, such as the technician carrying out the assay or the lot or age of the reagents.
3. Identify any differences between your protocol and that in the package insert. Test whether the protocol is involved by repeating the assays with a high ED_{50} using the primary protocol in the package insert.
4. Check the control values. If they are severely biased, the calibrators may have been incorrectly reconstituted.
5. Consult the manufacturer for advice, asking for the expected binding for the lot and age of the reagents used and whether other laboratories have reported similar findings.

Checklist of Useful Information

It may be useful to have the following information available:

- reagent lot numbers and the dates the assays were performed;
- when the problem started;
- details of any other coincident changes in performance, e.g. within-assay precision;
- QC charts;
- assay printout(s) or relevant calibration data;

- details of any differences between the protocol used and the protocol in the package insert;
- details of all the equipment used;
- information on any coincident change in the protocol, equipment, or technician.

Case Study

QC charts for an immunoassay kit showed an increase in ED_{50}. (Cause – the change in ED_{50} corresponded to the introduction of a new incubator which was set at 30 °C instead of 37 °C.)

HIGH NON-SPECIFIC BINDING

In a *competitive* immunoassay, NSB is measured by carrying out an assay without the addition of antiserum. If the antiserum is normally bound to a solid phase, e.g. particles, wells, tubes, or beads, a 'blank' antiserum reagent should ideally be used, which consists of the normal solid phase without antibody.

In an *immunometric* immunoassay, NSB is measured by using zero concentration samples (if available), samples treated to remove analyte, or buffer. The NSB is therefore a measurement of the binding of the labeled antibody to the solid phase in the absence of analyte.

Possible Causes

Poor assay design or optimization – check that the high NSB is consistent. If the product is poorly designed it should be apparent from the data used to validate the assay originally. Check with the manufacturer.

Faulty tracer – if the high NSB coincides with the introduction of a new lot of tracer, try replacing with an alternative lot.

Faulty separation system reagent – if the high NSB coincides with the introduction of a new lot of separation system reagent, try replacing with an alternative lot. If the separation system is faulty, within-assay precision is also likely to be poor.

Change in calibrator matrix – in competitive assays, try measuring the NSB using patient samples instead of calibrator. In immunometric assays, try using an alternative zero concentration material or leave out the sample completely to find out the effect on NSB.

Premature deterioration of reagents within shelf-life or exposure of reagent to extreme temperature during transport – if reagents are deteriorating during their shelf-life the NSB and imprecision will increase with the age of the reagent. If the high NSB is due to extreme temperature during shipment, other kits or reagent packs in the same shipment may also be affected. If it is likely that extreme temperatures were encountered, ask the manufacturer whether the reagents are sensitive to freezing or heating. Check by replacing with new reagent from a different shipment.

Binding of tracer to plastic tube surface – less likely in solid-phase assays. In liquid-phase assays, e.g. with PEG-assisted separations, try using different types of plastic and glass tubes, to find one that causes less interference.

Binding of tracer to wells, tubes, beads, particles, or other solid phase used to determine NSB – in solid-phase assays, NSB is sometimes measured by using a sample of solid phase which has not been coated with antiserum, but has been coated with a blocking material, e.g. bovine serum albumin. However, this can cause artifacts, and the results are not necessarily representative of the situation in a normal assay.

Problems with sample used to measure NSB, including interference due to the sample collection device – this is a problem with some immunometric assays. To identify the cause of the interference or to rule out collection device interference or the individual sample as the cause, try using samples from a variety of sample collection methods. Sample collection device interference can be minimized by careful assay design, the use of correction factors or 'decoys', or improvement in the design of the collection device itself.

Incorrect storage of reagents or faulty refrigerator – check that the storage temperature is correct.

Use of reagents after expiry date – check the date of the assay against the expiry date.

Contamination with antiserum (competitive assays) – check the experimental procedure.

Contamination with analyte (immunometric assays) – check that the sample used had a concentration of zero. Try to use an alternative, zero concentration sample to check the consistency of the results.

Contamination with tracer, e.g. on the outside of tube or well (especially immunometric assays) – likely to be inconsistent between duplicates. The high NSB replicates will be reduced if the outside of the tube or well is wiped with a tissue and recounted.

Incubation temperature too high – check the protocol against the package insert.

Protocol used was different from that in the package insert, especially at the washing stage – check the protocol against the package insert.

Coated tubes or tubes containing PEG or magnetizable cellulose separation systems not tapped hard enough (manual assays) – check the tubes for residual droplets of moisture.

Faulty washer – clean the washer thoroughly, using wire to clean the dispensing nozzles and flushing through with copious amounts of distilled water. Test the washer by dispensing intensely colored dye into the tubes or wells and monitoring the removal of the dye during washing.

Suggested Course of Action

1. Analyze the data from more than one assay for the consistency of high NSB. Check that the method used to estimate NSB is valid. If a method is not given in the package insert, ask the manufacturer for advice. Also check the age of the reagents used.
2. Check whether the high NSB occurs with a variety of samples or only with one type.
3. Identify differences between your protocol and that in the package insert. Test whether the protocol is involved by repeating an assay with high NSB using the primary protocol in the package insert.
4. Check the performance of the equipment, particularly the washer, if one is used.
5. Consult the manufacturer for advice. Ask if any other users have reported high NSB.

Checklist of Useful Information

It may be useful to have the following information available:

- reagent lot numbers;
- when the problem started;
- description of the method and samples used to determine NSB;
- details of any other coincident changes in performance noticed, e.g. change in binding at zero dose (B_0), curve shape, ED_{50}, or control values;
- assay printout(s) or relevant calibration data;
- details of differences between the protocol used and the protocol in the package insert;
- details of all the equipment, particularly the washer, if one is used;
- details of the tubes or wells used, unless supplied by the manufacturer.

Case Study

Staff in a laboratory were concerned about high NSB in a β-human chorionic gonadotropin enzyme immunometric assay, i.e. high signal at zero concentration. (Cause – the washer was faulty and was not removing all the unbound conjugate at the final washing stage.)

POOR LINEARITY OF DILUTION

Possible Causes

Poor product design or optimization – check that the non-linearity is consistent by running serial dilutions of several high-concentration samples. If the product is poorly designed it should be apparent from the dilution data used to validate the assay originally. Check with the manufacturer.

Poor tracer, antiserum, or separation system – if the manufacturer believes that the non-linearity is atypical, ask for different lots of the reagents and repeat the dilution study with a range of high-concentration samples and the highest concentration calibrators.

High-concentration hook effect in immunometric assays – check by testing a range of analyte concentrations, added to a low-concentration sample or the zero calibrator, up to the maximum concentration ever likely to be present in a patient sample. Plot the calculated concentration of added analyte against the analyte concentration as measured in the assay and compare any non-linearity with the dilution study results.

Error in calibrator concentration – if this is the cause, control bias and curve-fitting error may also be apparent.

Error in stored calibration curve or manufacturer's calibration curve – recalibrate or use reagents from a different lot.

Antigen variation in samples, e.g. in patients with tumors or to whom a purified form of the analyte has been administered – antigen variation between patients can cause poor linearity of dilution. Many proteins, especially glycoproteins, are heterogeneous and they may be present in unusual forms during illness. Assays vary in their susceptibility to this type of effect. Check the clinical records of the patient(s) concerned for the disease history. Check also whether any purified analyte has been administered during the course of treatment. The manufacturer should be consulted about the capability of the assay for monitoring analyte concentration in these clinical situations.

Difference between immunoassay behavior with respect to patient sample analyte and an international standard, e.g. due to antiserum specificity – this is most likely in assays for proteins, where international standards sometimes contain purified analyte with a slightly different structure. If this is the cause, recovery and dilution experiments will only give good results when the international standard is used as the source of analyte.

Lipemic or hemolyzed patient samples or sample collection device causing interference – check the appearance of the sample(s). If sample collection device interference is suspected, check with the source of the samples and find out which type of collection device was used. Other reports may have been received by the manufacturer. For further information *see* UNEXPECTED OR INCONSISTENT CLINICAL CLASSIFICATION.

Use of old samples – if the samples have been stored for a long time, the analyte may have deteriorated. Many analytes are unstable if stored for long periods at −20 °C. Repeat the experiment with fresh samples.

Use of high concentration calibrator to measure dilution properties of assay, if calibrator is in non-human matrix or contains analyte in a different form to that present in patient samples – repeat the experiment with a high-concentration patient sample or ask the manufacturer for advice.

Poor within-assay precision (*see* POOR WITHIN-ASSAY PRECISION for possible causes) – check the agreement between duplicates, particularly at high concentrations in competitive assays. Repeat the dilution experiment on more than one occasion, and include dilutions between the neat sample and 1/2, to give a clearer picture of the dilution profile.

Protocol used was different from that in package insert – major differences between your protocol and that in the package insert should be tested for their effects on dilution by repeating the experiment using the exact protocol in the package insert.

Measurement of dilution of low-concentration samples, which is subject to much greater error than dilution of high-concentration samples – it is difficult to achieve linear dilution at very low concentrations, e.g. between zero and the adjacent calibrator. This requires a high degree of precision and accuracy, careful technique, properly maintained equipment, an exceptionally good curve-fitting program, and accurate assignment of values to the calibrators. Any slight imperfections in the assay, e.g. in the areas of NSB, washing, or signal measurement, are magnified and may not give a representative picture of assay performance across a wider concentration range.

Use of inappropriate diluent – check the suitability of the diluent fluid used with the manufacturer.

Incubator or analyzer not allowed sufficient time to warm up – check the manufacturer's instructions for the warm-up time.

Faulty equipment – check for other signs of poor assay performance such as poor precision, control bias, or low binding. These may give clues as to the likely cause. Test by substituting different equipment if possible, or service and test the equipment according to the manufacturer's instructions.

Bias due to curve-fitting error at high or low concentrations – if this is the cause, differences between *fitted* and *actual* calibrator values should reflect the bias of points from a linear dilution plot in the dilution experiment.

Suggested Course of Action

1. Repeat the dilution experiment with several high concentration patient samples. All the dilutions should be run in duplicate. Check that the precision of the duplicates is satisfactory. Investigate any obvious inconsistencies in the shape of the dilution curve, e.g. due to incorrect values assigned to calibrators.
2. Identify any differences between your protocol and that in the package insert. Test whether the protocol is involved by repeating the dilution test using the primary protocol in the package insert.
3. Check the quality of the curve-fit (by comparing *fitted* and *actual* calibrator values).
4. Consult the manufacturer for advice. Check that the diluent used is suitable. Ask if any other users have reported high NSB.

Checklist of Useful Information

It may be useful to have the following information available:

- reagent lot numbers;
- description of the samples and diluent used;
- assay printout(s) or relevant calibration data;
- details of differences between the protocol used and the protocol in the package insert;
- details of all the equipment used.

Case Study

Staff in a laboratory in a hospital specializing in the treatment of patients suffering from cancer reported poor dilution of prolactinoma patient samples in a prolactin assay, although some samples diluted more linearly than others. (Cause – prolactin arising from tumors is often different in structure from prolactin in non-tumor samples. In addition the international standard for prolactin, on which the product standardization was based, is slightly different to prolactin in samples. It was a combination of these two effects that caused the poor linearity of dilution with prolactinoma samples.)

UNEXPECTED OR INCONSISTENT CLINICAL CLASSIFICATION

Possible Causes

Poor assay design or optimization – first check that sample determinations are consistent. If the assay is poorly designed it should be apparent from the data used to validate the assay originally. Either the data will show the same problem or there will be insufficient data for the clinical categories involved. Check clinical trial reports if available.

In immunometric assays, patient samples with zero or very low analyte concentrations may give apparently elevated concentrations if the capture and labeled antibodies are linked together (e.g. by human anti-mouse activity (HAMA)), unless this is suppressed by additives in the reagents such as mouse serum. The same apparent type of effect occurs if the labeled antibody binds non-specifically to the solid phase (particles, well, bead, or tube). This may vary according to the exact protocol used. Test the sample in an alternative assay for the same analyte, to see how the results compare.

Reagent lot at fault – any change from a normal to an abnormal situation may be evident in your QC data. If it is due to the calibration of the calibrators, the change should also be evident in the manufacturer's own QC charts. Other reports may have been received by the manufacturer. Ask the manufacturer for reagents from a different lot to replace the one that caused the problem.

Error or instability in stored calibration curve or manufacturer's calibration curve – all samples and controls with a similar concentration are likely to be affected. Recalibrate and repeat the determination. If the manufacturer's calibration curve is at fault, other users of the same reagent lot will also be affected.

Your perception of the assay's application is different to that intended by the manufacturer – differences should become apparent when checking for your application in the package insert. If the assay does not meet your needs, seek a more suitable alternative.

The immunoassay test correctly indicates patient states, and the interpretation of other clinical information is incorrect – run the sample in an alternative assay for the same analyte, to check the test method being used. Check the reasons for the clinical assessment.

Unusual type of patient sample – this is characterized by specific patient samples that give the same clinically inappropriate result on more than one occasion. Run the sample(s) in an alternative assay for the same analyte to check the test method being used. Many types of interference can occur and some of the more common ones are often listed in package inserts, such as hemolyzed, icteric, lipemic, or turbid samples.

Immunometric assays may be affected by the presence of cross-linking immunoglobulins (e.g. HAMA) or rheumatoid factor.

Competitive assays may be affected by the presence of endogenous antibodies for the analyte being measured.

Some types of assays are affected by the presence of particulate matter in samples.

Check that the type of sample concerned is validated for the assay (e.g. specific plasma collection method).

Samples may be diluted if taken from a catheter, reducing the analyte concentration.

Samples taken from patients on heparin can continue to produce fibrin after clotting.

Some unusual effects are caused by variants of normal proteins produced by tumors, which may be detected to a greater or lesser degree by different immunoassays for the same analyte.

Some analytes are affected by the presence of specific substances, e.g. digoxin-like immunoreactive substances (*see* DIGOXIN). Check package insert or contact the manufacturer for advice.

The action required depends on the frequency of occurrence and the nature of the cause. It may be necessary to use a confirmatory test for values outside the reference interval.

See SUBJECT PREPARATION, SAMPLE COLLECTION AND HANDLING for more information.

High concentration hook effect in immunometric assay – this can, if extreme, bring very high concentration patient samples into the assay range, giving them apparent analyte concentrations much lower than they really are.

Interference by cross-reactants, e.g. drug metabolites – run the sample(s) in an alternative assay for the same analyte to check the test method being used. The problem is likely to be limited to individual patients or types of patients. Drug-related problems may become apparent from the patient histories/medication records. The action required depends on the frequency of the problem and how it affects interpretation of the results.

Interference due to sample collection device – run the sample(s) in an alternative assay for the same analyte, to check the test method being used. If sample collection device interference is suspected, check with the source of the samples and find out which type of collection device was used. Several similar reports are likely to have been received by the manufacturer. Investigation of these types of problems involves incubating samples with different parts of the device to identify the source of the interference, and collaboration with the device manufacturer to prevent it. This is best carried out by the assay manufacturer. Changing the type of collection device will solve the problem but this is sometimes outside the control of the testing laboratory.

Incomplete mixing or warming up of reagents – if this is the cause, inappropriate sample measurements will not be repeatable in a second assay. Control values may also be affected. Test and resolve by ensuring that reagents are mixed and allowed to reach ambient temperature before retesting samples. Check that magnetic particle suspensions are adequately mixed, particularly when kept on board automated analyzers.

Protocol used was different from that in package insert – major differences between your protocol and that in the package insert should be tested for their effects on patient sample values by repeating tests on the samples using the exact protocol in the package insert.

Faulty pipette or automatic sampler/dispenser; dirty pipette tips; contaminated sample or reagent probe; poor connection between pipette tip and pipette; poor connection between probe and tubing; leaking gasket; carryover of analyte from

high-concentration sample to low-concentration sample; carryover of reagent by reagent-dispensing probe; splashing of reagents; tubing constricted – if any of these is the cause, inappropriate sample measurements will not be repeatable in a second assay. Control values will also be affected. Test by substituting alternative dispensing equipment and repeating the assay.

Insufficient penetration of sample by sampling probe – observe the sampling process and look for bubbles. The probe level detection system may require adjustment or replacement. Affected patient sample values will be biased low.

Dilution of sample by water in tubing – check the equipment by using control samples with known concentrations.

Penetration of clot by sample dispenser when using primary sample tubes – if a primary sample tube does not contain serum above the clot or if the liquid sensor is not working correctly, the sample dispenser may disturb the clot and the sample. Repeat the test with serum decanted from the clot into another tube.

Sample dispenser blocked by rubber, film, or foil during penetration of cover on primary sample container – check the probe.

Interference by coating on pipette tip – try using an alternative pipette tip.

Faulty mixing of sample and reagents – inappropriate sample measurements will not be repeatable in a second assay. Control values may also be affected and within-assay precision will be poor. Duplicate values for the same sample will be in poor agreement.

Faulty incubator – inconsistent temperature control could cause positional effects. Samples reassayed in different positions would give different results. Test by using a different incubation system if possible.

Incubator not allowed sufficient time to warm up – check the manufacturer's instructions for warm-up time. Also check the control values.

Faulty magnet or magnet not in close enough proximity to paramagnetic particles during separation – check the separation module of analyzer or the method of use of manual separation device.

Faulty washer – if a washer is not functioning adequately, certain samples may show different values to those expected. Normally this is evident in immunometric assays if patient sample values that should be close to zero show up with significant analyte concentrations apparent. This may be aggravated by the presence of substances in the sample that reduce washing efficiency or attract the labeled antibody to the solid phase, preventing its complete removal. Washers may be tested by washing out wells or tubes containing a colored dye. Check wash probes for obstructions.

Faulty or contaminated counter/signal reader – duplicate values for the same sample will be in poor agreement. Values for the sample run in repeat assays will be different from the original results. Test the counter or signal reader by running multiple measurements on the same sample tube and checking backgrounds.

Contamination of reusable reagent or distilled water container by dirt, disinfectant, or detergent – check the appearance of the container and the cleaning procedure.

Assay drift – run the sample at the beginning and the end of an assay or over a suitable period of time on a random-access analyzer. If the results are different *see* ASSAY DRIFT.

Curve-fit error – test by looking at the difference between *fitted* and *actual* calibrator values on the assay printout. If this is the cause of error the magnitude of the differences should reflect the biases in the values of clinical samples from those expected.

Use of incorrect reference interval – check against the reference interval in the package insert.

Use of incorrect units – check the reference interval given in the package insert for the units being used.

Incorrect storage of samples or reagents, or use of reagents beyond the expiry date – check the method of storage against package insert instructions. Your refrigerator or freezer may not be at the correct temperature. Some analytes are more prone to instability than others (ask the manufacturer for advice). Sample evaporation can cause a significant increase in analyte concentration after 2 h if the sample container is left unstoppered.

Reagents deteriorating while on board analyzer – a gradual change in sample and control values may be apparent from the first use of a new reagent pack until its last use, followed by a return to the initial values with the new pack. Check that the manufacturer's instructions are being followed.

Wrong sample tested or wrong label applied to sample tube – request another sample from the patient if this is suspected.

Reference method at fault – check by using a third method, or alternative source of the reference method.

Insufficient reagent in reagent container – check the reagent containers.

Defective values in reagent supply information – this could be either due to detector error or because the reagent container was not full when loaded onto the analyzer. Check the reagent containers.

Sample or reagent in incorrect position – in some multianalyte analyzers the test carried out on the sample is determined by the position of the sample relative to the position of the reagents. Check the sample and reagent positions and retest the sample.

Suggested Course of Action

1. Repeat the test.
2. If the immunoassay is being used for clinical diagnosis, inform the laboratory manager and the manufacturer immediately.
3. Analyze the information and write down a concise description of the problem.
4. If there have been similar reports before, analyze them to see if the problem is the same.
5. Identify differences between your protocol and that in the package insert.
6. Find out if the manufacturer has ever seen similar results in prelaunch trials or if similar observations have been reported by other users.
7. Keep relevant samples in a safe place. If requested send aliquots of the samples to the manufacturer.
8. If the samples concerned are from unusual types of patients or the assay is known to give rise to false positives or negatives, check the clinical interpretation of the results with the manufacturer. Your interpretation may be different from that intended by the manufacturer.
9. If you or the manufacturer have conducted correlation studies with another method, look for individual points that were out of consensus with the others. They may have been incorrectly assumed to be part of the normal statistical population. Were those samples retested?
10. If samples are retested by the manufacturer and give acceptable results, the cause of the problem may be specific to your laboratory. Look at the assay printouts (or calibration and control data) for evidence of control bias, imprecision, poor calibration curve shape, or curve-fit error (by comparing *fitted* with *actual* calibrator values).
11. If your results are confirmed by the manufacturer and the product is shown not to be performing as claimed, copies of the data should be handed over to the manufacturer for a full investigation.

Checklist of Useful Information

It may be useful to have the following information available:

- reagent lot number(s);
- when the problem started;
- patient sample values – document full details of the extent of the problem including the frequency of misclassification of each type of patient sample concerned and the number of times each sample was assayed;
- the clinical status and drug therapy of the patient(s);
- relevant results obtained using other assays, either for the same analyte or related analytes;
- assay printout(s) or relevant calibration data;
- details of differences between the protocol used and the protocol in the package insert;
- details of all the equipment used.

Case Study

A thyrotoxic patient with a highly elevated free thyroxine (T_4) concentration had a thyrotropin (TSH) value of 0.25 μU/mL in a high-sensitivity TSH IRMA assay. The biochemist expected the value to be <0.04 μU/mL, the assay detection limit. (Cause – further assays showed the TSH concentration to be approximately 0.06 μU/mL in two different tests, below the reference intervals for each assay. The washing module on the analyzer was faulty and this gave rise to contamination of some of the assay tubes with unbound tracer, increasing the signal and hence the apparent concentration.)

Further Reading

Tietz, N.W. (ed), *Clinical Guide to Laboratory Tests* 3rd edn, (AACC Press, Washington, 1995)

APPARENT SHIFT IN REFERENCE INTERVAL (NORMAL RANGE)

Possible Causes

Poorly calibrated or unstable calibrators – this should be clearly visible by a corresponding change in QC control values, coincident with a change in the calibrator lot or with the age of the reagents.

Bias due to interaction of set of calibrators with curve-fit method used – if individual calibrators in a set have been miscalibrated, laboratories using spline fits may see more bias than those who use more 'smoothed' fits. With most curve-fit programs, bias in individual calibrators is normally apparent as a difference between the *fitted* and *actual* values in the calibration curve part of the assay printout. Bias in high or low-concentration calibrators may cause problems with logit-log curve-fitting, biasing the fitted line from most of the calibrator points. This should be apparent from excessive bias between fitted and actual calibrator values in the printouts. Curve-fit problems due to miscalibrated calibrators can be distinguished from other curve-fitting problems by substituting another calibrator lot.

Error or instability in stored calibration curve or manufacturer's calibration curve – the change in the reference interval should either coincide with the introduction of the new calibration curve or, if due to instability, be corrected by recalibration. If the manufacturer's calibration curve is at fault, other users of the same kit lot will also be affected.

Reagent lot (other than calibrator) – likely to be mirrored by QC control values but not necessarily so. If this is the cause, other users are likely to have reported a change to the manufacturer, or the manufacturer's QC charts may show corresponding changes. Identify reagent lot changes coincident with the change in reference interval, to pinpoint the reagent at fault. If a reagent lot caused the change, ask the manufacturer for an alternative lot.

Change in assay protocol – the protocol details need to be checked against the package insert, before and after the change occurred.

Change in sample collection method (if method sensitive to such a change) – check to see if this has occurred. Several similar reports are likely to have been received by the manufacturer. Investigation of these types of problems involves incubating samples with different parts of the device to identify the source of the interference, and collaboration with the device manufacturer to prevent the interference. This is best carried out by the assay manufacturer. Sometimes changes in the formulation of the immunoassay reagents can reduce or eliminate these effects. Changing the type of collection device will solve the problem but this is often outside the control of the testing laboratory.

Change in sample storage method – check to see if this has occurred. Check the package insert for storage instructions. Ask whether there have been any delays to samples sent by mail or collected by a delivery service.

Change in origin or 'mix' of samples sent to the testing laboratory (e.g. more samples from geriatric patients), change in method of patient selection or criteria for classifying patients as 'normal' – this is often difficult to confirm. However, such a change should be suspected if all other causes have been ruled out.

Equipment fault or change in the type of equipment used – other effects should be apparent, e.g. a coincident change in curve shape or a deterioration in within-assay precision. *See* POOR WITHIN-ASSAY PRECISION and sections on changes in curve shape. Check to see if the change occurred after service or repair of any of the automated equipment used in the assay.

Change in laboratory temperature or humidity – other effects should be apparent such as a coincident change in curve shape. Recent extremes of laboratory temperature may have occurred. They are clearly most likely to affect assays with room temperature incubations. Exceptionally dry atmospheric conditions can cause beads, tubes, or wells to dry out after aspiration, before the next reagent is added. If this is the cause, controls and within-assay precision will also be affected.

Change in curve-fit program – this is self-evident if printouts are available from assays before and after the change occurred. Changes in values for patient samples and controls should correlate with changes in the amount of curve-fit bias. Test for curve-fit bias by comparing *fitted* and *actual* calibrator concentrations on the assay printout.

Change in technician – a change in patient sample values may be accompanied by changes in curve shape or between-assay precision. Check to see if the technician change was coincident with the change in the reference interval. If technician effects are suspected, check the way the technicians carry out the assay.

Use of reagents beyond expiry date – check the dates of the assays against the age or expiry date of the reagent lot(s) involved.

Suggested Course of Action

1. If available, retest patient samples that were tested before the change was noticed and compare the values obtained.
2. If the immunoassay is being used for clinical diagnosis and the change is significant clinically, inform the laboratory manager and the manufacturer immediately.
3. Analyze the information and write down a concise description of the problem.
4. If there have been similar reports before, analyze them to see if the problem is the same.
5. Identify differences between your protocol and that in the package insert.
6. Look for changes in the control values. If coincident changes are apparent, compare your control data with the manufacturer's and identify similarities and discrepancies.
7. Check assay printouts for curve-fit bias (by comparing *fitted* and *actual* calibrator values in the calibration curve), poor precision, or any major changes in the curve shape when the change in reference interval appears to have occurred.
8. Find out whether the patient selection criteria could have changed.
9. Find out whether the method of sample collection or origin of the samples has changed.
10. If your results are confirmed by the manufacturer, copies of the data should be handed over to the manufacturer for a full investigation.

Checklist of Useful Information

It may be useful to have the following information available:

- reagent lot number(s);
- when the problem started;
- details of patient sample distributions, daily patient sample means or medians before and after the change;

- results from individual patient samples tested before and after the apparent change;
- control values before and after the change;
- assay printout(s) or relevant calibration data;
- details of differences between the protocol used and the protocol in the package insert;
- details of all the equipment used.

Case Study

A laboratory manager noticed over several months that there was a significant increase in the number of patient samples with TSH values just above the upper limit of the TSH reference interval, which had been established over the previous year. He did not know whether patients above the reference interval were euthyroid or subclinical hypothyroids. (Cause – closure of another laboratory had resulted in an increase in the number of geriatric patient samples sent to this laboratory. Elevated TSH values, with no clinical symptoms, are more common in older patients.)

NEGATIVE PATIENT SAMPLE CONCENTRATIONS

Possible Causes

Patient sample contains little or no analyte and negative values are due to statistical variation around zero – patient sample measurement will not be significantly different to zero concentration. Check to see if the same patient sample gives a negative value on a number of occasions. If less than half the values are negative, statistical variation is likely to be the cause. Calculate the confidence interval for the values and check whether there is a significant difference from zero.

Zero calibrator contains analyte – there will be a significant difference between the signal level of the zero calibrator and that of the sample. In an immunometric assay the signal level will be higher in the calibrator; in a competitive assay it will be lower. However, other causes, such as matrix effects, could also produce negative values. If the assay is immunometric, the presence of analyte may be detected by pre-incubating a sample of the zero calibrator with a low concentration of antibody. If analyte is present, the binding in the assay will be reduced. This experiment requires a source of antiserum and is therefore best carried out by the manufacturer. It may be possible to measure analyte concentration in the zero calibrator using a more sensitive assay or a method less prone to matrix interference (e.g. GC–MS). The presence of analyte in zero calibrator may vary from lot to lot. If the incidence of negative values varies with the lot of calibrators, it is likely that the calibrators contain analyte. If the magnitude or incidence of the negative values changes according to the lot of one of the other reagents, the cause is likely to be due to matrix differences between the calibrators and the samples. Presence of significant levels of analyte in the zero calibrator may cause non-linearity at the low-concentration end of logit-log curve-fits. However, this can also be due to matrix effects. It is difficult to make zero calibrators for some analytes on a large scale and the manufacturer may have decided to provide the 'zero' calibrator with a measurable but low level of analyte, where this does not affect the use of the assay in normal clinical situations.

Zero calibrator unstable – instability may be detected by comparing the incidence of the negative values with the age of the lot of calibrators used. This may occur in liquid calibrators or in freeze-dried calibrators after reconstitution. It is most likely in competitive assays, and is due to changes in the calibrator matrix affecting binding at zero concentration.

Matrix effect between zero calibrator and patient samples – this is most likely in competitive assays, especially when a human serum zero calibrator has been highly processed to remove analyte, or if the calibrators are made up in buffer. This affects the binding at zero concentration causing positive or negative values for patient samples not containing analyte. It can only be corrected by the manufacturer. Individual samples may themselves cause artifacts in the assay, especially in competitive assays, if they are hemolyzed, lipemic, or icteric. The appearance of samples giving negative values should be checked.

Error in stored calibration curve or manufacturer's calibration curve – the negative values should either coincide with the introduction of the new calibration curve or, if due to instability, be corrected by recalibration. If the manufacturer's calibration curve is at fault, other users of the same reagent lot will also be affected.

Assay drift – suspect assay drift if negative samples occur at the end of a manual assay or on the first occasion an analyzer is used after a prolonged period of inactivity. Retest the samples positioned next to the calibrators in the assay. If the values become normal *see* ASSAY DRIFT.

Coated bead, well, or tube variation within assay – if this is the cause replicates will be inconsistent, and only certain replicates for one sample will give negative values. Check by running the same sample several times in one assay.

Contamination of zero calibrator with analyte during assay, e.g. by contaminated sample dispenser or probe, or splashing – if the zero calibrator is run in duplicate, check the replication. If the first replicate tends to be higher in concentration than the second, it is likely that contamination is occurring when dispensing calibrators. This has the effect of increasing the value of the zero calibrator, causing zero concentration patient samples to give negative values. If any of the laboratory staff have been handling pure analyte in the laboratory, this can, in certain instances, cause contamination of assays carried out in the same laboratory.

Contamination of zero calibrator with labeled reagent – this is characterized by poor replication of duplicates. Check that the dispensing equipment does not splash reagent onto the outside of the tubes or wells.

Sample collection or storage method affects apparent value in assay – look for a pattern in the origin or treatment of the samples that give negative values.

Poor within- or between-assay precision – *see* POOR WITHIN-ASSAY PRECISION or POOR BETWEEN-ASSAY PRECISION for possible causes if there is evidence of imprecision in the assay data.

Curve-fit error – check the value of the zero calibrator when the signal is reread from the fitted calibration curve as an unknown. If the curve-fit program is causing the bias, the zero calibrator will also have a negative value and the signal level (e.g. light units, optical density (OD), or counts per minute (cpm)) of the sample will not be significantly different from that of the zero calibrator.

Inconsistent reagent dispenser – if this is the cause the same negative value will not be obtained when the assay is repeated. Duplicate precision will also be poor. Check by substituting a different dispenser.

Temperature fluctuations in incubator – small variations in temperature can affect the rate of binding between antigen and antibody. This can give differences in binding between the samples and calibrators. This is especially true of large airflow incubators used for microtitration plate assays, where tubes or wells in the center of a tray can take longer to warm up. Problems with incubators can be tested for and rectified by using water-baths or special high-performance microtitration plate incubators.

Incubator not allowed sufficient time to warm up – check the manufacturer's instructions for the warm-up time.

Incomplete or variable centrifugation – if the centrifugation speed or time is insufficient or if braking is too severe, poor within-assay precision will occur and the sample will not give the same negative value in repeat assays with the brake left off.

Inconsistent washer – duplicates for the zero calibrator will be in poor agreement, and signal (e.g. light units, ODs, or cpm) will be higher than normal. Wash using alternative equipment or service the washer. Other duplicates will be poor if the washer is faulty.

Contaminated multiwell counter or signal reader – duplicates for the zero calibrator or sample will be in poor agreement. In immunometric assays signal (e.g. light units, ODs, or cpm) for the zero calibrator will be higher than normal. Check the background counts or clean the reader, counting trays, or holders.

Use of unvalidated protocol – major differences between your protocol and that in the package insert should be tested for their effect on patient sample values.

Suggested Course of Action

1. Repeat the test.
2. Analyze the information and write down a concise description of the problem. Examine the assay printouts (or calibration and control data), check that the calibrator values have been entered correctly and check that light units, OD, or cpm for the negative patient samples are higher than the zero calibrator (competitive assays) or lower (immunometric assays), to exclude curve-fitting bias near zero. Check that the negative values are statistically significantly different from zero.
3. If there have been similar reports before, analyze them to see if the problem is the same.
4. Identify any differences between your protocol and that in the package insert.
5. If possible, compare your QC and curve shape data with the manufacturer's data for the same lot and identify similarities and discrepancies. Check to see if you are obtaining similar light units, OD, or cpm for the zero calibrator to the manufacturer.
6. Find out whether the manufacturer observed negative values for patient samples during clinical trials of the product.
7. Ask the manufacturer for advice on the possible cause.
8. If convenient, send some of the negative patient samples to the manufacturer for reassay.

Checklist of Useful Information

- reagent lot number(s);
- when the problem started;
- full details, including the proportion of assays showing the problem, the types of samples involved, and the actual results;
- results obtained using alternative assays for the same analyte;
- information on the method of sample collection and storage for the patient samples involved;
- assay printout(s) or relevant calibration data;
- details of differences between the protocol used and the protocol in the package insert;
- details of all the equipment used.

Case Study

Several users of a human chorionic gonadotropin (hCG) immunoassay reported a low but significant incidence of negative hCG values for women who were not pregnant after a specific kit lot had been introduced.

(Cause – the 'zero' calibrator introduced in this kit lot was contaminated with a small amount of hCG.)

POOR RECOVERY

In a recovery experiment, the analyte concentration in a sample is assayed before and after the addition of a known quantity of analyte. The increase in concentration, as measured in the immunoassay, is compared to the quantity of analyte added. For example, if the original sample has 20 units/mL in the assay, and 100 units are added (or 'spiked') into 1 mL of the sample, the final concentration should be 120 units/mL. If this is the concentration obtained in the assay the recovery is 100%. If the measured concentration is 100 units/mL, the recovered amount is 80 units/mL, and recovery is 80%. This is 'low' recovery or 'under-recovery'. If the measured concentration is 130 units/mL, the recovered amount is 110 units/mL, and recovery is 110%. This is 'high recovery' or 'over-recovery'. Normally a recovery of 90–110% is considered acceptable.

Recovery is a fundamental check of an immunoassay's ability to measure an analyte accurately.

Unless the volume of analyte solution added is very small (i.e. less than 2% of the volume of sample to which it is added), an allowance must be made for the effect of this volume on the concentration of the original analyte in the sample, which will now be slightly lower.

Possible Causes

Immunoassay incorrectly standardized – check for any bias between the assay and other tests for the same analyte in proficiency testing/external quality assessment schemes. Consult the manufacturer for advice.

Calibrator lot incorrectly calibrated – if this is the cause, the control values obtained in the kit would have changed when the calibrator lot was introduced. Curve-fit bias may also be apparent.

Deterioration of manufacturer's reference standards – check whether the kit has gradually drifted away from other methods for the same analyte in external quality assessment schemes. The control values may show long-term stepwise changes coincident with the introduction of successive calibrator lots.

Deterioration of calibrator lot before use – control values would also be biased. Check by using fresh calibrators.

Error or instability in stored calibration curve or manufacturer's calibration curve – controls are also likely to be affected. The recovery should improve if the immunoassay is recalibrated, unless the fault is in the manufacturer's calibration curve.

Reagent lot at fault – a change in control values should clearly coincide with the change in reagent lot number.

Assay drift – try running the samples from the recovery experiment immediately after the calibrators. *See* ASSAY DRIFT.

Incorrect reconstitution of analyte used to spike sample in recovery experiment, particularly reference preparations in ampoules – accurate reconstitution of international standards in ampoules takes skill and practice. Check the technique used. The recoveries as assessed from two different ampoules should be in close agreement.

Instability of analyte before or after reconstitution, or use of impure analyte – the source of the analyte should be checked. Either fresh material or a preparation with a validated shelf-life should be used for the experiment. International Standard preparations should always be used if available. Following reconstitution, aliquots of International Standards should be stored at or below −60 °C. Repeated freezing and thawing of standards should be avoided.

Use of different matrix for reconstitution and/or dilution of analyte to that used by the manufacturer – consult the supplier of the International Standard for the correct diluent for reconstitution. Ask the manufacturer for information about further dilution for use in the assay. Some steroids and drugs are initially dissolved in ethanol. However, the final concentration of ethanol in the sample during the assay must be negligible.

Weighing or volumetric error – check for the correct operation of balances, pipettes, and volumetric flasks. Check the technique used.

Dilution of analyte in sample (when spiked) not allowed for in calculation – check the volumes used and the calculation. Recalculate recovery, adjusting for the change in base concentration after addition of the analyte solution.

Mathematical error in calculation of recovery – check the calculations.

Poor precision at high concentration if spiked sample near upper limit of assay range – check the duplicates. Repeat the experiment on more than one occasion to check the consistency of the results. Try using lower concentrations of analyte for spiking and samples with lower base concentrations.

Measurement of spiked sample outside assay range by extrapolation or dilution – repeat the experiment using lower concentrations of analyte for spiking and samples with lower base concentrations.

Use of different units or International Standard to product standardization – check the package insert or consult the manufacturer.

Only one assay performed – repeat the recovery experiment to check the consistency of the results.

Unusual, old, lipemic, icteric, hemolyzed, or turbid sample used in recovery experiment – check the appearance and origin of the sample. If the sample is suspect, repeat the recovery experiment using fresh sample. Ideally several samples should be used.

Use of expired or poor-quality reagents (due to incorrect storage conditions, faulty refrigerator, or contamination) – check the condition of the reagent, the expiry date, and the storage temperature.

Incomplete mixing of reagents or calibrators – check the technique used for mixing. The precision (if duplicates run) may be poor, depending on the stage where the poor mixing occurred.

Protocol used was different from that in package insert – major differences between your protocol and that in the package insert should be tested for their effects on recovery by repeating the experiment using the exact protocol in the package insert.

Room temperature incubation carried out at temperature outside allowed range – if extreme temperatures have been experienced, check whether the ambient temperature was within the range specified in the package insert.

Incubator not allowed sufficient time to warm up – check manufacturer's instructions for warm-up time.

Equipment fault – check for other signs of poor assay performance such as poor precision, control bias, or low binding. These may give clues as to the likely cause. Test by substituting different equipment, if possible, or service and test the equipment according to the manufacturer's instructions.

Poor decanting technique or insufficient draining time (manual separations) – the precision of the duplicates may also be poor. Check the technique used.

Overaggressive tapping of microparticle-based assays after decanting (manual separations) – the precision of the duplicates may also be poor. Check the technique used.

Coated tubes or tubes containing PEG or magnetizable cellulose separation systems not tapped hard enough – check the tubes for residual droplets of moisture.

Tubes or microtiter plates are being decanted, placed the right way up, and then reinverted – the precision of the duplicates may also be poor. Check the technique used.

Contamination of outside of tube or well by tracer (especially immunometric assays) – this is characterized by individual replicates with higher than expected signal levels that become normal if the tube or well is wiped with a tissue and recounted.

Contamination by dust or fiber particles (especially enzyme immunoassays at the stage of signal development) – the precision of the duplicates is also likely to be poor. Check the environment. If tissues are used for tapping, check that they are not shedding fibers that interfere with the assay. Some latex gloves shed material that can interfere with enzyme-based signal generation systems.

Faulty or contaminated signal reader – check by reading the signal in an empty tube (or run a zero concentration TSH (thyrotropin) calibrator in a TSH immunometric assay and check how close signal is to zero).

Insufficient counting or signal reading time (manual counters or signal readers) – check whether sufficient signal was generated.

Curve-fitting error – if this is the cause, differences between *fitted* and *actual* calibrator values should be of a similar magnitude to the bias from 100% recovery. Note that curve-fitting error can affect the apparent concentration of the base or spiked samples, or both.

Technician error – ask another technician to repeat the entire experiment and check the consistency of the results.

If the within-assay precision is poor *see* POOR WITHIN-ASSAY PRECISION.

If the results from different recovery experiments are not consistent *see* POOR BETWEEN-ASSAY PRECISION.

If the signal level or curve shape is abnormal, see the relevant section.

Suggested Course of Action

1. Repeat the recovery experiment with several base samples and different amounts of added analyte. Compare the results with the previous recovery experiment. Check that the precision of the duplicates, curve shape, and control values are satisfactory. Investigate any obvious inconsistencies in the results.
2. Check the assay printouts for curve-fit bias near the concentrations of the sample before and after analyte was added (by comparing *fitted* and *actual* calibrator values).
3. Check that the calculation of recovery allowed for the dilution of the analyte in the base sample when the concentrated solution of analyte was added.
4. Identify differences between your protocol and that in the package insert. Test whether the protocol is involved by repeating the recovery experiment using the primary protocol in the package insert.
5. Consult the manufacturer for advice. Ask if any other users have reported poor recovery.

Checklist of Useful Information

It may be useful to have the following information available:

- reagent lot number(s);
- when the problem started;
- details of the analyte used, including the source of the analyte, the method of reconstitution, the storage temperature, and the age of analyte solution used;
- details of the calculation used to determine the recovery;
- assay printout(s) or relevant calibration data;
- details of differences between the protocol used and the protocol in the package insert.

POOR ASSAY SENSITIVITY

Assay sensitivity is the precision of measurement at zero concentration. (A related, but clinically more useful measurement is the between-assay precision of a control with a very low, but detectable, concentration of analyte.) Assay sensitivity should not be confused with clinical sensitivity, which is a measure of an assay's ability to detect a clinical condition, or functional sensitivity, which establishes the lowest concentration of analyte that can be determined at a given level of precision. Although assay sensitivity is an unreliable measure of the analyte detection limit capability of an assay, it is a very useful troubleshooting parameter as it is normally easy to measure and can differentiate between different possible causes of precision problems. Assay sensitivity is unique in that it measures the precision of the signal in the absence of any analyte.

Possible Causes

Reagent lot at fault – check the assay sensitivity of different reagent lots. Check whether the signal levels, control values, and curve shape are similar to those in previous assays and in the package insert. The sensitivity should be measured on more than one occasion with each lot. *F*-tests can be performed to determine whether differences between lots are statistically significant.

Poor immunoassay design or optimization – check the package insert or consult with the manufacturer to find out the claimed assay sensitivity. Ask the manufacturer for advice, other users may have commented on the assay sensitivity.

Variation of coated wells, beads, or tubes – this is difficult to differentiate from other possible causes. It can only be confirmed by running replicates from the affected lot alongside a good lot at the same time. Equipment should be thoroughly checked first.

Premature deterioration of reagent during shelf-life or exposure of reagent to extremes of temperature during transport – if reagents are deteriorating during their shelf-life the sensitivity will worsen as the reagents near expiry. If poor sensitivity is due to extreme temperatures during shipment, other reagents in the same shipment may also be affected. If it is likely that extreme temperatures were encountered, ask the manufacturer whether the reagents are sensitive to freezing or heating.

Insufficient slope of curve at very low concentrations – if the shape of the curve is flat at low concentrations, ask the manufacturer for advice. This may be due to bias in the curve-fit program being used.

Incomplete mixing of zero calibrator – check by mixing the zero calibrator before dispensing.

Dirty, shiny, or siliconized polystyrene assay tubes or cuvettes or use of polypropylene tubes with microparticle or liquid-phase assays – check by using an alternative source of polystyrene assay tubes or cuvettes.

Reagent not allowed to reach room temperature before use – ensure that reagents are allowed to reach room temperature before use. Volumes in excess of 100 mL may take more than an hour to reach ambient temperature.

Incomplete mixing of reagent (if freeze-dried or suspension) – check the appearance of the reagents for even mixing. Follow the instructions in the package insert.

Pipette, dispenser, or probe not primed – check the procedure.

Defective pipettes or automatic sampler/dispenser; dirty pipette tips; contaminated sample or reagent probe; poor connection between pipette and pipette tips; poor connection between probe and tubing; carryover of analyte from high-concentration sample; carryover of reagent by reagent-dispensing probe; splashing of sample or reagent – these are often characterized by worse precision in the first few replicates of the zero calibrator in a sensitivity test.

Use of reduced volumes or other difference from the normal protocol – major differences between your protocol and that in the package insert should be tested for their effects on precision.

Insufficient or overaggressive mixing of sample with reagents – in manual assays test by vortex mixing for longer at a lower speed.

Use of air incubator (except purpose-designed incubators) – large air incubators are unsuitable for many immunoassays. Test by incubating in a water-bath.

Incubator not allowed sufficient time to warm up – check the manufacturer's instructions for the warm-up time.

Insufficient centrifugation (time or *g* force) or excessive centrifuge braking – check the *g* force from the speed and radius of the centrifuge rotor using a nomogram. Try using the centrifuge with the brake off.

Faulty magnet or magnet not in close enough proximity to paramagnetic particles during separation – check separation module of analyzer or method of use of manual separation device.

Faulty washer – clean the washer thoroughly, using wire to clean the dispensing nozzles and flushing through with copious amounts of distilled water. Test the washer by dispensing intensely colored dye into the tubes or wells and monitoring the removal of the dye during washing.

Poor decanting technique or insufficient draining time (manual separations) – if this is the cause, sensitivity may vary between technicians.

Overaggressive tapping of microparticle-based tubes after decanting (manual separations) – check the technique used.

Coated tubes or tubes containing PEG or magnetizable cellulose separation systems not tapped hard enough – check the tubes for residual droplets of moisture.

Tubes or microtiter plates are being decanted, placed the right way up, and then reinverted – check the technique used.

Contamination of outside of assay reaction cuvette, well, or tube, or inside of incubator, by tracer (especially immunometric assays) – this is characterized by individual replicates with higher than expected signal levels that become normal if the cuvette, well, or tube is wiped with a tissue and the signal level re-measured.

Contamination by dust or fiber particles (especially enzyme assays at the stage of signal development) – check the environment. If tissues are used for tapping, check that they are not shedding fibers that interfere with the assay. Some latex gloves shed material that can interfere with enzyme-based signal generation systems.

Faulty or contaminated signal reader – check by reading the signal in an empty tube (or run a zero concentration TSH (thyrotropin) calibrator and check how close signal is to zero).

Insufficient counting or signal reading time (manual counters or signal readers) – check whether sufficient signal was generated.

Use of expired or poor-quality reagents (due to incorrect storage conditions, faulty refrigerator, or contamination) – check the condition of the reagent, the expiry date, and the storage temperature.

Suggested Course of Action

1. Ensure that the sensitivity has been calculated correctly. Carefully check the within-assay precision for low-concentration controls and the sensitivity from several assays for patterns. If duplicates are not normally run, temporarily include them to provide more data. If sensitivity or precision has deteriorated, when did the problem start? Check whether the incidence of the problem varies between technicians, indicating that technique is involved.
2. Try running replicates of the zero calibrator in different positions in the assay. Poor replicates in the corners of microtitration plates may indicate uneven warming in an incubator. Poor replicates at the beginning of an assay or when an analyzer is first used after a period of inactivity may be due to insufficient pipette or dispenser priming, or carryover on a probe.
3. Check for obvious causes such as insufficient draining or contamination with tracer or signal reagent. In immunometric assays, check the quality of the washing procedure by observing a dye being washed from the cuvettes, wells, or tubes. Clean the washer carefully.
4. Check the signal level or count rate, and the curve shape, to see if they are normal. If they are not, check the expiry date, appearance, and odor of the reagents.
5. Identify differences between your protocol and that in the package insert. Test whether the protocol is involved by repeating an assay with poor sensitivity using the primary protocol in the package insert.
6. Try using a completely different set of equipment, if possible, to test for equipment problems. If alternatives are not available, clean and check the equipment according to the manufacturer's instructions.
7. Consult the manufacturer for advice. Check that they calculate sensitivity in the same way. Ask if any other users have reported poor sensitivity.

Checklist of Useful Information

It may be useful to have the following information available:

- reagent lot number(s);
- when the problem started;
- details of any other coincident changes in performance noted, e.g. within-assay precision, change in binding at zero dose (%B_0), curve shape, ED_{50}, or NSB;
- assay printout(s) and sensitivity calculations;
- details of differences between the protocol used and the protocol in the package insert;
- details of all the equipment used;
- information on any coincident change in protocol, equipment, or technician.

Case Study

A laboratory reported that the error in an immunometric assay for thyrotropin (TSH) was such that hyperthyroid samples (low TSH) could not be distinguished from normal (euthyroid) patient samples with certainty. The error was also apparent as poor agreement between the duplicate points of the zero calibrator. (Cause – due to a faulty reagent dispenser, causing splashing, the signal reader in the analyzer was being contaminated with tracer.)

POOR CORRELATION BETWEEN TWO IMMUNOASSAYS

Possible Causes

Either immunoassay incorrectly standardized when developed – consult the manufacturers for advice. Proficiency testing (external quality assessment) schemes are a useful source of information about the relative biases of individual methods from the overall laboratory mean and help to identify methods that are out of consensus.

Reagent lot at fault (either immunoassay) – any change from a normal to an abnormal situation may be evident in your QC data. If it is due to the calibration of a specific lot of calibrators, the change will also be evident in the manufacturer's own QC charts.

Instability of manufacturer's reference standards (either immunoassay) – this will result in a stepwise shift in assay standardization, noticeable with the introduction of each new lot of calibrators.

Error or instability in stored calibration curve or manufacturer's calibration curve – the poor correlation should be corrected by recalibration.

Use of different specificity antibodies in each immunoassay – consult the manufacturers for advice. This is particularly noticeable with some immunometric assays calibrated against international standards.

Use of old samples from a sample library, that have deteriorated – repeat the correlation with fresh samples.

Limited range of concentrations used – this causes considerable statistical error in correlations. Check the confidence intervals for the slope and intercept and repeat the correlation with samples with a wider range of concentrations.

Limited number of patient samples used for study – if only a few samples are available but they cover a wide range of concentrations, repeat the correlation with the same samples to increase the statistical confidence in the result.

Only one or two assays performed – ideally, correlation studies require a number of assays to be performed to be valid because of assay-to-assay variation.

Curve-fit bias with either or both immunoassays – test by looking at the difference between *fitted* and *actual* calibrator values on the assay printouts. If this is the cause of the poor correlation the magnitude of the difference should reflect the difference between the results obtained with the two immunoassays, at each part of the concentration range.

Use of protocol different from that in the package insert (either immunoassay) – major differences should be tested for their effects on patient sample values by repeat testing of the samples with the exact protocol in the package insert.

Room temperature incubation carried out at temperature outside allowed range – repeat the test of the samples at the correct ambient temperature.

Equipment fault – control values and within-assay precision may also be affected. Test by substituting alternative equipment.

Use of different units or International Standards – check package inserts.

Misinterpretation of statistical validity of correlation – check the confidence intervals for the slope and intercept.

Interference due to sample collection device – this should be suspected if correlation is consistently poor for specific samples. Check the method of treatment of the samples.

Suggested Course of Action

1. If only one assay was carried out with either of the products or the samples were run on just one occasion on an analyzer, repeat the assay and compare with the original values.
2. Look at the printout from a proficiency testing (external quality assessment) scheme to find out if the differences between the products are typical.
3. Identify any differences between your protocols and those in the package inserts.
4. Examine recent control QC data for changes in mean control values between lots.
5. Check for certain analyte concentrations giving more discrepant results than others.
6. Check assay printouts for curve-fit bias (by comparing *fitted* and *actual* calibrator values in the calibration curve).
7. Check the range of sample concentrations; it may be that only a limited range has been covered.
8. Check the age and storage temperature of the samples used.
9. Check that the products are calibrated using the same units.

Checklist of Useful Information

It may be useful to have the following information available:

- reagent lot numbers;
- details of the types of controls and samples used;
- plots of the correlation with statistical calculations of correlation coefficient, slope (with confidence interval), and intercept (with confidence interval);
- details of any differences between the protocol used and the protocol in the package insert (either product);
- details of all the equipment used;
- details of control values in recent assays.

Case Study

Customers reported that a new cortisol immunoassay with a protocol for measuring cortisol in urine gave lower values than a previously established product. (Cause – the antiserum in the new product was more specific for cortisol and less affected by cross-reacting metabolites of cortisol in urine.)

Part 4

Applications

45 Thyroid

Rhys John, Carole A. Spencer, Nic Christofides
and David Wild

NORMAL THYROID FUNCTION

The thyroid gland lies in the anterior neck, just below the thyroid cartilage. The principle hormone arising from the thyroid is thyroxine (T_4), which is considered as a prohormone for the more metabolically active thyroid hormone triiodothyronine (T_3). Normally, approximately 20% of T_3 in the circulation originates directly from thyroid gland secretion. Most T_3 in blood is produced extrathyroidally from T_4 by a tightly regulated monodeiodination enzymic process, catalyzed by 5′-deiodinase enzymes. Normal thyroid hormone status is critical for the health of both children and adults. In infancy and childhood, thyroid hormone is essential for normal physical and mental development as well as growth. In the adult, thyroid hormones exert a wide range of actions that control protein synthesis, oxygen consumption, heat generation, and overall metabolic activity. It follows that thyroid hormone excess (hyperthyroidism) causes a state of hyperactivity, whereas thyroid hormone deficiency (hypothyroidism) is characterized by symptoms of slow metabolic activity such as tiredness, lethargy, and cold intolerance. However, since the signs and symptoms of both hyper- and hypothyroidism are nonspecific and often attributed to the aging process, biochemical testing is essential to determine whether any thyroid dysfunction is present.

All aspects of thyroid gland function are controlled by thyrotropin, also known as thyroid-stimulating hormone (TSH). TSH is secreted by the anterior pituitary gland in response to the hypothalamic tripeptide, TRH. Homeostasis of the hypothalamic-pituitary-thyroid axis is achieved through negative feedback inhibition by circulating thyroid hormone at both the hypothalamic and pituitary level (see Figure 45.1).

Iodine is an essential element for thyroid hormone production. An adequate supply of iodine in the diet (~100 μg/day) is therefore necessary for normal thyroid function. When dietary iodine is deficient, there is a compensatory thyroid gland hypertrophy (goiter). TSH stimulates the thyroid follicular cell to trap iodine from the circulation. The iodine is then activated and incorporated

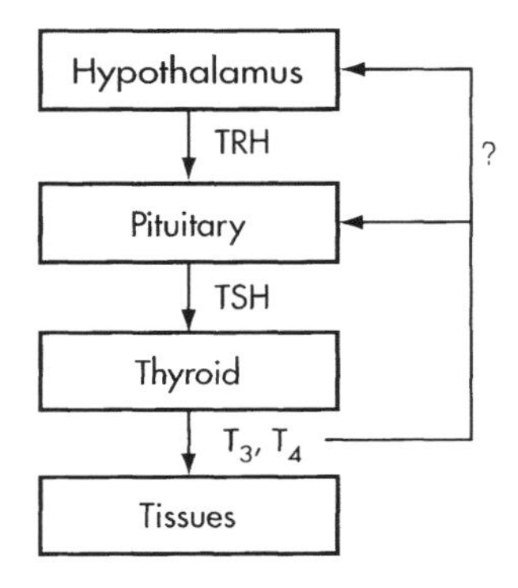

Fig. 45.1 Control of thyroid hormone secretion (simplified).

into the tyrosine residues of the large glycoprotein precursor molecule, thyroglobulin (Tg) by an enzymic process catalyzed by the thyroid peroxidase enzyme, which is located at the apical membrane of the thyroid follicular cell. Iodinated Tg is stored in the colloid of thyroid follicles and serves as a storage reservoir of thyroid hormone. Under the influence of TSH, Tg is endocytosed and digested by lyzosomal enzymes to release the thyroid hormones and some undigested Tg into the circulation.

Calcitonin is also synthesized in the thyroid gland. *See* BONE METABOLISM.

CLINICAL DISORDERS

People with normal thyroid function are described as **euthyroid**.

HYPOTHYROIDISM (THYROID HORMONE DEFICIENCY)

Congenital hypothyroidism presents at birth and unless it is diagnosed and treated within the first few weeks of life, it causes growth retardation and irreversible brain damage. The most common cause is due to a **thyroid dysgenesis** but there are also many rare inherited enzyme deficiencies, which are collectively described as **dyshormonogenesis**. The incidence of congenital hypothyroidism is about 1 in 3500 newborns. It is detected in many countries by screening infants a few days after birth by measuring TSH or T_4 in a blood sample obtained from a heel prick and absorbed onto a filter paper disk. TSH increases immediately after birth, but falls to normal within 24–48 h. In infants with congenital primary hypothyroidism, blood spot TSH concentrations are usually increased well above those in normal infants, so that screening with TSH is a more sensitive and specific test than screening with T_4. However, the rare infant with hypothyroidism due to hypothalamic or pituitary disease (central hypothyroidism) is missed using TSH screening. Since some infants have transient hypothyroidism, follow-up confirmatory tests are always necessary.

Childhood hypothyroidism. The most common feature of hypothyroidism in children and adolescents is failure in growth. There is a moderate increase in weight and bone retardation occurs. In children over 4 years of age, the most common cause of hypothyroidism is Hashimoto's thyroiditis and occurs most commonly in association with deficiency of other endocrine glands: pancreas, adrenals, gonads, and parathyroids. In places without a congenital hypothyroid screening program, failure of hypoplastic or ectopic thyroid tissue, as growth occurs, results in cases of childhood hypothyroidism.

Primary hypothyroidism. Most cases of hypothyroidism in adults arise from spontaneous thyroid failure caused by autoimmune thyroid disease (**Hashimoto's thyroiditis**) which progressively destroys thyroid gland function over a period of years. The prevalence in the general population approximates to 2%. Typically, hypothyroidism presents in differing degrees ranging from mild (subclinical) to overt. The prevalence of this condition increases with age and is fourfold greater in women as compared to men. Thyroid failure also frequently occurs in the months or years following therapies for hyperthyroidism (surgery, radioiodine ablation, or overtreatment with antithyroid drugs). The biochemical profile of overt hypothyroidism is typically marked by an elevated TSH associated with a subnormal serum free T_4. The classic clinical presentation includes clinical symptoms such as tiredness and lethargy, slow mentation, intolerance of cold, increased menstruation, constipation, weight gain, dry skin, edema, deafness, and loss of balance. Thyroid autoantibodies, especially thyroid peroxidase antibody (TPOAb), are usually detected and are a good marker for the underlying autoimmune process.

Mild (subclinical) hypothyroidism. This condition is characterized by an elevated serum TSH associated with a normal range free T_4. Studies suggest that the prevalence of mild hypothyroidism may be as much as 15–20% in elderly female patients. This TSH abnormality is now considered as an indicator of thyroxine deficiency, as judged by studies demonstrating abnormalities in tissue markers of thyroid hormone action and psychometric testing. Hence, the American Thyroid Association is now recommending that the term 'subclinical hypothyroidism' be replaced by the term 'mild hypothyroidism'. Since most cases of mild hypothyroidism (high TSH/normal T_4) are secondary to Hashimoto's thyroiditis and progress to overt hypothyroidism (high TSH/low T_4) at a rate of 5% per year, physicians increasingly recognize the value of diagnosing and treating this condition. A TPOAb measurement is an indicator of the severity of the underlying autoimmune condition, since the higher the TPOAb the more rapidly the thyroid failure progresses. Both overt and subclinical hypothyroidism are often missed in the older woman, since the symptoms of thyroxine deficiency have an insidious onset and are nonspecific, such that progressive thyroid failure is often attributed to the aging process.

Secondary hypothyroidism, caused by the failure of hypothalamic TRH and/or pituitary TSH to stimulate the thyroid gland, accounts for $<1\%$ of all cases of thyroid failure. In some cases of secondary hypothyroidism, serum TSH is not necessarily low but is often paradoxically normal or slightly elevated (5–10 mU/L) range. It is now known that the biological activity of the TSH secreted by a damaged or understimulated pituitary is impaired and this explains the inappropriately normal TSH. Secondary hypothyroidism may result from a variety of disease conditions or arise from hypothalamic or pituitary trauma. Usually, pituitary dysfunction is not confined to the thyroid axis but is evident from deficiencies in adrenal and gonadal functions. An impaired or delayed TSH response to TRH stimulation ($<$twofold rise) is an indicator of secondary thyroid failure.

Both primary and secondary hypothyroidism are effectively treated by exogenous L-T_4 therapy, which serves as an adequate source of substrate for T_3 production in extrathyroidal tissues (*see* L-THYROXINE REPLACEMENT THERAPY).

HYPERTHYROIDISM

There are three main causes of hyperthyroidism, namely, Graves' disease, toxic multinodular goiter, and toxic adenoma, and there are other rarer causes. In the UK, Graves' disease is the most common, but in iodine-deficient areas, toxic adenoma is more common. Hyperthyroidism is approximately 10 times more common in women than men, and in women has a prevalence of 2.5–4.7 per 1000. Graves' disease occurs most commonly in women between 20 and 50 years of age, but toxic multinodular goiter occurs more commonly in the over 60-year age group.

Graves' disease is caused by thyroid-stimulating autoantibodies (TSAb) which bind to the TSH receptor and mimic the action of TSH on thyroid cells.

The metabolic effects of thyroid hormone excess produce tiredness, excess sweating, and dyspnea on exertion. Most patients notice weight loss and decreased muscle mass, despite an increased appetite. Additional characteristic features include opthalmopathic manifestations including protruding, staring eyes (exophthalmos) and an enlarged thyroid gland.

TSAb stimulation causes excess T_3 and T_4 secretion, which suppresses pituitary TSH secretion to $<1/100$ of euthyroid levels (<0.02 mU/L) which can now be detected by the current third generation TSH immunometric assays.

Graves' disease can be treated with antithyroid drugs, such as carbimazole, propylthiouracil, or methimazole, which are typically taken for 12–24 months. If the disease recurs after cessation of the drug treatment, it may be treated by radioiodine ablation, or surgical removal of the thyroid gland.

Hyperthyroidism due to a **toxic multinodular goiter** (**TMNG**) is a condition where thyroid hypertrophy and nodularity develop, usually over a long period of time (years) often in response to iodine deficiency. As the thyroid enlarges, autonomy develops and thyroid hormone concentrations increase. There is typically an inverse relationship between thyroid size and serum TSH. However, the degree of thyroid hormone excess and the magnitude of TSH suppression are usually less than in Graves' disease. Despite a milder degree of thyrotoxicosis, patients with TMNG tend to be older than Graves' patients and more prone to atrial fibrillation, secondary to the exacerbation of intrinsic cardiac disease by thyroid hormone excess.

A **toxic adenoma** may exist for some time before hyperthyroidism develops, with a corresponding decrease in TSH levels. Recent studies suggest that the autonomy in such nodules represents constitutive activation of the TSH receptor secondary to a genetic mutation. In some cases the tumor secretes T_3 predominantly causing **T_3-toxicosis**. In this condition free T_3 is elevated but free T_4 is normal. It should be suspected when clinical symptoms suggest hyperthyroidism but free T_4 is within the reference interval and serum TSH is suppressed. **T_4-toxicosis**, where free T_4 is elevated but free T_3 is normal, can also occur. Toxic adenoma is most effectively treated by surgical removal. **Thyroid carcinoma** is a very rare cause of hyperthyroidism, as is hyperthyroidism caused by excessive secretion of TSH from the pituitary. These latter patients have normal or increased TSH concentrations in the presence of increased free thyroid hormone concentrations, leading to the term 'inappropriate TSH secretion' to describe this condition. These result from either a TSH-secreting pituitary adenoma or nontumorous TSH hypersecretion due to thyroid hormone resistance. Thyroid hormones may be secreted by ectopic thyroid tissue in an ovarian teratoma (**struma ovarii**). The concentrations of thyroid hormones may increase and TSH decrease, as in thyroid tumors.

The symptoms of hyperthyroidism can result from a hypothyroid patient receiving too high a dose of T_4 therapy. **Thyrotoxicosis factitia** is the name given to a condition in which a formerly euthyroid person takes excessive amounts of thyroid hormones, for example, in a misguided attempt to lose weight. Serum T_4 and T_3 concentrations are elevated, unless the patient has taken T_3, in which case T_4 is subnormal and T_3 is increased.

Hyperthyroidism can be induced in patients with goiters if iodine is administered. This condition is known as **iodine-induced hyperthyroidism** or the **Jod–Basedow** phenomenon.

Hyperthyroidism may be caused by **de Quervain's subacute thyroiditis** (also known as **granulomatous thyroiditis** or **silent thyroiditis**), which is probably due to a viral infection and causes inflammation of the thyroid. The inflammation causes leakage of excess thyroid hormones into the blood from the reservoir of hormone normally held in the thyroid gland. This can result in a moderate elevation in serum thyroid hormone concentrations. Autoimmunity may play a role in the etiology of subacute thyroiditis and thyroid autoantibodies can sometimes be detected. Patients either return to normal or pass through a euthyroid state temporarily, becoming hypothyroid for several months before finally returning to normal thyroid function. A few of these patients become hypothyroid years later. The **postpartum thyroiditis syndrome** has a similar pattern to subacute thyroiditis. Transient hyperthyroidism occurs after delivery of the baby, followed by a period of hypothyroidism for several months and, usually, by a return to the euthyroid state. Most patients have low positive TPO antibody levels. Postpartum thyroiditis occurs in 5–10% of women after giving birth.

NONTOXIC GOITER

Simple or nontoxic goiter is the term used to describe thyroid enlargement without hypo- or hyperthyroidism and not due to inflammation or a tumor.

NONTHYROIDAL ILLNESS/SICK EUTHYROID SYNDROME

Severe illness or injury can induce changes in thyroid hormone concentrations. The conversion of T_4 to T_3 in the tissues is inhibited, causing a decrease in T_3 concentration. Free thyroxine index (FTI) and labeled analog free T_4 tests are unreliable in this situation, tending to give an underestimate of the true value. The latest labeled antibody tests for free T_4 are designed to provide unbiased free T_4 values in nonthyroidal illness (NTI). In seriously ill patients, TSH, T_3, and T_4 may all decrease, and a severe decrease is associated with a poor prognosis. As patients with less severe NTI recover, T_3 gradually returns to normal, with a transient increase in TSH, which also returns to normal.

L-THYROXINE REPLACEMENT THERAPY

A normal serum TSH (~1–2 mU/L) is now the recognized therapeutic endpoint for levothyroxine replacement therapy. This is usually accomplished with a dose of 1.6 μg/kg/day. Normally, about one-fifth of T_3, the active thyroid hormone, originates directly from thyroid secretion. The remainder is produced in peripheral tissues from T_4 by monodeiodination. In hypothyroidism the majority of T_3 must be produced extrathyroidally. In hypothyroidism the endogenous T_3 component is lost and it is necessary to maintain free T_4 in the upper half of the normal range for TSH to be normalized. Since it takes time for the pituitary to re-equilibrate to any change in free thyroxine status, it is necessary to wait for 6–8 weeks after initiating L-T_4 therapy or changing the L-T_4 dose before checking the TSH status. Patients who are noncompliant may have normalized their serum T_4 by taking a large dose prior to their clinic visit. Noncompliance is the main reason for a TSH/free T_4 discordance in an L-T_4 treated patient.

ENDOGENOUS ANTIBODIES TO T_3 AND T_4

A few patients with silent thyroiditis, primary hypothyroidism or Graves' disease have antibodies to T_3, T_4 or both hormones in their blood. They tend to occur in patients who also have increased concentrations of thyroglobulin antibodies. If present, they can affect immunoassays for total and free hormones by increasing the quantity of T_3 or T_4 antibodies. This changes the apparent concentration of hormone in the assay because competitive immunoassays rely on a constant, limited amount of antibody being present in both calibrator and sample tubes. The latest labeled antibody free T_4 tests have been designed to be resistant to interference by T_4 antibodies. Endogenous antibodies to thyroid hormones can also cause an increase in total thyroid hormone concentration while the free hormones remain normal. Endogenous antibodies should be suspected when clinical symptoms do not match the laboratory test results.

FAMILIAL DYSALBUMINEMIC HYPERTHYROXINEMIA

Familial dysalbuminemic hyperthyroxinemia (**FDH**) is a rare inherited autosomal dominant condition in which the albumin is in an altered form that has an increased affinity for T_4 (and sometimes T_3). In this condition the concentration of total T_4 is elevated, although free T_4 is normal. Thus, FTI determinations and T_4/thyroxine-binding globulin (TBG) ratios are increased and inappropriately high for their TSH concentration. Whilst earlier analog free T_4 gave erroneously increased results in this condition, current labeled antibody free T_4 assays give normal results.

ANALYTES

THYROTROPIN

Thyrotropin, or TSH, is a 29 kDa glycoprotein hormone consisting of two subunits, designated α and β. The α subunit is common to TSH, luteinizing hormone (LH), follicle-stimulating hormone (FSH) and human chorionic gonadotropin (hCG). Although there are some similarities between the β subunits of these hormones, the differences between the amino acid sequences and oligosaccharides give them unique biological activities.

Function

TSH is secreted from the anterior pituitary, and it acts upon the thyroid gland, stimulating the production of T_4 and T_3. TSH is controlled by a negative feedback system that maintains a constant concentration of free thyroid hormone in serum.

Reference Interval

0.465–4.68 mIU/L 2nd IRP 90/558

(Vitros™ ECi TSH)

Clinical Applications

In primary hypothyroidism, the thyroid gland fails to produce sufficient T_4, so that TSH increases, and in severe cases by several orders of magnitude. The congenital hypothyroidism of infants can be diagnosed by measuring TSH in blood spots taken soon after birth (generally after day 3).

In secondary hypothyroidism, TSH concentration can be low but is often normal or even slightly increased, but it is associated with a low free T_4 concentration.

In hyperthyroidism, T_4 and T_3 are increased causing TSH secretion to be suppressed. This decrease is detectable in ultrasensitive immunometric assays for TSH (see Figure 45.2).

TSH immunometric assays vary in their ability to discriminate between normal and suppressed TSH concentrations. The assay sensitivity of a TSH assay gives only an indication of the assay's ability to discriminate hyperthyroid from normal. Sensitivity is a measurement of the within-assay error at a concentration of zero, normally quoted as the concentration of TSH that is two standard deviations above zero. The minimum detection limit is a more relevant parameter than the sensitivity, being the concentration of TSH that is statistically significant from zero, using the between-assay precision profile. Another useful measurement is

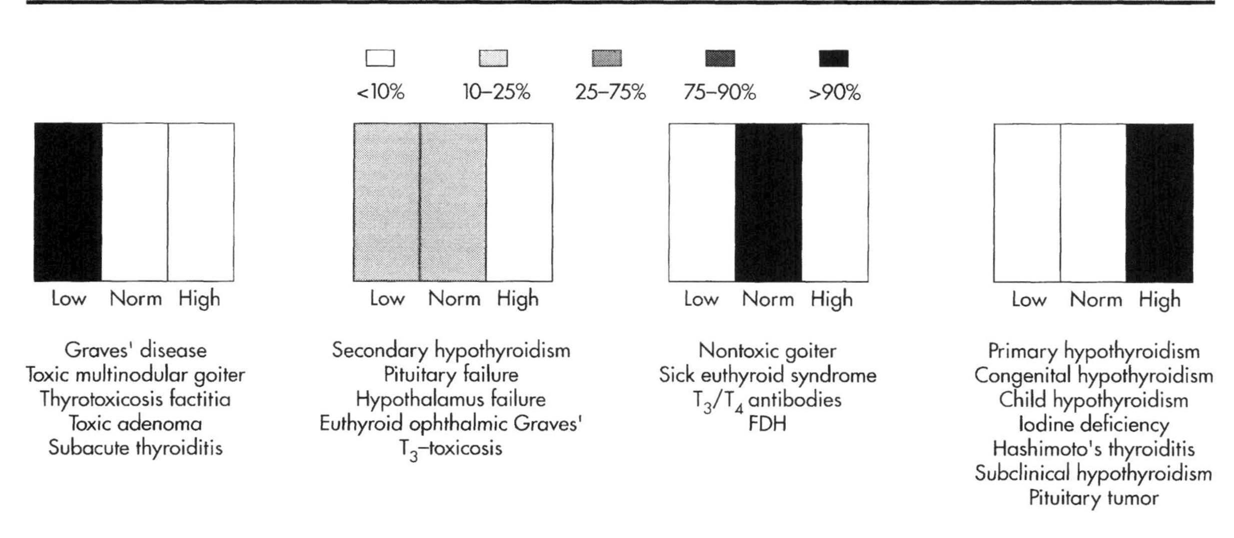

Fig. 45.2 High-sensitivity thyrotropin (hs-TSH).

the percent coefficient of variation (%CV) of a control or patient sample pool with a TSH concentration near the lower end of the reference interval. This should be determined over a number of assays (ideally at least 20) and using several batches of reagents. However, the measure that has received widespread support is the functional sensitivity, which is the concentration at which the %CV is at a certain level, e.g. 20%. It is determined by plotting precision (%CV) vs. concentration. Klee and Hay (1989) proposed five criteria that TSH assays should fulfill if they are to be used as single tests for front-line screening for thyroid disorders (*see* FURTHER READING). However, many laboratories are now using TSH and free T_4 together in front-line screening as neither test is considered reliable enough to be used alone in every clinical situation.

Limitations

In screening a well population for thyroid disorders, TSH is the most sensitive indicator of hypo- or hyperthyroidism, and when combined with FT_4 and/or FT_3, allows all subjects to be categorized as hypo-, eu-, or hyperthyroid. However, in ill patients, TSH concentrations as well as FT_4 and FT_3 may be affected by the illness or the drugs the patients are taking.

- In severe NTI, TSH can be suppressed below the reference range. However, using third generation TSH assays, the suppression due to NTI is not as great as that seen due to hyperthyroidism. In less sensitive assays, there may be overlap in the lowered TSH concentrations due to NTI and the TSH concentration due to hyperthyroidism.
- Both dopamine and glucocorticoid, used in the emergency room for medical treatment, can lower TSH concentrations.
- TSH may be reduced below the reference interval due to diurnal rhythm, starvation, or depression, and in the first and second trimester of pregnancy.
- TSH on its own will miss cases of secondary hypothyroidism as TSH may often be within the reference interval or even slightly increased (due to the secretion of a less biologically active TSH).
- TSH tends to react to T_4 treatment for hypothyroidism more slowly than free T_4, sometimes lagging behind by several months. TSH in patients successfully treated with T_4 replacement therapy tends to be lower than normal. This is because the ratio of T_4 to T_3 is increased compared to the normal situation where T_4 and T_3 are both secreted by the thyroid. Because T_3 is the most active hormone in the tissues, the overall concentration of thyroid hormones needs to be higher, and this depresses TSH secretion.
- As with all two-site immunometric assays, heterophilic antibodies can interfere with TSH assays, generally causing an erroneously increased TSH result by cross-linking the capture and labeled antibodies, but occasionally producing an artifactually low TSH result by preferentially binding labeled antibody. Manufacturers minimize this interference by the addition of nonimmune serum to the assay systems, but it can still present a problem in the occasional patient.

Assay Technology

Most laboratories now use immunometric assays for TSH, as competitive immunoassays are insufficiently sensitive to distinguish between normal and hyperthyroid patients. Because of the very wide range of concentrations that need to be measured, there are several critical aspects of assay design that have to be optimized correctly, and there are significant differences in quality between the many products on the market.

Most TSH assays utilize a capture antibody immobilized onto beads, tubes, wells, or microparticles. Some assays have a single incubation format in which sample and labeled antibody are incubated together with the immobilized capture antibody. Many have a dual incubation format in which the sample is incubated with the capture antibody first, followed by a washing stage before the labeled antibody is added. The principles of immunometric assays are explained in Part 1 and many of the common products are described in detail in Part 2 of this book.

Products commonly used to measure TSH include the AxSYM™ (microparticle enzyme immunoassay) and Centaur™ (acridinium ester-based luminescence).

Signal measurement and separation equipment such as counters and washers must be of a high specification and be kept properly maintained for TSH assays. Other sources of error in TSH assays are carry-over in sample dispensers, calibration error in very low-concentration calibrators, the presence of TSH in the 'zero' calibrator, and bias in the curve-fit at low concentrations. The solid phase must have a high capacity for the capture antibody and the tracer must have a high specific activity, to give a good signal-to-noise ratio and a linear relationship between signal and TSH concentration.

Some TSH immunometric assays are affected by interfering factors in a small proportion of patient samples. These can nonspecifically cross-link the capture and labeled antibodies. An example is 'anti-mouse' activity which can cross-link mouse monoclonal antibodies. This particular form of interference can be suppressed by the presence of mouse serum in one of the assay reagents.

One of the controls included in routine assays should have a concentration close to the lower limit of the reference interval, to check the clinical discrimination between hyperthyroid and normal patient samples.

Types of Sample

Serum or plasma. Some assays are restricted to serum samples. Screening for congenital hypothyroidism in infants is carried out using blood spots absorbed onto filter paper.

Frequency of Use

Very common.

THYROXINE

Thyroxine (T_4) is produced in the thyroid gland by the coupling of two diiodinated tyrosine molecules (see Figure 45.3). The full name is tetraiodothyronine as it contains four atoms of iodine.

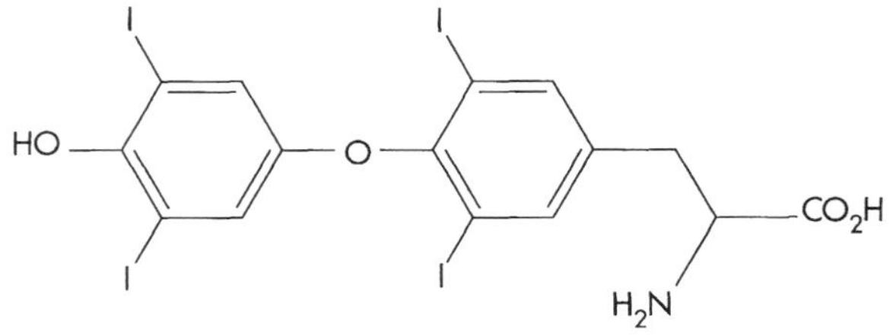

Fig. 45.3 Structure of thyroxine (T_4).

Function

T_4 is the thyroid hormone with the highest concentration in the body. It is normally present in blood at a concentration that is 50–100 times greater than that of T_3. However, T_3 is several times more biologically active than T_4. The most important function of T_4 is to serve as the precursor and primary source of T_3 in the tissues through deiodination. For the effects of thyroid hormones *see* NORMAL THYROID FUNCTION.

Approximately 99.98% of T_4 and 99.7% of T_3 are bound to specific binding proteins in blood, but it is the minute fraction of free, unbound hormone that is biologically active, as only the free hormone can pass into the cells in the target tissues. Total hormone concentrations in blood not only reflect the balance between output of the thyroid gland and metabolism, but also depend on the concentration of specific binding proteins. As there are a number of situations where abnormal concentrations of binding proteins occur, total hormone concentrations also increase or decrease. However, the free hormone concentration that controls the negative feedback mechanism is normal, reflecting the euthyroid state in these subjects.

Reference Interval

71–141 nmol/L (Vitros™ ECi)

5.5–11.0 μg/100 mL (Vitros™ ECi)

Clinical Applications

The concentration of T_4 in serum or plasma is an indicator of thyroid status. In most cases, T_4 concentration is reduced in hypothyroidism and increased in hyperthyroidism (see Figure 45.4).

Limitations

- Abnormal amounts of thyroid hormone binding proteins result in changes in the total T_4 concentration in euthyroid people, because it is the free, unbound fraction of T_4 that is maintained constant. Pregnancy, oral contraceptives, and various drug treatments cause increased concentrations of binding proteins. Decreases may occur in NTI, and as a side-effect of certain drug

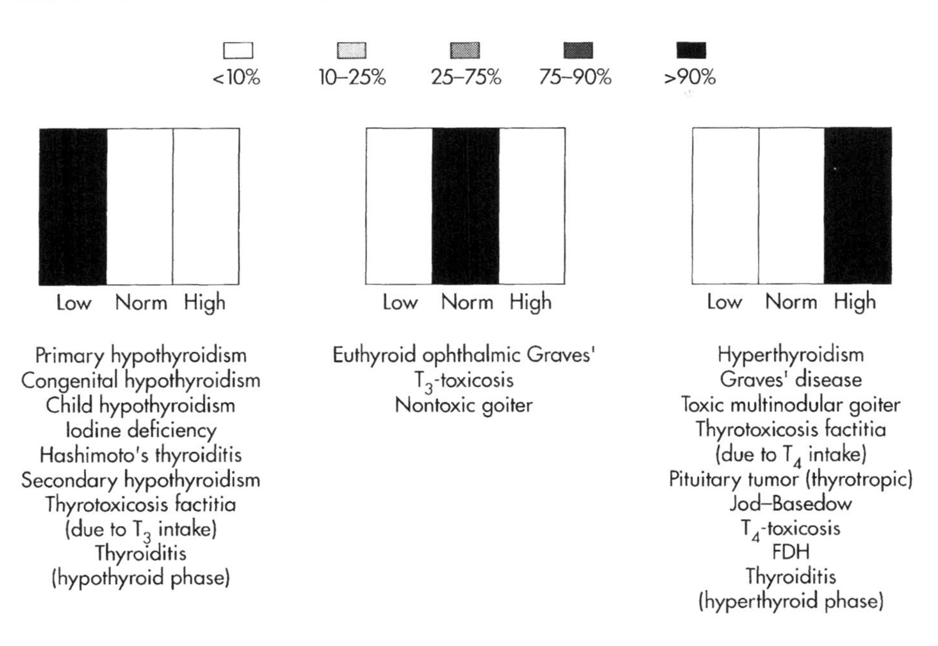

Fig. 45.4 Thyroxine (T_4).

treatments. For these reasons, most laboratories either measure free T_4 directly or calculate the FTI, using total T_4 in conjunction with T_3 uptake, which tests thyroid hormone binding capacity in the serum. A few laboratories measure TBG and T_4 and calculate the T_4/TBG ratio.

- Some drugs displace thyroid hormones from their binding sites, resulting in a reduction in the total T_4 hormone concentration while the free hormones remain normal, e.g. phenytoin, phenylbutazone, and salicylates.
- Amiodarone treatment results in an increase in serum total T_4 concentration.
- T_4 is normal in T_3-toxicosis.
- In FDH the proportion of T_4 that is bound by albumin is greatly increased, resulting in an increased total T_4 concentration. The free T_4 will be at a normal level. (Some free T_4 assays also give artifactually high values in FDH.)
- T_4 immunoassays are affected by endogenous antibodies to T_4 if they are present in patient samples. In most types of assays the apparent concentration of T_4 is increased. In addition, patients with T_4 antibodies may have a higher level of total T_4 but a normal free T_4 concentration.

Assay Technology

The most common technology used to measure T_4 in the USA and Japan is fluorescence polarization immunoassay (AxSYM™).

Total T_4 assays contain one or more decoupling agents, such as 8-anilino-1-naphthalene sulfonic acid (ANS) or salicylate, to displace the T_4 from the binding proteins in the serum.

Types of Sample

Serum or plasma, except in screening newborn infants for congenital hypothyroidism, for which blood spots dried onto filter paper are used.

Frequency of Use

Very common.

T_3 OR T UPTAKE AND CALCULATION OF FREE THYROXINE INDEX

As explained above, the free T_4 concentration is a more relevant indicator of thyroid function than total T_4. The concentration of free T_4 in serum depends on the relative concentrations of total T_4 and the three binding proteins: TBG, albumin, and thyroxine-binding prealbumin (TBPA). About 99.98% of T_4 is bound (70% to TBG, 20% to albumin, and 10% to TBPA). Thyroid hormone uptake tests measure the thyroid hormone binding capacity of the serum so that the FTI can be determined, using the thyroid hormone uptake and total T_4 test results.

Clinically, T_3 and T uptake tests are indistinguishable. The difference is in the methodology (*see* ASSAY TECHNOLOGY).

Reference Interval

23.5–40.6% (Vitros™ ECi)

1.40–0.75 T_3 uptake units (Vitros™ ECi)

Clinical Applications

Thyroid hormone uptake results are used in conjunction with total T_4 to calculate the FTI, using the following equation:

$$\text{Free } T_4 \text{ index} = \text{total } T_4 \times \% \text{ uptake}/100$$

The FTI is used as an indicator of thyroid status. In nearly all cases it is increased in hyperthyroidism and decreased in hypothyroidism. The FTI is a better measure of thyroid status than total T_4 alone as it estimates the amount of free T_4 activity. The free T_3 index can also be calculated using T_3 and T_3 uptake (see Figure 45.5).

Limitations

- In euthyroid patients with circulating T_4 antibody, total T_4 increases but the free T_4 remains normal. The FTI is elevated because of the increased total T_4 in serum. In addition, the antibodies interfere directly with T_4 immunoassays, which depend on a constant amount of antibody being added to each sample.
- In FDH the proportion of T_4 bound by albumin is increased, which increases the total T_4 concentration. The free T_4 concentration is normal. This results in an increase in FTI.
- In patients with NTI, the serum is thought to contain a factor that releases T_4 from the binding proteins preferentially over T_3. In this situation, the FTI underestimates the free T_4 concentration.
- Some drugs displace thyroid hormones from their binding sites, eventually causing a reduction in the total T_4 hormone concentration while the free hormones remain normal. This results in a lower FTI value, e.g. phenytoin, phenylbutazone, and salicylates.
- Some drugs cause an increase in serum total T_4 concentration, and hence in the FTI value, e.g. amiodarone.
- FTI is normal in T_3-toxicosis.
- FTI is depressed in very sick patients.
- FTI may be abnormal in euthyroid patients with familial high or low TBG concentrations.

Assay Technology

Traditional radioactive methods for determining T_3 uptake measure the proportion of ^{125}I-labeled T_3 bound to a secondary binder (such as an ion-exchange resin, macroaggregated albumin, or antibody to T_3) in the presence of the sample. A high percentage uptake in such tests means that the sample has a low concentration of unoccupied binding sites in the serum. The TDx, IMx, and AxSYM tests for T uptake measure the binding of a T_4-fluorescein tracer to thyroid hormone binding proteins

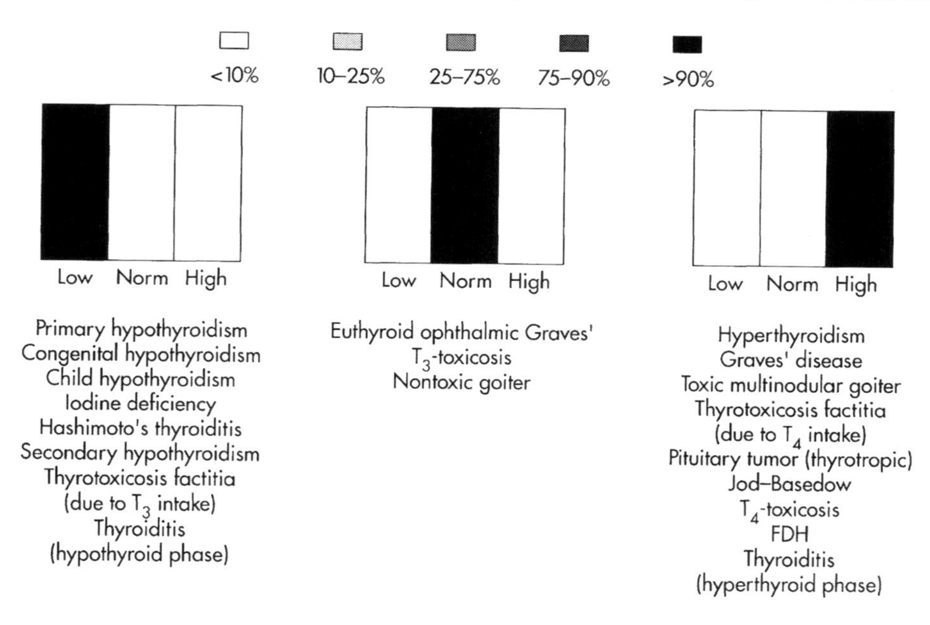

Fig. 45.5 Free thyroxine index (FTI).

directly. In this test a normal thyroid hormone binding capacity is given a value of 1. In contrast to the percentage uptake scale, lower values correspond to lower binding capacities in the sample.

Labeled T_3 is used in preference to T_4 in most assay designs because T_3 has a less intense binding to serum proteins, so that signal levels bound to the solid phase are higher, reducing counting times. This is satisfactory for determining the FTI, as T_3 and T_4 bind to the same sites on TBG, the principal binding protein (see Figure 45.6).

Several homogeneous assays are commonly used. These include CEDIA™ and EMIT™ format assays used on clinical chemistry analyzers, and assays for the aca™ and Dimension™.

Most thyroid hormone uptake tests are not immunoassays as only a few employ antibodies in the technology. However, they are described here because they are similar in design to immunoassays and have similar protocols. As in immunoassays, thyroid hormone uptake tests use some form of labeled tracer to generate a test signal that is proportional or inversely proportional to the concentration of thyroid hormone binding sites in the blood sample.

The technology used to test thyroid hormone uptake mirrors that used for total T_4, as laboratories tend to carry out both tests together. One of the most common technologies used to measure thyroid hormone uptake is fluorescence polarization (e.g. AxSYM). This assay measures the amount of fluorescein-labeled T_4 that binds to thyroid hormone binding proteins in the serum.

Other radioassays use alternative secondary binders to measure the amount of ^{125}I-labeled T_3 that remains unbound after incubation with the sample, such as macroaggregated albumin or resin.

Types of Sample

Serum or plasma.

Frequency of Use

The use of FTI is declining as more laboratories change over to direct free T_4 assays.

FREE T_4

The concentration of free T_4 is independent of the concentration of binding proteins and is the biologically active fraction of T_4. Free T_4 assays have to determine concentrations of just a few pg/mL in the presence of bound T_4, which circulates at 5000 times the free T_4 concentration. A conventional competitive immunoassay cannot be used, as some of the labeled T_4 would bind to the binding proteins as well as to the antibody, so that the final measurement would depend on the concentration of the binding proteins. Also, antibodies with a high affinity for T_4 could remove some of the T_4 from the binding proteins, increasing the apparent free hormone concentration.

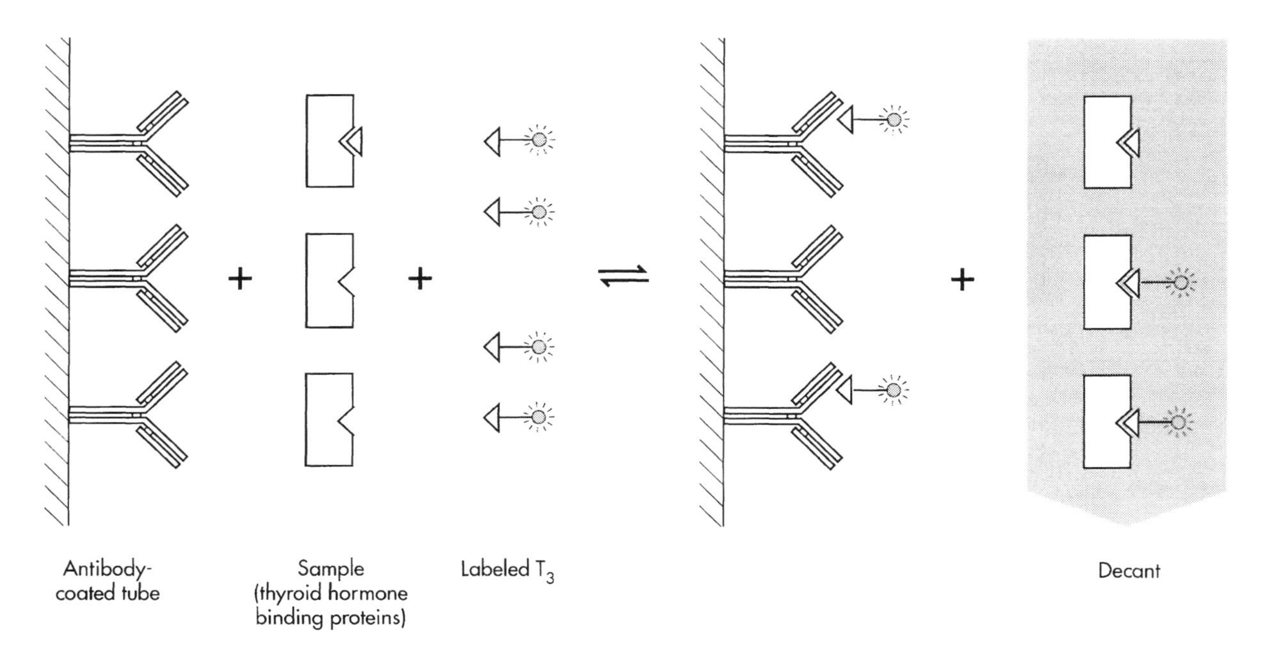

Fig. 45.6 Principle of T_3 uptake test.

Most free T_4 assays do not measure free T_4 directly, but are indirect methods. The principles are described in FREE ANALYTE IMMUNOASSAY. Standardization of free T_4 assays is against the more direct methods of equilibrium dialysis or ultrafiltration.

Free T_4 assays have the practical advantage that only one initial test of thyroid function is needed instead of two (T_4 and T_3 uptake). More recent free T_4 assays overcome most of the limitations of FTI measurements mentioned previously.

Reference Interval

10–28 pmol/L (Vitros™ ECi)

0.78–2.19 ng/100 mL (Vitros™ ECi)

Clinical Applications

As for T_4, free T_4 is important in the investigation of all patients suspected of thyroid disorders and can be of use in monitoring patients undergoing treatment. In most cases, free T_4 is increased in hyperthyroidism and decreased in hypothyroidism (see Figure 45.7).

Limitations

- As discussed in the chapter FREE ANALYTE IMMUNOASSAY, there are significant methodological differences in FT_4 assays resulting in numerical disagreement between methods. These differences are especially evident in sera with low T_4 binding capacities.
- Some immunoassays may yield artifactually high FT_4 concentrations in patients with FDH. In these patients the mutant albumin has an affinity towards T_4 which is approximately 80 times higher than normal albumin. The artifactual elevation of FT_4 may be the result of interaction of this albumin with the immunoassay reagents (e.g. analog tracer) or due to the presence of specific buffer ions (e.g. chloride ions) that cause dissociation of T_4 from the mutant albumin.
- In patients successfully treated with T_4 replacement therapy, free T_4 tends to be about 20% higher than normal. This is because the ratio of T_4 to T_3 is increased, compared to the normal situation where T_4 and T_3 are both secreted by the thyroid, and T_3 is the most active hormone in the tissues. Normal levels of free T_4 may be associated with undertreated patients.
- Transient hormonal abnormalities may be found in systemically (nonthyroidal) ill patients.
- Patients receiving heparin can have biased free T_4 results as heparin stimulates the production of nonesterified fatty acids, which displace T_4 from albumin. Heparin can also arise from indwelling cannulas containing a heparin-saline solution.
- Free T_4 is normal in T_3-toxicosis.
- Free T_4 may be increased in patients receiving treatment with amiodarone.
- Transient elevations in FT_4 concentrations may be observed following administration of drugs (e.g. furosemide, ketoprofen, phenylbutazone, mefenamic acid, probenecid, sulindac, and fenclofenac) that displace T_4 from their binding proteins.

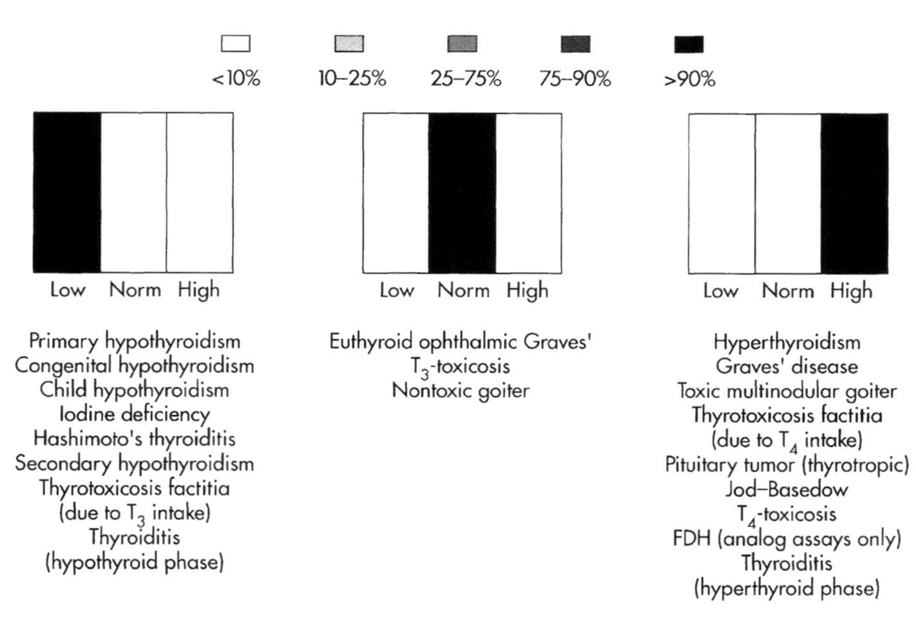

Fig. 45.7 Free thyroxine (free T_4).

- FT_4 concentrations fall during the second and third trimester of pregnancy and thus the use of pregnancy-specific euthyroid ranges should be established and used.
- Some patient sera containing avid autoantibodies to thyroid hormones may interfere in FT_4 methods based on one-step methodologies.
- Although manufacturers include materials to minimize interference by heterophilic antibodies, there is no guarantee that all such sera will be fully corrected.

Assay Technology

The physicochemical principles governing the measurement of free T_4 have been discussed in the previous chapter on FREE ANALYTE IMMUNOASSAY.

This section provides a summary of some of the FT_4 methodologies that are in routine use. The 1998 College of American Pathologist (CAP) scheme survey indicated that approximately 1800 laboratories use FT_4 assays obtained from 19 vendors. It is interesting to note that the number of TT_4 users (approximately 2000) is higher than the number using direct FT_4 assays. Seventy-five percent of the TT_4 users do, however, also use T- or T_3-uptake tests to derive an indirect measure of the free T_4 concentration. Thus, currently only approximately 10% of users rely solely on TT_4 measurement. This number is lower than the corresponding number obtained in 1995 (approximately 20%), indicating that an increasing number of laboratories have discovered that measurement of FT_4 is a superior diagnostic test than TT_4. Over this period (1995–1998) an increasing number of laboratories appear to have switched from the indirect measurement of FT_4 (i.e. TT_4 and T-uptake) to direct FT_4 methods (the number of direct FT_4 users, relative to the number of users of indirect FT_4 methods, increased from 43% in 1995 to 54% in 1998). The shift towards the direct measurement of FT_4 is expected to increase further in the future.

A brief description of some of commercial FT_4 assays (information as provided by company representatives or obtained from product literature) is given below.

Nichols Institute ED

This is the only commercially available (from the Nichols Institute Diagnostics, San Juan Capistrano, CA, USA) direct equilibrium dialysis (ED) FT_4 method. The development and validation of this method has been described by Nelson and Tomei (1988). This method is obtainable from the Nichols Institute Diagnostics, San Clemente, CA, USA. The method involves dialysis of the FT_4 from 200 μL of undiluted serum into 2.4 mL of dialysis buffer (at 37 °C, for 16–18 h). This method has been used as the yardstick with which other commercial assays are compared (e.g. Nelson *et al.*, 1994a,b; Wong *et al.*, 1992; Christofides *et al.*, 1999).

AxSYM FT_4

In the AxSYM FT_4 assay (Abbott Laboratories, Diagnostics Division, Abbott Park, IL, USA), the sample and anti-T_4 coated microparticles are pipetted into a well. During this first incubation the antibody binds to serum T_4. An aliquot of the reaction mixture is then transferred to the matrix cell, where the microparticles bind irreversibly to a glass-fiber matrix. The matrix cell is then washed to remove unbound material, and T_3-alkaline phosphatase conjugate is added. During this second incubation, the conjugate binds to unoccupied (by T_4) antibody binding sites. The matrix is then washed and the substrate (4-methylumbelliferyl phosphate) is added. The resultant fluorescent product is measured by the microparticle enzyme immunoassay (MEIA) optical assembly.

Elecsys™ FT_4

In the Elecsys FT_4 assay (Roche Diagnostic Corporation, Indianapolis, IN, USA) the sample is incubated with an anti-T_4 antibody labeled with ruthenium complex. During this period endogenous T_4 is bound to the labeled antibody. Biotinylated T_4 and streptavidin-coated microparticles are then added to the reaction mixture and the assay incubated further. During this second period the biotinylated T_4 binds to the still-free binding sites of the labeled antibody and the entire antibody–antigen complex gets bound to the solid phase via interaction of biotin and streptavidin. The reaction mixture is then aspirated into the measuring cell where the microparticles are magnetically captured onto the surface of the electrode and unbound substances removed. Application of a voltage to the electrode then induces chemiluminescent emission, which is measured by a photomultiplier. The assay uses 15 μL sample in a total reaction volume of 165 μL. The total duration of the assay is 18 min.

Bayer ADVIA® Centaur™ FrT_4

The Bayer ADVIA® Centaur™ FrT_4 assay (Bayer Corporation, Norwood, MA) is a competitive analog tracer method. The FT_4 in the sample (25 μL) competes with acridinium ester-labeled T_4 in the Lite Reagent (100 μL) for a limited amount of polyclonal rabbit anti-T_4 antibody, which is covalently coupled to paramagnetic particles in the solid phase (volume 450 μL). Incubation is at 37 °C for 7.5 min. The solid phase is then washed and 300 μL each of acid and base reagents are added to initiate the chemiluminescent reaction.

Vitros ECi Free T_4

The Vitros ECi FT_4 assay (Ortho-Clinical Diagnostics, Amersham, UK) is an enzyme immunoassay based on the labeled antibody procedure. FT_4 present in the sample (25 μL) competes with ligand on the modified well surface for a limited number of binding sites on a horseradish peroxidate (HRP)-labeled antibody conjugate (sheep anti-T_4, 100 μL). Incubation time is 16 min at 37 °C. The well surface has been modified to act as a ligand for uncombined conjugate. Unbound materials are removed by washing. The bound HRP conjugate is measured by luminescent reaction. A reagent containing luminogenic substrates (a luminol derivative and a peracid salt) and an electron transfer agent is added to the wells. The HRP in the bound conjugate catalyses the oxidation of the luminol derivative, producing light. The electron transfer agent (a substituted acetanilide) increases the level of light produced and prolongs its emission.

Types of Sample

Serum or plasma.

Frequency of Use

Common.

TRIIODOTHYRONINE

Triiodothyronine (T_3) differs from T_4 in that it has three iodine atoms instead of four. It is derived primarily from T_4 through deiodination outside the thyroid gland (see Figure 45.8).

Function

Although T_3 has only 1–2% of the concentration of T_4 in serum, less is bound to proteins in the serum. The concentration of free T_3 is about one-quarter of the concentration of free T_4. However, because T_3 has about four times the biological activity of T_4, T_3 and T_4 represent similar levels of thyroid hormone activity in the blood. In tissues, primarily the liver, T_4 is converted to T_3, so T_3 is normally considered to be the primary hormone, and T_4 is, in some respects, a prohormone. T_3 has a shorter half-life in the blood (1 day) than T_4 (7 days).

T_3, like T_4, increases the rate of metabolic activity in many tissues in the body. It is essential for normal physical and mental development throughout childhood and for controlling the rate of metabolic activity in adults. For a description of the effects of thyroid hormones *see* NORMAL THYROID FUNCTION.

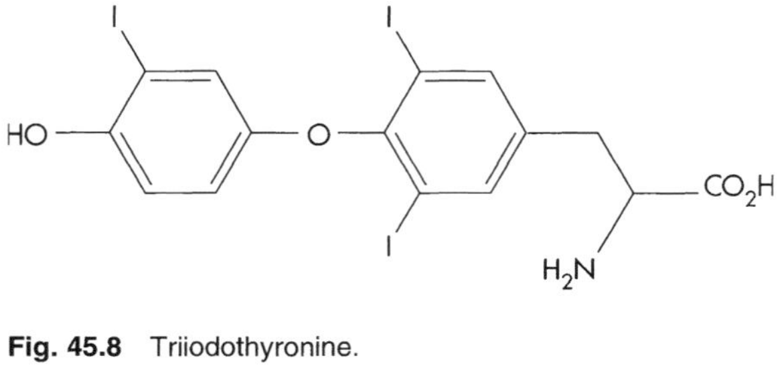

Fig. 45.8 Triiodothyronine.

About 99.7% of T_3 is bound to proteins in the blood but it is the minute fraction of free, unbound T_3 that is biologically active as only the free hormone can pass into the cells in the target tissues.

Reference Interval

1.5–2.6 nmol/L (Vitros ECi)

97–169 ng/100 mL (Vitros ECi)

Clinical Applications

Normally, TSH, free T_4, or FTI tests are used to screen the large number of patients for whom thyroid function tests are requested. T_3 is an important follow-up test for the investigation of thyroid status in patients who appear to be hyperthyroid in the screening test. T_3 is also useful for detecting hyperthyroidism where clinical symptoms indicate this condition but free T_4 or FTI tests are normal. This is because, in T_3-toxicosis, T_3 is elevated whereas T_4 is not. Although T_3 is the most metabolically active hormone, it is unsuitable as an analyte to test for hypothyroidism. As hypothyroidism develops, the T_4 concentration falls and TSH concentration increases. However, T_3 often remains normal, as the thyroid gland preferentially secretes T_3 in this situation (see Figure 45.9).

Limitations

- T_3 is unsuitable as a test for hypothyroidism.
- Abnormal levels of thyroid hormone binding proteins result in increases or decreases in total T_3 concentration in euthyroid people because it is the free, unbound fraction of T_3 that is biologically active. Pregnancy, oral contraceptives, and various drug treatments cause increased concentrations of binding proteins. Decreases occur in many cases of NTI, and as a side-effect of certain drug treatments. Some laboratories either measure free T_3 directly or calculate the free T_3 index, using total T_3 in conjunction with T_3 uptake or T uptake, which are tests for thyroid hormone binding capacity.
- Some drugs displace thyroid hormones from their binding sites, resulting in a reduction in the total thyroid hormone concentration while the free hormones remain normal, e.g. phenytoin, phenylbutazone, and salicylates.
- Amiodarone treatment can cause reduced T_3 concentrations.
- T_3 may be normal in T_4-toxicosis.
- T_3 immunoassays are affected by endogenous antibodies to T_3 if present in patient samples. In most types of assays the apparent concentration is increased. In addition, patients with T_3 antibodies may have a higher level of total T_3 but a normal free T_3 concentration.

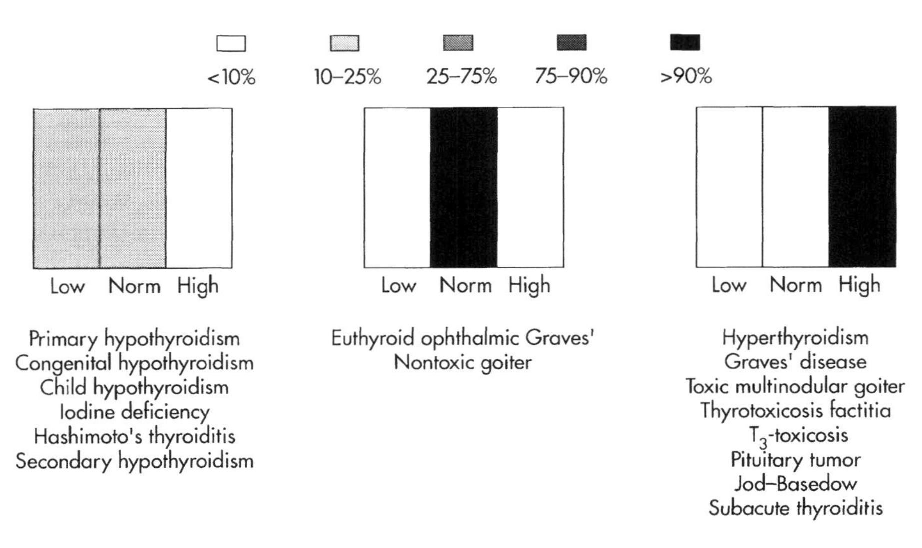

Fig. 45.9 Triiodothyronine (T_3).

Assay Technology

The most common methods at the time of writing are AxSYM (fluorescence polarization immunoassay) and Centaur™ (acridinium ester luminescence). Total T_3 assay reagents include one or more decoupling agents, such as 8-anilino-1-naphthalene sulfonic acid (ANS) or salicylate, to displace the T_3 from the binding proteins in the serum.

Types of Sample

Serum or plasma.

Frequency of Use

Common, but much less used than T_4.

FREE T_3

Free T_3 is a more relevant indicator of thyroid function than total T_3 but until recently free T_3 was difficult to measure directly. The 1998 College of American Pathologist (CAP) scheme survey indicated that 237 laboratories use FT_3 assays obtained from seven vendors. It is interesting to note that the number of laboratories using total T_3 is more than fivefold higher (1232 laboratories). Thus, in contrast to the situation seen with FT_4, the expected switch from TT_3 to FT_3 has been very slow in coming. The concentration of free T_3 in the blood depends on the concentrations of total T_3 and the thyroid hormone binding proteins, TBG, TBPA, and albumin. About 99.7% of T_3 is bound (80% to TBG, 10% to TBPA, and 10% to albumin). The technical problems associated with developing tests for free T_3 are similar to those described for free T_4 (*see* chapter on FREE ANALYTE IMMUNOASSAY).

Reference Interval

4.3–8.1 pmol/L (Vitros™ ECi)

2.8–5.3 pg/mL (Vitros™ ECi)

Clinical Applications

As for T_3, free T_3 is an important test for diagnosing or monitoring hyperthyroidism. Free T_3 is not suitable for diagnosing hypothyroidism as some hypothyroid patients have reduced free T_4 and elevated TSH, but a normal free T_3. However, clinical studies have shown that free T_3 is a better indicator of hyperthyroidism than total T_3, as it is independent of thyroid hormone binding protein concentrations (see Figure 45.10).

Limitations

- Free T_3 is unsuitable as a test for hypothyroidism (approximately 50% of hypothyroid patients have FT_3 levels that lie within the euthyroid range).
- Free T_3 may be normal in T_4-toxicosis.
- Some patient sera containing avid auto-antibodies to thyroid hormones may interfere in FT_3 methods based on one-step methodologies.
- Low concentrations of albumin cause an artificial decrease in free T_3 in some labeled analog assays.
- Free T_3 may be decreased in very sick patients.
- Patients receiving heparin can have biased free T_3 results as heparin stimulates the production of nonesterified fatty acids, which displace T_3 from albumin. Heparin can also arise from indwelling cannulas containing a heparin-saline solution.
- Amiodarone treatment can cause reduced free T_3 concentrations.

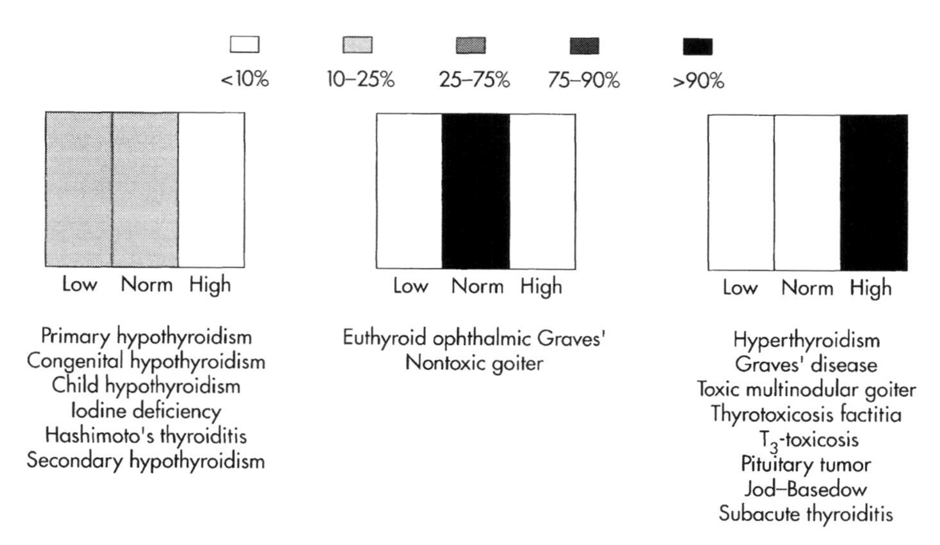

Fig. 45.10 Free T_3.

- Transient elevations in FT_3 concentrations may be observed following administration of drugs (e.g. furosemide, ketoprofen, phenylbutazone, mefenamic acid, probenecid, sulindac, and fenclofenac) that displace T_3 from their binding proteins.
- Although manufacturers include materials to minimize interference by heterophilic antibodies, there is no guarantee that all such sera will be fully corrected.

Assay Technology

The physicochemical principles describing the assay design of most commercial FT_3 assays have been discussed in the chapter on FREE ANALYTE IMMUNOASSAY. A brief description of some of the commercial FT_3 assays (information as provided by company representatives or obtained from product literature) is given below.

AxSYM FT_3

In the AxSYM FT_3 assay (Abbott Laboratories, Diagnostics Division, Abbott Park, IL, USA), the sample and anti-T_3 coated microparticles are pipetted into a well to form an antibody–antigen complex. An aliquot of the reaction mixture is then transferred to the matrix cell, where the microparticles bind irreversibly to a glass-fiber matrix. The T_3-alkaline phosphatase conjugate is dispensed onto the matrix cell and binds to the available sites on the anti-T_3 coated microparticles. The matrix is then washed and the substrate (4-methylumbelliferyl phosphate) is added. The resultant fluorescent product is measured by the MEIA optical assembly.

Bayer ADVIA® Centaur™ FT_3

The Bayer ADVIA® Centaur™ FT_3 assay (Bayer Corporation, Norwood, MA) is a competitive labeled antibody method using chemiluminescent technology. Fifty microliters of sample are incubated with 100 μL of Lite Reagent (acridinium ester (AE)-labeled monoclonal mouse anti-T_3 antibody) for 5 min at 37 °C. Four hundred and fifty microliters of solid phase (T_3 analog covalently coupled to paramagnetic particles) are then added and allowed to incubate for a further 2.5 min at 37 °C. After aspiration of the liquid and washing of the paramagnetic particles, 300 μL each of acid reagent and base reagent are added to initiate the chemiluminescent reaction.

Elecsys FT_3

In the Elecsys FT_3 assay (Roche Diagnostic Corporation, Indianapolis, IN, USA) the sample (30 μL for the 2020 system, and 20 μL for the 1010 system) is incubated with an anti-T_3 antibody labeled with ruthenium complex. During this period endogenous T_3 gets bound to the labeled antibody. Biotinylated T_3 and streptavidin-coated microparticles are then added to the reaction mixture and the assay incubated further. During this second period the biotinylated T_3 binds to the still-free binding sites of the labeled antibody and the entire antibody–antigen complex is bound to the solid phase via interaction of biotin and streptavidin. The reaction mixture is then aspirated into the measuring cell where the microparticles are magnetically captured onto the surface of the electrode and unbound substances removed. Application of a voltage to the electrode then induces chemiluminescent emissions that are measured by a photomultiplier. The total duration of the assay is 18 min.

Vitros Immunodiagnostic Products Free T_3

The Vitros FT_3 assay (Ortho-Clinical Diagnostics, Amersham, UK) is an enzyme immunoassay based on the labeled antibody procedure. FT_3 present in the sample (25 μL) competes with ligand on the modified well surface for a limited number of binding sites on an HRP-labeled antibody conjugate (sheep anti-T_3, 100 μL). Incubation time is 16 min at 37 °C. The well surface has been modified to act as a ligand for uncombined conjugate. Unbound materials are removed by washing. The bound HRP conjugate is measured by luminescent reaction. A reagent containing luminogenic substrates (a luminol derivative and a peracid salt) and an electron transfer agent is added to the wells. The HRP in the bound conjugate catalyzes the oxidation of the luminol derivative, producing light. The electron transfer agent (a substituted acetanilide) increases the level of light produced and prolongs its emission.

Types of Sample

Serum or plasma.

Frequency of Use

Fairly common in the UK, Japan, and parts of Europe, where it is replacing total T_3. Much less common than free T_4.

THYROXINE-BINDING GLOBULIN

TBG is a glycoprotein with a molecular mass of about 60 kDa. It has a single high-affinity binding site for T_3 and T_4.

Function

TBG provides the main reservoir of thyroid hormones in the body outside the thyroid gland. Approximately 80% of T_4 and 90% of T_3 circulating in the blood is bound to TBG. This is because TBG has a much stronger affinity than the other binding proteins for T_4 and T_3. In normal people, about one-third of the available binding capacity is utilized, but the proportion of free binding sites decreases in hyperthyroidism and increases in hypothyroidism. TBG is synthesized in the liver and has a biological half-life of approximately 5 days.

Reference Interval

TBG

13.3–28.3 mg/L (Amerlite TBG)

T_4/TBG ratio (T_4 in μg/100 mL, TBG in mg/100 mL)

2.59–5.91 (Amerlite TBG and T_4)

NOTE: Amerlite is no longer available, but these values give an indication of reference intervals. They should not be used for other methods.

Clinical Applications

As explained above, it is the free, unbound T_3 and T_4 that are metabolically active and therefore most relevant to the clinician. The concentrations of free T_3 and free T_4 depend on the total hormones and their binding protein concentrations in serum. Variation in TBG, which is the principal thyroid hormone binding protein, has the greatest effect. TBG concentrations vary widely under the influence of hormones and drugs, in disease states, and because of genetic factors. In these situations, the body maintains constant free hormone concentrations by increasing or decreasing the supply of T_3 and T_4. Measurement of TBG provides a corrected T_4 (the T_4/TBG ratio) which allows for the effect of the variation in TBG concentration.

The T_4/TBG ratio provides a more accurate differentiation of euthyroid, hypothyroid, and hyperthyroid metabolic states in most patients than total T_4 measurement (see Figure 45.11).

Limitations

- The T_4/TBG ratio only allows for the effects of abnormal TBG concentrations. It does not allow for the effects of abnormal albumin or TBPA concentrations.
- TBG is not an appropriate test for patients who do not have TBG in their blood owing to an inherited metabolic disorder.
- In euthyroid patients with circulating T_4 antibody, total T_4 increases but the free T_4 remains normal. The T_4/TBG ratio is elevated because of the increased total T_4 in the serum. In addition, the antibodies interfere directly with T_4 immunoassays, which depend on a constant amount of antibody being added to each sample.
- In FDH the proportion of T_4 that is bound by albumin is greatly increased and the total T_4 concentration is elevated to maintain a normal free T_4. This results in an increase in the T_4/TBG ratio.
- In patients with NTI, the serum is thought to contain a factor that releases T_4 from the binding proteins preferentially over T_3. In this situation, the T_4/TBG ratio underestimates the free T_4 concentration.
- Some drugs displace thyroid hormones from their binding sites, eventually causing a reduction in the total T_4 hormone concentration while the free hormones remain normal. This results in a lower T_4/TBG ratio.
- T_4/TBG ratios are normal in T_3-toxicosis.

Assay Technology

A variety of assay formats are available for TBG measurement, including competitive radioimmunoassay.

Types of Sample

Serum or plasma.

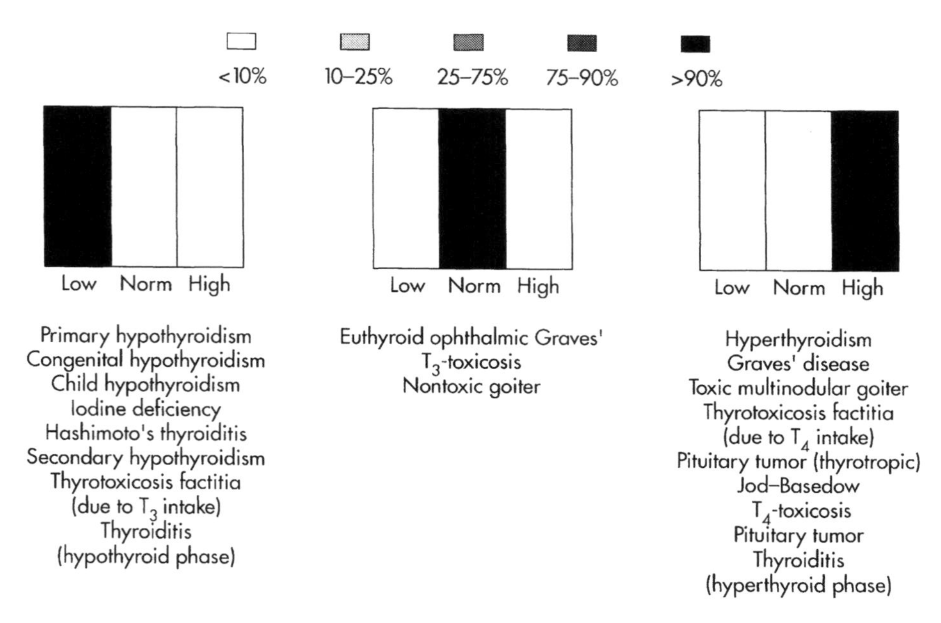

Fig. 45.11 T_4/TBG ratio.

Frequency of Use

TBG is used infrequently compared with other thyroid tests.

THYROTROPIN RECEPTOR ANTIBODIES

The thyrotropin receptor (TSH-R) on thyroid follicular cells is the target of autoantibodies that can either stimulate (TSAb) or inhibit (TBAb) thyroid hormone secretion, causing Graves' disease or idiopathic myxedema, respectively. In the case of Graves' disease, stimulation of the thyrotropin receptor by TSAb causes thyroid hyperplasia and hyperthyroidism, whereas in myxedema TBAb competes with TSH for the TSH-R and blocks the biological effect of TSH.

In autoimmune thyroid disease, both autoantibodies can be present, but in the case of Graves' disease, TSAb predominates, and in the case of myxedema, TBAb predominates. The specific measurement of stimulating and blocking activity requires a bioassay and the presence of TSH-R antibodies in serum has been identified routinely in assays using the inhibition of ^{125}I-labeled TSH binding to thyroid cell membranes. Antibodies detected by the binding assays are termed thyrotrophin-binding inhibitory immunoglobulins (TBII). These assays are unable to distinguish between stimulating and blocking antibody activity. Through the use of recombinant TSH receptor, newer assays that measure stimulating and blocking activity are being developed and may play an increasing role in the diagnosis and management of thyroid disorders. Most studies of thyrotropin receptor antibodies (TRAb) have used the generally available assay by competitive inhibition (TBII).

Reference Interval

$<$ 1.0 U/L (Brahms Diagnostica TRAK™ assay).

Clinical Applications

Currently, TRAb assays have only a very limited application in clinical practice. The autoimmune basis of Graves' disease can be established by the more easily performed and inexpensive TPO-Ab assay in the majority of patients. TRAb assays can be useful for special problems, as in patients who present with initial unilateral exophthalmos, or sometimes with subclinical hyperthyroidism, in hyperemesis with thyrotoxicosis or in the differential diagnosis of postpartum painless thyroiditis.

In pregnancy, in women with Graves' disease, transplacental passage of TSAb from mother to fetus can result in transient neonatal hyperthyroidism. This only affects 2–10% of women with Graves, but the severity of neonatal hyperthyroidism is related to high concentrations of TSAb in the mother's serum, measured late in pregnancy. TSH-receptor antibodies should be measured in the last trimester and if high, careful evaluation of the neonate is required to detect hyperthyroidism. Very rarely, in Hashimoto's disease blocking antibodies may cross the placenta in pregnant women and lead to transient neonatal hypothyroidism.

The measurement of TRAb has not been found to be sufficiently sensitive to predict relapse or remission in the treatment of Graves' disease (see Figure 45.12).

Limitations

The TSH-binding inhibition immunoglobulin (TBII) test nonselectively detects antibodies that bind to cell membrane receptors for TSH, whereas Graves' disease is caused by antibodies that not only bind but also stimulate the TSH receptors. However, more than 98% of patients with Graves' disease give positive results in this assay.

Assay Technology

Assays for the measurement of TSAb fall into two categories:

- bioassays that measure the stimulatory activity;
- competitive binding assays that rely on the ability of TSAb or TBAb to displace ^{125}I-labeled TSH from solubilized TSH-R.

Antibodies detected by the binding assays are termed TBII, and have been the assays most widely used.

Type of Sample

Serum.

Frequency of Use

Uncommon. This test is normally only carried out in specialist centers.

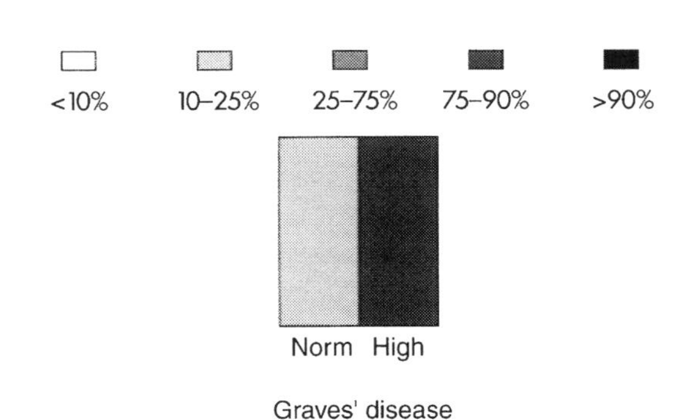

Fig. 45.12 Thyrotropin receptor antibody (TRAb).

THYROID PEROXIDASE ANTIBODIES

The autoimmune thyroid diseases (AITD) include Graves' disease, lymphocytic thyroiditis (Hashimoto's thyroiditis and primary myxedema) and postpartum thyroid dysfunction and are all diseases in which the immune system is primarily responsible for the disease process (although not necessarily the initiator). The thyroid autoantibodies to thyroid peroxidase (TPOAb) and thyroglobulin (TgAb) are a secondary response to thyroid injury. Both are polyclonal IgG antibodies and the amounts present correlate with lymphocytic infiltration of the thyroid, have complement-fixing cytotoxic activity and TPO autoantibodies correlate with thyroidal damage. In non-immune thyroid disease, e.g. thyroiditis (subacute or de Quervain's and Riedel's) and nontoxic goiter, although immunological disturbances occur, these are secondary to the disease process.

Reference Interval

In the Roche Elecsys Anti TPO assay 95% of normals had values less than 34 IU/mL (National Institute for Biological Standards and Central Reference Preparation MRC 66/387).

Clinical Applications

The main indication to measure TPOAb is to establish autoimmunity as the basis of any thyroid dysfunction, either Hashimoto's or Graves' disease, so as to distinguish it from other forms of thyroid failure. Assays for TPOAb are more sensitive for the diagnosis of autoimmune thyroiditis than TgAb, but with quantitative sensitive assay one or both antibodies are found in almost 100% of patients. TPOAb is of higher affinity and is usually present in higher concentrations than TgAb. High concentrations of thyroid antibodies confirm the diagnosis of primary autoimmune disease in patients in whom the clinical picture is unclear. An increased TPOAb concentration may be an indication to start treatment with thyroxine in those patients who are subclinically hypothyroid with an elevated TSH and normal FT_4, as progression to overt thyroid failure occurs at the rate of 5% per year. In women, there is a prevalence of postpartum thyroid disease of about 6% occurring around 6 months after delivery. Approximately 50% of women who are TPOAb positive early in pregnancy go on to develop some form of postpartum thyroid dysfunction, usually of a transient nature. High concentrations of TPOAb identify those women who are at risk of developing postpartum thyroid dysfunction, which may be severe enough in some women to warrant treatment. As AITD occurs commonly with other autoimmune diseases, such as insulin-dependent diabetes mellitus, which places them at increased risk of developing AITD, the presence of TPOAb is helpful in selecting which patients require monitoring of their thyroid function (see Figure 45.13).

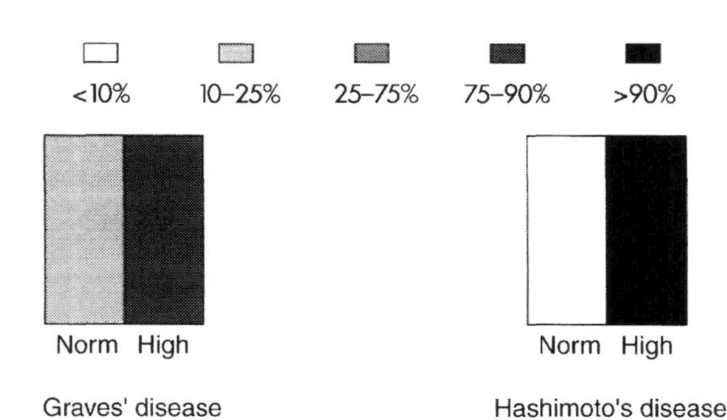

Fig. 45.13 TPO antibodies.

Limitations

- In the normal population, the prevalence of thyroid autoantibodies depends on the method used for their analysis. TPOAb is common in the general population and occurs more frequently in women than men, and in the elderly.
- TPOAb may be found in low concentrations in some healthy individuals with completely normal thyroid function. The significance of this is uncertain.
- In a small percentage of patients with autoimmune disease, TPOAb may not be detected in their sera.

Assay Technology

Older methods such as hemagglutination, which has low sensitivity and is operator dependent, and immunofluorescence are now obsolete. These assays have been replaced by quantitative assays with better sensitivity and specificity, which are more precise. They use a variety of techniques including radioimmunoassay, immunometric assay, and enzyme-linked immunoassays. These assays are available on most automated immunoassay analyzers, so can be done with FT_4/TSH as part of thyroid function testing strategy.

Type of Sample

Serum.

Frequency of Use

Widely available.

THYROGLOBULIN ANTIBODIES

In Hashimoto's thyroiditis, immunocytes invade the thyroid gland and synthesize antibody to thyroglobulin. Thyroglobulin is a large protein and is the site of synthesis of T_3 and T_4. The autoantibodies gradually destroy the thyroid tissue and prevent the production of thyroid hormones, causing hypothyroidism.

Reference Interval

In the Roche Enzymun assay, 95% of normals had values < 115 IU/mL WHO 1st IRP 1971, MRC 65/93. This assay has been replaced by a similar test on the Elecsys systems.

Clinical Applications

As with tests for TPO antibodies, the presence of thyroglobulin antibodies indicates the autoimmune basis of Hashimoto's thyroiditis or Graves' disease. However, they are less often present and less pathogenic than TPO antibodies, so this test is not as useful. If a sensitive immunoassay for TPO-Ab is available, only TPO-Ab need be done. As TgAb can interfere in the assay for thyroglobulin used for monitoring thyroid cancer patients, there is a requirement to measure TgAb in these patients and, if present, to establish whether it interferes in the assay for Tg by doing a recovery experiment (*see* THYROGLOBULIN ASSAY TECHNOLOGY) (see Figure 45.14).

Assay Technology

TgAb is measured by the same techniques as TPOAb (see previously) with hemagglutination assays being replaced by quantitative immunoassays.

Type of Sample

Serum.

Frequency of Use

Decreasing need to measure this.

THYROGLOBULIN

Thyroglobulin (Tg) is a large (660 kDa) glycoprotein, which is the site of synthesis of T_3 and T_4. It also acts as a large storage reservoir of thyroid hormones. Nearly all of the protein is present within the thyroid but a small amount is normally detectable in blood.

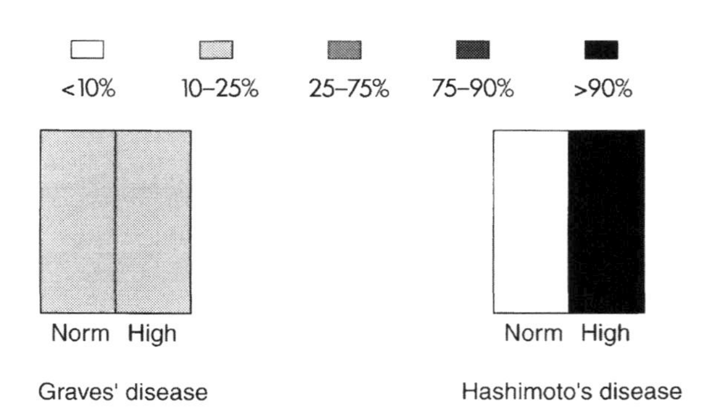

Fig. 45.14 Tg antibodies.

Reference Interval

4–40 ng/mL in normal subjects (Nichols Advantage). Undetectable in treated thyroid cancer patients.

Clinical Applications

Thyroglobulin assays are now in widespread use as a tumor marker for monitoring patients with differentiated thyroid carcinoma. In patients with papillary or follicular carcinoma, and following total thyroidectomy or radioactive ablation to remove the tumor, previously increased Tg concentrations will reduce to very low or undetectable levels. An increasing Tg concentration, when on a suppressive dose of thyroxine, indicates the recurrence of tumor or metastatic spread. Patients should be monitored every 6–12 months, but at more regular intervals if a detectable Tg concentration is found.

In comparing results from Tg measurements with whole body scans, it is important to realize that metastatic cells are unable to take up iodine and give a negative scan result, but are capable of releasing Tg into the circulation which gives a positive Tg result.

Tg assays have no place in the initial diagnosis of thyroid carcinoma.

Assay Technology

Thyroglobulin is measured by competitive and immunometric assays. Because thyroglobulin antibodies (Tg-Ab) can interfere with Tg assays, the choice of Tg assay is critical, and the effect of Tg-Abs on the assay in a particular patient sample needs to be known. For this reason, laboratories assay for Tg-Ab in every sample for which a Tg assay is requested, by a sensitive two-site assay, and perform a recovery experiment in all samples containing Tg-Ab.

Interference from Tg-Ab can be minimized by careful choice of reagent. The immunoradiometric Tg assay from Sanofi Pasteur uses a mixture of four monoclonal antibodies and a fifth which is radiolabeled, all chosen as being directed against epitopes not recognized by the majority of Tg-Abs. This reduces the incidence of Tg-Ab interference in the Tg assay, and ensures a more robust Tg assay for monitoring patients.

Type of Sample

Serum.

Frequency of Use

Uncommon. This test is normally only carried out in specialist centers.

THYROID TESTING STRATEGIES

Most clinical biochemistry laboratories have experienced an inexorable increase in requests for thyroid function

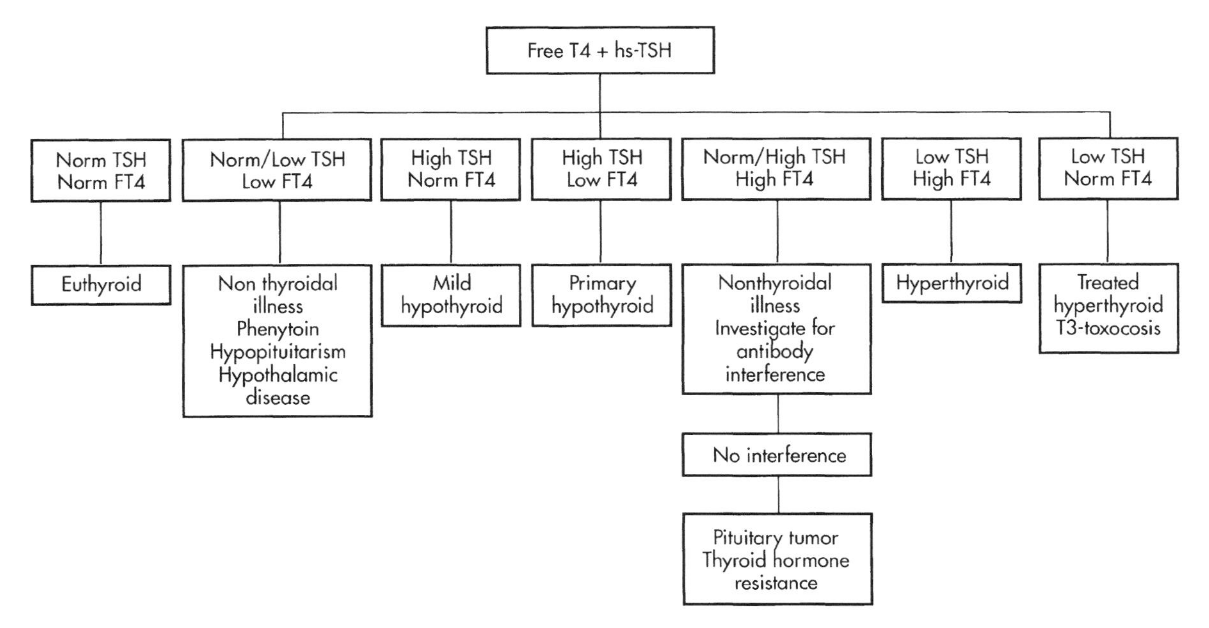

Fig. 45.15 Typical thyroid function screening strategy (first phase).

tests. To meet this demand, laboratories have instituted guidelines for appropriate requesting of thyroid function tests, introduced automated procedures and adopted testing strategies. For most samples received, there is a low suspicion of a thyroid disorder and the request merely requires any thyroid dysfunction as a cause of a patient's illness to be ruled out. In this situation, the ideal initial screening test would be sensitive enough to detect any degree of thyroid disorder, give a normal result when thyroid disease was not present and not be affected by a patient's illness or prescribed drugs. There is no single test of thyroid function which is totally reliable in all these situations, but a number of laboratories operate a policy of screening by means of a single test, either free T_4 or TSH. Both have their limitations: free T_4 can be lowered in NTI, and TSH is normal in secondary hypothyroidism. Both need to be followed up with additional tests, either TSH if free T_4 is used as a frontline screen, or free T_4 if TSH is used as a frontline screen. In a small number of samples, free T_3 may also be required. For completeness, the autoimmune basis of any abnormal thyroid dysfunction can be documented by assaying for TPO-Ab and TRAb. The newer automated immunoassay analyzers have the ability to reflex test from one analyte to another, after appropriate decision limits have been set. This ensures that a range of thyroid function tests are performed before samples are removed from the analyzer. Using the Bayer Centaur as an example, and TSH as the frontline screen, the time to a TSH result is 19 min, reflex testing to a free T_4 takes 37 min and to a free T_3 takes a total of 52 min.

Considerable cost and time savings are made by introducing a reflex testing strategy. Other laboratories carry out free T_4 and TSH tests on all incoming samples, and would follow this up with free T_3 and thyroid antibody assays in certain selected samples.

Whichever screening strategy is adopted, consideration needs to be given to a patient's previous clinical history, treatment, and drug use to be able to interpret thyroid function test results in all situations. Most assays for free T_4, free T_3, and TSH are sufficiently sensitive, precise, and robust to accurately predict the thyroid status in the majority of patients. However, there are exceptions, where seemingly anomalous results are obtained, and these require careful follow-up. These conditions include antibody interference in any of the commonly performed assays for free T_4, free T_3 or TSH, thyroid hormone resistance syndromes, TSH resistance syndromes, and pituitary TSH-secreting tumors. All laboratories should have a protocol for dealing with these rare occurrences. In the case of antibody interference, re-assay by a different methodology, preferably one which is little affected by antibody interference is often the first investigation, although other methods may also be affected. Dilutions of the sample and re-assay as well as polyethylene glycol precipitation of the antibody fraction and re-assay may also be helpful. For follow-up of the other rare conditions listed above, referral to a specialist endocrine center is required, where appropriate testing can be carried out. *See* TSH, LIMITATIONS and FREE T_4, LIMITATIONS.

A typical initial thyroid function testing strategy is shown in Figure 45.15. Although not shown here, T_3 or free T_3 tests are used for following up suspected cases of hyperthyroidism, including patients with clinical symptoms of hyperthyroidism but normal free T_4 and TSH results. Whichever laboratory testing strategy is used, an overall picture of thyroid hormone levels, consideration of the patient's symptoms and drug use, and a record of previous treatment are needed before the thyroid status of a patient can be reliably determined.

REFERENCES AND FURTHER READING

Braverman, L.E. and Utigar, R.D. (eds), *Werner & Ingbar's The Thyroid. A Fundamental and Clinical Text*, 6th edn (Lippincott, Philadelphia, PA, 1991).

Christofides, N.D., Sheehan, C.P. and Midgley, J.E.M. One-step, labeled-antibody assay for measuring free thyroxin. I Assay development and validation. *Clin. Chem.* **38**, 11–18 (1992).

Christofides, N.D., Wilkinson, E., Stoddart, M., Ray, D.C. and Beckett, G.J. Assessment of serum thyroxine binding capacity-dependent biases in free thyroxine assays. *Clin. Chem.* **45**, 520–525 (1999).

Demers, L.M. Thyroid function testing and automation. *J. Clin. Ligand Assay* **22**, 38–41 (1999).

Feldt-Rasmussen, U. Analytical and clinical performance goals for testing autoantibodies to thyroperoxidase, thyroglobulin, and thyrotropin receptor. *Clin. Chem.* **42**, 160–163 (1996).

Hall, R. In: *Fundamentals of Clinical Endocrinology*, 4th edn (eds Hall, R. and Besser, M.), (Churchill Livingstone, Edinburgh, 1989).

Hay, I.D., Bayer, M.F., Kaplan, M.M., Klee, G.G., Larsen, P.R., Spencer, C.A. for the Committee on Nomenclature of the American Thyroid Association. American Thyroid Association assessment of current free thyroid hormone and thyrotropin measurements and guidelines for future clinical assays. *Clin. Chem.* **37**, 2002–2008 (1991).

Klee, G.G. and Hay, I.D. Assessment of sensitive thyrotropin assays for an expanded role in thyroid function testing: proposed criteria for analytic performance and clinical utility. *Clin. Chem.* **64**, 461–471 (1989).

Larsen, P.R. and Ingbar, S.H. In: *Williams' Textbook of Endocrinology* (eds Wilson, J.D. and Foster, D.W.), (Saunders, Philadelphia, PA, 1992).

Laurberg, P., Nygaard, B., Glinoer, D., Grussendorf, M. and Orgiazzi, J. Guidelines for TSH-receptor antibody measurement in pregnancy: results of an evidence-based symposium organized by the European Thyroid Association. *Eur. J. Endocrinol.* **139**, 584–586 (1998).

Nelson, J.C., Nayak, S.S. and Wilcox, R.B. Variable underestimates by serum free thyroxine (T_4) immunoassays of free T_4 concentrations in simple solutions. *J. Clin. Endocrinol. Metab.* **79**, 1373–1375 (1994a).

Nelson, J.C. and Tomei, R.T. Direct determination of free thyroxine in undiluted serum by equilibrium dialysis/radioimmunoassay. *Clin. Chem.* **34**, 1737–1744 (1988).

Nelson, J.C., Weiss, R.M. and Wilcox, R.B. Underestimates of serum free thyroxine (T_4) concentrations by free T_4 immunoassays. *J. Clin. Endocrinol. Metab.* **79**, 76–79 (1994).

Sheehan, C.P. and Christofides, N.D. One-step, labeled-antibody assay for measuring free thyroxin. II. Performance in a multi-center trial. *Clin. Chem.* **38**, 19–25 (1992).

Wong, T.K., Pekary, A.E., Hoo, G.S., Bradley, M.E. and Hershman, J.M. Comparison of methods for measuring free thyroxine in nonthyroidal illness. *Clin. Chem.* **38**, 720–724 (1992).

46 The Adrenal Cortex

Sami Medbak

NORMAL ADRENOCORTICAL FUNCTION

The adrenal glands, situated one above each kidney, are composed of an outer cortex and an inner medulla. The adrenal cortex uses cholesterol to synthesize a variety of steroid hormones. Histologically, the cortex is composed of three layers. The outer layer, zona glomerulosa, predominantly secretes the mineralocorticoid hormone, aldosterone. The middle and inner zones, zona fasciculata, and zona reticularis, secrete glucocorticoids and androgens although cortisol, the main glucocorticoid, is mostly secreted by the zona fasciculata. Cortisol and aldosterone are the most important physiological products of the adrenal cortex. They are collectively known as corticosteroids.

Over 90% of cortisol in plasma is protein-bound to an α_2-globulin, known as cortisol-binding globulin (CBG) or transcortin, but only the nonprotein-bound 'free' hormone is biologically active. Cortisol is mainly metabolized in the liver into a number of inactive compounds; small amounts of free cortisol are excreted in urine.

Corticosteroids exert their cellular effects through specific receptors in the cell cytosol, producing vital and widespread effects on various tissues in the body including actions on carbohydrate, protein and fat metabolism, water metabolism, hematopoiesis, and hemostasis. They affect the gastrointestinal, cardiovascular, skeletal, neuromuscular, and immunological systems and possess anti-inflammatory activity.

Cortisol secretion is controlled by the anterior pituitary hormone, **adrenocorticotropic hormone (ACTH)**. ACTH is in turn controlled by the hypothalamic peptide **corticotropin-releasing hormone (CRH)**. CRH is released from the median eminence of the hypothalamus into the hypophyseal portal blood and acts on the corticotrophs of the anterior pituitary to cause the synthesis and release of ACTH. Vasopressin, from the posterior pituitary, acts in a synergistic way with CRH to promote ACTH secretion. Other factors, including amines and neuropeptides, are also important in the control of ACTH secretion (see Figure 46.1).

Three major factors are involved in the control of the hypothalamic-pituitary-adrenocortical (HPA) axis: **circadian rhythm, negative feedback**, and **stress mechanisms**. Cortisol secretion in man is periodic, with increased secretion mainly in the early morning and decreased secretion in the late evening. Healthy individuals have undetectable serum cortisol levels at the beginning of sleep (usually at midnight). Patients with hyperfunction of the adrenal cortex lose this cortisol circadian rhythm and have detectable midnight cortisol levels during sleep (see Figure 46.2).

Cortisol exerts a negative feedback effect whereby high concentrations of serum cortisol suppress CRH and ACTH production and, consequently, cortisol secretion is reduced. Conversely, CRH and ACTH production is increased when serum cortisol levels become too low. Stress is the third control mechanism and overrides

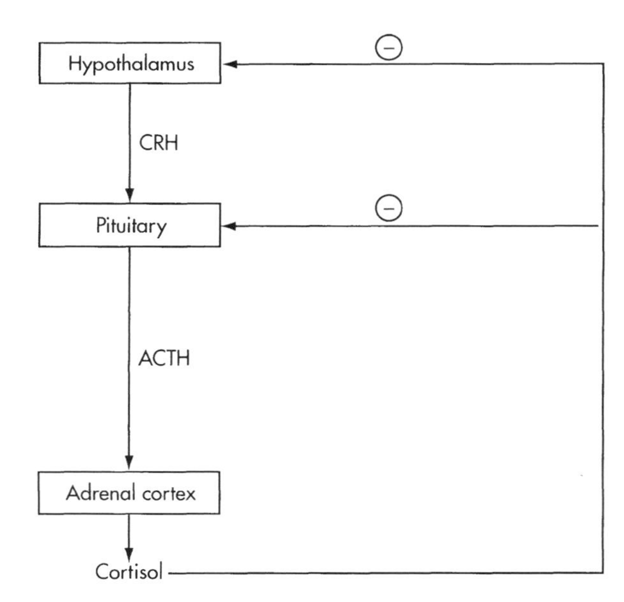

Fig. 46.1 Control of cortisol secretion (simplified).

The Immunoassay Handbook. *Third Edition*
D. Wild (Ed.)

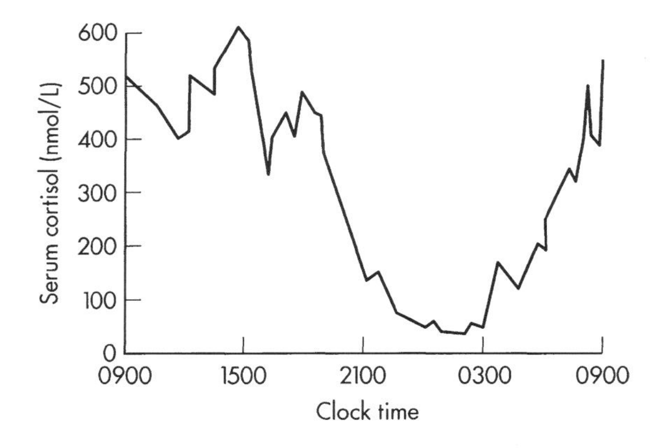

Fig. 46.2 Periodicity of cortisol secretion in normal people.

the other two. Thus, during any stressful condition (such as illness, accidents, operations or psychological stress), the HPA axis is activated, leading to increased secretion of ACTH and cortisol, which can reach extremely high concentrations.

CLINICAL DISORDERS

The main disorders of the adrenal cortex are those relating to abnormal cortisol secretion. Rarely, abnormalities in aldosterone or androgen secretion are encountered.

HYPERCORTISOLEMIA

The condition that results from the excessive secretion of cortisol is termed **Cushing's syndrome**. There are many causes of Cushing's syndrome and they are traditionally subdivided into:

- those conditions that are not due to excess ACTH secretion, the so-called **ACTH-independent** causes, which mainly comprise **adrenal tumors** (both adenomas and carcinomas) and exogenous administration of glucocorticoids;
- those conditions characterized by hypersecretion of ACTH, i.e. **ACTH-dependent Cushing's syndrome**, such as **pituitary tumors** (also referred to as **Cushing's disease**), and **ectopic ACTH syndrome** (where ACTH is secreted by nonendocrine tumors, such as lung cancer).

The exogenous administration of ACTH in the treatment of certain conditions is another but rare causative factor. Cushing's syndrome can also occur in patients with severe depression or after excessive drinking of alcohol (**pseudo-Cushing's** states).

Cushing's syndrome is more common in women, and the chronic excess of cortisol that characterizes the illness leads to a number of symptoms and signs including obesity (mainly of the face and trunk), easy bruising, purple abdominal striae, hirsutism, acne and greasy skin, hypertension, muscular weakness, menstrual disturbances, depression, and osteoporosis.

The diagnosis of Cushing's syndrome depends on:

- demonstrating the presence of excessive cortisol secretion;
- establishing the cause of the hypercortisolemia.

This diagnostic process often involves difficult and prolonged investigations, including the measurement of serum cortisol and plasma ACTH, first under basal conditions (circadian rhythm studies), and also after a number of dynamic tests of both stimulation and suppression, such as the dexamethasone suppression test (DST), insulin-induced hypoglycemia, CRH stimulation test, and venous catheter studies. A number of radiological imaging procedures are also required. It is therefore essential that such patients are cared for in medical centers with special expertise in dealing with this condition.

Treatment of Cushing's syndrome depends on the cause. Hypercortisolemia can be effectively controlled using medical therapy, surgery, or radiotherapy.

HYPOCORTISOLEMIA

Hypocortisolemia is caused by adrenocortical insufficiency which is either **primary** (disease of the adrenal cortex) or **secondary** (due to pituitary or hypothalamic lesions). A life-threatening acute form of primary adrenocortical insufficiency may occur following operations or overwhelming systemic infections with hemorrhage. However, it is more commonly chronic and is termed **Addison's disease**. The commonest cause of Addison's disease in the western world is autoimmunity. Other causative factors include sarcoidosis, tuberculosis, amyloidosis, hemochromatosis, and secondary malignant deposits.

Patients with Addison's disease suffer from generalized tiredness, anorexia and nausea, pigmentation or white patches in the skin, weight loss, dizziness and hypotension, and slow recovery from illness. Most of these features are due to the deficiency of cortisol. Excessive skin pigmentation, which may be the earliest sign of the disease and is characteristic of this condition, is thought to be due to the high levels of plasma ACTH (or its related peptides). In secondary hypoadrenalism there is no skin pigmentation and little change in blood pressure.

Diagnosis of Addison's disease relies on the demonstration of low or undetectable serum cortisol levels that fail to rise following stimulation with ACTH (tetracosactrin). The basal plasma ACTH concentrations in these patients are typically very elevated. In contrast, patients with secondary adrenocortical insufficiency have low or undetectable serum cortisol with low levels of plasma ACTH.

Treatment of adrenocortical insufficiency is with hydrocortisone or prednisolone (a synthetic glucocorticoid) given in dosage forms so as to mimic the normal circadian rhythm of serum cortisol. A typical dose would be 20 mg hydrocortisone in the morning and 10 mg in the evening although in some patients glucocorticoids given three times a day is a more appropriate treatment.

Other diseases of the adrenal cortex that are much rarer include **congenital adrenal hyperplasia** and **primary hyperaldosteronism**. In congenital adrenal hyperplasia, there is a congenital enzymatic defect in the pathway of cortisol synthesis. The commonest deficiency is that of 21-hydroxylase enzyme leading to hypocortisolemia with elevated ACTH levels. The latter leads to bilateral adrenal enlargement (and hyperplasia). There is a concomitant deficiency of aldosterone in over 60% of cases resulting in salt loss. The enzyme block, coupled with excess ACTH, leads to increased androgen secretion by the adrenals resulting in virilization. The diagnosis relies on finding:

1. Low serum cortisol.
2. Elevated levels of 17α-hydroxyprogesterone (a precursor of cortisol before the enzyme block) either basally or following stimulation with ACTH. Androstenedione and ACTH levels are also raised.

Early diagnosis and treatment with cortisol (and fludrocortisone, a synthetic mineralocorticoid) can be life saving. In primary hyperaldosteronism, which is usually due to an adrenal adenoma and known as **Conn's syndrome**, there are high levels of aldosterone in plasma resulting in hypokalemic alkalosis with muscle weakness, polyuria and polydypsia, and high blood pressure.

ANALYTES

CORTISOL

Cortisol is the most important steroid hormone. It is produced by the adrenal cortex from cholesterol by several enzymatic reactions (see Figure 46.3).

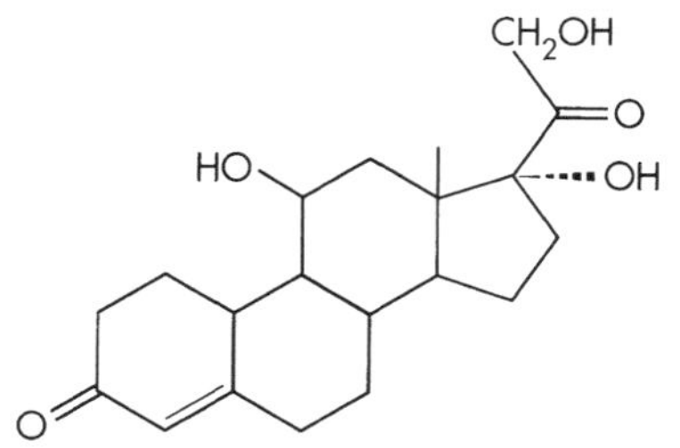

Fig. 46.3 Cortisol.

Function

Cortisol is an essential hormone and has a wide variety of effects on most tissues in the body (*see* NORMAL ADRENOCORTICAL FUNCTION). It is particularly important in coping with situations of mental or physical stress such as infections and operations.

Most of the cortisol circulates bound to protein in plasma with small amounts circulating in the free, biologically active form. Most routinely used assays measure the total cortisol in serum and thus the levels are artifactually affected in situations where there are abnormal concentrations of CBG, such as pregnancy and estrogen therapy, where high levels are recorded.

Reference Interval

220–720 nmol/L at 9 a.m.
8.0–26.0 μg/100 mL at 9 a.m.

Clinical Applications

The measurement of cortisol in serum is used in the diagnosis of disorders of the HPA axis. In contrast to other endocrine disorders where assay of single or basal samples is sufficient for diagnosis, dynamic tests (stimulation and/or suppression tests) are almost always required in the diagnosis of disorders of the HPA axis.

Cushing's Syndrome

In Cushing's syndrome, where there is an excessive secretion of cortisol, the following diagnostic strategy is required:

1. Confirm over-production of cortisol by:
   (a) screening tests (out-patients); any of the following:
      - overnight (1 mg single dose) DST: healthy individuals suppress their serum cortisol to undetectable levels following the ingestion of dexamethasone, a synthetic glucocorticoid; patients with Cushing's syndrome do not;
      - low-dose DST (0.5 mg 6-hourly for 48 h): again, healthy individuals suppress their cortisol to undetectable levels; patients with Cushing's syndrome do not (see high-dose DST below);
      - 24-h urinary free cortisol measurement: elevated in patients with Cushing's syndrome;
      - insulin-induced hypoglycemia test: patients with Cushing's syndrome show no cortisol rise during hypoglycemia. This is a valuable test for excluding depression as the cause of hypercortisolemia (see Figure 46.4).

   (b) in-patient tests:
      - confirm over-production of cortisol as above;
      - circadian-rhythm study: circadian rhythm of cortisol is lost in Cushing's syndrome,

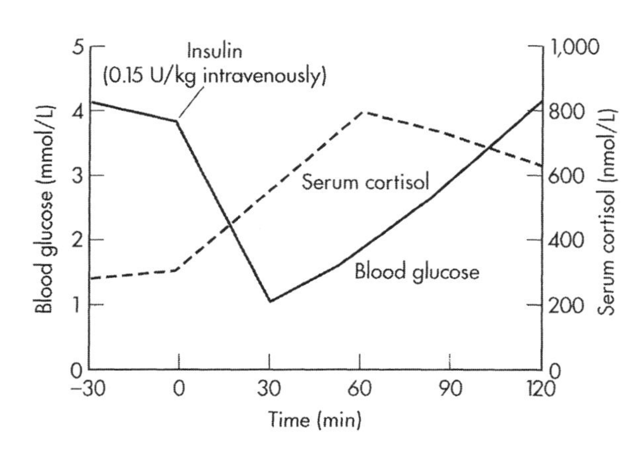

Fig. 46.4 Insulin hypoglycemia test (normals).

i.e. serum cortisol levels at midnight are detectable (see Figure 46.5).

2. Establish the cause of Cushing's syndrome:

   (a) plasma ACTH levels: if ACTH is undetectable an adrenal tumor is indicated; detectable ACTH levels could be due to pituitary or ectopic sources;
   (b) high-dose DST: 2 mg dexamethasone orally 6-hourly for 48 h; 50% or more suppression of serum cortisol indicates pituitary disease. This test is often combined with the low-dose DST (see Figure 46.6);
   (c) CRH test: 100 μg human CRH-41 injected intravenously and serum cortisol (and ACTH) measured. An exaggerated response (rise in cortisol to above 600 nmol/L) indicates pituitary disease. In contrast, patients with the ectopic ACTH syndrome do not respond to tests (b) and (c) (see Figure 46.7);
   (d) venous catheter studies: *see* ACTH, CLINICAL APPLICATIONS.

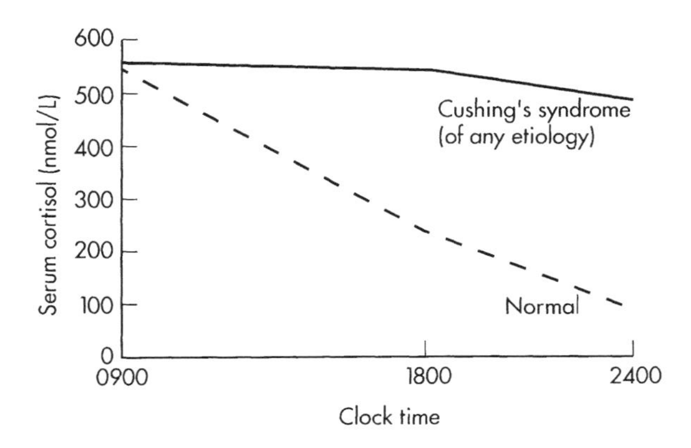

Fig. 46.5 Circadian rhythm studies.

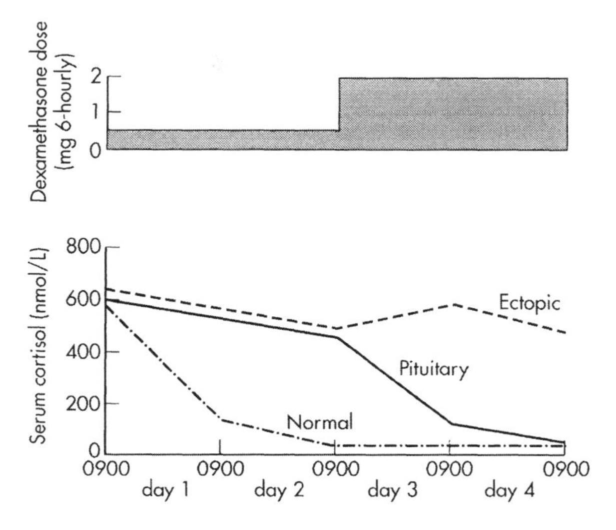

Fig. 46.6 Examples of responses in dexamethasone suppression test.

Addison's Disease

1. Morning (9 a.m.) serum cortisol: low or undetectable.
2. ACTH (tetracosactrin; Synacthen) stimulation test: short and long Synacthen tests. In healthy individuals serum cortisol levels rise to 500–600 nmol/L or more 30–60 min after 0.25 mg Synacthen is given intramuscularly. No cortisol rise is seen in Addison's disease. A delayed rise (6–24 h) is seen in secondary hypoadrenocorticism (due to pituitary or hypothalamic causes).

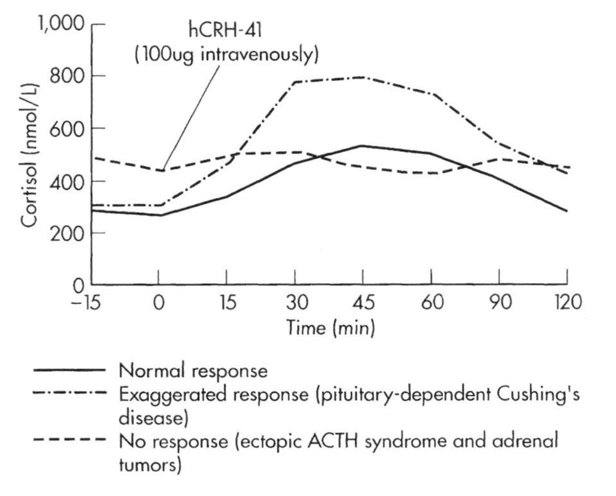

Fig. 46.7 Examples of cortisol response in CRH test.

Limitations

- Serum cortisol levels are affected by changes in the concentrations of CBG, thus cortisol is artifactually elevated in pregnancy and in females on estrogen-containing oral contraceptive pills, where CBG is increased.
- Cortisol measurements can be affected by one or more of the synthetic or natural steroids and so it is unreliable in patients taking prednisolone, a commonly prescribed glucocorticoid. Some immunoassay methods for cortisol cross-react significantly with 11-deoxycortisol, the immediate precursor of cortisol. This compound is excessively elevated in patients with Cushing's syndrome on metyrapone treatment (metyrapone inhibits the conversion of 11-deoxycortisol to cortisol). This is a serious drawback because this cross-reactivity results in apparently normal or even elevated cortisol levels in patients who have been rendered hypoadrenal. Such assays must therefore be used with great care.
- Some methods for cortisol measurement that use nonisotopic immunoassays have a serious problem of nonspecific interference, often associated with reagents used to denature cortisol-binding proteins before assay. Such assays must therefore be viewed with caution.
- Some immunoassays have inadequate sensitivity that could result in false negatives during the low-dose DST, which is a fundamental diagnostic test. This is a major problem that can be avoided by using kits with a lower detection limit of no more than 50 nmol/L.

Many laboratories nowadays use automated immunoassay systems. These tend to possess the necessary sensitivity and specificity.

Assay Technology

The commonly used methods for cortisol measurement include Chiron ACS™: 180, Abbott TDX™/FLX™, Bayer Immuno-1™, DPC IMMULITE™, DELFIA™, DPC Coat-a-Count™, and Roche Diagnostics Enzymun™. The trend now is towards the use of nonradioactive assays.

Desirable Assay Performance Characteristics

Knowledge of the degree of cross-reactivity with other steroids, especially 11-deoxycortisol, prednisolone, and dexamethasone, is vital. A clinical sensitivity of 25 nmol/L is required. An analyzer should generate a result within 60 min (for urgent cases, e.g. some patients in intensive care units with occult adrenocortical insufficiency).

Types of Sample

Serum or plasma. Ideally, specimen collection should be done at 8–10 a.m. since reference ranges are usually established for this time of the day.

Frequency of Use

Very common.

ADRENOCORTICOTROPIC HORMONE

ACTH (corticotropin) is a 39 amino acid peptide with a relative molecular mass of 4500 (4.5 kDa) that is derived from a larger (31 kDa) precursor, pro-opiomelanocortin (POMC), which is secreted by the corticotroph cells of the anterior pituitary. ACTH exists in the circulation in different molecular forms, some of which may be inactive biologically and undetected by immunoassays.

Function

ACTH is secreted by the anterior pituitary to stimulate the adrenal cortex directly, to synthesize and secrete glucocorticoids, the most important of which is cortisol. The bioactivity of ACTH resides in its first 24 amino-terminal amino acids. Serum cortisol exerts a negative feedback effect on the hypothalamus and pituitary to control the level of ACTH in the circulation.

Reference Interval

10–60 ng/L at 9 a.m.
2.2–13.2 pmol/L at 9 a.m.

Clinical Applications

Cushing's Syndrome

- *Basal ACTH*. Measurement of ACTH at 9 a.m. helps to establish the cause of Cushing's syndrome. Peripheral plasma ACTH levels are suppressed to below detection if the underlying pathology is an adrenal tumor (both adenomas and carcinomas) because these tumors secrete cortisol independently of ACTH and lead to suppression of plasma ACTH levels through the negative feedback mechanism. In contrast, if plasma ACTH concentrations are either detectable or high then a diagnosis of adrenal tumor is excluded and the underlying pathology is most likely to be either a pituitary or nonpituitary (ectopic) tumor such as lung cancer. Excessively elevated levels (above 500 ng/L) are highly indicative of an ectopic source of ACTH.
- *CRH test*. The test involves the intravenous administration of 100 μg human CRH-41 and the measurement of serum cortisol and plasma ACTH thereafter. Patients with pituitary disease respond by elevating their ACTH levels to above 100 ng/L whereas patients with the ectopic ACTH syndrome do not show a response.
- *Venous catheter studies*. ACTH measurements during venous catheter studies are extremely valuable in the localization of tumors responsible for Cushing's syndrome. Thus, if a pituitary cause is suspected, the inferior petrosal sinuses (veins that drain the pituitary) can be

cannulated and blood obtained for ACTH measurements before and after an injection of CRH. This procedure helps in the diagnosis of pituitary-dependent Cushing's disease and in localizing the tumor within the pituitary itself. Plasma ACTH measurement also helps in the follow-up of patients treated for Cushing's disease.

If the cause of Cushing's syndrome is suspected to be an ectopic (nonpituitary) tumor then a whole-body catheter study may need to be done where different veins in the chest and abdomen are cannulated and blood drawn for ACTH measurement. High concentrations of ACTH will be demonstrable in the veins draining the tumor and thus aid in tumor localization. *See* MANAGEMENT OF PATIENTS WITH CUSHING'S SYNDROME later in chapter for management of Cushing's syndrome.

Measurement of plasma ACTH concentrations is extremely valuable in the diagnosis and follow-up of patients with **Nelson's syndrome**, a rare but serious ACTH-secreting pituitary tumor that arises as a complication of bilateral adrenalectomy, a surgical procedure employed in difficult cases of Cushing's disease.

Adrenocortical Insufficiency

- Plasma ACTH levels are very useful in discerning the etiology of adrenocortical failure. ACTH levels are elevated (sometimes grossly so) in primary adrenocortical insufficiency (Addison's disease), whereas in secondary adrenocortical insufficiency (due to pituitary or hypothalamic causes) ACTH concentrations are low or undetectable.
- CRH test: this test occasionally will help differentiate pituitary from hypothalamic causes of adrenocortical insufficiency. Thus, patients with hypothalamic lesions will have raised ACTH levels after CRH injection whereas patients with pituitary disease show no response.

Limitations

- ACTH is an unstable peptide in plasma and therefore processing of the blood specimen must be done quickly with the need for a refrigerated centrifuge and solid CO_2 to freeze the plasma immediately.
- The assay is useful for differential diagnosis, but not for diagnosis of Cushing's syndrome where only cortisol measurement is required.
- The absolute levels of ACTH are not very useful in distinguishing pituitary from ectopic causes of Cushing's syndrome. In the majority of such cases, the ACTH concentrations lie between 30 and 200 ng/L. Thus, other ways to differentiate these two entities are often required.
- Not all of the ACTH assays available are sensitive enough to detect levels in all healthy individuals and are therefore not useful in differentiating low normal levels from the suppressed levels recorded in patients with adrenal tumors.

Assay Technology

Until recently, most assays used for the measurement of ACTH were competitive radioimmunoassays that used a plasma extraction procedure, such as the use of vycor glass to adsorb the peptide. Despite the fact that these extraction assays are time-consuming and labor-intensive with low sample throughput, they remain reliable and informative across the whole spectrum of diseases of the HPA axis.

However, most laboratories nowadays use immunometric assay methods and these appear to achieve the required sensitivity and are quick and convenient. Nonetheless, it must be remembered that ACTH immunometric assays require special expertise to set up because of the very low concentrations of ACTH circulating in the blood and the need to produce suitable antisera to achieve the required sensitivity and specificity. In addition, specificity can be a problem as some immunometric assays are so specific for ACTH 1–39 that they do not detect the larger forms of ACTH or its smaller fragments. These forms may be the only ones circulating in patients with the ectopic ACTH syndrome and would therefore be reported as undetectable instead of grossly elevated. This could lead to disastrous errors in diagnosis. It is, therefore, mandatory to check for the specificities of ACTH assays in relation to their clinical performance before they can be used for patient samples.

Most smaller laboratories send specimens to specialized centers for ACTH measurement. The popular methods used include Nichols Allegro™, Nichols ICMA™, DPC IMMULITE™, and Euro-Diagnostica IRMA in addition to a variety of in-house radioimmuno- and immunometric assays.

Desirable Assay Performance Characteristics

Clinical assay sensitivity of 5–10 ng/L is necessary. Detailed knowledge of assay specificity is mandatory to avoid errors in diagnosis. High-dose hook effects can be observed in immunometric assays when specimens from patients with Nelson's syndrome are analyzed.

Type of Sample

Plasma. Blood is collected into plastic tubes containing heparin or EDTA; glass tubes must not be used for blood collection as ACTH will be adsorbed to glass and lost. Owing to the instability of the peptide, blood needs to be centrifuged at 4 °C immediately after venepuncture and plasma decanted, flash frozen, and stored at − 20 °C until assay (stable for 6 months). Ideally, sample collection should be performed at 8–10 a.m. since reference ranges are usually established for this time of day.

Frequency of Use

Uncommon.

MANAGEMENT OF PATIENTS WITH CUSHING'S SYNDROME

PRE-ADMISSION

- Diagnosis of Cushing's syndrome suspected on clinical grounds.
- Diagnosis of Cushing's syndrome established biochemically by low-dose DST or by urinary free cortisol. Ideally, obtain a plasma sample for ACTH.

ADMISSION TO ENDOCRINE WARD (E.G. ON MONDAY)

Monday

Routine chemistry (plasma potassium, bicarbonate, and glucose in particular), hematology and chest X-ray. Blood sample for plasma ACTH.

Tuesday

CRH test: give 100 μg hCRH i.v. and measure cortisol and ACTH.

Wednesday

Circadian rhythm study: samples for cortisol at 9 a.m., 6 p.m., and midnight (whilst asleep).

Thursday to Sunday

Low dose followed by high-dose DST: 0.5 mg dexamethasone orally 6 hourly for 48 h followed by 2 mg 6 hourly for a further 48 h.

By following Monday

All results of tests are available and provide provisional diagnoses:

- No suppression on low-dose DST = Cushing's syndrome. If pseudo-Cushing's due to depression is suspected, perform an insulin hypoglycemia test.
- Plasma ACTH undetectable: diagnostic of adrenal tumors. Proceed to radioimaging of abdomen and adrenalectomy.
- Plasma ACTH detectable but not very grossly elevated: pituitary or ectopic ACTH-producing tumor:

  (a) Serum cortisol rises during CRH test and suppresses during high-dose DST = pituitary tumor. Proceed to pituitary radioimaging, petrosal sinus catheter study, and hypophysectomy.
  (b) Serum cortisol does not rise during CRH test and does not suppress on high-dose DST = ectopic ACTH-producing tumor. Plasma ACTH may be grossly elevated (greater than 300 ng/L) plus hypokalemia, alkalosis, and hyperglycemia. Proceed to radioimaging studies (and venous catheters) of chest and abdomen to localize tumor followed by its surgical removal.
  (c) If results of CRH and high-dose DST are discrepant, i.e. do not point to a uniform diagnosis as in (a) or (b), it becomes imperative to proceed to petrosal catheter study since pituitary disease is much more common than ectopic ACTH-producing lesions. If petrosal catheterization is not possible, patient needs to be referred to a specialist center capable of performing the procedure.

FURTHER READING

Clark, P.M., Neylon, I., Raggatt, P.R., Sheppard, M.C. and Stewart, P.M. Defining the normal cortisol response to the short Synacthen test: implications for the investigation of hypothalamic-pituitary disorders. *Clin. Endocrinol.* **49**, 287–292 (1998).

De Brabandere, V.I., Thienpont, L.M., De Stockl, D. and Leenheer, A.P. Three routine methods for serum cortisol evaluated by comparison with an isotope dilution gas chromatography-mass–spectrometry method. *Clin. Chem.* **41**, 1781–1783 (1995).

Erturk, E., Jaffe, C.A. and Barkan, A.L. Evaluation of the integrity of the hypothalamic–pituitary–adrenal axis by insulin hypoglycemia test. *J. Clin. Endocrinol. Metab.* **83**, 2350–2354 (1998).

Gibson, S., Pollock, A., Littley, M., Shalet, S. and White, A. Advantages of IRMA over RIA in the measurement of ACTH. *Ann. Clin. Biochem.* **26**, 500–507 (1989).

Grinspoon, S.K. and Biller, B.M.K. Laboratory assessment of adrenal insufficiency. *J. Clin. Endocrinol. Metab.* **79**, 923–931 (1994).

Grossman, A.B., Howlett, T.A., Perry, L., Coy, H., Savage, M.O., Lavender, P., Rees, L.H. and Besser, G.M. Corticotropin-releasing hormone in the differential diagnosis of Cushing's syndrome: a comparison with the dexamethasone suppression test. *Clin. Endocrinol.* **29**, 167–178 (1988).

Horrocks, P.M., Jones, A.F., Ratcliffe, W.A., Holder, G., White, A., Holder, R., Ratcliffe, J.G. and London, D.R. Patterns of ACTH and cortisol pulsatility over twenty-four hours in normal males and females. *Clin. Endocrinol.* **32**, 127–134 (1990).

Hurel, S.J., Thompson, C.J., Watson, M.J., Baylis, P.H. and Kendall-Taylor, P. The short synacthen and insulin stress tests in the assessment of the hypothalamic–pituitary–adrenal axis. *Clin. Endocrinol.* **44**, 141–146 (1996).

Lamberts, S.W.J., Bruining, H.A. and De Jong, F.H. Corticosteroid therapy in severe illness. *N. Engl. J. Med.* **337**, 1285–1293 (1997).

Mayenknecht, J., Diederich, S., Bahr, V., Plockinger, U. and Oelkers, W. Comparison of low and high dose corticotropin stimulation tests in patients with pituitary disease. *J. Clin. Endocrinol. Metab.* **83**, 1558–1562 (1998).

Mukherjee, J.J., Jacome de Castro, J., Kaltsas, G., Afshar, F., Grossman, A.B., Wass, J.A.H. and Besser, G.M. A comparison of the insulin tolerance/glucagon test with the short ACTH stimulation test in the assessment of the hypothalamo-pituitary-adrenal axis in the early post-operative period after hypophysectomy. *Clin. Endocrinol.* **47**, 51–60 (1997).

Newell-Price, J. The desmopressin test and Cushing's syndrome: current state of play. *Clin. Endocrinol.* **47**, 173–174 (1997).

Newell-Price, J., Trainer, P.J., Perry, L.A., Wass, J.A.H., Grossman, A.B. and Besser, G.M. A sleeping midnight cortisol has 100% sensitivity for the diagnosis of Cushing's syndrome. *Clin. Endocrinol.* **43**, 545–550 (1995).

Oldfield, E.H., Doppman, J.L., Nieman, L.K., Chrousos, G.P., Miller, D.L., Katz, D.A., Cutler, G.B. Jr. and Loriaux, D.L. Petrosal sinus sampling with and without corticotropin-releasing hormone for the differential diagnosis of Cushing's syndrome. *N. Engl. J. Med.* **325**, 897–905 (1991).

Orth, D.M. Differential diagnosis of Cushing's syndrome. *N. Engl. J. Med.* **325**, 957–959 (1991).

Papanicolaou, D.A., Yanovski, J.A., Cutler, G.B. Jr., Chrousos, G.P. and Nieman, L.K. A single midnight serum cortisol measurement distinguishes Cushing's syndrome from pseudo-Cushing's states. *J. Clin. Endocrinol. Metab.* **83**, 1163–1167 (1998).

Perry, L.A. and Grossman, A.B. The role of the laboratory in the diagnosis of Cushing's syndrome. *Ann. Clin. Biochem.* **34**, 345–359 (1997).

Raff, H. and Findling, J.W. A new immunoradiometric assay for corticotropin evaluated in normal subjects and patients with Cushing's syndrome. *Clin. Chem.* **35**, 595–600 (1989).

Ross, R.J.M. and Trainer, P.J. Endocrine investigation: Cushing's syndrome. *Clin. Endocrinol.* **49**, 153–155 (1998).

Streeten, D.H., Anderson, G.H. Jr., Brennan, S. and Jones, C. Suppressibility of plasma adrenocorticotropin by hydrocortisone: potential usefulness in the diagnosis of Cushing's disease. *J. Clin. Endocrinol. Metab.* **83**, 1114–1120 (1998).

Streeton, D.H.P., Anderson, G.H. and Bonaventura, M.M. The potential for serious consequences from misinterpreting normal responses to the rapid adrenocorticotropin test. *J. Clin. Endocrinol. Metab.* **81**, 285–290 (1996).

Von Werder, K. and Muller, O.A. Cushing's syndrome. In: *Clinical Endocrinology* (ed Grossman, A.), 442–456 (Blackwell Scientific Publications, Oxford, 1992).

Weintrob, N., Sprecher, E., Josefsberg, Z., Weininger, C., Aurbach-Klipper, Y., Lazard, D., Karp, M. and Pertzelan, A. Standard and low-dose short adrenocorticotropin test compared with insulin-induced hypoglycemia for assessment of the hypothalamic-pituitary-adrenal axis in children with idiopathic pituitary hormone deficiencies. *J. Clin. Endocrinol. Metab.* **83**, 88–92 (1998).

White, A. and Gibson, S. ACTH precursors: biological significance and clinical relevance. *Clin. Endocrinol.* **48**, 251–255 (1998).

47 Bone Metabolism

Kay W. Colston and John C. Stevenson

NORMAL CALCIUM METABOLISM

Calcium is a divalent cation that is involved in neuromuscular function, including cardiac muscle. Its level in extracellular fluid is tightly controlled by regulatory hormones. Calcium ions have important intracellular functions but the purpose of this chapter is to review the regulation of extracellular calcium.

A normal adult contains between 700 and 1000 g of calcium of which over 99% is in the skeleton. Approximately 50% of circulating calcium exists as free calcium ions. The remainder is bound to albumin or complexed to anions such as citrate, pyruvate, and lactate.

The extracellular calcium concentration (2.1–2.6 mmol/L) is relatively high compared to the intracellular concentration of free calcium (usually less than 1 μmol/L). The large mass of calcium in bones may not be immediately available but it can still be reabsorbed and thus, in the long term, influence the soluble calcium pools.

Calcium absorption, excretion, and mobilization from bone are closely regulated. The hormones central to this control include vitamin D metabolites, parathyroid hormone (PTH), and calcitonin. Of these, PTH and 1,25-dihydroxyvitamin D ($1,25(OH)_2D$) have the strongest effect on plasma calcium and their interactions are illustrated in Figure 47.1.

Both hormones act to raise plasma calcium levels. $1,25(OH)_2D$ acts predominantly to increase calcium absorption from the gut and to increase its resorption from bone. PTH produces its major regulatory effect on calcium by increasing its resorption in the kidney; at supraphysiological levels it causes enhanced bone resorption. Effects of calcitonin on changes in plasma calcium levels are minimal under normal physiological circumstances. Calcitonin may play a greater role in bone metabolism but its importance in overall calcium homeostasis is still not determined.

For diagnostic purposes, changes in plasma calcium should be interpreted in the context of a thorough medical history and examination. A change in plasma calcium should be corrected for any change in plasma albumin. A general rule of thumb is that for any 10 g/L alteration in albumin from 40 g/L, the plasma calcium should be adjusted by 0.17 mmol in the opposite direction.

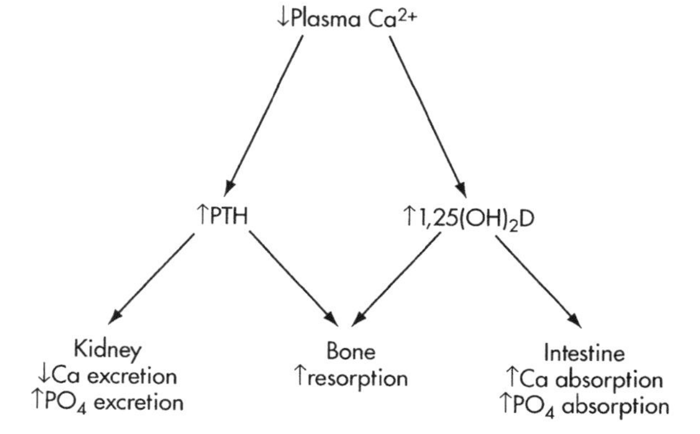

Fig. 47.1 Effects of PTH and $1,25(OH)_2D$ on Ca^{2+}.

CLINICAL DISORDERS

HYPERCALCEMIA

Hypercalcemia is often asymptomatic but can cause thirst, polyuria, dehydration, constipation, abdominal pain, renal calculi, nephrocalcinosis, and pancreatitis. With very high levels of calcium, cardiac dysrhythmias may occur; there is impairment of consciousness, coma and eventual death (Table 47.1).

Differentiation between the various causes of hypercalcemia is not difficult if the clinical setting is obvious and reliable diagnostic tests are available. If the patient is otherwise well and there is documentation of an elevated serum calcium concentration several years previously, this suggests primary hyperparathyroidism (see below). This diagnosis is particularly likely if hypophosphatemia

The Immunoassay Handbook. *Third Edition*
D. Wild (Ed.)

Table 47.1. Causes of hypercalcemia.

1 Primary hyperthyroidism		
2 Malignant disease:	Metastases in bone	e.g. carcinoma of the breast
	Humoral hypercalcemia	e.g. squamous cell carcinoma of the lung
	Hematological malignancies	e.g. multiple myeloma
3 Sarcoidosis and other granulomatous diseases		
4 Vitamin D intoxication		
5 Renal disease:	Diuretic phase of acute tubular necrosis	
	Parathyroid hyperplasia in chronic renal failure	
6 Familial hypocalciuric hypercalcemia		
7 Milk alkali syndrome		
8 Idiopathic hypercalcemia of infancy		
9 Thyrotoxicosis		
10 Thiazide diuretics (transient)		
11 Immobilization:	Paget's disease	
	Quadriplegia	
12 Addison's disease		

has been recorded on any occasion. In **hypercalcemia of malignancy**, imaging, biochemical and hematological tests may suggest the diagnosis, which should ideally be confirmed by subsequent histology. Steroid suppression tests are unreliable for distinguishing between hyperparathyroidism and other causes of hypercalcemia because both false positives and false negatives may occur. Most patients with hypercalcemia have either primary hyperparathyroidism or malignancy. Other causes are extremely rare.

The intact PTH assay is used to distinguish patients with primary hyperparathyroidism from those with hypercalcemia of malignancy. The term **humoral hypercalcemia of malignancy** has been used to refer to those patients with malignancy-associated hypercalcemia who display biochemical features similar to those seen in primary hyperparathyroidism.

PTH-related protein (PTHrP) is thought to be the humoral factor responsible for hypercalcemia associated with cancer. As with excess secretion of PTH, PTHrP production by tumors results in hypercalcemia because of increased bone resorption and reduced urinary calcium excretion. The protein also promotes renal phosphate excretion and increases nephrogenous cyclic AMP. However, whereas a mild hyperchloremic acidosis is often a feature in primary hyperparathyroidism, hypokalemic alkalosis is frequently found in humoral hypercalcemia of malignancy. The basis for these differences remains to be established.

PARATHYROID DISORDERS

Primary hyperparathyroidism is a common disorder occurring in approximately 1 in 1000 of the general population but the incidence is reported to be higher among postmenopausal women. The condition is most commonly associated with the presence of a solitary **parathyroid adenoma** autonomously secreting PTH. Primary hyperparathyroidism may also develop when certain undefined pathological processes lead to hyperplasia of the parathyroid gland. This may occur sporadically or form part of the syndrome of **multiple endocrine neoplasia**.

Patients with primary hyperparathyroidism are usually asymptomatic, but may present with renal stones or nephrocalcinosis. In elderly patients the presentation may be that of **acute hypercalcemic crisis** characterized by confusion and dehydration. Other forms of presentation include non-specific gastrointestinal symptoms, polyuria, polydypsia and occasionally psychiatric disorders. Overt evidence of bone disease is seen in about 10% of patients. Frequently, however, hypercalcemia is an incidental finding during routine biochemical testing for the investigation of some unrelated condition. Serum calcium is usually but not invariably raised. Serum phosphate concentration is sometimes low or below normal. Plasma chloride is usually in the upper normal range or clearly elevated with reverse changes in serum bicarbonate. Serum alkaline phosphatase activity is usually normal but may be raised when there is bone disease. Circulating levels of PTH are raised or inappropriately high for the level of serum calcium. The development of assays that measure the intact hormone has improved the usefulness of this test.

Decreased secretion of PTH occurs either because of surgical damage to the parathyroid gland or because of spontaneous failure of parathyroid gland function, usually due to autoimmune factors. The clinical features result from acute or chronic effects of hypocalcemia including neuromuscular hyperexcitability, carpal or pedal spasm, non-specific psychological changes and cataracts.

Hypoparathyroidism is diagnosed from the combination of hypocalcemia and hyperphosphatemia with low or undetectable PTH values. Hypocalcemia and hyperphosphatemia also occur in chronic renal failure and **pseudohypoparathyroidism** but circulating PTH concentrations are high in these conditions.

HYPOCALCEMIA

In **hypocalcemia** a decrease in the concentration in extra-cellular calcium ions causes hyperexcitability of the nervous system. The clinical manifestations are tetany, either latent or overt, proximal weakness, and ECG abnormalities. Other effects of long-standing hypocalcemia are cataracts, cutaneous candidiasis, calcification of the basal ganglion and psychiatric disorders, particularly mental depression (Table 47.2).

VITAMIN D DISORDERS

Vitamin D intoxication presents clinically as hypercalcemia with its attendant features. **Vitamin D deficiency** manifests itself predominantly as skeletal defects resulting from impaired mineralization of bone. It is now uncommon in the UK but may be seen in growing children or in the elderly, particularly those who are housebound. It is more common in Asian communities, mainly because of traditional habits in diet and dress. In growing children, the disease may manifest as bone pain, tenderness and deformities with expanded epiphyses, muscle weakness, and even symptomatic hypocalcemia; this is the clinical picture of **rickets**. In adults, the condition is known as **osteomalacia** which may manifest as bone pains and tenderness, fractures, proximal myopathy, and symptomatic hypocalcemia.

Table 47.2. Major causes of hypocalcemia.

Major causes of hypocalcemia
Protracted
Privational vitamin D deficiency
Hypoparathyroidism
Pseudohypoparathyroidism
Chronic renal disease
Malabsorption syndrome
Vitamin D-dependent rickets
Transient
Postoperative:
Thyroid
Parathyroid
Other neck surgery
Acute pancreatitis
Hypomagnesemia

Biochemical Changes

Plasma calcium may be subnormal in advanced cases but initially the tendency to hypocalcemia is corrected by increased parathyroid activity so that a normal plasma calcium concentration is not uncommon. Plasma phosphate concentration is usually in the low normal or subnormal range. There is generally an increase in the activity of the bone isoenzyme of alkaline phosphatase because of increased osteoblastic activity. Some patients do not have increased plasma total alkaline phosphatase so that a normal result does not exclude osteomalacia. A common biochemical feature is the presence of secondary hyperparathyroidism. Urinary calcium excretion is usually low as a result of increased PTH secretion.

The development of **vitamin D insufficiency**, due in part to defective 1α-hydroxylation of 25-hydroxyvitamin D, occurs in chronic renal failure. A wide spectrum of biochemical abnormalities is present because of the combined effects of secondary hyperparathyroidism, osteomalacia, and renal failure.

Serum 25-hydroxyvitamin D concentrations may be normal or low while $1,25(OH)_2D$ concentrations are decreased. PTH concentration may be overestimated by immunoassays that do not measure the whole molecule, because of retention of PTH and its fragments due to impaired metabolism and renal excretion.

It is well recognized that patients on long-term anticonvulsant therapy have an increased prevalence of osteomalacia. Such patients have low plasma 25-hydroxyvitamin D concentrations thought to result from induction of hepatic microsomal enzymes, leading to increased catabolism and a decreased half-life of vitamin D and its metabolites. Anti-tuberculosis drugs may similarly cause osteomalacia.

Both hereditary and acquired forms of rickets and osteomalacia occur due to defective 1,25-dihydroxyvitamin D synthesis. **Vitamin D-dependent rickets type I** is a rare disorder inherited as an autosomal recessive trait. The disorder is thought to be a hereditary defect in the renal 1α-hydroxylase enzyme responsible for the production of 1,25-dihydroxyvitamin D. Severe rickets usually manifests before the age of 6 months. Plasma biochemistry shows the typical changes of vitamin D deficiency: hypocalcemia, hypophosphatemia, secondary hyperparathyroidism and increased alkaline phosphatase activity. The rickets can be healed by treatment with small doses of 1,25-dihydroxyvitamin D but very large doses of the parent vitamin (1000–3000 μg per day) are needed.

Vitamin D-dependent rickets type II is a disorder characterized by rickets or osteomalacia that is present despite marked increases in circulating 1,25-dihydroxyvitamin D. The disorder appears to be one of target organ resistance to the hormonal form of vitamin D. It is likely that structural abnormalities in the receptor account for the defective response of target organs to 1,25-dihydroxyvitamin D.

Because the renal activation of vitamin D is regulated by the vitamin D status, high doses of vitamin D are needed to produce hypercalcemia. **Vitamin D intoxication** may occur as a complication of therapy in hypoparathyroidism, but occasionally vitamin D ingestion is surreptitious. Hypercalcemia due to intoxication with vitamin D itself is often protracted, lasting weeks or occasionally months, unless treated with glucocorticoids. Intoxication with 1,25-dihydroxyvitamin D_3 (calcitriol) also produces hypercalcemia, but the duration is much shorter (a few days), because of its shorter biological half-life. Assay of 25-hydroxyvitamin D (in the case of vitamin D intoxication) or 1,25-dihydroxyvitamin D (in the case of 1,25-dihydroxyvitamin D intoxication) will reveal the diagnosis if this is not clinically obvious.

About 10% of patients with sarcoidosis exhibit hypercalcemia, and a much higher proportion have hypercalciuria (about 50%). The hypercalcemia is associated with increased absorption of calcium in the intestine, and it has recently been demonstrated that it is associated with extra-renal production of 1,25-dihydroxyvitamin D by sarcoid tissue producing inappropriately increased circulating 1,25-dihydroxyvitamin D concentration. Other granulomatous diseases such as beryllîosis and tuberculosis can also be associated with hypercalcemia.

In **lymphoma** and in acute and chronic **leukemia**, hypercalcemia is sometimes associated with inappropriately high 1,25-dihydroxyvitamin D concentrations suggesting that the tumor cells may be the site of unregulated hydroxylation of 25-hydroxyvitamin D.

MEDULLARY CARCINOMA OF THE THYROID

Medullary carcinoma of the thyroid, which may present as a lump in the neck, is associated with hypersecretion of calcitonin. Families in whom this condition is an inherited trait require screening for excessive calcitonin secretion.

PAGET'S DISEASE

In **Paget's disease** there is excessive focal bone resorption by osteoclasts with a compensatory but disordered increase in osteoblastic formation. Affected bones become expanded and deformed, and may be more liable to fracture. A number of other complications can occur. Serum alkaline phosphatase is clearly elevated in this condition.

OSTEOPOROSIS

Osteoporosis commonly occurs in postmenopausal women. There is a loss of bone density due to increased bone turnover, predominantly resorption, of bone tissue, both mineral and matrix. This leads to an increased risk of fracture.

ANALYTES

In all immunometric assays, the following are important. Initial extraction procedures should be efficient and reproducible. The tracer should be stable and of sufficiently high specific activity, and the solid-phase employed for the capture antibody should have a high capacity, giving a linear relationship between signal and analyte concentration. Efficient separation of bound from free should also be achieved. Quality control samples should be included in every assay; these should have hormone concentrations covering the usual assay range.

VITAMIN D METABOLITES: 25-HYDROXYVITAMIN D, 1,25-DIHYDROXYVITAMIN D

Vitamin D metabolites consist of a steroid nucleus with the important structural variations of the different metabolites being related to the number and position of hydroxyl groups.

The synthesis of vitamin D and its metabolites involves metabolic pathways in a number of different tissues. Vitamin D_3 (cholecalciferol) is produced in the skin by the action of sunlight on the precursor molecule 7-dehydrocholesterol. Ultraviolet irradiation leads to fission of the ring between carbons 9 and 10 to produce vitamin D_3. Chronic lack of sunlight may result in a deficiency of this vitamin. Vitamin D is converted in the liver to 25-hydroxyvitamin D (25OHD). A second hydroxylation step takes place in the kidney to form 1α,25-dihydroxyvitamin D ($1,25(OH)_2D$), the biologically active metabolite. Disturbed tissue function at either of these sites has the potential to alter vitamin D levels and thus calcium metabolism. An alternative hydroxylation in the 24 position produces 24,25-dihydroxyvitamin D ($24,25(OH)_2D$), a relatively inert metabolite (Table 47.3).

With normal sunlight exposure, dietary sources of vitamin D are of marginal significance. Plasma transport of vitamin D metabolites is on a specific carrier protein, vitamin D binding protein. A second form of vitamin D, vitamin D_2 (ergocalciferol), can be produced by ultra violet irradiation of the fungal sterol, ergosterol.

Table 47.3. Vitamin D.

Diet	
	Vitamin D_3 (cholecalciferol)
	Vitamin D_2 (ergocalciferol)
Skin	
	Ultraviolet light
	7-dehydrocholesterol
	Vitamin D_3
Liver	
	25-hydroxyvitamin D_3 (25OHD_3)
	(25-hydroxyvitamin D_2)
Kidney	
	1,25-dihydroxyvitamin D_3 (1,25$(OH)_2D_3$)
	or 24,25-dihydroxyvitamin D_3 (24,25$(OH)_2D_3$)
	(1,25-dihydroxyvitamin D_2 (1,25$(OH)_2D_2$) or 24,25-dihydroxyvitamin D_2)

Its side chain structure differs slightly from that of vitamin D_3 but otherwise it appears to have equivalent biological metabolism in humans.

Metabolism

Recent interest in the catabolism of 1,25$(OH)_2$D has led to the elucidation of a pathway involving C24 oxidation and side chain cleavage with ultimate formation of calcitroic acid. Metabolites containing 24 hydroxylations have very little biological activity, although they may form part of the controlling mechanism for the regulation of production of metabolites.

Function

The major target organ for 1,25$(OH)_2$D is the small intestine where it acts to increase active calcium absorption. This action involves vitamin D-induced changes in the calcium transport system, of which an increase in the concentration of a specific calcium binding protein, calbindin, is an important part. Vitamin D metabolites also act on bone to increase calcium resorption by an indirect effect on osteoclasts, and act directly on osteoblasts to stimulate their activity, resulting in increased alkaline phosphatase levels in plasma. However, the predominant role of vitamin D in bone metabolism is to promote mineralization and thus retention of calcium in bone by raising extracellular calcium levels.

Although in the past vitamin D has been considered solely as a calcium-regulatory hormone, there are now suggestions that it may have more of a general role in cellular metabolism *via* modulation of intracellular calcium transport. It is now clear that vitamin D has many different effects in various tissues, and its receptors are widely distributed throughout the body.

Reference Intervals

25OHD:

20–100 nmol/L (winter)
25–150 nmol/L (summer)
(commercially available RIA method)

1,25$(OH)_2$D:

40–120 pmol/L (adults)
(RIA, Hollis, 1997)

Clinical Applications

The measurement of vitamin D metabolites is useful in the investigation of hypo- and hypercalcemia. Levels of 25OHD are elevated in vitamin D intoxication and reduced in vitamin D deficiency. Measurement of vitamin D metabolites is useful in the investigation of patients with hyper- and hypocalcemia. Levels of 25OHD are elevated in vitamin D intoxication and reduced in vitamin D deficiency. Cutaneous production of vitamin D is related to intensity of sunlight exposure and decreases with advancing age. In elderly subjects living in Northern Europe, vitamin D supplementation has been recommended during the winter months in order to maintain adequate 25OHD concentrations. Serum concentrations of 25OHD show a seasonal variation with highest levels seen in the late summer months and a nadir in late winter/early spring. In general, a serum concentration of 25OHD in autumn of less than 40 nmol/L may indicate insufficient body stores to maintain levels during the winter months. Serum levels of 25OHD less than 10 nmol/L indicate severe deficiency associated with osteomalacia.

In chronic renal failure, serum 25OHD concentrations may be normal or low while 1,25$(OH)_2$D concentrations are decreased. Patients on long-term anticonvulsant therapy or taking anti-tuberculous drugs have an increased prevalence of osteomalacia, with low plasma 25OHD concentrations.

Vitamin D dependent rickets type I, also known as **pseudo-vitamin D deficiency rickets** (**PDDR**) is an inborn error in conversion of 25OHD TO 1,25$(OH)_2D_3$ due to deficiency of the 1α-hydroxylase. Serum levels of 1,25$(OH)_2D_3$ are low or undetectable, and patients with this condition can be successfully treated with physiological doses of the metabolite. Vitamin D dependent rickets type II **(hereditary vitamin D resistant rickets, HVDRR)** is a disorder characterized by early onset rickets with marked increases in circulating 1,25$(OH)_2$D, which is attributable to end organ resistance to the hormone due to a receptor defect. In lymphoma and in acute and chronic leukemia, hypercalcemia is sometimes

Table 47.4. 25-Hydroxyvitamin D concentrations in various conditions associated with disturbances of calcium metabolism.

Low	Normal	High
Privational osteomalacia (diet, sunlight)	Sarcoidosis Vitamin D -dependent rickets	Vitamin D intoxication
Cirrhosis		
Anti-convulsant therapy		
Rickets/osteomalacia		

associated with inappropriately high $1,25(OH)_2D$ concentrations. In patients with sarcoidosis, hypercalcemia may be due to synthesis of $1,25(OH)_2D_3$ by sarcoid tissue. Levels of $1,25(OH)_2D_3$ may also be elevated in a subgroup of stone forming patients who have absorptive hypercalciuria and normal parathyroid function (Bataille *et al.*, 1987). (Tables 47.4 and 47.5).

Limitations

- Low 25OHD levels indicate vitamin D deficiency only. Diagnosis of osteomalacia may be made by histomorphometry.
- Monitoring vitamin D replacement therapy by measurement of $1,25(OH)_2D$ levels is only useful in patients treated with alfacalcidol (1α hydroxyvitamin D) or calcitriol ($1,25(OH)_2D_3$).

Table 47.5. Serum 1,25-dihydroxyvitamin D concentrations.

Causes of increased 1,25-dihydroxyvitamin D levels	
1 Physiological:	Growth pregnancy lactation
2 Hyperparathyroidism	
3 Sarcoidosis	
4 Acromegaly	
5 Hypothyroidism	
6 Vitamin D-dependent rickets type II	
Causes of decreased 1,25-dihydroxyvitamin D levels	
1 Renal failure	
2 Vitamin D deficiency	
3 Hypoparathyroidism	
4 Vitamin D-dependent rickets type I	
5 Hyperthyroidism	

Assay Technology

Anti-rachitic potency used to be assessed by the 'line test'. Nutritional rickets is induced in test rats and the degree of linear calcification in the radial epiphysis in response to vitamin D or an unknown sample is quantified. The sensitivity of this bioassay is about one international unit of vitamin D_2 or vitamin D_3. More sensitive assays for metabolites of vitamin D were subsequently developed that took advantage of the high affinity and selectivity of the target tissue receptors and serum transport proteins. These have now been largely superseded by radioimmunoassays (RIAs).

25-Hydroxyvitamin D

25-hydroxyvitamin D (25OHD) is the major circulating form of vitamin D in serum, accounting for more than 80–90% of total vitamin D metabolites. Measurement of serum 25OHD concentration is a useful index of overall body stores of vitamin D. A competitive protein-binding assay for 25OHD was developed in the early 1970s, which gained widespread use. Initially, vitamin D binding protein present in rat serum or kidney cytosol was utilized as a specific binding agent (Haddad and Chyu, 1971). The procedure involved organic extraction of the serum sample with ethanol, acetonitrile or other solvents followed by sample pre-purification using Sephadex or silica columns or Sep–Pak cartridges. These methods used $[^3H]$ $25OHD_3$ as tracer and required individual sample recovery estimation to correct for endogenous losses of 25OHD during extraction and chromatographic steps. This method measures both $25OHD_3$ and $25OHD_2$. The first valid direct UV quantitative HPLC assay was introduced in 1977 (Eisman *et al.*, 1976). HPLC detection provided the advantage of being able to separately quantitate $25OHD_3$ and $25OHD_2$. The disadvantages of this assay include the requirement for expensive equipment, a large sample size and the considerable expertise required. Hence these approaches to determine circulating 25OHD levels have largely been superseded by RIA. This has eliminated the requirement for sample chromatography, although a solvent extraction step is still required. The development of ^{125}I-25OHD tracer has further increased the performance of this assay (Hollis *et al.*, 1993).

1,25-Dihydroxyvitamin D

In 1978, the first RIA for $1,25(OH)_2D_3$ was introduced. This assay was relatively non-specific and lengthy sample preparative steps were still required. In 1984, a new approach to determining circulating concentrations of $1,25(OH)_2D_3$ was introduced involving more rapid and simple methodology.

However this assay still had the disadvantage of utilizing $[^3H]1,25(OH)_2D_3$ as tracer and as a result Hollis *et al.* (1996) reported the development of a new RIA using an ^{125}I-based tracer, as well as standards in an equivalent serum matrix so individual sample recoveries are not required. Development of RIAs for quantitation of circulating $1,25(OH)_2D_3$ has been hampered by the relatively poor specificity of the antibodies which have been generated. Even the best antibodies still show a cross-reactivity of approximately 1% with the non-1-hydroxylated metabolites which circulate at concentrations of more than 1000 times greater than $1,25(OH)_2D_3$. In the method developed by Hollis *et al.*, sample pre-treatment with sodium periodate is incorporated in order to convert interfering non-1-hydroxylated metabolites (chiefly $24,25(OH)_2D_3$ and $25,26(OH)_2D_3$) to their respective aldehyde and ketone forms, which are then removed by the subsequent chromatographic steps. Use of this purification procedure reduced cross-reactivity of non-1-hydroxylated vitamin D compounds to insignificant levels (Hollis, 1997). A second ^{125}I based RIA for $1,25(OH)_2D_3$, which is commercially available, involves the immunoextraction of $1,25(OH)_2D_3$ from serum or plasma samples with a specific monoclonal antibody bound to solid support. This antibody is directed to the 1α-hydroxylated A ring of $1,25(OH)_2D_3$, and interference from other 1-hydroxylated metabolites such as $1,25(OH)_2D_3$-26,23-lactone has been reported (Hollis, 1997).

Type of Sample

Serum, store frozen.

Frequency of Use

Moderate. Used in specialized centers.

PARATHYROID HORMONE

PTH (parathyrin) is a single-chain polypeptide composed of 84 amino acids, produced by two discrete pairs of parathyroid glands located at the upper and lower poles of the thyroid gland in the neck. Like many peptide hormones, PTH is synthesized as a larger precursor molecule, preproparathyroid hormone (a 114 amino acid form) and this peptide is cleaved to form proparathyroid hormone. The pro sequence of six amino acids is cleaved to form the 84 amino acid active peptide, which is stored in secretory granules of the parathyroid cells.

Function

In the circulation, intact PTH has a short half-life (less than 5 min) compared to the biologically inactive middle and carboxy-terminal fragments. The main actions of PTH are on bone, where it stimulates bone turnover, and the kidneys, where it acts directly to promote calcium reabsorption (and promote phosphate excretion), and indirectly *via* its ability to increase the activity of the 1α-hydroxylase enzyme controlling $1,25(OH)_2D$ synthesis. The major factor regulating the release of PTH is the extracellular calcium concentration: secretion is markedly increased by minor falls in plasma calcium just below the reference interval. Other factors have also been shown to influence PTH release, including $1,25(OH)_2D$, which is reported to suppress PTH gene expression.

Reference Interval

1–6 pmol/L
(RIA, whole molecule)

Clinical Applications

In primary hyperparathyroidism, circulating levels of PTH are raised or inappropriately high for the level of serum calcium. The development of assays that measure the intact hormone has improved the usefulness of this test.

Decreased secretion of PTH occurs either because of surgical damage to the parathyroid gland or because of spontaneous failure of parathyroid gland function, usually due to autoimmune factors. Hypoparathyroidism is diagnosed from the combination of hypocalcemia and hyperphosphatemia with low or undetectable PTH values. Hypocalcemia and hyperphosphatemia also occur in chronic renal failure and pseudohypoparathyroidism, but circulating PTH concentrations are high in these conditions.

Limitations

Use of antisera directed towards the *C*-terminal region of the molecule detects biologically inactive fragments (*see* ASSAY TECHNOLOGY).

Assay Technology

RIA is the commonest method used for routine measurement of circulating PTH. A major problem is the variety of immunological species of the hormone present in plasma. The major immunoreactive species is the biologically inactive *C*-terminal fragment. Intact PTH and its fragments are cleared from the circulation both by the kidneys and liver. The clearance of *C*-terminal fragments is slower than that of the intact hormone. It is also more dependent upon renal mechanisms, and consequently renal impairment leads to greater accumulation of this fragment. Most of the early RIAs for PTH were directed against the *C*-terminal region. Although these assays show raised values in the majority of patients subsequently shown to have primary hyperparathyroidism, a variable proportion have normal values, and patients with renal impairment have raised concentrations regardless of parathyroid secretion rate because of impaired clearance of the *C*-terminal fragments. The early assays were also based on antisera, calibrators and tracer from non-human species, which contributed to the insensitivity of the assay and nonparallel dilution characteristics.

Recent developments have improved the sensitivity and reproducibility of the measurement of PTH, and the two-site assay measuring intact hormone is becoming the universally accepted method. One commercially available intact PTH immunoradiometric assay (IRMA) utilizes two different polyclonal antibodies. One, directed against amino acids 39–84, is bound to a solid phase. The second antibody, which recognizes the first 34 amino acids, is labeled with ^{125}Iodine. Samples are incubated simultaneously with both antibodies followed by a washing procedure to remove any unbound labeled antibody. This method eliminates interference by *C*-terminal and mid-region fragments. Specificity, as assessed by the lack of interference on addition of high concentrations of PTH fragments, is good, with mid-molecule and *C*-terminal fragments exhibiting $<0.2\%$ cross-reactivity. Inter-assay coefficient of variation (CV) is reported to be $<7\%$.

Types of Sample

Because of the instability of the intact hormone, blood samples should be taken onto ice, separated without delay, and serum or plasma frozen immediately. We have found that plasma is more stable than serum.

Frequency of Use

Moderate. Used in specialized centers.

PARATHYROID HORMONE-RELATED PROTEIN

A humoral factor responsible for hypercalcemia and associated with cancer has been isolated from several solid tumors. The gene for this factor has been cloned leading to the recognition of the protein now known as parathyroid hormone-related protein (PTHrP, parathyrin-related protein). On the basis of analysis of messenger RNA from tumor tissue, at least three polypeptides of different lengths have been predicted. The amino-terminal portion of PTHrP has close sequence homology with PTH.

Mechanism of Action

The mechanism by which PTHrP induces hypercalcemia is by interaction with receptors that also bind PTH. As with excess secretion of PTH, PTHrP production by tumors results in hypercalcemia because of increased bone resorption and reduced urinary calcium excretion. The protein also promotes renal phosphate excretion and increases nephrogenous cyclic AMP. However, whereas a mild hyperchloremic acidosis is often a feature in primary hyperparathyroidism, hypokalemic alkalosis is frequently found in humoral hypercalcemia of malignancy.

Assay Technology

RIAs for PTHrP have been of limited utility thus far because of poor sensitivity and the requirement in early assays for sample extraction. Ratcliffe *et al.* (1991) reported the development and validation of an IRMA for PTHrP in unextracted plasma. The assay involves a polyclonal antibody to amino acids 1–34 of PTHrP coupled to cellulose particles as the capture antibody and a rabbit anti-PTHrP (37–67) as radiolabeled antibody. Ratcliffe and colleagues reported raised PTHrP concentrations in 95% of patients studied with hypercalcemia of malignancy. Plasma samples from normal subjects and patients with primary hyperparathyroidism had undetectable levels (detection limit of assay 0.23 pmol/L). Pandian *et al.* (1992) reported the development of a modified IRMA for PTHrP that uses affinity-purified polyclonal immunoglobins. Antibodies recognizing PTHrP 37–74 are immobilized onto polystyrene beads, and antibodies to epitopes within the 1–36 amino acid region of PTHrP are labeled with ^{125}I. The detection limit of this assay is reported to be 0.1 pmol/L, and low but detectable concentrations of PTHrP were reported in some normal individuals. In the study, 91% of patients with hypercalcemia associated with non-hematological malignancies had raised levels of PTHrP.

Type of Sample

Plasma. Protease inhibitors must be added quickly to maintain stability at 4 °C.

Frequency of Use

Rare at present.

CALCITONIN

Calcitonin is a 32 amino acid peptide hormone produced by the C cells located predominantly in the thyroid, but also present in the parathyroid, thymus and lung. As with many other peptide hormones, calcitonin is derived from a larger precursor molecule with post-translational modification cleaving both *N*- and *C*-terminal segments. As well as existing in plasma as both mono- and dimeric forms, alternative splicing of mRNA results in heterogeneous circulating peptides with both immunological and biological activity.

Function

The major action of calcitonin is on bone, where it inhibits bone resorption. This is achieved by inhibition of osteoclastic activity and, in the longer term, a reduction in the number of osteoclasts. Acute intravenous administration lowers plasma calcium but physiologically calcitonin is not thought to play a major role in the control of plasma calcium concentrations. A role for calcitonin in minimizing postprandial rises in serum calcium has been suggested. Elevated levels are seen in pregnancy suggesting that a physiological role may be that of skeletal protection.

Reference Interval

There is a marked sex difference in calcitonin concentrations with males having higher values than females.

Males: <120 ng/L
Females: <60 ng/L
Pregnancy: <120 ng/L
(Radioimmunoassay)

Clinical Applications

Plasma calcitonin is considered important clinically in situations where there is hypersecretion. The classic example is medullary carcinoma of the thyroid. Families in whom this condition is an inherited trait require screening for excessive calcitonin secretion. Increased calcitonin levels following provocative tests of intravenous calcium, pentagastrin or oral alcohol are used to confirm the diagnosis, although occasional false negatives occur. Plasma calcitonin may be useful as a tumor marker in a variety of different conditions.

It is not clear whether calcitonin deficiency *per se* results in any clinical deficit, although the development of osteoporosis may be enhanced.

Limitations

- Lack of specificity; most assays use polyclonal antibodies.
- Non-specific interference with binding from other serum constituents.
- Inadequate sensitivity; extraction procedures or long incubations are necessary.

Assay Technology

Early RIAs required extraction and concentration techniques to detect the low circulating levels of calcitonin. The assay of Hillyard *et al.* (1977) involved an initial extraction using Sepharose beads and elution of calcitonin with acetone. A rabbit antibody was used with ^{131}I calcitonin as tracer and, after a 24 h incubation, separation was achieved using charcoal. The sensitivity was 4–8 ng/L. Inter- and intra-assay variations were <10 and <14%, respectively. Although the sensitivity can be improved to 2 pg per tube with a 7 day incubation, this is of little use clinically because the main indication for calcitonin measurement is as a tumor marker, where high levels are present. For the same reason, in the assay of Body and Heath (1983), which used a silica extraction method and a goat antibody, long pre-incubation and incubation periods were involved and sensitivity was <1 ng/L.

Other techniques, including a bioassay and competitive RIA binding to cell membranes, are largely of historical interest.

Current two-site immunometric assays have satisfactory precision and sensitivity for all clinical uses.

Type of Sample

Plasma samples should be placed on ice immediately and stored frozen at −20 °C.

BIOCHEMICAL MARKERS OF BONE TURNOVER

The activity of cells that regulate bone remodeling, the osteoblasts and osteoclasts, is reflected in serum concentration of elaborated cellular products. Over recent years there has been an extensive development of biochemical assays that measure indices of bone formation and resorption. During these phases of bone activity, osteoblastic and osteoclastic enzymes and other proteins are secreted together with release of components of the organic extracellular matrix. Measurement of these markers of bone turnover in conjunction with the calciotropic hormones provides an important adjunct to imaging procedures for clinical assessment of the skeleton.

MARKERS OF BONE FORMATION

Alkaline Phosphatase

Alkaline phosphatase (EC3.1.3.1) is a marker of bone formation since this enzyme is present in the osteoblast membrane and appears to play a role in phosphate acquisition in the formation of the hydroxyapatite complex. Alkaline phosphatase occurs in the body as four isoenzymes: placental, intestinal, germ cell and liver/bone/kidney—the latter being the predominant form in serum. The isoforms from bone, liver, and kidney are encoded by the same gene but post-translational modification gives rise to differences, which can be detected electrophoretically. Liver and bone isoforms are most commonly associated with increased activity in serum, although in some patients with cancer, placental or germ cell isoenzymes may be responsible. Because of lack of tissue specificity, the usefulness of total serum alkaline phosphatase activity is of limited value. However, in diseases where there is significant skeletal involvement, such as Paget's disease of bone, alkaline phosphatase remains a clinically useful test. In patients with less dramatic biochemical changes, such as in osteoporosis, any changes in bone alkaline phosphatase are obscured by the small contribution that they make to the circulating pool of the enzyme. Many assay procedures have been developed to improve identification of bone alkaline phosphatase in serum. These have relied mainly on electrophoretic characteristics, however resolution has been improved by heat inactivation, lecithin precipitation and more recently by immunoassay.

Osteocalcin

Two further serum markers used as indicators of bone formation are osteocalcin and procollagen I extension peptides, both of which are bone specific (Delmas and Garnero, 1998). Osteocalcin, also known as bone γ-carboxy glutamic acid protein (bone gla protein) is the most abundant non-collagenous bone matrix protein, comprising 1–2% of total bone protein. Initially synthesized by osteoblasts and odontoblasts as pro-osteocalcin, a 75 amino acid peptide, the secreted osteocalcin peptide consists of 49 amino acids and is unique in having three glutamic acid residues in the central region of the molecule that are carboxylated by a vitamin K dependent process (Eriksen *et al.*, 1995). Synthesis of osteocalcin is dependent on the actions of $1{,}25(OH)_2D_3$, which promotes transcriptional activation of the osteocalcin gene. Osteocalcin detected in serum derives almost exclusively from synthesis by osteoblasts since very little is released during bone resorption. In addition to the intact molecule, a large *N*-terminal 'mid fragment' of 43 amino acids, as well as smaller fragments, has been identified in serum. Osteocalcin is cleared by the kidneys and consequently circulating concentrations are affected by impaired renal function. The serum half-life is 15–70 min and there is a pronounced circadian variation with levels peaking during the night and a nadir in the afternoon. Serum concentrations of osteocalcin may be measured by RIA or by enzyme linked immunosorbent assay (ELISA). Variability between assays can be attributed to differing antibody specificities to fragments and the intact molecule. In postmenopausal women serum osteocalcin is 10–30% higher compared to levels in pre-menopausal women. In osteoporosis the concentration may be normal or slightly raised above the expected postmenopausal range (Price and Thompson, 1995). This reflects the variable bone turnover states observed in this condition and the fact that patients may have high or low osteoblastic activity. However, bone formation is invariably reduced relative to levels of resorption. Serum osteocalcin concentration is increased in most conditions associated with bone mineralization (see summary table near end of chapter) but the concentrations do not always parallel those seen with bone alkaline phosphatase. Diseases characterized by increased concentrations of circulating osteocalcin include Paget's disease of bone (although this is variable), hyperparathyroidism, hyperthyroidism, osteomalacia, renal osteodystrophy, and acromegaly. Decreased osteocalcin levels have been reported in hypothyroidism, hypoparathyroidism, growth hormone deficiency, and pregnancy. Although there is a need for consensus regarding standardization, the majority of clinical studies correlating circulating osteocalcin with other biochemical or bone histomorphometric measurements of bone turnover have shown that osteocalcin is a useful marker of bone formation.

Reference Interval

Premenopausal females 3.0–7.4 ng/mL
Males 2.3–5.4 ng/mL

(Commercially available IRMA kit which detects 1–49 intact osteocalcin (carboxylated and uncarboxylated) and the 1–43 peptide fragment).

Procollagen I Extension Peptides

Collagen is the major bone protein and over 90% of bone collagen is type I. It is synthesized by osteoblasts as a precursor molecule, which has a central triple helix domain and carboxy- and amino-terminal extension peptides. These extension peptides are cleaved before collagen becomes incorporated into the bone matrix. Measurement of circulating levels of these peptides by immunoassay can provide an indication of rate of collagen type I synthesis. The procollagen I carboxypeptide (PICP) has a molecular weight of approximately 100 kDa and therefore is not subject to excretion by glomerular filtration. It can be detected in the circulation by RIA. The majority of studies of this new marker of bone formation have been undertaken with an RIA utilizing an antibody raised against the carboxy-terminus of the propeptide. Increases in PICP in the serum are seen in conditions associated with cancellous bone formation that correlate with other indices such as bone histomorphology and whole body calcium kinetics where there is co-existing matrix formation and mineralization (Eriksen *et al.*, 1993). However, when these are uncoupled the correlation is not apparent (Price and Thompson, 1995).

Reference Interval

50–170 μg/L
(from Eriksen *et al.*, 1993)

MARKERS OF BONE RESORPTION

Collagen Cross-Link Molecules (Pyridinoline and Deoxypyridinoline)

The extracellular matrix is stabilized by the formation of covalent cross-links between adjacent collagen chains. There are two major cross-link molecules: hydroxylysyl pyridinoline (pyrydinoline) and lysyl pyridinoline (deoxypyridinoline, DPD). These molecules form small

Table 47.6. Conditions associated with changes in bone markers.

	bALP	Osteocalcin	PICP	DpD	Telopeptides
Paget's disease	+++	++	++	++	++
Rickets/osteomalacia	+++	+++	++		
Osteoporosis	+	+		++	++
CRF	+++	++	++	++	++
Bone metastases				++	+++
HPTH	+	+	+	++	++

CRF, chronic renal failure; HPTH, primary hyperparathyroidism

non-reducible cross-links that stabilize the collagen fibrils. Pyridinoline is mainly present in cartilage with only a small amount in bone. DPD is less abundant than pyridinoline and is almost exclusively found in bone and dentine. The two pyridinoline compounds are not degraded on resorption or metabolized *in vivo* and are excreted in urine as free (40%) or peptide-bound (60%) forms. The fact that the cross-link molecules are only found in mature collagen means that the excretion in urine only reflects degradation of mature collagen and does not include collagen which has been synthesized but not incorporated into collagen fibrils. As the great majority of cross-links in urine are bone derived, there is good correlation between cross-link excretion and bone resorption (Delmas and Garnero, 1998). As with osteocalcin, there is a pronounced circadian variation with lowest urinary excretion of DPD observed in the afternoon. Indeed, cross-link excretion has been shown to decrease by approximately 30% between 8.00 and 11.00 am and thus standardization of sampling time is of critical importance for serial measurements.

Initial assays of pyridinoline and DPD were performed by HPLC with fluorescent detection following hydrolytic sample derivatization. More recently a variety of commercial ELISAs have become available (Eriksen *et al.*, 1995).

Sample

For measurement of urinary free DPD a preservative free random urine sample (first or second morning void collection) is required. Samples may be stored at − 20 °C. Exposure to ultra violet light should be avoided.

DPD excretion is corrected for urinary concentration by calculating the urine DPD:creatinine ratio [ref range to follow].

Pyridinium Cross-linked Carboxyterminal Telopeptides

During collagen type I degradation, the carboxy-terminal cross-linked telopeptide region is liberated into the

Table 47.7. Expected analyte changes in hypercalcemia and hypocalcemia.

	Ca^{2+}	PO_4^{2-}	ALP	PTH	25OHD	1,25D
Hypercalcemia						
1^y HPTH	⇑ / ⇑⇑	–/⇓	–/⇑	⇑/⇑⇑	–/⇓	⇑
HCOM	⇑⇑ / ⇑⇑⇑	–	–/⇑	⇓⇓	–	–/⇑ *
Vit D intoxication	⇑⇑ / ⇑⇑⇑	–/⇑	–	⇓⇓	⇑⇑/⇑⇑⇑	⇓
Sarcoidosis	⇑	–	–/⇑	⇓⇓	–	⇑⇑/⇑⇑⇑
Hypocalcemia						
Hypoparathyroidism	⇓/⇓⇓	⇑	–	⇓⇓	–	–/⇑
2^y HPTH	–/⇓	⇓/⇓⇓	⇑⇑	⇑⇑	–/⇓	⇑/⇑⇑
Pseudohypopara	⇓/⇓⇓	⇑	–	⇑⇑	–	–/⇓
CRF	⇓/⇓⇓	⇑⇑	⇑⇑	⇑⇑	–/⇓	⇓⇓

1^y HPTH = primary hyperparathyroidism; HCOM = hypercalcemia of malignancy (* = reticuloendothelial malignancy); 2^y HPTH = secondary hypothyroidism; pseudohypopara = pseudohyoparathyroidism; CRF = chronic renal failure circulation as an immunologically intact fragment that resists further degradation. This trimeric antigen has been isolated and used as an antigen to raise antibodies for immunoassay (Delmas and Garnero, 1998). Carboxyterminal telopeptide concentrations in serum show a significant correlation with rate of bone resorption as indicated by histomorphometry (Eriksen *et al.*, 1993). In a study of postmenopausal women with mild osteoporosis treated with hormone replacement, levels of this bone marker in serum were not found to be a sensitive indicator of changes in bone resorption induced by therapy (Hassager *et al.*, 1994). However, newer assays under development may well prove to be better in terms of clinical applications. Measurement of the cross-linked *N*-telopeptide of type I collagen has also been reported to be a sensitive and specific marker of bone resorption.

SERUM TARTRATE RESISTANT ACID PHOSPHATASE (TRAP)

Osteoclasts contain an isoenzyme of acid phosphatase that can be distinguished from prostatic acid phosphatase because it is tartrate resistant. The measurement of tartrate resistant acid phosphatase (TRAP) activity in serum has been used as an index of osteoclast activity. RIAs for serum TRAP are also being developed.

CONDITIONS ASSOCIATED WITH CHANGES IN BONE MARKERS

The conditions are summarized in Tables 47.6 and 47.7.

REFERENCES

Bataille, P., Bouillon, R., Fournier, A., Renaud, H., Gueris, J. and Idrissi, A. Increased plasma concentration of total and free $1,25(OH)_2D_3$ in calcium stone formers with idiopathic hypercalciuria. *Contrib. Nephrol.* **58**, 137–142 (1987).

Body, J.J. and Heath, H. Estimates of circulating monomeric calcitonin: physiological studies in normal and thyroidectomized man. *J. Clin. Endocrinol. Metab.* **57**, 897–903 (1983).

Delmas, P.D. and Garnero, P. Biochemical markers of bone turnover in osteoprosis. In: *Osteoporosis* (eds Stevenson, J.C. and Lindsay, R.), 117–136 (Chapman & Hall, London, 1998).

Eisman, J.A., Hamstra, A.J., Cream, B.E. and DeLuca, H.F. A sensitive, precise and convenient method for determination of 1,25-dihydroxyvitamin D in human plasma. *Arch. Biochem. Biophys.* **178**, 235–243 (1976).

Eisman, J.A., Sheppard, R.M. and DeLuca, H.F. Determination of 25-hydroxyvitamin D_2 and 25-hydroxyvitamin D_3 in human plasma using high-pressure liquid chromatography. *Anal. Biochem.* **80**, 298–305 (1977).

Eriksen, E.F., Charles, P., Melsen, F., Mosekilde, L., Risteli, L. and Risteli, J. Serum markers of type I collagen formation and degradation in metabolic bone disease; correlation with bone histomorphology. *J. Bone Min. Res.* **8**, 127–132 (1993).

Eriksen, F., Brixen, K. and Charles, P. New markers of bone metabolism: clinical use in metabolic bone disease. *Eur. J. Endocrinol.* **132**, 251–263 (1995).

Habener, J.F., Rosenblatt, M. and Potts, J.T. Parathyroid hormone: biochemical aspects of biosynthesis, secretion, action and metabolism. *Physiol. Rev.* **64**, 985–1053 (1984).

Haddad, J.G. and Chyu, K.J. Competitive protein binding radioassay for 25-hydroxycholecalciferol. *J. Clin. Endocrinol. Metab.* **33**, 992–995 (1971).

Hassager, C., Jenson, L.T., Podenphant, J., Thomsen, K. and Christiansen, C. The carboxyterminal pyridinoline cross-linked teleopeptide of type I collagen in serum as a marker of bone resorption: the effect of nandrolone decanoate and hormone replacement therapy. *Calcif. Tissue Int.* **54**, 30–33 (1994).

Hillyard, C.J., Cooke, T.J.C., Coombes, R.C., Evans, I.M.A. and MacInytre, I. Normal plasma calcitonin: circadian variation and response to stimuli. *Clin. Endocrinol.* **6**, 291–298 (1977).

Hollis, B.W. Detection of vitamin D and its major metabolites. In: *Vitamin D* (ed Feldman, D.), 587–606 (Academic Press, San Diego, 1997).

Hollis, B.W., Kamerud, J.Q., Selvaag, S.R., Lorenz, J.D. and Napoli, J.L. Determination of vitamin D status by radioimmunoassay with an ^{125}I-labelled tracer. *Clin. Chem.* **42**, 586–592 (1993).

Hollis, B.W., Kamerud, J.Q., Kurkowski, A., Beaulieu, J. and Napoli, J.L. Quantification of circulating 1,25-dihydroxyvitamin D by radioimmunoassay with an ^{125}I-labelled tracer. *Clin. Chem.* **42**, 586–592 (1996).

Johansen, J.S., Riis, B.J., Delmas, P.D. and Christiansen, C. Plasma BGP. An indicator of spontaneous bone loss and of the effect of oestrogen treatment in postmenopausal women. *Eur. J. Chem. Invest.* **18**, 191–195 (1988).

Pandian, M.R., Morgan, C.H., Carlton, E. and Segre, G.V. Modified immunoradiometric assay of parathyroid hormone-related protein: chemical application in the differential diagnosis of hypercalcemia. *Clin. Chem.* **38**, 282–288 (1992).

Price, C.P. and Thompson, P.W. The role of biochemical tests in the screening and monitoring of osteoporosis. *Ann. Clin. Biochem.* **32**, 244–260 (1995).

Ratcliffe, W.A., Norbury, S., Heath, D.A. and Ratcliffe, J.G. Development and validation of an immunoradiometric assay of parathyrin-related protein in unextracted plasma. *Clin. Chem.* **37**, 678–685 (1991).

Reinhardt, T.A., Horst, R.L., Orf, J.W. and Hollis, B.W. A microassay for 1,25-dihydroxyvitamin D not requiring high performance liquid chromatography: application to clinical studies. *Endocrinol. Metab.* **58**, 91–98 (1984).

Stevenson, J.C. Calcium. *Br. J. Hosp. Med.* **32**, 71–76 (1984).

Stevenson, J.C. (ed), *New Techniques in Metabolic Bone Disease* (Wright, London, 1990).

48 Infertility

Michael J. Wheeler

Fertility in males and females is controlled by a homeostatic balance between the hypothalamus, pituitary, and gonads (see Figure 48.1).

Gonadotropin releasing hormone (GnRH), formerly known as luteinizing hormone releasing hormone (LHRH), is secreted from the arcuate nucleus of the hypothalamus and passed down neurofibrils to the median eminence. Here it is released into the portal system and taken to the anterior pituitary where it binds to specific receptors on the gonadotroph cells. GnRH stimulates both the synthesis and the release of luteinizing hormone (lutropin, LH) and follicle stimulating hormone (follitropin, FSH). These hormones are also known as gonadotropins. To be effective, GnRH must be secreted in a series of pulses with a pulse frequency of 90–120 min. This is reflected in the pulsed secretion of the gonadotropins, LH in particular. LH and FSH are carried in the bloodstream to their target organs, the gonads.

In females, the secretion of the gonadotropins is both tonic and cyclic. As in men, the tonic secretion regulates the minute-by-minute secretion of the gonadal steroids and in turn is modified by the steroidal feedback on the hypothalamus and pituitary. The cyclic secretion, present only in women, governs the female menstrual cycle (see Figure 48.2).

At the beginning of the menstrual cycle, FSH stimulates the development of ovarian primordial follicles, one of which becomes dominant and matures into the Graafian follicle. The remaining follicles that had begun to develop degenerate. As the follicle matures (for which both LH and FSH are required) it secretes an increasing amount of estradiol. This steroid has a negative feedback on the pituitary during the early and mid-follicular phase. Thus, during this stage of the menstrual cycle there is a fall in FSH levels in particular. Another recently isolated non-steroidal hormone, inhibin, also changes during the menstrual cycle. Inhibin circulates as two biologically active forms, inhibin A and inhibin B. Concentrations of inhibin B rise at the time of menses, are maximal during the early and mid-follicular phases of the cycle and fall during the late follicular phase. Although it has been suggested that inhibin B suppresses FSH secretion in the mid-follicular phase, direct evidence for this is lacking. At about day 12, estradiol feedback changes. The mechanism is still unclear but the steroidal feedback becomes positive causing a sharp surge of LH and FSH. This in turn stimulates ovulation, i.e. rupture of the Graafian follicle and release of the ovum into the fallopian tube. There is also a sharp rise in inhibin concentrations at this time.

The remaining granulosa and thecal cells of the follicle develop into the corpus luteum. LH is required for the maintenance of this structure, which secretes progesterone and estradiol.

Therefore, during the luteal phase there is a small rise in estradiol and a large increase in progesterone concentrations. During the luteal phase concentrations of inhibin B remain at low levels. Inhibin A concentrations, which are low during the follicular phase, begin to increase at mid-cycle reaching peak levels at the mid-luteal phase. Levels therefore reflect growth of the corpus luteum from

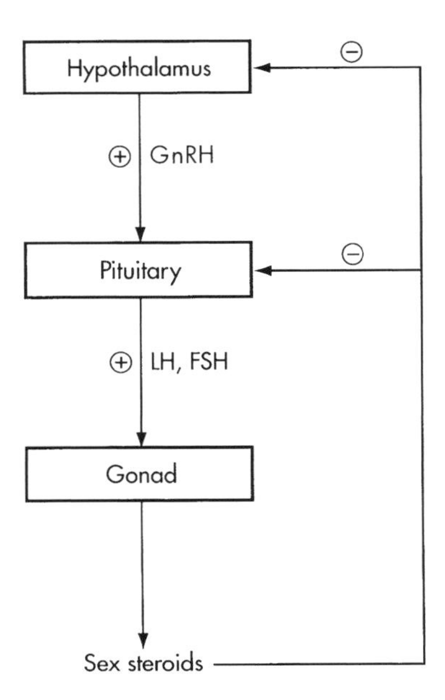

Fig. 48.1 Control of sex hormone secretion (simplified).

The Immunoassay Handbook. *Third Edition*
D. Wild (Ed.)

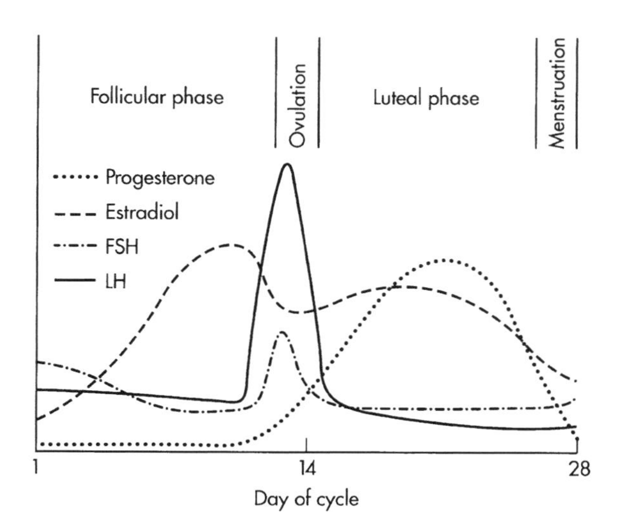

Fig. 48.2 Serum hormone changes during the menstrual cycle.

which this hormone is secreted. Changes in inhibin A concentrations closely follow changes in progesterone concentration. If no fertilization of the ovum takes place there is then a fall in progesterone levels. This, in turn, leads to necrotic changes in the endometrium with loss of the bulk of the tissue accompanied by the menstrual bleed. Inhibin A concentrations also fall at the end of the luteal phase and are associated with an increase in FSH concentrations suggesting that inhibin A suppresses FSH secretion during the luteal phase.

Eventually, the supply of oocytes declines so that in the late fifth and early sixth decade of life the woman enters the perimenopause. During this time there may still be cyclic secretion of the reproductive hormones with irregular menstrual cycles. Although gonadotropin levels are higher overall, they may also be in the normal range for the follicular phase. Therefore, normal LH and FSH concentrations do not exclude the perimenopause in a woman over 40 years of age with irregular menstrual cycles. Surprisingly, estradiol concentrations tend to be between 200 and 400 pmol/L rather than low (see Prior, 1998 for review). Presumably the raised FSH concentration is due to low inhibin levels at this time. Eventually, menstruation and ovarian function cease completely and the woman enters the menopause. Gonadotropin concentrations are initially very high, thereafter declining slowly with age; estrogen levels are very low.

In males, as in the female, LH and FSH concentrations are low before puberty. At puberty, surges of LH, FSH, and testosterone occur during the night. As puberty proceeds there is an overall increase in gonadotropin secretion with the gradual loss of circadian secretion.

LH stimulates the Leydig or interstitial cells of the testis to secrete the male sex hormone, testosterone. The main site of action of FSH is in the Sertoli cells in the seminiferous tubules. These cells secrete the peptides inhibin B and androgen-binding protein (no inhibin A is detectable in the serum of men). Androgen-binding protein is involved in the transport of the androgens, testosterone and 5α-dihydrotestosterone (DHT), to the developing sperm cells, whose development they support. Inhibin B secretion is stimulated by FSH but there appears to be an additional FSH-independent component to its secretion.

Testosterone and inhibin B are carried in the blood stream back to the hypothalamus and pituitary where they exert a negative feedback. Several studies have confirmed that inhibin B seems to influence the secretion of FSH, in particular.

A male climacteric has been proposed. However, many studies suggest that testosterone concentrations do not fall until about the age of 60 years. Free testosterone (non-protein bound) probably falls earlier than this due to a rise in sex hormone-binding globulin (SHBG) which starts at approximately 50 years of age. There is a corresponding gradual increase in gonadotropin levels.

CLINICAL DISORDERS

PRIMARY HYPOGONADISM IN THE FEMALE

Just under half of the women who present with infertility have ovarian dysfunction. An organized investigative approach is likely to lead to a faster diagnosis with a lower cost and less inconvenience to the patient. Breckwoldt *et al.* (1993) suggested that a detailed medical history and a meticulous physical examination, with special attention to the target organs for sex steroids, frequently provided better information than a battery of uncoordinated laboratory tests. A number of algorithms have been proposed which help in a logical approach to female infertility. The classification proposed in 1976, by the WHO Scientific Group on Agents Stimulating Gonadal Function in the Human, has been widely used and other algorithms are largely based on this classification.

Gonadal dysgenesis is associated with failure of the gonads to develop properly. They are present, if at all, as streaks of tissue. Two classes of patient may be conveniently recognized. First, those in whom development is associated with an abnormality of the sex chromosome, namely **Turner's syndrome** and its variants. In the typical condition the karyotype is 45XO and is associated with short stature, sexual infantilism, and several somatic abnormalities. All the patients are female. A partial abnormality of the second sex chromosome is associated with fewer abnormalities. Gonadotropins are elevated as expected in postpubertal patients. The second class of patients has a normal or near normal karyotype, 46XX or 46XY, but the gonads are absent. Usually, the somatic abnormalities of Turner's syndrome are absent and patients are of normal or tall height. Affected men have variable sexual development, presumably dependent upon the time at which the gonads degenerated. If this occurred before 6 weeks of fetal life the patient will have a complete female phenotype. If later, there will be variable development of the genital ducts and ambiguous genitalia. Gonadotropin concentrations will be increased in all adult patients.

Gonadotropin levels are also raised in women who have entered the **menopause**. A very high FSH concentration is usually diagnostic. (An exception would be if a very rare gonadotropin-secreting tumor was present.) Laparoscopy studies in some women, who appeared to have entered an early menopause, showed them to have normal ovaries. These studies suggested that the ovaries were insensitive to the elevated gonadotropins (**resistant ovary syndrome**). This syndrome may be associated with **primary amenorrhea** (failure of menstruation to commence at puberty) or **secondary amenorrhea** (absence of menstruation postpuberty). Again, levels of gonadotropins in the blood are increased, LH more significantly than FSH.

SECONDARY HYPOGONADISM IN THE FEMALE

Secondary hypogonadism may result from dysfunction of the pituitary, hypothalamus or higher brain centers; gonadotropin levels may be within or below the normal follicular range. For instance, tumors of the hypothalamus (**craniopharyngioma**) and pituitary (**prolactinomas**, **chromophobe adenomas**) may be associated with normal or low LH and FSH concentrations, whereas hypothalamic dysfunction (**hypogonadotropic hypogonadism**) is associated with low levels. Low gonadotropin concentrations are found in cases of **anorexia nervosa**, and after **irradiation** to the pituitary area, which can lead to either hypothalamic or pituitary dysfunction.

Oligomenorrhea and amenorrhea associated with hirsutism are discussed in the HIRSUTISM AND VIRILIZATION IN THE FEMALE chapter.

INFERTILITY AND NORMAL MENSTRUAL FUNCTION

Cyclic activity of the hypothalamus, pituitary, and ovary is required for cyclic hormonal changes. However, it is still possible for there to be some endocrine deficiency. Some women have a high incidence of anovulatory cycles or poor development of the corpus luteum with inadequate secretion of progesterone. Careful monitoring of basal body temperature (BBT) and the measurement of luteal progesterone are helpful in determining the problem.

If there is cyclic hormonal secretion with amenorrhea, degeneration of the endometrium or adhesions within the uterus are indicated. Once associated with tuberculosis, this is now rare.

Examination of the male partner is indicated if there is normal endocrine function and menstruation in the woman. In some cases where both partners appear endocrinologically normal, and if hostility of the secretions of the female tract has also been excluded, sex counseling may be required.

PRIMARY HYPOGONADISM IN THE MALE

LH and FSH concentrations are increased because of reduced feedback by testosterone and inhibin. Testosterone may be very low or only slightly reduced.

Men with **Klinefelter's syndrome** have an extra X chromosome. The classical form is XXY although cases with further additional X chromosomes have occurred. Affected men typically have small, firm testes, and a tall stature with eunuchoid appearance and gynecomastia. Social maladjustment and low IQ are also reported. Testosterone levels range from very low to just within the normal reference range. Gonadotropin concentrations are elevated. The syndrome has an incidence of about 0.2% and diagnosis is confirmed by chromosome analysis.

Patients with **male Turner's syndrome** have features similar to women with Turner's syndrome but their chromosome pattern is normal. Affected individuals have small, soft testes, low testosterone, and increased gonadotropin levels. In **anorchia** (absence of the testes), patients remain prepubertal. In the **Sertoli-cell-only syndrome** the testes are almost normal in size although the germ cells are completely absent. The seminiferous tubules are diagnostically not hyalinized as in the other conditions above. Because Leydig cell function is normal, LH and testosterone levels are normal. The azospermia is associated with increased FSH concentrations.

Viral orchitis is the most common type of acquired testicular failure. About 30% of men who develop orchitis after puberty experience testicular atrophy. Other causes of acquired testicular failure are **trauma**, **radiation** and **drugs**. Damage by radiation is related to duration and dose. Spironolactone, ketoconazole, and ethanol lower testosterone levels by inhibiting its synthesis.

Primary testicular dysfunction is also associated with a number of systemic diseases such as renal failure and cirrhosis of the liver. It is difficult to determine whether the effect on the testes is due to the disease or concomitant malnutrition.

Androgen resistance (**pseudohermaphrodism**: **Reifenstein's syndrome**, **testicular feminization**) is characterized by a complete or incomplete female phenotype. Unlike primary hypogonadism, testosterone concentration is normal or even elevated in the presence of increased LH and FSH values.

SECONDARY HYPOGONADISM IN THE MALE

As in the female, hypogonadism can result from dysfunction of the pituitary, hypothalamus or higher brain centers. Isolated gonadotropin deficiency may occur as in **Kallman's syndrome** or be part of a more generalized pituitary failure. In the former condition, individuals do not have a normal pubertal development, and typically have **anosmia** (absence of a sense of smell). This and other forms of **hypogonadotropic hypogonadism** (e.g. **Prader–Willi** and **Bardet–Biedl syndromes**) are a result of a hypothalamic defect. This is indicated by the efficacy of repeated administrations of GnRH therapy, which can eventually stimulate LH and FSH secretion by the pituitary.

Hyperprolactinemia due to a pituitary adenoma causes infertility and azospermia. Men usually present with headaches and visual problems, caused by a large pituitary tumor, rather than with infertility. Prolactin seems to interfere with the normal synthesis of LH, FSH, and testosterone.

Men with **anorexia nervosa** and with **psychogenic infertility** show gonadotropin secretion similar to women. Psychiatric treatment, rather than hormone replacement, is required and if successful can restore normal hormone secretion. Frequently psychogenic infertility is accompanied by impotence but in most cases gonadotropin and testosterone concentrations remain within the normal range.

IMPAIRED SPERM TRANSPORT AND SPERM VIABILITY

Infertility can occasionally result from an obstruction in the reproductive tract; this may be unilateral or bilateral. Hormone concentrations are normal, and diagnosis is made from seminal analysis, vasograms and, if necessary, testicular biopsy. Poor sperm viability is another cause of infertility. There may be oligospermia, a greater than normal percentage of abnormal forms, poor motility or the presence of anti-sperm antibodies. An increase in FSH concentrations is usually the only hormone abnormality although the levels of this hormone are frequently normal. Treatment with testosterone or gonadotropins is frequently unsatisfactory.

ANALYTES

LUTEINIZING HORMONE (LUTROPIN)

LH is structurally similar to FSH, TSH, and hCG. They are all glycoproteins consisting of two subunits. The α subunit is similar in all four hormones but the β subunit is unique to each of them. The latter subunit confers biological activity to the hormone by governing receptor-binding specificity. The LH molecule is about 30 kDa with 89 amino acids in the α chain and 129 amino acids in the β chain. The carbohydrate content is between 15 and 30%.

Function

LH is both stored and secreted by the anterior pituitary. In men, it acts on the Leydig cells of the testes, stimulating the synthesis and secretion of testosterone. In women, LH is involved in the development of the follicle, ovulation at mid-cycle, and maintenance of the corpus luteum. It is cleared by the liver and has a half-life of about 30 min.

Synthesis and secretion of LH is controlled by the stimulation of the gonadotroph cells of the anterior pituitary by GnRH from the hypothalamus. GnRH is secreted as pulses with a frequency of about 90 min. This is reflected in the episodic secretion of LH. The amount of GnRH and LH secreted is, in turn, controlled by the negative feedback of the sex steroids released from the gonads. Thus, high steroid levels suppress LH secretion and low levels increase it. However, in the middle of the menstrual cycle, estradiol exerts a positive feedback, stimulating LH secretion.

Reference Interval

Reference ranges for LH depend on the methodology used, especially whether the assay is competitive or immunometric. Immunometric assays give values that are about 30% lower. This is because monoclonal antibodies used in immunometric assays do not recognize subunits or all of the circulating LH epitopes (see Table 48.1).

Clinical Application

LH levels are increased in patients who have suffered primary gonadal failure. Conversely, where there is hypothalamic dysfunction (hypogonadotropic hypogonadism, anorexia nervosa) the concentration of LH is low. It may also be low in any cause of hypopituitarism; for example it may be low or normal when a pituitary tumor is present. Therefore, if the concentration of LH is low, other pituitary hormones should be investigated.

Table 48.1. Reference intervals for LH.

		Bayer Advia: Centaur® (IU/L, IS 80/552)
Women	Follicular phase	1.9–12.5
	Mid-cycle	8.7–76.3
	Luteal phase	0.5–16.9
	Postmenopausal	15.9–54.0
Men		1.5–9.3

LH secretion is very variable in the perimenopausal period and a single measurement is often unhelpful or misleading. Three measurements of LH, 1 week apart, will give a clearer clinical picture. LH levels are normal in many cases of infertility, and other hormone assays are required. The pituitary reserve of LH can be examined with a GnRH test. This may be particularly helpful after pituitary surgery or in the investigation of hypogonadotropic hypogonadism.

Limitations

- The technical limitations are the same as those described for TSH (*see* THYROID).
- LH is suppressed by estrogen, but in women taking oral contraceptives the concentration of LH may be normal or low.
- Excessive dieting and weight loss may lead to low gonadotropin levels.
- It is not uncommon for amenorrhea to be investigated in a woman who presents with amenorrhea but is, in fact, pregnant. Most current immunometric assays have little or no cross-reaction with hCG, and both LH and FSH concentrations will be very low or undetectable. Vivekanandan and Andrew (2002) have reported that the DPC Immulite method has sufficient cross-reaction with hCG to give a falsely detectable LH concentration. The UK NEQAS scheme has shown that an hCG concentration of 16 kIU/L results in an LH result of between 7 and 9 IU/L in the Bayer Immuno 1 and the DPC Immulite LH assays. Estradiol, testosterone, SHBG, and prolactin concentrations will be elevated. In these circumstances the hCG concentration should be measured in the same serum sample.
- Some pairs of monoclonal antibodies in immunometric assays fail to recognize certain epitopes of LH that seem to be biologically active. This is discussed more fully in ASSAY TECHNOLOGY below.
- The presence of heterophilic antibodies in serum may lead to erroneous results, which may be higher or lower than the true value. This problem may be encountered in any immunometric assay using two monoclonal antibodies. Ismail *et al.* (2002) examined the TSH, LH, and FSH results of 5310 patients. They reported that 0.53% of results were analytically incorrect due to interference.
- LH is released in a pulsed manner. The value in a single blood specimen may not be representative of the 24 h secretory mean level.

GnRH Stimulation Test

GnRH is a decapeptide secreted by the hypothalamus. It has been synthesized and is available commercially. An intravenous injection of 100 μg GnRH is given and blood is taken at 0, 20, and 60 min. A normal response shows an increase in LH values of 5–10 fold at 20 min over the basal level. At 60 min, LH levels usually decrease again, but may still increase slightly in a small proportion of patients (see Figure 48.3).

An exaggerated response is found in patients with primary hypogonadism and is also typical of the polycystic ovary syndrome (*see* HIRSUTISM AND VIRILIZATION IN THE FEMALE). However, these observations are rarely of help in diagnosis. Both hypothalamic and pituitary dysfunction may lead to little or no response to a GnRH test, and therefore a lack of response does not indicate pituitary dysfunction *per se*. When hypothalamic dysfunction is present, repeated injections or pulsed administration of GnRH will lead to increased secretion of the gonadotropins. A GnRH test is also helpful in establishing whether gonadotroph function is normal after pituitary surgery.

Assay Technology

Competitive radioimmunoassays have essentially been replaced by immunometric assays. In 1991, there were over 44 different kits of which 26 were non-radioactive immunometric methods. Today, the majority of kits are immunometric. Most clinical laboratories use non-isotopic immunometric assays run on fully automated, random access analyzers. The immunometric technique has the

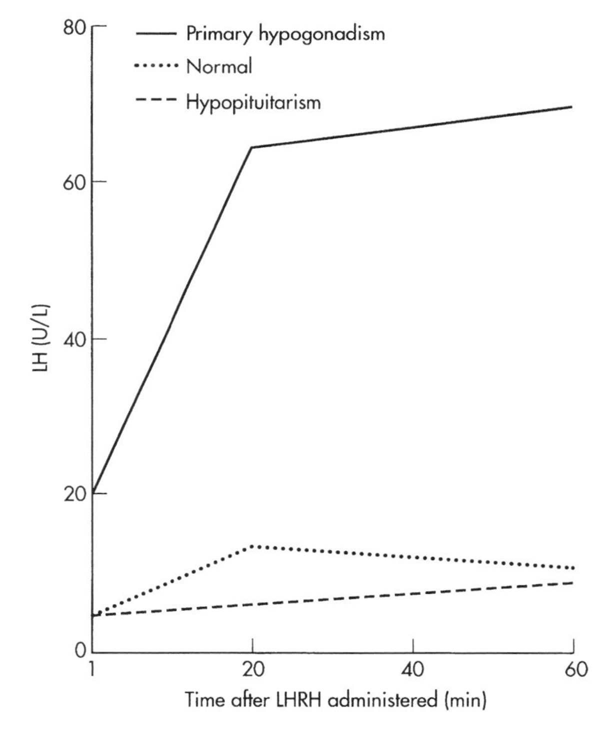

Fig. 48.3 Examples of LH responses in LHRH stimulation tests. LH concentrations after injection of 100 μg LHRH in normal subjects, patients with hypopituitarism and patients with primary hypogonadism.

benefits of faster assays, a wider working range and frequently greater sensitivity. Some individuals have anti-mouse antibodies in their serum which interfere with antibody binding in these assays. Also when analyte concentrations are very high in patient sera, suppression of tracer binding may occur, leading to falsely low results (the 'high-dose hook' effect). A knowledge of the limitations of each assay is therefore very important when choosing which one to use.

Two monoclonal antibodies are usually used in immunometric assays. The epitopes of LH measured in a method depend on the pair of monoclonal antibodies used. In some patient sera, significant differences in LH results are found between kits. It is now becoming clear that some women produce a genetic variant of LH, which is biologically active but not recognized by some monoclonal antibodies. Undetectable LH values are recorded in these patients, who have normal cyclical activity of their steroids.

Some immunometric assays have a polyclonal capture antibody and a monoclonal labeled antibody for signal generation, an arrangement that overcomes most of the problems associated with a dual monoclonal antibody system.

Totally automated random access analyzers are now used to perform this assay. Such assays have a sensitivity of about 0.1 IU/L and may have a total imprecision of <3% over the normal reference range.

Types of Sample

Normally serum or plasma. However, some kits are only suitable for serum.

Frequency of Use

Common.

FOLLICLE STIMULATING HORMONE (FOLLITROPIN)

As with LH, FSH has two subunits, α and β. The β subunit differs between the two hormones and bestows their independent biological activities. FSH is cleared more slowly from the circulation than LH. Its control by the hypothalamus is similar to that already described for LH but, because of its longer half-life, episodic secretion is less obvious. Stimulation of the pituitary with exogenous GnRH, either by a bolus injection or by pulsed infusion, stimulates FSH synthesis and secretion. FSH acts on the granulosa cell of the ovary, stimulating steroidogenesis. It is also thought to stimulate the development of the next cohort of follicles at the beginning of each menstrual cycle. It does not seem to be essential for either ovulation or the maintenance of the corpus luteum.

In men, FSH acts on the Sertoli cells of the testis, stimulating the synthesis of inhibin and androgen-binding protein. Thus, it indirectly supports spermatogenesis. The negative feedback of testosterone on the pituitary and hypothalamus controls FSH in the same way as LH. However, inhibin seems specifically to inhibit FSH secretion. In cases of azospermia, inhibin concentrations are low and FSH levels are specifically increased.

Reference Interval

Generally, there is little difference between competitive and immunometric assay techniques (see Table 48.2).

Clinical Application

FSH is diagnostically the best hormone for confirming primary hypogonadism in women. In general, changes in FSH and LH concentrations are concurrent in women. However, in men FSH concentrations are increased when there is a spermatogenic defect although the LH and testosterone concentrations may be normal. If there is accompanying dysfunction of the Leydig cells, LH levels will also be increased.

Limitations

Technical limitations are the same as described for TSH (*see* THYROID) and LH (above). The problem encountered with the measurement of LH epitopes by immunometric assays has not been reported for FSH methods at present. The effects of heterophilic antibodies on an immunometric assay system are the same as for LH.

Assay Technology

Methods for LH and FSH have been developed in parallel since these hormones are usually measured at the same time in most laboratories. Competitive and immunometric assays generally give similar results. This assay is also carried out on automated immunoassay analyzers with sensitivity and imprecision similar to the LH assays.

Types of Sample

Serum or plasma.

Table 48.2. Reference intervals for FSH.

		Bayer Advia: Centaur® (IU/L, IS 94/632)
Women	Follicular phase	2.5–10.2
	Mid-cycle	3.4–33.4
	Luteal phase	1.5–9.1
	Postmenopausal	23–116
Men		1.4–18.1

Frequency of Use

Common.

PROLACTIN

Prolactin is a protein hormone secreted from the lactotrophs of the anterior pituitary. It consists of a single polypeptide chain of about 200 amino acids. Its main action is on the mammary gland where it is involved in the growth of the gland and in the induction and maintenance of milk production. There is evidence that it may be involved in steroidogenesis in the gonad, acting synergistically with LH. Certainly very high levels of prolactin seem to inhibit steroidogenesis as well as having a local effect on the pituitary, inhibiting LH and FSH production.

Prolactin shows a noticeable circadian rhythm, being elevated during sleep. It rises rapidly after conception and continues to rise until the third trimester. After delivery, levels in the suckling mother rapidly fall, but may be maintained for several months at concentrations just above the reference range. Suckling itself stimulates the release of prolactin.

Reference Interval

See Table 48.3.

Clinical Application

Prolactin measurement is used for the diagnosis of hyperprolactinemia and for monitoring the effectiveness of subsequent treatment. Microadenomas can be treated with the dopamine agonists, such as bromocriptine, carbergoline, and pergolide. Bromocriptine therapy usually reduces the size of even large prolactin-secreting pituitary tumors, and therefore pituitary surgery is only rarely required nowadays.

Limitations

- Because prolactin is raised during sleep, particularly just before waking, it has been recommended that blood for prolactin measurement should be taken 2 h after sleep.
- Moderate increases of up to twice the upper limit of the normal range can occur in stressed patients and after mild exercise (e.g. climbing several flights of stairs). Therefore, it is important that patients are well rested before blood is taken. Multiple venepuncture may also increase prolactin levels. In particularly anxious patients, especially when a slightly raised prolactin level has been previously reported, some clinicians make use of a butterfly needle. After insertion of the needle, the patient is left to rest for about 20 min before collecting blood. Others prefer to take blood every 10 min for 30 min. A fall in prolactin is observed if the raised prolactin level was due to stress.
- Prolactin is elevated in about 30% of acromegalics, in hypothyroidism (in concert with TSH) and in polycystic ovarian disease. A moderate increase is reported in epileptic patients immediately following a seizure.
- Rarely after pituitary surgery, very high concentrations of prolactin may be found, with a steady fall in levels over the following months. Presumably this represents residual pituitary tissue, containing lactotrophs, no longer under hypothalamic control, which gradually regresses.
- Prolactin secretion is increased by drugs such as reserpine, methyldopa, morphine, metoclopromide, and the psychotropic drugs. These impair the action of dopamine from the hypothalamus which inhibits prolactin secretion. A careful drug history and clinical examination is therefore required before hyperprolactinemia due to a prolactinoma can be diagnosed.
- Prolactin is also increased in hyperthyroidism due to stimulation by the increased levels of TRH.
- In some patients prolactin circulates bound to immunoglobulin. This form is called macroprolactin, is not thought to be biologically active and can lead to falsely high results. The degree to which assays detect macroprolactin varies. Most laboratories now screen for the presence of macroprolactin, in specimens with an elevated prolactin concentration, using a simple PEG precipitation method (see Fahie-Wilson and Soule, 1997; Leslie *et al.*, 2001).

Table 48.3. Reference intervals for prolactin.

Women	2.8–29.2 ng/mL	59–619 mIU/L (3rd IRP 84/500)
Men	2.1–17.7 ng/mL	45–375 mIU/L (3rd IRP 84/500)

Bayer Advia: Centaur®.

Some workers have advocated the use of a TRH test to distinguish hyperprolactinemia due to a prolactinoma from other causes. A normal response shows a marked increase of prolactin at 20 min with prolactin levels falling at 60 min. Patients with a prolactinoma have a flat response. However, many have found the test to be unreliable and it is little used today.

TRH Test

An intravenous, bolus injection of 200 μg is given. Blood is taken at 0, 20, and 60 min. A normal response is an increase of 100% over the basal concentration or a threefold increase of the basal concentration (see Carmina and Lobo, 1997). Patients with a prolactinoma frequently show a suppressed or flat response.

Assay Technology

As with LH and FSH, radioimmunoassays have been replaced by immunometric assays, which are mostly non-radioactive methods. Despite the presence of dimeric as well as monomeric prolactin there is good agreement between assays. Routine measurement of prolactin is carried out on fully automated immunoassay analyzers. These assays have sensitivities down to about 10 mIU/L, and a total imprecision of $<3\%$ over the reference range.

Types of Sample

Serum or plasma.

Frequency of Use

Common.

INHIBIN

The inhibins are glycoproteins that are heterodimers. They comprise of a common alpha subunit and one of two beta subunits. In women, inhibin B is produced by the developing follicles while inhibin A is produced by the corpus luteum. In men, only inhibin B is secreted from the Sertoli cells of the testes.

Reference Interval

Using the most commonly reported assay developed by Groome, concentrations of inhibin B in normal men are reported as 187 ± 28 ng/L.

Inhibin A and B concentrations change markedly during the menstrual cycle. Peak levels of inhibin B occur in the early follicular phase when concentrations are reported as 86.8 ± 13.8 ng/L whereas inhibin A concentrations peak in the mid-luteal phase to 59.5 ± 15 ng/L. The surge of inhibin A at mid-cycle reaches concentrations of 39.3 ± 11.0 ng/L.

Clinical Application

Although a clinical role for the measurement of inhibins has been suggested in IVF treatment and as tumor markers, no clear application has yet been established. Recently, it has been shown that the measurement of inhibin A helps in assessing the risk of Down's syndrome in pregnancy, and with AFP, βhCG and estriol, has led to the quadruple test for Down's syndrome screening (see Wald, *et al.*, 2003a). A recent study (Wald *et al.*, 2003b) indicated that this is the best test for women who present in the second trimester of pregnancy. For women who present in the first trimester of pregnancy the integrated test, which combines markers from the first and second trimesters, is the test of choice.

Assay Technology

The first assay to be developed for the measurement of inhibin was a radioimmunoassay and is referred to as the Monash assay. Although it was thought that this assay measured dimeric inhibin it was later found to cross-react with the circulating α-subunit and other non-biologically active forms of inhibin. Specific ELISAs for inhibin A and B, developed by Professor Nigel Groome at the Oxford Brooke University, UK, have been used in a large number of studies. Other groups are now developing their own in-house assays but commercial assays using Professor Groome's reagents are available from DSL Ltd.

Types of Sample

Serum.

Frequency of Use

Very rare.

ESTRADIOL

The major estrogens in man are estradiol, estrone, and estriol, of which estradiol is biologically the most active (see Figure 48.4).

The estrogens are characterized by a phenolic A ring. A small amount of estradiol, which may constitute as much as 30% of the total estrogen production in some men, is produced by the testes. The ovaries are the main source of estrogen in the non-pregnant woman. During the follicular phase of the menstrual cycle there is a steady increase in the concentration of estradiol, which reflects the follicular growth in the ovary. Over the first 10 or 11 days of the cycle, estradiol has a negative feedback on the pituitary. This is reflected by a small fall in FSH concentration over this period. At about day 12 the feedback of estradiol becomes positive and stimulates the mid-cycle surge of LH and FSH at day 14. There is a fall in estradiol levels at mid-cycle but the concentrations increase again in the luteal phase, peaking at mid-cycle.

In postmenopausal women, most of the estrogen produced is from the peripheral conversion of androstenedione to estrone (15–60 μg/day). About 95% of the androstenedione production in the postmenopausal

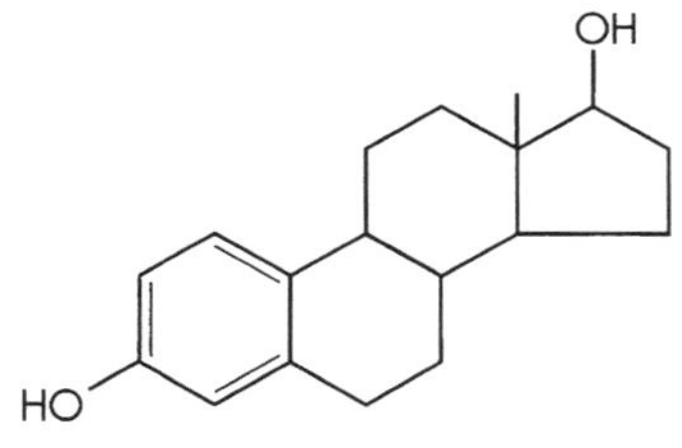

OH

HO

Fig. 48.4 Estradiol.

woman is from the adrenal glands. Androstenedione is converted to estrone in peripheral body fat. Therefore the more body fat there is, the higher the estrone concentration.

Estradiol circulates in the blood bound to SHBG and albumin. Only 1–3% of the total circulating estradiol remains unbound. It is a matter of great controversy at the moment whether any or all of the bound steroid is available to tissues (*see* FURTHER READING).

Estradiol is the main reproductive hormone in women and is involved in the development and maintenance of the reproductive organs. It is responsible for the development of the reproductive tract in the fetus and the female habitus at puberty. In the adult it is involved in the maturation and maintenance of the uterus during the menstrual cycle and the control of female reproductive behavior. It is also essential for the development of the mammary gland and lactation.

Reference Interval

See Table 48.4.

Clinical Application

Although estradiol estimation is one of the most frequently requested hormone tests, its usefulness in the investigation of infertility in women is generally considered to be limited (*see* FURTHER READING). A progestogen challenge is more helpful in establishing estrogenization of the uterus. A single intramuscular injection of 100 mg progesterone results in uterine bleeding within the next 7 days in women who have sufficient estrogen to produce endometrial proliferation. It also demonstrates that the ovary is responding to LH and FSH secretion from the pituitary and that the endometrium is responsive to estrogen and progesterone.

Estradiol is not helpful in establishing whether a woman has entered the menopause. The measurement of FSH is the most useful test in this case.

There is a large variation in the estradiol concentration both within and between women receiving hormone replacement therapy. This is especially so in women who are taking tablets or who have had implants. Levels are more constant when women have patches containing estradiol. However, menopausal symptoms frequently do not correlate with the estradiol concentration, so again the measurement of total estradiol is unhelpful.

Limitations

- There is a large overlap of the reference intervals during the menstrual cycle and for the menopause. This can make the interpretation of estradiol results difficult.
- Drugs can affect the final estradiol result either by cross-reacting in the assays, e.g. Danazol metabolites, or by altering SHBG concentrations, e.g. androgens and anticonvulsants.
- Estradiol results in women on oral contraceptives are unreliable due to the variable cross-reaction of synthetic and horse estrogens, in these preparations, with the estradiol antibodies used in different assays.

Assay Technology

For the purist, the only accurate way to measure steroids is by solvent extraction and column chromatography followed by radioimmunoassay. Chromatography may be required to remove estrone, which can have a high cross-reaction with the antiserum to estradiol. However for routine purposes, most laboratories now use commercial non-extraction methods.

A working range of about 50–2000 pmol/L is required for the investigation of infertility whereas for IVF purposes a range of 150–15,000 pmol/L is needed. No single assay can adequately cover the complete range of 50–15,000 pmol/L. In fact, some manufacturers have developed two quite different assays for the two areas of work. For instance, DPC have a 'second antibody' method for infertility work and a 'coated-tube' method for IVF monitoring. Medgenix gives two protocols for the same kit reagents.

Estradiol assays are available on all the current automated immunoassay analyzers but they generally have a functional sensitivity of about 150 pmol/L. Recently launched assays are showing slightly improved functional sensitivity.

Types of Sample

Serum or plasma.

Table 48.4. Reference intervals for estradiol.

		Bayer ACS: Centaur (pg/mL)	Bayer Advia: Centaur® (pmol/L)
Women	Follicular phase	11–69	40–253
	Mid-cycle	146–526	536–1930
	Luteal phase	68–196	250–719
	Postmenopausal	ND–37	ND–136

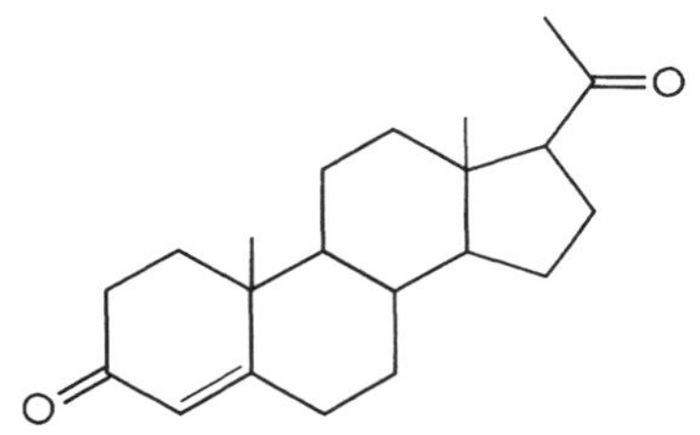

Fig. 48.5 Progesterone.

Frequency of Use

Common.

PROGESTERONE

Progesterone is produced from pregnenolone in all steroid-producing cells. It can then be further synthesized to 17α-hydroxyprogesterone or androstenedione. Large amounts (up to 30 mg/day) are produced by the corpus luteum, and by the placenta (between 250 and 500 mg/day in late pregnancy) (see Figure 48.5).

The main site of action of progesterone is on the uterus where, during the luteal phase, it increases the vascular bed, the tortuosity of the glands and glandular secretion, and reduces myometrial activity. In this way it prepares the uterus for implantation and supports the developing fetus. During pregnancy, progesterone is required for the maintenance of the placenta.

Reference Interval

See Table 48.5.

Clinical Application

The measurement of progesterone concentrations is used to show that ovulation has occurred and the corpus luteum is functioning normally. Three samples of blood may be taken around the mid-luteal phase to determine whether secretion of progesterone is adequate.

Progesterone measurement is also used in some centers during IVF therapy. Concentrations are monitored just before ovulation is due, when rising levels indicate that ovulation has or is about to occur.

Limitations

A single blood specimen taken during the luteal phase is frequently inadequate as an indicator of normal luteal function. Three specimens, on days 19, 21, and 23, are more informative.

BBT rises when progesterone levels begin to increase at mid-cycle. However, BBT is quite variable between cycles and can be difficult to interpret. A rise in BBT will usually indicate that ovulation has occurred but does not suggest that corpus luteum function or progesterone concentrations are normal.

Assay Technology

Very few laboratories use methods with solvent extraction before immunoassay; most use commercial radioimmunoassays. Companies have found it difficult to achieve precision at low concentrations, in non-radioactive assays, but several assays are now available.

Types of Sample

Serum or plasma.

Frequency of Use

Common.

TESTOSTERONE

During fetal life, at approximately 12 weeks, there is a rise in testosterone concentration in the male fetus due to stimulus of the Leydig cells in the developing testes by hCG. Testosterone falls to low levels in the third trimester of pregnancy, but there is another increase in the male neonate after about 3 weeks of life, reaching a maximum at about 2 months. Concentrations may be almost into the adult normal range at peak secretion. After about 6 months the concentration falls to less than 1.0 nmol/L (0.3 ng/L) and remains at low levels until puberty.

Testosterone is the main male sex hormone and is secreted by the Leydig or interstitial cells of the testes. Small amounts of testosterone are secreted by the ovary and the adrenal but about 50% of the testosterone production in women is derived from androstenedione by peripheral conversion.

Table 48.5. Reference intervals for progesterone.

		Bayer Advia: Centaur® (ng/mL)	Bayer Advia: Centaur® (nmol/L)
Women	Follicular phase	0.15–1.40	0.48–4.45
	Mid-luteal phase	4.4–28.0	14.1–89.1
	Postmenopausal	ND–0.73	ND–2.32

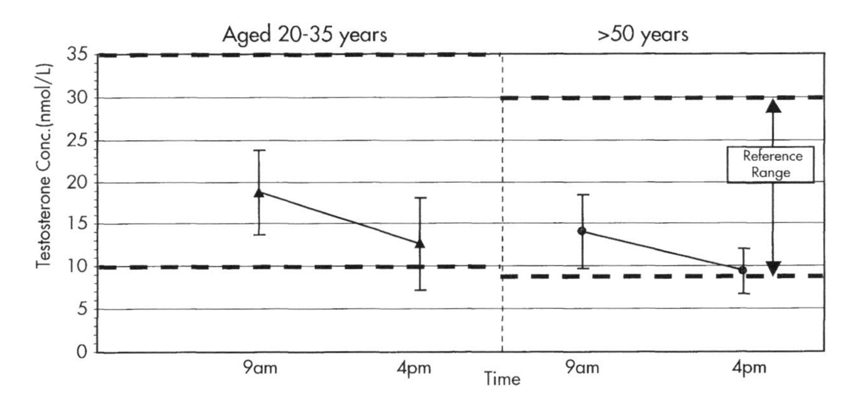

Fig. 48.6 Testosterone concentrations in men at 9 a.m. and 4 p.m.

Concentrations are less than 1 nmol/L (0.3 ng/mL) prepubertally. There is a gradual rise to adult levels during puberty in the male. Initially, an increase in concentration occurs at night but as puberty progresses, daytime levels also increase. In the adult, secretion of testosterone is episodic. There is a small circadian rhythm which, until recently, has been regarded as clinically insignificant. Recently we have shown that if testosterone is measured in the serum of normal men during the afternoon, concentrations may be below the reference range. In such cases, normal men may be wrongly diagnosed as hypogonadal (Figure 48.6).

Testosterone in the male has a negative feedback on the hypothalamus and pituitary. Most of the testosterone circulates in blood bound to SHBG and albumin. About 2% of the total plasma testosterone in men is non-protein bound; about 1% in women. It has been suggested that salivary testosterone represents this 'free' fraction.

Testosterone has a variety of actions in the body. In both sexes it stimulates secondary sexual hair growth, alters the concentration of several enzymes of the kidney, stimulates erythropoiesis and increases libido, competitiveness and aggression. In men, it is responsible for the change in voice at puberty and promotes growth and development of the sex glands and organs (Figure 48.7).

Reference Interval

See Table 48.6.

Fig. 48.7 Testosterone.

Clinical Application

The measurement of testosterone for the investigation of hirsutism and virilization in the female is explained in chapter 50.

Testosterone may be measured in boys with delayed puberty. Increases in the secretion of the gonadotropins and testosterone, particularly during sleep, indicate that puberty is progressing. Random blood samples may be taken at 3–6 month intervals to monitor testosterone levels; an increase in concentration may precede definite clinical evidence of puberty.

Infertile men may have normal testosterone concentrations. In these cases seminology tests should be done. Low libido may be related to low testosterone concentrations, but impotence is more commonly associated with neuropathic, vascular, or psychogenic causes with normal testosterone levels.

Low testosterone concentrations may be due to primary or secondary hypogonadism. Measurement of the gonadotropins aids the diagnosis; increased gonadotropins indicate primary hypogonadism. The capacity of the testes to secrete testosterone can be determined from an hCG test.

Treatment of infertility in men is often unsuccessful although pulsed GnRH therapy has been found effective in patients with hypogonadotropic hypogonadism and in some cases the GnRH may be supplemented with FSH to achieve complete spermatogenesis. Injections of hCG are also used to stimulate testosterone production and again FSH may be used as a supplement to therapy to

Table 48.6. Reference intervals for testosterone.

Women	14–76 ng/dL	0.5–2.6 mmol/L
Men	241–827 ng/dL	8.4–28.7 mmol/L

Bayer Advia: Centaur®.

Table 48.7. Regimen for hCG test.

Day 1	Blood sample	Inject 1500 IU hCG (intramuscularly)
Day 2	Blood sample	Inject 1500 IU hCG
Day 3	Blood sample	Inject 1500 IU hCG
Day 4	Blood sample	

achieve spermatogenesis. In primary hypogonadism, one of several injectable analogs can be used to maintain libido and androgenization. In these cases there is usually no treatment for infertility.

hCG Test

There are a variety of regimens in use. The one given here is our standard protocol to investigate the ability of the adult testes to secrete testosterone. Blood specimens taken on day 1 and day 4 are frequently adequate for clinical diagnosis (see Table 48.7 and Figure 48.8).

Limitations

- Total testosterone measurement is greatly influenced by the level of SHBG, which is increased by estrogens and anticonvulsants, in cirrhosis of the liver and some cases of hypothyroidism. In these situations, the concentration of free testosterone may be low despite a normal total testosterone level. Anticonvulsant therapy is associated with primary hypogonadism with elevated LH and FSH concentrations, but this is masked by the increased SHBG concentrations. Low testosterone values are found in patients with a variety of systemic diseases. Many centers measure serum SHBG concentrations in addition to total testosterone to provide an indication of free testosterone levels (*see* HIRSUTISM AND VIRILIZATION).
- Testosterone is released in a pulsed manner. The value in a single blood specimen may not be representative of the 24 h secretory mean level. There is a small, but clinically insignificant, circadian rhythm in adults.

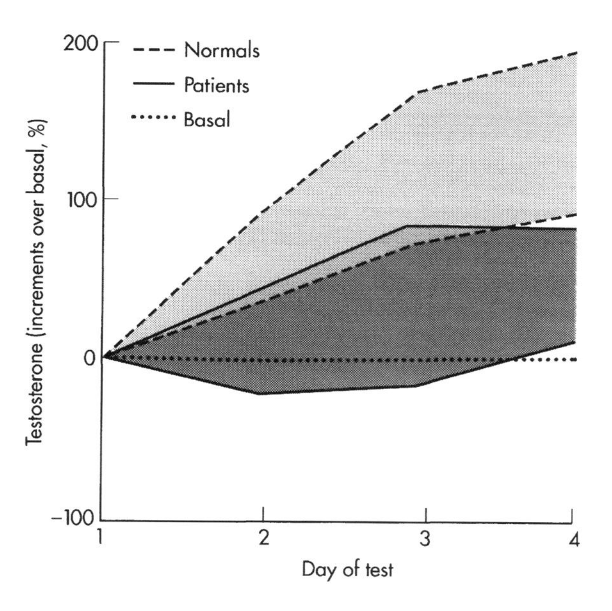

Fig. 48.8 Percentage change in testosterone concentration after 1500 IU hCG after 1, 2 and 3 days in normals (dashed lines) and men with primary hypogonadism (solid lines).

Assay Technology

Direct testosterone assays are now available on most automated immunoassay analyzers. The performance of the methods varies widely as does the agreement. Some methods appear to suffer from either interference or cross-reaction from unidentified substances in serum that leads to spuriously high results. It has not been possible to associate these results with any particular clinical condition. As well as interference these methods also have poor sensitivity, the functional sensitivity being no more than 1.0 nmol/L. Methods that employ solvent extraction before immunoassay have greater sensitivity. These methods have a functional sensitivity of 0.1 nmol/L and can be further modified so that the low levels of testosterone in saliva and hair can be measured.

Free testosterone radioimmunoassay kits are marketed by DPC and DSL. Good correlation between results from the DPC kit and total testosterone, free testosterone and the free androgen index has been reported. However, it is also reported that the results from this kit are about half those measured by equilibrium dialysis and steady-state gel filtration and it is suggested that the kit does not measure free testosterone but simply a fraction of testosterone (see Rosner, 2001).

Types of Sample

Serum and plasma. Some centers measure testosterone levels in saliva for research purposes.

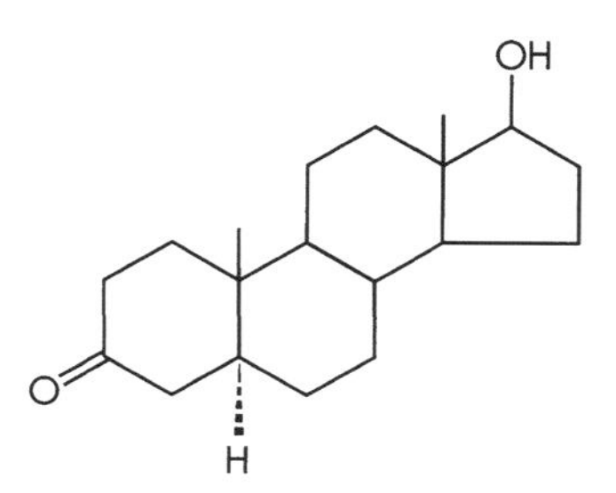

Fig. 48.9 5α-Dihydrotestosterone.

Table 48.8. Reference intervals for DHT.

	RIA after extraction and HPLC	
	(ng/mL)	(nmol/L)
Women	0.12–0.43	0.4–1.5
Men	0.38–0.72	1.3–2.5

Frequency of Use

Common.

DIHYDROTESTOSTERONE

Testosterone is converted in many tissues to 5α-dihydrotestosterone by the enzyme 5α-reductase. In most bioassays DHT is more active than testosterone.

DHT is required for the normal development of the prostate, and the urogenital sinus that forms the penis and scrotum. DHT is also produced in skin, brain, lung, salivary glands, and heart muscle (Figure 48.9).

Reference Interval

See Table 48.8.

Clinical Application

The measurement of DHT is used in the investigation of neonates with ambiguous genitalia, which could result from a deficiency of the enzyme 5α-reductase. Normal men have a testosterone:DHT ratio of 10:1. This is greatly increased in 5α-reductase deficiency.

Adult men with this deficiency present with poor masculinization and a microphallus. Where diagnosis is uncertain an hCG test may uncover the deficiency. Normally, there is a proportional increase in testosterone and DHT following stimulation of the testes with hCG, but in 5α-reductase deficiency the increase in the secretion of DHT is greatly reduced.

Limitations

The concentration of DHT is very low both in men and in women and a sensitive assay is required. In addition, antisera raised against DHT have a significant cross-reaction with testosterone which has a concentration up to 10 times that of DHT. Thus, testosterone must be removed from specimens before DHT is measured. This can be achieved by column chromatography, high-performance liquid chromatography (HPLC), or the oxidation of testosterone with potassium permanganate. The latter method was adopted by Nycomed Amersham (now GE Medical) to produce a research kit for the measurement of testosterone and DHT.

The above methods use solvent extraction and are technically difficult. Therefore, the measurement of DHT is not suited to the routine laboratory.

Types of Sample

Serum or plasma.

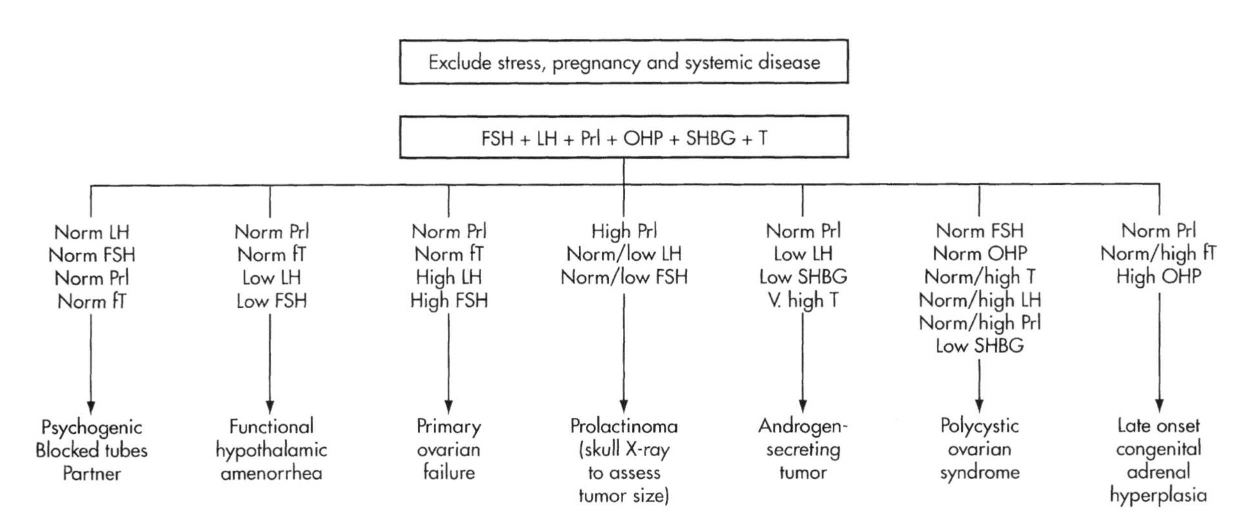

Fig. 48.10 Simplified scheme for investigation of female infertility. Prl, prolactin; T, testosterone; FT, 'free' testosterone; OHP, 17α-hydroxyprogesterone; SHBG, sex hormone-binding globulin.

Frequency of Use

Infrequent.

TEST STRATEGY FOR INFERTILITY IN WOMEN

Many different strategies are used for investigating infertility. As an example, a simplified scheme for the investigation of infertility in women is shown in Figure 48.10.

ACKNOWLEDGEMENTS

My sincere thanks to Dr Andrew Macleod, Department of Endocrinology and Chemical Pathology, St Thomas' Hospital, for his critical reading and comments on the clinical aspects of this chapter.

REFERENCES AND FURTHER READING

Breckwoldt, M., Zahradrick, H.P. and Neulin, J. Classification and diagnosis of ovarian insufficiency. In: *Infertility: Male and Female* (eds Insler, V. and Lunenfeld, B.), 229–251 (Churchill Livingstone, Edinburgh, 1993).

Carmina, A. and Lobo, R.A. Dynamic tests for hormone evaluation. In: *Infertility, Contraception and Reproductive Endocrinology* (eds Lobo, R.A., Mishell, D.R., Paulsen, R.J. and Shoupe, D.), 471–483 (Blackwell Science, Massachusetts, 1997).

Fahie-Wilson, M.N. and Soule, S.G. Macroprolactin: contribution to hyperprolactinaemia in a district general hospital and evaluation of a screening test based on precipitation with polyethylene glycol. *Ann. Clin. Biochem.* **34**, 252–258 (1997).

Glezerman, M. and Lunenfeld, B. Diagnosis of male infertility. In: *Infertility – Male and Female* (eds Insler, V. and Lunenfeld, B.), 229–251 (Churchill Livingstone, Edinburgh, 1993).

Greene, S., Zachmann, M., Manella, B., Hesse, V., Hoepffner, W., Willgerodt, H. and Prader, A. Comparison of two tests to recognize or exclude 5α-reductase deficiency in prepubertal childhood. *Acta Endocrinol.* **114**, 113–117 (1987).

Griffin, J.E. and Wilson, J.D. Disorders of the testes and male reproductive tract. In: *William's Textbook of Endocrinology* (eds Wilson, J.D., Foster, D.W., Kronenburg, H.M. and Reed Larsen, P.), 819–875 (Saunders, Philadelphia, 1998).

Groome, N.P., Illingworth, P.J., O'Brien, M., Pai, R., Rodger, F.E., Mather, J.P. and McNeilly, A.S. Measurement of dimeric inhibin B throughout the human menstrual cycle. *J. Clin. Endocrinol. Metab.* **81**, 1401–1405 (1996).

Hayes, F.J., Hall, J.E., Boepple, P.A. and Crowley, W.F. Differential control of gonadotropin secretion in the human: endocrine role of inhibin (clinical review). *J. Clin. Endocrinol. Metab.* **83**, 1835–1841 (1998).

Howles, C.M. and Macnamee, M.C. Endocrine monitoring for assisted human conception. *Br. Med. Bull.* **47**, 616–627 (1991).

Illingworth, P.J., Groome, N.P., Byrd, W., Rainey, W.E., McNeilly, A.S., Mather, J.P. and Bremner, W.J. Inhibin-B: a likely candidate for the physiologically important form of inhibin in men. *J. Clin. Endocrinol. Metab.* **81**, 1321–1325 (1996).

Imperato-McGinley, J., Gautier, T., Pichardo, M. and Shackleton, C. The diagnosis of 5α-reductase deficiency in infancy. *J. Clin. Endocrinol. Metab.* **63**, 1313–1318 (1986).

Ismail, A.A.A., Walker, P.L., Barth, J.H., Lewandowski, K.C., Jones, R. and Burr, W.A. Wrong biochemistry results: two case reports and observational study in 5310 patients on potentially misleading thyroid-stimulating hormone and gonadotropin immunoassay results. *Clin. Chem.* **48**, 2023–2029 (2002).

Leslie, H., Courtney, C.H., Bell, P.M., Hadden, D.R., McCance, D.R., Ellis, P.K., Sheridan, B. and Atkinson, A.B. Laboratory and clinical experience in 55 patients with macroprolactinemia identified by a simple polyethylene glycol precipitation method. *J. Clin. Endocrinol. Metab.* **86**, 2743–2746 (2001).

Odame, I., Donaldson, M.C.D., Wallace, A.M., Cochran, W. and Smith, P.J. Early diagnosis and management of 5α-reductase deficiency. *Arch. Dis. Child.* **67**, 720–727 (1992).

Prior, J.C. Perimenopause: the complex endocrinology of the menopause transition. *Endocrinol. Rev.* **19**, 397–428 (1998).

Rosner, W. An extraordinary inaccurate assay for free testosterone is still with us. *Clin. Endocrinol. Metab. (Letter)* **86**, 2903 (2001).

Ross, G.T. Disorders of the ovary and female reproductive tract. In: *William's Textbook of Endocrinology* (eds Wilson, J.D. and Foster, D.W.), 206–258 (Saunders, Philadelphia, 1985).

Shoupe, D., Brenner, P.F. and Mishell, D.R. Menopause. In: *Infertility, Contraception and Reproductive Endocrinology* (eds Lobo, R.A., Mishell, D.R., Paulsen, R.J. and

Shoupe, D.), 415–448 (Blackwell Science, Massachusetts, 1997).

Vivekanandan, S. and Andrew, C.E. Cross-reaction of human chorionic gonadotropin in Immulite 2000 luteinizing hormone assay. *Clin. Chem.* **39**, 318–319 (2002).

Wald, N., Huttley, W. and Hackshaw, A.K. Antenatal screening for Down's syndrome with the quadruple test: 5 year results from a screening programme. *Lancet* **361**, 835–836 (2003a).

Wald, N.J., Rodeck, C., Hackshaw, A.K., Walters, J., Chitty, L. and Mackinson, A.M. First and second trimester antenatal screening for Down's syndrome: the results of the serum, urine and ultrasound study (SURUSS). *J. Med. Screen.* **10**, 56–104 (2003b).

Yen, S.S.C. and Jaffe, R.B. (eds), *Reproductive Endocrinology: Physiology, Pathophysiology and Clinical Management* (Saunders, Philadelphia, 2004).

49 *In Vitro* Fertilization and Embryo Transfer (IVF–ET)

August C. Olivar

In 1978, Drs Patrick Steptoe and Robert Edwards announced the birth of Louise Brown, the first human baby conceived by IVF–ET, in England. Merit should be given to Dr M.C. Chang from the Worcester Foundation in Massachusetts for developing the first successful *In Vitro* Fertilization and Embryo Transfer in the animal model (rabbit). After 1978, clinics in Australia and the United States continued the work of Steptoe and Edwards, trying to improve the methods and pregnancy success rates, which at that time were low. The delivery of baby Brown was the product of many attempts by Steptoe and Edwards' team, using unstimulated natural cycles, and laparoscopy for oocyte retrieval.

Among the pioneers to improve pregnancy success rates in the US were Howard and Georgeanna Jones from Eastern Virginia Medical School in Norfolk, Virginia (now the Jones' Institute) and Dr Alan Trounsen from Melbourne, Australia, who made the procedure more efficient by using Human Menopausal Gonadotropins (Pergonal, Serono Laboratories, Milan, Italy, and Braintree, Massachusetts, USA). Using menotropins, more oocytes were recovered, and as a corollary, more embryos could be transferred (Jones, 1984). It was clear that the number of embryos transferred into the uterus had a significant effect on the pregnancy success rates.

The problem with transferring too many embryos, however, is the increased incidence of multiple pregnancies because of prematurity and high-perinatal mortality, and the prolonged stay of these babies in the intensive care nursery.

To avoid the high incidence of multiple pregnancy, many countries, including England, Germany, and Australia, have regulated the number of embryos transferred to a maximum of three in most cases. In the United States, the American Society of Reproductive Medicine has published guidelines that recommend the transfer of three embryos, unless the patient is over 39 years old, in which case four embryos may be transferred. (editor's note: new guidelines have been issued in 2004)

The use of vaginal ultrasonography (initially transabdominal), has made oocyte retrieval much easier, due to the new high-resolution transvaginal probes of 5.0 and 7.5 MGHz. The procedure is performed in most clinics as an office procedure, decreasing the cost and improving efficiency, since there is no need to transport the freshly recovered oocytes from the operating theater to the embryology laboratory.

As a result of these improvements, the pregnancy success rates continue to rise every year (see Table 49.1) in the United States, Europe, and most other parts of the world where Assisted Reproduction is routinely performed. In order to have uniform reports of pregnancy success rates, in 1992 the US Congress passed the Fertility Clinic Success Rate and Certification Act, which requires the Centers for Disease Control and Prevention (CDC) to

Table 49.1. Assisted reproductive technology in the United States and Canada, results generated from the ASRM/SART registry (1985–1995).

TYPE	1985	1986	1987	1988	1989	1990	1991	1992	1993	1994	1995
IVF-ET	14(10)	17(13)	16(12)	15(12)	18(14)	18(14)	21(15)	24(16)	25(18)	26(20)	27(22)
GIFT			23(16)	26(21)	30(23)	29(22)	33(26)	33(26)	35(28)	35(28)	34(27)
ZIFT				27(20)	21(17)	21(16)	29(22)	32(22)	28(24)	35(29)	35(28)

Numbers outside the brackets refer to the percentage of pregnancy rates. Numbers inside the brackets refer to the percentage of live births.

The Immunoassay Handbook. *Third Edition*
D. Wild (Ed.)

publish pregnancy success rates for fertility clinics in the United States. The first report was published in December 1997, with the most recent data compiled in 1995 and submitted from 281 fertility clinics performing Assisted Reproductive Procedures in the United States. This data will be published annually by the CDC with the cooperation of the American Society of Reproductive Medicine and the Society of Assisted Reproductive Technology (SART).

In Vitro Fertilization and Embryo Transfer has four well-defined stages:

STAGE I: INDUCTION OF OVULATION

The purpose of this stage is to stimulate the development of several mature Graafian follicles in the ovaries. The patient starts the cycle in the mid-luteal phase (day 21 or 22 of the cycle). She is instructed to take subcutaneous injections of gonadotropin releasing hormone (GnRH) agonist (luprolide acetate, Lupron, TAP Pharmaceuticals, Inc., Deerfield, IL) or a nasal spray form. This drug produces down regulation of FSH and LH receptors in the pituitary gland, preventing secretion of these gonadotropins.

The GnRH agonist inhibits a spontaneous surge of LH, minimizing the chances for early luteinization, which is detrimental to oocyte fertilization. In the past, spontaneous LH surge was the most common cause of cycle cancellation.

GnRH agonist is administered usually for 10 days, followed by the administration of a mixture of FSH and LH (Pergonal™, Serono Laboratories, Inc., Norwell, MA, USA; Humegon™, Organon, Inc., W. Orange, NJ, USA) or 'Pure FSH' extracted from the urine of post-menopausal women (Fertinex™, Serono Laboratories, Braintree, MA) or Pure Recombinant FSH (Follistim™, Organon Diagnostics or Gonal-F™, Serono Laboratories). The administration of gonadotropins is monitored closely by vaginal ultrasound and serum estradiol-17β, almost on a daily basis.

When the ovarian follicles have reached a size consistent with maturity an injection of human chorionic gonadotropin (hCG), 10,000 IU, is given I.M.

STAGE II: OOCYTE RETRIEVAL

Thirty-four hours after the injection of hCG, oocyte retrieval is performed transvaginally, with a transvaginal probe with a guided needle. Every ovarian follicle of at least 8–9 mm is aspirated. The follicular fluid is examined under a dissecting microscope to search for the oocyte.

STAGE III: OOCYTE INSEMINATION AND INCUBATION

Insemination of the oocytes is usually performed 3–4 h after the oocyte recovery. The preparation of the sperm prior to insemination is accomplished by the 'swim-up' method or in a way by which the sperm can be isolated, free of debris or any other contaminants left as a residual from any glandular secretion contributing to the ejaculate.

Approximately 200,000 sperm are placed per oocyte or 500,000 per 3–4 oocytes in the Petri dish and placed in the incubator. After 18–21 h, the oocytes are viewed under the microscope to determine the presence of two polar bodies, the first indication of fertilization. At present, most laboratories place the gametes in culture media devoid of serum, but enriched with other sources of protein, which must be free of any possible contaminants or infections.

The embryos remain in the incubator for between 48 and 72 h, before transfer. The difference in culture time depends on the guidelines for each assisted reproductive technology (ART) clinic. With better culture media, there is a tendency for most clinics to transfer the embryos later, at a more advanced stage of development.

STAGE IV: EMBRYO TRANSFER

Approximately 66–70 h after insemination the embryos are at about the 6–10 cell stage. For the transfer the patient is placed in the transfer room (next to the IVF laboratory), usually in dorso-lithotomy position. Some clinics will transfer the embryos in genu-pectoral position if the patient has a marked anteverted, anteflexed uterus.

The transfer is performed with a semi-rigid, plastic catheter, which is passed into the uterine cavity in an atraumatic manner with a guide. Once in position the embryologist fills the inner catheter with 3–4 embryos and very small amount of culture medium.

The number of embryos transferred to the uterus depends on the clinic guidelines. Most clinics transfer up to four embryos, freezing the rest of them. In England, and some other countries, the number of transferred embryos can only be up to three in order to minimize the chances of multiple pregnancies.

OTHER ASSISTED REPRODUCTIVE TECHNOLOGY PROCEDURES

GAMETE INTRA-FALLOPIAN TUBE TRANSFER

Ricardo Asch *et al.* (1986) reported his preliminary experiences with **gamete intra-Fallopian tube transfer (GIFT)**. Asch's idea was to facilitate the fertilization process by placing the gametes where they are supposed to meet, and in this manner make the process more physiologic, since the embryos would reach the endometrial cavity in about 3 days. Under normal circumstances the embryos usually reach the uterine cavity after 3 days when they are at the blastocyst stage and can, after

2 or 3 days, hatch and implant. This brilliant idea works pretty well. In this procedure, the first stage, induction of ovulation, is performed in the same manner as in IVF; however, at the time of oocyte retrieval, there is a double set-up for the laparoscopy in the operating room. Once the oocytes are isolated, laparoscopy is performed. After properly identifying the oviducts, usually two oocytes with approximately 500,000 sperm are transferred into the ampullary portion of each oviduct.

This procedure is performed under general anesthesia and requires expertise to transfer the oocytes and the sperm through a fine catheter of approximately 1.5–2 mm in diameter into the ampullary portion of each oviduct.

Overall the average pregnancy success rate for this procedure is about 2–4% better than IVF-ET (Table 37.1), however, because of the laparoscopy it is not so popular and only represents approximately 6% of the total number of ART procedures performed in the US.

ZYGOTE INTRA-FALLOPIAN TRANSFER

ZIFT is the acronym for **zygote intra-Fallopian transfer (ZIFT)**. This procedure is performed the day following oocyte retrieval, after fertilization has been accomplished.

Approximately 18 h after oocyte insemination, the oocytes are inspected under the microscope for the first sign of fertilization, which is the presence of two pronuclei (a male and a female pronucleus).

These zygotes are then transferred in the operating room via laparoscopy in the ampullary portion of each oviduct. The usual practice is to place two zygotes in each oviduct. Pregnancy success rates with this procedure are shown in Table 49.1.

MICROMANIPULATION OF GAMETES

ASSISTED HATCHING

In some cases, despite successful induction of ovulation, satisfactory oocyte retrieval, good fertilization rates, and adequate transfer, pregnancy is not accomplished. When there are definite signs of thick zona pellucida or in the case of advanced maternal age, a defect or 'hole' in the zona pellucida can be made using an acidic solution (tyrode's) utilizing micropipettes under a contrast microscope (Cohen *et al.*, 1990).

Before the widespread use of tyrode's solution, **assisted hatching** was tried in several ways. Initially, a defect in the zona was produced using the sharp edge of a micropipette. This technique was difficult to master, however, and not always replicable. Another method was the use of a laser in order to vaporize a small hole in the zona pellucida. A concern about this technique is that it is difficult to know the detrimental effects of ionizing radiation produced by the laser and the high cost of acquiring a laser machine must also be considered.

In some centers, assisted hatching is used in all embryos before transfer. This technique achieves better pregnancy rates according to some of these centers (Schoolcraft *et al.*, 1994) and the technique has many followers among members of the Society for Assisted Reproductive Technologies (SART).

INTRACYTOPLASMIC SPERM INJECTION

Palermo *et al.* (1992) and Van Steirteghen *et al.* (1993) were the first to report the successful use of **intracytoplasmic sperm injection** (**ICSI**). Under a contrast microscope the oocyte is positioned with a holding micropipette, and a single spermatozoa is injected through the zona pellucida, directly into the cytoplasm of the oocyte. This technique raised many eyebrows in the beginning, due to the fact that theoretically there was no time for 'sperm capacitation', causing concern about poor fertilization and pregnancy success rates, with the possibilities of chromosomally abnormal concepti. Fortunately, these fears have not proven to be the case and at present, ICSI seems to be the best answer to treat male infertility. Treatment of severe oligozoospermia, asthenozoospermia, or teratozoospermia yields good results (Table 49.2). Even with azoospermia but with sperm in the testicle, pregnancy is now possible (Silber *et al.*, 1994).

Currently, ICSI has made past procedures such as **subzona sperm injection** (**SUSI**) obsolete, because the pregnancy success rates with ICSI are superior and more consistent.

ANALYTES

Except for vaginal ultrasound, the determination of estradiol-17β levels in serum is the most important parameter to monitor the growth and maturity of the ovarian follicles.

ESTRADIOL

Function

There are many estrogens in the body, some more biologically potent than others. The most potent estrogen in the body is estradiol-17β (known as E_2). Other known

Table 49.2. Outcome of intra-cytoplasmic sperm injection.

Data from	Number of cycles	Percentage live births
Van Steirtenghen, 1995, American Society of Reproductive Medicine, Seattle, WA	1275	28.4
Centers for Disease Control and Prevention 1995 (11% of ART in the USA)	5049*	27.2**

*Approximate.
**Live birth/retrieval.

estrogens such as estrone (E_1) or estriol (E_3) have the same function as E_2, however, are less potent, especially E_3. Their function is to stimulate proliferation of the endometrium, which then becomes primed for the action of progesterone, which will be produced by the corpus luteum in preparation for implantation of the embryo.

In the normal menstrual cycle, estrogens stimulate the cervical mucus, making it abundant, clear, low, or absent from endocervical cells, and causing a high content of sodium chloride (which will produce the typical 'fern pattern' when dried on the slide). These characteristics of cervical mucus allow directional and easy penetration by the sperm.

Estrogens also promote stimulation of the oviductal cilia which have adrenergic inervation in order to beat toward the ostium allowing an adequate timed journey of the fertilized oocyte. Besides the action at the target organ, the uterus, estrogens have somatic effects on many other organs of the body (development of secondary sexual characteristics, maintenance of the reproductive tract in women, etc.)

Estradiol-17β has a phenolic A ring with two hydroxyl groups in the 3 and 17 positions (Figure 49.1).

Reference Interval

See Table 49.3.

Clinical Application

Serum estradiol determination is the most important laboratory hormonal assay for *in vitro* fertilization, as it indicates the size of the Graafian follicle and maturity of the oocyte. Serum estradiol is ordered at the initiation of the induction of ovulation to provide a baseline. The value should be low (usually lower than 30 pg/mL) because down regulation of the pituitary gland is accomplished with the use of GnRH analog (luprolide acetate: Lupron) for approximately 10 days preceding the administration of gonadotropins. Even if GnRH analog is not used, serum estradiol should be low, marking the beginning of the menstrual cycle in the early follicular phase.

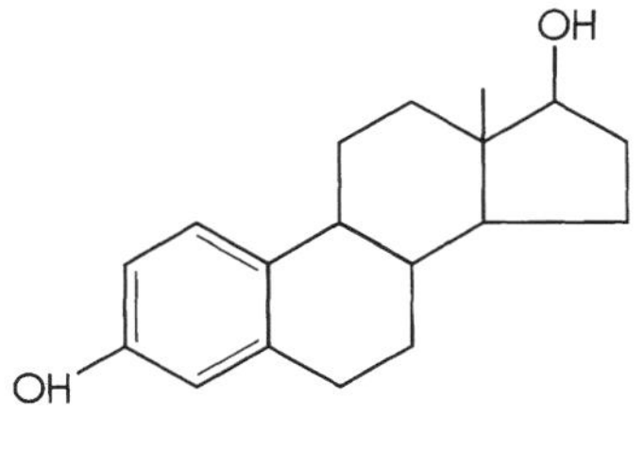

Most fertility clinics then allow the patient to take gonadotropins for 3 or 4 days, before ordering another estradiol determination. From thereon, daily levels of estradiol are tested and correlated with the size of the follicles measured by ultrasound. Good response to the gonadotropins is manifested by increasing levels of estradiol, anywhere from 33 to 75%. Usually, if estradiol levels plateau or decrease the cycle is cancelled. After the injection of hCG, estradiol levels should continue to rise or plateau. If levels decrease (more than 10%) the cycle is cancelled by most clinics.

The expected estradiol production by each mature follicle is approximately 150–200 pg/mL, however, in some cases some of the quasi mature follicles can produce substantial amounts of estrogens, making it more difficult to predict maturity in these smaller follicles.

OH

OH

Fig. 49.1 Estradiol-17β.

Limitations

There is some difficulty predicting the amount of estradiol produced by each follicle, due to the different states (sizes)

Table 49.3. Example reference intervals for estradiol.

Group	Common units (pg/mL)	SI units (pmol/L)
Adult female		
Follicular	26–158	95–158
Mid-cycle	69–364	253–1336
Luteal	51–219	187–804
Postmenopausal	0–47	0–172

of development. There is no specified amount of estradiol produced by a follicle measuring 10 mm in diameter, for example.

In general, it is adequate to start induction of ovulation with very low levels of estradiol, however there is no exact amount which would allow us to withhold the cycle. Some drugs, such as antibiotics (e.g. ampicillin) may decrease the production of estrogens by interfering with bacterial action on the production of estrogens. Other drugs, such as anticonvulsants or antimycotic agents, may also produce the same effect.

Assay Technology

See INFERTILITY – ESTRADIOL – ASSAY TECHNOLOGY.

Desirable Assay Performance Characteristics

Serum estradiol is an analyte with exceptional performance requirements in IVF, because most of the clinical decisions made for the induction of ovulation, such as whether to proceed or stop the cycle, are based on the daily levels of this hormone in response to the gonadotropins administered.

Imprecision

The Immuno-1 assay (Bayer, Tarrytown, NY, USA) represents an acceptable assay with intra- and inter-assay precision in the range of 2.8–9.7%.

Correlation Data

A wide range of patient samples should be compared with a well-established method, in order to check for satisfactory correlation. The level of agreement between different methods for serum estradiol is in most cases very adequate, considering the degree of sensitivity that all these assays provide.

Sensitivity

The minimum detectable concentration of estradiol should be around 10 pg/mL (37 pmol/L) or better. Most assays are able to detect this low concentration of estradiol which is adequate to diagnose any clinical condition requiring this assay.

Specificity

Method literature should be checked for sample interferences. For example in the Immuno-1 estradiol assay, the use of lipemic (up to 1300 mg/dL urea nitrogen), and hemolized (up to 1000 mg/dL of hemoglobin) samples should not affect method performance. Abnormal levels of albumin, immunoglobulin A, immunoglobulin G, and immunoglobulin M have no effect on method performance. Icteric samples (>4.0 mg/dL bilirubin) may give abnormal results in some cases and should be avoided.

Cross-reactivity has been tested with several compounds that could interfere in the assay. The potential cross-reactant is spiked into base serum and tested with the chosen method.

Types of sample

Serum or plasma.

Frequency of use

Extremely common.

ACKNOWLEDGEMENT

I would like to thank Ms Hope Keener Olivar for her expert technical assistance on the preparation of this manuscript.

REFERENCES AND FURTHER READING

Asch, R.H., Balmaceda, J.P., Elsworth, L.R. and Wong, P.C. Preliminary experience with gamete intra Fallopian transfer (GIFT). *Fertil. Steril.* **45**, 366–369 (1986).

Bayer Corporation. Publication No. DA4-1214K95. Tarrytown, New York (1995).

Centers for Disease Control and Prevention. Assisted Reproductive Technology Success Rates. National Summary and Fertility Clinic Reports. US Department of Health and Human Services (1995).

Cohen, J., Wright, G., Malter, H., Massey, J., Mayer, M.P. and Wiemer, K. Impairment of the hatching process following *in vitro* fertilization in the human and improvement of implantation by assisted hatching using micromanipulation. *Hum. Reprod.* **5**, 7–13 (1990).

Davis, O.K. and Rosenwaks, Z. In: *In Vitro Fertilization. Reproductive Endocrinology, Surgery and Technology*, vol. 2 (eds Adashi, E.Y., Rock, J.A. and Rosenwaks, Z.), 2330 (Lippincott-Raven, Philadelphia, 1996).

Jones, G.S. Update in *in vitro* fertilization. *Endocr. Rev.* **5**, 62–65 (1984).

Palermo, G., Joris, H., Devroey, P. and Van Steirteghen, A.C. Pregnancies after intracytoplasmic injection of a single spermatozoon into an oocyte. *Lancet* **340**, 17–18 (1992).

Practice Committee of the Society for Assisted Reproductive Technology and the American Society for Reproductive Medicine. Guidelines on the number of embryos transfered. *Fert. and Sterility*, **82**, 773–774 (2004).

Schoolcraft, W.B., Schlenker, T., Gee, M., Jones, G.S. and Jones, H.W. Jr. Assisted hatching in the treatment of poor prognosis *in vitro* fertilization candidates. *Fertil. Steril.* **62**, 551–554 (1994).

Silber, S.J., Devroey, P., Nagy, Z., Liu, J., Tournaye, H. and Van Steirteghen, A.C. *ICSI with Testicular and Epididymal Sperm*, 36–41 (ESHRE Workshop, Brussels, Belgium, 1994).

Speroff, L., Glass, R.H. and Kase, N.G. Clinical assays, In: *Clinical Gynecologic Endocrinology and Infertility*, 5th edn, 967–976 (Williams & Wilkins, Baltimore, MD, 1994a).

Speroff, L., Glass, R.H. and Kase, N.G. Hormone biosynthesis, metabolism and mechanism of action. In: *Clinical Gynecologic Endocrinology and Infertility*, 5th edn, 31–86 (Williams & Wilkins, Baltimore, MD, 1994b).

Steptoe, P.C. and Edwards, R.G. Birth after reimplantation of a human embryo. *Lancet* **2**, 366–368 (1978).

Van Steirteghen, A., Liu, J., Nagy, Z., Joris, H., Tournaye, H., Liebaers, I. and Devroey, P. Use of assisted fertilization. *Hum. Reprod.* **8**, 1984 (1993).

50 Hirsutism and Virilization in the Female

Michael J. Wheeler

The androgens have a wide variety of actions in both sexes at every stage of development. In women, the ovaries and adrenal glands secrete mostly weak androgens: dehydroepiandrosterone (DHEA) and androstenedione. These weak androgens are then converted to the more potent androgen, testosterone, in peripheral tissue. Testosterone is subsequently converted to 5α-dihydrotestosterone (DHT) in many androgen-sensitive tissues, e.g. hair follicles. The presence of increased levels of testosterone and DHT in the female causes hirsutism and virilization. **Hirsutism** is the term for increased hair growth, usually on the face, but sometimes on the thighs, arms, breasts, and supra-pubic areas. **Virilization** describes the appearance of male characteristics such as clitoromegaly, deepening of the voice, and increased muscle mass.

Both testosterone and DHT are required for the development of the male sex organs. Absence, or reduced synthesis, of these steroids or their receptors during fetal life leads to feminization. Conversely, exposure of the female fetus to high concentrations of androgens leads to virilization and masculinization.

Androgens (produced by the adrenals during the adrenarche) are required for the development of secondary sex hair (pubis, axilla) in both sexes. This is demonstrated by the development of pubic and axillary hair in agonadal men. The additional requirement for androgens to be converted to testosterone and then to DHT is shown by the lack of secondary sexual hair in the complete form of **testicular feminization** where there is an absence of the receptors for DHT. Asexual hair is also responsive to the action of androgens. Hence, men develop coarse terminal hair on the face, chest, and limbs and, in women, hirsutism is a common clinical problem. Racial and genetic factors have to be taken into consideration when assessing excess hair growth in women. For instance, noticeable hair growth on the face of a woman of Mediterranean origin may be expected but the same hair growth on the face of a Chinese woman would be regarded as abnormal.

Excessive secretion of androgens in the female leads to masculinization. If this occurs during fetal life, the child is born with ambiguous genitalia. There is an enlargement of the clitoris to form a pseudo-penis and labial fusion. The extent of the masculinization is variable but complete labial fusion with a terminal urethral opening has been reported. Masculinization of the adult female involves increase in muscle mass, clitoromegaly, deepening of the voice, development of a male escutcheon, and frontal balding. The degree of virilization or masculinization is positively correlated with the concentration of circulating androgen.

CLINICAL DISORDERS

POLYCYSTIC OVARIAN SYNDROME

In 1935, Stein and Leventhal described a syndrome of obesity, hirsutism, menstrual disturbance, and polycystic ovaries: the **Stein–Leventhal syndrome**. We now know that this represents just one subgroup of the **polycystic ovarian syndrome** (**PCOS**), which is typified by the presence of enlarged cystic ovaries that are covered with a pale, thickened, glistening capsule. The cysts are numerous, measuring 2–8 mm in diameter and arranged around the periphery of the ovary. The stroma is reported as enlarged. In a study of 1741 patients with polycystic ovaries, Balen *et al.* (1995) reported that 29.7% had a normal menstrual cycle, 47.0% had oligomenorrhea, 19.2% had amenorrhea, 2.7% had polymenorrhea, and 1.4% had menorrhagia. In addition, 66.2% had hirsutism; mild in 20.6%, moderate in 40.7%, and severe in 4.9%, while 34.7% of the patients had acne and 2.5% had acanthosis nigricans. PCOS is therefore a spectrum of disease.

Testosterone concentrations are frequently above the reference interval, androstenedione concentrations occasionally so. Free testosterone concentrations may

The Immunoassay Handbook. *Third Edition*
D. Wild (Ed.)

be increased proportionately more than total testosterone concentrations because of the lowered sex hormone-binding globulin (SHBG) concentrations. The level of SHBG has been shown to be inversely correlated with weight but this seems in turn to be related to insulin resistance in these patients. A large number of studies have now shown that insulin appears to diminish SHBG levels and hence increases free testosterone levels. In most patients the ovaries are the major source of the increased androgen secretion. Up to 10% of patients with PCOS may have increased dehydroepiandrosterone sulfate (DHEAS) concentrations in the serum, indicating increased androgen secretion from the adrenal. Several groups have investigated adrenal function in PCO patients and subtle defects in adrenal steroidogenesis have been reported in 12–40% of these patients although no increase in adrenocorticotropic hormone (ACTH) secretion has been found.

Polycystic ovaries are frequently present in other clinical disorders that are associated with increased secretion of androgens from the adrenal (congenital adrenal hyperplasia (CAH), *see* below; and Cushing's syndrome, *see* ADRENAL CORTEX). Because about 50% of patients with PCOS are hirsute and obese, it may be difficult to decide whether the patient has Cushing's syndrome or PCOS. Appropriate tests can quickly exclude or confirm Cushing's syndrome.

Luteinizing hormone (LH) levels in PCOS may be increased above or at the top of the reference interval for the follicular phase. Follicle-stimulating hormone (FSH) concentrations are usually normal. The ratio of LH:FSH has been used to indicate the presence of PCOS and a ratio of greater than 3 was said to be diagnostic. Although this was a helpful diagnostic tool when radioimmunoassays were used to measure the gonadotropins, current immunometric methods for LH give lower values and this ratio is no longer valid. Although some groups have tried to re-establish a diagnostic ratio, it has generally been found to be unhelpful and the most useful indicator of polycystic ovaries is an LH concentration greater than 10 IU/L.

Patients may present with hirsutism or infertility, usually due to oligomenorrhea or amenorrhea. The use of anti-androgens, for example, cyproterone acetate or spironolactone, is the most effective treatment for hirsutism. A number of different procedures have been carried out to achieve conception. These include clomiphene citrate, conventional gonadotropin therapy, pulsatile GnRH, GnRH agonist with gonadotropin therapy, and diathermy. Although for most of these treatments a conception rate of 50–80% has been reported, spontaneous abortion has been reported to be as high as 40% in some studies. Those obese women with PCOS who are able to lose weight often resume normal menstrual function.

Insulin resistance is present in about 40% of patients with PCOS. It occurs both in obese and nonobese PCOS patients although obesity further exacerbates insulin resistance. Hyperinsulinemia is thought to play a key role in PCOS. Reported actions include increase of hyperandrogenemia, decrease in SHBG concentrations, increase in LH secretion from the pituitary, and increased estrogen production by the ovaries. The key role is demonstrated by the fact that suppression of androgen concentrations does not lead to normal insulin sensitivity (*see* FURTHER READING for reviews). Harbourne *et al.* (2003) reported that treatment with the anti-hyperglycemic drug, metformin, reduced androgen slightly but showed a greater reduction in the Ferriman Galwey score when compared with Dianette. Patients perceived greater reduction of hirsutism with metformin treatment. In support of these findings, most studies examining the effect of metformin treatment of PCOS patients find a modest reduction in circulating free androgens.

Idiopathic hirsutism

Some women have hirsutism without the other features of PCOS being present. In some cases the hirsutism may be due to genetic or racial factors. For instance, it is well recognized that women of Mediterranean origin are more hirsute, and dark-haired women may be more conscious of visible hair on the upper lip in particular. Social pressures can also have an important role to play. In these women androgen levels are normal.

However, there is a separate group of women who have a marked increase of hair on the face, which may be present on other areas of the body. In the strict definition of the term androgen levels are within the reference interval. An index of the free testosterone concentration shows this parameter to be increased but no pathology is obvious with no ultrasound evidence of polycystic ovaries. A number of androgen indicators have been used to disclose an abnormality of androgen synthesis or secretion and these are discussed further under the appropriate analyte. Some investigators have seen this condition as one end of the spectrum of the polycystic ovary syndrome and would so classify these patients.

A number of drugs have been used to treat hirsutism, including estrogens, dexamethasone, and anti-androgens (spironolactone, cyproterone acetate).

ANDROGEN-SECRETING TUMORS OF THE OVARY

Androgen-secreting tumors of the ovary are very rare. Several different types of tumor can occur, depending on the main type of cell present. Testosterone concentrations are typically greater than 1.7 ng/mL (6 nmol/L). Androstenedione may also be elevated in concentration. Patients may present with hirsutism or with evidence of virilization. Hirsutism of sudden onset and of increasing severity should alert the clinician to the possibility of an androgen-secreting tumor.

CONGENITAL ADRENAL HYPERPLASIA

CAH is due to an enzyme deficiency in steroid biosynthesis. The consequent fall in cortisol secretion leads to increased secretion of ACTH. This, in turn, increases the stimulus of this hormone to the adrenal and hence causes hyperplasia of the adrenal cortical cells. A number of enzyme deficiencies have been identified but about 95% of the cases are due to a deficiency of the 21-hydroxylase enzyme (see Figure 50.1).

In this particular condition there is increased secretion of 17α-hydroxyprogesterone (17-OHP), the steroid immediately before the enzyme deficiency. At the same time more 17-OHP is converted to the androgens and the increased concentration of these steroids causes virilization of the female fetus. Patients are treated with exogenous glucocorticoids but, as up to 75% of the individuals also suffer salt loss, exogenous mineralocorticoid must be given.

Diagnosis of CAH is made by measuring the steroid immediately before the deficient enzyme. Therefore, nearly all cases may be diagnosed by measuring 17-OHP in the serum or saliva of the neonate.

Effective treatment minimizes further virilization and allows puberty to proceed at a more normal rate. Therefore, it is most important to control androgen levels. Although testosterone has been used to monitor treatment, androstenedione may be viewed as a more appropriate androgen. When androstenedione is suppressed well into the reference interval, the level of 17-OHP may be as much as four times the upper limit of its reference interval. If 17-OHP levels are within the reference interval, it is likely that the patient is being overtreated and growth will be retarded as well as other side-effects of glucocorticoid excess being manifest. Therefore, it is inappropriate to monitor treatment with 17-OHP measurement alone. The classical form of CAH is due to an autosomal recessive gene, closely linked to the human leucocyte antigen-B (HLA-B) locus. HLA typing can be carried out in families with an affected individual to assess the carrier status of relatives. Individuals who are heterozygous for the gene mutation also show abnormalities of steroid synthesis. 17-OHP may be slightly above or just within the reference interval but stimulation of steroidogenesis with ACTH results in a greater increase of 17-OHP secretion than seen in normal subjects. This abnormality may be further exposed by calculating the 17-OHP:cortisol ratio. Patients usually have no physical abnormality.

Another group of patients has been shown to have a milder 21-hydroxylase deficiency due to a different gene abnormality. No sign of abnormal steroid secretion occurs until after puberty and these patients are said to have **late-onset CAH (LOCAH)**. Patients have hirsutism, oligomenorrhea, and infertility, and since they also have polycystic ovaries (as do most patients with CAH), they may be wrongly diagnosed as having PCOS if appropriate biochemical tests are not carried out. 17-OHP levels are above the reference interval for the follicular phase of the menstrual cycle and give an exaggerated response to ACTH stimulation. Patients may be treated with exogenous glucocorticoids although anti-androgens may also be required for effective treatment of hirsutism.

Most other cases of CAH are due to a deficiency of the 11β-hydroxylase enzyme. There is increased production of 11-deoxycortisol, the steroid produced immediately before the enzyme block (*see* CONGENITAL ADRENAL HYPERPLASIA) and diagnosis is made by measurement of this steroid in serum. There is also increased secretion of 11-deoxycorticosterone and the androgens. Patients develop hypertension and affected girls have some degree of masculinization. Other enzyme deficiencies have been described but these are very rare.

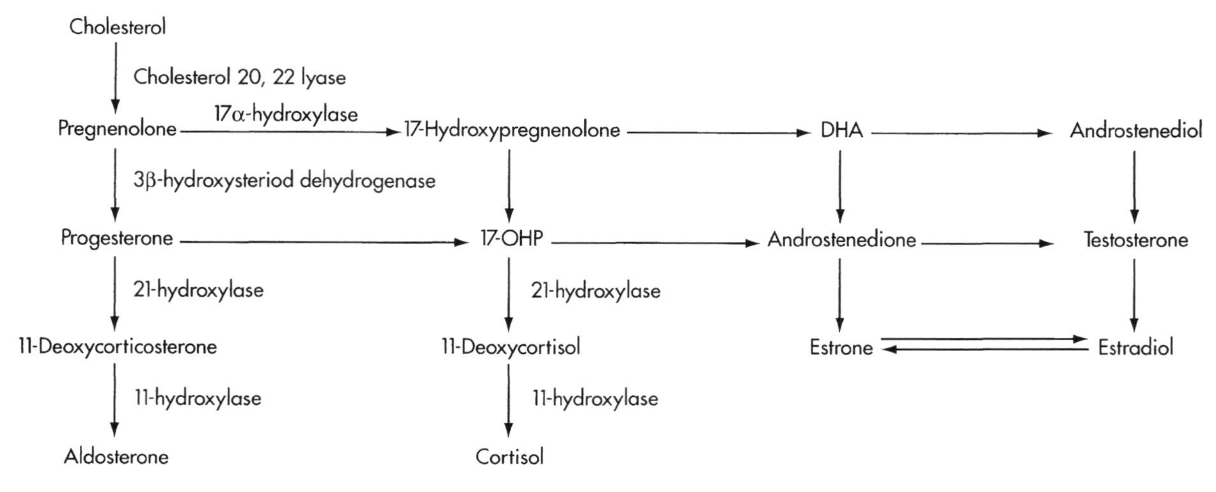

Fig. 50.1 Steroid pathways.

CUSHING'S SYNDROME

See also ADRENAL CORTEX.

Cushing's syndrome results from an overproduction of cortisol. It may be the result of a pituitary adenoma (**Cushing's disease**), an adrenal adenoma, adrenal carcinoma, or an ectopic source of ACTH. The disease is nine times more common in women. Increased steroidogenesis leads to increased secretion of androgens. The resulting hirsutism is generally mild in Cushing's disease but as adrenal tumors secrete greater quantities of androgens, severe hirsutism, clitoromegaly, and deepening of the voice may occur. Rarely, cortisol and androgen secretion is greatly increased when an adrenal carcinoma is present but in the case of adrenal adenomas DHEAS is usually below the reference interval. Treatment is appropriate for the abnormality and includes pituitary surgery, adrenalectomy, or removal of an ACTH-secreting tumor. Further treatment by pituitary irradiation or chemotherapy may be required. Careful assessment of pituitary and adrenal function is required after surgery and exogenous glucocorticoid given if necessary.

ANALYTES

LUTEINIZING HORMONE AND FOLLICLE-STIMULATING HORMONE

See also INFERTILITY.

The following section describes the measurement of these hormones in the diagnosis of androgen disorders.

Clinical Application

The secretion of LH and FSH is suppressed by very high levels of androgen. Thus, LH and FSH levels are usually low, normal, or below the reference interval in LOCAH and Cushing's syndrome. Although the concentration of FSH is normal in patients with PCOS, LH concentrations are frequently above the reference range and may be up to twice the upper limit of the reference interval. An LH concentration greater than 10 IU/L with a normal FSH level is suggestive of polycystic ovaries for a serum sample taken day 2–5 of the menstrual cycle or in an amenorrheic woman.

Limitations

- The measurement of LH and FSH is not helpful in the diagnosis of Cushing's syndrome, CAH, or androgen-secreting tumors.
- Many publications have stated that an LH:FSH ratio greater than 3 is diagnostic of PCOS. This was generally accepted when RIAs were used to measure the gonadotropins. Because immunometric assays yield lower results for LH than RIAs, it is no longer possible to state such a figure. It has been questioned whether LH and FSH assays have a role in the diagnosis of PCOS when good ultrasound facilities are available.

Frequency of Use

Common.

TESTOSTERONE

See also INFERTILITY.

The following section describes the use of testosterone measurement in the investigation of hirsutism and virilization.

Clinical Applications

A testosterone measurement is usually requested in every case of hirsutism. The primary use of this parameter is to diagnose an androgen-secreting tumor when values are typically greater than 1.7 ng/mL (6 nmol/L). The clinician should be alerted to the possibility of a tumor when hirsutism is of sudden onset with increasing severity.

Testosterone levels are usually raised in patients with PCOS. Discrimination between patients with PCOS, idiopathic hirsutism, and normal women is enhanced by including the measurement of SHBG. From these two parameters an indication of the free testosterone level may be achieved. Testosterone may be normal, but if the SHBG is low, the level of free testosterone may be abnormally high. It is questionable whether knowing the result of any of these three parameters will change the clinician's management of the patient with mild hirsutism. However, when there is a mixed ethnic population, an indication of the circulating free testosterone concentration may help to distinguish racial or genetic causes of hirsutism from the abnormal pathology. In addition, they may be helpful in confirming suppression or compliance during treatment with suppressive therapy.

Testosterone concentrations are higher in patients with CAH, LOCAH, and Cushing's syndrome but the measurement of this hormone has no place in the diagnosis of these diseases. However, the measurement of testosterone has been used to monitor the treatment of patients with CAH.

Limitations

- Total testosterone measurement is greatly influenced by SHBG levels. SHBG concentrations are increased by estrogens and anti-convulsants, and in cirrhosis of the liver and some cases of hypothyroidism. In hirsutism SHBG is often low, resulting in an elevated concentration of nonprotein-bound testosterone. This may be a result of obesity or accompanying insulin resistance in these patients. In these situations, measurement of total testosterone alone can be particularly misleading. Free testosterone levels may be estimated by using an SHBG measurement to derive a free androgen index.

- The measurement of testosterone gives no indication of the source of increased androgen secretion in women. The testosterone level does not predict the efficacy of any treatment instituted to treat hirsutism, apart from when an androgen-secreting tumor is present.
- As about half the circulating testosterone in women is derived from the peripheral conversion of the weaker androgens secreted by both the adrenal and ovary, testosterone is likely to be less sensitive than androstenedione in monitoring the treatment of CAH.
- It has been reported that unidentified steroid metabolites, which cross-react in direct assays, may occasionally be secreted in large amounts. A falsely high testosterone concentration will be achieved and any results from a direct assay that are unexpectedly above 1.7 ng/mL (6 nmol/L) should be confirmed by an extraction assay. In the UK, this assay is available through the Supraregional Assay Service.
- Testosterone exhibits diurnal variation, being highest in the early morning and falling by 25–30% to a minimum in the early evening. Normal males can have a testosterone concentration in the late afternoon at the bottom or just below the 9 a.m. reference range. Testosterone concentrations close to the lower limit of the reference range, in samples taken in the afternoon, should always be checked with a 9 a.m. sample.
- Testosterone levels may be elevated owing to alcohol abuse, stress, or hard physical exercise of short duration. It has been shown that extended exercise such as marathon running lowers testosterone concentrations.

Frequency of Use

Common.

SEX HORMONE-BINDING GLOBULIN

SHBG is a β-globulin, has a molecular mass of about 52 kDa, and is secreted by the liver. Secretion of this protein is stimulated by estrogens and suppressed by androgens. This results in a sex difference in plasma concentrations. In the circulation, SHBG binds several steroids. It has a high avidity for testosterone and DHT (about 1.5×10^9 mol/L) but a lower avidity for estradiol (5.0×10^8 mol/L). Nevertheless, even the binding with estradiol is considerably higher than the binding of these steroids to albumin ($3.7–6.4 \times 10^4$ mol/L).

An inverse correlation between SHBG and obesity and SHBG and insulin resistance has been demonstrated – situations which are found in patients with PCOS.

Measurement of SHBG and testosterone levels enables the concentration of biologically active testosterone to be estimated.

Function

There is still much debate about the precise role of SHBG. It has been suggested that SHBG dampens any large fluctuations in steroid levels, thus maintaining a fairly constant level of unbound steroid available to the tissues. Considering the marked episodic secretion of testosterone, this would seem a reasonable hypothesis. However, other experiments suggest that, if this is true, it is an oversimplification of the function of SHBG. Of greatest interest are the recent studies that have shown the presence of cell membrane receptors for SHBG, as well as evidence for internalization of the protein. Hence, the whole role of SHBG at the cellular level is yet to be elucidated.

Reference Interval

Men: 10–60 nmol/L
Women: 47–110 nmol/L
(Orion SHBG IRMA)

Clinical Application

The measurement of SHBG is used in the investigation of hirsutism. When combined with a testosterone test, the free testosterone index (or free androgen index) can be calculated (see below). This is particularly useful when the total testosterone concentration is normal but the SHBG is low, resulting in an elevated concentration of nonprotein-bound testosterone. An undetectable SHBG concentration has been reported but is very rare.

During treatment of hirsutism with the combined treatment of anti-androgen and estrogen, SHBG concentrations are usually above, often twice, the upper limit of the reference interval. Therefore, the measurement of SHBG levels can be used to confirm or refute compliance in patients who are not responding to therapy.

Limitations

- The concentration of SHBG is increased by estrogens (oral contraceptives, pregnancy) and decreased by androgens. It is also altered in a number of clinical conditions and by several drugs, e.g. anti-epileptics and barbiturates.
- SHBG levels are increased in hepatic cirrhosis, thyrotoxicosis, testicular feminization, and hypogonadism in the male.
- SHBG levels are decreased in myxedema, Cushing's syndrome, CAH, and acromegaly.
- Exogenous T_3 increases the secretion of SHBG, as do most of the anti-convulsant drugs. Dexamethasone is reported to cause a small increase in SHBG levels.

Assay Technology

Early methods were based on the binding of tritiated testosterone to SHBG after endogenous steroids had been removed from the serum with charcoal. Tritiated DHT soon replaced tritiated testosterone because its higher binding affinity for SHBG meant that removal of endogenous steroids was not required. Some laboratories still use a protein-binding method to measure SHBG routinely. With the lack of a reference preparation, these methods may also be used to assign values to standards used in IRMA methods. However, a possible reference preparation is being investigated at the National Institute of Biological Standards and Control, South Mimms, UK.

Another procedure uses Concanavalin A-Sepharose 4B to bind SHBG before addition of tritiated DHT, whereas Iqbal and Johnson used a column of Cibacron-blue Sepharose 4B layered on top of LH-20.

The above methods are used by a minority of laboratories. The most common method is now immunometric assay, which has better sensitivity and precision, and a simpler protocol. There are a number of commercial kits available, e.g. an immunoradiometric assay (IRMA) from Orion and a chemiluminescence-based method for the IMMULITE™ automated immunoassay analyzer from DPC.

Types of Sample

Serum is preferred for protein binding and chromatographic techniques. Serum or plasma may be used for IRMA assays.

Frequency of Use

Common.

FREE TESTOSTERONE

Only about 1% of the testosterone in the blood of women remains unbound to protein; in men, about 2% remains unbound. This 'free' fraction has been traditionally regarded as the biologically active portion. However, over the last few years this view has been challenged. Some workers suggest that both the free and albumin-bound fractions should be determined as the biologically important portions, whereas, more controversially, the SHBG-bound part has been suggested as the fraction available to tissues, *see* FURTHER READING. Nevertheless, there is little doubt that the nonprotein-bound steroid correlates well with the clinical state.

Free testosterone concentrations are often measured routinely in the investigation of hirsutism. An indication of free testosterone levels can be achieved from the ratio of testosterone and SHBG concentrations – the free testosterone or free androgen index. A closer approximation to the free testosterone concentration can be calculated (derived free testosterone) from the testosterone, SHBG and albumin concentrations and a variety of calculations have been published using some or all of these parameters.

Reference Interval

Women: 2.88–14.4 pg/mL (10–50 pmol/L)
Men: 38.6–243 pg/mL (134–844 pmol/L)
(Determined by micro-steady-state gel filtration in the author's laboratory)

Clinical Application

An estimate of the free testosterone concentration may influence the treatment of hirsutism in women who have a normal total testosterone level. Treatment may be more aggressive in women where an underlying androgen abnormality is revealed.

Limitations

In general terms, the levels of free and total testosterone are positively correlated with the degree of hirsutism and virilization in women. Nevertheless, there is considerable interindividual variation; some women have increased androgen concentrations but little or no hirsutism, whereas others have normal androgen results but quite marked hirsutism.

Clinicians disagree about the importance of a free testosterone measurement. Some clinicians are guided in their treatment by the result, although most argue that they treat the hirsutism, and whether androgen concentrations are normal or slightly elevated is irrelevant.

Assay Technology

The percentage of free testosterone in serum can be determined by equilibrium dialysis, steady-state gel filtration, ultracentrifugation, and microfiltration (*see* FURTHER READING). These methods are lengthy, tedious, and require good technical skill and experience. They are unsuited to routine clinical use.

An estimate of the free testosterone concentration may be made from the total testosterone and SHBG results. There is a good correlation between the percent free testosterone and the SHBG concentration (see Figure 50.2).

An equation describing this correlation can be used to calculate the percent free testosterone from the SHBG result. The free testosterone concentration can then be calculated from the total testosterone result. Some laboratories calculate a free testosterone index, or free androgen index, from the ratio of the total testosterone and SHBG results. Laboratories should establish their own reference intervals because of differences between methods.

It has been reported that albumin-bound testosterone diffuses readily from the circulation into tissues. It has

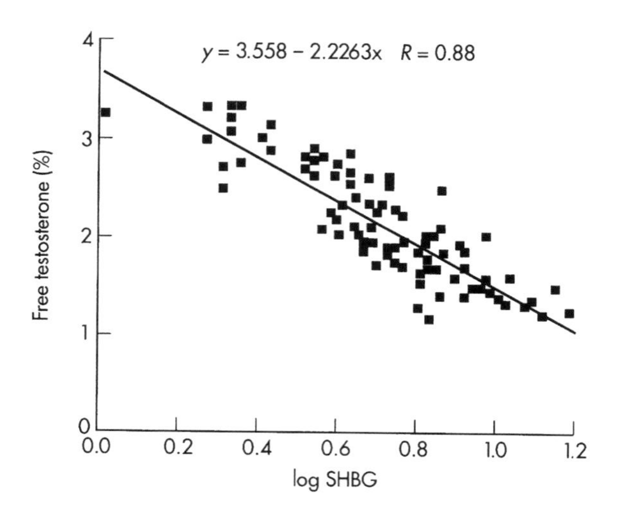

Fig. 50.2 Percentage of free testosterone against log SHBG. The correlation between the percentage of free testosterone, measured by steady-state gel filtration, and the log SHBG concentration measured by IRMA.

been proposed, therefore, that the free testosterone plus the albumin-bound testosterone (the non-SHBG bound testosterone) represent the fraction of testosterone available to target cells. The measurement of this fraction can be carried out using a simple ammonium precipitation of the SHBG-bound testosterone before immunoassay.

A number of mathematical models have been devised to calculate the non-SHBG-bound testosterone. Their complexity depends on the number of different parameters (e.g. SHBG, albumin, testosterone, cortisol-binding globulin) that are included in the calculation.

A radioimmunoassay kit is available from DPC. Studies that have examined the reliability of this kit have shown good correlations with total testosterone, free testosterone, and the androgen index, although values about half those obtained by equilibrium dialysis have been reported. It has been suggested that rather than measuring the free testosterone concentration, the method measures a constant proportion of the total testosterone (*see* FURTHER READING). Rosner (2001) has been particularly critical of this assay.

Vermeulen *et al.* (1999) evaluated the methods for estimating free and non-SHBG-bound testosterone, comparing results with equilibrium dialysis measurement of the free testosterone. They concluded that neither the free testosterone measured by the DPC kit nor the free androgen index was a reliable parameter of free testosterone concentration. The free testosterone calculated from the testosterone and SHBG measurement was rapid, simple and a reliable index of bioavailable testosterone. It was comparable to the results from equilibrium dialysis and suitable for clinical use, except in pregnancy. Calculated nonspecifically bound testosterone reliably reflected the measured non-SHBG-bound testosterone concentration.

Type of Sample

Serum.

Frequency of Use

Direct measurement by the more demanding methods is restricted to research studies but some laboratories use the DPC kit despite the many criticisms (see above). A derived free testosterone concentration by calculation or an androgen index is now common.

ANDROSTENEDIONE

Androstenedione is secreted by the testis, ovary, and adrenal. It is a weak androgen that is converted in peripheral tissue to the more potent testosterone. Therefore, increased secretion of androstenedione gives rise to hirsutism through this conversion (see Figure 50.3).

Reference Interval

Pre-pubertal children: <106 ng/dL (<3.7 nmol/L)
Women: 115–292 ng/dL (4.0–10.2 nmol/L)
Men: 126–304 ng/dL (4.4–10.6 nmol/L)
(Determined by extraction method with tritium label in the author's laboratory)

Clinical Applications

- Most requests for the determination of androstenedione levels are for the investigation of hirsutism. However, there is a disagreement among clinicians about the usefulness of this measurement. Our own investigations suggest that only 3% of hirsute women with PCOS have an increased androstenedione concentration alone (see Figure 50.4).
- Apart from salt loss, the major problem of CAH is virilization. Treatment of CAH should aim at maintaining androgen levels at physiological levels in order to minimize further virilization and to allow puberty to develop more normally. Although the measurement of 17-OHP, androstenedione, and total testosterone have all been used to monitor the treatment of patients with

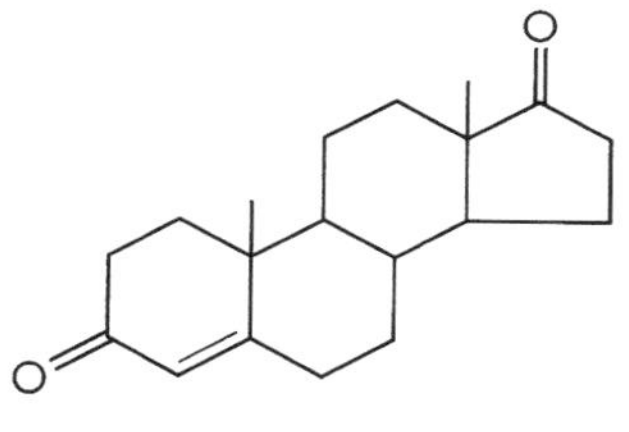

Fig. 50.3 Androstenedione.

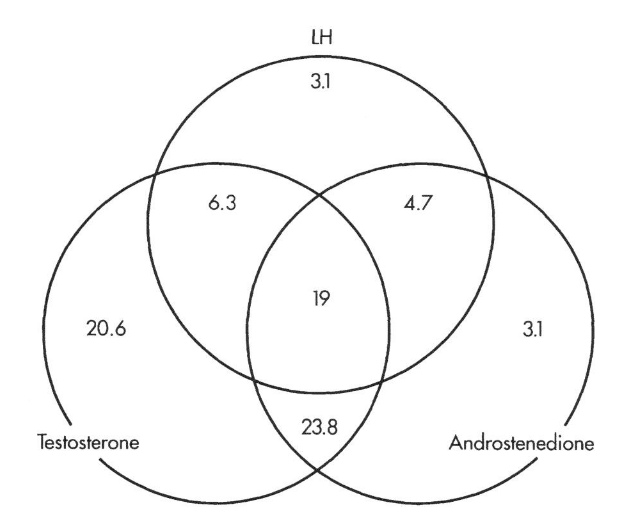

Fig. 50.4 A Venn diagram showing the percentage of patients with polycystic ovarian syndrome who had one or more abnormal androgen results. In only 3.1% of patients was androstenedione the only abnormal result, whereas 20.6% had a raised testosterone level. No hormonal abnormality, 14.2%; not every test done, 4.7%.

CAH, androstenedione seems the most appropriate. Its main source in CAH is from the adrenal, whereas testosterone is from peripheral conversion of androstenedione. It is well established that when androstenedione levels are suppressed to physiological levels, 17-OHP remains at levels up to five times the upper limit of its reference interval in men, and in women for the follicular phase. Thus, suppression of 17-OHP into the normal range leads to overtreatment with exogenous glucocorticoid.
- A deficiency of the 17β-hydroxysteroid dehydrogenase enzyme results in reduced testosterone synthesis by the testis. This is a rare condition associated with incomplete masculinization in the male. Nearly all affected neonates have been assigned female status. Theoretically, androstenedione secretion by the testes will be increased but, practically, a human chorionic gonadotropin (hCG) test (*see* INFERTILITY) is required to uncover the defect. An increased ratio of androstenedione to testosterone is diagnostic.

Limitations

- An androstenedione result provides little additional information in the investigation of hirsutism. Androstenedione is secreted by both the ovary and the adrenal, and the measurement of this steroid does not establish the source of increased androgen secretion.
- Concentrations in the neonate and the child are $<$0.9 ng/mL ($<$0.3 nmol/L), which is below the sensitivity of many assays. In addition, it is difficult to establish reference intervals for neonates and few laboratories have experience at accurately diagnosing 17β-hydroxysteroid dehydrogenase deficiency. Such investigations should be referred to an expert center.

Assay Technology

Many laboratories use RIA after solvent extraction of the steroid although commercial kits are now available. DPC manufactures a kit involving an extraction procedure, and a direct (nonextraction) method is available from DSL. There is good agreement between 'in-house' methods and commercial kits.

Types of Sample

Plasma or serum.

Frequency of Use

Moderate.

DEHYDROEPIANDROSTERONE SULFATE

Essentially all of the DHEAS in the circulation is derived from direct secretion by the adrenal glands. Because it is sulfated, it has a long half-life and hence lacks a circadian rhythm (see Figure 50.5).

Reference Interval

The reference interval changes with age, peaking just after puberty and gradually falling after the third decade of life (see Figure 50.6 and Table 50.1).

Clinical Application

About 10% of patients with PCOS have a concentration of DHEAS above the reference interval although some studies have reported up to 42% of patients. Because this suggests that the adrenal is the major source of increased androgen secretion, exogenous glucocorticoid may be used initially to treat such patients. Response to this treatment is variable.

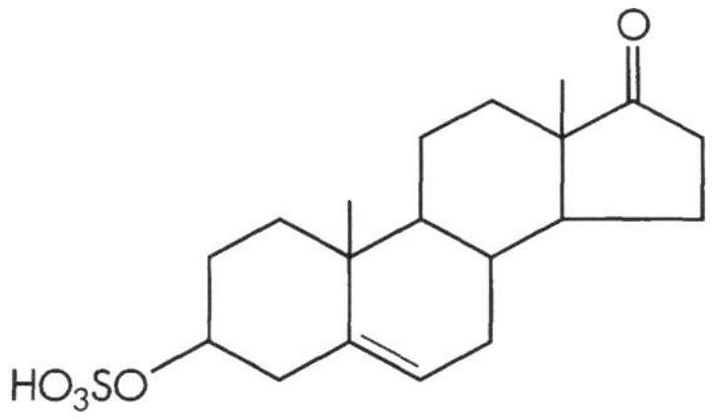

HO_3SO

Fig. 50.5 Dehydroepiandrosterone sulfate (DHEAS).

Fig. 50.6 Diagrammatic representation of the changes in DHEAS levels throughout life.

Patients with Cushing's syndrome due to an adrenal adenoma frequently have a DHEAS concentration below the reference interval.

Limitations

- Although DHEAS levels may be increased in hirsute patients, this does not exclude increased androgen secretion by the ovary. In fact, studies have shown that it is very rare to have increased androgen secretion from the adrenal alone.
- A normal DHEAS concentration in a patient with Cushing's syndrome does not exclude an adrenal adenoma.

Assay Technology

Because the concentration of DHEAS is 1000-fold greater than the other androgens, specimens are usually diluted by at least 100-fold before assay. Therefore, assays do not suffer from serum effects. Iodinated DHEAS is available and several simple nonextraction assays have been developed commercially.

It has been reported that some assays appear to cross-react with as yet unidentified metabolites, leading to spuriously high results. No specific clinical condition has been associated with these metabolites.

Coated tube RIA kits are available from DPC and DSL. Several companies have now developed methods for their automated immunoassay analyzers. These include DPC for the Immulite and Immulite 2000 and Nichols for the Advantage™.

Table 50.1. Reference intervals for DHEAS.

Approximate ranges for post-puberty	
15–30 years	0.7–4.5 μg/mL (1.7–11.5 μmol/L)
30–40 years	1.2–4.2 μg/mL (3.1–10.8 μmol/L)
40–50 years	0.8–4.0 μg/mL (2.0–10.2 μmol/L)
50–60 years	0.3–2.7 μg/mL (0.8–6.9 μmol/L)
>60 years	0.2–1.8 μg/mL (0.4–4.7 μmol/L)

The graph and ranges are based on data from the author's laboratory.
The conversion factor used (2.56) is based on the use of the sodium salt, which is used as a standard.

Type of Sample

Serum.

Frequency of Use

Infrequent.

17α-HYDROXYPROGESTERONE

Gonadal and adrenal tissues are able to synthesize 17-OHP (see Figure 50.7) from progesterone. 17-OHP is readily converted to 11-deoxycortisol in the adrenal and to androstenedione in the gonads and adrenal. Concentrations of the steroid rise in the late follicular phase and peak at the same time as estradiol. A second increase occurs in the luteal phase that is similar to progesterone. Thus, both the follicle and the corpus luteum appear to secrete 17-OHP during the menstrual cycle.

Reference Interval

Follicular phase: 0.55–1.84 ng/mL (1.7–5.7 nmol/L)
Luteal phase: 0.55–6.31 ng/mL (1.7–19.6 nmol/L)

Clinical Application

The measurement of serum 17-OHP is used for the diagnosis of CAH caused by a deficiency of the 21-hydroxylase enzyme. Affected infants have a concentration greater than 9.7 ng/mL (30 nmol/L). Heterozygote individuals also have increased 17-OHP levels compared

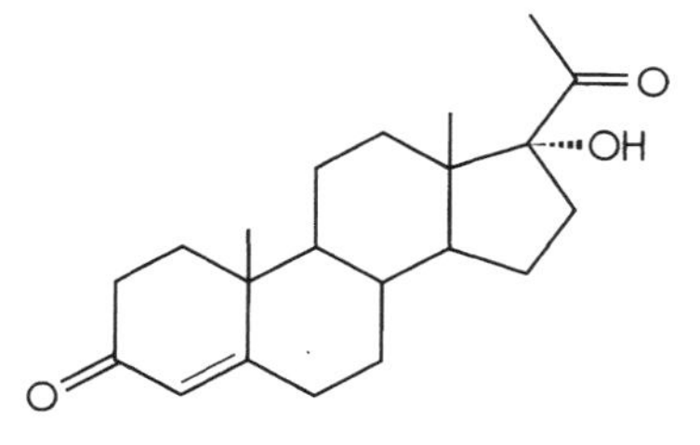

Fig. 50.7 17α-Hydroxyprogesterone.

with the normal population. Identification of heterozygotes is important in genetic counseling.

Patients with LOCAH usually have 17-OHP concentrations above the reference interval for the follicular phase of the menstrual cycle. A synacthen test will identify those patients whose results are equivocal. In affected patients there is an exaggerated increase in the secretion of 17-OHP (more than three times the basal level).

Limitations

- 17-OHP is mildly elevated in those cases of CAH that are due to an 11β-hydroxylase deficiency. This enzyme abnormality should always be considered when hypertension is present and the concentration of 17-OHP is only moderately increased.
- Interpretation of results is complicated in the first 24 h of life because of maternal steroids still present in the neonate's circulation.
- 17-OHP should not be used to monitor the effectiveness of treatment for CAH. If 17-OHP is suppressed into the normal range, it is likely that the patient is being overtreated. Androstenedione is the most appropriate analyte for monitoring treatment for CAH.

Assay Technology

'In-house' extraction assays are common. DPC manufactures a direct kit that has a good correlation with the extraction assay for 17-OHP concentrations in the adult. However, when using this kit, serum from neonates and children less than 6 months old should be extracted before the measurement of 17-OHP concentrations due to the cross-reaction with other steroids present in the serum.

Cross-reaction with other steroids can confuse the interpretation of results when using immunoassay methods. In the UK, a GC–MS is available through the Supra Regional Assay Laboratory at the University College London Hospitals.

The investigation of CAH is usually made in neonates and prepubertal children where the collection of blood in sufficient quantity is difficult. Methods have been developed to measure 17-OHP in saliva (*see* FURTHER READING) and from blood spots.

Types of Sample

Serum, plasma, saliva or blood spot.

Frequency of Use

Moderate.

DIHYDROTESTOSTERONE

DHT is formed in peripheral tissue from testosterone by the enzyme 5α-reductase. Its biological activity, relative to testosterone, depends on the bioassay and animal species used. However, it is generally regarded as the more potent androgen with a higher binding affinity for SHBG. Testosterone is converted to DHT in the testes, skin, brain, salivary glands, lung, heart and pectoral muscle.

Reference Interval

Women: 377–725 pg/mL (1.3–2.5 nmol/L)
Men: 116–435 pg/mL (0.4–1.5 nmol/L)
(Derived by an RIA method after HPLC separation of DHT from the other androgens)

Clinical Application

Some studies suggested that the measurement of DHT might be helpful in the investigation of hirsutism. However, because DHT is produced in many tissues, it is not a sensitive indicator of DHT production in the skin.

The concentration of DHT is low in both sexes and it is difficult for most assays to achieve the required sensitivity. This is because it is difficult to avoid losses during the separation of DHT from testosterone and other androgens.

Assay Technology

Antisera raised against DHT conjugates have a high cross-reaction with testosterone. Because attempts to produce a specific monoclonal antibody have been unsuccessful, testosterone must be removed from patient sera before DHT can be measured.

DHT has usually been separated from testosterone by a chromatographic procedure after an initial solvent extraction. These methods have the advantage of separating out all the androgens, which can then be measured by RIA. However, these procedures are usually limited to research studies.

A slightly simpler method uses potassium permanganate to oxidize the testosterone. Several companies now sell commercial kits. These include DSL, Research Diagnostics Inc., New Jersey, USA, and IBL-Hamburg, Germany. The ELISA from IBL-Hamburg is a simple method with no oxidation or extraction steps. To achieve good precision and reliability in the extraction methods, a high degree of experience and skill is required.

Types of Sample

Serum or plasma.

Frequency of Use

Uncommon.

ANDROSTANEDIOLS AND THEIR GLUCURONIDES

DHT is rapidly converted to a variety of other compounds that include androsterone and the androstanediols, 5α-androstane-3α, 17β-diol (3αdiol) and 5α-androstane-3β,17β-diol (3β-diol), by a number of tissues. These include the accessory sex glands, skin, brain, salivary glands, and heart muscle. These steroids are further metabolized to their glucuronides and sulfates. All these steroids have been investigated as markers of hirsutism. The most abundant of these steroid metabolites is androsterone sulfate. However, Zwicker and Rittmaster (1993) showed that this steroid was not suitable as a marker of hirsutism. It correlated poorly with other androgens and concentrations were not elevated in hirsute women. It was shown that the steroid is derived almost entirely from the adrenal glands. Little or no 3β-diol is formed in the human and so studies have focused on 3α-diol and its glucuronide (3α-diol G). The latter has been measured in both urine and serum although recent studies have measured 3α-diol G in serum. Many of these studies have reported that 3α-diol and 3α-diol G have a role in the investigation of hirsutism.

Reference Interval

3α-Androstanediol glucuronide

Adult males: 3.4–22.0 ng/mL
Adult females
 Premenopausal: 0.5–5.4 ng/mL
 Postmenopausal: 0.1–6.0 ng/mL
 Hirsute: 1.3–9.4 ng/mL

(ARUP Laboratories RIA)

Clinical Application

Studies have shown that DHT levels are infrequently increased in idiopathic hirsutism. The 3α-diol was investigated as a better indicator of increased androgen activity in peripheral tissue as it is the end metabolite of androgen metabolism in the skin. Findings have been variable. One group of investigators reported increased 3α-diol concentrations in 93% of their patients with idiopathic hirsutism, whereas other workers have found normal results.

Interest switched to 3α-diol glucuronide (3α-diol G) when it was reported that this metabolite is predominantly formed in sexual tissue and skin. Early studies showed that 3α-diol G concentration was greatly increased in hirsute women whereas testosterone was only modestly increased. It was later shown that methodological deficiencies in the early method caused erroneously elevated results. More recent studies have shown that 3α-diol G is only modestly increased in hirsutism. In a detailed review Rittmaster (1991) examined studies that had looked at the origin and clinical use of 3α-diol and 3α-diol G. He concluded that the primary source of precursors for androgen conjugates appeared to be the adrenal glands and that the liver may be the major source of glucuronide conjugation. Therefore, he concluded that the clinical utility of 3α-diol G in the investigation of hirsutism was limited and androgen-dependent hirsutism is best assessed by examining the distribution of body hair. Nevertheless, there continue to be many studies of PCOS and hirsutism where 3α-diol and 3α-diol G have been measured. Chen *et al.* (2002) discuss these studies and their varied outcomes in a more recent review.

Limitations

Results from different studies have been inconsistent, casting doubt on the usefulness of these assays. They probably add little to the information achieved from the measurement of other androgens. Azziz *et al.* (2000) concluded that the routine measurement of serum 3α-diol G is not recommended in the evaluation of idiopathic hirsutism or in other hirsute patients. The assays are not routinely available.

Assay Technology

Commercial assays are available from DSL Inc., Webster, TX 77598, USA and from ARUP Laboratories, Salt Lake City, UT 84108-1221, USA.

Type of Sample

Serum.

Frequency of Use

Rare.

FURTHER READING

Azziz, R., Black, V.Y., Knochenhauer, G.A., Hines, G.A. and Boots, L.R. Ovulation after glucocorticoid suppression of adrenal androgens in polycystic ovary syndrome is not predicted by the basal dehydroepiandrosterone sulfate level. *J. Clin. Endocrinol. Metab.* **84**, 946–950 (1999).

Balen, A.H., Conway, G.S., Kaltsas, G., Techatraisak, K., Manning, P.J., West, C. and Jacobs, H.S. Polycystic ovary syndrome: the spectrum of the disorder in 1,741 patients. *Hum. Reprod.* **10**, 2107–2111 (1995).

Chen, W., Thiboutot, D. and Zouboulis, C.C. Cutaneous androgen metabolism: basic research and clinical perspectives. *J. Invest. Dermatol.* **119**, 992–1007 (2002).

Dunaif, A. (ed) Polycystic ovary syndrome. *Endocrinol. Metab. Clin. N. Am.* **28**, (1999).

Ekins, R. Measurement of free hormones in blood. *Endocr. Rev.* **11**, 5–46 (1990).

Fears, T.R., Ziegler, R.G., Donaldson, J.L., Falk, R.T., Hoover, R.N., Stanczyk, F.Z., Vaught, J.B. and Gail, M.H. Reproducibility studies and interlaboratory concordance for androgen assays in female plasma. *Cancer Epidemiol. Biomarkers Prev.* **9**, 403–412 (2000).

Franks, S. Polycystic ovary syndrome: a changing perspective. *Clin. Endocrinol.* **31**, 87–120 (1989).

Goudas, V.T. and Dumesic, D.A. Polycystic ovary syndrome. *Endocrinol. Metab. Clin. N. Am.* **26**, 893–912 (1997).

Gower, D.B. Extraction, purification and estimation of the androgens and their derivatives. In: *Steroid Analysis* (els Makin, H.L.J., Gower, D.B. and Kirk, D.N.), (Blackie Academic and Professional, London, 1995).

Harbourne, L., Fleming, R., Lyall, H., Sattar, N. and Norman, J. Metformin or antiandrogen in the treatment of hirsutism in polycystic ovary syndrome. *J. Clin. Endocrinol. Metab.* **88**, 4116–4123 (2003).

Horton, R. and Loban, R.A. (eds), Androgen metabolism in hirsute and normal females, *Clinics in Endocrinology and Metabolism*, vol. 15 (W.B. Saunders, London, 1986).

Ibanez, L., Potau, N. and Carrascosa, A. Insulin resistance, premature adrenarche, and a risk factor of the polycystic ovary syndrome (PCOS). *TEM* **9**, 72–77 (1998).

Jacobs, H.S. (ed) Polycystic ovary syndrome. In: *Bailliere's Clinical Endocrinology and Metabolism*, vol. 10 (2). (Bailliere Tindall, London, 1996).

Jeffcoate, W. The treatment of women with hirsutism. *Clin. Endocrinol.* **39**, 143–150 (1993).

New, M.L. and Speiser, P.W. Genetics of adrenal steroid 21-hydroxylase deficiency. *Endocr. Rev.* **7**, 331–349 (1986).

Pucci, E. and Petraglia, F. Treatment of androgen excess in females: yesterday, today and tomorrow. *Gynaecol. Endocrinol.* **11**, 411–433 (1997).

Riad-Fahmy, D., Read, G.F., Walker, R.F. and Griffiths, K. Steroids in saliva for assessing endocrine function. *Endocr. Rev.* **3**, 367–395 (1982).

Rittmaster, R.S. Androgen conjugates: physiology and clinical significance. *Endocr. Rev.* **14**, 121–132 (1993).

Rittmaster, R.S. Medical treatment of androgen-dependent hirsutism: clinical review. *J. Clin. Endocrinol. Metab.* **80**, 2559–2563 (1995).

Rosner, W. An extraordinary inaccurate assay for free testosterone is still with us (Letter). *J. Clin. Endocrinol. Metab.* **86**, 2903 (2001).

Speiser, P.W. and White, P.C. Congenital adrenal hyperplasia due to steroid 21-hydroxylase deficiency. *Clin. Endocrinol.* **48**, 411–417 (1998).

Stewart, P.M., Penn, R., Holder, R., Parton, A., Ratcliffe, J. and London, D.R. The hypothalamo–pituitary–adrenal axis across the normal menstrual cycle and in polycystic ovary syndrome. *Clin. Endocrinol.* **38**, 387–391 (1993).

Vermeulen, A., Verdonck, L. and Kaufman, J.M. A critical evaluation of simple methods for the estimation of free testosterone in serum. *J. Clin. Endocrinol. Metab.* **84**, 3666–3672 (1999).

Wallace, A.M. Analytical support for the detection and treatment of congenital adrenal hyperplasia. *Ann. Clin. Biochem.* **32**, 9–27 (1995).

Wheeler, M.J. The determination of bio-available testosterone. *Ann. Clin. Biochem.* **32**, 345–357 (1995).

Winters, S.J., Kelley, D.E. and Goodpaster, B. The analog free testosterone assay. Are the results useful?. *Clin. Chem.* **44**, 2178–2182 (1998).

Zwicker, H. and Rittmaster, R.S. Androsterone sulphate: physiology and clinical significance in hirsute women. *J. Clin. Endocrinol. Metab.* **76**, 112–116 (1993).

51 Pregnancy

Tim Chard
(1937–2003)

This chapter is dedicated to the memory of Tim Chard who died on 20th July 2003. He was one of the first authors of a textbook on immunoassay: An Introduction to Radioimmunoassay and related techniques (Elsevier), now into its 5th edition. It was translated, among other languages, into Russian and Japanese. He will be sadly missed.

From the earliest stages of gestation the human fetus and placenta synthesize a number of compounds that are considered to be qualitatively and quantitatively specific to pregnancy. Measurement of these compounds is widely used in the diagnosis of various pregnancy disorders. In addition, maternal tissues such as the uterine epithelium (endometrium) and ovary (corpus luteum) secrete materials that may be of diagnostic value (see Table 51.1).

Many of these materials circulate at remarkably high levels. For example, the serum concentration of estrogens near term is 100 times greater than the highest level found in a woman who is not pregnant. Still more surprisingly, the functions of many of these materials are unknown and may even be nonexistent. Rare pregnancies in which estriol is almost totally absent from maternal blood are normal in every other respect.

As with functions, there is a similar dearth of information on control mechanisms. The type of feedback relationship that exists in the physiology of the normal adult is not found with the placental hormones or similar compounds. The circulating levels of most of the compounds listed in Table 51.1 show no diurnal rhythm or change with physiological events such as sleep, exercise, or meals. The rate of synthesis of placental products seems to relate solely to the mass of the tissue of origin (the placental trophoblast) and to the rate of uteroplacental blood flow. For most clinical purposes this is, of course, advantageous; a single measurement of a fetoplacental product can be taken

Table 51.1. Fetal, placental, and maternal products that are found at relatively high levels in a normal pregnancy.

Fetal	Placental	Maternal
Alphafetoprotein (AFP)	Heat-stable alkaline phosphatase (HSAP)	Estrogen*
Oxytocin	Cystine aminopeptidase (CAP; oxytocinase)	Progesterone*
	Estrogens	Relaxin*
	Progesterone	Oxytocin
	Human chorionic gonadotropin (hCG)	IGFBP-1**
	Human placental lactogen (hPL)	PP14**
	Placental growth hormone (pGH)	
	Schwangerschaftsprotein 1 (SP1)	
	Releasing hormones	
	Placental protein 5 (PP5)	
	Pregnancy-associated plasma protein-A (PAPP-A)	

*From the ovary.

**From the endometrium/decidua.

IGFBP-1, Insulin-like growth factor binding protein 1; PP14, placental protein 14.

without concern for the physiological status of the mother.

One special feature of the measurement of fetoplacental and related products is the noticeable change in levels in relation to the stage of pregnancy. For this reason it is not possible to quote a single reference range for a given analyte; instead, a separate reference range has to be used for each week of pregnancy. This has two important implications. First, large numbers of samples are required to establish a normal reference range. Second, it is almost impossible for the clinician or analyst to recall the range for each week without some sort of prompt. As a partial solution of this problem it is now common to quote levels as '**multiples of the median**' (**MOM**). Thus, when the action line for alphafetoprotein (AFP) is stated as being '2.5 times the median' or '2.5 MOM' this will be the same regardless of the week of pregnancy. The median will, of course, change but the action line expressed as a MOM does not.

The relation between analyte levels and gestational age places a heavy demand on the accuracy of gestational dating. A given AFP value could be normal for 16 weeks but above the reference interval 1 week earlier. In clinical practice, dating errors of up to 4 weeks or more are relatively common. Dating is usually based on the first day of the last menstrual period (LMP). This can be confirmed by ultrasound measurements of the fetus (crown-rump length (CRL) before 13 weeks of pregnancy; biparietal diameter (BPD) thereafter). If the LMP is in doubt, or there is a gross discrepancy with ultrasound then the ultrasound date is accepted as the 'gold standard'.

The pattern of change of maternal circulating levels of fetoplacental products varies between materials. By way of simplification, the types of pattern are summarized in Table 51.2.

Table 51.2. The pattern of maternal levels of fetoplacental products during pregnancy. In all cases there is a progressive rise from 4 to 8 weeks after the last menstrual period to a peak or plateau at the times shown.

Pattern	Analytes
Peak at 8–10 weeks	hCG, PP14, relaxin
Plateau at 12–16 weeks	IGFBP-1, binding proteins (e.g. SHBG)
Peak at 32 weeks	AFP
Plateau at 36 weeks*	hPL, progesterone, estriol, SP1, PAPP-A, PP5

*The general pattern of these analytes throughout pregnancy is a sigmoid curve.

One unusual feature of specific biochemical tests in pregnancy is that several different analytes may be used for the same clinical purpose. For example, all of the placental materials shown in Table 51.1 were at various times advocated for the diagnosis of placental insufficiency and all probably gave very similar information. Custom rather than any very fundamental advantage usually dictated which test was most widely used.

CLINICAL DISORDERS

DETECTION OF EARLY PREGNANCY

Though not strictly a 'disorder', detection of early gestation by measurement of hCG is probably the commonest immunoassay test in pregnancy. The indication is almost always maternal concern (which may reflect a positive or a negative frame of mind). In a woman without symptoms there are very few medical indications for a pregnancy test. However, it is by far the commonest and most popular of all 'home-use' tests, a marketplace supplied by highly convenient dipstick procedures.

THREATENED ABORTION

Threatened abortion can occur at any time during the first 24 weeks of pregnancy. The woman presents with clinical features of an abortion process (pain and bleeding) but the outcome is uncertain. Many cases of this type will proceed normally to term. There is no specific treatment for threatened abortion but if the fetus is dead the contents of the uterus are often surgically evacuated (dilatation and curettage).

ECTOPIC PREGNANCY

Fertilization takes place at the outer end of the Fallopian tube. The fertilized ovum then passes along the tube and implants in the uterus 6–7 days later. If the process is interrupted then implantation takes place in the tube itself. This is known as **ectopic pregnancy**. It is often followed by rupture of the tube and hemorrhage into the abdomen, a life-threatening emergency that may become apparent at any time in the first 8 weeks of pregnancy. The treatment is laparotomy or laparoscopy and removal of the ectopic pregnancy.

CHROMOSOME DEFECTS OF THE FETUS

The best-known defect is **Down's syndrome** or **trisomy 21** (three copies of chromosome 21 in place of the usual two copies). The child is born with the appearance that used to be described as a 'mongol'. Associated defects are common, especially congenital heart disease. Most children with Down's syndrome are mentally retarded, some severely so. Advances in medical treatment have increased the median life expectancy of Down's syndrome individuals to about 50 years, although with a significant risk of Alzheimer's disease in later life. Prenatal detection and the offer of termination of the pregnancy is the only way to reduce the incidence of this condition. The condition occurs in 1 in 600 pregnancies and is common among older mothers (1 in 385 at age 35; 1 in 110 at age 40). Screening for Down's syndrome is almost universal in obstetric practice. Originally, this was based solely on maternal age (35 or over). In recent years various biochemical tests have been introduced (see below) and in the very recent past ultrasound tests have been described, including measurement of nuchal translucency. The latter, together with some biochemical tests, enables detection in the late first and early second trimester (10–13 weeks of pregnancy). Those identified as being at risk by one or more of the screening tests are offered amniocentesis (collection of amniotic fluid through a needle), usually done at 16–18 weeks. The fetal skin cells in the amniotic fluid can be examined by standard cytogenetic techniques to make a firm diagnosis. However, amniocentesis should not be carried out in all women because the procedure itself carries a risk (now approximately 1 in 200) of causing an abortion.

There are numerous chromosome defects apart from trisomy 21. Some affect the sex chromosomes (**Turner's syndrome**, XO; **Klinefelter's syndrome**, XXY); these are relatively benign conditions which are not usually the subject of targeted diagnosis. Others affect the autosomes (**trisomy 13**, **trisomy 18**); these are occasionally found as a coincident result of Down's syndrome (see below).

NEURAL TUBE DEFECTS OF THE FETUS

The term **neural tube defect** (NTD) is used to describe various degrees of failure of formation of the central nervous system (brain and spinal cord) and its coverings. There are two main types: **anencephaly** and **spina bifida**. In anencephaly most of the brain is absent, together with the cranial bones. This is incompatible with extrauterine life and the child dies either before or immediately after delivery. In spina bifida there are defects of the lower spinal nerve roots and their coverings; spinal membranes bulge through the defect as a myelomeningocele. Around one-third of these children survive, often helped by surgery. However, the quality of life is poor: they are usually paralyzed from the waist down (paraplegia) and incontinent. After detection of NTDs, most women choose to have the pregnancy terminated before 24 weeks.

Screening for NTDs has been available since the early 1970s, based on measurement of maternal serum AFP. More recently, detection and diagnosis by ultrasound have become the norm.

Women who have had a previous child with a NTD have a very high recurrence risk (about 1 in 30). This risk is greatly reduced by folic acid supplements.

PREMATURE LABOR

Labor occurring after 24 and before 37 weeks is referred to as **premature labor**. The child is small (typically less than 2500 g) and at risk of potentially lethal complications such as the respiratory distress syndrome. In the longer term they may suffer permanent brain damage (mental retardation and cerebral palsy). Premature labor is treated (often unsuccessfully) with drugs that inhibit uterine contractions. If the child is delivered it may need to be transferred to a neonatal intensive care unit. Despite some promising leads, there is no predictive test for premature labor, biochemical or otherwise.

PLACENTAL INSUFFICIENCY

Placental insufficiency is a condition of usually ill-defined cause and pathology in which there is partial failure of placental transfer (nutrients to the fetus and waste-product removal). This can lead to fetal growth retardation, fetal distress or fetal death. The failure is also reflected by reduced secretion of the specific products of the placenta (see Table 51.1), the so-called 'placental function tests'.

Fetal growth retardation is the situation in which, as a result of placental insufficiency, the child is born at less than its expected weight. Typically, a neonate is defined as 'growth retarded' or 'small for dates' if birthweight is less than the 10th centile (some units use the 5th centile) for the relevant week of gestation. The child may display features of poor nutrition (increased ratio of head to abdominal circumference) and there is an association with hypoxia during delivery and also long-term brain damage. The diagnosis is based on clinical and ultrasound examination. At one time biochemical tests were also used (see below) but these have been largely replaced by the biophysical procedures. The treatment, once the condition has been identified, is early delivery (induction of labor or, if not possible, by Cesarean section).

Fetal death *in utero* in the last 12 weeks of pregnancy is referred to as **stillbirth**. The cause is sometimes placental insufficiency but in many cases no specific reason is found. Identification and management of the risk are the same as that of growth retardation.

Fetal distress: as a result of placental insufficiency, or sometimes because of mechanical problems in the delivery process, the fetus may become hypoxic during labor. Clinical signs include slowing of the fetal heart and passage of feces by the fetus (meconium). Death or brain damage may result if the child is not delivered rapidly.

PRE-ECLAMPSIA

In the condition known as **pre-eclampsia**, the mother has hypertension and/or proteinuria. Very rarely these signs precede the convulsive state known as **eclampsia**, a major medical emergency. In most cases the concern is the association with placental insufficiency and its potential consequences for the fetus: growth retardation, hypoxia, brain damage and death. There is no specific biochemical test for pre-eclampsia, apart from those for placental insufficiency. Treatment, in severe cases, includes reduction of blood pressure and early delivery.

MISCELLANEOUS DISORDERS

In **placental abruption**, part of the placenta separates with the formation of a large clot between the placenta and maternal tissues. This is sometimes associated with pre-eclampsia. It can lead to severe external bleeding, to a hypercoagulable state, to premature labor and to placental insufficiency (see above). If severe, the treatment is urgent delivery.

Uncontrolled **maternal diabetes** leads to overgrowth of both fetus and placenta, with elevated levels of fetoplacental products. If untreated, fetal death is likely in the last 4–6 weeks of pregnancy. Some cases of diabetes appear only during pregnancy (**gestational diabetes**). Today, with careful control of the mother's metabolic state, the diabetic pregnancy is virtually normal from every biological and clinical standpoint.

Rhesus (Rh) disease. The Rh-positive fetus can sensitize the Rh-negative mother. In subsequent pregnancies, maternal antibodies cross the placenta causing a hemolytic anemia in the fetus. If severe this can lead to stillbirth of an edematous fetus ('fetal hydrops'). If less severe the fetus survives but the newborn may develop severe jaundice and brain damage. The condition is associated with placental hypertrophy and elevated levels of fetoplacental products. Treatment is by exchange transfusion (either *in utero* or after birth) and early delivery.

ANALYTES

ALPHAFETOPROTEIN

AFP is composed of a single polypeptide chain with a relative molecular mass of 69,000. It consists of 590 amino acids and 39% of the amino acid sequence corresponds to that of albumin. Unlike albumin it is a glycoprotein with 4% of carbohydrate residues. In amniotic fluid there are two types of AFP, differing in the structure of their carbohydrate chains and therefore in their ability to bind to concanavalin A.

There are low levels of AFP in normal subjects. In the fetus, AFP is synthesized by the liver, the gastrointestinal tract, and the yolk sac. The concentration of AFP in fetal serum rises rapidly to reach a peak at 12–14 weeks' gestation. Thereafter it falls until term; the fall continues in the newborn with a half-life of 3–4 days during the first weeks of life, eventually reaching adult levels at 8 months. In amniotic fluid there is a steady fall in AFP levels from a peak in the first trimester; the source is complex and includes maternal blood, fetal urine and the yolk sac. In the mother, circulating AFP levels rise progressively to reach a peak at 32 weeks, then decrease towards term.

There is no information on mechanisms that may control the synthesis of AFP by the fetus and thus determine the concentrations in maternal and fetal fluids. The only physiological factor that influences AFP levels is fetal sex, the levels in umbilical blood at term being substantially higher if the child is a boy. The functions of AFP are also unknown. Probably, the most convincing suggestion is that it serves a function similar to that of albumin. Other possibilities include estrogen binding, though the activity of human AFP in this respect is far less than that of the more specific sex hormone-binding globulin. It has also been suggested that AFP has immunosuppressive activity and thus plays a role in the maternal–fetal graft relationship (see Tables 51.3 and 51.4).

Reference Interval

The ranges of AFP levels in maternal serum and amniotic fluid at 15–20 weeks' gestation are shown as originally determined with the Ortho-Clinical Diagnostics Amerlex-M AFP kit (no longer available). Ranges have been established for other weeks of pregnancy but have little or no clinical application (see Figures 51.1 and 51.2).

Table 51.3. Median values of maternal serum AFP in normal pregnancies.

Gestation (weeks)	Median AFP (IU/mL)	95th centile AFP (IU/mL)
15	29.8	56.0
16	32.2	60.6
17	37.1	69.8
18	42.3	79.6
19	51.0	96.0
20	60.5	114.0

Table 51.4. Median values of amniotic fluid AFP in normal pregnancies.

Gestation (weeks)	Median AFP (kIU/mL)	99th centile AFP (kIU/mL)
15	11.8	30.2
16	10.5	26.8
17	9.31	23.7
18	8.25	21.0
19	7.36	18.8
20	6.51	16.6

Clinical Applications

Neural tube defects. As a result of the defect in the fetal body surface, fetal AFP leaks across exposed capillaries into amniotic fluid, and from there across the membranes into the maternal circulation. Measurement of AFP in maternal serum (MSAFP), therefore, provides a screening test for NTD. The test is most effective when applied at 16, 17, and 18 weeks; the predictive value is less before 15 weeks and after 20 weeks. A positive result of the screening test (defined as more than two MOMs in some units and 2.5 MOMs in others) leads to a series of follow-up investigations. The end result of the whole process is detection and termination of 80–90% of spina bifida and 90–100% of anencephaly.

There is considerable overlap in MSAFP levels between normal and NTD pregnancies. Other causes of elevated AFP are listed in Table 51.5.

A woman with an elevated MSAFP level is subjected to a careful ultrasound examination which can identify most of the alternative diagnoses shown, including many cases of NTD. Ultrasound can miss some cases of spina bifida, especially low lumbo-sacral defects. If nothing is seen in a woman with a raised MSAFP level, then amniocentesis should be carried out. If amniotic fluid AFP is less than two MOMs, no further action is taken. If it is greater than two MOMs, acetylcholinesterase should be measured. Positive findings on ultrasound or raised levels of both amniotic fluid AFP and acetylcholinesterase, are an indication for termination of the pregnancy (subject, of course, to full counseling of the woman).

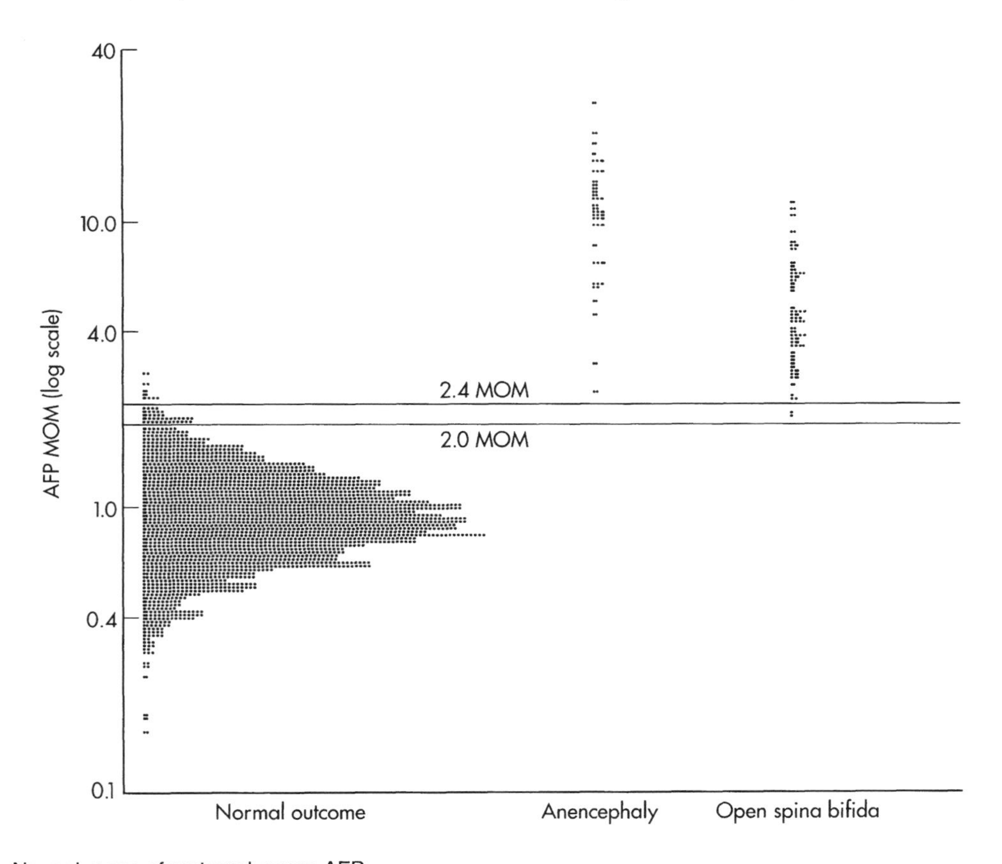

Fig. 51.1 Normal range of maternal serum AFP.

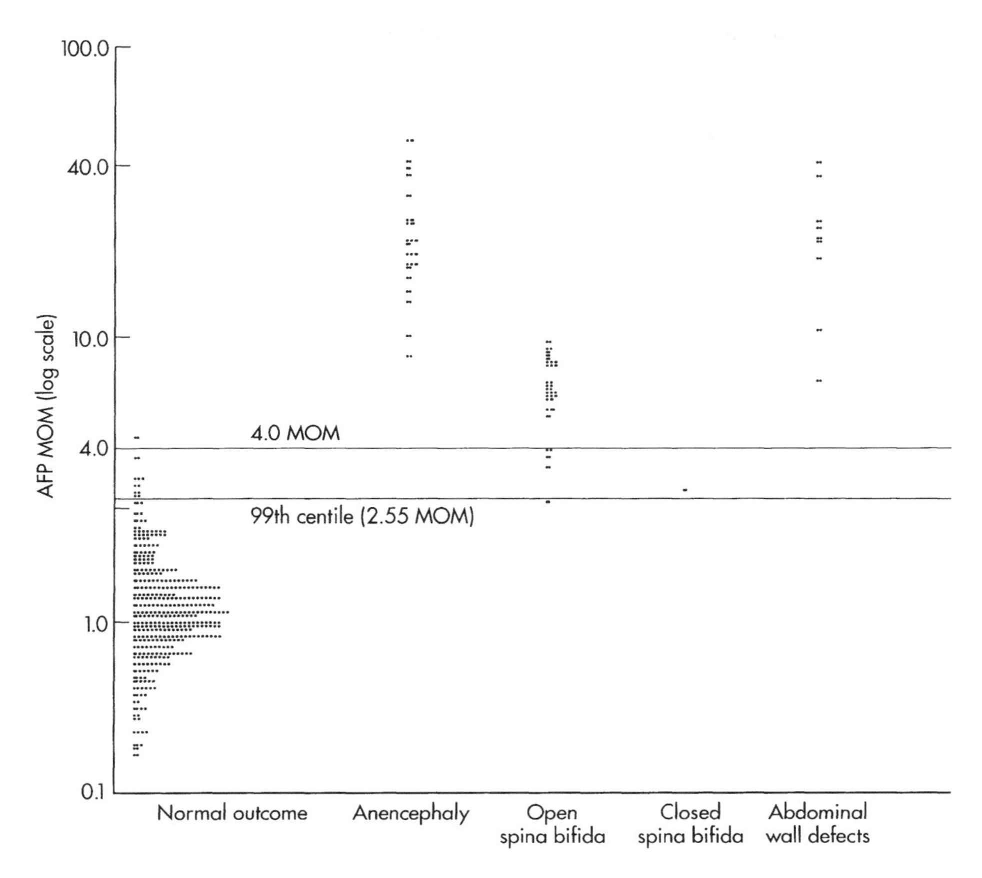

Fig. 51.2 Normal range of amniotic fluid AFP.

Down's syndrome. The use of low levels of MSAFP is discussed later in the chapter (*see* SCREENING FOR DOWN'S SYNDROME).

Placental insufficiency. In some cases with elevated MSAFP levels but no NTD there is evidence of increased fetal risk, especially low birth weight, later in the pregnancy. A woman with raised MSAFP levels should always be regarded as 'at risk'.

Assay Technology

A wide range of immunoassay techniques is available for AFP, including competitive and immunometric assays with a variety of signal generation systems. Apart from factors common to all immunoassays there are no very special demands on an assay for AFP. Specificity is not an issue in practice because there are no known interfering compounds. In amniotic fluid there are variant forms of AFP differing in the extent of glycosylation. These forms can be separated by lectin-affinity chromatography. Although some of these forms are considered to be characteristic of NTD, this detailed analysis is rarely done.

Extreme sensitivity is not required; all current assays are capable of measuring the levels present in pregnancy. Two features of the assay that should be considered are:

- wide operating range: this should embrace both the low normal levels found in Down's syndrome and the high

Table 51.5. Differential diagnosis of raised MSAFP levels.

1. Normal pregnancy (upper end of normal range)
2. NTDs
3. Other congenital abnormalities (notably exomphalos)
4. Multiple pregnancy
5. Fetal death (due to the associated fetomaternal hemorrhage)
6. Underestimation of stage of gestation
7. Tumors (hepatoma, gonadal teratomas)

normal levels associated with NTD. The need for precision over a wide range gives a theoretical advantage to labeled antibody (immunometric) assays, though this advantage has yet to be confirmed in clinical practice;

- convenient automation: AFP is a screening test and samples are usually assayed in large batches. Some degree of automation is usually essential.

Types of Sample

Serum, plasma or amniotic fluid. As with most immunoassays, serum is generally preferred to plasma because it is less liable to interference from anticoagulants, precipitates formed on storage, etc.

Frequency of Use

Common.

HUMAN CHORIONIC GONADOTROPIN

Human chorionic gonadotropin (hCG) is one of a family of glycoprotein hormones, the other members being luteinizing hormone (LH), follicle stimulating hormone (FSH), and thyrotropin (TSH). Each of these consists of two subunits: an α subunit (92 amino acids) which is virtually identical in all four; and a β subunit which is characteristic of the individual hormone. The β subunit of hCG is a single chain of 145 amino acids. The first 121 *N*-terminal amino acids share 80% sequence homology with β-LH; the *C* terminus of β-hCG has a 24 amino acid extension not present in β-LH. Both subunits of the molecule are needed for biological activity but the β subunit determines the specificity of the action.

Chorionic gonadotropin is produced by the syncytiotrophoblast (and possibly also by the cytotrophoblast). It appears in maternal blood shortly after implantation and then rises rapidly until 8 weeks' gestation. Levels show little change at 8–12 weeks, then decline until 18 weeks and remain fairly constant until term. The pattern of hCG in fetal blood is similar to that in the mother but at 2–3% of the concentration. At term the levels in the female fetus are substantially higher than those in the male. The levels and pattern of hCG in amniotic fluid are similar to those in blood. The half-life of the hormone shows multiple components, the initial faster phase being 6 h. In urine some hCG is in the form of intact hormone (20–25%) whereas the remainder consists of free β subunit and, in particular, a fragment known as 'β core' which is synthesized from hCG in the kidney.

The mechanisms that determine the levels of hCG in maternal blood are unknown. Unlike hPL and the steroid hormones, there is no relationship to the mass of tissue of origin.

In the first trimester hCG may be the major luteotrophic factor from the implanting embryo ensuring maintenance of the corpus luteum. In the second trimester it has been postulated that hCG is the stimulus for testosterone synthesis by the fetal testis.

Reference Interval

The range of hCG levels in the first half of pregnancy is shown as originally determined using the Ortho-Clinical Diagnostics Amerlex-M kit, which is no longer available (see Figure 51.3 and Tables 51.6 and 51.7.)

Clinical Applications

Detection of early pregnancy. This is the commonest use of hCG immunoassays. The result is qualitative (i.e. yes or no). Most commercial tests have a detection limit of 25 IU/L. This permits detection of pregnancy in the majority of subjects at approximately 7 days post-implantation (i.e. around the time of the missed period). More sensitive research assays can detect pregnancy as early as 1–2 days after implantation. At one time there was much concern about false-negative and false-positive results. With current technology the chance of either of

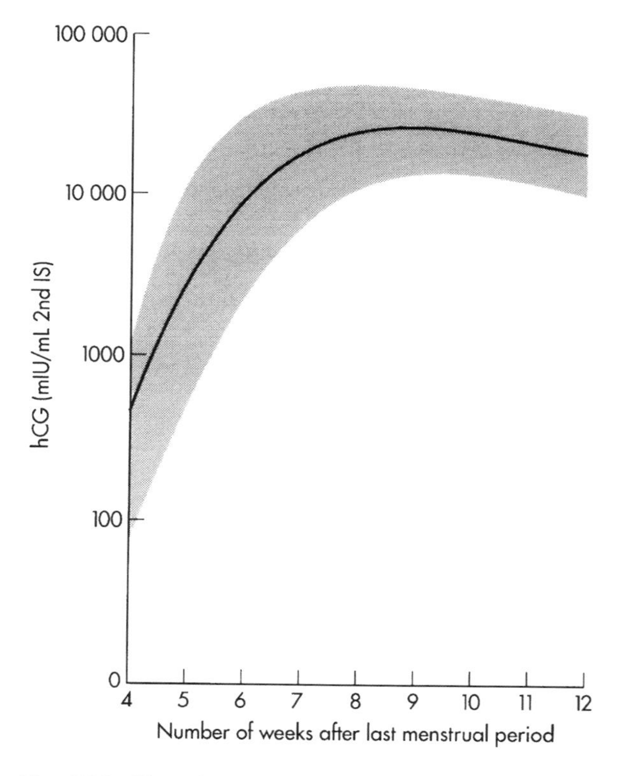

Fig. 51.3 Normal range of hCG in the first trimester of pregnancy.

Table 51.6. Mean values of serum hCG from normal pregnancies (4–12 weeks).

Gestation (weeks)	Mean hCG concentration mIU/mL 2nd IS	Range
Nonpregnant	<10	
4	415	70–1350
5	5300	560–22,600
6	14,900	2600–50,200
7	29,800	10,000–73,700
8	46,100	17,500–93,900
9	44,900	20,400–81,200
10	42,900	14,800–80,000
11	34,100	6400–71,300
12	30,700	16,000–60,400
2nd trimester	12,900	530–34,100
3rd trimester	9800	1400–23,900

these errors by 7 days after the missed period must be virtually zero.

Threatened abortion. The hCG levels are generally low in those cases in which the pregnancy will be lost and normal in those which proceed satisfactorily. However, there is considerable overlap between the two groups and a useful distinction is not possible before 8 weeks gestation. By this stage the fetal heart can be seen on ultrasound, and the presence or absence of the fetal heart provides a more definitive test. Some workers have advocated serial measurements of hCG; because the doubling-time in early pregnancy is 2 days, failure to increase over a period of 4 days or more is an unfavorable sign.

Ectopic pregnancy. The hCG levels are positive in all cases of ectopic pregnancy. The clinical importance of this is as follows. Many women present with lower abdominal pain of unknown cause. If hCG is negative, then a pregnancy-related condition can safely be excluded. If the test is positive, and even if there is no sign of pregnancy on ultrasound, then further investigation must focus on the possibility of an ectopic pregnancy (ultrasound and laparoscopy).

Table 51.7. The median values of serum hCG from normal pregnancies (15–20 weeks).

Gestation (weeks)	Median* IU/mL 3rd IS	Range 5–95th centile IU/mL 3rd IS
15	29.9	12.2–73.5
16	27.9	11.4–68.4
17	21.6	8.8–52.9
18	19.6	8.0–48.0
19	18.4	7.5–45.1
20	19.5	7.9–47.7

*The first International Reference preparation (1st IRP) (3rd IS) for hCG (75/37).

Down's syndrome. The use of elevated levels of hCG is discussed later in the chapter (*see* SCREENING FOR DOWN'S SYNDROME).

Late pregnancy. Low levels are found in placental insufficiency and high levels in pre-eclampsia. The test is rarely if ever used clinically.

Assay Technology

Because of the very similar nature of the β subunits of LH and hCG, early immunoassays showed considerable cross-reaction between the two. Unambiguous detection of early pregnancy was therefore not possible. The introduction of assays with an antibody directed to the β subunit of hCG yielded assays that were considerably more specific, and a further enhancement resulted from the use of immunometric assays and monoclonal antibodies. For practical purposes, most current hCG assays can be regarded as entirely specific to this analyte; interference by LH is so unlikely that it can safely be ignored.

An important and on-going problem of hCG assays arises from the fact that blood (and especially urine) contains a mixture of intact hCG, free β subunit and free α subunit. Individual assays may respond quite differently to each of these: for example, some competitive assays are much more sensitive to free subunit than to intact hormone, whereas some immunometric assays react with intact hormone but not at all with free subunit. Because the intact/subunit composition of calibrators is usually unlike that of samples, different assays can give very different results on the same sample.

The major practical effect of this is the chaotic picture that emerges from external quality control schemes. Fortunately, however, there is no evidence that the effect causes a problem in clinical practice. The ratio of intact hormone to free subunit does not vary systematically between individuals. Provided that a laboratory uses one assay, and adheres to the reference intervals derived from that assay, no problems should arise. However, there is now good evidence that an assay specific to the free β subunit (i.e. not reacting with intact hCG) is more efficient than nonspecific assays in screening for Down's syndrome.

As already noted, qualitative (yes/no) immunoassays for hCG are widely used in clinical practice. Some of these are based on agglutination, others on enzyme labels with a color endpoint. Ease of use is essential because the tests will often be used by untrained clinicians in emergency circumstances. The type of pregnancy test sold over-the-counter to the general public is perfectly adequate for this purpose (e.g. Carter–Wallace First Response).

Types of Sample

Serum, plasma, or urine. Urine is widely used for qualitative pregnancy tests and gives results virtually identical to blood. Contrary to popular myth, an early morning specimen does not contain the highest concentration of hCG.

Frequency of Use

Common.

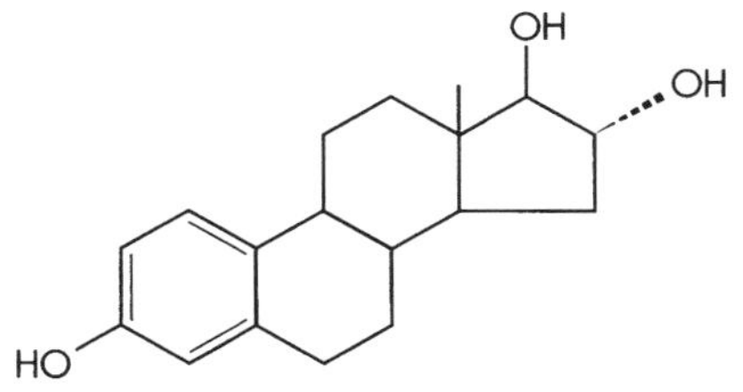

Fig. 51.4 Estriol.

ESTRIOL

Estriol is one of the three classical ovarian estrogens (estrone, estradiol, and estriol); it has the characteristic aromatic A-ring of all estrogens, together with hydroxyl groups at the 3, 15, and 16 positions (Figure 51.4). The unique feature of estriol is that it is produced by the syncytiotrophoblast from fetal precursors. The fetal adrenal produces dehydroepiandrosterone (DHEA) and the fetal liver converts this to 16-hydroxy-DHEA. Both of these compounds circulate in the fetus as the sulfate conjugate. In the placenta, the 16-hydroxy-DHEA-S is deconjugated by a sulfatase: the A-ring of the molecule is acted on by an aromatase to yield the form characteristic of estrogens. Estriol is then secreted into the maternal circulation. The pattern in maternal blood is a progressive rise throughout gestation.

The factors determining the levels of estriol in the maternal circulation are the supply of fetal precursors, and the conversion of these to estriol in the placenta. The latter is the rate-limiting step and the time-to-time control of estriol secretion is the same as that of products such as hPL, namely the mass of the trophoblast and the rate of uteroplacental blood flow. Because of the unique synthetic pathway of estriol, there are a number of unusual observations in experimental and pathological situations. These include:

- Estriol levels fall after administration of corticosteroids to the mother. These compounds cross the placenta, suppress the fetal pituitary-adrenal axis, and hence the production of DHEA.
- Anencephaly (with an inactive fetal pituitary) and absence or hypoplasia of the fetal adrenal gland are associated with very low levels of estriol.
- Low levels of estriol are associated with the rare condition of placental sulfatase deficiency in which the precursor (16-hydroxy-DHEA-S) cannot be deconjugated and hence is not available for conversion in the placenta.

As with many placental products, there is no evidence for a specific physiological role of estriol in pregnancy. Pregnancies in which estriol is deficient (see above) appear to be normal in every other respect.

Reference Interval

The range of serum unconjugated estriol levels in early and late pregnancy is shown as originally determined using the Ortho-Clinical Diagnostics Amerlex Specific Estriol (Unconjugated) kit, which is no longer available (see Table 51.8 and Figure 51.5).

Table 51.8. Median values of unconjugated serum estriol in normal pregnancies.

Gestation (weeks)	Median nmol/L unconjugated estriol (ng/mL)	Range 5–95th centile, nmol/L unconjugated estriol (ng/mL)
15	3.98 (1.12)	2.21–6.82 (0.64–1.96)
16	4.47 (1.29)	2.54–7.85 (0.73–2.26)
17	5.07 (1.46)	2.89–8.91 (0.83–2.57)
18	6.45 (1.86)	3.67–11.33 (1.06–3.26)
19	7.46 (2.15)	4.25–13.10 (1.22–3.77)
20	8.87 (2.55)	5.05–15.58 (1.45–4.49)

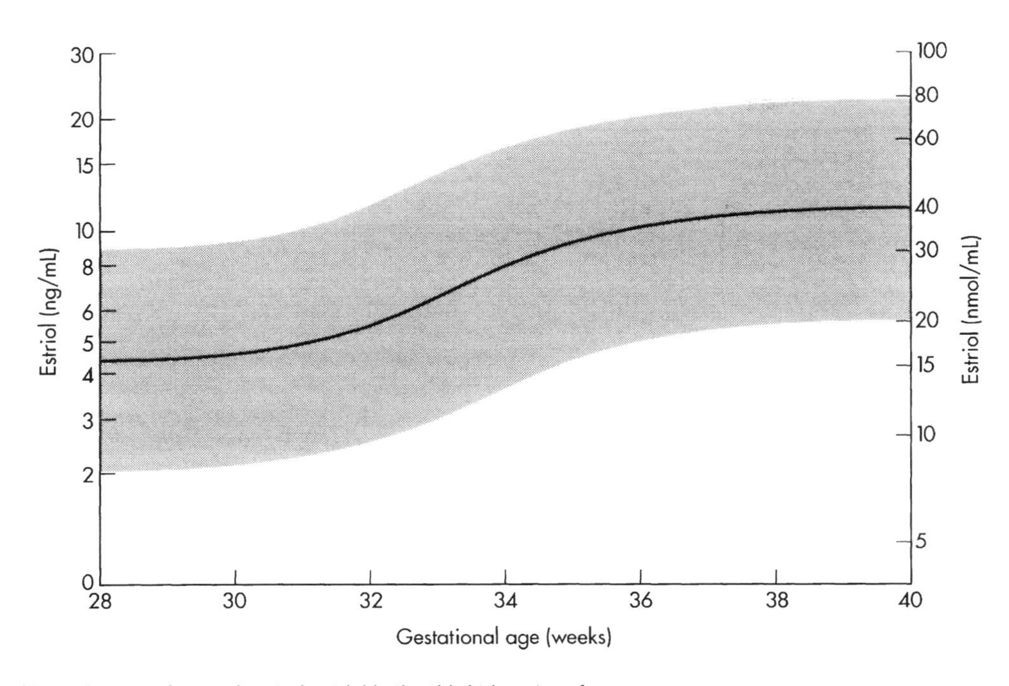

Fig. 51.5 Normal range of unconjugated estriol in the third trimester of pregnancy.

Clinical Applications

Down's syndrome. The use of low levels of unconjugated estriol is discussed later in the chapter (*see* SCREENING FOR DOWN'S SYNDROME).

Placental insufficiency. In the 1960s and 1970s, measurement of urine estriol was the most widely used special test of fetal well-being. In the late 1970s and 1980s the urine sample was replaced by blood. In the later 1980s and 1990s estriol measurement was replaced by biophysical techniques. However, there is no doubt that placental insufficiency is reflected by reduced levels of estriol. The test was superseded not because it had no predictive value, but because other tests are believed, sometimes on doubtful grounds, to be more effective.

Assay Technology

All immunoassays for estriol are competitive in design. Sensitivity is not a requirement because the levels in pregnancy are very high. Specificity is not an issue because estriol is the predominant estrogen in pregnancy blood; interference by other estrogens would have only minimal effects. Some assays include an enzyme that cleaves estriol conjugates so that all estriol is measured (total estriol). There are also assays with very specific antisera that only measure unconjugated estriol. However, for clinical purposes there is little or nothing to choose between the different forms.

Type of Sample

Serum is preferable for the only common current application: Down's syndrome screening.

Frequency of Use

Fairly common.

PREGNANCY-ASSOCIATED PLASMA PROTEIN-A

Pregnancy-associated plasma protein-A (PAPP-A) is a glycoprotein that has been described as having both anti-complementary and anti-proteolytic functions. It has also been suggested that low levels are predictive of unsatisfactory outcome in cases of threatened abortion in which fetal heart action can be seen by ultrasound. Levels of other trophoblast proteins (hCG, etc.) are normal under these circumstances.

In pregnancies with a Down's syndrome fetus, maternal levels of PAPP-A are substantially reduced in the first trimester but become normal in the second trimester. This estimation has therefore become the linchpin, together with free β hCG, of screening for Downs in the first trimester.

Reference Interval

The range of serum PAPP-A levels in early pregnancy is shown as originally determined using the Ortho-Clinical

Table 51.9. Median values of serum PAPP-A (8–14 weeks).

Gestation (weeks)	Median PAPP-A concentration (mg/L)	25–75 centile
8	1.86	0.89–3.15
9	3.07	1.66–6.33
10	5.56	2.98–10.1
11	9.86	5.74–15.0
12	14.5	9.90–27.5
13	23.4	14.1–34.6
14	29.1	19.5–38.4

Diagnostics Amerlex-M PAPP-A kit, which is no longer available, in Table 51.9.

Clinical Applications

Down's syndrome. The use of low levels of PAPP-A is discussed later in the chapter.

Assay Technology

A number of both isotopic and nonisotopic immunoassays have been described. Earlier problems of specificity appear to have been satisfactorily solved.

Type of Sample

Serum is preferable.

Frequency of Use

Fairly common.

OTHER ANALYTES

This category includes all those analytes listed in the first table that are not in widespread clinical use. For this reason reference intervals and assay technology are not discussed. In some cases the analyte might provide perfectly valid clinical information: for example, SP1 gives very similar results to hCG as a marker of early pregnancy, but custom dictates that hCG is the preeminent test.

Human Placental Lactogen

Human placental lactogen (hPL) is closely related, both chemically and biologically, to pituitary prolactin and growth hormone. It is secreted by the placental syncytiotrophoblast throughout pregnancy. The levels in maternal blood follow a sigmoid curve with the maximum rate of increase in the latter part of the second trimester and a plateau after 36 weeks.

The principal factors determining the levels of hPL in the circulation are the mass of functioning trophoblast and the rate of uteroplacental blood flow. This explains why there is no systematic (as opposed to random) time-to-time variation in hPL levels. In the case of hPL and other placental products, there does not seem to be any of the feedback control that characterizes most endocrine systems. The absence of classic feedback control mechanisms is accompanied by an apparent absence of biological function. There is a rare (1 in 3000) 'experiment of nature' in which there is total deletion of the hPL gene and therefore no hPL in maternal blood; these pregnancies and the resulting neonates are entirely normal in every clinical respect.

At term there is a clear-cut relationship between hPL levels and the weight of both the fetus and placenta.

Inhibins

The inhibins are members of a family of growth factors. Inhibin is produced by sertoli cells of the testis and granulosa cells of the ovary in the nonpregnant state and, during pregnancy, from the placenta. The inhibins are heterodimeric peptides consisting of one α subunit and one of two different β subunits. As with many other pregnancy products, the exact function is unknown. Measurement of inhibin A (α-βA) shows elevated maternal levels in the second trimester in Down's syndrome pregnancies.

Progesterone

Progesterone is produced by the corpus luteum in the first 6 weeks of pregnancy and by the placenta thereafter. It is believed to be the hormone principally responsible for the maintenance of early pregnancy. Though measurement of progesterone levels in both early and late pregnancy gives results similar to those of hPL, it has been rarely used in obstetric practice (but is very widely used in gynecological endocrinology).

Schwangerschaftsprotein 1

Schwangerschaftsprotein 1 (SP1) is a glycoprotein of the same family as carcino-embryonic antigen (CEA), but produced specifically by the placental trophoblast. The function is totally unknown. Measurement of SP1, gives clinical results similar to hPL and hCG. Because of its very high levels it has been advocated as an alternative to hCG for early pregnancy testing, especially in women who have received hCG therapeutically.

Placental Protein 5

Placental Protein 5 (PP5) has the biological function of antithrombin III, and could play a key role in

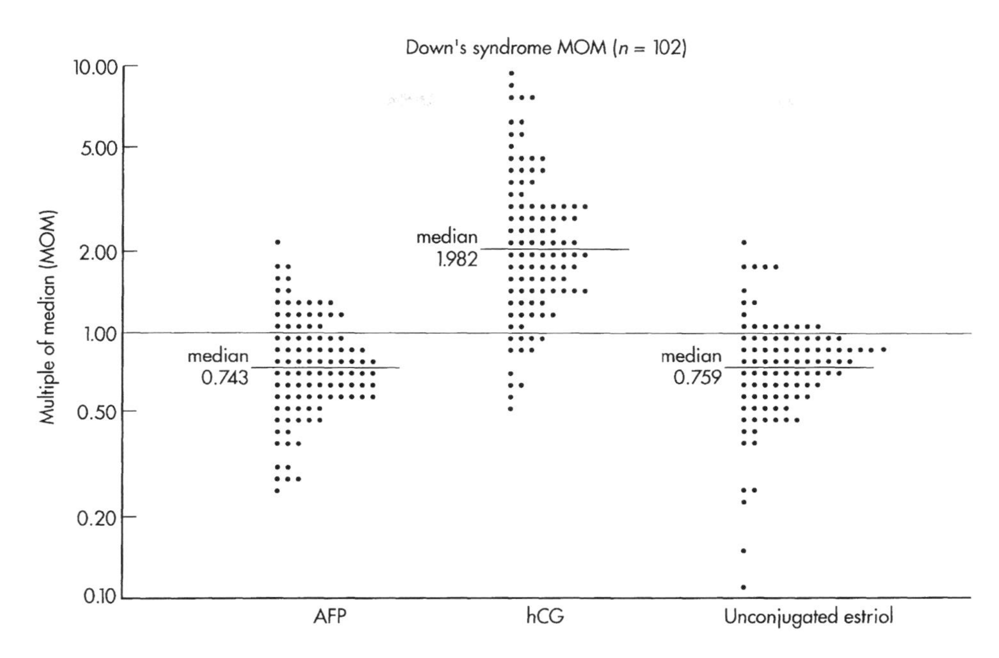

Fig. 51.6 Maternal serum AFP, hCG and unconjugated estriol for Down's syndrome-affected pregnancies.

preventing coagulation in the maternal circulation of the placenta (intervillous space). Both *in vitro* and *in vivo*, levels are greatly increased in the presence of heparin. Elevated levels are found in severe pre-eclampsia and placental abruption: it is notable that both of these conditions are associated with placental thrombosis and infarction, and disseminated intravascular coagulation (DIC).

Placental Protein 12

PP12 is an endometrial protein that has proved to be identical to insulin-like growth factor binding protein 1 (IGFBP-1). In the third trimester, the levels of IGFBP-1 are elevated in both the maternal and fetal circulation in association with intrauterine growth retardation and fetal hypoxia.

Placental Protein 14

Placental Protein 14 (PP14) (also known as glycodelin) is the principal product of the endometrial epithelium of early pregnancy (decidua). Chemically, it is analogous to the β-lactoglobulins. It may play a role in fertilization and implantation. Synthesis is dependent on an ovarian factor, possibly relaxin. The pattern of serum levels is identical to that of hCG. It is likely that the clinical application, if any, would be as a test of endometrial function.

Relaxin

Relaxin is a peptide of the insulin family that is produced by the ovary, and possibly the placenta, from an early stage of pregnancy. Functions might include softening of the pelvic ligaments and dilation of the uterine cervix near term. Measurement of relaxin might be clinically useful, but this will not be known until reagents for its immunoassay are more widely available.

Oxytocin

Oxytocin is a small peptide (eight amino acids) that is secreted by the posterior pituitary gland. There is a substantial increase in maternal circulating levels in association with labor, and an even more noticeable increase in fetal blood. The release takes the form of a series of 'spurts'. It is still uncertain whether these phenomena are a cause of, or a result of, the labor process. The immunoassay of oxytocin is technically demanding and no routine clinical application has ever been proposed.

SCREENING FOR DOWN'S SYNDROME

Down's syndrome is associated with elevated levels of hCG (especially free β subunit) and reduced levels of AFP and estriol in blood in mid-trimester. The mechanism is unknown but might represent a shift in balance between placental (hCG) and fetal (AFP, estriol) products. This is

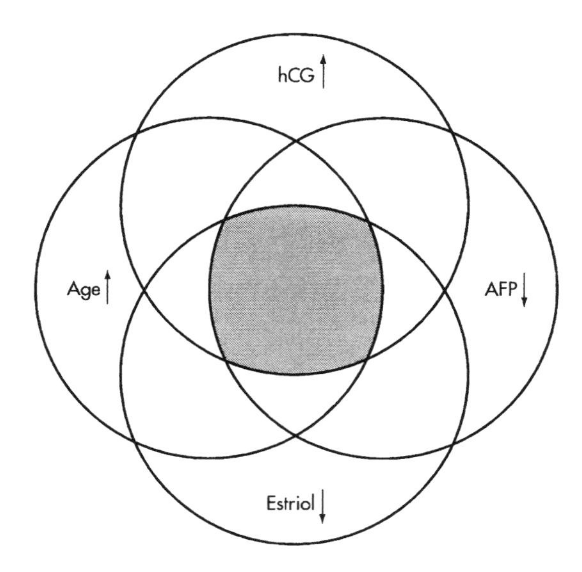

Fig. 51.7 Diagram to illustrate the principle of risk assessment for Down's syndrome using maternal age and maternal serum AFP, hCG, and estriol. In an individual woman, the probabilities given by each test are combined to yield a single overall probability indicated by the shaded area. The probabilities are based on analysis of the overlapping Gaussian curves for normal and Down's syndrome pregnancies. A computer program is essential for this calculation and a number of such programs are now available.

supported by the fact that other placental products (SP1 and progesterone) are also elevated. The test is done at 16–18 weeks of pregnancy. Earlier testing (10–14 weeks) is also feasible using measurement of PAPP-A and the free β subunit of hCG (see Figure 51.6). Many other analytes have been examined for their possible role in Down's screening. Of these, dimeric inhibin A appears especially promising in the second trimester.

There is substantial overlap between the values in Down's syndrome and normal pregnancy. Consequently, it is not possible, as it is with amniotic fluid AFP and NTD, to select a single cutoff point above or below which action is taken. Instead, the probabilities of Down's syndrome indicated by the result for each analyte, together with the probability for the maternal age, are combined to give a single overall probability. The principle of this is illustrated in Figure 51.7. If the overall probability exceeds some predetermined level (typically between 1:200 and 1:300) then the woman is counseled to have an amniocentesis and karyotyping.

Numerous studies indicate that this screening process can detect 60% or more of Down's syndrome for a 5% amniocentesis rate. [Editor's note: it is essential that screening algorithms are validated for the specific immunoassay tests being used.]

Assay Technology

The assays of hCG, AFP and estriol have already been discussed.

FURTHER READING

Canick, J.A., Saller, D.N. Jr. and Lambert-Messerlian, G.M. Prenatal screening for Down syndrome: current and future methods. *Clin. Lab. Med.* **23**, 395–411 (2003).

Chapman, M., Grudzinskas, J.G. and Chard, T. (eds), *The Embryo* (Springer, London, 1991).

Chard, T. What has happened to placental function tests? *Ann. Clin. Biochem.* **24**, 435–439 (1987).

Chard, T. *Radioimmunoassay and Related Techniques* 5th edn (Elsevier, Amsterdam, 1995).

Chard, T. and Lilford, R.J. *Basic Sciences for Obstetrics and Gynaecology* 5th edn (Springer, London, 1997).

Grudzinskas, J.G. and Ward, R.H.T. (eds), *Screening for Down's syndrome in the first trimester* (RCOG Press, London, 1997).

52 Growth and Growth Hormone Deficiency

Michael Preece and Jane Pringle

NORMAL CHILDHOOD GROWTH

The human being takes between 15 and 20 years to complete growth to mature size. During this process the progress of growth is influenced by a multitude of factors including genetics, nutrition, the general environment, and internal hormone secretion. These factors bear upon the timing and magnitude of various parts of the growth process, such as the inter-relationship between prepubertal growth and the adolescent growth spurt. There are three main phases of growth: infancy, which is predominantly dependent on nutrition; childhood, which is dominated by growth hormone (GH); and puberty, where the gonadotropin and sex steroids have a major role.

Thyroid hormones, the gonadotropins, and the sex and adrenal steroids are dealt with elsewhere (*see* THYROID, INFERTILITY, and ADRENAL CORTEX). This chapter is restricted to GH, its binding protein and the dependent insulin-like growth factors (IGFs) and their binding proteins.

The pituitary gland secretes GH under the joint control of the hypothalamic peptides: growth hormone-releasing hormone (GHRH) and growth hormone-release inhibiting hormone or somatostatin. The exact mechanisms are at present unclear, but it seems most likely that there is a continuous interplay between these hormones leading to intermittent, but well regulated, pulsed release of growth hormone at about 200-min intervals, mostly by night. There is a weak association between the first attainment of deep sleep and the major GH peak of the night.

In serum, GH is associated with specific binding proteins. A high-affinity binding protein (GHBP) is present in quantities approximately equimolar with inter-pulse levels of GH and may bind about 30% of serum GH at these levels. It is homologous with the extracellular domain of the membrane-bound GH receptor and may be derived by proteolytic cleavage during cellular processing of the receptor. Levels of the binding protein might therefore reflect expression of the GH receptor and the biological activity of GH. There is a striking age-related increase in GHBP, being very low in the neonatal period and rising through childhood.

Growth hormone exerts its further effects through two different pathways. It has potent metabolic effects that appear to act directly through specific GH receptors, particularly in adipocytes and muscle. These effects are directly opposed to those of insulin and form one of the major counter-regulatory mechanisms to balance insulin action. In contrast, the growth-promoting actions of GH are mediated through the intermediate IGFs. These are two in number (IGF-I and IGF-II) and they act in a paracrine/autocrine fashion, being locally produced in many tissues. There is, however, a considerable quantity of both IGF-I and -II in the circulation bound to a variety of binding proteins. It is not clear whether these circulating forms have a more general endocrine function.

IGF-I is thought to be an important signal for many cell types integrating information on GH secretion and nutritional status. Appropriate levels of IGF-I allow proliferating cells to proceed through the cell cycle, to differentiate, and to perform specific functions such as production of extracellular matrix compounds (e.g. the proteoglycans in cartilage). These effects are particularly important in the growing organism. The physiological importance of IGF-II is far less clear.

The IGFs in the circulation are tightly bound to one of a number of specific binding proteins; only about 1% of circulating IGF is in the free form. At least six classes of IGF binding protein (IGFBP) can be distinguished on the basis of their amino acid structure. They are designated IGF-BP1 to IGF-BP6.

CLINICAL DISORDERS

GROWTH HORMONE DEFICIENCY OR INSUFFICIENCY

Severe or **total GH deficiency** occurs relatively rarely after major structural lesions of the hypothalamus and pituitary gland, following surgery for their treatment, or as a result of deletion of the structural gene for GH. The latter is a rare autosomal recessive condition. Much more commonly there is a variable severity of **GH insufficiency** that is clinically almost indistinguishable from the severe forms at one end and normality at the other. Diagnosis of these conditions is largely based on clinical assessment of growth, usually coupled with some form of provocation test of GH secretion. Because of the pulsatile nature of GH secretion, basal levels are useless; measurements of GH concentrations following frequent sampling throughout 24-h periods have been used but are essentially a research tool because of the resource implications.

Hypopituitarism or **multiple pituitary hormone deficiencies** may be idiopathic, genetic or secondary to structural lesions. GH is by far the most commonly affected hormone, with the gonadotropins, thyrotropin (TSH), and adrenocorticotropin (ACTH) being affected in turn. The GH aspects of diagnosis are unaltered except that the clinical picture is usually more severe.

Growth hormone deficiency is treated by replacement with recombinant human growth hormone, given subcutaneously. Over recent years various regimens have been explored, but it is now more or less universally accepted that daily injections are required at doses between 5 and 10 mg/m^2 body surface area (pure somatropin has a bioactivity of 3 IU/mg).

GROWTH HORMONE RESISTANCE

Growth hormone receptor deficiency or **Laron-type dwarfism** clinically resembles severe GH deficiency, but is usually characterized by normal or excessively high serum concentrations of GH and very low concentrations of IGF-I. There is, usually, a very low concentration of GHBP and absent cell membrane binding of GH leading to the assumption that there is a deficiency of the GH receptor; it follows an autosomal recessive pattern of inheritance. A nearly indistinguishable variant has been described where normal circulating GHBP is found; it is believed that the GH receptor is present but that post-ligand binding events are abnormal.

Growth hormone antibodies are a rare cause of GH resistance. They do not occur spontaneously but may rarely develop while a patient is on GH therapy. This is probably confined to the few patients with severe gene-deletion GH deficiency where the subjects' immune system has never been exposed to growth hormone and fails to recognize it as self when administered therapeutically. When antibodies do occur, they effectively block the effects of the growth hormone. There is no known treatment for growth hormone antibodies.

EXCESSIVE GROWTH HORMONE SECRETION

Pituitary gigantism occurs with pituitary adenomas that produce excessive GH secretion during childhood. There is excessive skeletal and visceral growth often to a marked degree; it is extremely rare. Circulating GH concentrations may not be strikingly elevated on basal estimation but they fail to be suppressed after a glucose load.

Acromegaly is a far commoner condition that follows the development of a pituitary adenoma in adults when growth has ceased. There is still bony overgrowth but it mostly affects membrane bones because of the fusion of the growth plates. As with pituitary gigantism there are a number of important metabolic sequelae of which the most important is diabetes mellitus.

ANALYTES

GROWTH HORMONE

Growth hormone (GH, somatotropin) is a heterogeneous protein secreted predominantly as its 191 amino acid form, often referred to as 22 kDa GH. There are several other variants of GH found in the circulation, produced as a result of posttranscriptional, posttranslational and postsecretory events. This processing results in variants of GH of differing length such as the 20 kDa form (178 amino acids) or with structural changes caused by, for example, deamidation or glycosylation. Aggregation of the monomeric variants can result in dimeric and oligomeric forms, all of which can interact in an assay to a greater or lesser degree.

There is little evidence to date on the role of these variants and whether the proportion is consistent or alters under particular circumstances. This has implications for the measurement of GH by immunoassay (*see* ASSAY TECHNOLOGY).

It should also be noted that there is a second GH gene, which is only expressed in the placenta. This particular GH is found in late pregnancy serum and may also cross-react in certain GH assays. There is considerable structural variability between different species.

Function

Growth hormone is essential for growth, particularly during childhood. It is also important for certain metabolic functions throughout life.

Reference Interval

Because of the pulsatile nature of GH secretion, reference intervals for basal values are without meaning. The heterogenic nature of GH described above also makes the assignment of expected values, other than for the specific assay being used, unhelpful. Laboratories offering GH assays, particularly for the diagnosis of GH insufficiency states, need to define their own reference intervals for the assay method in use. There is considerable overlap between the concentrations seen in normal individuals and in those with hypo- or hypersecretion of GH.

The reference standard for somatotropin, issued by the National Institute for Biological Standards and Control (South Mimms, UK), is IS 80/505 and is pituitary derived. It has an assigned biological potency of 2 IU/mg and is 95% monomeric GH. More recently, a new standard preparation of recombinant human GH (2nd IS 98/574 r-hGH) has been distributed with a potency of 3.0 IU/mg r-hGH. This standard is 100% 22 kDa GH. The relative purity of IS 80/505 for monomeric 22 kDa GH and the absolute purity of IRP 98/574 have led to some problems. The specificity of some of the antibodies used in assays is targeted against specific epitopes of the hGH molecule that may be found on many of the GH forms. The monomeric nature of the standard can lead to incomplete recognition of the circulating GH components and, consequently, to lower readings. It is unlikely that the recombinant standard will be adopted in the foreseeable future. Details can be found at http://www.nibsc.ac.uk/catalog/standards/ifu. The heterogeneity of hGH contributes to the range of product bias against the 'All-Lab Trimmed Mean' (ALTM) seen in the UK National External Quality Assurance Scheme (NEQAS), which can be anything from −40 to +25%.

As a result of the huge differences between assays described above, it is impossible to establish a common reference value following an appropriate provocation test (*see* DYNAMIC TESTS) above which GH insufficiency can be excluded. For example, in one assay the cut-off for clinically proven GH insufficiency is 13.5 mU/L but in another it is 35 mU/L. Particular care is required around the time of onset of puberty, particularly when delayed, as artificially low responses may be seen. Diagnosis of excess GH secretion depends upon the failure to suppress circulating levels following a glucose load.

Clinical Applications

Measurements of basal levels of GH have no clinical use because of the pulsatile nature of GH secretion. Frequent sampling throughout a 24-h period is seldom used because of the time taken and the expense. Diagnostic information is normally sought by using dynamic tests that stimulate or suppress GH secretion.

Dynamic Tests for Growth Hormone Insufficiency

Provocation tests for GH insufficiency have been legion, indicating the lack of a clearly ideal test. The following are the most important at this time. The normal responses should be assessed in each laboratory.

Insulin Tolerance or Stress Test (ITT or IST)

Soluble insulin (0.05–0.15 U/kg) is given intravenously following an overnight fast. The induced hypoglycemia acts as a stimulus to GH release, which is measured at 15-min intervals for 90 min. This is the test with which the greatest experience has been gained and it has become something of a 'gold standard'. However, it has a significant mortality if used in an inappropriate way and it is doubtful whether it should be considered as the first choice test any longer. The risks lie in the development of severe hypoglycemia and then the overzealous correction with hyperosmolar glucose solutions leading to hyperosmolar coma and death.

Glucagon Tolerance Test

Intramuscular glucagon (0.1 mg/kg) is administered after an overnight fast and serum GH measured at 30-min intervals for 180 min. The reliability of the test is about the same as the insulin tolerance test, but it does take considerably longer. It is important that a meal is given and retained after completion of the test as prolonged hypoglycemia may occur otherwise.

Clonidine Test

Clonidine is a selective α-receptor agonist, stimulating GH release through GHRH secretion; it is administered orally in a dose of 0.15 mg/m^2. Blood is taken every 30 min for 150 min. It seems to be an entirely safe test although unpleasant postural hypotension and drowsiness may occur.

Arginine Test

Arginine is one of the basic amino acids that, when administered intravenously, stimulates the release of GH; the mechanism is unclear. The dose is 500 mg/kg and blood is taken at 30-min intervals for 150 min. It is slightly less reliable than the previous tests in the sense that there is a greater frequency of false-positive results.

Growth Hormone-Releasing Hormone

The use of GHRH as a diagnostic test for GH insufficiency is still somewhat controversial as the discriminatory power is disputed. It is probably best considered as a research tool at the present time.

Dynamic Tests for Growth Hormone Excess

Glucose Tolerance Test

Following an overnight fast, oral glucose (1.75 g/kg) is ingested. Serum GH concentrations are measured every 30 min for 180 min. The failure of GH levels to fall below 2 mIU/L within 60–120 min suggests excess GH secretion.

Dynamic Tests for Growth Hormone Resistance

See IGF-I.

Limitations

- ITT and glucagon tests both depend on manipulation of glucose homeostasis and can be dangerous.
- All tests have a relatively high rate of false positives (apparently low GH response in an individual who is subsequently seen to be normal) ranging between 15 and 25%.

Assay Technology

There are 12 different methods for the measurement of GH registered in the UK NEQAS scheme, most of which are immunometric in type. They vary in the combination of monoclonal or polyclonal antibodies used and in the type of solid-phase element. Some use nonradioisotopic tracers and many assays are now performed on automated systems.

The heterogeneity of GH means that for any two antibodies used in combination there are differences in the recognition of the growth hormone variants. Some assays only recognize the 22 kDa variant whilst others cross-react to varying degrees with other GH variants and their aggregates. If the proportion of variants changes after stimulation then, potentially the amount of GH recognized changes as well. This leads to huge differences in the absolute number obtained for analysis on any sample. The situation is compounded by the different calibrants that are available, some of which are pituitary derived and some of which are generated from recombinant GH.

There are methods available to measure variants of GH other than 22 kDa, either directly or indirectly. One is a non-22 kDa GH exclusion assay which is an indirect measure of GH variants other than 22 kDa: the 22 kDa GH is removed by a specific antibody and the remaining GH is measured by an immunoassay that recognizes all GH forms. Direct methods for measuring 20 kDa GH are also available. Both these assays are mainly for research at present but may have a role to play when the function of these variants is better understood.

Conventional immunoassays only provide information related to the *structure* of the analyte and provide no insight into *functional activity*. The immunofunctional GH assay was constructed to look at the structural integrity of the GH molecule and therefore its potential to be functionally active. It is not a bioassay but a sophisticated immunoassay that relies on the ability of the GH to dimerize the GH receptor, a process required for GH action. This method is available commercially but it remains to be seen whether it is more useful than a conventional immunoassay.

Note that a significant proportion of GH is bound to GHBP in the circulation. For the most part GHBP does not cause interference, probably because the antibodies used have a higher affinity for GH. However, some assays, particularly those with short incubation times, may be affected by the amount of GHBP present.

Urinary Growth Hormone

One attempt to circumvent the problems of the pulsatile secretion of serum GH has been the development of assays for urinary GH. There are major problems because of the very small percentage of circulating GH that is excreted in urine and considerable variation in day-to-day GH excretion within individuals. Also, it only gives an integrated GH concentration relative to the duration of the urine collection and says nothing about the pulsatile nature of GH. It may have a role to play in the monitoring of treatment in the management of GH excess.

GROWTH HORMONE BINDING PROTEIN

There is an approximately equimolar relationship between GH binding protein (GHBP) and interpulse levels of GH. Levels of the binding protein may reflect expression of the GH receptor and the biological activity of GH. There is a striking age-related increase in GHBP, being very low in the neonatal period and rising through childhood.

This remains predominantly a research tool and is only available in a number of specialist laboratories. Measurement still depends upon the quantitative separation of bound and free radiolabeled GH following incubation with an aliquot of the serum sample of interest. There are no international standards.

INSULIN-LIKE GROWTH FACTORS: IGF-I (SOMATOMEDIN C) AND IGF-II

IGF-I and IGF-II are peptides of 70 and 67 amino acids, respectively, and share considerable structural homology with each other and pro-insulin.

Function

IGFs are the mediators of the growth-promoting activity of GH. The mechanism of action of IGF-I is much better understood than that of IGF-II, which has an uncertain role at present. Approximately 99% of the IGFs in the circulation are bound to one of a group of IGFBPs.

Reference Interval

Reference intervals for IGF-I are age dependent, ranging from 15–75 ng/mL in infancy to 185–540 ng/mL at 13–15 years. After puberty there is a gradual decline to 37–144 ng/mL in the seventh decade. IGF-II is less age dependent with an increase from 132–430 ng/mL at birth to 330–767 ng/mL at 5 years with no significant change thereafter.

There is an International Reference Preparation for IGF-I (IRP 87/518) distributed by the National Institute for Biological Standards and Control (South Mimms, UK).

Clinical Applications

IGF-I assays are available commercially and are part of the repertoire of some automated analyzers. Their clinical use is restricted mainly because of difficulties of interpretation. Low IGF-I levels (with or without IGFBP-3 levels) are not helpful in determining if a child is GH deficient although normal levels suggest other causes of short stature. There is, however, a value in the use of serum IGF-I measurements in the diagnosis and management of conditions of GH excess, particularly acromegaly where there is an established experience. More recently, IGF-I measurements have been used to titrate and monitor recombinant hGH replacement therapy particularly in adults, but increasingly in children.

The clinical value of IGF-II is unclear at present, except possibly in the investigation of hypoglycemia, particularly if thought to be tumor related.

Insulin-Like Growth Factor I Generation Test (Dynamic Test for Growth Hormone Resistance)

The IGF-I generation test depends upon the measurement of IGF-I and IGFBP-3 concentrations in response to a dose of GH, usually given intramuscularly over a period of 3–5 days. The protocols are variable and this should be considered as a research tool for the time being.

Limitations

The value of IGF-I measurement in the diagnosis of GH insufficiency is limited because of the considerable overlap between the reference interval and values seen in GH-insufficient states, especially in infancy and young childhood. Other problems are the very low values found in some normal young children and the need for age- and puberty-matched reference intervals. IGF-II has a very unclear clinical significance.

Assay Technology

Many assays for IGF-I have been developed in-house although reagents are available commercially, including fully automated methods. The number of immunometric methods is increasing and both monoclonal and polyclonal antisera exist; the most useful have a low cross-reactivity with IGF-II (less than 1%). The presence of a number of IGFBPs interferes with the assay and some form of serum extraction procedure is desirable. The most widely used involves the extraction of the IGFs into acid–ethanol, sometimes followed by chromatography. It is also possible to mop up the binding proteins after acidification of the serum using excess IGF-II and then measuring the IGF-I using a highly specific antibody (no cross-reactivity with IGF-II). These procedures give a measure of the total IGF-I.

Free IGF-I assays also exist. These are mainly a research tool at present although they may prove to be more useful than total IGF-1.

There are fewer assays available for the measurement of IGF-II, although the necessary reagents are available commercially. Again IGFBPs can cause interference and extraction procedures comparable to those performed for IGF-I are required. There is no international standard. At present it must be considered as a research tool only.

INSULIN-LIKE GROWTH FACTOR BINDING PROTEINS

There are currently six IGFBPs identified although other putative binding proteins have been reported. They all have different affinities for IGF-I and -II and the role of many of these binding proteins has yet to be fully elucidated.

IGF-BP1 is a single-chain polypeptide with a molecular mass of 25.7 kDa. Serum levels are high in newborns and decline through childhood. It has a marked circadian rhythm, with high levels at night, and has an inverse relationship with food intake and insulin. Phosphorylated and lesser-phosphorylated forms of IGF-BP1 have been reported: the lesser-phosphorylated forms appear to have a reduced affinity for IGF-1, which may increase IGF-1 bioavailability.

The most abundant IGFBP in the circulation after the neonatal period is IGF-BP3, binding both IGF-I and -II with similar affinities. It is substantially glycosylated with a total, estimated, molecular mass of 42 kDa. IGF-I or -II forms a high molecular mass ternary (three-component) complex (120–150 kDa) with IGF-BP3 and an acid-labile subunit (ALS). ALS may be involved in transporting of IGFs to the binding protein.

Immunoassays, including some commercial kits, exist for IGFBP-1 (nonphosphorylated and phosphorylated), BP-3 and ALS but their significance in clinical endocrinology is controversial. They should be used with caution due to difficulty in interpreting results. Assays also exist for BP-2 and BPs 4–6 but these are purely a research tool.

GENERAL STRATEGY

The principal tool for the diagnosis of growth disorders remains the careful clinical assessment of the growth pattern. A low growth velocity is always suspicious and justifies further investigation. This should start with the exclusion of general medical problems that may influence growth. This done, specific endocrine tests are appropriate. In a child who is growing at a normal rate, but below the third centile for height, it is important to take into account parental heights and a measure of skeletal maturity, normally bone age. The commonest cause is simple constitutional growth delay.

The biochemical tests must only be considered as adjuncts, except in some highly specific situations. Disorders of GH excess are probably an exception, where the measurement of circulating GH levels (and possibly IGF-I) in carefully controlled circumstances is of primary importance.

FURTHER READING

Buchanan, C.R. and Preece, M.A. Hormonal control of bone growth. In: *Bone Growth* (ed Hall, B.K.), 53–89 (CRC Press, Boca Raton, 1992).

Preece, M.A. Principles of normal growth: auxology and endocrinology. In: *Clinical Endocrinology*, 2nd edn (ed Grossman, A.), (Blackwell Scientific Publishers, Oxford, 1998).

Pringle, P.J., Jones, J., Hindmarsh, P.C., Preece, M.A. and Brook, C.G.D. Performance of proficiency survey samples in two immunoradiometric assays of human growth hormone and comparison with patients' samples. *Clin. Chem.* **38**, 553–557 (1992).

Ranke, M.B. (ed), *Functional Endocrinologic Disorders in Children and Adolescents* (J & J Verlag, Mannheim, 1992).

Rosen, C.J. and Pollack, M. New perspectives for a new century. *Trends Endocrinol. Metab.* **10**, 136–141 (1999).

53 Diabetes Mellitus

Penny Clark

The concentration of glucose in blood is closely controlled in healthy individuals, regardless of the quantity of carbohydrate ingested in the diet. The hormone insulin plays a central role in the regulation of blood glucose concentrations by influencing tissue glucose uptake and the metabolic pathways of gluconeogenesis, glycogenolysis, and lipolysis. It is produced by the β cells of the islets of Langerhans in the pancreas and secreted into the circulation within minutes if the concentration of glucose in the blood increases. The secretion or administration of insulin causes a fall in the concentration of blood glucose. A concentration of blood glucose below that found in healthy fasting individuals is known as **hypoglycemia**. Measured glucose concentrations are dependent on sample type (arterial/venous/capillary, whole blood or plasma).

In **diabetes mellitus** the concentration of glucose in the blood is abnormally high (**hyperglycemia**) either due to insufficient insulin secretion or due to an inability of insulin to act at a cellular level. The hyperglycemia may occur when the patient is fasting and/or post-prandially. Other abnormalities of intermediary metabolism such as lipid metabolism are found. Diabetes mellitus is treated by diet, and administering insulin or drugs that enhance insulin secretion, or drugs that increase sensitivity to insulin. However, there are long-term complications of the disease: microvascular complications involving the kidney (nephropathy) and retina (retinopathy); macrovascular complications such as coronary heart disease, peripheral vascular disease, and also the neuropathies.

There are two main types of diabetes mellitus, which are classified based on the pathophysiology of the disease rather than on treatment (Table 53.1). **Type I** (previously **insulin-dependent diabetes mellitus**, **IDDM**) has a varied incidence of 1–29 per 100,000 of the population. Those affected have a significantly reduced secretion of insulin due to autoimmune destruction of the pancreatic beta cells. Type I diabetes is usually diagnosed in childhood or early adulthood, with the patient presenting with polyuria (excretion of an increased volume of urine), excessive thirst, weight loss, and **ketosis**, which is characterized by drowsiness, headache, and deep respiration, due to an accumulation of ketone bodies, β-hydroxybutyric acid, acetoacetic acid, and acetone in the blood. Diabetic ketoacidosis is a medical emergency. Type I diabetes is primarily treated with insulin.

Type II diabetes is associated with insulin resistance and obesity. Obesity itself will also contribute to the resistance to insulin. There is some evidence to suggest that insulin deficiency may also be a factor. Type II diabetes has a prevalence of about 3% and this may be higher in certain populations such as the Pima Indians in the United States. The prevalence of Type II diabetes is increasing both in adults and in children around the world and has been associated with the increase in obesity and changes in lifestyle (Ehtisham and Barrett, 2004). Type II diabetes is generally treated in the first instance by weight loss following dietary advice, drugs that will stimulate the secretion of insulin (oral hypoglycemics) or drugs that improve the sensitivity to insulin. Treatment with insulin may be necessary.

A number of diseases can cause secondary diabetes and there are also several rarer conditions associated with it.

The diagnosis of diabetes mellitus is made by measuring the concentration of glucose in plasma or blood. The most recent criteria introduced for the diagnosis of diabetes include a random venous plasma glucose (>11.1 mmol/L or 200 mg/dL) or a fasting venous plasma glucose (>7.0 mmol/L or 126 mg/dL) in the presence of symptoms of diabetes mellitus. With no symptoms, the diagnosis should not be based on a single glucose determination—further determinations are required. Two other groups of 'impaired glucose tolerance' and 'impaired fasting glycemia' are now defined (Table 53.2). It is recommended that those with impaired fasting glycemia should have an oral glucose tolerance test (OGTT) (American Diabetes Association, 2004a). Different criteria have been defined for gestational diabetes. In the OGTT glucose is administered orally and the blood glucose concentration is measured at timed intervals. The test must be standardized as a number of factors, such as the type of sample collected, the glucose dose and the preceding diet, all have a significant effect on the results of the test.

The Immunoassay Handbook. *Third Edition*
D. Wild (Ed.)

Table 53.1. Classification of disorders of glycemic control.

(1) *Type I diabetes mellitus*
  β-cell destruction leading to insulin deficiency)
  Autoimmune
  Idiopathic
(2) *Type II diabetes mellitus*
  (ranges from insulin resistance/relative insulin deficiency with or without secretory defect to secretory defect with or without insulin resistance)
(3) *Other types*
  Genetic defects of β-cell function, e.g. MODY
  Genetic defects in insulin action, e.g. Type A insulin resistance
  Diseases of the exocrine pancreas, e.g. cystic fibrosis, pancreatitis
  Endocrinopathies, e.g. acromegaly, Cushing's syndrome
  Drug/chemical induced
  Infection, e.g. CMV, congenital rubella
  Uncommon immune mediated, e.g. antibodies to insulin
  Associated with other genetic syndromes, e.g. Turner's, Wolfram's
(4) *Gestational diabetes mellitus*

The measurement of plasma insulin is not necessary for the diagnosis of diabetes mellitus. An understanding of the dynamics and processes of insulin secretion are needed to unravel the pathophysiological mechanisms of the disease. The importance of insulin assays is underlined by the fact that the first radioimmunoassay to be described (Yalow and Berson, 1959) was for plasma insulin.

Control of blood glucose within certain limits by adjusting treatment is advocated for patients with Type I and Type II diabetes, for both immediate metabolic control and to prevent or delay the onset of complications. (The Diabetic Control and Complication Trials Research Group, 1993; The United Kingdom Prospective Diabetes Study Group, 1998; Sacks *et al*., 2002). Central to good control of blood glucose in clinical practice is self-monitoring of blood glucose, the measurement of markers of glycemic control such as glycohemoglobin and markers of diabetic complications such as microalbumin. Immunoassays exist for many of these analytes.

ANALYTES

INSULIN, PROINSULIN(S) AND C-PEPTIDE

Insulin, like many polypeptide hormones, is synthesized as a precursor molecule: proinsulin. Limited proteolysis of proinsulin at sites marked by pairs of basic amino acids produces insulin by way of a number of partially processed forms. The C-peptide is a fragment that connects the A and B chains of insulin in the proinsulin molecule (Figure 53.1).

Insulin is secreted in a pulsatile manner and, with C-peptide, is secreted in equimolar concentrations into the portal circulation. The first pass through the liver clears approximately 50% of this insulin, whereas C-peptide is excreted mainly through the kidneys. Because C-peptide has a longer biological half-life than

Table 53.2. Glucose values (mmol/L) for the diagnosis of diabetes mellitus and hyperglycemia (American Diabetes Association 2004a).

	Whole blood (venous)	Whole blood (capillary)	Plasma (venous)
Diabetes mellitus			
Fasting	≥6.1	≥6.1	≥7.0
or			
2-h post glucose load	≥10.0	≥11.1	≥11.1
or both			
Impaired glucose tolerance			
Fasting	<6.1	<6.1	<7.0
and			
2-h post glucose load	≥6.7 and <10.0	≥7.8 and <11.1	≥7.8 and <11.1
Impaired fasting glycemia (IFG)			
Fasting	≥5.6 and <6.1	≥5.6 and <6.1	≥6.1 and <7.0
2-h post glucose load (if measured)	<6.7	<7.8	<7.8

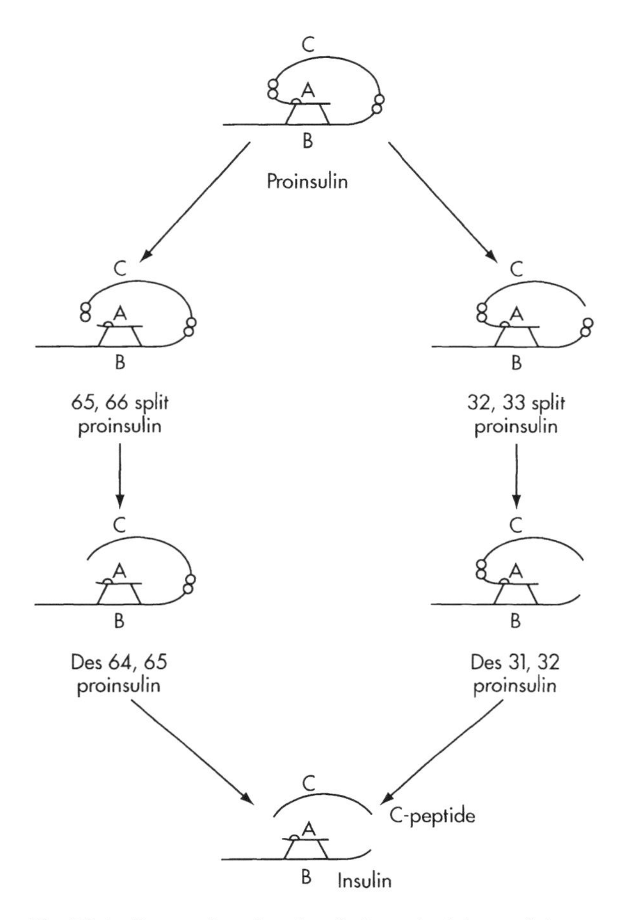

Fig. 53.1 Processing of proinsulin to major intermediates and to insulin.

insulin, its measurement in a body fluid may be used as an indicator of insulin secretion. Circulating concentrations of C-peptide will be elevated in renal failure. Proinsulin and its partially processed forms are thought to have little biological activity in comparison to insulin.

The amino acid composition of insulin is conserved among many species except for residues 4, 8, 9, and 10 of the A chain and 1, 2, 3, 27, 29, and 30 of the B chain. There is structural homology with insulin-like growth factors I and II.

Glucose is the primary stimulator of insulin secretion in mammals though secretion is also stimulated by the amino acids leucine and arginine and inhibited by the hormone somatostatin.

Insulin has a number of important metabolic actions central to the control of plasma glucose. It suppresses the endogenous production of glucose. Insulin inhibits the breakdown of glycogen to glucose in the liver and stimulates the uptake, storage and use of glucose in other tissues such as muscle and adipose tissue. Insulin also increases the synthesis of fatty acids, triglycerides, and proteins, and suppresses proteolysis and lipolysis. The counter-regulatory hormones to insulin, which increase the concentration of glucose in the blood, include glucagon, epinephrine (adrenaline), and to a lesser extent growth hormone and cortisol.

Reference Intervals

A number of factors influence the plasma concentration of these hormones, most notably the plasma glucose concentration, obesity (generally expressed as a body mass index, kg/m^2), age and ethnic group, and the assay used. The insulin/glucose relationship in neonates differs to that in adults. Fasting plasma insulin is increased at the time of puberty and insulin resistance is found in pregnancy.

The upper limit of the reference range for insulin for euglycaemic subjects is quoted as being between 90 and 195 pmol/L and the lower limit between 0 and 35 pmol/L. These differences may reflect differences in assay sensitivity and specificity, and also differences in the populations studied (Wark, 2003).

Fasting plasma proinsulin concentrations are normally <7 pmol/L though there is variation between assays and the numbers of subjects studied are few. The des 64,65 split proinsulin (i.e. proinsulin cleaved between amino acids 65 and 66 with subsequent loss of the pair of basic amino acids) is usually undetectable in fasting human plasma (less than 5 pmol/L) and the des 31,32 split proinsulin may be in the range of 1–20 pmol/L depending on body mass.

The upper limit of the reference range for C-peptide for euglycaemic subjects is quoted as being between 700 and 2300 pmol/L and the lower limit between 100 and 400 pmol/L. These differences may reflect differences in assay specificity and also differences in the populations studied (Wark, 2003).

In the investigation of hypoglycemia and the interpretation of the response to the 72 h fast a variety of figures are quoted as the 'normal' response for plasma insulin, C-peptide and proinsulin, e.g. <30, <300, <20 pmol/L, respectively, (Marks and Teale, 1993) and <36 pmol/L, <200 pmol/L, <5.0 pmol/L, respectively (Service, 1995).

Clinical Applications

The concentrations of plasma insulin, proinsulin, and C-peptide are normally measured to investigate the causes of hypoglycemia (plasma glucose <2.2 mmol/L, 40 mg per 100 mL) (Table 53.3) and in particular to diagnose insulinoma or nesidioblastosis where the plasma insulin (and/or proinsulin) is inappropriately high. Blood samples may be collected after an overnight or prolonged fast. Suppression tests are rarely used now. Clamping techniques and selective venous localization may be done in specialist centers and require technical expertise, special equipment, and medical supervision. Contra-

Table 53.3. Causes of hypoglycemia in adults.

In the 'well' patient
Drugs, e.g. oral hypoglycemic agents, salicylate
Ethanol
Insulinoma
Factitious induced by insulin
Ketotic hypoglycemia
Intense exercise
In the 'ill' patient
Drugs, e.g. oral hypoglycemic agents, salicylate, insulin, quinine, quinidine
Liver disease
Renal disease
Congestive cardiac failure
Non-β cell tumors, e.g. mesenchymal
Endocrine deficiency, e.g. cortisol, growth hormone
Sepsis
Surgical removal of pheochromocytoma
Insulin-antibody induced
Starvation
Shock
Total parenteral nutrition and insulin therapy

Service, 1995.

indications to the performance of any test should be considered. If exogenous insulin administration is suspected, the specificity of the immunoassay for different animal and human insulins should be taken into account (Owen and Roberts, 2004) and C-peptide should be measured. A drug screen for oral hypoglycemic agents should be performed.

In specialized laboratories these assays may be used in investigations into the pathophysiology of diabetes mellitus. A number of dynamic tests may be used, such as the OGTT, the intravenous glucose tolerance test (IVGTT) and various clamping techniques. These tests measure the insulin response to a glucose load and determine the degree of insulin resistance and/or secretion. It has been suggested that the loss of the first phase insulin response in the IVGTT can predict the development of Type I diabetes. A number of mathematical models and transformations have been used to derive functions of insulin sensitivity and beta cell function, for example homeostatic model assessment (HOMA) (Matthews *et al.*, 1985) and the insulin/glucose ratio (Katz *et al.*, 2000). The methods for some of these tests, and their interpretations, vary from center to center. They are applicable in research and epidemiological studies rather than to individuals. However, whichever method is used, the clinical diagnosis of diabetes mellitus is still determined from the plasma glucose concentration and clinical features. It remains to be established whether formal assessment of insulin resistance will become established in the assessment of those patients to be treated with drugs that improve tissue sensitivity to insulin.

It has also been suggested that the response of C-peptide to intravenous glucagon can predict the need for insulin treatment in Type II diabetes (Binder, 1991). In practice this is not used routinely, the decision on treatment protocols being made on clinical grounds, and the plasma glucose profile and glycohemoglobin results. However, the C-peptide response to glucagon may be used to assess beta cell function in pancreatic/islet cell transplantation.

Limitations

- Immunoassays for insulin, proinsulin(s) and C-peptide vary in their sensitivities and specificities. These factors should be taken into account when interpreting results. A number of assays for insulin measure proinsulin(s) and when these are elevated, as in Type II diabetes, the immunoassay may overestimate the plasma insulin concentration.
- Some assays may lack the sensitivity to detect suppressed insulin concentrations that are found in non-insulinoma hypoglycemia.
- Hemolysis may result in a fall in the measured hormone concentration and hemolyzed samples should not be assayed. The effects of other possible interferences should also be noted.
- It is recommended that samples be collected on ice and transported to the laboratory for immediate separation of plasma/serum, which should be stored frozen. Data suggest that it is C-peptide which is the least stable and that the degree of instability detected is assay dependent. Repeated freezing and thawing should be avoided.
- Some patients, particularly Type I diabetics treated with animal insulins, those with other autoimmune disease, or treated with sulfidryl-containing drugs such as methimazole and penicillamine, may develop antibodies to the insulin. These antibodies may interfere in assays for insulin, proinsulin, and C-peptide unless the sample is pre-treated, for example with polyethylene glycol. This procedure must be validated. Insulin autoantibodies have also been described in a number of non-diabetic subjects.
- Preparations of the split proinsulin(s) for use as standards are not widely available. There is an international reference reagent (coded IRR 84/611) for proinsulin and an international reference reagent (coded IRR 66/304) for insulin. Although there are commercially available quality assurance materials for insulin these are not always suitable for assays specific for human insulin. Quality assurance materials for the proinsulins are not widely available. The external quality assurance schemes are limited to insulin and C-peptide.

- There are few reported assays for the partially processed proinsulins.
- Cross-reactivity with synthetic/modified insulins should be known.
- Many drugs can raise or lower insulin concentrations through *in vivo* effects.
- Immunoassays for these hormones have been compared with high performance liquid chromatography and stable isotope dilution assays. These methods are technically demanding and use large volumes of sample and hence are unsuitable for routine use. However, they may be used to validate immunoassays.

Assay Technology

Conventional assays for insulin, proinsulin, and C-peptide were competitive radioimmunoassays (RIA). Care is required in reading the literature to determine the specificity and calibration of the assays used. More recently the development of immunometric assays, using monoclonal antibodies or both mono- and polyclonal antibodies in excess concentrations, has led to improvements in the sensitivity and specificity of these assays (Clark, 1999). A number of non-isotopic assays have also been described and are available commercially on automated immunoassay analyzers.

Insulin

Immunometric assays for insulin are available using an isotopic endpoint or more commonly horseradish peroxidase (HRP), alkaline phosphatase (ALP) or other non-isotopic signal generation systems such as chemiluminescence. The majority of these assays use a microtiter® plate as the solid phase or are available on automated immunoassay analyzers. Commercially available competitive assays are generally based on ^{125}I as a label, and PEG/second antibody separation or coated tube technology.

Proinsulin

A commercially radioimmunoassay is available for proinsulin, the majority of assays being immunometric assays with a microtiter plate solid phase. HRP and ALP (with enzyme amplification) have been used as the label in assays for intact and total proinsulin(s) as has chemiluminescence.

C-peptide

Many commercially available assays for C-peptide are classical radioimmunoassays though some competitive assays use HRP or alkaline phosphatase as label. Immunometric assays for C-peptide are available and have been automated.

Desirable Assay Performance Characteristics

Insulin

Competitive immunoassays for insulin tend to demonstrate significant cross-reactivity with intact proinsulin and the partially processed proinsulins whilst immunometric assays may show greater specificity. For the routine investigation of hypoglycemia the use of a non-specific assay may be preferable as some insulinomas are known to secrete significant amounts of proinsulin(s). In contrast, for research purposes, a more specific assay may be required in order to determine more accurately insulin secretion in the presence of increased concentrations of proinsulin(s). Some commercially available immunoassays, whether competitive or immunometric, lack the sensitivity to determine insulin concentrations at the lower limit of the reference range.

Proinsulin(s)

Immunoassays for intact and partially processed proinsulins vary in their degree of specificity. As a minimum full cross-reactivity data should be quoted. The sensitivity of many assays is insufficient to measure concentrations at the lower limit of the reference range.

C-peptide

C-peptide is unstable in whole blood, on storage and on repeated freeze–thaw cycles and the degree of instability has been shown to depend on the antiserum used. Stability data should be quoted.

Types of Sample

Blood should be collected and immediately taken on ice to the laboratory for separation. Plasma or serum may be used though reference ranges may differ and manufacturers' instructions need to be followed. Samples should be stored frozen. There is little published information on the effects of gel tubes and clot accelerators and this information should be sought from manufacturers. C-peptide is sometimes measured in urine. Insulin can be measured in blood-spot samples though such samples should be stored frozen (Butter *et al.*, 2001). A sample for the measurement of plasma glucose should be collected at the same time.

Frequency of Use

Mainly limited to specialized laboratories. Uncommon.

GLYCOHEMOGLOBIN (GHb)

More tests are done to monitor the health of diabetic patients than to diagnose the condition itself. This is true of both types of diabetes and careful, continuous management of the patient can lead to a full and normal life for several decades. The findings and recommendations of recent major studies (American Diabetes Association, 2004b) have emphasized the importance and consequent benefits of rigorous monitoring and especially the role of glycohemoglobin.

Two types of assays are used for monitoring diabetic patients: those that check the degree of metabolic control, and those that detect the appearance and progression of the complications that can accompany prolonged diabetes. In neither case are the analytes determined of primary significance to diabetes but rather they are secondary indicators of the short- and long-term effects of the disease.

The primary indicator of metabolic control in diabetes is the mean concentration of blood glucose, which is maintained by the patient through a combination of diet and drug therapy. Because of the rapid fluctuations in blood glucose concentration in response to diet and exercise, a random determination may be of limited use. Blood glucose meters do allow daily monitoring but there are limitations to the frequency with which measurements can be made. A means of making such a retrospective measurement is provided, however, by the ability of glucose to react directly with exposed amino groups of proteins in a non-enzymic glycation reaction. All proteins with susceptible residues will react if exposed to glucose and the extent to which they are glycated reflects three parameters: their intrinsic reactivity, their half-life and the mean glucose concentration. For any given protein, the first two factors are normally constant and hence the extent of protein glycation is a reliable indirect measure of the time-averaged glucose concentration experienced by the protein throughout its life. Results of these measurements are expressed as the percentage of analyte that is glycated.

Many proteins, particularly plasma constituents, have been studied in this context, but the one most useful for routine monitoring is glycohemoglobin. Hemoglobin is an $\alpha_2\beta_2$ tetramer of relative molecular mass 64.5 kDa, but its structure and function as an oxygen carrier are merely incidental to its use as an analyte for monitoring glycemic control in diabetes. What is important is the half-life of the molecule, which is determined in turn by the lifespan of the red blood cell of about 120 days. The molecule of hemoglobin is glycated at several surface lysine residues, but also possesses a uniquely reactive site, the amino-group of the *N*-terminal valine of the β chain. Glycation at this residue sufficiently alters the pK_a of the amino group to impart a relative negative charge to the molecule at neutral pH. It was on this basis that a glycated form of the molecule was first identified and separated by conventional biochemical techniques.

The reaction between the open chain form of glucose and the β chain valine residue is shown in Figure 53.2. The reaction proceeds in two stages, the first to form an unstable aldimine (or Schiff base) often termed pre-HbA_{1c}, which reflects the prevailing glucose concentration, and a second, slower and effectively

Val-NH$_2$ + HC=O–HCOH–HOCH–HCOH–HCOH–CH$_2$OH ⇌ HC=N-Val–HCOH–HOCH–HCOH–HCOH–CH$_2$OH ⇌ CH$_2$-NH-Val–HC=O–HOCH–HCOH–HCOH–CH$_2$OH

HbA_0 + Glucose ⇌ (Rapid) Pre-HbA_{1c} ⇌ (Slow, Amadori) HbA_{1c}

Aldimine Schiff base; Ketoamine

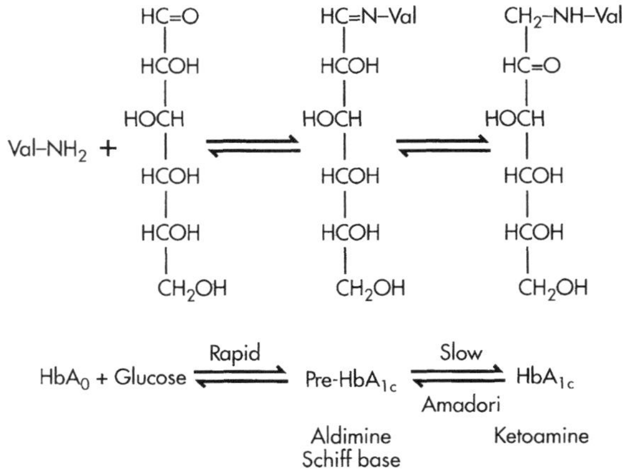

Fig. 53.2 The formation of HbA_{ic} from glucose and hemoglobin via the Schiff base pre-HbA_{ic}.

irreversible rearrangement that produces a stable ketoamine, HbA_{1c}. Other reactive hexose-phosphates (glucose-6-phosphate and fructose-1,6-bisphosphate) react with the same residue to form derivatives that also differ by charge and form fractions of the fast HbA_1 group. A comparable reaction sequence involving the ε-amino-group of lysine also occurs at a number of surface residues, to produce a heterogeneous fraction termed glycohemoglobin (GHb).

The analysis of GHb is complex because of the subtle chemical differences between the variously related derivatives of hemoglobin and the different subfractions that are measured. GHb is an unusual analyte, in that it can be determined by a multiplicity of fundamentally different methods that are not limited by considerations of sensitivity, since the analyte is present in high concentrations (John, 1997). Methods are based on differences in the antigenicity, charge or chemical reactivity of the glycosylated compared to the non-glycosylated forms. Moreover, the particular method used effectively defines the exact molecular species measured and, although all methods correlate reasonably well, the differences between methods may be significant and are not always fully appreciated.

In recent years these methods have been added to, and considerable effort has been made to develop a consensus view and especially an absolute reference method. Immunoassay methods, both enzymeimmunoassays (EIAs) and immunoturbidimetric methods, represent only part of the methodology available. Older methods, such as gel electrophoresis and chemical affinity (on boronate columns) have generally been replaced and improved upon by developments such as capillary electrophoresis, labeled boronate ligands and improvements to HPLC ion-exchange methods (Kobold *et al.*, 1997). Some of the chromatographic methods have now been automated in a form with a throughput of less than 10 min and of small enough size for use in clinics or satellite laboratories. More recently, electrospray mass spectrometry has been proposed as an absolute reference method (Roberts *et al.*, 1997; Peterson *et al.*, 1998). With such a variety of methods available, it is not surprising that standardization has proved difficult. Immunoassay methods, although generally precise and free from most interferences, are not necessarily superior to other methods and frequently the choice is dictated by other considerations including cost and instrumentation.

Reference Intervals

UK national guidelines (NICE, 2002a) for the treatment of Type II diabetes recommend that for an individual, a target HbA_{1c} (DCCT aligned) should be between 6.5 and 7.5%, dependant on the risk of micro- and macrovascular complications. The higher limit may be preferred in those at risk of iatrogenic hypoglycemia. *See* LIMITATIONS for a discussion of HbA_{1c} standardization and the relevance to reference intervals.

Separate reference ranges may be required for pregnancy (Hartland *et al.*, 1999).

Clinical Applications

Measurements of glycohemoglobin, whether using immunoassays or other technology, are used virtually exclusively for the monitoring of glycemic control as part of the routine management of diabetes. Hemoglobin has proved to be the most applicable analyte for this purpose, partly because of the relative ease with which the sample can be taken, but chiefly because the life span of the erythrocyte means that the retrospective window is about 60–120 days. Shorter time frames have not proved as useful except in the close monitoring of 'brittle diabetics', newly diagnosed or pregnant patients for which measurement of other shorter-lived glycated plasma proteins are appropriate (see later). Measurements of glycated proteins are not accepted as providing a diagnosis of diabetes, which remains based on the measurement of plasma glucose.

Limitations

- In general, immunoassays have high precision and good specificity, which is not always true of the other methods. Although most other technologies determine secondary changes, such as the charge or chemical affinity of the analyte, immunoassays use antibodies that react directly to the glycated sequence of the β chain. Depending on how much more of the sequence is required by the antibody for this recognition, an immunoassay may respond to some specific hemoglobin variants and not others. Of most relevance are HbS and HbC, which both have single amino acid substitutions at position 6 of the β chain. Although glycation of these variants does not give rise to HbA_{1c} as such, the glycated variants nevertheless still reflect the glycemic state of the patient. As long as there is no change in the percentage of the variant, and the effect of the variant in the assay system is not changed, it is theoretically possible to establish baseline values for a given patient. However, the usual reference ranges will not apply and this approach cannot be advocated.
- Physiological (e.g. pregnancy) and pathological conditions (e.g. renal failure, hemolytic anemias) that lead to decreased red cell survival, result in decreased values for glycohemoglobin, which do not reflect the degree of metabolic control. Blood glucose monitoring or measurement of glycated serum protein or glycated serum albumin may be used.
- The lack of agreement of GHb methods has led to international efforts to develop reference methods and standards. Clinicians should be made aware that ranges indicating good and poor glycemic control vary between assays.

In the mid-nineties the National Glycohemoglobin Standardization Program (http://www.ngsp.org) established a procedure such that any given assay could be traced to the DCCT reference method (DCCT aligned). Subsequently the International Federation of Clinical Chemistry published a reference method based on enzymatic cleavage of hemoglobin followed by reverse-phase high performance liquid chromatography and electron-spray ionization mass spectrometry or capillary electrophoresis (Jeppsson *et al.*, 2002). Most notably the reference ranges determined using the IFCC reference method are approximately 2% lower than those from DCCT aligned methods. This difference is clinically significant. Other national bodies had established reference methods. Further standardization with the preparation of appropriate and available standards and controls will be required.

Assay Technology

With increasing demand for the measurement of HbA_{1c} most assays provided by clinical laboratories will be fully automated, whether the assays are based on affinity/liquid chromatography or less commonly immunoassay. Automated methods are based on immunoturbidimetry and, because they determine an effective concentration of HbA_{1c}, also require a separate measurement of the total hemoglobin concentration, so that the result can be expressed as the conventional $\%HbA_{1c}$. A latex agglutination method that uses a monoclonal antibody raised against a glycated peptide corresponding to the *N*-terminal sequence of HbA_{1c} is widely used. The antibody agglutinates a suspension of latex particles that are coupled to this peptide. HbA_{1c} in the sample competes for the antibody thus inhibiting the agglutination and produces a type of competitive immunoassay.

The method also performs a simultaneous determination of the total hemoglobin concentration of the sample and these measurements are made, together with the necessary calibration, in a disposable cassette, which is read in a dedicated reader. Suitable instrumentation is available for use in diabetic clinics, using fingerprick samples and with results being available in less than 10 min.

Similarly a homogeneous, competitive immunoturbidimetry assay, together with a photometric total hemoglobin method, has been used on bench-top and larger analyzers. A polyclonal antibody recognizes the first four amino acids of the glycated β chain of hemoglobin and forms a soluble immunocomplex with HbA_{1c} in the analyte. Excess unreacted antibody then forms an immune complex with a poly-hapten form of the peptide coupled to a dextran carrier; aggregation is measured by increasing turbidimetry at 340 nm.

There are additionally a number of Point of Care devices, based on dry chemistries but which do require a degree of technical skill.

Desirable Assay Performance Characteristics

The National Glycohemoglobin Standardization Program recommends a target precision of CV $\leq$ 5%. With respect to bias the same group recommends that the 95% confidence intervals of differences between test methods and the Secondary Reference Laboratory (SRL) should be within $\pm$1% GHb of the SRL. The International Federation of Clinical Chemistry and American Association of Clinical Chemistry recommend that between run coefficients of variation should be less than 5%, though CVs of less than 3% are more clinically useful.

Types of Sample

These assays use whole blood as the sample, typically small volumes of around 100 μL. In order to liberate hemoglobin from the erythrocytes, part of the process requires the sample to be lyzed. Commonly available collection methods and anticoagulants (lithium-heparin, EDTA, fluoride–oxalate) can all be used. In addition, several attempts have been made to use hemoglobin eluted from dried blood spots, although this method is not in common use.

Frequency of Use

The management of diabetes varies both between and within countries, but most patients in the developed world are monitored at least annually and in some cases as often as four times per year. The National Institute of Clinical Excellence (2002a) has recommended that HbA_{1c} should be monitored at 2–6 month intervals. The interval depends on the acceptable level of control and stability of blood glucose control and/or changes in blood glucose concentrations and/or changes in therapy.

OTHER GLYCATED PROTEINS

Although a large number of proteins, both plasma and structural, may be glycated in diabetes besides hemoglobin, the only other glycated protein to be used as an index of glycemic control is glycated albumin. This is sometimes determined as a specific analyte, e.g. by immunoassay or as a major component in the glycated proteins of plasma or serum, e.g. as fructosamine. As with hemoglobin, the structure of albumin and its function as a transport protein are peripheral in this context and the most significant property of the molecule is its half-life, which, at about 20 days, is much shorter than that of hemoglobin. Thus albumin offers a measure of glycemic control intermediate between hemoglobin and the direct and immediate glucose determination.

Glucose reacts with albumin in the same way as with hemoglobin, but no reaction has been detected at the amino group of the *N*-terminal asparagine. All the glycated sites are reactive lysines, and there is no fraction of albumin analogous to HbA_{1c} that can be separated on the basis of charge. Glycated albumin is, therefore, a heterogeneous fraction of molecules modified at several sites; Lys 525 has been shown to be the most reactive with nearly half the non-enzymic glycation occurring at this residue.

Reference Interval

Glycated albumin measurements are not sufficiently common for reference intervals to be widely reported. The suggested upper limit of the non-diabetic range can vary between 2.4 and 3.0%.

Clinical Applications

Immunoassays for glycated albumin may be regarded as more specific alternatives to fructosamine and can be applied where short-term monitoring is important. However, of the two fructosamine is the more commonly used assay.

Limitations

Glycated albumin determinations either as fructosamine or specifically by immunoassay are only of use for short-term monitoring of diabetes and are not likely to replace glycated hemoglobin as the analyte of choice for routine long-term management. Results may not reflect glycemic control accurately in those patients where serum albumin concentrations are low.

Assay Technology

As with GHb, there are a number of simple chemical technologies available for the determination of glycated albumin and immunoassays have only recently become available. The ability of the ketoamine derivative to reduce nitro-blue tetrazolium (NBT) at alkaline pH forms the basis for the fructosamine method, which effectively determines the concentration of this derivative in unfractionated serum or plasma. Affinity chromatography on phenyl-boronate columns has also been applied to the measurement of glycated albumin using simple manual protocols derived from the more widely used method for glycated hemoglobin.

A number of immunoassays have been developed based both on monoclonal antibodies and on polyclonal antisera. The raising of these antibodies is critical to the specificity of the test. Some success has been achieved in producing antibodies specific to the glycated lysine residue alone which then have broad specificities for glycated proteins. In a different approach, others have raised antibodies to the glucitol-lysine derivative that is formed by the reduction of the ketoamine. This produces a more immunogenic molecule, but any assay must subsequently incorporate the same reduction step into its protocol.

One commercially available assay kit uses a 50 μL sample with a working range of up to 160 μg. A measurement of total plasma albumin is required to obtain the percentage of glycated albumin, the clinically relevant parameter.

Desirable Assay Performance Characteristics

Though desirable and detailed performance characteristics may not have been reported there is guidance on quality specifications (Fraser and Petersen, 1999). For long-term monitoring of glycemic control long-term precision will be critical.

Types of Sample

Both serum and plasma have been used for the determination of glycated albumin.

Frequency of Use

Rarely used.

MICROALBUMIN

The term 'microalbuminuria', i.e. the presence of low concentrations of albumin in the urine, is strictly a misnomer as it suggests the presence of small molecules of albumin. The term has become firmly established and will be used here. The determination of microalbuminuria is widely used as a predictor of microvascular disease in diabetes but is also a risk factor for cardiovascular disease and to a lesser extent end-stage renal disease. Elevations in urine microalbumin in patients with Type II diabetes with retinopathy, suggest that there is likely to be diabetic renal disease. If there is no retinopathy, the probability of other renal disease is increased.

Reference Interval

There is considerable variation in the values used to define proteinuria and microalbuminuria, which may relate in part to different analytical methods, different urine collection procedures and population differences. Those used by NICE (2002b) are given in Table 53.4.

Clinical Applications

Classic proteinuria, of which albumin is the major component, is associated with the late and irreversible stages of renal disease, and only comparatively recently have the earlier, incipient phases been unequivocally correlated with the lower urine concentrations of albumin, known as microalbuminuria. As diabetic nephropathy develops, there is a loss both in the size and in the charge selectivity of the basement membrane, and the excretion of albumin increases in absolute terms and also as a proportion of the total excreted protein.

The early detection of microalbuminuria, a risk factor for subsequent renal impairment and other complications, is critically important to its treatment, and once persistent proteinuria has developed, the decline of kidney function seems to be irreversible. At this stage conventional tests of renal function are used. There is much data to support the early treatment of hypertension in the diabetic, even when mild, in order

Table 53.4. Definitions of urine albumin excretion (NICE, 2002b).

Parameter	Normal	Higher-risk urine albumin excretion	
		Microalbuminuria	**Proteinuria**
Concentration (μg/mL or mg/L)	<20	20–200	>200
Albumin/creatinine ratio (ACR) (mg/mmol)	<3	Women ≥3.5 Men ≥2.5	≥30

to prevent and/or delay the development of nephropathy and other complications. Patients with diabetes are routinely monitored for microalbuminuria.

Limitations

- Physiological variations in excretion mean that care must be taken with the type of sample collected (see below).
- The range of concentrations of albumin in the urine in this patient group will be wide, and procedures should be in place to avoid the high dose hook effect.
- Differences in calibration of assays, particularly when the patient is monitored by both clinic and laboratory methods, may result in changes in measured analyte concentrations that are clinically significant resulting in inappropriate changes to treatment.
- There is a requirement for long-term stability of assays as they will be used for life-time monitoring of patients.
- Urine albumin is stable for 7–14 days at room temperature and at +4 °C (in the absence of urinary tract infection). Stability at −20 °C has been questioned, though lower temperatures are thought to be suitable.

Assay Technology

Overt proteinuria does not require a sensitive assay, and simple dipstick/Point of Care tests have been based on the spectral changes caused by the binding of albumin to an organic dye. Automated chemical methods are used to quantitate overt proteinuria. Point of Care devices should be evaluated to ensure that there is adequate sensitivity for the detection of microalbuminuria and appropriate quality assurance procedures put in place.

To detect albumin concentrations 100 times lower, however, more sensitive immunoassay methods have been developed including RIAs, immunoturbidometric and nephelometric assays and ELISAs, all of conventional design. The requirements for high sample throughput and fast assays means that automation is essential.

Desirable Assay Performance Characteristics

Because the measurement of microalbuminuria is used to monitor patients over many years long-term stability of the assays is required. It has been suggested that a method should have an interassay coefficient of variation of less than 12% over the concentration range 5–200 mg/L and that the assay should be able to detect changes of 10 mg/L in the range 5–35 mg/L (Rowe *et al.*, 1990). Criteria of adequate performance should be established for the albumin–creatinine ratio.

Types of Sample

The rate of albumin excretion is subject to several factors that influence renal function, but are unconnected with any pathological state. These include posture, rest or exercise, and time of day; these considerations are reflected in the type of urine sample collected and whether the results are expressed as a simple concentration or a rate of excretion. There are four ways in which urine is commonly collected:

1. **Timed overnight**. Allows the overnight AER to be determined; this is regarded as the most reliable index. Allows more stringent criteria to be applied and measures a lower AER because of overnight inactivity.
2. **First morning urine**. More convenient for the patient and retaining most of the advantages of a timed overnight sample, but only able to determine a single albumin concentration, rather than the AER.
3. **Random**. Most convenient for the patient, but single measurements of randomly taken urine are of little or no use because of the uncontrolled factors already mentioned.
4. **24 h**. Allows the average albumin excretion rate (AER) to be determined in a 24-h period, but is least convenient for the patient and subject to errors of collection.

It is common for additional measurements to be made to eliminate variations in renal function. Creatinine is the normal choice of analyte, and albumin concentrations are then expressed as an albumin/creatinine ratio (ACR).

Frequency of Use

It has been suggested that urine microalbumin should be measured at diagnosis and annually thereafter. If microalbuminuria or proteinuria is found, this should be repeated on at least two occasions within 1 month. More regular monitoring may be necessary to monitor treatment.

AUTOIMMUNE AND OTHER ASSAYS

Recent advances in the understanding of the pathogenesis of diabetes, especially of the autoimmune mechanism of Type I diabetes, have led to the identification of a number of immunological markers associated with this condition. These include autoantibodies to islet cell cytoplasm (ICA), endogenous insulin (IAA), two tyrosine phosphatases (IA-2A and IA 2βA) and to glutamic acid decarboxylase (GAD 65A).

The islet cell cytoplasmic antibody (ICA) is usually measured in peripheral blood by indirect immunofluorescence methods using frozen sections of human pancreas. Results are compared to a standard serum and reported as Juvenile Diabetes Foundation (JDF) units. Generally, results of 10 JDF units on two separate occasions or a single result ≥20 JDF units are considered as significant.

ICA has been detected in roughly 3% of a background population and in 15–30% of patients with Type I diabetes, though this rises to 70–80% in patients at the time of diagnosis. The antibodies may be detected temporarily in healthy subjects or following viral illness. The assessment of the risk of an individual developing Type I diabetes is therefore difficult and methodology must be accurate.

Antibodies to the protein tyrosine phosphatase IA–2A, IA2 beta A and GAD 65A have also been associated with Type 1 diabetes, though the radioligand binding assays used to detect the antibodies are technically demanding (Borg *et al.*, 1997). Radioligand assays using ^{35}S-labeled GAD showed high sensitivity but are technically demanding and subsequently commercial assays using ^{125}I have become available.

In addition to insulin antibodies detected in patients treated with exogenous insulin (normally polyclonal IgGs), insulin autoantibodies (IAA) have been identified in untreated patients who are either newly diagnosed with Type I diabetes or shown to be at high risk of developing it. These are measured by radioisotopic methods based on the displacement of insulin radioligand after the addition of excess non-labeled insulin. Whether these antibodies are a direct causal factor or a consequence of β-cell destruction is uncertain. The importance of IAA thus remains controversial and there are many conflicting reports in the literature, some of which have been attributed to shortcomings in the assays used (Greenham and Palmer, 1991).

None of the antibody assays mentioned above can be regarded as diagnostic nor of routine use in screening or monitoring. They can be considered as research tools, and the assays are not sufficiently well defined to give detailed performance characteristics. Nevertheless, the hope and expectation is that there will be major advances in the prediction of those at risk of developing Type I diabetes which will allow the development of therapeutic strategies for the prevention of this disease.

Obesity and insulin resistance are clearly associated with the development of Type II diabetes and the metabolic syndrome. Adipose tissue is now recognized as an endocrine tissue that plays a role in the control of energy homeostasis. Adipose tissue secretes a number of bioactive proteins, named adipocytokines which include leptin, adiponectin, tumor necrosis factor (TNF) alpha and resistin.

Leptin is a 16 kDa adipocyte derived protein that circulates in free and bound form (Considine *et al.*, 1996). Leptin acts by binding to specific receptors in the hypothalamus to affect energy intake and expenditure. Mutations in genes for both leptin and its receptor have been described in humans and are associated with gross obesity and serum leptin concentrations inappropriate for the degree of obesity. Generally though serum concentrations are directly related to body mass index. Leptin itself may suppress insulin secretion. A number of immunoassays are available for the measurement of circulating concentrations of leptin, including assays for serum free and bound leptin (Lewandoski *et al.*, 1999).

Human adiponectin is a 28 kDa collagen-like protein secreted by white adipose tissue. It has four domains including a carboxyterminal globular domain. It circulates as an oligomer formed by the association of three monomers. Four to six trimers may then associate to form oligomers in the peripheral circulation (Chandran *et al.*, 2003). This complex structure has important consequences for the design of immunoassays for the protein. Some of the differences in reported concentrations may reflect differences in antibody specificity. A number of commercial research assays are available. Although secreted by adipocytes, paradoxically circulating concentrations are higher in lean individuals than in obese. Adiponectin may have a causative role in insulin resistance and equally circulating concentrations may be influenced by hyperinsulinaemia. Antiinflammatory properties have also been described.

Resistin is an adipocyte-derived polypeptide that has been described in rodents with human resistin showing only 59% homology to the mouse form of the polypeptide. Study of mRNA expression, circulating tissue concentrations in both animals and humans has yielded conflicting data concerning the role of the protein in insulin resistance and obesity. The biological role of resistin remains to be established (Pittas *et al.*, 2004). Both competitive and immunometric assays for resistin are available (Pfutzner *et al.*, 2003). Their use is chiefly in the research laboratory.

REFERENCES

American Diabetes Association Position Statement, Diagnosis and classification of diabetes mellitus. *Diabetes Care* **27**, S5–S10 (2004a).

American Diabetes Association Position Statement, Tests of glycemia in diabetes. *Diabetes Care* **27** (Suppl 1), S91–S93 (2004b).

Binder, C. C-peptide and β-cell function in diabetes mellitus. In: *Textbook of Diabetes* (eds Pickup, J. and Williams, G.), 348–354 (Blackwell, Oxford, 1991).

Borg, H., Fernlund, P. and Sundkvist, G. Measurement of antibodies against glutamic acid decarboxylase 65 (GADA): two new ^{125}I assays compared with [^{35}S] GAD 65-ligand binding assay. *Clin. Chem.* **43**, 779–785 (1997).

Butter, N.L., Hattersley, A.T. and Clark, P.M. Development of a bloodspot assay for insulin. *Clin. Chim. Acta* **310**, 141–150 (2001).

Chandran, M., Phillips, S.A., Ciaraldi, T. and Henry, R.R. Adiponectin: more than just another fat cell hormone? *Diabetes Care* **26**, 2442–2450 (2003).

Clark, P.M. Assays for insulin, proinsulin(s) and C-peptide. *Ann. Clin. Biochem.* **36**, 541–564 (1999).

Considine, R.V., Sinha, M.K., Heiman, M.L. *et al.* Serum immunoreactive-leptin concentrations in normal-weight and obese humans. *N. Engl. J. Med.* **334**, 292–295 (1996).

Ehtisham, S. and Barrett, T.G. Emergence of type 2 diabetes in childhood. *Ann. Clin. Biochem.* **41**, 10–16 (2004).

Fraser, C.G. and Petersen, P.H. Analytical performance characteristics should be judged against objective quality specifications. *Clin. Chem.* **45**, 321–323 (1999).

Greenham, C.J. and Palmer, J.P. Insulin antibodies and autoantibodies. *Diabetes Med.* **8**, 97–105 (1991).

Hartland, A.J., Smith, J.M., Clark, P.M.S., Webber, J., Chowdhury, T. and Dunne, F. Establishing trimester-related and ethnic group-related reference ranges for fructosamine and HBA_{1c} in non-diabetic pregnant women. *Ann. Clin. Biochem.* **36**, 235–237 (1999).

Jeppsson, J.O., Kobold, U., Barr, J. *et al.* Approved IFCC reference method for the measurement of HbA1c in human blood. *Clin. Chem. Lab. Med.* **40**, 78–89 (2002).

John, W.G. Glycated haemoglobin analysis. *Ann. Clin. Biochem.* **34**, 17–31 (1997).

Katz, A., Nambi, S.R., Mather, K. *et al.* Quantitative insulin sensitivity index: a simple, accurate method for assessing insulin sensitivity in humans. *J. Clin. Endocrinol. Metab.* **85**, 2402–2410 (2000).

Kobold, U., Jeppsson, J.O., Dulffer, T. *et al.* Candidate reference method for hemoglobin A_{1c} based on peptide mapping. *Clin. Chem.* **43**, 1944–1951 (1997).

Lewandoski, K., Horn, R., O'Callahan, C.J. *et al.* Free leptin, bound leptin and soluble leptin receptor in normal and diabetic pregnancies. *J. Clin. Endocrinol. Metab.* **84**, 300–306 (1999).

Marks, V. and Teale, J.D. Hypoglycaemia in the adult. In: *Hypoglycaemia* (eds Gregory J.W. and Aynsley-Green A.), *Baillieres Clin. Endocrinol. Metab.* **7**, 705–729 (1993).

Matthews, D.R., Hosker, J.P., Rudenski, A.S. *et al.* Homeostatic model assessment: insulin resistance and beta-cell function from fasting plasma glucose and insulin concentrations in man. *Diabetologia* **28**, 412–419 (1985).

National Glycohemoglobin Standardization Program: http://www.missouri.edu/~diabetes/ngsp/

National Institute of Clinical Excellence. Management of Type 2 diabetes. Management of blood glucose Guideline G. London (2002a). www.nice.org.uk (Accessed April 2004).

National Institute of Clinical Excellence. Management of Type 2 diabetes. Renal disease: prevention and early management Guideline F. London (2002b). www.nice.org.uk (Accessed April 2004).

Owen, W.E. and Roberts, W.L. Cross-reactivity of three recombinant insulin analogs with five commercial insulin immunoassays. *Clin. Chem.* **50**, 257–259 (2004).

Peterson, K.P. *et al.* What is hemoglobin A_{1c}? An analysis of glycated hemoglobins by electrospray ionization mass spectrometry. *Clin. Chem.* **44**, 1951–1958 (1998).

Pittas, A.G., Joseph, N.A. and Greenberg, A.S. Adipocytokines and insulin resistance. *J. Clin. Endocrinol. Metab.* **89**, 447–452 (2004).

Pfutzner, A., Langenfeld, M., Kunt, T., Lobig, M. and Forst, T. Evaluation of human resistin assays with serum from patients with Type 2 diabetes and different degrees of insulin resistance. *Clin. Lab.* **49**, 571–576 (2003).

Roberts, N.B. *et al.* Potential of electrospray mass spectrometry for quantifying glycohemoglobin. *Clin. Chem.* **43**, 771–778 (1997).

Rowe, D.J.F., Dawnay, A. and Watts, G.F. Microalbuminuria in diabetes mellitus: review and recommendations for the measurement of albumin in urine. *Ann. Clin. Biochem.* **27**, 297–312 (1990).

Sacks, D.B., Bruns, D.E., Goldstein, D.E. *et al.* Guidelines and recommendations for laboratory analysis in the diagnosis and management of diabetes mellitus. *Clin. Chem.* **48**, 436–472 (2002).

Service, F.J. Hypoglycaemic disorders. *N. Engl. J. Med.* **332**, 1144–1152 (1995).

The Diabetic Control and Complication Trials Research Group, The effect of intensive treatment of diabetes on the development and progression of long-term complications in insulin-dependent diabetes mellitus. *N. Engl. J. Med.* **329**, 977–986 (1993).

UK Prospective Diabetes Study (UKPDS) Group, Intensive blood glucose control with sulphonylureas or insulin compared with conventional treatment and risk of complications in patients with Type 2 diabetes (UKPDS 33). *Lancet* **352**, 837–853 (1998).

Wark, G. UKNEQAS Guildford Peptide Hormones Annual Review. (2003). www.surrey.ac.uk/SBMS/eqas/gph.html (Accessed April 2004).

Yalow, R.S. and Berson, S.A. Assay of plasma insulin in human subjects by immunological methods. *Nature* **184**, 1648–1649 (1959).

54 Hematology

Derek Dawson, Harry Waters and John Ardern

NORMAL BLOOD FUNCTION

Blood is responsible for the transport of oxygen from the lungs to the tissues, and carbon dioxide from the tissues to the lungs. Blood also carries hormones, absorbed food materials, useful metabolites and waste products to the appropriate tissues and organs. Blood is used to regulate water balance and body temperature, and to protect the body against infection. It has a sophisticated mechanism that allows clotting at sites of injury, yet maintains fluidity in the rest of the circulation.

The manufacture of blood in the bone marrow is known as **hemopoiesis**, the production of red blood cells is **erythropoiesis**. DNA synthesis in the nuclei of maturing cell precursors depends upon a supply of two vitamins: vitamin B_{12} and folate. Vitamin B_{12} (properly called cyanocobalamin) is a member of the cobalamin family. However, unless a specific application is referred to, the term vitamin B_{12} will be used here since this is the stable form to which all other cobalamins are converted for reagent and testing purposes. Absorption of vitamin B_{12} from the normal diet depends upon the secretion of a carrier, **intrinsic factor**, by the parietal cells of the gastric mucosa. The B_{12}/intrinsic factor complex is absorbed in the terminal small bowel. Folate and iron, in contrast, are absorbed in the proximal small bowel.

The **erythron** is the total erythropoietic organ, including bone marrow and circulating red cells. The production of **hemoglobin**, the carrier of oxygen within the red cell, increases as the precursor cell matures. Hemoglobin, which gives the red cell its color, consists of iron atoms in heme molecules inserted into a protein shell (globin) that is essential for the proper control of the attachment and release of oxygen from the iron. Most of the iron in the body is in the circulating hemoglobin (1 mg in about 2 mL of blood) and much of the rest (500–1000 mg) is in the storage reticuloendothelial cells of the bone marrow, spleen and liver. The hemoglobin level is maintained by utilization of recycled iron from hemoglobin and, when necessary, by iron from the stores. The main storage protein for iron is **ferritin**, and the level of ferritin in the blood gives a reliable indication of the iron stores. In plasma, iron is bound to a β-globulin, **transferrin**.

Other cells in the circulation include the white blood cells, **leucocytes**, consisting of several types. **Polymorph neutrophils**, made in the bone marrow, respond to many acute bacterial infections whereas **lymphocytes** are concerned with defense against viruses and fungi, and with developing and controlling the immune response. A major component of this immune response is the synthesis of antibodies on exposure to foreign molecules. Re-exposure stimulates a much greater response. The lymphocytes originate in the lymphoid tissues of the bone marrow, thymus, spleen, and lymph nodes. A third group of circulating cells are the **platelets**, which are constantly used to repair minor capillary defects and also, when necessary, participate in the clotting of blood.

Plasma factors are very much concerned with **hemostasis**, the name given to a complex system of activating and feedback mechanisms that has evolved to provide not only a rapid response to tissue injury, with its potential blood loss, but also mechanisms to prevent unwanted thrombus formation. It depends upon the interaction of many factors, involving the endothelium of the vessel wall, platelets, coagulation factors in the plasma, naturally occurring anticoagulants (e.g. protein C, protein S, antithrombin) and **fibrinolysis** (clot digestion). When vascular damage occurs, platelets aggregate on the endothelium and a series of coagulation factors, known by their Roman numerals, are activated. Von Willebrand factor (vWF) supports platelet interaction with exposed endothelium and also acts as a carrier of factor VIII. The coagulation pathways lead to the splitting of fibrinogen by thrombin to give a fibrin gel. During this sequence proteases are generated that not only activate, but also degrade the coagulation factors. The deposition of fibrin and its removal are under the control of the fibrinolytic system consisting of a proenzyme, **plasminogen**, with its associated activators and inhibitors. Plasminogen is mainly activated by a tissue protease, **tissue-plasminogen activator** (t-PA), at the same time as the coagulation sequence is started.

Plasmin, the end point of the fibrinolytic system, cleaves fibrinogen and fibrin into degradation products (FDPs) which include fibrinopeptide A and D-Dimer. Plasminogen activator inhibitor-1 (PAI-1) is produced mainly by endothelial cells and is a powerful inhibitor of t-PA.

CLINICAL DISORDERS

ANEMIA

A shortage of red blood cells, and hence a low level of hemoglobin, results in insufficient oxygen being transported to the tissues. This condition is known as **anemia.** The hematocrit (packed cell volume) and the red cell count are used as indicators of red cell mass. The symptoms of anemia are the same, whatever its cause, and include breathlessness on exertion, and fatigue.

The causes of anemia are

- **failure of red cell production**:

  (a) decrease in erythropoietic mass, **aplastic anemia**, caused by failure of the bone marrow. This is a very serious disease, which is sometimes treated in young patients by bone marrow transplantation from a matched donor;

  (b) faulty cell formation:
  (i) nuclear defect, e.g. vitamin B_{12} or folate deficiency, which causes **megaloblastic anemia**, characterized by the presence of large, nucleated primitive bone marrow cells. An example is **pernicious anemia**;
  (ii) heme defect, e.g. **iron-deficiency anemia**, depletion of the body's iron stores. This is the commonest cause of anemia;
  (iii) where the rate of erythropoiesis exceeds the supply of iron to the erythron, irrespective of the level of storage iron present, a **functional iron deficiency** can occur (Cavill *et al.*, 1997). Many advances have been incorporated into new generation blood count analyzers. The use of flow cytometry has brought about a re-evaluation of some routine, traditional parameters (Waters and Seal, 2001). With some instruments it is possible to determine the hemoglobin content of individual red cells. This measurement reflects the iron content of circulating red cells and provides a direct indication of iron delivery to the bone marrow (Macdougall *et al.*, 1992). Accurate counts of circulating reticulocytes (juvenile red cells) can be used to indicate the amount of RNA present and hence their maturity. The presence and number of their precursor nucleated red cells in the circulation can also be indicated. These parameters are useful indicators of erythropoiesis, which can contribute to the differential diagnosis of the various forms of anemia;
  (iv) globin defect, **hemoglobinopathies.** The commonest group of conditions, **thalassemia**, involves a reduced rate of synthesis of one or more of the globin chains;

- **hemorrhage**;
- **hemolysis**, **hemolytic anemia**: increased red cell destruction within the circulation or reticuloendothelial system brought about by stress situations e.g. sickle-cell anemia, characterized by severe anemia with sudden exacerbations.

POLYCYTHEMIA

This term describes a group of conditions that have an **increase in red cell production** (**erythrocytosis**) as their common essential feature (Messinezy and Pearson, 1999). In contrast to anemia the hematocrit is raised in these conditions. The erythrocytosis may be **primary** due to an intrinsic bone marrow defect (polycythemia vera) or **secondary** due to increased erythropoietin (EPO) secretion. Both primary and secondary polycythemias can be congenital or acquired. A third, heterogeneous, group has been termed **idiopathic erythrocytosis** (Messinezy and Pearson, 1999). In addition to full blood count and hematocrit, direct radioisotopic measurement of the patient's red cell mass and plasma volume together with assay of the circulating EPO level are usually required for the differential diagnosis of polycythemia.

IRON OVERLOAD AND HEMOCHROMATOSIS

There is no active system for excreting excess iron from the body. No harm results when the iron is confined to macrophages but ultimately there is deposition in tissues, especially the liver, with resulting cirrhosis, diabetes and skin pigmentation. This state is called **hemochromatosis**. Genetic hemochromatosis is an inherited condition resulting from excessive amounts of iron being absorbed from the diet. It is one of the most common recessive disorders in Northern Europe. Measurement of transferrin saturation is the 'best' test for indicating potential or early iron overload and the degree of overload can be confirmed and monitored using the serum ferritin concentration. A diagnosis of hemochromatosis should be confirmed by demonstrating liver iron overload. When follow up and family studies are carried out genetic testing is valuable (Worwood, 1998). Commonly known mutations of the *HFE* gene are C282Y and, to a lesser extent, H63D. **Iron overload** can also be secondary to a prolonged inability to use iron that has entered the body in excessive amounts. This is typically seen in thalassemia when the anemia is treated by blood transfusions.

HEMOSTATIC DISORDERS

Hemorrhage occurs when there are disturbances of the capillary/platelet system (often spontaneous bleeding) or of the coagulation process (excessive bleeding following injury). **Thrombocytopenia**, a deficiency of platelets, is common and usually due to infection or autoimmune disease.

Coagulation defects may be due to an inherited deficiency of a plasma factor. **Hemophilia**, factor VIII deficiency, is the classic example. More commonly they are due to an acquired deficiency of several factors resulting, for example, from decreased production (as in hepatic disease) or due to excessive utilization as in **disseminated intravascular coagulation** (**DIC**), a not uncommon result of severe infection or tissue trauma. The fibrinolytic system and other plasma proteins that degrade activated coagulation factors limit coagulation to the site of injury. **Thrombosis**, intravascular coagulation of the blood with vessel occlusion, may occur spontaneously when these factors are lacking. Factor V Leiden is a variant coagulation factor, inherited as an autosomal dominant trait, resistant to degradation by the anticoagulant protein C.

The lupus anticoagulant is found in a variety of autoimmune diseases and occasionally in otherwise healthy persons. Contrary to what its name suggests, its presence is associated with a thrombotic predisposition.

The ideal requirement in clinical practice is for reliable estimates of the number and aggregability of platelets, the functional activity of the coagulation factors, and tests for evidence of hypercoagulability and aberrant coagulation. In general, immunologic techniques are of greater value in conditions of hypercoagulability than in bleeding disorders.

LEUKEMIA AND LYMPHOMA

Leukemia is a malignant proliferation of precursor (acute type) or mature (chronic type) cells of the leukocyte series, with an abnormal increase in the number of white blood cells in the circulation. **Lymphoma** is a malignant state of the lymphoid tissue, usually producing enlarged lymph nodes. **Hodgkin's disease** is the most common form.

ANALYTES

ANEMIA

Vitamin B_{12} and folate

These two vitamins are considered together because of their close interaction and the identical hemopoietic changes that result from their deficiency.

The cobalamins are found in nature in several forms. Vitamin B_{12} is one of these (cyanocobalamin). Their common structure consists of two parts: a planar group and a nucleotide set at right angles to it. The planar group is a corrin ring with a central cobalt atom, the nucleotide consists of a 5,6-dimethyl benzimidazole and a phosphorylated sugar. The functional group is attached directly to the cobalt atom; the most common form is deoxyadenosyl, found in the liver and most tissues. Other forms are methylcobalamin, the predominant type in plasma, and hydroxycobalamin, to which the other forms convert on exposure to light. The vitamin was first crystallized as cyanocobalamin (molecular mass 1355) and was designated vitamin B_{12}. This does not occur naturally. However, it is the stable form to which the functional groups are converted in the assay. Corrinoids are analogs of the vitamin lacking the specific nucleotide.

Folic acid, pteroylglutamic acid (PGA, molecular mass 441), consists of a pterin that is attached to glutamic acid through a *para*-aminobenzoic group. In nature it is found in reduced forms with an additional single carbon unit (e.g. methyl, methylene, formyl) and a variable number of glutamate radicals. Folates act as coenzymes in the transfer of single carbon units in amino acid metabolism. Because of the many functional forms, and the rarity of PGA in nature, the term folate is used to cover all forms. Plasma folate is a monoglutamate whereas red cells contain predominantly penta- and hexa-glutamates.

Only two reactions are known in man that require cobalamin. In one, deoxyadenosyl cobalamin is required as coenzyme in the conversion of methylmalonyl coenzyme A (CoA) to succinyl CoA. With cobalamin deficiency the serum level and the urinary excretion of methylmalonic acid increase. These changes can be used to demonstrate deficiency though this reaction is not connected with the defect leading to the anemia. In the other reaction, methylation of homocysteine to methionine requires both methylcobalamin and reduced methylfolate. Reduced folates, other than the methyl form, are built up into polyglutamates to keep the folates within the cells. 5,10-methylene tetrahydrofolate polyglutamate is the enzyme form (shown as Polyglutamate) required in the synthesis of thymidylate (dTMP) for DNA (see Figure 54.1).

The anemia due to deficiency of these vitamins is recognized by morphological (megaloblastic) changes in the nuclei of the bone marrow precursor cells. Cobalamin deficiency causes a build up of methylfolate and lack of

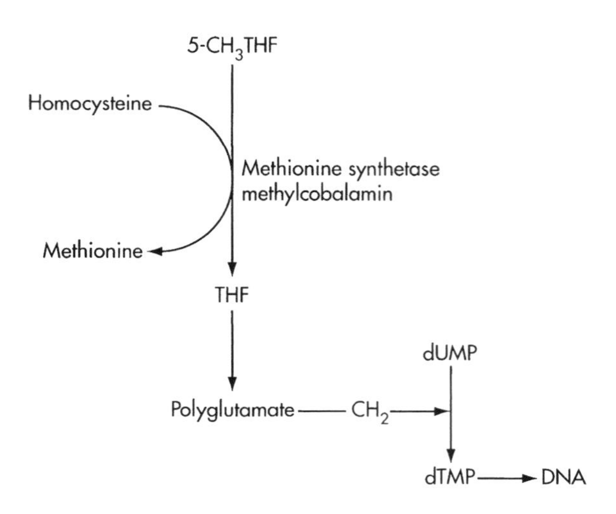

Fig. 54.1 Interaction of cobalamin and folate in DNA synthesis.

folate polyglutamate whereas folate deficiency leads to a direct lack of the latter. Synthesis of the pyrimidine is essential in man whereas it is available from a salvage pathway in other species. Hence, the characteristic megaloblastic anemia is seen only in man. The cause of the demyelination of axons with resulting nerve damage that occurs with cobalamin deficiency is unknown.

Higher animals and plants are unable to synthesize the vitamin and rely on the activity of microorganisms for their provision. The only source for man is food of animal origin, and it is present in virtually all animal tissues, eggs, and dairy products. Folates are found in most foods. Adults require a daily intake of at least 1 mg cobalamin and 100 mg folate. Dietary insufficiency is worldwide (Dawson and Waters, 1994). The main stores of both are in the liver. Vitamin B_{12} (cobalamin) deficiency is usually due to malabsorption or, in vegans and severe lacto-vegetarians, malnutrition.

Vitamin B_{12} malabsorption of **cobalamin** results from:

- deficiency of intrinsic factor (IF) in the gastric juice. Physiological amounts of the vitamin cannot be absorbed unless bound to IF. Pernicious anemia is due to an autoimmune gastritis in which the IF-producing parietal cells are destroyed. Resection of the stomach has the same effect;
- lack of gastric acid, achlorhydria, may prevent proper digestion of animal foods, and the vitamin may not be released to combine with IF;
- disease or resection of the terminal small bowel, the site of absorption of the B_{12}-IF complex. Crohn's disease, ulcerative colitis, tuberculosis and severe celiac disease are the usual causes;
- utilization of the vitamin by excessive bacterial flora in the gut, usually due to jejunal diverticulitis.

Deficiency of **folate** usually results from

- *malnutrition*. This is by far the most common cause worldwide;
- *malabsorption*, due to disease of the proximal small bowel. Celiac disease nearly always causes folate deficiency;
- *increased requirement*, as in pregnancy and chronic hemolytic states (e.g. sickle-cell anemia).

Reference Intervals

For vitamin B_{12}, a reference interval of 200–900 ng/L (148–664 pmol/L) embraces most assays though some have a lower limit of 150 ng/L. Values below but within 20% of the lower end of the interval are usually considered to be in an indeterminate area (British Committee for Standards in Haematology (BCSH), 1994a).

The interval for serum folate is 2–15 μg/L (5–34 nmol/L) and for red cell folate 160–600 μg/L (362–1360 nmol/L). For red cell folate determination, a whole blood hemolysate is assayed, followed by calculation of the red cell folate using the hematocrit. This conversion is needed to allow for the effect of anemia. There is considerable variation in the assay results of both serum and whole blood folate by different methods. Up to 9-fold differences in folate concentration, especially marked at low levels, with both serum and whole blood have been noted (Gunter *et al.*, 1996) and this is reflected to a certain extent in the variability of the reference intervals offered by manufacturers. This may be partly due to variation in the population upon which the reference interval has been based but is also due to assay methodology, especially with regard to sample preparation. All manufacturers state that their reference intervals are for guidance only and that users should determine their own. Because of these differences the result of the assay needs to be interpreted, whether normal, indeterminate or reduced, both for the clinician and for external quality assurance programs.

Clinical Applications

- **Macrocytic anemia**. Deficiency of either vitamin leads to the production of erythrocytes with an increased mean corpuscular volume (MCV). Anemia develops later. When the particular nuclear abnormalities have been demonstrated the term **megaloblastic anemia** is applied and vitamin deficiency is indicated.

With severe deficiency the leukocyte and platelet count also fall.

- **Neuropathy**. Damage to nerves, peripheral neuropathy, occurs with B_{12} deficiency. With progressive deficiency, damage to the spinal cord occurs (subacute combined degeneration of the cord).
- **Psychiatric changes**, **mental impairment** and even **dementia** may occur with B_{12} deficiency, and **depression** may accompany folate deficiency.
- **Infertility** can occur with either deficiency.
- **Intestinal investigations**. Because of the particular sites for the absorption of the vitamins, assessment of the patient's vitamin status may help in the differential diagnosis and management of intestinal disorders.
- **Homocysteinemia**. Deficiency of either vitamin may lead to an increase in plasma homocysteine. Folic acid can reduce the plasma level even when it is normal (Daly *et al.*, 2002). Hyperhomocysteinemia is associated with thrombotic and vascular disease (Scott and Weir, 1996; Quinlivan *et al.*, 2002) and assessment of the folate status may need wider application than in the investigation of blood disorders.

Limitations

- The serum B_{12} level depends not only upon the amount in the stores but also upon the level and turnover of the serum binders, the transcobalamins. A low serum B_{12} level may be found without tissue depletion in folate deficiency, pregnancy and the malignant bone disease, myelomatosis. It is, therefore, necessary to review the results of vitamin B_{12} and folate assays together. In folate deficiency the folate level falls much more than the B_{12}, whereas in B_{12} deficiency the low serum B_{12} is usually accompanied by a normal serum folate and a red cell folate that is normal or only mildly reduced (see Table 54.1).
- A 'falsely' normal or raised vitamin B_{12} level may occur in myeloproliferative diseases and with a marked leucocytosis.
- Methotrexate and folinic acid cross-react with folic acid and can interfere with the assay of serum folate.

Table 54.1. Comparison of typical results of vitamin B_{12} and folate assays in vitamin B_{12} and folate deficiency.

Deficiency	Serum vitamin B_{12}	Folate	Red cell folate
Vitamin B_{12}	↓↓(↓)	N(↑)	N(↓)
Folate	N(↓)	↓↓	↓↓(↓)

Arrows in brackets indicate less common finding; N = normal.

- Human anti-mouse antibodies, generated by therapy or diagnostic procedures, can interfere with the immunologic stage of some assays, depending on the preliminary treatment of the sample.
- A single serum folate measurement must be interpreted with care since the level can vary under the influence of physiological factors such as diet. In addition, a state of negative folate balance, without tissue depletion, is common in acute illnesses, and a low serum folate may be found in 30% or more of hospital patients.
- A 'falsely' low red cell folate is common in B_{12} deficiency since this is required to build up the intracellular folates.
- Young red cells have a higher folate content than older cells and an increase in reticulocytes may cause the red cell level to be normal when there is tissue deficiency.
- A blood transfusion may give a falsely normal result.
- The red cell level may be normal with acute folate deficiency. This is particularly seen in intensive care units.
- A case can be made for either the serum or whole blood folate to be assayed but, because of the limitations of each, it would seem reasonable to assay both or to assay the whole blood whenever the serum folate is low. The practice of assaying whole blood and converting to red cell folate using the hematocrit without correction for the serum folate level can give falsely elevated results in patients with tissue depletion who have recently started folate therapy.

Assay Technology

Vitamin B_{12}

This is measured using IF in place of the antibody used in an immunoassay.

The type of assay is referred to as competitive protein binding (CPB). Semi and fully automated non-radioisotopic systems have largely superseded the original radioisotopic manual kits though the principle of the assay remains the same. In these assays, a known amount of labeled vitamin B_{12} is diluted with the B_{12} in the test sample, which is first extracted from the serum binders. A volume of the mixture is bound to the specific binding protein, IF, which is added in an amount insufficient to bind all the labeled B_{12}. The bound is separated from the free and its label measured. The reaction signal is inversely related to the vitamin B_{12} concentration in the test material. True vitamin B_{12} radioimmunoassays (O'Sullivan *et al.*, 1992) have been described but are not available commercially. In assays employing IF as the binding agent an immunologic step is commonly used in setting up the competitive binding conditions whereby an anti-IF is attached to a solid phase *via* an anti-Ig antibody (Wallac, Bayer, Beckman). The Wallac AutoDelfia assay system may be taken as an example. In this, the microplate wall is coated with anti-mouse IgG. In the first incubation anti-IF IgG binds to the coated wall. In the second, vitamin B_{12}

labeled with tracer (europium) and that in the sample bind to added IF, and the B_{12}–IF complexes compete for sites on the solid-phase anti-IF antibody. Unbound B_{12} is removed by washing the plates. Enhancement solution releases the bound lanthanide ions to chelate with an organic ligand to give a fluorescent signal (see Figure 54.2). An advantage of this type of system is that anti-IF antibodies that may be present in the sample do not interfere with such an assay.

The extraction of vitamin B_{12} from its binders

The binders can be denatured by heat at an alkaline pH or by raising the pH to 12.0 or higher. Human anti-mouse antisera are also denatured at this high pH. The denatured protein will give a varying amount of non-specific binding (NSB) which may be excessive with sera in certain diseases such as chronic granulocytic leukemia. In solid-phase systems this would yield a falsely low result. Dithiothreitol (DTT) reduces NSB. Conversion of all cobalamins to the cyano (vitamin B_{12}) form is usually undertaken at this stage since the affinities of the co-enzymes for IF differ.

Binding agent

Porcine IF is used in purified form or its specificity achieved by attachment to IF antibody or addition of cobinamide, a cobalamin analog, which preferentially binds to non-IF cobalamin binders rather than to IF. The subsequent adjustment of the pH to that appropriate for the later IF-binding stage is especially important. In solid-phase systems the IF can be carried on the wall of the reaction tube, microparticles, beads, paramagnetic particles, or on a glass fiber matrix. Recombinant human IF has been produced (Fedosov *et al.*, 2003). The key properties of this preparation were identical to those of native IF. This offers the potential for large scale production of human IF for analytical and therapeutic purposes.

Label

The label is either isotopic, CN(^{57}Co)cobalamin, or chemiluminescent (acridinium ester, or alkaline phosphatase/dioxetane-phosphate) or fluorescent (alkaline phosphatase/4-methylumbelliferyl phosphate or europium).

Calibrators

The calibrators are made using cyanocobalamin and range from 0 to 2000 ng/L (1476 pmol/L). With automated systems a master rather than a full set of calibrators may be used for individual runs. The zero calibrator corrects for NSB. It is important that the protein content of the calibrators should be as similar to that of the test sample as possible. A WHO B_{12} standard is available from the National Institute for Biological Standards and Control (NIBSC – *see* CALIBRATORS AND STANDARDIZATION for address and web site). Its use should reduce differences between methods in their reference intervals. However, a multi-laboratory study showed that all methods then in use, although giving a wide scatter of values when assaying normal and pernicious anemia sera, were able to give meaningful clinical results (International Committee for Standardisation in Haematology (ICSH), 1986). A method for the preparation of calibrators, approved by the International Committee for Standardisation in Hematology, has been published (Nexo *et al.*, 1989).

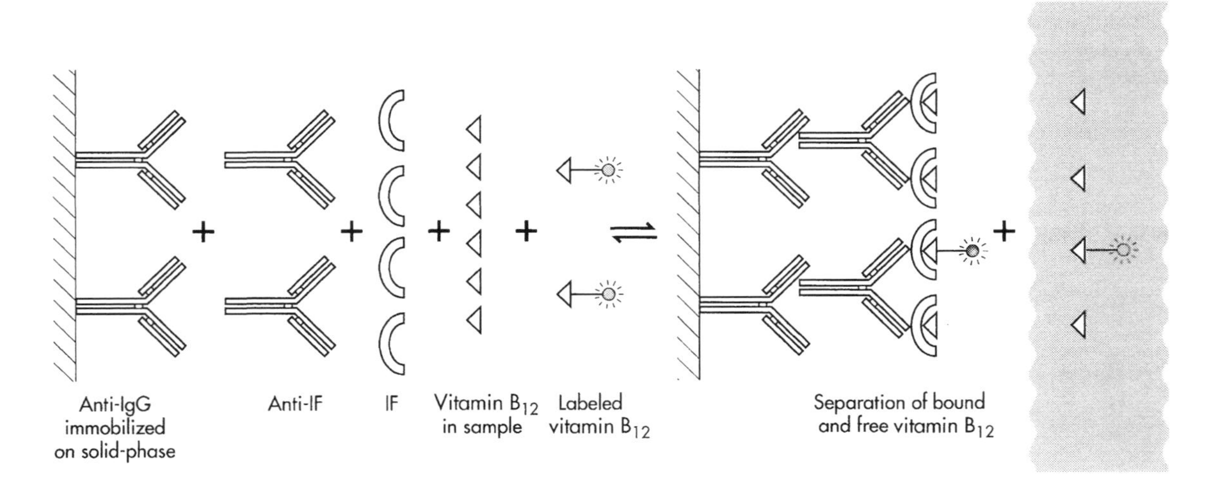

Fig. 54.2 CPB assay for vitamin B_{12}.

Separation of free and bound B_{12}

In semi and fully automated systems, the vitamin B_{12} bound to IF is separated from the free B_{12} by washing or magnetization of the solid phase. In liquid systems, charcoal or cellulose is used to remove unbound B_{12}. The amount of labeled vitamin B_{12} that is bound is usually estimated and this is inversely related to the concentration in the sample.

Serum Folate

The principle and procedure of the assay is the same as for vitamin B_{12}. Anti-mouse IgG may be used to attach anti-folate binder IgG to a solid phase (e.g. Beckman, Wallac, Bayer). Ascorbic acid or DTT are used to maintain the folate in the reduced, stable form. β-Lactoglobulin from cows' milk is the binding agent commonly used and is more reliable than porcine serum. The correct pH is important at the binding stage. At pH 9.3 the affinity of folic acid (calibrator) and methyl folate (serum) for the binder is the same. Non-radioactive labels are in common use. The tracer in radioactive kits is ^{125}I folic acid. Because the emissions of this and of ^{57}Co can be readily distinguished, and because the pH for folate binding is satisfactory for IF-B_{12} binding, dual assays for the simultaneous determination of cobalamin and folate have been popular.

Calibrators

Most methods use folic acid calibrators since methylfolate is less stable. This may cause some underestimation of the serum level. An International whole blood standard is available from NIBSC.

Red Cell Folate

The affinity of the binder for folates varies with the number of glutamate residues. It is necessary, therefore, to hemolyze the sample to release the folates and convert them to one form. For reproducible results, complete conversion to a monoglutamate is required. Adequate dilution of the red cells, a pH between 3 and 6, ascorbic acid to preserve the reduced folate and a deconjugase, provided by the plasma in the diluted whole blood sample, are needed for conversion, reduction, and preservation of the folate. Inadequate lysis and deconjugation give falsely low results. Dilutions of 1 in 20 or greater in fresh ascorbic acid solution and incubation in the dark at room temperature for 30–60 min, are suitable. The final ascorbic acid concentration recommended by manufacturers varies from 0.2 to 1.1%. The higher concentration is the more appropriate and kits using the weaker dilutions are generally associated with higher results. It is essential to the accuracy of the final result that the manufacturer's instructions for hemolysate preparation and storage are followed fully. The hemolysate is assayed in the same way as a serum sample, correcting the result for the dilution serum folate level and hematocrit. With severe anemia adjustment for the low hematocrit can contribute to imprecision.

Desirable Assay Performance Characteristics

Precision for within batch assay should be <5% and for between batches <10%. The best precision is required at the lower end of the reference interval, and the amount of labeled vitamin B_{12} added should be such that about 40–50% of it is bound at this level. Accuracy should be controlled by the standardization of calibrators against the vitamin B_{12} and whole blood folate International Standards. It has been shown previously that, for vitamin B_{12}, the mean of all methods in external proficiency programs is close to the true concentration (Dawson *et al.*, 1987) though, with changing technology, this can no longer be presumed. Reagent stability over a reasonable period is important for the laboratory carrying out small numbers of tests.

With fully automated analyzers errors are most likely to be limited to the pre- and post-analytical stages. These may be dilutional, for example, in the preparation of the red cell lysate, or transcription errors. Consideration, therefore, needs to be given to sample labeling and tracing features. The use of barcoded labels reduces the possibility of transcription faults in addition to faster patient data input. Adequate interfacing of analyzers to a host computer should allow bi-directional transfer and matching of patient identification and results. Linkage to an electronic transfer system reduces the chance of error at the request receipt and result delivery stages. In addition automation allows faster throughput of specimens and in some systems allows both continual and random access.

Types of Sample

Serum is usually used for vitamin B_{12}. Plasma from EDTA-anticoagulated blood may be used, though with liquid systems the level may be slightly lower than in serum. Heparinized plasma and plasma containing ascorbate and fluoride should be avoided.

Serum or plasma may be used for folate determinations. Anticoagulated blood is used for red cell folate.

Frequency of Use

Assay of serum B_{12} and folate are common and less so for red cell folate. These, together with the serum ferritin assay (below) are the most commonly used assays in hematology. This has resulted largely from the change in clinical practice from requesting these tests only when relevant to the clinical features and investigation to a non-investigative profiling approach (Waters and Seal, 2001).

INTRINSIC FACTOR ANTIBODY

Pernicious anemia is an autoimmune disease. Circulating antibodies to IF can be detected in the majority of patients. Two types have been described. Type I (blocking antibody) prevents the attachment of B_{12} to IF and type II prevents the attachment of IF or the IF-B_{12} complex to the ileal receptor. Antibodies to gastric parietal cells are found in the sera of most patients but are non-specific.

Reference Interval

Methods are essentially qualitative, giving a positive or negative result, or at best are semi-quantitative. However, they are mentioned here because of the unique relationship between measurement of serum B_{12} and detection of these autoantibodies.

Clinical Applications

Antibodies to IF are found in 60–75% of patients with pernicious anemia, though not in those with the juvenile form. The incidence increases with the duration of the disease. Together with a low serum B_{12} level they can be taken to indicate pernicious anemia. Most importantly, finding this combination obviates the need for further investigation, in particular vitamin B_{12} absorption studies, which are inconvenient for the patient (collection of full 24 h urine), involve administration of radioisotope (^{57}Co oral dose) and are expensive.

Limitations

- Intrinsic factor antibodies can rarely be found in patients with thyroid disease, diabetes mellitus and myasthenia.
- Serum from patients receiving vitamin B_{12} therapy may give false positive results when testing for type I antibody.

Assay Technology

The detection of type I antibodies is based upon the inhibition of vitamin B_{12} binding to IF by the test serum. The sera are incubated with IF immobilized on to a solid phase. The IF-B_{12} complex is then separated and Cn(^{57}Co) cobalamin added. If present, type I antibodies bind to the IF and reduce the radioactivity taken up by the solid phase. The results with test sera are compared with negative and positive sera provided by the manufacturers. Radio and enzyme immunoassay methods capable of detecting both type I and type II antibody (Conn, 1986; Waters *et al.*, 1989), or type II alone (Sourial, 1988) have also been described. The recombinant IF (Fedosov *et al.*, 2003) mentioned above has reagent potential, either as binder or label, in the demonstration of IF antibodies.

Desirable Assay Performance Characteristics

A balance between sensitivity and specificity has to be found. The former may be enhanced by increasing the concentration of serum but this may lessen specificity unless prior removal or neutralization of B_{12} binders in the sample is done (Nimo and Carmel, 1987). However, even in the most optimized conditions, antibodies have not been found in more than 75% of patients with pernicious anemia.

The incidence of type II antibodies is at least as high as that of type I (Conn, 1986; Waters *et al.*, 1993). The antibodies do not always coexist, so a system capable of detecting both antibodies is desirable.

Type of Sample

Serum.

Frequency of Use

Infrequent but increasing. Increasing difficulties are experienced with carrying out differential *in vivo* radioactive methods for determining the patient's ability to absorb vitamin B_{12}. Consequently the use of this simple, non-invasive, diagnostic test for pernicious anemia, despite its limitations, has gained popularity. Addition of IF antibody detection to the repertoire of immunoassay analyzers capable of measuring vitamin B_{12} is a logical development.

FERRITIN

Ferritin is the main iron-storage protein, providing a reserve of iron for heme (hemoglobin) synthesis. It consists of a 450 kDa protein, forming a hollow shell into which iron atoms may come and go. It is found in all tissues, with high concentrations in the liver, spleen and bone marrow. Ferritins from different tissues show heterogeneity (isoferritins). Those of the liver and spleen are basic and those of other tissues and of some tumors are acidic. In contrast to tissue ferritin, that found in serum consists of a number of glycosylated isoferritins of relatively low iron content. However, the concentration in serum may be directly related to body iron stores. This was first demonstrated by Addison *et al.* (1972).

One third of the stores is made up of another protein–iron complex: hemosiderin. This is composed of aggregates of ferritin molecules where the iron cores have become detached from their protein shells. Hemosiderin is not water soluble and is demonstrated by cytochemical techniques. The hemoglobin level is maintained by utilization of recycled iron from effete red cells and, when necessary, from the stores. Iron may be mobilized from both storage forms, but is more accessible from ferritin. Iron-deficiency anemia develops when the stores are exhausted.

Reference Interval

Most results from adult males fall within an interval of 15–300 μg/L. Those from females vary more with age, with premenopausal levels being lower than postmenopausal. Children tend to have lower concentrations than adults and the detection of childhood iron deficiency is more difficult. The lower end of the intervals for males proposed by manufacturers vary from 10–25 μg/L and for females from 3–17 μg/L. Nearly all manufacturers state that their reference ranges are for guidance only and that users should determine their own. Blood donation and iron supplementation need to be considered when doing this (Ledue *et al.*, 1994). The variation in reference intervals has been lessened by wider use of the WHO International Standard available from NIBSC.

Clinical Applications

The iron status of an individual may be assessed by a complete blood count to determine whether anemia is present and, if confirmed and there is doubt whether iron deficiency is responsible, by determining the level of iron stores. This can be done by aspirating bone marrow and staining it for hemosiderin. However, it is more practicable and less invasive to determine the serum ferritin concentration because this usually relates to the tissue stores. Each μg/L of serum ferritin is roughly equivalent to 8 mg of stored iron. Ferritin is, therefore, useful in detecting iron deficiency (low levels) and iron overload, e.g. due to hemochromatosis (high levels).

Limitations

- Although assay results below 10 μg/L indicate iron deficiency, there are a number of situations where the serum level does not equate so directly with the stores.
- Because ferritin is an acute-phase protein the level can be raised in inflammatory disorders and figures up to 50 μg/L may be encountered in these diseases, even when there is iron deficiency.
- Levels above 300 μg/L suggest iron overload but, in addition to inflammatory disorders, artificially high levels may be encountered in acute and chronic liver disease and in malignant diseases.
- In symptomatic idiopathic hemochromatosis, levels between 2000 and 4000 μg/L may be encountered. However, in the early stages the serum ferritin may be normal, when evidence of iron accumulation can be observed by an increase in the serum iron concentration and transferrin saturation. This is particularly important to remember when screening relatives for this condition.
- Current iron therapy gives falsely raised results.
- Blood transfusion, due to the introduction of hemoglobin iron raises the serum ferritin level.

Assay Technology

Competitive binding and two-site immunometric assays using radioactive labels have been largely superseded by immunometric assays using non-radioactive labels, primarily based on fluorescence or luminescence. These involve sandwiching the ferritin in the sample between two layers of antibody. Solid-phase immobilized anti-ferritin antibodies react with the ferritin and are then washed. Subsequent addition of excess-labeled antibody allows binding of this tracer in direct proportion to the ferritin content of the sample. The solid phase may be the reaction tube itself, beads, glass fiber carriers or paramagnetic particles. In these assays, the antigen–antibody complex is identified by measuring the activity of peroxidase or alkaline phosphatase. Addition of substrate leads to color, fluorescence or luminescence proportional to the antigen concentration of the sample.

Immunometric assays are sensitive and can detect levels of 0.2 μg/L ferritin. Serum proteins, in particular anti-immunoglobulin antibodies, which are to be found in 10% of individuals, may inhibit the binding of ferritin to the solid phase and affect systems using labeled antibodies. This can be a source of serious error producing falsely lowered or elevated ferritin values. Some manufacturers incorporate species immunoglobulins in the conjugate preparation to block any cross-reaction. The kit protocol should be scrutinized for information concerning cross-reaction. In practice it is very difficult to identify samples from patients falling into this category but cross-reaction should be considered when an unexpectedly high or low result is obtained.

In competitive binding radioimmunoassays the ferritin in the sample competes with ^{125}I-labeled ferritin for a limited number of antibody binding sites. The level of radioactivity bound is inversely related to the concentration of antigen in the sample. After incubation the bound and free ferritin are separated by precipitation of the antigen–antibody complex with a second antibody, an anti-globulin. The sensitivity of RIA is about 6–10 μg/L. In general competitive radioimmunoassays give lower results than immunometric assays.

The ferritin used as immunogen and for the calibrators should be of the same isoferritin and subunit composition as serum ferritin. Liver ferritin is almost identical and gives practically 100% cross-reactivity, whereas spleen ferritin is less so, with about 90% cross-reactivity. Some kits use monoclonal antibodies, which further enhance specificity. Standardization should be against the WHO standard.

A 'high-dose hook' effect, in which high concentrations of ferritin give the same signal as much lower concentrations, may occur; especially in single-incubation assays. Where an assay is known to be affected in this way samples should be tested at two dilutions. However, in practice, it is sufficient for high concentration samples, or samples with results out of accord with the clinical expectation, to be repeated at a higher dilution. This does not appear to be a problem with most of the present systems.

Desirable Assay Performance Characteristics

Because serum ferritin assays are required to reflect an extremely wide range of concentrations, careful consideration of the relationship between the analytical error involved and clinical classification is important. A dose response carrying maximum sensitivity and minimum error over the range of concentrations relevant to the purpose for which the assay is intended is desirable. Precision, within batch and between batch, should be 5% or less.

Absence of any 'high-dose hook' effect (see above) is important.

Types of Sample

Serum. Anticoagulated plasma may also be used with some systems but care must be taken and the manufacturer's protocol carefully scrutinized for information on this.

Frequency of Use

Common, and increasing due to the profiling approach mentioned above (*see* FREQUENCY OF USE OF VITAMIN B_{12} AND FOLATE ASSAYS).

TRANSFERRIN

In the plasma, iron is bound to a β-globulin, transferrin, that is indicative of the total plasma iron binding capacity. One milligram of transferrin binds 1.4 mg of iron.

Reference Interval

2.0–3.0 g/L (radial immunodiffusion).

Clinical Applications

Transferrin is normally one-third saturated with iron. In iron deficiency, the saturation is decreased and the transferrin concentration elevated before the appearance of anemia. The measurement of the serum iron and transferrin levels, and the derivation of the percentage saturation of the latter, are of value in screening for hemochromatosis (*see* FERRITIN above). In the early stages of iron overload a raised saturation of >50% in women and of >60% in men is to be found.

Limitations

- Estrogens (pregnancy, oral contraceptives) cause an increase in the transferrin concentration.
- The transferrin level and its saturation are reduced in inflammatory and malignant diseases. In hypoproteinemia both the level and its saturation are reduced.

Assay Technology

The measurement of transferrin is conventionally carried out by immunological techniques such as radial immunodiffusion and nephelometry. There is generally a good correlation between the results of chemical and immunological assays.

Desirable Assay Performance Characteristics

The usefulness of the assay is reduced by the lack of uniformity in calibration.

Type of Sample

Serum.

Frequency of Use

Infrequent as an individual measurement.

TRANSFERRIN RECEPTORS

The uptake of iron by immature red cells, and many other proliferating cells, involves the binding of transferrin to specific receptors on the cell surface (TfR). Transferrin delivers its iron to these receptors. Since the major use of iron is for hemoglobin production most TfR molecules are found on the erythroid precursor cells. During the iron transfer process TfR molecules are cleaved from the cell surface to join a serum pool (sTfR). Having the highest concentration of TfR, erythropoietic cells are the main contributors to this pool. Assay of sTfR, therefore, provides a indirect measure of total TfR and reflects either the cellular need for iron or the rate of erythropoiesis.

Reference Interval

Soluble receptor 8.7–28.1 nmol/L (Klemow *et al.*, 1990).

Altitude and ethnic origin can affect the reference interval (Allen *et al.*, 1998). These authors have reviewed calibration and reference intervals.

Clinical Applications

In theory, the estimation of the reticulocyte count together with the sTfR level should give the best guide to the level of effective erythropoiesis. As iron stores are exhausted the TfR concentration rises as the erythropoietic system responds. Unlike serum ferritin, sTfR levels are not influenced by inflammation or infection. sTfR assay should, therefore, be considered in anemic patients whose differential diagnosis includes iron deficiency and anemia of chronic disease, being raised in the former but normal or only slightly raised in the latter. Raised levels are also found in other causes of hyperplastic erythropoiesis (e.g. hemolytic anemia, β-thalassemia, polycythemia) whereas reduced levels occur in hypoplastic conditions (e.g. aplastic anemia, chronic renal failure, post-transplant anemia). It has been suggested that the sTfR assay may lack specificity, however, when expressed in combination with the serum ferritin concentration the results are more informative (Punnonen *et al.*, 1997; Means *et al.*, 1999; Flowers and Cook, 1999; Suominen *et al.*, 2000).

Limitations

Malignant lymphoid disorders may markedly increase the serum level.

Assay Technology

Circulating transferrin receptors may be detected by enzyme-linked and chemiluminescence immunoassay using dual monoclonal antibodies.

Desirable Assay Performance Characteristics

Antibodies may react differently to free and transferrin-bound receptors, giving widely differing values (Trowbridge, 1989). Convergence in antibody action would be of benefit.

Type of Sample

Serum. Anticoagulated plasma may also be used with some systems but care must be taken and the manufacturer's protocol carefully scrutinized for information on this.

Frequency of Use

Infrequent but likely to increase. As more immunoassay systems offer this assay its application will rise, particularly where the instrument software provides the option of linking the sTfR and serum ferritin levels, thus facilitating the adoption of an algorithm approach.

ERYTHROPOIETIN

Maintenance of the red cell volume is required to ensure an adequate oxygen supply to the tissues. The mechanism linking tissue oxygen delivery with red cell production is controlled mainly by erythropoietin (EPO). EPO is a glycoprotein with a molecular mass of 30.4 kDa (Kendall, 2001). It is secreted by the liver in the early stages of fetal development and thereafter 90% of production is taken up by the kidneys according to the concentration of oxygen in the blood. Lower oxygen levels result in increased secretion and *vice versa*. Underproduction of EPO will, therefore, result in anemia (e.g. in end stage renal failure) whereas over-production produces a polycythemia. The pharmacological use of recombinant human EPO (rHuEPO) in the management of patients with renal disease is now well established (Kendall, 2001).

Reference Interval

The range encountered in normal adult subjects lies between 10 and 30 mu/mL. Babies less than 3 months old have lower EPO levels than adults. The EPO response to hypoxia is lowest in the most immature neonates. Normally a stable level is achieved by 3 months of age and through to adulthood. No sex related differences have been demonstrated and levels are unaffected by the menstrual cycle. There is a diurnal variation with the highest levels at night. In pregnancy the level progressively increases (Kendall, 2001).

Clinical Applications

The level of EPO is raised in anemia, including aplastic, iron-deficiency and hemolytic anemias. EPO levels are depressed in patients with kidney disease, causing anemia, and low concentrations may give an early warning of kidney transplant rejection. Some tumors produce EPO and, in these cases, the concentration may be used as a tumor marker to monitor the effectiveness of treatment. EPO can be used to monitor AIDS patients undergoing Zidovudine (AZT) therapy. An increased concentration confirms that the anemia associated with AZT therapy is due to red cell hypoplasia or aplasia.

The EPO assay is useful when investigating polycythemia. In the primary form (polycythemia vera) overproduction of red blood cells occurs in the presence of EPO levels which are normal or significantly lower than in the secondary forms. The secondary polycythemias may be divided into those producing a raised EPO appropriate to the level of hypoxia (e.g. cardiac or pulmonary insufficiency or high affinity hemoglobins) and those with an inappropriate rise (e.g. hyper secretion from tumors or cysts). That 93% of primary and secondary polycythemias can be differentiated using measurement of EPO illustrates the value of the assay (Kendall, 2001).

Limitations

- EPO levels are increased by anabolic steroids, during pregnancy and in otherwise clinically normal persons following hemorrhage.
- EPO levels may be reduced in chronic infections, rheumatoid arthritis, AIDS and in the anemias associated with prematurity and hypothyroidism.

Assay Technology

Radioimmunoassays using purified human urinary EPO were developed originally. The introduction of rHuEPO obviated the need to use the native form for antibody and tracer production since results obtained from assays using the recombinant source were identical to those from established methods. Relatively few non-isotopic methods have been developed. Those which have use either monoclonal or affinity purified polyclonal anti-EPO antibody and are reported to be highly sensitive and quicker than radioassay (Kendall, 2001).

Desirable Assay Performance Characteristics

The assay should be standardized against one of the international reference preparations available from NIBSC.

Type of Sample

Serum. EDTA plasma may give slightly lower results.

Frequency of Use

Usually carried out at specialist centers.

THROMBOSIS AND HEMOSTASIS

Many immunological tests, especially ELISA methods, for the determination of platelet and coagulation factors, activators and inhibitors of the hemostatic system, and fibrinolytic products are now available. The diagnosis of coagulation factor deficiencies is conventionally made by clotting tests because functional activity, which is the important measurement, and antigen concentration do not always parallel each other. An ELISA has, however, been developed for determining the functional activity of vWF. In general, immunologic techniques have been of greater value in conditions of hypercoagulability than in bleeding disorders. Those for routine use include tests for a hereditary thrombotic state or predisposition: proteins C and S, and antithrombin; and tests for evidence of DIC such as FDPs, D-dimer and fibrinogen assays. The other tests below are mainly for specialized centers and research. The assay technology for each is listed in Table 54.2.

Kits and reagents for the following assays are available from several companies, including Diagnostica Stago, Immuno, Dade and Sysmex. It is essential to follow their instructions carefully. Because of the dynamic state of the hemostatic system the blood sample should be taken, without stasis or frothing on collection, into a plastic container and it should never be taken from indwelling catheters.

A list of reference materials, including International Standards, for many of the coagulation factors and the following assays are available from NIBSC (*see* CALIBRATION AND STANDARDIZATION).

THROMBOPHILIA

PROTEINS C AND S

Proteins C and S are vitamin K-dependent physiological anticoagulants produced in the liver. They act together as one of the main regulators of hemostasis by neutralizing activated coagulation factors V and VIII. Deficiency of Protein C accounts for 5–10% of venous thromboses in younger persons, deficiency of Protein S about half this number. Most cases are heterozygotes. Massive thromboses and skin necrosis may occur in the homozygous neonate. Factor V Leiden is probably an even more common cause of hypercoagulability but as yet there is no immunological test for it. Screening for thrombophilia has been reviewed recently by Jennings and Cooper (2003) and guidelines on testing are also available (Walker *et al.*, 2001).

Table 54.2. Immunological tests in hemostatic disorders.

ELISA	PF4, factors VII, VIII, IX, X, vWF, proteins C, S, β-TG, fibrinopeptide A, plasminogen, tPA, D-dimer, heparin cofactor II, antithrombin
Immunodiffusion	Factors VII, X, XI, prothrombin, fibrinogen, proteins C, S
Electroimmunodiffusion	Factor IX, fibrinogen, vWF, proteins C, S. antithrombin
Immunoelectrophoresis	Factors X, XII, XIII, fibrinogen, vWF, plasminogen, FDPs, proteins C, S
IRMA	PF4, factors VIII, IX, vWF
RIA	β-TG, PF4, vWF, fibrinogen, antithrombin, proteins C, S
Particle agglutination	β-TG, tPA, FDPs, D-dimer, antithrombin

Reference Interval

Protein C and S, 70–140% of average normal (Protein C: 4 mg/L, Protein S: 35 mg/L), by functional (coagulation) assays. Reference intervals appropriate to the local method and population should be determined. The effect of gender, age and pre-analytical variables such as anticoagulant therapy (for both), pregnancy (for protein S) and oral contraceptive use (for protein S) should be taken into account (Jennings and Cooper, 2003).

Limitations

Protein C functional activity and antigen concentration can differ. An immunologic assay is required to identify variant proteins when the functional test gives a low result. About 60% of Protein S is bound in the plasma and has to be freed before immunoassay to give the relevant measurement. Their levels are reduced by oral anticoagulant therapy (which has to be changed to heparin before assay), liver disease, DIC and post-surgery. The level of Protein S is about half normal in pregnancy and about one-fifth in neonates. Individuals with thrombotic disease may have levels only just below the reference interval.

Assay Technology

Early immunoelectrophoretic methods have been superseded by more sensitive ELISA and radioimmunoassay techniques (Table 54.2) based on monoclonal antibodies.

Desirable Assay Performance Characteristics

International Standards (NIBSC 86/622 – protein C and 93/590 – protein S) should be used to reduce differences in results from different methods.

Type of Sample

Citrated plasma.

Frequency of Use

Common.

ANTITHROMBIN

Antithrombin (previously known as antithrombin III) is a glycoprotein which primarily inhibits thrombin but also inhibits several activated coagulation factors. Its action is enormously enhanced by heparin. Deficiency can be congenital or acquired. Congenital deficiency accounts for about 3% of venous thrombotic disease in young persons, when a level of 40–50% of normal may be expected. The level in neonates, normally about 60–80% of the adult value, can be 30% or less when affected. Two types of congenital deficiency have been demonstrated. Functional activity and antigen levels are reduced in parallel in type I (quantitative) whereas the two levels are discrepant in type II (qualitative). Acquired deficiency can be caused by thrombosis, liver disease, sepsis, nephrotic syndrome, heparin therapy and treatment with L-asparaginase (Jennings and Cooper, 2003).

Reference Interval

0.125–0.39 g/L (70–130% of average normal).

Limitations

The antigen level may be normal when function is reduced (type II congenital deficiency). Therefore, the functional assay is more important and must always be available. High values are seen after myocardial infarction.

Assay Technology

Antithrombin can be measured by ELISA, radioimmunoassay, electroimmunodiffusion techniques and latex immunoassay (Table 54.2). Crossed immunoelectrophoresis can be used to identify molecular variants (Jennings and Cooper, 2003).

Desirable Assay Performance Characteristics

The assay should be standardized against an International Standard (NIBSC 93/768 – human plasma or 96/520– concentrate).

Type of Sample

Citrated plasma.

Frequency of Use

Uncommon but increasing.

COAGULATION

COAGULATION FACTORS

As mentioned above, functional activity of the clotting factors is of primary importance. However, the divergence between activity and antigen level can help in the differential diagnosis of two hereditary conditions, hemophilia, in which factor VIII activity is low and the antigen level is normal, and von Willebrand's disease, in which activity and antigen level decline in parallel (type I) or are discrepant (type II).

Reference Interval

A range of 50–200% of the reference population covers the normal range of most coagulation factors.

Limitations

Factor VIII and vWF are acute phase proteins and may be transiently raised into the normal range by stress, exercise, pregnancy, and estrogen-containing contraceptives.

Because of the wide reference interval it is essential to pool many normal plasmas (at least six) for the control sample or to use a commercial preparation for this purpose.

Desirable Assay Performance Characteristics

An International Standard for vWF is available from NIBSC (00/514). NIBSC also supplies standards for most other coagulation factors.

Type of Sample

Citrated plasma.

Frequency of Use

Functional assays are common; antigen level uncommon except in specialized centers.

FIBRINOGEN

Fibrinogen (Factor I) is a dimeric glycoprotein of molecular weight 340 kDa. The end product of the coagulation cascade, thrombin, is a powerful proteolytic enzyme that transforms fibrinogen into fibrin monomers by removing amino acids from the ends of two of its chains. The cleaved products are called fibrinopeptides and the large residual portions are fibrin monomers which combine and give a visible, strong fibrin polymer.

Reference Interval

2.0–4.0 g/L

Clinical Applications

Hypofibrinogenemia commonly occurs as part of a wider coagulation abnormality. Deficiency or dysfibrinogenemia (production of an abnormal form) are rare congenital events. An immunologic test is required when a functional test shows a low level; this may show a normal or even raised level in dysfibrinogenemia. The association of high levels with cardiovascular disease (Meade, 1997) may be an important reason for its estimation.

Limitations

Fibrinogen is an acute phase protein elevated by stress, infections, etc.; these factors need to be avoided when samples for study of vascular disease are taken.

Immunological measurements are influenced by FDPs in patients with increased fibrinolysis and their presence needs to be determined before interpreting the fibrinogen assay result.

Desirable Assay Performance Characteristics

There is still a wide variation in the reference intervals used in different laboratories. International Standards (NIBSC 98/614 – concentrate, human; 98/612 – plasma) are available and should help to resolve this problem.

Type of Sample

Citrated, platelet-poor plasma.

Frequency of Use

Occasional.

EVIDENCE OF DISSEMINATED INTRAVASCULAR COAGULATION

FIBRINOGEN/FIBRIN DEGRADATION PRODUCTS

Activation of the coagulation pathway results in the generation of thrombin, the enzyme that converts fibrinogen to fibrin. Cross-linked fibrin monomers are the main clot component. Fibrinolysis, the physiological mechanism for clot degradation, is activated by fibrin formation. The fibrin clot is degraded by plasmin (the enzyme end point of fibrinolysis) resulting in the release of fibrinogen/fibrin degradation products (FDPs) into the circulation. The test for FDPs is used to detect DIC in patients with hemostatic failure. Raised levels are also found in thrombotic diseases and with severe tissue damage (pneumonia, surgery).

Reference Interval

Normal: <10 mg/L (1 in 5 dilution, latex agglutination)

Assay Technology

In the latex agglutination test, latex particles are sensitized with antibodies to fibrinogen and fibrin degradation products D and E. A suspension is mixed on a glass slide with diluted serum, and aggregation indicates their presence. Serial dilution gives a semi-quantitative result.

Type of Sample

Serum. Whole blood is collected into a special tube that contains thrombin and an antifibrinolytic agent, and allowed to clot at 37 °C for 30 min.

Frequency of Use

Common.

D-DIMER TEST

The D-Dimer test is for fibrin derivatives containing linked D fragments, specific products of the lysis of fibrin clots by plasmin. The plasmin degradation of fibrin releases FDPs (see above), which contain D-Dimers. Since the presence of D-Dimer in the circulation results from fibrinolysis in response to activation of the coagulation system the detection is an indirect measure of thrombin generation and subsequent clot formation. D-Dimer is raised in DIC, deep vein thrombosis and pulmonary embolism. The level is not influenced by fibrinogen and the test is more specific than FDP for intravascular coagulation.

Reference Interval

Normal: <200 μg/L (latex agglutination)*
<400 μg/L (ELISA)*

Limitations

Methods lack standardization and the local cut-off level* must be determined prior to use. D-Dimers are also found in pregnancy and old age, and as a result of inflammation.

Assay Technology

Latex agglutination or ELISA can be used (Table 54.2). Assays are based on the monoclonal antibody recognition of D-Dimer epitopes. ELISA is the gold standard method but the latex agglutination test, which is similar to that for FDPs, is more popular for practical considerations.

Desirable Assay Performance Characteristics

The assay must be specific, showing no cross-reaction with fibrinogen, FDPs or fibrin monomers, otherwise an overestimation of the D-Dimer level may result.

Type of Sample

Citrated plasma or serum (follow manufacturers' instructions).

Frequency of Use

Common.

INFREQUENTLY USED AND RESEARCH ASSAYS

β-THROMBOGLOBULIN

In response to blood vessel injury, platelets aggregate at the site and release intracellular constituents, one of which is β-thromboglobulin (β-TG). The function of this globulin is not understood. The level of β-TG in the circulation depends upon the rate of its release and clearance by the kidneys.

Reference Interval

Normal: <50 μg/L

Clinical Applications

The release of β-TG may be rapid, due to thrombus (clot) formation, or steadier in conditions with either damaged vascular endothelium or non-endothelial surfaces (prosthetic valves, Dacron grafts). The level is raised in both venous and arterial thromboembolism as well as in renal failure. It has been found elevated in pre-eclampsia where platelet thrombi occur in the placenta.

Limitations

Many infections can raise the level of β-TG. Certain drugs in high dosage, including aspirin and opiates, may interfere with the assay. Raised β-TG and platelet factor 4 (PF4) levels lack specificity for thrombosis and hypercoagulability and are of limited value in individual patients.

Desirable Assay Performance Characteristics

The assay should be standardized against the first International Standard for β-TG (NIBSC 83/501).

Type of Sample

Plasma is used. The careful collection of the blood sample and its subsequent handling are most important. Free-flowing blood, taken without stasis, is withdrawn and the first 3 mL discarded. Blood is collected in a tube, provided with the kit, which is chilled and kept on melting ice. The tube contains an inhibitor of platelet activation and a calcium chelator. This procedure is essential to prevent artifactual release of platelet factors.

Measurement of both β-TG and PF4 is recommended to distinguish genuine from falsely elevated levels.

The ratio of β-TG to PF4 is usually more than 5:1. With faulty sample handling the ratio can be less than 2:1. The normally high ratio is due to the very rapid clearance *in vivo* of PF4.

Frequency of Use

Uncommon. This assay is a research tool for the investigation of platelet function, particularly in arterial disease.

PLATELET FACTOR 4

PF4 neutralizes heparinoids. The causes of raised levels and the precautions required in sample handling are the same as for β-TG. PF4 is cleared from the plasma very rapidly by vascular endothelium from which it may be displaced by heparin.

Reference Interval

< 10 μg/L (Abbott RIA)
0–5 IU/mL (Diagnostica Stago)

Limitations

High levels may be encountered with heparin therapy.

Desirable Assay Performance Characteristics

The assay should be standardized against the first International Standard for PF4 (NIBSC 83/505).

Type of Sample

Plasma, as for β-TG.

Frequency of Use

Uncommon. This assay is a research tool for the investigation of platelet function, particularly in arterial disease.

HEPARIN COFACTOR II

This is a specific inhibitor of thrombin in plasma. The significance of a deficiency is unclear since a reduced level has been found in both patients with a thrombotic tendency and in healthy individuals. The level is reduced in congenital deficiency, liver disease and DIC.

Reference Interval

45–120 mg/L

Limitations

The plasma concentration may be raised by oral contraceptives.

Type of Sample

Citrated plasma.

Frequency of Use

Research only.

PLASMINOGEN

Plasminogen is a single chain polypeptide of molecular weight 92 kDa. Four of its five looped structures possess a 'lysine binding site' through which it interacts with lysine residues in its substrates. Its native form (glu-plasminogen) has a *N*-terminal glutamic acid residue. Tissue activators (t-PA) cleave the chain to give glu-plasmin, and further, autocatalytic, cleavage yields lys-plasmin with exposure of the lysine binding sites that markedly enhances its interaction with fibrin. Binding to fibrin localizes its lytic activity. A series of proteolytic steps results in the production of FDPs. Initial removal of several peptides (fragments A, B, C) leaves a large portion, containing fibrinopeptide A, before further digestion cleaves another fragment, D.

Reference Interval

800–1200 U/L (Functional assay)

Clinical Applications

Deficiency may account for about 3% of thrombotic disease in young persons. The level is reduced in neonates, liver disease, DIC and with thrombolytic therapy. It is often raised in infections and malignant disease.

Desirable Assay Characteristics

A British standard for glu-type Human Plasminogen is available from NIBSC (78/646).

Type of Sample

Citrated plasma.

Frequency of Use

Uncommon.

TISSUE PLASMINOGEN ACTIVATOR

Congenital deficiency of tissue plasminogen activator (t-PA) is responsible for a small number of thromboses in young persons. The level is increased by venous stasis and exercise.

Reference Interval

~ 5 μg/L

Desirable Assay Performance Characteristics

Human recombinant tissue plasminogen activator is available from NIBSC as the third International Standard (98/714).

Frequency of Use

Rare.

IMMUNODETECTION METHODS

LEUKEMIA AND LYMPHOMA ANALYTES

Normal and neoplastic cells may be identified by the presence of antigens on the cell membrane and in the cytoplasm, and by intranuclear enzymes. When the nature of a neoplastic cell cannot be determined from morphology and cytochemistry it is essential to employ immunophenotypic analysis (European Working Group on Clinical Cell Analysis, 1996).

Clinical Applications

There are numerous monoclonal antibodies available, but only those that recognize the different types of acute leukemia and the nature of a chronic lymphoid proliferation are required. A suitable panel may, therefore, include the following cluster designation (CD) reagents for acute leukemia: Anti-TdT for a nuclear enzyme marker; for myeloid markers: CD33, CD13 (cytoplasmic) and antimyeloperoxidase; and for lymphoid markers: B cell – CD10, CD19, CD22 (cytoplasmic); T cell – CD2, CD7, CD3 (cytoplasmic).

For a lymphoproliferative disorder, markers are needed to distinguish between B and T cell disease and between B cell chronic leukemia and other B cell disorders. An appropriate panel would include the cell markers for CD22, CD5, CD19, CD20; markers (preferably polyclonal) to assess clonality – anti-k and anti-λ; and markers to determine the kind of B cell disease – CD22, CD23, FMC7.

A panel is required to recognize antigens from the earlier stages of cell development e.g. CD3, 13, 22 (cytoplasmic) and to clarify the overlap in expression of some antigens in the different leukemias. A suitable panel would include the monoclonal antibodies CD 2, 3, 4, 5, 7, 8, 10, 13, 14, 19, 22, 33, 41, 61, anti-IgM, and anti-TdT. Becton Dickinson, Beckman Coulter, Dakocytomation and Serotec provide these and other reagents with instructions for their use. The initial selection is usually based on those suggested by routine hematological tests e.g. CBC and blood film morphology.

Assay Technology

Two techniques are used to label and identify the cells: immunofluorescence and immuno-enzymatic labeling (British Committee for Standards in Haematology. General Haematology Task Force, 1994b,c).

Immunofluorescence on Cell Suspensions

The separated cells are resuspended and incubated with cell culture media to remove cytophilic antibodies. They are then incubated with an optimal dilution of monoclonal antibody followed by incubation with fluorescein isothiocyanate (FITC)-conjugated anti-mouse immunoglobulin. The cells are washed after each step. For microscopy the cells are resuspended, mounted on a glass slide and sealed. Alternatively the cells are resuspended in diluent and examined in a flow cytometer (FACS, EPICS).

Immunoenzymatic Labeling on Slides

The slide is first incubated with the monoclonal antibody, then with anti-mouse immunoglobulin, and finally with the immunoalkaline phosphatase anti-alkaline phosphatase (APAAP). The cell spread is developed in a solution of naphthol As-Mx phosphate, levamisole, *N,N*-dimethylformamide and FAST red TR salt. The nuclei are counterstained with hematoxylin.

Limitations

Cells are generally viable for up to 24 h at 4 °C except when the cell count is high. Cells in cerebrospinal fluid (CSF) are more labile. The effect of different anticoagulants on some antigens is uncertain. Immunoenzymatic tests may lose antigens through fixation, fewer cells can be examined, and dual labeling is not possible. Immunofluorescence does not normally detect antigens in cytoplasm (where they are first expressed) and retrospective examination is not possible. Negative and positive controls are essential with both techniques, particularly to check for NSB.

Desirable Assay Performance Characteristics

Antibodies with the same CD number are not necessarily of equal specificity and stability. Advice in their selection should be sought from a specialized laboratory.

Types of Sample

Whole blood and bone marrow samples collected in anticoagulant. Preservative-free heparin permits other studies (e.g. cytogenetics) to be made. Cell suspensions, CSF.

Frequency of Use

Common in specialized centers. Use extending in general hospitals.

MALARIAL PARASITES

Clinical Applications

Modern methods of travel and increased migration enable malaria to be detected in patients anywhere in the world. In addition to the clinical implications for the individual patient there is a risk of transmitting malaria by blood transfusion. In the case of *Plasmodium falciparum* infection the parasites should be quantified by morphological examination of thick and thin blood films. This may be supplemented by immunological, polymerase chain reaction (PCR), or fluorescence methods (British Committee for Standards in Haematology. General Haematology Task Force, 1997).

Assay Technology

Methods are available for direct detection of antigen or indirect detection of antibodies to malarial parasites. The most sensitive of the direct methods are based on PCR, which detects the parasite DNA. These suffer from high expense and complexity. Immuno chromatographic and non-chromatographic antigen detection methods that allow both the rapid diagnosis of malaria and differentiation between species are also available.

Indirect methods for the detection of anti-malarial antibodies can only be used to confirm the diagnosis since the antibodies are not detectable for some days following infection. Antigen-coated-solid phase combined with a detection system of enzyme labeled antigen is typical of the methods used.

Desirable Assay Performance Characteristics

Since the introduction of a single malarial parasite to a host can lead to the disease, testing is required in both endemic and non-endemic countries. Ideally tests need to be of high sensitivity and specificity, easy to perform and inexpensive.

Type of Sample

Undiluted blood or plasma.

Frequency of Use

Variable, depending on geographic location and population, and for the above reasons all laboratories should be able to detect parasites or forward the sample to a reference center. It is recommended that all positive tests should be sent to a reference center for verification (British Committee for Standards in Haematology. General Haematology Task Force, 1997).

HEMOGLOBINOPATHIES

Newborn and antenatal hemoglobinopathy screening programmes have been established and are well developed. Newborn screening is a qualitative process whereas measurement of various hemoglobin levels, in particular A_2 and F is required for antenatal screening. The programmes are almost exclusively based on high performance liquid chromatography (HPLC) and electrophoretic techniques. Hemoglobin F can also be measured using radial immunodiffusion (HbF QUIPlate Kit, Helena Laboratories) but this is used infrequently. Antibodies to hemoglobin variants exist but are used mainly for research purposes.

REFERENCES AND FURTHER READING

Addison, G.M., Beamish, M.R., Hales, C.N., Hodgkins, M., Jacobs, A. and Llewellin, P. An immunoradiometric assay for ferritin in the serum of normal subjects and patients with iron deficiency and iron overload. *J. Clin. Pathol.* **25**, 326–329 (1972).

Allen, J., Backstrom, K., Cooper, J., Cooper, M., Detwiler, T., Essex, D.W., Fritz, R., Means, R.T., Meier, P.B., Pearlman, S., Roitman-Johnson, B. and Seligman, P.A. Measurement of soluble transferrin receptor in serum of healthy adults. *Clin. Chem.* **44**, 35–39 (1998).

Bloom, A.L., Forbes, C.D., Duncan, P.T. and Tuddenham, E.G.D. (eds), *Haemostasis and Thrombosis* 3rd edn (Churchill Livingstone, Edinburgh, 1994).

British Committee for Standards in Haematology, General Haematology Task Force, Guidelines on the investigation and diagnosis of cobalamin and folate deficiencies. *Clin. Lab. Haem.* **16**, 101–115 (1994a).

British Committee for Standards in Haematology. General Haematology Task Force, Immunophenotyping in the diagnosis of acute leukemia. *J. Clin. Pathol.* **47**, 777–781 (1994b).

British Committee for Standards in Haematology. General Haematology Task Force, Immunophenotyping in the diagnosis of chronic lymphoproliferative disorders. *J. Clin. Pathol.* **44**, 871–875 (1994c).

British Committee for Standards in Haematology. General Haematology Task Force, Malaria Working Party. The laboratory diagnosis of malaria. *Clin. Lab. Haem.* **19**, 165–170 (1997).

Cavill, I., Macdougall, I.C., Gokal, R., Beguin, Y., Peters, T., Pippard, M., Schaefer, R. and Winearls, C. Iron management in patients on rHuEPO. *Br. J. Renal. Med.* **2**, 6–8 (1997).

Chanarin, I., *The Megaloblastic Anaemias* 3rd edn (Blackwell Scientific Publications, Oxford, 1990).

Conn, D.A. Intrinsic factor antibody detection and quantitation. *Med. Lab. Sci.* **43**, 48–52 (1986).

Dacie, J.V. and Lewis, S.M. *Practical Haematology*, 9th edn, (eds Lewis, S.M., Bain, B.J. and Bates, I.), (Churchill Livingstone, Edinburgh, 2001).

Daly, S., Mills, J.L., Molloy, A.M., Conley, M., McPartlin, J., Lee, Y.J., Young, P.B., Kirke, P.N., Weir, D.G. and Scott, J.M. Low-dose folic acid lowers plasma homocysteine levels in women of child-bearing age. *Q. J. Med.* **95**, 733–740 (2002).

Dawson, D.W. and Waters, H.M. Malnutrition: folate and cobalamin deficiency. *Br. J. Biomed. Sci.* **51**, 221–227 (1994).

Dawson, D.W., Fish, D.I., Frew, I.D.O., Roome, T. and Tilston, I. Laboratory diagnosis of megaloblastic anaemia: current methods assessed by external quality assurance trials. *J. Clin. Pathol.* **40**, 393–397 (1987).

European Working Group on Clinical Cell Analysis (EWGCCA), Consensus document on leukemia immunophenotyping. *Leukemia* **10**, 877–895 (1996).

Fedosov, S.N., Laursen, N.B., Nexo, E., Moestrup, S.K., Petersen, T.E., Jensen, E. and Berglund, L. Human intrinsic factor expressed in the plant *Arabidopsis thaliana*. *Eur. J. Biochem.* **270**, 3362–3367 (2003).

Flowers, C.H. and Cook, J.D. Dried plasma spot measurements for ferritin and transferrin receptor for assessing iron status. *Clin. Chem.* **45**, 1826–1832 (1999).

Gunter, E.W., Bowman, B.A., Caudill, S.P., Twite, D.B., Adams, M.J. and Sampson, E.J. Results of an international round robin for serum and whole-blood folate. *Clin. Chem.* **42**, 1689–1694 (1996).

Harford, J.B., Rouault, T.A., Huebers, H.A. and Klausner, R.D. Molecular mechanisms of iron metabolism. In: *The Molecular Basis of Blood Disorders* (eds Stamatoyannopoulos, G., Nienhuis, A.W., Majerus, P.W. and Varmus, H.), 351–378 (Saunders, Philadelphia, 1994).

Hershko, C. (ed), *Clinical Disorders of Iron Metabolism*, vol. 7/4, Clinical Haematology, (Balliere Tindall, London, 1994).

International Committee for Standardization in Haematology, Proposed serum standard for human serum vitamin B_{12} assay. *Br. J. Haematol.* **64**, 809–811 (1986).

Jennings, I. and Cooper, P. Screening for thrombophilia: a laboratory perspective. *Br. J. Biomed. Sci.* **60**, 39–51 (2003).

Kendall, R.G. Erythropoietin. *Clin. Lab. Haem.* **23**, 71–80 (2001).

Klemow, D., Einsphar, D., Brown, T.A., Flowers, C.H. and Skikne, B.S. Serum transferrin receptor measurements in hematologic malignancies. *Am. J. Hematol.* **34**, 193–198 (1990).

Ledue, T.B., Craig, W.Y., Ritchie, R.F. and Haddow, J.E. Influence of blood donation and iron supplementation on indicators of iron status. *Clin. Chem.* **40**, 1345–1346 (1994).

Macdougall, I.C., Cavill, I., Hulme, B., Bain, B., McGregor, E., McKay, P., Sanders, E., Coles, G.A. and Williams, J.D. Detection of functional iron deficiency during erythropoietin treatment: a new approach. *Br. Med. J.* **304**, 225–226 (1992).

Meade, T.W. Fibrinogen and cardiovascular disease. *J. Clin. Pathol.* **50**, 13–15 (1997).

Means, R.T. Jr., Allen, J., Sears, D.A. and Schuster, S.J. Serum soluble transferrin receptor and the prediction of marrow aspirate iron results in a heterogeneous group of patients. *Clin. Lab. Haem.* **21**, 161–167 (1999).

Messinezy, M. and Pearson, T.C. The classification and diagnostic criteria of the erythrocytoses (polycythaemias). *Clin. Lab. Haem.* **21**, 309–316 (1999).

Nexo, E., Tibbing, G., Weber, T., Hansen, H. and Grasbeck, R. Preparation of calibration standards for quantitation of cobalamins in plasma, a NORDKEM project. *Scan. J. Clin. Invest.* **49** (Suppl. 194), 35–36 (1989).

Nimo, R.E. and Carmel, R. Increased sensitivity of detection of the blocking (type I) anti-intrinsic factor antibody. *Am. J. Clin. Pathol.* **88**, 729–733 (1987).

O'Sullivan, J.J., Leeming, R.J., Lynch, S.S. and Pollock, A. Radioimmunoassay that measures serum vitamin B_{12}. *J. Clin. Pathol.* **45**, 328–331 (1992).

Punnonen, K., Irjala, K. and Rajamaki, A. Serum transferrin receptor and its ratio to serum ferritin in the diagnosis of iron deficiency. *Blood* **89**, 1052–1057 (1997).

Quinlivan, E.P., McPartlin, J., McNulty, H., Ward, M., Strain, J.J., Weir, D.G. and Scott, J.M. Importance of both folic acid and vitamin B_{12} in reduction of risk of vascular disease. *Lancet* **359**, 227–228 (2002).

Roberts, B.E. (ed), *Standard Haematology Practice/2* (Blackwell Science, Oxford, 1994).

Scott, J. and Weir, D. Homocysteine and cardiovascular disease. *Q. J. Med.* **89**, 561–563 (1996).

Sourial, N.A. Rapid protein A assay for intrinsic factor and its binding antibody. *J. Clin. Pathol.* **41**, 568–572 (1988).

Suominen, P., Möttönen, T., Rajamäki, A. and Irjala, K. Single values of serum transferrin receptor and transferrin receptor ferritin index can be used to detect true and functional iron deficiency in rheumatoid arthritis patients with anemia. *Arthrit. Rheumat.* **43**, 1016–1020 (2000).

Trowbridge, I.S. Immunoassay of serum transferrin receptors: clinical implications. *J. Lab. Clin. Med.* **114**, 336–337 (1989).

Walker, I.D., Greaves, M. and Preston, F.E. Investigation and management of heritable thrombophilia. *Br. J. Haematol.* **114**, 512–518 (2001).

Waters, H.M. and Seal, L.H. A systematic approach to the assessment of erythropoiesis. *Clin. Lab. Haem.* **23**, 271–283 (2001).

Waters, H.M., Smith, C., Howarth, J.E., Dawson, D.W. and Delamore, I.W. New enzyme immunoassay for detecting total, type I, and type II intrinsic factor antibodies. *J. Clin. Pathol.* **42**, 307–312 (1989).

Waters, H.M., Dawson, D.W., Howarth, J.E. and Geary, C.G. High incidence of type II autoantibodies in pernicious anaemia. *J. Clin. Pathol.* **46**, 45–47 (1993).

Wickramasinghe, S.N. (ed), *Megaloblastic Anaemia*, vol. 8/3, Clinical Haematology, (Bailliere Tindall, London, 1995).

Worwood, M. Haemochromatosis. *Clin. Lab. Haem.* **20**, 65–75 (1998).

55 Cardiac Markers

Alan H.B. Wu

NORMAL HEART FUNCTION

The heart is a muscular pump that ejects blood into the lung and systemic circulation. For it to function continuously, the heart muscle needs an uninterrupted and plentiful supply of oxygenated blood. This is supplied from the coronary arteries, which leave the ascending aorta soon after exiting the left ventricle. The main coronary artery divides into right, left, and circumflex arteries, which run over the surface of the upper part of the heart. These arteries in turn divide into finer arteries and eventually penetrate the whole of the myocardium ensuring that all the individual myocardial cells receive adequate oxygen and nutrients. The blood eventually collects into the coronary sinus, which drains into the right atrium before traversing to the pulmonary circulation for further oxygenation.

Much of the oxygen delivered to the myocardium is bound to myoglobin, a heme protein that gives cardiac muscle its deep red color and that is located inside the muscle fibers. The major requirement for energy provision in the myocardial cells is for adenosine triphosphate (ATP), which is used to produce mechanical energy at the level of the contractile myofibrils. Most high-energy phosphate in the heart is stored not as ATP but as creatine phosphate. The cytosolic enzyme creatine (phospho) kinase (CK) catalyzes the transfer of high-energy phosphate from creatine phosphate to adenosine diphosphate (ADP) to give adequate supplies of ATP. The ATP is used for a wide variety of reactions, but the most important is the interaction of myosin and actin in the myofibril thick and thin filaments. During this reaction, the chemical energy of ATP is transduced into mechanical energy, causing thick and thin filaments to slide past each other, making the muscle contract. This process is controlled by the level of intracellular free calcium ions, which act on a series of regulatory proteins (the troponin–tropomyosin complex) that, in turn, allow the thick and thin filaments to interact. To summarize, the cardiac pump function requires a continuous supply of oxygenated blood, ATP production from creatine phosphate, and the interaction of the myofibrillar proteins.

CLINICAL DISORDERS

Although there are a number of different disorders that affect the biochemical, physiological, and mechanical functions of the heart, by far the most common is the spectrum of changes that comes under the general umbrella of coronary artery disease (CAD).

CORONARY ARTERY DISEASE

Coronary artery disease is a continuous spectrum of disorders that for many, begins at an early age, and progresses until death. The major events are atherosclerosis, stable angina, unstable angina, and acute myocardial infarction (AMI), and cardiac death. Individuals who survive episodes of myocardial infarction develop heart failure (CHF). As there are other causes of heart failure, this disease is listed separately.

Atherosclerosis

The initial stages of CAD are characterized by **atherosclerosis**, which is the buildup and deposition of plaques within the coronary arteries. During this period, the individual is asymptomatic and essentially disease free. The progression of atherosclerosis, and the subsequent risk for CAD, is dependent on numerous risk factors such as abnormally high concentrations of total- and low-density lipoprotein (LDL) cholesterol, smoking, hypertension, diabetes, being a male or a postmenopausal female, or having an immediate family with a history of premature (men < 45 years, women < 55 years) heart disease. In addition to cholesterol, LDL, and high-density lipoprotein (HDL) cholesterol, new biochemical markers have been investigated as additional risk factors for CAD. Some of these new markers include the apolipoproteins A-I and B-100, Lp(a), homocysteine, high-sensitivity C-reactive protein, coagulation factors such as fibrinogen, factor VII, tissue plasminogen activator antigen, and plasminogen activator inhibitor (PAI-1), nitrous oxide, and oxidized LDL.

The Immunoassay Handbook. *Third Edition*
D. Wild (Ed.)

Stable Angina Pectoris and Silent Ischemia

Stable angina pectoris is defined as episodes of chest pain precipitated by physiologic situations of increased oxygen demands to the heart. It occurs most commonly during or immediately after exercise. In patients with atherosclerosis, angina is caused by the narrowing of coronary arteries to the point where there is insufficient delivery of blood and oxygen to actively respiring myocardial tissue. The affected areas of the heart are said to be **ischemic**, i.e. they are in danger for permanent myocardial damage, but myocardial *necrosis* does not occur in this condition, therefore most biomarkers in blood are present within the normal range. However, there may be release of cardiac troponin, providing the basis for risk stratification. The atherosclerotic plaques are stable, in that they have a thick fibrous cap, and are not in any immediate risk to rupture (Figure 55.1a). The chest pain is relieved by rest or by medications such as nitroglycerin, which functions to diminish the oxygen demands of the heart.

Patients with silent ischemia have episodes of reduced blood flow that can lead to both myocardial ischemia and cell necrosis. Unlike patients with stable angina and acute coronary syndromes, those with silent ischemia do not present with any symptoms or knowledge that an ischemic event has taken place. Biochemical markers might be increased in the blood of patients suffering an event. However, in the absence of symptoms, the patient would not know to present to a hospital for blood collection. The diagnosis is made on the basis of stress testing, whereby ischemic episodes are induced under controlled conditions and monitored by continuous electrocardiographic recordings (ECG ST-segment depression of >1 mm) or nuclear imaging techniques (reduced coronary artery blood flow).

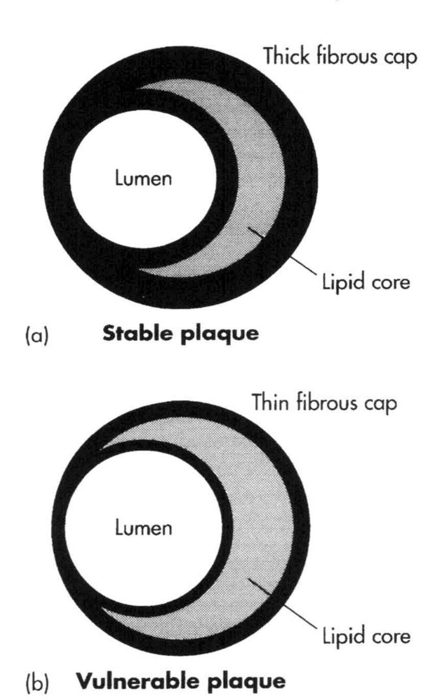

Fig. 55.1 Schematic cross-section of a coronary artery (a) Stable atherosclerotic plaque with a thick fibrous cap (b) Unstable plaque vulnerable to rupture.

Acute Coronary Syndromes

Patients who present with chest pain at rest have **unstable angina pectoris** or **acute myocardial infarction**, collectively known as **acute coronary syndromes**. In these patients, atherosclerotic plaques have become vulnerable due to the thinning of the fibrous cap (Figure 55.1b). Under the shear stress of circulating blood, there is rupture of the plaque, which leads to the exposure of the lipid filled core, causing an aggregation of platelets and the formation of a **thrombus**. When the clot is subocclusive, the patient has unstable angina pectoris. This condition is associated with little or no myocardial damage, and the ECG may be normal. Patients with unstable angina are treated with anti-inflammatory drugs (e.g. salicylates), β-adrenergic blockers, anti-thrombotic drugs (e.g. unfractionated or low molecular weight heparin), and anti-platelet medications (e.g. glycoprotein IIb/III receptor inhibitors).

When the clot from a ruptured plaque completely blocks coronary artery blood flow, the patient has suffered an AMI. This leads to myocardial cell death and release of intracellular components into the circulation. The World Health Organization criteria for AMI is a triad that includes a clinical presentation of chest pain, ST-segment elevations (>1 mm) on ECG, and increases in serum enzymes. The diagnosis of AMI is given when two of these criteria are fulfilled. In 2000, the definition of myocardial infarction was redefined by a joint committee of the European Society of Cardiology (ESC) and American College of Cardiology (ACC). The current criteria for AMI are increased concentrations of a cardiac marker with evidence of myocardial ischemia (chest pain, ECG changes, angiographic, and autopsy documentation).

The common biochemical markers of AMI include total creatine kinase (CK), the CK-MB isoenzyme, lactate dehydrogenase (LD) isoenzymes, myoglobin, and cardiac troponin T and I. ECG-documented AMI patients are acutely treated with intravenous thrombolytic agents such as streptokinase and tissue plasminogen activators, or revascularization by coronary angioplasty. In many cases, the non-Q-wave AMI is indistinguishable from unstable angina, and the therapy is similar.

A significant number of AMI patients will die within minutes after suffering AMI, before emergency treatments can be rendered. These patients die of uncontrolled cardiac arrhythmias. For those who survive, the risk for future myocardial infarctions is high.

HEART FAILURE

A clinical syndrome whereby the heart is unable to generate sufficient cardiac output to meet the body's demands defines **heart failure (HF)**. This is characterized by intravascular and interstitial volume overload, including shortness of breath, dyspnea, rales, and edema, or conditions of inadequate tissue perfusion, including fatigue and exercise intolerance. Heart failure is caused by a primary disorder of the heart muscle (cardiomyopathy), valvular disease, infiltrative processes, and ischemic heart disease. One measure of left ventricular function is the ejection fraction, determined by the echocardiogram. Heart failure occurs when there is over-stimulation of the renin–angiotensin–aldosterone axis that leads to volume overload and remodeling (enlargement) of the heart. Aldosterone stimulates vasoconstriction, hypernatremia, and fluid retention. As a compensatory mechanism, the natriuretic peptides are released in an attempt to reverse these changes. These include A-type natriuretic peptide (ANP) that is found in high concentrations in the atrium as granules, and B-type natriuretic peptide (BNP) that is upregulated and produced in the ventricles.

HYPERTENSION

Hypertension is defined when the systolic blood pressure exceeds 140 mm Hg and the diastolic blood pressure exceeds 90 mm Hg. It can be the result of a fault in the complex physiological, biochemical, and endocrine control mechanisms. One important factor is the level of activity in the renin–angiotensin system, which generates angiotensin-II, a powerful vasoconstrictor that gives rise to elevated blood pressure. Direct or indirect measurements of elevated plasma renin activity can be used to give an indication of hypertension. Other factors are the natriuretic peptide system (notably atrial and brain natriuretic peptide (BNP)), which controls fluid, and electrolyte balance.

ANALYTES

CREATINE KINASE AND THE MB ISOENZYME

Creatine kinase is a dimeric protein composed of two enzymatically active M (muscle-type) and B (brain-type) subunits. There are three major cytosolic isoenzymes. The CK-MM isoenzyme is the major form in skeletal and cardiac muscles whereas CK-BB is found in brain and other organs. CK-MB is found predominantly in cardiac muscle where it accounts for up to 30% of the total CK present, the rest being CK-MM. Small amounts of CK-MB, normally less than 1% of the total, are found in skeletal muscle. There are also two major mitochondrial isoenzymes of CK that can be released into blood after injury. These isoenzymes are unstable as they lose their enzymatic activity rapidly after they appear in blood.

Function

Mitochondrial CK functions to convert ATP generated from the electron transport system to creatine phosphate, which diffuses into the cytoplasm and acts as a reservoir of high-energy phosphate bonds. When energy is required for active-muscle contraction, cytosolic CK catalyzes the production of ATP from creatine phosphate stores. CK is also found in other non-contractile tissue such as the distal nephron. The role of CK in other tissues may be to shuttle high-energy phosphate groups for other ATP-dependent functions such as the sodium–potassium membrane pump. Total CK and CK isoenzymes are cleared from the blood by the reticuloendothelial system of the cell.

Reference Interval

Total CK: males 0–200 U/L, females 0–160 U/L
CK-MB: <5 ng/mL
Relative index: <2.5%

Clinical Applications

Total CK is not specific to cardiac damage, as it is also increased in patients with skeletal muscle disease or injury. Measurement of the CK-MB isoenzyme improves the specificity of the assay. Both total CK and CK-MB begin to increase in the blood of patients with AMI within the first 4–6 h after the onset of chest pain, peak at 18–24 h, and return to normal with 72 h (Figure 55.2).

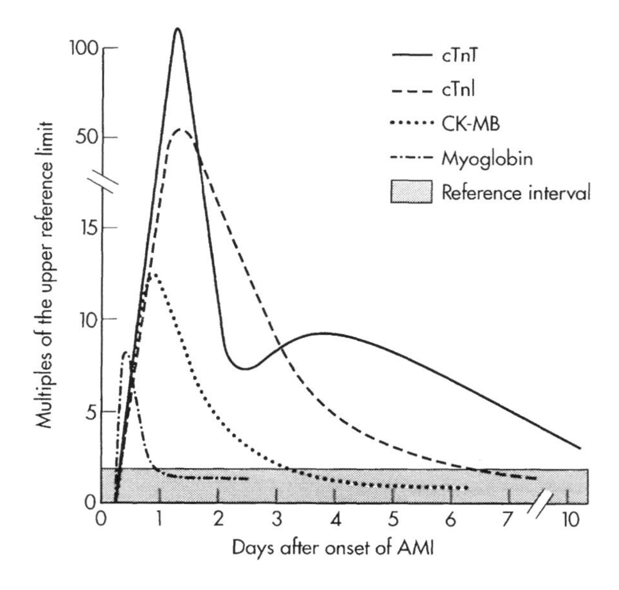

Fig. 55.2 Cardiac marker release patterns vs. time after onset of chest pain.

If there are no further episodes of ischemic injury, the activity declines in a mono-exponential fashion.

Limitations

The major limitation of CK-MB determination as a cardiac-specific marker of heart damage stems from the low levels of CK-MB present in skeletal muscle. In cases where severe skeletal muscle trauma may occur, such as in patients with polymyositis, muscular dystrophy, severe muscle trauma, and after prolonged physical exercise, CK-MB levels can be greatly increased above the absolute cut-off concentration indicative of myocardial infarction. In these cases, it is necessary to calculate a relative index of CK-MB (in ng/mL)/total CK (in U/L). An abnormal concentration of CK-MB with a normal relative index indicates a non-cardiac source of CK-MB. Neither the relative index nor the absolute concentration of CK-MB can be used to determine the presence of myocardial damage when skeletal muscle disease is also present.

Due to the lack of myocardial tissue specificity, CK-MB levels cannot be used to measure minor myocardial injury. To avoid a high number of false positive results, the cut-off concentration used for CK-MB is preset to differentiate AMI from non-AMI disease. Therefore, CK-MB results are often within the normal range in patients with low levels of myocardial necrosis, such as in patients with myocarditis, heart failure, and myocardial ischemia. In this respect, there is no clear cut-off but usually one might not expect to see the normal time profile of elevated CK-MB levels typical of mild to severe infarction. Single elevated samples should therefore be treated with caution. Typically, a sample would be taken on admission and a minimum of once or twice daily for the subsequent 2–3 days. More frequent sampling not only aids in positive diagnosis, but also may provide a semi-quantitative measure of the extent of damage. Numerous studies have shown that the cumulative release of CK-MB is well correlated with the extent of myocardial necrosis (infarct sizing), when compared to the amount of tissue damaged, when anatomically determined during an autopsy.

Assay Technology

Either the activity or the mass concentration of CK-MB can be measured. Activity measurements include electrophoresis and immunoinhibition. However, these assays have largely been replaced by immunoenzymetric mass assays, using a combination of monoclonal or polyclonal antibodies, with one raised against CK-MB, and the other against an individual subunit, usually the B subunit. Most commercial assays make use of the monoclonal antibody licensed from the Department of Laboratory Medicine, Washington University (St Louis, MO).

Desirable Assay Performance Characteristics

The assay for CK-MB should be sensitive to below 1 ng/mL, and have a dynamic range of at least 100 ng/mL. The acceptable precision is about 10% at the cut-off concentration.

Type of Sample

Serum or heparinized plasma.

Frequency of Use

Currently considered the 'gold standard' for diagnosis of AMI.

MYOGLOBIN

Function

Myoglobin is an iron-containing protein with a molecular mass of 18 kDa. It resembles hemoglobin but binds one rather than four molecules of oxygen. Its binding characteristics are such that it takes up oxygen from hemoglobin in the blood and releases it for use into the mitochondria where oxidative reactions occur.

Reference Interval

Myoglobin: 25–90 ng/mL

Clinical Applications

After AMI, when the cardiac myocyte becomes necrotic, myoglobin is released through the damaged cell membrane where it becomes detectable in the circulation. Elevated levels of myoglobin are normally detectable between 2 and 6 h after infarction peaking within 5–18 h. It is generally detectable before CK-MB, because of its low-molecular weight, it is readily filtered by the glomerulus, and cleared by the kidneys. Myoglobin levels return to normal within 24 h after injury (Figure 55.2). Because of its rapid return to baseline concentrations, myoglobin can also be used to detect new myocardial injury, such as a reinfarction.

Limitations

Myoglobin is found in all oxidative muscle fibers, including skeletal muscle, so injury to such tissues will give rise to elevated myoglobin levels. This can occur after skeletal muscle trauma, extreme physical pain, skeletal myopathy, or rhabdomyosarcoma. Myoglobin concentrations are also increased in patients with acute or chronic renal failure due to reduced clearance of the protein. Cardiac related conditions in which myoglobin elevations may be seen include heart failure, tachyarrhythmias, ischemia, and some cardiomyopathies.

Assay Technology

The original myoglobin assays were based on radioimmunoassays. Subsequently, immunoturbidimetric assays were developed. These have all been replaced by non-isotopic immunoassays with enzyme, fluorescence, or chemiluminescence signal generation systems. Assays usually use antibodies to human myoglobin raised in rabbits. Both competitive and non-competitive assay formats are available. The solid phase may be microtiter plates, or polystyrene particles in the case of some turbidimetric assays. In all other technical respects, assays for this protein are straightforward.

Desirable Assay Performance Characteristics

The myoglobin assay should have a sensitivity of <5 ng/mL and a dynamic range of at least 500 ng/mL. The acceptable precision is about 10% at the cut-off concentration.

Types of Sample

Serum, plasma, or urine.

Frequency of Use

Myoglobin is used in hospitals that have rapid rule-out protocols for chest pain patients (e.g. 'emergency room chest pain centers'). In these centers, triaging decisions are made within the first 3–6 h after admission. Myoglobin is useful because other cardiac markers such as CK-MB and troponin T or I are not reliably increased during this time interval to be useful for AMI rule-out. For hospitals that take a more conservative approach and admit the majority of patients with chest pain, the measurement of myoglobin becomes unnecessary, as more definitive markers such as cardiac troponin T or I can be used.

CARDIAC TROPONIN (T AND I)

Function

The thin filament of contractile muscle contains actin, myosin, and troponin, a complex of three proteins. Cardiac troponin T (cTnT) functions to bind the troponin complex to tropomyosin and has a molecular mass of 37 kDa. Cardiac troponin I (cTnI) functions to inhibit calcium-dependent ATPase, and has a mass of 24 kDa. Troponin C is so named because it has four binding sites for calcium, and weighs 18 kDa. When bound to calcium, this complex undergoes a stearic rearrangement enabling the thin filament to slide past the thick myosin filament when a signal for muscle contraction is received. As with other myofibrillar proteins, troponin T and I exist in more than one isoform, most of which have characteristic tissue distributions. The cardiac isoforms of T and I are specific to heart tissue. The cardiac isoform of C is identical to the skeletal muscle form. The majority of cTnT and cTnI are myofibrillar-bound. However, there is a small free cytosolic pool of cTnT (6%) and cTnI (2%), possibly as a precursor for incorporation into the myofibril.

Reference Interval

The cut-off concentrations for most cardiac markers, such as CK-MB and myoglobin, are set such that they differentiate between non-AMI cardiac diseases such as unstable angina and AMI. These cut-offs are generally higher than the upper limit of normal and are necessary because some healthy subjects can have high concentrations of these markers due to normal skeletal muscle turnover. For cTnT and cTnI, there is no contribution from skeletal muscle troponin and there is only a small amount of cardiac troponin in healthy individuals. Therefore, low cut-off concentrations are used to detect minor myocardial damage. The ESC/ACC has recommended use of a cut-off at the 99th percentile of the normal range, and has an assay coefficient of variance (CV) of 10% or less. Currently, there are no commercial assays that can meet these criteria. Therefore, the current recommendation is to use the lowest troponin concentration that can be measured with a CV of at least 10%. Figure 55.3 shows typical precision results for the 95th, 99th, and 10% imprecision levels.

cTnT: 0–0.03 ng/mL (10% CV); cTnI: varies with manufacturer.
AMI cut-off: not established for cTnT, varies with manufacturer for cTnI.

Clinical Applications

After AMI, necrosis of cardiac myocytes allows intracellular components such as cTnT and cTnI to leak into the circulation. The initial release of proteins is due to the free cytosolic pool, and abnormal concentrations of cTnT are detected in blood within the first 6–12 h after the onset of chest pain. Despite the release of free cytosolic cTnI, there is very little free form found in serum after AMI, because this form is hydrophobic and presumably binds to other proteins. Over the subsequent days, there is degradation of the myofibrillar elements and release of the ternary troponin complex. Much of this protein degrades in blood to the troponin I-C binary complex and free troponin T. The concentrations of cardiac troponin T and I are increased for 7–10 days after AMI, depending on the marker and the reperfusion status. For patients with successful reperfusion (by thrombolytic therapy or spontaneously), cTnT exhibits a biphasic release pattern (Figure 55.2). For patients with unsuccessful reperfusion and for cTnI, a monophasic release pattern is observed. In most cases, cTnI clears from the blood sooner than cTnT.

A major application of assays for cTnT and cTnI is in the risk stratification of patients with unstable angina. Roughly, one-third of patients with a diagnosis of unstable

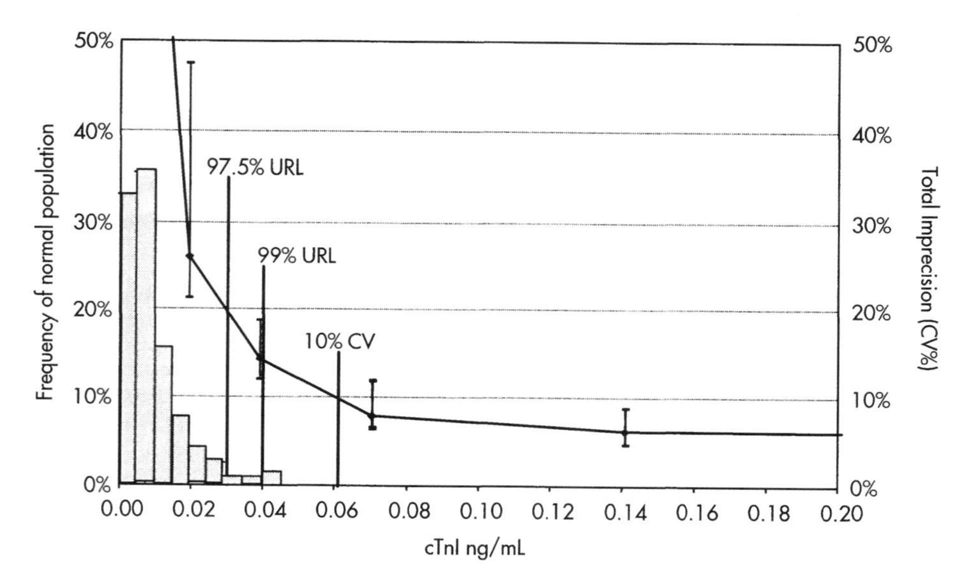

Fig. 55.3 Assay precision vs. troponin I concentration curve. The frequency of cTnI concentrations is given on the left *y*-axis (bars) and the assay precision on the right *y*-axis (curve). The troponin I concentrations for the 97.5, 99 and 10% CV cutpoints are shown on the *x*-axis.

angina will have measurable concentrations of cTnT or cTnI, in the presence of a normal concentration of CK-MB. These patients have suffered minor myocardial injury. In prospective clinical trials, UA patients with an abnormal troponin will have a fivefold higher incidence of cardiac death or AMI within the following 4 weeks than a matched group of UA patients who have normal troponin. These data indicate that measurement of cTnT or cTnI is useful for risk stratification. These markers may be used to triage patients to receive new anti-thrombotic and anti-platelet drugs. Those with high concentrations of cardiac troponin will benefit most from these new therapies.

Limitations

Both cTnT and cTnI have been shown to be highly specific for cardiac injury. Assays have no cross-reactivity toward skeletal muscle forms. There are increases in both cTnT and cTnI in a minority of patients with chronic renal failure, with a higher percentage of abnormal results occurring for cTnT. These results are due to the presence of true myocardial injury, which is common in patients with chronic renal failure. There is no release of cTnT or cTnI in any other non-cardiac diseases.

Assay Technology

All troponin assays use a two-site non-isotopic immunometric assay format with enzyme, fluorescence, or chemiluminescence detection. Due to patent restrictions, all cTnT assays are produced by the same manufacturer and are calibrated to the same reference material. Assays for cTnI are available through many manufacturers. These cTnI assays have significant biases between them. However, a cTnI reference standard has become commercially available. Manufacturers will be gradually standardizing their assays to this material. Qualitative and quantitative point-of-care testing devices are also available using an immunochromatography format. In some cases, multiple analytes are available on the same strip, e.g. cTnI, myoglobin, and CK-MB (e.g. Biosite Diagnostics and First Medical).

Desirable Assay Performance Characteristics

Assays should exhibit no cross-reactivity towards skeletal muscle troponin. Cardiac troponin T should have a sensitivity to detect the 99th percentile of a healthy population with an assay precision of 10% or less.

Type of Sample

Serum or heparinized plasma.

Frequency of Use

Troponin has become the standard biomarker for use in acute coronary syndromes.

FREE FATTY ACID BINDING PROTEINS AND CARBONIC ANHYDRASE III ISOENZYME

Function

Fatty acid binding protein functions (FABP) as a long-chain fatty acid carrier in blood. It plays an important role in lipid metabolism. The heart type isoenzyme is found in the heart and skeletal muscles. It has a molecular weight of 14–15 kDa. Carbonic anhydrase III (CAIII) is a 28 kDa cytoplasmic protein that is found in the skeletal muscles but not the heart. It functions to convert carbonic acid to carbon dioxide and water.

Reference Interval

CAIII: <56 ng/mL[1]
Myoglobin/CAIII: <1.9[1]
FABP: <12 ng/mL[1]
Myoglobin/FABP: <9.0[1]

Clinical Application

The clinical performance of free fatty acid binding protein is similar to that of myoglobin. Both are low molecular weight proteins that are released within 3–6 h after the onset of chest pain, and both are increased in skeletal muscle injury or disease. CAIII concentrations are normal in patients with AMI. Therefore, measurement of either FABP or CAIII alone offers little clinical advantage over myoglobin. However, when myoglobin concentrations are increased, measurement of the ratio of myoglobin to FABP, or of myoglobin to CAIII may be useful in determining the source of myoglobin release. These ratios are significantly different when the injury originates from the skeletal muscles, as opposed to myocardial injury. Therefore, the additional measurement of either FABP or CAIII (but not both) increases the specificity of myoglobin.

Limitations

The addition of either FABP or CAIII to myoglobin doubles the amount of laboratory work and expenses for such testing.

Assay Technology

All assays for FABP and CAIII are based on a two-site immunometric assay format.

Frequency of Use

Commercial assays for FABP and CAIII are not currently available.

[1] Provisional cut-offs.

APOLIPOPROTEINS AI, AII, AND B

Function

The protein part of the lipoproteins is composed of components termed apolipoproteins. Each type of lipoprotein has a specific and relatively constant composition of apolipoprotein. Apolipoprotein A (Apo A) is the major component of the HDLs and is in turn composed of two major subcomponents: Apo AI and Apo AII. Apo AI constitutes about 60–75% of the Apo A in HDL and is necessary for HDL formation. Apo AI is also responsible for the activation of lecithin cholesterol acyl transferase, which catalyzes cholesterol esterification before catabolism of cholesterol and liver excretion. HDL formation helps remove cholesterol from blood vessels and as such is anti-atherogenic.

Apo B is the major protein constituent of LDL (about 90–95%) and also makes up about 40% of the very LDLs and chylomicrons. Apo B is a mixed group of proteins, the two major components being Apo B-100 and Apo B-48. Apo B is of major functional importance because it is recognized by the receptors of LDL and therefore plays a crucial role in LDL catabolism, transporting cholesterol to cells for deposition.

Reference Interval

Sigma Apo AI: 112–152 mg/dL
Immuno Apo AI: >115 mg/dL
Sigma Apo B: 70–96 mg/dL

Clinical Applications

Patients with atherosclerosis and associated CAD have consistently lower levels of HDL and consequently Apo AI and AII. This leads to increased blood levels of the LDL, which in turn results in cholesterol deposition in blood vessels and atherosclerotic plaques. Apo AI levels show a significant correlation with HDL-cholesterol although it has been shown that Apo AI levels are a better predictor of CAD than HDL or HDL-cholesterol.

The indications for Apo B levels in atherosclerosis are directly counter to those for Apo A with increased levels of Apo B being observed consistently in CAD. From this, the ratio of Apo B/Apo AI is being used as a positive predictor of coronary atherosclerosis.

Limitations

The apolipoproteins have not been studied as extensively as the more traditional total, HDL, and LDL cholesterol for risk assessment in CAD, because these markers are new, there is less epidemiologic data available. Moreover, studies on the effect of lipid lowering drugs have largely focused on these traditional lipids and not on the apolipoproteins.

Assay Technology

Many different assay formats for detection of Apo A (usually Apo AI) levels are available. These range from immunoturbidimetric (Immuno; Sigma; Boehringer) to radioimmunoassay (Pharmacia). There are no particularly unusual features in these assays. The assay from Pharmacia is a competitive RIA using specific monoclonal antibodies attached to solid-phase micro-Sepharose®.

Type of Sample

Serum.

Frequency of Use

Moderate.

LIPOPROTEIN (A)

Function

Lipoprotein (a) (Lp(a)), a unique lipoprotein molecule, is a spherical lipid particle very similar in structure to LDL, but with an additional apolipoprotein (a) component. This is covalently bound by disulfide bridges to Apo B-100. In consequence it differs from LDL in molecular mass, protein: lipid ratio and electrophoretic mobility. Lp(a) has a core rich in cholesterol esters.

Reference Interval

Lp(a) $>$ 30 mg/dL = borderline risk
Lp(a) $>$ 50 mg/dL = high risk

Clinical Applications

Several studies have shown a positive correlation between raised plasma Lp(a) levels and the risk of CAD. When the level of Lp(a) is elevated above 30 mg/dL, the risk of atherosclerosis is doubled. This is thought to be due to the large degree of homology between Lp(a) and plasminogen. Lp(a) may inhibit fibrinolysis by competing for plasminogen receptors on the epothelial surface for binding to fibrin. Lp(a) promotes vascular smooth muscle cell proliferation by inhibiting the activation of growth factor. It is taken up by macrophages and may contribute to the formation of foam cells. Unlike other markers of cardiovascular risk, the concentration of Lp(a) in plasma is genetically determined and is independent of diet and exercise. Recent clinical studies have shown, however, the plasma concentrations of Lp(a) can be lowered by niacin therapy. Whether reducing Lp(a) concentrations in blood equates to a lowering of cardiovascular risks is as yet unknown.

Limitations

The apo (a) structure of Lp(a) exists in polymorphic forms, with a variable number of multiple repeats within the kringle 4 protein domain. These apo(a) isoforms have a range of molecular weights from 280–800 kDa. Because reporting units are on a mass basis (mg/dL) rather than a molar basis, some individuals will have a higher Lp(a) result if they normally produce the higher molecular weight form. Moreover, immunoassays that make use of antibodies directed against the variable portions of the apo(a) molecule will have different immunoreactivities to these forms. Therefore, it is difficult if not impossible to standardize immunoassay results from different manufacturers, unless they all agree to use antibodies directed that are not sensitive to the variable regions of the protein.

Assay Technology

Assays for Lp(a) are mainly of the ELISA or immunoturbidimetric type. The ELISA from Immuno Ltd (Enzyquick) is a two-site assay with polyclonal anti-Lp(a) coated on the solid phase and a peroxidase-conjugated Fab fragment of anti-Lp(a) for signal generation. In contrast, Terumo Medical Corporation uses a monoclonal anti-Lp(a) antibody on the solid phase and peroxidase-conjugated polyclonal anti-Lp(a) antibody for signal generation. In both cases, the first antibody is coated on to plastic microtiter strips. An immunoturbidimetric assay from Immuno uses monospecific polyclonal Lp(a) antibody.

The assay for apolipoprotein (a) from Pharmacia is a two-site immunoradiometric assay using two different monoclonal antibodies in excess with the same solid phase as above. Recently, electrophoretic assays have been developed that isolate the Lp(a) fraction from the other lipid fractions, thereby enabling a measurement of the cholesterol content of Lp(a). This assay is not subject to polymorphic heterogeneity.

Frequency of Use

Assays for Lp(a) have been commercially available for many years; in the US, they have only recently become approved by the FDA.

HOMOCYSTEINE

Function

Homocysteine is a sulfhydryl-containing amino acid formed by the demethylation of methionine (Figure 55.4). It consists of a mixture of homocysteine, homocystine, and homocysteine–cysteine mixed disulfides. Under normal conditions, it is metabolized to cysteine through the transsulfuration pathway, or re-methylated back to methionine through the transmethylation pathway. In patients with hereditary homocystinuria, a deficiency in a metabolic enzyme such as cystathione β-synthase

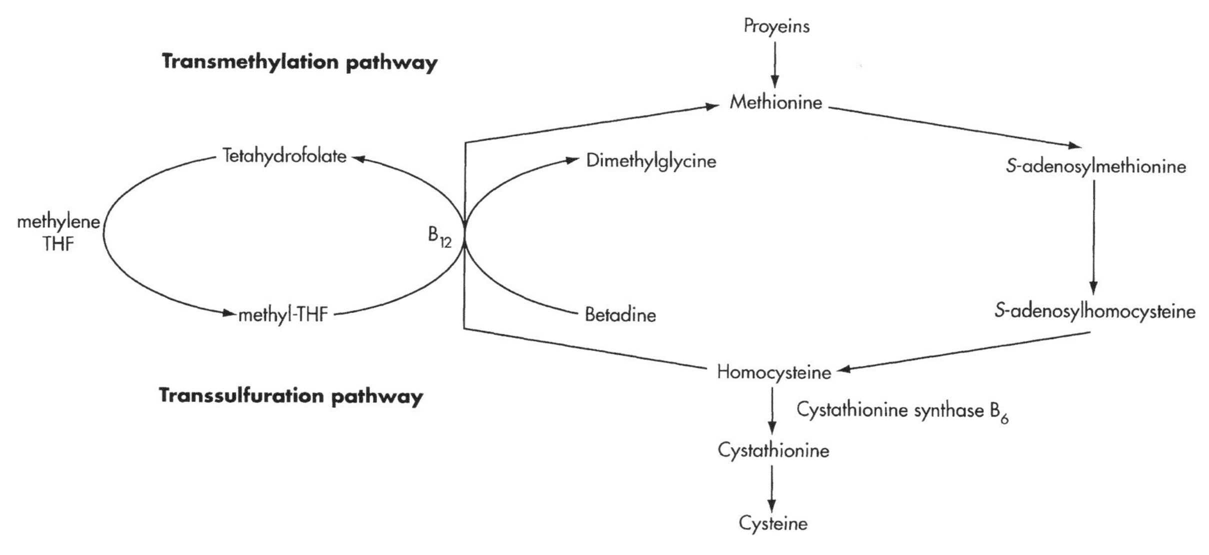

Fig. 55.4 Normal metabolism of homocysteine.

causes severe elevations of plasma and urinary homocysteine. In individuals who have deficiencies in vitamin B_6, B_{12}, and folate, homocysteine accumulates in blood because B_6 is a necessary cofactor for the transsulfation pathway, and B_{12} and folate are necessary for the transmethylation pathway.

Reference Interval

Homocysteine: <15 μmol/L

Clinical Applications

Homocysteinemia has been identified as a cardiovascular risk factor. The original observation was made in untreated homocystinuric children who died of stroke and AMI before adulthood. Despite normal levels of cholesterol, examination of their coronary arteries revealed extensive atherosclerosis, similar in presentation to adults with CAD. Subsequent studies have shown that adult patients with high plasma homocysteine concentrations are at greater risk for cardiovascular disease than age-matched controls. Homocysteinemia is also associated with peripheral arterial and venous occlusive diseases. The cardiovascular risk can be reduced by vitamin supplementation. The current US recommended daily allowance is 2 mg for B_6, 5 μg for B_{12}, and 200 μg for folic acid. To increase the folate concentration in the general population, grain products are supplemented with folate in the US.

Assay Technology

Homocysteine can be measured by high-performance liquid chromatography and immunoassay. Samples must be treated to free protein-bound homocysteine, and to reduce homocystine and the mixed disulfides to free homocysteine. HPLC eluates can be measured using electrochemical detectors, or fluorometrically following derivatization. Commercial immunoassays are currently being developed. In the Abbott AxSYM assay, bound and oxidized forms are reduced by dithiothreitol, converted to *S*-adenosyl-L-homocysteine (SAH) by SAH hydrolase and excess adenosine, and measured by fluorescence polarization immunoassay.

C-REACTIVE PROTEIN

Function

C-reactive protein was first described in 1930. It is an acute-phase protein that is released into the circulation during acute and chronic inflammation. It functions to bind to the polysaccharide component of bacteria, fungi, and parasites. Once bound, CRP promotes opsonization, phagocytosis, and lysis via activation of the classical complement pathway. CRP is a pentamer of identical subunits with a combined molecular weight of 118 kDa. In patients suffering acute inflammation, CRP is released into blood within 24 h to exceptionally high levels (>1000 times the normal range).

Reference Interval

Normal sensitivity assay (10 mg/L) and High-sensitivity assay (<1.0 mg/L) indicates the cardiovascular lowest risk group (Figure 55.5).

(a)

1. Determine TC:HDLC Quintile
2. Determine hs-CRP Quintile
3. Assess Cardiovascular Relative Risk

(b)

Quintile	hs-CRP (mg/L)	TC:HDLC (Women)	TC:HDLC (Men)
1	0.1 - 0.7	< 3.4	< 3.4
2	0.7 - 1.1	3.4 - 4.1	3.4 - 4.0
3	1.2 - 1.9	4.1 - 4.7	4.0 - 4.7
4	2.0 - 3.8	4.7 - 5.8	4.7 - 5.5
5	3.9 - 15.0	> 5.8	> 5.5

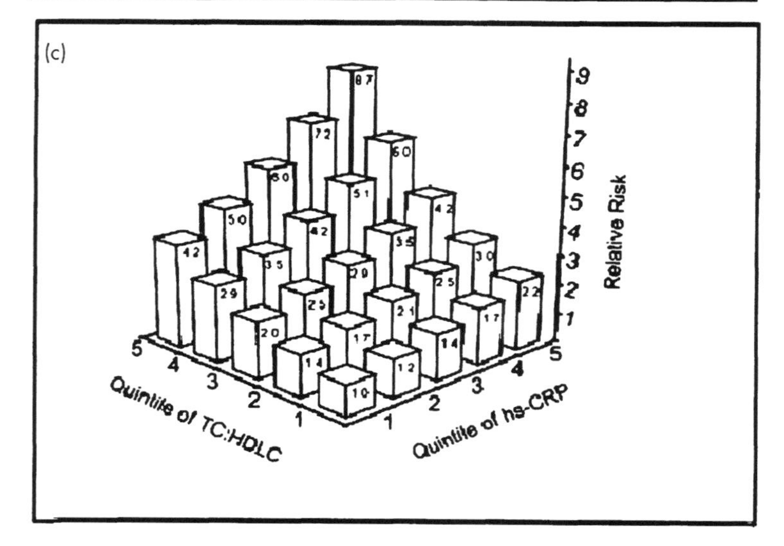

Fig. 55.5 Primary cardiovascular risk assessment using a combination of hsCRP and total cholesterol: HDL cholesterol. Cut-offs are broken down into quintiles. Used with permission from the American Association for Clinical Chemistry (*Clin. Chem.* **47**: 28–30 (2001)).

Clinical Applications

Monitoring CRP concentrations in blood has been used as a marker for acute inflammation for many years. CRP has been shown to be more sensitive and specific than the erythrocyte sedimentation rate. Recently, the role of inflammation has been clarified in the pathophysiology of acute coronary syndromes. As such, minor increases in CRP have been linked to higher risk for cardiovascular disease. Use of CRP as a predictor of cardiac events requires the use of a high-sensitivity assay, as increased risk is shown for CRP concentrations that are within the normal range of conventional (low-sensitivity) assays. The predictive value of high-sensitivity CRP assay is independent to total cholesterol and HDL measurements. The combination of hsCRP and lipids produces more precise risk stratification information that use of lipids and lipoproteins alone.

Limitations

If CRP is to be used for acute detection of inflammation, an assay with low sensitivity is sufficient for clinical use. However, for cardiovascular disease risk, high-sensitivity CRP assays are necessary. The limit of quantitation necessary for this use is ≤ 0.3 mg/L. If the CRP assay is performed by light-scattering immunoassays, the phenomenon of 'prozoning' or antigen excess can occur. As shown in Figure 55.6, the antigen–antibody complex is insoluble, and a linear relationship is initially observed between the light-scattering curve and the antigen concentration. At high-antigen concentration, however, a soluble complex is formed, resulting in a reduction in the analytical signal back to the baseline. As a result, two different antigen concentrations can produce the same analytical signal. To prevent the reporting of the wrong antigen concentration, the sample must be diluted. A reduction in the antigen concentration upon dilution

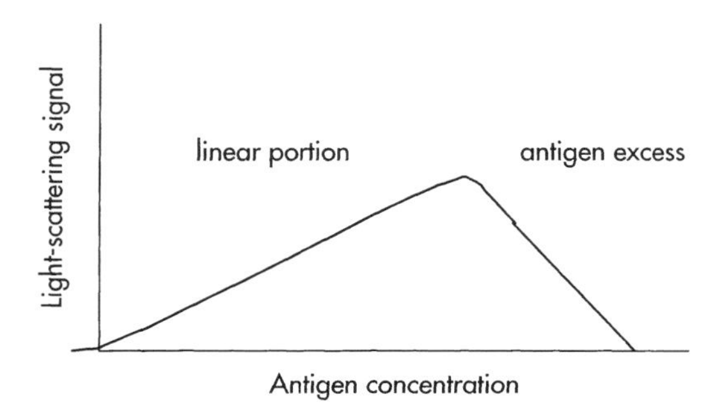

Fig. 55.6 Light-scattering curve vs. antigen concentration in nephelometric assays, demonstrating the antigen excess zone.

suggests that the initial sample was within the acceptable linear portion of the curve. A high antigen concentration recovery upon dilution suggests that the sample CRP concentration is too high. Heterogeneous immunometric assays with other signal generation systems do not suffer from the prozone effect, but may suffer from the hook effect common to these types of assays. The reporting units for the conventional (low-sensitivity) CRP assay have been in mg/dL. This can produce considerable confusion if the recommended mg/L units are used for the high-sensitivity assay.

Assay Technology

C-reactive protein can be measured by rate nephelometry and turbidimetry. For high-sensitivity determinations, heterogeneous immunoassays are preferred.

Types of Sample

Serum and heparinized plasma.

BRAIN NATRIURETIC PEPTIDE AND NT-PRO-BNP

Function

The natriuretic peptides are a collection of hormones that regulate body fluid homeostasis and blood pressure by diuresis, natriuresis, vasorelaxation, and inhibition of the renin–aldosterone axis. Atrial natriuretic peptide is secreted from the atrium and is a 28 amino acid peptide. BNP is found in both the brain and the ventricles of the heart, and is a 32 amino acid peptide. C-type natriuretic peptide is a 22 amino acid peptide and is though to originate in the endothelial cells. BNP and the inactive metabolite, NT-proBNP are widely used as markers for heart failure. Unlike lipid markers that participate in the disease process of atherosclerosis, or cardiac markers that are released as a consequence of cardiac injury, BNP is released as a *compensatory* mechanism to the failing heart to improve left ventricular function by reducing cardiovascular load.

Within the myocyte, Figure 55.7 shows that BNP and NT-proBNP are derived from preproBNP, a 134 amino acid peptide. This peptide is cleaved to proBNP (108 amino acids) and a signal peptide. Under conditions of ventricular stretch, proBNP is cleaved to BNP, the biologically active hormone (32 amino acids), and the inactive *N*-terminal fragment (NT-proBNP, 76 amino acids). Both of these peptides circulate in blood.

Reference Interval

The concentrations for BNP and NT-proBNP from healthy individuals increase with age. Healthy women have higher values than men. Cut-off concentrations for use in heart failure are:

BNP: <100 pg/mL.
NT-proBNP: <125 pg/mL for <75 years; <450 pg/mL for ≥ 75 years

Clinical Applications

The New York Heart Association (NYHA) and other profession groups have established four classifications of heart failure based on subjective criteria of functional capacities. These are listed in Table 55.1. Clinical studies have shown that there is an incremental rise in blood concentrations of BNP within the increasing severity of NYHA classes. In patients with systolic dysfunction (ventricular ejection defect), the natriuretic peptide concentrations are inversely correlated to the ejection fraction as measured by echocardiographic analysis. There is no correlation with BNP and the LVEF in diastolic heart failure (i.e. ventricular filling defect). These clinical studies suggest that BNP may be useful as a diagnostic marker of CHF. BNP might also be useful in monitoring the success of drug therapy given to these patients. For risk stratification, increased BNP is correlated

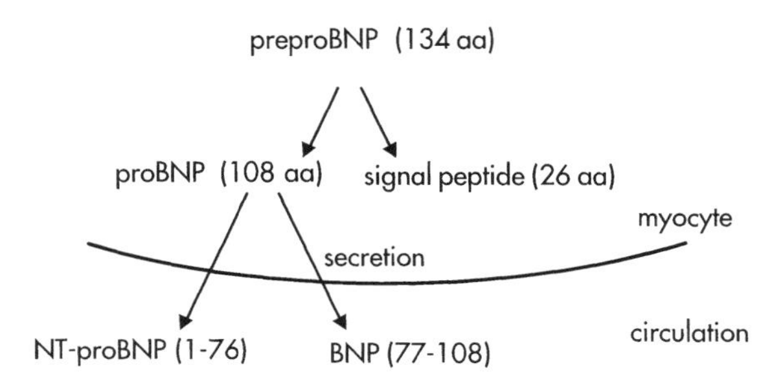

Fig. 55.7 Derivation and secretion into blood of BNP and NT-proBNP from myocytes.

Table 55.1. New York Heart Association Classification of heart failure.

Functional assessment	
Class I	Patients with cardiac disease but without resulting limitation of physical activity. Ordinary physical activity does not cause undue fatigue, palpitation, dyspnea, or anginal pain
Class II	Cardiac disease resulting in slight limitation of physical activity. They are comfortable at rest. Ordinary physical activity results in fatigue, palpitation, dyspnea, or anginal pain
Class III	Marked limitation of physical activity. They are comfortable at rest. Less than ordinary activity causes fatigue, palpitation, dyspnea, or anginal pain
Class IV	Cardiac disease resulting in inability to carry on any physical activity without discomfort. Symptoms of heart failure or of the anginal syndrome may be present even at rest. If any physical activity is undertaken, discomfort is increased
Objective assessment	
A	No objective evidence of cardiovascular disease
B	Objective evidence of minimal cardiovascular disease
C	Objective evidence of moderately severe cardiovascular disease, and
D	Objective evidence of severe cardiovascular disease

Functional capacity and objective assessment. In: Dolgin (1994).

with a higher risk for death and AMI for patients who present with acute coronary syndromes or CHF. In this regard, BNP provides complementary information to hsCRP and cardiac troponin.

Limitations

BNP is not stable beyond 24 h at 4 °C. NT-proBNP is stable for 72 h at 4 °C. Both analytes are increased in patients with chronic renal failure due to volume overload.

Assay Technology

Prototypic assays for BNP are based on two-site sandwich immunoassays using monoclonal antibodies. Assays are available on a point-of-care platform, using immunochromatography, and on automated immunochemistry analyzers.

Types of Sample

BNP: whole blood or plasma collected in EDTA only.
NT-proBNP: serum, heparin, or EDTA plasma.

Frequency of Use

BNP and NT-proBNP are widely used for diagnosis and management of heart failure. These markers have been incorporated by the ESC into clinical practice guidelines.

PLASMA RENIN

Function

Angiotensin I is formed from angiotensinogen by the action of renin, which is released from the kidney juxtaglomerular cells. Angiotensin I is further cleaved to angiotensin II by angiotensin converting enzyme. Angiotensin II is an extremely potent vasoconstrictor but has a very short half-life, being degraded by angiotensinases to inactive fragments. Because angiotensin II is difficult to measure, angiotensin I levels are determined as a measure of renin activity under conditions in which angiotensin breakdown is blocked by inhibiting plasma-converting enzyme and proteolysis by angiotensinases.

Reference Interval

Plasma renin activity (pH 6.0):

renin (supine): 0.2–2.8 ng/mL/h
renin (upright): 1.5–5.7 ng/mL/h.

The above values are only indicative for a sodium intake of 100–150 mEq per 24 h.

Clinical Applications

The determination of plasma renin activity has been widely used to evaluate the renin–angiotensin system in disease states. In patients with hypertension due to primary hyperaldosteronism, plasma renin activity is reduced. In contrast, in patients with renovascular hypertension, plasma renin activity, and aldosterone secretion are both elevated. Measurement of plasma renin activity has therefore been suggested as an important aid in the differential diagnosis of primary and secondary aldosteronism. A differentiation between low- and high-renin hypertensive states may also be helpful in selecting anti-hypertensive medications (e.g. angiotensin converting enzyme inhibitors vs. diuretics).

Limitations

Because the analyte in this case is generated in the assay, the specificity of the antiserum in question is not critical because sample blanks are deducted from the final results. The interference from some sample proteins may limit the assay. In addition, prorenin, the inactive precursor of renin, is cryoactivated to renin when the sample is exposed to 4 °C temperatures for extended time, leading to falsely high results.

Assay Technology

The common assay format is a competitive radioimmunoassay using ^{125}I-labeled angiotensin I. Assays from Du-Pont (RIANEN^{TM} assay system) and CIS both use this methodology. In the former case, the antigen–antibody reaction is conducted in free solution with subsequent addition of second antibody and centrifugation to separate free from bound label. In the latter system, the competition for the fixed number of binding sites is done with primary antibody-coated tubes. Nichols Institute Diagnostics have developed a two-site immunoradiometric sandwich assay making use of one antibody coupled to biotin and the second radiolabeled renin antibody.

Type of Sample

Plasma, using EDTA as anticoagulant, as heparin interferes with angiotensin I production.

PLASMINOGEN ACTIVATORS AND INHIBITORS

Function

Tissue plasminogen activator (t-PA), a 70 kDa serine protease, is responsible for the conversion of plasminogen to plasmin, which in turn lyzes fibrin. The activity of t-PA is enhanced 100-fold if fibrin is present, leading to further dissolution of thrombi. Plasminogen is activated on the surface of fibrin by t-PA from endothelial cells. Although it is not normally detectable in blood, t-PA is present after stimulation by activated protein C, or due to stress, physical exertion, or venous occlusion. As t-PA activity is limited to the local area of the fibrin clot, the plasmin produced is therefore protected from the inhibitory action of anti-plasmin. Inhibitors of t-PA, called plasminogen-activator inhibitors (PAIs) are also present in the circulation. The activities of t-PA and PAIs *in vivo* are therefore good indicators of the level of activity of the fibrinolytic system and in turn of the risk of thromboembolism leading to CAD.

Reference Interval

t-PA: 1.3–10.4 ng/mL
PAI: not available.

Clinical Applications

Myocardial infarction is just one of the number of conditions that can arise as a result of thromboembolism. The risk of thrombotic complications is associated with t-PA deficiency, which can arise as a result of reduced synthesis or release of t-PA or by increased concentrations of t-PA inhibitors. Determination of the levels of either t-PA and/or PAIs can give an indication of the activity of the fibrinolytic system.

Limitations

Plasma must not be used where it has been collected with the aid of fibrinolysis inhibitors.

Assay Technology

Common assay protocols for t-PA use a non-competitive two-site ELISA with monoclonal anti-t-PA bound to the solid phase (e.g. Innogenetics). The t-PA concentration is determined by routine enzyme-labeled second-site monoclonal antibody and colorimetric determination.

Likewise, ELISAs for PAIs (Innogenetics) also use the same two-site protocols with bound and enzyme-labeled monoclonal antibodies.

Types of Sample

Fresh citrated plasma for t-PA and fresh venous plasma samples for PAIs.

REFERENCES

Adams, J.E., Bodor, G.S., Davila-Roman, V.G., Delmez, J.A., Apple, F.S., Ladenson, J.H. and Jaffe, A.S. Cardiac troponin I. A marker with high specificity for cardiac injury. *Circulation* **88**, 101–106 (1993).

Apple, F.S. Acute myocardial infarction and coronary reperfusion. Serum cardiac markers for the 1990s. *Am. J. Clin. Pathol.* **97**, 217–226 (1992).

Apple, F.S. and Wu, A.H.B. Myocardial infarction redefined: role of cardiac troponin testing [Editorial]. *Clin. Chem.* **47**, 377–379 (2001).

Arakawa, N., Nakamura, M., Aoki, H. and Hiramori, K. Plasma brain natriuretic peptide concentrations predict survival after acute myocardial infarction. *J. Am. Coll. Cardiol.* **27**, 1656–1661 (1996).

Bessman, S.P. and Carpenter, C.L. The creatine–creatine phosphate energy shuttle. *Ann. Rev. Histochem.* **54**, 831–862 (1985).

Bodor, G.S., Porter, S., Landt, Y. and Ladenson, J.H. Development of monoclonal antibodies for the assay of cardiac troponin I and preliminary results in suspected cases of myocardial infarction. *Clin. Chem.* **38**, 2203–2214 (1992).

Booth, N.A., Simpson, A.J., Croll, A., Bennett, B. and MacGregor, I.R. Plasminogen activator inhibitor-1 in plasma and platelets. *J. Hematol.* **70**, 327–333 (1988).

Chan, K.M., Ladenson, J.H., Pierce, G.F. and Jaffe, A.S. Increased creatine kinase MB in the absence of acute myocardial infarction. *Clin. Chem.* **32**, 2044–2051 (1986).

Chapelle, J.P. and Heusghem, C. Semiquantitative estimation of serum myoglobin by a rapid agglutination method: an emergency screening test for acute myocardial infarction. *Clin. Chim. Acta* **145**, 143–150 (1985).

Functional capacity and objective assessment. In: *Nomenclature and Criteria for Diagnosis of Diseases of the Heart and Great Vessels* 9th edn (ed Dolgin, M.), 253–255 (Little, Brown and Co., New York, 1994).

Feng, Y.G., Chen, C., Fallon, J.T., Lai, T., Chen, L., Knibbs, D.R., Waters, D.D. and Wu, A.H.B. Comparison of cardiac troponin I, creatine kinase-MB and myoglobin for detection of acute ischemic myocardial injury in a swine model. *Am. J. Clin. Pathol.* **110**, 70–77 (1998).

Fuster, V., Badimon, L., Badimon, J.J. and Chesebro, J.H. The pathogenesis of coronary artery disease and the acute coronary syndromes. *N. Engl. J. Med.* **326**, 242–250 (1992).

Genest, J.J., McNamara, J.R., Salem, D.N., Wilson, P.W.F., Schaefer, E.J. and Malinow, M.R. Plasma homocyst(e)ine levels in men with premature coronary artery disease. *J. Am. Coll. Cardiol.* **16**, 1114–1119 (1990).

Hamm, C.W., Ravkilde, J., Gerhardt, W., Jorgensen, P., Peheim, E., Ljungdahl, L., Goldmann, B. and Katus, H.A. The prognostic value of serum troponin T in unstable angina. *N. Engl. J. Med.* **327**, 146–150 (1992).

Hedges, J.R., Young, G.P., Henkel, G.F., Gibler, W.B., Green, T.R. and Swanson, J.R. Early CK-MB elevations predict ischemic events in stable chest pain patients. *Acad. Emerg. Med.* **1**, 9–16 (1994).

Joint European Society of Cardiology/American College of Cardiology Committee, Myocardial infarction redefined – a consensus document of the joint European Society of Cardiology/American College of Cardiology Committee for the redefinition of myocardial infarction. *J. Am. Coll. Cardiol.* **36**, 959–969 (2000).

Konstam, M.A., Dracup, K., Baker, D.W. *et al. Heart failure: Evaluation and Care of Patients with Left-ventricular Systolic Dysfunction* (US Department of Health Human Services, Rockville, MD, 1994).

Mair, J., Morndell, D., Genser, N., Lechleitner, P., Dienstl, F. and Puschendorf, B. Equivalent early sensitivities of myoglobin, creatine kinase MB mass, creatine kinase isoform ratios and cardiac troponins I and T for acute myocardial infaction. *Clin. Chem.* **41**, 1266–1277 (1995).

McCullough, P.A., Duc, P., Omland, T., *et al.* For the BNP Multinational Study Investigators B-Type natriuretic peptide and renal function in the diagnosis of heart failure: an analysis from the breathing not properly multinational study. *Am. J. Kidney Dis.* **41**, 571–579 (2003).

Miles, L.A., Fless, G.M., Levin, E.G., Scanu, A.M. and Plow, E.F. A potential basis for the thrombotic risks associated with lipoprotein(a). *Nature* **339**, 301–303 (1989).

Muller-Bardorff, M., Hallermayer, K., Schroder, A., Ebert, C., Borgya, A., Gerhardt, W., Remppis, A., Zehelein, J. and Katus, H.A. Improved troponin T ELISA specific for cardiac troponin T isoform: assay development and analytical and clinical validation. *Clin. Chem.* **43**, 458–466 (1997).

Redfield, M.M., Rodeheffer, R.J., Jacobsen, S.J., Mahoney, D.W., Bailey, K.R. and Burnett, J.C. Jr. Plasma brain natriuretic peptide concentration: impact of age and gender. *J. Am. Coll. Cardiol.* **40**, 976–982 (2002).

Reinhart, R.A., Gani, K., Arndt, M.R. and Broste, S.K. Apolipoproteins A-I and B as predictors of angiographically defined coronary heart disease. *Arch. Intern. Med.* **150**, 1629–1633 (1990).

Remme, W.J. and Swedberg, K. Guidelines for the diagnosis and treatment of chronic heart failure. Task Force for the Diagnosis and Treatment of Chronic Heart Failure, European Society of Cardiology. *Eur. Heart J.* **22**, 1527–1560 (2001).

Ridker, P.M., Cushman, M., Stampfer, M.J., Tracy, R.P. and Hennekens, C.H. Inflammation, aspirin and the risk of cardiovascular disease in apparently healthy men. *N. Engl. J. Med.* **336**, 973–979 (1997).

Rifai, N. and Ridker, P.M. Proposed cardiovascular risk assessment algorithm using high-sensitivity C-reactive protein and lipid screening. *Clin. Chem.* **47**, 28–30 (2001).

Sealey, J.E. Plasma renin activity and plasma prorenin assays. *Clin. Chem.* **37**, 1811–1819 (1991).

Stampfer, J.J., Sacks, F.M., Salvini, S., Willett, W.C. and Henneckens, C.H. A prospective study of cholesterol, apolipoproteins and the risk of myocardial infarction. *N. Engl. J. Med.* **325**, 373–381 (1991).

Stampfer, J.J., Malinow, M.R., Willett, W.C., Newcomer, L.M., Upson, B., Ullmann, D., Tishler, P.V. and Hennekens, C.H. A prospective study of plasma homocyst(e)ine and risk of myocardial infarction in US Physicians. *JAMA* **268**, 877–881 (1992).

Vaidya, H.C. Myoglobin: an early biochemical marker for the diagnosis of acute myocardial infarction. *J. Clin. Immunoassay* **17**, 35–39 (1994).

Vaidya, H.C., Maynard, Y., Dietzler, D.N. and Ladenson, J.H. Direct measurement of creatine kinase-MB activity in serum after extraction with a monoclonal antibody specific to the MB isoenzyme. *Clin. Chem.* **32**, 657–663 (1986).

Van Leeuwen, M.A. and Van Rijswijk, M.H. Acute phase proteins in the monitoring of inflammatory disorders. *Baillieres Clin. Rheumatol.* **8**, 531–552 (1994).

Van Nieuwenhoben, F.A., Kleine, A.H., Wodzig, K.W.H., Hermens, W.T., Kragten, H.A., Maessen, J.G., Punt, C.D., Van Dieijen, P., Van der Vusse, and Glatz, J.F.C. Discrimination between myocardial and skeletal muscle injury by assessment of the plasma ratio of myoglobin over fatty acid-binding protein. *Circulation* **92**, 2848–2854 (1995).

Vuori, J., Syrjala, H. and Vaananen, K. Myoglobin carbonic anhydrase III ratio: highly specific and sensitive early indicator for myocardial infarction. *Clin. Chem.* **36**, 107–109 (1991).

Wieczorek, S.J., Wu, A.H.B., Christenson, R., Rosano, T., Hager, D., Bailly, K., Dahlen, J., Chambers, B.S. and Maisel, A. A rapid B-type natriuretic peptide (BNP) assay accurately diagnoses left ventricular dysfunction and heart failure – a multi-center evaluation. *Am. Heart J.* **144**, 834–839 (2002).

Wei, C.M., Heublein, D.M., Perrella, M.A., Lerman, A., Rodeheffer, R.J., McGregor, C.G.A., Edwards, W.D., Schaff, H.V. and Burnett, J.C. Natriuretic peptide system in human heart failure. *Circulation* **88**, 1004–1009 (1993).

Wu, A.H.B. (ed), *Cardiac Markers* 2nd edn (Humana, Totowa, NJ, 2003).

Yun, D.D. and Alpert, J.S. Acute coronary syndromes. *Cardiology* **88**, 223–237 (1997).

Zuo, W.M., Pratt, R.E., Heusser, C.H., Bews, J.P.A., de Gasparo, M. and Dzau, V.J. Characterization of a monoclonal antibody specific for human active renin. *Hypertension* **19**, 249–254 (1992).

56 Cancer Markers

Mavanur R. Suresh

INTRODUCTION

During normal growth and development, tissues and organs develop originally from a single cell by the process of differentiation. Although the processes involved are poorly understood, they are clearly complex and highly regulated. In cancer, cell division goes out of control, often as a result of the gradual accumulation of multiple mutations over a prolonged period. Small growths sometimes occur that are harmless and not cancerous; these are said to be **benign** and are mostly well differentiated. Their close similarity to normal tissues makes detection difficult. In cancer, the dangerous growths are described as **malignant** and are predominantly less differentiated. The resulting growth and eventual systematic spread or **metastasis** of the disease often kills the patient. **Carcinoma** refers to tumors of skin or mucous membranes, **sarcoma** to tumors of connective tissue. Abnormal cancerous growth produces abnormal types and levels of substances we now recognize as cancer markers or tumor markers. These markers appear in blood and other body fluids due to the loss of polarity and/or anatomic damage from the tumor. The properties of an ideal tumor marker are summarized in Table 56.1, although no tumor marker exhibits all of these properties.

Specific Bence-Jones proteins were discovered in the urine of some cancer patients as long ago as 1846. During the next 100 years, ectopic hormones and isoenzymes were identified as cancer markers. The last few decades of research have resulted in a virtual explosion in the discovery, validation, and clinical application of these analytes in cancer patient management, largely due to the development of monoclonal antibody and recombinant DNA technology (Oppenheimer, 1985; Virji *et al.*, 1988; Rittenhouse *et al.*, 1985; Sell, 1990). This chapter attempts to provide an overview of cancer marker immunoassays with special reference to those analytes now available for routine clinical use. It is pertinent to mention at this juncture that various countries have different modes of regulating the use of tumor marker assays. It ranges from minimal to moderate regulatory constraints in Canada and Western Europe, to substantive requirements in the USA and Japan. Until 1996, a new tumor marker assay to be introduced in the USA was regulated as a Class III device requiring retrospective and prospective clinical trial as part of the pre-market application or PMA. Now, tumor markers are classified as Class II medical devices in the USA requiring a 510k application, a simplified process that generally requires 90–120 days for approval.

A definition of tumor markers was adopted at the 5th International Conference on Human Tumor Markers held in Stockholm, Sweden in 1988. It states that

> Biochemical tumor markers are substances developed in tumor cells and secreted into body fluids in which they can be quantitated by non-invasive analyses. Because of a correlation between marker concentration and active tumor mass, tumor markers are useful in the management of cancer patients. Markers, which are available for most cancer cases, are additional, valuable tools in patient prognosis, surveillance and therapy monitoring, whereas they are presently not applicable for screening. Serodiagnostic measurements of markers should emphasize relative trends instead of absolute values and cut-off levels.

The potential clinical applications of tumor marker assays are listed below.

Screening

Some tumor markers have been utilized in mass screening programs of asymptomatic individuals, with limited success, in high-risk sectors of the population. However, it is to be emphasized that no biochemical tumor marker is yet specific and sensitive enough to be recommended as a definitive screening test for cancer. In some countries, screening programs have been conducted for certain cancers highly prevalent in those regions. In China, AFP measurements have been used for hepatocellular carcinoma screening. Individuals with a previous history of hepatitis infection or liver cirrhosis are at a higher risk for developing liver cancer. In Japan, screening for neuroblastoma in children <1 year has been conducted by measuring urinary vanillylmandelic acid and homovanillic acid. Other screening markers tested

The Immunoassay Handbook. *Third Edition*
D. Wild (Ed.)

Table 56.1. Properties of an ideal tumor marker.

- High clinical sensitivity
- High clinical specificity
- Tumor marker levels proportional to tumor volume
- Short half life to rapidly mirror treatment schedules
- Reflect tumor heterogeneity
- Low levels in healthy population
- Low levels in benign diseases
- Discriminatory to identify tumor and metastasis from benign to healthy states
- Provide adequate lead times for early diagnosis and early treatment
- Assay sensitivity to detect stage I cancers

include fecal occult blood/hemoglobin for colorectal cancer, CA125 for ovarian cancer, prostate-specific antigen (PSA) for prostate cancer and p21 *ras* oncoprotein. Prostate cancer screening using PSA and digital rectal examination was recommended by the American Cancer Society in November 1992 for men over 50 years of age. However, it remains controversial, and in spite of this strong endorsement, a significant number of oncologists question the merits of screening as related to treatment options and survival benefit. Another prominent marker worth mentioning is the identification of mutated forms of hereditary cancer susceptibility BRCA1 and BRCA2 genes or gene products, which identifies high-risk individuals.

In 2003, a new two-in-one test was approved by the FDA for the primary screening of cervical cancer. It includes a sandwich immunoassay test with the traditional 50 year old *PAP* smear for women over the age of 30. The PAP test is widely used as a screening test that is done annually to detect the presence of abnormal cellular markers indicative of cancerous or precancerous conditions. This was developed by George Papanicolaou and involves the collection of vaginal fluid or cells scraped from the cervix to predict or detect cervical cancer. The new component of the combined test (Digene Inc., USA) is the genetic test for 13 strains of the human papilloma viruses (HPV), which are largely sexually transmitted and the likely cause of 99% of cervical cancers. While millions of women are infected with HPV, only those above the age of 30 and having persistent infections are at high risk. The Hybrid Capture® 2 DNA test uses a two RNA probe mixture to distinguish between the carcinogenic and low-risk HPV types. The company offers a specific specimen collection device to accompany the combined test procedure although it is similar to the traditional sample collection method. The principle of the hybrid capture method is essentially a sandwich immunoassay that detects the specific DNA–RNA hybrid as the antigen. The specimen is dissolved in a base solution to dissociate the various components including the target viral DNA. The specific HPV RNA probe is added to form a DNA–RNA hybrid, which is selectively captured by a solid phase coated with an antibody with specificity to the hybrid nucleic acids. The captured DNA–RNA hybrid is detected by an alkaline phosphatase-labeled antibody and detected by chemiluminescent dioxetane substrate. The combination test has been found to have better sensitivity than the use of either test alone. In some countries outside the USA, the HPV test is also used as a primary screening test alone or in conjunction with the *PAP* test.

Diagnosis

Almost every cancer marker has been investigated for its suitability as a primary diagnostic test for cancer in symptomatic individuals. However, sufficient false positives and false negatives have been encountered with every marker so far discovered to preclude their use in distinguishing malignant and non-malignant conditions. The ultimate goal of identifying tumor-specific antigens has so far eluded oncologists because most tumor markers have been found in some normal tissues and the serum of some non-cancerous individuals and in many benign diseases. For this reason, these antigens are often referred to as tumor-associated antigens. Nevertheless, a number of cancer markers have proved to be useful in confirming diagnosis, often in conjunction with a battery of other clinical methods. Another approach attempts to use multiple tumor markers to diagnose tumors and to identify the primary origin of metastic disease (Wu and Nakamura, 1997; Hanausek and Walaszek, 1998).

Differential Diagnosis and Classification

Immunoassays for some cancer markers are used in clinics to distinguish between clinical conditions with similar symptoms, where one or both could be cancerous. For example, the measurement of neuron-specific enolase levels allows differentiation between neuroblastoma and Wilm's tumor when a child presents with a palpable abdominal mass. Similarly, PSA and prostatic acid phosphatase (PAP) can distinguish prostate cancer metastasis from other secondary tumors whose primary origin is not the prostate gland. Antibody probes specific for B or T cells can establish the lineage and classify leukemias and lymphomas in an immunohistochemical assay (*see* HEMATOLOGY). Antibodies specific to lymphoid malignancies can distinguish between non-Hodgkin's lymphoma and undifferentiated cancer of non-hematopoietic origin.

Staging and Grading

The degree of elevation in the concentration of several tumor markers can help to **stage** tumors. In general the mean circulating levels of these tumor markers increase with the stage of the cancer. In contrast, placental alkaline phosphatase (ALP) is a tumor marker related to the **grade** of cancer, and serum levels of this analyte are higher in grade 1 and 2 tumors than in grade 3 ovarian carcinomas.

Prognosis

Prognosis is the probability of cure of a cancer patient. Positive lymph node detection is a classical method of determining prognosis invasively. The magnitude of tumor marker levels in several cancers corresponds to the mass of tumor. Moderate elevations are suggestive of better prognosis than persistent high levels. An important prognostic factor in ovarian and breast cancers is the amplification of the c-*erb*B-2 gene (HER-2/*neu*) and protein. Tumor aggressiveness resulting in widespread metastasis precipitates very high serum tumor marker levels indicating poor prognosis. Generally, well-differentiated tumors tend to be less aggressive than undifferentiated or anaplastic tumors. While most tumor marker overexpression indicates poor prognosis, the increased levels of progesterone and estrogen receptors (ER) in breast cancers determine the type of treatment (hormone) as well as good prognosis.

Monitoring and Recurrence

The profile of tumor marker concentration against time can mirror the condition of patients diagnosed to have cancer, for example indicating whether therapy has been successful, or if remission has occurred (Figure 56.1). This is one area where tumor markers are most useful (Pannall and Kotasek, 1997; Suresh, 1996).

Tumor marker profiles usually reflect one of the following classical patterns:

- A rapid decline in tumor marker level to normal concentrations following surgery or other forms of first-line therapy suggests that treatment has been successful.
- The lack of a decline to basal levels following first-line therapy may indicate that treatment has only been partially successful.
- Continued low levels of the tumor marker indicate that remission has been maintained as a result of treatment.
- A subsequent rise in the concentration of the tumor marker (from the basal level) suggests a recurrence of the disease. Tumor markers can warn of renewed tumor growth or recurrence 3–12 months before other methods provide confirmation.

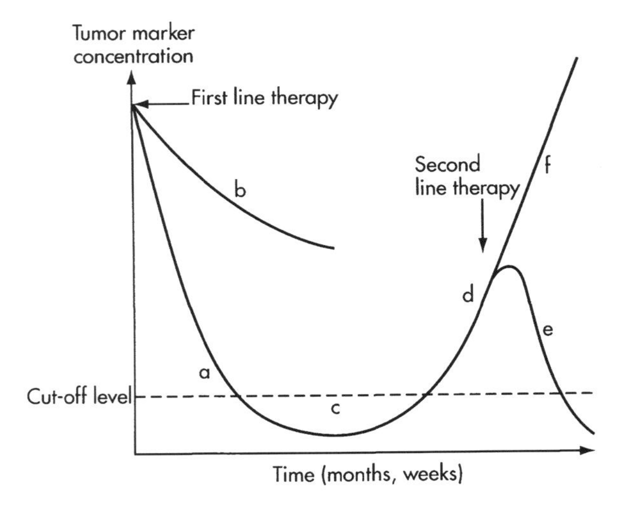

Fig. 56.1 Classical trends in tumor marker profiles: (a) successful first line therapy with reduction to normal levels; (b) unsuccessful first-line therapy or partial response; (c) continued clinical remission; (d) recurrence of cancer; (e) response to second line of therapy; (f) failure of or resistance to therapy, with poor prognosis.

- Decline of the marker levels after an increase has been associated with a recurrence, is suggestive of the responsiveness of a tumor to second line or subsequent treatment.
- If tumor marker concentrations remain elevated after treatment, the tumor may be resistant to the therapeutic method employed and the prognosis of the patient is poor unless alternative therapeutic modalities are available.

These characteristic profiles can be observed for many tumor markers, e.g. carcino-embryonic antigen (CEA) in colorectal cancers, cancer antigen 125 (CA 125) in ovarian cancers, or PSA in prostatic cancer.

Although these classical patterns in tumor marker profiles are seen in the majority of patients, they do not reflect the clinical status of every patient. Hence some oncologists recommend the estimation of more than one marker (Wu and Nakamura, 1997). For example, in pancreatic cancer, carbohydrate antigens 19-9 (CA 19-9), 50 (CA 50) and CEA may all be elevated. However, CA 19-9 is positive in 75% of the patients, whereas CEA is positive in less than 50%. In certain germ-cell tumors the combined measurement of human chorionic gonadotropin (hCG) and AFP is desirable to confirm diagnosis and manage patients. The association of multiple tumor markers with a range of cancers is shown in Figure 56.2.

With the advent of automated multi-analyte analyzers for tumor markers, the assay of more than one marker is desirable and straightforward to carry out. This provides greater confidence in establishing clinical status and may

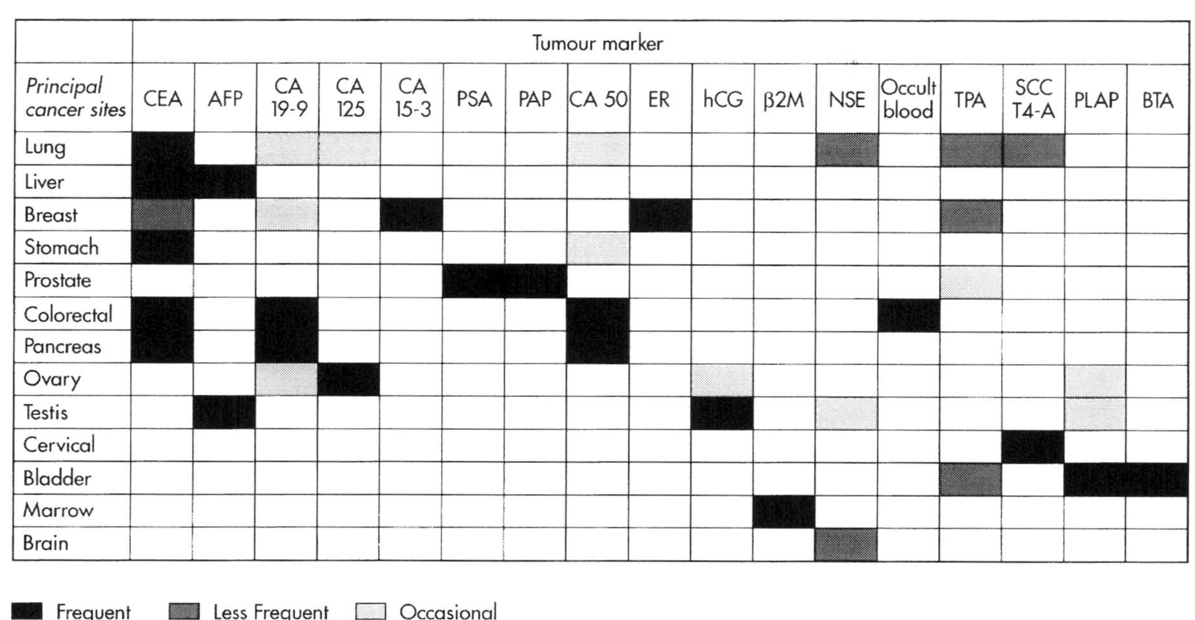

Fig. 56.2 The association of tumor markers with different cancer sites.

become a feature of the use of tumor markers in oncology in the future.

A welcome adjunct is the availability of tumor marker reference controls for several analytes like CA 19-9, CA 125, CA 15-3, CEA, tissue polypeptide antigen (TPA) and squamous cell carcinoma antigen (SCC) from Bioref in Germany and Biorad in North America. Ciba Corning has also developed a tumor marker control containing 20 cancer-related markers including a comprehensive list of well known markers. This provides two levels of the constituents, one in the normal and the other in the abnormal range. In addition, the donors for the product have also been screened for hepatitis C virus. Widespread use of these controls would aid in the standardization of tumor marker analysis across different clinical laboratories.

HISTORY AND CLASSIFICATION

The history of the discovery of biochemical tumor markers starts with the description of a urinary substance in 1846 which is now known to be excessive secretion of the immunoglobin light chain in multiple myeloma (8). The next hundred years saw sporadic description of hormones (hCG, ACTH), enzymes and isoenzymes (AP, PAP) and cytokeratins (TPA) as potential tumor markers. The development of key techniques in the latter half of the 20th century was important for the rapid discovery and development of new tumor markers and their immunoassays. Central to the theme of this entire book is the development of the immunoassay concept in the 1950s by Yalow and Berson involving the application of antibodies as reagents to measure specific substances in complex mixtures. Numerous immunoassays emerged using polyclonal antibodies subsequent to this period, although most were to measure non-cancerous analytes. It was in the early 1970s that the CEA immunoassay was introduced as a commercial test for cancer. Monoclonal antibody techniques introduced in 1975, and the development of the immunometric (sandwich) immunoassay format in 1982, revolutionized the field of tumor markers. This resulted in a virtual explosion in the discovery of new tumor antigens, and the introduction of several among them as immunoassays for routine clinical use. Recombinant antibody techniques also provided invaluable insights into the understanding of the structure and putative functions of tumor markers. Molecular biology techniques were key to the recent discovery of several emerging tumor markers belonging to the class of oncogenes, tumor suppresser genes and a host of other molecules including angiogenic factors, cyclins, nuclear matrix proteins, cell adhesion factors, heat shock proteins, growth factors and their receptors, and telomerase (Wu and Nakamura, 1997; Suresh, 1996).

Almost every new major, and many minor, molecule discovered in the last 10 years has been investigated as a potential new tumor marker. Hence, it is beyond the scope of this chapter to describe all of these new tumor markers and the reader is directed to an excellent recent book devoted to this topic (Wu and Nakamura, 1997).

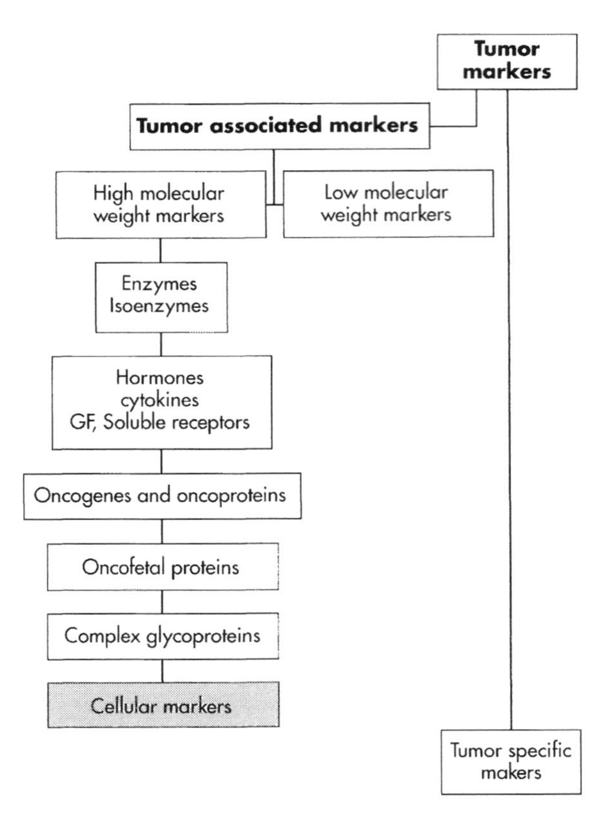

Fig. 56.3 Classification of tumor markers.

A classification of tumor markers is provided in Figure 56.3. Almost all tumor markers are now considered tumor-*associated* antigens, due to their expression to some extent in some non-cancerous tissues. However, rare examples of highly specific tumor antigens can be recognized as a separate subgroup. These include the B-cell tumor immunoglobulin idiotype (the unique paratope or variable region of the specific immunoglobulin expressed on the surface or as a secreted myeloma protein), T-cell receptor of T-cell leukemia, mutated forms of oncogenes and tumor suppressor genes, and several virus induced antigens found predominantly in non-human cancers. Tumor associated markers can be classified into two sub-categories based on the size of the molecules. The macromolecular markers have several categories including proteins, genes, chromosomes and histologically identifiable cellular markers.

NOMENCLATURE AND IDIOSYNCRASIES OF GLYCOPROTEIN TUMOR ANTIGENS

The discovery of several new large glycoprotein tumor markers has been fuelled by monoclonal antibody technology. However, a word of caution is relevant at this juncture to illustrate the misuse of monoclonal antibody (mAb) technology. Claims of discovery of a new tumor marker are abundant, based mainly on the development of a new monoclonal and its putative recognition of a new tumor antigen. Several monoclonals have been developed measuring complex and large glycoproteins elevated in breast, ovarian, pancreatic, gastric, lung, and colorectal cancers. The nomenclature of many of these cancer antigens (CA) has been derived from the arbitrary clonal designations of the various mAbs. For example, the mAb OC125 was developed by immunizing mice with human ovarian cancer cells (Bast *et al.*, 1981). The antigen identified by this mAb is now recognized as CA125. Similarly CA19.9 antigen was originally identified by the NS19-9 mAb. Some confusion has been introduced in the tumor marker literature, due to the development of several different mAbs to overlapping or distinct epitopes on the large glycoprotein antigens with claims of identifying new markers with increased clinical sensitivity/specificity of one over the other. Adding to this confusion is the question of what is measured by these glycoprotein tumor marker assays—epitope, antigenic determinant, domain or the antigen. It is sufficient to state that every tumor marker assay measures antigens albeit *via* unique mAbs binding to unique epitopes. New tumor marker antigens/epitopes have been described purely based on the development of a new mAb without making serious attempts to compare and contrast it with pre-existing mAbs and antigens. This important issue is not easy to resolve due to the idiosyncrasies of the large cancer glycoproteins. Unlike traditional analytes, these large cancer glycoproteins and mucins have special features (Suresh, 1991). The precise estimation of these glycoproteins is influenced by pH, valency, and distribution or density of the epitope and serum anti-carbohydrate antibodies. An anomalous feature unique to these analytes is that many cancer serum samples exhibit increased recovery/estimation of the antigen upon dilution. It is not often appreciated that human serum has a substantive amount of anti-carbohydrate IgM and IgG antibodies that can interfere with glycoprotein immunoassays. It appears that in serum, these large cancer glycoproteins can exist as supramacromolecular complexes promoted partly due to glycan–glycan interactions and partly as a result of weak cross-linking by anti-carbohydrate antibodies. This hypothesis (Suresh, 1991) explains why upon dilution or lowering the pH of serum during assay (e.g. CA19-9 kit, Fujirebio—previously Centocor), one can recover higher amounts of the antigen in an immunoassay, presumably due to disassociation of the large complexes. This concept also can explain the often observed phenomenon of co-expression of CA125, CA19.9, CA15.3, sialyl Lewis^X, Lewis^X, Lewis^Y and other antigens.

An attempt to critically study some of these complex issues has been initiated by the International Society For Oncodevelopmental Biology and Medicine (ISOBM).

They initiated tissue differentiation (TD) workshops a few years ago, analogous to the CD workshops to classify leukocyte antigens. These workshops have been conducted for several tumor markers such as CEA (Hammarstrom *et al.*, 1989; Nap *et al.*, 1992), CA125 (Nustad *et al.*, 1996; Nap *et al.*, 1996), AFP (Alpert and Abelev, 1998), PSA (Stenman *et al.*, 1999), MUC-1 (Price *et al.*, 1998), cytokeratins (Stigbrand *et al.*, 1998), and sialyl Lewis A (Rye *et al.*, 1998), primarily to standardize and compare the numerous mAbs described by different groups putatively identifying the same antigen. For example, the TD-4 workshop compared the 56 different mAbs against the MUC-1 mucin (the international antigen designation for CA15.3 or CA27.29). The majority of the antibodies (34/56) apparently react with the 20 amino acid tandem repeat sequence of the core peptide of MUC-1 mucin (TAPPAHGVTSAPDTRPAPGS). Many of the remaining antibodies react with carbohydrate epitopes. This type of analysis at least takes the first step towards a blind comparison of the various antibodies, all of which are putatively measuring the same antigen with various affinities and overlapping epitope specificities. It is not surprising that the clinical sensitivities and specificities of breast immunoassays constructed using the various antibodies are essentially similar, but some may exhibit unique or subtle abilities to be clinically more useful than others. An ISOBM-TD Workshop web site provides information about participation and additional information (www.uio.no/dnr/ISOBM/Tdbackground.html).

In the remainder of this chapter, the analytes generally regarded as being the most clinically useful are discussed in detail. Unfortunately, an exhaustive survey of all the cancer markers utilized in various continents is not possible in this review as also the numerous tumor markers and their assays described in the research literature (see references). A host of new analytes has appeared on the commercial scene in the last few years and the clinical utility of many of these needs to be further explored.

NEW DEVELOPMENTS

Two new developments that are likely to open new vistas in cancer research need to be addressed. One is the emerging fields of genomics and proteomics that have ushered in a new era of diagnostic possibilities. Several new cancer associated/specific genes and corresponding assays have been developed. A new paradigm of understanding the total proteome fingerprint patterns is underway and serum/plasma is the largest repository of the low and high abundant proteins. Human serum is estimated to contain 30,000 proteins and six proteins: albumin, immunoglobulin G, α-1-antitrypsin, transferrin, immunoglobulin A and haptoglobulin account for ~85% of the bulk. New techniques are being developed including gene and protein array technology coupled with sophisticated mass spectrometry procedures to detect the disease specific/associated markers. A new ovarian cancer test based on an algorithm of reading proteomic fragmentation patterns has been described recently with the claim of 100% accuracy. This has generated both excitement and controversy and will be the subject of much debate in the next few years as this nascent technology evolves. In 2003, the FDA approved the first commercially available bladder cancer detection blood test based on proteomics technology called the BladderChek™. Several new groups and companies are developing multiplexed immunoassays based on microarray and nanoarray platforms. It is likely that such assays for multiple analytes may be approved for routine use in the future.

Lastly, several new cancer therapeutic biotechnology medicines have been approved based on the understanding of cancer markers. These include monoclonals as cancer therapeutics such as Herceptin® for breast cancer associated marker HER2/neu, Rituximab® for B-cell cancers and the intense research and development on CA based therapeutics, often referred to as cancer vaccines. Cancer markers have come a long way since their first discovery nearly 150 years ago and could be the basis of dominant treatment strategies in the future.

ANALYTES

CARCINO-EMBRYONIC ANTIGEN

The discovery of CEA and α-fetoprotein (AFP) forty years ago ushered in a renewed interest in human tumor markers. The development of radioimmunoassay technology for insulin a few years earlier had set the stage for the emergence of new non-invasive techniques to aid the cancer patient. Thus, CEA and AFP are considered classical tumor markers. Gold and Freedman (1965), in their landmark experiment, immunized rabbits with extracts of human colon cancer tissue. The resulting rabbit antiserum was absorbed with extracts from normal human gut tissue and the enriched polyclonal antibody obtained reacted specifically to cancer tissues and their extracts. Because the antigen identified was also found in embryonic tissues, the term CEA was introduced. Both CEA and AFP are members of the family of oncofetal antigens that are normally only expressed in any quantity in embryonic development but are also found in adult neoplastic tissues. CEA is one of the most widely used tumor marker immunoassays with sales of $50–100 million worldwide.

CEA is a heavily glycosylated cell surface glycoprotein and is one of a large family of related molecules belonging to what is now fashionably called a superfamily, which also includes immunoglobulins. This general classification is based on the degree of similarity between the domains of different proteins. Nearly 36 different glycoproteins have been identified in the CEA family and they appear to be derived from 10 genes localized on chromosome 19 in two clusters. CEA is a non-mucinous, 180 kDa glycoprotein secreted by the epithelial cells of the digestive tract in the normal fetus and in adult cancers. It exhibits β-electrophoretic mobility and contains 60% carbohydrate by weight, constituting *N*-acetylglucosamine, mannose, fucose, galactose and sialic acid. The oligosaccharide chains are approximately 80 in number, linked to the polypeptide core by asparagine-*N*-acetyl glucosamine linkages (this is termed an *N*-linked oligosaccharide core in contrast to the serine or threonine-*O*-linked core typical of mucins) with a high proportion of branched oligosaccharide chains. Although CEA has a high carbohydrate content, due to the composition of the sugars and the predominant *N*-linked oligosaccharide chains, it is not considered a typical mucin like CA 19-9 or CA 15-3. Considerable heterogeneity exists in CEA preparations from various sources and this is probably due to variation in the oligosaccharide chains.

In contrast, the polypeptide chain is fairly consistent between different preparations of CEA. The single protein chain consists of approximately 829 amino acids with several intrachain disulfide bonds. Monclonal antibodies utilized in CEA assays bind primarily to the protein chain, rather than to the oligosaccharides.

A number of molecules with structures similar to CEA have been discovered. These include normal cross-reactive antigens (NCA 1 and NCA 2), tumor-extracted related antigen (TEX), normal fecal antigens (NFA 1 and 2), meconium antigen (MA) and biliary glycoprotein 1 (BGP-1). Higher molecular mass forms of CEA have been reported from some colon tumor extracts. The concept of organ-specific CEA has also been proposed.

Function

As with many tumor markers, the function of CEA, and the reason for its appearance in the serum of patients with cancer, is largely a mystery. As with mucins, the association of CEA with the epithelial cells of the digestive tract may suggest a protective role in the turnover of the digestive epithelium. It is estimated that 70 mg/day of CEA is normally secreted into the digestive lumen to eventually end in feces and its appearance in blood is presumed to be due to a reversal or loss of the normal polar secretory function of the epithelial cells (Pannall and Kotasek, 1997).

Reference Interval

Nearly 99% of non-smokers have less than 5 ng/mL and 99% of smokers have less than 10 ng/mL in the Abbott CEA assays. Some assays set the upper limit for normals at 2.5 ng/mL.

The World Health Organization (WHO) has made available the first International Reference Preparation for CEA (73/601). One International Unit of this standard is equivalent to 100 ng of CEA glycoprotein.

Clinical Applications

CEA is one of the most widely used tumor markers in oncology today, now surpassed only by PSA. Despite its widespread use, it is not suitable as a screening test for asymptomatic people, nor is it a reliable diagnostic test in patients with symptoms that may be due to cancer. This is because of the considerable incidence of false positives and false negatives. However, the presence of carcinoma is strongly indicated in patients with elevated values and CEA is often a very useful test as part of the multiparametric diagnosis of cancer. The most significant use of CEA assays is in the management of cancer patients by serial monitoring to determine:

- the recurrence or metastatic spread of cancer after first-line therapy;
- the presence of residual or occult metastatic cancer;
- the effectiveness of therapy;
- the prognosis and staging of patients, when used with other additional information in colorectal and lung cancer. Most colorectal patients with preoperative CEA in excess of 20 ng/mL would manifest recurrence within 14 months after surgery (Wu and Nakamura, 1997).

Although CEA is primarily associated with colorectal cancers, other malignancies that can cause elevated CEA concentrations are those arising from the lung, breast, stomach, ovary, pancreas, and other organs. A number of benign conditions may also be responsible for CEA levels significantly higher than normal. These include inflammatory diseases of the lung and gastrointestinal (GI) tract and benign liver disease. Heavy smokers, as a group, also have an elevated range of CEA values.

However, the most useful clinical application of CEA analysis is as a non-invasive test for the recurrence of colorectal cancer. This is particularly diagnostic in patients whose postoperative levels initially decrease to a normal level within six weeks. CEA concentrations are significantly elevated when the liver is the metastatic site for a primary colorectal cancer. Patients with elevated preoperative levels of CEA that fail to reduce to normal after first-line therapy are suspected as having residual disease or occult cancer. In all these patients the rise or fall of CEA values generally reflects progression or regression of disease as a function of the therapeutic treatment.

CEA can be used to stage disease and estimate the prognosis. A good correlation exists between preoperative CEA values and increased risk of recurrence of disease, particularly in Dukes' C stage of colorectal cancer. Fully differentiated colorectal cancer tends to secrete CEA copiously compared to undifferentiated tumors, which are associated with low levels or do not express the antigen.

Elevated CEA levels are also common in breast and lung cancer patients with disseminated disease. Increases in CEA values are usually not apparent in localized or primary disease. The profiles of CEA values in patients being treated for metastatic breast cancer appear to correlate well with the therapeutic effectiveness.

Limitations

- Most of the immunoassay tests for CEA are reasonably well correlated. However it is important to note that the different antibodies utilized have subtle differences in their affinity for CEA, and in their cross-reactivities to CEA-like material. Hence the switching of immunoassay tests during the course of monitoring a single patient is usually not recommended.
- CEA has a low clinical sensitivity and specificity as a tumor marker and is hence not recommended for screening. Clinically an elevated CEA value is in itself not of diagnostic value as a test for cancer and this parameter should only be used in conjunction with other clinical observations and diagnostic parameters. Some patients with colorectal cancer do not exhibit elevated CEA values and elevated CEA levels in some patients do not change in accordance with progression or regression of disease. CEA values can be elevated in a number of benign conditions.
- Smokers constitute a distinct group with a higher range of baseline values.

Assay Technology

The commonly used Abbott assays for CEA employ the sandwich (immunometric) assay principle with either an enzyme or radioactive label as the signal generation method. Older methods of CEA estimation involved pretreatment of the sample, for example using perchloric acid extraction or heat treatment. Most immunoassays for CEA currently utilize a pair of monoclonal antibodies or a combination of a polyclonal capture antibody with a monoclonal labeled antibody.

Types of Sample

Serum or EDTA plasma for Abbott kits.

Frequency of Use

Very common.

α-FETOPROTEIN

The Russian scientific group led by Abelev in 1963 discovered the presence of AFP in adult mice with hepatomas (liver cancer). The protein is abundantly present in the fetus and levels decline rapidly after birth. This important milestone in the history of tumor marker oncology resulted in the emergence of the concept of oncodevelopmental or oncofetal antigens as possible tumor markers. The notion of regarding a tumor or cancerous state as simulating the fetal or ontogenic phenotype subsequently emerged. Many other oncofetal tumor markers have since been described including CEA and other cell surface glycoconjugates. The dedifferentiation of the adult cells and tissues expressing these early embryonic antigens suggests that they may be involved in cell division and the regulation of growth.

AFP is a 70 kDa glycoprotein with a single polypeptide chain. It is similar to serum albumin in size, structure and amino acid composition but has distinct immunological properties. Unlike AFP, however, albumin is not a glycoprotein. A fucosylated form of AFP has been identified as being associated with liver cancer but not in benign liver diseases. AFP is synthesized by the liver, yolk sac and GI tract of the fetus reaching a peak serum concentration of up to 10 mg/mL at 12 weeks of gestation. This peak level gradually decreases and, one year after the birth of the newborn, the serum levels decrease to less than 25 ng/mL. Albumin becomes the major serum component in adult serum with concentrations up to 60 mg/mL.

Functions

AFP is one of the major components of fetal serum and is replaced by albumin postpartum. Both of these proteins are known to be responsible for the maintenance of serum osmotic pressure and to have various transport functions.

Reference Interval

100% of healthy males and 97% of healthy non-pregnant females have AFP values less than 15 ng/mL (Abbott AFP EIA).

Clinical Applications

AFP determinations are used primarily in two areas. The application of this test in the detection of open neural tube defects in the fetus is described elsewhere (*see* PREGNANCY). In the Orient, due to the high prevalence of liver cancer, particularly in the hepatitis and cirrhosis groups, AFP has been successfully used in screening applications.

The other common application in the field of cancer is in the management of germ-cell tumors and hepatomas. The most common testicular malignancies derive from seminiferous tubules and germ cells. They are classified into two groups, namely seminomas and non-seminomas. The non-seminomatous group includes the embryonal carcinomas, teratocarcinomas, and choriocarcinomas.

The magnitude of AFP elevation has been found to correlate with the stage of non-seminomatous testicular cancers, particularly in the embryonal carcinoma group. For example, continued AFP elevations following orchidectomy (removal of a testis) suggest that the disease is at stage II or beyond. When sequential monitoring gives rise to a profile of continued elevations or a rise in AFP levels, residual disease or a recurrence is strongly suggested.

Decreases in AFP levels in patients are associated with clinical remission. hCG is often used in conjunction with AFP to monitor testicular cancers, particularly non-seminomatous tumors of the choriocarcinoma type (*see* HUMAN CHORIONIC GONADOTROPIN).

AFP levels are elevated in more than 60% of liver cancers. Hepatomas are not common in the western world but are more prevalent in Africa and Asia. A strong etiological association between hepatomas and viral hepatitis, other infections and aflatoxin poison ingestion has been observed. Clinically, the response to treatment and hence prognosis of patients with liver cancer has been poor to date. Nevertheless, AFP determinations have been useful in monitoring these patients during the course of their treatment. A few AFP screening programs for hepatomas have been conducted in high-risk populations with considerable success.

Limitations

- AFP levels are known to be elevated in a number of benign diseases and conditions including pregnancy and non-malignant liver diseases such as hepatitis and cirrhosis.
- Although limited screening studies have been conducted with promising results, AFP is neither recommended as a screening test n or as a diagnostic test.

Assay Technology

Competitive and immunometric assays are used for AFP determinations with a trend towards immunometric assays for oncology applications. These normally use a pair of monoclonal antibodies or a polyclonal capture antibody with a labeled monoclonal antibody.

Types of Sample

Serum or plasma (and amniotic fluid for pregnancy applications).

Frequency of Use

Common.

CARBOHYDRATE ANTIGEN 19-9 (SIALYL LEWIS[a])

Carbohydrate antigen 19-9 (CA 19-9) or sialyl Lewis[a] is a tumor marker predominantly associated with pancreatic, gall bladder, gastric, and colorectal cancers, which are collectively classified as GI malignancies. The term GI cancer-associated antigen (GICA) has also been used, though less frequently, to identify the same antigen. The antigen was originally described as a cell surface monosialoganglioside isolated from the SW1116 human colorectal carcinoma cell line grown *in vitro*. The antigen has the chemical structure shown in Figure 56.4 along with other related tumor antigens.

The original hybridoma secreting the monoclonal antibody 1116 NS-19.9 used to characterize the ganglioside antigen was developed by immunizing mice with the SW1116 human cancer cells. The minimal structure recognized by this antibody, and several other antibodies developed subsequently, appears to be the terminal tetrasaccharide of the CA 19-9 antigen. Deletion of the sialic acid moiety or the fucose residue abolishes or greatly reduces the antigen-antibody interaction. The first comparative study (TD-6 Workshop) of 20 monoclonal antibodies against sialyl Lewis[a] and related antigens was completed recently (Rye *et al.*, 1998). Cross-reactivities to closely related oligosaccharides such as sialyl Lewis[x], Lewis[a], Lewis[x], LSTa (CA50) and others (Figure 56.4) were studied. Most antibodies reacted to the sialyl Lewis[a] antigen and exhibited varying degrees of cross-reactivities to related structures. The subtle differences in their cross-reactivities and affinities, together with the class of the antibody (bivalent IgG versus decavalent IgM measuring polyepitopic mucinous antigens) could explain the spectrum of clinical results obtained. Claims of superior sensitivities and specificities have been made for the various immunoassays even though these are recommended only for monitoring and not for screening or diagnosis.

The CA 19-9 antigen was initially found to be present in serum from patients with GI malignancies, but not in

Common name	Structure
Sialyl Lewis a	Gal β 1 ---> 3GlcNAc β-O-R ↑ 2,3 (NeuAc α) ↑ 1,4 (Fuc α)
Lewis a	Gal β 1 ---> 3GlcNAc β-O-R ↑ 1,4 (Fuc α)
CA50	Gal β 1 ---> 3GlcNAc β-O-R ↑ 2,3 (NeuAc α)
Sialyl Tn	NeuAc α 2 ---> 6GalNAc β-O-Ser/Thr
Sialyl Lewis X	Gal β 1 ---> 4GlcNAc β-O-R ↑ 2,3 (NeuAc α) ↑ 1,4 (Fuc α)

Fig. 56.4 Carbohydrate tumor and related antigens.

normal sera. Based upon these findings it was hypothesized that the ganglioside antigen was shed into the serum. However, a more detailed investigation revealed that the circulating antigen was a high molecular mass mucinous antigen. Virtually no ganglioside antigen was found in the serum of cancer patients. Several other forms of sialyl Lewis^a antigen have been described from seminal plasma, normal saliva and human milk. A sialyl Lewis^a hexasaccharide with a reducing end (i.e. without the ceramide of the CA 19-9 ganglioside) has been purified from human milk. Thus it is now established that multiple species of CA 19-9 antigens exist with univalent (monosialoganglioside, hexasaccharide), oligovalent or polyvalent (glycoproteins and mucins) expression of sialyl Lewis^a residues.

The mucinous form of the antigen has been further characterized following purification. The subunit structure of the mucin appears to be a 210 kDa glycoprotein which, in the absence of detergents or other dissociating conditions, aggregates to form higher relative molecular mass species in the range of 600–2000 kDa. More than 85% of the glycoprotein is carbohydrate by weight. About 35% of the core protein is composed of serine, threonine, and proline, a feature typical of epithelial tumor-associated mucin antigens.

Function

The CA 19-9 antigen is a sialated derivative of the Lewis^a blood group antigen. The specific function of such antigens is largely unknown although a number of theories have been put forward. Gangliosides are thought to be involved in interactions between cells. Mucins derived from epithelial cells may have a protective role. Milk oligosaccharides appear to have a bacteriostatic effect.

Reference Interval

Ninety-nine per cent of apparently normal blood bank donors have serum levels that are less than 37 U/mL (CA 19-9 RIA, Centocor, USA). The arbitrary unit of antigen has been given a gravimetric value of 0.89 ng.

Clinical Applications

In a significant number of GI malignancies, CA 19-9 levels are elevated above the 37 U/mL level. This is particularly pronounced in pancreatic and gall bladder cancer patients, followed by gastric and colorectal cancer patients. Like all tumor marker assays currently available commercially, the clinical sensitivity of the CA 19-9 marker is moderate in early-stage disease. The important feature of CA 19-9 assays is the high specificity of the test. Less than 1% of apparently healthy blood donors exhibit elevated values. A number of benign conditions related to GI disease have been tested for the presence of the antigen and although levels are higher than the healthy blood donor group, they tend to be much less elevated than those samples from cancer patients that give a positive result in the test.

CA 19-9 levels are found to be remarkably high in symptomatic pancreatic and gall bladder cancers. The mean serum level for these cancers is 10–100 times higher than those for gastric and colorectal cancers. Because benign gall bladder disease, pancreatitis and benign hepatobiliary conditions can frequently cause CA 19-9 levels above the range for healthy individuals it is helpful to use an elevated cutoff level for cancers of the pancreas and gall bladder.

The main clinical application of CA 19-9 determinations is in the monitoring of pancreatic, colorectal and gastric cancer patients. It has also been suggested that CA 19-9 could be used for diagnosing the presence of cancer of the pancreas and gall bladder. The use of this tumor marker in conjunction with others can increase the predictive value of the test.

Limitations

- One of the most important limitations of CA 19-9 determinations is the particular sensitivity of the assay and the tumor marker to viral and bacterial neuraminidases resulting in false negatives. Serum samples should be carefully prepared to avoid bacterial contamination.
- The Lewis blood group antigens are classified into Lewis^a (approximately 40%), Lewis^b (40%), Lewis^{ab} (15%) and Lewis$^{a\text{-}b\text{-}}$ (5%). CA 19-9 antigen is not synthesized in individuals who are genotypically Lewis$^{a\text{-}b\text{-}}$ because of the lack of the enzyme fucosyl transferase.
- The distribution of antigen levels in normal donors may vary because of the population distribution of Lewis genotypes in a given geographical area. Most manufacturers recommend that establishment of cut-off values is determined by the clinical laboratory.
- Non-linear dilution, with increased recovery of the antigen, is common in immunoassays for mucins. This is probably due to a variety of factors such as the presence of high levels of anti-carbohydrate antibodies in serum, which generate complexes, the inherent property of mucins to aggregate and disaggregate into a range of molecular species, and other matrix-related effects (Suresh, 1991).
- Elevated levels of CA 19-9 can be found in some benign conditions such as cirrhosis and other liver diseases, gall bladder disease, pancreatitis and cystic fibrosis, thus limiting the diagnostic utility of the marker.

Assay Technology

Most of the kits developed for CA 19-9 utilize immunometric assay methodology although one kit (TRU-QUANT® GI™ RIA) is based on competitive inhibition assay. The sandwich assay format utilizes the same antibody for capture and signal generation. Thus the CA

19-9 species measured by this homo-sandwich technology needs to be oligovalent or polyvalent for sialyl Lewis[a]. The competitive assay for CA 19-9 uses a solid-phase coated with CA 19-9 and has the potential to measure all species of CA 19-9 irrespective of the valency for sialyl Lewis[a].

The range for CA 19-9 sandwich assays is generally 0–240 U/mL. The competitive assay has a higher range up to 600 U/mL, in a coated-tube format.

Types of Sample

Serum or plasma. Some assays are validated only for serum samples.

Frequency of Use

Very common in Japan, and common in western Europe. Not yet approved for clinical use by the FDA in the USA.

CANCER ANTIGEN 125 (Muc-16)

CA 125 is the most important cancer-associated marker for the management of ovarian cancer. It was discovered using a monoclonal antibody, OC125, generated by immunizing a mouse with a human ovarian cystadenocarcinoma cell line. This antibody exhibits specificity for staining epithelial ovarian carcinoma cell lines and tumor tissues. The CA 125 antigen is also expressed in a number of gynecological, non-ovarian and normal tissues of Mullerian origin. Several other monoclonal antibodies have been developed subsequently to measure CA 125 antigen and a comparative blind evaluation was the subject of the TD 1 workshop. These CA 125 antibodies appears to cluster into two major epitope groups, namely 'OC125' like and 'M-11' like (Nustad *et al.*, 1996). Two new antigens CA 130 and CA 602 have been described which appear to be CA 125-like. The CA 130 antigen employs 130-22 as the solid-phase antibody and OC125 as the tracer, while the CA 602 antigen is measured using the two anti-clear cell ovarian cancer Mabs MA602-1 and MA602-6. The CA 125-II immunoassay is a second generation heterosandwich (immunometric) assay combining the M-11 (solid-phase) and OC125 antibodies.

The CA 125 antigen has been characterized as a high molecular mass glycoprotein aggregate. Size exclusion chromatography of native CA 125 material from either body fluids or *in vitro* culture supernatants of ovarian cancer cells results in at least two broad peaks of antigen reactivity with approximate relative molecular masses of 400 and >2000 kDa. The minimal subunit molecular mass of the antigen obtained under strong denaturing conditions was found to be 210 kDa with a 24% carbohydrate composition by weight. The relatively low carbohydrate content, the low buoyant density, and the presence of both *N*-linked and *O*-linked oligosaccharides suggests that CA 125 is not a typical mucin. The data currently available indicates that the CA 125 antigen is a glycoprotein aggregate of at least two molecular species whose minimal subunit molecular mass is 210 kDa. Although this general structure appears to be consistent with the antigen isolated from a variety of sources, size heterogeneity of the denatured subunits and the native aggregated states has been reported. One report describes the isolation and identification of a 40 kDa subunit associated only with tumor-associated CA 125 and not with CA 125 antigen found in benign or normal tissues and fluids. Another study describes that under strong denaturing conditions only one 55 kDa protein band is seen, suggesting that the 205 kDa species could be a tetramer. Recently, the CA 125 has been cloned and is designated Muc-16.

Function

The CA 125 antigen is expressed in copious amounts in the tissues and serum of epithelial ovarian carcinoma patients. The antigen is not present in normal serum, or in adult or fetal ovaries. Using immunohistochemical techniques, CA 125 reactive material has been detected in some normal tissues such as adult pleura, pericardium, peritoneum, Fallopian tubes, endometrium and endocervix. The antigen is also found in the chorionic membrane, extracts of maternal decidua and, abundantly, in the amniotic fluid. Amniotic fluid CA 125 has two dissimilar subunits and is not derived from the fetus. The function of CA 125 antigen is largely unknown and one hypothesis associates the antigen with Mullerian differentiation.

Reference Interval

Homosandwich immunoassay kits employing OC125 antibody adopt a 35 U/mL discrimination value that encompasses 99% of healthy donors. The TRUQUANT® OV2™ immunoassay employs the B27.1/B43.13 antibodies and utilizes a discrimination value of 45 U/mL. The highly purified CA 125 has a specific activity of 317 units/μg protein.

Clinical Applications

The measurement of CA 125 antigen is very helpful in the management of serous ovarian carcinomas. Epithelial ovarian carcinomas frequently metastasize into the peritoneal cavity on the serosal surfaces, often producing ascites. Primary ovarian cancer is usually treated by surgically removing the ovaries and giving the patient chemotherapy to ablate any residual disease. CA 125 antigen measurement is used to monitor residual tumor burden in patients who have undergone such therapy. Antigen levels above the normal range are usually predictive of residual or recurrent ovarian carcinoma if other causes of CA 125 elevations can be eliminated (*see* LIMITATIONS). This intended clinical use of the CA 125 antigen assay was approved by the US FDA in 1986 and its routine use has had a strong positive impact in the management of epithelial ovarian cancers. Subsequently numerous clinical reports have appeared in the literature suggesting the extension of the use of CA 125 immunoassays for a variety of other oncological applications such as limited diagnosis of ovarian cancer, monitoring of ovarian, lung and breast cancer patients, and in applications involving endometrial and Fallopian tube cancer. Some studies have attempted to establish that CA 125 could be used to screen for ovarian cancer, using a higher cutoff value than the usual 99% confidence interval for normals. Although the studies looked promising, consensus opinion is that further work is necessary in this regard.

Limitations

- CA 125 antigen levels are elevated above the recommended cutoff value in 1% of normals, 5% of benign diseases and 28% of non-gynecological cancers. The benign conditions associated with increased CA 125 in serum include ovarian cysts, severe endometriosis, menstruation, first trimester of pregnancy, cirrhosis and pericarditis.
- Higher CA 125 antigen levels are also found in non-ovarian tumors such as those originating in the breast, lung, uterus, endometrium, pancreas and liver. Some early attempts were made to use CA 125 to classify unknown cancers as being from an ovarian primary. However, the presence of elevated CA 125 antigen levels in non-ovarian cancers limits the potential of such an application.
- Increased CA 125 antigen levels are found in a number of ascites fluids and pleural effusions in both malignant and benign conditions. In our experience we have found CA 125 levels in these fluids in the range of 2000–500,000 U/mL.
- Radiolabeled OC125 antibody has been used to identify cancer sites *in vivo* often missed by other diagnostic methods. The injection of the mouse antibody into humans elicits a human anti-mouse antibody (anti-isotypic and anti-idotypic HAMA) response, capable of increasing the apparent CA 125 concentration in subsequent serum samples tested in immunometric assays. Assays employing alternative antibodies such as TRUQUANT OV2™ may be used in these situations.
- As with CA 19-9, increased recovery of antigen can occur in dilution experiments. Considerable caution should be exercised when carrying out comparisons between methods or changing from one method to another.

Assay Technology

CA 125 assays are based on immunometric assay methodology. A number of homo-sandwich RIAs and EIAs have been developed using the original OC125 antibody with a range of 0–500 U/mL of CA 125 antigen. Higher ranges up to 2000 U/mL and lack of HAMA interference have been achieved by a two-step assay employing B27.1 and B43.13 antibodies. The CA125-II immunoassay incorporates an M11 monoclonal antibody on the solid-phase, with the OC125 tracer antibody. (Adequate overlapping determinations are recommended when changing test).

Types of Sample

Serum or plasma. Some assays are validated only for serum samples. Ascites and pleural effusions should not be used as they have higher antigen levels than are found in serum.

Frequency of Use

Common.

CANCER ANTIGEN 15-3

The breast CA 15-3 is a large mucinous glycoprotein with a native molecular mass in excess of 400 kDa. The antigen is identified using a sandwich assay employing two monoclonal antibodies. The solid-phase mAb, 115D8, was generated by immunizing mice with defatted human milk fat globule (HMFG) antigens. The tracer mAb, DF3, was developed against enriched antigens from the membrane of human breast carcinoma metastasis. MAb DF3 is more specific for cancers than 115D8. The sandwich assay that uses these two antibodies detects antigens that have been variously described as MAM6, milk mucin, human mammary epithelial antigen, HMFG antigen, and polymorphic epithelial mucin. The first international workshop on cancer-associated mucins has assigned the name MUC-1 to the breast cancer-associated mucin. The antigen identified by the DF3 antibody in human milk consists of a single high molecular mass species, whereas in breast carcinomas, the antibody binds to two glycoproteins with molecular masses of 330 and 450 kDa. Approximately 50% of the composition by mass is carbohydrate. The antigenic site identified by DF3 appears to be sensitive to neuraminidase, alkaline borohydride treatment and proteases, suggesting that it is a combined sialyl oligosaccharide and peptide on the CA 15-3 antigen. Microheterogeneity and genetic polymorphism are observed in these epithelial antigens causing considerable variation in the size of native oligomers, subunits and core proteins from different sources. Recently, a 309 base pair cDNA encoding the sequence for the DF3 antigen has been isolated and using this probe it was demonstrated that the polymorphism of these mucins is a reflection of the variations in the size of the alleles. The conserved sequence is rich in guanine and cytosine with a 60 base pair tandem repeat encoding a serine, threonine and proline rich polypeptide. The number of tandem repeats of this 20 amino acid sequence (PDTRPAPGSTAPPAHGVTSA) is thought to be the basis for the polymorphism in these mucins. This peptide has been synthesized without any oligosaccharide chains and at high concentrations it can block DF3 binding to solid-phase mucin antigen.

A number of other antibodies have been prepared that apparently react with the same family of polymorphic epithelial mucins (e.g. CA 27.29, CA 549, MCA) etc. These 56 mAbs were investigated in the TD-4 workshop by 16 international groups (Price *et al.* 1998). Most of the mAbs [34/56] were mapped within the immunodominant 20 amino acid tandem repeat domain. The bulk of the remaining antibodies appear to recognize carbohydrate incorporating epitopes.

Function

CA 15-3 is found in normal and cancerous epithelial cells and, as a mucin, it is often assumed to play a protective role. The antigen constitutes approximately 15% of the total membrane protein of HMFGs. The quantity of DF3 antigen expressed appears to correlate with the degree of breast cancer differentiation. Because human milk also contains the antigen, the DF3 antigen is considered a marker of differentiation of mammary epithelial cells.

Reference Interval

Ninety-nine per cent of healthy individuals have less than 30 U/mL in the Centocor CA 15-3 RIA kit.

Clinical Applications

The CA 15-3 antigen is an epithelial membrane antigen expressed on normal cells and found in serum. Elevated levels of this antigen are found in about 60% of preoperative breast cancer and 80% of advanced metastatic breast cancer. Breast cancer is one of the most common cancers in women in the western world and the CA 15-3 assay has proved to be helpful in patient monitoring, with better clinical sensitivity than CEA assays. An advantage over CEA is that the antigen levels are not abnormally elevated in smokers. The CA 15-3 assay is not suitable as a diagnostic test because of its low sensitivity in stage I and II disease but, in advanced mammary carcinomas, trends in the antigen levels provide a useful non-invasive indicator of early recurrence, presence of residual disease, and continued remission or poor prognosis. Combined use of CA 15-3 and CEA does not appear to give any improved clinical information.

Limitations

- CA 15-3 is sensitive to proteases and neuraminidases and hence it is important to prepare and store samples with great care to avoid microbial contamination.
- Elevated values are seen in less than 10% of benign diseases of liver, breast, ovary, GI tract, and lung.
- The polymorphic nature of CA 15-3 and its complex glycoprotein structure presents similar assay problems (e.g. dilution non-linearity) to those described for other mucins such as CA 19-9.

Assay Technology

The TRUQUANT BR RIA is a competitive assay with a mucin coated solid-phase and was the first breast cancer MUC-1 marker test approved by the FDA in 1995. Subsequently the reclassification of tumor markers as class II devices by FDA allowed other similar immunoassays to be approved as well. The CA 15-3 assay is an immunometric assay now approved by the FDA (Centocor Diagnostics, PA, USA), utilizing two different monoclonal antibodies. MAb 115D8 is used as the solid-phase capture antibody, and labeled DF3 is employed as the signal generation mAb. The tracer appears to have a more restricted antigen specificity than the capture mAb. An FDA approved, automated, 15 min, luminescence-based assay for CA27.29 has also been developed (Bayer (previously Chiron) Diagnostics, USA).

Types of Sample

Serum or plasma. Some assays may only be suitable for serum sample.

Frequency of Use

Common in Japan, Europe, and the USA.

ESTROGEN RECEPTOR AND PROGESTERONE RECEPTOR

Estrogens are female sex hormones synthesized by the ovary and adrenal cortex. The hormonal action of estrogens is mediated by an ER protein called estrophilin, which is present in the nuclei of target cells. It was originally believed that β-estradiol, the major estrogen, was bound by a cytosolic ER which subsequently underwent macromolecular size alterations prior to translocation into the nucleus to regulate gene expression as a transcription factor. It is now understood that most, if not all, of the ER is a nuclear protein with a high affinity for estradiol. The dissociation constant (k_d) is in the range of 10^{-10} to 10^{-9} M. This 66 kDa protein has a steroid-binding site as well as a DNA-binding site. Upon binding of the steroid, the complex binds to DNA and regulates gene expression. The ERs in human breast tissue generally decrease during the development and onset of mammary carcinoma. The estimation of ER in breast cancer tissues is an important aid in deciding the course of treatment.

Progesterone is a steroid hormone that influences the endometrium to allow implantation of the fertilized ovum and its gestation. It is biochemically also a precursor of adrenal corticosteroids, estrogens, and androgens. The cellular progesterone receptor (PR) has two molecular components with molecular masses of 120 and 95 kDa. Estrogen modulates the appearance of the PR and its analysis complements the information derived from ER assays.

Function

The function of ER is to act as the second messenger of estrogen action by regulating gene expression in the nucleus. The estrogen–ER complex is capable of stimulating gene expression by acting on nuclear DNA. Progesterone upon binding to PR promotes binding and activates the hormone specific genes.

Reference Interval

A cutoff of 10 fmol/mg cytosol protein is recommended for the Abbott ER enzyme immunoassay. The PR enzyme immunoassay has a cut off level of 15 fmol/mg cytosol protein.

Clinical Applications

Approximately two-thirds of endometrial and breast cancers are positive for ER. At least 50% of ER-positive breast cancer patients respond favorably to endocrine therapy while less than 10% of ER-negative patients show such a good clinical response. Hence, estimation of ER levels has become fairly routine in determining the choice of therapy for breast cancer patients. The level of ER also has a prognostic value as there is a good correlation between breast cancer patients who benefit from endocrine therapy and the amount of the receptor present in the sample.

In addition to quantitative analysis of ER in breast cancer tissues, several oncologists promote the use of direct visualization of the receptor in tissues by immunohistology or immunocytology. This method involves sectioning of tissue or smearing a fine-needle aspirate on a slide followed by staining ER-containing cells with a specific probe such as an anti-ER antibody. Immunocytochemical methods reveal the heterogeneity in the tumor with regard to ER status and make obvious the contribution, if any, from normal tissue. With this method it is possible to distinguish, for example between one patient exhibiting ER-positivity in homogenates due to high receptor content in a small proportion of the tumor and another with moderate amounts of ER in most of the tumor cells. The latter patient is likely to respond better to endocrine therapy.

Combination of ER and PR estimations appears to increase the predictive value of those patients likely to respond to endocrine therapy.

Limitations

- A biopsy sample is required for the assay.
- Immunoassays for ER (and PR) analysis measure both the unbound and hormone-bound protein forms, unlike steroid-binding assays, which use radiolabeled estrogen or progesterone as tracers. Hence, some discrepancies may be observed between the two types of assays.

Assay Technology

The first assay method used to identify and measure receptors depended on the binding of ^{3}H-labeled steroids. Such assays are affected by endogenous steroids, which block the binding sites causing an underestimate of the receptor content. The immunometric assays using monoclonal antibodies to measure ER and PR, introduced by Abbott Laboratories, appear to be unaffected by endogenous hormones.

Types of Sample

Homogenate of a biopsy tissue prepared carefully to avoid heat stress, which destroys the receptors.

For immunocytochemical analysis, fresh or frozen specimens or fine-needle aspirates are required. Paraffin sections are less desirable for the Abbott ER assay.

Frequency of Use

Common for ER, but uncommon for PR assays.

FECAL OCCULT BLOOD

A number of colorectal disorders, benign and malignant, precipitate the rupture of tissues and blood vessels, resulting in the presence of blood in the lumen of the colon and the rectum. Some of these blood components are found in the feces, and chemical and immunochemical methods have been developed to detect their presence. All the commercial tests available are designed to detect hemoglobin.

The chemical tests for fecal occult blood detection are popularly known as guaiac tests and are based on the pseudoperoxidase activity of heme in hemoglobin. In the presence of suitable substrates such as gum guaiac (a natural resin from the wood of *Guaiacum officinale* containing α-guaiaconic acid) and hydrogen peroxide, the heme catalyzes a peroxidation reaction generating a blue quinone product. Exploiting this principle a variety of tests are available based on guaiac-impregnated paper or tape which can be used to detect fecal occult blood in a laboratory, physician's office or now even as a home test. The presence of blood in the stool is a diagnostic aid in the detection of a number of colorectal disorders including colorectal cancer.

The immunochemical test for fecal occult blood detection utilizes a monoclonal antibody that is highly specific for human hemoglobin and has a low cross-reactivity with hemoglobins from common dietary meat products. This assay appears to possess better sensitivity and specificity than chemical tests.

Reference Interval

Fecal occult blood tests are qualitative tests with a positive or negative end point.

Clinical Applications

The detection of occult blood in human feces gives a general indication of disorders in the colon and rectum and is not specific for colorectal cancer. Non-cancerous conditions showing a positive fecal blood test include peptic ulcer, ulcerative colitis and iron-deficiency anemia. Despite these limitations, the qualitative assay for detecting fecal blood was the first test used in the western world as a cancer screening test, with limited success.

Generally the fecal blood test is recommended as a diagnostic aid during routine physical examinations of people above the age of 50. The American Cancer Society recommends serial testing for three consecutive days to minimize false-negative results. A special diet is recommended for at least two days prior to the chemical test to avoid false positivity due to any consumption of red meat or peroxidase-rich vegetables and fruits. Large amounts of vitamin C in the diet can cause false-negative results.

The immunochemical fecal blood test does not require patient compliance to the special diet. In an asymptomatic group, about 2–3% of individuals score positive for fecal occult blood, of which the incidences of adenomatous polyps and colorectal cancer are 1 and 0.2%, respectively. The polyps are often considered a precancerous condition. When used in conjunction with sigmoidoscopy, colonoscopy and barium enema, the fecal blood test is a useful and simple initial test for the detection of colorectal diseases, including cancer.

Limitations

- The chemical tests that detect the pseudoperoxidase activity of hemoglobin are plagued by a variety of dietary factors. Red meat and peroxidase-rich vegetables and fruits generate false positives while vitamin C can cause false negatives. Patient compliance to a restricted diet is essential to increase the utility of the test. The immunochemical test does not appear to be sensitive to the above factors.
- Intermittent bleeding and a lack of homogeneity in the distribution of blood in the feces can cause a wide variation in results. Serial testing is therefore often recommended.

Assay Technology

The chemical test (Hemeoccult®) is based on the generation of a blue product when hydrogen peroxide is added to guaiac-impregnated paper and when the fecal smear has traces of hemoglobin. The development of a monoclonal antibody reactive only to human hemoglobin has resulted in an immunochemical assay (Hemeselect™), with increased sensitivity and specificity.

Type of Sample

Feces.

Frequency of Use

Common.

PROSTATE-SPECIFIC ANTIGEN

PSA is a glycoprotein with a molecular mass of 34 kDa with a single polypeptide chain. Immunologically and biochemically, PSA is distinct from PAP. PSA is a serine protease (the active site of the enzyme has a serine residue) and its labile nature could be partially attributable to its autocatalytic activity. Human seminal plasma is a rich source of PSA, and histologically it is restricted to the cytoplasm of the acinar cells and ductal epithelium of the prostate gland. PSA derives its name from the observation that it is a normal antigen of the prostate but is not found in other normal or malignant tissues, although PSA like material has been recently described in the breast tissue. The antigen is present in benign, malignant and metastatic prostate cancer, and immunohistochemical analysis of distant metastasis for PSA can usually identify whether the primary origin of the cancer is from the prostate.

In serum, at least three forms of complexed PSA have been identified in addition to free PSA. One is bound to α_2-macroglobulin, and it appears that the PSA epitopes are covered and inaccessible for measurement by current assays. The second major species is the PSA–ACT (α1,anti-chymotrypsin) complex. The third complex is with α1-protease inhibitor (PSA-API). The discrepancy between the various immunoassays for PSA could be due to their epitope specificity and the relative ability of measuring the various species by the different mAbs employed. Recently, immunoassays measuring total PSA (often referred to as equimolar assays measuring PSA and PSA–ACT complex equally well) and free PSA have been introduced by several diagnostic companies. The research literature documents numerous novel immunoassays for the measurement of PSA (Kreutz and Suresh, 1997) and the recent development of two ultrasensitive PSA immunoassays, deserves special mention (Yu *et al.*, 1997; Ellis *et al.*, 1997). Both of these novel immunoassays demonstrate utility by monitoring very low PSA antigen <0.1 ng/mL, which is the limit of most the current clinical assays. In the serum of patients who have undergone radical prostatectomy, theoretically, PSA levels should be zero or very close to it after a few weeks of surgery. The use of ultrasensitive immunoassays for monitoring very early recurrence of metastatic disease, and potential early second line treatment, is an exciting possibility not only for prostate cancer, but for the whole range of clinically relevant tumor markers.

An international workshop (TD-3) on the comparative properties of 82 PSA antibodies was organized recently (Stenman *et al.*, 1999). A significant finding was that nearly 17 of these cross-reacted with human glandular kallikrein (hK2), which shares considerable homology with PSA (see below). In the light of these observations, studies on the estimation of PSA in serum, or claims of identifying PSA in non-prostate tissues, are only credible if the assays employ monoclonal antibodies specific to unique PSA epitopes not shared by hK2 and other kallikreins. For a comprehensive summary of the biology and the clinical applications of PSA and kallikreins, the reader is directed to a recent critical review (Rittenhouse *et al.*, 1998).

Function

PSA is a protease whose role in the prostate gland, prostatic fluid and seminal plasma is largely unknown. It is related to the serine proteases of the Kallikrein family with 62% homology to human pancreatric/renal kallikrein and 80% homology to human glandular/serum kallikreins. PSA may be involved in the thinning of seminal clots by cleaving the predominant protein in the seminal vesicle fluid.

Reference Interval

Ninety-nine per cent of apparently healthy donors have total PSA levels of <4 ng/mL (Hybritech Tandem-R® PSA assay). Values in benign prostate hypertrophy (BPH) are generally in the range of 4–10 ng/mL, which overlaps with the levels also seen in malignancy. Total PSA values >10 mg/mL are however more likely due to malignancy. Some authors have suggested that the ratio of free and bound PSA be used to discriminate between BPH and prostate cancer. The PSA–ACT fraction is higher in cancer than in BPH.

Clinical Applications

Prostate cancer is the second most prevalent form of male malignancy and early diagnosis is the key to a potential cure. The diagnosis of prostatic carcinoma, like all other cancers, is done by carrying out a number of procedures in combination, such as rectal examination, fine-needle biopsy, chest X-ray, bone scan and serum PAP tests. The development of immunoassays to measure serum PSA has provided a valuable adjunct to the diagnosis and management of patients with prostatic cancer. The American Cancer Society in 1992 recommended the use of annual PSA tests for screening prostate cancer in conjunction with digital rectal examination in males above the age of 50. While this has lead to an enormous interest in the development of PSA immunoassays for screening applications, the oncology community is split on the value of such mass applications in asymptomatic people. Serum PSA has been found to be more useful than PAP because of increased clinical sensitivity. However about 5% of patients have increased PAP but normal PSA levels. For this reason some experts recommend that a combined PSA and PAP measurement is more useful than either one in isolation.

Elevations of serum PSA concentrations above 4 ng/mL are found not only in prostate cancer but also in benign prostatic hypertrophy. The magnitude of the serum PSA elevation is progressive with the stage of the disease, and the highest levels are seen in stage D prostatic cancer with metastatic involvement. PSA is not useful as a specific diagnostic test for prostate cancer because of the elevated values in benign prostatic hypertrophy and attempts have been made to achieve discrimination based on proportion and ratios of free: bound PSA. Nevertheless, PSA is now a routine test in the management of patients who have been confirmed to have prostate cancer. In this clinical application for monitoring prostate cancer, PSA is superior to PAP as a reliable tumor marker. Changes in tumor marker levels correspond to classical trends (*see* CANCER MARKERS, INTRODUCTION) in most cases of prostate cancer. PSA is a good marker for establishing prognosis in prostate cancer.

Limitations

- Elevation of PSA above 4 ng/mL is not diagnostic of prostate cancer because benign prostatic hypertrophy and some benign genitourinary diseases also result in elevated values.
- Massaging the prostate prior to blood sample collection can result in transient PSA increases.
- About 5% of patients with prostate cancer have elevated PAP, but normal PSA concentrations.

Assay Technology

Both competitive and immunometric assays, using monoclonal antibodies, are available. The immunometric assays have largely superceded competitive assays. The Tandem-R® total PSA assay has a range of 0–100 ng/mL. Free PSA assays are also available from Hybritech (now a subsidiary of Beckman Coulter Inc., San Diego, USA) as are semi-quantitative or dip stick formats (VEDA labs, France). The latter test is designed for screening applications to measure if serum PSA levels are more than 4 ng/mL. An automated assay estimating all PSA complexes (cPSA) has been developed (Bayer, Germany). Research assays for PSA–AMG and PSA–API have been described. The ratio of PSA–AMG to total PSA was significantly higher in BPH than prostate cancer while the proportion of PSA–API was lower in prostate cancer than BPH.

Types of Sample

Serum or plasma depending on the assay used.

Frequency of Use

Common in the western world and may overtake CEA as the most used tumor marker, with an estimated worldwide market of ~ $200 million.

PROSTATIC ACID PHOSPHATASE

PAP is a lysosomal enzyme optimally active at acidic or low pH. In contrast, ALPs are optimally active at moderately high pH values. The principal PAP isoenzyme has a molecular mass of 102 kDa, 13% being carbohydrate. Unlike PSA, this protein is a dimer with an acidic isoelectric point. The enzyme cleaves the phosphoric monoester bond and its activity varies depending on the substrate used.

The PAP glycoprotein is secreted by the prostate gland and is found in seminal plasma and urine. Acid phosphatase isoenzymes in serum are derived from various tissues including the prostate gland.

Function

PAP is a lysosomal hydrolytic enzyme and is most probably involved in the catabolism of organic phosphate esters. The precise physiological role, beyond understanding its phosphatase activity, is largely speculative. The enzyme is thought to be involved in hydrolyzing phosphorylcholine and other phospholipids in seminal fluid.

Reference Interval

Of healthy males, 97.5% have <3 ng/mL PAP in the Tandem-R PAP assay.

Clinical Applications

The clinical significance of PAP in oncology was first reported in 1936 by Gutman and co-workers who observed increased phosphatase activity in bone metastases of prostate cancer. Serum PAP levels were found to be elevated in patients with prostatic carcinoma, particularly those in the stage C and D groups. PAP was the tumor marker of choice for the management of prostate cancers prior to the introduction of PSA. The specificity of the test is high for prostate cancer but sensitivity is low compared to the PSA assay. In one comparative study of 143 prostate cancer patients, the clinical specificities of the PSA and PAP immunoassays were 96–97%. However, PAP had a clinical sensitivity of 57% compared to the high value of 93% for PSA. The PAP test is not suitable as a diagnostic or screening test for cancer although the trends in the levels of the tumor marker usually mirror the progression or regression of the disease. Some experts advocate using both PAP and PSA to increase the overall predictive value.

Limitations

- Benign prostatic hyperplasia and other genitourinary diseases can cause high serum PAP values.
- The biochemical and immunochemical activities of PAP are sensitive to pH 8 and above, and care should be taken during sample collection and assay to avoid inaccurate results.
- Rectal examination or any other manipulation of the prostate gland leads to short-term PAP elevations in serum.

Assay Technology

The first assays for PAP were based on its enzyme activity. Interference by other isoenzymes in this assay was a major problem. Immunoassays avoid this problem by employing antibodies with minimal cross-reactivity with acid phosphatases of erythrocytes, platelets, leukocytes, liver, spleen, and other tissues. Competitive and immunometric assay procedures are available commercially. Most PAP assays have a range of up to 30–40 ng/mL and one kit from Diagnostic Products Corporation has a range up to 150 ng/mL.

Types of Sample

Serum or plasma depending on the assay used.

Frequency

Common.

β_2-MICROGLOBULIN

β_2-Microglobulin (β_2M) is a single-chain aglycosyl protein composed of 100 amino acids. Its molecular mass is 11.8 kDa and it is now known to be the light-chain component of the histocompatibility antigens (HLA). It is therefore found on all nucleated cells and is present in high concentrations on the lymphocyte cell surface. This small protein bears sequence homology with immunoglobulins and is hence classified as belonging to the superfamily of immunoglobulins.

β_2M, being a small protein, escapes the glomerular filtration network of the kidneys. Most of what passes the glomeruli is reabsorbed and catabolized by the cells of the proximal tubules. A small amount of the protein is detected in normal urine with elevated levels in patients with proximal tubular dysfunction.

The normal serum levels of β_2M are primarily a reflection of HLA metabolism and turnover. It is estimated that on a daily basis about 150 mg of free β_2M protein is secreted into the body fluids. The serum levels are altered in various benign and malignant conditions and hence β_2M is a non-specific tumor marker.

Function

β_2M is an integral component of the HLA antigen system and is similar in structure to immunoglobulins. The specific role of the protein is not yet defined but as part of the histocompatibility complex it is thought to be involved in molecular recognition, particularly in distinguishing between self and non-self. The molecule also appears to stabilize the heavy-chain conformation of the HLA class I molecule which may be important in immune recognition and restriction.

Reference Interval

In the Abbott RIA kit the 95th percentile is 2 mg/mL for normal serum and 160 μg/mL for urine.

Clinical Applications

Serum β_2M levels are elevated in the presence of a number of solid tumors and lymphomas. However, a variety of non-malignant conditions such as rheumatoid arthritis, AIDS, lupus, Crohn's disease, and renal tubular dysfunction cause elevated levels of the marker. The level of serum β_2M also appears to be an indicator of acute renal transplant rejection.

The role of β_2M levels is less certain for solid tumors, either in monitoring the disease or as an indicator of prognosis. There appears to be a use for this marker in the lymphoid malignancies such as Hodgkin's and non-Hodgkin's lymphoma, multiple myeloma and chronic lymphocytic leukemia. A high initial level of β_2M is an indicator of poor prognosis and an advanced stage of the disease. It is also useful for monitoring the course of the disease in these cancers, particularly in multiple myelomas.

Limitations

- β_2M elevations are not diagnostic of cancer, as a number of non-malignant conditions also give rise to elevated concentrations. The changes in β_2M levels found in non-cancer conditions include inflammatory disorders such as rheumatoid arthritis, Crohn's disease, lupus, AIDS, renal tubular dysfunction, and renal transplant rejection.
- Although β_2M levels are elevated in some solid tumors, the marker is not useful in prognosis or in monitoring the disease state in these situations.

Assay Technology

Most kits are competitive in design because of the relatively small size of the antigen. Solid-phase polyclonal or monoclonal antibody-based assays are available with enzyme or radiolabeled β_2M as the tracer.

Types of Sample

Serum, plasma, and urine can be used in the Abbott RIA kit.

Frequency of Use

Not very common, except in Japan.

NEURON-SPECIFIC ENOLASE

Enolase is a ubiquitous glycolytic enzyme which catalyzes the conversion of 2-phosphoglycerate to phosphoenolpyruvate (Figure 56.5).

The enzyme enolase is also referred to as 2-phospho-D-glycerate hydrolase or phosphopyruvate hydratase. It is a dimer which can be composed of three different types of subunit, namely α, β or γ. The αα isoenzyme dimer is synthesized by most of the cells in the body and by glial cells in the brain. This form is sometimes referred to as non-neuronal enolase or NNE. The β enolase appears to be specific to muscle tissue. The γγ and αγ isoenzymes are collectively referred to as neuron-specific enolase or NSE. NSE is produced by nerve cells or neurons, and neuroendocrine cells, particularly the cells of the amine precursor uptake and decarboxylation lineage. NSE is an acidic protein with a native molecular mass of 78 kDa and a subunit molecular mass of 39 kDa. NSE and NNE are immunologically distinct and have different sensitivities to chloride ions and temperature.

Function

NSE is a glycolytic enzyme involved in the energy-generating process of the cell. Ontogenetically the NSE isoenzyme appears in the final stages of neuronal differentiation and is hence a good nerve cell maturation marker.

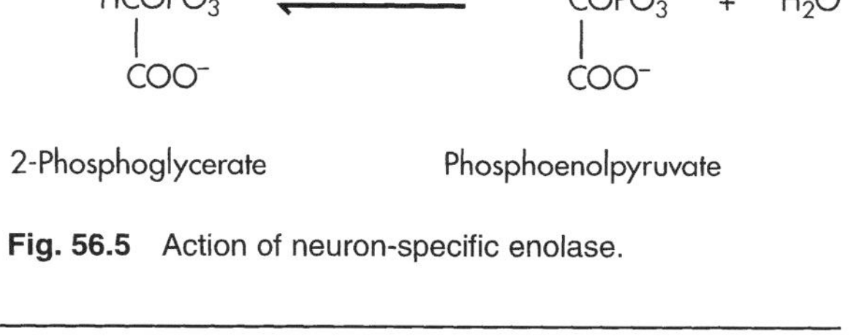

$$\begin{array}{c} CH_2OH \\ | \\ HCOPO_3^{2-} \\ | \\ COO^- \end{array} \underset{}{\overset{NSE}{\rightleftharpoons}} \begin{array}{c} CH_2 \\ | \\ COPO_3^{2-} \\ | \\ COO^- \end{array} + H_2O$$

2-Phosphoglycerate   Phosphoenolpyruvate

Fig. 56.5 Action of neuron-specific enolase.

Reference Interval

In the Pharmacia NSE RIA assay, the upper limit for normal sera is 12.5 ng/mL which represents the 95th percentile value.

Clinical Applications

NSE is found elevated primarily in small cell lung cancer (SCLC) and neuroblastomas. Other neuroendocrine tumors with elevated levels of NSE include insulinomas, medullary thyroid carcinomas, phaeochromocytoma and gut carcinoids. The main clinical application of NSE is in the monitoring of these tumors for response to chemotherapy or to detect early relapse.

SCLC is the most aggressive of the lung cancers and most of the patients have already progressed to metastasis by the time of diagnosis. However, SCLC responds particularly well to chemotherapy compared to other lung cancers. NSE levels can help to classify the type of lung cancer when used in conjunction with histology, enabling the appropriate course of therapy to be initiated. Monitoring NSE levels can also assist in determining the effectiveness of chemotherapy and to predict relapse of the disease.

Neuroblastoma is a common childhood cancer which is often malignant. In addition to monitoring, NSE levels help to differentiate between neuroblastomas and Wilm's tumor, which originates in the kidney. Both these conditions may present as a palpable abdominal mass and elevated levels of NSE are suggestive of neuroblastoma.

Limitations

- Careful specimen handling is essential for this immunoassay because of the presence of NSE in erythrocytes and other blood cells. The Pharmacia NSE RIA kit package insert warns that samples that are hemolyzed and exhibiting an absorbance of at least 0.3 at 500 nm should not be used.

Assay Technology

A competitive assay with radiolabeled NSE and polyclonal anti-NSE is commercially available from Pharmacia and Eiken. A polyclonal–monoclonal immunometric assay is available from Roche. An interesting immunoassay method called immunocapture assay has been described in the literature but not yet commercialized. In this format the monoclonal antibody captures NSE which is subsequently quantified by a bioluminescent assay exploiting the enzymatic activity of the protein.

Type of Sample

Non-hemolyzed serum samples are preferred and repeated freezing and thawing are to be avoided.

Frequency of Use

Uncommon.

CARBOHYDRATE ANTIGEN 50

A novel glycolipid antigen termed carbohydrate antigen 50 (CA 50) was isolated from the Colo 205 colorectal carcinoma cell line. This cell line was used as an immunogen to generate the C-50 monoclonal antibody.

This sialyllacto-*N*-tetraosyl ceramide structure is similar to the CA 19-9 glycolipid described earlier except for the lack of the α-L-fucose residue. It is important to understand that although the C-50 antibody identified the CA 50 glycolipid as a new tumor-associated antigen, this mAb exhibits significant cross-reactivity with CA 19-9 antigen or sialyl Lewis^{a}. Consequently immunoassays developed utilizing this antibody measure any one or both antigens in a given sample. CA 50 assays thus have a broader cancer specificity than CA 19-9 assays. Individuals with Lewis^{a-b-} antigen status who are unable to synthesize CA 19-9 can, however, express CA 50 (Figure 56.6).

The serum CA 50 antigen is mainly a large molecular mass mucinous structure similar to the CA 19-9 antigen. Both glycolipid and glycoprotein antigens are found on the cell surface of Colo 205 cells but apparently only the glycoprotein antigens are measured in serum. CA 50 antigen elevations are found in colorectal, gastric, pancreatic, gynecological, and lung cancers.

Function

The role of CA 50 antigen, either as a glycolipid or glycoprotein, is largely unknown.

Reference Interval

About 95% of healthy subjects have CA 50 values less than 24 U/mL in the CA 50 IRMA kit from Stena Diagnostics (now Pharmacia).

Clinical Applications

Considerable overlap exists between cancer patients positive for CA 50 and those identified by the CA 19-9 assay. This is because of the structural similarity between the two antigens and the cross-reactivity of the C-50 mAb towards the CA 19-9 antigen. Thus, a significant percentage of patients with pancreatic, GI and colorectal cancer exhibit elevated CA 50 levels. Unlike the CA 19-9 marker, the CA 50 antigen is elevated in hepatocellular carcinomas (60–70%) and in a higher proportion of carcinomas of the lung, uterus, prostate, ovary, kidney,

Neu5Acα2→3Galβ1→GlcNAcβ1→3Gal1→4Glcβ1→1Ceramide

Fig. 56.6 Structure of CA5O glycolipid.

breast, and cholangium. The CA 50 assay has been found useful in monitoring the course of disease in various cancers.

Limitations

- CA 50 antigen is elevated in Crohn's disease which is an inflammatory bowel condition. The levels appear not to correspond to the clinical activity.
- Because of the strong cross-reactivity of the C-50 monoclonal antibody to the CA 19-9 antigen there is a considerable overlap between CA 50 and CA 19-9 assays. Further work is necessary to resolve the antigens identified by the CA 50 assay in various cancers and if indeed the CA50 glycolipid or glycoproteins are elevated in body fluids. The development of a Mab reactive only to CA 50 would be essential to clarify the physical and clinical properties of the antigen.
- As with CA 19-9 assays, the CA 50 assay is sensitive to microbial neuraminidases and hence it is crucial to avoid sample contamination.
- CA 50 levels are also elevated in ulcerative colitis, some autoimmune diseases and liver cirrhosis as well as melanomas and lymphomas.

Assay Technology

The first assay developed for CA 50 was a competitive inhibition assay utilizing a CA 50 ganglioside-coated solid-phase. The more sensitive homo-sandwich (immunometric) assays were introduced later. One format uses a radioactive label and the other is based on time-resolved fluorescence using the DELFIA system (delayed emission lanthanide fluorescent immunoassay).

Types of Sample

Serum or plasma.

Frequency of Use

Not very common.

SQUAMOUS CELL CARCINOMA ANTIGEN

SCC antigen is a 48 kDa glycoprotein, originally isolated from a squamous cancer of the uterine cervix. This antigen is a subfraction of TA-4, a tumor-associated antigen that is so designated because it is isolated during the 4th purification step of the antigen. Resolution of this fraction by isoelectric focusing shows 14 bands with a range of molecular masses between 42 and 48 kDa. The currently available Abbott SCC RIA can measure all of these subfractions. Immunohistochemically SCC was found to be a cytoplasmic protein of normal and cancerous squamous cells.

Although not approved by the FDA for routine clinical use, the development of a tumor marker assay for squamous cancers of the head, neck, lung and cervix heralds an important step in the management of these cancers.

Function

Not known.

Reference Interval

The Abbott SCC RIA does not recommend a reference interval. In one study 95% of healthy subjects had SCC values below 3.9 ng/mL. The 95th percentile for females was 5 ng/mL and for males, 3.3 ng/mL.

Clinical Applications

SCC antigen is the first commercially available tumor marker for squamous cancers. The serum levels of SCC antigen are elevated in a significant percentage of patients with squamous cancers of the cervix, head, neck and lung, and the level of the tumor marker increases with the stage of the disease. The specificity appears to be good for squamous cancers, and adenocarcinomas do not give rise to abnormal concentrations of this marker. Some benign gynecological and pulmonary diseases are responsible for higher SCC values than the normal reference interval. SCC antigen levels tend to be normal in early stage squamous cancers. About 40% of stage III and 60% of stage IV head and neck cancers exhibit SCC antigen levels above the reference interval for normals. Cervical squamous carcinomas in similar stages have shown a higher proportion (80%) of patients with elevated antigen levels. The degree of differentiation of the tumor does not appear to be related to the level of SCC antigen. Monitoring patients with squamous cancers has demonstrated that the assay can detect recurrence and provide a prognosis.

Assay Technology

Currently the assay methodology employs radiolabeled SCC antigen and polyclonal antibody in a competitive assay format. The range of the assay is 0–150 ng/mL of antigen.

Type of Sample

Serum.

Frequency of Use

Uncommon.

TISSUE POLYPEPTIDE ANTIGEN, TISSUE POLYPEPTIDE SPECIFIC ANTIGEN AND CYFRA 21-1

TPA is a pan-carcinoma marker. This antigen was discovered in 1957 as an insoluble residue from human carcinomas. TPA is now known to belong to a class of cytoskeletal proteins called cytokeratins or intermediate filaments. Cytokeratins 8, 18, and 19 react with anti-TPA antibodies. These cytokeratins are cytoplasmic proteins and are found in all normal epithelial cells, and cells lining the ducts and their sacs. Thus various tumors arising from different organ sites are known to express TPA, which is also released into the serum by cell destruction. TPA assays represent the first generation cytokeratin tumor marker tests. Tissue polypeptide specific antigen (TPS) and CYFRA 21-1 are second generation monoclonal immunoassays measuring specific fragments or components of the TPA Cytokeratin family (Van Dalen, 1996). TPS measures fragments of cytokeratin 18, CYFRA detects 21-1 fragments of Cytokeratin 19, and the TPA (Cyk) test estimates cytokeratin 8 and 18.

Function

The cytoskeleton is responsible for the physical three-dimensional architecture of the cell. During cell division the cytoskeleton assumes a crucial, dynamic, functional role. The precise function of individual cytokeratins is yet to be fully understood but as an intermediate filament it has an obvious role in defining the structure of the cytoskeleton and its dynamics during cell division.

Reference Interval

Ninety-five per cent of apparently healthy individuals have TPA values of less than 55 U/L in the Prolifigen® TPA IRMA assay. The CYFRA 21-1 RIA (Centocor, USA) and EIA (Boehringer Mannheim, Germany) are available and the 95th percentile for the latter is 1.2 ng/mL. The Prolifigen-TPS assay (Beki Diagnostics, Sweden) has a 95th percentile of 95 mU/mL.

Clinical Applications

Cytokeratin markers are indicators of cell proliferation. Many carcinoma patients have elevated TPA/TPS levels in their serum and the magnitude of the elevation correlates with tumor progression. The widespread distribution of this marker is in some respects similar to CEA. Some experts recommend that these two analytes be measured in combination. Among the types of cancer that show increases in TPA levels are breast, digestive tract, lung, prostate, and ovarian. TPA is particularly useful as a very sensitive marker for confirmation of the diagnosis of transition cell carcinoma of the bladder in its early stages. TPS appears to be more sensitive and specific for breast cancer than CEA and CA 15-3.

TPA has a half-life of 7 days in circulation and a stable level is reached in 3–4 weeks after treatment of the cancer. Serum TPA levels are altered in relation to the proliferation of tumors. Thus it is likely that a tumor without significant cell division and growth should not result in an increased level of TPA in the serum.

CYFRA 21-1, which measures cytokeratin 19, is elevated in most lung tumors of the non-small cell category, with the highest sensitivity in lung squamous cell cancers.

Limitations

- Cytokeratin markers are not suitable for diagnosis of carcinoma but are used to monitor patients, often along with other organ-specific tumor markers.
- Elevations in TPA are seen in the last trimester of pregnancy and in various benign diseases of the lung, liver, stomach, and pancreas.
- Monitoring of patients during therapy with cytokeratin markers is more complex than using other markers. Further work is needed to resolve the nature of these soluble serum fragments of cytokeratin parent molecules, which are more insoluble by nature.

Assay Technology

A polyclonal competitive inhibition assay, employing radiolabeled TPA, was the first assay to be developed. A more recent IRMA assay has also been developed by Sangtec Medical.

The range of the assay is 25–2000 U/L for the IRMA format. The TPS assay uses the labeled M3 monoclonal antibody, which appears to measure cytokeratin 18. Two monoclonals are employed in the CYFRA 21-1 assay.

Type of Sample

Serum should be collected prior to treatment because transient increases of TPA can occur in response to therapy. It is also recommended that stored samples are thoroughly vortexed prior to assay.

Frequency of Use

Common in western Europe.

PLACENTAL ALKALINE PHOSPHATASE

ALPs are a family of ubiquitous membrane-bound glycoprotein enzymes with optimal activity at an alkaline pH. This is in contrast to acid phosphatases, which have optimal activity in an acidic environment. Elevated ALP levels in serum are found in patients with bone and liver metastasis of a variety of cancers. One ancestral gene is thought to have given rise to the four ALP genes coding for the isoenzymes. One of the isoenzymes is placental alkaline phosphatase (PLAP), which is a homo-dimer with good thermostability. PLAP, unlike other forms, is not denatured when heated to 65 °C for 5 min, and is not inhibited by urea.

Function

PLAP is a hydrolytic enzyme and, like its other isoforms, is thought to be involved in the catabolism of organic phosphates. ALPs also catalyze phosphotransferase reactions. A role in the transport of phosphate has also been attributed to it.

Reference Interval

Ninety-five per cent of healthy blood donors have serum PLAP concentrations of less than 4 nKat/L in the Prolifigen PLAP enzyme-immunoassay.

Clinical Applications

Patients with various gynecological cancers have elevated levels of PLAP in their serum. PLAP concentrations have been observed to vary with the grade of differentiation of ovarian tumor but not with the stage of the disease. Higher PLAP values are found in grade 1 and 2 non-mucinous ovarian cancers than in grade 3 cancers. PLAP is similar to CA 125 in that serum elevations are seen in serous but not mucinous ovarian tumors.

Trophoblastic cancers and germ-cell tumors of the male reproductive organs show high PLAP elevations. A number of other malignancies such as those of lung, breast, colon, kidney, stomach, pancreas, and bladder also have elevated PLAP levels.

Limitations

- In some benign conditions, pregnancy and in smokers, elevated serum levels of the enzyme are seen.
- The specificity of the assay can be influenced by the antibody reactivity to other isoenzyme forms.

Assay Technology

The Prolifigen PLAP EIA is an immunoenzymetric assay. It utilizes a solid-phase monoclonal antibody with high specificity to PLAP and minimal reactivity to the isoenzyme in smokers. The PLAP captured by the monoclonal is estimated by measuring enzyme activity spectrophotometrically.

Type of Sample

Serum should be obtained prior to any treatment. Phosphatase activity is inhibited by anticoagulants such as fluoride, citrate and EDTA.

Frequency of Use

Not very common.

HUMAN CHORIONIC GONADOTROPIN

See PREGNANCY: HUMAN CHORIONIC GONADOTROPIN for further information on this marker.

Reference Interval

The reference value for pregnancy is >25 mIU/mL in the Ortho-Clinical Diagnostics Amerlite HCG-60 assay (no longer available). However, for oncology applications a cutoff of 5 mIU/mL is recommended.

Clinical Applications

hCG is a major analyte in the diagnosis of pregnancy and this aspect is covered in a separate chapter. Choriocarcinomas and male germ-cell tumors are characterized by elevated levels of hCG and its subunits. Increases in hCG levels have also been found in cancers of the breast, lung, and small intestine and in some prostate cancers. The combined measurement of hCG and AFP levels has been shown to be superior in confirming diagnosis and managing non-seminomatous germ-cell tumors. Monitoring of germ-cell tumors is effective with these markers, which mirror the clinical progression or regression of the disease.

Cancers secreting hCG often produce abnormal forms of the molecule. These include altered glycosylation of the peptide, and secretion of α-chains. Some scientists have attempted to develop an hCG assay that is specific for cancerous conditions by exploiting these anomalous features. This type of assay has not yet become commercially available.

Limitations

See PREGNANCY.

Assay Technology

See PREGNANCY.

Types of Sample

Serum or plasma.

Frequency of Use

Not very common as a tumor marker.

CATHEPSIN D

Cathepsin D is classified as an aspartyl protease due to the presence of aspartic acid in its active site. Other types of proteases include serine proteases (PSA), cystenyl protease (cathepsin B) and metalloproteases (collagenase). Various proteases have come into prominence along with angiogenesis factors as essential prerequisites to metastatic spread. Proteolytic cleavage of the tissue matrix and basement membrane around the primary tumor site is generally regarded as an early event in the metastic process. Hence, tumor-derived and tumor-associated proteases could be important markers correlating with the progression of cancer and targets for treatment.

Function

Cathepsin D is an intracellular enzyme found in lysosomes, golgi, and endosomes. Its acidic proteolytic activity and localization suggest that it is one of the key catabolic enzymes in cells. Many tumors with an increased expression of protease are likely to have a higher onotogenetic advantage for metastatic spread.

Reference Interval

A cathepsin D level of 30 nmol/g cytosolic protein is considered as being favorable for prognosis. Levels above 60 nmol/g cytosolic protein are associated with a favorable prognosis.

Clinical Applications

Cathepsin D is more useful as a prognostic marker rather than for routine monitoring. In mode-negative breast cancer patients the level of this enzyme appears to predict recurrence of disease. Lower cathepsin D levels are related to longer survival.

Limitations

A biopsy sample is usually required.

Assay Technology

Sandwich (immunometric) immunoassay.

Frequency of Use

Not yet in extensive use.

INTERLEUKIN-2 RECEPTOR

The high affinity Interleukin-2 receptor is a hetero-dimer with 75 kDa α-and 55 kDa β-chains. Resting T cells express limited $\propto$-chains only. Activated T cells over-express the β-chain compared to the $\propto$-chain. The IL-2β-chain is a membrane glycoprotein and the extra-cellular domain of this molecule of approximately 19 kDa appears in serum as the soluble receptor fragment (sIL-2R). The presence of the sIL-2R is presumably due to catabolism or metabolic turnover generated by a protease.

Function

The T-cells are normally quiescent and when an antigen is presented by the antigen presenting cell (APC) the specific population(s) of T-cells are activated. A key manifestation of this activation is the appearance of high affinity IL-2R in its dimeric form. The individual $\propto$- and β-chains have moderate and low affinities to the cytokine IL-2. The growth promoting cytokine IL-2 binds to IL-2R and promotes clonal expansion of the unique population of T-cells, either in an autocrine or paracrine fashion. Both IL-2 and its receptor are transiently produced in the presence of appropriate antigen and APC as a natural mechanism to regulate T-cell function.

Clinical Applications

The measurement of sIL-2R in the serum of lymphoid malignancies can assist in the monitoring of patients undergoing treatment. The main application is in the management of T-cell leukemias and hairy cell leukemias. The FDA has approved the use of sIL-2R assay recently.

Frequency of Use

Mostly in the western countries.

C-ERB B-2 (HER-2/NEU) ONCOPROTEIN

It is now well accepted that the development of cancer is a complex and long drawn out multistep process. Exciting new research in the area of oncogenes, tumor supressor genes, DNA repair genes and apoptosis (programmed cell death) have lead to a better understanding of normal cell division and potential events leading to abnormal cell division, and eventually cancer. Oncogenes are the mutated versions of genes that regulate normal cell division and growth called proto-oncogenes. A variety of proto-oncogenes have been described which include growth factors and their receptors, proteins involved in signal transduction, transcription factors and regulators of apoptosis.

The c-*erb* B-2 gene encodes a 185 kDa transmembrane receptor associated with tyrosine protein kinase. Like the interleukin-2 receptor, the extra cellular domain of the c-*erb* B-2 protein could be clipped and shed into circulation.

Function

The c-*erb* B-2 protein has considerable homology with the epidermal growth factor receptor. As a member of the growth factor family, c-*erb*-B-2 protein is involved in the early events of signal transduction, from the external mitogen to the intracellular millieu, *via* tyrosine kinase dependent cascade events.

Reference Interval

Favorable prognosis is indicated if levels of the p185 protein are less than 10 U/mL cytosol.

Clinical Applications

Mutations in the c-*erb* B-2 proto-oncogene and its amplification are frequently found in breast and ovarian cancers. A gene-based test (Ventana Medical Systems, Arizona, USA) for predicting breast cancer recurrence and aggressiveness is approved in Europe and by the FDA. Nearly a third of breast cancer patients positive for gene amplification died within five years after preliminary surgical treatment. In the negative group 97% survived a minimum of 5 years. The measurement of the levels of the c-*erb*-B-2 protein expressed by the gene is also an important marker in the prognosis of breast cancer patients. As with many other prognostic tumor markers, increased levels indicate poor prognosis, early relapse and shorter duration of survival.

Assay Technology

Sandwich (immunometric) EIA.

Frequency of Use

This marker is one of the new tumor markers and its clinical use is likely to increase in the future.

P53

The P53 gene is a negative regulator of cell growth and development and an important tumor suppressor gene. In the normal cell, several genes are thought to regulate cell division, including some that suppress the entry to—or initiation of—the mitotic cycle. Mutations or aberrations in these cell-cycle suppressor genes result in an unregulated cell division cycle that can ultimately lead to the development of tumors. Oncogene amplifications or increased levels of oncoprotein expression are more readily identified, unlike tumor suppressor genes whose detection is difficult due to their absence (Pannall and Kotasek, 1997). Mutations in the p53 gene are very common in human cancers. Further, mutations in the germ line p53 appear to increase the propensity of these individuals to develop cancers. The polypeptide encoded by the p53 gene is a 53 kDa nuclear phosphoprotein.

Function

The p53 nucleophosphoprotein functions as a key transcription factor, suppressing cell growth and proliferation, particularly in cells with damage to DNA. The native p53 protein can induce apoptosis or programmed cell death, and hence check the onset of uncontrolled growth in damaged cells. Its mechanism of action is to activate a number of genes *via* its DNA binding domain, including p21 (*ras*) gene expression, as well as inhibiting G1 cyclin-dependent kinases. Mutations in the DNA binding domain of p53 are seen in most cancers. One hypothesis speculates that apoptotic events in the vicinity of a primary tumor result in the release of proteases and other lytic enzymes capable of initiating invasiveness and metastasis (Pannall and Kotasek, 1997).

Clinical Applications

Widespread detection of p53 mutations in a number of cancers has resulted in the mutated p53 protein being considered as an important tumor marker. One could even regard this as a tumor-specific marker. Mutated p53 gene and protein have been detected in 50% of all types of cancer, including adenocarcinomas as well as SCCs. Altered p53 has also been reported as a marker of cancer aggressiveness and in some cancers as a prognostic indicator.

Assay Technology

A sandwich EIA is available to measure mutant p53. Alternative qualitative technologies include PCR amplification and sequencing, or immunohistochemical detection of the mutant nuclear p53 protein. Another new ELISA has been developed to measure p53 autoantibodies (Dianova, Hamburg, Germany) which is reported to be associated with more aggressive tumors and hence, poor survival. In this assay, wild type p53 protein is the solid-phase antigen capturing human anti-p53 and detected by enzyme labeled anti-human antibody.

Frequency of Use

A new marker that holds much promise in the years to come. Current use appears to be mostly in the western world.

BLADDER TUMOR ANTIGEN

Bladder cancer is the sixth most frequent cancer in women and the fourth most common in men. Men are three times more likely to get bladder carcinoma than women. Smoking and exposure to chemicals appear to be risk factors. Most bladder cancers (90%) are transitional cell carcinomas of epithelial origin. The rest are either squamous cell cancers, adenocarcinomas, or undifferentiated carcinomas. Traditional detection methods are cytoscopy and urinary cytology, analogous to the PAP smear tests for cervical cancer. As early as 1945, Papanicolaou and Marshall described the detection of cancer cells of the urinary tract by urinary cytospin sediments. Several bladder tumor marker tests have been developed including bladder tumor antigen (BTA), urinary cytokeratin, NMP22 (nuclear matrix protein) and fibrin/fibrinogen degradation products (FDP).

The BTA antigen appears to be the complement factor H or a closely related protein. The apparent molecular weight of the BTA antigen is predominantly 150 kDa, although some degraded fragments have also been identified. Several different cancer cell lines have been shown to secrete BTA into the medium.

Function

The BTA antigen has a complement factor C3b binding site and it degrades C3b in the presence of complement factor I. The structural relationship to serum complement factor H (hCFH) was deduced by partial sequence analysis. It is speculated that the secretion of complement factor-like activities may confer a selective *in situ* growth advantage to cancer cells by blocking complement mediated lytic activity. The function of hCFH is to interact with complement factor C3b and inhibit the formation of membrane attack complex, thus preventing cell lysis. The hCFH appears to have a role in the regulation of the alternate complement pathway.

Reference Interval

A cut off value of 14 U/mL has been suggested, based on the mean plus 3 SD, derived from urine of healthy individuals.

Clinical Applications

The BTA antigen is useful in monitoring transitional cell carcinomas which represent the bulk of bladder cancers. The sensitivity is excellent for non-invasive high-grade tumors and relatively high for low grade and *in situ* tumors. The specificity of the assay was high both in healthy individuals and in patients with other cancers and benign disorders. The BTA quantitative assay is superior to traditional methods of bladder cancer detection such as voided urine cytology for monitoring recurrent bladder cancers. A significant number of patients with no evidence of disease were positive in one study, and further work is needed to establish if this represents false positives or detection of early disease recurrence.

Limitations

- Urine samples (fresh, refrigerated, or frozen) are used in this assay and care should be taken in serial measurements due to variable amounts of voided urine.

Assay Technology

The BTA TRAK enzyme immunoassay is a dual monoclonal sandwich assay. The same pair of monoclonals are used in the new qualitative BTA Stat test, which was recently approved by the FDA as the first home use device for cancer marker recurrence. This qualitative test is similar to the lateral flow dipstick pregnancy test based on the principle of immunochromatography. This test is an adjunct to cytoscopy, and based on the results of the BTA Stat test, the urologist has the choice to use either a rigid or flexible cytoscope. If the results are positive, the rigid cytoscope could be used under general anesthesia to remove the recurrent tumor during examination. With a negative BTA Stat result, the flexible exploratory cytoscope could be used with only a local anesthetic.

Frequency of Use

The BTA TRACK™ EIA and BTA Stat tests (Bard Diagnostic Sciences, Redmond, WA, USA) are relatively new assays and hence their use is currently limited. The regulatory endorsement of the qualitative test may increase its home use in the future.

IMMUNOCHROMATOGRAPHY ASSAYS FOR TUMOR MARKERS

In recent years there has been a growing number of qualitative or semi-quantitative cancer marker assays developed on the user-friendly lateral flow, immunochromatography format. This format can accommodate immunometric or competitive immunoassays. In the immunometric format, one of the antibodies is labeled with either colloidal gold, colored latex or colloidal carbon to generate the pink, blue or black lines respectively. In the competitive assay format, it is the analyte that is labeled. The sample moves along the device by capillary flow, solubilizing the label and carrying it across a band of immobilized antibody, on a color-contrasted white membrane. A detailed description of several of these types of assays (for other analyte applications) can be found in the PRODUCT TECHNOLOGY chapters (Part 2). While most of these types of tests provide qualitative information on the presence or absence of a given analyte, some degree of quantitative information is now possible with the recent development of several instruments developed to measure the intensity of the end-point band. In the future it is likely that such tests would be performed in primary health care centers, such as the physician's office, to initiate appropriate therapeutic courses of action, or referral to secondary and tertiary health care centers for subsequent follow up.

Currently a few tumor marker assays are available in the immunochromatography format. They are:

- Ideal Rapid UBC™ test (IDL Biotech, Sweden)—A one step self-testing assay for urinary bladder cancer detecting cytokeratin fragments. A quantitative ELISA assay format is also available from the same source.
- BTA Stat™ test (Bard Diagnostics, Redmond, Washington, USA)—An FDA approved home test for monitoring recurrence of urinary bladder cancer. This assay measures the presence of urinary complement factor H and related proteins.
- One-Step FOB™ test (TECO Diagnostics, California, USA)—Fecal occult blood test detects the presence of human hemoglobin in feces. This test utilizes monoclonal and polyclonal antibodies specific to human hemoglobin and hence has less interference from dietary source hemoglobin, vitamin C, or iron.
- AFP Card™ test (TECO Diagnostics, California, USA)—This is a one step assay detecting AFP and could have applications in hepatomas.
- PSA test (VEDA Labs, France)—The 1992 recommendation of the American Cancer Society to use an annual PSA test in conjunction with digital rectal examination for every male above the age of 50 years has likely resulted in the development of several rapid PSA immunochromatography tests. The test from VEDA Labs claims to rapidly discriminate PSA values above or below 4 ng/mL, the accepted cut-off for apparently healthy males.

FREE LIGHT CHAIN (FLC) ASSAYS

The first biochemical cancer marker to be discovered was the Bence Jones proteins in urine. These are the mostly monoclonal homogeneous kappa or lambda light chains of immunoglobulins derived from the malignant B-cells. Until now, the traditional method of detecting FLCs has been either by electrophoresis of proteins or immunofixation electrophoresis of urine. The urinary tests have several limitations such as low sensitivity, need for stringent urine collection over 24 h, subsequent concentration prior to assay and the metabolism of the light chains by the proximal tubules, thus potentially masking the malignant condition to some extent. More recently, serum FLC assays have been developed that are more sensitive and can be automated (Bradwell, 2003). FDA approved immunoassay tests are now available as an adjunct for the diagnosis and monitoring of multiple myeloma. The kits are also available for use on two automated systems (Beckman Coulter IMMAGE® and Dade-Behring BNII).

Function

The light chain is part of the two-chain immunoglobulin molecule and is found in all types of immunoglobulin with some rare exceptions such as camel immunoglobulins, which have only one chain. The light chain has a constant region and three variable regions which have unique amino acid sequences in different antibodies. Together with the three variable regions of the immunoglobulin heavy chain, the six unique domains of every immunoglobulin combine to form the specific antigen combining site called the paratope. The paratope is the complementary face that binds to the specific epitope (antigenic determinant) face on the antigen.

Clinical Applications

Serum FLC levels are elevated in 85% of patients with non-secretory multiple myelomas, 95% of intact immunoglobulin multiple myeloma and 100% of light chain multiple myeloma. The most important application of the FLC assays is in the detection of non-secretory multiple myelomas when serum or urine electrophoretic tests for monoclonal proteins are in the normal range. The half-life of FLC in serum is ~2–4 h compared to intact IgG. Hence serial measurements of the FLC can be a quicker indicator of the outcomes of a therapeutic regimen. The serum levels of FLC are also seen in amyloid disease wherein the light chains can form polymeric deposits. Decreasing levels of serum FLC as a result of chemotherapy is a very good indicator of long term survival.

Assay Technology

The immunoassays are based on the development of antibodies specific to masked unique light chain epitopes. The developed kits using these antibodies are based on latex-enhanced nephelometric and turbidimetric methods. Unlike the classical electrophoretic tests, concentration of sample is not necessary.

Limitations

The FLC are not specific to cancer and only 20% of B-cell chronic lymphocytic leukemias have abnormal levels. Most monoclonal gammopathies are of undetermined significance and 60% have elevated FLC.

Frequency of Use

This is a relatively new test and in view of its advantages of sensitivity and in measuring clinically useful levels in serum, it is likely to replace the urine based methods in the future.

NOVEL EXPERIMENTAL AND OTHER MINOR MARKERS

This section includes a brief overview of recent immunoassays for tumor markers that have not yet achieved widespread acceptance. Many of the antigens have been identified by the reactivity of monoclonal antibodies developed by various investigators. It appears that a number of these immunoassays determine the levels of novel mucinous antigens. Much remains to be learned about the biochemistry of these complex molecules, and their utility in a clinical setting is yet to be established. Nevertheless some of the early results show promise in the monitoring of cancer patients. An entirely new class of tumor marker gene based assays is emerging, which is beyond the scope of this chapter and book. Several minor tumor markers e.g. des-gamma-prothrombin, calcitonin, ACTH, TA-4, creatine kinase ββ, inhibition, LDH, TSH, catecholamines - are used in specific clinical niches and are summarized in Table 56.2.

CANCER ANTIGEN 195 (CA 195)

This immunoassay was developed by Hybritech and utilizes the homo-sandwich (immunometric) assay principle employing a monoclonal antibody with reactivity to both sialyl Lewis[a] (CA 19-9). and Lewis[a] (Figure 56.4) CA 195 appears similar to CA 19-9 with higher serum values in pancreatic, gastric, and colorectal cancer.

TUMOR-ASSOCIATED GLYCOPROTEIN 72.4 (TAG 72.4)

Centocor developed this novel panadenocarcinoma mucinous marker utilizing a pair of monoclonal antibodies CC-49 and B72.3. MAb B72.3 reacts with sialyl 2 ⇒ 6GalNAc-*0*-Serine/threonine (Sialyl Tn) which is considered another oncofetal antigen (Figure 56.4). This assay appears to be useful in gastric carcinoma.

S-100 ANTIGEN

The S-100 antigen of neuroendocrine origin is an acidic protein and is a homo or heterodimer of A and B subunits. The S-100 B serum assay (Sangtec Medical AB, Sweden) is a melanoma marker useful in monitoring of patients and as a prognostic marker. Elevation of this antigen in surgically treated and disease free melanoma patients suggests early recurrence.

SIALYL LEWIS^X (SLX-I)

Otsuka (Japan) introduced this homo-sandwich immunoassay utilizing the antibody FH6, which binds to the sialated derivative of the stage-specific embryonic

Table 56.2. Minor tumor markers useful in niche clinical applications.

Tumor markers	Clinical application
Des-gamma-carboxyprothrombin	Hepatoma versus cirrhosis
Calcitonin	Bone metastasis, meduallary carcinoma of thyroid
Adrenocorticotropic hormone (ACTH)	Neuroendocrine tumors
Creatine kinase BB	Neuroendocrine tumors
TA-4	Squamous cell carcinoma
Cancer associated serum antigen (CASA)	Mucinous ovarian cancer
Inhibin	Granulosas and mucinous cystadeno-carcinoma
Lactate dehydrogenase (LDH)	Germ cell tumors
Thyroid stimulating hormone and thyroglobulin	Thyroid cancer
Catecholamines	Neuroblastomas, pheochromocytoma, carcinoid tumors
Tumor associated trypsin inhibitor (TATI)	Renal and gastric carcinomas
Neopterin	Prognostic in myeloma and hematological malignancies
Epidermal growth factor receptor (EGFR)	SCC and prognostic in breast cancer
Ferritin	Advanced adenocarcinomas
5-Hydroxyindoleacetic acid (5-HIAA)	Carcinoid tumors
Lipid associated sialic acid	Non-specific marker in several cancers
Parathyroid hormone-related peptide (PTH-RP)	Tumors with hypercalecemia
Terminal deoxynucleotidyl transferase	Classification of leukemias
Urinary gonadotropin peptide (UGP)	Ovarian cancer

antigen (Figure 56.4). The antigen appears to be elevated in a number of cancers and most notably in lung cancers.

MUCIN-LIKE CARCINOMA-ASSOCIATED ANTIGEN (MCA)

Roche launched a two-step enzyme immunoassay for the management of patients with breast cancer. The antigen appears to belong to the group of mammary mucins and gives results similar to other tests measuring breast cancer mucins. The sensitivity in stages I and II of breast cancer is lower than for more advanced stages.

CANCER-ASSOCIATED ANTIGEN 549 (CA 549)

Hybritech developed the CA 549 sandwich assay (which uses two antibodies) for the management of patients with metastatic breast cancer. One antibody was developed against a breast cancer cell line and the other developed against HMFG membrane. The various immunoassays for breast cancer, namely CA 15-3, CA 27.29, MCA, and CA 549, all appear to measure breast cancer mucins with comparable clinical specificities and sensitivities.

BONE ALKALINE PHOSPHATASE

Metra Biosystems (Mountain View, California, USA) has introduced a new assay to measure bone alkaline phosphatase (ALP). This assay appears to be more specific in identifying bone metastasis, instead of measuring total ALP which is also elevated in liver metastasis, hepatomas, and prostate cancer.

B/T GENE REARRANGEMENT TEST

This test is not an immunoassay but deserves special mention for its place in cancer diagnosis and management. The FDA approved DNA probe test, developed by Oncor (USA), is useful for the diagnosis of leukemia and lymphomas. The test allows the determination of clonality, and can distinguish between T and B cells, and between lymphoid malignancies and benign non-lymphoid cancer. In addition, gene rearrangement is detected to allow the oncologist to alter the course of treatment.

DNA probes represent an exciting new development that is likely to aid the oncologist to diagnose and manage the cancer patient more effectively.

BRCA1 AND BRCA2

These two genetic tumor markers are also not detected by classical immunoassay technology. Yet they represent a growing number of tumor susceptibility genes as emerging tumor markers. Mutations in these genes increase the susceptibility to develop breast cancer. The BRCA1 gene was the first cancer susceptibility gene identified in familial breast and ovarian cancer. The BRCA2 gene appears to be associated predominantly with breast cancer. Mutations in these two genes are identified by gene sequence analysis.

P21 OR *RAS*

Mutations in the *ras* proto-oncogene are seen in about a fifth of all human cancers. The p 21 *ras* protein belongs to the family of signal transducers. Point mutations in the *ras* proto-oncogene have been identified in colorectal cancers and adenomas and may even be prognostically useful. An ELISA is marketed for qualitative determinations by Oncogene Science for the detection of mutated forms of p 21.

DR-70

This is a new cancer marker polyclonal sandwich EIA assay (AMDL, Tustin, California, USA) measuring the pan tumor marker DR-70 antigen, found in lung, stomach, breast, and colorectal cancers.

90K/MAC-2 BP

A 90 kDa human serum glycoprotein identified as the Mac-2BP, the ligand for the lactose/galactose specific s-lectin, has been shown to be elevated in breast, ovarian, colorectal and non-hodgkin lymphomas. An ELISA assay for measurement of this antigen (Diesse Diagnostica Senese, Italy) is available and a high level of this antigen is indicative of poor prognosis.

BCL-2

A quantitative ELISA is available (Bender MedSystems, Vienna, Austria) to estimate the anti-apoptotic protein bcl-2 in whole cell lysates. The Bcl-2 gene is another important member of the Bcl-2 oncogene family along with the p 53 gene. The quantitative estimation of the Bcl-2 oncoprotein has a prognostic significance in non-hodgkin's lymphoma, SCCs, multiple myeloma and carcinomas of breast and gall bladder.

NMP-22

The nuclear matrix protein (NMP) network is the residual framework seen in the nucleus after exhaustive extraction to remove membrane, chromatin, and cytoskeletal proteins. The NMP-22 assay (Matritech, Massachusetts, USA) for bladder cancer, identifies the nuclear mitotic apparatus protein associated with the mitotic spindle during mitotis.

MMP ASSAYS

Several ELISAs for quantitative detection of matrix metalloproteinases (MMP) have been introduced and may have applications in cancer invasion and metastasis. These are zinc containing enzymes that can dissolve adjacent connective tissue and are thought to be one of the factors in metastatic spread of cancer as well as tumor induced angiogenesis. Several MMP enzymatic activity assays and immunoassays are available (MMP-1, 2, 3, and 8, Chemicon, California, USA; MMP-9, Amersham Pharmacia Biotech, Amersham, UK) employing specific fluorescent substrates or antibodies. These antigens belong to the group of emerging tumor markers that could have applications in the early detection of invasion and metastasis. To date nearly 26 MMPs have been described including a distinct class of membrane bound forms.

TAG-12

The ELISA for estimation of tumor associated glycoprotein (TAG-12, Medac GmbH, Hamburg, Germany) in breast cancer employs two monoclonals. The BM2 monoclonal is specific for the peptide epitope while the solid-phase antibody BM7 binds to a carbohydrate epitope. The TAG12 assay correlates well with the CA 15-3 assay although the antibodies recognize different epitopes of the same MUC-1 antigen.

CA 1-18

The CA 1-18 is a serum tumor marker (IMI, Mississauga, Ont., Canada) found in patients with colorectal cancer. The test appears to detect early stage colorectal cancer.

ANGIOGENIC FACTORS

Angiogenesis is the development of new blood vessels, and it is essential for the growth of tumors beyond a certain size (~2 mm). Both positive (fibroblast growth factor and vascular endothelial growth factor) and negative (endostatin) regulators have been reported. Endostatin is a fragment of collagen XVIII with anti-angiogenic properties. Enzyme immunoassays are now available (Tri Delta Developments, Ireland) for endostatin, basic fibroblast growth factor and vascular endothelial growth factor. Measurement of these regulators may provide useful information guiding the future course of therapeutic intervention.

CHROMOGRANIN A

Chromogranin A (CgA) is a component of the neuroendocrine granules. It is an acidic protein of 45 kDa elevated in pheochromocytomas, GI tumors and pancreatic endocrine tumors. A dual monoclonal sandwich assay is available to measure CgA (Cis Bio International, France).

The RIA test for CgA in plasma (Euro-Diagnostica BV, Netherlands) is a competitive assay using polyclonal antibodies that measures both whole and fragments of the antigen.

TELOMERASE

Telomeres are the specialized nucleoprotein ends of all eukaryotic chromosomes, composed of tandem repeats of the nucleotide sequence TTAGGG. Progressive cell divisions shorten the telomere length, which is thought to be the biological clock that triggers senescence. However in stem cells and cancer cells this is regenerated by the enzyme telomerase. Cancer cells appear to escape this replicative senescence by upregulating telomerase activity. In most cancer cells the telomerase can be demonstrated, unlike in normal cells, and hence this is emerging as an exciting new marker. A new PCR-ELISA has been developed (Roche Molecular Biochemicals) to measure telomerase activity. A biotin labeled specific primer is elongated by telomerase and subsequently amplified by PCR, hybridized to the telomerase repeat-specific detection probe. A streptavidin-coated microtiter plate is used in the final ELISA detection step.

URINARY BLADDER CANCER (UBC) ANTIGEN

The quantitative UBC antigen is a new test that is useful in the management of urinary bladder cancer. Unlike most other tumor markers whose levels in body fluids reflect the tumor burden, the UBC antigen levels are indicative of the course of the disease measured in terms of tumor cell activity. Hence, this test is indicated for monitoring tumor recurrences and its levels are correlated with the stage and grade of cancer. Both IRMA and ELISA versions are available for diagnosis and follow up (IDL Biotech, Sweden).

HHV-8 ANTIBODY

The Kaposi's sarcoma associated herpes virus, also called the human herpes virus 8, is also associated with some primary effusion lymphomas, non-hodgkins lymphomas and multiple myelomas. Two new immunofluorescence assays have been developed (Stellar Bio Systems, USA) for the detection of antibodies to the lytic and latent antigens of HHV-8.

NOR-/METANEPHRINE RIA

Normetanephrine and metanephrine are *O*-methylated metabolites of the catecholamines noradrenaline and adrenaline respectively. These metabolites are elevated in pheochromocytoma, ganglioneuroma and other neurogenic tumors. An RIA for the detection of these two metabolites in heparin plasma has been developed (DLD Diagnostika GmbH, Germany).

HER2/NEU

Herceptin® is an approved monoclonal therapeutic for the treatment of metastatic breast cancer patients with overexpression of the HER2/neu growth factor antigen. A serum assay has been developed (Bayer Diagnostics, UK) to identify the subset of patients with elevated serum antigen levels who would benefit from Herceptin treatment. An additional use of the assay is to predict the patient's response to hormonal treatment. The assay measures the shed extracellular domain of the growth factor receptor using two monoclonal antibodies. No interferences are seen in the presence of therapeutic doses of Herceptin.

PML PROTEIN

A new cancer marker for acute promyelocytic leukemia (APL) has been identified along with the development of a monoclonal probe (Dako Inc., Denmark) for identification of the promyelocytic leukemia protein (PML). In APL cells the nuclei exhibit a microgranular pattern with numerous small dots while the normal hematopoietic cells exhibit the speckled pattern with 5–10 nuclear dots. The architecture of the PML nuclear granules is distorted in APL cells bearing the characteristic reciprocal 15:17 chromosomal translocation. It is important to diagnose APL from the other types since this leukemia subtype is successfully treated often with chemotherapy and all-*trans*-retinoic acid.

ACKNOWLEDGEMENTS

The author would like to acknowledge the various immunoassay manufacturers and their valuable product brochures for key information cited in this chapter. The author also thanks CIHR and Biomira for Chair support and NSERC for grant funding.

This chapter is dedicated to Professor Frank M. Huennekens, of Scripps Clinic and Research Foundation, La Jolla, California, in whose laboratory the author was initiated into the field of cancer research.

REFERENCES AND FURTHER READING

Abelev, G.I. Production of embryonal serum alpha globulin by hepatomas. Review of experimental and clinical data. *Cancer Res.* **28**, 1344–1350 (1968).

Alpert, E. and Abelev, G.I. Summary report: epitope analysis of human alpha-fetoprotein. *Tumor Biol.* **19**, 290–292 (1998).

Bast, R.C., Feeny, M., Lazarus, H., Nadler, L.M., Colvin, R.B. and Knapp, R.C. Reactivity of a monoclonal antibody with human ovarian carcinoma. *J. Clin. Invest.* **68**, 1331–1337 (1981).

Bradwell, A.R. *Serum Free Light Chain Assays* (Binding Site Ltd, Birmingham, UK, 2003).

Ellis, W.J., Vessella, R.L., Noteboom, J.L., Lange, P.H., Wolfert, R.L. and Rittenhouse, H.G. Early detection of recurrent prostate cancer with an ultrasensitive chemiluminescent prostate-specific antigen assay. *Urology* **50**, 573–579 (1997).

Gold, P. and Freedom, S.O. Demonstration of tumor specific antigen in human colonic carcinoma by immunological tolerance and absorption techniques. *Exp. Med.* **121**, 439–461 (1965).

Hammarstrom, S., Shively, J.E., Paxton, R.J. *et al.* Antigenic sites in carcino embryonic antigen. *Cancer Res.* **49**, 4852–4858 (1989).

Hanausek, P. and Walaszek, Z. (eds), *Tumor Marker Protocols* (Humana Press, New Jersey, 1998).

Kreutz, F.T. and Suresh, M.R. Novel bispecific immunoprobe for rapid sensitive detection of prostate-specific antigen. *Clin. Chem.* **43**, 649–656 (1997).

Nap, M., Hammarstrom, M.L., Bormer, O. *et al.* Specificity and affinity of monoclonal antibodies against carcino embryonic antigen. *Cancer Res.* **52**, 2329–2339 (1992).

Nap, M., Vitali, A., Nustad, K. *et al.* Immunohistochemical characterization of 22 monoclonal antibodies against the CA 125 antigen: 2nd report from the ISOBM TD-1 workshop. *Tumor Biol.* **17**, 325–331 (1996).

Nustad, K., Bast, R.C. Jr., Brien, T.J. *et al.* Specificity and affinity of 26 monoclonal antibodies against the CA 125 antigen: first report from the ISOBM TD-1 workshop. *Tumor Biol.* **17**, 196–219 (1996).

Oppenheimer, S.B. *Cancer, a Biological and Clinical Introduction*, 2nd edn (Jones and Bartlett, Boston, 1985).

Pannall, P. and Kotasek, D. *Cancer and Clinical Biochemistry* (Piggott Printers, Cambridge, UK, 1997).

Price, M.R., Rye, P.D., Petrakou, E. *et al.* Summary report on the ISOBM TD-4 workshop: analysis of 56 monoclonal antibodies against MUC mucin. *Tumor Biol.* **19** (S1), 1–152 (1998).

Rittenhouse, H.G. Early detection of recurrent prostate cancer with an ultrasensitive chemiluminescent prostate-specific antigen assay. *Urology* **50**, 573–579 (1997).

Rittenhouse, H.G., Manderino, G.L. and Hass, G.M. Mucin-type glycoproteins as tumor markers by laboratory. *Medicine* **16**, 556–560 (1985).

Rittenhouse, H.G., Finlay, J.A., Mikolajczky, S.D. and Partin, A.W. Human kallikrein 2 (hk2) and prostate-specific antigen (PSA): two closely related but distinct kallikreins in the prostate. *Crit. Rev. Clin. Lab. Sci.* **35**, 275–368 (1998).

Rye, P., Bovin, N.V., Vlasova, E.V. *et al.* Summary report on the ISOBM TD-6 workshop: analysis of 20 monoclonal antibodies against Sialyl Lewis[a] and related antigens. *Tumor Biol.* **19**, 390–420 (1998).

Sell, S. Cancer markers of the 1990s. *Clin. Lab. Med.* **10**, 1–37 (1990).

Stenman, U.H., Paus, E., Allard, W.J. *et al.* Summary report of the TD-3 workshops: characterization of 83 antibodies against prostate-specific antigen. *Tumour Biol.* **20** Supp. 1, 1–12 (1999).

Stigbrand, T., Andres, C., Bellanger, L. *et al.* Epitope specificity of 30 monoclonal antibodies against cytokeratin antigens: the ISOBM TDS-1 workshop. *Tumor Biol.* **19**, 132–152 (1998).

Suresh, M.R. Immunoassays for cancer-associated carbohydrate antigens. In: *Seminars in Cancer Biology—Glycosylation Changes Associated with Malignant Change*, (ed Longenecker, B.M.) vol. 2, 367–377 (W.D. Saunders Press, London, 1991).

Suresh, M.R. Classification of tumor markers. *Anticancer Res.* **15**, 1–6 (1996).

Van Dalen, A. Significance of cytokeratin markers TPA, TPA (cyk) TPS and CYFRA 21.1 in metastic disease. *Anticancer Res.* **16**, 2345–2350 (1996).

Virji, M.A., Mercer, D.W. and Herberman, R.B. Tumor markers in cancer diagnosis and prognosis. *Ca: Cancer J. Clin.* **38**, 42–63 (1988).

Wu, J. and Nakamura, R. *Human Circulating Tumor Markers: Current Concepts and Clinical Applications* (American Society of Clinical Pathologists Press, Chicago, 1997).

Yu, H., Diamandis, E.P., Wong, P-Y., Nam, R. and Trachtenberg, J. Detection of prostate cancer relapse with prostate specific antigen monitoring at levels of 0.001 to 0.1 μg/L. *J. Urol.* **157**, 913–918 (1997).

57 Allergy

Lars Yman

ALLERGIC DISEASE

Allergy is not an easy disease to define. In 1901, von Pirquet described it as an uncommon sensitivity against substances completely harmless to the majority. This hypersensitivity to common factors in the environment or food can cause a range of symptoms, such as inflammation of the nasal membranes (**rhinitis**), **asthma**, or **dermatitis**. Allergic diseases belong to a large group of inflammatory conditions involving several interacting systems in the human body, sometimes described simply as type I, II, III, or IV (Coombs and Gell, 1968). Figure 57.1 summarizes some of the major components that also act as the body's defense against harmful foreign substances or organisms and direct physical damage. This powerful system can be triggered by a wide range of external and internal events. The ability of some of the systems to react to a variety of stimuli, and the high degree of communication between the systems, can lead to the development of disorders showing the same symptoms regardless of the trigger.

The mast cell plays a central role in allergy, releasing mediators that stimulate glandular secretions and smooth muscle contraction, and increase the permeability of the small blood vessels. These typical features of the immediate phase of an asthma attack or an outbreak of rhinitis are often followed by a late phase, initiated by the release of chemotactic factors from the mast cell. These activate cells, such as eosinophil leucocytes, to release substances like eosinophil cationic protein (ECP), damaging the tissues (Svensson *et al.*, 1990).

A characteristic of mast cells is that they carry immunoglobulin molecules of the IgE class on their surface. Allergic individuals produce excessive amounts of IgE antibodies when exposed to certain foreign macromolecules by inhalation, ingestion, or injection and develop allergic symptoms when re-exposed to the same material. The foreign materials that precipitate IgE production are referred to as **allergens** and the affected individuals as **atopic**. The allergens penetrate mucosal surfaces and bind to IgE antibodies on the surface of mast cells triggering potentially life-threatening immediate and late allergic reactions.

DIAGNOSIS AND THERAPY

In diagnosis, atopic IgE-mediated allergy must be differentiated from conditions initiated by other mechanisms, both immunological and non-immunological (see Figure 57.2). Normally this involves testing for total IgE in serum. A raised level confirms that the allergy is atopic. In order to treat the disease, the allergen specificity of the IgE antibodies needs to be identified. IgE-mediated sensitivity was traditionally detected by skin testing, injecting solutions of suspected allergens into the dermis. However, immunoassays are now used extensively to detect and quantify IgE antibodies in the blood and other body fluids. The IgE antibodies in the body fluids are in equilibrium with the antibodies bound to cells and can be measured quantitatively with high sensitivity and specificity. (For review, *see* Johansson and Yman, 1988.)

Allergy can be treated by eliminating exposure to the sensitizing allergen, injecting the allergen (specific immunotherapy), or administering drugs. These inhibit the release of mast cell mediators or block their effects. The effectiveness of the therapy can be monitored by assaying factors involved in the inflammatory reactions.

ALLERGENS

Substances capable of eliciting an allergic reaction are generally referred to as allergens, regardless of the mechanism involved. This chapter is limited to allergens that can stimulate the synthesis of IgE antibodies, causing atopic allergy. Identification of the offending allergen can be difficult. Although the timing of a seasonal allergy may give clues to the cause, most affected patients are sensitized by allergens from many sources with overlapping seasons of exposure and often with overlapping immunological specificity.

Most allergens are biological materials which, when inhaled or ingested, dissolve in mucosal secretions. They also tend to be comparatively easy to extract and dissolve in buffered solutions. Typically the allergy source releases a complex mixture of potentially allergenic macromolecules such as polypeptides, glycoproteins, and peptidoglycans. Atopic individuals may produce IgE antibodies to

The Immunoassay Handbook. *Third Edition*
D. Wild (Ed.)

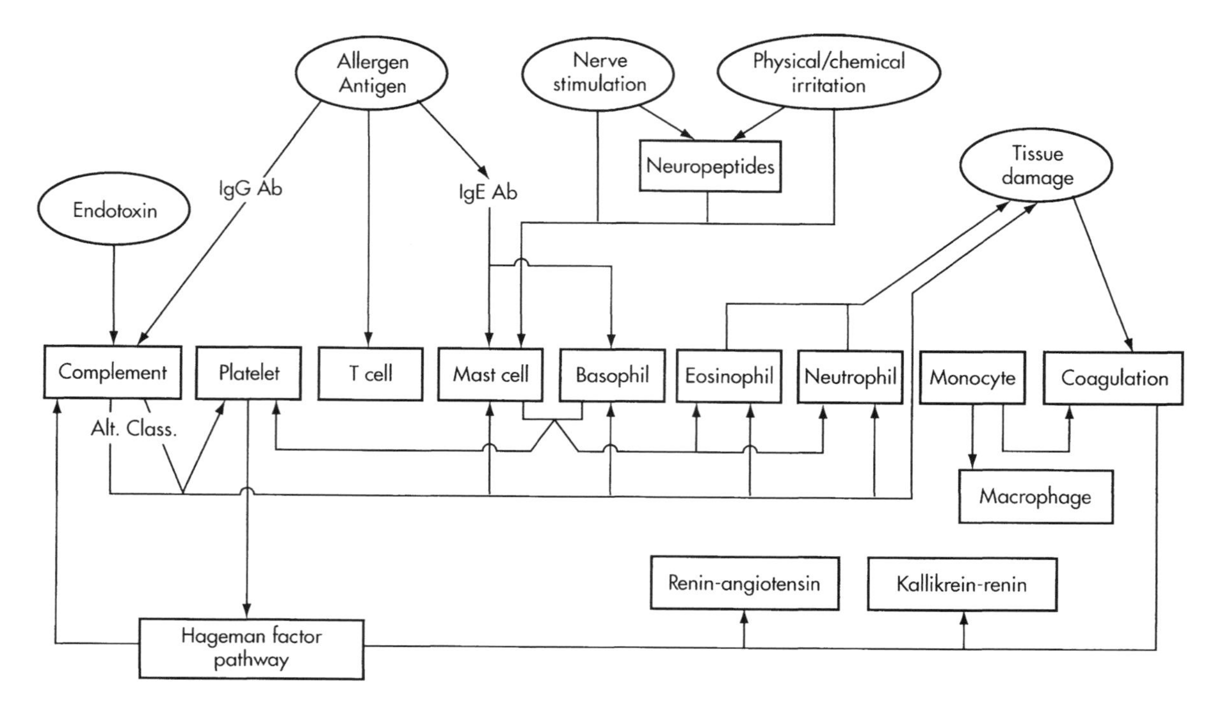

Fig. 57.1 Systems involved in allergic reactions. The diagram summarizes the interactions between systems potentially involved in inflammatory reactions that may lead to allergic symptoms. Allergens reacting with IgE antibodies (IgE Ab) on the surface of mast cells are major, but not sole triggers of such symptoms. Differentiation is essential for adequate therapy. (Alt., alternative; Class., classic pathway).

one or many of the allergenic components. Thus every allergic patient has a range of unique antibodies to a given allergen source. Because some allergenic components have a higher relative concentration and greater immunogenic activity than others, it is possible to test allergic sensitivity qualitatively using selective combinations of a limited range of allergens. However, quantitative measurement of allergen-specific IgE antibodies against a specified allergen source, to indicate the degree of sensitization, requires the presence of an excess of all the allergenic components in the test (see Figure 57.3). The number of allergenic components identified increases rapidly and nomenclature rules were proposed to minimize confusion (King *et al.*, 1995).

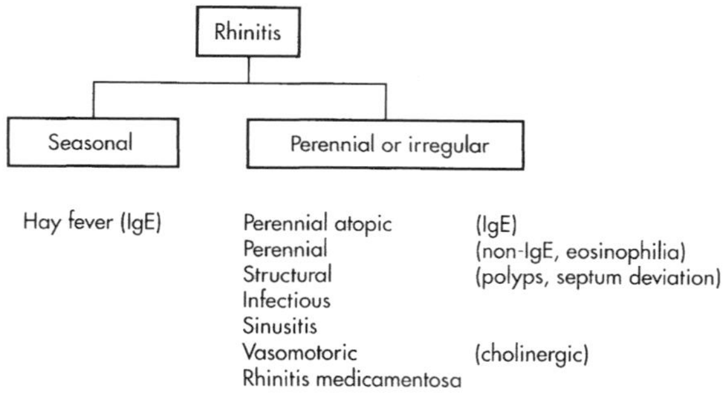

Rhinitis

Seasonal

Hay fever (IgE)

Perennial or irregular

Perennial atopic (IgE)
Perennial (non-IgE, eosinophilia)
Structural (polyps, septum deviation)
Infectious
Sinusitis
Vasomotoric (cholinergic)
Rhinitis medicamentosa

Fig. 57.2 Allergy-like symptoms. The majority of cases have perennial or irregular symptoms that may be IgE mediated or triggered by other conditions. Differentiation mostly requires extensive testing.

ANALYTES

TOTAL SERUM IgE

Immunoassays for total serum IgE are used to test for elevated IgE concentration, confirming atopic disease. They use a wide range of assay principles. Competitive double-antibody solid-phase RIAs co-exist with immunometric (sandwich) enzyme immunoassays. Solid phases in common use include microparticles, polystyrene beads and tubes, and cellulose foam. The signals measured are radioactivity, color, fluorescence, and luminescence. Concentrations are expressed as kU/L defined by the WHO International Reference.

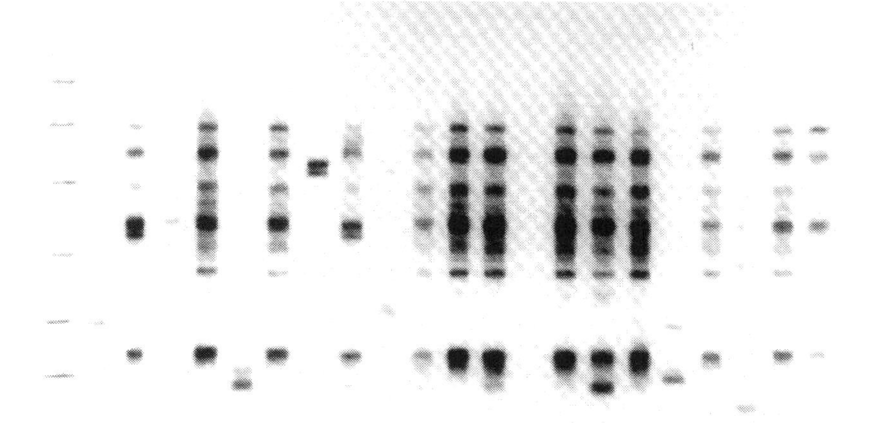

Fig. 57.3 Diversity of IgE antibody patterns against soy bean allergen in sera from 21 individuals. The allergenic components were separated by SDS-g-polyacrylamide gel electrophoresis and blotted onto nitrocellulose. Allergen-IgE complexes were detected by radiolabeled anti-IgE and autoradiography (Perborn, 1990, unpublished data).

Clinical Applications

The serum total IgE level is generally raised in atopic disease, although an IgE concentration within the normal range does not rule out IgE-mediated disease, especially if the disease is due to a single allergen.

Total IgE measurements have been used for differentiation of atopic and non-atopic diseases and for prediction of allergy in children by measurement in cord blood or in newborns.

The serum IgE concentration is increased in atopic allergy, especially in **eczema** patients who may exceed 20,000 kU/L. The total serum IgE concentration is highly correlated with the number of allergens involved. IgE is also raised in **parasitosis** and some immunodeficiencies.

Reference Intervals

The serum IgE level in healthy individuals increases during childhood from <1 kU/L at birth to an adult level from the age of about 10 years. The geometric mean of truly healthy non-smoking adults is about 10 kU/L and the mean + 2SD is slightly above 100 kU/L. It is recommended that allergy investigation is done on all children with levels higher than the mean for the age + 1SD and in adults if the level is higher than 100 kU/L (Johansson and Yman, 1988).

The validity of the original adult reference values, geometric mean = 13.2 kU/L, mean + 2SD = 114 kU/L (Zetterström and Johansson, 1981), was recently confirmed with UniCAP, testing 63 healthy blood donors without known allergy. Geometric mean = 17.4 kU/L, mean + 2SD = 112.9 kU/L (Persson and Yman, 1997).

Limitations

Total IgE levels are less closely linked to allergic disease in areas with high incidence of parasitosis.

Assay Technology

A wide variety of immunoassay procedures are in routine use. The bulk of testing is performed in laboratory systems with varying degrees of automation.

Desirable Assay Performance Characteristics

An assay covering the range 0.5–5000 kU/L would satisfy the practical clinical needs for IgE determinations from cord serum measurements on allergy risk babies to testing of subjects with atopic dermatitis or parasitosis. To keep imprecision (repeatability and reproducibility) as low as possible this range is typically split into a routine range (preferably 2–2000 kU/L) and a low range for primarily pediatric use. UniCAP low-measuring range is equivalent to the range covered by the specific IgE assay, i.e. 0.35–100 kU/L. Although some other commercial assays include diluents as zero calibrators the detection limits are always higher. An option to store calibration curves is desirable.

Types of Sample

Assays for total and specific IgE are generally validated for serum and plasma. Measurements in tears (Kari *et al.*, 1985; Sainte Laudy *et al.*, 1994), nasal secretion (Deuschl and Johansson, 1977; Sensi *et al.*, 1994), and feces (Sasai *et al.*, 1992) have been reported. Applications outside those described in the manufacturer's instructions for use must be validated by the investigator.

ALLERGEN-SPECIFIC IgE ANTIBODY

Allergen-specific antibodies of a single immunoglobulin class have to be measured in the presence of other antibodies of the same class and antibodies of other classes specific for the same allergen. This requires specific recognition of the antigen-binding sites (Fab) and the class-specific epitopes (Fc) in the same assay. In the case of IgE the concentration of specific antibody should be measurable down to picograms per milliliter in samples that may contain total IgE and competing antibodies of other classes well into the microgram per milliliter range. The first test designed for this purpose was the Phadebas radioallergosorbent test (RAST) (see Figure 57.4).

Assay Characteristics

IgE assays use a solid phase coated with allergen to bind the IgE antibody in the sample. There is a critical washing step after the initial allergen–IgE antibody reaction and before the incubation with labeled anti-IgE. The solid phase may alternatively be coated with ligands that catch allergen–IgE complexes before the washing step. A high-capacity solid phase, like the chemically activated foam of the Pharmacia CAP System and UniCAP, provides a large excess of allergen that maximizes the binding of the IgE antibody. This makes the assay less sensitive to competing antibodies of other classes. It also means that the bound IgE antibody can be detected with less amplification. The resulting lower concentration of anti-IgE, less-intensive labeling, and lower substrate concentration minimize several of the classic contributions to high and variable non-specific background.

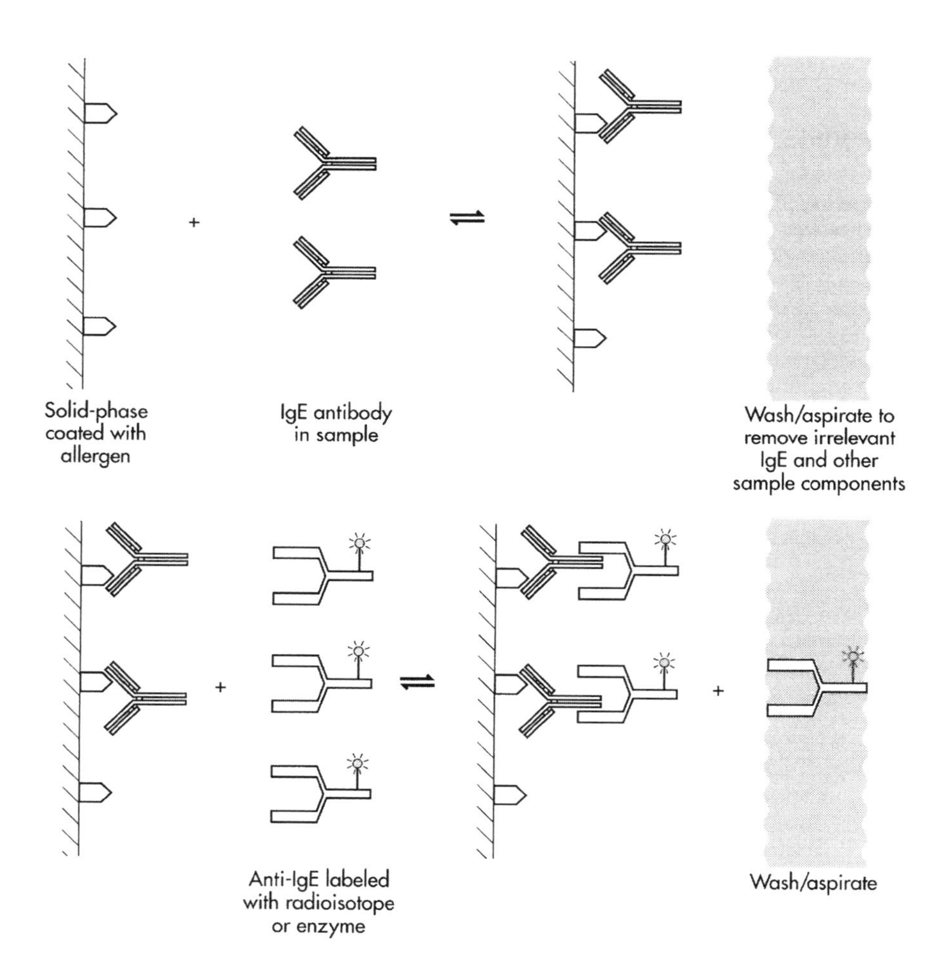

Fig. 57.4 Principle of Phadebas RAST®, Pharmacia CAP System™ RAST®, and UniCAP® specific IgE.

Allergen in Excess

A high-capacity solid phase provides conditions for quantitation of the sum of all antibodies regardless of specificity and affinity. The Law of Mass Action, applied to heterogeneous solid-phase immunoassays, predicts that when the available allergen concentration is increased to a level where K (the affinity constant) multiplied by the allergen concentration (equation (57.3)) reaches ≥ 10, the proportion of IgE antibody bound to the solid phase will be $\geq 90\%$ and essentially independent of the affinity between the antibody and the allergen (Peterman, 1991; Yman, 1994).

Law of Mass Action

$$\text{Association Constant} = \frac{\text{AB}}{\text{A} \times \text{B}} \quad (57.1)$$

$$K = \frac{[\text{Allergen IgE Ab complex}]}{[\text{Allergen}] \times [\text{IgE Ab}]} \quad (57.2)$$

$$\frac{[\text{IgE Ab}_{\text{bound}}]}{[\text{IgE Ab}_{\text{total}}]} = \frac{K[\text{Allergen}_{\text{avail}}]}{1 + K[\text{Allergen}_{\text{avail}}]} \quad (57.3)$$

When the assay is essentially affinity independent, i.e. when more than approximately 90% of the IgE antibody is bound, a further increase of the allergen concentration will have negligible effect on the measured value (Figure 57.5).

Anti-IgE

Anti-IgE preparations must be IgE Fc specific and are preferentially combinations of, e.g. monoclonal antibodies with specificity against more than one epitope on the Fc fragment and with complementary dose–response characteristics.

Labeled anti-IgE has been prepared using several signal generation systems, in each case aiming at a high sensitivity.

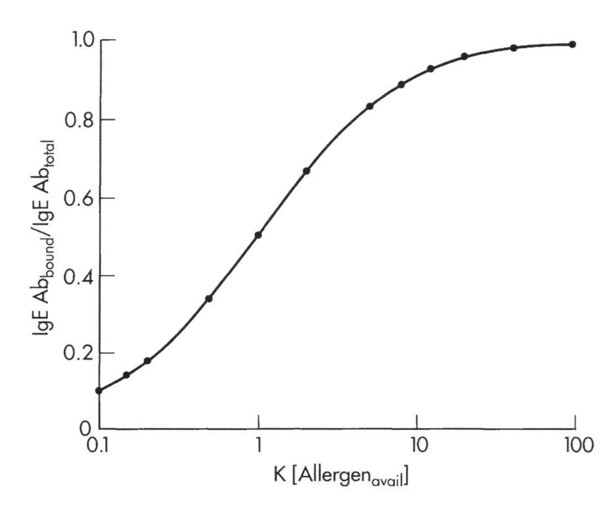

Fig. 57.5 Effect of increased allergen concentration on the proportion of the added IgE antibody that is bound to the solid-phase allergen.

Calibration

Calibrators for both specific and total IgE measurements should be traceable to the WHO International Reference Preparation for human IgE (75/502). This allows true quantitation of the specific antibodies provided that the allergen is in excess (see further discussion under QUANTITATIVE MEASUREMENTS). The calibrator range of UniCAP-specific IgE is 0.35–100 kU/L.

The design of the assay should permit storage of calibration curves up to 1 month using curve controls in each assay to verify the validity of the stored calibration curve.

Test Selection, Use, and Interpretation

Allergy tests should be used selectively, taking into account the local distribution of the allergens and the clinical history of the individual patient. Unless there is a clear link between the appearance of symptoms and occasions of exposure to one allergen, multisensitivity should always be suspected.

Pollen Allergens

Most of the allergen-inducing pollens emanate from wind-pollinated plants in the temperate zones of the northern and southern hemispheres. Trees and grasses pollinating in spring and weeds flowering in summer and autumn are responsible for most of the seasonal allergies. As part of its natural function a pollen grain rapidly releases species-specific recognition substances and enzymes when it comes into contact with a moist surface. Pollens from related plants tend to release substances that cross-react and this should be taken into account when interpreting the results of allergy tests. The degree of cross-reactivity and the probability of multiple sensitivity caused by pollens can be estimated from the botanical relationships of the plants (Weber and Nelson, 1985; Yman, 1982; Lewis and Elvin-Lewis, 1977). An example of the relationships between trees pollinating in spring is given in Figure 57.6.

Epithelial Allergens

Airborne exposure to allergens from cats and dogs commonly gives rise to allergic reactions. Those who work with other species may have similar problems. Cross-reactivity of allergens from different species exists (Ohman *et al.*, 1976), especially with serum-derived

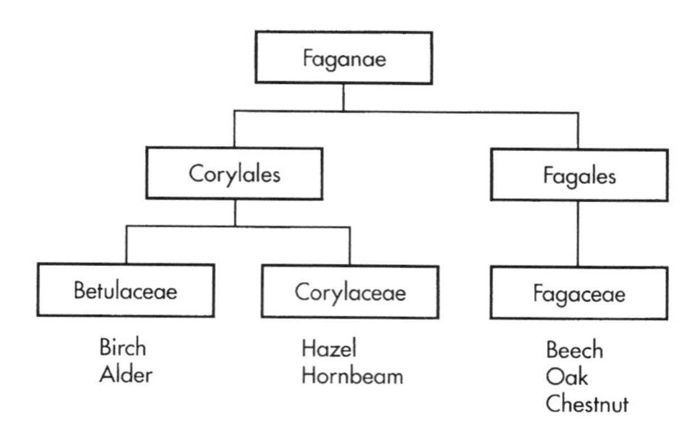

Fig. 57.6 The order Fagales includes three families. Pollen allergens from the seven major genera show a high degree of immunological cross-reactivity within a family and moderate cross-reactivity between families.

substances such as albumin (Brandt and Yman, 1980; Sabbah *et al.*, 1994). Other epithelial components are species-specific but there is no evidence for breed-specific allergens (Wüthrich *et al.*, 1985). Allergens have been identified in saliva (Anderson and Baer, 1981) and urine (Lutsky *et al.*, 1980) but dander appears to be the most common source of airborne epithelial allergens (Vanto *et al.*, 1982).

House Dust

Mites belonging to the genus *Dermatophagoides* are often the cause of allergy to house dust, although other mites can also be allergenic. They are often referred to as storage mites (Yman *et al.*, 1988a,b). Other common house dust allergens arise from cats and dogs (Ammann and Wüthrich, 1985), cockroaches (Kang and Chang, 1985), moths and midges (Komase *et al.*, 1997), molds (Gravesen, 1972), and other microorganisms (Bernstein and Safferman, 1973). House dust is impossible to standardize in allergy tests for obvious reasons and is best replaced by a defined combination of allergens or a range of individual allergen tests. A rapidly increasing number of homogeneous allergenic protein preparations are made available by purification or recombinant DNA techniques (Chapman *et al.*, 1997). Some of these may eventually prove useful in clinical routine diagnostic investigations.

Occupational Dusts

Allergens that are normally rare may be a prime cause of allergy in a particular occupational environment. Sensitivity to many occupational dusts can be identified by *in vitro* tests (Johansson and Yman, 1988).

Molds

The air that we breathe contains a variety of fungi as spores, mycelial fragments, and mold degradation products. The major mold allergens emanate from the mycelium rather than the spores. Some mold allergens are carbohydrate-rich glycoproteins. Immunochemical studies have shown that only the protein part of the molecule is capable of binding IgE antibodies, whereas the carbohydrate parts seem to trigger an IgG antibody response (Swärd-Nordmo *et al.*, 1989).

It is normally difficult to predict mold allergy from a patient's clinical history. Hardly any environment is free of fungal growth from a number of species. *Alternaria*, *Cladosporium*, *Penicillium*, *Candida*, and *Aspergillus* are frequently used for testing. The clinical sensitivity of *in vitro* testing is high if high-quality allergens are applied in a proper assay (Bahna and Yee, 1986). Molds like *Botrytis* and *Helminthosporium* are also potential mold allergens (Beaumont *et al.*, 1985; Jonsson *et al.*, 1987) in certain environments.

A yeast-like fungus, *Pityrosporum*, is commonly present on the human skin and sometimes causes sensitization that contributes further to already existing problems in **atopic dermatitis** (Ring *et al.*, 1992).

Food

Intact macromolecules from partially digested food pass through the intestinal mucosa into the circulation. The IgE antibody patterns of food-sensitized patients are often even more complex than those seen for inhalant allergens. Egg white and cow's milk are the major causes of allergy in early childhood. A high IgE antibody concentration becomes apparent before the age of 1 year. Two or three years later the antibodies and symptoms may decline and eventually disappear (Sigurs *et al.*, 1988), at least if the relevant food is eliminated from the diet.

Sensitivity to nuts and peanut is surprisingly common in small children even without known exposure. IgE antibodies against peanut do not decrease as children grow older and sensitivity seems to remain for life (Zimmerman *et al.*, 1989). Inadvertent consumption of peanut or related legume seeds may be responsible for this. Common thickening and stabilizing agents in industrially prepared food contain related proteins from other members of the legume family (Yman *et al.*, 1988a,b), and may also provide a continuing source of low-dose exposure.

Seeds

Legume seeds

Hidden exposure to legume seeds (peanut, soy, pea) accounts for almost half of the unexpected allergic reactions to food reported in recent surveys in Sweden (Figure 57.7; Malmheden Yman, 1997). Several fatal anaphylactic reactions have been reported lately.

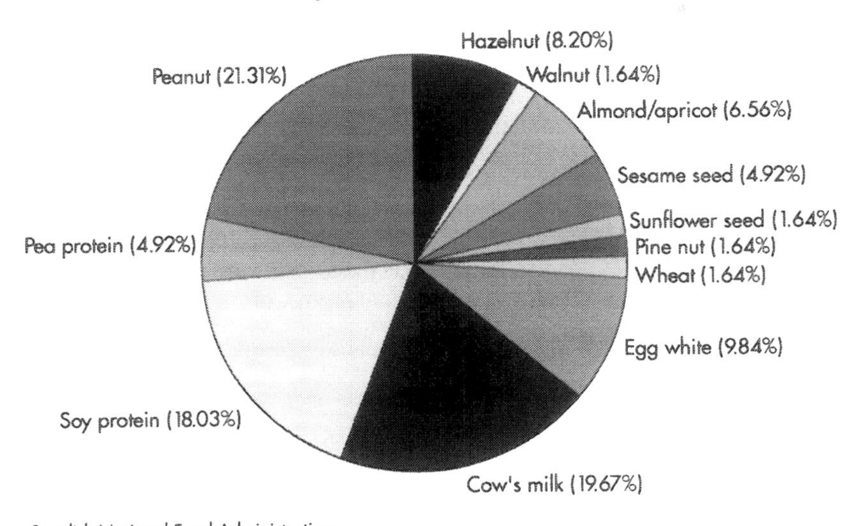

Fig. 57.7 Unexpected exposure to peanut, soy, and pea was responsible for 44% of reported severe reactions to food (information from Swedish National Food Administration).

The typical case is of a school age child with known asthma and allergy to peanut. The patient has high anti-peanut IgE and low to moderate anti-soy IgE. The severe reaction occurs up to an hour after consumption of, e.g. a meat product containing soy protein. Exercise or low temperature is often involved (Foucard *et al.*, 1997).

Nuts

Several frequently consumed seeds are commonly referred to as nuts or tree nuts although only a few qualify for this botanical term. There are relations not only within this group, e.g. between walnut and pecan nut (Juglandaceae), but also between the food and inhaled allergens, as for hazel nut and birch pollen (Fagales). Unexpected cross-reactivity that cannot easily be explained by a close botanical relationship may also occur in some patients. The reactivity to birch pollen, hazel nut, and apple is a classic example (de Groot *et al.*, 1996) explained by structural similarities between the major birch allergen, Bet v 1, and similar proteins present in others.

Cereals

Cereal grains are often involved in the development of gastrointestinal and skin problems. Measurement of IgE antibodies can be especially helpful in cases with a late onset of allergic symptoms (Wraith *et al.*, 1979). Wheat allergens cross-react with grain allergens from rye, barley, and oats, and to a minor extent with grass pollen allergens.

Egg, Milk, and Meat

Egg white is a common allergen in early childhood (*see* also below under MULTIFOOD ALLERGENS and PREDICTION AND MONITORING OF THE DEVELOPMENT OF ALLERGIC DISEASE) but decreases in importance with increasing age. Some individuals, however, retain their sensitivity to egg as adults. Studies on a group of patients failing to develop tolerance showed exposure to cage birds to be the probable reason (Añibarro Basuela *et al.*, 1991). Such patients usually have IgE antibodies and clinical sensitivity to egg yolk and chicken meat rather than to egg white.

Cow's milk allergy is also important in childhood. Care should be taken not to introduce cow's milk too early to children at hereditary risk (Saarinen and Kajosaari, 1995).

Allergy to meat is more frequent than commonly expected. A prevalence of about 9% has been observed in atopic children (Restani *et al.*, 1997). Many of the children studied showed sensitivity to meat from more than one animal species due to the chemical and immunological similarities between their serum albumins. Serum albumin is one of the allergens in all kinds of milk and meat.

Seafood

Fish may, from an allergy perspective, be divided into at least four types: sharks, e.g. dogfish; the order Gadiformes like codfish and hake; the Scombroid fishes, e.g. tuna and mackerels; and the order Pleuronectiformes (flatfish) such as megrim and sole. The overlap of allergen specificity between the types varies from moderate to small (Bernhisel-Broadbent *et al.*, 1992; de Martino *et al.*, 1990; Pascual *et al.*, 1988, 1992; Helbling *et al.*, 1996). Furthermore, the order Clupeiformes, e.g. herrings and sardines (Alonso *et al.*, 1993), and the family Xiphiidae, swordfish (Kelso *et al.*, 1996), may constitute allergenically distinct groups.

Generally speaking, skin test and serum-specific IgE antibody measurements show good agreement in allergy to seafood. Exceptions have been observed, especially for scombroid fish and other species where histamine is rapidly formed upon storage and false positive skin test results are common (Williams *et al.*, 1992).

Specific IgE antibodies against a marine fish parasite, *Anisakis*, have been demonstrated in patients with urticaria and gastrointestinal reactions (Audicana, 1995). Adverse reactions after fish consumption can be an IgE-mediated allergy to the parasite instead of to the fish (Lindqvist *et al.*, 1993; see Figure 57.8). The parasite is common and the presence of IgE antibodies to the parasite has diagnostic importance in reactions to seafood. Several other parasites, especially helminths, cause synthesis of IgE antibodies against a wide range of proteins released by the organism during the human stage of their life cycles. Presence of cross-reacting allergens cannot be excluded. However, studies on patient populations with suspected reactions to fish (Montoro *et al.*, 1996; del Pozo *et al.*, 1996) showed low or no cross-reactivity to *Ascaris* and *Toxocara* with Pharmacia CAP System.

Latex

IgE antibodies were shown to be involved in adverse reactions to natural rubber latex (Turjanmaa *et al.*, 1996). A range of IgE-binding proteins have been identified and their role in allergy to natural rubber latex, latex products, and to more or less related food products are under study (Levy and Leynadier, 1997). Allergen-specific IgE antibodies can be quantitated in serum, and testing should always be considered in the investigation of suspected latex allergy, in food allergy in children, and in urticaria and anaphylaxis of unknown cause.

Allergens of banana and avocado carry epitopes homologous to latex allergens and are most probably different proteins with partial sequence homologies, rather than a universal fruit allergen (Beezhold *et al.*, 1996). This is supported by the results of measurements of IgE antibodies to latex and banana in 19 latex allergic patients (Delbourg *et al.*, 1996) showing no correlation (Figure 57.9).

The clinical efficiency of specific IgE antibody measurement (Pharmacia CAP System) is high in children and adults, with and without occupational exposure, and in spina bifida patients. Compiled data (Yman and Lundberg, 1997) from 11 published studies show agreement with clinical judgment in 483/524 cases (92.2%).

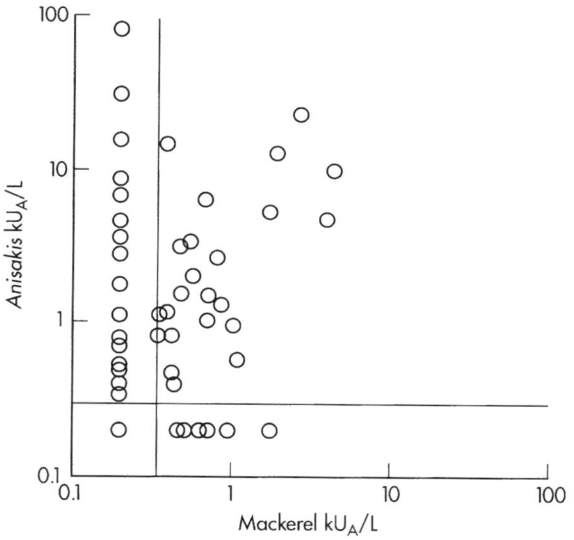

Fig. 57.8 IgE antibody concentrations against mackerel and the fish parasite *Anisakis* in 138 atopic dermatitis patients from Japan (Lindqvist *et al.*, 1993). Lines at 0.35 kU_A/L indicate low end of measuring range.

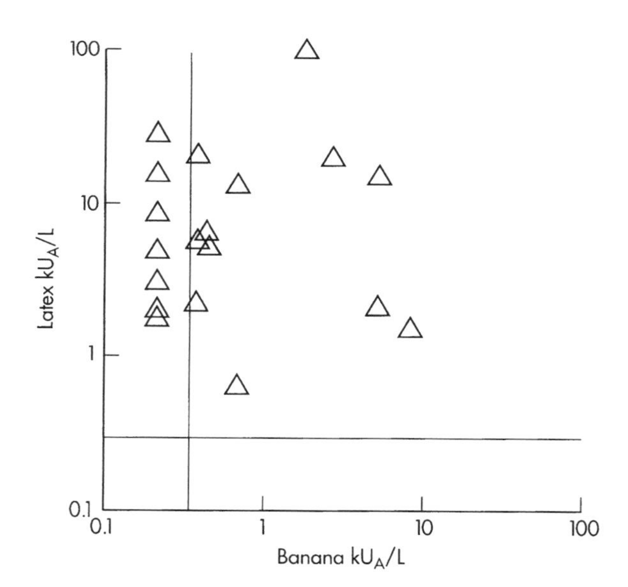

Fig. 57.9 Measured IgE antibody concentrations against natural rubber latex and banana in a group of latex sensitive patients (data from Delbourg *et al.*, 1996). Lines at 0.35 kU_A/L indicate low end of measuring range.

Chemicals and Drugs

Isocyanates (Butcher *et al.*, 1980) and anhydrides (Maccia *et al.*, 1976) are examples of reactive chemicals that, if inhaled, may bind to carrier proteins in the body and act as haptens, stimulating an IgE antibody response. The β-lactam structure of penicillin also spontaneously reacts with proteins such as albumin and forms an immunogenic compound. (For further information on drug allergens, see Baldo and Harle, 1990.)

IgE Antibodies in Penicillin Allergy

The early work of Levine and Zolov (1969) on the allergenic properties of penicillins and their metabolites, and the studies of IgE antibody measurements in the diagnosis of penicillin allergy by Kraft *et al.* (1977) have been followed by numerous immunological and clinical studies on penicillins and other β-lactam antibiotics.

In cases of adverse reactions to drugs the prevalence is low, the cause of the reactions extremely complex, and therefore the true clinical diagnosis less obvious. Less than 10% of all adverse reactions to drugs are considered to have immunological (allergic) background (Boguniewicz, 1995), and only a part of those are IgE mediated. Extensive prospective studies on consecutive cases with suspected allergy have therefore never been reported for drug allergens. Sensitivity data reported in the literature are, generally speaking, based on agreement of test (*in vivo* or *in vitro*) with the physician's interpretation of historical data or agreement between skin test and IgE antibody measurement in samples collected and stored over long periods of time.

True specificity data are not available because very few positive test reactions, *in vivo* or *in vitro*, are confirmed or disqualified by drug challenge.

Considering this situation and the fact that standardized skin testing reagents and procedures do not exist and that challenge and even skin testing often are avoided because of the risk, it is not surprising that reported data on performance of *in vitro* tests are variable and difficult to interpret. However, in a study during 1983–1990 (Surtees *et al.*, 1991) there were 175 patients with a history of immediate type reaction to penicillin referred to an allergy clinic by general practitioners. Of these patients, 132 were tested and 4 were found to have penicillin-specific IgE antibodies. The 128 patients without detectable IgE antibodies were challenged with 250 mg phenoximethylpenicillin orally. None of them had any reaction. The clinical sensitivity of the IgE antibody measurement was high, i.e. no patients (0/128) were falsely classified as negative.

Multiallergens

Combinations of allergens, e.g. representing common molds, or the summer pollens of a certain area, are useful if the clinical history is inconclusive.

Phadiatop®

Before trying to identify offending allergens, patients with allergy symptoms should first be tested to confirm that the disease is atopic. If not, further searches for offending allergens are rarely successful. Testing with Phadiatop detects at least 90% of patients with atopic inhalant allergy (Eriksson, 1990). Neither Phadiatop nor any other commercial test with this type of indication gives quantitative results.

Phadiatop was designed to be a test for atopic allergy with its main use for patients with respiratory symptoms. This means that common inhaled allergens form the base. Furthermore, it means that cases primarily sensitized by food or insect sting, occupational exposure, or by keeping more or less exotic pets, like birds, guinea pigs, etc. cannot be detected unless the patients are also sensitive to common inhalants.

The allergens selected for Phadiatop, and the way they are combined, are aimed at achieving an approximately 90% probability of correct classification of atopics and non-atopics. The combined effects of exposure, multisensitization, and cross-reactivities between allergens give a high cumulative efficiency already for a relatively limited combination of species. The order of the allergens in terms of contribution to the cumulative efficiency varies between populations depending on relative intensity of exposure but it has been possible to select three combinations that covered the needs of Europe, North America, and Japan. Equivalent function of all three reagent versions (Europe, North America, and East Asia) in the latest system version, UniCAP, was recently confirmed in new prospective studies. The data from the European studies (Paganelli *et al.*, 1998) are summarized in Table 57.1. Recent data from the US on 142 patients, age 6–18 years, showed 95% sensitivity and 97% specificity (Portnoy *et al.*, 1997). Comparative studies in Japan showed full agreement between UniCAP and Pharmacia CAP System.

Multifood Allergens (fx5)

It is well documented that small children start with allergy to food allergens. Egg is the most frequent sensitizer and

Table 57.1. UniCAP Phadiatop.

Clinical diagnosis	Positive	Negative	Total
Atopic	483	38	521
Non-atopic	34	281	315
Total	517	319	836

Results of clinical trials in six European clinics. Clinical sensitivity and specificity were 93 and 89%, respectively.

most of the symptoms involve the gastrointestinal tract and the skin. This situation changes with increasing age (Figure 57.10). The frequency and intensity of food sensitivity decreases and inhalant allergy (asthma and later rhinitis) to mites, animals, and pollen takes over (Sigurs *et al.*, 1988).

Allergy to inhalant allergens is rare in the first year, uncommon before the age of three, and becomes dominant at school age. Small children therefore rarely have any IgE antibodies against inhalants. Phadiatop, designed to be a test for the identification of atopic sensitivity to common inhalants, therefore should be expected to show increasing sensitivity with increasing age over the first years of life (Zimmerman and Forsyth, 1988).

The documentation on Phadiatop clearly supports its use as a sensitive test for atopy in patients older than 3 years suffering from respiratory symptoms. The Zimmerman studies and a range of subsequent investigations also showed that a multifood allergen (fx5), including egg, milk, fish, wheat, soy, and peanut, was an effective test for atopy in children younger than about 3 years (Figure 57.11).

The combination of the two tests correctly assigned the presence or absence of atopy with a sensitivity of 98% and a specificity of 98%, compared with a full laboratory evaluation in a group of 109 children with a median age of 3 years.

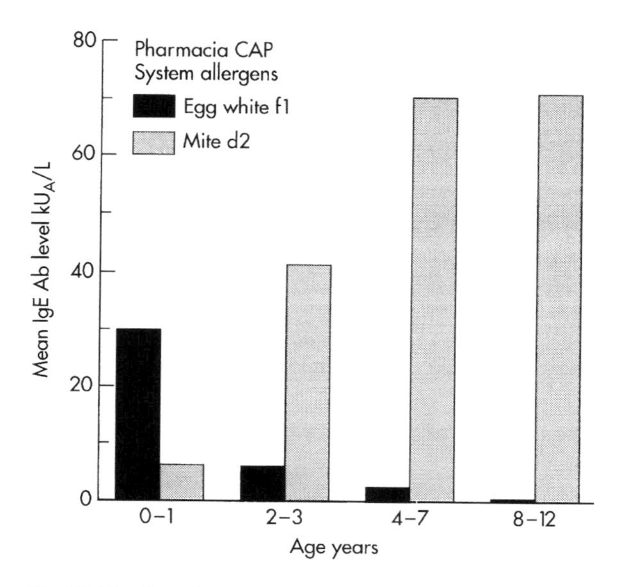

Fig. 57.10 Specific IgE antibodies against food and inhalants in Japanese children (data extracted from Okudaira *et al.*, 1991).

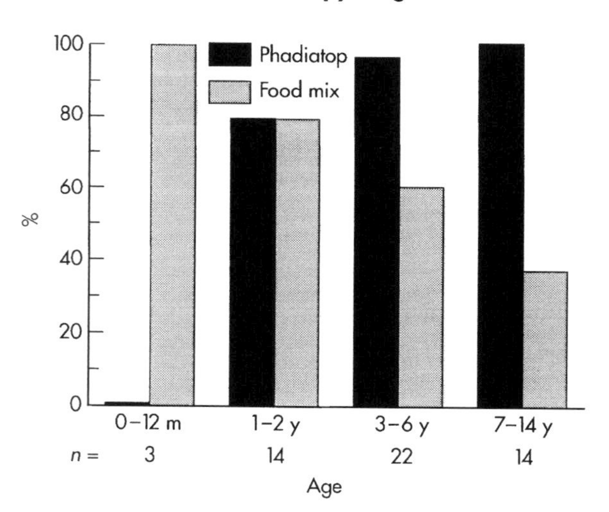

Fig. 57.11 The sensitivity of Phadiatop and the multifood allergen (fx5) for the detection of atopic allergy in childhood as a function of age (Zimmerman *et al.*, 1988).

IgG ANTIBODY ASSAYS

Tests for the measurement of allergen-specific IgG and IgG_4 antibodies are commercially available. Measurements of elevation of IgG antibody concentrations are not only used primarily for monitoring insect venom immunotherapy but may also be helpful in investigations of allergic diseases and adverse reactions to food as markers of allergen exposure. Gliadin, milk proteins, and mold antigens are of particular interest in this respect (Hoest *et al.*, 1992; Leser *et al.*, 1992).

IgG antibodies to human proteins, nucleic acids, etc. are sometimes formed and involved in so-called autoimmune diseases. Measurement of antibodies against various autoantigens is part of the diagnostic investigation of these special forms of allergic diseases. UniCAP and ELISA are major assay techniques.

IgA ANTIBODY ASSAY

The presence of IgA antibodies against wheat gliadins in serum is one of the characteristics of celiac disease. Measurement of serum concentration is a major diagnostic method and an effective tool for monitoring of compliance to the gluten-free diet. Total avoidance of wheat and related cereals results in healing of the gut mucosa and disappearance of gliadin-specific IgA antibodies from serum (Chartrand *et al.*, 1997).

UniCAP Gliadin IgA correlates with clinical disease in children and adults (Ribes-Koninckx, 1998).

MARKERS OF CELL ACTIVATION

Several of the mediators released during allergic reactions can be measured by immunoassay. Tests for the mast-cell products, tryptase and histamine (as methyl histamine), and ECP are examples of commercially available assays. Tryptase is of particular interest in investigations of anaphylactic reactions caused by, e.g. drugs or food (Hogan and Schwartz, 1997). ECP is released by activated eosinophil leucocytes and is a marker for ongoing inflammation in asthmatic patients (Kristjánsson *et al.*, 1996).

STANDARDIZATION AND EVALUATION

Allergy tests are used to help in the diagnosis and monitoring of allergic patients. Assays need to be sensitive and specific and permit quantitative measurements over a wide range (Bernstein, 1988). As with all immunoassays, there is a need for common standards and definitions of performance parameters. In the case of total serum IgE, the WHO standard serves its purpose, but for the measurement of allergen-specific antibodies there is no official standard collection of units and there are no uniform definitions.

In reality, however, Phadebas RAST, first available in 1974, became a working standard against which all subsequent commercial tests were compared. Phadebas RAST includes a reference consisting of several dilutions of a serum containing IgE antibodies specific for birch pollen allergen. The dilutions were selected to permit the construction of a calibration curve and evaluation of test results in arbitrary units (PRU/mL) or semi-quantitative classes. This standard was internally calibrated against the WHO IgE standard (Lundkvist, 1975). In the new generations of RAST (Pharmacia CAP System and UniCAP) the birch reference was replaced by a direct substandard to the WHO standard 75/502 for IgE (Yman, 1990) covering the range 0.35–100 kU/L. Concentrations lower than the 0.35 kU/L calibrator are considered to be undetectable. Tests for all allergens are evaluated against this IgE standard.

All other commercial test kits largely adhere to this principle. The apparent agreement in terms of classes or units is, however, usually not supported by data defining the true concentrations of the calibrators. Direct comparisons are therefore difficult.

Modified scoring systems are used routinely by laboratories and physicians. Recent instrument and microprocessor technology development permits the storage of calibration curves for each reagent batch, although control sera should always be run with the patient samples under test. Multiallergen reagents, such as group-specific combinations or Phadiatop, are not designed to be quantitative and test results should be interpreted as detectable or undetectable by means of a clearly defined cutoff.

QUANTITATIVE MEASUREMENT OF ALLERGEN-SPECIFIC IgE ANTIBODIES

The IgE molecule has a large number of epitopes on the Fc part. Combinations of capture-anti-IgE and conjugate-anti-IgE can be found that make the calibrator IgE and allergen-bound IgE appear identical. To what extent this applies to a certain assay can only be judged on the basis of experimental data and not on general assumptions. In the case of the Pharmacia CAP System it has been shown in model experiments with several allergens that the measurement of allergen-bound IgE is equivalent to IgE protein (Figure 57.12).

This is possible because practically all the IgE antibodies of the sample are bound to the solid phase (Borgå *et al.*, 1992). For instance, in the case of the timothy ImmunoCAP, the capacity to bind IgE antibodies was shown to be 25 times greater than is required to bind

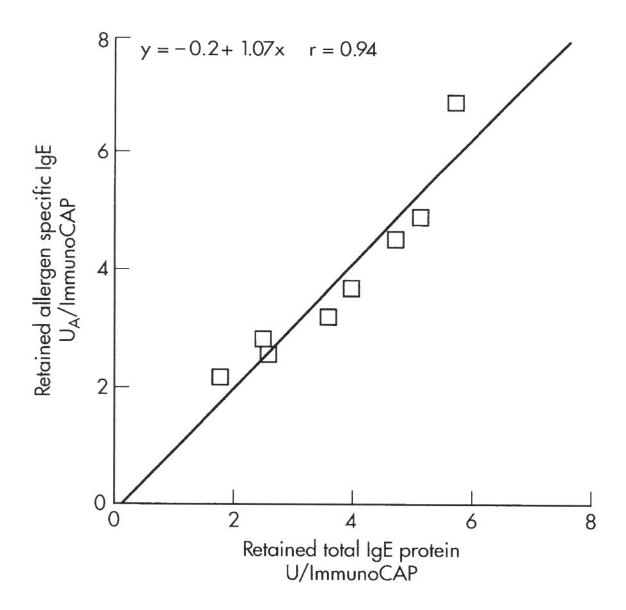

Fig. 57.12 Measurement of the change of concentration of IgE antibody and IgE protein after stepwise depletion of specific IgE antibody from serum. The allergen specific antibody unit (U_A) is equivalent to the international IgE unit (U) (Lindqvist *et al.*, 1995).

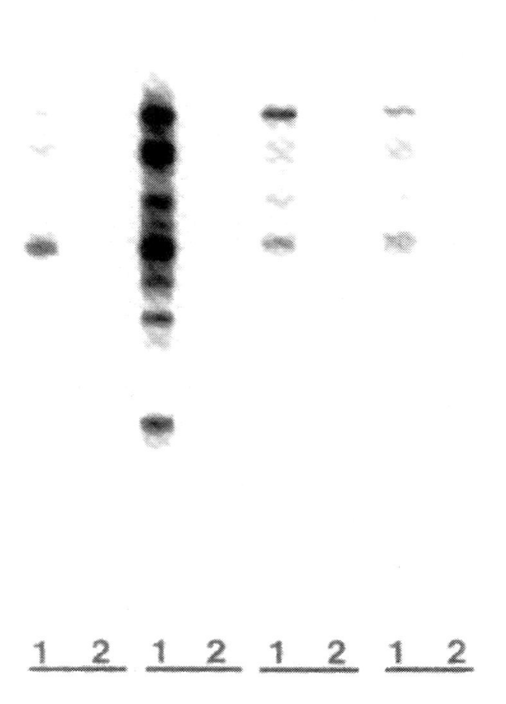

Fig. 57.13 Immunoblotting analysis of four sera with wheat-specific IgE antibodies before (1) and after (2) contact with a wheat ImmunoCAP. Antibodies of all subspecificities were bound to the solid-phase allergens.

all IgE antibodies at a concentration corresponding to the high end of the measuring range (100 kU_A/L). This is a large excess and referring to the Law of Mass Action it is clear that formation of allergen–IgE complex is favored. As mentioned above, the excess is large enough to make the antibody binding independent of the association constant of the interaction (the affinity), regardless of the IgE antibody concentration. Furthermore, if all the relevant allergenic components of the source are present on the solid phase and shown to be capable of binding IgE antibodies (Figure 57.13), the system is capable of measuring the sum of allergen-specific antibodies and satisfies the requirements of a quantitative assay (Perborn *et al.*, 1995).

REFERENCE VALUES

According to the rules of WHO for establishment of standards for biological materials an International Reference Preparation for IgE was established in 1968. The preparation was assigned arbitrary mass units (International Units) and submitted to collaborative trials. Later the International Unit was shown to correspond to 2.42 ng pure IgE by parallel chemical and immunological measurement (Bazaral and Hamburger, 1972). The International Unit is well established for total IgE measurements and it is evident that measurements of specific IgE antibodies also should move towards increasing standardization and that the antibody concentrations ultimately should be expressed in the same mass unit of IgE protein.

Measurable concentrations of specific IgE antibodies are not present in non-atopic individuals. Antibodies indicate sensitization and may be found in asymptomatic individuals who have shown symptoms earlier or who may develop them in the future. The serum concentration of allergen-specific IgE antibodies is correlated to the severity of the symptoms and to skin test sensitivity (Lichtenstein *et al.*, 1973). Antibodies to one allergen may constitute a major part of the total IgE in monosensitized patients or in the early phase of the development of severe disease (Gleich and Jacob, 1975).

Allergen-specific IgG antibodies are also present at variable levels in healthy individuals. Reference values are therefore not established. The concentration increases as a result of treatment in patients undergoing allergen immunotherapy.

Reference values for gliadin-specific IgA and IgG antibodies are currently method and age dependent. A standardization according to the principles established for IgE antibodies was recently introduced also for IgA and IgG antibodies (UniCAP).

QUALITY CONTROL

Adequate controls should be included in every assay. Universally available IgE antibody control sera with defined concentration and allergen specificity are preferred over in-house pools. National and international quality control programs provide regular assessment of reagent and laboratory performance (Fifield *et al.*, 1987).

The most common causes of error in RAST-type assays are inadequate washing and failure to remove washing solution effectively and evenly from all the reaction vessels. Measurement in fluids other than those specified in the directions for use should be avoided if possible. However, if this is essential, then proper negative and positive controls should be formulated and run, and dilution and recovery should be confirmed.

The quality documentation that should be produced for an IgE antibody assay by the manufacturer includes:

- quality and reproducibility of allergen source materials;
- specificity and dose–response characteristics of labeled anti-IgE;
- capacity and reproducibility of solid-phase binding;
- calibration;

- linearity and parallelism of calibrator and sample dilutions regardless of allergen;
- recovery;
- signal-to-noise ratio for the lowest calibrator;
- freedom of interference from non-specific IgE and competing IgG antibodies;
- high-dose hook data;
- stability of reagents and reagent combinations;
- reproducibility;
- correlation to reference method;
- clinical sensitivity and specificity calculated on relevant patient populations.

CLINICAL APPLICATIONS

DIAGNOSIS OF ATOPIC DISEASE AND IDENTIFICATION OF OFFENDING ALLERGEN

The clinical sensitivity and specificity of IgE antibody measurements should be judged in relation to the allergy specialist's diagnosis of groups of consecutive patients. The groups must be representative of the populations for which the test is intended in terms of prevalence of disease and degree of sensitization. Extensive clinical trials in accordance with this principle on more than 2000 individuals showed the average diagnostic efficiency of the Pharmacia CAP System to be about 90% (Yman, 1990). A new immunoassay system (UniCAP) was recently evaluated. Six clinical studies included 894 consecutive patients. The clinical sensitivity and specificity of UniCAP-specific IgE derived from 5170 comparisons to clinical diagnosis were 89 and 91%, respectively (Paganelli *et al.*, 1998).

A few examples of clinical applications are mentioned earlier in this chapter in relation to some important allergens.

PREDICTION AND MONITORING OF THE DEVELOPMENT OF ALLERGIC DISEASE

The severity of an atopic patient's symptoms is dependent on the degree of sensitization, the intensity of allergen exposure, the sensitivity of the target organs to the mediators at the time, medication, and a number of other factors. IgE antibody concentration in serum has been shown to correlate with the severity of the symptoms and the sensitivity to allergen challenge in the skin and nasal mucosa. A quantitative measure of IgE antibody provides a measure of the degree of sensitization independent of the short-term fluctuations of the patient's status. Studies with the Pharmacia CAP System suggest that the wide measuring range further increases the value of IgE antibody measurements by providing a basis for the choice of therapy depending on the IgE concentration (Lindholm and Bjärneman, 1988).

The early appearance of food-specific IgE in very young children and the relation to skin problems is well documented (Rowntree *et al.*, 1985; Saarinen and Kajosaari, 1995). Recent studies (Sasai *et al.*, 1996) confirm this and show that measurable (>0.35 kU_A/L) egg white-specific IgE at 6 months is an effective predictor of development of allergic disease to common Japanese inhalant allergens, e.g. the house dust mite, during the first 5 years of life. The positive and negative predictive values are both 85%. Attempts to measure below 0.35 kU_A/L did not contribute further information of value.

Elimination of egg and milk in children with atopic dermatitis resulted in decreases of allergen-specific IgE antibody as well as the symptoms. Specific IgE antibodies to food were shown to be useful indexes of the effect of elimination diets in food-sensitive patients (Agata *et al.*, 1993).

ECP, finally, was proposed for monitoring the effectiveness of environmental control in the homes of allergic patients (Boner *et al.*, 1993).

REFERENCES

Agata, H., Kondo, N., Fukutomi, O., Shinoda, S. and Orii, T. Effect of elimination diets on food-specific IgE antibodies and lymphocyte proliferative responses to food antigens in atopic dermatitis patients exhibiting sensitivity to food allergens. *J. Allergy Clin. Immunol.* **91**, 668–679 (1993).

Alonso, M.D., Dávila, I., Conde Salazar, L., Cuevas, M., Martin, J.A., Guimaraens, M.D. and Losada, J.A. Occupational protein contact dermatitis from herring. *Allergy* **48**, 349–352 (1993).

Ammann, B. and Wüthrich, B. Bedeutung der tierepithelien als 'Hausstauballergene'. *Dtsch. Med. Wochenschr.* **110**, 1239–1245 (1985).

Anderson, M.C. and Baer, H. Allergenically active components of cat allergen extracts. *J. Immunol.* **127**, 972–975 (1981).

Añibarro Basuela, B., Martín Esteban, M., Martínez Alsamora, F., Pascual Marcos, C. and Ojeda Casas, J.A. Egg protein sensitization in patients with bird feather allergy. *Allergy* **46**, 614–618 (1991).

Audicana, M.T., Fernández de Corres, L., Muños, D., Fernández, E., Navarro, J.A. and del Pozo, M.D. Recurrent anaphylaxis caused by *Anisakis simplex* parasitizing fish. *J. Allergy Clin. Immunol.* **96**, 558–560 (1995).

Bahna, S.L. and Yee, E.H. Comparison of RAST by different manufactures for IgE to molds. *J. Allergy Clin. Immunol.* **77**, 242 (1986).

Baldo, B.A. and Harle, D.G. Drug allergenic determinants. *Monogr. Allergy* **28**, 11–51 (1990).

Bazaral, M. and Hamburger, R.N. Standardization and stability of immunoglobulin E (IgE). *J. Allergy Clin. Immunol.* **49**, 189–191 (1972).

Beaumont, F., Kauffman, H.F., de Monchy, J.G.R., Sluiter, H.J. and de Vries, K. Volumetric aerobiological survey of conidial fungi in the north-east Netherlands. *Allergy* **40**, 181–186 (1985).

Beezhold, D.H., Sussman, G.L., Liss, G.M. and Chang, N.S. Latex allergy can induce clinical reactions to specific foods. *Clin. Exp. Allergy* **26**, 416–422 (1996).

Bernhisel-Broadbent, J., Scanlon, S.M. and Sampson, H.A. Fish hypersensitivity. *J. Allergy Clin. Immunol.* **89**, 730–737 (1992).

Bernstein, I.L. Proceedings of the task force on guidelines for standardizing old and new technologies used for diagnosis and treatment of allergic diseases. *J. Allergy Clin. Immunol.* **82**, 487–526 (1988).

Bernstein, I.L. and Safferman, R. Clinical sensitivity to green algae demonstrated by nasal challenge and *in vitro* tests of immediate hypersensitivity. *J. Allergy Clin. Immunol.* **51**, 22–28 (1973).

Boguniewicz, M. Adverse reactions to antibiotics. Is the patient really allergic? *Drug Saf.* **13**, 273–280 (1995).

Boner, A.L., Peroni, D.G., Piacentini, G.L. and Venge, P. Influence of allergen avoidance at high altitude on serum markers of eosinophil activation in children with allergic asthma. *Clin. Exp. Allergy* **23**, 1021–1026 (1993).

Borgå, Å., Karlsson, T., Perborn, H., Lindqvist, A. and Yman, L. Standardization of timothy, peanut and *Aspergillus fumigatus* allergens for immunoassay of specific IgE (Pharmacia CAP System). *J. Allergy Clin. Immunol.* **89** (No. 1, Part 2), 148 (1992).

Brandt, R. and Yman, L. Dog dander allergens. Specificity studies based on the Radioallergosorbent technique. *Int. Arch. Allergy Appl. Immunol.* **61**, 361–370 (1980).

Butcher, B.T., O'Neil, C.E., Reed, M.A. and Salvaggio, J.E. Radioallergosorbent testing of toluene di-isocyanate reactive individuals using *p*-tolyl isocyanate antigen. *J. Allergy Clin. Immunol.* **66**, 213–216 (1980).

Chapman, M.D., Smith, A.M., Vailes, L.D. and Arruda, L.K. Defined epitopes: *in vivo* and *in vitro* studies using recombinant allergens. *Int. Arch. Allergy Immunol.* **113**, 102–104 (1997).

Chartrand, L.J., Agulnik, J., Vanounou, T., Russo, P.A., Baehler, P. and Seidman, E.G. Effectiveness of antigliadin antibodies as a screening test for celiac disease in children. *CMAJ* **157**, 527–533 (1997).

Coombs, R.R.A. and Gell, P.G.H. Classification of allergic reactions responsible for clinical hypersensitivity and disease. In: *Clinical Aspects of Immunology* (eds Gell, P.G.H., Coombs, R.R.A. and Lachman, P.J.), 317 (Blackwell, Oxford, 1968).

de Groot, H., de Jong, N.W., Wuijk, M.H. and van Wijk, G. Birch pollinosis and atopy caused by apple, peach and hazelnut; comparison of three extraction procedures with two apple strains. *Allergy* **51**, 712–718 (1996).

Delbourg, M.F., Guilloux, L., Moneret-Vautrin, D.A. and Ville, G. Hypersensitivity to banana in latex-allergic patients. Identification of two major banana allergens of 33 and 37 kD. *Ann. Allergy Asthma Immunol.* **76**, 321–326 (1996).

del Pozo, M.D., Moneo, I., Fernández de Corres, L., Audicana, M.T., Muños, D., Fernández, E., Navarro, J.A. and Garcia, M. Laboratory determinations in *Anisakis simplex* allergy. *J. Allergy Clin. Immunol.* **97**, 977–984 (1996).

de Martino, M., Novembre, E., Galli, L., de Marco, A., Botarelli, P., Marano, E. and Vierucci, A. Allergy to different fish species in cod-allergic children: *in vivo* and *in vitro* studies. *J. Allergy Clin. Immunol.* **86**, 909–914 (1990).

Deuschl, H. and Johansson, S.G.O. Specific IgE antibodies in nasal secretions from patients with allergic rhinitis and with negative or weakly positive RAST in the serum. *Clin. Allergy* **7**, 195–202 (1977).

Eriksson, N.E. Allergy screening with Phadiatop® and CAP Phadiatop® in combination with a questionnaire in adults with asthma and rhinitis. *Allergy* **45**, 285–292 (1990).

Fifield, R., Bird, A.G., Carter, R.H., Ward, A.M. and Whicher, J.T. Total IgE and allergen-specific IgE assays: guidelines for the provision of a laboratory service. *Ann. Clin. Biochem.* **24**, 232–245 (1987).

Foucard, T., Edberg, U. and Malmheden Yman, I. Fatal and severe food hypersensitivity. Peanut and soya underestimated allergens. *Läkartidningen* **94**, 2635–2638 (1997).

Gleich, G.J. and Jacob, G.L. Immunoglobulin E antibodies to pollen allergens account for high percentages of total immunoglobulin E protein. *Science* **190**, 1106–1108 (1975).

Gravesen, S. Identification and quantitation of indoor airborne micro-fungi during 12 months from 44 Danish homes. *Acta Allergol.* **27**, 337–354 (1972).

Helbling, A., McCants, M.L., Musmand, J.J., Schwarts, H.J. and Lehrer, S.B. Immunopathogenesis of fish allergy: identification of fish-allergic adults by skin test and radioallergosorbent test. *Ann. Allergy* **77**, 48–54 (1996).

Hoest, A., Husby, S., Gjesing, B., Larsen, J.N. and Loewenstein, H. Prospective estimation of IgG. IgG subclass and IgE antibodies to dietary proteins in infants with cow milk allergy – levels of antibodies to whole milk protein, BLG and ovalbumin in relation to repeated milk challenge and clinical course of cow milk allergy. *Allergy* **47**, 218–229 (1992).

Hogan, A.D. and Schwartz, L.B. Markers of mast cell degranulation. *Methods* **13**, 43–52 (1997).

Johansson, S.G.O. and Yman, L. *In vitro* assays for immunoglobulin E. *Clin. Rev. Allergy* **6**, 93–139 (1988).

Jonsson, P., Rolfsen, W. and Yman, L. Mesure des IgE spécifiques dirigées contre 16 genres de moisissures très répandus. Amélioration de l'efficacité du diagnostic de l'allergie aux moisissures. *Allergie Immunol.* **19**, 407–409 (1987).

Kang, B. and Chang, J.L. Allergenic impact of inhaled arthropod material. *Clin. Rev. Allergy* **3**, 363–375 (1985).

Kari, O., Salo, O.P., Björksten, F. and Backman, A. Allergic conjunctivitis, total and specific IgE in the tear fluid. *Acta Ophthalmol.* **63**, 97–99 (1985).

Kelso, J.M., Jones, R.T. and Yunginger, J.W. Monospecific allergy to swordfish. *Ann. Allergy Asthma Immunol.* **77**, 227–228 (1996).

King, T.P., Hoffman, D., Loewenstein, H., Marsch, D.G. and Platts-Mills, T.A.E. Allergen nomenclature. *J. Allergy Clin. Immunol.* **96**, 5–14 (1995).

Komase, Y., Sakata, M., Azuma, T., Tanaka, A. and Nakagawa, T. IgE antibodies against midge and moth found in Japanese asthmatic subjects and comparison of allergenicity between these insects. *Allergy* **52**, 75–81 (1997).

Kraft, D., Roth, A., Mischer, P., Pichler, H. and Ebner, H. Specific and total serum IgE measurements in the diagnosis of penicillin allergy. A long term follow up study. *Clin. Allergy* **7**, 21–28 (1977).

Kristjánsson, S., Strannegård, I.L. and Wennergren, G. Inflammatory markers in childhood asthma. *Ann. Med.* **28**, 395–399 (1996).

Leser, C., Kauffman, H., Virchow, C. and Menz, G. Specific serum immunopatterns in clinical phases of allergic bronchopulmonary aspergillosis. *J. Allergy Clin. Immunol.* **90**, 589–599 (1992).

Levine, B.B. and Zolov, D.M. Prediction of penicillin allergy by immunological tests. *J. Allergy Clin. Immunol.* **43**, 231–244 (1969).

Levy, D. and Leynadier, F. Latex and food allergy. *Rev. Fr. Allergol.* **37**, 1188–1194 (1997).

Lewis, W.H. and Elvin-Lewis, M.P.F. *Medical Botany: Plants Affecting Man's Health* (Wiley-Interscience, New York, 1977).

Lichtenstein, L.M., Ishizaka, K., Norman, P.S., Sobotka, A.K. and Hill, B.M. IgE antibody measurements in ragweed hayfever. Relationship to clinical severity and the result of immunotherapy. *J. Clin. Invest.* **52**, 472–482 (1973).

Lindholm, N.B. and Bjärneman, P. Severity of birch allergy in relation to specific IgE measured with a new test. *N. Engl. Reg. Allergy Proc.* **9**, 422 (1988).

Lindqvist, A., Ikezawa, Z., Tanaka, A. and Yman, L. Seafood specific IgE in atopic dermatitis. *Ann. Allergy* **70**, 58 (1993).

Lindqvist, A., Maaninen, E., Zimmerman, K., Rimland, A., Andersson, O., Holmquist, I., Karlsson, T. and Yman, L. Quantitative measurement of allergen specific IgE antibodies applied in a new immunoassay system, UniCAP. In: *XVI European Congress of Allergology and Clinical Immunology ECACI 95* (eds Basomba, A., Hernandez, F. and de Rojas, M.D.), 195–200 (Monduzzi Editore, Bologna, 1995).

Lundkvist, U. Research and development of the RAST technology. In: *Advances in Diagnosis of Allergy: RAST* (ed Evans, R.), 85–99 (Symposia Specialists, Miami, 1975).

Lutsky, I., Fink, J.A., Hoffman, D.R. and Morouse, M. Skin test reactivity to dog-derived antigens. *Clin. Allergy* **10**, 331–340 (1980).

Maccia, C.A., Bernstein, I.L., Emmett, E.A. and Brooks, S.M. *In vitro* demonstration of specific IgE in phthalic anhydride hypersensitivity. *Am. Rev. Respir. Dis.* **113**, 701–704 (1976).

Malmheden Yman, I. Se upp med sojaprotein och jordnötter! Allvarliga reaktioner på livsmedel (Watch out for soy protein and peanuts! Severe reactions to food). *Vår Föda* **5**, 8–9 (1997).

Montoro, A., Perteguer, M.J., Chivato, T., Laguna, R. and Cuéllar, C. Acute recidivant urticaria caused by *Anisakis simplex*. *Allergy* **51**, 27 (1996).

Ohman, J.L. Jr., Bloch, K.J. and Kendall, S. Allergens of mammalian origin. IV. Evidence for common allergens in cat and dog serum. *J. Allergy Clin. Immunol.* **57**, 560–568 (1976).

Okudaira, H., Ito, K., Miyamoto, T., Wagatsuma, Y., Matsuyama, R., Kobayashi, S., Nakazawa, T., Okuda, M., Otsuka, H., Baba, M., Iwasaki, E., Takahashi, T., Adachi, M., Kokubu, F., Nishima, S., Shibata, R., Yoshida, H. and Maeda, K. Evaluation of a new system for the detection of IgE antibodies (CAP) in atopic disease. *Arerugi* **40**, 544–554 (1991).

Paganelli, R., Ansotegui, I.J., Sastre, J., Lange, C.-E., Roovers, M.H.W.M., de Groot, H., Lindholm, N.B. and Ewan, P.W. Specific IgE antibodies in the diagnosis of atopic disease. Clinical evaluation of a new *in vitro* test system, UniCAP, in six European allergy clinics. *Allergy* **53**, 763–768 (1998).

Pascual, C., Larramendi, C.H., Martín Esteban, M., Fiandor, A. and Ojeda, J.A. Fish allergy and fish allergens. *J. Allergy Clin. Immunol.* **81**, 264 (1988).

Pascual, C., Martín Esteban, M. and Fernandez Crespo, J. Fish allergy: evaluation of the importance of cross-reactivity. *J. Pediatr.* **121**, S29–S34 (1992).

Perborn, H. *et al.* Standardization of allergen reagents for immunoassay of allergen-specific IgE antibodies (UniCAP). Allergen excess and component specificity. In: *XVI European Congress of Allergology and Clinical Immunology ECACI'95* (eds Basomba, A., Hernandez, F. and de Rojas, M.D.), 191–194 (Monduzzi Editore, Bologna, 1995).

Persson, E. and Yman, L. Unpublished data (1997).

Peterman, J.H. Immunochemical considerations in the analysis of data from non-competitive solid-phase immunoassays. In: *Immunochemistry of Solid-Phase Immunoassay* (ed Butler, J.E.), 47–65 (CRC Press, Boca Raton, FL, 1991).

Portnoy, J.M., Williams, B.P. and Siegel, C.J. Personal communication (1997).

Restani, P., Fiocchi, A., Beretta, B., Velonà, T., Giovannini, M. and Galli, C.L. Meat allergy: III – proteins involved and cross-reactivity between different animal species. *J. Am. Coll. Nutr.* **16**, 383–389 (1997).

Ribes-Koninckx, C. Personal communication (1998).

Ring, J., Abeck, D. and Neuber, K. Atopic eczema – role of microorganisms on the skin surface. *Allergy* **47**, 265–269 (1992).

Rowntree, S., Cogswell, J.J., Platts-Mills, T.A.E. and Mitchell, E.B. Development of IgE and IgG antibodies to food and inhalant allergens in children at risk of allergic disease. *Arch. Dis. Child.* **60**, 727–735 (1985).

Saarinen, U.M. and Kajosaari, M. Breastfeeding as prophylaxis against atopic disease: prospective follow up study until 17 years old. *Lancet* **346**, 1065–1069 (1995).

Sabbah, A., Lauret, M.G., Chene, J., Boutet, S. and Drouet, M. The pork–cat syndrome or crossed allergy between pork meat and cat epithelia. *Allerg. Immunol. Paris* **26**, 177–180 (1994).

Sainte Laudy, J., Couturier, P. and Basset-Sthème, D. Intérêt des dosage lacrymaux (IgE totales, IgE spécifiques et albumine) pour l'exploration des conjuctivites allergiques. *Allerg. Immunol.* **26**, 95–96 (1994).

Sasai, K., Furukawa, S., Sugawara, T., Kaneko, K., Baba, M. and Yabuta, K. IgE levels in faecal extracts of patients with food allergy. *Allergy* **47**, 594–598 (1992).

Sasai, K., Furukawa, S., Muto, T., Baba, M., Yabuta, K. and Fukuwatari, Y. Early detection of specific IgE antibody against house dust mite in children at risk of allergic disease. *J. Pediatr.* **128**, 834–840 (1996).

Sensi, L.G., Piacentini, E., Nobile, E., Ghebregzagher, M., Brunori, R., Zanolla, L., Boner, A.L. and Marcucci, F. Changes in nasal specific IgE to mites after periods of allergen exposure-avoidance: a comparison with serum levels. *Clin. Exp. Allergy* **24**, 377–382 (1994).

Sigurs, N., Hildebrand, H., Hultkvist, K., Litwin, E., Malmqvist, L.-Å., Sandahl, G., Lothe, L. and Svenonius, E. Atopy in childhood identified by Phadebas RAST, serum IgE and Phadiatop. *Allergy* **43** (Suppl. 7), 8 (1988).

Surtees, S.J., Stockton, M.G. and Gietzen, T.W. Allergy to penicillin: fable or fact? *Br. Med. J.* **302**, 1051–1052 (1991).

Svensson, C., Andersson, M., Persson, C.G.A., Venge, P., Alkner, V. and Pipkorn, U. Albumin, bradykinins and eosinophil cationic protein on the nasal mucosal surface in patients with hay fever during natural allergen exposure. *J. Allergy Clin. Immunol.* **85**, 828–833 (1990).

Swärd-Nordmo, M., Smedstad Paulsen, B. and Wold, J.K. Immunological studies of the glycoprotein allergen Ag-54 (Cla h II) in *Cladosporium herbarum* with special attention to the carbohydrate and protein moieties. *Int. Arch. Allergy Appl. Immun.* **90**, 155–161 (1989).

Turjanmaa, K., Alenius, H., Mäkinen-Kiljunen, S., Reunala, T. and Palosuo, T. Natural rubber latex allergy. *Allergy* **51**, 593–602 (1996).

Vanto, T., Viander, M., Koiviko, A., Schwartz, B. and Loewenstein, H. RAST in the diagnosis of dog dander allergy. *Allergy* **37**, 75–85 (1982).

Weber, R.W. and Nelson, H.S. Pollen allergens and their interrelationships. *Clin. Rev. Allergy* **3**, 291–318 (1985).

Williams, P.B., Nolte, H., Dolen, W.K. *et al.* The histamine content of allergen extracts. *J. Allergy Clin. Immunol.* **89**, 738–745 (1992).

Wraith, D., Merrett, J., Roth, A., Yman, L. and Merrett, T. The recognition of food allergic patients and their allergens by the RAST technique and clinical investigation. *Clin. Allergy* **9**, 25–36 (1979).

Wüthrich, B., Guerin, B. and Hewitt, B. Cross-allergenicity between extracts of hair from different dog breeds and cat fur. *Clin. Allergy* **15**, 87–93 (1985).

Yman, L. *Botanical Relations and Immunological Cross-reactions in Pollen Allergy* (Pharmacia Diagnostics, Uppsala, 1982).

Yman, L. Die neue generation der allergie-testung: Pharmacia CAP system. *In-vitro Diagn. Spec.* **1**, 18–22 (1990).

Yman, L. Quantitative measurement of allergen-specific IgE antibody. A theoretical model supported by Pharmacia CAP System data. *Allergy Clin. Immunol. News* (Suppl 2), 495 (1994).

Yman, L. and Lundberg, M. Serological aspects of latex allergy. Some recent developments. *Rev. Fr. Allergol.* **37**, 1195–1200 (1997).

Yman, L., Borgå, Å. and Bengtsson, A. *Dermatophagoides*-specific IgE antibodies. In: *International Symposium on Mite and Midge Allergy* (Ministry of Education, Science and Culture, Tokyo, 1988a).

Yman, L., Rolfsen, W. and Malmheden-Yman, I. Food additives from the legume family (Leguminosae/ Fabaceae). A potential allergy risk. *Allergy* **43**, 81 (1988b).

Zetterström, O. and Johansson, S.G.O. IgE concentrations measured by PRIST® in serum of healthy adults and in patients with respiratory allergy. A diagnostic approach. *Allergy* **36**, 537–547 (1981).

Zimmerman, B. and Forsyth, S. Diagnosis of allergy in different age groups of children: use of mixed allergen RAST discs. Phadiatop and Paediatric mix. *Clin. Allergy* **18**, 581–587 (1988).

Zimmerman, B., Forsyth, S. and Gold, M. Highly atopic children: formation of IgE antibody to food protein, especially peanut. *J. Allergy Clin. Immunol.* **83**, 764–770 (1989).

58 Autoimmune Disease

David F. Keren

The term 'autoimmune disease' implies the presence of a well-defined entity where an aberrant reaction of the immune system with a specific organ or tissues has resulted in the development of a disease. This is rarely true. In reality, the clinical conditions of individuals with autoimmune diseases are extremely heterogeneous. Although they do get lumped together in chapters such as this, they have little in common except that one is able to demonstrate some sort of autoreactive phenomenon – usually autoantibodies in their serum. Indeed, a clear-cut causal relationship between the autoreactive phenomenon and the pathogenesis of the disease is uncommon. Such a relationship is found in Goodpasture's syndrome, where an autoantibody directed to the glomerular basement membrane is logically related to the glomerular disease, and perhaps to the respiratory disease, that these patients experience. However, in the vast majority of autoimmune diseases, the autoantibodies demonstrated in serum are only indirectly, or not at all, related to the pathologic lesions. A good example is primary biliary cirrhosis, where patients suffer progressive damage to their bile ducts that eventuates in liver failure. Yet, their serum does not contain antibody or cellular reactivity against bile ducts. It contains antibodies that react with mitochondria (anti-mitochondrial antibody – AMA). These antibodies are not specific for mitochondria in bile ducts. Because mitochondria are not on the cell surface, they are resistant to exposure to these antibodies. Therefore, although the presence of AMA is useful to the clinician in confirming the diagnosis of primary biliary cirrhosis, it does not provide information about the pathogenesis of the disease, and serves as an example of the inappropriate assumptions that are inherent in the use of the term 'autoimmune disease'. Is primary biliary cirrhosis caused by the immune system reacting with the bile ducts? It is not known. It is not even likely. Yet, because of the presence of AMA in the serum, primary biliary cirrhosis is classified as an autoimmune disease.

Because the body is not expected to develop immune reactivity to its own components, a wide variety of theories and models have been developed to explain how this may come about. This is not a new concept. Indeed, Paul Erlich referred to the ability of the immune system to react against an individual's own tissues as 'horror autotoxicus'. With our more recent information about the subtypes of cells within the immune system, a better understanding has evolved about how such self-reactivity may occur. The immune system has built-in checks and balances between the T and B lymphocytes by which the body develops tolerance to its own antigens. Some of the T cells are suppressor cells that inhibit the formation of immune reactivity against a variety of antigens, including self-antigens. The concepts of how the body's immune system is able to escape from this 'tolerance' to self-antigens and produce antibodies and/or cell-mediated immunity against its own cells are wide-ranging and beyond the scope of this chapter. Some recent review articles are recommended for those interested in more details (*see* FURTHER READING).

However, for the purposes of using the presence of autoantibodies for clinical diagnosis, it is not necessary to know why they are formed, nor is it necessary to assume that they are involved in the pathogenesis of the disease. They are merely serving as markers of the diseases described in this chapter. But even as markers of disease, there is considerable variation in the use of autoantibody assays. This is because a certain amount (albeit very low levels) of autoreactivity exists in all of us. Perhaps antibodies against nuclear antigens, for example, are a normal part of tissue repair, aging, cell renewal, or a reaction to microbial DNA. In any event, if we had very sensitive assays, we would detect the presence of low levels of many types of autoantibodies in clinically well individuals. For such autoantibodies to be useful in detecting diseases, they need to be present in relatively high titers compared to the control populations.

When reviewing the various tests listed below, it is necessary to pay particular attention to the antigen preparation used. For instance, it is possible to speak glibly about anti-nuclear antibody (ANA) reactivity, but what does it really mean? There are a plethora of proteins and nucleic acids in the nucleus that can serve as antigens in the ANA test. Therefore, a positive ANA means reactivity against something in the nucleus, but does not give the antigen specificity of the autoantibody. For this reason, the ANA test is a general screen. When the assay is

The Immunoassay Handbook. *Third Edition*
D. Wild (Ed.)

employed on symptomatic patients with an appropriate dilution of serum (1:80 in our laboratory), a negative result makes it highly unlikely that the patient has systemic lupus erythematosus (SLE) (95–98% of patients with SLE have a positive ANA). Therefore, the test has good sensitivity. But when used on all patients with minor aches and pains, or with serum that is not diluted sufficiently, a large number of normal people have a positive ANA. Therefore, the ANA test does not have good specificity. A lack of understanding about the limitations of tests such as ANA leads to misunderstanding about its utility, and a global distrust of the laboratory.

Lastly, all ANA screening tests are not identical. When frozen sections of animal tissue substrates are used, different results can be obtained than when tissue culture monolayers are used for these indirect immunofluorescence (IIF) assays. Furthermore, the type of fixative used affects the results. Results obtained from IIF assays may differ from those obtained with the new enzyme immunoassay (EIA) ANA screens that employ mixtures of several nuclear antigens. Results obtained from one manufacturer's EIA ANA screen differ from those of another manufacturer who uses a different mixture. Yet, no matter what method is used, the final result usually simply appears on the patient's chart as 'ANA positive' or 'ANA negative'. It is very difficult for the clinician to be aware of all the nuances of the above testing variations. Therefore, a laboratory must be prepared to advise clinicians about appropriate screening tests, and to answer questions about possible limitations and helpful further testing. We keep a knowledgeable pathologist available 24 h a day to help the clinician interpret this information, and to guide further testing, if it is needed.

ANALYTES

ANTI-NUCLEAR ANTIBODIES

Anti-nuclear antibodies (ANA) are extremely heterogeneous antibodies that are directed against components of the nucleus. Detection of these antibodies is used as a screen for **systemic lupus erythematosus** (SLE) and other connective tissue (rheumatoid) diseases.

Reference Interval

For an IIF autoantibody test, the ideal reference interval is no detectable nuclear reactivity at the titer chosen by the individual laboratory as its cut-off. That cut-off varies from laboratory to laboratory and from one substrate to another. Some laboratories use the substrate manufacturer's suggested cut-off, typically 1:20 or 1:40. However, a standard cut-off for one kit does not take into consideration variables inherent to each laboratory. For instance, the type of microscope, its light source, optics, and condenser play some role in the detection of ANA. Other laboratories establish their cut-off by studying normal individuals of appropriate age and sex for the population tested. A 5–10% positive rate among control individuals is chosen as the typical cut-off. In our laboratory, for instance, we have established a cut-off of 1:80. The disparity from one laboratory to another has made comparisons of samples from the same patient performed at different laboratories difficult and laboratorians need to be aware of it.

For the EIA ANA screening methods, the reference interval is less than a defined optical density (OD) for a particular lot of a specific company's kit. Although some studies have found that results from EIA screening kits compare well with standard IIF ANA kits, there is no standard EIA antigen preparation and, therefore, OD numbers vary from laboratory to laboratory and kit to kit. Further, little independent information is available in the literature about lot-to-lot variation. The ease in automation of this technique may encourage manufacturers to standardize the antigen preparations.

Clinical Applications

Typically, the ANA test is used to screen for the presence of SLE, but antibodies that react with the nucleus are present in a wide variety of other conditions including: mixed connective tissue disease, scleroderma, Sjögren's syndrome, drug-induced lupus, polymyositis, dermatomyositis, Raynaud's phenomenon, and in as many as 10% of normal individuals. The absence of ANA is useful in ruling out SLE because it is found in 95–98% of patients with that disease. However, its presence cannot by itself establish that diagnosis. Multiple specific clinical features must be present along with specific antibodies to define the presence of SLE in an individual. As described below, antibodies to many of the different antigenic components of the nucleus can now be distinguished with specific tests. Some of them, such as antibodies against Sm and DNA, are highly specific for SLE. When the IIF test for ANA is used, a laboratory usually reports a pattern of reactivity in addition to its titer (Table 58.1). Before the advent of specific antigenic testing, the pattern of reactivity was used as a determinant of subsequent testing. Now the presence of specific antibody tests is

Table 58.1. Reactivity of ANA on HEp-2 substrate.

Pattern	Specificities
Homogeneous/ peripheral	dsDNA, ssDNA, histone, DNP
Speckled	Sm, RNP, SSA/Ro, SSB/La, Scl-70
Centromere	Centromere
Nucleolar	Nucleolar RNA

the focus of the serologic evaluation of patients with positive ANA testing.

Limitations

The ANA screen available today has replaced the older lupus erythematosus (LE) cell preparation. Although ANA is quite sensitive for detecting SLE, it is highly nonspecific. There does not seem to be any difference in specificity between the immunofluorescence method and the EIA method. Many normal individuals have a positive ANA. ANA reactivity has also been found in individuals with nonspecific rheumatologic complaints. Sometimes, the titer of control individuals can be relatively high (such as 1:640 with a cut-off of 1:80). Therefore, patient selection prior to testing (i.e. symptomatic, not general screening) and subsequent specific antibody testing are critical to its effective use.

Assay Technology

Most laboratories use an IIF assay (Figure 58.1). IIF involves incubation of the patient's serum on a substrate that contains nuclei. There are many substrates that are used, but they fall into two basic categories: frozen sections of rodent tissues and tissue culture cells. The tissue culture substrates (especially the HEp-2 cell line) are the most sensitive and in the past 10 years have become the choice of most laboratories. Frozen section substrates derive from mouse kidney or rat liver. The frozen section substrates do not represent the centromere antigens to advantage because they contain few dividing cells, nor do they preserve the important SSA/Ro antigen as well as HEp-2 substrates. With either substrate, following incubation with the patient's serum, the substrate is washed thoroughly with buffer and then

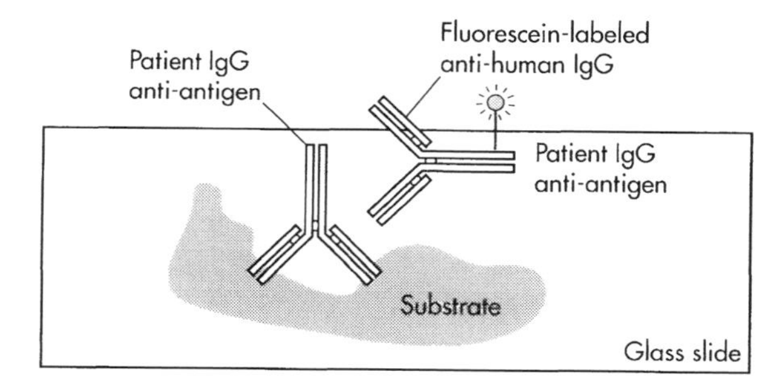

Fig. 58.1 Indirect immunofluorescence technique. A substrate containing the antigen of interest is placed on a glass slide. The patient's antibodies react with the antigen. After a wash step, a reagent (often mouse or rabbit) anti-human IgG that is labeled with fluorescein is added. Following another wash step, a coverslip is applied to the substrate and the slide is examined with a microscope equipped for darkfield fluorescence.

covered with fluorescein-conjugated antibody against human IgG. This binds with the patient's IgG that has bound nuclear antigens. After this incubation, the substrate is washed with buffer, a coverslip is applied to the substrate on the slide, and it is examined by fluorescence microscopy. The presence of reactivity against the nuclear antigen is recorded along with the pattern of fluorescence and the strength of the reaction (indicated by the reciprocal of the highest dilution of the patient's serum that gives reactivity). One variation of this technique employs an enzyme-conjugated anti-human immunoglobulin that gives a colored reaction product. While this obviates the need for a fluorescence microscope, it requires an additional substrate step that slows the overall process. Since most laboratories that perform the ANA test have fluorescence microscopes for other assays and are familiar with its use, this technique is not widely used.

The pattern of nuclear fluorescence has traditionally been used to indicate the specificity of antibody present. Homogeneous and peripheral reactivity with the nuclei are the patterns most strongly associated with SLE because they can be a result of antibody directed against double-stranded DNA. Unfortunately, the specificity of this reactivity is often overstated. Antibodies against other nuclear antigens, such as histones, can produce this pattern. Because anti-histone may be associated with transient phenomena like drug-induced lupus, one should not ascribe too much importance to pattern detection. The nucleolar pattern has been weakly associated with scleroderma, but there are much better specific tests such as Scl-70 and centromere antibodies for this disease (see below). The fourth pattern is the speckled pattern. Attempts to subdivide the speckled pattern into fine speckles, coarse speckles, or other such descriptions have been impossible to reproduce due to interpretive subjectivity. Now that specific antibody tests are available to identify different clinically significant antibodies that can result in the speckled pattern, the pattern of reactivity is much less important.

EIA as an ANA screen is relatively new. An antigen preparation that includes clinically relevant nuclear components such as DNA, Sm, RNP, SSA/Ro, SSB/La, among others is used to coat a solid surface (usually a microtiter well). The antigen preparations vary from one company to another in concentrations of each of the above and are, therefore, not standardized. The EIA procedure is a standard, easily automated serology technique. Serum diluted in buffer reacts with the antigen-coated solid surface followed by a wash step. An enzyme-conjugated anti-human immunoglobulin preparation reacts with any immunoglobulin bound to the surface of the well by the nuclear antigens (Figure 58.2). Following another wash step, the chromogenic (color producing) substrate for the enzyme is added and the color produced is compared to that of standard positive and negative solutions provided.

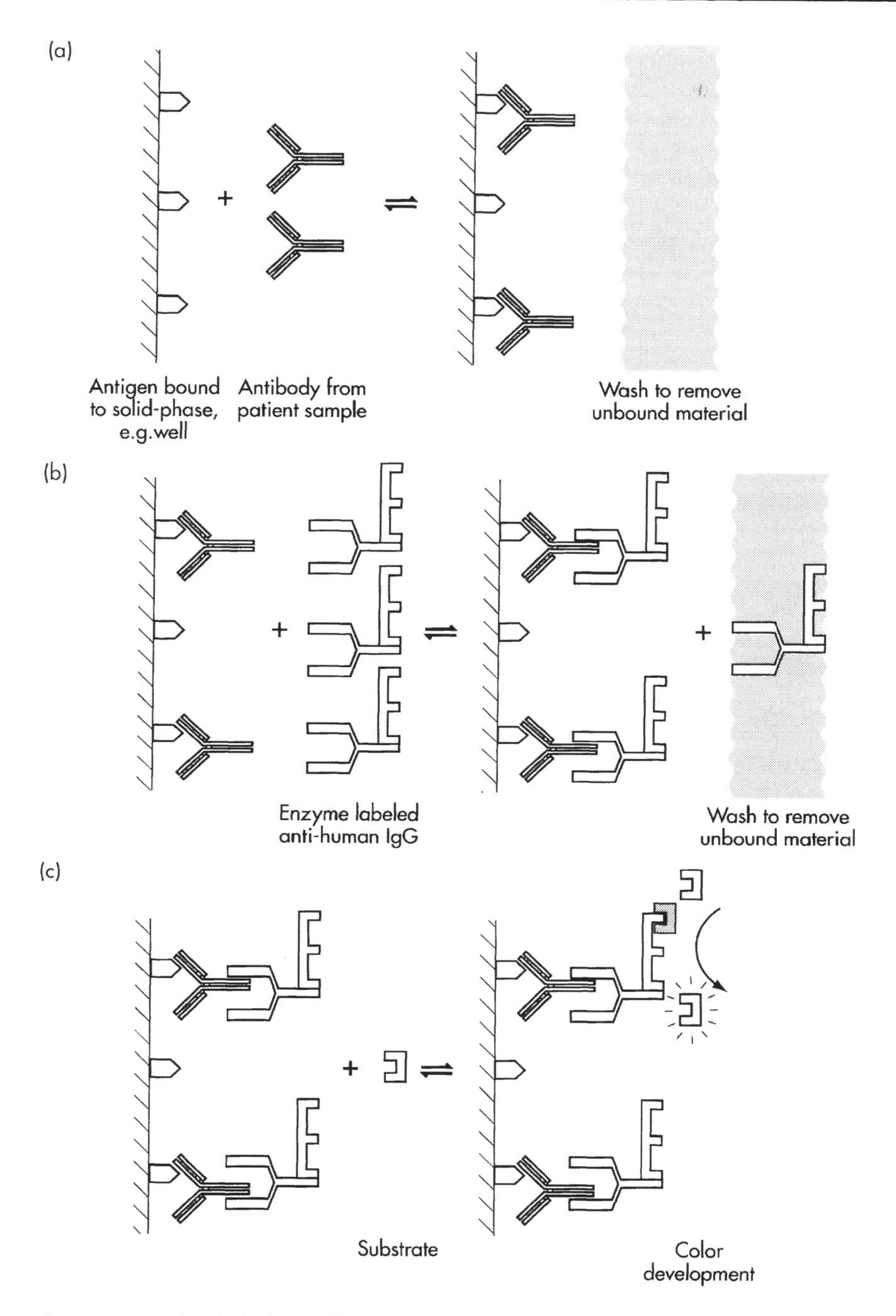

Fig. 58.2 Enzyme immunoassay (EIA). Antigen of interest is coated to a solid surface (microtiter well or a bead is typically used). The patient's antibodies react with the antigen. After a wash step, a reagent (often mouse) anti-human IgG that is labeled with an enzyme is added. Following another wash step, the substrate for the enzyme produces a color reaction.

Type of Sample

Serum or plasma.

ANTI-DOUBLE-STRANDED (DS) DNA

Anti-dsDNA antibodies are found mainly in patients with SLE. They usually produce a peripheral or homogeneous pattern on either the frozen section or HEp-2 cell line preparations used in the IIF assay.

Reference Interval

Anti-dsDNA should be less than the cut-off. With the IIF test, it is undetectable at the titer used for the *Crithidia luciliae* test. With the EIA, anti-dsDNA gives an OD less than that of the cut-off established by the kit manufacturer.

Clinical Applications

Anti-dsDNA is one of the most specific laboratory tests for SLE. It correlates with the presence of renal disease in affected patients, one of the most common causes of death in these patients. The glomeruli of patients with SLE who suffer from renal disease usually contain complexes of DNA with anti-DNA antibodies and complement. The titer of anti-dsDNA may correlate with disease activity. Anti-single-stranded (ss) DNA is also present in patients with SLE as well as in patients with drug-induced lupus. It does not have the same specificity as anti-dsDNA for renal disease or disease activity.

Limitations

Anti-dsDNA is not an effective screening test, because patients with SLE may lack these antibodies, and contain other antibodies, such as Sm or SSA/Ro. Therefore, the absence of these antibodies does not exclude SLE. However, the presence of anti-dsDNA is strong evidence supporting the diagnosis of SLE. Although the assays claim to be specific for anti-dsDNA, many of these patients have antibodies that react with either single-stranded (ss) or dsDNA. Anti-ssDNA has less specificity for SLE, but is often present in the serum from these patients. Some patients with drug-induced lupus may have anti-ssDNA.

Assay Technology

The older Farr assay is rarely used today. It is a radioimmunoassay (RIA) where radiolabeled DNA is allowed to react with anti-dsDNA in serum. The complexes are precipitated from solution by either specific antibodies or an ammonium sulfate solution.

EIA techniques use preparations of dsDNA coated on a solid surface (usually a microtiter well). Serum is incubated on these coated wells. Following a wash step with buffer, enzyme-conjugated anti-human immunoglobulin is incubated with the well. After another wash step, the chromogenic substrate for the enzyme results in a color-producing reaction. This color is compared to known reactivity in positive and negative samples provided by the kit manufacturer.

IIF is another method used to identify anti-dsDNA. The substrate is the hemoflagellate *Crithidia luciliae*. This test takes advantage of the fact that the kinetoplast of these microorganisms contains dsDNA with no ssDNA. It should be noted that the kinetoplast also contains a small amount of histone which may result in a false-positive response with strong anti-histone antibodies in some patients with drug-induced lupus.

Type of Sample

Serum or plasma.

ANTI-SM

Anti-Sm (short for Smith – the first person described who had this type of antibody reactivity detected) results in a speckled pattern by IIF ANA assays. Sm is a 95 kDa protein that exists as a nuclear protein–RNA complex as part of the ribonucleoprotein (RNP) antigen. Sm and RNP are part of the ribonucleoproteins that function to splice transcriptional mRNA. The antibodies against Sm react with several U1 RNP antigens: B, B′, D, and E.

Reference Interval

Most laboratories use either EIA or a gel diffusion technique to detect anti-Sm. The reference interval in the EIA is reactivity below the cut-off established by the manufacturer. In the gel diffusion technique the reference interval is the *lack* of a precipitin band that has identity with control positive serum.

Clinical Applications

Anti-Sm reactivity is present only in 20–30% of patients with SLE. However, its presence is strong evidence for the presence of SLE. Unlike the ANA test which is very sensitive, but not specific, the anti-Sm test is very specific for SLE.

Limitations

Only 20–30% of patients with SLE will have anti-Sm. Because of cross-reactivity with RNP antibodies, stringent conditions must be used to avoid false positives with that antibody.

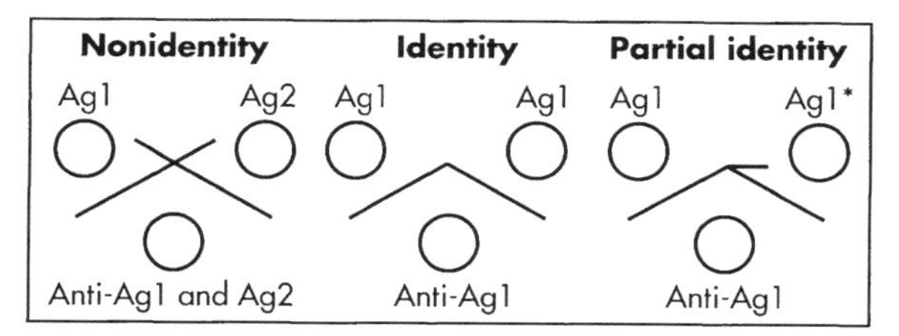

Fig. 58.3 Ouchterlony technique. For a nonidentity reaction, antibodies are present in one well that react with two different antigens in two different wells. This produces two precipitin bands that cross each other. For an identity reaction, only one type of antigen is present in two adjacent wells. The antibody that reacts with this antigen is present in the third well. This produces two precipitin bands that meet each other but do not cross each other. For a partial identity reaction, there are two different, but related (antigenically similar) molecules. The precipitin bands meet, and often one line spurs toward the well of one of the antigens.

Assay Technology

Gel diffusion (Ouchterlony) plates are standard double-diffusion in two-dimension assays (Figure 58.3). Sm antigen is placed in a central well in an agarose-coated surface. Surrounding wells contain patient sera alternated with known anti-Sm positive sera for comparison. After diffusion is complete, precipitin bands resulting from known anti-Sm positive sera are compared to the adjacent patient reactions. If the patient's serum contains an antibody that produces a band giving an identity line with the known positive control, it is said to contain anti-Sm.

EIA technology uses antigen preparations that contain Sm. After a standard EIA reaction with the patient's serum, the reactivity is compared to known positive and negative controls.

Immunoblotting analysis may also be used to determine the presence of anti-Sm reactivity. However, this technique is not readily automated and is beyond the degree of complexity offered by most clinical laboratories.

Type of Sample

Serum or plasma.

ANTI-RNP

Anti-ribonucleoprotein (RNP) results in a speckled pattern on IIF ANA screening tests. Anti-RNP reacts with a 70 kDa nuclear matrix antigen and epitopes A and C on U1 RNP. It is present in high titers in patients with **mixed connective tissue disease** (**MCTD**).

Reference Interval

Most laboratories use either an EIA or a gel diffusion technique to detect anti-RNP. The reference interval in the EIA is reactivity below the cut-off established by the manufacturer. In the gel diffusion technique the reference interval is the *lack* of a precipitin band that has identity with control positive serum. In the immunoblot assay, lack of band reactivity is the reference interval.

Clinical Applications

Anti-RNP is present in low titer in many autoimmune diseases, including SLE – where it occurs in about 25% of cases. In MCTD, anti-RNP is often the only, or major reactivity present. About 95% of patients with MCTD have anti-RNP in their serum. Identification of patients with MCTD is useful because it is usually a milder form of autoimmune disease than SLE. Patients with MCTD often present with Raynaud's phenomenon, but lack involvement of the kidneys, lungs, or heart.

Limitations

Although this test may be useful to distinguish patients with SLE from those with MCTD, about a third of patients with SLE also have anti-RNP.

Assay Technology

Gel diffusion (Ouchterlony) plates are standard double-diffusion in two-dimension assays. RNP antigen is placed in a central well on an agarose-coated plate. Surrounding wells are prefilled with patient's sera alternating with known positive sera for comparison. After diffusion is complete, precipitin bands resulting from known anti-RNP positive samples are compared to the adjacent patient reactions. If the patient's serum contains an antibody that produces a band giving an identity line with the known positive control, it is assumed to contain anti-RNP.

EIA technology uses antigen preparations that contain RNP. After a standard EIA reaction with the patient's serum, the reactivity is compared to known positive and negative controls provided by the manufacturer.

Immunoblotting analysis may also be used to determine the presence of anti-RNP reactivity. However, it is not readily automated and is beyond the degree of complexity offered by most clinical laboratories.

Type of Sample

Serum or plasma.

ANTI-SSA/RO

Anti-SSA/Ro produces a speckled pattern on most IIF ANA screening tests. These antibodies were first discovered in patients with Sjögren's syndrome (SS), but are also present in patients with SLE and occasionally in patients with rheumatoid arthritis (RA). The antibody recognizes a 52 and 60 kDa nuclear protein. However, this antigen is easily extracted by the buffer wash steps in IIF techniques and, therefore, may be missed unless the manufacturer has used special measures to fix it on the substrate. Indeed, before the widespread use of newer HEp-2 cell lines, SLE patients with mainly or only anti-SSA/Ro reactivity produced negative ANA screening tests.

Reference Interval

Laboratories use an EIA, a gel diffusion technique (Ouchterlony or counter-immunoelectrophoresis – CIE), or an immunoblot assay to detect anti-SSA/Ro. The reference interval in the EIA is reactivity below the cut-off established by the manufacturer. In the gel diffusion and immunoblot techniques the reference interval is the *lack* of a precipitin band that has identity with control positive serum, or an appropriate molecular weight.

Clinical Applications

About 75% of patients with primary **Sjögren's syndrome** (protean symptoms include dry eyes, dry mouth, and arthritis) have anti-SSA/Ro. A smaller percentage (10–25%) of patients with secondary Sjögren's syndrome (the same symptoms together with another autoimmune disease) have these antibodies. Anti-SSA/Ro is also present in about 25% of patients with SLE. These antibodies are strongly associated with a subgroup of SLE patients who present with photosensitive skin rash and Sjögren's syndrome (subacute cutaneous lupus erythematosus). Some patients with drug-induced lupus (especially those associated with toxicity to D-penicillamine) have this antibody, as well. SLE and neonatal lupus often have anti-SSA/Ro activity in the serum. Lastly, anti-SSA/Ro is a marker antibody for **neonatal lupus**. Often, it is the only autoantibody present in the serum of these babies.

Limitations

When the tissue section immunofluorescence test is used as a screen for ANA, the SSA/Ro antigen may be removed during the wash steps because it is easily eluted from the nuclei in the section. Using these older substrates, about 60% of ANA negative cases who had SLE were found to have anti-SSA/Ro in their serum. Newer substrates, especially tissue culture substrates such as HEp-2, are better at preserving this antigen.

Assay Technology

Gel diffusion (Ouchterlony) plates are standard double-diffusion in two-dimension assays. A preparation containing SSA/Ro antigen is placed in a central well on an agarose-coated slide. Surrounding wells contain patient's sera alternating with known positive sera for comparison. After diffusion is complete, precipitin bands resulting from the known anti-SSA/Ro positive are compared to the adjacent patient reactions. If the patient's serum contains an antibody that produces a band giving an identity line with the known positive control, it is said to contain anti-SSA/Ro. Similarly in CIE, known positive SSA/Ro antisera are placed in wells adjacent to wells containing patient's serum for comparison following electrophoresis that propels the antibodies toward wells containing the SSA/Ro antigen preparation.

EIA technology uses antigen preparations that contain SSA/Ro. After a standard EIA reaction with the patient's serum, the reactivity is compared to known positive and negative controls provided by the manufacturer.

Immunoblotting analysis may also be used to determine the presence of anti-SSA/Ro reactivity. However, it is not readily automated and is beyond the degree of complexity offered by most clinical laboratories.

Type of Sample

Serum or plasma.

ANTI-SSB/LA

Anti-SSB/La produces a speckled pattern on most IIF ANA screening tests. As with anti-SSA/Ro, these antibodies were first discovered in patients with Sjögren's syndrome (SS). SSB/La antigen is a 47 kDa protein that associates with RNA polymerase III transcripts. SSB/La is present in much greater concentration than SSA/Ro in substrates for IIF ANA screening.

Reference Interval

Laboratories use an EIA, a gel diffusion technique (Ouchterlony or counter-immunoelectrophoresis – CIE), or an immunoblot assay to detect anti-SSB/La. The reference interval in the EIA is reactivity below the cut-off established by the manufacturer. In the gel diffusion and immunoblot techniques the reference interval is the *lack* of a precipitin band that has identity with control positive serum, or an appropriate molecular weight.

Clinical Applications

About 40% of patients with primary Sjögren's syndrome have anti-SSA/Ro. Anti-SSB/La is also found in SLE, but unlike SSA/Ro is neither a marker for neonatal lupus, nor subacute cutaneous lupus erythematosus.

Limitations

Although anti-SSB/La is found in many individuals with both SLE and Sjögren's syndrome, it has not proven to be a particularly useful marker of any specific subset of these diseases.

Assay Technology

The assay technology is identical to that of SSA/Ro (see above).

Type of Sample

Serum or plasma.

ANTI-HISTONE

Antibodies against the basic nuclear proteins, histones, usually give a homogeneous pattern on the immunofluorescence ANA screen. These antibodies may be present in patients with SLE or in patients with drug-induced lupus. Based on immunologic studies, histones are subdivided into five groups: H1, H2, H2B, H3, and H4.

Reference Interval

Anti-histone may be detected by IIF on substrates, by EIA or immunoblot. The reference interval is absence of detectable antibody.

Clinical Applications

It is common for patients with SLE to have anti-histone antibodies along with more specific antibodies such as anti-Sm or anti-dsDNA (see above). Because they are not specific for this disease, there is no need to test for these antibodies unless drug-induced lupus is suspected. In patients with SLE, the anti-histone antibodies are usually directed against H1 and H2B. Anti-histone antibodies are also present in about 10% of patients with rheumatoid arthritis (RA). Testing for these antibodies in patients with RA does not seem to add to the useful information. The vast majority of patients with **drug-induced lupus** have anti-histone antibodies. Patients with procainamide-induced lupus typically develop antibodies that react with H2A–H2B complex, whereas patients with hydralazine-induced lupus develop antibodies against H3 and H4.

Limitations

The presence of anti-histone does not mean that the patient has drug-induced lupus. The antibodies are often present in asymptomatic individuals receiving one of the above medications. Further, in patients who do have symptoms of drug-induced lupus, cessation of the drug usually results in a reversal of the symptoms in weeks, although the antibodies may persist for months or years; recall that the half-life of IgG is about 21 days.

Assay Technology

EIA and immunoblot technology are similar to those described above.

IIF has been used in the past to detect these antibodies. However, IIF involves extracting and then reconstituting histones in the tissues, which is clumsy and insensitive. Therefore, it is not recommended.

Type of Sample

Serum or plasma.

ANTI-DNP

The lupus erythematosus (LE) cell test is the oldest autoantibody test for SLE. It is no longer used because it is insensitive, laborious, and subjective. However, the antibody responsible for the LE cell phenomenon is anti-deoxyribonucleoprotein. Simple latex agglutination and EIA tests for DNP are now available that serve as rapid (albeit insensitive and relatively nonspecific) screening tests for SLE.

Reference Interval

The reference interval is *lack* of agglutination in the latex agglutination test. For the EIA test, reference interval is reactivity below the cut-off established by the manufacturer.

Clinical Applications

About half of the patients with SLE have anti-DNP antibodies. When present, they are often in high titer and, therefore, easily detected with a simple screening test.

Limitations

The test is not specific for SLE, nor is it as good a screen as the ANA test (see above). It is also positive in patients with scleroderma, polymyositis, rheumatoid arthritis, and even chronic active hepatitis.

Assay Technology

EIA technology is similar to the EIA technique described in previous sections.

Latex agglutination uses latex beads coated with DNP antigen that result in visible changes in clarity of the mixture due to clumping of the latex beads. Both screening reaction and titers of the patient's serum can be performed.

Type of Sample

Serum or plasma.

ANTI-CENTROMERE

This antibody directed against centromere proteins is useful in the diagnosis of the CREST syndrome (see below). On IIF, anti-centromere gives a characteristic staining pattern (Table 58.2). Uniform-sized speckles are present in resting cells and in dividing cells; speckles conform to and are present on chromosomes.

Reference Interval

Anti-centromere may be detected by IIF or by EIA. The reference interval is *absence* of detectable antibody.

Clinical Applications

Most patients with **CREST syndrome** have autoantibodies that react with chromosomal centromeres and lack antibodies that react with Scl-70 (topoisomerase I – see below). CREST stands for autoimmune disease with the following manifestations: calcinosis (i.e. rock-like calcification within soft tissues), Raynaud's phenomenon (cutaneous cyanosis), esophageal dysmotility (difficulty swallowing), and telangiectasias (spider-like red patches on skin or mucosal surfaces). It is an important subset of scleroderma because these individuals usually lack deep organ involvement and, therefore, have a much better prognosis.

Limitations

Although anti-centromere is present in about 95% of patients with CREST syndrome, it is present in a small number (about 10%) of patients with diffuse scleroderma. Further, it is present in about 25% of patients with isolated Raynaud's phenomenon who lack any other features of CREST.

Assay Technology

EIA technology is similar to those described above.

IIF on HEp-2 or other tissue culture substrate is an excellent technique to detect these antibodies. They are not definable, however, on frozen section substrates because the nuclei of these normal rat livers or kidneys lack large numbers of dividing cells that produce and concentrate centromere proteins.

Table 58.2. Pattern of ANA in progressive systemic sclerosis.

Pattern	Antigen	Significance
Nucleolar	RNA	60% of PSS
Centromere	Centromere protein	80% of CREST patients
Fine Speckled	Topoisomerase (Scl-70)	20% of PSS
Coarse Speckled	Unknown	Unknown

PSS, progressive systemic sclerosis (scleroderma).

Type of Sample

Serum or plasma.

ANTI-SCL-70

Anti-Scl-70 also called the scleroderma antibody, or Scl-1, is an antibody against a 70 kDa antigen present in DNA topoisomerase I. On IIF, anti-Scl-70 gives a speckled pattern with fine, almost homogeneous speckles.

Reference Interval

Anti-Scl-70 may be detected with EIA or immunoblot. The reference interval is *absence* of detectable antibody.

Clinical Applications

Anti-Scl-70 identifies about 75% of patients with **progressive systemic sclerosis** (**PSS** or **scleroderma**). It is extremely uncommon for this antibody to be present in cases of CREST syndrome subset of PSS. There has been an association between the subsequent development of cancer in scleroderma patients with the presence of anti-Scl-70.

Limitations

Anti-Scl-70 is also present in about 40% of patients with acrosclerosis, 10% of patients with primary Raynaud's phenomenon, and about 5% of patients with Sjögren's syndrome.

Assay Technology

EIA, gel diffusion, and immunoblot technology are similar to those described above.

Type of Sample

Serum or plasma.

RHEUMATOID FACTOR

Rheumatoid factor is an antibody (usually IgM) directed against the Fc portion of human IgG.

Reference Interval

The reference interval for rheumatoid factor is undetectable, or detectable below a cut-off value provided by the manufacturer's kit.

Clinical Applications

More than 90% of patients with **rheumatoid arthritis** have rheumatoid factor present in high titer in their serum. Although often considered to be a disease of the small joints, rheumatoid arthritis is a systemic autoimmune disease. Systemic manifestations include a mild fever (especially early on during the course of the disease), cutaneous nodules as a result of vascular damage, and rarely, inflammation of the lungs, liver, and heart. Women are affected twice as often as men. It usually begins after the third decade. Patients have involvement mainly of small joints symmetrically. Cutaneous rheumatoid nodules are often present on the extensor surfaces of the elbows, and at sites of repeated trauma.

Limitations

Most normal individuals develop rheumatoid factor during acute infectious diseases. Thus, the presence of rheumatoid factor may be confusing. Because of this, the specificity of rheumatoid factor is only about 72% in recent studies. The combined use of anti-CCP has been reported to be beneficial in improving the specificity of a positive rheumatoid factor assay. Since the early stages of rheumatic fever may present with systemic manifestations and fever, repeat testing when the acute period has ended helps to confirm the diagnosis of RA. Of course a negative result does not rule out the disease. The titer of rheumatoid factor will fluctuate, and repeat testing is warranted in cases with strong clinical suspicion despite the absence of detectable rheumatoid factor on a single sample, especially early in the disease. This suggests that rheumatoid factor is not the etiologic agent of RA, but only a byproduct of the inflammatory process that is useful for diagnosis.

Assay Technology

The agglutination and precipitation tests that employed erythrocytes or latex beads coated with immunoglobulin are being replaced by more reproducible nephelometric tests. The older procedures are subjective and suffer from poor reproducibility.

In the past 10 years, nephelometric assays with objective rate units have become the mainstay to detect rheumatoid factor.

Type of Sample

Serum or plasma.

ANTI-CYCLIC CITRULLINATED (ANTI-CCP) PEPTIDE

Anti-CCP is an antibody that recognizes citrullinated forms of native proteins and which is directed against an antigenic determinant containing the deimidated form of arginine (called citrulline).

Reference Interval

The reference interval for anti-CCP is undetectable, or detectable below a cut-off value provided by the manufacturer's kit.

Clinical Applications

The presence of anti-CCP antibodies has a specificity of >90% for rheumatoid arthritis. They may be detected years before the onset of clinical symptoms of rheumatoid arthritis. Its presence is also associated with the severity of the disease at the time of diagnosis. Since the presence of anti-CCP early in the disease is associated with an increased risk of joint erosion, more aggressive therapy may be indicated than in anti-CCP negative cases.

Limitations

While anti-CCP is more specific than the rheumatoid factor assay (above) for distinguishing patients with rheumatoid arthritis from controls, the sensitivity of anti-CCP is less than that of the rheumatoid factor test. Because of this, some workers have suggested that the combination of anti-CCP EIA with IgM anti-RF significantly improves the sensitivity of detection of rheumatoid arthritis.

Assay Technology

EIA technology is similar to those described above. The most recent generation of commercially available kits has achieved a specificity of >90% in some studies, although the sensitivity still lags at about 66%.

Type of Sample

Serum.

ANTI-NEUTROPHIL CYTOPLASMIC ANTIBODIES (C-ANCA, P-ANCA)

Cytoplasmic (C) and perinuclear (P) ANCA are found in patients with Wegener's granulomatosis and microscopic (i.e. small vessel) vasculitis, respectively. They are usually, but not always, directed against serine protease 3 (C-ANCA) and myeloperoxidase (P-ANCA) in the neutrophil granules. This is one of the most specific of all autoimmune tests and one of the few that should be offered on a STAT basis. It should be considered for a STAT assay in a patient suspected of having Wegener's granulomatosis or crescentic glomerulonephritis. Treatment before the creatinine is markedly elevated can prevent chronic renal failure and may reverse lung collapse.

Reference Interval

C-ANCA and P-ANCA may be detected by IIF and/or by EIA. Reference interval is *absence* of antibody reactivity.

Clinical Applications

Wegener's granulomatosis is an uncommon condition wherein small blood vessels in the kidney, lungs, and upper respiratory tract are damaged by necrotizing inflammation. Because its course is highly variable and the histologic evidence is difficult to obtain on one biopsy, the development of a highly specific and highly sensitive autoantibody test is of great help. Because C-ANCA is present in the majority of patients with Wegener's granulomatosis, but only uncommonly in other patients or controls, it is the test of choice for diagnosing this condition. The presence of this antibody in an appropriate clinical setting results in chemotherapy being administered that may restore renal or pulmonary function.

P-ANCA are less specific than C-ANCA. They are often seen in patients with **microscopic vasculitis**, but are not as specific as C-ANCA.

Limitations

Although C-ANCA is highly specific for Wegener's granulomatosis, its absence cannot exclude the disease. The titer of reactivity waxes and wanes with the course of the disease and in some patients it may not be detectable. The presence of granular cytoplasmic staining on the ethanol-fixed neutrophils (i.e. the hallmark of C-ANCA) must be confirmed by repeating the study on formalin-fixed neutrophils. With true anti-serine protease 3, this also produces a granular cytoplasmic pattern, typically stronger in reactivity than the ethanol-fixed slides. Performing a final identification step using an EIA kit (with the purified protease 3 antigen) is also recommended. The EIA kits for anti-serine protease 3 have not been standardized with each other and disparate results have been found between different kits. This antigen deteriorates with time, and reactivity must be verified on each run.

P-ANCA is much less specific than C-ANCA. If IIF on ethanol-fixed neutrophils is used, it is necessary to be aware that ANA will give a false-positive reactivity. It is recommended to react the antisera on formalin-fixed neutrophils where the ANA reactivity usually disappears and anti-MPO activity gives a cytoplasmic rather than perinuclear staining pattern.

Further, it is recommended that all C-ANCA and P-ANCA reactivity be confirmed on EIA kits with the specific antigen. If discrepant reactivity is seen between the two types of assays (IIF and EIA), the clinician should be informed of the atypical reaction. Because the kits are not well standardized, the results may reflect a particular kit's antigen solution. In addition, other cytoplasmic antigens such as cathepsin and lactoferrin have been described that also give ANCA patterns in patients with microscopic vasculitis.

Assay Technology

IIF is the screening procedure. Neutrophils from a control individual are cytocentrifuged onto glass slides. Some of the slides are fixed with ethanol, others with formalin. Serum (usually diluted 1:20 in buffer) is allowed to react with each fixed substrate for about 30 min. Following a triple wash in buffer, fluorescein-conjugated antibody against human immunoglobulin is layered onto the slides. After another 30 min incubation and wash, the slides are examined under a fluorescence microscope. The presence of granular cytoplasmic fluorescence on both the ethanol-fixed and formalin-fixed slides indicates the presence of C-ANCA. The presence of perinuclear staining on the ethanol-fixed slides and cytoplasmic granular staining on the formalin-fixed slides indicates the presence of P-ANCA.

EIA is performed to confirm the above pattern specificity. Kits with serine protease 3 and MPO-coated wells are available but not well standardized from manufacturer to manufacturer. Despite this, we confirm all positive C-ANCA and P-ANCA results. In the uncommon instances where they do not correlate, we do not dismiss the results. The clinician is informed of the discrepancy. Other antigens such as lactoferrin and cathepsin can give reactivity and such antibody reactivity has been described in patients with microscopic vasculitis.

P-ANCA reactivity has also been described in patients with inflammatory bowel disease, especially ulcerative colitis, and in patients with primary sclerosing cholangitis. The relatives of patients who have inflammatory bowel disease may also demonstrate P-ANCA. This implies the presence of a genetic predisposition for forming such antibodies.

Type of Sample

Serum or plasma.

ANTI-GLOMERULAR BASEMENT MEMBRANE (GBM)

Anti-GBM is an antibody that reacts with glomerular basement membranes and which may cross-react with alveolar basement membranes. It is useful in the diagnosis of Goodpasture's syndrome.

Reference Interval

Anti-GBM may be detected by IIF, direct immunofluorescence, EIA or Western blot. The reference interval is *absence* of detectable antibody.

Clinical Applications

Patients with **Goodpasture's syndrome** suffer from a necrotizing crescentic glomerulonephritis and pulmonary hemorrhage. These clinical features are associated with autoantibodies directed against their glomerular basement membranes. The presence of these antibodies is highly specific for the diagnosis of Goodpasture's syndrome when found in the appropriate clinical setting. These antibodies react with type IV collagen, a key component in the basement membrane.

Limitations

Although the Western blot technology is highly specific, it is not available in kit form for general laboratory use. The indirect and direct immunofluorescence tests require experience to prevent overinterpretation. The available EIA kit (see below) may not be as sensitive as some Western blots.

Assay Technology

Western blotting is available only through some commercial or research laboratories that have developed that test.

Renal biopsies from a patient may be stained with fluorescein-conjugated anti-human IgG to demonstrate the linear fluorescence characteristic of anti-GBM (i.e. direct immunofluorescence).

Serum followed by fluorescein-conjugated anti-human IgG may be used to stain glomeruli from kits that use (mainly) monkey kidney sections (i.e. IIF). This technique is subjective and may not be as sensitive as the Western blot methodology.

EIA kits have been produced that use a subunit of type IV collagen (alpha-3) that has been implicated as the specific antigen for anti-GBM. However, both false positives and negatives have been reported. EIA may be useful to determine the strength of reactivity of a sample with a known positive Western blot.

Type of Sample

Serum or plasma.

ANTI-JO-1

Anti-Jo-1 is an antibody that reacts with the enzyme histidyl-tRNA synthetase and has been used as a marker for **myositis**, an idiopathic immune or inflammatory disease of skeletal muscle.

Reference Interval

The reference interval is *absence* of detectable antibody.

Clinical Applications

Anti-Jo-1 has been used to identify patients with myositis who may benefit from steroid therapy.

Limitations

Although relatively sensitive for myositis, a negative anti-Jo-1 cannot rule out myositis, polymyositis, or dermatomyositis. Anti-Jo-1 alone may not give a positive ANA. Therefore, patients with symptoms of myositis may benefit from an anti-Jo-1 test even when their ANA screen is negative.

Assay Technology

EIA or Ouchterlony double-diffusion are the techniques used to detect anti-Jo-1. The methodology is the same as described above.

Type of Sample

Serum or plasma.

ANTI-MICROSOMAL (THYROID PEROXIDASE – TPO)

Antibodies against the microsomal fraction of thyroid epithelium are present in patients with autoimmune (Hashimoto's) thyroiditis and in those with Grave's disease. Recently, it has been found that the specific reactivity is against thyroid peroxidase (TPO), a cytoplasmic enzyme. Therefore, one may see either the term anti-microsomal or anti-TPO in the literature. They refer to the same antibody.

Reference Interval

The reference interval is *absence* of detectable antibody.

Clinical Applications

Patients with either **Grave's disease** (hyperthyroidism) or **Hashimoto's thyroiditis** (usually hypothyroid, although they may be euthyroid or even hyperthyroid early in the course), both have anti-microsomal antibodies. The presence of these antibodies and the appropriate clinical and laboratory features help to secure the diagnosis. Anti-TPO fix complement, damage thyroid tissue and lead to lymphocytic infiltration of its parenchyma. Other autoantibodies are also present in these conditions. Thyroid-stimulating immunoglobulins (TSI) may react with the TSH receptors on thyroid epithelium and mimic the stimulatory effect of TSH resulting in Grave's disease (hyperthyroidism). Doing further testing for TSI with the typical case of Grave's disease or Hashimoto's thyroiditis is not recommended. Another autoantibody usually found in these cases is anti-thyroglobulin. Because anti-TPO is almost always present in these cases, the presence or absence of anti-thyroglobulin does not add useful information to the diagnosis and using it as a routine test except when anti-TPO is negative is not recommended.

Limitations

Rarely, anti-TPO antibodies may be absent in a case of Hashimoto's thyroiditis or Grave's disease. In those cases, an anti-thyroglobulin test may be useful. Since the antibodies do not respond to therapy with levothyroxine, there is no need to monitor the anti-MPO levels after the diagnosis.

Assay Technology

Most laboratories use EIA kits for TPO to look for microsomal antibodies. The technique is as described above. Latex agglutination techniques that use beads coated with TPO are also available.

The IIF test uses tissue sections of thyroid glands from rats or mice. These sections are reacted with the patient's serum as above for the ANA test. Anti-microsomal antibodies react with the cytoplasm of the gland, while anti-thyroglobulin reacts with the colloid.

Type of Sample

Serum or plasma.

ISLET CELL AUTOANTIBODIES (ICA)

ICA are often present early on in patients with diabetes mellitus type 1.

Reference Interval

The reference interval is *absence* of detectable antibody.

Clinical Applications

At the present time, the ICA test should be used on a research basis only. It should not be considered as a part of the routine diagnosis of diabetes mellitus. In the future, if research studies show the possibility of early intervention in relatives or patients with these antibodies, the test may be of considerable clinical use.

Limitations

The test is not invariably positive in type 1 diabetes. Because it is present only transiently, its diagnostic utility is very limited.

Assay Technology

The assay is available from research laboratories mainly as an IIF assay.

Type of Sample

Serum or plasma.

ANTI-ADRENAL CORTICAL ANTIBODIES

Antibodies that react with the adrenal cortex are usually present in the serum from patients with autoimmune adrenal insufficiency (Addison's disease).

Reference Interval

The reference interval is *absence* of detectable antibody.

Clinical Applications

Patients with idiopathic **Addison's disease** usually have anti-adrenal antibodies in their serum. The demonstration of these antibodies is useful to distinguish idiopathic (autoimmune) Addison's disease from other causes of adrenal insufficiency, such as tuberculosis. As with most other autoimmune disease testing, patient selection prior to testing is key to optimizing the utility of the test.

Limitations

Anti-adrenal cortical antibodies may be found in normal individuals. Although this may indicate a predisposition to Addison's disease, the significance of these antibodies in normal individuals is unclear. About 25% of these patients also have other autoimmune diseases including autoimmune thyroiditis and pernicious anemia, but have normal adrenal cortical function. Tests for autoantibodies associated with those conditions (i.e. anti-TPO, intrinsic factor, and parietal cell antibodies) may be helpful.

Assay Technology

Both IIF and radioimmunoassay have been used to detect anti-adrenal antibodies.

Type of Sample

Serum or plasma.

ANTI-PARIETAL CELL ANTIBODIES (PCA)

Antibodies against the parietal cells (the cells that make hydrochloric acid) in the stomach are present in most patients with pernicious anemia.

Reference Interval

The reference interval is *absence* of detectable antibody.

Clinical Applications

About 90% of patients with **pernicious anemia** have anti-parietal cell antibodies in their serum. When present in the context of an older, usually female patient with a macrocytic anemia, fatigue, dyspepsia and possibly symptoms of neuropathy, it is a useful confirmation of the presence of this autoimmune disease. These individuals often have antibodies against intrinsic factor, but testing for these antibodies is only useful if the PCA test is negative.

Limitations

False positives may occur with serum from patients with AMA because parietal cells of all species contain abundant mitochondria. A technique should be employed for detection of PCA that allows for the simultaneous detection of AMA. The presence of these PCA in otherwise normal individuals may have no significance.

Assay Technology

IIF is usually used for this test. The typical substrate employs rodent stomach wrapped around rodent kidney. The presence of reactivity in both the stomach and the kidney tubule epithelium is evidence of AMA. PCA give reactivity in the parietal cells of the stomach, but not in the kidney tubules.

Type of Sample

Serum or plasma.

ANTI-MITOCHONDRIAL ANTIBODIES (AMA, OR M2)

Antibodies that react with mitochondria have served as a useful marker for the liver disease primary biliary cirrhosis. Although nine different mitochondrial antigens have been described, the M2 antigens are the most specific for the diagnosis of primary biliary cirrhosis. These antigens are part of the 2-oxo-acid dehydrogenase complex in mitochondria.

Reference Interval

The reference interval is *absence* of detectable antibody.

Clinical Applications

Patients with **primary biliary cirrhosis** often have an insidious onset of their disease. Early symptoms of fatigue are often gradual and inconsistent. The patient typically seeks medical attention when her/his pruritus becomes a problem or when the sclera becomes noticeably yellow. The presence of an elevation of alkaline phosphatase in the presence of normal ALT and AST is the usual pattern of serum enzymes in the disease. AMA are found in about 90% of patients with primary biliary cirrhosis. In some of the patients, anti-smooth muscle antibodies may be present. This is not usually a problem because serum from patients with autoimmune chronic active hepatitis gives a much different serum enzyme pattern than that from patients with primary biliary cirrhosis.

Recently, patients with the clinical picture of primary biliary cirrhosis who do not have AMA in their serum, but who often do have ANA reactivity have been described. Although some authors refer to these patients as a distinct subset called **autoimmune cholangitis**, clinically, they do not seem to differ significantly from primary biliary cirrhosis.

Limitations

The titer of AMA does not reflect the severity of the disease in an individual. The degree of liver damage is determined by use of clinical information and biopsy. Because some cases of primary biliary cirrhosis fall into the subset called autoimmune cholangitis, patients who are AMA negative should be tested for the presence of ANA and anti-smooth muscle antibodies.

Antibodies against liver–kidney microsomal (LKM) fraction can mimic AMA. The LKM antibodies are associated with autoimmune hepatitis and stain the proximal tubules, but not the distal tubules in the kidney in IIF techniques. However, LKM do not stain the gastric epithelium that AMA always do (see below).

Assay Technology

Both IIF and EIA tests are available for AMA detection.

The IIF test employs tissue sections of kidney encircled by stomach from rats or mice. These sections are reacted with the patient's serum as above for the ANA test. AMA give granular reactivity with the cytoplasm of the kidney tubules and cells within the gastric mucosa.

EIA tests have been developed that can distinguish between different types of mitochondrial antigens and may provide better specificity in the diagnosis of primary biliary cirrhosis. Although there have been nine mitochondrial antigens that have been characterized, the M2 antigen is the one that has been most closely associated with the diagnosis of primary biliary cirrhosis.

Type of Sample

Serum or plasma.

ANTI-SMOOTH MUSCLE ANTIBODIES (SMA)

SMA are antibodies that have as their antigen F-actin. Typically, smooth muscle from rat stomach and/or kidney is used to detect these antibodies.

Reference Interval

The reference interval is *absence* of detectable antibody.

Clinical Applications

Although SMA have been used to detect patients with **autoimmune hepatitis**, they are also present in patients with conditions such as primary biliary cirrhosis and even in normal individuals. Patients with autoimmune hepatitis frequently have other antibodies as well, including ANA and AMA. Autoimmune hepatitis is found most often in young to middle-aged women. The autoimmune nature of the condition is speculative, but the patients usually respond well to steroid therapy.

Limitations

Smooth muscle antibodies are not specific for autoimmune hepatitis. Indeed, it is important to use standard serologic tests for hepatitis B and C to rule out infectious etiologies for the condition. Other hepatic diseases such as Wilson's disease and primary biliary cirrhosis also need to be excluded. Usually, the history and serum enzyme pattern help to distinguish primary biliary cirrhosis from autoimmune hepatitis. Because of the nonspecificity of SMA, laboratories usually perform the assay using two or three dilutions of the patient's serum. Only reactivity at the higher dilutions (typically 1:80 or 1:160) have reasonable predictive value.

Assay Technology

SMA are detected by immunofluorescence on rat stomach and/or kidney sections. The staining pattern on the stomach involves both the muscularis propria and the muscularis mucosa. SMA react mainly with the muscular layers of the arteries in the kidney sections.

Type of Sample

Serum or plasma.

ANTI-LIVER–KIDNEY MICROSOMAL (LKM)

Anti-LKM antibodies react with cytochrome P450. Western blotting studies have found the reactivity to reside mainly within a 50 kDa band.

Reference Interval

The reference interval is *absence* of detectable antibody.

Clinical Applications

Anti-LKM is a helpful marker for a subgroup of patients, mainly children, with autoimmune hepatitis. In the United States, less than 5% of cases of autoimmune hepatitis characterized by the presence of anti-LKM reactivity are seen in adults. Interestingly, in Europe, as many as 20% of such cases occur among adults.

Limitations

These patients are said to be more likely to have a wide variety of autoantibodies than patients with the usual SMA-associated autoimmune hepatitis. Antibodies against thyroid microsomal and parietal cells are common. The latter may cause confusion with AMA if stomach and renal substrates are not used concomitantly.

Assay Technology

Most laboratories use IIF to detect LKM. The antibodies react with the proximal convoluted tubules in the kidney, but unlike AMA, they do not react with the distal tubules or the loop of Henle. Further, in gastric substrates, the gastric parietal cells are usually negative. Some patients have unexpected staining of the gastric parietal cells.

Type of Sample

Serum or plasma.

IgA ANTI-ENDOMYSIUM

IgA anti-endomysium antibodies react with smooth muscle endomysium (cell membrane) in frozen sections of monkey kidney or human umbilical cord. It is present in the vast majority of patients with celiac disease (sprue) and is uncommon in other gastrointestinal conditions that may be in the differential diagnosis of malabsorption.

Reference Interval

The reference interval is *absence* of detectable antibody.

Clinical Applications

IgA anti-endomysium is used to detect patients with **celiac disease** (**sprue**). These patients have severe malabsorption due to a reaction to gluten in their diet. Detection of IgA anti-endomysium may be helpful in selecting patients who would benefit from a biopsy of the small intestine – currently required for definitive diagnosis. By following the titer of IgA anti-endomysium, one may be able to determine if the patient is maintaining a gluten-free diet without a biopsy of the small intestine. Because IgA anti-endomysium persists longer than anti-gliadin, some authors have suggested that anti-gliadin is a better test to follow therapy.

IgA anti-endomysium is a more specific antibody test than the older IgG anti-reticulin. The latter has similar diagnostic uses, but has largely and justifiably been replaced.

Limitations

IgA anti-endomysium is occasionally found in individuals who have normal small bowel biopsies. Therefore, one should confirm the diagnosis of celiac disease by biopsy before submitting the patient to a lifelong gluten-free diet (which requires abstinence from all starches except rice). IgA anti-endomysium may be absent in patients with typical celiac disease. Therefore, in the presence of strong clinical suspicion of celiac disease, a negative study does not preclude biopsy.

Assay Technology

IIF using frozen sections of monkey esophagus or human umbilical cord is the most common technique to detect IgA anti-endomysium. For this technique to be effective, two dilutions of the patient serum are employed: 1:5 and 1:50. Because some of the patients have other autoantibodies such as SMA that obscure detection of IgA anti-endomysium, the larger dilution is more specific for celiac disease. The more concentrated serum, however, provides better sensitivity in the majority of sera lacking interfering antibodies.

Type of Sample

Serum or plasma.

IgA ANTI-TISSUE TRANSGLUTAMINASE (ANTI-tTG)

IgA anti-tTG antibodies react with the major autoantigen in smooth muscle endomysium that results in the positive staining on frozen sections of monkey kidney or human umbilical cord. Not surprisingly, as with IgA anti-endomysium, IgA anti-tTG is present in the vast majority of patients with celiac disease (sprue) and is uncommonly present in other gastrointestinal conditions that may be in the differential diagnosis of malabsorption.

Reference Interval

The reference interval is *absence* of detectable antibody.

Clinical Applications

IgA anti-tTG is used to detect patients with celiac disease. It has been used instead of the IgA anti-endomysium to select patients who would benefit from a biopsy of the small intestine. One function of tTG is to deamidate glutamine into glutamic acid which changes the charge on gluten peptides. This action of tTG on gluten results in a heightened immunoreactivity of T cells for gluten in individuals with a genetic proclivity to develop celiac disease.

Limitations

IgA anti-tTG is occasionally found in individuals who have normal small bowel biopsies. Therefore, the diagnosis of celiac disease should be confirmed by biopsy before submitting the patient to a lifelong gluten-free diet (which requires abstinence from all starches except rice). Uncommonly, IgA anti-tTG may be absent in patients with typical celiac disease. Of course, no individual who is IgA deficient will be able to manufacture IgA anti-tTG. For those individuals, the finding of a positive IgG anti-gliadin (see below) in an appropriate clinical setting would encourage biopsy proof of celiac disease.

Assay Technology

EIA tests and dot blot assays are available commercially that have performed similarly to the IgA anti-endomysium test in detecting patients with celiac disease.

Type of Sample

Serum or plasma.

IgG AND IgA ANTI-GLIADIN

These antibodies react with the alcohol-soluble fraction of gluten that is responsible for celiac disease.

Reference Interval

The reference interval is absence of detectable antibody.

Clinical Applications

IgG and IgA anti-gliadin are elevated in most patients with celiac disease. Although the IgA anti-gliadin seems to be more specific for celiac disease than IgG anti-gliadin, it is a less sensitive test. Although neither is as specific and sensitive as IgA anti-endomysium, it is often useful to use both the anti-gliadin and anti-endomysium tests for celiac disease. Because celiac disease occurs with increased frequency among children who are deficient in production of IgA, redundant testing with IgG anti-gliadin may be helpful. IgG anti-gliadin has been useful to follow patients' adherence to their unpalatable gluten-free diets.

Limitations

Anti-gliadin is neither as specific, nor as sensitive as IgA anti-endomysium for diagnosing celiac disease.

Assay Technology

EIA kits are available for detection of IgG and IgA anti-gliadin. They are, however, not well standardized from one manufacturer to another.

Type of Sample

Serum or plasma.

ANTI-ACETYLCHOLINE RECEPTOR (ACHR)

ACHR reacts with the acetylcholine receptor at the neuromuscular postsynaptic junction. They are one of the few autoantibodies whose specificity is directly related to the clinical symptoms involved with their associated neuromuscular disease – myasthenia gravis.

Reference Interval

The reference interval is *absence* of detectable antibody.

Clinical Applications

In patients who suffer from **myasthenia gravis**, the normal transmission of acetylcholine across the tiny gap present between the nerve and muscle is blocked by antibodies that react with or near that receptor. This results in weakness in the muscles that are used most frequently. Typically this involves the eyelids (**ptosis**), ocular muscles (**diplopia**), and respiratory muscles, which leads to difficulty in breathing. As with many autoimmune diseases, women are affected twice as often as men. Different fractions of antibodies that bind to this receptor have been described: binding antibodies, blocking antibodies, and modulating antibodies. One or more of these antibodies are detectable in about 90% of patients with generalized myasthenia gravis.

In addition to the neuromuscular problem, about half of the cases in adolescents or young adults have an associated thymic hyperplasia, and about 10% of cases in middle-aged or older adults have a thymoma.

Limitations

The absence of ACHR does not rule out myasthenia gravis. The presence or absence of the antibodies does not identify whether a patient has a thymoma or thymic hyperplasia.

Assay Technology

A complex radioimmunoassay is used mainly in research or commercial laboratories to detect the antibodies associated with acetylcholine receptors. The technique takes advantage of the binding of the snake venom, α-bungarotoxin, to acetylcholine receptors. For the assay, α-bungarotoxin is labeled with ^{125}I. This then reacts with and binds irreversibly to preparations of human acetylcholine receptors prepared from human neural tissues. The patient's serum is allowed to react with preparations of the human acetylcholine receptor complexes – labeled α-bungarotoxin. If the serum contains ACHR, it binds to this labeled complex and precipitates. The amount of radioactivity in the precipitate correlates with the concentration of ACHR. The modulating antibody is detected by the binding of antibodies to the surface of cultured muscle cells also using the radiolabeled α-bungarotoxin.

Modulating antibodies can be detected by the amount that they reduce the binding sites available to the labeled snake venom.

ACHR blocking antibodies are antibodies that bind near the receptor. They are detected by the inhibition of binding of a known positive ACHR sample to the labeled receptors.

Type of Sample

Serum or plasma.

STRIATIONAL ANTIBODIES

Striational antibodies react with many skeletal muscle antigens including actin, myosin, and ryanodine receptor.

Reference Interval

The reference interval is *absence* of detectable antibody.

Clinical Applications

A small number of patients who have symptoms of myasthenia gravis lack antibodies to ACHR, ACHR modulating antibodies, and ACHR blocking antibodies. Some of these patients have striational antibodies.

Limitations

Because most patients with myasthenia gravis have ACHR antibodies, assay for striational antibodies is not the first screening test of choice. Striational antibodies are usually not present in adolescents and children with myasthenia gravis. These antibodies are also found in about 5% of patients with the clinically related Eaton–Lambert syndrome. Striational antibodies are also present in about 5% of patients with lung cancer, and in many patients with autoimmune hepatitis.

Assay Technology

An EIA with antigen preparations from homogenates of skeletal muscle is the technique used in research and commercial laboratories for detecting striational antibodies.

Type of Sample

Serum or plasma.

CALCIUM CHANNEL ANTIBODIES

Calcium channel antibodies react with plasma membrane proteins that are involved with initiating release of acetylcholine and neurotransmission. Interference with the function of the calcium channels by these antibodies can produce the Lambert–Eaton syndrome.

Reference Interval

The reference interval is absence of detectable antibody.

Clinical Applications

Lambert–Eaton syndrome can have symptoms that mimic myasthenia gravis. In addition to the weakness, however, they also suffer from xerostomia, xerophthalmia, and other autonomic nervous system impairments. More than 90% of these patients have antibodies that react with calcium channel peptide antigens, which makes this a sensitive assay for this unusual condition.

Limitations

As with many autoantibody tests described above, one cannot exclude the diagnosis of Lambert–Eaton syndrome with a negative result. Positive results are present in a small percentage of controls (less than 5%), but are found more frequently in patients with neuropathies associated with epithelial neoplasms including those of breast, ovary, or lung origin.

Assay Technology

The assay for these antibodies is a research radioimmunoassay procedure using antigen from human brain. Radiolabeled calcium channel peptides are used in a radioimmunoprecipitation procedure similar to that of ACHR binding antibodies. The radiolabeled synthetic calcium channel peptides are added to preparations of human brain. This is added to the patient's serum. If antibodies are present that bind to the labeled calcium channel peptides (that are attached to high affinity calcium channel receptors in the human brain), they will react with the brain-antigen preparation. Then, anti-human immunoglobulin is used to precipitate the patient's immunoglobulins. If it has bound to the radiolabeled complexes, radioactivity will be detectable in the immunoprecipitate in proportion to the immune reactivity.

Type of Sample

Serum or plasma.

ANTI-CARDIOLIPIN ANTIBODIES

Antibodies against cardiolipin may actually react with beta-2 glycoprotein I that in turn binds to cardiolipin.

Reference Interval

The reference interval is *absence* of detectable antibody.

Clinical Applications

Cardiolipin antibodies of the IgG and IgM class have been associated with the **phospholipid antibody syndrome**. Patients with the phospholipid antibody syndrome have significant problems with both arterial and venous thromboses, thrombocytopenia and recurrent fetal loss. The vascular thromboses may lead to life-threatening cerebrovascular accidents. In addition, other patients suffer from myocardial infarcts, endocarditis, pulmonary hypertension, and pulmonary infarcts. There is a moderate concordance between the cardiolipin immunoassays and the lupus anticoagulant functional hematologic assay. Patients with the phospholipid syndrome may have any or none of these antibodies.

Limitations

Many individuals develop antibodies against cardiolipin transiently during infectious diseases. Cardiolipin is a component of the VDRL test for syphilis. Therefore, some individuals with syphilis display positive cardiolipin antibody results.

Assay Technology

EIA with cardiolipin-coated microtiter wells has become the standard laboratory test for anti-cardiolipin antibodies. The reagents should be able to react with IgG and IgM. Some studies suggest that it may be useful to look at IgA anti-cardiolipin, but the data indicate that individuals with only IgA against cardiolipin are vanishingly rare.

Recently, some studies have suggested that anti-cardiolipin antibodies may really have specificity for beta-2 glycoprotein I, which then have an affinity for phospholipids, especially cardiolipin. It will be worthwhile to watch as the literature evolves to see if tests for antibodies against beta-2 glycoprotein I and cardiolipin are looking at the same population of antibodies.

Type of Sample

Serum or plasma.

ANTI-MYELIN OLIGODENDROCYTE GLYCOPROTEIN (MOG) AND ANTI-MYELIN BASIC PROTEIN (MBP) ANTIBODIES

Anti-MOG and anti-MBP antibodies are found in the serum of patients with multiple sclerosis.

Reference Interval

The reference interval is *absence* of detectable antibody.

Clinical Applications

Multiple sclerosis is a disease that presents with varied and nonspecific symptoms. The best laboratory test to confirm that one has this disease is the oligoclonal band test performed on cerebrospinal fluid. However, when this test and the clinical picture have confirmed the presence of multiple sclerosis, there is a wide variation in the progression to clinically definite multiple sclerosis. Because immunosuppressive therapy is now available and thought to be most effective when given early in the disease, tests that help predict which patients will progress are being sought. Recent data indicate that anti-MOG and anti-MBP, when present early on in the course of multiple sclerosis, predict an early transition to clinically definite multiple sclerosis.

Limitations

Currently, the assay is available only by Western blotting that is performed mainly in research laboratories. While helpful in predicting progression from early multiple sclerosis to definite multiple sclerosis, anti-MOG and anti-MBP have not been recommended for the initial diagnosis of multiple sclerosis. Further, although the vast majority of patients who give a positive test undergo more rapid progression of the disease than those who are seronegative for anti-MOG and anti-MBP, when both results are reactive, the positive predictive value of the test is only 76.5%.

Assay Technology

Western blotting is used to detect anti-MOG and anti-MBP. The test is not yet well standardized and is available mainly through research laboratories. However, if the assay proves to be useful for detecting likelihood of progression of early stage multiple sclerosis, it is likely that new EIA techniques will be developed for more general use. The reader is advised to consult the current literature on this topic.

Type of Sample

Serum.

FURTHER READING

Basso, D., Gallo, N., Guariso, G., Pittoni, M., Piva, M.G. and Plebani, M. Role of anti-transglutaminase (anti-tTG), anti-gliadin, and anti-endomysium serum antibodies in diagnosing celiac disease: a comparison of four different commercial kits for anti-tTG determination. *J. Clin. Lab. Anal.* **15**, 112–115 (2001).

Berger, T., Rubner, P., Schautzer, F., Egg, R., Ulmer, H., Mayringer, I., Dilitz, E., Deisenhammer, F. and Reindl, M. Antimyelin antibodies as a predictor of clinically definite multiple sclerosis after a first demyelinating event. *N. Engl. J. Med.* **349**, 139–145 (2003).

Bodil, E., Roth, K.S. and Stenberg, P. Biochemical and immuno-pathological aspects of tissue transglutaminase in celiac disease. *Autoimmunity* **36**, 221–226 (2003).

Bonamico, M., Tasore-Quartino, A., Mariani, P., Scartezzini, P., Cerruti, P., Tozzi, M.C., Cingolani, M. and Gemme, G. Down syndrome and coeliac disease: usefulness of antigliadin and antiendomysium antibodies. *Acta Paediatr.* **85**, 1503–1505 (1996).

Bylund, D.J. and McHutchison, J. Autoimmune liver diseases, In: *Progress and Controversies in Autoimmune Disease Testing. Clinics in Laboratory Medicine*, vol. **17** (eds Keren, D.F. and Nakamura, R.), 483–498 (W.B. Saunders, Philadelphia, PA, 1997).

Caja, A.J. Diagnosis and therapy of autoimmune liver disease. Management of chronic liver disease. *Med. Clin. N. Am.* **80**, 973–994 (1996).

Carey, J.L. III and Keren, D.F. Autoimmune disease and serology. In: *Laboratory Medicine* (ed McClatchey, K.D.), 1599–1634 (W.B. Saunders, Philadelphia, PA, 1994), Chapter 62.

Collins, A.B. and Colvin, R.B. Kidney and lung disease mediated by glomerular basement membrane antibodies: detection by Western blot analysis. In: *Manual of Clinical Laboratory Immunology*, 5th edn (eds Rose, N.R., Conway de Macario, E., Folds, J.D., Lane, H.C. and

Nakamura, R.M.), 1008–1012 (ASM Press, Washington, DC, 1997).

Esdaile, J.M., Abrahamowicz, M., Joseph, L., MacKenzie, T., Li, Y. and Danoff, D. Laboratory tests as predictors of disease exacerbations in systemic lupus erythematosus. Why some tests fail. *Arthritis Rheum.* **39**, 370–378 (1996).

Falk, R.J., Hogan, S., Carey, T.S. and Jennette, C. Clinical course of anti-neutrophil cytoplasmic autoantibody-associated glomerulonephritis. *Ann. Intern. Med.* **113**, 656–663 (1990).

Feltkamp, T.E.W. Antinuclear antibody determination in a routine laboratory. *Ann. Rheum. Dis.* **55**, 723–727 (1996).

Gniewek, R.A., Sandbulte, C. and Fox, P.C. Comparison of antinuclear antibody testing methods by ROC analysis with reference to disease diagnosis. *Clin. Chem.* **43**, 1987–1989 (1997).

Griesmann, G.E. and Lennon, V.A. Detection of autoantibodies in myasthenia gravis and Lambert–Eaton myasthenic syndrome. In: *Manual of Clinical Laboratory Immunology* (ed Rose, N.), 983–988 (ASM Press, Washington, DC, 1997).

Homberg, J.C., Abuaf, N. and Bernard, O. Chronic active hepatitis associated with anti-liver/kidney microsome antibody type 1: a second type of autoimmune hepatitis. *Hepatology* **7**, 1333–1339 (1987).

Huston, D.P. The biology of the immune system. *JAMA* **278**, 1804–1814 (1997).

Itoh, Y., Hamada, H., Imai, T., Seki, T., Igarashi, T., Yuge, K., Fukunaga, Y. and Yamamoto, M. Antinuclear antibodies in children with chronic nonspecific complaints. *Autoimmunity* **25**, 243–250 (1997).

Jarzabek-Chorzelska, M., Blaszczyk, M., Jablonska, S., Chorzelski, T., Kumar, V. and Beutner, E.H. Scl-70 antibody – a specific marker of systemic sclerosis. *Br. J. Dermatol.* **115**, 393–401 (1986).

Jitsukawa, T., Nakajima, S., Junka, U. and Watanabe, H. Detection of anti-nuclear antibodies from patients with systemic rheumatic diseases by ELISA using HEp-2 cell nuclei. *J. Clin. Lab. Anal.* **5**, 49–53 (1991).

Kavanaugh, A., Tomar, R., Reveille, J., Solomon, D.H. and Homburger, H.A. Guidelines for clinical use of the antinuclear antibody test and tests for specific autoantibodies to nuclear antigens. *Arch. Pathol. Lab. Med.* **124**, 71–81 (2000).

Keren, D.F. Antinuclear antibody testing, In: *Test Selection Strategies. Clinics in Laboratory Medicine*, vol. **22** (ed Burke, D.), 447–474 (W.B. Saunders, Philadelphia, PA, 2002).

Keren, D.F. and Goeken, J.A. Autoimmune reactivity in inflammatory bowel disease, In: *Progress and Controversies in Autoimmune Disease Testing. Clinics in Laboratory Medicine*, vol. **17** (eds Keren, D.F. and Nakamura, R.M.), 465–482 (W.B. Saunders, Philadelphia, PA, 1997).

Kohda, S., Kanayama, Y., Okamura, M., Amatsu, K., Negoro, N., Takeda, T. and Inoue, T. Clinical significance of antibodies to histones in systemic lupus erythematosus. *J. Rheumatol.* **16**, 24–28 (1989).

Kotzin, B.L. Systemic lupus erythematosus. *Cell* **85**, 303–306 (1996).

Lennon, V.A., Kryzer, T.J., Griesmann, G.E., O'Suilleabhain, P.E., Windebank, A.J., Woppmann, A., Miljanich, G.P. and Lambert, E.H. Calcium-channel antibodies in the Lambert–Eaton syndrome and other paraneoplastic syndromes. *N. Engl. J. Med.* **332**, 1467–1474 (1995).

Levine, S.R., Deegan, M.J., Futrell, N. and Welch, K.M.A. Cerebrovascular and neurologic disease associated with antiphospholipid antibodies: 48 cases. *Neurology* **40**, 1181–1189 (1990).

Malberg, K., Malfertheiner, P., Bannert, N. and Gunther, T. IgA-tissue transglutaminase (tTG) antibodies are highly sensitive serum markers for celiac disease. *Am. J. Gastroenterol.* **94**, 3079–3080 (1999).

Malleson, P.N., Sailer, M. and Mackinnon, M.J. Usefulness of antinuclear antibody testing to screen for rheumatic diseases. *Arch. Dis. Child.* **77**, 299–304 (1997).

Meini, A., Pillan, N.M., Villanacci, V., Monafo, V., Ugazio, A.G. and Plebani, A. Prevalence and diagnosis of celiac disease in IgA-deficient children. *Ann. Allergy Asthma Immunol.* **77**, 333–336 (1996).

Miller, M.H., Littlejohn, G.O., Davidson, A., Jones, B. and Toplis, D.J. The clinical significance of the anticentromere antibody. *Br. J. Rheumatol.* **26**, 17–21 (1987).

Mond, C.B., Peterson, M.G.E. and Rothfield, N.F. Correlation of anti-Ro antibody with photosensitivity rash in systemic lupus erythematosus patients. *Arthritis Rheum.* **32**, 202–204 (1989).

Nakamura, R.M. and Binder, W.L. Current concepts and diagnostic evaluation of autoimmune disease. *Arch. Pathol. Lab. Med.* **112**, 869–877 (1988).

Nakamura, R.M., Keren, D.F. and Bylund, D.J. *Clinical and Laboratory Evaluation of Human Autoimmune Diseases* (ASCP Press, Chicago, IL, 2002).

Pearce, E.N., Farwell, A.P. and Braverman, L.E. Thyroiditis. *N. Engl. J. Med.* **348**, 2646–2655 (2003).

Pisetsky, D.S. Antibody responses to DNA in normal immunity and aberrant immunity. *Clin. Diagn. Lab. Immunol.* **5**, 1–6 (1998).

Pisetsky, D.S., Gilkeson, G. and St. Clair, E.W. Systemic lupus erythematosus diagnosis and treatment. *Adv. Rheumatol. Med. Clin. N. Am.* **81**, 113–128 (1997).

Rader, M.D., Codding, C. and Reichlin, M. Differences in the fine specificity of anti-Ro (SS-A) in relation to the presence of other precipitating autoantibodies. *Arthritis Rheum.* **32**, 1563–1571 (1989).

Rote, N.S., Dostal-Johnson, D. and Branch, D.W. Antiphospholipid antibodies and recurrent pregnancy loss: correlation between the activated partial thromboplastin time and antibodies against phosphatidylserine and cardiolipin. *Am. J. Obstet. Gynecol.* **163**, 575–584 (1990).

Rothfield, N., Kurtzman, S., Vazques-Abad, D., Charron, C., Daniels, L. and Greenberg, B. Association of anti-topoisomerase 1 with cancer. *Arthritis Rheum.* **35**, 724 (1992).

Rutgers, A., Heeringa, P., Damoiseaux, J.G. and Cohen Tervaert, J.W. ANCA and anti-GBM antibodies in diagnosis and follow-up of vasculitic disease. *Eur. J. Intern. Med.* **14**, 287–295 (2003).

Sun, D., Martinez, A., Sullivan, K.F., Sharp, G.C. and Hoch, S.O. Detection of anticentromere antibodies using recombinant human CENP-A protein. *Arthritis Rheum.* **39**, 863–867 (1996).

Totoritis, M.C., Tan, E.M., McNally, E.M. and Rubin, R.L. Association of antibody to histone complex H2A–H2B with symptomatic procainamide-induced lupus. *N. Engl. J. Med.* **318**, 1431–1436 (1988).

Weigle, W.O. Advances in basic concepts of autoimmune disease, In: *Progress and Controversies in Autoimmune Disease Testing. Clinics in Laboratory Medicine*, vol. **17** (eds Keren, D.F. and Nakamura, R.), 329–340 (W.B. Saunders, Philadelphia, PA, 1997).

Wesierska-Gadek, J., Penner, E., Lindner, H., Hitchman, E. and Sauermann, G. Autoantibodies against different histone H1 subtypes in systemic lupus erythematosus sera. *Arthritis Rheum.* **33**, 1273–1278 (1990).

59 Sexually Transmitted Diseases

Bruce J. Dille, Alan S. Armstrong and Isa K. Mushahwar

NEISSERIA GONORRHOEAE

Etiologic Agent

Neisseria gonorrhoeae, the causative agent of **gonorrhea**, is a Gram-negative diplococcal bacterium. It is a fastidious organism requiring complex media for growth. The medium has to contain appropriate antibiotics to inhibit selectively any undesirable normal flora. Gonococci are antigenically diverse, sharing protein and lipopolysaccharide determinants with other pathogenic and nonpathogenic *Neisseria* species (Mardh and Danielsson, 1990).

Gonorrhea is a major cause of sexually transmitted disease (STD) worldwide. In the USA, the reported incidence of gonorrhea rapidly increased during the 1960s and early 1970s. The annual incidence remained constant (1 million cases) until 1982 followed by a gradually decreasing trend to approximately 326,000 civilian cases in 1996 (Centers for Disease Control, 1997). There are estimated to be 62,000,000 infections annually worldwide.

Pathogenesis

Gonorrhea is an infection of the genitourinary tract in both males and females. In males, following an incubation period of 2–5 days, acute infection usually causes anterior **urethritis** accompanied by purulent discharge (Hook and Handsfield, 1990). Typically, gonococci are present in the exudate within polymorphonuclear leucocytes. In the absence of complications, acute urethritis usually subsides within 2–4 weeks. Symptomatic men seek treatment; however, there are asymptomatic males who remain carriers for months. Complications of gonorrhea in men include chronic urethritis, prostatitis, and epididymitis. The latter infrequently leads to infertility.

In females, the primary site of urogenital infection is the endocervix. The incubation period is generally difficult to define because of vague and nonspecific symptoms. Usually, there is a mild and transient urethritis, which may be accompanied by a purulent endocervical discharge. However, less than 25% of females with gonorrhea will have an easily discernable cervical infection. About 10–15% of infected women develop **gonococcal pelvic inflammatory disease** which usually involves damage to the Fallopian tubes and subsequently may ascend to the ovaries and into the peritoneal cavity. This condition can lead to sterility and, in some cases, is life threatening.

In less than 1% of individuals with gonorrhea, gonococci become disseminated from the mucosa into the blood, usually causing diverse clinical manifestations such as fever, skin lesions, and septic arthritis.

Newborns and children also can be infected with *N. gonorrhoeae*. In newborns, conjunctivitis is the most common clinical manifestation which, if left untreated, may lead to blindness due to corneal scarring. Silver nitrate instilled into the conjunctivae at birth is prophylactic. **Gonococcal vaginitis** is the most common form of gonorrhea in prepubertal girls. In young males, **gonococcal urethritis** is less common.

For three decades, penicillin was the recommended treatment for gonorrhea. However, in 1976, the first penicillin-resistant gonococcus was identified in the USA. Subsequently, infections caused by antimicrobial-resistant gonococci have become an increasing problem. Extrachromosomal plasmids are responsible for resistance to penicillin and tetracycline. In addition, resistance has been found to be chromosomally mediated. In 1989, over 50,000 gonococcal infections in the United States were due to resistant organisms. Therefore, the current Centers for Disease Control recommended therapy for uncomplicated gonorrhea, established in 1993, is a single oral dose of ciprofloxacin (500 mg) or ofloxacin (400 mg) (Krieger, 1995). Treatment failures using this regimen have been reported in the United Kingdom, Australia, Canada, Hong Kong, and the United States. The alternate recommended treatment is then 125 mg ceftiaxone, intramuscularly, or 400 mg cefixime, orally. Different treatment modalities are used for other types of gonococcal infections, but these are not described here.

The Immunoassay Handbook. *Third Edition*
D. Wild (Ed.)

Diagnosis and Assay Technology

Diagnosis of gonorrhea in the symptomatic male with urethritis can be made by Gram-staining a smear of the exudate. The presence of Gram-negative intracellular diplococci is considered diagnostic for gonorrhea. However, in females, although the Gram stain can be used, the interpretation is more difficult. The sensitivity and specificity of a Gram-stained male urethral smear are 75–98% and 95–99%, respectively. In females, a Gram-stained endocervical smear is reported to have a sensitivity of 23–65% and a specificity of 88–100%. The accuracy of interpretation of smears can vary greatly depending on the skill of the microscopist (Goodhart *et al.*, 1982).

To confirm the diagnosis of gonorrhea, particularly in females, a swab specimen should be cultured for the presence of *N. gonorrhoeae*. Although culture is considered the 'gold standard', it has limitations. The sensitivity of a single cervical culture may range from 70 to 90%. Presumptively identified (oxidase positive, Gram-negative diplococci) gonococcal colonies may be confirmed as *N. gonorrhoeae* using the traditional carbohydrate degradation method. However, problems associated with this method have led to the development of nonculture confirmatory tests. These assays utilize specific monoclonal antibodies. The reagents contain either fluorescein-labeled antibody, requiring a fluorescent microscope, or immunoglobulin-coated particles agglutinating in the presence of gonococci. The antibody-based confirmatory tests generally are accurate, demonstrating a sensitivity and specificity of approximately 99% (Ison, 1990; Judson, 1978).

The direct detection of *N. gonorrhoeae* in clinical specimens is accomplished using an enzyme-linked immunosorbent assay (ELISA). The ELISA has the advantage of being able to identify *N. gonorrhoeae* that fail to grow on standard selective culture media. The gonococci do not need to remain viable during transport, and properly transported and stored samples can be used up to 7 days after collection. Antibiotic sensitivity cannot be determined by ELISA, and inadequate sample collection can result in too few organisms on the swab to give accurate results in the assay (Martin *et al.*, 1984; Stamm *et al.*, 1984; Thomas *et al.*, 1986; Woods, 1995).

Direct detection of *N. gonorrhoeae* can also be accomplished using a nonisotopic DNA probe assay. Instead of using antibodies to gonococci antigens, this type of assay uses specific DNA sequences to unique regions of the gonococci DNA or RNA to identify organisms in a sample (Panke *et al.*, 1991). Limitations of this assay format are similar to those of an ELISA.

The number of organisms on the swab is less of a problem for DNA amplification assays, which can detect fewer than 100 organisms. Assays based on polymerase chain reaction (PCR) or ligase chain reaction (LCR) are able to amplify a specific region of DNA unique to *N. gonorrhoeae* to copy numbers easily detectable by nonisotopic instrument-based detection methods (Buimer *et al.*, 1996; Crotchfelt *et al.*, 1997; DiDomenico *et al.*, 1996; Hook *et al.*, 1997). The amplification assays are also able to detect nonviable organisms. Because of their ability to amplify so few organisms to detectable levels, these assays necessitate a reevaluation of culture as the gold standard for gonococci detection. Resolution of discordant samples between these assays and culture requires an independent, but comparable, assay to the amplification assay being evaluated. Biased sensitivity and specificity figures can result if proper care is not taken in analyzing the results (Buimer *et al.*, 1996). Amplification assays are not able to determine antibiotic sensitivity.

Serology is not recommended for the diagnosis of gonorrhea because current antibody tests are unable to distinguish acute from past infections.

REFERENCES

Buimer, M., Van Doornum, G.J.J., Ching, S., Peerbooms, P.G.H., Plier, P.K., Ram, D. and Lee, H.H. Detection of *Chlamydia trachomatis* and *Neisseria gonorrhoeae* by ligase chain reaction-based assays with clinical specimens from various sites: implications for diagnostic testing and screening. *J. Clin. Microbiol.* **34**, 2395–2400 (1996).

Centers for Disease Control, Summary of notifiable diseases, United States 1996. *Morb. Mortal. Wkly Rep.* **45**, 74 (1997).

Crotchfelt, K.A., Welsh, L.E., DeBonville, D., Rosenstraus, M. and Quinn, T.C. Detection of *Neisseria gonorrhoeae* and *Chlamydia trachomatis* in genitourinary specimens from men and women by a coamplification PCR assay. *J. Clin. Microbiol.* **35**, 1536–1540 (1997).

DiDomenico, N., Link, H., Knobel, R., Caratsch, T., Weschler, W., Loewy, Z.G. and Rosenstraus, M. Cobas Amplicor: fully automated RNA and DNA amplification and detection system for routine diagnostic PCR. *Clin. Chem.* **42**, 1915–1923 (1996).

Goodhart, M., Ogden, J., Zaidi, A. and Kraus, S. Factors affecting the performance of smear and culture tests for the detection of *Neisseria gonorrhoeae*. *Sex. Transm. Dis.* **9**, 63–69 (1982).

Hook, E. and Handsfield, H. Gonococcal infections in adults. In: *Sexually Transmitted Diseases* 2nd edn. (eds Holmes, K., Mardh, P., Sparling, F., Wiesner, P., Cates, W., Lemon, S. and Stamm, W.), 149–165 (McGraw-Hill, New York, 1990).

Hook, E.W., Ching, S.F., Stephens, J., Hardy, K.F., Smith, K.R. and Lee, H.H. Diagnosis of Neisseria gonorrhoeae infections in women by using the ligase chain reaction on patient-obtained vaginal swabs. *J. Clin. Microbiol.* **35**, 2129–2132 (1997).

Ison, C. Methods of diagnosing gonorrhoea. *Genitourin. Med.* **66**, 453–459 (1990).

Judson, F. A clinic-based system for monitoring the quality of techniques for the diagnosis of gonorrhea. *Sex. Transm. Dis.* **5**, 141–145 (1978).

Krieger, J.N. New sexually transmitted diseases treatment guidelines. *J. Urol.* **154**, 209–213 (1995).

Mardh, P. and Danielsson, D. *Neisseria gonorrhoeae*. In: *Sexually Transmitted Diseases* 2nd edn. (eds Holmes, K., Mardh, P., Sparling, F., Wiesner, P., Cates, W., Lemon, S. and Stamm, W.), 903–916 (McGraw-Hill, New York, 1990).

Martin, R., Wentworth, B., Coopes, S. and Larson, E. Comparison of Transgrow and Gonozyme for the detection of *Neisseria gonorrhoeae* in mailed specimens. *J. Clin. Microbiol.* **19**, 893–895 (1984).

Panke, E.S., Yang, L.I., Leist, P.A., Magevney, P., Fry, R.J. and Lee, R.F. Comparison of Gen-Probe DNA probe test and culture for the detection of *Neisseria gonorrhoeae* in endocervical specimens. *J. Clin. Microbiol.* **29**, 883–888 (1991).

Stamm, W.E., Cole, B., Fennel, C., Bonin, P., Armstrong, A., Herrmann, J. and Holmes, K. Antigen detection for the diagnosis of gonorrhea. *J. Clin. Microbiol.* **19**, 399–403 (1984).

Thomas, E., Scott, S., Grefkees, I., Hession, G., Pollack, R., Martin, T. and Albritton, W. Validity and cost-effectiveness of the Gonozyme test in the diagnosis of gonorrhea. *Can. Med. Assoc. J.* **134**, 121–124 (1986).

Woods, G.L. Update on laboratory diagnosis of sexually transmitted diseases. *Gynecol. Pathol.* **15**, 665–684 (1995).

CHLAMYDIA TRACHOMATIS

Etiologic Agent

Chlamydia trachomatis is a Gram-negative obligate intracellular bacterium. It multiplies within the cytoplasm of host cells and is able to synthesize both DNA and RNA. Although able to perform most metabolic functions, *C. trachomatis* is dependent on host ATP for energy. To multiply, the chlamydia infectious particle (elementary body, EB) attaches to the host's cell surface. After phagocytosis, the EB is internalized and forms a reticulate body that replicates. The newly formed elements condense to form EBs which at the end of the cycle are released by the host cell to infect new cells. *C. trachomatis* contains a group (genus)-specific carbohydrate which is composed of a lipopolysaccharide (LPS) complex. Species-specific protein antigens (major outer membrane proteins, MOMP) are also present in the cell wall, along with 15 type-specific protein antigens. Serotypes A, B, Ba and C are associated with trachoma. Lymphogranuloma venereum (LGV) is caused by types L1, L2, and L3. Chlamydia-associated nongonococcal urethritis in males and **cervicitis** in women are caused by serotypes D-K.

C. trachomatis is recognized as causing more STDs than *N. gonorrhoeae* and is more common in patients with gonorrhea than those who are not infected. Chlamydia infections became nationally notifiable in the United States in 1995 with nearly 500,000 cases reported in 1996 (Centers for Disease Control, 1997). Nevertheless, there are believed to be over 4 million cases per year in the United States and 3 million cases annually in Europe (Peeling and Brunham, 1996).

Pathogenesis

In women, generally the endocervix is the primary site of infection. Unfortunately, many of those infected are asymptomatic and do not seek medical help. When symptoms such as discharge are present, they are frequently mild and vague, making a diagnosis on clinical information difficult. If these infections remain untreated, they can progress to **pelvic inflammatory disease (PID)** and other complications including premature delivery and prenatal mortality. Infertility and ectopic pregnancy are two long-term sequelae of persistent chlamydia infections. Repeated episodes can lead to scarred Fallopian tubes resulting in sterility in over 50% of infected patients.

In males, approximately 50% of **nongonococcal urethritis** is caused by *C. trachomatis*, and in men with **postgonococcal urethritis** this percentage is higher. The incubation period is usually 7–14 days. Similar to gonorrhea, chlamydia infection in men is manifest by dysuria and discharge; however, the symptoms are usually milder. In fact, even in many chlamydia-infected symptomatic men, the discharge is very scant. Compared to men with gonorrhea, patients with chlamydia urethritis wait much longer before seeking help. In sexually active young men, *C. trachomatis* appears to be the major cause of **epididymitis. Chlamydia proctitis** usually is found only in homosexual men (Martin, 1990; Stamm and Holmes, 1990).

C. trachomatis may be transferred from vaginally infected mothers during birth to as many as 50% of newborns. In addition, the organism may cause **ophthalmic neonatorum** or **conjunctivitis** in up to 50% of infants exposed during delivery. In these infections, blindness is a rare occurrence. Nasopharyngeal infection occurs in 15–20% of the infants, and 3–18% develop pneumonia (Harrison and Alexander, 1990).

Although rare in the USA, *C. trachomatis* causes **endemic trachoma** in North Africa, the Middle East, and South East Asia. Infection may lead to corneal scarring and blindness, which currently is the most common cause of preventable blindness in the world. **LGV**, a lymphoadenopathy of the inguinal region, caused by *C. trachomatis*, is infrequently found in the USA but is prevalent in parts of Africa, Asia, and South America. Acute infections are more easily detected in males than females and, as a result, women usually suffer more severe complications (Perine and Osoba, 1990).

For the treatment of uncomplicated urethral, endocervical or rectal *C. trachomatis* infections, doxycycline is the drug of choice for most infections (Krieger, 1995). Azithromycin may also be used. Both sexual partners should be treated even when asymptomatic. For pregnant women, erythromycin is recommended. Tetracycline and erythromycin are effective in eradicating LGV. Instillation of one prophylactic agent (silver nitrate, erythromycin, or tetracycline) into the eyes of all newborn infants is required to prevent gonococcal ophthalmia neonatorum. However, the reason for their efficacy in preventing chlamydia eye infection is not clear, and further studies are required. To date, significant antimicrobial resistance has not been observed although resistance to tetracycline has been reported recently. Treatment failures have not been observed.

Diagnosis and Assay Technology

The gold standard for the diagnosis of chlamydia infection is the identification of the organism in tissue culture. The sensitivity of this method is about 80% under optimum conditions. This method requires that tissue cells, usually McCoy, be maintained in the laboratory. Clinical specimens (swabs) bathed in appropriate transport media are inoculated onto pretreated cell culture monolayers which are incubated at 37 °C for at least 48 h. After incubation, the monolayer is stained, usually with iodine, Giemsa or fluorescent antibody. Iodine staining is the most commonly used technique for detection of *C. trachomatis* and is based on the reaction of iodine with glycogen formed during the growth of chlamydia. The immunofluorescent monoclonal antibody method, which requires a fluorescence microscope, is reported to be more sensitive than iodine or Giemsa particularly as the number of chlamydia per cell decreases. The sensitivity of the chemical staining techniques is reported to be relatively low, ranging from 20% with male urethral specimens to 40% with samples from the endocervix. Sensitivity may be improved by secondary passage onto fresh cell monolayers (Barnes, 1989).

There are several commercial ELISAs available for the direct detection of *C. trachomatis* in clinical specimens. The immunoassays are capable of detecting chlamydia antigen from urogenital, endocervical, conjunctival, nasopharyngeal and, in some cases, urine specimens, providing results in less than 4 h. As with the direct detection assays of *N. gonorrhoeae*, the organism does not need to be viable, but does need to be present in sufficient numbers to be accurately detected (Black, 1997; Gann *et al.*, 1990).

Specificity of the ELISA can be improved by incorporating a monoclonal antibody to block specific *C. trachomatis* capture. The sensitivity of culture for detecting cervical infection is estimated to be 70–80%. The evaluation of the specificity of a direct antigen test is complicated by apparent false-positive results, due to failure of culture methods to detect all infections. In the blocking assay, a monoclonal antibody is added to the rabbit anti-*C. trachomatis* antibody which specifically blocks the reaction of the assay with *C. trachomatis*. In contrast, in the absence of antigen, the monoclonal antibody will not block a positive result. The specificity of the assay can be improved from 96.9% compared to culture alone or 99.2% compared to culture and immunofluorescence assay to 99.9%. Confirmation of positive samples is particularly important in low-prevalence populations.

Direct specimen detection can be accomplished using a fluorescein-labeled monoclonal antibody specific for the MOMP. The antibody reacts with chlamydia elementary bodies (infectious stage) which appear as pinpoints of apple-green fluorescence when smears are viewed under a fluorescence microscope. The assay can be used with urogenital, rectal, conjunctival, and nasopharyngeal specimens. The sensitivity and specificity using these kinds of specimens are 92–100% and 98–100%, respectively. A critical step in performing this assay is the collection of a specimen having sufficient host cells attached to the slide. The direct specimen test requires less than 1 h to complete and is best utilized when the specimen load is relatively small. Smears stored at 2–8 °C can be held for 7 days before staining. As with any direct fluorescent antibody test, critical parameters are good smear and staining techniques, a properly maintained microscope and a skilled microscopist who can correctly interpret the smear results.

Recently, to improve health care by allowing greater patient access to rapidly advancing immunodiagnostic technology, manufacturers of diagnostic tests for *C. trachomatis* have developed less complex rapid visual readout ELISAs. These tests are instrument-independent and can be performed at or near the site of collection and provide results within 20–30 min. The rapid tests are an advantage where laboratory facilities are limited, or where diagnosis before a patient leaves the office or clinic is desirable so that appropriate treatment can be initiated without having to worry about situations of poor patient compliance and follow-up.

Direct detection of *C. trachomatis* can be accomplished using a nonisotopic DNA probe assay. The assay utilizes a chemiluminescent probe to the 16S ribosomal RNA (rRNA) of chlamydia, and by capitalizing on the 10^4 copies of rRNA per cell achieves detection of about

10-fold fewer EB's than an ELISA (Beebe *et al.*, 1997; Limberger *et al.*, 1992; Pasternack *et al.*, 1997a).

DNA amplification tests can detect between 10 and 100 organisms (Dille *et al.*, 1993). This level of detection from either the PCR or LCR based assays allows for detection of chlamydia in urine samples permitting noninvasive sample collection. Viability of the chlamydia does not need to be maintained during transport, improving that aspect of sample handling. Extreme care must be taken in the laboratory when using amplification tests in order to prevent cross-contamination between samples and introduction of amplified product into a sample being prepared for amplification. The tests employ inactivation schemes to destroy amplified product after the assay is complete, but prevention of contamination before inactivation is still a risk that the laboratory must control (Black, 1997; Mouton *et al.*, 1997; Pasternack *et al.*, 1997b; Peterson *et al.*, 1997).

Assessment of amplification assay performance has similar limitations to those for the *N. gonorrhea* amplification tests. The ability to amplify very few organisms to detectable levels, and lack of reliance on viability means that these assays can have better sensitivity for detection of confirmed infection than culture. At least two different methods need to be used as the standard to confirm infection in order to assess the performance of any assay. Using this expanded gold standard, culture has a sensitivity, depending on the laboratory, of 60–80% which is similar to sensitivities obtained by ELISAs. Amplification assays have percent sensitivities into the 90s with percent specificities into the high 90s.

The leukocyte esterase test (LET) has been used to screen for chlamydia infections. The test detects the presence of polymorphonuclear leukocytes in urine through the presence of an esterase that has been released. The LET performs reasonably as a predictor of urinary tract infection, but poorly as an assay to detect any one specific infection such as chlamydia or gonorrhea (Chow *et al.*, 1996).

Serology has not played an important role in the immunodiagnosis of *C. trachomatis* infections. Clearly, antibody can be induced by infection. However, as many as 30% of a normal population may contain chlamydia antibody. Therefore, the value of serology probably is appreciated best in patients suspected of having systemic chlamydia infections such as PID. The relationship between antibody and infection needs to be explored further before serology becomes a reliable and important diagnostic tool (Woods, 1995).

REFERENCES

Barnes, R. Laboratory diagnosis of human chlamydial infections. *Clin. Microbiol. Rev.* **2**, 119–136 (1989).

Beebe, J.L., Sharpton, T.R., Zanto, S.N., Steece, R.S., Rogers, C. and Mottice, S.L. Performance characteristics of the Gen-Probe probe competition assay used as a supplementary test for the Gen-Probe PACE 2 and 2C assays for detection of *Chlamydia trachomatis*. *J. Clin. Microbiol.* **35**, 477–478 (1997).

Black, C.M. Current methods of laboratory diagnosis of *Chlamydia trachomatis* infections. *Clin. Microbiol. Rev.* **10**, 160–184 (1997).

Centers for Disease Control, Summary of notifiable diseases, United States 1996. *Morb. Mortal. Wkly Rep.* **45**, 74 (1997).

Chow, J.M., Moncada, J., Brooks, D., Bolan, G., Shaw, H. and Schachter, J. Is urine leukocyte esterase test a useful screening method to predict *Chlamydia trachomatis* infection is women? *J. Clin. Microbiol.* **34**, 534–536 (1996).

Dille, B.J., Butzen, C.C. and Birkenmeyer, L.G. Amplification of *Chlamydia trachomatis* DNA by ligase chain reaction. *J. Clin. Microbiol.* **31**, 729–731 (1993).

Gann, P., Herrmann, J., Candib, L. and Hudson, R. Accuracy of *Chlamydia trachomatis* antigen detection methods in a low prevalence population in a primary care setting. *J. Clin. Microbiol.* **28**, 1580–1585 (1990).

Harrison, H. and Alexander, E. Chlamydial infections in infants and children. In: *Sexually Transmitted Diseases* 2nd edn, (eds Holmes, K., Mardh, P., Sparling, F., Wiesner, P., Cates, W., Lemon, S. and Stamm, W.), 811–820 (McGraw-Hill, New York, 1990).

Krieger, J.N. New sexually transmitted diseases treatment guidelines. *J. Urol.* **154**, 209–213 (1995).

Limberger, R.J., Biega, R., Evancoe, A., McCarthy, L., Sliviendki, L. and Kirkwood, M. Evaluation of culture and Gen-Probe PACE 2 assay for detection of *Neisseria gonorrhoeae* and *Chlamydia trachomatis* in endocervical specimens transported to a state health laboratory. *J. Clin. Microbiol.* **30**, 1162–1166 (1992).

Martin, O. Chlamydial infections. *Med. Clin. N. Am.* **74**, 1367–1387 (1990).

Moncada, J., Schachter, J., Bolan, G., Engelman, J., Howard, L., Mushahwar, I., Ridgway, G., Mumtaz, G., Stamm, W. and Clark, A. Confirmatory assay increases specificity of Chlamydiazyme test for *Chlamydia trachomatis* infections of the cervix. *J. Clin. Microbiol.* **28**, 1770–1773 (1990).

Mouton, J.W., Verkooyen, R., van der Meijden, W.I., van Rijsoort-Vos, T.H., Goessens, W.H.F., Kluytmans, J.A.J.W., Deelen, S.D.A., Luijendijk, A. and Verbrugh, H.A. Detection of *Chlamydia trachomatis* in male and female urine specimens by using the amplified *Chlamydia trachomatis* test. *J. Clin. Microbiol.* **35**, 1369–1372 (1997).

Pasternack, R., Vuorinen, P. and Miettinen, A. Evaluation of the Gen-Probe *Chlamydia trachomatis* transcription-mediated amplification assay with urine specimens from women. *J. Clin. Microbiol.* **35**, 676–678 (1997a).

Pasternack, R., Vuorinen, P., Pitkäjärvi, T., Koskela, M. and Miettinen, A. Comparison of manual Amplicor PCR, Cobas Amplicor PCR and LCx assays for detection of *Chlamydia trachomatis* infection in women by using urine specimens. *J. Clin. Microbiol.* **35**, 402–405 (1997b).

Peeling, R.W. and Brunham, R.C. Chlamydia as pathogens: new species and new issues. *Emerg. Infect. Dis.* **2**, 307–319 (1996).

Perine, P. and Osoba, A. Lymphogranuloma venereum. In: *Sexually Transmitted Diseases* 2nd edn, (eds Holmes, K., Mardh, P., Sparling, F., Wiesner, P., Cates, W., Lemon, S. and Stamm, W.), 195–204 (McGraw-Hill, New York, 1990).

Peterson, E.M., Darrow, V., Blanding, J., Aarnaes, S. and de la Maza, L.M. Reproducibility problems with Amplicor PCR *Chlamydia trachomatis* test. *J. Clin. Microbiol.* **35**, 957–959 (1997).

Stamm, W. and Holmes, K. *Chlamydia trachomatis* infections of the adult. In: *Sexually Transmitted Diseases* 2nd edn, (eds Holmes, K., Mardh, P., Sparling, F., Wiesner, P., Cates, W., Lemon, S. and Stamm, W.), 181–193 (McGraw-Hill, New York, 1990).

Stamm, W. and Mardh, P. *Chlamydia trachomatis*. In: *Sexually Transmitted Diseases* 2nd edn, (eds Holmes, K., Mardh, P., Sparling, F., Wiesner, P., Cates, W., Lemon, S. and Stamm, W.), 917–925 (McGraw-Hill, New York, 1990).

Woods, G.L. Update on laboratory diagnosis of sexually transmitted diseases. *Gynecol. Pathol.* **15**, 665–684 (1995).

TRICHOMONAS VAGINALIS

Etiologic Agent

Trichomonas vaginalis is a sexually transmitted protozoan residing in the genitourinary tracts of men and women. The organism is usually ovoid in shape with an anterior blunt end. *T. vaginalis* is an extracellular motile parasite having four anterior flagella which help in locomotion. The organism divides by longitudinal binary fission and can be propagated in complex laboratory media. Extensive immunological heterogeneity exists between isolates of *T. vaginalis* which can undergo phase variation during growth and replication. Several protein and glycoprotein antigens have been identified, some of which appear to be species-specific and which may be useful in the development of diagnostic immunoassays.

Infection with *T. vaginalis* is believed to be the most common STD in the world. It has been estimated that, in the United States, as many as 6 million females may be infected each year. Worldwide, over 180 million infections are thought to occur annually. The prevalence of **trichomoniasis** appears to correlate with the level of sexual activity. For example, *T. vaginalis* can be recovered from about 5% of those attending family planning clinics whereas up to 75% of prostitutes may be infected. As many as 50% of patients attending STD clinics harbor the organism. In addition, gonorrhea occurs twice as frequently in women with trichomoniasis compared with women not infected. Women with disease can infect 30–40% of their sexual partners. However, the prevalence in men generally has not been well documented (Rein and Muller, 1990; Sobel, 1990; Thomason and Gelbart, 1989).

Pathogenesis

In women the incubation period for *T. vaginalis* infection is believed to be in the range of 3–28 days. Vaginal discharge is prominent in about 50–75% of those infected. Other symptoms such as dysuria and mild dyspareuria may be present. In the majority of cases, trichomoniasis does not produce a characteristic discharge. In addition, as many as 50% of those infected may be asymptomatic. In general, these infections do not lead to adverse sequelae and usually are self limiting.

T. vaginalis has been recovered in 5–20% of men having **nongonococcal urethritis**. The majority of infected men are asymptomatic, and only about 50% of symptomatic men have mild discharge. Complications are rare in men, although in acute disease the prostate is often infected.

The recommended treatment for *T. vaginalis* infection is metronidazole, 2 g orally in one dose. Treatment failures do occur and may result from reinfection of the patient by untreated partners. Therefore, sexual partners should be treated. Because of the mutagenic and possibly teratogenic effects, metronidazole should not be prescribed during early pregnancy (Krieger, 1995).

Diagnosis and Assay Technology

The clinical features of trichomoniasis alone are not sufficiently specific to make an accurate diagnosis. Typically, the diagnosis is made by microscopic examination of vaginal or urethral fluid alone although trichomonads can be cultured in the laboratory. In the office setting, microscopy is useful; however, the accuracy is only 60–80% compared to culture. The wet mount is examined for the presence of motile organisms of the typical appearance of *T. vaginalis*. Before a preparation is deemed negative, 10–20 fields at 400 × magnification should be viewed, which may take 10 min. In addition, the pH of the fluid sample should be determined. A pH of at least 4.5 is indicative of infection. To make an accurate diagnosis, the fluid must be fresh so as to observe motile organisms which, if nonmotile, may be confused with host white cells. The failure to see motile trichomonads is a common error which may lead to misdiagnosis (Fouts and Kraus, 1980).

Culture is the most accurate method to diagnose infection. There are several commercial broth media available for the isolation of *T. vaginalis* from clinical specimens. The inoculated broth must be incubated at 37 °C for at least 1 week before infection can be ruled out. During this period, the broth should be regularly inspected microscopically for the presence of motile trichomonads. For maximum recovery, the media must be warmed prior to inoculation. In women, the sensitivity of culture approaches 95%, whereas in men, the culture is about 60% sensitive (Bickley *et al.*, 1989).

Although laboratory diagnostic tests for *T. vaginalis* are relatively simple and fast, most physicians ignore them and treat empirically using the 'shotgun' approach to therapy. Immunoassays have been developed to replace or serve as adjuncts to current diagnostic methods, but, to date, none are in wide use. These assays are immunometric, or 'sandwich' assays, using rabbit antibody-coated solid-phase to capture the antigen. An enzyme-labeled rabbit polyclonal antibody is used to detect the bound antigen. These assays can be completed within 1–4 h. Compared to culture, the sensitivity of these assays varies from 77 to 93%, and the specificity ranges from 97.5 to 100%. In contrast, the conventional wet mount is much less sensitive (39–72%) (Lisa *et al.*, 1988; Yule *et al.*, 1987).

A 2-min latex agglutination test utilizes latex particles coated with rabbit antibodies to *T. vaginalis*. Agglutination of the test sample, but not the control, is indicative of the presence of *T. vaginalis*. Limited results indicate that, compared to culture, the latex test demonstrates a sensitivity of 95% and a specificity of 99%. Experience in reading agglutination patterns is required to differentiate correctly between specific and nonspecific agglutination of the latex particles (Carney *et al.*, 1988).

Currently, many clinicians believe that their interpretation of the wet mount is highly accurate. In addition, they feel that the wet mount, because of its simplicity and low cost, is adequate for the diagnosis of *T. vaginalis* infection. However, it is clear that the vaginal fluid wet mount has inadequate diagnostic sensitivity, and this needs to be convincingly demonstrated to physicians.

To date, although several serological tests have been described for the detection of antibody to *T. vaginalis*, none have been widely accepted. Usually these tests are unreliable for a number of reasons, which limits their clinical utility.

REFERENCES

Bickley, L., Krisher, K., Punsalang, A., Trupei, M., Reichman, R. and Menegus, M. Comparison of direct fluorescent antibody, acridine orange, wet mount, and culture for detection of *Trichomonas vaginalis* in women attending a Public Sexually Transmitted Diseases Clinic. *Sex. Transm. Dis.* **16**, 127–131 (1989).

Carney, J., Unadkat, P., Yule, A., Rajakumar, R., Lacey, C. and Ackers, J. New rapid latex agglutination test for diagnosing *Trichomonas vaginalis* infection. *J. Clin. Pathol.* **41**, 806–808 (1988).

Fouts, A. and Kraus, S. *Trichomonas vaginalis*: reevaluation of its clinical presentation and laboratory diagnosis. *J. Infect. Dis.* **141**, 137–143 (1980).

Krieger, J.N. New sexually transmitted diseases treatment guidelines. *J. Urol.* **154**, 209–213 (1995).

Lisa, P., Dondero, R., Kwiatkoski, D., Spence, M., Rein, M. and Alderete, J. Monoclonal-antibody-based enzyme-linked immunosorbent assay for *Trichomonas vaginalis*. *J. Clin. Microbiol.* **26**, 1684–1686 (1988).

Rein, M. and Muller, M. *Trichomonas vaginalis* and trichomoniasis. In: *Sexually Transmitted Diseases* 2nd edn. (eds Holmes, K., Mardh, P., Sparling, F., Wiesner, P., Cates, W., Lemon, S. and Stamm, W.), 481–492 (McGraw-Hill, New York, 1990).

Sobel, J. Vaginal infections in adult women. *Med. Clin. N. Am.* **74**, 1573–1602 (1990).

Spence, M., Hollander, D., Smith, J., McCaig, L., Sewell, D. and Brockman, M. The clinical and laboratory diagnosis of *Trichomonas vaginalis* infection. *Sex. Transm. Dis.* **7**, 168–171 (1980).

Thomason, J. and Gelbart, S. *Trichomonas vaginalis*. *Obstet. Gynecol.* **74**, 536–541 (1989).

Watt, R., Philip, A., Wos, S. and Sam, G. Rapid assay for immunological detection of *Trichomonas vaginalis*. *J. Clin. Microbiol.* **24**, 551–555 (1986).

Yule, A., Gellan, M., Oriel, J. and Ackers, J. Detection of *Trichomonas vaginalis* antigen in women by enzyme-immunoassay. *J. Clin. Pathol.* **40**, 566–568 (1987).

CANDIDA ALBICANS – VAGINITIS

Also *see* MISCELLANEOUS INFECTIOUS DISEASES: *CANDIDA ALBICANS*, INVASIVE CANDIDIASIS.

Etiologic Agent

Vulvovaginal candidiasis is caused by the *Candida* species, of which at least 90% are *C. albicans*. The vegetative yeast cells are Gram-positive, round or oval in morphology (2.5–6 mm in diameter) and reproduce asexually by budding. Filamentous forms, indicative of infection, may be present in clinical specimens.

Mannan, a well characterized large protein–carbohydrate complex, is a major surface antigen which is important in antigen detection and serology. In addition, less characterized and defined protein antigens present in the cytoplasm have also been utilized in immunoassays (Sobel, 1990b).

C. albicans is the most frequently isolated fungal pathogen and is responsible for approximately 25% of all vaginal infections. **C. vulvovaginitis** is much more prevalent than vaginitis due to *T. vaginalis*. It is difficult to determine the number of women with vaginal candidiasis because it is not a reportable disease. On the basis of estimates, approximately 75% of all women will get at least one infection in their lifetime and many will suffer multiple infections. Less than 5% will be subject to recurrent and chronic infections. About 20% of healthy women will carry the organism without symptoms for various periods of time before eliminating the yeast (Sobel, 1990a).

Pathogenesis

Factors determining the virulence of *C. albicans* are just beginning to be studied. Under some conditions, normal carriers of *C. albicans* become symptomatic and seek medical help. It is clear that, during estrogen therapy and pregnancy, women become more susceptible to vaginal candidiasis. Women undergoing prolonged antibiotic therapy, diabetics, and immunocompromised patients are also subject to candida vulvovaginitis.

In women, intense itching and vaginal discharge are common complaints, but these are neither specific nor consistently associated with vaginitis. In symptomatic women, vulvar itching is the most frequent symptom. The discharge has been described as being like cottage cheese in nature with varying viscosity. Typically, candida vaginitis is benign and self-limiting. After treatment, the infection usually is eradicated. *C. albicans* does not appear to reproduce in the male urogenital tract. However, about 20% of males of infected partners are colonized but show no symptoms. The possibility of sexual transmission of *C. albicans* to partners is open to conjecture. Concurrent cases of candidal vaginitis and bacterial vaginosis have been reported, further complicating diagnosis (Rajakumar *et al.*, 1987).

The recommended treatment for candida vulvovaginitis is butoconazole or clotrimazole for 3 or 7–14 days, respectively. Single dose treatment using clotrimazole vaginal tables or fluconazole orally are also approved. The efficacious treatment of patients with recurrent and chronic vulvovaginitis remains an important clinical challenge (Krieger, 1995).

Diagnosis and Assay Technology

Diagnosis and treatment of vulvovaginal candidiasis are usually based on clinical signs and symptoms. For diagnosis, a wet mount is routinely done which has a sensitivity of 40–60% compared to culture. To improve the accuracy, 10% KOH is added to the wet mount to enhance the observation of germinating yeast forms. During infection vaginal pH is 4–4.5. Although vaginal culture is the most sensitive method available for the detection of *C. albicans*, a positive culture does not unequivocally indicate that symptoms result from the yeast. A significant number (50%) of women may carry low levels of *C. albicans* in the absence of symptoms. However, there are also symptomatic women carrying small numbers of organisms. Several agar media are available, such as Sabouraud dextrose agar, for the laboratory isolation of *C. albicans*. The culture is usually incubated for 7 days at 25–30 °C before a specimen is considered negative. If necessary, standard laboratory methods are available for specific identification of the colonies (Sobel, 1990b).

There are relatively few immunoassays available for the identification of vulvovaginitis candidiasis. Slide latex agglutination tests have been developed for detection of *C. albicans* in vaginal specimens. The presence of *C. albicans* produces visible agglutination of the test latex but not the control latex. The overall sensitivity of the latex assays has been reported to be 71–91% and the specificity 96–98% (Evans *et al.*, 1986; Hopwood *et al.*, 1987; Rajakumar *et al.*, 1987).

Some experience is required to interpret results accurately, particularly when observing a weakly positive agglutination reaction. The simple and rapid latex tests are particularly valuable when microscopy is unavailable, and the immunoassays that can be carried out in the course of the patient examination have merit as a diagnostic tool. However, most physicians believe that the conventional wet mount is a reliable and inexpensive method to diagnose vulvovaginal candidiasis. Therefore, relatively few of them routinely utilize slide agglutination tests.

REFERENCES

Evans, E., Lacey, C. and Carney, J. Criteria for the diagnosis of vaginal candidosis: evaluation of a new latex agglutination test. *Eur. J. Obstet. Gynecol. Reprod. Biol.* **22**, 365–371 (1986).

Hopwood, V., Warnock, D., Milne, J., Crowley, T., Horrocks, C. and Taylor, P. Evaluation of a new slide latex agglutination test for diagnosis of vaginal candidosis. *Eur. J. Clin. Microbiol.* **6**, 392–394 (1987).

Krieger, J.N. New sexually transmitted diseases treatment guidelines. *J. Urol.* **154**, 209–213 (1995).

Rajakumar, R., Lacey, C., Evans, E. and Carney, J. Use of slide latex agglutination test for rapid diagnosis of vaginal candidosis. *Genitourin. Med.* **63**, 192–195 (1987).

Redondo-Lopez, V., Meuwether, C., Schmitt, C., Opitz, M., Cook, R. and Sobel, J. Vulvovaginal candidiasis complicating recurrent bacterial vaginosis. *Sex. Transm. Dis.* **17**, 51–53 (1990).

Sobel, J. Vaginal infections in adult women. *Med. Clin. N. Am.* **74**, 1573–1602 (1990a).

Sobel, J. Vulvovaginal candidiasis. In: *Sexually Transmitted Diseases* 2nd edn, (eds Holmes, K., Mardh, P., Sparling, F., Wiesner, P., Cates, W., Lemon, S. and Stamm, W.), 515–523 (McGraw-Hill, New York, 1990b).

GARDNERELLA VAGINALIS

Etiologic Agent

Historically, *Gardnerella vaginalis*, a Gram-variable coccobacillus bacterium, was considered to be the sole causative agent of **nonspecific vaginitis** in contrast to vaginitis associated specifically with the presence of *C. albicans* or *T. vaginalis*. *G. vaginalis* can be isolated virtually from all women with nonspecific vaginitis and is the most common organism present. However, additional studies demonstrated the presence of *G. vaginalis* in up to 50% of females without nonspecific vaginitis. In symptomatic women, the concentration of *G. vaginalis* is 100–1000-fold higher than in asymptomatic women. Recently, because of better culture methods, other bacteria such as *Mobiluncus* and *Bacteroides* species have been isolated from patients with this clinical syndrome. Therefore, the term **bacterial vaginosis** (**BV**) was introduced to include other bacterial agents as well as *G. vaginalis*. BV is the most common vaginal infection in premenopausal women, and those attending STD clinics have the highest prevalence of BV (Holst *et al.*, 1987; Sobel, 1990).

Pathogenesis

Bacterial vaginosis occurs when normal vaginal flora, predominantly lactobacilli, are replaced with a mixed flora composed of *G. vaginalis*, anaerobes and *Mycoplasma hominis*. Further research is required to understand fully the modification of normal vaginal flora. Bacterial vaginosis may be present in up to 50% of females showing no symptoms. Complaints typical of other vaginal infections, such as pruritis, dysuria, and dyspareuria are not usually present. Generally a thin water-like grayish-white adherent discharge is present on the external vaginal area. Additional clinical abnormalities are not apparent. Historically, BV has been considered to be a female annoyance. However, recently this syndrome has been associated with preterm membrane rupture, low birth weight, postpartum endometritis, and amniotic fluid infection. In males, these organisms do not appear to cause infection, and the role played by men in the sexual transmission of BV is not clear (Hillier and Holmes, 1990).

Metronidazole has been the most effective drug in the treatment of BV. Either a single large dose or smaller multiple doses yield cure rates of at least 90%. Recent studies indicate that the antibiotic clindomycin offers comparable efficacy. It has not been a practice to treat asymptomatic females because patients eventually improve. Unfortunately, as many as 30% of treated infections may recur within a few months after treatment. Most investigations have determined that treatment of sexual partners does not reduce the recurrence (Krieger, 1995; Lossick, 1990).

Diagnosis and Assay Technology

Women with BV do not present with characteristic signs and clinical symptoms, making diagnosis on these grounds alone unreliable. Culture is the most sensitive method. However, isolation of *G. vaginalis* is not specific because as many as 50% of women without BV may be colonized with low levels of *G. vaginalis* (Eschenbach *et al.*, 1988; Nugent *et al.*, 1991; Woods, 1995).

Typically, the diagnosis of BV is made when at least three of the following four criteria are met:

1. a homogeneous adherent discharge;
2. vaginal pH greater than 4.5;
3. addition of 10% KOH to secretions produces a fishy odor;
4. presence of clue cells (exfoliated squamous cells covered with *G. vaginalis*).

Although these diagnostic criteria appear to be straightforward, they are highly dependent on clinical experience and acuity. As a result, it is difficult to determine accurately their sensitivity and specificity. Unfortunately, many clinicians do not utilize these simple tools and rely on 'shotgun' therapy to eliminate symptoms (Sobel, 1997).

Recent evidence indicates that the Gram stain may be a valuable diagnostic tool. Other laboratory methods, such as those based on monoclonal antibodies and DNA probes, have been described but, to date, none appear to offer broad clinical utility. The utility of serology as a diagnostic approach for BV remains to be investigated.

REFERENCES

Eschenbach, D., Hillier, S., Critchlow, C., Stevens, C., DeRouen, T. and Holmes, K. Diagnosis and clinical manifestations of bacterial vaginosis. *Am. J. Obstet. Gynecol.* **158**, 819–828 (1988).

Hillier, S. and Holmes, K. Bacterial vaginosis. In: *Sexually Transmitted Diseases* 2nd edn, (eds Holmes, K., Mardh, P., Sparling, F., Wiesner, P., Cates, W., Lemon, S. and Stamm, W.), 547–559 (McGraw-Hill, New York, 1990).

Holst, E., Wathne, B., Hovelius, B. and Mardh, P. Bacterial vaginosis: microbiological and clinical findings. *Eur. J. Clin. Microbiol.* **6**, 536–541 (1987).

Krieger, J.N. New sexually transmitted diseases treatment guidelines. *J. Urol.* **154**, 209–213 (1995).

Lossick, J. Treatment of sexually transmitted vaginosis/vaginitis. *Rev. Infect. Dis.* **12**, S665–S681 (1990).

Nugent, R., Krohn, M. and Hillier, S. Reliability of diagnosing bacterial vaginosis is improved by a standardized method of Gram stain interpretation. *J. Clin. Microbiol.* **29**, 297–301 (1991).

Sobel, J. Vaginal infections in adult women. *Med. Clin. N. Am.* **74**, 1573–1602 (1990).

Sobel, J.D. Vaginitis. *N. Engl. J. Med.* **337**, 1896–1903 (1997).

Woods, G.L. Update on laboratory diagnosis of sexually transmitted diseases. *Gynecol. Pathol.* **15**, 665–684 (1995).

HERPES SIMPLEX VIRUS TYPE 2

Etiologic Agent

Herpes simplex virus type 2 (HSV-2) shares about 50% base sequence homology with HSV type 1, and almost all of the virus structural proteins exhibit antigenic homology. The HSV-2 double-stranded DNA has a relative molecular mass of just under 10^8, with a coding capacity for nearly 80 proteins, of which the function of only a handful is understood (Roizman and Sears, 1990). Upon entry into a cell the virus takes over the biosynthetic pathways of the cell, and produces viral proteins in a tightly regulated cascade starting with the **immediate early** (α) proteins. Messenger RNA for this set of regulatory proteins is transcribed directly from the infecting virus DNA in the absence of new DNA synthesis, and some of these proteins direct the synthesis of the **early** (β) proteins. The early proteins shut off α protein synthesis and direct the production of DNA and the **late** or **structural** proteins.

Virion assembly then takes place in the cell nucleus. The virus buds from the inner nuclear membrane to acquire its envelope with at least eight virus-coded glycoproteins. Under conditions that are not yet understood, productive virus infection does not occur in some cells and the virus establishes latency. During the latent state the viral DNA remains in the host cell with little or no protein synthesis and no progeny virus is produced. The molecular changes that release the virus from this latent state and allow for virus replication are also a mystery.

Pathogenesis

Genital herpes has been recognized as an STD since the mid-1700s. HSV-2 causes 80–90% of all primary genital herpes infections and greater than 95% of all recurrent infections. It is estimated that over 500,000 new cases of genital herpes occur each year in the USA. Recent serology studies in the United States have found that 21.9% of the general population over age 12 show evidence of previous HSV-2 exposure (Flemming *et al.*, 1997). The seroprevalence increased 30% from 1988 to 1994 with a 5 × increase occurring in white teenagers. Of major concern are reports that over 90% of those with serological evidence of previous exposure to HSV-2 give no history of infection and are unaware of their exposure. There is, therefore, a large population of infected people who can unknowingly transmit virus.

HSV-2 is acquired through direct contact with virus shed from another person. Infections acquired in the absence of HSV-1 antibody are usually more severe. The onset of symptoms occurs 7–10 days after exposure and the symptoms generally last 7–14 days. A typical clinical sequence starts with redness at the lesion site (macular stage) which becomes raised (papular) before forming a clear blister (vesicular stage) (Sunstrum, 1989). Multiple blisters may be present in primary infections. The blisters become pustular before breaking to begin the ulcerative stage. The ulcer then heals, becoming crusted, and the lesion clears without a scar. During primary infection secondary blisters may form as the first lesion begins to heal. The virus establishes latency in the sacral ganglia (for genital infections) very early in the course of the infection.

Reactivation can be triggered by stress, hormone changes, trauma to the lesion site, etc. Recurrences occur at a site involved in the original infection, and usually follow the same sequence of stages as a primary infection but last 3–5 days. Very short or absent vesicular stages and atypical ulcers are not uncommon.

Asymptomatic genital infections are a significant problem. HSV-2 infection can be acquired asymptomatically, particularly if antibody from a prior HSV-1 infection is present (Koutsky *et al.*, 1990; Nahmias *et al.*, 1990). In addition, 2–5% of all infected individuals are shedding virus asymptomatically at any given time. It is estimated that 70% of HSV-2 transmission occurs during periods of asymptomatic shedding.

This asymptomatic transmission has alarming epidemiological implications and makes control of virus transmission particularly difficult. Patient education and proper screening methods are essential prerequisites for effective control. Treatment of primary infection is with topical or oral acyclovir, which significantly reduces the severity of the infection but does not prevent the establishment of latency. Frequent or severe recurrences are treated with oral acyclovir or valaciclovir (Fife *et al.*, 1997; Stone and Whittington, 1990).

The most serious consequence of genital herpes infection is to the newborn at the time of delivery (Whitley *et al.*, 1988). Untreated neonatal herpes has a very high mortality rate (80%) with less than 10% of the survivors developing normally. An estimated 2500 cases of neonatal herpes occur each year in the USA. If the risk for infection is known from the history of the mother, and infection is identified early and treated promptly with acyclovir, the prognosis for the newborn is significantly improved.

Diagnosis and Assay Technology

Cell culture is the 'gold standard' for HSV diagnosis (Sunstrum, 1989). A number of culture confirmation schemes have been introduced to support the interpretation of cytopathic effects (CPE) in the cells. Typically, a sample is collected by rubbing the base of a lesion with a cotton or dacron swab (calcium alginate swabs inactivate the virus) and placing the swab in transport medium (buffered salts or culture medium with gelatin or bovine serum albumin). Samples should be transported and stored short-term at 4 °C or long-term at −60 °C. Storage at −20 °C should be avoided.

The best cultures for clinical specimens are primary cell lines such as primary rabbit kidney, mink lung, or human embryonic lung, foreskin, or tonsil. Confirmation of culture using commercial polyclonal or monoclonal antibodies can be carried out at various times after infection. (Monoclonal antibodies can be used for typing.) Screening at 2 days postinfection can be used for a faster turn-around time with another culture incubating for 7 days. Confirmation after CPE is observed, or after 7 days in culture, is used to verify microscopic findings. This can be performed using either a fluorescein conjugate, which requires an ultraviolet microscope, or an enzyme-linked conjugate followed by the addition of an appropriate substrate, with observation using a light microscope.

A number of rapid, nonculture tests are available in a variety of formats. No direct antigen detection test yet has the sensitivity to assess acceptably the asymptomatic shedding of virus because of the low numbers of virus present (Mertz, 1990). Immunofluorescence (IFA) tests can be used directly on patient samples or on cells in culture. By employing monoclonal antibodies, typing of virus isolates is possible. The direct assay is very dependent on the quality of the sample to ensure that an adequate number of cells are collected. This method also requires investment in an ultraviolet microscope and expertise to use it properly.

Enzyme-immunoassays (ELISA) capture antigen in the sample onto a solid support, then anti-HSV antibody labeled with enzyme is added. Substrate is introduced and color read by an appropriate instrument. These assays provide an objective print-out of the results, but not typing of the virus. Published sensitivities for ELISA range from mid-60% to low-90% with percent specificity generally in the 90s (Corey, 1986).

Latex agglutination tests are also rapid to perform, but have very subjective interpretation. Reported sensitivities for this type of assay range down to under 50% (Cruse and Jenkins, 1987). *In situ* hybridization using a HSV-specific DNA probe is also available for direct testing on patient samples or for culture confirmation. Like the IFA, the direct assay is dependent on an adequate sample to provide enough cells.

A variety of IFA and EIA tests are available to detect antibody to HSV. A true primary infection can be diagnosed by the appearance of HSV-specific antibody in convalescent serum, but only if acute serum was negative for HSV antibody. Antibody titers may remain unchanged during recurrent episodes. Research assays have established that the HSV type-specific glycoprotein G (gG) is capable of distinguishing previous exposure to HSV-1 and HSV-2 serologically (Dille *et al.*, 1990; Lee *et al.*, 1985; Spiezia *et al.*, 1990; Sullender *et al.*, 1988). Commercial assays based on gG are becoming available. No assay which uses whole virus or infected cell lysate preparations is able to accurately distinguish HSV-1 from HSV-2.

REFERENCES

Corey, L. Laboratory diagnosis of herpes simplex virus infection. *Diagn. Microbiol. Infect. Dis.* **4**, 111S–119S (1986).

Cruse, M.A. and Jenkins, S.G. Detection of herpes simplex virus in clinical specimens by latex agglutination and ELISA techniques. *87th Annual Meeting of American Society for Microbiology* C-141 (1987)

Dille, B.J., Spiezia, K.V., Kifle, L. and Mushahwar, I.K. Evaluation of a recombinant herpes simplex type 2 glycoprotein in an enzyme-linked immunoassay for the detection of specific HSV-2 antibodies. *30th Annual Interscience Conference on Antimicrobial Agents and Chemotherapy* No. 195 (1990)

Fife, K.H., Barbarash, R.A., Rudolph, T., Degregorio, B. and Roth, R. Valaciclovir versus acyclovir in the treatment of first-episode genital herpes infection. Results of an international, multicenter, double-blind, randomized clinical trial. The Valaciclovir International Herpes Simplex Study Group. *Sex. Transm. Dis.* **24**, 481–486 (1997).

Flemming, D.T., McQuillan, G.M., Johnson, R.E., Nahmias, A.J., Aral, S.O., Lee, F.K. and St. Louis, M.E. Herpes simplex virus type 2 in the United States, 1976 to 1994. *N. Engl. J. Med.* **337**, 1105–1111 (1997).

Koutsky, L.A., Ashley, R.L., Holmes, K.K., Stevens, C.E., Critchlow, C.W., Kiviat, N., Lipinski, C.M., Wülner-Hanssen, P. and Corey, L. The frequency of unrecognized type 2 herpes simplex infection among women. *Sex. Transm. Dis.* **17**, 90–94 (1990).

Lee, F.K., Coleman, R.M., Pereira, L., Bailey, P., Tatsuno, M. and Nahmias, A.J. Detection of herpes simplex virus type 2 specific antibodies with glycoprotein G. *J. Clin. Microbiol.* **22**, 641–644 (1985).

Mertz, G.J. Genital herpes simplex virus infections. *Med. Clin. N. Am.* **74**, 1433–1454 (1990).

Nahmias, A.J., Lee, F.K. and Beckman-Nahmias, S. Sero-epidemiological and sociological patterns of herpes simplex virus infection in the world. *Scand. J. Infect. Dis.* **69** (Suppl.), 19–36 (1990).

Roizman, B. and Sears, A.E. Herpes simplex viruses and their replication. In: *Virology* (eds Fields, B.N. and Knipe, D.M.), 1795–1841 (Raven Press, New York, 1990).

Spiezia, K.V., Dille, B.J., Mushahwar, I.K., Kifle, L. and Okasinski, G.F. Prevalence of specific antibodies to herpes simplex virus type 2 as revealed by an enzyme-linked immunoassay and Western blot analysis. *Adv. Exp. Med. Biol.* **278**, 231–242 (1990).

Stone, K.M. and Whittington, W.L. Treatment of genital herpes. *Rev. Infect. Dis.* **12**, S610–S619 (1990).

Sullender, W.M., Yasudawa, L.L., Schwartz, M., Pereira, L., Hensleigh, P.A., Prober, C.G. and Arvin, A.M. Type-specific antibodies to herpes simplex virus type 2 (HSV-2) glycoprotein G in pregnant women, infants exposed to maternal HSV-2 infection at delivery, and infants with neonatal herpes. *J. Infect. Dis.* **157**, 164–171 (1988).

Sunstrum, J. Herpes simplex infections: a review. *J. Clin. Immunoassay* **12**, 175–178 (1989).

Whitley, R.J., Corey, L., Arvin, A., Lakeman, F.O., Sumaya, C.V., Wright, P.F., Dunkle, L.M., Steele, R.W., Soong, S., Nahmias, A.J., Alford, C.A., Powell, D.A. and San Joaquin, V. NIAID Collaborative Antiviral Study Group and Changing presentation of herpes simplex virus infection in neonates. *J. Infect. Dis.* **158**, 109–116 (1988).

SYPHILIS (*TREPONEMA PALLIDUM*)

Etiologic Agent

The causative agent of syphilis is *Treponema pallidum* (*Spirocheta pallida*), a delicate corkscrew-shaped spirochete of 6–15 μm in length that possesses 5–20 rigid and regular spirals. Attempts (over the past 90 years) to cultivate this spirochete have been unsuccessful owing to the inability to identify a culture medium capable of supporting its growth. The surface of the spirochete has been found to be antigenically inert, hence the inability of the afflicted patient to develop a protective immune response, resulting in a prolonged infection.

Pathogenesis

Syphilis is a STD of humans, with an incubation period of 10–90 days. The period of communicability is during the primary and secondary stages of the disease and also during mucocutaneous relapses. Congenital transmission occurs throughout the child-bearing period. Transfusion-acquired syphilis is very rare because the spirochete cannot survive in citrated blood stored at 4 °C beyond 72 h (Guthrie, 1951). The clinical manifestations of the disease are classified by the stage of infection as primary, secondary, and tertiary syphilis.

The chancre (syphilitic ulcer) of primary syphilis usually appears within 2–9 weeks of infection of the genital organ. The secondary stage of the disease is characterized by a generalized skin rash, particularly palmar and plantar rashes, accompanied by low-grade fever, malaise and lymphadenopathy. The tertiary stage of the disease is characterized by cardiac and central nervous system involvement that eventually leads to the death of the host (Termini and Music, 1972).

Three pathogenic mechanisms in syphilis have recently been identified: dissemination of *T. pallidum* throughout the body and passage through intercellular junctions of endothelial cells; avoidance of the host's immune system due to the antigenic inertness of the spirochete membrane; and disease manifestation which is accelerated by the immune response of the host to material associated with the outer membrane of the spirochete (Thomas *et al.*, 1988).

Diagnosis and Assay Technology

The serologic tests for syphilis are classified into preliminary or **nontreponemal** tests and confirmatory or **treponemal** tests. The nontreponemal tests measure both IgM and IgG class antibodies to lipids released from damaged host cells because of treponemal infections. These tests are nonspecific and generally used for screening. All reactive sera have to be confirmed by specific treponemal tests. Currently, there are four nontreponemal tests. These include Venereal Disease Research Laboratory (VDRL), unheated serum reagin (USR), rapid plasma reagin (RPR), and the reagin screen test (RST).

The treponemal tests detect specific antibodies to the antigenic proteins of *T. pallidum*. Currently there are four such tests. These are the fluorescent treponemal antibody absorption test (FTA-ABS); fluorescent treponemal antibody absorption double staining (FTA-ABS DS); microhemagglutination assay for antibodies to *T. pallidum*

(MHA-TP), and the hemagglutination treponemal test (HATTS) (Hunter, 1971; Wentworth *et al.*, 1978).

Treponemal antibodies generally appear early in the course of syphilis, peak during the secondary stage, and often remain high for many years. Reagents for both nontreponemal and treponemal tests are available in kit form from many commercial manufacturers (Cummings *et al.*, 1996; Larsen *et al.*, 1995).

REFERENCES

Cummings, M.C., Lukehart, S.A., Marra, C., Smith, B.L., Shaffer, J., Demeo, L.R., Castro, C. and McCormack, W.M. Comparison of methods for the detection of *Treponema pallidum* in lesions of early syphilis. *Sex. Transm. Dis.* **23**, 366-369 (1996).

Guthrie, N. Failure of stored syphilitic blood to transmit syphilis: a case report. *J. Ven. Dis. Infect.* **32**, 243–246 (1951).

Hare, M.J. Serological tests for treponemal disease in pregnancy. *J. Obstet. Gynaecol. Br. Common.* **80**, 515–519 (1973).

Hunter, E.F. Characteristics of patient sera, conjugates and antigens used in FTA-ABS tests. *Ann. N.Y. Acad. Sci.* **177**, 48–53 (1971).

Larsen, S.A., Steiner, B.M. and Rudolph, A.H. Laboratory diagnosis and interpretation of tests for syphilis. *Clin. Microbiol. Rev.* **8**, 1–21 (1995).

Termini, B.A. and Music, S.I. The natural history of syphilis. *South. Med. J.* **65**, 241–245 (1972).

Thomas, D.D., Navab, H. and Fogelmen, D.A. *Treponema pallidum* invades intracellular junctions of endothelial cell monolayers. *Proc. Natl Acad. Sci. USA* **85**, 3608–3612 (1988).

Wentworth, B.B., Thompson, M.A., Peter, C.R., Bawdon, R.E. and Wilson, D.L. Comparison of a hemagglutination treponemal test for syphilis (HATTS) with other serologic methods for the diagnosis of syphilis. *Sex. Transm. Dis.* **5**, 103–111 (1978).

60 Congenital Diseases of Microbiological Origin

Bruce J. Dille, John W. Safford Jr. and Isa K. Mushahwar

CYTOMEGALOVIRUS

Etiologic Agent

Human cytomegalovirus (CMV), a member of the Herpesviridae family, is extremely complex, containing a linear double-stranded DNA of about 230 kb and over 200 open reading frames (Chee *et al.*, 1990). There are over 30 structural proteins in the virus, which consists of a 110 nm capsid surrounded by an amorphous tegument, which in turn is enclosed in an envelope containing surface spikes of viral glycoprotein. The slowly replicating virus characteristically causes cell enlargement and intranuclear inclusions.

Pathogenesis

CMV is endemic to populations throughout the world, infecting 50–80% of the adults in developed nations. The virus is transmitted from person to person primarily via oropharyngeal secretions under conditions of close personal contact, but can also be transmitted vertically by intrauterine transmission, by blood transfusion, and by bone marrow and solid organ transplantation (Bruggeman, 1993). For instance, children in day care centers may acquire the virus and spread it to their mothers and other family members. In the healthy child or adult, primary infection with CMV is usually subclinical or results in a very mild disease which resembles infectious mononucleosis. During the course of the infection, virus is found in the blood, throat, and urine for weeks to months. Acute primary infection is followed by establishment of a latent infection which can recur asymptomatically under certain conditions with renewed viral shedding. Reinfection with antigenically diverse exogenous virus is not uncommon.

CMV infection causes serious disease in a number of situations in which the host is immunologically deficient or receives large doses of the virus. Active maternal infection during pregnancy leads to infection of the immunologically immature fetus with devastating effects. Immunosuppressed allograft recipients and patients with acquired immunodeficiency syndrome (AIDS) frequently develop severe CMV infections. Both primary and reactivated latent infections during pregnancy can lead to the infection of the fetus with CMV. The rate of infection of the primary infection is much greater, at 40%, than the 0.2–1.8% occurring during reactivation. CMV is the most common cause of congenital infection in the United States with an estimated occurrence between 0.5 and 2.2% of all live births (Nelson and Demmler, 1997). Approximately, 10% of neonates infected *in utero* will be symptomatic at birth while 90% will be chronically infected but asymptomatic. The most common signs of congenital CMV among the symptomatic group are petechiae (small, hemorrhagic spots), hepatosplenomegaly, jaundice, and microcephaly. Mortality in this group is high (10–20%) (Litwin and Hill, 1997), and those surviving are likely to suffer mental retardation (70%), blindness (20%), or deafness (50%). Approximately, 10–15% of the asymptomatic neonate group will develop intelligence defects and deafness later in life. Both groups will shed high concentrations of virus in urine and the nasopharynx for years.

Allograft recipients undergoing immunosuppression are subject to CMV disease which can be initiated in several ways (Kanj *et al.*, 1996). Primary infection, in which CMV is transmitted to a seronegative recipient from a seropositive organ donor, is the most frequent cause of significant CMV disease in allograft recipients. Patients seropositive at the time of transplantation may develop overt CMV disease by reactivation of latent virus or by reinfection with virus from the organ donor. Fever and respiratory difficulty are usually the first signs of CMV infection. This may be followed by pneumonia, which is particularly serious if bacterial or fungal coinfection is present. Hepatitis, chorioretinitis, and gastroenteritis are also observed. Graft rejection is higher in CMV-infected

The Immunoassay Handbook. *Third Edition*
D. Wild (Ed.)

recipients; however, a causal relationship between CMV and graft rejection has not yet been proved.

CMV is the major opportunistic infection in AIDS patients (Dal Monte *et al.*, 1996); virtually all homosexual men have evidence of recent infection, either acquired or reactivated. Viremia or shedding of virus in urine or semen is observed in at least 50% of AIDS patients. CMV is responsible for, or at least contributes to, several of the diseases associated with AIDS including retinitis, gastroenteritis, and pneumonitis (Hennis *et al.*, 1989; Klotman and Hamilton, 1987).

CMV is transmitted by blood transfusion but poses very little risk except in certain situations. Seronegative blood should be used for seronegative, pregnant patients, seronegative, low birth weight infants, and seronegative recipients of organs from seronegative donors.

Diagnosis and Assay Technology

Cell culture is the 'gold standard' for diagnosis of CMV disease, detecting virus in samples such as urine, blood, throat washings, and bronchoalveolar lavage. Cytopathic effects (CPE) can be observed in high titered samples within 3–7 days, but most cultures are held up to 3 weeks or longer to ensure that low virus titers have a chance for CPE to develop (Ehrnst, 1996). Using the shell vial tissue culture technique allows results of CMV culture to be reported within 18–48 h. Enzyme immunoassays (ELISA) have been used to detect CMV antigens in urine but with poor sensitivity due to the fact that β_2-microglobulin binds to the virus and inhibits the ELISA. The polymerase chain reaction (PCR) has successfully detected CMV in urine from congenitally infected infants and may become more widely used as standardized tests become readily available.

Serological assays for IgG, IgM, or total (IgG, IgM, and IgA) antibody to CMV are useful in screening, determining susceptibility to primary infection, providing serological evidence of recent infection, and in certain situations, differentiating between primary and reactivated infection or reinfection. Many immunoassay technologies have been used for the detection of antibody to CMV, including complement fixation (CF), indirect hemagglutination (IHA), indirect fluorescent antibody assay (IFA), radioimmunoassay (RIA), latex agglutination (LA), ELISA and the recently introduced microparticle enzyme-immunoassay (MEIA). To date, no internationally recognized CMV antibody standard has been developed to which assays can be standardized. Of the many technologies available, three of them, LA, ELISA, and MEIA, are becoming the most widely used. LA is simple and quick to perform and is scored subjectively. ELISAs may be configured to detect IgG antibody, IgM antibody, or total antibody and, depending on the system, are easy to perform and provide objective results. The MEIA assays are specific for either IgG or IgM antibody and provide objective results. MEIA is simple to perform, automated and rapid, with 24 samples processed in about 35 min.

Assay protocols and manufacturers' instructions for sample preparation and storage should be closely followed. Because CMV infection is often accompanied by significant rheumatoid factor (RF), it is particularly important that tests used to detect IgM antibody should include reagents and procedures to neutralize RF.

The selection of seronegative blood units and organ donors can be critical, as discussed above, but this can be accomplished using current assays for IgG-specific or total antibody. It is also possible to identify seronegative women who are at risk of primary infection during pregnancy and who could be informed of ways to avoid situations presenting high risk of exposure to CMV. Recent infection (primary, reactivated or reinfection) is indicated by increasing levels of IgG antibody when comparing paired sera. Seroconversion from negative to positive confirms that the infection acquired during the interval between the first and second serum draw is primary. A positive IgM antibody result suggests a recent infection with CMV but must be interpreted cautiously. Primary infection is accompanied by the appearance of IgM antibody to CMV; however, reactivated infection and reinfection with exogenous virus may also induce detectable IgM antibody levels. In addition, IgM antibody is not a clear indicator of recent infection because prolonged IgM responses are not unusual, particularly in chronic CMV infections. Detection of IgM antibody to CMV in cord blood is diagnostic for congenital infection. However, a negative result cannot rule out congenital CMV disease. Studies have shown that the assays fail to detect IgM antibody in 20–30% of cases in which cell culture has confirmed congenital CMV (Numazaki and Chiba, 1997).

REFERENCES

Bruggeman, C. Cytomegalovirus and latency: an overview. *Virchows Arch. B Cell Pathol.* **64**, 323–333 (1993).

Chee, M., Bankier, A., Beck, S., Bohni, R., Brown, C., Cerny, R., Horsnell, T., Hutchinson, C., Kouzarides, T., Martignetti, J., Preddie, E., Satchwell, S., Tomlinson, P., Weston, K. and Barrell, B. Analysis of the protein-coding content of the sequence of human cytomegalovirus strain AD169. *Curr. Top. Microbiol. Immunol.* **154**, 125–169 (1990).

Dal Monte, P., Lazzarotto, T., Ripalti, A. and Landini, M. Human cytomegalovirus infection: a complex diagnostic problem in which molecular biology has induced a rapid evolution. *Intervirology* **39**, 193–203 (1996).

Ehrnst, A. The clinical relevance of different laboratory tests in CMV diagnosis. *Scand. J. Infect. Dis. Suppl.* **100**, 64–71 (1996).

Hennis, H., Scott, A. and Apple, D. Cytomegalovirus retinitis. *Surv. Ophthalmol.* **34**, 193–203 (1989).

Kanj, S., Sharara, S., Clavien, P. and Hamilton, J. Cytomegalovirus infection following liver transplantation: review of the literature. *Clin. Infect. Dis.* **22**, 537–549 (1996).

Klotman, M. and Hamilton, J. Cytomegalovirus pneumonia. *Sem. Resp. Infect.* **2**, 95–103 (1987).

Litwin, C. and Hill, H. Serologic and DNA-based testing for congenital and perinatal infections. *Pediatr. Infect. Dis. J.* **16**, 1166–1175 (1997).

Nelson, C. and Demmler, G. Cytomegalovirus infection in the pregnant mother, fetus, and newborn infant. *Infect. Perinatol.* **24**, 151–160 (1997).

Numazaki, K. and Chiba, S. Current aspects of diagnosis and treatment of cytomegalovirus infections in infants. *Clin. Diagn. Virol.* **8**, 169–181 (1997).

RUBELLA VIRUS

Etiologic Agent

Rubella virus is the sole member of the genus Rubivirus within the Togaviridae family. The virion is a spherical particle, about 60–70 nm in diameter. A single copy of the message sense, single-stranded RNA genome is contained within a 30 nm nucleocapsid which is surrounded by a lipid envelope derived from the host cell. Two glycosylated structural proteins are inserted in the envelope, E1 (58–62 kDa) and E2 (42–54 kDa). A nonglycosylated structural protein, C (32–38 kDa), is used in the construction of the nucleocapsid (Wolinsky, 1990).

The virus multiplies in the cytoplasm of the infected cells and is assembled at either intracellular or plasma membrane sites that have been modified by insertion of the envelope glycoproteins. The virus acquires the lipid envelope and is released from the cell through a process called budding. Although experimental infections can be induced in a variety of laboratory animals, man is the only known natural reservoir for rubella virus.

Pathogenesis

Postnatal primary infections, acquired by inhalation of aerosols, are nearly always mild. Within a week after exposure, virus can be found in both the blood and nasopharynx; the latter being responsible for the spread of the virus from one individual to another. Symptoms of primary rubella infection may include a rash (the hallmark of clinical rubella), low-grade fever, lymphadenopathy, sore throat, conjunctivitis, and arthralgia. Joint involvement is very common in adults, particularly among women. Serious sequelae are rare but central nervous system involvement (0.02%) and thrombocytopenia (0.03%) have been reported (see Figure 60.1).

Serology plays a critical role in the diagnosis of rubella for two main reasons. First, 30% of primary infections are subclinical; and second, other rash-inducing illnesses may confuse a diagnosis of acute, primary rubella infection. A primary immune response, consisting both of IgG and of IgM antibodies, can be detected at about the same time as the rash. IgM antibody usually falls to undetectable or very low levels within 4–6 months; however, in some cases, it may remain at detectable levels for a year or more. IgG antibody may eventually decline to a low level but lasts indefinitely.

Life-long immunity is nearly always conferred on the individual following infection. Reinfection of immune individuals is subclinical, without viremia, although a low level of nasopharyngeal virus may be detected. A secondary immune response induced by reinfection may be demonstrated by rising levels of IgG antibody with little, if any, IgM antibody.

Although most postnatal infections are subclinical to mild and of little consequence, primary rubella infection during pregnancy can be devastating. Viremia associated with primary infection leads to infection of the placenta and fetus which can lead to fetal death. Those infants surviving fetal infection may exhibit one or more of a variety of symptoms known collectively as **congenital rubella syndrome** (**CRS**), which include low birth weight, deafness, eye disease, mental retardation, cardiac abnormalities, hepatomegaly, splenomegaly, and thrombocytopenia (Cochi *et al.*, 1989). Virus may be isolated from virtually every organ and is shed in urine and nasopharyngeal secretions; a few infants will continue to shed virus for well over a year.

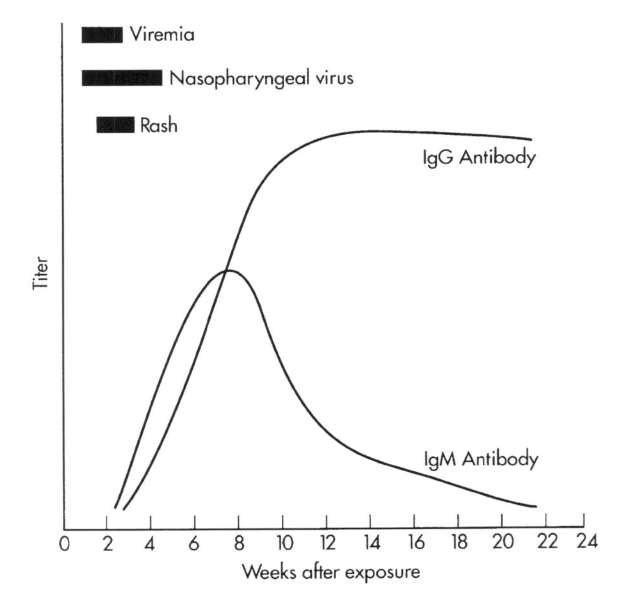

Fig. 60.1 Primary rubella virus infection.

Infants born without apparent symptoms may develop late-onset disease, months to years later, which includes hearing loss, mental retardation, retinopathy, and diabetes mellitus. Rare cases of late-onset progressive **rubella panencephalitis** have been reported. The risk of CRS or late-onset disease is as high as 85% if infection occurs during the first trimester, falls to 10–24% during the 13th–16th weeks and approaches 0% beyond the 20th week of gestation. No treatment, other than symptomatic, is available and the incidence of CRS must be controlled by the immunization of susceptible individuals. A rubella vaccine was introduced in the USA in 1969 during which 57,686 cases of postnatal rubella were reported; 28 years later the reported cases had fallen to 161. The number of confirmed and compatible cases of CRS has been dramatically reduced from a high of 77 in 1970 to only 4 in 1997 (Centers for Disease Control, 1998).

Diagnosis and Assay Technology

Diagnosis of rubella infection and the determination of an individual's immune status are accomplished with serological assays detecting antibody in serum samples (Litwin and Hill, 1997). Hemagglutination inhibition (HAI) was the first serological assay to find general use in clinical laboratories and remains the standard against which many of the newer assays are compared. HAI has largely been replaced by passive hemagglutination (PHA), latex agglutination, ELISA, and MEIA (Herrmann, 1985).

ELISA and MEIA have the advantages of differentiating between IgG and IgM antibody classes and of providing objective data not subject to variations in interpretation. Some ELISAs and the IgG-specific MEIA provide quantitative results expressed in International Units (IU) of antibody per mL referenced to the World Health Organization International Standard for Anti-Rubella Serum. The MEIA has the added advantages of automation and providing results for 24 sera in 35 min (Abbott *et al.*, 1990; Fiore *et al.*, 1988). Assay protocols and manufacturers' instructions for sample preparation and storage should be closely followed.

Use of the appropriate assay provides data required in the three principal areas of rubella serodiagnosis: immune status screening, detection of recent primary infection, and diagnosis of congenital rubella syndrome.

Immune Status Screening

It is generally agreed that levels of antibody detectable by HAI, about 10–15 IU/mL, are protective. Antibodies at or above these levels are sufficient to protect an individual from clinical reinfection; levels of antibody below 10–15 IU/mL may not prevent clinical rubella after exposure. Instances of CRS in babies born to mothers who had detectable antibody prior to conception have been reported but are very rare. Qualitative immune or nonimmune results are provided by LA and PHA, both of which are subjectively interpreted by the technician. Objective results are obtained using IgG-specific ELISA or MEIA assays.

Serodiagnosis of Recent Primary Infection

The first method routinely available to clinical laboratories to serodiagnose recent exposure to the virus involved testing paired sera by HAI. The first sample was drawn from the patient during the acute stage of the disease and a second, the convalescent sample, about 10–14 days later. A fourfold, or greater, rise in HAI titer indicated recent exposure. IgG-specific ELISAs available today may be used in a similar manner if the equivalent of a fourfold rise in HAI titer for that assay has been defined. However, paired sera testing only differentiates primary from secondary rubella infections if the acute sample is seronegative and the convalescent sample is positive, demonstrating seroconversion. This distinction is of particular importance during pregnancy because secondary maternal infections present little threat to the fetus whereas primary maternal infection can have devastating effects.

Primary infection induces both an IgG and a significant IgM antibody response. The secondary immune response induced by secondary rubella infection is characterized by rising IgG antibody in the absence of significant IgM antibody. IgM-specific ELISA or the IgM MEIA may be used specifically to detect only that class of antibody directed against rubella virus. The presence of clinically significant levels of IgM antibody is serodiagnostic evidence of recent, primary rubella infection. Because of the relatively short duration of IgM antibody, the time at which a sample is drawn after exposure is critical. The optimum time to obtain the sample for IgM antibody assays is from 1 to 3 weeks after onset of symptoms (typically the rash), because sera taken before or after that interval may not contain clinically significant levels of IgM antibody. In IgM assays, RF can cause false-positive reactions in the presence of IgG antibody to rubella virus requiring that the IgM antibody assay incorporate some means to neutralize RF to ensure a correct result.

Serodiagnosis of Congenital Rubella Syndrome

Maternal IgM antibody induced by primary rubella infection does not cross the intact placenta as does maternal IgG. The infected fetus develops IgM antibody to the virus and detection of that antibody in the neonate aids in the diagnosis of CRS. A sample should be drawn from the neonate as soon as possible following parturition. Comparison of IgG antibody in the neonate at the time of birth with that 6 months later also aids in the diagnosis of CRS. A significant drop in antibody over that time interval suggests decreasing maternal antibody in the absence of perinatal or prenatal infection. Pre- or perinatal infection should be suspected if stable or increasing IgG antibody levels are detected.

REFERENCES

Abbott, G.G., Safford, J.W., MacDonald, R.G., Craine, M. and Applegren, R.R. Development of automated immunoassays for immune status screening and serodiagnosis of rubella infection. *J. Virol. Methods* **27**, 227–240 (1990).

Centers for Disease Control, Provisional cases of selected notifiable diseases preventable by vaccination, United States, weeks ending December 27, 1997, and December 28, 1996 (52nd week). *Morbid. Mortal. Weekly Rep.* **46**, 1259–1269 (1998).

Cochi, S.L., Edmonds, L.E., Dyer, K., Greaves, W.L., Marks, J.S., Rovira, E.Z., Preblud, S.R. and Orenstein, W.A. Congenital rubella syndrome in the USA, 1970–1985. On the verge of elimination. *Am. J. Epidemiol.* **129**, 349–361 (1989).

Fiore, M., Mitchell, J., Doan, T., Nelson, R., Winter, G., Grandone, C., Zeng, K., Haraden, R., Smith, J., Harris, K., Leszczynski, J., Berry, D., Safford, S., Barnes, G., Scholnick, A. and Ludington, K. The Abbott IMx™ automated benchtop immunochemistry analyzer system. *Clin. Chem.* **34**, 1726–1732 (1988).

Herrmann, K.L. Available rubella serologic tests. *Rev. Infect. Dis.* **7**, S108–S112 (1985).

Litwin, C. and Hill, H. Serologic and DNA-based testing for congenital and perinatal infections. *Pediatr. Infect. Dis. J.* **16**, 1166–1175 (1997).

Wolinsky, J.S. Rubella. In: *Fields Virology*, vol. 1 (eds Fields, B.N. and Knipe, D.M.), 815–838 (Raven Press, New York, 1990).

TOXOPLASMA GONDII

Etiologic Agent

Toxoplasma gondii is an obligate intracellular parasite infecting a wide variety of birds and mammals including man. The sexual cycle of *T. gondii* occurs in the intestine of cats, the definitive host, after ingestion of oocysts or infected animals. Ultimately, oocysts are excreted in the feces and ingested by the wide variety of intermediate hosts (Lynfield and Guerina, 1997).

The rapidly growing form of *T. gondii*, the tachyzoite or trophozoite, develops in tissues of the host followed by a slower-growing form, the bradyzoite, which becomes encysted. The tachyzoite is an oval- to crescent-shaped organism about 2–4 μm wide and 4–8 μm long, rounded at one end and pointed at the other. The tachyzoite attaches to the host cell membrane and is able to modify the membrane by means of a penetration-enhancing factor. Entry into the cell is followed by rapid multiplication and cell lysis, which releases organisms that then repeat the cycle. In intermediate hosts, bradyzoites develop within cells which can contain from a few to several thousand organisms. They can persist for the life of the host in virtually every organ, but most commonly this occurs in the brain, heart, and skeletal muscle. It is thought that the immune system of healthy individuals holds the encysted bradyzoites in check. This form of the parasite is relatively resistant to digestive fluids and is responsible for the spread of infection among carnivorous animals. Infection of herbivores occurs by the ingestion of oocysts excreted in cat feces.

Pathogenesis

Toxoplasmosis is acquired by man via the ingestion of cat feces or undercooked meats (most often lamb). Serological surveys show that exposure to *T. gondii* varies widely from one geographical area to another and ranges from about 20 to 90% of the population. Infection of the normal adult is usually asymptomatic. In those cases with clinical manifestations the most common symptom is lymphadenopathy, which can be accompanied by an array of other symptoms which can make differential diagnosis difficult. Severe to fatal infections can occur in adults immunocompromised by cancer chemotherapy or by the immunosuppressive treatment that accompanies organ transplantation. *T. gondii* has been recognized as the most frequent cause of encephalitis in AIDS patients. Infections in immunocompromised adults are thought to be reactivated latent infections and usually involve the central nervous system.

Transplacental transmission of the parasite, resulting in **congenital toxoplasmosis**, can occur during acute acquired maternal infection. Approximately, 2–6 cases of congenital toxoplasmosis occur for every 1000 live births (Litwin and Hill, 1997). On average, a woman who seroconverts during pregnancy has a 40% chance of transmitting the parasite to the fetus if left untreated (Holliman *et al.*, 1991). The individual risk and outcome of fetal infection is a function of the time at which maternal infection is acquired during pregnancy. Maternal infections acquired before conception present very little, if any, risk to the fetus. The incidence of congenital toxoplasmosis increases as pregnancy progresses; conversely the severity of congenital toxoplasmosis is greatest when maternal infection is acquired early in pregnancy. The majority of infants infected *in utero* are asymptomatic at birth, particularly, if infection occurred during the third trimester, with sequelae appearing later in life – from 1 month to 9 years. Congenital toxoplasmosis results in severe generalized or neurologic disease in about 20–30% of infants infected *in utero*; approximately 10% exhibit ocular involvement only, and the remaining 60–70% are asymptomatic at birth. Up to 85% of the asymptomatic congenitally infected infants will go on to develop sequelae including chorioretinitis, severe neurologic disability, and mental retardation

(Koppe and Rothova, 1989; Wilson *et al.*, 1980). Subclinical infection may result in premature delivery and subsequent neurologic, intelligence, and audiologic defects.

Diagnosis and Assay Technology

Serological assays used to aid in the diagnosis of toxoplasmosis include the Sabin–Feldman dye test, indirect hemagglutination assays (IHA), IFA, immunosorbent agglutination assay (ISAGA), ELISA, and MEIA (Alger, 1997).

The dye test, although of remarkable specificity and widely considered the standard of toxoplasma serology, is not routinely used. This is because the laboratory must maintain a source of viable tachyzoites, and a high degree of technical expertise is required to perform the test and interpret the results.

The IHA results must be interpreted with caution because antibody becomes detectable only after weeks to months postexposure; false-positive reactions with 'natural' IgM antibodies have been reported.

IFA can be adapted to detect specific IgG or IgM antibodies and provides good results in the hands of skilled technicians able to interpret fluorescent reaction patterns correctly. IgM-specific IFA is subject to false-positive reactions with antinuclear antibodies and false-negative results may occur in the presence of high levels of IgG antibody. More accurate IgM IFA results are obtained if IgM is separated from IgG on small columns provided for that purpose.

ISAGA IgM antibody results are subjectively interpreted by scoring the agglutination of tachyzoites bound to IgM antibody captured by an anti-IgM-coated solid phase.

ELISA and MEIA technologies are becoming the most widely used methods because they provide accurate, objective results for both IgG and IgM antibodies and require minimal technical expertise. MEIA is fully automated and can process 24 samples in about 35 min (Safford *et al.*, 1991).

The dye test, IFA, ELISA, and MEIA detect antibody within 1–2 weeks after toxoplasma infection and are therefore suitable as screening tests. Tests used to detect IgM antibody should include reagents and procedures to neutralize RF to ensure accurate results. Assay protocols and manufacturers' instructions for sample preparation and storage should be closely followed.

Perhaps the most important use of toxoplasma serology is for screening women during pregnancy to determine if acute acquired infection occurred after conception. Studies have shown that prenatal diagnosis of infection followed by prenatal therapy reduces the frequency and severity of congenital toxoplasmosis. A negative result identifies women at risk who should be monitored during pregnancy. Seroconversion from negative to positive gives an accurate estimation of the time of infection. Rising titers of antibody or the presence of IgM antibody are indicative of recent exposure.

Determination of the date of infection based solely on detectable IgM antibody to *T. gondii* is not recommended but should include the clinical history and the previous serology, because low levels of IgM antibody may persist for up to 12 years (Bobic *et al.*, 1991). Acute acquired toxoplasmosis in normal adults is difficult to diagnose and may present only as mild lymphadenopathy. The demonstration of rising antibody titers or the presence of IgM antibody is a valuable aid in the diagnosis of this disease with nonspecific symptoms.

Toxoplasma serology has not proved to be useful in diagnosing toxoplasmic encephalitis in the AIDS patient. However, a negative result may be helpful to rule out toxoplasmic encephalitis, which is usually a reactivated latent infection. AIDS patients positive for antibody to *T. gondii* are at risk of developing toxoplasmic encephalitis. It has been reported that the presence of antibody in cerebrospinal fluid is diagnostic of toxoplasmic encephalitis.

REFERENCES

Alger, L. Toxoplasmosis and parvovirus B19. *Infect. Obstet.* **11**, 55–75 (1997).

Bobic, B., Sibalic, D. and Djurkovic-Djakovic, O. High levels of IgM antibodies specific for *Toxoplasma gondii* in pregnancy 12 years after primary toxoplasma infection. *Gynecol. Obstet. Invest.* **31**, 182–184 (1991).

Holliman, R., Johnson, J. and Constantine, G. Difficulties in the diagnosis of congenital toxoplasmosis by cordocentesis: case report. *Br. J. Obstet. Gynaecol.* **98**, 832–834 (1991).

Koppe, J. and Rothova, A. Congenital toxoplasmosis. A long-term follow-up of 20 years. *Int. Ophthalmol.* **13**, 387–390 (1989).

Litwin, C. and Hill, H. Serologic and DNA-based testing for congenital and perinatal infections. *Pediatr. Infect. Dis. J.* **16**, 1166–1175 (1997).

Lynfield, R. and Guerina, N. Toxoplasmosis. *Pediatr. Rev.* **18**, 75–83 (1997).

Safford, J.W., Abbott, G.G., Craine, M.C. and MacDonald, R.G. Automated microparticle enzyme immunoassays for IgG and IgM antibodies to *Toxoplasma gondii*. *J. Clin. Pathol.* **44**, 238–242 (1991).

Wilson, C., Remington, J., Stagno, S. and Reynolds, D. Development of adverse sequelae in children born with subclinical congenital *Toxoplasma* infection. *Pediatrics* **66**, 767–774 (1980).

IgG AVIDITY TESTS

It is well known that IgM antibodies to rubella, CMV, and toxoplasma infections are long lasting. In many cases, IgM antibodies last for several months after a primary infection with these infectious agents. Because of this, it is very difficult to differentiate between recent (acute) and remote infection utilizing currently available IgM immunoassays. It has been reported several years ago (Inouye *et al.*, 1984) that the avidity of specific IgG antibody is low in primary acute viral infections and it increases with time. Thus, the IgG avidity index is useful in distinguishing acute from remote infection. This has been demonstrated to be the case for rubella (Hedman and Rousseau, 1989), CMV (Grangeot-Keros *et al.*, 1997), *Toxoplasma* (Roberts *et al.*, 2001), Hanta (Hedman *et al.*, 1991), Epstein–Barr (Gray, 1995), and herpes 6 (Ward *et al.*, 1993) infections. The avidity index is calculated by measuring absorbancies in the presence and absence of a denaturing agent such as urea, SDS, and ethanolamine. The avidity index is calculated as follows:

$$\frac{\text{Absorbancy value (assay with denaturing agent)}}{\text{Absorbancy value (assay without denaturing agent)}} \times 100$$

Avidity indices of 50% or less are considered low avidity indices.

REFERENCES

Grangeot-Keros, L., Mayaux, M., Lebon, P., Freymuth, F., Eugene, G., Stricker, R. and Dussaix, E. *J. Infect. Dis.* **175**, 944–946 (1997).

Gray, J. Avidity of EBV VCA-specific IgG antibodies: distinction between recent primary infection, past infection and reactivation. *J. Virol. Methods* **52**, 95–104 (1995).

Hedman, K. and Rousseau, S. Measurement of avidity of specific IgG for verification of recent primary rubella. *J. Med. Virol.* **27**, 288–292 (1989).

Hedman, K., Vaheri, A. and Brummer-Korvenkontio, M. Rapid diagnosis of hantavirus disease with an IgG-avidity assay. *Lancet* **338**, 1353–1356 (1991).

Inouye, S., Hasegawa, A., Matsuno, S. and Katow, S. Changes in antibody avidity after virus infectious: detection by an immunosorbent assay in which a mild protein-denaturing agent is employed. *J. Clin. Microbiol.* **20**, 525–529 (1984).

Roberts, A., Hedman, K., Luyasu, V., Zufferey, J., Bessieres, M., Blatz, R., Candolfi, E., Decoster, A., Enders, G., Gross, U., Guy, E., Hayde, M., Ho-Yen, D., Johnson, J., Lecolier, B., Naessens, A., Pelloux, H., Thulliez, P. and Petersen, E. Multicenter evaluation of strategies for serodiagnosis of primary infection with *Toxoplasma gondii*. *Eur. J. Clin. Microbiol. Infect. Dis.* **20**, 467–474 (2001).

Ward, K., Gray, J., Joslin, M. and Sheldon, M. Avidity of IgG antibodies to human herpesvirus-6 distinguishes primary from recurrent infection in organ transplant recipients and excludes cross-reactivity with other herpesviruses. *J. Med. Virol.* **39**, 44–49 (1993).

PARVOVIRUS B19

Etiologic Agent

Parvovirus B19 is an autonomously replicating member of the genus parvovirus, family Parvoviridae. B19 is a nonenveloped icosohedral virus 18–26 nm in diameter, containing a linear single-stranded DNA. The viral genome has been fully sequenced (Cotmore and Tatersall, 1984) and is approximately 5500 nucleotides in length, with a 300 nucleotide palindromic terminal repeat at each end. The genome contains two major open reading frames (Carper and Kurtzman, 1996). Open reading frame 1 (ORF-1) (map position 5–40%) encodes one and possibly two nonstructural proteins. ORF-2 encodes the two structural proteins VP1 (approximately 83 kDa) and VP2 (approximately 60 kDa). VP2 is the major structural protein, forming capsids in an approximate ratio of 6:1 with VP1. The start site for VP2 is within the VP1 coding region and the amino acid sequence of VP2 is fully contained within VP1.

Pathogenesis

Parvovirus B19 was first described morphologically in 1975 (Cossart *et al.*, 1975), but was not conclusively linked to human disease until 1981, when the virus was implicated as a causative agent for hypoplastic crises in patients with **sickle cell anemia** (Pattison *et al.*, 1981). Evidence for this association has been strengthened, and it now appears that at least 90% of sickle cell aplastic crises are caused by B19 infection. The virus has also been linked to a variety of other hemolytic conditions, including **chronic anemia** in patients with immunodeficiencies (Centers for Disease Control, 1989).

Parvovirus B19 is a common human infection with seroprevalence in developed countries from 10% in children under 5 to over 85% in those over 70 (Kerr, 1996). Volunteer challenge studies indicate that the main route of infection is respiratory, and the primary site of infection is the erythroid precursor cell. The lytic nature of the viral life cycle results in a significant decline in circulating reticulocytes, and a subsequent fall in the level of hemoglobin. Healthy subjects with long-lived red blood cells (RBCs) show little effect. However, individuals with reticuloendothelial disorders are unable to compensate for this reticulocytopenia, and a severe anemic crisis often occurs. In healthy individuals, the transient effect on the reticuloendothelial system is often followed by the symptomatic appearance of the erythematous rash, fever, and arthropathy described below.

Anderson *et al.* (1984) demonstrated that B19 was the causative agent for the disease **erythema infectiosum**

(EI), also known as **Fifth's disease**. This is a common, often subclinical disease, primarily in children. When symptoms occur, the disease is characterized by an intense erythematous facial rash (**slapped cheek syndrome**), fever, and occasionally **polyarthralgia**.

Studies of B19 outbreaks and volunteer challenge studies indicate that the disease progresses in two stages. The first stage is an intense viremia that generally occurs 5–10 days postinfection and is sometimes associated with nonspecific symptoms such as fever and chills. This stage also correlates with reticulocytopenia and detection of IgM in sera. If symptoms appear, the classic manifestations of EI (fever, rash, polyarthralgia) generally appear after the viremic phase, between days 14 and 18. IgG titers begin to rise in this time frame as well. The illness is generally short-lived and innocuous.

The most significant complication is joint involvement, which is rare in children, but is often the primary symptom and sequelae in adults. Patients infected with B19 may present to rheumatology clinics with symptoms consistent with early benign rheumatoid arthritis but test as RF negative (Reid *et al.*, 1985). This polyarthralgia generally resolves within a few weeks, but can persist for significantly longer in some cases (see below).

A third significant association with human illness is the linkage with fetal infection and fetal loss. Spontaneous abortion associated with hydrops fetalis and proven presence of B19 viral nucleic acids in fetal tissues were first described by Anand *et al.* (1987). Epidemiologic studies indicate that maternal infection is not always transmitted, and fetal infection is usually resolved without significant morbidity or mortality (Anand *et al.*, 1987). However, the Centers for Disease Control (CDC) estimate that the risk of fetal death after exposure to B19 is between 1.5 and 2.5%. These significant statistics indicate the need for additional research and investigation into whether infection control methods are useful and necessary.

Human Parvovirus B19 and Rheumatoid Arthritis

The etiology of rheumatoid arthritis is unknown, however, many studies have indicated the role of a virus such as rubella, cytomegalovirus, human parvovirus B19, and human T cell leukemia virus as causative agents of this classic inflammatory synovitis disease. Recently, Takahashi *et al.* (1998) showed unequivocally that human parvovirus B19 is a causative agent for rheumatoid arthritis. These authors had shown that parvovirus B19 DNA was detected in the synovial tissue in 30 of 39 patients with rheumatoid arthritis and infrequently in those with osteoarthritis and traumatic joints. They also showed that the expression of the B19 antigen VP-1 was specific (27/27) in rheumatoid arthritis with active synovial lesions, but not in osteoarthritis and controls. The target cells of B19 were macrophages, follicular dendritic cells, T cells, and B cells, but not synovial lining cells in the synovium. They also showed that the expression of VP-1 and the production of interleukin 6 and tumor necrosis factor α were significantly inhibited by the addition of neutralizing antibody for B19, suggesting that B19 detected in rheumatoid synovial cells is infective. They concluded that B19 is involved in the initiation and perpetuation of rheumatoid synovitis leading to joint lesions.

Diagnosis and Assay Technology

Because the virus can only be propagated in primary bone marrow culture, standard methods of viral growth and isolation cannot be used. However, in the acute phase of infection, virus can be readily detected by RIA or ELISA, because the virus often reaches titers of 10^{11} per mL. During this phase of infection, characteristic hematological changes are also detectable, in conjunction with the signs and symptoms described above.

IgM immune response can be detected within 2–3 days of infection and generally lasts 2–3 weeks. IgG response follows rapidly (within 5 days of the IgM response) and appears to persist for life. Care must be taken in interpreting assay results when used to determine when infection occurred since specific IgM has been detected up to 10 months after exposure, and some patients never mount an IgG response (Alger, 1997; Serjeant *et al.*, 1993).

The virus has also been detected by direct DNA hybridization to viral sequences present in serum or tissue (Kajigaya *et al.*, 1991), and by PCR amplification of viral nucleic acids (Anderson *et al.*, 1985). Amplification offers the most sensitive means of viral detection, with a theoretical limit of a single viral genome. With the advent of more 'user-friendly' procedures for sample preparation and amplification, this could become the method of choice for highly sensitive detection of the virus. During acute infection, however, the viremia only lasts 2–4 days and is usually gone by the time the patient presents for diagnosis with the rash (Brown and Young, 1997).

REFERENCES

Alger, L. Toxoplasmosis and parvovirus B19. *Infect. Obstet.* **11**, 55–75 (1997).

Anand, A., Gray, E.S., Brown, T., Clewley, J.P. and Cohen, B.J. Human parvovirus infection in pregnancy and hydrops fetalis. *N. Engl. J. Med.* **316**, 183–186 (1987).

Anderson, M.J., Lewis, E., Kidd, I.M., Hall, S.M. and Cohen, B.J. An outbreak of erythema infectiosum associated with human parvovirus infection. *J. Hyg.* **93**, 85–93 (1984).

Anderson, M.J., Jones, S.E. and Minson, A.C. Diagnosis of human parvovirus infection by dot-blot hybridization using cloned viral DNA. *J. Med. Virol.* **15**, 163–172 (1985).

Brown, K. and Young, N. Parvovirus B19 in human disease. *Annu. Rev. Med.* **48**, 59–67 (1997).

Carper, E. and Kurtzman, G. Human parvovirus B19 infection. *Curr. Opin. Hematol.* **3**, 111–117 (1996).

Centers for Disease Control, Risks associated with human parvovirus B19 infection. *Morbid. Mortal. Weekly Rep.* **38**, 81–88; 93–97 (1989).

Cossart, Y.E., Field, A.M., Cant, B. and Widdows, D. Parvovirus-like particles in human sera. *Lancet* **i**, 72–73 (1975).

Cotmore, S.F. and Tatersall, P. Characterization and molecular cloning of a human parvovirus genome. *Science* **226**, 1161–1165 (1984).

Kajigaya, S., Fujii, H., Field, A., Anderson, S., Rosenfeld, S., Anderson, L.J., Shimada, T. and Young, N.S. Self-assembled B19 Parvovirus capsids, produced in a baculovirus system, are antigenically and immunogenically similar to native virions. *Proc. Natl Acad. Sci. USA* **88**, 4646–4650 (1991).

Kerr, J. Parvovirus B19 infection. *Eur. J. Clin. Microbiol. Infect. Dis.* **15**, 10–29 (1996).

Kinney, J.S., Anderson, L.J., Farrar, J., Strikas, R.A., Kumar, M.L., Kleigman, R.M., Sever, J.L., Hurwitz, E.S. and Sikes, R.K. Risk of adverse outcomes of pregnancy after human parvovirus B19 infection. *J. Infect. Dis.* **157**, 663–667 (1988).

Pattison, J.R., Jones, S.E., Hodgson, J., Davis, L.R., White, J.M., Stroud, C.E. and Murtaza, L. Parvovirus infections and hypoplastic crises in sickle cell anemia. *Lancet* **i**, 664–665 (1981).

Reid, D.M., Reid, T.M.S., Brown, T., Rennie, J.A.N. and Eastmond, C.J. Human parvovirus-associated arthritis: a clinical and laboratory description. *Lancet* **ii**, 422–425 (1985).

Serjeant, G., Serjeant, B., Thomas, P., Anderson, M., Patou, G. and Pattison, J. Human parvovirus infection in homozygous sickle cell disease. *Lancet* **342**, 1237–1240 (1993).

Takahashi, Y., Murai, C., Shibata, S., Munakata, Y., Ishii, T., Ishii, K., Saitoh, T., Sawai, T., Sugamura, K. and Sasaki, T. *Proc. Natl Acad. Sci. USA* **95**, 8227–8232 (1998).

GROUP B STREPTOCOCCI

Etiologic Agent

Streptococcus agalactiae is commonly called group B streptococci (GBS), a major cause of perinatal morbidity and mortality which is capable of being spread venereally. Although the organism was isolated in the 1930s, from postpartum women with fever, the importance as a pathogen causing early- and late-onset neonatal infections was not recognized until later. The organism is a Gram-positive coccus which can be cultivated on routine media (i.e. blood agar). GBS have been subclassified into five serotypes (Ia, Ib, Ic, II, III) based on the composition of cell wall polysaccharides. Type III strains appear to be associated with the bulk of late-onset infections.

Based on estimates, GBS is the leading cause of neonatal bacterial sepsis in the USA, accounting for about 12,000 cases annually. Of those, about 9000 are early-onset and 3000 are late-onset (Schuchat, 1998). In the USA, over 300 infants die each year from this infection.

Pathogenesis

Two distinct forms of neonatal disease can be found: an early-onset and a late-onset syndrome. Clinically, early-onset GBS infection is manifested usually within 72 h after birth. Usually, the organism is transmitted from the colonized mother to the infant at or near birth. Over 80% of the cases occur within the first 3 days postpartum. In most cases the respiratory tract is involved and organisms can be isolated from blood. Often respiratory stress is evident. About 1–2% of infants born of colonized mothers become diseased (Alkalay *et al.*, 1996). However, recent studies suggest that the rate may be higher, particularly when the mother is heavily colonized. Neonates with this form of disease are often premature, having a low birth weight. The mortality of early-onset disease ranges from 30 to 50%.

Late-onset infection usually affects normal-appearing infants 2–4 weeks of age, and generally is less fulminant than early-onset disease. About 50% of the late-onset infections result from colonized mothers. In about 80% of the cases, the major clinical manifestation is associated with meningitis; however, some studies show sepsis to be a major presenting manifestation. Usually infection cannot be traced to obstetric problems in the mother. Thus, infant colonization may come from other sources, as well as from the mother. The infant mortality rate is 15–25% but significant sequelae, such as nerve deficits and blindness, can occur in up to half of the survivors.

Maternal carriage of GBS may be chronic, but usually it is intermittent or transient. About 70% of prenatal carriers remain positive at delivery. In contrast, less than 10% of women become culture positive at the time of delivery.

In vitro, GBS are sensitive to penicillin and ampicillin, as well as other antibiotics. However, in spite of appropriate sensitivity, treatment failures and recurrences in neonates occur with penicillin. A few studies indicate that therapy employing antibiotic combinations may be effective, but additional studies are needed to confirm these results. Prophylactic antibiotic treatment of colonized mothers has not been found to be effective at preventing infection of the newborn. Administration of penicillin during labor reduces neonatal colonization and sepsis in infants of GBS carriers (Gilbert *et al.*, 1995). Additional controlled studies are needed to confirm the efficacy of this approach. Because the gastrointestinal tract is thought likely to be an important reservoir of infection, therapeutic elimination of the carrier state may be difficult to accomplish.

Both the American Academy of Pediatrics (AAP) and the American College of Obstetricians and Gynecologists (ACOG) issued recommendations for the prevention of early-onset neonatal GBS infection (American Academy of Pediatrics, 1992; American College of Obstetricians and Gynecologists, 1992, 1993). The recommendations at best are controversial as to the routine screening of all pregnant women for the detection of GBS and the timing of intrapartum antibiotic treatment (Towers, 1995). They agree, however, as to the timing of intrapartum antibiotic treatment of women with high risk factors associated with GBS. These include preterm labor or preterm rupture of the membranes before 37 weeks gestation, fever in labor, multiple births, rupture of membranes for more than 18 h at any gestational age, or a previous affected child.

Diagnosis and Assay Technology

In neonates with respiratory distress, having characteristic GBS infection, the Gram stain of respiratory or tracheal aspirations can be very helpful in narrowing the range of possible etiologic agents. Routinely, specimens of body fluids suspected of harboring GBS are cultured on blood agar and/or selective broth (Watson and Fenton, 1995). Colonies typically appear pigmented (orange-yellow) and most show a narrow zone of weak beta (clear) hemolysis surrounding the colony. Presumptively identified colonies should be subjected to standard procedures such as sensitivity to bacitracin, bile–esculin reaction, and hydrolysis of hippurate.

Several direct antigen-detection systems have been developed to facilitate the diagnosis of GBS infections. Methods, including counter immunoelectrophoresis, coagglutination, and latex particle agglutination (LPA), have been devised to detect antigen in body fluids such as blood, tracheal aspirates, cerebrospinal fluid, and urine (Gibbs *et al.*, 1992; Walker *et al.*, 1992). Of the three, the LPA format appears to be the most popular. These assays have been reported to be 88–100% sensitive and 81–100% specific, and utilize latex beads coated with polyclonal rabbit anti-GBS polysaccharide antibodies. Control particles are provided and should be used with each specimen. Sensitivities are reported to be about a nanogram of GBS polysaccharides, detecting a minimum of 10^5 organisms per mL. Thus, the use of LPA tests with patients having invasive GBS disease may produce false-negative results.

Serology has not been used as a method for the diagnosis of GBS infections. The presence of anti-GBS antibodies in normal pregnant females may be the result of a previous GBS infection, and thus has no diagnostic significance.

REFERENCES

Alkalay, A., Brunell, P., Greenspan, J. and Pomerance, J. Management of neonates born to mothers with group B streptococcus colonization. *J. Perinatol.* **16**, 470–477 (1996).

American Academy of Pediatrics, Guidelines for prevention of group B streptococcal infection by chemoprophylaxis. Committee on infectious diseases and committee on fetus and newborn. *Pediatrics* **90**, 775–778 (1992).

American College of Obstetricians and Gynecologists, Group B streptococcal infections in pregnancy. *ACOG Tech. Bull.* **170**, 1–5 (1992).

American College of Obstetricians and Gynecologists, Group B streptococcal infections in pregnancy. ACOG recommendations. *ACOG Newslett.* **37**, 2 (1993).

Gibbs, R., Hall, R., Yow, M., McCracken, G. and Nelson, J. Consensus: perinatal prophylaxis for group B streptococcal infection. *Pediatr. Infect. Dis. J.* **11**, 179–183 (1992).

Gilbert, G., Isaacs, D., Burgess, M., Garland, S., Grimwood, K., Hogg, G. and McIntyre, P. Prevention of neonatal group B streptococcal sepsis: is routine antenatal screening appropriate. *Aust. NZ J. Obstet. Gynaecol.* **35**, 120–126 (1995).

Schuchat, A. Epidemiology of group B streptococcal disease in the United States shifting paradigms. *Clin. Microbiol. Rev.* **11**, 497–513 (1998).

Towers, C.V. Group B streptococcus: the US controversy. *Lancet* **346**, 197–198 (1995).

Walker, C., Crombleholme, W., Ohm-Smith, M. and Sweet, R. Comparison of rapid tests for detection of group B streptococcal colonization. *Am. J. Perinatol.* **9**, 304–308 (1992).

Watson, W. and Fenton, L. Group B streptococcal infections in the perinatal period: current approaches. *SDJ Med.* **48**, 149–153 (1995).

61 Hepatitis

Isa K. Mushahwar

HEPATITIS A VIRUS, ANTI-HAV IgM, ANTI-HAV IgG

Etiologic Agent

The hepatitis A virus (HAV) is a positive-sense, 27 nm, single-stranded non-enveloped RNA virus of the family Picornavirdae. It does not appear to be highly defective, and there is no immunologic similarity to any of the other known hepatitis virus antigens. During the course of HAV disease, there is a short period of viremia, but there is no evidence of a long-term carrier state. During the early incubation period and prior to the onset of clinical symptoms, the virus is excreted in relatively large quantities in the feces.

Pathogenesis

The term **hepatitis** means inflammation of the liver. The liver has a wide range of metabolic functions and the symptoms of hepatitis are correspondingly generalized. Early symptoms are similar to influenza: general fatigue, pain in the joints and muscles, and loss of appetite. Nausea, vomiting, and diarrhea or constipation may follow with an increase in temperature. The liver may gradually become enlarged and tender. Bilirubin (a waste product from the breakdown of red blood cells) may accumulate in the blood, the skin, and the eyes causing **jaundice** (**icterus**). Symptoms vary in severity between patients infected with HAV, with about one-third remaining asymptomatic. Only about 10% of patients infected with HAV become jaundiced. Symptomatic patients suffer from an abrupt onset of the disease, which lasts for 1–3 weeks.

HAV is transmitted to humans via the fecal-oral (enteric) route from contaminated food, water-borne vectors, or close personal contact. For this reason, HAV infection is often called **infectious hepatitis**. The virus is excreted into the feces during the incubation period of 2–6 weeks, with the highest concentration before the onset of acute clinical symptoms. Virus excretion may persist at detectable levels for about a week after onset of symptoms.

Diagnosis and Assay Technology

Abnormal liver function is detected by clinical chemistry tests for two specific liver enzymes: alanine aminotransferase (ALT, SGPT) and aspartate aminotransferase (AST, SGOT), and to a lesser extent by the level of bilirubin. In cases of hepatitis, the concentrations of these analytes are greatly elevated.

HAV in stools is one of the first markers of HAV infection. Viremia in HAV disease is generally of short duration and not easily demonstrated. However, many patients no longer shed the virus in their feces by the time the symptoms appear. Diagnosis of hepatitis A through HAV in stool or serum is, therefore, difficult and unreliable. A virus specific antibody of IgM class has proven to be very useful for diagnosis of HAV infection during acute illness. Anti-HAV IgM appears concurrently with onset of clinical symptoms in over 99% of individuals infected with HAV and persists during early convalescence as illustrated graphically in Figure 61.1. After the appearance of anti-HAV IgM, anti-HAV IgG begin to reach significant levels, associated with recovery and persisting immunity.

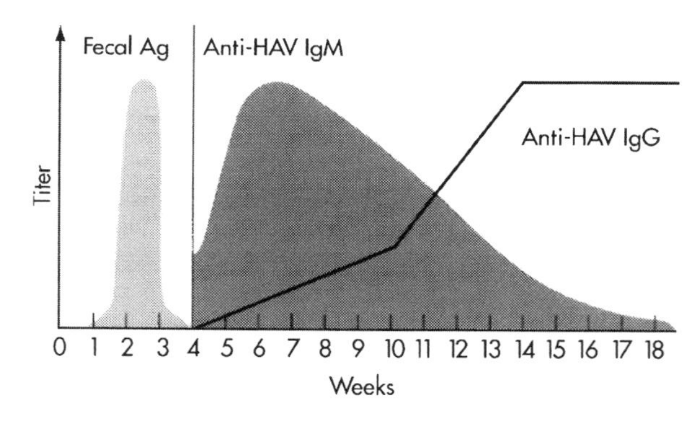

Fig. 61.1 Serological profile of HAV infection.

The Immunoassay Handbook. *Third Edition*
D. Wild (Ed.)

The four markers of HAV infection are fecal HAV, anti-HAV IgM, serum HAV RNA, and anti-HAV IgG. They have been measured by a variety of direct and indirect (competitive) solid-phase immunoassays and by the polymerase chain reaction (PCR). The test for anti-HAV IgG employs solid-phase HAV and labeled antibody. Serum anti-HAV IgG is detected by competition with the labeled antibody for solid-phase HAV. This competitive procedure does not distinguish IgM and IgG classes of immunoglobulins. Paired serum specimens are needed for diagnosis of HAV infection in order to demonstrate an increase in titer of anti-HAV by immunoassay.

A reliable and reproducible specific immunoassay for the detection of anti-HAV IgM is demonstrated in Figure 61.2, which is based on the following reactions:

1. Solid-phase anti-IgM antibody + IgM → solid-phase Ab-IgM
2. Solid-phase Ab-IgM + HAV → solid-phase Ab-IgM-HAV
3. Solid-phase Ab-IgM-HAV + *Ab-HAV → solid-phase Ab-IgM-HAV-*Ab-HAV

The solid-phase anti-IgM is a polystyrene surface coated with μ chain-specific goat anti-human antibody. In step (1) if anti-HAV IgM is present in a patient's serum, it is bound by the μ chain-specific solid-phase antibody. In step (2) HAV becomes attached to it to form the complex solid-phase Ab-IgM-HAV. This complex is then detected in step (3) by incubation with the probe antibody (*Ab-HAV), a radiolabeled or enzyme-labeled human anti-HAV IgG. The resulting count rate or absorbency of the multiple-layer product, solid-phase Ab-IgM-HAV-*AbHAV is in proportion to anti-HAv IgM concentration in the patient's serum.

The utility of serum HAV RNA for the detection of recent HAV during the early convalescent phase of acute HAV has recently been examined. Two hundred sera from 38 patients with type A acute hepatitis and 20 patients with non-A, non-B, non-C acute hepatitis were examined for the presence of HAV RNA. HAV RNA was detected by nested reverse transcription-PCR (RT-PCR) with primers located at the 5′ non-translated region of HAV. HAV was detected in 35 of 38 (92%) type A hepatitis patients and in none of 44 serum specimens from 20 non-A, non-B, non-C acute hepatitis patients.

The immunoassays for anti-HAV IgG and IgM are not only useful in the differential diagnosis of hepatitis, but are

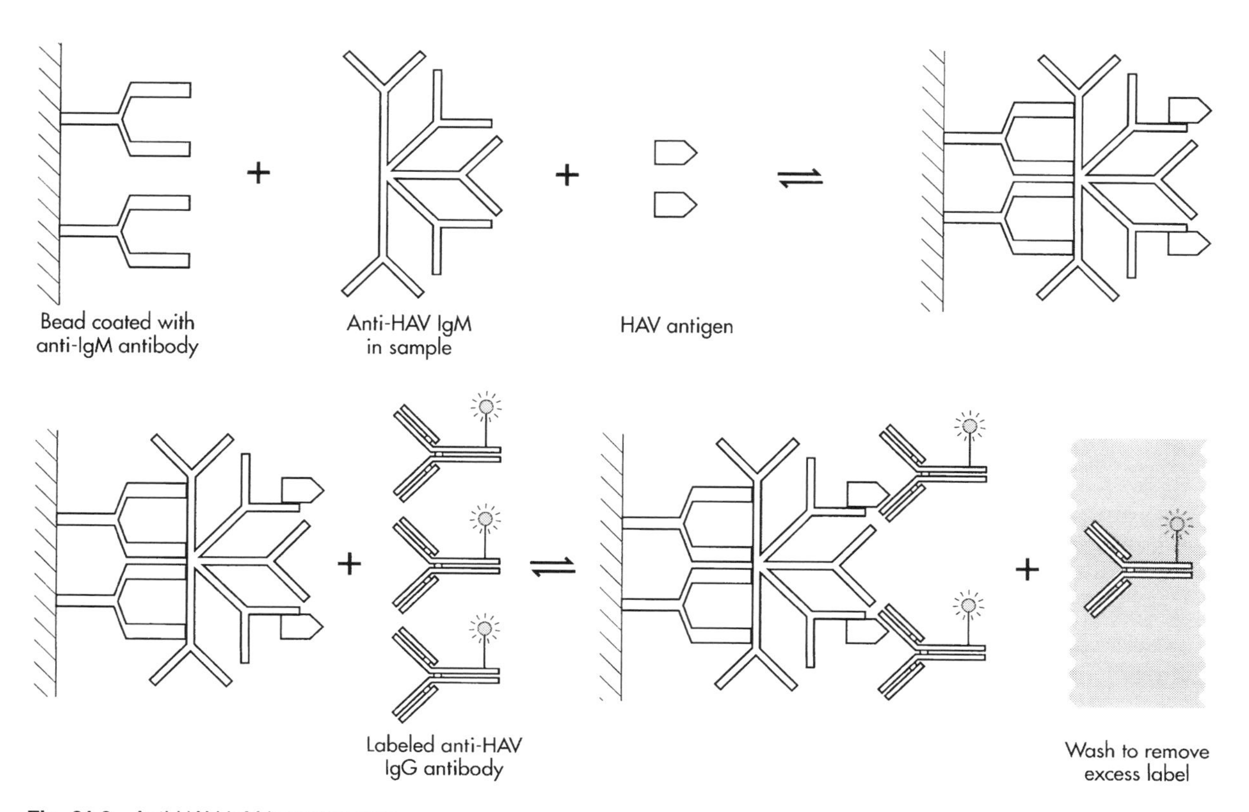

Fig. 61.2 Anti-HAV IgM immunoassay.

also useful for measurement of anti-HAV IgG titer in blood, blood products, and sera of vaccinated individuals with HAV vaccine; for understanding the epidemiology of the disease; for the determination of the immune status of individuals; and for controlling HAV infection in institutions and groups with high risks of HAV transmission.

FURTHER READING

Birkenmeyer, L.G. and Mushahwar, I.K. Detection of hepatitis A, B and D virus by the polymerase chain reaction. *J. Virol. Methods* **49**, 101–112 (1994).

Chernesky, M.A., Gretch, D., Mushahwar, I.K., Swenson, B.D. and Yarbough, P.O. *Laboratory Diagnosis of Hepatitis Viruses*, Cumitech Series, (ed Young, S.), (American Society of Microbiology, Washington, DC, 1998), in press.

Fujiwara, K., Yokosuka, O., Ehata, T., Imazeki, F., Saisho, H., Miki, M. and Omata, M. Frequent detection of hepatitis A viral RNA in serum during the early convalescent phase of acute hepatitis A. *Hepatology* **26**, 1634–1639 (1997).

Koff, R.S. Hepatitis A. *Lancet* **351**, 1643–1649 (1998).

Lemon, S.M. Type A viral hepatitis: epidemiology, diagnosis, and prevention. *Clin. Chem.* **43**, 1494–1499 (1997).

HEPATITIS B VIRUS, HBsAg, HBcAg, HBeAg, ANTI-HBs, ANTI-HBc, ANTI-HBc IgM, ANTI-HBe

Etiologic Agent

Hepatitis B virus (HBV) is a large 42 nm double-shelled spherical particle known as the Dane particle. It contains circular, partially double-stranded DNA and DNA polymerase. The HBV DNA is approximately 1.6×10^5 Da in molecular weight and the circular molecule is single stranded for 15–45% of its length. The virion HBV DNA polymerase converts the virion DNA into a fully double-stranded form with a size of approximately 2.1×10^6 Da or 3200 base pairs. Disruption of HBV by non-ionic detergent in the presence of a reducing agent results in the removal of the glycoprotein outer coat of the virion. The outer coat of HBV is the hepatitis B surface antigen, HBsAg, a defective 22 nm particle without nucleic acid. The detergent treatment of HBV exposes the nucleocapsid, a 27 nm spherical particle containing the hepatitis B core antigen (HBcAg). HBcAg is not found free in serum, but free core particles are observed in nuclei of infected hepatocytes. Another associated antigen, hepatitis B e antigen (HBeAg), occurs in soluble form in serum during HBsAg antigenemia.

The outer coat of HBV consists of three envelope proteins. These are the large (L, pre S1), middle (M pre S2), and small (S, HBsAg). The L protein contains the pre S1, pre S2, and S regions. The M protein contains the pre S2 and S regions. Spherical and tubular forms of HBsAg are found in the blood of HBV-infected patients. HBsAg is a complex lipid and glycoprotein particle consisting of a group specific determinant, 'a', and two pairs of type-specific determinants, '*d*/*y*' and '*w*/*r*'. HBV strains have been classified into nine different subtypes, *ayw*1, *ayw*2, *ayw*3, *ayw*4, *ayr*, *adw*2, *adw*4q^-, *adr*q^+, and *adr*q^-, according to the antigenic determinants and subdeterminants of their HBsAg. HBV has also been classified into six genotypes, A–F, on the basis of nucleotide sequence differences. These HBV genotypes correlate with geographic origin (Table 61.1).

Antibodies to the above three antigenic components of HBV are induced during the course of clinical and subclinical disease and recovery. These are antibody to HBsAg (anti-HBs), antibody to HBcAg (anti-HBc), and antibody to HBeAg (anti-HBe).

Pathogenesis

HBV (once referred to as **Australia antigen**) is a widespread cause of liver disease. In 1988, it was estimated that there were 300 million hepatitis B carriers worldwide. Only about half of HBV-infected patients have clinical symptoms, which are similar to those of hepatitis A infection (*see* HEPATITIS A). However, the onset of the

Table 61.1. Geographic distribution of HBV genomic groups and subtypes.

Genomic group	Subtype	Area of prevalence
A	*adw*2	Northwestern Europe
	*adyw*1	Central Africa
B	*adw*2	Indonesia, China
	*ayw*1	Vietnam
C	*adw*2	East Africa
	adrq^+	Korea, China, Japan
	adrq^-	Polynesia
	ayr	Vietnam
D	*ayw*2	Mediterranean area
	*ayw*3	India
E	*ayw*4	West Africa
F	*adw*4q^-	South America, Polynesia

Personal communication from L.O. Magnius.

symptoms is slower and it may take several months before patients with **acute hepatitis** feel well again. About 20% of HBV-infected individuals have jaundice. The mortality of HBV is much greater than that of hepatitis A. Approximately 1% of HBV patients die as a result of liver failure.

A significant proportion (10–15%) of infected patients go on to develop **chronic hepatitis**. These individuals may be asymptomatic although they can still transmit the disease to others. The chronic state may last for several years and can result in damage to the liver. HBV DNA becomes integrated into the chromosomes of the host liver cells, where it can subsequently cause **hepatic cancer**.

HBV is found in the serum of patients during acute and chronic infection. It is transmitted by direct percutaneous inoculation of blood or blood products and also by close physical contact with carriers of the virus, presumably by the passage of bodily fluids through cutaneous breaks or through oral and genital membranes. Hemodialysis patients, residents of mental institutions, users of illegal parenteral drugs, homosexually active men and prostitutes have an increased risk of contracting the disease. Medical staff are also at risk. All blood products derived from humans must be tested for HBsAg before use, including the serum, plasma, and blood proteins used in immunoassay calibrators and reagents.

In countries with a high incidence of hepatitis B, and a correspondingly high incidence of liver cancer, national vaccination programs have been initiated. Hospital workers are also routinely vaccinated against HBV in many countries.

Clinical Applications of HBV Marker Analytes

Acute HBV Infection

There is a characteristic sequence in the appearance of HBV markers following infection. HBsAg and HBeAg become detectable first, with a progressive increase in concentration. In acute HBV infection, HBeAg declines before HBsAg, then is replaced by anti-HBe. As anti-HBe increases, HBsAg starts to decline. Anti-HBs appears some months after the disappearance of HBsAg. Anti-HBc also appears early on in the disease and all three antibodies remain in the blood for many years (see Figure 61.3).

Chronic HBV Infection

Chronic persistent HBs antigenemia may take two different courses. In both, HBsAg remains in the blood for years, but in chronic HBV infection with late seroconversion, anti-HBe is eventually detectable, preceded by a decline in HBeAg. In chronic HBV infection without seroconversion, anti-HBe is not detectable, and the level of HBeAg remains high. In both types of chronic illness anti-HBc IgM is produced early on, followed by anti-HBc IgG (see Figures 61.4 and 61.5).

Vaccination

Anti-HBs tests are used to check the immune response of vaccinated individuals, usually 1 month after the final dose of vaccine, and then every few years, to check whether a booster dose is required. A level of 10 mIU/mL is typically used as being indicative of protective immunity.

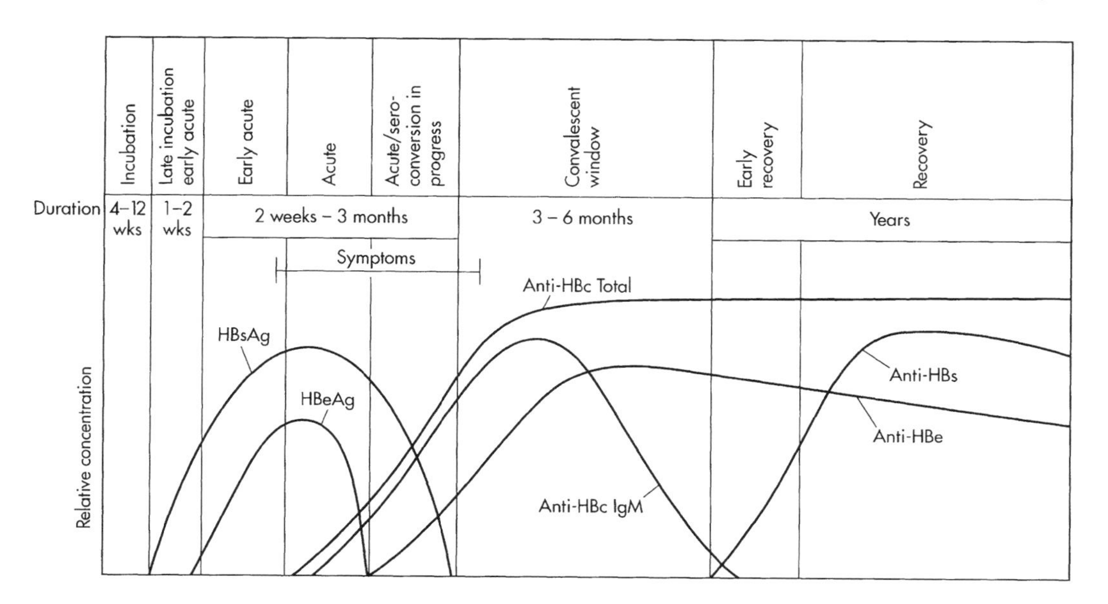

Fig. 61.3 Hepatitis B diagnostic profile; serological profile in 75–85% of patients with acute type B hepatitis. Reproduced with permission from *Serodiagnostic Assessment of Acute Viral Hepatitis*, Abbott Diagnostics, Abbott Park, Illinois.

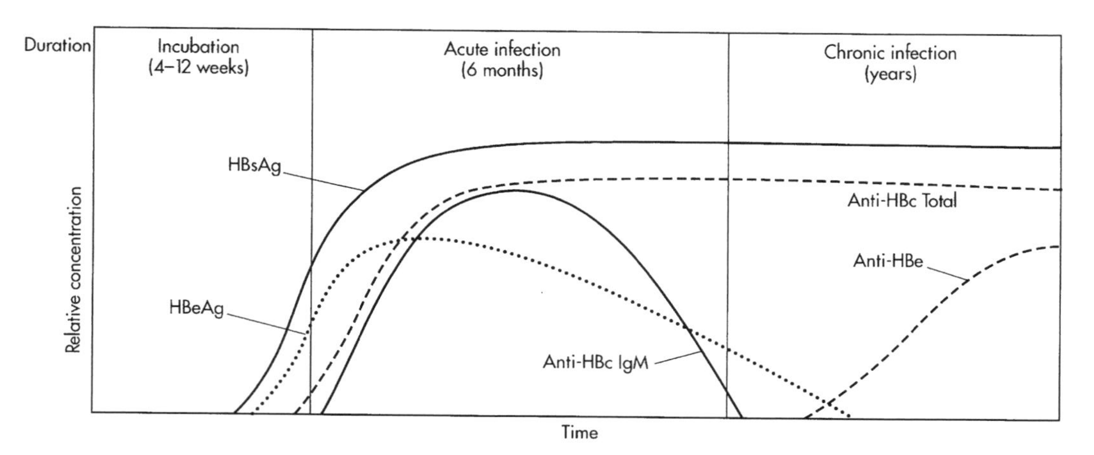

Fig. 61.4 Chronic HBV infection with late seroconversion. Redrawn, with permission, from Coslett, G.D. *Hepatitis Learning Guide*, Abbott Diagnostics, Abbott Park, Illinois, 1988.

HBsAg

HBsAg is the commonest hepatitis B test. The presence of HBsAg in serum indicates that the patient has an active HBV infection. There are several applications:

- screening to identify those at risk of spreading the disease, e.g. blood donors, pregnant women, intravenous drug abusers, health care workers, institutionalized people, transplant donors and recipients, and donors of semen for artificial insemination;
- differentiation of hepatitis B from other causes of hepatitis;
- monitoring disease progression;
- management of treatment of chronic hepatitis patients;
- management of dialysis units (screening of patients and staff).

HBsAg screening assays are normally supported by confirmatory tests, which are used to confirm repeatably reactive (positive) results. Typically, the confirmatory test involves neutralization of the HBsAg in the sample using a specific antibody. On reassay, the HBsAg level should decrease by at least 50%.

Anti-HBc and Anti-HBc IgM

Anti-HBc IgM is the first antibody to be detectable following infection. There is a period of time immediately after acute infection when HBsAg and anti-HBs are both undetectable, and anti-HBc or anti-HBc IgM kits may be used to monitor the course of the disease. Anti-HBc (total) kits measure IgM and IgG antibodies.

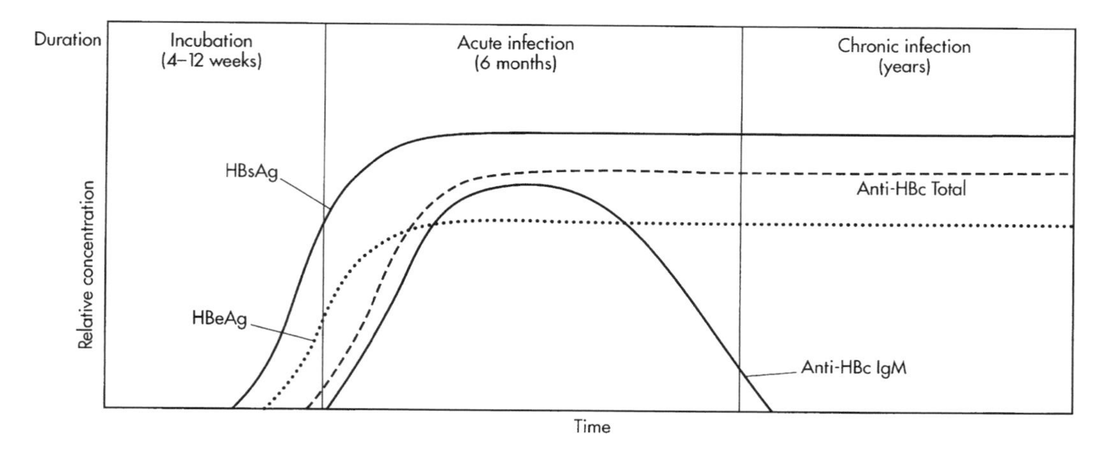

Fig. 61.5 Chronic HBV infection without seroconversion. Redrawn, with permission, from Caslett, G.D. *Hepatitis Learning Guide*, Abbott Diagnostics, Abbott Park, Illinois, 1988.

Anti-HBs

The anti-HBs test is used to confirm recovery and immunity in patients with acute hepatitis. It is also used to check that vaccination has been effective.

HBeAg and Anti-HBe

HBeAg is found during the early phase of HBV infection, soon after HBsAg is first detected. The presence of HBeAg correlates with an increased number of HBV particles and indicates that patients are at increased risk of transmitting the virus to their contacts. Persistence of HBeAg in HBV carriers is often associated with chronic active hepatitis. The presence of anti-HBe, following seroconversion, corresponds to a reduced level of infectivity.

Assay Technology

HBsAg Assay

This is a direct solid-phase immunoassay illustrated in Figure 61.6. Hyperimmune anti-HBs antibodies are attached to the surface of a solid support such as microparticles, polystyrene beads, test tubes, or microtiter wells. The solid-phase antibody is then utilized as a two-step reaction for HBsAg determination in serum or other fluids. The initial step involves incubation of a specimen with the anti-HBs-coated solid-phase. After incubation and washing, if HBsAg is present in the serum, it is bound via the specific antibody to the solid phase. In the second step of the procedure a highly purified labeled probe, usually non-radioactive, using colorimetric, fluorometric, or luminescent signal generation systems based on enzyme labels, is used in an

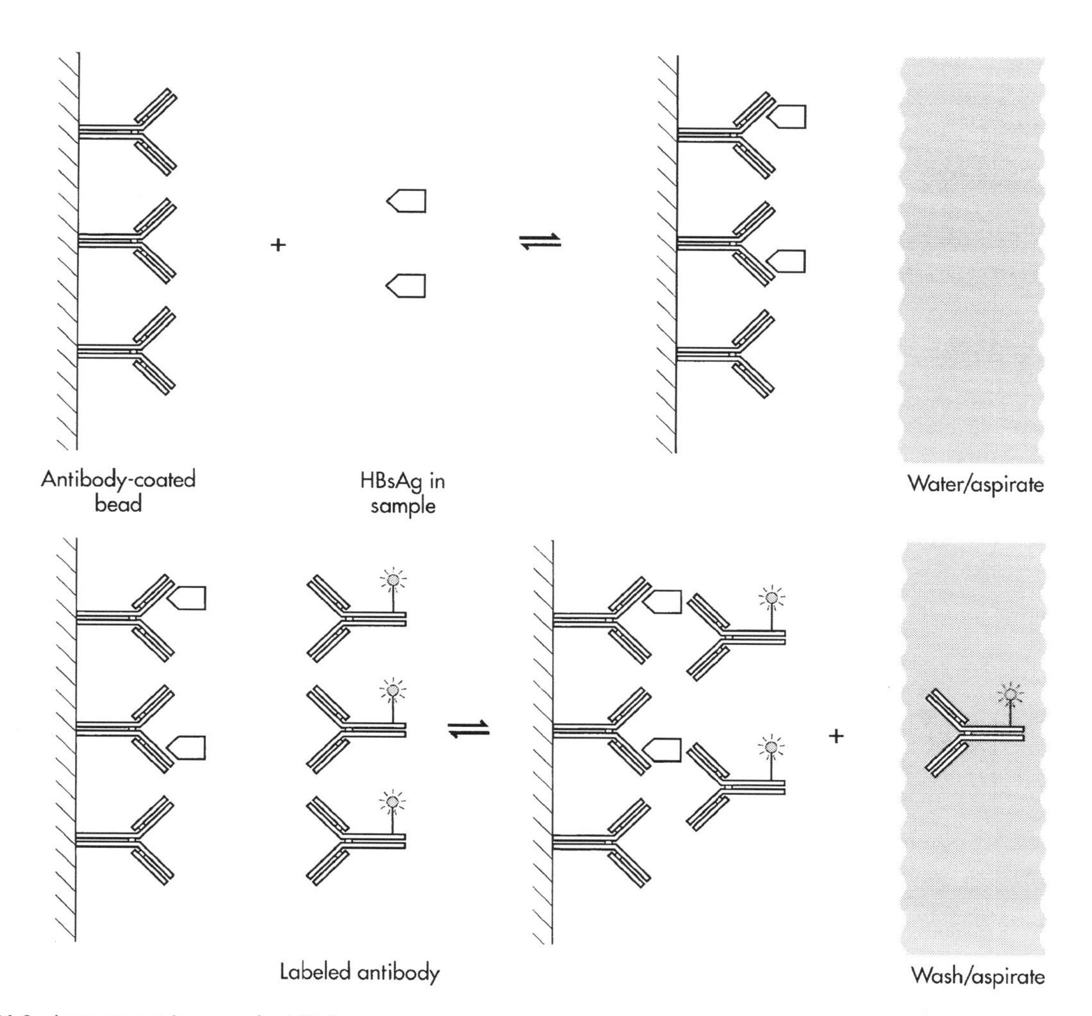

Fig. 61.6 Immunometric assay for HBsAg.

'inquiry' reaction whereby labeled antibody will bind to the available free antigenic sites on any captured antigen. Upon removal of the excess labeled antibody by thorough washing, the amount of bound labeled antibody is measured and it is directly proportional to the amount of antigen in the original serum specimen. This direct solid-phase sandwich technique is generally applicable to all macromolecules and complex structures possessing two or more antibody binding sites. Similar assays have been constructed for HBcAg and HBeAg determinations.

Confirmatory HBsAg Assays

False-positive HBsAg are infrequently encountered when performing solid-phase immunoassays for HBsAg, which is generally confirmed by well-established confirmatory tests utilizing high-titered human anti-HBs for this purpose. This is accomplished by repeating the HBsAg screening in the presence of anti-HBs. Two basic confirmation procedures, namely, a first-step neutralization and a second-step neutralization are generally used. Both procedures are effective for confirmation purposes; however, the first-step neutralization procedure fails to confirm high-titered HBsAg samples (prozone effect) when these samples are appropriately diluted. The second-step neutralization technique does not require dilutions of any test samples.

Anti-HBs Assay

A similar direct solid-phase sandwich immunoassay for anti-HBs detection and quantification is illustrated in Figure 61.7. In this assay, the serum is incubated with a solid phase that has been coated with highly purified recombinant HBsAg. If anti-HBs is present in the serum, it forms a complex with the solid-phase antigen. In a second

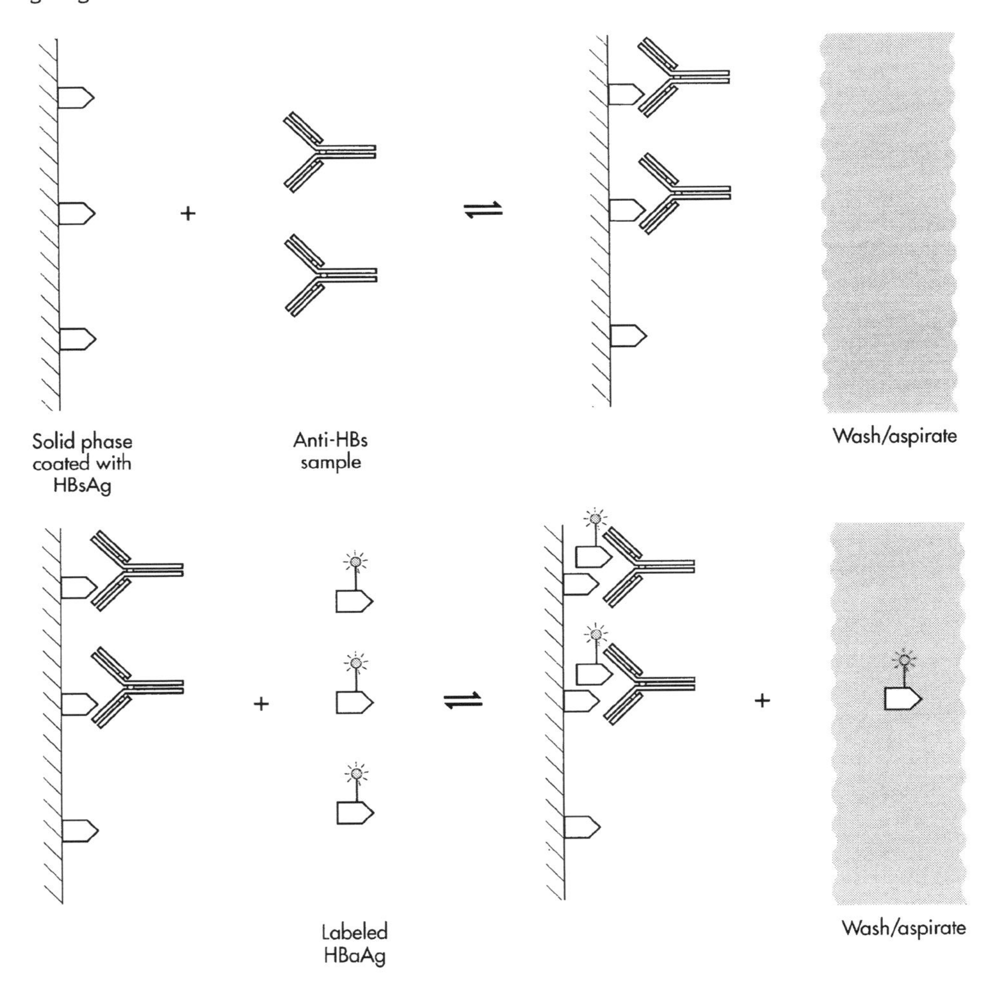

Fig. 61.7 Immunometric assay for anti-HBs.

step, labeled HBsAg is added and reacts with the immobilized antibody. The amount of the antibody in the original serum is in direct proportion to the amount of label fixed to the solid phase.

Indirect (Competitive) Immunoassays for HBV Antibodies

The generic indirect competitive solid-phase immunoassay for antibody detection is illustrated in Figure 61.8. In this assay, labeled antibodies are utilized to assay for serum antibodies. The solid phase is coated with an appropriate antigen. If antibodies are present in the specimen during the first step of the reaction, they are captured by the solid-phase antigen. The labeled antibodies present in the second inquiry step of the reaction will have fewer antigen-binding sites for reaction, and the final signal of the solid phase will be inversely proportional to the amount of antibody in the assay specimen.

Anti-HBc Assay

The solid-phase reagent in this assay is a recombinant produced HBcAg. In the reaction, this capture reagent is first reacted with an unknown serum containing anti-HBc. The resulting complex is challenged with a constant amount of labeled anti-HBc IgG in a second step. The degree of competition of unknown with labeled probe is an indication of the amount of anti-HBc. Fifty percent inhibition when compared to a negative serum was selected as a basis of anti-HBc positivity.

Anti-HBe Assay

In this assay, the unknown sample is mixed with an equal volume of standardized HBeAg positive serum containing a predetermined quantity of HBeAg termed the neutralizing reagent. The mixture is then incubated overnight at

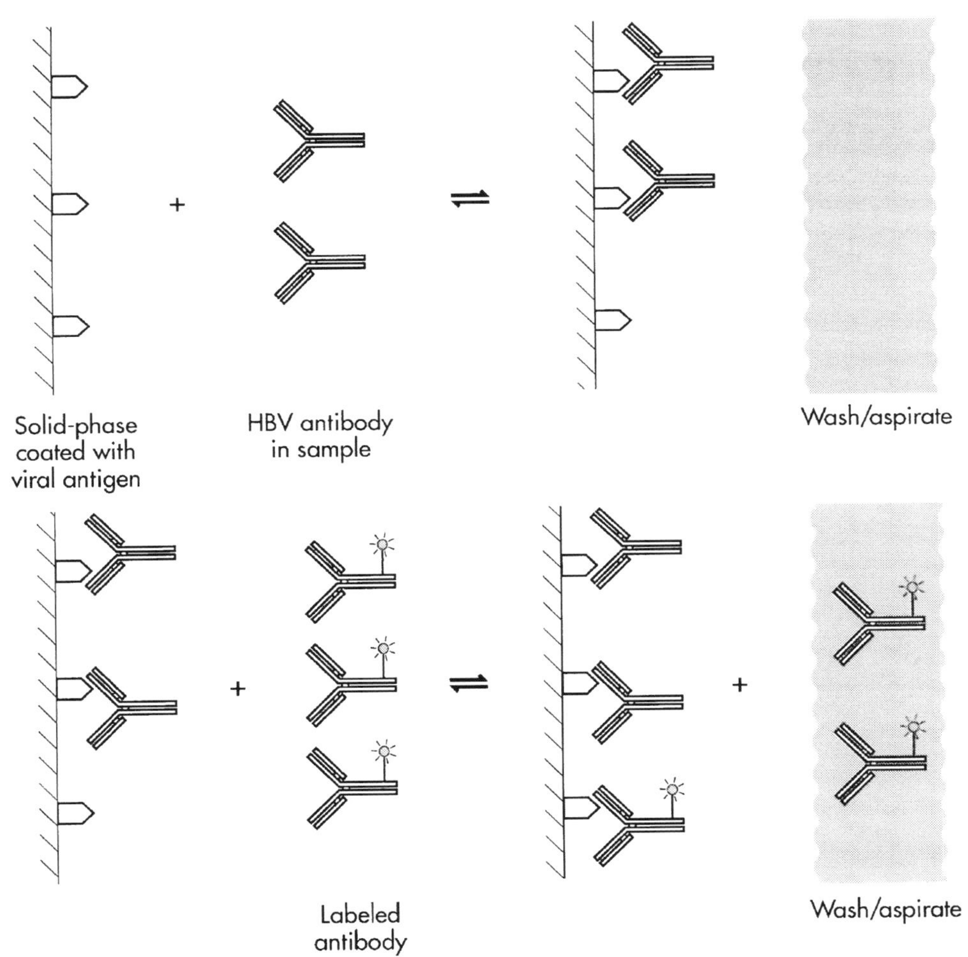

Fig. 61.8 Competitive assay for HBV antibodies.

room temperature with a solid phase coated with high-titered human anti-HBe. The solid phase is then incubated with labeled anti-HBe, and a significant (50%) reduction of signal indicates the presence of anti-HBe in the test sample.

Interpretation of HBV Serologic Markers

Unlike many other viral diseases, HBV infection is characterized by several distinctive serologic and immunologic responses. The temporal profiles of HBV infection can serve as useful guides to monitor the course of the disease and to provide a serologic correlation with the disease's progress. When only a single serum is available, diagnostic accuracy and efficiency are improved by reference to the complete battery of HBV markers. The particular combination of markers can be of prognostic value in assessing future clinical courses of the disease and level of infectivity.

FURTHER READING

Chernesky, M.A., Gretch, D., Mushahwar, I.K., Swenson, B.D. and Yarbough, P.O. *Laboratory Diagnosis of Hepatitis Viruses*, Cumitech Series, (ed Young, S.), (American Society of Microbiology, Washington, DC, 1998), in press.

Ganem, D. and Varmus, H.E. The molecular biology of hepatitis B virus. *Annu. Rev. Biochem.* **56**, 651–693 (1987).

Gritlin, N. Hepatitis B: diagnosis, prevention, and treatment. *Clin. Chem.* **43**, 1500–1506 (1997).

Lee, W.M. Hepatitis B virus infection. *N. Engl. J. Med.* **337**, 1733 (1997).

Mushahwar, I.K., Dienstag, J.L., Polesky, H.F., McGrath, L.C., Decker, R.H. and Overby, L.R. Interpretation of various serological profiles of hepatitis B virus infection. *Am. J. Clin. Pathol.* **76**, 773–777 (1981).

HEPATITIS C VIRUS

Etiologic Agent

The etiologic agent of parenterally transmitted non-A, non-B hepatitis is an enveloped virus that has been named hepatitis C virus (HCV). It is a single-stranded, positive sense RNA genome of approximately 9500 nucleotides. The genome possesses a large open reading frame encoding a polyprotein of about 3000 amino acids. It is a member of the Flaviviridae family of viruses that possesses similar gene products. Its structural proteins, namely, core, E1, and E2 are located at the 5′ end, and the non-structural (NS) proteins, namely, NS2, NS3, NS4, and NS5 are located in the remaining portion of the protein. Various HCV strains have been identified. These are divided into six major genotypes, provisionally classed as separate species. Within these genotypes are more than 100 subtypes and within them, the virus mutates into isolates that are specific to individual carriers, creating many quasispecies or swarms of closely related but different viruses.

Pathogenesis

The incubation period of HCV varies widely and evidence of infection may take several weeks to develop. About 80–90% of afflicted individuals with acute HCV infection progress to chronicity. In about 20–30% of HCV chronic carriers, hepatitis progresses to cirrhosis of the liver within 15–20 years of exposure. Twenty percent of these patients with cirrhosis develop fatal primary hepatocellular carcinoma.

Diagnosis and Assay Technology

The anti-HCV first generation assays were indirect enzyme-linked immunoassays (ELISAs) utilizing the C 100-3 protein of HCV as the solid-phase antigen. The C 100-3 protein utilized in this assay represented only 12% of the encoding capacity of the virus, and obviously, the first generation assay had its limitations. First, it was not an adequate assay for detecting all stages of HCV infection. Second, the test's inability to flag all infectious blood units hinted at its poor sensitivity, especially among acutely infected individuals. Third, delay in anti-HCV appearance suggested that some blood donors capable of transmitting HCV infection are not detected by this assay. For these reasons, recombinant antigens from different regions on the HCV genome were expressed, purified, and utilized to develop a second generation assay.

The antigens utilized in the second generation assay include a structural antigen encoded by sequences at the 5′ end of HCV (core region), and non-structural antigens encoded by NS3 and NS4 regions of HCV using epitopes of 33C and C100. These antigens (representing approximately 27% of the encoding capacity of HCV) are bound to a solid-phase (polystyrene, microparticles, microtiter plates, etc.) by adsorption. The serum to be tested for anti-HCV is incubated with the antigen coated in the solid phase. The horseradish peroxidase labeled probe used in a second step is a highly purified goat anti-human IgG. The resulting absorbency at 492 nm of the multiple layered solid-phase after incubation with an appropriate enzyme–substrate is then in direct proportion to anti-HCV in the serum sample.

The second generation assay was shown to be highly sensitive and specific. It detected anti-HCV antibodies in 95% of post-transfusional non-A, non-B hepatitis cases. Furthermore, additional anti-HCV positive cases were detected among patients with post-transfusional hepatitis, hemophiliacs, intravenous drug abusers, and patients on maintenance hemodialysis. In serial bleeds from patients with HCV infection as well as chimpanzees experimentally infected with the Hutchinson's strain of HCV, the second generation assay detected anti-HCV at least 2–3 weeks earlier than the first generation anti-HCV assay. No doubt, the improvement in sensitivity of the second generation anti-HCV assay has increased the safety of the blood supply worldwide. The current risk of transfusing an anti-HCV positive unit in the United States is now estimated at 1 in 103,000 donations.

Soon after the introduction of second generation anti-HCV assays, third generation assays were also introduced. In addition to the HCV proteins used in the second generation assays, an antigen from the NS5 region was included. In comparison with the second generation assay, it is claimed that the third generation assay increased the reactivity of the NS3 antigen. However, the addition of the NS5 antigen has resulted in a variety of false-positives.

Several anti-HCV confirmatory assays have been introduced. Among the currently used ones are: Abbott's Matrix HCV Assay, Chiron's RIBA Third Generation Assay, and Murex HCV Western Blot Assay. The third generation confirmatory assays have resolved a number of indeterminate samples, although indeterminate samples do still exist. These samples are resolvable only by a specific HCV-PCR assay. Currently, it is believed that the diagnosis of HCV infection in clinical laboratories, followed by confirmation of positive or weakly positive ELISAs with immunoblot-based confirmatory assays is no longer needed. HCV RNA detection by PCR helps to resolve weakly positive or negative ELISA results when the clinical context is compatible with an HCV infection.

A specific IgM solid-phase enzyme-linked immunoassay for the diagnosis of a recent HCV infection has been developed. The assay utilizes a structural antigen encoded by sequences at the 5′ end of HCV (core antigen) and non-structural antigens encoded by the NS3 (33C) and NS4 (C 100-3) regions of the HCV genome. Anti-HCV IgM was frequently but transiently detected in serial serum samples from several patients with clinically diagnosed post-transfusion non-A, non-B hepatitis (PT-NANBH). In individuals who resolved their HCV infection or progressed to chronicity, anti-HCV IgM was produced transiently at or near the onset of clinically diagnosed hepatitis.

FURTHER READING

Alter, M.J., Mast, E.E., Moyer, L.A. and Margolis, H.S. Hepatitis C. *Infect. Dis. Clin. N. Am.* **12**, 13–26 (1998).

Chau, K.H., Dawson, G.J., Mushahwar, I.K., Gutierrez, R.A., Johnson, R.G., Lesniewski, R.R., Mattsson, L. and Weiland, O. IgM-antibody response to hepatitis C virus antigens in acute and chronic post-transfusion non-A, non-B hepatitis. *J. Virol. Methods* **35**, 343–352 (1991).

Chernesky, M.A., Grech, D., Mushahwar, I.K., Swenson, B.D. and Yarbough, P.O. *Laboratory Diagnosis of Hepatitis Viruses*, Cumitech Series, (ed Young, S.), (American Society of Microbiology, Washington, DC, 1998), in press.

Clarke, B. Molecular virology of hepatitis C virus. *J. Gen. Virol.* **78**, 2397–2410 (1997).

Courouсé, A.-M. Development of screening and confirmation tests for antibodies to hepatitis C virus. In: *Hepatitis C Virus* (ed Reesink, H.W.), 64–75 (Karger, New York, NY, 1998).

Major, E.M. and Feinstone, S.M. The molecular virology of hepatitis C. *Hepatology* **25**, 1527–1538 (1997).

Mushahwar, I.K. Recent progress on anti-HCV screening. *Can. Dis. Weekly Rep.* **1755**, 41–43 (1991).

Schlauder, G.G. and Mushahwar, I.K. Detection of hepatitis C and E virus by the polymerase chain reaction. *J. Virol. Methods* **47**, 243–253 (1994).

Simmonds, P. Clinical relevance of hepatitis C virus genotypes. *Gut* **40**, 291–293 (1997).

Vrielink, H., Reesink, H.W., van den Burg, P.J.M., Zaaijer, H.L., Cuypers, H.T.M., Lelie, P.N. and van der Poel, C.L. Performance of three generations of anti-hepatitis C virus enzyme-linked immunosorbent assays in donors and patients. *Transfusion* **37**, 845–849 (1997).

HEPATITIS D VIRUS

Etiologic agent

The hepatitis D virus (HDV) is a transmissible animal viroid-like agent that depends upon HBV for its expression and replication. It is a 36 nm particle that consists of hepatitis D antigen (HDAg), a protein encapsulated with a low molecular weight (1.7 kb) circular single-stranded RNA in an intermediate size envelope protein (the HBsAg particle of HBV). Two forms of HDAg have been characterized. These are the small antigen (delta-Ag-S) which is a nuclear phosphoprotein, and the large antigen (delta-Ag-L) which is essential for virus assembly. The delta antigen has been detected in serum and also in the livers of patients with chronic HBV infection. It localizes mainly in the nuclei of hepatocytes. Three genotypes (I, II, and III) of HDV have been identified, each with different geographic distribution and disease association. Genotype I has a wide geographic distribution that predominates in Europe and the United States, while genotype II is predominant in Taiwan and Japan.

Pathogenesis

The delta agent is pathogenic and can cause either acute or chronic hepatitis. It is an important cause of severe and progressive liver disease. The course of HDV infection depends primarily on the presence of acute or chronic HBV infection. Simultaneous infection with HBV and HDV (coinfection) generally leads to an acute self-limited episode with transient HBs- and HD-antigenemias and eventual recovery. However, the liver damage of each agent is likely to be additive. It has been reported, but not confirmed, that fulminant hepatitis occurs more frequently in patients with either acute delta coinfection with HBV, or acute delta superinfection with HBV carriers, than in patients with acute HBV infection alone. The mortality rate for acute delta hepatitis ranges from 2 to 20% compared to less than 1% for acute HBV infection. Infection of chronic HBV carriers with HDV (superinfection) often leads to severe chronic disease with HDV cirrhosis in about 75% of patients. Chronic delta hepatitis is more severe than other types of viral hepatitis. Hepatitis delta virus infection is also associated with a wide range of different autoantibodies. The humoral immune response in chronic HDV infection is directed against the cytoskeleton, the nucleus, the nuclear lamina, and the endoplasmic reticulum. Autoantibodies directed against antigens of the endoplasmic reticulum (LKM) are also common in chronic HDV infection.

Diagnosis and Assay Technology

The molecular and serologic markers of HDV infection are serum HDV RNA, serum and liver HDAg, serum anti-HD IgM, and serum anti-HD IgG. Immunofluorescence, immunoperoxidase and ELISA methods have been used for the localization of HDAg in serum and liver tissue. Competitive ELISA is the current test for the determination of anti-HD IgG. In this test, non-labeled anti-delta from the patient's serum competes with a constant amount of human anti-delta labeled IgG for a limited number of binding sites on solid phase coated with HDAg. Thus, the proportion of labeled anti-HD bound to the solid phase is inversely proportional to the concentration of anti-delta in the test specimen. Anti-HD IgM is carried out in a similar fashion to the anti-HAV IgM test described under the HAV section.

In acute coinfection with HBV and HDV, the following sequential markers appear in the serum: HBsAg, HDV RNA, HDAg, anti-HD IgM, and anti-HD IgG. Both antibodies are generally of low concentration and of short duration. In superinfected patients, there is a persistence of HBs- and HD-antigenemia and HDV RNA in the serum and liver of infected individuals. Generally, anti-HD IgM is short-lived, while anti-HD IgG is present at higher titer. In interferon-treated patients, anti-HD IgM disappears in patients who respond to therapy with a persistent normalization of serum aminotransferase level and concomitant clearance of serum HBsAg and HDV RNA. It is believed that anti-HD IgM is elevated in response to HDV-induced damage and thus represents a valid surrogate marker of liver damage that is immunopathologically related to HDV infection. Thus, besides providing diagnostic information, serum anti-HD IgM is the best predictor of impending resolution of chronic HDV disease, whether spontaneous or interferon-induced.

FURTHER READING

Bichko, V., Netter, H.J., Wu, T.T. and Taylor, J. Pathogenesis associated with replication of hepatitis delta virus. *Infect. Agents Dis.* **3**, 94–97 (1994).

Birkenmeyer, L.G. and Mushahwar, I.K. Detection of hepatitis A, B, and D virus by the polymerase chain reaction. *J. Virol. Methods* **49**, 1–12 (1994).

Borghesio, E., Rosina, F., Smedile, A., Lagget, M., Niro, M.G., Marinucci, G. and Rizzetto, M. Serum immunoglobulin M antibody to hepatitis D as a surrogate marker of hepatitis D in interferon-treated patients and in patients who underwent liver transplantation. *Hepatology* **27**, 873–876 (1998).

Hadziyannis, S.J. Review: hepatitis delta. *J. Gastroenterol. Hepatol.* **12**, 289–298 (1997).

Mushahwar, I.K., Gerin, J.L., Dienstag, J.L., Decker, R.H., Smedile, A. and Rizzetto, M.D. Interpretation of hepatitis B virus and hepatitis delta virus serologic profiles. *Pathologist* **38**, 648–650 (1984).

Negro, F. and Rizzetto, M. Diagnosis of hepatitis delta virus infection. *J. Hepatol.* **22**, 136–139 (1995).

Phillip, T., Obermayer-Straub, P. and Manns, M.P. Autoantibodies in hepatitis delta. *Biomed. Pharmacother.* **49**, 344–349 (1995).

Rizzetto, M. Hepatitis delta virus (HDV) infection and disease. *Ric. Clin. Lab.* **19**, 11–26 (1989).

HEPATITIS E VIRUS

Etiologic Agent

The hepatitis E virus (HEV) is a spherical, non-enveloped particle. The virus particle measures 32–34 nm in diameter. The mature, complete virion isopycnically bands at 1.29 g/cm^3 in potassium tartrate–glycerol buoyant density gradients, and has a sedimentation coefficient of 183S in sucrose. The viral genome consists of a positive-sense single-stranded poly-adenylated RNA molecule of approximately 7.5 kb. It possesses three discontinuous open reading frames (ORF) that encode polyproteins of 1693 amino acids (ORF 1), 660 amino acids (ORF 2), and

123 amino acids (ORF 3). The genome is organized with the structural genes located at the 3′ end and the non-structural genes at the 5′ end. Analysis of the nucleotides reveals that ORF 1 contains several conserved motifs such as an RNA helicase, and methyl transferase, a papain-like proteinase, and an RNA-dependent RNA polymerase. ORF 2 encodes for the capsid protein, while the function of ORF 3 is unknown. Both ORF 2 and ORF 3 proteins, however, are highly antigenic.

Pathogenesis

The incubation period ranges between 2 and 9 weeks. The infection causes symptoms of a self-limiting, acute, icteric disease. Chronic liver disease or persistent viremia have not been observed. Among HEV infected individuals, the most common complaints are fatigue accompanied by gastrointestinal disorders such as anorexia, nausea, vomiting, and abdominal discomfort. Diarrhea, fever, and epistaxis (bleeding from the nose) are less common symptoms. Most patients become jaundiced. Some patients progress to fulminant hepatitis with a high mortality rate, especially among pregnant women who acquire HEV infection during the third trimester. The clinical manifestations of both HAV and HEV are practically indistinguishable. The main points of difference for HEV are: the incubation period is longer, cholestasis is more predominant, and the mortality rate during pregnancy is high, especially if HEV is acquired during the late pregnancy period. The highest attack rate is observed in young to middle-aged adults. The occurrence of subclinical cases in younger individuals is suspected, but not yet documented.

Epidemiology

HEV has a worldwide distribution. A number of epidemics have been reported on the Indian subcontinent, in Central and in Southeast Asia as well as in Africa and Mexico. Recent surveys have shown that the sporadic form of HEV is also spread globally, and that outbreaks and clusters of cases can occur in any situation where drinking water may be fecally contaminated. In industrialized countries HEV infection occurs occasionally as imported sporadic cases. Recent data indicate that HEV is a zoonotic virus, and that swine, chickens, and rats are among its natural hosts. The high prevalence of infection in these animals suggests that they may serve as a reservoir for HEV in non-industrialized countries. Recently, a novel virus, designated swine hepatitis E virus (swine HEV) was identified in pig herds from the Midwestern United States, and a new US HEV strain has been characterized.

Clinical Course

The clinical and serological course of an HEV infection in cynomolgus monkeys has been established by several investigators. The studies show that the incubation in monkeys is short, averaging 24 days, and ranging from 21 to 45 days. The first marker of early infection appears in feces and bile. Bile samples collected between days 7 and 41 are positive for HEV RNA by PCR. Similar results are observed for fecal samples. The virus is first detected in serum on day 9, and appears to be present at higher titer by PCR from days 14 to 23, then disappears by day 28. Anti-HEV IgG first appears in serum on day 27, at the height of ALT elevation.

Similar studies in humans have also been conducted. The icteric phase of infection (first 37 days) was characterized by anorexia, epigastric pain, and discolored urine (day 30). The icteric phase extended from day 38 to day 120 after infection and was characterized by elevation in ALT levels, with peak levels noted on day 46. HEV RNA was detected via RT-PCR in serum on day 22 and disappeared by day 46. HEV RNA was likewise detected in stools between days 38 and 46. Antibodies to HEV were first detected on day 41 and have persisted for at least 2 years. The similarities in this study to the above-mentioned monkey studies validated the latter as a non-human primate for HEV. Other studies in humans have shown that HEV RNA has regularly been found in serum by RT-PCR in virtually all cases studied within 2 weeks after onset of illness. Prolonged periods of HEV RNA positivity in serum ranging from 4 to 16 weeks have also been reported.

Diagnosis and Assay Technology

Hepatitis E isolated from geographically distinct regions of the globe has been shown by immune electron microscopy (IEM) to possess at least one major cross-reactive epitope. Acute and convalescent phase sera from well-documented cases of HEV occurring in China, Somalia, Burma, Mexico, Russia, Borneo, India, Pakistan, and Greece also have been shown to block antibody binding in a fluorescent antibody (FA) test that utilizes an FITC-labeled IgG probe (Mexico) and liver section from HEV-Burma infected monkey, thus indicating that one virus or class of serologically related viruses is responsible for HEV infections worldwide.

Reliable techniques for anti-HEV detection such as immunofluorescence and IEM have been utilized and developed. These techniques, however, require special labor-intensive procedures that are not widely available to many laboratories. The availability of highly purified recombinant HEV antigens and synthetic peptides enables scientists to develop highly sensitive and specific ELISA for the detection of both IgM and IgG-class antibodies to HEV. Analysis of serum specimens collected from HEV infected patients during the incubation period, acute phase, convalescence, and recovery phase of disease has shown the order of appearance of various diagnostic molecules as shown in Figure 61.9. The usual sequential order is serum HEV RNA, IgA/IgM anti-HEV, and anti-HEV IgG. Serum HEV RNA appeared at the peak of ALT elevation and disappeared in all cases

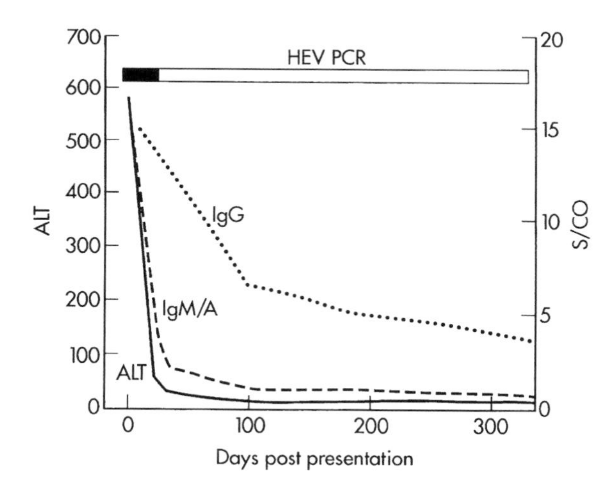

Fig. 61.9 HEV antibody response.

studied 10 days later. Anti-HEV IgM was also detected in all cases at the peak of ALT elevation and disappeared 20 days later. In contrast, anti-HEV IgG was detected at high levels (S/CO $\geq$ 3.0) in all patients and continued to be detected at declining levels approximately a year later. Generally, the IgG response is much stronger in individuals with a recent HEV infection.

The complete set of diagnostic markers for HEV provides rapid diagnosis of this disease from a single serum specimen and also by exclusion, identifies acute episodes of non-A, non-B, non-C (non-A–C) hepatitis. Acute HEV infection is readily identified by the presence in some cases of anti-HEV IgM and/or anti-HEV IgA.

HEV diagnostic molecules enable us to differentiate several consequences of infection. These include:

- acute disease markers: HEV, HEV RNA, anti-HEV IgA, anti-HEV IgM, anti-HEV IgG titer;
- infectious virus markers: HEV and HEV RNA;
- epidemiological marker: anti-HEV IgG and genetic diversity;
- immune status marker: anti-HEV IgG.

Generally, the intact virion (HEV) is excreted in the feces during the incubation period and prior to the onset of symptoms. Analysis of fecal antigen by RT-PCR is more practical than by IEM. However, collection of fecal material prior to a knowledge of the onset of symptoms or exposure is not commonly applicable to clinical practice.

Isolates of HEV have been described from a number of geographic locations. These include Mexico, Burma, China, Pakistan, and United States. For epidemiological studies, it is of importance to test for genetic diversity of HEV by cDNA HEV sequencing in order to identify correctly the source of infection and outbreaks of HEV.

FURTHER READING

Balayan, M.S. Epidemiology of hepatitis E virus infection. *J. Viral. Hepat.* **4**, 155–165 (1997).

Krawczynski, K., McCaustland, K., Mast, E., Yarbough, P.O., Purdy, M., Favorou, M.O. and Spellbring, J. Elements of pathogenesis of HEV infection in man and experimentally infected primates. In: *Enterically-Transmitted Hepatitis Viruses* (eds Buisson, Y., Coursaget, P. and Kane, M.), 317–328 (La Simarre, Tours, 1996).

Meng, X.J., Purcell, R.H., Halbur, P.G., Lehman, J.R., Webb, D.M., Tsareva, T.S., Haynes, J.S., Thacker, B.J. and Emerson, S.U. A novel virus in swine is closely related to the human hepatitis E virus. *Proc. Natl Acad. Sci. USA* **94**, 9860–9865 (1997).

Mushahwar, I.K. and Dawson, G.J. *Hepatitis E virus: epidemiology, molecular biology and diagnosis*, Prospectus: Molecular Medical Series, (eds Zuckerman, A.J. and Harrison, T.J.), 33–43 (Wiley, Chichester, 1997).

Mushahwar, I.K., Dawson, G.J., Bile, K. and Magnius, L.O. Serological studies of an enterically transmitted non-A, non-B hepatitis in Somalia. *J. Med. Virol.* **20**, 218–221 (1993).

Mushahwar, I.K., Dawson, G.J. and Reyes, G.R. Hepatitis E virus: molecular biology and diagnosis. *Eur. J. Gastroenterol. Hepatol.* **8**, 312–318 (1996).

Mushahwar, I.K., Schlauder, G.G. and Dawson, G.J. Hepatitis E virus diagnostic molecules. In: *Viral Hepatitis and Liver Disease* (eds Rizzetto, M., Purcell, R.H., Gerin, J.L. and Verme, G.), 317–320 (Edzioni Minerva Medica, Turin, 1997).

Schlauder, G.G., Dawson, G.J., Erker, J.C., Kwo, P.Y., Knigge, M.F., Smalley, D.L., Rosenblatt, J.E., Desai, S.M. and Mushahwar, I.K. The sequence and phylogenetic analysis of a novel hepatitis E virus isolated from a patient with acute hepatitis reported in the United States. *J. Gen. Virol.* **79**, 447–456 (1998).

Tong, M.J., Roué, R., Nahor, M., Gwara, N., Desrame, J., Buisson, Y. and Coursaget, P. Clinical aspects of hepatitis E and hepatitis A: a comparison. In: *Enterically-Transmitted Hepatitis Viruses* (eds Buisson, Y., Coursaget, P. and Kane, M.), 1–10 (La Simarre, Tours, 1996).

GB VIRUSES

The discovery of three novel Flavi-like viruses, namely GB virus A (GBV-A), GB virus B (GBV-B), and GB virus C (GBV-C) was reported in 1995. Each of these viruses contains a single-stranded, positive sense RNA genome of approximately 9000 nucleotides that possesses a continuous, translational, single open reading frame that encodes a polyprotein of about 3000 amino acids. The genomes are organized with structural genes located at the 5′ end and non-structural genes at the 3′ end. Analysis of the nucleotides reveals a pattern of serine protease, conserved helicase sequences, and coding for an RNA dependent-RNA polymerase. Recent evidence showed that GBV-A and GBV-B are non-human primate viruses. The sequence and genomic organization of a variant of GBV-A isolated from a captive tamarin has been recently reported. GBV-C, however, has been proven to be a human virus.

GB VIRUS C

Etiologic Agent

GBV-C is a positive strand RNA virus particle that is enveloped with no nucleocapsid. It measures 50–100 nm in diameter with a buoyant density in sucrose gradients of 1.08–1.13 g/mL. The viral particles are highly associated with lipoproteins with moderate carbohydrate moieties on their surfaces. The genome is single stranded of approximately 8600 nucleotides. It possesses a long open reading frame that encodes a polyprotein of about 2900 amino acids that cleaves into a series of structural and non-structural peptides during infection. Genotypes of GBV-C have already been identified by analysis of the entire 5′-NTR sequence. Phylogenetic analysis of these sequences demonstrated the presence of three major types of GBV-C isolates (i.e. 1, 2, and 3) as well as subtypes which correlated with geographic origin. Hepatitis G virus (HGV) is actually a member of one of these genotypes and is indeed a different isolate of GBV-C. This conclusion is based on the alignment of the 5′-UTR nucleotide sequences of HGV with that of GBV-C. The alignment demonstrated amino acid sequence identity at 95%.

Pathogenesis

Most GBV-C infections appear to be asymptomatic, transient, and self-limiting, with slight or no elevation of amino-transferase levels. Most of GBV-C subclinical cases resolve after loss of serum GBV-C RNA with a concomitant appearance of antibody to the envelope E2 of GBV-C (anti-GBV-C E2). GBV-C is capable of inducing persistent infection in about 10% of GBV-C infected individuals. This persistent infection lasts for more than 15 years. The importance of GBV-C persistency is very crucial for our understanding of the pathogenesis of GBV-C. It has been reported by many investigators that GBV-C infections account for 30–50% of fulminant hepatitis in non-A–E hepatitis cases. The role of GBV-C in the etiology of fulminant hepatitis is not yet fully established. The spectrum of liver disease associated with GBV-C infection is wide, with a variety of histological liver lesions ranging from steatosis to fibrosis and cirrhosis in particular, non-specific inflammatory bile duct lesions. Numerous studies have shown that in HCV coinfected individuals, GBV-C does not seem to affect HCV replication, HCV RNA concentration, and liver disease. Other studies showed that although coinfection with GBV-C did not alter the biochemical and virological profile of patients with HCV hepatitis, there was an association, however, between GBV-C and HCV viremia and portal and periportal inflammation. It was also noted that the duration of HCV/GBV-C coinfection may be an important factor in progression of liver disease. Furthermore, it was observed that inflammation with necrosis in the portal and periportal tracts was significantly higher in patients with combined viremia compared to those with HCV infection alone. These findings suggest that GBV-C in patients with HCV infection might accelerate liver injury.

Epidemiology

GBV-C RNA is present in sera obtained from a variety of sources and global regions. These include sera from hemophiliacs, thalassemic patients, intravenous drug abusers, multiply transfused individuals, transfusion associated hepatitis cases, volunteer blood donors with both normal and elevated serum transaminase values, chronic HBV and HCV carriers, and sera from both acute and chronic non-A–E hepatitis patients, kidney transplant recipients, liver transplant recipients, fulminant hepatitis in the United States, Japan, Germany, and England, and in patients treated by maintenance hemodialysis. These data provide evidence that GBV-C is transmitted by blood and blood products and is globally distributed. Examination of serum specimens collected from healthy volunteer blood donors from different parts of the world confirmed the presence of GBV-C RNA in a remarkable 1–4% of the specimens.

Diagnosis and Assay Technology

The laboratory diagnosis of GBV-C infection is accomplished by testing for GBV-C RNA in serum. Serological immunoassays for the detection of antibodies to GBV-C have proven to be inadequate because only about 20–25% of GBV-C infected individuals develop antibodies to some poor immunogenic regions of the GBV-C protein. An ELISA for the detection of antibodies to the E2 structural protein of GBV-C has proven to be useful for assessing the recovery from GBV-C infection. On testing various populations for GBV-C E2 antibody, it was found that none of the serum specimens that were positive for GBV-C E2 antibodies were positive for GBV-C RNA. Of 100 volunteer blood donors from the United States, one was GBV-C RNA positive and three were GBV-C E2 antibody positive for a total incidence of 4%. In contrast, the exposure in commercial blood donors was 39%, and in intravenous drug abusers was 60%. The data suggest that exposure to GBV-C is much higher than in GBV-RNA studies alone. Other studies have shown that GBV-C E2 antibodies are long lived with a fairly constant titer, and that they are associated not only with clearance of GBV-C infection but also with protective immunity.

Because of the failure to identify an immunodominant epitope useful for serological diagnosis of an ongoing GBV-C infection, RT-PCR assays were developed to detect GBV-C RNA in human biological specimens. Many assays have been described in the literature. The best sensitive and specific assays are the ones that utilize oligonucleotide primers to amplify the 5′-NTR domain. These assays effectively detect GBV-C RNA from GBV-C infected individuals. Because testing a large number of samples by manual RT-PCR is tedious, labor intensive, and time consuming, an automated RT-PCR assay format with oligomer hybridization for the detection of sequence within the 5′-NTR was developed. This assay is a single tube assay that requires only the addition of serum-derived nucleic acids to a pre-aliquoted reaction vial. Reverse transcription, PCR amplification, and oligomer hybridization occur in the same tube containing recombinant *Thermus thermophilus* polymerase, adamantene-labeled sense and anti-sense oligonucleotide primers for the 5′-NTR of GBV-C, and carbazole-labeled probe. Detection of GBV-C product employs the automated LCx detection system (Abbott Laboratories, Abbott Park, IL) utilizing a microparticle enzyme immunoassay.

FURTHER READING

Dille, B.J., Surowy, T.K., Gutierrez, R.A., Coleman, P.F., Knigge, M.F., Carrick, R.J., Aach, R.D., Hollinger, F.B., Stevens, C.E., Barbosa, L.H., Nemo, G.J., Mosley, J.W., Dawson, G.J. and Mushahwar, I.K. An ELISA for detection of antibodies to the E2 protein of GB virus C. *J. Infect. Dis.* **176**, 458–461 (1997).

Erker, J.C., Desai, S.M., Leary, T.P., Chalmers, M.L., Montes, C.C. and Mushahwar, I.K. Genomic analysis of two GB virus A variants isolated from captive monkeys. *J. Gen. Virol.* **79**, 41–45 (1998).

Erker, J.C., Desai, S.M. and Mushahwar, I.K. Rapid detection of GB virus C RNA by reverse transcription-polymerase chain reaction (RT-PCR) using primers derived from the 5′-nontranslated region. *J. Virol. Methods* **70**, 1–5 (1998).

Fiordalisi, G., Zanella, I., Mantero, G., Bettinardi, A., Stellini, R., Paraninfo, G., Cadeo, G. and Primi, D. High prevalence of GB virus C infection in a group of Italian patients with hepatitis of unknown etiology. *J. Infect. Dis.* **174**, 181–183 (1996).

Gutierrez, R.A., Dawson, G.J., Knigge, M.F., Melvin, S.L., Heynen, C.A., Kyrk, C.R., Young, C., Carrick, R.J., Schlauder, G.G., Surowy, T.K., Dille, B.J., Coleman, P.F., Thiele, D.L., Lantino, J.R., Pachuki, C. and Mushahwar, I.K. Seroprevalence of GB virus C and persistence of RNA and antibody. *J. Med. Virol.* **53**, 167–173 (1997).

Gutierrez, R.A., Dawson, G.J. and Mushahwar, I.K. ELISA for detection of antibody to the E2 protein of GB virus C. *J. Virol. Methods* **69**, 1–6 (1997).

Leary, T.P., Desai, S.M., Yamaguchi, J., Chalmers, M.L., Schlauder, G.G., Dawson, G.J. and Mushahwar, I.K. Species-specific variants of GB virus A (GBV-A) in captive monkeys. *J. Virol.* **70**, 9028–9030 (1996).

Manolakopoulos, S., Morris, A., Davies, S., Brown, D., Hajat, S. and Dusheiko, G. Influence of GB virus C viremia on the clinical, virological, and histological features of early hepatitis C-related hepatic disease. *J. Hepatol.* **28**, 173–178 (1998).

Marshall, R.L., Cockerill, J., Friedman, P., Hayden, M., Hodges, S., Holas, C., Jennings, C., Jou, C.K., Kratochvil, J., Laffler, T., Lewis, N., Scheffel, C., Traylor, D., Wang, L. and Solomon, N. Detection of GB virus C by the RT-PCR LCx system. *J. Virol. Methods* **73**, 99–107 (1998).

Muerhoff, A.S., Simons, J.N., Erker, J.C., Chalmers, M.C., Desai, S.M. and Mushahwar, I.K. Conserved nucleotides sequences within the GB virus 5′-untranslated region: design of PCR primers for detection of viral RNA. *J. Virol. Methods* **62**, 55–62 (1996).

Muerhoff, A.S., Smith, D.B., Leary, T.P., Erker, J.C., Desai, S.M. and Mushahwar, I.K. Identification of GB virus C variants by phylogenetic analysis of 5′-untranslated and coding region sequences. *J. Virol.* **71**, 6501–6508 (1997).

Mushahwar, I.K. and Zuckerman, J.N. Clinical implications of GB virus C. *J. Med. Virol.* **56**, 1–3 (1998).

Ross, R.S., Viazov, S., Kruppenbacher, J.P., Elaner, S., Sarr, S., Lange, R., Eigler, F.-W. and Roggendorf, M. GB virus C infection in patients who underwent liver transplantation. *Liver* **17**, 238–243 (1997).

Simons, J.N., Leary, T.P., Dawson, G.J., Pilot-Matias, T.J., Muerhoff, A.S., Schlauder, G.G., Desai, S.M. and Mushahwar, I.K. Isolation of novel flavivirus-like sequences associated with human hepatitis. *Nat. Med.* **1**, 564–569 (1995).

Simons, J.N., Pilot-Matias, T.J., Dawson, G.J., Schlauder, G.G., Desai, S.M., Leary, T.P., Muerhoff, A.S., Erker, J.C., Buijk, S.L., Chalmers, M.L., Van Sant, C.L. and Mushahwar, I.K. Identification of two flavivirus-like genomes in the GB hepatitis agent. *Proc. Natl Acad. Sci. USA* **92**, 3401–3405 (1995).

Yoshiba, M., Okamoto, H. and Mishiro, S. Detection of GBV-C hepatitis virus genome in serum from patients with fulminant hepatitis. *Lancet* **346**, 1131–1132 (1995).

Zuckerman, A.J. Alphabet of hepatitis viruses. *Lancet* **347**, 558–559 (1996).

62 Human Retroviruses

George J. Dawson and Isa K. Mushahwar

ETIOLOGIC AGENTS

In 1980, the first human retrovirus {human T-cell lymphotropic virus type I (HTLV-I)} was identified in the lymphocytes of a patient with cutaneous T-cell lymphoma (Poiesz *et al.*, 1980). This virus and its close relative, HTLV-II, are classified as members of the bovine leukemia virus/HTLV genus of the *retroviridae*. In 1983, the human immunodeficiency virus type I (HIV-1) was identified in the T-cell lymphocytes of a patient with **acquired immunodeficiency syndrome (AIDS)** (Barré-Sinoussi *et al.*, 1983). HIV-1 and its close relative HIV-2 are classified as members of the Lentivirus genus of the *retroviridae*. Over the past three decades, there have been numerous reports of retrovirus particles associated with human tissues, but only these four retroviruses to date have been demonstrated to be both exogenous and transmissible from human to human.

The *retroviridae* are spherical, enveloped virions (80–120 nm in diameter) and consist of at least three major types of gene products including *gag* proteins (internal structural proteins), *pol* proteins {RNA-dependent DNA polymerase (reverse transcriptase – RT)} and *env* proteins (envelope proteins). The viral genomes of the *retroviridae* contain positive-sense, single-stranded RNA (7–10 kilobases) composed of two identical subunits. During its replication cycle, retroviruses utilize RT, a unique enzyme required to generate a DNA sequence complimentary to the viral RNA genome. The human retroviruses have several common features. The genomes of HTLV-I/-II and HIV-1/-2, unlike the majority of other *retroviridae*, code for regulatory proteins that upregulate transcription or enhance expression of viral proteins post-transcriptionally (reviewed by Haseltine, 1991). Another common feature of the human retroviruses is that they typically infect human lymphocytes with HIV-1/-2 infecting CD4-bearing T-cells and HTLV-I/-II generally infecting CD8-bearing T-cells.

The human retroviruses establish long-term persistent infections in their hosts. HIV-1 appears to be continuously replicating at any point in time after infection as evidenced by the ease of detecting viral RNA and/or viral antigens in serum following infection. For HTLV-I/-II, the level of viral replication is much reduced relative to HIV-1/-2, making it difficult to detect viral RNA or viral antigens in serum. Exposure to HTLV-I/-II is best detected by antibody response to viral proteins or by detecting viral DNA in lymphocytes. Since the human retroviruses are parenterally transmitted and can cause serious disease, blood screening tests have been implemented to reduce their spread to recipients of blood or blood products.

Unlike the HTLV viruses which appear to have little sequence variability, the genetic variation of HIV-1 is quite high, due to a combination of factors including the rapid turnover of HIV and the high viral load observed in infected individuals (Hu *et al.*, 1996). As HIV infection continues, the presence of 'quasispecies' (a heterogeneous group of related but distinct viruses present in an infected individual) of HIV is common, especially as the disease progresses.

HTLV-I/-II

Pathogenesis

Although it appears that HTLV-I/-II enters susceptible cells via specific cell receptors, the receptor has not been identified. Following the entry of the virus into susceptible cells, RT converts genomic RNA to DNA, followed by integration of the genome into the host cell genome. In general, it is believed that HTLV replication does not produce a large yield of virions in infected cells *in vivo* and that this low level of replication may permit the virus to evade immune elimination by the infected host.

HTLV-I virions were identified in an established T-lymphoblastoid cell line (HUT-102) originating from a patient with cutaneous T-cell lymphoma (Poiesz *et al.*, 1980). A second research group demonstrated that the virus was also present in another cell line derived from a patient with **Adult T-cell leukemia (ATL)**. Additional studies indicated that individuals diagnosed with ATL produce antibodies that react with HTLV-I antigens produced by the cell line, establishing that HTLV-I causes

ATL (Hinuma *et al.*, 1981). It has been shown that the proviral genome (double-stranded DNA of the viral RNA genome) becomes integrated into the host cell genome and can immortalize lymphocytes, resulting in tumors that are usually monoclonal in nature. Thus, malignancy may develop from clonal expansion of a single immortalized cell. There appear to be only low levels of viral expression in HTLV-I-transformed cells obtained *in vivo*. In general, it is believed that only about 1% of the HTLV-I-infected individuals will develop a viral-induced tumor after infection, and in most cases, this event occurs 20–30 years after infection. Once a diagnosis of ATL is made, the mean survival time is about 6–11 months. Other hematologic disorders that appear to be attributed to HTLV-I infection include cases of T-cell non-Hodgkins lymphoma, mycosis fungoides, small cell carcinoma, and large granular lymphocytic leukemia; further studies are needed to confirm these preliminary results (reviewed by Uchiyama, 1997).

In 1985, a study conducted in the Caribbean indicated that antibodies to HTLV-I were frequently detected in patients with a neurological disease known as **tropical spastic paraparesis (TSP)** (Gessain *et al.*, 1985). At the same time, a report from Japan described elevated antibody titers in patients with a slowly progressing myelopathy and pyramidal disturbances, termed **HTLV-I-associated myelopathy (HAM)**, very similar to TSP in patient presentation and outcome (Osame *et al.*, 1986). Patients presenting with TSP/HAM usually have weakness and spasticity of the extremities, hyperfelexia, and urinal and fecal incontinence. Autopsy studies indicate that there is HTLV-I-associated demyelineation of spinal cord tissues. Frequently, specific antibodies directed against HTLV-I antigens are detected in the cerebrospinal fluid. In general, the antibody titers to HTLV-I proteins detected in TSP/HAM patients are much higher than that noted in patients with ATL or asymptomatic patients, suggesting that humoral antibodies may play a role in inducing HAM/TSP. As with ATL, only about 1% of HTLV-I-infected individuals develop HAM/TSP over a lifetime of infection.

HTLV-II was first identified in 1982, in an established T-cell line derived from an individual diagnosed with hairy cell leukemia. HTLV-II is closely related to HTLV-I (approximately 60% nucleotide identity across the viral genome), and like HTLV-I, is able to immortalize T-cells *in vitro*. There have been interesting reports linking HTLV-II infection with T-cell malignancies and with various neuropathologic syndromes. However, there has been no consistent disease link of HTLV-II with any human malignancy or with neurologic disorders.

Transmission/Epidemiology

Unlike HIV, HTLV-I can only be transmitted by the introduction of infected lymphocytes into a recipient. Transmission is believed to occur by three major routes. First, HTLV-I can be transmitted from mother to child primarily by ingestion of breast milk containing HTLV-I infected lymphocytes. Secondly, sexual transmission may occur from male to female via infected lymphocytes in semen. Thirdly, HTLV-I can be transmitted by infected blood or blood products that are comprised at least in part of infected lymphocytes. In the US, all donors of blood and blood products have been screened for antibodies to HTLV-I since 1988. HTLV-II is also believed to be parenterally transmitted as is apparent from the high rate of HTLV-II infection among intravenous drug users.

Worldwide there are 10–20 million individuals infected with HTLV-I, with at least one million of these individuals residing in Japan (Cann and Chen, 1996). HTLV-I infection is quite common in southwestern Japan (the islands of Okinawa, Kyusu, and Shikoku) and in the Caribbean. HTLV-II is more commonly found in some of the Native American populations in southwestern US and in South America. HTLV-II infection is common among intravenous drug users.

Diagnostic and Immunoassay Technology

Antibody tests based on ELISAs employing purified virions are the most common methods utilized to detect exposure to HTLV-I. These ELISAs are utilized to screen blood and blood products and to diagnose infection with HTLV in diagnostic laboratories. Most of the ELISAs rely on purified HTLV-I coated on the solid phase (Gallo *et al.*, 1996). Specimens that are reactive in the ELISA are tested in a second (supplemental) assay to confirm the results of the first assay. The most commonly utilized supplemental assay is the Western blot utilizing purified viral lysates. Public health agencies such as the World Health Organization have recommended that the serologic confirmation of HTLV-I/-II infection requires the detection of antibodies to at least one *gag* and one *envelope* protein (Varma *et al.*, 1995).

In general, the immunoassays are believed to be adequately sensitive to prevent most HTLV infections via blood or blood products. The seroprevalence of HTLV in volunteer blood donors is estimated to be between 0.009 and 0.043% (Sandler *et al.*, 1990). There are approximately equal numbers of HTLV-I and HTLV-II-infected blood donors in the US. The estimated 'window period' (time between exposure to virus and development of antibodies) for HTLV-I and -II is 51 days with a range of 36–72 days (Manns *et al.*, 1992). In 1992, the calculated risk of transfusion and transmission of HTLV was 1 in 50,000 (Dodd, 1992). Since the likelihood that a serious disease may occur after transmission of HTLV is believed to be about 5%, the risk of HTLV-related disease in recipients of blood and blood products is believed to be relatively low (Table 62.1).

There have been several recent reports indicating that some individuals whose lymphocytes have been found

Table 62.1. Risk of adverse outcomes for HIV/HTLV infections acquired by allogeneic blood transfusion.

Virus	Seroprevalence in blood donors (%)	Risk of estimated infection per unit (12.057 million unit)	Risk of infection per recipient (5 units/recipient)	Approximate risk of serious disease developing after infection (%)	Risk that recipients will develop chronic disease or die as a result of transfusion (likelihood)*
HIV-1/-2	0.012–0.041	1:450,000	1:90,000	100	1:128,000
HTLV-I/-II	0.009–0.043	1:50,000	1:10,000	5	1:285,000

* Values are estimates that are corrected to exclude about 30% of recipients who are likely to die of underlying disease or trauma in the first 2 years after transfusion.

by PCR to contain HTLV-I/-II DNA are seronegative in the existing tests (Zucker-Franklin *et al.*, 1997; Zehender *et al.*, 1996). Some of these seronegative individuals are infected with HTLV-II. These data suggest either that immunoassays need to be improved or that in some cases, an assay detecting HTLV nucleic acids may be needed to ensure a safe blood supply.

Although there is significant serologic cross-reactivity between HTLV-I and -II, 3–5% of the HTLV-II seropositive donors are not detected with the ELISAs employing HTLV-I proteins (Gallo *et al.*, 1996). Recently, an immunoassay that employs both HTLV-I and HTLV-II proteins has been licensed for screening in the US, resulting in increased antibody detection among HTLV-II carriers (Roberston *et al.*, 1994).

Summary

Both HTLV-I and HTLV-II can be transmitted by transfusions. HTLV-I infections result in serious disease (ATL or TSP) over a lifetime in less than 5% of the HTLV-I-infected individuals. There is no clear link between HTLV-II infection and human disease. Blood screening tests are utilized to detect HTLV-infected individuals, and have been successfully utilized to reduce the spread of HTLV by blood transfusion.

REFERENCES

Barré-Sinoussi, F., Chermann, J.C., Rey, F., Nugeyre, M.T., Charmaret, S., Gruest, J., Dauget, C., Axler-Blin, C., Vézinet-Brun, F., Rouzioux, C., Rozenbaum, W. and Montagnier, L. Isolation of a T-lymphotropic retrovirus from a patient at risk for AIDS. *Science* **220**, 868–870 (1983).

Cann, A.J. and Chen, S.Y. Human T-cell leukemia virus types I and II. In: *Fields Virology* (eds Fields, B.N., Knipe, D.M., Howley, P.M., Chanock, R.M., Melnick, J.L., Monath, T.P., Roizman, B. and Straus, S.E.), 1849–1881 (Lippincott-Raven, Philadelphia, PA, 1996), Chapter 59.

Dodd, R.Y. The risk of transfusion-transmitted infection. *N. Engl. J. Med.* **327**, 419–421 (1992).

Gallo, D., Yeh, E.T., Moore, E.S. and Hanson, C.V. Comparison of four enzyme immunoassays for detection of human T-cell lymphotropic virus type 2 antibodies. *J. Clin. Microbiol.* **34**, 213–215 (1996).

Gessain, A., Barin, F., Vernant, J.C., Gout, O., Maurs, L., Calendar, A. and de The, G. Antibodies to human T-cell lymphotropic virus type-I in patients with tropical spastic paraparesis. *Lancet* **2**, 407–410 (1985).

Haseltine, W.A. Molecular biology of the human immunodeficiency virus type 1. *FASEB J.* **5**, 2349–2360 (1991).

Hinuma, Y., Nagata, K., Hanakoa, M. *et al.* Adult T-cell leukemia: antigen in an ATL cell line and detection of antibodies to the antigen in human sera. *Proc. Natl. Acad. Sci. USA* **78**, 6476–6480 (1981).

Hu, D.J., Dondero, T.J., Rayfield, M.A., George, J.R., Schochetman, G., Jaffe, H.W., Luo, C.-C., Kalish, M.L., Weniger, B.G., Pau, C.-P., Schable, C.A. and Curran, J.W. The emerging genetic diversity of HIV. *JAMA* **275**, 210–216 (1996).

Manns, A. *et al.* Prospective study of transmission by transfusion of HTLV-I and risk factors associated with seroconversion. *Int. J. Cancer* **51**, 886–891 (1992).

Osame, M., Usuku, K., Izumo, S., Iijichi, N., Amitani, H., Igata, A., Matsumoto, M. and Tara, M. HTLV-I associated myelopathy, a new clinical entity. *Lancet* **i**, 1031–1032 (1986).

Poiesz, B.J., Ruscetti, F.W., Gazdar, A.F., Bunn, P.A., Minna, J.D. and Gallo, R.C. Detection and isolation of a type C retrovirus particles from fresh and cultured lymphocytes of a patient with cutaneous T-cell lymphoma. *Proc. Natl. Acad. Sci. USA* **77**, 7415–7419 (1980).

Robertson, E., Stephens, J., Chan, E., Motley, C., Butyendorp, M., Prillaman, J., Coleman, C., Garrett, P., Weiblan, B., Contoreggi, C., Hoffman, W. and Phelps, B. Evaluation of an HTLV-II risk population with an HTLV-I/HTLV-II combination immunoassay and an HTLV-I/-II western immunoblot. *AIDS Res. Hum. Retrovir.* **10**, 497 (1994), abstract.

Sandler, S.G., Fang, C. and Williams, A. Retroviral infections transmitted by blood transfusion. *N. Engl. J. Med.* **316**, 918–922 (1990).

Uchiyama, T. Human T cell leukemia virus type I (HTLV-I) and human diseases. *Ann. Rev. Immunol.* 15–37 (1997).

Varma, M., Rudolph, D.L., Knuchel, M., Switzer, W.M., Hadlock, K.G., Velligan, M., Chan, L., Fouing, S.K.H. and Lal, R.B. Enhanced specificity of truncated transmembrane protein for serologic confirmation of human T-cell lymphotropic virus type 1 (HTLV-1) and HTLV-2 infections by western blot (immunoblot) assay containing recombinant envelope glycoproteins. *J. Clin. Microbiol.* **33**, 3239–3244 (1995).

Zehender, G., Girotto, M., De Maddalena, C., Francisco, G., Moroni, M. and Galli, M. HTLV infection in ELISA-negative blood donors. *AIDS Res. Hum. Retrovir.* **12**, 737–740 (1996).

Zucker-Franklin, D., Pancake, B.A., Marmor, M. and Legler, P.M. Reexamination of human T cell lymphotropic virus (HTLV-I/-II) prevalence. *Proc. Natl. Acad. Sci. USA* **94**, 6403–6407 (1997).

HIV-1/-2

Pathogenesis

The natural history of HIV-1 infection involves a continuous process wherein immune dysfunction and the loss of CD4-bearing cells begin at infection and progressively increase over time. Eventually, the immune system is impaired and opportunistic infections ensue. The mean time from infection to the development of frank AIDS is approximately 8–10 years.

HIV enters susceptible cells by the specific, high-affinity interaction between the outer envelope glycoprotein (gp) 120 of HIV and the CD4 molecule present on the surface both of T helper/inducer lymphocytes and of monocytes. Alternatively, HIV may gain entry into cells by phagocytosis, independent of CD4 on the cell surface. Following entry, RT catalyzes the synthesis of double-stranded HIV DNA, which enters the nucleus and integrates into the host cell DNA. The virus may then initiate replication or, alternatively, assume a latent state. Latently infected cells may later be activated in the presence of certain cytokines.

The first HIV genes to be expressed are those that encode for regulatory functions that enhance RNA transcription and affect the transport of spliced transcripts into the cytoplasm (for review see Greene, 1991). These transcripts encode for the HIV structural and enzymatic proteins necessary for the production of infectious progeny. As the replication cycle is completed, the HIV-infected cell may be destroyed as a direct result of the viral products, or because of the effects of viral budding on the cellular membrane, complexing of gp120 and CD4 molecules intracellularly, or the accumulation of toxic viral products such as unintegrated viral DNA (Rosenberg and Fauci, 1991; Panteleo and Fauci, 1996). Alternatively, $CD4^+$ cells may be killed indirectly by antibody-dependent cellular cytotoxicity, complement-mediated cell lysis, or via syncytial formation with other CD4-bearing cells (for review see Rosenberg and Fauci, 1991).

Among the factors that allow HIV to persist are the heterogeneity of the virus, both within a given individual and between populations of infected individuals (Goudsmit *et al.*, 1991) and the ability of HIV to evade the host immune system by establishing a latent infection (Levy *et al.*, 1987) or by replicating in monocyte/macrophage cells after being opsonized (Robinson *et al.*, 1988).

The key event in the initiation and progression of HIV disease is the selective destruction of CD4-bearing T-cells (T helper/inducer cells). The T helper/inducer $CD4^+$ lymphocyte population may be reduced from approximately 1000×10^6 cells/L to approximately $200–400 \times 10^6$ cells/L (Panteleo and Fauci, 1996). As the numbers of T helper cells decrease, an immune impairment ensues that renders the host susceptible to diseases caused by organisms usually held in check by the host's immune system. Most of the illnesses and deaths attributed to HIV infections are due to opportunistic infections by organisms such as *Pneumocystis carinii* (pneumonia), *Toxoplasma gondii* (toxoplasmosis – brain dysfunction), *Cryptococcus neoformans* (cryptococcal meningitis), *Mycobacterim avium* (pneumonia, lymphadenopathy, diarrhea), cytomegalovirus (retinitis, esophagitis, interstitial pneumonitis), *Candida* species (thrush, esophagitis), and *Histoplasma capsulatum* (histoplasmosis – pulmonary infection, disseminated infections).

Infection with HIV-2 may also lead to the development of AIDS (Clavel, 1987). However, the immunologic impairment in HIV-2-infected individuals appears to be less severe than with HIV-1, and the disease progression is much slower.

Transmission/Epidemiology

Transmission of HIV occurs by sexual intercourse (via semen or vaginal secretions), by parenteral routes (e.g. blood transfusions, intravenous drug use, or infusion of blood products by hemophiliacs), and by perinatal exposure (*in utero* or via colostrum). The time from exposure to the virus until the development of a detectable antibody response is generally believed to be 6–12 weeks. Following primary HIV-1 infection, individuals may either remain asymptomatic or develop a mononucleosis type illness (Panteleo and Fauci, 1996). During this initial phase of infection, HIV-1 frequently produces a viremia resulting in the detection of antigenemia (Paul *et al.*, 1987), concomitant with or preceding the development of IgG class, and sometimes IgM class antibodies against HIV proteins.

Recent estimates indicate that over 30 million persons have been infected with HIV-1 and that 28 million persons worldwide are currently living with HIV. HIV-1 is more prevalent than HIV-2 worldwide, and has been more extensively studied. HIV-2 infections are more prevalent in western Africa, most notably in Senegal, Guinea-Bissau, and The Gambia (Clavel, 1987). Except where noted in this chapter, the term HIV will be used to indicate HIV-1.

Diagnosis and Assay Technology

A direct method for determining exposure to HIV is to culture the virus from plasma or from the peripheral blood mononuclear cells (PBMCs) of infected individuals (Gallo *et al.*, 1984; Jackson *et al.*, 1990). It was reported that among 409 HIV-seropositive individuals, HIV could be isolated from each of 56 patients with frank AIDS, from 87 of 88 patients with ARC (AIDS-related complex) and from 259 of 265 asymptomatic patients (Jackson *et al.*, 1990). Although this is a very effective procedure, it requires dedicated facilities, highly trained personnel, and several weeks to obtain reliable results.

The discovery that interleukin-2 (IL-2) activates T-cells and permits their sustained growth *in vitro* (Poiesz *et al.*, 1980), provided needed factors for propagating HIV-1 *in vitro* (Barré-Sinoussi *et al.*, 1983). The ability to propagate HIV-1 *in vitro* was a major breakthrough permitting researchers to produce HIV-1 at a large scale (Popovic *et al.*, 1984) resulting in the development of ELISAs and Western blots for the detection of antibodies to HIV. The current tests most commonly utilized to detect exposure to HIV-1/-2 include antibody tests (ELISAs, agglutination tests, etc.), detection of viral antigens in serum, and detection of viral RNA or DNA in serum or in blood cells. For the most part, antibodies can be detected quite readily with assays employing at least one antigen from the envelope of the virus and a second antigen from the *gag* or *pol* gene products. Unlike most viral infections, nearly 100% of individuals seropositive for both the screening and confirmatory antibody tests continue to harbor HIV as detected by cell culture or polymerase chain reaction (PCR) (Jackson *et al.*, 1990). Thus the presence of antibody does not indicate clearance of the virus, but instead is a reliable indicator that an individual is infected with HIV.

The early serological techniques utilized infected cells (immunofluorescence) or purified virus (ELISA tests, Western blots) to determine exposure to HIV. Confirmatory tests that use recombinant antigens have also been useful (Dawson *et al.*, 1988). Since the mid-1990s, most of the HIV antibody tests have been based on recombinant antigens. These antibody tests are utilized to determine exposure to HIV among donors of blood products and in diagnostic laboratories.

With the discovery of HIV-2 in western Africa and its subsequent appearance in Europe and the US, many of the commercial tests were redesigned to include one or more HIV-2 proteins. The earlier HIV-1 tests, which did not include HIV-2-specific antigens, detected 60–90% of the HIV-2-infected individuals. Approximately 30 individuals in the US have been identified as having been infected with HIV-2, with most of these individuals having migrated from West Africa. Most were not blood donors. In fact, only two HIV-2 positive units have been identified among 60 million units screened in the US. Even though infection with HIV-2 appears to be rare in the US, the FDA has recommended that all antibody tests for HIV include HIV-2 proteins as a safety measure in the event of increases in HIV-2 seroprevalence over time.

Phylogenetic analysis of the HIV-1 virus strains indicates that most of the virus isolates are classified within the major (M) group of viruses, which contains 10 subtypes. In the late 1980s a genetically distinct outlier (O) of HIV-1 was identified and is now referred to as HIV-1 Group O. In studying the worldwide distribution of HIV it can been seen that some of the subtypes (B, C, and F) are found in diverse geographical regions, while the Group O isolate is rarely detected outside of Central Africa. Even within Central Africa, infection with HIV Group O accounts for less than 10% of the total HIV infections (Hu *et al.*, 1996). Nevertheless, cases of Group O have been identified in Europe. In July 1996, the Centers for Diseases Control reported on an HIV-1 Group O infected person residing in the US, who had migrated from Central Africa. Although the existing assays detect about 80% of the individuals infected with HIV-1 Group O, manufacturers are already modifying their tests to increase detection of antibodies in individuals infected with Group O (Jongerius *et al.*, 1997). Although this report shows enhanced detection of antibodies in individuals infected with Group O, reports have not yet been published which utilize HIV-1 Group O specific antigens. However, preliminary data indicates that the inclusion of HIV-1 Group O recombinant proteins in immunoassays will enhance the detection of antibodies in cases of Group O infected individuals (Hunt *et al.*, 1997). Immunoassays employing Group O antigens are now available in the market.

As with HTLV, two types of antibody tests are utilized for detection of exposure to HIV. Primary tests are used to screen large numbers of samples, and secondary tests, which are often more expensive or difficult to perform, are used to further examine samples that are repeatedly positive in the primary test. Most primary tests employ a solid phase, coated with native viral antigens or recombinant antigens to detect antibodies to HIV and are designed to give maximal sensitivity. Although these tests are sensitive and relatively specific, specimens may be falsely reactive. All samples that are repeatedly reacting in the primary test are tested with a supplemental test such as Western blot, in order to identify samples that are truly positives.

The seroprevalence rate for HIV infection among US blood donors was between 0.009 and 0.041% in 1986–1987 (United States General Accounting Office, 1997). Although the latest generation of antibody tests provides improved sensitivity for detecting antibodies to HIV, the window to seroconversion remains at about 22–25 days. Since it has been known that viral antigens can be detected in serum prior to seroconversion (Paul *et al.*, 1987) and that transmission of HIV by antibody-negative, antigen-positive blood donors has been reported (Gilcher *et al.*, 1990; Irani *et al.*, 1991), the Federal Drug Administration has licensed a new HIV-1 test to detect the p24 antigen. The window is reduced from 22–25 days with antibody testing alone to about 16–19 days, by using both an antibody test and the antigen test. The FDA has recommended that all blood units be tested for HIV antigen beginning in June 1996. Newly developed antigen detection tests have been developed which show increased sensitivity and allow detection of HIV antigens derived from HIV-1 Groups M and O, as well as for HIV-2 (Fransen *et al.*, 1997).

The estimated risk of HIV-1 infection among recipients of blood screened for HIV-1 ranges from 1:6800 to 1:493,000 (United States General Accounting Office, 1997; Table 62.1). It has been estimated that the window period would be reduced an additional 5 days if plasma from blood donors were tested for HIV RNA (Hewlett and Epstein, 1997). It is unlikely however, that nucleic acid probes technologies such as PCR would be rapidly licensed for screening blood for HIV RNA due to the lack of an inexpensive, highly reproducible, user-friendly assay that has the appropriate quality control measures in place. It is likely that as manufacturers continue their work on nucleic acid probe-based tests, these obstacles will be overcome. In the near future it is more likely that HIV nucleic acid tests such as RT-PCR will be utilized in specialized situations such as monitoring of patients on antiviral therapy, detection of HIV infection in newborns, as a prognostic indicator for disease progression, and possibly as a confirmatory assay for cases of indeterminate serology.

Tests that Predict Prognosis of HIV Disease Progression

The earliest method for identifying the onset of frank AIDS was the demonstration of T4-cell depletion as determined by flow cytometry of PBMCs (Lane and Fauci, 1985). Several other serological tests have been utilized in order to predict which individuals are most likely to develop AIDS (Lange *et al.*, 1989). Among these is the disappearance of antibodies to p24 (Goudsmit *et al.*, 1987; Lange *et al.*, 1987) and to p17 (Mehta *et al.*, 1990). Often the disappearance of these antibodies precedes or coincides with the detection of HIV p24 antigenemia (Paul *et al.*, 1987). Other markers that have been used to stage HIV disease with varying degrees of success include elevated serum β_2 microglobulin and elevated serum and urine neopterin levels (for review see Lange *et al.*, 1989). Newer concepts in staging HIV disease have focused on determining the number and percentage of HIV-infected $CD4^+$ cells among asymptomatic individuals followed over a period of time. In general, patients with increasing frequencies of HIV-infected cells develop HIV-related symptoms more rapidly than individuals who retain stable frequencies of HIV-infected cells (Schnittman *et al.*, 1990). Further studies on viral load indicate that increased levels of HIV RNA are detected in the serum of individuals who more rapidly develop AIDS.

Vaccination and Therapy

For an HIV vaccine to be effective it must protect against infection both by cell-free virions and by infected cells. In addition, the vaccines must be broad enough in reactivity to prevent infection with a virus that is known to be extremely heterogeneous. Although there have been numerous attempts at developing a vaccine that prevents transmission of HIV-1 to vaccinated individuals at risk of infections, there is no viable candidate vaccine at the present time (Heyward *et al.*, 1997).

Considerable progress has been made in developing antiviral substances that interfere with the replication of HIV *in vitro*. One of the antiviral drugs, 3′-azido-2′,3′-dideoxythymidine (AZT or zidovudine), currently available as an FDA-approved drug, inhibits RT activity, and has been effective in both prolonging the survival time of treated patients and reducing the severity of symptoms associated with HIV disease. However, the virus eventually develops resistance to AZT through the appearance of mutated forms of RT, and disease progression continues. Recently, a protease inhibitor drug has been designed that inhibits HIV replication and appears to dramatically reduce the viral load in treated patients (Kempf, 1994). This compound has also been approved by the FDA and is currently available. Although there appear to be some benefits from the protease inhibitor, HIV is not eradicated from the infected individual and the virus does develop resistance.

There remains optimism that in the future, combinations of drugs designed to interrupt HIV replication at various stages of the viral life cycle may provide greater benefits to HIV-infected individuals. A greater understanding of the viral life cycle, including the interplay between viral regulatory genes, replication, and pathogenesis, is likely to be central to the development of efficacious vaccines and antiviral therapies.

Summary

Currently, two immunoassays (an antibody test and an antigen test) are performed concomitantly to identify individuals infected with HIV-1 and/or HIV-2. The 'window period' between exposure to the virus and detection of antibodies or antigens is about 15 days. The risk of a recipient developing disease as a result of transfusion is about 1:128,000. Further improvements in screening tests may require the use of nucleic acid probe technology to detect viral nucleic acids in the serum or cells of infected individuals.

REFERENCES

Barré-Sinoussi, F., Chermann, J.C., Rey, F., Nugeyre, M.T., Charmaret, S., Gruest, J., Dauget, C., Axler-Blin, C., Vézinet-Brun, F., Rouzioux, C., Rozenbaum, W. and Montagnier, L. Isolation of a T-lymphotropic retrovirus from a patient at risk for AIDS. *Science* **220**, 868–870 (1983).

Clavel, F. HIV-2, the West African AIDS virus. *AIDS* **1**, 135–140 (1987).

Dawson, G.J., Heller, J.S., Wood, C.A., Gutierrez, R.A., Webber, J.S., Hunt, J.C., Hojvat, S.A., Senn, D., Devare, S.G. and Decker, R.H. Reliable detection of individuals seropositive for human immunodeficiency virus by competitive immunoassays using *Escherichia coli*-expressed HIV structural proteins. *J. Infect. Dis.* **157**, 149–155 (1988).

Fransen, K., Mertens, G., Stynen, D., Goris, A., Nys, P., Nkengasong, J., Heyndrickx, L., Janssens, W. and van der Groen, G. Evaluation of a newly developed HIV antigen test. *J. Med. Virol.* **53**, 31–35 (1997).

Gallo, R.C., Salahuddin, S.Z., Popovic, M., Shearer, G.M., Kaplan, M., Hayner, B.F., Palker, T.J., Redfield, R., Oleske, J., Safai, B., White, G., Foster, P. and Markham, P.D. Frequent detection and isolation of cytopathic retroviruses (HTLV-III) from patients with AIDS and at risk for AIDS. *Science* **224**, 500–502 (1984).

Gilcher, R.O., Smith, J., Thompson, S., Chandler, L., Epstein, J. and Axelrod, F. Transfusion associated HIV from anti-HIV non-reactive, antigen-reactive donor blood. In: *Abstracts of the International Society of Blood Transfusion/American Association of Blood Banks Joint Congress, Los Angeles, November 10–15, 1990* (American Association of Blood Banks, Arlington, VA, 1990).

Goudsmit, J., Lange, J.M.A., Paul, D.A. and Dawson, G.J. Antigen and antibody titers to core and envelope antigens in AIDS, AIDS-related complex and subclinical human immunodeficiency virus infection. *J. Infect. Dis.* **155**, 598–660 (1987).

Goudsmit, J., Back, N.K.T. and Nara, P. Genomic diversity and antigenic variation of HIV-I: links between pathogenesis, epidemiology and vaccine development. *FASEB J.* **5**, 2427–2436 (1991).

Greene, W.C. The molecular biology of human immunodeficiency virus type 1 infection. *N. Engl. J. Med.* **324**, 308–317 (1991).

Hewlett, I.K. and Epstein, J.S. Food and Drug Administration conference on the feasibility of genetic technology to close the HIV window in donor screening. *Transfusion* **37**, 346–351 (1997).

Heyward, W.L., MacQueen, K.M. and Jaffe, H.W. Obstacles and progress towards development of a preventive HIV vaccine. *J. Int. Assoc. Physicians AIDS Care* **3**, 28–34 (1997).

Hu, D.J., Dondero, T.J., Rayfield, M.A., George, J.R., Schochetman, G., Jaffe, H.W., Luo, C.-C., Kalish, M.L., Weniger, B.G., Pau, C.-P., Schable, C.A. and Curran, J.W. The emerging genetic diversity of HIV. *JAMA* **275**, 210–216 (1996).

Hunt, J.C., Golden, A.M., Lund, J.K., Gurtler, L.G., Zekeng, L., Obiang, J., Kapute, L., Hampl, H., Vallari, A. and Devare, S.G. Envelope sequence variability and serologic characterization of HIV type I Group O isolates from equatorial Guinea. *AIDS Res. Hum. Retrovir.* **13**, 995–1005 (1997).

Irani, M.S., Dudley, A.W. Jr. and Luccu, L.J. Case of HIV-I transmission by antigen-positive antibody-negative blood. *N. Engl. J. Med.* **325**, 1174–1175 (1991).

Jackson, J.B., Kwok, S.Y., Sninsky, J.J., Hopsicker, J.S., Sannerud, K.J., Rhame, F.S., Henry, K., Simpson, M. and Balfour, H.H. Jr. Human immunodeficiency virus type 1 detected in all seropositive symptomatic and asymptomatic individuals. *J. Clin. Microbiol.* **28**, 20–26 (1990).

Jongerius, J.M., van der Poel, A.M., van Loon, R., van den Akker, R., Schaasberg, W. and van Leeuwen, E.F. Human immunodeficiency virus (HIV) antibodies detected by new assays that are enhanced for HIV-1 subtype O. *Transfusion* **37**, 841–844 (1997).

Kempf, D.J. Design of symmetry-based peptidomimetic inhibitors of human immunodeficiency virus protease. In: *Retroviral Proteases* (eds Kuo, L.C. and Shafer, J.A.), 334–354 (Academic Press, San Diego, CA, 1994).

Lane, H.C. and Fauci, A.S. Immunologic abnormalities in the acquired immunodeficiency syndrome. *Ann. Rev. Immunol.* **3**, 477–500 (1985).

Lange, J.M.A., deWolf, F., Krone, W.J.A., Danner, S.A., Coutinho, R.A. and Goudsmit, J. Decline of antibody to outer core viral protein p17 is an earlier serological marker of disease progression in human immunodeficiency virus infection than anti-p24 decline. *AIDS* **1**, 155–159 (1987).

Lange, M.A., deWolf, F. and Goudsmit, J. Markers for progression in HIV infection. *AIDS* **3** (Suppl 1), S153–S160 (1989).

Levy, J.A., Evans, L., Cheng-Mayer, C., Pan, L.Z., Lane, A., Staben, C., Dina, D., Wiley, C. and Nelson, J. Biologic and molecular properties of the AIDS-associated retrovirus that affect antiviral therapy. *Ann. Inst. Pasteur* **138**, 101–111 (1987).

Mehta, S.U., Rupprecht, K.R., Hunt, J.C., Kramer, D.E., McRae, B.J., Allen, R.G., Dawson, G.J. and Devare, S.G. Prevalence of antibodies to core protein p17, a serological marker during HIV infection. *AIDS Res. Hum. Retrovir.* **6**, 443–454 (1990).

Panteleo, G. and Fauci, A.S. Immunopathogenesis of HIV infection. *Annu. Rev. Microbiol.* **50**, 825–854 (1996).

Paul, D.A., Falk, L.A., Kessler, H.A., Chase, R.M. and Blauuw, B. Correlation of serum HIV antigen and antibody with clinical status in HIV-infected patients. *J. Med. Virol.* **22**, 357–363 (1987).

Poiesz, B.J., Ruscetti, F.W., Gazdar, A.F., Bunn, P.A., Minna, J.D. and Gallo, R.C. Detection and isolation of type C retrovirus particles from fresh and cultured lymphocytes of a patient with cutaneous T-cell lymphoma. *Proc. Natl. Acad. Sci. USA* **77**, 7415–7419 (1980).

Popovic, M., Sarangadharan, M.G., Read, E. and Gallo, R.C. Detection, isolation and continuous production of cytopathic retroviruses (HTLV-III) from patients with AIDS and pre-AIDS. *Science* **224**, 497–500 (1984).

Robinson, W.E. Jr., Montefiori, D.C. and Mitchell, W.M. Antibody-dependent enhancement of human immunodeficiency virus type 1 infection. *Lancet* **i**, 790–794 (1988).

Rosenberg, Z.F. and Fauci, A.S. Immunopathogenesis of HIV infection. *FASEB J.* **5**, 2382–2390 (1991).

Schnittman, S.M., Greenhouse, J.J., Psallidopoulos, M.C., Baseler, M., Salzman, N.P., Fauci, A.S. and Lane, H.C. Increasing viral burden in $CD4^+$ T cells in patients with human immunodeficiency virus (HIV) infection reflects rapidly progressive immunosuppression and clinical disease. *Ann. Intern. Med.* **113**, 438–443 (1990).

United States General Accounting Office. Blood supply: transfusion associated risks. In: *Blood Supply: FDA Oversight of Remaining Issues of Safety* (February 1997).

63 Dengue Virus Infections

Bruce J. Dille and Isa K. Mushahwar

Descriptions of illnesses consistent with that caused by dengue virus can be found as early as the Chin Dynasty in China around 400 AD. Major epidemics occurred in Asia, Africa, and North America in 1779 and 1780. Despite this long history of causing disease, dengue fever, and the associated dengue hemorrhagic fever (DHF) and dengue shock syndrome (DSS), are on all lists of emerging infectious diseases. Multiple demographic, cultural, and economic factors have put more than 40% of the world's population at risk for this re-emerging disease.

Etiologic Agent

The four serotypes of dengue virus are members of the Flaviviridae family. Other members of this family of viruses include yellow fever virus, hepatitis C virus (*see* HEPATITIS – HEPATITIS C VIRUS), and the newly discovered GB viruses (*see* HEPATITIS – GB VIRUSES).

The spherical virus particles are 40–50 nm with a lipid envelope that has virus encoded glycoproteins on the surface. The single-stranded RNA genome is positive sense with approximately 11,000 bases coding for the three structural and seven nonstructural proteins. The RNA is translated into a single polyprotein, which is cleaved by a combination of host and virus coded proteases to produce the individual virus proteins.

The four dengue virus serotypes have type-common epitopes present on the viral glycoproteins, making accurate serologic diagnosis difficult. Infection with one serotype does provide lifelong immunity to reinfection with that same serotype, but does not provide cross-protective immunity to the other three serotypes.

Pathogenesis

The dengue viruses can be transmitted to humans via a number of mosquito species in the genus *Aedes* (Gubler, 1998). The principal vector is *Aedes aegypti*, a small, highly domesticated tropical mosquito that lays its eggs in artificial containers found in and around homes. These mosquitoes prefer to dwell indoors, and often feed on multiple individuals in the course of a single blood meal. Infected mosquitoes can also transmit the virus vertically to their offspring (Figure 63.1).

Uncomplicated, classic dengue fever develops 3–14 days after being bitten by an infected mosquito. The disease is characterized by an acute onset of fever, with headache, malaise, lumbosacral aching, and generalized muscle, joint, and bone pain (Kalayanarooj *et al.*, 1997). After a brief period of improvement, fever can recur and a measles-like rash may appear. Primary dengue infections are always self-limiting and the disease resolves within 2 weeks of onset. Dengue fever is primarily observed in older children and adults.

DHF and DSS, in contrast, are predominantly infections seen in children under 15 years of age (Dietz *et al.*, 1996). The acute phase of DHF is almost indistinguishable from the acute phase of dengue fever, or the acute phase of any of a number of illnesses including measles, rubella, influenza, typhoid, leptospirosis, or malaria, which are all common in dengue endemic regions. The critical time for DHF is when the fever starts to fall. A blood test usually shows that the patient has thrombocytopenia with a platelet count less than 100,000 mm^{-3}. Hemorrhagic manifestations begin to appear including petechiae, purpuric lesions, and ecchymoses. In severe cases epistaxis, bleeding gums, gastrointestinal hemorrhage, and hematuria occur. In contrast to dengue fever, which is a self-limiting mild infection, DHF and DSS can progress rapidly, resulting in death within 8–24 h of onset of symptoms. If left untreated, the fatality rate can exceed 20% (Monath, 1994). With early recognition of infection and aggressive supportive therapy, particularly with fluid and electrolyte replacement, starting with saline then on to plasma and plasma expanders in more severe cases, the fatality rate can be reduced to less than 1%. Once the shock is overcome, recovery is rapid, usually within 2–3 days.

The majority of cases of DHF appear to be the result of a secondary infection with a dengue strain different than that which caused the primary dengue fever infection (Halstead, 1988). While protective antibody was made to the first virus, cross-reactive antibodies do not neutralize other strains of dengue, thus permitting infection of the second strain. The cross-reactive antibody binds to the virus enhancing uptake into leukocytes through normal Fc

The Immunoassay Handbook. *Third Edition*
D. Wild (Ed.)

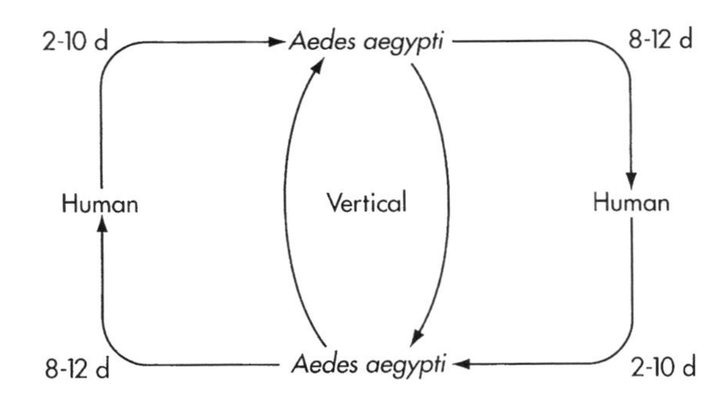

Fig. 63.1 Typical urban transmission cycle.

receptor binding and uptake functions. This antibody-dependent enhancement heightens the infection and replication of the virus in mononuclear cells. These cells then produce and secrete vasoactive mediators in response to the dengue infection causing increased vascular permeability leading to hypovolemia and shock (Kurane *et al.*, 1994).

Not all cases of DHF can be accounted for by reinfection. These cases, which account for about 1% of all DHF cases, may result from genetic changes which lead to increased virus replication and greater severity of disease (Rosen, 1977).

Epidemiology

Dengue virus causes more illness and death than any other arbovirus disease in humans. Dengue fever is now endemic in more than 100 countries throughout the tropical regions of Africa, the Americas, the Eastern Mediterranean, Southeast Asia, and the Western Pacific (Figure 63.2). There are an estimated 100 million cases of dengue fever occurring throughout the world each year (Monath, 1994). At least 41 countries have experienced DHF epidemics, up from only nine countries in 1970. These DHF epidemics account for 500,000 hospitalizations annually, mostly in children under 15 years of age. Worldwide, about 25,000 people die annually as a result of DHF, and DHF is the leading cause of hospitalization and death among children in many countries.

The principal vector of dengue infection, *A. aegypti*, can be found throughout the tropical and sub-tropical regions of the world, bringing the potential of dengue virus infection to 2.5 billion people (Figure 63.2). If the geographic distribution of the secondary vector *Aedes albopictus* is included, nearly two-thirds of the world's population is at risk for infection (Moore and Mitchell, 1997; Pinheiro and Corber, 1997). The importation of *A. albopictus* from its native Asia to the Americas is a prime example of the infectious disease impact of global trade practices (Reiter and Sprenger, 1987). In the early 1980s,

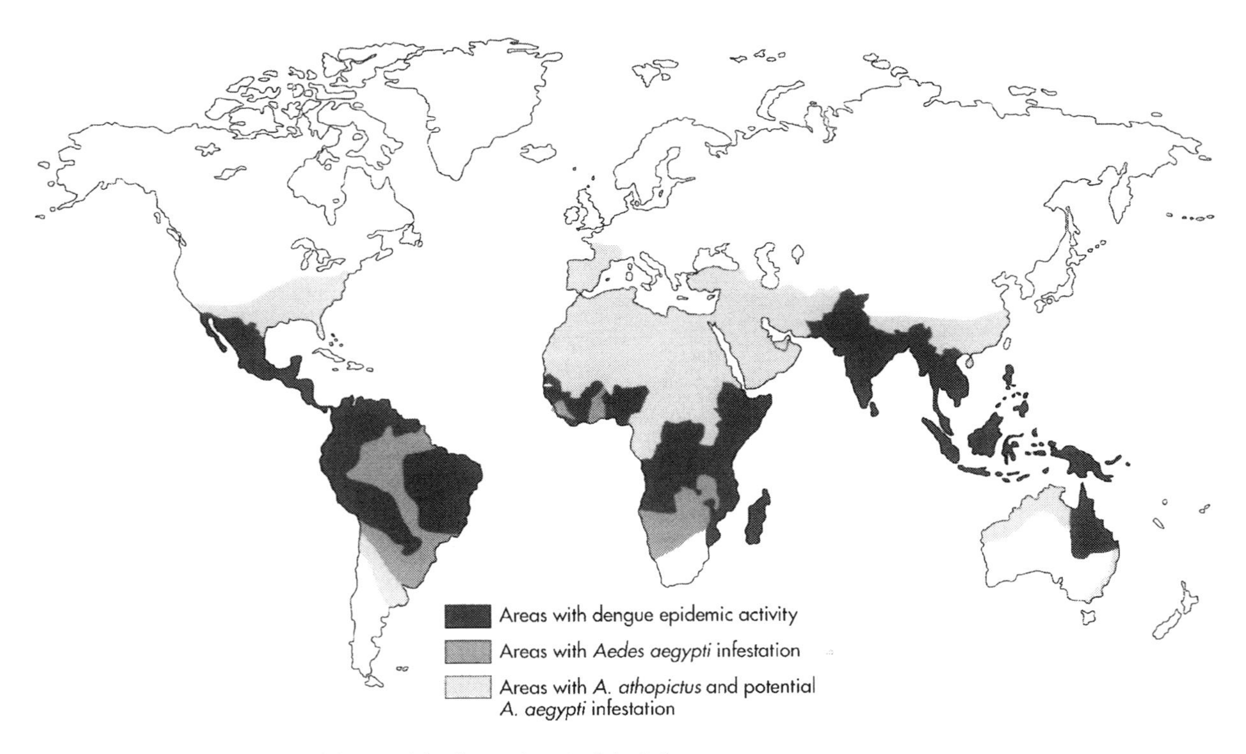

Fig. 63.2 World distribution of dengue infection and vector infestation.

companies in the United States were importing over one million used truck tires per year from Asia for the purposes of recapping and reselling. The tires were stored before shipping in the open in Asia where they collected rainwater, and subsequently the eggs of *A. albopictus*. On arrival in the United States, about 20% of the tires were deemed unusable and were discarded into the environment. The eggs subsequently hatched and the winter-hardy species rapidly spread throughout the eastern part of the United States. A similar import and expansion of *A. albopictus* occurred in South America at about the same time. *A. albopictus* has so far been shown to harbor eastern equine encephalitis and dengue virus.

The re-emergence of dengue infections, particularly in the Americas demonstrates the interplay between ecological adaptation of the mosquito vector and dramatic socioeconomic changes in humans (Monath, 1994). The main vector, *A. aegypti*, has evolved in Africa from a forest dwelling mosquito using tree holes containing rainwater for egg and larva development to an urban dweller that utilizes artificial containers (tires, vases, cisterns, barrels, etc.) for its eggs. The fully domesticated mosquito thrives indoors, and has adopted feeding habits to successfully elude the swats from the bitten human host. As human populations, particularly in Asia and the Americas, have become more urban and mobile, the *A. aegypti* population has increased and spread.

This mobility of the human population bringing the mosquito along leads to the introduction of new serotypes into new regions of the world. During the 1960s, for example, only dengue virus types 2 and 3 were present in the Americas. Dengue type 1 was introduced in 1977 and rapidly spread and became endemic in the region (Gubler, 1989). These hyperendemic regions, where multiple strains of dengue virus are circulating, have led to the emergence and spread of DHF. In the 1970s, only Puerto Rico had reported any cases of DHF. In 1981, a dengue epidemic occurred in Cuba with 344,203 cases, 116,000 people hospitalized, and 10,312 classified as DHF (Kourí *et al.*, 1998). Over the next decade, 11 countries in the Americas reported DHF outbreaks.

A. aegypti, in addition to urbanizing along with humans to facilitate the spread of infection, is also inadvertently contributing to increasing the virulence of the dengue virus. The mosquito is an inefficient vector of infection having both a low susceptibility to oral infection and a low rate of transovarial infection. The virus titer in blood must be more than 10^5-10^7 infectious virus per milliliter for the mosquito to be infected. Thus, only aggressive virus strains that replicate efficiently to high titers are able to be maintained in the population.

An additional human factor is contributing to the spread of *A. aegypti*. Mosquito eradication programs that were begun in the late 1940s and were successful in many countries by the late 1960s and early 1970s have now mostly been abandoned. As a result of abandoning these programs, the mosquitoes have rapidly reinvested the previously clear regions leading to explosive epidemic outbreaks (Gubler and Clark, 1995; Halstead, 1984). Rio de Janeiro had been freed of *A. aegypti*, but after reintroduction of the mosquito an epidemic of dengue type 1 occurred in 1986 involving over one million people. The introduction of dengue type 2 in 1990 led to outbreaks of DHF.

In the United States, dengue infections have so far been limited to several hundred cases per year among travelers returning from dengue endemic regions (Rigau-Perez *et al.*, 1994). There have been four incidents of secondary transmission from imported cases since 1980. Each of these outbreaks ended quickly, but given the widespread presence of dengue vectors, these incidents underscore the potential for more widespread outbreaks occurring in the United States.

Diagnosis and Assay Technology

Traditional serologic methods of hemagglutination inhibition (HI), complement fixation, and neutralization have been used to determine immunologic evidence of dengue virus infection. These tests are now giving way to a variety of enzyme immunoassay (ELISA) formats (Gubler, 1998; Guzman and Kourí, 1996; Talarmin *et al.*, 1998). One of the more widely used formats is an IgM capture ELISA. IgM is present in 80% of patients' sera by day 5 of illness, increasing to 93% by day 10 and 99% by day 20. For this assay, IgM from the patient sera is captured on the solid support by anti-IgM antibodies coated to the support. Once captured, dengue antigen is added with enzyme-labeled monoclonal antibody to the antigen. If dengue antigen-specific IgM is present, it will capture the antigen-labeled antibody complex resulting in a positive signal after substrate is added.

The IgM assay has an advantage over HI in that a single sample can be used to diagnose acute infection since IgM rarely persists more than 90 days after infection. The HI needs acute and convalescent samples to show a rise in antibody titer to distinguish acute infections from past infections.

ELISAs to detect IgG have also been developed and can be used to assess previous exposure to dengue virus. Similar to the HI tests, IgG testing needs acute and convalescent serum samples and must demonstrate a significant increase in antibody titer to be useful for acute infection diagnosis.

In addition to the standard microtiter format for these ELISAs, rapid (less than 7 min) chromatographic assays have been developed (Lam and Devine, 1998; Vaughn *et al.*, 1998). For this assay, a drop of patient serum migrates along a nitrocellulose strip and the IgG and IgM are captured on separate bands where anti-human IgG or anti-IgM antibodies have been applied. Gold labeled anti-dengue monoclonal antibody–dengue antigen complex is then applied, and visible bands appear where human anti-dengue antibody has been captured causing the gold-labeled conjugate to accumulate. Diagnosis of a primary dengue infection is made if a band appears for the IgM line, but not the IgG line. Secondary dengue infection is defined by the presence of IgG with or without IgM. There are certain limitations to this approach. The sample during primary infection must be late enough after infection to have detectable IgM, but early enough not to have detectable IgG, a window of about 5–10 days. The presence of IgG must also be viewed cautiously. The IgG is indicative of previous exposure to dengue virus, but its presence does not necessarily prove that the set of symptoms the patient is presenting with is a secondary dengue infection. As discussed earlier, a host of other infectious diseases, all with similar symptoms, must be ruled out before a secondary dengue infection diagnosis can be definitively made.

Another limitation of the serology based diagnostic assays for dengue virus is their inability to distinguish virus serotypes due to the extensive antigenic cross-reactivity between serotypes. In hyperendemic regions, where multiple serotypes are circulating, identification of the serotype responsible for a person's primary infection is advantageous for determining risk for infection with a second serotype that may lead to DHF. Serotyping infections can also alert health care workers that a new serotype has been introduced into a population, increasing the risk of DHF in that population.

The definitive diagnosis of dengue infection is isolation of the dengue virus from an infected individual. The dengue virus can then be positively identified using direct or indirect immunofluorescence and the serotype determined using monoclonal antibodies. Unfortunately, the virus grows poorly in mammalian cell culture, which is the easiest culture type for laboratories to work with.

The most sensitive virus isolation technique is direct inoculation of mosquitoes (Rosen and Gubler, 1974). The virus replicates to high titers within 4–5 days. Detection of virus is performed by direct immunofluorescence on mosquito tissue, usually brain or salivary glands. This method is not practical for most diagnostic labs, however.

The primary choice for virus isolation is mosquito cell culture. Although not quite as sensitive as infecting mosquitoes directly, the procedure is considerably more practical for most laboratories with cell culture facilities (Kuno *et al.*, 1985). The presence of virus in the sample can be detected by the development of cytopathic effects in the cells. More definitive diagnosis is through the use of direct or indirect immunofluorescence with anti-dengue antibodies, where the serotype of the virus can also be determined.

Reverse transcriptase polymerase chain reaction (RT-PCR) is a very rapid and sensitive method for detecting dengue virus RNA directly from a clinical sample (Morita *et al.*, 1994). Proper selection of primers also allows for serotyping the virus at the same time (Lanciotti *et al.*, 1992). Due to the high sensitivity of RT-PCR, extreme care must be used in the laboratory with proper controls for each assay to prevent false-positive results from amplicon contamination of the laboratory.

Control and Prevention

A number of different vaccine strategies are being pursued including attenuated vaccine viruses, inactivated whole virus, synthetic peptides, subunit vaccines, vector expression, recombinant live vector systems, infectious cDNA clone-derived vaccines, and naked DNA (Gubler, 1998). Most of these vaccines are still in the early stages of development, and none are likely to be in use soon. One of the problems with any dengue vaccine is that it must provide complete protective immunity to all four serotypes. Providing only partial protection will actually increase the risk for the more severe DHF to develop in vaccines.

The most direct method of dengue control is controlling the mosquito vector (Gubler, 1989; Gubler and Clark, 1994). Community-wide efforts to eliminate or clean water-holding containers are seen as the most likely means to achieve sustainable *A. aegypti* control (World Health Organization, 1998). Unfortunately, limited health care resources significantly hamper any control efforts.

REFERENCES

Dietz, B., Gubler, D., Ortiz, S., Kuno, G., Casta-Valez, A., Sather, G., Gomez, I. and Vergne, E. The 1986 dengue and dengue hemorrhagic fever epidemic in Puerto Rico: epidemiologic and clinical observations. *P.R. Health Sci. J.* **15**, 201–210 (1996).

Gubler, D. *Aedes aegypti* and *Aedes aegypti*-borne disease control in the 1990s: top down or bottom up. *Am. J. Trop. Med. Hyg.* **40**, 571–578 (1989).

Gubler, D. Dengue and dengue hemorrhagic fever. *Clin. Microbiol. Rev.* **11**, 480–496 (1998).

Gubler, D. and Clark, G. Community-based integrated control of *Aedes aegypti*: a brief overview of current programs. *Am. J. Trop. Med. Hyg.* **50**, 50–60 (1994).

Gubler, D. and Clark, G. Dengue/dengue hemorrhagic fever: the emergence of a global health problem. *Emerg. Infect. Dis.* **1**, 55–57 (1995).

Guzman, M. and Kourí, G. Advances in dengue diagnosis. *Clin. Diagn. Lab. Immunol.* **3**, 621–627 (1996).

Halstead, S. Selective primary health care: strategies for control of disease in the developing world. XI. Dengue. *Rev. Infect. Dis.* **6**, 251–264 (1984).

Halstead, S. Pathogenesis of dengue: challenges to molecular biology. *Science* **239**, 476–481 (1988).

Kalayanarooj, S., Vaughn, D., Nimmannitya, S., Green, S., Suntayakorn, S., Kunentrasai, N., Viramitrachai, W., Ratanachu-eke, S., Kiatpolpoj, S., Innis, B., Rothman, A., Nisalak, A. and Ennis, F. Early clinical and laboratory indicators of acute dengue illness. *J. Infect. Dis.* **176**, 313–321 (1997).

Kourí, G., Guzman, M., Valdés, L., Carbonel, I., Rosario, D., Vazquez, S., Laferté, J., Delgado, J. and Cabrera, M. Reemergence of dengue in Cuba: a 1997 epidemic in Santiago de Cuba. *Emerg. Infect. Dis.* **4**, 89–92 (1998).

Kuno, G., Gubler, D., Valez, M. and Oliver, A. Comparative sensitivity of three mosquito cell lines for isolation of dengue viruses. *Bull. WHO* **63**, 279–286 (1985).

Kurane, I., Rothman, A., Livingston, P., Green, S., Gagnon, S., Janus, J., Innis, B., Nimmannitya, S., Nisalak, A. and Ennis, F. Immunopathologic mechanisms of dengue hemorrhagic fever and dengue shock syndrome. *Arch. Virol.* (Suppl) **9**, 59–64 (1994).

Lam, S. and Devine, P. Evaluation of capture ELISA and rapid immunochromatographic test for the determination of IgM and IgG antibodies produced during dengue infection. *Clin. Diagn. Virol.* **10**, 75–81 (1998).

Lanciotti, R., Calisher, C., Gubler, D., Chang, G. and Vorndam, A. Rapid detection and typing of viruses from clinical samples by using reverse transcriptase-polymerase chain reaction. *J. Clin. Microbiol.* **30**, 545–551 (1992).

Monath, T. Dengue: the risk to developed and developing countries. *Proc. Natl Acad. Sci. USA* **91**, 2395–2400 (1994).

Moore, C. and Mitchell, C. *Aedes albopictus* in the United States: ten-year presence and public health implications. *Emerg. Infect. Dis.* **3**, 329–334 (1997).

Morita, K., Maemoto, T., Honda, S., Onishi, K., Murata, M., Tanaka, M. and Igarashi, A. Rapid detection of virus genome from imported dengue fever and dengue hemorrhagic fever patients by direct polymerase chain reaction. *J. Med. Virol.* **44**, 54–58 (1994).

Pinheiro, F. and Corber, S. Global situation of dengue and dengue hemorrhagic fever, and its emergence in the Americas. *World Health Stat. Q.* **50**, 161–169 (1997).

Reiter, P. and Sprenger, D. The used tire trade: a mechanism for the worldwide dispersal of container breeding mosquitoes. *J. Am. Mosq. Control Assoc.* **3**, 494–501 (1987).

Rigau-Perez, J., Gubler, D., Vorndam, A. and Clark, G. Dengue surveillance – United States, 1986–1992. *Mor. Mortal. Wkly Rep. CDC Surveill. Summ.* **43**, 7–19 (1994).

Rosen, L. The Emperor's new clothes revisited, or reflections on the pathogenesis of dengue hemorrhagic fever. *Am. J. Trop. Med. Hyg.* **26**, 337–343 (1977).

Rosen, L. and Gubler, D. The use of mosquitoes to detect and propagate dengue viruses. *Am. J. Trop. Med. Hyg.* **21**, 1153–1160 (1974).

Talarmin, A., Labeau, B., Lelarge, J. and Sarthou, J. Immunoglobulin A-specific capture enzyme-linked immunosorbent assay for diagnosis of dengue fever. *J. Clin. Microbiol.* **36**, 1189–1192 (1998).

Vaughn, D., Nisalak, A., Kalayanarooj, S., Solomon, T., Dung, N., Cuzzubbo, A. and Devine, P. Evaluation of a rapid immunochromatographic test for diagnosis of dengue virus infection. *J. Clin. Microbiol.* **36**, 234–238 (1998).

World Health Organization, Dengue. *Wkly Epidem. Rec.* **73**, 185–186 (1998).

64 Miscellaneous Diseases of Microbiological Origin

Bruce J. Dille and Isa K. Mushahwar

HERPES SIMPLEX VIRUS TYPE 1

Etiologic Agent

Herpes simplex virus type 1 (HSV-1) consists of a central protein-DNA core surrounded by a lipid envelope derived from the host cell nuclear membrane. The envelope contains at least eight different viral glycoproteins. The linear double-stranded DNA has about 152,000 base pairs divided into a unique long and a unique short region, each bounded by terminal repeat regions (Roizman and Sears, 1990). HSV-1 is closely related to HSV type 2, with which it shows considerable antigenic homology.

Upon entering a host cell, the virus shuts off host DNA, RNA, and protein synthesis. The synthesis of viral proteins is tightly regulated, and proceeds in a sequential fashion. The **immediate early proteins** are made in the absence of DNA synthesis, and direct the synthesis of the **early proteins**. The early proteins shut off the immediate early proteins and direct the synthesis of viral DNA and the **late** or **structural proteins**. Virion assembly then takes place in the cell nucleus. In certain host cells this sequence of protein synthesis and virion production is circumvented by mechanisms that are not understood. A latent infection is established where the DNA resides in the cell with minimal, if any, protein synthesis, and without virus production. The molecular trigger that releases the virus from the latent state, resulting in production of the virus, is also a mystery.

Pathogenesis

Infection caused by HSV-1 has been recognized for centuries with the name 'herpes' being used some 2000 years ago. HSV-1 primarily causes orolabial lesions (cold sores) and can also be found in the mouth (gingivostomatitis), pharynx, and esophagus.

Recurrent infection in the eye is very serious and can lead to blindness. HSV-1 causes about 10–20% of primary genital herpes, but is rarely responsible for recurrent genital lesions. Herpes encephalitis in adults is usually caused by HSV-1, but is rare with about three cases per million people per year. Extensive serology studies show the incidence of HSV-1 in adults worldwide at about 80%, ranging from about 40% for first-year college students to nearly 100% in some developing countries (Nahmias *et al.*, 1990). In the USA the prevalence increases with age, from around 30% in children under 10 years old to over 80%, by the age of 60.

The virus is spread by direct contact with virus shed from an infected person, who may be either symptomatic or asymptomatic. Virus infection tends to favor mucus membranes and is facilitated by breaks in the skin. Primary infections in childhood are usually asymptomatic or very mild, but occasionally symptomatic infections can occur that are quite severe. Symptomatic infections occur within 7–10 days of infection and start in the vesicular stage with a clear blister or group of blisters (Sunstrum, 1989). The vesicles become pustular then break to start the ulcerative stage. The ulcer becomes crusted and then heals without a scar. This sequence can take 7–14 days with new vesicles appearing as the original lesion begins to heal. The virus establishes latency in the trigeminal ganglia (for infections of the mouth and eyes) very early in the infection, possibly even before the lesion is apparent.

Reactivation of virus occurs in only 20–40% of infected people and can be triggered by stress, hormone changes, ultraviolet light, or trauma to the lesion site. Virus always recurs at an original site of infection, although that site may have been inapparent at the time. Recurrences often start with a tingling sensation at the lesion site called the prodrome, followed by redness (macular stage) which becomes raised (papular stage) before forming a vesicle. The pustular, ulcerative, and crusted stages follow. Recurrences usually resolve in 3–5 days without secondary lesion formation.

Primary herpes infections can be treated with topical or oral acyclovir to shorten the lesion duration, but treatment

The Immunoassay Handbook. *Third Edition*
D. Wild (Ed.)

does not affect the establishment of latency. Severe or frequent recurrences can be suppressed with oral acyclovir (Stone and Whittington, 1990).

Diagnosis and Assay Technology

In spite of the introduction in recent years of immunoassays for herpes simplex, the 'gold standard' for HSV diagnosis remains cell culture (Sunstrum, 1989). Typically a sample is collected using a cotton- or dacron-tipped swab (calcium alginate must be avoided) which is then inserted into a tube with transport medium (typically culture media or buffered salts with a protein stabilizer like gelatin or bovine serum albumin). Samples should be held at 4 °C short term and below −60 °C long term. Storage at −20 °C should be avoided. HSV-1 infects virtually all cell culture lines, however, the best results for clinical specimens are with primary cell lines such as primary rabbit kidney, human embryonic lung, tonsil or foreskin, mink lung, etc. Culture results can be confirmed directly on the cells by fixing the culture and probing with commercial polyclonal or monoclonal anti-herpes antibody followed by a fluorescein or enzyme (peroxidase or alkaline phosphatase) conjugate. The fluorescein conjugate is visualized using an ultraviolet microscope. Substrate is added to the enzyme conjugate which can then be observed with a light microscope.

A variety of rapid, non-culture tests have been introduced. Enzyme immunoassays (ELISA) capture antigen in the sample onto a solid support then anti-HSV antibody labeled with enzyme is added. Substrate is introduced and color read by an appropriate instrument. Published sensitivities for these assays range from mid-60% to low-90% with percent specificity in the 90s (Corey, 1986). An immunofluorescence (IFA) test is available and can be used either directly on a patient specimen or on the cells in culture after incubation with the sample. The direct method is very dependent on adequate sample collection with as many as one-third of samples being rejected as inadequate.

Latex agglutination tests are rapid to perform but the interpretation of the results can be very subjective. Reported sensitivities range down to under 50% for this type of assay (Cruse and Jenkins, 1987). *In situ* hybridization using an HSV-specific DNA probe is also available. Like the IFA, adequate sampling is essential to collect enough cells. This assay can also be used to confirm culture results.

A number of tests are available to detect antibody to HSV. Acute and convalescent testing, to detect a rise in titer, is effective only during a true primary infection when no HSV antibody is present in acute-phase serum. Antibody titers may remain unchanged during recurrent episodes (Mann and Hilty, 1982).

Commercially available HSV antibody tests that are based on infected cell lysate or whole virus preparations are not capable of accurately distinguishing HSV-1 from HSV-2 due to the extensive antigenic cross-reactivity between the two viruses (Lee *et al.*, 1985). Research has established that the HSV type-specific glycoprotein G (gG) is capable of distinguishing virus types (Dille *et al.*, 1990; Lee *et al.*, 1986; Sullender *et al.*, 1988). Commercial assays based on purified or recombinant gG are becoming available.

REFERENCES

Corey, L. Laboratory diagnosis of herpes simplex virus infection. *Diagn. Microbiol. Infect. Dis.* **4**, 111S–119S (1986).

Cruse, M.A. and Jenkins, S.G. Detection of herpes simplex virus in clinical specimens by latex agglutination and ELISA techniques. *87th Annual Meeting of American Society for Microbiology* C-141 (1987).

Dille, B.J., Spiezia, K.V., Kifle, L. and Mushahwar, I.K. Evaluation of a recombinant herpes simplex type 2 glycoprotein in an enzyme-linked immunoassay for the detection of specific HSV-2 antibodies. *30th Annual Interscience Conference on Antimicrobial Agents and Chemotherapy* No. 195 (1990).

Lee, F.K., Coleman, R.M., Pereira, L., Bailey, P., Tatsuno, M. and Nahmias, A.J. Detection of herpes simplex virus type 2 specific antibodies with glycoprotein G. *J. Clin. Microbiol.* **22**, 641–644 (1985).

Lee, F.K., Pereira, L., Griffin, C., Reid, E. and Nahmias, A. A novel glycoprotein for detection of herpes simplex virus type 1-specific antibodies. *J. Virol. Methods* **14**, 111–118 (1986).

Mann, D.R. and Hilty, M.D. Antibody response to herpes simplex virus type 1 polypeptides and glycoproteins in primary and recurrent infection. *Pediatr. Res.* **16**, 176–180 (1982).

Nahmias, A.J., Lee, F.K. and Beckman-Nahmias, S. Sero-epidemiological and sociological patterns of herpes simplex virus infection in the world. *Scand. J. Infect. Dis.* **69** (Suppl), 19–36 (1990).

Roizman, B. and Sears, A.E. Herpes simplex viruses and their replication. In: *Virology* (eds Fields, B.N. and Knipe, D.M.), 1795–1841 (Raven Press, New York, 1990).

Stone, K.M. and Whittington, W.L. Treatment of genital herpes. *Rev. Infect. Dis.* **12**, S610–S619 (1990).

Sullender, W.M., Yasudawa, L.L., Schwartz, M., Pereira, L., Hensleigh, P.A., Prober, C.G. and Arvin, A.M. Type-specific antibodies to herpes simplex virus type 2 (HSV-2) glycoprotein G in pregnant women, infants exposed to maternal HSV-2 infection at delivery, and infants with neonatal herpes. *J. Infect. Dis.* **157**, 164–171 (1988).

Sunstrum, J. Herpes simplex infections: a review. *J. Clin. Immunoassay* **12**, 175–178 (1989).

EPSTEIN–BARR VIRUS

Etiologic Agent

Epstein–Barr virus (EBV) was discovered in 1964 and is a member of the herpesvirus family. The icosahedral structure of the viral capsid contains 162 capsomeres and is surrounded by a lipid envelope, which contains virus-coded glycoproteins. The double-stranded DNA has over 172,000 base pairs with a coding capacity for over 80 proteins, few of which are understood (Baer *et al.*, 1984). EBV grows very poorly in culture. Only human and some primate lymphoid cells have been shown to be susceptible. Infection of peripheral lymphocytes results in transformation into immature blastoid cells capable of growth in culture. Virus production must be stimulated with phorbol esters to achieve even marginal yields.

Pathogenesis

Over 95% of the world's population have been infected with EBV, making it the most ubiquitous virus known. Most people are infected as infants and children in whom the infection may cause a mild sore throat and minor fever for a few days. Transmission of the virus requires direct contact with a person shedding virus, which can be found in saliva (Chang and Le, 1984).

The virus is thought to infect the epithelial cells of the throat where it gains access to the lymphocytes in the blood. EBV establishes a latent infection in the B-lymphocytes, where it remains for life. The infected lymphocytes are transformed and will replicate if not kept suppressed by cytotoxic T cells.

Reactivation of the virus and shedding in the saliva occur asymptomatically. If an individual escapes childhood without getting infected (generally only in developed countries), the primary infection as an adolescent or adult can be more severe (Henle and Henle, 1982). **Infectious mononucleosis (IM)** was first recognized as a distinct clinical condition in 1888, but it was not until 1969 that EBV was identified as the cause. About one-third of primary infections in adults result in IM, with clinically apparent IM occurring with a frequency of about 45 per 100,000.

Clinical symptoms include fever, malaise, swollen lymph nodes, and sometimes splenomegaly. Usually IM is self-limiting, but occasionally (about 1%) the disease has neurological complications. The situation is very different in individuals who have genetic or acquired immunosuppression (transplant recipients, AIDS, X-linked lymphoproliferative syndrome) (Okano *et al.*, 1988). EBV infection (either primary or reactivation) is not controlled properly by the immune system and the polyclonal B cell growth is almost always severe and can be fatal.

There has been a debate over the last few years about a chronic mononucleosis-like disease. The name 'chronic mononucleosis' was proposed and serological 'evidence' linking it to EBV was presented (Jones *et al.*, 1985). However, the link between EBV and the condition now called **chronic fatigue syndrome (CFS)** did not hold up under close scrutiny (Holmes *et al.*, 1988). EBV does cause a rare **chronic mononucleosis syndrome**, but is generally not involved with the more common CFS (Straus, 1988).

EBV is closely associated with two malignancies. **Burkitt's lymphoma** occurs in about 10 per 100,000 children in equatorial Africa and New Guinea. **Nasopharyngeal carcinoma (NPC)** constitutes 20% of all cancers in southeast China. Eskimo populations descendant from that region of China also show a higher incidence than found in the general population, where NPC is rare.

Diagnosis and Assay Technology

IM must be distinguished from other more serious conditions that produce similar symptoms. These conditions include streptococcal exudative tonsillitis, infectious hepatitis, cytomegalovirus (CMV) mononucleosis, Lyme disease, toxoplasmosis, and chronic lymphatic diseases such as Hodgkin's disease, lymphoma, and leukemia (Okano *et al.*, 1988).

Traditionally, a first step in the diagnosis of IM is hematology. Lymphocytosis occurs with an increase in lymphocytes and monocytes to 50% or more and atypical lymphocytes represent 10–20% of the lymphocyte and monocyte population.

In 1932, Paul and Bunnell discovered that heterophile antibodies in serum from IM patients caused sheep red blood cells to agglutinate (Paul and Bunnell, 1932). Inclusion of an absorption step using guinea pig kidney a few years later improved specificity. Even though heterophile antibodies have nothing to do with EBV, the cause of IM, heterophile antibody tests remain the principal method to diagnose IM. Numerous versions of heterophile antibody tests are available commercially, most using agglutination of red cells as the end point. Latex agglutination versions and a few enzyme immunoassays (ELISA) are also available.

False-positive heterophile antibody tests have been reported, including several reports of false-positive tests during primary human immunodeficiency virus infection (Van Essen *et al.*, 1988; Dubois, 1988).

After EBV was shown to cause IM in 1969, extensive serological studies were conducted. Four groups of antigens have been found to have diagnostic significance in EBV infections (Henle *et al.*, 1974). These antigens are available commercially either as part of diagnostic kits, which are usually immunofluorescence assays, or on slides for which anti-human IgM or IgG fluorescein conjugates can be purchased separately. The first group of at least 30 antigens constitutes all of the virus structural antigens and is called **virus capsid antigen** (VCA). Everyone infected with EBV develops an IgM response to VCA and all have lifelong IgG titers.

The **early antigens** (EA) are divided into two groups based on fluorescence patterns observed in cells: a **diffuse** pattern (EA-D) and a **restricted** pattern (EA-R). The EA-R antigens are also denatured by methanol fixation. IgG to EA-D is found in about 70% of acute IM, but rarely seen after recovery. IgG to EA-R is found only in silent primary infections in infants and in Burkitt's lymphoma patients.

The final group of antigens is the **nuclear antigens** (EBNA), and these may be involved in the establishment and maintenance of latency or cell transformation. These antigens cannot be detected by conventional IFA, and anticomplement immunofluorescence procedures are needed. This requirement may be due to the small amount of EBNA present in the cells used in the tests. Anti-EBNA IgG, as measured by anti-complement IFA, appears months after primary infection and persists for life in 100% of the people infected.

Comparison studies between heterophile antibody tests and IFA to specific EBV antigens reveal that less than 90% of adults with IM produce heterophile antibodies (Henle and Henle, 1982). In children, heterophile antibodies are present in less than 50% of primary EBV infections. In spite of the low sensitivity, not being specific for EBV, and only being a presumptive diagnostic test, the heterophile antibody test remains the principal method of diagnosis of IM. This is partly because these tests are fast and inexpensive, and partly due to the difficulty of the IFA test, which requires an ultraviolet microscope and expertise to interpret the results. However, detection of IgM to VCA by IFA is the definitive diagnostic test for IM (Okano *et al.*, 1988).

A number of enzyme immunoassays (ELISA) have been introduced using a variety of antigen strategies. There are several assays that use purified heterophile antigens on the solid phase and detect the presence of heterophile antibodies for a presumptive infectious mononucleosis diagnosis (Elgh and Linderholm, 1996). ELISAs that specifically detect antibodies to EBV antigens rely on purified or recombinant VCA, EBNA, or EA. These tests are designed to assess IgM for acute infection diagnosis or IgG for past infection (Gutiérrez *et al.*, 1997; Svahn *et al.*, 1997; Weber *et al.*, 1996). Performance was variable among different test formats with IgM detection using VCA antigens comparing the best to reference assays for acute IM diagnosis.

Virus detection is difficult and rarely done and has little diagnostic value. Because previously infected people can periodically shed virus, it is impossible to determine from the presence of virus alone whether the patient's symptoms are due to a primary EBV infection or another cause (i.e. CMV or hepatitis), with EBV present merely incidentally (Okano *et al.*, 1988). EBV antigen detection, virus detection, and serology are all of little value for the diagnosis of CFS (Holmes *et al.*, 1988).

REFERENCES

Baer, R., Bankier, A.T., Biggin, M.D., Deininger, P.L., Farrell, P.J., Gibson, T.J., Hatfull, G., Hudson, G.S., Satchwell, S.C., Seguin, C., Tuffnell, P.S. and Barrell, B.G. DNA sequence and expression of the B95-8 Epstein–Barr virus genome. *Nature* **310**, 207–211 (1984).

Chang, R.S. and Le, C.T. Failure to acquire Epstein–Barr virus infection after intimate exposure to the virus. *Am. J. Epidemiol.* **119**, 392–395 (1984).

Dubois, R.E. Delayed male-to-female transmission of human immunodeficiency virus, with a false-positive, rapid heterophile test. *J. Infect. Dis.* **157**, 208 (1988).

Elgh, F. and Linderholm, M. Evaluation of six commercially available kits using purified heterophile antigen for the rapid diagnosis of infectious mononucleosis compared with Epstein–Barr virus-specific serology. *Clin. Diagn. Virol.* **7**, 17–21 (1996).

Gutiérrez, J., Rodríguez, M., Maroto, C. and Piédrola, G. Reliability of four methods for the diagnosis of acute infection by Epstein–Barr virus. *J. Clin. Lab. Anal.* **11**, 78–81 (1997).

Henle, W. and Henle, G. Epstein–Barr virus and infectious mononucleosis. In: *Human Herpesvirus Infections* (eds Glaser, R. and Gotlieb-Stematsky, T.), 151–167 (Marcel Dekker, New York, 1982).

Henle, W., Henle, G. and Horwitz, C.A. Epstein–Barr virus specific diagnostic tests in infectious mononucleosis. *Hum. Pathol.* **5**, 551–565 (1974).

Holmes, G.P., Kaplan, J.E., Gantz, N.M., Kamaroff, A.L., Schonberger, L.B., Strauss, S.E., Jones, J.F., Dubois, R.E., Cunningham-Rundles, C., Pahuva, S., Tosato, G., Zegans, L.S., Purtilo, D.T., Brown, N., Schooley, R.T. and Brus, I. Chronic fatigue syndrome: a working case definition. *Ann. Intern. Med.* **108**, 387–389 (1988).

Jones, J.F., Ray, C.G., Minnish, L.L., Hicks, M.J., Kibler, R. and Lucas, D.O. Evidence for active Epstein–Barr virus infection in patients with persistent, unexplained illnesses: elevated anti-early antigen antibodies. *Ann. Intern. Med.* **102**, 1–7 (1985).

Okano, M., Thiele, G.M., Davis, J.R., Grierson, H.L. and Purtilo, D.T. Epstein–Barr virus and human diseases: recent advances in diagnosis. *Clin. Microbiol. Rev.* **1**, 300–312 (1988).

Paul, J.R. and Bunnell, W.W. The presence of heterophile antibodies in infectious mononucleosis. *Am. J. Med. Sci.* **183**, 90–104 (1932).

Smith, R.S., Rhodes, G., Vaughn, J.H., Horwitz, C.A., Getiosky, J.E. and Whalley, A.S. A synthetic peptide for detecting antibodies to Epstein–Barr virus nuclear antigen in sera from patients with infectious mononucleosis. *J. Infect. Dis.* **154**, 885–889 (1986).

Straus, S.E. The chronic mononucleosis syndrome. *J. Infect. Dis.* **157**, 405–412 (1988).

Svahn, A., Magnusson, M., Jägdahl, L., Schloss, L., Kahlmeter, G. and Linde, A. Evaluation of three commercial enzyme-linked immunosorbent assays and two latex agglutination assays for diagnosis of primary Epstein–Barr virus infection. *J. Clin. Microbiol.* **35**, 2728–2732 (1997).

Van Essen, G.G., Lieverse, A.G., Sprenger, H.G., Schirm, J. and Weitz, J. False-positive Paul–Bunnell test in HIV seroconversion. *Lancet* **ii**, 747–748 (1988).

Weber, B., Brunner, M., Preiser, W. and Doerr, H.W. Evaluation of 11 enzyme immunoassays for the detection of immunoglobulin M antibodies to Epstein–Barr virus. *J. Virol. Methods* **57**, 87–93 (1996).

CHAGAS' DISEASE

Etiologic Agent

American **trypanosomiasis** or Chagas' disease is transmitted by the parasite *Trypanosoma cruzi* (*T. cruzi*). These parasites are characterized by the presence of one flagellum that runs alongside the body of the protozoan, a spindle-shaped or U- and C-shaped, 15–20 μm parasite. The common mode of transmission to mammals is through the bloodsucking triatomine bugs or reduviids. Generally, these bugs are infected with viable trypomastigotes (extracellular non-dividing forms) from humans or animals that have been already infected with *T. cruzi*. In the insect's midgut, the trypomastigotes are transformed into epimastigotes, then 3–4 weeks later, the epimastigotes transform into trypomastigotes in the insect's hind gut. The trypomastigotes infect mammals through the bite wound in infected bug feces. The trypomastigotes penetrate mammalian cells directly where they transform into intracellular amastigotes. Mammalian cells harboring the parasites release them upon rupturing and a new cycle of infection and multiplication will then be initiated.

Chagas' disease is prevalent in Latin America. Approximately 20 million people are currently infected with the disease and it is estimated that 90 million people are considered at risk.

The predominant route of transmission in humans occurs after the bite of the infected triatomine bug is contaminated with the insect's contagious feces. However, other routes of transmission such as through blood transfusion, organ transplantation, sexual activity, breast milk, intravenous drug use, and congenital transmission through the placenta have been reported.

Pathogenesis

Three clinical manifestations have been reported to occur in humans during Chagas' infection. These are acute disease, late or indeterminate phase, and chronic disease. As reported by many investigators (Kirchhoff and Neva, 1985; Neva, 1988; Maguire and Hoff, 1989) and elegantly described by Guidino and Linares (1990) and Tanowitz *et al.* (1992), acute Chagas' disease, characterized by a high level of parasitemia, occurs mostly in children and is subclinical and self-limiting in more than 90% of cases. The incubation period varies from 20 to 40 days. A local area of erythema and induration (chagoma) may develop occasionally at the parasite's entry site. A periorbital swelling (Romaña sign) with painless conjunctivitis and periauricular lymphadenopathy can be seen when the route of infection is the conjunctiva.

The indeterminate phase generally follows acute infection and then is usually followed by an apparent complete clinical recovery. It is characterized by the presence in serum of *T. cruzi* antibodies. Only 20–30% of the indeterminate cases progress to chronic Chagas' disease (Kirchhoff and Neva, 1985; Neva, 1988; Maguire and Hoff, 1989).

According to Neva (1988), the most common manifestation of chronic Chagas' disease is cardiac disease, where cardiac arrhythmias are frequent. Another common chronic manifestation is mega disease of the esophagus or colon. Patients with megacolon suffer from chronic constipation, abdominal pain with volvulus obstruction and perforation of the bowel (Gudino and Linares, 1990).

Mega disease of the esophagus and the colon may be present in the same patient, and cardiomyopathy can be present with either form of the disease (Neva, 1988).

Diagnosis and Assay Technology

Several detection systems such as indirect hemagglutination, direct and indirect immunofluorescence, complement fixation, enzyme-linked immunoassay (ELISA), and polymerase chain reaction (PCR) have been developed to diagnose Chagas' infection during the acute and chronic phases of the disease. In the acute phase of *T. cruzi* infection, the high level of parasitemia in the blood stream facilitates diagnosis by direct microscopic examination of fresh blood, cerebrospinal, and lymph fluids. During the chronic phase of infection, the microscopic detection of *T. cruzi* is limited due to the low level of parasitemia. However, high titers of antibodies to *T. cruzi* (Israelski *et al.*, 1988) allow detection by a variety of sensitive complement fixation and indirect immunofluorescence tests. These serologic tests have been shown to suffer from poor specificity (false-positive reactions).

Xenodiagnosis (Segura, 1987) is another procedure that has been described in which triatomine bugs are allowed to feed on the blood of a Chagas' patient. Upon subsequent dissection of the insects several weeks later, a direct microscopic examination of the insects' intestinal content to verify the presence of *T. cruzi* is performed. The test is reported to be highly specific as expected. However, like the other tests, it suffers from very poor sensitivity since parasites are detected in only 50% of known infected individuals.

Because of the poor sensitivities and specificities of the above-described procedures, alternative tests have been devised such as an ELISA antibody test and a PCR DNA test for the accurate diagnosis of Chagas' disease.

Brashear *et al.* (1993) evaluated the sensitivity and specificity of an ELISA for the detection of antibodies to *T. cruzi* among blood donors in the southwestern and western United States. The authors reported that their test had a specificity of 99.98% with negative samples and a sensitivity of 100% in specimens positive by consensus (i.e. reactive in ELISA, indirect hemagglutination assay, and immunofluorescence assays).

Supplemental ELISAs and radioimmunoprecipitation assays for confirmation of the above-mentioned antibody assay have been described (Winkler *et al.*, 1995). These assays utilize purified *T. cruzi* antigens GP90, GP 60/50, and lipophosphonopeptidoglycan (LPPG).

A sensitive PCR assay for the detection of the parasite has been described by Moser *et al.* (1989). The authors reported the detectability of one parasite in 20 mL of blood from an infected individual. They also reported the inability to detect any parasite in two samples of blood from patients chronically infected with *T. cruzi*. This was attributed to known intermittent parasitemia in such patients. Nevertheless, such amplification assays when automated can be utilized in the evaluation of large numbers of samples in both insects and mammalian hosts.

Treatment

No vaccines are available at this time. However, two drugs are currently available for the treatment of Chagas' disease. These are nifurtimo X (Lampit; Bayer 2502) and benzinidazole (Radamil; Roche 7-1051). These drugs are effective if taken for an extended period of time during the acute phase of the disease, but have no beneficial effect on the chronic phase.

REFERENCES

Brashear, R.J., Winkler, M.A., Schur, J.D., Lee, H., Burczak, J.D., Hall, H.J. and Pan, A.A. Detection of antibodies to *Trypanosoma cruzi* among blood donors in the southwestern and western United States. 1. Evaluation of the sensitivity and specificity of an enzyme immunoassay for detecting antibodies to *T. cruzi*. *Transfusion* **35**, 213–218 (1993).

Gudino, M.D. and Linares, J. Chagas' disease and blood transfusion. In: *Emerging Global Patterns in Transfusion-Transmitted Infections* (eds Westphal, R.G., Carlson, K.B. and Ture, J.M.), 65–86 (American Association of Blood Banks, Arlington, VA, 1990).

Israelski, D.M., Sadler, R. and Araujo, F.G. Antibody response and antigen recognition in human infection with *Trypanosoma cruzi*. *Am. J. Trop. Med. Hyg.* **39**, 445–455 (1988).

Kirchhoff, L.V. and Neva, F.A. Trypanosoma species (Chagas' disease). In: *Mandell, Douglas and Bennett's Principles and Practice of Infectious Diseases* 4th edn (eds Mandell, G., Bennett, J.E. and Dolin, R.), 1531–1537 (Churchill Livingstone, New York, 1985).

Maguire, J.H. and Hoff, R. American trypanosomiasis. In: *Infectious Diseases* (eds Hoepnicke, P.D. and Jordan, M.C.), 1257–1266 (Lippincott, Philadelphia, PA, 1989).

Moser, D.R., Kirchhoff, L.V. and Donelson, J.E. Detection of *Trypanosoma cruzi* by DNA amplification using the polymerase chain reaction. *J. Clin. Microbiol.* **27**, 1477–1482 (1989).

Neva, F.A. American Trypanosomiasis (Chagas' disease). In: *Cecil's Textbook of Medicine* (eds Wyngaarden, J.B. and Smith, L.H.), 1865–1869 (W.B. Saunders, Philadelphia, PA, 1988).

Segura, E.L. Xenodiagnosis. In: *Chagas' Disease Vectors. Vol. 1. Taxonomic, Ecologic, and Epidemiological Aspects* (eds Brenner, R., Stoka, R. and de la Merced, A.), 41–45 (CRC Press, Boca Raton, FL, 1987).

Tanowitz, H.B., Kirchhoff, L.V., Simon, D., Morris, S.A., Weiss, L.M. and Wittner, M. Chagas' disease. *Clin. Microbiol. Rev.* **5**, 400–419 (1992).

Winkler, M.A., Brashear, R.J., Hall, H.J., Schur, J.D. and Pann, A.A. Detection of antibodies to *Trypanosoma cruzi* among blood donors in the southwestern and western United States. II. Evaluation of a supplemental enzyme immunoassay and radioimmunoprecipitation assay for confirmation of seroreactivity. *Transfusion* **35**, 219–225 (1995).

CANDIDA ALBICANS – INVASIVE CANDIDIASIS

Pathogenesis

Also see SEXUALLY TRANSMITTED DISEASES: *CANDIDA ALBICANS* – VAGINITIS.

In addition to causing superficial vulvovaginitis, *Candida albicans* can become invasive, infecting tissues below the epithelial surfaces. Patients whose body fluids or tissues harbor *C. albicans* are considered to have **invasive candidiasis**. Systemic infections have become an important cause of morbidity and mortality in patients undergoing chemotherapy, and transplant recipients. **Chronic mucocutaneous candidiasis** is common in patients with acquired immunodeficiency syndrome. Burn patients are particularly susceptible to disseminated candidiasis. In patients with invasive candidiasis, the mortality rate may be as high as 70%. It has been estimated that approximately 5% of hospitalized patients acquire infections during their stay, and 5% of these patients become infected with *Candida* species.

Amphotericin B remains the mainstay for treatment of invasive candidiasis. However, treatment failures and toxic side effects are numerous. Dosage and duration of treatment are not well documented. Increased resistance to the drug has been reported.

Diagnosis and Assay Technology

Culture is the standard method for identifying patients with invasive candidiasis. Blood agar is a satisfactory medium for the isolation of *C. albicans* from body fluids. However, blood and other tissues require special culture techniques to isolate *Candida* species. Unfortunately, the diagnostic efficiency of these techniques is low. Invasive candidiasis can be caused by species other than *C. albicans*. It is therefore important to identify the species in order to identify the source of the infection.

There are many publications describing immunoassays for the detection of antibodies and antigens in patients with invasive candidiasis (Bennett, 1987; Lew, 1989; Ness *et al.*, 1989; Phillips *et al.*, 1990). Unfortunately, most of the earlier serological assays utilized crude antigens making interpretation of titers difficult. Recently, better-defined antigens have been used in RIA and ELISA. However, primarily because of poor performance, none of these immunoassays have been commercialized. At this time, serology does not appear to be a useful tool for the diagnosis of severe *Candida* infections (De Repentigny, 1989).

Enzyme immunoassay and latex tests have been described for the detection of candidiasis. Reported sensitivities of various assay kits are low, ranging from 33 to 82% (Fujita and Hashimoto, 1992; Morhart *et al.*, 1994). One of the major problems associated with some antigen-detection assays has been the presence of antibody or other proteins blocking the detection of antigen (Jones, 1990). Thus, sera must be treated to eliminate these antigen-binding proteins.

REFERENCES

Bennett, J. Rapid diagnosis of candidiasis and aspergillosis. *Rev. Infect. Dis.* **9**, 398–402 (1987).

De Repentigny, L. Serological techniques for diagnosis of fungal infections. *Eur. J. Clin. Microbiol. Infect. Dis.* **8**, 362–375 (1989).

Fujita, S. and Hashimoto, T. Detection of serum *Candida* antigens by enzyme-linked immunosorbent assay and latex agglutination test with anti-*Candida albicans* and anti-*Candida krusei* antibodies. *J. Clin. Microbiol.* **30**, 3132–3137 (1992).

Jones, J. Laboratory diagnosis of invasive candidiasis. *Clin. Microbiol. Rev.* **3**, 32–45 (1990).

Lew, M. Diagnosis of systemic *Candida* infections. *Annu. Rev. Med.* **40**, 87–97 (1989).

Morhart, M., Rennie, R., Ziola, B., Bow, E. and Louie, T.J. Evaluation of enzyme immunoassay for *Candida* cytoplasmic antigens in neutropenic cancer patients. *J. Clin. Microbiol.* **32**, 766–776 (1994).

Ness, M., Vaughan, W. and Woods, G. *Candida* antigen latex test for detection of invasive candidiasis in immunocompromised patients. *J. Infect. Dis.* **159**, 495–502 (1989).

Phillips, P., Dowd, A., Jewesson, P., Radigan, G., Tweeddale, M., Clarke, A., Geere, I. and Kelly, M. Nonvalue of antigen detection of immunoassays for diagnosis of candidemia. *J. Clin. Microbiol.* **28**, 2320–2326 (1990).

BORRELIA BURGDORFERI

Etiologic Agent

Lyme disease is caused by *Borrelia burgdorferi*, a member of the Spirochaetaceae family (Burgdorfer *et al.*, 1982). It is transmitted from infected deer, mice, raccoons or people by the deer tick of the genus *Ixodes*. The ticks have a 2-year life cycle. The adults feed and mate on large animals in the fall through early spring, when the females lay their eggs in the ground. The eggs hatch by summer into larvae, which feed on mice and other small mammals. The larvae are inactive during winter, then molt into nymphs in spring. The nymphs, which are less than 2 mm in size, feed on small rodents and birds, then molt into adults in the fall, thus completing the cycle. The larvae and nymphs become infected by feeding on small mammals, particularly the white-footed mouse, and the bacteria remain with the ticks through adulthood.

Pathogenesis

B. burgdorferi is transmitted to mammals, including humans, in the course of normal feeding behavior of the nymph or adult tick. Most transmission is suspected to be by the nymph, which is most active during the summer months when people are most likely to be outdoors, and is smaller and more likely to feed undiscovered for the 2 or more days required for optimum inoculation of the *B. burgdorferi*.

Lyme disease is a chronic disease with acute exacerbations and is characterized in a general way by the presence of a 'Bull's-eye' rash or **erythema migrans**. Other signs of disease include a flu-like illness, with fever, headache, extreme fatigue, and stiff neck and joints. If untreated, peripheral nerve damage, numbness, facial paralysis, seizures, and cardiac damage can develop (Cooke and Dattwyler, 1992).

The incubation period is from a few days to a month after the tick bite. The period of communicability is for as long as the spirochete exists in the circulating blood in sufficient quantities. While transfusion transmission is theoretically possible, no documented transfusion-transmitted cases have occurred. Susceptibility appears to be universal, and exposure does not guarantee immunity.

Lyme disease is endemic in northern Europe, Russia, and regionally in the Untied States, where it is the most prevalent vector-borne disease (Centers for Disease Control and Prevention, 1995; Dekonenko *et al.*, 1988). Over 70,000 cases have been reported in the United States since national surveillance began in 1982. Regions of the United States with highest incidence are the Northeast, from Massachusetts to Maryland; North central, especially Wisconsin and Minnesota; and the West coast, particularly northern California.

Treatment and Prevention

Prevention strategies should be used in endemic areas, particularly during the late spring to early summer when the nymphal stage is prevalent. Recommendations include avoiding tick habitats, such as leaf litter, overgrown grassy areas, and low-lying vegetation, wearing protective clothing, using insect repellents containing DEET on clothes and skin or permethrine (solid aspermanone) on clothing, inspecting for and promptly removing ticks, and including the inspection of household cats and dogs for ticks.

In December 1998, the United States Food and Drug Administration licensed a Lyme disease vaccine. This vaccine contains a lipidated recombinant outer surface protein A (OspA) derived from the *B. burgdorferi* strain responsible for Lyme disease in the United States (Centers for Disease Control and Prevention, 1999). The vaccine, for use in persons 15–70 years of age, requires two doses 1 month apart followed by a third dose 12 months after the first. Vaccine efficacy at preventing Lyme disease is 50% after two doses and 78% after the third dose (Steere *et al.*, 1998). The duration of immunity and the need for booster doses are not known.

The principal candidates for vaccination are individuals whose activities place them in prolonged or frequent exposure to ticks in areas endemic for Lyme disease. The benefits of vaccinating people with limited exposure to ticks have not been determined.

Treatment of Lyme disease is usually successful following standard antibiotic regimens (Nadelman and Wormser, 1998). Regimens for early disease states such as erythema migrans include 14 days of doxycycline, amoxicillin, cefuroxime axetil, phenoxymethylpenicillin, tetracycline (except pregnant patients and children under 9 years), or azithromycin. There is no evidence to support the use of quinolones, sulfa drugs, rifampicin, or first generation cephalosporins; nor combination or pulse antibiotic therapies. Patients exhibiting later stages of infection including encephalomyelitis or chronic encephalopathy may need 14–28 days of ceftriaxone, cefotaxime, or penicillin G. Patients presenting with arthritis should receive 28 days of doxycycline or amoxicillin or 14 days of ceftriaxone.

Diagnosis and Assay Technology

The Centers for Disease Control and Prevention in the United States and the American College of Physicians (ACP) have developed guidelines for the diagnosis of Lyme disease (Centers for Disease Control and Prevention, 1997; Tugwell *et al.*, 1997a). The CDC guidelines, which were developed for surveillance case definition not clinical diagnosis, consider a case as confirmed if erythema migrans is present or at least one of the late manifestations occurs and it is laboratory confirmed. The late manifestations include the musculoskeletal system with recurrent, brief attacks of objective joint swelling sometimes followed by chronic arthritis in one or a few joints; the nervous system with lymphocytic meningitis, cranial neuritis, particularly facial palsy, radiculoneuropathy, or, rarely, encephalomyelitis; and the cardiovascular system with acute onset high-grade (2nd or 3rd degree) atrioventricular conduction defects that are sometimes associated with myocarditis.

Laboratory guidelines from the ACP state that isolation of the organism in culture is the best available evidence of causality. Although *B. burgdorferi* grows well in culture, collecting an adequate sample is difficult. The leading edge of the suspected erythema migrans lesion is the best source to obtain organisms, but only 60–80% of Lyme patients present with erythema migrans, and organism is only successfully recovered in 60–80% of them.

Serologic tests for *B. burgdorferi* have been developed with variable success (Karlsson and Granstrom, 1989; Karlsson *et al.*, 1989; Stiernstedt *et al.*, 1984). Both sensitivity and specificity issues exist for these assays. In addition, the nature of the antibody response makes using serology for acute infection diagnosis less accurate than most infections. Both IgG and IgM to *B. burgdorferi* can persist for a considerable time (years) even after treatment and may not be reliable for monitoring progression of disease or response to therapy (Hilton *et al.*, 1997).

The ACP guidelines recommend serologic testing only if the pretest probability of Lyme disease in the community is 0.20–0.80. With a probability less than 0.20, testing will result in more false-positive results than true positive results (Tugwell *et al.*, 1997b). A two test protocol is recommended for patients within the probability range; first an ELISA or immunofluorescence assay to detect IgG or IgM to *B. burgdorferi* antigens, followed by Western blot of all specimens found to be indeterminate. A positive result, combined with objective clinical signs of Lyme disease, is indicative of disease, and the patient should have antibiotic therapy. Negative results are useful for ruling out disease.

Although PCR is an extremely sensitive technique, the sample site and timing of sample collection are critical factors. Negative PCR results do not exclude the possibility of disease in that an inappropriate or inadequate sample may have been taken. A positive PCR result after treatment is not necessarily an indication of treatment failure due to persistence of PCR reactivity for weeks after treatment (Schmidt, 1997).

The complex nature of *B. burgdorferi* sampling, persistence, and treatment failures needs more investigation before testing strategies can give reliable results that allow accurate assessment of disease state and prognosis.

REFERENCES

Burgdorfer, W., Barbour, A.G., Hayes, S.F., Benach, J.L., Grunwaldt, E. and Davis, J.P. Lyme disease – a tick-borne spirochetosis? *Science* **216**, 1317–1319 (1982).

Centers for Disease Control and Prevention, Lyme disease – United States. *Morb. Mortal. Wkly Rep.* **44**, 590–591 (1995).

Centers for Disease Control and Prevention, Case definitions for infectious conditions under public health surveillance. *Morb. Mortal. Wkly Rep.* **46** (1997).

Centers for Disease Control and Prevention, Availability of Lyme disease vaccine. *Morb. Mortal. Wkly Rep.* **48**, 35–43 (1999).

Cooke, W.D. and Dattwyler, R.J. Complications of Lyme borreliosis. *Annu. Rev. Med.* **41**, 93–103 (1992).

Dekonenko, E.J., Steere, A.C., Berardi, B.P. and Kravchuk, L.N. Lyme borreliosis in the Soviet Union: a cooperative US–USSR report. *J. Infect. Dis.* **158**, 748–753 (1988).

Hilton, E., Tramontano, A., DeVoti, J. and Sood, S.K. Temporal study of immunoglobulin M seroreactivity to *Borrelia burgdorferi* in patients treated for lyme borreliosis. *J. Clin. Microbiol.* **35**, 774–776 (1997).

Karlsson, M. and Granstrom, M. An IgM-antibody capture enzyme immunoassay for serodiagnosis of Lyme borreliosis. *Serodiagn. Immunother. Infect. Dis.* **3**, 413–421 (1989).

Karlsson, M., Mollegard, I., Stiernstedt, G. and Wretlind, B. Comparison of western blot and enzyme-linked immunosorbent assay for diagnosis of Lyme borreliosis. *Eur. J. Clin. Microbiol. Infect. Dis.* **8**, 871–877 (1989).

Nadelman, R.B. and Wormser, G.P. Lyme borreliosis. *Lancet* **352**, 557–565 (1998).

Nielsen, S.L., Young, K.K.Y. and Barbour, A.G. Detection of *Borrelia burgdorferi* DNA by the polymerase chain reaction. *Mol. Cell. Probes* **4**, 73–79 (1990).

Schmidt, B.L. PCR in laboratory diagnosis of human *Borrelia burgdorferi* infections. *Clin. Microbiol. Rev.* **10**, 185–201 (1997).

Steere, A.C., Sikand, V.K., Meurice, F., Parenti, D.L., Fikrig, E., Schoen, R.T., Nowakowski, J., Schmid, C.H., Laukamp, S., Buscarino, C. and Krause, D.S. Vaccination against Lyme disease with recombinant *Borrelia burgdorferi* outer-surface lipoprotein A with adjuvant. Lyme disease vaccine study group. *N. Engl. J. Med.* **339**, 209–215 (1998).

Stiernstedt, G.T., Granstrom, M., Hedersledt, B. and Skoldenberg, B. Enzyme-linked immunosorbent assay and indirect immunofluorescence assay for Lyme disease. *J. Infect. Dis.* **149**, 465–470 (1984).

Tugwell, P., Dennis, D.T., Weinstein, A., Wells, G., Nichol, G., Shea, B., Hayward, R., Lightfoot, R., Baker, P. and Steere, A.C. Clinical guidelines, part 1: guidelines for laboratory evaluation in the diagnosis of Lyme disease. *Ann. Intern. Med.* **127**, 1106–1108 (1997a).

Tugwell, P., Dennis, D.T., Weinstein, A., Wells, G., Shea, B., Nichol, G., Hayward, R., Lightfoot, R., Baker, P. and Steere, A.C. Clinical guidelines, part 2: laboratory evaluation in the diagnosis of Lyme disease. *Ann. Intern. Med.* **127**, 1109–1123 (1997b).

HELICOBACTER PYLORI

Etiologic Agent

Helicobacter pylori (previously known as *Campylobacter pylori*) is a spiral shaped Gram-negative bacterium. It is microaerophilic, which means that although the species requires oxygen, it grows best at low levels. The bacterium moves by means of a single, thick flagellum. It is a parasite, the only organism to colonize the acid-secreting stomach of humans.

Pathogenesis

H. pylori is strongly associated with **gastritis** (inflammation of the stomach). The mode of transmission remains unclear, but once acquired, the organism may persist in the gastric mucosa for many years (Cave, 1997). The mode of action of the bacterium also is uncertain, one theory is that the organism secretes ammonia through urease activity, leading to overproduction of gastrin, which results in excess acid secretion by gastric mucosal cells. In addition, *H. pylori* protease and lipase degrade gastric mucus allowing for cell injury from back diffusion of gastric acid (Smoot, 1997). *H. pylori* infection may be the major cause of **peptic ulcers**, which affect 7–10% of the population at some time in their lives. Peptic ulcers can occur in the duodenum (**duodenal ulcer**) or stomach (**gastric ulcer**). *H. pylori* are present in virtually all duodenal ulcer cases, except those linked to non-steroidal anti-inflammatory drugs (NSAIDs) or aspirin. Depending on the location, between 30 and 70% of the adult population have antibodies to *H. pylori*.

There is also a strong association between chronic *H. pylori* infection and **gastric cancer**. Chronic inflammation and impaired host defense mechanisms are being investigated as factors contributing to the development of gastric cancer (Asaka *et al.*, 1997).

Anti-microbial therapy has recently been used to eradicate the organism in many patients, although not all respond to the treatment and recurrences are common. It is thought that effective treatment of *H. pylori* infections could much reduce the incidence of gastritis and peptic ulcer disease, and it may even prevent stomach cancer, which has a poor prognosis, once detected. Prophylactic eradication of *H. pylori* is not being recommended except for infected family members of patients with gastric cancer (Lee and O'Morain, 1997). Further research is needed to define the benefits of treatment in the variety of conditions currently associated with *H. pylori* infection.

Diagnosis and Assay Technology

Test methods for *H. pylori* can be divided into two broad areas: invasive sampling requiring endoscopy and non-invasive (including serology) (Dunn *et al.*, 1997). During endoscopic examination, biopsies are taken for culture, rapid urease testing or PCR. The distribution of *H. pylori* in the gastric mucosa can be patchy, leading to possible false negative biopsy results due to inadequate sample.

The immune response to *H. pylori* at the mucosal lining of the stomach is predominantly IgA. IgG is readily detectable in the serum of infected individuals. In contrast to the initial antibody response to many acute microbial infections, IgM antibodies against *H. pylori* are seldom detectable, so cannot be used to differentiate between infected individuals and those who have been successfully treated. A small percentage of individuals infected with *H. pylori* fail to produce IgG antibodies and in these patients only IgA may be detected.

Tests for antibodies to *H. pylori* are used to reduce the need for endoscopy in patients presenting with **dyspepsia** (indigestion) and/or abdominal pain. In one study only patients who were sero-positive for *H. pylori*, taking NSAIDs or over 45 years of age underwent endoscopy. This is particularly useful in avoiding unnecessary endoscopy in children. However, serological assays for *H. pylori* are not 100% reliable and sole use of these tests for diagnosis is risky and can lead to errors (Marcheldon *et al.*, 1996; von Wulffen, 1992).

Other non-invasive testing includes the urea breath test (UBT) which utilizes [^{13}C] or [^{14}C] urea and PCR of saliva. In the UBT the *H. pylori*, if present, produce urease which hydrolyzes the urea into bicarbonate, which in turn is exhaled as labeled CO_2 (Faigel *et al.*, 1996). Detection of released isotope in the collected CO_2 is indicative of the presence of *H. pylori*. Sampling of urine or blood for by-products of the *H. pylori* urease is also being investigated. When PCR was used to detect *H. pylori* DNA in saliva, 57 of 68 (84%) confirmed infections were detected (Li *et al.*, 1996).

REFERENCES

Asaka, M., Takeda, H., Sugiyama, T. and Kato, M. What role does *Helicobacter pylori* play in gastric cancer? *Gastroenterology* **113**, S56–S60 (1997).

Cave, D.R. Epidemiology and transmission of *Helicobacter pylori* infection. *Gastroenterology* **113**, S9–S14 (1997).

Dunn, B., Cohen, H. and Blaser, M. *Helicobacter pylori*. *Clin. Microbiol. Rev.* **10**, 720–741 (1997).

Faigel, D.O., Childs, M., Furth, E.E., Alavi, A. and Metz, D.C. New noninvasive tests for *Helicobacter pylori* gastritis. *Dig. Dis. Sci.* **41**, 740–748 (1996).

Lee, J. and O'Morain, C. Who should be treated for *Helicobacter pylori* infection? A review of consensus conferences and guidelines. *Gastroenterology* **113**, S99–S106 (1997).

Li, C., Ha, T., Ferguson, D.A., Chi, D.S., Zhao, R., Patel, N.R., Krishnaswamy, G. and Thomas, E. A newly developed PCR assay of *H. pylori* in gastric biopsy, saliva, and feces. *Dig. Dis. Sci.* **41**, 2142–2149 (1996).

Marcheldon, P., Ciota, L., Zamaniyan, F., Peacock, J. and Graham, D. Evaluation of three commercial enzyme immunoassays compared with the ^{13}C urea breath test for detection of *Helicobacter pylori* infection. *J. Clin. Microbiol.* **34**, 1147–1152 (1996).

Smoot, D.T. Pathogenesis of *Helicobacter pylori* infection. *Gastroenterology* **113**, S31–S34 (1997).

Von Wulffen, H. An assessment of serological tests for detection of *Helicobacter pylori*. *Eur. J. Clin. Microbiol. Infect. Dis.* **11**, 577–582 (1992).

GROUP A STREPTOCOCCUS

Etiologic Agent

Group A streptococcus (*Streptococcus pyogenes*) is the most common bacterial agent associated with upper respiratory tract infections. These pathogenic bacteria produce an extracellular enzyme that lyses red blood cells (hemolysin). Colonies on agar medium containing whole blood are surrounded by clear zones in which the erythrocytes have been destroyed. This mode of action is known as β-hemolysis. The pyogenic streptococci are sensitive to extremes of temperature, being unable to grow at 10 or 40 °C.

Streptococci from group A are characterized by the presence of group A carbohydrate. They can be differentiated from other β-hemolytic streptococci by use of Bacitracin disks. The sensitivity of a β-hemolytic, Gram-positive *Streptococcus* to Bacitracin is diagnostic.

Pathogenesis

The most common type of streptococcal infection is primary pharyngeal infection with Group A β-hemolytic streptococci, normally resulting in sore throat, tonsillitis, etc. Streptococcal pharyngitis is more prevalent in schoolchildren, particularly in winter and spring months. If left untreated, the infection may lead to more serious complications, such as rheumatic fever and acute glomerulonephritis. *S. pyogenes* produces several toxic substances, in addition to hemolysin. These include fibrinolysin, an enzyme that dissolves clots of fibrin; hyaluronidase, which attacks hyaluronic acid in connective tissue; an erythrogenic toxin, responsible for the rash in scarlet fever; and leucocidin, a substance that destroys white blood cells.

Strep A infections respond rapidly to treatment with antibiotics. Correct differentiation of Strep A from viral infections at an early stage allows antibiotics to be administered only to those who will benefit from them.

In hospitalized patients, the diagnosis of Strep A infection is particularly important, especially in the case of wound infections. Outbreaks in hospitals need prompt attention to prevent the disease spreading.

Diagnosis and Assay Technology

Bacteriological testing involves an overnight incubation of swabs for the primary isolation. Full identification requires a second isolation of β-hemolytic colonies, which are spread onto a blood-agar plate. The characteristic clear zones around the colonies appear colorless or pale pink. Group A streptococci are differentiated by use of Bacitracin disks (Facklam and Carey, 1985).

Non-culture testing is available for group A strep. Test formats include latex agglutination and instrument-independent enzyme immunoassays. Both test formats are very rapid and provide visual results. Research results reporting the performance of these assays give quite variable performance numbers (Baker *et al.*, 1995; Heiter and Bourbeau, 1995; Laubscher *et al.*, 1995; Roe *et al.*, 1995). In general, the latex tests are less sensitive and more difficult to read than the ELISA tests, which are less sensitive than culture but have good specificity. Therefore, positive results are quite reliable, but negative results should be backed up by culture (Roe *et al.*, 1995).

REFERENCES

Baker, D., Cooper, R., Rhodes, C., Weymouth, L. and Dalton, H. Superiority of conventional culture technique over rapid detection of group A *Streptococcus* by optical immunoassay. *Diagn. Microbiol. Infect. Dis.* **21**, 61–64 (1995).

Facklam, R.R. and Carey, R.B. Streptococci and Aerococci. In: *Manual of Clinical Microbiology*, 4th edn (ed Lennette, E.H.), 154–175 (American Society for Microbiology, Washington, DC, 1985).

Heiter, B. and Bourbeau, P. Comparison of two rapid streptococcal antigen detection assays with culture for diagnosis of streptococcal pharyngitis. *J. Clin. Microbiol.* **33**, 1408–1410 (1995).

Laubscher, B., van Melle, G., Dreyfuss, N. and de Crousaz, H. Evaluation of a new immunologic test kit for rapid detection of group A streptococci, the Abbott TestPack Strep A plus. *J. Clin. Microbiol.* **33**, 260–261 (1995).

Roe, M., Kishiyama, C., Davidson, K., Schaefer, L. and Todd, J. Comparison of BioStar Strep A OIA optical immunoassay, Abbott TestPack Plus Strep A, and culture with selective media for diagnosis of group A streptococcal pharyngitis. *J. Clin. Microbiol.* **33**, 1551–1553 (1995).

HUMAN HERPESVIRUS 6

Etiologic Agent

Human herpesvirus 6 (HHV-6) was first isolated from patients with lymphoproliferative diseases in 1986 and was provisionally named human B-lymphotropic herpesvirus due to an observed tropism for B cells. The virus was subsequently found to predominate in $CD4^+$ cells and renamed HHV-6. Studies of numerous isolates by restriction endonuclease analysis, cell tropism, and monoclonal antibody reactivity have lead to the identification of two groups, designated group A and group B (Braun *et al.*, 1997).

HHV-6A and HHV-6B, as with all herpesviruses, contain a large double-stranded DNA core surrounded by an icosahedral capsid. The outer lipid envelope contains virus coded glycoproteins and appears to be derived by budding into cytoplasmic vesicles after the capsid-containing nuclear tegusome fuses with the nuclear membrane releasing tegumented capsids into the cytoplasm (Roffman *et al.*, 1990). Mature virions, which are approximately 200 nm in diameter, are released by exocytosis.

The structure of the viral DNA, which has been completely sequenced, is unusual compared to other members of the human herpesviruses. The full-length DNA is approximately 162–170 kb composed of a 143 kb unique long segment bracketed by 8–13 kb direct repeat regions. There is no unique short region as is found in the previous five human herpesviruses (Gompels *et al.*, 1995).

Pathogenesis

HHV-6 is acquired very early in life. Seroprevalence by age 13 months ranges from 64 to 83% in locations throughout the world; 10% of infants less than 1 month old were found to be PCR positive (Hall *et al.*, 1994). Between 85 and 95% of the adult population show evidence of exposure to HHV-6, making it one of the most ubiquitous human viruses known.

HHV-6A and HHV-6B have very distinct pathogenic profiles. HHV-6A has primarily been found in adults, but has not been definitively linked to any disease. HHV-6B causes a majority of the cases of the childhood disease **exanthem subitum** (also known as **roseola infantum**) (Yamanishi *et al.*, 1988). The disease generally starts with a fever that may exceed 40 °C and lasts from 3 to 5 days. As the temperature returns to normal, a maculopapular rash develops first on the trunk then spreading to the extremities, including the neck and face. HHV-6B DNA has been found in saliva, and the salivary gland is a possible reservoir of persistent or latent viral infection. Saliva may, therefore, act as a vehicle for transmission, particularly from mother to infant (Mukai *et al.*, 1994).

Primary infections with HHV-6B in adults are rare, but usually have more severe symptoms including a mononucleosis-like syndrome, prolonged lymphadenopathy, and even fulminant hepatitis (Sobue *et al.*, 1991). HHV-6A has been primarily found in immunocompromised or chronically ill adults; however, no widespread assessment of the virus incidence has been conducted in the general population.

The potential role of HHV-6 in the pathogenesis of multiple sclerosis (MS) is being investigated (Wilborn *et al.*, 1994). Reports have stated that higher HHV-6 antibody titers are found in MS patients than controls; HHV-6 DNA has been found in the central nervous system of MS patients but not controls; and HHV-6 has been found in oligodendrocytes near plaques in brain tissue of MS patients (Challoner *et al.*, 1995). Conversely, some groups have found no associations between HHV-6 and MS (Merelli *et al.*, 1997). Proving a cause-and-effect relationship between HHV-6 and MS given the very high prevalence of HHV-6 in the general population could be a lively research topic for years to come.

The immunosuppression induced to prevent graft rejection after transplantation can lead to serious complications due to reactivation of latent HHV-6 infection. Bone marrow transplant patients have been the most extensively studied group for HHV-6 involvement in post-transplant diseases. While proof of direct cause-and-effect relationship between HHV-6 and post-transplantation diseases has been elusive, substantial circumstantial evidence exists, implicating the virus in a variety of conditions including graft versus host disease, pneumonitis, sinusitis, febrile episodes, rash, and suppression of graft outgrowth. Additional research has focused on renal, liver, and cardiac transplants. No clear link between HHV-6 replication and organ rejection has yet been found; however, episodes of fever and rash may result from HHV-6 infection.

The acquired immunodeficiency that results from an HIV infection leads to significant disease complications due to reactivation of latent herpesvirus infections. HHV-6 has been of particular interest because it infects and replicates in $CD4^+$ cells, the same target as HIV-1. No clear role of HHV-6 as a cofactor in the progression of HIV infection to AIDS has been identified. HHV-6 has also been implicated in specific conditions such as retinitis and pneumonitis.

Diagnosis and Assay Technology

Diagnostic procedures for acute HHV-6 infection include cell culture, antigen capture ELISA, PCR, and IFA. Serodiagnosis is accomplished by standard techniques involving IFA and ELISA (Sloots *et al.*, 1996). IgM response can be measured by ELISA, but interpretation of the results is complicated by unusual immune response characteristics. While IgM to HHV-6 can usually be found during primary infection, it can also be present during reactivation. Studies have found 5% of infected adults are IgM positive (Suga *et al.*, 1992). The same study found culture-positive children often do not have a detectable IgM response.

Cell culture involves culturing the patient's lymphocytes alone or with fresh lymphocytes and activating with phytohemagglutinin or antibody to CD3 in the presence of IL-2. Culture detection is most successful during the febrile, prerash phase of exanthem subitum in children.

REFERENCES

Braun, D., Donimguez, G. and Pellett, P. Human herpesvirus 6. *Clin. Microbiol. Rev.* **10**, 521–567 (1997).

Challoner, P., Smith, K., Parker, J., MacLeod, D., Coulter, S., Rose, T., Schultz, E., Bennett, R., Garber, R., Chang, M., Schad, P., Stewart, P., Nowinski, R., Brown, J. and Burmer, G. Plaque-associated expression of human herpesvirus 6 in multiple sclerosis. *Proc. Natl Acad. Sci. USA* **92**, 7440–7444 (1995).

Gompels, U., Nicholas, J., Lawrence, G., Jones, M., Thomson, B., Martin, M., Efstanthiou, S., Craxton, M. and Macaulay, H. The DNA sequence of human herpesvirus-6: structure, coding content, and genome evolution. *Virology* **209**, 29–51 (1995).

Hall, C., Long, C., Schabel, K., Caserta, M., McIntyre, K., Costanzo, M., Knott, A., Dewhurst, S., Insel, R. and Epstein, L. Human herpesvirus-6 infection in children. A prospective study of complications and reactivation. *N. Engl. J. Med.* **331**, 432–438 (1994).

Merelli, E., Bedin, R., Sola, P., Barozzi, P., Mancardi, G., Ficarra, G. and Franchini, G. Human herpesvirus 6 and human herpesvirus 8 DNA sequences in brains of multiple sclerosis patients, normal adults and children. *J. Neurol.* **244**, 450–454 (1997).

Mukai, T., Yamamoto, T., Kondo, T., Kondo, K., Okuno, T., Kosuge, H. and Yamanishi, K. Molecular epidemiological studies of human herpesvirus 6 in families. *J. Med. Virol.* **42**, 224–227 (1994).

Roffman, E., Albert, J., Goff, J. and Frenkel, N. Putative site for the acquisition of human herpesvirus 6 virion tegument. *J. Virol.* **64**, 6308–6313 (1990).

Sloots, T., Kapeleris, J., MacKay, I., Batham, M. and Devine, P. Evaluation of a commercial enzyme-linked immunosorbent assay for detection of serum immunoglobulin G response to human herpesvirus 6. *J. Clin. Microbiol.* **34**, 675–679 (1996).

Sobue, R., Miyazaki, H., Okamoto, M., Hirano, M., Yoshikawa, T., Suga, S. and Asano, Y. Fulminant hepatitis in primary human herpesvirus-6 infection. *N. Engl. J. Med.* **324**, 1290 (1991).

Suga, S., Yoshikawa, T., Asano, Y., Nakashima, T., Yazaki, T., Fukuda, M., Kojima, S., Matsuyama, T., Ono, Y. and Oshima, S. IgM neutralizing antibody responses to human herpesvirus-6 in patients with exanthem subitum or organ transplantation. *Microbiol. Immunol.* **36**, 495–506 (1992).

Wilborn, F., Schmidt, C., Brinkmann, V., Jenroska, K., Oettle, H. and Siegert, W. A potential role for human herpesvirus type 6 in nervous system disease. *J. Neuroimmunol.* **49**, 213–214 (1994).

Yamanishi, K., Okuno, T., Shiraki, K., Takahashi, M., Kondo, T., Asano, Y. and Kurata, T. Identification of human herpesvirus-6 as a causal agent for exanthem subitum. *Lancet* **i**, 1065–1067 (1988).

HUMAN HERPESVIRUS 8 (KAPOSI'S SARCOMA HERPESVIRUS)

Etiologic Agent

In December 1994, a new virus was identified from **Kaposi's sarcoma** (KS) tumors using the state-of-the-art molecular biology technique, **representational difference analysis** (Chang *et al.*, 1994). This virus was identified as a member of the Herpesviridae family and named Kaposi's sarcoma herpesvirus (KSHV) or human herpesvirus 8 (HHV-8). HHV-8 is closely related to the gammaherpesviruses including EBV.

The double-stranded DNA viral genome has been cloned and sequenced and found to be 165 kb with 81 open reading frames. A long unique DNA region of 140.5 kb is flanked by high G + C terminal repeats (Sarid *et al.*, 1998).

Virus has been recovered from KS tissue, peripheral blood B lymphocytes, and primary effusion lymphomas, also called body cavity-based lymphomas (Boshoff and Weiss, 1997). Cell lines derived from virus-infected tissue can be induced to produce infectious virus particles using phorbol esters. Raji cells and fetal cord blood lymphocytes have been infected in culture with HHV-8, but transformation similar to that seen for EBV has not been observed.

Pathogenesis

Evidence of HHV-8 infection has been found in over 95% of KS patients (Lennette *et al.*, 1996). The same study found less than 25% of the general adult population seropositive. Other studies have put the incidence in the general population at less than 3% (Boshoff and Weiss, 1997; Simpson *et al.*, 1996). Direct examination of KS lesions has found HHV-8 DNA in greater than 95%, with less than 2% of control tissues having amplifiable HHV-8 DNA (Chang and Moore, 1996). Further epidemiology studies have found that 100% of patients with African endemic KS are seropositive for HHV-8.

Epidemiologic evidence developed to date is consistent with HHV-8 being sexually transmitted (Boshoff and Weiss, 1997). The detection of HHV-8 DNA in semen, and the very low incidence of HHV-8 and KS in hemophiliac patients and injection drug users, further support the sexual transmission theory.

Diagnosis and Assay Technology

Diagnostic methods for HHV-8 are still in the early stages of development as evidenced by the wide disparity in incidence rates reported in the literature. PCR is the only definitive test for acute diagnosis and has been instrumental in associating KS lesions with HHV-8.

Serologic studies using immunoblotting or immunofluorescence have been based on lytic antigens derived from cells induced with phorbol esters or latent antigens isolated from the nuclei of latently infected B cells. These antigen preparations have tended to be crude making comparisons between studies difficult. Recombinant antigens are becoming available and should improve reproducibility and provide a better source of antigen for ELISA assays (Simpson *et al.*, 1996).

REFERENCES

Boshoff, C. and Weiss, R. Kaposi's sarcoma-associated herpesvirus. *Curr. Opin. Infect. Dis.* **10**, 26–31 (1997).

Chang, Y. and Moore, P. Kaposi's sarcoma (KS)-associated herpesvirus and its role in KS. *Infect. Agents Dis.* **5**, 215–222 (1996).

Chang, Y., Cesarman, E., Pessin, S., Lee, F., Culpepper, J., Knowles, D. and Moore, P. Identification of herpesvirus-like DNA sequences in AIDS-associated Kaposi's sarcoma. *Science* **266**, 1865–1869 (1994).

Lennette, E., Blackbourn, D. and Levy, J. Antibodies to human herpesvirus type 8 in the general population and in Kaposi's sarcoma patients. *Lancet* **348**, 858–861 (1996).

Sarid, R., Ornella, F., Bohenzky, R., Chang, Y. and Moore, P. Transcription mapping of the Kaposi's sarcoma-associated herpesvirus (human herpesvirus 8) genome in a body cavity-based lymphoma cell line (BC-1). *J. Virol.* **72**, 1005–1012 (1998).

Simpson, G., Schulz, T., Whitby, D., Cook, P., Boshoff, C., Rainbow, L., Howard, M., Gao, S., Bohenzky, R., Simmonds, P., Lee, C., de Ruiter, A., Hatzakis, A., Tedder, R., Weller, I., Weiss, R. and Moore, P. Prevalence of Kaposi's sarcoma associated herpesvirus infection measured by antibodies to recombinant capsid protein and latent immunofluorescence antigen. *Lancet* **349**, 1133–1138 (1996).

65 Therapeutic Drug Monitoring (TDM)

Philip A. Routledge and Alun D. Hutchings

INTRODUCTION

Although drug treatments are prescribed in terms of the dose to be administered, it is the subsequent concentration of active drug in the patient's blood plasma that governs its effectiveness. The relationship between dose and plasma concentration can be highly variable between patients. Measurement of the concentration of the drug in serum or plasma enables the dose administered to be adjusted to the optimum level.

Plasma drug concentrations are monitored for three main reasons:

- to ensure that the drug concentration is high enough to be therapeutically effective;
- to minimize dose-related (type A or toxic) side-effects of the drug;
- to check for patient noncompliance with the prescribed therapy.

Monitoring is only of value when the wanted or unwanted effects of the drug are related to the concentration in plasma. It is particularly useful when there is a wide variation in the rate of metabolism or excretion of the drug between individuals, resulting in marked differences in the plasma concentration at any given dose. Monitoring of serum or plasma concentrations is also particularly useful for those drugs in which the therapeutic concentration is close to the toxic concentration (low therapeutic ratio) such as digoxin and theophylline. Similarly, it is useful in monitoring those drugs (e.g. phenytoin) where the rate of metabolism can fall with even small increases of dose over the therapeutic range of drug concentration so that their concentrations are not linearly related to dose.

ASSAY TECHNOLOGY

There are a number of methods for determining drug concentrations in serum or plasma. Excellent analytical methods are available that rely on extraction and chromatographic separations, e.g. high-performance liquid chromatography (HPLC). However, immunoassay is the commonest method for routine therapeutic drug monitoring because of its simplicity and reliability.

Radioimmunoassay

In radioimmunoassay (RIA), a biological sample is mixed with a radiolabeled compound of interest and an immobilized antibody specific to that agent. Measured radioactivity bound to the immobilized antibody is inversely proportional to the agent of interest. The technique can be extremely sensitive and has been used widely for a number of agents found in very low concentrations (cannabis, lysergic acid diethylamide, digoxin, and paraquat). The disadvantages of this assay include the problems arising from the use of radiolabeled compounds (e.g. the need for expensive reagents and equipment, and the inconveniences associated with the safe disposal of radioactive material), the availability of those radiolabels and the specificity of the antibody (compared to methods that involve separation such as HPLC).

Nonisotopic Immunoassay

Nonisotopic immunoassay methods use inexpensive, simple and rapid measurements with similar specificity and sensitivity to RIA. Immunoassay methodology may be broadly separated into homogeneous and heterogeneous assays. Heterogeneous assays include the classical RIA technique in that labeled antigen competes with unlabeled antigen for limited antibody binding sites. Separation of the free antigen from the bound variety and the measurement of enzyme activity in either fraction correlates with the concentration of free antigen. Homogeneous immunoassay, on the other hand, does not rely on this separation step, enabling simpler, faster protocols, so it is particularly suited to random-access clinical

The Immunoassay Handbook. *Third Edition*
D. Wild (Ed.)

chemistry analyzers. Although most homogeneous assays are less sensitive than heterogeneous assays, the majority of therapeutic drugs tend to circulate at concentrations well within the working ranges of these assays, making them ideally suited for therapeutic drug monitoring.

Enzyme-Multiplied Immunoassay Technique (Emit®)

The enzyme-multiplied immunoassay technique (Emit) is a simple, rapid homogeneous method now commonly employed to measure a wide range of substances (particularly drugs). The technique works on the basis that the drug present is proportional to the inhibition of an enzyme substrate reaction. A known quantity of drug is chemically labeled with an enzyme (e.g. glucose-6-phosphate dehydrogenase) and antibodies specific to that drug bind that drug–enzyme complex, so reducing enzyme activity. The introduction of a biological sample containing the same drug releases the enzyme-labeled drug from the antibody complex, thereby increasing enzymatic activity. Enzyme activity correlates with drug concentration in the specimen as measured by absorbance changes resulting from the activity of the enzyme on a particular substrate.

Fluoroimmunoassay

Fluoroimmunoassay offers an alternative signal generation and detection system to radioimmunoassay with potentially greater sensitivity. Fluoroimmunoassays may also be categorized as homogeneous and heterogeneous. Heterogeneous methods require a separation step since the activity of the fluorophor is unaffected by its antibody binding. In contrast, the fluorophor in the homogenous assay is affected by its attachment to the antibody and therefore does not require a separation step, making it particularly useful for clinical chemistry systems and enabling faster assay times.

Fluorescence enhancement and **quenching assays** are based on the principle that the fluorescence properties of a fluorescent-labeled antigen are altered once bound to an antibody. These assays do not require a separation step, and unlike many immunoassay techniques do not rely on specialist equipment, using conventional fluorimeters for their measurements.

Polarization fluoroimmunoassay is based on the behavior of a fluorophor-labeled antigen when subjected to plane-polarized light. Preferential excitement occurs when the axes of the molecules lie parallel to the light plane, and rotation of the molecule results in depolarization of that light. Since the degree of depolarization is dependent on the size of the molecule (large molecules rotate more slowly), the binding of the complex to a large antibody significantly reduces this rotation, resulting in a reduction of depolarization. In an immunoassay of this type, unlabeled antigen competes with the fluorescent-labeled antigen for binding sites on the antibody, and the depolarization of the emitted light is proportional to the concentration of unlabeled antigen.

Nonisotopic immunoassays are largely replacing radioimmunoassay (RIA) approaches due to their simplicity. As with RIA no sample pretreatment (e.g. solvent extraction) is required so the technique is ideally suited to the emergency situation. Like RIA, however, the technique often suffers from a lack of specificity, and cross-reactivity with other substances, such as metabolites, is common. All immunoassays depend on the availability of suitable antibodies, which may not always be available. For example, it has not been possible to produce an antibody for amiodarone that does not cross-react with triiodothyronine and thyroxine. Nevertheless, the ease with which such assays may be performed have in many cases reduced the need for specialist laboratory services and placed such assays within the remit of the general hospital biochemistry laboratory, thereby making such measurements more widely available.

MEASUREMENT OF FREE DRUG CONCENTRATION

A basic principle of therapeutic drug monitoring is that the circulating drug concentration in blood or plasma is related to the concentration of the drug at its site of action and therefore correlates with the magnitude of drug effect. Most drug assays measure the total concentration (bound plus free) of the substance in blood, and in most cases this is an adequate reflection of drug concentration at the site of action. Certain drugs are highly bound in plasma to albumin, alpha-1-acid glycoprotein, lipoproteins and other proteins and it is the free (unbound) drug concentration that reflects the concentration at the site of action (Routledge, 1986). Provided there is a reasonable correlation between the free and total concentration, measurement of the latter is a useful surrogate measure for the former. Any marked variability in the plasma protein binding in a population reduces the validity of total concentration monitoring. Such variability is relatively low if the drug is predominantly bound to albumin and the patient group is healthy (Ebden *et al.*, 1984; Rimmer *et al.*, 1984). Conditions causing low albumin concentrations (e.g. renal or liver disease or malnutrition), or displacement of the drug from binding sites (e.g. by other drugs, nonesterified fatty acids or in uremia) mean that the total drug concentration may underestimate the free (active) drug concentration in plasma. Measurement of free drug concentration may therefore be useful in such circumstances.

Alpha-1-acid glycoprotein (AAG) is an acute-phase protein that may vary markedly in its plasma concentration, both within and between individuals. Concentrations rise after burns, trauma, inflammation, and acute myocardial infarction. Lower than normal concentrations may occur in neonates, chronic liver disease, nephrotic syndrome, and during pregnancy. A number of basic drugs (e.g. lidocaine and quinidine) are bound to a significant degree to this protein and the total drug concentration may not adequately reflect the free drug concentration in these circumstances (Routledge *et al.*, 1985). Even in healthy individuals, the plasma AAG concentration may vary with time, possibly because of intercurrent viral or other infections, resulting in changes in the plasma protein binding of some (often basic) drugs.

Despite these considerations, total drug concentrations correlate well with free drug concentrations of commonly measured drugs (e.g. phenytoin or theophylline) in most general patient populations, so that routine measurement of free drug concentrations has not become common. Nevertheless, the availability of rapid, reliable and inexpensive techniques to measure free drug concentration would be useful in circumstances where variability in protein binding of certain drugs might be greater than normal.

PRACTICAL ASPECTS OF TDM

Since drugs are normally given at fixed intervals, the plasma concentration varies between doses during the processes of absorption, distribution, metabolism, and excretion. The greatest variability is in rate of absorption so samples taken when absorption is relatively complete are therefore more reflective of the average (steady-state) concentration between doses. Therefore, samples should be taken at least 8 h after digoxin administration and around 12 h after lithium. In certain cases, peak levels are more important in assessing efficacy (e.g. with aminoglycoside antibiotics) and since they occur around 30–60 min after an intramuscular injection or immediately at the end of an intravenous infusion, samples should ideally be taken at those times. Peak levels of orally administered drugs are achieved at 30–180 min after conventional formulations and later after modified-release preparations. For many drugs, a sample taken just before the next dose is due (predose concentration) will correlate best with the average (steady-state) concentration, although it will of course be lower.

It takes around five half-lives of a drug before the plasma concentrations reach their maximum, steady-state level. Sampling before this time has elapsed is therefore most usually only considered if toxicity of the drug due to excessive accumulation is suspected or anticipated. In other circumstances, five half-lives should be allowed to elapse before steady-state plasma concentrations are measured. This may be a considerable time in certain cases (e.g. 9 months in the case of amiodarone, which has a half-life of 45 days). Details of sampling time relative to dose, time of last dosage change and present daily dose schedule should always be stated on the assay request form to aid in the clinical interpretation of the plasma concentration.

Finally, the concentration response relationship is a continuous one and adequate efficacy may be seen in some patients when the plasma drug concentration is below the accepted therapeutic range. The ranges should therefore be considered as guidelines based on the average patient group, and the best dose of drug for an individual patient is the lowest one consistent with adequate efficacy and minimum toxicity.

IMPORTANT NOTE: All doses shown refer to adults and are for general guidance; they are not intended for clinical use.

ANTIARRHYTHMIC DRUGS

These agents are generally basic compounds, some of which (e.g. lidocaine) are used predominantly for the control of ventricular arrhythmias, whereas others (e.g. disopyramide) are used to treat arrhythmias arising either from ventricular or supraventricular sites.

ACECAINIDE (*N*-ACETYLPROCAINAMIDE)

Clinical Applications

Acecainide (*N*-acetylprocainamide) is used in some countries in the treatment of ventricular arrhythmias, in particular life-threatening arrhythmias in patients with procainamide-induced lupus erythematosus. It is the major metabolite of procainamide.

Dose and Mode of Administration

Orally, 1.5–10 g daily in divided doses.

Pharmacological Effects

The drug increases the effective refractory period with a selective lengthening of the cardiac potential by lengthening repolarization. Since this is accomplished without an effect on depolarization, the drug is classified as a class III antiarrhythmic agent (procainamide is a class 1A agent).

Therapeutic Range

6–20 mg/L
6–20 μg/mL

Potentially Toxic Concentration

> 20 mg/L
> 20 μg/mL

Toxic Side-Effects

Common adverse effects include light-headedness, insomnia, nausea, stomach upset, and blurred vision.

It is thought to be unlikely to be associated with the production of drug-induced lupus.

Type of Sample

Serum or plasma.

DISOPYRAMIDE

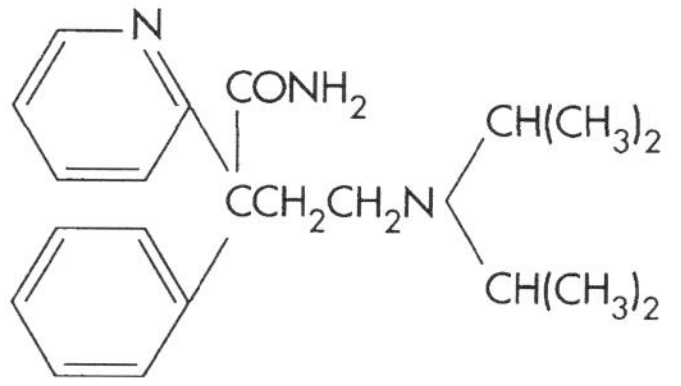

Fig. 65.1 Disopyramide.

Clinical Applications

Disopyramide is used for the treatment of ventricular and supraventricular arrhythmias, the former especially after acute myocardial infarction.

Dose and Mode of Administration

Orally, 300–800 mg daily in divided doses. The drug may also be given intravenously at a different dose.

Pharmacological Effects

Disopyramide has a group 1A antiarrhythmic activity with a pharmacological action similar to quinidine and procainamide. It prolongs the effective refractory period in the atria, atrioventricular node, and ventricles by decreasing the rate of diastolic depolarization.

Disopyramide also has anticholinergic (muscarinic) effects. A major metabolite, *N*-monodesisopropyldisopyramide is produced, which has around half the antiarrhythmic activity of disopyramide.

Therapeutic Range

2–5 mg/L
2–5 µg/mL

Potentially Toxic Concentration

$>$ 5 mg/L
$>$ 5 µg/mL

Toxic Side-Effects

Anticholinergic effects such as dry mouth, retention of urine, blurred vision, retention of urine, and reduction in rate and force of contraction of the heart leading to hypotension. Ventricular arrhythmias, including ventricular tachycardia, ventricular fibrillation or torsades de pointes and atrioventricular block have also been reported.

Type of Sample

Serum or plasma.

LIDOCAINE

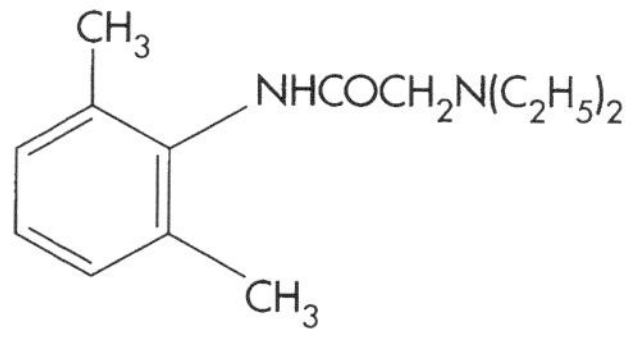

Fig. 65.2 Lidocaine.

Clinical Applications

The treatment and prevention of ventricular arrhythmias, particularly those occurring after acute myocardial infarction.

Dose and Mode of Administration

Normally, 1.5–3 mg of lidocaine ('lignocaine' in Europe) hydrochloride per minute by intravenous infusion. A loading dose is recommended to achieve early therapeutic concentrations. Reduced dose is required in cardiac failure or liver cirrhosis.

Pharmacological Effects

Lidocaine reduces the rate of phase 4 depolarization and depresses spontaneous automaticity of heart muscle. It is a class 1B agent and although it decreases action potential duration, it increases the effective refractory period relative to the action potential duration. It has less effect on the force of contraction of the heart than some class 1A agents.

Therapeutic Range

1.5–5 mg/L
1.5–5 µg/mL

Potentially Toxic Concentration

$>$ 5 mg/L
$>$ 5 µg/mL

Toxic Side-Effects

Predominantly on the central nervous system including confusion, coma, and convulsions. Dizziness, paresthesiae, hypotension, and bradycardia (sometimes severe) may also occur.

Type of Sample

Serum or plasma.

PROCAINAMIDE

$CONHCH_2CH_2N(CH_2CH_3)_2 \cdot HCl$

NH_2

Fig. 65.3 Procainamide hydrochloride.

Clinical Applications

Procainamide is used in the treatment of ventricular arrhythmias (especially after myocardial infarction) and atrial tachycardia.

Dose and Modes of Administration

For ventricular arrhythmias, procainamide can be given orally at doses up to 50 mg/kg/day in divided doses, preferably controlled by monitoring serum-procainamide concentration (dosage intervals can range from 3 to 6 h). For atrial arrhythmias, higher doses may be required.

Procainamide can be administered by slow intravenous injection, at a rate not exceeding 50 mg/min, 100 mg with ECG monitoring, repeated at 5-min intervals until arrhythmia is controlled: maximum 1 g.

It can also be administered by intravenous infusion: 500–600 mg over 25–30 min with ECG monitoring, followed by maintenance at a rate of 2–6 mg/min; then if necessary oral treatment as above, starting 3–4 h after infusion.

Note: serum-procainamide concentration for optimum response 3–10 mg/L.

Pharmacological Effects

Procainamide is a class 1A antiarrhythmic that prolongs the action potential duration and the effective refractory period. It can also reduce the force of contraction of the heart.

Therapeutic Range

4–10 mg/L
4–10 μg/mL

Potentially Toxic Concentration

$>$ 8 mg/L
$>$ 8 μg/mL

Toxic Side-Effects

Procainamide can reduce blood pressure by causing myocardial depression as well as nausea, vomiting, diarrhea and on occasion, psychosis. It may cause drug-induced lupus erythematosus in as many as 30% of those patients taking procainamide for 6 months or longer. This condition does not appear to be caused by *N*-acetylprocainamide, the major metabolite of procainamide, which has pharmacological activity in its own right (*see* ACECAINIDE).

Type of Sample

Serum or plasma.

PROPRANOLOL

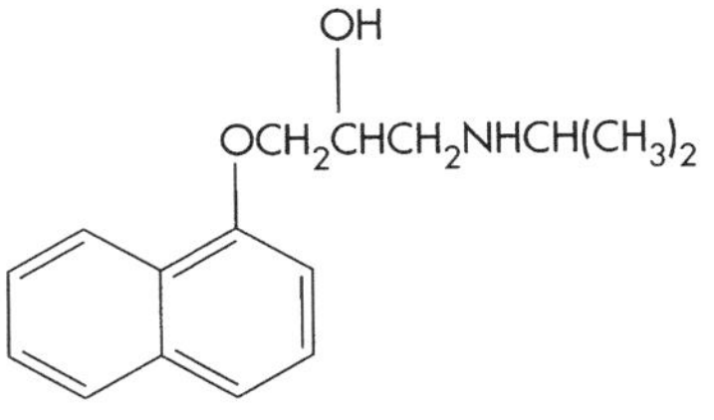

OH

$OCH_2CHCH_2NHCH(CH_3)_2$

Fig. 65.4 Propranolol.

Clinical Applications

Propranolol is a beta adrenoceptor-blocking drug used in the treatment of hypertension, angina, supraventricular and ventricular arrhythmias and in the secondary prevention of mortality after myocardial infarction. It is also used in the treatment of portal hypertension, in anxiety with symptoms such as palpitations, sweating, tremor, and in the treatment of essential tremor and thyrotoxic crisis, as well as migraine prophylaxis.

Dose and Modes of Administration

Oral 80–160 mg b.d., depending on indication. The drug undergoes variable and extensive presystemic (first-pass) metabolism in the liver. The intravenous dose is 1 mg/min up to a total dose of 10 mg.

Pharmacological Effects

Propranolol blocks both β_1 and β_2 receptors and therefore slows the rate and force of conduction of the heart (β_1 receptor blockade) and reduces tremor (β_2 blockade).

Therapeutic Range

Effective concentration: >50 μg/L (>50 ng/mL). Measurement of concentration is only of value in assessing compliance.

Potentially Toxic Concentration

Highly variable.

Toxic Side-Effects

By blocking β_2 receptors in the bronchi, propranolol may worsen asthma or bronchitis and should never be used in these conditions. It may also cause peripheral vasoconstriction. By blockade of β_1 receptors, propranolol may induce or worsen heart failure in patients with poor cardiac function and cause hypotension or heart block. Gastrointestinal disturbances, fatigue, sleep disturbances, skin rash and dry eyes (reversible on stopping the drug), and worsening of already existing psoriasis have also been reported.

Type of Sample

Serum or plasma.

QUINIDINE

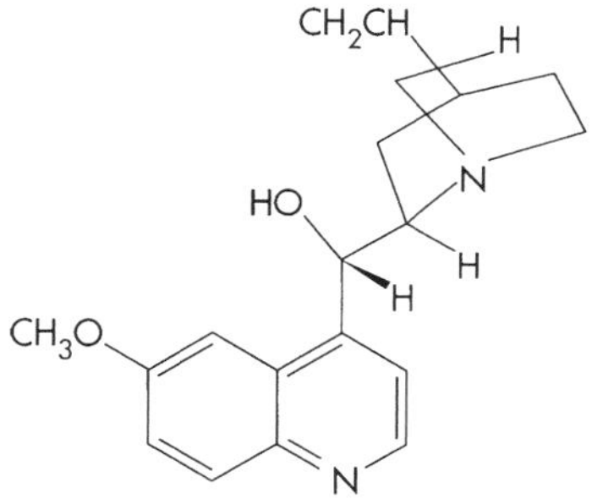

CH_2CH H HO N H H CH_3O N

Fig. 65.5 Quinidine.

Clinical Applications

Quinidine is used for the treatment and prevention of both supraventricular and ventricular arrhythmias.

Dose and Mode of Administration

Orally, 600–1600 mg of quinidine sulfate daily in 3–4 divided doses.

Pharmacological Effects

Quinidine is a class 1A agent that increases the action potential duration and the effective refractory period. It also reduces the force of contraction of the heart and possesses anticholinergic activity.

Therapeutic Range

2–5 mg/L
2–5 μg/mL

Potentially Toxic Concentration

> 5 mg/L
> 5 μg/mL

Toxic Side-Effects

Quinidine may cause nausea and vomiting and the syndrome of 'cinchonism' (tinnitus, dizziness) and it may also cause hypotension or serious ventricular arrhythmias. In addition, quinidine may produce various reactions related to hypersensitivity (not dose related).

Assay Limitations

Some immunoassays suffer from cross-reactivity from dihydroquinidine (drug impurity) and quinidine metabolites.

Type of Sample

Serum or plasma.

AMIODARONE

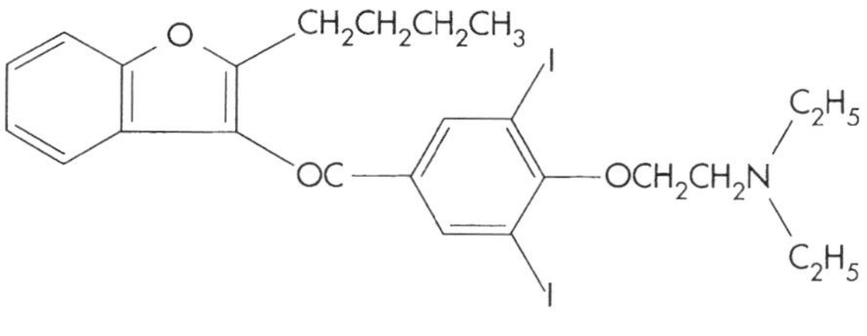

O $CH_2CH_2CH_2CH_3$ I C_2H_5 OC OCH_2CH_2N C_2H_5 I

Fig. 65.6 Amiodarone.

Clinical Applications

Amiodarone is used in the treatment of ventricular arrhythmias. It is effective in the treatment of supraventricular arrhythmias, particularly those in association with Wolff–Parkinson–White syndrome.

Dose and Modes of Administration

Maintenance dose as low as possible consistent with efficacy but often around 200 mg daily, orally. Can also be given intravenously.

Pharmacological Effects

Amiodarone is a class 3 antiarrhythmic that slows the action potential duration.

Therapeutic Range

0.5–2 mg/L
0.5–2 μg/mL
(Approximate only)

Potentially Toxic Concentration

> 2 mg/L
> 2 μg/mL

Toxic Side-Effects

Toxic effects include hypo- or hyperthyroidism, hepatitis, peripheral neuropathy, and diffuse lung infiltration (alveolitis). These effects are not clearly related to the concentration of the drug in the plasma. Some of the toxicity may be related to the principal metabolite desethyl-amiodarone (DEA).

Assay Technology

Amiodarone concentration is usually measured by HPLC. It is included in this book because it can interfere with thyroid function (*see* THYROID).

FLECAINIDE

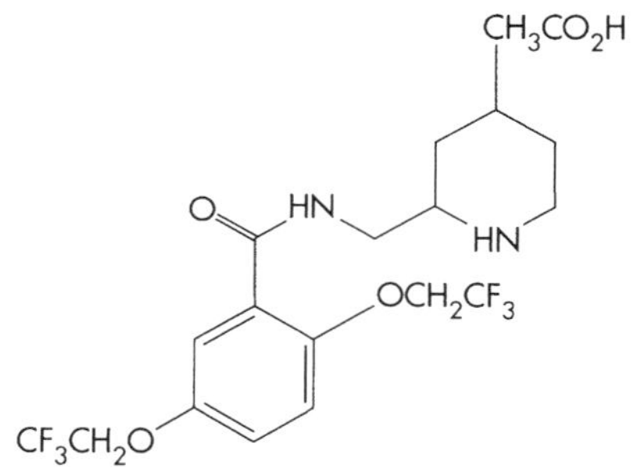

Fig. 65.7 Flecainide.

Clinical Applications

Flecainide is used for the treatment of serious supraventricular and ventricular arrhythmias, particularly in patients resistant to (or intolerant of) other treatments.

Dose and Mode of Administration

Orally, 100–400 mg daily, in divided doses. Can also be given by slow intravenous injection or infusion.

Pharmacological Effects

Flecainide is a class 1C agent that prolongs the effect of the action potential duration. It may also reduce the force of contraction of the heart.

Therapeutic Range

0.2–1 mg/L
0.2–1 μg/mL

Potentially Toxic Concentration

> 1 mg/L
> 1 μg/mL

Toxic Side-Effects

Dizziness and hypotension, nausea and vomiting, visual disturbance (e.g. corneal deposits), precipitation of heart failure, provocation of severe ventricular arrhythmias. Although photosensitivity, reversible increases in liver enzymes, jaundice, ataxia, peripheral neuropathy, pulmonary fibrosis, pneumonitis are occasionally reported, it is unclear if these are dose-related (toxic) side-effects.

Type of Sample

Serum or plasma.

ANTIBIOTICS

Many of the antibiotics (e.g. penicillins and cephalosporins) have a wide safety margin and plasma monitoring is therefore not required. Monitoring is of greater value with the aminoglycoside antibiotics (e.g. amikacin, kanamycin, gentamicin, and tobramycin) which in excessive dose may cause kidney damage or damage to the VIII nerve, which serves hearing and balance.

AMIKACIN

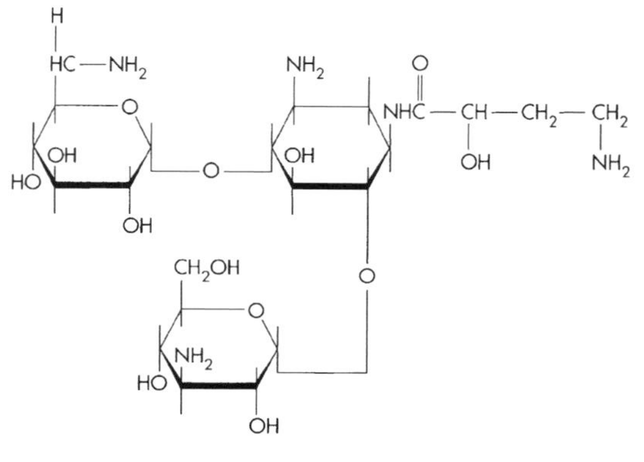

Fig. 65.8 Amikacin.

Clinical Applications

Amikacin is a semisynthetic aminoglycoside antibiotic used in the treatment of Gram-negative infections.

Dose and Modes of Administration

Amikacin is not absorbed orally and must be given by either intramuscular or slow intravenous injection at a dose of up to 15 mg/kg daily (in 2–3 divided doses). REDUCE DOSE IN RENAL IMPAIRMENT.

Pharmacological Effects

Like all aminoglycoside antibiotics, amikacin blocks the production of protein by inhibiting messenger RNA in the bacterial cell. A 'therapeutic range' is not applicable to antibiotics, in which the aim is to achieve a concentration above the minimum inhibitory concentration for the bacteria while ensuring that trough levels are sufficiently low to reduce the risk of damage to susceptible organs.

Target Peak Concentration

Varies according to severity of infection. Moderate to severe Gram-negative infection 20–25 mg/L (20–25 μg/mL), life-threatening Gram-negative infection 25–30 mg/L (25–30 μg/mL).

Target Trough Concentration

Moderate to severe Gram-negative infection 1–4 mg/L (1–4 μg/mL), life-threatening Gram-negative infections 4–8 mg/L (4–8 μg/mL).

Potentially Toxic Concentration and Side-Effects

Above target peak or trough concentrations, there is an increased risk of kidney damage or damage to the VIII nerve, particularly the vestibular branch, which mediates balance (but also the auditory branch, which subserves hearing).

Type of Sample

Serum or plasma.

KANAMYCIN

See Figure 65.9.

Clinical Applications

Kanamycin is similar to amikacin and has the same indications. In some countries it is considered to have been superseded by other aminoglycosides.

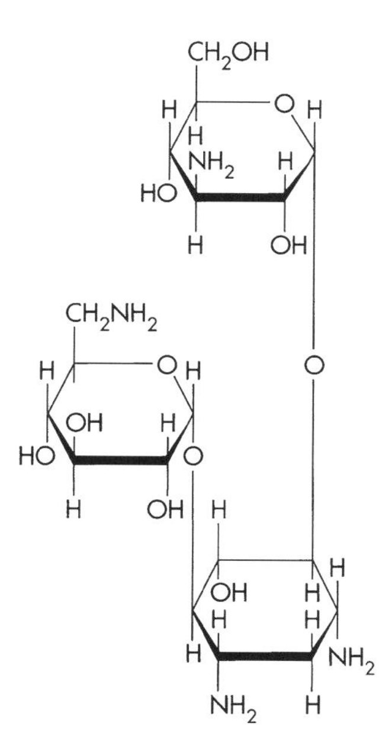

Fig. 65.9 Kanamycin.

Dose and Modes of Administration

Intramuscular injection, 250 mg every 6 h or 500 mg every 12 h. May also be given by slow intravenous infusion (15–30 mg/kg daily in divided doses every 8–12 h). REDUCE DOSE IN RENAL IMPAIRMENT.

Pharmacological Effects

See AMIKACIN.

Target Peak Concentration

See AMIKACIN.

Target Trough Concentration

See AMIKACIN.

Potentially Toxic Concentration and Side-Effects

See AMIKACIN.

Type of Sample

Serum or plasma.

GENTAMICIN

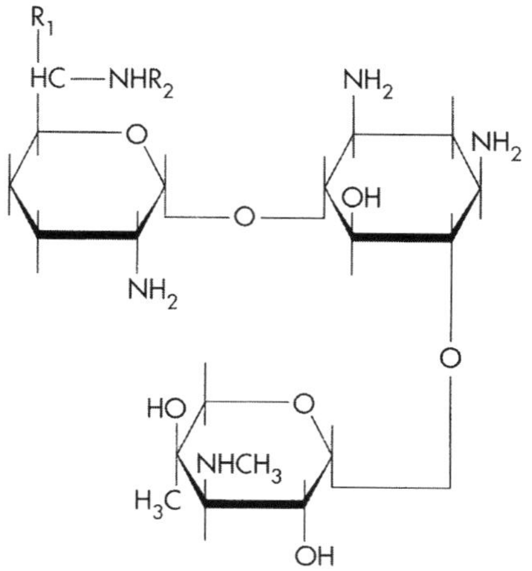

Fig. 65.10 Gentamicin: C_1, $R_1 = R_2 = CH_3$; C_2, $R_1 = CH_3$, $R_2 = H$; $R_1 = R_2 = H$.

Clinical Applications

Gentamicin is an aminoglycoside antibiotic used in the treatment of severe infections, particularly those caused by Gram-negative organisms (e.g. septicemia and neonatal sepsis; meningitis and other CNS infections; biliary-tract infection, acute pyelonephritis or prostatitis, endocarditis caused by viridans streptococci or *Enterococcus faecalis* (with a penicillin); pneumonia in hospital patients and as adjunctive therapy in Listerial meningitis).

Dose and Mode of Administration

Must be given by intramuscular injection (2–5 mg/kg daily in divided doses every 8 h) or by slow intravenous injection or infusion.

Pharmacological Effects

See AMIKACIN.

Target Peak Concentration

6–8 mg/L (6–8 μg/mL)
8–10 mg/L (8–10 μg/mL)
For life-threatening infections.

Target Trough Concentration

0.5–1.5 mg/L (0.5–1.5 μg/mL)
1–2 mg/L (1–2 μg/mL)
For life-threatening infections.

Potentially Toxic Concentration

Above target peak or trough concentrations.

Toxic Side-Effects

Eighth cranial nerve (vestibular and auditory) damage and nephrotoxicity. Hypomagnesemia has been rarely reported after prolonged therapy. Antibiotic-associated colitis may occur and nausea, vomiting, and rash may also be seen.

Type of Sample

Serum or plasma.

TOBRAMYCIN

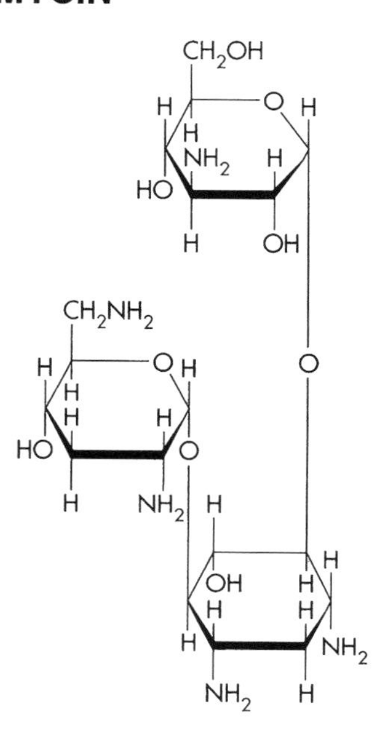

Fig. 65.11 Tobramycin.

Clinical Applications

Tobramycin is very similar to gentamicin. For data (except dose) *see* GENTAMICIN.

Dose and Modes of Administration

Intramuscular injection, 3–5 mg/kg daily in divided doses every 8 h. REDUCE DOSE IN RENAL IMPAIRMENT.

NETILMICIN

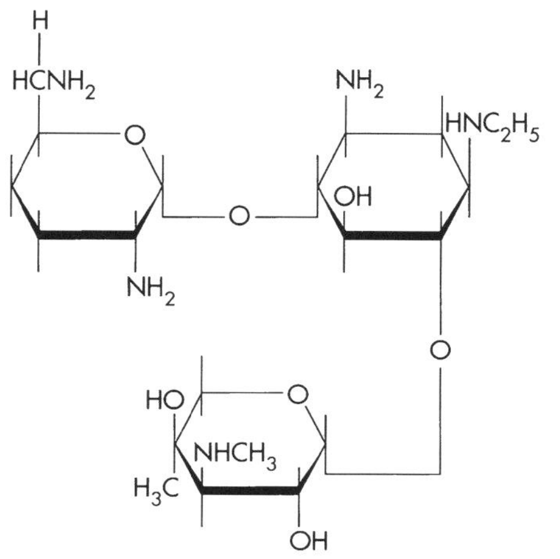

Fig. 65.12 Netilmicin.

Clinical Applications

Serious Gram-negative infections resistant to gentamicin and in the treatment of gonorrhea (single dose).

Dose and Modes of Administration

Intramuscular or intravenous injection or infusion, 4–6 mg/kg daily, in divided doses every 8–12 h. An increased dose is necessary in the first 48 h of treatment of severe infections. REDUCE DOSE IN RENAL IMPAIRMENT.

Pharmacological Effects

See AMIKACIN.

Target Peak Concentration

6–8 mg/L (6–8 μg/mL)
8–10 mg/L (8–10 μg/mL)
For life-threatening infections.

Target Trough Concentration

0.5–1.5 mg/L (0.5–1.5 μg/mL)
1–2 mg/L (1–2 μg/mL)
For life-threatening infections.

Potentially Toxic Concentration

Above target peak or trough concentrations.

Toxic Side-Effects

See AMIKACIN. But may cause less ototoxicity than other aminoglycosides when given for more than 10 days.

Type of Sample

Serum or plasma.

CHLORAMPHENICOL

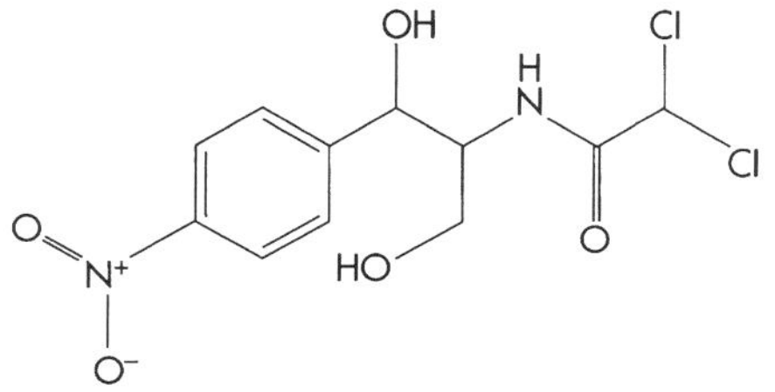

Fig. 65.13 Chloramphenicol.

Clinical Applications

Although chloramphenicol is an effective broad-spectrum antibiotic, it is associated with serious hematological side-effects when given systemically. It is therefore only indicated for the treatment of life-threatening bacterial infections, particularly those caused by *Haemophilus influenzae*, and for typhoid fever. To avoid certain serious toxic side-effects, measurement of plasma concentration must be performed in neonates and is also preferred in children aged less than 4 years.

Dose and Modes of Administration

Orally, or by intravenous injection or infusion, normally 50 mg/kg daily in four divided doses.

Pharmacological Effects

Chloramphenicol probably acts primarily on the 50S subunit of the bacterial 70S ribosome and inhibits peptide bond formation by suppressing peptidyl transferase activity. It is bacteriostatic rather than bactericidal.

Target Peak Concentration

15–25 mg/L
15–25 μg/mL
(Measured 1 h after intravenous injection or infusion)

Target Trough Concentration

< 15 mg/L
< 15 μg/mL

Potentially Toxic Concentration

Above target peak or trough concentrations.

Toxic Side-Effects

Blood dyscrasias, including reversible (and concentration-related) as well as irreversible aplastic anemia, peripheral and optic neuritis, nausea, vomiting, diarrhea, stomatitis, glossitis, and nocturnal hemoglobinuria have all been reported. The 'gray baby' syndrome (abdominal distension, pallor, cyanosis, and circulatory collapse) may follow excessive doses in neonates with immature hepatic metabolism (glucuronide conjugation).

Type of Sample

Serum or plasma.

VANCOMYCIN

Clinical Applications

The amphoteric glycopeptide antibiotic, vancomycin, is used systemically in the prophylaxis and treatment of endocarditis and other serious infections caused by Gram-positive cocci including multiresistant staphylococci. Vancomycin (added to dialysis fluid) is also used in the treatment of peritoneal dialysis-associated peritonitis. Given orally (when it is not significantly absorbed) it is effective in the treatment of antibiotic-associated (pseudomembranous) colitis.

Dose and Modes of Administration

By intravenous infusion, 500 mg over at least 60 min every 6 h or 1 g over at least 100 min every 12 h.

Pharmacological Effects

Vancomycin inhibits Gram-positive bacterial cell wall synthesis at a different site from the beta lactam antibiotics (cross-resistance therefore does not occur) and is normally bactericidal. It cannot penetrate the outer membrane of most Gram-negative organisms and so has limited activity.

Target Peak Concentration

30 mg/L
30 μg/mL
(Measured 2 h after intravenous injection or infusion)

Target Trough Concentration

< 10 mg/L
< 10 μg/mL

Potentially Toxic Concentration

Above target peak or trough concentrations.

Toxic Side-Effects

After parenteral administration, vancomycin may be associated with renal damage and ototoxicity (discontinue the drug if the patient experiences tinnitus). If the infusion is given rapidly, severe hypotension (including shock and cardiac arrest), wheezing, dyspnea, urticaria, pruritus, flushing of the upper body ('red man' syndrome), pain and muscle spasm in the back and chest have been reported.

Type of Sample

Serum or plasma.

ANTICONVULSANTS

Anticonvulsant monitoring is used both to reduce the risk of drug toxicity and to indicate likely therapeutic concentrations because the clinical end point (absence of seizures) can only be assessed retrospectively.

PHENYTOIN

Fig. 65.14 Phenytoin.

Clinical Applications

Phenytoin is used in the treatment and prevention of epilepsy, particularly generalized seizures and *status epilepticus* (but not absence seizures). It is also used to treat trigeminal neuralgia.

Dose and Modes of Administration

Normally 200–500 mg daily by mouth, or (in emergency) by slow intravenous injection. Phenytoin is poorly and variably absorbed after intramuscular administration and should not be given by this route.

Pharmacological Effects

The mode of action is unknown but phenytoin appears to reduce the spread of epileptic discharges.

Therapeutic Range

10–20 mg/L
10–20 μg/mL

Potentially Toxic Concentration

> 20 mg/L
> 20 μg/mL

Toxic Side-Effects

Phenytoin can cause nystagmus (abnormal jerky movements of the eyes), nausea, vomiting, confusion, tremor, insomnia, nervousness, drowsiness, and coma. Long-term use of the drug may cause GI symptoms, acne, gingival hyperplasia, and hirsutism. Hypersensitivity (nondose related) reactions (e.g. leucopenia and severe skin rashes) are rare but may occur.

Type of Sample

Serum or plasma.

PHENOBARBITAL (PHENOBARBITONE)

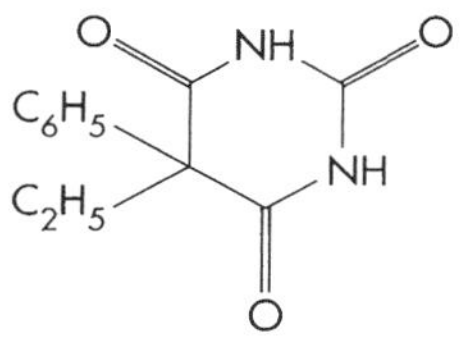

Fig. 65.15 Phenobarbital.

Clinical Applications

Phenobarbital is used in the treatment of all forms of epilepsy except absence seizures and in the treatment of *status epilepticus*.

Dose and Mode of Administration

Orally, 60–180 mg once daily (usually at night). By intramuscular injection, 200 mg, repeated after 6 h if necessary. In *status epilepticus*, by intravenous injection (dilute injection 1 in 10 with water for injections), 10 mg/kg at a rate of not more than 100 mg/min: max. 1 g.

Pharmacological Effects

Phenobarbital increases the seizure threshold and reduces the spread of discharge from an epileptic focus.

Therapeutic Range

15–40 mg/L
15–40 μg/mL

Potentially Toxic Concentration

> 40 mg/L
> 40 μg/mL

Toxic Side-Effects

Sedation, lethargy, depression, unsteadiness, stupor, and coma. Paradoxical excitement, restlessness, and confusion may occur in the elderly and hyperkinesia may be a problem in children. Megaloblastic anemia (sometimes responsive to folic acid) may occur after prolonged use.

Type of Sample

Serum or plasma.

PRIMIDONE

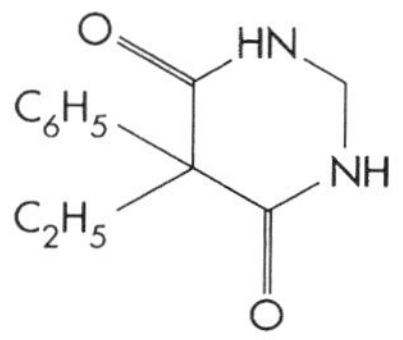

Fig. 65.16 Primidone.

Clinical Applications

Primidone is used in the treatment of all forms of epilepsy except absence seizures. It is also used to treat essential tremor. Primidone is largely inactive but can be metabolized to two major active metabolites, phenylethylmalonamide and (more slowly) to phenobarbital (phenobarbitone).

Dose and Mode of Administration

Orally, 500–1500 mg daily in two divided doses.

Pharmacological Effects

See PHENOBARBITAL.

Therapeutic Range

5–12 mg/L
5–12 μg/mL

Normally measured as phenobarbital (*see* PHENOBARBITAL).

Potentially Toxic Concentration

> 15 mg/L
> 15 μg/mL

Normally measured as phenobarbital. For toxic side-effects *see* PHENOBARBITAL.

Type of Sample

Serum or plasma.

CARBAMAZEPINE

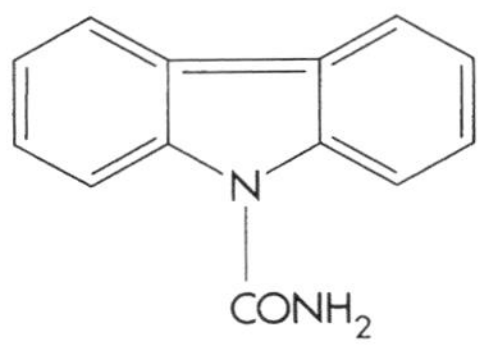

$$\text{(dibenzazepine ring)}-N-CONH_2$$

Fig. 65.17 Carbamazepine.

Clinical Applications

Carbamazepine is an anticonvulsant used in the treatment of partial and secondary generalized tonic–clonic seizures, but not in primary generalized seizures. It is also used to treat trigeminal neuralgia and in the prevention of bipolar affective disorder (manic-depressive psychosis) unresponsive to lithium.

Dose and Mode of Administration

Dose is normally given orally at a dose of 800–1200 mg daily. The drug may be given by suppository.

Pharmacological Effects

Carbamazepine reduces the spread of impulses from epileptic foci. It also increases the amount of cytochrome P-450 in the liver and may reduce the plasma concentration of drugs metabolized by this route (e.g. estrogens in oral contraceptives) by causing 'enzyme induction'. The principal metabolite carbamazepine 10–11-epoxide has around one-third of the anticonvulsant activity of the parent drug.

Therapeutic Range

4–12 mg/L
4–12 μg/mL

Potentially Toxic Concentration

> 12 mg/L
> 12 μg/mL

Toxic Side-Effects

Unsteadiness, drowsiness, headache, confusion, agitation (particularly in the elderly) double vision, drowsiness, nausea, vomiting, loss of appetite, constipation, and diarrhea have been reported. Hyponatremia is also reported. Hypersensitivity reactions (e.g. rash, liver or renal damage) may occur rarely. Suppositories may sometimes cause rectal irritation.

Type of Sample

Serum or plasma.

VALPROIC ACID

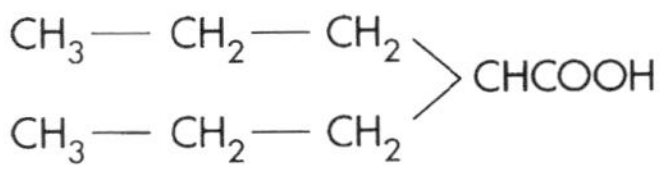

$$CH_3-CH_2-CH_2 \quad CH_3-CH_2-CH_2 > CHCOOH$$

Fig. 65.18 Valproic acid.

Clinical Applications

Valproic acid is effective against a wide variety of seizure types including generalized (tonic–clonic) seizures and partial seizures.

Dose and Mode of Administration

Oral and intravenous administration. Oral dose 1–2.5 g daily.

Pharmacological Effects

Valproic acid may act by increasing the brain concentrations of gamma-aminobutyric acid (GABA) which is an inhibitory neurotransmitter at the GABA–benzodiazepine complex in the central nervous system.

Therapeutic Range

50–100 mg/L
50–100 μg/mL

Potentially Toxic Concentration

> 100 mg/L
> 100 μg/mL

Toxic Side-Effects

Nausea and vomiting, drowsiness, confusion, ataxia and tremor, transient alopecia (with subsequent growth of curly hair), increased appetite, weight gain, and edema have been reported. Hypersensitivity reactions such as a reduction in platelet count and liver damage can occur. The latter is potentially serious and may proceed to liver failure.

Type of Sample

Serum or plasma.

ETHOSUXIMIDE

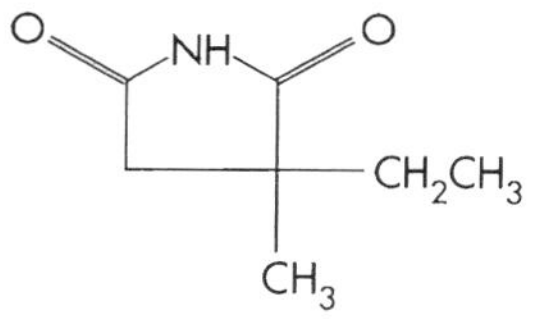

Fig. 65.19 Ethosuximide.

Clinical Applications

Ethosuximide is the drug of choice in simple absence seizures. It may also be used in myoclonic seizures and in atypical absence, atonic and tonic seizures.

Dose and Mode of Administration

Orally, 1–2 g.

Pharmacological Effects

Mode of action unknown but may act on low-frequency inhibitory pathways to reduce duration of the spike and wave discharges (3 Hz) seen in absence seizures.

Therapeutic Range

40–100 mg/L
40–100 μg/mL

Potentially Toxic Concentration

> 100 mg/L
> 100 μg/mL

Toxic Side-Effects

Nausea and vomiting, drowsiness, unsteadiness, abnormal movements, hiccup, headache, photophobia, and depression. Slight euphoria and, in some cases, aggression and frank psychosis have been reported. Agranulocytosis is a very serious but rare (and probably nondose related) side-effect. Others include severe rash and hepatotoxicity.

Type of Sample

Serum or plasma.

CHEMOTHERAPEUTIC AGENTS

METHOTREXATE

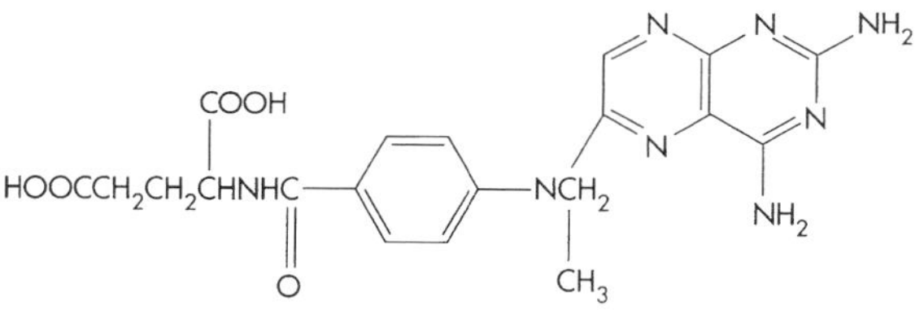

Fig. 65.20 Methotrexate.

Clinical Applications

Methotrexate is used for the maintenance therapy of childhood acute lymphoblastic leukemia and other malignancies. It is also used in the treatment of severe psoriasis and rheumatoid arthritis.

Dose and Modes of Administration

Methotrexate can be given orally, intravenously, intramuscularly, or intrathecally. The oral dose for leukemia in children is 15 mg/m^2 (body surface area) each week, in combination with other drugs. For psoriasis in adults it is 10–25 mg orally each week and for rheumatoid arthritis in adults up to 15 mg weekly, orally.

Pharmacological Effects

Methotrexate inhibits the enzyme dihydrofolate reductase and the synthesis of purines and pyrimidines necessary for nucleic acid synthesis.

Therapeutic Range

The minimum cytotoxic concentration is 0.01 μmol/L.

Potentially Toxic Concentration

Similar to the minimum cytotoxic concentration. Toxicity is dependent on duration of exposure.

Toxic Side-Effects

Bone-marrow suppression, ulceration of mucus membranes and, more rarely, lung infiltration or liver damage.

Type of Sample

Serum or plasma.

MISCELLANEOUS

ACETAMINOPHEN (PARACETAMOL)

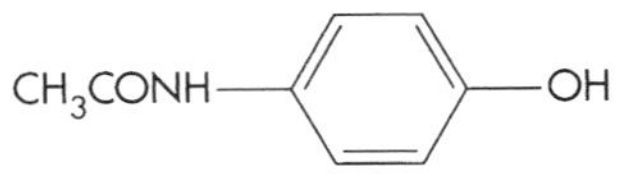

CH_3CONH — OH

Fig. 65.21 Acetaminophen.

Clinical Applications

Acetaminophen is used in the treatment of mild to moderate pain and in the relief of pyrexia.

Dose and Mode of Administration

Orally, 0.5–1 g every 4–6 h (maximum daily dose 4 g).

Pharmacological Effects

The mode of action of acetaminophen is unknown. It may involve central inhibition of prostaglandin-mediated pathways.

Therapeutic Range

10–30 mg/L
10–30 μg/mL

Potentially Toxic Concentration

$>$ 100 mg/L
$>$ 100 μg/mL
(4 h after ingestion)

The risk of toxicity appears to be related to the plasma concentrations at any given time after overdose, as well as to the nutritional state, chronic alcohol misuse and concomitant drug therapy. Above the appropriate recommended treatment line, antidotal treatment (e.g. *N*-acetylcysteine) should be used. Such treatment may be effective in reducing the risk or severity of liver damage after overdose. The recommended treatment lines differ in different countries (Routledge *et al.*, 1998).

Toxic Side-Effects

Liver damage and, more rarely, renal damage.

Type of Sample

Serum or plasma.

TRICYCLIC ANTIDEPRESSANTS

Clinical Applications

Tricyclic antidepressants are most effective in treating moderate to severe endogenous depression but there may be an interval of 2–4 weeks before beneficial effects are seen. They are also effective in the management of panic disorder. Although many related compounds are available, amitriptyline, imipramine, and desipramine are well-established compounds for which TDM may occasionally be helpful.

Dose and Mode of Administration

Amitriptyline: maximum 50–200 mg daily, orally. Imipramine: 50–200 mg daily, orally.

Pharmacological Effects

The mode of action is unknown but tricyclic antidepressants possess peripheral and central anticholinergic properties.

Therapeutic Range

Amitriptyline: 150–250 μg/L (150–250 ng/mL)
Imipramine: 200–300 μg/L (200–300 ng/mL)
Desipramine: 70–200 μg/L (70–200 mg/mL)

Toxic Side-Effects

Dry mouth, sedation (especially with amitriptyline), blurred vision, constipation, difficulty in passing urine. Tachycardia and postural fall in blood pressure are also common.

Type of Sample

Serum or plasma.

CYCLOSPORIN

Cyclosporin is a cyclic polypeptide consisting of 11 amino acids. It was first isolated as a metabolite from two strains of fungi and is a nonmyelotoxic potent immunosuppressant.

Clinical Applications

Cyclosporin is an immunosuppressant used in the field of organ and tissue (bone marrow, kidney, liver, pancreas, heart, and heart–lung) transplantation to prevent graft rejection and for the prophylaxis of graft-versus-host disease. It is also used in severe atopic dermatitis, psoriasis, and rheumatoid arthritis.

Dose and Modes of Administration

The drug can be given orally or by intravenous infusion. Specialist advice is required for doses in particular situations.

Pharmacological Effects

Inhibits activation of T (helper) cells and reduces production of lymphokines especially interleukin-2 (IL-2).

Therapeutic Range

100–300 μg/L
100–300 ng/mL

Dependent on the specificity of the technique used. (Some methods also measure a number of metabolites.)

Potentially Toxic Concentration

> 400 μg/L
> 400 ng/mL

Toxic Side-Effects

Cyclosporin may cause kidney damage in up to a third of individuals. Tremor, hypertension, hepatic dysfunction, nausea and vomiting, increased hair growth, and overgrowth of the gums (gingival hyperplasia) may also occur.

Type of Sample

Whole blood, serum, or plasma.

DIGOXIN

Digoxin is one of the cardiac glycosides, which are characterized by a complex multiring structure.

Clinical Applications

Digoxin is used in the treatment of heart failure, to prevent supraventricular arrhythmias, and to control heart rate in atrial fibrillation.

Dose and Modes of Administration

Orally, 62.5–500 μg daily. Digoxin can also be given by intravenous injection, preferably by slow infusion.

Pharmacological Effects

Digoxin increases the force of contraction of the heart. It also slows conduction through the atrioventricular node and works in this way to terminate or reduce the risk of recurrence of supraventricular tachycardia.

Therapeutic Range

Approx. 0.8–2.0 μg/L (0.8–2.0 ng/mL).

Sample taken at least 6 h after dose.

Potentially Toxic Concentration

> 2.4 μg/L (>2.4 ng/mL)

Lower if predisposing factors are present (e.g. low blood potassium).

Toxic Side-Effects

Loss of appetite, nausea and vomiting, diarrhea, abdominal pain, slowing of the heart and cardiac arrhythmias (both abnormally fast and abnormally slow). Effects on the central nervous system include visual disturbances, headache, fatigue, drowsiness, confusion, delirium, and hallucinations.

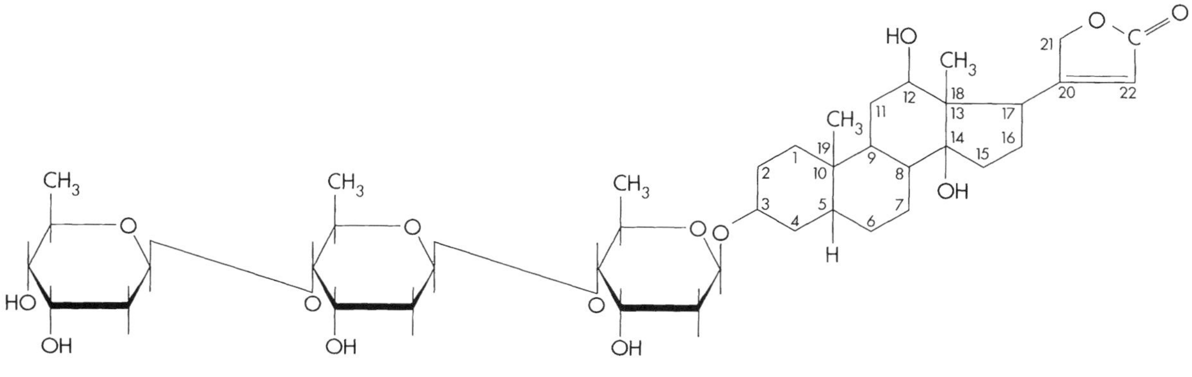

Fig. 65.22 Digoxin.

Type of Sample

Serum or plasma.

Digoxin-Like Factors

Some endogenous substances other than digoxin are detected by many digoxin immunoassays. They are known collectively as digoxin-like factors (DLF), digoxin-like immunoreactive factors (DLIF) or digoxin-like immunoreactive substances (DLIS). The extent of the interference varies from assay to assay, primarily due to the specificity of the antisera. As a result of DLF, 'digoxin' concentrations have been found in patients who have never received the drug, primarily neonates, pregnant women, and patients with renal or hepatic failure. DLF may be the natural ligands for digoxin receptors in the body but their exact identity is as yet unknown. In some affected assays, extending the assay incubation time can reduce the interference. Pre-ultrafiltration is an effective method for reducing DLF in samples.

Effect of Digibind

Another source of interference in digoxin immunoassays is the drug 'Digibind', made from Fab fragments of digoxin antibodies and used to treat life-threatening digoxin overdoses. The antibody fragments mop up the free digoxin, rendering it inactive in the patient. However, the extra digoxin antibody present in the sample can cause misleading results in immunoassays. For example, in a solid-phase competitive assay, the labeled digoxin may be bound by the antibody in the sample, leading to a low signal and apparently high concentration. In a polyethylene glycol-precipitation assay, this labeled digoxin–Fab complex is precipitated and the apparent concentration of digoxin may be zero. Using an antibody with a relatively higher affinity for digoxin than the Digibind antibodies may reduce the interference. Alternatively, a pretreatment stage that removes proteins may be effective.

DIGITOXIN

Digitoxin has a similar structure to digoxin.

Clinical Applications

See DIGOXIN.

Dose and Mode of Administration

Orally, 50–200 mg daily.

Pharmacological Effects

See DIGOXIN. However, digitoxin has a longer half-life and is primarily excreted via the liver.

Therapeutic Range

15–30 μg/L
15–30 ng/mL

Potentially Toxic Concentration

$>$ 40 μg/L ($>$40 ng/mL)

Lower if predisposing factors are present (e.g. low blood potassium).

Toxic Side-Effects

See DIGOXIN.

Type of Sample

Serum or plasma.

THEOPHYLLINE

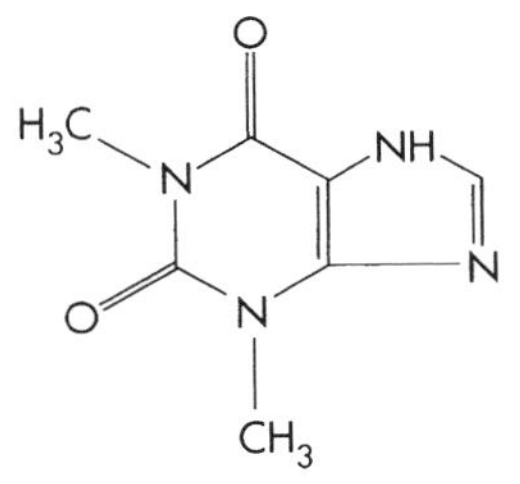

Fig. 65.23 Theophylline.

Clinical Applications

Theophylline is used in the treatment of patients with reversible airways obstruction (asthma and chronic airways obstruction disease).

Dose and Modes of Administration

Orally, 500–1200 mg per day, depending on plasma concentrations achieved. Theophylline can also be given in the form of an ethylenediamine salt (aminophylline) orally or rarely intravenously (must be given slowly). Theophylline (80 mg) is equivalent to aminophylline (100 mg).

Pharmacological Effects

The mode of action of theophylline is unknown, though it may act in part by blocking adenosine receptors.

Therapeutic Range

10–20 mg/L
10–20 μg/mL

Potentially Toxic Concentration

> 20 mg/L (>20 μg/mL)

There is a narrow margin between therapeutic and toxic dose.

Toxic Side-Effects

Nausea, vomiting, tachycardia, palpitations, arrhythmias, headache, insomnia, and convulsions.

Type of Sample

Serum or plasma.

FURTHER READING

Ebden, P., Leopold, D., Buss, D., Smith, A.P. and Routledge, P.A. Free and total plasma theophylline concentrations in chronic airflow obstruction. *Thorax* **39**, 352–355 (1984).

Flanagan, R.J. Guidelines for the interpretation of analytical toxicology results and unit of measurement conversion factors. *Ann. Clin. Biochem.* **35**, 261–267 (1998).

Hammett-Stabler, C.A. and Johns, T. Laboratory guidelines for monitoring of antimicrobial drugs. National Academy of Clinical Biochemistry. *Clin. Chem.* **44**, 1129–1140 (1998).

Holford, N.H.G. and Tett, S. Therapeutic drug monitoring: the strategy of target concentration intervention. In: *Avery's Drug Treatment* (eds Speight, T.M. and Holford, N.H.G.), 225–259 (Adis, Auckland, 1997).

Linder, M.W. and Keck, P.E. Jr. Standards of laboratory practice: antidepressant drug monitoring. National Academy of Clinical Biochemistry. *Clin. Chem.* **44**, 1073–1084 (1998).

Rimmer, E.M., Buss, D.C., Routledge, P.A. and Richens, A. Should we routinely measure free plasma phenytoin concentration? *Br. J. Clin. Pharmacol.* **17**, 99–102 (1984).

Rolan, P.E. Plasma protein binding displacement interactions – why are they still regarded as clinically important? *Br. J. Clin. Pharmacol.* **37**, 125–128 (1994).

Routledge, P.A. The plasma protein binding of basic drugs. *Br. J. Clin. Pharmacol.* **22**, 499–506 (1986).

Routledge, P.A., Lazar, J.D., Barchowsky, A., Stargel, W.W., Wagner, G.S. and Shand, D.G. A free lignocaine index as a guide to unbound drug concentration. *Br. J. Clin. Pharmacol.* **20**, 695–698 (1985).

Routledge, P.A., Vale, J.A. *et al.* Paracetamol (acetaminophen) poisoning. No need to change current guidelines to accident departments. *BMJ* **317**, 1609–1610 (1998).

Valdes, R. Jr., Jortani, S.A. and Gheorghiade, M. Standards of laboratory practice: cardiac drug monitoring. National Academy of Clinical Biochemistry. *Clin. Chem.* **44**, 1096–1109 (1998).

White, S. and Wong, S.H.Y. Standards of laboratory practice: analgesic drug monitoring. National Academy of Clinical Biochemistry. *Clin. Chem.* **44**, 1110–1123 (1998).

66 Drugs of Abuse☆

Brian Widdop

The nonmedical use of drugs such as heroin, cocaine, and amphetamines remains epidemic and affects all aspects of social and economic life.

The recent establishment of a link between intravenous drug misuse and the spread of AIDS has fueled concern, particularly in Western countries. Legislative measures, no matter how punitive, have had marginal impact and it is now recognized that these must be accompanied by education, treatment, and rehabilitation programs if the problem is to be brought under control. Diagnostic tests carried out on biological samples are an integral feature of these programs and those based on immunological principles are widely applied.

Many of those who advocate a more radical approach to the problem, such as the decriminalization of cannabis use, often point to nicotine and alcohol as examples of substances that are much more harmful and yet, in most countries, are quite legal and widely available. Irrespective of the merits of this argument, there is no doubt that nicotine is highly addictive and that regular tobacco use can lead to a number of severe medical disorders. By the same token, chronic alcohol abuse has serious physical and neurological consequences as well as being a major cause of domestic upheaval and loss of livelihood. Smokers and alcohol abusers undergoing treatment are notorious, when faced with medical interrogation, for either claiming complete abstinence from the habit or grossly underreporting their consumption. Again, immunoassays can provide an objective measure of the true pattern of use.

Self-administration of performance-enhancing drugs by high-profile athletes receives a great deal of media coverage. It is known that the practice now extends to local competitions, e.g. school and county championships, and immunoassays have a role in detecting this type of drug misuse. Anabolic steroids taken in regular doses throughout an intercompetition training period help to build muscle and can boost athletic performance well beyond an individual's natural capability. Apart from the fact that an unfair advantage may be gained, there is great concern about the subsequent physical and mental health of young people who embark on such a course. Anabolic steroid administration in order to increase the performance of racehorses and other animals featured in sport is also prohibited and testing of biological samples for this group of substances is carried out on a regular basis using immunoassays. In competitive human sport, however, chromatographic and mass spectrometric techniques now take precedence.

APPLICATIONS

DRUG DEPENDENCE TREATMENT CENTERS

These centers may be either out-patient or in-patient units and are designed to deal with the more severely affected patients. New patients undergo a medical examination, are questioned on their drug abuse history, and a urine sample is requested for analysis. The history volunteered is often unreliable; many drug abusers are unaware of the composition of 'street' preparations; some report heroin abuse but omit to mention regular intake of cannabis, benzodiazepines, etc.; others may be drug free and attempting to obtain a prescription for drugs for subsequent illegal sale. The results of the urine analysis therefore provide the only objective evidence of the pattern of drug abuse and are crucial to the initial diagnosis.

The ideal treatment aim is total and continued abstinence from drug abuse. For some patients who are heavily dependent on opiate drugs such as heroin, this is an unrealistic goal and it is common practice to prescribe the safer heroin substitute methadone. After a while, a sizeable proportion of these patients are unable to resist the craving for heroin and other psychoactive drugs and relapse into multiple drug abuse. The only reliable means

☆Some of the product performance data reproduced here may be from a previous version of the product. Please contact manufacturer or check pack insert for latest performance information.

The Immunoassay Handbook. *Third Edition*
D. Wild (Ed.)

of monitoring compliance with prescribed medication in patients of this type is by regular urine analyses.

Other patients may be successfully detoxified by controlled withdrawal of drugs and then embark on a course of rehabilitation. This invariably includes routine urine checks at appropriate intervals to ensure continued abstinence from any form of nonmedical drug taking.

PSYCHIATRIC CLINICS

Abused drugs act on the central nervous system (CNS) and can provoke signs of mental illness. For example, drowsiness and slurred speech might suggest sedative or cannabis abuse. Stimulant drugs (amphetamines, cocaine) induce excessive agitation, and prolonged and heavy abuse can lead to psychosis. The hallucinogenic properties of LSD and phencyclidine are well documented. Urine tests for the presence of these drugs help to differentiate between endogenous and drug-induced mental disorders.

MEDICAL–LEGAL APPLICATIONS

Children of drug-abusing parents are often at high risk of mental and physical neglect. In certain circumstances, the child may be taken into care until such time that the mother or father can demonstrate prolonged abstinence from drug abuse. A decision to renew custody of the child may rest on the results of urine drug analyses on samples collected from the parent at weekly intervals for up to 3 months.

Occasionally, drug abusers dose their own children, either to gain respite from the child's demands or out of malice. Analytical tests on samples from such children are crucial in establishing a diagnosis of poisoning and securing their future safety by legal means.

Drug abuse carries the danger of overdosage, particularly among intravenous users. This may be deliberate, as in a suicide attempt, or unintentional, for example when an illicit supply that is far more concentrated than that previously tolerated is injected. The practice of smuggling packages of drugs across international boundaries by concealment in the rectum or vagina or by swallowing them ('body-packing') can cause severe and even fatal poisoning if the packages leak. Tests for drugs of abuse on biological samples form part of the clinical diagnosis. If the outcome is fatal, the results of these tests are used by the pathologist to establish the cause of death.

Many countries legislate against driving under the influence of drugs, and analyses of blood or urine specimens form part of the evidence put forward to secure a conviction. A more recent development is roadside testing by police officers equipped with hand-held devices that give presumptive results within 5 min. Clearly, sample collection has to be noninvasive and must not infringe privacy. For these reasons, saliva is the favored sample and the results of early trials look promising. Offenders with a history of drug abuse applying for renewal of a driving license may be required to undergo periodic urine tests before this is granted.

DRUG ABUSE IN THE WORKPLACE

Employees who misuse drugs are more likely to make bad decisions, manufacture faulty goods, and cause accidents. In the USA random urine testing for drug abuse is mandatory in sensitive government posts, the armed forces, and the transport industries. Many other industries test job applicants before employment as a routine exercise. This approach has become widely practiced in several European countries and, because a high proportion of samples are negative, immunoassays, which are rapid and readily automated, are much favored by the testing laboratories. It must be emphasized, however, that great care is needed to avoid false accusation of an innocent employee. A positive immunoassay result must always be substantiated by a second confirmatory analysis by gas chromatography/mass spectrometry (GC/MS) before any action is taken.

IMMUNOASSAYS FOR DRUGS OF ABUSE

Quantitative immunoassays for abused drugs in serum, plasma, whole blood, or saliva can be used to diagnose clinical overdose and fatal cases. Saliva tests for roadside testing after an accident are being seriously evaluated in Europe. Sweat patches that can soak up drugs excreted by the sweat glands over several days are now used in the USA to monitor prisoners for drug abstention during periods of parole. For the other situations described above, semiquantitative tests on urine remain most prevalent. This chapter therefore deals mainly with immunoassays for detection of the drugs in urine.

A distinction must be made between the sensitivity of the test, which can be defined as the lowest concentration of analyte that can be detected reliably in a given matrix, and the threshold (or 'cutoff') concentration. Urine is a complex and variable matrix that exacerbates the problem of distinguishing a signal due to the presence of a drug from that of background instrument noise. To overcome this, commercial assays are assigned a threshold concentration that exceeds the detection limit by several times. The threshold adopted must also be practical in that it allows recent drug abuse to be detected. Once the threshold limit has been selected, test samples are assigned positive or negative status by comparison with the threshold (cutoff) response level. Selection of threshold levels is also governed by the views of national bodies, for example, in the USA the Substance Abuse and Mental Health Service Administration (SAMHSA) lays down its own guidelines on these for workplace testing (DHHS/SAMHSA, 1994). In the European Union, recommended threshold levels for workplace testing were put forward by an international panel in 1996 (Killander *et al.*, 1997), but there is as yet no legislation to enforce them. Examples of common assay cutoff values are shown in Table 66.1. Note that for substances which are extensively metabolized before reaching the urine (e.g. cocaine), cutoff values are assigned to the most prevalent urinary metabolite.

Although radioimmunoassay (using ^{125}I labeling) still has a place, particularly when testing for drugs present in very low concentrations, homogeneous nonisotopic immunoassays such as the enzyme-multiplied immunoassay technique (EMIT), the cloned enzyme donor immunoassay (CEDIA), fluorescence polarization immunoassay (FPIA), microparticle-based immunoassays such as OnLine, and heterogeneous assays such as enzyme-linked immunoabsorbent assay (ELISA) have proved the most popular in routine use. There has always been pressure on the manufacturers to develop very simple 'dipstick' tests which can be used with little or no training by doctors, nurses, policemen and the like to give on-the-spot answers. The early devices were rather unreliable, but some of the recent products have to be taken seriously and will gain increasing favor over the next few years.

The cross-reacting characteristics of the various antisera used are crucial in this area and these are presented in detail for some of the main kits commercially available under the appropriate sections. Space does not allow all products to be dealt with here in such detail, and the relevant information should be sought from the companies themselves.

Although the manufacturers take great pains to evaluate the cross-reactivity of as many potentially interfering substances as possible, their lists are not exhaustive. Although possible cross-reactivity can sometimes be forecast on the basis of a similarity in chemical structure, unrelated substances can cause false-positive results. Moreover, lack of interference in the assays by parent compounds does not always rule out interference by their urinary metabolites. Sometimes a compound can affect the whole assay range as in the example of the antibacterial drug ciprofloxacin, which produces high absorbance readings in EMIT assays. Interference from natural products is very rare, although there have been reports of problems with FPIAs in subjects taking abnormally high quantities of riboflavin to treat migraine.

It is most important for the analyst to be aware that adulteration of urine samples by donors wishing to avoid detection is a common practice. Immunoassays are vulnerable to changes in the sample matrix, pH, and ionic strength brought about by adding sodium chloride, sodium bicarbonate, bleach, and disinfectant. Other household products such as liquid detergent and hand-soap also disrupt immunoassays. In theory, tightly controlled sample collection procedures should eliminate this problem, but it is wise to examine samples carefully for abnormalities before analysis. Apart from visual abnormalities (odd colors, soap bubbles, undissolved solids) many laboratories check the pH, creatinine level, specific gravity, and osmolality for additional signs of tampering.

Finally, the choice of assay system depends on several factors such as sensitivity required, specificity, sample numbers, equipment available, ease of use and, not least, cost. For most systems there are at least two suppliers and one may have the edge over the others when these factors are taken into account by an individual laboratory. In the sections that follow, an example of each system is included and one of the most important characteristics, i.e. specificity, is considered in detail. However, inclusion of an assay here does not imply superiority over those produced by alternative diagnostic companies.

Table 66.1. Common assay cutoff values for drugs of abuse immunoassays.

Compound	Cutoff values (ng/mL)
Amphetamines	300, 1000
Benzodiazepines*	300
Barbiturates	200, 300
Cannabis metabolite**	20, 50, 100
Cocaine metabolite†	300
Methadone	300
Opiates‡	300
Phencyclidine	25, 75

*Usually as oxazepam. **11-nor-δ-9-tetrahydrocannabinol-9-carboxylic acid. †Benzoyl ecgonine. ‡Morphine.

AMPHETAMINE

Structure

See Figure 66.1.

Dose and Modes of Administration

Orally or intravenously, 10–30 mg; prolonged abuse leads to tolerance and dosage may exceed 200 mg daily.

Pharmacological Effects

Amphetamine is a potent sympathomimetic amine with respect to stimulation of the CNS. Effective doses produce elevation of mood, increased alertness, self-confidence,

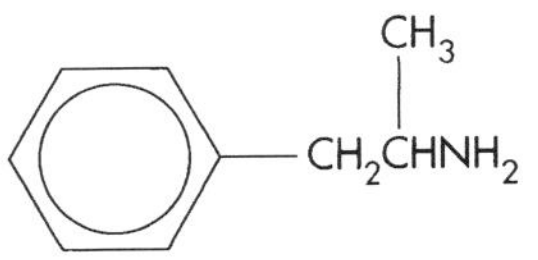

CH_3 / CH_2CHNH_2

Fig. 66.1 Amphetamine.

and ability to concentrate. Concomitant stimulation of the peripheral nervous system improves physical performance. The D-isomer (dextroamphetamine) is four times as potent as the L-isomer. Its use as an anorexiant in treating obesity had little success and has long been abandoned.

Toxic Effects

Chronic abuse of high doses leads to weight loss, hallucinations, and paranoid psychosis. Acute overdose causes agitation, hyperthermia, convulsions, coma, and respiratory and/or cardiac failure.

Assay Technology

See METHYLENEDIOXYMETHAMPHETAMINE.

METHAMPHETAMINE

Structure

See Figure 66.2.

Dose and Modes of Administration

Orally, 2.5–15 mg. The D-isomer is abused intravenously by addicts in doses of up to 200 mg daily. Insufflation of the free base is also practiced.

Pharmacological Effects

These are identical to those of amphetamine. The L-isomer has weaker central, but greater peripheral sympathomimetic activity and is used in some nonprescription inhalers as a decongestant. About 5% of a dose of methamphetamine is excreted as amphetamine in the urine.

Toxic Effects

These are similar to those of amphetamine.

Assay Technology

See METHYLENEDIOXYMETHAMPHETAMINE.

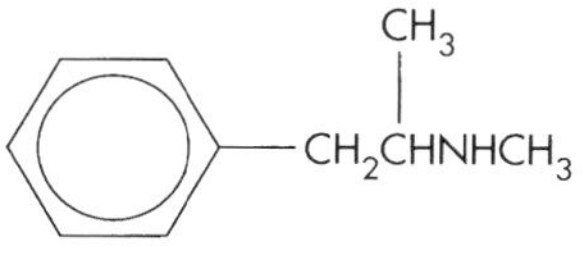

Fig. 66.2 Methamphetamine.

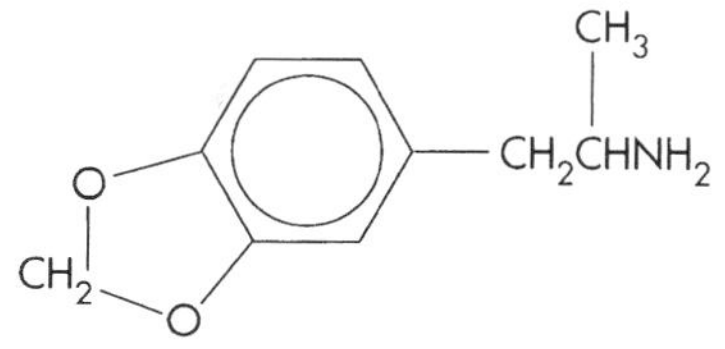

Fig. 66.3 Methylenedioxyamphetamine.

METHYLENEDIOXYAMPHETAMINE

Structure

See Figure 66.3.

Dose and Modes of Administration

Methylenedioxyamphetamine (MDA) is a ring-substituted derivative of amphetamine and is a member of the group of drugs popularly known as 'Ecstacy' which also includes methylenedioxymethylamphetamine (MDMA) and methylenedioxyethylamphetamine (MDEA). MDA is administered both orally and intravenously in doses of 50–250 mg as an illicit drug.

Pharmacological Effects

This drug has, in the main, central stimulant properties and large doses induce hallucinations.

Toxic Effects

Overdosage causes agitation, tremor, tachycardia, hyperthermia, muscular rigidity, hyperventilation, and coma.

Assay Technology

See METHYLENEDIOXYMETHAMPHETAMINE.

METHYLENEDIOXY-METHAMPHETAMINE

Structure

See Figure 66.4.

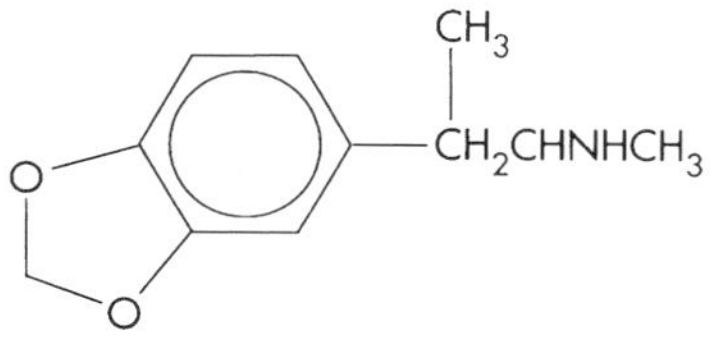

Fig. 66.4 Methylenedioxymethamphetamine.

Dose and Modes of Administration

Methylenedioxymethamphetamine (MDMA) was previously used as an adjunct to psychotherapy. Drug abusers take oral doses of 100–150 mg. It is metabolized to MDA.

Pharmacological Effects

MDMA has both central and peripheral sympathomimetic activity and doses of around 200 mg can lead to visual, auditory, and tactile hallucinations.

Toxic Effects

MDMA abuse is associated with acid-house parties and deaths have been reported that were due to hyperthermia and to cardiovascular causes after 'normal' dosage.

Assay Technology (All Amphetamines)

Commercial immunoassay kits for detecting amphetamines in urine specimens have been available for many years. The antibodies used in these individual systems exhibit a wide range of cross-reactivities, particularly towards other drugs derived from phenylethylamine. Some of these, e.g. ephedrine and phenylpropanolamine, are present in numerous over-the-counter (OTC) medicines and normal use can trigger a positive response in certain amphetamine assays. Wide cross-reactivity can be an advantage if other related and abused drugs such as phentermine, mephentermine, MDA, MDMA, and other members of the 'Ecstacy' group of drugs are to be detected. Some of the 'Ecstacy' compounds, e.g. *N*-methyl-1-(3,4,-methylenedioxyphenyl)-2-butanamine (MBDB), are missed by the commercial immunoassays and if misuse of these substances spreads an effort will be needed to develop more sensitive tests for this group.

The existence of isomeric forms of amphetamine and methamphetamine is a further complication. For example, L-methamphetamine, which is a much less potent CNS stimulant than D-methamphetamine, is used in nonprescription anticongestant inhalers. Some assays are designed to avoid significant cross-reactivity towards the L-species.

The manufacturers are at pains to test antibody specificity towards a comprehensive range of common drugs, but their data are not exhaustive. On occasions they overlook the fact that a nonreactive parent drug will be metabolized to a reactive product, e.g. the anorectic drugs diethyl propion, clobenox, fenproporex have no cross-reactivity, but are metabolized to amphetamine. Cross-reactivity of other drugs, such as labetolol in the EMIT polyclonal assay, is unpredictable and any findings of this nature should be reported immediately to the manufacturer.

Finally, amphetamine immunoassays are only preliminary tests and positive results should always be confirmed by a chromatographic procedure such as gas chromatography or, for medical–legal purposes, gas chromatography/mass spectrometry (GC/MS).

Radioimmunoassays

Roche Abuscreen™

The Roche High Specificity Amphetamine Radioimmunoassay uses the double-antibody separation technique. The kit detects D-amphetamine at 5 ng/mL and is very selective for this isomer (see Table 66.2).

Diagnostic Products Corporation (DPC)

The double-antibody radioimmunoassay produced by DPC is designed both for qualitative and for quantitative measurements and is highly selective for amphetamine (see Table 66.3).

DPC also produces a solid-phase RIA (Coat-a-Count™) for methamphetamine designed specifically to detect the D-isomer. The following structurally related drugs are not detected at levels of 100,000 ng/mL: L-pseudoephedrine, beta-phenethylamine D,L-normetanephrine, dopamine, D,L-metonephrine, tyramine, L-norepinephrine, D-ephedrine, L-norpseudoephedrine. The cross-reactivity data for other structurally related compounds are shown in Table 66.4.

Enzyme-multiplied immunoassay technique

EMIT technology was developed in the 1970s by the Syva company and since that time several other manufacturers have adapted the principle to produce tests which differ in cross-reactivity, sensitivity, and ease of use. Here the EMIT assays produced by the originators of the system are presented in detail. Similar data for alternative enzyme immunoassays can be obtained from the respective companies.

Table 66.2. Drugs found not to cross-react with the Roche High Specificity Amphetamine Radioimmunoassay at a concentration of 10,000 ng/mL.

Compound	Concentration found (ng/mL)
L-amphetamine	771
D-methamphetamine	100
L-methamphetamine	49
Methylenedioxymethamphetamine (MDMA)	90
Phenylethylamine	258
Phenylpropanolamine	13
Tyramine	146

Table 66.3. Amphetamine-related drugs found not to produce a false positive (cutoff level 1000 ng/mL) with the DPC double-antibody assay.

Drug/concentration (ng/mL)	Concentration found (ng/mL)
Dopamine (10,000)	18
Hydroxyamphetamine (1000)	294
D,L-methamphetamine (10,000)	29
3,4-Methylenedioxy-methamphetamine (10,000)	23
Norpseudoephedrine (10,000)	5
Phenylethylamine (1000)	281
Phenylpropanolamine (10,000)	74

EMIT polyclonal assay

The EMIT polyclonal assay screens for the entire class of amphetamine compounds. Cross-reactivity data are shown in Table 66.5.

The antibody cross-reactivity towards L-methamphetamine has not been reported, but is probably equivalent to that for D-methamphetamine. The L-isomer of methamphetamine is 10 times less potent than the D-form and is used in the USA in common cold decongestants.

Table 66.5. Amphetamine compounds producing a positive result with the Syva EMIT polyclonal assay (cutoff level 300 ng/mL D,L-amphetamine).

Drug	Concentration (ng/mL)
D-amphetamine	300
D,L-amphetamine	300
D-methamphetamine	1000
Methylenedioxy-amphetamine (MDA)	10,000
Methylenedioxymeth-amphetamine (MDMA)	10,000

To reduce the incidence of false positives, a monoclonal assay kit was developed with much less sensitivity towards the L-isomers of amphetamine and methamphetamine. This kit is also more sensitive towards MDA and MDMA (see Table 66.6).

Both assays respond to phenethylamine compounds, some of which are present in proprietary cold cures, but the monoclonal kit is less prone to interference (see Table 66.7).

An amphetamine confirmation kit is available which contains sodium periodate as an oxidant. Compounds with hydroxyl groups proximal to the amino group (e.g. ephedrine, phenylpropanolamine) undergo carbon–carbon bond cleavage at an aliphatic chain or oxidative deamination. Compounds not susceptible to the reaction

Table 66.4. Cross-reactivity of amphetamine-related drugs with the DPC methamphetamine (Cost-a-Count) RIA.

Compound	Concentration added (ng/mL)	Concentration found (ng/mL)	Cross-reactivity (%)
L-methamphetamine	10,000	396	3.9
	100,000	3323	3.3
D,L-methamphetamine	500	381	76
	1000	668	67
3,4-methylenedioxymethamphetamine (MDMA)	500	1446	289
	100	438	438
D-amphetamine	100,000	319	0.3
D,L-amphetamine	100,000	276	0.3
D-pseudoephedrine	100,000	332	0.3
L-ephedrine	100,000	884	0.9
Benzphetamine	100,000	315	0.3
Apomorphine	100,000	700	0.7
Hydroxyamphetamine	100,000	236	0.2
3-Methoxy-4,5-methylenedioxyamphetamine	100,000	426	0.4
Phenylpropanolamine	100,000	93	0.1
Ranitidine	100,000	1215	1.2

Table 66.6. Amphetamine compounds producing a positive result with the Syva EMIT monoclonal assay (cutoff level 1000 ng/mL D-methamphetamine).

Drug	Concentration (ng/mL)
D-amphetamine	≤400
D,L-amphetamine	1000
L-amphetamine	10,000
D-methamphetamine	1000
L-methamphetamine	12,000
Methylenedioxy-amphetamine (MDA)	1000
Methylenedioxymeth-amphetamine (MDMA)	3000

(e.g. isoxsuprine, phentermine) still interfere. For medical–legal purposes, chromatographic confirmation procedures are essential. Other drugs known to cross-react in these assays after therapeutic dosage are listed in Table 66.8.

Cloned enzyme donor immunoassay (CEDIA)™

This technology has been around since 1986 but its commercial development is relatively recent. In the assay, enzyme donor units react with enzyme acceptor units to form a fully active tetrameric molecule of β-galactosidase, which then reacts with a substrate (galactopyranoside) to give a colored product. Competitive protein binding means that active enzyme formation and the amount of product depend on the concentration of the analyte present in the same way as for EMIT. Cross-reactivity data for the CEDIA amphetamine assay are given in Table 66.9.

Table 66.7. Urine concentrations (ng/mL) of amphetamine-like compounds above which positive results may occur.

Drug	Polyclonal assay	Monoclonal assay
Ephedrine	1000	50,000
Fenfluramine	–*	10,000
Mephentermine	500	10,000
Phendimetrazine	–*	100,000
Phenethylamine	–*	10,000
Phenmetrazine	1000	100,000
Phenylephrine	–*	200,000
Phenylpropanolamine	1000	75,000

*No data available.

Table 66.8. Drugs known to produce false-positive results with Syva EMIT amphetamine assays after therapeutic dosage.

Polyclonal assay	Monoclonal assay
Labetalol	Chlorpromazine
N,N-dibenzylethylenediamine	Chloroquine
Phenelzine	*N*-acetyl procainamide
	Procainamide
	Quinacrine
	Ranitidine

Although the CEDIA system detects fewer compounds from proprietary medicines, the number of false positives is still substantial. Producing antibodies which bind solely amphetamine and methamphetamine is probably an unrealistic goal and, as stated previously, wider specificity can be advantageous when related illicit compounds such as the those of the 'Ecstasy' group (MDMA, MDA) are of concern.

The technology is designed for use on high throughput clinical analyzers and various advantages in use are claimed. For example, reconstitution of the dried reagents in a volume of buffer is less critical compared to EMIT. One of the reagents (enzyme donor) is colored red and not likely to be confused with the EMIT noncolored enzyme acceptor reagent (both of the EMIT reagents are colorless).

Fluorescence Polarization Immunoassay

The Abbott TDx is the most widely used FPIA and has equal reactivity towards amphetamine and methamphetamine

Table 66.9. Cross-reactivity of amphetamine-related drugs in the CEDIA amphetamine assay.

Compound	Cross-reactivity (%)
D,L-methamphetamine	67
L-ephedrine	0.5
D,L-amphetamine	52
3,4-Methylenedioxy-amphetamine (MDA)	2.2
3,4-Methylenedioxymethyl-amphetamine (MDMA)	70
Phentermine	1.9
D-phenylpropanolamine	<0.1
D-pseudoephedrine	0.6

Table 66.10. Urine concentrations of amphetamine-like compounds above which positive results occur with the Abbott TDx amphetamine assay based on a D-amphetamine cutoff level of 1000 ng/mL.

Drug	Concentration (ng/mL)
Fenfluramine	10,000
Isometheptene	50,000
Labetalol	100,000
Mephentermine	50,000
Methylenedioxy-amphetamine (MDA)	1000
Methylenedioxymeth-amphetamine (MDMA)	1000
Phenethylamine	100,000
Phentermine	10,000
Propylhexedrine	10,000
Tyramine	100,000

with a detection sensitivity of 300 ng/mL. No significant cross-reactivity occurs with ephedrine or phenylpropanolamine, two drugs present in many OTC preparations, although some other related compounds show significant cross-reactivity (see Table 66.10).

Abuscreen OnLine

In the OnLine series of assays drug-microparticle conjugates react in solution with free antibodies and this causes aggregation, the rate of which can be measured by a change in light absorbance. In the presence of a test analyte, competitive antibody binding brings about a slowing down in the aggregation process that is proportional to the concentration of analyte present. Again, this type of technology is aimed for mass screening using automated clinical analyzers. Cross-reactivity data for the OnLine amphetamine assay are listed in Table 66.11.

Other structurally related substances such as ephedrine, mephentermine, and psuedoephedrine showed hardly any cross-reactivity with the assay, even at concentrations greater than 100,000 ng/mL. No other drugs have been found to have any significant interference.

This amphetamine assay has an extensive dynamic range and good precision at the cutoff, which improves the ability to discriminate between a negative sample and the cutoff and therefore provides a more specific test for amphetamine and methamphetamine. The OnLine assay also contains two monoclonal antibodies for amphetamine and methamphetamine. Methamphetamine is metabolized to amphetamine and a positive urine sample should contain both compounds. At the 1000 ng/mL cutoff the methamphetamine antibody has very low response to methamphetamine, which means that there is little cross-reactivity towards OTC preparations. If amphetamine is present, albeit in low concentrations, the response is enhanced and the assay will report positive any sample containing 200 ng/mL amphetamine and 500 ng/mL methamphetamine. The objective of this design is to reduce the number of samples requiring confirmatory analysis by GC–MS.

Enzyme-Linked Immunoabsorbent Assay

COZART Bioscience has developed a simple ELISA procedure which can be used on urine samples, serum and plasma, saliva, sweat and also on more difficult

Table 66.11. Cross-reactivity of amphetamine-related drugs in the Abuscreen OnLine amphetamine assay.

Compound	Cross-reactivity (%)* 500 ng/mL cutoff	1000 ng/mL cutoff
D,L-amphetamine	62	56
p-hydroxyamphetamine	25	9
L-amphetamine	5	7
D-methamphetamine	98	0.5
D,L-methamphetamine	47	0.2
3,4-methylenedioxyamphetamine (MDA)	35	35
3,4-methylenedioxymethylamphetamine (MDMA)	30	0.2
Phentermine	0.1	<0.2
D-phenylpropanolamine	0.1	<0.2
β-Phenethylamine	1.4	2.3

*Data derived by generating from inhibition curves for each compound and determining for each one the amount equivalent to the 500 and 1000 ng/mL D-amphetamine cutoffs.

Table 66.12. Cross-reactivity of amphetamine-related compounds in the Cozart methamphetamine microplate EIA assay.

Compound	Concentration added (ng/mL)	Concentration found (ng/mL)	Cross-reactivity (%)
D-amphetamine	1000	100	10
	10,000	200	2.0
	100,000	600	0.6
β-Phenylethylamine	10,000	<25	–
	100,000	79	0.08
	250,000	170	0.07
L-phenylalanine	100,000	<25	<0.025
L-ephedrine	10,000	290	2.9
	100,000	>500	–
Pseudoephedrine	10,000	25	2.5
	100,000	>500	–
Phenylpropanolamine	10,000	80	0.8
	100,000	190	0.19
Phentermine	10,000	60	0.6
	100,000	100	0.1
	50,000	311	0.62
MDEA	1000	10	1.0
	5000	50	1.0
	10,000	100	1.0
MDA	10,000	360	3.6
	100,000	>500	–
MDMA	10	125	1250
	25	202	808
	50	379	758
	100	>500	–

matrices such as whole blood and hair extracts. The assay is based on horseradish peroxidase-labeled enzyme and antibody immobilized on the wall of a 96-well microplate. Sample and enzyme conjugate are incubated in the well for 30 min and after washing the plate the substrate (3,3′,5,5′-tetramethylbenzidine) is added. After a further 30 min of incubation, the reaction is stopped by adding sulfuric acid and the absorbance measured at 460 nm within half an hour (Table 66.12).

The specific amphetamine assay has far less cross-reactivity towards drugs commonly used in cold cures, but at the same time is less sensitive than the methamphetamine kit towards MDEA and MDMA. However, the specific amphetamine test may still pick up MDMA use due to the strong affinity of the antibody for the MDA metabolite (Table 66.13).

BARBITURATES

Structure

See Figure 66.5.

Dose and Modes of Administration

A variety of these barbituric acid derivatives, about 12 in all, are used as sedatives, hypnotics, anesthetics, and antiepileptic drugs. Short-acting barbiturates, which have effects lasting up to 3 h, are the most commonly abused and include pentobarbitone and seconal (quinalbarbitone). These are taken mainly by the oral route in doses of up 200 mg with much larger doses taken by tolerant abusers. One of the long-acting barbiturates, phenobarbitone, has on occasions been used as a heroin adulterant.

Pharmacological Effects

Barbiturates act as depressants on the CNS to produce drowsiness and sedation, which is often accompanied by a decrease in mental agility. With increasing dosage the speech becomes slurred and ataxia develops.

Toxic Effects

Overdose leads to dramatic falls in blood pressure and body temperature, depressed respiration, and coma.

Table 66.13. Cross-reactivity of amphetamine-related drugs with the Cozart amphetamine specific microplate EIA.

Compound	Concentration added (ng/mL)	Concentration found (ng/mL)	Cross-reactivity (%)
L-phenylalanine	100,000	<25	<0.025
L-ephedrine	100,000	<25	<0.025
L-methamphetamine	100,000	<25	<0.025
Pseudoephedrine	100,000	<25	<0.025
Phenylpropanolamine	100,000	<25	<0.025
β-Phenethylamine	5000	33	0.66
	10,000	134	1.3
	100,000	442	0.44
Fenfluramine	100,000	<25	<0.025
Phentermine	1000	28	2.8
	10,000	134	1.3
	50,000	311	0.62
	100,000	442	0.44
MDEA	100,000	160	160
MDA	10	21	213
MDMA	100,000	77	0.07

The clinical syndrome resembles opiate poisoning. Continued use of barbiturates leads to marked tolerance and withdrawal can be hazardous; patients have suffered fatal grand mal seizures 2–3 days into withdrawal.

Assay Technology

The short-acting barbiturates are extensively metabolized by the liver to more polar and pharmacologically inactive hydroxylated compounds and only a very small proportion of the parent compound (<0.2%) appears in the 24-h urine. However, with the large doses involved there is usually sufficient of the parent compound present to give an adequate response in the immunoassays and some of the hydroxylated metabolites also cross-react. Secobarbital (quinalbarbitone) is the most commonly used target analyte and calibrator.

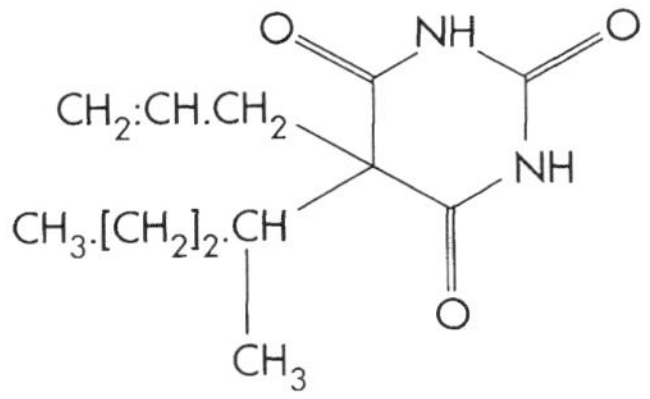

Fig. 66.5. Secobarbital.

BENZODIAZEPINES

Structure (Oxazepam)

See Figure 66.6.

Dose and Modes of Administration

The benzodiazepines are the most widely prescribed sedative/hypnotic drugs and over 20 congeners are marketed. Because of the huge variation in potency, doses range from 1 to 200 mg. Intravenous abuse of liquid encapsulated forms (e.g. temazepam) occurs, but the usual route is oral.

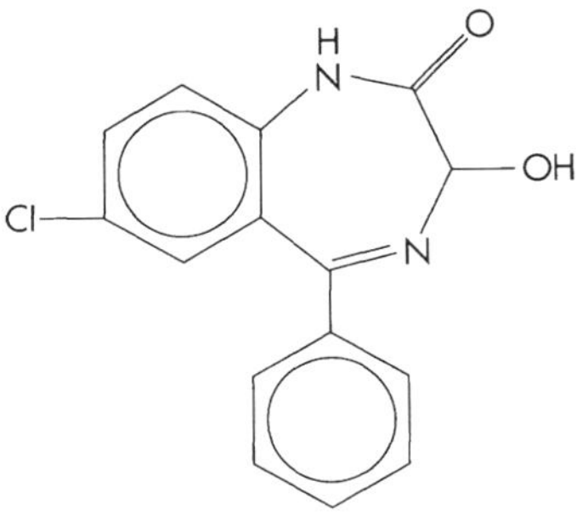

Fig. 66.6 Oxazepam.

Pharmacological Effects

The most prominent effects of benzodiazepines on the CNS are sedation, hypnosis, decreased anxiety, and anticonvulsant activity. There are virtually no effects on the peripheral tissues even in overdose.

Toxic Effects

Chronic abuse leads to blurred vision, confusion, slow reflexes, slurred speech, and hypotension. Benzodiazepines are relatively safe drugs in overdose and deaths are usually the result of concomitant ingestion of ethanol or other drugs.

Assay Technology

Benzodiazepines are extensively metabolized by the liver by processes of *N*-dealkylation and hydroxylation. Only trace amounts of the parent compounds appear in the urine. Hydroxylated metabolites of benzodiazepines and those with a hydroxyl group in the C3 position are conjugated to glucuronic acid; these conjugates account for the major proportion of the dose eliminated in the urine.

The diversity of ring substituents and metabolites is immense and no immunoassay yet devised can claim to cover all members of the group. The compromise has been to raise antibodies towards the most common metabolites encountered (in particular, oxazepam and nordiazepam), and hope that there is sufficient cross-reactivity towards other products to widen the scope of the assay. Where the therapeutic dose is very low, as in the case of the so-called 'date-rape' drug flunitrazepam, immunoassays can give false negatives. Hydrolyzing samples with β-glucuronidase can increase detection rates and some manufacturers incorporate this enzyme in the reagent. For some low-dose benzodiazepines such as alprazolam and triazolam, further improvement can be made by lowering the cutoff level. A comprehensive account of the problems of benzodiazepine detection by immunoassays can be found in Fraser and Meatherall (1996).

Radioimmunoassay

DPC

DPC produces a double-antibody benzodiazepine kit capable of detecting oxazepam in urine at a sensitivity of 0.6 ng/mL. The antibodies presumably have a high affinity towards oxazepam glucuronide, although this metabolite has not been evaluated. Table 66.14 gives cross-reactivity data for a series of benzodiazepines and benzodiazepine metabolites at varying concentrations.

Some of the data in Table 66.14 underestimate the capability of the kit. For example, chlordiazepoxide, clorazepate, and medazepam are extensively metabolized to oxazepam glucuronide and therefore their abuse should be easily detected. It is clear, however, that abuse of more recent benzodiazepine drugs such as flurazepam, lorazepam, and triazolam would not be detected by this assay.

Table 66.14. DPC benzodiazepines RIA; percentage cross-reactivity (on a weight-for-weight basis) relative to oxazepam.

	10,000 ng/mL	1000 ng/mL	500 ng/mL	100 ng/mL	50 ng/mL
Alprazolam	+	+	+	330	354
α-OH alprazolam	+	54	69	87	90
Bromazepam	+	20	20	19	16
Chlordiazepoxide	+	15	13	5	6
Clorazepate	8	11	11	9	8
Clonazepam	1	1	1	4	2
Demoxepam	+	47	60	92	94
Desmethyldiazepam	+	58	44	54	20
Diazepam	+	+	+	39	302
Flurazepam	1	1	2	2	3
Flunitrazepam	+	32	35	39	43
Halazepam	4	9	11	16	14
Lorazepam	1	2	2	4	4
Medazepam	9	6	6	5	4
Nitrazepam	+	32	43	75	96
Prazepam	3	7	7	8	8
Temazepam	+	+	+	420	352
Triazolam	2	2	4	7	6

A plus sign (+) indicates an apparent oxazepam result greater than 1000 ng/mL. All values are expressed in percentage.

Table 66.15. Concentrations of benzodiazepine compounds showing a positive response in the EMIT d.a.u. benzodiazepine assay. Cross-reactivity data relative to oxazepam are listed in Table 66.16.

Compound	Concentration (ng/mL)
Chlordiazepoxide	3000
Clonazepam	2000
Demoxepam	2000
Desalkyflurazepam	2000
N-desmethyldiazepam	2000
Diazepam	2000
Flunitrazepam	2000
Flurazepam	2000
Lorazepam	3000
Nitrazepam	2000
Oxazepam	300

No significant interference from nonbenzodiazepine compounds in the DPC assay has been reported.

Enzyme-Multiplied Immunoassay Technique

The EMIT d.a.u. benzodiazepine assay detects primarily those drugs that have oxazepam glucuronide as a major urinary metabolite. More recent benzodiazepine drugs such as alprazolam and midazolam are also readily detected by this assay. A positive result is based on a response greater than that of a 300 ng/mL oxazepam calibrator (see Table 66.15).

Cross-reactivity data relative to oxazepam are listed in Table 66.16.

The EMIT assay is unlikely to detect the use of flurazepam, flunitrazepam, or triazolam because of poor cross-reactivity and low concentrations of the urinary metabolites.

False-positive results have occurred in samples containing the nonsteroidal anti-inflammatory drug, oxaprozin, but no major interference from other non-benzodiazepine drugs has been reported.

Table 66.16. Relative cross-reactivity of benzodiazepine derivatives to oxazepam in the EMIT d.a.u. benzodiazepine assay.

Compound	Relative cross-reactivity*
Chlordiazepoxide	0.03–0.33
Clidinium bromide	0.07
Clonazepam	0.15
Clorazepate	0.25
Demoxepam	0.15
N-desalkylflurazepam	0.14–1.00
Diazepam	0.15–0.63
Flurazepam	>0.01–0.23
Hydroxyethylflurazepam	>0.1
3-Hydroxydesalkyl-flurazepam	0.50
Lorazepam	>0.01–0.23
Medazepam	0.06
Nitrazepam	0.15–0.35
Norchlordiazepoxide	0.17
Nordiazepam	0.15–1.11
Oxazepam	1.00
Temazepam (3-hydroxydiazepam)	0.45

*Relative cross-reactivity is defined as the test concentration of oxazepam divided by the concentration of cross-reacting compound required to give an equal response.

Table 66.17. Cross-reactivity of benzodiazepines in the CEDIA benzodiazepine assay.

Compound	Cross-reactivity (%)
Nitrazepam	100
Alprazolam	205
Bromazepam	110
Chlordiazepoxide	13
Clobazam	62
Clonazepam	140
Diazepam	247
Flunitrazepam	135
Flurazepam	190
Lorazepam	122
Medazepam	135
Oxazepam	107
Nordiazepam	210
Temazepam	144
Triazolam	191

Cloned Enzyme Donor Immunoassay

The CEDIA assays have greater expected rate differences between the zero and cutoff calibrators than, say, EMIT II. This gives better discrimination between blank urine samples and those with drugs at cutoff concentrations. The high sensitivity (HS) protocol benzodiazepine assay also improves the rate of positive detection by an enhancement of the sensitivity towards benzodiazepine glucuronides. Cross-reactivity data for the CEDIA benzodiazepine assay are listed in Table 66.17. False-positive results have been noted in samples containing metabolites of the antidepressant drug sertraline, but changing the antibodies has now eliminated the problem. In other reports, some interference was noted in samples containing the antihistamine embramine and, more recently, the nonsteroidal anti-inflammatory drug, oxaprozin, has been shown to cause false positives.

Fluorescence Polarization Immunoassay

The TDx kit uses antiserum raised against nordiazepam and has wider cross-reactivity towards benzodiazepines than other commercial products. The set threshold level is 200 ng/mL, i.e. samples giving a response equal to or greater than this cutoff are taken as positive (see Table 66.18).

False-positive results have been given by samples containing the nonsteroidal anti-inflammatory drug, oxaprozin. No other drugs or metabolites, either structurally related or dissimilar to benzodiazepines, are known to give false-positive results.

Abuscreen OnLine

The Abuscreen OnLine benzodiazepine system is very sensitive and is claimed to detect at least 6 ng/mL of nordiazepam. Manufacturers have been criticized in the past for publishing cross-reactivity data only for the parent drugs, but no such accusation can apply with this assay where a long list of benzodiazepines have been evaluated together with, in many cases, their metabolites (see Table 66.19).

Enzyme-Linked Immunoadsorbent Assay (ELISA)

The Cozart benzodiazepines microplate EIA is designed for use on serum or whole blood and can give a semiquantitative result. The calibrators consist of a protein matrix with temazepam at concentrations of 0, 1, 10, and 100 ng/mL. (For a description of the kit *see* AMPHETAMINES). No major interferences have been reported and a list of benzodiazepine cross-reactivity data is given in Table 66.20.

CANNABIS

Structure (Tetrahydrocannabinol)

See Figure 66.7.

Dose and Modes of Administration

The *Cannabis sativa* plant produces a resinous substance containing various cannabinoids. The most potent principle is δ-9-tetrahydrocannabinol (THC). The resin (hashish) contains 3–6% THC and the flowering tops of the female plant (marijuana) 1–3%. Hashish oil is a much more potent form, which

Table 66.18. Cross-reactivity of benzodiazepines and metabolites in the TDx benzodiazepines assay.

Compound	Concentration added (ng/mL)	Concentration found (ng/mL)	Cross-reactivity (%)
Alprazolam	200	233	116.5
Bromazepam	800	222	27.8
Chlordiazepoxide	2400	161	6.7
Clonazepam	1200	283.1	23.6
Demoxepam	2400	301.4	12.6
Desalkylflurazepam	400	226.4	56.6
Diazepam	200	288.1	144.1
Flunitrazepam	400	209.3	52.3
Flurazepam	400	241.4	60.4
1-*N*-hydroxyethylflurazepam	400	335.9	84
Lorazepam	800	253	31.6
Medazepam	400	368.5	92.1
Midazolam hydrochloride	400	295.3	73.8
Nitrazepam	400	263.6	65.9
Norchlordiazepoxide	800	249	31.1
Oxazepam	400	344.6	86.2
Prazepam	200	237.2	118.6
Temazepam	400	406.4	101.6
Triazolam	400	261.9	65.5

Table 66.19. Cross-reactivity of benzodiazepine compounds in the Abuscreen OnLine benzodiazepine assay.

Compound	Concentration equivalent to 100 ng/mL nordiazepam	Cross-reactivity (%)
Alprazolam	112	89
α-Hydroxyprazolam	114	88
4-Hydroxyprazolam	116	86
Bromazepam	135	74
Chlordiazepoxide	172	58
Desmethylchlordiazepoxide	179	56
Clonazepam	167	60
Demoxepam	128	78
Diazepam	118	85
Oxazepam	139	72
N-Methyloxazepam	127	79
Flunitrazepam	182	55
Desmethylflunitrazepam	169	59
3-Hydroxyflunitrazepam	385	26
Flurazepam	164	61
Desalkylflurazepam	175	57
Didesethylflurazepam	125	80
Hydroxyethylflurazepam	123	81
Lorazepam	169	59
Medazepam	345	29
Desmethylmedazepam	286	35
Midazolam	130	29
Nitrazepam	133	75
7-Acetamidonitrazepam	62,500	0.2
7-Aminonitrazepam	189	53
Pinazepam	127	79
Prazepam	139	72
Triazolam	127	79
α-Hydroxytriazolam	115	87
4-Hydroxytriazolam	196	51

No interference from other drugs has been reported.

contains 30–50% THC. Synthetic THC (dronabinol) is marketed by Roxane as an antiemetic under the trade name Marinol. Lilly produces a synthetic cannabinoid (nabilone) with the trade name Cesamet that is also used as an antiemetic. Cannabis is usually smoked in cigarettes or pipes; oral use in cakes and confectionery is also popular. Although rare, intravenous injection of hashish oil has occurred, often with fatal consequences. An effective smoked dose is about 10 mg of THC.

Pharmacological Effects

Crude plant extracts, synthetic THC, and other cannabinoids have been used in the treatment of various medical conditions (glaucoma, asthma, multiple sclerosis) and as antinauseants in cancer chemotherapy. THC has sedative and euphoric properties and distorts the sense of space and time.

Toxic Effects

High and prolonged use has been linked to psychosis. Conjunctivitis is a frequent symptom and deleterious effects on the cardiovascular and respiratory systems can occur. Acute overdose can cause hallucinations, coma, and death.

Assay Technology

Only a small fraction of a dose of THC is excreted in the urine and therefore immunoassays are designed to detect a major inactive oxidation product, 11-nor-δ-9-tetrahydrocannabinol-9-carboxylic acid (11-COOH-THC).

Table 66.20. Cross-reactivity of benzodiazepines in the Cozart microplate benzodiazepine assay.

Compound	Concentration (ng/mL)	Cross-reactivity (%)
Temazepam	1, 10, 100	100
Alprazolam	1	100
	10	40
	100	60
Nordiazepam	1, 10	10
	100	5
	1000	2
Oxazepam	10	5
	100	0.5
	500	1.0
Triazolam	100	2
	1000	0.4
	10,000	0.08
Nitrazepam	100	2
	1000	0.4
	10,000	0.7
Diazepam	1	1
	10	10
	100	100
Flunitrazepam	10, 100	5
	1000	2
Clobazam	10	50
	100	21
	1000	6.5

Radioimmunoassays

Abuscreen

This is a specific and sensitive test that detects 11-COOH-THC at 5 ng/mL. A wide range of drugs has been tested against the assay at urine concentrations of 10,000 ng/mL and no positive values greater or equal to the detection sensitivity of 5 ng/mL occur. Some of the drugs tested are listed in Table 66.21.

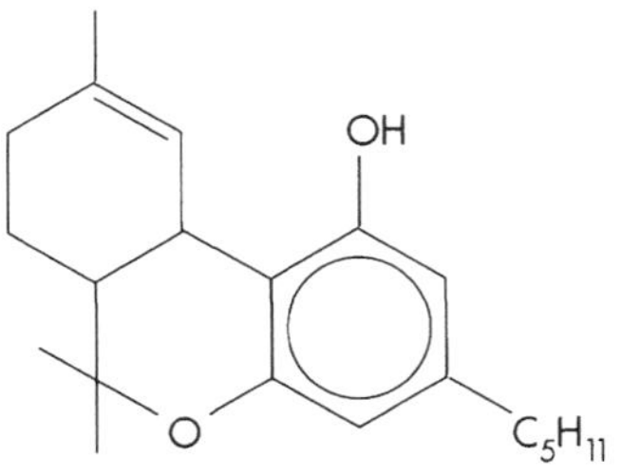

Fig. 66.7 Tetrahydrocannabinol.

Table 66.21. Drugs shown not to interfere with the Roche Abuscreen RIA for cannabinoids at an added concentration of 10,000 ng/mL.

Amitriptyline	Diphenylhydantoin
Amylobarbitone	Fenoprofen
Amphetamine	Hydrochlorothiazide
Ampicillin	Ibuprofen
Aspirin	Imipramine
Atropine	MDMA
Benzocaine	Methadone
Benzoylecgonine	Methamphetamine
Butobarbitone	Methaqualone
Caffeine	Morphine
Chlordiazepoxide	Naproxen
Chloroquine	Oxazepam
Chlorpheniramine	Penicillin G
Chlorpromazine	Pentobarbitone
Cocaine	Phencyclidine
Codeine	Phenobarbitone
Dextromethorphan	Phenothiazine
Dextropropoxyphene	Phenylpropanolamine
Diazepam	

DPC

The DPC radioimmunoassay for cannabinoids uses the double-antibody technique and has a detection limit of approximately 2 ng/mL.

No interference has been found with the drugs listed in Table 66.22.

The cross-reactivity of other cannabinoids in the assay is shown in Table 66.23.

Table 66.22. Drugs not detected in urine by the DPC cannabinoid RIA when present in concentrations of 100,000 ng/mL.

Acetaminophen	Fentanyl
Amphetamine	Furosemide
Aspirin	Ibuprofen
Benzoylecgonine	Lidocaine
Buspirone	Methadone
Caffeine	Methamphetamine
Clonazepam	Methaqualone
Cocaine	Morphine
Codeine	Morphine 3-glucuronide
Cotinine	Nalorphine
Diazepam	Normorphine
Ecgonine	Phenobarbitone
Ethylmorphine	D-proxyphene
Fenfluramine	Secobarbitone

Table 66.23. Cross-reactivity of cannabinoids in the DPC RIA.

Compound	Concentration added (ng/mL)	Concentration found (ng/mL)	Cross-reactivity (%)
11-Nor-δ-9-THC-9-carboxylic acid	100	100	100
Cannabinol	10,000	3.4	0.03
	100,000	24	0.02
Cannabidiol	10,000	5.5	0.06
	100,000	21	0.02
δ-8-THC	100	>250	>100
	1000	>250	>100
δ-8-THC-11-oic acid	10	9.7	97
	100	125	125
	1000	>250	>100

Enzyme-Multiplied Immunoassay Technique

The EMIT d.a.u. is available in three kits with cutoff levels of 20, 50, and 100 ng/mL.

The assay detects the major metabolites of THC in urine (see Tables 66.24 and 66.25).

Cloned Enzyme Donor Immunoassay

The CEDIA d.a.u. multilevel THC system is very similar in performance to EMIT in terms of sensitivity and specificity. Like most nonisotopic assays, it detects about 10% fewer positives than the RIA techniques in comparative trials, but most of these false-negative samples are found to contain very low concentrations of the target metabolite, 11-nor-Δ^9-tetrahydrocannabinol-9-carboxylic acid (11-COOH-THC). Cross-reactivity data are listed in Table 66.26.

Table 66.24. Concentrations of THC metabolites showing a positive response in the EMIT d.a.u. cannabinoid 50 ng assay.

Compound	Concentration (ng/mL)
8-β-11-Dihydroxy-δ-9-THC	1000
8-β-Hydroxy-δ-9-THC	1000
11-Hydroxy-δ-8-THC	1000
11-Hydroxy-δ-9-THC	1000
11-Nor-δ-9-THC-9-carboxylic acid	50

Table 66.25. Compounds shown to give a negative response with the Syva EMIT d.a.u. cannabinoid 50 ng assay.

Compound	Concentration tested (μg/mL)
Acetylsalicylic acid	1000
Amitriptyline	1000
Amphetamine	100
Benzoylecgonine	400
Chlorpromazine	12*
Meperidine	1000
Methaqualone	500
Morphine	200
Oxazepam	300
Phencyclidine	1000
Promethazine	125
Propoxyphene	100
Secobarbitone	1000

*Solubility limit for chlorpromazine under assay conditions.

The assay can also be extended to blood samples by first extracting with acetone, evaporating the extract and then reconstituting the residue in a mixture of methanol and buffer. The same process can be used to adapt the FPIA system described below for blood analyses.

Fluorescence Polarization Immunoassay

The TDx kit for cannabinoids has a stated sensitivity for 11-COOH-THC of 10 ng/mL in urine, although some reports claim a better sensitivity of about 2 ng/mL. The assay detects a range of THC metabolites in urine (see Table 66.27).

Table 66.26. Cross-reactivity of cannabinoids in the CEDIA THC assay.

Compound	Concentration (ng/mL)	Cross-reactivity (%)
11-Nor-Δ^9-THC-COOH	50	100
11-Nor-Δ^8-THC-COOH	40	125
Δ^9-THC	500	10.4
11-OH-11-Δ^9-THC	125	43
8-β-OH-11-Δ^9-THC	1000	2.8
8,11-di-OH-11-Δ^9-THC	500	8.4
1-Δ^9-THC-glucuronide	62	72
Cannabinol	1000	2.9
Cannabidiol	1000	<0.1

Table 66.27. Cross-reactivity of cannabinoids in the Abbott TDx FPIA.

Compound	Concentration added (ng/mL)	Concentration found (ng/mL)	Cross-reactivity (%)
11-Nor-δ-8-THC-9-carboxylic acid	100	109	100
1-Hydroxy-δ-9-THC	200	96.8	48.4
	100	56.7	56.7
	50	31.5	62.9
8-β-Hydroxy-δ-9-THC	200	57.5	28.8
	100	50.6	50.6
	50	40.3	80.6
8-β-Dihydroxy-δ-9-THC	200	58.5	29.3
	100	52.8	52.8
	50	41.8	83.6
Cannabidiol	200	Not detected	
Cannabinol	200	15.8	7.9
	100	Not detected	

Well over 200 drugs and other compounds have been evaluated for interference in the assay and no cross-reactivity within the sensitivity of the assay (10 ng/mL) has been observed.

Abuscreen OnLine

The Abuscreen OnLine THC system has been evaluated on more than one occasion against the more sensitive RIA methods. Initially, the performance was equivalent to RIA, but in a later trial it moved into line with other nonisotopic assays in giving a false-negative rate of around 10% in samples containing cannabinoids. This was put down to a change in the calibrator material from a racemic mixture of the D- and L-isomers of 11-nor-Δ^9-tetrahydrocannabinol-9-carboxylic acid to the naturally occurring L-isomer. Be that as it may, the test is still adequate for screening purposes and the false-negative samples were usually at the low end of the 11-THC-COOH concentration range. A list of cross-reacting cannabinoids is given in Table 66.28.

More than 100 common drug compounds have been tested against the assay at concentrations of 100,000 ng/mL and none gave values in excess of the assay sensitivity level of 5 ng/mL.

Table 66.28. Cross-reactivity of cannabinoids in the Abuscreen OnLine THC assay.

Compound	Equivalent concentration (ng/mL)*	Cross-reactivity (%)
8-α-Hydroxy-Δ^9-THC	227	22
11-Hydroxy-Δ^9-THC	278	18
Δ^9-THC	455	11
8-β-11-Dihydroxy-Δ^9-THC	500	10
11-Hydroxycannabinol	1000	5
Cannabinol	2500	2

*Represents the approximate concentration of each compound equivalent in assay reactivity to a 50 ng/mL 11-Δ^9-THC-COOH assay cutoff.

Enzyme-Linked Immunoadsorbent Assay

Designed for use in whole blood or serum, the Cozart microplate EIA has calibrators containing 0, 2, 10, and 50 ng/mL of 11-nor-Δ^9-THC-COOH made up in a stabilized protein matrix. The cross-reactivity data for cannabinoids are shown in Table 66.29.

COCAINE

Structure

See Figure 66.8.

Dose and Modes of Administration

Cocaine is rapidly hydrolyzed and inactivated in the stomach and is therefore abused intravenously or by nasal insufflation as the hydrochloride salt. Free-base forms (e.g. 'crack') are very potent and give an immediate 'high' when smoked. Doses range from 10 to 120 mg. Heavy abusers can consume up to 4 g daily.

Table 66.29. Cross-reactivity of cannabinoids in the Cozart microplate EIA.

Compound	Concentration added (ng/mL)	Cross-reactivity (%)
Cannabinol	10	42
	100	34.3
Δ^8-THC	5	62
	10	42
	100	9.2
Δ^9-THC	5	114
	10	100
	100	135
Δ^9-THC-glucuronide	5	228
	10	163
	25	174

Table 66.30. Related substances found to cross-react in the Roche Abuscreen cocaine RIA method.

Compound concentration (ng/mL)	Cross-reactivity (counts/min)
Cocaine	
1000	12
10,000	118
100,000	719
Ecgonine HCl	
1000	29
10,000	203
100,000	876
Ecgonine methyl ester HCl	
1000	2
10,000	9
100,000	69

Pharmacological Effects

Cocaine stimulates the CNS, has local anesthetic properties, and also increases blood pressure, heart rate, and body temperature. The euphoric effects show rapid onset but wear off within the hour, leaving anxiety, fatigue, and depression.

Toxic Effects

Cocaine has similar actions to amphetamine, and chronic abuse can lead to psychosis. Myocardial infarction, cardiac dysrhythmias, and cerebrovascular accident can occur with both chronic abuse and overdose.

Assay Technology

Cocaine is rapidly hydrolyzed by blood cholinesterase to ecgonine methyl ester. Spontaneous chemical hydrolysis to benzoylecgonine also occurs and cocaine is therefore unstable in aqueous solution (and urine) at pH values above neutrality.

Immunoassays are targeted at the polar benzoylecgonine breakdown product, which accounts for 30–40% of the dose eliminated in the urine.

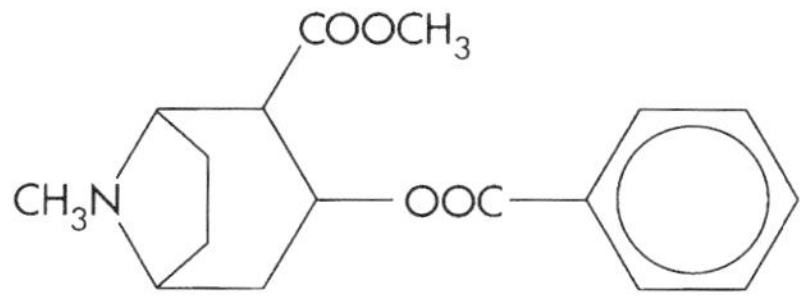

Fig. 66.8 Cocaine.

Radioimmunoassays

Abuscreen

This kit has a detection sensitivity for benzoylecgonine of 5 ng/mL. There is some cross-reactivity with cocaine and its other metabolites (*see* Table 66.30), but no unrelated compounds are known to interfere with the assay.

DPC

This is a solid-phase (Coat-a-Count) RIA which can be used for both qualitative and quantitative measurement of benzoylecgonine in urine with a sensitivity of 3 ng/mL (see Tables 66.31 and 66.32).

Table 66.31. Drugs shown not to be detected by the DPC (Coat-a-Count) cocaine metabolite RIA.

Acetaminophen	Lidocaine
Acetylsalicylic acid	Mepivacaine
D,L-amphetamine	Methadone
Atropine (D-hyoscyamine)	Methaqualone
Benzocaine	Morphine
Caffeine	Phenazocine
Codeine	Phencyclidine
Cotinine (nicotine metabolite)	Phenobarbitone
Dextropropoxyphene	Secobarbitone
L-hyoscyamine HCl	

Table 66.32. Cross-reactivity of various compounds in the DPC cocaine metabolite RIA.

Compound	Concentration added (ng/mL)	Concentration found (ng/mL)	Cross-reactivity (%)
Tetracaine HCl	10,000	ND	ND
	100,000	24	0.02
D,L-homotropine HBr	10,000	ND	ND
	100,000	43	0.04
Procainamide	10,000	6	0.06
Ecgonine (cocaine metabolite)	100,000	70	0.07
Procaine HCl	10,000	20	0.20
Dibucaine	10,000	40	0.40
	100,000	472	0.47
Tropacocaine	10,000	2763	28
Ecgonine methyl ester	1000	486	49
Benzoylecgonine	900	900	100
Cocaine	125	5400	>100

ND, not determined.

Enzyme-Multiplied Immunoassay Technique

The EMIT d.a.u. cocaine metabolite kit is the most frequently used testing method and has a benzoylecgonine detection sensitivity of 300 ng/mL. The assay also detects cocaine and ecgonine at levels greater than 25,000 and 5000 ng/mL, respectively.

Concentrations of compounds showing a negative response in the assay are shown in Table 66.33.

No substances producing false-positive results with the assay have yet been reported.

Table 66.33. Concentrations of compounds with a negative response to the Syva EMIT d.a.u. cocaine metabolite assay.

Compound	Concentration tested (μg/mL)
Acetaminophen	1000
Acetylsalicylic acid	1000
Amitriptyline	100
Amphetamine	500
Chlorpromazine	12*
Cocaine	25
Codeine	500
Dextromethorphan	175
Ecgonine	50
Methadone	500
Methaqualone	100
Monoethylglycinexylidide	1000
Morphine	200
Oxazepam	250
p-Aminobenzoic acid	1000
Phencyclidine	750
Procainamide	1000
Propoxyphene	500
Secobarbitone	1000

*Solubility limit for chlorpromazine under assay conditions.

Cloned Enzyme Donor Immunoassay

This assay compares very favorably with both RIA and other nonisotopic methods such as EMIT in terms of reliability, sensitivity, and specificity. Like the other systems, the target analyte is benzoylecgonine. Cross-reactivity characteristics are listed in Table 66.34.

Like most other immunoassays for cocaine detection, the specificity is extremely high and no interfering compounds have been reported.

Table 66.34. Cross-reactivity of cocaine and metabolites in the CEDIA cocaine assay.

Compound	Concentration (ng/mL)	Cross-reactivity (%)
Benzoylecgonine	300	100
Cocaethylene	312	57
Cocaine	312	54
Ecgonine	10,000	1.1
Ecgonine-methyl ester	10,000	<0.1

Table 66.35. Cross-reactivity of cocaine and metabolites in the Abbott TDx FPIA cocaine metabolite assay.

Compound	Concentration (ng/mL)	Cross-reactivity (%)
Cocaine	100	0.79
	10	0.4
	1	None
Ecgonine	10	0.3
	1	None
Ecgonine methyl ester	100	0.03
	10	None

Fluorescence Polarization Immunoassay

The TDx FPIA cocaine metabolite assay has a benzoylecgonine detection limit in urine of 300 ng/mL. Cocaine and its other common metabolites cross-react as shown in Table 66.35.

No compounds have been found as yet to give rise to false-positive results in this assay.

Abuscreen OnLine

This system performs as well as any other immunoassay for cocaine detection and has very high specificity for the benzoylecgonine metabolite (Table 66.36).

Enzyme-Linked Immunoadsorbent Assay

The Cozart microplate cocaine kit is designed for use in diluted serum or whole blood to detect the major cocaine metabolite benzoylecgonine. It also shows good cross-reactivity towards cocaine itself and reasonable cross-reactivity for the metabolic product formed by the simultaneous abuse of cocaine and alcohol, cocaethylene (Table 66.37).

Table 66.36. Cross-reactivity of cocaine and metabolites in the Abuscreen OnLine assay.

Compound	Concentration*	Cross-reactivity (%)
Ecgonine	25,000	1.2
Cocaine	30,928	0.97
Ecgonine methyl ester	96,774	0.31

*Represents the approximate concentration of each compound equivalent in assay reactivity to a 300 ng/mL benzoylecgonine assay cutoff.

Table 66.37. Cross-reactivity data of cocaine and metabolites in the Cozart microplate cocaine metabolite assay.

Compound	Concentration added (ng/mL)	Cross-reactivity (%)
Cocaine	10	110
	100	101
	1000	>30
Cocaethylene	25	64
	50	56
	100	41
	500	24

FENTANYL

Structure

See Figure 66.9.

Dose and Mode of Administration

Fentanyl and its congeners (alfentanil, sufentanil) are synthetic opioids related in structure to pethidine. They are given by parenteral injection in single doses of 25–100 μg, usually as the citrate salt. Methylfentanyl (α-methylfentanyl; 'China White') and the much more potent 3-methylfentanyl are abused in various parts of the USA and these and several other derivatives are classified as 'designer drugs'. They have been smoked and 'snorted' as well as abused intravenously. A typical intravenous dose of methylfentanyl is 5–10 μg.

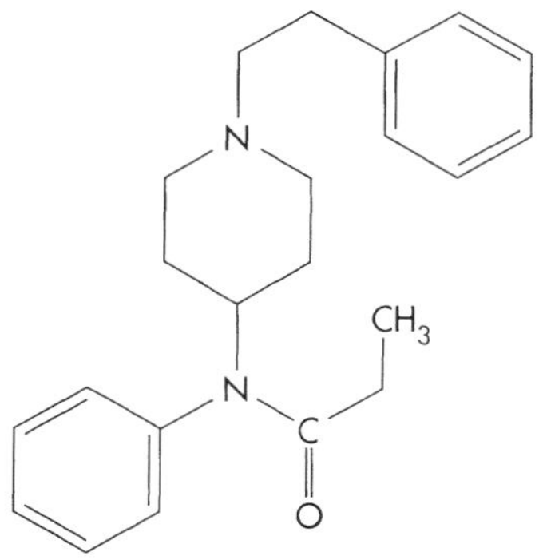

Fig. 66.9 Fentanyl.

Pharmacological Effects

Fentanyl is used mainly to produce surgical anesthesia and is short acting after single doses, although the elimination half-life is prolonged due to rapid distribution into the tissues. When high doses are given to maintain cardiac stability, assisted ventilation is needed to combat respiratory depression.

Toxic Effects

Fentanyl and its 'designer drug' derivatives are highly potent and death from respiratory depression can occur within minutes after intravenous use.

Assay Technology

A substantial proportion of an injected fentanyl dose is metabolized by the liver and very little of the parent drug appears in the urine. The most abundant urinary metabolite is norfentanyl together with smaller amounts of despropionylfentanyl. Due to the extremely low concentrations of fentanyl and its metabolites, radioimmunoassay has for long been the preferred technique for their detection in either blood or urine, although ELISA technology is now available and may well gain ground if the incidence of abuse of these drugs increases.

Radioimmunoassay-DPC Kit

The DPC Coat-a-Count kit can detect as little as 0.1 ng/mL of fentanyl and for qualitative purposes a threshold (cutoff) concentration of 0.5 ng/mL is used. High specificity for fentanyl and its metabolites is claimed (Table 66.38) and the cross-reactivity towards similar drugs (alfentanil, carfentanil, lofentanil, sufentanil) is surprisingly low with no reaction seen with any of these at 50 ng/mL.

Table 66.38. Cross-reactivity data for structurally related substances in the DPC Coat-a-Count fentanyl radioimmunoassay.

Compound	Concentration (ng/mL)	Cross-reactivity (%)
p-Fluorofentanyl	50	28
	25	28
	5	32
Thienylfentanyl	50	16
	25	19
	5	26
3-Methylfentanyl	50	15
	25	15
	5	22
trans-3-Methylfentanyl	10	58
	5	32
cis-3-Methylfentanyl	10	27
	5	4.8
α-Methylfentanyl	50	5
	25	6
	5	12
α-Methylthiofentanyl	100	9.7
	10	6.5
	5	7.3
2-Hydroxyfentanyl	100	8.4
	10	6.6
	5	8.1
Norfentanyl	100	7.6
	10	16.1
	5	1.0
Benzfentanyl	10,000	0.39

LYSERGIC ACID DIETHYLAMIDE

Structure

See Figure 66.10.

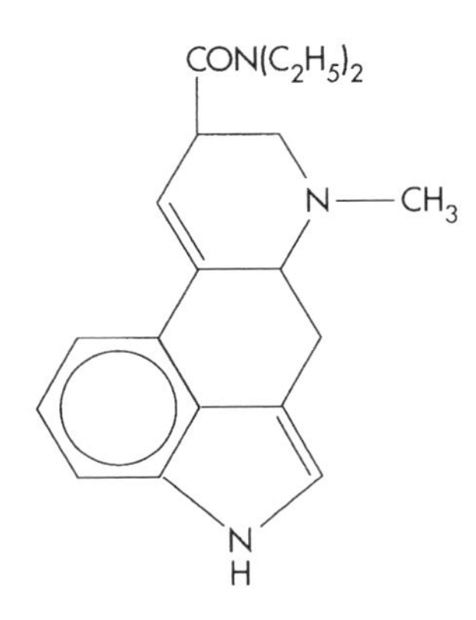

Fig. 66.10 Lysergic acid diethylamide.

Dose and Modes of Administration

Lysergic acid diethylamide (LSD) is taken orally as the tartrate salt. Doses ranging from 100 to 200 μg were common for many years, but recently the tendency has been to take less (30–50 μg).

Pharmacological Effects

The D-isomer of LSD is one of the most potent hallucinogenic agents known (the L-isomer is inactive). This drug belongs to the serotonin class of psychedelics, which disrupts the function of the brain 5-hydroxytryptamine systems. LSD is structurally related to naturally occurring ergot alkaloids. Disorientation, euphoria, and hallucination are the prevalent features that occur after a 'trip' dose and these subside after 12 h.

Toxic Effects

Recurrent symptoms without further dosage for several months ('flashback') have been reported. Overdose can cause delirium and coma, but no deaths have been reported.

Assay Technology

LSD is extensively metabolized by *N*-demethylation, *N*-deethylation, and hydroxylation to inactive metabolites, and only trace amounts of unchanged drug are excreted in the urine. This, together with the minute dosage, makes detection in urine difficult, and for a long time this was feasible only by the most sensitive method, radioimmunoassay. Over the last 3 years nonisotopic methods have emerged and are gradually gaining favor. In a number of comparative trials, including some where the LSD samples were urines collected from primates dosed with LSD, there have been significant discrepancies between isotopic and nonisotopic immunoassays, and GC–MS. Usually, the immunoassays have given a higher positive rate than GC–MS and this could well be due to cross-reactivity with LSD metabolites, which persist in urine for much longer than unchanged LSD. This may prove advantageous in future if the antibodies can be used to isolate these metabolites and facilitate their identification.

Table 66.39. Ergot derivatives not detectable by the DPC Coat-a-Count LSD assay at added concentrations of 100,000 ng/mL.

5-Hydroxytryptamine
Dihydroergotamine
Ergokryptine
Ergotamine
Lysergic acid

Radioimmunoassay

The DPC LSD assay (Coat-a-Count) is a solid-phase method which has a detection limit for LSD in urine of 20 pg/mL and can be used quantitatively. For qualitative purposes a threshold concentration of 500 pg/mL has been selected.

The antiserum used is highly specific for LSD and probably its metabolites, although these have not been characterized separately. Cross-reactivity toward the pharmacologically inactive L-isomer has not been reported.

The ergot derivatives in Table 66.39 were not detectable by the assay. Cross-reactivity data for other substances are given in Table 66.40.

Ergot alkaloids and their derivatives are used clinically, mainly in treating postpartum hemorrhage and in the relief of migraine. With a test of this sensitivity, use of medications containing ergonovine, dihydroergokryptine, or methysergide might yield false-positive reactions.

Table 66.40. Cross-reactivity of various substances in the DPC Coat-a-Count LSD assay.

Compound	Concentration added (ng/mL)	Concentration found (ng/mL)	Cross-reactivity (%)
Lysergic acid	1000	15,428	1.5
Methylpropylamide	100	5627	5.6
Lysergic acid	100,000	1903	0.002
Hydroxyethylamide	10,000	328	0.003
Lysergol	100,000	163	<0.001
Ergonovine	100,000	1549	0.002
	10,000	248	0.002
Dihydroergokryptine	100,000	106	<0.001
Methysergide	100,000	1607	0.002
	10,000	329	0.003
Methysergide maleate	100,000	1686	0.002
	10,000	199	<0.001
Ecgonine	100,000	433	<0.001
Ecgonine methyl ester	100,000	356	<0.001

Table 66.41. Concentrations of substances which gave a response equivalent to 0.5 ng/mL in the EMIT LSD assay.

Compound	Concentration (μg/mL)
D-amphetamine	500
Ergonovine	1
Methadone	400
D-methamphetamine	100
Methylsergide	3
Phencyclidine	30
Propoxyphene	1000

Heavy abuse of cocaine with consequent excretion of large concentrations of ecgonine and ecgonine methyl ester might also cause false positives.

No significant cross-reactivity has been reported with any other substances.

Enzyme-Multiplied Immunoassay Technique

This assay is based on monoclonal antibodies raised against lysergic acid diethylamide and has a recommended cutoff level of 0.5 ng/mL. Calibrators containing 0, 0.5, 1.5, and 2.5 ng/mL are supplied, and low and high controls are available. Cross-reactivities for some other abused drugs are given in Table 66.41.

These are very high urine concentrations and are unlikely to be found in real samples.

The manufacturers warn that samples from patients taking chlorpromazine may yield false-positive results with this assay. Some workers have reported a false-positive rate of more than 10% in samples from psychiatric patients receiving antipsychotic, antidepressant, and anxiolytic drugs and a much lower and acceptable rate in the general hospital population of about 0.5%. This illustrates the difficulties of evaluating the specificity of an assay which may be applied to samples containing a huge range of other pharmaceuticals and their metabolites, especially when the concentration of analyte sought is so low.

Cloned Enzyme Donor Immunoassay

This assay has a range of 0–4 ng/mL at a cutoff of 0.5 ng/mL in urine and is said to have an intra-assay precision of around 15% and a specificity of 99.9%. Cross-reactivity data for the assay towards other ergotamine-type compounds are given in Table 66.42. There is a report of false-positive results in samples from patients on the mucolytic drug ambroxol, even though there is no similarity between this and LSD in chemical structure.

Table 66.42. Cross-reactivity of compounds structurally related to LSD in the CEDIA LSD assay.

Compound	Concentration (ng/mL)	Cross-reactivity (%)
Dihydroergotamine	125,000	<0.001
Alpha ergotamine	500,000	<0.001
Lysergic acid	100,000	<0.001
Lysergol	50,000	<0.0001
Ecgonine	100,000	<0.001
Serotonin	1,000,000	<0.001
Ecgonine methyl ester	100,000	<0.001
Psilcybin	10,000	<0.001
Psylocyn	10,000	<0.001

Like several other immunoassays for LSD, cross-reactivity towards metabolites is very likely, but this has not yet been fully categorized.

Abuscreen OnLine

This assay performs slightly better than some of the other nonisotopic LSD immunoassays and has good precision around the cutoff concentration of 0.5 ng/mL. The antibody is raised against LSD itself, but there is clear evidence of binding to LSD metabolites such as nor-LSD, which might explain why it can perform better than GC–MS in controlled trials (Table 66.43).

Enzyme-Linked Immunoadsorbent Assay

The Cozart microplate assay is designed to detect LSD in urine samples. Precision testing at the 0.5 ng/mL cutoff gave a coefficient of variation of 6%. The assay stands up well against RIA methods and may have slightly enhanced sensitivity. There is significant cross-reaction with nor-LSD (16–28% depending on the concentration), which is advantageous in a screening assay (Table 66.44). The assay has also been adapted for use in serum or whole blood samples.

No significant interference was found for a whole range of substances similar in structure to LSD (Table 66.45). When these were tested at concentrations of 10,000 ng/mL, the apparent LSD concentration was less than 0.5 ng/mL in all cases (% cross-reactivity <0.05).

METHADONE

Structure

See Figure 66.11.

Table 66.43. Cross-reactivity of LSD metabolites and compounds structurally related to LSD in the Abuscreen OnLine assay.

Compound	Concentration (ng/mL) equivalent to 0.5 ng/mL LSD	Cross-reactivity (%)
2-Bromo-α-ergocryptine	>20,000	<0.0025
iso-LSD	21	2.2
Lysergic acid *N*-(methylpropyl) amide	3.5	24
Lysergic acid *N*-(hydroxyethyl) amide	1560	0.034
D-lysergic acid	>20,000	0.0025
N-demethyl-LSD (nor-LSD)	1.4	44
Methylsergide maleate	5500	<0.024
α-Ergocryptine	>20,000	<0.0025
Ergotamine tartrate	13,300	0.0039
Ergonovine maleate	13,000	0.0063

Dose and Route of Administration

Methadone is a favored heroin substitute in opiate withdrawal programs and tolerant patients are given the drug orally as a linctus in doses of 40–100 mg daily. It is also available as a racemic mixture of the hydrochloride salt, as tablets of 5–10 mg and as a 10 mg/L solution for parenteral injection.

Pharmacological Effects

Methadone has analgesic properties similar to morphine, but has marked sedative effects due to drug accumulation. Although given as D,L-methadone, only the L-isomer is active.

Toxic Effects

Methadone overdose causes stupor, respiratory depression, hypotension, coma, and circulatory collapse. Doses of 50 mg can prove fatal in nontolerant adults.

Table 66.44. Cross-reactivity of nor-LSD in the Cozart microplate LSD (urine) assay.

Compound	Concentration (ng/mL)	Cross-reactivity (%)
Nor-LSD	1	25
	2.5	28
	5	20
	10	16
	25	>20

Assay Technology

Methadone is metabolized by mono- and di-*N*-demethylation to unstable metabolites that cyclize spontaneously to give 2-ethylidene-1,5-dimethyl-3,3-diphenylpyrrolidine (EDDP) and 2-ethyl-5-methyl-3,3-diphenylpyrroline (EMDP). These, together with methadone, are the main urinary excretion products. The commercial immunoassays detect methadone and cross-react poorly or not at all with EDDP or EMDP. This can be a disadvantage when testing urine samples taken a long time after the last dose, since the metabolites may still be present when methadone itself is below the detection limit. Since methadone is prescribed on a huge scale, there is a danger of supplying the street market if drug abusers obtain doses under false pretences by spiking their urine samples with the drug. Most immunoassays fail to distinguish these as adulterated samples. This has stimulated the development of assays that are aimed at picking up the metabolites and have very low cross-reactivity towards methadone itself. It is likely that these will be used more widely in future years.

Table 66.45. Compounds similar in structure to LSD which have insignificant cross-reactivity in the Cozart microplate LSD (urine) assay.

Dihydroergocristine	Ergosine (base)
Dihydroergotamine	Ergosinine (base)
Ergocornine	Ergotamine tartrate
Ergocryptine	Lysergic acid
Ergocristine	Serotonin
Ergometrinine (base)	L-trytophan

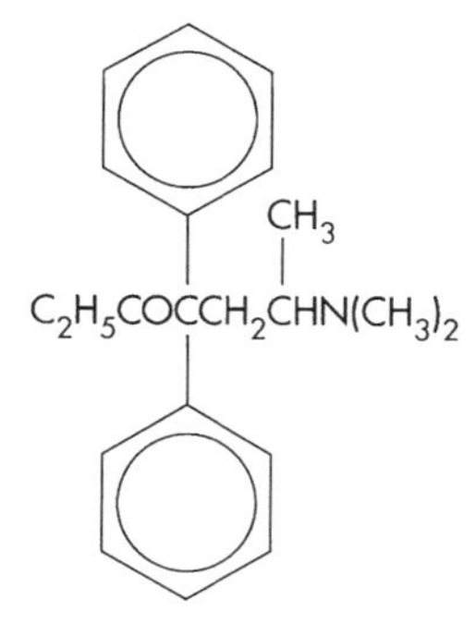

Fig. 66.11 Methadone.

Radioimmunoassay

DPC

The DPC Coat-a-Count methadone assay is a solid-phase RIA with a detection limit in urine of around 0.2 ng/mL.

The antiserum is highly specific for methadone and no significant cross-reactivity with structurally related or dissimilar drugs has been observed (see Table 66.46).

Many other compounds have been tested for interference at added concentrations of 100 μg/mL and no response higher than the assay sensitivity (100 ng/mL) has been observed. A positive reaction to L-α-acetylmethadol (LAAM), a long-acting methadone derivative, may well occur following high dosage.

Enzyme-Multiplied Immunoassay Technique

The EMIT d.a.u. methadone assay is set to detect methadone in urine at a threshold or 'cutoff' concentration of 300 ng/mL. High concentrations of the antihistamine drugs, doxylamine and diphenhydramine, are also detected, but no interference from other compounds has been reported (see Table 66.47).

Table 66.47. Concentrations of compounds showing a negative response in the EMIT d.a.u. methadone assay.

Compound	Concentration tested (ng/mL)
Acetaminophen	1000
Acetylsalicylic acid	1000
Amitriptyline	50
Amphetamine	500
Benzodiazepine	400
Chlorpromazine	12*
Codeine	500
Dextromethorphan	300
Diphenhydramine	100
Meperidine	200
Methaqualone	100
Morphine	200
Naloxone	500
Oxazepam	250
Phencyclidine	500
Promethazine	75
Propoxyphene	300
Secobarbitone	100

*Solubility limit for chlorpromazine under assay conditions.

Cloned Enzyme Donor Immunoassay

The CEDIA methadone assay recommends a cutoff of 150 ng/mL. Again there is poor cross-reactivity with the methadone metabolites (Table 66.48).

Table 66.46. Drugs shown not to be detected by the DPC methadone assay.

Compound	Concentration added (ng/mL)	Concentration found (ng/mL)	Cross-reactivity (%)
Disopyramide	10,000	0	0
	100,000	3	0.003
Dihydrocodeine	100,000	3	0.003
EDDP*	100,000	5	0.005
Meperidine	100,000	8	0.008
Nortriptyline	100,000	8	0.008
Doxylamine	100,000	10	0.01
Diphenhydramine	10,000	20	0.02
D-propoxyphene	100,000	19	0.02
Imipramine	100,000	16	0.02
Chlorpromazine	100,000	34	0.03

*2-Ethylidene-1,5-dimethyl-3,3-diphenylpyrrolidine.

Table 66.48. Cross-reactivity of methadone-related substances in the CEDIA methadone assay.

Compound	Concentration (ng/mL)	Cross-reactivity (%)
Methadone	300	100
α-Methadol	33,333	2.65
EDDP	500,000	0.02
EMDP	100,000	0.03
LAAM	20,000	1.48
Methadol	25,000	1.50
Morphine-3-glucuronide	100,000	0.01
Norpropoxyphene	500,000	0.03
Propoxyphene	500,000	0.03

Fluorescence Polarization Immunoassay

The TDx methadone assay is set to a threshold concentration of 250 ng/mL. The sensitivity of the assay is 100 ng/mL. Cross-reactivity towards the major urinary methadone metabolites (EDDP and EMDP) is negligible, but a positive response to L-α-acetylmethadol (LAAM), a long-acting methadone derivative, could occur after high dosage (see Table 66.49).

Table 66.49. Cross-reactivity of metabolites of methadone and L-α-acetylmethadol in the TDx methadone assay.

Compound	Concentration added (ng/mL)	Concentration found (ng/mL)	Cross-reactivity (%)
EDDP	100,000	0	0
EDMP	100,000	0	0
L-α-acetyl-methadol (LAAM)	1000	210	21.0
Nor-LAAM	4000	110	2.8
Dinor-LAAM	4000	0	0
L-α-methadol	4000	810	20.3
	1000	370	37.0
	500	230	46.0

Abuscreen OnLine

This assay uses a cutoff value for methadone in urine of 300 ng/mL and the polyclonal antibodies are raised against the parent drug. Table 66.50 gives the cross-reactivity data for related compounds and other common drugs. Like most methadone immunoassays, there is the disadvantage of very little reaction with the methadone metabolites EDDP and EMDP. Detection of the methadone alternative, L-α-acetylmethadol (LAAM), is possible at high doses.

Table 66.50. Cross-reactivity of various substances in the Abuscreen OnLine methadone assay.

Compound	Concentration* (ng/mL)	Cross-reactivity (%)
Methadol	250	120
Hydroxymethadone	577	52
L-α-acetylmethadol (LAAM)	1000	30
Promethazine	12,000	2.5
Diphenhydramine	60,000	0.50
Amitriptyline	91,000	0.33
Chlorpheniramine	91,000	0.33
Doxylamine	100,000	0.30
Imipramine	100,000	0.30
D-propoxyphene	107,000	0.28
Benzphetamine	130,000	0.23
Meperidine (pethidine)	136,000	0.22
Dextromethorphan	150,000	0.20
Chlorpromazine	250,000	0.12
EDDP	273,000	0.11
EMDP	333,000	0.09

*Represents the concentration needed to give a reaction equivalent to the methadone 300 ng/mL cutoff.

Enzyme-Linked Immunoadsorbent Assay

The Cozart microplate assay for diluted serum or whole blood has methadone calibrators of 5, 25, and 100 ng/mL. Methadone toxicity can develop at blood concentration of about 0.2 mg/L (200 ng/mL) which leaves plenty of scope for dilution of forensic samples into the analytical range. There is significant cross-reactivity with the methadone substitute L-α-acetylmethadol, but in a forensic case this would be discovered by additional and more discriminating tests. Cross-reactivity data are presented in Table 66.51.

OPIATES

Structure

See Figures 66.12–66.14.

Table 66.51. Cross-reactivity of substances related to methadone in the Cozart microplate methadone assay.

Compound	Concentration added (ng/mL)	Concentration found (ng/mL)	Cross-reactivity (%)
L-α acetyl-methadol (LAAM)	5	3.3	67
	10	10.1	101
	25	31.7	127
	1000	191.6	19.1
EDDP	100	0.7	0.69
	10,000	2.6	0.03
	100,000	9.2	0.01
EMDP	100	0.6	0.55
	10,000	2.0	0.02
	100,000	13.0	0.01

Dose and Modes of Administration

The opium poppy contains two narcotic analgesics, namely, morphine and codeine. Synthetic derivatives such as diacetyl morphine (heroin), hydrocodone, and dihydrocodeine are more properly classified as opioids, but their detection is also considered in this section.

Heroin is the most commonly abused derivative and tolerant addicts take up to 200 mg daily either by intravenous injection or nasal insufflation ('snorting'). Inhalation of vaporized heroin, colloquially known as 'chasing the dragon', is also practiced.

Pharmacological Effects

These drugs are potent depressants of the CNS, causing analgesia, euphoria, and narcosis. Regular use leads to tolerance and physical dependence.

Toxic Effects

Opiate poisoning is characterized by pinpoint pupils, respiratory depression, and deep coma. Following overdosage by intravenous injection, death can occur within a few minutes.

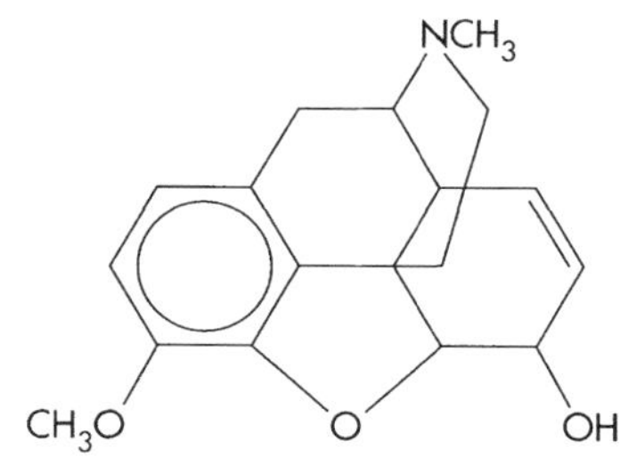

Fig. 66.13 Codeine.

Assay Technology

Heroin is rapidly metabolized to morphine, which is excreted in the urine mainly as glucuronide conjugates. The immunoassays are therefore directed towards morphine, which is also a metabolite of codeine. Other phenanthrene narcotics such as codeine itself, dihydrocodeine, hydromorphine and levorphanol and their metabolites also cross-react to varying degrees.

Pholcodine (β-morpholinylethylmorphine) is widely used in cough remedies in Europe and gives a strong positive response to some immunoassays. Foods containing poppy seeds, which are a source of morphine and codeine, can also yield positive urine samples by immunoassay.

It is important, therefore, to reanalyze urine samples that are positive by specific chromatographic methods to discern which opioid congeners are present.

Fully synthetic opioids such as meperidine (pethidine), propoxyphene, and methadone show little or no cross-reactivity in these immunoassays.

Radioimmunoassays

Abuscreen

The kit has a detection sensitivity for morphine of 10 ng/mL. There is strong cross-reactivity towards codeine (see Table 66.52).

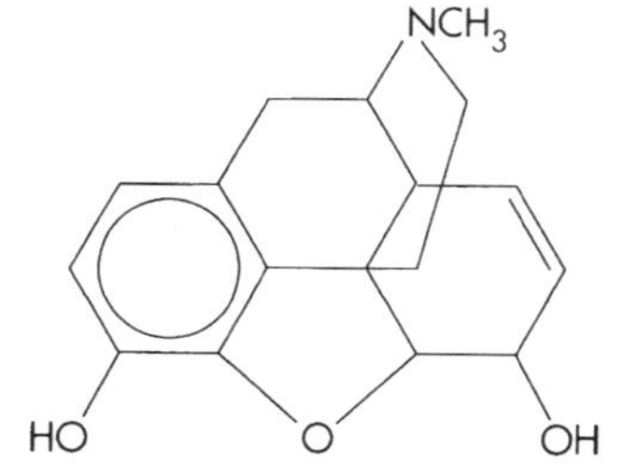

Fig. 66.12 Morphine.

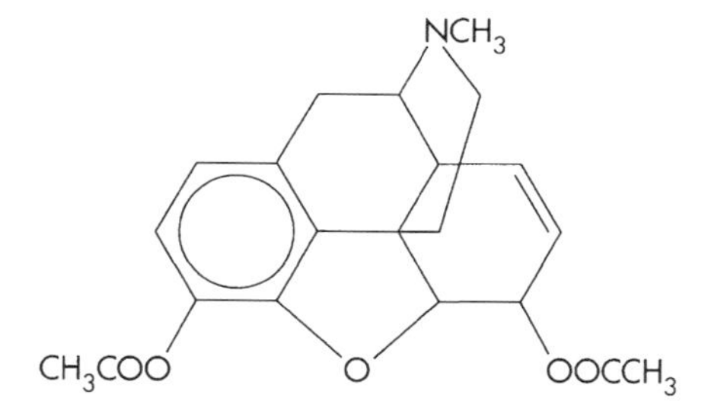

Fig. 66.14 Heroin.

Table 66.52. Cross-reactivity of opiate drugs in the Roche Abuscreen opiate RIA.

Compound	Concentration (ng/mL)	Cross-reactivity (%)
Codeine	222	135
Dihydrocodeine	1007	30
Ethylmorphine	219	137
Hydrocodone	1634	18
Morphine 3-glucoronide	784	38
Morphine	300	100

The compounds in Table 66.53 cross-react at significantly higher concentrations.

DPC

DPC markets three solid-phase RIA (Coat-a-Count) kits: an opiate screen with broad specificity and designed only for veterinary use, and two much more specific tests for morphine in serum (quantitative) and in urine.

The serum morphine kit is equipped with five calibrators with morphine concentrations ranging from 2.5 to 250 ng/mL, and can detect 0.8 ng/mL. The urine morphine kit has calibrators containing from 2.5 to 500 ng/mL and can detect as little as 0.3 ng/mL. The antiserum used in both kits is highly specific for free (unconjugated) morphine with low cross-reactivity towards morphine-3-glucuronide and codeine (see Table 66.54).

Enzyme-Multiplied Immunoassay Technique

The EMIT d.a.u. kit detects morphine, morphine-3-glucuronide, and codeine in urine. Related synthetic opiates such as dihydrocodeine, levorphanol, and pholcodine are also detected. High concentrations of meperidine (pethidine) and the narcotic analgesic, nalorphine, can give a positive response, but no significant interference from other compounds structurally unrelated to morphine has been reported. The kit is supplied with low and medium calibrators containing 300 and 1000 ng/mL of morphine, respectively. A positive result is based on a response equal to or greater than that of the low calibrator (see Table 66.55).

The compounds listed in Table 66.56 gave negative results.

Table 66.53. Apparent morphine concentrations (ng/mL) of opioid drugs in the Roche Abuscreen opiate RIA.

Compound	Concentration 1000 ng/mL	10,000 ng/mL
Dihydromorphine	160	903
Hydromorphine	144	720
Meperidine (pethidine)	8	18
6-Monoacetyl-morphine	145	1000
Oxycodone	15	46
n-Norcodeine	29	233
Thebaine	44	190

Table 66.54. Cross-reactivity of opiate compounds in the DPC (Coat-a-Count) urine morphine assay.

Compound	Concentration added (ng/mL)	Concentration found (ng/mL)	Cross-reactivity (%)
Nalorphine	500	135	27
	100	28	28
	50	13	26
	10	2.4	24
Normorphine	1000	91	9.1
	500	48	9.6
	100	9	9.0
Morphine-3-glucuronide	10,000	3.1	0.03
	5000	1.9	0.04
Codeine	10,000	6.1	0.06
	5000	2.8	0.06
Hydromorphine	10,000	75.1	0.75
	5000	45.2	0.90
	1000	11.6	1.16
	500	5.9	1.19
	100	1.5	1.50
Dihydrocodeine	10,000	3.9	0.04
	5000	2.4	0.05
Morphine-6-glucoronide	10,000	15.3	0.15
	5000	8.5	0.17

Cloned Enzyme Donor Immunoassay

The CEDIA d.a.u. opiate kit has more sensitivity than some of the other nonisotopic assays for the target drugs morphine and codeine due to its greater rate separation (milliabsorbance units change per minute) between the negative and the 300 ng/mL cutoff calibrator. It is also more selective towards morphine and codeine, but like most other immunoassays for this class of drugs, there is considerable cross-reactivity with other opiate drugs and their metabolites, the exceptions being oxycodone and its metabolite, oxymorphone (Table 66.57).

Table 66.55. Concentration of compounds showing a positive response in the Syva EMIT d.a.u. opiate assay.

Compound	Concentration (ng/mL)
Codeine	1000
Dihydrocodeine	260
Hydrocodone	1000
Hydromorphine	3000
Levallorphan	1000
Levorphanol	300
Monoacetylmorphine	460
Morphine	300
Morphine-3-glucuronide	3000
Oxycodone	50,000
Norlevorphanol	23,000
Oxymorphone	82,000

Fluorescence Polarization Immunoassay

The TDx assay has a test sensitivity for morphine of 25 ng/mL, although detection down to 15 ng/mL is technically feasible. The system is designed to operate at a cutoff level for morphine of 200 ng/mL. A number of morphine derivatives show significant cross-reactivity (see Table 66.58).

No significant interference has been demonstrated in tests involving over 120 other drugs.

Table 66.56. Compounds showing a negative response in the Syva EMIT d.a.u. opiate assay.

Compound	Concentration (μg/mL)
Alphaprodine	74
Amphetamine	1000
Benzoylecgonine	1000
Buprenorphine	1000
Butorphanol	1000
Chlorpromazine	12
Dextromethorphan	175
Doxylamine	1000
Meperidine	20
Meptazinol	100
Methadone	500
Nalbuphine	1000
Nalorphine	20
Naltrexone	5000
Naloxone	150
Norpropoxyphene	1000
Oxazepam	250
Papaverine	1000
Pentazocine	1000
Phencyclidine	1000
Propoxyphene	1000
Secobarbitone	1000

Table 66.57. Cross-reactivity of opiate drugs in the CEDIA opiate assay.

Compound	Concentration (ng/mL)	Cross-reactivity (%)
Morphine	300	100
Codeine	300	125
Diacetylmorphine	300	53
Dihydrocodeine	300	50
Hydrocodone	300	48
Morphine-3-glucuronide	300	81
Morphine-6-glucuronide	300	47
6-Monoacetyl-morphine	300	81
Oxymorphone	20,000	1.9
Oxycodone	10,000	3.1

Abuscreen OnLine

The OnLine opiates assay compares well with RIA and has somewhat better precision around the cutoff point (300 ng/mL) than, for example, EMIT II. It also has a more stable calibration curve which can reduce reagent and time costs in laboratories carrying out large numbers of assays routinely. Some cross-reactivity data for opiates are given in Table 66.59.

Enzyme-Linked Immunoadsorbent Assay

Cozart Bioscience produces two microplate opiate assays, both of which are for use in diluted serum or whole blood. One is for general opiate detection and picks up codeine and several other opiate drugs and the other is designed for the specific quantification of morphine and has calibrators of 0, 5, 10, 25, 50, and 100 ng of morphine (Tables 66.60 and 66.61).

PHENCYCLIDINE

Structure

See Figure 66.15.

Table 66.58. Cross-reactivity of opioid drugs in the Abbott TDx opiates FPIA.

Compound	Concentration added (ng/mL)	Concentration found (ng/mL)	Cross-reactivity (%)
Codeine	500	633	126.6
Diacetylmorphine	1000	359	35.9
Dihydrocodeine	1000	476	47.6
Dihydromorphine	1000	468	46.8
Ethylmorphine	1000	949	94.9
Hydrocodone	1000	466	46.6
Levorphanol	1000	324	32.4
6-Monoacetylmorphine	1000	402	40.2
Morphine-3-glucuronide	1000	466	46.6
N-normorphine	1000	41	4.1
N-norcodeine	1000	67	6.7
Noroxymorphone	100,000	51	0.05
Oxycodone	1000	105	10.5
Oxymorphone	1000	60	6.0
Thebaine	1000	240	24.0

Dose and Modes of Administration

Phencyclidine (PCP) is used in doses ranging from 2 to 6 mg by smoking with tobacco, nasal insufflation ('snorting'), intravenous injection, and oral ingestion. Its abuse is virtually unknown outside the USA.

Pharmacological Effects

Phencyclidine has a legitimate use as a veterinary tranquilizer and was developed for human use as an intravenous anesthetic agent, being similar in structure to ketamine. It no longer has any therapeutic uses in humans because of its pronounced hallucinogenic properties.

Table 66.59. Cross-reactivity of opiates in the Abuscreen OnLine opiates assay.

Compound	Concentration (ng/mL)*	Cross-reactivity (%)
Codeine	225	134
Ethyl morphine	265	113
6-Monoacetylmorphine	311	97
Dihydrocodeine	317	95
Thebaine	351	85
Dihydromorphine	371	81
Hydrocodone	479	63
Morphine-3-glucuronide	480	62
Hydromorphone	620	48
Norcodeine	11,744	3
Oxycodone	23,166	1

*Represents the approximate concentration equivalent in assay reactivity to the 300 ng/mL morphine cutoff calibrator.

Table 66.60. Cross-reactivity of opiates in the Cozart microplate opiates assay.

Compound	Concentration (ng/mL)	Cross-reactivity (%)
Codeine	10	577
Morphine-3-glucuronide	10	17.6
	100	6.0
	1000	3.0
6-Monoacetylmorphine	10	62.4
	100	28.1
	1000	34.7
Normorphine	10	23
	100	2.9
	1000	0.7
Nalorphine	10	12.5
	100	3.0
	1000	0.8
Diacetylmorphine	10	45.4
	100	30.0
	1000	22.7
Hydromorphone	10	45.8
	100	20.1
	1000	18.7
Hydrocodone	10	262
	100	102

Table 66.61. Cross-reactivity data for the Cozart specific microplate morphine assay.

Compound	Concentration (ng/mL)	Cross-reactivity (%)
Codeine	1000	<1
Morphine-3-glucuronide	1000	<1
	10,000	0.62
Diacetylmorphine	1000	<1
	10,000	0.2
Normorphine	10	40
	100	30
Nalorphine	10	490
6-Monoacetylmorphine	100	<10
	1000	<1
Hydromorphone	100	<10
	1000	<1

Toxic Effects

Abuse of phencyclidine causes lethargy, hallucinations, and loss of coordination. Signs of intoxication include hypertension, seizures, violent behavior, coma, and respiratory depression. Doses of 100–120 mg can cause death from respiratory failure. Chronic abuse can lead to memory loss, inarticulation, depression and a PCP psychosis that recurs on exposure to the drug.

Assay Technology

Phencyclidine undergoes oxidative metabolism to at least two inactive metabolites, which appear in the urine as glucuronide conjugates. The immunoassays are relatively specific for phencyclidine and its metabolites.

Radioimmunoassays

Abuscreen

This RIA is designed specifically to detect the presence of phencyclidine and its monohydroxylated metabolites, but detects several other metabolites as well. The sensitivity of the test is 2.5 ng/mL. No cross-reactivity with drugs other than close structural analogs of PCP has been reported.

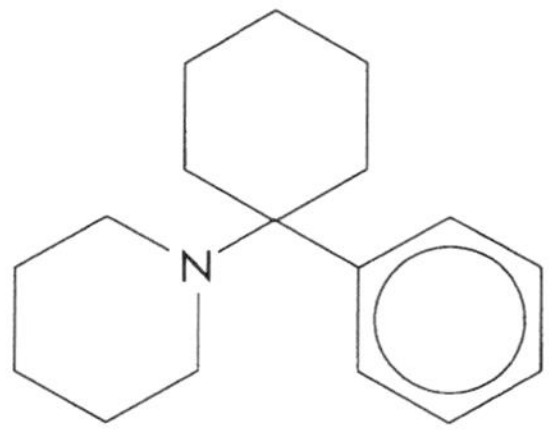

Fig. 66.15 Phencyclidine.

DPC

The DPC solid-phase RIA (Coat-a-Count) kit is both qualitative and quantitative for PCP in urine and has a detection limit of 0.2 ng/mL. Little or no cross-reactivity occurs with other compounds.

Enzyme-Multiplied Immunoassay Technique

The EMIT d.a.u. phencyclidine assay has a cutoff of 25 ng/mL. The compounds in Table 66.62 gave negative results against the low calibrator when tested at the specified levels.

Cloned Enzyme Donor Immunoassay

The CEDIA PCP assay has an average limit of detection in urine of 0.6 ng/mL. Apart from one report of a false-positive finding in a sample containing diphenhydramine,

Table 66.62. Compounds found to give negative results with the Syva EMIT d.a.u. PCP assay.

Drug	Concentration (μg/mL)
Acetaminophen	1000
Albuterol	1000
Benzoic acid	32
Buspirone	914
Cimetidine	1000
Dextrorphan	100
Diclofenac	1000
Diphenhydramine	1000
Fentanyl	100
Haloperidol	1000
Hydroxyzine	50 mg dose
Ketoprofen	1000
Levallorphan	10
Mesoridazine	10
Methenamine	162
Norlevorphanol	1000
Normeperidine	1000
Norpropoxyphene	1000
Orphenadrine	1000
Oxycodone	1000
Pentazocine	1000
Phenytoin	30
Salicylamide	65
Sodium salicylate	97
Terfenadine	1000
Thioridazine	1000
Tripelennamine	1000

Table 66.63. Cross-reactivity data for the CEDIA phencyclidine assay.

Compound	Concentration (ng/mL)	Cross-reactivity (%)
1-Phenylcyclohexyl-4-hydroxypiperidine	32	106
4-Phenyl-4-piperidine-cyclohexanol	1000	2.5
Phenylcyclohexylpyrrolidine	25	68
5-(1-Phenylcyclohexylamine)	1000	0.6
1-[1-(2-Thienyl)cyclohexyl]-piperidine	100	31
Phenylcyclohexylamine	100	37

the test is highly specific, and cross-reactivity data for various metabolites and related substances are listed in Table 66.63.

Fluorescence Polarization Immunoassay

The TDx system is a semiquantitative assay for PCP in urine which has a detection sensitivity of 5 ng/mL. Positive samples are taken as those giving a response greater than the 25 ng/mL cutoff value. Although several PCP analogs cross-react significantly, no other compounds are known to give false-positive results.

Abuscreen OnLine

The Abuscreen OnLine assay has analytical sensitivity of around 5 ng/mL. A limited amount of cross-reactivity data is available for this assay, but no interference from other substances has yet been reported (Table 66.64).

PROPOXYPHENE

Structure

See Figure 66.16.

Dose and Mode of Administration

Dextropropoxyphene is marketed as an oral preparation either as the hydrochloride (32 or 65 mg) or as the napsylate (50 or 100 mg) and used in daily doses of around 400 mg. It is often formulated in combination with aspirin or paracetamol and prescribed on a large scale for the relief of chronic pain. There are many reports of its successful use in heroin maintenance or withdrawal treatment programs in doses of 800–1400 mg per day.

Table 66.64. Cross-reactivity data for the Abuscreen OnLine PCP assay.

Compound	Concentration*	Cross-reactivity (%)
Dextromethorphan	272,000	0.01
Thienylcyclohexyl-piperidine	31	80

*Approximate concentration equivalent to the 25 ng/mL phencyclidine cutoff.

Pharmacological Effects

Dextropropoxyphene is a mild narcotic analgesic, less potent than codeine and closely related to methadone in chemical structure. The L-isomer has virtually no narcotic effects and is used as an antitussive.

Toxic Effects

These are similar to methadone in overdose with symptoms of respiratory depression, stupor, hypotension, coma, and circulatory collapse. It has a long half-life and therefore the toxic effects are prolonged. There are many reports of fatal overdose and the minimum lethal dose in adults is estimated to be between 500 and 800 mg.

Assay Technology

Most commercial immunoassay systems have propoxyphene in the portfolio and are targeted at the parent compound. There is usually good cross-reactivity towards the demethylated metabolite, norpropoxyphene, but little response towards methadone despite the similarity in chemical structure.

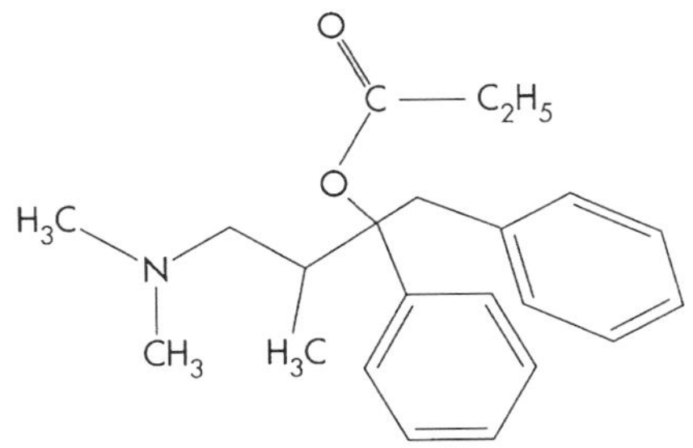

O
C — C_2H_5
O
H_3C
N
CH_3
H_3C

Fig. 66.16 Propoxyphene.

LEGAL ADDICTIVE SUBSTANCES

ALCOHOL (ETHYL ALCOHOL) AND THE USE OF CARBOHYDRATE-DEFICIENT TRANSFERRIN

Toxic Effects of Alcohol

The occasional bout of heavy social drinking usually causes nothing more serious than the morning-after hangover effects of headache, dehydration, and gastric upset and most people recover as the day moves on. Deaths from overindulgence do occur, however, for example when an individual loses consciousness and aspirates vomit into the lungs. Continual heavy intake (alcoholism) leads to chronic poisoning and some of the features are liver disease (cirrhosis), cardiac abnormalities, nerve degeneration, and loss of mental function. Those who abuse alcohol are notoriously reluctant to admit how much they imbibe, even when the information is asked for by medical personnel trying to evaluate their health. Laboratories have been able to measure blood alcohol levels for many years and, for a long time, simple methods of measuring alcohol in saliva or breath samples have been around. The drawback of this approach is that if the abuser is given notice of the date and time of the test he/she may well refrain from drinking for a day or so and the alcohol result will be negative. Even a positive alcohol measurement is only a snapshot of the situation at the time and is no indicator of how substantial and regular the abuse is, or how long it has been going on. This has led to a search for biological markers of alcohol abuse and one of these, carbohydrate-deficient transferrin (CDT), is measured in serum by immunoassay techniques. This is now accepted as the most sensitive and specific test for diagnosing heavy alcohol consumption, but there are some flaws in that it is more specific in men than women, and less sensitive when applied to nonhospital populations such as university students. Like many other diagnostic biological tests, it is best combined with measurement of other parameters such as γ-glutamyltransferase and erythrocyte aldehyde dehydrogenase (ADH).

Carbohydrate-Deficient Transferrin

The glycoprotein, transferrin, takes part in the delivery of iron to the body tissues and is synthesized in liver cells. Chronic exposure to alcohol disrupts this process, possibly by acetaldehyde-mediated inhibition of glycosyl transfer, and a proportion of the resulting transferrin is missing some carbohydrate terminal chains (e.g. sialic acid, galactose, *N*-acetylglucosamine), hence, the term carbohydrate-deficient transferrin. At least 1 week of heavy drinking is needed to give raised serum CDT levels and these normalize slowly with a half-life of about 15 days after drinking stops.

Assay Technology

The first methods concentrated on a combination of isoelectric focusing with immunofixation techniques and were soon superceded by isocratic anion-exchange chromatography. In this procedure iron-saturated serum is passed through a microcolumn and the separated isotransferrins quantified by radioimmunoassay. The first commercial kit used a similar principle of separating the isoforms on a microcolumn and quantifying the eluted deficient transferrins by a double-antibody immunoassay. In a second kit, serum transferrin is radiolabeled using antibody fragments and the carbohydrate-deficient form separated on an ion-exchange chromatography minicolumn. The relative amount of CDT eluted is measured by counting the radioactivity, and any elevation above 2.5% is considered diagnostic of alcohol abuse. In comparative trials the first kit was found to be more sensitive, but alterations in serum transferrin concentration markedly reduced the specificity of the assay. More recently, commercial competitive-binding enzyme immunoassays have been introduced. Again the transferrin isoforms are separated on an ion-exchange microcolumn and the eluted CDT then quantified by ELISA technology.

An alternative approach has been to go back to isoelectric focusing, use direct immunofixation by a specific antibody and quantify by computerized scanning densitometry. This is quite an easy method to carry out, is inexpensive, and has a reputed specificity of 100% (defined as the ability to give normal results when there is no heavy alcohol use) and a sensitivity of about 95% (defined as the ability to detect heavy alcohol consumption when it really exists). The method has been described in detail by Dumon *et al.* (1996). The technology has now been adapted to evaluate CDT derived from dried blood spots.

NICOTINE AND THE USE OF COTININE

Structure

See Figure 66.17.

Dose and Mode of Administration

Tobacco is the most abundant source of nicotine and the leaves contain 1–6% by weight. Smoking is by far

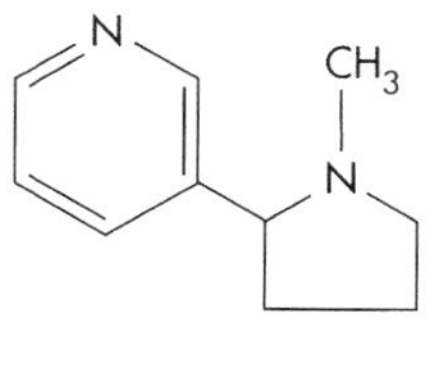

N
CH_3
N

Fig. 66.17 Nicotine.

the most popular means of absorbing nicotine, although chewing tobacco or snorting powdered tobacco (snuff) is still practiced on a small scale. The average cigarette contains 13–19 mg of nicotine and cigars have 15–40 mg. During smoking a significant amount of nicotine is lost by combustion and in the side-stream smoke. Depending on the depth of inhalation, cigarette smokers take in 0.2–2.5 mg through the lungs. Cigar and pipe smokers absorb nicotine predominantly by the mouth and take in 10–50% via this route. In recent years the dangers of nonsmokers being exposed to nicotine by inhaling smoke (passive inhalation) have been emphasized. Nicotine can also be absorbed through the skin and this property has been exploited by designing skin patches impregnated with nicotine that continually deliver small amounts to the blood stream to help stave off the craving. Nicotine patches contain up to 50 mg of the drug and deliver doses of 5–21 mg. An alternative formulation is nicotine-impregnated chewing gum and each tablet contains 2–4 mg.

Pharmacological Effects

Nicotine causes stimulation of the autonomic ganglia and the CNS.

Toxic Effects

Nicotine is an extremely toxic substance and it has been estimated that as little as 40 mg can be lethal in an adult. A lethal dose causes paralysis of the CNS, including the respiratory center, hypotension, tachycardia, muscle paralysis, convulsions and death within a few minutes to 1 h after ingestion. Fortunately, acute nicotine poisoning is rare and is usually associated with accidental exposure to concentrated nicotine insecticide solutions. There is far more concern about the pronounced carcinogenic effects of other components of tobacco smoke and also of the contribution of the smoking habit to heart disease.

Assay Technology

Nicotine is extensively metabolized by the liver to form largely inactive products. One of the main urinary excretion products is the oxidized form (cotinine) and this has been the favored antigen in developing tests for smoking patterns. A substantial amount of quantitative data has been accumulated that can be used to judge whether an individual is an active or passive smoker and how much tobacco is being consumed. From an analytical viewpoint, cotinine measurements have the advantage of being able to discount problems of equipment and biological samples being contaminated by environmental nicotine. To obtain reliable data for nicotine itself calls for the analyses to be carried out under almost aseptic conditions.

Radioimmunoassays are still the most widely used to measure both nicotine and cotinine in either serum or urine samples, but the results of comparative trials against gas-chromatographic methods such as that described by Biber *et al.* (1987) show large discrepancies in the quantitative data. This is less pronounced for cotinine in serum samples than in urine samples where RIA tends to give much higher concentrations. There are also reports of huge variations in nicotine and cotinine concentrations in serum and urine samples reported by different laboratories applying RIA to actual and spiked samples. This is most likely due to problems of varying nonspecific binding, although it should be pointed out that in some of these trials, there have also been substantial interlaboratory discrepancies amongst the gas-chromatography groups. However, the difference in serum cotinine concentrations between samples from nonsmokers exposed to environmental tobacco smoke and people who smoke actively is substantial, and both types of assay have been shown capable of making a clear distinction. From a practical standpoint this is often the only information required.

STEROID ABUSE IN SPORT

ANABOLIC ANDROGENIC STEROIDS

Anabolic androgenic steroids (AAS) are based on the naturally occurring testosterone and numerous forms are available for legitimate use in medicine to treat a variety of conditions ranging from osteoporosis, disseminated breast cancer, protein deficiency states through to anemia and post-traumatic catabolism. Abuse by athletes, in particular weight lifters, throwers (discus, shot), and sprinters, wishing to increase muscle bulk has been a cause of great concern in sports organizations for many years and detection of this practice presents a major challenge to analysts working in the sports testing field. This includes analysts concerned with the integrity of animal sports such as horse racing. AAS are extensively metabolized and the parent drugs are detectable only for short periods after administration. As a consequence, analytical strategies are directed towards identifying the metabolites to extend the detection time period. Schanzer (1996) has given a very comprehensive account of the structures and metabolism of these compounds. A major problem is that endogenous steroids such as testosterone and dihydrotestosterone, which is an active metabolite of testosterone, are the most often used and finding an unusually high urine concentration of these is not sufficient evidence to confirm abuse. Some headway has been made in recent years by measuring the ratios in urine samples of testosterone and dihydrotestosterone to other endogenous steroids such as epitestosterone, and

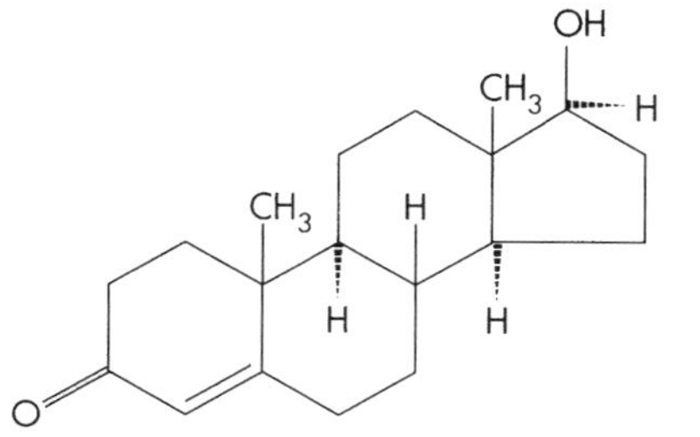

Fig. 66.18 Testosterone.

Southan *et al.* (1992) have published a review of these and other possible markers of administration.

The exogenous anabolic steroids tend to present less of a problem analytically and nandralone decanoate, which is the one most widely abused, serves as an example of the group.

Structures of Testosterone and Nandralone

See Figures 66.18 and 66.19.

Dose and Mode of Administration

Nandralone decanoate is administered for medical purposes by intramuscular injection in doses of 50 mg. Athletes may take much larger and more frequent doses.

Pharmacological Effects

Nandralone is primarily an anabolic substance and has very little androgenic effect, although long-term use can produce acne, hirsutism, and deepening of the voice in women. There is conflicting evidence for a direct effect on muscle development in athletes by anabolic steroids, but it is accepted that, coupled with hard and continuous exercise and protein supplements, they can increase muscle strength. Many now take the view that there is an indirect effect due to the action of the steroids on the CNS, which causes increased aggression and competitiveness. Since the aggression may not always be channeled into sporting activities, this is another reason for concern.

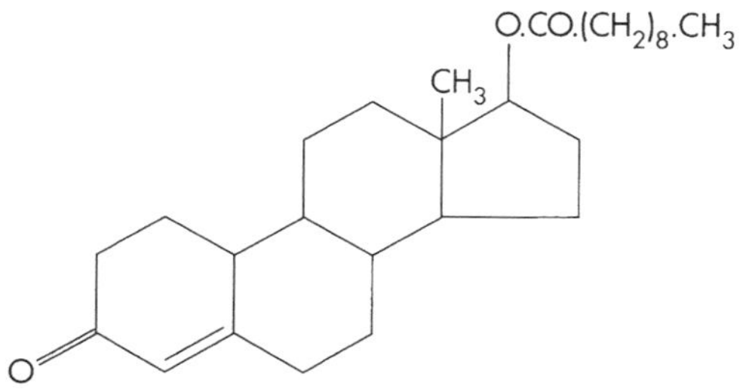

Fig. 66.19 Nandralone decanoate.

Toxic Effects

In athletes anabolic steroid abuse has been linked to cardiovascular disorders such as arteriosclerosis with consequent coronary heart disease, fatal hypotension, carcinomas of the liver and kidney and, in males, drastic reductions in sperm counts.

Assay Technology

Radioimmunoassays offered the first opportunity to screen for anabolic androgens during the Summer Olympic Games of 1976, but because of an unacceptable number of false negatives and the need for specific measurement of testosterone and epitestosterone, the technique was virtually abandoned in 1980. Those laboratories which were accredited for testing by the International Olympic Committee scheme moved on to chromatographic/mass spectrometric screening methods for this group of compounds, although they still fall back on immunoassays for those such as trebolone which are difficult to detect by these techniques. It is fair to say that there is now far more interest in using immunoaffinity chromatography to isolate the anabolic steroids from the complex biological matrix prior to analysis by mass spectrometry or to apply the immunoassay to HPLC fractions.

The use and development of immunoassays for anabolic steroids are still very much in vogue in the animal sports testing laboratories, and ELISA technology has largely replaced RIA in this area. The more advanced laboratories often develop and manufacture 'in-house' ELISA kits for anabolic steroids, either when these are not available commercially, to ensure continuity of supply, or as a cost-efficiency exercise. There will most likely be an increasing development of specific ELISA kits over the next few years, applying phage display antibody technology, which has been described in detail by Dorsam *et al.* (1997). It can also be predicted that the human sports testing laboratories will follow this lead, given the increasing pressure for more intensive inter-competition monitoring.

Finally, there is the less exciting, though important, problem of monitoring the illegal use of anabolic steroids, such as 19-nortestosterone, in meat-producing animals. In the veterinary testing laboratories immunoassays are an integral part of the routine operation. Here again, ELISA is likely to become the favored technique.

RAPID IMMUNOASSAY TESTS FOR DRUGS OF ABUSE

Doctors have long been attracted to the idea of being able to evaluate a patient's drug status on the spot rather than waiting until the results come back from a laboratory. Even if the laboratory is local, there are inevitable delays in sample transport and processing which mean that treatment decisions are delayed, often for a day or more. The growth of drug testing in the workplace, prisons, schools and in road traffic offences, has stimulated the development of simple tests which need very little skill, give an almost immediate answer and allow action to be taken straight away. Not surprisingly, alcohol tests for use at the roadside were the first to be introduced and were based on the colorimetric reaction of alcohol in the breath with potassium dichromate. Since then these rather crude 'blow-tubes' have been replaced by hand-held devices which rely on fuel cell technology. These give a reasonable estimate of the breath alcohol level and select out those who are obliged either to undergo a more accurate evidential breath test in the police station or to opt for laboratory analysis of a blood or urine sample. Attempts to apply wet chemistry methods to detect abused drugs in urine samples were doomed to failure. Most of the tests that did reach the marketing stage were very unreliable and false-positive results were commonplace. The emergence of dry reagent immunology systems for various biochemical markers of disease has opened up this field significantly and several products are now available. The most popular versions use gold-labeled immunotechnology and are either in the form of a small hand-held disposable device that detects several drugs in the same urine sample, or as a paper test strip similar in appearance to biochemical dipsticks with separate test strips for each drug. The Triage™ (from Biosite Inc.) is a multitest device for seven classes of drugs (amphetamines, barbiturates, benzodiazepines, cocaine metabolite, cannabinoids, opiates, and phencyclidine). Figure 66.20 gives a top view of the Triage.

The reaction well at the top of the device contains three reagent beads. The antibodies bead contains monoclonal antibodies for all the seven drug classes. The conjugate bead contains a representative drug from each class, bound to colloidal gold particles. The third bead contains a buffer. When a urine sample is added (0.14 mL) to the reagent well, the beads are reconstituted and the mixture is left to incubate for 10 min. If the urine sample contains one or more of the target drugs at or above the threshold concentrations, the antibodies bind both free and conjugated drug to leave some unbound conjugates. The reaction mixture is transferred to the detection area which has monoclonal antibodies immobilized on a nylon membrane in seven discrete areas. The detection area also has two other discrete regions, one impregnated with negative control material and the other with a positive control. After the mixture has soaked into the membrane, a wash solution is added to remove any free drug-colloidal gold conjugates. Free gold conjugate binds to the immobilized antibodies producing a red bar opposite the name of whichever drug is present. For a test to be valid, the positive control must give a red bar and the negative should be blank.

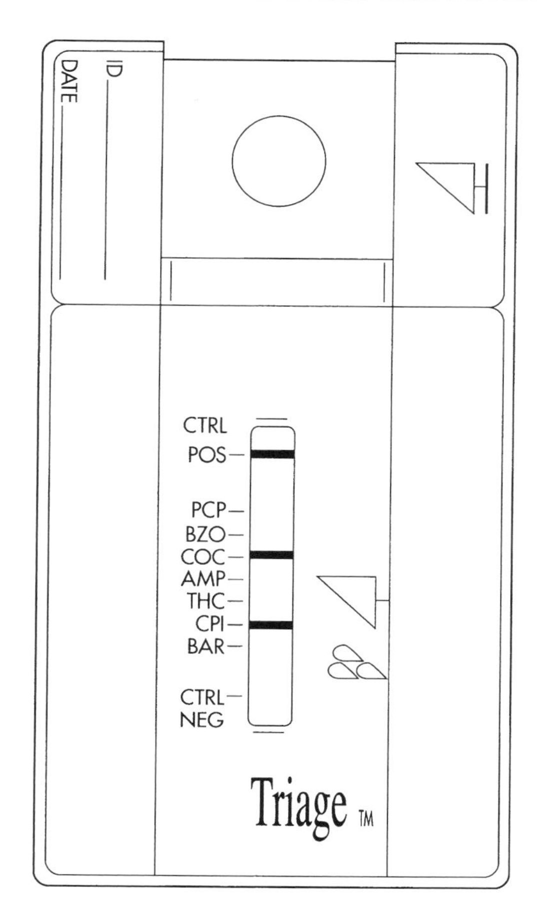

Fig. 66.20 Top view of the Triage device.

The thresholds (or cutoffs) are similar to those recommended by SAMHSA for workplace testing. The kit has performed well in trials against laboratory immunoassay techniques and chromatographic methods. Opinions vary on how easy it is to handle by first-time users. Some reports say that it is difficult to make mistakes, whereas others point out the dangers of losing one or more of the beads as the lid of the reaction cup is opened. People with no laboratory training might ignore this, and the test would be invalid. There can be false-positive amphetamine results due to reactions with other amphetamine-like substances (e.g. phenylalkylamines in postmortem urine), but this is not uncommon with laboratory drugs of abuse immunoassay systems. At low drug concentrations, the bands formed can sometimes be faint and may be due to the presence of artifacts rather than drugs.

Another product, OnTrak Frontline® (Roche Diagnostics), is a test strip that uses immunochromatography. The basic design is shown in Figure 66.21.

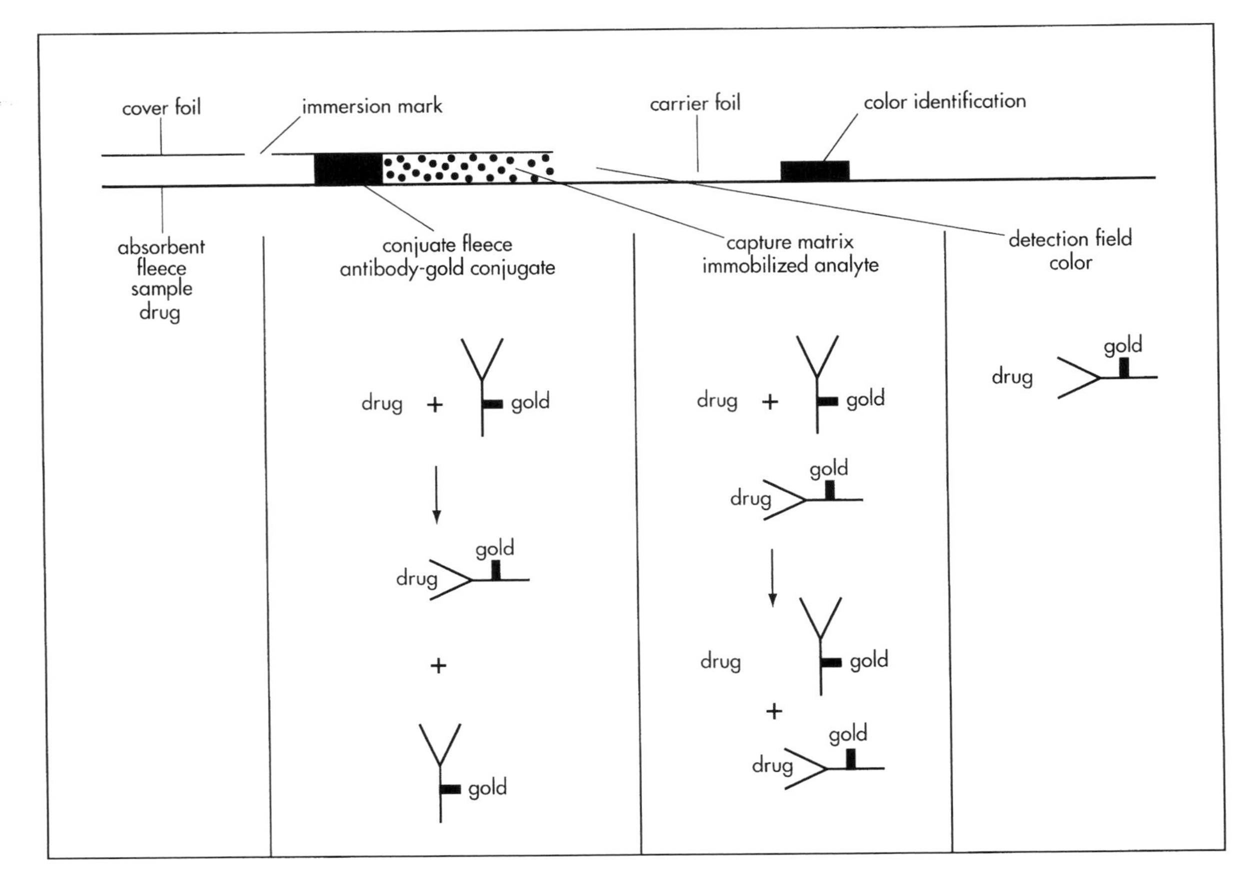

Fig. 66.21 Principle of the FRONTLINE rapid test.

The strip is dipped into the sample for a few seconds and the absorbent fleece takes up a uniform volume of urine. This is then drawn along the strip by capillary action and reaches an area of fleece containing antibodies conjugated to gold particles. These are eluted completely as the urine passes through. Any target drug in the urine sample binds with the conjugate and this, together with unbound conjugate moves into the next (capture) zone. This is coated with an immobilized polyhapten of the target drug that binds any free antibody-gold conjugate, allowing only the drug-bound conjugate to pass through to the detection zone. Here, the drug-gold conjugate spreads homogeneously into a cellulose fleece and the intensity of the red color is proportional to the amount of drug in the original sample. The intensity is compared visually to a color scale on the vial containing the test strips, and rough estimates of drug concentration can be made. Again, the threshold concentrations for detecting positives are linked to the SAMSHA recommendations. In practical terms this test could not be simpler to use, although there have been problems in interpreting the significance of reaction colors by novice users, particularly for the cocaine and cannabis tests. This resulted in a high number of false positives, but the problem disappeared with more experience. The strips have been evaluated very thoroughly in a multicenter trial against the conventional laboratory techniques of Emit and FPIA, and appear to be quite reliable. The only discrepancies noted were false positives due to cross-reactivity of the cocaine test strip with the methadone metabolite EDDP and also with metabolites of the antipsychotic drug clozapine.

Cozart Bioscience have developed a more sophisticated device that uses saliva (RapiScan) and which is currently being evaluated for roadside testing and clinical use. This is similar in weight to a mobile telephone and fits into the palm of the hand. A fixed volume of saliva (1 mL) is collected onto a swab, which is then placed into a disposable immunoassay cartridge in which the immunochemical reaction takes place. The cartridge is inserted into the RapiScan and after 5 min a digital display of the results appears. This is a multitesting device and the analyte menu so far consists of amphetamines, benzodiazepines, cannabinoids, and opiates with a test for methadone on the way. An obvious advantage of this sort of technology is the ease of sample collection and the fact that this can be closely observed without infringing the individual's privacy.

These rapid tests should prove a boon to the drug dependency clinics where they are used in the context of improving patient care. However, in other areas where punitive measures may follow a positive result, the tests should be regarded as presumptive. For example, while it may be considered acceptable for an employee that tests positive to be suspended from work temporarily, any further action should be withheld until a laboratory has carried out confirmatory analyses by gas chromatography–mass spectrometry.

REFERENCES

Biber, A., Scherer, G., Hoepfner, I. *et al.* Determination of nicotine and cotinine in human serum and urine: an interlaboratory study. *Toxicol. Lett.* **35**, 45–52 (1987).

DHHS/SAMHSA, Mandatory guidelines for federal workplace drug testing programs; notice. *Fed. Regist.* **59**, 29908–29931 (1994).

Dorsam, H., Rohrbach, P., Kurschner, T. *et al.* Antibodies to steroids from a small human naive IgM library. *FEBS Lett.* **414**, 7–13 (1997).

Dumon, M.F., Nau, A., Hervouet, J. *et al.* Isoelectric focusing (IEF) and immunofixation for determination of disialotransferrin. *Clin. Biochem.* **29**, 549–554 (1996).

Fraser, A.D. and Meatherall, R. Comparative evaluation of five immunoassays for the analysis of alprazolam and triazolam metabolites in urine: effects of lowering the screening and GC–MS cutoff values. *J. Anal. Toxicol.* **20**, 217–223 (1996).

Killander, J., De la Torre, R., Segura, J. *et al.* Recommendations for the reliable detection of illicit drugs in urine, with special attention to drugs in the work-place, in the European Union. *Scand. J. Clin. Lab. Invest.* **57**, 97–104 (1997).

Schanzer, W. Metabolism of anabolic androgenic steroids. *Clin. Chem.* **42**, 1001–1020 (1996).

Southan, G.J., Brooks, R.V., Cowan, D.A., Kicman, A.T., Unnadkat, N. and Walker, C.J. Possible indices for the detection of the administration of dihydrotestosterone to athletes. *J. Steroid Biochem. Mol. Biol.* **42**, 87–94 (1992).

FURTHER READING

Baselt, R.C. Urine drug screening by immunoassay: interpretation of results. In: *Advances in Analytical Toxicology*, vol. 1, (ed Baselt, R.C.), 81–123 (Biomedical Publications, Foster City, CA, 1984).

Segura, J. and de la Torre, R. *Current Issues of Drug Abuse Testing. First International Symposium* (CRC Press, Boca Raton, FL, 1992).

Tan, K. and Marks, V. Use of immunoassays to detect drugs in body fluids. In: *The Analysis of Drugs of Abuse* (ed Gough, T.A.), 311–335 (Wiley, New York, 1991).

Warner, A. Interference of common household chemicals in immunoassay methods for drugs of abuse. *Clin. Chem.* **35**, 648–651 (1989).

67 Assays for Drug-screening Applications and Research

Jeffrey K. Horton, Stephen J. Capper, Molly J. Price Jones and Kelvin T. Hughes

Immunoassay techniques are widely used in drug-screening and life-science research because of their sensitivity and specificity. Immunoassays are simpler, and often more rapid, for large numbers of samples than many of the other analytical techniques used in scientific research. They have proved their value in the elucidation of the physiological roles and pathological significance of many biologically active compounds.

Immunoassays have had their greatest impact in clinical research, contributing to the understanding of diseases and, in many cases, leading to the development of useful new diagnostic tests. More basic biochemical research has also benefited from the quantification of agents, e.g. the cyclic nucleotides and kinases, involved in signal transduction and other cell regulatory processes.

Apart from academia, the major immunoassay users in life-science research are in the pharmaceutical industry. Immunoassays are widely used in all areas of drug discovery and development process. During screening for potential therapeutics the measurement, e.g. of the arachidonic acid metabolite prostaglandin E_2, is a useful end-point monitor for potential cyclo-oxygenase inhibitors, such as new non-steroidal anti-inflammatory drugs (NSAIDs). Also cyclic AMP (cAMP) determinations are commonly used to monitor G-protein-associated receptor activation. The requirements for a high-throughput drug-screening assay are quite different from those used in clinical trials, and other basic research studies, and this subject will be discussed later in the chapter, with particular reference to cAMP.

Another application of immunoassay in the pharmaceutical industry is in metabolism and toxicology studies. Defining effects of potential therapeutics on the endocrine system is required to satisfy authorities for safety compliance, and tests are often carried out on animal models (with, e.g. rat hormone immunoassay). Likewise numerous immunoassays are performed on animals and human subjects during clinical trials to assess longer-term effects on multiple endogenous measures.

Some of the compounds that are measured using commercially available assay kits suitable for pharmaceutical screening or academic research are listed in Table 67.1. The list is not exhaustive and immunoassays have been developed for many hundreds if not thousands of analytes.

This chapter describes well-established and also novel immunoassay techniques used by scientists from academia and those carrying out pharmaceutical research and development. The functions of some of the more interesting analytes will be described together with some examples of the applications of immunoassay techniques in drug-screening and academic research.

Techniques used by the researcher differ from those of a diagnostic laboratory. Immunoassay kits very much dominate the diagnostic laboratory, often dedicated to specific instrumentation. In contrast, the researcher is much more likely to develop his or her own assay from bought-in components, or materials obtained from colleagues. The reason for this 'in-house' preference in research laboratories is varied. The higher cost of kits is a key factor, especially in laboratories where the throughput is high, although if analytes are assayed infrequently the convenience of kits outweighs the cost. In addition, kits are purchased if there are time constraints inappropriate for the development and validation of a new assay. Another factor is the availability of kits. Owing to the nature of most researchers' work it is unlikely that a commercial kit will be available for speculative new molecules or specialist compounds of limited interest.

One of the difficulties facing the manufacturer of assay kits for the scientific community is the extent of assay validation to perform. The intended use of a diagnostic kit is well defined and usually confined to one or two sample types and a limited number of pathological indications. In contrast, the potential use of kits by the research community

The Immunoassay Handbook. *Third Edition*
D. Wild (Ed.)

Table 67.1. Some commercially available life-science research immunoassays.

Research area	Analyte	Technique
Cytokines, chemokines, and growth factors	Eotaxin	ELISA
	GM-CSF	ELISA
	IFNα	ELISA
	IFNγ	ELISA
	IL-1α	ELISA
	IL-1β	ELISA
	IL-2	ELISA
	IL-4	ELISA
	IL-5	ELISA
	IL-6	ELISA
	IL-8	ELISA
	IL-10	ELISA
	IL-12, p40/p70	ELISA
	IL-12, p70 specific	ELISA
	IL-16	ELISA
	MCP-1	ELISA
	MIP-1α	ELISA
	RANTES	ELISA
	TGFβ_1	ELISA
	TNFα	ELISA
Eicosanoids	6-Keto-prostaglandin $F_{1\alpha}$	EIA
	6-Keto-prostaglandin $F_{1\alpha}$ [^{125}I]	RIA
	Leukotriene B_4	EIA
	Leukotriene B_4 [^3H]	RIA
	Leukotriene C_4 [^3H]	RIA
	Leukotriene $C_4/D_4/E_4$	EIA
	Prostaglandin E_2	EIA
	Prostaglandin E_2 [^{125}I]	RIA
	Prostaglandin $F_{2\alpha}$ [^3H]	RIA
	Thromboxane B_2	EIA
	Thromboxane B_2 [^{125}I]	RIA
Matrix metalloproteinases (MMPs)	MMP-1	ELISA/AA
	MMP-2	ELISA/AA
	MMP-3	ELISA/AA
	MMP-7	ELISA
	MMP-8	ELISA/AA
	MMP-9	ELISA/AA
	MMP-13	ELISA/AA
	MMP-14	AA
	TIMP-1	ELISA
	TIMP-2	ELISA
	TIMP-2 (rabbit)	ELISA
	TIMP-2 (mouse)	ELISA
	MMP-3 (rabbit)	ELISA
Metabolism and toxicology (rat)	Corticosterone [^{125}I]	RIA
	FSH	EIA
	FSH [^{125}I]	RIA
	Growth hormone	EIA
	Growth hormone [^{125}I]	RIA
	Insulin	EIA
	Insulin [^{125}I]	RIA
	Luteinizing hormone	EIA

Table 67.1. Continued

Research area	Analyte	Technique
	Luteinizing hormone [^{125}I]	RIA
	Prolactin	EIA
	Prolactin [^{125}I]	RIA
	TSH	EIA
	TSH [^{125}I]	RIA
Signal transduction	cAMP	EIA/FPIA
	cAMP [^{3}H]	RIA
	cAMP [^{125}I]	RIA
	cGMP	EIA
	cGMP [^{3}H]	RIA
	cGMP [^{125}I]	RIA
(Receptor assay)	Inositol trisphosphate [^{3}H]	RR
Neurodegeneration	β-Amyloid 1–40	ELISA
	β-Amyloid 1–40	ELISA
	α-Synuclein	ELISA
Cardiovascular	Atrial natriuretic peptide [^{125}I]	RIA
	Complement C3a [^{125}I]	RIA
	Complement C4a [^{125}I]	RIA
	Complement C5a [^{125}I]	RIA
	Endothelin-1	ELISA
	Endothelin-1,2 [^{125}I]	RIA
	Endothelin 1–21 [^{125}I]	RIA

The assays referred to are human assays; assays for other animal species are available. High-sensitivity versions of some of these assays are also available.

is extensive and they are often used as an end-point to a complex *in vitro* pre-study. This makes full validation that covers all possible applications virtually impossible and only when a defined market need is known will specific support studies be performed. Otherwise the accountability for validating the kit in a specific experimental design is the responsibility solely of the investigator. Immunoassay kits for the scientific community do not require licensing by the FDA or other national bodies. The best mark of quality assurance is to use kits from manufacturers that are accredited to international quality systems such as ISO 9000.

Another difference from the diagnostics industry is the great diversity of immunoassay techniques in use. Traditional radioimmunoassay (RIA) techniques using tritiated tracers and charcoal separation of bound from free analyte are still widely used. Conservatism and a general reluctance to change from existing methods, unless there are clear benefits in assay performance, are typical of many scientific staff. The incentive to change to faster, more convenient systems is difficult to justify if a complicated and time-consuming pre-assay sample treatment is obligatory. The move towards simpler protocols using scintillation proximity assay (SPA), fluorescence polarization immunoassay (FPIA) and other non-isotopic assays has often been slow in areas where RIA is well established. It is only in the emerging scientific fields such as cytokines and chemokines that non-isotopic methods have become well established, and in pharmaceutical screening, where high assay throughput using homogeneous methodologies such as SPA and FPIA has become paramount.

However, the trend towards non-isotopic methodologies is gaining momentum in traditional RIA areas. This has been brought about not just by environmental and safety considerations but by the acceptance that assay performance of these methods is equivalent or superior to RIA, and by the promise of technological advances such as multiplexing. Although radioisotopic technology has matured, many non-isotopic techniques hold considerable promise in terms of speed, sensitivity, and convenience.

ASSAY TECHNOLOGY

TRITIUM RADIOIMMUNOASSAYS

Tritium RIAs are still widely used in research. This is because wide ranges of tritium compounds and antisera are available from suppliers, enabling a cheap assay for an unusual analyte to be developed quickly. Tritium tracers are also chemically identical to the analytes under test, which sometimes makes it easier to optimize assay performance. The long half-life of tritium is also an advantage although reagent shelf-life is still limited by chemical stability. However, tritium assays are very labor intensive, as the supernatants have to be transferred to scintillation vials, mixed with scintillant, and counted in a liquid scintillation counter.

Assay Protocol

An example of a tritium-based assay for leukotriene B_4 is shown below.

A single lyophilized leukotriene B_4 calibrator is provided which is diluted by the user to prepare a calibration curve. The tracer is [5,6,8,9,11,12,14,15(n)-^{3}H]leukotriene B_4 in methanol:water:acetic acid (60:40:0.01) at pH 5.6. The assay uses a charcoal separation system. All the reagents require reconstitution or dilution before use and the assay buffer and charcoal need mixing using a magnetic stirrer and bead.

The assay is a conventional competitive immunoassay with a 2 h incubation at 25 °C or an alternative overnight incubation at 4 °C. The charcoal separation stage requires a 5 min incubation followed by centrifugation at 4 °C. The supernatants, which contain the bound fraction, are carefully decanted into scintillation vials and 10 mL scintillant added to each. The counting time required is 4 min.

IODINE-125 RADIOIMMUNOASSAYS

The time-consuming counting stage of tritium RIA can be speeded up by using iodine-125 tracers. In addition, other time-saving developments from the clinical field, such as magnetic separations, have been used in research assays.

The following example protocol describes a method for the measurement of prostaglandin E_2 (PGE_2). The PGE_2 is first stabilized by converting to the methyl oximate. An antibody directed to the methyl oximate is used together with a ^{125}I-labeled prostaglandin E_2 proline–tyrosine conjugate.

This is done as follows:

1. Reconstitute the prostaglandin E_2 fraction from a solid-phase extraction column with 100 μL phosphate-buffered gelatin saline, pH 7.0.
2. Add 100 μL of the methyl oximation reagent. Vortex mix the resulting solution and incubate at 60 °C for 1 h to allow methyl oximation of the sample. Alternatively, derivatization can be achieved using an overnight incubation at room temperature.
3. Following methyl oximation, dilute to a final volume of 500 μL with the phosphate-buffered gelatin saline. Samples prepared in this way can be assayed directly or can be stored at −20 °C for up to 6 days before analysis.

Assay Protocol

A single lyophilized PGE_2 calibrator is provided which is diluted by the user to prepare a calibration curve. The tracer is the ^{125}I-labeled prostaglandin E_2 proline–tyrosine conjugate (methyl oximate derivative). The assay uses magnetizable latex polymer particles coated with donkey anti-rabbit antiserum as a separation system. Some of the reagents require reconstitution or dilution before use.

The assay is a conventional competitive immunoassay with a 2 h incubation at 25 °C or an alternative overnight incubation at room temperature. The magnetic separation takes 15 min. Following decantation, counting for 1 min is required.

SCINTILLATION PROXIMITY ASSAY

Although there have been many advances in immunoassay design, most have been confined to the area of clinical biochemistry. Improvements to assays used in research laboratories have mainly been due to superior quality antibodies, which have improved specificity and sensitivity. Little has been done to simplify the protocols of research assays. However, the introduction of the SPA has greatly simplified research RIAs, allowing full automation for the first time.

SPA is based on the principle that relatively weak β emitters, such as ^{3}H β particles and ^{125}I Auger electrons, need to be close to scintillant molecules to produce light; otherwise the energy is dissipated and lost to the solvent. This concept has been used by Amersham Biosciences™ to develop a range of homogeneous immunoassay products by coupling second antibodies or Protein A to fluomicrospheres. Fluomicrospheres are solid-phase support particles or beads impregnated with substances that fluoresce when excited by radioactive energy.

When added to an antibody/antigen mixture the antibody is captured on the fluomicrosphere, bringing any bound radiolabeled antigen close enough to allow the radiation energy emitted to activate the fluorescent compound and emit light energy.

If the concentration of fluomicrospheres is optimized, only signal from the radiolabeled ligand bound to antibody is detected, eliminating the need for any separation of bound and free ligand. The level of light energy emitted, which is indicative of the extent to which the ligand is bound to the antibody, may be measured in a liquid scintillation counter.

A further advantage of tritium-based SPAs is that liquid scintillant addition can be avoided, thus reducing the costs and hazards associated with solvent-based scintillants.

Scintillation proximity technology is applicable both to ^{3}H- and ^{125}I-based RIA systems. The range of assays that has been developed using this technique is shown in Table 67.2.

With a few exceptions, SPAs can be developed by simple substitution of the existing separation system with the appropriate second antibody or Protein A-coupled fluomicrospheres, available from Amersham Biosciences.

Table 67.2. Scintillation proximity assay analytes.

Tritium	**Iodine-125**
Arachidonic acid metabolites and lipids	
Thromboxane B_2	Thromboxane B_2
6-Keto-prostaglandin $F_{1\alpha}$	6-Keto-prostaglandin $F_{1\alpha}$
Bicyclic prostaglandin E_2	11-Dehydro-thromboxane B_2
Prostaglandin D_2	Prostaglandin E_2
Prostaglandin E_2	
Leukotriene B_4	
Leukotriene $C_4/D_4/E_4$	
Platelet activating factor	
Steroids	
Progesterone	
Testosterone	
Cortisol	
Aldosterone	
Androstenedione	
Dehydroepiandrosterone	
11β-Hydroxyandrostenedione	
1α,25-Dihydroxy vitamin D	
Cyclic nucleotides	
Cyclic AMP	Cyclic AMP
Cyclic GMP	Cyclic GMP
Peptides/proteins/growth factors	
	Atrial natriuretic peptide
	Insulin
	Interleukins-1 and -2
	Tumor necrosis factor
	Platelet-derived growth factor
	Substance P
	Neuropeptide Y
	Cholecystokinin
	Albumin
	Cholesteryl ester transfer protein
Other analytes	
Abscisic acid	
Leukocyte antigen-related phosphatase	
Nuclear factor-kappa B (NF-κB)	
Iloprost	
Ranitidine	
Acyclovir	
Lacidipine	
Fucosyl-GM	

Assay Protocol

The simplicity of the protocol compared with traditional research assays can be seen from the example of an assay for the measurement of 6-keto-$PGF_{1\alpha}$, the stable metabolite of prostacyclin.

The choice of protocol is determined by the individual assay reagents, conditions, and laboratory convenience. In general, equilibrium is reached faster if sample, tracer, and first antibody are preincubated before the addition of the SPA reagent. This allows same-day counting in many assays. However, single incubation of all reagents is more suited to laboratories wishing to automate assay processes.

Speed of reaction is markedly increased if the tubes containing the SPA reagents are agitated to keep the fluomicrospheres in suspension. The use of orbital shakers is recommended for shorter incubation times.

Assays can be performed in any flat- or round-bottomed tubes compatible with direct counting in a liquid scintillation counter. Insufficient agitation is achieved in conical-shaped tubes. Further simplification is achieved by carrying out assays in 96 or 384-well microtiter plates that can then be counted in the Wallac MicroBeta™ or Packard TopCount™ instruments as an alternative to tubes. These new generation liquid scintillation counters use multiple detectors enabling the simultaneous determination of count rates in six or more samples with the resulting benefit of increased assay throughput.

Assay Performance

In general, assay sensitivity and dose–response curves are similar to those of heterogeneous assay systems. However, there have been a few examples where steric effects associated with the solid-phase antibody–antigen reaction reduce the sensitivity.

The separation procedure of any heterogeneous immunoassay normally has a greater influence on assay precision than any other aspect of the protocol. Consequently SPAs, in common with other homogeneous methodologies, exhibit better precision than heterogeneous methods.

A potential drawback of homogeneous assays is the need for antigen–antibody reaction equilibrium before end-point determination. This is particularly important in assays where the end-point determination is relatively slow. SPAs are dynamic because the reaction is not halted by the separation of bound from free antibody. Fast incubation times are therefore difficult to achieve, and overnight incubations are normally preferred.

THE APPLICATION OF SPA TECHNOLOGY TO THE MEASUREMENT OF PROTEIN: PROTEIN INTERACTIONS USING AN ANTIBODY CAPTURE FORMAT

Interactions between proteins are a key feature of many biochemical processes, e.g. cell signaling. In the absence of any enzymatic activity, measurement of protein–protein interactions has presented problems. SPA technology (Cook, 1996) permits the direct measurement of binding of one protein to another. In addition, in cases where the dissociation rate of the interaction is high, SPA provides a means of equilibrium counting in a homogeneous assay format. The first SPA assay for studying protein–protein interactions involved measurement of the direct binding of the GTP-ase activating protein, neurofibromin-1 (NF1) to an oncogenic Ras protein (Leu-61 Ras) (Skinner *et al.*, 1994). The following example highlights the versatility of SPA technology.

Antibody Capture Format

There are several alternative assay concepts potentially applicable to a protein–protein interaction assay. The NF1–Ras SPA format exploited the fact that the NF1 protein was expressed as a GST (*Schistosoma japonicum* glutathione-*S*-transferase) fusion protein, i.e. GST–NF1. The GST–NF1 fusion protein was coupled to a polyvinyltoluene (PVT) Protein A SPA bead via a commercially available anti-GST antibody, as shown schematically in Figure 67.1.

An obvious alternative bead for this type of format would be one of the anti-species antibody-binding beads, e.g. anti-rabbit, anti-mouse, or anti-sheep, according to the source of the capture antibody. The specificity, affinity, and purity of the antibody will effect the maximum level of signal achievable and choice of bead. This format offers a generic design for the capture of any protein when a suitable antibody exists. An SPA assay utilizing Protein A beads and an anti-GST antibody has recently been published for the measurement of the binding of GST-Raf (a serine/threonine kinase) to Ras (Gorman *et al.*, 1996).

Factors affecting the signal:noise ratio will be the relative amounts of each reagent present, the quality and affinity of the capture antibody, and the affinity of the two proteins participating in the interaction.

Antibody

For all assays using an antibody capture format, the purity and affinity of the antibody will greatly influence the amount of antibody (and hence the amount of bead) required for maximal signal. It will be essential that the antibody:protein interaction does not interfere with the protein:protein interaction.

For the NF1–Ras assay, the initial experiment assessed the titer of the anti-GST antibody required. The anti-GST antibody was diluted in assay buffer and a constant volume of each dilution added to each assay well. Maximal binding was achieved using a dilution of 1:20 (Figure 67.2).

Assay Validation

The specificity of the NF1–Ras interaction was determined using unlabeled GTP, ATP, and GDP over a range of concentrations in a SPA competition binding assay (see Figure 67.3). GTP appeared to have an inhibitory effect on the NF1–Ras interaction under normal assay conditions with IC_{50} values of 0.1–1.0 mM being observed.

Skinner *et al.*, 1994 used a neutralizing anti-Ras monoclonal antibody Y13-259 and a detergent *N*-dodecyl maltoside (a specific inhibitor of NF1 catalytic activity) to

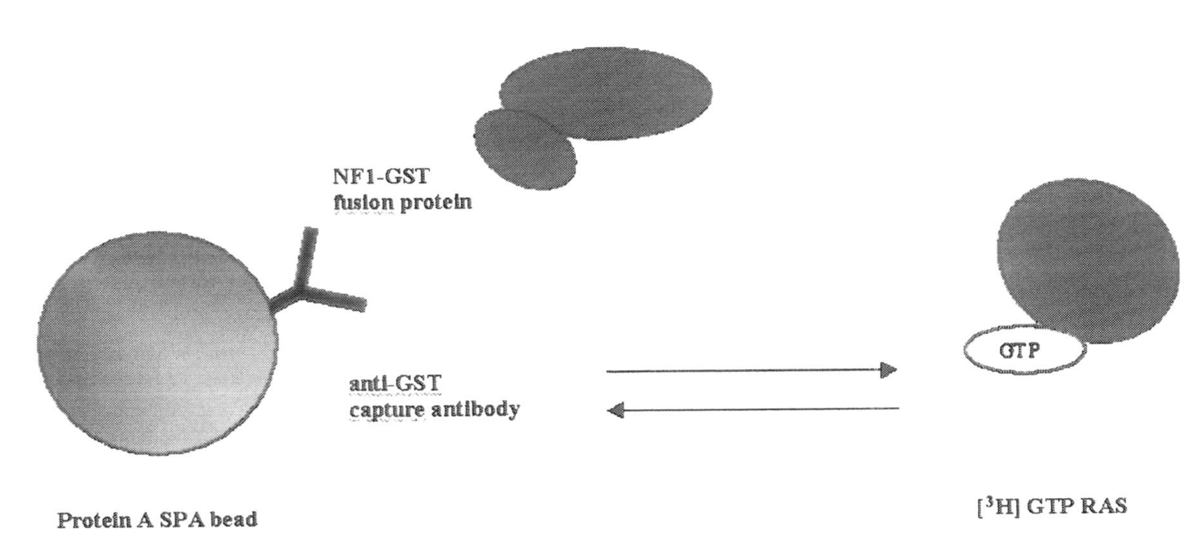

Fig. 67.1 Schematic representation of the NF1–Ras SPA utilizing Protein A SPA beads and anti-GST antibody.

demonstrate specific binding. Both reagents abolished the signal from the NF1/Ras SPA but neither affected the signal from a control SPA in which a [^{3}H]GTP·GST–Ras fusion protein was bound to Protein A beads.

The interaction between Ras and NF1 is known to show a high degree of specificity for the GTP-bound form of Ras. This was demonstrated using both Ras [^{3}H]GTP and Ras [^{3}H]GDP (Figure 67.3) where no binding of Ras [^{3}H]GTP to GST.NF1 was detectable.

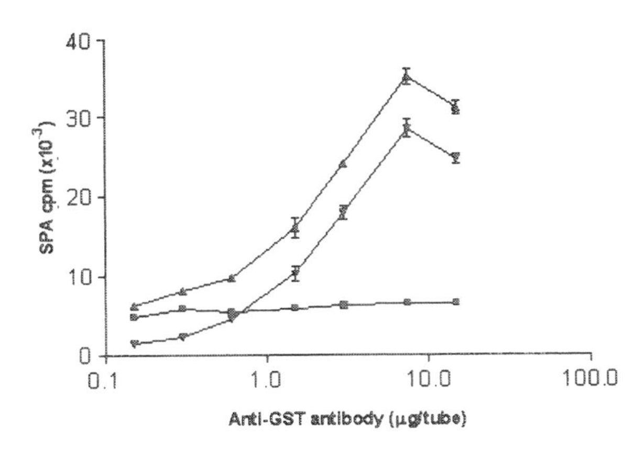

Fig. 67.2 The effect of anti-GST antibody dilution on the binding of H-RasL61·GTP to GST–NF1 in the presence of 1 mg Protein A bead in 50 mM Tris–HCl, pH 7.5, 2 mM dithiothreitol. ■, SPA cpm bound in the absence of GST–NF1; ▲, SPA cpm bound in the presence of 30 pmol/well GST–NF1; ▼, specific SPA cpm bound. (Results are means ± SEM, n = 3.)

The kinetics of the NF1–Ras interaction and also the Ras–Raf interaction have been studied (Skinner *et al.*, 1994; Gorman *et al.*, 1996; Sermon *et al.*, 1996). The homogeneous nature of the SPA technology permits the kinetic analysis of such interactions despite the very fast dissociation rates (>1 min^{-1}). Skinner *et al.*, 1994 were able to calculate the affinity of the NF1–Ras interaction, giving a K_d of approximately 40 nM.

In summary, the measurement of protein:protein interactions in the absence of enzymatic activity is feasible using SPA technology. The NF1–Ras SPA format demonstrates that interactions of micromolar affinity and

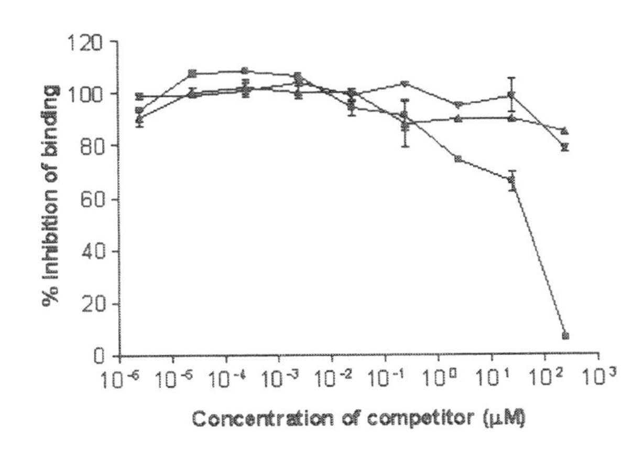

Fig. 67.3 Inhibition of H-RasL61·[^{3}H]GTP to GST–NF1 by GTP (■), ATP (▼), and GDP (▲). Assay conditions as for Figure 67.2 using 10 pmol H-RasL61·[^{3}H]GTP and 40 pmol GST–NF1. (Results are means ± SD, n = 2.)

fast dissociation rates can be measured using SPA, permitting kinetic analysis and inhibitor screening.

Indeed, SPA greatly simplifies RIA protocols. Only four pipetting steps are required, thus reducing assay time. Assays are carried out in one tube and the addition of liquid scintillant is not required. Assays can be automated, something difficult to achieve with traditional heterogeneous assays. Furthermore, when used in conjunction with imaging devices and technologies, SPA can greatly facilitate assay throughput, thereby significantly assisting with the discovery of new pharmacologically active compounds during the drug discovery process (see next section).

IMAGING TECHNOLOGIES AND INSTRUMENTATION FOR ULTRA HIGH-THROUGHPUT DRUG-SCREENING

In recent years, there has been a demand from the pharmaceutical industry to increase both the number of high-throughput screens run and the number of compounds being put through the screen. This is because of the increased numbers of drug targets that have been identified from the sequencing of the human genome and the increased rate of synthesis of novel chemical compounds. The rate-limiting factor for throughput had been the instrumentation for reading microplates, particularly with radioactive assays, where a maximum of 12 wells per 96 or 384-well plate could be read at any one time, taking up to 32 min per plate using a 1 min count per well. The solution to this problem has been to use a charge-coupled device (CCD) camera imaging system, which can image a complete microplate in one read. An example of this type of system is the LEADseeker™ multimodality imaging system (Figures 67.4 and 67.5).

The light source, filters, and dichroic mirrors in the instrument make it capable of reading microplates using five different assay modalities – radiometric, luminescent, steady-state fluorescence, time-resolved fluorescence, and fluorescence polarization. It can read any microplate that conforms to the standard footprint and so can be used to read 96, 384, or 1536-well microplates. The telecentric lens enables it to gather a consistent image of all the wells of the microplate. The CCD chip in the camera is cooled to $-100\,°C$ to eliminate any dark current and give maximum sensitivity of detection.

An increase in well density per plate means that more samples can be run at any one time and assay volumes can be significantly reduced, thereby reducing the amount of reagent required per assay point. Typical assay volumes are 200 μL for 96 well, 50 μL for 384 well and 5 μL for 1536-well plates. Microplates are imaged for between 20 sec and 5 min depending on the assay modality. The throughput requires that all the assays are carried out robotically and the instruments are designed to be built in to integrated robotic screening systems.

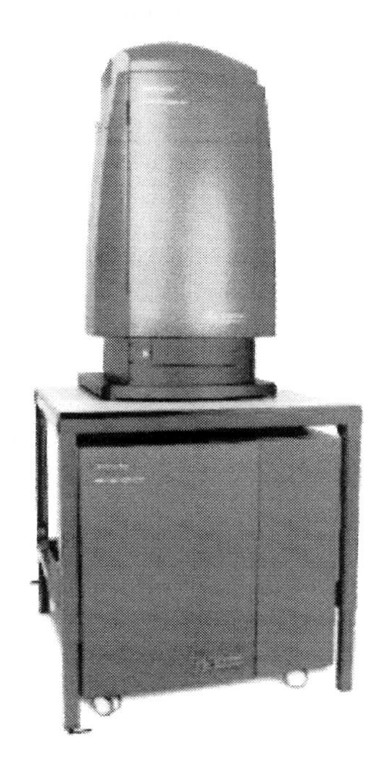

Fig. 67.4 The LEADseeker multimodality system.

The 'chips' in CCD cameras have maximal sensitivity to light in the red region of the spectrum (~600 nm), so detection reagents used in imaging assays need to emit light in this region. For the LEADseeker multimodality imaging system these is a proprietary set of reagents including SPA imaging beads, and for non-radioactive detection, CyDye™ fluorescent dyes and europium terpyridine-bis(methyl-enamine)tetra acetic acid (TMT) time-resolved fluorescent chelates.

One example of an immunoassay which has been run on the LEADseeker multimodality imaging system is the radiometric assay for measuring cAMP (*see* SIGNAL TRANSDUCTION), a technique commonly used for the discovery of new pharmacologically active compounds active at G-protein-coupled receptors. This was carried out in a 384-well microplate in a final volume of 80 μL. The assay was set up as previously described except that the SPA beads, which emit light at 420 nm and are designed for use in detectors with photomultiplier tubes, were replaced by SPA imaging beads. These were Protein A-coated polystyrene beads with a 2 μm diameter that contain a proprietary europium chelate. When the bead is stimulated to emit light by Auger electrons from the [^{125}I]cAMP, the light emitted has a wavelength of 615 nm which can then be detected by the CCD camera. A false color image of a uniform plate from the camera is shown in Figure 67.6. The image is

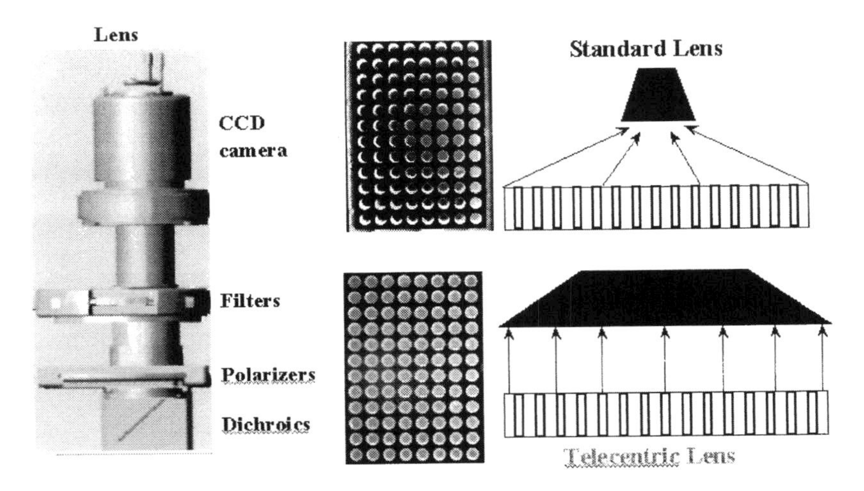

Fig. 67.5 The LEADseeker multimodality imaging system showing telecentric lens.

converted to numerical data, which are expressed as integrated optical density (IOD) units.

A typical standard curve is shown in Figure 67.7.

ENZYME IMMUNOASSAYS

RIA is highly sensitive, accurate, and precise but it does have several significant disadvantages, such as the short isotopic half-life of iodine-125. In addition there are growing concerns about the safety and disposal of radioactive compounds. As in the clinical field, this has stimulated an intensive search for practical alternatives.

Many different non-isotopic signal generation systems have been tried over the last 20 years. However, very few of these have approached RIA in terms of sensitivity, convenience, and precision. One of the more popular approaches has been the use of enzyme labels which, when substrate is added, generate color that can be measured by a spectrophotometer. Some of these tracers, particularly those based on horseradish peroxidase (HRP), have provided assays with excellent sensitivity. The convenience of these enzyme immunoassays (EIA) has also been markedly improved by the combination of microtiter® plates and automated multiscan spectrophotometers. Microtiter plates provide a very convenient format for the separation and washing stages necessary in sensitive heterogeneous assays.

The move from radioactivity to non-isotopic signal generation systems has been much slower in life-science research than in clinical laboratories. It is only in the newer areas of research, perhaps where RIA methodology is less entrenched, that non-isotopic methods have gained wide acceptance. There is now, however, an increasing movement away from radioactive techniques.

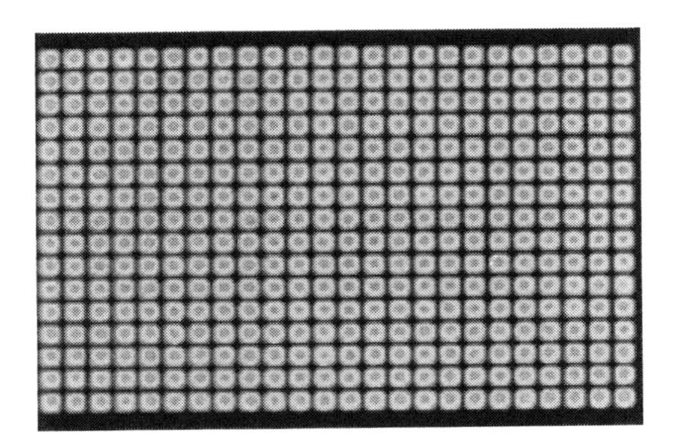

Fig. 67.6 Image of a 384-well microplate.

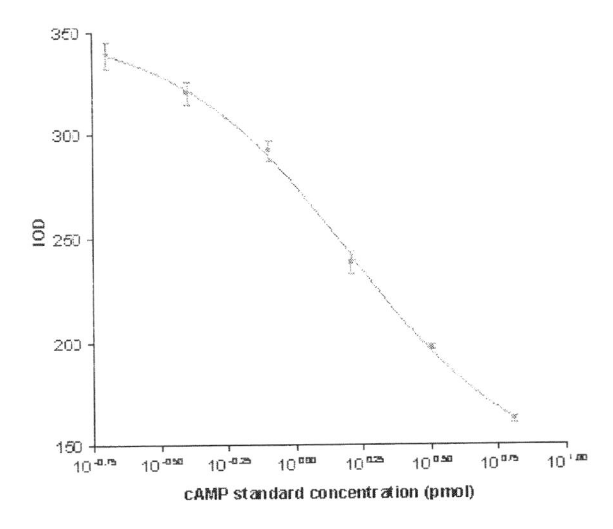

Fig. 67.7 Typical standard curve for imaging cAMP assay.

One reason for the slow acceptance of EIA in research laboratories is the general perception that enzyme-based assays are less sensitive and less precise than tube-based RIAs. In our laboratories we have found sensitivity to be enhanced in EIAs for haptens by as much as 10-fold without compromising precision. This has been achieved by careful optimization of assay parameters. In addition, assays can be carried out and results processed much more rapidly. The simple protocol for thromboxane B_2 (TxB_2) serves as a good example of a competitive EIA. The assay for interleukin-2 (IL-2) is also described to illustrate an immunometric EIA.

Assay Protocol

Competitive

This competitive assay uses pre-coated anti-rabbit IgG microtiter strips, which fit into a standard 96-well plate holder. A specific rabbit primary antibody is immobilized onto the second antibody-coated plate enabling TxB_2 and TxB_2 labeled with peroxidase to compete for binding sites. The amount of peroxidase-labeled TxB_2 that binds to the antibody is inversely proportional to the concentration of native TxB_2 in the well. After incubation for 1 h at room temperature, unbound TxB_2 is removed by washing the wells four times. Substrate is added, enabling color development in proportion to the amount of peroxidase-labeled TxB_2 bound to the well. The concentration of TxB_2 in a sample can be determined by interpolation from a calibration curve.

Immunometric

Immunometric assays are suitable for compounds of high molecular mass that contain more than one epitope. IL-2 in the sample is captured onto microtiter wells pre-coated with a specific antibody to one epitope of IL-2. The plate is washed, and an excess of a second antibody specific to a second epitope of IL-2 and labeled with peroxidase is added. The plate is incubated and any unbound antibody removed by washing the plate four times. Substrate solution is added and the color development is proportional to the amount of peroxidase-labeled anti-IL-2 bound, which reflects the amount of IL-2 in the sample.

REFERENCES AND FURTHER READING

Baxendale, P.M., Martin, R.C., Hughes, K.T., Lee, D.Y. and Waters, N.M. Development of scintillation proximity assays for prostaglandins and related compounds. *Adv. Prost. Thromb. Leuk. Res.* **21A**, 303–306 (1990).

Benedetto, C., McDonald-Gibson, R.G., Nigam, and Slater, T.F. *Prostaglandins and Related Substances, A Practical Approach* (IRL Press, Oxford, 1987).

Bosworth, N. and Towers, P. Scintillation proximity assay. *Nature* **341**, 167–168 (1989).

Cook, N.D. Scintillation proximity assay: a versatile high-throughput screening technology. *DDT* **1**, 287–294 (1996).

Gorman, C., Skinner, R.H., Skelly, J.V., Neidle, S. and Lowe, P.N. Equilibrium and kinetic measurements reveal rapidly reversible binding of Ras to Raf. *J. Biol. Chem.* **271**, 6713–6719 (1996).

Kelly, R.W., Graham, B.J.M. and O'Sullivan, M.J. Measurement of PGE_2 as the methyl oxime by radioimmunoassay using a novel iodinated label. *Prost. Leuk. Essential Fatty Acids* **27**, 187–191 (1989).

O'Sullivan, M.J. Enzymeimmunoassay. In: *Practical Immunoassay* (ed Butt, W.R.), 37–69 (Marcel Dekker, New York, 1984).

Powell, W.S. Rapid extraction of metabolites of arachidonic acid from biological samples using octadecylsilyl silica. *Prostaglandins* **25**, 947–957 (1980).

Sermon, B.A., Eccleston, J.F., Skinner, R.H. and Lowe, P.N. Mechanism of inhibition of arachidonic acid of the catalytic activity of Ras GTP-ase-activating proteins. *J. Biol. Chem.* **271**, 1566–1572 (1996).

Skinner, R.H., Picardo, M., Gane, N.M., Cook, N.D., Morgan, L., Rowedder, J. and Lowe, P.N. Direct measurement of the binding of RAS to neurofibromin using a scintillation proximity assay. *Anal. Biochem.* **223**, 259–265 (1994).

ANALYTES

PROSTAGLANDINS

Background

Arachidonic acid metabolism via the cyclo-oxygenase pathway leads to the formation of prostaglandin H_2 (PGH_2). A number of biologically active compounds including prostacyclin, thromboxane A_2 (TxA_2), prostaglandin D_2, prostaglandin E_2 (PGE_2), and prostaglandin $F_{2\alpha}$ ($PGF_{2\alpha}$) are formed from PGH_2.

These prostaglandins have a broad range of biological effects and have been implicated in a number of disease states such as atherosclerosis, cancer, and inflammatory disorders. Due to their instability *in vivo*, immunoassays have generally been developed for the stable metabolite of the active compound. Another approach has been to rapidly stabilize the compound *in vitro* and assay the derivative. It is also important to include a cyclo-oxygenase inhibitor during sample collection to prevent any artifactual prostaglandin formation prior to assay.

Useful reviews of methods of handling prostaglandins and leukotrienes are given by Benedetto *et al.* (1987) (see previous section for reference).

Thromboxane B_2

Thromboxane A_2 (TxA_2) is a labile bicyclic compound formed from prostaglandin endoperoxides. It has a half-life of about 30 sec and is rapidly hydrolyzed to its stable, biologically inactive metabolite, thromboxane B_2 (TxB_2).

The short half-life of TxA_2 impedes its measurement at physiological concentrations. Consequently, the measurement of TxB_2 is generally accepted as an indicator of TxA_2 production. This compound has a physiological half-life of 20–30 min and, more importantly, it can be readily quantified by immunoassay. More recently it has been suggested that 11-dehydro-thromboxane B_2 is a better index of TxA_2 production, because it is less likely to be found as a sampling artifact.

TxA_2 is a potent inducer of platelet aggregation and it also acts as a vasoconstrictor. The most active antagonist of TxA_2-induced platelet aggregation is prostacyclin (PGI_2). These two compounds act on platelets by influencing the activity of adenylate cyclase. PGI_2 activates and TxA_2 inhibits this enzyme. The opposing effects of TxA_2 and PGI_2 on adenylate cyclase have fostered the theory that platelet homeostasis is dependent on a 'reciprocal regulation' of cAMP levels by PGI_2 and TxA_2.

Platelet activity may influence atherosclerotic cardiovascular disease, so researchers are very interested in estimating TxA_2 production through the use of an immunoassay for TxB_2.

6-Keto-prostaglandin F1α

Prostacyclin, also known as prostaglandin I_2 (PGI_2), is an unstable vinyl ether formed from the prostaglandin endoperoxide, prostaglandin H_2. This conversion of PGH_2 to prostacyclin is catalyzed by prostacyclin synthetase. The two primary sites of synthesis are the veins and arteries. Prostacyclin has biological properties opposing the effect of thromboxane A_2. Prostacyclin is a vasodilator and a potent inhibitor of platelet aggregation whereas thromboxane A_2 is a vasoconstrictor and a promoter of platelet aggregation. A physiological balance between the activities of these two effectors is probably important in maintaining a healthy blood supply.

Prostacyclin is unstable and undergoes a spontaneous hydrolysis to 6-keto-prostaglandin F1α (6-keto-PGF1α). Study of this reaction *in vitro* established that prostacyclin has a half-life of about 3 min. This half-life increases to 9–23 min in platelet-poor plasma. Because of this spontaneous hydrolysis of prostacyclin, the quantification of 6-keto-PGF1α is accepted by many researchers as a measure of prostacyclin formation.

Prostaglandin E_2

PGE_2 has numerous biological activities including involvement in human parturition, respiratory function, protection of the gastrointestinal tract, blood pressure regulation, and macrophage function. It has a short half-life in the bloodstream and is rapidly metabolized to 13,14-dihydro-15-keto-PGE_2. However, the use of this dihydro metabolite to monitor PGE_2 formation is complicated by subsequent chemical reactions. The metabolite readily dehydrates to give PGA_2, which can bind covalently to albumin. Fortunately all three metabolites can be converted by alkaline treatment to bicyclic PGE_2, which can be readily quantified by RIA.

In certain situations it may still be desirable to measure PGE_2 itself but this can present difficulties because PGE_2 is readily converted to the dehydration artifacts, PGA_2 and PGB_2. The development of a PGE_2 immunoassay involves several steps where dehydration can occur, such as during coupling to protein to produce an antigenic hapten and during antiserum production in the animal. In addition, PGE_2 in biological samples can degrade during storage. These problems can be circumvented by forming a methyl oxime of PGE_2 by a simple oximation procedure. Antiserum has been raised against PGE_2-9-methyl oxime acid; the calibrators and tracers are similarly methyl oximated.

More recently, Amersham Biosciences has introduced a direct EIA for measurement of PGE_2 without the need for a meth-oximation step. The method involves the use of a highly specific monoclonal antibody to PGE_2. The assay is commonly used in drug discovery laboratories in the quest for new NSAIDs, where minimal sample preparation and throughput is vital. The sensitivity of the assay is equivalent to, if not better than, the corresponding radioactive assay.

Sample Preparation

Prostaglandin E_2 is present in a wide variety of biological material. Samples should be processed immediately after collection and assayed as soon as possible. Although prostaglandin E_2 is reported to be stable in urine if kept frozen, the level of prostaglandin E_2 in plasma decreases significantly if stored at $-20\,°C$ for 1 week. Samples should be purified and derivatized to the methyl oximate as quickly as possible and stored at $-20\,°C$ in their derivatized state. Samples are stable for up to 6 days under these conditions.

Blood samples should be collected into a tube containing an anticoagulant (e.g. 3.8% w/v citrate in 0.9% w/v NaCl solution) and kept on ice. The blood should be centrifuged immediately at 2500*g* for 10 min at 4 °C. If blood samples cannot be rapidly processed, then the addition of indomethacin or aspirin (1% w/v final concentration) to the anticoagulant is recommended. Either of these compounds will inhibit the subsequent metabolism of arachidonic acid to prostaglandins.

Sample Purification

Non-esterified fatty acids can interfere with prostaglandin assays. Several methods are available for purifying PGE_2. An example is the solid-phase extraction procedure developed by Kelly *et al.* (1989) (see references in previous section), as follows.

Pipette 0.5 mL plasma sample, 0.5 mL of 1:4 water: ethanol, and 10 μL glacial acetic acid into a 1.5 mL microfuge polypropylene tube. Gently mix and leave at room temperature for 5 min. Centrifuge at 2500*g* for 2 min. Remove the supernatant and apply to an Amprep® C18 (100 mg size, product code RPN1900) minicolumn which has been primed with 2 column volumes of 10% ethanol.

Wash the column with 1 column volume of distilled water and 1 column volume of hexane. Elute the prostaglandin E_2 twice with 0.75 mL ethyl acetate. Collect the ethyl acetate fractions and evaporate to dryness under nitrogen. Using this method, levels of prostaglandin E_2 in normal plasma have been reported at approximately 5 pg/mL. This value agrees with previously reported levels of prostaglandin E_2 in plasma of between 3 and 15 pg/mL.

More recently, Amersham Biosciences have introduced a number of cell lysis reagents that assist the processing and assay of PGE_2 generated from cultured cells. These reagents are combined with the EIA technology described above.

LEUKOTRIENES

Background

A diverse array of mammalian cells oxidize arachidonic acid to physiologically active compounds including thromboxanes, prostacyclin, prostaglandins, and leukotrienes. The biological properties of leukotrienes suggest that they are potent mediators in inflammatory and allergic disease. The enzyme 5-lipoxygenase can convert arachidonic acid into 5-hydroperoxyeicosatetraenoic acid (5-HPETE), which in turn can be converted to leukotriene A_4 (LTA_4), the parent leukotriene. Subsequently LTA_4 can be converted to leukotriene B_4 (LTB_4) by LTA_4 epoxide hydrolase or to leukotriene C_4 (LTC_4) by glutathione-*S*-transferase. LTC_4 can be further metabolized to leukotriene D_4 (LTD_4) and leukotriene E_4 (LTE_4).

Leukotriene B_4

It is widely considered that LTB_4 is a mediator involved in chronic inflammation. It is a potent stimulator both of random and directional movement of polymorphonuclear leukocytes (PMNs). It induces adhesion of leukocytes to the endothelial cells of postcapillary venules and aggregation of PMNs. It also causes degranulation and the release of lysosomal enzymes from PMNs.

Peptido-leukotrienes

The peptido-leukotrienes (LTC_4, LTD_4, and LTE_4) have been characterized as the slow-reacting substances of anaphylaxis and are considered to be potent mediators of hypersensitivity reactions. They cause bronchoconstriction, vascular and non-vascular smooth-muscle contraction, increased vascular permeability, and epithelial mucous secretion. It is believed that the peptido-leukotrienes are mediators involved in asthma and considerable effort has been spent on developing potent antagonists as potential therapeutic agents.

Assay Technology

Immunoassays have been widely used in an attempt to elucidate the physiological role and pathological significance of leukotrienes. However, considerable methodological hurdles have had to be overcome. These include the short biological half-life of some of these compounds, their chemical instability, their cost and availability, the low concentration in some biological samples, and their structural similarity to many endogenous compounds. Many of these problems have now been overcome by improved synthetic procedures, by the availability of high specific activity tracers, and the introduction of highly specific, high-avidity antisera.

Assay Specificity

Antisera are generally screened in an attempt to identify those antibodies that recognize the active compound rather than structurally related inactive compounds. For instance, most LTB_4 antibodies are chosen to be specific for LTB_4 rather than its inactive metabolites 20-OH-LTB_4 or 20-COOH-LTB_4. However, absolute specificity may not always be appropriate. LTC_4 is rapidly converted, in some situations, to LTD_4 and LTE_4. Both of these compounds are biologically active and can have their own specific receptors. Peptido-leukotriene assays have been designed that are specific for LTC_4. Alternatively, antibodies that cross-react with all three peptido-leukotrienes have also proved useful. These have been used to measure total peptido-leukotriene formation. There are sometimes situations in which it may be advantageous to measure inactive metabolites rather than the parent compound. This will be discussed in a later section. Clearly the antiserum specificity should always be appropriate to the analytical problem.

Sample Purification

Leukotriene RIAs have been applied to many diverse types of samples, components of which can potentially interfere with the assay. The degree of purification required before the assay depends on the sample. In some situations it may be possible to assay leukotrienes such as LTB_4 without prior extraction and chromatography. For example, LTB_4 in supernatants derived from incubation of human PMNs with the calcium ionophore A23187 can be determined directly. Alternatively, solid-phase extraction techniques using commercially available reverse-phase silica minicolumns have proved very popular. The procedure described by Powell (1980) (see references in previous section) is widely used. It is strongly recommended that all new sample types should be validated before RIA results are accepted. Typical schemes should include recovery and sample dilution experiments. High extraction efficiency of tritium-labeled compounds is not proof of a valid procedure for sample purification. A very valuable additional test is to apply the partially purified sample to an appropriate high-performance liquid chromatography (HPLC) system. All immunoreactivity recovered from the column should co-elute with pure material. Leukotrienes are typically measured by competitive RIA or EIA.

RAT HORMONE ASSAYS

Background

The measurement of pituitary hormones is becoming increasingly common in toxicology laboratories around the globe. These hormones have been recognized as vital markers in studying the toxicology of potential therapeutic agents. Generally studies are carried out using control and treated animals, and then comparing levels found. However, as in any study involving animals many other factors may come into play. This is a brief attempt to examine the factors that should be borne in mind when measuring pituitary hormone levels in the rat. Most of this work has been abstracted from the literature whilst some has been carried out at our own laboratories.

Follicle-Stimulating Hormone

Rat follicle-stimulating hormone (FSH) is a glycoprotein of molecular weight approximately 36 kDa, comprising two subunits. Both FSH and luteinizing hormone (LH) are termed gonadotropins. The basophilic cells of the anterior pituitary secrete these hormones. FSH acts in females to stimulate the development of ovarian follicles, while in males it stimulates spermatogenesis. The role of the adrenal gland in the regulation of FSH, LH, and prolactin in the lactating rat has been studied. The pituitary–adrenal system seems capable of influencing the maintenance of normal secretion of gonadotropins and prolactin as well as the maintenance of ovarian function in the rat.

Growth Hormone

Rat growth hormone (GH), also called somatotropin, is a pituitary hormone of molecular weight approximately 22 kDa. It is a simple polypeptide hormone with no carbohydrate or lipid moieties. As its name implies this hormone stimulates growth. Unlike most of the pituitary hormones GH does not have a single target gland. It can promote growth in bone, soft tissues, and viscera. GH has also been shown to play an auxiliary role in lactation, which may be important in times of prolactin insufficiency.

Luteinizing Hormone

Rat luteinizing hormone (LH) is a glycoprotein of molecular weight approximately 36 kDa, comprising two subunits. Both FSH and LH are termed gonadotropins. The basophilic cells of the anterior pituitary secrete these hormones. LH acts in females to induce ovulation and then maintain the secretory function of the corpus luteum. In males LH acts by stimulating the Leydig cells of the testes to produce testosterone. The secretion of LH and FSH is regulated by a single gonadotropin-releasing hormone produced in the hypothalamus. In turn the secretions of the hypothalamic-pituitary unit are regulated by a negative feedback mechanism involving steroid hormones, gonadotropins, and inhibin.

Prolactin

Prolactin is a pituitary hormone with a molecular weight of approximately 23 kDa. It is a single polypeptide of about 200 amino acids with three disulfide bonds. In mammals prolactin has been claimed to exert a wide range of physiological effects. These include stimulation of mammary gland development and lactation, hair maturation, synergism with androgen in male sex accessory growth, and maintenance and secretion in the corpus luteum. Prolactin is predominantly under inhibitory control by the hypothalamus. Stimulation of prolactin can be mediated by dopamine and thyrotropin-releasing factor.

Thyroid-Stimulating Hormone

Thyroid-stimulating hormone (TSH), often referred to as thyrotropin, is a glycoprotein hormone of molecular weight approximately 28 kDa, comprising two linked subunits (alpha and beta). The structure of the alpha subunit is similar for TSH, LH, and FSH; however, the beta subunit is different in each hormone. TSH is synthesized and secreted by specific basophil cells of the anterior pituitary called thyrotrophs. TSH stimulates hormonogenesis in the thyroid gland and release of the thyroid hormones, thyroxine (T_4) and triiodothyronine (T_3). T_4 and T_3 levels exert negative feedback on the pituitary to decrease the secretion of TSH. Numerous sex-related differences in thyroid functions have been reported in the rat. In particular, lower plasma T_4 and TSH concentrations are associated with a higher plasma level of T_3 in the female than in the male.

Factors Affecting Sample Values

The hormones of the anterior pituitary in the rat have been extensively studied in the past, and many authors have noted that each of the hormones exists as a heterogeneous mixture. This heterogeneity can be both pre- and post-translational. The charge heterogeneity has been studied extensively using isoelectric focusing. In addition to charge heterogeneity there have been extensive reports of differences in glycosylation patterns. These issues have been noted widely in human as well.

In addition to variations in hormone composition the laboratory rat itself can provide the researcher with challenges. For example, it is widely documented that stress will profoundly affect the levels of the pituitary hormones, especially prolactin. For this reason when commencing studies on hormone levels it is essential to accustom the animals to any procedure which may produce stress. For example if the subject animals are to be bled on multiple occasions the animals should be introduced to such a procedure before beginning the study. Animals that are not used to manual handling will also produce high levels of hormone; for this reason many laboratories use only animals that have been introduced to manual handling.

It is also important to recognize that the site and method of blood sampling can profoundly affect the results. In some circumstances where multiple blood samples are taken cannulation may be preferable to venous puncture. Similarly if the animals are to be exsanguinated the type of anesthetic can affect the levels. Also the choice of bleeding procedure will definitely affect the levels in serum.

Whilst stress can have a massive effect on hormone levels it is clear that other factors can also affect the levels. Amongst the most important of these are the environmental conditions under which the animals are maintained. It is possible to show that not only does the light–dark cycle affect the hormone levels but also the 'temperature' of the light. Similar though less pronounced effects can be observed with the type of diet used. Similarly the production of some hormones is dependent on the point in the diurnal cycle. For example GH is produced by the male rat in fairly large pulses about every 3 h; however, the female rat seems to produce smaller pulses of GH production but more frequently.

When using female rats it is also critical to remember that the hormone levels will vary enormously during the estrus cycle. Obviously the use of pregnant animals in the study should be carefully monitored.

In addition to the above factors many investigators recognize the significance of the exact strain of rat used in any study. For many authors the typical inbred strain of rat may be the Sprague–Dawley. However, some drug companies have become so concerned at variation in this area that they have developed their own in-house breeding program to produce a company-specific rat.

One final significant factor to bear in mind when designing a study is the age of the animals. Several studies have shown that hormone levels vary significantly with the age of the animal. Similarly if ageing animals are to be used then behavioral problems leading to fighting and stress should be borne in mind.

The conclusion from these observations is that any study using rats will require internal control. Most groups now use control and treated animals, and these should be treated in an identical manner, including admission of suitable placebos to the control group. Careful note should be taken of environmental conditions in both groups. Stress seems to affect all the pituitary hormones, and handling the animals regularly can help overcome this. However, it is also important to recognize that the scientist or technician responsible for handling the animals or taking blood samples has been thoroughly trained. We have observed significant differences between serum levels of hormones in samples taken by experienced versus inexperienced staff (see Table 67.3).

Assay Technology

Traditional RIAs have been widely used in this area. In general iodine tracers have been used in combination with second antibody separations. The commercially available kits from Amersham Biosciences use magnetizable beads to separate the bound and free fractions. Amersham Biosciences have a range of EIAs, and these non-radioactive assays complement the existing range of RIAs. Each rat hormone EIA is based on the competition between unlabeled hormone and a fixed quantity of biotin-labeled hormone for a limited amount of hormone-specific antibody. The labeled ligand that is bound to the antibody is immobilized on precoated microtiter wells. After washing, Amdex™ amplification reagent is added. Any immobilized labeled ligand will bind Amdex. Following a wash to remove any unbound amplification reagent, a substrate is added to the wells. Color develops in proportion to the amount of hormone present in the initial step. The Amdex amplification reagent is a high-performance conjugate, based on an exceptional chemistry, that utilizes a hydrophilic straight chain dextran backbone to which many hundreds of HRP molecules are covalently coupled, together with,

Table 67.3. Summary of factors that may affect hormone levels.

Stress	Experience of staff
	Manual handling
Rat	Strain
	Age
Environmental conditions	Lighting type
	Lighting cycle
	Temperature
	Diet
Blood sampling	Time of day
	Method of bleeding
	Anesthetic used
Miscellaneous	Estrus cycle

on average, 10 streptavidin molecules. The result is a multifunctional conjugate with a significantly enhanced activity and well-controlled non-specific binding properties.

REFERENCES AND FURTHER READING

Baldwin, D.M. *et al.* An *in vitro* study of LH release, synthesis and heterogeneity in pituitaries from proestrous and short-term ovariectomized rats. *Biology Reprod.* **34**, 304–315 (1986).

Boockfoor, F.R. *et al.* Cultures of GH3 cells are functionally heterogeneous: thyrotropin-releasing hormone, estradiol and cortisol cause reciprocal shifts in the proportions of growth hormone and prolactin secretors. *Endocrinology* **117**, 418–420 (1985).

Brelje, T.C. *et al.* Regulation of islet beta-cell proliferation by prolactin in rat islets. *Diabetes* **43**, 263–273 (1994).

Chi, H.J. and Shin, S.H. The effect of exposure to ether on prolactin secretion and the half-life of endogenous prolactin in normal and castrated male rats. *Neuroendocrinology* **26**, 193–201 (1978).

Laakso, M.-L. *et al.* Spectral properties of light affect plasma and pituitary gonadotropins in male rats. *Hormone Res.* **27**, 84–94 (1987).

Mattheij, J.A. and van Pijkeren, T.A. Plasma prolactin in undisturbed cannulated male rats; effects of perphenazine, frequent sampling, stress and castration plus oestrone treatment. *Acta Endocrinol.* **84**, 51–61 (1971).

Ozawa, K. and Wakabayashi, K. Dynamic change in charge heterogeneity of pituitary FSH throughout the estrous cycle in female rats. *Endocrinol. Jpn.* **35**, 321–332 (1988).

Szkudlinski, M.W. *et al.* Asparagine-linked oligosaccharide structures determine clearance and organ distribution of pituitary and recombinant thyrotropin. *Endocrinology* **136**, 3325–3330 (1995).

Wakabayashi, K. Heterogeneity of rat luteinizing hormone revealed by radioimmunoassay and electrofocusing studies. *Endocrinol. Jpn.* **24**, 473–485 (1977).

CYTOKINES

Background

Cytokines can be described as secreted regulatory proteins or glycoproteins that act as chemical communicators between cells. They include growth factors, colony-stimulating factors, monokines, lymphokines, chemokines, interferons, and interleukins. Cytokines bind to specific receptors on the surface of target cells that are coupled to intracellular signal transduction pathways. Most cytokines are growth and differentiation factors that generally act on cells within the hematopoietic system (red cells, platelets, phagocytic cells, and lymphocytes). They are often involved in immune and inflammatory responses.

Before the advent of molecular biology cloning techniques this whole area was very confused, consisting mainly of observations of a large number of disparate biological activities present in complex biological samples. With the availability of pure proteins, it became possible to correlate these activities with individual proteins and to begin to understand the function of cytokines. Even so the situation is still difficult to understand, as cytokines often have many different biological activities and the same biological response can be mediated by several different cytokines.

The cytokine area is a very active field of research. An appropriate immune response is essential for resolving infectious disease. An inappropriate response can lead to chronic disease, inflammation, and allergy. The main aim of research in this area is to manipulate the immune response so as to inhibit its harmful aspects – as seen in inflammatory diseases such as rheumatoid arthritis and allergies such as asthma – while maximizing its defensive properties against pathogenic organisms.

Given the complexity of the field, it is difficult to gain an overview of the importance of individual or groups of cytokines. Descriptions of the cytokines often include large lists of *in vitro* activities that may or may not be important *in vivo*. Attempts have been made to classify cytokines in terms of three-dimensional structure and receptor interactions. An alternative classification is based on function. The following functional classification is intended for guidance rather than as a rigid system.

Pro-inflammatory Cytokines

A number of cytokines are important in the acute inflammatory response to infection or injury. Interleukin-1 (IL-1) and tumor necrosis factor-α (TNFα) are very important pro-inflammatory mediators that orchestrate the immune response. They are rapidly produced by macrophages following infection or trauma, producing both local and systematic inflammatory effects. Their effects are often mediated through other cytokines.

Tumor Necrosis Factor-α

TNFα can exert effects on most cell types. It is a key regulator of other pro-inflammatory cytokines such as IL-1β, IL-6, and IL-8. Low levels of TNFα confer protection against infectious agents, tumors, and tissue damage. Overproduction can lead to autoimmunity or immunopathological disease. The systematic overproduction of TNFα during infection with Gram-negative bacteria causes widespread tissue damage and is a common cause of death due to circulatory shock.

Interleukin-1

IL-1 is a highly inflammatory cytokine similar to TNFα in many respects. It affects diverse cell types and has numerous biological effects including fever, hypotension, increased synthesis of acute phase proteins, increased resistance to infection, increased cytokine secretion, and increased eicosanoid production. There are three members of the IL-1 gene family: IL-1α, IL-1β, and IL-1 receptor antagonist. IL-1RA may function to modulate the activity of IL-1. There is considerable interest in identifying potential anti-inflammatory agents that reduce the production or activity of TNFα and IL-1.

Interleukin-6

IL-6 was first identified as a T-cell-derived factor acting on B-cells to induce immunoglobulin secretion. Its activities also include production of acute-phase proteins by the liver, formation of platelets, osteoclast (a macrophage-like cell in bone tissue) activation, and fever. Its production is induced by TNFα and IL-1. Although generally considered to be a pro-inflammatory cytokine, recent findings have called this classification into question.

Interleukin-8

IL-8 is an inflammatory cytokine that is also classified as a chemokine. It plays a key role in the accumulation of neutrophils at the sites of inflammation. IL-8 has been shown to induce liposomal enzyme release, oxidative burst by neutrophils, and other pro-inflammatory effects. It is produced in response to IL-1, TNF, and exogenous irritants. Its normal role is in the resolution of bacterial infections but elevated levels have been reported in diseases like rheumatoid arthritis.

Interleukin-12

IL-12 is a key cytokine in initiating resistance to infection and for promoting cell-mediated immunity. Certain infectious agents induce IL-12 synthesis by macrophages. This IL-12 acts upon natural killer (NK) cells, stimulating them to produce interferon-γ (IFNγ). IFNγ activates macrophages, enabling them to destroy invading microorganisms. IL-12 also acts upon T cells to promote differentiation towards the Th-1 phenotype (see next section). Th-1 cells are believed to be responsible for initiating an effective cell-mediated immune response. IL-12 is unusual in that it is a disulfide-linked heterodimer (termed p70) composed of a heavy chain (termed p40) and a light chain (termed p35). The p40 form is produced in large excess over p70 and it specifically inhibits binding of the biologically active p70 to its receptor. It has been proposed that p40 is a natural IL-12 antagonist. Assays are available that measure either p70, or p40 plus p70.

Th-1 and Th-2 Cytokines

These mainly lymphocyte-derived cytokines can be thought of as agents that act to fine-tune the immune response. There is considerable evidence for the existence of subsets of T helper (Th) cells called Th-1 and Th-2 that secrete distinct sets of cytokines. These two cytokine groups have differential patterns of activity. A Th-1 type response is associated with macrophage activation and the production of IgG subclasses that are effective at activating complement and stimulating phagocytosis. This results in enhanced cell-mediated immunity and the destruction of mainly intracellular microbes. A Th-2 type response is associated with IgE production and activation of eosinophils and mast cells. This humoral response is generally thought to be protective against larger extracellular organisms such as helminthic (worm-like) parasites.

It is considered that the resolution of infections produced by pathogenic organisms depends upon an appropriate balance between Th-1 and Th-2 type responses, and that balance will depend upon the nature of the invading pathogen. The response is normally tightly controlled. Unfortunately many organisms have developed mechanisms for subverting the balance of this response. Control also breaks down in inflammatory conditions and allergies. Manipulation of the Th-1/Th-2 response may be one way of treating these conditions, which accounts for the drug industry's interest in the area.

Th-1 Cytokines

Interferon-γ

IFNγ has potent immunoregulatory effects on a wide range of cells. It activates macrophages and NK cells, enhances production of complement, binding IgG_2a, blocks Th-2 cell formation, enhances the cytotoxic activity of TNF, and the antiviral activity of IFNα and IFNβ. It plays a vital role in defensive mechanisms against infectious diseases. Macrophages are the central element in host defenses against intracellular parasites. They are avidly phagocytic and are capable of secreting a wide variety of toxic compounds such as hydrolases, proteases, lipases, DNAses, and reactive oxygen intermediates such as H_2O_2. These agents are effective in killing many types of bacteria. However, macrophages, when inappropriately stimulated, can make a major contribution to chronic destructive or autoimmune diseases such as multiple sclerosis or rheumatoid arthritis.

Tumor Necrosis Factor-β

TNFβ, also known as lymphotoxin, is closely related to TNFα and shares many of its activities. It binds to the two types of TNF receptors with comparable affinities to that of TNFα. TNFβ has a broad spectrum of activity reflecting the presence of its receptors on most cell types. It has wide-ranging pro-inflammatory activities. Its activities include cytolytic activity towards transformed and virus infected cells, activation of granulocytes and macrophages, induction of IL-2 and IL-6 synthesis, induction of adhesion molecule expression, and inhibition of collagen synthesis.

Interleukin-2

IL-2 synthesis and secretion is triggered by antigen-induced activation of T-lymphocytes. Subsequent binding of IL-2 to its receptor induces expansion of antigen-specific T cells, an essential step in the generation of an antigen-specific immune response. It also stimulates B-cell growth and differentiation, and activates NK cells and macrophages. There are three IL-2 receptor subunits. The IL-2Rα subunit is undetectable in resting T cells, but is induced on T cell activation. However, IL-2Rα is expressed in certain abnormal cells associated with leukemias, autoimmune diseases, and allograft (genetically non-identical grafts from the same species) rejection. In such situations the serum concentration of a soluble form of the subunit, released from activated cells, is elevated. It has been suggested that sIL-2R may down-regulate cellular immune responses by competing with transmembrane IL-2R for IL-2.

Th-2 Cytokines

Interleukin-4

IL-4 is important in directing the immune system towards a humoral response. It causes a strong polarization of T cells towards the Th-2 phenotype and inhibits Th-1 cell formation by preventing the synthesis, and antagonizing the action, of IFNγ. IL-4 blocks the synthesis of pro-inflammatory cytokines by macrophages. It causes B-cell class switching to promote the synthesis of IgG1 and IgE, immunoglobulins involved in allergic responses. It also stimulates mast cell growth, cells that again are implicated in the allergic response. On binding IgE, mast cells can degranulate releasing a host of biologically active molecules such as histamine, proteases, chemotactic factors for eosinophils, neutrophils, and monocytes, prostaglandin D_2, and platelet-activating factor. These agents are thought to be generally protective against parasitic infections, damaging and expelling the invading organism from the body. However, they will also damage bystander cells and their prolonged activation in allergies such as asthma is definitely harmful.

Interleukin-5

IL-5 promotes eosinophil growth and differentiation, activates mature eosinophils, and increases their survival in parasitic infections. Eosinophils have the ability to release a wide variety of toxic substances against non-phagocytosable surfaces and are so able to kill certain metazoan parasites. However, these substances are also highly toxic to host cells and tissues. For instance, they cause destruction of the tracheal epithelium in bronchial asthma. Once again, appropriate regulation of this powerful effector arm of the immune system is essential to protect the host against invading parasites without causing excessive bystander tissue damage.

Interleukin-10

IL-10 inhibits the synthesis of cytokine production by Th-1 cells through an indirect effect on macrophages. It also inhibits the formation of IFNγ and TNFα by NK cells and some additional aspects of macrophage activity. Together with IL-4 it acts as a stimulator of mast cell growth. These activities are consistent with an inhibition of cell-mediated and an activation of humoral responses. The Epstein Barr virus synthesizes a protein that has considerable homology with IL-10 and shares its macrophage-inhibiting activity, which may give a selective advantage to the virus by suppressing aspects of the host's immune system. IL-10 may find use in the amelioration of inflammatory biological responses.

Interleukin-13

IL-13 shares many activities with IL-4. Like IL-4, it inhibits the synthesis of pro-inflammatory cytokines such as IL-1 and TNFα by macrophages. It also induces proliferation of human B-cells and a switch to the synthesis of IgE. Unlike IL-4, it does not appear to act upon T cells.

Interferons

The interferons were originally defined by their ability to inhibit viral replication in uninfected cells. This is still thought to be the primary role of IFNα and IFNβ, but IFNγ has a much wider range of activities and has been discussed as a Th-1 cytokine.

The IFNα family is secreted by leukocytes and contains a large number of sub-types. IFNβ is secreted by fibroblasts and has about 30% amino acid homology to IFNα. They both share the same receptor and have similar biological activities. These interferons are produced in response to viral infections. They have potent antiviral activity but at higher concentrations have antiproliferative activity against both normal and tumor cells. They can also enhance the expression of class 1 major histocompatibility complex (MHC) gene product, which enhances the ability of T-cytotoxic lymphocytes to kill cells infected by virus. Interferons have been used in the treatment of various cancers and viral diseases. Their effectiveness may be improved by incorporation into multi-drug therapy regimes.

Chemokines

The chemokine superfamily contains approximately 50 members and more are likely to be discovered. The superfamily is currently considered to contain four structural branches, CXC, CC, CX3C, and C. Chemokines have three highly conserved cysteine residues. The CXC chemokine family has the first two NH_2-terminal cysteines separated by one non-conserved amino acid residue. In the CC family these cysteines are placed alongside each other. The current single C family member has only one NH_2-terminal cysteine residue. The current sole CX3C family member (fractakline) is a transmembrane protein. Many new chemokines and chemokine receptors have been recently identified, mainly from genomic databases. The functions of most of these remain to be established.

Chemokines are involved in host defense mechanisms against invading pathogens. They attract leukocytes to areas of inflammation, where these cells destroy the infectious agent. On occasions these defensive mechanisms go astray with subsequent destruction of host tissue. For this reason chemokines have become prime targets for therapeutic intervention in chronic inflammation.

A family of G-protein-coupled receptors mediates the specific effects of chemokines on leukocytes. Around 10 receptors had been identified by 1997. Chemokines often share receptors, and cells can express multiple receptor types. The chemokines therefore exhibit multiple overlapping activities. Interest in the area has been stimulated by the discovery that chemokine receptors are implicated in the entry of the HIV virus into susceptible cells.

CXC Chemokines (α-Chemokines), e.g. IL-8

IL-8 is considered to be the best-characterized CXC chemokine. Its biological role has been discussed in PRO-INFLAMMATORY CYTOKINES. Other CXC chemokines include epithelial neutrophil-activating protein-78 (ENA-78), three closely related proteins, growth-related oncogene-α (GRO-α), GRO-β, and GRO-γ, and granulocyte chemotactic protein-2. Two receptors for IL-8 have been characterized, one of which is selective for IL-8, whereas the other has high affinity for several CXC chemokines. The main site of chemokine interaction with this receptor is the *N*-terminal domain, which must contain the sequence Glu-Leu-Arg (ELR) preceding the first cysteine. The ELR motif is common to all CXC chemokines that activate neutrophils.

CC Chemokines (β-Chemokines), e.g. MCP-1 and Eotaxin

Members of this family include monocyte chemotactic protein (MCP-1 to MCP-4), macrophage inflammatory protein (MIP-1α and MIP-1β), RANTES, and eotaxin. This family was originally considered to act specifically on macrophages, but it is now known that they also act on other cell types, in particular, basophils and eosinophils. MCP-1 is the best-studied CC chemokine. It is a potent chemoattractant for monocytes and it has been suggested that MCP-1 could play a significant role in the tissue infiltration of monocytes as seen in many inflammatory lesions. Recently eotaxin has attracted considerable interest. Unlike other chemokines it appears to be a specific chemoattractant for eosinophils. Eotaxin and its receptor CCR3 have become prime targets in anti-inflammatory research into asthma and other disorders characterized by tissue eosinophilia.

Chemokines and the HIV Virus

Interest in chemokines has increased considerably since the finding that certain chemokines can inhibit the growth of HIV in cell culture. Considerable progress has been made in elucidating the mechanisms of this inhibition. It had long been known that HIV uses a T-lymphocyte receptor called CD4 to infect cells, but it was also clear that an additional factor was required. This factor has now been identified as a chemokine receptor. It is also now known that HIVs that cause the initial infection predominantly use the receptor CCR5. This receptor normally binds RANTES, MIP-1α, and MIP-1β. The HIVs that predominate in the final stages of the disease bind to CXCR4, the receptor for SDF-1. Pharmaceutical companies are now racing to develop antagonists to these receptors.

Cytokines in Hematopoiesis

A number of cytokines are involved in the maturation of blood cells. These are considered to act in a stepwise fashion. IL-3 stimulates stem cells to develop into many different cell types such as early erythrocytes, neutrophils, eosinophils, basophils, macrophages, and megakaryocytes (platelet precursors). Granulocyte macrophage colony-stimulating factor (GM-CSF) acts somewhat later than IL-3 and gives rise mainly to neutrophils, macrophages, and eosinophils.

A number of other cytokines are committed to actions on cells of particular lineages. Granulocyte colony-stimulating factor (G-CSF) favors the development of neutrophils while macrophage colony-stimulating factor (M-CSF) enhances the development of macrophages. IL-4 and IL-5 stimulate the production of eosinophils and mast cells. In the presence of the hormone erythropoietin, red blood cells and megakaryocytes develop. IL-9 and IL-11 appear to be additional differentiation factors for red cells and megakaryocytes, respectively.

The Transforming Growth Factor-β Family

TGFβ is a multi-functional cytokine involved in tissue remodeling, wound repair, development, modulation of the immune response, and hematopoiesis. It is the prototype member of a family of secreted, disulfide-linked homodimeric polypeptides. Three closely related proteins have been identified. TGFβ is assembled as a four-protein complex, with the *C*-terminal homodimer interacting non-covalently with the two pro-segments. This complex is considered latent, as it is unable to react with cell surface receptors. Particularly in platelets, the complex is associated covalently with an additional protein termed

TGFβ-binding protein. *In vitro*, latent TGFβ can be released from the complex in an active state by acid treatment. *In vivo* activation is thought to occur by proteolytic digestion of the pro-segments.

TGFβ is perhaps the cytokine with the widest range of biological activities. Its activities may also be isotype specific. It is important in embryonic development, inhibiting the growth of epithelial cells and stimulating the growth of fibroblasts. In wound healing it decreases levels of proteolytic enzymes, increases synthesis of matrix proteins and metalloproteinase inhibitors. These activities promote wound healing, but excess TGFβ can lead to scar formation.

TGFβ has potent anti-inflammatory activity, inhibiting production of IFNα and IL-2 while promoting IL-10 production, suppressing macrophage function. It appears to shift immune reactions to a Th-2 response. TGFβ1 gene knockout mice are characterized as suffering from multifocal inflammatory disease, massive lymphocytic infiltration, and early demise of newborn mice.

TGFβ is an important tumor suppressor that normally acts to restrain cell proliferation. However, the type 11 TGFβ receptor is very susceptible to mutation. Cells containing non-functional type 11 receptors lose their sensitivity to TGFβ growth inhibition. In fact many tumors over-express TGFβ. In the absence of a functional receptor, TGFβ may provide a growth advantage through suppression of the immune system and by promoting new blood vessel formation.

Assay Technology

Cytokines are almost invariably measured by two-site ELISAs, either in cell culture supernatants, serum, or plasma. Levels in healthy humans tend to be low, usually in the low pg/mL range. In disease states, levels are often elevated 10 to 100-fold. A range of fully configured assay kits is offered by Amersham Biosciences, along with products for measurement of cytokines in rodent species. High-sensitivity versions are also available that utilize the AMDEX technology.

REFERENCES AND FURTHER READING

Abbas, A.K. Functional diversity of T lymphocytes. *Nature* **383**, 787–793 (1996).

Bacon, K.B. Chemokines and chemokine receptors: therapeutic targets in inflammation and infectious diseases. *Drug News Perspect.* **10**, 133–143 (1997).

Belardi, F. Role of interferons and other cytokines in the regulation of the immune response. *APMIS* **103**, 161–179 (1995).

Billiau, A. Interferon-130 (1996).

Callard, R. and Gearing, A. *The Cytokine Facts Book* (Academic Press, London, 1994).

Cohen, J. Exploiting the HIV–chemokine nexus. *Science* **275**, 1261–1264 (1997).

Cohen, M.C. and Cohen, S. Cytokine function, a study in biologic diversity. *Am. J. Clin. Pathol.* **105**, 589–598 (1996).

Devos, R., Plaetinck, G., Cornelis, S., Guisez, Y., van der Heyden, J. and Tavernier, J. Interleukin-5 and its receptor: a drug target for eosinophilia associated with chronic allergic disease. *J. Leuk. Biol.* **57**, 813–819 (1995).

de Vries, J.E. Immunosuppressive and anti-inflammatory properties of Interleukin-10. *Ann. Med.* **27**, 537–541 (1995).

Dinarello, C.A. Biologic basis for interleukin-1 in disease. *Blood* **87**, 2095–2147 (1996).

Gazzinelli, R.T. Molecular and cellular basis of interleukin 12 activity in prophylaxis and therapy against infectious diseases. *Mol. Med. Today* **2**, 258–267 (1996).

Hamblin, A.S. *Cytokines and Cytokine Receptors* (IRL Press, Oxford, 1993).

Harada, A., Mukaida, N. and Matsushiman, K. Interleukin 8 as a novel target for intervention in acute inflammatory diseases. *Mol. Med. Today* **2**, 482–489 (1996).

Jose, P.J., Adcock, I.M., Griffiths-Johnson, D.A., Berkman, N., Wells, T.N., Williams, T.J. and Powers, C.A. Eotaxin: cloning of an eosinophil chemoattractant cytokine and increased mRNA expression in allergen-challenged guinea-pig lungs. *Biochem. Biophys. Res. Commun.* **205**, 788–794 (1994).

Kekow, J. and Wiedemann, G.J. Transforming growth factor 182 (1995).

Marriott, J.B., Westby, M. and Dalgleish, A.G. Therapeutic potential of TNF-α inhibitors old and new. *Drug Discovery Today* **2**, 273–282 (1997).

Miller, E.J., Nagao, S., Carr, F.K., Noble, J.M. and Cohen, A.B. Interleukin-8 (IL-8) is a major neutrophil chemotaxin from human alveolar macrophages stimulated with staphylococcal entertoxin A (SEA). *Inflamm. Res.* **45**, 386–392.

Nicola, N.A. (ed), *Guidebook to Cytokines and Their Receptors* (Oxford University Press, Oxford, 1994).

Omori, F., Okamura, S., Shimoda, K., Otsuka, T., Harada, M. and Niho, Y. Levels of human serum granulocytes colony-stimulating factor and granulocyte-macrophage colony-stimulating factor under pathological conditions. *Biotherapy* **4**, 147–153 (1992).

Otsuka, T., Miyajima, A., Brown, N. *et al.* Solubilisation and characterisation of an expressible cDNA encoding human IL-3. *J. Immunol.* **140**, 2288–2295 (1988).

Roberts, A.B. Transforming growth factor-β: activity and efficacy in animal models of wound healing. *Wound Rep. Regen.* **3**, 408–418 (1995).

Rollins, B.J. Monocyte chemoattractant protein 1: a potential regulator of monocyte recruitment in inflammatory disease. *Mol. Med. Today* **2**, 198–204 (1996).

Strieter, R.M., Standiford, T.J., Huffnagle, G.B., Colletti, L.M., Lukacs, N.W. and Kunkel, S.L. The good, the bad, and the ugly. The role of chemokines in models of human disease. *J. Immunol.* **156**, 3583–3586 (1996).

Tagboto, S.K. Interleukin-5, eosinophils and the control of helminth infections in man and laboratory animals. *J. Helminthol.* **69**, 271–278 (1995).

Theze, J., Alzari, P.M. and Bertoglio, J. Interleukin 2 and its receptor: recent advances and new immunological functions. *Immunol. Today* **17**, 481–486 (1996).

Tilg, H., Dinarello, C.A. and Weir, J.W. IL-6 and APPs: anti-inflammatory and immunosuppressive mediators. *Immunol. Today* **18**, 428–432 (1997).

Vaddi, K., Keller, M. and Newton, R.C. *The Chemokine Facts Book* (Academic Press, London, 1997).

Zurawski, G. and de Vries, J. Interleukin-13 elicits a subset of the activities of its close relative interlekin-4. *Stem Cells* **12**, 169–174 (1994a).

Zurawski, G. and de Vries, J. Interleukin-13, an interleukin-4 like cytokine that acts on monocytes and B cells, but not on T cells. *Immunol. Today* **15**, 19–26 (1994b).

SIGNAL TRANSDUCTION ASSAYS

Cyclic Nucleotides: cAMP and cGMP

Adenosine 3′,5′ cyclic monophosphate (cAMP) and guanosine 3′,5′ cyclic monophosphate (cGMP) are involved in a myriad of normal and pathological processes. Indeed, these cyclic nucleotides serve as second messengers for the action of a variety of endogenous and exogenous agents ranging from bacteria to humans. **Second messenger** is a term used to describe molecules that transmit intracellular signals from other molecules that cannot themselves enter target cells. The universal nature of cAMP and cGMP has made their measurement essential to the study of numerous hormones, local mediators, neurotransmitters, pharmacological agents, and toxins. Typically, with cAMP, the binding of e.g. a hormone, agonist, or neuromodulator to its receptor is followed by activation or inhibition of a G-protein which, in turn, activates the effector adenylate cyclase, evoking the generation of cAMP from ATP. The activation of protein kinase A by cAMP results in the phosphorylation of specific substrates, which include enzymes, ion channels, and transcription factors. Because cAMP can activate a cascade of reactions, the involvement of cAMP greatly amplifies the cellular response to a variety of drug and hormonal stimuli.

cGMP has also been shown to be widely distributed, occurring at low concentrations (1–10% of cAMP levels) in most tissues. Since its discovery, a number of roles have been suggested for cGMP based on changes in levels after challenge of tissues with various ligands. For example, acetylcholine, oxytocin, insulin, serotonin, and histamine cause an increase in intracellular cGMP levels. Vasodilators such as nitroprusside, nitroglycerine, sodium nitrate, and nitric oxide also increase cGMP levels via soluble guanylate cyclase. Both cAMP and cGMP are present at very low concentrations in biological fluids, and methods developed for measurement must contend with high concentrations of interfering non-cyclic nucleotide substances. There are several rapid, selective, and highly sensitive methods for measuring cAMP and cGMP by competitive immunoassays. Technologies include:

- RIA, using a high specific activity adenosine or guanosine 3′,5′-cyclic phosphoric acid 2′-*O*-succinyl-3-[^{125}I] iodotyrosine methyl ester together with a second antibody bound to magnetic polymer particles
- EIA, involving the linking of succinyl cAMP or cGMP to HRP, and combining this with stable second antibody-coated microtiter plates (Horton *et al.*, 1992).
- homogeneous RIA (non-separation), using SPA technology (Horton and Baxendale, 1995).
- homogeneous FPIA (non-separation), using cyanine dye (e.g. Cy3B™) technology.

D-myo-Inositol 1,4,5-Trisphosphate (IP_3)

The phosphoinositide cycle is a major pathway by which a variety of hormones and other signaling molecules can generate second messengers and thus affect the intracellular milieu. Key amongst these intermediates is inositol 1,4,5-trisphosphate (IP_3). It is now well established that IP_3 acts as a second messenger for calcium in a variety of cell types. IP_3 may be measured by competitive protein-binding assay, an approach that is analogous to a typical RIA.

Second Messenger (Cyclic Nucleotide) Assays

Preparation of Antisera and Tracers

Antibodies to the cyclic nucleotides are raised in rabbits after immunization with antigen, in which a 2′-*O*-succinyl derivative of the cyclic nucleotide is conjugated to protein (Horton and Baxendale, 1995). Adenosine or guanosine 3′,5′-cyclic phosphoric acid 2′-*O*-succinyl-3-[^{125}I] iodotyrosine methyl ester is prepared whereby 2′-*O*-monosuccinyl adenosine or guanosine 3′,5′-cyclic monophosphoric acid tyrosine methyl ester is reacted with sodium [^{125}I] iodide and chloramine-T to give a specific activity of 2000 Ci/mmol. The product is purified by HPLC.

For the EIA, cAMP and cGMP are conjugated to HRP through an activated *N*-hydroxysuccinimide ester (Erlanger, 1973). The ester is prepared by reaction of the carboxylic acid group of 2′-*O*-monosuccinyl adenosine or guanosine 3′,5′-cyclic monophosphate with *N*-hydroxysuccinimide in the presence of 1-ethyl-3 (3-dimethylaminopropyl) carbodiimide hydrochloride. HRP is dissolved in sodium phosphate buffer pH 8.0, the activated cyclic nucleotide added, and the mixture incubated overnight at room temperature. Low molecular weight components of the reaction mixture are removed by gel filtration.

For the FPIA, cAMP was coupled to Cy3B through a carbamate linker. The product was purified by reverse-phase HPLC on C-18 columns. Derivatization of the ribose ring via a carbamate rather than the more commonly used ester grouping resulted in a final labeled conjugate with significantly increased hydrolytic activity.

Sample Preparation of Cell Culture Samples

Established methods for preparing cellular extracts for cyclic measurements usually involve tedious processes, such as acid or solvent extraction, in order to obtain samples in a suitable form for subsequent assay. The traditional techniques involve cell lysis, removal of extracting agent, and assay. These steps must be carried out sequentially, thereby adding to the time and cost of each assay. Furthermore, none of these methods are very suitable for applications where large numbers of samples are required to be processed. Amersham Biosciences have developed a direct assay technique eliminating cell culture sample preparation methods. The procedure enables culture of cells, followed by the direct extraction and assay of cyclic nucleotides, to be achieved with little technical intervention. Novel cell lysis reagents are included in the method facilitating a simple and rapid extraction of intracellular cyclic nucleotides, obviating the need for removal of extracting reagents prior to measurement, whilst ensuring cyclic nucleotides are available for subsequent analysis.

Assay Procedures

cAMP or cGMP in test samples are analyzed by either acetylation or non-acetylation assay procedures. Preliminary acetylation (Harper and Brooker, 1975) of samples and standards markedly increases the detection limit of the cyclic nucleotide immunoassays.

RIA

Non-acetylation method

Standards are prepared in assay buffer (0.05 M acetate, pH 5.8) with cAMP or cGMP diluted over the range of 0.25–16 pmol/mL. Diluted sample or standard (100 μL) is incubated with specific antisera (100 μL, diluted 1/11,000) and [^{125}I]cAMP or cGMP (100 μL, 20,000–30,000 cpm) for 3 h at 4 °C. Non-specific binding is determined in the absence of specific rabbit antisera. Amerlex-M™ second antibody reagent (500 μL) is added to each tube except the total counts, and incubated for 10 min at room temperature. The antibody-bound fraction is separated by magnetic separation or centrifugation with decantation of the supernatant. The radioactivity present in each tube is determined by counting in a multiwell γ spectrophotometer.

Acetylation method

Standards are prepared in assay buffer with cAMP or cGMP diluted over the range of 0.04–2.56 pmol/mL. The zero standard tube contains assay buffer without standard. Standards and samples (500 μL) are acetylated by the addition of 25 μL of a 2:1 (v/v) mixture of triethylamine and acetic anhydride. For the assay, an aliquot (100 μL) of acetylated samples or standards is incubated with antisera (100 μL) and [^{125}I] cAMP or cGMP overnight at 4 °C. Non-specific binding is determined in the absence of specific rabbit antisera. The antibody-bound fraction is separated using magnetic separation or centrifugation.

EIA

For full details of this method, the reader is referred to Horton *et al.* (1992). The acetylation step is carried out essentially as described for the RIA. Sample or standard (100 μL) was incubated with specific antisera (100 μL) on donkey anti-rabbit coated microtiter plates for 2 h at 4 °C before addition of HRP-succinyl cAMP or cGMP conjugate (100 μL; approximately 0.004 U enzyme). The microtiter plate is incubated for a further 1 h at 4 °C and thoroughly washed with 0.01 M sodium phosphate buffer, pH 7.5, containing 0.05% (v/v) Tween-20. Activity of enzyme bound to the solid phase is determined by the addition of 0.03% (w/v) tetramethylbenzidine hydrochloride in 10 mM citric acid, pH 3.5, containing 0.01% (v/v) hydrogen peroxide. The substrate is incubated for 1 h at room temperature, and the reaction terminated with 1.0 M sulfuric acid (100 μL/well). The absorbance generated in the presence of bound conjugate is measured at 450 nm using a microtiter plate spectrophotometer. Standard curves are constructed in assay buffer with cAMP or cGMP diluted over the range of 0.125–32 pmol/mL (non-acetylation) and 0.04–2.56 pmol/mL (acetylation assay).

SPA

Details of the SPA immunoassay method for cyclic nucleotide measurement are as follows. Assays may be carried out in tubes (100 μL reagent volumes) or in microplate wells (50 μL reagent volumes). Sample

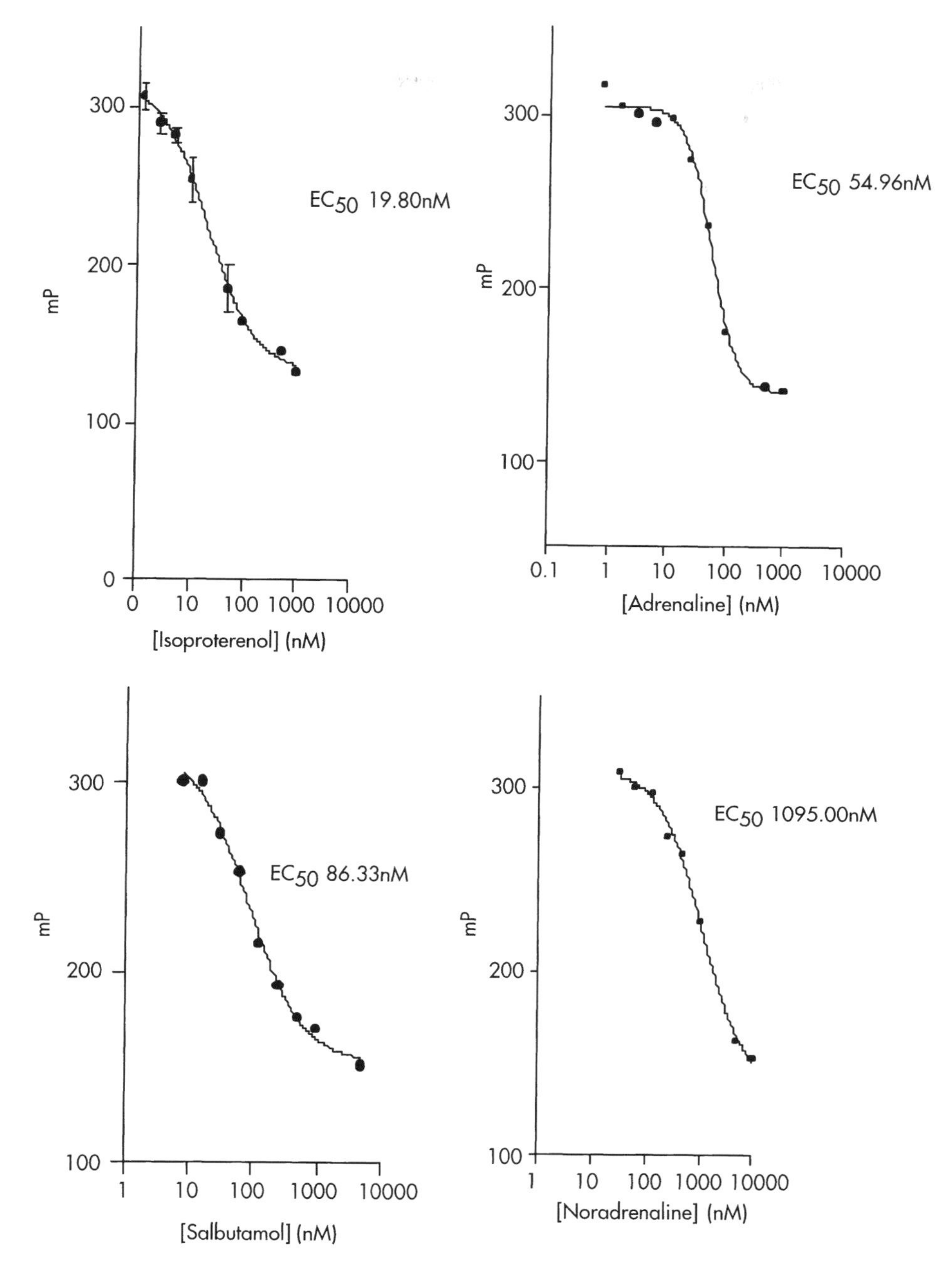

Fig. 67.8 Agonist-induced cAMP generation from A431 cells.

or standard is added to the specific antisera and [^{125}I] cAMP or cGMP. Non-specific binding is determined, as before, in the absence of specific rabbit antisera. The SPA reagent is thoroughly mixed before adding, to ensure a homogenous suspension. All tubes contain 400 μL, all wells 200 μL. The samples are incubated overnight, before counting in a β-scintillation counter for 2 min. Standard curves are prepared in assay buffer with cAMP or cGMP diluted over the range 2–128 pmol/mL (non-acetylation) and 0.04–2.56 pmol/mL (acetylation).

FPIA

The cAMP Fluorescence Polarization (FP) Biotrak™ Immunoassay System from Amersham Biosciences is a convenient approach to detect and quantify cAMP in a wide variety of samples. The method is available in 96-well or 384-well formats and uses the bright polarization dye, Cy3B, for fluorescent detection. Cy3B is well suited to fluorescence polarization measurements, due to its high quantum yield, fluorescence lifetime, and therefore a wide assay window.

Traditionally, cAMP requires extraction from cell culture prior to assay. Proprietary reagents (see above) from Amersham Biosciences lyse the cells and release the cAMP for simple, direct, measurement reducing hands-on time and offering a simple one-stage protocol.

Quantification of cAMP is measured by competitive displacement of Cy3B-cAMP from anti-cAMP binding. The signal decrease is proportional to the cAMP concentration. Known concentrations of cAMP are provided to prepare a standard curve. The method is ideal for compound screening employed during the drug discovery process.

Measurement of intracellular cAMP generation has become an established means of screening for antagonists and agonists of receptors linked to adenylate cyclase via either inhibitory or stimulatory G-proteins. Indeed, there is evidence that functional assay technologies for screening at GPCRs will lead to the identification of new compounds not detected in ligand-binding screens. Fluorescence polarization may be measured on a microplate fluorescence reader or using a multimodality imaging system (e.g. LEADseeker, Amersham Biosciences).

Using the FPIA approach, pharmacological studies may be carried out as follows. Chinese hamster ovary (CHO), NIH 3T3, or A431 cells are seeded into 96- or 384-well cluster plates suitable for cell culture and fluorescence detection. Cells are stimulated (20 min, 37 °C, 5% CO_2, 95% humidity) with forskolin (1–100 μM) or with the β adrenoreceptor agonists, isoproterenol, adrenaline, salbutamol, and noradrenaline (1.25–1000 nM) or pre-incubated for 30 min with the non-selective β adrenoreceptor antagonist propranolol (0.001–5000 nM) before addition of isoproterenol (50 nM). The culture supernatant is gently aspirated and cells lyzed with 0.25% (w/v) dodecyltrimethylammonium bromide. Working standards (0.2–51.2 pmol/well) are pipetted into empty wells on the culture plate. Rabbit antisera and Cy3B-linked cAMP are added to the wells containing the standards and the samples. The plates are incubated overnight at room temperature and fluorescence polarization measured. A431 cells exhibit low levels of basal cAMP production in the absence of agonist. Under normal assay conditions, there is a dose-dependent increase in cAMP generation with the β adrenoreceptor agonists, isoproterenol, adrenaline, salbutamol, and noradrenaline (Figure 67.8). The response of A431 cells to isoproterenol is clearly inhibited by the β adrenoreceptor antagonist propranolol (Figure 67.9).

The following procedure describes a general approach that may be used in a pharmacological screen using cultured cells. A431 cells are stimulated for 30 min with each agent obtained from a Library of Pharmacologically Active Compounds (LOPAC™, Sigma), in an agonist screen. Cells are lyzed (see above) and levels of cAMP measured. The mean cAMP levels ±3 standard deviations are derived (expressed as millipolarization units) from the negative control compounds. A 'hit' from a test compound is considered, when there is a significant increase in cAMP levels greater than 3 standard deviations above the mean (Figure 67.10). The test compounds were used at 1 μM.

Unlike diagnostic or clinical research immunoassays, assay performance in drug-screening assays is monitored by the use of Z', as described by Zhang *et al.* (1999). Z' values exceeding 0.5 are acceptable for a screen, those values approaching 1.0 are ideal for drug-screening applications.

Inositol 1,4,5-trisphosphate (IP_3) assay

Here unlabeled IP_3 competes with a fixed amount of $[^3H]$-labeled IP_3 from a receptor preparation obtained from bovine adrenal cortex. The bound IP_3 is separated from free IP_3 by centrifugation, a step that sediments the receptor preparation. The free IP_3 is discarded by simple decantation, leaving the bound fraction at the base of the tube. Measurement of the radioactivity in the tube enables the amount of unlabeled IP_3 in the sample to be determined by interpolation from a standard curve.

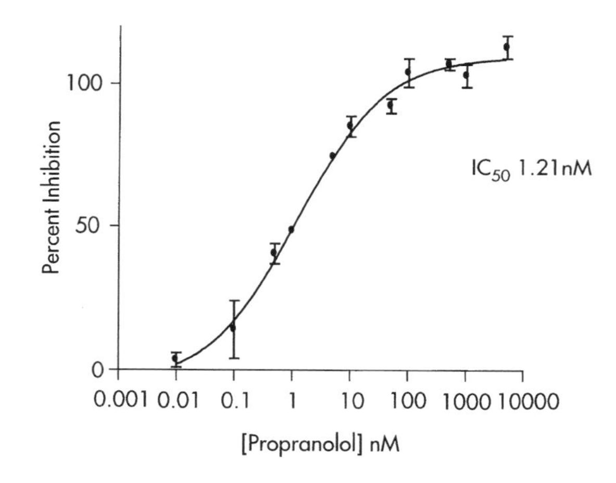

Fig. 67.9 Effect of propranolol on isoproterenol-induced cAMP generation (n = 3).

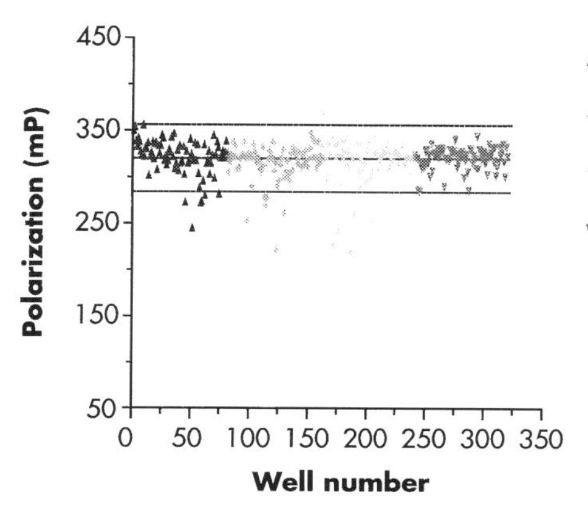

Fig. 67.10 Data from A431 cells stimulated with 320 compounds from a LOPAC library (Sigma) and measured with the cAMP FPIA, showing zero false positives. Three standard deviations are shown as a dashed line.

REFERENCES AND FURTHER READING

Erlanger, G.F. Principles and methods for the preparation of drug-protein conjugates for immunological studies. *Pharmacol. Rev.* **25**, 271–280 (1973).

Harper, J.F. and Brooker, G. Femtomole sensitive radioimmunoassay for cyclic AMP and cyclic GMP after 2′-O-acetylation by acetic anhydride in aqueous solution. *J. Cyclic Nucleotide Res.* **1**, 207–218 (1975).

Horton, J.K. and Baxendale, P.M. Mass measurements of cyclic AMP formation by radioimmunoassay, enzymeimmunoassay and scintillation proximity assay. *Methods Mol. Biol.* **41**, 91–105 (1995).

Horton, J.K., Martin, R.C., Kalinka, S., Cushing, A., Kitcher, J.P., O'Sullivan, M.J. and Baxendale, P.M. Enzyme immunoassays for the estimation of adenosine 3′5′-cyclic monophosphate and guanosine 3′5′-cyclic monophosphate in biological fluids. *J. Immunol. Methods* **155**, 31–40 (1992).

Palmer, S., Hughes, K.T., Lee, D.Y. and Wakelam, M.J.O. Development of a novel Ins(1,4,5)P_3-specific binding assay. *Cell Signalling* **1**, 147–153 (1989).

Zhang, J.H., Chung, T.D.Y. and Oldenburg, K.R. A simple statistical parameter for use in evaluation and validation of high throughput screening assays. *J. Biomol. Screen* **4**, 67–73 (1999).

CELL PROLIFERATION IMMUNOASSAY

Traditionally, measurement of cell proliferation is determined by counting cells directly, the determination of the mitotic index, or by performing a clonogenic assay. All these methods are labor-intensive and are impractical for evaluating large numbers of samples. Alternatively, as an indirect measure of viable cell number, metabolic activity may be measured with tetrazolium salts. Furthermore, measurement of cell proliferation has involved the use of [^{3}H]-thymidine to allow monitoring of DNA synthesis. More recently, alternative non-radioactive techniques have been developed in which a thymidine analog, 5-bromo-2′-deoxyuridine (BrdU), is incorporated into replicating DNA and subsequently localized using a specific monoclonal antibody.

The principle of the assay involves culturing cells in the presence of the test substances in a 96-well microtiter plate at 37 °C at 1–5 days. BrdU is added to the cells and these are re-incubated (usually 2–24 h). During this labeling period, the pyrimidine analogue BrdU is incorporated into the DNA of proliferating cells. After removing the culture medium, the cells are fixed and the DNA denatured (the denaturation of DNA is necessary to improve the accessibility of the incorporated BrdU for detection by the antibody). The peroxidase-labeled anti-BrdU binds to the BrdU incorporated in newly synthesized, cellular DNA. The immune complexes are detected by the subsequent substrate reaction, and the absorbance measured at 450 nm in a microtiter plate spectrophotometer. The absorbance values correlate directly to the amount of DNA synthesis and thereby to the number of proliferating cells in culture.

REFERENCE

Porstmann, T., Ternynck, T. and Avrameas, S. Quantitation of 5-bromo-2′-deoxyuridine incorporation into DNA. An enzyme immunoassay for the assessment of the lymphoid cell proliferative response. *J. Immunol. Methods* **82**, 169–180 (1980).

MATRIX METALLOPROTEINASES (MMPs)

The Family of MMPs and Their Role

Many normal physiological and pathological processes such as embryogenesis, morphological growth changes, ovulation and pregnancy, wound healing, atherosclerosis, inflammation, tumor invasion, and metastasis involve breakdown and remodeling of the extracellular matrix. This degradation is due to the family of important enzymes known as the matrix metalloproteinases (MMPs).

There are now many MMPs identified (Table 67.4). They are quite a homologous family due to the high degree of similarity in their domain structures. They can be further classified into substrate-specific groups although the distinctions are becoming less clear as more becomes known about them. The main types of MMPs can be categorized as collagenases, stromelysins, gelatinases, and membrane-type or MT-MMPs.

Several collagenases have now been identified – the more ubiquitous fibroblast collagenase (MMP-1), neutrophil collagenase (MMP-8), and more recently MMP-13 which may be involved in breast tumors and may also represent the principal murine collagenase. They are unique in their ability to degrade collagens.

Stromelysins have a broader substrate specificity, including proteoglycans collagen type IV and IX, fibronectin, and laminin, and they can activate MMP-1. The archetypal enzyme is stromelysin-1 (MMP-3) which has strong homology to stromelysin-2 (MMP-10). Also included in this group is matrilysin (MMP-7) which has a quite distinctive truncated domain to the other MMPs and may represent the most evolutionary primitive MMP.

The gelatinases are 72 kDa gelatinase (gelatinase A, MMP-2) and 92 kDa gelatinase (gelatinase B, MMP-9). They degrade denatured collagens (gelatins), type IV basement membrane collagen, elastin, and fibronectin.

MT-MMPs have an additional transmembrane sequence to anchor them to the cell and are thought to be involved in the activation of other MMPs.

These very potent enzymes have their activity tightly controlled at the transcription and activation stages to prevent inappropriate matrix degradation. Furthermore, specific inhibitors, the tissue inhibitors of metalloproteinases (TIMPs), bind to the active MMPs in a 1:1 ratio to form inactive complexes. So far, four TIMPs have been identified, with again, a high degree of homology. TIMP-1 and 2 form complexes with the active MMP and, as far as is known, all active MMPs are capable of forming complexes with all the TIMPs. There are two particular additions to this situation in that proMMP-9 and proMMP-2 (latent zymogens) can form complexes with TIMP-1 and TIMP-2, respectively. It is thought that TIMPs may not only play a role in stabilizing the proenzyme to autoactivation but also participate in the activation process. In addition, the non-specific protease inhibitor α_2-macroglobulin (α_2M) can also entrap active enzyme providing a further inhibitory mechanism. These all serve to tightly regulate MMP activity.

MMPs need to be activated to exert their effects. The enzymes are secreted as inactive (latent) proMMPs, which require another agent to cleave a propeptide sequence and/or perturb the conformation leading to an unstable intermediate. This autocatalytically processes itself to the fully active enzyme by further propeptide sequence cleavage. Residues such as cysteine that are important in bonding the active site Zn^{2+} are removed during this process to reveal the active catalytic site. The fully active

Table 67.4. Members of the MMP family.

MMP no.	Human enzyme	Latent MW	Active MW	Substrates
MMP-1	Interstitial collagenase-1	55,000	45,000	Collagens, gelatin, proteoglycan, etc.
MMP-8	Neutrophil collagenase	75,000	58,000	Collagens, gelatin, proteoglycan, etc.
MMP-13	Collagenase-3	60,000	48,000	Collagens, gelatin, PAI2, aggrecan
MMP-18	Xenopus collagenase	55,000	?	
MMP-3	Stromelysin-1	57,000	45,000	Collagens, gelatin, aggrecan, fibronectin, laminin, casein, MMPs
MMP-10	Stromelysin-2	57,000	44,000	Collagens, gelatin, aggrecan, fibronectin, laminin, casein, MMPs
MMP-11	Stromelysin-3	51,000	44,000	α_1-Proteinase inhibitor
MMP-19	RASI, stromelysin-4	57,000	?	Aggrecan, COMP
MMP-7	Matrilysin	28,000	19,000	As MMP-2
MMP-26	Matrilysin-2	30,000	?	As MMP-2
MMP-2	72 kDa gelatinase	72,000	66,000	Gelatin, collagens, fibronectin, MMPs
MMP-9	92 kDa gelatinase	92,000	86,000	Gelatin, collagens, fibronectin
MMP-14	MT-MMP-1	66,000	56,000	Collagens, casein, fibronectin, MMP-2,13

enzyme is then capable of degrading the appropriate matrix substrates. However, it is then also open to inhibition by high-affinity (10^{-10} M) binding of TIMPs.

MMPs can be activated *in vitro* by proteinases (e.g. trypsin and plasmin), mercurials (e.g. 4-aminophenylmercuric acetate (APMA)), and *in vivo* by proteases such as plasmin and other MMPs.

Due to their implication in many disease states the MMPs are a potential target for therapeutic intervention and this is currently an active area of drug development for several pharmaceutical companies. Most approaches are based on the use of low molecular weight inhibitors of the activity of the MMP. The balance between MMPs and their inhibitors is thought to be altered in disease states and the ability to measure their relative levels is crucial to the investigation of their role in these conditions.

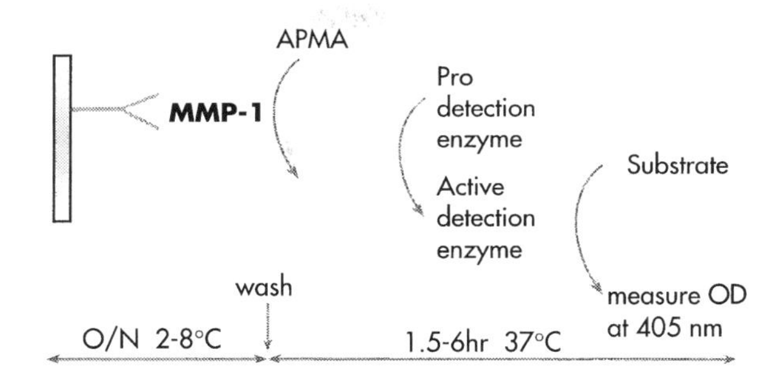

Fig. 67.11 A schematic of an activity assay protocol using the MMP-1 assay as an example.

Measurement Methodology

It is only in the last decade that immunoassays have been applied to the measurement of MMPs. Typically they have been measured by substrate cleavage assays and/or zymography, which rely on the enzyme's biological activity in degradation. However, these techniques do have several disadvantages. Substrate cleavage assays, e.g. using radiolabeled collagen as substrate, can lack convenience, specificity, and sensitivity, and are prone to interference from TIMPs, which, by definition, will inhibit enzyme cleavage. Zymography can also be variable and labor-intensive, and is only semi-quantitative.

ELISA

At Amersham Biosciences we have developed the first commercially available immunoassays to a range of MMPs and TIMPs. The assay methodology is conventional 'sandwich' ELISA using HRP as detection enzyme and a colorimetric signal. Some of these assays have been developed in collaboration with Fuji Chemical Industries of Japan.

An ELISA is clearly measuring something different to that which is detected in 'bioassays'. An ELISA is a quantitative measure of the amount of MMP or TIMP present and may not provide information on how much 'activity' there is. In order to address this question we have developed a series of activity assays for the MMPs that allow for a quantitative and specific measure of the MMP present in either the latent or active form (where latent can be regarded as the potential for activity and is probably equal to the pro form of enzyme).

Activity Assays

By modifying the plasma-specific activation site of the enzyme pro-urokinase into a site that is only recognized by MMPs, we obtained a general MMP-specific substrate-proenzyme that upon activation by MMPs can easily be assayed using a chromogenic substrate. We have adapted the assay from a general MMP activity assay into a specific assay by introducing specific antibodies that capture the appropriate MMP from a biological fluid (Figure 67.11).

The assay technology, termed Quickzyme™, uses the pro form of a modified pro-urokinase where the activation sequence, normally recognized by plasmin (Pro-Arg-Phe-Lys⇩Ile-Ile-Gly-Gly), has been replaced by a sequence that is specifically recognized by MMPs (Arg-Pro-Leu-Gly⇩Ile-Ile-Gly-Gly). Captured active MMP cleaves this sequence through a single proteolytic event to yield the active form of urokinase detection enzyme. MMP-activated detection enzyme can then be measured using a specific chromogenic peptide substrate S-2444™ (Figure 67.11). Standards and samples are incubated in microtiter wells precoated with a specific anti-MMP antibody. Any of the appropriate MMP present will be bound to the wells, other components of the sample being removed by washing and aspiration. Either the endogenous levels of active MMP or the total level (pro and active) can be detected. In order to measure the total MMP content, any bound MMP in its pro form is activated using *p*-APMA. Active MMP is detected by using buffer blank in place of the APMA. The standard is in the pro form too, and is activated in parallel for both types (active and total) of sample. The captured active MMP is then detected through activation of the pro detection enzyme and the subsequent cleavage of its chromogenic peptide substrate. The resultant color is read at 405 nm in a microtiter plate reader.

Specificity and Sample Considerations

The expression and activation of MMPs is complex with several different MMPs being produced at the same site or by the same cell. The MMPs may interact to activate each other as well as displaying their matrix substrate specificities. For example, MMP-3 can activate MMP-1 to a highly activated form. In addition, the inhibitory TIMPs are found in close proximity and again can be produced by the same cells. Therefore it is necessary to specifically identify the activity of each MMP and its inhibitors that may be present in a mixture of MMPs in order to fully understand their action.

Furthermore, the complicated activation and subsequent inhibition mechanisms of the MMPs can cause several forms of these enzymes to be present in a single sample. As a result, researchers studying MMPs are often perplexed as to what to measure and how best to do it.

One of the main challenges for manufacturers/developers of kits and for customers starting to use these assays is to define what is being measured.

So, before considering an assay system to measure MMP, one should ask:

- In what form are the MMPs present in the sample?
- Is the assay's specificity appropriate for this form?

There are potentially several forms in which a given MMP can exist given its activation and inhibition cascade. However, the cross-reactivity profile of an assay can be defined quite accurately for a given assay. It is very important to define the cross-reactivity for these forms and not just related MMPs. An example of the cross-reactivity profile for the Amersham Biosciences MMP-3 ELISA assay is shown as an example (Table 67.5). In this particular case the assay cross-reacts virtually equally with pro, active and complexed active forms of MMP-3, i.e. it can be regarded as measuring 'total' MMP-3 levels. Another particularly common profile found is for an assay to measure the pro form only, with no cross-reaction with active MMP. TIMPs are interesting in that potentially their ELISA reactivity when complexed to some MMPs may not be equal to that when they are free, posing the question as to which form is actually the most significant in a given sample.

Activity assays have slightly different considerations, as by definition they need to be able to bind, and therefore can potentially measure, both pro and active forms of the MMP. It is a consequence of active MMPs (and some of the pro forms too) being bound to TIMPs that they will be present with the captured analyte and are often not dissociated by antibody capture and subsequent assay procedures. Therefore they will be present and can inhibit the MMP activation of the urokinase detection enzyme. This means that the cross-reaction of activity assays with TIMP-complexed forms of MMPs is inevitably less than can be found in the equivalent ELISA, for example. One way of interpreting the signal obtained with an activity assay in the light of this is to regard it as a measure of the active potential of the MMP.

Therefore the other element required judging assay suitability is knowledge of what form of MMP may be present in a sample. ELISAs themselves have helped answer this. For example, it has been shown that MMP-9 is elevated in the plasma of hepatocellular carcinoma (HCC) patients. By using gel filtration and a variety of assays and antibodies it was shown that the predominant form of plasma MMP-9 immunoreactivity is proMMP-9 complexed to TIMP-1. In this case the Amersham Biosciences ELISA measures proMMP-9 and so would be appropriate for such samples.

Table 67.5. Cross-reactivity of MMP-3 ELISA (AB).

Compound	Cross-reactivity (%)
MMP-3 standard	100.0
Activated MMP-3*	96.45
MMP-1	<0.118
MMP-2	<0.118
MMP-9	<0.118
TIMP-1	<0.118
TIMP-2	<0.118
MMP-3/TIMP-1 complex	112.68
MMP-3/TIMP-2 complex	90.10

*APMA activated.
For certain compounds the measured concentrations were below the sensitivity (detection limit) of the assay. Hence the assay sensitivity value was used in the calculation of cross-reactivity.

Sample Levels

Research assays are applied to many more sample types than clinical assays. These assays have been validated for use in serum and plasma as they are commonly found sample types. Typically for MMPs, this has involved testing for linearity of dilution and recovery values and a small-scale estimation of the levels likely to be encountered in normals. One note of caution about recovery testing should be that researchers who wish to do this should ensure that, for spiking, the proMMP is as pure as possible. Active MMPs are likely to be trapped by α_2-macroglobulin which is present in great excess in serum. Even using pure proMMP is no guarantee of successful recovery because serum proteases may well be present that can activate the MMP enzyme to labile forms.

Normal levels of MMPs and TIMPS in serum and plasma have been determined as mentioned above. The results obtained are for guidance only and are shown in Table 67.6.

There is little indication at this stage that any sample type should be avoided for the MMPs. Serum MMP-8 is higher than plasma due to neutrophil release during clotting. For TIMP-1 the serum level may be higher and more variable due to platelet lysis than that found in plasma. It has also been reported that TIMP-2 levels are elevated in heparin plasma although the reason for this is unclear at present.

When activity assays have been applied to serum and plasma samples the results show a similar response to those found with ELISA tests on the same samples. However, due to the presence of complexed TIMP, results are generally lower where samples contain significant

Table 67.6. Ranges of normal levels of MMPs and TIMPs in serum and plasma measured by ELISA (values as ng/mL).

	Sample type			
MMP	**Serum**	**EDTA plasma**	**Citrate plasma**	**Heparin plasma**
MMP-1	ND	ND	ND	ND
MMP-2	470–800	365–649	NT	NT
MMP-3	28–99	17–41	23–79	25–136
MMP-9	ND	ND	ND	ND
MMP-7	ND	ND	ND	ND
MMP-8	40–57	1.4–10.3	1.0–1.8	1.4–4.7
MMP-13	ND	ND	ND	ND
TIMP-1	280–520	142–198	64–100	104–162
TIMP-2	29–108	21–108	NT	NT

ND, not detectable; NT, not tested.

amounts of this form of the MMP (e.g. MMP-2, 9), for the reasons explained above for cross-reactivity. We have found that all serum and plasma types can be used with activity assays with the exception of EDTA plasma. This sample type is clearly not recommended as EDTA will be able to chelate the Zn^{2+} at the MMP active site which has been shown to inactivate these enzymes.

REFERENCES AND FURTHER READING

Birkedal-Hansen, H. Proteolytic remodelling of extracellular matrix. *Curr. Opin. Cell Biol.* **7**, 728–735 (1995).

Cawston, T. Matrix metalloproteinases and TIMPs: properties and implications for the rheumatic diseases. *Mol. Med. Today*, 130–137 (1998).

Fujimoto, N., Hosokawa, N., Iwata, K., Shinya, T., Okada, Y. and Hayakawa, T. A one-step sandwich enzyme immunoassay for inactive precursor and complexed forms of human matrix metalloproteinase 9 (92 kDa gelatinase/type IV collagenase, gelatinase B) using monoclonal antibodies. *Clin. Chim. Acta* **231**, 79–88 (1994).

Jung, K., Nowak, L., Lein, M., Henke, W., Schnorr, D. and Loening, S.A. Role of specimen collection in preanalytical variation of metalloproteinases and their inhibitors in blood. *Clin. Chem.* **42**, 2043–2045 (1996).

Jung, K., Laube, C., Lein, M., Lichtinghagen, R., Tschesche, H., Schnorr, D. and Loening, S.A. Kind of sample as preanalytical determinant of matrix metalloproteinases 2 and 9 (MMP-2; MMP-9) and tissue inhibitor of metalloproteinases 2 (TIMP-2) in blood. *Clin. Chem.* **44**, 1060–1062 (1998).

Verheijen, J., Nieuwenbroek, N., Beekman, B., Hanemaaijer, R., Verspaget, H., Ronday, H. and Bakker, A. Modified proenzymes as artificial substrates for proteolytic enzymes; colorimetric assay of bacterial collagenase and matrix metalloproteinase activity using modified pro-urokinase. *Biochem. J.* **323**, 603–609 (1997).

CARDIOVASCULAR PEPTIDES

Endothelins

Endothelin-1 (ET-1) is a potent vasoconstrictor peptide produced by vascular endothelial cells. It is an acidic 21 amino acid peptide with a molecular weight of 2492 Da, and contains two sets of intrachain disulfide bonds, an unusual feature for a mammalian endogenous peptide, but a configuration often found in many peptide toxins. In fact, ET-1 shows a striking similarity to a group of peptide toxins from snake venom.

ET-1 is produced in vascular endothelial cells from a larger prepro-peptide that requires an unusual proteolytic processing between a tryptophan and a valine residue of a 38-residue intermediate (big endothelin). ET was originally purified from porcine aortic endothelial cells and was later found to be identical to human ET. Rat ET was sequenced and found to be homologous to porcine ET. Since then, this family has been expanded and renamed after the discovery of three ET genes in humans, of which porcine ET is ET-1, [Trp^6,Leu^7] ET is ET-2, and rat ET ([Thr^2,Phe^4,Thr^5,Tyr^6,Lys^7,Tyr^{14}]ET) is ET-3.

ET-1 is the most potent vasoconstrictor known to date, causing a strong and sustained vasoconstrictive response in most arteries and veins of many mammalian species and exhibiting extremely long-lasting pressor activities *in vivo*. This activity is mediated by an increase in the intracellular concentration of Ca^{2+}, by influx of extracellular Ca^{2+} through plasma membrane channels, and/or mobilization of intracellular Ca^{2+} by phospholipase C-stimulated inositol trisphosphate formation. However, it also has an extensive range of binding sites, not confined to vascular tissue, suggesting a wider range of activities than simply vasoconstriction. In fact, from recent *in vitro* experiments, ET-1 has been reported to stimulate the release of several hormones including atrial natriuretic peptide (ANP) from rat cardiac myocytes, eicosanoids, and endothelium-derived relaxing factor (EDRF) from vascular beds, and to modulate the release of noradrenaline from sympathetic termini. It also has effects on kidney cells, including the stimulation of mitogenesis in rat glomerular mesangial cells, the inhibition of renin release from rat glomerulus, and acute renal failure when perfused through isolated rat kidneys. Finally it stimulates the proliferation of vascular smooth muscle cells and contracts both airway and intestinal smooth muscle.

ET-like immunoreactivity has been identified in the plasma of normal and hypertensive subjects and has been shown to be elevated in hemodialysis patients. It has been demonstrated that plasma ET-like immunoreactivity consists of both ET-1 and its precursor, big ET. As big ET is thought to be inactive, it is important to determine if the contributions of active and inactive ET forms to plasma immunoreactivity vary independently in any situation.

Assay Technology

The ETs have been mostly assayed by RIA using ^{125}I-ET-1 as tracer with a phase-separation step using a magnetic or precipitating second-antibody. The antibodies used have been mostly polyclonal. Obviously, the antibody is a key feature of any competitive immunoassay but for assays of ETs there are two particular features that require selection. First, the antibodies selected must be of high affinity. ET-1, e.g. is found in the circulation at very low levels (<5 fmol/mL) and so it is crucial to have a very sensitive assay. Secondly, the specificity is very important due to the various iso forms of the active ETs and also the inactive pro forms. In order partly to address both these issues over recent years 2-site sandwich ELISAs have also been introduced, e.g. for the assay of ET-1 and big ET-1.

Specificity

There is a high degree of homology in sequence between ET-1, 2, and 3. This has made it difficult to develop specific antibodies that can discriminate between them. However, it is possible to distinguish ET-1 and 3, e.g. with the right antibody. It is much harder to be as specific for ET-1 as 2 but in practice this is not an issue as ET-2 is thought to be poorly expressed, if ever. A bigger concern has been the development of assays that discriminate the active ET-1 from its inactive pre-cursor big ET. It has been possible to develop RIAs that measure ET-1 to 21 forms distinct from big ET as well as high-sensitivity RIAs that will measure ET-1, 2 (total) forms distinct from ET-3, for example. Now ELISA has also done this.

Sample Preparation

Small peptides such as ET generally require extraction from samples such as serum and plasma prior to assay. It is not clear why this is, although the presence of binding proteins and the obvious potential for serum proteases to degrade these labile peptides, which are present at low concentrations, are thought to be the reasons. Typically solid-phase column extraction, e.g. Amprep™ or Sep-pak™, is the method of choice. The sample (1–2 mL) is loaded onto a pre-equilibrated column and following wash steps, e.g. 5 mL 0.1% TFA, the peptide is eluted in an acidified organic solvent, e.g. 80% methanol/0.1% TFA. The sample can then be dried under nitrogen or under centrifugal evaporation prior to reconstitution and assay.

There are obvious disadvantages with solid-phase extraction such as the requirement for larger (1 mL) sample volumes and the inconvenience of the extraction process. However, it appears to be a necessary prerequisite of ET assays.

Atrial Natriuretic Peptide

ANP can be produced by the atria following volume loading and high plasma sodium concentration. It has potent natriuretic, diuretic, and hypotensive effects. The active 28 amino acid peptide (αANP) which is the main circulating form is produced from a 126 amino acid prehormone precursor. This can further undergo degradation to various metabolites, some of which can be inactive. Some of these metabolites include the family of atriopeptins (I–III) first identified in rat, which were originally thought to be the active form of ANP.

The functions of ANP include natriuresis, diuresis, kaliuresis, and hypotensive effects; vasodilation; inhibition of angiotensin, norepinephrine, and potassium-induced vasoconstriction; positive modulation of renal gluconeogenesis; inhibition of adenylate cyclase activity and aldosterone production.

Plasma levels of ANP have been shown to be elevated in cases of chronic renal failure, paroxysmal atrial tachycardia, heart failure, and during pregnancy.

Assays

Several competitive RIAs for αANP have been described using ^{125}I-tracer and second-antibody separation methods. Similar to ET the antibody plays a crucial role in determining the two important features of any ANP immunoassay – sensitivity and specificity. Normal levels of αANP are <10 fmol/mL and so assay sensitivity is vital. It is not so much of a problem in specifically distinguishing the active αANP from its very much larger inactive precursor as it is with ETs. However, poorly active or inactive metabolites are much more of a problem, and antibodies which can distinguish the intact 28 amino acid forms particularly from C-terminally cleared peptides are thought to be very useful in ensuring accurate ANP measurement.

Sample Preparation

As for ETs sample extraction, typically by solid-phase column, appears to be mandatory for ANP. Platelets seem to release ANP, and so problems can occur in plasma that is not properly separated.

REFERENCES AND FURTHER READING

Capper, S.J., Smith, S.W., Spensley, C.A. and Whateley, J.G. Specificities compared for a radioreceptor assay and a radioimmunoassay of atrial natriuretic peptide. *Clin. Chem.* **36**, 656–658 (1990).

Capper, S.J., Ella, S.J. and Kalinka, S. Development and application of three radioimmunoassays of endothelin of varying specificities. *J. Cardiovasc. Pharmacol.* **17** (Suppl 7), S425–S426 (1991).

NEURODEGENERATION ASSAYS

Dementia

Alzheimer's Disease (**AD**) and **Dementia with Lewy Bodies** (**DLB**) are the most common forms of dementia, and are one of the largest health-care problems in developed countries. AD is considered to be the most important form of all neurodegenerative disorders with devastating consequences for the affected patients and their families. AD is characterized clinically by a progressive loss of memory and cognition, often with additional psychiatric manifestations, that gradually lead to a profound mental deterioration, and ultimately, death. AD is characterized pathologically by the presence of extracellular senile plaques and intraneuronal neurofibrillary tangles in the brain. Current treatments for AD, whilst slowing the rate of decline of patients, are considered to be palliative, and there is presently no cure.

Beta-amyloid Peptides

The major protein component of senile plaques is a β-amyloid peptide, a 40–43 amino acid peptide cleaved from amyloid precursor protein (APP) by β-secretase. The aggregation and deposition of β-amyloid is thought to play an early, and perhaps causative, role in AD. It has been reported that β-amyloid 1–42 is deposited preferentially in the cerebral cortex causing senile plaque formation, whereas β-amyloid 1–40 is mainly deposited in the cerebral blood vessels resulting in angiopathy.

Synuclein

Studies of familial **Parkinson's Disease** (**PD**) have identified the **alpha synuclein** gene as a cause of PD in rare families. Alpha-synuclein is, however, a key component of **Lewy bodies**, the major pathological hallmark of PD, and a short peptide fragment of alpha synuclein termed NACP, is also found in senile plaques in AD. Alpha synuclein is a major synaptic protein and it has been suggested that alterations in synaptic proteins, synaptic loss, and injury are the primary causes of impaired cognitive function in AD and DLB.

Assay Procedures

Immunoassays are available using two-site sandwich ELISA approaches, with monoclonal antibody-coated wells and polyclonal second antibodies. In all cases standards are prepared with synthetic peptides (synuclein, amino acid residues 117–131). Detection is with tetramethylbenzidine.

Levels of beta amyloid 1–40 and 1–42 are increased in human brain tissue samples obtained at autopsy from patients with Alzheimer's disease (Figure 67.12).

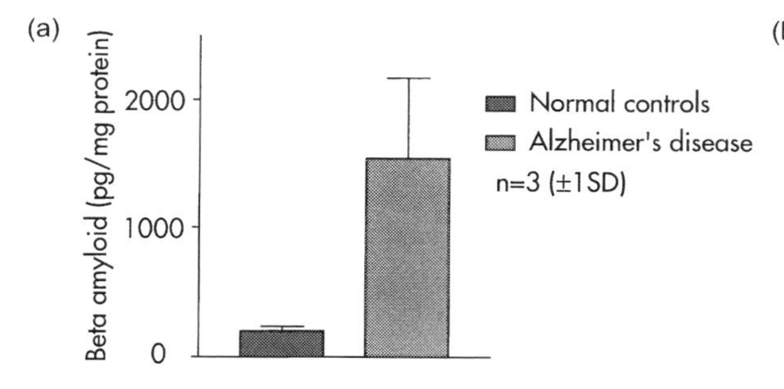

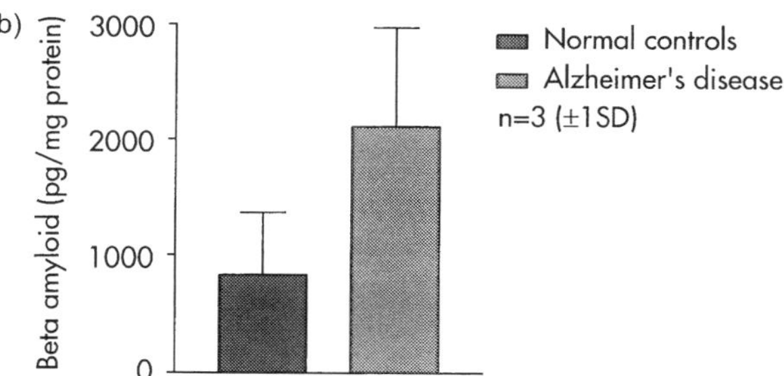

Figure 67.12 Measurement of beta amyloid (a) 1–40 and (b) 1–42 extracted from brain tissue from patients with AD and normal controls. Data is shown as mean ± 1 standard deviation (n = 3).

All the assays are simple, precise, highly specific, and easy to perform, with colorimetric end-points. The sandwich ELISA developed for the beta amyloid peptide and alpha synuclein have been validated using a number of sample types. The methods are ideally suited for the discovery of new compounds for the treatment of Alzheimer's and Parkinson's Diseases.

REFERENCES AND FURTHER READING

DeKosky, S.T. and Orgogozo, J.M. Alzheimer's disease; diagnosis, costs and dimensions of treatment. *Alzheimer Dis. Assoc. Disord.* **15** (Suppl 1), S3–S7 (2001).

Hardy, J.A. and Higgins, G.A. Alzheimer's disease: the amyloid cascade hypothesis. *Science* **256**, 184–185 (1992).

Ince, P.G., Perry, E.K. and Morris, C.M. Dementia with Lewy bodies. A distinct non-Alzheimer dementia syndrome? *Brain Pathol.* **8**, 299–324 (1998).

Polymeropoulos, M.H., Lavedan, C., Leroy, E., Ide, S.E., Dehjia, A., Dutra, A., Pike, B., Root, H., Rubenstein, J., Boyer, R., Stenroos, E.S., Chandraseharappa, S., Athanassiadou, A., Papapetropoulos, T., Johnson, W.G., Lazzarini, A.M., Duvoisin, R.C., Di Iorio, G., Golbe, L.I. and Nussbaum, R.L. Mutation in the alpha-synuclein gene identified in families with Parkinson's disease. *Science* **276**, 2045–2047 (1997).

Terry, R.D., Masliah, E., Salmon, D.P., Butters, N., DeTeresa, R., Hill, R., Hansen, L.A. and Katzman, R. Physical basis of cognitive alterations in Alzheimer's disease: synapse loss is the major correlate of cognitive impairment. *Ann. Neurol.* **30**, 572–580 (1991).

Vassar, R., Bennett, B.D., Babu-Khan, S., Kahn, S., Mendiaz, E.A., Denis, P., Teplow, D.B., Ross, S., Amarante, P., Loeloff, R., Luo, Y., Fisher, S., Fuller, J., Edenson, S., Lile, J., Jarosinski, M.A., Biere, A.L., Curran, E., Burgess, T., Louis, J.C., Collins, F., Treanor, J., Rogers, G. and Citron, M. Beta-secretase cleavage of Alzheimer's amyloid precursor protein by the transmembrane aspartic protease BACE. *Science* **286**, 735–741 (1999).

68 Immunoassay Applications in Veterinary Diagnostics

Erwin Workman

Diagnosis is defined as the art or act of identifying disease from its signs and symptoms. The veterinarian, as the physician, uses a variety of tools in practicing this art. One such tool that continues to increase in importance is the immunodiagnostic test. Increased usage of this technology in veterinary medicine reflects the tremendous strides that have been made in assay development, driven by improvements and advances in instrumentation, solid-phase chemistry, assay device technology, detection technology, and biotechnology.

The needs of the veterinary diagnostic laboratory for diagnostic tests may be divided into three categories. First, and most important, is the need to diagnose disease. This usually involves testing for the presence of viral or bacterial antigens, or antibodies to them. Next, there is a need to assess immune status to certain diseases as a function of vaccination or previous exposure. For this type of determination, tests should detect the presence of antibodies to viral or bacterial antigens and provide semiquantitative results. The last category is the need to assess reproductive and metabolic status. Diagnostics in this category are primarily quantitative tests for various hormones.

For years these types of tests relied on traditional methodologies such as viral isolation, virus neutralization, plate agglutination, hemagglutination inhibition, immunodiffusion, classical microbiological culture techniques, high-pressure liquid chromatography, and thin layer chromatography. However, during the 1980s and 1990s, improved immunoassay-based technologies have evolved which have all but replaced the more traditional methods, resulting in improved accuracy and precision while requiring less time and labor.

Enzyme, fluorescence and colloidal particle-based immunoassay systems are replacing radioimmunoassays and various agglutination and tissue culture methodologies. The use of monoclonal antibodies as well as recombinant and synthetic peptide antigens has given rise to improvements in sensitivity, specificity, and ease of use. The availability of inexpensive and visually read single-use delivery systems has made it possible for veterinarians to perform tests in the clinic or on the farm that once could be done only in reference laboratories.

This chapter provides an overview of the various immunoassays that are commercially relevant and available to meet the needs of veterinary medicine for diagnostics. The information is grouped by category (infectious disease diagnosis/assessment of immune status and assessment of reproductive/metabolic status) and by species (feline, canine, porcine, equine, bovine, avian, and murine).

INFECTIOUS DISEASE DIAGNOSIS/ASSESSMENT OF IMMUNE STATUS

FELINE

Feline Leukemia Virus

Classification of Organism/Pathogenesis

The feline leukemia viruses (FeLV) are oncogenic retroviruses that are transmitted horizontally by close contact or by *in utero* transmission. A variety of neoplastic and non-neoplastic diseases are caused by FeLV infection, including lymphosarcoma, myelogenous leukemia, thymic degenerative disease, panleukopenia-like diseases, and non-regenerative anemias. Furthermore, because FeLV infection is immunosuppressive, infected cats are susceptible to a variety of secondary and opportunistic infections.

Test Analyte

FeLV group-specific antigen (p27).

Types of Sample

Whole blood, serum, plasma, saliva, or tears.

The Immunoassay Handbook. *Third Edition*
D. Wild (Ed.)

Assay Technology

Most laboratories and veterinary clinics use some form of enzyme immunoassay (EIA) or colloidal particle-based immunoassay (CPIA) to detect the presence of p27, although some reference laboratories still offer the less sensitive immunofluorescence assay. All EIAs and CPIAs utilize a capture antibody immobilized onto a solid phase (wells, particles, dipsticks, etc.) and an enzyme or colloidal particle-labeled antibody which is incubated simultaneously or sequentially with the capture antibody and sample. In general, the reference laboratories utilize EIAs in microwell formats because of ease of batching and cost considerations while the clinics prefer EIAs or CPIAs in the single-use immunoassay delivery formats.

IDEXX has been the market leader since 1987, 1 year after introducing its concentration immunoassay technology (CITE®), which utilized membrane filtration technology. In the CITE format, three different reagents were prespotted onto separate areas of a fiber membrane in a triangular pattern. The sample spot contained monoclonal antibodies to p27 and the two control spots contained reagents to represent positive and negative controls. A prefilter fitted on top of the CITE device removed insoluble material from the sample. The device was designed to ensure an even and regulated flow of materials across reactive membrane surfaces. The test sample was first mixed with an enzyme-labeled monoclonal antibody to p27 (the conjugate). Any p27 antigen present became bound to the conjugate. Then the entire solution was applied to the CITE device. The membrane-bound p27 antibody in the sample spot captured the p27-conjugate complex. Unbound conjugate and other substances were washed away and an enzyme substrate/chromogen solution was added. Subsequent color development in the sample spot indicated the presence of p27 and the intensity of the color was proportional to p27 concentration. Color also developed in the positive control spot, indicating that the reagents were active. The negative control spot, containing mouse antibodies with no reactivity to p27, developed color only if there was non-specific reactivity in the sample (Figure 68.1).

IDEXX over the last 10 years has introduced second and third generation immunoassay devices, CITE PROBE and SNAP, respectively. CITE PROBE was similar to CITE, but eliminated reagent bottles, the reagents being packaged in a sealed tray (see Figure 68.2). SNAP, the current market leader, has on board reagents for color development and requires less hands on time than CITE or CITE PROBE. In the case of SNAP (see Figure 68.3), an enzyme labeled conjugate is mixed with serum, plasma, or whole blood and immediately added to the sample well of the SNAP device. When the sample reaches the activate circle (30–60 sec) having flowed past the reactive spots, the device is activated. Activation results in a reversing of flow and release of wash and substrate. Results are complete within 10 min.

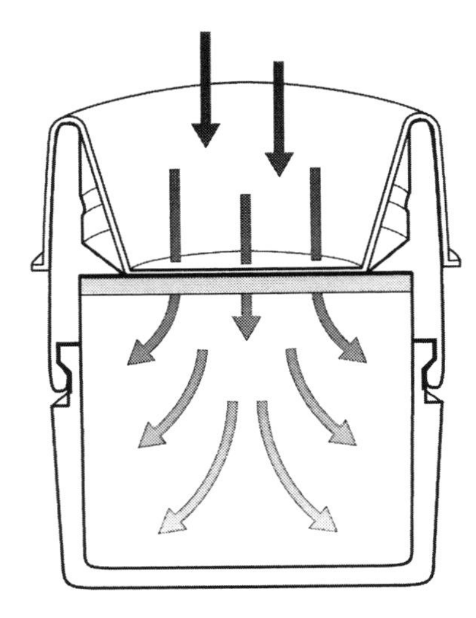

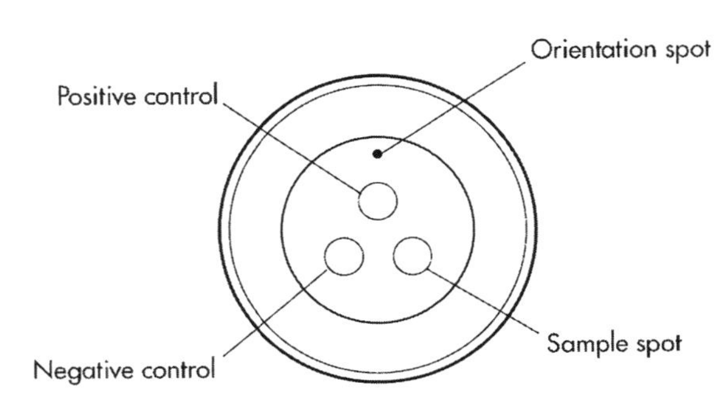

Fig. 68.1 CITE®-FeLV (cross-section).

It should be noted that most FeLV tests are performed as combination tests with feline immunodeficiency virus in the SNAP format. This test will be described in a later section.

In addition to the SNAP-FeLV and SNAP-FeLV/FIV, other products are available for the detection of p27 antigen in the USA and Europe. Synbiotics markets a dipstick-based test (Assure™ FeLV), a lateral flow

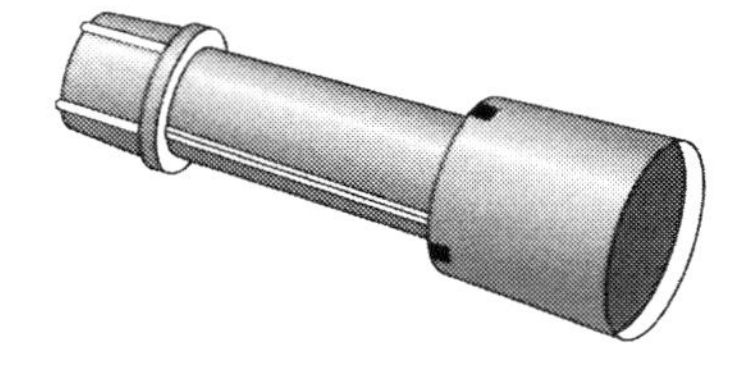

Fig. 68.2 Cite® Probe. The device is incubated with sample in the first well, and then moved sequentially through conjugate, wash and substrate wells. Color develops on the tip in a pattern similar to Cite®.

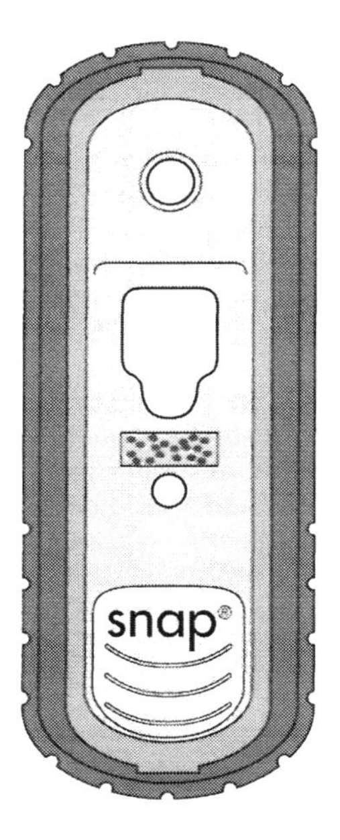

Fig. 68.3 SNAP device.

immunochromatography device (WITNESS™), as well as a microwell test (ViraCHEK™/FeLV). The WITNESS product is a CPIA which utilizes colloidal gold as the label. IDEXX also has a microwell product with a confirmatory protocol (neutralization). Other manufacturers such as Bio Veto Test (BVT) of France sell only outside the United States, and market CPIA products.

Frequency of Use

Very common.

Feline Immunodeficiency Virus

Classification of Organism/Pathogenesis

The feline immunodeficiency virus (FIV; formerly feline T-lymphotropic lentivirus) is a recently described feline-specific retrovirus that can produce chronic immunodeficiency-like disorders in cats. The FIV agent has a strong tropism for feline T-lymphocyte cells. The cytopathic effect exhibited following infection of these cells may be responsible for the immunosuppressive nature of the virus. Viral particle morphology and the Mg^{2+} requirement of the viral reverse transcriptase are characteristic of lentiviruses. Members of the lentivirus subgroup include the human immunodeficiency viruses (Pederson *et al.*, 1987).

Clinical signs most commonly observed in infected cats include chronic rhinitis, chronic gingivitis and periodontitis, anemia, diarrhea, pustular dermatitis, and generalized lymphadenopathy. Symptoms vary from animal to animal and may persist for several years. FIV is infectious within cat populations and is spread by intimate or prolonged contact.

Test Analyte

An assessment of exposure to FIV via natural infection can be made by measurement of specific antibody titer to FIV in feline serum, plasma, or whole blood. In addition, measurement of the predominant group-associated antigen of FIV (p26) can be used to demonstrate the presence of the virus in tissue culture medium following culturing of lymphocytes from feline blood or tissue extracts.

Types of Sample

Serum, plasma, whole blood, or tissue extracts, (*see* above).

Assay Technology

EIA is the predominant technology used for the diagnosis of FIV exposure or infection. At present IDEXX is the only manufacturer of FIV diagnostics in the USA, supplying two antibody detection systems and one antigen detection system. One of the antibody tests is in the SNAP format (SNAP COMBO-FeLV/FIV) while the second is configured as a microwell product. (For a description of the SNAP technology *see* FeLV, ASSAY TECHNOLOGY). The antigen test is available only in a microwell format. In both antibody tests the solid phase contains antigen and the conjugate consists of labeled antigen rather than a labeled anti-species antibody, which results in increased sensitivity and specificity.

Other FIV diagnostics are available outside the USA. Synbiotics markets WITNESS FIV, a CIPA for use in the veterinary clinic, and ViraCHEK FIV, a microwell product. Both utilize reagents designed to detect antibodies to gp40. BVT of France also sells an in-clinic product in the CIPA format.

Frequency of Use

Very common.

Feline Infectious Peritonitis Virus

Classification of Organism/Pathogenesis

Feline infectious peritonitis virus (FIPV) is a coronavirus that is antigenically related to canine coronavirus, human bronchitis virus 229E, porcine transmissible gastroenteritis virus, and feline enteric coronavirus (FeCV). The mode of natural infection is as yet unknown (Aiello and Mays, 1998). Experimental studies have shown that regardless of the initial site of replication, once the virus enters macrophages, virus replication begins in earnest. These cells also carry the virus to the various target tissues such as the serosal membranes lining the abdominal and pleural cavities and organs, meninges and ependyma of brain and spinal cord. Two forms of feline infectious peritonitis (FIP) may occur: the dry form, characterized by the development of a disseminated granulomatous disease, and the wet form, characterized by the appearance of abdominal and/or pleural effusions. Mortality among clinically ill animals is essentially 100%.

Test Analyte

FIPV antibody.

Type of Sample

Serum.

Assay Technology

At present three test kits are commercially available for FIP testing. The first is manufactured by IDEXX and is a competitive enzyme-immunoassay. It utilizes a monoclonal antibody specific for the E1 envelope antigen to compete with any E1-specific antibody in the test serum. The plastic microwells in the kit have been precoated with inactivated, purified FIPV. Cat serum is incubated simultaneously in the well with the monoclonal antibody to which horseradish peroxidase (HRPO) has been conjugated. Color formation is inversely proportional to the amount of anti-E1 present in the sample.

This product is intended for the diagnosis and management of FIP, but it is not totally specific for FIPV as there is cross-reactivity with anti-FeCV. The second product is manufactured by Biogal-Galed Labs in Israel. It is designed to determine the serum IgG antibody titers for FIP using the ImmunoComb, a plastic card shaped like a comb on which FIP antigens are coated. The conjugate is an enzyme-labeled anti-feline IgG antibody. The assay is performed by incubating the comb with sample which allows specific antibody to bind to coated antigens. After a wash step, the comb is incubated with conjugate. After a second wash, the comb is placed into substrate, generating a color if antibodies are present. This product also is not specific for FIP.

A third product for the detection of antibodies to FeCV is manufactured by BVT of France and is available in the CIPA format.

Frequency of Use

Occasional, because of lack of specificity of the current products.

CANINE

Dirofilaria immitis (Heartworm)

Classification of Organism/Pathogenesis

Dirofilaria immitis (*D. immitis*) is a filarial nematode that causes heartworm disease complex (dirofilariasis) in dogs and sometimes in cats. The insect vector for *D. immitis* is the mosquito. Adult worms reach a size of 20–36 cm in length and 3 mm in diameter. They inhabit the blood and vascular tissue, especially the heart and adjacent blood vessels. *D. immitis* often interferes with heart function and blood circulation and may damage other vital organs. Fertile female worms release microfilariae into the circulation, and microscopic detection of this larval stage in the blood is the most common method used to diagnose heartworm disease. The accuracy and sensitivity of microscopic diagnostic methods are limited by several factors. First, *D. immitis* infections may occur without subsequent production of microfilariae. These occult infections may be caused by immune-mediated or drug-induced worm sterility, as well as single-sex, prepatent or ectopic infections. In addition, the presence of morphologically similar microfilariae (*Dipetalonema reconditum*) may result in misdiagnosis. Furthermore, microfilarial counts are influenced by the diurnal rhythm of the microfilariae, the volume of blood examined, and the skill of the examiner.

Accurate diagnosis of heartworm infections is important. Inappropriate administration of heartworm preventatives to an infected animal may result in severe or fatal reaction. Failure to eliminate adult worms can result in impaired heart function and circulation, and progressive deterioration of other organs and vascular tissue. However, proper treatment of adult worms in the early stages of infection can result in complete recovery.

Test Analyte

D. immitis antigen.

Types of Sample

Serum, plasma, or whole blood.

Assay Technology

The technologies utilized for the detection of heartworm antigen are similar to those used for FeLV detection. Commercially available tests utilize EIA, CPIA, or agglutination technologies. These include microwells, dipsticks, membrane filtration systems, immunochromatography devices, and lateral flow systems. Prior to 1995 most tests required some form of sample pretreatment to break down immune complexes typical in occult infections. At present, with few exceptions, the protocols are pretreatment free. Manufacturers include Synbiotics with microwell, agglutination, and immunochromatography devices, Heska with a CPIA based product, IDEXX with a microwell test, the SNAP Canine Heartworm test, and the SNAP Canine Heartworm Antigen/*Borrelia burgdorferi*/*Ehrlichia canis* Antibody Test kit (3DX), a three way combination product. All tests utilize monoclonal antibodies, a combination of monoclonal and polyclonal antibodies, or polyclonal antibodies.

The semi-quantitative SNAP Canine Heartworm PF test is the market leader on a worldwide basis. In the SNAP format, there are four different spots of reagent on a porous polyethylene matrix. Two contain different, calibrated levels of antibody raised against the *D. immitis* antigen, and the remaining spots act as positive and negative controls. The conjugate and test sample are mixed and added to the device and allowed to pass through the reagent spots. *D. immitis* antigen, if present, is captured by the enzyme-conjugated anti-heartworm antibodies and the immobilized anti-heartworm antibodies forming an antibody–antigen–antibody sandwich. The two calibrated spots on the matrix vary in their capacity to bind *D. immitis* antigen. After the conjugate and sample mixture passes through the reagent area, the device is activated thus allowing wash and substrate to flow sequentially into the matrix. Subsequent color development on the calibrated spots is proportional to the *D. immitis* antigen concentration in the sample, indicating low to high antigen levels (see Figure 68.4).

In an independent field trial with 55 heartworm-infected dogs, antigen levels were found to correspond to overall worm burden, with low burden defined as up to 1.5 g of worm per animal and high burden as greater than 1.5 g. Note that one gravid female worm weighs approximately 0.1 g.

The IDEXX 3DX product provides simultaneous detection of *D. immitis* antigen, antibody to *B. burgdorferi*, and antibody to *E. canis* in canine whole blood, serum, or plasma. Sensitivity and specificity for *D. immitis* antigen are similar to the SNAP Canine Heartworm test.

Frequency of Use

Very common, particularly in late winter and during the spring.

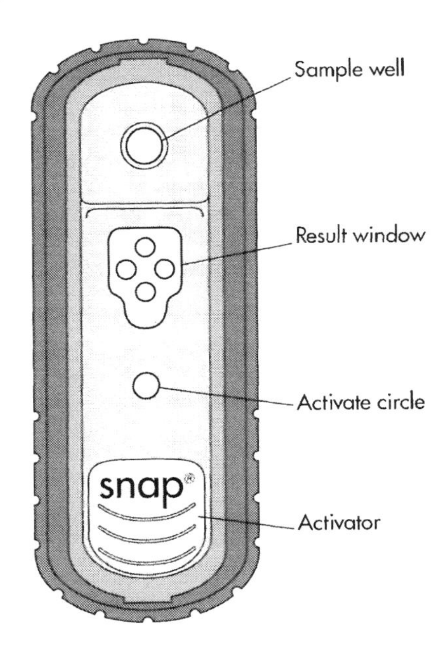

Fig. 68.4 Quantitative SNAP-format.

Canine Parvovirus

Classification of Organism/Pathogenesis

Canine parvovirus is an autonomous parvovirus similar to feline panleukopenia virus and mink enteritis virus that causes an enteritis of acute onset with varying morbidity and mortality. Ingestion of fecal material from infected animals is the major route of infection. After ingestion, tonsillar crypts and Peyer's patches are infected. Subsequent to infection of lymphatic tissues, infection of intestinal crypts occurs resulting in loss of intestinal villi. The virus causes two disease forms in dogs: myocarditis and enteritis. Maternal immunity has all but eliminated the incidence of myocarditis, while enteritis, although reduced, continues to infect dogs of all ages. Enteritis is characterized by acute, severe diarrhea, vomiting, panleukopenia, and rapid dehydration. The disease is most severe in puppies and can be fatal.

Test Analyte

Canine parvovirus antigen.

Type of Sample

Because large numbers of viral particles are shed in the feces, fecal samples are appropriate for the detection of acute, transmissible infections.

Assay Technology

Rapid, accurate diagnosis of canine parvoviral infection allows the initiation of prompt treatment and quarantine of infected dogs. EIA technology allows rapid and accurate diagnosis. IDEXX's quick and convenient CITE-probe and SNAP-based tests are most commonly used in clinics. Synbiotics offers a dipstick product, ASSURE-PARVO™, and Nippon Zenyaku offers a monoclonal antibody based product.

Similar to other SNAP products (*see* FeLV, ASSAY TECHNOLOGY), the Canine Parvovirus test is a monoclonal antibody-based enzyme immunoassay incorporating positive and negative controls. Total assay time is under 10 min.

Frequency of Use

Occasional.

Borrelia burgdorferi (Lyme Disease)

Classification of Organism/Pathogenesis

The causative agent of Lyme disease has been identified as the tick-borne spirochete *B. burgdorferi* (Steere, 1989). This organism is known to infect a wide variety of mammals and birds. Typical signs of the disease in dogs include skin lesions, fever, lethargy, anorexia, depression, generalized joint pain or arthritis and intermittent joint lameness.

Test Analyte

B. burgdorferi antibody.

Types of Sample

Serum, plasma, or whole blood.

Assay Technology

An enzyme immunoassay is available from IDEXX Laboratories in the SNAP format as a combination product with *D. immitis* antigen and *E. canis* antibody. The SNAP test utilizes *B. burgdorferi* antigen on the solid phase for antibody capture, but the conjugate is inactivated *B. burgdorferi* antigen conjugated to HRPO. The SNAP Lyme Antibody test has been designed to demonstrate exposure to *B. burgdorferi*. Most Lyme-vaccinated dogs will be non-reactive on this test.

Frequency of Use

Frequent during heartworm season.

Leishmania infantum/Leishmania donovani

Classification of Organism/Pathogenesis

Leishmaniasis, caused by parasites of the genus Leishmania, has wide distribution throughout the Mediterranean Basin. The parasites are transmitted by sand flies to many mammals, including humans and canines. Dogs are likely an important reservoir for both the human and canine diseases. Leishmania primarily infects white blood cells leading to impaired immune function, blood disorders, and various visceral and/or skin lesions (Georgi and Georgi, 1992). Clinical signs of visceral leishmaniasis include weight loss, muscle atrophy, dermatitis, and lymphadenitis. In dogs, the diagnosis of visceral leishmaniasis can be made following observation of typical clinical signs and the measurement of a significant antibody titer to Leishmania.

Test Analyte

Antibody to Leishmania.

Types of Samples

Serum, plasma, or whole blood.

Assay Technology

Products on the market or soon to be on the market utilize CPIA or EIA technologies. Both Heska and BVT of France market colloidal gold immunochromatography products. Both utilize processed or recombinant antigens for detection of specific antibodies. IDEXX has introduced SNAP Leishmania, an EIA formatted in similar fashion to the SNAP products described in previous sections. The assay is constructed such that specific antibody is sandwiched between two different antigens, one immobilized and one labeled with HRPO. Positive and negative controls are integrated into the assay. A comparison of the three products in terms of sensitivity and specificity may be seen in Table 68.1. Sample populations were obtained in Europe and were made up of 27 negatives and 38 positives (as determined by IFA, currently the gold standard). All three assays had similar sensitivity on positives, ranging from 95 to 97%. SNAP seemed to have the edge in specificity. It should be noted that the populations are small.

Table 68.1. Leishmania sensitivity/specificity comparison.

	Negatives	Positives
SNAP	27/27, 100%	37/38, 97%
BVT	21/27, 78%	36/38, 95%
Heska	24/27, 89%	37/38, 97%

Note: Negative population was obtained in France; positive population from Spain. Assays were performed per manufacturer's instructions.

Frequency of Use

Frequently in endemic areas of Europe.

Ehrlichia canis

Classification of Organism/Pathogenesis

Canine ehrlichiosis is a tick-borne disease of dogs caused by the rickettsial parasite, *E. canis*. Replication of the organism occurs within infected mononuclear cells and spreads to organs containing mononuclear phagocytes. Infection can result in thrombocytopenia, leukopenia, and/or anemia. Clinical signs of infection include fever, dyspnea, weight loss, hemorrhages, and epistaxis. In dogs, diagnosis of Ehrlichiosis has been made following observations of typical clinical signs and by the measurement of a significant antibody titer to *E. canis*.

Test Analyte

Antibody to *E. canis*.

Types of Sample

Serum, plasma, or whole blood.

Assay Technology

At present, it appears that EIAs are available from Biogal-Galed Labs of Israel in their ImmunoComb format, and from IDEXX in the SNAP format. The SNAP technology has been described previously. Note that the test for *E. canis* is incorporated in a combination assay for

D. immitis antigen and antibody to *B. burgdorferi*. CPIA formatted products are available from BVT and Mega Cor Diagnostik. The protocol for ImmunoComb has been described previously.

Frequency of Use

Frequent during heartworm season.

PORCINE

Pseudorabies Virus

Classification of Organism/Pathogenesis

Pseudorabies, or Aujesky's disease, is caused by a type 1 porcine herpesvirus (pseudorabies virus, PRV). Typically, the virus is taken into the oro-nasal passages and initiates infection in a variety of cells, including olfactory nerve cells, and progresses into the brain and other nervous tissue. Glomerulonephritis develops and is followed by meningitis, panencephalitis, and acute pneumonia.

Infections with the highest mortality rate are those affecting suckling pigs born to a susceptible sow. Baby pigs in the fatal course of the disease exhibit difficulty in breathing, fever, hypersalivation, anorexia, vomiting, diarrhea, trembling, and depression. Within this age group, the final stages of infection are commonly characterized by ataxia, nystagmus, running fits, intermittent convulsions, coma and death. Death usually occurs within 24–48 h of the appearance of clinical symptoms.

Test Analyte

Pseudorabies antibody (*see* ASSAY TECHNOLOGY below).

Type of Sample

Serum.

Assay Technology

The EIA has all but replaced serum neutralization testing since IDEXX introduced the screening and verification microwell products in the 1980s. These are standard antibody tests where purified antigen is coated on the solid phase and serves to sequester specific antibodies to PRV from the sample. The bound antibodies are then detected using an anti-porcine antibody conjugated to HRPO. The only difference between the screening test and the verification test is the incorporation of a normal host cell (NHC) control configured in alternating rows on the microwell plate. The extent of host cell contribution to the total signal is assessed by relating PRV activity to NHC reactivity. Total assay time for each assay is approximately 30 min.

In addition to the screening tests, which are designed to detect total antibody against PRV, immunoassays are now available that detect antibodies to certain viral proteins, the DNA for which has been deleted from companion genetically engineered vaccines. This allows for serological differentiation of vaccinated animals from animals exposed to field strains of PRV, because only animals exposed to field strains develop antibodies to the 'deleted' protein. Examples include IDEXX's anti-PRV-gpX test (complements the SyntroVet gpX-deleted PRV vaccine), IDEXX's anti-PRV-g1 test (complements the Boehringer Ingelheim, Pfizer, and other g1-deleted PRV vaccines around the world). A number of other anti-PRV-g1 products are manufactured in Europe by firms such as Synbiotics, Svanova, and Bomelli Laboratories. All of these products utilize EIA technology and microwell formats. In terms of performance, differential tests should have sensitivity for field strain detection equivalent to that of the whole-virus assays, and should ignore all antibodies produced in the animal as a function of vaccination (see Figure 68.5).

Frequency of Use

Pseudorabies tests are performed frequently in official laboratories around the world.

Porcine Reproductive and Respiratory Syndrome (PRRS)

Classification of Organism/Pathogenesis

A new swine disease causing reproductive problems, respiratory disease and mild neurologic signs was first reported in 1987. Due to the general clinical symptoms presented in most cases, diagnosis was often confused with swine influenza, pseudorabies, hog cholera, parvovirus, encephalomyocarditis, chlamydia, and mycoplasma (Collins *et al.*, 1992). A major component of the syndrome is reproductive failure resulting in premature births, late term abortions, pigs born weak, increased stillbirths, mummified fetuses, decreased farrowing rates, and delayed return to estrus. These aspects of the syndrome have been observed to last 1–3 months. Respiratory disease is another significant feature of the disease that most affects pigs less than 3–4 weeks of age. Respiratory signs can occur in most stages of the production cycle.

European and North American scientists have successfully isolated and characterized the agent responsible for this disease (Wensvoort *et al.*, 1992; Benfield *et al.*, 1992; Terpstra *et al.*, 1991). The etiologic agent has been described as a virus similar to equine arteritis virus and the lactate dehydrogenase-elevating virus. Researchers in the Netherlands have proposed that these viruses be grouped in a new family, Arteriviridae, based on genomic sequencing information (Muelenberg *et al.*, 1992).

Test Analyte

PRRS-specific antibodies.

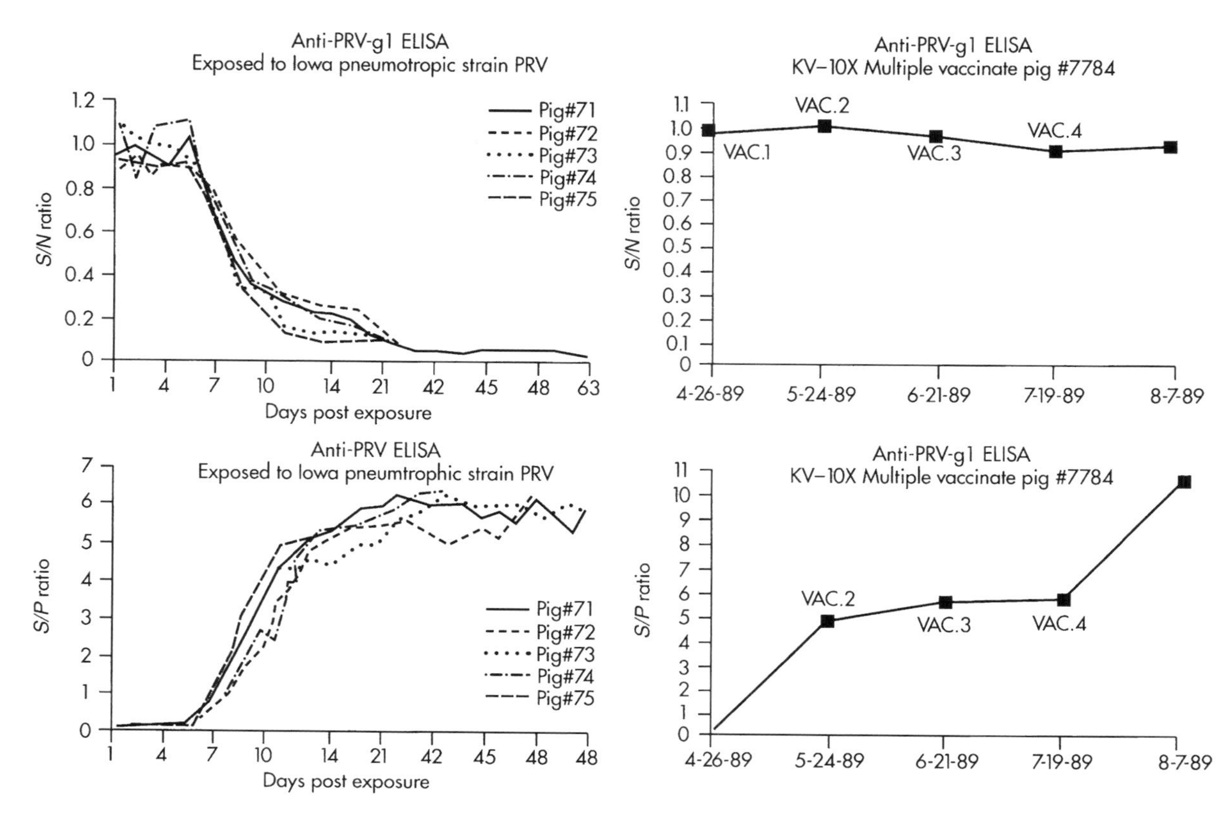

Fig. 68.5 Left, the performance of the IDEXX anti-PRV-g1 test on field strain infected animals compared to that of the IDEXX whole-virus assay. Note that the anti-g1 test is configured as a competitive E1A, thus a sample/negative (S/N) ratio of 1.0 is equal to the negative control; samples with S/Ns $\leq$ 0.7 are considered positive. Seroconversion is detected on or about day 7 for both assays indicating equivalent sensitivity (S/P, sample/positive). Right, the performance of the same test on an animal hyperimmunized with the Boehringer Ingleheim gl-detected vaccine. The whole-virus test indicates the presence of substantial titer against PRV, while the g1 test remains negative over the course of the experiment.

Type of Sample

Serum.

Assay Technology

As assessment of exposure to the PRRS virus as a result of natural infection is facilitated by a measurement of antibodies in the serum, it is important that the assay measuring the antibodies be sensitive to the several strains of the virus. The first product available to the market was the IDEXX HerdChek™ PRRS assay, an EIA designed to detect the presence of antibody to PRRS in swine serum. The assay is configured in a way similar to the PRV verification assay described above (*see* PSEUDORABIES VIRUS – ASSAY TECHNOLOGY) and utilizes PRRS antigens from different strains as well as NHC antigens coated on alternating columns of a microwell plate. Total time for the assay is approximately 75 min.

Within the last few years Bomelli Laboratories has introduced an EIA for the detection of antibody to PRRS virus and is similar in configuration to other microwell assays provided by Bomelli Laboratories.

Frequency of Use

PRRS tests are performed frequently in corporate and government laboratories around the world.

Classical Swine Fever Virus (Hog Cholera Virus)

Classification of Organism/Pathogenesis

Classical Swine Fever Virus (CSFV) is a small, enveloped RNA virus of the family Flaviviridae. This pestivirus is antigenically similar to Bovine Viral Diarrhea Virus and Border Disease Virus. CSFV causes serious losses in the pig industry since it is highly pathogenic and can cause widespread deaths. Pigs infected with highly virulent CSFV strains may shed a large amount of virus before showing clinical signs of the disease such as fever, depression, and loss of appetite. The primary lesion is generalized vasculitis (Kosmidou *et al.*, 1995) manifested in live pigs as hemorrhages in the skin. Animals that survive an acute or subacute infection develop antibodies and will no longer spread the virus. Moderately virulent, less pathogenic strains may lead to chronic infection, when pigs excrete the virus continuously or intermittently until death. Congenital infection can result in abortion, fetal mummification, stillborn and/or weak piglets or embryonic malformations.

Test Analyte

Antigen or antibody specific for CSFV.

Type of Sample

Peripheral blood leukocytes, whole blood, and tissue (antigen test); serum or plasma (antibody test).

Assay Technology

EIAs are available for both antibody and antigen detection and all are configured in microwell formats. In general, the antigen tests are sandwich assays utilizing polyclonal/ monoclonal combinations or dual monoclonals. The antibody tests are typically of the competitive type and utilize native or recombinant antigen on the solid phase and enzyme-labeled monoclonal antibody as the conjugate. Products for the diagnosis of classical swine fever are available from IDEXX, Synbiotics, and Bomelli Laboratories.

Frequency of Use

Such tests are performed frequently in government laboratories in countries where disease outbreaks occur.

EQUINE

Equine Infectious Anemia Virus

Classification of Organism/Pathogenesis

Equine infectious anemia virus (EIAV) is an equine-specific retrovirus (lentivirus) that causes equine infectious anemia in horses around the world. The virus persists in the white blood cells of infected horses for life. Transmission occurs by transfer of blood cells from an infected horse, for example, via insertion and withdrawal of contaminated hypodermic needles. The primary pathology of the disease is the immune-mediated destruction of red blood cells.

Test Analyte

EIAV-specific antibodies.

Type of Sample

Serum.

Assay Technology

At present, both agar gel immunodiffusion (AGID) and EIA technologies are being used to detect antibodies against p26 antigen, the major group-specific antigen found internally in the virus. AGID tests are available from IDEXX and Synbiotics. The oldest commercially available EIA is manufactured by IDEXX in a competitive microwell format. The kit contains microwells precoated with monoclonal antibody specific for p26. The conjugated p26 antigen has been chemically linked to HRPO. Horse serum is incubated with conjugated p26 antigen in the well. Serum antibodies specific for p26 compete with the bound anti-p26 monoclonal antibodies for the HRPO-p26. Other EIAs are available from Centaur, Inc. and Synbiotics. The latter utilizes recombinant p26 antigen on the microwell and a HRPO-linked p26 antigen as a conjugate reagent. If antibodies are present in the sample, they bind to the microwell and to the HRPO-linked antigen forming a complex. Subsequent to washing, color development is proportional to the amount of antibody present.

Frequency of Use

Used often in official testing laboratories around the world.

Foal Immunoglobulin G

Equines are born with little or no circulating immunoglobulin. Neonatal immunity to infectious agents requires the uptake and absorption of maternal antibodies from colostrum. Failure of this passive transfer can occur as a result of premature lactation, deficient suckling, malabsorption, or low levels of immunoglobulin G (IgG) in colostrum. Partial or complete failure of immune transfer occurs in 10–25% of all foals, and these animals are at a high risk of serious illness or death.

Several studies have identified specific serum concentrations of IgG as indicators of the success of immune transfer. Greater than 800 mg IgG per 100 mL serum is considered as an adequate level of immunity. Levels of 400–800 mg/dL may be adequate, but foals at this level are possibly at risk. IgG levels between 200 and 400 mg/dL reflect a partial failure of immune transfer, while concentrations of less than 200 mg/dL suggest a total failure.

Rapid identification of low IgG levels is essential to the early initiation of treatment of immunodeficient foals. Furthermore, posttreatment testing allows a timely evaluation of the success of IgG supplementation.

Test Analyte

IgG, semiquantitative measurement.

Types of Sample

Serum, plasma, or whole blood.

Assay Technology

IDEXX's Foal IgG is an enzyme immunoassay and now utilizes SNAP immunoassay device technology (*see* FeLV, ASSAY TECHNOLOGY). In the SNAP format, polyclonal antibodies to equine IgG, as well as calibration levels of equine IgG, have been spotted separately onto the device matrix. This assay utilizes a modified SNAP protocol. When sample is applied directly to the sample spots on the device any equine IgG present is captured by the immobilized anti-IgG on the sample spot. Enzyme-conjugated polyclonal antibodies are then added to the well of the device, which bind to the captured equine IgG forming an antibody–equine-IgG–antibody sandwich. Subsequent to device activation, unbound material is washed from the matrix, and enzyme substrate solution is added. Subsequent color development is proportional to the concentration of equine IgG captured.

Color also develops in the IgG calibration spots. These spots contain equine IgG corresponding to serum IgG levels of 400 and 800 mg/dL. Because these spots are calibrated with specific levels of IgG, a comparison of color intensity between the sample and the controls allows an assessment of IgG level in the sample. In addition to their calibration function, these spots indicate that the assay reagents are active.

BOVINE

Brucella abortus (Brucellosis)

Classification of Organism/Pathogenesis

Brucellosis in cattle is a disease caused by *B. abortus*, a facultative, intracellular bacterium. This organism is able to survive and multiply within the reticuloendothelial system. The major mode of disease transmission is by ingestion of *B. abortus* organisms that may be present in tissues of aborted fetuses, fetal membranes, and uterine fluids. In addition, infection may occur as a result of cattle ingesting feed or water contaminated with *B. abortus*. Infection in cows has also occurred through venereal transmission of the organisms by infected bulls (Davis *et al.*, 1980).

Abortion is the most outstanding clinical feature of the disease. If a carrier state develops in the majority of infected cows in a herd, the clinical manifestations may be reduced milk production, dead calves at term, and/or a higher frequency of retained placenta. Disease in the bull may produce infections of the seminal vesicles and testicles resulting in shedding of the organisms in semen.

Test Analyte

B. abortus antibody.

Types of Sample

Serum, plasma, or milk.

Assay Technology

Diagnosis is based on serological and/or bacteriological procedures. Although a positive bacteriological finding provides the only definitive diagnosis, several weeks may be required to obtain final culture results. The success of disease eradication is dependent upon the accurate identification and elimination of *B. abortus* reactors in a herd. Reliable serological techniques are commercially available and provide a rapid and accurate assessment of natural infection or response to vaccine by a measurement of antibodies to *B. abortus* in the serum.

The predominant assay technology that is commercially available is IDEXX's PCFIA (particle concentration fluorescence immunoassay) technology. PCFIA is a fluorescence immunoassay technique which utilizes sub-micron polystyrene particles as the solid phase onto which *B. abortus* antigens are attached. The conjugate consists of fluorescein-labeled *B. abortus* antibodies.

In the assay, the test sample and coated particles are initially mixed and incubated in a specially designed 96-well plate. After this initial incubation period, conjugate is added, and allowed to react, then the reaction mixture is filtered through the membrane at the bottom of each well. The particles, being too large to pass through the membrane, are retained on its surface. Each well is washed to remove unbound conjugate and antibody, and then moved into position below a front-surface fluorimetric reading system. The amount of particle-bound fluorescence is measured as counts from the membrane surface. A longitudinal cross-section of a number of wells is shown in Figure 68.6.

The system reads fluorescence in two channels, a sample channel (A) and a reference channel (B). The ratio of the counts in each channel normalizes any variation due to pipetting of particles.

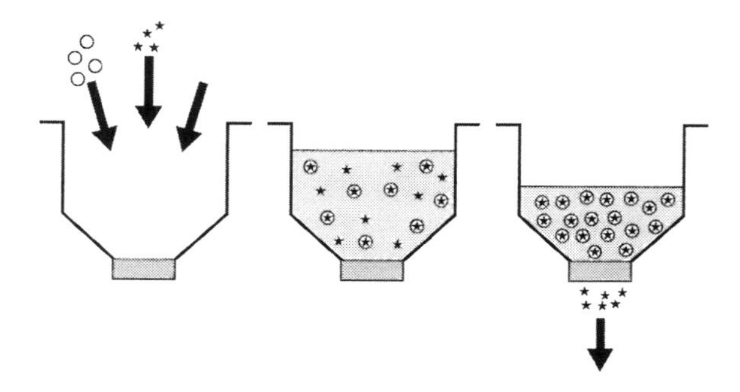

Fig. 68.6 Principles of PCFIA. After the reaction is complete, the mixture is filtered, washed, and concentrated at the bottom of the well. The total particle-bound fluorescence is determined by front-surface fluorimetry.

The anti-*B. abortus* PCFIA is a competitive immunoassay for antigen-specific antibodies. During the reaction, specific antibodies in the sample compete with the conjugate for antigen binding sites on the solid phase. The amount of conjugate bound to the solid phase decreases as the specific antibody concentration in the sample increases, thus providing an inverse measurement of specific antibody. Binding of a non-specific antibody to the solid phase does not affect the competition between specific antibody in the sample and the conjugate, and thus does not affect the resulting signal. Non-specific binding to the antigen on the solid phase is displaced by either the higher affinity conjugate or specific antibody in the sample and, again, does not affect the resulting signal.

Results are expressed as an *S/N* value which is the ratio of sample or positive control signal to the negative control signal. The presence or absence of antibody to *B. abortus* is determined on the basis of the *S/N* value. Serum samples with a *S/N* value greater than 0.70 are considered negative for *B. abortus* antibody; samples with *S/N* values less than or equal to 0.70 are considered positive.

The specificity of a test may be defined as the number of test negatives divided by the number of true or confirmed negatives. In the absence of a gold-standard test for true negativity, an effort was made to test populations in areas classified as free of brucellosis. Assuming that all cattle tested are true negatives, one may obtain a 'worst case' estimate of the specificity of the *B. abortus* PCFIA. For example, a total of 52,856 samples were tested at the Pennsylvania laboratory. A frequency distribution is presented in the figure. At a cutoff of 0.7 the false-positive rate was 0.21%, or a specificity of 99.79% (see Figure 68.7).

The sensitivity of a test may be defined as the number of test positives divided by the number of true or confirmed positives. Serum samples from 505 culture-positive animals were obtained from field trial sites and tested by PCFIA. A frequency distribution is presented in the figure. At a cutoff of 0.7 the false-negative rate was 0.79%, or a sensitivity of 99.21% (see Figure 68.8).

In Europe, indirect ELISA tests are available from a number of firms including IDEXX, Bomelli Laboratories, Synbiotics, Svanova, and Institute Pourquier. However, the bulk of the testing is done using various agglutination technologies.

Frequency of Use

Very common.

Bovine Herpesvirus 1

Classification of Organism/Pathogenesis

This herpesvirus is known to cause a variety of syndromes in cattle, most notably infectious bovine rhinotracheitis (IBR). In addition to causing respiratory disease, bovine herpesvirus 1 (BHV-1) can cause conjunctivitis, vulvovaginitis, abortions, encephalitis, and generalized systemic infections (Wyler *et al.*, 1990). The virus is present in a variety of secretions including ocular, nasal and vaginal, and may be found in aborted fetuses.

Although clinical findings may be highly suggestive of IBR, no real pathopneumonic signs are exclusive to IBR. Therefore, laboratory testing is necessary in order to confirm BHV-1 infection.

Test Analyte

Measurement of circulating antibody is required to confirm natural exposure to BHV-1 or vaccination.

Types of Sample

Serum or milk.

Assay Technology

EIA technology dominates the BHV-1 testing markets, which to date is primarily in Europe. In most assays, wells coated with antigen bind antibody in the sample. The presence of bound antibody is detected with an enzyme-labeled anti-bovine antibody. Such assays are

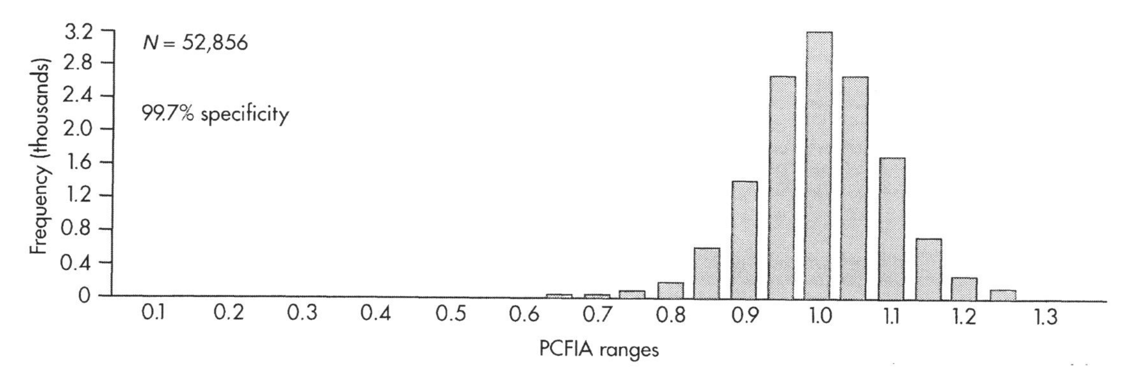

Fig. 68.7 Specificity of PCFIA.

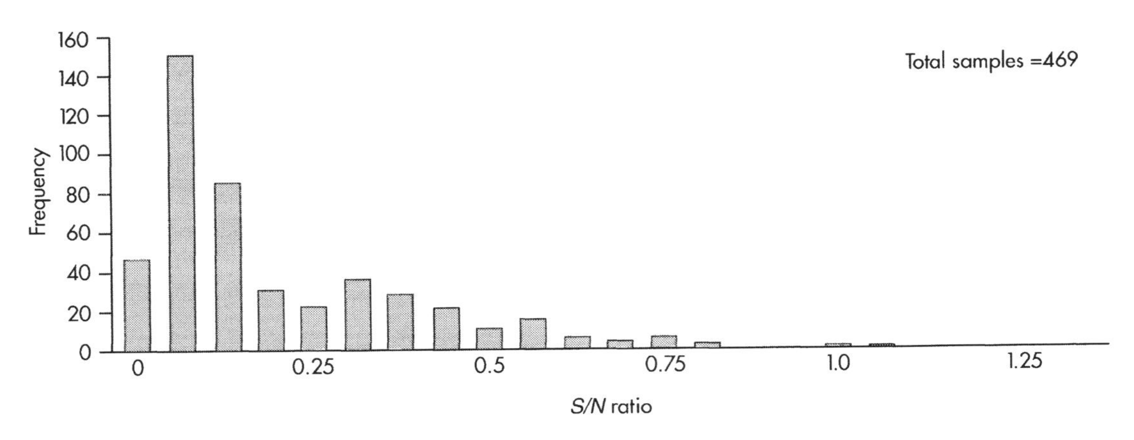

Fig. 68.8 Culture-positive *Brucella* samples. By PCFIA, 99.4% are positive.

available from IDEXX, Bomelli Laboratories, Pourquier, Svanova, etc.

As with Pseudorabies, there are tests available that allow differentiation of field infection from vaccination. A test is available from IDEXX and is designed to detect antibodies to the gE antigen of BHV-1. The test is compatible with vaccines from Bayer and Hoechst Roussel Vet which utilize virus from which gE has been deleted. The assay is configured in a competitive format in which an anti-BHV-1-gE monoclonal labeled with HRPO competes with antibodies to gE in the sample for binding to a BHV-1 antigen coated microwell. The presence of anti-gE BHV-1 antibodies indicates a previous exposure to a field strain or application of conventional gE-positive modified-live or killed virus vaccines. The presence of BHV-1 antibodies detected by a standard screening test, but absence of antibodies to gE antigen, indicates a response to a gE-deleted vaccine.

Frequency of Use

Very common in Europe due to eradication efforts.

Bovine Leukemia Virus

Classification of Organism/Pathogenesis

The bovine leukemia virus (BLV), a retrovirus, is the causative agent of enzootic bovine leukosis in cattle. This is a highly fatal neoplasia of cattle characterized by the aggregation of neoplastic lymphocytes in lymph nodes. Clinical signs most commonly associated with infection include weight loss, decreased milk production, lymphadenopathy, and posterior paresis. Transmission of BLV occurs through the transfer of infected lymphocytes from one animal to another. Once acquired, viral infection is life long. An assessment of exposure to BLV via natural infection can be made by the measurement of specific antibody titer to BLV. A positive antibody titer to BLV indicates that the animal has been exposed to BLV and may be persistently and chronically infected (Johnson and Kaneene, 1991; Miller *et al.*, 1981).

Test Analyte

BLV antibody.

Types of Sample

Serum or plasma.

Assay Technology

As with BHV-1, EIA is the dominant assay technology. Due to the existence of eradication programs, the bulk of the testing done worldwide occurs in Europe. A variety of assays are commercially available, primarily in the format described for BHV-1. Firms currently supplying kits include IDEXX, Bomelli Laboratories, Synbiotics, Svanova, and Pourquier.

Frequency of Use

Very common in Europe.

Mycobacterium paratuberculosis

Classification of Organism/Pathogenesis

Paratuberculosis or Johne's disease is a chronic, debilitating enteritis of ruminants caused by *M. paratuberculosis*. Infection is caused by ingestion of the organism. Calves generally contract infection from the environment shortly after birth, but clinical manifestations may not appear for a number of years. During this incubation period small numbers of organisms may be excreted in the feces. Although some animals recover at this point, many develop clinical disease with characteristic symptoms including chronic diarrhea and extreme weight loss. The organism may be isolated from feces as well as a number of tissues including the lymph nodes, intestinal wall and reproductive tract (Amstutz, 1984).

Diagnosis of Johne's disease can be made only by identification of the organism by microbiological culture or by using the IDEXX *M. paratuberculosis* DNA Probe Test Kit. Exposure to the organism may be assessed via a determination of T-cell sensitization or by the measurement of specific antibodies.

Test Analyte

See above.

Types of Sample

Serum or whole blood for the antibody and T-cell activation tests, respectively.

Assay Technology

During the active stage of infection and prior to the onset of clinical disease, cattle generally develop antibodies to *M. paratuberculosis* antigens. Uninfected cattle lack specific antibodies to the organism but may have cross-reacting antibodies to other mycobacteria. These cross-reacting antibodies can be removed by absorption from the serum or plasma sample using *Mycobacterium phlei* before starting a standard ELISA test in which *M. paratuberculosis* antigens are coated onto the solid phase. Such tests are available from IDEXX, Svanova, and Commonwealth Serum Laboratories in Australia.

Animals that have been exposed to or infected with *M. paratuberculosis* have circulating lymphocytes sensitized to mycobacterial antigens. A delayed-type hypersensitive (DTH) response is observed *in vivo* following intradermal injection of *M. avium* or *M. paratuberculosis* in an infected animal. Using the Commonwealth Serum Laboratories' γ-interferon test protocol this DTH response is mimicked *in vitro* when sensitized lymphocytes from infected cattle are exposed to *M. avium* purified protein derivative (PPD). In response to this exposure, the sensitized lymphocytes release γ-interferon which can be detected by the bovine γ-interferon test. Lymphocytes from cattle not previously exposed to these antigens do not release γ-interferon. Therefore, γ-interferon correlates with exposure to mycobacteria of the *M. paratuberculosis–M. avium* complex, because *M. paratuberculosis* and *M. avium* are antigenically indistinguishable. The Johne's γ-INF test is an enzyme immunoassay designed to detect the presence of bovine γ-interferon (γ-INF) in plasma. The double monoclonal-based EIA provides a qualitative assessment of γ-INF levels in approximately 2 h. For each animal three samples of heparinized bovine blood are incubated for 24 h: one in the absence of antigen (control); one in the presence of *M. avium* PPD; and one in the presence of *M. bovis* PPD. Cells are then separated from plasma by centrifugation and the three plasma samples are individually assayed for the presence of bovine γ-INF. A direct comparison of γ-INF levels in each is then made to determine if exposure to mycobacteria belonging to the *M. paratuberculosis–M. avium* complex has occurred.

Frequency of Use

Very common in the USA and Europe.

Bovine Viral Diarrhea Virus

Classification of Organism/Pathogenesis

Bovine viral diarrhea virus (BVDV) is a pestivirus of the family Flaviviridae and is related antigenically to CSFV. Replication occurs in lymphoid cells and may cause immunosuppression. It is one of the most important pathogenic viruses in cattle, causing considerable losses in the dairy and beef industries worldwide. The virus crosses the placenta in infected, pregnant cows causing reproductive losses due to abortions, stillborn calves or calves that die early in life. Some calves that survive are immunotolerant to the virus and these animals excrete large amounts of infectious virus for the rest of their lives. It is important to identify these animals to break the cycle of infections in herds. As a consequence of the *in utero* infection, BVDV is a frequent contaminant of biological products, such as vaccines and pharmaceuticals.

Test Analyte

BVDV specific antibodies or antigens.

Types of Sample

Serum, plasma, milk for antibody tests; peripheral blood leukocytes, whole blood, tissue samples, nasal swabs, and cell cultures for antigen tests.

Assay Technology

Commercial products are available from a number of manufacturers including IDEXX, Svanova, Bomelli Laboratories, and Synbiotics. All assays are EIAs and are similar to tests for specific antibodies and antigens previously described. In general, the antibody tests are competitive assays utilizing specific monoclonal conjugates, while the antigen tests utilize monoclonals in a sandwich assay configuration.

Frequency of Use

Frequently in Europe.

Neospora caninum

Classification of Organism/Pathogenesis

N. caninum is a recently discovered obligate intracellular protozoal (Apicomplexan) parasite which can cause abortion and neonatal morbidity and mortality in cattle, sheep, goats, and horses (Dubey and Linsay, 1993).

It has been reported in California that infection with *Neospora* is the most important diagnosed cause of abortion in dairy cattle (Anderson *et al.*, 1991, 1995; Pare *et al.*, 1996). Currently, it is not known how the organism is introduced into herds or whether there are means of transmitting the organism between cattle other than by congenital transmission (Holmdahl *et al.*, 1995). The complete life cycle of this organism remains to be determined.

Test Analyte

Antibody to *N. caninum*.

Type of Sample

Serum.

Assay Technology

Indirect EIAs are available from IDEXX and Bomelli Laboratories in microwell formats. *N. caninum* antigens are coated on the solid phase, and bound bovine antibodies are detected using an anti-bovine HRPO conjugate. A competitive EIA is available from VMRD, Inc.

Frequency of Use

Currently used infrequently, but growing for prepurchase testing, prebreeding testing, and control measures (culling).

Foot-and-Mouth Disease Virus

Classification of Organism/Pathogenesis

Foot-and-mouth disease (FMD) is a highly infectious viral infection caused by a virus of the family Picornaviridae. The virus infects cloven-hoofed animals including cattle, sheep and swine, causing high fever, anorexia and vesicle or blister development in and around the mouth and on the feet. Transmission occurs via contact with infected animals, as they shed virus in all excretions and secretions, but virus may be spread via infected/contaminated milk, meat, and feed. The mucosa of the respiratory tract is the primary site of infection and replication (Aiello and Mays, 1998).

There are seven antigenically distinct serotypes: A, O, C, Asia 1, and SAT (South African Territories) 1, 2, 3. The existence of these distinct serotypes complicates vaccination and serological diagnosis. Antibodies against structural proteins, which are essential for protection, exhibit little cross-reactivity among the various serotypes, indicating the need for multiple vaccines. On the other hand, antibodies against non-structural proteins are highly cross-reactive among serotypes and have been used as diagnostic reagents.

The generally accepted method of FMD control involves eradicating all infected animals as well as animals in surrounding areas where exposure may have occurred. Vaccination is often used on a limited basis to help control the spread, but the practice may lead to trade issues due to the presence of antibodies to the virus potentially indicating exposure or infection. Thus, vaccinated animals are ultimately destroyed as well.

Test Analyte

Antibody to FMD viral antigens.

Types of Sample

Serum or plasma.

Assay Technology

The gold standard for diagnosis of FMD remains virus isolation from tissue culture followed by immunological characterization as to serotype via EIA using serotype-specific reagents. In order to take full advantage of vaccination to help control the disease, there must be a diagnostic test that can distinguish between antibodies caused by infection versus vaccination. At present there are two commercially available EIAs that can differentiate if used in conjunction with modern FMD vaccines that incorporate highly purified structural proteins essential for protection.

Bomelli Laboratories and United Biomedical, Inc. (UBI) each produce test kits that utilize non-structural proteins on the microwell to capture antibodies to FMD virus. Bomelli utilizes a recombinant non-structural protein, 3ABC, while UBI incorporates a synthetic peptide from non-structural protein, 3B.

As with other indirect EIAs, antibody in the sample specific to the immobilized antigen binds and, subsequent to a wash step, is detected via a HRPO conjugate reagent that binds to the captured antibody from the sample. Both companies claim excellent sensitivity to all serotypes and excellent specificity on samples from negative and vaccinated animals.

These assays seem appropriate for use as herd management and surveillance tools, but should not be used as standalone individual animal diagnostics.

Frequency of Use

Potentially high during outbreaks but likely dependent on acceptance of vaccination on a broader scale.

Bovine Spongiform Encephalopathy (BSE)

Classification of Organism/Pathogenesis

Bovine spongiform encephalopathy (BSE) is a progressive neurodegenerative disease of cattle and one of a number of **transmissible spongiform encephalopathies** (**TSE**) that affect animals and humans. In addition to BSE, **Scrapie** in sheep and **chronic wasting disease** (**CWD**) in deer and elk are becoming significant issues in veterinary medicine. Clinical manifestations include nervousness, lack of coordination, loss of body weight, and ultimately death. These degenerative disorders of the brain and central nervous system are believed to be caused not by viruses or bacteria, but by infectious proteins called **prions**. Transmission in cattle is believed to occur via ingestion of food containing infectious prions followed by replication and subsequent migration to the central nervous system.

Prion protein (**PrP**) is present in all vertebrates although no physiologic function for this cell surface protein has been identified. PrP can exist in multiple conformations; normal **cellular PrP** is referred to as **PrPc**, while the abnormal PrP causing disease is referred to as **PrPsc**. It is believed that PrPsc having an altered conformation is able to transform PrPc upon association, thereby multiplying the pathogenic prion. PrPsc molecules tend to aggregate and form plaques in the brain resulting in neuron destruction and vacuole formation. It is important to note that PrPsc having an altered conformation with an increased level of β-pleated-sheet folding is less susceptible to protease degradation.

Test Analyte

Abnormal prion protein, PrPsc.

Type of Sample

Bovine brain tissue.

Assay Technology

Currently, there are no commercially available tests for BSE in live cattle; only postmortem tests are available. Furthermore, these tests provide little assurance that 'negative' animals are truly negative and not in an early incubation stage when PrPsc is undetectable in the brain. Although the gold standard for postmortem testing remains histopathology, there are a number of immunoassays on the market for the detection of PrPsc in brain tissue. At present, most products rely on the protease resistance of PrPsc to eliminate cross-reactivity or interference by PrPc, which is present in high concentrations relative to PrPsc in all brain samples. In effect, the PrPc is eliminated during a sample pretreatment step subsequent to homogenization and extraction via digestion with an optimized concentration of proteinase-K, leaving only PrPsc. The remaining PrPsc is detected via several immunoassay technologies depending on the commercial product.

One of the first products developed in Europe for BSE testing was the Prionics-Check Western, a western blot assay utilizing proteinase-K treated samples and a monoclonal antibody, Prionics Antibody 6H4 for detection of PrPsc on the membrane subsequent to transfer from the electrophoresis gel. A second antibody, goat anti-mouse IgG coupled to alkaline phosphatase, and a chemiluminescence substrate allow imaging of the PrPsc electrophoretic pattern on the membrane. Prionics AG, a Swiss company, has also developed an antigen capture luminescence immunoassay: Prionics Check LIA™. This product also requires proteinase K-treated samples, but employs an antibody coated microwell plate for capture. Prepared samples are treated with an enzyme–antibody conjugate prior to incubation on the plate. Both products have been approved by the EU.

There are two additional EU approved immunoassays on the market. Enfer Scientific Ltd. has developed a chemiluminescence-based EIA and Bio-Rad Laboratories markets an EIA, Platelia™ BSE, both of which are antigen capture assays utilizing enzyme-labeled antibodies for detection of bound PrPsc. The Bio-Rad product was initially developed by CEA, the French Atomic Energy Commission. Both assays require proteinase K digestion to remove PrPc from the sample prior to analysis.

Several assays are in the approval process in the EU. Among those are assays from InPro Biotechnology, Inc. and IDEXX Laboratories, Inc., both of which have unique features.

InPro has developed what is called a **Conformation-Dependent Immunoassay** (**CDI**), InPro CDI-5. In this method, the sample is treated briefly with proteinase K, subsequent to homogenization and precipitation with phosphotungstate to concentrate the PrPsc. Half of the sample is denatured at high temperature. Samples, both native and denatured, are introduced into microwells coated with capture antibody. Antibody binding sites are hidden in native PrPsc but exposed in PrPc. Denaturation exposes all antibody binding sites in both forms. Subsequent to capture, detection antibody labeled with europium is introduced that binds to all PrP present. Unbound antibody is washed away and europium is released for detection via time-resolved fluorescence measurement. Comparing fluorescence in the native and denatured samples allows determination of whether the sample contained PrPsc.

IDEXX Labs, Inc. has developed a TSE diagnostic assay using Seprion affinity capture technology applied to a microwell plate format. The plate is coated with a non-biological PrPsc-specific ligand that binds only PrPsc even in the presence of an excess of PrPc. The captured PrPsc is detected by incubation with an anti-PrP antibody–enzyme conjugate and subsequent color development. In this method, there is no need for proteinase K treatment of samples and sample preparation consists of simple homogenization of tissue and addition of dilution buffer. This assay has been approved by the USDA for CWD testing in the US.

Frequency of Use

High frequency of use in Europe.

AVIAN

Avian Encephalomyelitis Virus

Classification of Organism/Pathogenesis

Avian encephalomyelitis (AE) is caused by a picornavirus (avian encephalomyelitis virus, AEV) that is widespread and affects young chickens. The disease is characterized by a variety of neurological signs, including lack of coordination, ataxia and tremors of the head and neck. Vaccines have been successfully developed for this disease.

Test Analyte

AEV antibody for assessment of immune status or identification of AEV infection.

Type of Sample

Serum.

Assay Technology

An enzyme immunoassay is available from IDEXX that is designed to measure the relative level of antibody to AE in chicken serum. In the test, antibody in the sample binds to purified AEV coated on 96-well plates, and is subsequently probed using an anti-chicken/HRPO conjugate. The presence or absence of antibody to AEV is determined by relating the optical density of the unknown to the optical density of the positive control. The positive control has been standardized and represents significant antibody levels to AEV in chicken serum. The relative level of antibody in the unknown can be determined by calculating the sample to positive (*S/P*) ratio.

The immune status of a flock is best assessed by monitoring and recording antibody titers in representative samples as a function of time. The resulting flock profiles allow an assessment of the distribution of antibody titers and an analysis of changes in titer over time. Pullets are typically checked pre- and postvaccination to ensure the presence of uniform titers before going 'into lay'.

Frequency of Use

Occasional.

Avian Leukosis Virus

Classification of Organism/Pathogenesis

Lymphoid leukosis is the most common manifestation of the avian leukosis/sarcoma group of viruses, which produces a variety of neoplastic diseases including erythroblastosis, myelocyomatosis, myeloblastosis and others (Purchase and Payne, 1984; Purchase and Fadley, 1983). Although not all infected animals develop tumors, nearly every commercial flock contains infected birds and sporadic occurrence of tumors may result. Infection occurs horizontally, by direct or indirect contact between birds, or vertically, from an infected hen to her eggs, as virus is shed into the albumin of the egg. In addition, vertical transmission may occur from virus incorporated into the DNA of germ cells. Viremia in the hen is strongly associated with the transmission of virus congenitally.

An important subgroup for this virus is subgroup J. This virus was first isolated in meat-type chickens in the late 1980s and designated as a unique subgroup partly based on the envelope glycoprotein, gp85 (Payne *et al.*, 1991). Clinically, ALV-J causes predominantly myeloid leukosis, with variable tumor frequency across chicken lines (Payne *et al.*, 1991; Payne and Fadly, 1997). As with other avian leukosis viruses, ALV-J is transmitted both vertically and horizontally.

Test Analyte

Avian leukosis viral antigen, p27, for detection of all subgroups; antibodies specific for ALV-J gp85 for detection of ALV-J.

Types of Sample

Serum or egg albumin.

Assay Technology

Enzyme immunoassays in microwell formats for ALV p27 antigen are available from Synbiotics, Euro-Diagnostica, and IDEXX. These are sandwich assays that utilize either monoclonal or polyclonal antibodies for both capture and enzyme-labeled conjugate.

The ALV-J antibody test is offered by IDEXX and is configured as an indirect EIA. ALV-J gp85 antigen is coated onto the solid phase. Antibody specific to this antigen is bound and subsequently detected with an anti-chicken HRPO conjugate.

Frequency of Use

Very common.

Avian Reovirus

Classification of Organism/Pathogenesis

Avian reoviruses are ubiquitous among poultry populations and have been reported to be responsible for viral arthritis (tenosynovitis), respiratory infections and cloacal pasting in chicks. Symptoms are most apparent in older birds but respiratory signs may be seen in young chicks. The incidence of reovirus infection in older birds is high, but clinical symptoms are not seen in most birds.

Test Analyte

Both assessment of immune status and serologic identification of avian reovirus require a measurement of antibody to reovirus.

Type of Sample

Serum.

Assay Technology

Standard enzyme-immunoassays, utilizing microwells as the solid phase, are available from Synbiotics, Guildhay, and IDEXX. In all tests purified virus is immobilized on the solid phase; conjugates are anti-chicken coupled to HRPO.

Frequency of Use

Very common.

Infectious Bronchitis Virus

Classification of Organism/Pathogenesis

Infectious bronchitis is a highly contagious viral disease of chickens which is usually manifested as a respiratory condition and may cause high mortality. The causative agent is a coronavirus (infectious bronchitis virus, IBV) which may be spread via aerosols or contaminated equipment.

Both live and killed vaccines are available. Proper management of flocks requires an assessment of immune status pre- and postvaccination.

Test Analyte

IBV antibody.

Type of Sample

Serum.

Assay Technology

Standard indirect enzyme-immunoassays, utilizing microwells as the solid phase (*see* AVIAN REOVIRUS, ASSAY TECHNOLOGY) are available from Synbiotics, Guildhay, and IDEXX.

Frequency of Use

Very common.

Infectious Bursal Disease Virus

Classification of Organism/Pathogenesis

Infectious bursal disease (IBD) is a highly contagious viral disease affecting young chickens 3–6 weeks of age. The bursa becomes swollen and the immune system is suppressed, often resulting in concurrent or secondary infections by other organisms. Symptoms include anorexia, lack of coordination, and depression. Also characteristic of the disease are edema and swelling of the cloacal bursa, the organ from which the infectious bursal disease virus (IBDV) may be readily isolated. The virus is shed in the feces, and because of its stability may be very difficult to eradicate from living quarters. Losses may approach 20% in an infected flock. Both assessment of immune status and serologic identification of IBD require measurement of antibody to the virus in serum.

Although there is no treatment, effective vaccines are available. Monitoring of titers in breeder flocks is particularly important because high titers are required to provide adequate levels of parental immunity to offspring during the first few weeks of life.

Test Analyte

IBDV antibody.

Type of Sample

Serum.

Assay Technology

Standard indirect enzyme-immunoassays utilizing microwells as the solid phase (*see* AVIAN REOVIRUS, ASSAY TECHNOLOGY) are available from Synbiotics, Guildhay, and IDEXX.

Frequency of Use

Very common.

Mycoplasma gallisepticum

Classification of Organism/Pathogenesis

M. gallisepticum is a fastidious organism which causes respiratory disease in chickens. The usual types of infection from mycoplasma are chronic respiratory disease, airsaculitis, sinusitis, and synovitis (Yoder, 1984).

Infections typically occur in the epithelium of the air passages, but the organism may be found in many tissues during acute infection. Symptoms include difficulty in breathing, coughing, and sneezing. In many cases, however, the infection may be identified only through serological and culture methods.

Test Analyte

Chicken flocks are monitored for exposure by testing for mycoplasma antibody.

Type of Sample

Serum.

Assay Technology

Standard indirect enzyme-immunoassays utilizing microwells as the solid phase are available from Synbiotics, Guildhay, and IDEXX (*see* AVIAN REOVIRUS, ASSAY TECHNOLOGY). Also available from IDEXX is a polymerase chain reaction (PCR)-amplified DNA probe assay which utilizes tracheal swabs for direct detection of the organism.

Newcastle Disease Virus

Classification of Organism/Pathogenesis

Newcastle disease is a highly contagious and sometimes fatal illness affecting poultry. The causal agent of the disease is a hemagglutinating paramyxovirus (Newcastle disease virus, NDV). The severity of the disease is a function of the virulence of the infecting viral strain. Mild strains (mesogenic) infect the trachea, lungs and air sacs, and interfere with egg production. Severe strains (velogenic) are manifested through lack of coordination, paralysis, swelling of tissue around the eyes, diarrhea, and eventual death (Miers *et al.*, 1983).

Test Analyte

Both assessment of immune status and serological identification require testing for NDV antibody.

Type of Sample

Serum.

Assay Technology

Standard indirect enzyme-immunoassays utilizing microwells as the solid phase are available from Synbiotics, Guildhay, and IDEXX (*see* AVIAN REOVIRUS, ASSAY TECHNOLOGY).

Pasturella multocida

Classification of Organism/Pathogenesis

Fowl cholera, caused by *P. multocida* infection, is a commonly occurring disease of birds. It is caused by a small, Gram-negative rod bacterium. In the acute form, its usual symptom is septicemia with associated high morbidity and mortality. Chronic localized infections can also occur, following either an acute exposure or infection with organisms of low virulence. Clinical signs of acute infections are typical of bacterial septicemia, whereas the signs of chronic disease are typically related to the anatomic location of the infection (Rhoades and Heddleston, 1980).

Test Analyte

Both the assessment of immune status and the serological identification of *P. multocida* require a test for antibody to *P. multocida* (Snyder *et al.*, 1984).

Type of Sample

Serum.

Assay Technology

Standard indirect enzyme-immunoassays, utilizing microwells as the solid phase, are available from IDEXX and Synbiotics (*see* AVIAN REOVIRUS, ASSAY TECHNOLOGY).

Reticuloendotheliosis Virus

Classification of Organism/Pathogenesis

Reticuloendotheliosis is caused by a retrovirus which is morphologically similar to, but genetically and antigenically distinct from, the avian leukosis/sarcoma viruses. Runting disease has been reported in chickens through the application of vaccines contaminated with reticuloendotheliosis virus (REV). In addition, REV can produce disease in experimentally infected chickens pathologically indistinguishable from lymphoid leukosis (Witter and Calnek, 1984).

Test Analyte

An assessment of exposure to REV requires a measurement of antibody to REV.

Type of Sample

Serum.

Assay Technology

A standard indirect enzyme-immunoassay utilizing microwells as the solid phase is available from IDEXX (*see* AVIAN REOVIRUS, ASSAY TECHNOLOGY).

Frequency of Use

Occasional.

Chicken Anemia Virus

Classification of Organism/Pathogenesis

Chicken anemia virus (CAV) is an important pathogen of poultry and has been found in broilers, breeders, and SPF flocks (McNulty, 1991; Pope, 1991). This single-stranded DNA virus is a member of the family Circoviridae and may be transmitted both vertically and horizontally. Infection by this virus causes anemia, lymphoid depletion and hemorrhaging. Outbreaks of the clinical disease are fairly rare, since most breeder flocks have already developed immunity when infected during rearing. Seroconversion at this time appears to prevent subsequent vertical transmission.

Test Analyte

Antibodies specific for CAV.

Type of Sample

Serum.

Assay Technology

Standard indirect EIAs utilizing microwells as the solid phase (*see* AVIAN REOVIRUS, ASSAY TECHNOLOGY) are available from IDEXX, Bomelli, Synbiotics, and Guildhay.

Frequency of Use

Occasional.

Salmonella enteritidis

Classification of Organism/Pathogenesis

S. enteritidis is a pathogen of poultry and has been isolated from broilers, breeders, and commercial egg laying flocks (McIlroy *et al.*, 1989). It is a member of the bacterial genus *Salmonella*, which is a member of the family Enterobacteriaceae. It is of particular concern because it can infect other species such as man. Bacteriological identification of positive birds is difficult due to the intermittent shedding of the organism. Symptoms such as depression, poor growth, weakness, diarrhea, and dehydration in laying hens can be insignificant, thus, infection may not be detected until postmortem examination. Testing is performed to monitor exposure and is a tool to reduce the prevalence of the organism.

Test Analyte

Antibodies specific for *S. enteritidis.*

Type of Sample

Serum or egg yolk.

Assay Technology

Enzyme immunoassays are available in competitive and indirect formats utilizing microwells as the solid phase from IDEXX, Guildhay, Synbiotics, etc.

Frequency of Use

Occasional.

MURINE

Disease prevention and monitoring programs are important in the proper management of mouse and rat colonies. New animals should be free of common pathogens and care should be taken to prevent the accidental introduction of several pathogens. A number of enzyme immunoassays are available for detecting antibody to various rodent pathogens. Suppliers include Organon Teknika, Charles River Laboratories, and M.A. Bioproducts. Below are brief descriptions of the pathogens for which at least one enzyme immunoassay is available commercially.

Mycoplasma pulmonis

This organism is believed to be the principal agent responsible for chronic respiratory syndrome, and constitutes one of the more important health problems affecting laboratory rodents. It is characterized by inflammation of the respiratory tract and middle ear, with clinical symptoms including nasal discharge, sneezing, lack of coordination, weight loss, and ocular discharge. The available enzyme immunoassays provide the appropriate level of sensitivity and specificity when screening for evidence of *M. pulmonis* infection.

Test Analyte

M. pulmonis antibody.

Type of Sample

Serum.

Assay Technology

Standard indirect enzyme-immunoassays utilizing various solid phases and anti-mouse/rat conjugates are available (*see* AVIAN REOVIRUS, ASSAY TECHNOLOGY).

Ectromelia Virus (Mousepox)

This virus causes a serious disease in laboratory mice resulting in various lesions of the extremities and certain organs and ultimately in death. Vaccination with vaccinia virus may control some outbreaks. Often overt signs of the disease are not present even though the virus has been present in colonies for long periods of time.

Test Analyte

Mousepox antibody.

Type of Sample

Serum.

Assay Technology

Standard indirect enzyme-immunoassays utilizing various solid phases and anti-mouse conjugates are available (*see* AVIAN REOVIRUS, ASSAY TECHNOLOGY).

Mouse Hepatitis Virus

This coronaviral infection is highly contagious and primarily affects the intestinal tract. Clinical symptoms include diarrhea, jaundice, and weight loss.

Test Analyte

Mouse hepatitis virus antibody.

Type of Sample

Serum.

Assay Technology

Standard indirect enzyme-immunoassays utilizing various solid phases and anti-mouse conjugates are available (*see* AVIAN REOVIRUS, ASSAY TECHNOLOGY).

Sendai Virus

Infection with Sendai virus may be the major cause of respiratory disease in mice. Symptoms are pneumonia-like.

Test Analyte

Sendai virus antibody.

Type of Sample

Serum.

Assay Technology

Standard indirect enzyme-immunoassays utilizing various solid phases and anti-mouse/rat conjugate are available (*see* AVIAN REOVIRUS, ASSAY TECHNOLOGY).

Other Mouse/Rat Pathogens

Enzyme immunoassays available for other mouse/rat pathogens include rat coronavirus, mouse pneumonia virus, mouse choriomeningitis virus and reovirus.

ASSESSMENT OF REPRODUCTIVE/METABOLIC STATUS

FELINE/CANINE

Thyroxine

Disorders of the thyroid gland are common in both cats and dogs. The thyroid gland produces several hormones that exert broad biological effects by controlling many aspects of cellular metabolism (Feldman and Nelson, 1987). Secretion of these hormones from the thyroid is controlled both by the hypothalamus and by the pituitary gland, with thyrotropin (TSH) from the pituitary being the principal regulator.

Because of the broad metabolic influence of thyroid hormones, clinical signs of thyroid dysfunction are polysystemic, highly variable and non-specific. As a result, a definitive diagnosis of thyroid disease by clinical signs alone is difficult (Nelson and Ihle, 1987).

Thyroxine (T_4), an iodinated derivative of the amino acid tyrosine, is the most abundant of the hormones produced by the thyroid. The measurement of T_4 levels provides the most useful indication of overall status of thyroid function (McIlroy *et al.*, 1989). T_4 levels, in combination with clinical observations, are diagnostic of thyroid disorders.

Hypothyroidism (too little thyroid hormone) and hyperthyroidism (too much thyroid hormone) are conditions that occur both in dogs and cats.

Hypothyroidism, particularly common among middle-aged dogs, is generally due to dysfunction of the thyroid gland itself (primary hypothyroidism). Clinical signs reflect a generally reduced metabolism: lethargy, weight gain, mental dullness and intolerance to exercise and cold. In addition, dermatological symptoms such as alopecia, hyperkeritosis, myxedema, pyoderma, and seborrhea are generally presented. The polysystemic, non-specific nature of symptoms makes diagnosis difficult. The reported reference interval for dogs is approximately 10–40 ng/mL. Serum or plasma T_4 concentrations of 10 ng/mL or less are consistent with hypothyroidism.

Hyperthyroidism, most commonly encountered in older cats, reflects the general stimulatory effect of excess thyroid hormone secretion. It is characterized by weight loss, polyphagia, tachycardia, polyurial/polydipsia, vomiting, diarrhea, and general hyperactivity. The reported reference interval for T_4 in cats is approximately 10–40 ng/mL. Serum or plasma total T_4 concentrations of greater than 40 ng/mL are consistent with hyperthyroidism.

Types of Sample

Serum or plasma.

Assay Technology

The only commercial assay specifically designed for measuring total T_4 in dogs and cats is the SNAP T_4 product produced by IDEXX. The SNAP T_4 test uses a competitive EIA format. In the SNAP T_4 device, two stripes of reagents are contained on the solid phase; the bottom stripe is the T_4 test reagent and the top stripe is a reference reagent. In the test procedure, the serum sample is first incubated with an anti-T_4 antibody–HRPO conjugate. During incubation, T_4 present in the serum sample binds to the conjugate. The mixture is then added to the SNAP device. Any unbound conjugate binds to the T_4 test reagent stripe. The color developed is, therefore, inversely proportional to the amount of T_4 present in the serum. The T_4 concentration is then calculated from the ratio of T_4 test stripe color to reference stripe color.

Frequency of Use

Common.

Cortisol

Cortisol is an important glucocorticoid hormone secreted by the adrenal gland. Canine hyperadrenocorticism (Cushing's syndrome) is associated with chronic excessive serum cortisol levels. The excess serum cortisol has several possible pathophysiologic origins: pituitary tumor or hyperplasia leading to over production of adrenocorticotropic hormone (ACTH) by the pituitary gland, adrenocortical carcinoma or adenoma, or excessive administration of glucocorticoids or ACTH (Feldman and Nelson, 1996; Feldman, 1995).

Clinical signs of Cushing's syndrome progress slowly, are highly variable and are non-specific. As a result, a definitive diagnosis of Cushing's by clinical signs alone is difficult. Laboratory confirmation of a diagnosis of Cushing's syndrome is generally accomplished by manipulation of the pituitary-adrenocortisol axis via low dose dexamethasone suppression or ACTH stimulation testing.

Spontaneous hypoadrenocorticism (Addison's disease) is associated with low cortisol production by the adrenal cortex, most often caused by immune mediated disease or drug therapy. The test of choice to diagnose Addison's disease is the ACTH stimulation test (Hardy, 1995).

Types of Sample

Serum.

Assay Technology

IDEXX produces a commercial assay for measuring cortisol levels in dog serum. The assay is similar to the SNAP T_4 test described in the previous section.

Frequency of Use

Occasional.

Other Feline/Canine Reproductive/ Metabolic Markers

Enzyme immunoassays and radioimmunoassays are available for a number of infrequently measured analytes such as progesterone, prolactin, testosterone, and luteinizing hormone. Diagnostic Products Corporation is a major supplier.

EQUINE

Progesterone

Progesterone is a steroid hormone important in the regulation of reproductive function in the mare. This hormone functions to regulate uterine activity and plays an essential role in the coordination of the estrus cycle. In addition, progesterone is essential to the survival of the embryo in pregnant mares. The hormone is produced by the corpus luteum during the estrus cycle and early pregnancy, and by the placenta later in pregnancy. In the mare, serum levels of progesterone are low (<1 ng/mL) during estrus (heat) and rise rapidly to 10–20 ng/mL during the luteal phase. If pregnancy does not occur, progesterone levels drop again to below 1 ng/mL and the estrus cycle begins again. In the event of pregnancy, the progesterone concentration remains high throughout gestation.

Accurate measurement of progesterone levels in the mare provides valuable information for equine reproductive management. Specific applications include:

- Confirmation of adequate progesterone levels to maintain pregnancy (progesterone >4 ng/mL).
- Detection or confirmation of estrus (progesterone <1 ng/mL).
- Monitoring ovarian function either during the normal cycle or in the event of persistent luteal phase syndrome.
- Evaluation of progesterone levels in conjunction with: (1) prostaglandin therapy; and (2) embryo transfer practices.

Types of Sample

Equine serum or plasma.

Assay Technology

A number of enzyme immunoassays are available for the determination of progesterone in equine serum or plasma. All are competitive EIAs. Manufacturers include Synbiotics and Vetoquinol.

Frequency of Use

Occasional.

BOVINE

Progesterone

The utility of progesterone testing in cows is similar to that in horses. Potential uses include verification of estrus, monitoring of the estrus cycle, and screening for ovarian dysfunction. In addition, an indication of pregnancy status may be obtained, in that low values (<2 ng/mL) suggest with a high degree of accuracy that the cow is open.

Types of Sample

Bovine milk, serum or plasma, depending on the assay configuration.

Assay Technology

As with equine progesterone, enzyme immunoassays are commercially available for the determination of bovine progesterone levels. All are in competitive assay formats. Assays are available from Synbiotics and Vetoquinol.

Frequency of Use

Occasional.

CONCLUSION

Immunoassays have become an important diagnostic tool in veterinary medicine. Their rise in importance has been driven both by technological advances, making them more accurate and easier to use, and by their increased availability through commercial channels.

Currently, immunoassays are commercially available for many of the important infectious diseases of companion and food animals. There is every indication that the list will continue to grow given the level of commitment of a number of companies in the USA and Europe to veterinary diagnostics.

In contrast to the long list of infectious disease products, there are few commercially available tests for metabolic and reproductive functions. The number of products in this category is sure to grow as university researchers and diagnostic companies continue to apply new technology to meet the needs of the veterinarian. Additional thyroid tests, new species-specific pregnancy markers and various other hormone tests are likely in the near future.

REFERENCES

Aiello, S.E. and Mays, A. *The Merck Veterinary Manual*, 8th edn (Merck, Whitehouse Station, New Jersey, 1998).

Amstutz, H.E. Bovine paratuberculosis: an update. *Mod. Vet. Pract.* **65**, 134–135 (1984).

Anderson, M.L., Blanchard, P.C., Barr, B.C. *et al. Neospora*-like protozoan infection as a major cause of abortion in California dairy cattle. *J. Am. Vet. Med. Assoc.* **198**, 241–244 (1991).

Anderson, M.L., Palmer, C.W., Thurmond, M.C. *et al.* Evaluation of abortions in cattle attributable to Neosporosis in selected dairy herds in California. *J. Am. Vet. Med. Assoc.* **207**, 1206–1210 (1995).

Benfield, D.A., Nelson, E., Collins, J.E. *et al.* Characterization of swine infertility and respiratory syndrome (SIRS) virus (isolate ATCC VC-2332). *J. Vet. Diagn. Invest.* **4**, 127–133 (1992).

Collins, J.E., Benfield, D.A., Christianson, W.T. *et al.* Isolation of swine infertility and respiratory syndrome virus (isolate ATCC VR2332). *J. Vet. Diagn. Invest.* **4**, 117–126 (1992).

Davis, B.R., Dulbecco, R., Eisen, H.N. and Ginsberg, H. *Microbiology*, 2nd edn 686–690 (Harper & Row, New York, 1980).

Dubey, J.P. and Linsay, D.S. Neosporosis. *Parasitol. Today* **9**, 452–458 (1993).

Feldman, E.C. Hyperadrenocorticism. In: *The Textbook of Veterinary Internal Medicine*, 4th edn 1538–1578 (W.B. Saunders, Philadelphia, 1995).

Feldman, E.C. and Nelson, R.W. *Canine and Feline Endocrinology and Reproduction*, 55–136 (W.B. Saunders, Philadelphia, 1987).

Feldman, E.C. and Nelson, R.W. Disorders of growth hormone. In: *Canine and Feline Endocrinology and Reproduction*, vol. 2, 3rd edn 55–136 (W.B. Saunders, Philadelphia, 1996).

Georgi, J.R. and Georgi, M.E. *Canine Clinical Parasitology* (Lea & Febiger, Philadelphia, 1992).

Hardy, R.M. Hypoadrenal gland disease. In: *The Textbook of Veterinary Internal Medicine*, 4th edn 1579–1593 (W.B. Saunders, Philadelphia, 1995).

Holmdahl, O.J.M., Bjorkman, C. and Uggla, A. A case of *Neospora* associated bovine abortion in Sweden. *ACTA Vet. Scand.* **36**, 279–281 (1995).

Johnson, R. and Kaneene, J.B. Bovine leukemia virus. Part 1. Descriptive epidemiology, clinical manifestations, and diagnostic tests – the compendium. *Food Anim.* **13**, 315–325 (1991).

Kosmidou, A., Ahl, R., Thiel, H.-J. and Weiland, E. Differentiation of classical swine fever virus (CSFV) strains using monoclonal antibodies against structural glycoproteins. *Vet. Microbiol.* **47**, 111–118 (1995).

McIlroy, S.G., McCracken, R.M., Neill, S.D. and O'Brien, J.J. *Vet. Rec.* **125**, 545–548 (1989).

McNulty, M.S. Chicken anemia agent: a review. *Avian Pathol.* **20**, 186–203 (1991).

Miers, L.A., Bankowski, R.A. and Zee, Y.C. Optimizing the enzyme-linked immunosorbent assay for evaluating immunity of chickens to Newcastle Disease. *Avian Dis.* **27**, 1112–1125 (1983).

Miller, J.M., Schmerr, M.J. and Van der Maaten, M.J. Comparisons of four serologic tests for the detection of antibodies to bovine leukemia virus. *Am. J. Vet. Res.* **42**, 5–8 (1981).

Muelenberg, J.J.M., Hulst, M.M., de Meijer, E.J. *et al.* Lelystad virus, the causative agent of porcine epidemic abortion and respiratory syndrome (PEARS), is related to LDV and EAV. *Virology* **192**, 62–72 (1992).

Nelson, R.W. and Ihle, S.L. Hypothyroidism in dogs and cats: a difficult deficiency to diagnose. *Vet. Med.* **82**, 60–70 (1987).

Pare, J., Thurmond, M.C. and Hietala, S.K. Congenital *Neospora caninum* infection in dairy cattle and associated calfhood mortality. *Can. J. Vet. Res.* **60**, 133–139 (1996).

Payne, L.N. and Fadly, A.M. Neoplastic diseases/leukosis/sarcoma Group. In: *Diseases of Poultry*, 10th edn (ed. Calnek, B.W.), 414–466 (ISU Press, Ames, IA, 1997).

Payne, L.N., Brown, S.R., Bumstead, N., Howes, K., Frazier, J.A. and Thouless, M.E. A novel subgroup of exogenous avian leukosis virus in chickens. *J. Gen. Virol.* **72**, 801–807 (1991).

Pederson, N.C., Ho, E.W., Brown, M.L. and Yamamoto, J.K. Isolation of a T-lymphotrophic virus from domestic cats with an immunodeficiency-like syndrome. *Science* **235**, 790–793 (1987).

Pope, C.R. Chicken anemia agent. *Vet. Immunol. Immunopathol.* **30**, 51–65 (1991).

Purchase, H.G. and Fadley, A.M. Leukosis and sarcomas. In: *Avian Disease Manual* (eds Purchase, H.G. and Fadley, A.M.), 54–58 (University of Pennsylvania, Kennett Square, PA, 1983).

Purchase, H.G. and Payne, L.N. Leukosis/sarcoma group. In: *Diseases of Poultry*, 8th edn (ed. Hofstad, M.S.), 360–405 (Iowa State University Press, Ames, IA, 1984).

Rhoades, K.R. and Heddleston, K.L. Pasteurellosis and pseudotuberculosis. In: *Pasteurellosis in Isolation and Identification of Avian Pathogens*, 2nd edn (eds Hitchner, S.B., Domermuth, C.H., Purchase, H.G. and Williams, J.E.), 11–15 (Kendall/Hunt Publishing Co., Dubuque, IA, 1980).

Snyder, D.B., Marquardt, W.W., Mallinson, E.T., Savage, P.K. and Allen, D.C. Rapid serological profiling by enzyme-linked immunosorbent assay. III. Simultaneous measurements of antibody titers to infectious bronchitis. Infectious bursal disease and Newcastle disease virus in a single serum dilution. *Avian Dis.* **28**, 12–24 (1984).

Steere, A.C. Lyme disease. *N. Engl. J. Med.* **321**, 586–596 (1989).

Terpstra, C., Wensvoort, G. and Pol, J.M. Experimental reproduction of porcine epidemic abortion and respiratory syndrome (mystery swine disease) by infection with Lelystad virus: Koch's postulates fulfilled. *Vet. Q.* **13**, 121–130 (1991).

Wensvoort, G., de Kluyver, E.P., Pol, J.M. *et al.* Lelystad virus the cause of porcine epidemic abortion and respiratory syndrome: a review of mystery swine disease research at Lelystad. *Vet. Microbiol.* **33**, 185–193 (1992).

Witter, R.L. and Calnek, B.W. Marek's disease. In: *Diseases of Poultry*, 8th edn (ed. Hofstad, M.S.), 406–417 (Iowa State University Press, Ames, IA, 1984).

Wyler, R. *et al.* Infectious bovine rhinotracheitis/vulvovaginitis (BHV-1). In: *Herpesvirus of Cattle, Horses and Pigs* (ed. Whittmann, G.), 1–72 (Kluwer, Boston, MA, 1990).

Yoder, H.W. Jr. *Mycoplasma gallisepticum* infection. In: *Diseases of Poultry*, 8th edn (eds Hofstaf, M.S., Barnes, H.F., Calnek, B.W., Reid, W.M. and Yoper, H.W. Jr.), 190–202, 212–220. (Iowa State University Press, Ames, IA, 1984).

INDEX

A

D

E

F

G

H

I

M

N

O

P

Q

R

S

T

U

V

W

Z